Be sure to keep up-to-date in 2011!

Register now and we'll notify you by email each quarter when the newest cost index is available free online.

Please provide your name, address, and email below and return this card by mail or Fax: 1-800-632-6732.

To register online:
www.reedconstructiondata.com/rsmeans/chgnotice/842339

Name _____

email _____

Title _____

Company _____

Street _____

City/Town _____ State/Prov. _____ Zip/Postal Code _____

RSMeans

D1710582

RSMeans
Quarterly Updates
700 Longwater Drive
Suite 100
Norwell, MA 02061
USA

74.95 per copy (in United States)
ice is subject to change without prior notice.

021

RSMeans
Mechanical
Cost Data

34TH ANNUAL EDITION

2011

RSMeans
A division of Reed Construction Data
Construction Publishers & Consultants
700 Longwater Drive
Norwell, MA 02061
USA
1-800-334-3509 ·

Printed in the United States of America
ISSN 0748-2698
ISBN 978-1-936335-14-5

Senior Editor
Melville J. Mossman, PE

Contributing Editors
Christopher Babbitt
Ted Baker
Barbara Balboni
Robert A. Bastoni
John H. Chiang, PE
Gary W. Christensen
David G. Drain, PE
Cheryl Elsmore
Robert J. Kuchta
Robert C. McNichols
Jeannene D. Murphy
Stephen C. Plotner
Eugene R. Spencer
Marshall J. Stetson
Phillip R. Waier, PE

Cover Design
Paul Robertson
Wayne Engebretson

President & CEO
Iain Melville

Vice President, Product & Development
Richard Remington

Vice President of Finance & Operations
David C. Walsh

Vice President, Data Aquisition & Fulfillment
Jane Rafeedie

Chief Customer Officer
Lisa Fiondella

Director, RSMeans
Daimon Bridge

Vice President, Marketing & Business
Steven Ritchie

Engineering Manager
Bob Mewis, CCC

Product Manager
Andrea Sillah

Production Manager
Michael Kokernak

Production Coordinator
Jill Goodman

Production
Paula Reale-Camelio
Sheryl A. Rose
Jonathan Forgit
Debbie Panarelli
Mary Lou Geary
Caroline Sullivan

Technical Support
Chris Anderson
Tom Dion
Roger Hancock
Gary L. Hoitt
Genevieve Medeiros
Kathryn S. Rodriguez
Sharon Proulx

This book is printed using soy-based printing ink.
This book is recyclable.

Related RSMeans Products and Services

The engineers at RSMeans suggest the following products and services as companion information resources to *RSMeans Mechanical Cost Data:*

Construction Cost Data Books
Assemblies Cost Data 2011
Square Foot Costs 2011

Reference Books
ADA Compliance Pricing Guide
Estimating Building Costs
RSMeans Estimating Handbook
Green Building: Project Planning & Estimating
How to Estimate with Means Data and CostWorks
Plan Reading & Material Takeoff
Project Scheduling & Management for Construction
Repair & Remodeling Estimating Methods
Universal Design

Seminars and In-House Training
Unit Price Estimating
Means CostWorks® Training
Means Data for Job Order Contracting (JOC)
Practical Project Management for Construction Professionals
Scheduling with MSProject for Construction Professionals
Mechanical & Electrical Estimating

RSMeans on the Internet
Visit RSMeans at **www.reedconstructiondata.com/rsmeans** for a one-stop portal for the most reliable and current resources available on the market. Here you can request or download **FREE** estimating software demos, visit our bookstore for convenient ordering, and learn more about new publications and companion products.
Also visit **www.reedconstructiondata.com** for RSMeans business solutions, seminars, training, and more.

RSMeans Electronic Data
Get the information found in RSMeans cost books electronically on *Means CostWorks®* CD or on the Web at **MeansCostWorks.com**.

RSMeans Business Solutions
Cost engineers and analysts conduct facility life cycle and benchmark research studies and predictive cost modeling, as well as offer consultation services for conceptual estimating, real property management, and job order contracting. Research studies are designed with the application of proprietary cost and project data from Reed Construction Data and RSMeans' extensive North American databases. Analysts offer building product manufacturers qualitative and quantitative market research, as well as time/motion studies for new products, and Web-based dashboards for market opportunity analysis. Clients are from both the public and private sectors, including federal, state, and municipal agencies, corporations, institutions, construction management firms, hospitals, and associations.

RSMeans for Job Order Contracting (JOC)
Best practice job order contracting (JOC) services support the owner in the development of their program. RSMeans develops all required contracting documents, works with stakeholders, and develops qualified and experienced contractor lists. Means JOCWorks™ software is the best of class in the functional tools that meet JOC specific requirements. RSMeans data is organized to meet JOC cost estimating requirements.

Construction Costs for Software Applications
Over 25 unit price and assemblies cost databases are available through a number of leading estimating and facilities management software providers (listed below). For more information see the "Other RSMeans Products" pages at the back of this publication.
MeansData™ is also available to federal, state, and local government agencies as multi-year, multi-seat licenses.

- 4Clicks-Solutions, LLC
- ArenaSoft Estimating
- Beck Technology
- BSD – Building Systems Design, Inc.
- CMS – Construction Management Software
- Corecon Technologies, Inc.
- CorVet Systems
- Hard Dollar Corporation
- MC^2
- Parsons Corporation
- Sage Timberline Office
- UDA Technologies, Inc.
- US Cost, Inc.
- VFA – Vanderweil Facility Advisers
- WinEstimator, Inc.
- R & K Solutions
- Maxwell Systems
- Tririga

Table of Contents

Foreword

Our Mission

Since 1942, RSMeans has been actively engaged in construction cost publishing and consulting throughout North America.

Today, more than 65 years after RSMeans began, our primary objective remains the same: to provide you, the construction and facilities professional, with the most current and comprehensive construction cost data possible.

Whether you are a contractor, owner, architect, engineer, facilities manager, or anyone else who needs a reliable construction cost estimate, you'll find this publication to be a highly useful and necessary tool.

With the constant flow of new construction methods and materials today, it's difficult to find the time to look at and evaluate all the different construction cost possibilities. In addition, because labor and material costs keep changing, last year's cost information is not a reliable basis for today's estimate or budget.

That's why so many construction professionals turn to RSMeans. We keep track of the costs for you, along with a wide range of other key information, from city cost indexes . . . to productivity rates . . . to crew composition . . . to contractor's overhead and profit rates.

RSMeans performs these functions by collecting data from all facets of the industry and organizing it in a format that is instantly accessible to you. From the preliminary budget to the detailed unit price estimate, you'll find the data in this book useful for all phases of construction cost determination.

The Staff, the Organization, and Our Services

When you purchase one of RSMeans' publications, you are, in effect, hiring the services of a full-time staff of construction and engineering professionals.

Our thoroughly experienced and highly qualified staff works daily at collecting, analyzing, and disseminating comprehensive cost information for your needs. These staff members have years of practical construction experience and engineering training prior to joining the firm. As a result, you can count on them not only for accurate cost figures, but also for additional background reference information that will help you create a realistic estimate.

The RSMeans organization is always prepared to help you solve construction problems through its variety of data solutions, including online, CD, and print book formats, as well as cost estimating expertise available via our business solutions, training, and seminars.

Besides a full array of construction cost estimating books, RSMeans also publishes a number of other reference works for the construction industry. Subjects include construction estimating and project and business management, special topics such as green building and job order contracting, and a library of facility management references.

In addition, you can access all of our construction cost data electronically using *Means CostWorks*® CD or on the Web using **MeansCostWorks.com.**

What's more, you can increase your knowledge and improve your construction estimating and management performance with an RSMeans construction seminar or in-house training program. These two-day seminar programs offer unparalleled opportunities for everyone in your organization to become updated on a wide variety of construction-related issues.

RSMeans is also a worldwide provider of construction cost management and analysis services for commercial and government owners.

In short, RSMeans can provide you with the tools and expertise for constructing accurate and dependable construction estimates and budgets in a variety of ways.

Robert Snow Means Established a Tradition of Quality That Continues Today

Robert Snow Means spent years building RSMeans, making certain he always delivered a quality product.

Today, at RSMeans, we do more than talk about the quality of our data and the usefulness of our books. We stand behind all of our data, from historical cost indexes to construction materials and techniques to current costs.

If you have any questions about our products or services, please call us toll-free at 1-800-334-3509. Our customer service representatives will be happy to assist you. You can also visit our Web site at **www.rsmeans.com**

How the Book is Built: An Overview

The Construction Specifications Institute (CSI) and Construction Specifications Canada (CSC) have produced the 2010 edition of MasterFormat™, a system of titles and numbers used extensively to organize construction information.

All unit price data in the RSMeans cost data books is now arranged in the 50-division MasterFormat™ 2010 system.

A Powerful Construction Tool

You have in your hands one of the most powerful construction tools available today. A successful project is built on the foundation of an accurate and dependable estimate. This book will enable you to construct just such an estimate.

For the casual user the book is designed to be:

- quickly and easily understood so you can get right to your estimate.
- filled with valuable information so you can understand the necessary factors that go into the cost estimate.

For the regular user, the book is designed to be:

- a handy desk reference that can be quickly referred to for key costs.
- a comprehensive, fully reliable source of current construction costs and productivity rates so you'll be prepared to estimate any project.
- a source book for preliminary project cost, product selections, and alternate materials and methods.

To meet all of these requirements, we have organized the book into the following clearly defined sections.

New: Quick Start

See our new "Quick Start" instructions on the following page to get started right away.

Estimating with RSMeans Unit Price Cost Data

Please refer to these steps for guidance on completing an estimate using RSMeans unit price cost data.

How To Use the Book: The Details

This section contains an in-depth explanation of how the book is arranged . . . and how you can use it to determine a reliable construction cost estimate. It includes information about how we develop our cost figures and how to completely prepare your estimate.

Unit Price Section

All cost data has been divided into the 50 divisions according to the MasterFormat system of classification and numbering. For a listing of these divisions and an outline of their subdivisions, see the Unit Price Section Table of Contents.

Estimating tips are included at the beginning of each division.

Assemblies Section

The cost data in this section has been organized in an "Assemblies" format. These assemblies are the functional elements of a building and are arranged according to the 7 elements of the UNIFORMAT II classification system. For a complete explanation of a typical "Assemblies" page, see "How To Use the Assemblies Cost Tables."

Reference Section

This section includes information on Equipment Rental Costs, Crew Listings, Historical Cost Indexes, City Cost Indexes, Location Factors, Reference Tables, Change Orders, Square Foot Costs, and a listing of Abbreviations.

Equipment Rental Costs: This section contains the average costs to rent and operate hundreds of pieces of construction equipment.

Crew Listings: This section lists all of the crews referenced in the book. For the purposes of this book, a crew is composed of more than one trade classification and/or the addition of power equipment to any trade classification. Power equipment is included in the cost of the crew. Costs are shown both with bare labor rates and with the installing contractor's overhead and profit added. For each, the total crew cost per eight-hour day and the composite cost per labor-hour are listed.

Historical Cost Indexes: These indexes provide you with data to adjust construction costs over time.

City Cost Indexes: All costs in this book are U.S. national averages. Costs vary because of the regional economy. You can adjust costs by CSI Division to over 700 locations throughout the U.S. and Canada by using the data in this section.

Location Factors: You can adjust total project costs to over 900 locations throughout the U.S. and Canada by using the data in this section.

Reference Tables: At the beginning of selected major classifications in the Unit Price section are reference numbers shown in a shaded box. These numbers refer you to related information in the Reference Section. In this section, you'll find reference tables, explanations, estimating information that support how we develop the unit price data, technical data, and estimating procedures.

Change Orders: This section includes information on the factors that influence the pricing of change orders.

Square Foot Costs: This section contains costs for 59 different building types that allow you to make a rough estimate for the overall cost of a project or its major components.

Abbreviations: A listing of abbreviations used throughout this book, along with the terms they represent, is included in this section.

Index

A comprehensive listing of all terms and subjects in this book will help you quickly find what you need when you are not sure where it occurs in MasterFormat.

The Scope of This Book

This book is designed to be as comprehensive and as easy to use as possible. To that end we have made certain assumptions and limited its scope in two key ways:

1. We have established material prices based on a national average.
2. We have computed labor costs based on a 30-city national average of union wage rates.

For a more detailed explanation of how the cost data is developed, see "How To Use the Book: The Details."

Project Size/Type

The material prices in Means data cost books are "contractor's prices." They are the prices that contractors can expect to pay at the lumberyards, suppliers/distributers warehouses, etc. Small orders of speciality items would be higher than the costs shown, while very large orders, such as truckload lots, would be less. The variation would depend on the size, timing, and negotiating power of the contractor. The labor costs are primarily for new construction or major renovation rather than repairs or minor alterations.

With reasonable exercise of judgment, the figures can be used for any building work.

Absolute Essentials for a Quick Start

If you feel you are ready to use this book and don't think you will need the detailed instructions that begin on the following page, this Absolute Essentials for a Quick Start page is for you. These steps will allow you to get started estimating in a matter of minutes.

1 Scope

Think through the project that you will be estimating, and identify the many individual work tasks that will need to be covered in your estimate.

2 Quantify

Determine the number of units that will be required for each work task that you identified.

3 Pricing

Locate individual Unit Price line items that match the work tasks you identified. The Unit Price Section Table of Contents that begins on page 1 and the Index in the back of the book will help you find these line items.

4 Multiply

Multiply the Total Incl O&P cost for a Unit Price line item in the book by your quantity for that item. The price you calculate will be an estimate for a completed item of work performed by a subcontractor. Keep adding line items in this manner to build your estimate.

5 Project Overhead

Include project overhead items in your estimate. These items are needed to make the job run and are typically, but not always, provided by the General Contractor. They can be found in Division 1. An alternate method of estimating project overhead costs is to apply a percentage of the total project cost.

Include rented tools not included in crews, waste, rubbish handling, and cleanup.

6 Estimate Summary

Include General Contractor's markup on subcontractors, General Contractor's office overhead and profit, and sales tax on materials and equipment.

Adjust your estimate to the project's location by using the City Cost Indexes or Location Factors found in the Reference Section.

Editors' Note: We urge you to spend time reading and understanding the supporting material in the front of this book. An accurate estimate requires experience, knowledge, and careful calculation. The more you know about how we at RSMeans developed the data, the more accurate your estimate will be. In addition, it is important to take into consideration the reference material in the back of the book such as Equipment Listings, Crew Listings, City Cost Indexes, Location Factors, and Reference Numbers.

Estimating with RSMeans Unit Price Cost Data

ollowing these steps will allow you to omplete an accurate estimate using RSMeans nit Price cost data.

Scope Out the Project

Identify the individual work tasks that will need to be covered in your estimate.

The Unit Price data in this book has been divided into 50 Divisions according to the CSI MasterFormat 2010—their titles are listed on the back cover of your book.

Think through the project and identify those CSI Divisions needed in your estimate.

The Unit Price Section Table of Contents on page 1 may also be helpful when scoping out your project.

Experienced estimators find it helpful to begin with Division 2 and continue through completion. Division 1 can be estimated after the full project scope is known.

2 Quantify

Determine the number of units required for each work task that you identified.

Experienced estimators include an allowance for waste in their quantities. (Waste is not included in RSMeans Unit Price line items unless so stated.)

3 Price the Quantities

- Use the Unit Price Table of Contents, and the Index, to locate individual Unit Price line items for your estimate.

- Reference Numbers indicated within a Unit Price section refer to additional information that you may find useful.

- The crew indicates who is performing the work for that task. Crew codes are expanded in the Crew Listings in the Reference Section to include all trades and equipment that comprise the crew.

- The Daily Output is the amount of work the crew is expected to do in one day.

- The Labor-Hours value is the amount of time it will take for the crew to install one unit of work.

- The abbreviated Unit designation indicates the unit of measure upon which the crew, productivity, and prices are based.

- Bare Costs are shown for materials, labor, and equipment needed to complete the Unit Price line item. Bare costs do not include waste, project overhead, payroll insurance, payroll taxes, main office overhead, or profit.

- The Total Incl O&P cost is the billing rate or invoice amount of the installing contractor or subcontractor who performs the work for the Unit Price line item.

4 Multiply

- Multiply the total number of units needed for your project by the Total Incl O&P cost for each Unit Price line item.

- Be careful that your take off unit of measure matches the unit of measure in the Unit column.

- The price you calculate is an estimate for a completed item of work.

- Keep scoping individual tasks, determining the number of units required for those tasks, matching each task with individual Unit Price line items in the book, and multiply quantities by Total Incl O&P costs.

- An estimate completed in this manner is priced as if a subcontractor, or set of subcontractors, is performing the work. The estimate does not yet include Project Overhead or Estimate Summary components such as General Contractor markups on subcontracted work, General Contractor office overhead and profit, contingency, and location factor.

5 Project Overhead

- Include project overhead items from Division 1 – General Requirements.

- These items are needed to make the job run. They are typically, but not always, provided by the General Contractor. Items include, but are not limited to, field personnel, insurance, performance bond, permits, testing, temporary utilities, field office and storage facilities, temporary scaffolding and platforms, equipment mobilization and demobilization, temporary roads and sidewalks, winter protection, temporary barricades and fencing, temporary security, temporary signs, field engineering and layout, final cleaning and commissioning.

- Each item should be quantified, and matched to individual Unit Price line items in Division 1, and then priced and added to your estimate.

- An alternate method of estimating project overhead costs is to apply a percentage of the total project cost, usually 5% to 15% with an average of 10% (see General Conditions, page ix).

- Include other project related expenses in your estimate such as:
  - Rented equipment not itemized in the Crew Listings
  - Rubbish handling throughout the project (see 02 41 19.20)

6 Estimate Summary

- Include sales tax as required by laws of your state or country.
- Include the General Contractor's markup on self-performed work, usually 5% to 15% with an average of 10%.
- Include the General Contractor's markup on subcontracted work, usually 5% to 15% with an average of 10%.
- Include General Contractor's main office overhead and profit.
- RSMeans gives general guidelines on the General Contractor's main office overhead (see section 01 31 13.50 and Reference Number R013113-50).
- RSMeans gives no guidance on the General Contractor's profit.
- Markups will depend on the size of the General Contractor's operations, his projected annual revenue, the level of risk he is taking on, and on the level of competition in the local area and for this project in particular.
- Include a contingency, usually 3% to 5%, if appropriate.

- Adjust your estimate to the project's location by using the City Cost Indexes or the Location Factors in the Reference Section.
- Look at the rules on the pages for How to Use the City Cost Indexes to see how to apply the Indexes for your location.
- When the proper Index or Factor has been identified for the project's location, convert it to a multiplier by dividing it by 100, and then multiply that multiplier by your estimated total cost. The original estimated total cost will now be adjusted up or down from the national average to a total that is appropriate for your location.

Editors' Notes:

1) *We urge you to spend time reading and understanding the supporting material in the front of this book. An accurate estimate requires experience, knowledge, and careful calculation. The more you know about how we at RSMeans developed the data, the more accurate your estimate will be. In addition, it is important to take into consideration the reference material* in the back of the book such as Equipment Listings, Crew Listings, City Cost Indexes, Location Factors, and Reference Numbers.

2) *Contractors who are bidding or are involved in JOC, DOC, SABER, or IDIQ type contracts are cautioned that workers' compensation insurance, federal and state payroll taxes, waste, project supervision, project overhead, main office overhead, and profit are not included in bare costs. The coefficient or multiplier must cover these costs.*

How to Use the Book: The Details

What's Behind the Numbers? The Development of Cost Data

The staff at RSMeans continually monitors developments in the construction industry in order to ensure reliable, thorough, and up-to-date cost information. While overall construction costs may vary relative to general economic conditions, price fluctuations within the industry are dependent upon many factors. Individual price variations may, in fact, be opposite to overall economic trends. Therefore, costs are constantly tracked and complete updates are published yearly. Also, new items are frequently added in response to changes in materials and methods.

Costs—$ (U.S.)

All costs represent U.S. national averages and are given in U.S. dollars. The RSMeans City Cost Indexes can be used to adjust costs to a particular location. The City Cost Indexes for Canada can be used to adjust U.S. national averages to local costs in Canadian dollars. No exchange rate conversion is necessary.

G The processes or products identified by the green symbol in our publications have been determined to be environmentally responsible and/or resource-efficient solely by the RSMeans engineering staff. The inclusion of the green symbol does not represent compliance with any specific industry association or standard.

Material Costs

The RSMeans staff contacts manufacturers, dealers, distributors, and contractors all across the U.S. and Canada to determine national average material costs. If you have access to current material costs for your specific location, you may wish to make adjustments to reflect differences from the national average. Included within material costs are fasteners for a normal installation. RSMeans engineers use manufacturers' recommendations, written specifications, and/or standard construction practice for size and spacing of fasteners. Adjustments to material costs may be required for your specific application or location. Material costs do not include sales tax.

Labor Costs

Labor costs are based on the average of wage rates from 30 major U.S. cities. Rates are determined from labor union agreements or prevailing wages for construction trades for the current year. Rates, along with overhead and profit markups, are listed on the inside back cover of this book.

- If wage rates in your area vary from those used in this book, or if rate increases are expected within a given year, labor costs should be adjusted accordingly.

Labor costs reflect productivity based on actual working conditions. In addition to actual installation, these figures include time spent during a normal weekday on tasks such as, material receiving and handling, mobilization at site, site movement, breaks, and cleanup. Productivity data is developed over an extended period so as not to be influenced by abnormal variations and reflects a typical average.

Equipment Costs

Equipment costs include not only rental, but also operating costs for equipment under normal use. The operating costs include parts and labor for routine servicing such as repair and replacement of pumps, filters, and worn lines. Normal operating expendables, such as fuel, lubricants, tires, and electricity (where applicable), are also included. Extraordinary operating expendables with highly variable wear patterns, such as diamond bits and blades, are excluded. These costs are included under materials. Equipment rental rates are obtained from industry sources throughout North America—contractors, suppliers, dealers, manufacturers, and distributors.

Rental rates can also be treated as reimbursement costs for contractor-owned equipment. Owned equipment costs include depreciation, loan payments, interest, taxes, insurance, storage and major repairs.

Equipment costs do not include operators' wages; nor do they include the cost to move equipment to a job site (mobilization) or from a job site (demobilization).

Equipment Cost/Day——The cost of power equipment required for each crew is included in the Crew Listings in the Reference Section (small tools that are considered as essential everyday tools are not listed out separately). The Crew Listings itemize specialized tools and heavy equipment along with labor trades. The daily cost of itemized equipment included in a crew is based on dividing the weekly bare rental rate by 5 (number of working days per week) and then adding the hourly operating cost times 8 (the number of hours per day). This Equipment Cost/Day is shown in the last column of the Equipment Rental Cost pages in the Reference Section.

Mobilization/Demobilization——The cost to move construction equipment from an equipment yard or rental company to the job site and back again is not included in equipment costs. Mobilization (to the site) and demobilization (from the site) costs can be found in the Unit Price Section. If a piece of equipment is already at the job site, it is not appropriate to utilize mob./demob. costs again in an estimate.

Overhead and Profit

Total Cost including O&P for the *Installing Contractor* is shown in the last column on the Unit Price and/or the Assemblies pages of this book. This figure is the sum of the bare material cost plus 10% for profit, the bare labor cost plus total overhead and profit, and the bare equipment cost plus 10% for profit. Details for the calculation of Overhead and Profit on labor are shown on the inside back cover and in the Reference Section of this book. (See the "How to Use the Unit Price Pages" for an example of this calculation.)

General Conditions

Cost data in this book is presented in two ways: Bare Costs and Total Cost including O&P (Overhead and Profit). General Conditions, or General Requirements, of the contract should also be added to the Total Cost including O&P when applicable. Costs for General Conditions are listed in Division 1 of the Unit Price Section and the Reference

Section of this book. General Conditions for the *Installing Contractor* may range from 0% to 10% of the Total Cost including O&P. For the *General* or *Prime Contractor*, costs for General Conditions may range from 5% to 15% of the Total Cost including O&P, with a figure of 10% as the most typical allowance. If applicable, the Assemblies and Models sections of this book use costs that include the installing contractor's overhead and profit (O&P).

Factors Affecting Costs

Costs can vary depending upon a number of variables. Here's how we have handled the main factors affecting costs.

Quality——The prices for materials and the workmanship upon which productivity is based represent sound construction work. They are also in line with U.S. government specifications.

Overtime——We have made no allowance for overtime. If you anticipate premium time or work beyond normal working hours, be sure to make an appropriate adjustment to your labor costs.

Productivity——The productivity, daily output, and labor-hour figures for each line item are based on working an eight-hour day in daylight hours in moderate temperatures. For work that extends beyond normal work hours or is performed under adverse conditions, productivity may decrease. (See the section in "How To Use the Unit Price Pages" for more on productivity.)

Size of Project——The size, scope of work, and type of construction project will have a significant impact on cost. Economies of scale can reduce costs for large projects. Unit costs can often run higher for small projects.

Location——Material prices in this book are for metropolitan areas. However, in dense urban areas, traffic and site storage limitations may increase costs. Beyond a 20-mile radius of large cities, extra trucking or transportation charges may also increase the material costs slightly. On the other hand, lower wage rates may be in effect. Be sure to consider both of these factors when preparing an estimate, particularly if the job site is located in a central city or remote rural location. In addition, highly specialized subcontract items may require travel and per-diem expenses for mechanics.

Other Factors—

- season of year
- contractor management
- weather conditions
- local union restrictions
- building code requirements
- availability of:
  - adequate energy
  - skilled labor
  - building materials
- owner's special requirements/restrictions
- safety requirements
- environmental considerations

Unpredictable Factors——General business conditions influence "in-place" costs of all items. Substitute materials and construction methods may have to be employed. These may affect the installed cost and/or life cycle costs. Such factors may be difficult to evaluate and cannot necessarily be predicted on the basis of the job's location in a particular section of the country. Thus, where these factors apply, you may find significant but unavoidable cost variations for which you will have to apply a measure of judgment to your estimate.

Rounding of Costs

In general, all unit prices in excess of $5.00 have been rounded to make them easier to use and still maintain adequate precision of the results. The rounding rules we have chosen are in the following table.

Prices from . . .	Rounded to the nearest . . .
$.01 to $5.00	$.01
$5.01 to $20.00	$.05
$20.01 to $100.00	$.50
$100.01 to $300.00	$1.00
$300.01 to $1,000.00	$5.00
$1,000.01 to $10,000.00	$25.00
$10,000.01 to $50,000.00	$100.00
$50,000.01 and above	$500.00

How Subcontracted Items Affect Costs

A considerable portion of all large construction jobs is usually subcontracted. In fact, the percentage done by subcontractors is constantly increasing and may run over 90%. Since the workers employed by these companies do nothing else but install their particular product, they soon become expert in that line. The result is, installation by these firms is accomplished so efficiently that the total in-place cost, even adding the general contractor's overhead and profit, is no more, and often less, than if the principal contractor had handled the installation himself/herself. Companies that deal with construction specialties are anxious to have their product perform well and, consequently, the installation will be the best possible.

Contingencies

The allowance for contingencies generally provides for unforeseen construction difficulties. On alterations or repair jobs, 20% is not too much. If drawings are final and only field contingencies are being considered, 2% or 3% is probably sufficient, and often nothing need be added. Contractually, changes in plans will be covered by extras. The contractor should consider inflationary price trends and possible material shortages during the course of the job. These escalation factors are dependent upon both economic conditions and the anticipated time between the estimate and actual construction. If drawings are not complete or approved, or a budget cost is wanted, it is wise to add 5% to 10%. Contingencies, then, are a matter of judgment. Additional allowances for contingencies are shown in Division 1.

Important Estimating Considerations

The "productivity," or daily output, of each craftsman includes mobilization and cleanup time, break time and plan layout time, as well as an allowance to carry stock from the storage trailer or location on the job site up to 200' into the building and to the first or second floor. If material has to be transported over greater distances or to higher floors, an additional allowance should be considered by the estimator. An allowance has also been included in the piping and fittings installation time for leak checking and minor tightening. Equipment installation time includes the following applicable items: positioning, leveling and securing the unit in place, connecting all associated piping, ducts, vents, etc., which shall have been estimated separately, connecting to an adjacent power source, filling/bleeding, startup, adjusting the controls up and down to ensure proper response, setting the integral controls/valves/regulators/thermostats for proper operation (does not include external building type control systems, DDC systems, etc.), explaining/training owner's operator, and warranty. A reasonable breakdown of the labor costs is as follows:

1. Movement into building, installation/setting of equipment 35%
2. Connection to piping/duct/power, etc. 25%
3. Filling/flushing/cleaning/touchup, etc. 15%
4. Startup/running adjustments 5%
5. Training owner's representative 5%
6. Warranty/call back/service 15%

Note that cranes or other lifting equipment are not included on any lines in the Mechanical divisions. For example, if a crane is required to lift a heavy piece of pipe into place high above a gym floor, or to put a rooftop unit on the roof of a four-story building, etc., it must be added. Due to the potential for extreme variation—from nothing additional required to a major crane or helicopter—we feel that including a nominal amount for "lifting contingency" would be useless and detract from the accuracy of the estimate. When using equipment rental from RSMeans, remember to include the cost of the operator(s).

Estimating Labor-Hours

The labor-hours expressed in this publication are based on Average Installation time, using an efficiency level of approximately 60%–65% (see item 7), which has been found reasonable and acceptable by many contractors. The book uses this national efficiency average to establish a consistent benchmark. For bid situations, adjustments to this efficiency level should be the responsibility of the contractor bidding the project. The unit labor-hour is divided in the following manner. A typical day for a journeyman might be:

1.	Study Plans	3%	14.4 min.
2.	Material Procurement	3%	14.4 min.
3.	Receiving and Storing	3%	14.4 min
4.	Mobilization	5%	24.0 min.
5.	Site Movement	5%	24.0 min.
6.	Layout and Marking	8%	38.4 min.
7.	Actual Installation	64%	307.2 min.
8.	Cleanup	3%	14.4 min.
9.	Breaks, Nonproductive	6%	28.8 min.
		100%	480.0 min.

If any of the percentages expressed in this breakdown do not apply to the particular work or project situation, then that percentage or a portion of it may be deducted from or added to labor-hours.

Final Checklist

Estimating can be a straightforward process provided you remember the basics. Here's a checklist of some of the steps you should remember to complete before finalizing your estimate. Did you remember to . . .

- factor in the City Cost Index for your locale?
- take into consideration which items have been marked up and by how much?
- mark up the entire estimate sufficiently for your purposes?
- read the background information on techniques and technical matters that could impact your project time span and cost?
- include all components of your project in the final estimate?
- double check your figures for accuracy?
- call RSMeans if you have any questions about your estimate or the data you've found in our publications? Remember, RSMeans stands behind its publications. If you have any questions about your estimate . . . about the costs you've used from our books . . . or even about the technical aspects of the job that may affect your estimate, feel free to call the RSMeans editors at 1-800-334-3509.

Unit Price Section

Table of Contents

1

Table of Contents (cont.)

How to Use the Unit Price Pages

The following is a detailed explanation of a sample entry in the Unit Price Section. Next to each bold number below is the described item with the appropriate component of the sample entry following in parentheses. Some prices are listed as bare costs; others as costs that include overhead and profit of the installing contractor. In most cases, if the work is to be subcontracted, the general contractor will need to add an additional markup (RSMeans suggests using 10%) to the figures in the column "Total Incl. O&P."

1 Section Number/Title (03 30/Cast-In-Place Concrete)

Use the Unit Price Section Table of Contents to locate specific Level 2 Sections. The sections are classified according to the CSI MasterFormat 2004 system.

2 Line Numbers (03 30 53.40 3920)

Each unit price line item has been assigned a unique 12-digit code based on the CSI MasterFormat classification.

- MasterFormat Level 1 (Division 03)
- MasterFormat Level 2 (03 30)
- MasterFormat Level 3 (03 30 53)

03 30 53.40 3920

- MasterFormat Level 4 (03 30 53.40)
- RSMeans 12-Digit Line Number

3 Description (Concrete In Place, etc.)

Each line item is described in detail. Sub-items and additional sizes are indented beneath the appropriate line items. The first line or two after the main item (in boldface) may contain descriptive information that pertains to all line items beneath this boldface listing.

4 Reference Number Information

R033105 -20 You'll see reference numbers shown in shaded boxes at the beginning of some sections. These refer to related items in the Reference Section, visually identified by a vertical gray bar on the page edges.

The relation may be an estimating procedure that should be read before estimating, or technical information.

The "R" designates the Reference Section. The numbers refer to the MasterFormat 2004 classification system.

It is strongly recommended that you review all reference numbers that appear within the section in which you are working.

03 30 Cast-In-Place Concrete

03 30 53 – Miscellaneous Cast-In-Place Concrete

03 30 53.40 Concrete In Place		Crew	Daily Output	Labor-Hours	Unit	Material	2011 Bare Costs Labor	Equipment	Total	Total Incl O&P
0010	**CONCRETE IN PLACE**	R033105-10								
0020	Including forms (4 uses), reinforcing steel, concrete, placement,	R033105-20								
0050	and finishing unless otherwise indicated	R033105-50								
0300	Beams (3500 psi), 5 kip per L.F., 10' span	R033105-65	C-14A	15.62	12.804	C.Y.	291	555	49	895	1,225
3825	1 C.Y. to 5 C.Y.			43	2.605		185	108	.55	293.55	370
3850	Over 5 C.Y.			75	1.493		171	61.50	.31	232.81	283
3900	Footings, strip (3000 psi), 18" x 9", unreinforced		C-14L	40	2.400		119	96.50	.58	216.08	279
3920	18" x 9", reinforced		C-14C	35	3.200		141	132	.67	273.67	360
3925	20" x 10", unreinforced		C-14L	45	2.133		116	85.50	.51	202.01	260
3930	20" x 10", reinforced		C-14C	40	2.800		134	116	.59	250.59	325
3935	24" x 12", unreinforced		C-14L	55	1.745		114	70	.42	184.42	233
3940	24" x 12", reinforced		C-14C	48	2.333		132	96.50	.49	228.99	294

5 Crew (C-14C)

The "Crew" column designates the typical trade or crew used to install the item. If an installation can be accomplished by one trade and requires no power equipment, that trade and the number of workers are listed (for example, "2 Carpenters"). If an installation requires a composite crew, a crew code designation is listed (for example, "C-14C"). You'll find full details on all composite crews in the Crew Listings.

* For a complete list of all trades utilized in this book and their abbreviations, see the inside back cover.

Crews

Crew No.	Bare Costs		Incl. Subs O&P		Cost Per Labor-Hour	
Crew C-14C	Hr.	Daily	Hr.	Daily	Bare Costs	Incl. O&P
1 Carpenter Foreman (out)	$45.05	$360.40	$69.10	$552.80	$41.31	$63.60
6 Carpenters	43.05	2066.40	66.05	3170.40		
2 Rodmen (reinf.)	48.40	774.40	77.35	1237.60		
4 Laborers	34.35	1099.20	52.70	1686.40		
1 Cement Finisher	40.85	326.80	59.50	476.00		
1 Gas Engine Vibrator		23.20		25.52	.21	.23
112 L.H., Daily Totals		$4650.40		$7148.72	$41.52	$63.83

6 Productivity: Daily Output (35.0)/ Labor-Hours (3.20)

The "Daily Output" represents the typical number of units the designated crew will install in a normal 8-hour day. To find out the number of days the given crew would require to complete the installation, divide your quantity by the daily output. For example:

Quantity	÷	Daily Output	=	Duration
100 C.Y.	÷	35.0/ Crew Day	=	2.86 Crew Days

The "Labor-Hours" figure represents the number of labor-hours required to install one unit of work. To find out the number of labor-hours required for your particular task, multiply the quantity of the item times the number of labor-hours shown. For example:

Quantity	x	Productivity Rate	=	Duration
100 C.Y.	x	3.20 Labor-Hours/ C.Y.	=	320 Labor-Hours

7 Unit (C.Y.)

The abbreviated designation indicates the unit of measure upon which the price, production, and crew are based (C.Y. = Cubic Yard). For a complete listing of abbreviations, refer to the Abbreviations Listing in the Reference Section of this book.

8 Bare Costs:
Mat. (Bare Material Cost) (141)

The unit material cost is the "bare" material cost with no overhead or profit included. Costs shown reflect national average material prices for January of the current year and include delivery to the job site. No sales taxes are included.

Labor (132)

The unit labor cost is derived by multiplying bare labor-hour costs for Crew C-14C by labor-hour units. The bare labor-hour cost is found in the Crew Section under C-14C. (If a trade is listed, the hourly labor cost—the wage rate—is found on the inside back cover.)

Labor-Hour Cost Crew C-14C	x	Labor-Hour Units	=	Labor
$41.31	x	3.20	=	$132.00

Equip. (Equipment) (.67)

Costs for equipment included in each crew are listed in the Crew Listings in the Reference Section. Costs for small tools that are considered as essential everyday tools are covered in the "Overhead" column on the "Installing Contractor's Overhead and Profit" table that lists labor trades, base rates and markups. The unit equipment cost displayed here is derived by multiplying the bare equipment hourly cost in the Crew Listing by the unit Labor-Hours value.

Equipment Cost Crew C-14C	x	Labor-Hour Units	=	Equip.
$.21	x	3.20	=	$.67

Total (273.67)

The total of the bare costs is the arithmetic total of the three previous columns: mat., labor, and equip.

Material	+	Labor	+	Equip.	=	Total
$141	+	$132	+	$.67	=	$273.67

9 Total Costs Including O & P

This figure is the sum of the bare material cost plus 10% for profit; the bare labor cost plus total overhead and profit (per the inside back cover or, if a crew is listed, from the crew listings); and the bare equipment cost plus 10% for profit.

Material is Bare Material Cost + 10% = 141 + 14.1	=	$155.10
Labor for Crew C-14C = Labor-Hour Cost (63.60) x Labor-Hour Units (3.20)	=	$203.50
Equip. is Bare Equip. Cost + 10% = .67 + .07	=	$ .74
Total (Rounded)	=	$ 360

Estimating Tips

01 20 00 Price and Payment Procedures

- Allowances that should be added to estimates to cover contingencies and job conditions that are not included in the national average material and labor costs are shown in section 01 21.

- When estimating historic preservation projects (depending on the condition of the existing structure and the owner's requirements), a 15%–20% contingency or allowance is recommended, regardless of the stage of the drawings.

01 30 00 Administrative Requirements

- Before determining a final cost estimate, it is a good practice to review all the items listed in Subdivisions 01 31 and 01 32 to make final adjustments for items that may need customizing to specific job conditions.

- Requirements for initial and periodic submittals can represent a significant cost to the General Requirements of a job. Thoroughly check the submittal specifications when estimating a project to determine any costs that should be included.

01 40 00 Quality Requirements

- All projects will require some degree of Quality Control. This cost is not included in the unit cost of construction listed in each division. Depending upon the terms of the contract, the various costs of inspection and testing can be the responsibility of either the owner or the contractor. Be sure to include the required costs in your estimate.

01 50 00 Temporary Facilities and Controls

- Barricades, access roads, safety nets, scaffolding, security, and many more requirements for the execution of a safe project are elements of direct cost. These costs can easily be overlooked when preparing an estimate. When looking through the major classifications of this subdivision, determine which items apply to each division in your estimate.

- Construction Equipment Rental Costs can be found in the Reference Section in section 01 54 33. Operators' wages are not included in equipment rental costs.

- Equipment mobilization and demobilization costs are not included in equipment rental costs and must be considered separately in section 01 54 36.50.

- The cost of small tools provided by the installing contractor for his workers is covered in the "Overhead" column on the "Installing Contractor's Overhead and Profit" table that lists labor trades, base rates and markups and, therefore, is included in the "Total Incl. O&P" cost of any Unit Price line item. For those users who are constrained by contract terms to use only bare costs, there are two line items in section 01 54 39.70 for small tools as a percentage of bare labor cost. If some of those users are further constrained by contract terms to refrain from using line items from Division 1, you are advised to cover the cost of small tools within your coefficient or multiplier.

01 70 00 Execution and Closeout Requirements

- When preparing an estimate, thoroughly read the specifications to determine the requirements for Contract Closeout. Final cleaning, record documentation, operation and maintenance data, warranties and bonds, and spare parts and maintenance materials can all be elements of cost for the completion of a contract. Do not overlook these in your estimate.

Reference Numbers

Reference numbers are shown in shaded boxes at the beginning of some major classifications. These numbers refer to related items in the Reference Section. The reference information may be an estimating procedure, an alternate pricing method, or technical information.

Note: Not all subdivisions listed here necessarily appear in this publication.

Division 1 - General Requirements

01 11 Summary of Work

01 11 31 – Professional Consultants

01 11 31.10 Architectural Fees

		Crew	Daily Output	Labor-Hours	Unit	Material	2011 Bare Costs Labor	Equipment	Total	Total Incl O&P
0010	**ARCHITECTURAL FEES** R011110-10									
2000	For "Greening" of building [G]				Project					3%

01 11 31.30 Engineering Fees

		Crew	Daily Output	Labor-Hours	Unit	Material	2011 Bare Costs Labor	Equipment	Total	Total Incl O&P
0010	**ENGINEERING FEES** R011110-30									
0020	Educational planning consultant, minimum				Project					.50%
0100	Maximum				"					2.50%
0200	Electrical, minimum				Contrct					4.10%
0300	Maximum									10.10%
0400	Elevator & conveying systems, minimum									2.50%
0500	Maximum									5%
0600	Food service & kitchen equipment, minimum									8%
0700	Maximum									12%
1000	Mechanical (plumbing & HVAC), minimum									4.10%
1100	Maximum									10.10%

01 21 Allowances

01 21 16 – Contingency Allowances

01 21 16.50 Contingencies

		Crew	Daily Output	Labor-Hours	Unit	Material	2011 Bare Costs Labor	Equipment	Total	Total Incl O&P
0010	**CONTINGENCIES**, Add to estimate									
0020	Conceptual stage				Project					20%
0050	Schematic stage									15%
0100	Preliminary working drawing stage (Design Dev.)									10%
0150	Final working drawing stage									3%

01 21 53 – Factors Allowance

01 21 53.50 Factors

		Crew	Daily Output	Labor-Hours	Unit	Material	2011 Bare Costs Labor	Equipment	Total	Total Incl O&P
0010	**FACTORS** Cost adjustments R012153-10									
0100	Add to construction costs for particular job requirements									
0500	Cut & patch to match existing construction, add, minimum				Costs	2%	3%			
0550	Maximum					5%	9%			
0800	Dust protection, add, minimum					1%	2%			
0850	Maximum					4%	11%			
1100	Equipment usage curtailment, add, minimum					1%	1%			
1150	Maximum					3%	10%			
1400	Material handling & storage limitation, add, minimum					1%	1%			
1450	Maximum					6%	7%			
1700	Protection of existing work, add, minimum					2%	2%			
1750	Maximum					5%	7%			
2000	Shift work requirements, add, minimum						5%			
2050	Maximum						30%			
2300	Temporary shoring and bracing, add, minimum					2%	5%			
2350	Maximum					5%	12%			
2400	Work inside prisons and high security areas, add, minimum						30%			
2450	Maximum						50%			

01 21 55 – Job Conditions Allowance

01 21 55.50 Job Conditions

		Crew	Daily Output	Labor-Hours	Unit	Material	2011 Bare Costs Labor	Equipment	Total	Total Incl O&P
0010	**JOB CONDITIONS** Modifications to applicable									
0020	cost summaries									
0100	Economic conditions, favorable, deduct				Project				2%	2%
0200	Unfavorable, add								5%	5%
0300	Hoisting conditions, favorable, deduct								2%	2%

01 21 Allowances

01 21 55 – Job Conditions Allowance

01 21 55.50 Job Conditions	Crew	Daily Output	Labor-Hours	Unit	Material	2011 Bare Costs Labor	Equipment	Total	Total Incl O&P
0400 Unfavorable, add				Project				5%	5%
0700 Labor availability, surplus, deduct								1%	1%
0800 Shortage, add								10%	10%
0900 Material storage area, available, deduct								1%	1%
1000 Not available, add								2%	2%
1100 Subcontractor availability, surplus, deduct								5%	5%
1200 Shortage, add								12%	12%
1300 Work space, available, deduct								2%	2%
1400 Not available, add								5%	5%

01 21 57 – Overtime Allowance

01 21 57.50 Overtime

		Crew	Daily Output	Labor-Hours	Unit	Material	2011 Bare Costs Labor	Equipment	Total	Total Incl O&P
0010	**OVERTIME** for early completion of projects or where	R012909-90								
0020	labor shortages exist, add to usual labor, up to				Costs		100%			

01 21 61 – Cost Indexes

01 21 61.10 Construction Cost Index

		Crew	Daily Output	Labor-Hours	Unit	Material	2011 Bare Costs Labor	Equipment	Total	Total Incl O&P
0010	**CONSTRUCTION COST INDEX** (Reference) over 930 zip code locations in									
0020	the U.S. and Canada, total bldg. cost, min. (Guymon, OK)				%					67.30%
0050	Average									100%
0100	Maximum (New York, NY)									133%

01 21 61.20 Historical Cost Indexes

0010	**HISTORICAL COST INDEXES** (See Reference Section)									

01 21 61.30 Labor Index

		Crew	Daily Output	Labor-Hours	Unit	Material	2011 Bare Costs Labor	Equipment	Total	Total Incl O&P
0010	**LABOR INDEX** (Reference) For over 930 zip code locations in									
0020	the U.S. and Canada, minimum (Guymon, OK)				%		31.10%			
0050	Average						100%			
0100	Maximum (New York, NY)						166.30%			

01 21 61.50 Material Index

		Crew	Daily Output	Labor-Hours	Unit	Material	2011 Bare Costs Labor	Equipment	Total	Total Incl O&P
0011	**MATERIAL INDEX** For over 930 zip code locations in									
0020	the U.S. and Canada, minimum (Elizabethtown, KY)				%	91%				
0040	Average					100%				
0060	Maximum (Ketchikan, AK)					136.90%				

01 21 63 – Taxes

01 21 63.10 Taxes

		Crew	Daily Output	Labor-Hours	Unit	Material	2011 Bare Costs Labor	Equipment	Total	Total Incl O&P
0010	**TAXES**	R012909-80								
0020	Sales tax, State, average				%	5.04%				
0050	Maximum	R012909-85				8.25%				
0200	Social Security, on first $106,800 of wages						7.65%			
0300	Unemployment, combined Federal and State, minimum						.80%			
0350	Average						6.20%			
0400	Maximum						13.16%			

01 31 Project Management and Coordination

01 31 13 – Project Coordination

01 31 13.20 Field Personnel

		Crew	Daily Output	Labor-Hours	Unit	Material	2011 Bare Costs Labor	2011 Bare Costs Equipment	Total	Total Incl O&P
0010	**FIELD PERSONNEL**									
0020	Clerk, average				Week		410		410	630
0100	Field engineer, minimum						975		975	1,500
0120	Average						1,265		1,265	1,950
0140	Maximum						1,450		1,450	2,225
0160	General purpose laborer, average						1,350		1,350	2,075
0180	Project manager, minimum						1,800		1,800	2,750
0200	Average						2,075		2,075	3,175
0220	Maximum						2,375		2,375	3,650
0240	Superintendent, minimum						1,750		1,750	2,675
0260	Average						1,925		1,925	2,950
0280	Maximum						2,200		2,200	3,375
0290	Timekeeper, average				↓		1,130		1,130	1,725

01 31 13.30 Insurance

			Crew	Daily Output	Labor-Hours	Unit	Material	2011 Bare Costs Labor	2011 Bare Costs Equipment	Total	Total Incl O&P
0010	**INSURANCE**	R013113-40									
0020	Builders risk, standard, minimum					Job					.24%
0050	Maximum	R013113-50									.64%
0200	All-risk type, minimum										.25%
0250	Maximum	R013113-60				↓					.62%
0400	Contractor's equipment floater, minimum					Value					.50%
0450	Maximum					"					1.50%
0600	Public liability, average					Job					2.02%
0800	Workers' compensation & employer's liability, average										
0850	by trade, carpentry, general					Payroll		16.07%			
1000	Electrical							5.81%			
1150	Insulation							12.30%			
1450	Plumbing							7.04%			
1550	Sheet metal work (HVAC)					↓		8.94%			

01 31 13.50 General Contractor's Mark-Up

		Crew	Daily Output	Labor-Hours	Unit	Material	2011 Bare Costs Labor	2011 Bare Costs Equipment	Total	Total Incl O&P
0010	**GENERAL CONTRACTOR'S MARK-UP** on Change Orders									
0200	Extra work, by subcontractors, add				%					10%
0250	By General Contractor, add									15%
0400	Omitted work, by subcontractors, deduct all but									5%
0450	By General Contractor, deduct all but									7.50%
0600	Overtime work, by subcontractors, add									15%
0650	By General Contractor, add				↓					10%

01 31 13.60 Installing Contractor's Main Office Overhead

		Crew	Daily Output	Labor-Hours	Unit	Material	2011 Bare Costs Labor	2011 Bare Costs Equipment	Total	Total Incl O&P
0010	**INSTALLING CONTRACTOR'S MAIN OFFICE OVERHEAD**									
0020	As percent of direct costs, minimum				%				5%	
0050	Average								13%	
0100	Maximum				↓				30%	

01 31 13.80 Overhead and Profit

		Crew	Daily Output	Labor-Hours	Unit	Material	2011 Bare Costs Labor	2011 Bare Costs Equipment	Total	Total Incl O&P
0010	**OVERHEAD & PROFIT** Allowance to add to items in this									
0020	book that do not include Subs O&P, average				%				25%	
0100	Allowance to add to items in this book that									
0110	do include Subs O&P, minimum				%					5%
0150	Average									10%
0200	Maximum									15%
0300	Typical, by size of project, under $100,000								30%	
0350	$500,000 project								25%	
0400	$2,000,000 project								20%	
0450	Over $10,000,000 project				↓				15%	

01 31 Project Management and Coordination

01 31 13 – Project Coordination

01 31 13.90 Performance Bond		Crew	Daily Output	Labor-Hours	Unit	Material	2011 Bare Costs Labor	Equipment	Total	Total Incl O&P
0010	**PERFORMANCE BOND**	R013113-80								
0020	For buildings, minimum					Job				.60%
0100	Maximum					"				2.50%

01 32 Construction Progress Documentation

01 32 33 – Photographic Documentation

01 32 33.50 Photographs

		Crew	Daily Output	Labor-Hours	Unit	Material	2011 Bare Costs Labor	Equipment	Total	Total Incl O&P
0010	**PHOTOGRAPHS**									
0020	8" x 10", 4 shots, 2 prints ea., std. mounting				Set	475			475	520
0100	Hinged linen mounts					530			530	580
0200	8" x 10", 4 shots, 2 prints each, in color					415			415	460
0300	For I.D. slugs, add to all above					5.30			5.30	5.85
1500	Time lapse equipment, camera and projector, buy					845			845	930
1550	Rent per month					260			260	286
1700	Cameraman and film, including processing, B.&W.				Day	1,375			1,375	1,525
1720	Color				"	1,375			1,375	1,525

01 41 Regulatory Requirements

01 41 26 – Permit Requirements

01 41 26.50 Permits

		Crew	Daily Output	Labor-Hours	Unit	Material	2011 Bare Costs Labor	Equipment	Total	Total Incl O&P
0010	**PERMITS**									
0020	Rule of thumb, most cities, minimum				Job					.50%
0100	Maximum				"					2%

01 45 Quality Control

01 45 23 – Testing and Inspecting Services

01 45 23.50 Testing

		Crew	Daily Output	Labor-Hours	Unit	Material	2011 Bare Costs Labor	Equipment	Total	Total Incl O&P
0010	**TESTING** and Inspecting Services									
5820	Non-destructive metal testing, dye penetrant				Day				310	340
5840	Magnetic particle								310	340
5860	Radiography								450	495
5880	Ultrasonic								310	340
9000	Thermographic testing, for bldg envelope heat loss, average 2,000 S.F. [G]				Ea.					500

01 51 Temporary Utilities

01 51 13 – Temporary Electricity

01 51 13.80 Temporary Utilities

		Crew	Daily Output	Labor-Hours	Unit	Material	2011 Bare Costs Labor	Equipment	Total	Total Incl O&P
0010	**TEMPORARY UTILITIES**									
0100	Heat, incl. fuel and operation, per week, 12 hrs. per day	1 Skwk	100	.080	CSF Flr	27.50	3.55		31.05	36
0200	24 hrs. per day	"	60	.133		52.50	5.90		58.40	67
0350	Lighting, incl. service lamps, wiring & outlets, minimum	1 Elec	34	.235		2.73	11.85		14.58	20.50
0360	Maximum	"	17	.471		5.80	23.50		29.30	41.50
0400	Power for temp lighting only, per month, min/month 6.6 KWH								.92	1.01
0430	Average/month 11.8 KWH								1.65	1.82
0450	Maximum/month 23.6 KWH								3.30	3.63

01 51 Temporary Utilities

01 51 13 – Temporary Electricity

01 51 13.80 Temporary Utilities	Crew	Daily Output	Labor-Hours	Unit	Material	2011 Bare Costs Labor	Equipment	Total	Total Incl O&P
0600	Power for job duration incl. elevator, etc., minimum				CSF Flr				47	51.50
0650	Maximum				↓				110	121
1000	Toilet, portable, see Equip. Rental 01 54 33 in Reference Section									

01 52 Construction Facilities

01 52 13 – Field Offices and Sheds

01 52 13.20 Office and Storage Space

		Crew	Daily Output	Labor-Hours	Unit	Material	2011 Bare Costs Labor	Equipment	Total	Total Incl O&P
0010	**OFFICE AND STORAGE SPACE**									
0020	Trailer, furnished, no hookups, 20' x 8', buy	2 Skwk	1	16	Ea.	8,550	710		9,260	10,500
0250	Rent per month					146			146	160
0300	32' x 8', buy	2 Skwk	.70	22.857		12,300	1,025		13,325	15,100
0350	Rent per month					193			193	213
0400	50' x 10', buy	2 Skwk	.60	26.667		23,300	1,175		24,475	27,500
0450	Rent per month					288			288	315
0500	50' x 12', buy	2 Skwk	.50	32		26,000	1,425		27,425	30,800
0550	Rent per month					360			360	395
0700	For air conditioning, rent per month, add				↓	41.50			41.50	45.50
0800	For delivery, add per mile				Mile	4.60			4.60	5.05
1000	Portable buildings, prefab, on skids, economy, 8' x 8'	2 Carp	265	.060	S.F.	85	2.60		87.60	97.50
1100	Deluxe, 8' x 12'	"	150	.107	"	95	4.59		99.59	112
1200	Storage boxes, 20' x 8', buy	2 Skwk	1.80	8.889	Ea.	2,175	395		2,570	2,975
1250	Rent per month					71.50			71.50	78.50
1300	40' x 8', buy	2 Skwk	1.40	11.429		3,025	505		3,530	4,100
1350	Rent per month				↓	94			94	103

01 52 13.40 Field Office Expense

		Crew	Daily Output	Labor-Hours	Unit	Material	2011 Bare Costs Labor	Equipment	Total	Total Incl O&P
0010	**FIELD OFFICE EXPENSE**									
0100	Office equipment rental average				Month	200			200	220
0120	Office supplies, average				"	86			86	94.50
0125	Office trailer rental, see Section 01 52 13.20									
0140	Telephone bill; avg. bill/month incl. long dist.				Month	81			81	89
0160	Lights & HVAC				"	152			152	167

01 54 Construction Aids

01 54 16 – Temporary Hoists

01 54 16.50 Weekly Forklift Crew

		Crew	Daily Output	Labor-Hours	Unit	Material	2011 Bare Costs Labor	Equipment	Total	Total Incl O&P
0010	**WEEKLY FORKLIFT CREW**									
0100	All-terrain forklift, 45' lift, 35' reach, 9000 lb. capacity	A-3P	.20	40	Week		1,750	2,225	3,975	5,050

01 54 19 – Temporary Cranes

01 54 19.50 Daily Crane Crews

		Crew	Daily Output	Labor-Hours	Unit	Material	2011 Bare Costs Labor	Equipment	Total	Total Incl O&P
0010	**DAILY CRANE CREWS** for small jobs, portal to portal									
0100	12-ton truck-mounted hydraulic crane	A-3H	1	8	Day		370	870	1,240	1,525
0200	25-ton	A-3I	1	8			370	1,025	1,395	1,675
0300	40-ton	A-3J	1	8			370	1,250	1,620	1,925
0400	55-ton	A-3K	1	16			695	1,750	2,445	2,950
0500	80-ton	A-3L	1	16			695	2,425	3,120	3,700
0600	100-ton	A-3M	1	16	↓		695	2,400	3,095	3,650

01 54 Construction Aids

01 54 19 – Temporary Cranes

01 54 19.60 Monthly Tower Crane Crew	Crew	Daily Output	Labor-Hours	Unit	Material	2011 Bare Costs Labor	Equipment	Total	Total Incl O&P
0010 **MONTHLY TOWER CRANE CREW**, excludes concrete footing									
0100 Static tower crane, 130' high, 106' jib, 6200 lb. capacity	A-3N	.05	176	Month		8,175	21,700	29,875	36,100

01 54 23 – Temporary Scaffolding and Platforms

01 54 23.70 Scaffolding

		Crew	Daily Output	Labor-Hours	Unit	Material	Labor	Equipment	Total	Total Incl O&P
0010	**SCAFFOLDING** R015423-10									
0015	Steel tube, regular, no plank, labor only to erect & dismantle									
0090	Building exterior, wall face, 1 to 5 stories, 6'-4" x 5' frames	3 Carp	8	3	C.S.F.		129		129	198
0200	6 to 12 stories	4 Carp	8	4			172		172	264
0301	13 to 20 stories	5 Clab	8	5			172		172	264
0460	Building interior, wall face area, up to 16' high	3 Carp	12	2			86		86	132
0560	16' to 40' high		10	2.400			103		103	159
0800	Building interior floor area, up to 30' high		150	.160	C.C.F.		6.90		6.90	10.55
0900	Over 30' high	4 Carp	160	.200	"		8.60		8.60	13.20
0906	Complete system for face of walls, no plank, material only rent/mo				C.S.F.	36			36	39.50
0908	Interior spaces, no plank, material only rent/mo				C.C.F.	3.42			3.42	3.76
0910	Steel tubular, heavy duty shoring, buy									
0920	Frames 5' high 2' wide				Ea.	93			93	102
0925	5' high 4' wide					105			105	116
0930	6' high 2' wide					106			106	117
0935	6' high 4' wide					124			124	136
0940	Accessories									
0945	Cross braces				Ea.	18			18	19.80
0950	U-head, 8" x 8"					21.50			21.50	24
0955	J-head, 4" x 8"					15.80			15.80	17.40
0960	Base plate, 8" x 8"					17.60			17.60	19.35
0965	Leveling jack					38			38	41.50
1000	Steel tubular, regular, buy									
1100	Frames 3' high 5' wide				Ea.	71			71	78
1150	5' high 5' wide					83			83	91.50
1200	6'-4" high 5' wide					104			104	114
1350	7'-6" high 6' wide					179			179	197
1500	Accessories cross braces					20.50			20.50	22.50
1550	Guardrail post					18.60			18.60	20.50
1600	Guardrail 7' section					9.90			9.90	10.90
1650	Screw jacks & plates					27			27	29.50
1700	Sidearm brackets					43			43	47.50
1750	8" casters					35.50			35.50	39
1800	Plank 2" x 10" x 16'-0"					51			51	56
1900	Stairway section					325			325	360
1910	Stairway starter bar					36.50			36.50	40
1920	Stairway inside handrail					66			66	72.50
1930	Stairway outside handrail					98			98	108
1940	Walk-thru frame guardrail					47			47	51.50
2000	Steel tubular, regular, rent/mo.									
2100	Frames 3' high 5' wide				Ea.	5			5	5.50
2150	5' high 5' wide					5			5	5.50
2200	6'-4" high 5' wide					5.15			5.15	5.65
2250	7'-6" high 6' wide					7			7	7.70
2500	Accessories, cross braces					1			1	1.10
2550	Guardrail post					1			1	1.10
2600	Guardrail 7' section					1			1	1.10
2650	Screw jacks & plates					2			2	2.20

13

01 54 23 – Temporary Scaffolding and Platforms

01 54 23.70 Scaffolding		Crew	Daily Output	Labor-Hours	Unit	Material	2011 Bare Costs Labor	Equipment	Total	Total Incl O&P
2700	Sidearm brackets				Ea.	2			2	2.20
2750	8" casters					8			8	8.80
2800	Outrigger for rolling tower					3			3	3.30
2850	Plank 2" x 10" x 16'-0"					6			6	6.60
2900	Stairway section					40			40	44
2940	Walk-thru frame guardrail					2.50			2.50	2.75
3000	Steel tubular, heavy duty shoring, rent/mo.									
3250	5' high 2' & 4' wide				Ea.	5			5	5.50
3300	6' high 2' & 4' wide					5			5	5.50
3500	Accessories, cross braces					1			1	1.10
3600	U - head, 8" x 8"					1			1	1.10
3650	J - head, 4" x 8"					1			1	1.10
3700	Base plate, 8" x 8"					1			1	1.10
3750	Leveling jack					2			2	2.20
5700	Planks, 2" x 10" x 16'-0", labor only to erect & remove to 50' H	3 Carp	72	.333			14.35		14.35	22
5800	Over 50' high	4 Carp	80	.400			17.20		17.20	26.50
6000	Heavy duty shoring for elevated slab forms to 8'-2" high, floor area									
6100	Labor only to erect & dismantle	4 Carp	16	2	C.S.F.		86		86	132
6110	Materials only, rent.mo				"	29.50			29.50	32.50
6500	To 14'-8" high									
6600	Labor only to erect & dismantle	4 Carp	10	3.200	C.S.F.		138		138	211
6610	Materials only, rent/mo				"	43			43	47.50

01 54 23.75 Scaffolding Specialties		Crew	Daily Output	Labor-Hours	Unit	Material	2011 Bare Costs Labor	Equipment	Total	Total Incl O&P
0010	**SCAFFOLDING SPECIALTIES**									
1200	Sidewalk bridge, heavy duty steel posts & beams, including									
1210	parapet protection & waterproofing									
1220	8' to 10' wide, 2 posts	3 Carp	15	1.600	L.F.	35	69		104	144
1230	3 posts	"	10	2.400	"	53	103		156	218
1500	Sidewalk bridge using tubular steel									
1510	scaffold frames, including planking	3 Carp	45	.533	L.F.	5.60	23		28.60	41
1600	For 2 uses per month, deduct from all above					50%				
1700	For 1 use every 2 months, add to all above					100%				
1900	Catwalks, 20" wide, no guardrails, 7' span, buy				Ea.	145			145	160
2000	10' span, buy					203			203	223
3720	Putlog, standard, 8' span, with hangers, buy					75			75	82.50
3730	Rent per month					10			10	11
3750	12' span, buy					113			113	124
3755	Rent per month					15			15	16.50
3760	Trussed type, 16' span, buy					260			260	286
3770	Rent per month					20			20	22
3790	22' span, buy					310			310	345
3795	Rent per month					30			30	33
3800	Rolling ladders with handrails, 30" wide, buy, 2 step					237			237	261
4000	7 step					725			725	800
4050	10 step					1,000			1,000	1,100
4100	Rolling towers, buy, 5' wide, 7' long, 10' high					1,350			1,350	1,500
4200	For 5' high added sections, to buy, add					207			207	228
4300	Complete incl. wheels, railings, outriggers,									
4350	21' high, to buy				Ea.	2,250			2,250	2,475
4400	Rent/month = 5% of purchase cost				"	113			113	124

01 54 Construction Aids
01 54 36 – Equipment Mobilization

01 54 36.50 Mobilization		Crew	Daily Output	Labor-Hours	Unit	Material	2011 Bare Costs Labor	2011 Bare Costs Equipment	Total	Total Incl O&P
0010	**MOBILIZATION** (Use line item again for demobilization) R015433-10									
0015	Up to 25 mi haul dist (50 mi RT for mob/demob crew)									
0020	Dozer, loader, backhoe, excav., grader, paver, roller, 70 to 150 H.P.	B-34N	4	2	Ea.		69	116	185	233
0100	Above 150 HP	B-34K	3	2.667			91.50	259	350.50	425
0900	Shovel or dragline, 3/4 C.Y.	"	3.60	2.222			76.50	216	292.50	355
1100	Small equipment, placed in rear of, or towed by pickup truck	A-3A	8	1			33.50	14.60	48.10	67
1150	Equip up to 70 HP, on flatbed trailer behind pickup truck	A-3D	4	2			67	57	124	165
2000	Crane, truck-mounted, up to 75 ton, (driver only, one-way)	1 Eqhv	7.20	1.111			51.50		51.50	77
2100	Crane, truck-mounted, over 75 ton	A-3E	2.50	6.400			259	46.50	305.50	440
2200	Crawler-mounted, up to 75 ton	A-3F	2	8			325	400	725	920
2300	Over 75 ton	A-3G	1.50	10.667	↓		430	595	1,025	1,300
2500	For each additional 5 miles haul distance, add						10%	10%		
3000	For large pieces of equipment, allow for assembly/knockdown									
3100	For mob/demob of micro-tunneling equip, see Section 33 05 23.19									

01 54 39 – Construction Equipment
01 54 39.70 Small Tools

01 54 39.70 Small Tools		Crew	Daily Output	Labor-Hours	Unit	Material	2011 Bare Costs Labor	2011 Bare Costs Equipment	Total	Total Incl O&P
0010	**SMALL TOOLS** R013113-50									
0020	As % of contractor's bare labor cost for project, minimum				Total		.50%			
0100	Maximum				"		2%			

01 55 Vehicular Access and Parking
01 55 23 – Temporary Roads
01 55 23.50 Roads and Sidewalks

01 55 23.50 Roads and Sidewalks		Crew	Daily Output	Labor-Hours	Unit	Material	2011 Bare Costs Labor	2011 Bare Costs Equipment	Total	Total Incl O&P
0010	**ROADS AND SIDEWALKS** Temporary									
0050	Roads, gravel fill, no surfacing, 4" gravel depth	B-14	715	.067	S.Y.	4	2.43	.45	6.88	8.60
0100	8" gravel depth	"	615	.078	"	8	2.83	.52	11.35	13.70
1000	Ramp, 3/4" plywood on 2" x 6" joists, 16" O.C.	2 Carp	300	.053	S.F.	1.34	2.30		3.64	5
1100	On 2" x 10" joists, 16" O.C.	"	275	.058	"	1.92	2.50		4.42	5.95

01 56 Temporary Barriers and Enclosures
01 56 13 – Temporary Air Barriers
01 56 13.60 Tarpaulins

01 56 13.60 Tarpaulins		Crew	Daily Output	Labor-Hours	Unit	Material	2011 Bare Costs Labor	2011 Bare Costs Equipment	Total	Total Incl O&P
0010	**TARPAULINS**									
0020	Cotton duck, 10 oz. to 13.13 oz. per S.Y., minimum				S.F.	.48			.48	.53
0050	Maximum					.60			.60	.66
0100	Polyvinyl coated nylon, 14 oz. to 18 oz., minimum					.48			.48	.53
0150	Maximum					.68			.68	.75
0200	Reinforced polyethylene 3 mils thick, white					.15			.15	.17
0300	4 mils thick, white, clear or black					.20			.20	.22
0400	5.5 mils thick, clear					.30			.30	.33
0500	White, fire retardant					.35			.35	.39
0600	7.5 mils, oil resistant, fire retardant					.40			.40	.44
0700	8.5 mils, black					.53			.53	.58
0710	Woven polyethylene, 6 mils thick					.35			.35	.39
0730	Polyester reinforced w/integral fastening system 11 mils thick					1.07			1.07	1.18
0740	Mylar polyester, non-reinforced, 7 mils thick				↓	1.17			1.17	1.29

01 56 Temporary Barriers and Enclosures

01 56 13 – Temporary Air Barriers

01 56 13.90 Winter Protection

		Crew	Daily Output	Labor-Hours	Unit	Material	2011 Bare Costs Labor	Equipment	Total	Total Incl O&P
0010	**WINTER PROTECTION**									
0100	Framing to close openings	2 Clab	750	.021	S.F.	.39	.73		1.12	1.55
0200	Tarpaulins hung over scaffolding, 8 uses, not incl. scaffolding		1500	.011		.25	.37		.62	.84
0250	Tarpaulin polyester reinf. w/integral fastening system 11 mils thick		1600	.010		.80	.34		1.14	1.41
0300	Prefab fiberglass panels, steel frame, 8 uses	↓	1200	.013	↓	.85	.46		1.31	1.64

01 56 16 – Temporary Dust Barriers

01 56 16.10 Dust Barriers, Temporary

		Crew	Daily Output	Labor-Hours	Unit	Material	2011 Bare Costs Labor	Equipment	Total	Total Incl O&P
0010	**DUST BARRIERS, TEMPORARY**									
0020	Spring loaded telescoping pole & head, to 12', erect and dismantle	1 Clab	240	.033	Ea.		1.14		1.14	1.76
0025	Cost per day (based upon 250 days)				Day	.22			.22	.24
0030	To 21', erect and dismantle	1 Clab	240	.033	Ea.		1.14		1.14	1.76
0035	Cost per day (based upon 250 days)				Day	.39			.39	.42
0040	Accessories, caution tape reel, erect and dismantle	1 Clab	480	.017	Ea.		.57		.57	.88
0045	Cost per day (based upon 250 days)				Day	.05			.05	.05
0060	Foam rail and connector, erect and dismantle	1 Clab	240	.033	Ea.		1.14		1.14	1.76
0065	Cost per day (based upon 250 days)				Day	.10			.10	.11
0070	Caution tape	1 Clab	384	.021	C.L.F.	2.67	.72		3.39	4.03
0080	Zipper, standard duty		60	.133	Ea.	8	4.58		12.58	15.85
0090	Heavy duty		48	.167	"	9.50	5.75		15.25	19.25
0100	Polyethylene sheet, 4 mil		37	.216	Sq.	3.34	7.45		10.79	15.05
0110	6 mil	↓	37	.216	"	5.25	7.45		12.70	17.15
1000	Dust partition, 6 mil polyethylene, 1" x 3" frame	2 Carp	2000	.008	S.F.	.28	.34		.62	.84
1080	2" x 4" frame	"	2000	.008	"	.30	.34		.64	.86

01 56 23 – Temporary Barricades

01 56 23.10 Barricades

		Crew	Daily Output	Labor-Hours	Unit	Material	2011 Bare Costs Labor	Equipment	Total	Total Incl O&P
0010	**BARRICADES**									
0020	5' high, 3 rail @ 2" x 8", fixed	2 Carp	20	.800	L.F.	4.93	34.50		39.43	58.50
0150	Movable		30	.533		4.10	23		27.10	39.50
1000	Guardrail, wooden, 3' high, 1" x 6", on 2" x 4" posts		200	.080		1.11	3.44		4.55	6.50
1100	2" x 6", on 4" x 4" posts	↓	165	.097		2.08	4.17		6.25	8.70
1200	Portable metal with base pads, buy					19.30			19.30	21
1250	Typical installation, assume 10 reuses	2 Carp	600	.027	↓	2.50	1.15		3.65	4.51
1300	Barricade tape, polyethylene, 7 mil, 3" wide x 500' long roll				Ea.	25			25	27.50

01 56 26 – Temporary Fencing

01 56 26.50 Temporary Fencing

		Crew	Daily Output	Labor-Hours	Unit	Material	2011 Bare Costs Labor	Equipment	Total	Total Incl O&P
0010	**TEMPORARY FENCING**									
0020	Chain link, 11 ga, 5' high	2 Clab	400	.040	L.F.	4.50	1.37		5.87	7.05
0100	6' high		300	.053		5.30	1.83		7.13	8.65
0200	Rented chain link, 6' high, to 1000' (up to 12 mo.)		400	.040		6.30	1.37		7.67	9.05
0250	Over 1000' (up to 12 mo.)	↓	300	.053		3.29	1.83		5.12	6.45
0350	Plywood, painted, 2" x 4" frame, 4' high	A-4	135	.178		5.60	7.35		12.95	17.30
0400	4" x 4" frame, 8' high	"	110	.218		10.45	9		19.45	25
0500	Wire mesh on 4" x 4" posts, 4' high	2 Carp	100	.160		8.55	6.90		15.45	19.95
0550	8' high	"	80	.200	↓	13	8.60		21.60	27.50

01 56 29 – Temporary Protective Walkways

01 56 29.50 Protection

		Crew	Daily Output	Labor-Hours	Unit	Material	2011 Bare Costs Labor	Equipment	Total	Total Incl O&P
0010	**PROTECTION**									
0020	Stair tread, 2" x 12" planks, 1 use	1 Carp	75	.107	Tread	4.08	4.59		8.67	11.55
0100	Exterior plywood, 1/2" thick, 1 use		65	.123		1.74	5.30		7.04	10.05
0200	3/4" thick, 1 use	↓	60	.133	↓	2.45	5.75		8.20	11.50

01 56 Temporary Barriers and Enclosures

01 56 29 – Temporary Protective Walkways

01 56 29.50 Protection	Crew	Daily Output	Labor-Hours	Unit	Material	2011 Bare Costs Labor	Equipment	Total	Total Incl O&P
2200	Sidewalks, 2" x 12" planks, 2 uses	1 Carp	350	.023	S.F.	.68	.98		1.66	2.26
2300	Exterior plywood, 2 uses, 1/2" thick		750	.011		.29	.46		.75	1.02
2400	5/8" thick		650	.012		.35	.53		.88	1.19
2500	3/4" thick		600	.013		.41	.57		.98	1.33

01 58 Project Identification

01 58 13 – Temporary Project Signage

01 58 13.50 Signs

		Crew	Daily Output	Labor-Hours	Unit	Material	Labor	Equipment	Total	Total Incl O&P
0010	**SIGNS**									
0020	High intensity reflectorized, no posts, buy				S.F.	26.50			26.50	29.50

01 74 Cleaning and Waste Management

01 74 13 – Progress Cleaning

01 74 13.20 Cleaning Up

		Crew	Daily Output	Labor-Hours	Unit	Material	Labor	Equipment	Total	Total Incl O&P
0010	**CLEANING UP**									
0020	After job completion, allow, minimum				Job					.30%
0040	Maximum				"					1%
0050	Cleanup of floor area, continuous, per day, during const.	A-5	24	.750	M.S.F.	1.70	25.50	1.79	28.99	43.50
0100	Final by GC at end of job	"	11.50	1.565	"	2.71	53.50	3.73	59.94	89
1000	Mechanical demolition, see Section 02 41 19									

01 91 Commissioning

01 91 13 – General Commissioning Requirements

01 91 13.50 Building Commissioning

		Crew	Daily Output	Labor-Hours	Unit	Material	Labor	Equipment	Total	Total Incl O&P
0010	**BUILDING COMMISSIONING**									
0100	Basic building commissioning, minimum				%					.25%
0150	Maximum									.50%
0200	Enhanced building commissioning, minimum									.50%
0250	Maximum									1%

01 93 Facility Maintenance

01 93 13 – Facility Maintenance Procedures

01 93 13.15 Mechanical Facilities Maintenance

		Crew	Daily Output	Labor-Hours	Unit	Material	Labor	Equipment	Total	Total Incl O&P
0010	**MECHANICAL FACILITIES MAINTENANCE**									
0100	Air conditioning system maintenance									
0130	Belt, replace	1 Stpi	15	.533	Ea.		29		29	43
0170	Fan, clean		16	.500			27		27	40.50
0180	Filter, remove, clean, replace		12	.667			36		36	54
0190	Flexible coupling alignment, inspect		40	.200			10.80		10.80	16.15
0200	Gas leak locate and repair		4	2			108		108	162
0250	Pump packing gland, remove and replace		11	.727			39.50		39.50	58.50
0270	Tighten		32	.250			13.55		13.55	20
0290	Pump disassemble and assemble		4	2			108		108	162
0300	Air pressure regulator, disassemble, clean, assemble	1 Skwk	4	2			89		89	136
0310	Repair or replace part	"	6	1.333			59		59	91
0320	Purging system	1 Stpi	16	.500			27		27	40.50

01 93 13.15 Mechanical Facilities Maintenance		Crew	Daily Output	Labor-Hours	Unit	Material	2011 Bare Costs Labor	2011 Bare Costs Equipment	Total	Total Incl O&P
0400	Compressor, air, remove or install fan wheel	1 Skwk	20	.400	Ea.		17.75		17.75	27
0410	Disassemble or assemble 2 cylinder, 2 stage		4	2			89		89	136
0420	4 cylinder, 4 stage		1	8			355		355	545
0430	Repair or replace part		2	4			178		178	272
0700	Demolition, for mech. demolition see Section 23 05 05.10 or 22 05 05.10									
0800	Ductwork, clean									
0810	Rectangular									
0820	6" [G]	1 Shee	187.50	.043	L.F.		2.20		2.20	3.33
0830	8" [G]		140.63	.057			2.94		2.94	4.44
0840	10" [G]		112.50	.071			3.67		3.67	5.55
0850	12" [G]		93.75	.085			4.41		4.41	6.65
0860	14" [G]		80.36	.100			5.15		5.15	7.75
0870	16" [G]		70.31	.114			5.90		5.90	8.90
0900	Round									
0910	4" [G]	1 Shee	358.10	.022	L.F.		1.15		1.15	1.74
0920	6" [G]		238.73	.034			1.73		1.73	2.62
0930	8" [G]		179.05	.045			2.31		2.31	3.49
0940	10" [G]		143.24	.056			2.88		2.88	4.36
0950	12" [G]		119.37	.067			3.46		3.46	5.25
0960	16" [G]		89.52	.089			4.62		4.62	7
1000	Expansion joint, not screwed, install or remove	1 Stpi	3	2.667	Ea.		144		144	215
1010	Repack	"	6	1.333	"		72		72	108
1200	Fire protection equipment									
1220	Fire hydrant, replace	Q-1	3	5.333	Ea.		257		257	385
1230	Service, lubricate, inspect, flush, clean	1 Plum	7	1.143			61.50		61.50	91.50
1240	Test	"	11	.727			39		39	58
1310	Inspect valves, pressure, nozzle,	1 Spri	4	2	System		106		106	159
1500	Instrumentation preventive maintenance, rule of thumb									
1510	Annual cost per instrument for calibration & trouble shooting									
1520	Includes flow meters, analytical transmitters, flow control									
1530	chemical feed systems, recorders and distrib. control sys.				Ea.					325
1800	Plumbing fixtures, for installation see Section 22 41 00									
1801	Plumbing fixtures									
1840	Open drain with snake	1 Plum	13	.615	Ea.		33		33	49
1850	Open drain with toilet auger		16	.500			27		27	40
1860	Grease trap, clean		2	4			214		214	320
1870	Plaster trap, clean		6	1.333			71.50		71.50	107
1881	Clean commode	1 Clab	20	.400		.47	13.75		14.22	21.50
1882	Clean commode seat		48	.167		.24	5.75		5.99	9.05
1883	Clean single sink		28	.286		.31	9.80		10.11	15.40
1884	Clean double sink		20	.400		.47	13.75		14.22	21.50
1885	Clean lavatory		28	.286		.31	9.80		10.11	15.40
1886	Clean bathtub		12	.667		.95	23		23.95	36
1887	Clean fiberglass shower		12	.667		.95	23		23.95	36
1888	Clean fiberglass tub/shower		8	1		1.18	34.50		35.68	54
1889	Clean faucet set		32	.250		.19	8.60		8.79	13.40
1891	Clean shower head		80	.100		.02	3.44		3.46	5.30
1893	Clean water heater		16	.500		1.65	17.20		18.85	28.50
1900	Relief valve, test and adjust	1 Stpi	20	.400			21.50		21.50	32.50
1910	Clean pump, heater, or motor for whirlpool	1 Clab	16	.500		.24	17.20		17.44	27
1920	Clean thermal cover for whirlpool	"	16	.500		2.36	17.20		19.56	29
2000	Repair or replace-steam trap	1 Stpi	8	1			54		54	81
2020	Y-type or bell strainer		6	1.333			72		72	108

01 93 Facility Maintenance

01 93 13 – Facility Maintenance Procedures

01 93 13.15 Mechanical Facilities Maintenance		Crew	Daily Output	Labor-Hours	Unit	Material	2011 Bare Costs Labor	Equipment	Total	Total Incl O&P
2040	Water trap or vacuum breaker, screwed joints	1 Stpi	13	.615	Ea.		33.50		33.50	49.50
2100	Steam specialties, clean									
2120	Air separator with automatic trap, 1" fittings	1 Stpi	12	.667	Ea.		36		36	54
2130	Bucket trap, 2" pipe		7	1.143			62		62	92.50
2150	Drip leg, 2" fitting		45	.178			9.60		9.60	14.35
2200	Thermodynamic trap, 1" fittings		50	.160			8.65		8.65	12.90
2210	Thermostatic		65	.123			6.65		6.65	9.95
2240	Screen and seat in y-type strainer, plug type.		25	.320			17.30		17.30	26
2242	Screen and seat in y-type strainer, flange type.		12	.667			36		36	54
2500	Valve, replace broken handwheel	↓	24	.333	↓		18.05		18.05	27
3000	Valve, overhaul, regulator, relief, flushometer, mixing									
3040	Cold water, gas	1 Stpi	5	1.600	Ea.		86.50		86.50	129
3050	Hot water, steam		3	2.667			144		144	215
3080	Globe, gate, check up to 4" cold water, gas		10	.800			43.50		43.50	64.50
3090	Hot water, steam		5	1.600			86.50		86.50	129
3100	Over 4" ID hot or cold line	↓	1.40	5.714			310		310	460
3120	Remove and replace, gate, globe or check up to 4"	Q-5	6	2.667			130		130	194
3130	Over 4"	"	2	8			390		390	580
3150	Repack up to 4"	1 Stpi	13	.615			33.50		33.50	49.50
3160	Over 4"	"	4	2	↓		108		108	162

Division Notes

	CREW	DAILY OUTPUT	LABOR-HOURS	UNIT	2011 BARE COSTS				TOTAL INCL O&P
					MAT.	LABOR	EQUIP.	TOTAL	

Estimating Tips

02 30 00 Subsurface Investigation

In preparing estimates on structures involving earthwork or foundations, all information concerning soil characteristics should be obtained. Look particularly for hazardous waste, evidence of prior dumping of debris, and previous stream beds.

02 41 00 Demolition and Structure Moving

The costs shown for selective demolition do not include rubbish handling or disposal. These items should be estimated separately using RSMeans data or other sources.

- Historic preservation often requires that the contractor remove materials from the existing structure, rehab them, and replace them. The estimator must be aware of any related measures and precautions that must be taken when doing selective demolition and cutting and patching. Requirements may include special handling and storage, as well as security.

- In addition to Subdivision 02 41 00, you can find selective demolition items in each division. Example: Roofing demolition is in Division 7.

02 42 10 Building Deconstruction

This section provides costs for the careful dismantling and recycling of most of low-rise building materials.

02 56 13 Containment of Hazardous Waste

This section addresses on-site hazardous waste disposal costs.

02 80 00 Hazardous Material Disposal/ Remediation

This subdivision includes information on hazardous waste handling, asbestos remediation, lead remediation, and mold remediation. See reference R028213-20 and R028319-60 for further guidance in using these unit price lines.

02 91 10 Monitoring Chemical Sampling, Testing Analysis

This section provides costs for on-site sampling and testing hazardous waste.

Reference Numbers

Reference numbers are shown in shaded boxes at the beginning of some major classifications. These numbers refer to related items in the Reference Section. The reference information may be an estimating procedure, an alternate pricing method, or technical information.

Note: Not all subdivisions listed here necessarily appear in this publication.

Division 2 - Existing Conditions

02 41 Demolition

02 41 13 – Selective Site Demolition

02 41 13.17 Demolish, Remove Pavement and Curb		Crew	Daily Output	Labor-Hours	Unit	Material	2011 Bare Costs		Total	Total Incl O&P
							Labor	Equipment		
0010	**DEMOLISH, REMOVE PAVEMENT AND CURB**	R024119-10								
5010	Pavement removal, bituminous roads, 3" thick	B-38	690	.058	S.Y.		2.25	1.65	3.90	5.20
5050	4" to 6" thick		420	.095			3.70	2.71	6.41	8.60
5100	Bituminous driveways		640	.063			2.43	1.78	4.21	5.65
5200	Concrete to 6" thick, hydraulic hammer, mesh reinforced		255	.157			6.10	4.46	10.56	14.10
5300	Rod reinforced		200	.200	↓		7.75	5.70	13.45	18
5400	Concrete, 7" to 24" thick, plain		33	1.212	C.Y.		47	34.50	81.50	110
5500	Reinforced	↓	24	1.667	"		64.50	47.50	112	150
5600	With hand held air equipment, bituminous, to 6" thick	B-39	1900	.025	S.F.		.92	.10	1.02	1.51
5700	Concrete to 6" thick, no reinforcing		1600	.030			1.09	.12	1.21	1.79
5800	Mesh reinforced		1400	.034			1.24	.14	1.38	2.05
5900	Rod reinforced	↓	765	.063	↓		2.27	.25	2.52	3.75

02 41 13.33 Minor Site Demolition

		Crew	Daily Output	Labor-Hours	Unit	Material	Labor	Equipment	Total	Total Incl O&P
0010	**MINOR SITE DEMOLITION**	R024119-10								
0015	No hauling, abandon catch basin or manhole	B-6	7	3.429	Ea.		128	46	174	246
0020	Remove existing catch basin or manhole, masonry		4	6			225	80	305	430
0030	Catch basin or manhole frames and covers, stored		13	1.846			69	24.50	93.50	132
0040	Remove and reset	↓	7	3.429			128	46	174	246
0900	Hydrants, fire, remove only	B-21A	5	8			340	87	427	610
0950	Remove and reset	"	2	20	↓		855	217	1,072	1,550
4000	Sidewalk removal, bituminous, 2-1/2" thick	B-6	325	.074	S.Y.		2.76	.99	3.75	5.30
4050	Brick, set in mortar		185	.130			4.86	1.73	6.59	9.30
4100	Concrete, plain, 4"		160	.150			5.60	2	7.60	10.75
4200	Mesh reinforced	↓	150	.160	↓		6	2.14	8.14	11.45

02 41 19 – Selective Demolition

02 41 19.13 Selective Building Demolition

		Crew	Daily Output	Labor-Hours	Unit	Material	Labor	Equipment	Total	Total Incl O&P
0010	**SELECTIVE BUILDING DEMOLITION**									
0020	Costs related to selective demolition of specific building components									
0025	are included under Common Work Results (XX 05)									
0030	in the component's appropriate division.									

02 41 19.19 Selective Demolition, Dump Charges

		Crew	Daily Output	Labor-Hours	Unit	Material	Labor	Equipment	Total	Total Incl O&P
0010	**SELECTIVE DEMOLITION, DUMP CHARGES**	R024119-10								
0020	Dump charges, typical urban city, tipping fees only									
0100	Building construction materials				Ton	82			82	90
0200	Trees, brush, lumber					70			70	77
0300	Rubbish only					70			70	77
0500	Reclamation station, usual charge				↓	82			82	90

02 41 19.23 Selective Demolition, Rubbish Handling

		Crew	Daily Output	Labor-Hours	Unit	Material	Labor	Equipment	Total	Total Incl O&P
0010	**SELECTIVE DEMOLITION, RUBBISH HANDLING**	R024119-10								
0020	The following are to be added to the demolition prices									
0400	Chute, circular, prefabricated steel, 18" diameter	B-1	40	.600	L.F.	53	21		74	90.50
0440	30" diameter	"	30	.800	"	47.50	28		75.50	95.50
0600	Dumpster, weekly rental, 1 dump/week, 6 C.Y. capacity (2 Tons)				Week	300			300	330
0700	10 C.Y. capacity (4 Tons)					475			475	525
0725	20 C.Y. capacity (8 Tons)	R024119-20				630			630	695
0800	30 C.Y. capacity (10 Tons)					750			750	825
0840	40 C.Y. capacity (13 Tons)				↓	875			875	965
2000	Load, haul, dump and return, 50' haul, hand carried	2 Clab	24	.667	C.Y.		23		23	35
2005	Wheeled		37	.432			14.85		14.85	23
2040	51' to 100' haul, hand carried		16.50	.970			33.50		33.50	51
2045	Wheeled	↓	25	.640	↓		22		22	33.50

02 41 Demolition

02 41 19 – Selective Demolition

02 41 19.23 Selective Demolition, Rubbish Handling

		Crew	Daily Output	Labor-Hours	Unit	Material	2011 Bare Costs Labor	2011 Bare Costs Equipment	Total	Total Incl O&P
2080	Over 100' haul, add per 100 L.F., hand carried	2 Clab	35.50	.451	C.Y.		15.50		15.50	24
2085	Wheeled		54	.296			10.20		10.20	15.60
2120	In elevators, per 10 floors, add		140	.114			3.93		3.93	6
2130	Load, haul, dump and return, up to 50' haul, incl. up to 5 rsr stair, hand		23	.696			24		24	36.50
2135	Wheeled		35	.457			15.70		15.70	24
2140	6 – 10 riser stairs, hand carried		22	.727			25		25	38.50
2145	Wheeled		34	.471			16.15		16.15	25
2150	11 – 20 riser stairs, hand carried		20	.800			27.50		27.50	42
2155	Wheeled		31	.516			17.75		17.75	27
2160	21 – 40 riser stairs, hand carried		16	1			34.50		34.50	52.50
2165	Wheeled		24	.667			23		23	35
2170	100' haul, incl. 5 riser stair, hand carried		15	1.067			36.50		36.50	56
2175	Wheeled		23	.696			24		24	36.50
2180	6 – 10 riser stair, hand carried		14	1.143			39.50		39.50	60
2185	Wheeled		21	.762			26		26	40
2190	11 – 20 riser stair, hand carried		12	1.333			46		46	70.50
2195	Wheeled		18	.889			30.50		30.50	47
2200	21 – 40 riser stair, hand carried		8	2			68.50		68.50	105
2205	Wheeled		12	1.333			46		46	70.50
2210	Over 100' haul, add per 100 L.F., hand carried		35.50	.451			15.50		15.50	24
2215	Wheeled		54	.296			10.20		10.20	15.60
2220	For each additional flight of stairs, up to 5 risers, add		550	.029	Flight		1		1	1.53
2225	6 – 10 risers, add		275	.058			2		2	3.07
2230	11 – 20 risers, add		138	.116			3.98		3.98	6.10
2235	21 – 40 risers, add		69	.232			7.95		7.95	12.20
3000	Loading & trucking, including 2 mile haul, chute loaded	B-16	45	.711	C.Y.		25	11.25	36.25	50.50
3040	Hand loading truck, 50' haul	"	48	.667			23	10.55	33.55	47
3080	Machine loading truck	B-17	120	.267			9.80	5.30	15.10	20.50
5000	Haul, per mile, up to 8 C.Y. truck	B-34B	1165	.007			.24	.43	.67	.84
5100	Over 8 C.Y. truck	"	1550	.005			.18	.33	.51	.63

02 41 19.27 Selective Demolition, Torch Cutting

		Crew	Daily Output	Labor-Hours	Unit	Material	2011 Bare Costs Labor	2011 Bare Costs Equipment	Total	Total Incl O&P
0010	**SELECTIVE DEMOLITION, TORCH CUTTING** R024119-10									
0020	Steel, 1" thick plate	1 Clab	360	.022	L.F.	.24	.76		1	1.44
0040	1" diameter bar	"	210	.038	Ea.		1.31		1.31	2.01
1000	Oxygen lance cutting, reinforced concrete walls									
1040	12" to 16" thick walls	1 Clab	10	.800	L.F.		27.50		27.50	42
1080	24" thick walls	"	6	1.333	"		46		46	70.50

02 65 Underground Storage Tank Removal

02 65 10 – Underground Tank and Contaminated Soil Removal

02 65 10.30 Removal of Underground Storage Tanks

			Crew	Daily Output	Labor-Hours	Unit	Material	2011 Bare Costs Labor	2011 Bare Costs Equipment	Total	Total Incl O&P
0010	**REMOVAL OF UNDERGROUND STORAGE TANKS** R026510-20										
0011	Petroleum storage tanks, non-leaking										
0100	Excavate & load onto trailer										
0110	3000 gal. to 5000 gal. tank	G	B-14	4	12	Ea.		435	80	515	755
0120	6000 gal. to 8000 gal. tank	G	B-3A	3	13.333			485	315	800	1,100
0130	9000 gal. to 12000 gal. tank	G	"	2	20			730	475	1,205	1,650
0190	Known leaking tank, add					%				100%	100%
0200	Remove sludge, water and remaining product from tank bottom										
0201	of tank with vacuum truck										
0300	3000 gal. to 5000 gal. tank	G	A-13	5	1.600	Ea.		70	147	217	266

02 65 Underground Storage Tank Removal

02 65 10 – Underground Tank and Contaminated Soil Removal

02 65 10.30 Removal of Underground Storage Tanks	Crew	Daily Output	Labor-Hours	Unit	Material	2011 Bare Costs Labor	Equipment	Total	Total Incl O&P
0310	6000 gal. to 8000 gal. tank G	A-13	4	2	Ea.		87	184	271	330
0320	9000 gal. to 12000 gal. tank G	↓	3	2.667	↓		116	245	361	445
0390	Dispose of sludge off-site, average				Gal.				6	6.60
0400	Insert inert solid CO_2 "dry ice" into tank									
0401	For cleaning/transporting tanks (1.5 lb./100 gal. cap) G	1 Clab	500	.016	Lb.	1.87	.55		2.42	2.90
1020	Haul tank to certified salvage dump, 100 miles round trip									
1023	3000 gal. to 5000 gal. tank				Ea.				710	780
1026	6000 gal. to 8000 gal. tank								810	890
1029	9,000 gal. to 12,000 gal. tank				↓				950	1,050
1100	Disposal of contaminated soil to landfill									
1110	Minimum				C.Y.				125	140
1111	Maximum				"				400	440
1120	Disposal of contaminated soil to									
1121	bituminous concrete batch plant									
1130	Minimum				C.Y.				70	77
1131	Maximum				"				110	120
2010	Decontamination of soil on site incl poly tarp on top/bottom									
2011	Soil containment berm, and chemical treatment									
2020	Minimum G	B-11C	100	.160	C.Y.	7.05	6.40	3.20	16.65	21
2021	Maximum G	"	100	.160	↓	9.15	6.40	3.20	18.75	23.50
2050	Disposal of decontaminated soil, minimum								125	140
2055	Maximum				↓				400	440

02 82 Asbestos Remediation

02 82 13 – Asbestos Abatement

02 82 13.39 Asbestos Remediation Plans and Methods

		Crew	Daily Output	Labor-Hours	Unit	Material	2011 Bare Costs Labor	Equipment	Total	Total Incl O&P
0010	**ASBESTOS REMEDIATION PLANS AND METHODS**									
0100	Building Survey-Commercial Building				Ea.				2,000	2,200
0200	Asbestos Abatement Remediation Plan				"				1,200	1,325

02 82 13.41 Asbestos Abatement Equip.

		Crew	Daily Output	Labor-Hours	Unit	Material	2011 Bare Costs Labor	Equipment	Total	Total Incl O&P
0010	**ASBESTOS ABATEMENT EQUIP.** R028213-20									
0011	Equipment and supplies, buy									
0200	Air filtration device, 2000 C.F.M.				Ea.	870			870	955
0250	Large volume air sampling pump, minimum					355			355	390
0260	Maximum					520			520	570
0300	Airless sprayer unit, 2 gun					4,800			4,800	5,300
0350	Light stand, 500 watt				↓	46			46	50.50
0400	Personal respirators									
0410	Negative pressure, 1/2 face, dual operation, min.				Ea.	25.50			25.50	28
0420	Maximum					31.50			31.50	35
0450	P.A.P.R., full face, minimum					107			107	118
0460	Maximum					195			195	215
0470	Supplied air, full face, incl. air line, minimum					187			187	206
0480	Maximum					385			385	425
0500	Personnel sampling pump, minimum					221			221	244
1500	Power panel, 20 unit, incl. G.F.I.					550			550	610
1600	Shower unit, including pump and filters					1,175			1,175	1,275
1700	Supplied air system (type C)					5,325			5,325	5,875
1750	Vacuum cleaner, HEPA, 16 gal., stainless steel, wet/dry					1,175			1,175	1,300
1760	55 gallon					1,925			1,925	2,100
1800	Vacuum loader, 9 – 18 ton/hr.				↓	90,000			90,000	99,000

02 82 Asbestos Remediation

02 82 13 – Asbestos Abatement

02 82 13.41 Asbestos Abatement Equip.

		Crew	Daily Output	Labor-Hours	Unit	Material	2011 Bare Costs Labor	Equipment	Total	Total Incl O&P
1900	Water atomizer unit, including 55 gal. drum				Ea.	230			230	253
2000	Worker protection, whole body, foot, head cover & gloves, plastic					8.70			8.70	9.60
2500	Respirator, single use					29			29	32
2550	Cartridge for respirator					6.85			6.85	7.55
2570	Glove bag, 7 mil, 50" x 64"					17.25			17.25	19
2580	10 mil, 44" x 60"					5.40			5.40	5.95
2590	6 mil, 44" x 60"					5.40			5.40	5.95
3000	HEPA vacuum for work area, minimum					1,375			1,375	1,500
3050	Maximum					1,425			1,425	1,575
6000	Disposable polyethylene bags, 6 mil, 3 C.F.					2.30			2.30	2.53
6300	Disposable fiber drums, 3 C.F.					16.90			16.90	18.60
6400	Pressure sensitive caution labels, 3" x 5"					3.16			3.16	3.48
6450	11" x 17"					7.05			7.05	7.75
6500	Negative air machine, 1800 C.F.M.					820			820	905

02 82 13.42 Preparation of Asbestos Containment Area

		Crew	Daily Output	Labor-Hours	Unit	Material	2011 Bare Costs Labor	Equipment	Total	Total Incl O&P
0010	**PREPARATION OF ASBESTOS CONTAINMENT AREA** R028213-20									
0100	Pre-cleaning, HEPA vacuum and wet wipe, flat surfaces	A-9	12000	.005	S.F.	.02	.25		.27	.41
0200	Protect carpeted area, 2 layers 6 mil poly on 3/4" plywood	"	1000	.064		2	3.01		5.01	6.85
0300	Separation barrier, 2" x 4" @ 16", 1/2" plywood ea. side, 8' high	2 Carp	400	.040		3	1.72		4.72	5.95
0310	12' high		320	.050		3	2.15		5.15	6.60
0320	16' high		200	.080		2	3.44		5.44	7.50
0400	Personnel decontam. chamber, 2" x 4" @ 16", 3/4" ply ea. side		280	.057		4	2.46		6.46	8.15
0450	Waste decontam. chamber, 2" x 4" studs @ 16", 3/4" ply ea. side		360	.044		4	1.91		5.91	7.35
0500	Cover surfaces with polyethylene sheeting									
0501	Including glue and tape									
0550	Floors, each layer, 6 mil	A-9	8000	.008	S.F.	.05	.38		.43	.64
0551	4 mil		9000	.007		.04	.33		.37	.56
0560	Walls, each layer, 6 mil		6000	.011		.05	.50		.55	.83
0561	4 mil		7000	.009		.04	.43		.47	.70
0570	For heights above 12', add						20%			
0575	For heights above 20', add						30%			
0580	For fire retardant poly, add					100%				
0590	For large open areas, deduct					10%	20%			
0600	Seal floor penetrations with foam firestop to 36 sq. in.	2 Carp	200	.080	Ea.	7.60	3.44		11.04	13.70
0610	36 sq. in. to 72 sq. in.		125	.128		15.25	5.50		20.75	25
0615	72 sq. in. to 144 sq. in.		80	.200		30.50	8.60		39.10	46.50
0620	Wall penetrations, to 36 square inches		180	.089		7.60	3.83		11.43	14.25
0630	36 sq. in. to 72 sq. in.		100	.160		15.25	6.90		22.15	27.50
0640	72 sq. in. to 144 sq. in.		60	.267		30.50	11.50		42	51
0800	Caulk seams with latex	1 Carp	230	.035	L.F.	.15	1.50		1.65	2.47
0900	Set up neg. air machine, 1-2k C.F.M./25 M.C.F. volume	1 Asbe	4.30	1.860	Ea.		87.50		87.50	135

02 82 13.43 Bulk Asbestos Removal

		Crew	Daily Output	Labor-Hours	Unit	Material	2011 Bare Costs Labor	Equipment	Total	Total Incl O&P
0010	**BULK ASBESTOS REMOVAL**									
0020	Includes disposable tools and 2 suits and 1 respirator filter/day/worker									
0200	Boiler insulation	A-9	480	.133	S.F.	.48	6.25		6.73	10.20
0210	With metal lath, add				%		50%			
0300	Boiler breeching or flue insulation	A-9	520	.123	S.F.	.37	5.80		6.17	9.35
0310	For active boiler, add				%		100%			
0400	Duct or AHU insulation	A-10B	440	.073	S.F.	.22	3.42		3.64	5.55
0500	Duct vibration isolation joints, up to 24 sq. in. duct	A-9	56	1.143	Ea.	3.47	53.50		56.97	87
0520	25 sq. in. to 48 sq. in. duct		48	1.333		4.05	62.50		66.55	101
0530	49 sq. in. to 76 sq. in. duct		40	1.600		4.86	75		79.86	121

02 82 Asbestos Remediation

02 82 13 – Asbestos Abatement

02 82 13.43 Bulk Asbestos Removal

		Crew	Daily Output	Labor-Hours	Unit	Material	2011 Bare Costs Labor	Equipment	Total	Total Incl O&P
0600	Pipe insulation, air cell type, up to 4" diameter pipe	A-9	900	.071	L.F.	.22	3.34		3.56	5.40
0610	4" to 8" diameter pipe		800	.080		.24	3.76		4	6.05
0620	10" to 12" diameter pipe		700	.091		.28	4.29		4.57	6.95
0630	14" to 16" diameter pipe		550	.116		.35	5.45		5.80	8.85
0650	Over 16" diameter pipe		650	.098	S.F.	.30	4.62		4.92	7.50
0700	With glove bag up to 3" diameter pipe		200	.320	L.F.	6.05	15.05		21.10	29.50
1000	Pipe fitting insulation up to 4" diameter pipe		320	.200	Ea.	.61	9.40		10.01	15.15
1100	6" to 8" diameter pipe		304	.211		.64	9.90		10.54	16
1110	10" to 12" diameter pipe		192	.333		1.01	15.65		16.66	25
1120	14" to 16" diameter pipe		128	.500		1.52	23.50		25.02	38
1130	Over 16" diameter pipe		176	.364	S.F.	1.11	17.10		18.21	27.50
1200	With glove bag, up to 8" diameter pipe		75	.853	L.F.	6.55	40		46.55	69
2000	Scrape foam fireproofing from flat surface		2400	.027	S.F.	.08	1.25		1.33	2.03
2100	Irregular surfaces		1200	.053		.16	2.50		2.66	4.05
3000	Remove cementitious material from flat surface		1800	.036		.11	1.67		1.78	2.70
3100	Irregular surface		1000	.064		.14	3.01		3.15	4.80
6000	Remove contaminated soil from crawl space by hand		400	.160	C.F.	.49	7.50		7.99	12.15
6100	With large production vacuum loader	A-12	700	.091	"	.28	4.29	1.05	5.62	8.10
7000	Radiator backing, not including radiator removal	A-9	1200	.053	S.F.	.16	2.50		2.66	4.05
9000	For type B (supplied air) respirator equipment, add				%					10%

02 82 13.44 Demolition In Asbestos Contaminated Area

		Crew	Daily Output	Labor-Hours	Unit	Material	2011 Bare Costs Labor	Equipment	Total	Total Incl O&P
0010	**DEMOLITION IN ASBESTOS CONTAMINATED AREA**									
0200	Ceiling, including suspension system, plaster and lath	A-9	2100	.030	S.F.	.09	1.43		1.52	2.31
0210	Finished plaster, leaving wire lath		585	.109		.33	5.15		5.48	8.30
0220	Suspended acoustical tile		3500	.018		.06	.86		.92	1.39
0230	Concealed tile grid system		3000	.021		.06	1		1.06	1.62
0240	Metal pan grid system		1500	.043		.13	2		2.13	3.24
0250	Gypsum board		2500	.026		.08	1.20		1.28	1.95
0260	Lighting fixtures up to 2' x 4'		72	.889	Ea.	2.70	41.50		44.20	67.50
0400	Partitions, non load bearing									
0410	Plaster, lath, and studs	A-9	690	.093	S.F.	.85	4.36		5.21	7.70
0450	Gypsum board and studs	"	1390	.046	"	.14	2.16		2.30	3.49
9000	For type B (supplied air) respirator equipment, add				%					10%

02 82 13.45 OSHA Testing

		Crew	Daily Output	Labor-Hours	Unit	Material	2011 Bare Costs Labor	Equipment	Total	Total Incl O&P
0010	**OSHA TESTING**									
0100	Certified technician, minimum				Day				200	220
0110	Maximum								300	330
0120	Industrial hygienist, minimum								300	330
0130	Maximum								400	440
0200	Asbestos sampling and PCM analysis, NIOSH 7400, minimum	1 Asbe	8	1	Ea.	2.75	47		49.75	75.50
0210	Maximum		4	2		3	94		97	148
1000	Cleaned area samples		8	1		2.63	47		49.63	75.50
1100	PCM air sample analysis, NIOSH 7400, minimum		8	1		30.50	47		77.50	107
1110	Maximum		4	2		3.20	94		97.20	149
1200	TEM air sample analysis, NIOSH 7402, minimum								90	120
1210	Maximum								360	450

02 82 13.46 Decontamination of Asbestos Containment Area

		Crew	Daily Output	Labor-Hours	Unit	Material	2011 Bare Costs Labor	Equipment	Total	Total Incl O&P
0010	**DECONTAMINATION OF ASBESTOS CONTAINMENT AREA**									
0100	Spray exposed substrate with surfactant (bridging)									
0200	Flat surfaces	A-9	6000	.011	S.F.	.36	.50		.86	1.17
0250	Irregular surfaces		4000	.016	"	.31	.75		1.06	1.50
0300	Pipes, beams, and columns		2000	.032	L.F.	.56	1.50		2.06	2.94

02 82 Asbestos Remediation

02 82 13 – Asbestos Abatement

02 82 13.46 Decontamination of Asbestos Containment Area

		Crew	Daily Output	Labor-Hours	Unit	Material	2011 Bare Costs Labor	Equipment	Total	Total Incl O&P
1000	Spray encapsulate polyethylene sheeting	A-9	8000	.008	S.F.	.31	.38		.69	.92
1100	Roll down polyethylene sheeting		8000	.008	"		.38		.38	.58
1500	Bag polyethylene sheeting		400	.160	Ea.	.77	7.50		8.27	12.45
2000	Fine clean exposed substrate, with nylon brush		2400	.027	S.F.		1.25		1.25	1.94
2500	Wet wipe substrate		4800	.013			.63		.63	.97
2600	Vacuum surfaces, fine brush		6400	.010			.47		.47	.73
3000	Structural demolition									
3100	Wood stud walls	A-9	2800	.023	S.F.		1.07		1.07	1.66
3500	Window manifolds, not incl. window replacement		4200	.015			.72		.72	1.11
3600	Plywood carpet protection		2000	.032			1.50		1.50	2.32
4000	Remove custom decontamination facility	A-10A	8	3	Ea.	15	141		156	235
4100	Remove portable decontamination facility	3 Asbe	12	2	"	12.75	94		106.75	159
5000	HEPA vacuum, shampoo carpeting	A-9	4800	.013	S.F.	.05	.63		.68	1.03
9000	Final cleaning of protected surfaces	A-10A	8000	.003	"		.14		.14	.22

02 82 13.47 Asbestos Waste Pkg., Handling, and Disp.

		Crew	Daily Output	Labor-Hours	Unit	Material	2011 Bare Costs Labor	Equipment	Total	Total Incl O&P
0010	**ASBESTOS WASTE PACKAGING, HANDLING, AND DISPOSAL**									
0100	Collect and bag bulk material, 3 C.F. bags, by hand	A-9	400	.160	Ea.	2.30	7.50		9.80	14.15
0200	Large production vacuum loader	A-12	880	.073		.80	3.42	.83	5.05	7.10
1000	Double bag and decontaminate	A-9	960	.067		2.30	3.13		5.43	7.35
2000	Containerize bagged material in drums, per 3 C.F. drum	"	800	.080		16.90	3.76		20.66	24.50
3000	Cart bags 50' to dumpster	2 Asbe	400	.040			1.88		1.88	2.90
5000	Disposal charges, not including haul, minimum				C.Y.				60	66
5020	Maximum				"				350	385
5100	Remove refrigerant from system	1 Stpi	40	.200	Lb.		10.80		10.80	16.15
9000	For type B (supplied air) respirator equipment, add				%					10%

02 82 13.48 Asbestos Encapsulation With Sealants

		Crew	Daily Output	Labor-Hours	Unit	Material	2011 Bare Costs Labor	Equipment	Total	Total Incl O&P
0010	**ASBESTOS ENCAPSULATION WITH SEALANTS**									
0100	Ceilings and walls, minimum	A-9	21000	.003	S.F.	.27	.14		.41	.52
0110	Maximum		10600	.006	"	.42	.28		.70	.90
0300	Pipes to 12" diameter including minor repairs, minimum		800	.080	L.F.	.37	3.76		4.13	6.20
0310	Maximum		400	.160	"	1.04	7.50		8.54	12.75

02 85 Mold Remediation

02 85 16 – Mold Remediation Preparation and Containment

02 85 16.50 Preparation of Mold Containment Area

		Crew	Daily Output	Labor-Hours	Unit	Material	2011 Bare Costs Labor	Equipment	Total	Total Incl O&P
0010	**PREPARATION OF MOLD CONTAINMENT AREA** R028213-20									
0100	Pre-cleaning, HEPA vacuum and wet wipe, flat surfaces	A-9	12000	.005	S.F.	.02	.25		.27	.41
0300	Separation barrier, 2" x 4" @ 16", 1/2" plywood ea. side, 8' high	2 Carp	400	.040		3	1.72		4.72	5.95
0310	12' high		320	.050		3	2.15		5.15	6.60
0320	16' high		200	.080		2	3.44		5.44	7.50
0400	Personnel decontam. chamber, 2" x 4" @ 16", 3/4" ply ea. side		280	.057		4	2.46		6.46	8.15
0450	Waste decontam. chamber, 2" x 4" studs @ 16", 3/4" ply each side		360	.044		4	1.91		5.91	7.35
0500	Cover surfaces with polyethylene sheeting									
0501	Including glue and tape									
0550	Floors, each layer, 6 mil	A-9	8000	.008	S.F.	.06	.38		.44	.65
0551	4 mil		9000	.007		.04	.33		.37	.57
0560	Walls, each layer, 6 mil		6000	.011		.09	.50		.59	.87
0561	4 mil		7000	.009		.07	.43		.50	.74
0570	For heights above 12', add						20%			
0575	For heights above 20', add						30%			

02 85 Mold Remediation

02 85 16 – Mold Remediation Preparation and Containment

02 85 16.50 Preparation of Mold Containment Area	Crew	Daily Output	Labor-Hours	Unit	Material	2011 Bare Costs Labor	Equipment	Total	Total Incl O&P
0580	For fire retardant poly, add					100%				
0590	For large open areas, deduct					10%	20%			
0600	Seal floor penetrations with foam firestop to 36 sq. in.	2 Carp	200	.080	Ea.	7.60	3.44		11.04	13.70
0610	36 sq. in. to 72 sq. in.		125	.128		15.25	5.50		20.75	25
0615	72 sq. in. to 144 sq. in.		80	.200		30.50	8.60		39.10	46.50
0620	Wall penetrations, to 36 square inches		180	.089		7.60	3.83		11.43	14.25
0630	36 sq. in. to 72 sq. in.		100	.160		15.25	6.90		22.15	27.50
0640	72 sq. in. to 144 sq. in.		60	.267		30.50	11.50		42	51
0800	Caulk seams with latex caulk	1 Carp	230	.035	L.F.	.15	1.50		1.65	2.47
0900	Set up neg. air machine, 1-2k C.F.M./25 M.C.F. volume	1 Asbe	4.30	1.860	Ea.		87.50		87.50	135

02 85 33 – Removal and Disposal of Materials with Mold

02 85 33.50 Demolition in Mold Contaminated Area	Crew	Daily Output	Labor-Hours	Unit	Material	2011 Bare Costs Labor	Equipment	Total	Total Incl O&P
0010	**DEMOLITION IN MOLD CONTAMINATED AREA**									
0200	Ceiling, including suspension system, plaster and lath	A-9	2100	.030	S.F.	.09	1.43		1.52	2.31
0210	Finished plaster, leaving wire lath		585	.109		.33	5.15		5.48	8.30
0220	Suspended acoustical tile		3500	.018		.06	.86		.92	1.39
0230	Concealed tile grid system		3000	.021		.06	1		1.06	1.62
0240	Metal pan grid system		1500	.043		.13	2		2.13	3.24
0250	Gypsum board		2500	.026		.08	1.20		1.28	1.95
0255	Plywood		2500	.026		.08	1.20		1.28	1.95
0260	Lighting fixtures up to 2' x 4'		72	.889	Ea.	2.70	41.50		44.20	67.50
0400	Partitions, non load bearing									
0410	Plaster, lath, and studs	A-9	690	.093	S.F.	.85	4.36		5.21	7.70
0450	Gypsum board and studs		1390	.046		.14	2.16		2.30	3.49
0465	Carpet & pad		1390	.046		.14	2.16		2.30	3.49
0600	Pipe insulation, air cell type, up to 4" diameter pipe		900	.071	L.F.	.22	3.34		3.56	5.40
0610	4" to 8" diameter pipe		800	.080		.24	3.76		4	6.05
0620	10" to 12" diameter pipe		700	.091		.28	4.29		4.57	6.95
0630	14" to 16" diameter pipe		550	.116		.35	5.45		5.80	8.85
0650	Over 16" diameter pipe		650	.098	S.F.	.30	4.62		4.92	7.50
9000	For type B (supplied air) respirator equipment, add				%					10%

Estimating Tips

General

- Carefully check all the plans and specifications. Concrete often appears on drawings other than structural drawings, including mechanical and electrical drawings for equipment pads. The cost of cutting and patching is often difficult to estimate. See Subdivision 03 81 for Concrete Cutting, Subdivision 02 41 19.16 for Cutout Demolition, Subdivision 03 05 05.10 for Concrete Demolition, and Subdivision 02 41 19.23 for Rubbish Handling (handling, loading and hauling of debris).

- Always obtain concrete prices from suppliers near the job site. A volume discount can often be negotiated, depending upon competition in the area. Remember to add for waste, particularly for slabs and footings on grade.

03 10 00 Concrete Forming and Accessories

- A primary cost for concrete construction is forming. Most jobs today are constructed with prefabricated forms. The selection of the forms best suited for the job and the total square feet of forms required for efficient concrete forming and placing are key elements in estimating concrete construction. Enough forms must be available for erection to make efficient use of the concrete placing equipment and crew.

- Concrete accessories for forming and placing depend upon the systems used. Study the plans and specifications to ensure that all special accessory requirements have been included in the cost estimate, such as anchor bolts, inserts, and hangers.

- Included within costs for forms-in-place are all necessary bracing and shoring.

03 20 00 Concrete Reinforcing

- Ascertain that the reinforcing steel supplier has included all accessories, cutting, bending, and an allowance for lapping, splicing, and waste. A good rule of thumb is 10% for lapping, splicing, and waste. Also, 10% waste should be allowed for welded wire fabric.

- The unit price items in the subdivision for Reinforcing In Place include the labor to install accessories such as beam and slab bolsters, high chairs, and bar ties and tie wire. The material cost for these accessories is not included; they may be obtained from the Accessories Division.

03 30 00 Cast-In-Place Concrete

- When estimating structural concrete, pay particular attention to requirements for concrete additives, curing methods, and surface treatments. Special consideration for climate, hot or cold, must be included in your estimate. Be sure to include requirements for concrete placing equipment, and concrete finishing.

- For accurate concrete estimating, the estimator must consider each of the following major components individually: forms, reinforcing steel, ready-mix concrete, placement of the concrete, and finishing of the top surface. For faster estimating, Subdivision 03 30 53.40 for Concrete-In-Place can be used; here, various items of concrete work are presented that include the costs of all five major components (unless specifically stated otherwise).

03 40 00 Precast Concrete
03 50 00 Cast Decks and Underlayment

- The cost of hauling precast concrete structural members is often an important factor. For this reason, it is important to get a quote from the nearest supplier. It may become economically feasible to set up precasting beds on the site if the hauling costs are prohibitive.

Reference Numbers

Reference numbers are shown in shaded boxes at the beginning of some major classifications. These numbers refer to related items in the Reference Section. The reference information may be an estimating procedure, an alternate pricing method, or technical information.

Note: Not all subdivisions listed here necessarily appear in this publication.

Division 3 - Concrete

03 01 Maintenance of Concrete

03 01 30 – Maintenance of Cast-In-Place Concrete

03 01 30.62 Concrete Patching		Crew	Daily Output	Labor-Hours	Unit	Material	2011 Bare Costs Labor	Equipment	Total	Total Incl O&P
0010	**CONCRETE PATCHING**									
0100	Floors, 1/4" thick, small areas, regular grout	1 Cefi	170	.047	S.F.	1.15	1.92		3.07	4.06
0150	Epoxy grout	"	100	.080	"	7.65	3.27		10.92	13.15
2000	Walls, including chipping, cleaning and epoxy grout									
2100	1/4" deep	1 Cefi	65	.123	S.F.	7.50	5.05		12.55	15.55
2150	1/2" deep		50	.160		15.05	6.55		21.60	26
2200	3/4" deep		40	.200		22.50	8.15		30.65	37

03 11 Concrete Forming

03 11 13 – Structural Cast-In-Place Concrete Forming

03 11 13.40 Forms In Place, Equipment Foundations

		Crew	Daily Output	Labor-Hours	Unit	Material	2011 Bare Costs Labor	Equipment	Total	Total Incl O&P
0010	**FORMS IN PLACE, EQUIPMENT FOUNDATIONS**									
0020	1 use	C-2	160	.300	SFCA	2.66	12.60		15.26	22
0050	2 use		190	.253		1.47	10.60		12.07	17.85
0100	3 use		200	.240		1.07	10.05		11.12	16.60
0150	4 use		205	.234		.87	9.80		10.67	16

03 11 13.45 Forms In Place, Footings

		Crew	Daily Output	Labor-Hours	Unit	Material	2011 Bare Costs Labor	Equipment	Total	Total Incl O&P
0010	**FORMS IN PLACE, FOOTINGS**									
0020	Continuous wall, plywood, 1 use	C-1	375	.085	SFCA	5	3.49		8.49	10.85
0050	2 use		440	.073		2.76	2.97		5.73	7.60
0100	3 use		470	.068		2	2.78		4.78	6.45
0150	4 use		485	.066		1.63	2.70		4.33	5.95
5000	Spread footings, job-built lumber, 1 use		305	.105		1.79	4.29		6.08	8.55
5050	2 use		371	.086		1	3.53		4.53	6.50
5100	3 use		401	.080		.72	3.26		3.98	5.80
5150	4 use		414	.077		.58	3.16		3.74	5.50

03 11 13.65 Forms In Place, Slab On Grade

		Crew	Daily Output	Labor-Hours	Unit	Material	2011 Bare Costs Labor	Equipment	Total	Total Incl O&P
0010	**FORMS IN PLACE, SLAB ON GRADE**									
3000	Edge forms, wood, 4 use, on grade, to 6" high	C-1	600	.053	L.F.	.27	2.18		2.45	3.64
6000	Trench forms in floor, wood, 1 use		160	.200	SFCA	1.55	8.20		9.75	14.25
6050	2 use		175	.183		.85	7.50		8.35	12.40
6100	3 use		180	.178		.62	7.25		7.87	11.85
6150	4 use		185	.173		.50	7.05		7.55	11.40
8760	Void form, corrugated fiberboard, 4" x 12", 4' long	G	3000	.011	S.F.	2.39	.44		2.83	3.29
8770	6" x 12", 4' long		3000	.011		2.81	.44		3.25	3.76
8780	1/4" thick hardboard protective cover for void form	2 Carp	1500	.011		.57	.46		1.03	1.33

03 15 Concrete Accessories

03 15 05 – Concrete Forming Accessories

03 15 05.75 Sleeves and Chases

		Crew	Daily Output	Labor-Hours	Unit	Material	2011 Bare Costs Labor	Equipment	Total	Total Incl O&P
0010	**SLEEVES AND CHASES**									
0100	Plastic, 1 use, 12" long, 2" diameter	1 Carp	100	.080	Ea.	1.74	3.44		5.18	7.20
0150	4" diameter		90	.089		5.10	3.83		8.93	11.45
0200	6" diameter		75	.107		9	4.59		13.59	16.95
0250	12" diameter		60	.133		25.50	5.75		31.25	37
5000	Sheet metal, 2" diameter	G	100	.080		1.42	3.44		4.86	6.85
5100	4" diameter	G	90	.089		1.77	3.83		5.60	7.80
5150	6" diameter	G	75	.107		1.77	4.59		6.36	9
5200	12" diameter	G	60	.133		3.53	5.75		9.28	12.70

03 15 Concrete Accessories

03 15 05 – Concrete Forming Accessories

03 15 05.75 Sleeves and Chases

		Crew	Daily Output	Labor-Hours	Unit	Material	2011 Bare Costs Labor	Equipment	Total	Total Incl O&P
6000	Steel pipe, 2" diameter	G 1 Carp	100	.080	Ea.	3.49	3.44		6.93	9.15
6100	4" diameter	G	90	.089		10.30	3.83		14.13	17.20
6150	6" diameter	G	75	.107		18.20	4.59		22.79	27
6200	12" diameter	G	60	.133		60.50	5.75		66.25	76

03 21 Reinforcing Steel

03 21 10 – Uncoated Reinforcing Steel

03 21 10.60 Reinforcing In Place

		Crew	Daily Output	Labor-Hours	Unit	Material	2011 Bare Costs Labor	Equipment	Total	Total Incl O&P
0015	**REINFORCING IN PLACE**, 50-60 ton lots, A615 Grade 60									
0020	Includes labor, but not material cost, to install accessories									
0030	Made from recycled materials	G								
0502	Footings, #4 to #7	G 4 Rodm	4200	.008	Lb.	.47	.37		.84	1.10
0552	#8 to #18	G	7200	.004		.47	.21		.68	.85
0602	Slab on grade, #3 to #7	G	4200	.008		.45	.37		.82	1.09
0900	For other than 50 – 60 ton lots									
1000	Under 10 ton job, #3 to #7, add				Ton	25%	10%			
1010	#8 to #18, add					20%	10%			
1050	10 – 50 ton job, #3 to #7, add					10%				
1060	#8 to #18, add					5%				
1100	60 – 100 ton job, #3 to #7, deduct					5%				
1110	#8 to #18, deduct					10%				
1150	Over 100 ton job, #3 to #7, deduct					10%				
1160	#8 to #18, deduct					15%				

03 22 Welded Wire Fabric Reinforcing

03 22 05 – Uncoated Welded Wire Fabric

03 22 05.50 Welded Wire Fabric

		Crew	Daily Output	Labor-Hours	Unit	Material	2011 Bare Costs Labor	Equipment	Total	Total Incl O&P
0010	**WELDED WIRE FABRIC** ASTM A185									
0030	Made from recycled materials	G								
0050	Sheets									
0100	6 x 6 - W1.4 x W1.4 (10 x 10) 21 lb. per C.S.F.	G 2 Rodm	35	.457	C.S.F.	12.50	22		34.50	49.50

03 30 Cast-In-Place Concrete

03 30 53 – Miscellaneous Cast-In-Place Concrete

03 30 53.40 Concrete In Place

		Crew	Daily Output	Labor-Hours	Unit	Material	2011 Bare Costs Labor	Equipment	Total	Total Incl O&P
0010	**CONCRETE IN PLACE**									
0020	Including forms (4 uses), reinforcing steel, concrete, placement,									
0050	and finishing unless otherwise indicated									
0500	Chimney foundations (5000 psi), industrial, minimum	C-14C	32.22	3.476	C.Y.	144	144	.73	288.73	380
0510	Maximum	"	23.71	4.724	"	168	195	.99	363.99	485
3540	Equipment pad (3000 psi), 3' x 3' x 6" thick	C-14H	45	1.067	Ea.	40.50	45.50	.51	86.51	115
3550	4' x 4' x 6" thick		30	1.600		62	68	.77	130.77	173
3560	5' x 5' x 8" thick		18	2.667		111	113	1.28	225.28	297
3570	6' x 6' x 8" thick		14	3.429		150	146	1.65	297.65	390
3580	8' x 8' x 10" thick		8	6		320	255	2.88	577.88	750
3590	10' x 10' x 12" thick		5	9.600		550	410	4.61	964.61	1,225
3800	Footings (3000 psi), spread under 1 C.Y.	C-14C	28	4	C.Y.	158	165	.84	323.84	430

03 30 Cast-In-Place Concrete

03 30 53 – Miscellaneous Cast-In-Place Concrete

03 30 53.40 Concrete In Place		Crew	Daily Output	Labor-Hours	Unit	Material	2011 Bare Costs Labor	Equipment	Total	Total Incl O&P
3825	1 C.Y. to 5 C.Y.	C-14C	43	2.605	C.Y.	185	108	.55	293.55	370
3850	Over 5 C.Y.	↓	75	1.493		171	61.50	.31	232.81	283
3900	Footings, strip (3000 psi), 18" x 9", unreinforced	C-14L	40	2.400		119	96.50	.58	216.08	279
3920	18" x 9", reinforced	C-14C	35	3.200		141	132	.67	273.67	360
3925	20" x 10", unreinforced	C-14L	45	2.133		116	85.50	.51	202.01	260
3930	20" x 10", reinforced	C-14C	40	2.800		134	116	.59	250.59	325
3935	24" x 12", unreinforced	C-14L	55	1.745		114	70	.42	184.42	233
3940	24" x 12", reinforced	C-14C	48	2.333		132	96.50	.49	228.99	294
3945	36" x 12", unreinforced	C-14L	70	1.371		111	55	.33	166.33	206
3950	36" x 12", reinforced	C-14C	60	1.867		127	77	.39	204.39	259
4000	Foundation mat (3000 psi), under 10 C.Y.		38.67	2.896		192	120	.61	312.61	395
4050	Over 20 C.Y.	↓	56.40	1.986		169	82	.42	251.42	310
4650	Slab on grade (3500 psi), not including finish, 4" thick	C-14E	60.75	1.449		117	61.50	.38	178.88	225
4700	6" thick	"	92	.957	↓	113	41	.25	154.25	188
4701	Thickened slab edge (3500 psi), for slab on grade poured									
4702	monolithically with slab; depth is in addition to slab thickness;									
4703	formed vertical outside edge, earthen bottom and inside slope									
4705	8" deep x 8" wide bottom, unreinforced	C-14L	2190	.044	L.F.	3.18	1.76	.01	4.95	6.20
4710	8" x 8", reinforced	C-14C	1670	.067		5.30	2.77	.01	8.08	10.15
4715	12" deep x 12" wide bottom, unreinforced	C-14L	1800	.053		6.55	2.14	.01	8.70	10.50
4720	12" x 12", reinforced	C-14C	1310	.086		10.40	3.53	.02	13.95	16.90
4725	16" deep x 16" wide bottom, unreinforced	C-14L	1440	.067		11.10	2.68	.02	13.80	16.30
4730	16" x 16", reinforced	C-14C	1120	.100		15.70	4.13	.02	19.85	23.50
4735	20" deep x 20" wide bottom, unreinforced	C-14L	1150	.083		16.85	3.35	.02	20.22	23.50
4740	20" x 20", reinforced	C-14C	920	.122		22.50	5.05	.03	27.58	33
4745	24" deep x 24" wide bottom, unreinforced	C-14L	930	.103		24	4.14	.02	28.16	32.50
4750	24" x 24", reinforced	C-14C	740	.151	↓	31.50	6.25	.03	37.78	44
4751	Slab on grade (3500 psi), incl. troweled finish, not incl. forms									
4760	or reinforcing, over 10,000 S.F., 4" thick	C-14F	3425	.021	S.F.	1.29	.82	.01	2.12	2.64
4820	6" thick		3350	.021		1.89	.84	.01	2.74	3.33
4840	8" thick		3184	.023		2.59	.88	.01	3.48	4.16
4900	12" thick		2734	.026		3.88	1.02	.01	4.91	5.80
4950	15" thick	↓	2505	.029	↓	4.88	1.12	.01	6.01	7

03 31 Structural Concrete

03 31 05 – Normal Weight Structural Concrete

03 31 05.35 Normal Weight Concrete, Ready Mix

		Crew	Daily Output	Labor-Hours	Unit	Material	2011 Bare Costs Labor	Equipment	Total	Total Incl O&P
0010	**NORMAL WEIGHT CONCRETE, READY MIX**, delivered									
0012	Includes local aggregate, sand, Portland cement, and water									
0015	Excludes all additives and treatments									
0020	2000 psi				C.Y.	91.50			91.50	101
0100	2500 psi					94			94	103
0150	3000 psi					99			99	109
0200	3500 psi					99.50			99.50	110
0300	4000 psi					103			103	113
0350	4500 psi					106			106	116
0400	5000 psi					111			111	122
0411	6000 psi					127			127	139
0412	8000 psi					206			206	227
0413	10,000 psi					293			293	320
0414	12,000 psi				↓	355			355	390

03 31 Structural Concrete

03 31 05 – Normal Weight Structural Concrete

03 31 05.35 Normal Weight Concrete, Ready Mix

03 31 05.35 Normal Weight Concrete, Ready Mix	Crew	Daily Output	Labor-Hours	Unit	Material	2011 Bare Costs Labor	Equipment	Total	Total Incl O&P
1000 For high early strength cement, add				C.Y.	10%				
1300 For winter concrete (hot water), add					4.25			4.25	4.68
1400 For hot weather concrete (ice), add					9.35			9.35	10.25
1410 For mid-range water reducer, add					4.13			4.13	4.54
1420 For high-range water reducer/superplasticizer, add					6.35			6.35	6.95
1430 For retarder, add					2.71			2.71	2.98
1440 For non-Chloride accelerator, add					4.83			4.83	5.30
1450 For Chloride accelerator, per 1%, add					3.28			3.28	3.61
1460 For fiber reinforcing, synthetic (1 lb./C.Y.), add					6.65			6.65	7.30
1500 For Saturday delivery, add					8.85			8.85	9.70
1510 For truck holding/waiting time past 1st hour per load, add				Hr.	87.50			87.50	96.50
1520 For short load (less than 4 C.Y.), add per load				Ea.	112			112	124
2000 For all lightweight aggregate, add				C.Y.	45%				

03 31 05.70 Placing Concrete

	Crew	Daily Output	Labor-Hours	Unit	Material	2011 Bare Costs Labor	Equipment	Total	Total Incl O&P
0010 **PLACING CONCRETE**									
0020 Includes labor and equipment to place, strike off and consolidate									
1900 Footings, continuous, shallow, direct chute	C-6	120	.400	C.Y.		14.30	.39	14.69	22
1950 Pumped	C-20	150	.427			15.70	5.25	20.95	30
2000 With crane and bucket	C-7	90	.800			29.50	14.20	43.70	60.50
2100 Footings, continuous, deep, direct chute	C-6	140	.343			12.25	.33	12.58	19
2150 Pumped	C-20	160	.400			14.70	4.92	19.62	28
2200 With crane and bucket	C-7	110	.655			24.50	11.60	36.10	50
2400 Footings, spread, under 1 C.Y., direct chute	C-6	55	.873			31	.85	31.85	48.50
2450 Pumped	C-20	65	.985			36	12.10	48.10	68.50
2500 With crane and bucket	C-7	45	1.600			59.50	28.50	88	121
2600 Over 5 C.Y., direct chute	C-6	120	.400			14.30	.39	14.69	22
2650 Pumped	C-20	150	.427			15.70	5.25	20.95	30
2700 With crane and bucket	C-7	100	.720			27	12.75	39.75	54.50
2900 Foundation mats, over 20 C.Y., direct chute	C-6	350	.137			4.91	.13	5.04	7.60
2950 Pumped	C-20	400	.160			5.90	1.97	7.87	11.10
3000 With crane and bucket	C-7	300	.240			8.90	4.26	13.16	18.20

03 35 Concrete Finishing

03 35 29 – Tooled Concrete Finishing

03 35 29.30 Finishing Floors

	Crew	Daily Output	Labor-Hours	Unit	Material	2011 Bare Costs Labor	Equipment	Total	Total Incl O&P
0010 **FINISHING FLOORS**									
0012 Finishing requires that concrete first be placed, struck off & consolidated									
0015 Basic finishing for various unspecified flatwork									
0100 Bull float only	C-10	4000	.006	S.F.		.23		.23	.34
0125 Bull float & manual float		2000	.012			.46		.46	.69
0150 Bull float, manual float, & broom finish, w/edging & joints		1850	.013			.50		.50	.74
0200 Bull float, manual float & manual steel trowel		1265	.019			.73		.73	1.09
0210 For specified Random Access Floors in ACI Classes 1, 2, 3 and 4 to achieve									
0215 Composite Overall Floor Flatness and Levelness values up to F35/F25									
0250 Bull float, machine float & machine trowel (walk-behind)	C-10C	1715	.014	S.F.		.54	.02	.56	.82
0300 Power screed, bull float, machine float & trowel (walk-behind)	C-10D	2400	.010			.39	.04	.43	.62
0350 Power screed, bull float, machine float & trowel (ride-on)	C-10E	4000	.006			.23	.06	.29	.40
0352 For specified Random Access Floors in ACI Classes 5, 6, 7 and 8 to achieve									
0354 Composite Overall Floor Flatness and Levelness values up to F50/F50									
0356 Add for two-dimensional restraightening after power float	C-10	6000	.004	S.F.		.15		.15	.23
0358 For specified Random or Defined Access Floors in ACI Class 9 to achieve									

03 35 Concrete Finishing

03 35 29 – Tooled Concrete Finishing

03 35 29.30 Finishing Floors

		Crew	Daily Output	Labor-Hours	Unit	Material	2011 Bare Costs Labor	Equipment	Total	Total Incl O&P
0360	Composite Overall Floor Flatness and Levelness values up to F100/F100									
0362	Add for two-dimensional restraightening after bull float & power float	C-10	3000	.008	S.F.		.31		.31	.46
0364	For specified Superflat Defined Access Floors in ACI Class 9 to achieve									
0366	Minimum Floor Flatness and Levelness values of F100/F100									
0368	Add for 2-dim'l restraightening after bull float, power float, power trowel	C-10	2000	.012	S.F.		.46		.46	.69

03 35 29.35 Control Joints, Saw Cut

		Crew	Daily Output	Labor-Hours	Unit	Material	2011 Bare Costs Labor	Equipment	Total	Total Incl O&P
0010	**CONTROL JOINTS, SAW CUT**									
0100	Sawcut control joints in green concrete									
0120	1" depth	C-27	2000	.008	L.F.	.05	.33	.07	.45	.60
0140	1-1/2" depth		1800	.009		.07	.36	.07	.50	.69
0160	2" depth	↓	1600	.010	↓	.10	.41	.08	.59	.80
0180	Sawcut joint reservoir in cured concrete									
0182	3/8" wide x 3/4" deep, with single saw blade	C-27	1000	.016	L.F.	.07	.65	.13	.85	1.18
0184	1/2" wide x 1" deep, with double saw blades		900	.018		.14	.73	.15	1.02	1.38
0186	3/4" wide x 1-1/2" deep, with double saw blades	↓	800	.020		.30	.82	.17	1.29	1.70
0190	Water blast joint to wash away laitance, 2 passes	C-29	2500	.003			.11	.02	.13	.20
0200	Air blast joint to blow out debris and air dry, 2 passes	C-28	2000	.004	↓		.16	.01	.17	.25
0300	For backer rod, see Section 07 91 23.10									
0340	For joint sealant, see Section 07 92 13.20									

03 54 Cast Underlayment

03 54 16 – Hydraulic Cement Underlayment

03 54 16.50 Cement Underlayment

		Crew	Daily Output	Labor-Hours	Unit	Material	2011 Bare Costs Labor	Equipment	Total	Total Incl O&P
0010	**CEMENT UNDERLAYMENT**									
2510	Underlayment, P.C based self-leveling, 4100 psi, pumped, 1/4"	C-8	20000	.003	S.F.	1.35	.11	.04	1.50	1.69
2520	1/2"		19000	.003		2.71	.11	.04	2.86	3.19
2530	3/4"		18000	.003		4.06	.12	.04	4.22	4.69
2540	1"		17000	.003		5.40	.13	.04	5.57	6.20
2550	1-1/2"	↓	15000	.004		8.10	.14	.05	8.29	9.20
2560	Hand placed, 1/2"	C-18	450	.020		2.71	.69	.12	3.52	4.17
2610	Topping, P.C. based self-level/dry 6100 psi, pumped, 1/4"	C-8	20000	.003		2.03	.11	.04	2.18	2.44
2620	1/2"		19000	.003		4.06	.11	.04	4.21	4.68
2630	3/4"		18000	.003		6.10	.12	.04	6.26	6.95
2660	1"		17000	.003		8.15	.13	.04	8.32	9.20
2670	1-1/2"	↓	15000	.004		12.20	.14	.05	12.39	13.65
2680	Hand placed, 1/2"	C-18	450	.020	↓	4.06	.69	.12	4.87	5.65

03 63 Epoxy Grouting

03 63 05 – Grouting of Dowels and Fasteners

03 63 05.10 Epoxy Only

		Crew	Daily Output	Labor-Hours	Unit	Material	2011 Bare Costs Labor	Equipment	Total	Total Incl O&P
0010	**EPOXY ONLY**									
1500	Chemical anchoring, epoxy cartridge, excludes layout, drilling, fastener									
1530	For fastener 3/4" diam. x 6" embedment	2 Skwk	72	.222	Ea.	4.95	9.85		14.80	20.50
1535	1" diam. x 8" embedment		66	.242		7.45	10.75		18.20	24.50
1540	1-1/4" diam. x 10" embedment		60	.267		14.85	11.85		26.70	34.50
1545	1-3/4" diam. x 12" embedment		54	.296		25	13.15		38.15	47
1550	14" embedment		48	.333		29.50	14.80		44.30	55
1555	2" diam. x 12" embedment	↓	42	.381	↓	39.50	16.90		56.40	69.50

03 63 Epoxy Grouting

03 63 05 - Grouting of Dowels and Fasteners

03 63 05.10 Epoxy Only		Crew	Daily Output	Labor-Hours	Unit	Material	2011 Bare Costs Labor	Equipment	Total	Total Incl O&P
1560	18" embedment	2 Skwk	32	.500	Ea.	49.50	22		71.50	88.50

03 82 Concrete Boring

03 82 13 - Concrete Core Drilling

03 82 13.10 Core Drilling

	03 82 13.10 Core Drilling	Crew	Daily Output	Labor-Hours	Unit	Material	2011 Bare Costs Labor	Equipment	Total	Total Incl O&P
0010	**CORE DRILLING**									
0020	Reinf. conc slab, up to 6" thick, incl. bit, layout & set up									
0100	1" diameter core	B-89A	17	.941	Ea.	3.66	37	6.45	47.11	68
0150	Each added inch thick in same hole, add		1440	.011		.69	.44	.08	1.21	1.51
0300	3" diameter core		16	1		12.20	39.50	6.85	58.55	81.50
0350	Each added inch thick in same hole, add		720	.022		1.41	.88	.15	2.44	3.06
0500	4" diameter core		15	1.067		15.45	42	7.30	64.75	89.50
0550	Each added inch thick in same hole, add		480	.033		1.73	1.31	.23	3.27	4.16
0700	6" diameter core		14	1.143		20	45	7.85	72.85	99.50
0750	Each added inch thick in same hole, add		360	.044		2.74	1.75	.30	4.79	6.05
0900	8" diameter core		13	1.231		33	48.50	8.45	89.95	120
0950	Each added inch thick in same hole, add		288	.056		3.76	2.19	.38	6.33	7.90
1100	10" diameter core		12	1.333		26.50	52.50	9.15	88.15	120
1150	Each added inch thick in same hole, add		240	.067		3.76	2.63	.46	6.85	8.65
1300	12" diameter core		11	1.455		31	57.50	10	98.50	133
1350	Each added inch thick in same hole, add		206	.078		4.56	3.06	.53	8.15	10.30
1500	14" diameter core		10	1.600		37	63	11	111	149
1550	Each added inch thick in same hole, add		180	.089		6.20	3.50	.61	10.31	12.85
1700	18" diameter core		9	1.778		45.50	70	12.20	127.70	170
1750	Each added inch thick in same hole, add		144	.111		8.80	4.38	.76	13.94	17.25
1760	For horizontal holes, add to above						20%	20%		
1770	Prestressed hollow core plank, 8" thick									
1780	1" diameter core	B-89A	17.50	.914	Ea.	2.38	36	6.25	44.63	64.50
1790	Each added inch thick in same hole, add		3840	.004		.48	.16	.03	.67	.81
1800	3" diameter core		17	.941		4.96	37	6.45	48.41	69.50
1810	Each added inch thick in same hole, add		1920	.008		.86	.33	.06	1.25	1.51
1820	4" diameter core		16.50	.970		6.80	38	6.65	51.45	73.50
1830	Each added inch thick in same hole, add		1280	.013		1.26	.49	.09	1.84	2.24
1840	6" diameter core		15.50	1.032		8.35	40.50	7.10	55.95	79.50
1850	Each added inch thick in same hole, add		960	.017		1.43	.66	.11	2.20	2.71
1860	8" diameter core		15	1.067		11.85	42	7.30	61.15	85.50
1870	Each added inch thick in same hole, add		768	.021		2	.82	.14	2.96	3.62
1880	10" diameter core		14	1.143		15.70	45	7.85	68.55	95
1890	Each added inch thick in same hole, add		640	.025		2.08	.98	.17	3.23	3.99
1900	12" diameter core		13.50	1.185		18.35	46.50	8.15	73	100
1910	Each added inch thick in same hole, add		548	.029		3.33	1.15	.20	4.68	5.65
1999	Drilling, core, minimum labor/equipment charge		4	4	Job		158	27.50	185.50	272

03 82 16 - Concrete Drilling

03 82 16.10 Concrete Impact Drilling

	03 82 16.10 Concrete Impact Drilling	Crew	Daily Output	Labor-Hours	Unit	Material	2011 Bare Costs Labor	Equipment	Total	Total Incl O&P
0010	**CONCRETE IMPACT DRILLING**									
0050	Up to 4" deep in conc/brick floor/wall, incl. bit & layout, no anchor									
0100	Holes, 1/4" diameter	1 Carp	75	.107	Ea.	.05	4.59		4.64	7.10
0150	For each additional inch of depth, add		430	.019		.01	.80		.81	1.24
0200	3/8" diameter		63	.127		.05	5.45		5.50	8.45
0250	For each additional inch of depth, add		340	.024		.01	1.01		1.02	1.56

03 82 Concrete Boring

03 82 16 – Concrete Drilling

03 82 16.10 Concrete Impact Drilling		Crew	Daily Output	Labor-Hours	Unit	Material	2011 Bare Costs Labor	Equipment	Total	Total Incl O&P
0300	1/2" diameter	1 Carp	50	.160	Ea.	.05	6.90		6.95	10.60
0350	For each additional inch of depth, add		250	.032		.01	1.38		1.39	2.12
0400	5/8" diameter		48	.167		.07	7.20		7.27	11.10
0450	For each additional inch of depth, add		240	.033		.02	1.43		1.45	2.22
0500	3/4" diameter		45	.178		.10	7.65		7.75	11.85
0550	For each additional inch of depth, add		220	.036		.02	1.57		1.59	2.43
0600	7/8" diameter		43	.186		.13	8		8.13	12.45
0650	For each additional inch of depth, add		210	.038		.03	1.64		1.67	2.56
0700	1" diameter		40	.200		.14	8.60		8.74	13.35
0750	For each additional inch of depth, add		190	.042		.04	1.81		1.85	2.82
0800	1-1/4" diameter		38	.211		.22	9.05		9.27	14.15
0850	For each additional inch of depth, add		180	.044		.06	1.91		1.97	3
0900	1-1/2" diameter		35	.229		.34	9.85		10.19	15.45
0950	For each additional inch of depth, add		165	.048		.08	2.09		2.17	3.29
1000	For ceiling installations, add						40%			

Estimating Tips

04 05 00 Common Work Results for Masonry

- The terms *mortar* and *grout* are often used interchangeably, and incorrectly. Mortar is used to bed masonry units, seal the entry of air and moisture, provide architectural appearance, and allow for size variations in the units. Grout is used primarily in reinforced masonry construction and is used to bond the masonry to the reinforcing steel. Common mortar types are M(2500 psi), S(1800 psi), N(750 psi), and O(350 psi), and conform to ASTM C270. Grout is either fine or coarse and conforms to ASTM C476, and in-place strengths generally exceed 2500 psi. Mortar and grout are different components of masonry construction and are placed by entirely different methods. An estimator should be aware of their unique uses and costs.

- Mortar is included in all assembled masonry line items. The mortar cost, part of the assembled masonry material cost, includes all ingredients, all labor, and all equipment required. Please see reference number R040513-10.

- Waste, specifically the loss/droppings of mortar and the breakage of brick and block, is included in all masonry assemblies in this division. A factor of 25% is added for mortar and 3% for brick and concrete masonry units.

- Scaffolding or staging is not included in any of the Division 4 costs. Refer to Subdivision 01 54 23 for scaffolding and staging costs.

04 20 00 Unit Masonry

- The most common types of unit masonry are brick and concrete masonry. The major classifications of brick are building brick (ASTM C62), facing brick (ASTM C216), glazed brick, fire brick, and pavers. Many varieties of texture and appearance can exist within these classifications, and the estimator would be wise to check local custom and availability within the project area. For repair and remodeling jobs, matching the existing brick may be the most important criteria.

- Brick and concrete block are priced by the piece and then converted into a price per square foot of wall. Openings less than two square feet are generally ignored by the estimator because any savings in units used is offset by the cutting and trimming required.

- It is often difficult and expensive to find and purchase small lots of historic brick. Costs can vary widely. Many design issues affect costs, selection of mortar mix, and repairs or replacement of masonry materials. Cleaning techniques must be reflected in the estimate.

- All masonry walls, whether interior or exterior, require bracing. The cost of bracing walls during construction should be included by the estimator, and this bracing must remain in place until permanent bracing is complete. Permanent bracing of masonry walls is accomplished by masonry itself, in the form of pilasters or abutting wall corners, or by anchoring the walls to the structural frame. Accessories in the form of anchors, anchor slots, and ties are used, but their supply and installation can be by different trades. For instance, anchor slots on spandrel beams and columns are supplied and welded in place by the steel fabricator, but the ties from the slots into the masonry are installed by the bricklayer. Regardless of the installation method, the estimator must be certain that these accessories are accounted for in pricing.

Reference Numbers

Reference numbers are shown in shaded boxes at the beginning of some major classifications. These numbers refer to related items in the Reference Section. The reference information may be an estimating procedure, an alternate pricing method, or technical information.

Note: Not all subdivisions listed here necessarily appear in this publication.

04 05 Common Work Results for Masonry

04 05 23 – Masonry Accessories

04 05 23.19 Masonry Cavity Drainage, Weepholes, and Vents	Crew	Daily Output	Labor-Hours	Unit	Material	2011 Bare Costs Labor	Equipment	Total	Total Incl O&P
0010 **MASONRY CAVITY DRAINAGE, WEEPHOLES, AND VENTS**									
0020 Extruded aluminum, 4" deep, 2-3/8" x 8-1/8"	1 Bric	30	.267	Ea.	30.50	11.50		42	51
0050 5" x 8-1/8"		25	.320		40	13.80		53.80	64.50
0100 2-1/4" x 25"		25	.320		69.50	13.80		83.30	97
0150 5" x 16-1/2"		22	.364		58.50	15.70		74.20	87.50
0200 6" x 16-1/2"		22	.364		81	15.70		96.70	113
0250 7-3/4" x 16-1/2"		20	.400		69	17.25		86.25	102
0400 For baked enamel finish, add					35%				
0500 For cast aluminum, painted, add					60%				
1000 Stainless steel ventilators, 6" x 6"	1 Bric	25	.320		126	13.80		139.80	159
1050 8" x 8"		24	.333		132	14.40		146.40	168
1100 12" x 12"		23	.348		153	15		168	191
1150 12" x 6"		24	.333		134	14.40		148.40	169
1200 Foundation block vent, galv., 1-1/4" thk, 8" high, 16" long, no damper		30	.267		24.50	11.50		36	44.50
1250 For damper, add					8.20			8.20	9

04 21 Clay Unit Masonry

04 21 13 – Brick Masonry

04 21 13.15 Chimney

		Crew	Daily Output	Labor-Hours	Unit	Material	2011 Bare Costs Labor	Equipment	Total	Total Incl O&P
0010	**CHIMNEY**, excludes foundation, scaffolding, grout and reinforcing									
0100	Brick, 16" x 16", 8" flue	D-1	18.20	.879	V.L.F.	21.50	34.50		56	75
0150	16" x 20" with one 8" x 12" flue		16	1		34	39		73	96
0200	16" x 24" with two 8" x 8" flues		14	1.143		50	44.50		94.50	122
0250	20" x 20" with one 12" x 12" flue		13.70	1.168		38.50	45.50		84	111
0300	20" x 24" with two 8" x 12" flues		12	1.333		57	52		109	141
0350	20" x 32" with two 12" x 12" flues		10	1.600		68	62.50		130.50	169
1800	Metal, high temp. steel jacket, factory lining, 24" diam.	E-2	65	.862		210	41	25	276	330
1900	60" diameter	"	30	1.867		765	88.50	54	907.50	1,050
2100	Poured concrete, brick lining, 200' high x 10' diam.					7,275			7,275	8,000
2800	500' x 20' diameter					12,700			12,700	14,000

Estimating Tips

05 05 00 Common Work Results for Metals

- Nuts, bolts, washers, connection angles, and plates can add a significant amount to both the tonnage of a structural steel job and the estimated cost. As a rule of thumb, add 10% to the total weight to account for these accessories.

- Type 2 steel construction, commonly referred to as "simple construction," consists generally of field-bolted connections with lateral bracing supplied by other elements of the building, such as masonry walls or x-bracing. The estimator should be aware, however, that shop connections may be accomplished by welding or bolting. The method may be particular to the fabrication shop and may have an impact on the estimated cost.

05 12 23 Structural Steel

- Steel items can be obtained from two sources: a fabrication shop or a metals service center. Fabrication shops can fabricate items under more controlled conditions than can crews in the field. They are also more efficient and can produce items more economically. Metal service centers serve as a source of long mill shapes to both fabrication shops and contractors.

- Most line items in this structural steel subdivision, and most items in 05 50 00 Metal Fabrications, are indicated as being shop fabricated. The bare material cost for these shop fabricated items is the "Invoice Cost" from the shop and includes the mill base price of steel plus mill extras, transportation to the shop, shop drawings and detailing where warranted, shop fabrication and handling, sandblasting and a shop coat of primer paint, all necessary structural bolts, and delivery to the job site. The bare labor cost and bare equipment cost for these shop fabricated items is for field installation or erection.

- Line items in Subdivision 05 12 23.40 Lightweight Framing, and other items scattered in Division 5, are indicated as being field fabricated. The bare material cost for these field fabricated items is the "Invoice Cost" from the metals service center and includes the mill base price of steel plus mill extras, transportation to the metals service center, material handling, and delivery of long lengths of mill shapes to the job site. Material costs for structural bolts and welding rods should be added to the estimate. The bare labor cost and bare equipment cost for these items is for both field fabrication and field installation or erection, and include time for cutting, welding and drilling in the fabricated metal items. Drilling into concrete and fasteners to fasten field fabricated items to other work are not included and should be added to the estimate.

05 20 00 Steel Joist Framing

- In any given project the total weight of open web steel joists is determined by the loads to be supported and the design. However, economies can be realized in minimizing the amount of labor used to place the joists. This is done by maximizing the joist spacing, and therefore minimizing the number of joists required to be installed on the job. Certain spacings and locations may be required by the design, but in other cases maximizing the spacing and keeping it as uniform as possible will keep the costs down.

05 30 00 Steel Decking

- The takeoff and estimating of metal deck involves more than simply the area of the floor or roof and the type of deck specified or shown on the drawings. Many different sizes and types of openings may exist. Small openings for individual pipes or conduits may be drilled after the floor/roof is installed, but larger openings may require special deck lengths as well as reinforcing or structural support. The estimator should determine who will be supplying this reinforcing. Additionally, some deck terminations are part of the deck package, such as screed angles and pour stops, and others will be part of the steel contract, such as angles attached to structural members and cast-in-place angles and plates. The estimator must ensure that all pieces are accounted for in the complete estimate.

05 50 00 Metal Fabrications

- The most economical steel stairs are those that use common materials, standard details, and most importantly, a uniform and relatively simple method of field assembly. Commonly available A36 channels and plates are very good choices for the main stringers of the stairs, as are angles and tees for the carrier members. Risers and treads are usually made by specialty shops, and it is most economical to use a typical detail in as many places as possible. The stairs should be pre-assembled and shipped directly to the site. The field connections should be simple and straightforward to be accomplished efficiently, and with minimum equipment and labor.

Reference Numbers

Reference numbers are shown in shaded boxes at the beginning of some major classifications. These numbers refer to related items in the Reference Section. The reference information may be an estimating procedure, an alternate pricing method, or technical information. *Note:* Not all subdivisions listed here necessarily appear in this publication.

05 05 21 – Fastening Methods for Metal

05 05 21.15 Drilling Steel

		Crew	Daily Output	Labor-Hours	Unit	Material	2011 Bare Costs Labor	Equipment	Total	Total Incl O&P
0010	**DRILLING STEEL**									
1910	Drilling & layout for steel, up to 1/4" deep, no anchor									
1920	Holes, 1/4" diameter	1 Sswk	112	.071	Ea.	.07	3.47		3.54	6.20
1925	For each additional 1/4" depth, add		336	.024		.07	1.16		1.23	2.11
1930	3/8" diameter		104	.077		.08	3.73		3.81	6.70
1935	For each additional 1/4" depth, add		312	.026		.08	1.24		1.32	2.29
1940	1/2" diameter		96	.083		.08	4.05		4.13	7.25
1945	For each additional 1/4" depth, add		288	.028		.08	1.35		1.43	2.47
1950	5/8" diameter		88	.091		.12	4.41		4.53	7.95
1955	For each additional 1/4" depth, add		264	.030		.12	1.47		1.59	2.73
1960	3/4" diameter		80	.100		.15	4.86		5.01	8.75
1965	For each additional 1/4" depth, add		240	.033		.15	1.62		1.77	3.02
1970	7/8" diameter		72	.111		.19	5.40		5.59	9.75
1975	For each additional 1/4" depth, add		216	.037		.19	1.80		1.99	3.39
1980	1" diameter		64	.125		.20	6.05		6.25	10.90
1985	For each additional 1/4" depth, add		192	.042		.20	2.02		2.22	3.79
1990	For drilling up, add						40%			

05 05 23 – Metal Fastenings

05 05 23.05 Anchor Bolts

			Crew	Daily Output	Labor-Hours	Unit	Material	2011 Bare Costs Labor	Equipment	Total	Total Incl O&P
0010	**ANCHOR BOLTS**										
0015	Made from recycled materials	G									
0025	Single bolts installed in fresh concrete, no templates										
0030	Hooked w/nut and washer, 1/2" diameter, 8" long	G	1 Carp	132	.061	Ea.	1.36	2.61		3.97	5.50
0040	12" long	G		131	.061		1.51	2.63		4.14	5.70
0050	5/8" diameter, 8" long	G		129	.062		2.96	2.67		5.63	7.35
0060	12" long	G		127	.063		3.65	2.71		6.36	8.15
0070	3/4" diameter, 8" long	G		127	.063		3.65	2.71		6.36	8.15
0080	12" long	G		125	.064		4.56	2.76		7.32	9.25
0090	2-bolt pattern, including job-built 2-hole template, per set										
0100	J-type, incl. hex nut & washer, 1/2" diameter x 6" long	G	1 Carp	21	.381	Set	4.50	16.40		20.90	30
0110	12" long	G		21	.381		5.10	16.40		21.50	30.50
0120	18" long	G		21	.381		6	16.40		22.40	31.50
0130	3/4" diameter x 8" long	G		20	.400		9.40	17.20		26.60	37
0140	12" long	G		20	.400		11.20	17.20		28.40	39
0150	18" long	G		20	.400		13.95	17.20		31.15	42
0160	1" diameter x 12" long	G		19	.421		19.65	18.15		37.80	49.50
0170	18" long	G		19	.421		23.50	18.15		41.65	53.50
0180	24" long	G		19	.421		28	18.15		46.15	58.50
0190	36" long	G		18	.444		37.50	19.15		56.65	71
0200	1-1/2" diameter x 18" long	G		17	.471		60	20.50		80.50	97
0210	24" long	G		16	.500		71.50	21.50		93	112
0300	L-type, incl. hex nut & washer, 3/4" diameter x 12" long	G		20	.400		10.70	17.20		27.90	38.50
0310	18" long	G		20	.400		13.10	17.20		30.30	41
0320	24" long	G		20	.400		15.50	17.20		32.70	43.50
0330	30" long	G		20	.400		19.15	17.20		36.35	47.50
0340	36" long	G		20	.400		21.50	17.20		38.70	50
0350	1" diameter x 12" long	G		19	.421		16.75	18.15		34.90	46.50
0360	18" long	G		19	.421		20.50	18.15		38.65	50.50
0370	24" long	G		19	.421		24.50	18.15		42.65	55
0380	30" long	G		19	.421		28.50	18.15		46.65	59.50
0390	36" long	G		18	.444		32.50	19.15		51.65	65.50
0400	42" long	G		18	.444		39	19.15		58.15	72.50

05 05 23 – Metal Fastenings

05 05 23.05 Anchor Bolts			Crew	Daily Output	Labor-Hours	Unit	Material	2011 Bare Costs Labor	Equipment	Total	Total Incl O&P
0410	48" long	G	1 Carp	18	.444	Set	43.50	19.15		62.65	77.50
0420	1-1/4" diameter x 18" long	G		18	.444		30	19.15		49.15	62.50
0430	24" long	G		18	.444		35	19.15		54.15	68.50
0440	30" long	G		17	.471		40.50	20.50		61	75.50
0450	36" long	G		17	.471		45.50	20.50		66	81.50
0460	42" long	G	2 Carp	32	.500		51.50	21.50		73	89.50
0470	48" long	G		32	.500		58.50	21.50		80	97
0480	54" long	G		31	.516		68.50	22		90.50	109
0490	60" long	G		31	.516		75	22		97	117
0500	1-1/2" diameter x 18" long	G		33	.485		43.50	21		64.50	80
0510	24" long	G		32	.500		50.50	21.50		72	88.50
0520	30" long	G		31	.516		57	22		79	96.50
0540	42" long	G		30	.533		74	23		97	116
0550	48" long	G		29	.552		82.50	24		106.50	128
0560	54" long	G		28	.571		100	24.50		124.50	148
0570	60" long	G		28	.571		110	24.50		134.50	159
0580	1-3/4" diameter x 18" long	G		31	.516		61.50	22		83.50	102
0590	24" long	G		30	.533		72	23		95	114
0600	30" long	G		29	.552		83.50	24		107.50	128
0610	36" long	G		28	.571		95	24.50		119.50	142
0620	42" long	G		27	.593		106	25.50		131.50	156
0630	48" long	G		26	.615		116	26.50		142.50	169
0640	54" long	G		26	.615		144	26.50		170.50	199
0650	60" long	G		25	.640		155	27.50		182.50	214
0660	2" diameter x 24" long	G		27	.593		96.50	25.50		122	145
0670	30" long	G		27	.593		108	25.50		133.50	158
0680	36" long	G		26	.615		119	26.50		145.50	172
0690	42" long	G		25	.640		132	27.50		159.50	189
0700	48" long	G		24	.667		152	28.50		180.50	211
0710	54" long	G		23	.696		180	30		210	244
0720	60" long	G		23	.696		194	30		224	259
0730	66" long	G		22	.727		207	31.50		238.50	276
0740	72" long	G		21	.762		227	33		260	300
1000	4-bolt pattern, including job-built 4-hole template, per set										
1100	J-type, incl. hex nut & washer, 1/2" diameter x 6" long	G	1 Carp	19	.421	Set	6.90	18.15		25.05	35.50
1110	12" long	G		19	.421		8.15	18.15		26.30	37
1120	18" long	G		18	.444		9.95	19.15		29.10	40.50
1130	3/4" diameter x 8" long	G		17	.471		16.70	20.50		37.20	49.50
1140	12" long	G		17	.471		20.50	20.50		41	53.50
1150	18" long	G		17	.471		26	20.50		46.50	59.50
1160	1" diameter x 12" long	G		16	.500		37.50	21.50		59	74
1170	18" long	G		15	.533		44.50	23		67.50	84
1180	24" long	G		15	.533		54	23		77	94
1190	36" long	G		15	.533		73	23		96	115
1200	1-1/2" diameter x 18" long	G		13	.615		118	26.50		144.50	171
1210	24" long	G		12	.667		140	28.50		168.50	199
1300	L-type, incl. hex nut & washer, 3/4" diameter x 12" long	G		17	.471		19.25	20.50		39.75	52
1310	18" long	G		17	.471		24	20.50		44.50	57.50
1320	24" long	G		17	.471		29	20.50		49.50	63
1330	30" long	G		16	.500		36	21.50		57.50	73
1340	36" long	G		16	.500		41	21.50		62.50	78
1350	1" diameter x 12" long	G		16	.500		31.50	21.50		53	67.50
1360	18" long	G		15	.533		38.50	23		61.50	77.50

05 05 23 – Metal Fastenings

05 05 23.05 Anchor Bolts		Crew	Daily Output	Labor-Hours	Unit	Material	2011 Bare Costs Labor	Equipment	Total	Total Incl O&P
1370	24" long	G 1 Carp	15	.533	Set	47	23		70	87
1380	30" long	G	15	.533		55.50	23		78.50	96
1390	36" long	G	15	.533		63	23		86	104
1400	42" long	G	14	.571		76	24.50		100.50	121
1410	48" long	G	14	.571		85	24.50		109.50	131
1420	1-1/4" diameter x 18" long	G	14	.571		58	24.50		82.50	101
1430	24" long	G	14	.571		68.50	24.50		93	113
1440	30" long	G	13	.615		79	26.50		105.50	127
1450	36" long	G	13	.615		89.50	26.50		116	139
1460	42" long	G 2 Carp	25	.640		101	27.50		128.50	154
1470	48" long	G	24	.667		115	28.50		143.50	170
1480	54" long	G	23	.696		135	30		165	194
1490	60" long	G	23	.696		148	30		178	208
1500	1-1/2" diameter x 18" long	G	25	.640		85	27.50		112.50	136
1510	24" long	G	24	.667		99	28.50		127.50	153
1520	30" long	G	23	.696		112	30		142	169
1540	42" long	G	22	.727		146	31.50		177.50	208
1550	48" long	G	21	.762		163	33		196	230
1560	54" long	G	20	.800		198	34.50		232.50	271
1570	60" long	G	20	.800		217	34.50		251.50	292
1580	1-3/4" diameter x 18" long	G	22	.727		121	31.50		152.50	181
1590	24" long	G	21	.762		142	33		175	207
1600	30" long	G	21	.762		165	33		198	232
1610	36" long	G	20	.800		187	34.50		221.50	259
1620	42" long	G	19	.842		210	36.50		246.50	287
1630	48" long	G	18	.889		231	38.50		269.50	315
1640	54" long	G	18	.889		286	38.50		324.50	375
1650	60" long	G	17	.941		310	40.50		350.50	400
1660	2" diameter x 24" long	G	19	.842		191	36.50		227.50	266
1670	30" long	G	18	.889		215	38.50		253.50	295
1680	36" long	G	18	.889		236	38.50		274.50	320
1690	42" long	G	17	.941		263	40.50		303.50	350
1700	48" long	G	16	1		300	43		343	395
1710	54" long	G	15	1.067		360	46		406	465
1720	60" long	G	15	1.067		385	46		431	495
1730	66" long	G	14	1.143		410	49		459	530
1740	72" long	G	14	1.143		450	49		499	570
1990	For galvanized, add				Ea.	75%				

05 05 23.15 Chemical Anchors

		Crew	Daily Output	Labor-Hours	Unit	Material	Labor	Equipment	Total	Total Incl O&P
0010	**CHEMICAL ANCHORS**									
0020	Includes layout & drilling									
1430	Chemical anchor, w/rod & epoxy cartridge, 3/4" diam. x 9-1/2" long	B-89A	27	.593	Ea.	8.25	23.50	4.07	35.82	49.50
1435	1" diameter x 11-3/4" long		24	.667		16.05	26.50	4.57	47.12	63
1440	1-1/4" diameter x 14" long		21	.762		29.50	30	5.25	64.75	84.50
1445	1-3/4" diameter x 15" long		20	.800		58	31.50	5.50	95	118
1450	18" long		17	.941		69.50	37	6.45	112.95	141
1455	2" diameter x 18" long		16	1		84	39.50	6.85	130.35	161
1460	24" long		15	1.067		109	42	7.30	158.30	193

05 05 23.20 Expansion Anchors

		Crew	Daily Output	Labor-Hours	Unit	Material	Labor	Equipment	Total	Total Incl O&P
0010	**EXPANSION ANCHORS**									
0100	Anchors for concrete, brick or stone, no layout and drilling									
0200	Expansion shields, zinc, 1/4" diameter, 1-5/16" long, single	G 1 Carp	90	.089	Ea.	.42	3.83		4.25	6.30

05 05 23 – Metal Fastenings

05 05 23.20 Expansion Anchors		Crew	Daily Output	Labor-Hours	Unit	Material	2011 Bare Costs Labor	Equipment	Total	Total Incl O&P
0300	1-3/8" long, double	G	1 Carp	85	.094	Ea.	.50	4.05		4.55	6.75
0400	3/8" diameter, 1-1/2" long, single	G		85	.094		.72	4.05		4.77	7
0500	2" long, double	G		80	.100		1.08	4.31		5.39	7.80
0600	1/2" diameter, 2-1/16" long, single	G		80	.100		1.36	4.31		5.67	8.10
0700	2-1/2" long, double	G		75	.107		1.65	4.59		6.24	8.85
0800	5/8" diameter, 2-5/8" long, single	G		75	.107		2.46	4.59		7.05	9.75
0900	2-3/4" long, double	G		70	.114		2.39	4.92		7.31	10.20
1000	3/4" diameter, 2-3/4" long, single	G		70	.114		2.74	4.92		7.66	10.55
1100	3-15/16" long, double	G	▼	65	.123	▼	4.65	5.30		9.95	13.25
2100	Hollow wall anchors for gypsum wall board, plaster or tile										
2300	1/8" diameter, short	G	1 Carp	160	.050	Ea.	.19	2.15		2.34	3.51
2400	Long	G		150	.053		.22	2.30		2.52	3.76
2500	3/16" diameter, short	G		150	.053		.34	2.30		2.64	3.89
2600	Long	G		140	.057		.48	2.46		2.94	4.30
2700	1/4" diameter, short	G		140	.057		.56	2.46		3.02	4.39
2800	Long	G		130	.062		.52	2.65		3.17	4.63
3000	Toggle bolts, bright steel, 1/8" diameter, 2" long	G		85	.094		.18	4.05		4.23	6.40
3100	4" long	G		80	.100		.23	4.31		4.54	6.85
3200	3/16" diameter, 3" long	G		80	.100		.21	4.31		4.52	6.85
3300	6" long	G		75	.107		.31	4.59		4.90	7.40
3400	1/4" diameter, 3" long	G		75	.107		.27	4.59		4.86	7.35
3500	6" long	G		70	.114		.45	4.92		5.37	8.05
3600	3/8" diameter, 3" long	G		70	.114		.67	4.92		5.59	8.30
3700	6" long	G		60	.133		1.13	5.75		6.88	10.05
3800	1/2" diameter, 4" long	G		60	.133		1.67	5.75		7.42	10.65
3900	6" long	G	▼	50	.160	▼	2.30	6.90		9.20	13.10
5000	Screw anchors for concrete, masonry,										
5100	stone & tile, no layout or drilling included										
5700	Lag screw shields, 1/4" diameter, short	G	1 Carp	90	.089	Ea.	.31	3.83		4.14	6.20
5800	Long	G		85	.094		.40	4.05		4.45	6.65
5900	3/8" diameter, short	G		85	.094		.59	4.05		4.64	6.85
6000	Long	G		80	.100		.73	4.31		5.04	7.40
6100	1/2" diameter, short	G		80	.100		.99	4.31		5.30	7.70
6200	Long	G		75	.107		1.35	4.59		5.94	8.55
6300	5/8" diameter, short	G		70	.114		1.58	4.92		6.50	9.30
6400	Long	G		65	.123		2.12	5.30		7.42	10.50
6600	Lead, #6 & #8, 3/4" long	G		260	.031		.16	1.32		1.48	2.21
6700	#10 - #14, 1-1/2" long	G		200	.040		.26	1.72		1.98	2.93
6800	#16 & #18, 1-1/2" long	G		160	.050		.32	2.15		2.47	3.65
6900	Plastic, #6 & #8, 3/4" long			260	.031		.04	1.32		1.36	2.07
7000	#8 & #10, 7/8" long			240	.033		.04	1.43		1.47	2.24
7100	#10 & #12, 1" long			220	.036		.05	1.57		1.62	2.46
7200	#14 & #16, 1-1/2" long		▼	160	.050	▼	.07	2.15		2.22	3.38
8000	Wedge anchors, not including layout or drilling										
8050	Carbon steel, 1/4" diameter, 1-3/4" long	G	1 Carp	150	.053	Ea.	.32	2.30		2.62	3.87
8100	3-1/4" long	G		140	.057		.42	2.46		2.88	4.23
8150	3/8" diameter, 2-1/4" long	G		145	.055		.38	2.38		2.76	4.06
8200	5" long	G		140	.057		.67	2.46		3.13	4.51
8250	1/2" diameter, 2-3/4" long	G		140	.057		.79	2.46		3.25	4.64
8300	7" long	G		125	.064		1.35	2.76		4.11	5.70
8350	5/8" diameter, 3-1/2" long	G		130	.062		1.40	2.65		4.05	5.60
8400	8-1/2" long	G		115	.070		2.97	3		5.97	7.85
8450	3/4" diameter, 4-1/4" long	G	▼	115	.070		2.09	3		5.09	6.90

05 05 23 – Metal Fastenings

05 05 23.20 Expansion Anchors		Crew	Daily Output	Labor-Hours	Unit	Material	2011 Bare Costs Labor	Equipment	Total	Total Incl O&P
8500	10" long	G	1 Carp	95	.084	Ea.	4.75	3.63		8.38	10.80
8550	1" diameter, 6" long	G		100	.080		7.65	3.44		11.09	13.70
8575	9" long	G		85	.094		9.95	4.05		14	17.10
8600	12" long	G		75	.107		10.70	4.59		15.29	18.85
8650	1-1/4" diameter, 9" long	G		70	.114		21.50	4.92		26.42	31
8700	12" long	G		60	.133		27.50	5.75		33.25	39
8750	For type 303 stainless steel, add						350%				
8800	For type 316 stainless steel, add						450%				
8950	Self-drilling concrete screw, hex washer head, 3/16" diam. x 1-3/4" long	G	1 Carp	300	.027	Ea.	.18	1.15		1.33	1.96
8960	2-1/4" long	G		250	.032		.20	1.38		1.58	2.33
8970	Phillips flat head, 3/16" diam. x 1-3/4" long	G		300	.027		.18	1.15		1.33	1.96
8980	2-1/4" long	G		250	.032		.19	1.38		1.57	2.32

05 05 23.30 Lag Screws

			Crew	Daily Output	Labor-Hours	Unit	Material	Labor	Equipment	Total	Total Incl O&P
0010	**LAG SCREWS**										
0020	Steel, 1/4" diameter, 2" long	G	1 Carp	200	.040	Ea.	.09	1.72		1.81	2.74
0100	3/8" diameter, 3" long	G		150	.053		.26	2.30		2.56	3.81
0200	1/2" diameter, 3" long	G		130	.062		.48	2.65		3.13	4.59
0300	5/8" diameter, 3" long	G		120	.067		.99	2.87		3.86	5.50

05 05 23.35 Machine Screws

			Crew	Daily Output	Labor-Hours	Unit	Material	Labor	Equipment	Total	Total Incl O&P
0010	**MACHINE SCREWS**										
0020	Steel, round head, #8 x 1" long	G	1 Carp	4.80	1.667	C	3.73	72		75.73	114
0110	#8 x 2" long	G		2.40	3.333		5.75	144		149.75	226
0200	#10 x 1" long	G		4	2		5.80	86		91.80	138
0300	#10 x 2" long	G		2	4		7.80	172		179.80	273

05 05 23.40 Machinery Anchors

			Crew	Daily Output	Labor-Hours	Unit	Material	Labor	Equipment	Total	Total Incl O&P
0010	**MACHINERY ANCHORS**, heavy duty, incl. sleeve, floating base nut,										
0020	lower stud & coupling nut, fiber plug, connecting stud, washer & nut.										
0030	For flush mounted embedment in poured concrete heavy equip. pads.										
0200	Stud & bolt, 1/2" diameter	G	E-16	40	.400	Ea.	68.50	19.80	2.72	91.02	114
0300	5/8" diameter	G		35	.457		76	22.50	3.11	101.61	127
0500	3/4" diameter	G		30	.533		87.50	26.50	3.63	117.63	147
0600	7/8" diameter	G		25	.640		95.50	31.50	4.36	131.36	166
0800	1" diameter	G		20	.800		100	39.50	5.45	144.95	186
0900	1-1/4" diameter	G		15	1.067		133	53	7.25	193.25	249

05 05 23.50 Powder Actuated Tools and Fasteners

			Crew	Daily Output	Labor-Hours	Unit	Material	Labor	Equipment	Total	Total Incl O&P
0010	**POWDER ACTUATED TOOLS & FASTENERS**										
0020	Stud driver, .22 caliber, buy, minimum					Ea.	104			104	114
0100	Maximum					"	370			370	405
0300	Powder charges for above, low velocity					C	7			7	7.70
0400	Standard velocity						7.70			7.70	8.50
0600	Drive pins & studs, 1/4" & 3/8" diam., to 3" long, minimum	G	1 Carp	4.80	1.667		4.02	72		76.02	114
0700	Maximum	G	"	4	2		12.85	86		98.85	146

05 05 23.55 Rivets

			Crew	Daily Output	Labor-Hours	Unit	Material	Labor	Equipment	Total	Total Incl O&P
0010	**RIVETS**										
0100	Aluminum rivet & mandrel, 1/2" grip length x 1/8" diameter	G	1 Carp	4.80	1.667	C	6.60	72		78.60	117
0200	3/16" diameter	G		4	2		11.20	86		97.20	144
0300	Aluminum rivet, steel mandrel, 1/8" diameter	G		4.80	1.667		8.75	72		80.75	120
0400	3/16" diameter	G		4	2		14	86		100	147
0500	Copper rivet, steel mandrel, 1/8" diameter	G		4.80	1.667		7.70	72		79.70	119
1200	Steel rivet and mandrel, 1/8" diameter	G		4.80	1.667		7.50	72		79.50	118
1300	3/16" diameter	G		4	2		11.10	86		97.10	144

05 05 Common Work Results for Metals

05 05 23 – Metal Fastenings

05 05 23.55 Rivets		Crew	Daily Output	Labor-Hours	Unit	Material	2011 Bare Costs Labor	Equipment	Total	Total Incl O&P
1400	Hand riveting tool, minimum				Ea.	55			55	60.50
1500	Maximum					257			257	283
1600	Power riveting tool, minimum					450			450	495
1700	Maximum					2,450			2,450	2,700

05 05 23.70 Structural Blind Bolts										
0010	**STRUCTURAL BLIND BOLTS**									
0100	1/4" diameter x 1/4" grip	G	1 Sswk	240	.033	Ea.	1.24	1.62		2.86	4.22
0150	1/2" grip	G		216	.037		.95	1.80		2.75	4.23
0200	3/8" diameter x 1/2" grip	G		232	.034		1.75	1.67		3.42	4.89
0250	3/4" grip	G		208	.038		1.83	1.87		3.70	5.30
0300	1/2" diameter x 1/2" grip	G		224	.036		5.15	1.73		6.88	8.70
0350	3/4" grip	G		200	.040		5.60	1.94		7.54	9.60
0400	5/8" diameter x 3/4" grip	G		216	.037		8.25	1.80		10.05	12.30
0450	1" grip	G		192	.042		9.05	2.02		11.07	13.50

05 12 Structural Steel Framing

05 12 23 – Structural Steel for Buildings

05 12 23.40 Lightweight Framing		Crew	Daily Output	Labor-Hours	Unit	Material	2011 Bare Costs Labor	Equipment	Total	Total Incl O&P
0010	**LIGHTWEIGHT FRAMING**									
0015	Made from recycled materials	G								
0400	Angle framing, field fabricated, 4" and larger	G	E-3	440	.055	Lb.	.65	2.69	.25	3.59	5.75
0450	Less than 4" angles	G		265	.091		.68	4.46	.41	5.55	9.05
0600	Channel framing, field fabricated, 8" and larger	G		500	.048		.68	2.36	.22	3.26	5.15
0650	Less than 8" channels	G		335	.072		.68	3.53	.33	4.54	7.35
1000	Continuous slotted channel framing system, shop fab, minimum	G	2 Sswk	2400	.007		3.49	.32		3.81	4.41
1200	Maximum	G	"	1600	.010		3.94	.49		4.43	5.20
1250	Plate & bar stock for reinforcing beams and trusses	G					1.24			1.24	1.36
1300	Cross bracing, rods, shop fabricated, 3/4" diameter	G	E-3	700	.034		1.35	1.69	.16	3.20	4.64
1310	7/8" diameter	G		850	.028		1.35	1.39	.13	2.87	4.08
1320	1" diameter	G		1000	.024		1.35	1.18	.11	2.64	3.70
1330	Angle, 5" x 5" x 3/8"	G		2800	.009		1.35	.42	.04	1.81	2.27
1350	Hanging lintels, shop fabricated, average	G		850	.028		1.35	1.39	.13	2.87	4.08
1380	Roof frames, shop fabricated, 3'-0" square, 5' span	G	E-2	4200	.013		1.35	.63	.39	2.37	2.98
1400	Tie rod, not upset, 1-1/2" to 4" diameter, with turnbuckle	G	2 Sswk	800	.020		1.46	.97		2.43	3.33
1420	No turnbuckle	G		700	.023		1.41	1.11		2.52	3.51
1500	Upset, 1-3/4" to 4" diameter, with turnbuckle	G		800	.020		1.46	.97		2.43	3.33
1520	No turnbuckle	G		700	.023		1.41	1.11		2.52	3.51

05 12 23.60 Pipe Support Framing										
0010	**PIPE SUPPORT FRAMING**									
0020	Under 10#/L.F., shop fabricated	G	E-4	3900	.008	Lb.	1.51	.40	.03	1.94	2.40
0200	10.1 to 15#/L.F.	G		4300	.007		1.49	.36	.03	1.88	2.30
0400	15.1 to 20#/L.F.	G		4800	.007		1.46	.33	.02	1.81	2.22
0600	Over 20#/L.F.	G		5400	.006		1.44	.29	.02	1.75	2.11

Division Notes

	CREW	DAILY OUTPUT	LABOR-HOURS	UNIT	2011 BARE COSTS				TOTAL INCL O&P
					MAT.	LABOR	EQUIP.	TOTAL	

Estimating Tips

06 05 00 Common Work Results for Wood, Plastics, and Composites

- Common to any wood-framed structure are the accessory connector items such as screws, nails, adhesives, hangers, connector plates, straps, angles, and hold-downs. For typical wood-framed buildings, such as residential projects, the aggregate total for these items can be significant, especially in areas where seismic loading is a concern. For floor and wall framing, the material cost is based on 10 to 25 lbs. per MBF. Hold-downs, hangers, and other connectors should be taken off by the piece.

  Included with material costs are fasteners for a normal installation. RSMeans engineers use manufacturer's recommendations, written specifications, and/or standard construction practice for size and spacing of fasteners. Prices for various fasteners are shown for informational purposes only. Adjustments should be made if unusual fastening conditions exist.

06 10 00 Carpentry

- Lumber is a traded commodity and therefore sensitive to supply and demand in the marketplace. Even in "budgetary" estimating of wood-framed projects, it is advisable to call local suppliers for the latest market pricing.

- Common quantity units for wood-framed projects are "thousand board feet" (MBF). A board foot is a volume of wood, 1" x 1' x 1', or 144 cubic inches. Board-foot quantities are generally calculated using nominal material dimensions—dressed sizes are ignored. Board foot per lineal foot of any stick of lumber can be calculated by dividing the nominal cross-sectional area by 12. As an example, 2,000 lineal feet of 2 x 12 equates to 4 MBF by dividing the nominal area,2 x 12, by 12, which equals 2, and multiplying by 2,000 to give 4,000 board feet. This simple rule applies to all nominal dimensioned lumber.

- Waste is an issue of concern at the quantity takeoff for any area of construction. Framing lumber is sold in even foot lengths, i.e., 10', 12', 14', 16', and depending on spans, wall heights, and the grade of lumber, waste is inevitable. A rule of thumb for lumber waste is 5%–10% depending on material quality and the complexity of the framing.

- Wood in various forms and shapes is used in many projects, even where the main structural framing is steel, concrete, or masonry. Plywood as a back-up partition material and 2x boards used as blocking and cant strips around roof edges are two common examples. The estimator should ensure that the costs of all wood materials are included in the final estimate.

06 20 00 Finish Carpentry

- It is necessary to consider the grade of workmanship when estimating labor costs for erecting millwork and interior finish. In practice, there are three grades: premium, custom, and economy. The RSMeans daily output for base and case moldings is in the range of 200 to 250 L.F. per carpenter per day. This is appropriate for most average custom-grade projects. For premium projects, an adjustment to productivity of 25%–50% should be made, depending on the complexity of the job.

Reference Numbers

Reference numbers are shown in shaded boxes at the beginning of some major classifications. These numbers refer to related items in the Reference Section. The reference information may be an estimating procedure, an alternate pricing method, or technical information.

Note: Not all subdivisions listed here necessarily appear in this publication.

06 05 Common Work Results for Wood, Plastics, and Composites

06 05 23 – Wood, Plastic, and Composite Fastenings

06 05 23.10 Nails

		Crew	Daily Output	Labor-Hours	Unit	Material	2011 Bare Costs Labor	Equipment	Total	Total Incl O&P
0010	**NAILS**, material only, based upon 50# box purchase									
0020	Copper nails, plain				Lb.	10			10	11
0400	Stainless steel, plain					7.80			7.80	8.60
0500	Box, 3d to 20d, bright					1.25			1.25	1.38
0520	Galvanized					1.77			1.77	1.95
0600	Common, 3d to 60d, plain					1.30			1.30	1.43
0700	Galvanized					1.56			1.56	1.72
0800	Aluminum					5.80			5.80	6.40
1000	Annular or spiral thread, 4d to 60d, plain					1.80			1.80	1.98
1200	Galvanized					3			3	3.30
1400	Drywall nails, plain					3.59			3.59	3.95
1600	Galvanized					2.19			2.19	2.41
1800	Finish nails, 4d to 10d, plain					1.36			1.36	1.50
2000	Galvanized					1.79			1.79	1.97
2100	Aluminum					5.80			5.80	6.40
2300	Flooring nails, hardened steel, 2d to 10d, plain					3.10			3.10	3.41
2400	Galvanized					4			4	4.40
2500	Gypsum lath nails, 1-1/8", 13 ga. flathead, blued					3.59			3.59	3.95
2600	Masonry nails, hardened steel, 3/4" to 3" long, plain					2.20			2.20	2.42
2700	Galvanized					3.75			3.75	4.13
5000	Add to prices above for cement coating					.11			.11	.12
5200	Zinc or tin plating					.14			.14	.15
5500	Vinyl coated sinkers, 8d to 16d					.65			.65	.72

06 05 23.40 Sheet Metal Screws

		Crew	Daily Output	Labor-Hours	Unit	Material	2011 Bare Costs Labor	Equipment	Total	Total Incl O&P
0010	**SHEET METAL SCREWS**									
0020	Steel, standard, #8 x 3/4", plain				C	2.92			2.92	3.21
0100	Galvanized					2.92			2.92	3.21
0300	#10 x 1", plain					4.03			4.03	4.43
0400	Galvanized					4.03			4.03	4.43
0600	With washers, #14 x 1", plain					12.35			12.35	13.60
0700	Galvanized					12.35			12.35	13.60
0900	#14 x 2", plain					16.90			16.90	18.60
1000	Galvanized					16.90			16.90	18.60
1500	Self-drilling, with washers, (pinch point) #8 x 3/4", plain					7.80			7.80	8.55
1600	Galvanized					7.80			7.80	8.55
1800	#10 x 3/4", plain					8.90			8.90	9.80
1900	Galvanized					8.90			8.90	9.80
3000	Stainless steel w/aluminum or neoprene washers, #14 x 1", plain					30			30	33
3100	#14 x 2", plain					40			40	44

06 05 23.50 Wood Screws

		Crew	Daily Output	Labor-Hours	Unit	Material	2011 Bare Costs Labor	Equipment	Total	Total Incl O&P
0010	**WOOD SCREWS**									
0020	#8 x 1" long, steel				C	3.41			3.41	3.75
0100	Brass					10.50			10.50	11.55
0200	#8, 2" long, steel					5.25			5.25	5.80
0300	Brass					19.95			19.95	22
0400	#10, 1" long, steel					3.65			3.65	4.02
0500	Brass					14.15			14.15	15.60
0600	#10, 2" long, steel					5.85			5.85	6.45
0700	Brass					24.50			24.50	27
0800	#10, 3" long, steel					9.80			9.80	10.75
1000	#12, 2" long, steel					8.05			8.05	8.85
1100	Brass					29.50			29.50	32

06 05 Common Work Results for Wood, Plastics, and Composites

06 05 23 – Wood, Plastic, and Composite Fastenings

06 05 23.50 Wood Screws	Crew	Daily Output	Labor-Hours	Unit	Material	Labor	Equipment	Total	Total Incl O&P
1500	#12, 3" long, steel				C	11.65			11.65	12.80
2000	#12, 4" long, steel					21.50			21.50	23.50

06 16 Sheathing

06 16 36 – Wood Panel Product Sheathing

06 16 36.10 Sheathing

		Crew	Daily Output	Labor-Hours	Unit	Material	Labor	Equipment	Total	Total Incl O&P
0010	**SHEATHING**									
0012	Plywood on roofs, CDX									
0030	5/16" thick	2 Carp	1600	.010	S.F.	.55	.43		.98	1.26
0035	Pneumatic nailed		1952	.008		.55	.35		.90	1.14
0050	3/8" thick		1525	.010		.47	.45		.92	1.21
0055	Pneumatic nailed		1860	.009		.47	.37		.84	1.09
0100	1/2" thick		1400	.011		.58	.49		1.07	1.40
0105	Pneumatic nailed		1708	.009		.58	.40		.98	1.26
0200	5/8" thick		1300	.012		.70	.53		1.23	1.58
0205	Pneumatic nailed		1586	.010		.70	.43		1.13	1.44
0300	3/4" thick		1200	.013		.82	.57		1.39	1.78
0305	Pneumatic nailed		1464	.011		.82	.47		1.29	1.62
0500	Plywood on walls with exterior CDX, 3/8" thick		1200	.013		.47	.57		1.04	1.40
0505	Pneumatic nailed		1488	.011		.47	.46		.93	1.23
0600	1/2" thick		1125	.014		.58	.61		1.19	1.58
0605	Pneumatic nailed		1395	.011		.58	.49		1.07	1.40
0700	5/8" thick		1050	.015		.70	.66		1.36	1.78
0705	Pneumatic nailed		1302	.012		.70	.53		1.23	1.58
0800	3/4" thick		975	.016		.82	.71		1.53	1.98
0805	Pneumatic nailed		1209	.013		.82	.57		1.39	1.77

Division Notes

	CREW	DAILY OUTPUT	LABOR-HOURS	UNIT	2011 BARE COSTS				TOTAL INCL O&P
					MAT.	LABOR	EQUIP.	TOTAL	

Estimating Tips

07 10 00 Dampproofing and Waterproofing

- Be sure of the job specifications before pricing this subdivision. The difference in cost between waterproofing and dampproofing can be great. Waterproofing will hold back standing water. Dampproofing prevents the transmission of water vapor. Also included in this section are vapor retarding membranes.

07 20 00 Thermal Protection

- Insulation and fireproofing products are measured by area, thickness, volume or R-value. Specifications may give only what the specific R-value should be in a certain situation. The estimator may need to choose the type of insulation to meet that R-value.

07 30 00 Steep Slope Roofing
07 40 00 Roofing and Siding Panels

- Many roofing and siding products are bought and sold by the square. One square is equal to an area that measures 100 square feet.

This simple change in unit of measure could create a large error if the estimator is not observant. Accessories necessary for a complete installation must be figured into any calculations for both material and labor.

07 50 00 Membrane Roofing
07 60 00 Flashing and Sheet Metal
07 70 00 Roofing and Wall Specialties and Accessories

- The items in these subdivisions compose a roofing system. No one component completes the installation, and all must be estimated. Built-up or single-ply membrane roofing systems are made up of many products and installation trades. Wood blocking at roof perimeters or penetrations, parapet coverings, reglets, roof drains, gutters, downspouts, sheet metal flashing, skylights, smoke vents, and roof hatches all need to be considered along with the roofing material. Several different installation trades will need to work together on the roofing system. Inherent difficulties in the scheduling and coordination of various trades must be accounted for when estimating labor costs.

07 90 00 Joint Protection

- To complete the weather-tight shell, the sealants and caulkings must be estimated. Where different materials meet—at expansion joints, at flashing penetrations, and at hundreds of other locations throughout a construction project—they provide another line of defense against water penetration. Often, an entire system is based on the proper location and placement of caulking or sealants. The detailed drawings that are included as part of a set of architectural plans show typical locations for these materials. When caulking or sealants are shown at typical locations, this means the estimator must include them for all the locations where this detail is applicable. Be careful to keep different types of sealants separate, and remember to consider backer rods and primers if necessary.

Reference Numbers

Reference numbers are shown in shaded boxes at the beginning of some major classifications. These numbers refer to related items in the Reference Section. The reference information may be an estimating procedure, an alternate pricing method, or technical information.

Note: Not all subdivisions listed here necessarily appear in this publication.

07 65 Flexible Flashing

07 65 10 – Sheet Metal Flashing

07 65 10.10 Sheet Metal Flashing and Counter Flashing	Crew	Daily Output	Labor-Hours	Unit	Material	2011 Bare Costs Labor	Equipment	Total	Total Incl O&P
0010 **SHEET METAL FLASHING AND COUNTER FLASHING**									
0011 Including up to 4 bends									
0020 Aluminum, mill finish, .013" thick	1 Rofc	145	.055	S.F.	.73	2.02		2.75	4.16
0030 .016" thick		145	.055		.84	2.02		2.86	4.28
0060 .019" thick		145	.055		1.07	2.02		3.09	4.54
0100 .032" thick		145	.055		1.25	2.02		3.27	4.74
0200 .040" thick		145	.055		2.14	2.02		4.16	5.70
0300 .050" thick		145	.055		2.25	2.02		4.27	5.85
0400 Painted finish, add					.34			.34	.37
1600 Copper, 16 oz, sheets, under 1000 lb.	1 Rofc	115	.070		5.90	2.54		8.44	10.75
1700 Over 4000 lb.		155	.052		5.55	1.89		7.44	9.25
1900 20 oz sheets, under 1000 lb.		110	.073		7.35	2.66		10.01	12.55
2000 Over 4000 lb.		145	.055		7	2.02		9.02	11.05
2200 24 oz sheets, under 1000 lb.		105	.076		9.25	2.78		12.03	14.85
2300 Over 4000 lb.		135	.059		8.80	2.17		10.97	13.25
2500 32 oz sheets, under 1000 lb.		100	.080		11.10	2.92		14.02	17.05
2600 Over 4000 lb.		130	.062		10.55	2.25		12.80	15.35
5800 Lead, 2.5 lb. per S.F., up to 12" wide		135	.059		5.40	2.17		7.57	9.55
5900 Over 12" wide		135	.059		4.43	2.17		6.60	8.50
8900 Stainless steel sheets, 32 ga, .010" thick		155	.052		4	1.89		5.89	7.55
9000 28 ga, .015" thick		155	.052		5.65	1.89		7.54	9.35
9100 26 ga, .018" thick		155	.052		7.25	1.89		9.14	11.15
9200 24 ga, .025" thick		155	.052		8.90	1.89		10.79	12.95
9400 Terne coated stainless steel, .015" thick, 28 ga		155	.052		7.35	1.89		9.24	11.25
9500 .018" thick, 26 ga		155	.052		8.30	1.89		10.19	12.25
9600 Zinc and copper alloy (brass), .020" thick		155	.052		11.10	1.89		12.99	15.35
9700 .027" thick		155	.052		15.60	1.89		17.49	20.50
9800 .032" thick		155	.052		18.80	1.89		20.69	23.50
9900 .040" thick		155	.052		21	1.89		22.89	26

07 65 13 – Laminated Sheet Flashing

07 65 13.10 Laminated Sheet Flashing	Crew	Daily Output	Labor-Hours	Unit	Material	2011 Bare Costs Labor	Equipment	Total	Total Incl O&P
0010 **LAMINATED SHEET FLASHING**, Including up to 4 bends									
4300 Copper-clad stainless steel, .015" thick, under 500 lb.	1 Rofc	115	.070	S.F.	5.80	2.54		8.34	10.65
4400 Over 2000 lb.		155	.052		5.55	1.89		7.44	9.25
4600 .018" thick, under 500 lb.		100	.080		6.55	2.92		9.47	12.05
4700 Over 2000 lb.		145	.055		6.20	2.02		8.22	10.20
8550 3 ply copper and fabric, 3 oz		155	.052		2.32	1.89		4.21	5.70
8600 7 oz		155	.052		3.25	1.89		5.14	6.70
8700 Lead on copper and fabric, 5 oz		155	.052		2.78	1.89		4.67	6.20
8800 7 oz		155	.052		5.25	1.89		7.14	8.95

07 65 19 – Plastic Sheet Flashing

07 65 19.10 Plastic Sheet Flashing and Counter Flashing	Crew	Daily Output	Labor-Hours	Unit	Material	2011 Bare Costs Labor	Equipment	Total	Total Incl O&P
0010 **PLASTIC SHEET FLASHING AND COUNTER FLASHING**									
7300 Polyvinyl chloride, black, .010" thick	1 Rofc	285	.028	S.F.	.21	1.03		1.24	1.94
7400 .020" thick		285	.028		.30	1.03		1.33	2.04
7600 .030" thick		285	.028		.39	1.03		1.42	2.14
7700 .056" thick		285	.028		.94	1.03		1.97	2.74
7900 Black or white for exposed roofs, .060" thick		285	.028		2	1.03		3.03	3.91
8060 PVC tape, 5" x 45 mils, for joint covers, 100 L.F./roll				Ea.	128			128	141
8850 Polyvinyl chloride, .030" thick	1 Rofc	160	.050	S.F.	.46	1.83		2.29	3.55

07 71 Roof Specialties

07 71 23 – Manufactured Gutters and Downspouts

07 71 23.10 Downspouts		Crew	Daily Output	Labor-Hours	Unit	Material	2011 Bare Costs Labor	Equipment	Total	Total Incl O&P
0010	**DOWNSPOUTS**									
0020	Aluminum 2" x 3", .020" thick, embossed	1 Shee	190	.042	L.F.	1.03	2.18		3.21	4.42
0100	Enameled		190	.042		1.26	2.18		3.44	4.68
4800	Steel, galvanized, round, corrugated, 2" or 3" diameter, 28 gauge		190	.042		1.72	2.18		3.90	5.20
4900	4" diameter, 28 gauge		145	.055		2.35	2.85		5.20	6.90
5100	5" diameter, 28 gauge		130	.062		3.48	3.18		6.66	8.65
5200	26 gauge		130	.062		3.60	3.18		6.78	8.75
5400	6" diameter, 28 gauge		105	.076		4	3.94		7.94	10.35
5500	26 gauge		105	.076		5.20	3.94		9.14	11.70

07 72 Roof Accessories

07 72 26 – Ridge Vents

07 72 26.10 Ridge Vents and Accessories

		Crew	Daily Output	Labor-Hours	Unit	Material	2011 Bare Costs Labor	Equipment	Total	Total Incl O&P
0010	**RIDGE VENTS AND ACCESSORIES**									
2300	Ridge vent strip, mill finish	1 Shee	155	.052	L.F.	2.73	2.67		5.40	7.05

07 72 53 – Snow Guards

07 72 53.10 Snow Guard Options

		Crew	Daily Output	Labor-Hours	Unit	Material	2011 Bare Costs Labor	Equipment	Total	Total Incl O&P
0010	**SNOW GUARD OPTIONS**									
0100	Slate & asphalt shingle roofs, fastened with nails	1 Rofc	160	.050	Ea.	9.40	1.83		11.23	13.40
0200	Standing seam metal roofs, fastened with set screws		48	.167		14.60	6.10		20.70	26
0300	Surface mount for metal roofs, fastened with solder		48	.167		6.60	6.10		12.70	17.40
0400	Double rail pipe type, including pipe		130	.062	L.F.	22.50	2.25		24.75	28.50

07 72 73 – Pitch Pockets

07 72 73.10 Pitch Pockets, Variable Sizes

		Crew	Daily Output	Labor-Hours	Unit	Material	2011 Bare Costs Labor	Equipment	Total	Total Incl O&P
0010	**PITCH POCKETS, VARIABLE SIZES**									
0100	Adjustable, 4" to 7", welded corners, 4" deep	1 Rofc	48	.167	Ea.	12.90	6.10		19	24.50
0200	Side extenders, 6"	"	240	.033	"	1.97	1.22		3.19	4.20

07 84 Firestopping

07 84 13 – Penetration Firestopping

07 84 13.10 Firestopping

			Crew	Daily Output	Labor-Hours	Unit	Material	2011 Bare Costs Labor	Equipment	Total	Total Incl O&P
0010	**FIRESTOPPING**	R078413-30									
0100	Metallic piping, non insulated										
0110	Through walls, 2" diameter		1 Carp	16	.500	Ea.	13.75	21.50		35.25	48
0120	4" diameter			14	.571		21	24.50		45.50	60.50
0130	6" diameter			12	.667		28.50	28.50		57	75
0140	12" diameter			10	.800		50	34.50		84.50	108
0150	Through floors, 2" diameter			32	.250		8.35	10.75		19.10	25.50
0160	4" diameter			28	.286		12	12.30		24.30	32
0170	6" diameter			24	.333		15.75	14.35		30.10	39.50
0180	12" diameter			20	.400		26.50	17.20		43.70	56
0190	Metallic piping, insulated										
0200	Through walls, 2" diameter		1 Carp	16	.500	Ea.	19.50	21.50		41	54.50
0210	4" diameter			14	.571		26.50	24.50		51	67
0220	6" diameter			12	.667		34	28.50		62.50	81.50
0230	12" diameter			10	.800		55.50	34.50		90	114
0240	Through floors, 2" diameter			32	.250		14.10	10.75		24.85	32
0250	4" diameter			28	.286		17.70	12.30		30	38.50

07 84 13 – Penetration Firestopping

07 84 13.10 Firestopping		Crew	Daily Output	Labor-Hours	Unit	Material	2011 Bare Costs Labor	Equipment	Total	Total Incl O&P
0260	6" diameter	1 Carp	24	.333	Ea.	21.50	14.35		35.85	45.50
0270	12" diameter	↓	20	.400	↓	26.50	17.20		43.70	56
0280	Non metallic piping, non insulated									
0290	Through walls, 2" diameter	1 Carp	12	.667	Ea.	56.50	28.50		85	107
0300	4" diameter		10	.800		71	34.50		105.50	131
0310	6" diameter		8	1		99	43		142	175
0330	Through floors, 2" diameter		16	.500		44.50	21.50		66	82
0340	4" diameter		6	1.333		55	57.50		112.50	149
0350	6" diameter	↓	6	1.333	↓	66	57.50		123.50	161
0370	Ductwork, insulated & non insulated, round									
0380	Through walls, 6" diameter	1 Carp	12	.667	Ea.	29	28.50		57.50	76
0390	12" diameter		10	.800		57	34.50		91.50	116
0400	18" diameter		8	1		93	43		136	169
0410	Through floors, 6" diameter		16	.500		15.75	21.50		37.25	50.50
0420	12" diameter		14	.571		29	24.50		53.50	69.50
0430	18" diameter	↓	12	.667	↓	50	28.50		78.50	99
0440	Ductwork, insulated & non insulated, rectangular									
0450	With stiffener/closure angle, through walls, 6" x 12"	1 Carp	8	1	Ea.	24	43		67	92
0460	12" x 24"		6	1.333		31.50	57.50		89	123
0470	24" x 48"		4	2		90.50	86		176.50	232
0480	With stiffener/closure angle, through floors, 6" x 12"		10	.800		12.95	34.50		47.45	67
0490	12" x 24"		8	1		23.50	43		66.50	91.50
0500	24" x 48"	↓	6	1.333	↓	45.50	57.50		103	138
0510	Multi trade openings									
0520	Through walls, 6" x 12"	1 Carp	2	4	Ea.	50	172		222	320
0530	12" x 24"	"	1	8		202	345		547	750
0540	24" x 48"	2 Carp	1	16		810	690		1,500	1,950
0550	48" x 96"	"	.75	21.333		3,250	920		4,170	4,975
0560	Through floors, 6" x 12"	1 Carp	2	4		50	172		222	320
0570	12" x 24"	"	1	8		202	345		547	750
0580	24" x 48"	2 Carp	.75	21.333		810	920		1,730	2,300
0590	48" x 96"	"	.50	32	↓	3,250	1,375		4,625	5,700
0600	Structural penetrations, through walls									
0610	Steel beams, W8 x 10	1 Carp	8	1	Ea.	31.50	43		74.50	101
0620	W12 x 14		6	1.333		50	57.50		107.50	143
0630	W21 x 44		5	1.600		100	69		169	217
0640	W36 x 135		3	2.667		244	115		359	445
0650	Bar joists, 18" deep		6	1.333		46	57.50		103.50	139
0660	24" deep		6	1.333		57	57.50		114.50	151
0670	36" deep		5	1.600		86	69		155	201
0680	48" deep	↓	4	2	↓	100	86		186	243
0690	Construction joints, floor slab at exterior wall									
0700	Precast, brick, block or drywall exterior									
0710	2" wide joint	1 Carp	125	.064	L.F.	7.15	2.76		9.91	12.15
0720	4" wide joint	"	75	.107	"	14.30	4.59		18.89	23
0730	Metal panel, glass or curtain wall exterior									
0740	2" wide joint	1 Carp	40	.200	L.F.	17	8.60		25.60	32
0750	4" wide joint	"	25	.320	"	23.50	13.80		37.30	46.50
0760	Floor slab to drywall partition									
0770	Flat joint	1 Carp	100	.080	L.F.	7.05	3.44		10.49	13.05
0780	Fluted joint		50	.160		14.30	6.90		21.20	26.50
0790	Etched fluted joint	↓	75	.107	↓	9.30	4.59		13.89	17.30
0800	Floor slab to concrete/masonry partition									

07 84 Firestopping

07 84 13 – Penetration Firestopping

07 84 13.10 Firestopping	Crew	Daily Output	Labor-Hours	Unit	Material	2011 Bare Costs Labor	Equipment	Total	Total Incl O&P
0810	Flat joint	1 Carp	75	.107	L.F.	15.75	4.59		20.34	24.50
0820	Fluted joint	"	50	.160	"	18.65	6.90		25.55	31
0830	Concrete/CMU wall joints									
0840	1" wide	1 Carp	100	.080	L.F.	8.60	3.44		12.04	14.75
0850	2" wide		75	.107		15.75	4.59		20.34	24.50
0860	4" wide	↓	50	.160	↓	30	6.90		36.90	43.50
0870	Concrete/CMU floor joints									
0880	1" wide	1 Carp	200	.040	L.F.	4.31	1.72		6.03	7.40
0890	2" wide		150	.053		7.90	2.30		10.20	12.15
0900	4" wide	↓	100	.080	↓	15.05	3.44		18.49	22

07 91 Preformed Joint Seals

07 91 13 – Compression Seals

07 91 13.10 Compression Seals

		Crew	Daily Output	Labor-Hours	Unit	Material	2011 Bare Costs Labor	Equipment	Total	Total Incl O&P
0010	COMPRESSION SEALS									
4900	O-ring type cord, 1/4"	1 Bric	472	.017	L.F.	.47	.73		1.20	1.62
4910	1/2"		440	.018		1.78	.78		2.56	3.14
4920	3/4"		424	.019		3.61	.81		4.42	5.20
4930	1"		408	.020		6.95	.85		7.80	8.90
4940	1-1/4"		384	.021		13.60	.90		14.50	16.30
4950	1-1/2"		368	.022		17.20	.94		18.14	20.50
4960	1-3/4"		352	.023		36	.98		36.98	41
4970	2"	↓	344	.023	↓	63.50	1		64.50	71.50

07 91 16 – Joint Gaskets

07 91 16.10 Joint Gaskets

		Crew	Daily Output	Labor-Hours	Unit	Material	2011 Bare Costs Labor	Equipment	Total	Total Incl O&P
0010	JOINT GASKETS									
4400	Joint gaskets, neoprene, closed cell w/adh, 1/8" x 3/8"	1 Bric	240	.033	L.F.	.26	1.44		1.70	2.45
4500	1/4" x 3/4"		215	.037		.58	1.61		2.19	3.05
4700	1/2" x 1"		200	.040		1.30	1.73		3.03	4.02
4800	3/4" x 1-1/2"	↓	165	.048	↓	1.44	2.09		3.53	4.72

07 91 23 – Backer Rods

07 91 23.10 Backer Rods

		Crew	Daily Output	Labor-Hours	Unit	Material	2011 Bare Costs Labor	Equipment	Total	Total Incl O&P
0010	BACKER RODS									
0030	Backer rod, polyethylene, 1/4" diameter	1 Bric	4.60	1.739	C.L.F.	2.83	75		77.83	116
0050	1/2" diameter		4.60	1.739		6.60	75		81.60	120
0070	3/4" diameter		4.60	1.739		10.10	75		85.10	124
0090	1" diameter	↓	4.60	1.739	↓	18.70	75		93.70	134

07 91 26 – Joint Fillers

07 91 26.10 Joint Fillers

		Crew	Daily Output	Labor-Hours	Unit	Material	2011 Bare Costs Labor	Equipment	Total	Total Incl O&P
0010	JOINT FILLERS									
4360	Butyl rubber filler, 1/4" x 1/4"	1 Bric	290	.028	L.F.	.14	1.19		1.33	1.95
4365	1/2" x 1/2"		250	.032		.57	1.38		1.95	2.69
4370	1/2" x 3/4"		210	.038		.85	1.64		2.49	3.41
4375	3/4" x 3/4"		230	.035		1.28	1.50		2.78	3.66
4380	1" x 1"	↓	180	.044	↓	1.71	1.92		3.63	4.76
4390	For coloring, add					12%				
4980	Polyethylene joint backing, 1/4" x 2"	1 Bric	2.08	3.846	C.L.F.	12	166		178	262
4990	1/4" x 6"		1.28	6.250	"	36	270		306	445
5600	Silicone, room temp vulcanizing foam seal, 1/4" x 1/2"	↓	1312	.006	L.F.	.29	.26		.55	.72

07 91 Preformed Joint Seals

07 91 26 – Joint Fillers

07 91 26.10 Joint Fillers		Crew	Daily Output	Labor-Hours	Unit	Material	2011 Bare Costs Labor	Equipment	Total	Total Incl O&P
5610	1/2" x 1/2"	1 Bric	656	.012	L.F.	.57	.53		1.10	1.42
5620	1/2" x 3/4"		442	.018		.86	.78		1.64	2.12
5630	3/4" x 3/4"		328	.024		1.29	1.05		2.34	3
5640	1/8" x 1"		1312	.006		.29	.26		.55	.72
5650	1/8" x 3"		442	.018		.86	.78		1.64	2.12
5670	1/4" x 3"		295	.027		1.72	1.17		2.89	3.65
5680	1/4" x 6"		148	.054		3.44	2.33		5.77	7.30
5690	1/2" x 6"		82	.098		6.90	4.21		11.11	13.85
5700	1/2" x 9"		52.50	.152		10.30	6.60		16.90	21
5710	1/2" x 12"		33	.242		13.75	10.45		24.20	31

07 92 Joint Sealants

07 92 13 – Elastomeric Joint Sealants

07 92 13.20 Caulking and Sealant Options		Crew	Daily Output	Labor-Hours	Unit	Material	2011 Bare Costs Labor	Equipment	Total	Total Incl O&P
0010	**CAULKING AND SEALANT OPTIONS**									
0050	Latex acrylic based, bulk				Gal.	27			27	29.50
0055	Bulk in place 1/4" x 1/4" bead	1 Bric	300	.027	L.F.	.09	1.15		1.24	1.82
0060	1/4" x 3/8"		294	.027		.14	1.17		1.31	1.92
0065	1/4" x 1/2"		288	.028		.19	1.20		1.39	2.01
0075	3/8" x 3/8"		284	.028		.21	1.22		1.43	2.05
0080	3/8" x 1/2"		280	.029		.28	1.23		1.51	2.16
0085	3/8" x 5/8"		276	.029		.35	1.25		1.60	2.27
0095	3/8" x 3/4"		272	.029		.42	1.27		1.69	2.37
0100	1/2" x 1/2"		275	.029		.38	1.26		1.64	2.30
0105	1/2" x 5/8"		269	.030		.47	1.28		1.75	2.45
0110	1/2" x 3/4"		263	.030		.57	1.31		1.88	2.60
0115	1/2" x 7/8"		256	.031		.66	1.35		2.01	2.75
0120	1/2" x 1"		250	.032		.76	1.38		2.14	2.90
0125	3/4" x 3/4"		244	.033		.85	1.41		2.26	3.06
0130	3/4" x 1"		225	.036		1.14	1.53		2.67	3.55
0135	1" x 1"		200	.040		1.52	1.73		3.25	4.26
0190	Cartridges				Gal.	22.50			22.50	24.50
0200	11 fl. oz cartridge				Ea.	1.93			1.93	2.12
0500	1/4" x 1/2"	1 Bric	288	.028	L.F.	.16	1.20		1.36	1.97
0600	1/2" x 1/2"		275	.029		.32	1.26		1.58	2.23
0800	3/4" x 3/4"		244	.033		.71	1.41		2.12	2.90
0900	3/4" x 1"		225	.036		.95	1.53		2.48	3.34
1000	1" x 1"		200	.040		1.18	1.73		2.91	3.89
1400	Butyl based, bulk				Gal.	26			26	28.50
1500	Cartridges				"	32			32	35
1700	1/4" x 1/2", 154 L.F./gal.	1 Bric	288	.028	L.F.	.17	1.20		1.37	1.99
1800	1/2" x 1/2", 77 L.F./gal.	"	275	.029	"	.34	1.26		1.60	2.25
2300	Polysulfide compounds, 1 component, bulk				Gal.	56			56	61.50
2600	1 or 2 component, in place, 1/4" x 1/4", 308 L.F./gal.	1 Bric	300	.027	L.F.	.18	1.15		1.33	1.93
2700	1/2" x 1/4", 154 L.F./gal.		288	.028		.36	1.20		1.56	2.20
2900	3/4" x 3/8", 68 L.F./gal.		272	.029		.82	1.27		2.09	2.81
3000	1" x 1/2", 38 L.F./gal.		250	.032		1.47	1.38		2.85	3.69
3200	Polyurethane, 1 or 2 component				Gal.	50			50	55
3300	Cartridges				"	70			70	77
3500	Bulk, in place, 1/4" x 1/4"	1 Bric	300	.027	L.F.	.16	1.15		1.31	1.91
3655	1/2" x 1/4"		288	.028		.32	1.20		1.52	2.16

07 92 Joint Sealants

07 92 13 – Elastomeric Joint Sealants

07 92 13.20 Caulking and Sealant Options	Crew	Daily Output	Labor-Hours	Unit	Material	2011 Bare Costs Labor	Equipment	Total	Total Incl O&P
3800 3/4" x 3/8"	1 Bric	272	.029	L.F.	.74	1.27		2.01	2.71
3900 1" x 1/2"	↓	250	.032	↓	1.30	1.38		2.68	3.50
4100 Silicone rubber, bulk				Gal.	41			41	45
4200 Cartridges				"	41			41	45

07 92 16 – Rigid Joint Sealants

07 92 16.10 Rigid Joint Sealants

	Crew	Daily Output	Labor-Hours	Unit	Material	2011 Bare Costs Labor	Equipment	Total	Total Incl O&P
0010 **RIGID JOINT SEALANTS**									
5800 Tapes, sealant, P.V.C. foam adhesive, 1/16" x 1/4"				C.L.F.	5.65			5.65	6.20
5900 1/16" x 1/2"					8.35			8.35	9.20
5950 1/16" x 1"					13.85			13.85	15.25
6000 1/8" x 1/2"				↓	9.35			9.35	10.30

07 92 19 – Acoustical Joint Sealants

07 92 19.10 Acoustical Sealant

	Crew	Daily Output	Labor-Hours	Unit	Material	2011 Bare Costs Labor	Equipment	Total	Total Incl O&P
0010 **ACOUSTICAL SEALANT**									
0020 Acoustical sealant, elastomeric, cartridges				Ea.	2.30			2.30	2.53
0025 In place, 1/4" x 1/4"	1 Bric	300	.027	L.F.	.09	1.15		1.24	1.83
0030 1/4" x 1/2"		288	.028		.19	1.20		1.39	2.01
0035 1/2" x 1/2"		275	.029		.38	1.26		1.64	2.29
0040 1/2" x 3/4"		263	.030		.56	1.31		1.87	2.59
0045 3/4" x 3/4"		244	.033		.85	1.41		2.26	3.05
0050 1" x 1"	↓	200	.040	↓	1.50	1.73		3.23	4.24

Division Notes

	CREW	DAILY OUTPUT	LABOR-HOURS	UNIT	2011 BARE COSTS				TOTAL INCL O&P
					MAT.	LABOR	EQUIP.	TOTAL	

Estimating Tips

08 10 00 Doors and Frames

All exterior doors should be addressed for their energy conservation.

- Most metal doors and frames look alike, but there may be significant differences among them. When estimating these items, be sure to choose the line item that most closely compares to the specification or door schedule requirements regarding:
  - type of metal
  - metal gauge
  - door core material
  - fire rating
  - finish
- Wood and plastic doors vary considerably in price. The primary determinant is the veneer material. Lauan, birch, and oak are the most common veneers. Other variables include the following:
  - hollow or solid core
  - fire rating
  - flush or raised panel
  - finish

08 30 00 Specialty Doors and Frames

- There are many varieties of special doors, and they are usually priced per each. Add frames, hardware, or operators required for a complete installation.

08 40 00 Entrances, Storefronts, and Curtain Walls

- Glazed curtain walls consist of the metal tube framing and the glazing material. The cost data in this subdivision is presented for the metal tube framing alone or the composite wall. If your estimate requires a detailed takeoff of the framing, be sure to add the glazing cost.

08 50 00 Windows

- Most metal windows are delivered preglazed. However, some metal windows are priced without glass. Refer to 08 80 00 Glazing for glass pricing. The grade C indicates commercial grade windows, usually ASTM C-35.
- All wood windows and vinyl are priced preglazed. The glazing is insulating glass. Some wood windows may have single pane float glass. Add the cost of screens and grills if required, and not already included.

08 70 00 Hardware

- Hardware costs add considerably to the cost of a door. The most efficient method to determine the hardware requirements for a project is to review the door schedule.

- Door hinges are priced by the pair, with most doors requiring 1-1/2 pairs per door. The hinge prices do not include installation labor because it is included in door installation. Hinges are classified according to the frequency of use.

08 80 00 Glazing

- Different openings require different types of glass. The three most common types are:
  - float
  - tempered
  - insulating
  - impact-resistant
  - ballistic-resistant
- Most exterior windows are glazed with insulating glass. Entrance doors and window walls, where the glass is less than 18" from the floor, are generally glazed with tempered glass. Interior windows and some residential windows are glazed with float glass.
- Coastal communities are requiring the use of impact-resistant glass.
- The insulation or 'u' value is a strong consideration, along with solar heat gain.

Reference Numbers

Reference numbers are shown in shaded boxes at the beginning of some major classifications. These numbers refer to related items in the Reference Section. The reference information may be an estimating procedure, an alternate pricing method, or technical information.

Note: Not all subdivisions listed here necessarily appear in this publication.

08 34 13.10 Doors for Cold Area Storage	Crew	Daily Output	Labor-Hours	Unit	Material	2011 Bare Costs Labor	Equipment	Total	Total Incl O&P
0010 **DOORS FOR COLD AREA STORAGE**									
0020 Single, 20 ga. galvanized steel									
0300 Horizontal sliding, 5' x 7', manual operation, 3.5" thick	2 Carp	2	8	Ea.	3,175	345		3,520	4,000
0400 4" thick		2	8		3,150	345		3,495	3,975
0500 6" thick		2	8		3,300	345		3,645	4,150
0800 5' x 7', power operation, 2" thick		1.90	8.421		5,375	365		5,740	6,450
0900 4" thick		1.90	8.421		5,500	365		5,865	6,600
1000 6" thick		1.90	8.421		6,275	365		6,640	7,450
1300 9' x 10', manual operation, 2" insulation		1.70	9.412		4,375	405		4,780	5,425
1400 4" insulation		1.70	9.412		4,475	405		4,880	5,550
1500 6" insulation		1.70	9.412		5,375	405		5,780	6,550
1800 Power operation, 2" insulation		1.60	10		7,400	430		7,830	8,800
1900 4" insulation		1.60	10		7,650	430		8,080	9,050
2000 6" insulation	↓	1.70	9.412	↓	8,600	405		9,005	10,100
2300 For stainless steel face, add					22%				
3000 Hinged, lightweight, 3' x 7'-0", 2" thick	2 Carp	2	8	Ea.	1,375	345		1,720	2,025
3050 4" thick		1.90	8.421		1,700	365		2,065	2,425
3300 Aluminum doors, 3' x 7'-0", 4" thick		1.90	8.421		1,300	365		1,665	1,975
3350 6" thick		1.40	11.429		2,275	490		2,765	3,250
3600 Stainless steel, 3' x 7'-0", 4" thick		1.90	8.421		1,675	365		2,040	2,375
3650 6" thick		1.40	11.429		2,850	490		3,340	3,875
3900 Painted, 3' x 7'-0", 4" thick		1.90	8.421		1,225	365		1,590	1,900
3950 6" thick	↓	1.40	11.429	↓	2,275	490		2,765	3,250
5000 Bi-parting, electric operated									
5010 6' x 8' opening, galv. faces, 4" thick for cooler	2 Carp	.80	20	Opng.	7,000	860		7,860	9,025
5050 For freezer, 4" thick		.80	20		7,725	860		8,585	9,825
5300 For door buck framing and door protection, add		2.50	6.400		620	276		896	1,100
6000 Galvanized batten door, galvanized hinges, 4' x 7'		2	8		1,750	345		2,095	2,450
6050 6' x 8'		1.80	8.889		2,400	385		2,785	3,200
6500 Fire door, 3 hr., 6' x 8', single slide		.80	20		8,025	860		8,885	10,200
6550 Double, bi-parting	↓	.70	22.857	↓	12,200	985		13,185	14,900

08 91 19.10 Aluminum Louvers	Crew	Daily Output	Labor-Hours	Unit	Material	2011 Bare Costs Labor	Equipment	Total	Total Incl O&P
0010 **ALUMINUM LOUVERS**									
0020 Aluminum with screen, residential, 8" x 8"	1 Carp	38	.211	Ea.	10.55	9.05		19.60	25.50
0100 12" x 12"		38	.211		15	9.05		24.05	30.50
0200 12" x 18"		35	.229		18.85	9.85		28.70	35.50
0250 14" x 24"		30	.267		26	11.50		37.50	46
0300 18" x 24"		27	.296		29	12.75		41.75	51.50
0500 24" x 30"		24	.333		32	14.35		46.35	57
0700 Triangle, adjustable, small		20	.400		32	17.20		49.20	61.50
0800 Large	↓	15	.533	↓	41	23		64	80
1200 Extruded aluminum, see Section 23 37 15.40									
2100 Midget, aluminum, 3/4" deep, 1" diameter	1 Carp	85	.094	Ea.	.69	4.05		4.74	6.95
2150 3" diameter		60	.133		1.87	5.75		7.62	10.85
2200 4" diameter		50	.160		3.64	6.90		10.54	14.55
2250 6" diameter	↓	30	.267		3.74	11.50		15.24	21.50

08 95 Vents

08 95 13 – Soffit Vents

08 95 13.10 Wall Louvers

		Crew	Daily Output	Labor-Hours	Unit	Material	2011 Bare Costs Labor	Equipment	Total	Total Incl O&P
0010	**WALL LOUVERS**									
2330	Soffit vent, continuous, 3" wide, aluminum, mill finish	1 Carp	200	.040	L.F.	.48	1.72		2.20	3.17
2340	Baked enamel finish		200	.040	"	.52	1.72		2.24	3.21
2400	Under eaves vent, aluminum, mill finish, 16" x 4"		48	.167	Ea.	1.37	7.20		8.57	12.50
2500	16" x 8"		48	.167	"	1.93	7.20		9.13	13.10

08 95 16 – Wall Vents

08 95 16.10 Louvers

		Crew	Daily Output	Labor-Hours	Unit	Material	2011 Bare Costs Labor	Equipment	Total	Total Incl O&P
0010	**LOUVERS**									
0020	Redwood, 2'-0" diameter, full circle	1 Carp	16	.500	Ea.	169	21.50		190.50	219
0100	Half circle		16	.500		189	21.50		210.50	241
0200	Octagonal		16	.500		129	21.50		150.50	175
0300	Triangular, 5/12 pitch, 5'-0" at base		16	.500		189	21.50		210.50	241
1100	Rectangular, 1'-4" x 1'-8"		16	.500		25.50	21.50		47	61
1200	1'-4" x 2'-2"		15	.533		29.50	23		52.50	67
1300	1'-9" x 2'-2"		15	.533		35	23		58	73.50
1400	2'-3" x 2'-2"		14	.571		47	24.50		71.50	89
1700	2'-4" x 2'-11"		13	.615		47	26.50		73.50	92
2000	Aluminum, 12" x 16"		25	.320		13.45	13.80		27.25	36
2010	16" x 20"		25	.320		22.50	13.80		36.30	45.50
2020	24" x 30"		25	.320		50.50	13.80		64.30	76.50
2100	6' triangle		12	.667		162	28.50		190.50	222
7000	Vinyl gable vent, 8" x 8"		38	.211		11.55	9.05		20.60	26.50
7020	12" x 12"		38	.211		22.50	9.05		31.55	39
7080	12" x 18"		35	.229		29.50	9.85		39.35	47.50
7200	18" x 24"		30	.267		36.50	11.50		48	57.50

61

Division Notes

	CREW	DAILY OUTPUT	LABOR-HOURS	UNIT	2011 BARE COSTS				TOTAL INCL O&P
					MAT.	LABOR	EQUIP.	TOTAL	

Estimating Tips

General

- Room Finish Schedule: A complete set of plans should contain a room finish schedule. If one is not available, it would be well worth the time and effort to obtain one.

09 20 00 Plaster and Gypsum Board

- Lath is estimated by the square yard plus a 5% allowance for waste. Furring, channels, and accessories are measured by the linear foot. An extra foot should be allowed for each accessory miter or stop.

- Plaster is also estimated by the square yard. Deductions for openings vary by preference, from zero deduction to 50% of all openings over 2 feet in width. The estimator should allow one extra square foot for each linear foot of horizontal interior or exterior angle located below the ceiling level. Also, double the areas of small radius work.

- Drywall accessories, studs, track, and acoustical caulking are all measured by the linear foot. Drywall taping is figured by the square foot. Gypsum wallboard is estimated by the square foot. No material deductions should be made for door or window openings under 32 S.F.

09 60 00 Flooring

- Tile and terrazzo areas are taken off on a square foot basis. Trim and base materials are measured by the linear foot. Accent tiles are listed per each. Two basic methods of installation are used. Mud set is approximately 30% more expensive than thin set. In terrazzo work, be sure to include the linear footage of embedded decorative strips, grounds, machine rubbing, and power cleanup.

- Wood flooring is available in strip, parquet, or block configuration. The latter two types are set in adhesives with quantities estimated by the square foot. The laying pattern will influence labor costs and material waste. In addition to the material and labor for laying wood floors, the estimator must make allowances for sanding and finishing these areas unless the flooring is prefinished.

- Sheet flooring is measured by the square yard. Roll widths vary, so consideration should be given to use the most economical width, as waste must be figured into the total quantity. Consider also the installation methods available, direct glue down or stretched.

09 70 00 Wall Finishes

- Wall coverings are estimated by the square foot. The area to be covered is measured, length by height of wall above baseboards, to calculate the square footage of each wall. This figure is divided by the number of square feet in the single roll which is being used. Deduct, in full, the areas of openings such as doors and windows. Where a pattern match is required allow 25%-30% waste.

09 80 00 Acoustic Treatment

- Acoustical systems fall into several categories. The takeoff of these materials should be by the square foot of area with a 5% allowance for waste. Do not forget about scaffolding, if applicable, when estimating these systems.

09 90 00 Painting and Coating

- A major portion of the work in painting involves surface preparation. Be sure to include cleaning, sanding, filling, and masking costs in the estimate.

- Protection of adjacent surfaces is not included in painting costs. When considering the method of paint application, an important factor is the amount of protection and masking required. These must be estimated separately and may be the determining factor in choosing the method of application.

Reference Numbers

Reference numbers are shown in shaded boxes at the beginning of some major classifications. These numbers refer to related items in the Reference Section. The reference information may be an estimating procedure, an alternate pricing method, or technical information.

Note: Not all subdivisions listed here necessarily appear in this publication.

09 22 Supports for Plaster and Gypsum Board

09 22 03 – Fastening Methods for Finishes

09 22 03.20 Drilling Plaster/Drywall		Crew	Daily Output	Labor-Hours	Unit	Material	2011 Bare Costs Labor	Equipment	Total	Total Incl O&P
0010	**DRILLING PLASTER/DRYWALL**									
1100	Drilling & layout for drywall/plaster walls, up to 1" deep, no anchor									
1200	Holes, 1/4" diameter	1 Carp	150	.053	Ea.	.01	2.30		2.31	3.53
1300	3/8" diameter		140	.057		.01	2.46		2.47	3.78
1400	1/2" diameter		130	.062		.01	2.65		2.66	4.07
1500	3/4" diameter		120	.067		.01	2.87		2.88	4.41
1600	1" diameter		110	.073		.02	3.13		3.15	4.82
1700	1-1/4" diameter		100	.080		.03	3.44		3.47	5.35
1800	1-1/2" diameter		90	.089		.04	3.83		3.87	5.90
1900	For ceiling installations, add						40%			

09 69 Access Flooring

09 69 13 – Rigid-Grid Access Flooring

09 69 13.10 Access Floors		Crew	Daily Output	Labor-Hours	Unit	Material	2011 Bare Costs Labor	Equipment	Total	Total Incl O&P
0010	**ACCESS FLOORS**									
0015	Access floor package including panel, pedestal, stringers & laminate cover									
0100	Computer room, greater than 6,000 S.F.	4 Carp	750	.043	S.F.	7	1.84		8.84	10.50
0110	Less than 6,000 S.F.	2 Carp	375	.043		7.65	1.84		9.49	11.20
0120	Office, greater than 6,000 S.F.	4 Carp	1050	.030		5	1.31		6.31	7.50
0250	Panels, particle board or steel, 1250# load, no covering, under 6,000 S.F.	2 Carp	600	.027		3.50	1.15		4.65	5.60
0300	Over 6,000 S.F.		640	.025		3	1.08		4.08	4.95
0400	Aluminum, 24" panels		500	.032		31.50	1.38		32.88	37
0600	For carpet covering, add					8.55			8.55	9.40
0700	For vinyl floor covering, add					6.70			6.70	7.35
0900	For high pressure laminate covering, add					5.45			5.45	6
0910	For snap on stringer system, add	2 Carp	1000	.016		1.44	.69		2.13	2.64
0950	Office applications, steel or concrete panels,									
0960	no covering, over 6,000 S.F.	2 Carp	960	.017	S.F.	10.10	.72		10.82	12.20
1000	Machine cutouts after initial installation	1 Carp	50	.160	Ea.	5.05	6.90		11.95	16.10
1050	Pedestals, 6" to 12"	2 Carp	85	.188		7.95	8.10		16.05	21
1100	Air conditioning grilles, 4" x 12"	1 Carp	17	.471		66	20.50		86.50	104
1150	4" x 18"	"	14	.571		90.50	24.50		115	137
1200	Approach ramps, minimum	2 Carp	60	.267	S.F.	24.50	11.50		36	44.50
1300	Maximum	"	40	.400	"	33.50	17.20		50.70	63
1500	Handrail, 2 rail, aluminum	1 Carp	15	.533	L.F.	94.50	23		117.50	139

09 91 Painting

09 91 23 – Interior Painting

09 91 23.52 Miscellaneous, Interior		Crew	Daily Output	Labor-Hours	Unit	Material	2011 Bare Costs Labor	Equipment	Total	Total Incl O&P
0010	**MISCELLANEOUS, INTERIOR** R099100-10									
3800	Grilles, per side, oil base, primer coat, brushwork	1 Pord	520	.015	S.F.	.12	.58		.70	.99
3850	Spray		1140	.007		.13	.26		.39	.53
3880	Paint 1 coat, brushwork		520	.015		.19	.58		.77	1.07
3900	Spray		1140	.007		.21	.26		.47	.62
3920	Paint 2 coats, brushwork		325	.025		.37	.92		1.29	1.78
3940	Spray		650	.012		.42	.46		.88	1.16
3950	Prime & paint 1 coat		325	.025		.31	.92		1.23	1.71
3960	Prime & paint 2 coats		270	.030		.31	1.11		1.42	1.99
4500	Louvers, one side, primer, brushwork		524	.015		.06	.57		.63	.92

09 91 Painting

09 91 23 – Interior Painting

09 91 23.52 Miscellaneous, Interior	Crew	Daily Output	Labor-Hours	Unit	Material	2011 Bare Costs Labor	Equipment	Total	Total Incl O&P
4520	Paint one coat, brushwork	1 Pord	520	.015	S.F.	.09	.58		.67	.96
4530	Spray		1140	.007		.10	.26		.36	.50
4540	Paint two coats, brushwork		325	.025		.17	.92		1.09	1.56
4550	Spray		650	.012		.19	.46		.65	.90
4560	Paint three coats, brushwork		270	.030		.25	1.11		1.36	1.93
4570	Spray		500	.016		.28	.60		.88	1.20
5000	Pipe, 1" - 4" diameter, primer or sealer coat, oil base, brushwork	2 Pord	1250	.013	L.F.	.06	.48		.54	.78
5100	Spray		2165	.007		.06	.28		.34	.48
5200	Paint 1 coat, brushwork		1250	.013		.10	.48		.58	.82
5300	Spray		2165	.007		.09	.28		.37	.51
5350	Paint 2 coats, brushwork		775	.021		.18	.78		.96	1.35
5400	Spray		1240	.013		.20	.48		.68	.94
5420	Paint 3 coats, brushwork		775	.021		.27	.78		1.05	1.44
5450	5" - 8" diameter, primer or sealer coat, brushwork		620	.026		.13	.97		1.10	1.58
5500	Spray		1085	.015		.21	.55		.76	1.05
5550	Paint 1 coat, brushwork		620	.026		.27	.97		1.24	1.74
5600	Spray		1085	.015		.30	.55		.85	1.15
5650	Paint 2 coats, brushwork		385	.042		.36	1.56		1.92	2.70
5700	Spray		620	.026		.40	.97		1.37	1.88
5720	Paint 3 coats, brushwork		385	.042		.53	1.56		2.09	2.90
5750	9" - 12" diameter, primer or sealer coat, brushwork		415	.039		.19	1.45		1.64	2.36
5800	Spray		725	.022		.33	.83		1.16	1.59
5850	Paint 1 coat, brushwork		415	.039		.28	1.45		1.73	2.45
6000	Spray		725	.022		.31	.83		1.14	1.57
6200	Paint 2 coats, brushwork		260	.062		.54	2.31		2.85	4.02
6250	Spray		415	.039		.60	1.45		2.05	2.81
6270	Paint 3 coats, brushwork		260	.062		.80	2.31		3.11	4.31
6300	13" - 16" diameter, primer or sealer coat, brushwork		310	.052		.26	1.94		2.20	3.16
6350	Spray		540	.030		.29	1.11		1.40	1.97
6400	Paint 1 coat, brushwork		310	.052		.37	1.94		2.31	3.28
6450	Spray		540	.030		.41	1.11		1.52	2.10
6500	Paint 2 coats, brushwork		195	.082		.72	3.08		3.80	5.35
6550	Spray		310	.052		.80	1.94		2.74	3.75
6600	Radiators, per side, primer, brushwork	1 Pord	520	.015	S.F.	.06	.58		.64	.93
6620	Paint, one coat		520	.015		.06	.58		.64	.92
6640	Two coats		340	.024		.17	.88		1.05	1.50
6660	Three coats		283	.028		.25	1.06		1.31	1.85

Division Notes

	CREW	DAILY OUTPUT	LABOR-HOURS	UNIT	2011 BARE COSTS				TOTAL INCL O&P
					MAT.	LABOR	EQUIP.	TOTAL	

Estimating Tips

General

- The items in this division are usually priced per square foot or each.

- Many items in Division 10 require some type of support system or special anchors that are not usually furnished with the item. The required anchors must be added to the estimate in the appropriate division.

- Some items in Division 10, such as lockers, may require assembly before installation. Verify the amount of assembly required. Assembly can often exceed installation time.

10 20 00 Interior Specialties

- Support angles and blocking are not included in the installation of toilet compartments, shower/dressing compartments, or cubicles. Appropriate line items from Divisions 5 or 6 may need to be added to support the installations.

- Toilet partitions are priced by the stall. A stall consists of a side wall, pilaster, and door with hardware. Toilet tissue holders and grab bars are extra.

- The required acoustical rating of a folding partition can have a significant impact on costs. Verify the sound transmission coefficient rating of the panel priced to the specification requirements.

- Grab bar installation does not include supplemental blocking or backing to support the required load. When grab bars are installed at an existing facility, provisions must be made to attach the grab bars to solid structure.

Reference Numbers

Reference numbers are shown in shaded boxes at the beginning of some major classifications. These numbers refer to related items in the Reference Section. The reference information may be an estimating procedure, an alternate pricing method, or technical information.

Note: Not all subdivisions listed here necessarily appear in this publication.

Division 10 - Specialties

10 31 Manufactured Fireplaces

10 31 13 – Manufactured Fireplace Chimneys

10 31 13.10 Fireplace Chimneys

		Crew	Daily Output	Labor-Hours	Unit	Material	2011 Bare Costs Labor	Equipment	Total	Total Incl O&P
0010	**FIREPLACE CHIMNEYS**									
0500	Chimney dbl. wall, all stainless, over 8'-6", 7" diam., add to fireplace	1 Carp	33	.242	V.L.F.	72	10.45		82.45	95
0600	10" diameter, add to fireplace		32	.250		106	10.75		116.75	134
0700	12" diameter, add to fireplace		31	.258		139	11.10		150.10	169
0800	14" diameter, add to fireplace		30	.267		183	11.50		194.50	219
1000	Simulated brick chimney top, 4' high, 16" x 16"		10	.800	Ea.	258	34.50		292.50	335
1100	24" x 24"		7	1.143	"	480	49		529	605

10 31 23 – Prefabricated Fireplaces

10 31 23.10 Fireplace, Prefabricated

		Crew	Daily Output	Labor-Hours	Unit	Material	2011 Bare Costs Labor	Equipment	Total	Total Incl O&P
0010	**FIREPLACE, PREFABRICATED**, free standing or wall hung									
0100	With hood & screen, minimum	1 Carp	1.30	6.154	Ea.	1,200	265		1,465	1,700
0150	Average		1	8		1,675	345		2,020	2,375
0200	Maximum		.90	8.889		3,475	385		3,860	4,400
1500	Simulated logs, gas fired, 40,000 BTU, 2' long, minimum		7	1.143	Set	445	49		494	565
1600	Maximum		6	1.333		1,025	57.50		1,082.50	1,225
1700	Electric, 1,500 BTU, 1'-6" long, minimum		7	1.143		175	49		224	269
1800	11,500 BTU, maximum		6	1.333		395	57.50		452.50	520

10 35 Stoves

10 35 13 – Heating Stoves

10 35 13.10 Woodburning Stoves

		Crew	Daily Output	Labor-Hours	Unit	Material	2011 Bare Costs Labor	Equipment	Total	Total Incl O&P
0010	**WOODBURNING STOVES**									
0015	Cast iron, minimum	2 Carp	1.30	12.308	Ea.	1,200	530		1,730	2,150
0020	Average		1	16		1,600	690		2,290	2,800
0030	Maximum		.80	20		2,875	860		3,735	4,475
0050	For gas log lighter, add					42			42	46.50

Estimating Tips

General

- The items in this division are usually priced per square foot or each. Many of these items are purchased by the owner for installation by the contractor. Check the specifications for responsibilities and include time for receiving, storage, installation, and mechanical and electrical hookups in the appropriate divisions.

- Many items in Division 11 require some type of support system that is not usually furnished with the item. Examples of these systems include blocking for the attachment of casework and support angles for ceiling-hung projection screens. The required blocking or supports must be added to the estimate in the appropriate division.

- Some items in Division 11 may require assembly or electrical hookups. Verify the amount of assembly required or the need for a hard electrical connection and add the appropriate costs.

Reference Numbers

Reference numbers are shown in shaded boxes at the beginning of some major classifications. These numbers refer to related items in the Reference Section. The reference information may be an estimating procedure, an alternate pricing method, or technical information.

Note: Not all subdivisions listed here necessarily appear in this publication.

11 11 Vehicle Service Equipment

11 11 13 – Compressed-Air Vehicle Service Equipment

11 11 13.10 Compressed Air Equipment		Crew	Daily Output	Labor-Hours	Unit	Material	2011 Bare Costs Labor	Equipment	Total	Total Incl O&P
0010	**COMPRESSED AIR EQUIPMENT**									
0030	Compressors, electric, 1-1/2 H.P., standard controls	L-4	1.50	16	Ea.	410	650		1,060	1,450
0550	Dual controls		1.50	16		740	650		1,390	1,800
0600	5 H.P., 115/230 volt, standard controls		1	24		2,175	970		3,145	3,900
0650	Dual controls		1	24		2,350	970		3,320	4,075

11 11 19 – Vehicle Lubrication Equipment

11 11 19.10 Lubrication Equipment

		Crew	Daily Output	Labor-Hours	Unit	Material	Labor	Equipment	Total	Total Incl O&P
0010	**LUBRICATION EQUIPMENT**									
3000	Lube equipment, 3 reel type, with pumps, not including piping	L-4	.50	48	Set	7,900	1,950		9,850	11,700
3700	Pump lubrication, pneumatic, not incl. air compressor									
3710	Oil/gear lube	Q-1	9.60	1.667	Ea.	910	80.50		990.50	1,125
3720	Grease	"	9.60	1.667	"	945	80.50		1,025.50	1,175

11 11 33 – Vehicle Spray Painting Equipment

11 11 33.10 Spray Painting Equipment

		Crew	Daily Output	Labor-Hours	Unit	Material	Labor	Equipment	Total	Total Incl O&P
0010	**SPRAY PAINTING EQUIPMENT**									
4000	Spray painting booth, 26' long, complete	L-4	.40	60	Ea.	15,900	2,425		18,325	21,100

11 23 Commercial Laundry and Dry Cleaning Equipment

11 23 16 – Drying and Conditioning Equipment

11 23 16.13 Dryers

		Crew	Daily Output	Labor-Hours	Unit	Material	Labor	Equipment	Total	Total Incl O&P
0010	**DRYERS**, Not including rough-in									
1500	Industrial, 30 lb. capacity	1 Plum	2	4	Ea.	2,975	214		3,189	3,600
1600	50 lb. capacity	"	1.70	4.706	"	3,200	252		3,452	3,900

11 23 26 – Commercial Washers and Extractors

11 23 26.13 Washers and Extractors

		Crew	Daily Output	Labor-Hours	Unit	Material	Labor	Equipment	Total	Total Incl O&P
0010	**WASHERS AND EXTRACTORS**, not including rough-in									
6000	Combination washer/extractor, 20 lb. capacity	L-6	1.50	8	Ea.	5,400	420		5,820	6,575
6100	30 lb. capacity		.80	15		8,550	790		9,340	10,600
6200	50 lb. capacity		.68	17.647		10,000	925		10,925	12,400
6300	75 lb. capacity		.30	40		18,700	2,100		20,800	23,700
6350	125 lb. capacity		.16	75		24,900	3,950		28,850	33,300

11 23 33 – Coin-Operated Laundry Equipment

11 23 33.13 Coin Operated Washers and Dryers

		Crew	Daily Output	Labor-Hours	Unit	Material	Labor	Equipment	Total	Total Incl O&P
0010	**COIN OPERATED WASHERS AND DRYERS**									
0990	Dryer, gas fired									
1000	Commercial, 30 lb. capacity, coin operated, single	1 Plum	3	2.667	Ea.	3,100	143		3,243	3,625
1100	Double stacked	"	2	4	"	6,275	214		6,489	7,225
5290	Clothes washer									
5300	Commercial, coin operated, average	1 Plum	3	2.667	Ea.	1,125	143		1,268	1,475

11 24 Maintenance Equipment

11 24 19 – Vacuum Cleaning Systems

11 24 19.10 Vacuum Cleaning		Crew	Daily Output	Labor-Hours	Unit	Material	2011 Bare Costs Labor	Equipment	Total	Total Incl O&P
0010	**VACUUM CLEANING**									
0020	Central, 3 inlet, residential	1 Skwk	.90	8.889	Total	1,025	395		1,420	1,750
0200	Commercial		.70	11.429		1,225	505		1,730	2,125
0400	5 inlet system, residential		.50	16		1,450	710		2,160	2,700
0600	7 inlet system, commercial		.40	20		1,675	890		2,565	3,200
0800	9 inlet system, residential	↓	.30	26.667		3,600	1,175		4,775	5,775
4010	Rule of thumb: First 1200 S.F., installed				↓				1,325	1,450
4020	For each additional S.F., add				S.F.					.24

11 26 Unit Kitchens

11 26 13 – Metal Unit Kitchens

11 26 13.10 Commercial Unit Kitchens

11 26 13.10 Commercial Unit Kitchens		Crew	Daily Output	Labor-Hours	Unit	Material	2011 Bare Costs Labor	Equipment	Total	Total Incl O&P
0010	**COMMERCIAL UNIT KITCHENS**									
1500	Combination range, refrigerator and sink, 30" wide, minimum	L-1	2	8	Ea.	925	415		1,340	1,650
1550	Maximum		1	16		3,675	830		4,505	5,275
1570	60" wide, average		1.40	11.429		3,225	595		3,820	4,400
1590	72" wide, average		1.20	13.333		4,200	695		4,895	5,650
1600	Office model, 48" wide		2	8		2,650	415		3,065	3,525
1620	Refrigerator and sink only	↓	2.40	6.667	↓	2,725	345		3,070	3,500
1640	Combination range, refrigerator, sink, microwave									
1660	Oven and ice maker	L-1	.80	20	Ea.	4,450	1,050		5,500	6,450

11 27 Photographic Processing Equipment

11 27 13 – Darkroom Processing Equipment

11 27 13.10 Darkroom Equipment

11 27 13.10 Darkroom Equipment		Crew	Daily Output	Labor-Hours	Unit	Material	2011 Bare Costs Labor	Equipment	Total	Total Incl O&P
0010	**DARKROOM EQUIPMENT**									
0020	Developing sink, 5" deep, 24" x 48"	Q-1	2	8	Ea.	5,625	385		6,010	6,750
0050	48" x 52"		1.70	9.412		5,675	455		6,130	6,900
0200	10" deep, 24" x 48"		1.70	9.412		7,775	455		8,230	9,225
0250	24" x 108"		1.50	10.667		10,200	515		10,715	12,100
3500	Washers, round, minimum sheet 11" x 14"		2	8		3,400	385		3,785	4,325
3550	Maximum sheet 20" x 24"		1	16		3,475	770		4,245	4,975
3800	Square, minimum sheet 20" x 24"		1	16		3,125	770		3,895	4,575
3900	Maximum sheet 50" x 56"	↓	.80	20	↓	4,850	965		5,815	6,775
4500	Combination tank sink, tray sink, washers, with									
4510	Dry side tables, average	Q-1	.45	35.556	Ea.	10,600	1,725		12,325	14,300

11 31 Residential Appliances

11 31 13 – Residential Kitchen Appliances

11 31 13.13 Cooking Equipment

11 31 13.13 Cooking Equipment		Crew	Daily Output	Labor-Hours	Unit	Material	2011 Bare Costs Labor	Equipment	Total	Total Incl O&P
0010	**COOKING EQUIPMENT**									
0020	Cooking range, 30" free standing, 1 oven, minimum	2 Clab	10	1.600	Ea.	505	55		560	640
0050	Maximum		4	4		1,700	137		1,837	2,075
0150	2 oven, minimum		10	1.600		1,050	55		1,105	1,225
0200	Maximum	↓	10	1.600	↓	1,100	55		1,155	1,275

11 31 Residential Appliances

11 31 13 – Residential Kitchen Appliances

11 31 13.23 Refrigeration Equipment	Crew	Daily Output	Labor-Hours	Unit	Material	2011 Bare Costs Labor	Equipment	Total	Total Incl O&P
0010 **REFRIGERATION EQUIPMENT**									
5200　Icemaker, automatic, 20 lb. per day	1 Plum	7	1.143	Ea.	880	61.50		941.50	1,050
5350　　51 lb. per day	"	2	4	"	1,250	214		1,464	1,700

11 31 13.33 Kitchen Cleaning Equipment

0010 **KITCHEN CLEANING EQUIPMENT**									
2750　Dishwasher, built-in, 2 cycles, minimum	L-1	4	4	Ea.	221	208		429	555
2800　　Maximum		2	8		400	415		815	1,050
2950　　4 or more cycles, minimum		4	4		325	208		533	665
2960　　　Average		4	4		430	208		638	785
3000　　　Maximum		2	8		1,075	415		1,490	1,825

11 31 13.43 Waste Disposal Equipment

0010 **WASTE DISPOSAL EQUIPMENT**									
3300　Garbage disposal, sink type, minimum	L-1	10	1.600	Ea.	72.50	83		155.50	204
3350　　Maximum	"	10	1.600	"	191	83		274	335

11 31 13.53 Kitchen Ventilation Equipment

0010 **KITCHEN VENTILATION EQUIPMENT**									
4150　Hood for range, 2 speed, vented, 30" wide, minimum	L-3	5	3.200	Ea.	62.50	150		212.50	297
4200　　Maximum		3	5.333		1,000	251		1,251	1,475
4300　　42" wide, minimum		5	3.200		188	150		338	435
4330　　　Custom		5	3.200		1,650	150		1,800	2,050
4350　　　Maximum		3	5.333		2,025	251		2,276	2,600
4500　For ventless hood, 2 speed, add					17.05			17.05	18.75
4650　For vented 1 speed, deduct from maximum					45.50			45.50	50

11 31 23 – Residential Laundry Appliances

11 31 23.13 Washers

0010 **WASHERS**									
5000　Residential, 4 cycle, average	1 Plum	3	2.667	Ea.	800	143		943	1,100
6650　Washing machine, automatic, minimum		3	2.667		400	143		543	655
6700　　Maximum		1	8		1,300	430		1,730	2,075

11 31 23.23 Dryers

0010 **DRYERS**									
0500　Gas fired residential, 16 lb. capacity, average	1 Plum	3	2.667	Ea.	635	143		778	915
7450　Vent kits for dryers	1 Carp	10	.800	"	37.50	34.50		72	94

11 31 33 – Miscellaneous Residential Appliances

11 31 33.13 Sump Pumps

0010 **SUMP PUMPS**									
6400　Cellar drainer, pedestal, 1/3 H.P., molded PVC base	1 Plum	3	2.667	Ea.	119	143		262	345
6450　　Solid brass	"	2	4	"	263	214		477	610
6460　Sump pump, see also Section 22 14 29.16									

11 31 33.23 Water Heaters

0010 **WATER HEATERS**									
6900　Electric, glass lined, 30 gallon, minimum	L-1	5	3.200	Ea.	520	166		686	815
6950　　Maximum		3	5.333		720	277		997	1,200
7100　　80 gallon, minimum		2	8		935	415		1,350	1,650
7150　　　Maximum		1	16		1,300	830		2,130	2,650
7180　Gas, glass lined, 30 gallon, minimum	2 Plum	5	3.200		805	172		977	1,150
7220　　Maximum		3	5.333		1,125	286		1,411	1,650
7260　　50 gallon, minimum		2.50	6.400		845	345		1,190	1,425
7300　　　Maximum		1.50	10.667		1,175	570		1,745	2,125

11 31 Residential Appliances

11 31 33 – Miscellaneous Residential Appliances

11 31 33.43 Air Quality	Crew	Daily Output	Labor-Hours	Unit	Material	2011 Bare Costs Labor	Equipment	Total	Total Incl O&P
0010 **AIR QUALITY**									
2450 Dehumidifier, portable, automatic, 15 pint				Ea.	154			154	170
2550 40 pint					171			171	188
3550 Heater, electric, built-in, 1250 watt, ceiling type, minimum	1 Elec	4	2		98.50	101		199.50	258
3600 Maximum		3	2.667		161	134		295	375
3700 Wall type, minimum		4	2		159	101		260	325
3750 Maximum		3	2.667		171	134		305	385
3900 1500 watt wall type, with blower		4	2		159	101		260	325
3950 3000 watt		3	2.667		325	134		459	555
4850 Humidifier, portable, 8 gallons per day					165			165	182
5000 15 gallons per day					198			198	218

11 41 Foodservice Storage Equipment

11 41 13 – Refrigerated Food Storage Cases

11 41 13.10 Refrigerated Food Cases

	Crew	Daily Output	Labor-Hours	Unit	Material	2011 Bare Costs Labor	Equipment	Total	Total Incl O&P
0010 **REFRIGERATED FOOD CASES**									
0030 Dairy, multi-deck, 12' long	Q-5	3	5.333	Ea.	10,700	260		10,960	12,200
0100 For rear sliding doors, add					1,525			1,525	1,700
0200 Delicatessen case, service deli, 12' long, single deck	Q-5	3.90	4.103		7,225	200		7,425	8,250
0300 Multi-deck, 18 S.F. shelf display		3	5.333		6,450	260		6,710	7,475
0400 Freezer, self-contained, chest-type, 30 C.F.		3.90	4.103		7,400	200		7,600	8,450
0500 Glass door, upright, 78 C.F.		3.30	4.848		10,000	236		10,236	11,400
0600 Frozen food, chest type, 12' long		3.30	4.848		7,250	236		7,486	8,325
0700 Glass door, reach-in, 5 door		3	5.333		13,900	260		14,160	15,700
0800 Island case, 12' long, single deck		3.30	4.848		8,225	236		8,461	9,375
0900 Multi-deck		3	5.333		17,400	260		17,660	19,500
1000 Meat case, 12' long, single deck		3.30	4.848		5,975	236		6,211	6,900
1050 Multi-deck		3.10	5.161		10,300	251		10,551	11,700
1100 Produce, 12' long, single deck		3.30	4.848		7,900	236		8,136	9,050
1200 Multi-deck		3.10	5.161		8,625	251		8,876	9,850

11 41 13.20 Refrigerated Food Storage Equipment

	Crew	Daily Output	Labor-Hours	Unit	Material	2011 Bare Costs Labor	Equipment	Total	Total Incl O&P
0010 **REFRIGERATED FOOD STORAGE EQUIPMENT**									
2350 Cooler, reach-in, beverage, 6' long	Q-1	6	2.667	Ea.	4,625	129		4,754	5,275
4300 Freezers, reach-in, 44 C.F.		4	4		2,700	193		2,893	3,275
4500 68 C.F.		3	5.333		3,650	257		3,907	4,375

11 44 Food Cooking Equipment

11 44 13 – Commercial Ranges

11 44 13.10 Cooking Equipment

	Crew	Daily Output	Labor-Hours	Unit	Material	2011 Bare Costs Labor	Equipment	Total	Total Incl O&P
0010 **COOKING EQUIPMENT**									
0020 Bake oven, gas, one section	Q-1	8	2	Ea.	4,650	96.50		4,746.50	5,275
0300 Two sections		7	2.286		9,300	110		9,410	10,400
0600 Three sections		6	2.667		13,500	129		13,629	15,000
0900 Electric convection, single deck	L-7	4	7		6,475	291		6,766	7,575
6350 Kettle, w/steam jacket, tilting, w/positive lock, SS, 20 gallons		7	4		7,400	166		7,566	8,375
6600 60 gallons		6	4.667		9,800	194		9,994	11,100

11 47 Ice Machines

11 47 10 – Commercial Ice Machines

11 47 10.10 Commercial Ice Equipment		Crew	Daily Output	Labor-Hours	Unit	Material	2011 Bare Costs Labor	Equipment	Total	Total Incl O&P
0010	**COMMERCIAL ICE EQUIPMENT**									
5800	Ice cube maker, 50 pounds per day	Q-1	6	2.667	Ea.	1,475	129		1,604	1,825
6050	500 pounds per day	"	4	4	"	2,825	193		3,018	3,400

11 48 Cleaning and Disposal Equipment

11 48 13 – Commercial Dishwashers

11 48 13.10 Dishwashers

		Crew	Daily Output	Labor-Hours	Unit	Material	Labor	Equipment	Total	Total Incl O&P
0010	**DISHWASHERS**									
2700	Dishwasher, commercial, rack type									
2720	10 to 12 racks per hour	Q-1	3.20	5	Ea.	4,300	241		4,541	5,075
2730	Energy star rated, 35 to 40 racks/hour G		1.30	12.308		4,650	595		5,245	5,975
2740	50 to 60 racks/hour G		1.30	12.308		10,100	595		10,695	12,000
2800	Automatic, 190 to 230 racks per hour	L-6	.35	34.286		14,300	1,800		16,100	18,400
2820	235 to 275 racks per hour		.25	48		30,300	2,525		32,825	37,100
2840	8,750 to 12,500 dishes per hour		.10	120		49,500	6,300		55,800	64,000

11 53 Laboratory Equipment

11 53 13 – Laboratory Fume Hoods

11 53 13.13 Recirculating Laboratory Fume Hoods

		Crew	Daily Output	Labor-Hours	Unit	Material	Labor	Equipment	Total	Total Incl O&P
0010	**RECIRCULATING LABORATORY FUME HOODS**									
0600	Fume hood, with countertop & base, not including HVAC									
0610	Simple, minimum	2 Carp	5.40	2.963	L.F.	635	128		763	895
0620	Complex, including fixtures		2.40	6.667		1,800	287		2,087	2,425
0630	Special, maximum		1.70	9.412		1,825	405		2,230	2,625
0670	Service fixtures, average				Ea.	287			287	315
0680	For sink assembly with hot and cold water, add	1 Plum	1.40	5.714	"	705	305		1,010	1,225

11 53 13.23 Exhaust Hoods

		Crew	Daily Output	Labor-Hours	Unit	Material	Labor	Equipment	Total	Total Incl O&P
0010	**EXHAUST HOODS**									
0650	Ductwork, minimum	2 Shee	1	16	Hood	3,625	825		4,450	5,250
0660	Maximum	"	.50	32	"	5,575	1,650		7,225	8,650

11 53 19 – Laboratory Sterilizers

11 53 19.13 Sterilizers

		Crew	Daily Output	Labor-Hours	Unit	Material	Labor	Equipment	Total	Total Incl O&P
0010	**STERILIZERS**									
0700	Glassware washer, undercounter, minimum	L-1	1.80	8.889	Ea.	5,850	460		6,310	7,125
0710	Maximum	"	1	16		11,700	830		12,530	14,000
1850	Utensil washer-sanitizer	1 Plum	2	4		10,900	214		11,114	12,300

11 53 33 – Emergency Safety Appliances

11 53 33.13 Emergency Equipment

		Crew	Daily Output	Labor-Hours	Unit	Material	Labor	Equipment	Total	Total Incl O&P
0010	**EMERGENCY EQUIPMENT**									
1400	Safety equipment, eye wash, hand held				Ea.	405			405	445
1450	Deluge shower				"	740			740	815

11 53 43 – Service Fittings and Accessories

11 53 43.13 Fittings

		Crew	Daily Output	Labor-Hours	Unit	Material	Labor	Equipment	Total	Total Incl O&P
0010	**FITTINGS**									
1600	Sink, one piece plastic, flask wash, hose, free standing	1 Plum	1.60	5	Ea.	1,800	268		2,068	2,375
1610	Epoxy resin sink, 25" x 16" x 10"	"	2	4	"	198	214		412	540
8000	Alternate pricing method: as percent of lab furniture									

11 53 Laboratory Equipment

11 53 43 – Service Fittings and Accessories

11 53 43.13 Fittings

11 53 43.13 Fittings		Crew	Daily Output	Labor-Hours	Unit	Material	2011 Bare Costs Labor	Equipment	Total	Total Incl O&P
8050	Installation, not incl. plumbing & duct work				% Furn.				20%	22%
8100	Plumbing, final connections, simple system								9.09%	10%
8110	Moderately complex system								13.64%	15%
8120	Complex system								18.18%	20%
8150	Electrical, simple system								9.09%	10%
8160	Moderately complex system								18.18%	20%
8170	Complex system								31.80%	35%

11 71 Medical Sterilizing Equipment

11 71 10 – Medical Sterilizers & Distillers

11 71 10.10 Sterilizers and Distillers

11 71 10.10 Sterilizers and Distillers		Crew	Daily Output	Labor-Hours	Unit	Material	2011 Bare Costs Labor	Equipment	Total	Total Incl O&P
0010	**STERILIZERS AND DISTILLERS**									
0700	Distiller, water, steam heated, 50 gal. capacity	1 Plum	1.40	5.714	Ea.	18,500	305		18,805	20,800
5600	Sterilizers, floor loading, 26" x 62" x 42", single door, steam					158,500			158,500	174,500
5650	Double door, steam					204,000			204,000	224,000
5800	General purpose, 20" x 20" x 38", single door					15,300			15,300	16,800
6000	Portable, counter top, steam, minimum					3,800			3,800	4,175
6020	Maximum					5,950			5,950	6,550
6050	Portable, counter top, gas, 17" x 15" x 32-1/2"					39,300			39,300	43,200
6150	Manual washer/sterilizer, 16" x 16" x 26"	1 Plum	2	4		54,000	214		54,214	59,500
6200	Steam generators, electric 10 kW to 180 kW, freestanding									
6250	Minimum	1 Elec	3	2.667	Ea.	8,000	134		8,134	9,000
6300	Maximum	"	.70	11.429		28,800	575		29,375	32,600
8200	Bed pan washer-sanitizer	1 Plum	2	4		7,125	214		7,339	8,175

11 73 Patient Care Equipment

11 73 10 – Patient Treatment Equipment

11 73 10.10 Treatment Equipment

11 73 10.10 Treatment Equipment		Crew	Daily Output	Labor-Hours	Unit	Material	2011 Bare Costs Labor	Equipment	Total	Total Incl O&P
0010	**TREATMENT EQUIPMENT**									
1800	Heat therapy unit, humidified, 26" x 78" x 28"				Ea.	3,400			3,400	3,725
8400	Whirlpool bath, mobile, sst, 18" x 24" x 60"					4,500			4,500	4,950
8450	Fixed, incl. mixing valves	1 Plum	2	4		9,200	214		9,414	10,400

11 74 Dental Equipment

11 74 10 – Dental Office Equipment

11 74 10.10 Diagnostic and Treatment Equipment

11 74 10.10 Diagnostic and Treatment Equipment		Crew	Daily Output	Labor-Hours	Unit	Material	2011 Bare Costs Labor	Equipment	Total	Total Incl O&P
0010	**DIAGNOSTIC AND TREATMENT EQUIPMENT**									
0020	Central suction system, minimum	1 Plum	1.20	6.667	Ea.	1,550	355		1,905	2,225
0100	Maximum	"	.90	8.889		4,250	475		4,725	5,375
0600	Chair, electric or hydraulic, minimum	1 Skwk	.50	16		2,625	710		3,335	3,975
0700	Maximum		.25	32		14,400	1,425		15,825	18,100
2000	Light, ceiling mounted, minimum		8	1		1,225	44.50		1,269.50	1,425
2100	Maximum		8	1		1,625	44.50		1,669.50	1,875
2200	Unit light, minimum	2 Skwk	5.33	3.002		600	133		733	865
2210	Maximum		5.33	3.002		1,625	133		1,758	2,000
2220	Track light, minimum		3.20	5		1,525	222		1,747	2,025
2230	Maximum		3.20	5		3,725	222		3,947	4,450

11 74 Dental Equipment

11 74 10 – Dental Office Equipment

11 74 10.10 Diagnostic and Treatment Equipment	Crew	Daily Output	Labor-Hours	Unit	Material	2011 Bare Costs Labor	Equipment	Total	Total Incl O&P
2300	Sterilizers, steam portable, minimum				Ea.	1,850			1,850	2,050
2350	Maximum					10,400			10,400	11,400
2600	Steam, institutional					4,250			4,250	4,675
2650	Dry heat, electric, portable, 3 trays					1,225			1,225	1,325

11 78 Mortuary Equipment

11 78 13 – Mortuary Refrigerators

11 78 13.10 Mortuary and Autopsy Equipment

		Crew	Daily Output	Labor-Hours	Unit	Material	Labor	Equipment	Total	Total Incl O&P
0010	**MORTUARY AND AUTOPSY EQUIPMENT**									
0015	Autopsy table, standard	1 Plum	1	8	Ea.	8,850	430		9,280	10,400
0020	Deluxe	"	.60	13.333		15,600	715		16,315	18,300
3200	Mortuary refrigerator, end operated, 2 capacity					12,800			12,800	14,100
3300	6 capacity					23,900			23,900	26,300

11 78 16 – Crematorium Equipment

11 78 16.10 Crematory

		Crew	Daily Output	Labor-Hours	Unit	Material	Labor	Equipment	Total	Total Incl O&P
0010	**CREMATORY**									
1500	Crematory, not including building, 1 place	Q-3	.20	160	Ea.	63,000	8,175		71,175	81,500
1750	2 place	"	.10	320	"	90,000	16,300		106,300	123,500

11 82 Solid Waste Handling Equipment

11 82 19 – Packaged Incinerators

11 82 19.10 Packaged Gas Fired Incinerators

		Crew	Daily Output	Labor-Hours	Unit	Material	Labor	Equipment	Total	Total Incl O&P
0010	**PACKAGED GAS FIRED INCINERATORS**									
4400	Incinerator, gas, not incl. chimney, elec. or pipe, 50#/hr., minimum	Q-3	.80	40	Ea.	11,500	2,050		13,550	15,800
4420	Maximum		.70	45.714		37,100	2,325		39,425	44,300
4440	200 lb. per hr., minimum (batch type)		.60	53.333		63,500	2,725		66,225	74,000
4460	Maximum (with feeder)		.50	64		90,000	3,275		93,275	104,000
4480	400 lb. per hr., minimum (batch type)		.30	106		70,500	5,450		75,950	85,500
4500	Maximum (with feeder)		.25	128		95,500	6,525		102,025	115,000
4520	800 lb. per hr., with feeder, minimum		.20	160		115,500	8,175		123,675	139,000
4540	Maximum		.17	188		159,000	9,600		168,600	189,500
4560	1,200 lb. per hr., with feeder, minimum		.15	213		155,000	10,900		165,900	187,000
4580	Maximum		.11	290		185,000	14,900		199,900	225,500
4600	2,000 lb. per hr., with feeder, minimum		.10	320		379,000	16,300		395,300	441,500
4620	Maximum		.05	640		519,000	32,700		551,700	620,000
4700	For heat recovery system, add, minimum		.25	128		73,000	6,525		79,525	90,500
4710	Add, maximum		.11	290		234,000	14,900		248,900	279,500
4720	For automatic ash conveyer, add		.50	64		30,700	3,275		33,975	38,700
4750	Large municipal incinerators, incl. stack, minimum		.25	128	Ton/day	18,700	6,525		25,225	30,400
4850	Maximum		.10	320	"	49,700	16,300		66,000	79,000

11 82 26 – Facility Waste Compactors

11 82 26.10 Compactors

		Crew	Daily Output	Labor-Hours	Unit	Material	Labor	Equipment	Total	Total Incl O&P
0010	**COMPACTORS**									
0020	Compactors, 115 volt, 250#/hr., chute fed	L-4	1	24	Ea.	11,600	970		12,570	14,200
0100	Hand fed		2.40	10		8,450	405		8,855	9,925
0300	Multi-bag, 230 volt, 600#/hr, chute fed		1	24		10,500	970		11,470	13,100
0400	Hand fed		1	24		9,025	970		9,995	11,400
0500	Containerized, hand fed, 2 to 6 C.Y. containers, 250#/hr.		1	24		11,100	970		12,070	13,800

11 82 Solid Waste Handling Equipment

11 82 26 – Facility Waste Compactors

11 82 26.10 Compactors	Crew	Daily Output	Labor-Hours	Unit	Material	2011 Bare Costs Labor	Equipment	Total	Total Incl O&P
0550 For chute fed, add per floor	L-4	1	24	Ea.	1,275	970		2,245	2,900
1000 Heavy duty industrial compactor, 0.5 C.Y. capacity		1	24		7,525	970		8,495	9,775
1050 1.0 C.Y. capacity		1	24		11,500	970		12,470	14,100
1100 3.0 C.Y. capacity		.50	48		15,700	1,950		17,650	20,200
1150 5.0 C.Y. capacity		.50	48		19,400	1,950		21,350	24,300
1200 Combination shredder/compactor (5,000 lb./hr.)		.50	48		38,300	1,950		40,250	45,200
1400 For handling hazardous waste materials, 55 gallon drum packer, std.					17,800			17,800	19,600
1410 55 gallon drum packer w/HEPA filter					22,200			22,200	24,400
1420 55 gallon drum packer w/charcoal & HEPA filter					29,700			29,700	32,600
1430 All of the above made explosion proof, add					13,500			13,500	14,900
5800 Shredder, industrial, minimum					21,900			21,900	24,100
5850 Maximum					117,500			117,500	129,000
5900 Baler, industrial, minimum					8,775			8,775	9,650
5950 Maximum					512,000			512,000	563,500

Division Notes

	CREW	DAILY OUTPUT	LABOR-HOURS	UNIT	2011 BARE COSTS				TOTAL INCL O&P
					MAT.	LABOR	EQUIP.	TOTAL	

Estimating Tips

General

- The items and systems in this division are usually estimated, purchased, supplied, and installed as a unit by one or more subcontractors. The estimator must ensure that all parties are operating from the same set of specifications and assumptions, and that all necessary items are estimated and will be provided. Many times the complex items and systems are covered, but the more common ones, such as excavation or a crane, are overlooked for the very reason that everyone assumes nobody could miss them. The estimator should be the central focus and be able to ensure that all systems are complete.
- Another area where problems can develop in this division is at the interface between systems. The estimator must ensure, for instance, that anchor bolts, nuts, and washers are estimated and included for the air-supported structures and pre-engineered buildings to be bolted to their foundations. Utility supply is a common area where essential items or pieces of equipment can be missed or overlooked due to the fact that each subcontractor may feel it is another's responsibility. The estimator should also be aware of certain items which may be supplied as part of a package but installed by others, and ensure that the installing contractor's estimate includes the cost of installation. Conversely, the estimator must also ensure that items are not costed by two different subcontractors, resulting in an inflated overall estimate.

13 30 00 Special Structures

- The foundations and floor slab, as well as rough mechanical and electrical, should be estimated, as this work is required for the assembly and erection of the structure. Generally, as noted in the book, the pre-engineered building comes as a shell. Pricing is based on the size and structural design parameters stated in the reference section. Additional features, such as windows and doors with their related structural framing, must also be included by the estimator. Here again, the estimator must have a clear understanding of the scope of each portion of the work and all the necessary interfaces.

Reference Numbers

Reference numbers are shown in shaded boxes at the beginning of some major classifications. These numbers refer to related items in the Reference Section. The reference information may be an estimating procedure, an alternate pricing method, or technical information.

Note: Not all subdivisions listed here necessarily appear in this publication.

13 11 Swimming Pools

13 11 13 – Below-Grade Swimming Pools

13 11 13.50 Swimming Pools	Crew	Daily Output	Labor-Hours	Unit	Material	2011 Bare Costs Labor	2011 Bare Costs Equipment	Total	Total Incl O&P
0010 **SWIMMING POOLS** Residential in-ground, vinyl lined, concrete									
0020 Sides including equipment, sand bottom	B-52	300	.187	SF Surf	15.95	7.45	1.68	25.08	31
0100 Metal or polystyrene sides	B-14	410	.117		13.35	4.24	.78	18.37	22
0200 Add for vermiculite bottom				↓	1.02			1.02	1.12
0500 Gunite bottom and sides, white plaster finish									
0600 12' x 30' pool	B-52	145	.386	SF Surf	29.50	15.40	3.48	48.38	60
0720 16' x 32' pool		155	.361		27	14.40	3.25	44.65	55
0750 20' x 40' pool	↓	250	.224	↓	24	8.90	2.02	34.92	42.50
0810 Concrete bottom and sides, tile finish									
0820 12' x 30' pool	B-52	80	.700	SF Surf	30	28	6.30	64.30	82.50
0830 16' x 32' pool		95	.589		25	23.50	5.30	53.80	69.50
0840 20' x 40' pool	↓	130	.431	↓	19.70	17.15	3.88	40.73	52
1100 Motel, gunite with plaster finish, incl. medium									
1150 capacity filtration & chlorination	B-52	115	.487	SF Surf	36.50	19.40	4.38	60.28	74.50
1200 Municipal, gunite with plaster finish, incl. high									
1250 capacity filtration & chlorination	B-52	100	.560	SF Surf	47.50	22.50	5.05	75.05	91.50
1350 Add for formed gutters				L.F.	69.50			69.50	76.50
1360 Add for stainless steel gutters				"	206			206	226
1700 Filtration and deck equipment only, as % of total				Total				20%	20%
1800 Deck equipment, rule of thumb, 20' x 40' pool				SF Pool				1.18	1.30
1900 5000 S.F. pool				"				1.73	1.90
3000 Painting pools, preparation + 3 coats, 20' x 40' pool, epoxy	2 Pord	.33	48.485	Total	1,700	1,825		3,525	4,550
3100 Rubber base paint, 18 gallons	"	.33	48.485		1,100	1,825		2,925	3,900
3500 42' x 82' pool, 75 gallons, epoxy paint	3 Pord	.14	171		7,150	6,425		13,575	17,400
3600 Rubber base paint	"	.14	171	↓	4,500	6,425		10,925	14,500

13 11 46 – Swimming Pool Accessories

13 11 46.50 Swimming Pool Equipment

	Crew	Daily Output	Labor-Hours	Unit	Material	2011 Bare Costs Labor	2011 Bare Costs Equipment	Total	Total Incl O&P
0010 **SWIMMING POOL EQUIPMENT**									
0020 Diving stand, stainless steel, 3 meter	2 Carp	.40	40	Ea.	11,000	1,725		12,725	14,800
0300 1 meter	"	2.70	5.926	"	6,375	255		6,630	7,400
1100 Gutter system, stainless steel, with grating, stock,									
1110 contains supply and drainage system	E-1	20	1.200	L.F.	268	57	5.45	330.45	395
1120 Integral gutter and 5' high wall system, stainless steel	"	10	2.400	"	360	114	10.90	484.90	600
2100 Lights, underwater, 12 volt, with transformer, 300 watt	1 Elec	1	8	Ea.	263	400		663	885
2200 110 volt, 500 watt, standard		1	8		209	400		609	825
2400 Low water cutoff type	↓	1	8	↓	254	400		654	875
2800 Heaters, see Section 23 52 28.10									

13 18 Ice Rinks

13 18 13 – Ice Rink Floor Systems

13 18 13.50 Ice Skating

	Crew	Daily Output	Labor-Hours	Unit	Material	2011 Bare Costs Labor	2011 Bare Costs Equipment	Total	Total Incl O&P
0010 **ICE SKATING** Equipment incl. refrigeration, plumbing & cooling									
0020 coils & concrete slab, 85' x 200' rink									
0300 55° system, 5 mos., 100 ton				Total	575,000			575,000	632,500
0700 90° system, 12 mos., 135 ton				"	600,000			600,000	660,000
1200 Subsoil heating system (recycled from compressor), 85' x 200'	Q-7	.27	118	Ea.	25,000	6,100		31,100	36,600
1300 Subsoil insulation, 2 lb. polystyrene with vapor barrier, 85' x 200'	2 Carp	.14	114	"	30,000	4,925		34,925	40,600

13 18 Ice Rinks

13 18 16 – Ice Rink Dasher Boards

13 18 16.50 Ice Rink Dasher Boards	Crew	Daily Output	Labor-Hours	Unit	Material	2011 Bare Costs Labor	Equipment	Total	Total Incl O&P
0010 **ICE RINK DASHER BOARDS**									
1000 Dasher boards, 1/2" H.D. polyethylene faced steel frame, 3' acrylic									
1020 screen at sides, 5' acrylic ends, 85' x 200'	F-5	.06	533	Ea.	125,000	23,200		148,200	173,000
1100 Fiberglass & aluminum construction, same sides and ends	"	.06	533	"	125,000	23,200		148,200	173,000

13 21 Controlled Environment Rooms

13 21 13 – Clean Rooms

13 21 13.50 Clean Room Components

	Crew	Daily Output	Labor-Hours	Unit	Material	2011 Bare Costs Labor	Equipment	Total	Total Incl O&P
0010 **CLEAN ROOM COMPONENTS**									
1100 Clean room, soft wall, 12' x 12', Class 100	1 Carp	.18	44.444	Ea.	15,500	1,925		17,425	20,000
1110 Class 1,000		.18	44.444		12,600	1,925		14,525	16,800
1120 Class 10,000		.21	38.095		11,100	1,650		12,750	14,700
1130 Class 100,000		.21	38.095		10,400	1,650		12,050	14,000
2800 Ceiling grid support, slotted channel struts 4'-0" O.C., ea. way				S.F.				5.90	6.50
3000 Ceiling panel, vinyl coated foil on mineral substrate									
3020 Sealed, non-perforated				S.F.				1.27	1.40
4000 Ceiling panel seal, silicone sealant, 150 L.F./gal.	1 Carp	150	.053	L.F.	.30	2.30		2.60	3.85
4100 Two sided adhesive tape	"	240	.033	"	.13	1.43		1.56	2.34
4200 Clips, one per panel				Ea.	1.04			1.04	1.14
6000 HEPA filter, 2' x 4', 99.97% eff., 3" dp beveled frame (silicone seal)					365			365	400
6040 6" deep skirted frame (channel seal)					390			390	430
6100 99.99% efficient, 3" deep beveled frame (silicone seal)					380			380	420
6140 6" deep skirted frame (channel seal)					405			405	445
6200 99.999% efficient, 3" deep beveled frame (silicone seal)					395			395	435
6240 6" deep skirted frame (channel seal)					420			420	465
7000 Wall panel systems, including channel strut framing									
7020 Polyester coated aluminum, particle board				S.F.				18.20	20
7100 Porcelain coated aluminum, particle board								32	35
7400 Wall panel support, slotted channel struts, to 12' high								16.35	18

13 21 26 – Cold Storage Rooms

13 21 26.50 Refrigeration

	Crew	Daily Output	Labor-Hours	Unit	Material	2011 Bare Costs Labor	Equipment	Total	Total Incl O&P
0010 **REFRIGERATION**									
0020 Curbs, 12" high, 4" thick, concrete	2 Carp	58	.276	L.F.	4.37	11.90		16.27	23
1000 Doors, see Section 08 34 13.10									
2400 Finishes, 2 coat portland cement plaster, 1/2" thick	1 Plas	48	.167	S.F.	1.20	6.55		7.75	11.05
2500 For galvanized reinforcing mesh, add	1 Lath	335	.024		.80	.92		1.72	2.22
2700 3/16" thick latex cement	1 Plas	88	.091		2.05	3.58		5.63	7.60
2900 For glass cloth reinforced ceilings, add	"	450	.018		.49	.70		1.19	1.58
3100 Fiberglass panels, 1/8" thick	1 Carp	149.45	.054		2.71	2.30		5.01	6.50
3200 Polystyrene, plastic finish ceiling, 1" thick		274	.029		2.50	1.26		3.76	4.68
3400 2" thick		274	.029		2.85	1.26		4.11	5.05
3500 4" thick		219	.037		3.17	1.57		4.74	5.90
3800 Floors, concrete, 4" thick	1 Cefi	93	.086		1.29	3.51		4.80	6.50
3900 6" thick	"	85	.094		2.01	3.84		5.85	7.80
4000 Insulation, 1" to 6" thick, cork				B.F.	1.16			1.16	1.28
4100 Urethane					.51			.51	.56
4300 Polystyrene, regular					.50			.50	.55
4400 Bead board					.29			.29	.32
4600 Installation of above, add per layer	2 Carp	657.60	.024	S.F.	.38	1.05		1.43	2.03
4700 Wall and ceiling juncture		298.90	.054	L.F.	1.87	2.30		4.17	5.60

13 21 Controlled Environment Rooms

13 21 26 – Cold Storage Rooms

13 21 26.50 Refrigeration

	13 21 26.50 Refrigeration	Crew	Daily Output	Labor-Hours	Unit	Material	2011 Bare Costs Labor	Equipment	Total	Total Incl O&P
4900	Partitions, galvanized sandwich panels, 4" thick, stock	2 Carp	219.20	.073	S.F.	7.85	3.14		10.99	13.40
5000	Aluminum or fiberglass	↓	219.20	.073	"	8.55	3.14		11.69	14.25
5200	Prefab walk-in, 7'-6" high, aluminum, incl. door & floors,									
5210	not incl. partitions or refrigeration, 6' x 6' O.D. nominal	2 Carp	54.80	.292	SF Flr.	140	12.55		152.55	173
5500	10' x 10' O.D. nominal		82.20	.195		113	8.40		121.40	137
5700	12' x 14' O.D. nominal		109.60	.146		101	6.30		107.30	121
5800	12' x 20' O.D. nominal	↓	109.60	.146		88	6.30		94.30	107
6100	For 8'-6" high, add					5%				
6300	Rule of thumb for complete units, w/o doors & refrigeration, cooler	2 Carp	146	.110		127	4.72		131.72	147
6400	Freezer		109.60	.146	↓	150	6.30		156.30	175
6600	Shelving, plated or galvanized, steel wire type		360	.044	SF Hor.	11.10	1.91		13.01	15.20
6700	Slat shelf type	↓	375	.043		13.70	1.84		15.54	17.90
6900	For stainless steel shelving, add				↓	300%				
7000	Vapor barrier, on wood walls	2 Carp	1644	.010	S.F.	.16	.42		.58	.82
7200	On masonry walls	"	1315	.012	"	.42	.52		.94	1.26
7500	For air curtain doors, see Section 23 34 33.10									

13 24 Special Activity Rooms

13 24 16 – Saunas

13 24 16.50 Saunas and Heaters

	13 24 16.50 Saunas and Heaters	Crew	Daily Output	Labor-Hours	Unit	Material	2011 Bare Costs Labor	Equipment	Total	Total Incl O&P
0010	**SAUNAS AND HEATERS**									
0020	Prefabricated, incl. heater & controls, 7' high, 6' x 4', C/C	L-7	2.20	12.727	Ea.	5,125	530		5,655	6,450
0050	6' x 4', C/P		2	14		4,300	580		4,880	5,625
0400	6' x 5', C/C		2	14		5,225	580		5,805	6,650
0450	6' x 5', C/P		2	14		4,450	580		5,030	5,800
0600	6' x 6', C/C		1.80	15.556		5,550	645		6,195	7,075
0650	6' x 6', C/P		1.80	15.556		4,750	645		5,395	6,200
0800	6' x 9', C/C		1.60	17.500		6,875	730		7,605	8,650
0850	6' x 9', C/P		1.60	17.500		6,425	730		7,155	8,175
1000	8' x 12', C/C		1.10	25.455		11,200	1,050		12,250	13,900
1050	8' x 12', C/P		1.10	25.455		9,900	1,050		10,950	12,500
1200	8' x 8', C/C		1.40	20		8,525	830		9,355	10,700
1250	8' x 8', C/P		1.40	20		7,750	830		8,580	9,800
1400	8' x 10', C/C		1.20	23.333		9,450	970		10,420	11,900
1450	8' x 10', C/P		1.20	23.333		8,500	970		9,470	10,800
1600	10' x 12', C/C		1	28		11,800	1,175		12,975	14,800
1650	10' x 12', C/P	↓	1	28		10,300	1,175		11,475	13,200
2500	Heaters only (incl. above), wall mounted, to 200 C.F.					615			615	675
2750	To 300 C.F.					795			795	875
3000	Floor standing, to 720 C.F., 10,000 watts, w/controls	1 Elec	3	2.667		2,100	134		2,234	2,500
3250	To 1,000 C.F., 16,000 watts	"	3	2.667	↓	2,700	134		2,834	3,150

13 24 26 – Steam Baths

13 24 26.50 Steam Baths and Components

	13 24 26.50 Steam Baths and Components	Crew	Daily Output	Labor-Hours	Unit	Material	2011 Bare Costs Labor	Equipment	Total	Total Incl O&P
0010	**STEAM BATHS AND COMPONENTS**									
0020	Heater, timer & head, single, to 140 C.F.	1 Plum	1.20	6.667	Ea.	2,100	355		2,455	2,825
0500	To 300 C.F.		1.10	7.273		2,425	390		2,815	3,250
1000	Commercial size, with blow-down assembly, to 800 C.F.		.90	8.889		7,800	475		8,275	9,275
1500	To 2500 C.F.	↓	.80	10		8,775	535		9,310	10,500
2000	Multiple, motels, apts., 2 baths, w/blow-down assm., 500 C.F.	Q-1	1.30	12.308		5,900	595		6,495	7,375
2500	4 baths	"	.70	22.857	↓	7,450	1,100		8,550	9,850

13 24 Special Activity Rooms

13 24 26 – Steam Baths

13 24 26.50 Steam Baths and Components	Crew	Daily Output	Labor-Hours	Unit	Material	2011 Bare Costs Labor	Equipment	Total	Total Incl O&P
2700	Conversion unit for residential tub, including door				Ea.	4,100			4,100	4,500

13 34 Fabricated Engineered Structures

13 34 23 – Fabricated Structures

13 34 23.10 Comfort Stations

		Crew	Daily Output	Labor-Hours	Unit	Material	2011 Bare Costs Labor	Equipment	Total	Total Incl O&P
0010	**COMFORT STATIONS** Prefab., stock, w/doors, windows & fixt.									
0100	Not incl. interior finish or electrical									
0300	Mobile, on steel frame, minimum				S.F.	46.50			46.50	51
0350	Maximum					75			75	82.50
0400	Permanent, including concrete slab, minimum	B-12J	50	.320		199	12.95	17.15	229.10	257
0500	Maximum	"	43	.372		288	15.05	19.95	323	360
0600	Alternate pricing method, mobile, minimum				Fixture	2,150			2,150	2,375
0650	Maximum					3,175			3,175	3,500
0700	Permanent, minimum	B-12J	.70	22.857		12,200	925	1,225	14,350	16,300
0750	Maximum	"	.50	32		20,400	1,300	1,725	23,425	26,200

13 34 23.16 Fabricated Control Booths

		Crew	Daily Output	Labor-Hours	Unit	Material	2011 Bare Costs Labor	Equipment	Total	Total Incl O&P
0010	**FABRICATED CONTROL BOOTHS**									
0100	Guard House, prefab conc w/bullet resistant doors & windows, roof & wiring									
0110	8' x 8', Level III	L-10	1	24	Ea.	24,200	1,175	655	26,030	29,300
0120	8' x 8', Level IV	"	1	24	"	30,300	1,175	655	32,130	36,000

Division Notes

		CREW	DAILY OUTPUT	LABOR-HOURS	UNIT	2011 BARE COSTS				TOTAL INCL O&P
						MAT.	LABOR	EQUIP.	TOTAL	

Estimating Tips

General

- Many products in Division 14 will require some type of support or blocking for installation not included with the item itself. Examples are supports for conveyors or tube systems, attachment points for lifts, and footings for hoists or cranes. Add these supports in the appropriate division.

14 10 00 Dumbwaiters
14 20 00 Elevators

- Dumbwaiters and elevators are estimated and purchased in a method similar to buying a car. The manufacturer has a base unit with standard features. Added to this base unit price will be whatever options the owner or specifications require. Increased load capacity, additional vertical travel, additional stops, higher speed, and cab finish options are items to be considered. When developing an estimate for dumbwaiters and elevators, remember that some items needed by the installers may have to be included as part of the general contract.

Examples are:
— shaftway
— rail support brackets
— machine room
— electrical supply
— sill angles
— electrical connections
— pits
— roof penthouses
— pit ladders

Check the job specifications and drawings before pricing.

- Installation of elevators and handicapped lifts in historic structures can require significant additional costs. The associated structural requirements may involve cutting into and repairing finishes, mouldings, flooring, etc. The estimator must account for these special conditions.

14 30 00 Escalators and Moving Walks

- Escalators and moving walks are specialty items installed by specialty contractors. There are numerous options associated with these items. For specific options, contact a manufacturer or contractor. In a method similar to estimating dumbwaiters and elevators, you should verify the extent of general contract work and add items as necessary.

14 40 00 Lifts
14 90 00 Other Conveying Equipment

- Products such as correspondence lifts, chutes, and pneumatic tube systems, as well as other items specified in this subdivision, may require trained installers. The general contractor might not have any choice as to who will perform the installation or when it will be performed. Long lead times are often required for these products, making early decisions in scheduling necessary.

Reference Numbers

Reference numbers are shown in shaded boxes at the beginning of some major classifications. These numbers refer to related items in the Reference Section. The reference information may be an estimating procedure, an alternate pricing method, or technical information.

Note: Not all subdivisions listed here necessarily appear in this publication.

Division 14 - Conveying Equipment

14 91 Facility Chutes

14 91 33 – Laundry and Linen Chutes

14 91 33.10 Chutes

		Crew	Daily Output	Labor-Hours	Unit	Material	2011 Bare Costs Labor	Equipment	Total	Total Incl O&P
0011	**CHUTES**, linen, trash or refuse									
0050	Aluminized steel, 16 ga., 18" diameter	2 Shee	3.50	4.571	Floor	1,625	236		1,861	2,150
0100	24" diameter		3.20	5		1,725	258		1,983	2,275
0200	30" diameter		3	5.333		2,050	275		2,325	2,675
0300	36" diameter		2.80	5.714		2,475	295		2,770	3,175
0400	Galvanized steel, 16 ga., 18" diameter		3.50	4.571		1,000	236		1,236	1,450
0500	24" diameter		3.20	5		1,125	258		1,383	1,650
0600	30" diameter		3	5.333		1,275	275		1,550	1,825
0700	36" diameter		2.80	5.714		1,500	295		1,795	2,100
0800	Stainless steel, 18" diameter		3.50	4.571		2,725	236		2,961	3,350
0900	24" diameter		3.20	5		2,950	258		3,208	3,650
1000	30" diameter		3	5.333		3,525	275		3,800	4,300
1005	36" diameter		2.80	5.714		3,825	295		4,120	4,650
1200	Linen chute bottom collector, aluminized steel		4	4	Ea.	1,400	207		1,607	1,850
1300	Stainless steel		4	4		1,800	207		2,007	2,275
1500	Refuse, bottom hopper, aluminized steel, 18" diameter		3	5.333		1,025	275		1,300	1,550
1600	24" diameter		3	5.333		1,275	275		1,550	1,825
1800	36" diameter		3	5.333		2,375	275		2,650	3,025

14 91 82 – Trash Chutes

14 91 82.10 Trash Chutes and Accessories

		Crew	Daily Output	Labor-Hours	Unit	Material	2011 Bare Costs Labor	Equipment	Total	Total Incl O&P
0010	**TRASH CHUTES AND ACCESSORIES**									
2900	Package chutes, spiral type, minimum	2 Shee	4.50	3.556	Floor	2,425	184		2,609	2,950
3000	Maximum	"	1.50	10.667	"	6,325	550		6,875	7,800

14 92 Pneumatic Tube Systems

14 92 10 – Conventional, Automatic and Computer Controlled Pneumatic Tube Systems

14 92 10.10 Pneumatic Tube Systems

		Crew	Daily Output	Labor-Hours	Unit	Material	2011 Bare Costs Labor	Equipment	Total	Total Incl O&P
0010	**PNEUMATIC TUBE SYSTEMS**									
0020	100' long, single tube, 2 stations, stock									
0100	3" diameter	2 Stpi	.12	133	Total	6,600	7,225		13,825	18,100
0300	4" diameter	"	.09	177	"	7,450	9,625		17,075	22,600
0400	Twin tube, two stations or more, conventional system									
0600	2-1/2" round	2 Stpi	62.50	.256	L.F.	13.30	13.85		27.15	35
0700	3" round		46	.348		15.25	18.80		34.05	45
0900	4" round		49.60	.323		16.90	17.45		34.35	44.50
1000	4" x 7" oval		37.60	.426		24.50	23		47.50	61.50
1050	Add for blower		2	8	System	5,200	435		5,635	6,375
1110	Plus for each round station, add		7.50	2.133	Ea.	585	115		700	815
1150	Plus for each oval station, add		7.50	2.133	"	585	115		700	815
1200	Alternate pricing method: base cost, minimum		.75	21.333	Total	6,375	1,150		7,525	8,750
1300	Maximum		.25	64	"	12,700	3,450		16,150	19,100
1500	Plus total system length, add, minimum		93.40	.171	L.F.	8.20	9.25		17.45	23
1600	Maximum		37.60	.426	"	24.50	23		47.50	61
1800	Completely automatic system, 4" round, 15 to 50 stations		.29	55.172	Station	20,100	2,975		23,075	26,600
2200	51 to 144 stations		.32	50		15,600	2,700		18,300	21,300
2400	6" round or 4" x 7" oval, 15 to 50 stations		.24	66.667		25,200	3,600		28,800	33,100
2800	51 to 144 stations		.23	69.565		21,100	3,775		24,875	28,800

Estimating Tips

22 10 00 Plumbing Piping and Pumps

This subdivision is primarily basic pipe and related materials. The pipe may be used by any of the mechanical disciplines, i.e., plumbing, fire protection, heating, and air conditioning.

- The labor adjustment factors listed in Subdivision 22 01 02.20 apply throughout Divisions 21, 22, and 23. CAUTION: the correct percentage may vary for the same items. For example, the percentage add for the basic pipe installation should be based on the maximum height that the craftsman must install for that particular section. If the pipe is to be located 14' above the floor but it is suspended on threaded rod from beams, the bottom flange of which is 18' high (4' rods), then the height is actually 18' and the add is 20%. The pipe coverer, however, does not have to go above the 14', and so his or her add should be 10%.

- Most pipe is priced first as straight pipe with a joint (coupling, weld, etc.) every 10' and a hanger usually every 10'.

There are exceptions with hanger spacing such as for cast iron pipe (5') and plastic pipe (3 per 10'). Following each type of pipe there are several lines listing sizes and the amount to be subtracted to delete couplings and hangers. This is for pipe that is to be buried or supported together on trapeze hangers. The reason that the couplings are deleted is that these runs are usually long, and frequently longer lengths of pipe are used. By deleting the couplings, the estimator is expected to look up and add back the correct reduced number of couplings.

- When preparing an estimate, it may be necessary to approximate the fittings. Fittings usually run between 25% and 50% of the cost of the pipe. The lower percentage is for simpler runs, and the higher number is for complex areas, such as mechanical rooms.

- For historic restoration projects, the systems must be as invisible as possible, and pathways must be sought for pipes, conduit, and ductwork. While installations in accessible spaces (such as basements and attics) are relatively straightforward to estimate, labor costs may be more difficult to determine when delivery systems must be concealed.

22 40 00 Plumbing Fixtures

- Plumbing fixture costs usually require two lines: the fixture itself and its "rough-in, supply, and waste."

- In the Assemblies Section (Plumbing D2010) for the desired fixture, the System Components Group at the center of the page shows the fixture on the first line. The rest of the list (fittings, pipe, tubing, etc.) will total up to what we refer to in the Unit Price section as "Rough-in, supply, waste, and vent." Note that for most fixtures we allow a nominal 5' of tubing to reach from the fixture to a main or riser.

- Remember that gas- and oil-fired units need venting.

Reference Numbers

Reference numbers are shown in shaded boxes at the beginning of some major classifications. These numbers refer to related items in the Reference Section. The reference information may be an estimating procedure, an alternate pricing method, or technical information.

Note: Not all subdivisions listed here necessarily appear in this publication.

Note: **Trade Service,** *in part, has been used as a reference source for some of the material prices used in Division 22.*

22 01 02.10 Boilers, General		Crew	Daily Output	Labor-Hours	Unit	Material	2011 Bare Costs Labor	Equipment	Total	Total Incl O&P
0010	**BOILERS, GENERAL**, Prices do not include flue piping, elec. wiring,									
0020	gas or oil piping, boiler base, pad, or tankless unless noted									
0100	Boiler H.P.: 10 KW = 34 lb./steam/hr. = 33,475 BTU/hr.	R235000-50								
0150	To convert SFR to BTU rating: Hot water, 150 x SFR;									
0160	Forced hot water, 180 x SFR; steam, 240 x SFR									

22 01 02.20 Labor Adjustment Factors		Crew	Daily Output	Labor-Hours	Unit	Material	2011 Bare Costs Labor	Equipment	Total	Total Incl O&P
0010	**LABOR ADJUSTMENT FACTORS**, (For Div. 21, 22 and 23)									
1000	Add to labor for elevated installation (Above floor level)									
1080	10' to 14.5' high	R221113-70					10%			
1100	15' to 19.5' high						20%			
1120	20' to 24.5' high						25%			
1140	25' to 29.5' high						35%			
1160	30' to 34.5' high						40%			
1180	35' to 39.5' high						50%			
1200	40' and higher						55%			
2000	Add to labor for crawl space									
2100	3' high						40%			
2140	4' high						30%			
3000	Add to labor for multi-story building									
3100	Add per floor for floors 3 thru 19						2%			
3140	Add per floor for floors 20 and up						4%			
4000	Add to labor for working in existing occupied buildings									
4100	Hospital						35%			
4140	Office building						25%			
4180	School						20%			
4220	Factory or warehouse						15%			
4260	Multi dwelling						15%			
5000	Add to labor, miscellaneous									
5100	Cramped shaft						35%			
5140	Congested area						15%			
5180	Excessive heat or cold						30%			
9000	Labor factors, The above are reasonable suggestions, however									
9010	each project should be evaluated for its own peculiarities.									
9100	Other factors to be considered are:									
9140	Movement of material and equipment through finished areas									
9180	Equipment room									
9220	Attic space									
9260	No service road									
9300	Poor unloading/storage area									
9340	Congested site area/heavy traffic									

22 05 Common Work Results for Plumbing

22 05 05 – Selective Plumbing Demolition

22 05 05.10 Plumbing Demolition		Crew	Daily Output	Labor-Hours	Unit	Material	2011 Bare Costs Labor	Equipment	Total	Total Incl O&P
0010	**PLUMBING DEMOLITION** R220105-10									
0400	Air compressor, up thru 2 H.P.	Q-1	10	1.600	Ea.		77		77	115
0410	3 H.P. thru 7-1/2 H.P.		5.60	2.857			138		138	206
0420	10 H.P. thru 15 H.P.	↓	1.40	11.429			550		550	825
0430	20 H.P. thru 30 H.P.	Q-2	1.30	18.462			925		925	1,375
0500	Backflow preventer, up thru 2" diameter	1 Plum	17	.471			25		25	37.50
0510	2-1/2" thru 3" diameter	Q-1	10	1.600			77		77	115
0520	4" thru 6" diameter	"	5	3.200			154		154	231
0530	8" thru 10" diameter	Q-2	3	8	↓		400		400	595
0700	Carriers and supports									
0710	Fountains, sinks, lavatories and urinals	1 Plum	14	.571	Ea.		30.50		30.50	45.50
0720	Water closets	"	12	.667	"		35.50		35.50	53.50
0730	Grinder pump or sewage ejector system									
0732	Simplex	Q-1	7	2.286	Ea.		110		110	165
0734	Duplex	"	2.80	5.714			276		276	410
0738	Hot water dispenser	1 Plum	36	.222			11.90		11.90	17.80
0740	Hydrant, wall		26	.308			16.50		16.50	24.50
0744	Ground		12	.667			35.50		35.50	53.50
0760	Cleanouts and drains, up thru 4" pipe diameter	↓	10	.800			43		43	64
0764	5" thru 8" pipe diameter	Q-1	10	1.600			77		77	115
0780	Industrial safety fixtures	1 Plum	8	1	↓		53.50		53.50	80
1020	Fixtures, including 10' piping									
1100	Bathtubs, cast iron	1 Plum	4	2	Ea.		107		107	160
1120	Fiberglass		6	1.333			71.50		71.50	107
1140	Steel	↓	5	1.600			86		86	128
1150	Bidet	Q-1	7	2.286			110		110	165
1200	Lavatory, wall hung	1 Plum	10	.800			43		43	64
1220	Counter top		8	1			53.50		53.50	80
1300	Sink, single compartment		8	1			53.50		53.50	80
1320	Double compartment	↓	7	1.143			61.50		61.50	91.50
1340	Shower, stall and receptor	Q-1	6	2.667			129		129	192
1350	Group	"	7	2.286			110		110	165
1400	Water closet, floor mounted	1 Plum	8	1			53.50		53.50	80
1420	Wall mounted	"	7	1.143			61.50		61.50	91.50
1440	Wash fountain, 36" diameter	Q-2	8	3			150		150	224
1442	54" diameter	"	7	3.429			172		172	256
1500	Urinal, floor mounted	1 Plum	4	2			107		107	160
1520	Wall mounted	"	7	1.143			61.50		61.50	91.50
1590	Whirl pool or hot tub	Q-1	2.60	6.154			297		297	445
1600	Water fountains, free standing	1 Plum	8	1			53.50		53.50	80
1620	Wall or deck mounted		6	1.333			71.50		71.50	107
1800	Medical gas specialties		8	1			53.50		53.50	80
1810	Plumbing demo, Floor drain, Remove		12	.667			35.50		35.50	53.50
1900	Piping fittings, single connection, up thru 1-1/2" diameter		30	.267			14.30		14.30	21.50
1910	2" thru 4" diameter		14	.571			30.50		30.50	45.50
1980	Pipe hanger/support removal		80	.100	↓		5.35		5.35	8
1990	Glass pipe with fittings, 1" thru 3" diameter		200	.040	L.F.		2.14		2.14	3.20
1992	4" thru 6" diameter		150	.053			2.86		2.86	4.27
2000	Piping, metal, up thru 1-1/2" diameter		200	.040			2.14		2.14	3.20
2050	2" thru 3-1/2" diameter	↓	150	.053			2.86		2.86	4.27
2100	4" thru 6" diameter	2 Plum	100	.160			8.60		8.60	12.80
2150	8" thru 14" diameter	"	60	.267			14.30		14.30	21.50
2153	16" thru 20" diameter	Q-18	70	.343	↓		17.30	.46	17.76	26.50

22 05 05.10 Plumbing Demolition		Crew	Daily Output	Labor-Hours	Unit	Material	2011 Bare Costs Labor	Equipment	Total	Total Incl O&P
2155	24" thru 26" diameter	Q-18	55	.436	L.F.		22	.59	22.59	33.50
2156	30" thru 36" diameter	↓	40	.600			30.50	.81	31.31	46
2160	Plastic pipe with fittings, up thru 1-1/2" diameter	1 Plum	250	.032			1.72		1.72	2.56
2162	2" thru 3" diameter	"	200	.040			2.14		2.14	3.20
2164	4" thru 6" diameter	Q-1	200	.080			3.86		3.86	5.75
2166	8" thru 14" diameter		150	.107			5.15		5.15	7.70
2168	16" diameter		100	.160	↓		7.70		7.70	11.50
2170	Prison fixtures, lavatory or sink		18	.889	Ea.		43		43	64
2172	Shower		5.60	2.857			138		138	206
2174	Urinal or water closet		13	1.231			59.50		59.50	88.50
2180	Pumps, all fractional horse-power		12	1.333			64.50		64.50	96
2184	1 H.P. thru 5 H.P.		6	2.667			129		129	192
2186	7-1/2 H.P. thru 15 H.P.	↓	2.50	6.400			310		310	460
2188	20 H.P. thru 25 H.P.	Q-2	4	6			300		300	450
2190	30 H.P. thru 60 H.P.		.80	30			1,500		1,500	2,250
2192	75 H.P. thru 100 H.P.		.60	40			2,000		2,000	2,975
2194	150 H.P.	↓	.50	48			2,400		2,400	3,575
2198	Pump, sump or submersible	1 Plum	12	.667			35.50		35.50	53.50
2200	Receptors and interceptors, up thru 20 GPM	"	8	1			53.50		53.50	80
2204	25 thru 100 GPM	Q-1	6	2.667			129		129	192
2208	125 thru 300 GPM	"	2.40	6.667			320		320	480
2211	325 thru 500 GPM	Q-2	2.60	9.231	↓		460		460	690
2212	Deduct for salvage, aluminum scrap				Ton				700	770
2214	Brass scrap								2,450	2,675
2216	Copper scrap								3,200	3,525
2218	Lead scrap								520	570
2220	Steel scrap				↓				180	200
2230	Temperature maintenance cable	1 Plum	1200	.007	L.F.		.36		.36	.53
2250	Water heater, 40 gal.	"	6	1.333	Ea.		71.50		71.50	107
3100	Tanks, water heaters and liquid containers									
3110	Up thru 45 gallons	Q-1	22	.727	Ea.		35		35	52.50
3120	50 thru 120 gallons		14	1.143			55		55	82.50
3130	130 thru 240 gallons		7.60	2.105			102		102	152
3140	250 thru 500 gallons	↓	5.40	2.963			143		143	213
3150	600 thru 1000 gallons	Q-2	1.60	15			750		750	1,125
3160	1100 thru 2000 gallons		.70	34.286			1,725		1,725	2,550
3170	2100 thru 4000 gallons	↓	.50	48			2,400		2,400	3,575
6000	Remove and reset fixtures, minimum	1 Plum	6	1.333			71.50		71.50	107
6100	Maximum	"	4	2			107		107	160
9100	Valve, metal valves, strainers and similar, up thru 1-1/2" diameter	1 Stpi	28	.286			15.45		15.45	23
9110	2" thru 3" diameter	Q-1	11	1.455			70		70	105
9120	4" thru 6" diameter	"	8	2			96.50		96.50	144
9130	8" thru 14" diameter	Q-2	8	3			150		150	224
9140	16" thru 20" diameter		2	12			600		600	895
9150	24" diameter	↓	1.20	20			1,000		1,000	1,500
9200	Valve, plastic, up thru 1-1/2" diameter	1 Plum	42	.190			10.20		10.20	15.25
9210	2" thru 3" diameter		15	.533			28.50		28.50	42.50
9220	4" thru 6" diameter		12	.667			35.50		35.50	53.50
9300	Vent flashing and caps	↓	55	.145			7.80		7.80	11.65
9350	Water filter, commercial, 1" thru 1-1/2"	Q-1	2	8			385		385	575
9360	2" thru 2-1/2"	"	1.60	10	↓		485		485	720
9400	Water heaters									
9410	Up thru 245 GPH	Q-1	2.40	6.667	Ea.		320		320	480

22 05 Common Work Results for Plumbing

22 05 05 – Selective Plumbing Demolition

22 05 05.10 Plumbing Demolition

		Crew	Daily Output	Labor-Hours	Unit	Material	2011 Bare Costs Labor	2011 Bare Costs Equipment	Total	Total Incl O&P
9420	250 thru 756 GPH	Q-1	1.60	10	Ea.		485		485	720
9430	775 thru 1640 GPH	↓	.80	20			965		965	1,450
9440	1650 thru 4000 GPH	Q-2	.50	48			2,400		2,400	3,575
9470	Water softener	Q-1	2	8	↓		385		385	575

22 05 23 – General-Duty Valves for Plumbing Piping

22 05 23.10 Valves, Brass

		Crew	Daily Output	Labor-Hours	Unit	Material	2011 Bare Costs Labor	2011 Bare Costs Equipment	Total	Total Incl O&P
0010	**VALVES, BRASS**									
0032	For motorized valves, see Section 23 09 53.10									
0500	Gas cocks, threaded									
0510	1/4"	1 Plum	26	.308	Ea.	9.25	16.50		25.75	34.50
0520	3/8"		24	.333		9.25	17.85		27.10	36.50
0530	1/2"		24	.333		8.90	17.85		26.75	36.50
0540	3/4"		22	.364		11.20	19.50		30.70	41.50
0550	1"		19	.421		17.65	22.50		40.15	53
0560	1-1/4"		15	.533		38	28.50		66.50	84.50
0570	1-1/2"		13	.615		51	33		84	105
0580	2"	↓	11	.727	↓	75	39		114	141
0672	For larger sizes use lubricated plug valve, Section 23 05 23.70									

22 05 23.20 Valves, Bronze

			Crew	Daily Output	Labor-Hours	Unit	Material	2011 Bare Costs Labor	2011 Bare Costs Equipment	Total	Total Incl O&P
0010	**VALVES, BRONZE**	R220523-90									
1020	Angle, 150 lb., rising stem, threaded										
1030	1/8"		1 Plum	24	.333	Ea.	102	17.85		119.85	139
1040	1/4"			24	.333		102	17.85		119.85	139
1050	3/8"			24	.333		102	17.85		119.85	139
1060	1/2"			22	.364		102	19.50		121.50	141
1070	3/4"			20	.400		139	21.50		160.50	185
1080	1"			19	.421		200	22.50		222.50	254
1090	1-1/4"			15	.533		257	28.50		285.50	325
1100	1-1/2"			13	.615		335	33		368	420
1110	2"		↓	11	.727	↓	540	39		579	655
1380	Ball, 150 psi, threaded										
1400	1/4"		1 Plum	24	.333	Ea.	11.15	17.85		29	39
1430	3/8"			24	.333		11.15	17.85		29	39
1450	1/2"			22	.364		11.30	19.50		30.80	41.50
1460	3/4"			20	.400		18.60	21.50		40.10	52.50
1470	1"			19	.421		23.50	22.50		46	59.50
1480	1-1/4"			15	.533		39.50	28.50		68	86
1490	1-1/2"			13	.615		55	33		88	110
1500	2"			11	.727		66.50	39		105.50	131
1510	2-1/2"			9	.889		209	47.50		256.50	300
1520	3"		↓	8	1	↓	340	53.50		393.50	450
1600	Butterfly, 175 psi, full port, solder or threaded ends										
1610	Stainless steel disc and stem										
1620	1/4"		1 Plum	24	.333	Ea.	12.45	17.85		30.30	40
1630	3/8"			24	.333		12.45	17.85		30.30	40
1640	1/2"			22	.364		13.15	19.50		32.65	43.50
1650	3/4"			20	.400		21	21.50		42.50	55.50
1660	1"			19	.421		26	22.50		48.50	62
1670	1-1/4"			15	.533		42	28.50		70.50	88.50
1680	1-1/2"			13	.615		54	33		87	109
1690	2"		↓	11	.727	↓	67.50	39		106.50	132
1750	Check, swing, class 150, regrinding disc, threaded										

22 05 Common Work Results for Plumbing

22 05 23 – General-Duty Valves for Plumbing Piping

22 05 23.20 Valves, Bronze		Crew	Daily Output	Labor-Hours	Unit	Material	2011 Bare Costs Labor	Equipment	Total	Total Incl O&P
1800	1/8"	1 Plum	24	.333	Ea.	47.50	17.85		65.35	78.50
1830	1/4"		24	.333		47.50	17.85		65.35	78.50
1840	3/8"		24	.333		47.50	17.85		65.35	78.50
1850	1/2"		24	.333		49.50	17.85		67.35	81
1860	3/4"		20	.400		72	21.50		93.50	111
1870	1"		19	.421		104	22.50		126.50	148
1880	1-1/4"		15	.533		149	28.50		177.50	207
1890	1-1/2"		13	.615		174	33		207	240
1900	2"		11	.727		256	39		295	340
1910	2-1/2"	Q-1	15	1.067		575	51.50		626.50	705
1920	3"	"	13	1.231		765	59.50		824.50	930
2000	For 200 lb, add					5%	10%			
2040	For 300 lb, add					15%	15%			
2060	Check swing, 300#, sweat, 3/8" size	1 Plum	24	.333	Ea.	34	17.85		51.85	63.50
2070	1/2"		24	.333		34	17.85		51.85	63.50
2080	3/4"		20	.400		43	21.50		64.50	79.50
2090	1"		19	.421		60	22.50		82.50	99.50
2100	1-1/4"		15	.533		77.50	28.50		106	128
2110	1-1/2"		13	.615		103	33		136	162
2120	2"		11	.727		156	39		195	230
2130	2-1/2"	Q-1	15	1.067		365	51.50		416.50	475
2140	3"	"	13	1.231		490	59.50		549.50	630
2350	Check, lift, class 150, horizontal composition disc, threaded									
2430	1/4"	1 Plum	24	.333	Ea.	128	17.85		145.85	168
2440	3/8"		24	.333		163	17.85		180.85	206
2450	1/2"		24	.333		142	17.85		159.85	183
2460	3/4"		20	.400		173	21.50		194.50	222
2470	1"		19	.421		246	22.50		268.50	305
2480	1-1/4"		15	.533		330	28.50		358.50	410
2490	1-1/2"		13	.615		395	33		428	485
2500	2"		11	.727		655	39		694	785
2850	Gate, N.R.S., soldered, 125 psi									
2900	3/8"	1 Plum	24	.333	Ea.	42	17.85		59.85	72.50
2920	1/2"		24	.333		35	17.85		52.85	65
2940	3/4"		20	.400		40.50	21.50		62	76.50
2950	1"		19	.421		57	22.50		79.50	96
2960	1-1/4"		15	.533		91	28.50		119.50	143
2970	1-1/2"		13	.615		97	33		130	156
2980	2"		11	.727		138	39		177	210
2990	2-1/2"	Q-1	15	1.067		345	51.50		396.50	455
3000	3"	"	13	1.231		450	59.50		509.50	585
3350	Threaded, class 150									
3410	1/4"	1 Plum	24	.333	Ea.	59	17.85		76.85	91.50
3420	3/8"		24	.333		59	17.85		76.85	91.50
3430	1/2"		24	.333		55.50	17.85		73.35	87.50
3440	3/4"		20	.400		63.50	21.50		85	102
3450	1"		19	.421		80.50	22.50		103	122
3460	1-1/4" size		15	.533		109	28.50		137.50	163
3470	1-1/2"		13	.615		156	33		189	220
3480	2"		11	.727		186	39		225	262
3490	2-1/2"	Q-1	15	1.067		435	51.50		486.50	550
3500	3" size	"	13	1.231		615	59.50		674.50	765
3600	Gate, flanged, 150 lb.									

22 05 23 – General-Duty Valves for Plumbing Piping

22 05 23.20 Valves, Bronze		Crew	Daily Output	Labor-Hours	Unit	Material	2011 Bare Costs Labor	Equipment	Total	Total Incl O&P
3610	1"	1 Plum	7	1.143	Ea.	1,050	61.50		1,111.50	1,250
3620	1-1/2"		6	1.333		1,325	71.50		1,396.50	1,550
3630	2"		5	1.600		1,925	86		2,011	2,250
3634	2-1/2"	Q-1	5	3.200		3,000	154		3,154	3,500
3640	3"	"	4.50	3.556		3,475	172		3,647	4,075
3850	Rising stem, soldered, 300 psi									
3900	3/8"	1 Plum	24	.333	Ea.	86	17.85		103.85	121
3920	1/2"		24	.333		86	17.85		103.85	121
3940	3/4"		20	.400		93.50	21.50		115	135
3950	1"		19	.421		136	22.50		158.50	184
3960	1-1/4"		15	.533		189	28.50		217.50	250
3970	1-1/2"		13	.615		222	33		255	293
3980	2"		11	.727		360	39		399	455
3990	2-1/2"	Q-1	15	1.067		805	51.50		856.50	960
4000	3"	"	13	1.231		1,150	59.50		1,209.50	1,350
4250	Threaded, class 150									
4310	1/4"	1 Plum	24	.333	Ea.	53.50	17.85		71.35	85
4320	3/8"		24	.333		53.50	17.85		71.35	85
4330	1/2"		24	.333		49.50	17.85		67.35	81
4340	3/4"		20	.400		57.50	21.50		79	95.50
4350	1"		19	.421		77	22.50		99.50	118
4360	1-1/4"		15	.533		104	28.50		132.50	158
4370	1-1/2"		13	.615		132	33		165	194
4380	2"		11	.727		174	39		213	250
4390	2-1/2"	Q-1	15	1.067		415	51.50		466.50	530
4400	3"	"	13	1.231		580	59.50		639.50	725
4500	For 300 psi, threaded, add					100%	15%			
4540	For chain operated type, add					15%				
4850	Globe, class 150, rising stem, threaded									
4920	1/4"	1 Plum	24	.333	Ea.	76	17.85		93.85	110
4940	3/8" size		24	.333		76	17.85		93.85	110
4950	1/2"		24	.333		76	17.85		93.85	110
4960	3/4"		20	.400		102	21.50		123.50	144
4970	1"		19	.421		160	22.50		182.50	210
4980	1-1/4"		15	.533		254	28.50		282.50	325
4990	1-1/2"		13	.615		310	33		343	390
5000	2"		11	.727		465	39		504	570
5010	2-1/2"	Q-1	15	1.067		935	51.50		986.50	1,100
5020	3"	"	13	1.231		1,325	59.50		1,384.50	1,550
5120	For 300 lb threaded, add					50%	15%			
5130	Globe, 300 lb., sweat, 3/8" size	1 Plum	24	.333	Ea.	95	17.85		112.85	132
5140	1/2"		24	.333		97	17.85		114.85	134
5150	3/4"		20	.400		132	21.50		153.50	177
5160	1"		19	.421		183	22.50		205.50	235
5170	1-1/4"		15	.533		289	28.50		317.50	365
5180	1-1/2"		13	.615		350	33		383	435
5190	2"		11	.727		515	39		554	630
5200	2-1/2"	Q-1	15	1.067		1,000	51.50		1,051.50	1,175
5210	3"	"	13	1.231		1,300	59.50		1,359.50	1,525
5600	Relief, pressure & temperature, self-closing, ASME, threaded									
5640	3/4"	1 Plum	28	.286	Ea.	132	15.30		147.30	169
5650	1"		24	.333		202	17.85		219.85	249
5660	1-1/4"		20	.400		390	21.50		411.50	455

22 05 23.20 Valves, Bronze		Crew	Daily Output	Labor-Hours	Unit	Material	2011 Bare Costs Labor	Equipment	Total	Total Incl O&P
5670	1-1/2"	1 Plum	18	.444	Ea.	745	24		769	850
5680	2"	↓	16	.500	↓	790	27		817	910
5950	Pressure, poppet type, threaded									
6000	1/2"	1 Plum	30	.267	Ea.	35	14.30		49.30	60
6040	3/4"	"	28	.286	"	37.50	15.30		52.80	64
6400	Pressure, water, ASME, threaded									
6440	3/4"	1 Plum	28	.286	Ea.	74.50	15.30		89.80	105
6450	1"		24	.333		158	17.85		175.85	201
6460	1-1/4"		20	.400		251	21.50		272.50	310
6470	1-1/2"		18	.444		345	24		369	415
6480	2"		16	.500		500	27		527	590
6490	2-1/2"	↓	15	.533	↓	1,900	28.50		1,928.50	2,150
6900	Reducing, water pressure									
6920	300 psi to 25-75 psi, threaded or sweat									
6940	1/2"	1 Plum	24	.333	Ea.	239	17.85		256.85	290
6950	3/4"		20	.400		239	21.50		260.50	295
6960	1"		19	.421		370	22.50		392.50	445
6970	1-1/4"		15	.533		665	28.50		693.50	775
6980	1-1/2"		13	.615		1,000	33		1,033	1,150
6990	2"	↓	11	.727		1,450	39		1,489	1,650
7100	For built-in by-pass or 10-35 psi, add				↓	22.50			22.50	25
7700	High capacity, 250 psi to 25-75 psi, threaded									
7740	1/2"	1 Plum	24	.333	Ea.	299	17.85		316.85	355
7780	3/4"		20	.400		299	21.50		320.50	360
7790	1"		19	.421		465	22.50		487.50	545
7800	1-1/4"		15	.533		830	28.50		858.50	955
7810	1-1/2"		13	.615		1,250	33		1,283	1,425
7820	2"		11	.727		1,800	39		1,839	2,050
7830	2-1/2"		9	.889		2,150	47.50		2,197.50	2,425
7840	3"	↓	8	1		2,200	53.50		2,253.50	2,500
7850	3" flanged (iron body)	Q-1	10	1.600		2,925	77		3,002	3,350
7860	4" flanged (iron body)	"	8	2	↓	3,850	96.50		3,946.50	4,375
7920	For higher pressure, add					25%				
8000	Silent check, bronze trim									
8010	Compact wafer type, for 125 or 150 lb. flanges									
8020	1-1/2"	1 Plum	11	.727	Ea.	297	39		336	385
8021	2"	"	9	.889		315	47.50		362.50	415
8022	2-1/2"	Q-1	9	1.778		330	86		416	495
8023	3"		8	2		370	96.50		466.50	550
8024	4"	↓	5	3.200		630	154		784	925
8025	5"	Q-2	6	4		835	200		1,035	1,225
8026	6"		5	4.800		985	240		1,225	1,425
8027	8"		4.50	5.333		2,100	267		2,367	2,725
8028	10"		4	6		3,850	300		4,150	4,675
8029	12"	↓	3	8	↓	8,700	400		9,100	10,200
8050	For 250 or 300 lb flanges, thru 6" no change									
8051	For 8" and 10", add				Ea.	40%	10%			
8060	Full flange wafer type, 150 lb.									
8061	1"	1 Plum	14	.571	Ea.	310	30.50		340.50	390
8062	1-1/4"		12	.667		325	35.50		360.50	410
8063	1-1/2"		11	.727		390	39		429	490
8064	2"	↓	9	.889		525	47.50		572.50	650
8065	2-1/2"	Q-1	9	1.778	↓	595	86		681	785

22 05 Common Work Results for Plumbing

22 05 23 – General-Duty Valves for Plumbing Piping

22 05 23.20 Valves, Bronze		Crew	Daily Output	Labor-Hours	Unit	Material	2011 Bare Costs Labor	Equipment	Total	Total Incl O&P
8066	3"	Q-1	8	2	Ea.	660	96.50		756.50	870
8067	4"	↓	5	3.200		970	154		1,124	1,300
8068	5"	Q-2	6	4		1,250	200		1,450	1,675
8069	6"	"	5	4.800		1,650	240		1,890	2,175
8080	For 300 lb., add				↓	40%	10%			
8100	Globe type, 150 lb.									
8110	2"	1 Plum	9	.889	Ea.	615	47.50		662.50	745
8111	2-1/2"	Q-1	9	1.778		760	86		846	965
8112	3"		8	2		910	96.50		1,006.50	1,150
8113	4"	↓	5	3.200		1,250	154		1,404	1,600
8114	5"	Q-2	6	4		1,550	200		1,750	2,000
8115	6"	"	5	4.800	↓	2,150	240		2,390	2,700
8130	For 300 lb., add					20%	10%			
8140	Screwed end type, 250 lb.									
8141	1/2"	1 Plum	24	.333	Ea.	55	17.85		72.85	87
8142	3/4"		20	.400		55	21.50		76.50	92.50
8143	1"		19	.421		63	22.50		85.50	103
8144	1-1/4"		15	.533		87	28.50		115.50	139
8145	1-1/2"		13	.615		97.50	33		130.50	156
8146	2"	↓	11	.727	↓	132	39		171	203
8350	Tempering, water, sweat connections									
8400	1/2"	1 Plum	24	.333	Ea.	98	17.85		115.85	135
8440	3/4"	"	20	.400	"	114	21.50		135.50	157
8650	Threaded connections									
8700	1/2"	1 Plum	24	.333	Ea.	141	17.85		158.85	182
8740	3/4"		20	.400		445	21.50		466.50	515
8750	1"		19	.421		500	22.50		522.50	585
8760	1-1/4"		15	.533		770	28.50		798.50	890
8770	1-1/2"		13	.615		845	33		878	975
8780	2"	↓	11	.727	↓	1,275	39		1,314	1,450

22 05 23.60 Valves, Plastic

	22 05 23.60 Valves, Plastic	Crew	Daily Output	Labor-Hours	Unit	Material	2011 Bare Costs Labor	Equipment	Total	Total Incl O&P
0010	**VALVES, PLASTIC** R220523-90									
1100	Angle, PVC, threaded									
1110	1/4"	1 Plum	26	.308	Ea.	69.50	16.50		86	101
1120	1/2"		26	.308		69.50	16.50		86	101
1130	3/4"		25	.320		82.50	17.15		99.65	117
1140	1"	↓	23	.348	↓	100	18.65		118.65	138
1150	Ball, PVC, socket or threaded, single union									
1230	1/2"	1 Plum	26	.308	Ea.	26.50	16.50		43	53.50
1240	3/4"		25	.320		30	17.15		47.15	58.50
1250	1"		23	.348		38	18.65		56.65	69.50
1260	1-1/4"		21	.381		51	20.50		71.50	86.50
1270	1-1/2"		20	.400		60.50	21.50		82	98.50
1280	2"	↓	17	.471		87.50	25		112.50	134
1290	2-1/2"	Q-1	26	.615		370	29.50		399.50	450
1300	3"		24	.667		450	32		482	545
1310	4"	↓	20	.800		760	38.50		798.50	895
1360	For PVC, flanged, add					100%	15%			
1450	Double union 1/2"	1 Plum	26	.308		32	16.50		48.50	59.50
1460	3/4"		25	.320		38	17.15		55.15	67
1470	1"		23	.348		45	18.65		63.65	77.50
1480	1-1/4"	↓	21	.381		62.50	20.50		83	99

22 05 23.60 Valves, Plastic		Crew	Daily Output	Labor-Hours	Unit	Material	2011 Bare Costs Labor	2011 Bare Costs Equipment	Total	Total Incl O&P
1490	1-1/2"	1 Plum	20	.400	Ea.	75	21.50		96.50	115
1500	2"	↓	17	.471	↓	99	25		124	147
1650	CPVC, socket or threaded, single union									
1700	1/2"	1 Plum	26	.308	Ea.	50	16.50		66.50	79.50
1720	3/4"		25	.320		62.50	17.15		79.65	94.50
1730	1"		23	.348		75	18.65		93.65	111
1750	1-1/4"		21	.381		120	20.50		140.50	162
1760	1-1/2"		20	.400		120	21.50		141.50	163
1770	2"	↓	17	.471		165	25		190	219
1780	3"	Q-1	24	.667		455	32		487	550
1840	For CPVC, flanged, add					65%	15%			
1880	For true union, socket or threaded, add				↓	50%	5%			
2050	Polypropylene, threaded									
2100	1/4"	1 Plum	26	.308	Ea.	38.50	16.50		55	66.50
2120	3/8"		26	.308		38.50	16.50		55	66.50
2130	1/2"		26	.308		38.50	16.50		55	66.50
2140	3/4"		25	.320		46.50	17.15		63.65	76.50
2150	1"		23	.348		53.50	18.65		72.15	87
2160	1-1/4"		21	.381		72	20.50		92.50	110
2170	1-1/2"		20	.400		88	21.50		109.50	129
2180	2"	↓	17	.471		119	25		144	169
2190	3"	Q-1	24	.667		270	32		302	345
2200	4"	"	20	.800	↓	475	38.50		513.50	585
2550	PVC, three way, socket or threaded									
2600	1/2"	1 Plum	26	.308	Ea.	59	16.50		75.50	89
2640	3/4"		25	.320		67	17.15		84.15	99
2650	1"		23	.348		72	18.65		90.65	107
2660	1-1/2"		20	.400		170	21.50		191.50	219
2670	2"	↓	17	.471		194	25		219	252
2680	3"	Q-1	24	.667		425	32		457	515
2740	For flanged, add				↓	60%	15%			
3150	Ball check, PVC, socket or threaded									
3200	1/4"	1 Plum	26	.308	Ea.	39	16.50		55.50	67.50
3220	3/8"		26	.308		39	16.50		55.50	67.50
3240	1/2"		26	.308		39	16.50		55.50	67.50
3250	3/4"		25	.320		44	17.15		61.15	73.50
3260	1"		23	.348		55	18.65		73.65	88.50
3270	1-1/4"		21	.381		92	20.50		112.50	132
3280	1-1/2"		20	.400		92	21.50		113.50	133
3290	2"	↓	17	.471		125	25		150	176
3310	3"	Q-1	24	.667		345	32		377	430
3320	4"	"	20	.800		490	38.50		528.50	600
3360	For PVC, flanged, add				↓	50%	15%			
3750	CPVC, socket or threaded									
3800	1/2"	1 Plum	26	.308	Ea.	60.50	16.50		77	91
3840	3/4"		25	.320		72	17.15		89.15	105
3850	1"		23	.348		85	18.65		103.65	122
3860	1-1/2"		20	.400		149	21.50		170.50	196
3870	2"	↓	17	.471		197	25		222	255
3880	3"	Q-1	24	.667		535	32		567	635
3920	4"	"	20	.800		715	38.50		753.50	845
3930	For CPVC, flanged, add				↓	40%	15%			
4340	Polypropylene, threaded									

22 05 23.60 Valves, Plastic		Crew	Daily Output	Labor-Hours	Unit	Material	2011 Bare Costs Labor	Equipment	Total	Total Incl O&P
4360	1/2"	1 Plum	26	.308	Ea.	38.50	16.50		55	66.50
4400	3/4"		25	.320		45	17.15		62.15	75
4440	1"		23	.348		57	18.65		75.65	90.50
4450	1-1/2"		20	.400		110	21.50		131.50	153
4460	2"		17	.471		138	25		163	190
4500	For polypropylene flanged, add					200%	15%			
4850	Foot valve, PVC, socket or threaded									
4900	1/2"	1 Plum	34	.235	Ea.	66.50	12.60		79.10	92
4930	3/4"		32	.250		75	13.40		88.40	103
4940	1"		28	.286		97.50	15.30		112.80	130
4950	1-1/4"		27	.296		187	15.90		202.90	230
4960	1-1/2"		26	.308		187	16.50		203.50	231
4970	2"		24	.333		218	17.85		235.85	266
4980	3"		20	.400		520	21.50		541.50	600
4990	4"		18	.444		915	24		939	1,025
5000	For flanged, add					25%	10%			
5050	CPVC, socket or threaded									
5060	1/2"	1 Plum	34	.235	Ea.	68	12.60		80.60	94
5070	3/4"		32	.250		78.50	13.40		91.90	107
5080	1"		28	.286		96.50	15.30		111.80	129
5090	1-1/4"		27	.296		155	15.90		170.90	194
5100	1-1/2"		26	.308		155	16.50		171.50	195
5110	2"		24	.333		199	17.85		216.85	246
5120	3"		20	.400		405	21.50		426.50	475
5130	4"		18	.444		735	24		759	845
5140	For flanged, add					25%	10%			
5280	Needle valve, PVC, threaded									
5300	1/4"	1 Plum	26	.308	Ea.	44.50	16.50		61	73.50
5340	3/8"		26	.308		51.50	16.50		68	81
5360	1/2"		26	.308		51.50	16.50		68	81
5380	For polypropylene, add					10%				
5800	Y check, PVC, socket or threaded									
5820	1/2"	1 Plum	26	.308	Ea.	77.50	16.50		94	110
5840	3/4"		25	.320		84	17.15		101.15	118
5850	1"		23	.348		91	18.65		109.65	128
5860	1-1/4"		21	.381		143	20.50		163.50	188
5870	1-1/2"		20	.400		156	21.50		177.50	203
5880	2"		17	.471		194	25		219	251
5890	2-1/2"		15	.533		410	28.50		438.50	495
5900	3"	Q-1	24	.667		410	32		442	500
5910	4"	"	20	.800		670	38.50		708.50	795
5960	For PVC flanged, add					45%	15%			
6350	Y sediment strainer, PVC, socket or threaded									
6400	1/2"	1 Plum	26	.308	Ea.	44	16.50		60.50	73
6440	3/4"		24	.333		47	17.85		64.85	78.50
6450	1"		23	.348		57	18.65		75.65	90.50
6460	1-1/4"		21	.381		94.50	20.50		115	135
6470	1-1/2"		20	.400		94.50	21.50		116	136
6480	2"		17	.471		116	25		141	166
6490	2-1/2"		15	.533		284	28.50		312.50	355
6500	3"	Q-1	24	.667		284	32		316	360
6510	4"	"	20	.800		470	38.50		508.50	580
6560	For PVC flanged, add					55%	15%			

22 05 29 – Hangers and Supports for Plumbing Piping and Equipment

22 05 29.10 Hangers & Supp. for Plumb'g/HVAC Pipe/Equip.	Crew	Daily Output	Labor-Hours	Unit	Material	2011 Bare Costs Labor	Equipment	Total	Total Incl O&P
0010 **HANGERS AND SUPPORTS FOR PLUMB'G/HVAC PIPE/EQUIP.**									
0011 TYPE numbers per MSS-SP58									
0050 Brackets									
0060 Beam side or wall, malleable iron, TYPE 34									
0070 3/8" threaded rod size	1 Plum	48	.167	Ea.	3.34	8.95		12.29	17
0080 1/2" threaded rod size		48	.167		5.80	8.95		14.75	19.75
0090 5/8" threaded rod size		48	.167		9.10	8.95		18.05	23.50
0100 3/4" threaded rod size		48	.167		10.95	8.95		19.90	25.50
0110 7/8" threaded rod size		48	.167		12.85	8.95		21.80	27.50
0120 For concrete installation, add						30%			
0150 Wall, welded steel, medium, TYPE 32									
0160 0 size, 12" wide, 18" deep	1 Plum	34	.235	Ea.	234	12.60		246.60	276
0170 1 size, 18" wide, 24" deep		34	.235		278	12.60		290.60	325
0180 2 size, 24" wide, 30" deep		34	.235		365	12.60		377.60	425
0200 Beam attachment, welded, TYPE 22									
0202 3/8"	Q-15	80	.200	Ea.	8.15	9.65	.41	18.21	24
0203 1/2"		76	.211		8.15	10.15	.43	18.73	24.50
0204 5/8"		72	.222		8.60	10.70	.45	19.75	26
0205 3/4"		68	.235		9.40	11.35	.48	21.23	28
0206 7/8"		64	.250		13.60	12.05	.51	26.16	33.50
0207 1"		56	.286		19.45	13.80	.58	33.83	42.50
0300 Clamps									
0310 C-clamp, for mounting on steel beam flange, w/locknut, TYPE 23									
0320 3/8" threaded rod size	1 Plum	160	.050	Ea.	3.15	2.68		5.83	7.45
0330 1/2" threaded rod size		160	.050		3.91	2.68		6.59	8.30
0340 5/8" threaded rod size		160	.050		6	2.68		8.68	10.60
0350 3/4" threaded rod size		160	.050		8.05	2.68		10.73	12.90
0352 7/8" threaded rod size		140	.057		25.50	3.06		28.56	33
0400 High temperature to 1050°F, alloy steel									
0410 4" pipe size	Q-1	106	.151	Ea.	26.50	7.30		33.80	40
0420 6" pipe size		106	.151		44.50	7.30		51.80	60
0430 8" pipe size		97	.165		50.50	7.95		58.45	67.50
0440 10" pipe size		84	.190		79.50	9.20		88.70	101
0450 12" pipe size		72	.222		92.50	10.70		103.20	118
0460 14" pipe size		64	.250		243	12.05		255.05	285
0470 16" pipe size		56	.286		254	13.80		267.80	300
0480 Beam clamp, flange type, TYPE 25									
0482 For 3/8" bolt	1 Plum	48	.167	Ea.	5.40	8.95		14.35	19.30
0483 For 1/2" bolt		44	.182		7.90	9.75		17.65	23
0484 For 5/8" bolt		40	.200		7.90	10.70		18.60	24.50
0485 For 3/4" bolt		36	.222		7.90	11.90		19.80	26.50
0486 For 1" bolt		32	.250		7.90	13.40		21.30	28.50
0500 I-beam, for mounting on bottom flange, strap iron, TYPE 21									
0510 2" flange size	1 Plum	96	.083	Ea.	14	4.47		18.47	22
0520 3" flange size		95	.084		16.60	4.51		21.11	25
0530 4" flange size		93	.086		7.90	4.61		12.51	15.55
0540 5" flange size		92	.087		8.70	4.66		13.36	16.50
0550 6" flange size		90	.089		9.95	4.76		14.71	18.05
0560 7" flange size		88	.091		10.95	4.87		15.82	19.30
0570 8" flange size		86	.093		11.65	4.99		16.64	20.50
0600 One hole, vertical mounting, malleable iron									
0610 1/2" pipe size	1 Plum	160	.050	Ea.	1.14	2.68		3.82	5.25

22 05 29 – Hangers and Supports for Plumbing Piping and Equipment

22 05 29.10 Hangers & Supp. for Plumb'g/HVAC Pipe/Equip.	Crew	Daily Output	Labor-Hours	Unit	Material	2011 Bare Costs Labor	Equipment	Total	Total Incl O&P
0620	3/4" pipe size	1 Plum	145	.055	Ea.	1.39	2.96		4.35	5.95
0630	1" pipe size		136	.059		1.92	3.15		5.07	6.80
0640	1-1/4" pipe size		128	.063		2.30	3.35		5.65	7.55
0650	1-1/2" pipe size		120	.067		2.56	3.57		6.13	8.15
0660	2" pipe size		112	.071		3.72	3.83		7.55	9.80
0670	2-1/2" pipe size		104	.077		7.30	4.12		11.42	14.15
0680	3" pipe size		96	.083		10.35	4.47		14.82	18.05
0690	3-1/2" pipe size		90	.089		12.80	4.76		17.56	21
0700	4" pipe size		84	.095		14	5.10		19.10	23
0750	Riser or extension pipe, carbon steel, TYPE 8									
0756	1/2" pipe size	1 Plum	52	.154	Ea.	4.42	8.25		12.67	17.15
0760	3/4" pipe size		48	.167		4.42	8.95		13.37	18.20
0770	1" pipe size		47	.170		4.70	9.10		13.80	18.75
0780	1-1/4" pipe size		46	.174		5.80	9.30		15.10	20.50
0790	1-1/2" pipe size		45	.178		6.25	9.55		15.80	21
0800	2" pipe size		43	.186		6.45	9.95		16.40	22
0810	2-1/2" pipe size		41	.195		7.05	10.45		17.50	23.50
0820	3" pipe size		40	.200		7.80	10.70		18.50	24.50
0830	3-1/2" pipe size		39	.205		9.55	11		20.55	27
0840	4" pipe size		38	.211		9.75	11.30		21.05	27.50
0850	5" pipe size		37	.216		13.20	11.60		24.80	32
0860	6" pipe size		36	.222		16.35	11.90		28.25	36
0870	8" pipe size		34	.235		25.50	12.60		38.10	47.50
0880	10" pipe size		32	.250		38	13.40		51.40	62
0890	12" pipe size		28	.286		54	15.30		69.30	82.50
0900	For plastic coating 3/4" to 4", add					190%				
0910	For copper plating 3/4" to 4", add					58%				
0950	Two piece, complete, carbon steel, medium weight, TYPE 4									
0960	1/2" pipe size	Q-1	137	.117	Ea.	3.38	5.65		9.03	12.10
0970	3/4" pipe size		134	.119		3.38	5.75		9.13	12.30
0980	1" pipe size		132	.121		3.40	5.85		9.25	12.50
0990	1-1/4" pipe size		130	.123		4.48	5.95		10.43	13.80
1000	1-1/2" pipe size		126	.127		4.48	6.15		10.63	14.10
1010	2" pipe size		124	.129		5.20	6.25		11.45	15
1020	2-1/2" pipe size		120	.133		5.55	6.45		12	15.70
1030	3" pipe size		117	.137		6.20	6.60		12.80	16.65
1040	3-1/2" pipe size		114	.140		8.75	6.75		15.50	19.75
1050	4" pipe size		110	.145		8.75	7		15.75	20
1060	5" pipe size		106	.151		16.85	7.30		24.15	29.50
1070	6" pipe size		104	.154		21.50	7.40		28.90	34.50
1080	8" pipe size		100	.160		25.50	7.70		33.20	39.50
1090	10" pipe size		96	.167		46	8.05		54.05	62.50
1100	12" pipe size		89	.180		59	8.65		67.65	78
1110	14" pipe size		82	.195		113	9.40		122.40	139
1120	16" pipe size		68	.235		123	11.35		134.35	152
1130	For galvanized, add					45%				
1150	Insert, concrete									
1160	Wedge type, carbon steel body, malleable iron nut									
1170	1/4" threaded rod size	1 Plum	96	.083	Ea.	5.25	4.47		9.72	12.45
1180	3/8" threaded rod size		96	.083		7.70	4.47		12.17	15.15
1190	1/2" threaded rod size		96	.083		8	4.47		12.47	15.45
1200	5/8" threaded rod size		96	.083		8.40	4.47		12.87	15.90
1210	3/4" threaded rod size		96	.083		8.90	4.47		13.37	16.45

22 05 Common Work Results for Plumbing

22 05 29 – Hangers and Supports for Plumbing Piping and Equipment

	22 05 29.10 Hangers & Supp. for Plumb'g/HVAC Pipe/Equip.	Crew	Daily Output	Labor-Hours	Unit	Material	2011 Bare Costs Labor	Equipment	Total	Total Incl O&P
1220	7/8" threaded rod size	1 Plum	96	.083	Ea.	12.90	4.47		17.37	21
1250	Pipe guide sized for insulation									
1260	No. 1, 1" pipe size, 1" thick insulation	1 Stpi	26	.308	Ea.	152	16.65		168.65	192
1270	No. 2, 1-1/4"-2" pipe size, 1" thick insulation		23	.348		179	18.80		197.80	225
1280	No. 3, 1-1/4"-2" pipe size, 1-1/2" thick insulation		21	.381		179	20.50		199.50	228
1290	No. 4, 2-1/2"-3-1/2" pipe size, 1-1/2" thick insulation		18	.444		179	24		203	233
1300	No. 5, 4"-5" pipe size, 1-1/2" thick insulation		16	.500		199	27		226	260
1310	No. 6, 5"-6" pipe size, 2" thick insulation	Q-5	21	.762		221	37		258	299
1320	No. 7, 8" pipe size, 2" thick insulation		16	1		305	48.50		353.50	410
1330	No. 8, 10" pipe size, 2" thick insulation		12	1.333		505	65		570	650
1340	No. 9, 12" pipe size, 2" thick insulation	Q-6	17	1.412		505	71.50		576.50	660
1350	No. 10, 12"-14" pipe size, 2-1/2" thick insulation		16	1.500		560	76		636	735
1360	No. 11, 16" pipe size, 2-1/2" thick insulation		10.50	2.286		560	115		675	790
1370	No. 12, 16"-18" pipe size, 3" thick insulation		9	2.667		790	135		925	1,075
1380	No. 13, 20" pipe size, 3" thick insulation		7.50	3.200		790	162		952	1,100
1390	No. 14, 24" pipe size, 3" thick insulation		7	3.429		1,075	173		1,248	1,425
1400	Bands									
1410	Adjustable band, carbon steel, for non-insulated pipe, TYPE 7									
1420	1/2" pipe size	Q-1	142	.113	Ea.	.72	5.45		6.17	8.90
1430	3/4" pipe size		140	.114		.72	5.50		6.22	9.05
1440	1" pipe size		137	.117		.72	5.65		6.37	9.20
1450	1-1/4" pipe size		134	.119		.78	5.75		6.53	9.45
1460	1-1/2" pipe size		131	.122		.79	5.90		6.69	9.65
1470	2" pipe size		129	.124		.79	6		6.79	9.80
1480	2-1/2" pipe size		125	.128		1.42	6.20		7.62	10.75
1490	3" pipe size		122	.131		1.54	6.35		7.89	11.15
1500	3-1/2" pipe size		119	.134		2.27	6.50		8.77	12.20
1510	4" pipe size		114	.140		2.49	6.75		9.24	12.85
1520	5" pipe size		110	.145		4.19	7		11.19	15.10
1530	6" pipe size		108	.148		4.67	7.15		11.82	15.80
1540	8" pipe size		104	.154		6.30	7.40		13.70	18.05
1550	For copper plated, add					50%				
1560	For galvanized, add					30%				
1570	For plastic coating, add					30%				
1600	Adjusting nut malleable iron, steel band, TYPE 9									
1610	1/2" pipe size, galvanized band	Q-1	137	.117	Ea.	1.31	5.65		6.96	9.85
1620	3/4" pipe size, galvanized band		135	.119		1.44	5.70		7.14	10.15
1630	1" pipe size, galvanized band		132	.121		1.44	5.85		7.29	10.35
1640	1-1/4" pipe size, galvanized band		129	.124		1.58	6		7.58	10.70
1650	1-1/2" pipe size, galvanized band		126	.127		1.58	6.15		7.73	10.90
1660	2" pipe size, galvanized band		124	.129		1.67	6.25		7.92	11.15
1670	2-1/2" pipe size, galvanized band		120	.133		2.70	6.45		9.15	12.55
1680	3" pipe size, galvanized band		117	.137		2.84	6.60		9.44	12.95
1690	3-1/2" pipe size, galvanized band		114	.140		3.78	6.75		10.53	14.25
1700	4" pipe size, cadmium plated band		110	.145		3.78	7		10.78	14.65
1740	For plastic coated band, add					35%				
1750	For completely copper coated, add					45%				
1800	Clevis, adjustable, carbon steel, for non-insulated pipe, TYPE 1									
1810	1/2" pipe size	Q-1	137	.117	Ea.	1.67	5.65		7.32	10.25
1820	3/4" pipe size		135	.119		1.67	5.70		7.37	10.40
1830	1" pipe size		132	.121		1.70	5.85		7.55	10.60
1840	1-1/4" pipe size		129	.124		1.89	6		7.89	11.05
1850	1-1/2" pipe size		126	.127		1.89	6.15		8.04	11.25

22 05 Common Work Results for Plumbing

22 05 29 – Hangers and Supports for Plumbing Piping and Equipment

22 05 29.10 Hangers & Supp. for Plumb'g/HVAC Pipe/Equip.		Crew	Daily Output	Labor-Hours	Unit	Material	2011 Bare Costs Labor	Equipment	Total	Total Incl O&P
1860	2" pipe size	Q-1	124	.129	Ea.	2.33	6.25		8.58	11.85
1870	2-1/2" pipe size		120	.133		3.66	6.45		10.11	13.65
1880	3" pipe size		117	.137		4.38	6.60		10.98	14.65
1890	3-1/2" pipe size		114	.140		4.79	6.75		11.54	15.35
1900	4" pipe size		110	.145		5.50	7		12.50	16.55
1910	5" pipe size		106	.151		7.30	7.30		14.60	18.85
1920	6" pipe size		104	.154		9.70	7.40		17.10	22
1930	8" pipe size		100	.160		15.05	7.70		22.75	28
1940	10" pipe size		96	.167		27.50	8.05		35.55	42.50
1950	12" pipe size		89	.180		36	8.65		44.65	52.50
1960	14" pipe size		82	.195		47	9.40		56.40	65.50
1970	16" pipe size		68	.235		71	11.35		82.35	95
1971	18" pipe size		54	.296		85.50	14.30		99.80	116
1972	20" pipe size	▼	38	.421	▼	162	20.50		182.50	209
1980	For galvanized, add					66%				
1990	For copper plated 1/2" to 4", add					77%				
2000	For light weight 1/2" to 4", deduct					13%				
2010	Insulated pipe type, 3/4" to 12" pipe, add					180%				
2020	Insulated pipe type, chrome-moly U-strap, add					530%				
2250	Split ring, malleable iron, for non-insulated pipe, TYPE 11									
2260	1/2" pipe size	Q-1	137	.117	Ea.	5.95	5.65		11.60	14.90
2270	3/4" pipe size		135	.119		6	5.70		11.70	15.15
2280	1" pipe size		132	.121		6.35	5.85		12.20	15.70
2290	1-1/4" pipe size		129	.124		7.85	6		13.85	17.60
2300	1-1/2" pipe size		126	.127		9.55	6.15		15.70	19.65
2310	2" pipe size		124	.129		10.80	6.25		17.05	21
2320	2-1/2" pipe size		120	.133		15.60	6.45		22.05	27
2330	3" pipe size		117	.137		19.50	6.60		26.10	31.50
2340	3-1/2" pipe size		114	.140		20.50	6.75		27.25	32.50
2350	4" pipe size		110	.145		20.50	7		27.50	33
2360	5" pipe size		106	.151		27	7.30		34.30	40.50
2370	6" pipe size		104	.154		60.50	7.40		67.90	77.50
2380	8" pipe size	▼	100	.160	▼	99.50	7.70		107.20	121
2390	For copper plated, add					8%				
2500	Washer, flat steel									
2502	3/8"	1 Plum	240	.033	Ea.	.06	1.79		1.85	2.74
2503	1/2"		220	.036		.13	1.95		2.08	3.05
2504	5/8"		200	.040		.23	2.14		2.37	3.45
2505	3/4"		180	.044		.34	2.38		2.72	3.93
2506	7/8"		160	.050		.47	2.68		3.15	4.52
2507	1"		140	.057		.56	3.06		3.62	5.20
2508	1-1/4"	▼	120	.067	▼	.81	3.57		4.38	6.25
2520	Nut, steel, hex									
2522	3/8"	1 Plum	200	.040	Ea.	.85	2.14		2.99	4.14
2523	1/2"		180	.044		1.64	2.38		4.02	5.35
2524	5/8"		160	.050		2.77	2.68		5.45	7.05
2525	3/4"		140	.057		4.19	3.06		7.25	9.20
2526	7/8"		120	.067		6.65	3.57		10.22	12.70
2527	1"		100	.080		9.65	4.29		13.94	17
2528	1-1/4"	▼	80	.100	▼	12.35	5.35		17.70	21.50
2532	Turnbuckle, TYPE 13									
2534	3/8"	1 Plum	80	.100	Ea.	3.30	5.35		8.65	11.65
2535	1/2"	▼	72	.111	▼	8	5.95		13.95	17.70

22 05 29.10 Hangers & Supp. for Plumb'g/HVAC Pipe/Equip.	Crew	Daily Output	Labor-Hours	Unit	Material	2011 Bare Costs Labor	Equipment	Total	Total Incl O&P
2536	5/8"	1 Plum	64	.125	Ea.	10.40	6.70		17.10	21.50
2537	3/4"		56	.143		17.95	7.65		25.60	31
2538	7/8"		48	.167		22.50	8.95		31.45	38.50
2539	1"		40	.200		35.50	10.70		46.20	55
2540	1-1/4"		32	.250		58	13.40		71.40	83.50
2650	Rods, carbon steel									
2660	Continuous thread									
2670	1/4" thread size	1 Plum	144	.056	L.F.	1.35	2.98		4.33	5.95
2680	3/8" thread size		144	.056		1.44	2.98		4.42	6
2690	1/2" thread size		144	.056		2.26	2.98		5.24	6.95
2700	5/8" thread size		144	.056		3.21	2.98		6.19	7.95
2710	3/4" thread size		144	.056		5.65	2.98		8.63	10.70
2720	7/8" thread size		144	.056		7.10	2.98		10.08	12.25
2721	1" thread size	Q-1	160	.100		12.10	4.83		16.93	20.50
2722	1-1/8" thread size	"	120	.133		17.15	6.45		23.60	28.50
2725	1/4" thread size, bright finish	1 Plum	144	.056		.85	2.98		3.83	5.40
2726	1/2" thread size, bright finish	"	144	.056		2.92	2.98		5.90	7.65
2730	For galvanized, add					40%				
2750	Both ends machine threaded 18" length									
2760	3/8" thread size	1 Plum	240	.033	Ea.	5.75	1.79		7.54	9
2770	1/2" thread size		240	.033		9.25	1.79		11.04	12.85
2780	5/8" thread size		240	.033		13.35	1.79		15.14	17.30
2790	3/4" thread size		240	.033		21	1.79		22.79	25.50
2800	7/8" thread size		240	.033		29	1.79		30.79	34
2810	1" thread size		240	.033		34	1.79		35.79	40
2820	Rod couplings									
2821	3/8"	1 Plum	60	.133	Ea.	1.83	7.15		8.98	12.65
2822	1/2"		54	.148		2.43	7.95		10.38	14.50
2823	5/8"		48	.167		2.81	8.95		11.76	16.45
2824	3/4"		44	.182		3.85	9.75		13.60	18.80
2825	7/8"		40	.200		6.45	10.70		17.15	23
2826	1"		34	.235		8.65	12.60		21.25	28.50
2827	1-1/8"		30	.267		39.50	14.30		53.80	65
2860	Pipe hanger assy, adj. clevis, saddle, rod, clamp, insul. allowance									
2864	1/2" pipe size	Q-5	35	.457	Ea.	39.50	22.50		62	76.50
2866	3/4" pipe size		34.80	.460		41	22.50		63.50	78.50
2868	1" pipe size		34.60	.462		35	22.50		57.50	72
2869	1-1/4" pipe size		34.30	.466		35.50	22.50		58	73
2870	1-1/2" pipe size		33.90	.472		36.50	23		59.50	74.50
2872	2" pipe size		33.30	.480		41.50	23.50		65	80.50
2874	2-1/2" pipe size		32.30	.495		49.50	24		73.50	90.50
2876	3" pipe size		31.20	.513		57.50	25		82.50	101
2880	4" pipe size		30.70	.521		60.50	25.50		86	105
2884	6" pipe size		29.80	.537		77	26		103	124
2888	8" pipe size		28	.571		109	28		137	162
2892	10" pipe size		25.20	.635		124	31		155	182
2896	12" pipe size		23.20	.690		182	33.50		215.50	250
2900	Rolls									
2910	Adjustable yoke, carbon steel with CI roll, TYPE 43									
2918	2" pipe size	Q-1	140	.114	Ea.	16.25	5.50		21.75	26
2920	2-1/2" pipe size		137	.117		16.25	5.65		21.90	26.50
2930	3" pipe size		131	.122		18.10	5.90		24	28.50
2940	3-1/2" pipe size		124	.129		24	6.25		30.25	36

22 05 29 – Hangers and Supports for Plumbing Piping and Equipment

22 05 29.10 Hangers & Supp. for Plumb'g/HVAC Pipe/Equip.		Crew	Daily Output	Labor-Hours	Unit	Material	2011 Bare Costs Labor	Equipment	Total	Total Incl O&P
2950	4" pipe size	Q-1	117	.137	Ea.	24	6.60		30.60	36.50
2960	5" pipe size		110	.145		28	7		35	41.50
2970	6" pipe size		104	.154		37	7.40		44.40	52
2980	8" pipe size		96	.167		52	8.05		60.05	69.50
2990	10" pipe size		80	.200		66.50	9.65		76.15	87.50
3000	12" pipe size		68	.235		97.50	11.35		108.85	124
3010	14" pipe size		56	.286		243	13.80		256.80	289
3020	16" pipe size		48	.333		300	16.10		316.10	355
3050	Chair, carbon steel with CI roll									
3060	2" pipe size	1 Plum	68	.118	Ea.	21.50	6.30		27.80	33
3070	2-1/2" pipe size		65	.123		23	6.60		29.60	35.50
3080	3" pipe size		62	.129		23.50	6.90		30.40	36.50
3090	3-1/2" pipe size		60	.133		30	7.15		37.15	44
3100	4" pipe size		58	.138		32.50	7.40		39.90	46.50
3110	5" pipe size		56	.143		35	7.65		42.65	50.50
3120	6" pipe size		53	.151		49	8.10		57.10	66
3130	8" pipe size		50	.160		66.50	8.60		75.10	86
3140	10" pipe size		48	.167		83	8.95		91.95	105
3150	12" pipe size		46	.174		121	9.30		130.30	147
3170	Single pipe roll, (see line 2650 for rods), TYPE 41, 1" pipe size	Q-1	137	.117		14.65	5.65		20.30	24.50
3180	1-1/4" pipe size		131	.122		15.15	5.90		21.05	25.50
3190	1-1/2" pipe size		129	.124		15.65	6		21.65	26
3200	2" pipe size		124	.129		16.20	6.25		22.45	27
3210	2-1/2" pipe size		118	.136		17.40	6.55		23.95	29
3220	3" pipe size		115	.139		18.40	6.70		25.10	30
3230	3-1/2" pipe size		113	.142		20.50	6.85		27.35	32.50
3240	4" pipe size		112	.143		22	6.90		28.90	34.50
3250	5" pipe size		110	.145		26	7		33	39
3260	6" pipe size		101	.158		37.50	7.65		45.15	52.50
3270	8" pipe size		90	.178		52	8.60		60.60	70.50
3280	10" pipe size		80	.200		61.50	9.65		71.15	82.50
3290	12" pipe size		68	.235		95	11.35		106.35	122
3291	14" pipe size		56	.286		148	13.80		161.80	183
3292	16" pipe size		48	.333		201	16.10		217.10	245
3293	18" pipe size		40	.400		216	19.30		235.30	266
3294	20" pipe size		35	.457		295	22		317	360
3296	24" pipe size		30	.533		465	25.50		490.50	555
3297	30" pipe size		25	.640		920	31		951	1,075
3298	36" pipe size		20	.800		1,125	38.50		1,163.50	1,300
3300	Saddles (add vertical pipe riser, usually 3" diameter)									
3310	Pipe support, complete, adjust., CI saddle, TYPE 36									
3320	2-1/2" pipe size	1 Plum	96	.083	Ea.	123	4.47		127.47	142
3330	3" pipe size		88	.091		124	4.87		128.87	144
3340	3-1/2" pipe size		79	.101		125	5.45		130.45	146
3350	4" pipe size		68	.118		180	6.30		186.30	207
3360	5" pipe size		64	.125		183	6.70		189.70	211
3370	6" pipe size		59	.136		191	7.25		198.25	221
3380	8" pipe size		53	.151		209	8.10		217.10	242
3390	10" pipe size		50	.160		231	8.60		239.60	267
3400	12" pipe size		48	.167		239	8.95		247.95	276
3450	For standard pipe support, one piece, CI deduct					34%				
3460	For stanchion support, CI with steel yoke, deduct					60%				
3550	Insulation shield 1" thick, 1/2" pipe size, TYPE 40	1 Asbe	100	.080	Ea.	6.10	3.75		9.85	12.50

22 05 29.10 Hangers & Supp. for Plumb'g/HVAC Pipe/Equip.	Crew	Daily Output	Labor-Hours	Unit	Material	2011 Bare Costs Labor	Equipment	Total	Total Incl O&P
3560	3/4" pipe size	1 Asbe	100	.080	Ea.	7	3.75		10.75	13.50
3570	1" pipe size		98	.082		7.55	3.83	-	11.38	14.20
3580	1-1/4" pipe size		98	.082		8.05	3.83		11.88	14.80
3590	1-1/2" pipe size		96	.083		8.05	3.91		11.96	14.95
3600	2" pipe size		96	.083		8.35	3.91		12.26	15.25
3610	2-1/2" pipe size		94	.085		8.65	3.99		12.64	15.65
3620	3" pipe size		94	.085		9.25	3.99		13.24	16.30
3630	2" thick, 3-1/2" pipe size		92	.087		10.35	4.08		14.43	17.65
3640	4" pipe size		92	.087		14.75	4.08		18.83	22.50
3650	5" pipe size		90	.089		21	4.17		25.17	29.50
3660	6" pipe size		90	.089		22	4.17		26.17	30.50
3670	8" pipe size		88	.091		33	4.26		37.26	42.50
3680	10" pipe size		88	.091		44.50	4.26		48.76	55.50
3690	12" pipe size		86	.093		51	4.36		55.36	63
3700	14" pipe size		86	.093		98	4.36		102.36	115
3710	16" pipe size		84	.095		104	4.47		108.47	122
3720	18" pipe size		84	.095		116	4.47		120.47	134
3730	20" pipe size		82	.098		127	4.58		131.58	147
3732	24" pipe size		80	.100		154	4.69		158.69	177
3733	30" pipe size		78	.103		475	4.81		479.81	525
3735	36" pipe size	↓	75	.107	↓	715	5		720	795
3750	Covering protection saddle, TYPE 39									
3760	1" covering size									
3770	3/4" pipe size	1 Plum	68	.118	Ea.	11	6.30		17.30	21.50
3780	1" pipe size		68	.118		11	6.30		17.30	21.50
3790	1-1/4" pipe size		68	.118		11	6.30		17.30	21.50
3800	1-1/2" pipe size		66	.121		11.95	6.50		18.45	23
3810	2" pipe size		66	.121		11.95	6.50		18.45	23
3820	2-1/2" pipe size		64	.125		11.95	6.70		18.65	23
3830	3" pipe size		64	.125		16.85	6.70		23.55	28.50
3840	3-1/2" pipe size		62	.129		17.80	6.90		24.70	30
3850	4" pipe size		62	.129		17.80	6.90		24.70	30
3860	5" pipe size		60	.133		17.80	7.15		24.95	30
3870	6" pipe size	↓	60	.133	↓	21.50	7.15		28.65	34
3900	1-1/2" covering size									
3910	3/4" pipe size	1 Plum	68	.118	Ea.	13.40	6.30		19.70	24
3920	1" pipe size		68	.118		13.40	6.30		19.70	24
3930	1-1/4" pipe size		68	.118		13.40	6.30		19.70	24
3940	1-1/2" pipe size		66	.121		13.40	6.50		19.90	24.50
3950	2" pipe size		66	.121		14.55	6.50		21.05	25.50
3960	2-1/2" pipe size		64	.125		17.20	6.70		23.90	29
3970	3" pipe size		64	.125		17.20	6.70		23.90	29
3980	3-1/2" pipe size		62	.129		17.40	6.90		24.30	29.50
3990	4" pipe size		62	.129		17.40	6.90		24.30	29.50
4000	5" pipe size		60	.133		17.40	7.15		24.55	30
4010	6" pipe size		60	.133		25	7.15		32.15	38
4020	8" pipe size		58	.138		31.50	7.40		38.90	45.50
4022	10" pipe size		56	.143		31.50	7.65		39.15	46
4024	12" pipe size	↓	54	.148	↓	56	7.95		63.95	73.50
4028	2" covering size									
4029	2-1/2" pipe size	1 Plum	62	.129	Ea.	16.45	6.90		23.35	28.50
4032	3" pipe size		60	.133		19.35	7.15		26.50	32
4033	4" pipe size	↓	58	.138	↓	19.35	7.40		26.75	32.50

22 05 29 – Hangers and Supports for Plumbing Piping and Equipment

22 05 29.10 Hangers & Supp. for Plumb'g/HVAC Pipe/Equip.	Crew	Daily Output	Labor-Hours	Unit	Material	2011 Bare Costs Labor	2011 Bare Costs Equipment	Total	Total Incl O&P
4034	6" pipe size	1 Plum	56	.143	Ea.	28	7.65		35.65	42
4035	8" pipe size		54	.148		32.50	7.95		40.45	48
4080	10" pipe size		58	.138		35	7.40		42.40	49.50
4090	12" pipe size		56	.143		63	7.65		70.65	80.50
4100	14" pipe size		56	.143		63	7.65		70.65	80.50
4110	16" pipe size		54	.148		90.50	7.95		98.45	111
4120	18" pipe size		54	.148		90.50	7.95		98.45	111
4130	20" pipe size		52	.154		101	8.25		109.25	123
4150	24" pipe size		50	.160		118	8.60		126.60	143
4160	30" pipe size		48	.167		129	8.95		137.95	155
4180	36" pipe size		45	.178		143	9.55		152.55	171
4186	2-1/2" covering size									
4187	3" pipe size	1 Plum	58	.138	Ea.	18.65	7.40		26.05	31.50
4188	4" pipe size		56	.143		20.50	7.65		28.15	34
4189	6" pipe size		52	.154		31	8.25		39.25	46.50
4190	8" pipe size		48	.167		38	8.95		46.95	55
4191	10" pipe size		44	.182		41.50	9.75		51.25	60
4192	12" pipe size		40	.200		68	10.70		78.70	91
4193	14" pipe size		36	.222		68	11.90		79.90	93
4194	16" pipe size		32	.250		95	13.40		108.40	125
4195	18" pipe size		28	.286		106	15.30		121.30	140
4200	Sockets									
4210	Rod end, malleable iron, TYPE 16									
4220	1/4" thread size	1 Plum	240	.033	Ea.	1.86	1.79		3.65	4.72
4230	3/8" thread size		240	.033		2.02	1.79		3.81	4.89
4240	1/2" thread size		230	.035		2.40	1.86		4.26	5.40
4250	5/8" thread size		225	.036		4.73	1.91		6.64	8.05
4260	3/4" thread size		220	.036		7.75	1.95		9.70	11.40
4270	7/8" thread size		210	.038		14.05	2.04		16.09	18.55
4290	Strap, 1/2" pipe size, TYPE 26	Q-1	142	.113		4.23	5.45		9.68	12.75
4300	3/4" pipe size		140	.114		4.32	5.50		9.82	13
4310	1" pipe size		137	.117		4.51	5.65		10.16	13.35
4320	1-1/4" pipe size		134	.119		4.98	5.75		10.73	14.10
4330	1-1/2" pipe size		131	.122		6.30	5.90		12.20	15.70
4340	2" pipe size		129	.124		6.45	6		12.45	16.05
4350	2-1/2" pipe size		125	.128		11.65	6.20		17.85	22
4360	3" pipe size		122	.131		12.25	6.35		18.60	23
4370	3-1/2" pipe size		119	.134		14.55	6.50		21.05	26
4380	4" pipe size		114	.140		15.45	6.75		22.20	27
4400	U-bolt, carbon steel									
4410	Standard, with nuts, TYPE 42									
4420	1/2" pipe size	1 Plum	160	.050	Ea.	1.86	2.68		4.54	6.05
4430	3/4" pipe size		158	.051		1.86	2.71		4.57	6.10
4450	1" pipe size		152	.053		1.92	2.82		4.74	6.30
4460	1-1/4" pipe size		148	.054		2.30	2.90		5.20	6.85
4470	1-1/2" pipe size		143	.056		2.46	3		5.46	7.20
4480	2" pipe size		139	.058		2.68	3.08		5.76	7.55
4490	2-1/2" pipe size		134	.060		4.45	3.20		7.65	9.70
4500	3" pipe size		128	.063		4.67	3.35		8.02	10.15
4510	3-1/2" pipe size		122	.066		4.92	3.51		8.43	10.65
4520	4" pipe size		117	.068		5	3.67		8.67	10.95
4530	5" pipe size		114	.070		5.05	3.76		8.81	11.15
4540	6" pipe size		111	.072		10.35	3.86		14.21	17.15

22 05 29.10 Hangers & Supp. for Plumb'g/HVAC Pipe/Equip.		Crew	Daily Output	Labor-Hours	Unit	Material	2011 Bare Costs Labor	Equipment	Total	Total Incl O&P
4550	8" pipe size	1 Plum	109	.073	Ea.	12.30	3.93		16.23	19.40
4560	10" pipe size		107	.075		22	4.01		26.01	30
4570	12" pipe size		104	.077		30	4.12		34.12	39
4580	For plastic coating on 1/2" thru 6" size, add					150%				
4700	U-hook, carbon steel, requires mounting screws or bolts									
4710	3/4" thru 2" pipe size									
4720	6" long	1 Plum	96	.083	Ea.	.68	4.47		5.15	7.40
4730	8" long		96	.083		.86	4.47		5.33	7.60
4740	10" long		96	.083		.98	4.47		5.45	7.75
4750	12" long		96	.083		1.07	4.47		5.54	7.85
4760	For copper plated, add					50%				
8000	Pipe clamp, plastic, 1/2" CTS	1 Plum	80	.100		.22	5.35		5.57	8.25
8010	3/4" CTS		73	.110		.23	5.85		6.08	9
8020	1" CTS		68	.118		.52	6.30		6.82	9.95
8080	Economy clamp, 1/4" CTS		175	.046		.05	2.45		2.50	3.72
8090	3/8" CTS		168	.048		.07	2.55		2.62	3.89
8100	1/2" CTS		160	.050		.07	2.68		2.75	4.08
8110	3/4" CTS		145	.055		.14	2.96		3.10	4.56
8200	Half clamp, 1/2" CTS		80	.100		.08	5.35		5.43	8.10
8210	3/4" CTS		73	.110		.12	5.85		5.97	8.90
8300	Suspension clamp, 1/2" CTS		80	.100		.22	5.35		5.57	8.25
8310	3/4" CTS		73	.110		.23	5.85		6.08	9
8320	1" CTS		68	.118		.52	6.30		6.82	9.95
8400	Insulator, 1/2" CTS		80	.100		.36	5.35		5.71	8.40
8410	3/4" CTS		73	.110		.37	5.85		6.22	9.15
8420	1" CTS		68	.118		.38	6.30		6.68	9.80
8500	J hook clamp with nail 1/2" CTS		240	.033		.13	1.79		1.92	2.81
8501	3/4" CTS		240	.033		.13	1.79		1.92	2.81
8800	Wire cable support system									
8810	Cable with hook terminal and locking device									
8830	2 mm, (.079") dia cable, (100 lb. cap.)									
8840	1 m, (3.3') length, with hook	1 Shee	96	.083	Ea.	2.86	4.30		7.16	9.65
8850	2 m, (6.6') length, with hook		84	.095		3.30	4.92		8.22	11.10
8860	3 m, (9.9') length, with hook		72	.111		3.74	5.75		9.49	12.80
8870	5 m, (16.4') length, with hook	Q-9	60	.267		4.69	12.40		17.09	24
8880	10 m, (32.8') length, with hook	"	30	.533		6.80	25		31.80	45
8900	3mm, (.118") dia cable, (200 lb. cap.)									
8910	1 m, (3.3') length, with hook	1 Shee	96	.083	Ea.	3.66	4.30		7.96	10.55
8920	2 m, (6.6') length, with hook		84	.095		4.11	4.92		9.03	11.95
8930	3 m, (9.9') length, with hook		72	.111		4.54	5.75		10.29	13.70
8940	5 m, (16.4') length, with hook	Q-9	60	.267		5.60	12.40		18	25
8950	10 m, (32.8') length, with hook	"	30	.533		8.10	25		33.10	46.50
9000	Cable system accessories									
9010	Anchor bolt, 3/8", with nut	1 Shee	140	.057	Ea.	1.10	2.95		4.05	5.65
9020	Air duct corner protector		160	.050		.49	2.58		3.07	4.45
9030	Air duct support attachment		140	.057		.94	2.95		3.89	5.50
9040	Flange clip, hammer-on style									
9044	For flange thickness 3/32" - 9/64", 160 lb. cap.	1 Shee	180	.044	Ea.	.25	2.30		2.55	3.75
9048	For flange thickness 1/8" - 1/4", 200 lb. cap.		160	.050		.25	2.58		2.83	4.19
9052	For flange thickness 5/16" - 1/2", 200 lb. cap.		150	.053		.52	2.75		3.27	4.74
9056	For flange thickness 9/16" - 3/4", 200 lb. cap.		140	.057		.71	2.95		3.66	5.25
9060	Wire insulation protection tube		180	.044	L.F.	.32	2.30		2.62	3.82
9070	Wire cutter				Ea.	30.50			30.50	33.50

22 05 Common Work Results for Plumbing

22 05 48 – Vibration and Seismic Controls for Plumbing Piping and Equipment

22 05 48.40 Vibration Absorbers

		Crew	Daily Output	Labor-Hours	Unit	Material	2011 Bare Costs Labor	Equipment	Total	Total Incl O&P
0010	**VIBRATION ABSORBERS**									
0100	Hangers, neoprene flex									
0200	10 – 120 lb. capacity				Ea.	22.50			22.50	25
0220	75 – 550 lb. capacity					30.50			30.50	33.50
0240	250 – 1100 lb. capacity					63			63	69.50
0260	1000 – 4000 lb. capacity					100			100	110
0500	Spring flex, 60 lb. capacity					31.50			31.50	35
0520	450 lb. capacity					41.50			41.50	46
0540	900 lb. capacity					55.50			55.50	61
0560	1100 – 1300 lb. capacity					55.50			55.50	61
0600	Rubber in shear									
0610	45 – 340 lb., up to 1/2" rod size	1 Stpi	22	.364	Ea.	14.30	19.65		33.95	45.50
0620	130 – 700 lb., up to 3/4" rod size		20	.400		31	21.50		52.50	66.50
0630	50 – 1000 lb., up to 3/4" rod size		18	.444		36.50	24		60.50	76
1000	Mounts, neoprene, 45 – 380 lb. capacity					15.35			15.35	16.85
1020	250 – 1100 lb. capacity					40.50			40.50	44.50
1040	1000 – 4000 lb. capacity					90.50			90.50	99.50
1100	Spring flex, 60 lb. capacity					62.50			62.50	69
1120	165 lb. capacity					62.50			62.50	69
1140	260 lb. capacity					62.50			62.50	69
1160	450 lb. capacity					83.50			83.50	91.50
1180	600 lb. capacity					83.50			83.50	91.50
1200	750 lb. capacity					83.50			83.50	91.50
1220	900 lb. capacity					86.50			86.50	95
1240	1100 lb. capacity					92			92	101
1260	1300 lb. capacity					92			92	101
1280	1500 lb. capacity					136			136	150
1300	1800 lb. capacity					146			146	161
1320	2200 lb. capacity					146			146	161
1340	2600 lb. capacity					146			146	161
1399	Spring type									
1400	2 Piece									
1410	50 – 1000 lb.	1 Stpi	12	.667	Ea.	130	36		166	197
1420	1100 – 1600 lb.	"	12	.667	"	141	36		177	209
1500	Double spring open									
1510	150 – 450 lb.	1 Stpi	24	.333	Ea.	89	18.05		107.05	125
1520	500 – 1000 lb.		24	.333		89	18.05		107.05	125
1530	1100 – 1600 lb.		24	.333		89	18.05		107.05	125
1540	1700 – 2400 lb.		24	.333		97	18.05		115.05	133
1550	2500 – 3400 lb.		24	.333		165	18.05		183.05	209
2000	Pads, cork rib, 18" x 18" x 1", 10-50 psi					146			146	160
2020	18" x 36" x 1", 10-50 psi					292			292	320
2100	Shear flexible pads, 18" x 18" x 3/8", 20-70 psi					68.50			68.50	75.50
2120	18" x 36" x 3/8", 20-70 psi					146			146	160
2150	Laminated neoprene and cork									
2160	1" thick	1 Stpi	16	.500	S.F.	72.50	27		99.50	121
3000	Note overlap in capacities due to deflections									

22 05 53 – Identification for Plumbing Piping and Equipment

22 05 53.10 Piping System Identification Labels

0010	**PIPING SYSTEM IDENTIFICATION LABELS**									
0100	Indicate contents and flow direction									
0106	Pipe markers									

22 05 Common Work Results for Plumbing

22 05 53 – Identification for Plumbing Piping and Equipment

22 05 53.10 Piping System Identification Labels	Crew	Daily Output	Labor-Hours	Unit	Material	2011 Bare Costs Labor	Equipment	Total	Total Incl O&P
0110	Plastic snap around									
0114	1/2" pipe	1 Plum	80	.100	Ea.	4.10	5.35		9.45	12.50
0116	3/4" pipe		80	.100		4.10	5.35		9.45	12.50
0118	1" pipe		80	.100		4.10	5.35		9.45	12.50
0120	2" pipe		75	.107		5.60	5.70		11.30	14.70
0122	3" pipe		70	.114		10.40	6.15		16.55	20.50
0124	4" pipe		60	.133		10.40	7.15		17.55	22
0126	6" pipe		60	.133		10.40	7.15		17.55	22
0128	8" pipe		56	.143		15.80	7.65		23.45	29
0130	10" pipe		56	.143		15.80	7.65		23.45	29
0200	Over 10" pipe size		50	.160		19.20	8.60		27.80	34
1110	Self adhesive									
1114	1" pipe	1 Plum	80	.100	Ea.	3.20	5.35		8.55	11.50
1116	2" pipe		75	.107		3.60	5.70		9.30	12.50
1118	3" pipe		70	.114		4.90	6.15		11.05	14.55
1120	4" pipe		60	.133		4.90	7.15		12.05	16.05
1122	6" pipe		60	.133		4.90	7.15		12.05	16.05
1124	8" pipe		56	.143		4.90	7.65		12.55	16.85
1126	10" pipe		56	.143		9.90	7.65		17.55	22.50
1200	Over 10" pipe size		50	.160		9.90	8.60		18.50	23.50
2000	Valve tags									
2010	Numbered plus identifying legend									
2100	Brass, 2" diameter	1 Plum	40	.200	Ea.	2.95	10.70		13.65	19.25
2200	Plastic, 1-1/2" Dia.	"	40	.200	"	2.70	10.70		13.40	18.95

22 07 Plumbing Insulation

22 07 16 – Plumbing Equipment Insulation

22 07 16.10 Insulation for Plumbing Equipment

		Crew	Daily Output	Labor-Hours	Unit	Material	Labor	Equipment	Total	Total Incl O&P
0010	**INSULATION FOR PLUMBING EQUIPMENT**									
2900	Domestic water heater wrap kit									
2920	1-1/2" with vinyl jacket, 20-60 gal.　G	1 Plum	8	1	Ea.	17	53.50		70.50	98.50

22 07 19 – Plumbing Piping Insulation

22 07 19.10 Piping Insulation

		Crew	Daily Output	Labor-Hours	Unit	Material	Labor	Equipment	Total	Total Incl O&P
0010	**PIPING INSULATION**									
0100	Rule of thumb, as a percentage of total mechanical costs				Job				10%	
0110	Insulation req'd is based on the surface size/area to be covered									
2930	Insulated protectors, (ADA)									
2935	For exposed piping under sinks or lavatories									
2940	Vinyl coated foam, velcro tabs									
2945	P Trap, 1-1/4" or 1-1/2"	1 Plum	32	.250	Ea.	19.80	13.40		33.20	42
2960	Valve and supply cover									
2965	1/2", 3/8", and 7/16" pipe size	1 Plum	32	.250	Ea.	19.35	13.40		32.75	41.50
2970	Extension drain cover									
2980	Tailpiece offset (wheelchair)									
2985	1-1/4" pipe size	1 Plum	32	.250	Ea.	16.05	13.40		29.45	37.50
4000	Pipe covering (price copper tube one size less than IPS)									
4280	Cellular glass, closed cell foam, all service jacket, sealant,									
4281	working temp. (-450°F to +900°F), 0 water vapor transmission									
4284	1" wall,									
4286	1/2" iron pipe size　G	Q-14	120	.133	L.F.	2.01	5.65		7.66	10.90

22 07 19 – Plumbing Piping Insulation

22 07 19.10 Piping Insulation		Crew	Daily Output	Labor-Hours	Unit	Material	2011 Bare Costs Labor	Equipment	Total	Total Incl O&P
4300	1-1/2" wall,									
4301	1" iron pipe size	G Q-14	105	.152	L.F.	3.22	6.45		9.67	13.50
4304	2-1/2" iron pipe size	G	90	.178		5.40	7.50		12.90	17.55
4306	3" iron pipe size	G	85	.188		5.50	7.95		13.45	18.35
4308	4" iron pipe size	G	70	.229		7.30	9.65		16.95	23
4310	5" iron pipe size	G	65	.246		8.40	10.40		18.80	25.50
4320	2" wall,									
4322	1" iron pipe size	G Q-14	100	.160	L.F.	4.80	6.75		11.55	15.75
4324	2-1/2" iron pipe size	G	85	.188		7.10	7.95		15.05	20
4326	3" iron pipe size	G	80	.200		7.30	8.45		15.75	21
4328	4" iron pipe size	G	65	.246		8.20	10.40		18.60	25
4330	5" iron pipe size	G	60	.267		9.80	11.25		21.05	28
4332	6" iron pipe size	G	50	.320		11.60	13.50		25.10	34
4336	8" iron pipe size	G	40	.400		15	16.90		31.90	42.50
4338	10" iron pipe size	G	35	.457		15.50	19.30		34.80	47
4350	2-1/2" wall,									
4360	12" iron pipe size	G Q-14	32	.500	L.F.	24.50	21		45.50	59.50
4362	14" iron pipe size	G "	28	.571	"	26.50	24		50.50	66.50
4370	3" wall,									
4378	6" iron pipe size	G Q-14	48	.333	L.F.	15.50	14.05		29.55	39
4380	8" iron pipe size	G	38	.421		18.10	17.75		35.85	47.50
4382	10" iron pipe size	G	33	.485		24	20.50		44.50	58
4384	16" iron pipe size	G	25	.640		31	27		58	76
4386	18" iron pipe size	G	22	.727		34	30.50		64.50	84.50
4388	20" iron pipe size	G	20	.800		40	34		74	95.50
4400	3-1/2" wall,									
4412	12" iron pipe size	G Q-14	27	.593	L.F.	35	25		60	77
4414	14" iron pipe size	G "	25	.640	"	37	27		64	82.50
4430	4" wall,									
4446	16" iron pipe size	G Q-14	22	.727	L.F.	37.50	30.50		68	88.50
4448	18" iron pipe size	G	20	.800		48.50	34		82.50	106
4450	20" iron pipe size	G	18	.889		50.50	37.50		88	114
4480	Fittings, average with fabric and mastic									
4484	1" wall,									
4486	1/2" iron pipe size	G 1 Asbe	40	.200	Ea.	3.56	9.40		12.96	18.40
4500	1-1/2" wall,									
4502	1" iron pipe size	G 1 Asbe	38	.211	Ea.	6.15	9.85		16	22
4504	2-1/2" iron pipe size	G	32	.250		7.65	11.75		19.40	26.50
4506	3" iron pipe size	G	30	.267		8.40	12.50		20.90	28.50
4508	4" iron pipe size	G	28	.286		12.95	13.40		26.35	34.50
4510	5" iron pipe size	G	24	.333		14.25	15.65		29.90	39.50
4520	2" wall,									
4522	1" iron pipe size	G 1 Asbe	36	.222	Ea.	7.10	10.40		17.50	24
4524	2-1/2" iron pipe size	G	30	.267		8.40	12.50		20.90	28.50
4526	3" iron pipe size	G	28	.286		10.10	13.40		23.50	31.50
4528	4" iron pipe size	G	24	.333		13.55	15.65		29.20	39
4530	5" iron pipe size	G	22	.364		17.10	17.05		34.15	45.50
4532	6" iron pipe size	G	20	.400		20	18.75		38.75	51
4536	8" iron pipe size	G	12	.667		30.50	31.50		62	82
4538	10" iron pipe size	G	8	1		37.50	47		84.50	114
4550	2-1/2" wall,									
4560	12" iron pipe size	G 1 Asbe	6	1.333	Ea.	64.50	62.50		127	168
4562	14" iron pipe size	G "	4	2	"	75	94		169	228

22 07 19.10 Piping Insulation		Crew	Daily Output	Labor-Hours	Unit	Material	2011 Bare Costs Labor	Equipment	Total	Total Incl O&P
4570	3" wall,									
4578	6" iron pipe size	G	1 Asbe	16	.500	Ea.	23	23.50		46.50	62
4580	8" iron pipe size	G		10	.800		34	37.50		71.50	95.50
4582	10" iron pipe size	G		6	1.333		48.50	62.50		111	150
4584	16" iron pipe size	G		2	4		116	188		304	420
4586	18" iron pipe size	G		2	4		150	188		338	455
4588	20" iron pipe size	G	↓	2	4	↓	177	188		365	485
4600	3-1/2" wall,									
4612	12" iron pipe size	G	1 Asbe	4	2	Ea.	84	94		178	238
4614	14" iron pipe size	G	"	2	4	"	93	188		281	390
4630	4" wall,									
4646	16" iron pipe size	G	1 Asbe	2	4	Ea.	139	188		327	445
4648	18" iron pipe size	G		2	4		170	188		358	475
4650	20" iron pipe size	G	↓	2	4	↓	199	188		387	510
4900	Calcium silicate, with cover									
5100	1" wall, 1/2" iron pipe size	G	Q-14	170	.094	L.F.	2.82	3.97		6.79	9.25
5130	3/4" iron pipe size	G		170	.094		2.84	3.97		6.81	9.25
5140	1" iron pipe size	G		170	.094		2.75	3.97		6.72	9.20
5150	1-1/4" iron pipe size	G		165	.097		2.82	4.09		6.91	9.45
5160	1-1/2" iron pipe size	G		165	.097		2.86	4.09		6.95	9.50
5170	2" iron pipe size	G		160	.100		3.28	4.22		7.50	10.15
5180	2-1/2" iron pipe size	G		160	.100		3.49	4.22		7.71	10.40
5190	3" iron pipe size	G		150	.107		3.77	4.50		8.27	11.10
5200	4" iron pipe size	G		140	.114		4.68	4.82		9.50	12.60
5210	5" iron pipe size	G		135	.119		4.93	5		9.93	13.15
5220	6" iron pipe size	G		130.	.123		5.25	5.20		10.45	13.80
5280	1-1/2" wall, 1/2" iron pipe size	G		150	.107		3.05	4.50		7.55	10.30
5310	3/4" iron pipe size	G		150	.107		3.10	4.50		7.60	10.35
5320	1" iron pipe size	G		150	.107		3.38	4.50		7.88	10.65
5330	1-1/4" iron pipe size	G		145	.110		3.63	4.66		8.29	11.20
5340	1-1/2" iron pipe size	G		145	.110		3.89	4.66		8.55	11.50
5350	2" iron pipe size	G		140	.114		4.29	4.82		9.11	12.15
5360	2-1/2" iron pipe size	G		140	.114		4.68	4.82		9.50	12.60
5370	3" iron pipe size	G		135	.119		4.92	5		9.92	13.15
5380	4" iron pipe size	G		125	.128		5.70	5.40		11.10	14.60
5390	5" iron pipe size	G		120	.133		6.40	5.65		12.05	15.75
5400	6" iron pipe size	G		110	.145		6.60	6.15		12.75	16.80
5402	8" iron pipe size	G		95	.168		8.70	7.10		15.80	20.50
5404	10" iron pipe size	G		85	.188		11.35	7.95		19.30	25
5406	12" iron pipe size	G		80	.200		13.40	8.45		21.85	28
5408	14" iron pipe size	G		75	.213		15.10	9		24.10	30.50
5410	16" iron pipe size	G		70	.229		16.90	9.65		26.55	33.50
5412	18" iron pipe size	G		65	.246		18.65	10.40		29.05	36.50
5460	2" wall, 1/2" iron pipe size	G		135	.119		4.70	5		9.70	12.90
5490	3/4" iron pipe size	G		135	.119		4.94	5		9.94	13.20
5500	1" iron pipe size	G		135	.119		5.20	5		10.20	13.45
5510	1-1/4" iron pipe size	G		130	.123		5.60	5.20		10.80	14.20
5520	1-1/2" iron pipe size	G		130	.123		5.80	5.20		11	14.45
5530	2" iron pipe size	G		125	.128		6.10	5.40		11.50	15.10
5540	2-1/2" iron pipe size	G		125	.128		7.30	5.40		12.70	16.35
5550	3" iron pipe size	G		120	.133		7.35	5.65		13	16.80
5560	4" iron pipe size	G		115	.139		8.50	5.85		14.35	18.45
5570	5" iron pipe size	G		110	.145		9.70	6.15		15.85	20

22 07 19.10 Piping Insulation		Crew	Daily Output	Labor-Hours	Unit	Material	2011 Bare Costs Labor	Equipment	Total	Total Incl O&P
5580	6" iron pipe size [G]	Q-14	105	.152	L.F.	10.60	6.45		17.05	21.50
5581	8" iron pipe size [G]		95	.168		12.60	7.10		19.70	25
5582	10" iron pipe size [G]		85	.188		15.75	7.95		23.70	29.50
5583	12" iron pipe size [G]		80	.200		17.50	8.45		25.95	32.50
5584	14" iron pipe size [G]		75	.213		19.25	9		28.25	35
5585	16" iron pipe size [G]		70	.229		21	9.65		30.65	38.50
5586	18" iron pipe size [G]		65	.246		23	10.40		33.40	41.50
5587	3" wall, 1-1/4" iron pipe size [G]		100	.160		8.60	6.75		15.35	19.95
5588	1-1/2" iron pipe size [G]		100	.160		8.70	6.75		15.45	20
5589	2" iron pipe size [G]		95	.168		9	7.10		16.10	21
5590	2-1/2" iron pipe size [G]		95	.168		10.50	7.10		17.60	22.50
5591	3" iron pipe size [G]		90	.178		10.65	7.50		18.15	23.50
5592	4" iron pipe size [G]		85	.188		13.45	7.95		21.40	27
5593	6" iron pipe size [G]		80	.200		16.35	8.45		24.80	31
5594	8" iron pipe size [G]		75	.213		19.45	9		28.45	35.50
5595	10" iron pipe size [G]		65	.246		23.50	10.40		33.90	42
5596	12" iron pipe size [G]		60	.267		26	11.25		37.25	46
5597	14" iron pipe size [G]		55	.291		29	12.30		41.30	51
5598	16" iron pipe size [G]		50	.320		32	13.50		45.50	56
5599	18" iron pipe size [G]		45	.356		35	15		50	61.50
5600	Calcium silicate, no cover									
5720	1" wall, 1/2" iron pipe size [G]	Q-14	180	.089	L.F.	2.56	3.75		6.31	8.60
5740	3/4" iron pipe size [G]		180	.089		2.56	3.75		6.31	8.60
5750	1" iron pipe size [G]		180	.089		2.45	3.75		6.20	8.50
5760	1-1/4" iron pipe size [G]		175	.091		2.50	3.86		6.36	8.70
5770	1-1/2" iron pipe size [G]		175	.091		2.52	3.86		6.38	8.70
5780	2" iron pipe size [G]		170	.094		2.90	3.97		6.87	9.35
5790	2-1/2" iron pipe size [G]		170	.094		3.07	3.97		7.04	9.55
5800	3" iron pipe size [G]		160	.100		3.30	4.22		7.52	10.20
5810	4" iron pipe size [G]		150	.107		4.13	4.50		8.63	11.50
5820	5" iron pipe size [G]		145	.110		4.29	4.66		8.95	11.90
5830	6" iron pipe size [G]		140	.114		4.51	4.82		9.33	12.40
5900	1-1/2" wall, 1/2" iron pipe size [G]		160	.100		2.71	4.22		6.93	9.55
5920	3/4" iron pipe size [G]		160	.100		2.75	4.22		6.97	9.60
5930	1" iron pipe size [G]		160	.100		3	4.22		7.22	9.85
5940	1-1/4" iron pipe size [G]		155	.103		3.23	4.36		7.59	10.30
5950	1-1/2" iron pipe size [G]		155	.103		3.48	4.36		7.84	10.60
5960	2" iron pipe size [G]		150	.107		3.83	4.50		8.33	11.15
5970	2-1/2" iron pipe size [G]		150	.107		4.18	4.50		8.68	11.55
5980	3" iron pipe size [G]		145	.110		4.37	4.66		9.03	12
5990	4" iron pipe size [G]		135	.119		5.05	5		10.05	13.30
6000	5" iron pipe size [G]		130	.123		5.70	5.20		10.90	14.30
6010	6" iron pipe size [G]		120	.133		5.80	5.65		11.45	15.10
6020	7" iron pipe size [G]		115	.139		6.85	5.85		12.70	16.60
6030	8" iron pipe size [G]		105	.152		7.70	6.45		14.15	18.45
6040	9" iron pipe size [G]		100	.160		9.25	6.75		16	20.50
6050	10" iron pipe size [G]		95	.168		10.20	7.10		17.30	22.50
6060	12" iron pipe size [G]		90	.178		12.10	7.50		19.60	25
6070	14" iron pipe size [G]		85	.188		13.70	7.95		21.65	27.50
6080	16" iron pipe size [G]		80	.200		15.30	8.45		23.75	30
6090	18" iron pipe size [G]		75	.213		16.95	9		25.95	32.50
6120	2" wall, 1/2" iron pipe size [G]		145	.110		4.28	4.66		8.94	11.90
6140	3/4" iron pipe size [G]		145	.110		4.51	4.66		9.17	12.15

22 07 19.10 Piping Insulation		Crew	Daily Output	Labor-Hours	Unit	Material	2011 Bare Costs Labor	Equipment	Total	Total Incl O&P
6150	1" iron pipe size	G Q-14	145	.110	L.F.	4.75	4.66		9.41	12.45
6160	1-1/4" iron pipe size	G	140	.114		5.10	4.82		9.92	13.05
6170	1-1/2" iron pipe size	G	140	.114		5.30	4.82		10.12	13.30
6180	2" iron pipe size	G	135	.119		5.60	5		10.60	13.90
6190	2-1/2" iron pipe size	G	135	.119		6.70	5		11.70	15.15
6200	3" iron pipe size	G	130	.123		6.75	5.20		11.95	15.45
6210	4" iron pipe size	G	125	.128		7.80	5.40		13.20	16.90
6220	5" iron pipe size	G	120	.133		8.90	5.65		14.55	18.50
6230	6" iron pipe size	G	115	.139		9.70	5.85		15.55	19.80
6240	7" iron pipe size	G	110	.145		10.55	6.15		16.70	21
6250	8" iron pipe size	G	105	.152		11.55	6.45		18	22.50
6260	9" iron pipe size	G	100	.160		13.10	6.75		19.85	25
6270	10" iron pipe size	G	95	.168		14.55	7.10		21.65	27
6280	12" iron pipe size	G	90	.178		16.15	7.50		23.65	29.50
6290	14" iron pipe size	G	85	.188		17.75	7.95		25.70	32
6300	16" iron pipe size	G	80	.200		19.55	8.45		28	34.50
6310	18" iron pipe size	G	75	.213		21.50	9		30.50	37.50
6320	20" iron pipe size	G	65	.246		26.50	10.40		36.90	45
6330	22" iron pipe size	G	60	.267		29	11.25		40.25	49.50
6340	24" iron pipe size	G	55	.291		30	12.30		42.30	51.50
6360	3" wall, 1/2" iron pipe size	G	115	.139		7.75	5.85		13.60	17.65
6380	3/4" iron pipe size	G	115	.139		7.80	5.85		13.65	17.70
6390	1" iron pipe size	G	115	.139		7.85	5.85		13.70	17.75
6400	1-1/4" iron pipe size	G	110	.145		8	6.15		14.15	18.30
6410	1-1/2" iron pipe size	G	110	.145		8.05	6.15		14.20	18.35
6420	2" iron pipe size	G	105	.152		8.30	6.45		14.75	19.05
6430	2-1/2" iron pipe size	G	105	.152		9.75	6.45		16.20	20.50
6440	3" iron pipe size	G	100	.160		9.85	6.75		16.60	21.50
6450	4" iron pipe size	G	95	.168		12.55	7.10		19.65	25
6460	5" iron pipe size	G	90	.178		13.60	7.50		21.10	26.50
6470	6" iron pipe size	G	90	.178		15.30	7.50		22.80	28.50
6480	7" iron pipe size	G	85	.188		17.05	7.95		25	31
6490	8" iron pipe size	G	85	.188		18.25	7.95		26.20	32.50
6500	9" iron pipe size	G	80	.200		20.50	8.45		28.95	36
6510	10" iron pipe size	G	75	.213		22	9		31	38
6520	12" iron pipe size	G	70	.229		24.50	9.65		34.15	42
6530	14" iron pipe size	G	65	.246		27.50	10.40		37.90	46
6540	16" iron pipe size	G	60	.267		30	11.25		41.25	50.50
6550	18" iron pipe size	G	55	.291		33	12.30		45.30	55.50
6560	20" iron pipe size	G	50	.320		39.50	13.50		53	64.50
6570	22" iron pipe size	G	45	.356		42.50	15		57.50	70
6580	24" iron pipe size	G	40	.400		46	16.90		62.90	76.50
6600	Fiberglass, with all service jacket									
6640	1/2" wall, 1/2" iron pipe size	G Q-14	250	.064	L.F.	.79	2.70		3.49	5.05
6660	3/4" iron pipe size	G	240	.067		.89	2.81		3.70	5.35
6670	1" iron pipe size	G	230	.070		.92	2.94		3.86	5.55
6680	1-1/4" iron pipe size	G	220	.073		.98	3.07		4.05	5.85
6690	1-1/2" iron pipe size	G	220	.073		1.13	3.07		4.20	6
6700	2" iron pipe size	G	210	.076		1.21	3.22		4.43	6.30
6710	2-1/2" iron pipe size	G	200	.080		1.26	3.38		4.64	6.60
6840	1" wall, 1/2" iron pipe size	G	240	.067		.95	2.81		3.76	5.40
6860	3/4" iron pipe size	G	230	.070		1.04	2.94		3.98	5.70
6870	1" iron pipe size	G	220	.073		1.12	3.07		4.19	6

22 07 Plumbing Insulation

22 07 19 – Plumbing Piping Insulation

22 07 19.10 Piping Insulation		Crew	Daily Output	Labor-Hours	Unit	Material	2011 Bare Costs Labor	Equipment	Total	Total Incl O&P
6880	1-1/4" iron pipe size	G	Q-14	210	.076	L.F.	1.21	3.22		4.43	6.30
6890	1-1/2" iron pipe size	G		210	.076		1.31	3.22		4.53	6.40
6900	2" iron pipe size	G		200	.080		1.41	3.38		4.79	6.75
6910	2-1/2" iron pipe size	G		190	.084		1.61	3.55		5.16	7.25
6920	3" iron pipe size	G		180	.089		1.72	3.75		5.47	7.70
6930	3-1/2" iron pipe size	G		170	.094		1.86	3.97		5.83	8.20
6940	4" iron pipe size	G		150	.107		2.27	4.50		6.77	9.45
6950	5" iron pipe size	G		140	.114		2.57	4.82		7.39	10.30
6960	6" iron pipe size	G		120	.133		2.72	5.65		8.37	11.70
6970	7" iron pipe size	G		110	.145		3.24	6.15		9.39	13.05
6980	8" iron pipe size	G		100	.160		4.45	6.75		11.20	15.35
6990	9" iron pipe size	G		90	.178		4.69	7.50		12.19	16.75
7000	10" iron pipe size	G		90	.178		4.72	7.50		12.22	16.80
7010	12" iron pipe size	G		80	.200		5.15	8.45		13.60	18.70
7020	14" iron pipe size	G		80	.200		6.15	8.45		14.60	19.85
7030	16" iron pipe size	G		70	.229		7.85	9.65		17.50	23.50
7040	18" iron pipe size	G		70	.229		8.70	9.65		18.35	24.50
7050	20" iron pipe size	G		60	.267		9.70	11.25		20.95	28
7060	24" iron pipe size	G		60	.267		11.80	11.25		23.05	30.50
7080	1-1/2" wall, 1/2" iron pipe size	G		230	.070		1.81	2.94		4.75	6.55
7100	3/4" iron pipe size	G		220	.073		1.82	3.07		4.89	6.75
7110	1" iron pipe size	G		210	.076		1.94	3.22		5.16	7.10
7120	1-1/4" iron pipe size	G		200	.080		2.12	3.38		5.50	7.55
7130	1-1/2" iron pipe size	G		200	.080		2.21	3.38		5.59	7.65
7140	2" iron pipe size	G		190	.084		2.44	3.55		5.99	8.20
7150	2-1/2" iron pipe size	G		180	.089		2.62	3.75		6.37	8.70
7160	3" iron pipe size	G		170	.094		2.74	3.97		6.71	9.15
7170	3-1/2" iron pipe size	G		160	.100		3	4.22		7.22	9.85
7180	4" iron pipe size	G		140	.114		3.11	4.82		7.93	10.85
7190	5" iron pipe size	G		130	.123		3.49	5.20		8.69	11.90
7200	6" iron pipe size	G		110	.145		3.68	6.15		9.83	13.55
7210	7" iron pipe size	G		100	.160		4.09	6.75		10.84	14.95
7220	8" iron pipe size	G		90	.178		5.15	7.50		12.65	17.30
7230	9" iron pipe size	G		85	.188		5.40	7.95		13.35	18.20
7240	10" iron pipe size	G		80	.200		5.55	8.45		14	19.20
7250	12" iron pipe size	G		75	.213		6.30	9		15.30	21
7260	14" iron pipe size	G		70	.229		7.55	9.65		17.20	23
7270	16" iron pipe size	G		65	.246		9.80	10.40		20.20	27
7280	18" iron pipe size	G		60	.267		11.05	11.25		22.30	29.50
7290	20" iron pipe size	G		55	.291		11.25	12.30		23.55	31.50
7300	24" iron pipe size	G		50	.320		13.80	13.50		27.30	36
7320	2" wall, 1/2" iron pipe size	G		220	.073		2.81	3.07		5.88	7.85
7340	3/4" iron pipe size	G		210	.076		2.90	3.22		6.12	8.15
7350	1" iron pipe size	G		200	.080		3.08	3.38		6.46	8.60
7360	1-1/4" iron pipe size	G		190	.084		3.26	3.55		6.81	9.10
7370	1-1/2" iron pipe size	G		190	.084		3.40	3.55		6.95	9.25
7380	2" iron pipe size	G		180	.089		3.58	3.75		7.33	9.75
7390	2-1/2" iron pipe size	G		170	.094		3.86	3.97		7.83	10.40
7400	3" iron pipe size	G		160	.100		4.09	4.22		8.31	11.05
7410	3-1/2" iron pipe size	G		150	.107		4.43	4.50		8.93	11.80
7420	4" iron pipe size	G		130	.123		4.77	5.20		9.97	13.30
7430	5" iron pipe size	G		120	.133		5.45	5.65		11.10	14.70
7440	6" iron pipe size	G		100	.160		5.65	6.75		12.40	16.65

22 07 19 – Plumbing Piping Insulation

22 07 19.10 Piping Insulation		Crew	Daily Output	Labor-Hours	Unit	Material	2011 Bare Costs Labor	Equipment	Total	Total Incl O&P
7450	7" iron pipe size	[G] Q-14	90	.178	L.F.	6.40	7.50		13.90	18.65
7460	8" iron pipe size	[G]	80	.200		6.85	8.45		15.30	20.50
7470	9" iron pipe size	[G]	75	.213		7.50	9		16.50	22
7480	10" iron pipe size	[G]	70	.229		8.20	9.65		17.85	24
7490	12" iron pipe size	[G]	65	.246		9.20	10.40		19.60	26
7500	14" iron pipe size	[G]	60	.267		11.80	11.25		23.05	30.50
7510	16" iron pipe size	[G]	55	.291		12.95	12.30		25.25	33.50
7520	18" iron pipe size	[G]	50	.320		14.50	13.50		28	37
7530	20" iron pipe size	[G]	45	.356		16.45	15		31.45	41
7540	24" iron pipe size	[G]	40	.400		17.65	16.90		34.55	45.50
7560	2-1/2" wall, 1/2" iron pipe size	[G]	210	.076		3.32	3.22		6.54	8.60
7562	3/4" iron pipe size	[G]	200	.080		3.47	3.38		6.85	9
7564	1" iron pipe size	[G]	190	.084		3.62	3.55		7.17	9.50
7566	1-1/4" iron pipe size	[G]	185	.086		3.75	3.65		7.40	9.80
7568	1-1/2" iron pipe size	[G]	180	.089		3.95	3.75		7.70	10.15
7570	2" iron pipe size	[G]	170	.094		4.14	3.97		8.11	10.70
7572	2-1/2" iron pipe size	[G]	160	.100		4.76	4.22		8.98	11.80
7574	3" iron pipe size	[G]	150	.107		5	4.50		9.50	12.45
7576	3-1/2" iron pipe size	[G]	140	.114		5.45	4.82		10.27	13.45
7578	4" iron pipe size	[G]	120	.133		5.75	5.65		11.40	15.05
7580	5" iron pipe size	[G]	110	.145		6.85	6.15		13	17.05
7582	6" iron pipe size	[G]	90	.178		8.25	7.50		15.75	20.50
7584	7" iron pipe size	[G]	80	.200		8.30	8.45		16.75	22
7586	8" iron pipe size	[G]	70	.229		8.80	9.65		18.45	24.50
7588	9" iron pipe size	[G]	65	.246		9.65	10.40		20.05	26.50
7590	10" iron pipe size	[G]	60	.267		10.50	11.25		21.75	29
7592	12" iron pipe size	[G]	55	.291		12.75	12.30		25.05	33
7594	14" iron pipe size	[G]	50	.320		14.90	13.50		28.40	37.50
7596	16" iron pipe size	[G]	45	.356		17.10	15		32.10	42
7598	18" iron pipe size	[G]	40	.400		18.55	16.90		35.45	46.50
7602	24" iron pipe size	[G]	30	.533		24.50	22.50		47	62
7620	3" wall, 1/2" iron pipe size	[G]	200	.080		4.25	3.38		7.63	9.90
7622	3/4" iron pipe size	[G]	190	.084		4.50	3.55		8.05	10.45
7624	1" iron pipe size	[G]	180	.089		4.76	3.75		8.51	11.05
7626	1-1/4" iron pipe size	[G]	175	.091		4.86	3.86		8.72	11.30
7628	1-1/2" iron pipe size	[G]	170	.094		5.10	3.97		9.07	11.75
7630	2" iron pipe size	[G]	160	.100		5.50	4.22		9.72	12.60
7632	2-1/2" iron pipe size	[G]	150	.107		5.70	4.50		10.20	13.25
7634	3" iron pipe size	[G]	140	.114		6.10	4.82		10.92	14.15
7636	3-1/2" iron pipe size	[G]	130	.123		6.70	5.20		11.90	15.45
7638	4" iron pipe size	[G]	110	.145		7.20	6.15		13.35	17.45
7640	5" iron pipe size	[G]	100	.160		8.15	6.75		14.90	19.45
7642	6" iron pipe size	[G]	80	.200		8.70	8.45		17.15	22.50
7644	7" iron pipe size	[G]	70	.229		10	9.65		19.65	26
7646	8" iron pipe size	[G]	60	.267		10.85	11.25		22.10	29.50
7648	9" iron pipe size	[G]	55	.291		11.70	12.30		24	32
7650	10" iron pipe size	[G]	50	.320		12.55	13.50		26.05	35
7652	12" iron pipe size	[G]	45	.356		15.65	15		30.65	40.50
7654	14" iron pipe size	[G]	40	.400		18.05	16.90		34.95	46
7656	16" iron pipe size	[G]	35	.457		20.50	19.30		39.80	52.50
7658	18" iron pipe size	[G]	32	.500		22	21		43	56.50
7660	20" iron pipe size	[G]	30	.533		24	22.50		46.50	61
7662	24" iron pipe size	[G]	28	.571		30	24		54	70.50

22 07 Plumbing Insulation

22 07 19 – Plumbing Piping Insulation

22 07 19.10 Piping Insulation		Crew	Daily Output	Labor-Hours	Unit	Material	2011 Bare Costs Labor	Equipment	Total	Total Incl O&P
7664	26" iron pipe size	G	Q-14	26	.615	L.F.	33.50	26		59.50	76.50
7666	30" iron pipe size	G	↓	24	.667	↓	38	28		66	85
7800	For fiberglass with standard canvas jacket, deduct						5%				
7802	For fittings, add 3 L.F. for each fitting										
7804	plus 4 L.F. for each flange of the fitting										
7810	Finishes										
7814	For single layer of felt, add						10%	10%			
7816	For roofing paper, 45 lb to 55 lb, add						25%	10%			
7820	Polyethylene tubing flexible closed cell foam, UV resistant										
7828	Standard temperature (-90°F to +212°F)										
7830	3/8" wall, 1/8" iron pipe size	G	1 Asbe	130	.062	L.F.	.19	2.89		3.08	4.67
7831	1/4" iron pipe size	G		130	.062		.20	2.89		3.09	4.68
7832	3/8" iron pipe size	G		130	.062		.22	2.89		3.11	4.70
7833	1/2" iron pipe size	G		126	.063		.25	2.98		3.23	4.88
7834	3/4" iron pipe size	G		122	.066		.29	3.08		3.37	5.05
7835	1" iron pipe size	G		120	.067		.33	3.13		3.46	5.20
7836	1-1/4" iron pipe size	G		118	.068		.41	3.18		3.59	5.35
7837	1-1/2" iron pipe size	G		118	.068		.50	3.18		3.68	5.45
7838	2" iron pipe size	G		116	.069		.60	3.23		3.83	5.65
7839	2-1/2" iron pipe size	G		114	.070		.82	3.29		4.11	6
7840	3" iron pipe size	G		112	.071		1.35	3.35		4.70	6.70
7842	1/2" wall, 1/8" iron pipe size	G		120	.067		.28	3.13		3.41	5.15
7843	1/4" iron pipe size	G		120	.067		.30	3.13		3.43	5.15
7844	3/8" iron pipe size	G		120	.067		.33	3.13		3.46	5.20
7845	1/2" iron pipe size	G		118	.068		.37	3.18		3.55	5.35
7846	3/4" iron pipe size	G		116	.069		.42	3.23		3.65	5.45
7847	1" iron pipe size	G		114	.070		.47	3.29		3.76	5.60
7848	1-1/4" iron pipe size	G		112	.071		.55	3.35		3.90	5.80
7849	1-1/2" iron pipe size	G		110	.073		.79	3.41		4.20	6.10
7850	2" iron pipe size	G		108	.074		.95	3.47		4.42	6.40
7851	2-1/2" iron pipe size	G		106	.075		1.07	3.54		4.61	6.65
7852	3" iron pipe size	G		104	.077		1.42	3.61		5.03	7.15
7853	3-1/2" iron pipe size	G		102	.078		1.63	3.68		5.31	7.50
7854	4" iron pipe size	G		100	.080		1.87	3.75		5.62	7.85
7855	3/4" wall, 1/8" iron pipe size	G		110	.073		.42	3.41		3.83	5.70
7856	1/4" iron pipe size	G		110	.073		.45	3.41		3.86	5.75
7857	3/8" iron pipe size	G		108	.074		.53	3.47		4	5.95
7858	1/2" iron pipe size	G		106	.075		.63	3.54		4.17	6.15
7859	3/4" iron pipe size	G		104	.077		.76	3.61		4.37	6.45
7860	1" iron pipe size	G		102	.078		.87	3.68		4.55	6.65
7861	1-1/4" iron pipe size	G		100	.080		1.18	3.75		4.93	7.10
7862	1-1/2" iron pipe size	G		100	.080		1.55	3.75		5.30	7.50
7863	2" iron pipe size	G		98	.082		1.72	3.83		5.55	7.80
7864	2-1/2" iron pipe size	G		96	.083		1.83	3.91		5.74	8.05
7865	3" iron pipe size	G		94	.085		2.54	3.99		6.53	8.95
7866	3-1/2" iron pipe size	G		92	.087		2.78	4.08		6.86	9.35
7867	4" iron pipe size	G		90	.089		3.12	4.17		7.29	9.90
7868	1" wall, 1/4" iron pipe size	G		100	.080		.93	3.75		4.68	6.80
7869	3/8" iron pipe size	G		98	.082		1.04	3.83		4.87	7.05
7870	1/2" iron pipe size	G		96	.083		1.16	3.91		5.07	7.35
7871	3/4" iron pipe size	G		94	.085		1.36	3.99		5.35	7.65
7872	1" iron pipe size	G		92	.087		1.66	4.08		5.74	8.15
7873	1-1/4" iron pipe size	G		90	.089		1.90	4.17		6.07	8.55

22 07 19 – Plumbing Piping Insulation

22 07 19.10 Piping Insulation		Crew	Daily Output	Labor-Hours	Unit	Material	2011 Bare Costs Labor	Equipment	Total	Total Incl O&P
7874	1-1/2" iron pipe size	G	1 Asbe	90	.089	L.F.	2.38	4.17		6.55	9.05
7875	2" iron pipe size	G		88	.091		2.85	4.26		7.11	9.75
7876	2-1/2" iron pipe size	G		86	.093		3.22	4.36		7.58	10.30
7877	3" iron pipe size	G		84	.095		4.06	4.47		8.53	11.35
7878	Contact cement, quart can	G				Ea.	9.85			9.85	10.85
7879	Rubber tubing, flexible closed cell foam										
7880	3/8" wall, 1/4" iron pipe size	G	1 Asbe	120	.067	L.F.	.55	3.13		3.68	5.45
7900	3/8" iron pipe size	G		120	.067		.68	3.13		3.81	5.60
7910	1/2" iron pipe size	G		115	.070		.70	3.26		3.96	5.80
7920	3/4" iron pipe size	G		115	.070		.85	3.26		4.11	6
7930	1" iron pipe size	G		110	.073		.98	3.41		4.39	6.35
7940	1-1/4" iron pipe size	G		110	.073		1.20	3.41		4.61	6.55
7950	1-1/2" iron pipe size	G		110	.073		1.30	3.41		4.71	6.70
8100	1/2" wall, 1/4" iron pipe size	G		90	.089		.50	4.17		4.67	7
8120	3/8" iron pipe size	G		90	.089		.55	4.17		4.72	7.05
8130	1/2" iron pipe size	G		89	.090		.61	4.22		4.83	7.15
8140	3/4" iron pipe size	G		89	.090		.68	4.22		4.90	7.25
8150	1" iron pipe size	G		88	.091		.75	4.26		5.01	7.45
8160	1-1/4" iron pipe size	G		87	.092		.87	4.31		5.18	7.60
8170	1-1/2" iron pipe size	G		87	.092		1.06	4.31		5.37	7.80
8180	2" iron pipe size	G		86	.093		1.38	4.36		5.74	8.25
8190	2-1/2" iron pipe size	G		86	.093		1.74	4.36		6.10	8.65
8200	3" iron pipe size	G		85	.094		1.95	4.41		6.36	8.95
8210	3-1/2" iron pipe size	G		85	.094		2.70	4.41		7.11	9.75
8220	4" iron pipe size	G		80	.100		2.89	4.69		7.58	10.45
8230	5" iron pipe size	G		80	.100		3.97	4.69		8.66	11.60
8240	6" iron pipe size	G		75	.107		3.97	5		8.97	12.10
8300	3/4" wall, 1/4" iron pipe size	G		90	.089		.77	4.17		4.94	7.30
8320	3/8" iron pipe size	G		90	.089		.84	4.17		5.01	7.35
8330	1/2" iron pipe size	G		89	.090		1.01	4.22		5.23	7.60
8340	3/4" iron pipe size	G		89	.090		1.23	4.22		5.45	7.85
8350	1" iron pipe size	G		88	.091		1.41	4.26		5.67	8.15
8360	1-1/4" iron pipe size	G		87	.092		1.90	4.31		6.21	8.75
8370	1-1/2" iron pipe size	G		87	.092		2.15	4.31		6.46	9
8380	2" iron pipe size	G		86	.093		2.52	4.36		6.88	9.50
8390	2-1/2" iron pipe size	G		86	.093		3.38	4.36		7.74	10.45
8400	3" iron pipe size	G		85	.094		3.85	4.41		8.26	11.05
8410	3-1/2" iron pipe size	G		85	.094		4.58	4.41		8.99	11.85
8420	4" iron pipe size	G		80	.100		4.88	4.69		9.57	12.60
8430	5" iron pipe size	G		80	.100		5.90	4.69		10.59	13.70
8440	6" iron pipe size	G		80	.100		7.10	4.69		11.79	15.05
8444	1" wall, 1/2" iron pipe size	G		86	.093		1.93	4.36		6.29	8.85
8445	3/4" iron pipe size	G		84	.095		2.34	4.47		6.81	9.45
8446	1" iron pipe size	G		84	.095		2.72	4.47		7.19	9.90
8447	1-1/4" iron pipe size	G		82	.098		3.08	4.58		7.66	10.45
8448	1-1/2" iron pipe size	G		82	.098		3.58	4.58		8.16	11
8449	2" iron pipe size	G		80	.100		4.78	4.69		9.47	12.50
8450	2-1/2" iron pipe size	G		80	.100		6.25	4.69		10.94	14.10
8456	Rubber insulation tape, 1/8" x 2" x 30'	G				Ea.	11.40			11.40	12.50
8460	Polyolefin tubing, flexible closed cell foam, UV stabilized, work										
8462	temp.-165°F to +210°F, 0 water vapor transmission										
8464	3/8" wall, 1/8" iron pipe size	G	1 Asbe	140	.057	L.F.	.37	2.68		3.05	4.55
8466	1/4" iron pipe size	G		140	.057		.39	2.68		3.07	4.57

22 07 Plumbing Insulation

22 07 19 – Plumbing Piping Insulation

22 07 19.10 Piping Insulation		Crew	Daily Output	Labor-Hours	Unit	Material	2011 Bare Costs Labor	Equipment	Total	Total Incl O&P
8468	3/8" iron pipe size [G]	1 Asbe	140	.057	L.F.	.42	2.68		3.10	4.60
8470	1/2" iron pipe size [G]		136	.059		.48	2.76		3.24	4.79
8472	3/4" iron pipe size [G]		132	.061		.55	2.84		3.39	5
8474	1" iron pipe size [G]		130	.062		.60	2.89		3.49	5.10
8476	1-1/4" iron pipe size [G]		128	.063		.78	2.93		3.71	5.40
8478	1-1/2" iron pipe size [G]		128	.063		.96	2.93		3.89	5.60
8480	2" iron pipe size [G]		126	.063		1.07	2.98		4.05	5.80
8482	2-1/2" iron pipe size [G]		123	.065		1.56	3.05		4.61	6.45
8484	3" iron pipe size [G]		121	.066		2.03	3.10		5.13	7
8486	4" iron pipe size [G]		118	.068		3.34	3.18		6.52	8.60
8500	1/2" wall, 1/8" iron pipe size [G]		130	.062		.54	2.89		3.43	5.05
8502	1/4" iron pipe size [G]		130	.062		.58	2.89		3.47	5.10
8504	3/8" iron pipe size [G]		130	.062		.63	2.89		3.52	5.15
8506	1/2" iron pipe size [G]		128	.063		.68	2.93		3.61	5.30
8508	3/4" iron pipe size [G]		126	.063		.79	2.98		3.77	5.45
8510	1" iron pipe size [G]		123	.065		.86	3.05		3.91	5.65
8512	1-1/4" iron pipe size [G]		121	.066		1.07	3.10		4.17	5.95
8514	1-1/2" iron pipe size [G]		119	.067		1.29	3.15		4.44	6.30
8516	2" iron pipe size [G]		117	.068		1.50	3.21		4.71	6.60
8518	2-1/2" iron pipe size [G]		114	.070		2.04	3.29		5.33	7.35
8520	3" iron pipe size [G]		112	.071		2.90	3.35		6.25	8.40
8522	4" iron pipe size [G]		110	.073		3.60	3.41		7.01	9.20
8534	3/4" wall, 1/8" iron pipe size [G]		120	.067		.83	3.13		3.96	5.75
8536	1/4" iron pipe size [G]		120	.067		.88	3.13		4.01	5.80
8538	3/8" iron pipe size [G]		117	.068		1.01	3.21		4.22	6.05
8540	1/2" iron pipe size [G]		114	.070		1.12	3.29		4.41	6.35
8542	3/4" iron pipe size [G]		112	.071		1.45	3.35		4.80	6.80
8544	1" iron pipe size [G]		110	.073		1.67	3.41		5.08	7.10
8546	1-1/4" iron pipe size [G]		108	.074		2.26	3.47		5.73	7.85
8548	1-1/2" iron pipe size [G]		108	.074		2.62	3.47		6.09	8.25
8550	2" iron pipe size [G]		106	.075		3.10	3.54		6.64	8.85
8552	2-1/2" iron pipe size [G]		104	.077		3.90	3.61		7.51	9.90
8554	3" iron pipe size [G]		102	.078		4.44	3.68		8.12	10.60
8556	4" iron pipe size [G]		100	.080		6.30	3.75		10.05	12.75
8570	1" wall, 1/8" iron pipe size [G]		110	.073		1.53	3.41		4.94	6.95
8572	1/4" iron pipe size [G]		108	.074		1.60	3.47		5.07	7.10
8574	3/8" iron pipe size [G]		106	.075		1.69	3.54		5.23	7.30
8576	1/2" iron pipe size [G]		104	.077		1.96	3.61		5.57	7.75
8578	3/4" iron pipe size [G]		102	.078		2.18	3.68		5.86	8.10
8580	1" iron pipe size [G]		100	.080		2.34	3.75		6.09	8.35
8582	1-1/4" iron pipe size [G]		97	.082		2.52	3.87		6.39	8.75
8584	1-1/2" iron pipe size [G]		97	.082		3.27	3.87		7.14	9.60
8586	2" iron pipe size [G]		95	.084		4.29	3.95		8.24	10.80
8588	2-1/2" iron pipe size [G]		93	.086		5.60	4.03		9.63	12.40
8590	3" iron pipe size [G]		91	.088		6.70	4.12		10.82	13.70
8606	Contact adhesive (R-320) [G]				Qt.	16.70			16.70	18.40
8608	Contact adhesive (R-320) [G]				Gal.	53.50			53.50	58.50
8610	NOTE: Preslit/preglued vs unslit, same price									

22 07 19.30 Piping Insulation Protective Jacketing, PVC

0010	**PIPING INSULATION PROTECTIVE JACKETING, PVC**									
0100	PVC, white, 48" lengths cut from roll goods									
0120	20 mil thick									

22 07 19.30 Piping Insulation Protective Jacketing, PVC	Crew	Daily Output	Labor-Hours	Unit	Material	2011 Bare Costs Labor	Equipment	Total	Total Incl O&P
0140	Size based on OD of insulation								
0150	1-1/2" ID	Q-14	270	.059	L.F.	.26	2.50		2.76	4.16
0152	2" ID		260	.062		.32	2.60		2.92	4.37
0154	2-1/2" ID		250	.064		.40	2.70		3.10	4.62
0156	3" ID		240	.067		.47	2.81		3.28	4.87
0158	3-1/2" ID		230	.070		.54	2.94		3.48	5.15
0160	4" ID		220	.073		.60	3.07		3.67	5.40
0162	4-1/2" ID		210	.076		.67	3.22		3.89	5.70
0164	5" ID		200	.080		.75	3.38		4.13	6.05
0166	5-1/2" ID		190	.084		.82	3.55		4.37	6.40
0168	6" ID		180	.089		.88	3.75		4.63	6.75
0170	6-1/2" ID		175	.091		.95	3.86		4.81	7
0172	7" ID		170	.094		1.03	3.97		5	7.30
0174	7-1/2" ID		164	.098		1.10	4.12		5.22	7.55
0176	8" ID		161	.099		1.18	4.19		5.37	7.80
0178	8-1/2" ID		158	.101		1.24	4.27		5.51	7.95
0180	9" ID		155	.103		1.31	4.36		5.67	8.20
0182	9-1/2" ID		152	.105		1.38	4.44		5.82	8.35
0184	10" ID		149	.107		1.46	4.53		5.99	8.60
0186	10-1/2" ID		146	.110		1.52	4.62		6.14	8.80
0188	11" ID		143	.112		1.59	4.72		6.31	9.05
0190	11-1/2" ID		140	.114		1.66	4.82		6.48	9.30
0192	12" ID		137	.117		1.74	4.93		6.67	9.50
0194	12-1/2" ID		134	.119		1.80	5.05		6.85	9.80
0195	13" ID		132	.121		1.88	5.10		6.98	9.95
0196	13-1/2" ID		132	.121		1.95	5.10		7.05	10.05
0198	14" ID		130	.123		2.02	5.20		7.22	10.25
0200	15" ID		128	.125		2.15	5.30		7.45	10.50
0202	16" ID		126	.127		2.30	5.35		7.65	10.85
0204	17" ID		124	.129		2.43	5.45		7.88	11.05
0206	18" ID		122	.131		2.58	5.55		8.13	11.40
0208	19" ID		120	.133		2.72	5.65		8.37	11.70
0210	20" ID		118	.136		2.86	5.70		8.56	12
0212	21" ID		116	.138		3	5.80		8.80	12.30
0214	22" ID		114	.140		3.14	5.90		9.04	12.60
0216	23" ID		112	.143		3.28	6.05		9.33	12.90
0218	24" ID		110	.145		3.43	6.15		9.58	13.25
0220	25" ID		108	.148		3.56	6.25		9.81	13.55
0222	26" ID		106	.151		3.71	6.35		10.06	13.95
0224	27" ID		104	.154		3.84	6.50		10.34	14.25
0226	28" ID		102	.157		3.99	6.60		10.59	14.65
0228	29" ID		100	.160		4.13	6.75		10.88	15
0230	30" ID		98	.163		4.27	6.90		11.17	15.35
0300	For colors, add				Ea.	10%				
1000	30 mil thick									
1010	Size based on OD of insulation									
1020	2" ID	Q-14	260	.062	L.F.	.50	2.60		3.10	4.57
1022	2-1/2" ID		250	.064		.60	2.70		3.30	4.84
1024	3" ID		240	.067		.72	2.81		3.53	5.15
1026	3-1/2" ID		230	.070		.82	2.94		3.76	5.45
1028	4" ID		220	.073		.92	3.07		3.99	5.75
1030	4-1/2" ID		210	.076		1.03	3.22		4.25	6.10
1032	5" ID		200	.080		1.13	3.38		4.51	6.45

22 07 19.30 Piping Insulation Protective Jacketing, PVC	Crew	Daily Output	Labor-Hours	Unit	Material	2011 Bare Costs Labor	2011 Bare Costs Equipment	Total	Total Incl O&P
1034	5-1/2" ID	Q-14	190	.084	L.F.	1.24	3.55		4.79	6.85
1036	6" ID		180	.089		1.34	3.75		5.09	7.25
1038	6-1/2" ID		175	.091		1.46	3.86		5.32	7.55
1040	7" ID		170	.094		1.56	3.97		5.53	7.85
1042	7-1/2" ID		164	.098		1.68	4.12		5.80	8.20
1044	8" ID		161	.099		1.78	4.19		5.97	8.45
1046	8-1/2" ID		158	.101		1.89	4.27		6.16	8.70
1048	9" ID		155	.103		1.99	4.36		6.35	8.95
1050	9-1/2" ID		152	.105		2.11	4.44		6.55	9.15
1052	10" ID		149	.107		2.21	4.53		6.74	9.45
1054	10-1/2" ID		146	.110		2.32	4.62		6.94	9.70
1056	11" ID		143	.112		2.42	4.72		7.14	9.95
1058	11-1/2" ID		140	.114		2.53	4.82		7.35	10.25
1060	12" ID		137	.117		2.63	4.93		7.56	10.50
1062	12-1/2" ID		134	.119		2.74	5.05		7.79	10.80
1063	13" ID		132	.121		2.86	5.10		7.96	11.05
1064	13-1/2" ID		132	.121		2.96	5.10		8.06	11.15
1066	14" ID		130	.123		3.07	5.20		8.27	11.45
1068	15" ID		128	.125		3.28	5.30		8.58	11.75
1070	16" ID		126	.127		3.49	5.35		8.84	12.15
1072	17" ID		124	.129		3.71	5.45		9.16	12.50
1074	18" ID		122	.131		3.92	5.55		9.47	12.85
1076	19" ID		120	.133		4.13	5.65		9.78	13.25
1078	20" ID		118	.136		4.35	5.70		10.05	13.65
1080	21" ID		116	.138		4.56	5.80		10.36	14
1082	22" ID		114	.140		4.77	5.90		10.67	14.40
1084	23" ID		112	.143		4.98	6.05		11.03	14.80
1086	24" ID		110	.145		5.20	6.15		11.35	15.25
1088	25" ID		108	.148		5.40	6.25		11.65	15.60
1090	26" ID		106	.151		5.65	6.35		12	16.05
1092	27" ID		104	.154		5.85	6.50		12.35	16.50
1094	28" ID		102	.157		6.05	6.60		12.65	16.90
1096	29" ID		100	.160		6.30	6.75		13.05	17.35
1098	30" ID		98	.163		6.50	6.90		13.40	17.80
1300	For colors, add				Ea.	10%				
2000	PVC, white, fitting covers									
2020	Fiberglass insulation inserts included with sizes 1-3/4" thru 9-3/4"									
2030	Size is based on OD of insulation									
2040	90° Elbow fitting									
2060	1-3/4"	Q-14	135	.119	Ea.	.49	5		5.49	8.30
2062	2"		130	.123		.60	5.20		5.80	8.70
2064	2-1/4"		128	.125		.66	5.30		5.96	8.90
2068	2-1/2"		126	.127		.74	5.35		6.09	9.10
2070	2-3/4"		123	.130		.82	5.50		6.32	9.40
2072	3"		120	.133		.83	5.65		6.48	9.60
2074	3-3/8"		116	.138		.95	5.80		6.75	10.05
2076	3-3/4"		113	.142		1.08	6		7.08	10.45
2078	4-1/8"		110	.145		1.34	6.15		7.49	10.95
2080	4-3/4"		105	.152		1.61	6.45		8.06	11.70
2082	5-1/4"		100	.160		1.88	6.75		8.63	12.50
2084	5-3/4"		95	.168		2.52	7.10		9.62	13.75
2086	6-1/4"		90	.178		4.24	7.50		11.74	16.25
2088	6-3/4"		87	.184		4.49	7.75		12.24	16.95

22 07 19.30 Piping Insulation Protective Jacketing, PVC		Crew	Daily Output	Labor-Hours	Unit	Material	2011 Bare Costs		Total	Total Incl O&P
							Labor	Equipment		
2090	7-1/4"	Q-14	85	.188	Ea.	5.60	7.95		13.55	18.50
2092	7-3/4"		83	.193		5.80	8.15		13.95	19
2094	8-3/4"		80	.200		7.60	8.45		16.05	21.50
2096	9-3/4"		77	.208		10.15	8.75		18.90	24.50
2098	10-7/8"		74	.216		11.30	9.10		20.40	26.50
2100	11-7/8"		71	.225		12.95	9.50		22.45	29
2102	12-7/8"		68	.235		16.70	9.95		26.65	34
2104	14-1/8"		66	.242		18.55	10.25		28.80	36.50
2106	15-1/8"		64	.250		20	10.55		30.55	38.50
2108	16-1/8"		63	.254		22	10.70		32.70	40.50
2110	17-1/8"		62	.258		24.50	10.90		35.40	44
2112	18-1/8"		61	.262		30.50	11.05		41.55	50.50
2114	19-1/8"		60	.267		42.50	11.25		53.75	64.50
2116	20-1/8"	▼	59	.271	▼	55	11.45		66.45	78
2200	45° Elbow fitting									
2220	1-3/4" thru 9-3/4" same price as 90° Elbow fitting									
2320	10-7/8"	Q-14	74	.216	Ea.	11.30	9.10		20.40	26.50
2322	11-7/8"		71	.225		12.50	9.50		22	28.50
2324	12-7/8"		68	.235		13.90	9.95		23.85	30.50
2326	14-1/8"		66	.242		15.85	10.25		26.10	33
2328	15-1/8"		64	.250		17.05	10.55		27.60	35
2330	16-1/8"		63	.254		19.50	10.70		30.20	38
2332	17-1/8"		62	.258		22	10.90		32.90	41
2334	18-1/8"		61	.262		27	11.05		38.05	46.50
2336	19-1/8"		60	.267		36.50	11.25		47.75	57.50
2338	20-1/8"	▼	59	.271	▼	41.50	11.45		52.95	63
2400	Tee fitting									
2410	1-3/4"	Q-14	96	.167	Ea.	.93	7.05		7.98	11.90
2412	2"		94	.170		1.03	7.20		8.23	12.25
2414	2-1/4"		91	.176		1.15	7.40		8.55	12.70
2416	2-1/2"		88	.182		1.25	7.65		8.90	13.25
2418	2-3/4"		85	.188		1.37	7.95		9.32	13.80
2420	3"		82	.195		1.49	8.25		9.74	14.40
2422	3-3/8"		79	.203		1.71	8.55		10.26	15.10
2424	3-3/4"		76	.211		1.96	8.90		10.86	15.90
2426	4-1/8"		73	.219		2.30	9.25		11.55	16.85
2428	4-3/4"		70	.229		2.88	9.65		12.53	18.05
2430	5-1/4"		67	.239		3.45	10.10		13.55	19.40
2432	5-3/4"		63	.254		4.57	10.70		15.27	21.50
2434	6-1/4"		60	.267		6	11.25		17.25	24
2436	6-3/4"		59	.271		7.45	11.45		18.90	26
2438	7-1/4"		57	.281		12.05	11.85		23.90	31.50
2440	7-3/4"		54	.296		13.20	12.50		25.70	34
2442	8-3/4"		52	.308		16.05	13		29.05	37.50
2444	9-3/4"		50	.320		18.95	13.50		32.45	42
2446	10-7/8"		48	.333		19.15	14.05		33.20	43
2448	11-7/8"		47	.340		21.50	14.35		35.85	45.50
2450	12-7/8"		46	.348		23.50	14.70		38.20	48.50
2452	14-1/8"		45	.356		25.50	15		40.50	51
2454	15-1/8"		44	.364		27.50	15.35		42.85	54
2456	16-1/8"		43	.372		30	15.70		45.70	57.50
2458	17-1/8"		42	.381		32	16.10		48.10	60
2460	18-1/8"	▼	41	.390		35	16.45		51.45	64

22 07 19.30 Piping Insulation Protective Jacketing, PVC		Crew	Daily Output	Labor-Hours	Unit	Material	2011 Bare Costs Labor	Equipment	Total	Total Incl O&P
2462	19-1/8"	Q-14	40	.400	Ea.	38.50	16.90		55.40	68
2464	20-1/8"	↓	39	.410	↓	41.50	17.30		58.80	72.50
4000	Mechanical grooved fitting cover, including insert									
4020	90° Elbow fitting									
4030	3/4" & 1"	Q-14	140	.114	Ea.	6.75	4.82		11.57	14.90
4040	1-1/4" & 1-1/2"		135	.119		7.05	5		12.05	15.50
4042	2"		130	.123		7.75	5.20		12.95	16.55
4044	2-1/2"		125	.128		8.60	5.40		14	17.80
4046	3"		120	.133		9.65	5.65		15.30	19.35
4048	3-1/2"		115	.139		11.20	5.85		17.05	21.50
4050	4"		110	.145		12.50	6.15		18.65	23.50
4052	5"		100	.160		15.45	6.75		22.20	27.50
4054	6"		90	.178		22	7.50		29.50	35.50
4056	8"		80	.200		25	8.45		33.45	40.50
4058	10"		75	.213		32	9		41	49
4060	12"		68	.235		46	9.95		55.95	66
4062	14"		65	.246		57	10.40		67.40	78.50
4064	16"		63	.254		77.50	10.70		88.20	102
4066	18"	↓	61	.262	↓	106	11.05		117.05	134
4100	45° Elbow fitting									
4120	3/4" & 1"	Q-14	140	.114	Ea.	6.45	4.82		11.27	14.55
4130	1-1/4" & 1-1/2"		135	.119		6.75	5		11.75	15.20
4140	2"		130	.123		7.10	5.20		12.30	15.85
4142	2-1/2"		125	.128		7.85	5.40		13.25	17
4144	3"		120	.133		8.65	5.65		14.30	18.25
4146	3-1/2"		115	.139		10.10	5.85		15.95	20.50
4148	4"		110	.145		10.45	6.15		16.60	21
4150	5"		100	.160		13.95	6.75		20.70	26
4152	6"		90	.178		19.40	7.50		26.90	33
4154	8"		80	.200		22	8.45		30.45	37
4156	10"		75	.213		28	9		37	44.50
4158	12"		68	.235		39.50	9.95		49.45	59
4160	14"		65	.246		44.50	10.40		54.90	65
4162	16"		63	.254		65.50	10.70		76.20	89
4164	18"	↓	61	.262	↓	94	11.05		105.05	120
4200	Tee fitting									
4220	3/4" & 1"	Q-14	93	.172	Ea.	9.30	7.25		16.55	21.50
4230	1-1/4" & 1-1/2"		90	.178		9.90	7.50		17.40	22.50
4240	2"		87	.184		10.95	7.75		18.70	24
4242	2-1/2"		84	.190		12.20	8.05		20.25	26
4244	3"		80	.200		13.10	8.45		21.55	27.50
4246	3-1/2"		77	.208		15.25	8.75		24	30.50
4248	4"		73	.219		17.95	9.25		27.20	34
4250	5"		67	.239		21	10.10		31.10	38.50
4252	6"		60	.267		25	11.25		36.25	45
4254	8"		54	.296		31	12.50		43.50	53.50
4256	10"		50	.320		39.50	13.50		53	64.50
4258	12"		46	.348		52.50	14.70		67.20	80.50
4260	14"		43	.372		67	15.70		82.70	98
4262	16"		42	.381		89	16.10		105.10	123
4264	18"	↓	41	.390	↓	118	16.45		134.45	155

22 07 Plumbing Insulation

22 07 19 – Plumbing Piping Insulation

22 07 19.40 Pipe Insulation Protective Jacketing, Aluminum	Crew	Daily Output	Labor-Hours	Unit	Material	2011 Bare Costs Labor	Equipment	Total	Total Incl O&P
0010 **PIPE INSULATION PROTECTIVE JACKETING, ALUMINUM**									
0100 Metal roll jacketing									
0120 Aluminum with polykraft moisture barrier									
0140 Smooth, based on OD of insulation, .016" thick									
0180 1/2" ID	Q-14	220	.073	L.F.	.24	3.07		3.31	5
0190 3/4" ID		215	.074		.32	3.14		3.46	5.20
0200 1" ID		210	.076		.38	3.22		3.60	5.40
0210 1-1/4" ID		205	.078		.46	3.29		3.75	5.60
0220 1-1/2" ID		202	.079		.53	3.34		3.87	5.75
0230 1-3/4" ID		199	.080		.61	3.39		4	5.90
0240 2" ID		195	.082		.68	3.46		4.14	6.10
0250 2-1/4" ID		191	.084		.75	3.54		4.29	6.30
0260 2-1/2" ID		187	.086		.83	3.61		4.44	6.50
0270 2-3/4" ID		184	.087		.90	3.67		4.57	6.65
0280 3" ID		180	.089		.97	3.75		4.72	6.85
0290 3-1/4" ID		176	.091		1.04	3.84		4.88	7.10
0300 3-1/2" ID		172	.093		1.12	3.93		5.05	7.30
0310 3-3/4" ID		169	.095		1.19	4		5.19	7.50
0320 4" ID		165	.097		1.26	4.09		5.35	7.75
0330 4-1/4" ID		161	.099		1.34	4.19		5.53	7.95
0340 4-1/2" ID		157	.102		1.41	4.30		5.71	8.20
0350 4-3/4" ID		154	.104		1.49	4.38		5.87	8.45
0360 5" ID		150	.107		1.55	4.50		6.05	8.65
0370 5-1/4" ID		146	.110		1.63	4.62		6.25	8.95
0380 5-1/2" ID		143	.112		1.70	4.72		6.42	9.15
0390 5-3/4" ID		139	.115		1.78	4.86		6.64	9.45
0400 6" ID		135	.119		1.85	5		6.85	9.80
0410 6-1/4" ID		133	.120		1.92	5.10		7.02	9.95
0420 6-1/2" ID		131	.122		2	5.15		7.15	10.15
0430 7" ID		128	.125		2.14	5.30		7.44	10.50
0440 7-1/4" ID		125	.128		2.21	5.40		7.61	10.80
0450 7-1/2" ID		123	.130		2.29	5.50		7.79	11
0460 8" ID		121	.132		2.43	5.60		8.03	11.30
0470 8-1/2" ID		119	.134		2.58	5.65		8.23	11.60
0480 9" ID		116	.138		2.72	5.80		8.52	12
0490 9-1/2" ID		114	.140		2.87	5.90		8.77	12.30
0500 10" ID		112	.143		3.02	6.05		9.07	12.60
0510 10-1/2" ID		110	.145		3.17	6.15		9.32	13
0520 11" ID		107	.150		3.32	6.30		9.62	13.40
0530 11-1/2" ID		105	.152		3.46	6.45		9.91	13.75
0540 12" ID		103	.155		3.61	6.55		10.16	14.10
0550 12-1/2" ID		100	.160		3.75	6.75		10.50	14.60
0560 13" ID		99	.162		3.89	6.80		10.69	14.85
0570 14" ID		98	.163		4.19	6.90		11.09	15.25
0580 15" ID		96	.167		4.48	7.05		11.53	15.85
0590 16" ID		95	.168		4.78	7.10		11.88	16.25
0600 17" ID		93	.172		5.05	7.25		12.30	16.85
0610 18" ID		92	.174		5.25	7.35		12.60	17.15
0620 19" ID		90	.178		5.65	7.50		13.15	17.85
0630 20" ID		89	.180		5.95	7.60		13.55	18.30
0640 21" ID		87	.184		6.25	7.75		14	18.85
0650 22" ID		86	.186		6.55	7.85		14.40	19.35

22 07 19 – Plumbing Piping Insulation

22 07 19.40 Pipe Insulation Protective Jacketing, Aluminum	Crew	Daily Output	Labor-Hours	Unit	Material	2011 Bare Costs Labor	Equipment	Total	Total Incl O&P
0660	23" ID	Q-14	84	.190	L.F.	6.85	8.05		14.90	19.95
0670	24" ID	↓	83	.193		7.10	8.15		15.25	20.50
0710	For smooth .020" thick, add					27%	10%			
0720	For smooth .024" thick, add					52%	20%			
0730	For smooth .032" thick, add					104%	33%			
0800	For stucco embossed, add					1%				
0820	For corrugated, add					2.50%				
0900	White aluminum with polysurlyn moisture barrier									
0910	Smooth, % is an add to polykraft lines of same thickness									
0940	For smooth .016" thick, add				L.F.	35%				
0960	For smooth .024" thick, add				"	22%				
1000	Aluminum fitting covers									
1010	Size is based on OD of insulation									
1020	90° LR elbow, 2 piece									
1100	1-1/2"	Q-14	140	.114	Ea.	3.05	4.82		7.87	10.80
1110	1-3/4"		135	.119		3.05	5		8.05	11.10
1120	2"		130	.123		3.83	5.20		9.03	12.25
1130	2-1/4"		128	.125		3.83	5.30		9.13	12.35
1140	2-1/2"		126	.127		3.83	5.35		9.18	12.50
1150	2-3/4"		123	.130		3.83	5.50		9.33	12.70
1160	3"		120	.133		4.16	5.65		9.81	13.30
1170	3-1/4"		117	.137		4.16	5.75		9.91	13.50
1180	3-1/2"		115	.139		4.91	5.85		10.76	14.50
1190	3-3/4"		113	.142		5.05	6		11.05	14.80
1200	4"		110	.145		5.25	6.15		11.40	15.30
1210	4-1/4"		108	.148		5.40	6.25		11.65	15.60
1220	4-1/2"		106	.151		5.40	6.35		11.75	15.80
1230	4-3/4"		104	.154		5.40	6.50		11.90	16
1240	5"		102	.157		6.10	6.60		12.70	16.95
1250	5-1/4"		100	.160		7.05	6.75		13.80	18.20
1260	5-1/2"		97	.165		8.80	6.95		15.75	20.50
1270	5-3/4"		95	.168		8.80	7.10		15.90	20.50
1280	6"		92	.174		7.50	7.35		14.85	19.60
1290	6-1/4"		90	.178		7.50	7.50		15	19.85
1300	6-1/2"		87	.184		9	7.75		16.75	22
1310	7"		85	.188		10.95	7.95		18.90	24.50
1320	7-1/4"		84	.190		10.95	8.05		19	24.50
1330	7-1/2"		83	.193		16.10	8.15		24.25	30.50
1340	8"		82	.195		12	8.25		20.25	26
1350	8-1/2"		80	.200		22	8.45		30.45	37.50
1360	9"		78	.205		22	8.65		30.65	38
1370	9-1/2"		77	.208		16.30	8.75		25.05	31.50
1380	10"		76	.211		16.30	8.90		25.20	31.50
1390	10-1/2"		75	.213		16.65	9		25.65	32.50
1400	11"		74	.216		16.65	9.10		25.75	32.50
1410	11-1/2"		72	.222		18.95	9.40		28.35	35.50
1420	12"		71	.225		18.95	9.50		28.45	35.50
1430	12-1/2"		69	.232		33.50	9.80		43.30	52
1440	13"		68	.235		33.50	9.95		43.45	52.50
1450	14"		66	.242		45	10.25		55.25	66
1460	15"		64	.250		47.50	10.55		58.05	68.50
1470	16"	↓	63	.254	↓	51.50	10.70		62.20	73.50
2000	45° Elbow, 2 piece									

22 07 19.40 Pipe Insulation Protective Jacketing, Aluminum		Crew	Daily Output	Labor-Hours	Unit	Material	2011 Bare Costs Labor	Equipment	Total	Total Incl O&P
2010	2-1/2"	Q-14	126	.127	Ea.	3.17	5.35		8.52	11.80
2020	2-3/4"		123	.130		3.17	5.50		8.67	12
2030	3"		120	.133		3.58	5.65		9.23	12.65
2040	3-1/4"		117	.137		3.58	5.75		9.33	12.85
2050	3-1/2"		115	.139		4.08	5.85		9.93	13.60
2060	3-3/4"		113	.142		4.08	6		10.08	13.75
2070	4"		110	.145		4.15	6.15		10.30	14.05
2080	4-1/4"		108	.148		4.73	6.25		10.98	14.85
2090	4-1/2"		106	.151		4.73	6.35		11.08	15.05
2100	4-3/4"		104	.154		4.73	6.50		11.23	15.25
2110	5"		102	.157		5.55	6.60		12.15	16.35
2120	5-1/4"		100	.160		5.55	6.75		12.30	16.55
2130	5-1/2"		97	.165		5.85	6.95		12.80	17.15
2140	6"		92	.174		5.85	7.35		13.20	17.75
2150	6-1/2"		87	.184		8.05	7.75		15.80	21
2160	7"		85	.188		8.05	7.95		16	21
2170	7-1/2"		83	.193		8.10	8.15		16.25	21.50
2180	8"		82	.195		8.10	8.25		16.35	21.50
2190	8-1/2"		80	.200		10.10	8.45		18.55	24
2200	9"		78	.205		10.10	8.65		18.75	24.50
2210	9-1/2"		77	.208		14.05	8.75		22.80	29
2220	10"		76	.211		14.05	8.90		22.95	29
2230	10-1/2"		75	.213		13.10	9		22.10	28.50
2240	11"		74	.216		13.10	9.10		22.20	28.50
2250	11-1/2"		72	.222		15.75	9.40		25.15	32
2260	12"		71	.225		15.75	9.50		25.25	32
2270	13"		68	.235		18.50	9.95		28.45	36
2280	14"		66	.242		22.50	10.25		32.75	41
2290	15"		64	.250		35	10.55		45.55	55
2300	16"		63	.254		38	10.70		48.70	58
2310	17"		62	.258		37	10.90		47.90	57.50
2320	18"		61	.262		42	11.05		53.05	63.50
2330	19"		60	.267		51.50	11.25		62.75	74
2340	20"		59	.271		50	11.45		61.45	72.50
2350	21"		58	.276		54	11.65		65.65	77
3000	Tee, 4 piece									
3010	2-1/2"	Q-14	88	.182	Ea.	24.50	7.65		32.15	39
3020	2-3/4"		86	.186		24.50	7.85		32.35	39
3030	3"		84	.190		28	8.05		36.05	43
3040	3-1/4"		82	.195		28	8.25		36.25	43.50
3050	3-1/2"		80	.200		29	8.45		37.45	45
3060	4"		78	.205		29.50	8.65		38.15	46
3070	4-1/4"		76	.211		31	8.90		39.90	48
3080	4-1/2"		74	.216		31	9.10		40.10	48
3090	4-3/4"		72	.222		31	9.40		40.40	48.50
3100	5"		70	.229		32	9.65		41.65	50.50
3110	5-1/4"		68	.235		32	9.95		41.95	51
3120	5-1/2"		66	.242		33.50	10.25		43.75	53
3130	6"		64	.250		33.50	10.55		44.05	53.50
3140	6-1/2"		60	.267		37	11.25		48.25	58
3150	7"		58	.276		37	11.65		48.65	58.50
3160	7-1/2"		56	.286		41	12.05		53.05	63.50
3170	8"		54	.296		41	12.50		53.50	64.50

22 07 19 – Plumbing Piping Insulation

22 07 19.40 Pipe Insulation Protective Jacketing, Aluminum	Crew	Daily Output	Labor-Hours	Unit	Material	2011 Bare Costs Labor	Equipment	Total	Total Incl O&P
3180	8-1/2"	Q-14	52	.308	Ea.	42	13		55	66
3190	9"		50	.320		42	13.50		55.50	67
3200	9-1/2"		49	.327		31	13.80		44.80	56
3210	10"		48	.333		31	14.05		45.05	56.50
3220	10-1/2"		47	.340		33	14.35		47.35	58.50
3230	11"		46	.348		33	14.70		47.70	59
3240	11-1/2"		45	.356		35	15		50	61.50
3250	12"		44	.364		35	15.35		50.35	62
3260	13"		43	.372		37	15.70		52.70	65
3270	14"		42	.381		38.50	16.10		54.60	67.50
3280	15"		41	.390		41.50	16.45		57.95	71.50
3290	16"		40	.400		43	16.90		59.90	73
3300	17"		39	.410		49	17.30		66.30	81
3310	18"		38	.421		51	17.75		68.75	83.50
3320	19"		37	.432		56.50	18.25		74.75	90
3330	20"		36	.444		57.50	18.75		76.25	92.50
3340	22"		35	.457		73.50	19.30		92.80	111
3350	23"		34	.471		75.50	19.85		95.35	114
3360	24"		31	.516		77.50	22		99.50	119

22 07 19.50 Pipe Insulation Protective Jacketing, St. Stl.

		Crew	Daily Output	Labor-Hours	Unit	Material	2011 Bare Costs Labor	Equipment	Total	Total Incl O&P
0010	**PIPE INSULATION PROTECTIVE JACKETING, STAINLESS STEEL**									
0100	Metal roll jacketing									
0120	Type 304 with moisture barrier									
0140	Smooth, based on OD of insulation, .010" thick									
0260	2-1/2" ID	Q-14	250	.064	L.F.	2.30	2.70		5	6.70
0270	2-3/4" ID		245	.065		2.50	2.76		5.26	7
0280	3" ID		240	.067		2.70	2.81		5.51	7.30
0290	3-1/4" ID		235	.068		2.91	2.87		5.78	7.65
0300	3-1/2" ID		230	.070		3.11	2.94		6.05	7.95
0310	3-3/4" ID		225	.071		3.32	3		6.32	8.30
0320	4" ID		220	.073		3.52	3.07		6.59	8.60
0330	4-1/4" ID		215	.074		3.72	3.14		6.86	8.95
0340	4-1/2" ID		210	.076		3.93	3.22		7.15	9.30
0350	5" ID		200	.080		4.34	3.38		7.72	9.95
0360	5-1/2" ID		190	.084		4.74	3.55		8.29	10.70
0370	6" ID		180	.089		5.15	3.75		8.90	11.45
0380	6-1/2" ID		175	.091		5.55	3.86		9.41	12.05
0390	7" ID		170	.094		5.95	3.97		9.92	12.70
0400	7-1/2" ID		164	.098		6.35	4.12		10.47	13.35
0410	8" ID		161	.099		6.80	4.19		10.99	13.95
0420	8-1/2" ID		158	.101		7.20	4.27		11.47	14.50
0430	9" ID		155	.103		7.60	4.36		11.96	15.10
0440	9-1/2" ID		152	.105		8	4.44		12.44	15.65
0450	10" ID		149	.107		8.40	4.53		12.93	16.25
0460	10-1/2" ID		146	.110		8.80	4.62		13.42	16.85
0470	11" ID		143	.112		9.25	4.72		13.97	17.45
0480	12" ID		137	.117		10.05	4.93		14.98	18.65
0490	13" ID		132	.121		10.85	5.10		15.95	19.85
0500	14" ID		130	.123		11.65	5.20		16.85	21
0700	For smooth .016" thick, add					45%	33%			
1000	Stainless steel, Type 316, fitting covers									
1010	Size is based on OD of insulation									

22 07 19.50 Pipe Insulation Protective Jacketing, St. Stl.	Crew	Daily Output	Labor-Hours	Unit	Material	2011 Bare Costs Labor	Equipment	Total	Total Incl O&P
1020	90° LR Elbow, 2 piece									
1100	1-1/2"	Q-14	126	.127	Ea.	19.55	5.35		24.90	30
1110	2-3/4"		123	.130		12.50	5.50		18	22.50
1120	3"		120	.133		13.05	5.65		18.70	23
1130	3-1/4"		117	.137		13.05	5.75		18.80	23.50
1140	3-1/2"		115	.139		13.75	5.85		19.60	24
1150	3-3/4"		113	.142		14.50	6		20.50	25
1160	4"		110	.145		15.95	6.15		22.10	27
1170	4-1/4"		108	.148		21	6.25		27.25	33
1180	4-1/2"		106	.151		21	6.35		27.35	33.50
1190	5"		102	.157		21.50	6.60		28.10	34.50
1200	5-1/2"		97	.165		32.50	6.95		39.45	46.50
1210	6"		92	.174		35.50	7.35		42.85	50.50
1220	6-1/2"		87	.184		49	7.75		56.75	65.50
1230	7"		85	.188		49	7.95		56.95	66
1240	7-1/2"		83	.193		57	8.15		65.15	75
1250	8"		80	.200		57	8.45		65.45	75.50
1260	8-1/2"		80	.200		59	8.45		67.45	78
1270	9"		78	.205		89.50	8.65		98.15	112
1280	9-1/2"		77	.208		88	8.75		96.75	111
1290	10"		76	.211		88	8.90		96.90	111
1300	10-1/2"		75	.213		105	9		114	130
1310	11"		74	.216		101	9.10		110.10	125
1320	12"		71	.225		114	9.50		123.50	141
1330	13"		68	.235		158	9.95		167.95	189
1340	14"		66	.242		159	10.25		169.25	191
2000	45° Elbow, 2 piece									
2010	2-1/2"	Q-14	126	.127	Ea.	10.40	5.35		15.75	19.75
2020	2-3/4"		123	.130		10.40	5.50		15.90	19.95
2030	3"		120	.133		11.20	5.65		16.85	21
2040	3-1/4"		117	.137		11.20	5.75		16.95	21.50
2050	3-1/2"		115	.139		11.35	5.85		17.20	21.50
2060	3-3/4"		113	.142		11.35	6		17.35	22
2070	4"		110	.145		14.60	6.15		20.75	25.50
2080	4-1/4"		108	.148		20.50	6.25		26.75	32
2090	4-1/2"		106	.151		20.50	6.35		26.85	32.50
2100	4-3/4"		104	.154		20.50	6.50		27	32.50
2110	5"		102	.157		21	6.60		27.60	33.50
2120	5-1/2"		97	.165		21	6.95		27.95	34
2130	6"		92	.174		24.50	7.35		31.85	38.50
2140	6-1/2"		87	.184		24.50	7.75		32.25	39
2150	7"		85	.188		42	7.95		49.95	58.50
2160	7-1/2"		83	.193		42.50	8.15		50.65	59.50
2170	8"		82	.195		42.50	8.25		50.75	60
2180	8-1/2"		80	.200		50	8.45		58.45	68
2190	9"		78	.205		50	8.65		58.65	68.50
2200	9-1/2"		77	.208		58.50	8.75		67.25	78
2210	10"		76	.211		58.50	8.90		67.40	78.50
2220	10-1/2"		75	.213		70.50	9		79.50	91.50
2230	11"		74	.216		70.50	9.10		79.60	91.50
2240	12"		71	.225		76.50	9.50		86	98.50
2250	13"		68	.235		90	9.95		99.95	114

22 11 13 – Facility Water Distribution Piping

22 11 13.14 Pipe, Brass

	Crew	Daily Output	Labor-Hours	Unit	Material	Labor	Equipment	Total	Total Incl O&P
0010 **PIPE, BRASS**, Plain end									
0900 Field threaded, coupling & clevis hanger assembly 10' O.C.									
0920 Regular weight									
1120 1/2" diameter	1 Plum	48	.167	L.F.	6.50	8.95		15.45	20.50
1140 3/4" diameter		46	.174		8.60	9.30		17.90	23.50
1160 1" diameter	↓	43	.186		12.50	9.95		22.45	28.50
1180 1-1/4" diameter	Q-1	72	.222		18.90	10.70		29.60	37
1200 1-1/2" diameter		65	.246		22.50	11.90		34.40	42.50
1220 2" diameter		53	.302		31	14.55		45.55	56
1240 2-1/2" diameter		41	.390		50.50	18.85		69.35	83.50
1260 3" diameter	↓	31	.516		66.50	25		91.50	110
1300 4" diameter	Q-2	37	.649	↓	116	32.50		148.50	176
1930 To delete coupling & hanger, subtract									
1940 1/2" diam.					14%	46%			
1950 3/4" diam. to 1-1/2" diam.					8%	47%			
1960 2" diam. to 4" diam.					10%	37%			

22 11 13.16 Pipe Fittings, Brass

	Crew	Daily Output	Labor-Hours	Unit	Material	Labor	Equipment	Total	Total Incl O&P
0010 **PIPE FITTINGS, BRASS**, Rough bronze, threaded.									
1000 Standard wt., 90° Elbow									
1040 1/8"	1 Plum	13	.615	Ea.	15.25	33		48.25	66
1060 1/4"		13	.615		15.25	33		48.25	66
1080 3/8"		13	.615		15.25	33		48.25	66
1100 1/2"		12	.667		15.25	35.50		50.75	70.50
1120 3/4"		11	.727		20.50	39		59.50	80.50
1140 1"	↓	10	.800		33	43		76	101
1160 1-1/4"	Q-1	17	.941		53.50	45.50		99	127
1180 1-1/2"		16	1		66.50	48.50		115	145
1200 2"		14	1.143		107	55		162	201
1220 2-1/2"		11	1.455		259	70		329	390
1240 3"	↓	8	2		395	96.50		491.50	580
1260 4"	Q-2	11	2.182		800	109		909	1,050
1280 5"		8	3		2,200	150		2,350	2,650
1300 6"	↓	7	3.429		3,275	172		3,447	3,850
1500 45° Elbow, 1/8"	1 Plum	13	.615		18.75	33		51.75	69.50
1540 1/4"		13	.615		18.75	33		51.75	69.50
1560 3/8"		13	.615		18.75	33		51.75	69.50
1580 1/2"		12	.667		18.75	35.50		54.25	74
1600 3/4"		11	.727		26.50	39		65.50	87.50
1620 1"	↓	10	.800		45.50	43		88.50	114
1640 1-1/4"	Q-1	17	.941		72	45.50		117.50	148
1660 1-1/2"		16	1		90.50	48.50		139	172
1680 2"		14	1.143		147	55		202	244
1700 2-1/2"		11	1.455		280	70		350	415
1720 3"	↓	8	2		430	96.50		526.50	620
1740 4"	Q-2	11	2.182		960	109		1,069	1,225
1760 5"		8	3		1,800	150		1,950	2,200
1780 6"	↓	7	3.429		2,500	172		2,672	3,000
2000 Tee, 1/8"	1 Plum	9	.889		17.90	47.50		65.40	90.50
2040 1/4"		9	.889		17.90	47.50		65.40	90.50
2060 3/8"		9	.889		17.90	47.50		65.40	90.50
2080 1/2"		8	1		17.90	53.50		71.40	99.50
2100 3/4"	↓	7	1.143		25.50	61.50		87	120

22 11 Facility Water Distribution

22 11 13 – Facility Water Distribution Piping

22 11 13.16 Pipe Fittings, Brass

		Crew	Daily Output	Labor-Hours	Unit	Material	2011 Bare Costs Labor	Equipment	Total	Total Incl O&P
2120	1"	1 Plum	6	1.333	Ea.	46	71.50		117.50	158
2140	1-1/4"	Q-1	10	1.600		79	77		156	202
2160	1-1/2"		9	1.778		89	86		175	226
2180	2"		8	2		148	96.50		244.50	305
2200	2-1/2"		7	2.286		350	110		460	550
2220	3"		5	3.200		540	154		694	820
2240	4"	Q-2	7	3.429		1,325	172		1,497	1,725
2260	5"		5	4.800		2,800	240		3,040	3,425
2280	6"		4	6		4,350	300		4,650	5,250
2500	Coupling, 1/8"	1 Plum	26	.308		12.75	16.50		29.25	38.50
2540	1/4"		22	.364		12.75	19.50		32.25	43
2560	3/8"		18	.444		12.75	24		36.75	49.50
2580	1/2"		15	.533		12.75	28.50		41.25	56.50
2600	3/4"		14	.571		17.90	30.50		48.40	65
2620	1"		13	.615		30.50	33		63.50	82.50
2640	1-1/4"	Q-1	22	.727		52.50	35		87.50	110
2660	1-1/2"		20	.800		66.50	38.50		105	131
2680	2"		18	.889		110	43		153	185
2700	2-1/2"		14	1.143		218	55		273	320
2720	3"		10	1.600		259	77		336	400
2740	4"	Q-2	12	2		535	100		635	735
2760	5"		10	2.400		1,000	120		1,120	1,275
2780	6"		9	2.667		1,425	133		1,558	1,750
3000	Union, 125 lb									
3020	1/8"	1 Plum	12	.667	Ea.	43.50	35.50		79	101
3040	1/4"		12	.667		43.50	35.50		79	101
3060	3/8"		12	.667		43.50	35.50		79	101
3080	1/2"		11	.727		43.50	39		82.50	106
3100	3/4"		10	.800		59.50	43		102.50	130
3120	1"		9	.889		90	47.50		137.50	170
3140	1-1/4"	Q-1	16	1		130	48.50		178.50	215
3160	1-1/2"		15	1.067		155	51.50		206.50	247
3180	2"		13	1.231		209	59.50		268.50	320
3200	2-1/2"		10	1.600		570	77		647	745
3220	3"		7	2.286		885	110		995	1,150
3240	4"	Q-2	10	2.400		2,775	120		2,895	3,225
3320	For 250 lb. (navy pattern), add					100%				

22 11 13.23 Pipe/Tube, Copper

		Crew	Daily Output	Labor-Hours	Unit	Material	2011 Bare Costs Labor	Equipment	Total	Total Incl O&P
0010	**PIPE/TUBE, COPPER**, Solder joints R221113-50									
0100	Solder									
0120	Solder, lead free, roll				Lb.	15.35			15.35	16.85
1000	Type K tubing, couplings & clevis hanger assemblies 10' O.C.									
1100	1/4" diameter	1 Plum	84	.095	L.F.	4.41	5.10		9.51	12.45
1120	3/8" diameter		82	.098		4.55	5.25		9.80	12.80
1140	1/2" diameter		78	.103		5.20	5.50		10.70	13.90
1160	5/8" diameter		77	.104		6.50	5.55		12.05	15.45
1180	3/4" diameter		74	.108		8.90	5.80		14.70	18.40
1200	1" diameter		66	.121		11.80	6.50		18.30	22.50
1220	1-1/4" diameter		56	.143		14.70	7.65		22.35	27.50
1240	1-1/2" diameter		50	.160		18.90	8.60		27.50	34
1260	2" diameter		40	.200		28.50	10.70		39.20	47.50
1280	2-1/2" diameter	Q-1	60	.267		43.50	12.85		56.35	67

22 11 Facility Water Distribution

22 11 13 – Facility Water Distribution Piping

22 11 13.23 Pipe/Tube, Copper		Crew	Daily Output	Labor-Hours	Unit	Material	2011 Bare Costs Labor	2011 Bare Costs Equipment	Total	Total Incl O&P
1300	3" diameter	Q-1	54	.296	L.F.	60.50	14.30		74.80	88
1320	3-1/2" diameter		42	.381		83	18.40		101.40	119
1330	4" diameter		38	.421		103	20.50		123.50	144
1340	5" diameter		32	.500		229	24		253	288
1360	6" diameter	Q-2	38	.632		340	31.50		371.50	415
1380	8" diameter	"	34	.706		575	35.50		610.50	685
1390	For other than full hard temper, add					13%				
1440	For silver solder, add						15%			
1800	For medical clean, (oxygen class), add					12%				
1950	To delete cplgs. & hngrs., 1/4"-1" pipe, subtract					27%	60%			
1960	1-1/4" -3" pipe, subtract					14%	52%			
1970	3-1/2"-5" pipe, subtract					10%	60%			
1980	6"-8" pipe, subtract					19%	53%			
2000	Type L tubing, couplings & clevis hanger assemblies 10' O.C.									
2100	1/4" diameter	1 Plum	88	.091	L.F.	2.71	4.87		7.58	10.25
2120	3/8" diameter		84	.095		3.65	5.10		8.75	11.60
2140	1/2" diameter		81	.099		3.97	5.30		9.27	12.25
2160	5/8" diameter		79	.101		5.85	5.45		11.30	14.55
2180	3/4" diameter		76	.105		5.95	5.65		11.60	14.95
2200	1" diameter		68	.118		9	6.30		15.30	19.35
2220	1-1/4" diameter		58	.138		12.45	7.40		19.85	24.50
2240	1-1/2" diameter		52	.154		15.70	8.25		23.95	29.50
2260	2" diameter		42	.190		24	10.20		34.20	42
2280	2-1/2" diameter	Q-1	62	.258		37	12.45		49.45	59.50
2300	3" diameter		56	.286		50	13.80		63.80	75.50
2320	3-1/2" diameter		43	.372		69.50	17.95		87.45	104
2340	4" diameter		39	.410		85.50	19.80		105.30	124
2360	5" diameter		34	.471		183	22.50		205.50	235
2380	6" diameter	Q-2	40	.600		258	30		288	330
2400	8" diameter	"	36	.667		430	33.50		463.50	520
2410	For other than full hard temper, add					21%				
2590	For silver solder, add						15%			
2900	For medical clean, (oxygen class), add					12%				
2940	To delete cplgs. & hngrs., 1/4"-1" pipe, subtract					37%	63%			
2960	1-1/4"-3" pipe, subtract					12%	53%			
2970	3-1/2"-5" pipe, subtract					12%	63%			
2980	6"-8" pipe, subtract					24%	55%			
3000	Type M tubing, couplings & clevis hanger assemblies 10' O.C.									
3100	1/4" diameter	1 Plum	90	.089	L.F.	3.01	4.76		7.77	10.40
3120	3/8" diameter		87	.092		3.06	4.93		7.99	10.70
3140	1/2" diameter		84	.095		3.09	5.10		8.19	11
3160	5/8" diameter		81	.099		4.72	5.30		10.02	13.10
3180	3/4" diameter		78	.103		4.56	5.50		10.06	13.20
3200	1" diameter		70	.114		7.30	6.15		13.45	17.20
3220	1-1/4" diameter		60	.133		10.40	7.15		17.55	22
3240	1-1/2" diameter		54	.148		13.90	7.95		21.85	27
3260	2" diameter		44	.182		21.50	9.75		31.25	38
3280	2-1/2" diameter	Q-1	64	.250		32.50	12.05		44.55	53.50
3300	3" diameter		58	.276		43.50	13.30		56.80	68
3320	3-1/2" diameter		45	.356		63	17.15		80.15	94.50
3340	4" diameter		40	.400		78.50	19.30		97.80	116
3360	5" diameter		36	.444		183	21.50		204.50	233
3370	6" diameter	Q-2	42	.571		258	28.50		286.50	325

22 11 13.23 Pipe/Tube, Copper

		Crew	Daily Output	Labor-Hours	Unit	Material	2011 Bare Costs Labor	Equipment	Total	Total Incl O&P
3380	8" diameter	Q-2	38	.632	L.F.	430	31.50		461.50	515
3440	For silver solder, add						15%			
3960	To delete cplgs. & hngrs., 1/4"-1" pipe, subtract					35%	65%			
3970	1-1/4"-3" pipe, subtract					19%	56%			
3980	3-1/2"-5" pipe, subtract					13%	65%			
3990	6"-8" pipe, subtract					28%	58%			
4000	Type DWV tubing, couplings & clevis hanger assemblies 10' O.C.									
4100	1-1/4" diameter	1 Plum	60	.133	L.F.	11.05	7.15		18.20	23
4120	1-1/2" diameter		54	.148		13.65	7.95		21.60	27
4140	2" diameter	↓	44	.182		18.30	9.75		28.05	34.50
4160	3" diameter	Q-1	58	.276		33.50	13.30		46.80	57
4180	4" diameter		40	.400		59.50	19.30		78.80	94.50
4200	5" diameter	↓	36	.444		167	21.50		188.50	215
4220	6" diameter	Q-2	42	.571		239	28.50		267.50	305
4240	8" diameter	"	38	.632	↓	575	31.50		606.50	675
4730	To delete cplgs. & hngrs., 1-1/4"-2" pipe, subtract					16%	53%			
4740	3"-4" pipe, subtract					13%	60%			
4750	5"-8" pipe, subtract					23%	58%			
5200	ACR tubing, type L, hard temper, cleaned and									
5220	capped, no couplings or hangers									
5240	3/8" OD				L.F.	1.89			1.89	2.08
5250	1/2" OD					2.87			2.87	3.16
5260	5/8" OD					3.49			3.49	3.84
5270	3/4" OD					4.94			4.94	5.45
5280	7/8" OD					5.50			5.50	6.05
5290	1-1/8" OD					7.95			7.95	8.75
5300	1-3/8" OD					10.65			10.65	11.75
5310	1-5/8" OD					13.65			13.65	15.05
5320	2-1/8" OD					21.50			21.50	23.50
5330	2-5/8" OD					32			32	35
5340	3-1/8" OD					43			43	47
5350	3-5/8" OD					56			56	61.50
5360	4-1/8" OD				↓	71.50			71.50	78.50
5380	ACR tubing, type L, hard, cleaned and capped									
5381	No couplings or hangers									
5384	3/8"	1 Stpi	160	.050	L.F.	1.89	2.71		4.60	6.10
5385	1/2"		160	.050		2.87	2.71		5.58	7.20
5386	5/8"		160	.050		3.49	2.71		6.20	7.90
5387	3/4"		130	.062		4.94	3.33		8.27	10.40
5388	7/8"		130	.062		5.50	3.33		8.83	11
5389	1-1/8"		115	.070		7.95	3.76		11.71	14.35
5390	1-3/8"		100	.080		10.65	4.33		14.98	18.20
5391	1-5/8"		90	.089		13.65	4.81		18.46	22.50
5392	2-1/8"	↓	80	.100		21.50	5.40		26.90	31.50
5393	2-5/8"	Q-5	125	.128		32	6.25		38.25	44.50
5394	3-1/8"		105	.152		43	7.40		50.40	58
5395	4-1/8"	↓	95	.168	↓	71.50	8.20		79.70	91
5800	Refrigeration tubing, dryseal, 50' coils									
5840	1/8" OD				Coil	38.50			38.50	42.50
5850	3/16" OD					45			45	49.50
5860	1/4" OD					52			52	57
5870	5/16" OD					69.50			69.50	76.50
5880	3/8" OD				↓	74			74	81.50

22 11 Facility Water Distribution

22 11 13 – Facility Water Distribution Piping

22 11 13.23 Pipe/Tube, Copper

		Crew	Daily Output	Labor-Hours	Unit	Material	2011 Bare Costs Labor	Equipment	Total	Total Incl O&P
5890	1/2" OD				Coil	104			104	115
5900	5/8" OD					142			142	156
5910	3/4" OD					170			170	187
5920	7/8" OD					254			254	279
5930	1-1/8" OD					365			365	400
5940	1-3/8" OD					575			575	630
5950	1-5/8" OD					725			725	800
9400	Sub assemblies used in assembly systems									
9410	Chilled water unit, coil connections per unit under 10 ton	Q-5	.80	20	System	1,150	975		2,125	2,725
9420	Chilled water unit, coil connections per unit 10 ton and up		1	16		1,750	780		2,530	3,100
9430	Chilled water dist. piping per ton, less than 61 ton systems		26	.615		16.15	30		46.15	62.50
9440	Chilled water dist. piping per ton, 61 through 120 ton systems	Q-6	31	.774		31	39		70	93
9450	Chilled water dist. piping/ton, 135 ton systems and up	Q-8	25.40	1.260		42	65	1.27	108.27	145
9510	Refrigerant piping/ton of cooling for remote condensers	Q-5	2	8		480	390		870	1,100
9520	Refrigerant piping per ton up to 10 ton w/remote condensing unit		2.40	6.667		203	325		528	710
9530	Refrigerant piping per ton, 20 ton w/remote condensing unit		2	8		305	390		695	915
9540	Refrigerant piping per ton, 40 ton w/remote condensing unit		1.90	8.421		405	410		815	1,050
9550	Refrigerant piping per ton, 75-80 ton w/remote condensing unit	Q-6	2.40	10		605	505		1,110	1,425
9560	Refrigerant piping per ton, 100 ton w/remote condensing unit	"	2.20	10.909		800	550		1,350	1,700

22 11 13.25 Pipe/Tube Fittings, Copper

		Crew	Daily Output	Labor-Hours	Unit	Material	2011 Bare Costs Labor	Equipment	Total	Total Incl O&P
0010	**PIPE/TUBE FITTINGS, COPPER**, Wrought unless otherwise noted									
0040	Solder joints, copper x copper									
0070	90° elbow, 1/4"	1 Plum	22	.364	Ea.	6.40	19.50		25.90	36
0090	3/8"		22	.364		6.10	19.50		25.60	35.50
0100	1/2"		20	.400		2.03	21.50		23.53	34
0110	5/8"		19	.421		6.80	22.50		29.30	41
0120	3/4"		19	.421		4.19	22.50		26.69	38
0130	1"		16	.500		11.20	27		38.20	52.50
0140	1-1/4"		15	.533		16.90	28.50		45.40	61
0150	1-1/2"		13	.615		26.50	33		59.50	78
0160	2"		11	.727		48	39		87	111
0170	2-1/2"	Q-1	13	1.231		96	59.50		155.50	195
0180	3"		11	1.455		129	70		199	246
0190	3-1/2"		10	1.600		450	77		527	610
0200	4"		9	1.778		330	86		416	490
0210	5"		6	2.667		1,375	129		1,504	1,725
0220	6"	Q-2	9	2.667		1,850	133		1,983	2,225
0230	8"	"	8	3		6,800	150		6,950	7,700
0250	45° elbow, 1/4"	1 Plum	22	.364		11.40	19.50		30.90	41.50
0270	3/8"		22	.364		9.75	19.50		29.25	40
0280	1/2"		20	.400		3.73	21.50		25.23	36
0290	5/8"		19	.421		17.70	22.50		40.20	53
0300	3/4"		19	.421		6.50	22.50		29	40.50
0310	1"		16	.500		16.40	27		43.40	58
0320	1-1/4"		15	.533		22	28.50		50.50	67
0330	1-1/2"		13	.615		26.50	33		59.50	78.50
0340	2"		11	.727		44.50	39		83.50	107
0350	2-1/2"	Q-1	13	1.231		95	59.50		154.50	194
0360	3"		13	1.231		141	59.50		200.50	244
0370	3-1/2"		10	1.600		237	77		314	375
0380	4"		9	1.778		300	86		386	460
0390	5"		6	2.667		1,175	129		1,304	1,475

22 11 13.25 Pipe/Tube Fittings, Copper		Crew	Daily Output	Labor-Hours	Unit	Material	2011 Bare Costs Labor	Equipment	Total	Total Incl O&P
0400	6"	Q-2	9	2.667	Ea.	1,825	133		1,958	2,225
0410	8"	"	8	3		6,225	150		6,375	7,075
0450	Tee, 1/4"	1 Plum	14	.571		12.80	30.50		43.30	59.50
0470	3/8"		14	.571		10.25	30.50		40.75	57
0480	1/2"		13	.615		3.47	33		36.47	53
0490	5/8"		12	.667		22.50	35.50		58	78
0500	3/4"		12	.667		8	35.50		43.50	62.50
0510	1"		10	.800		26	43		69	92.50
0520	1-1/4"		9	.889		35.50	47.50		83	110
0530	1-1/2"		8	1		53	53.50		106.50	139
0540	2"		7	1.143		85	61.50		146.50	185
0550	2-1/2"	Q-1	8	2		171	96.50		267.50	335
0560	3"		7	2.286		262	110		372	455
0570	3-1/2"		6	2.667		760	129		889	1,025
0580	4"		5	3.200		635	154		789	925
0590	5"		4	4		2,075	193		2,268	2,575
0600	6"	Q-2	6	4		2,825	200		3,025	3,425
0610	8"	"	5	4.800		10,900	240		11,140	12,400
0612	Tee, reducing on the outlet, 1/4"	1 Plum	15	.533		21.50	28.50		50	66.50
0613	3/8"		15	.533		20.50	28.50		49	65
0614	1/2"		14	.571		17.90	30.50		48.40	65
0615	5/8"		13	.615		34.50	33		67.50	87
0616	3/4"		12	.667		9.40	35.50		44.90	64
0617	1"		11	.727		29.50	39		68.50	90
0618	1-1/4"		10	.800		38	43		81	106
0619	1-1/2"		9	.889		40.50	47.50		88	116
0620	2"		8	1		59	53.50		112.50	145
0621	2-1/2"	Q-1	9	1.778		184	86		270	330
0622	3"		8	2		220	96.50		316.50	385
0623	4"		6	2.667		450	129		579	685
0624	5"		5	3.200		2,075	154		2,229	2,500
0625	6"	Q-2	7	3.429		2,825	172		2,997	3,375
0626	8"	"	6	4		10,900	200		11,100	12,300
0630	Tee, reducing on the run, 1/4"	1 Plum	15	.533		27	28.50		55.50	72.50
0631	3/8"		15	.533		36.50	28.50		65	82.50
0632	1/2"		14	.571		24.50	30.50		55	72.50
0633	5/8"		13	.615		40	33		73	93
0634	3/4"		12	.667		9.85	35.50		45.35	64.50
0635	1"		11	.727		31	39		70	92
0636	1-1/4"		10	.800		50	43		93	119
0637	1-1/2"		9	.889		81.50	47.50		129	161
0638	2"		8	1		110	53.50		163.50	202
0639	2-1/2"	Q-1	9	1.778		269	86		355	425
0640	3"		8	2		380	96.50		476.50	565
0641	4"		6	2.667		845	129		974	1,125
0642	5"		5	3.200		1,975	154		2,129	2,400
0643	6"	Q-2	7	3.429		3,000	172		3,172	3,525
0644	8"	"	6	4		10,900	200		11,100	12,300
0650	Coupling, 1/4"	1 Plum	24	.333		1.48	17.85		19.33	28
0670	3/8"		24	.333		1.97	17.85		19.82	28.50
0680	1/2"		22	.364		1.54	19.50		21.04	30.50
0690	5/8"		21	.381		4.64	20.50		25.14	35.50
0700	3/4"		21	.381		3.09	20.50		23.59	34

22 11 13.25 Pipe/Tube Fittings, Copper		Crew	Daily Output	Labor-Hours	Unit	Material	2011 Bare Costs Labor	Equipment	Total	Total Incl O&P
0710	1"	1 Plum	18	.444	Ea.	6.15	24		30.15	42.50
0715	1-1/4"		17	.471		10.95	25		35.95	49.50
0716	1-1/2"		15	.533		14.45	28.50		42.95	58.50
0718	2"		13	.615		24	33		57	75.50
0721	2-1/2"	Q-1	15	1.067		51.50	51.50		103	134
0722	3"		13	1.231		77	59.50		136.50	174
0724	3-1/2"		8	2		148	96.50		244.50	305
0726	4"		7	2.286		162	110		272	345
0728	5"		6	2.667		400	129		529	630
0731	6"	Q-2	8	3		660	150		810	950
0732	8"	"	7	3.429		2,150	172		2,322	2,600
0741	Coupling, reducing, concentric									
0743	1/2"	1 Plum	23	.348	Ea.	4.02	18.65		22.67	32.50
0745	3/4"		21.50	.372		7.20	19.95		27.15	38
0747	1"		19.50	.410		11	22		33	45
0748	1-1/4"		18	.444		15.55	24		39.55	52.50
0749	1-1/2"		16	.500		20.50	27		47.50	62.50
0751	2"		14	.571		42.50	30.50		73	92.50
0752	2-1/2"		13	.615		89	33		122	147
0753	3"	Q-1	14	1.143		107	55		162	200
0755	4"	"	8	2		225	96.50		321.50	390
0757	5"	Q-2	7.50	3.200		1,200	160		1,360	1,550
0759	6"		7	3.429		1,900	172		2,072	2,325
0761	8"		6.50	3.692		3,925	185		4,110	4,600
0771	Cap, sweat									
0773	1/2"	1 Plum	40	.200	Ea.	1.48	10.70		12.18	17.65
0775	3/4"		38	.211		2.75	11.30		14.05	19.90
0777	1"		32	.250		6.50	13.40		19.90	27
0778	1-1/4"		29	.276		8.75	14.80		23.55	31.50
0779	1-1/2"		26	.308		12.65	16.50		29.15	38.50
0781	2"		22	.364		23.50	19.50		43	55
0791	Flange, sweat									
0793	3"	Q-1	22	.727	Ea.	286	35		321	370
0795	4"		18	.889		400	43		443	505
0797	5"		12	1.333		745	64.50		809.50	915
0799	6"	Q-2	18	1.333		780	66.50		846.50	960
0801	8"	"	16	1.500		2,150	75		2,225	2,450
0850	Unions, 1/4"	1 Plum	21	.381		43.50	20.50		64	78.50
0870	3/8"		21	.381		44	20.50		64.50	78.50
0880	1/2"		19	.421		23.50	22.50		46	59
0890	5/8"		18	.444		78.50	24		102.50	122
0900	3/4"		18	.444		29.50	24		53.50	67.50
0910	1"		15	.533		50.50	28.50		79	98
0920	1-1/4"		14	.571		87.50	30.50		118	142
0930	1-1/2"		12	.667		115	35.50		150.50	181
0940	2"		10	.800		196	43		239	280
0950	2-1/2"	Q-1	12	1.333		430	64.50		494.50	565
0960	3"	"	10	1.600		1,100	77		1,177	1,350
0980	Adapter, copper x male IPS, 1/4"	1 Plum	20	.400		21	21.50		42.50	55.50
0990	3/8"		20	.400		10.50	21.50		32	43.50
1000	1/2"		18	.444		4.24	24		28.24	40
1010	3/4"		17	.471		7.15	25		32.15	45.50
1020	1"		15	.533		18.45	28.50		46.95	63

22 11 13 – Facility Water Distribution Piping

22 11 13.25 Pipe/Tube Fittings, Copper		Crew	Daily Output	Labor-Hours	Unit	Material	2011 Bare Costs Labor	Equipment	Total	Total Incl O&P
1030	1-1/4"	1 Plum	13	.615	Ea.	27.50	33		60.50	79
1040	1-1/2"		12	.667		31.50	35.50		67	88
1050	2"	↓	11	.727		53	39		92	117
1060	2-1/2"	Q-1	10.50	1.524		203	73.50		276.50	335
1070	3"		10	1.600		269	77		346	410
1080	3-1/2"		9	1.778		330	86		416	495
1090	4"		8	2		365	96.50		461.50	545
1200	5"	↓	6	2.667		1,725	129		1,854	2,075
1210	6"	Q-2	8.50	2.824	↓	2,100	141		2,241	2,500
1214	Adapter, copper x female IPS									
1216	1/2"	1 Plum	18	.444	Ea.	6.75	24		30.75	43
1218	3/4"		17	.471		9.25	25		34.25	47.50
1220	1"		15	.533		21.50	28.50		50	66
1221	1-1/4"		13	.615		31.50	33		64.50	84
1222	1-1/2"		12	.667		49	35.50		84.50	108
1224	2"		11	.727		66.50	39		105.50	132
1250	Cross, 1/2"		10	.800		31.50	43		74.50	99
1260	3/4"		9.50	.842		61.50	45		106.50	135
1270	1"		8	1		104	53.50		157.50	195
1280	1-1/4"		7.50	1.067		150	57		207	251
1290	1-1/2"		6.50	1.231		214	66		280	335
1300	2"	↓	5.50	1.455		405	78		483	560
1310	2-1/2"	Q-1	6.50	2.462		930	119		1,049	1,200
1320	3"	"	5.50	2.909	↓	1,175	140		1,315	1,500
1500	Tee fitting, mechanically formed,(Type 1, 'branch sizes up to 2 in.')									
1520	1/2" run size, 3/8" to 1/2" branch size	1 Plum	80	.100	Ea.		5.35		5.35	8
1530	3/4" run size, 3/8" to 3/4" branch size		60	.133			7.15		7.15	10.65
1540	1" run size, 3/8" to 1" branch size		54	.148			7.95		7.95	11.85
1550	1-1/4" run size, 3/8" to 1-1/4" branch size		48	.167			8.95		8.95	13.35
1560	1-1/2" run size, 3/8" to 1-1/2" branch size		40	.200			10.70		10.70	16
1570	2" run size, 3/8" to 2" branch size		35	.229			12.25		12.25	18.30
1580	2-1/2" run size, 1/2" to 2" branch size		32	.250			13.40		13.40	20
1590	3" run size, 1" to 2" branch size		26	.308			16.50		16.50	24.50
1600	4" run size, 1" to 2" branch size	↓	24	.333			17.85		17.85	26.50
1640	Tee fitting, mechanically formed, (Type 2, branches 2-1/2" thru 4")									
1650	2-1/2" run size, 2-1/2" branch size	1 Plum	12.50	.640	Ea.		34.50		34.50	51
1660	3" run size, 2-1/2" to 3" branch size		12	.667			35.50		35.50	53.50
1670	3-1/2" run size, 2-1/2" to 3-1/2" branch size		11	.727			39		39	58
1680	4" run size, 2-1/2" to 4" branch size		10.50	.762			41		41	61
1698	5" run size, 2" to 4" branch size		9.50	.842			45		45	67.50
1700	6" run size, 2" to 4" branch size		8.50	.941			50.50		50.50	75.50
1710	8" run size, 2" to 4" branch size	↓	7	1.143	↓		61.50		61.50	91.50
1800	ACR fittings, OD size									
1802	Tee, straight									
1808	5/8"	1 Stpi	12	.667	Ea.	3.47	36		39.47	58
1810	3/4"		12	.667		22.50	36		58.50	78.50
1812	7/8"		10	.800		8.35	43.50		51.85	73.50
1813	1"		10	.800		60.50	43.50		104	131
1814	1-1/8"		10	.800		26	43.50		69.50	93
1816	1-3/8"		9	.889		35.50	48		83.50	111
1818	1-5/8"		8	1		54.50	54		108.50	141
1820	2-1/8"	↓	7	1.143		85	62		147	186
1822	2-5/8"	Q-5	8	2		171	97.50		268.50	335

22 11 13 – Facility Water Distribution Piping

22 11 13.25 Pipe/Tube Fittings, Copper		Crew	Daily Output	Labor-Hours	Unit	Material	2011 Bare Costs Labor	Equipment	Total	Total Incl O&P
1824	3-1/8"	Q-5	7	2.286	Ea.	262	111		373	455
1826	4-1/8"	↓	5	3.200	↓	635	156		791	930
1830	90° elbow									
1836	5/8"	1 Stpi	19	.421	Ea.	6.80	23		29.80	41.50
1838	3/4"	19	19	.421		12.35	23		35.35	47.50
1840	7/8"		16	.500		12.35	27		39.35	54
1842	1-1/8"		16	.500		16.45	27		43.45	58.50
1844	1-3/8"		15	.533		16.90	29		45.90	61.50
1846	1-5/8"		13	.615		26.50	33.50		60	78.50
1848	2-1/8"		11	.727		48	39.50		87.50	112
1850	2-5/8"	Q-5	13	1.231		96	60		156	196
1852	3-1/8"	↓	11	1.455		129	71		200	247
1854	4-1/8"	↓	9	1.778	↓	330	86.50		416.50	490
1860	Coupling									
1866	5/8"	1 Stpi	21	.381	Ea.	1.54	20.50		22.04	32.50
1868	3/4"		21	.381		4.41	20.50		24.91	36
1870	7/8"		18	.444		3.12	24		27.12	39.50
1871	1"		18	.444		10.50	24		34.50	47.50
1872	1-1/8"		18	.444		6.25	24		30.25	43
1874	1-3/8"		17	.471		10.95	25.50		36.45	50
1876	1-5/8"		15	.533		14.45	29		43.45	59
1878	2-1/8"	↓	13	.615		24	33.50		57.50	76
1880	2-5/8"	Q-5	15	1.067		54.50	52		106.50	138
1882	3-1/8"	↓	13	1.231		79	60		139	176
1884	4-1/8"	↓	7	2.286	↓	179	111		290	365
2000	DWV, solder joints, copper x copper									
2030	90° Elbow, 1-1/4"	1 Plum	13	.615	Ea.	19.65	33		52.65	70.50
2050	1-1/2"		12	.667		29.50	35.50		65	86
2070	2"	↓	10	.800		38.50	43		81.50	107
2090	3"	Q-1	10	1.600		95.50	77		172.50	220
2100	4"	"	9	1.778		545	86		631	730
2150	45° Elbow, 1-1/4"	1 Plum	13	.615		18.10	33		51.10	69
2170	1-1/2"		12	.667		14.95	35.50		50.45	70
2180	2"	↓	10	.800		34.50	43		77.50	102
2190	3"	Q-1	10	1.600		73	77		150	196
2200	4"	"	9	1.778		350	86		436	515
2250	Tee, Sanitary, 1-1/4"	1 Plum	9	.889		39	47.50		86.50	114
2270	1-1/2"		8	1		48.50	53.50		102	133
2290	2"	↓	7	1.143		56.50	61.50		118	154
2310	3"	Q-1	7	2.286		229	110		339	415
2330	4"	"	6	2.667		585	129		714	830
2400	Coupling, 1-1/4"	1 Plum	14	.571		9.20	30.50		39.70	55.50
2420	1-1/2"		13	.615		11.40	33		44.40	61.50
2440	2"	↓	11	.727		15.85	39		54.85	75.50
2460	3"	Q-1	11	1.455		30.50	70		100.50	139
2480	4"	"	10	1.600	↓	98	77		175	223
2602	Traps, see Section 22 13 16.60									
3500	Compression joint fittings									
3510	As used for plumbing and oil burner work									
3520	Fitting price includes nuts and sleeves									
3540	Sleeve, 1/8"					Ea.	.19		.19	.21
3550	3/16"						.19		.19	.21
3560	1/4"				↓	.07			.07	.08

22 11 13.25 Pipe/Tube Fittings, Copper	Crew	Daily Output	Labor-Hours	Unit	Material	2011 Bare Costs Labor	Equipment	Total	Total Incl O&P
3570	5/16"				Ea.	.20			.20	.22
3580	3/8"					.39			.39	.43
3600	1/2"					.52			.52	.57
3620	Nut, 1/8"					.27			.27	.30
3630	3/16"					.29			.29	.32
3640	1/4"					.28			.28	.31
3650	5/16"					.42			.42	.46
3660	3/8"					.46			.46	.51
3670	1/2"					.81			.81	.89
3710	Union, 1/8"	1 Plum	26	.308		1.70	16.50		18.20	26.50
3720	3/16"		24	.333		1.63	17.85		19.48	28.50
3730	1/4"		24	.333		1.71	17.85		19.56	28.50
3740	5/16"		23	.348		2.19	18.65		20.84	30.50
3750	3/8"		22	.364		2.21	19.50		21.71	31.50
3760	1/2"		22	.364		3.40	19.50		22.90	32.50
3780	5/8"		21	.381		3.77	20.50		24.27	34.50
3820	Union tee, 1/8"		17	.471		6.50	25		31.50	44.50
3830	3/16"		16	.500		3.77	27		30.77	44
3840	1/4"		15	.533		3.40	28.50		31.90	46
3850	5/16"		15	.533		4.78	28.50		33.28	48
3860	3/8"		15	.533		4.86	28.50		33.36	48
3870	1/2"		15	.533		8.20	28.50		36.70	51.50
3910	Union elbow, 1/4"		24	.333		2.09	17.85		19.94	29
3920	5/16"		23	.348		3.69	18.65		22.34	32
3930	3/8"		22	.364		3.56	19.50		23.06	33
3940	1/2"		22	.364		5.80	19.50		25.30	35.50
3980	Female connector, 1/8"		26	.308		1.32	16.50		17.82	26
4000	3/16" x 1/8"		24	.333		1.68	17.85		19.53	28.50
4010	1/4" x 1/8"		24	.333		1.49	17.85		19.34	28
4020	1/4"		24	.333		2	17.85		19.85	28.50
4030	3/8" x 1/4"		22	.364		2	19.50		21.50	31
4040	1/2" x 3/8"		22	.364		3.19	19.50		22.69	32.50
4050	5/8" x 1/2"		21	.381		4.49	20.50		24.99	35.50
4090	Male connector, 1/8"		26	.308		1.32	16.50		17.82	26
4100	3/16" x 1/8"		24	.333		1.10	17.85		18.95	27.50
4110	1/4" x 1/8"		24	.333		1.14	17.85		18.99	28
4120	1/4"		24	.333		1.14	17.85		18.99	28
4130	5/16" x 1/8"		23	.348		1.34	18.65		19.99	29.50
4140	5/16" x 1/4"		23	.348		1.51	18.65		20.16	29.50
4150	3/8" x 1/8"		22	.364		1.59	19.50		21.09	31
4160	3/8" x 1/4"		22	.364		1.83	19.50		21.33	31
4170	3/8"		22	.364		1.83	19.50		21.33	31
4180	3/8" x 1/2"		22	.364		2.51	19.50		22.01	32
4190	1/2" x 3/8"		22	.364		2.43	19.50		21.93	31.50
4200	1/2"		22	.364		2.73	19.50		22.23	32
4210	5/8" x 1/2"		21	.381		3.15	20.50		23.65	34
4240	Male elbow, 1/8"		26	.308		2.19	16.50		18.69	27
4250	3/16" x 1/8"		24	.333		2.41	17.85		20.26	29
4260	1/4" x 1/8"		24	.333		1.83	17.85		19.68	28.50
4270	1/4"		24	.333		2.08	17.85		19.93	29
4280	3/8" x 1/4"		22	.364		2.64	19.50		22.14	32
4290	3/8"		22	.364		4.53	19.50		24.03	34
4300	1/2" x 1/4"		22	.364		4.37	19.50		23.87	34

22 11 13 – Facility Water Distribution Piping

22 11 13.25 Pipe/Tube Fittings, Copper		Crew	Daily Output	Labor-Hours	Unit	Material	2011 Bare Costs Labor	Equipment	Total	Total Incl O&P
4310	1/2" x 3/8"	1 Plum	22	.364	Ea.	6.90	19.50		26.40	36.50
4340	Female elbow, 1/8"		26	.308		4.70	16.50		21.20	29.50
4350	1/4" x 1/8"		24	.333		2.67	17.85		20.52	29.50
4360	1/4"		24	.333		3.51	17.85		21.36	30.50
4370	3/8" x 1/4"		22	.364		3.46	19.50		22.96	33
4380	1/2" x 3/8"		22	.364		5.90	19.50		25.40	35.50
4390	1/2"		22	.364		7.95	19.50		27.45	38
4420	Male run tee, 1/4" x 1/8"		15	.533		3.28	28.50		31.78	46
4430	5/16" x 1/8"		15	.533		6.95	28.50		35.45	50
4440	3/8" x 1/4"		15	.533		5.05	28.50		33.55	48
4480	Male branch tee, 1/4" x 1/8"		15	.533		3.80	28.50		32.30	46.50
4490	1/4"		15	.533		4.09	28.50		32.59	47
4500	3/8" x 1/4"		15	.533		4.76	28.50		33.26	48
4510	1/2" x 3/8"		15	.533		7.40	28.50		35.90	50.50
4520	1/2"		15	.533		11.85	28.50		40.35	55.50
4800	Flare joint fittings									
4810	Refrigeration fittings									
4820	Flare joint nuts and labor not incl. in price. Add 1 nut per jnt.									
4830	90° Elbow, 1/4"				Ea.	2.58			2.58	2.84
4840	3/8"					2.81			2.81	3.09
4850	1/2"					3.62			3.62	3.98
4860	5/8"					4.97			4.97	5.45
4870	3/4"					12.65			12.65	13.90
5030	Tee, 1/4"					3.10			3.10	3.41
5040	5/16"					3.97			3.97	4.37
5050	3/8"					4.37			4.37	4.81
5060	1/2"					6.70			6.70	7.35
5070	5/8"					14.50			14.50	15.95
5080	3/4"					16.50			16.50	18.15
5140	Union, 3/16"					1.16			1.16	1.28
5150	1/4"					.80			.80	.88
5160	5/16"					1.45			1.45	1.60
5170	3/8"					1.34			1.34	1.47
5180	1/2"					2			2	2.20
5190	5/8"					3.21			3.21	3.53
5200	3/4"					8.95			8.95	9.85
5260	Long flare nut, 3/16"	1 Stpi	42	.190		2.09	10.30		12.39	17.70
5270	1/4"		41	.195		.82	10.55		11.37	16.65
5280	5/16"		40	.200		1.48	10.80		12.28	17.80
5290	3/8"		39	.205		1.78	11.10		12.88	18.50
5300	1/2"		38	.211		2.84	11.40		14.24	20
5310	5/8"		37	.216		4.96	11.70		16.66	23
5320	3/4"		34	.235		14.50	12.75		27.25	35
5380	Short flare nut, 3/16"		42	.190		.48	10.30		10.78	15.95
5390	1/4"		41	.195		.58	10.55		11.13	16.40
5400	5/16"		40	.200		.69	10.80		11.49	16.90
5410	3/8"		39	.205		.90	11.10		12	17.55
5420	1/2"		38	.211		1.34	11.40		12.74	18.45
5430	5/8"		36	.222		1.87	12		13.87	20
5440	3/4"		34	.235		6.25	12.75		19	26
5500	90° Elbow flare by MIPS, 1/4"					2.45			2.45	2.70
5510	3/8"					2.64			2.64	2.90
5520	1/2"					4.22			4.22	4.64

22 11 Facility Water Distribution

22 11 13 – Facility Water Distribution Piping

22 11 13.25 Pipe/Tube Fittings, Copper	Crew	Daily Output	Labor-Hours	Unit	Material	2011 Bare Costs Labor	Equipment	Total	Total Incl O&P
5530	5/8"				Ea.	7.55			7.55	8.30
5540	3/4"					12.25			12.25	13.50
5600	Flare by FIPS, 1/4"					4.01			4.01	4.41
5610	3/8"					4.35			4.35	4.79
5620	1/2"					6.05			6.05	6.65
5670	Flare by sweat, 1/4"					6.60			6.60	7.25
5680	3/8"					7.35			7.35	8.10
5690	1/2"					12.80			12.80	14.05
5700	5/8"					9.65			9.65	10.60
5760	Tee flare by IPS, 1/4"					4.01			4.01	4.41
5770	3/8"					5.40			5.40	5.95
5780	1/2"					9.45			9.45	10.40
5790	5/8"					6.95			6.95	7.60
5850	Connector, 1/4"					.87			.87	.96
5860	3/8"					1.30			1.30	1.43
5870	1/2"					2.12			2.12	2.33
5880	5/8"					2.15			2.15	2.37
5890	3/4"					6.45			6.45	7.10
5950	Seal cap, 1/4"					.48			.48	.53
5960	3/8"					.95			.95	1.05
5970	1/2"					1.07			1.07	1.18
5980	5/8"					2.23			2.23	2.45
5990	3/4"					6.50			6.50	7.15
6000	Water service fittings									
6010	Flare joints nut and labor are included in the fitting price.									
6020	90° Elbow, C x C, 3/8"	1 Plum	19	.421	Ea.	78.50	22.50		101	120
6030	1/2"		18	.444		78.50	24		102.50	122
6040	3/4"		16	.500		99.50	27		126.50	150
6050	1"		15	.533		189	28.50		217.50	251
6080	2"		10	.800		650	43		693	780
6090	90° Elbow, C x MPT, 3/8"		19	.421		56.50	22.50		79	95.50
6100	1/2"		18	.444		56.50	24		80.50	97.50
6110	3/4"		16	.500		65.50	27		92.50	112
6120	1"		15	.533		174	28.50		202.50	235
6130	1-1/4"		13	.615		365	33		398	455
6140	1-1/2"		12	.667		365	35.50		400.50	460
6150	2"		10	.800		480	43		523	590
6160	90° Elbow, C x FPT, 3/8"		19	.421		60.50	22.50		83	100
6170	1/2"		18	.444		60.50	24		84.50	102
6180	3/4"		16	.500		75	27		102	123
6190	1"		15	.533		173	28.50		201.50	233
6200	1-1/4"		13	.615		193	33		226	261
6210	1-1/2"		12	.667		740	35.50		775.50	870
6220	2"		10	.800		1,075	43		1,118	1,250
6230	Tee, C x C x C, 3/8"		13	.615		114	33		147	175
6240	1/2"		12	.667		114	35.50		149.50	180
6250	3/4"		11	.727		134	39		173	206
6260	1"		10	.800		238	43		281	325
6330	Tube nut, C x nut seat, 3/8"		40	.200		25.50	10.70		36.20	44
6340	1/2"		38	.211		25.50	11.30		36.80	45
6350	3/4"		34	.235		25.50	12.60		38.10	47
6360	1"		32	.250		42	13.40		55.40	66.50
6380	Coupling, C x C, 3/8"		19	.421		70	22.50		92.50	111

22 11 Facility Water Distribution

22 11 13 – Facility Water Distribution Piping

22 11 13.25 Pipe/Tube Fittings, Copper

		Crew	Daily Output	Labor-Hours	Unit	Material	2011 Bare Costs Labor	Equipment	Total	Total Incl O&P
6390	1/2"	1 Plum	18	.444	Ea.	70	24		94	113
6400	3/4"		16	.500		89	27		116	138
6410	1"		15	.533		164	28.50		192.50	223
6420	1-1/4"		12	.667		266	35.50		301.50	345
6430	1-1/2"		12	.667		400	35.50		435.50	500
6440	2"		10	.800		605	43		648	730
6450	Adapter, C x FPT, 3/8"		19	.421		53	22.50		75.50	92
6460	1/2"		18	.444		53	24		77	94
6470	3/4"		16	.500		64	27		91	111
6480	1"		15	.533		137	28.50		165.50	193
6490	1-1/4"		13	.615		292	33		325	370
6500	1-1/2"		12	.667		292	35.50		327.50	375
6510	2"		10	.800		380	43		423	480
6520	Adapter, C x MPT, 3/8"		19	.421		47.50	22.50		70	85.50
6530	1/2"		18	.444		47.50	24		71.50	87.50
6540	3/4"		16	.500		65.50	27		92.50	112
6550	1"		15	.533		115	28.50		143.50	170
6560	1-1/4"		13	.615		289	33		322	365
6570	1-1/2"		12	.667		289	35.50		324.50	370
6580	2"		10	.800		390	43		433	495
6992	Polybutylene/polyethylene pipe, See Section 22 11 13.76 for plastic ftng.									
7000	Insert type Brass/copper, 100 psi @ 180°F, CTS									
7010	Adapter MPT 3/8" x 3/8" CTS	1 Plum	29	.276	Ea.	1.65	14.80		16.45	24
7020	1/2" x 1/2"		26	.308		1.65	16.50		18.15	26.50
7030	3/4" x 1/2"		26	.308		2.82	16.50		19.32	27.50
7040	3/4" x 3/4"		25	.320		2.43	17.15		19.58	28
7050	Adapter CTS 1/2" x 1/2" sweat		24	.333		.62	17.85		18.47	27
7060	3/4" x 3/4" sweat		22	.364		.78	19.50		20.28	30
7070	Coupler center set 3/8" CTS		25	.320		.80	17.15		17.95	26.50
7080	1/2" CTS		23	.348		.62	18.65		19.27	28.50
7090	3/4" CTS		22	.364		.78	19.50		20.28	30
7100	Elbow 90°, copper 3/8"		25	.320		1.77	17.15		18.92	27.50
7110	1/2" CTS		23	.348		1.13	18.65		19.78	29
7120	3/4" CTS		22	.364		1.37	19.50		20.87	30.50
7130	Tee copper 3/8" CTS		17	.471		2.10	25		27.10	40
7140	1/2" CTS		15	.533		1.45	28.50		29.95	44
7150	3/4" CTS		14	.571		2.22	30.50		32.72	48
7160	3/8" x 3/8" x 1/2"		16	.500		2.22	27		29.22	42.50
7170	1/2" x 3/8" x 1/2"		15	.533		1.44	28.50		29.94	44
7180	3/4" x 1/2" x 3/4"		14	.571		2.22	30.50		32.72	48

22 11 13.27 Pipe/Tube, Grooved Joint for Copper

		Crew	Daily Output	Labor-Hours	Unit	Material	2011 Bare Costs Labor	Equipment	Total	Total Incl O&P
0010	**PIPE/TUBE, GROOVED JOINT FOR COPPER**									
4000	Fittings: coupling material required at joints not incl. in fitting price.									
4001	Add 1 selected coupling, material only, per joint for installed price.									
4010	Coupling, rigid style									
4018	2" diameter	1 Plum	50	.160	Ea.	14.40	8.60		23	28.50
4020	2-1/2" diameter	Q-1	80	.200		16.40	9.65		26.05	32.50
4022	3" diameter		67	.239		18.20	11.50		29.70	37
4024	4" diameter		50	.320		28	15.45		43.45	53.50
4026	5" diameter		40	.400		47	19.30		66.30	80.50
4028	6" diameter	Q-2	50	.480		62	24		86	105
4100	Elbow, 90° or 45°									

139

22 11 13.27 Pipe/Tube, Grooved Joint for Copper

		Crew	Daily Output	Labor-Hours	Unit	Material	2011 Bare Costs Labor	Equipment	Total	Total Incl O&P
4108	2" diameter	1 Plum	25	.320	Ea.	26	17.15		43.15	54
4110	2-1/2" diameter	Q-1	40	.400		28	19.30		47.30	60
4112	3" diameter		33	.485		39.50	23.50		63	78.50
4114	4" diameter		25	.640		91	31		122	146
4116	5" diameter		20	.800		259	38.50		297.50	345
4118	6" diameter	Q-2	25	.960		415	48		463	525
4200	Tee									
4208	2" diameter	1 Plum	17	.471	Ea.	42.50	25		67.50	84.50
4210	2-1/2" diameter	Q-1	27	.593		45.50	28.50		74	92.50
4212	3" diameter		22	.727		67.50	35		102.50	127
4214	4" diameter		17	.941		149	45.50		194.50	231
4216	5" diameter		13	1.231		415	59.50		474.50	550
4218	6" diameter	Q-2	17	1.412		515	70.50		585.50	670
4300	Reducer, concentric									
4310	3" x 2-1/2" diameter	Q-1	35	.457	Ea.	37	22		59	74
4312	4" x 2-1/2" diameter		32	.500		76.50	24		100.50	121
4314	4" x 3" diameter		29	.552		76.50	26.50		103	124
4316	5" x 3" diameter		25	.640		216	31		247	284
4318	5" x 4" diameter		22	.727		216	35		251	291
4320	6" x 3" diameter	Q-2	28	.857		234	43		277	320
4322	6" x 4" diameter		26	.923		234	46		280	325
4324	6" x 5" diameter		24	1		234	50		284	335
4350	Flange, w/groove gasket									
4351	ANSI class 125 and 150									
4355	2" diameter	1 Plum	23	.348	Ea.	113	18.65		131.65	152
4356	2-1/2" diameter	Q-1	37	.432		118	21		139	160
4358	3" diameter		31	.516		123	25		148	173
4360	4" diameter		23	.696		135	33.50		168.50	199
4362	5" diameter		19	.842		182	40.50		222.50	261
4364	6" diameter	Q-2	23	1.043		198	52		250	296

22 11 13.44 Pipe, Steel

		Crew	Daily Output	Labor-Hours	Unit	Material	2011 Bare Costs Labor	Equipment	Total	Total Incl O&P
0010	**PIPE, STEEL**	R221113-50								
0020	All pipe sizes are to Spec. A-53 unless noted otherwise	R221113-70								
0032	Schedule 10, see Line 22 11 13.48 0500									
0050	Schedule 40, threaded, with couplings, and clevis hanger									
0060	assemblies sized for covering, 10' O.C.									
0540	Black, 1/4" diameter	1 Plum	66	.121	L.F.	6.50	6.50		13	16.85
0550	3/8" diameter		65	.123		7.25	6.60		13.85	17.85
0560	1/2" diameter		63	.127		3.80	6.80		10.60	14.35
0570	3/4" diameter		61	.131		4.37	7.05		11.42	15.30
0580	1" diameter		53	.151		6.15	8.10		14.25	18.85
0590	1-1/4" diameter	Q-1	89	.180		7.55	8.65		16.20	21.50
0600	1-1/2" diameter		80	.200		8.70	9.65		18.35	24
0610	2" diameter		64	.250		11.45	12.05		23.50	30.50
0620	2-1/2" diameter		50	.320		17.30	15.45		32.75	42
0630	3" diameter		43	.372		22	17.95		39.95	51
0640	3-1/2" diameter		40	.400		30	19.30		49.30	61.50
0650	4" diameter		36	.444		32.50	21.50		54	67.50
0809	A-106, gr. A/B, seamless w/cplgs. & clevis hanger assemblies									
0811	1/4" diameter	1 Plum	66	.121	L.F.	8.20	6.50		14.70	18.75
0812	3/8" diameter		65	.123		8.20	6.60		14.80	18.85
0813	1/2" diameter		63	.127		8.40	6.80		15.20	19.40

22 11 Facility Water Distribution

22 11 13 – Facility Water Distribution Piping

22 11 13.44 Pipe, Steel		Crew	Daily Output	Labor-Hours	Unit	Material	2011 Bare Costs Labor	Equipment	Total	Total Incl O&P
0814	3/4" diameter	1 Plum	61	.131	L.F.	10.10	7.05		17.15	21.50
0815	1" diameter	↓	53	.151		11.10	8.10		19.20	24.50
0816	1-1/4" diameter	Q-1	89	.180		13.25	8.65		21.90	27.50
0817	1-1/2" diameter		80	.200		22.50	9.65		32.15	39
0819	2" diameter		64	.250		23.50	12.05		35.55	43.50
0821	2-1/2" diameter		50	.320		27	15.45		42.45	53
0822	3" diameter		43	.372		29	17.95		46.95	59
0823	4" diameter	↓	36	.444	↓	44.50	21.50		66	80.50
1220	To delete coupling & hanger, subtract									
1230	1/4" diam. to 3/4" diam.					31%	56%			
1240	1" diam. to 1-1/2" diam.					23%	51%			
1250	2" diam. to 4" diam.					23%	41%			
1280	All pipe sizes are to Spec. A-53 unless noted otherwise									
1281	Schedule 40, threaded, with couplings and clevis hanger									
1282	assemblies sized for covering, 10' O. C.									
1290	Galvanized, 1/4" diameter	1 Plum	66	.121	L.F.	8.60	6.50		15.10	19.20
1300	3/8" diameter		65	.123		9.45	6.60		16.05	20.50
1310	1/2" diameter		63	.127		5.15	6.80		11.95	15.80
1320	3/4" diameter		61	.131		10.25	7.05		17.30	22
1330	1" diameter	↓	53	.151		8.10	8.10		16.20	21
1340	1-1/4" diameter	Q-1	89	.180		10.15	8.65		18.80	24
1350	1-1/2" diameter		80	.200		11.80	9.65		21.45	27.50
1360	2" diameter		64	.250		15.60	12.05		27.65	35
1370	2-1/2" diameter		50	.320		25	15.45		40.45	51
1380	3" diameter		43	.372		32	17.95		49.95	62
1390	3-1/2" diameter		40	.400		40.50	19.30		59.80	73.50
1400	4" diameter	↓	36	.444	↓	46	21.50		67.50	82.50
1750	To delete coupling & hanger, subtract									
1760	1/4" diam. to 3/4" diam.					31%	56%			
1770	1" diam. to 1-1/2" diam.					23%	51%			
1780	2" diam. to 4" diam.					23%	41%			
2000	Welded, sch. 40, on yoke & roll hanger assy's, sized for covering, 10' O.C.									
2040	Black, 1" diameter	Q-15	93	.172	L.F.	4.64	8.30	.35	13.29	17.90
2050	1-1/4" diameter		84	.190		5.85	9.20	.39	15.44	20.50
2060	1-1/2" diameter		76	.211		6.30	10.15	.43	16.88	22.50
2070	2" diameter		61	.262		7.20	12.65	.53	20.38	27.50
2080	2-1/2" diameter		47	.340		9.70	16.45	.69	26.84	36
2090	3" diameter		43	.372		11.40	17.95	.76	30.11	40.50
2100	3-1/2" diameter		39	.410		13.75	19.80	.83	34.38	45.50
2110	4" diameter		37	.432		15.80	21	.88	37.68	49.50
2120	5" diameter	↓	32	.500		21	24	1.02	46.02	60
2130	6" diameter	Q-16	36	.667		25	33.50	.90	59.40	78.50
2140	8" diameter		29	.828		36.50	41.50	1.12	79.12	103
2150	10" diameter		24	1		51	50	1.35	102.35	132
2160	12" diameter		19	1.263		87.50	63	1.71	152.21	192
2170	14" diameter, (two rod roll type hanger for 14" dia. and up)		15	1.600		97	80	2.16	179.16	228
2180	16" diameter, (two rod roll type hanger)		13	1.846		107	92.50	2.49	201.99	259
2190	18" diameter, (two rod roll type hanger)		11	2.182		124	109	2.95	235.95	300
2200	20" diameter, (two rod roll type hanger)		9	2.667		152	133	3.60	288.60	370
2220	24" diameter, (two rod roll type hanger)	↓	8	3	↓	217	150	4.05	371.05	465
2345	Sch. 40, A-53, gr. A/B, ERW, welded w/hngrs.									
2346	2" diameter	Q-15	61	.262	L.F.	6.70	12.65	.53	19.88	27
2347	2-1/2" diameter	↓	47	.340	↓	8.85	16.45	.69	25.99	35

22 11 13.44 Pipe, Steel		Crew	Daily Output	Labor-Hours	Unit	Material	2011 Bare Costs Labor	Equipment	Total	Total Incl O&P
2348	3" diameter	Q-15	43	.372	L.F.	11.15	17.95	.76	29.86	40
2349	4" diameter	↓	38	.421		14.80	20.50	.85	36.15	47.50
2350	6" diameter	Q-16	37	.649	↓	56	32.50	.88	89.38	111
2351	8" diameter		29	.828		78	41.50	1.12	120.62	149
2352	10" diameter	↓	24	1	↓	95.50	50	1.35	146.85	181
2363	.375" wall, A-53, gr. A/B, ERW, welded w/hngrs.									
2364	12" diameter	Q-16	20	1.200	L.F.	116	60	1.62	177.62	218
2365	14" diameter	↓	17	1.412	↓	166	70.50	1.91	238.41	289
2366	16" diameter	↓	14	1.714	↓	350	86	2.31	438.31	515
2560	To delete hanger, subtract									
2570	1" diam. to 1-1/2" diam.					15%	34%			
2580	2" diam. to 3-1/2" diam.					9%	21%			
2590	4" diam. to 12" diam.					5%	12%			
2596	14" diam. to 36" diam.					3%	10%			
2640	Galvanized, 1" diameter	Q-15	93	.172	L.F.	8.60	8.30	.35	17.25	22
2650	1-1/4" diameter		84	.190		10.85	9.20	.39	20.44	26
2660	1-1/2" diameter		76	.211		12.25	10.15	.43	22.83	29
2670	2" diameter		61	.262		15.45	12.65	.53	28.63	36.50
2680	2-1/2" diameter		47	.340		22	16.45	.69	39.14	50
2690	3" diameter		43	.372		28	17.95	.76	46.71	58.50
2700	3-1/2" diameter		39	.410		36	19.80	.83	56.63	70.50
2710	4" diameter		37	.432		39	21	.88	60.88	75
2720	5" diameter	↓	32	.500		45.50	24	1.02	70.52	87
2730	6" diameter	Q-16	36	.667		50.50	33.50	.90	84.90	106
2740	8" diameter		29	.828		56.50	41.50	1.12	99.12	125
2750	10" diameter		24	1		90	50	1.35	141.35	175
2760	12" diameter	↓	19	1.263	↓	112	63	1.71	176.71	220
3160	To delete hanger, subtract									
3170	1" diam. to 1-1/2" diam.					30%	34%			
3180	2" diam. to 3-1/2" diam.					19%	21%			
3190	4" diam. to 12" diam.					10%	12%			
3250	Flanged, 150 lb. weld neck, on yoke & roll hangers									
3260	sized for covering, 10' O.C.									
3290	Black, 1" diameter	Q-15	70	.229	L.F.	9.90	11.05	.46	21.41	28
3300	1-1/4" diameter		64	.250		11.10	12.05	.51	23.66	31
3310	1-1/2" diameter		58	.276		11.55	13.30	.56	25.41	33
3320	2" diameter		45	.356		13.25	17.15	.72	31.12	41
3330	2-1/2" diameter		36	.444		16.60	21.50	.90	39	51
3340	3" diameter		32	.500		18.70	24	1.02	43.72	57.50
3350	3-1/2" diameter		29	.552		23	26.50	1.12	50.62	65.50
3360	4" diameter		26	.615		25	29.50	1.25	55.75	73.50
3370	5" diameter	↓	21	.762		35.50	37	1.55	74.05	95.50
3380	6" diameter	Q-16	25	.960		41	48	1.30	90.30	118
3390	8" diameter		19	1.263		63.50	63	1.71	128.21	166
3400	10" diameter		16	1.500		105	75	2.03	182.03	230
3410	12" diameter	↓	14	1.714		149	86	2.31	237.31	295
3470	For 300 lb. flanges, add					63%				
3480	For 600 lb. flanges, add				↓	310%				
3960	To delete flanges & hanger, subtract									
3970	1" diam. to 2" diam.					76%	65%			
3980	2-1/2" diam. to 4" diam.					62%	59%			
3990	5" diam. to 12" diam.					60%	46%			
4040	Galvanized, 1" diameter	Q-15	70	.229	L.F.	13.85	11.05	.46	25.36	32

22 11 13.44 Pipe, Steel	Crew	Daily Output	Labor-Hours	Unit	Material	2011 Bare Costs Labor	Equipment	Total	Total Incl O&P
4050 1-1/4" diameter	Q-15	64	.250	L.F.	16.10	12.05	.51	28.66	36.50
4060 1-1/2" diameter		58	.276		17.50	13.30	.56	31.36	39.50
4070 2" diameter		45	.356		21	17.15	.72	38.87	49.50
4080 2-1/2" diameter		36	.444		29	21.50	.90	51.40	65
4090 3" diameter		32	.500		35	24	1.02	60.02	75.50
4100 3-1/2" diameter		29	.552		44	26.50	1.12	71.62	88.50
4110 4" diameter		26	.615		48	29.50	1.25	78.75	99
4120 5" diameter		21	.762		60	37	1.55	98.55	123
4130 6" diameter	Q-16	25	.960		65.50	48	1.30	114.80	145
4140 8" diameter		19	1.263		70.50	63	1.71	135.21	174
4150 10" diameter		16	1.500		116	75	2.03	193.03	242
4160 12" diameter		14	1.714		144	86	2.31	232.31	289
4220 For 300 lb. flanges, add					65%				
4240 For 600 lb. flanges, add					350%				
4660 To delete flanges & hanger, subtract									
4670 1" diam. to 2" diam.					73%	65%			
4680 2-1/2" diam. to 4" diam.					58%	59%			
4690 5" diam. to 12" diam.					43%	46%			
4750 Schedule 80, threaded, with couplings, and clevis hanger assemblies									
4760 sized for covering, 10' O.C.									
4790 Black, 1/4" diameter	1 Plum	54	.148	L.F.	6.05	7.95		14	18.50
4800 3/8" diameter		53	.151		7.60	8.10		15.70	20.50
4810 1/2" diameter		52	.154		5.35	8.25		13.60	18.20
4820 3/4" diameter		50	.160		6.40	8.60		15	19.80
4830 1" diameter		45	.178		8.50	9.55		18.05	23.50
4840 1-1/4" diameter	Q-1	75	.213		11	10.30		21.30	27.50
4850 1-1/2" diameter		69	.232		12.75	11.20		23.95	30.50
4860 2" diameter		56	.286		17.05	13.80		30.85	39.50
4870 2-1/2" diameter		44	.364		26	17.55		43.55	54.50
4880 3" diameter		38	.421		34	20.50		54.50	68
4890 3-1/2" diameter		35	.457		41.50	22		63.50	78.50
4900 4" diameter		32	.500		25	24		49	63.50
5061 A-106, gr. A/B seamless with cplgs. & clevis hanger assemblies, 1/4" dia.	1 Plum	63	.127		8.55	6.80		15.35	19.55
5062 3/8" diameter		62	.129		9.90	6.90		16.80	21
5063 1/2" diameter		61	.131		8.65	7.05		15.70	20
5064 3/4" diameter		57	.140		9.95	7.50		17.45	22
5065 1" diameter		51	.157		11.75	8.40		20.15	25.50
5066 1-1/4" diameter	Q-1	85	.188		15.45	9.10		24.55	30.50
5067 1-1/2" diameter		77	.208		27	10.05		37.05	44.50
5071 2" diameter		61	.262		33.50	12.65		46.15	56
5072 2-1/2" diameter		48	.333		27	16.10		43.10	54
5073 3" diameter		41	.390		35.50	18.85		54.35	67
5074 4" diameter		35	.457		52	22		74	90
5430 To delete coupling & hanger, subtract									
5440 1/4" diam. to 1/2" diam.					31%	54%			
5450 3/4" diam. to 1-1/2" diam.					28%	49%			
5460 2" diam. to 4" diam.					21%	40%			
5510 Galvanized, 1/4" diameter	1 Plum	54	.148	L.F.	9.20	7.95		17.15	22
5520 3/8" diameter		53	.151		7.65	8.10		15.75	20.50
5530 1/2" diameter		52	.154		7.20	8.25		15.45	20
5540 3/4" diameter		50	.160		8.60	8.60		17.20	22.50
5550 1" diameter		45	.178		11.70	9.55		21.25	27
5560 1-1/4" diameter	Q-1	75	.213		15.40	10.30		25.70	32.50

22 11 13.44 Pipe, Steel		Crew	Daily Output	Labor-Hours	Unit	Material	2011 Bare Costs			Total	Total Incl O&P
							Labor	Equipment		Total	
5570	1-1/2" diameter	Q-1	69	.232	L.F.	18.05	11.20			29.25	36.50
5580	2" diameter		56	.286		24.50	13.80			38.30	47.50
5590	2-1/2" diameter		44	.364		38	17.55			55.55	67.50
5600	3" diameter		38	.421		49.50	20.50			70	85
5610	3-1/2" diameter		35	.457		59	22			81	98
5620	4" diameter	▼	32	.500	▼	70.50	24			94.50	114
5930	To delete coupling & hanger, subtract										
5940	1/4" diam. to 1/2" diam.					31%	54%				
5950	3/4" diam. to 1-1/2" diam.					28%	49%				
5960	2" diam. to 4" diam.					21%	40%				
6000	Welded, on yoke & roller hangers										
6010	sized for covering, 10' O.C.										
6040	Black, 1" diameter	Q-15	85	.188	L.F.	9.25	9.10	.38		18.73	24
6050	1-1/4" diameter		79	.203		11.95	9.75	.41		22.11	28
6060	1-1/2" diameter		72	.222		13.70	10.70	.45		24.85	31.50
6070	2" diameter		57	.281		17.40	13.55	.57		31.52	40
6080	2-1/2" diameter		44	.364		26	17.55	.74		44.29	55.50
6090	3" diameter		40	.400		33.50	19.30	.81		53.61	67
6100	3-1/2" diameter		34	.471		40.50	22.50	.96		63.96	80
6110	4" diameter		33	.485		23.50	23.50	.98		47.98	62
6120	5" diameter	▼	26	.615		39	29.50	1.25		69.75	88.50
6130	6" diameter	Q-16	30	.800		46	40	1.08		87.08	111
6140	8" diameter		25	.960		109	48	1.30		158.30	192
6150	10" diameter		20	1.200		202	60	1.62		263.62	315
6160	12" diameter	▼	15	1.600	▼	395	80	2.16		477.16	555
6540	To delete hanger, subtract										
6550	1" diam. to 1-1/2" diam.					30%	14%				
6560	2" diam. to 3" diam.					23%	9%				
6570	3-1/2" diam. to 5" diam.					12%	6%				
6580	6" diam. to 12" diam.					10%	4%				
6610	Galvanized, 1" diameter	Q-15	85	.188	L.F.	12.50	9.10	.38		21.98	27.50
6650	1-1/4" diameter		79	.203		16.35	9.75	.41		26.51	33
6660	1-1/2" diameter		72	.222		19	10.70	.45		30.15	37.50
6670	2" diameter		57	.281		25	13.55	.57		39.12	48
6680	2-1/2" diameter		44	.364		37.50	17.55	.74		55.79	68.50
6690	3" diameter		40	.400		49	19.30	.81		69.11	84
6700	3-1/2" diameter		34	.471		58.50	22.50	.96		81.96	99
6710	4" diameter		33	.485		69	23.50	.98		93.48	112
6720	5" diameter	▼	26	.615		72.50	29.50	1.25		103.25	126
6730	6" diameter	Q-16	30	.800		97.50	40	1.08		138.58	169
6740	8" diameter		25	.960		117	48	1.30		166.30	202
6750	10" diameter		20	1.200		202	60	1.62		263.62	315
6760	12" diameter	▼	15	1.600	▼	246	80	2.16		328.16	390
7150	To delete hanger, subtract										
7160	1" diam. to 1-1/2" diam.					26%	14%				
7170	2" diam. to 3" diam.					16%	9%				
7180	3-1/2" diam. to 5" diam.					10%	6%				
7190	6" diam. to 12" diam.					7%	4%				
7250	Flanged, 300 lb. weld neck, on yoke & roll hangers										
7260	sized for covering, 10' O.C.										
7290	Black, 1" diameter	Q-15	66	.242	L.F.	15.70	11.70	.49		27.89	35
7300	1-1/4" diameter		61	.262	▼	18.40	12.65	.53		31.58	39.50
7310	1-1/2" diameter	▼	54	.296		20	14.30	.60		34.90	44

22 11 Facility Water Distribution

22 11 13 – Facility Water Distribution Piping

22 11 13.44 Pipe, Steel		Crew	Daily Output	Labor-Hours	Unit	Material	2011 Bare Costs Labor	Equipment	Total	Total Incl O&P
7320	2" diameter	Q-15	42	.381	L.F.	25.50	18.40	.77	44.67	56.50
7330	2-1/2" diameter		33	.485		35.50	23.50	.98	59.98	75
7340	3" diameter		29	.552		43	26.50	1.12	70.62	87.50
7350	3-1/2" diameter		24	.667		56	32	1.35	89.35	111
7360	4" diameter		23	.696		39	33.50	1.41	73.91	94.50
7370	5" diameter		19	.842		62.50	40.50	1.71	104.71	131
7380	6" diameter	Q-16	23	1.043		68	52	1.41	121.41	155
7390	8" diameter		17	1.412		147	70.50	1.91	219.41	269
7400	10" diameter		14	1.714		276	86	2.31	364.31	435
7410	12" diameter		12	2		490	100	2.70	592.70	690
7470	For 600 lb. flanges, add					100%				
7940	To delete flanges & hanger, subtract									
7950	1" diam. to 1-1/2" diam.					75%	66%			
7960	2" diam. to 3" diam.					62%	60%			
7970	3-1/2" diam. to 5" diam.					54%	66%			
7980	6" diam. to 12" diam.					55%	62%			
8040	Galvanized, 1" diameter	Q-15	66	.242	L.F.	18.90	11.70	.49	31.09	39
8050	1-1/4" diameter		61	.262		23	12.65	.53	36.18	44.50
8060	1-1/2" diameter		54	.296		25.50	14.30	.60	40.40	50
8070	2" diameter		42	.381		33	18.40	.77	52.17	65
8080	2-1/2" diameter		33	.485		47	23.50	.98	71.48	88
8090	3" diameter		29	.552		58.50	26.50	1.12	86.12	105
8100	3-1/2" diameter		24	.667		73.50	32	1.35	106.85	130
8110	4" diameter		23	.696		84.50	33.50	1.41	119.41	145
8120	5" diameter		19	.842		96.50	40.50	1.71	138.71	168
8130	6" diameter	Q-16	23	1.043		120	52	1.41	173.41	212
8140	8" diameter		17	1.412		156	70.50	1.91	228.41	278
8150	10" diameter		14	1.714		276	86	2.31	364.31	435
8160	12" diameter		12	2		340	100	2.70	442.70	525
8240	For 600 lb. flanges, add					100%				
8900	To delete flanges & hangers, subtract									
8910	1" diam. to 1-1/2" diam.					72%	66%			
8920	2" diam. to 3" diam.					59%	60%			
8930	3-1/2" diam. to 5" diam.					51%	66%			
8940	6" diam. to 12" diam.					49%	62%			
9000	Threading pipe labor, one end, all schedules through 80									
9010	1/4" through 3/4" pipe size	1 Plum	80	.100	Ea.		5.35		5.35	8
9020	1" through 2" pipe size		73	.110			5.85		5.85	8.75
9030	2-1/2" pipe size		53	.151			8.10		8.10	12.10
9040	3" pipe size		50	.160			8.60		8.60	12.80
9050	3-1/2" pipe size	Q-1	89	.180			8.65		8.65	12.95
9060	4" pipe size		73	.219			10.60		10.60	15.80
9070	5" pipe size		53	.302			14.55		14.55	22
9080	6" pipe size		46	.348			16.80		16.80	25
9090	8" pipe size		29	.552			26.50		26.50	39.50
9100	10" pipe size		21	.762			37		37	55
9110	12" pipe size		13	1.231			59.50		59.50	88.50
9200	Welding labor per joint									
9210	Schedule 40,									
9230	1/2" pipe size	Q-15	32	.500	Ea.		24	1.02	25.02	37
9240	3/4" pipe size		27	.593			28.50	1.20	29.70	44
9250	1" pipe size		23	.696			33.50	1.41	34.91	51.50
9260	1-1/4" pipe size		20	.800			38.50	1.62	40.12	59.50

22 11 13.44 Pipe, Steel

		Crew	Daily Output	Labor-Hours	Unit	Material	2011 Bare Costs Labor	Equipment	Total	Total Incl O&P
9270	1-1/2" pipe size	Q-15	19	.842	Ea.		40.50	1.71	42.21	62.50
9280	2" pipe size		16	1			48.50	2.03	50.53	74
9290	2-1/2" pipe size		13	1.231			59.50	2.50	62	91
9300	3" pipe size		12	1.333			64.50	2.71	67.21	99
9310	4" pipe size		10	1.600			77	3.25	80.25	119
9320	5" pipe size		9	1.778			86	3.61	89.61	132
9330	6" pipe size		8	2			96.50	4.06	100.56	148
9340	8" pipe size		5	3.200			154	6.50	160.50	238
9350	10" pipe size		4	4			193	8.10	201.10	297
9360	12" pipe size		3	5.333			257	10.85	267.85	395
9370	14" pipe size		2.60	6.154			297	12.50	309.50	460
9380	16" pipe size		2.20	7.273			350	14.75	364.75	540
9390	18" pipe size		2	8			385	16.25	401.25	595
9400	20" pipe size		1.80	8.889			430	18.05	448.05	660
9410	22" pipe size		1.70	9.412			455	19.10	474.10	700
9420	24" pipe size	▼	1.50	10.667	▼		515	21.50	536.50	795
9450	Schedule 80,									
9460	1/2" pipe size	Q-15	27	.593	Ea.		28.50	1.20	29.70	44
9470	3/4" pipe size		23	.696			33.50	1.41	34.91	51.50
9480	1" pipe size		20	.800			38.50	1.62	40.12	59.50
9490	1-1/4" pipe size		19	.842			40.50	1.71	42.21	62.50
9500	1-1/2" pipe size		18	.889			43	1.80	44.80	66
9510	2" pipe size		15	1.067			51.50	2.17	53.67	79.50
9520	2-1/2" pipe size		12	1.333			64.50	2.71	67.21	99
9530	3" pipe size		11	1.455			70	2.95	72.95	108
9540	4" pipe size		8	2			96.50	4.06	100.56	148
9550	5" pipe size		6	2.667			129	5.40	134.40	198
9560	6" pipe size		5	3.200			154	6.50	160.50	238
9570	8" pipe size		4	4			193	8.10	201.10	297
9580	10" pipe size		3	5.333			257	10.85	267.85	395
9590	12" pipe size	▼	2	8			385	16.25	401.25	595
9600	14" pipe size	Q-16	2.60	9.231			460	12.45	472.45	705
9610	16" pipe size		2.30	10.435			520	14.10	534.10	795
9620	18" pipe size		2	12			600	16.20	616.20	915
9630	20" pipe size		1.80	13.333			665	18	683	1,025
9640	22" pipe size		1.60	15			750	20.50	770.50	1,150
9650	24" pipe size	▼	1.50	16	▼		800	21.50	821.50	1,225

22 11 13.45 Pipe Fittings, Steel, Threaded

		Crew	Daily Output	Labor-Hours	Unit	Material	2011 Bare Costs Labor	Equipment	Total	Total Incl O&P
0010	**PIPE FITTINGS, STEEL, THREADED**									
0020	Cast Iron									
0040	Standard weight, black									
0060	90° Elbow, straight									
0070	1/4"	1 Plum	16	.500	Ea.	7.40	27		34.40	48
0080	3/8"		16	.500		10.65	27		37.65	52
0090	1/2"		15	.533		4.68	28.50		33.18	47.50
0100	3/4"		14	.571		4.88	30.50		35.38	51
0110	1"	▼	13	.615		5.75	33		38.75	55.50
0120	1-1/4"	Q-1	22	.727		8.20	35		43.20	61.50
0130	1-1/2"		20	.800		11.35	38.50		49.85	70
0140	2"		18	.889		17.70	43		60.70	83.50
0150	2-1/2"		14	1.143		42.50	55		97.50	129
0160	3"	▼	10	1.600	▼	69.50	77		146.50	192

22 11 13 – Facility Water Distribution Piping

22 11 13.45 Pipe Fittings, Steel, Threaded	Crew	Daily Output	Labor-Hours	Unit	Material	2011 Bare Costs Labor	Equipment	Total	Total Incl O&P
0170	3-1/2"	Q-1	8	2	Ea.	189	96.50		285.50	350
0180	4"	↓	6	2.667	↓	129	129		258	335
0250	45° Elbow, straight									
0260	1/4"	1 Plum	16	.500	Ea.	9.10	27		36.10	50
0270	3/8"		16	.500		9.75	27		36.75	51
0280	1/2"		15	.533		7.15	28.50		35.65	50.50
0300	3/4"		14	.571		7.20	30.50		37.70	53.50
0320	1"	↓	13	.615		8.45	33		41.45	58.50
0330	1-1/4"	Q-1	22	.727		11.35	35		46.35	65
0340	1-1/2"		20	.800		18.80	38.50		57.30	78
0350	2"		18	.889		21.50	43		64.50	88
0360	2-1/2"		14	1.143		56.50	55		111.50	145
0370	3"		10	1.600		89.50	77		166.50	214
0380	3-1/2"		8	2		204	96.50		300.50	370
0400	4"	↓	6	2.667	↓	186	129		315	395
0500	Tee, straight									
0510	1/4"	1 Plum	10	.800	Ea.	11.55	43		54.55	76.50
0520	3/8"		10	.800		11.20	43		54.20	76.50
0530	1/2"		9	.889		7.30	47.50		54.80	79
0540	3/4"		9	.889		8.50	47.50		56	80.50
0550	1"	↓	8	1		7.55	53.50		61.05	88.50
0560	1-1/4"	Q-1	14	1.143		13.80	55		68.80	97.50
0570	1-1/2"		13	1.231		17.95	59.50		77.45	108
0580	2"		11	1.455		25	70		95	133
0590	2-1/2"		9	1.778		65	86		151	200
0600	3"		6	2.667		99.50	129		228.50	300
0610	3-1/2"		5	3.200		201	154		355	450
0620	4"	↓	4	4	↓	194	193		387	500
0660	Tee, reducing, run or outlet									
0661	1/2"	1 Plum	9	.889	Ea.	18	47.50		65.50	91
0662	3/4"		9	.889		20	47.50		67.50	93
0663	1"	↓	8	1		17.75	53.50		71.25	99.50
0664	1-1/4"	Q-1	14	1.143		25	55		80	110
0665	1-1/2"		13	1.231		29.50	59.50		89	121
0666	2"		11	1.455		44.50	70		114.50	154
0667	2-1/2"		9	1.778		85.50	86		171.50	222
0668	3"		6	2.667		174	129		303	385
0669	3-1/2"		5	3.200		300	154		454	560
0670	4"	↓	4	4	↓	315	193		508	635
0674	Reducer, concentric									
0675	3/4"	1 Plum	18	.444	Ea.	13.05	24		37.05	50
0676	1"	"	15	.533		8.85	28.50		37.35	52.50
0677	1-1/4"	Q-1	26	.615		26.50	29.50		56	73.50
0678	1-1/2"		24	.667		41.50	32		73.50	93.50
0679	2"		21	.762		47	37		84	107
0680	2-1/2"		18	.889		67	43		110	138
0681	3"		14	1.143		96	55		151	188
0682	3-1/2"		12	1.333		181	64.50		245.50	295
0683	4"	↓	10	1.600	↓	196	77		273	330
0687	Reducer, eccentric									
0688	3/4"	1 Plum	16	.500	Ea.	27.50	27		54.50	70.50
0689	1"	"	14	.571		29.50	30.50		60	78
0690	1-1/4"	Q-1	25	.640		48	31		79	98.50

22 11 13.45 Pipe Fittings, Steel, Threaded		Crew	Daily Output	Labor-Hours	Unit	Material	2011 Bare Costs Labor	Equipment	Total	Total Incl O&P
0691	1-1/2"	Q-1	22	.727	Ea.	62.50	35		97.50	122
0692	2"		20	.800		90.50	38.50		129	157
0693	2-1/2"		16	1		124	48.50		172.50	209
0694	3"		12	1.333		196	64.50		260.50	310
0695	3-1/2"		10	1.600		273	77		350	415
0696	4"	▼	9	1.778	▼	355	86		441	520
0700	Standard weight, galvanized cast iron									
0720	90° Elbow, straight									
0730	1/4"	1 Plum	16	.500	Ea.	12.40	27		39.40	53.50
0740	3/8"		16	.500		12.40	27		39.40	53.50
0750	1/2"		15	.533		14.15	28.50		42.65	58
0760	3/4"		14	.571		14.60	30.50		45.10	61.50
0770	1"	▼	13	.615		16.05	33		49.05	66.50
0780	1-1/4"	Q-1	22	.727		25	35		60	80
0790	1-1/2"		20	.800		34.50	38.50		73	95
0800	2"		18	.889		50.50	43		93.50	120
0810	2-1/2"		14	1.143		104	55		159	197
0820	3"		10	1.600		158	77		235	289
0830	3-1/2"		8	2		252	96.50		348.50	420
0840	4"	▼	6	2.667	▼	290	129		419	510
0900	45° Elbow, straight									
0910	1/4"	1 Plum	16	.500	Ea.	14.20	27		41.20	55.50
0920	3/8"		16	.500		14.20	27		41.20	55.50
0930	1/2"		15	.533		14.20	28.50		42.70	58
0940	3/4"		14	.571		16.20	30.50		46.70	63.50
0950	1"	▼	13	.615		15.90	33		48.90	66.50
0960	1-1/4"	Q-1	22	.727		21.50	35		56.50	76
0970	1-1/2"		20	.800		39.50	38.50		78	101
0980	2"		18	.889		55.50	43		98.50	125
0990	2-1/2"		14	1.143		109	55		164	203
1000	3"		10	1.600		174	77		251	305
1010	3-1/2"		8	2		278	96.50		374.50	450
1020	4"	▼	6	2.667	▼	320	129		449	540
1100	Tee, straight									
1110	1/4"	1 Plum	10	.800	Ea.	16.15	43		59.15	82
1120	3/8"		10	.800		16.15	43		59.15	82
1130	1/2"		9	.889		14.45	47.50		61.95	87
1140	3/4"		9	.889		19.95	47.50		67.45	93
1150	1"	▼	8	1		21.50	53.50		75	104
1160	1-1/4"	Q-1	14	1.143		38	55		93	124
1170	1-1/2"		13	1.231		50	59.50		109.50	144
1180	2"		11	1.455		62.50	70		132.50	174
1190	2-1/2"		9	1.778		135	86		221	276
1200	3"		6	2.667		350	129		479	575
1210	3-1/2"		5	3.200		360	154		514	625
1220	4"	▼	4	4	▼	400	193		593	730
1300	Extra heavy weight, black									
1310	Couplings, steel straight									
1320	1/4"	1 Plum	19	.421	Ea.	3.08	22.50		25.58	37
1330	3/8"		19	.421		3.35	22.50		25.85	37
1340	1/2"		19	.421		4.53	22.50		27.03	38.50
1350	3/4"		18	.444		4.84	24		28.84	41
1360	1"	▼	15	.533		6.10	28.50		34.60	49

22 11 Facility Water Distribution

22 11 13 – Facility Water Distribution Piping

22 11 13.45 Pipe Fittings, Steel, Threaded	Crew	Daily Output	Labor-Hours	Unit	Material	2011 Bare Costs Labor	Equipment	Total	Total Incl O&P
1370	1-1/4"	Q-1	26	.615	Ea.	9.90	29.50		39.40	55.50
1380	1-1/2"		24	.667		9.90	32		41.90	59
1390	2"		21	.762		15.05	37		52.05	71.50
1400	2-1/2"		18	.889		22.50	43		65.50	88.50
1410	3"		14	1.143		26.50	55		81.50	112
1420	3-1/2"		12	1.333		35.50	64.50		100	136
1430	4"		10	1.600		42	77		119	161
1510	90° Elbow, straight									
1520	1/2"	1 Plum	15	.533	Ea.	23.50	28.50		52	68.50
1530	3/4"		14	.571		24.50	30.50		55	72.50
1540	1"		13	.615		29.50	33		62.50	81.50
1550	1-1/4"	Q-1	22	.727		44	35		79	101
1560	1-1/2"		20	.800		55	38.50		93.50	118
1580	2"		18	.889		67.50	43		110.50	139
1590	2-1/2"		14	1.143		162	55		217	262
1600	3"		10	1.600		215	77		292	350
1610	4"		6	2.667		450	129		579	685
1650	45° Elbow, straight									
1660	1/2"	1 Plum	15	.533	Ea.	34	28.50		62.50	80
1670	3/4"		14	.571		33	30.50		63.50	81.50
1680	1"		13	.615		39.50	33		72.50	92.50
1690	1-1/4"	Q-1	22	.727		65	35		100	124
1700	1-1/2"		20	.800		71.50	38.50		110	136
1710	2"		18	.889		102	43		145	176
1720	2-1/2"		14	1.143		175	55		230	276
1800	Tee, straight									
1810	1/2"	1 Plum	9	.889	Ea.	37.50	47.50		85	113
1820	3/4"		9	.889		37.50	47.50		85	113
1830	1"		8	1		45.50	53.50		99	130
1840	1-1/4"	Q-1	14	1.143		67.50	55		122.50	157
1850	1-1/2"		13	1.231		87	59.50		146.50	185
1860	2"		11	1.455		108	70		178	224
1870	2-1/2"		9	1.778		231	86		317	380
1880	3"		6	2.667		315	129		444	535
1890	4"		4	4		610	193		803	965
4000	Standard weight, black									
4010	Couplings, steel straight, merchants									
4030	1/4"	1 Plum	19	.421	Ea.	.60	22.50		23.10	34
4040	3/8"		19	.421		.99	22.50		23.49	34.50
4050	1/2"		19	.421		1.05	22.50		23.55	34.50
4060	3/4"		18	.444		1.35	24		25.35	37
4070	1"		15	.533		1.88	28.50		30.38	44.50
4080	1-1/4"	Q-1	26	.615		2.41	29.50		31.91	47
4090	1-1/2"		24	.667		3.05	32		35.05	51.50
4100	2"		21	.762		4.36	37		41.36	60
4110	2-1/2"		18	.889		12.40	43		55.40	77.50
4120	3"		14	1.143		17.45	55		72.45	102
4130	3-1/2"		12	1.333		31	64.50		95.50	130
4140	4"		10	1.600		31	77		108	149
4166	Plug, 1/4"	1 Plum	38	.211		2.08	11.30		13.38	19.15
4167	3/8"		38	.211		2.08	11.30		13.38	19.15
4168	1/2"		38	.211		2.08	11.30		13.38	19.15
4169	3/4"		32	.250		6.30	13.40		19.70	27

149

22 11 13.45 Pipe Fittings, Steel, Threaded	Crew	Daily Output	Labor-Hours	Unit	Material	2011 Bare Costs Labor	Equipment	Total	Total Incl O&P
4170	1"	1 Plum	30	.267	Ea.	6.70	14.30		21	29
4171	1-1/4"	Q-1	52	.308		7.70	14.85		22.55	30.50
4172	1-1/2"		48	.333		10.90	16.10		27	36
4173	2"		42	.381		14.05	18.40		32.45	43
4176	2-1/2"		36	.444		21	21.50		42.50	55
4180	4"		20	.800		36	38.50		74.50	97
4200	Standard weight, galvanized									
4210	Couplings, steel straight, merchants									
4230	1/4"	1 Plum	19	.421	Ea.	.95	22.50		23.45	34.50
4240	3/8"		19	.421		1.22	22.50		23.72	35
4250	1/2"		19	.421		1.30	22.50		23.80	35
4260	3/4"		18	.444		1.62	24		25.62	37.50
4270	1"		15	.533		2.26	28.50		30.76	45
4280	1-1/4"	Q-1	26	.615		2.90	29.50		32.40	47.50
4290	1-1/2"		24	.667		3.60	32		35.60	52
4300	2"		21	.762		5.40	37		42.40	61
4310	2-1/2"		18	.889		15.25	43		58.25	81
4320	3"		14	1.143		20.50	55		75.50	105
4330	3-1/2"		12	1.333		35.50	64.50		100	136
4340	4"		10	1.600		35.50	77		112.50	155
4370	Plug, galvanized, square head									
4374	1/2"	1 Plum	38	.211	Ea.	5	11.30		16.30	22.50
4375	3/4"		32	.250		5	13.40		18.40	25.50
4376	1"		30	.267		5	14.30		19.30	27
4377	1-1/4"	Q-1	52	.308		8.40	14.85		23.25	31
4378	1-1/2"		48	.333		11.30	16.10		27.40	36.50
4379	2"		42	.381		14.05	18.40		32.45	43
4380	2-1/2"		36	.444		29.50	21.50		51	64.50
4381	3"		28	.571		38	27.50		65.50	83
4382	4"		20	.800		104	38.50		142.50	173
4700	Nipple, black									
4710	1/2" x 4" long	1 Plum	19	.421	Ea.	1.77	22.50		24.27	35.50
4712	3/4" x 4" long		18	.444		2.16	24		26.16	38
4714	1" x 4" long		15	.533		3	28.50		31.50	46
4716	1-1/4" x 4" long	Q-1	26	.615		3.76	29.50		33.26	48.50
4718	1-1/2" x 4" long		24	.667		4.44	32		36.44	53
4720	2" x 4" long		21	.762		6.20	37		43.20	62
4722	2-1/2" x 4" long		18	.889		16.90	43		59.90	82.50
4724	3" x 4" long		14	1.143		21.50	55		76.50	106
4726	4" x 4" long		10	1.600		29	77		106	147
4800	Nipple, galvanized									
4810	1/2" x 4" long	1 Plum	19	.421	Ea.	2.19	22.50		24.69	36
4812	3/4" x 4" long		18	.444		2.75	24		26.75	38.50
4814	1" x 4" long		15	.533		3.66	28.50		32.16	46.50
4816	1-1/4" x 4" long	Q-1	26	.615		4.49	29.50		33.99	49.50
4818	1-1/2" x 4" long		24	.667		5.70	32		37.70	54.50
4820	2" x 4" long		21	.762		7.25	37		44.25	63
4822	2-1/2" x 4" long		18	.889		18.55	43		61.55	84.50
4824	3" x 4" long		14	1.143		25	55		80	111
4826	4" x 4" long		10	1.600		34	77		111	153
5000	Malleable iron, 150 lb.									
5020	Black									
5040	90° elbow, straight									

22 11 13 – Facility Water Distribution Piping

22 11 13.45 Pipe Fittings, Steel, Threaded		Crew	Daily Output	Labor-Hours	Unit	Material	2011 Bare Costs Labor	Equipment	Total	Total Incl O&P
5060	1/4"	1 Plum	16	.500	Ea.	3.63	27		30.63	44
5070	3/8"		16	.500		3.63	27		30.63	44
5080	1/2"		15	.533		3.39	28.50		31.89	46
5090	3/4"		14	.571		4.16	30.50		34.66	50
5100	1"		13	.615		7.70	33		40.70	57.50
5110	1-1/4"	Q-1	22	.727		12.15	35		47.15	66
5120	1-1/2"		20	.800		15.90	38.50		54.40	75
5130	2"		18	.889		26	43		69	92.50
5140	2-1/2"		14	1.143		44	55		99	131
5150	3"		10	1.600		64.50	77		141.50	186
5160	3-1/2"		8	2		177	96.50		273.50	340
5170	4"		6	2.667		138	129		267	345
5250	45° elbow, straight									
5270	1/4"	1 Plum	16	.500	Ea.	5.50	27		32.50	46
5280	3/8"		16	.500		5.50	27		32.50	46
5290	1/2"		15	.533		4.15	28.50		32.65	47
5300	3/4"		14	.571		5.20	30.50		35.70	51.50
5310	1"		13	.615		6.45	33		39.45	56
5320	1-1/4"	Q-1	22	.727		11.45	35		46.45	65
5330	1-1/2"		20	.800		14.15	38.50		52.65	73
5340	2"		18	.889		21.50	43		64.50	87.50
5350	2-1/2"		14	1.143		62	55		117	151
5360	3"		10	1.600		80.50	77		157.50	204
5370	3-1/2"		8	2		158	96.50		254.50	320
5380	4"		6	2.667		158	129		287	365
5450	Tee, straight									
5470	1/4"	1 Plum	10	.800	Ea.	5.30	43		48.30	70
5480	3/8"		10	.800		5.30	43		48.30	70
5490	1/2"		9	.889		3.36	47.50		50.86	74.50
5500	3/4"		9	.889		4.82	47.50		52.32	76.50
5510	1"		8	1		8.25	53.50		61.75	89
5520	1-1/4"	Q-1	14	1.143		13.40	55		68.40	97.50
5530	1-1/2"		13	1.231		16.65	59.50		76.15	107
5540	2"		11	1.455		28.50	70		98.50	136
5550	2-1/2"		9	1.778		61	86		147	196
5560	3"		6	2.667		90	129		219	291
5570	3-1/2"		5	3.200		209	154		363	460
5580	4"		4	4		218	193		411	530
5601	Tee, reducing, on outlet									
5602	1/2"	1 Plum	9	.889	Ea.	8.95	47.50		56.45	81
5603	3/4"		9	.889		7.65	47.50		55.15	79.50
5604	1"		8	1		13.25	53.50		66.75	94.50
5605	1-1/4"	Q-1	14	1.143		23.50	55		78.50	108
5606	1-1/2"		13	1.231		24	59.50		83.50	115
5607	2"		11	1.455		32	70		102	141
5608	2-1/2"		9	1.778		89.50	86		175.50	227
5609	3"		6	2.667		128	129		257	330
5610	3-1/2"		5	3.200		289	154		443	550
5611	4"		4	4		265	193		458	580
5650	Coupling									
5670	1/4"	1 Plum	19	.421	Ea.	4.51	22.50		27.01	38.50
5680	3/8"		19	.421		4.51	22.50		27.01	38.50
5690	1/2"		19	.421		3.47	22.50		25.97	37.50

22 11 13.45 Pipe Fittings, Steel, Threaded		Crew	Daily Output	Labor-Hours	Unit	Material	2011 Bare Costs Labor	Equipment	Total	Total Incl O&P
5700	3/4"	1 Plum	18	.444	Ea.	4.07	24		28.07	40
5710	1"	↓	15	.533		6.10	28.50		34.60	49
5720	1-1/4"	Q-1	26	.615		7.90	29.50		37.40	53
5730	1-1/2"		24	.667		10.65	32		42.65	60
5740	2"		21	.762		15.80	37		52.80	72.50
5750	2-1/2"		18	.889		43.50	43		86.50	112
5760	3"		14	1.143		59	55		114	148
5770	3-1/2"		12	1.333		126	64.50		190.50	235
5780	4"	↓	10	1.600		119	77		196	245
5840	Reducer, concentric, 1/4"	1 Plum	19	.421		4.75	22.50		27.25	39
5850	3/8"		19	.421		6.10	22.50		28.60	40
5860	1/2"		19	.421		4.81	22.50		27.31	39
5870	3/4"		16	.500		5.70	27		32.70	46.50
5880	1"	↓	15	.533		9.60	28.50		38.10	53
5890	1-1/4"	Q-1	26	.615		10.75	29.50		40.25	56.50
5900	1-1/2"		24	.667		15.40	32		47.40	65
5910	2"		21	.762		22.50	37		59.50	79.50
5911	2-1/2"		18	.889		49.50	43		92.50	119
5912	3"		14	1.143		62.50	55		117.50	151
5913	3-1/2"		12	1.333		190	64.50		254.50	305
5914	4"	↓	10	1.600		145	77		222	274
5981	Bushing, 1/4"	1 Plum	19	.421		3.16	22.50		25.66	37
5982	3/8"		19	.421		4.45	22.50		26.95	38.50
5983	1/2"		19	.421		6.50	22.50		29	40.50
5984	3/4"		16	.500		4.08	27		31.08	44.50
5985	1"	↓	15	.533		6.65	28.50		35.15	50
5986	1-1/4"	Q-1	26	.615		8.25	29.50		37.75	53.50
5987	1-1/2"		24	.667		7.10	32		39.10	56
5988	2"	↓	21	.762		8.95	37		45.95	65
5989	Cap, 1/4"	1 Plum	38	.211		4.07	11.30		15.37	21.50
5991	3/8"		38	.211		2.65	11.30		13.95	19.75
5992	1/2"		38	.211		2.40	11.30		13.70	19.50
5993	3/4"		32	.250		3.45	13.40		16.85	24
5994	1"	↓	30	.267		4.17	14.30		18.47	26
5995	1-1/4"	Q-1	52	.308		5.50	14.85		20.35	28
5996	1-1/2"		48	.333		7.55	16.10		23.65	32.50
5997	2"	↓	42	.381		11	18.40		29.40	39.50
6000	For galvanized elbows, tees, and couplings add				↓	20%				
6058	For galvanized reducers, caps and bushings add					20%				
6100	90° Elbow, galvanized, 150 lb., reducing									
6110	3/4" x 1/2"	1 Plum	15.40	.519	Ea.	7.25	28		35.25	49.50
6112	1" x 3/4"		14	.571		9.30	30.50		39.80	55.50
6114	1" x 1/2"	↓	14.50	.552		9.85	29.50		39.35	55
6116	1-1/4" x 1"	Q-1	24.20	.661		15.55	32		47.55	64.50
6118	1-1/4" x 3/4"		25.40	.630		18.70	30.50		49.20	66
6120	1-1/4" x 1/2"		26.20	.611		19.80	29.50		49.30	66
6122	1-1/2" x 1-1/4"		21.60	.741		25	35.50		60.50	81
6124	1-1/2" x 1"		23.50	.681		25	33		58	76.50
6126	1-1/2" x 3/4"		24.60	.650		25	31.50		56.50	74.50
6128	2" x 1-1/2"		20.50	.780		29	37.50		66.50	88
6130	2" x 1-1/4"		21	.762		33.50	37		70.50	91.50
6132	2" x 1"		22.80	.702		34.50	34		68.50	88.50
6134	2" x 3/4"	↓	23.90	.669		35.50	32.50		68	87

22 11 13 – Facility Water Distribution Piping

22 11 13.45 Pipe Fittings, Steel, Threaded	Crew	Daily Output	Labor-Hours	Unit	Material	2011 Bare Costs Labor	Equipment	Total	Total Incl O&P
6136	2-1/2" x 2"	Q-1	12.30	1.301	Ea.	98.50	63		161.50	202
6138	2-1/2" x 1-1/2"		12.50	1.280		110	62		172	213
6140	3" x 2-1/2"		8.60	1.860		179	90		269	330
6142	3" x 2"		11.80	1.356		155	65.50		220.50	269
6144	4" x 3"		8.20	1.951		405	94		499	585
6160	90° Elbow, black, 150 lb., reducing									
6170	1" x 3/4"	1 Plum	14	.571	Ea.	6.35	30.50		36.85	52.50
6174	1-1/2" x 1"	Q-1	23.50	.681		15	33		48	65.50
6178	1-1/2" x 3/4"		24.60	.650		17.20	31.50		48.70	66
6182	2" x 1-1/2"		20.50	.780		22	37.50		59.50	80
6186	2" x 1"		22.80	.702		24.50	34		58.50	77.50
6190	2" x 3/4"		23.90	.669		26	32.50		58.50	77
6194	2-1/2" x 2"		12.30	1.301		59.50	63		122.50	159
7000	Union, with brass seat									
7010	1/4"	1 Plum	15	.533	Ea.	17.80	28.50		46.30	62
7020	3/8"		15	.533		12.25	28.50		40.75	56
7030	1/2"		14	.571		10.80	30.50		41.30	57.50
7040	3/4"		13	.615		12.75	33		45.75	63
7050	1"		12	.667		16.65	35.50		52.15	72
7060	1-1/4"	Q-1	21	.762		24	37		61	81.50
7070	1-1/2"		19	.842		30	40.50		70.50	93.50
7080	2"		17	.941		34.50	45.50		80	106
7090	2-1/2"		13	1.231		103	59.50		162.50	203
7100	3"		9	1.778		124	86		210	265
7120	Union, galvanized									
7124	1/2"	1 Plum	14	.571	Ea.	14.75	30.50		45.25	62
7125	3/4"		13	.615		16.95	33		49.95	67.50
7126	1"		12	.667		22	35.50		57.50	78
7127	1-1/4"	Q-1	21	.762		32	37		69	90.50
7128	1-1/2"		19	.842		39	40.50		79.50	104
7129	2"		17	.941		44.50	45.50		90	117
7130	2-1/2"		13	1.231		155	59.50		214.50	260
7131	3"		9	1.778		217	86		303	365
7500	Malleable iron, 300 lb									
7520	Black									
7540	90° Elbow, straight, 1/4"	1 Plum	16	.500	Ea.	12.25	27		39.25	53.50
7560	3/8"		16	.500		11.30	27		38.30	52.50
7570	1/2"		15	.533		14.65	28.50		43.15	58.50
7580	3/4"		14	.571		16.60	30.50		47.10	64
7590	1"		13	.615		21	33		54	72.50
7600	1-1/4"	Q-1	22	.727		31	35		66	86.50
7610	1-1/2"		20	.800		36.50	38.50		75	97.50
7620	2"		18	.889		52.50	43		95.50	122
7630	2-1/2"		14	1.143		136	55		191	232
7640	3"		10	1.600		156	77		233	287
7650	4"		6	2.667		365	129		494	595
7700	45° Elbow, straight, 1/4"	1 Plum	16	.500		18.95	27		45.95	61
7720	3/8"		16	.500		18.95	27		45.95	61
7730	1/2"		15	.533		21	28.50		49.50	65.50
7740	3/4"		14	.571		23.50	30.50		54	71.50
7750	1"		13	.615		26	33		59	77.50
7760	1-1/4"	Q-1	22	.727		41.50	35		76.50	98.50
7770	1-1/2"		20	.800		54.50	38.50		93	117

22 11 13 – Facility Water Distribution Piping

22 11 13.45 Pipe Fittings, Steel, Threaded	Crew	Daily Output	Labor-Hours	Unit	Material	2011 Bare Costs Labor	Equipment	Total	Total Incl O&P
7780	2"	Q-1	18	.889	Ea.	82	43		125	154
7790	2-1/2"		14	1.143		183	55		238	284
7800	3"		10	1.600		241	77		318	380
7810	4"	▼	6	2.667		545	129		674	790
7850	Tee, straight, 1/4"	1 Plum	10	.800		16.05	43		59.05	81.50
7870	3/8"		10	.800		17	43		60	82.50
7880	1/2"		9	.889		21.50	47.50		69	95
7890	3/4"		9	.889		23.50	47.50		71	96.50
7900	1"	▼	8	1		28	53.50		81.50	111
7910	1-1/4"	Q-1	14	1.143		40.50	55		95.50	127
7920	1-1/2"		13	1.231		47	59.50		106.50	140
7930	2"		11	1.455		69.50	70		139.50	181
7940	2-1/2"		9	1.778		168	86		254	315
7950	3"		6	2.667		240	129		369	455
7960	4"	▼	4	4		685	193		878	1,050
8050	Couplings, straight, 1/4"	1 Plum	19	.421		11.50	22.50		34	46
8070	3/8"		19	.421		11.50	22.50		34	46
8080	1/2"		19	.421		12.80	22.50		35.30	47.50
8090	3/4"		18	.444		14.75	24		38.75	52
8100	1"	▼	15	.533		16.85	28.50		45.35	61
8110	1-1/4"	Q-1	26	.615		20	29.50		49.50	66.50
8120	1-1/2"		24	.667		30	32		62	81
8130	2"		21	.762		43	37		80	103
8140	2-1/2"		18	.889		79.50	43		122.50	152
8150	3"		14	1.143		115	55		170	209
8160	4"	▼	10	1.600	▼	144	77		221	273
8200	Galvanized									
8220	90° Elbow, straight, 1/4"	1 Plum	16	.500	Ea.	23	27		50	65
8222	3/8"		16	.500		23.50	27		50.50	65.50
8224	1/2"		15	.533		27.50	28.50		56	73
8226	3/4"		14	.571		31	30.50		61.50	79.50
8228	1"	▼	13	.615		40	33		73	93
8230	1-1/4"	Q-1	22	.727		64	35		99	123
8232	1-1/2"		20	.800		68	38.50		106.50	132
8234	2"		18	.889		114	43		157	190
8236	2-1/2"		14	1.143		252	55		307	360
8238	3"		10	1.600		300	77		377	445
8240	4"	▼	6	2.667	▼	895	129		1,024	1,175
8280	45° Elbow, straight									
8282	1/2"	1 Plum	15	.533	Ea.	41	28.50		69.50	88
8284	3/4"		14	.571		47.50	30.50		78	98
8286	1"	▼	13	.615		52	33		85	106
8288	1-1/4"	Q-1	22	.727		80.50	35		115.50	142
8290	1-1/2"		20	.800		104	38.50		142.50	173
8292	2"	▼	18	.889		145	43		188	223
8310	Tee, straight, 1/4"	1 Plum	10	.800		32.50	43		75.50	99.50
8312	3/8"		10	.800		33	43		76	101
8314	1/2"		9	.889		41.50	47.50		89	117
8316	3/4"		9	.889		45.50	47.50		93	121
8318	1"	▼	8	1		56	53.50		109.50	142
8320	1-1/4"	Q-1	14	1.143		80	55		135	171
8322	1-1/2"		13	1.231		83.50	59.50		143	181
8324	2"	▼	11	1.455	▼	133	70		203	252

22 11 13 – Facility Water Distribution Piping

22 11 13.45 Pipe Fittings, Steel, Threaded		Crew	Daily Output	Labor-Hours	Unit	Material	2011 Bare Costs Labor	Equipment	Total	Total Incl O&P
8326	2-1/2"	Q-1	9	1.778	Ea.	420	86		506	595
8328	3"		6	2.667		480	129		609	715
8330	4"	▼	4	4		1,225	193		1,418	1,650
8380	Couplings, straight, 1/4"	1 Plum	19	.421		14.95	22.50		37.45	50
8382	3/8"		19	.421		21.50	22.50		44	57
8384	1/2"		19	.421		22	22.50		44.50	58
8386	3/4"		18	.444		25	24		49	63
8388	1"	▼	15	.533		33	28.50		61.50	79
8390	1-1/4"	Q-1	26	.615		43	29.50		72.50	92
8392	1-1/2"		24	.667		60.50	32		92.50	115
8394	2"		21	.762		73.50	37		110.50	136
8396	2-1/2"		18	.889		193	43		236	277
8398	3"		14	1.143		230	55		285	335
8399	4"	▼	10	1.600	▼	240	77		317	380
8529	Black									
8530	Reducer, concentric, 1/4"	1 Plum	19	.421	Ea.	15.20	22.50		37.70	50
8531	3/8"		19	.421		15.20	22.50		37.70	50
8532	1/2"		17	.471		18.30	25		43.30	57.50
8533	3/4"		16	.500		26	27		53	68.50
8534	1"	▼	15	.533		31.50	28.50		60	77
8535	1-1/4"	Q-1	26	.615		46.50	29.50		76	96
8536	1-1/2"		24	.667		48.50	32		80.50	102
8537	2"	▼	21	.762		70	37		107	132
8550	Cap, 1/4"	1 Plum	38	.211		11.80	11.30		23.10	30
8551	3/8"		38	.211		11.80	11.30		23.10	30
8552	1/2"		34	.235		11.70	12.60		24.30	31.50
8553	3/4"		32	.250		14.75	13.40		28.15	36.50
8554	1"	▼	30	.267		19.35	14.30		33.65	43
8555	1-1/4"	Q-1	52	.308		21.50	14.85		36.35	45.50
8556	1-1/2"		48	.333		31.50	16.10		47.60	58.50
8557	2"	▼	42	.381		43.50	18.40		61.90	75.50
8570	Plug, 1/4"	1 Plum	38	.211		2.93	11.30		14.23	20
8571	3/8"		38	.211		3.07	11.30		14.37	20
8572	1/2"		34	.235		3.22	12.60		15.82	22.50
8573	3/4"		32	.250		4.01	13.40		17.41	24.50
8574	1"	▼	30	.267		6.15	14.30		20.45	28.50
8575	1-1/4"	Q-1	52	.308		12.65	14.85		27.50	36
8576	1-1/2"		48	.333		14.45	16.10		30.55	40
8577	2"	▼	42	.381		23	18.40		41.40	52.50
9500	Union with brass seat, 1/4"	1 Plum	15	.533		26	28.50		54.50	71
9530	3/8"		15	.533		26.50	28.50		55	71.50
9540	1/2"		14	.571		21.50	30.50		52	69
9550	3/4"		13	.615		23.50	33		56.50	75
9560	1"	▼	12	.667		31	35.50		66.50	87.50
9570	1-1/4"	Q-1	21	.762		50.50	37		87.50	111
9580	1-1/2"		19	.842		52.50	40.50		93	119
9590	2"		17	.941		65	45.50		110.50	140
9600	2-1/2"		13	1.231		210	59.50		269.50	320
9610	3"		9	1.778		272	86		358	425
9620	4"	▼	5	3.200		865	154		1,019	1,175
9630	Union, all iron, 1/4"	1 Plum	15	.533		40.50	28.50		69	87
9650	3/8"		15	.533		40.50	28.50		69	87
9660	1/2"	▼	14	.571	▼	42.50	30.50		73	92

22 11 13.45 Pipe Fittings, Steel, Threaded		Crew	Daily Output	Labor-Hours	Unit	Material	2011 Bare Costs Labor	Equipment	Total	Total Incl O&P
9670	3/4"	1 Plum	13	.615	Ea.	45.50	33		78.50	99
9680	1"	↓	12	.667		56	35.50		91.50	115
9690	1-1/4"	Q-1	21	.762		86	37		123	150
9700	1-1/2"		19	.842		105	40.50		145.50	176
9710	2"		17	.941		133	45.50		178.50	214
9720	2-1/2"		13	1.231		235	59.50		294.50	345
9730	3"	↓	9	1.778		445	86		531	620
9750	For galvanized unions, add				↓	15%				
9757	Forged steel, 3000 lb.									
9758	Black									
9760	90° Elbow, 1/4"	1 Plum	16	.500	Ea.	17.65	27		44.65	59.50
9761	3/8"		16	.500		17.65	27		44.65	59.50
9762	1/2"		15	.533		13.55	28.50		42.05	57.50
9763	3/4"		14	.571		17	30.50		47.50	64
9764	1"	↓	13	.615		25	33		58	77
9765	1-1/4"	Q-1	22	.727		48	35		83	105
9766	1-1/2"		20	.800		62	38.50		100.50	126
9767	2"	↓	18	.889		75	43		118	147
9780	45° Elbow 1/4"	1 Plum	16	.500		22.50	27		49.50	64.50
9781	3/8"		16	.500		22.50	27		49.50	64.50
9782	1/2"		15	.533		22	28.50		50.50	67
9783	3/4"		14	.571		25.50	30.50		56	73.50
9784	1"	↓	13	.615		35	33		68	87.50
9785	1-1/4"	Q-1	22	.727		48	35		83	106
9786	1-1/2"		20	.800		69	38.50		107.50	134
9787	2"	↓	18	.889		95	43		138	169
9800	Tee, 1/4"	1 Plum	10	.800		21.50	43		64.50	87.50
9801	3/8"		10	.800		21.50	43		64.50	87.50
9802	1/2"		9	.889		19.30	47.50		66.80	92
9803	3/4"		9	.889		26	47.50		73.50	99.50
9804	1"	↓	8	1		34.50	53.50		88	118
9805	1-1/4"	Q-1	14	1.143		66.50	55		121.50	156
9806	1-1/2"		13	1.231		78	59.50		137.50	174
9807	2"	↓	11	1.455		101	70		171	216
9820	Reducer, concentric, 1/4"	1 Plum	19	.421		10.55	22.50		33.05	45
9821	3/8"		19	.421		11	22.50		33.50	45.50
9822	1/2"		17	.471		11	25		36	49.50
9823	3/4"		16	.500		13.05	27		40.05	54.50
9824	1"	↓	15	.533		17	28.50		45.50	61
9825	1-1/4"	Q-1	26	.615		30	29.50		59.50	77.50
9826	1-1/2"		24	.667		32.50	32		64.50	83.50
9827	2"	↓	21	.762		47	37		84	107
9840	Cap, 1/4"	1 Plum	38	.211		6.50	11.30		17.80	24
9841	3/8"		38	.211		6.50	11.30		17.80	24
9842	1/2"		34	.235		6.30	12.60		18.90	26
9843	3/4"		32	.250		8.80	13.40		22.20	29.50
9844	1"	↓	30	.267		13.50	14.30		27.80	36.50
9845	1-1/4"	Q-1	52	.308		21.50	14.85		36.35	46
9846	1-1/2"		48	.333		26	16.10		42.10	52.50
9847	2"	↓	42	.381		37.50	18.40		55.90	68.50
9860	Plug, 1/4"	1 Plum	38	.211		2.93	11.30		14.23	20
9861	3/8"		38	.211		3.07	11.30		14.37	20
9862	1/2"	↓	34	.235		3.22	12.60		15.82	22.50

22 11 13 – Facility Water Distribution Piping

22 11 13.45 Pipe Fittings, Steel, Threaded

		Crew	Daily Output	Labor-Hours	Unit	Material	2011 Bare Costs Labor	Equipment	Total	Total Incl O&P
9863	3/4"	1 Plum	32	.250	Ea.	4.01	13.40		17.41	24.50
9864	1"	▼	30	.267		6.15	14.30		20.45	28.50
9865	1-1/4"	Q-1	52	.308		12.65	14.85		27.50	36
9866	1-1/2"		48	.333		14.45	16.10		30.55	40
9867	2"	▼	42	.381		22.50	18.40		40.90	52.50
9880	Union, bronze seat, 1/4"	1 Plum	15	.533		48	28.50		76.50	95
9881	3/8"		15	.533		48	28.50		76.50	95
9882	1/2"		14	.571		46	30.50		76.50	96
9883	3/4"		13	.615		61.50	33		94.50	117
9884	1"	▼	12	.667		69.50	35.50		105	130
9885	1-1/4"	Q-1	21	.762		148	37		185	218
9886	1-1/2"		19	.842		139	40.50		179.50	214
9887	2"	▼	17	.941		164	45.50		209.50	249
9900	Coupling, 1/4"	1 Plum	19	.421		6.40	22.50		28.90	40.50
9901	3/8"		19	.421		6.40	22.50		28.90	40.50
9902	1/2"		17	.471		5.55	25		30.55	43.50
9903	3/4"		16	.500		7.20	27		34.20	48
9904	1"	▼	15	.533		12.75	28.50		41.25	56.50
9905	1-1/4"	Q-1	26	.615		21	29.50		50.50	67.50
9906	1-1/2"		24	.667		27	32		59	77.50
9907	2"	▼	21	.762	▼	33.50	37		70.50	92

22 11 13.47 Pipe Fittings, Steel

		Crew	Daily Output	Labor-Hours	Unit	Material	2011 Bare Costs Labor	Equipment	Total	Total Incl O&P
0010	**PIPE FITTINGS, STEEL**, Flanged, Welded & Special									
0020	Flanged joints, C.I., standard weight, black. One gasket & bolt									
0040	set, mat'l only, required at each joint, not included (see line 0620)									
0060	90° Elbow, straight, 1-1/2" pipe size	Q-1	14	1.143	Ea.	415	55		470	540
0080	2" pipe size		13	1.231		246	59.50		305.50	360
0090	2-1/2" pipe size		12	1.333		266	64.50		330.50	390
0100	3" pipe size		11	1.455		222	70		292	350
0110	4" pipe size		8	2		275	96.50		371.50	445
0120	5" pipe size	▼	7	2.286		650	110		760	880
0130	6" pipe size	Q-2	9	2.667		430	133		563	675
0140	8" pipe size		8	3		740	150		890	1,050
0150	10" pipe size		7	3.429		1,625	172		1,797	2,050
0160	12" pipe size	▼	6	4	▼	3,275	200		3,475	3,900
0171	90° Elbow, reducing									
0172	2-1/2" by 2" pipe size	Q-1	12	1.333	Ea.	805	64.50		869.50	980
0173	3" by 2-1/2" pipe size		11	1.455		825	70		895	1,000
0174	4" by 3" pipe size		8	2		610	96.50		706.50	815
0175	5" by 3" pipe size	▼	7	2.286		1,300	110		1,410	1,600
0176	6" by 4" pipe size	Q-2	9	2.667		755	133		888	1,025
0177	8" by 6" pipe size		8	3		1,100	150		1,250	1,425
0178	10" by 8" pipe size		7	3.429		2,100	172		2,272	2,550
0179	12" by 10" pipe size	▼	6	4		4,025	200		4,225	4,725
0200	45° Elbow, straight, 1-1/2" pipe size	Q-1	14	1.143		500	55		555	635
0220	2" pipe size		13	1.231		355	59.50		414.50	480
0230	2-1/2" pipe size		12	1.333		380	64.50		444.50	515
0240	3" pipe size		11	1.455		370	70		440	510
0250	4" pipe size		8	2		415	96.50		511.50	605
0260	5" pipe size	▼	7	2.286		995	110		1,105	1,275
0270	6" pipe size	Q-2	9	2.667		680	133		813	950
0280	8" pipe size	▼	8	3		995	150		1,145	1,325

22 11 Facility Water Distribution

22 11 13 – Facility Water Distribution Piping

22 11 13.47 Pipe Fittings, Steel		Crew	Daily Output	Labor-Hours	Unit	Material	2011 Bare Costs Labor	Equipment	Total	Total Incl O&P
0290	10" pipe size	Q-2	7	3.429	Ea.	2,100	172		2,272	2,550
0300	12" pipe size	↓	6	4	↓	3,200	200		3,400	3,825
0310	Cross, straight									
0311	2-1/2" pipe size	Q-1	6	2.667	Ea.	780	129		909	1,050
0312	3" pipe size		5	3.200		820	154		974	1,125
0313	4" pipe size		4	4		1,075	193		1,268	1,475
0314	5" pipe size	↓	3	5.333		2,250	257		2,507	2,850
0315	6" pipe size	Q-2	5	4.800		2,250	240		2,490	2,825
0316	8" pipe size		4	6		3,625	300		3,925	4,425
0317	10" pipe size		3	8		4,300	400		4,700	5,325
0318	12" pipe size	↓	2	12		6,550	600		7,150	8,100
0350	Tee, straight, 1-1/2" pipe size	Q-1	10	1.600		475	77		552	640
0370	2" pipe size		9	1.778		269	86		355	425
0380	2-1/2" pipe size		8	2		390	96.50		486.50	575
0390	3" pipe size		7	2.286		275	110		385	465
0400	4" pipe size		5	3.200		420	154		574	690
0410	5" pipe size	↓	4	4		1,125	193		1,318	1,525
0420	6" pipe size	Q-2	6	4		605	200		805	965
0430	8" pipe size		5	4.800		1,025	240		1,265	1,500
0440	10" pipe size		4	6		2,800	300		3,100	3,525
0450	12" pipe size	↓	3	8	↓	4,375	400		4,775	5,425
0459	Tee, reducing on outlet									
0460	2-1/2" by 2" pipe size	Q-1	8	2	Ea.	740	96.50		836.50	960
0461	3" by 2-1/2" pipe size		7	2.286		815	110		925	1,075
0462	4" by 3" pipe size		5	3.200		860	154		1,014	1,175
0463	5" by 4" pipe size	↓	4	4		1,850	193		2,043	2,350
0464	6" by 4" pipe size	Q-2	6	4		830	200		1,030	1,200
0465	8" by 6" pipe size		5	4.800		1,300	240		1,540	1,775
0466	10" by 8" pipe size		4	6		2,950	300		3,250	3,700
0467	12" by 10" pipe size	↓	3	8	↓	5,125	400		5,525	6,225
0476	Reducer, concentric									
0477	3" by 2-1/2"	Q-1	12	1.333	Ea.	495	64.50		559.50	635
0478	4" by 3"		9	1.778		660	86		746	855
0479	5" by 4"	↓	8	2		855	96.50		951.50	1,075
0480	6" by 4"	Q-2	10	2.400		740	120		860	995
0481	8" by 6"		9	2.667		935	133		1,068	1,225
0482	10" by 8"		8	3		1,925	150		2,075	2,350
0483	12" by 10"	↓	7	3.429	↓	3,325	172		3,497	3,900
0492	Reducer, eccentric									
0493	4" by 3"	Q-1	8	2	Ea.	900	96.50		996.50	1,125
0494	5" by 4"	"	7	2.286		1,375	110		1,485	1,675
0495	6" by 4"	Q-2	9	2.667		850	133		983	1,125
0496	8" by 6"		8	3		1,075	150		1,225	1,400
0497	10" by 8"		7	3.429		2,750	172		2,922	3,275
0498	12" by 10"	↓	6	4		3,700	200		3,900	4,375
0500	For galvanized elbows and tees, add					100%				
0520	For extra heavy weight elbows and tees, add					140%				
0620	Gasket and bolt set, 150#, 1/2" pipe size	1 Plum	20	.400		2.84	21.50		24.34	35
0622	3/4" pipe size		19	.421		3.10	22.50		25.60	37
0624	1" pipe size		18	.444		3.29	24		27.29	39
0626	1-1/4" pipe size		17	.471		2.67	25		27.67	40.50
0628	1-1/2" pipe size		15	.533		2.84	28.50		31.34	45.50
0630	2" pipe size	↓	13	.615	↓	5.15	33		38.15	54.50

158

22 11 13 – Facility Water Distribution Piping

22 11 13.47 Pipe Fittings, Steel	Crew	Daily Output	Labor-Hours	Unit	Material	2011 Bare Costs Labor	Equipment	Total	Total Incl O&P
0640	2-1/2" pipe size	1 Plum	12	.667	Ea.	7.10	35.50		42.60	61.50
0650	3" pipe size		11	.727		5.50	39		44.50	64
0660	3-1/2" pipe size		9	.889		8.70	47.50		56.20	80.50
0670	4" pipe size		8	1		10.10	53.50		63.60	91
0680	5" pipe size		7	1.143		16.30	61.50		77.80	109
0690	6" pipe size		6	1.333		17.50	71.50		89	126
0700	8" pipe size		5	1.600		20.50	86		106.50	151
0710	10" pipe size		4.50	1.778		33.50	95.50		129	179
0720	12" pipe size		4.20	1.905		36	102		138	192
0730	14" pipe size		4	2		38	107		145	202
0740	16" pipe size		3	2.667		39.50	143		182.50	257
0750	18" pipe size		2.70	2.963		77	159		236	320
0760	20" pipe size		2.30	3.478		125	186		311	415
0780	24" pipe size		1.90	4.211		156	226		382	505
0790	26" pipe size		1.60	5		211	268		479	630
0810	30" pipe size		1.40	5.714		410	305		715	910
0830	36" pipe size		1.10	7.273		770	390		1,160	1,425
0850	For 300 lb gasket set, add					40%				
2000	Flanged unions, 125 lb., black, 1/2" pipe size	1 Plum	17	.471	Ea.	62.50	25		87.50	107
2040	3/4" pipe size		17	.471		82.50	25		107.50	129
2050	1" pipe size		16	.500		80	27		107	128
2060	1-1/4" pipe size	Q-1	28	.571		95	27.50		122.50	145
2070	1-1/2" pipe size		27	.593		87	28.50		115.50	138
2080	2" pipe size		26	.615		103	29.50		132.50	158
2090	2-1/2" pipe size		24	.667		139	32		171	201
2100	3" pipe size		22	.727		158	35		193	227
2110	3-1/2" pipe size		18	.889		269	43		312	360
2120	4" pipe size		16	1		215	48.50		263.50	310
2130	5" pipe size		14	1.143		500	55		555	635
2140	6" pipe size	Q-2	19	1.263		470	63		533	610
2150	8" pipe size	"	16	1.500		1,075	75		1,150	1,275
2200	For galvanized unions, add					150%				
2290	Threaded flange									
2300	Cast iron									
2310	Black, 125 lb., per flange									
2320	1" pipe size	1 Plum	27	.296	Ea.	36	15.90		51.90	63
2330	1-1/4" pipe size	Q-1	44	.364		43	17.55		60.55	73
2340	1-1/2" pipe size		40	.400		39.50	19.30		58.80	72.50
2350	2" pipe size		36	.444		39.50	21.50		61	75.50
2360	2-1/2" pipe size		28	.571		46.50	27.50		74	92
2370	3" pipe size		20	.800		59.50	38.50		98	123
2380	3-1/2" pipe size		16	1		83.50	48.50		132	164
2390	4" pipe size		12	1.333		80.50	64.50		145	185
2400	5" pipe size		10	1.600		114	77		191	240
2410	6" pipe size	Q-2	14	1.714		129	86		215	269
2420	8" pipe size		12	2		203	100		303	370
2430	10" pipe size		10	2.400		360	120		480	575
2440	12" pipe size		8	3		790	150		940	1,100
2460	For galvanized flanges, add					95%				
2490	Blind flange									
2492	Cast iron									
2494	Black, 125 lb., per flange									
2496	1" pipe size	1 Plum	27	.296	Ea.	57	15.90		72.90	86.50

22 11 Facility Water Distribution

22 11 13 – Facility Water Distribution Piping

22 11 13.47 Pipe Fittings, Steel		Crew	Daily Output	Labor-Hours	Unit	Material	2011 Bare Costs Labor	Equipment	Total	Total Incl O&P
2500	1-1/2" pipe size	Q-1	40	.400	Ea.	64	19.30		83.30	99.50
2502	2" pipe size		36	.444		72	21.50		93.50	112
2504	2-1/2" pipe size		28	.571		79.50	27.50		107	129
2506	3" pipe size		20	.800		95.50	38.50		134	163
2508	4" pipe size		12	1.333		122	64.50		186.50	230
2510	5" pipe size		10	1.600		198	77		275	335
2512	6" pipe size	Q-2	14	1.714		216	86		302	365
2514	8" pipe size		12	2		340	100		440	525
2516	10" pipe size		10	2.400		500	120		620	730
2518	12" pipe size		8	3		940	150		1,090	1,250
2520	For galvanized flanges, add					80%				
2570	Threaded flange									
2580	Forged steel,									
2590	Black 150 lb., per flange									
2600	1/2" pipe size	1 Plum	30	.267	Ea.	26.50	14.30		40.80	50.50
2610	3/4" pipe size		28	.286		26.50	15.30		41.80	52
2620	1" pipe size		27	.296		26.50	15.90		42.40	52.50
2630	1-1/4" pipe size	Q-1	44	.364		26.50	17.55		44.05	55
2640	1-1/2" pipe size		40	.400		26.50	19.30		45.80	58
2650	2" pipe size		36	.444		29.50	21.50		51	64.50
2660	2-1/2" pipe size		28	.571		36	27.50		63.50	81
2670	3" pipe size		20	.800		36.50	38.50		75	97.50
2690	4" pipe size		12	1.333		42.50	64.50		107	143
2700	5" pipe size		10	1.600		66.50	77		143.50	189
2710	6" pipe size	Q-2	14	1.714		73	86		159	209
2720	8" pipe size		12	2		126	100		226	287
2730	10" pipe size		10	2.400		228	120		348	430
2860	Black 300 lb., per flange									
2870	1/2" pipe size	1 Plum	30	.267	Ea.	29	14.30		43.30	53.50
2880	3/4" pipe size		28	.286		29	15.30		44.30	55
2890	1" pipe size		27	.296		29	15.90		44.90	55.50
2900	1-1/4" pipe size	Q-1	44	.364		29	17.55		46.55	58
2910	1-1/2" pipe size		40	.400		29	19.30		48.30	61
2920	2" pipe size		36	.444		34	21.50		55.50	69.50
2930	2-1/2" pipe size		28	.571		47.50	27.50		75	93
2940	3" pipe size		20	.800		50.50	38.50		89	113
2960	4" pipe size		12	1.333		73	64.50		137.50	176
2970	6" pipe size	Q-2	14	1.714		136	86		222	278
3000	Weld joint, butt, carbon steel, standard weight									
3040	90° elbow, long radius									
3050	1/2" pipe size	Q-15	16	1	Ea.	45	48.50	2.03	95.53	124
3060	3/4" pipe size		16	1		45	48.50	2.03	95.53	124
3070	1" pipe size		16	1		21	48.50	2.03	71.53	97
3080	1-1/4" pipe size		14	1.143		21	55	2.32	78.32	108
3090	1-1/2" pipe size		13	1.231		21	59.50	2.50	83	114
3100	2" pipe size		10	1.600		22.50	77	3.25	102.75	144
3110	2-1/2" pipe size		8	2		27.50	96.50	4.06	128.06	179
3120	3" pipe size		7	2.286		29	110	4.64	143.64	202
3130	4" pipe size		5	3.200		48	154	6.50	208.50	291
3136	5" pipe size		4	4		103	193	8.10	304.10	410
3140	6" pipe size	Q-16	5	4.800		106	240	6.50	352.50	485
3150	8" pipe size		3.75	6.400		200	320	8.65	528.65	710
3160	10" pipe size		3	8		400	400	10.80	810.80	1,050

22 11 13.47 Pipe Fittings, Steel		Crew	Daily Output	Labor-Hours	Unit	Material	2011 Bare Costs Labor	Equipment	Total	Total Incl O&P
3170	12" pipe size	Q-16	2.50	9.600	Ea.	590	480	12.95	1,082.95	1,375
3180	14" pipe size		2	12		970	600	16.20	1,586.20	2,000
3190	16" pipe size		1.50	16		1,325	800	21.50	2,146.50	2,700
3191	18" pipe size		1.25	19.200		1,525	960	26	2,511	3,125
3192	20" pipe size		1.15	20.870		2,450	1,050	28	3,528	4,275
3194	24" pipe size		1.02	23.529		3,450	1,175	32	4,657	5,575
3200	45° Elbow, long									
3210	1/2" pipe size	Q-15	16	1	Ea.	62.50	48.50	2.03	113.03	143
3220	3/4" pipe size		16	1		62.50	48.50	2.03	113.03	143
3230	1" pipe size		16	1		22	48.50	2.03	72.53	98.50
3240	1-1/4" pipe size		14	1.143		22	55	2.32	79.32	110
3250	1-1/2" pipe size		13	1.231		22	59.50	2.50	84	116
3260	2" pipe size		10	1.600		22	77	3.25	102.25	143
3270	2-1/2" pipe size		8	2		26.50	96.50	4.06	127.06	177
3280	3" pipe size		7	2.286		27.50	110	4.64	142.14	201
3290	4" pipe size		5	3.200		49.50	154	6.50	210	293
3296	5" pipe size		4	4		75	193	8.10	276.10	380
3300	6" pipe size	Q-16	5	4.800		97.50	240	6.50	344	475
3310	8" pipe size		3.75	6.400		163	320	8.65	491.65	670
3320	10" pipe size		3	8		325	400	10.80	735.80	960
3330	12" pipe size		2.50	9.600		460	480	12.95	952.95	1,225
3340	14" pipe size		2	12		615	600	16.20	1,231.20	1,600
3341	16" pipe size		1.50	16		1,100	800	21.50	1,921.50	2,450
3342	18" pipe size		1.25	19.200		1,550	960	26	2,536	3,175
3343	20" pipe size		1.15	20.870		1,625	1,050	28	2,703	3,350
3345	24" pipe size		1.05	22.857		2,300	1,150	31	3,481	4,250
3346	26" pipe size		.85	28.235		2,700	1,425	38	4,163	5,125
3347	30" pipe size		.45	53.333		2,975	2,675	72	5,722	7,325
3349	36" pipe size		.38	63.158		3,275	3,150	85.50	6,510.50	8,450
3350	Tee, straight									
3352	For reducing tees and concentrics see starting line 4600									
3360	1/2" pipe size	Q-15	10	1.600	Ea.	110	77	3.25	190.25	240
3370	3/4" pipe size		10	1.600		110	77	3.25	190.25	240
3380	1" pipe size		10	1.600		54.50	77	3.25	134.75	179
3390	1-1/4" pipe size		9	1.778		68.50	86	3.61	158.11	207
3400	1-1/2" pipe size		8	2		68.50	96.50	4.06	169.06	224
3410	2" pipe size		6	2.667		54.50	129	5.40	188.90	258
3420	2-1/2" pipe size		5	3.200		75.50	154	6.50	236	320
3430	3" pipe size		4	4		84	193	8.10	285.10	390
3440	4" pipe size		3	5.333		118	257	10.85	385.85	525
3446	5" pipe size		2.50	6.400		195	310	13	518	690
3450	6" pipe size	Q-16	3	8		203	400	10.80	613.80	830
3460	8" pipe size		2.50	9.600		355	480	12.95	847.95	1,125
3470	10" pipe size		2	12		695	600	16.20	1,311.20	1,675
3480	12" pipe size		1.60	15		975	750	20.50	1,745.50	2,225
3481	14" pipe size		1.30	18.462		1,700	925	25	2,650	3,275
3482	16" pipe size		1	24		1,900	1,200	32.50	3,132.50	3,925
3483	18" pipe size		.80	30		3,025	1,500	40.50	4,565.50	5,625
3484	20" pipe size		.75	32		4,750	1,600	43	6,393	7,675
3486	24" pipe size		.70	34.286		6,150	1,725	46.50	7,921.50	9,350
3487	26" pipe size		.55	43.636		6,775	2,175	59	9,009	10,800
3488	30" pipe size		.30	80		7,425	4,000	108	11,533	14,200
3490	36" pipe size		.25	96		8,200	4,800	130	13,130	16,300

22 11 13.47 Pipe Fittings, Steel		Crew	Daily Output	Labor-Hours	Unit	Material	2011 Bare Costs		Total	Total Incl O&P
							Labor	Equipment		
3491	Eccentric reducer, 1-1/2" pipe size	Q-15	14	1.143	Ea.	30.50	55	2.32	87.82	119
3492	2" pipe size		11	1.455		43	70	2.95	115.95	156
3493	2-1/2" pipe size		9	1.778		50	86	3.61	139.61	187
3494	3" pipe size		8	2		59	96.50	4.06	159.56	213
3495	4" pipe size		6	2.667		75.50	129	5.40	209.90	281
3496	6" pipe size	Q-16	5	4.800		232	240	6.50	478.50	620
3497	8" pipe size		4	6		345	300	8.10	653.10	840
3498	10" pipe size		3	8		435	400	10.80	845.80	1,075
3499	12" pipe size		2.50	9.600		385	480	12.95	877.95	1,150
3501	Cap, 1-1/2" pipe size	Q-15	28	.571		25.50	27.50	1.16	54.16	70.50
3502	2" pipe size		22	.727		25.50	35	1.48	61.98	82
3503	2-1/2" pipe size		18	.889		26.50	43	1.80	71.30	95
3504	3" pipe size		16	1		26.50	48.50	2.03	77.03	103
3505	4" pipe size		12	1.333		38.50	64.50	2.71	105.71	141
3506	6" pipe size	Q-16	10	2.400		65.50	120	3.24	188.74	255
3507	8" pipe size		8	3		99	150	4.05	253.05	335
3508	10" pipe size		6	4		180	200	5.40	385.40	505
3509	12" pipe size		5	4.800		272	240	6.50	518.50	665
3511	14" pipe size		4	6		293	300	8.10	601.10	785
3512	16" pipe size		4	6		455	300	8.10	763.10	960
3513	18" pipe size		3	8		575	400	10.80	985.80	1,250
3517	Weld joint, butt, carbon steel, extra strong									
3519	90° elbow, long									
3520	1/2" pipe size	Q-15	13	1.231	Ea.	55.50	59.50	2.50	117.50	153
3530	3/4" pipe size		12	1.333		55.50	64.50	2.71	122.71	160
3540	1" pipe size		11	1.455		27	70	2.95	99.95	138
3550	1-1/4" pipe size		10	1.600		27	77	3.25	107.25	149
3560	1-1/2" pipe size		9	1.778		27	86	3.61	116.61	162
3570	2" pipe size		8	2		27.50	96.50	4.06	128.06	179
3580	2-1/2" pipe size		7	2.286		38.50	110	4.64	153.14	213
3590	3" pipe size		6	2.667		49.50	129	5.40	183.90	252
3600	4" pipe size		4	4		81.50	193	8.10	282.60	385
3606	5" pipe size		3.50	4.571		195	221	9.30	425.30	555
3610	6" pipe size	Q-16	4.50	5.333		206	267	7.20	480.20	635
3620	8" pipe size		3.50	6.857		395	345	9.25	749.25	950
3630	10" pipe size		2.50	9.600		830	480	12.95	1,322.95	1,650
3640	12" pipe size		2.25	10.667		1,025	535	14.40	1,574.40	1,925
3650	45° Elbow, long									
3660	1/2" pipe size	Q-15	13	1.231	Ea.	62	59.50	2.50	124	159
3670	3/4" pipe size		12	1.333		62	64.50	2.71	129.21	167
3680	1" pipe size		11	1.455		29	70	2.95	101.95	140
3690	1-1/4" pipe size		10	1.600		29	77	3.25	109.25	150
3700	1-1/2" pipe size		9	1.778		29	86	3.61	118.61	163
3710	2" pipe size		8	2		29	96.50	4.06	129.56	180
3720	2-1/2" pipe size		7	2.286		64	110	4.64	178.64	241
3730	3" pipe size		6	2.667		37	129	5.40	171.40	239
3740	4" pipe size		4	4		58.50	193	8.10	259.60	360
3746	5" pipe size		3.50	4.571		139	221	9.30	369.30	495
3750	6" pipe size	Q-16	4.50	5.333		160	267	7.20	434.20	585
3760	8" pipe size		3.50	6.857		276	345	9.25	630.25	825
3770	10" pipe size		2.50	9.600		540	480	12.95	1,032.95	1,325
3780	12" pipe size		2.25	10.667		790	535	14.40	1,339.40	1,675
3800	Tee, straight									

22 11 13 – Facility Water Distribution Piping

22 11 13.47 Pipe Fittings, Steel	Crew	Daily Output	Labor-Hours	Unit	Material	2011 Bare Costs Labor	Equipment	Total	Total Incl O&P
3810	1/2" pipe size	Q-15	9	1.778	Ea.	136	86	3.61	225.61	281
3820	3/4" pipe size		8.50	1.882		131	91	3.82	225.82	284
3830	1" pipe size		8	2		72.50	96.50	4.06	173.06	228
3840	1-1/4" pipe size		7	2.286		67	110	4.64	181.64	244
3850	1-1/2" pipe size		6	2.667		81.50	129	5.40	215.90	288
3860	2" pipe size		5	3.200		68.50	154	6.50	229	315
3870	2-1/2" pipe size		4	4		114	193	8.10	315.10	420
3880	3" pipe size		3.50	4.571		106	221	9.30	336.30	455
3890	4" pipe size		2.50	6.400		171	310	13	494	660
3896	5" pipe size		2.25	7.111		375	345	14.45	734.45	935
3900	6" pipe size	Q-16	2.25	10.667		288	535	14.40	837.40	1,125
3910	8" pipe size		2	12		545	600	16.20	1,161.20	1,525
3920	10" pipe size		1.75	13.714		965	685	18.50	1,668.50	2,125
3930	12" pipe size		1.50	16		1,400	800	21.50	2,221.50	2,775
4000	Eccentric reducer, 1-1/2" pipe size	Q-15	10	1.600		17.30	77	3.25	97.55	138
4010	2" pipe size		9	1.778		47.50	86	3.61	137.11	184
4020	2-1/2" pipe size		8	2		71.50	96.50	4.06	172.06	227
4030	3" pipe size		7	2.286		59	110	4.64	173.64	235
4040	4" pipe size		5	3.200		95.50	154	6.50	256	345
4046	5" pipe size		4.70	3.404		245	164	6.90	415.90	520
4050	6" pipe size	Q-16	4.50	5.333		259	267	7.20	533.20	695
4060	8" pipe size		3.50	6.857		390	345	9.25	744.25	950
4070	10" pipe size		2.50	9.600		665	480	12.95	1,157.95	1,450
4080	12" pipe size		2.25	10.667		855	535	14.40	1,404.40	1,750
4090	14" pipe size		2.10	11.429		1,575	570	15.45	2,160.45	2,600
4100	16" pipe size		1.90	12.632		2,025	630	17.05	2,672.05	3,200
4151	Cap, 1-1/2" pipe size	Q-15	24	.667		31.50	32	1.35	64.85	84.50
4152	2" pipe size		18	.889		25.50	43	1.80	70.30	94
4153	2-1/2" pipe size		16	1		34.50	48.50	2.03	85.03	112
4154	3" pipe size		14	1.143		38.50	55	2.32	95.82	127
4155	4" pipe size		10	1.600		89	77	3.25	169.25	217
4156	6" pipe size	Q-16	9	2.667		107	133	3.60	243.60	320
4157	8" pipe size		7	3.429		160	172	4.63	336.63	435
4158	10" pipe size		5	4.800		245	240	6.50	491.50	635
4159	12" pipe size		4	6		330	300	8.10	638.10	820
4190	Weld fittings, reducing, standard weight									
4200	Welding ring w/spacer pins, 2" pipe size				Ea.	3.57			3.57	3.93
4210	2-1/2" pipe size					4.32			4.32	4.75
4220	3" pipe size					4.47			4.47	4.92
4230	4" pipe size					4.65			4.65	5.10
4236	5" pipe size					5.45			5.45	6
4240	6" pipe size					5.45			5.45	6
4250	8" pipe size					6.45			6.45	7.10
4260	10" pipe size					7.90			7.90	8.70
4270	12" pipe size					9.10			9.10	10
4280	14" pipe size					10.55			10.55	11.60
4290	16" pipe size					11.90			11.90	13.10
4300	18" pipe size					13.75			13.75	15.10
4310	20" pipe size					14.55			14.55	16.05
4330	24" pipe size					16.95			16.95	18.65
4340	26" pipe size					22.50			22.50	25
4350	30" pipe size					22.50			22.50	25
4370	36" pipe size					25.50			25.50	28

22 11 13.47 Pipe Fittings, Steel	Crew	Daily Output	Labor-Hours	Unit	Material	2011 Bare Costs Labor	Equipment	Total	Total Incl O&P
4600	Tee, reducing on outlet								
4601	2-1/2" x 2" pipe size	Q-15	5	3.200	Ea.	98	154	6.50	258.50	345
4602	3" x 2-1/2" pipe size		4	4		118	193	8.10	319.10	425
4604	4" x 3" pipe size		3	5.333		124	257	10.85	391.85	535
4605	5" x 4" pipe size		2.50	6.400		268	310	13	591	770
4606	6" x 5" pipe size	Q-16	3	8		370	400	10.80	780.80	1,025
4607	8" x 6" pipe size		2.50	9.600		460	480	12.95	952.95	1,225
4608	10" x 8" pipe size		2	12		850	600	16.20	1,466.20	1,850
4609	12" x 10" pipe size		1.60	15		1,250	750	20.50	2,020.50	2,525
4610	16" x 12" pipe size		1.50	16		2,000	800	21.50	2,821.50	3,425
4611	14" x 12" pipe size		1.52	15.789		1,900	790	21.50	2,711.50	3,275
4618	Reducer, concentric								
4619	2-1/2" by 2" pipe size	Q-15	10	1.600	Ea.	42.50	77	3.25	122.75	166
4620	3" by 2-1/2" pipe size		9	1.778		35	86	3.61	124.61	170
4621	3-1/2" by 3" pipe size		8	2		89.50	96.50	4.06	190.06	246
4622	4" by 2-1/2" pipe size		7	2.286		54.50	110	4.64	169.14	230
4623	5" by 3" pipe size		7	2.286		114	110	4.64	228.64	295
4624	6" by 4" pipe size	Q-16	6	4		94	200	5.40	299.40	410
4625	8" by 6" pipe size		5	4.800		124	240	6.50	370.50	505
4626	10" by 8" pipe size		4	6		256	300	8.10	564.10	740
4627	12" by 10" pipe size		3	8		296	400	10.80	706.80	930
4660	Reducer, eccentric								
4662	3" x 2" pipe size	Q-15	8	2	Ea.	59	96.50	4.06	159.56	213
4664	4" x 3" pipe size		6	2.667		71.50	129	5.40	205.90	277
4666	4" x 2" pipe size		6	2.667		87.50	129	5.40	221.90	294
4670	6" x 4" pipe size	Q-16	5	4.800		148	240	6.50	394.50	530
4672	6" x 3" pipe size		5	4.800		232	240	6.50	478.50	620
4676	8" x 6" pipe size		4	6		188	300	8.10	496.10	665
4678	8" x 4" pipe size		4	6		345	300	8.10	653.10	840
4682	10" x 8" pipe size		3	8		247	400	10.80	657.80	880
4684	10" x 6" pipe size		3	8		435	400	10.80	845.80	1,075
4688	12" x 10" pipe size		2.50	9.600		385	480	12.95	877.95	1,150
4690	12" x 8" pipe size		2.50	9.600		605	480	12.95	1,097.95	1,400
4691	14" x 12" pipe size		2.20	10.909		675	545	14.75	1,234.75	1,575
4693	16" x 14" pipe size		1.80	13.333		1,025	665	18	1,708	2,150
4694	16" x 12" pipe size		2	12		1,375	600	16.20	1,991.20	2,425
4696	18" x 16" pipe size		1.60	15		1,075	750	20.50	2,645.50	3,225
5000	Weld joint, socket, forged steel, 3000 lb., schedule 40 pipe								
5010	90° elbow, straight								
5020	1/4" pipe size	Q-15	22	.727	Ea.	21.50	35	1.48	57.98	78
5030	3/8" pipe size		22	.727		21.50	35	1.48	57.98	78
5040	1/2" pipe size		20	.800		12.05	38.50	1.62	52.17	72.50
5050	3/4" pipe size		20	.800		12.55	38.50	1.62	52.67	73
5060	1" pipe size		20	.800		16.10	38.50	1.62	56.22	77
5070	1-1/4" pipe size		18	.889		31	43	1.80	75.80	100
5080	1-1/2" pipe size		16	1		36	48.50	2.03	86.53	114
5090	2" pipe size		12	1.333		53	64.50	2.71	120.21	157
5100	2-1/2" pipe size		10	1.600		150	77	3.25	230.25	284
5110	3" pipe size		8	2		259	96.50	4.06	359.56	435
5120	4" pipe size		6	2.667		685	129	5.40	819.40	955
5130	45° Elbow, straight								
5134	1/4" pipe size	Q-15	22	.727	Ea.	21.50	35	1.48	57.98	78
5135	3/8" pipe size		22	.727		21.50	35	1.48	57.98	78

22 11 Facility Water Distribution

22 11 13 – Facility Water Distribution Piping

22 11 13.47 Pipe Fittings, Steel	Crew	Daily Output	Labor-Hours	Unit	Material	2011 Bare Costs Labor	Equipment	Total	Total Incl O&P
5136	1/2" pipe size	Q-15	20	.800	Ea.	16.05	38.50	1.62	56.17	77
5137	3/4" pipe size		20	.800		18.65	38.50	1.62	58.77	80
5140	1" pipe size		20	.800		24.50	38.50	1.62	64.62	86.50
5150	1-1/4" pipe size		18	.889		33.50	43	1.80	78.30	103
5160	1-1/2" pipe size		16	1		41	48.50	2.03	91.53	119
5170	2" pipe size		12	1.333		66	64.50	2.71	133.21	171
5180	2-1/2" pipe size		10	1.600		173	77	3.25	253.25	310
5190	3" pipe size		8	2		288	96.50	4.06	388.56	465
5200	4" pipe size		6	2.667		565	129	5.40	699.40	825
5250	Tee, straight									
5254	1/4" pipe size	Q-15	15	1.067	Ea.	24	51.50	2.17	77.67	105
5255	3/8" pipe size		15	1.067		24	51.50	2.17	77.67	105
5256	1/2" pipe size		13	1.231		14.90	59.50	2.50	76.90	108
5257	3/4" pipe size		13	1.231		18.25	59.50	2.50	80.25	111
5260	1" pipe size		13	1.231		25	59.50	2.50	87	119
5270	1-1/4" pipe size		12	1.333		38	64.50	2.71	105.21	141
5280	1-1/2" pipe size		11	1.455		50.50	70	2.95	123.45	164
5290	2" pipe size		8	2		73	96.50	4.06	173.56	228
5300	2-1/2" pipe size		6	2.667		209	129	5.40	343.40	430
5310	3" pipe size		5	3.200		500	154	6.50	660.50	790
5320	4" pipe size		4	4		805	193	8.10	1,006.10	1,175
5350	For reducing sizes, add					60%				
5450	Couplings									
5451	1/4" pipe size	Q-15	23	.696	Ea.	13.40	33.50	1.41	48.31	66.50
5452	3/8" pipe size		23	.696		13.40	33.50	1.41	48.31	66.50
5453	1/2" pipe size		21	.762		6.20	37	1.55	44.75	63.50
5454	3/4" pipe size		21	.762		8	37	1.55	46.55	65.50
5460	1" pipe size		20	.800		8.85	38.50	1.62	48.97	69
5470	1-1/4" pipe size		20	.800		15.65	38.50	1.62	55.77	76.50
5480	1-1/2" pipe size		18	.889		17.35	43	1.80	62.15	85
5490	2" pipe size		14	1.143		27.50	55	2.32	84.82	116
5500	2-1/2" pipe size		12	1.333		62	64.50	2.71	129.21	167
5510	3" pipe size		9	1.778		139	86	3.61	228.61	285
5520	4" pipe size		7	2.286		209	110	4.64	323.64	400
5570	Union, 1/4" pipe size		21	.762		29	37	1.55	67.55	88
5571	3/8" pipe size		21	.762		29	37	1.55	67.55	88
5572	1/2" pipe size		19	.842		24	40.50	1.71	66.21	88.50
5573	3/4" pipe size		19	.842		28	40.50	1.71	70.21	93.50
5574	1" pipe size		19	.842		36	40.50	1.71	78.21	102
5575	1-1/4" pipe size		17	.941		58.50	45.50	1.91	105.91	135
5576	1-1/2" pipe size		15	1.067		63.50	51.50	2.17	117.17	149
5577	2" pipe size		11	1.455		91	70	2.95	163.95	208
5600	Reducer, 1/4" pipe size		23	.696		31.50	33.50	1.41	66.41	86
5601	3/8" pipe size		23	.696		32.50	33.50	1.41	67.41	87
5602	1/2" pipe size		21	.762		22	37	1.55	60.55	80.50
5603	3/4" pipe size		21	.762		22	37	1.55	60.55	80.50
5604	1" pipe size		21	.762		25.50	37	1.55	64.05	84.50
5605	1-1/4" pipe size		19	.842		39	40.50	1.71	81.21	105
5607	1-1/2" pipe size		17	.941		42	45.50	1.91	89.41	117
5608	2" pipe size		13	1.231		46.50	59.50	2.50	108.50	142
5612	Cap, 1/4" pipe size		46	.348		12.85	16.80	.71	30.36	40
5613	3/8" pipe size		46	.348		12.85	16.80	.71	30.36	40
5614	1/2" pipe size		42	.381		7.80	18.40	.77	26.97	37

22 11 13 – Facility Water Distribution Piping

22 11 13.47 Pipe Fittings, Steel	Crew	Daily Output	Labor-Hours	Unit	Material	2011 Bare Costs Labor	Equipment	Total	Total Incl O&P
5615	3/4" pipe size	Q-15	42	.381	Ea.	9.10	18.40	.77	28.27	38.50
5616	1" pipe size		42	.381		13.80	18.40	.77	32.97	43.50
5617	1-1/4" pipe size		38	.421		16.25	20.50	.85	37.60	49.50
5618	1-1/2" pipe size		34	.471		22.50	22.50	.96	45.96	60
5619	2" pipe size		26	.615		35	29.50	1.25	65.75	84.50
5630	T-O-L, 1/4" pipe size, nozzle		23	.696		6.70	33.50	1.41	41.61	59
5631	3/8" pipe size, nozzle		23	.696		6.80	33.50	1.41	41.71	59
5632	1/2" pipe size, nozzle		22	.727		6.70	35	1.48	43.18	61.50
5633	3/4" pipe size, nozzle		21	.762		7.70	37	1.55	46.25	65
5634	1" pipe size, nozzle		20	.800		9	38.50	1.62	49.12	69
5635	1-1/4" pipe size, nozzle		18	.889		13.85	43	1.80	58.65	81
5636	1-1/2" pipe size, nozzle		16	1		14.25	48.50	2.03	64.78	90
5637	2" pipe size, nozzle		12	1.333		14.25	64.50	2.71	81.46	115
5638	2-1/2" pipe size, nozzle		10	1.600		54.50	77	3.25	134.75	179
5639	4" pipe size, nozzle		6	2.667		123	129	5.40	257.40	335
5640	W-O-L, 1/4" pipe size, nozzle		23	.696		15.85	33.50	1.41	50.76	69
5641	3/8" pipe size, nozzle		23	.696		14.95	33.50	1.41	49.86	68
5642	1/2" pipe size, nozzle		22	.727		13.90	35	1.48	50.38	69.50
5643	3/4" pipe size, nozzle		21	.762		14.65	37	1.55	53.20	73
5644	1" pipe size, nozzle		20	.800		15.30	38.50	1.62	55.42	76
5645	1-1/4" pipe size, nozzle		18	.889		18.15	43	1.80	62.95	86
5646	1-1/2" pipe size, nozzle		16	1		18.15	48.50	2.03	68.68	94
5647	2" pipe size, nozzle		12	1.333		18.35	64.50	2.71	85.56	119
5648	2-1/2" pipe size, nozzle		10	1.600		42	77	3.25	122.25	165
5649	3" pipe size, nozzle		8	2		46	96.50	4.06	146.56	199
5650	4" pipe size, nozzle		6	2.667		58	129	5.40	192.40	262
5651	5" pipe size, nozzle		5	3.200		143	154	6.50	303.50	395
5652	6" pipe size, nozzle		4	4		164	193	8.10	365.10	480
5653	8" pipe size, nozzle		3	5.333		315	257	10.85	582.85	740
5654	10" pipe size, nozzle		2.60	6.154		455	297	12.50	764.50	960
5655	12" pipe size, nozzle		2.20	7.273		865	350	14.75	1,229.75	1,500
5656	14" pipe size, nozzle		2	8		2,800	385	16.25	3,201.25	3,675
5674	S-O-L, 1/4" pipe size, outlet		23	.696		8.90	33.50	1.41	43.81	61.50
5675	3/8" pipe size, outlet		23	.696		8.90	33.50	1.41	43.81	61.50
5676	1/2" pipe size, outlet		22	.727		8.30	35	1.48	44.78	63.50
5677	3/4" pipe size, outlet		21	.762		8.45	37	1.55	47	66
5678	1" pipe size, outlet		20	.800		9.35	38.50	1.62	49.47	69.50
5679	1-1/4" pipe size, outlet		18	.889		15.60	43	1.80	60.40	83
5680	1-1/2" pipe size, outlet		16	1		15.60	48.50	2.03	66.13	91.50
5681	2" pipe size, outlet	↓	12	1.333	↓	17.75	64.50	2.71	84.96	119
6000	Weld-on flange, forged steel									
6020	Slip-on, 150 lb. flange (welded front and back)									
6050	1/2" pipe size	Q-15	18	.889	Ea.	17.50	43	1.80	62.30	85
6060	3/4" pipe size		18	.889		17.50	43	1.80	62.30	85
6070	1" pipe size		17	.941		17.50	45.50	1.91	64.91	89.50
6080	1-1/4" pipe size		16	1		17.50	48.50	2.03	68.03	93.50
6090	1-1/2" pipe size		15	1.067		17.50	51.50	2.17	71.17	98.50
6100	2" pipe size		12	1.333		19.45	64.50	2.71	86.66	120
6110	2-1/2" pipe size		10	1.600		33.50	77	3.25	113.75	155
6120	3" pipe size		9	1.778		26.50	86	3.61	116.11	161
6130	3-1/2" pipe size		7	2.286		33	110	4.64	147.64	207
6140	4" pipe size		6	2.667		33	129	5.40	167.40	234
6150	5" pipe size	↓	5	3.200	↓	57.50	154	6.50	218	300

22 11 Facility Water Distribution

22 11 13 – Facility Water Distribution Piping

22 11 13.47 Pipe Fittings, Steel	Crew	Daily Output	Labor-Hours	Unit	Material	2011 Bare Costs Labor	Equipment	Total	Total Incl O&P
6160	6" pipe size	Q-16	6	4	Ea.	54.50	200	5.40	259.90	365
6170	8" pipe size		5	4.800		82.50	240	6.50	329	460
6180	10" pipe size		4	6		143	300	8.10	451.10	615
6190	12" pipe size		3	8		212	400	10.80	622.80	840
6191	14" pipe size		2.50	9.600		281	480	12.95	773.95	1,050
6192	16" pipe size		1.80	13.333		440	665	18	1,123	1,500
6200	300 lb. flange									
6210	1/2" pipe size	Q-15	17	.941	Ea.	23.50	45.50	1.91	70.91	95.50
6220	3/4" pipe size		17	.941		23.50	45.50	1.91	70.91	95.50
6230	1" pipe size		16	1		23.50	48.50	2.03	74.03	99.50
6240	1-1/4" pipe size		13	1.231		23.50	59.50	2.50	85.50	117
6250	1-1/2" pipe size		12	1.333		23.50	64.50	2.71	90.71	124
6260	2" pipe size		11	1.455		31	70	2.95	103.95	142
6270	2-1/2" pipe size		9	1.778		35	86	3.61	124.61	170
6280	3" pipe size		7	2.286		38	110	4.64	152.64	212
6290	4" pipe size		6	2.667		55.50	129	5.40	189.90	259
6300	5" pipe size		4	4		95	193	8.10	296.10	400
6310	6" pipe size	Q-16	5	4.800		95.50	240	6.50	342	470
6320	8" pipe size		4	6		163	300	8.10	471.10	640
6330	10" pipe size		3.40	7.059		278	355	9.55	642.55	840
6340	12" pipe size		2.80	8.571		340	430	11.55	781.55	1,025
6400	Welding neck, 150 lb. flange									
6410	1/2" pipe size	Q-15	40	.400	Ea.	25.50	19.30	.81	45.61	58
6420	3/4" pipe size		36	.444		25.50	21.50	.90	47.90	61
6430	1" pipe size		32	.500		25.50	24	1.02	50.52	65
6440	1-1/4" pipe size		29	.552		25.50	26.50	1.12	53.12	68.50
6450	1-1/2" pipe size		26	.615		25.50	29.50	1.25	56.25	74
6460	2" pipe size		20	.800		29.50	38.50	1.62	69.62	91.50
6470	2-1/2" pipe size		16	1		32	48.50	2.03	82.53	109
6480	3" pipe size		14	1.143		35	55	2.32	92.32	124
6500	4" pipe size		10	1.600		42.50	77	3.25	122.75	165
6510	5" pipe size		8	2		66.50	96.50	4.06	167.06	222
6520	6" pipe size	Q-16	10	2.400		64	120	3.24	187.24	253
6530	8" pipe size		7	3.429		113	172	4.63	289.63	385
6540	10" pipe size		6	4		180	200	5.40	385.40	505
6550	12" pipe size		5	4.800		262	240	6.50	508.50	655
6551	14" pipe size		4.50	5.333		380	267	7.20	654.20	825
6552	16" pipe size		3	8		570	400	10.80	980.80	1,225
6553	18" pipe size		2.50	9.600		800	480	12.95	1,292.95	1,600
6554	20" pipe size		2.30	10.435		975	520	14.10	1,509.10	1,875
6556	24" pipe size		2	12		1,300	600	16.20	1,916.20	2,375
6557	26" pipe size		1.70	14.118		1,425	705	19.05	2,149.05	2,650
6558	30" pipe size		.90	26.667		1,650	1,325	36	3,011	3,850
6559	36" pipe size		.75	32		1,900	1,600	43	3,543	4,525
6560	300 lb. flange									
6570	1/2" pipe size	Q-15	36	.444	Ea.	31	21.50	.90	53.40	67
6580	3/4" pipe size		34	.471		31	22.50	.96	54.46	69
6590	1" pipe size		30	.533		31	25.50	1.08	57.58	73.50
6600	1-1/4" pipe size		28	.571		31	27.50	1.16	59.66	76.50
6610	1-1/2" pipe size		24	.667		31	32	1.35	64.35	83.50
6620	2" pipe size		18	.889		38.50	43	1.80	83.30	108
6630	2-1/2" pipe size		14	1.143		44	55	2.32	101.32	134
6640	3" pipe size		12	1.333		44.50	64.50	2.71	111.71	148

22 11 13.47 Pipe Fittings, Steel		Crew	Daily Output	Labor-Hours	Unit	Material	2011 Bare Costs			Total	Total Incl O&P
							Labor	Equipment			
6650	4" pipe size	Q-15	8	2	Ea.	73	96.50	4.06	173.56	228
6660	5" pipe size	↓	7	2.286		111	110	4.64	225.64	292
6670	6" pipe size	Q-16	9	2.667		113	133	3.60	249.60	330
6680	8" pipe size		6	4		195	200	5.40	400.40	520
6690	10" pipe size		5	4.800		355	240	6.50	601.50	755
6700	12" pipe size		4	6		455	300	8.10	763.10	960
6710	14" pipe size		3.50	6.857		980	345	9.25	1,334.25	1,600
6720	16" pipe size	↓	2	12	↓	1,150	600	16.20	1,766.20	2,200
7740	Plain ends for plain end pipe, mechanically coupled									
7750	Cplg. & labor required at joints not included, add 1 per									
7760	joint for installed price, see line 9180									
7770	Malleable iron, painted, unless noted otherwise									
7800	90° Elbow 1"				Ea.	81			81	89
7810	1-1/2"					96			96	105
7820	2"					143			143	157
7830	2-1/2"					167			167	184
7840	3"					173			173	191
7860	4"					197			197	216
7870	5" welded steel					226			226	248
7880	6"					286			286	315
7890	8" welded steel					535			535	590
7900	10" welded steel					710			710	780
7910	12" welded steel					790			790	870
7970	45° Elbow 1"					48			48	53
7980	1-1/2"					72			72	79
7990	2"					151			151	166
8000	2-1/2"					151			151	166
8010	3"					173			173	191
8030	4"					181			181	199
8040	5" welded steel					226			226	248
8050	6"					259			259	285
8060	8"					305			305	335
8070	10" welded steel					325			325	360
8080	12" welded steel					590			590	650
8140	Tee, straight 1"					90.50			90.50	99.50
8150	1-1/2"					116			116	128
8160	2"					116			116	128
8170	2-1/2"					152			152	167
8180	3"					234			234	257
8200	4"					335			335	370
8210	5" welded steel					460			460	505
8220	6"					400			400	440
8230	8" welded steel					575			575	630
8240	10" welded steel					900			900	990
8250	12" welded steel				↓	1,100			1,100	1,200
8340	Segmentally welded steel, painted									
8390	Wye 2"				Ea.	145			145	160
8400	2-1/2"					145			145	160
8410	3"					163			163	179
8430	4"					243			243	267
8440	5"					295			295	325
8450	6"					405			405	445
8460	8"				↓	525			525	580

22 11 13 – Facility Water Distribution Piping

22 11 13.47 Pipe Fittings, Steel	Crew	Daily Output	Labor-Hours	Unit	Material	2011 Bare Costs Labor	Equipment	Total	Total Incl O&P
8470	10"				Ea.	775			775	855
8480	12"					1,200			1,200	1,325
8540	Wye, lateral 2"					157			157	173
8550	2-1/2"					181			181	199
8560	3"					215			215	237
8580	4"					296			296	325
8590	5"					505			505	555
8600	6"					520			520	570
8610	8"					870			870	960
8620	10"					905			905	995
8630	12"					1,675			1,675	1,825
8690	Cross, 2"					154			154	169
8700	2-1/2"					154			154	169
8710	3"					183			183	202
8730	4"					251			251	276
8740	5"					355			355	390
8750	6"					470			470	515
8760	8"					600			600	660
8770	10"					880			880	965
8780	12"					1,275			1,275	1,400
8800	Tees, reducing 2" x 1"					99.50			99.50	109
8810	2" x 1-1/2"					99.50			99.50	109
8820	3" x 1"					99.50			99.50	109
8830	3" x 1-1/2"					99.50			99.50	109
8840	3" x 2"					99.50			99.50	109
8850	4" x 1"					145			145	159
8860	4" x 1-1/2"					145			145	159
8870	4" x 2"					145			145	159
8880	4" x 2-1/2"					148			148	163
8890	4" x 3"					148			148	163
8900	6" x 2"					214			214	235
8910	6" x 3"					230			230	253
8920	6" x 4"					280			280	310
8930	8" x 2"					292			292	320
8940	8" x 3"					288			288	315
8950	8" x 4"					288			288	315
8960	8" x 5"					310			310	340
8970	8" x 6"					335			335	370
8980	10" x 4"					440			440	485
8990	10" x 6"					440			440	485
9000	10" x 8"					470			470	515
9010	12" x 6"					680			680	750
9020	12" x 8"					680			680	750
9030	12" x 10"				▼	705			705	780
9080	Adapter nipples 3" long									
9090	1"				Ea.	13.60			13.60	14.95
9100	1-1/2"					13.60			13.60	14.95
9110	2"					13.60			13.60	14.95
9120	2-1/2"					15.80			15.80	17.40
9130	3"					19.20			19.20	21
9140	4"					32			32	35
9150	6"				▼	109			109	120
9180	Coupling, mechanical, plain end pipe to plain end pipe or fitting									

22 11 13 – Facility Water Distribution Piping

22 11 13.47 Pipe Fittings, Steel

		Crew	Daily Output	Labor-Hours	Unit	Material	2011 Bare Costs Labor	Equipment	Total	Total Incl O&P
9190	1"	Q-1	29	.552	Ea.	48.50	26.50		75	93
9200	1-1/2"		28	.571		48.50	27.50		76	94.50
9210	2"		27	.593		48.50	28.50		77	96
9220	2-1/2"		26	.615		48.50	29.50		78	98
9230	3"		25	.640		71	31		102	125
9240	3-1/2"		24	.667		82.50	32		114.50	139
9250	4"		22	.727		82.50	35		117.50	144
9260	5"	Q-2	28	.857		118	43		161	193
9270	6"		24	1		143	50		193	233
9280	8"		19	1.263		253	63		316	375
9290	10"		16	1.500		330	75		405	470
9300	12"		12	2		415	100		515	605
9310	Outlets for precut holes through pipe wall									
9331	Strapless type, with gasket									
9332	4" to 8" pipe x 1/2"	1 Plum	13	.615	Ea.	53	33		86	108
9333	4" to 8" pipe x 3/4"		13	.615		57	33		90	112
9334	10" pipe and larger x 1/2"		11	.727		53	39		92	117
9335	10" pipe and larger x 3/4"		11	.727		57	39		96	121
9341	Thermometer wells with gasket									
9342	4" to 8" pipe, 6" stem	1 Plum	14	.571	Ea.	80	30.50		110.50	134
9343	8" pipe and larger, 6" stem	"	13	.615	"	80	33		113	137
9940	For galvanized fittings for plain end pipe, add					20%				

22 11 13.48 Pipe, Fittings and Valves, Steel, Grooved-Joint

		Crew	Daily Output	Labor-Hours	Unit	Material	2011 Bare Costs Labor	Equipment	Total	Total Incl O&P
0010	**PIPE, FITTINGS AND VALVES, STEEL, GROOVED-JOINT**									
0012	Fittings are ductile iron. Steel fittings noted.									
0020	Pipe includes coupling & clevis type hanger assemblies, 10' O.C.									
0500	Schedule 10, black									
0550	2" diameter	1 Plum	43	.186	L.F.	6.10	9.95		16.05	21.50
0560	2-1/2" diameter	Q-1	61	.262		7.70	12.65		20.35	27.50
0570	3" diameter		55	.291		8.90	14.05		22.95	31
0580	3-1/2" diameter		53	.302		10.30	14.55		24.85	33.50
0590	4" diameter		49	.327		11.75	15.75		27.50	36.50
0600	5" diameter		40	.400		13.80	19.30		33.10	44
0610	6" diameter	Q-2	46	.522		17.85	26		43.85	58.50
0620	8" diameter	"	41	.585		24	29.50		53.50	70
0700	To delete couplings & hangers, subtract									
0710	2" diam. to 5" diam.					25%	20%			
0720	6" diam. to 8" diam.					27%	15%			
1000	Schedule 40, black									
1040	3/4" diameter	1 Plum	71	.113	L.F.	5.40	6.05		11.45	14.95
1050	1" diameter		63	.127		4.09	6.80		10.89	14.65
1060	1-1/4" diameter		58	.138		5.10	7.40		12.50	16.70
1070	1-1/2" diameter		51	.157		5.75	8.40		14.15	18.85
1080	2" diameter		40	.200		6.60	10.70		17.30	23.50
1090	2-1/2" diameter	Q-1	57	.281		9.25	13.55		22.80	30
1100	3" diameter		50	.320		11.30	15.45		26.75	35.50
1110	4" diameter		45	.356		15.75	17.15		32.90	43
1120	5" diameter		37	.432		21.50	21		42.50	54.50
1130	6" diameter	Q-2	42	.571		27.50	28.50		56	72.50
1140	8" diameter		37	.649		41	32.50		73.50	94
1150	10" diameter		31	.774		61.50	38.50		100	126
1160	12" diameter		27	.889		86.50	44.50		131	162

22 11 13.48 Pipe, Fittings and Valves, Steel, Grooved-Joint	Crew	Daily Output	Labor-Hours	Unit	Material	2011 Bare Costs Labor	Equipment	Total	Total Incl O&P
1170	14" diameter	Q-2	20	1.200	L.F.	93.50	60		153.50	193
1180	16" diameter		17	1.412		108	70.50		178.50	224
1190	18" diameter		14	1.714		123	86		209	264
1200	20" diameter		12	2		151	100		251	315
1210	24" diameter		10	2.400		182	120		302	380
1740	To delete coupling & hanger, subtract									
1750	3/4" diam. to 2" diam.					65%	27%			
1760	2-1/2" diam. to 5" diam.					41%	18%			
1770	6" diam. to 12" diam.					31%	13%			
1780	14" diam. to 24" diam.					35%	10%			
1800	Galvanized									
1840	3/4" diameter	1 Plum	71	.113	L.F.	11.25	6.05		17.30	21.50
1850	1" diameter		63	.127		8.05	6.80		14.85	19
1860	1-1/4" diameter		58	.138		10.15	7.40		17.55	22
1870	1-1/2" diameter		51	.157		11.70	8.40		20.10	25.50
1880	2" diameter		40	.200		14.85	10.70		25.55	32.50
1890	2-1/2" diameter	Q-1	57	.281		22	13.55		35.55	44.50
1900	3" diameter		50	.320		28	15.45		43.45	54
1910	4" diameter		45	.356		39	17.15		56.15	68.50
1920	5" diameter		37	.432		46	21		67	81.50
1930	6" diameter	Q-2	42	.571		51.50	28.50		80	99
1940	8" diameter		37	.649		58	32.50		90.50	113
1950	10" diameter		31	.774		86	38.50		124.50	153
1960	12" diameter		27	.889		111	44.50		155.50	189
2540	To delete coupling & hanger, subtract									
2550	3/4" diam. to 2" diam.					36%	27%			
2560	2-1/2" diam. to 5" diam.					19%	18%			
2570	6" diam. to 12" diam.					14%	13%			
2600	Schedule 80, black									
2610	3/4" diameter	1 Plum	65	.123	L.F.	7.05	6.60		13.65	17.65
2650	1" diameter		61	.131		8.85	7.05		15.90	20.50
2660	1-1/4" diameter		55	.145		11.40	7.80		19.20	24
2670	1-1/2" diameter		49	.163		13.25	8.75		22	27.50
2680	2" diameter		38	.211		17.10	11.30		28.40	35.50
2690	2-1/2" diameter	Q-1	54	.296		25.50	14.30		39.80	49.50
2700	3" diameter		48	.333		33.50	16.10		49.60	61
2710	4" diameter		44	.364		24	17.55		41.55	52.50
2720	5" diameter		35	.457		40	22		62	77
2730	6" diameter	Q-2	40	.600		44.50	30		74.50	94
2740	8" diameter		35	.686		108	34.50		142.50	170
2750	10" diameter		29	.828		198	41.50		239.50	280
2760	12" diameter		24	1		390	50		440	505
3240	To delete coupling & hanger, subtract									
3250	3/4" diam. to 2" diam.					30%	25%			
3260	2-1/2" diam. to 5" diam.					14%	17%			
3270	6" diam. to 12" diam.					12%	12%			
3300	Galvanized									
3310	3/4" diameter	1 Plum	65	.123	L.F.	9.30	6.60		15.90	20
3350	1" diameter		61	.131		12.10	7.05		19.15	24
3360	1-1/4" diameter		55	.145		15.80	7.80		23.60	29
3370	1-1/2" diameter		46	.174		18.60	9.30		27.90	34.50
3380	2" diameter		38	.211		24.50	11.30		35.80	44
3390	2-1/2" diameter	Q-1	54	.296		37.50	14.30		51.80	62.50

22 11 Facility Water Distribution

22 11 13 – Facility Water Distribution Piping

22 11 13.48 Pipe, Fittings and Valves, Steel, Grooved-Joint	Crew	Daily Output	Labor-Hours	Unit	Material	2011 Bare Costs Labor	Equipment	Total	Total Incl O&P
3400	3" diameter	Q-1	48	.333	L.F.	49.50	16.10		65.60	78
3410	4" diameter		44	.364		69.50	17.55		87.05	102
3420	5" diameter	▼	35	.457		73	22		95	113
3430	6" diameter	Q-2	40	.600		96	30		126	150
3440	8" diameter		35	.686		116	34.50		150.50	179
3450	10" diameter		29	.828		197	41.50		238.50	278
3460	12" diameter	▼	24	1	▼	237	50		287	335
3920	To delete coupling & hanger, subtract									
3930	3/4" diam. to 2" diam.					30%	25%			
3940	2-1/2" diam. to 5" diam.					15%	17%			
3950	6" diam. to 12" diam.					11%	12%			
3990	Fittings: coupling material required at joints not incl. in fitting price.									
3994	Add 1 selected coupling, material only, per joint for installed price.									
4000	Elbow, 90° or 45°, painted									
4030	3/4" diameter	1 Plum	50	.160	Ea.	43	8.60		51.60	60.50
4040	1" diameter		50	.160		23	8.60		31.60	38
4050	1-1/4" diameter		40	.200		23	10.70		33.70	41
4060	1-1/2" diameter		33	.242		23	13		36	44.50
4070	2" diameter	▼	25	.320		23	17.15		40.15	50.50
4080	2-1/2" diameter	Q-1	40	.400		23	19.30		42.30	54
4090	3" diameter		33	.485		40.50	23.50		64	79.50
4100	4" diameter		25	.640		44	31		75	94.50
4110	5" diameter	▼	20	.800		105	38.50		143.50	174
4120	6" diameter	Q-2	25	.960		124	48		172	208
4130	8" diameter		21	1.143		258	57		315	370
4140	10" diameter		18	1.333		470	66.50		536.50	615
4150	12" diameter		15	1.600		750	80		830	945
4170	14" diameter		12	2		835	100		935	1,075
4180	16" diameter	▼	11	2.182		1,075	109		1,184	1,375
4190	18" diameter	Q-3	15	2.133		1,375	109		1,484	1,700
4200	20" diameter		13	2.462		1,825	126		1,951	2,200
4210	24" diameter	▼	11	2.909		2,625	149		2,774	3,125
4250	For galvanized elbows, add				▼	26%				
4690	Tee, painted									
4700	3/4" diameter	1 Plum	38	.211	Ea.	46	11.30		57.30	68
4740	1" diameter		33	.242		35.50	13		48.50	58.50
4750	1-1/4" diameter		27	.296		35.50	15.90		51.40	62.50
4760	1-1/2" diameter		22	.364		35.50	19.50		55	68
4770	2" diameter	▼	17	.471		35.50	25		60.50	76.50
4780	2-1/2" diameter	Q-1	27	.593		35.50	28.50		64	81.50
4790	3" diameter		22	.727		49	35		84	106
4800	4" diameter		17	.941		74.50	45.50		120	150
4810	5" diameter	▼	13	1.231		174	59.50		233.50	280
4820	6" diameter	Q-2	17	1.412		201	70.50		271.50	325
4830	8" diameter		14	1.714		440	86		526	615
4840	10" diameter		12	2		915	100		1,015	1,150
4850	12" diameter		10	2.400		1,275	120		1,395	1,575
4851	14" diameter		9	2.667		1,125	133		1,258	1,425
4852	16" diameter	▼	8	3		1,275	150		1,425	1,625
4853	18" diameter	Q-3	11	2.909		1,575	149		1,724	1,950
4854	20" diameter		10	3.200		2,250	163		2,413	2,750
4855	24" diameter	▼	8	4		3,450	204		3,654	4,100
4900	For galvanized tees, add				▼	24%				

172

22 11 Facility Water Distribution

22 11 13 – Facility Water Distribution Piping

22 11 13.48 Pipe, Fittings and Valves, Steel, Grooved-Joint	Crew	Daily Output	Labor-Hours	Unit	Material	2011 Bare Costs Labor	Equipment	Total	Total Incl O&P
4906	Couplings, rigid style, painted									
4908	1" diameter	1 Plum	100	.080	Ea.	17.20	4.29		21.49	25.50
4909	1-1/4" diameter		100	.080		17.20	4.29		21.49	25.50
4910	1-1/2" diameter		67	.119		17.20	6.40		23.60	28.50
4912	2" diameter		50	.160		18.80	8.60		27.40	33.50
4914	2-1/2" diameter	Q-1	80	.200		22	9.65		31.65	38.50
4916	3" diameter		67	.239		25.50	11.50		37	45
4918	4" diameter		50	.320		35.50	15.45		50.95	62
4920	5" diameter		40	.400		46	19.30		65.30	80
4922	6" diameter	Q-2	50	.480		61	24		85	103
4924	8" diameter		42	.571		96	28.50		124.50	149
4926	10" diameter		35	.686		159	34.50		193.50	225
4928	12" diameter		32	.750		177	37.50		214.50	251
4930	14" diameter		24	1		204	50		254	300
4931	16" diameter		20	1.200		266	60		326	385
4932	18" diameter		18	1.333		310	66.50		376.50	440
4933	20" diameter		16	1.500		420	75		495	570
4934	24" diameter		13	1.846		540	92.50		632.50	735
4940	Flexible, standard, painted									
4950	3/4" diameter	1 Plum	100	.080	Ea.	12.40	4.29		16.69	20
4960	1" diameter		100	.080		12.40	4.29		16.69	20
4970	1-1/4" diameter		80	.100		16.40	5.35		21.75	26
4980	1-1/2" diameter		67	.119		18	6.40		24.40	29.50
4990	2" diameter		50	.160		19	8.60		27.60	34
5000	2-1/2" diameter	Q-1	80	.200		22	9.65		31.65	39
5010	3" diameter		67	.239		25	11.50		36.50	44.50
5020	3-1/2" diameter		57	.281		36	13.55		49.55	59.50
5030	4" diameter		50	.320		36	15.45		51.45	62.50
5040	5" diameter		40	.400		54.50	19.30		73.80	89
5050	6" diameter	Q-2	50	.480		64	24		88	107
5070	8" diameter		42	.571		104	28.50		132.50	158
5090	10" diameter		35	.686		173	34.50		207.50	241
5110	12" diameter		32	.750		197	37.50		234.50	272
5120	14" diameter		24	1		211	50		261	305
5130	16" diameter		20	1.200		276	60		336	395
5140	18" diameter		18	1.333		325	66.50		391.50	455
5150	20" diameter		16	1.500		430	75		505	585
5160	24" diameter		13	1.846		555	92.50		647.50	755
5176	Lightweight style, painted									
5178	1-1/2" diameter	1 Plum	67	.119	Ea.	15.80	6.40		22.20	27
5180	2" diameter	"	50	.160		16	8.60		24.60	30.50
5182	2-1/2" diameter	Q-1	80	.200		18.60	9.65		28.25	35
5184	3" diameter		67	.239		21.50	11.50		33	41
5186	3-1/2" diameter		57	.281		30	13.55		43.55	53
5188	4" diameter		50	.320		30	15.45		45.45	56
5190	5" diameter		40	.400		43	19.30		62.30	76.50
5192	6" diameter	Q-2	50	.480		51.50	24		75.50	93
5194	8" diameter		42	.571		81	28.50		109.50	132
5196	10" diameter		35	.686		216	34.50		250.50	288
5198	12" diameter		32	.750		241	37.50		278.50	320
5200	For galvanized couplings, add					33%				
5220	Tee, reducing, painted									
5225	2" x 1-1/2" diameter	Q-1	38	.421	Ea.	75	20.50		95.50	113

173

22 11 Facility Water Distribution

22 11 13 – Facility Water Distribution Piping

22 11 13.48 Pipe, Fittings and Valves, Steel, Grooved-Joint	Crew	Daily Output	Labor-Hours	Unit	Material	2011 Bare Costs Labor	Equipment	Total	Total Incl O&P
5226	2-1/2" x 2" diameter	Q-1	28	.571	Ea.	75	27.50		102.50	124
5227	3" x 2-1/2" diameter		23	.696		66.50	33.50		100	123
5228	4" x 3" diameter		18	.889		89.50	43		132.50	163
5229	5" x 4" diameter		15	1.067		192	51.50		243.50	288
5230	6" x 4" diameter	Q-2	18	1.333		211	66.50		277.50	330
5231	8" x 6" diameter		15	1.600		440	80		520	605
5232	10" x 8" diameter		13	1.846		575	92.50		667.50	775
5233	12" x 10" diameter		11	2.182		880	109		989	1,125
5234	14" x 12" diameter		10	2.400		815	120		935	1,075
5235	16" x 12" diameter		9	2.667		1,000	133		1,133	1,300
5236	18" x 12" diameter	Q-3	12	2.667		1,200	136		1,336	1,525
5237	18" x 16" diameter		11	2.909		1,525	149		1,674	1,900
5238	20" x 16" diameter		10	3.200		2,000	163		2,163	2,450
5239	24" x 20" diameter		9	3.556		3,150	182		3,332	3,750
5240	Reducer, concentric, painted									
5241	2-1/2" x 2" diameter	Q-1	43	.372	Ea.	27	17.95		44.95	56.50
5242	3" x 2-1/2" diameter		35	.457		32	22		54	68.50
5243	4" x 3" diameter		29	.552		39	26.50		65.50	82.50
5244	5" x 4" diameter		22	.727		54	35		89	112
5245	6" x 4" diameter	Q-2	26	.923		62.50	46		108.50	138
5246	8" x 6" diameter		23	1.043		162	52		214	256
5247	10" x 8" diameter		20	1.200		330	60		390	455
5248	12" x 10" diameter		16	1.500		590	75		665	760
5255	Eccentric, painted									
5256	2-1/2" x 2" diameter	Q-1	42	.381	Ea.	56	18.40		74.40	89
5257	3" x 2-1/2" diameter		34	.471		64	22.50		86.50	105
5258	4" x 3" diameter		28	.571		78	27.50		105.50	127
5259	5" x 4" diameter		21	.762		106	37		143	172
5260	6" x 4" diameter	Q-2	25	.960		124	48		172	208
5261	8" x 6" diameter		22	1.091		250	54.50		304.50	355
5262	10" x 8" diameter		19	1.263		685	63		748	850
5263	12" x 10" diameter		15	1.600		945	80		1,025	1,175
5270	Coupling, reducing, painted									
5272	2" x 1-1/2" diameter	1 Plum	52	.154	Ea.	27	8.25		35.25	42
5274	2-1/2" x 2" diameter	Q-1	82	.195		35	9.40		44.40	52.50
5276	3" x 2" diameter		69	.232		40	11.20		51.20	60.50
5278	4" x 2" diameter		52	.308		63	14.85		77.85	91
5280	5" x 4" diameter		42	.381		71	18.40		89.40	106
5282	6" x 4" diameter	Q-2	52	.462		107	23		130	153
5284	8" x 6" diameter	"	44	.545		160	27.50		187.50	217
5290	Outlet coupling, painted									
5294	1-1/2" x 1" pipe size	1 Plum	65	.123	Ea.	34	6.60		40.60	47.50
5296	2" x 1" pipe size	"	48	.167		35	8.95		43.95	52
5298	2-1/2" x 1" pipe size	Q-1	78	.205		53.50	9.90		63.40	74
5300	2-1/2" x 1-1/4" pipe size	1 Plum	70	.114		60.50	6.15		66.65	75.50
5302	3" x 1" pipe size	Q-1	65	.246		68.50	11.90		80.40	93
5304	4" x 3/4" pipe size		48	.333		76	16.10		92.10	108
5306	4" x 1-1/2" pipe size		46	.348		108	16.80		124.80	144
5308	6" x 1-1/2" pipe size	Q-2	44	.545		153	27.50		180.50	209
5750	Flange, w/groove gasket, black steel									
5754	See Line 22 11 13.47 0620 for gasket & bolt set									
5760	ANSI class 125 and 150, painted									
5780	2" pipe size	1 Plum	23	.348	Ea.	76.50	18.65		95.15	112

22 11 Facility Water Distribution

22 11 13 – Facility Water Distribution Piping

22 11 13.48 Pipe, Fittings and Valves, Steel, Grooved-Joint	Crew	Daily Output	Labor-Hours	Unit	Material	2011 Bare Costs Labor	Equipment	Total	Total Incl O&P
5790	2-1/2" pipe size	Q-1	37	.432	Ea.	94.50	21		115.50	135
5800	3" pipe size		31	.516		102	25		127	149
5820	4" pipe size		23	.696		137	33.50		170.50	200
5830	5" pipe size		19	.842		159	40.50		199.50	235
5840	6" pipe size	Q-2	23	1.043		173	52		225	268
5850	8" pipe size		17	1.412		195	70.50		265.50	320
5860	10" pipe size		14	1.714		310	86		396	470
5870	12" pipe size		12	2		405	100		505	595
5880	14" pipe size		10	2.400		820	120		940	1,075
5890	16" pipe size		9	2.667		950	133		1,083	1,250
5900	18" pipe size		6	4		1,175	200		1,375	1,575
5910	20" pipe size		5	4.800		1,400	240		1,640	1,900
5920	24" pipe size		4.50	5.333		1,800	267		2,067	2,375
5940	ANSI class 350, painted									
5946	2" pipe size	1 Plum	23	.348	Ea.	95.50	18.65		114.15	133
5948	2-1/2" pipe size	Q-1	37	.432		110	21		131	152
5950	3" pipe size		31	.516		151	25		176	203
5952	4" pipe size		23	.696		201	33.50		234.50	271
5954	5" pipe size		19	.842		229	40.50		269.50	310
5956	6" pipe size	Q-2	23	1.043		266	52		318	370
5958	8" pipe size		17	1.412		305	70.50		375.50	440
5960	10" pipe size		14	1.714		485	86		571	665
5962	12" pipe size	1 Plum	12	.667		520	35.50		555.50	625
6100	Cross, painted									
6110	2" diameter	1 Plum	12.50	.640	Ea.	63.50	34.50		98	121
6112	2-1/2" diameter	Q-1	20	.800		63.50	38.50		102	128
6114	3" diameter		15.50	1.032		113	50		163	200
6116	4" diameter		12.50	1.280		188	62		250	299
6118	6" diameter	Q-2	12.50	1.920		495	96		591	685
6120	8" diameter		10.50	2.286		650	114		764	885
6122	10" diameter		9	2.667		1,100	133		1,233	1,425
6124	12" diameter		7.50	3.200		1,600	160		1,760	2,000
7400	Suction diffuser									
7402	Grooved end inlet x flanged outlet									
7410	3" x 3"	Q-1	50	.320	Ea.	620	15.45		635.45	705
7412	4" x 4"		38	.421		840	20.50		860.50	955
7414	5" x 5"		30	.533		985	25.50		1,010.50	1,125
7416	6" x 6"	Q-2	38	.632		1,250	31.50		1,281.50	1,425
7418	8" x 8"		27	.889		2,300	44.50		2,344.50	2,625
7420	10" x 10"		20	1.200		3,150	60		3,210	3,550
7422	12" x 12"		16	1.500		5,175	75		5,250	5,775
7424	14" x 14"		15	1.600		6,025	80		6,105	6,775
7426	16" x 14"		14	1.714		6,225	86		6,311	6,950
7500	Strainer, tee type, painted									
7506	2" pipe size	1 Plum	38	.211	Ea.	420	11.30		431.30	475
7508	2-1/2" pipe size	Q-1	62	.258		440	12.45		452.45	505
7510	3" pipe size		50	.320		495	15.45		510.45	565
7512	4" pipe size		38	.421		560	20.50		580.50	645
7514	5" pipe size		30	.533		810	25.50		835.50	930
7516	6" pipe size	Q-2	38	.632		875	31.50		906.50	1,000
7518	8" pipe size		27	.889		1,350	44.50		1,394.50	1,550
7520	10" pipe size		20	1.200		1,975	60		2,035	2,275
7522	12" pipe size		16	1.500		2,550	75		2,625	2,900

22 11 13 – Facility Water Distribution Piping

22 11 13.48 Pipe, Fittings and Valves, Steel, Grooved-Joint		Crew	Daily Output	Labor-Hours	Unit	Material	2011 Bare Costs Labor	Equipment	Total	Total Incl O&P
7524	14" pipe size	Q-2	15	1.600	Ea.	8,150	80		8,230	9,075
7526	16" pipe size	↓	14	1.714	↓	10,100	86		10,186	11,200
7570	Expansion joint, max. 3" travel									
7572	2" diameter	1 Plum	38	.211	Ea.	520	11.30		531.30	590
7574	3" diameter	Q-1	50	.320		600	15.45		615.45	685
7576	4" diameter	"	38	.421		790	20.50		810.50	900
7578	6" diameter	Q-2	38	.632	↓	1,225	31.50		1,256.50	1,400
7800	Ball valve w/handle, carbon steel trim									
7810	1-1/2" pipe size	1 Plum	50	.160	Ea.	119	8.60		127.60	144
7812	2" pipe size	"	38	.211		132	11.30		143.30	162
7814	2-1/2" pipe size	Q-1	62	.258		288	12.45		300.45	335
7816	3" pipe size		50	.320		465	15.45		480.45	540
7818	4" pipe size	↓	38	.421		720	20.50		740.50	825
7820	6" pipe size	Q-2	30	.800		2,225	40		2,265	2,475
7830	With gear operator									
7834	2-1/2" pipe size	Q-1	62	.258	Ea.	585	12.45		597.45	660
7836	3" pipe size		50	.320		825	15.45		840.45	935
7838	4" pipe size	↓	38	.421		1,050	20.50		1,070.50	1,175
7840	6" pipe size	Q-2	30	.800	↓	2,475	40		2,515	2,775
7870	Check valve									
7874	2-1/2" pipe size	Q-1	62	.258	Ea.	199	12.45		211.45	238
7876	3" pipe size		50	.320		236	15.45		251.45	282
7878	4" pipe size		38	.421		249	20.50		269.50	305
7880	5" pipe size	↓	30	.533		415	25.50		440.50	495
7882	6" pipe size	Q-2	38	.632		490	31.50		521.50	585
7884	8" pipe size		27	.889		670	44.50		714.50	805
7886	10" pipe size		20	1.200		1,950	60		2,010	2,225
7888	12" pipe size	↓	16	1.500	↓	2,300	75		2,375	2,625
7900	Plug valve, balancing, w/lever operator									
7906	3" pipe size	Q-1	50	.320	Ea.	495	15.45		510.45	570
7908	4" pipe size	"	38	.421		610	20.50		630.50	700
7909	6" pipe size	Q-2	30	.800	↓	935	40		975	1,075
7916	With gear operator									
7920	3" pipe size	Q-1	50	.320	Ea.	940	15.45		955.45	1,050
7922	4" pipe size	"	38	.421		975	20.50		995.50	1,100
7924	6" pipe size	Q-2	38	.632		1,300	31.50		1,331.50	1,500
7926	8" pipe size		27	.889		2,025	44.50		2,069.50	2,300
7928	10" pipe size		20	1.200		2,750	60		2,810	3,125
7930	12" pipe size	↓	16	1.500	↓	4,200	75		4,275	4,700
8000	Butterfly valve, 2 position handle, with standard trim									
8010	1-1/2" pipe size	1 Plum	50	.160	Ea.	188	8.60		196.60	219
8020	2" pipe size	"	38	.211		188	11.30		199.30	223
8030	3" pipe size	Q-1	50	.320		238	15.45		253.45	285
8050	4" pipe size	"	38	.421		296	20.50		316.50	355
8070	6" pipe size	Q-2	38	.632		600	31.50		631.50	700
8080	8" pipe size		27	.889		805	44.50		849.50	950
8090	10" pipe size	↓	20	1.200	↓	1,425	60		1,485	1,675
8200	With stainless steel trim									
8240	1-1/2" pipe size	1 Plum	50	.160	Ea.	238	8.60		246.60	275
8250	2" pipe size	"	38	.211		238	11.30		249.30	279
8270	3" pipe size	Q-1	50	.320		320	15.45		335.45	375
8280	4" pipe size	"	38	.421		345	20.50		365.50	410
8300	6" pipe size	Q-2	38	.632	↓	650	31.50		681.50	760

22 11 13 – Facility Water Distribution Piping

22 11 13.48 Pipe, Fittings and Valves, Steel, Grooved-Joint	Crew	Daily Output	Labor-Hours	Unit	Material	2011 Bare Costs Labor	Equipment	Total	Total Incl O&P
8310	8" pipe size	Q-2	27	.889	Ea.	1,600	44.50		1,644.50	1,850
8320	10" pipe size		20	1.200		2,900	60		2,960	3,275
8322	12" pipe size		16	1.500		3,325	75		3,400	3,750
8324	14" pipe size		15	1.600		4,025	80		4,105	4,550
8326	16" pipe size		14	1.714		5,825	86		5,911	6,550
8328	18" pipe size	Q-3	12	2.667		7,175	136		7,311	8,100
8330	20" pipe size		11	2.909		8,825	149		8,974	9,950
8332	24" pipe size		10	3.200		11,800	163		11,963	13,200
8336	Note: sizes 8" up w/manual gear operator									
9000	Cut one groove, labor									
9010	3/4" pipe size	Q-1	152	.105	Ea.		5.10		5.10	7.60
9020	1" pipe size		140	.114			5.50		5.50	8.25
9030	1-1/4" pipe size		124	.129			6.25		6.25	9.30
9040	1-1/2" pipe size		114	.140			6.75		6.75	10.10
9050	2" pipe size		104	.154			7.40		7.40	11.10
9060	2-1/2" pipe size		96	.167			8.05		8.05	12
9070	3" pipe size		88	.182			8.75		8.75	13.10
9080	3-1/2" pipe size		83	.193			9.30		9.30	13.90
9090	4" pipe size		78	.205			9.90		9.90	14.80
9100	5" pipe size		72	.222			10.70		10.70	16
9110	6" pipe size		70	.229			11.05		11.05	16.45
9120	8" pipe size		54	.296			14.30		14.30	21.50
9130	10" pipe size		38	.421			20.50		20.50	30.50
9140	12" pipe size		30	.533			25.50		25.50	38.50
9150	14" pipe size		20	.800			38.50		38.50	57.50
9160	16" pipe size		19	.842			40.50		40.50	60.50
9170	18" pipe size		18	.889			43		43	64
9180	20" pipe size		17	.941			45.50		45.50	68
9190	24" pipe size		15	1.067			51.50		51.50	77
9210	Roll one groove									
9220	3/4" pipe size	Q-1	266	.060	Ea.		2.90		2.90	4.33
9230	1" pipe size		228	.070			3.39		3.39	5.05
9240	1-1/4" pipe size		200	.080			3.86		3.86	5.75
9250	1-1/2" pipe size		178	.090			4.34		4.34	6.45
9260	2" pipe size		116	.138			6.65		6.65	9.95
9270	2-1/2" pipe size		110	.145			7		7	10.50
9280	3" pipe size		100	.160			7.70		7.70	11.50
9290	3-1/2" pipe size		94	.170			8.20		8.20	12.25
9300	4" pipe size		86	.186			9		9	13.40
9310	5" pipe size		84	.190			9.20		9.20	13.70
9320	6" pipe size		80	.200			9.65		9.65	14.40
9330	8" pipe size		66	.242			11.70		11.70	17.45
9340	10" pipe size		58	.276			13.30		13.30	19.85
9350	12" pipe size		46	.348			16.80		16.80	25
9360	14" pipe size		30	.533			25.50		25.50	38.50
9370	16" pipe size		28	.571			27.50		27.50	41
9380	18" pipe size		27	.593			28.50		28.50	42.50
9390	20" pipe size		25	.640			31		31	46
9400	24" pipe size		23	.696			33.50		33.50	50

22 11 13.60 Tubing, Stainless Steel

| 0010 | **TUBING, STAINLESS STEEL** | | | | | | | | | |
| 5010 | Type 304, no joints, no hangers | | | | | | | | | |

22 11 13.60 Tubing, Stainless Steel

		Crew	Daily Output	Labor-Hours	Unit	Material	2011 Bare Costs Labor	Equipment	Total	Total Incl O&P
5020	.035 wall									
5021	1/4"	1 Plum	160	.050	L.F.	2.89	2.68		5.57	7.20
5022	3/8"		160	.050		3.64	2.68		6.32	8
5023	1/2"		160	.050		4.69	2.68		7.37	9.15
5024	5/8"		160	.050		5.75	2.68		8.43	10.35
5025	3/4"		133	.060		6.95	3.22		10.17	12.40
5026	7/8"		133	.060		7.25	3.22		10.47	12.75
5027	1"		114	.070		8.05	3.76		11.81	14.45
5040	.049 wall									
5041	1/4"	1 Plum	160	.050	L.F.	3.39	2.68		6.07	7.75
5042	3/8"		160	.050		4.69	2.68		7.37	9.15
5043	1/2"		160	.050		5.20	2.68		7.88	9.70
5044	5/8"		160	.050		7.05	2.68		9.73	11.75
5045	3/4"		133	.060		7.50	3.22		10.72	13.05
5046	7/8"		133	.060		8.75	3.22		11.97	14.45
5047	1"		114	.070		9.35	3.76		13.11	15.85
5060	.065 wall									
5061	1/4"	1 Plum	160	.050	L.F.	3.80	2.68		6.48	8.20
5062	3/8"		160	.050		5.25	2.68		7.93	9.75
5063	1/2"		160	.050		6.35	2.68		9.03	10.95
5064	5/8"		160	.050		7.70	2.68		10.38	12.50
5065	3/4"		133	.060		8.20	3.22		11.42	13.80
5066	7/8"		133	.060		10.55	3.22		13.77	16.40
5067	1"		114	.070		11.35	3.76		15.11	18.05
5210	Type 316									
5220	.035 wall									
5221	1/4"	1 Plum	160	.050	L.F.	4.50	2.68		7.18	8.95
5222	3/8"		160	.050		4.98	2.68		7.66	9.50
5223	1/2"		160	.050		7.40	2.68		10.08	12.15
5224	5/8"		160	.050		10.20	2.68		12.88	15.20
5225	3/4"		133	.060		9.05	3.22		12.27	14.75
5226	7/8"		133	.060		16.80	3.22		20.02	23.50
5227	1"		114	.070		11	3.76		14.76	17.70
5240	.049 wall									
5241	1/4"	1 Plum	160	.050	L.F.	5.45	2.68		8.13	10
5242	3/8"		160	.050		6.25	2.68		8.93	10.90
5243	1/2"		160	.050		7.50	2.68		10.18	12.25
5244	5/8"		160	.050		10.55	2.68		13.23	15.65
5245	3/4"		133	.060		10.30	3.22		13.52	16.15
5246	7/8"		133	.060		14.05	3.22		17.27	20.50
5247	1"		114	.070		14.15	3.76		17.91	21
5260	.065 wall									
5261	1/4"	1 Plum	160	.050	L.F.	5.05	2.68		7.73	9.55
5262	3/8"		160	.050		9.20	2.68		11.88	14.10
5263	1/2"		160	.050		10.35	2.68		13.03	15.35
5264	5/8"		160	.050		12.25	2.68		14.93	17.50
5265	3/4"		133	.060		13.10	3.22		16.32	19.20
5266	7/8"		133	.060		16.45	3.22		19.67	23
5267	1"		114	.070		15.65	3.76		19.41	23

22 11 13.61 Tubing Fittings, Stainless Steel

0010	**TUBING FITTINGS, STAINLESS STEEL**									
8200	Tube fittings, compression type									

22 11 Facility Water Distribution

22 11 13 – Facility Water Distribution Piping

22 11 13.61 Tubing Fittings, Stainless Steel	Crew	Daily Output	Labor-Hours	Unit	Material	2011 Bare Costs Labor	Equipment	Total	Total Incl O&P
8202 Type 316									
8204 90° elbow									
8206 1/4"	1 Plum	24	.333	Ea.	14.40	17.85		32.25	42.50
8207 3/8"		22	.364		17.60	19.50		37.10	48.50
8208 1/2"		22	.364		29	19.50		48.50	61
8209 5/8"		21	.381		32.50	20.50		53	66.50
8210 3/4"		21	.381		51.50	20.50		72	87
8211 7/8"		20	.400		78.50	21.50		100	119
8212 1"		20	.400		98	21.50		119.50	140
8220 Union tee									
8222 1/4"	1 Plum	15	.533	Ea.	20.50	28.50		49	65
8224 3/8"		15	.533		26	28.50		54.50	71
8225 1/2"		15	.533		40.50	28.50		69	87
8226 5/8"		14	.571		45	30.50		75.50	95
8227 3/4"		14	.571		60.50	30.50		91	112
8228 7/8"		13	.615		108	33		141	168
8229 1"		13	.615		129	33		162	191
8234 Union									
8236 1/4"	1 Plum	24	.333	Ea.	9.95	17.85		27.80	37.50
8237 3/8"		22	.364		14.15	19.50		33.65	44.50
8238 1/2"		22	.364		21	19.50		40.50	52
8239 5/8"		21	.381		27.50	20.50		48	60.50
8240 3/4"		21	.381		34	20.50		54.50	68
8241 7/8"		20	.400		56	21.50		77.50	93.50
8242 1"		20	.400		59	21.50		80.50	97
8250 Male connector									
8252 1/4" x 1/4"	1 Plum	24	.333	Ea.	6.40	17.85		24.25	33.50
8253 3/8" x 3/8"		22	.364		9.95	19.50		29.45	40
8254 1/2" x 1/2"		22	.364		14.70	19.50		34.20	45
8256 3/4" x 3/4"		21	.381		22.50	20.50		43	55.50
8258 1" x 1"		20	.400		39.50	21.50		61	75.50

22 11 13.64 Pipe, Stainless Steel

	Crew	Daily Output	Labor-Hours	Unit	Material	2011 Bare Costs Labor	Equipment	Total	Total Incl O&P
0010 **PIPE, STAINLESS STEEL**									
0020 Welded, with clevis type hanger assemblies, 10' O.C.									
0500 Schedule 5, type 304									
0540 1/2" diameter	Q-15	128	.125	L.F.	7.50	6.05	.25	13.80	17.55
0550 3/4" diameter		116	.138		9.35	6.65	.28	16.28	20.50
0560 1" diameter		103	.155		11.50	7.50	.32	19.32	24
0570 1-1/4" diameter		93	.172		15.80	8.30	.35	24.45	30
0580 1-1/2" diameter		85	.188		21	9.10	.38	30.48	37
0590 2" diameter		69	.232		34.50	11.20	.47	46.17	55
0600 2-1/2" diameter		53	.302		38	14.55	.61	53.16	64.50
0610 3" diameter		48	.333		46	16.10	.68	62.78	75
0620 4" diameter		44	.364		58.50	17.55	.74	76.79	91
0630 5" diameter		36	.444		119	21.50	.90	141.40	163
0640 6" diameter	Q-16	42	.571		110	28.50	.77	139.27	163
0650 8" diameter		34	.706		168	35.50	.95	204.45	239
0660 10" diameter		26	.923		233	46	1.25	280.25	325
0670 12" diameter		21	1.143		305	57	1.54	363.54	425
0700 To delete hangers, subtract									
0710 1/2" diam. to 1-1/2" diam.					8%	19%			
0720 2" diam. to 5" diam.					4%	9%			

22 11 Facility Water Distribution

22 11 13 – Facility Water Distribution Piping

22 11 13.64 Pipe, Stainless Steel	Crew	Daily Output	Labor-Hours	Unit	Material	2011 Bare Costs Labor	Equipment	Total	Total Incl O&P
0730	6" diam. to 12" diam.					3%	4%		
0750	For small quantities, add				L.F.	10%			
1250	Schedule 5, type 316								
1290	1/2" diameter	Q-15	128	.125	L.F.	11.50	6.05	.25	17.80	22
1300	3/4" diameter		116	.138		14.25	6.65	.28	21.18	26
1310	1" diameter		103	.155		16.80	7.50	.32	24.62	30
1320	1-1/4" diameter		93	.172		20	8.30	.35	28.65	35
1330	1-1/2" diameter		85	.188		28.50	9.10	.38	37.98	45
1340	2" diameter		69	.232		48.50	11.20	.47	60.17	70.50
1350	2-1/2" diameter		53	.302		59	14.55	.61	74.16	87.50
1360	3" diameter		48	.333		75	16.10	.68	91.78	107
1370	4" diameter		44	.364		93	17.55	.74	111.29	129
1380	5" diameter	▼	36	.444		177	21.50	.90	199.40	228
1390	6" diameter	Q-16	42	.571		173	28.50	.77	202.27	233
1400	8" diameter		34	.706		260	35.50	.95	296.45	340
1410	10" diameter		26	.923		370	46	1.25	417.25	480
1420	12" diameter	▼	21	1.143		480	57	1.54	538.54	610
1490	For small quantities, add				▼	10%			
1940	To delete hanger, subtract								
1950	1/2" diam. to 1-1/2" diam.					5%	19%		
1960	2" diam. to 5" diam.					3%	9%		
1970	6" diam. to 12" diam.					2%	4%		
2000	Schedule 10, type 304								
2040	1/4" diameter	Q-15	131	.122	L.F.	4.70	5.90	.25	10.85	14.20
2050	3/8" diameter		128	.125		6.15	6.05	.25	12.45	16.05
2060	1/2" diameter		125	.128		6.85	6.20	.26	13.31	17.05
2070	3/4" diameter		113	.142		7.45	6.85	.29	14.59	18.70
2080	1" diameter		100	.160		11.50	7.70	.32	19.52	24.50
2090	1-1/4" diameter		91	.176		11.80	8.50	.36	20.66	26
2100	1-1/2" diameter		83	.193		12.85	9.30	.39	22.54	28.50
2110	2" diameter		67	.239		15.40	11.50	.48	27.38	34.50
2120	2-1/2" diameter		51	.314		20.50	15.15	.64	36.29	45.50
2130	3" diameter		46	.348		25	16.80	.71	42.51	53
2140	4" diameter		42	.381		36.50	18.40	.77	55.67	68.50
2150	5" diameter	▼	35	.457		47	22	.93	69.93	86
2160	6" diameter	Q-16	40	.600		51.50	30	.81	82.31	102
2170	8" diameter		33	.727		89.50	36.50	.98	126.98	154
2180	10" diameter		25	.960		125	48	1.30	174.30	210
2190	12" diameter	▼	21	1.143		155	57	1.54	213.54	258
2250	For small quantities, add				▼	10%			
2650	To delete hanger, subtract								
2660	1/4" diam. to 3/4" diam.					9%	22%		
2670	1" diam. to 2" diam.					4%	15%		
2680	2-1/2" diam. to 5" diam.					3%	8%		
2690	6" diam. to 12" diam.					3%	4%		
2750	Schedule 10, type 316								
2790	1/4" diameter	Q-15	131	.122	L.F.	5.95	5.90	.25	12.10	15.60
2800	3/8" diameter		128	.125		7.55	6.05	.25	13.85	17.60
2810	1/2" diameter		125	.128		8	6.20	.26	14.46	18.30
2820	3/4" diameter		113	.142		10.15	6.85	.29	17.29	21.50
2830	1" diameter		100	.160		11.50	7.70	.32	19.52	24.50
2840	1-1/4" diameter		91	.176		17.10	8.50	.36	25.96	32
2850	1-1/2" diameter	▼	83	.193	▼	23	9.30	.39	32.69	39.50

22 11 13 – Facility Water Distribution Piping

22 11 13.64 Pipe, Stainless Steel

		Crew	Daily Output	Labor-Hours	Unit	Material	2011 Bare Costs Labor	Equipment	Total	Total Incl O&P
2860	2" diameter	Q-15	67	.239	L.F.	28	11.50	.48	39.98	48.50
2870	2-1/2" diameter		51	.314		30.50	15.15	.64	46.29	57
2880	3" diameter		46	.348		37	16.80	.71	54.51	67
2890	4" diameter		42	.381		44	18.40	.77	63.17	77
2900	5" diameter		35	.457		55.50	22	.93	78.43	95
2910	6" diameter	Q-16	40	.600		70.50	30	.81	101.31	123
2920	8" diameter		33	.727		128	36.50	.98	165.48	197
2930	10" diameter		25	.960		169	48	1.30	218.30	259
2940	12" diameter		21	1.143		212	57	1.54	270.54	320
2990	For small quantities, add					10%				
3430	To delete hanger, subtract									
3440	1/4" diam. to 3/4" diam.					6%	22%			
3450	1" diam. to 2" diam.					3%	15%			
3460	2-1/2" diam. to 5" diam.					2%	8%			
3470	6" diam. to 12" diam.					2%	4%			
3500	Threaded, couplings and clevis hanger assemblies, 10' O.C.									
3520	Schedule 40, type 304									
3540	1/4" diameter	1 Plum	54	.148	L.F.	6.45	7.95		14.40	18.95
3550	3/8" diameter		53	.151		8.15	8.10		16.25	21
3560	1/2" diameter		52	.154		9.60	8.25		17.85	23
3570	3/4" diameter		51	.157		11.40	8.40		19.80	25
3580	1" diameter		45	.178		15.65	9.55		25.20	31.50
3590	1-1/4" diameter	Q-1	76	.211		21	10.15		31.15	38.50
3600	1-1/2" diameter		69	.232		24.50	11.20		35.70	43.50
3610	2" diameter		57	.281		34	13.55		47.55	57.50
3620	2-1/2" diameter		44	.364		58.50	17.55		76.05	90.50
3630	3" diameter		38	.421		80.50	20.50		101	119
3640	4" diameter	Q-2	51	.471		97.50	23.50		121	142
3740	For small quantities, add					10%				
4200	To delete couplings & hangers, subtract									
4210	1/4" diam. to 3/4" diam.					15%	56%			
4220	1" diam. to 2" diam.					18%	49%			
4230	2-1/2" diam. to 4" diam.					34%	40%			
4250	Schedule 40, type 316									
4290	1/4" diameter	1 Plum	54	.148	L.F.	7.75	7.95		15.70	20.50
4300	3/8" diameter		53	.151		8.75	8.10		16.85	21.50
4310	1/2" diameter		52	.154		11.65	8.25		19.90	25
4320	3/4" diameter		51	.157		14.20	8.40		22.60	28
4330	1" diameter		45	.178		20	9.55		29.55	36
4340	1-1/4" diameter	Q-1	76	.211		28	10.15		38.15	45.50
4350	1-1/2" diameter		69	.232		31	11.20		42.20	51
4360	2" diameter		57	.281		44.50	13.55		58.05	68.50
4370	2-1/2" diameter		44	.364		78.50	17.55		96.05	112
4380	3" diameter		38	.421		103	20.50		123.50	144
4390	4" diameter	Q-2	51	.471		145	23.50		168.50	194
4490	For small quantities, add					10%				
4900	To delete couplings & hangers, subtract									
4910	1/4" diam. to 3/4" diam.					12%	56%			
4920	1" diam. to 2" diam.					14%	49%			
4930	2-1/2" diam. to 4" diam.					27%	40%			
5000	Schedule 80, type 304									
5040	1/4" diameter	1 Plum	53	.151	L.F.	17.45	8.10		25.55	31.50
5050	3/8" diameter		52	.154		18.85	8.25		27.10	33

22 11 13.64 Pipe, Stainless Steel		Crew	Daily Output	Labor-Hours	Unit	Material	2011 Bare Costs Labor	Equipment	Total	Total Incl O&P
5060	1/2" diameter	1 Plum	51	.157	L.F.	24.50	8.40		32.90	39
5070	3/4" diameter		48	.167		28.50	8.95		37.45	44.50
5080	1" diameter	↓	43	.186		34	9.95		43.95	52
5090	1-1/4" diameter	Q-1	73	.219		44	10.60		54.60	64
5100	1-1/2" diameter		67	.239		55	11.50		66.50	77.50
5110	2" diameter	↓	54	.296		72	14.30		86.30	101
5190	For small quantities, add				↓	10%				
5700	To delete couplings & hangers, subtract									
5710	1/4" diam. to 3/4" diam.					10%	53%			
5720	1" diam. to 2" diam.					14%	47%			
5750	Schedule 80, type 316									
5790	1/4" diameter	1 Plum	53	.151	L.F.	20	8.10		28.10	34
5800	3/8" diameter		52	.154		21.50	8.25		29.75	36
5810	1/2" diameter		51	.157		28.50	8.40		36.90	44
5820	3/4" diameter		48	.167		38	8.95		46.95	55.50
5830	1" diameter	↓	43	.186		45	9.95		54.95	64.50
5840	1-1/4" diameter	Q-1	73	.219		64	10.60		74.60	86
5850	1-1/2" diameter		67	.239		70.50	11.50		82	94.50
5860	2" diameter	↓	54	.296		100	14.30		114.30	132
5950	For small quantities, add				↓	10%				
7000	To delete couplings & hangers, subtract									
7010	1/4" diam. to 3/4" diam.					9%	53%			
7020	1" diam. to 2" diam.					14%	47%			
8000	Weld joints with clevis type hanger assemblies, 10' O.C.									
8010	Schedule 40, type 304									
8050	1/8" pipe size	Q-15	126	.127	L.F.	4.90	6.15	.26	11.31	14.85
8060	1/4" pipe size		125	.128		5.25	6.20	.26	11.71	15.30
8070	3/8" pipe size		122	.131		6.70	6.35	.27	13.32	17.10
8080	1/2" pipe size		118	.136		7.70	6.55	.28	14.53	18.50
8090	3/4" pipe size		109	.147		8.85	7.10	.30	16.25	20.50
8100	1" pipe size		95	.168		11.50	8.15	.34	19.99	25
8110	1-1/4" pipe size		86	.186		14.50	9	.38	23.88	29.50
8120	1-1/2" pipe size		78	.205		16.95	9.90	.42	27.27	34
8130	2" pipe size		62	.258		21.50	12.45	.52	34.47	42.50
8140	2-1/2" pipe size		49	.327		30	15.75	.66	46.41	57
8150	3" pipe size		44	.364		41.50	17.55	.74	59.79	72.50
8160	3-1/2" pipe size		44	.364		42.50	17.55	.74	60.79	73.50
8170	4" pipe size		39	.410		42.50	19.80	.83	63.13	77.50
8180	5" pipe size	↓	32	.500		60.50	24	1.02	85.52	104
8190	6" pipe size	Q-16	37	.649		72	32.50	.88	105.38	128
8200	8" pipe size		29	.828		144	41.50	1.12	186.62	221
8210	10" pipe size		24	1		243	50	1.35	294.35	345
8220	12" pipe size	↓	20	1.200	↓	275	60	1.62	336.62	390
8300	Schedule 40, type 316									
8310	1/8" pipe size	Q-15	126	.127	L.F.	6.80	6.15	.26	13.21	16.95
8320	1/4" pipe size		125	.128		6.30	6.20	.26	12.76	16.45
8330	3/8" pipe size		122	.131		7	6.35	.27	13.62	17.45
8340	1/2" pipe size		118	.136		9.35	6.55	.28	16.18	20.50
8350	3/4" pipe size		109	.147		11.10	7.10	.30	18.50	23
8360	1" pipe size		95	.168		15.15	8.15	.34	23.64	29
8370	1-1/4" pipe size		86	.186		19.90	9	.38	29.28	36
8380	1-1/2" pipe size		78	.205		22.50	9.90	.42	32.82	40
8390	2" pipe size	↓	62	.258		29.50	12.45	.52	42.47	51.50

22 11 13 – Facility Water Distribution Piping

22 11 13.64 Pipe, Stainless Steel		Crew	Daily Output	Labor-Hours	Unit	Material	2011 Bare Costs Labor	Equipment	Total	Total Incl O&P
8400	2-1/2" pipe size	Q-15	49	.327	L.F.	44	15.75	.66	60.41	72.50
8410	3" pipe size		44	.364		55.50	17.55	.74	73.79	88
8420	3-1/2" pipe size		44	.364		69.50	17.55	.74	87.79	103
8430	4" pipe size		39	.410		79	19.80	.83	99.63	117
8440	5" pipe size		32	.500		120	24	1.02	145.02	169
8450	6" pipe size	Q-16	37	.649		136	32.50	.88	169.38	198
8460	8" pipe size		29	.828		262	41.50	1.12	304.62	350
8470	10" pipe size		24	1		350	50	1.35	401.35	460
8480	12" pipe size		20	1.200		475	60	1.62	536.62	615
8500	Schedule 80, type 304									
8510	1/4" pipe size	Q-15	110	.145	L.F.	16.05	7	.30	23.35	28.50
8520	3/8" pipe size		109	.147		17.20	7.10	.30	24.60	30
8530	1/2" pipe size		106	.151		22.50	7.30	.31	30.11	35.50
8540	3/4" pipe size		96	.167		25.50	8.05	.34	33.89	41
8550	1" pipe size		87	.184		29.50	8.85	.37	38.72	46
8560	1-1/4" pipe size		81	.198		33	9.55	.40	42.95	51
8570	1-1/2" pipe size		74	.216		42.50	10.45	.44	53.39	62.50
8580	2" pipe size		58	.276		55.50	13.30	.56	69.36	81.50
8590	2-1/2" pipe size		46	.348		114	16.80	.71	131.51	151
8600	3" pipe size		41	.390		152	18.85	.79	171.64	196
8610	4" pipe size		33	.485		238	23.50	.98	262.48	298
8630	6" pipe size	Q-16	30	.800		525	40	1.08	566.08	635
8640	Schedule 80, type 316									
8650	1/4" pipe size	Q-15	110	.145	L.F.	18.60	7	.30	25.90	31.50
8660	3/8" pipe size		109	.147		19.80	7.10	.30	27.20	33
8670	1/2" pipe size		106	.151		26	7.30	.31	33.61	39.50
8680	3/4" pipe size		96	.167		34.50	8.05	.34	42.89	50.50
8690	1" pipe size		87	.184		38.50	8.85	.37	47.72	56
8700	1-1/4" pipe size		81	.198		50	9.55	.40	59.95	69.50
8710	1-1/2" pipe size		74	.216		54.50	10.45	.44	65.39	76
8720	2" pipe size		58	.276		78	13.30	.56	91.86	106
8730	2-1/2" pipe size		46	.348		208	16.80	.71	225.51	255
8740	3" pipe size		41	.390		243	18.85	.79	262.64	296
8760	4" pipe size		33	.485		340	23.50	.98	364.48	405
8770	6" pipe size	Q-16	30	.800		910	40	1.08	951.08	1,050
9100	Threading pipe labor, sst, one end, schedules 40 & 80									
9110	1/4" through 3/4" pipe size	1 Plum	61.50	.130	Ea.		6.95		6.95	10.40
9120	1" through 2" pipe size		55.90	.143			7.65		7.65	11.45
9130	2-1/2" pipe size		41.50	.193			10.35		10.35	15.40
9140	3" pipe size		38.50	.208			11.15		11.15	16.60
9150	3-1/2" pipe size	Q-1	68.40	.234			11.30		11.30	16.85
9160	4" pipe size		73	.219			10.60		10.60	15.80
9170	5" pipe size		40.70	.393			18.95		18.95	28.50
9180	6" pipe size		35.40	.452			22		22	32.50
9190	8" pipe size		22.30	.717			34.50		34.50	51.50
9200	10" pipe size		16.10	.994			48		48	71.50
9210	12" pipe size		12.30	1.301			63		63	93.50
9250	Welding labor per joint for stainless steel									
9260	Schedule 5 and 10									
9270	1/4" pipe size	Q-15	36	.444	Ea.		21.50	.90	22.40	33
9280	3/8" pipe size		35	.457			22	.93	22.93	34
9290	1/2" pipe size		35	.457			22	.93	22.93	34
9300	3/4" pipe size		28	.571			27.50	1.16	28.66	42.50

22 11 13.64 Pipe, Stainless Steel	Crew	Daily Output	Labor-Hours	Unit	Material	2011 Bare Costs Labor	2011 Bare Costs Equipment	Total	Total Incl O&P
9310	1" pipe size	Q-15	25	.640	Ea.		31	1.30	32.30	47.50
9320	1-1/4" pipe size		22	.727			35	1.48	36.48	54
9330	1-1/2" pipe size		21	.762			37	1.55	38.55	56.50
9340	2" pipe size		18	.889			43	1.80	44.80	66
9350	2-1/2" pipe size		12	1.333			64.50	2.71	67.21	99
9360	3" pipe size		9.73	1.644			79.50	3.34	82.84	122
9370	4" pipe size		7.37	2.171			105	4.41	109.41	161
9380	5" pipe size		6.15	2.602			126	5.30	131.30	193
9390	6" pipe size		5.71	2.802			135	5.70	140.70	208
9400	8" pipe size		3.69	4.336			209	8.80	217.80	320
9410	10" pipe size		2.91	5.498			265	11.15	276.15	405
9420	12" pipe size		2.31	6.926			335	14.05	349.05	515
9500	Schedule 40									
9510	1/4" pipe size	Q-15	28	.571	Ea.		27.50	1.16	28.66	42.50
9520	3/8" pipe size		27	.593			28.50	1.20	29.70	44
9530	1/2" pipe size		25.40	.630			30.50	1.28	31.78	47
9540	3/4" pipe size		22.22	.720			34.50	1.46	35.96	53.50
9550	1" pipe size		20.25	.790			38	1.60	39.60	59
9560	1-1/4" pipe size		18.82	.850			41	1.73	42.73	63
9570	1-1/2" pipe size		17.78	.900			43.50	1.83	45.33	67
9580	2" pipe size		15.09	1.060			51	2.15	53.15	79
9590	2-1/2" pipe size		7.96	2.010			97	4.08	101.08	149
9600	3" pipe size		6.43	2.488			120	5.05	125.05	185
9610	4" pipe size		4.88	3.279			158	6.65	164.65	243
9620	5" pipe size		4.26	3.756			181	7.60	188.60	279
9630	6" pipe size		3.77	4.244			205	8.60	213.60	315
9640	8" pipe size		2.44	6.557			315	13.30	328.30	485
9650	10" pipe size		1.92	8.333			400	16.90	416.90	620
9660	12" pipe size		1.52	10.526			510	21.50	531.50	785
9750	Schedule 80									
9760	1/4" pipe size	Q-15	21.55	.742	Ea.		36	1.51	37.51	55
9770	3/8" pipe size		20.75	.771			37	1.57	38.57	57
9780	1/2" pipe size		19.54	.819			39.50	1.66	41.16	61
9790	3/4" pipe size		17.09	.936			45	1.90	46.90	69.50
9800	1" pipe size		15.58	1.027			49.50	2.08	51.58	76.50
9810	1-1/4" pipe size		14.48	1.105			53.50	2.24	55.74	82
9820	1-1/2" pipe size		13.68	1.170			56.50	2.37	58.87	87
9830	2" pipe size		11.61	1.378			66.50	2.80	69.30	103
9840	2-1/2" pipe size		6.12	2.614			126	5.30	131.30	194
9850	3" pipe size		4.94	3.239			156	6.55	162.55	240
9860	4" pipe size		3.75	4.267			206	8.65	214.65	315
9870	5" pipe size		3.27	4.893			236	9.95	245.95	360
9880	6" pipe size		2.90	5.517			266	11.20	277.20	405
9890	8" pipe size		1.87	8.556			415	17.35	432.35	635
9900	10" pipe size		1.48	10.811			520	22	542	805
9910	12" pipe size		1.17	13.675			660	28	688	1,025
9920	Schedule 160, 1/2" pipe size		17	.941			45.50	1.91	47.41	70
9930	3/4" pipe size		14.81	1.080			52	2.19	54.19	80.50
9940	1" pipe size		13.50	1.185			57	2.41	59.41	88
9950	1-1/4" pipe size		12.55	1.275			61.50	2.59	64.09	95
9960	1-1/2" pipe size		11.85	1.350			65	2.74	67.74	101
9970	2" pipe size		10	1.600			77	3.25	80.25	119
9980	3" pipe size		4.28	3.738			180	7.60	187.60	277

22 11 13 – Facility Water Distribution Piping

22 11 13.64 Pipe, Stainless Steel	Crew	Daily Output	Labor-Hours	Unit	Material	2011 Bare Costs Labor	Equipment	Total	Total Incl O&P
9990	4" pipe size	Q-15	3.25	4.923	Ea.		238	10	248	365

22 11 13.66 Pipe Fittings, Stainless Steel

		Crew	Daily Output	Labor-Hours	Unit	Material	2011 Bare Costs Labor	Equipment	Total	Total Incl O&P
0010	**PIPE FITTINGS, STAINLESS STEEL**									
0100	Butt weld joint, schedule 5, type 304									
0120	90° Elbow, long									
0140	1/2"	Q-15	17.50	.914	Ea.	19.80	44	1.86	65.66	90
0150	3/4"		14	1.143		19.80	55	2.32	77.12	107
0160	1"		12.50	1.280		21	62	2.60	85.60	118
0170	1-1/4"		11	1.455		28.50	70	2.95	101.45	140
0180	1-1/2"		10.50	1.524		24	73.50	3.09	100.59	140
0190	2"		9	1.778		28.50	86	3.61	118.11	163
0200	2-1/2"		6	2.667		66	129	5.40	200.40	270
0210	3"		4.86	3.292		72.50	159	6.70	238.20	325
0220	3-1/2"		4.27	3.747		182	181	7.60	370.60	480
0230	4"		3.69	4.336		99	209	8.80	316.80	430
0240	5"	▼	3.08	5.195		385	251	10.55	646.55	810
0250	6"	Q-16	4.29	5.594		297	280	7.55	584.55	755
0260	8"		2.76	8.696		625	435	11.75	1,071.75	1,350
0270	10"		2.18	11.009		970	550	14.85	1,534.85	1,900
0280	12"	▼	1.73	13.873		1,375	695	18.75	2,088.75	2,575
0320	For schedule 5, type 316, add				▼	30%				
0600	45° Elbow, long									
0620	1/2"	Q-15	17.50	.914	Ea.	19.80	44	1.86	65.66	90
0630	3/4"		14	1.143		19.80	55	2.32	77.12	107
0640	1"		12.50	1.280		21	62	2.60	85.60	118
0650	1-1/4"		11	1.455		28.50	70	2.95	101.45	140
0660	1-1/2"		10.50	1.524		24	73.50	3.09	100.59	140
0670	2"		9	1.778		28.50	86	3.61	118.11	163
0680	2-1/2"		6	2.667		66	129	5.40	200.40	270
0690	3"		4.86	3.292		72.50	159	6.70	238.20	325
0700	3-1/2"		4.27	3.747		182	181	7.60	370.60	480
0710	4"		3.69	4.336		83.50	209	8.80	301.30	410
0720	5"	▼	3.08	5.195		310	251	10.55	571.55	725
0730	6"	Q-16	4.29	5.594		209	280	7.55	496.55	660
0740	8"		2.76	8.696		440	435	11.75	886.75	1,150
0750	10"		2.18	11.009		770	550	14.85	1,334.85	1,675
0760	12"	▼	1.73	13.873		965	695	18.75	1,678.75	2,100
0800	For schedule 5, type 316, add				▼	25%				
1100	Tee, straight									
1130	1/2"	Q-15	11.66	1.372	Ea.	59.50	66	2.79	128.29	168
1140	3/4"		9.33	1.715		59.50	82.50	3.48	145.48	193
1150	1"		8.33	1.921		62.50	92.50	3.90	158.90	211
1160	1-1/4"		7.33	2.183		50.50	105	4.43	159.93	217
1170	1-1/2"		7	2.286		49.50	110	4.64	164.14	225
1180	2"		6	2.667		51.50	129	5.40	185.90	255
1190	2-1/2"		4	4		124	193	8.10	325.10	435
1200	3"		3.24	4.938		198	238	10	446	585
1210	3-1/2"		2.85	5.614		260	271	11.40	542.40	705
1220	4"		2.46	6.504		141	315	13.20	469.20	640
1230	5"		2	8		445	385	16.25	846.25	1,075
1240	6"	Q-16	2.85	8.421		350	420	11.35	781.35	1,025
1250	8"	▼	1.84	13.043	▼	750	655	17.60	1,422.60	1,825

22 11 13.66 Pipe Fittings, Stainless Steel		Crew	Daily Output	Labor-Hours	Unit	Material	2011 Bare Costs Labor	Equipment	Total	Total Incl O&P
1260	10"	Q-16	1.45	16.552	Ea.	1,200	830	22.50	2,052.50	2,575
1270	12"		1.15	20.870		1,675	1,050	28	2,753	3,425
1320	For schedule 5, type 316, add					25%				
2000	Butt weld joint, schedule 10, type 304									
2020	90° elbow, long									
2040	1/2"	Q-15	17	.941	Ea.	14.85	45.50	1.91	62.26	86.50
2050	3/4"		14	1.143		16.35	55	2.32	73.67	103
2060	1"		12.50	1.280		14.25	62	2.60	78.85	111
2070	1-1/4"		11	1.455		21.50	70	2.95	94.45	132
2080	1-1/2"		10.50	1.524		16.50	73.50	3.09	93.09	132
2090	2"		9	1.778		19.50	86	3.61	109.11	153
2100	2-1/2"		6	2.667		49.50	129	5.40	183.90	252
2110	3"		4.86	3.292		46	159	6.70	211.70	295
2120	3-1/2"		4.27	3.747		136	181	7.60	324.60	430
2130	4"		3.69	4.336		70.50	209	8.80	288.30	395
2140	5"		3.08	5.195		289	251	10.55	550.55	705
2150	6"	Q-16	4.29	5.594		203	280	7.55	490.55	650
2160	8"		2.76	8.696		430	435	11.75	876.75	1,125
2170	10"		2.18	11.009		665	550	14.85	1,229.85	1,575
2180	12"		1.73	13.873		940	695	18.75	1,653.75	2,075
2500	45° elbow, long									
2520	1/2"	Q-15	17.50	.914	Ea.	14.85	44	1.86	60.71	84.50
2530	3/4"		14	1.143		14.85	55	2.32	72.17	101
2540	1"		12.50	1.280		14.25	62	2.60	78.85	111
2550	1-1/4"		11	1.455		21.50	70	2.95	94.45	132
2560	1-1/2"		10.50	1.524		16.50	73.50	3.09	93.09	132
2570	2"		9	1.778		19.50	86	3.61	109.11	153
2580	2-1/2"		6	2.667		49.50	129	5.40	183.90	252
2590	3"		4.86	3.292		37	159	6.70	202.70	285
2600	3-1/2"		4.27	3.747		136	181	7.60	324.60	430
2610	4"		3.69	4.336		57	209	8.80	274.80	380
2620	5"		3.08	5.195		231	251	10.55	492.55	640
2630	6"	Q-16	4.29	5.594		143	280	7.55	430.55	585
2640	8"		2.76	8.696		300	435	11.75	746.75	995
2650	10"		2.18	11.009		525	550	14.85	1,089.85	1,425
2660	12"		1.73	13.873		655	695	18.75	1,368.75	1,775
2670	Reducer, concentric									
2674	1" x 3/4"	Q-15	13.25	1.208	Ea.	29.50	58.50	2.45	90.45	122
2676	2" x 1-1/2"	"	9.75	1.641		24	79	3.33	106.33	148
2678	6" x 4"	Q-16	4.91	4.888		72	245	6.60	323.60	450
2680	Caps									
2682	1"	Q-15	25	.640	Ea.	30	31	1.30	62.30	80.50
2684	1-1/2"		21	.762		36	37	1.55	74.55	96
2685	2"		18	.889		34	43	1.80	78.80	103
2686	4"		7.38	2.168		54	105	4.40	163.40	220
2687	6"	Q-16	8.58	2.797		86.50	140	3.78	230.28	310
3000	Tee, straight									
3030	1/2"	Q-15	11.66	1.372	Ea.	44.50	66	2.79	113.29	151
3040	3/4"		9.33	1.715		44.50	82.50	3.48	130.48	177
3050	1"		8.33	1.921		43	92.50	3.90	139.40	189
3060	1-1/4"		7.33	2.183		57.50	105	4.43	166.93	225
3070	1-1/2"		7	2.286		34	110	4.64	148.64	207
3080	2"		6	2.667		35.50	129	5.40	169.90	237

22 11 13 – Facility Water Distribution Piping

22 11 13.66 Pipe Fittings, Stainless Steel		Crew	Daily Output	Labor-Hours	Unit	Material	2011 Bare Costs Labor	Equipment	Total	Total Incl O&P
3090	2-1/2"	Q-15	4	4	Ea.	93	193	8.10	294.10	400
3100	3"		3.24	4.938		73.50	238	10	321.50	445
3110	3-1/2"		2.85	5.614		108	271	11.40	390.40	535
3120	4"		2.46	6.504		96	315	13.20	424.20	590
3130	5"		2	8		335	385	16.25	736.25	965
3140	6"	Q-16	2.85	8.421		240	420	11.35	671.35	905
3150	8"		1.84	13.043		510	655	17.60	1,182.60	1,550
3151	10"		1.45	16.552		825	830	22.50	1,677.50	2,150
3152	12"		1.15	20.870		1,150	1,050	28	2,228	2,825
3154	For schedule 10, type 316, add					25%				
3281	Butt weld joint, schedule 40, type 304									
3284	90° Elbow, long, 1/2"	Q-15	12.70	1.260	Ea.	17.35	61	2.56	80.91	113
3288	3/4"		11.10	1.441		17.35	69.50	2.93	89.78	126
3289	1"		10.13	1.579		18.15	76	3.21	97.36	137
3290	1-1/4"		9.40	1.702		24	82	3.46	109.46	153
3300	1-1/2"		8.89	1.800		19	87	3.65	109.65	155
3310	2"		7.55	2.119		27	102	4.30	133.30	188
3320	2-1/2"		3.98	4.020		52	194	8.15	254.15	355
3330	3"		3.21	4.984		67	241	10.10	318.10	445
3340	3-1/2"		2.83	5.654		248	273	11.50	532.50	690
3350	4"		2.44	6.557		116	315	13.30	444.30	610
3360	5"		2.13	7.512		390	360	15.25	765.25	980
3370	6"	Q-16	2.83	8.481		340	425	11.45	776.45	1,025
3380	8"		1.83	13.115		615	655	17.70	1,287.70	1,675
3390	10"		1.44	16.667		1,300	835	22.50	2,157.50	2,700
3400	12"		1.14	21.053		1,650	1,050	28.50	2,728.50	3,425
3410	For schedule 40, type 316, add					25%				
3460	45° Elbow, long, 1/2"	Q-15	12.70	1.260	Ea.	17.35	61	2.56	80.91	113
3470	3/4"		11.10	1.441		17.35	69.50	2.93	89.78	126
3480	1"		10.13	1.579		18.15	76	3.21	97.36	137
3490	1-1/4"		9.40	1.702		24	82	3.46	109.46	153
3500	1-1/2"		8.89	1.800		19	87	3.65	109.65	155
3510	2"		7.55	2.119		27	102	4.30	133.30	188
3520	2-1/2"		3.98	4.020		52	194	8.15	254.15	355
3530	3"		3.21	4.984		53	241	10.10	304.10	430
3540	3-1/2"		2.83	5.654		248	273	11.50	532.50	690
3550	4"		2.44	6.557		82.50	315	13.30	410.80	575
3560	5"		2.13	7.512		272	360	15.25	647.25	855
3570	6"	Q-16	2.83	8.481		239	425	11.45	675.45	910
3580	8"		1.83	13.115		430	655	17.70	1,102.70	1,475
3590	10"		1.44	16.667		910	835	22.50	1,767.50	2,275
3600	12"		1.14	21.053		1,150	1,050	28.50	2,228.50	2,875
3610	For schedule 40, type 316, add					25%				
3660	Tee, straight 1/2"	Q-15	8.46	1.891	Ea.	44.50	91.50	3.84	139.84	189
3670	3/4"		7.40	2.162		44.50	104	4.39	152.89	210
3680	1"		6.74	2.374		47	115	4.82	166.82	228
3690	1-1/4"		6.27	2.552		110	123	5.20	238.20	310
3700	1-1/2"		5.92	2.703		50.50	130	5.50	186	257
3710	2"		5.03	3.181		75	153	6.45	234.45	320
3720	2-1/2"		2.65	6.038		95	291	12.25	398.25	550
3730	3"		2.14	7.477		95.50	360	15.20	470.70	660
3740	3-1/2"		1.88	8.511		181	410	17.30	608.30	835
3750	4"		1.62	9.877		181	475	20	676	930

22 11 13.66 Pipe Fittings, Stainless Steel	Crew	Daily Output	Labor-Hours	Unit	Material	2011 Bare Costs Labor	Equipment	Total	Total Incl O&P
3760	5"	Q-15	1.42	11.268	Ea.	415	545	23	983	1,300
3770	6"	Q-16	1.88	12.766		370	640	17.25	1,027.25	1,375
3780	8"		1.22	19.672		735	985	26.50	1,746.50	2,325
3790	10"		.96	25		1,425	1,250	34	2,709	3,500
3800	12"		.76	31.579		1,900	1,575	42.50	3,517.50	4,500
3810	For schedule 40, type 316, add					25%				
3820	Tee, reducing on outlet, 3/4" x 1/2"	Q-15	7.73	2.070	Ea.	59.50	100	4.20	163.70	219
3822	1" x 1/2"		7.24	2.210		64.50	107	4.49	175.99	235
3824	1" x 3/4"		6.96	2.299		59.50	111	4.67	175.17	237
3826	1-1/4" x 1"		6.43	2.488		145	120	5.05	270.05	345
3828	1-1/2" x 1/2"		6.58	2.432		87.50	117	4.94	209.44	276
3830	1-1/2" x 3/4"		6.35	2.520		81.50	122	5.10	208.60	277
3832	1-1/2" x 1"		6.18	2.589		62.50	125	5.25	192.75	261
3834	2" x 1"		5.50	2.909		112	140	5.90	257.90	340
3836	2" x 1-1/2"		5.30	3.019		94	146	6.15	246.15	325
3838	2-1/2" x 2"		3.15	5.079		142	245	10.30	397.30	530
3840	3" x 1-1/2"		2.72	5.882		144	284	11.95	439.95	595
3842	3" x 2"		2.65	6.038		120	291	12.25	423.25	580
3844	4" x 2"		2.10	7.619		299	370	15.45	684.45	895
3846	4" x 3"		1.77	9.040		217	435	18.35	670.35	910
3848	5" x 4"		1.48	10.811		500	520	22	1,042	1,350
3850	6" x 3"	Q-16	2.19	10.959		555	550	14.80	1,119.80	1,450
3852	6" x 4"		2.04	11.765		480	590	15.90	1,085.90	1,425
3854	8" x 4"		1.46	16.438		1,125	820	22	1,967	2,475
3856	10" x 8"		.69	34.783		1,875	1,750	47	3,672	4,725
3858	12" x 10"		.55	43.636		2,525	2,175	59	4,759	6,100
3950	Reducer, concentric, 3/4" x 1/2"	Q-15	11.85	1.350		33	65	2.74	100.74	137
3952	1" x 3/4"		10.60	1.509		34.50	73	3.06	110.56	150
3954	1-1/4" x 3/4"		10.19	1.570		85	76	3.19	164.19	210
3956	1-1/4" x 1"		9.76	1.639		43	79	3.33	125.33	169
3958	1-1/2" x 3/4"		9.88	1.619		72	78	3.29	153.29	200
3960	1-1/2" x 1"		9.47	1.690		58	81.50	3.43	142.93	189
3962	2" x 1"		8.65	1.850		25	89.50	3.75	118.25	164
3964	2" x 1-1/2"		8.16	1.961		29	94.50	3.98	127.48	177
3966	2-1/2" x 1"		5.71	2.802		125	135	5.70	265.70	345
3968	2-1/2" x 2"		5.21	3.071		63.50	148	6.25	217.75	298
3970	3" x 1"		4.88	3.279		86	158	6.65	250.65	340
3972	3" x 1-1/2"		4.72	3.390		43.50	164	6.90	214.40	300
3974	3" x 2"		4.51	3.548		41.50	171	7.20	219.70	310
3976	4" x 2"		3.69	4.336		58.50	209	8.80	276.30	385
3978	4" x 3"		2.77	5.776		43.50	279	11.75	334.25	475
3980	5" x 3"		2.56	6.250		261	300	12.70	573.70	750
3982	5" x 4"		2.27	7.048		210	340	14.30	564.30	755
3984	6" x 3"	Q-16	3.57	6.723		134	335	9.10	478.10	660
3986	6" x 4"		3.19	7.524		111	375	10.15	496.15	695
3988	8" x 4"		2.44	9.836		315	490	13.30	818.30	1,100
3990	8" x 6"		2.22	10.811		236	540	14.60	790.60	1,075
3992	10" x 6"		1.91	12.565		440	630	16.95	1,086.95	1,450
3994	10" x 8"		1.61	14.907		365	745	20	1,130	1,550
3995	12" x 6"		1.63	14.724		745	735	19.90	1,499.90	1,925
3996	12" x 8"		1.41	17.021		675	850	23	1,548	2,050
3997	12" x 10"		1.27	18.898		435	945	25.50	1,405.50	1,900
4000	Socket weld joint, 3000 lb., type 304									

22 11 13.66 Pipe Fittings, Stainless Steel		Crew	Daily Output	Labor-Hours	Unit	Material	2011 Bare Costs Labor	Equipment	Total	Total Incl O&P
4100	90° Elbow									
4140	1/4"	Q-15	13.47	1.188	Ea.	44	57.50	2.41	103.91	137
4150	3/8"		12.97	1.234		57	59.50	2.50	119	155
4160	1/2"		12.21	1.310		62.50	63	2.66	128.16	166
4170	3/4"		10.68	1.498		72	72.50	3.04	147.54	190
4180	1"		9.74	1.643		108	79.50	3.33	190.83	241
4190	1-1/4"		9.05	1.768		189	85.50	3.59	278.09	340
4200	1-1/2"		8.55	1.871		229	90.50	3.80	323.30	390
4210	2"		7.26	2.204		370	106	4.47	480.47	570
4300	45° Elbow									
4340	1/4"	Q-15	13.47	1.188	Ea.	83	57.50	2.41	142.91	179
4350	3/8"		12.97	1.234		83	59.50	2.50	145	183
4360	1/2"		12.21	1.310		83	63	2.66	148.66	188
4370	3/4"		10.68	1.498		94	72.50	3.04	169.54	214
4380	1"		9.74	1.643		136	79.50	3.33	218.83	272
4390	1-1/4"		9.05	1.768		224	85.50	3.59	313.09	375
4400	1-1/2"		8.55	1.871		225	90.50	3.80	319.30	385
4410	2"		7.26	2.204		410	106	4.47	520.47	615
4500	Tee									
4540	1/4"	Q-15	8.97	1.784	Ea.	72.50	86	3.62	162.12	211
4550	3/8"		8.64	1.852		72.50	89.50	3.76	165.76	217
4560	1/2"		8.13	1.968		89	95	4	188	244
4570	3/4"		7.12	2.247		103	108	4.56	215.56	280
4580	1"		6.48	2.469		138	119	5	262	335
4590	1-1/4"		6.03	2.653		245	128	5.40	378.40	465
4600	1-1/2"		5.69	2.812		350	136	5.70	491.70	595
4610	2"		4.83	3.313		530	160	6.70	696.70	830
5000	Socket weld joint, 3000 lb., type 316									
5100	90° Elbow									
5140	1/4"	Q-15	13.47	1.188	Ea.	54	57.50	2.41	113.91	147
5150	3/8"		12.97	1.234		63	59.50	2.50	125	161
5160	1/2"		12.21	1.310		76	63	2.66	141.66	181
5170	3/4"		10.68	1.498		100	72.50	3.04	175.54	221
5180	1"		9.74	1.643		142	79.50	3.33	224.83	278
5190	1-1/4"		9.05	1.768		253	85.50	3.59	342.09	410
5200	1-1/2"		8.55	1.871		286	90.50	3.80	380.30	455
5210	2"		7.26	2.204		485	106	4.47	595.47	700
5300	45° Elbow									
5340	1/4"	Q-15	13.47	1.188	Ea.	104	57.50	2.41	163.91	202
5350	3/8"		12.97	1.234		104	59.50	2.50	166	206
5360	1/2"		12.21	1.310		104	63	2.66	169.66	211
5370	3/4"		10.68	1.498		121	72.50	3.04	196.54	244
5380	1"		9.74	1.643		182	79.50	3.33	264.83	320
5390	1-1/4"		9.05	1.768		258	85.50	3.59	347.09	415
5400	1-1/2"		8.55	1.871		292	90.50	3.80	386.30	460
5410	2"		7.26	2.204		435	106	4.47	545.47	645
5500	Tee									
5540	1/4"	Q-15	8.97	1.784	Ea.	73	86	3.62	162.62	212
5550	3/8"		8.64	1.852		89.50	89.50	3.76	182.76	235
5560	1/2"		8.13	1.968		99.50	95	4	198.50	255
5570	3/4"		7.12	2.247		123	108	4.56	235.56	300
5580	1"		6.48	2.469		188	119	5	312	390
5590	1-1/4"		6.03	2.653		299	128	5.40	432.40	525

22 11 13.66 Pipe Fittings, Stainless Steel		Crew	Daily Output	Labor-Hours	Unit	Material	2011 Bare Costs Labor	Equipment	Total	Total Incl O&P
5600	1-1/2"	Q-15	5.69	2.812	Ea.	420	136	5.70	561.70	675
5610	2"	↓	4.83	3.313		670	160	6.70	836.70	985
5700	For socket weld joint, 6000 lb., type 304 and 316, add				↓	100%				
6000	Threaded companion flange									
6010	Stainless steel, 150 lb., type 304									
6020	1/2" diam.	1 Plum	30	.267	Ea.	35	14.30		49.30	60
6030	3/4" diam.		28	.286		39	15.30		54.30	66
6040	1" diam.	↓	27	.296		43	15.90		58.90	71
6050	1-1/4" diam.	Q-1	44	.364		55	17.55		72.55	86.50
6060	1-1/2" diam.		40	.400		55	19.30		74.30	89.50
6070	2" diam.		36	.444		72	21.50		93.50	111
6080	2-1/2" diam.		28	.571		100	27.50		127.50	151
6090	3" diam.		20	.800		105	38.50		143.50	174
6110	4" diam.	↓	12	1.333		144	64.50		208.50	254
6130	6" diam.	Q-2	14	1.714		259	86		345	415
6140	8" diam.	"	12	2	↓	500	100		600	700
6150	For type 316 add					40%				
6260	Weld flanges, stainless steel, type 304									
6270	Slip on, 150 lb. (welded, front and back)									
6280	1/2" diam.	Q-15	12.70	1.260	Ea.	31	61	2.56	94.56	128
6290	3/4" diam.		11.11	1.440		32	69.50	2.92	104.42	142
6300	1" diam.		10.13	1.579		43	76	3.21	122.21	165
6310	1-1/4" diam.		9.41	1.700		55	82	3.45	140.45	186
6320	1-1/2" diam.		8.89	1.800		55	87	3.65	145.65	195
6330	2" diam.		7.55	2.119		72	102	4.30	178.30	237
6340	2-1/2" diam.		3.98	4.020		100	194	8.15	302.15	410
6350	3" diam.		3.21	4.984		105	241	10.10	356.10	485
6370	4" diam.	↓	2.44	6.557		144	315	13.30	472.30	645
6390	6" diam.	Q-16	1.89	12.698		259	635	17.15	911.15	1,250
6400	8" diam.	"	1.22	19.672	↓	500	985	26.50	1,511.50	2,050
6410	For type 316, add					40%				
6530	Weld neck 150 lb.									
6540	1/2" diam.	Q-15	25.40	.630	Ea.	15.10	30.50	1.28	46.88	63.50
6550	3/4" diam.		22.22	.720		19.10	34.50	1.46	55.06	74.50
6560	1" diam.		20.25	.790		24	38	1.60	63.60	85.50
6570	1-1/4" diam		18.82	.850		36	41	1.73	78.73	102
6580	1-1/2" diam.		17.78	.900		31.50	43.50	1.83	76.83	102
6590	2" diam.		15.09	1.060		30	51	2.15	83.15	112
6600	2-1/2" diam.		7.96	2.010		49	97	4.08	150.08	203
6610	3" diam.		6.43	2.488		48.50	120	5.05	173.55	238
6630	4" diam.		4.88	3.279		84	158	6.65	248.65	335
6640	5" diam.		4.26	3.756		104	181	7.60	292.60	395
6650	6" diam.	Q-16	5.66	4.240		115	212	5.70	332.70	450
6652	8" diam.		3.65	6.575		229	330	8.90	567.90	750
6654	10" diam.		2.88	8.333		325	415	11.25	751.25	985
6656	12" diam.	↓	2.28	10.526	↓	505	525	14.20	1,044.20	1,350
6670	For type 316 add					23%				
7000	Threaded joint, 150 lb., type 304									
7030	90° elbow									
7040	1/8"	1 Plum	13	.615	Ea.	25.50	33		58.50	77
7050	1/4"		13	.615		25.50	33		58.50	77
7070	3/8"		13	.615		28	33		61	79.50
7080	1/2"	↓	12	.667	↓	25	35.50		60.50	81

22 11 13 – Facility Water Distribution Piping

22 11 13.66 Pipe Fittings, Stainless Steel	Crew	Daily Output	Labor-Hours	Unit	Material	2011 Bare Costs Labor	Equipment	Total	Total Incl O&P
7090	3/4"	1 Plum	11	.727	Ea.	29.50	39		68.50	90.50
7100	1"		10	.800		39.50	43		82.50	108
7110	1-1/4"	Q-1	17	.941		61	45.50		106.50	135
7120	1-1/2"		16	1		70.50	48.50		119	150
7130	2"		14	1.143		102	55		157	195
7140	2-1/2"		11	1.455		255	70		325	385
7150	3"		8	2		370	96.50		466.50	555
7160	4"	Q-2	11	2.182		630	109		739	860
7180	45° elbow									
7190	1/8"	1 Plum	13	.615	Ea.	37	33		70	90
7200	1/4"		13	.615		37	33		70	90
7210	3/8"		13	.615		35	33		68	87.50
7220	1/2"		12	.667		35.50	35.50		71	92.50
7230	3/4"		11	.727		38.50	39		77.50	101
7240	1"		10	.800		44.50	43		87.50	113
7250	1-1/4"	Q-1	17	.941		60.50	45.50		106	135
7260	1-1/2"		16	1		78	48.50		126.50	158
7270	2"		14	1.143		111	55		166	205
7280	2-1/2"		11	1.455		360	70		430	500
7290	3"		8	2		530	96.50		626.50	725
7300	4"	Q-2	11	2.182		955	109		1,064	1,225
7320	Tee, straight									
7330	1/8"	1 Plum	9	.889	Ea.	39	47.50		86.50	114
7340	1/4"		9	.889		39	47.50		86.50	114
7350	3/8"		9	.889		39.50	47.50		87	114
7360	1/2"		8	1		37.50	53.50		91	121
7370	3/4"		7	1.143		40	61.50		101.50	136
7380	1"		6.50	1.231		49.50	66		115.50	153
7390	1-1/4"	Q-1	11	1.455		83	70		153	197
7400	1-1/2"		10	1.600		106	77		183	232
7410	2"		9	1.778		132	86		218	273
7420	2-1/2"		7	2.286		365	110		475	570
7430	3"		5	3.200		560	154		714	845
7440	4"	Q-2	7	3.429		1,375	172		1,547	1,775
7460	Coupling, straight									
7470	1/8"	1 Plum	19	.421	Ea.	10.25	22.50		32.75	45
7480	1/4"		19	.421		12.10	22.50		34.60	47
7490	3/8"		19	.421		14.50	22.50		37	49.50
7500	1/2"		19	.421		19.30	22.50		41.80	54.50
7510	3/4"		18	.444		25.50	24		49.50	64
7520	1"		15	.533		41.50	28.50		70	88
7530	1-1/4"	Q-1	26	.615		66.50	29.50		96	118
7540	1-1/2"		24	.667		74.50	32		106.50	130
7550	2"		21	.762		123	37		160	191
7560	2-1/2"		18	.889		287	43		330	380
7570	3"		14	1.143		390	55		445	515
7580	4"	Q-2	16	1.500		550	75		625	715
7600	Reducer, concentric, 1/2"	1 Plum	12	.667		20.50	35.50		56	76.50
7610	3/4"		11	.727		26.50	39		65.50	87.50
7612	1"		10	.800		44.50	43		87.50	113
7614	1-1/4"	Q-1	17	.941		94.50	45.50		140	172
7616	1-1/2"		16	1		103	48.50		151.50	185
7618	2"		14	1.143		161	55		216	260

22 11 13.66 Pipe Fittings, Stainless Steel		Crew	Daily Output	Labor-Hours	Unit	Material	2011 Bare Costs Labor	Equipment	Total	Total Incl O&P
7620	2-1/2"	Q-1	11	1.455	Ea.	400	70		470	545
7622	3"	↓	8	2		445	96.50		541.50	635
7624	4"	Q-2	11	2.182	↓	735	109		844	975
7710	Union									
7720	1/8"	1 Plum	12	.667	Ea.	47.50	35.50		83	106
7730	1/4"		12	.667		47.50	35.50		83	106
7740	3/8"		12	.667		55.50	35.50		91	115
7750	1/2"		11	.727		69.50	39		108.50	135
7760	3/4"		10	.800		96	43		139	169
7770	1"	↓	9	.889		138	47.50		185.50	223
7780	1-1/4"	Q-1	16	1		320	48.50		368.50	425
7790	1-1/2"		15	1.067		350	51.50		401.50	460
7800	2"		13	1.231		440	59.50		499.50	575
7810	2-1/2"		10	1.600		830	77		907	1,025
7820	3"	↓	7	2.286		1,100	110		1,210	1,400
7830	4"	Q-2	10	2.400	↓	1,525	120		1,645	1,850
7838	Caps									
7840	1/2"	1 Plum	24	.333	Ea.	12.95	17.85		30.80	41
7841	3/4"		22	.364		18.80	19.50		38.30	49.50
7842	1"	↓	20	.400		30	21.50		51.50	65
7843	1-1/2"	Q-1	32	.500		71.50	24		95.50	115
7844	2"	"	28	.571		90.50	27.50		118	141
7845	4"	Q-2	22	1.091		355	54.50		409.50	470
7850	For 150 lb., type 316, add				↓	25%				
8750	Threaded joint, 2000 lb., type 304									
8770	90° Elbow									
8780	1/8"	1 Plum	13	.615	Ea.	32.50	33		65.50	84.50
8790	1/4"		13	.615		32.50	33		65.50	84.50
8800	3/8"		13	.615		40	33		73	93
8810	1/2"		12	.667		54.50	35.50		90	113
8820	3/4"		11	.727		63	39		102	128
8830	1"	↓	10	.800		83.50	43		126.50	156
8840	1-1/4"	Q-1	17	.941		136	45.50		181.50	218
8850	1-1/2"		16	1		218	48.50		266.50	310
8860	2"	↓	14	1.143	↓	305	55		360	420
8880	45° Elbow									
8890	1/8"	1 Plum	13	.615	Ea.	67.50	33		100.50	124
8900	1/4"		13	.615		67.50	33		100.50	124
8910	3/8"		13	.615		84.50	33		117.50	142
8920	1/2"		12	.667		84.50	35.50		120	147
8930	3/4"		11	.727		93	39		132	160
8940	1"	↓	10	.800		111	43		154	186
8950	1-1/4"	Q-1	17	.941		194	45.50		239.50	281
8960	1-1/2"		16	1		260	48.50		308.50	360
8970	2"	↓	14	1.143	↓	299	55		354	415
8990	Tee, straight									
9000	1/8"	1 Plum	9	.889	Ea.	42	47.50		89.50	117
9010	1/4"		9	.889		42	47.50		89.50	117
9020	3/8"		9	.889		54	47.50		101.50	130
9030	1/2"		8	1		69	53.50		122.50	156
9040	3/4"		7	1.143		78.50	61.50		140	178
9050	1"	↓	6.50	1.231		108	66		174	218
9060	1-1/4"	Q-1	11	1.455	↓	182	70		252	305

22 11 13 – Facility Water Distribution Piping

22 11 13.66 Pipe Fittings, Stainless Steel		Crew	Daily Output	Labor-Hours	Unit	Material	2011 Bare Costs Labor	Equipment	Total	Total Incl O&P
9070	1-1/2"	Q-1	10	1.600	Ea.	305	77		382	450
9080	2"	↓	9	1.778	↓	420	86		506	595
9100	For couplings and unions use 3000 lb., type 304									
9120	2000 lb., type 316									
9130	90° Elbow									
9140	1/8"	1 Plum	13	.615	Ea.	40.50	33		73.50	93.50
9150	1/4"		13	.615		40.50	33		73.50	93.50
9160	3/8"		13	.615		47	33		80	101
9170	1/2"		12	.667		61	35.50		96.50	121
9180	3/4"		11	.727		75	39		114	141
9190	1"	↓	10	.800		111	43		154	187
9200	1-1/4"	Q-1	17	.941		202	45.50		247.50	290
9210	1-1/2"		16	1		233	48.50		281.50	330
9220	2"	↓	14	1.143	↓	395	55		450	520
9240	45° Elbow									
9250	1/8"	1 Plum	13	.615	Ea.	85.50	33		118.50	143
9260	1/4"		13	.615		85.50	33		118.50	143
9270	3/8"		13	.615		85.50	33		118.50	143
9280	1/2"		12	.667		85.50	35.50		121	148
9300	3/4"		11	.727		96.50	39		135.50	164
9310	1"	↓	10	.800		146	43		189	224
9320	1-1/4"	Q-1	17	.941		208	45.50		253.50	297
9330	1-1/2"		16	1		285	48.50		333.50	385
9340	2"	↓	14	1.143	↓	350	55		405	470
9360	Tee, straight									
9370	1/8"	1 Plum	9	.889	Ea.	51.50	47.50		99	128
9380	1/4"		9	.889		51.50	47.50		99	128
9390	3/8"		9	.889		63	47.50		110.50	140
9400	1/2"		8	1		78	53.50		131.50	166
9410	3/4"		7	1.143		96.50	61.50		158	198
9420	1"	↓	6.50	1.231		147	66		213	261
9430	1-1/4"	Q-1	11	1.455		239	70		309	370
9440	1-1/2"		10	1.600		345	77		422	490
9450	2"	↓	9	1.778	↓	545	86		631	730
9470	For couplings and unions use 3000 lb., type 316									
9490	3000 lb., type 304									
9510	Coupling									
9520	1/8"	1 Plum	19	.421	Ea.	12.80	22.50		35.30	47.50
9530	1/4"		19	.421		13.90	22.50		36.40	49
9540	3/8"		19	.421		16.20	22.50		38.70	51.50
9550	1/2"		19	.421		19.50	22.50		42	55
9560	3/4"		18	.444		26	24		50	64
9570	1"	↓	15	.533		44	28.50		72.50	91
9580	1-1/4"	Q-1	26	.615		107	29.50		136.50	162
9590	1-1/2"		24	.667		125	32		157	186
9600	2"	↓	21	.762	↓	165	37		202	236
9620	Union									
9630	1/8"	1 Plum	12	.667	Ea.	98	35.50		133.50	162
9640	1/4"		12	.667		98	35.50		133.50	162
9650	3/8"		12	.667		105	35.50		140.50	169
9660	1/2"		11	.727		105	39		144	173
9670	3/4"		10	.800		128	43		171	205
9680	1"	↓	9	.889	↓	198	47.50		245.50	288

22 11 13.66 Pipe Fittings, Stainless Steel

		Crew	Daily Output	Labor-Hours	Unit	Material	2011 Bare Costs Labor	Equipment	Total	Total Incl O&P
9690	1-1/4"	Q-1	16	1	Ea.	350	48.50		398.50	455
9700	1-1/2"		15	1.067		410	51.50		461.50	530
9710	2"	↓	13	1.231	↓	550	59.50		609.50	695
9730	3000 lb., type 316									
9750	Coupling									
9770	1/8"	1 Plum	19	.421	Ea.	14.40	22.50		36.90	49.50
9780	1/4"		19	.421		16.25	22.50		38.75	51.50
9790	3/8"		19	.421		17.40	22.50		39.90	52.50
9800	1/2"		19	.421		24.50	22.50		47	60.50
9810	3/4"		18	.444		36.50	24		60.50	75.50
9820	1"	↓	15	.533		60.50	28.50		89	110
9830	1-1/4"	Q-1	26	.615		136	29.50		165.50	195
9840	1-1/2"		24	.667		159	32		191	223
9850	2"	↓	21	.762	↓	219	37		256	296
9870	Union									
9880	1/8"	1 Plum	12	.667	Ea.	108	35.50		143.50	172
9890	1/4"		12	.667		108	35.50		143.50	172
9900	3/8"		12	.667		125	35.50		160.50	191
9910	1/2"		11	.727		125	39		164	195
9920	3/4"		10	.800		163	43		206	243
9930	1"	↓	9	.889		257	47.50		304.50	355
9940	1-1/4"	Q-1	16	1		450	48.50		498.50	565
9950	1-1/2"		15	1.067		555	51.50		606.50	685
9960	2"	↓	13	1.231	↓	705	59.50		764.50	865

22 11 13.74 Pipe, Plastic

		Crew	Daily Output	Labor-Hours	Unit	Material	2011 Bare Costs Labor	Equipment	Total	Total Incl O&P
0010	**PIPE, PLASTIC**									
0020	Fiberglass reinforced, couplings 10' O.C., clevis hanger assy's, 3 per 10'									
0080	General service									
0120	2" diameter	Q-1	59	.271	L.F.	9.15	13.10		22.25	29.50
0140	3" diameter		52	.308		12	14.85		26.85	35
0150	4" diameter		48	.333		15.65	16.10		31.75	41
0160	6" diameter	↓	39	.410		27	19.80		46.80	59
0170	8" diameter	Q-2	49	.490		42.50	24.50		67	83.50
0180	10" diameter		41	.585		62.50	29.50		92	112
0190	12" diameter	↓	36	.667	↓	80	33.50		113.50	138
0600	PVC, high impact/pressure, cplgs. 10' O.C., clevis hanger assy's, 3 per 10'									
1020	Schedule 80									
1070	1/2" diameter	1 Plum	50	.160	L.F.	5.25	8.60		13.85	18.60
1080	3/4" diameter		47	.170		6.15	9.10		15.25	20.50
1090	1" diameter		43	.186		8.35	9.95		18.30	24
1100	1-1/4" diameter		39	.205		10.45	11		21.45	28
1110	1-1/2" diameter	↓	34	.235		11.65	12.60		24.25	31.50
1120	2" diameter	Q-1	55	.291		14.65	14.05		28.70	37
1140	3" diameter		50	.320		28	15.45		43.45	53.50
1150	4" diameter		46	.348		39	16.80		55.80	68
1170	6" diameter	↓	38	.421	↓	81	20.50		101.50	120
1730	To delete coupling & hangers, subtract									
1740	1/4" diam. to 1/2" diam.					62%	80%			
1750	3/4" diam. to 1-1/4" diam.					58%	73%			
1760	1-1/2" diam. to 6" diam.					40%	57%			
1770	8" diam. to 12" diam.					34%	50%			
1800	PVC, couplings 10' O.C., clevis hanger assemblies, 3 per 10'									

22 11 13.74 Pipe, Plastic		Crew	Daily Output	Labor-Hours	Unit	Material	2011 Bare Costs Labor	Equipment	Total	Total Incl O&P
1820	Schedule 40									
1860	1/2" diameter	1 Plum	54	.148	L.F.	3.53	7.95		11.48	15.75
1870	3/4" diameter		51	.157		3.81	8.40		12.21	16.75
1880	1" diameter		46	.174		4.89	9.30		14.19	19.30
1890	1-1/4" diameter		42	.190		5.70	10.20		15.90	21.50
1900	1-1/2" diameter		36	.222		5.75	11.90		17.65	24
1910	2" diameter	Q-1	59	.271		6.75	13.10		19.85	27
1920	2-1/2" diameter		56	.286		9.05	13.80		22.85	30.50
1930	3" diameter		53	.302		11.10	14.55		25.65	34
1940	4" diameter		48	.333		14.20	16.10		30.30	39.50
1950	5" diameter		43	.372		19.20	17.95		37.15	48
1960	6" diameter		39	.410		24.50	19.80		44.30	56.50
1970	8" diameter	Q-2	48	.500		35.50	25		60.50	76.50
1980	10" diameter		43	.558		69.50	28		97.50	118
1990	12" diameter		42	.571		83.50	28.50		112	135
2000	14" diameter		31	.774		133	38.50		171.50	204
2010	16" diameter		23	1.043		190	52		242	287
2340	To delete coupling & hangers, subtract									
2360	1/2" diam. to 1-1/4" diam.					65%	74%			
2370	1-1/2" diam. to 6" diam.					44%	57%			
2380	8" diam. to 12" diam.					41%	53%			
2390	14" diam. to 16" diam.					48%	45%			
2420	Schedule 80									
2440	1/4" diameter	1 Plum	58	.138	L.F.	3.09	7.40		10.49	14.45
2450	3/8" diameter		55	.145		3.09	7.80		10.89	15.05
2460	1/2" diameter		50	.160		3.65	8.60		12.25	16.80
2470	3/4" diameter		47	.170		3.96	9.10		13.06	17.95
2480	1" diameter		43	.186		5.15	9.95		15.10	20.50
2490	1-1/4" diameter		39	.205		6.05	11		17.05	23
2500	1-1/2" diameter		34	.235		6.15	12.60		18.75	25.50
2510	2" diameter	Q-1	55	.291		7	14.05		21.05	29
2520	2-1/2" diameter		52	.308		8.85	14.85		23.70	32
2530	3" diameter		50	.320		12.40	15.45		27.85	36.50
2540	4" diameter		46	.348		16.10	16.80		32.90	42.50
2550	5" diameter		42	.381		22.50	18.40		40.90	52
2560	6" diameter		38	.421		33.50	20.50		54	67
2570	8" diameter	Q-2	47	.511		42	25.50		67.50	84.50
2580	10" diameter		42	.571		80.50	28.50		109	131
2590	12" diameter		38	.632		99.50	31.50		131	157
2830	To delete coupling & hangers, subtract									
2840	1/4" diam. to 1/2" diam.					66%	80%			
2850	3/4" diam. to 1-1/4" diam.					61%	73%			
2860	1-1/2" diam. to 6" diam.					41%	57%			
2870	8" diam. to 12" diam.					31%	50%			
2900	Schedule 120									
2910	1/2" diameter	1 Plum	50	.160	L.F.	3.77	8.60		12.37	16.95
2950	3/4" diameter		47	.170		4.13	9.10		13.23	18.15
2960	1" diameter		43	.186		5.35	9.95		15.30	21
2970	1-1/4" diameter		39	.205		6.35	11		17.35	23.50
2980	1-1/2" diameter		33	.242		6.65	13		19.65	26.50
2990	2" diameter	Q-1	54	.296		7.75	14.30		22.05	30
3000	2-1/2" diameter		52	.308		11.75	14.85		26.60	35
3010	3" diameter		49	.327		13.85	15.75		29.60	39

22 11 13 – Facility Water Distribution Piping

22 11 13.74 Pipe, Plastic		Crew	Daily Output	Labor-Hours	Unit	Material	2011 Bare Costs Labor	Equipment	Total	Total Incl O&P
3020	4" diameter	Q-1	45	.356	L.F.	19.35	17.15		36.50	47
3030	6" diameter	↓	37	.432	↓	40	21		61	75
3240	To delete coupling & hangers, subtract									
3250	1/2" diam. to 1-1/4" diam.					52%	74%			
3260	1-1/2" diam. to 4" diam.					30%	57%			
3270	6" diam.					17%	50%			
3300	PVC, pressure, couplings 10' O.C., clevis hanger assy's, 3 per 10'									
3310	SDR 26, 160 psi									
3350	1-1/4" diameter	1 Plum	42	.190	L.F.	5.15	10.20		15.35	21
3360	1-1/2" diameter	"	36	.222		5.15	11.90		17.05	23.50
3370	2" diameter	Q-1	59	.271		5.75	13.10		18.85	26
3380	2-1/2" diameter		56	.286		7.75	13.80		21.55	29
3390	3" diameter		53	.302		9.85	14.55		24.40	33
3400	4" diameter		48	.333		12.90	16.10		29	38
3420	6" diameter	↓	39	.410		24	19.80		43.80	55.50
3430	8" diameter	Q-2	48	.500	↓	35.50	25		60.50	76.50
3660	To delete coupling & clevis hanger assy's, subtract									
3670	1-1/4" diam.					63%	68%			
3680	1-1/2" diam. to 4" diam.					48%	57%			
3690	6" diam. to 8" diam.					60%	54%			
3720	SDR 21, 200 psi, 1/2" diameter	1 Plum	54	.148	L.F.	3.34	7.95		11.29	15.55
3740	3/4" diameter		51	.157		3.58	8.40		11.98	16.50
3750	1" diameter		46	.174		4.49	9.30		13.79	18.85
3760	1-1/4" diameter		42	.190		5.25	10.20		15.45	21
3770	1-1/2" diameter	↓	36	.222		5.30	11.90		17.20	23.50
3780	2" diameter	Q-1	59	.271		5.95	13.10		19.05	26
3790	2-1/2" diameter		56	.286		9.40	13.80		23.20	31
3800	3" diameter		53	.302		10.30	14.55		24.85	33.50
3810	4" diameter		48	.333		15.10	16.10		31.20	40.50
3830	6" diameter	↓	39	.410		25.50	19.80		45.30	58
3840	8" diameter	Q-2	48	.500	↓	38.50	25		63.50	80
4000	To delete coupling & hangers, subtract									
4010	1/2" diam. to 3/4" diam.					71%	77%			
4020	1" diam. to 1-1/4" diam.					63%	70%			
4030	1-1/2" diam. to 6" diam.					44%	57%			
4040	8" diam.					46%	54%			
4100	DWV type, schedule 40, couplings 10' O.C., clevis hanger assy's, 3 per 10'									
4210	ABS, schedule 40, foam core type									
4212	Plain end black									
4214	1-1/2" diameter	1 Plum	39	.205	L.F.	4.48	11		15.48	21.50
4216	2" diameter	Q-1	62	.258		4.93	12.45		17.38	24
4218	3" diameter		56	.286		7.45	13.80		21.25	28.50
4220	4" diameter		51	.314		9.70	15.15		24.85	33
4222	6" diameter	↓	42	.381	↓	17.85	18.40		36.25	47
4240	To delete coupling & hangers, subtract									
4244	1-1/2" diam. to 6" diam.					43%	48%			
4400	PVC									
4410	1-1/4" diameter	1 Plum	42	.190	L.F.	4.65	10.20		14.85	20.50
4420	1-1/2" diameter	"	36	.222		4.53	11.90		16.43	23
4460	2" diameter	Q-1	59	.271		4.97	13.10		18.07	25
4470	3" diameter		53	.302		7.50	14.55		22.05	30.50
4480	4" diameter		48	.333		9.50	16.10		25.60	34.50
4490	6" diameter	↓	39	.410		16.10	19.80		35.90	47

22 11 Facility Water Distribution

22 11 13 – Facility Water Distribution Piping

22 11 13.74 Pipe, Plastic	Crew	Daily Output	Labor-Hours	Unit	Material	2011 Bare Costs Labor	Equipment	Total	Total Incl O&P
4500 8" diameter	Q-2	48	.500	L.F.	23.50	25		48.50	63.50
4510 To delete coupling & hangers, subtract									
4520 1-1/4" diam. to 1-1/2" diam.					48%	60%			
4530 2" diam. to 8" diam.					42%	54%			
4550 PVC, schedule 40, foam core type									
4552 Plain end, white									
4554 1-1/2" diameter	i Plum	39	.205	L.F.	4.27	11		15.27	21
4556 2" diameter	Q-1	62	.258		4.65	12.45		17.10	23.50
4558 3" diameter		56	.286		6.90	13.80		20.70	28
4560 4" diameter		51	.314		8.60	15.15		23.75	32
4562 6" diameter		42	.381		14.25	18.40		32.65	43
4564 8" diameter	Q-2	51	.471		20.50	23.50		44	58
4568 10" diameter		48	.500		25.50	25		50.50	65.50
4570 12" diameter		46	.522		30	26		56	72
4580 To delete coupling & hangers, subtract									
4582 1-1/2" dia to 2" dia					58%	54%			
4584 3" dia to 12" dia					46%	42%			
4800 PVC, clear pipe, cplgs. 10' O.C., clevis hanger assy's 3 per 10', Sched. 40									
4840 1/4" diameter	1 Plum	59	.136	L.F.	3.59	7.25		10.84	14.80
4850 3/8" diameter		56	.143		3.89	7.65		11.54	15.70
4860 1/2" diameter		54	.148		4.47	7.95		12.42	16.75
4870 3/4" diameter		51	.157		5.10	8.40		13.50	18.15
4880 1" diameter		46	.174		7	9.30		16.30	21.50
4890 1-1/4" diameter		42	.190		8.35	10.20		18.55	24.50
4900 1-1/2" diameter		36	.222		9.25	11.90		21.15	28
4910 2" diameter	Q-1	59	.271		11.40	13.10		24.50	32
4920 2-1/2" diameter		56	.286		16.85	13.80		30.65	39
4930 3" diameter		53	.302		21	14.55		35.55	45
4940 3-1/2" diameter		50	.320		27	15.45		42.45	52.50
4950 4" diameter		48	.333		30	16.10		46.10	57
5250 To delete coupling & hangers, subtract									
5260 1/4" diam. to 3/8" diam.					60%	81%			
5270 1/2" diam. to 3/4" diam.					41%	77%			
5280 1" diam. to 1-1/2" diam.					26%	67%			
5290 2" diam. to 4" diam.					16%	58%			
5300 CPVC, socket joint, couplings 10' O.C., clevis hanger assemblies, 3 per 10'									
5302 Schedule 40									
5304 1/2" diameter	1 Plum	54	.148	L.F.	4.22	7.95		12.17	16.50
5305 3/4" diameter		51	.157		4.75	8.40		13.15	17.80
5306 1" diameter		46	.174		6.45	9.30		15.75	21
5307 1-1/4" diameter		42	.190		7.65	10.20		17.85	23.50
5308 1-1/2" diameter		36	.222		8.55	11.90		20.45	27
5309 2" diameter	Q-1	59	.271		10.05	13.10		23.15	30.50
5310 2-1/2" diameter		56	.286		15.85	13.80		29.65	38
5311 3" diameter		53	.302		18.65	14.55		33.20	42.50
5312 4" diameter		48	.333		27.50	16.10		43.60	54
5314 6" diameter		43	.372		46	17.95		63.95	77.50
5318 To delete coupling & hangers, subtract									
5319 1/2" diam. to 3/4" diam.					37%	77%			
5320 1" diam. to 1-1/4" diam.					27%	70%			
5321 1-1/2" diam. to 3" diam.					21%	57%			
5322 4" diam. to 6" diam.					16%	57%			
5324 Schedule 80									

22 11 13.74 Pipe, Plastic		Crew	Daily Output	Labor-Hours	Unit	Material	2011 Bare Costs Labor	Equipment	Total	Total Incl O&P
5325	1/2" diameter	1 Plum	50	.160	L.F.	4.41	8.60		13.01	17.65
5326	3/4" diameter		47	.170		5.05	9.10		14.15	19.15
5327	1" diameter		43	.186		6.85	9.95		16.80	22.50
5328	1-1/4" diameter		39	.205		8.30	11		19.30	25.50
5329	1-1/2" diameter		34	.235		9.40	12.60		22	29
5330	2" diameter	Q-1	55	.291		11.40	14.05		25.45	33.50
5331	2-1/2" diameter		52	.308		17.60	14.85		32.45	41.50
5332	3" diameter		50	.320		20	15.45		35.45	45
5333	4" diameter		46	.348		29	16.80		45.80	57
5334	6" diameter		38	.421		52.50	20.50		73	88
5335	8" diameter	Q-2	47	.511		101	25.50		126.50	149
5339	To delete couplings & hangers, subtract									
5340	1/2" diam. to 3/4" diam.					44%	77%			
5341	1" diam. to 1-1/4" diam.					32%	71%			
5342	1-1/2" diam. to 4" diam.					25%	58%			
5343	6" diam. to 8" diam.					20%	53%			
5360	CPVC, threaded, couplings 10' O.C., clevis hanger assemblies, 3 per 10'									
5380	Schedule 40									
5460	1/2" diameter	1 Plum	54	.148	L.F.	4.93	7.95		12.88	17.30
5470	3/4" diameter		51	.157		6	8.40		14.40	19.15
5480	1" diameter		46	.174		7.70	9.30		17	22.50
5490	1-1/4" diameter		42	.190		8.65	10.20		18.85	25
5500	1-1/2" diameter		36	.222		9.40	11.90		21.30	28
5510	2" diameter	Q-1	59	.271		11.15	13.10		24.25	32
5520	2-1/2" diameter		56	.286		16.95	13.80		30.75	39
5530	3" diameter		53	.302		20.50	14.55		35.05	44.50
5540	4" diameter		48	.333		33.50	16.10		49.60	61
5550	6" diameter		43	.372		49	17.95		66.95	81
5730	To delete coupling & hangers, subtract									
5740	1/2" diam. to 3/4" diam.					37%	77%			
5750	1" diam. to 1-1/4" diam.					27%	70%			
5760	1-1/2" diam. to 3" diam.					21%	57%			
5770	4" diam. to 6" diam.					16%	57%			
5800	Schedule 80									
5860	1/2" diameter	1 Plum	50	.160	L.F.	5.10	8.60		13.70	18.45
5870	3/4" diameter		47	.170		6.30	9.10		15.40	20.50
5880	1" diameter		43	.186		8.15	9.95		18.10	24
5890	1-1/4" diameter		39	.205		9.35	11		20.35	26.50
5900	1-1/2" diameter		34	.235		10.25	12.60		22.85	30
5910	2" diameter	Q-1	55	.291		12.50	14.05		26.55	35
5920	2-1/2" diameter		52	.308		18.70	14.85		33.55	42.50
5930	3" diameter		50	.320		21.50	15.45		36.95	47
5940	4" diameter		46	.348		35.50	16.80		52.30	64
5950	6" diameter		38	.421		55.50	20.50		76	91.50
5960	8" diameter	Q-2	47	.511		101	25.50		126.50	149
6060	To delete couplings & hangers, subtract									
6070	1/2" diam. to 3/4" diam.					44%	77%			
6080	1" diam. to 1-1/4" diam.					32%	71%			
6090	1-1/2" diam. to 4" diam.					25%	58%			
6100	6" diam. to 8" diam.					20%	53%			
6240	CTS, 1/2" diameter	1 Plum	54	.148	L.F.	3.16	7.95		11.11	15.35
6250	3/4" diameter		51	.157		4.47	8.40		12.87	17.45
6260	1" diameter		46	.174		5.80	9.30		15.10	20.50

22 11 13 – Facility Water Distribution Piping

22 11 13.74 Pipe, Plastic	Crew	Daily Output	Labor-Hours	Unit	Material	2011 Bare Costs Labor	Equipment	Total	Total Incl O&P
6270	1 1/4"	1 Plum	42	.190	L.F.	7.05	10.20		17.25	23
6280	1 1/2" diameter	↓	36	.222		8.30	11.90		20.20	27
6290	2" diameter	Q-1	59	.271	↓	11.95	13.10		25.05	32.50
6370	To delete coupling & hangers, subtract									
6380	1/2" diam.					51%	79%			
6390	3/4" diam.					40%	76%			
6392	1" thru 2" diam.					72%	68%			
7280	PEX, flexible, no couplings or hangers									
7282	Note: For labor costs add 25% to the couplings and fittings labor total.									
7300	Non-barrier type, hot/cold tubing rolls									
7310	1/4" diameter x 100'				L.F.	.46			.46	.51
7350	3/8" diameter x 100'					.51			.51	.56
7360	1/2" diameter x 100'					.57			.57	.63
7370	1/2" diameter x 500'					.57			.57	.63
7380	1/2" diameter x 1000'					.57			.57	.63
7400	3/4" diameter x 100'					1.05			1.05	1.16
7410	3/4" diameter x 500'					1.05			1.05	1.16
7420	3/4" diameter x 1000'					1.05			1.05	1.16
7460	1" diameter x 100'					1.80			1.80	1.98
7470	1" diameter x 300'					1.80			1.80	1.98
7480	1" diameter x 500'					1.78			1.78	1.96
7500	1-1/4" diameter x 100'					3.04			3.04	3.34
7510	1-1/4" diameter x 300'					3.04			3.04	3.34
7540	1-1/2" diameter x 100'					3.82			3.82	4.20
7550	1-1/2" diameter x 300'				↓	3.81			3.81	4.19
7596	Most sizes available in red or blue									
7700	Non-barrier type, hot/cold tubing straight lengths									
7710	1/2" diameter x 20'				L.F.	.57			.57	.63
7750	3/4" diameter x 20'					1.05			1.05	1.16
7760	1" diameter x 20'					1.80			1.80	1.98
7770	1-1/4" diameter x 20'					3.04			3.04	3.34
7780	1-1/2" diameter x 20'				↓	3.92			3.92	4.31
7796	Most sizes available in red or blue									
9000	Polypropylene pipe									
9002	For fusion weld fittings and accessories see 22 11 13.76 9400									
9004	Note: sizes 1/2" thru 4" use socket fusion									
9005	Sizes 6" thru 10" use butt fusion									
9010	SDR 7.4, (domestic hot water piping)									
9011	Enhanced to minimize thermal expansion and high temperature life									
9016	13' lengths, size is ID, includes joints 13' O.C. and hangers 3 per 10'									
9020	3/8" diameter	1 Plum	53	.151	L.F.	2.98	8.10		11.08	15.35
9022	1/2" diameter		52	.154		3.52	8.25		11.77	16.20
9024	3/4" diameter		50	.160		4.19	8.60		12.79	17.40
9026	1" diameter		45	.178		5.50	9.55		15.05	20.50
9028	1-1/4" diameter		40	.200		7.85	10.70		18.55	24.50
9030	1-1/2" diameter	↓	35	.229		11.25	12.25		23.50	30.50
9032	2" diameter	Q-1	58	.276		16	13.30		29.30	37.50
9034	2-1/2" diameter		55	.291		22	14.05		36.05	45.50
9036	3" diameter		52	.308		30.50	14.85		45.35	55.50
9038	3-1/2" diameter		49	.327		44	15.75		59.75	71.50
9040	4" diameter		46	.348		49.50	16.80		66.30	79.50
9042	6" diameter	↓	39	.410		60.50	19.80		80.30	96
9044	8" diameter	Q-2	48	.500	↓	109	25		134	157

22 11 13 – Facility Water Distribution Piping

22 11 13.74 Pipe, Plastic		Crew	Daily Output	Labor-Hours	Unit	Material	2011 Bare Costs Labor	Equipment	Total	Total Incl O&P
9046	10" diameter	Q-2	43	.558	L.F.	172	28		200	231
9050	To delete joint & hangers, subtract									
9052	3/8" diam. to 1" diam.					45%	65%			
9054	1-1/4" diam. to 4" diam.					15%	45%			
9056	6" diam. to 10" diam.					5%	24%			
9060	SDR 11, (domestic cold water piping)									
9062	13' lengths, size is ID, includes joints 13' O.C. and hangers 3 per 10'									
9064	1/2" diameter	1 Plum	57	.140	L.F.	3.10	7.50		10.60	14.65
9066	3/4" diameter		54	.148		3.52	7.95		11.47	15.70
9068	1" diameter		49	.163		4.31	8.75		13.06	17.80
9070	1-1/4" diameter		45	.178		6.10	9.55		15.65	21
9072	1-1/2" diameter		40	.200		8.15	10.70		18.85	25
9074	2" diameter	Q-1	62	.258		11.60	12.45		24.05	31.50
9076	2-1/2" diameter		59	.271		15.25	13.10		28.35	36.50
9078	3" diameter		56	.286		21.50	13.80		35.30	44
9080	3-1/2" diameter		53	.302		31	14.55		45.55	56.50
9082	4" diameter		50	.320		34.50	15.45		49.95	61
9084	6" diameter		47	.340		42.50	16.45		58.95	71.50
9086	8" diameter	Q-2	51	.471		74.50	23.50		98	117
9088	10" diameter	"	46	.522		114	26		140	164
9090	To delete joint & hangers, subtract									
9092	3/8" diam. to 1" diam.					45%	65%			
9094	1-1/4" diam. to 4" diam.					15%	45%			
9096	6" diam. to 10" diam.					5%	24%			

22 11 13.76 Pipe Fittings, Plastic										
0010	**PIPE FITTINGS, PLASTIC**									
0030	Epoxy resin, fiberglass reinforced, general service									
0090	Elbow, 90°, 2"	Q-1	33.10	.483	Ea.	93	23.50		116.50	137
0100	3"		20.80	.769		107	37		144	174
0110	4"		16.50	.970		109	47		156	190
0120	6"		10.10	1.584		214	76.50		290.50	350
0130	8"	Q-2	9.30	2.581		395	129		524	630
0140	10"		8.50	2.824		495	141		636	755
0150	12"		7.60	3.158		710	158		868	1,025
0160	45° Elbow, same as 90°									
0170	Elbow, 90°, flanged									
0172	2"	Q-1	23	.696	Ea.	133	33.50		166.50	196
0173	3"		16	1		152	48.50		200.50	239
0174	4"		13	1.231		198	59.50		257.50	305
0176	6"		8	2		360	96.50		456.50	540
0177	8"	Q-2	9	2.667		650	133		783	915
0178	10"		7	3.429		890	172		1,062	1,225
0179	12"		5	4.800		1,200	240		1,440	1,675
0186	Elbow, 45°, flanged									
0188	2"	Q-1	23	.696	Ea.	133	33.50		166.50	196
0189	3"		16	1		152	48.50		200.50	239
0190	4"		13	1.231		198	59.50		257.50	305
0192	6"		8	2		360	96.50		456.50	540
0193	8"	Q-2	9	2.667		615	133		748	880
0194	10"		7	3.429		785	172		957	1,125
0195	12"		5	4.800		1,000	240		1,240	1,450
0290	Tee, 2"	Q-1	20	.800		51.50	38.50		90	114

22 11 13 – Facility Water Distribution Piping

22 11 13.76 Pipe Fittings, Plastic	Crew	Daily Output	Labor-Hours	Unit	Material	2011 Bare Costs Labor	Equipment	Total	Total Incl O&P
0300	3"	Q-1	13.90	1.151	Ea.	62	55.50		117.50	151
0310	4"		11	1.455		84.50	70		154.50	198
0320	6"		6.70	2.388		214	115		329	405
0330	8"	Q-2	6.20	3.871		1,500	194		1,694	1,975
0340	10"		5.70	4.211		1,725	211		1,936	2,225
0350	12"		5.10	4.706		2,100	235		2,335	2,650
0352	Tee, flanged									
0354	2"	Q-1	17	.941	Ea.	180	45.50		225.50	266
0355	3"		10	1.600		238	77		315	375
0356	4"		8	2		267	96.50		363.50	435
0358	6"		5	3.200		460	154		614	735
0359	8"	Q-2	6	4		870	200		1,070	1,250
0360	10"		5	4.800		1,250	240		1,490	1,750
0361	12"		4	6		1,750	300		2,050	2,375
0365	Wye, flanged									
0367	2"	Q-1	17	.941	Ea.	355	45.50		400.50	460
0368	3"		10	1.600		490	77		567	655
0369	4"		8	2		655	96.50		751.50	865
0371	6"		5	3.200		920	154		1,074	1,225
0372	8"	Q-2	6	4		1,500	200		1,700	1,950
0373	10"		5	4.800		2,350	240		2,590	2,925
0374	12"		4	6		3,425	300		3,725	4,200
0380	Couplings									
0410	2"	Q-1	33.10	.483	Ea.	15.45	23.50		38.95	52
0420	3"		20.80	.769		16.55	37		53.55	73.50
0430	4"		16.50	.970		22	47		69	94.50
0440	6"		10.10	1.584		53	76.50		129.50	172
0450	8"	Q-2	9.30	2.581		90	129		219	292
0460	10"		8.50	2.824		132	141		273	355
0470	12"		7.60	3.158		176	158		334	430
0473	High corrosion resistant couplings, add					30%				
0474	Reducer, concentric, flanged									
0475	2" x 1-1/2"	Q-1	30	.533	Ea.	345	25.50		370.50	420
0476	3" x 2"		24	.667		375	32		407	465
0477	4" x 3"		19	.842		425	40.50		465.50	525
0479	6" x 4"		15	1.067		425	51.50		476.50	540
0480	8" x 6"	Q-2	16	1.500		695	75		770	875
0481	10" x 8"		13	1.846		645	92.50		737.50	850
0482	12" x 10"		11	2.182		995	109		1,104	1,275
0486	Adapter, bell x male or female									
0488	2"	Q-1	28	.571	Ea.	17.20	27.50		44.70	60
0489	3"		20	.800		25.50	38.50		64	85.50
0491	4"		17	.941		34	45.50		79.50	106
0492	6"		12	1.333		71.50	64.50		136	175
0493	8"	Q-2	15	1.600		104	80		184	233
0494	10"	"	11	2.182		150	109		259	330
0528	Flange									
0532	2"	Q-1	46	.348	Ea.	25.50	16.80		42.30	53
0533	3"		32	.500		31	24		55	70
0534	4"		26	.615		41	29.50		70.50	89.50
0536	6"		16	1		77	48.50		125.50	157
0537	8"	Q-2	18	1.333		118	66.50		184.50	230
0538	10"		14	1.714		168	86		254	310

22 11 13.76 **Pipe Fittings, Plastic**	Crew	Daily Output	Labor-Hours	Unit	Material	2011 Bare Costs Labor	Equipment	Total	Total Incl O&P
0539	12"	Q-2	10	2.400	Ea.	211	120		331	410
2100	PVC schedule 80, socket joint									
2110	90° elbow, 1/2"	1 Plum	30.30	.264	Ea.	2.21	14.15		16.36	23.50
2130	3/4"		26	.308		2.83	16.50		19.33	27.50
2140	1"		22.70	.352		4.55	18.90		23.45	33
2150	1-1/4"		20.20	.396		6.10	21		27.10	38
2160	1-1/2"		18.20	.440		6.55	23.50		30.05	42
2170	2"	Q-1	33.10	.483		7.90	23.50		31.40	43.50
2180	3"		20.80	.769		21	37		58	78.50
2190	4"		16.50	.970		31.50	47		78.50	105
2200	6"		10.10	1.584		90	76.50		166.50	213
2210	8"	Q-2	9.30	2.581		248	129		377	465
2250	45° elbow, 1/2"	1 Plum	30.30	.264		4.17	14.15		18.32	25.50
2270	3/4"		26	.308		6.35	16.50		22.85	31.50
2280	1"		22.70	.352		9.55	18.90		28.45	38.50
2290	1-1/4"		20.20	.396		12.15	21		33.15	45
2300	1-1/2"		18.20	.440		14.80	23.50		38.30	51.50
2310	2"	Q-1	33.10	.483		18.65	23.50		42.15	55.50
2320	3"		20.80	.769		47.50	37		84.50	108
2330	4"		16.50	.970		86	47		133	165
2340	6"		10.10	1.584		108	76.50		184.50	233
2350	8"	Q-2	9.30	2.581		234	129		363	450
2400	Tee, 1/2"	1 Plum	20.20	.396		6.25	21		27.25	38.50
2420	3/4"		17.30	.462		6.55	25		31.55	44
2430	1"		15.20	.526		8.20	28		36.20	51
2440	1-1/4"		13.50	.593		22.50	32		54.50	72
2450	1-1/2"		12.10	.661		22.50	35.50		58	77.50
2460	2"	Q-1	20	.800		28	38.50		66.50	88.50
2470	3"		13.90	1.151		38	55.50		93.50	125
2480	4"		11	1.455		44.50	70		114.50	154
2490	6"		6.70	2.388		156	115		271	345
2500	8"	Q-2	6.20	3.871		350	194		544	675
2510	Flange, socket, 150 lb., 1/2"	1 Plum	55.60	.144		12.15	7.70		19.85	25
2514	3/4"		47.60	.168		12.95	9		21.95	27.50
2518	1"		41.70	.192		14.45	10.30		24.75	31
2522	1-1/2"		33.30	.240		15.20	12.90		28.10	36
2526	2"	Q-1	60.60	.264		20	12.75		32.75	41
2530	4"		30.30	.528		43.50	25.50		69	86
2534	6"		18.50	.865		68.50	41.50		110	138
2538	8"	Q-2	17.10	1.404		123	70		193	240
2550	Coupling, 1/2"	1 Plum	30.30	.264		4.07	14.15		18.22	25.50
2570	3/4"		26	.308		5.40	16.50		21.90	30.50
2580	1"		22.70	.352		5.55	18.90		24.45	34
2590	1-1/4"		20.20	.396		9.95	21		30.95	42.50
2600	1-1/2"		18.20	.440		9.15	23.50		32.65	45
2610	2"	Q-1	33.10	.483		9.85	23.50		33.35	46
2620	3"		20.80	.769		28	37		65	86
2630	4"		16.50	.970		35	47		82	109
2640	6"		10.10	1.584		75	76.50		151.50	197
2650	8"	Q-2	9.30	2.581		102	129		231	305
2660	10"		8.50	2.824		350	141		491	595
2670	12"		7.60	3.158		405	158		563	680
2700	PVC (white), schedule 40, socket joints									

22 11 13 – Facility Water Distribution Piping

22 11 13.76 Pipe Fittings, Plastic	Crew	Daily Output	Labor-Hours	Unit	Material	2011 Bare Costs Labor	Equipment	Total	Total Incl O&P
2760	90° elbow, 1/2"	1 Plum	33.30	.240	Ea.	.43	12.90		13.33	19.65
2770	3/4"		28.60	.280		.48	15		15.48	23
2780	1"		25	.320		.87	17.15		18.02	26.50
2790	1-1/4"		22.20	.360		1.53	19.30		20.83	30.50
2800	1-1/2"		20	.400		1.65	21.50		23.15	34
2810	2"	Q-1	36.40	.440		2.58	21		23.58	34.50
2820	2-1/2"		26.70	.599		7.85	29		36.85	51.50
2830	3"		22.90	.699		9.40	33.50		42.90	61
2840	4"		18.20	.879		16.80	42.50		59.30	82
2850	5"		12.10	1.322		43.50	64		107.50	144
2860	6"		11.10	1.441		53.50	69.50		123	163
2870	8"	Q-2	10.30	2.330		138	117		255	325
2980	45° elbow, 1/2"	1 Plum	33.30	.240		.71	12.90		13.61	20
2990	3/4"		28.60	.280		1.10	15		16.10	23.50
3000	1"		25	.320		1.33	17.15		18.48	27
3010	1-1/4"		22.20	.360		1.85	19.30		21.15	31
3020	1-1/2"		20	.400		2.28	21.50		23.78	34.50
3030	2"	Q-1	36.40	.440		3.02	21		24.02	35
3040	2-1/2"		26.70	.599		7.85	29		36.85	51.50
3050	3"		22.90	.699		12.20	33.50		45.70	64
3060	4"		18.20	.879		22	42.50		64.50	87.50
3070	5"		12.10	1.322		43.50	64		107.50	144
3080	6"		11.10	1.441		54	69.50		123.50	164
3090	8"	Q-2	10.30	2.330		130	117		247	315
3180	Tee, 1/2"	1 Plum	22.20	.360		.53	19.30		19.83	29.50
3190	3/4"		19	.421		.62	22.50		23.12	34
3200	1"		16.70	.479		1.15	25.50		26.65	40
3210	1-1/4"		14.80	.541		1.80	29		30.80	45
3220	1-1/2"		13.30	.602		2.18	32		34.18	50.50
3230	2"	Q-1	24.20	.661		3.18	32		35.18	51
3240	2-1/2"		17.80	.899		10.50	43.50		54	76.50
3250	3"		15.20	1.053		13.80	51		64.80	91
3260	4"		12.10	1.322		25	64		89	123
3270	5"		8.10	1.975		60.50	95.50		156	209
3280	6"		7.40	2.162		84	104		188	249
3290	8"	Q-2	6.80	3.529		222	177		399	510
3380	Coupling, 1/2"	1 Plum	33.30	.240		.29	12.90		13.19	19.50
3390	3/4"		28.60	.280		.39	15		15.39	23
3400	1"		25	.320		.68	17.15		17.83	26.50
3410	1-1/4"		22.20	.360		.95	19.30		20.25	30
3420	1-1/2"		20	.400		1.01	21.50		22.51	33
3430	2"	Q-1	36.40	.440		1.54	21		22.54	33
3440	2-1/2"		26.70	.599		3.41	29		32.41	47
3450	3"		22.90	.699		5.35	33.50		38.85	56.50
3460	4"		18.20	.879		7.70	42.50		50.20	72
3470	5"		12.10	1.322		14.10	64		78.10	111
3480	6"		11.10	1.441		24.50	69.50		94	131
3490	8"	Q-2	10.30	2.330		45.50	117		162.50	224
3600	Cap, schedule 40, PVC socket 1/2"	1 Plum	60.60	.132		.39	7.10		7.49	11
3610	3/4"		51.90	.154		.45	8.25		8.70	12.85
3620	1"		45.50	.176		.71	9.40		10.11	14.85
3630	1-1/4"		40.40	.198		1.01	10.60		11.61	16.95
3640	1-1/2"		36.40	.220		1.10	11.80		12.90	18.80

22 11 13.76 Pipe Fittings, Plastic		Crew	Daily Output	Labor-Hours	Unit	Material	2011 Bare Costs Labor	Equipment	Total	Total Incl O&P
3650	2"	Q-1	66.10	.242	Ea.	1.33	11.70		13.03	18.90
3660	2-1/2"		48.50	.330		4.22	15.90		20.12	28.50
3670	3"		41.60	.385		4.62	18.55		23.17	32.50
3680	4"		33.10	.483		10.50	23.50		34	46.50
3690	6"		20.20	.792		25	38		63	84.50
3700	8"	Q-2	18.60	1.290		63	64.50		127.50	166
3710	Reducing insert, schedule 40, socket weld									
3712	3/4"	1 Plum	31.50	.254	Ea.	.45	13.60		14.05	21
3713	1"		27.50	.291		.82	15.60		16.42	24.50
3715	1-1/2"		22	.364		1.16	19.50		20.66	30.50
3716	2"	Q-1	40	.400		1.93	19.30		21.23	31
3717	4"		20	.800		10.20	38.50		48.70	68.50
3718	6"		12.20	1.311		25.50	63.50		89	123
3719	8"	Q-2	11.30	2.124		99.50	106		205.50	269
3730	Reducing insert, socket weld x female/male thread									
3732	1/2"	1 Plum	38.30	.209	Ea.	2.04	11.20		13.24	18.95
3733	3/4"		32.90	.243		1.26	13.05		14.31	21
3734	1"		28.80	.278		1.77	14.90		16.67	24
3736	1-1/2"		23	.348		3.18	18.65		21.83	31.50
3737	2"	Q-1	41.90	.382		3.41	18.40		21.81	31.50
3738	4"	"	20.90	.766		42	37		79	102
3742	Male adapter, socket weld x male thread									
3744	1/2"	1 Plum	38.30	.209	Ea.	.39	11.20		11.59	17.15
3745	3/4"		32.90	.243		.43	13.05		13.48	19.90
3746	1"		28.80	.278		.77	14.90		15.67	23
3748	1-1/2"		23	.348		1.26	18.65		19.91	29.50
3749	2"	Q-1	41.90	.382		1.66	18.40		20.06	29.50
3750	4"	"	20.90	.766		9.10	37		46.10	65
3754	Female adapter, socket weld x female thread									
3756	1/2"	1 Plum	38.30	.209	Ea.	.48	11.20		11.68	17.25
3757	3/4"		32.90	.243		.62	13.05		13.67	20
3758	1"		28.80	.278		.71	14.90		15.61	23
3760	1-1/2"		23	.348		1.26	18.65		19.91	29.50
3761	2"	Q-1	41.90	.382		1.70	18.40		20.10	29.50
3762	4"	"	20.90	.766		9.55	37		46.55	65.50
3800	PVC, schedule 80, socket joints									
3810	Reducing insert									
3812	3/4"	1 Plum	28.60	.280	Ea.	1.29	15		16.29	24
3813	1"		25	.320		3.70	17.15		20.85	29.50
3815	1-1/2"		20	.400		7.90	21.50		29.40	40.50
3816	2"	Q-1	36.40	.440		11.25	21		32.25	44
3817	4"		18.20	.879		43	42.50		85.50	111
3818	6"		11.10	1.441		57.50	69.50		127	167
3819	8"	Q-2	10.20	2.353		345	118		463	555
3830	Reducing insert, socket weld x female/male thread									
3832	1/2"	1 Plum	34.80	.230	Ea.	8.05	12.30		20.35	27.50
3833	3/4"		29.90	.268		4.96	14.35		19.31	27
3834	1"		26.10	.307		7.75	16.45		24.20	33
3836	1-1/2"		20.90	.383		9.75	20.50		30.25	41.50
3837	2"	Q-1	38	.421		14.30	20.50		34.80	46.50
3838	4"	"	19	.842		69.50	40.50		110	137
3844	Adapter, male socket x male thread									
3846	1/2"	1 Plum	34.80	.230	Ea.	3.12	12.30		15.42	22

22 11 Facility Water Distribution

22 11 13 – Facility Water Distribution Piping

22 11 13.76 Pipe Fittings, Plastic		Crew	Daily Output	Labor-Hours	Unit	Material	2011 Bare Costs Labor	Equipment	Total	Total Incl O&P
3847	3/4"	1 Plum	29.90	.268	Ea.	3.44	14.35		17.79	25.50
3848	1"		26.10	.307		5.95	16.45		22.40	31
3850	1-1/2"		20.90	.383		10	20.50		30.50	41.50
3851	2"	Q-1	38	.421		14.45	20.50		34.95	46.50
3852	4"	"	19	.842		32.50	40.50		73	96
3860	Adapter, female socket x female thread									
3862	1/2"	1 Plum	34.80	.230	Ea.	3.75	12.30		16.05	22.50
3863	3/4"		29.90	.268		5.55	14.35		19.90	27.50
3864	1"		26.10	.307		8.20	16.45		24.65	33.50
3866	1-1/2"		20.90	.383		16.30	20.50		36.80	48.50
3867	2"	Q-1	38	.421		28.50	20.50		49	62
3868	4"	"	19	.842		87	40.50		127.50	156
3872	Union, socket joints									
3874	1/2"	1 Plum	25.80	.310	Ea.	8.20	16.60		24.80	34
3875	3/4"		22.10	.362		10.45	19.40		29.85	40.50
3876	1"		19.30	.415		11.90	22		33.90	46
3878	1-1/2"		15.50	.516		27	27.50		54.50	71
3879	2"	Q-1	28.10	.569		36.50	27.50		64	81
3888	Cap									
3890	1/2"	1 Plum	54.50	.147	Ea.	3.92	7.85		11.77	16.05
3891	3/4"		46.70	.171		4.13	9.20		13.33	18.25
3892	1"		41	.195		7.35	10.45		17.80	23.50
3894	1-1/2"		32.80	.244		8.90	13.05		21.95	29.50
3895	2"	Q-1	59.50	.269		23.50	12.95		36.45	45
3896	4"		30	.533		70.50	25.50		96	116
3897	6"		18.20	.879		175	42.50		217.50	257
3898	8"	Q-2	16.70	1.437		225	72		297	355
4500	DWV, ABS, non pressure, socket joints									
4540	1/4 Bend, 1-1/4"	1 Plum	20.20	.396	Ea.	2.88	21		23.88	34.50
4560	1-1/2"	"	18.20	.440		2.19	23.50		25.69	37.50
4570	2"	Q-1	33.10	.483		3.15	23.50		26.65	38.50
4580	3"		20.80	.769		7.90	37		44.90	64
4590	4"		16.50	.970		15.90	47		62.90	87.50
4600	6"		10.10	1.584		75.50	76.50		152	197
4650	1/8 Bend, same as 1/4 Bend									
4800	Tee, sanitary									
4820	1-1/4"	1 Plum	13.50	.593	Ea.	3.44	32		35.44	51.50
4830	1-1/2"	"	12.10	.661		3.02	35.50		38.52	56.50
4840	2"	Q-1	20	.800		4.66	38.50		43.16	62.50
4850	3"		13.90	1.151		12.80	55.50		68.30	97
4860	4"		11	1.455		22.50	70		92.50	130
4862	Tee, sanitary, reducing, 2" x 1-1/2"		22	.727		4.71	35		39.71	57.50
4864	3" x 2"		15.30	1.046		9.30	50.50		59.80	86
4868	4" x 3"		12.10	1.322		22	64		86	120
4870	Combination Y and 1/8 bend									
4872	1-1/2"	1 Plum	12.10	.661	Ea.	7.20	35.50		42.70	61
4874	2"	Q-1	20	.800		8.10	38.50		46.60	66.50
4876	3"		13.90	1.151		18.45	55.50		73.95	104
4878	4"		11	1.455		36.50	70		106.50	146
4880	3" x 1-1/2"		15.50	1.032		18.65	50		68.65	95
4882	4" x 3"		12.10	1.322		29	64		93	127
4900	Wye, 1-1/4"	1 Plum	13.50	.593		3.97	32		35.97	52
4902	1-1/2"	"	12.10	.661		4.80	35.50		40.30	58.50

205

22 11 13 – Facility Water Distribution Piping

22 11 13.76 Pipe Fittings, Plastic		Crew	Daily Output	Labor-Hours	Unit	Material	2011 Bare Costs Labor	Equipment	Total	Total Incl O&P
4904	2"	Q-1	20	.800	Ea.	6	38.50		44.50	64
4906	3"		13.90	1.151		14.20	55.50		69.70	98.50
4908	4"		11	1.455		29	70		99	137
4910	6"		6.70	2.388		86	115		201	267
4918	3" x 1-1/2"		15.50	1.032		11.40	50		61.40	87
4920	4" x 3"		12.10	1.322		23	64		87	121
4922	6" x 4"		6.90	2.319		71	112		183	246
4930	Double Wye, 1-1/2"	1 Plum	9.10	.879		13.90	47		60.90	86
4932	2"	Q-1	16.60	.964		16.60	46.50		63.10	88
4934	3"		10.40	1.538		38	74		112	153
4936	4"		8.25	1.939		74	93.50		167.50	222
4940	2" x 1-1/2"		16.80	.952		14.50	46		60.50	84.50
4942	3" x 2"		10.60	1.509		28	73		101	140
4944	4" x 3"		8.45	1.893		58.50	91.50		150	201
4946	6" x 4"		7.25	2.207		90.50	106		196.50	259
4950	Reducer bushing, 2" x 1-1/2"		36.40	.440		1.61	21		22.61	33.50
4952	3" x 1-1/2"		27.30	.586		6.85	28.50		35.35	49.50
4954	4" x 2"		18.20	.879		13.15	42.50		55.65	78
4956	6" x 4"		11.10	1.441		35	69.50		104.50	143
4960	Couplings, 1-1/2"	1 Plum	18.20	.440		.99	23.50		24.49	36
4962	2"	Q-1	33.10	.483		1.32	23.50		24.82	36.50
4963	3"		20.80	.769		3.83	37		40.83	59.50
4964	4"		16.50	.970		7.25	47		54.25	78
4966	6"		10.10	1.584		29	76.50		105.50	146
4970	2" x 1-1/2"		33.30	.480		2.84	23		25.84	37.50
4972	3" x 1-1/2"		21	.762		7.90	37		44.90	63.50
4974	4" x 3"		16.70	.958		12.50	46		58.50	83
4978	Closet flange, 4"	1 Plum	32	.250		5.95	13.40		19.35	26.50
4980	4" x 3"	"	34	.235		7.85	12.60		20.45	27.50
5000	DWV, PVC, schedule 40, socket joints									
5040	1/4 bend, 1-1/4"	1 Plum	20.20	.396	Ea.	5.25	21		26.25	37.50
5060	1-1/2"	"	18.20	.440		1.44	23.50		24.94	36.50
5070	2"	Q-1	33.10	.483		2.27	23.50		25.77	37.50
5080	3"		20.80	.769		6.70	37		43.70	63
5090	4"		16.50	.970		13.20	47		60.20	84.50
5100	6"		10.10	1.584		46.50	76.50		123	165
5105	8"	Q-2	9.30	2.581		94	129		223	296
5106	10"	"	8.50	2.824		220	141		361	455
5110	1/4 bend, long sweep, 1-1/2"	1 Plum	18.20	.440		3.46	23.50		26.96	39
5112	2"	Q-1	33.10	.483		3.73	23.50		27.23	39
5114	3"		20.80	.769		8.80	37		45.80	65
5116	4"		16.50	.970		16.80	47		63.80	88.50
5150	1/8 bend, 1-1/4"	1 Plum	20.20	.396		3.44	21		24.44	35.50
5170	1-1/2"	"	18.20	.440		1.42	23.50		24.92	36.50
5180	2"	Q-1	33.10	.483		2.03	23.50		25.53	37
5190	3"		20.80	.769		6	37		43	62
5200	4"		16.50	.970		10.50	47		57.50	81.50
5210	6"		10.10	1.584		42.50	76.50		119	161
5215	8"	Q-2	9.30	2.581		74	129		203	274
5216	10"		8.50	2.824		184	141		325	415
5217	12"		7.60	3.158		340	158		498	610
5250	Tee, sanitary 1-1/4"	1 Plum	13.50	.593		5.25	32		37.25	53.50
5254	1-1/2"	"	12.10	.661		2.61	35.50		38.11	56

22 11 13 – Facility Water Distribution Piping

22 11 13.76 Pipe Fittings, Plastic		Crew	Daily Output	Labor-Hours	Unit	Material	2011 Bare Costs Labor	Equipment	Total	Total Incl O&P
5255	2"	Q-1	20	.800	Ea.	3.83	38.50		42.33	61.50
5256	3"		13.90	1.151		9.85	55.50		65.35	94
5257	4"		11	1.455		17.10	70		87.10	124
5259	6"		6.70	2.388		75	115		190	254
5261	8"	Q-2	6.20	3.871		249	194		443	560
5276	Tee, sanitary, reducing									
5281	2" x 1-1/2" x 1-1/2"	Q-1	23	.696	Ea.	3.38	33.50		36.88	53.50
5282	2" x 1-1/2" x 2"		22	.727		4.08	35		39.08	57
5283	2" x 2" x 1-1/2"		22	.727		3.39	35		38.39	56
5284	3" x 3" x 1-1/2"		15.50	1.032		7.15	50		57.15	82.50
5285	3" x 3" x 2"		15.30	1.046		7.85	50.50		58.35	84
5286	4" x 4" x 1-1/2"		12.30	1.301		19.30	63		82.30	115
5287	4" x 4" x 2"		12.20	1.311		16	63.50		79.50	112
5288	4" x 4" x 3"		12.10	1.322		22	64		86	120
5291	6" x 6" x 4"		6.90	2.319		72.50	112		184.50	247
5294	Tee, double sanitary									
5295	1-1/2"	1 Plum	9.10	.879	Ea.	5.80	47		52.80	77
5296	2"	Q-1	16.60	.964		7.85	46.50		54.35	78
5297	3"		10.40	1.538		22	74		96	135
5298	4"		8.25	1.939		35	93.50		128.50	179
5303	Wye, reducing									
5304	2" x 1-1/2" x 1-1/2"	Q-1	23	.696	Ea.	6.60	33.50		40.10	57.50
5305	2" x 2" x 1-1/2"		22	.727		5.75	35		40.75	59
5306	3" x 3" x 2"		15.30	1.046		18.70	50.50		69.20	96
5307	4" x 4" x 2"		12.20	1.311		13.80	63.50		77.30	110
5309	4" x 4" x 3"		12.10	1.322		18.60	64		82.60	116
5314	Combination Y & 1/8 bend, 1-1/2"	1 Plum	12.10	.661		6.30	35.50		41.80	60
5315	2"	Q-1	20	.800		7.50	38.50		46	66
5317	3"		13.90	1.151		18.65	55.50		74.15	104
5318	4"		11	1.455		37	70		107	146
5319	6"		6.70	2.388		122	115		237	305
5320	8"	Q-2	6.20	3.871		247	194		441	560
5324	Combination Y & 1/8 bend, reducing									
5325	2" x 2" x 1-1/2"	Q-1	22	.727	Ea.	8.85	35		43.85	62.50
5327	3" x 3" x 1-1/2"		15.50	1.032		15.35	50		65.35	91.50
5328	3" x 3" x 2"		15.30	1.046		11.10	50.50		61.60	87.50
5329	4" x 4" x 2"		12.20	1.311		18.05	63.50		81.55	114
5331	Wye, 1-1/4"	1 Plum	13.50	.593		6.70	32		38.70	55
5332	1-1/2"	"	12.10	.661		4.92	35.50		40.42	58.50
5333	2"	Q-1	20	.800		4.67	38.50		43.17	62.50
5334	3"		13.90	1.151		12.65	55.50		68.15	97
5335	4"		11	1.455		23	70		93	130
5336	6"		6.70	2.388		69	115		184	248
5337	8"	Q-2	6.20	3.871		203	194		397	510
5338	10"		5.70	4.211		435	211		646	795
5339	12"		5.10	4.706		710	235		945	1,125
5341	2" x 1-1/2"	Q-1	22	.727		5.75	35		40.75	59
5342	3" x 1-1/2"		15.50	1.032		8.25	50		58.25	83.50
5343	4" x 3"		12.10	1.322		18.60	64		82.60	116
5344	6" x 4"		6.90	2.319		51	112		163	223
5345	8" x 6"	Q-2	6.40	3.750		153	188		341	450
5347	Double wye, 1-1/2"	1 Plum	9.10	.879		10.65	47		57.65	82.50
5348	2"	Q-1	16.60	.964		11.95	46.50		58.45	82.50

22 11 Facility Water Distribution

22 11 13 – Facility Water Distribution Piping

22 11 13.76 Pipe Fittings, Plastic	Crew	Daily Output	Labor-Hours	Unit	Material	2011 Bare Costs Labor	Equipment	Total	Total Incl O&P
5349	3"	Q-1	10.40	1.538	Ea.	25	74		99	139
5350	4"	↓	8.25	1.939	↓	51	93.50		144.50	196
5353	Double wye, reducing									
5354	2" x 2" x 1-1/2" x 1-1/2"	Q-1	16.80	.952	Ea.	10.85	46		56.85	80.50
5355	3" x 3" x 2" x 2"		10.60	1.509		18.65	73		91.65	130
5356	4" x 4" x 3" x 3"		8.45	1.893		40	91.50		131.50	180
5357	6" x 6" x 4" x 4"	↓	7.25	2.207		131	106		237	305
5374	Coupling, 1-1/4"	1 Plum	20.20	.396		3.18	21		24.18	35
5376	1-1/2"	"	18.20	.440		.69	23.50		24.19	36
5378	2"	Q-1	33.10	.483		.93	23.50		24.43	36
5380	3"		20.80	.769		3.21	37		40.21	59
5390	4"		16.50	.970		5.70	47		52.70	76.50
5400	6"	↓	10.10	1.584		18.55	76.50		95.05	135
5402	8"	Q-2	9.30	2.581		45.50	129		174.50	243
5404	2" x 1-1/2"	Q-1	33.30	.480		2.08	23		25.08	37
5406	3" x 1-1/2"		21	.762		6.35	37		43.35	62
5408	4" x 3"		16.70	.958		10.30	46		56.30	80.50
5410	Reducer bushing, 2" x 1-1/4"		36.50	.438		3.14	21		24.14	35
5411	2" x 1-1/2"		36.40	.440		1.23	21		22.23	33
5412	3" x 1-1/2"		27.30	.586		5.90	28.50		34.40	48.50
5413	3" x 2"		27.10	.590		3.44	28.50		31.94	46.50
5414	4" x 2"		18.20	.879		11.45	42.50		53.95	76
5415	4" x 3"		16.70	.958		5.85	46		51.85	75.50
5416	6" x 4"	↓	11.10	1.441		30.50	69.50		100	138
5418	8" x 6"	Q-2	10.20	2.353		66	118		184	249
5425	Closet flange 4"	Q-1	32	.500		7.25	24		31.25	44
5426	4" x 3"	"	34	.471	↓	7.25	22.50		29.75	42
5450	Solvent cement for PVC, industrial grade, per quart				Qt.	19.90			19.90	22
5500	CPVC, Schedule 80, threaded joints									
5540	90° Elbow, 1/4"	1 Plum	32	.250	Ea.	10.85	13.40		24.25	32
5560	1/2"		30.30	.264		7.05	14.15		21.20	29
5570	3/4"		26	.308		9.40	16.50		25.90	35
5580	1"		22.70	.352		13.20	18.90		32.10	42.50
5590	1-1/4"		20.20	.396		25.50	21		46.50	59.50
5600	1-1/2"	↓	18.20	.440		27.50	23.50		51	65
5610	2"	Q-1	33.10	.483		36.50	23.50		60	75.50
5620	2-1/2"		24.20	.661		114	32		146	173
5630	3"		20.80	.769		122	37		159	190
5640	4"		16.50	.970		191	47		238	280
5650	6"	↓	10.10	1.584	↓	225	76.50		301.50	360
5660	45° Elbow same as 90° Elbow									
5700	Tee, 1/4"	1 Plum	22	.364	Ea.	21	19.50		40.50	52
5702	1/2"		20.20	.396		21	21		42	54.50
5704	3/4"		17.30	.462		30.50	25		55.50	70.50
5706	1"		15.20	.526		32.50	28		60.50	78
5708	1-1/4"		13.50	.593		33	32		65	83.50
5710	1-1/2"	↓	12.10	.661		34.50	35.50		70	91
5712	2"	Q-1	20	.800		38.50	38.50		77	99.50
5714	2-1/2"		16.20	.988		189	47.50		236.50	278
5716	3"		13.90	1.151		219	55.50		274.50	325
5718	4"		11	1.455		520	70		590	675
5720	6"	↓	6.70	2.388		620	115		735	850
5730	Coupling, 1/4"	1 Plum	32	.250	↓	13.60	13.40		27	35

22 11 13.76 Pipe Fittings, Plastic	Crew	Daily Output	Labor-Hours	Unit	Material	2011 Bare Costs Labor	Equipment	Total	Total Incl O&P
5732	1/2"	1 Plum	30.30	.264	Ea.	11.35	14.15		25.50	33.50
5734	3/4"		26	.308		18.35	16.50		34.85	44.50
5736	1"		22.70	.352		21	18.90		39.90	51
5738	1-1/4"		20.20	.396		22	21		43	56
5740	1-1/2"		18.20	.440		24	23.50		47.50	61
5742	2"	Q-1	33.10	.483		28	23.50		51.50	66
5744	2-1/2"		24.20	.661		50	32		82	103
5746	3"		20.80	.769		58.50	37		95.50	120
5748	4"		16.50	.970		119	47		166	201
5750	6"		10.10	1.584		162	76.50		238.50	293
5752	8"	Q-2	9.30	2.581		350	129		479	575
5900	CPVC, Schedule 80, socket joints									
5904	90° Elbow, 1/4"	1 Plum	32	.250	Ea.	10.30	13.40		23.70	31.50
5906	1/2"		30.30	.264		4.05	14.15		18.20	25.50
5908	3/4"		26	.308		5.15	16.50		21.65	30
5910	1"		22.70	.352		8.20	18.90		27.10	37
5912	1-1/4"		20.20	.396		17.75	21		38.75	51
5914	1-1/2"		18.20	.440		19.80	23.50		43.30	57
5916	2"	Q-1	33.10	.483		24	23.50		47.50	61.50
5918	2-1/2"		24.20	.661		55	32		87	108
5920	3"		20.80	.769		62.50	37		99.50	124
5922	4"		16.50	.970		112	47		159	193
5924	6"		10.10	1.584		225	76.50		301.50	360
5926	8"		9.30	1.720		550	83		633	730
5930	45° Elbow, 1/4"	1 Plum	32	.250		15.35	13.40		28.75	37
5932	1/2"		30.30	.264		4.95	14.15		19.10	26.50
5934	3/4"		26	.308		7.15	16.50		23.65	32.50
5936	1"		22.70	.352		11.35	18.90		30.25	40.50
5938	1-1/4"		20.20	.396		22.50	21		43.50	56
5940	1-1/2"		18.20	.440		23	23.50		46.50	60
5942	2"	Q-1	33.10	.483		25.50	23.50		49	63.50
5944	2-1/2"		24.20	.661		52.50	32		84.50	106
5946	3"		20.80	.769		67.50	37		104.50	130
5948	4"		16.50	.970		94	47		141	173
5950	6"		10.10	1.584		285	76.50		361.50	430
5952	8"		9.30	1.720		595	83		678	780
5960	Tee, 1/4"	1 Plum	22	.364		9.50	19.50		29	39.50
5962	1/2"		20.20	.396		9.50	21		30.50	42
5964	3/4"		17.30	.462		9.65	25		34.65	47.50
5966	1"		15.20	.526		11.85	28		39.85	55
5968	1-1/4"		13.50	.593		25	32		57	75
5970	1-1/2"		12.10	.661		28.50	35.50		64	84.50
5972	2"	Q-1	20	.800		32	38.50		70.50	92.50
5974	2-1/2"		16.20	.988		81	47.50		128.50	160
5976	3"		13.90	1.151		81	55.50		136.50	172
5978	4"		11	1.455		108	70		178	224
5980	6"		6.70	2.388		280	115		395	480
5982	8"	Q-2	6.20	3.871		805	194		999	1,175
5990	Coupling, 1/4"	1 Plum	32	.250		11	13.40		24.40	32
5992	1/2"		30.30	.264		4.28	14.15		18.43	25.50
5994	3/4"		26	.308		5.95	16.50		22.45	31
5996	1"		22.70	.352		8.05	18.90		26.95	37
5998	1-1/4"		20.20	.396		12.05	21		33.05	45

22 11 13 – Facility Water Distribution Piping

22 11 13.76 Pipe Fittings, Plastic		Crew	Daily Output	Labor-Hours	Unit	2011 Bare Costs				Total Incl O&P
						Material	Labor	Equipment	Total	
6000	1-1/2"	1 Plum	18.20	.440	Ea.	15.15	23.50		38.65	51.50
6002	2"	Q-1	33.10	.483		17.30	23.50		40.80	54
6004	2-1/2"		24.20	.661		39	32		71	90.50
6006	3"		20.80	.769		42.50	37		79.50	103
6008	4"		16.50	.970		56	47		103	132
6010	6"		10.10	1.584		132	76.50		208.50	259
6012	8"	Q-2	9.30	2.581		355	129		484	585
6200	CTS, 100 psi at 180°F, hot and cold water									
6230	90° Elbow, 1/2"	1 Plum	20	.400	Ea.	.32	21.50		21.82	32.50
6250	3/4"		19	.421		.53	22.50		23.03	34
6251	1"		16	.500		1.47	27		28.47	41.50
6252	1-1/4"		15	.533		2.52	28.50		31.02	45.50
6253	1-1/2"		14	.571		4.56	30.50		35.06	50.50
6254	2"	Q-1	23	.696		8.75	33.50		42.25	59.50
6260	45° Elbow, 1/2"	1 Plum	20	.400		.41	21.50		21.91	32.50
6280	3/4"		19	.421		.70	22.50		23.20	34.50
6281	1"		16	.500		1.47	27		28.47	41.50
6282	1-1/4"		15	.533		2.92	28.50		31.42	45.50
6283	1-1/2"		14	.571		4.23	30.50		34.73	50
6284	2"	Q-1	23	.696		10.85	33.50		44.35	62
6290	Tee, 1/2"	1 Plum	13	.615		.41	33		33.41	49.50
6310	3/4"		12	.667		.77	35.50		36.27	54.50
6311	1"		11	.727		3.32	39		42.32	61.50
6312	1-1/4"		10	.800		5.10	43		48.10	69.50
6313	1-1/2"		10	.800		6.65	43		49.65	71.50
6314	2"	Q-1	17	.941		10.75	45.50		56.25	80
6320	Coupling, 1/2"	1 Plum	22	.364		.26	19.50		19.76	29.50
6340	3/4"		21	.381		.35	20.50		20.85	31
6341	1"		18	.444		1.29	24		25.29	37
6342	1-1/4"		17	.471		1.56	25		26.56	39
6343	1-1/2"		16	.500		2.20	27		29.20	42.50
6344	2"	Q-1	28	.571		4.48	27.50		31.98	46
6360	Solvent cement for CPVC, commercial grade, per quart				Qt.	32.50			32.50	36
7340	PVC flange, slip-on, Sch 80 std., 1/2"	1 Plum	22	.364	Ea.	12.15	19.50		31.65	42.50
7350	3/4"		21	.381		12.95	20.50		33.45	45
7360	1"		18	.444		14.45	24		38.45	51.50
7370	1-1/4"		17	.471		14.85	25		39.85	54
7380	1-1/2"		16	.500		15.20	27		42.20	56.50
7390	2"	Q-1	26	.615		20	29.50		49.50	66.50
7400	2-1/2"		24	.667		31	32		63	82.50
7410	3"		18	.889		34.50	43		77.50	102
7420	4"		15	1.067		43.50	51.50		95	125
7430	6"		10	1.600		68.50	77		145.50	191
7440	8"	Q-2	11	2.182		123	109		232	298
7550	Union, schedule 40, socket joints, 1/2"	1 Plum	19	.421		4.22	22.50		26.72	38
7560	3/4"		18	.444		4.80	24		28.80	41
7570	1"		15	.533		4.95	28.50		33.45	48
7580	1-1/4"		14	.571		14.90	30.50		45.40	62
7590	1-1/2"		13	.615		16.80	33		49.80	67.50
7600	2"	Q-1	20	.800		22.50	38.50		61	82
7992	Polybutyl/polyethyl pipe, for copper fittings see Line 22 11 13.25 7000									
8000	Compression type, PVC, 160 psi cold water									
8010	Coupling, 3/4" CTS	1 Plum	21	.381	Ea.	3.20	20.50		23.70	34

22 11 Facility Water Distribution

22 11 13 – Facility Water Distribution Piping

22 11 13.76 Pipe Fittings, Plastic	Crew	Daily Output	Labor-Hours	Unit	Material	2011 Bare Costs Labor	2011 Bare Costs Equipment	Total	Total Incl O&P
8020	1" CTS	1 Plum	18	.444	Ea.	3.98	24		27.98	40
8030	1-1/4" CTS		17	.471		5.60	25		30.60	43.50
8040	1-1/2" CTS		16	.500		7.65	27		34.65	48.50
8050	2" CTS		15	.533		10.65	28.50		39.15	54
8060	Female adapter, 3/4" FPT x 3/4" CTS		23	.348		3.54	18.65		22.19	32
8070	3/4" FPT x 1" CTS		21	.381		4.32	20.50		24.82	35.50
8080	1" FPT x 1" CTS		20	.400		5.65	21.50		27.15	38.50
8090	1-1/4" FPT x 1-1/4" CTS		18	.444		7.40	24		31.40	43.50
8100	1-1/2" FPT x 1-1/2" CTS		16	.500		8.45	27		35.45	49.50
8110	2" FPT x 2" CTS		13	.615		10.25	33		43.25	60.50
8130	Male adapter, 3/4" MPT x 3/4" CTS		23	.348		3.42	18.65		22.07	32
8140	3/4" MPT x 1" CTS		21	.381		3.69	20.50		24.19	34.50
8150	1" MPT x 1" CTS		20	.400		4.78	21.50		26.28	37.50
8160	1-1/4" MPT x 1-1/4" CTS		18	.444		6.50	24		30.50	42.50
8170	1-1/2" MPT x 1-1/2" CTS		16	.500		7.85	27		34.85	48.50
8180	2" MPT x 2" CTS		13	.615		10.15	33		43.15	60
8200	Spigot adapter, 3/4" IPS x 3/4" CTS		23	.348		2.48	18.65		21.13	30.50
8210	3/4" IPS x 1" CTS		21	.381		3.05	20.50		23.55	34
8220	1" IPS x 1" CTS		20	.400		3.05	21.50		24.55	35.50
8230	1-1/4" IPS x 1-1/4" CTS		18	.444		4.59	24		28.59	40.50
8240	1-1/2" IPS x 1-1/2" CTS		16	.500		4.82	27		31.82	45.50
8250	2" IPS x 2" CTS	▼	13	.615	▼	5.95	33		38.95	55.50
8270	Price includes insert stiffeners									
8280	250 psi is same price as 160 psi									
8300	Insert type, nylon, 160 & 250 psi, cold water									
8310	Clamp ring stainless steel, 3/4" IPS	1 Plum	115	.070	Ea.	2.25	3.73		5.98	8.05
8320	1" IPS		107	.075		2.30	4.01		6.31	8.55
8330	1-1/4" IPS		101	.079		2.32	4.25		6.57	8.90
8340	1-1/2" IPS		95	.084		3.11	4.51		7.62	10.15
8350	2" IPS		85	.094		3.57	5.05		8.62	11.50
8370	Coupling, 3/4" IPS		22	.364		.79	19.50		20.29	30
8390	1-1/4" IPS		18	.444		1.22	24		25.22	37
8400	1-1/2" IPS		17	.471		1.45	25		26.45	39
8410	2" IPS		16	.500		2.74	27		29.74	43
8430	Elbow, 90°, 3/4" IPS		22	.364		1.58	19.50		21.08	30.50
8440	1" IPS		19	.421		1.75	22.50		24.25	35.50
8450	1-1/4" IPS		18	.444		1.95	24		25.95	37.50
8460	1-1/2" IPS		17	.471		2.30	25		27.30	40
8470	2" IPS		16	.500		3.22	27		30.22	43.50
8490	Male adapter, 3/4" IPS x 3/4" MPT		25	.320		.79	17.15		17.94	26.50
8500	1" IPS x 1" MPT		21	.381		.82	20.50		21.32	31.50
8510	1-1/4" IPS x 1-1/4" MPT		20	.400		1.29	21.50		22.79	33.50
8520	1-1/2" IPS x 1-1/2" MPT		18	.444		1.45	24		25.45	37
8530	2" IPS x 2" MPT		15	.533		2.77	28.50		31.27	45.50
8550	Tee, 3/4" IPS		14	.571		1.53	30.50		32.03	47
8560	1" IPS		13	.615		1.99	33		34.99	51
8570	1-1/4" IPS		12	.667		3.11	35.50		38.61	57
8580	1-1/2" IPS		11	.727		3.53	39		42.53	62
8590	2" IPS	▼	10	.800	▼	6.95	43		49.95	71.50
8610	Insert type, PVC, 100 psi @ 180°F, hot & cold water									
8620	Coupler, male, 3/8" CTS x 3/8" MPT	1 Plum	29	.276	Ea.	.65	14.80		15.45	22.50
8630	3/8" CTS x 1/2" MPT		28	.286		.65	15.30		15.95	23.50
8640	1/2" CTS x 1/2" MPT	▼	27	.296	▼	.65	15.90		16.55	24

22 11 13.76 Pipe Fittings, Plastic		Crew	Daily Output	Labor-Hours	Unit	Material	2011 Bare Costs Labor	Equipment	Total	Total Incl O&P
8650	1/2" CTS x 3/4" MPT	1 Plum	26	.308	Ea.	2.05	16.50		18.55	27
8660	3/4" CTS x 1/2" MPT		25	.320		1.85	17.15		19	27.50
8670	3/4" CTS x 3/4" MPT		25	.320		2.29	17.15		19.44	28
8700	Coupling, 3/8" CTS x 1/2" CTS		25	.320		3.37	17.15		20.52	29
8710	1/2" CTS		23	.348		3.94	18.65		22.59	32.50
8730	3/4" CTS		22	.364		6.20	19.50		25.70	36
8750	Elbow 90°, 3/8" CTS		25	.320		4	17.15		21.15	30
8760	1/2" CTS		23	.348		4.73	18.65		23.38	33
8770	3/4" CTS		22	.364		7.25	19.50		26.75	37
8800	Rings, crimp, copper, 3/8" CTS		120	.067		.14	3.57		3.71	5.50
8810	1/2" CTS		117	.068		.15	3.67		3.82	5.60
8820	3/4" CTS		115	.070		.20	3.73		3.93	5.75
8850	Reducer tee, 3/8" x 3/8" x 1/2" CTS		17	.471		1.22	25		26.22	39
8860	1/2" x 3/8" x 1/2" CTS		15	.533		1.22	28.50		29.72	44
8870	3/4" x 1/2" x 1/2" CTS		14	.571		1.88	30.50		32.38	47.50
8890	3/4" x 3/4" x 1/2" CTS		14	.571		1.88	30.50		32.38	47.50
8900	3/4" x 1/2" x 3/8" CTS		14	.571		1.88	30.50		32.38	47.50
8930	Tee, 3/8" CTS		17	.471		1.60	25		26.60	39.50
8940	1/2" CTS		15	.533		1.97	28.50		30.47	44.50
8950	3/4" CTS		14	.571		3.75	30.50		34.25	49.50
8960	Copper rings included in fitting price									
9000	Flare type, assembled, acetal, hot & cold water									
9010	Coupling, 1/4" & 3/8" CTS	1 Plum	24	.333	Ea.	3.36	17.85		21.21	30
9020	1/2" CTS		22	.364		3.84	19.50		23.34	33
9030	3/4" CTS		21	.381		5.70	20.50		26.20	37
9040	1" CTS		18	.444		7.25	24		31.25	43.50
9050	Elbow 90°, 1/4" CTS		26	.308		3.73	16.50		20.23	28.50
9060	3/8" CTS		24	.333		4	17.85		21.85	31
9070	1/2" CTS		22	.364		4.73	19.50		24.23	34
9080	3/4" CTS		21	.381		7.25	20.50		27.75	38.50
9090	1" CTS		18	.444		9.15	24		33.15	45.50
9110	Tee, 1/4" CTS		16	.500		4.09	27		31.09	44.50
9114	3/8" CTS		15	.533		4.13	28.50		32.63	47
9120	1/2" CTS		14	.571		5.25	30.50		35.75	51.50
9130	3/4" CTS		13	.615		8.25	33		41.25	58
9140	1" CTS		12	.667		11.05	35.50		46.55	65.50
9400	Polypropylene, fittings and accessories									
9404	Fittings fusion welded, sizes are I.D.									
9408	Note: sizes 1/2" thru 4" use socket fusion									
9410	Sizes 6" thru 10" use butt fusion									
9416	Coupling									
9420	3/8"	1 Plum	39	.205	Ea.	1.16	11		12.16	17.70
9422	1/2"		37.40	.214		1.51	11.45		12.96	18.75
9424	3/4"		35.40	.226		1.69	12.10		13.79	19.95
9426	1"		29.70	.269		2.23	14.45		16.68	24
9428	1-1/4"		27.60	.290		2.67	15.55		18.22	26
9430	1-1/2"		24.80	.323		6.05	17.30		23.35	32.50
9432	2"	Q-1	43	.372		11.20	17.95		29.15	39.50
9434	2-1/2"		35.60	.449		12.55	21.50		34.05	46.50
9436	3"		30.90	.518		27.50	25		52.50	68
9438	3-1/2"		27.80	.576		45	28		73	91
9440	4"		25	.640		59	31		90	111
9442	Reducing coupling, female to female									

22 11 13.76 Pipe Fittings, Plastic		Crew	Daily Output	Labor-Hours	Unit	Material	2011 Bare Costs Labor	Equipment	Total	Total Incl O&P
9446	2" to 1-1/2"	Q-1	49	.327	Ea.	16.15	15.75		31.90	41.50
9448	2-1/2" to 2"		41.20	.388		17.75	18.75		36.50	47.50
9450	3" to 2-1/2"		33.10	.483		24	23.50		47.50	61.50
9470	Reducing bushing, female to female									
9472	1/2" to 3/8"	1 Plum	38.20	.209	Ea.	1.51	11.20		12.71	18.40
9474	3/4" to 3/8" or 1/2"		36.50	.219		1.69	11.75		13.44	19.40
9476	1" to 3/4" or 1/2"		33.10	.242		2.26	12.95		15.21	22
9478	1-1/4" to 3/4" or 1"		28.70	.279		3.47	14.95		18.42	26.50
9480	1-1/2" to 1/2" thru 1-1/4"		26.20	.305		5.75	16.35		22.10	31
9482	2" to 1/2" thru 1-1/2"	Q-1	43	.372		11.50	17.95		29.45	39.50
9484	2-1/2" to 1/2" thru 2"		41.20	.388		12.85	18.75		31.60	42
9486	3" to 1-1/2" thru 2-1/2"		33.10	.483		28.50	23.50		52	66.50
9488	3-1/2" to 2" thru 3"		29.20	.548		46	26.50		72.50	90
9490	4" to 2-1/2" thru 3-1/2"		26.20	.611		71.50	29.50		101	123
9491	6" to 4" SDR 7.4		16.50	.970		85.50	47		132.50	164
9492	6" to 4" SDR 11		16.50	.970		85.50	47		132.50	164
9493	8" to 6" SDR 7.4		10.10	1.584		126	76.50		202.50	253
9494	8" to 6" SDR 11		10.10	1.584		116	76.50		192.50	241
9495	10" to 8" SDR 7.4		7.80	2.051		173	99		272	340
9496	10" to 8" SDR 11		7.80	2.051		151	99		250	315
9500	90° Elbow									
9504	3/8"	1 Plum	39	.205	Ea.	1.28	11		12.28	17.80
9506	1/2"		37.40	.214		1.62	11.45		13.07	18.90
9508	3/4"		35.40	.226		2.08	12.10		14.18	20.50
9510	1"		29.70	.269		3.01	14.45		17.46	25
9512	1-1/4"		27.60	.290		4.63	15.55		20.18	28
9514	1-1/2"		24.80	.323		9.95	17.30		27.25	37
9516	2"	Q-1	43	.372		15.30	17.95		33.25	44
9518	2-1/2"		35.60	.449		34	21.50		55.50	70
9520	3"		30.90	.518		62.50	25		87.50	107
9522	3-1/2"		27.80	.576		89	28		117	140
9524	4"		25	.640		137	31		168	197
9526	6" SDR 7.4		5.55	2.883		159	139		298	380
9528	6" SDR 11		5.55	2.883		131	139		270	350
9530	8" SDR 7.4	Q-2	8.10	2.963		550	148		698	825
9532	8" SDR 11		8.10	2.963		420	148		568	680
9534	10" SDR 7.4		7.50	3.200		770	160		930	1,100
9536	10" SDR 11		7.50	3.200		710	160		870	1,025
9551	45° Elbow									
9554	3/8"	1 Plum	39	.205	Ea.	1.24	11		12.24	17.75
9556	1/2"		37.40	.214		1.62	11.45		13.07	18.90
9558	3/4"		35.40	.226		2.08	12.10		14.18	20.50
9564	1"		29.70	.269		3.01	14.45		17.46	25
9566	1-1/4"		27.60	.290		4.63	15.55		20.18	28
9568	1-1/2"		24.80	.323		9.95	17.30		27.25	37
9570	2"	Q-1	43	.372		15.15	17.95		33.10	43.50
9572	2-1/2"		35.60	.449		33.50	21.50		55	69.50
9574	3"		30.90	.518		62	25		87	106
9576	3-1/2"		27.80	.576		88	28		116	139
9578	4"		25	.640		136	31		167	195
9580	6" SDR 7.4		5.55	2.883		157	139		296	380
9582	6" SDR 11		5.55	2.883		129	139		268	350
9584	8" SDR 7.4	Q-2	8.10	2.963		430	148		578	695

22 11 13.76 Pipe Fittings, Plastic		Crew	Daily Output	Labor-Hours	Unit	Material	2011 Bare Costs Labor	Equipment	Total	Total Incl O&P
9586	8" SDR 11	Q-2	8.10	2.963	Ea.	375	148		523	630
9588	10" SDR 7.4		7.50	3.200		690	160		850	1,000
9590	10" SDR 11	▼	7.50	3.200	▼	570	160		730	865
9600	Tee									
9604	3/8"	1 Plum	26	.308	Ea.	1.60	16.50		18.10	26.50
9606	1/2"		24.90	.321		2.19	17.20		19.39	28
9608	3/4"		23.70	.338		3.01	18.10		21.11	30.50
9610	1"		19.90	.402		3.83	21.50		25.33	36
9612	1-1/4"		18.50	.432		5.85	23		28.85	41
9614	1-1/2"	▼	16.60	.482		16.65	26		42.65	57
9616	2"	Q-1	26.80	.597		24	29		53	69.50
9618	2-1/2"		23.80	.672		40	32.50		72.50	92.50
9620	3"		20.60	.777		73.50	37.50		111	137
9622	3-1/2"		18.40	.870		115	42		157	189
9624	4"		16.70	.958		145	46		191	228
9626	6" SDR 7.4		3.70	4.324		170	209		379	495
9628	6" SDR 11	▼	3.70	4.324		129	209		338	450
9630	8" SDR 7.4	Q-2	5.40	4.444		465	222		687	845
9632	8" SDR 11		5.40	4.444		405	222		627	775
9634	10" SDR 7.4		5	4.800		805	240		1,045	1,250
9636	10" SDR 11	▼	5	4.800	▼	650	240		890	1,075
9638	For reducing tee use same tee price									
9660	End cap									
9662	3/8"	1 Plum	78	.103	Ea.	1.60	5.50		7.10	9.95
9664	1/2"		74.60	.107		2.19	5.75		7.94	11
9666	3/4"		71.40	.112		3.01	6		9.01	12.25
9668	1"		59.50	.134		3.65	7.20		10.85	14.75
9670	1-1/4"		55.60	.144		5.75	7.70		13.45	17.80
9672	1-1/2"	▼	49.50	.162		7.90	8.65		16.55	21.50
9674	2"	Q-1	80	.200		13.25	9.65		22.90	29
9676	2-1/2"		71.40	.224		19.20	10.80		30	37
9678	3"		61.70	.259		43.50	12.50		56	66.50
9680	3-1/2"		55.20	.290		52	14		66	78.50
9682	4"		50	.320		79.50	15.45		94.95	111
9684	6" SDR 7.4		26.50	.604		109	29		138	164
9686	6" SDR 11	▼	26.50	.604		109	29		138	164
9688	8" SDR 7.4	Q-2	16.30	1.472		109	73.50		182.50	230
9690	8" SDR 11		16.30	1.472		94.50	73.50		168	214
9692	10" SDR 7.4		14.90	1.611		164	80.50		244.50	300
9694	10" SDR 11	▼	14.90	1.611	▼	139	80.50		219.50	273
9800	Accessories and tools									
9802	Pipe clamps for suspension, not including rod or beam clamp									
9804	3/8"	1 Plum	74	.108	Ea.	2.81	5.80		8.61	11.75
9805	1/2"		70	.114		3.52	6.15		9.67	13
9806	3/4"		68	.118		4.06	6.30		10.36	13.85
9807	1"		66	.121		4.42	6.50		10.92	14.55
9808	1-1/4"		64	.125		4.54	6.70		11.24	15
9809	1-1/2"	▼	62	.129		4.95	6.90		11.85	15.75
9810	2"	Q-1	110	.145		6.15	7		13.15	17.25
9811	2-1/2"		104	.154		7.90	7.40		15.30	19.75
9812	3"		98	.163		8.50	7.90		16.40	21
9813	3-1/2"		92	.174		9.30	8.40		17.70	23
9814	4"	▼	86	.186	▼	10.45	9		19.45	25

22 11 Facility Water Distribution

22 11 13 – Facility Water Distribution Piping

22 11 13.76 Pipe Fittings, Plastic

		Crew	Daily Output	Labor-Hours	Unit	Material	2011 Bare Costs Labor	Equipment	Total	Total Incl O&P
9815	6"	Q-1	70	.229	Ea.	12.70	11.05		23.75	30.50
9816	8"	Q-2	100	.240		45.50	12		57.50	68
9817	10"	"	94	.255	↓	52	12.75		64.75	76
9820	Pipe cutter									
9822	For 3/8" thru 1-1/4"				Ea.	155			155	171
9824	For 1-1/2" thru 4"				"	410			410	450
9826	Note: Pipes may be cut with standard									
9827	iron saw with blades for plastic.									
9830	Manual welding device, 800 W, 110 V									
9831	For 3/8" thru 2"				Ea.	495			495	545
9834	Manual welding device, 1400 W, 110 V									
9835	For 1-1/2" thru 4"				Ea.	495			495	545
9838	Butt welding machine									
9839	For 6" thru 10"				Ea.	33,500			33,500	36,800
9982	For plastic hangers see Line 22 05 29.10 8000									
9986	For copper/brass fittings see Line 22 11 13.25 7000									

22 11 13.78 Pipe, High Density Polyethylene Plastic (HDPE)

		Crew	Daily Output	Labor-Hours	Unit	Material	2011 Bare Costs Labor	Equipment	Total	Total Incl O&P
0010	**PIPE, HIGH DENSITY POLYETHYLENE PLASTIC (HDPE)**									
0020	Not incl. hangers, trenching, backfill, hoisting or digging equipment.									
0030	Standard length is 40', add a weld for each joint									
0040	Single wall									
0050	Straight									
0054	1" diameter DR 11				L.F.	.63			.63	.69
0058	1-1/2" diameter DR 11					.79			.79	.87
0062	2" diameter DR 11					1.32			1.32	1.45
0066	3" diameter DR 11					1.59			1.59	1.75
0070	3" diameter DR 17					1.27			1.27	1.40
0074	4" diameter DR 11					2.65			2.65	2.92
0078	4" diameter DR 17					2.65			2.65	2.92
0082	6" diameter DR 11					6.60			6.60	7.30
0086	6" diameter DR 17					4.37			4.37	4.81
0090	8" diameter DR 11					11			11	12.10
0094	8" diameter DR 26					5.15			5.15	5.70
0098	10" diameter DR 11					17.20			17.20	18.90
0102	10" diameter DR 26					7.95			7.95	8.75
0106	12" diameter DR 11					25			25	27.50
0110	12" diameter DR 26					11.90			11.90	13.10
0114	16" diameter DR 11					38.50			38.50	42
0118	16" diameter DR 26					17.20			17.20	18.90
0122	18" diameter DR 11					49			49	54
0126	18" diameter DR 26					22.50			22.50	25
0130	20" diameter DR 11					59.50			59.50	65.50
0134	20" diameter DR 26					26.50			26.50	29
0138	22" diameter DR 11					73			73	80
0142	22" diameter DR 26					33			33	36.50
0146	24" diameter DR 11					86			86	94.50
0150	24" diameter DR 26					38.50			38.50	42
0154	28" diameter DR 17					79.50			79.50	87.50
0158	28" diameter DR 26					53			53	58.50
0162	30" diameter DR 21					74			74	81.50
0166	30" diameter DR 26					61			61	67
0170	36" diameter DR 26					61			61	67

22 11 13.78 Pipe, High Density Polyethylene Plastic (HDPE)	Crew	Daily Output	Labor-Hours	Unit	Material	2011 Bare Costs Labor	Equipment	Total	Total Incl O&P
0174	42" diameter DR 26				L.F.	118			118	130
0178	48" diameter DR 26					155			155	170
0182	54" diameter DR 26				↓	195			195	214
0300	90° Elbow									
0304	1" diameter DR 11				Ea.	4.85			4.85	5.35
0308	1-1/2" diameter DR 11					6.05			6.05	6.70
0312	2" diameter DR 11					6.05			6.05	6.70
0316	3" diameter DR 11					12.15			12.15	13.35
0320	3" diameter DR 17					12.15			12.15	13.35
0324	4" diameter DR 11					17			17	18.70
0328	4" diameter DR 17					17			17	18.70
0332	6" diameter DR 11					39			39	42.50
0336	6" diameter DR 17					39			39	42.50
0340	8" diameter DR 11					96			96	105
0344	8" diameter DR 26					85			85	93.50
0348	10" diameter DR 11					360			360	395
0352	10" diameter DR 26					330			330	365
0356	12" diameter DR 11					380			380	415
0360	12" diameter DR 26					340			340	375
0364	16" diameter DR 11					450			450	490
0368	16" diameter DR 26					440			440	485
0372	18" diameter DR 11					560			560	615
0376	18" diameter DR 26					530			530	580
0380	20" diameter DR 11					660			660	730
0384	20" diameter DR 26					640			640	705
0388	22" diameter DR 11					685			685	750
0392	22" diameter DR 26					660			660	730
0396	24" diameter DR 11					770			770	850
0400	24" diameter DR 26					750			750	825
0404	28" diameter DR 17					935			935	1,025
0408	28" diameter DR 26					880			880	970
0412	30" diameter DR 17					1,325			1,325	1,450
0416	30" diameter DR 26					1,225			1,225	1,325
0420	36" diameter DR 26					1,550			1,550	1,700
0424	42" diameter DR 26					1,975			1,975	2,175
0428	48" diameter DR 26					2,325			2,325	2,550
0432	54" diameter DR 26				↓	5,525			5,525	6,075
0500	45° Elbow									
0512	2" diameter DR 11				Ea.	4.85			4.85	5.35
0516	3" diameter DR 11					12.15			12.15	13.35
0520	3" diameter DR 17					12.15			12.15	13.35
0524	4" diameter DR 11					17			17	18.70
0528	4" diameter DR 17					17			17	18.70
0532	6" diameter DR 11					39			39	42.50
0536	6" diameter DR 17					39			39	42.50
0540	8" diameter DR 11					96			96	105
0544	8" diameter DR 26					56			56	61.50
0548	10" diameter DR 11					360			360	395
0552	10" diameter DR 26					330			330	365
0556	12" diameter DR 11					380			380	415
0560	12" diameter DR 26					340			340	375
0564	16" diameter DR 11					198			198	218
0568	16" diameter DR 26				↓	187			187	206

22 11 Facility Water Distribution

22 11 13 – Facility Water Distribution Piping

22 11 13.78 Pipe, High Density Polyethylene Plastic (HDPE)	Crew	Daily Output	Labor-Hours	Unit	Material	2011 Bare Costs Labor	Equipment	Total	Total Incl O&P
0572	18" diameter DR 11				Ea.	211			211	232
0576	18" diameter DR 26					193			193	212
0580	20" diameter DR 11					320			320	350
0584	20" diameter DR 26					305			305	335
0588	22" diameter DR 11					440			440	485
0592	22" diameter DR 26					420			420	460
0596	24" diameter DR 11					540			540	595
0600	24" diameter DR 26					520			520	570
0604	28" diameter DR 17					615			615	675
0608	28" diameter DR 26					595			595	655
0612	30" diameter DR 17					760			760	835
0616	30" diameter DR 26					740			740	815
0620	36" diameter DR 26					940			940	1,025
0624	42" diameter DR 26					1,200			1,200	1,325
0628	48" diameter DR 26					1,300			1,300	1,450
0632	54" diameter DR 26					1,750			1,750	1,925
0700	Tee									
0704	1" diameter DR 11				Ea.	6.35			6.35	6.95
0708	1-1/2" diameter DR 11					8.85			8.85	9.75
0712	2" diameter DR 11					7.60			7.60	8.35
0716	3" diameter DR 11					13.95			13.95	15.35
0720	3" diameter DR 17					13.95			13.95	15.35
0724	4" diameter DR 11					20.50			20.50	22.50
0728	4" diameter DR 17					20.50			20.50	22.50
0732	6" diameter DR 11					50.50			50.50	56
0736	6" diameter DR 17					50.50			50.50	56
0740	8" diameter DR 11					126			126	138
0744	8" diameter DR 17					126			126	138
0748	10" diameter DR 11					375			375	410
0752	10" diameter DR 17					375			375	410
0756	12" diameter DR 11					500			500	550
0760	12" diameter DR 17					500			500	550
0764	16" diameter DR 11					264			264	290
0768	16" diameter DR 17					217			217	239
0772	18" diameter DR 11					370			370	410
0776	18" diameter DR 17					305			305	335
0780	20" diameter DR 11					455			455	500
0784	20" diameter DR 17					370			370	410
0788	22" diameter DR 11					580			580	640
0792	22" diameter DR 17					460			460	505
0796	24" diameter DR 11					720			720	795
0800	24" diameter DR 17					610			610	670
0804	28" diameter DR 17					1,175			1,175	1,300
0812	30" diameter DR 17					1,350			1,350	1,475
0820	36" diameter DR 17					2,225			2,225	2,450
0824	42" diameter DR 26					2,475			2,475	2,725
0828	48" diameter DR 26					2,675			2,675	2,925
1000	Flange adptr, w/back-up ring and 1/2 cost of plated bolt set									
1004	1" diameter DR 11				Ea.	24			24	26.50
1008	1-1/2" diameter DR 11					24			24	26.50
1012	2" diameter DR 11					15.20			15.20	16.75
1016	3" diameter DR 11					17.75			17.75	19.55
1020	3" diameter DR 17					17.75			17.75	19.55

22 11 13.78 Pipe, High Density Polyethylene Plastic (HDPE)	Crew	Daily Output	Labor-Hours	Unit	Material	2011 Bare Costs Labor	Equipment	Total	Total Incl O&P
1024	4" diameter DR 11				Ea.	24			24	26.50
1028	4" diameter DR 17					24			24	26.50
1032	6" diameter DR 11					34			34	37.50
1036	6" diameter DR 17					34			34	37.50
1040	8" diameter DR 11					49.50			49.50	54.50
1044	8" diameter DR 26					49.50			49.50	54.50
1048	10" diameter DR 11					78.50			78.50	86.50
1052	10" diameter DR 26					78.50			78.50	86.50
1056	12" diameter DR 11					115			115	127
1060	12" diameter DR 26					115			115	127
1064	16" diameter DR 11					247			247	272
1068	16" diameter DR 26					247			247	272
1072	18" diameter DR 11					320			320	355
1076	18" diameter DR 26					320			320	355
1080	20" diameter DR 11					445			445	490
1084	20" diameter DR 26					445			445	490
1088	22" diameter DR 11					485			485	530
1092	22" diameter DR 26					480			480	530
1096	24" diameter DR 17					525			525	575
1100	24" diameter DR 32.5					525			525	575
1104	28" diameter DR 15.5					710			710	780
1108	28" diameter DR 32.5					710			710	780
1112	30" diameter DR 11					825			825	905
1116	30" diameter DR 21					825			825	905
1120	36" diameter DR 26					900			900	990
1124	42" diameter DR 26					1,025			1,025	1,125
1128	48" diameter DR 26					1,250			1,250	1,375
1132	54" diameter DR 26				▼	1,525			1,525	1,675
1200	Reducer									
1208	2" x 1-1/2" diameter DR 11				Ea.	7.30			7.30	8
1212	3" x 2" diameter DR 11					7.30			7.30	8
1216	4" x 2" diameter DR 11					8.50			8.50	9.35
1220	4" x 3" diameter DR 11					10.90			10.90	12
1224	6" x 4" diameter DR 11					25.50			25.50	28
1228	8" x 6" diameter DR 11					39			39	42.50
1232	10" x 8" diameter DR 11					66.50			66.50	73.50
1236	12" x 8" diameter DR 11					109			109	120
1240	12" x 10" diameter DR 11					87.50			87.50	96
1244	14" x 12" diameter DR 11					97			97	107
1248	16" x 14" diameter DR 11					124			124	136
1252	18" x 16" diameter DR 11					150			150	165
1256	20" x 18" diameter DR 11					298			298	330
1260	22" x 20" diameter DR 11					365			365	405
1264	24" x 22" diameter DR 11					410			410	455
1268	26" x 24" diameter DR 11					485			485	535
1272	28" x 24" diameter DR 11					620			620	680
1276	32" x 28" diameter DR 17					800			800	880
1280	36" x 32" diameter DR 17				▼	1,100			1,100	1,200
4000	Welding labor per joint, not including welding machine									
4010	Pipe joint size (cost based on thickest wall for each dia.)									
4030	1" pipe size	4 Skwk	273	.117	Ea.		5.20		5.20	8
4040	1-1/2" pipe size		175	.183			8.10		8.10	12.45
4050	2" pipe size	▼	128	.250	▼		11.10		11.10	17.05

22 11 13 – Facility Water Distribution Piping

22 11 13.78 Pipe, High Density Polyethylene Plastic (HDPE)	Crew	Daily Output	Labor-Hours	Unit	Material	2011 Bare Costs Labor	Equipment	Total	Total Incl O&P
4060	3" pipe size	4 Skwk	100	.320	Ea.		14.20		14.20	22
4070	4" pipe size	↓	77	.416			18.45		18.45	28.50
4080	6" pipe size	5 Skwk	63	.635			28		28	43
4090	8" pipe size		48	.833			37		37	57
4100	10" pipe size	↓	40	1			44.50		44.50	68
4110	12" pipe size	6 Skwk	41	1.171			52		52	79.50
4120	16" pipe size		34	1.412			62.50		62.50	96
4130	18" pipe size	↓	32	1.500			66.50		66.50	102
4140	20" pipe size	8 Skwk	37	1.730			77		77	118
4150	22" pipe size		35	1.829			81		81	125
4160	24" pipe size		34	1.882			83.50		83.50	128
4170	28" pipe size		33	1.939			86		86	132
4180	30" pipe size		32	2			89		89	136
4190	36" pipe size		31	2.065			91.50		91.50	141
4200	42" pipe size	↓	30	2.133			94.50		94.50	145
4210	48" pipe size	9 Skwk	33	2.182			97		97	149
4220	54" pipe size	"	31	2.323	↓		103		103	158
4300	Note: Cost for set up each time welder is moved.									
4301	Add 50% of a weld cost									
4310	Welder usually remains stationary with pipe moved through it.									
4340	Weld machine, rental per day based on dia. capacity									
4350	1" thru 2" diameter				Ea.		40.50		40.50	44.50
4360	3" thru 4" diameter						46		46	50.50
4370	6" thru 8" diameter						103		103	113
4380	10" thru 12" diameter						178		178	196
4390	16" thru 18" diameter						259		259	285
4400	20" thru 24" diameter						500		500	550
4410	28" thru 32" diameter						545		545	600
4420	36" diameter						570		570	625
4430	42" thru 54" diameter				↓		890		890	980
5000	Dual wall contained pipe									
5040	Straight									
5054	1" DR 11 x 3" DR 11				L.F.	5.85			5.85	6.45
5058	1" DR 11 x 4" DR 11					6.25			6.25	6.85
5062	1-1/2" DR 11 x 4" DR 17					6.45			6.45	7.05
5066	2" DR 11 x 4" DR 17					6.90			6.90	7.60
5070	2" DR 11 x 6" DR 17					10.35			10.35	11.40
5074	3" DR 11 x 6" DR 17					11.55			11.55	12.70
5078	3" DR 11 x 6" DR 26					9.95			9.95	10.90
5086	3" DR 17 x 8" DR 17					14.65			14.65	16.15
5090	4" DR 11 x 8" DR 17					17.15			17.15	18.90
5094	4" DR 17 x 8" DR 26					13.55			13.55	14.95
5098	6" DR 11 x 10" DR 17					26.50			26.50	29
5102	6" DR 17 x 10" DR 26					19.90			19.90	22
5106	6" DR 26 x 10" DR 26					18.90			18.90	21
5110	8" DR 17 x 12" DR 26					28.50			28.50	31
5114	8" DR 26 x 12" DR 32.5					24			24	26.50
5118	10" DR 17 x 14" DR 26					39.50			39.50	43.50
5122	10" DR 17 x 16" DR 26					43			43	47.50
5126	10" DR 26 x 16" DR 26					39			39	43
5130	12" DR 26 x 16" DR 26					44.50			44.50	49
5134	12" DR 17 x 18" DR 26					57			57	62.50
5138	12" DR 26 x 18" DR 26				↓	51			51	56

22 11 13.78 Pipe, High Density Polyethylene Plastic (HDPE)	Crew	Daily Output	Labor-Hours	Unit	Material	2011 Bare Costs Labor	Equipment	Total	Total Incl O&P
5142	14" DR 26 x 20" DR 32.5				L.F.	56			56	61.50
5146	16" DR 26 x 22" DR 32.5					84.50			84.50	92.50
5150	18" DR 26 x 24" DR 32.5					77.50			77.50	85
5154	20" DR 32.5 x 28" DR 32.5					100			100	110
5158	22" DR 32.5 x 30" DR 32.5					100			100	110
5162	24" DR 32.5 x 32" DR 32.5					128			128	141
5166	36" DR 32.5 x 42" DR 32.5				▼	216			216	238
5300	Force transfer coupling									
5354	1" DR 11 x 3" DR 11				Ea.	211			211	232
5358	1" DR 11 x 4" DR 17					221			221	244
5362	1-1/2" DR 11 x 4" DR 17					232			232	255
5366	2" DR 11 x 4" DR 17					243			243	267
5370	2" DR 11 x 6" DR 17					350			350	385
5374	3" DR 11 x 6" DR 17					350			350	385
5378	3" DR 11 x 6" DR 26					350			350	385
5382	3" DR 11 x 8" DR 11					370			370	410
5386	3" DR 11 x 8" DR 17					370			370	410
5390	4" DR 11 x 8" DR 17					370			370	410
5394	4" DR 17 x 8" DR 26					325			325	360
5398	6" DR 11 x 10" DR 17					470			470	520
5402	6" DR 17 x 10" DR 26					390			390	430
5406	6" DR 26 x 10" DR 26					390			390	430
5410	8" DR 17 x 12" DR 26					515			515	570
5414	8" DR 26 x 12" DR 32.5					415			415	460
5418	10" DR 17 x 14" DR 26					675			675	745
5422	10" DR 17 x 16" DR 26					675			675	745
5426	10" DR 26 x 16" DR 26					675			675	745
5430	12" DR 26 x 16" DR 26					850			850	935
5434	12" DR 17 x 18" DR 26					870			870	955
5438	12" DR 26 x 18" DR 26					870			870	955
5442	14" DR 26 x 20" DR 32.5					920			920	1,000
5446	16" DR 26 x 22" DR 32.5					945			945	1,050
5450	18" DR 26 x 24" DR 32.5					1,025			1,025	1,125
5454	20" DR 32.5 x 28" DR 32.5					1,250			1,250	1,375
5458	22" DR 32.5 x 30" DR 32.5					1,375			1,375	1,525
5462	24" DR 32.5 x 32" DR 32.5					1,500			1,500	1,650
5466	36" DR 32.5 x 42" DR 32.5				▼	2,475			2,475	2,725
5600	90° Elbow									
5654	1" DR 11 x 3" DR 11				Ea.	174			174	192
5658	1" DR 11 x 4" DR 17					167			167	184
5662	1-1/2" DR 11 x 4" DR 17					182			182	201
5666	2" DR 11 x 4" DR 17					193			193	213
5670	2" DR 11 x 6" DR 17					246			246	270
5674	3" DR 11 x 6" DR 17					282			282	310
5678	3" DR 17 x 6" DR 26					224			224	247
5682	3" DR 17 x 8" DR 11					455			455	500
5686	3" DR 17 x 8" DR 17					355			355	390
5690	4" DR 11 x 8" DR 17					415			415	455
5694	4" DR 17 x 8" DR 26					330			330	365
5698	6" DR 11 x 10" DR 17					555			555	610
5702	6" DR 17 x 10" DR 26					425			425	465
5706	6" DR 26 x 10" DR 26					395			395	435
5710	8" DR 17 x 12" DR 26				▼	630			630	695

22 11 13 – Facility Water Distribution Piping

22 11 13.78 Pipe, High Density Polyethylene Plastic (HDPE)	Crew	Daily Output	Labor-Hours	Unit	Material	2011 Bare Costs Labor	Equipment	Total	Total Incl O&P
5714	8" DR 26 x 12" DR 32.5				Ea.	585			585	640
5718	10" DR 17 x 14" DR 26					990			990	1,100
5722	10" DR 17 x 16" DR 26					950			950	1,050
5726	10" DR 26 x 16" DR 26					880			880	965
5730	12" DR 26 x 16" DR 26					975			975	1,075
5734	12" DR 17 x 18" DR 26					1,150			1,150	1,275
5738	12" DR 26 x 18" DR 26					1,075			1,075	1,175
5742	14" DR 26 x 20" DR 32.5					1,400			1,400	1,550
5746	16" DR 26 x 22" DR 32.5					1,425			1,425	1,550
5750	18" DR 26 x 24" DR 32.5					1,725			1,725	1,900
5754	20" DR 32.5 x 28" DR 32.5					2,000			2,000	2,200
5758	22" DR 32.5 x 30" DR 32.5					2,725			2,725	3,000
5762	24" DR 32.5 x 32" DR 32.5					2,650			2,650	2,925
5766	36" DR 32.5 x 42" DR 32.5					4,325			4,325	4,750
5800	45° Elbow									
5804	1" DR 11 x 3" DR 11				Ea.	104			104	115
5808	1" DR 11 x 4" DR 17					104			104	114
5812	1-1/2" DR 11 x 4" DR 17					111			111	122
5816	2" DR 11 x 4" DR 17					121			121	134
5820	2" DR 11 x 6" DR 17					153			153	168
5824	3" DR 11 x 6" DR 17					169			169	186
5828	3" DR 17 x 6" DR 26					140			140	154
5832	3" DR 17 x 8" DR 11					259			259	285
5836	3" DR 17 x 8" DR 17					207			207	228
5840	4" DR 11 x 8" DR 17					242			242	266
5844	4" DR 17 x 8" DR 26					197			197	217
5848	6" DR 11 x 10" DR 17					320			320	350
5852	6" DR 17 x 10" DR 26					250			250	275
5856	6" DR 26 x 10" DR 26					236			236	260
5860	8" DR 17 x 12" DR 26					380			380	420
5864	8" DR 26 x 12" DR 32.5					355			355	390
5868	10" DR 17 x 14" DR 26					595			595	655
5872	10" DR 17 x 16" DR 26					585			585	640
5876	10" DR 26 x 16" DR 26					545			545	600
5880	12" DR 26 x 16" DR 26					615			615	680
5884	12" DR 17 x 18" DR 26					725			725	800
5888	12" DR 26 x 18" DR 26					675			675	745
5892	14" DR 26 x 20" DR 32.5					875			875	960
5896	16" DR 26 x 22" DR 32.5					915			915	1,000
5900	18" DR 26 x 24" DR 32.5					1,075			1,075	1,200
5904	20" DR 32.5 x 28" DR 32.5					1,325			1,325	1,450
5908	22" DR 32.5 x 30" DR 32.5					1,675			1,675	1,850
5912	24" DR 32.5 x 32" DR 32.5					1,675			1,675	1,825
5916	36" DR 32.5 x 42" DR 32.5					2,600			2,600	2,850
6000	Access port with 4" riser									
6050	1" DR 11 x 4" DR 17				Ea.	228			228	251
6054	1-1/2" DR 11 x 4" DR 17					232			232	256
6058	2" DR 11 x 6" DR 17					284			284	310
6062	3" DR 11 x 6" DR 17					285			285	315
6066	3" DR 17 x 6" DR 26					280			280	310
6070	3" DR 17 x 8" DR 11					320			320	350
6074	3" DR 17 x 8" DR 17					305			305	335
6078	4" DR 11 x 8" DR 17					315			315	345

22 11 13.78 Pipe, High Density Polyethylene Plastic (HDPE)	Crew	Daily Output	Labor-Hours	Unit	Material	2011 Bare Costs Labor	Equipment	Total	Total Incl O&P
6082	4" DR 17 x 8" DR 26				Ea.	305			305	335
6086	6" DR 11 x 10" DR 17					395			395	435
6090	6" DR 17 x 10" DR 26					365			365	400
6094	6" DR 26 x 10" DR 26					355			355	390
6098	8" DR 17 x 12" DR 26					380			380	420
6102	8" DR 26 x 12" DR 32.5				▼	365			365	400
6200	End termination with vent plug									
6204	1" DR 11 x 3" DR 11				Ea.	189			189	207
6208	1" DR 11 x 4" DR 17					202			202	222
6212	1-1/2" DR 11 x 4" DR 17					202			202	222
6216	2" DR 11 x 4" DR 17					219			219	241
6220	2" DR 11 x 6" DR 17					250			250	275
6224	3" DR 11 x 6" DR 17					250			250	275
6228	3" DR 17 x 6" DR 26					250			250	275
6232	3" DR 17 x 8" DR 11					345			345	380
6236	3" DR 17 x 8" DR 17					345			345	380
6240	4" DR 11 x 8" DR 17					345			345	380
6244	4" DR 17 x 8" DR 26					300			300	330
6248	6" DR 11 x 10" DR 17					410			410	450
6252	6" DR 17 x 10" DR 26					335			335	370
6256	6" DR 26 x 10" DR 26					335			335	370
6260	8" DR 17 x 12" DR 26					500			500	550
6264	8" DR 26 x 12" DR 32.5					400			400	440
6268	10" DR 17 x 14" DR 26					520			520	570
6272	10" DR 17 x 16" DR 26					520			520	570
6276	10" DR 26 x 16" DR 26					520			520	570
6280	12" DR 26 x 16" DR 26					520			520	570
6284	12" DR 17 x 18" DR 26					685			685	750
6288	12" DR 26 x 18" DR 26					685			685	750
6292	14" DR 26 x 20" DR 32.5					695			695	765
6296	16" DR 26 x 22" DR 32.5					810			810	890
6300	18" DR 26 x 24" DR 32.5					835			835	915
6304	20" DR 32.5 x 28" DR 32.5					1,075			1,075	1,175
6308	22" DR 32.5 x 30" DR 32.5					1,175			1,175	1,300
6312	24" DR 32.5 x 32" DR 32.5					1,300			1,300	1,425
6316	36" DR 32.5 x 42" DR 32.5				▼	1,900			1,900	2,075
6600	Tee									
6604	1" DR 11 x 3" DR 11				Ea.	192			192	211
6608	1" DR 11 x 4" DR 17					238			238	262
6612	1-1/2" DR 11 x 4" DR 17					265			265	292
6616	2" DR 11 x 4" DR 17					290			290	320
6620	2" DR 11 x 6" DR 17					350			350	385
6624	3" DR 11 x 6" DR 17					405			405	445
6628	3" DR 17 x 6" DR 26					395			395	435
6632	3" DR 17 x 8" DR 11					475			475	525
6636	3" DR 17 x 8" DR 17					455			455	500
6640	4" DR 11 x 8" DR 17					495			495	545
6644	4" DR 17 x 8" DR 26					475			475	520
6648	6" DR 11 x 10" DR 17					620			620	680
6652	6" DR 17 x 10" DR 26					585			585	645
6656	6" DR 26 x 10" DR 26					580			580	640
6660	8" DR 17 x 12" DR 26					580			580	640
6664	8" DR 26 x 12" DR 32.5				▼	755			755	830

22 11 13 – Facility Water Distribution Piping

22 11 13.78 Pipe, High Density Polyethylene Plastic (HDPE)	Crew	Daily Output	Labor-Hours	Unit	Material	2011 Bare Costs Labor	Equipment	Total	Total Incl O&P
6668	10" DR 17 x 14" DR 26				Ea.	875			875	965
6672	10" DR 17 x 16" DR 26					920			920	1,000
6676	10" DR 26 x 16" DR 26					920			920	1,000
6680	12" DR 26 x 16" DR 26					990			990	1,100
6684	12" DR 17 x 18" DR 26					990			990	1,100
6688	12" DR 26 x 18" DR 26					1,100			1,100	1,200
6692	14" DR 26 x 20" DR 32.5					1,450			1,450	1,600
6696	16" DR 26 x 22" DR 32.5					1,775			1,775	1,950
6700	18" DR 26 x 24" DR 32.5					1,925			1,925	2,125
6704	20" DR 32.5 x 28" DR 32.5					2,450			2,450	2,675
6708	22" DR 32.5 x 30" DR 32.5					2,900			2,900	3,200
6712	24" DR 32.5 x 32" DR 32.5					3,475			3,475	3,825
6716	36" DR 32.5 x 42" DR 32.5					7,075			7,075	7,775
6800	Wye									
6816	2" DR 11 x 4" DR 17				Ea.	365			365	400
6820	2" DR 11 x 6" DR 17					415			415	455
6824	3" DR 11 x 6" DR 17					420			420	460
6828	3" DR 17 x 6" DR 26					405			405	445
6832	3" DR 17 x 8" DR 11					510			510	560
6836	3" DR 17 x 8" DR 17					475			475	525
6840	4" DR 11 x 8" DR 17					520			520	570
6844	4" DR 17 x 8" DR 26					495			495	540
6848	6" DR 11 x 10" DR 17					645			645	705
6852	6" DR 17 x 10" DR 26					595			595	655
6856	6" DR 26 x 10" DR 26					595			595	655
6860	8" DR 17 x 12" DR 26					850			850	935
6864	8" DR 26 x 12" DR 32.5					820			820	905
6868	10" DR 17 x 14" DR 26					1,025			1,025	1,125
6872	10" DR 17 x 16" DR 26					1,100			1,100	1,200
6876	10" DR 26 x 16" DR 26					1,075			1,075	1,175
6880	12" DR 26 x 16" DR 26					1,200			1,200	1,325
6884	12" DR 17 x 18" DR 26					1,350			1,350	1,500
6888	12" DR 26 x 18" DR 26					1,375			1,375	1,525
6892	14" DR 26 x 20" DR 32.5					1,675			1,675	1,825
6896	16" DR 26 x 22" DR 32.5					2,400			2,400	2,650
6900	18" DR 26 x 24" DR 32.5					2,825			2,825	3,125
6904	20" DR 32.5 x 28" DR 32.5					3,550			3,550	3,900
6908	22" DR 32.5 x 30" DR 32.5					4,050			4,050	4,450
6912	24" DR 32.5 x 32" DR 32.5					4,600			4,600	5,075
9000	Welding labor per joint, not including welding machine									
9010	Pipe joint size, outer pipe (cost based on the thickest walls)									
9020	Straight pipe									
9050	3" pipe size	4 Skwk	96	.333	Ea.		14.80		14.80	22.50
9060	4" pipe size	"	77	.416			18.45		18.45	28.50
9070	6" pipe size	5 Skwk	60	.667			29.50		29.50	45.50
9080	8" pipe size	"	40	1			44.50		44.50	68
9090	10" pipe size	6 Skwk	41	1.171			52		52	79.50
9100	12" pipe size		39	1.231			54.50		54.50	84
9110	14" pipe size		38	1.263			56		56	86
9120	16" pipe size		35	1.371			61		61	93.50
9130	18" pipe size	8 Skwk	45	1.422			63		63	97
9140	20" pipe size		42	1.524			67.50		67.50	104
9150	22" pipe size		40	1.600			71		71	109

22 11 Facility Water Distribution

22 11 13 – Facility Water Distribution Piping

22 11 13.78 Pipe, High Density Polyethylene Plastic (HDPE)	Crew	Daily Output	Labor-Hours	Unit	Material	2011 Bare Costs Labor	Equipment	Total	Total Incl O&P
9160	24" pipe size	8 Skwk	38	1.684	Ea.		75		75	115
9170	28" pipe size		37	1.730			77		77	118
9180	30" pipe size		36	1.778			79		79	121
9190	32" pipe size		35	1.829			81		81	125
9200	42" pipe size		32	2			89		89	136
9300	Note: Cost for set up each time welder is moved.									
9301	Add 100% of weld labor cost									
9310	For handling between fitting welds add 50% of weld labor cost									
9320	Welder usually remains stationary with pipe moved through it									
9360	Weld machine, rental per day based on dia. capacity									
9380	3" thru 4" diameter				Ea.			63.50	63.50	70
9390	6" thru 8" diameter							207	207	228
9400	10" thru 12" diameter							310	310	340
9410	14" thru 18" diameter							420	420	460
9420	20" thru 24" diameter							750	750	825
9430	28" thru 32" diameter							880	880	970
9440	42" thru 58" diameter							910	910	1,000

22 11 19 – Domestic Water Piping Specialties

22 11 19.10 Flexible Connectors

		Crew	Daily Output	Labor-Hours	Unit	Material	Labor	Equipment	Total	Total Incl O&P
0010	**FLEXIBLE CONNECTORS**, Corrugated, 7/8" O.D., 1/2" I.D.									
0050	Gas, seamless brass, steel fittings									
0200	12" long	1 Plum	36	.222	Ea.	15.10	11.90		27	34.50
0220	18" long		36	.222		18.70	11.90		30.60	38.50
0240	24" long		34	.235		22	12.60		34.60	43.50
0260	30" long		34	.235		24	12.60		36.60	45.50
0280	36" long		32	.250		26.50	13.40		39.90	49
0320	48" long		30	.267		33.50	14.30		47.80	58.50
0340	60" long		30	.267		40	14.30		54.30	65.50
0360	72" long		30	.267		46	14.30		60.30	72
2000	Water, copper tubing, dielectric separators									
2100	12" long	1 Plum	36	.222	Ea.	14	11.90		25.90	33
2220	15" long		36	.222		15.45	11.90		27.35	35
2240	18" long		36	.222		20.50	11.90		32.40	40.50
2260	24" long		34	.235		20.50	12.60		33.10	41.50

22 11 19.14 Flexible Metal Hose

		Crew	Daily Output	Labor-Hours	Unit	Material	Labor	Equipment	Total	Total Incl O&P
0010	**FLEXIBLE METAL HOSE**, Connectors, standard lengths									
0100	Bronze braided, bronze ends									
0120	3/8" diameter x 12"	1 Stpi	26	.308	Ea.	18.85	16.65		35.50	46
0140	1/2" diameter x 12"		24	.333		17.95	18.05		36	47
0160	3/4" diameter x 12"		20	.400		26.50	21.50		48	61.50
0180	1" diameter x 18"		19	.421		34	23		57	71.50
0200	1-1/2" diameter x 18"		13	.615		53.50	33.50		87	109
0220	2" diameter x 18"		11	.727		64.50	39.50		104	130
1000	Carbon steel ends									
1020	1/4" diameter x 12"	1 Stpi	28	.286	Ea.	13.60	15.45		29.05	38
1040	3/8" diameter x 12"		26	.308		13.95	16.65		30.60	40.50
1060	1/2" diameter x 12"		24	.333		15.40	18.05		33.45	44
1080	1/2" diameter x 24"		24	.333		39.50	18.05		57.55	70
1100	1/2" diameter x 36"		24	.333		60.50	18.05		78.55	93.50
1120	3/4" diameter x 12"		20	.400		25	21.50		46.50	60
1140	3/4" diameter x 24"		20	.400		48.50	21.50		70	86
1160	3/4" diameter x 36"		20	.400		57.50	21.50		79	96

22 11 19.14 Flexible Metal Hose		Crew	Daily Output	Labor-Hours	Unit	Material	2011 Bare Costs Labor	Equipment	Total	Total Incl O&P
1180	1" diameter x 18"	1 Stpi	19	.421	Ea.	31.50	23		54.50	68.50
1200	1" diameter x 30"		19	.421		56	23		79	95.50
1220	1" diameter x 36"		19	.421		67.50	23		90.50	109
1240	1-1/4" diameter x 18"		15	.533		43	29		72	90.50
1260	1-1/4" diameter x 36"		15	.533		90.50	29		119.50	143
1280	1-1/2" diameter x 18"		13	.615		61	33.50		94.50	117
1300	1-1/2" diameter x 36"		13	.615		98	33.50		131.50	158
1320	2" diameter x 24"		11	.727		88.50	39.50		128	156
1340	2" diameter x 36"		11	.727		121	39.50		160.50	192
1360	2-1/2" diameter x 24"		9	.889		202	48		250	295
1380	2-1/2" diameter x 36"		9	.889		271	48		319	370
1400	3" diameter x 24"		7	1.143		279	62		341	400
1420	3" diameter x 36"		7	1.143		385	62		447	520
2000	Carbon steel braid, carbon steel solid ends									
2100	1/2" diameter x 12"	1 Stpi	24	.333	Ea.	36	18.05		54.05	66.50
2120	3/4" diameter x 12"		20	.400		54.50	21.50		76	92
2140	1" diameter x 12"		19	.421		78	23		101	120
2160	1-1/4" diameter x 12"		15	.533		56.50	29		85.50	105
2180	1-1/2" diameter x 12"		13	.615		58.50	33.50		92	114
3000	Stainless steel braid, welded on carbon steel ends									
3100	1/2" diameter x 12"	1 Stpi	24	.333	Ea.	48.50	18.05		66.55	80
3120	3/4" diameter x 12"		20	.400		60	21.50		81.50	98.50
3140	3/4" diameter x 24"		20	.400		42	21.50		63.50	78.50
3160	3/4" diameter x 36"		20	.400		50.50	21.50		72	88
3180	1" diameter x 12"		19	.421		57.50	23		80.50	97
3200	1" diameter x 24"		19	.421		53	23		76	92.50
3220	1" diameter x 36"		19	.421		97.50	23		120.50	141
3240	1-1/4" diameter x 12"		15	.533		106	29		135	160
3260	1-1/4" diameter x 24"		15	.533		116	29		145	171
3280	1-1/4" diameter x 36"		15	.533		131	29		160	187
3300	1-1/2" diameter x 12"		13	.615		108	33.50		141.50	169
3320	1-1/2" diameter x 24"		13	.615		127	33.50		160.50	189
3340	1-1/2" diameter x 36"		13	.615		148	33.50		181.50	213
3400	Metal stainless steel braid, over corrugated stainless steel, flanged ends									
3410	150 PSI									
3420	1/2" diameter x 12"	1 Stpi	24	.333	Ea.	144	18.05		162.05	185
3430	1" diameter x 12"		20	.400		180	21.50		201.50	231
3440	1-1/2" diameter x 12"		15	.533		204	29		233	267
3450	2-1/2" diameter x 9"		12	.667		108	36		144	173
3460	3" diameter x 9"		9	.889		90	48		138	171
3470	4" diameter x 9"		7	1.143		110	62		172	214
3480	4" diameter x 30"		5	1.600		560	86.50		646.50	745
3490	4" diameter x 36"		4.80	1.667		580	90		670	775
3500	6" diameter x 11"		5	1.600		183	86.50		269.50	330
3510	6" diameter x 36"		3.80	2.105		685	114		799	920
3520	8" diameter x 12"		4	2		400	108		508	600
3530	10" diameter x 13"		3	2.667		610	144		754	885
3540	12" diameter x 14"	Q-5	4	4		925	195		1,120	1,325
6000	Molded rubber with helical wire reinforcement									
6010	150 PSI									
6020	1-1/2" diameter x 12"	1 Stpi	15	.533	Ea.	82	29		111	134
6030	2" diameter x 12"		12	.667		234	36		270	310
6040	3" diameter x 12"		8	1		244	54		298	350

22 11 Facility Water Distribution

22 11 19 – Domestic Water Piping Specialties

22 11 19.14 Flexible Metal Hose

		Crew	Daily Output	Labor-Hours	Unit	Material	2011 Bare Costs Labor	Equipment	Total	Total Incl O&P
6050	4" diameter x 12"	1 Stpi	6	1.333	Ea.	310	72		382	450
6060	6" diameter x 18"		4	2		465	108		573	670
6070	8" diameter x 24"		3	2.667		650	144		794	930
6080	10" diameter x 24"		2	4		790	216		1,006	1,200
6090	12" diameter x 24"	Q-5	3	5.333		900	260		1,160	1,375
7000	Molded teflon with stainless steel flanges									
7010	150 PSI									
7020	2-1/2" diameter x 3-3/16"	Q-1	7.80	2.051	Ea.	2,275	99		2,374	2,650
7030	3" diameter x 3-5/8"		6.50	2.462		1,575	119		1,694	1,925
7040	4" diameter x 3-5/8"		5	3.200		2,050	154		2,204	2,475
7050	6" diameter x 4"		4.30	3.721		2,875	180		3,055	3,450
7060	8" diameter x 6"		3.80	4.211		4,575	203		4,778	5,350

22 11 19.18 Mixing Valve

		Crew	Daily Output	Labor-Hours	Unit	Material	2011 Bare Costs Labor	Equipment	Total	Total Incl O&P
0010	**MIXING VALVE**, Automatic, water tempering.									
0040	1/2" size	1 Stpi	19	.421	Ea.	505	23		528	590
0050	3/4" size		18	.444		505	24		529	590
0100	1" size		16	.500		745	27		772	860
0120	1-1/4" size		13	.615		1,025	33.50		1,058.50	1,175
0140	1-1/2" size		10	.800		1,225	43.50		1,268.50	1,425
0160	2" size		8	1		1,525	54		1,579	1,750
0170	2-1/2" size		6	1.333		1,525	72		1,597	1,775
0180	3" size		4	2		3,600	108		3,708	4,100
0190	4" size		3	2.667		3,600	144		3,744	4,175

22 11 19.22 Pressure Reducing Valve

		Crew	Daily Output	Labor-Hours	Unit	Material	2011 Bare Costs Labor	Equipment	Total	Total Incl O&P
0010	**PRESSURE REDUCING VALVE**, Steam, pilot operated.									
0100	Threaded, iron body									
0200	1-1/2" size	1 Stpi	8	1	Ea.	1,700	54		1,754	1,950
0220	2" size	"	5	1.600	"	1,950	86.50		2,036.50	2,275
1000	Flanged, iron body, 125 lb. flanges									
1020	2" size	1 Stpi	8	1	Ea.	1,975	54		2,029	2,250
1040	2-1/2" size	"	4	2		2,375	108		2,483	2,775
1060	3" size	Q-5	4.50	3.556		2,875	173		3,048	3,425
1080	4" size	"	3	5.333		4,150	260		4,410	4,975
1500	For 250 lb. flanges, add					5%				

22 11 19.26 Pressure Regulators

		Crew	Daily Output	Labor-Hours	Unit	Material	2011 Bare Costs Labor	Equipment	Total	Total Incl O&P
0010	**PRESSURE REGULATORS**									
0100	Gas appliance regulators									
0106	Main burner and pilot applications									
0108	Rubber seat poppet type									
0109	1/8" pipe size	1 Stpi	24	.333	Ea.	14.10	18.05		32.15	42.50
0110	1/4" pipe size		24	.333		16.25	18.05		34.30	45
0112	3/8" pipe size		24	.333		17.30	18.05		35.35	46
0113	1/2" pipe size		24	.333		19.10	18.05		37.15	48
0114	3/4" pipe size		20	.400		22.50	21.50		44	57
0122	Lever action type									
0123	3/8" pipe size	1 Stpi	24	.333	Ea.	21.50	18.05		39.55	50.50
0124	1/2" pipe size		24	.333		21.50	18.05		39.55	50.50
0125	3/4" pipe size		20	.400		42.50	21.50		64	79
0126	1" pipe size		19	.421		42.50	23		65.50	80.50
0132	Double diaphragm type									
0133	3/8" pipe size	1 Stpi	24	.333	Ea.	31.50	18.05		49.55	62
0134	1/2" pipe size		24	.333		43.50	18.05		61.55	75

22 11 19.26 Pressure Regulators		Crew	Daily Output	Labor-Hours	Unit	Material	2011 Bare Costs Labor	Equipment	Total	Total Incl O&P
0135	3/4" pipe size	1 Stpi	20	.400	Ea.	73	21.50		94.50	113
0136	1" pipe size		19	.421		91	23		114	134
0137	1-1/4" pipe size		15	.533		310	29		339	385
0138	1-1/2" pipe size		13	.615		570	33.50		603.50	675
0139	2" pipe size		11	.727		570	39.50		609.50	685
0140	2-1/2" pipe size	Q-5	15	1.067		1,100	52		1,152	1,300
0141	3" pipe size		13	1.231		1,100	60		1,160	1,325
0142	4" pipe size (flanged)		8	2		1,975	97.50		2,072.50	2,325
0160	Main burner only									
0162	Straight-thru-flow design									
0163	1/2" pipe size	1 Stpi	24	.333	Ea.	35.50	18.05		53.55	66
0164	3/4" pipe size		20	.400		48.50	21.50		70	86
0165	1" pipe size		19	.421		69	23		92	110
0166	1-1/4" pipe size		15	.533		69	29		98	119
0200	Oil, light, hot water, ordinary steam, threaded									
0220	Bronze body, 1/4" size	1 Stpi	24	.333	Ea.	169	18.05		187.05	213
0230	3/8" size		24	.333		183	18.05		201.05	228
0240	1/2" size		24	.333		227	18.05		245.05	277
0250	3/4" size		20	.400		271	21.50		292.50	330
0260	1" size		19	.421		410	23		433	490
0270	1-1/4" size		15	.533		555	29		584	655
0280	1-1/2" size		13	.615		595	33.50		628.50	705
0290	2" size		11	.727		1,075	39.50		1,114.50	1,250
0320	Iron body, 1/4" size		24	.333		132	18.05		150.05	172
0330	3/8" size		24	.333		156	18.05		174.05	198
0340	1/2" size		24	.333		165	18.05		183.05	208
0350	3/4" size		20	.400		200	21.50		221.50	253
0360	1" size		19	.421		261	23		284	320
0370	1-1/4" size		15	.533		360	29		389	445
0380	1-1/2" size		13	.615		405	33.50		438.50	495
0390	2" size		11	.727		585	39.50		624.50	705
0500	Oil, heavy, viscous fluids, threaded									
0520	Bronze body, 3/8" size	1 Stpi	24	.333	Ea.	291	18.05		309.05	345
0530	1/2" size		24	.333		375	18.05		393.05	435
0540	3/4" size		20	.400		415	21.50		436.50	490
0550	1" size		19	.421		520	23		543	605
0560	1-1/4" size		15	.533		735	29		764	850
0570	1-1/2" size		13	.615		840	33.50		873.50	975
0600	Iron body, 3/8" size		24	.333		231	18.05		249.05	281
0620	1/2" size		24	.333		280	18.05		298.05	335
0630	3/4" size		20	.400		320	21.50		341.50	385
0640	1" size		19	.421		375	23		398	450
0650	1-1/4" size		15	.533		545	29		574	645
0660	1-1/2" size		13	.615		580	33.50		613.50	690
0800	Process steam, wet or super heated, monel trim, threaded									
0820	Bronze body, 1/4" size	1 Stpi	24	.333	Ea.	610	18.05		628.05	695
0830	3/8" size		24	.333		610	18.05		628.05	695
0840	1/2" size		24	.333		700	18.05		718.05	795
0850	3/4" size		20	.400		825	21.50		846.50	940
0860	1" size		19	.421		1,050	23		1,073	1,175
0870	1-1/4" size		15	.533		1,300	29		1,329	1,475
0880	1-1/2" size		13	.615		1,550	33.50		1,583.50	1,750
0920	Iron body, max 125 PSIG press out, 1/4" size		24	.333		132	18.05		150.05	172

22 11 19 – Domestic Water Piping Specialties

22 11 19.26 Pressure Regulators

		Crew	Daily Output	Labor-Hours	Unit	Material	2011 Bare Costs Labor	Equipment	Total	Total Incl O&P
0930	3/8" size	1 Stpi	24	.333	Ea.	156	18.05		174.05	198
0940	1/2" size		24	.333		165	18.05		183.05	208
0950	3/4" size		20	.400		200	21.50		221.50	253
0960	1" size		19	.421		261	23		284	320
0970	1-1/4" size		15	.533		360	29		389	445
0980	1-1/2" size		13	.615		405	33.50		438.50	495
0990	2" size	▼	11	.727	▼	585	39.50		624.50	705
1000	Flanged, Class 125									
1006	2-1/2" size	Q-5	10	1.600	Ea.	2,375	78		2,453	2,750
1008	3" size		9	1.778		2,875	86.50		2,961.50	3,300
1010	4" size		8	2		4,150	97.50		4,247.50	4,725
1020	6" size	▼	7	2.286	▼	10,200	111		10,311	11,400
3000	Steam, high capacity, bronze body, stainless steel trim									
3020	Threaded, 1/2" diameter	1 Stpi	24	.333	Ea.	1,450	18.05		1,468.05	1,625
3030	3/4" diameter		24	.333		1,450	18.05		1,468.05	1,625
3040	1" diameter		19	.421		1,625	23		1,648	1,800
3060	1-1/4" diameter		15	.533		1,800	29		1,829	2,025
3080	1-1/2" diameter		13	.615		2,050	33.50		2,083.50	2,300
3100	2" diameter	▼	11	.727		2,500	39.50		2,539.50	2,800
3120	2-1/2" diameter	Q-5	12	1.333		3,125	65		3,190	3,550
3140	3" diameter	"	11	1.455	▼	3,575	71		3,646	4,050
3500	Flanged connection, iron body, 125 lb. W.S.P.									
3520	3" diameter	Q-5	11	1.455	Ea.	3,925	71		3,996	4,425
3540	4" diameter	"	5	3.200	"	4,950	156		5,106	5,675
9002	For water pressure regulators, see Section 22 05 23.20									

22 11 19.30 Pressure and Temperature Safety Plug

		Crew	Daily Output	Labor-Hours	Unit	Material	2011 Bare Costs Labor	Equipment	Total	Total Incl O&P
0010	**PRESSURE & TEMPERATURE SAFETY PLUG**									
1000	3/4" external thread, 3/8" diam. element									
1020	Carbon steel									
1050	7-1/2" insertion	1 Stpi	32	.250	Ea.	87	13.55		100.55	116
1120	304 stainless steel									
1150	7-1/2" insertion	1 Stpi	32	.250	Ea.	81.50	13.55		95.05	110
1220	316 Stainless steel									
1250	7-1/2" insertion	1 Stpi	32	.250	Ea.	85.50	13.55		99.05	114

22 11 19.34 Sleeves and Escutcheons

		Crew	Daily Output	Labor-Hours	Unit	Material	2011 Bare Costs Labor	Equipment	Total	Total Incl O&P
0010	**SLEEVES & ESCUTCHEONS**									
0100	Pipe sleeve									
0110	Steel, w/water stop, 12" long, with link seal									
0120	2" diam. for 1/2" carrier pipe	1 Plum	8.40	.952	Ea.	45.50	51		96.50	126
0130	2-1/2" diam. for 3/4" carrier pipe		8	1		52	53.50		105.50	137
0140	2-1/2" diam. for 1" carrier pipe		8	1		49	53.50		102.50	134
0150	3" diam. for 1-1/4" carrier pipe		7.20	1.111		63.50	59.50		123	159
0160	3-1/2" diam. for 1-1/2" carrier pipe		6.80	1.176		58.50	63		121.50	159
0170	4" diam. for 2" carrier pipe		6	1.333		68.50	71.50		140	183
0180	4" diam. for 2-1/2" carrier pipe		6	1.333		69.50	71.50		141	184
0190	5" diam. for 3" carrier pipe		5.40	1.481		81.50	79.50		161	209
0200	6" diam. for 4" carrier pipe	▼	4.80	1.667		92.50	89.50		182	235
0210	10" diam. for 6" carrier pipe	Q-1	8	2		159	96.50		255.50	320
0220	12" diam. for 8" carrier pipe		7.20	2.222		222	107		329	405
0230	14" diam. for 10" carrier pipe		6.40	2.500		255	121		376	460
0240	16" diam. for 12" carrier pipe		5.80	2.759		292	133		425	520
0250	18" diam. for 14" carrier pipe	▼	5.20	3.077	▼	430	148		578	690

22 11 Facility Water Distribution

22 11 19 – Domestic Water Piping Specialties

22 11 19.34 Sleeves and Escutcheons		Crew	Daily Output	Labor-Hours	Unit	Material	2011 Bare Costs Labor	Equipment	Total	Total Incl O&P
0260	24" diam. for 18" carrier pipe	Q-1	4	4	Ea.	670	193		863	1,025
0270	24" diam. for 20" carrier pipe		4	4		580	193		773	930
0280	30" diam. for 24" carrier pipe	↓	3.20	5	↓	885	241		1,126	1,325
0500	Wall sleeve									
0510	Ductile iron with rubber gasket seal									
0520	3"	1 Plum	8.40	.952	Ea.	700	51		751	845
0530	4"		7.20	1.111		750	59.50		809.50	915
0540	6"		6	1.333		920	71.50		991.50	1,125
0550	8"		4	2		1,125	107		1,232	1,400
0560	10"		3	2.667		1,375	143		1,518	1,750
0570	12"	↓	2.40	3.333	↓	1,625	179		1,804	2,075
5000	Escutcheon									
5100	Split ring, pipe									
5110	Chrome plated									
5120	1/2"	1 Plum	160	.050	Ea.	1.05	2.68		3.73	5.15
5130	3/4"		160	.050		1.16	2.68		3.84	5.30
5140	1"		135	.059		1.17	3.18		4.35	6.05
5150	1-1/2"		115	.070		1.76	3.73		5.49	7.50
5160	2"		100	.080		1.96	4.29		6.25	8.55
5170	4"		80	.100		4.35	5.35		9.70	12.80
5180	6"	↓	68	.118	↓	7.25	6.30		13.55	17.40
5400	Shallow flange type									
5410	Chrome plated steel									
5420	1/2" CTS	1 Plum	180	.044	Ea.	.32	2.38		2.70	3.91
5430	3/4" CTS		180	.044		.29	2.38		2.67	3.88
5440	1/2" IPS		180	.044		.33	2.38		2.71	3.92
5450	3/4" IPS		180	.044		.28	2.38		2.66	3.87
5460	1" IPS		175	.046		.46	2.45		2.91	4.17
5470	1-1/2" IPS		170	.047		.72	2.52		3.24	4.55
5480	2" IPS	↓	160	.050	↓	.86	2.68		3.54	4.95

22 11 19.38 Water Supply Meters

22 11 19.38 Water Supply Meters		Crew	Daily Output	Labor-Hours	Unit	Material	2011 Bare Costs Labor	Equipment	Total	Total Incl O&P
0010	**WATER SUPPLY METERS**									
1000	Detector, serves dual systems such as fire and domestic or									
1020	process water, wide range cap., UL and FM approved									
1100	3" mainline x 2" by-pass, 400 GPM	Q-1	3.60	4.444	Ea.	6,500	214		6,714	7,475
1140	4" mainline x 2" by-pass, 700 GPM	"	2.50	6.400		6,500	310		6,810	7,600
1180	6" mainline x 3" by-pass, 1600 GPM	Q-2	2.60	9.231		9,950	460		10,410	11,600
1220	8" mainline x 4" by-pass, 2800 GPM		2.10	11.429		14,700	570		15,270	17,100
1260	10" mainline x 6" by-pass, 4400 GPM		2	12		21,100	600		21,700	24,100
1300	10" x 12" mainlines x 6" by-pass, 5400 GPM	↓	1.70	14.118	↓	28,600	705		29,305	32,600
2000	Domestic/commercial, bronze									
2020	Threaded									
2060	5/8" diameter, to 20 GPM	1 Plum	16	.500	Ea.	43.50	27		70.50	88
2080	3/4" diameter, to 30 GPM		14	.571		79.50	30.50		110	133
2100	1" diameter, to 50 GPM	↓	12	.667	↓	121	35.50		156.50	187
2300	Threaded/flanged									
2340	1-1/2" diameter, to 100 GPM	1 Plum	8	1	Ea.	295	53.50		348.50	405
2360	2" diameter, to 160 GPM	"	6	1.333	"	400	71.50		471.50	545
2600	Flanged, compound									
2640	3" diameter, 320 GPM	Q-1	3	5.333	Ea.	2,725	257		2,982	3,375
2660	4" diameter, to 500 GPM		1.50	10.667		4,350	515		4,865	5,575
2680	6" diameter, to 1,000 GPM	↓	1	16	↓	6,950	770		7,720	8,800

22 11 19 – Domestic Water Piping Specialties

22 11 19.38 Water Supply Meters		Crew	Daily Output	Labor-Hours	Unit	Material	Labor	2011 Bare Costs Equipment	Total	Total Incl O&P
2700	8" diameter, to 1,800 GPM	Q-1	.80	20	Ea.	10,900	965		11,865	13,500
7000	Turbine									
7260	Flanged									
7300	2" diameter, to 160 GPM	1 Plum	7	1.143	Ea.	530	61.50		591.50	675
7320	3" diameter, 450 GPM	Q-1	3.60	4.444		1,125	214		1,339	1,550
7340	4" diameter, 650 GPM	"	2.50	6.400		1,825	310		2,135	2,475
7360	6" diameter, 1800 GPM	Q-2	2.60	9.231		3,300	460		3,760	4,325
7380	8" diameter, to 2500 GPM		2.10	11.429		5,225	570		5,795	6,575
7400	10" diameter, to 5500 GPM	↓	1.70	14.118	↓	7,050	705		7,755	8,800

22 11 19.42 Backflow Preventers		Crew	Daily Output	Labor-Hours	Unit	Material	Labor	2011 Bare Costs Equipment	Total	Total Incl O&P
0010	**BACKFLOW PREVENTERS**, Includes valves									
0020	and four test cocks, corrosion resistant, automatic operation									
1000	Double check principle									
1010	Threaded, with ball valves									
1020	3/4" pipe size	1 Plum	16	.500	Ea.	228	27		255	291
1030	1" pipe size		14	.571		252	30.50		282.50	325
1040	1-1/2" pipe size		10	.800		485	43		528	600
1050	2" pipe size	↓	7	1.143	↓	560	61.50		621.50	710
1080	Threaded, with gate valves									
1100	3/4" pipe size	1 Plum	16	.500	Ea.	900	27		927	1,025
1120	1" pipe size		14	.571		910	30.50		940.50	1,050
1140	1-1/2" pipe size		10	.800		1,175	43		1,218	1,350
1160	2" pipe size	↓	7	1.143	↓	1,425	61.50		1,486.50	1,675
1200	Flanged, valves are gate									
1210	3" pipe size	Q-1	4.50	3.556	Ea.	2,050	172		2,222	2,525
1220	4" pipe size	"	3	5.333		2,525	257		2,782	3,150
1230	6" pipe size	Q-2	3	8		3,725	400		4,125	4,700
1240	8" pipe size		2	12		6,775	600		7,375	8,375
1250	10" pipe size	↓	1	24	↓	10,100	1,200		11,300	12,900
1300	Flanged, valves are OS&Y									
1370	1" pipe size	1 Plum	5	1.600	Ea.	990	86		1,076	1,200
1374	1-1/2" pipe size		5	1.600		1,250	86		1,336	1,500
1378	2" pipe size	↓	4.80	1.667		1,525	89.50		1,614.50	1,800
1380	3" pipe size	Q-1	4.50	3.556		4,325	172		4,497	5,000
1400	4" pipe size	"	3	5.333		4,800	257		5,057	5,675
1420	6" pipe size	Q-2	3	8		7,525	400		7,925	8,875
1430	8" pipe size	"	2	12	↓	14,900	600		15,500	17,300
4000	Reduced pressure principle									
4100	Threaded, bronze, valves are ball									
4120	3/4" pipe size	1 Plum	16	.500	Ea.	345	27		372	420
4140	1" pipe size		14	.571		375	30.50		405.50	455
4150	1-1/4" pipe size		12	.667		640	35.50		675.50	760
4160	1-1/2" pipe size	↓	10	.800		700	43		743	840
4180	2" pipe size	↓	7	1.143	↓	615	61.50		676.50	765
5000	Flanged, bronze, valves are OS&Y									
5060	2-1/2" pipe size	Q-1	5	3.200	Ea.	5,600	154		5,754	6,375
5080	3" pipe size		4.50	3.556		6,525	172		6,697	7,425
5100	4" pipe size	↓	3	5.333		7,650	257		7,907	8,800
5120	6" pipe size	Q-2	3	8	↓	12,200	400		12,600	14,000
5200	Flanged, iron, valves are gate									
5210	2-1/2" pipe size	Q-1	5	3.200	Ea.	2,325	154		2,479	2,775
5220	3" pipe size	↓	4.50	3.556	↓	2,475	172		2,647	2,975

22 11 19 – Domestic Water Piping Specialties

22 11 19.42 Backflow Preventers		Crew	Daily Output	Labor-Hours	Unit	Material	2011 Bare Costs Labor	Equipment	Total	Total Incl O&P
5230	4" pipe size	Q-1	3	5.333	Ea.	3,275	257		3,532	3,975
5240	6" pipe size	Q-2	3	8		4,600	400		5,000	5,675
5250	8" pipe size		2	12		8,275	600		8,875	10,000
5260	10" pipe size	↓	1	24	↓	11,700	1,200		12,900	14,600
5600	Flanged, iron, valves are OS&Y									
5660	2-1/2" pipe size	Q-1	5	3.200	Ea.	2,625	154		2,779	3,125
5680	3" pipe size		4.50	3.556		2,775	172		2,947	3,300
5700	4" pipe size	↓	3	5.333		3,475	257		3,732	4,200
5720	6" pipe size	Q-2	3	8		5,025	400		5,425	6,125
5740	8" pipe size		2	12		10,300	600		10,900	12,200
5760	10" pipe size	↓	1	24	↓	11,800	1,200		13,000	14,800

22 11 19.50 Vacuum Breakers

		Crew	Daily Output	Labor-Hours	Unit	Material	2011 Bare Costs Labor	Equipment	Total	Total Incl O&P
0010	**VACUUM BREAKERS**									
0013	See also backflow preventers Section 22 11 19.42									
1000	Anti-siphon continuous pressure type									
1010	Max. 150 PSI - 210°F									
1020	Bronze body									
1030	1/2" size	1 Stpi	24	.333	Ea.	160	18.05		178.05	203
1040	3/4" size		20	.400		160	21.50		181.50	209
1050	1" size		19	.421		165	23		188	216
1060	1-1/4" size		15	.533		325	29		354	405
1070	1-1/2" size		13	.615		400	33.50		433.50	490
1080	2" size	↓	11	.727	↓	415	39.50		454.50	515
1200	Max. 125 PSI with atmospheric vent									
1210	Brass, in-line construction									
1220	1/4" size	1 Stpi	24	.333	Ea.	63	18.05		81.05	96.50
1230	3/8" size	"	24	.333		63	18.05		81.05	96.50
1260	For polished chrome finish, add				↓	13%				
2000	Anti-siphon, non-continuous pressure type									
2010	Hot or cold water 125 PSI - 210°F									
2020	Bronze body									
2030	1/4" size	1 Stpi	24	.333	Ea.	40.50	18.05		58.55	71.50
2040	3/8" size		24	.333		40.50	18.05		58.55	71.50
2050	1/2" size		24	.333		46	18.05		64.05	77.50
2060	3/4" size		20	.400		54	21.50		75.50	92
2070	1" size		19	.421		84.50	23		107.50	127
2080	1-1/4" size		15	.533		148	29		177	206
2090	1-1/2" size		13	.615		174	33.50		207.50	241
2100	2" size		11	.727		271	39.50		310.50	355
2110	2-1/2" size		8	1		780	54		834	935
2120	3" size	↓	6	1.333	↓	1,025	72		1,097	1,250
2150	For polished chrome finish, add					50%				

22 11 19.54 Water Hammer Arresters/Shock Absorbers

		Crew	Daily Output	Labor-Hours	Unit	Material	2011 Bare Costs Labor	Equipment	Total	Total Incl O&P
0010	**WATER HAMMER ARRESTERS/SHOCK ABSORBERS**									
0490	Copper									
0500	3/4" male I.P.S. For 1 to 11 fixtures	1 Plum	12	.667	Ea.	20.50	35.50		56	76
0600	1" male I.P.S. For 12 to 32 fixtures		8	1		41.50	53.50		95	126
0700	1-1/4" male I.P.S. For 33 to 60 fixtures		8	1		44	53.50		97.50	129
0800	1-1/2" male I.P.S. For 61 to 113 fixtures		8	1		63	53.50		116.50	150
0900	2" male I.P.S. For 114 to 154 fixtures		8	1		92	53.50		145.50	181
1000	2-1/2" male I.P.S. For 155 to 330 fixtures	↓	4	2	↓	285	107		392	475
4000	Bellows type									

22 11 Facility Water Distribution

22 11 19 – Domestic Water Piping Specialties

22 11 19.54 Water Hammer Arresters/Shock Absorbers	Crew	Daily Output	Labor-Hours	Unit	Material	2011 Bare Costs Labor	Equipment	Total	Total Incl O&P
4010	3/4" FNPT, to 11 fixture units	1 Plum	10	.800	Ea.	205	43		248	290
4020	1" FNPT, to 32 fixture units		8.80	.909		415	48.50		463.50	530
4030	1" FNPT, to 60 fixture units		8.80	.909		620	48.50		668.50	760
4040	1" FNPT, to 113 fixture units		8.80	.909		1,550	48.50		1,598.50	1,800
4050	1" FNPT, to 154 fixture units		6.60	1.212		1,750	65		1,815	2,025
4060	1-1/2" FNPT, to 300 fixture units		5	1.600		2,150	86		2,236	2,500

22 11 19.64 Hydrants

		Crew	Daily Output	Labor-Hours	Unit	Material	Labor	Equipment	Total	Total Incl O&P
0010	**HYDRANTS**									
0050	Wall type, moderate climate, bronze, encased									
0200	3/4" IPS connection	1 Plum	16	.500	Ea.	600	27		627	700
0300	1" IPS connection	"	14	.571		690	30.50		720.50	805
0500	Anti-siphon type, 3/4" connection					520			520	570
1000	Non-freeze, bronze, exposed									
1100	3/4" IPS connection, 4" to 9" thick wall	1 Plum	14	.571	Ea.	405	30.50		435.50	490
1120	10" to 14" thick wall		12	.667		520	35.50		555.50	625
1140	15" to 19" thick wall		12	.667		520	35.50		555.50	625
1160	20" to 24" thick wall		10	.800		530	43		573	650
1200	For 1" IPS connection, add					15%	10%			
1240	For 3/4" adapter type vacuum breaker, add				Ea.	51			51	56
1280	For anti-siphon type, add				"	106			106	117
2000	Non-freeze bronze, encased, anti-siphon type									
2100	3/4" IPS connection, 5" to 9" thick wall	1 Plum	14	.571	Ea.	1,025	30.50		1,055.50	1,175
2120	10" to 14" thick wall		12	.667		1,050	35.50		1,085.50	1,225
2140	15" to 19" thick wall		12	.667		1,100	35.50		1,135.50	1,275
2160	20" to 24" thick wall		10	.800		1,150	43		1,193	1,350
2200	For 1" IPS connection, add					10%	10%			
3000	Ground box type, bronze frame, 3/4" IPS connection									
3080	Non-freeze, all bronze, polished face, set flush									
3100	2 feet depth of bury	1 Plum	8	1	Ea.	765	53.50		818.50	925
3120	3 feet depth of bury		8	1		825	53.50		878.50	985
3140	4 feet depth of bury		8	1		885	53.50		938.50	1,050
3160	5 feet depth of bury		7	1.143		945	61.50		1,006.50	1,150
3180	6 feet depth of bury		7	1.143		1,000	61.50		1,061.50	1,200
3200	7 feet depth of bury		6	1.333		1,050	71.50		1,121.50	1,275
3220	8 feet depth of bury		5	1.600		1,125	86		1,211	1,350
3240	9 feet depth of bury		4	2		1,175	107		1,282	1,450
3260	10 feet depth of bury		4	2		1,225	107		1,332	1,500
3400	For 1" IPS connection, add					15%	10%			
3450	For 1-1/4" IPS connection, add					325%	14%			
3500	For 1-1/2" connection, add					370%	18%			
3550	For 2" connection, add					445%	24%			
3600	For tapped drain port in box, add					69.50			69.50	76
4000	Non-freeze, CI body, bronze frame & scoriated cover									
4010	with hose storage									
4100	2 feet depth of bury	1 Plum	7	1.143	Ea.	1,425	61.50		1,486.50	1,650
4120	3 feet depth of bury		7	1.143		1,475	61.50		1,536.50	1,725
4140	4 feet depth of bury		7	1.143		1,525	61.50		1,586.50	1,775
4160	5 feet depth of bury		6.50	1.231		1,550	66		1,616	1,800
4180	6 feet depth of bury		6	1.333		1,575	71.50		1,646.50	1,825
4200	7 feet depth of bury		5.50	1.455		1,625	78		1,703	1,925
4220	8 feet depth of bury		5	1.600		1,700	86		1,786	1,975
4240	9 feet depth of bury		4.50	1.778		1,750	95.50		1,845.50	2,075

22 11 Facility Water Distribution

22 11 19 – Domestic Water Piping Specialties

22 11 19.64 Hydrants

		Crew	Daily Output	Labor-Hours	Unit	Material	2011 Bare Costs Labor	Equipment	Total	Total Incl O&P
4260	10 feet depth of bury	1 Plum	4	2	Ea.	1,800	107		1,907	2,125
4280	For 1" IPS connection, add					310			310	345
4300	For tapped drain port in box, add				↓	69.50			69.50	76
5000	Moderate climate, all bronze, polished face									
5020	and scoriated cover, set flush									
5100	3/4" IPS connection	1 Plum	16	.500	Ea.	530	27		557	625
5120	1" IPS connection	"	14	.571		655	30.50		685.50	765
5200	For tapped drain port in box, add				↓	69.50			69.50	76
6000	Ground post type, all non-freeze, all bronze, aluminum casing									
6010	guard, exposed head, 3/4" IPS connection									
6100	2 feet depth of bury	1 Plum	8	1	Ea.	750	53.50		803.50	905
6120	3 feet depth of bury		8	1		810	53.50		863.50	970
6140	4 feet depth of bury		8	1		875	53.50		928.50	1,050
6160	5 feet depth of bury		7	1.143		935	61.50		996.50	1,125
6180	6 feet depth of bury		7	1.143		1,000	61.50		1,061.50	1,200
6200	7 feet depth of bury		6	1.333		1,075	71.50		1,146.50	1,275
6220	8 feet depth of bury		5	1.600		1,125	86		1,211	1,375
6240	9 feet depth of bury		4	2		1,200	107		1,307	1,475
6260	10 feet depth of bury	↓	4	2	↓	1,275	107		1,382	1,550
6300	For 1" IPS connection, add					40%	10%			
6350	For 1-1/4" IPS connection, add					140%	14%			
6400	For 1-1/2" IPS connection, add					225%	18%			
6450	For 2" IPS connection, add					315%	24%			

22 11 23 – Domestic Water Pumps

22 11 23.10 General Utility Pumps

		Crew	Daily Output	Labor-Hours	Unit	Material	2011 Bare Costs Labor	Equipment	Total	Total Incl O&P
0010	**GENERAL UTILITY PUMPS**									
2000	Single stage									
3000	Double suction,									
3190	75 HP, to 2500 GPM	Q-3	.28	114	Ea.	16,800	5,825		22,625	27,200
3220	100 HP, to 3000 GPM		.26	123		21,200	6,275		27,475	32,800
3240	150 HP, to 4000 GPM	↓	.24	133	↓	32,800	6,800		39,600	46,300
4000	Centrifugal, end suction, mounted on base									
4010	Horizontal mounted, with drip proof motor, rated @ 100' head									
4020	Vertical split case, single stage									
4040	100 GPM, 5 HP, 1-1/2" discharge	Q-1	1.70	9.412	Ea.	3,250	455		3,705	4,250
4050	200 GPM, 10 HP, 2" discharge		1.30	12.308		3,675	595		4,270	4,900
4060	250 GPM, 10 HP, 3" discharge	↓	1.28	12.500		4,125	605		4,730	5,425
4070	300 GPM, 15 HP, 2" discharge	Q-2	1.56	15.385		4,625	770		5,395	6,225
4080	500 GPM, 20 HP, 4" discharge		1.44	16.667		5,200	835		6,035	6,975
4090	750 GPM, 30 HP, 4" discharge		1.20	20		5,250	1,000		6,250	7,275
4100	1050 GPM, 40 HP, 5" discharge		1	24		7,275	1,200		8,475	9,800
4110	1500 GPM, 60 HP, 6" discharge		.60	40		11,200	2,000		13,200	15,400
4120	2000 GPM, 75 HP, 6" discharge		.50	48		11,200	2,400		13,600	16,000
4130	3000 GPM, 100 HP, 8" discharge	↓	.40	60	↓	16,100	3,000		19,100	22,200
4200	Horizontal split case, single stage									
4210	100 GPM, 7.5 HP, 1-1/2" discharge	Q-1	1.70	9.412	Ea.	3,100	455		3,555	4,075
4220	250 GPM, 15 HP, 2-1/2" discharge	"	1.30	12.308		7,975	595		8,570	9,650
4230	500 GPM, 20 HP, 4" discharge	Q-2	1.60	15		8,500	750		9,250	10,500
4240	750 GPM, 25 HP, 5" discharge		1.54	15.584		8,775	780		9,555	10,900
4250	1000 GPM, 40 HP, 5" discharge	↓	1.20	20		11,600	1,000		12,600	14,300
4260	1500 GPM, 50 HP, 6" discharge	Q-3	1.42	22.535		17,600	1,150		18,750	21,000
4270	2000 GPM, 75 HP, 8" discharge	↓	1.14	28.070		19,500	1,425		20,925	23,600

22 11 Facility Water Distribution

22 11 23 – Domestic Water Pumps

22 11 23.10 General Utility Pumps	Crew	Daily Output	Labor-Hours	Unit	Material	2011 Bare Costs Labor	Equipment	Total	Total Incl O&P
4280	3000 GPM, 100 HP, 10" discharge	Q-3	.96	33.333	Ea.	21,500	1,700		23,200	26,300
4290	3500 GPM, 150 HP, 10" discharge		.86	37.209		26,300	1,900		28,200	31,800
4300	4000 GPM, 200 HP, 10" discharge	↓	.66	48.485	↓	35,400	2,475		37,875	42,600
4330	Horizontal split case, two stage, 500' head									
4340	100 GPM, 40 HP, 1-1/2" discharge	Q-2	1.70	14.118	Ea.	14,900	705		15,605	17,500
4350	200 GPM, 50 HP, 1-1/2" discharge	"	1.44	16.667		15,600	835		16,435	18,500
4360	300 GPM, 75 HP, 2" discharge	Q-3	1.57	20.382		21,500	1,050		22,550	25,200
4370	400 GPM, 100 HP, 3" discharge		1.14	28.070		25,400	1,425		26,825	30,100
4380	800 GPM, 200 HP, 4" discharge	↓	.86	37.209	↓	34,000	1,900		35,900	40,200
5000	Centrifugal, in-line									
5006	Vertical mount, iron body, 125 lb. flgd, 3550 RPM TEFC mtr									
5010	Single stage									
5020	50 GPM, 3 HP, 1-1/2" discharge	Q-1	2.30	6.957	Ea.	1,075	335		1,410	1,675
5030	75 GPM, 5 HP, 1-1/2" discharge		1.60	10		1,150	485		1,635	2,000
5040	100 GPM, 7.5 HP, 1-1/2" discharge	↓	1.30	12.308		1,425	595		2,020	2,450
5050	125 GPM, 10 HP, 1-1/2" discharge	Q-2	1.70	14.118		1,650	705		2,355	2,850
5060	150 GPM, 15 HP, 1-1/2" discharge		1.60	15		2,775	750		3,525	4,175
5080	250 GPM, 40 HP, 2" discharge		1.40	17.143		4,150	860		5,010	5,825
5090	300 GPM, 50 HP, 3" discharge	↓	1.30	18.462		4,425	925		5,350	6,250
5100	400 GPM, 75 HP, 3" discharge	Q-3	.60	53.333		5,625	2,725		8,350	10,300
5110	600 GPM, 100 HP, 3" discharge	"	.50	64	↓	5,900	3,275		9,175	11,400

22 11 23.11 Miscellaneous Pumps

		Crew	Daily Output	Labor-Hours	Unit	Material	Labor	Equipment	Total	Total Incl O&P
0010	**MISCELLANEOUS PUMPS**									
0020	Water pump, portable, gasoline powered									
0100	170 GPH, 2" discharge	Q-1	11	1.455	Ea.	715	70		785	890
0110	343 GPH, 3" discharge		10.50	1.524		980	73.50		1,053.50	1,175
0120	608 GPH, 4" discharge	↓	10	1.600	↓	1,375	77		1,452	1,625
0500	Pump, propylene body, housing and impeller									
0510	22 GPM, 1/3 HP, 40' HD	1 Plum	5	1.600	Ea.	510	86		596	690
0520	33 GPM, 1/2 HP, 40' HD	Q-1	5	3.200		550	154		704	835
0530	53 GPM, 3/4 HP, 40' HD	"	4	4	↓	900	193		1,093	1,275
0600	Rotary pump, CI									
0614	202 GPH, 3/4 HP, 1" discharge	Q-1	4.50	3.556	Ea.	960	172		1,132	1,300
0618	277 GPH, 1 HP, 1" discharge		4	4		960	193		1,153	1,350
0624	1100 GPH, 1.5 HP, 1-1/4" discharge	↓	3.60	4.444		2,400	214		2,614	2,950
0628	1900 GPH, 2 HP, 1-1/4" discharge	↓	3.20	5	↓	3,300	241		3,541	3,975
1000	Turbine pump, CI									
1010	50 GPM, 2 HP, 3" discharge	Q-1	.80	20	Ea.	3,325	965		4,290	5,100
1020	100 GPM, 3 HP, 4" discharge	Q-2	.96	25		5,225	1,250		6,475	7,625
1030	250 GPM, 15 HP, 6" discharge	"	.94	25.532		6,475	1,275		7,750	9,025
1040	500 GPM, 25 HP, 6" discharge	Q-3	1.22	26.230		6,525	1,350		7,875	9,175
1050	1000 GPM, 50 HP, 8" discharge		1.14	28.070		9,000	1,425		10,425	12,100
1060	2000 GPM, 100 HP, 10" discharge		1	32		10,700	1,625		12,325	14,200
1070	3000 GPM, 150 HP, 10" discharge		.80	40		17,200	2,050		19,250	22,000
1080	4000 GPM, 200 HP, 12" discharge		.70	45.714		18,800	2,325		21,125	24,200
1090	6000 GPM, 300 HP, 14" discharge		.60	53.333		28,600	2,725		31,325	35,600
1100	10,000 GPM, 300 HP, 18" discharge	↓	.58	55.172	↓	33,100	2,825		35,925	40,600
2000	Centrifugal stainless steel pumps									
2100	100 GPM, 100' TDH, 3 HP	Q-1	1.80	8.889	Ea.	9,450	430		9,880	11,000
2130	250 GPM, 100' TDH, 10 HP	"	1.28	12.500		13,200	605		13,805	15,400
2160	500 GPM, 100' TDH, 20 HP	Q-2	1.44	16.667	↓	15,100	835		15,935	17,900
2200	Vertical turbine stainless steel pumps									

234

22 11 Facility Water Distribution

22 11 23 – Domestic Water Pumps

22 11 23.11 Miscellaneous Pumps

22 11 23.11 Miscellaneous Pumps		Crew	Daily Output	Labor-Hours	Unit	Material	2011 Bare Costs Labor	2011 Bare Costs Equipment	Total	Total Incl O&P
2220	100 GPM, 100' TDH, 7.5 HP	Q-2	.95	25.263	Ea.	53,000	1,275		54,275	60,000
2240	250 GPM, 100' TDH, 15 HP		.94	25.532		58,500	1,275		59,775	66,500
2260	500 GPM, 100' TDH, 20 HP	↓	.93	25.806		64,000	1,300		65,300	72,500
2280	750 GPM, 100' TDH, 30 HP	Q-3	1.19	26.891		70,000	1,375		71,375	79,000
2300	1000 GPM, 100' TDH, 40 HP	"	1.16	27.586		75,500	1,400		76,900	85,000
2320	100 GPM, 200' TDH, 15 HP	Q-2	.94	25.532		81,000	1,275		82,275	91,500
2340	250 GPM, 200' TDH, 25 HP	Q-3	1.22	26.230		87,000	1,350		88,350	97,500
2360	500 GPM, 200' TDH, 40 HP		1.16	27.586		92,500	1,400		93,900	104,000
2380	750 GPM, 200' TDH, 60 HP		1.13	28.319		93,000	1,450		94,450	104,500
2400	1000 GPM, 200' TDH, 75 HP	↓	1.12	28.571	↓	96,000	1,450		97,450	108,000

22 11 23.13 Domestic-Water Packaged Booster Pumps

		Crew	Daily Output	Labor-Hours	Unit	Material	2011 Bare Costs Labor	2011 Bare Costs Equipment	Total	Total Incl O&P
0010	**DOMESTIC-WATER PACKAGED BOOSTER PUMPS**									
0200	Pump system, with diaphragm tank, control, press. switch									
0300	1 HP pump	Q-1	1.30	12.308	Ea.	5,800	595		6,395	7,250
0400	1-1/2 HP pump		1.25	12.800		5,850	620		6,470	7,375
0420	2 HP pump		1.20	13.333		6,000	645		6,645	7,550
0440	3 HP pump	↓	1.10	14.545		6,100	700		6,800	7,750
0460	5 HP pump	Q-2	1.50	16		6,750	800		7,550	8,625
0480	7-1/2 HP pump		1.42	16.901		7,500	845		8,345	9,500
0500	10 HP pump	↓	1.34	17.910	↓	7,850	895		8,745	10,000
2000	Pump system, variable speed, base, controls, starter									
2010	Duplex, 100' head									
2020	400 GPM, 7-1/2HP, 4" discharge	Q-2	.70	34.286	Ea.	30,000	1,725		31,725	35,600
2025	Triplex, 100' head									
2030	1000 GPM, 15HP, 6" discharge	Q-2	.50	48	Ea.	47,500	2,400		49,900	56,000
2040	1700 GPM, 30HP, 6" discharge	"	.30	80	"	58,000	4,000		62,000	70,000

22 12 Facility Potable-Water Storage Tanks

22 12 23 – Facility Indoor Potable-Water Storage Tanks

22 12 23.13 Facility Steel, Indoor Pot.-Water Storage Tanks

		Crew	Daily Output	Labor-Hours	Unit	Material	2011 Bare Costs Labor	2011 Bare Costs Equipment	Total	Total Incl O&P
0010	**FACILITY STEEL, INDOOR POT.-WATER STORAGE TANKS**									
2000	Galvanized steel, 15 gal., 14" diam., 26" LOA	1 Plum	12	.667	Ea.	1,250	35.50		1,285.50	1,425
2060	30 gal., 14" diam. x 49" LOA		11	.727		1,475	39		1,514	1,675
2080	80 gal., 20" diam. x 64" LOA		9	.889		2,300	47.50		2,347.50	2,600
2100	135 gal., 24" diam. x 75" LOA		6	1.333		3,350	71.50		3,421.50	3,775
2120	240 gal., 30" diam. x 86" LOA		4	2		6,550	107		6,657	7,350
2140	300 gal., 36" diam. x 76" LOA	↓	3	2.667		9,300	143		9,443	10,400
2160	400 gal., 36" diam. x 100" LOA	Q-1	4	4		11,400	193		11,593	12,900
2180	500 gal., 36" diam., x 126" LOA	"	3	5.333		14,100	257		14,357	16,000
3000	Glass lined, P.E., 80 gal., 20" diam. x 60" LOA	1 Plum	9	.889		2,475	47.50		2,522.50	2,800
3060	140 gal., 24" diam. x 80" LOA		6	1.333		3,425	71.50		3,496.50	3,875
3080	225 gal., 30" diam. x 78" LOA		4	2		4,675	107		4,782	5,300
3100	325 gal., 36" diam. x 81" LOA	↓	3	2.667		5,975	143		6,118	6,800
3120	460 gal., 42" diam. x 84" LOA	Q-1	4	4		6,300	193		6,493	7,225
3140	605 gal., 48" diam. x 87" LOA		3	5.333		12,000	257		12,257	13,600
3160	740 gal., 54" diam. x 91" LOA		3	5.333		13,500	257		13,757	15,300
3180	940 gal., 60" diam. x 93" LOA		2.50	6.400		15,600	310		15,910	17,700
3200	1330 gal., 66" diam. x 107" LOA		2	8		21,400	385		21,785	24,100
3220	1615 gal., 72" diam. x 110" LOA		1.50	10.667		24,300	515		24,815	27,600
3240	2285 gal., 84" diam. x 128" LOA	↓	1	16		30,100	770		30,870	34,300
3260	3440 gal., 96" diam. x 157" LOA	Q-2	1.50	16	↓	44,600	800		45,400	50,000

22 13 16.40 Pipe Fittings, Cast Iron for Drainage	Crew	Daily Output	Labor-Hours	Unit	Material	2011 Bare Costs Labor	2011 Bare Costs Equipment	Total	Total Incl O&P
0010 **PIPE FITTINGS, CAST IRON FOR DRAINAGE**, Special									
1000 Drip pan elbow, (safety valve discharge elbow)									
1010 Cast iron, threaded inlet									
1013 2"	Q-1	9	1.778	Ea.	98.50	86		184.50	236
1014 2-1/2"		8	2		122	96.50		218.50	278
1015 3"		6.40	2.500		162	121		283	360
1017 4"		4.80	3.333		243	161		404	505
1018 Cast iron, flanged inlet									
1019 6"	Q-2	3.60	6.667	Ea.	490	335		825	1,050
1020 8"	"	2.60	9.231	"	1,200	460		1,660	2,000

22 13 16.60 Traps	Crew	Daily Output	Labor-Hours	Unit	Material	2011 Bare Costs Labor	2011 Bare Costs Equipment	Total	Total Incl O&P
0010 **TRAPS**									
4700 Copper, drainage, drum trap									
4800 3" x 5" solid, 1-1/2" pipe size	1 Plum	16	.500	Ea.	177	27		204	234
4840 3" x 6" swivel, 1-1/2" pipe size	"	16	.500	"	280	27		307	350
5100 P trap, standard pattern									
5200 1-1/4" pipe size	1 Plum	18	.444	Ea.	130	24		154	179
5240 1-1/2" pipe size		17	.471		126	25		151	177
5260 2" pipe size		15	.533		195	28.50		223.50	257
5280 3" pipe size		11	.727		465	39		504	575
5340 With cleanout, swivel joint and slip joint									
5360 1-1/4" pipe size	1 Plum	18	.444	Ea.	92	24		116	137
5400 1-1/2" pipe size		17	.471		192	25		217	249
5420 2" pipe size		15	.533		310	28.50		338.50	385
6710 ABS DWV P trap, solvent weld joint									
6720 1-1/2" pipe size	1 Plum	18	.444	Ea.	5.85	24		29.85	42
6722 2" pipe size		17	.471		7.70	25		32.70	46
6724 3" pipe size		15	.533		30.50	28.50		59	76
6726 4" pipe size		14	.571		66.50	30.50		97	119
6732 PVC DWV P trap, solvent weld joint									
6733 1-1/2" pipe size	1 Plum	18	.444	Ea.	4.60	24		28.60	40.50
6734 2" pipe size		17	.471		6.20	25		31.20	44.50
6735 3" pipe size		15	.533		21	28.50		49.50	65.50
6736 4" pipe size		14	.571		48	30.50		78.50	98.50
6760 PP DWV, dilution trap, 1-1/2" pipe size		16	.500		186	27		213	245
6770 P trap, 1-1/2" pipe size		17	.471		46	25		71	88
6780 2" pipe size		16	.500		68	27		95	115
6790 3" pipe size		14	.571		125	30.50		155.50	184
6800 4" pipe size		13	.615		210	33		243	280
6830 S trap, 1-1/2" pipe size		16	.500		38	27		65	82
6840 2" pipe size		15	.533		57	28.50		85.50	105
6850 Universal trap, 1-1/2" pipe size		14	.571		75	30.50		105.50	128
6860 PVC DWV hub x hub, basin trap, 1-1/4" pipe size		18	.444		7.50	24		31.50	44
6870 Sink P trap, 1-1/2" pipe size		18	.444		7.50	24		31.50	44
6880 Tubular S trap, 1-1/2" pipe size		17	.471		13.90	25		38.90	53
6890 PVC sch. 40 DWV, drum trap									
6900 1-1/2" pipe size	1 Plum	16	.500	Ea.	20.50	27		47.50	63
6910 P trap, 1-1/2" pipe size		18	.444		5.10	24		29.10	41
6920 2" pipe size		17	.471		6.90	25		31.90	45
6930 3" pipe size		15	.533		23.50	28.50		52	68
6940 4" pipe size		14	.571		53.50	30.50		84	104
6950 P trap w/clean out, 1-1/2" pipe size		18	.444		8.45	24		32.45	45

22 13 Facility Sanitary Sewerage

22 13 16 – Sanitary Waste and Vent Piping

22 13 16.60 Traps	Crew	Daily Output	Labor-Hours	Unit	Material	2011 Bare Costs Labor	Equipment	Total	Total Incl O&P
6960	2" pipe size	1 Plum	17	.471	Ea.	14.30	25		39.30	53.50
6970	P trap adjustable, 1-1/2" pipe size		17	.471		6.55	25		31.55	44.50
6980	P trap adj. w/union & cleanout, 1-1/2" pipe size	↓	16	.500	↓	23	27		50	65.50

22 13 16.80 Vent Flashing and Caps

		Crew	Daily Output	Labor-Hours	Unit	Material	2011 Bare Costs Labor	Equipment	Total	Total Incl O&P
0010	**VENT FLASHING AND CAPS**									
0120	Vent caps									
0140	Cast iron									
0180	2-1/2" - 3-5/8" pipe	1 Plum	21	.381	Ea.	40	20.50		60.50	74.50
0190	4" - 4-1/8" pipe	"	19	.421	"	48	22.50		70.50	86.50
0900	Vent flashing									
1000	Aluminum with lead ring									
1020	1-1/4" pipe	1 Plum	20	.400	Ea.	10.70	21.50		32.20	44
1030	1-1/2" pipe		20	.400		10.95	21.50		32.45	44
1040	2" pipe		18	.444		11.45	24		35.45	48
1050	3" pipe		17	.471		12.70	25		37.70	51.50
1060	4" pipe	↓	16	.500	↓	15.30	27		42.30	57
1350	Copper with neoprene ring									
1400	1-1/4" pipe	1 Plum	20	.400	Ea.	19.95	21.50		41.45	54
1430	1-1/2" pipe		20	.400		19.95	21.50		41.45	54
1440	2" pipe		18	.444		21	24		45	58.50
1450	3" pipe		17	.471		24.50	25		49.50	64.50
1460	4" pipe	↓	16	.500	↓	27	27		54	70
2000	Galvanized with neoprene ring									
2020	1-1/4" pipe	1 Plum	20	.400	Ea.	14.20	21.50		35.70	47.50
2030	1-1/2" pipe		20	.400		14.20	21.50		35.70	47.50
2040	2" pipe		18	.444		14.70	24		38.70	51.50
2050	3" pipe		17	.471		15.40	25		40.40	54.50
2060	4" pipe	↓	16	.500	↓	17.90	27		44.90	59.50
2980	Neoprene, one piece									
3000	1-1/4" pipe	1 Plum	24	.333	Ea.	7.25	17.85		25.10	34.50
3030	1-1/2" pipe		24	.333		7.95	17.85		25.80	35.50
3040	2" pipe		23	.348		7.30	18.65		25.95	36
3050	3" pipe		21	.381		8.40	20.50		28.90	39.50
3060	4" pipe	↓	20	.400	↓	12.45	21.50		33.95	45.50
4000	Lead, 4#, 8" skirt, VTR									
4100	2" pipe	1 Plum	18	.444	Ea.	38	24		62	77
4110	3" pipe		17	.471		45	25		70	87
4120	4" pipe		16	.500		51.50	27		78.50	96.50
4130	6" pipe	↓	14	.571	↓	73.50	30.50		104	127

22 13 26 – Sanitary Waste Separators

22 13 26.10 Separators

		Crew	Daily Output	Labor-Hours	Unit	Material	2011 Bare Costs Labor	Equipment	Total	Total Incl O&P
0010	**SEPARATORS**, Entrainment eliminator, steel body, 150 PSIG.									
0100	1/4" size	1 Stpi	24	.333	Ea.	253	18.05		271.05	305
0120	1/2" size		24	.333		262	18.05		280.05	315
0140	3/4" size		20	.400		274	21.50		295.50	335
0160	1" size		19	.421		282	23		305	345
0180	1-1/4" size		15	.533		300	29		329	375
0200	1-1/2" size		13	.615		330	33.50		363.50	415
0220	2" size	↓	11	.727		370	39.50		409.50	465
0240	2-1/2" size	Q-5	15	1.067		1,825	52		1,877	2,075
0260	3" size		13	1.231		2,125	60		2,185	2,425
0280	4" size	↓	10	1.600	↓	2,350	78		2,428	2,700

22 13 Facility Sanitary Sewerage

22 13 26 – Sanitary Waste Separators

22 13 26.10 Separators	Crew	Daily Output	Labor-Hours	Unit	Material	2011 Bare Costs Labor	Equipment	Total	Total Incl O&P
0300	5" size	Q-5	6	2.667	Ea.	2,675	130		2,805	3,150
0320	6" size	↓	3	5.333		2,950	260		3,210	3,650
0340	8" size	Q-6	4.40	5.455		3,625	275		3,900	4,375
0360	10" size	"	4	6		5,825	305		6,130	6,850
1000	For 300 PSIG, add				↓	15%				

22 13 29 – Sanitary Sewerage Pumps

22 13 29.13 Wet-Pit-Mounted, Vertical Sewerage Pumps

		Crew	Daily Output	Labor-Hours	Unit	Material	2011 Bare Costs Labor	Equipment	Total	Total Incl O&P
0010	**WET-PIT-MOUNTED, VERTICAL SEWERAGE PUMPS**									
0020	Controls incl. alarm/disconnect panel w/wire. Excavation not included									
0260	Simplex, 9 GPM at 60 PSIG, 91 gal. tank				Ea.	2,975			2,975	3,275
0300	Unit with manway, 26" I.D., 18" high					3,375			3,375	3,700
0340	26" I.D., 36" high					3,500			3,500	3,850
0380	43" I.D., 4' high					3,575			3,575	3,925
0600	Simplex, 9 GPM at 60 PSIG, 150 gal. tank, indoor					3,150			3,150	3,475
0700	Unit with manway, 26" I.D., 36" high					3,925			3,925	4,325
0740	26" I.D., 4' high					4,000			4,000	4,400
2000	Duplex, 18 GPM at 60 PSIG, 150 gal. tank, indoor					6,125			6,125	6,725
2060	Unit with manway, 43" I.D., 4' high					7,350			7,350	8,100
2400	For core only				↓	1,600			1,600	1,775
3000	Indoor residential type installation									
3020	Simplex, 9 GPM at 60 PSIG, 91 gal. HDPE tank				Ea.	2,950			2,950	3,225

22 13 29.14 Sewage Ejector Pumps

		Crew	Daily Output	Labor-Hours	Unit	Material	2011 Bare Costs Labor	Equipment	Total	Total Incl O&P
0010	**SEWAGE EJECTOR PUMPS**, With operating and level controls									
0100	Simplex system incl. tank, cover, pump 15' head									
0500	37 gal. PE tank, 12 GPM, 1/2 HP, 2" discharge	Q-1	3.20	5	Ea.	440	241		681	845
0510	3" discharge		3.10	5.161		480	249		729	895
0530	87 GPM, .7 HP, 2" discharge		3.20	5		670	241		911	1,100
0540	3" discharge		3.10	5.161		725	249		974	1,175
0600	45 gal. coated stl tank, 12 GPM, 1/2 HP, 2" discharge		3	5.333		780	257		1,037	1,250
0610	3" discharge		2.90	5.517		810	266		1,076	1,275
0630	87 GPM, .7 HP, 2" discharge		3	5.333		1,000	257		1,257	1,475
0640	3" discharge		2.90	5.517		1,050	266		1,316	1,550
0660	134 GPM, 1 HP, 2" discharge		2.80	5.714		1,075	276		1,351	1,600
0680	3" discharge		2.70	5.926		1,150	286		1,436	1,675
0700	70 gal. PE tank, 12 GPM, 1/2 HP, 2" discharge		2.60	6.154		855	297		1,152	1,375
0710	3" discharge		2.40	6.667		910	320		1,230	1,475
0730	87 GPM, 0.7 HP, 2" discharge		2.50	6.400		1,100	310		1,410	1,650
0740	3" discharge		2.30	6.957		1,150	335		1,485	1,775
0760	134 GPM, 1 HP, 2" discharge		2.20	7.273		1,200	350		1,550	1,825
0770	3" discharge		2	8		1,275	385		1,660	1,975
0800	75 gal. coated stl. tank, 12 GPM, 1/2 HP, 2" discharge		2.40	6.667		935	320		1,255	1,500
0810	3" discharge		2.20	7.273		970	350		1,320	1,600
0830	87 GPM, .7 HP, 2" discharge		2.30	6.957		1,175	335		1,510	1,800
0840	3" discharge		2.10	7.619		1,225	370		1,595	1,900
0860	134 GPM, 1 HP, 2" discharge		2	8		1,250	385		1,635	1,950
0880	3" discharge	↓	1.80	8.889	↓	1,325	430		1,755	2,100
1040	Duplex system incl. tank, covers, pumps									
1060	110 gal. fiberglass tank, 24 GPM, 1/2 HP, 2" discharge	Q-1	1.60	10	Ea.	1,700	485		2,185	2,600
1080	3" discharge		1.40	11.429		1,800	550		2,350	2,800
1100	174 GPM, .7 HP, 2" discharge		1.50	10.667		2,200	515		2,715	3,175
1120	3" discharge		1.30	12.308		2,275	595		2,870	3,375
1140	268 GPM, 1 HP, 2" discharge	↓	1.20	13.333	↓	2,375	645		3,020	3,575

22 13 Facility Sanitary Sewerage

22 13 29 – Sanitary Sewerage Pumps

22 13 29.14 Sewage Ejector Pumps		Crew	Daily Output	Labor-Hours	Unit	Material	Labor	2011 Bare Costs Equipment	Total	Total Incl O&P
1160	3" discharge	Q-1	1	16	Ea.	2,475	770		3,245	3,850
1260	135 gal. coated stl. tank, 24 GPM, 1/2 HP, 2" discharge	Q-2	1.70	14.118		1,750	705		2,455	2,975
2000	3" discharge		1.60	15		1,850	750		2,600	3,150
2640	174 GPM, .7 HP, 2" discharge		1.60	15		2,275	750		3,025	3,625
2660	3" discharge		1.50	16		2,400	800		3,200	3,825
2700	268 GPM, 1 HP, 2" discharge		1.30	18.462		2,475	925		3,400	4,100
3040	3" discharge		1.10	21.818		2,625	1,100		3,725	4,525
3060	275 gal. coated stl. tank, 24 GPM, 1/2 HP, 2" discharge		1.50	16		2,175	800		2,975	3,600
3080	3" discharge		1.40	17.143		2,200	860		3,060	3,700
3100	174 GPM, .7 HP, 2" discharge		1.40	17.143		2,800	860		3,660	4,350
3120	3" discharge		1.30	18.462		2,975	925		3,900	4,650
3140	268 GPM, 1 HP, 2" discharge		1.10	21.818		3,075	1,100		4,175	5,000
3160	3" discharge		.90	26.667		3,250	1,325		4,575	5,575
3260	Pump system accessories, add									
3300	Alarm horn and lights, 115 V mercury switch	Q-1	8	2	Ea.	93	96.50		189.50	246
3340	Switch, mag. contactor, alarm bell, light, 3 level control		5	3.200		455	154		609	735
3380	Alternator, mercury switch activated		4	4		825	193		1,018	1,200

22 14 Facility Storm Drainage

22 14 29 – Sump Pumps

22 14 29.13 Wet-Pit-Mounted, Vertical Sump Pumps

		Crew	Daily Output	Labor-Hours	Unit	Material	Labor	2011 Bare Costs Equipment	Total	Total Incl O&P
0010	**WET-PIT-MOUNTED, VERTICAL SUMP PUMPS**									
0400	Molded PVC base, 21 GPM at 15' head, 1/3 HP	1 Plum	5	1.600	Ea.	119	86		205	259
0800	Iron base, 21 GPM at 15' head, 1/3 HP		5	1.600		150	86		236	293
1200	Solid brass, 21 GPM at 15' head, 1/3 HP		5	1.600		263	86		349	420

22 14 29.16 Submersible Sump Pumps

		Crew	Daily Output	Labor-Hours	Unit	Material	Labor	2011 Bare Costs Equipment	Total	Total Incl O&P
0010	**SUBMERSIBLE SUMP PUMPS**									
7000	Sump pump, automatic									
7100	Plastic, 1-1/4" discharge, 1/4 HP	1 Plum	6	1.333	Ea.	148	71.50		219.50	269
7140	1/3 HP		5	1.600		181	86		267	325
7160	1/2 HP		5	1.600		221	86		307	370
7180	1-1/2" discharge, 1/2 HP		4	2		246	107		353	430
7500	Cast iron, 1-1/4" discharge, 1/4 HP		6	1.333		170	71.50		241.50	294
7540	1/3 HP		6	1.333		200	71.50		271.50	325
7560	1/2 HP		5	1.600		242	86		328	395

22 15 General Service Compressed-Air Systems

22 15 13 – General Service Compressed-Air Piping

22 15 13.10 Compressor Accessories

		Crew	Daily Output	Labor-Hours	Unit	Material	Labor	2011 Bare Costs Equipment	Total	Total Incl O&P
0010	**COMPRESSOR ACCESSORIES**									
1700	Refrigerated air dryers with ambient air filters									
1710	10 CFM	Q-5	8	2	Ea.	675	97.50		772.50	890
1720	25 CFM		6.60	2.424		920	118		1,038	1,200
1730	50 CFM		6.20	2.581		1,300	126		1,426	1,625
1740	75 CFM		5.80	2.759		2,375	134		2,509	2,825
1750	100 CFM		5.60	2.857		2,700	139		2,839	3,175
4000	Couplers, air line, sleeve type									
4010	Female, connection size NPT									
4020	1/4"	1 Stpi	38	.211	Ea.	8.15	11.40		19.55	26

22 15 General Service Compressed-Air Systems

22 15 13 – General Service Compressed-Air Piping

22 15 13.10 Compressor Accessories		Crew	Daily Output	Labor-Hours	Unit	Material	2011 Bare Costs Labor	Equipment	Total	Total Incl O&P
4030	3/8"	1 Stpi	36	.222	Ea.	13.65	12		25.65	33
4040	1/2"		35	.229		19.85	12.35		32.20	40.50
4050	3/4"	↓	34	.235	↓	19.85	12.75		32.60	41
4100	Male									
4110	1/4"	1 Stpi	38	.211	Ea.	8.50	11.40		19.90	26.50
4120	3/8"		36	.222		9.35	12		21.35	28
4130	1/2"		35	.229		19.85	12.35		32.20	40.50
4140	3/4"	↓	34	.235		19.85	12.75		32.60	41
4150	Coupler, combined male and female halves									
4160	1/2"	1 Stpi	17	.471	Ea.	39.50	25.50		65	81.50
4170	3/4"	"	15	.533	"	39.50	29		68.50	86.50

22 15 19 – General Service Packaged Air Compressors and Receivers

22 15 19.10 Air Compressors

		Crew	Daily Output	Labor-Hours	Unit	Material	2011 Bare Costs Labor	Equipment	Total	Total Incl O&P
0010	**AIR COMPRESSORS**									
5250	Air, reciprocating air cooled, splash lubricated, tank mounted									
5300	Single stage, 1 phase, 140 psi									
5303	1/2 HP, 30 gal. tank	1 Stpi	3	2.667	Ea.	1,600	144		1,744	2,000
5305	3/4 HP, 30 gal. tank		2.60	3.077		1,650	166		1,816	2,075
5307	1 HP, 30 gal. tank	↓	2.20	3.636		2,000	197		2,197	2,500
5309	2 HP, 30 gal. tank	Q-5	4	4		2,475	195		2,670	3,025
5310	3 HP, 30 gal. tank		3.60	4.444		2,600	216		2,816	3,200
5314	3 HP, 60 gal. tank		3.50	4.571		3,300	223		3,523	3,950
5320	5 HP, 60 gal. tank		3.20	5		3,475	244		3,719	4,200
5330	5 HP, 80 gal. tank		3	5.333		3,725	260		3,985	4,500
5340	7.5 HP, 80 gal. tank	↓	2.60	6.154	↓	4,875	300		5,175	5,825
5600	2 stage pkg., 3 phase									
5650	6 CFM at 125 psi 1-1/2 HP, 60 gal. tank	Q-5	3	5.333	Ea.	2,950	260		3,210	3,650
5670	10.9 CFM at 125 psi, 3 HP, 80 gal. tank		1.50	10.667		3,500	520		4,020	4,625
5680	38.7 CFM at 125 psi, 10 HP, 120 gal. tank	↓	.60	26.667		5,925	1,300		7,225	8,475
5690	105 CFM at 125 psi, 25 HP, 250 gal. tank	Q-6	.60	40	↓	12,300	2,025		14,325	16,500
5800	With single stage pump									
5850	8.3 CFM at 125 psi, 2 HP, 80 gal. tank	Q-6	3.50	6.857	Ea.	3,425	345		3,770	4,300
5860	38.7 CFM at 125 psi, 10 HP, 120 gal. tank	"	.90	26.667	"	5,925	1,350		7,275	8,525
6000	Reciprocating, 2 stage, tank mtd, 3 Ph., Cap. rated @175 PSIG									
6050	Pressure lubricated, hvy duty, 9.7 CFM, 3 HP, 120 gal. tank	Q-5	1.30	12.308	Ea.	5,225	600		5,825	6,650
6054	5 CFM, 1-1/2 HP, 80 gal. tank		2.80	5.714		3,150	278		3,428	3,900
6056	6.4 CFM, 2 HP, 80 gal. tank		2	8		3,425	390		3,815	4,350
6058	8.1 CFM, 3 HP, 80 gal. tank		1.70	9.412		3,500	460		3,960	4,525
6059	14.8 CFM, 5 HP, 80 gal. tank		1	16		3,725	780		4,505	5,275
6060	16.5 CFM, 5 HP, 120 gal. tank		1	16		5,325	780		6,105	7,050
6063	13 CFM, 6 HP, 80 gal. tank		.90	17.778		5,000	865		5,865	6,775
6066	19.8 CFM, 7.5 HP, 80 gal. tank		.80	20		5,425	975		6,400	7,400
6070	25.8 CFM, 7-1/2 HP, 120 gal. tank		.80	20		7,800	975		8,775	10,000
6078	34.8 CFM, 10 HP, 80 gal. tank		.70	22.857		6,975	1,125		8,100	9,325
6080	34.8 CFM, 10 HP, 120 gal. tank	↓	.60	26.667		7,375	1,300		8,675	10,100
6090	53.7 CFM, 15 HP, 120 gal. tank	Q-6	.80	30		8,900	1,525		10,425	12,100
6100	76.7 CFM, 20 HP, 120 gal. tank		.70	34.286		11,700	1,725		13,425	15,500
6104	76.7 CFM, 20 HP, 240 gal. tank		.68	35.294		12,800	1,775		14,575	16,800
6110	90.1 CFM, 25 HP, 120 gal. tank		.63	38.095		12,200	1,925		14,125	16,400
6120	101 CFM, 30 HP, 120 gal. tank		.57	42.105		13,100	2,125		15,225	17,700
6130	101 CFM, 30 HP, 250 gal. tank	↓	.52	46.154		14,300	2,325		16,625	19,200
6200	Oil-less, 13.6 CFM, 5 HP, 120 gal. tank	Q-5	.88	18.182	↓	14,600	885		15,485	17,400

22 15 General Service Compressed-Air Systems

22 15 19 – General Service Packaged Air Compressors and Receivers

22 15 19.10 Air Compressors

	22 15 19.10 Air Compressors	Crew	Daily Output	Labor-Hours	Unit	Material	2011 Bare Costs Labor	Equipment	Total	Total Incl O&P
6210	13.6 CFM, 5 HP, 250 gal. tank	Q-5	.80	20	Ea.	16,000	975		16,975	19,100
6220	18.2 CFM, 7.5 HP, 120 gal. tank		.73	21.918		15,100	1,075		16,175	18,200
6230	18.2 CFM, 7.5 HP, 250 gal. tank		.67	23.881		16,200	1,175		17,375	19,500
6250	30.5 CFM, 10 HP, 120 gal. tank		.57	28.070		17,600	1,375		18,975	21,400
6260	30.5 CFM, 10 HP, 250 gal. tank		.53	30.189		18,700	1,475		20,175	22,800
6270	41.3 CFM, 15 HP, 120 gal. tank	Q-6	.70	34.286		19,100	1,725		20,825	23,600
6280	41.3 CFM, 15 HP, 250 gal. tank	"	.67	35.821		20,200	1,800		22,000	24,900

22 31 Domestic Water Softeners

22 31 13 – Residential Domestic Water Softeners

22 31 13.10 Residential Water Softeners

		Crew	Daily Output	Labor-Hours	Unit	Material	2011 Bare Costs Labor	Equipment	Total	Total Incl O&P
0010	**RESIDENTIAL WATER SOFTENERS**									
7350	Water softener, automatic, to 30 grains per gallon	2 Plum	5	3.200	Ea.	380	172		552	675
7400	To 100 grains per gallon	"	4	4	"	750	214		964	1,150

22 35 Domestic Water Heat Exchangers

22 35 30 – Water Heating by Steam

22 35 30.10 Water Heating Transfer Package

		Crew	Daily Output	Labor-Hours	Unit	Material	2011 Bare Costs Labor	Equipment	Total	Total Incl O&P
0010	**WATER HEATING TRANSFER PACKAGE**, Complete controls,									
0020	expansion tank, converter, air separator									
1000	Hot water, 180°F enter, 200°F leaving, 15# steam									
1010	One pump system, 28 GPM	Q-6	.75	32	Ea.	16,700	1,625		18,325	20,700
1020	35 GPM		.70	34.286		18,100	1,725		19,825	22,500
1040	55 GPM		.65	36.923		21,300	1,875		23,175	26,300
1060	130 GPM		.55	43.636		27,000	2,200		29,200	33,000
1080	255 GPM		.40	60		35,300	3,025		38,325	43,400
1100	550 GPM		.30	80		44,800	4,050		48,850	55,000
1120	800 GPM		.25	96		57,000	4,850		61,850	69,500
1220	Two pump system, 28 GPM		.70	34.286		22,600	1,725		24,325	27,400
1240	35 GPM		.65	36.923		26,300	1,875		28,175	31,700
1260	55 GPM		.60	40		28,300	2,025		30,325	34,200
1280	130 GPM		.50	48		37,200	2,425		39,625	44,500
1300	255 GPM		.35	68.571		47,600	3,475		51,075	57,500
1320	550 GPM		.25	96		59,500	4,850		64,350	72,500
1340	800 GPM		.20	120		77,000	6,050		83,050	93,500

22 41 Residential Plumbing Fixtures

22 41 39 – Residential Faucets, Supplies and Trim

22 41 39.10 Faucets and Fittings

		Crew	Daily Output	Labor-Hours	Unit	Material	2011 Bare Costs Labor	Equipment	Total	Total Incl O&P
0010	**FAUCETS AND FITTINGS**									
5000	Sillcock, compact, brass, IPS or copper to hose	1 Plum	24	.333	Ea.	10.60	17.85		28.45	38
6000	Stop and waste valves, bronze									
6100	Angle, solder end 1/2"	1 Plum	24	.333	Ea.	9.50	17.85		27.35	37
6110	3/4"		20	.400		11.10	21.50		32.60	44.50
6300	Straightway, solder end 3/8"		24	.333		6.15	17.85		24	33.50
6310	1/2"		24	.333		6.45	17.85		24.30	33.50
6320	3/4"		20	.400		3.48	21.50		24.98	36

22 41 Residential Plumbing Fixtures

22 41 39 – Residential Faucets, Supplies and Trim

22 41 39.10 Faucets and Fittings	Crew	Daily Output	Labor-Hours	Unit	Material	2011 Bare Costs Labor	Equipment	Total	Total Incl O&P
6410	Straightway, threaded 1/2"	1 Plum	24	.333	Ea.	3.82	17.85		21.67	30.50
6420	3/4"		20	.400		6.35	21.50		27.85	39
6430	1"		19	.421		12.60	22.50		35.10	47.50

22 51 Swimming Pool Plumbing Systems

22 51 19 – Swimming Pool Water Treatment Equipment

22 51 19.50 Swimming Pool Filtration Equipment

		Crew	Daily Output	Labor-Hours	Unit	Material	2011 Bare Costs Labor	Equipment	Total	Total Incl O&P
0010	**SWIMMING POOL FILTRATION EQUIPMENT**									
0900	Filter system, sand or diatomite type, incl. pump, 6,000 gal./hr.	2 Plum	1.80	8.889	Total	1,750	475		2,225	2,625
1020	Add for chlorination system, 800 S.F. pool		3	5.333	Ea.	103	286		389	540
1040	5,000 S.F. pool		3	5.333	"	1,875	286		2,161	2,500

22 52 Fountain Plumbing Systems

22 52 16 – Fountain Pumps

22 52 16.10 Fountain Water Pumps

		Crew	Daily Output	Labor-Hours	Unit	Material	2011 Bare Costs Labor	Equipment	Total	Total Incl O&P
0010	**FOUNTAIN WATER PUMPS**									
0100	Pump w/controls									
0200	Single phase, 100' cord, 1/2 H.P. pump	2 Skwk	4.40	3.636	Ea.	1,200	161		1,361	1,575
0300	3/4 H.P. pump		4.30	3.721		2,050	165		2,215	2,500
0400	1 H.P. pump		4.20	3.810		2,225	169		2,394	2,700
0500	1-1/2 H.P. pump		4.10	3.902		2,650	173		2,823	3,175
0600	2 H.P. pump		4	4		3,575	178		3,753	4,200
0700	Three phase, 200' cord, 5 H.P. pump		3.90	4.103		4,900	182		5,082	5,650
0800	7-1/2 H.P. pump		3.80	4.211		8,500	187		8,687	9,625
0900	10 H.P. pump		3.70	4.324		9,275	192		9,467	10,500
1000	15 H.P. pump		3.60	4.444		11,200	197		11,397	12,600
2000	DESIGN NOTE: Use two horsepower per surface acre.									

22 52 33 – Fountain Ancillary

22 52 33.10 Fountain Miscellaneous

		Crew	Daily Output	Labor-Hours	Unit	Material	2011 Bare Costs Labor	Equipment	Total	Total Incl O&P
0010	**FOUNTAIN MISCELLANEOUS**									
1100	Nozzles, minimum	2 Skwk	8	2	Ea.	145	89		234	296
1200	Maximum		8	2		310	89		399	475
1300	Lights w/mounting kits, 200 watt		18	.889		1,025	39.50		1,064.50	1,175
1400	300 watt		18	.889		1,225	39.50		1,264.50	1,400
1500	500 watt		18	.889		1,400	39.50		1,439.50	1,575
1600	Color blender		12	1.333		540	59		599	685

22 62 Vacuum Systems for Laboratory and Healthcare Facilities

22 62 19 – Vacuum Equipment for Laboratory and Healthcare Facilities

22 62 19.70 Healthcare Vacuum Equipment	Crew	Daily Output	Labor-Hours	Unit	Material	2011 Bare Costs Labor	Equipment	Total	Total Incl O&P
0010 **HEALTHCARE VACUUM EQUIPMENT**									
0100 Medical, with receiver									
0110 Duplex									
0120 20 SCFM	Q-1	1.14	14.035	Ea.	16,800	675		17,475	19,500
0130 60 SCFM, 10 HP	Q-2	1.20	20	"	30,500	1,000		31,500	35,000
0200 Triplex									
0220 180 SCFM	Q-2	.86	27.907	Ea.	47,300	1,400		48,700	54,000
0300 Dental oral									
0310 Duplex									
0330 165 SCFM with 77 gal. separator	Q-2	1.30	18.462	Ea.	49,400	925		50,325	56,000
1100 Vacuum system									
1110 Vacuum outlet alarm panel	1 Plum	3.20	2.500	Ea.	1,050	134		1,184	1,350

22 63 Gas Systems for Laboratory and Healthcare Facilities

22 63 13 – Gas Piping for Laboratory and Healthcare Facilities

22 63 13.70 Healthcare Gas Piping

	Crew	Daily Output	Labor-Hours	Unit	Material	2011 Bare Costs Labor	Equipment	Total	Total Incl O&P
0010 **HEALTHCARE GAS PIPING**									
1000 Nitrogen or oxygen system									
1010 Cylinder manifold									
1020 5 cylinder	1 Plum	.80	10	Ea.	5,675	535		6,210	7,025
1026 10 cylinder	"	.40	20	"	6,825	1,075		7,900	9,100
3000 Outlets and valves									
3010 Recessed, wall mounted									
3012 Single outlet	1 Plum	3.20	2.500	Ea.	68.50	134		202.50	275
3100 Ceiling outlet									
3200 Zone valve with box									
3210 2"	1 Plum	3.20	2.500	Ea.	630	134		764	895
4000 Alarm panel, medical gases and vacuum									
4010 Alarm panel	1 Plum	3.20	2.500	Ea.	1,050	134		1,184	1,350

Division Notes

	CREW	DAILY OUTPUT	LABOR-HOURS	UNIT	2011 BARE COSTS				TOTAL INCL O&P
					MAT.	LABOR	EQUIP.	TOTAL	

Estimating Tips

The labor adjustment factors listed in Subdivision 22 01 02.20 also apply to Division 23.

23 10 00 Facility Fuel Systems

- The prices in this subdivision for above- and below-ground storage tanks do not include foundations or hold-down slabs, unless noted. The estimator should refer to Divisions 3 and 31 for foundation system pricing. In addition to the foundations, required tank accessories, such as tank gauges, leak detection devices, and additional manholes and piping, must be added to the tank prices.

23 50 00 Central Heating Equipment

- When estimating the cost of an HVAC system, check to see who is responsible for providing and installing the temperature control system. It is possible to overlook controls, assuming that they would be included in the electrical estimate.

- When looking up a boiler, be careful on specified capacity. Some manufacturers rate their products on output while others use input.

- Include HVAC insulation for pipe, boiler, and duct (wrap and liner).

- Be careful when looking up mechanical items to get the correct pressure rating and connection type (thread, weld, flange).

23 70 00 Central HVAC Equipment

- Combination heating and cooling units are sized by the air conditioning requirements. (See Reference No. R236000-20 for preliminary sizing guide.)

- A ton of air conditioning is nominally 400 CFM.

- Rectangular duct is taken off by the linear foot for each size, but its cost is usually estimated by the pound. Remember that SMACNA standards now base duct on internal pressure.

- Prefabricated duct is estimated and purchased like pipe: straight sections and fittings.

- Note that cranes or other lifting equipment are not included on any lines in Division 23. For example, if a crane is required to lift a heavy piece of pipe into place high above a gym floor, or to put a rooftop unit on the roof of a four-story building, etc., it must be added. Due to the potential for extreme variation—from nothing additional required to a major crane or helicopter—we feel that including a nominal amount for "lifting contingency" would be useless and detract from the accuracy of the estimate. When using equipment rental cost data from RSMeans, do not forget to include the cost of the operator(s).

Reference Numbers

Reference numbers are shown in shaded boxes at the beginning of some major classifications. These numbers refer to related items in the Reference Section. The reference information may be an estimating procedure, an alternate pricing method, or technical information.

Note: Not all subdivisions listed here necessarily appear in this publication.

Note: **Trade Service,** *in part, has been used as a reference source for some of the material prices used in Division 23.*

23 05 Common Work Results for HVAC

23 05 02 – HVAC General

23 05 02.10 Air Conditioning, General	Crew	Daily Output	Labor-Hours	Unit	Material	2011 Bare Costs Labor	Equipment	Total	Total Incl O&P
0010	**AIR CONDITIONING, GENERAL** Prices are for standard efficiencies (SEER 13)									
0020	for upgrade to SEER 14 add					20%				

23 05 05 – Selective HVAC Demolition

23 05 05.10 HVAC Demolition

	23 05 05.10 HVAC Demolition	Crew	Daily Output	Labor-Hours	Unit	Material	Labor	Equipment	Total	Total Incl O&P
0010	**HVAC DEMOLITION** R220105-10									
0100	Air conditioner, split unit, 3 ton	Q-5	2	8	Ea.		390		390	580
0150	Package unit, 3 ton	Q-6	3	8			405		405	605
0190	Rooftop, self contained, up to 5 ton	1 Plum	1.20	6.667	↓		355		355	535
0250	Air curtain	Q-9	20	.800	L.F.		37		37	56
0254	Air filters, up thru 16,000 CFM		20	.800	Ea.		37		37	56
0256	20,000 thru 60,000 CFM		16	1			46.50		46.50	70.50
0297	Boiler blowdown	Q-5	8	2	↓		97.50		97.50	145
0298	Boilers									
0300	Electric, up thru 148 kW	Q-19	2	12	Ea.		590		590	880
0310	150 thru 518 kW	"	1	24			1,175		1,175	1,750
0320	550 thru 2000 kW	Q-21	.40	80			4,025		4,025	6,025
0330	2070 kW and up	"	.30	106			5,375		5,375	8,025
0340	Gas and/or oil, up thru 150 MBH	Q-7	2.20	14.545			750		750	1,125
0350	160 thru 2000 MBH		.80	40			2,050		2,050	3,075
0360	2100 thru 4500 MBH		.50	64			3,300		3,300	4,925
0370	4600 thru 7000 MBH		.30	106			5,500		5,500	8,200
0380	7100 thru 12,000 MBH		.16	200			10,300		10,300	15,400
0390	12,200 thru 25,000 MBH	↓	.12	266			13,700		13,700	20,500
0400	Central station air handler unit, up thru 15 ton	Q-5	1.60	10			485		485	725
0410	17.5 thru 30 ton	"	.80	20	↓		975		975	1,450
0430	Computer room unit									
0434	Air cooled split, up thru 10 ton	Q-5	.67	23.881	Ea.		1,175		1,175	1,725
0436	12 thru 23 ton		.53	30.189			1,475		1,475	2,200
0440	Chilled water, up thru 10 ton		1.30	12.308			600		600	895
0444	12 thru 23 ton		1	16			780		780	1,175
0450	Glycol system, up thru 10 ton		.53	30.189			1,475		1,475	2,200
0454	12 thru 23 ton		.40	40			1,950		1,950	2,900
0460	Water cooled, not including condenser, up thru 10 ton		.80	20			975		975	1,450
0464	12 thru 23 ton		.60	26.667			1,300		1,300	1,950
0600	Condenser, up thru 50 ton	↓	1	16			780		780	1,175
0610	51 thru 100 ton	Q-6	.80	30			1,525		1,525	2,250
0620	101 thru 1000 ton	"	.16	150			7,575		7,575	11,300
0660	Condensing unit, up thru 10 ton	Q-5	1.25	12.800			625		625	930
0670	11 thru 50 ton	"	.40	40			1,950		1,950	2,900
0680	60 thru 100 ton	Q-6	.30	80			4,050		4,050	6,025
0700	Cooling tower, up thru 400 ton		.80	30			1,525		1,525	2,250
0710	450 thru 600 ton		.53	45.283			2,275		2,275	3,425
0720	700 thru 1300 ton	↓	.40	60			3,025		3,025	4,525
0780	Dehumidifier, up thru 155 lb./hr.	Q-1	8	2			96.50		96.50	144
0790	240 lb./hr. and up	"	2	8	↓		385		385	575
1560	Ductwork									
1570	Metal, steel, sst, fabricated	Q-9	1000	.016	Lb.		.74		.74	1.12
1580	Aluminum, fabricated		485	.033	"		1.53		1.53	2.32
1590	Spiral, prefabricated		400	.040	L.F.		1.86		1.86	2.81
1600	Fiberglass, prefabricated		400	.040			1.86		1.86	2.81
1610	Flex, prefabricated		500	.032			1.49		1.49	2.25
1620	Glass fiber reinforced plastic, prefabricated	↓	280	.057	↓		2.66		2.66	4.02

23 05 05.10 HVAC Demolition	Crew	Daily Output	Labor-Hours	Unit	Material	2011 Bare Costs Labor	2011 Bare Costs Equipment	Total	Total Incl O&P
1630	Diffusers, registers or grills, up thru 20" max dimension	1 Shee	50	.160	Ea.		8.25		8.25	12.50
1640	21 thru 36" max dimension		36	.222			11.50		11.50	17.35
1650	Above 36" max dimension		30	.267			13.75		13.75	21
1700	Evaporator, up thru 12,000 BTUH	Q-5	5.30	3.019			147		147	219
1710	12,500 thru 30,000 BTUH	"	2.70	5.926			289		289	430
1720	31,000 BTUH and up	Q-6	1.50	16			810		810	1,200
1730	Evaporative cooler, up thru 5 H.P.	Q-9	2.70	5.926			275		275	415
1740	10 thru 30 H.P.	"	.67	23.881			1,100		1,100	1,675
1750	Exhaust systems									
1760	Exhaust components	1 Shee	8	1	System		51.50		51.50	78
1770	Weld fume hoods	"	20	.400	Ea.		20.50		20.50	31
2120	Fans, up thru 1 H.P. or 2000 CFM	Q-9	8	2			93		93	141
2124	1-1/2 thru 10 H.P. or 20,000 CFM		5.30	3.019			140		140	212
2128	15 thru 30 H.P. or above 20,000 CFM		4	4			186		186	281
2150	Fan coil air conditioner, chilled water, up thru 7.5 ton	Q-5	14	1.143			55.50		55.50	83
2154	Direct expansion, up thru 10 ton		8	2			97.50		97.50	145
2158	11 thru 30 ton		2	8			390		390	580
2170	Flue shutter damper	Q-9	8	2			93		93	141
2200	Furnace, electric	Q-20	2	10			470		470	710
2300	Gas or oil, under 120 MBH	Q-9	4	4			186		186	281
2340	Over 120 MBH	"	3	5.333			248		248	375
2730	Heating and ventilating unit	Q-5	2.70	5.926			289		289	430
2740	Heater, electric, wall, baseboard and quartz	1 Elec	10	.800			40		40	59.50
2750	Heater, electric, unit, cabinet, fan and convector	"	8	1			50.50		50.50	74.50
2760	Heat exchanger, shell and tube type	Q-5	1.60	10			485		485	725
2770	Plate type	Q-6	.60	40			2,025		2,025	3,025
2810	Heat pump									
2820	Air source, split, up thru 10 ton	Q-5	.90	17.778	Ea.		865		865	1,300
2830	15 thru 25 ton	Q-6	.80	30			1,525		1,525	2,250
2850	Single package, up thru 12 ton	Q-5	1	16			780		780	1,175
2860	Water source, up thru 15 ton	"	.90	17.778			865		865	1,300
2870	20 thru 50 ton	Q-6	.80	30			1,525		1,525	2,250
2910	Heat recovery package, up thru 20,000 CFM	Q-5	2	8			390		390	580
2920	25,000 CFM and up		1.20	13.333			650		650	970
2930	Heat transfer package, up thru 130 GPM		.80	20			975		975	1,450
2934	255 thru 800 GPM		.42	38.095			1,850		1,850	2,775
2940	Humidifier		10.60	1.509			73.50		73.50	110
2961	Hydronic unit heaters, up thru 200 MBH		14	1.143			55.50		55.50	83
2962	Above 200 MBH		8	2			97.50		97.50	145
2964	Valance units		32	.500			24.50		24.50	36.50
2966	Radiant floor heating									
2967	System valves, controls, manifolds	Q-5	16	1	Ea.		48.50		48.50	72.50
2968	Per room distribution		8	2			97.50		97.50	145
2970	Hydronic heating, baseboard radiation		16	1			48.50		48.50	72.50
2976	Convectors and free standing radiators		18	.889			43.50		43.50	64.50
2980	Induced draft fan, up thru 1 H.P.	Q-9	4.60	3.478			162		162	244
2984	1-1/2 H.P. thru 7-1/2 H.P.	"	2.20	7.273			340		340	510
2988	Infrared unit	Q-5	16	1			48.50		48.50	72.50
2992	Louvers	1 Shee	46	.174	S.F.		9		9	13.60
3000	Mechanical equipment, light items. Unit is weight, not cooling.	Q-5	.90	17.778	Ton		865		865	1,300
3600	Heavy items	"	1.10	14.545	"		710		710	1,050
3700	Deduct for salvage (when applicable), minimum				Job				73	80
3710	Maximum				"				455	500

23 05 05 – Selective HVAC Demolition

23 05 05.10 HVAC Demolition

		Crew	Daily Output	Labor-Hours	Unit	Material	2011 Bare Costs Labor	2011 Bare Costs Equipment	Total	Total Incl O&P
3720	Make up air unit, up thru 6000 CFM	Q-5	3	5.333	Ea.		260		260	390
3730	6500 thru 30,000 CFM	"	1.60	10			485		485	725
3740	35,000 thru 75,000 CFM	Q-6	1	24			1,200		1,200	1,800
3800	Mixing boxes, constant and VAV	Q-9	18	.889			41.50		41.50	62.50
4000	Packaged terminal air conditioner, up thru 18,000 BTUH	Q-5	8	2			97.50		97.50	145
4010	24,000 thru 48,000 BTUH	"	2.80	5.714			278		278	415
5000	Refrigerant compressor, reciprocating or scroll									
5010	Up thru 5 ton	1 Stpi	6	1.333	Ea.		72		72	108
5020	5.08 thru 10 ton	Q-5	6	2.667			130		130	194
5030	15 thru 50 ton	"	.40	40			1,950		1,950	2,900
5040	60 thru 130 ton	Q-6	.48	50			2,525		2,525	3,775
5100	Roof top air conditioner, up thru 10 ton	Q-5	1.40	11.429			555		555	830
5110	12 thru 40 ton	Q-6	1	24			1,200		1,200	1,800
5120	50 thru 140 ton		.50	48			2,425		2,425	3,625
5130	150 thru 300 ton		.30	80			4,050		4,050	6,025
6000	Self contained single package air conditioner, up thru 10 ton	Q-5	1.60	10			485		485	725
6010	15 thru 60 ton	Q-6	1.20	20			1,000		1,000	1,500
6100	Space heaters, up thru 200 MBH	Q-5	10	1.600			78		78	116
6110	Over 200 MBH		5	3.200			156		156	233
6200	Split ductless, both sections		8	2			97.50		97.50	145
6300	Steam condensate meter	1 Stpi	11	.727			39.50		39.50	58.50
6600	Thru-the-wall air conditioner	L-2	8	2			75.50		75.50	116
7000	Vent chimney, prefabricated, up thru 12" diameter	Q-9	94	.170	V.L.F.		7.90		7.90	11.95
7010	14" thru 36" diameter		40	.400			18.60		18.60	28
7020	38" thru 48" diameter		32	.500			23		23	35
7030	54" thru 60" diameter	Q-10	14	1.714			82.50		82.50	125
7400	Ventilators, up thru 14" neck diameter	Q-9	58	.276	Ea.		12.80		12.80	19.40
7410	16" thru 50" neck diameter		40	.400			18.60		18.60	28
7450	Relief vent, up thru 24" x 96"		22	.727			34		34	51
7460	48" x 60" thru 96" x 144"		10	1.600			74.50		74.50	112
8000	Water chiller up thru 10 ton	Q-5	2.50	6.400			310		310	465
8010	15 thru 100 ton	Q-6	.48	50			2,525		2,525	3,775
8020	110 thru 500 ton	Q-7	.29	110			5,675		5,675	8,475
8030	600 thru 1000 ton		.23	139			7,175		7,175	10,700
8040	1100 ton and up		.20	160			8,250		8,250	12,300
8400	Window air conditioner	1 Carp	16	.500			21.50		21.50	33

23 05 23 – General-Duty Valves for HVAC Piping

23 05 23.30 Valves, Iron Body

		Crew	Daily Output	Labor-Hours	Unit	Material	2011 Bare Costs Labor	2011 Bare Costs Equipment	Total	Total Incl O&P
0010	**VALVES, IRON BODY**	R220523-90								
0022	For grooved joint, see Section 22 11 13.48									
0100	Angle, 125 lb.									
0110	Flanged									
0116	2"	1 Plum	5	1.600	Ea.	1,000	86		1,086	1,225
0118	4"	Q-1	3	5.333		1,650	257		1,907	2,200
0120	6"	Q-2	3	8		3,225	400		3,625	4,150
0122	8"	"	2.50	9.600		5,800	480		6,280	7,100
0560	Butterfly, lug type, pneumatic operator, 2" size	1 Stpi	14	.571		201	31		232	267
0570	3"	Q-1	8	2		213	96.50		309.50	380
0580	4"	"	5	3.200		233	154		387	490
0590	6"	Q-2	5	4.800		320	240		560	715
0600	8"		4.50	5.333		415	267		682	855
0610	10"		4	6		520	300		820	1,025

23 05 23 – General-Duty Valves for HVAC Piping

23 05 23.30 Valves, Iron Body		Crew	Daily Output	Labor-Hours	Unit	Material	2011 Bare Costs Labor	Equipment	Total	Total Incl O&P
0620	12"	Q-2	3	8	Ea.	675	400		1,075	1,350
0630	14"		2.30	10.435		1,275	520		1,795	2,175
0640	18"		1.50	16		2,000	800		2,800	3,425
0650	20"		1	24		2,450	1,200		3,650	4,500
0790	Butterfly, lug type, gear operated, 2" size, 200 lb. except noted	1 Plum	14	.571		299	30.50		329.50	375
0800	2-1/2"	Q-1	9	1.778		300	86		386	460
0810	3"		8	2		315	96.50		411.50	490
0820	4"		5	3.200		355	154		509	625
0830	5"	Q-2	5	4.800		445	240		685	850
0840	6"		5	4.800		490	240		730	900
0850	8"		4.50	5.333		640	267		907	1,100
0860	10"		4	6		880	300		1,180	1,425
0870	12"		3	8		1,225	400		1,625	1,950
0880	14", 150 lb.		2.30	10.435		2,475	520		2,995	3,500
0890	16", 150 lb.		1.75	13.714		3,575	685		4,260	4,975
0900	18", 150 lb.		1.50	16		4,500	800		5,300	6,150
0910	20", 150 lb.		1	24		5,675	1,200		6,875	8,050
0930	24", 150 lb.		.75	32		10,200	1,600		11,800	13,600
1020	Butterfly, wafer type, gear actuator, 200 lb.									
1030	2"	1 Plum	14	.571	Ea.	167	30.50		197.50	229
1040	2-1/2"	Q-1	9	1.778		172	86		258	315
1050	3"		8	2		178	96.50		274.50	340
1060	4"		5	3.200		202	154		356	455
1070	5"	Q-2	5	4.800		269	240		509	655
1080	6"		5	4.800		305	240		545	695
1090	8"		4.50	5.333		405	267		672	845
1100	10"		4	6		505	300		805	1,000
1110	12"		3	8		705	400		1,105	1,375
1200	Wafer type, lever actuator, 200 lb.									
1220	2"	1 Plum	14	.571	Ea.	133	30.50		163.50	192
1230	2-1/2"	Q-1	9	1.778		137	86		223	278
1240	3"		8	2		145	96.50		241.50	305
1250	4"		5	3.200		176	154		330	425
1260	5"	Q-2	5	4.800		245	240		485	630
1270	6"		5	4.800		295	240		535	685
1280	8"		4.50	5.333		460	267		727	905
1290	10"		4	6		605	300		905	1,125
1300	12"		3	8		1,100	400		1,500	1,825
1600	Gate, threaded, 125 lb.									
1603	1-1/2"	1 Plum	13	.615	Ea.	247	33		280	320
1604	2"		11	.727		840	39		879	985
1605	2-1/2"		10	.800		1,100	43		1,143	1,275
1606	3"		8	1		1,100	53.50		1,153.50	1,275
1607	4"		5	1.600		1,525	86		1,611	1,825
1650	Gate, 125 lb., N.R.S.									
2150	Flanged									
2200	2"	1 Plum	5	1.600	Ea.	550	86		636	735
2240	2-1/2"	Q-1	5	3.200		565	154		719	850
2260	3"		4.50	3.556		630	172		802	950
2280	4"		3	5.333		905	257		1,162	1,375
2290	5"	Q-2	3.40	7.059		1,550	355		1,905	2,225
2300	6"		3	8		1,550	400		1,950	2,300
2320	8"		2.50	9.600		2,650	480		3,130	3,625

23 05 23.30 Valves, Iron Body		Crew	Daily Output	Labor-Hours	Unit	Material	2011 Bare Costs Labor	Equipment	Total	Total Incl O&P
2340	10"	Q-2	2.20	10.909	Ea.	4,650	545		5,195	5,925
2360	12"		1.70	14.118		6,375	705		7,080	8,075
2370	14"		1.30	18.462		9,000	925		9,925	11,300
2380	16"		1	24		12,500	1,200		13,700	15,600
2420	For 250 lb flanged, add					200%	10%			
3500	OS&Y, 125 lb., (225# noted), threaded									
3504	3/4" (225 lb.)	1 Plum	18	.444	Ea.	440	24		464	520
3505	1" (225 lb.)		16	.500		500	27		527	590
3506	1-1/2" (225 lb.)		13	.615		730	33		763	850
3507	2"		11	.727		900	39		939	1,050
3550	OS&Y, 125 lb., flanged									
3600	2"	1 Plum	5	1.600	Ea.	350	86		436	515
3640	2-1/2"	Q-1	5	3.200		360	154		514	625
3660	3"		4.50	3.556		395	172		567	690
3670	3-1/2"		3	5.333		520	257		777	955
3680	4"		3	5.333		565	257		822	1,000
3690	5"	Q-2	3.40	7.059		930	355		1,285	1,550
3700	6"		3	8		930	400		1,330	1,625
3720	8"		2.50	9.600		1,650	480		2,130	2,550
3740	10"		2.20	10.909		3,025	545		3,570	4,150
3760	12"		1.70	14.118		4,050	705		4,755	5,500
3770	14"		1.30	18.462		8,150	925		9,075	10,300
3780	16"		1	24		11,600	1,200		12,800	14,500
3790	18"		.80	30		14,900	1,500		16,400	18,700
3800	20"		.60	40		21,600	2,000		23,600	26,800
3830	24"		.50	48		30,900	2,400		33,300	37,500
3900	For 175 lb, flanged, add					200%	10%			
4350	Globe, OS&Y									
4540	Class 125, flanged									
4550	2"	1 Plum	5	1.600	Ea.	700	86		786	900
4560	2-1/2"	Q-1	5	3.200		705	154		859	1,000
4570	3"		4.50	3.556		855	172		1,027	1,200
4580	4"		3	5.333		1,225	257		1,482	1,725
4590	5"	Q-2	3.40	7.059		2,225	355		2,580	2,975
4600	6"		3	8		2,225	400		2,625	3,050
4610	8"		2.50	9.600		4,375	480		4,855	5,525
4612	10"		2.20	10.909		6,575	545		7,120	8,050
4614	12"		1.70	14.118		7,950	705		8,655	9,800
5040	Class 250, flanged									
5050	2"	1 Plum	4.50	1.778	Ea.	1,125	95.50		1,220.50	1,375
5060	2-1/2"	Q-1	4.50	3.556		1,450	172		1,622	1,850
5070	3"		4	4		1,500	193		1,693	1,950
5080	4"		2.70	5.926		2,200	286		2,486	2,850
5090	5"	Q-2	3	8		3,950	400		4,350	4,950
5100	6"		2.70	8.889		3,950	445		4,395	5,025
5110	8"		2.20	10.909		6,725	545		7,270	8,225
5120	10"		2	12		10,300	600		10,900	12,200
5130	12"		1.60	15		15,400	750		16,150	18,100
5240	Valve sprocket rim w/chain, for 2" valve	1 Stpi	30	.267		118	14.45		132.45	152
5250	2-1/2" valve		27	.296		118	16.05		134.05	154
5260	3-1/2" valve		25	.320		118	17.30		135.30	156
5270	6" valve		20	.400		118	21.50		139.50	163
5280	8" valve		18	.444		158	24		182	210

23 05 23.30 Valves, Iron Body		Crew	Daily Output	Labor-Hours	Unit	Material	2011 Bare Costs Labor	Equipment	Total	Total Incl O&P
5290	12" valve	1 Stpi	16	.500	Ea.	158	27		185	215
5300	16" valve		12	.667		218	36		254	294
5310	20" valve		10	.800		218	43.50		261.50	305
5320	36" valve		8	1		415	54		469	535
5450	Swing check, 125 lb., threaded									
5470	1"	1 Plum	13	.615	Ea.	575	33		608	685
5500	2"	"	11	.727		705	39		744	835
5540	2-1/2"	Q-1	15	1.067		675	51.50		726.50	820
5550	3"		13	1.231		750	59.50		809.50	915
5560	4"		10	1.600		1,300	77		1,377	1,550
5950	Flanged									
5994	1"	1 Plum	7	1.143	Ea.	161	61.50		222.50	269
5998	1-1/2"		6	1.333		223	71.50		294.50	355
6000	2"		5	1.600		284	86		370	440
6040	2-1/2"	Q-1	5	3.200		286	154		440	545
6050	3"		4.50	3.556		305	172		477	590
6060	4"		3	5.333		480	257		737	915
6070	6"	Q-2	3	8		820	400		1,220	1,500
6080	8"		2.50	9.600		1,550	480		2,030	2,425
6090	10"		2.20	10.909		2,625	545		3,170	3,725
6100	12"		1.70	14.118		4,275	705		4,980	5,750
6110	18"		1.30	18.462		12,400	925		13,325	15,000
6114	24"		.75	32		23,400	1,600		25,000	28,100
6160	For 250 lb flanged, add					200%	20%			
6600	Silent check, bronze trim									
6610	Compact wafer type, for 125 or 150 lb. flanges									
6630	1-1/2"	1 Plum	11	.727	Ea.	117	39		156	187
6640	2"	"	9	.889		140	47.50		187.50	225
6650	2-1/2"	Q-1	9	1.778		153	86		239	297
6660	3"		8	2		165	96.50		261.50	325
6670	4"		5	3.200		215	154		369	470
6680	5"	Q-2	6	4		274	200		474	600
6690	6"		6	4		370	200		570	705
6700	8"		4.50	5.333		655	267		922	1,125
6710	10"		4	6		1,125	300		1,425	1,700
6720	12"		3	8		2,175	400		2,575	2,975
6740	For 250 or 300 lb. flanges, thru 6" no change									
6741	For 8" and 10", add				Ea.	11%	10%			
6750	Twin disc									
6752	2"	1 Plum	9	.889	Ea.	243	47.50		290.50	340
6754	4"	Q-1	5	3.200		395	154		549	665
6756	6"	Q-2	5	4.800		600	240		840	1,025
6758	8"		4.50	5.333		905	267		1,172	1,400
6760	10"		4	6		1,450	300		1,750	2,050
6762	12"		3	8		1,875	400		2,275	2,675
6764	18"		1.50	16		8,100	800		8,900	10,100
6766	24"		.75	32		11,400	1,600		13,000	15,000
6800	Full flange type, 150 lb.									
6810	1"	1 Plum	14	.571	Ea.	137	30.50		167.50	197
6811	1-1/4"		12	.667		145	35.50		180.50	213
6812	1-1/2"		11	.727		168	39		207	242
6813	2"		9	.889		192	47.50		239.50	283
6814	2-1/2"	Q-1	9	1.778		233	86		319	385

23 05 23 – General-Duty Valves for HVAC Piping

23 05 23.30 Valves, Iron Body		Crew	Daily Output	Labor-Hours	Unit	Material	2011 Bare Costs Labor	Equipment	Total	Total Incl O&P
6815	3"	Q-1	8	2	Ea.	241	96.50		337.50	410
6816	4"	↓	5	3.200		299	154		453	560
6817	5"	Q-2	6	4		470	200		670	820
6818	6"		5	4.800		620	240		860	1,050
6819	8"		4.50	5.333		1,050	267		1,317	1,550
6820	10"	↓	4	6		1,325	300		1,625	1,900
6840	For 250 lb., add				↓	30%	10%			
6900	Globe type, 125 lb.									
6910	2"	1 Plum	9	.889	Ea.	315	47.50		362.50	415
6911	2-1/2"	Q-1	9	1.778		345	86		431	510
6912	3"		8	2		375	96.50		471.50	555
6913	4"	↓	5	3.200		505	154		659	785
6914	5"	Q-2	6	4		645	200		845	1,000
6915	6"		5	4.800		810	240		1,050	1,250
6916	8"		4.50	5.333		1,475	267		1,742	2,025
6917	10"		4	6		1,875	300		2,175	2,500
6918	12"		3	8		3,100	400		3,500	4,025
6919	14"		2.30	10.435		4,225	520		4,745	5,425
6920	16"		1.75	13.714		6,100	685		6,785	7,725
6921	18"		1.50	16		10,700	800		11,500	13,000
6922	20"		1	24		11,200	1,200		12,400	14,100
6923	24"	↓	.75	32		13,800	1,600		15,400	17,600
6940	For 250 lb., add				↓	40%	10%			
6980	Screwed end type, 125 lb.									
6981	1"	1 Plum	19	.421	Ea.	48.50	22.50		71	86.50
6982	1-1/4"		15	.533		63	28.50		91.50	112
6983	1-1/2"		13	.615		77.50	33		110.50	134
6984	2"	↓	11	.727		106	39		145	175

23 05 23.50 Valves, Multipurpose

		Crew	Daily Output	Labor-Hours	Unit	Material	2011 Bare Costs Labor	Equipment	Total	Total Incl O&P
0010	**VALVES, MULTIPURPOSE**									
0100	Functions as a shut off, balancing, check & metering valve									
1000	Cast iron body									
1010	Threaded									
1020	1-1/2" size	1 Stpi	11	.727	Ea.	340	39.50		379.50	435
1030	2" size	"	8	1		415	54		469	535
1040	2-1/2" size	Q-5	5	3.200		450	156		606	730
1050	3" size	"	4.50	3.556	↓	515	173		688	825
1200	Flanged									
1210	3" size	Q-5	4.20	3.810	Ea.	515	186		701	840
1220	4" size	"	3	5.333		1,050	260		1,310	1,550
1230	5" size	Q-6	3.80	6.316		1,225	320		1,545	1,825
1240	6" size		3	8		1,550	405		1,955	2,300
1250	8" size		2.50	9.600		2,675	485		3,160	3,650
1260	10" size		2.20	10.909		3,825	550		4,375	5,025
1270	12" size		2.10	11.429		6,900	575		7,475	8,425
1280	14" size	↓	2	12	↓	10,000	605		10,605	11,900

23 05 23.70 Valves, Semi-Steel

		Crew	Daily Output	Labor-Hours	Unit	Material	2011 Bare Costs Labor	Equipment	Total	Total Incl O&P
0010	**VALVES, SEMI-STEEL** R220523-90									
1020	Lubricated plug valve, threaded, 200 psi									
1030	1/2"	1 Plum	18	.444	Ea.	82	24		106	126
1040	3/4"		16	.500		82	27		109	130
1050	1"	↓	14	.571	↓	104	30.50		134.50	161

23 05 23 – General-Duty Valves for HVAC Piping

23 05 23.70 Valves, Semi-Steel		Crew	Daily Output	Labor-Hours	Unit	Material	2011 Bare Costs Labor	Equipment	Total	Total Incl O&P
1060	1-1/4"	1 Plum	12	.667	Ea.	124	35.50		159.50	190
1070	1-1/2"		11	.727		134	39		173	205
1080	2"		8	1		159	53.50		212.50	255
1090	2-1/2"	Q-1	5	3.200		246	154		400	500
1100	3"	"	4.50	3.556		300	172		472	585
6990	Flanged, 200 psi									
7000	2"	1 Plum	8	1	Ea.	197	53.50		250.50	297
7010	2-1/2"	Q-1	5	3.200		283	154		437	540
7020	3"		4.50	3.556		340	172		512	630
7030	4"		3	5.333		430	257		687	855
7036	5"		2.50	6.400		840	310		1,150	1,375
7040	6"	Q-2	3	8		705	400		1,105	1,375
7050	8"		2.50	9.600		1,375	480		1,855	2,225
7060	10"		2.20	10.909		2,175	545		2,720	3,225
7070	12"		1.70	14.118		4,700	705		5,405	6,225

23 05 23.80 Valves, Steel

23 05 23.80 Valves, Steel		Crew	Daily Output	Labor-Hours	Unit	Material	2011 Bare Costs Labor	Equipment	Total	Total Incl O&P
0010	**VALVES, STEEL** R220523-90									
0800	Cast									
1350	Check valve, swing type, 150 lb., flanged									
1370	1"	1 Plum	10	.800	Ea.	545	43		588	665
1400	2"	"	8	1		660	53.50		713.50	805
1440	2-1/2"	Q-1	5	3.200		765	154		919	1,075
1450	3"		4.50	3.556		780	172		952	1,100
1460	4"		3	5.333		1,125	257		1,382	1,600
1470	6"	Q-2	3	8		1,750	400		2,150	2,525
1480	8"		2.50	9.600		2,850	480		3,330	3,875
1490	10"		2.20	10.909		4,225	545		4,770	5,475
1500	12"		1.70	14.118		6,025	705		6,730	7,700
1510	14"		1.30	18.462		8,725	925		9,650	11,000
1520	16"		1	24		9,875	1,200		11,075	12,700
1548	For 600 lb., flanged, add					110%	20%			
1571	300 lb., 2"	1 Plum	7.40	1.081		700	58		758	855
1572	2-1/2"	Q-1	4.20	3.810		1,050	184		1,234	1,425
1573	3"		4	4		1,050	193		1,243	1,450
1574	4"		2.80	5.714		1,375	276		1,651	1,900
1575	6"	Q-2	2.90	8.276		2,650	415		3,065	3,550
1576	8"		2.40	10		3,825	500		4,325	4,975
1577	10"		2.10	11.429		5,675	570		6,245	7,100
1578	12"		1.60	15		7,950	750		8,700	9,850
1579	14"		1.20	20		12,000	1,000		13,000	14,700
1581	16"		.90	26.667		15,500	1,325		16,825	19,100
1950	Gate valve, 150 lb., flanged									
2000	2"	1 Plum	8	1	Ea.	710	53.50		763.50	860
2040	2-1/2"	Q-1	5	3.200		1,000	154		1,154	1,325
2050	3"		4.50	3.556		1,000	172		1,172	1,350
2060	4"		3	5.333		1,250	257		1,507	1,750
2070	6"	Q-2	3	8		1,850	400		2,250	2,625
2080	8"		2.50	9.600		1,925	480		2,405	2,850
2090	10"		2.20	10.909		4,400	545		4,945	5,650
2100	12"		1.70	14.118		5,975	705		6,680	7,625
2110	14"		1.30	18.462		9,500	925		10,425	11,900
2120	16"		1	24		12,300	1,200		13,500	15,300

23 05 23.80 Valves, Steel		Crew	Daily Output	Labor-Hours	Unit	Material	2011 Bare Costs Labor	Equipment	Total	Total Incl O&P
2130	18"	Q-2	.80	30	Ea.	14,900	1,500		16,400	18,700
2140	20"	↓	.60	40	↓	17,900	2,000		19,900	22,700
2650	300 lb., flanged									
2700	2"	1 Plum	7.40	1.081	Ea.	965	58		1,023	1,125
2740	2-1/2"	Q-1	4.20	3.810		1,300	184		1,484	1,700
2750	3"		4	4		1,300	193		1,493	1,725
2760	4"	↓	2.80	5.714		1,825	276		2,101	2,400
2770	6"	Q-2	2.90	8.276		3,075	415		3,490	4,000
2780	8"		2.40	10		4,800	500		5,300	6,025
2790	10"		2.10	11.429		6,450	570		7,020	7,950
2800	12"		1.60	15		9,300	750		10,050	11,300
2810	14"		1.20	20		17,600	1,000		18,600	20,900
2820	16"		.90	26.667		22,100	1,325		23,425	26,300
2830	18"		.70	34.286		31,500	1,725		33,225	37,300
2840	20"	↓	.50	48	↓	35,100	2,400		37,500	42,200
3650	Globe valve, 150 lb., flanged									
3700	2"	1 Plum	8	1	Ea.	890	53.50		943.50	1,050
3740	2-1/2"	Q-1	5	3.200		1,125	154		1,279	1,475
3750	3"		4.50	3.556		1,125	172		1,297	1,500
3760	4"	↓	3	5.333		1,650	257		1,907	2,200
3770	6"	Q-2	3	8		2,600	400		3,000	3,450
3780	8"		2.50	9.600		4,750	480		5,230	5,975
3790	10"		2.20	10.909		8,425	545		8,970	10,100
3800	12"	↓	1.70	14.118	↓	11,400	705		12,105	13,600
4080	300 lb., flanged									
4100	2"	1 Plum	7.40	1.081	Ea.	1,200	58		1,258	1,375
4140	2-1/2"	Q-1	4.20	3.810		1,625	184		1,809	2,050
4150	3"		4	4		1,625	193		1,818	2,075
4160	4"	↓	2.80	5.714		2,250	276		2,526	2,875
4170	6"	Q-2	2.90	8.276		4,050	415		4,465	5,075
4180	8"		2.40	10		6,625	500		7,125	8,025
4190	10"		2.10	11.429		12,500	570		13,070	14,600
4200	12"	↓	1.60	15	↓	14,500	750		15,250	17,000
4680	600 lb., flanged									
4700	2"	1 Plum	7	1.143	Ea.	1,650	61.50		1,711.50	1,925
4740	2-1/2"	Q-1	4	4		2,475	193		2,668	3,025
4750	3"		3.60	4.444		2,475	214		2,689	3,050
4760	4"	↓	2.50	6.400		4,000	310		4,310	4,850
4770	6"	Q-2	2.60	9.231		8,325	460		8,785	9,850
4780	8"	"	2.10	11.429	↓	13,500	570		14,070	15,800
4800	Silent check, 316 S.S. trim									
4810	Full flange type, 150 lb.									
4811	1"	1 Plum	14	.571	Ea.	485	30.50		515.50	580
4812	1-1/4"		12	.667		485	35.50		520.50	590
4813	1-1/2"		11	.727		530	39		569	645
4814	2"	↓	9	.889		640	47.50		687.50	775
4815	2-1/2"	Q-1	9	1.778		775	86		861	980
4816	3"		8	2		900	96.50		996.50	1,125
4817	4"	↓	5	3.200		1,225	154		1,379	1,575
4818	5"	Q-2	6	4		1,550	200		1,750	2,000
4819	6"		5	4.800		2,000	240		2,240	2,550
4820	8"		4.50	5.333		2,875	267		3,142	3,575
4821	10"	↓	4	6	↓	4,075	300		4,375	4,925

23 05 23.80 Valves, Steel		Crew	Daily Output	Labor-Hours	Unit	Material	2011 Bare Costs Labor	Equipment	Total	Total Incl O&P
4840	For 300 lb., add				Ea.	.20%	10%			
4860	For 600 lb., add					50%	15%			
4900	Globe type, 150 lb.									
4910	2"	1 Plum	9	.889	Ea.	765	47.50		812.50	910
4911	2-1/2"	Q-1	9	1.778		920	86		1,006	1,125
4912	3"		8	2		1,025	96.50		1,121.50	1,275
4913	4"		5	3.200		1,350	154		1,504	1,725
4914	5"	Q-2	6	4		1,700	200		1,900	2,175
4915	6"		5	4.800		2,100	240		2,340	2,675
4916	8"		4.50	5.333		2,950	267		3,217	3,650
4917	10"		4	6		5,100	300		5,400	6,075
4918	12"		3	8		6,275	400		6,675	7,500
4919	14"		2.30	10.435		9,475	520		9,995	11,200
4920	16"		1.75	13.714		13,200	685		13,885	15,600
4921	18"		1.50	16		13,900	800		14,700	16,500
4940	For 300 lb., add					30%	10%			
4960	For 600 lb., add					60%	15%			
5150	Forged									
5340	Ball valve, 1500 psi, threaded, 1/4" size	1 Plum	24	.333	Ea.	60	17.85		77.85	92.50
5350	3/8"		24	.333		60	17.85		77.85	92.50
5360	1/2"		24	.333		74.50	17.85		92.35	109
5370	3/4"		20	.400		99	21.50		120.50	141
5380	1"		19	.421		123	22.50		145.50	170
5390	1-1/4"		15	.533		165	28.50		193.50	224
5400	1-1/2"		13	.615		215	33		248	285
5410	2"		11	.727		284	39		323	375
5460	Ball valve, 800 lb., socket weld, 1/4" size	Q-15	19	.842		60	40.50	1.71	102.21	128
5470	3/8"		19	.842		60	40.50	1.71	102.21	128
5480	1/2"		19	.842		74.50	40.50	1.71	116.71	144
5490	3/4"		19	.842		99	40.50	1.71	141.21	171
5500	1"		15	1.067		123	51.50	2.17	176.67	215
5510	1-1/4"		13	1.231		165	59.50	2.50	227	272
5520	1-1/2"		11	1.455		215	70	2.95	287.95	345
5530	2"		8.50	1.882		284	91	3.82	378.82	455
5550	Ball valve, 150 lb., flanged									
5560	4"	Q-1	3	5.333	Ea.	725	257		982	1,175
5570	6"	Q-2	3	8		1,750	400		2,150	2,525
5580	8"	"	2.50	9.600		3,250	480		3,730	4,300
5650	Check valve, class 800, horizontal, socket									
5652	1/4"	Q-15	19	.842	Ea.	88	40.50	1.71	130.21	159
5654	3/8"		19	.842		88	40.50	1.71	130.21	159
5656	1/2"		19	.842		88	40.50	1.71	130.21	159
5658	3/4"		19	.842		94.50	40.50	1.71	136.71	166
5660	1"		15	1.067		111	51.50	2.17	164.67	201
5662	1-1/4"		13	1.231		217	59.50	2.50	279	330
5664	1-1/2"		11	1.455		217	70	2.95	289.95	345
5666	2"		8.50	1.882		305	91	3.82	399.82	475
5698	Threaded									
5700	1/4"	1 Plum	24	.333	Ea.	88	17.85		105.85	123
5720	3/8"		24	.333		88	17.85		105.85	123
5730	1/2"		24	.333		88	17.85		105.85	123
5740	3/4"		20	.400		94.50	21.50		116	136
5750	1"		19	.421		111	22.50		133.50	156

23 05 23 – General-Duty Valves for HVAC Piping

23 05 23.80 Valves, Steel

		Crew	Daily Output	Labor-Hours	Unit	Material	2011 Bare Costs Labor	Equipment	Total	Total Incl O&P
5760	1-1/4"	1 Plum	15	.533	Ea.	217	28.50		245.50	282
5770	1-1/2"		13	.615		217	33		250	288
5780	2"	↓	11	.727		305	39		344	395
5840	For class 150, flanged, add					100%	15%			
5860	For class 300, flanged, add				↓	120%	20%			
6100	Gate, class 800, OS&Y, socket									
6102	3/8"	Q-15	19	.842	Ea.	61	40.50	1.71	103.21	130
6103	1/2"		19	.842		61	40.50	1.71	103.21	130
6104	3/4"		19	.842		67	40.50	1.71	109.21	136
6105	1"		15	1.067		81.50	51.50	2.17	135.17	169
6106	1-1/4"		13	1.231		154	59.50	2.50	216	260
6107	1-1/2"		11	1.455		154	70	2.95	226.95	277
6108	2"	↓	8.50	1.882	↓	200	91	3.82	294.82	360
6118	Threaded									
6120	3/8"	1 Plum	24	.333	Ea.	61	17.85		78.85	94
6130	1/2"		24	.333		61	17.85		78.85	94
6140	3/4"		20	.400		67	21.50		88.50	106
6150	1"		19	.421		81.50	22.50		104	123
6160	1-1/4"		15	.533		154	28.50		182.50	212
6170	1-1/2"		13	.615		154	33		187	218
6180	2"	↓	11	.727		200	39		239	278
6260	For OS&Y, flanged, add				↓	100%	20%			
6700	Globe, OS&Y, class 800, socket									
6710	1/4"	Q-15	19	.842	Ea.	94.50	40.50	1.71	136.71	166
6720	3/8"		19	.842		94.50	40.50	1.71	136.71	166
6730	1/2"		19	.842		94.50	40.50	1.71	136.71	166
6740	3/4"		19	.842		108	40.50	1.71	150.21	181
6750	1"		15	1.067		141	51.50	2.17	194.67	234
6760	1-1/4"		13	1.231		276	59.50	2.50	338	395
6770	1-1/2"		11	1.455		276	70	2.95	348.95	415
6780	2"	↓	8.50	1.882		355	91	3.82	449.82	530
6860	For OS&Y, flanged, add				↓	300%	20%			
6880	Threaded									
6882	1/4"	1 Plum	24	.333	Ea.	94.50	17.85		112.35	131
6884	3/8"		24	.333		94.50	17.85		112.35	131
6886	1/2"		24	.333		94.50	17.85		112.35	131
6888	3/4"		20	.400		108	21.50		129.50	151
6890	1"		19	.421		141	22.50		163.50	189
6892	1-1/4"		15	.533		276	28.50		304.50	350
6894	1-1/2"		13	.615		276	33		309	355
6896	2"	↓	11	.727	↓	355	39		394	450

23 05 23.90 Valves, Stainless Steel

		Crew	Daily Output	Labor-Hours	Unit	Material	2011 Bare Costs Labor	Equipment	Total	Total Incl O&P
0010	**VALVES, STAINLESS STEEL** R220523-90									
1610	Ball, threaded 1/4"	1 Stpi	24	.333	Ea.	45	18.05		63.05	76.50
1620	3/8"		24	.333		45	18.05		63.05	76.50
1630	1/2"		22	.364		45	19.65		64.65	79
1640	3/4"		20	.400		74.50	21.50		96	114
1650	1"		19	.421		90.50	23		113.50	134
1660	1-1/4		15	.533		172	29		201	233
1670	1-1/2"		13	.615		180	33.50		213.50	248
1680	2"	↓	11	.727	↓	234	39.50		273.50	315
1700	Check, 200 lb., threaded									

23 05 23 – General-Duty Valves for HVAC Piping

23 05 23.90	Valves, Stainless Steel	Crew	Daily Output	Labor-Hours	Unit	Material	2011 Bare Costs Labor	Equipment	Total	Total Incl O&P
1710	1/4"	1 Plum	24	.333	Ea.	130	17.85		147.85	170
1720	1/2"		22	.364		130	19.50		149.50	172
1730	3/4"		20	.400		142	21.50		163.50	188
1750	1"		19	.421		182	22.50		204.50	234
1760	1-1/2"		13	.615		350	33		383	435
1770	2"		11	.727		585	39		624	705
1800	150 lb., flanged									
1810	2-1/2"	Q-1	5	3.200	Ea.	1,400	154		1,554	1,775
1820	3"		4.50	3.556		1,400	172		1,572	1,800
1830	4"		3	5.333		2,100	257		2,357	2,675
1840	6"	Q-2	3	8		3,725	400		4,125	4,700
1850	8"	"	2.50	9.600		6,250	480		6,730	7,600
2100	Gate, OS&Y, 150 lb., flanged									
2120	1/2"	1 Plum	18	.444	Ea.	415	24		439	490
2140	3/4"		16	.500		440	27		467	525
2150	1"		14	.571		540	30.50		570.50	640
2160	1-1/2"		11	.727		750	39		789	890
2170	2"		8	1		845	53.50		898.50	1,000
2180	2-1/2"	Q-1	5	3.200		1,400	154		1,554	1,750
2190	3"		4.50	3.556		2,050	172		2,222	2,500
2200	4"		3	5.333		2,050	257		2,307	2,625
2205	5"		2.80	5.714		3,550	276		3,826	4,325
2210	6"	Q-2	3	8		3,550	400		3,950	4,525
2220	8"		2.50	9.600		5,725	480		6,205	7,025
2230	10"		2.30	10.435		9,375	520		9,895	11,100
2240	12"		1.90	12.632		12,000	630		12,630	14,100
2260	For 300 lb., flanged, add					120%	15%			
2600	600 lb., flanged									
2620	1/2"	1 Plum	16	.500	Ea.	199	27		226	259
2640	3/4"		14	.571		215	30.50		245.50	283
2650	1"		12	.667		259	35.50		294.50	340
2660	1-1/2"		10	.800		415	43		458	520
2670	2"		7	1.143		570	61.50		631.50	715
2680	2-1/2"	Q-1	4	4		4,950	193		5,143	5,750
2690	3"	"	3.60	4.444		4,950	214		5,164	5,775
3100	Globe, OS&Y, 150 lb., flanged									
3120	1/2"	1 Plum	18	.444	Ea.	415	24		439	490
3140	3/4"		16	.500		460	27		487	545
3150	1"		14	.571		570	30.50		600.50	670
3160	1-1/2"		11	.727		845	39		884	990
3170	2"		8	1		1,025	53.50		1,078.50	1,200
3180	2-1/2"	Q-1	5	3.200		2,225	154		2,379	2,675
3190	3"		4.50	3.556		2,225	172		2,397	2,700
3200	4"		3	5.333		2,600	257		2,857	3,250
3210	6"	Q-2	3	8		4,575	400		4,975	5,650
5000	Silent check, 316 S.S. body and trim									
5010	Compact wafer type, 300 lb.									
5020	1"	1 Plum	14	.571	Ea.	805	30.50		835.50	930
5021	1-1/4"		12	.667		870	35.50		905.50	1,000
5022	1-1/2"		11	.727		900	39		939	1,050
5023	2"		9	.889		1,025	47.50		1,072.50	1,200
5024	2-1/2"	Q-1	9	1.778		1,400	86		1,486	1,675
5025	3"	"	8	2		1,625	96.50		1,721.50	1,950

23 05 Common Work Results for HVAC

23 05 23 – General-Duty Valves for HVAC Piping

	23 05 23.90 Valves, Stainless Steel	Crew	Daily Output	Labor-Hours	Unit	Material	2011 Bare Costs Labor	Equipment	Total	Total Incl O&P
5100	Full flange wafer type, 150 lb.									
5110	1"	1 Plum	14	.571	Ea.	565	30.50		595.50	665
5111	1-1/4"		12	.667		610	35.50		645.50	725
5112	1-1/2"		11	.727		695	39		734	825
5113	2"		9	.889		910	47.50		957.50	1,075
5114	2-1/2"	Q-1	9	1.778		1,075	86		1,161	1,300
5115	3"		8	2		1,225	96.50		1,321.50	1,500
5116	4"		5	3.200		1,800	154		1,954	2,225
5117	5"	Q-2	6	4		2,100	200		2,300	2,625
5118	6"		5	4.800		2,650	240		2,890	3,250
5119	8"		4.50	5.333		4,400	267		4,667	5,250
5120	10"		4	6		6,100	300		6,400	7,150
5200	Globe type, 300 lb.									
5210	2"	1 Plum	9	.889	Ea.	1,025	47.50		1,072.50	1,225
5211	2-1/2"	Q-1	9	1.778		1,275	86		1,361	1,525
5212	3"		8	2		1,575	96.50		1,671.50	1,875
5213	4"		5	3.200		2,175	154		2,329	2,625
5214	5"	Q-2	6	4		3,050	200		3,250	3,650
5215	6"		5	4.800		3,450	240		3,690	4,150
5216	8"		4.50	5.333		4,875	267		5,142	5,775
5217	10"		4	6		6,925	300		7,225	8,050
5300	Screwed end type, 300 lb.									
5310	1/2"	1 Plum	24	.333	Ea.	82	17.85		99.85	117
5311	3/4"		20	.400		97.50	21.50		119	139
5312	1"		19	.421		114	22.50		136.50	159
5313	1-1/4"		15	.533		150	28.50		178.50	208
5314	1-1/2"		13	.615		161	33		194	226
5315	2"		11	.727		201	39		240	279

23 05 93 – Testing, Adjusting, and Balancing for HVAC

	23 05 93.10 Balancing, Air	Crew	Daily Output	Labor-Hours	Unit	Material	2011 Bare Costs Labor	Equipment	Total	Total Incl O&P
0010	BALANCING, AIR (Subcontractor's quote incl. material and labor) R230500-10									
0900	Heating and ventilating equipment									
1000	Centrifugal fans, utility sets				Ea.				340	375
1100	Heating and ventilating unit								510	560
1200	In-line fan								510	560
1300	Propeller and wall fan								96.50	106
1400	Roof exhaust fan								227	250
2000	Air conditioning equipment, central station								740	810
2100	Built-up low pressure unit								680	750
2200	Built-up high pressure unit								795	875
2300	Built-up high pressure dual duct								1,250	1,375
2400	Built-up variable volume								1,475	1,625
2500	Multi-zone A.C. and heating unit								510	560
2600	For each zone over one, add								114	125
2700	Package A.C. unit								285	310
2800	Rooftop heating and cooling unit								400	440
3000	Supply, return, exhaust, registers & diffusers, avg. height ceiling								68	75
3100	High ceiling								102	112
3200	Floor height								57	62
3300	Off mixing box								45.50	50
3500	Induction unit								74	81
3600	Lab fume hood								340	375

23 05 Common Work Results for HVAC

23 05 93 – Testing, Adjusting, and Balancing for HVAC

23 05 93.10 Balancing, Air

		Crew	Daily Output	Labor-Hours	Unit	Material	2011 Bare Costs Labor	Equipment	Total	Total Incl O&P
3700	Linear supply				Ea.				171	187
3800	Linear supply high								199	219
4000	Linear return								57	62
4100	Light troffers								68	75
4200	Moduline - master								68	75
4300	Moduline - slaves								34	37.50
4400	Regenerators								455	500
4500	Taps into ceiling plenums								85.50	93.50
4600	Variable volume boxes								68	75

23 05 93.20 Balancing, Water

		Crew	Daily Output	Labor-Hours	Unit	Material	2011 Bare Costs Labor	Equipment	Total	Total Incl O&P
0010	**BALANCING, WATER** (Subcontractor's quote incl. material and labor) R230500-10									
0050	Air cooled condenser				Ea.				206	226
0080	Boiler								415	455
0100	Cabinet unit heater								70.50	77.50
0200	Chiller								500	550
0300	Convector								59	64.50
0400	Converter								294	325
0500	Cooling tower								380	420
0600	Fan coil unit, unit ventilator								106	116
0700	Fin tube and radiant panels								118	129
0800	Main and duct re-heat coils								109	120
0810	Heat exchanger								109	120
0900	Main balancing cocks								88	97
1000	Pumps								259	284
1100	Unit heater								82.50	90.50

23 05 93.50 Piping, Testing

		Crew	Daily Output	Labor-Hours	Unit	Material	2011 Bare Costs Labor	Equipment	Total	Total Incl O&P
0010	**PIPING, TESTING**									
0100	Nondestructive testing									
0110	Nondestructive hydraulic pressure test, isolate & 1 hr. hold									
0120	1" - 4" pipe									
0140	0 – 250 L.F.	1 Stpi	1.33	6.015	Ea.		325		325	485
0160	250 – 500 L.F.	"	.80	10			540		540	810
0180	500 – 1000 L.F.	Q-5	1.14	14.035			685		685	1,025
0200	1000 – 2000 L.F.	"	.80	20			975		975	1,450
0300	6" - 10" pipe									
0320	0 – 250 L.F.	Q-5	1	16	Ea.		780		780	1,175
0340	250 – 500 L.F.		.73	21.918			1,075		1,075	1,600
0360	500 – 1000 L.F.		.53	30.189			1,475		1,475	2,200
0380	1000 – 2000 L.F.		.38	42.105			2,050		2,050	3,050
1000	Pneumatic pressure test, includes soaping joints									
1120	1" - 4" pipe									
1140	0 – 250 L.F.	Q-5	2.67	5.993	Ea.	10.95	292		302.95	445
1160	250 – 500 L.F.		1.33	12.030		22	585		607	900
1180	500 – 1000 L.F.		.80	20		33	975		1,008	1,475
1200	1000 – 2000 L.F.		.50	32		44	1,550		1,594	2,375
1300	6" - 10" pipe									
1320	0 – 250 L.F.	Q-5	1.33	12.030	Ea.	10.95	585		595.95	885
1340	250 – 500 L.F.		.67	23.881		22	1,175		1,197	1,750
1360	500 – 1000 L.F.		.40	40		44	1,950		1,994	2,950
1380	1000 – 2000 L.F.		.25	64		55	3,125		3,180	4,700
2000	X-Ray of welds									
2110	2" diam.	1 Stpi	8	1	Ea.	11.60	54		65.60	94

23 05 Common Work Results for HVAC

23 05 93 – Testing, Adjusting, and Balancing for HVAC

23 05 93.50 Piping, Testing		Crew	Daily Output	Labor-Hours	Unit	Material	2011 Bare Costs Labor	Equipment	Total	Total Incl O&P
2120	3" diam.	1 Stpi	8	1	Ea.	11.60	54		65.60	94
2130	4" diam.		8	1		17.40	54		71.40	100
2140	6" diam.		8	1		17.40	54		71.40	100
2150	8" diam.		6.60	1.212		17.40	65.50		82.90	117
2160	10" diam.	↓	6	1.333	↓	23	72		95	134
3000	Liquid penetration of welds									
3110	2" diam.	1 Stpi	14	.571	Ea.	3.55	31		34.55	50
3120	3" diam.		13.60	.588		3.55	32		35.55	51.50
3130	4" diam.		13.40	.597		3.55	32.50		36.05	52
3140	6" diam.		13.20	.606		3.55	33		36.55	53
3150	8" diam.		13	.615		5.35	33.50		38.85	55.50
3160	10" diam.	↓	12.80	.625	↓	5.35	34		39.35	56.50

23 07 HVAC Insulation

23 07 13 – Duct Insulation

23 07 13.10 Duct Thermal Insulation

			Crew	Daily Output	Labor-Hours	Unit	Material	2011 Bare Costs Labor	Equipment	Total	Total Incl O&P
0010	**DUCT THERMAL INSULATION**										
0100	Rule of thumb, as a percentage of total mechanical costs					Job				10%	
0110	Insulation req'd is based on the surface size/area to be covered										
3000	Ductwork										
3020	Blanket type, fiberglass, flexible										
3140	FSK vapor barrier wrap, .75 lb. density										
3160	1" thick	G	Q-14	350	.046	S.F.	.20	1.93		2.13	3.20
3170	1-1/2" thick	G		320	.050		.23	2.11		2.34	3.51
3180	2" thick	G		300	.053		.28	2.25		2.53	3.79
3190	3" thick	G		260	.062		.40	2.60		3	4.46
3200	4" thick	G		242	.066		.60	2.79		3.39	4.97
3212	Vinyl Jacket, .75 lb. density, 1-1/2" thick	G	↓	320	.050	↓	.23	2.11		2.34	3.51
3280	Unfaced, 1 lb. density										
3310	1" thick	G	Q-14	360	.044	S.F.	.19	1.88		2.07	3.11
3320	1-1/2" thick	G		330	.048		.23	2.05		2.28	3.41
3330	2" thick	G	↓	310	.052	↓	.31	2.18		2.49	3.71
3400	FSK facing, 1 lb. density										
3420	1-1/2" thick	G	Q-14	310	.052	S.F.	.18	2.18		2.36	3.57
3430	2" thick	G	"	300	.053	"	.20	2.25		2.45	3.70
3450	FSK facing, 1.5 lb. density										
3470	1-1/2" thick	G	Q-14	300	.053	S.F.	.27	2.25		2.52	3.78
3480	2" thick	G	"	290	.055	"	.33	2.33		2.66	3.96
3730	Sheet insulation										
3760	Polyethylene foam, closed cell, UV resistant										
3770	Standard temperature (-90°F to +212°F)										
3771	1/4" thick	G	Q-14	450	.036	S.F.	1.38	1.50		2.88	3.84
3772	3/8" thick	G		440	.036		1.97	1.53		3.50	4.54
3773	1/2" thick	G		420	.038		2.42	1.61		4.03	5.15
3774	3/4" thick	G		400	.040		3.46	1.69		5.15	6.40
3775	1" thick	G		380	.042		4.67	1.78		6.45	7.90
3776	1-1/2" thick	G		360	.044		7.40	1.88		9.28	11.05
3777	2" thick	G		340	.047		9.90	1.99		11.89	13.95
3778	2-1/2" thick	G	↓	320	.050	↓	12.55	2.11		14.66	17.05
3779	Adhesive (see line 7878)										
3780	Foam, rubber										

23 07 HVAC Insulation

23 07 13 – Duct Insulation

23 07 13.10 Duct Thermal Insulation		Crew	Daily Output	Labor-Hours	Unit	Material	2011 Bare Costs Labor	Equipment	Total	Total Incl O&P
3782	1" thick	G	1 Stpi	50	.160	S.F.	4.15	8.65		12.80	17.45
3795	Finishes										
3800	Stainless steel woven mesh		Q-14	100	.160	S.F.	.84	6.75		7.59	11.35
3820	18 oz. fiberglass cloth, pasted on			170	.094		.77	3.97		4.74	7
3900	8 oz. canvas, pasted on		↓	180	.089	↓	.49	3.75		4.24	6.35
7878	Contact cement, quart can					Ea.	9.85			9.85	10.85

23 07 16 – HVAC Equipment Insulation

23 07 16.10 HVAC Equipment Thermal Insulation

			Crew	Daily Output	Labor-Hours	Unit	Material	2011 Bare Costs Labor	Equipment	Total	Total Incl O&P
0010	**HVAC EQUIPMENT THERMAL INSULATION**										
0100	Rule of thumb, as a percentage of total mechanical costs					Job				10%	
0110	Insulation req'd is based on the surface size/area to be covered										
1000	Boiler, 1-1/2" calcium silicate only	G	Q-14	110	.145	S.F.	2.92	6.15		9.07	12.70
1020	Plus 2" fiberglass	G	"	80	.200	"	3.82	8.45		12.27	17.25
2000	Breeching, 2" calcium silicate										
2020	Rectangular	G	Q-14	42	.381	S.F.	5.70	16.10		21.80	31.50
2040	Round	G	"	38.70	.413	"	5.95	17.45		23.40	33.50
2300	Calcium silicate block, + 200°F to + 1200°F										
2310	On irregular surfaces, valves and fittings										
2340	1" thick	G	Q-14	30	.533	S.F.	2.59	22.50		25.09	38
2360	1-1/2" thick	G		25	.640		2.92	27		29.92	45
2380	2" thick	G		22	.727		3.83	30.50		34.33	51.50
2400	3" thick	G	↓	18	.889	↓	5.90	37.50		43.40	64.50
2410	On plane surfaces										
2420	1" thick	G	Q-14	126	.127	S.F.	2.59	5.35		7.94	11.15
2430	1-1/2" thick	G		120	.133		2.92	5.65		8.57	11.90
2440	2" thick	G		100	.160		3.83	6.75		10.58	14.65
2450	3" thick	G	↓	70	.229	↓	5.90	9.65		15.55	21.50

23 09 Instrumentation and Control for HVAC

23 09 13 – Instrumentation and Control Devices for HVAC

23 09 13.60 Water Level Controls

		Crew	Daily Output	Labor-Hours	Unit	Material	2011 Bare Costs Labor	Equipment	Total	Total Incl O&P
0010	**WATER LEVEL CONTROLS**									
1000	Electric water feeder	1 Stpi	12	.667	Ea.	251	36		287	330
2000	Feeder cut-off combination									
2100	Steam system up to 5000 sq. ft.	1 Stpi	12	.667	Ea.	535	36		571	645
2200	Steam system above 5000 sq. ft.		12	.667		660	36		696	780
2300	Steam and hot water, high pressure	↓	10	.800	↓	775	43.50		818.50	920
3000	Low water cut-off for hot water boiler, 50 psi maximum									
3100	1" top & bottom equalizing pipes, manual reset	1 Stpi	14	.571	Ea.	300	31		331	375
3200	1" top & bottom equalizing pipes		14	.571		300	31		331	375
3300	2-1/2" side connection for nipple-to-boiler	↓	14	.571	↓	270	31		301	345
4000	Low water cut-off for low pressure steam with quick hook-up ftgs.									
4100	For installation in gauge glass tappings	1 Stpi	16	.500	Ea.	218	27		245	280
4200	Built-in type, 2-1/2" tap - 3-1/8" insertion		16	.500		180	27		207	239
4300	Built-in type, 2-1/2" tap - 1-3/4" insertion		16	.500		191	27		218	251
4400	Side connection to 2-1/2" tapping		16	.500		206	27		233	268
5000	Pump control, low water cut-off and alarm switch	↓	14	.571	↓	470	31		501	560
9000	Water gauges, complete									
9010	Rough brass, wheel type									
9020	125 PSI at 350°F									

23 09 Instrumentation and Control for HVAC

23 09 13 – Instrumentation and Control Devices for HVAC

23 09 13.60 Water Level Controls		Crew	Daily Output	Labor-Hours	Unit	Material	2011 Bare Costs Labor	Equipment	Total	Total Incl O&P
9030	3/8" pipe size	1 Stpi	11	.727	Ea.	55	39.50		94.50	119
9040	1/2" pipe size	"	10	.800	"	60	43.50		103.50	131
9060	200 PSI at 400°F									
9070	3/8" pipe size	1 Stpi	11	.727	Ea.	68.50	39.50		108	134
9080	1/2" pipe size		10	.800		68.50	43.50		112	140
9090	3/4" pipe size	↓	9	.889	↓	87.50	48		135.50	168
9130	Rough brass, chain lever type									
9140	250 PSI at 400°F									
9200	Polished brass, wheel type									
9210	200 PSI at 400°F									
9220	3/8" pipe size	1 Stpi	11	.727	Ea.	129	39.50		168.50	201
9230	1/2" pipe size		10	.800		131	43.50		174.50	209
9240	3/4" pipe size	↓	9	.889	↓	138	48		186	224
9260	Polished brass, chain lever type									
9270	250 PSI at 400°F									
9280	1/2" pipe size	1 Stpi	10	.800	Ea.	178	43.50		221.50	261
9290	3/4" pipe size	"	9	.889	"	194	48		242	285
9400	Bronze, high pressure, ASME									
9410	1/2" pipe size	1 Stpi	10	.800	Ea.	225	43.50		268.50	310
9420	3/4" pipe size	"	9	.889	"	238	48		286	335
9460	Chain lever type									
9470	1/2" pipe size	1 Stpi	10	.800	Ea.	300	43.50		343.50	395
9480	3/4" pipe size	"	9	.889	"	315	48		363	420
9500	316 stainless steel, high pressure, ASME									
9510	500 PSI at 450°F									
9520	1/2" pipe size	1 Stpi	10	.800	Ea.	720	43.50		763.50	860
9530	3/4" pipe size	"	9	.889	"	700	48		748	840

23 09 23 – Direct-Digital Control System for HVAC

23 09 23.10 Control Components/DDC Systems		Crew	Daily Output	Labor-Hours	Unit	Material	Labor	Equipment	Total	Total Incl O&P
0010	**CONTROL COMPONENTS/DDC SYSTEMS** (Sub's quote incl. M & L)									
0100	Analog inputs									
0110	Sensors (avg. 50' run in 1/2" EMT)									
0120	Duct temperature				Ea.				360	395
0130	Space temperature								575	635
0140	Duct humidity, +/- 3%								605	665
0150	Space humidity, +/- 2%								925	1,025
0160	Duct static pressure								490	540
0170	C.F.M./Transducer								660	730
0172	Water temp. (see Section 23 21 20 for well tap add)								570	625
0174	Water flow (see Section 23 21 20 for circuit sensor add)								2,075	2,300
0176	Water pressure differential (see Section 23 21 20 for tap add)								850	935
0177	Steam flow (see Section 23 21 20 for circuit sensor add)								2,075	2,300
0178	Steam pressure (see Section 23 21 20 for tap add)								890	980
0180	K.W./Transducer								1,175	1,300
0182	K.W.H. totalization (not incl. elec. meter pulse xmtr.)								545	600
0190	Space static pressure				↓				925	1,025
1000	Analog outputs (avg. 50' run in 1/2" EMT)									
1010	P/I Transducer				Ea.				550	605
1020	Analog output, matl. in MUX								265	291
1030	Pneumatic (not incl. control device)								560	615
1040	Electric (not incl control device)				↓				330	365
2000	Status (Alarms)									

23 09 23 – Direct-Digital Control System for HVAC

23 09 23.10 Control Components/DDC Systems	Crew	Daily Output	Labor-Hours	Unit	Material	2011 Bare Costs Labor	Equipment	Total	Total Incl O&P
2100 Digital inputs (avg. 50' run in 1/2" EMT)									
2110 Freeze				Ea.				380	415
2120 Fire								340	375
2130 Differential pressure, (air)								520	570
2140 Differential pressure, (water)								850	935
2150 Current sensor								380	415
2160 Duct high temperature thermostat								495	545
2170 Duct smoke detector				↓				615	675
2200 Digital output (avg. 50' run in 1/2" EMT)									
2210 Start/stop				Ea.				297	325
2220 On/off (maintained contact)				"				510	560
3000 Controller M.U.X. panel, incl. function boards									
3100 48 point				Ea.				4,575	5,050
3110 128 point				"				6,300	6,925
3200 D.D.C. controller (avg. 50' run in conduit)									
3210 Mechanical room									
3214 16 point controller (incl. 120 volt/1 phase power supply)				Ea.				1,925	3,125
3229 32 point controller (incl. 120 volt/1 phase power supply)				"				4,725	5,200
3230 Includes software programming and checkout									
3260 Space									
3266 V.A.V. terminal box (incl. space temp. sensor)				Ea.				735	805
3280 Host computer (avg. 50' run in conduit)									
3281 Package complete with PC, keyboard,									
3282 printer, color CRT, modem, basic software				Ea.				8,500	9,375
4000 Front end costs									
4100 Computer (P.C.)/software program				Ea.				5,675	6,250
4200 Color graphics software								3,400	3,750
4300 Color graphics slides								425	470
4350 Additional dot matrix printer				↓				850	935
4400 Communications trunk cable				L.F.				3.31	3.64
4500 Engineering labor, (not incl. dftg.)				Point				72	79
4600 Calibration labor								72	79
4700 Start-up, checkout labor				↓				109	120
4800 Drafting labor, as req'd									
5000 Communications bus (data transmission cable)									
5010 #18 twisted shielded pair in 1/2" EMT conduit				C.L.F.				330	365
8000 Applications software									
8050 Basic maintenance manager software (not incl. data base entry)				Ea.				1,700	1,875
8100 Time program				Point				5.95	6.55
8120 Duty cycle								11.85	13.05
8140 Optimum start/stop								36	39.50
8160 Demand limiting								17.80	19.55
8180 Enthalpy program				↓				36	39.50
8200 Boiler optimization				Ea.				1,075	1,175
8220 Chiller optimization				"				1,425	1,550
8240 Custom applications									
8260 Cost varies with complexity									

23 09 33 – Electric and Electronic Control System for HVAC

23 09 33.10 Electronic Control Systems

0010 **ELECTRONIC CONTROL SYSTEMS** R230500-10									
0020 For electronic costs, add to Section 23 09 43.10				Ea.					15%

23 09 Instrumentation and Control for HVAC

23 09 43 – Pneumatic Control System for HVAC

23 09 43.10 Pneumatic Control Systems

		Crew	Daily Output	Labor-Hours	Unit	Material	2011 Bare Costs Labor	Equipment	Total	Total Incl O&P
0010	**PNEUMATIC CONTROL SYSTEMS** R230500-10									
0011	Including a nominal 50 ft. of tubing. Add control panelboard if req'd.									
0100	Heating and ventilating, split system									
0200	Mixed air control, economizer cycle, panel readout, tubing									
0220	Up to 10 tons [G]	Q-19	.68	35.294	Ea.	3,675	1,750		5,425	6,625
0240	For 10 to 20 tons [G]		.63	37.915		3,925	1,875		5,800	7,100
0260	For over 20 tons [G]		.58	41.096		4,275	2,025		6,300	7,700
0270	Enthalpy cycle, up to 10 tons		.50	48.387		4,075	2,375		6,450	8,025
0280	For 10 to 20 tons		.46	52.174		4,375	2,575		6,950	8,650
0290	For over 20 tons		.42	56.604		4,750	2,775		7,525	9,375
0300	Heating coil, hot water, 3 way valve,									
0320	Freezestat, limit control on discharge, readout	Q-5	.69	23.088	Ea.	2,725	1,125		3,850	4,675
0500	Cooling coil, chilled water, room									
0520	Thermostat, 3 way valve	Q-5	2	8	Ea.	1,225	390		1,615	1,925
0600	Cooling tower, fan cycle, damper control,									
0620	Control system including water readout in/out at panel	Q-19	.67	35.821	Ea.	4,850	1,775		6,625	7,950
1000	Unit ventilator, day/night operation,									
1100	freezestat, ASHRAE, cycle 2	Q-19	.91	26.374	Ea.	2,675	1,300		3,975	4,875
2000	Compensated hot water from boiler, valve control,									
2100	readout and reset at panel, up to 60 GPM	Q-19	.55	43.956	Ea.	5,000	2,175		7,175	8,725
2120	For 120 GPM		.51	47.059		5,350	2,325		7,675	9,350
2140	For 240 GPM		.49	49.180		5,600	2,425		8,025	9,775
3000	Boiler room combustion air, damper to 5 S.F., controls		1.37	17.582		2,425	865		3,290	3,950
3500	Fan coil, heating and cooling valves, 4 pipe control system		3	8		1,100	395		1,495	1,775
3600	Heat exchanger system controls		.86	27.907		2,350	1,375		3,725	4,625
3900	Multizone control (one per zone), includes thermostat, damper									
3910	motor and reset of discharge temperature	Q-5	.51	31.373	Ea.	2,425	1,525		3,950	4,925
4000	Pneumatic thermostat, including controlling room radiator valve	"	2.43	6.593		730	320		1,050	1,275
4040	Program energy saving optimizer [G]	Q-19	1.21	19.786		6,050	975		7,025	8,100
4060	Pump control system	"	3	8		1,125	395		1,520	1,800
4080	Reheat coil control system, not incl coil	Q-5	2.43	6.593		945	320		1,265	1,525
4500	Air supply for pneumatic control system									
4600	Tank mounted duplex compressor, starter, alternator,									
4620	piping, dryer, PRV station and filter									
4630	1/2 HP	Q-19	.68	35.139	Ea.	8,975	1,725		10,700	12,500
4640	3/4 HP		.64	37.383		9,425	1,850		11,275	13,200
4650	1 HP		.61	39.539		10,300	1,950		12,250	14,200
4660	1-1/2 HP		.58	41.739		11,000	2,050		13,050	15,200
4680	3 HP		.55	43.956		14,900	2,175		17,075	19,600
4690	5 HP		.42	57.143		26,100	2,825		28,925	32,900
4800	Main air supply, includes 3/8" copper main and labor	Q-5	1.82	8.791	C.L.F.	300	430		730	975
4810	If poly tubing used, deduct									30%
7000	Static pressure control for air handling unit, includes pressure									
7010	sensor, receiver controller, readout and damper motors	Q-19	.64	37.383	Ea.	7,150	1,850		9,000	10,600
7020	If return air fan requires control, add									70%
8600	VAV boxes, incl. thermostat, damper motor, reheat coil & tubing	Q-5	1.46	10.989		1,125	535		1,660	2,025
8610	If no reheat coil, deduct									204
9400	Sub assemblies for assembly systems									

23 09 53 – Pneumatic and Electric Control System for HVAC

23 09 53.10 Control Components

		Crew	Daily Output	Labor-Hours	Unit	Material	2011 Bare Costs Labor	Equipment	Total	Total Incl O&P
0010	**CONTROL COMPONENTS**									
0500	Aquastats									

23 09 53 – Pneumatic and Electric Control System for HVAC

23 09 53.10 Control Components		Crew	Daily Output	Labor-Hours	Unit	Material	2011 Bare Costs Labor	Equipment	Total	Total Incl O&P
0508	Immersion type									
0510	High/low limit, breaks contact w/temp rise	1 Stpi	4	2	Ea.	365	108		473	560
0514	Sequencing, break 2 switches w/temp rise		4	2		261	108		369	450
0518	Circulating, makes contact w/temp rise	↓	4	2	↓	237	108		345	420
0600	Carbon monoxide detector system									
0606	Panel	1 Stpi	4	2	Ea.	1,350	108		1,458	1,650
0610	Sensor	"	7.30	1.096	"	785	59.50		844.50	955
0700	Controller, receiver									
0730	Pneumatic, panel mount, single input	1 Plum	8	1	Ea.	410	53.50		463.50	530
0740	With conversion mounting bracket		8	1		410	53.50		463.50	530
0750	Dual input, with control point adjustment	↓	7	1.143		565	61.50		626.50	710
0850	Electric, single snap switch	1 Elec	4	2		450	101		551	645
0860	Dual snap switches		3	2.667		605	134		739	865
0870	Humidity controller		8	1		210	50.50		260.50	305
0880	Load limiting controller		8	1		775	50.50		825.50	930
0890	Temperature controller	↓	8	1	↓	435	50.50		485.50	555
0900	Control panel readout									
0910	Panel up to 12 indicators	1 Stpi	1.20	6.667	Ea.	264	360		624	830
0914	Panel up to 24 indicators		.86	9.302		355	505		860	1,150
0918	Panel up to 48 indicators	↓	.50	16	↓	860	865		1,725	2,250
1000	Enthalpy control, boiler water temperature control									
1010	governed by outdoor temperature, with timer	1 Elec	3	2.667	Ea.	330	134		464	565
1300	Energy control/monitor									
1320	BTU computer meter and controller	1 Stpi	1	8	Ea.	1,425	435		1,860	2,200
1600	Flow meters									
1610	Gas	1 Stpi	4	2	Ea.	118	108		226	291
1620	Liquid	"	4	2	"	114	108		222	288
1640	Freezestat									
1644	20' sensing element, adjustable	1 Stpi	3.50	2.286	Ea.	64.50	124		188.50	256
2000	Gauges, pressure or vacuum									
2100	2" diameter dial	1 Stpi	32	.250	Ea.	14.30	13.55		27.85	35.50
2200	2-1/2" diameter dial		32	.250		15.65	13.55		29.20	37.50
2300	3-1/2" diameter dial		32	.250		35.50	13.55		49.05	59.50
2400	4-1/2" diameter dial	↓	32	.250	↓	52.50	13.55		66.05	78
2700	Flanged iron case, black ring									
2800	3-1/2" diameter dial	1 Stpi	32	.250	Ea.	102	13.55		115.55	132
2900	4-1/2" diameter dial		32	.250		105	13.55		118.55	136
3000	6" diameter dial	↓	32	.250	↓	167	13.55		180.55	203
3010	Steel case, 0 – 300 psi									
3012	2" diam. dial	1 Stpi	16	.500	Ea.	7.70	27		34.70	49
3014	4" diam. dial	"	16	.500	"	36.50	27		63.50	81
3020	Aluminum case, 0 – 300 psi									
3022	3-1/2" diam. dial	1 Stpi	16	.500	Ea.	33.50	27		60.50	77.50
3024	4-1/2" diam. dial		16	.500		101	27		128	152
3026	6" diam. dial		16	.500		165	27		192	223
3028	8-1/2" diam. dial	↓	16	.500	↓	281	27		308	350
3030	Brass case, 0 – 300 psi									
3032	2" diam. dial	1 Stpi	16	.500	Ea.	39.50	27		66.50	84
3034	4-1/2" diam. dial	"	16	.500	"	93.50	27		120.50	144
3040	Steel case, high pressure, 0 -10,000 psi									
3042	4-1/2" diam. dial	1 Stpi	16	.500	Ea.	118	27		145	170
3044	6-1/2" diam. dial		16	.500		228	27		255	292
3046	8-1/2" diam. dial	↓	16	.500	↓	330	27		357	405

23 09 53 – Pneumatic and Electric Control System for HVAC

23 09 53.10 Control Components	Crew	Daily Output	Labor-Hours	Unit	Material	2011 Bare Costs Labor	Equipment	Total	Total Incl O&P
3080	Pressure gauge, differential, magnehelic									
3084	0 – 2" W.C., with air filter kit	1 Stpi	6	1.333	Ea.	93.50	72		165.50	211
3300	For compound pressure-vacuum, add					18%				
3350	Humidistat									
3360	Pneumatic operation									
3361	Room humidistat, direct acting	1 Stpi	12	.667	Ea.	289	36		325	375
3362	Room humidistat, reverse acting		12	.667		289	36		325	375
3363	Room humidity transmitter		17	.471		310	25.50		335.50	380
3364	Duct mounted controller		12	.667		340	36		376	425
3365	Duct mounted transmitter		12	.667		310	36		346	395
3366	Humidity indicator, 3-1/2"		28	.286		123	15.45		138.45	158
3390	Electric operated	1 Shee	8	1		80	51.50		131.50	166
3400	Relays									
3430	Pneumatic/electric	1 Plum	16	.500	Ea.	310	27		337	380
3440	Pneumatic proportioning		8	1		209	53.50		262.50	310
3450	Pneumatic switching		12	.667		142	35.50		177.50	210
3460	Selector, 3 point		6	1.333		101	71.50		172.50	218
3470	Pneumatic time delay		8	1		264	53.50		317.50	370
3500	Sensor, air operated									
3520	Humidity	1 Plum	16	.500	Ea.	340	27		367	410
3540	Pressure		16	.500		58	27		85	104
3560	Temperature		12	.667		152	35.50		187.50	221
3600	Electric operated									
3620	Humidity	1 Elec	8	1	Ea.	89.50	50.50		140	173
3650	Pressure		8	1		1,500	50.50		1,550.50	1,725
3680	Temperature		10	.800		93.50	40		133.50	163
3700	Switches									
3710	Minimum position									
3720	Electrical	1 Stpi	4	2	Ea.	19.45	108		127.45	184
3730	Pneumatic	"	4	2	"	93.50	108		201.50	265
4000	Thermometers									
4100	Dial type, 3-1/2" diameter, vapor type, union connection	1 Stpi	32	.250	Ea.	227	13.55		240.55	270
4120	Liquid type, union connection		32	.250		440	13.55		453.55	505
4130	Remote reading, 15' capillary		32	.250		172	13.55		185.55	209
4500	Stem type, 6-1/2" case, 2" stem, 1/2" NPT		32	.250		60.50	13.55		74.05	86.50
4520	4" stem, 1/2" NPT		32	.250		74	13.55		87.55	102
4600	9" case, 3-1/2" stem, 3/4" NPT		28	.286		83	15.45		98.45	115
4620	6" stem, 3/4" NPT		28	.286		98	15.45		113.45	131
4640	8" stem, 3/4" NPT		28	.286		169	15.45		184.45	209
4660	12" stem, 1" NPT		26	.308		185	16.65		201.65	229
4670	Bi-metal, dial type, steel case brass stem									
4672	2" dial, 4" - 9" stem	1 Stpi	16	.500	Ea.	29.50	27		56.50	73
4673	2-1/2" dial, 4" - 9" stem		16	.500		29.50	27		56.50	73
4674	3-1/2" dial, 4" - 9" stem		16	.500		33	27		60	76.50
4680	Mercury filled, industrial, union connection type									
4682	Angle stem, 7" scale	1 Stpi	16	.500	Ea.	63	27		90	110
4683	9" scale		16	.500		169	27		196	227
4684	12" scale		16	.500		218	27		245	281
4686	Straight stem, 7" scale		16	.500		149	27		176	205
4687	9" scale		16	.500		151	27		178	207
4688	12" scale		16	.500		183	27		210	242
4690	Mercury filled, industrial, separable socket type									
4692	Angle stem, with socket, 7" scale	1 Stpi	16	.500	Ea.	59	27		86	105

23 09 53 – Pneumatic and Electric Control System for HVAC

23 09 53.10 Control Components	Crew	Daily Output	Labor-Hours	Unit	Material	2011 Bare Costs Labor	Equipment	Total	Total Incl O&P
4693	9" scale	1 Stpi	16	.500	Ea.	75.50	27		102.50	124
4694	12" scale		16	.500		139	27		166	194
4696	Straight stem, with socket, 7" scale		16	.500		29	27		56	72.50
4697	9" scale		16	.500		46	27		73	91
4698	12" scale		16	.500		98	27		125	149
5000	Thermostats									
5030	Manual	1 Stpi	8	1	Ea.	45	54		99	130
5040	1 set back, electric, timed G		8	1		30	54		84	114
5050	2 set back, electric, timed G		8	1		146	54		200	241
5100	Locking cover		20	.400		18.05	21.50		39.55	52.50
5200	24 hour, automatic, clock G	1 Shee	8	1		137	51.50		188.50	229
5220	Electric, low voltage, 2 wire	1 Elec	13	.615		32.50	31		63.50	81.50
5230	3 wire		10	.800		31.50	40		71.50	94
5236	Heating/cooling, low voltage, with clock		8	1		231	50.50		281.50	330
5240	Pneumatic									
5250	Single temp., single pressure	1 Stpi	8	1	Ea.	200	54		254	300
5251	Dual pressure		8	1		280	54		334	390
5252	Dual temp., dual pressure		8	1		257	54		311	365
5253	Reverse acting w/averaging element		8	1		198	54		252	299
5254	Heating-cooling w/deadband		8	1		445	54		499	570
5255	Integral w/piston top valve actuator		8	1		177	54		231	276
5256	Dual temp., dual pressure		8	1		181	54		235	280
5257	Low limit, 8' averaging element		8	1		171	54		225	269
5258	Room single temp. proportional		8	1		71.50	54		125.50	160
5260	Dual temp, direct acting for VAV fan		8	1		271	54		325	380
5262	Capillary tube type, 20', 30°F to 100°F		8	1		119	54		173	212
5300	Transmitter, pneumatic									
5320	Temperature averaging element	Q-1	8	2	Ea.	116	96.50		212.50	271
5350	Pressure differential	1 Plum	7	1.143		975	61.50		1,036.50	1,175
5370	Humidity, duct		8	1		310	53.50		363.50	420
5380	Room		12	.667		310	35.50		345.50	395
5390	Temperature, with averaging element		6	1.333		164	71.50		235.50	287
5500	Timer/time clocks									
5510	7 day, 12 hr. battery	1 Stpi	2.60	3.077	Ea.	186	166		352	455
5600	Recorders									
5610	Hydrograph humidity, wall mtd., aluminum case, grad. chart									
5612	8"	1 Stpi	1.60	5	Ea.	645	271		916	1,100
5614	10"	"	1.60	5	"	715	271		986	1,200
5620	Thermo-hydrograph, wall mtd., grad. chart, 5' SS probe									
5622	8"	1 Stpi	1.60	5	Ea.	565	271		836	1,025
5624	10"	"	1.60	5	"	1,125	271		1,396	1,650
5630	Time of operation, wall mtd., 1 pen cap, 24 hr. clock									
5632	6"	1 Stpi	1.60	5	Ea.	420	271		691	865
5634	8"	"	1.60	5	"	650	271		921	1,125
5640	Pressure & vacuum, bourdon tube type									
5642	Flush mount, 1 pen									
5644	4"	1 Stpi	1.60	5	Ea.	450	271		721	900
5645	6"		1.60	5		450	271		721	900
5646	8"		1.60	5		700	271		971	1,175
5648	10"		1.60	5		750	271		1,021	1,225
6000	Valves, motorized zone									
6100	Sweat connections, 1/2" C x C	1 Stpi	20	.400	Ea.	159	21.50		180.50	208
6110	3/4" C x C		20	.400		160	21.50		181.50	209

23 09 53 – Pneumatic and Electric Control System for HVAC

23 09 53.10 Control Components		Crew	Daily Output	Labor-Hours	Unit	Material	2011 Bare Costs Labor	2011 Bare Costs Equipment	Total	Total Incl O&P
6120	1" C x C	1 Stpi	19	.421	Ea.	208	23		231	263
6140	1/2" C x C, with end switch, 2 wire		20	.400		158	21.50		179.50	206
6150	3/4" C x C, with end switch, 2 wire		20	.400		169	21.50		190.50	218
6160	1" C x C, with end switch, 2 wire	↓	19	.421	↓	192	23		215	245
7090	Valves, motor controlled, including actuator									
7100	Electric motor actuated									
7200	Brass, two way, screwed									
7210	1/2" pipe size	L-6	36	.333	Ea.	208	17.50		225.50	255
7220	3/4" pipe size		30	.400		267	21		288	325
7230	1" pipe size		28	.429		310	22.50		332.50	380
7240	1-1/2" pipe size		19	.632		410	33		443	500
7250	2" pipe size	↓	16	.750	↓	680	39.50		719.50	805
7350	Brass, three way, screwed									
7360	1/2" pipe size	L-6	33	.364	Ea.	254	19.10		273.10	310
7370	3/4" pipe size		27	.444		296	23.50		319.50	360
7380	1" pipe size		25.50	.471		385	24.50		409.50	460
7384	1-1/4" pipe size		21	.571		490	30		520	585
7390	1-1/2" pipe size		17	.706		560	37		597	670
7400	2" pipe size	↓	14	.857	↓	745	45		790	885
7550	Iron body, two way, flanged									
7560	2-1/2" pipe size	L-6	4	3	Ea.	1,100	158		1,258	1,450
7570	3" pipe size		3	4		1,175	210		1,385	1,600
7580	4" pipe size	↓	2	6	↓	2,025	315		2,340	2,700
7850	Iron body, three way, flanged									
7860	2-1/2" pipe size	L-6	3	4	Ea.	1,125	210		1,335	1,550
7870	3" pipe size		2.50	4.800		1,250	252		1,502	1,775
7880	4" pipe size	↓	2	6	↓	1,625	315		1,940	2,250
8000	Pneumatic, air operated									
8050	Brass, two way, screwed									
8060	1/2" pipe size, class 250	1 Plum	24	.333	Ea.	164	17.85		181.85	207
8070	3/4" pipe size, class 250		20	.400		196	21.50		217.50	248
8080	1" pipe size, class 250		19	.421		229	22.50		251.50	286
8090	1-1/4" pipe size, class 125		15	.533		285	28.50		313.50	360
8100	1-1/2" pipe size, class 125		13	.615		365	33		398	450
8110	2" pipe size, class 125	↓	11	.727	↓	420	39		459	525
8180	Brass, three way, screwed									
8190	1/2" pipe size, class 250	1 Plum	22	.364	Ea.	180	19.50		199.50	227
8200	3/4" pipe size, class 250		18	.444		223	24		247	281
8210	1" pipe size, class 250		17	.471		262	25		287	325
8214	1-1/4" pipe size, class 250		14	.571		365	30.50		395.50	450
8220	1-1/2" pipe size, class 125		11	.727		435	39		474	540
8230	2" pipe size, class 125	↓	9	.889	↓	525	47.50		572.50	645
8450	Iron body, two way, flanged									
8460	2-1/2" pipe size, 250 lb. flanges	Q-1	5	3.200	Ea.	1,750	154		1,904	2,150
8470	3" pipe size, 250 lb. flanges		4.50	3.556		1,850	172		2,022	2,300
8480	4" pipe size, 250 lb. flanges		3	5.333		2,175	257		2,432	2,775
8510	2-1/2" pipe size, class 125		5	3.200		890	154		1,044	1,200
8520	3" pipe size, class 125		4.50	3.556		975	172		1,147	1,325
8530	4" pipe size, class 125	↓	3	5.333		1,450	257		1,707	1,975
8540	5" pipe size, class 125	Q-2	3.40	7.059		2,950	355		3,305	3,775
8550	6" pipe size, class 125	"	3	8		3,400	400		3,800	4,325
8560	Iron body, three way, flanged									
8570	2-1/2" pipe size, class 125	Q-1	4.50	3.556	Ea.	960	172		1,132	1,300

23 09 53 – Pneumatic and Electric Control System for HVAC

23 09 53.10 Control Components	Crew	Daily Output	Labor-Hours	Unit	Material	2011 Bare Costs Labor	Equipment	Total	Total Incl O&P
8580	3" pipe size, class 125	Q-1	4	4	Ea.	1,100	193		1,293	1,525
8590	4" pipe size, class 125		2.50	6.400		2,225	310		2,535	2,900
8600	6" pipe size, class 125	Q-2	3	8		3,400	400		3,800	4,325
9005	Pneumatic system misc. components									
9010	Adding/subtracting repeater	1 Stpi	20	.400	Ea.	156	21.50		177.50	205
9020	Adjustable ratio network		16	.500		147	27		174	203
9030	Comparator		20	.400		300	21.50		321.50	365
9040	Cumulator									
9041	Air switching	1 Stpi	20	.400	Ea.	142	21.50		163.50	189
9042	Averaging		20	.400		128	21.50		149.50	173
9043	2:1 ratio		20	.400		288	21.50		309.50	350
9044	Sequencing		20	.400		90.50	21.50		112	132
9045	Two-position		16	.500		177	27		204	235
9046	Two-position pilot		16	.500		340	27		367	410
9050	Damper actuator									
9051	For smoke control	1 Stpi	8	1	Ea.	134	54		188	228
9052	For ventilation	"	8	1	"	193	54		247	293
9053	Series duplex pedestal mounted									
9054	For inlet vanes on fans, compressors	1 Stpi	6	1.333	Ea.	1,475	72		1,547	1,725
9060	Enthalpy logic center		16	.500		287	27		314	355
9070	High/low pressure selector		22	.364		57	19.65		76.65	92.50
9080	Signal limiter		20	.400		90.50	21.50		112	132
9081	Transmitter		28	.286		34.50	15.45		49.95	61
9090	Optimal start									
9092	Mass temperature transmitter	1 Stpi	18	.444	Ea.	135	24		159	185
9100	Step controller with positioner, time delay restrictor									
9110	recycler air valve, switches and cam settings									
9112	With cabinet									
9113	6 points	1 Stpi	4	2	Ea.	885	108		993	1,125
9114	6 points, 6 manual sequences		2	4		775	216		991	1,175
9115	8 points		3	2.667		965	144		1,109	1,300
9200	Pressure controller and switches									
9210	High static pressure limit	1 Stpi	8	1	Ea.	202	54		256	305
9220	Pressure transmitter		8	1		213	54		267	315
9230	Differential pressure transmitter		8	1		835	54		889	1,000
9240	Static pressure transmitter		8	1		835	54		889	1,000
9250	Proportional-only control, single input		6	1.333		410	72		482	560
9260	Proportional plus integral, single input		6	1.333		610	72		682	780
9270	Proportional-only, dual input		6	1.333		565	72		637	730
9280	Proportional plus integral, dual input		6	1.333		745	72		817	930
9281	Time delay for above proportional units		16	.500		143	27		170	199
9290	Proportional/2 position controller units		6	1.333		430	72		502	585
9300	Differential pressure control direct/reverse		6	1.333		365	72		437	510
9310	Air pressure reducing valve									
9311	1/8" size	1 Stpi	18	.444	Ea.	31.50	24		55.50	70.50
9315	Precision valve, 1/4" size		17	.471		176	25.50		201.50	231
9320	Air flow controller		12	.667		164	36		200	234
9330	Booster relay volume amplifier		18	.444		97	24		121	143
9340	Series restrictor, straight		60	.133		4.99	7.20		12.19	16.25
9341	T-fitting		40	.200		7.25	10.80		18.05	24
9342	In-line adjustable		48	.167		32	9		41	49
9350	Diode tee		40	.200		16.70	10.80		27.50	34.50
9351	Restrictor tee		40	.200		16.70	10.80		27.50	34.50

23 09 53 – Pneumatic and Electric Control System for HVAC

23 09 53.10 Control Components		Crew	Daily Output	Labor-Hours	Unit	Material	2011 Bare Costs Labor	Equipment	Total	Total Incl O&P
9360	Pneumatic gradual switch	1 Stpi	8	1	Ea.	279	54		333	385
9361	Selector switch		8	1		101	54		155	192
9370	Electro-pneumatic motor driven servo		10	.800		880	43.50		923.50	1,025
9390	Fan control switch and mounting base		16	.500		97.50	27		124.50	148
9400	Circulating pump sequencer		14	.571		1,150	31		1,181	1,325
9410	Pneumatic tubing, fittings and accessories									
9414	Tubing, urethane									
9415	1/8" OD x 1/16" ID	1 Stpi	120	.067	L.F.	.66	3.61		4.27	6.15
9416	1/4" OD x 1/8" ID		115	.070		1.30	3.76		5.06	7.05
9417	5/32" OD x 3/32" ID		110	.073		.74	3.93		4.67	6.65
9420	Coupling, straight									
9422	Barb x barb									
9423	1/4" x 5/32"	1 Stpi	160	.050	Ea.	.74	2.71		3.45	4.85
9424	1/4" x 1/4"		158	.051		.38	2.74		3.12	4.51
9425	3/8" x 3/8"		154	.052		.51	2.81		3.32	4.76
9426	3/8" x 1/4"		150	.053		.51	2.89		3.40	4.87
9427	1/2" x 1/4"		148	.054		4.73	2.92		7.65	9.55
9428	1/2" x 3/8"		144	.056		.99	3.01		4	5.60
9429	1/2" x 1/2"		140	.057		.65	3.09		3.74	5.35
9440	Tube x tube									
9441	1/4" x 1/4"	1 Stpi	100	.080	Ea.	2.03	4.33		6.36	8.70
9442	3/8" x 1/4"		96	.083		3.77	4.51		8.28	10.90
9443	3/8" x 3/8"		92	.087		2.52	4.70		7.22	9.75
9444	1/2" x 3/8"		88	.091		4.05	4.92		8.97	11.80
9445	1/2" x 1/2"		84	.095		4.62	5.15		9.77	12.80
9450	Elbow coupling									
9452	Barb x barb									
9454	1/4" x 1/4"	1 Stpi	158	.051	Ea.	.87	2.74		3.61	5.05
9455	3/8" x 3/8"		154	.052		2.37	2.81		5.18	6.80
9456	1/2" x 1/2"		140	.057		1.32	3.09		4.41	6.05
9460	Tube x tube									
9462	1/2" x 1/2"	1 Stpi	84	.095	Ea.	6.40	5.15		11.55	14.75
9470	Tee coupling									
9472	Barb x barb x barb									
9474	1/4" x 1/4" x 5/32"	1 Stpi	108	.074	Ea.	1.31	4.01		5.32	7.45
9475	1/4" x 1/4" x 1/4"		104	.077		.89	4.16		5.05	7.20
9476	3/8" x 3/8" x 5/32"		102	.078		2.84	4.24		7.08	9.45
9477	3/8" x 3/8" x 1/4"		99	.081		1.31	4.37		5.68	8
9478	3/8" x 3/8" x 3/8"		98	.082		1.40	4.42		5.82	8.15
9479	1/2" x 1/2" x 1/4"		96	.083		3.57	4.51		8.08	10.70
9480	1/2" x 1/2" x 3/8"		95	.084		2.97	4.56		7.53	10.05
9481	1/2" x 1/2" x 1/2"		92	.087		1.59	4.70		6.29	8.75
9484	Tube x tube x tube									
9485	1/4" x 1/4" x 1/4"	1 Stpi	66	.121	Ea.	4.03	6.55		10.58	14.25
9486	3/8" x 1/4" x 1/4"		64	.125		12.30	6.75		19.05	23.50
9487	3/8" x 3/8" x 1/4"		61	.131		5.85	7.10		12.95	17
9488	3/8" x 3/8" x 3/8"		60	.133		6.10	7.20		13.30	17.50
9489	1/2" x 1/2" x 3/8"		58	.138		12.75	7.45		20.20	25
9490	1/2" x 1/2" x 1/2"		56	.143		8.60	7.75		16.35	21
9492	Needle valve									
9494	Tube x tube									
9495	1/4" x 1/4"	1 Stpi	86	.093	Ea.	7.15	5.05		12.20	15.40
9496	3/8" x 3/8"	"	82	.098	"	10.60	5.30		15.90	19.60

23 09 Instrumentation and Control for HVAC

23 09 53 – Pneumatic and Electric Control System for HVAC

23 09 53.10 Control Components	Crew	Daily Output	Labor-Hours	Unit	Material	2011 Bare Costs Labor	Equipment	Total	Total Incl O&P
9600	Electronic system misc. components									
9610	Electric motor damper actuator	1 Elec	8	1	Ea.	605	50.50		655.50	740
9620	Damper position indicator		10	.800		114	40		154	186
9630	Pneumatic-electronic transducer	↓	12	.667	↓	59.50	33.50		93	115
9700	Modulating step controller									
9701	2-5 steps	1 Elec	5.30	1.509	Ea.	810	76		886	1,000
9702	6-10 steps		3.20	2.500		1,025	126		1,151	1,300
9703	11-15 steps		2.70	2.963		1,475	149		1,624	1,850
9704	16-20 steps	↓	1.60	5	↓	2,175	252		2,427	2,750
9710	Staging network									
9740	Accessory unit regulator									
9750	Accessory power supply	1 Elec	18	.444	Ea.	175	22.50		197.50	225
9760	Step down transformer	"	16	.500	"	181	25		206	237

23 11 Facility Fuel Piping

23 11 13 – Facility Fuel-Oil Piping

23 11 13.10 Fuel Oil Specialties

		Crew	Daily Output	Labor-Hours	Unit	Material	2011 Bare Costs Labor	Equipment	Total	Total Incl O&P
0010	**FUEL OIL SPECIALTIES**									
0020	Foot valve, single poppet, metal to metal construction									
0040	Bevel seat, 1/2" diameter	1 Stpi	20	.400	Ea.	58	21.50		79.50	96.50
0060	3/4" diameter		18	.444		59	24		83	101
0080	1" diameter		16	.500		76	27		103	124
0100	1-1/4" diameter		15	.533		98.50	29		127.50	151
0120	1-1/2" diameter		13	.615		136	33.50		169.50	200
0140	2" diameter	↓	11	.727	↓	141	39.50		180.50	214
0160	Foot valve, double poppet, metal to metal construction									
0164	1" diameter	1 Stpi	15	.533	Ea.	152	29		181	210
0166	1-1/2" diameter	"	12	.667	"	176	36		212	248
0400	Fuel fill box, flush type									
0408	Nonlocking, watertight									
0410	1-1/2" diameter	1 Stpi	12	.667	Ea.	19.40	36		55.40	75.50
0440	2" diameter		10	.800		16.75	43.50		60.25	83
0450	2-1/2" diameter		9	.889		56	48		104	134
0460	3" diameter		7	1.143		54	62		116	152
0470	4" diameter	↓	5	1.600	↓	75	86.50		161.50	212
0500	Locking inner cover									
0510	2" diameter	1 Stpi	8	1	Ea.	82.50	54		136.50	172
0520	2-1/2" diameter		7	1.143		110	62		172	214
0530	3" diameter		5	1.600		118	86.50		204.50	259
0540	4" diameter	↓	4	2	↓	152	108		260	330
0600	Fuel system components									
0620	Spill container	1 Stpi	4	2	Ea.	635	108		743	860
0640	Fill adapter, 4", straight drop		8	1		72	54		126	160
0680	Fill cap, 4"		30	.267		35.50	14.45		49.95	61
0700	Extractor fitting, 4" x 1-1/2"		8	1		340	54		394	450
0740	Vapor hose adapter, 4"	↓	8	1		102	54		156	193
0760	Wood gage stick, 10'					14.55			14.55	16
1000	Oil filters, 3/8" IPT., 20 gal. per hour	1 Stpi	20	.400		19.25	21.50		40.75	53.50
1020	32 gal. per hour		18	.444		34.50	24		58.50	74
1040	40 gal. per hour		16	.500		37	27		64	81.50
1060	50 gal. per hour	↓	14	.571	↓	38.50	31		69.50	88

23 11 Facility Fuel Piping

23 11 13 – Facility Fuel-Oil Piping

23 11 13.10 Fuel Oil Specialties	Crew	Daily Output	Labor-Hours	Unit	Material	2011 Bare Costs Labor	Equipment	Total	Total Incl O&P
2000	Remote tank gauging system, self contained									
2100	Single tank kit/8 sensors inputs w/printer	1 Stpi	2.50	3.200	Ea.	3,500	173		3,673	4,125
2120	Two tank kit/8 sensors inputs w/printer		2	4		4,725	216		4,941	5,525
3000	Valve, ball check, globe type, 3/8" diameter		24	.333		12.80	18.05		30.85	41
3500	Fusible, 3/8" diameter		24	.333		13.10	18.05		31.15	41.50
3600	1/2" diameter		24	.333		32	18.05		50.05	62
3610	3/4" diameter		20	.400		70	21.50		91.50	110
3620	1" diameter		19	.421		205	23		228	260
4000	Nonfusible, 3/8" diameter		24	.333		22.50	18.05		40.55	51.50
4500	Shutoff, gate type, lever handle, spring-fusible kit									
4520	1/4" diameter	1 Stpi	14	.571	Ea.	34.50	31		65.50	84
4540	3/8" diameter		12	.667		33.50	36		69.50	91
4560	1/2" diameter		10	.800		49.50	43.50		93	119
4570	3/4" diameter		8	1		86	54		140	176
4580	Lever handle, requires weight and fusible kit									
4600	1" diameter	1 Stpi	9	.889	Ea.	218	48		266	310
4620	1-1/4" diameter		8	1		196	54		250	297
4640	1-1/2" diameter		7	1.143		252	62		314	370
4660	2" diameter		6	1.333		320	72		392	460
4680	For fusible link, weight and braided wire, add					5%				
5000	Vent alarm, whistling signal					26			26	28.50
5500	Vent protector/breather, 1-1/4" diameter	1 Stpi	32	.250		10.40	13.55		23.95	31.50
5520	1-1/2" diameter		32	.250		12	13.55		25.55	33
5540	2" diameter		32	.250		22.50	13.55		36.05	44.50
5560	3" diameter		28	.286		36	15.45		51.45	62.50
5580	4" diameter		24	.333		40	18.05		58.05	71
5600	Dust cap, breather, 2"		40	.200		11.70	10.80		22.50	29
8000	Fuel oil and tank heaters									
8020	Electric, capacity rated at 230 volts									
8040	Immersion element in steel manifold									
8060	96 GPH at 50°F rise	Q-5	6.40	2.500	Ea.	900	122		1,022	1,175
8070	128 GPH at 50°F rise		6.20	2.581		925	126		1,051	1,225
8080	160 GPH at 50°F rise		5.90	2.712		1,025	132		1,157	1,325
8090	192 GPH at 50°F rise		5.50	2.909		1,125	142		1,267	1,425
8100	240 GPH at 50°F rise		5.10	3.137		1,300	153		1,453	1,650
8110	288 GPH at 50°F rise		4.60	3.478		1,450	169		1,619	1,850
8120	384 GPH at 50°F rise		3.10	5.161		1,700	251		1,951	2,250
8130	480 GPH at 50°F rise		2.30	6.957		2,150	340		2,490	2,850
8140	576 GPH at 50°F rise		2.10	7.619		2,150	370		2,520	2,900
8300	Suction stub, immersion type									
8320	75" long, 750 watts	1 Stpi	14	.571	Ea.	610	31		641	715
8330	99" long, 2000 watts		12	.667		855	36		891	995
8340	123" long, 3000 watts		10	.800		920	43.50		963.50	1,100
8660	Steam, cross flow, rated at 5 PSIG									
8680	42 GPH	Q-5	7	2.286	Ea.	1,400	111		1,511	1,725
8690	73 GPH		6.70	2.388		1,650	116		1,766	1,975
8700	112 GPH		6.20	2.581		1,700	126		1,826	2,075
8710	158 GPH		5.80	2.759		1,850	134		1,984	2,250
8720	187 GPH		4	4		2,550	195		2,745	3,100
8730	270 GPH		3.60	4.444		2,750	216		2,966	3,350
8740	365 GPH	Q-6	4.90	4.898		3,225	247		3,472	3,925
8750	635 GPH		3.70	6.486		4,075	330		4,405	4,975
8760	845 GPH		2.50	9.600		5,950	485		6,435	7,275

23 11 Facility Fuel Piping

23 11 13 – Facility Fuel-Oil Piping

23 11 13.10 Fuel Oil Specialties		Crew	Daily Output	Labor-Hours	Unit	Material	2011 Bare Costs Labor	Equipment	Total	Total Incl O&P
8770	1420 GPH	Q-6	1.60	15	Ea.	9,700	760		10,460	11,800
8780	2100 GPH	↓	1.10	21.818	↓	13,300	1,100		14,400	16,400

23 11 23 – Facility Natural-Gas Piping

23 11 23.10 Gas Meters

		Crew	Daily Output	Labor-Hours	Unit	Material	2011 Bare Costs Labor	Equipment	Total	Total Incl O&P
0010	**GAS METERS**									
4000	Residential									
4010	Gas meter, residential, 3/4" pipe size	1 Plum	14	.571	Ea.	244	30.50		274.50	315
4020	Gas meter, residential, 1" pipe size		12	.667		244	35.50		279.50	320
4030	Gas meter, residential, 1-1/4" pipe size	↓	10	.800	↓	254	43		297	345

23 12 Facility Fuel Pumps

23 12 13 – Facility Fuel-Oil Pumps

23 12 13.10 Pump and Motor Sets

		Crew	Daily Output	Labor-Hours	Unit	Material	2011 Bare Costs Labor	Equipment	Total	Total Incl O&P
0010	**PUMP AND MOTOR SETS**									
1810	Light fuel and diesel oils									
1820	20 GPH 1/3 HP	Q-5	6	2.667	Ea.	830	130		960	1,100
1830	27 GPH, 1/3 HP		6	2.667		830	130		960	1,100
1840	80 GPH, 1/3 HP		5	3.200		830	156		986	1,150
1850	145 GPH, 1/2 HP		4	4		880	195		1,075	1,250
1860	277 GPH, 1 HP		4	4		980	195		1,175	1,375
1870	700 GPH, 1-1/2 HP		3	5.333		2,075	260		2,335	2,675
1880	1000 GPH, 2 HP		3	5.333		2,450	260		2,710	3,075
1890	1800 GPH, 5 HP	↓	1.80	8.889	↓	3,350	435		3,785	4,325

23 13 Facility Fuel-Storage Tanks

23 13 13 – Facility Underground Fuel-Oil, Storage Tanks

23 13 13.09 Single-Wall Steel Fuel-Oil Tanks

		Crew	Daily Output	Labor-Hours	Unit	Material	2011 Bare Costs Labor	Equipment	Total	Total Incl O&P
0010	**SINGLE-WALL STEEL FUEL-OIL TANKS**									
5000	Steel underground, sti-P3, set in place, not incl. hold-down bars.									
5500	Excavation, pad, pumps and piping not included									
5510	Single wall, 500 gallon capacity, 7 gauge shell	Q-5	2.70	5.926	Ea.	2,050	289		2,339	2,675
5520	1,000 gallon capacity, 7 gauge shell	"	2.50	6.400		3,150	310		3,460	3,950
5530	2,000 gallon capacity, 1/4" thick shell	Q-7	4.60	6.957		5,125	360		5,485	6,150
5535	2,500 gallon capacity, 7 gauge shell	Q-5	3	5.333		5,650	260		5,910	6,625
5540	5,000 gallon capacity, 1/4" thick shell	Q-7	3.20	10		7,300	515		7,815	8,825
5560	10,000 gallon capacity, 1/4" thick shell		2	16		10,200	825		11,025	12,400
5580	15,000 gallon capacity, 5/16" thick shell		1.70	18.824		17,400	970		18,370	20,600
5600	20,000 gallon capacity, 5/16" thick shell		1.50	21.333		23,000	1,100		24,100	27,000
5610	25,000 gallon capacity, 3/8" thick shell		1.30	24.615		31,100	1,275		32,375	36,100
5620	30,000 gallon capacity, 3/8" thick shell		1.10	29.091		31,700	1,500		33,200	37,100
5630	40,000 gallon capacity, 3/8" thick shell		.90	35.556		35,600	1,825		37,425	41,900
5640	50,000 gallon capacity, 3/8" thick shell	↓	.80	40	↓	39,500	2,050		41,550	46,600

23 13 13.13 Dbl-Wall Steel, Undrgrnd Fuel-Oil, Stor. Tanks

0010	**DOUBLE-WALL STEEL, UNDERGROUND FUEL-OIL, STORAGE TANKS**									
6200	Steel, underground, 360°, double wall, U.L. listed,									
6210	with sti-P3 corrosion protection,									
6220	(dielectric coating, cathodic protection, electrical									
6230	isolation) 30 year warranty,									

23 13 Facility Fuel-Storage Tanks

23 13 13 – Facility Underground Fuel-Oil, Storage Tanks

23 13 13.13 Dbl-Wall Steel, Undrgrnd Fuel-Oil, Stor. Tanks

		Crew	Daily Output	Labor-Hours	Unit	Material	2011 Bare Costs Labor	Equipment	Total	Total Incl O&P
6240	not incl. manholes or hold-downs.									
6250	500 gallon capacity	Q-5	2.40	6.667	Ea.	3,900	325		4,225	4,775
6260	1,000 gallon capactiy	"	2.25	7.111		5,700	345		6,045	6,800
6270	2,000 gallon capacity	Q-7	4.16	7.692		7,475	395		7,870	8,825
6280	3,000 gallon capacity		3.90	8.205		10,900	425		11,325	12,600
6290	4,000 gallon capacity		3.64	8.791		12,700	455		13,155	14,700
6300	5,000 gallon capacity		2.91	10.997		13,400	565		13,965	15,500
6310	6,000 gallon capacity		2.42	13.223		15,400	680		16,080	17,900
6320	8,000 gallon capacity		2.08	15.385		16,700	795		17,495	19,600
6330	10,000 gallon capacity		1.82	17.582		19,100	905		20,005	22,400
6340	12,000 gallon capacity		1.70	18.824		22,100	970		23,070	25,800
6350	15,000 gallon capacity		1.33	24.060		29,000	1,250		30,250	33,800
6360	20,000 gallon capacity		1.33	24.060		40,000	1,250		41,250	45,900
6370	25,000 gallon capacity		1.16	27.586		60,000	1,425		61,425	68,000
6380	30,000 gallon capacity		1.03	31.068		72,500	1,600		74,100	82,000
6390	40,000 gallon capacity		.80	40		93,000	2,050		95,050	105,000
6395	50,000 gallon capacity		.73	43.836		113,000	2,250		115,250	127,500
6400	For hold-downs 500-2000 gal., add		16	2	Set	188	103		291	360
6410	For hold-downs 3000-6000 gal., add		12	2.667		305	137		442	540
6420	For hold-downs 8000-12,000 gal., add		11	2.909		370	150		520	635
6430	For hold-downs 15,000 gal., add		9	3.556		490	183		673	815
6440	For hold-downs 20,000 gal., add		8	4		580	206		786	945
6450	For hold-downs 20,000 gal. plus, add		6	5.333		770	275		1,045	1,250
6500	For manways, add				Ea.	1,500			1,500	1,650
6600	In place with hold-downs									
6652	550 gallon capacity	Q-5	1.84	8.696	Ea.	4,100	425		4,525	5,125

23 13 13.23 Glass-Fiber-Reinfcd-Plastic, Fuel-Oil, Storage

		Crew	Daily Output	Labor-Hours	Unit	Material	2011 Bare Costs Labor	Equipment	Total	Total Incl O&P
0010	**GLASS-FIBER-REINFCD-PLASTIC, UNDERGRND FUEL-OIL, STORAGE**									
0210	Fiberglass, underground, single wall, U.L. listed, not including									
0220	manway or hold-down strap									
0225	550 gallon capacity	Q-5	2.67	5.993	Ea.	3,100	292		3,392	3,825
0230	1,000 gallon capacity	"	2.46	6.504		3,775	315		4,090	4,625
0240	2,000 gallon capacity	Q-7	4.57	7.002		6,075	360		6,435	7,250
0250	4,000 gallon capacity		3.55	9.014		8,650	465		9,115	10,200
0260	6,000 gallon capacity		2.67	11.985		9,600	620		10,220	11,500
0270	8,000 gallon capacity		2.29	13.974		11,400	720		12,120	13,600
0280	10,000 gallon capacity		2	16		12,900	825		13,725	15,400
0282	12,000 gallon capacity		1.88	17.021		21,800	875		22,675	25,200
0284	15,000 gallon capacity		1.68	19.048		23,900	980		24,880	27,800
0290	20,000 gallon capacity		1.45	22.069		25,800	1,125		26,925	30,100
0300	25,000 gallon capacity		1.28	25		38,200	1,300		39,500	43,900
0320	30,000 gallon capacity		1.14	28.070		46,100	1,450		47,550	52,500
0340	40,000 gallon capacity		.89	35.955		65,500	1,850		67,350	75,500
0360	48,000 gallon capacity		.81	39.506		77,500	2,025		79,525	88,000
0500	For manway, fittings and hold-downs, add					20%	15%			
0600	For manways, add					1,900			1,900	2,100
1000	For helical heating coil, add	Q-5	2.50	6.400		3,850	310		4,160	4,700
1020	Fiberglass, underground, double wall, U.L. listed									
1030	includes manways, not incl. hold-down straps									
1040	600 gallon capacity	Q-5	2.42	6.612	Ea.	6,925	320		7,245	8,100
1050	1,000 gallon capacity	"	2.25	7.111		8,650	345		8,995	10,000
1060	2,500 gallon capacity	Q-7	4.16	7.692		13,400	395		13,795	15,400

23 13 Facility Fuel-Storage Tanks

23 13 13 – Facility Underground Fuel-Oil, Storage Tanks

23 13 13.23 Glass-Fiber-Reinfcd-Plastic, Fuel-Oil, Storage

		Crew	Daily Output	Labor-Hours	Unit	Material	2011 Bare Costs Labor	Equipment	Total	Total Incl O&P
1070	3,000 gallon capacity	Q-7	3.90	8.205	Ea.	14,500	425		14,925	16,600
1080	4,000 gallon capacity		3.64	8.791		16,000	455		16,455	18,400
1090	6,000 gallon capacity		2.42	13.223		18,000	680		18,680	20,800
1100	8,000 gallon capacity		2.08	15.385		21,300	795		22,095	24,700
1110	10,000 gallon capacity		1.82	17.582		24,000	905		24,905	27,800
1120	12,000 gallon capacity		1.70	18.824		29,600	970		30,570	34,000
1122	15,000 gallon capacity		1.52	21.053		38,800	1,075		39,875	44,300
1124	20,000 gallon capacity		1.33	24.060		46,500	1,250		47,750	53,000
1126	25,000 gallon capacity		1.16	27.586		62,000	1,425		63,425	70,500
1128	30,000 gallon capacity		1.03	31.068		74,500	1,600		76,100	84,500
1140	For hold-down straps, add					2%	10%			
1150	For hold-downs 500-4000 gal., add	Q-7	16	2	Set	445	103		548	640
1160	For hold-downs 5000-15000 gal., add		8	4		885	206		1,091	1,275
1170	For hold-downs 20,000 gal., add		5.33	6.004		1,325	310		1,635	1,900
1180	For hold-downs 25,000 gal., add		4	8		1,775	410		2,185	2,575
1190	For hold-downs 30,000 gal., add		2.60	12.308		2,650	635		3,285	3,875
2210	Fiberglass, underground, single wall, U.L. listed, including									
2220	hold-down straps, no manways									
2225	550 gallon capacity	Q-5	2	8	Ea.	3,525	390		3,915	4,475
2230	1,000 gallon capacity	"	1.88	8.511		4,225	415		4,640	5,250
2240	2,000 gallon capacity	Q-7	3.55	9.014		6,525	465		6,990	7,875
2250	4,000 gallon capacity		2.90	11.034		9,100	570		9,670	10,900
2260	6,000 gallon capacity		2	16		10,500	825		11,325	12,700
2270	8,000 gallon capacity		1.78	17.978		12,300	925		13,225	14,900
2280	10,000 gallon capacity		1.60	20		13,800	1,025		14,825	16,800
2282	12,000 gallon capacity		1.52	21.053		22,700	1,075		23,775	26,500
2284	15,000 gallon capacity		1.39	23.022		24,800	1,175		25,975	29,100
2290	20,000 gallon capacity		1.14	28.070		27,100	1,450		28,550	32,100
2300	25,000 gallon capacity		.96	33.333		40,000	1,725		41,725	46,600
2320	30,000 gallon capacity		.80	40		48,700	2,050		50,750	56,500
3020	Fiberglass, underground, double wall, U.L. listed									
3030	includes manways and hold-down straps									
3040	600 gallon capacity	Q-5	1.86	8.602	Ea.	7,375	420		7,795	8,725
3050	1,000 gallon capacity	"	1.70	9.412		9,100	460		9,560	10,700
3060	2,500 gallon capacity	Q-7	3.29	9.726		13,900	500		14,400	16,100
3070	3,000 gallon capacity		3.13	10.224		15,000	525		15,525	17,300
3080	4,000 gallon capacity		2.93	10.922		16,500	565		17,065	18,900
3090	6,000 gallon capacity		1.86	17.204		18,900	885		19,785	22,100
3100	8,000 gallon capacity		1.65	19.394		22,200	1,000		23,200	25,900
3110	10,000 gallon capacity		1.48	21.622		24,900	1,125		26,025	29,100
3120	12,000 gallon capacity		1.40	22.857		30,500	1,175		31,675	35,300
3122	15,000 gallon capacity		1.28	25		39,700	1,300		41,000	45,600
3124	20,000 gallon capacity		1.06	30.189		47,900	1,550		49,450	55,000
3126	25,000 gallon capacity		.90	35.556		64,000	1,825		65,825	72,500
3128	30,000 gallon capacity		.74	43.243		77,000	2,225		79,225	88,500

23 13 23 – Facility Aboveground Fuel-Oil, Storage Tanks

23 13 23.13 Vertical, Steel, Abvground Fuel-Oil, Stor. Tanks

		Crew	Daily Output	Labor-Hours	Unit	Material	2011 Bare Costs Labor	Equipment	Total	Total Incl O&P
0010	**VERTICAL, STEEL, ABOVEGROUND FUEL-OIL, STORAGE TANKS**									
4000	Fixed roof oil storage tanks, steel, (1 BBL=42 gal. w/foundation 3'D x 1'W)									
4200	5,000 barrels				Ea.				194,000	213,500
4300	24,000 barrels								333,500	367,000
4500	56,000 barrels								729,000	802,000

23 13 Facility Fuel-Storage Tanks

23 13 23 – Facility Aboveground Fuel-Oil, Storage Tanks

23 13 23.13 Vertical, Steel, Abvground Fuel-Oil, Stor. Tanks	Crew	Daily Output	Labor-Hours	Unit	Material	2011 Bare Costs Labor	Equipment	Total	Total Incl O&P
4600	110,000 barrels				Ea.				1,060,000	1,166,000
4800	143,000 barrels								1,250,000	1,375,000
4900	224,000 barrels								1,360,000	1,496,000
5100	Floating roof gasoline tanks, steel, 5,000 barrels (w/foundation 3'D x 1'W)								204,000	225,000
5200	25,000 barrels								381,000	419,000
5400	55,000 barrels								839,000	923,000
5500	100,000 barrels								1,253,000	1,379,000
5700	150,000 barrels								1,532,000	1,685,000
5800	225,000 barrels								2,300,000	2,783,000

23 13 23.16 Horizontal, Stl, Abvgrd Fuel-Oil, Storage Tanks

		Crew	Daily Output	Labor-Hours	Unit	Material	2011 Bare Costs Labor	Equipment	Total	Total Incl O&P
0010	**HORIZONTAL, STEEL, ABOVEGROUND FUEL-OIL, STORAGE TANKS**									
3000	Steel, storage, above ground, including cradles, coating,									
3020	fittings, not including foundation, pumps or piping									
3040	Single wall, 275 gallon	Q-5	5	3.200	Ea.	405	156		561	680
3060	550 gallon	"	2.70	5.926		2,800	289		3,089	3,525
3080	1,000 gallon	Q-7	5	6.400		3,175	330		3,505	4,000
3100	1,500 gallon		4.75	6.737		5,025	345		5,370	6,075
3120	2,000 gallon		4.60	6.957		5,775	360		6,135	6,875
3140	5,000 gallon		3.20	10		6,075	515		6,590	7,475
3150	10,000 gallon		2	16		26,400	825		27,225	30,200
3160	15,000 gallon		1.70	18.824		24,800	970		25,770	28,800
3170	20,000 gallon		1.45	22.069		31,700	1,125		32,825	36,500
3180	25,000 gallon		1.30	24.615		32,500	1,275		33,775	37,600
3190	30,000 gallon		1.10	29.091		42,300	1,500		43,800	48,700
3320	Double wall, 500 gallon capacity	Q-5	2.40	6.667		2,200	325		2,525	2,900
3330	2000 gallon capacity	Q-7	4.15	7.711		8,250	395		8,645	9,675
3340	4000 gallon capacity		3.60	8.889		14,700	460		15,160	16,900
3350	6000 gallon capacity		2.40	13.333		17,400	685		18,085	20,100
3360	8000 gallon capacity		2	16		22,300	825		23,125	25,700
3370	10000 gallon capacity		1.80	17.778		25,000	915		25,915	28,900
3380	15000 gallon capacity		1.50	21.333		37,900	1,100		39,000	43,400
3390	20000 gallon capacity		1.30	24.615		43,200	1,275		44,475	49,400
3400	25000 gallon capacity		1.15	27.826		52,500	1,425		53,925	59,500
3410	30000 gallon capacity		1	32		57,500	1,650		59,150	66,000

23 21 Hydronic Piping and Pumps

23 21 20 – Hydronic HVAC Piping Specialties

23 21 20.10 Air Control

		Crew	Daily Output	Labor-Hours	Unit	Material	2011 Bare Costs Labor	Equipment	Total	Total Incl O&P
0010	**AIR CONTROL**									
0030	Air separator, with strainer									
0040	2" diameter	Q-5	6	2.667	Ea.	865	130		995	1,150
0080	2-1/2" diameter		5	3.200		980	156		1,136	1,300
0100	3" diameter		4	4		1,500	195		1,695	1,925
0120	4" diameter		3	5.333		2,150	260		2,410	2,775
0130	5" diameter	Q-6	3.60	6.667		2,750	335		3,085	3,525
0140	6" diameter		3.40	7.059		3,275	355		3,630	4,125
0160	8" diameter		3	8		4,925	405		5,330	6,000
0180	10" diameter		2.20	10.909		7,625	550		8,175	9,225
0200	12" diameter		1.70	14.118		12,800	715		13,515	15,200
0210	14" diameter		1.30	18.462		15,200	930		16,130	18,200
0220	16" diameter		1	24		22,900	1,200		24,100	27,000

Market Intelligence
driven by powerhouse databases

Reed Construction Data/RSMeans has the answers to your market intelligence questions – Want to know what vertical markets are poised for growth? Want to know effect of stimulus programs on renovation costs? Want to know total spend for HVAC systems for commercial buildings?

Check out our NEW Expansion Index at http://www.reedconstructiondata.com/rsmeans/expansion-index/

✓ Get answers to market size, market share and product demand
✓ RSMeans team of seasoned cost engineers and business analysts

What's different about our studies is that we're using REAL data, not forecasts. Call now for an exploratory discussion.

Contact Laura Dempsey at 404.433.3583

Looking for your next job to bid?

Get up-to-date construction project information, plans and specs with SmartProject News from Reed Construction Data.

- Find projects bidding in your area
- Browse full project details, plans and specs
- Identify key project participants
- Track projects with e-mail notifications

Register for free access at www.smartprojectnews.com.

SmartProject News
Reed Construction Data

23 21 20 – Hydronic HVAC Piping Specialties

23 21 20.10 Air Control		Crew	Daily Output	Labor-Hours	Unit	Material	2011 Bare Costs Labor	2011 Bare Costs Equipment	Total	Total Incl O&P
0230	18" diameter	Q-6	.80	30	Ea.	30,100	1,525		31,625	35,500
0240	20" diameter	↓	.60	40	↓	33,700	2,025		35,725	40,100
0300	Without strainer									
0310	2" diameter	Q-5	6	2.667	Ea.	670	130		800	930
0320	2-1/2" diameter		5	3.200		800	156		956	1,125
0330	3" diameter		4	4		1,125	195		1,320	1,525
0340	4" diameter		3	5.333		1,775	260		2,035	2,350
0350	5" diameter	↓	2.40	6.667		2,400	325		2,725	3,100
0360	6" diameter	Q-6	3.40	7.059		2,750	355		3,105	3,550
0370	8" diameter		3	8		3,775	405		4,180	4,750
0380	10" diameter		2.20	10.909		5,550	550		6,100	6,925
0390	12" diameter		1.70	14.118		8,550	715		9,265	10,500
0400	14" diameter		1.30	18.462		12,800	930		13,730	15,500
0410	16" diameter		1	24		17,900	1,200		19,100	21,500
0420	18" diameter		.80	30		23,300	1,525		24,825	27,900
0430	20" diameter	↓	.60	40	↓	27,900	2,025		29,925	33,700
1000	Micro-bubble separator for total air removal, closed loop system									
1010	Requires bladder type tank in system.									
1020	Water (hot or chilled) or glycol system									
1030	Threaded									
1040	3/4" diameter	1 Stpi	20	.400	Ea.	83.50	21.50		105	124
1050	1" diameter		19	.421		94	23		117	137
1060	1-1/4" diameter		16	.500		130	27		157	184
1070	1-1/2" diameter		13	.615		169	33.50		202.50	236
1080	2" diameter	↓	11	.727		820	39.50		859.50	960
1090	2-1/2" diameter	Q-5	15	1.067		910	52		962	1,075
1100	3" diameter		13	1.231		1,300	60		1,360	1,525
1110	4" diameter	↓	10	1.600	↓	1,400	78		1,478	1,675
1230	Flanged									
1250	2" diameter	1 Stpi	8	1	Ea.	1,050	54		1,104	1,225
1260	2-1/2" diameter	Q-5	5	3.200		1,125	156		1,281	1,475
1270	3" diameter	↓	4.50	3.556		1,525	173		1,698	1,925
1280	4" diameter	↓	3	5.333		1,675	260		1,935	2,250
1290	5" diameter	Q-6	3.40	7.059		2,700	355		3,055	3,500
1300	6" diameter	"	3	8	↓	3,275	405		3,680	4,200
1400	Larger sizes available									
1590	With extended tank dirt catcher									
1600	Flanged									
1620	2" diameter	1 Stpi	7.40	1.081	Ea.	1,375	58.50		1,433.50	1,600
1630	2-1/2" diameter	Q-5	4.20	3.810		1,450	186		1,636	1,875
1640	3" diameter	↓	4	4		1,900	195		2,095	2,375
1650	4" diameter	↓	2.80	5.714		2,050	278		2,328	2,675
1660	5" diameter	Q-6	3.10	7.742		3,025	390		3,415	3,900
1670	6" diameter	"	2.90	8.276	↓	3,725	420		4,145	4,700
1800	With drain/dismantle/cleaning access flange									
1810	Flanged									
1820	2" diameter	1 Stpi	7	1.143	Ea.	2,150	62		2,212	2,475
1830	2-1/2" diameter	Q-5	4	4		2,300	195		2,495	2,850
1840	3" diameter	↓	3.60	4.444		3,125	216		3,341	3,775
1850	4" diameter	↓	2.50	6.400		3,450	310		3,760	4,275
1860	5" diameter	Q-6	2.80	8.571		5,150	435		5,585	6,325
1870	6" diameter	"	2.60	9.231	↓	6,725	465		7,190	8,100
2000	Boiler air fitting, separator									

23 21 Hydronic Piping and Pumps

23 21 20 – Hydronic HVAC Piping Specialties

23 21 20.10 Air Control	Crew	Daily Output	Labor-Hours	Unit	Material	2011 Bare Costs Labor	Equipment	Total	Total Incl O&P
2010	1-1/4"	Q-5	22	.727	Ea.	193	35.50		228.50	265
2020	1-1/2"		20	.800		370	39		409	465
2021	2"	↓	18	.889	↓	635	43.50		678.50	760
2400	Compression tank air fitting									
2410	For tanks 9" to 24" diameter	1 Stpi	15	.533	Ea.	62.50	29		91.50	112
2420	For tanks 100 gallon or larger	"	12	.667	"	235	36		271	310

23 21 20.14 Air Purging Scoop

		Crew	Daily Output	Labor-Hours	Unit	Material	Labor	Equipment	Total	Total Incl O&P
0010	**AIR PURGING SCOOP**, With tappings.									
0020	For air vent and expansion tank connection									
0100	1" pipe size, threaded	1 Stpi	19	.421	Ea.	22	23		45	58.50
0110	1-1/4" pipe size, threaded		15	.533		22	29		51	67.50
0120	1-1/2" pipe size, threaded		13	.615		48.50	33.50		82	103
0130	2" pipe size, threaded	↓	11	.727		55.50	39.50		95	120
0140	2-1/2" pipe size, threaded	Q-5	15	1.067		118	52		170	208
0150	3" pipe size, threaded		13	1.231		154	60		214	260
0160	4" 150 lb. flanges	↓	3	5.333	↓	345	260		605	770

23 21 20.18 Automatic Air Vent

		Crew	Daily Output	Labor-Hours	Unit	Material	Labor	Equipment	Total	Total Incl O&P
0010	**AUTOMATIC AIR VENT**									
0020	Cast iron body, stainless steel internals, float type									
0060	1/2" NPT inlet, 300 psi	1 Stpi	12	.667	Ea.	90.50	36		126.50	154
0140	3/4" NPT inlet, 300 psi		12	.667		90.50	36		126.50	154
0180	1/2" NPT inlet, 250 psi		10	.800		289	43.50		332.50	380
0220	3/4" NPT inlet, 250 psi		10	.800		289	43.50		332.50	380
0260	1" NPT inlet, 250 psi	↓	10	.800		425	43.50		468.50	530
0340	1-1/2" NPT inlet, 250 psi	Q-5	12	1.333		895	65		960	1,075
0380	2" NPT inlet, 250 psi	"	12	1.333	↓	895	65		960	1,075
0600	Forged steel body, stainless steel internals, float type									
0640	1/2" NPT inlet, 750 psi	1 Stpi	12	.667	Ea.	945	36		981	1,100
0680	3/4" NPT inlet, 750 psi		12	.667		945	36		981	1,100
0760	3/4" NPT inlet, 1000 psi	↓	10	.800		1,425	43.50		1,468.50	1,625
0800	1" NPT inlet, 1000 psi	Q-5	12	1.333		1,425	65		1,490	1,650
0880	1-1/2" NPT inlet, 1000 psi		10	1.600		3,975	78		4,053	4,500
0920	2" NPT inlet, 1000 psi	↓	10	1.600	↓	3,975	78		4,053	4,500
1100	Formed steel body, noncorrosive									
1110	1/8" NPT inlet 150 psi	1 Stpi	32	.250	Ea.	12.25	13.55		25.80	33.50
1120	1/4" NPT inlet 150 psi		32	.250		38	13.55		51.55	62
1130	3/4" NPT inlet 150 psi	↓	32	.250	↓	38	13.55		51.55	62
1300	Chrome plated brass, automatic/manual, for radiators									
1310	1/8" NPT inlet, nickel plated brass	1 Stpi	32	.250	Ea.	6.50	13.55		20.05	27

23 21 20.22 Circuit Sensor

		Crew	Daily Output	Labor-Hours	Unit	Material	Labor	Equipment	Total	Total Incl O&P
0010	**CIRCUIT SENSOR**, Flow meter									
0020	Metering stations									
0040	Wafer orifice insert type									
0060	2-1/2" pipe size	Q-5	12	1.333	Ea.	182	65		247	297
0100	3" pipe size		11	1.455		203	71		274	330
0140	4" pipe size		8	2		235	97.50		332.50	405
0180	5" pipe size		7.30	2.192		298	107		405	485
0220	6" pipe size	↓	6.40	2.500		355	122		477	575
0260	8" pipe size	Q-6	5.30	4.528		495	229		724	885
0280	10" pipe size		4.60	5.217		565	263		828	1,025
0360	12" pipe size	↓	4.20	5.714	↓	940	289		1,229	1,450
2000	In-line probe type									

23 21 20.22 Circuit Sensor		Crew	Daily Output	Labor-Hours	Unit	Material	2011 Bare Costs Labor	Equipment	Total	Total Incl O&P
2200	Copper, soldered									
2210	1/2" size	1 Stpi	24	.333	Ea.	169	18.05		187.05	213
2220	3/4" size		20	.400		175	21.50		196.50	226
2230	1" size		19	.421		199	23		222	253
2240	1-1/4" size		15	.533		209	29		238	273
2250	1-1/2" size		13	.615		249	33.50		282.50	325
2260	2" size		11	.727		277	39.50		316.50	365
2270	2-1/2" size	Q-5	15	1.067		292	52		344	400
2280	3" size	"	13	1.231		315	60		375	440
2500	Brass, threaded									
2510	1" size	1 Stpi	19	.421	Ea.	209	23		232	264
2520	1-1/4" size		15	.533		231	29		260	297
2530	1-1/2" size		13	.615		277	33.50		310.50	355
2540	2" size		11	.727		292	39.50		331.50	380
2550	2-1/2" size	Q-5	15	1.067		315	52		367	430
2560	3" size	"	13	1.231		325	60		385	445
2700	Steel, threaded									
2710	1" size	1 Stpi	19	.421	Ea.	200	23		223	254
2720	1-1/4" size		15	.533		217	29		246	282
2730	1-1/2" size		13	.615		247	33.50		280.50	320
2740	2" size		11	.727		275	39.50		314.50	365
2750	2-1/2" size	Q-5	15	1.067		292	52		344	400
2760	3" size	"	13	1.231		320	60		380	440
3000	Pitot tube probe type									
3100	Weld-on mounting with probe									
3110	2" size	Q-17	18	.889	Ea.	226	43.50	1.80	271.30	315
3120	2-1/2" size		18	.889		231	43.50	1.80	276.30	320
3130	3" size		18	.889		244	43.50	1.80	289.30	335
3140	4" size		18	.889		325	43.50	1.80	370.30	420
3150	5" size		17	.941		365	46	1.91	412.91	470
3160	6" size		17	.941		470	46	1.91	517.91	590
3170	8" size		16	1		670	48.50	2.03	720.53	815
3180	10" size		15	1.067		725	52	2.17	779.17	880
3190	12" size		15	1.067		780	52	2.17	834.17	935
3200	14" size		14	1.143		890	55.50	2.32	947.82	1,075
3210	16" size		14	1.143		1,050	55.50	2.32	1,107.82	1,225
3220	18" size		13	1.231		1,125	60	2.50	1,187.50	1,325
3230	20" size		13	1.231		1,225	60	2.50	1,287.50	1,450
3240	24" size		12	1.333		1,300	65	2.71	1,367.71	1,525
3400	Weld-on wet tap with probe									
3410	2" size, 5/16" probe dia.	Q-17	14	1.143	Ea.	550	55.50	2.32	607.82	690
3420	2-1/2" size, 5/16" probe dia.		14	1.143		555	55.50	2.32	612.82	695
3430	3" size, 5/16" probe dia.		14	1.143		595	55.50	2.32	652.82	735
3440	4" size, 5/16" probe dia.		14	1.143		690	55.50	2.32	747.82	840
3450	5" size, 3/8" probe dia.		13	1.231		705	60	2.50	767.50	870
3460	6" size, 3/8" probe dia.		13	1.231		675	60	2.50	737.50	835
3470	8" size, 1/2" probe dia.		12	1.333		700	65	2.71	767.71	870
3480	10" size, 3/4" probe dia.		11	1.455		770	71	2.95	843.95	960
3490	12" size, 3/4" probe dia.		11	1.455		815	71	2.95	888.95	1,000
3500	14" size, 1" probe dia.		10	1.600		925	78	3.25	1,006.25	1,150
3510	16" size, 1" probe dia.		10	1.600		1,100	78	3.25	1,181.25	1,325
3520	18" size, 1" probe dia.		9	1.778		1,175	86.50	3.61	1,265.11	1,425
3530	20" size, 1" probe dia.		8	2		1,275	97.50	4.06	1,376.56	1,550

23 21 20.22 Circuit Sensor

		Crew	Daily Output	Labor-Hours	Unit	Material	2011 Bare Costs Labor	Equipment	Total	Total Incl O&P
3540	24" size, 1" probe dia.	Q-17	8	2	Ea.	1,425	97.50	4.06	1,526.56	1,725
3700	Clamp-on wet tap with probe									
3710	2" size, 5/16" probe dia.	Q-5	16	1	Ea.	790	48.50		838.50	945
3720	2-1/2" size, 5/16" probe dia.		16	1		685	48.50		733.50	830
3730	3" size, 5/16" probe dia.		15	1.067		685	52		737	835
3740	4" size, 5/16" probe dia.		15	1.067		795	52		847	955
3750	5" size, 3/8" probe dia.		14	1.143		840	55.50		895.50	1,000
3760	6" size, 3/8" probe dia.		14	1.143		890	55.50		945.50	1,075
3770	8" size, 1/2" probe dia.		13	1.231		980	60		1,040	1,175
3780	10" size, 3/4" probe dia.		12	1.333		1,025	65		1,090	1,225
3790	12" size, 3/4" probe dia.		12	1.333		1,175	65		1,240	1,400
3800	14" size, 1" probe dia.		11	1.455		1,525	71		1,596	1,800
3810	16" size, 1" probe dia.		11	1.455		1,775	71		1,846	2,075
3820	18" size, 1" probe dia.		10	1.600		1,925	78		2,003	2,250
3830	20" size, 1" probe dia		9	1.778		2,200	86.50		2,286.50	2,525
3840	24" size, 1" probe dia.	▼	9	1.778	▼	2,275	86.50		2,361.50	2,625
4000	Wet tap drills									
4010	3/8" dia. for 5/16" probe				Ea.	132			132	145
4020	1/2" dia. for 3/8" probe					136			136	150
4030	5/8" dia for 1/2" probe					162			162	178
4040	7/8" dia for 3/4" probe					211			211	232
4050	1-3/16" dia for 1" probe				▼	226			226	248
4100	Wet tap punch									
4110	5/16" dia for 5/16" probe				Ea.	136			136	149
4120	3/8" dia for 3/8" probe					146			146	161
4130	1/2" dia for 1/2" probe				▼	175			175	193
4150	Note: labor for wet tap drill or punch									
4160	is included with the mounting and probe assembly									
9000	Readout instruments									
9200	Gauges									
9220	GPM gauge				Ea.	2,225			2,225	2,425
9230	Dual gauge kit				"	2,300			2,300	2,550

23 21 20.26 Circuit Setter

		Crew	Daily Output	Labor-Hours	Unit	Material	2011 Bare Costs Labor	Equipment	Total	Total Incl O&P
0010	**CIRCUIT SETTER**, Balance valve									
0012	Bronze body, soldered									
0013	1/2" size	1 Stpi	24	.333	Ea.	66.50	18.05		84.55	100
0014	3/4" size	"	20	.400	"	70	21.50		91.50	110
0018	Threaded									
0019	1/2" pipe size	1 Stpi	22	.364	Ea.	68.50	19.65		88.15	105
0020	3/4" pipe size		20	.400		73.50	21.50		95	114
0040	1" pipe size		18	.444		95	24		119	141
0050	1-1/4" pipe size		15	.533		140	29		169	197
0060	1-1/2" pipe size		12	.667		165	36		201	235
0080	2" pipe size	▼	10	.800		236	43.50		279.50	325
0100	2-1/2" pipe size	Q-5	15	1.067		535	52		587	665
0120	3" pipe size	"	10	1.600	▼	750	78		828	940
0130	Cast iron body, flanged									
0136	3" pipe size, flanged	Q-5	4	4	Ea.	750	195		945	1,125
0140	4" pipe size	"	3	5.333		1,125	260		1,385	1,650
0200	For differential meter, accurate to 1%, add					865			865	955
0300	For differential meter 400 psi/200°F continuous, add				▼	2,150			2,150	2,350

23 21 20 – Hydronic HVAC Piping Specialties

23 21 20.30 Cocks, Drains and Specialties		Crew	Daily Output	Labor-Hours	Unit	Material	2011 Bare Costs Labor	Equipment	Total	Total Incl O&P
0010	**COCKS, DRAINS AND SPECIALTIES**									
1000	Boiler drain									
1010	Pipe thread to hose									
1020	Bronze									
1030	1/2" size	1 Stpi	36	.222	Ea.	9.35	12		21.35	28.50
1040	3/4" size	"	34	.235	"	9.45	12.75		22.20	29.50
1100	Solder to hose									
1110	Bronze									
1120	1/2" size	1 Stpi	46	.174	Ea.	7.60	9.40		17	22.50
1130	3/4" size	"	44	.182	"	8.30	9.85		18.15	24
1600	With built-in vacuum breaker									
1610	1/2" IP or solder	1 Stpi	36	.222	Ea.	29.50	12		41.50	50.50
1630	With tamper proof vacuum breaker									
1640	1/2" IP or solder	1 Stpi	36	.222	Ea.	33.50	12		45.50	55
1650	3/4" IP or solder	"	34	.235	"	34	12.75		46.75	56.50
3000	Cocks									
3010	Air, lever or tee handle									
3020	Bronze, single thread									
3030	1/8" size	1 Stpi	52	.154	Ea.	8.65	8.30		16.95	22
3040	1/4" size		46	.174		9.15	9.40		18.55	24
3050	3/8" size		40	.200		9.25	10.80		20.05	26.50
3060	1/2" size		36	.222		10.95	12		22.95	30
3100	Bronze, double thread									
3110	1/8" size	1 Stpi	26	.308	Ea.	10.90	16.65		27.55	37
3120	1/4" size		22	.364		15.65	19.65		35.30	46.50
3130	3/8" size		18	.444		11.80	24		35.80	49
3140	1/2" size		15	.533		14.55	29		43.55	59
4000	Steam									
4010	Bronze									
4020	1/8" size	1 Stpi	26	.308	Ea.	31.50	16.65		48.15	59.50
4030	1/4" size		22	.364		31.50	19.65		51.15	64
4040	3/8" size		18	.444		39.50	24		63.50	79.50
4050	1/2" size		15	.533		42	29		71	89
4060	3/4" size		14	.571		50	31		81	101
4070	1" size		13	.615		64.50	33.50		98	121
4080	1-1/4" size		11	.727		99	39.50		138.50	168
4090	1-1/2" size		10	.800		128	43.50		171.50	205
4100	2" size		9	.889		212	48		260	305
4300	Bronze, 3 way									
4320	1/4" size	1 Stpi	15	.533	Ea.	47	29		76	94.50
4330	3/8" size		12	.667		47	36		83	106
4340	1/2" size		10	.800		47	43.50		90.50	116
4350	3/4" size		9.50	.842		80.50	45.50		126	157
4360	1" size		8.50	.941		97	51		148	183
4370	1-1/4" size		7.50	1.067		145	57.50		202.50	246
4380	1-1/2" size		6.50	1.231		169	66.50		235.50	286
4390	2" size		6	1.333		250	72		322	385
4500	Gauge cock, brass									
4510	1/4" FPT	1 Stpi	24	.333	Ea.	9.15	18.05		27.20	37
4512	1/4" MPT	"	24	.333	"	13.55	18.05		31.60	42
4600	Pigtail, steam syphon									
4604	1/4"	1 Stpi	24	.333	Ea.	16.75	18.05		34.80	45.50

23 21 20.30 Cocks, Drains and Specialties	Crew	Daily Output	Labor-Hours	Unit	Material	2011 Bare Costs Labor	Equipment	Total	Total Incl O&P
4650	Snubber valve								
4654	1/4"	1 Stpi	22	.364	Ea.	12.15	19.65		31.80	43
4660	Nipple, black steel								
4664	1/4" x 3"	1 Stpi	37	.216	Ea.	1.99	11.70		13.69	19.65

23 21 20.34 Dielectric Unions

		Crew	Daily Output	Labor-Hours	Unit	Material	Labor	Equipment	Total	Total Incl O&P
0010	**DIELECTRIC UNIONS**, Standard gaskets for water and air									
0020	250 psi maximum pressure									
0280	Female IPT to sweat, straight									
0300	1/2" pipe size	1 Plum	24	.333	Ea.	4.79	17.85		22.64	32
0340	3/4" pipe size		20	.400		5.35	21.50		26.85	38
0360	1" pipe size		19	.421		7.35	22.50		29.85	41.50
0380	1-1/4" pipe size		15	.533		11.35	28.50		39.85	55
0400	1-1/2" pipe size		13	.615		17	33		50	67.50
0420	2" pipe size		11	.727		23	39		62	83.50
0580	Female IPT to brass pipe thread, straight									
0600	1/2" pipe size	1 Plum	24	.333	Ea.	10.40	17.85		28.25	38
0640	3/4" pipe size		20	.400		11.55	21.50		33.05	44.50
0660	1" pipe size		19	.421		21.50	22.50		44	57
0680	1-1/4" pipe size		15	.533		26	28.50		54.50	71
0700	1-1/2" pipe size		13	.615		38.50	33		71.50	91.50
0720	2" pipe size		11	.727		74.50	39		113.50	140
0780	Female IPT to female IPT, straight									
0800	1/2" pipe size	1 Plum	24	.333	Ea.	9.65	17.85		27.50	37
0840	3/4" pipe size		20	.400		10.85	21.50		32.35	44
0860	1" pipe size		19	.421		14.55	22.50		37.05	49.50
0880	1-1/4" pipe size		15	.533		19.70	28.50		48.20	64
0900	1-1/2" pipe size		13	.615		30	33		63	82
0920	2" pipe size		11	.727		44	39		83	107
2000	175 psi maximum pressure									
2180	Female IPT to sweat									
2240	2" pipe size	1 Plum	9	.889	Ea.	130	47.50		177.50	214
2260	2-1/2" pipe size	Q-1	15	1.067		141	51.50		192.50	232
2280	3" pipe size		14	1.143		193	55		248	296
2300	4" pipe size		11	1.455		515	70		585	670
2480	Female IPT to brass pipe									
2500	1-1/2" pipe size	1 Plum	11	.727	Ea.	162	39		201	236
2540	2" pipe size	"	9	.889		190	47.50		237.50	280
2560	2-1/2" pipe size	Q-1	15	1.067		279	51.50		330.50	380
2580	3" pipe size		14	1.143		325	55		380	440
2600	4" pipe size		11	1.455		585	70		655	745

23 21 20.38 Expansion Couplings

		Crew	Daily Output	Labor-Hours	Unit	Material	Labor	Equipment	Total	Total Incl O&P
0010	**EXPANSION COUPLINGS**, Hydronic									
0100	Copper to copper, sweat									
1000	Baseboard riser fitting, 5" stub by coupling 12" long									
1020	1/2" diameter	1 Stpi	24	.333	Ea.	16.35	18.05		34.40	45
1040	3/4" diameter		20	.400		23.50	21.50		45	58
1060	1" diameter		19	.421		32.50	23		55.50	69.50
1080	1-1/4" diameter		15	.533		42.50	29		71.50	89.50
1180	9" Stub by tubing 8" long									
1200	1/2" diameter	1 Stpi	24	.333	Ea.	14.70	18.05		32.75	43
1220	3/4" diameter		20	.400		19.45	21.50		40.95	54
1240	1" diameter		19	.421		27	23		50	63.50

23 21 20 – Hydronic HVAC Piping Specialties

23 21 20.38 Expansion Couplings

		Crew	Daily Output	Labor-Hours	Unit	Material	2011 Bare Costs Labor	2011 Bare Costs Equipment	Total	Total Incl O&P
1260	1-1/4" diameter	1 Stpi	15	.533	Ea.	38	29		67	84.50

23 21 20.42 Expansion Joints

		Crew	Daily Output	Labor-Hours	Unit	Material	2011 Bare Costs Labor	2011 Bare Costs Equipment	Total	Total Incl O&P
0010	**EXPANSION JOINTS**									
0100	Bellows type, neoprene cover, flanged spool									
0140	6" face to face, 1-1/4" diameter	1 Stpi	11	.727	Ea.	248	39.50		287.50	330
0160	1-1/2" diameter	"	10.60	.755		248	41		289	335
0180	2" diameter	Q-5	13.30	1.203		251	58.50		309.50	365
0190	2-1/2" diameter		12.40	1.290		260	63		323	380
0200	3" diameter		11.40	1.404		291	68.50		359.50	420
0210	4" diameter		8.40	1.905		315	93		408	485
0220	5" diameter		7.60	2.105		385	103		488	580
0230	6" diameter		6.80	2.353		395	115		510	605
0240	8" diameter		5.40	2.963		460	144		604	720
0250	10" diameter		5	3.200		635	156		791	935
0260	12" diameter		4.60	3.478		730	169		899	1,050
0480	10" face to face, 2" diameter		13	1.231		360	60		420	485
0500	2-1/2" diameter		12	1.333		380	65		445	515
0520	3" diameter		11	1.455		390	71		461	530
0540	4" diameter		8	2		440	97.50		537.50	630
0560	5" diameter		7	2.286		525	111		636	745
0580	6" diameter		6	2.667		545	130		675	795
0600	8" diameter		5	3.200		650	156		806	950
0620	10" diameter		4.60	3.478		720	169		889	1,050
0640	12" diameter		4	4		890	195		1,085	1,275
0660	14" diameter		3.80	4.211		1,100	205		1,305	1,500
0680	16" diameter		2.90	5.517		1,275	269		1,544	1,800
0700	18" diameter		2.50	6.400		1,425	310		1,735	2,050
0720	20" diameter		2.10	7.619		1,500	370		1,870	2,200
0740	24" diameter		1.80	8.889		1,750	435		2,185	2,575
0760	26" diameter		1.40	11.429		1,975	555		2,530	2,975
0780	30" diameter		1.20	13.333		2,200	650		2,850	3,400
0800	36" diameter	▼	1	16	▼	2,725	780		3,505	4,175
1000	Bellows with internal sleeves and external covers									
1010	Stainless steel, 150 lb.									
1020	With male threads									
1040	3/4" diameter	1 Stpi	20	.400	Ea.	132	21.50		153.50	179
1050	1" diameter		19	.421		140	23		163	188
1060	1-1/4" diameter		16	.500		146	27		173	202
1070	1-1/2" diameter		13	.615		178	33.50		211.50	246
1080	2" diameter	▼	11	.727	▼	228	39.50		267.50	310
1110	With flanged ends									
1120	1-1/4" diameter	1 Stpi	12	.667	Ea.	182	36		218	254
1130	1-1/2" diameter		11	.727		223	39.50		262.50	305
1140	2" diameter	▼	9	.889		283	48		331	380
1150	3" diameter	Q-5	8	2		495	97.50		592.50	690
1160	4" diameter	"	5	3.200		840	156		996	1,150
1170	5" diameter	Q-6	6	4		995	202		1,197	1,400
1180	6" diameter		5	4.800		1,175	242		1,417	1,650
1190	8" diameter		4.50	5.333		1,675	269		1,944	2,250
1200	10" diameter		4	6		2,025	305		2,330	2,675
1210	12" diameter	▼	3.60	6.667	▼	3,150	335		3,485	3,975

23 21 20.46 Expansion Tanks	Crew	Daily Output	Labor-Hours	Unit	Material	2011 Bare Costs Labor	2011 Bare Costs Equipment	Total	Total Incl O&P
0010 **EXPANSION TANKS**									
1400 Plastic, corrosion resistant, see Plumbing Cost Data									
1507 Underground fuel-oil storage tanks, see Section 23 13 13									
1512 Tank leak detection systems, see Section 28 33 33.50									
2000 Steel, liquid expansion, ASME, painted, 15 gallon capacity	Q-5	17	.941	Ea.	555	46		601	685
2020 24 gallon capacity		14	1.143		595	55.50		650.50	735
2040 30 gallon capacity		12	1.333		625	65		690	780
2060 40 gallon capacity		10	1.600		730	78		808	915
2080 60 gallon capacity		8	2		875	97.50		972.50	1,100
2100 80 gallon capacity		7	2.286		935	111		1,046	1,200
2120 100 gallon capacity		6	2.667		1,275	130		1,405	1,600
2130 120 gallon capacity		5	3.200		1,350	156		1,506	1,700
2140 135 gallon capacity		4.50	3.556		1,425	173		1,598	1,800
2150 175 gallon capacity		4	4		2,225	195		2,420	2,750
2160 220 gallon capacity		3.60	4.444		2,525	216		2,741	3,100
2170 240 gallon capacity		3.30	4.848		2,625	236		2,861	3,250
2180 305 gallon capacity		3	5.333		3,700	260		3,960	4,475
2190 400 gallon capacity		2.80	5.714		4,525	278		4,803	5,425
2360 Galvanized									
2370 15 gallon capacity	Q-5	17	.941	Ea.	980	46		1,026	1,150
2380 24 gallon capacity		14	1.143		1,075	55.50		1,130.50	1,250
2390 30 gallon capacity		12	1.333		1,175	65		1,240	1,375
2400 40 gallon capacity		10	1.600		1,375	78		1,453	1,650
2410 60 gallon capacity		8	2		1,575	97.50		1,672.50	1,900
2420 80 gallon capacity		7	2.286		1,825	111		1,936	2,175
2430 100 gallon capacity		6	2.667		2,375	130		2,505	2,800
2440 120 gallon capacity		5	3.200		2,525	156		2,681	3,000
2450 135 gallon capacity		4.50	3.556		2,675	173		2,848	3,175
2460 175 gallon capacity		4	4		4,325	195		4,520	5,075
2470 220 gallon capacity		3.60	4.444		4,950	216		5,166	5,775
2480 240 gallon capacity		3.30	4.848		5,175	236		5,411	6,050
2490 305 gallon capacity		3	5.333		7,525	260		7,785	8,675
2500 400 gallon capacity		2.80	5.714		9,250	278		9,528	10,600
3000 Steel ASME expansion, rubber diaphragm, 19 gal. cap. accept.		12	1.333		2,225	65		2,290	2,550
3020 31 gallon capacity		8	2		2,475	97.50		2,572.50	2,875
3040 61 gallon capacity		6	2.667		3,475	130		3,605	4,025
3060 79 gallon capacity		5	3.200		3,550	156		3,706	4,125
3080 119 gallon capacity		4	4		3,750	195		3,945	4,425
3100 158 gallon capacity		3.80	4.211		5,200	205		5,405	6,025
3120 211 gallon capacity		3.30	4.848		6,025	236		6,261	6,975
3140 317 gallon capacity		2.80	5.714		7,850	278		8,128	9,075
3160 422 gallon capacity		2.60	6.154		11,600	300		11,900	13,200
3180 528 gallon capacity		2.40	6.667		12,800	325		13,125	14,500

23 21 20.50 Float Valves

	Crew	Daily Output	Labor-Hours	Unit	Material	2011 Bare Costs Labor	2011 Bare Costs Equipment	Total	Total Incl O&P
0010 **FLOAT VALVES**									
0020 With ball and bracket									
0030 Single seat, threaded									
0040 Brass body									
0050 1/2"	1 Stpi	11	.727	Ea.	69	39.50		108.50	135
0060 3/4"		9	.889		78	48		126	158
0070 1"		7	1.143		103	62		165	207
0080 1-1/2"		4.50	1.778		157	96		253	315

23 21 Hydronic Piping and Pumps

23 21 20 – Hydronic HVAC Piping Specialties

23 21 20.50 Float Valves	Crew	Daily Output	Labor-Hours	Unit	Material	2011 Bare Costs Labor	2011 Bare Costs Equipment	Total	Total Incl O&P
0090	2"	1 Stpi	3.60	2.222	Ea.	160	120		280	355
0300	For condensate receivers, CI, in-line mount									
0320	1" inlet	1 Stpi	7	1.143	Ea.	113	62		175	218
0360	For condensate receiver, CI, external float, flanged tank mount									
0370	3/4" inlet	1 Stpi	5	1.600	Ea.	85	86.50		171.50	223

23 21 20.54 Flow Check Control

		Crew	Daily Output	Labor-Hours	Unit	Material	Labor	Equipment	Total	Total Incl O&P
0010	**FLOW CHECK CONTROL**									
0100	Bronze body, soldered									
0110	3/4" size	1 Stpi	20	.400	Ea.	56.50	21.50		78	94.50
0120	1" size	"	19	.421	"	67.50	23		90.50	108
0200	Cast iron body, threaded									
0210	3/4" size	1 Stpi	20	.400	Ea.	43.50	21.50		65	80
0220	1" size		19	.421		48.50	23		71.50	87
0230	1-1/4" size		15	.533		59.50	29		88.50	109
0240	1-1/2" size		13	.615		90.50	33.50		124	149
0250	2" size		11	.727		130	39.50		169.50	202
0300	Flanged inlet, threaded outlet									
0310	2-1/2" size	1 Stpi	8	1	Ea.	291	54		345	400
0320	3" size	Q-5	9	1.778		345	86.50		431.50	510
0330	4" size	"	7	2.286		675	111		786	910

23 21 20.58 Hydronic Heating Control Valves

		Crew	Daily Output	Labor-Hours	Unit	Material	Labor	Equipment	Total	Total Incl O&P
0010	**HYDRONIC HEATING CONTROL VALVES**									
0050	Hot water, nonelectric, thermostatic									
0100	Radiator supply, 1/2" diameter	1 Stpi	24	.333	Ea.	58	18.05		76.05	90.50
0120	3/4" diameter		20	.400		59.50	21.50		81	97.50
0140	1" diameter		19	.421		74.50	23		97.50	116
0160	1-1/4" diameter		15	.533		107	29		136	160
0500	For low pressure steam, add					25%				
1000	Manual, radiator supply									
1010	1/2" pipe size, angle union	1 Stpi	24	.333	Ea.	38	18.05		56.05	68.50
1020	3/4" pipe size, angle union		20	.400		47.50	21.50		69	85
1030	1" pipe size, angle union		19	.421		61.50	23		84.50	102
1100	Radiator, balancing, straight, sweat connections									
1110	1/2" pipe size	1 Stpi	24	.333	Ea.	14.40	18.05		32.45	43
1120	3/4" pipe size		20	.400		20	21.50		41.50	54.50
1130	1" pipe size		19	.421		32.50	23		55.50	70
1200	Steam, radiator, supply									
1210	1/2" pipe size, angle union	1 Stpi	24	.333	Ea.	37	18.05		55.05	67.50
1220	3/4" pipe size, angle union		20	.400		40	21.50		61.50	76.50
1230	1" pipe size, angle union		19	.421		46.50	23		69.50	85
1240	1-1/4" pipe size, angle union		15	.533		59.50	29		88.50	109
8000	System balancing and shut-off									
8020	Butterfly, quarter turn, calibrated, threaded or solder									
8040	Bronze, -30°F to +350°F, pressure to 175 psi									
8060	1/2" size	1 Stpi	22	.364	Ea.	13.15	19.65		32.80	44
8070	3/4" size		20	.400		21	21.50		42.50	56
8080	1" size		19	.421		26	23		49	62.50
8090	1-1/4" size		15	.533		42	29		71	89
8100	1-1/2" size		13	.615		54	33.50		87.50	109
8110	2" size		11	.727		67.50	39.50		107	133

23 21 20.62 Liquid Drainers	Crew	Daily Output	Labor-Hours	Unit	Material	2011 Bare Costs Labor	Equipment	Total	Total Incl O&P
0010 **LIQUID DRAINERS**									
0100 Guided lever type									
0110 Cast iron body, threaded									
0120 1/2" pipe size	1 Stpi	22	.364	Ea.	90.50	19.65		110.15	129
0130 3/4" pipe size		20	.400		241	21.50		262.50	298
0140 1" pipe size		19	.421		385	23		408	455
0150 1-1/2" pipe size		11	.727		835	39.50		874.50	980
0160 2" pipe size	↓	8	1	↓	835	54		889	1,000
0200 Forged steel body, threaded									
0210 1/2" pipe size	1 Stpi	18	.444	Ea.	890	24		914	1,025
0220 3/4" pipe size		16	.500		890	27		917	1,025
0230 1" pipe size	↓	14	.571		1,350	31		1,381	1,525
0240 1-1/2" pipe size	Q-5	10	1.600		3,900	78		3,978	4,400
0250 2" pipe size	"	8	2	↓	3,900	97.50		3,997.50	4,425
0300 Socket weld									
0310 1/2" pipe size	Q-17	16	1	Ea.	1,025	48.50	2.03	1,075.53	1,200
0320 3/4" pipe size		14	1.143		1,025	55.50	2.32	1,082.82	1,200
0330 1" pipe size		12	1.333		1,025	65	2.71	1,092.71	1,225
0340 1-1/2" pipe size		9	1.778		1,500	86.50	3.61	1,590.11	1,775
0350 2" pipe size	↓	7	2.286	↓	4,050	111	4.64	4,165.64	4,625
0400 Flanged									
0410 1/2" pipe size	1 Stpi	12	.667	Ea.	1,750	36		1,786	1,975
0420 3/4" pipe size		10	.800		1,750	43.50		1,793.50	2,000
0430 1" pipe size	↓	7	1.143		2,275	62		2,337	2,600
0440 1-1/2" pipe size	Q-5	9	1.778		5,050	86.50		5,136.50	5,675
0450 2" pipe size	"	7	2.286	↓	5,050	111		5,161	5,725
1000 Fixed lever and snap action type									
1100 Cast iron body, threaded									
1110 1/2" pipe size	1 Stpi	22	.364	Ea.	141	19.65		160.65	185
1120 3/4" pipe size		20	.400		370	21.50		391.50	445
1130 1" pipe size	↓	19	.421	↓	370	23		393	445
1200 Forged steel body, threaded									
1210 1/2" pipe size	1 Stpi	18	.444	Ea.	970	24		994	1,100
1220 3/4" pipe size		16	.500		970	27		997	1,125
1230 1" pipe size		14	.571		2,375	31		2,406	2,675
1240 1-1/4" pipe size	↓	12	.667	↓	2,375	36		2,411	2,675
2000 High pressure high leverage type									
2100 Forged steel body, threaded									
2110 1/2" pipe size	1 Stpi	18	.444	Ea.	1,900	24		1,924	2,100
2120 3/4" pipe size		16	.500		1,900	27		1,927	2,125
2130 1" pipe size		14	.571		3,175	31		3,206	3,550
2140 1-1/4" pipe size	↓	12	.667		3,175	36		3,211	3,550
2150 1-1/2" pipe size	Q-5	10	1.600		4,775	78		4,853	5,375
2160 2" pipe size	"	8	2	↓	4,775	97.50		4,872.50	5,400
2200 Socket weld									
2210 1/2" pipe size	Q-17	16	1	Ea.	2,050	48.50	2.03	2,100.53	2,325
2220 3/4" pipe size		14	1.143		2,050	55.50	2.32	2,107.82	2,325
2230 1" pipe size		12	1.333		3,325	65	2.71	3,392.71	3,775
2240 1-1/4" pipe size		11	1.455		3,325	71	2.95	3,398.95	3,775
2250 1-1/2" pipe size		9	1.778		4,925	86.50	3.61	5,015.11	5,550
2260 2" pipe size	↓	7	2.286	↓	4,925	111	4.64	5,040.64	5,600
2300 Flanged									

23 21 20 – Hydronic HVAC Piping Specialties

23 21 20.62 Liquid Drainers		Crew	Daily Output	Labor-Hours	Unit	Material	2011 Bare Costs Labor	Equipment	Total	Total Incl O&P
2310	1/2" pipe size	1 Stpi	12	.667	Ea.	2,750	36		2,786	3,075
2320	3/4" pipe size		10	.800		2,750	43.50		2,793.50	3,100
2330	1" pipe size		7	1.143		4,325	62		4,387	4,875
2340	1-1/4" pipe size		6	1.333		4,325	72		4,397	4,875
2350	1-1/2" pipe size	Q-5	9	1.778		5,925	86.50		6,011.50	6,650
2360	2" pipe size	"	7	2.286		5,925	111		6,036	6,700
3000	Guided lever dual gravity type									
3100	Cast iron body, threaded									
3110	1/2" pipe size	1 Stpi	22	.364	Ea.	345	19.65		364.65	410
3120	3/4" pipe size		20	.400		345	21.50		366.50	415
3130	1" pipe size		19	.421		505	23		528	590
3140	1-1/2" pipe size		11	.727		975	39.50		1,014.50	1,125
3150	2" pipe size		8	1		975	54		1,029	1,150
3200	Forged steel body, threaded									
3210	1/2" pipe size	1 Stpi	18	.444	Ea.	1,025	24		1,049	1,175
3220	3/4" pipe size		16	.500		1,025	27		1,052	1,200
3230	1" pipe size		14	.571		1,525	31		1,556	1,750
3240	1-1/2" pipe size	Q-5	10	1.600		4,100	78		4,178	4,625
3250	2" pipe size	"	8	2		4,100	97.50		4,197.50	4,650
3300	Socket weld									
3310	1/2" pipe size	Q-17	16	1	Ea.	1,175	48.50	2.03	1,225.53	1,350
3320	3/4" pipe size		14	1.143		1,175	55.50	2.32	1,232.82	1,350
3330	1" pipe size		12	1.333		1,700	65	2.71	1,767.71	1,950
3340	1-1/2" pipe size		9	1.778		4,250	86.50	3.61	4,340.11	4,800
3350	2" pipe size		7	2.286		4,250	111	4.64	4,365.64	4,850
3400	Flanged									
3410	1/2" pipe size	1 Stpi	12	.667	Ea.	1,875	36		1,911	2,125
3420	3/4" pipe size		10	.800		1,875	43.50		1,918.50	2,150
3430	1" pipe size		7	1.143		2,475	62		2,537	2,825
3440	1-1/2" pipe size	Q-5	9	1.778		5,225	86.50		5,311.50	5,875
3450	2" pipe size	"	7	2.286		5,225	111		5,336	5,925

23 21 20.66 Monoflow Tee Fitting

0010	**MONOFLOW TEE FITTING**	Crew	Daily Output	Labor-Hours	Unit	Material	Labor	Equipment	Total	Total Incl O&P
1100	For one pipe hydronic, supply and return									
1110	Copper, soldered									
1120	3/4" x 1/2" size	1 Stpi	13	.615	Ea.	17.65	33.50		51.15	69
1130	1" x 1/2" size		12	.667		17.65	36		53.65	73.50
1140	1" x 3/4" size		11	.727		17.65	39.50		57.15	78
1150	1-1/4" x 1/2" size		11	.727		18.40	39.50		57.90	78.50
1160	1-1/4" x 3/4" size		10	.800		18.40	43.50		61.90	84.50
1170	1-1/2" x 3/4" size		10	.800		26	43.50		69.50	93
1180	1-1/2" x 1" size		9	.889		26	48		74	101
1190	2" x 3/4" size		9	.889		40	48		88	116
1200	2" x 1" size		8	1		40	54		94	125

23 21 20.70 Steam Traps

0010	**STEAM TRAPS**	Crew	Daily Output	Labor-Hours	Unit	Material	Labor	Equipment	Total	Total Incl O&P
0030	Cast iron body, threaded									
0040	Inverted bucket									
0050	1/2" pipe size	1 Stpi	12	.667	Ea.	131	36		167	198
0070	3/4" pipe size		10	.800		228	43.50		271.50	315
0100	1" pipe size		9	.889		350	48		398	455
0120	1-1/4" pipe size		8	1		530	54		584	665

23 21 20.70 Steam Traps		Crew	Daily Output	Labor-Hours	Unit	Material	2011 Bare Costs Labor	Equipment	Total	Total Incl O&P
0130	1-1/2" pipe size	1 Stpi	7	1.143	Ea.	670	62		732	830
0140	2" pipe size	▼	6	1.333	▼	1,050	72		1,122	1,250
0200	With thermic vent & check valve									
0202	1/2" pipe size	1 Stpi	12	.667	Ea.	155	36		191	224
0204	3/4" pipe size		10	.800		261	43.50		304.50	350
0206	1" pipe size		9	.889		380	48		428	490
0208	1-1/4" pipe size		8	1		575	54		629	715
0210	1-1/2" pipe size		7	1.143		640	62		702	795
0212	2" pipe size	▼	6	1.333	▼	975	72		1,047	1,175
1000	Float & thermostatic, 15 psi									
1010	3/4" pipe size	1 Stpi	16	.500	Ea.	105	27		132	157
1020	1" pipe size		15	.533		127	29		156	182
1030	1-1/4" pipe size		13	.615		154	33.50		187.50	219
1040	1-1/2" pipe size		9	.889		224	48		272	320
1060	2" pipe size	▼	6	1.333		410	72		482	565
1100	For vacuum breaker, add				▼	56			56	61.50
1110	Differential condensate controller									
1120	With union and manual metering valve									
1130	1/2" pipe size	1 Stpi	24	.333	Ea.	515	18.05		533.05	595
1140	3/4" pipe size		20	.400		575	21.50		596.50	670
1150	1" pipe size		19	.421		770	23		793	885
1160	1-1/4" pipe size		15	.533		1,050	29		1,079	1,200
1170	1-1/2" pipe size		13	.615		1,250	33.50		1,283.50	1,425
1180	2" pipe size	▼	11	.727	▼	1,625	39.50		1,664.50	1,850
1290	Brass body, threaded									
1300	Thermostatic, angle union, 25 psi									
1310	1/2" pipe size	1 Stpi	24	.333	Ea.	54	18.05		72.05	86.50
1320	3/4" pipe size		20	.400		82	21.50		103.50	123
1330	1" pipe size	▼	19	.421	▼	122	23		145	168
1990	Carbon steel body, threaded									
2000	Controlled disc, integral strainer & blow down									
2010	Threaded									
2020	3/8" pipe size	1 Stpi	24	.333	Ea.	221	18.05		239.05	270
2030	1/2" pipe size		24	.333		221	18.05		239.05	270
2040	3/4" pipe size		20	.400		297	21.50		318.50	360
2050	1" pipe size	▼	19	.421	▼	380	23		403	455
2100	Socket weld									
2110	3/8" pipe size	Q-17	17	.941	Ea.	264	46	1.91	311.91	360
2120	1/2" pipe size		16	1		264	48.50	2.03	314.53	365
2130	3/4" pipe size		14	1.143		345	55.50	2.32	402.82	465
2140	1" pipe size	▼	12	1.333	▼	430	65	2.71	497.71	575
2170	Flanged, 600 lb. ASA									
2180	3/8" pipe size	1 Stpi	12	.667	Ea.	595	36		631	710
2190	1/2" pipe size		12	.667		595	36		631	710
2200	3/4" pipe size		10	.800		715	43.50		758.50	850
2210	1" pipe size	▼	7	1.143	▼	845	62		907	1,025
5000	Forged steel body									
5010	Inverted bucket									
5020	Threaded									
5030	1/2" pipe size	1 Stpi	12	.667	Ea.	395	36		431	490
5040	3/4" pipe size		10	.800		685	43.50		728.50	815
5050	1" pipe size		9	.889		1,025	48		1,073	1,200
5060	1-1/4" pipe size	▼	8	1		1,475	54		1,529	1,700

23 21 20 – Hydronic HVAC Piping Specialties

23 21 20.70 Steam Traps

		Crew	Daily Output	Labor-Hours	Unit	Material	2011 Bare Costs Labor	Equipment	Total	Total Incl O&P
5070	1-1/2" pipe size	1 Stpi	7	1.143	Ea.	2,100	62		2,162	2,425
5080	2" pipe size	↓	6	1.333	↓	3,700	72		3,772	4,175
5100	Socket weld									
5110	1/2" pipe size	Q-17	16	1	Ea.	475	48.50	2.03	525.53	600
5120	3/4" pipe size		14	1.143		765	55.50	2.32	822.82	925
5130	1" pipe size		12	1.333		1,100	65	2.71	1,167.71	1,325
5140	1-1/4" pipe size		10	1.600		1,575	78	3.25	1,656.25	1,850
5150	1-1/2" pipe size		9	1.778		2,200	86.50	3.61	2,290.11	2,525
5160	2" pipe size	↓	7	2.286	↓	3,800	111	4.64	3,915.64	4,350
5200	Flanged									
5210	1/2" pipe size	1 Stpi	12	.667	Ea.	730	36		766	860
5220	3/4" pipe size		10	.800		1,200	43.50		1,243.50	1,400
5230	1" pipe size		7	1.143		1,625	62		1,687	1,900
5240	1-1/4" pipe size	↓	6	1.333		2,200	72		2,272	2,525
5250	1-1/2" pipe size	Q-5	9	1.778		2,825	86.50		2,911.50	3,225
5260	2" pipe size	"	7	2.286	↓	4,425	111		4,536	5,050
6000	Stainless steel body									
6010	Inverted bucket									
6020	Threaded									
6030	1/2" pipe size	1 Stpi	12	.667	Ea.	131	36		167	198
6040	3/4" pipe size		10	.800		172	43.50		215.50	254
6050	1" pipe size	↓	9	.889	↓	610	48		658	745
6100	Threaded with thermic vent									
6110	1/2" pipe size	1 Stpi	12	.667	Ea.	150	36		186	218
6120	3/4" pipe size		10	.800		192	43.50		235.50	277
6130	1" pipe size	↓	9	.889	↓	635	48		683	770
6200	Threaded with check valve									
6210	1/2" pipe size	1 Stpi	12	.667	Ea.	187	36		223	260
6220	3/4" pipe size	"	10	.800	"	255	43.50		298.50	345
6300	Socket weld									
6310	1/2" pipe size	Q-17	16	1	Ea.	152	48.50	2.03	202.53	242
6320	3/4" pipe size	"	14	1.143	"	196	55.50	2.32	253.82	300

23 21 20.74 Strainers, Basket Type

		Crew	Daily Output	Labor-Hours	Unit	Material	2011 Bare Costs Labor	Equipment	Total	Total Incl O&P
0010	**STRAINERS, BASKET TYPE**, Perforated stainless steel basket									
0100	Brass or monel available									
2000	Simplex style									
2300	Bronze body									
2320	Screwed, 3/8" pipe size	1 Stpi	22	.364	Ea.	152	19.65		171.65	197
2340	1/2" pipe size		20	.400		244	21.50		265.50	300
2360	3/4" pipe size		17	.471		325	25.50		350.50	400
2380	1" pipe size		15	.533		325	29		354	405
2400	1-1/4" pipe size		13	.615		475	33.50		508.50	575
2420	1-1/2" pipe size		12	.667		470	36		506	570
2440	2" pipe size	↓	10	.800		695	43.50		738.50	830
2460	2-1/2" pipe size	Q-5	15	1.067		685	52		737	830
2480	3" pipe size	"	14	1.143		985	55.50		1,040.50	1,150
2600	Flanged, 2" pipe size	1 Stpi	6	1.333		725	72		797	910
2620	2-1/2" pipe size	Q-5	4.50	3.556		1,125	173		1,298	1,500
2640	3" pipe size		3.50	4.571		1,275	223		1,498	1,725
2660	4" pipe size	↓	3	5.333		2,175	260		2,435	2,775
2680	5" pipe size	Q-6	3.40	7.059		3,325	355		3,680	4,175
2700	6" pipe size	↓	3	8	↓	4,225	405		4,630	5,225

23 21 20.74 **Strainers, Basket Type**	Crew	Daily Output	Labor-Hours	Unit	Material	2011 Bare Costs Labor	Equipment	Total	Total Incl O&P
2710	8" pipe size	Q-6	2.50	9.600	Ea.	6,875	485		7,360	8,275
3600	Iron body									
3700	Screwed, 3/8" pipe size	1 Stpi	22	.364	Ea.	86	19.65		105.65	124
3720	1/2" pipe size		20	.400		88.50	21.50		110	130
3740	3/4" pipe size		17	.471		113	25.50		138.50	162
3760	1" pipe size		15	.533		115	29		144	169
3780	1-1/4" pipe size		13	.615		150	33.50		183.50	215
3800	1-1/2" pipe size		12	.667		166	36		202	236
3820	2" pipe size		10	.800		198	43.50		241.50	283
3840	2-1/2" pipe size	Q-5	15	1.067		265	52		317	370
3860	3" pipe size	"	14	1.143		320	55.50		375.50	440
4000	Flanged, 2" pipe size	1 Stpi	6	1.333		305	72		377	445
4020	2-1/2" pipe size	Q-5	4.50	3.556		415	173		588	715
4040	3" pipe size		3.50	4.571		435	223		658	805
4060	4" pipe size		3	5.333		665	260		925	1,125
4080	5" pipe size	Q-6	3.40	7.059		1,000	355		1,355	1,625
4100	6" pipe size		3	8		1,275	405		1,680	2,025
4120	8" pipe size		2.50	9.600		2,325	485		2,810	3,275
4140	10" pipe size		2.20	10.909		5,100	550		5,650	6,450
6000	Cast steel body									
6400	Screwed, 1" pipe size	1 Stpi	15	.533	Ea.	184	29		213	245
6410	1-1/4" pipe size		13	.615		278	33.50		311.50	355
6420	1-1/2" pipe size		12	.667		278	36		314	360
6440	2" pipe size		10	.800		380	43.50		423.50	485
6460	2-1/2" pipe size	Q-5	15	1.067		540	52		592	675
6480	3" pipe size	"	14	1.143		705	55.50		760.50	860
6560	Flanged, 2" pipe size	1 Stpi	6	1.333		615	72		687	790
6580	2-1/2" pipe size	Q-5	4.50	3.556		925	173		1,098	1,275
6600	3" pipe size		3.50	4.571		990	223		1,213	1,425
6620	4" pipe size		3	5.333		1,400	260		1,660	1,950
6640	6" pipe size	Q-6	3	8		2,575	405		2,980	3,425
6660	8" pipe size	"	2.50	9.600		4,150	485		4,635	5,300
7000	Stainless steel body									
7200	Screwed, 1" pipe size	1 Stpi	15	.533	Ea.	278	29		307	350
7210	1-1/4" pipe size		13	.615		430	33.50		463.50	525
7220	1-1/2" pipe size		12	.667		430	36		466	530
7240	2" pipe size		10	.800		645	43.50		688.50	770
7260	2-1/2" pipe size	Q-5	15	1.067		910	52		962	1,075
7280	3" pipe size	"	14	1.143		1,250	55.50		1,305.50	1,450
7400	Flanged, 2" pipe size	1 Stpi	6	1.333		1,025	72		1,097	1,225
7420	2-1/2" pipe size	Q-5	4.50	3.556		1,850	173		2,023	2,275
7440	3" pipe size		3.50	4.571		1,900	223		2,123	2,400
7460	4" pipe size		3	5.333		2,975	260		3,235	3,675
7480	6" pipe size	Q-6	3	8		5,175	405		5,580	6,275
7500	8" pipe size	"	2.50	9.600		7,525	485		8,010	9,000
8100	Duplex style									
8200	Bronze body									
8240	Screwed, 3/4" pipe size	1 Stpi	16	.500	Ea.	1,075	27		1,102	1,225
8260	1" pipe size		14	.571		1,075	31		1,106	1,225
8280	1-1/4" pipe size		12	.667		2,100	36		2,136	2,350
8300	1-1/2" pipe size		11	.727		2,100	39.50		2,139.50	2,350
8320	2" pipe size		9	.889		3,400	48		3,448	3,800
8340	2-1/2" pipe size	Q-5	14	1.143		4,400	55.50		4,455.50	4,900

23 21 Hydronic Piping and Pumps

23 21 20 – Hydronic HVAC Piping Specialties

23 21 20.74 Strainers, Basket Type

		Crew	Daily Output	Labor-Hours	Unit	Material	2011 Bare Costs Labor	Equipment	Total	Total Incl O&P
8420	Flanged, 2" pipe size	1 Stpi	6	1.333	Ea.	3,650	72		3,722	4,125
8440	2-1/2" pipe size	Q-5	4.50	3.556		5,050	173		5,223	5,800
8460	3" pipe size		3.50	4.571		5,500	223		5,723	6,375
8480	4" pipe size		3	5.333		7,900	260		8,160	9,075
8500	5" pipe size	Q-6	3.40	7.059		17,600	355		17,955	19,800
8520	6" pipe size	"	3	8		17,600	405		18,005	19,900
8700	Iron body									
8740	Screwed, 3/4" pipe size	1 Stpi	16	.500	Ea.	575	27		602	675
8760	1" pipe size		14	.571		575	31		606	680
8780	1-1/4" pipe size		12	.667		1,025	36		1,061	1,175
8800	1-1/2" pipe size		11	.727		1,025	39.50		1,064.50	1,175
8820	2" pipe size		9	.889		1,725	48		1,773	1,950
8840	2-1/2" pipe size	Q-5	14	1.143		1,900	55.50		1,955.50	2,150
9000	Flanged, 2" pipe size	1 Stpi	6	1.333		2,050	72		2,122	2,375
9020	2-1/2" pipe size	Q-5	4.50	3.556		1,950	173		2,123	2,400
9040	3" pipe size		3.50	4.571		2,125	223		2,348	2,675
9060	4" pipe size		3	5.333		3,600	260		3,860	4,350
9080	5" pipe size	Q-6	3.40	7.059		7,150	355		7,505	8,375
9100	6" pipe size		3	8		7,150	405		7,555	8,450
9120	8" pipe size		2.50	9.600		12,500	485		12,985	14,400
9140	10" pipe size		2.20	10.909		17,300	550		17,850	19,900
9160	12" pipe size		1.70	14.118		19,200	715		19,915	22,300
9170	14" pipe size		1.40	17.143		23,300	865		24,165	26,900
9180	16" pipe size		1	24		29,200	1,200		30,400	34,000
9300	Cast steel body									
9340	Screwed, 1" pipe size	1 Stpi	14	.571	Ea.	1,225	31		1,256	1,400
9360	1-1/2" pipe size		11	.727		1,975	39.50		2,014.50	2,225
9380	2" pipe size		9	.889		2,600	48		2,648	2,950
9460	Flanged, 2" pipe size		6	1.333		2,800	72		2,872	3,175
9480	2-1/2" pipe size	Q-5	4.50	3.556		4,500	173		4,673	5,225
9500	3" pipe size		3.50	4.571		4,925	223		5,148	5,725
9520	4" pipe size		3	5.333		6,075	260		6,335	7,075
9540	6" pipe size	Q-6	3	8		11,400	405		11,805	13,100
9560	8" pipe size	"	2.50	9.600		27,400	485		27,885	30,800
9700	Stainless steel body									
9740	Screwed, 1" pipe size	1 Stpi	14	.571	Ea.	1,575	31		1,606	1,800
9760	1-1/2" pipe size		11	.727		2,375	39.50		2,414.50	2,675
9780	2" pipe size		9	.889		3,325	48		3,373	3,725
9860	Flanged, 2" pipe size		6	1.333		3,625	72		3,697	4,100
9880	2-1/2" pipe size	Q-5	4.50	3.556		6,075	173		6,248	6,950
9900	3" pipe size		3.50	4.571		6,625	223		6,848	7,625
9920	4" pipe size		3	5.333		8,625	260		8,885	9,900
9940	6" pipe size	Q-6	3	8		13,200	405		13,605	15,100
9960	8" pipe size	"	2.50	9.600		36,800	485		37,285	41,200

23 21 20.76 Strainers, Y Type, Bronze Body

		Crew	Daily Output	Labor-Hours	Unit	Material	2011 Bare Costs Labor	Equipment	Total	Total Incl O&P
0010	**STRAINERS, Y TYPE, BRONZE BODY**									
0050	Screwed, 150 lb., 1/4" pipe size	1 Stpi	24	.333	Ea.	17.30	18.05		35.35	46
0070	3/8" pipe size		24	.333		23	18.05		41.05	52
0100	1/2" pipe size		20	.400		23	21.50		44.50	57.50
0120	3/4" pipe size		19	.421		26	23		49	62.50
0140	1" pipe size		17	.471		30.50	25.50		56	72
0150	1-1/4" pipe size		15	.533		62	29		91	111

23 21 20 – Hydronic HVAC Piping Specialties

23 21 20.76 Strainers, Y Type, Bronze Body

		Crew	Daily Output	Labor-Hours	Unit	Material	2011 Bare Costs Labor	Equipment	Total	Total Incl O&P
0160	1-1/2" pipe size	1 Stpi	14	.571	Ea.	66	31		97	119
0180	2" pipe size		13	.615		88.50	33.50		122	147
0182	3" pipe size		12	.667		670	36		706	790
0200	300 lb., 2-1/2" pipe size	Q-5	17	.941		420	46		466	535
0220	3" pipe size		16	1		835	48.50		883.50	995
0240	4" pipe size		15	1.067		1,900	52		1,952	2,175
0500	For 300 lb rating 1/4" thru 2", add					15%				
1000	Flanged, 150 lb., 1-1/2" pipe size	1 Stpi	11	.727	Ea.	435	39.50		474.50	540
1020	2" pipe size	"	8	1		590	54		644	730
1030	2-1/2" pipe size	Q-5	5	3.200		875	156		1,031	1,200
1040	3" pipe size		4.50	3.556		1,075	173		1,248	1,425
1060	4" pipe size		3	5.333		1,625	260		1,885	2,200
1080	5" pipe size	Q-6	3.40	7.059		1,625	355		1,980	2,325
1100	6" pipe size		3	8		3,125	405		3,530	4,025
1106	8" pipe size		2.60	9.231		3,425	465		3,890	4,475
1500	For 300 lb rating, add					40%				

23 21 20.78 Strainers, Y Type, Iron Body

		Crew	Daily Output	Labor-Hours	Unit	Material	2011 Bare Costs Labor	Equipment	Total	Total Incl O&P
0010	**STRAINERS, Y TYPE, IRON BODY**									
0050	Screwed, 250 lb., 1/4" pipe size	1 Stpi	20	.400	Ea.	9.60	21.50		31.10	43
0070	3/8" pipe size		20	.400		9.60	21.50		31.10	43
0100	1/2" pipe size		20	.400		9.60	21.50		31.10	43
0120	3/4" pipe size		18	.444		11.50	24		35.50	48.50
0140	1" pipe size		16	.500		15.85	27		42.85	58
0150	1-1/4" pipe size		15	.533		21	29		50	66
0160	1-1/2" pipe size		12	.667		26	36		62	82.50
0180	2" pipe size		8	1		39	54		93	124
0200	2-1/2" pipe size	Q-5	12	1.333		232	65		297	355
0220	3" pipe size		11	1.455		251	71		322	380
0240	4" pipe size		5	3.200		425	156		581	700
0500	For galvanized body, add					50%				
1000	Flanged, 125 lb., 1-1/2" pipe size	1 Stpi	11	.727	Ea.	135	39.50		174.50	207
1020	2" pipe size	"	8	1		142	54		196	237
1030	2-1/2" pipe size	Q-5	5	3.200		102	156		258	345
1040	3" pipe size		4.50	3.556		133	173		306	405
1060	4" pipe size		3	5.333		242	260		502	655
1080	5" pipe size	Q-6	3.40	7.059		380	355		735	945
1100	6" pipe size		3	8		460	405		865	1,125
1120	8" pipe size		2.50	9.600		775	485		1,260	1,575
1140	10" pipe size		2	12		1,475	605		2,080	2,525
1160	12" pipe size		1.70	14.118		2,225	715		2,940	3,525
1170	14" pipe size		1.30	18.462		4,125	930		5,055	5,925
1180	16" pipe size		1	24		5,825	1,200		7,025	8,200
1500	For 250 lb rating, add					20%				
2000	For galvanized body, add					50%				
2500	For steel body, add					40%				

23 21 20.80 Suction Diffusers

		Crew	Daily Output	Labor-Hours	Unit	Material	2011 Bare Costs Labor	Equipment	Total	Total Incl O&P
0010	**SUCTION DIFFUSERS**									
0100	Cast iron body with integral straightening vanes, strainer									
1000	Flanged									
1010	2" inlet, 1-1/2" pump side	1 Stpi	6	1.333	Ea.	310	72		382	450
1020	2" pump side	"	5	1.600		320	86.50		406.50	485
1030	3" inlet, 2" pump side	Q-5	6.50	2.462		350	120		470	565

23 21 20.80 Suction Diffusers		Crew	Daily Output	Labor-Hours	Unit	Material	2011 Bare Costs Labor	2011 Bare Costs Equipment	Total	Total Incl O&P
1040	2-1/2" pump side	Q-5	5.30	3.019	Ea.	580	147		727	855
1050	3" pump side		4.50	3.556		585	173		758	900
1060	4" inlet, 3" pump side		3.50	4.571		685	223		908	1,075
1070	4" pump side		3	5.333		770	260		1,030	1,225
1080	5" inlet, 4" pump side	Q-6	3.80	6.316		905	320		1,225	1,475
1090	5" pump side		3.40	7.059		1,075	355		1,430	1,700
1100	6" inlet, 4" pump side		3.30	7.273		940	365		1,305	1,575
1110	5" pump side		3.10	7.742		1,125	390		1,515	1,825
1120	6" pump side		3	8		1,200	405		1,605	1,900
1130	8" inlet, 6" pump side		2.70	8.889		1,275	450		1,725	2,075
1140	8" pump side		2.50	9.600		2,225	485		2,710	3,175
1150	10" inlet, 6" pump side		2.40	10		3,075	505		3,580	4,125
1160	8" pump side		2.30	10.435		3,450	525		3,975	4,575
1170	10" pump side		2.20	10.909		3,700	550		4,250	4,900
1180	12" inlet, 8" pump side		2	12		3,925	605		4,530	5,225
1190	10" pump side		1.80	13.333		4,350	675		5,025	5,775
1200	12" pump side		1.70	14.118		4,900	715		5,615	6,450
1210	14" inlet, 12" pump side		1.50	16		6,350	810		7,160	8,175
1220	14" pump side		1.30	18.462		6,475	930		7,405	8,525

23 21 20.84 Thermoflo Indicator

		Crew	Daily Output	Labor-Hours	Unit	Material	Labor	Equipment	Total	Total Incl O&P
0010	**THERMOFLO INDICATOR**, For balancing									
1000	Sweat connections, 1-1/4" pipe size	1 Stpi	12	.667	Ea.	555	36		591	665
1020	1-1/2" pipe size		10	.800		565	43.50		608.50	690
1040	2" pipe size		8	1		595	54		649	735
1060	2-1/2" pipe size		7	1.143		910	62		972	1,100
2000	Flange connections, 3" pipe size	Q-5	5	3.200		1,075	156		1,231	1,425
2020	4" pipe size		4	4		1,300	195		1,495	1,725
2030	5" pipe size		3.50	4.571		1,625	223		1,848	2,125
2040	6" pipe size		3	5.333		1,750	260		2,010	2,325
2060	8" pipe size		2	8		2,125	390		2,515	2,925

23 21 20.88 Venturi Flow

		Crew	Daily Output	Labor-Hours	Unit	Material	Labor	Equipment	Total	Total Incl O&P
0010	**VENTURI FLOW**, Measuring device									
0050	1/2" diameter	1 Stpi	24	.333	Ea.	281	18.05		299.05	335
0100	3/4" diameter		20	.400		257	21.50		278.50	315
0120	1" diameter		19	.421		276	23		299	340
0140	1-1/4" diameter		15	.533		340	29		369	420
0160	1-1/2" diameter		13	.615		355	33.50		388.50	440
0180	2" diameter		11	.727		365	39.50		404.50	465
0200	2-1/2" diameter	Q-5	16	1		500	48.50		548.50	625
0220	3" diameter		14	1.143		515	55.50		570.50	655
0240	4" diameter		11	1.455		775	71		846	955
0260	5" diameter	Q-6	4	6		1,025	305		1,330	1,575
0280	6" diameter		3.50	6.857		1,125	345		1,470	1,775
0300	8" diameter		3	8		1,450	405		1,855	2,200
0320	10" diameter		2	12		3,425	605		4,030	4,675
0330	12" diameter		1.80	13.333		4,950	675		5,625	6,450
0340	14" diameter		1.60	15		5,600	760		6,360	7,300
0350	16" diameter		1.40	17.143		6,250	865		7,115	8,175
0500	For meter, add					2,125			2,125	2,350

23 21 20.94 Weld End Ball Joints

		Crew	Daily Output	Labor-Hours	Unit	Material	Labor	Equipment	Total	Total Incl O&P
0010	**WELD END BALL JOINTS**, Steel									
0050	2-1/2" diameter	Q-17	13	1.231	Ea.	615	60	2.50	677.50	770

23 21 Hydronic Piping and Pumps

23 21 20 – Hydronic HVAC Piping Specialties

23 21 20.94 Weld End Ball Joints

		Crew	Daily Output	Labor-Hours	Unit	Material	2011 Bare Costs Labor	Equipment	Total	Total Incl O&P
0100	3" diameter	Q-17	12	1.333	Ea.	720	65	2.71	787.71	890
0120	4" diameter	↓	11	1.455		1,175	71	2.95	1,248.95	1,375
0140	5" diameter	Q-18	14	1.714		1,350	86.50	2.31	1,438.81	1,600
0160	6" diameter		12	2		2,000	101	2.70	2,103.70	2,350
0180	8" diameter		9	2.667		3,025	135	3.60	3,163.60	3,525
0200	10" diameter		8	3		4,075	152	4.05	4,231.05	4,700
0220	12" diameter	↓	6	4	↓	6,500	202	5.40	6,707.40	7,450

23 21 20.98 Zone Valves

		Crew	Daily Output	Labor-Hours	Unit	Material	2011 Bare Costs Labor	Equipment	Total	Total Incl O&P
0010	**ZONE VALVES**									
1000	Bronze body									
1010	2 way, 125 psi									
1020	Standard, 65' pump head									
1030	1/2" soldered	1 Stpi	24	.333	Ea.	112	18.05		130.05	150
1040	3/4" soldered		20	.400		112	21.50		133.50	156
1050	1" soldered		19	.421		120	23		143	166
1060	1-1/4" soldered	↓	15	.533	↓	141	29		170	198
1200	High head, 125' pump head									
1210	1/2" soldered	1 Stpi	24	.333	Ea.	153	18.05		171.05	195
1220	3/4" soldered		20	.400		158	21.50		179.50	207
1230	1" soldered	↓	19	.421	↓	172	23		195	223
1300	Geothermal, 150' pump head									
1310	3/4" soldered	1 Stpi	20	.400	Ea.	181	21.50		202.50	232
1320	1" soldered		19	.421		193	23		216	246
1330	3/4" threaded	↓	20	.400	↓	181	21.50		202.50	232
3000	3 way, 125 psi									
3200	Bypass, 65' pump head									
3210	1/2" soldered	1 Stpi	24	.333	Ea.	160	18.05		178.05	203
3220	3/4" soldered		20	.400		165	21.50		186.50	214
3230	1" soldered		19	.421		188	23		211	240
9000	Transformer, for up to 3 zone valves	↓	20	.400	↓	23	21.50		44.50	58

23 21 23 – Hydronic Pumps

23 21 23.13 In-Line Centrifugal Hydronic Pumps

		Crew	Daily Output	Labor-Hours	Unit	Material	2011 Bare Costs Labor	Equipment	Total	Total Incl O&P
0010	**IN-LINE CENTRIFUGAL HYDRONIC PUMPS**									
0600	Bronze, sweat connections, 1/40 HP, in line									
0640	3/4" size	Q-1	16	1	Ea.	192	48.50		240.50	283
1000	Flange connection, 3/4" to 1-1/2" size									
1040	1/12 HP	Q-1	6	2.667	Ea.	495	129		624	735
1060	1/8 HP		6	2.667		855	129		984	1,125
1100	1/3 HP		6	2.667		955	129		1,084	1,250
1140	2" size, 1/6 HP		5	3.200		1,025	154		1,179	1,350
1180	2-1/2" size, 1/4 HP		5	3.200		1,525	154		1,679	1,925
1220	3" size, 1/4 HP		4	4		1,600	193		1,793	2,075
1260	1/3 HP		4	4		1,975	193		2,168	2,475
1300	1/2 HP		4	4		2,000	193		2,193	2,500
1340	3/4 HP		4	4		2,200	193		2,393	2,700
1380	1 HP	↓	4	4	↓	3,550	193		3,743	4,200
2000	Cast iron, flange connection									
2040	3/4" to 1-1/2" size, in line, 1/12 HP	Q-1	6	2.667	Ea.	320	129		449	540
2060	1/8 HP		6	2.667		535	129		664	780
2100	1/3 HP		6	2.667		595	129		724	845
2140	2" size, 1/6 HP		5	3.200		655	154		809	950
2180	2-1/2" size, 1/4 HP	↓	5	3.200		830	154		984	1,150

23 21 Hydronic Piping and Pumps

23 21 23 – Hydronic Pumps

23 21 23.13 In-Line Centrifugal Hydronic Pumps	Crew	Daily Output	Labor-Hours	Unit	Material	2011 Bare Costs Labor	2011 Bare Costs Equipment	Total	Total Incl O&P
2220	3" size, 1/4 HP	Q-1	4	4	Ea.	840	193		1,033	1,225
2260	1/3 HP		4	4		1,150	193		1,343	1,550
2300	1/2 HP		4	4		1,175	193		1,368	1,600
2340	3/4 HP		4	4		1,375	193		1,568	1,800
2380	1 HP		4	4		1,975	193		2,168	2,475
2600	For non-ferrous impeller, add					3%				
3000	High head, bronze impeller									
3030	1-1/2" size 1/2 HP	Q-1	5	3.200	Ea.	1,025	154		1,179	1,375
3040	1-1/2" size 3/4 HP		5	3.200		1,125	154		1,279	1,475
3050	2" size 1 HP		4	4		1,375	193		1,568	1,800
3090	2" size 1-1/2 HP		4	4		1,700	193		1,893	2,175
4000	Close coupled, end suction, bronze impeller									
4040	1-1/2" size, 1-1/2 HP, to 40 GPM	Q-1	3	5.333	Ea.	1,875	257		2,132	2,450
4090	2" size, 2 HP, to 50 GPM		3	5.333		2,075	257		2,332	2,650
4100	2" size, 3 HP, to 90 GPM		2.30	6.957		2,225	335		2,560	2,950
4190	2-1/2" size, 3 HP, to 150 GPM		2	8		2,500	385		2,885	3,325
4300	3" size, 5 HP, to 225 GPM		1.80	8.889		2,875	430		3,305	3,800
4410	3" size, 10 HP, to 350 GPM		1.60	10		3,850	485		4,335	4,975
4420	4" size, 7-1/2 HP, to 350 GPM		1.60	10		3,750	485		4,235	4,850
4520	4" size, 10 HP, to 600 GPM	Q-2	1.70	14.118		4,275	705		4,980	5,750
4530	5" size, 15 HP, to 1000 GPM		1.70	14.118		4,300	705		5,005	5,775
4610	5" size, 20 HP, to 1350 GPM		1.50	16		4,550	800		5,350	6,225
4620	5" size, 25 HP, to 1550 GPM		1.50	16		6,225	800		7,025	8,050
5000	Base mounted, bronze impeller, coupling guard									
5040	1-1/2" size, 1-1/2 HP, to 40 GPM	Q-1	2.30	6.957	Ea.	5,100	335		5,435	6,125
5090	2" size, 2 HP, to 50 GPM		2.30	6.957		5,725	335		6,060	6,800
5100	2" size, 3 HP, to 90 GPM		2	8		6,300	385		6,685	7,500
5190	2-1/2" size, 3 HP, to 150 GPM		1.80	8.889		6,775	430		7,205	8,100
5300	3" size, 5 HP, to 225 GPM		1.60	10		7,775	485		8,260	9,300
5410	4" size, 5 HP, to 350 GPM		1.50	10.667		8,500	515		9,015	10,100
5420	4" size, 7-1/2 HP, to 350 GPM		1.50	10.667		9,550	515		10,065	11,300
5520	5" size, 10 HP, to 600 GPM	Q-2	1.60	15		12,400	750		13,150	14,700
5530	5" size, 15 HP, to 1000 GPM		1.60	15		12,600	750		13,350	14,900
5610	6" size, 20 HP, to 1350 GPM		1.40	17.143		15,000	860		15,860	17,800
5620	6" size, 25 HP, to 1550 GPM		1.40	17.143		16,500	860		17,360	19,400
5800	The above pump capacities are based on 1800 RPM,									
5810	at a 60 foot head. Increasing the RPM									
5820	or decreasing the head will increase the GPM.									

23 21 29 – Automatic Condensate Pump Units

23 21 29.10 Condensate Removal Pump System

0010	CONDENSATE REMOVAL PUMP SYSTEM									
0020	Pump with 1 gal. ABS tank	G								
0100	115 V	G								
0120	1/50 HP, 200 GPH	G	1 Stpi	12	.667	Ea.	176	36		212	248
0140	1/18 HP, 270 GPH	G		10	.800		183	43.50		226.50	266
0160	1/5 HP, 450 GPH	G		8	1		420	54		474	545
0200	230 V	G								
0240	1/18 HP, 270 GPH	G	1 Stpi	10	.800	Ea.	195	43.50		238.50	280
0260	1/5 HP, 450 GPH	G	"	8	1	"	465	54		519	595

23 22 Steam and Condensate Piping and Pumps

23 22 13 – Steam and Condensate Heating Piping

23 22 13.23 Aboveground Steam and Condensate Piping	Crew	Daily Output	Labor-Hours	Unit	Material	2011 Bare Costs Labor	Equipment	Total	Total Incl O&P
0010 **ABOVEGROUND STEAM AND CONDENSATE HEATING PIPING**									
0020 Condensate meter									
0100 500 lb. per hour	1 Stpi	14	.571	Ea.	2,350	31		2,381	2,650
0140 1500 lb. per hour		7	1.143		2,550	62		2,612	2,900
0160 3000 lb. per hour	↓	5	1.600		3,425	86.50		3,511.50	3,875
0200 12,000 lb. per hour	Q-5	3.50	4.571	↓	4,575	223		4,798	5,350

23 22 23 – Steam Condensate Pumps

23 22 23.10 Condensate Return System

	Crew	Daily Output	Labor-Hours	Unit	Material	2011 Bare Costs Labor	Equipment	Total	Total Incl O&P
0010 **CONDENSATE RETURN SYSTEM**									
2000 Simplex									
2010 With pump, motor, CI receiver, float switch									
2020 3/4 HP, 15 GPM	Q-1	1.80	8.889	Ea.	5,550	430		5,980	6,750
2100 Duplex									
2110 With 2 pumps and motors, CI receiver, float switch, alternator									
2120 3/4 HP, 15 GPM, 15 gal. CI rcvr	Q-1	1.40	11.429	Ea.	6,050	550		6,600	7,475
2130 1 HP, 25 GPM		1.20	13.333		7,250	645		7,895	8,925
2140 1-1/2 HP, 45 GPM		1	16		8,425	770		9,195	10,400
2150 1-1/2 HP, 60 GPM	↓	1	16	↓	9,450	770		10,220	11,600

23 22 23.23 Pumps, Pneumatic Ejector

	Crew	Daily Output	Labor-Hours	Unit	Material	2011 Bare Costs Labor	Equipment	Total	Total Incl O&P
0010 **PUMPS, PNEUMATIC EJECTOR**									
0020 With steel receiver, level controls, inlet/outlet gate/check valves									
0030 Cross connect. not incl. compressor, fittings or piping									
0040 Duplex									
0050 30 GPM	Q-2	1.70	14.118	Ea.	22,500	705		23,205	25,900
0060 50 GPM		1.56	15.385		26,400	770		27,170	30,300
0070 100 GPM		1.30	18.462		36,500	925		37,425	41,600
0080 150 GPM		1.10	21.818		67,000	1,100		68,100	75,000
0090 200 GPM	↓	.85	28.235		94,500	1,425		95,925	106,000
0100 250 GPM	Q-3	.91	35.165		108,000	1,800		109,800	121,000
0110 300 GPM	"	.57	56.140	↓	146,500	2,875		149,375	165,500

23 23 Refrigerant Piping

23 23 13 – Refrigerant Piping Valves

23 23 13.10 Valves

	Crew	Daily Output	Labor-Hours	Unit	Material	2011 Bare Costs Labor	Equipment	Total	Total Incl O&P
0010 **VALVES**									
8100 Check valve, soldered									
8110 5/8"	1 Stpi	36	.222	Ea.	115	12		127	144
8114 7/8"		26	.308		132	16.65		148.65	171
8118 1-1/8"		18	.444		158	24		182	209
8122 1-3/8"		14	.571		217	31		248	285
8126 1-5/8"		13	.615		279	33.50		312.50	355
8130 2-1/8"	↓	12	.667		310	36		346	395
8134 2-5/8"	Q-5	22	.727		465	35.50		500.50	565
8138 3-1/8"	"	20	.800	↓	565	39		604	685
8500 Refrigeration valve, packless, soldered									
8510 1/2"	1 Stpi	38	.211	Ea.	37.50	11.40		48.90	58
8514 5/8"		36	.222		39	12		51	60.50
8518 7/8"	↓	26	.308	↓	132	16.65		148.65	170
8520 Packed, soldered									
8522 1-1/8"	1 Stpi	18	.444	Ea.	161	24		185	213

23 23 13 – Refrigerant Piping Valves

23 23 13.10 Valves		Crew	Daily Output	Labor-Hours	Unit	Material	2011 Bare Costs Labor	Equipment	Total	Total Incl O&P
8526	1-3/8"	1 Stpi	14	.571	Ea.	284	31		315	355
8530	1-5/8"		13	.615		315	33.50		348.50	395
8534	2-1/8"		12	.667		500	36		536	605
8538	2-5/8"	Q-5	22	.727		680	35.50		715.50	800
8542	3-1/8"		20	.800		785	39		824	925
8546	4-1/8"		18	.889		1,375	43.50		1,418.50	1,600
8600	Solenoid valve, flange/solder									
8610	1/2"	1 Stpi	38	.211	Ea.	163	11.40		174.40	197
8614	5/8"		36	.222		205	12		217	243
8618	3/4"		30	.267		256	14.45		270.45	305
8622	7/8"		26	.308		320	16.65		336.65	380
8626	1-1/8"		18	.444		420	24		444	495
8630	1-3/8"		14	.571		510	31		541	605
8634	1-5/8"		13	.615		715	33.50		748.50	835
8638	2-1/8"		12	.667		785	36		821	920
8800	Thermostatic valve, flange/solder									
8810	1/2 – 3 ton, 3/8" x 5/8"	1 Stpi	9	.889	Ea.	131	48		179	216
8814	4 – 5 ton, 1/2" x 7/8"		7	1.143		151	62		213	259
8818	6 – 8 ton, 5/8" x 7/8"		5	1.600		151	86.50		237.50	295
8822	7 – 12 ton, 7/8" x 1-1/8"		4	2		176	108		284	355
8826	15 – 20 ton, 7/8" x 1-3/8"		3.20	2.500		176	135		311	395

23 23 16 – Refrigerant Piping Specialties

23 23 16.10 Refrigerant Piping Component Specialties

		Crew	Daily Output	Labor-Hours	Unit	Material	2011 Bare Costs Labor	Equipment	Total	Total Incl O&P
0010	**REFRIGERANT PIPING COMPONENT SPECIALTIES**									
0600	Accumulator									
0610	3/4"	1 Stpi	8.80	.909	Ea.	70.50	49		119.50	151
0614	7/8"		6.40	1.250		80.50	67.50		148	190
0618	1-1/8"		4.80	1.667		111	90		201	257
0622	1-3/8"		4	2		145	108		253	320
0626	1-5/8"		3.20	2.500		157	135		292	375
0630	2-1/8"		2.40	3.333		375	180		555	685
0700	Condensate drip/drain pan		24	.333		132	18.05		150.05	172
1000	Filter dryer									
1010	Replaceable core type, solder									
1020	1/2"	1 Stpi	20	.400	Ea.	150	21.50		171.50	198
1030	5/8"		19	.421		154	23		177	204
1040	7/8"		18	.444		154	24		178	206
1050	1-1/8"		15	.533		168	29		197	228
1060	1-3/8"		14	.571		168	31		199	231
1070	1-5/8"		12	.667		172	36		208	243
1080	2-1/8"		10	.800		181	43.50		224.50	264
1090	2-5/8"		9	.889		695	48		743	835
1100	3-1/8"		8	1		695	54		749	845
1200	Sealed in-line, solder									
1210	1/4", 3 cubic inches	1 Stpi	22	.364	Ea.	8.10	19.65		27.75	38.50
1220	3/8", 5 cubic inches		21	.381		7.40	20.50		27.90	39
1230	1/2", 9 cubic inches		20	.400		12	21.50		33.50	45.50
1240	1/2", 16 cubic inches		20	.400		15.60	21.50		37.10	49.50
1250	5/8", 16 cubic inches		19	.421		16.35	23		39.35	52
1260	5/8", 30 cubic inches		19	.421		26.50	23		49.50	63
1270	7/8", 30 cubic inches		18	.444		28	24		52	67
1280	7/8", 41 cubic inches		17	.471		37.50	25.50		63	79.50

23 23 Refrigerant Piping

23 23 16 – Refrigerant Piping Specialties

23 23 16.10 Refrigerant Piping Component Specialties	Crew	Daily Output	Labor-Hours	Unit	Material	2011 Bare Costs Labor	Equipment	Total	Total Incl O&P
1290	1-1/8", 60 cubic inches	1 Stpi	15	.533	Ea.	52	29		81	101
4000	P-Trap, suction line, solder									
4010	5/8"	1 Stpi	19	.421	Ea.	62	23		85	102
4020	3/4"		19	.421		92.50	23		115.50	136
4030	7/8"		18	.444		69	24		93	112
4040	1-1/8"		15	.533		105	29		134	158
4050	1-3/8"		14	.571		190	31		221	255
4060	1-5/8"		12	.667		289	36		325	375
4070	2-1/8"	↓	10	.800	↓	590	43.50		633.50	715
5000	Sightglass									
5010	Moisture and liquid indicator, solder									
5020	1/4"	1 Stpi	22	.364	Ea.	19.65	19.65		39.30	51
5030	3/8"		21	.381		20	20.50		40.50	53
5040	1/2"		20	.400		24	21.50		45.50	59
5050	5/8"		19	.421		25	23		48	61.50
5060	7/8"		18	.444		36.50	24		60.50	76.50
5070	1-1/8"		15	.533		39.50	29		68.50	86.50
5080	1-3/8"		14	.571		60	31		91	112
5090	1-5/8"		12	.667		68	36		104	129
5100	2-1/8"	↓	10	.800	↓	85.50	43.50		129	159
7400	Vacuum pump set									
7410	Two stage, high vacuum continuous duty	1 Stpi	8	1	Ea.	3,375	54		3,429	3,775

23 23 23 – Refrigerants

23 23 23.10 Anti-Freeze

		Crew	Daily Output	Labor-Hours	Unit	Material	2011 Bare Costs Labor	Equipment	Total	Total Incl O&P
0010	**ANTI-FREEZE**, Inhibited									
0900	Ethylene glycol concentrated									
1000	55 gallon drums, small quantities				Gal.	7.30			7.30	8.05
1200	Large quantities					7.55			7.55	8.30
2000	Propylene glycol, for solar heat, small quantities					12.05			12.05	13.25
2100	Large quantities				↓	13.75			13.75	15.15

23 23 23.20 Refrigerant

		Crew	Daily Output	Labor-Hours	Unit	Material	2011 Bare Costs Labor	Equipment	Total	Total Incl O&P
0010	**REFRIGERANT**									
4420	Refrigerant, R-22, 30 lb. disposable cylinder				Lb.	12.50			12.50	13.75
4428	Refrigerant, R-134A, 30 lb. disposable cylinder					7.50			7.50	8.25
4434	Refrigerant, R-407C, 30 lb. disposable cylinder					17.35			17.35	19.05
4440	Refrigerant, R-408A, 25 lb. disposable cylinder					22			22	24
4450	Refrigerant, R-410A, 25 lb. disposable cylinder					14.55			14.55	16
4470	Refrigerant, R-507, 25 lb. disposable cylinder				↓	16.95			16.95	18.65

23 31 HVAC Ducts and Casings

23 31 13 – Metal Ducts

23 31 13.13 Rectangular Metal Ducts

0010	**RECTANGULAR METAL DUCTS**	R233100-20								
0020	Fabricated rectangular, includes fittings, joints, supports,									
0030	allowance for flexible connections, no insulation	R233100-30								
0031	NOTE: Fabrication and installation are combined									
0040	as LABOR cost. Approx. 25% fittings assumed.	R233100-40								
0042	Fabrication/Inst. is to commercial quality standards									
0043	(SMACNA or equiv.) for structure, sealing, leak testing, etc.	R233100-50								
0050	Add to labor for elevated installation									

23 31 13.13 Rectangular Metal Ducts	Crew	Daily Output	Labor-Hours	Unit	Material	2011 Bare Costs Labor	Equipment	Total	Total Incl O&P
0051 of fabricated ductwork									
0052 10' to 15' high						6%			
0053 15' to 20' high						12%			
0054 20' to 25' high						15%			
0055 25' to 30' high						21%			
0056 30' to 35' high						24%			
0057 35' to 40' high						30%			
0058 Over 40' high						33%			
0072 For duct insulation see Line 23 07 13.10 3000									
0100 Aluminum, alloy 3003-H14, under 100 lb.	Q-10	75	.320	Lb.	3.51	15.40		18.91	27.50
0110 100 to 500 lb.		80	.300		2.31	14.45		16.76	24.50
0120 500 to 1,000 lb.		95	.253		2.19	12.20		14.39	21
0140 1,000 to 2,000 lb.		120	.200		2.11	9.65		11.76	16.90
0150 2,000 to 5,000 lb.		130	.185		2.11	8.90		11.01	15.75
0160 Over 5,000 lb.		145	.166		2.11	8		10.11	14.35
0500 Galvanized steel, under 200 lb.		235	.102		.66	4.92		5.58	8.20
0520 200 to 500 lb.		245	.098		.65	4.72		5.37	7.85
0540 500 to 1,000 lb.		255	.094		.64	4.54		5.18	7.55
0560 1,000 to 2,000 lb.		265	.091		.63	4.37		5	7.30
0570 2,000 to 5,000 lb.		275	.087		.62	4.21		4.83	7.05
0580 Over 5,000 lb.		285	.084		.62	4.06		4.68	6.85
0600 For large quantities special prices available from supplier									
1000 Stainless steel, type 304, under 100 lb.	Q-10	165	.145	Lb.	2.40	7		9.40	13.25
1020 100 to 500 lb.		175	.137		4.02	6.60		10.62	14.40
1030 500 to 1,000 lb.		190	.126		3.68	6.10		9.78	13.25
1040 1,000 to 2,000 lb.		200	.120		3.65	5.80		9.45	12.75
1050 2,000 to 5,000 lb.		225	.107		3.31	5.15		8.46	11.40
1060 Over 5,000 lb.		235	.102		3.26	4.92		8.18	11.05
1080 Note: Minimum order cost exceeds per lb. cost for min. wt.									
1100 For medium pressure ductwork, add				Lb.		15%			
1200 For high pressure ductwork, add						40%			
1210 For welded ductwork, add						85%			
1220 For 30% fittings, add						11%			
1224 For 40% fittings, add						34%			
1228 For 50% fittings, add						56%			
1232 For 60% fittings, add						79%			
1236 For 70% fittings, add						101%			
1240 For 80% fittings, add						124%			
1244 For 90% fittings, add						147%			
1248 For 100% fittings, add						169%			
1252 Note: Fittings add includes time for detailing and installation.									
1276 Duct tape, 2" x 60 yd roll				Ea.	5.20			5.20	5.70

23 31 13.16 Round and Flat-Oval Spiral Ducts

	Crew	Daily Output	Labor-Hours	Unit	Material	2011 Bare Costs Labor	Equipment	Total	Total Incl O&P
0010 **ROUND AND FLAT-OVAL SPIRAL DUCTS**									
1280 Add to labor for elevated installation									
1282 of prefabricated (purchased) ductwork									
1283 10' to 15' high						10%			
1284 15' to 20' high						20%			
1285 20' to 25' high						25%			
1286 25' to 30' high						35%			
1287 30' to 35' high						40%			
1288 35' to 40' high						50%			

23 31 13.16 Round and Flat-Oval Spiral Ducts	Crew	Daily Output	Labor-Hours	Unit	Material	2011 Bare Costs Labor	Equipment	Total	Total Incl O&P
1289 Over 40' high						55%			
5400 Spiral preformed, steel, galv., straight lengths, Max 10" spwg.									
5410 4" diameter, 26 ga.	Q-9	360	.044	L.F.	1.77	2.07		3.84	5.05
5416 5" diameter, 26 ga.		320	.050		1.77	2.32		4.09	5.45
5420 6" diameter, 26 ga.		280	.057		1.77	2.66		4.43	5.95
5425 7" diameter, 26 ga.		240	.067		2.26	3.10		5.36	7.20
5430 8" diameter, 26 ga.		200	.080		2.35	3.72		6.07	8.20
5440 10" diameter, 26 ga.		160	.100		2.94	4.65		7.59	10.30
5450 12" diameter, 26 ga.		120	.133		3.53	6.20		9.73	13.25
5460 14" diameter, 24 ga.		80	.200		4.62	9.30		13.92	19.15
5480 16" diameter, 24 ga.		60	.267		5.30	12.40		17.70	24.50
5490 18" diameter, 24 ga.	↓	50	.320		5.95	14.85		20.80	29
5500 20" diameter, 24 ga.	Q-10	65	.369		6.60	17.80		24.40	34.50
5510 22" diameter, 24 ga.		60	.400		7.25	19.30		26.55	37
5520 24" diameter, 24 ga.		55	.436		7.95	21		28.95	40.50
5540 30" diameter, 22 ga.		45	.533		11.85	25.50		37.35	52
5600 36" diameter, 22 ga.	↓	40	.600	↓	14.25	29		43.25	59
5800 Connector, 4" diameter	Q-9	100	.160	Ea.	2.15	7.45		9.60	13.60
5810 5" diameter		94	.170		2.31	7.90		10.21	14.50
5820 6" diameter		88	.182		2.42	8.45		10.87	15.45
5840 8" diameter		78	.205		2.70	9.55		12.25	17.35
5860 10" diameter		70	.229		3.25	10.60		13.85	19.65
5880 12" diameter		50	.320		3.80	14.85		18.65	26.50
5900 14" diameter		44	.364		4.02	16.90		20.92	30
5920 16" diameter		40	.400		4.40	18.60		23	33
5930 18" diameter		37	.432		4.68	20		24.68	35.50
5940 20" diameter		34	.471		5.50	22		27.50	39
5950 22" diameter		31	.516		6.65	24		30.65	44
5960 24" diameter		28	.571		7.10	26.50		33.60	48
5980 30" diameter		22	.727		8.35	34		42.35	60
6000 36" diameter		18	.889		10	41.50		51.50	73.50
6300 Elbow, 45°, 4" diameter		60	.267		6.65	12.40		19.05	26
6310 5" diameter		52	.308		6.90	14.30		21.20	29
6320 6" diameter		44	.364		7.35	16.90		24.25	33.50
6340 8" diameter		28	.571		8.90	26.50		35.40	50
6360 10" diameter		18	.889		9.35	41.50		50.85	73
6380 12" diameter		13	1.231		10.10	57		67.10	97.50
6400 14" diameter		11	1.455		14.05	67.50		81.55	117
6420 16" diameter		10	1.600		20	74.50		94.50	134
6430 18" diameter	↓	9.60	1.667		26.50	77.50		104	146
6440 20" diameter	Q-10	14	1.714		33.50	82.50		116	162
6450 22" diameter		13	1.846		56.50	89		145.50	198
6460 24" diameter		12	2		69.50	96.50		166	223
6480 30" diameter		9	2.667		101	129		230	305
6500 36" diameter	↓	7	3.429		122	165		287	385
6600 Elbow, 90°, 4" diameter	Q-9	60	.267		5.05	12.40		17.45	24.50
6610 5" diameter		52	.308		5.05	14.30		19.35	27
6620 6" diameter		44	.364		5.05	16.90		21.95	31
6625 7" diameter		36	.444		5.50	20.50		26	37
6630 8" diameter		28	.571		6.20	26.50		32.70	47
6640 10" diameter		18	.889		8.15	41.50		49.65	71.50
6650 12" diameter		13	1.231		10.20	57		67.20	97.50
6660 14" diameter	↓	11	1.455		17.45	67.50		84.95	121

23 31 HVAC Ducts and Casings

23 31 13 – Metal Ducts

23 31 13.16 Round and Flat-Oval Spiral Ducts

		Crew	Daily Output	Labor-Hours	Unit	Material	2011 Bare Costs Labor	Equipment	Total	Total Incl O&P
6670	16" diameter	Q-9	10	1.600	Ea.	25	74.50		99.50	140
6676	18" diameter	▼	9.60	1.667		33	77.50		110.50	154
6680	20" diameter	Q-10	14	1.714		42	82.50		124.50	171
6684	22" diameter		13	1.846		68	89		157	210
6690	24" diameter		12	2		83.50	96.50		180	238
6700	30" diameter		9	2.667		211	129		340	425
6710	36" diameter	▼	7	3.429	▼	253	165		418	530
6720	End cap									
6722	4"	Q-9	110	.145	Ea.	4.38	6.75		11.13	15
6724	6"		94	.170		4.38	7.90		12.28	16.75
6725	8"		88	.182		5.50	8.45		13.95	18.85
6726	10"		78	.205		6.85	9.55		16.40	22
6727	12"		70	.229		7.45	10.60		18.05	24.50
6728	14"		50	.320		10.45	14.85		25.30	34
6729	16"		44	.364		11.65	16.90		28.55	38.50
6730	18"		40	.400		16	18.60		34.60	45.50
6731	20"		37	.432		32.50	20		52.50	66.50
6732	22"		34	.471		36.50	22		58.50	73.50
6733	24"		30	.533		41	25		66	82.50
6800	Reducing coupling, 6" x 4"		46	.348		14.30	16.15		30.45	40.50
6820	8" x 6"		40	.400		13.40	18.60		32	43
6840	10" x 8"		32	.500		14.30	23		37.30	51
6860	12" x 10"		24	.667		16.25	31		47.25	65
6880	14" x 12"		20	.800		17.05	37		54.05	75
6900	16" x 14"		18	.889		22	41.50		63.50	86.50
6920	18" x 16"	▼	16	1		24	46.50		70.50	97
6940	20" x 18"	Q-10	24	1		38.50	48		86.50	116
6950	22" x 20"		23	1.043		44	50.50		94.50	125
6960	24" x 22"		22	1.091		49.50	52.50		102	134
6980	30" x 28"		18	1.333		124	64.50		188.50	233
7000	36" x 34"	▼	16	1.500		154	72.50		226.50	279
7100	Tee, 90°, 4" diameter	Q-9	40	.400		21	18.60		39.60	51
7110	5" diameter		34.60	.462		31.50	21.50		53	67.50
7120	6" diameter		30	.533		31.50	25		56.50	72.50
7130	8" diameter		19	.842		41	39		80	104
7140	10" diameter		12	1.333		56.50	62		118.50	156
7150	12" diameter		9	1.778		68.50	82.50		151	201
7160	14" diameter		7	2.286		104	106		210	275
7170	16" diameter		6.70	2.388		120	111		231	300
7176	18" diameter	▼	6	2.667		175	124		299	380
7180	20" diameter	Q-10	8.70	2.759		208	133		341	430
7190	24" diameter		7	3.429		325	165		490	610
7200	30" diameter		6	4		400	193		593	730
7210	36" diameter	▼	5	4.800	▼	510	231		741	910
7400	Steel, PVC coated both sides, straight lengths									
7410	4" diameter, 26 ga.	Q-9	360	.044	L.F.	4.11	2.07		6.18	7.65
7416	5" diameter, 26 ga.		240	.067		4.11	3.10		7.21	9.20
7420	6" diameter, 26 ga.		280	.057		4.11	2.66		6.77	8.55
7440	8" diameter, 26 ga.		200	.080		5.50	3.72		9.22	11.65
7460	10" diameter, 26 ga.		160	.100		6.85	4.65		11.50	14.55
7480	12" diameter, 26 ga.		120	.133		8.20	6.20		14.40	18.35
7500	14" diameter, 26 ga.		80	.200		9.60	9.30		18.90	24.50
7520	16" diameter, 24 ga.	▼	60	.267	▼	10.95	12.40		23.35	31

23 31 13.16 Round and Flat-Oval Spiral Ducts	Crew	Daily Output	Labor-Hours	Unit	Material	2011 Bare Costs Labor	Equipment	Total	Total Incl O&P
7540	18" diameter, 24 ga.	Q-9	45	.356	L.F.	12.30	16.55		28.85	38.50
7560	20" diameter, 24 ga.	Q-10	65	.369		13.65	17.80		31.45	42
7580	24" diameter, 24 ga.		55	.436		16.40	21		37.40	50
7600	30" diameter, 22 ga.		45	.533		25	25.50		50.50	66
7620	36" diameter, 22 ga.		40	.600		29.50	29		58.50	76
7890	Connector, 4" diameter	Q-9	100	.160	Ea.	2.94	7.45		10.39	14.50
7896	5" diameter		83	.193		3.39	8.95		12.34	17.30
7900	6" diameter		88	.182		3.73	8.45		12.18	16.90
7920	8" diameter		78	.205		4.41	9.55		13.96	19.25
7940	10" diameter		70	.229		5.10	10.60		15.70	21.50
7960	12" diameter		50	.320		6	14.85		20.85	29
7980	14" diameter		44	.364		6.35	16.90		23.25	32.50
8000	16" diameter		40	.400		6.90	18.60		25.50	35.50
8020	18" diameter		37	.432		7.70	20		27.70	39
8040	20" diameter		34	.471		10.05	22		32.05	44
8060	24" diameter		28	.571		12.45	26.50		38.95	53.50
8080	30" diameter		22	.727		16.60	34		50.60	69.50
8100	36" diameter		18	.889		20	41.50		61.50	85
8390	Elbow, 45°, 4" diameter		60	.267		10.70	12.40		23.10	30.50
8396	5" diameter		36	.444		11	20.50		31.50	43
8400	6" diameter		44	.364		11.35	16.90		28.25	38
8420	8" diameter		28	.571		12.30	26.50		38.80	53.50
8440	10" diameter		18	.889		13	41.50		54.50	77
8460	12" diameter		13	1.231		14.40	57		71.40	102
8480	14" diameter		11	1.455		15.50	67.50		83	119
8500	16" diameter		10	1.600		22.50	74.50		97	137
8520	18" diameter		9	1.778		28.50	82.50		111	156
8540	20" diameter	Q-10	13	1.846		64.50	89		153.50	206
8560	24" diameter		11	2.182		92	105		197	260
8580	30" diameter		9	2.667		146	129		275	355
8600	36" diameter		7	3.429		177	165		342	445
8800	Elbow, 90°, 4" diameter	Q-9	60	.267		15.10	12.40		27.50	35.50
8804	5" diameter		52	.308		15.70	14.30		30	39
8806	6" diameter		44	.364		15.75	16.90		32.65	43
8808	8" diameter		28	.571		16.45	26.50		42.95	58
8810	10" diameter		18	.889		17.10	41.50		58.60	81.50
8812	12" diameter		13	1.231		17.80	57		74.80	106
8814	14" diameter		11	1.455		24.50	67.50		92	129
8816	16" diameter		10	1.600		33	74.50		107.50	148
8818	18" diameter		9.60	1.667		40	77.50		117.50	161
8820	20" diameter	Q-10	14	1.714		50.50	82.50		133	181
8822	24" diameter		12	2		111	96.50		207.50	268
8824	30" diameter		9	2.667		292	129		421	515
8826	36" diameter		7	3.429		355	165		520	640
9000	Reducing coupling									
9040	6" x 4"	Q-9	46	.348	Ea.	12.65	16.15		28.80	38.50
9060	8" x 6"		40	.400		16.05	18.60		34.65	45.50
9080	10" x 8"		32	.500		18.75	23		41.75	55.50
9100	12" x 10"		24	.667		21	31		52	70.50
9120	14" x 12"		20	.800		24.50	37		61.50	83
9140	16" x 14"		18	.889		28.50	41.50		70	94
9160	18" x 16"		16	1		34	46.50		80.50	108
9180	20" x 18"	Q-10	24	1		41	48		89	118

23 31 HVAC Ducts and Casings

23 31 13 – Metal Ducts

23 31 13.16 Round and Flat-Oval Spiral Ducts

		Crew	Daily Output	Labor-Hours	Unit	Material	2011 Bare Costs Labor	Equipment	Total	Total Incl O&P
9200	24" x 22"	Q-10	22	1.091	Ea.	51	52.50		103.50	136
9220	30" x 28"		18	1.333		112	64.50		176.50	220
9240	36" x 34"		16	1.500		130	72.50		202.50	252
9800	Steel, stainless, straight lengths									
9810	3" diameter, 26 Ga.	Q-9	400	.040	L.F.	14.20	1.86		16.06	18.40
9820	4" diameter, 26 Ga.		360	.044		16.45	2.07		18.52	21
9830	5" diameter, 26 Ga.		320	.050		17.75	2.32		20.07	23
9840	6" diameter, 26 Ga.		280	.057		17.75	2.66		20.41	23.50
9850	7" diameter, 26 Ga.		240	.067		19.55	3.10		22.65	26
9860	8" diameter, 26 Ga.		200	.080		23	3.72		26.72	31
9870	9" diameter, 26 Ga.		180	.089		33.50	4.13		37.63	43.50
9880	10" diameter, 26 Ga.		160	.100		43.50	4.65		48.15	55
9890	12" diameter, 26 Ga.		120	.133		61.50	6.20		67.70	77
9900	14" diameter, 26 Ga..		80	.200		65	9.30		74.30	85.50

23 31 13.17 Round Grease Duct

		Crew	Daily Output	Labor-Hours	Unit	Material	2011 Bare Costs Labor	Equipment	Total	Total Incl O&P
0010	**ROUND GREASE DUCT**									
1280	Add to labor for elevated installation									
1282	of prefabricated (purchased) ductwork									
1283	10' to 15' high						10%			
1284	15' to 20' high						20%			
1285	20' to 25' high						25%			
1286	25' to 30' high						35%			
1287	30' to 35' high						40%			
1288	35' to 40' high						50%			
1289	Over 40' high						55%			
4020	Zero clearance, 2 hr. fire rated, UL listed, 3" double wall									
4030	304 interior, aluminized exterior, all sizes are ID									
4032	For 316 interior and aluminized exterior, add					6%				
4034	For 304 interior and 304 exterior, add					21%				
4036	For 316 interior and 316 exterior, add					40%				
4040	Straight									
4050	5"	Q-9	36.48	.439	L.F.	65	20.50		85.50	103
4052	6"		34.20	.468		82	22		104	123
4054	8"		29.64	.540		94	25		119	141
4056	10"		27.36	.585		106	27		133	158
4058	12"		25.10	.637		121	29.50		150.50	178
4060	14"		23.94	.668		139	31		170	200
4062	16"		22.80	.702		145	32.50		177.50	210
4064	18"		21.66	.739		179	34.50		213.50	249
4066	20"	Q-10	20.52	1.170		206	56.50		262.50	310
4068	22"		19.38	1.238		234	59.50		293.50	350
4070	24"		18.24	1.316		266	63.50		329.50	390
4072	26"		17.67	1.358		282	65.50		347.50	410
4074	28"		17.10	1.404		296	67.50		363.50	425
4076	30"		15.96	1.504		320	72.50		392.50	465
4078	32"		15.39	1.559		350	75		425	500
4080	36"		14.25	1.684		390	81		471	555
4100	Adjustable section, 30" long									
4102	5"	Q-9	18.24	.877	Ea.	204	41		245	286
4104	6"		17.10	.936		255	43.50		298.50	345
4106	8"		14.82	1.080		260	50		310	365
4108	10"		13.68	1.170		330	54.50		384.50	445

23 31 13.17 Round Grease Duct		Crew	Daily Output	Labor-Hours	Unit	Material	2011 Bare Costs		Total	Total Incl O&P
							Labor	Equipment		
4110	12"	Q-9	12.54	1.276	Ea.	380	59.50		439.50	510
4112	14"		11.97	1.337		435	62		497	575
4114	16"		11.40	1.404		455	65		520	600
4116	18"		10.83	1.477		560	68.50		628.50	720
4118	20"	Q-10	10.26	2.339		645	113		758	880
4120	22"		9.69	2.477		730	119		849	985
4122	24"		9.12	2.632		835	127		962	1,100
4124	26"		8.83	2.718		880	131		1,011	1,175
4126	28"		8.55	2.807		925	135		1,060	1,225
4128	30"		8.26	2.906		1,000	140		1,140	1,300
4130	32"		7.98	3.008		1,100	145		1,245	1,425
4132	36"		6.84	3.509		1,175	169		1,344	1,550
4140	Tee, 90°, grease									
4142	5"	Q-9	14.25	1.123	Ea.	540	52		592	670
4144	6"		13.68	1.170		635	54.50		689.50	775
4146	8"		12.54	1.276		730	59.50		789.50	890
4148	10"		11.97	1.337		850	62		912	1,025
4150	12"		11.40	1.404		995	65		1,060	1,200
4152	14"		10.26	1.559		1,175	72.50		1,247.50	1,400
4154	16"		9.12	1.754		1,325	81.50		1,406.50	1,575
4156	18"		7.98	2.005		1,550	93		1,643	1,875
4158	20"		9.69	1.651		1,750	77		1,827	2,050
4160	22"		8.26	1.937		1,950	90		2,040	2,275
4162	24"		6.84	2.339		2,150	109		2,259	2,550
4164	26"		6.55	2.443		2,450	114		2,564	2,850
4166	28"	Q-10	6.27	3.828		2,700	185		2,885	3,250
4168	30"		5.98	4.013		3,200	193		3,393	3,825
4170	32"		5.70	4.211		3,700	203		3,903	4,375
4172	36"		5.13	4.678		4,325	226		4,551	5,100
4180	Cleanout Tee Cap									
4182	5"	Q-9	22.80	.702	Ea.	159	32.50		191.50	224
4184	6"		21.09	.759		188	35.50		223.50	260
4186	8"		19.38	.826		190	38.50		228.50	267
4188	10"		18.24	.877		210	41		251	292
4190	12"		17.10	.936		219	43.50		262.50	305
4192	14"		15.96	1.003		251	46.50		297.50	345
4194	16"		14.25	1.123		275	52		327	385
4196	18"		13.68	1.170		310	54.50		364.50	425
4198	20"	Q-10	15.39	1.559		335	75		410	485
4200	22"		12.54	1.914		380	92.50		472.50	560
4202	24"		11.97	2.005		425	96.50		521.50	610
4204	26"		11.40	2.105		475	101		576	680
4206	28"		10.83	2.216		525	107		632	735
4208	30"		10.26	2.339		585	113		698	815
4210	32"		9.69	2.477		640	119		759	885
4212	36"		8.55	2.807		800	135		935	1,075
4220	Elbow, 45°									
4222	5"	Q-9	18.24	.877	Ea.	295	41		336	385
4224	6"		17.10	.936		345	43.50		388.50	445
4226	8"		14.82	1.080		395	50		445	510
4228	10"		13.68	1.170		440	54.50		494.50	565
4230	12"		12.54	1.276		500	59.50		559.50	645
4232	14"		11.97	1.337		580	62		642	730

23 31 13.17 Round Grease Duct		Crew	Daily Output	Labor-Hours	Unit	Material	2011 Bare Costs Labor	Equipment	Total	Total Incl O&P
4234	16"	Q-9	11.40	1.404	Ea.	650	65		715	815
4236	18"	▼	10.83	1.477		740	68.50		808.50	915
4238	20"	Q-10	10.26	2.339		845	113		958	1,100
4240	22"		9.69	2.477		950	119		1,069	1,225
4242	24"		9.12	2.632		1,075	127		1,202	1,400
4244	26"		8.83	2.718		1,150	131		1,281	1,475
4246	28"		8.55	2.807		1,225	135		1,360	1,550
4248	30"		8.26	2.906		1,425	140		1,565	1,775
4250	32"		7.98	3.008		1,650	145		1,795	2,025
4252	36"	▼	6.84	3.509	▼	2,325	169		2,494	2,800
4260	Elbow, 90°									
4262	5"	Q-9	18.24	.877	Ea.	470	41		511	575
4264	6"		17.10	.936		550	43.50		593.50	670
4266	8"		14.82	1.080		605	50		655	740
4268	10"		13.68	1.170		695	54.50		749.50	845
4270	12"		12.54	1.276		790	59.50		849.50	955
4272	14"		11.97	1.337		925	62		987	1,125
4274	16"		11.40	1.404		1,025	65		1,090	1,225
4276	18"	▼	10.83	1.477		1,200	68.50		1,268.50	1,425
4278	20"	Q-10	10.26	2.339		1,350	113		1,463	1,650
4280	22"		9.69	2.477		1,500	119		1,619	1,825
4282	24"		9.12	2.632		1,650	127		1,777	2,025
4284	26"		8.83	2.718		1,875	131		2,006	2,250
4286	28"		8.55	2.807		2,075	135		2,210	2,475
4288	30"		8.26	2.906		2,425	140		2,565	2,850
4290	32"		7.98	3.008		2,750	145		2,895	3,250
4292	36"	▼	6.84	3.509	▼	3,375	169		3,544	3,975
4300	Support strap									
4302	5"	Q-9	27	.593	Ea.	23	27.50		50.50	67
4304	6"		25.50	.627		23	29		52	69.50
4306	8"		22	.727		24.50	34		58.50	77.50
4308	10"		20.50	.780		24.50	36.50		61	81.50
4310	12"		18.80	.851		24.50	39.50		64	86.50
4312	14"		17.90	.894		24.50	41.50		66	89.50
4314	16"		17	.941		24.50	44		68.50	92.50
4316	18"	▼	16.20	.988		25.50	46		71.50	97.50
4318	20"	Q-10	15.40	1.558		26.50	75		101.50	144
4320	22"	"	14.50	1.655	▼	26.50	80		106.50	151
4340	Plate Support Assembly									
4342	5"	Q-9	17.10	.936	Ea.	110	43.50		153.50	187
4344	6"		14.82	1.080		128	50		178	217
4346	8"		12.54	1.276		145	59.50		204.50	249
4348	10"		11.40	1.404		159	65		224	273
4350	12"		10.26	1.559		164	72.50		236.50	291
4352	14"		9.69	1.651		200	77		277	335
4354	16"		9.12	1.754		210	81.50		291.50	355
4356	18"	▼	8.55	1.871		222	87		309	375
4358	20"	Q-10	9.12	2.632		234	127		361	450
4360	22"		8.55	2.807		241	135		376	470
4362	24"		7.98	3.008		248	145		393	490
4364	26"		7.69	3.121		305	150		455	560
4366	28"		7.41	3.239		345	156		501	615
4368	30"	▼	7.12	3.371	▼	350	162		512	630

23 31 13.17 Round Grease Duct		Crew	Daily Output	Labor-Hours	Unit	Material	2011 Bare Costs Labor	Equipment	Total	Total Incl O&P
4370	32"	Q-10	6.84	3.509	Ea.	395	169		564	690
4372	36"	▼	5.70	4.211	▼	490	203		693	845
4380	Wall Guide Assembly									
4382	5"	Q-9	20.20	.792	Ea.	128	37		165	197
4384	6"		19.10	.838		140	39		179	213
4386	8"		16.50	.970		147	45		192	230
4388	10"		15.30	1.046		154	48.50		202.50	243
4390	12"		14.10	1.135		161	52.50		213.50	257
4392	14"		13.40	1.194		174	55.50		229.50	275
4394	16"		12.75	1.255		176	58.50		234.50	282
4396	18"	▼	12.10	1.322		184	61.50		245.50	295
4398	20"	Q-10	11.50	2.087		193	101		294	365
4400	22"		10.90	2.202		199	106		305	380
4402	24"		10.20	2.353		200	113		313	390
4404	26"		9.90	2.424		210	117		327	405
4406	28"		9.60	2.500		225	121		346	430
4408	30"		9.30	2.581		234	124		358	445
4410	32"		9	2.667		244	129		373	465
4412	36"	▼	7.60	3.158	▼	266	152		418	525
4420	Hood Transition, Flanged or Unflanged									
4422	5"	Q-9	16.60	.964	Ea.	44	45		89	116
4424	6"		15.60	1.026		52	47.50		99.50	130
4426	8"		13.50	1.185		59	55		114	149
4428	10"		12.50	1.280		66	59.50		125.50	163
4430	12"		11.40	1.404		76.50	65		141.50	183
4432	14"		10.90	1.468		88	68		156	200
4434	16"		10.40	1.538		98.50	71.50		170	216
4436	18"	▼	9.80	1.633		113	76		189	240
4438	20"	Q-10	9.40	2.553		132	123		255	330
4440	22"		8.80	2.727		155	131		286	370
4442	24"		8.30	2.892		175	139		314	405
4444	26"		8.10	2.963		186	143		329	420
4446	28"		7.80	3.077		200	148		348	445
4448	30"		7.50	3.200		216	154		370	470
4450	32"		7.30	3.288		227	158		385	490
4452	36"	▼	6.20	3.871	▼	274	187		461	580
4460	Fan Adapter									
4462	5"	Q-9	16	1	Ea.	250	46.50		296.50	345
4464	6"		15	1.067		300	49.50		349.50	405
4466	8"		13	1.231		335	57		392	455
4468	10"		12	1.333		375	62		437	510
4470	12"		11	1.455		435	67.50		502.50	575
4472	14"		10.50	1.524		500	71		571	655
4474	16"		10	1.600		570	74.50		644.50	735
4476	18"	▼	9.50	1.684		650	78.50		728.50	835
4478	20"	Q-10	9	2.667		760	129		889	1,025
4480	22"		8.50	2.824		885	136		1,021	1,175
4482	24"		8	3		910	145		1,055	1,225
4484	26"		7.70	3.117		940	150		1,090	1,250
4486	28"		7.50	3.200		965	154		1,119	1,275
4488	30"		7.20	3.333		995	161		1,156	1,350
4490	32"		7	3.429		1,025	165		1,190	1,375
4492	36"	▼	6	4	▼	1,050	193		1,243	1,450

23 31 HVAC Ducts and Casings

23 31 13 – Metal Ducts

23 31 13.17 Round Grease Duct

		Crew	Daily Output	Labor-Hours	Unit	Material	2011 Bare Costs Labor	Equipment	Total	Total Incl O&P
4500	Tapered Increaser/Reducer									
4510	5" x 10" dia	Q-9	13.68	1.170	Ea.	485	54.50		539.50	610
4520	5" x 20" dia		10.30	1.553		755	72		827	940
4530	6" x 10" dia		13.40	1.194		485	55.50		540.50	615
4540	6" x 20" dia		10.10	1.584		795	73.50		868.50	985
4550	8" x 10" dia		13.20	1.212		455	56.50		511.50	585
4560	8" x 20" dia		9.90	1.616		800	75		875	990
4570	10" x 20" dia		10.60	1.509		855	70		925	1,050
4580	10" x 28" dia	Q-10	8.60	2.791		1,175	135		1,310	1,500
4590	12" x 20" dia		8.80	2.727		905	131		1,036	1,200
4600	14" x 28" dia		8.40	2.857		1,350	138		1,488	1,675
4610	16" x 20" dia		8.60	2.791		905	135		1,040	1,200
4620	16" x 30" dia		8.30	2.892		1,500	139		1,639	1,850
4630	18" x 24" dia		8.80	2.727		1,100	131		1,231	1,400
4640	18" x 30" dia		8.10	2.963		1,475	143		1,618	1,850
4650	20" x 24" dia		8.30	2.892		1,025	139		1,164	1,350
4660	20" x 32" dia		8	3		1,575	145		1,720	1,950
4670	24" x 26" dia		8	3		1,025	145		1,170	1,375
4680	24" x 36" dia		7.30	3.288		1,800	158		1,958	2,250
4690	28" x 30" dia		7.90	3.038		1,250	146		1,396	1,600
4700	28" x 36" dia		7.10	3.380		1,825	163		1,988	2,250
4710	30" x 36" dia		6.90	3.478		1,875	168		2,043	2,300
4730	Many intermediate standard sizes of tapered fittings are available									

23 31 13.19 Metal Duct Fittings

		Crew	Daily Output	Labor-Hours	Unit	Material	2011 Bare Costs Labor	Equipment	Total	Total Incl O&P
0010	**METAL DUCT FITTINGS**									
0050	Air extractors, 12" x 4"	1 Shee	24	.333	Ea.	16.10	17.20		33.30	44
0100	8" x 6"		22	.364		16.10	18.80		34.90	46.50
0120	12" x 6"		21	.381		22.50	19.70		42.20	54.50
0140	16" x 6"		20	.400		28	20.50		48.50	62
0160	24" x 6"		18	.444		46.50	23		69.50	86
0180	12" x 8"		20	.400		25	20.50		45.50	58.50
0200	20" x 8"		16	.500		36.50	26		62.50	79
0240	18" x 10"		14	.571		35.50	29.50		65	83.50
0260	24" x 10"		12	.667		46	34.50		80.50	103
0280	24" x 12"		10	.800		49.50	41.50		91	117
0300	30" x 12"		8	1		62	51.50		113.50	146
0350	Duct collar, spin-in type									
0352	Without damper									
0354	Round									
0356	4"	1 Shee	32	.250	Ea.	2.07	12.90		14.97	22
0357	5"		30	.267		2.07	13.75		15.82	23.50
0358	6"		28	.286		2.07	14.75		16.82	25
0359	7"		26	.308		2.38	15.90		18.28	26.50
0360	8"		24	.333		2.38	17.20		19.58	28.50
0361	9"		22	.364		3.60	18.80		22.40	32.50
0362	10"		20	.400		3.60	20.50		24.10	35
0363	12"		17	.471		4.61	24.50		29.11	42
2000	Fabrics for flexible connections, with metal edge		100	.080	L.F.	2.13	4.13		6.26	8.60
2100	Without metal edge		160	.050	"	1.53	2.58		4.11	5.60

23 31 HVAC Ducts and Casings

23 31 16 – Nonmetal Ducts

23 31 16.13 Fibrous-Glass Ducts

		Crew	Daily Output	Labor-Hours	Unit	Material	2011 Bare Costs Labor	Equipment	Total	Total Incl O&P
0010	**FIBROUS-GLASS DUCTS** R233100-20									
1280	Add to labor for elevated installation									
1282	of prefabricated (purchased) ductwork									
1283	10' to 15' high						10%			
1284	15' to 20' high						20%			
1285	20' to 25' high						25%			
1286	25' to 30' high						35%			
1287	30' to 35' high						40%			
1288	35' to 40' high						50%			
1289	Over 40' high						55%			
3490	Rigid fiberglass duct board, foil reinf. kraft facing									
3500	Rectangular, 1" thick, alum. faced, (FRK), std. weight	Q-10	350	.069	SF Surf	.77	3.31		4.08	5.85

23 31 16.16 Thermoset Fiberglass-Reinforced Plastic Ducts

		Crew	Daily Output	Labor-Hours	Unit	Material	2011 Bare Costs Labor	Equipment	Total	Total Incl O&P
0010	**THERMOSET FIBERGLASS-REINFORCED PLASTIC DUCTS** R233100-20									
1280	Add to labor for elevated installation									
1282	of prefabricated (purchased) ductwork									
1283	10' to 15' high						10%			
1284	15' to 20' high						20%			
1285	20' to 25' high						25%			
1286	25' to 30' high						35%			
1287	30' to 35' high						40%			
1288	35' to 40' high						50%			
1289	Over 40' high						55%			
3550	Rigid fiberglass reinforced plastic, FM approved									
3552	for acid fume and smoke exhaust system, nonflammable									
3554	Straight, 4" diameter	Q-9	200	.080	L.F.	10.60	3.72		14.32	17.25
3555	6" diameter		146	.110		13.90	5.10		19	23
3556	8" diameter		106	.151		16.85	7		23.85	29
3557	10" diameter		85	.188		20	8.75		28.75	35.50
3558	12" diameter		68	.235		23.50	10.95		34.45	42.50
3559	14" diameter		50	.320		35	14.85		49.85	61
3560	16" diameter		40	.400		39	18.60		57.60	71
3561	18" diameter		31	.516		43.50	24		67.50	84
3562	20" diameter	Q-10	44	.545		48	26.50		74.50	92.50
3563	22" diameter		40	.600		51.50	29		80.50	101
3564	24" diameter		37	.649		55.50	31.50		87	109
3565	26" diameter		34	.706		60	34		94	118
3566	28" diameter		28	.857		64.50	41.50		106	134
3567	30" diameter		24	1		68.50	48		116.50	149
3568	32" diameter		22.60	1.062		73	51		124	158
3569	34" diameter		21.60	1.111		77	53.50		130.50	166
3570	36" diameter		20.60	1.165		81.50	56		137.50	175
3571	38" diameter		19.80	1.212		125	58.50		183.50	227
3572	42" diameter		18.30	1.311		140	63		203	250
3573	46" diameter		17	1.412		155	68		223	273
3574	50" diameter		15.90	1.509		168	73		241	295
3575	54" diameter		14.90	1.611		185	77.50		262.50	320
3576	60" diameter		13.60	1.765		212	85		297	360
3577	64" diameter		12.80	1.875		310	90.50		400.50	475
3578	68" diameter		12.20	1.967		330	95		425	510
3579	72" diameter		11.90	2.017		355	97		452	535
3584	Note: joints are cemented with									

23 31 HVAC Ducts and Casings

23 31 16 – Nonmetal Ducts

23 31 16.16 Thermoset Fiberglass-Reinforced Plastic Ducts	Crew	Daily Output	Labor-Hours	Unit	Material	2011 Bare Costs Labor	Equipment	Total	Total Incl O&P
3586	fiberglass resin, included in material cost.								
3590	Elbow, 90°, 4" diameter	Q-9	20.30	.788	Ea.	51	36.50		87.50	112
3591	6" diameter		13.80	1.159		71.50	54		125.50	160
3592	8" diameter		10	1.600		79	74.50		153.50	199
3593	10" diameter		7.80	2.051		103	95.50		198.50	258
3594	12" diameter		6.50	2.462		131	114		245	315
3595	14" diameter		5.50	2.909		161	135		296	380
3596	16" diameter		4.80	3.333		191	155		346	445
3597	18" diameter		4.40	3.636		274	169		443	555
3598	20" diameter	Q-10	5.90	4.068		289	196		485	610
3599	22" diameter		5.40	4.444		355	214		569	715
3600	24" diameter		4.90	4.898		410	236		646	805
3601	26" diameter		4.60	5.217		415	251		666	835
3602	28" diameter		4.20	5.714		515	275		790	985
3603	30" diameter		3.90	6.154		575	297		872	1,075
3604	32" diameter		3.70	6.486		640	315		955	1,175
3605	34" diameter		3.45	6.957		755	335		1,090	1,325
3606	36" diameter		3.20	7.500		865	360		1,225	1,500
3607	38" diameter		3.07	7.818		1,225	375		1,600	1,925
3608	42" diameter		2.70	8.889		1,475	430		1,905	2,275
3609	46" diameter		2.54	9.449		1,725	455		2,180	2,600
3610	50" diameter		2.33	10.300		2,000	495		2,495	2,950
3611	54" diameter		2.14	11.215		2,300	540		2,840	3,350
3612	60" diameter		1.94	12.371		3,050	595		3,645	4,250
3613	64" diameter		1.82	13.187		4,425	635		5,060	5,825
3614	68" diameter		1.71	14.035		4,925	675		5,600	6,450
3615	72" diameter		1.62	14.815		5,450	715		6,165	7,050
3626	Elbow, 45°, 4" diameter	Q-9	22.30	.717		49	33.50		82.50	105
3627	6" diameter		15.20	1.053		68	49		117	149
3628	8" diameter		11	1.455		72.50	67.50		140	182
3629	10" diameter		8.60	1.860		76.50	86.50		163	215
3630	12" diameter		7.20	2.222		83.50	103		186.50	248
3631	14" diameter		6.10	2.623		102	122		224	297
3632	16" diameter		5.28	3.030		122	141		263	345
3633	18" diameter		4.84	3.306		143	154		297	390
3634	20" diameter	Q-10	6.50	3.692		165	178		343	450
3635	22" diameter		5.94	4.040		226	195		421	540
3636	24" diameter		5.40	4.444		257	214		471	610
3637	26" diameter		5.06	4.743		291	229		520	665
3638	28" diameter		4.62	5.195		315	250		565	725
3639	30" diameter		4.30	5.581		365	269		634	805
3640	32" diameter		4.07	5.897		405	284		689	875
3641	34" diameter		3.80	6.316		480	305		785	985
3642	36" diameter		3.52	6.818		550	330		880	1,100
3643	38" diameter		3.38	7.101		775	340		1,115	1,375
3644	42" diameter		2.97	8.081		925	390		1,315	1,625
3645	46" diameter		2.80	8.571		1,075	415		1,490	1,825
3646	50" diameter		2.56	9.375		1,250	450		1,700	2,050
3647	54" diameter		2.35	10.213		1,425	490		1,915	2,325
3648	60" diameter		2.13	11.268		1,925	545		2,470	2,925
3649	64" diameter		2	12		2,775	580		3,355	3,925
3650	68" diameter		1.88	12.766		3,075	615		3,690	4,300
3651	72" diameter		1.78	13.483		3,400	650		4,050	4,700

23 31 16.16 Thermoset Fiberglass-Reinforced Plastic Ducts	Crew	Daily Output	Labor-Hours	Unit	Material	2011 Bare Costs Labor	Equipment	Total	Total Incl O&P
3660	Tee, 90°, 4" diameter	Q-9	14.06	1.138	Ea.	28	53		81	111
3661	6" diameter		9.52	1.681		43	78		121	165
3662	8" diameter		6.98	2.292		58.50	107		165.50	226
3663	10" diameter		5.45	2.936		76.50	136		212.50	290
3664	12" diameter		4.60	3.478		95.50	162		257.50	350
3665	14" diameter		3.86	4.145		139	193		332	445
3666	16" diameter		3.40	4.706		165	219		384	510
3667	18" diameter		3.05	5.246		196	244		440	585
3668	20" diameter	Q-10	4.12	5.825		229	281		510	675
3669	22" diameter		3.78	6.349		262	305		567	755
3670	24" diameter		3.46	6.936		295	335		630	830
3671	26" diameter		3.20	7.500		335	360		695	915
3672	28" diameter		2.96	8.108		375	390		765	1,000
3673	30" diameter		2.75	8.727		420	420		840	1,100
3674	32" diameter		2.54	9.449		465	455		920	1,200
3675	34" diameter		2.40	10		505	480		985	1,300
3676	36" diameter		2.27	10.573		555	510		1,065	1,375
3677	38" diameter		2.15	11.163		835	540		1,375	1,725
3678	42" diameter		1.91	12.565		1,000	605		1,605	2,025
3679	46" diameter		1.78	13.483		1,175	650		1,825	2,275
3680	50" diameter		1.63	14.724		1,375	710		2,085	2,600
3681	54" diameter		1.50	16		1,600	770		2,370	2,925
3682	60" diameter		1.35	17.778		2,000	855		2,855	3,500
3683	64" diameter		1.27	18.898		3,000	910		3,910	4,675
3684	68" diameter		1.20	20		3,350	965		4,315	5,125
3685	72" diameter		1.13	21.239		3,750	1,025		4,775	5,675
3690	For Y @ 45°, add					22%				
3700	Blast gate, 4" diameter	Q-9	22.30	.717		136	33.50		169.50	200
3701	6" diameter		15.20	1.053		154	49		203	243
3702	8" diameter		11	1.455		207	67.50		274.50	330
3703	10" diameter		8.60	1.860		236	86.50		322.50	390
3704	12" diameter		7.20	2.222		289	103		392	470
3705	14" diameter		6.10	2.623		320	122		442	535
3706	16" diameter		5.28	3.030		355	141		496	605
3707	18" diameter		4.84	3.306		365	154		519	635
3708	20" diameter	Q-10	6.50	3.692		420	178		598	730
3709	22" diameter		5.94	4.040		455	195		650	795
3710	24" diameter		5.40	4.444		485	214		699	860
3711	26" diameter		5.06	4.743		520	229		749	920
3712	28" diameter		4.62	5.195		560	250		810	995
3713	30" diameter		4.30	5.581		595	269		864	1,050
3714	32" diameter		4.07	5.897		650	284		934	1,150
3715	34" diameter		3.80	6.316		710	305		1,015	1,250
3716	36" diameter		3.52	6.818		765	330		1,095	1,325
3717	38" diameter		3.38	7.101		1,025	340		1,365	1,650
3718	42" diameter		2.97	8.081		1,175	390		1,565	1,900
3719	46" diameter		2.80	8.571		1,350	415		1,765	2,125
3720	50" diameter		2.56	9.375		1,550	450		2,000	2,375
3721	54" diameter		2.35	10.213		1,750	490		2,240	2,675
3722	60" diameter		2.13	11.268		2,050	545		2,595	3,075
3723	64" diameter		2	12		2,275	580		2,855	3,375
3724	68" diameter		1.88	12.766		2,500	615		3,115	3,675
3725	72" diameter		1.78	13.483		2,750	650		3,400	4,000

23 31 16 – Nonmetal Ducts

23 31 16.16 Thermoset Fiberglass-Reinforced Plastic Ducts	Crew	Daily Output	Labor-Hours	Unit	Material	2011 Bare Costs Labor	Equipment	Total	Total Incl O&P
3740	Reducers, 4" diameter	Q-9	24.50	.653	Ea.	34	30.50		64.50	83.50
3741	6" diameter		16.70	.958		50.50	44.50		95	124
3742	8" diameter		12	1.333		68.50	62		130.50	169
3743	10" diameter		9.50	1.684		84.50	78.50		163	211
3744	12" diameter		7.90	2.025		101	94		195	254
3745	14" diameter		6.70	2.388		118	111		229	298
3746	16" diameter		5.80	2.759		135	128		263	340
3747	18" diameter		5.30	3.019		153	140		293	380
3748	20" diameter	Q-10	7.20	3.333		169	161		330	430
3749	22" diameter		6.50	3.692		273	178		451	570
3750	24" diameter		5.90	4.068		295	196		491	620
3751	26" diameter		5.60	4.286		320	207		527	660
3752	28" diameter		5.10	4.706		350	227		577	730
3753	30" diameter		4.70	5.106		375	246		621	780
3754	32" diameter		4.50	5.333		400	257		657	830
3755	34" diameter		4.20	5.714		425	275		700	880
3756	36" diameter		3.90	6.154		450	297		747	945
3757	38" diameter		3.70	6.486		680	315		995	1,225
3758	42" diameter		3.30	7.273		750	350		1,100	1,350
3759	46" diameter		3.10	7.742		820	375		1,195	1,475
3760	50" diameter		2.80	8.571		890	415		1,305	1,600
3761	54" diameter		2.60	9.231		965	445		1,410	1,725
3762	60" diameter		2.34	10.256		1,075	495		1,570	1,950
3763	64" diameter		2.20	10.909		1,675	525		2,200	2,650
3764	68" diameter		2.06	11.650		1,775	560		2,335	2,800
3765	72" diameter		1.96	12.245		1,900	590		2,490	2,975
3780	Flange, per each, 4" diameter	Q-9	40.60	.394		82	18.30		100.30	118
3781	6" diameter		27.70	.578		96	27		123	147
3782	8" diameter		20.50	.780		120	36.50		156.50	187
3783	10" diameter		16.30	.982		144	45.50		189.50	227
3784	12" diameter		13.90	1.151		165	53.50		218.50	263
3785	14" diameter		11.30	1.416		191	66		257	310
3786	16" diameter		9.70	1.649		215	76.50		291.50	350
3787	18" diameter		8.30	1.928		238	89.50		327.50	395
3788	20" diameter	Q-10	11.30	2.124		262	102		364	445
3789	22" diameter		10.40	2.308		286	111		397	485
3790	24" diameter		9.50	2.526		310	122		432	525
3791	26" diameter		8.80	2.727		335	131		466	570
3792	28" diameter		8.20	2.927		360	141		501	610
3793	30" diameter		7.60	3.158		380	152		532	645
3794	32" diameter		7.20	3.333		405	161		566	690
3795	34" diameter		6.80	3.529		430	170		600	725
3796	36" diameter		6.40	3.750		455	181		636	775
3797	38" diameter		6.10	3.934		480	190		670	810
3798	42" diameter		5.50	4.364		520	210		730	895
3799	46" diameter		5	4.800		570	231		801	980
3800	50" diameter		4.60	5.217		620	251		871	1,050
3801	54" diameter		4.20	5.714		665	275		940	1,150
3802	60" diameter		3.80	6.316		740	305		1,045	1,275
3803	64" diameter		3.60	6.667		785	320		1,105	1,350
3804	68" diameter		3.40	7.059		830	340		1,170	1,425
3805	72" diameter		3.20	7.500		880	360		1,240	1,525

23 31 16.19 PVC Ducts

		Crew	Daily Output	Labor-Hours	Unit	Material	2011 Bare Costs Labor	Equipment	Total	Total Incl O&P
0010	**PVC DUCTS**									
4000	Rigid plastic, corrosive fume resistant PVC									
4020	Straight, 6" diameter	Q-9	220	.073	L.F.	10.05	3.38		13.43	16.15
4040	8" diameter		160	.100		14.60	4.65		19.25	23
4060	10" diameter		120	.133		18.40	6.20		24.60	29.50
4070	12" diameter		100	.160		21.50	7.45		28.95	35
4080	14" diameter		70	.229		26.50	10.60		37.10	45.50
4090	16" diameter		62	.258		30.50	12		42.50	51.50
4100	18" diameter	▼	58	.276		42.50	12.80		55.30	66
4110	20" diameter	Q-10	75	.320		54.50	15.40		69.90	83.50
4130	24" diameter	"	55	.436	▼	59.50	21		80.50	97.50
4250	Coupling, 6" diameter	Q-9	88	.182	Ea.	19.70	8.45		28.15	34.50
4270	8" diameter		78	.205		25.50	9.55		35.05	42.50
4290	10" diameter		70	.229		29.50	10.60		40.10	48.50
4300	12" diameter		55	.291		32	13.50		45.50	55.50
4310	14" diameter		44	.364		46	16.90		62.90	76
4320	16" diameter		40	.400		48.50	18.60		67.10	81.50
4330	18" diameter	▼	38	.421		60	19.55		79.55	95.50
4340	20" diameter	Q-10	55	.436		73.50	21		94.50	113
4360	24" diameter	"	45	.533		85	25.50		110.50	133
4470	Elbow, 90°, 6" diameter	Q-9	44	.364		87	16.90		103.90	121
4490	8" diameter		28	.571		99.50	26.50		126	149
4510	10" diameter		18	.889		117	41.50		158.50	192
4520	12" diameter		15	1.067		135	49.50		184.50	224
4530	14" diameter		11	1.455		162	67.50		229.50	280
4540	16" diameter	▼	10	1.600		204	74.50		278.50	335
4550	18" diameter	Q-10	15	1.600		289	77		366	430
4560	20" diameter		14	1.714		430	82.50		512.50	600
4580	24" diameter	▼	12	2	▼	515	96.50		611.50	715
4750	Elbow 45°, use 90° and deduct					25%				
9990	Minimum labor/equipment charge	1 Shee	3	2.667	Job		138		138	208

23 31 16.21 Polypropylene Ducts

		Crew	Daily Output	Labor-Hours	Unit	Material	2011 Bare Costs Labor	Equipment	Total	Total Incl O&P
0010	**POLYPROPYLENE DUCTS**									
0020	Rigid plastic, corrosive fume resistant, flame retardant, PP									
0110	Straight, 4" diameter	Q-9	200	.080	L.F.	8.90	3.72		12.62	15.40
0120	6" diameter		146	.110		12.60	5.10		17.70	21.50
0140	8" diameter		106	.151		16.20	7		23.20	28.50
0160	10" diameter		85	.188		24	8.75		32.75	40
0170	12" diameter		68	.235		46	10.95		56.95	67
0180	14" diameter		50	.320		52	14.85		66.85	79.50
0190	16" diameter		40	.400		70	18.60		88.60	105
0200	18" diameter	▼	31	.516		114	24		138	162
0210	20" diameter	Q-10	44	.545		144	26.50		170.50	199
0220	22" diameter		40	.600		161	29		190	222
0230	24" diameter		37	.649		219	31.50		250.50	289
0250	28" diameter		28	.857		238	41.50		279.50	325
0270	30" diameter	▼	24	1	▼	335	48		383	440
0400	Coupling									
0410	4" diameter	Q-9	23	.696	Ea.	14.15	32.50		46.65	64.50
0420	6" diameter		16	1		18.40	46.50		64.90	90.50
0430	8" diameter		11.40	1.404		22.50	65		87.50	124
0440	10" diameter	▼	9	1.778	▼	28.50	82.50		111	156

23 31 HVAC Ducts and Casings

23 31 16 – Nonmetal Ducts

23 31 16.21 Polypropylene Ducts		Crew	Daily Output	Labor-Hours	Unit	Material	2011 Bare Costs Labor	Equipment	Total	Total Incl O&P
0450	12" diameter	Q-9	8	2	Ea.	34	93		127	179
0460	14" diameter		6.60	2.424		79.50	113		192.50	257
0470	16" diameter		5.70	2.807		88	130		218	294
0480	18" diameter		5.20	3.077		157	143		300	390
0490	20" diameter	Q-10	6.80	3.529		171	170		341	445
0500	22" diameter		6.20	3.871		208	187		395	510
0510	24" diameter		5.80	4.138		234	199		433	555
0800	Elbow, 90°									
0810	4" diameter	Q-9	22	.727	Ea.	29.50	34		63.50	83.50
0820	6" diameter		15.40	1.039		42.50	48.50		91	120
0830	8" diameter		11	1.455		51	67.50		118.50	158
0840	10" diameter		8.60	1.860		73.50	86.50		160	212
0850	12" diameter		7.70	2.078		191	96.50		287.50	355
0860	14" diameter		6.40	2.500		221	116		337	420
0870	16" diameter		5.40	2.963		232	138		370	465
0880	18" diameter	Q-10	7.50	3.200		480	154		634	765
0890	20" diameter		6.60	3.636		590	175		765	915
0900	24" diameter		5.60	4.286		945	207		1,152	1,350
1000	Elbow, 45°									
1020	4" diameter	Q-9	22	.727	Ea.	19.85	34		53.85	73
1030	6" diameter		15.40	1.039		35.50	48.50		84	112
1040	8" diameter		11	1.455		41	67.50		108.50	147
1050	10" diameter		8.60	1.860		55	86.50		141.50	192
1060	12" diameter		7.70	2.078		149	96.50		245.50	310
1070	14" diameter		6.40	2.500		166	116		282	360
1080	16" diameter		5.40	2.963		168	138		306	395
1090	18" diameter	Q-10	7.50	3.200		380	154		534	650
1100	20" diameter		6.60	3.636		490	175		665	805
1110	24" diameter		5.60	4.286		645	207		852	1,025
1200	Tee									
1210	4" diameter	Q-9	16.50	.970	Ea.	29.50	45		74.50	101
1220	6" diameter		11.60	1.379		156	64		220	268
1230	8" diameter		8.30	1.928		194	89.50		283.50	350
1240	10" diameter		6.50	2.462		266	114		380	465
1250	12" diameter		5.80	2.759		350	128		478	580
1260	14" diameter		4.80	3.333		460	155		615	740
1270	16" diameter		4.05	3.951		510	184		694	840
1280	18" diameter	Q-10	5.60	4.286		670	207		877	1,050
1290	20" diameter		4.90	4.898		725	236		961	1,150
1300	22" diameter		4.40	5.455		875	263		1,138	1,375
1310	24" diameter		4.20	5.714		1,100	275		1,375	1,625

23 33 13.13 Volume-Control Dampers	Crew	Daily Output	Labor-Hours	Unit	Material	2011 Bare Costs Labor	Equipment	Total	Total Incl O&P
0010 **VOLUME-CONTROL DAMPERS**									
5990 Multi-blade dampers, opposed blade, 8" x 6"	1 Shee	24	.333	Ea.	20	17.20		37.20	48
5994 8" x 8"		22	.364		21	18.80		39.80	51.50
5996 10" x 10"		21	.381		24.50	19.70		44.20	57
6000 12" x 12"		21	.381		27.50	19.70		47.20	60
6020 12" x 18"		18	.444		36.50	23		59.50	75
6030 14" x 10"		20	.400		26.50	20.50		47	60.50
6031 14" x 14"		17	.471		32.50	24.50		57	72.50
6032 16" x 10"		18	.444		29	23		52	66
6033 16" x 12"		17	.471		32.50	24.50		57	72.50
6035 16" x 16"		16	.500		40.50	26		66.50	83.50
6036 18" x 14"		16	.500		40.50	26		66.50	83.50
6037 18" x 16"		15	.533		44.50	27.50		72	90.50
6038 18" x 18"		15	.533		48	27.50		75.50	94.50
6040 18" x 24"		12	.667		62	34.50		96.50	120
6060 18" x 28"		10	.800		72	41.50		113.50	142
6068 20" x 6"		18	.444		26	23		49	63
6069 20" x 8"		16	.500		30	26		56	72.50
6070 20" x 16"		14	.571		48	29.50		77.50	97.50
6071 20" x 18"		13	.615		52	32		84	105
6072 20" x 20"		13	.615		57.50	32		89.50	112
6073 22" x 6"		20	.400		26.50	20.50		47	60.50
6074 22" x 18"		14	.571		57.50	29.50		87	108
6075 22" x 20"		12	.667		62.50	34.50		97	121
6076 24" x 16"		11	.727		57	37.50		94.50	120
6077 22" x 22"		10	.800		67	41.50		108.50	136
6078 24" x 20"		8	1		67	51.50		118.50	152
6080 24" x 24"		8	1		78.50	51.50		130	165
6100 24" x 28"		6	1.333		92	69		161	205
6110 26" x 26"		6	1.333		86.50	69		155.50	199
6120 28" x 28"	Q-9	11	1.455		98	67.50		165.50	210
6130 30" x 18"		10	1.600		83.50	74.50		158	204
6132 30" x 24"		7	2.286		104	106		210	276
6133 30" x 30"		6.60	2.424		124	113		237	305
6135 32" x 32"		6.40	2.500		144	116		260	335
6150 34" x 34"		6	2.667		161	124		285	365
6151 36" x 12"		10	1.600		65.50	74.50		140	184
6152 36" x 16"		8	2		89.50	93		182.50	239
6156 36" x 32"		6.30	2.540		161	118		279	355
6157 36" x 34"		6.20	2.581		168	120		288	365
6158 36" x 36"		6	2.667		176	124		300	380
6160 44" x 28"		5.80	2.759		181	128		309	395
6180 48" x 36"		5.60	2.857		237	133		370	460
6200 56" x 36"		5.40	2.963		287	138		425	525
6220 60" x 36"		5.20	3.077		305	143		448	550
6240 60" x 44"		5	3.200		360	149		509	620
7500 Variable volume modulating motorized damper, incl. elect. mtr.									
7504 8" x 6"	1 Shee	15	.533	Ea.	109	27.50		136.50	162
7506 10" x 6"		14	.571		109	29.50		138.50	165
7510 10" x 10"		13	.615		112	32		144	172
7520 12" x 12"		12	.667		117	34.50		151.50	181
7522 12" x 16"		11	.727		118	37.50		155.50	187
7524 16" x 10"		12	.667		115	34.50		149.50	179

23 33 13 – Dampers

23 33 13.13 Volume-Control Dampers	Crew	Daily Output	Labor-Hours	Unit	Material	2011 Bare Costs Labor	Equipment	Total	Total Incl O&P
7526	16" x 14"	1 Shee	10	.800	Ea.	120	41.50		161.50	194
7528	16" x 18"		9	.889		125	46		171	207
7540	18" x 12"		10	.800		120	41.50		161.50	195
7542	18" x 18"		8	1		127	51.50		178.50	218
7544	20" x 14"		8	1		130	51.50		181.50	221
7546	20" x 18"		7	1.143		136	59		195	240
7560	24" x 12"		8	1		132	51.50		183.50	223
7562	24" x 18"		7	1.143		143	59		202	247
7568	28" x 10"		7	1.143		132	59		191	235
7580	28" x 16"		6	1.333		149	69		218	268
7590	30" x 14"		5	1.600		140	82.50		222.50	279
7600	30" x 18"		4	2		180	103		283	355
7610	30" x 24"	↓	3.80	2.105		239	109		348	425
7690	48" x 48"	Q-9	6	2.667		760	124		884	1,025
7694	6' x 14' w/2 motors	"	3	5.333		3,250	248		3,498	3,950
7700	For thermostat, add	1 Shee	8	1		40.50	51.50		92	123
7800	For transformer 40 VA capacity, add	"	16	.500	↓	18	26		44	59
8000	Multi-blade dampers, parallel blade									
8100	8" x 8"	1 Shee	24	.333	Ea.	73	17.20		90.20	107
8120	12" x 8"		22	.364		73	18.80		91.80	109
8140	16" x 10"		20	.400		94	20.50		114.50	134
8160	18" x 12"		18	.444		107	23		130	153
8180	22" x 12"		15	.533		109	27.50		136.50	162
8200	24" x 16"		11	.727		120	37.50		157.50	188
8220	28" x 16"		10	.800		134	41.50		175.50	210
8240	30" x 16"		8	1		138	51.50		189.50	230
8260	30" x 18"	↓	7	1.143	↓	164	59		223	270
8400	Round damper, butterfly, vol. control w/lever lock reg.									
8410	6" diam.	1 Shee	22	.364	Ea.	23	18.80		41.80	54
8412	7" diam.		21	.381		24	19.70		43.70	56
8414	8" diam.		20	.400		24.50	20.50		45	58
8416	9" diam.		19	.421		25.50	22		47.50	61
8418	10" diam.		18	.444		28	23		51	65.50
8420	12" diam.		16	.500		29	26		55	70.50
8422	14" diam.		14	.571		37.50	29.50		67	85.50
8424	16" diam.		13	.615		40.50	32		72.50	92.50
8426	18" diam.		12	.667		49.50	34.50		84	107
8428	20" diam.		11	.727		60.50	37.50		98	124
8430	24" diam.		10	.800		75.50	41.50		117	146
8432	30" diam.		9	.889		106	46		152	186
8434	36" diam.	↓	8	1	↓	141	51.50		192.50	233
8500	Round motor operated damper									
8510	6" dia.	1 Shee	22	.364	Ea.	57	18.80		75.80	91
8512	7" dia.		21	.381		60	19.70		79.70	95.50
8514	8" dia.		20	.400		62	20.50		82.50	99
8516	10" dia.		18	.444		68.50	23		91.50	110
8518	12" dia.		16	.500		73.50	26		99.50	120
8520	14" dia.	↓	14	.571	↓	83.50	29.50		113	137

23 33 13.16 Fire Dampers

		Crew	Daily Output	Labor-Hours	Unit	Material	2011 Bare Costs Labor	Equipment	Total	Total Incl O&P
0010	**FIRE DAMPERS**									
3000	Fire damper, curtain type, 1-1/2 hr. rated, vertical, 6" x 6"	1 Shee	24	.333	Ea.	21.50	17.20		38.70	50
3020	8" x 6"	↓	22	.364	↓	21.50	18.80		40.30	52.50

23 33 13.16 Fire Dampers		Crew	Daily Output	Labor-Hours	Unit	Material	2011 Bare Costs Labor	Equipment	Total	Total Incl O&P
3021	8" x 8"	1 Shee	22	.364	Ea.	21.50	18.80		40.30	52.50
3030	10" x 10"		22	.364		21.50	18.80		40.30	52.50
3040	12" x 6"		22	.364		21.50	18.80		40.30	52.50
3044	12" x 12"		20	.400		21.50	20.50		42	55
3060	20" x 6"		18	.444		25	23		48	62
3080	12" x 8"		22	.364		25	18.80		43.80	56
3100	24" x 8"		16	.500		30	26		56	72
3120	12" x 10"		21	.381		21.50	19.70		41.20	54
3140	24" x 10"		15	.533		31.50	27.50		59	76.50
3160	36" x 10"		12	.667		41	34.50		75.50	97
3180	16" x 12"		20	.400		24	20.50		44.50	57
3200	24" x 12"		13	.615		33.50	32		65.50	85
3220	48" x 12"		10	.800		51	41.50		92.50	119
3238	14" x 14"		19	.421		29.50	22		51.50	65.50
3240	16" x 14"		18	.444		29.50	23		52.50	67
3260	24" x 14"		12	.667		36.50	34.50		71	92
3280	30" x 14"		11	.727		43	37.50		80.50	105
3298	16" x 16"		18	.444		33.50	23		56.50	71.50
3300	18" x 16"		17	.471		35.50	24.50		60	76.50
3320	24" x 16"		11	.727		38.50	37.50		76	99.50
3340	36" x 16"		10	.800		51.50	41.50		93	119
3356	18" x 18"		16	.500		37	26		63	79.50
3360	24" x 18"		10	.800		45	41.50		86.50	112
3380	48" x 18"		8	1		64.50	51.50		116	149
3398	20" x 20"		10	.800		38.50	41.50		80	105
3400	24" x 20"		8	1		42	51.50		93.50	124
3420	36" x 20"		7	1.143		55	59		114	150
3440	24" x 22"		7	1.143		43	59		102	137
3460	30" x 22"		6	1.333		46	69		115	155
3478	24" x 24"		8	1		44	51.50		95.50	127
3480	26" x 24"		7	1.143		47	59		106	141
3484	32" x 24"	Q-9	13	1.231		55	57		112	147
3500	48" x 24"		12	1.333		72	62		134	173
3520	28" x 26"		13	1.231		52.50	57		109.50	145
3540	30" x 28"		12	1.333		58.50	62		120.50	158
3560	48" x 30"		11	1.455		85	67.50		152.50	196
3580	48" x 36"		10	1.600		93	74.50		167.50	214
3590	32" x 40"		10	1.600		73.50	74.50		148	193
3600	44" x 44"		10	1.600		97	74.50		171.50	219
3620	48" x 48"		8	2		127	93		220	281
3700	U.L. label included in above									
3800	For horizontal operation, add				Ea.	20%				
3900	For cap for blades out of air stream, add					20%				
4000	For 10" 22 ga., U.L. approved sleeve, add					35%				
4100	For 10" 22 ga., U.L. sleeve, 100% free area, add					65%				
4200	For oversize openings group dampers									
4210	Fire damper, round, Type A									
4214	4" diam.	1 Shee	20	.400	Ea.	56.50	20.50		77	93.50
4216	5" diam.		16	.500		51	26		77	95
4218	6" diam.		14.30	.559		51	29		80	99.50
4220	7" diam.		13.30	.602		51	31		82	103
4222	8" diam.		11.40	.702		51	36.50		87.50	111
4224	9" diam.		10.80	.741		55.50	38.50		94	119

23 33 Air Duct Accessories

23 33 13 – Dampers

23 33 13.16 Fire Dampers

		Crew	Daily Output	Labor-Hours	Unit	Material	2011 Bare Costs Labor	Equipment	Total	Total Incl O&P
4226	10" diam.	1 Shee	10	.800	Ea.	58	41.50		99.50	126
4228	12" diam.		9.50	.842		63.50	43.50		107	136
4230	14" diam.		8.20	.976		80	50.50		130.50	164
4232	16" diam.		7.60	1.053		80	54.50		134.50	170
4234	18" diam.		7.20	1.111		95.50	57.50		153	192
4236	20" diam.		6.40	1.250		101	64.50		165.50	209
4238	22" diam.		6.10	1.311		109	67.50		176.50	222
4240	24" diam.		5.70	1.404		114	72.50		186.50	235
4242	26" diam.		5.50	1.455		128	75		203	255
4244	28" diam.		5.40	1.481		134	76.50		210.50	264
4246	30" diam.		5.30	1.509		140	78		218	272
4500	Fire/smoke combination damper, louver type, UL									
4506	6" x 6"	1 Shee	24	.333	Ea.	37	17.20		54.20	66.50
4510	8" x 8"		22	.364		130	18.80		148.80	172
4520	16" x 8"		20	.400		142	20.50		162.50	187
4540	18" x 8"		18	.444		142	23		165	191
4560	20" x 8"		16	.500		152	26		178	206
4580	10" x 10"		21	.381		147	19.70		166.70	192
4600	24" x 10"		15	.533		173	27.50		200.50	232
4620	30" x 10"		12	.667		179	34.50		213.50	249
4640	12" x 12"		20	.400		153	20.50		173.50	200
4660	18" x 12"		18	.444		156	23		179	207
4680	24" x 12"		13	.615		165	32		197	229
4700	30" x 12"		11	.727		171	37.50		208.50	245
4720	14" x 14"		17	.471		165	24.50		189.50	219
4740	16" x 14"		18	.444		170	23		193	222
4760	20" x 14"		14	.571		173	29.50		202.50	235
4780	24" x 14"		12	.667		177	34.50		211.50	247
4800	30" x 14"		10	.800		188	41.50		229.50	269
4820	16" x 16"		16	.500		183	26		209	240
4840	20" x 16"		14	.571		188	29.50		217.50	252
4860	24" x 16"		11	.727		194	37.50		231.50	270
4880	30" x 16"		8	1		202	51.50		253.50	300
4900	18" x 18"		15	.533		181	27.50		208.50	241
5000	24" x 18"		10	.800		193	41.50		234.50	276
5020	36" x 18"		7	1.143		212	59		271	325
5040	20" x 20"		13	.615		202	32		234	271
5060	24" x 20"		8	1		207	51.50		258.50	305
5080	30" x 20"		8	1		218	51.50		269.50	320
5100	36" x 20"		7	1.143		229	59		288	340
5120	24" x 24"		8	1		224	51.50		275.50	325
5130	30" x 24"		7.60	1.053		212	54.50		266.50	315
5140	36" x 24"	Q-9	12	1.333		244	62		306	365
5141	36" x 30"		10	1.600		258	74.50		332.50	395
5142	40" x 36"		8	2		460	93		553	645
5143	48" x 48"		4	4		895	186		1,081	1,275
5150	Damper operator motor, 24 or 120 volt	1 Shee	16	.500		245	26		271	310

23 33 13.25 Exhaust Vent Damper

		Crew	Daily Output	Labor-Hours	Unit	Material	2011 Bare Costs Labor	Equipment	Total	Total Incl O&P
0010	**EXHAUST VENT DAMPER**, Auto, OB with elect. actuator.									
1110	6" x 6"	1 Shee	24	.333	Ea.	59	17.20		76.20	91
1114	8" x 6"		23	.348		60.50	17.95		78.45	93.50
1116	8" x 8"		22	.364		62.50	18.80		81.30	97.50

23 33 Air Duct Accessories

23 33 13 – Dampers

23 33 13.25 Exhaust Vent Damper

		Crew	Daily Output	Labor- Hours	Unit	Material	2011 Bare Costs Labor	Equipment	Total	Total Incl O&P
1118	10" x 6"	1 Shee	23	.348	Ea.	62.50	17.95		80.45	96
1120	10" x 10"		22	.364		67	18.80		85.80	102
1122	12" x 6"		22	.364		63.50	18.80		82.30	98
1124	12" x 10"		21	.381		70.50	19.70		90.20	108
1126	12" x 12"		20	.400		74	20.50		94.50	113
1128	14" x 6"		21	.381		65.50	19.70		85.20	102
1130	14" x 10"		19	.421		74	22		96	115
1132	16" x 6"		21	.381		66	19.70		85.70	103
1134	16" x 8"		20	.400		73.50	20.50		94	112
1136	18" x 6"		20	.400		70	20.50		90.50	108
1138	18" x 8"		19	.421		74	22		96	115

23 33 13.28 Splitter Damper Assembly

		Crew	Daily Output	Labor- Hours	Unit	Material	2011 Bare Costs Labor	Equipment	Total	Total Incl O&P
0009	**SPLITTER DAMPER ASSEMBLY**									
0010	Self locking, 1' rod	1 Shee	24	.333	Ea.	21.50	17.20		38.70	50
7020	3' rod		22	.364		28	18.80		46.80	59.50
7040	4' rod		20	.400		31.50	20.50		52	65.50
7060	6' rod		18	.444		38	23		61	76.50

23 33 13.32 Relief Damper

		Crew	Daily Output	Labor- Hours	Unit	Material	2011 Bare Costs Labor	Equipment	Total	Total Incl O&P
0010	**RELIEF DAMPER**, Electronic bypass with tight seal									
8310	8" x 6"	1 Shee	22	.364	Ea.	163	18.80		181.80	209
8314	10" x 6"		22	.364		163	18.80		181.80	209
8318	10" x 10"		21	.381		170	19.70		189.70	217
8322	12" x 12"		20	.400		173	20.50		193.50	221
8326	12" x 16"		19	.421		180	22		202	231
8330	16" x 10"		20	.400		189	20.50		209.50	239
8334	16" x 14"		18	.444		194	23		217	248
8338	16" x 18"		17	.471		211	24.50		235.50	269
8342	18" x 12"		18	.444		211	23		234	267
8346	18" x 18"		15	.533		199	27.50		226.50	261
8350	20" x 14"		14	.571		196	29.50		225.50	261
8354	20" x 18"		13	.615		205	32		237	273
8358	24" x 12"		13	.615		200	32		232	268
8362	24" x 18"		10	.800		156	41.50		197.50	234
8363	24" x 24"		10	.800		218	41.50		259.50	305
8364	24" x 36"		9	.889		259	46		305	355
8365	24" x 48"		6	1.333		290	69		359	425
8366	28" x 10"		13	.615		144	32		176	206
8370	28" x 16"		10	.800		221	41.50		262.50	305
8374	30" x 14"		11	.727		151	37.50		188.50	223
8378	30" x 18"		8	1		228	51.50		279.50	330
8382	30" x 24"		6	1.333		242	69		311	370
8390	46" x 36"		4	2		320	103		423	510
8394	48" x 48"		3	2.667		350	138		488	595
8396	54" x 36"		2	4		370	207		577	715

23 33 19 – Duct Silencers

23 33 19.10 Duct Silencers

		Crew	Daily Output	Labor- Hours	Unit	Material	2011 Bare Costs Labor	Equipment	Total	Total Incl O&P
0009	**DUCT SILENCERS**									
0010	Silencers, noise control for air flow, duct				MCFM	55.50			55.50	61
9004	Duct sound trap, packaged									
9010	12" x 18" x 36", 2000 CFM	1 Shee	12	.667	Ea.	430	34.50		464.50	525
9011	24" x 18" x 36", 9000 CFM	Q-9	80	.200		495	9.30		504.30	560
9012	24" x 24" x 36", 9000 CFM		60	.267		745	12.40		757.40	835

23 33 Air Duct Accessories

23 33 19 – Duct Silencers

23 33 19.10 Duct Silencers

23 33 19.10 Duct Silencers		Crew	Daily Output	Labor-Hours	Unit	Material	2011 Bare Costs Labor	Equipment	Total	Total Incl O&P
9013	24" x 30" x 36", 9000 CFM	Q-9	50	.320	Ea.	795	14.85		809.85	895
9014	24" x 36" x 36", 9000 CFM		40	.400		990	18.60		1,008.60	1,125
9015	24" x 48" x 36", 9000 CFM		28	.571		1,300	26.50		1,326.50	1,475
9018	36" x 18" x 36", 6300 CFM		42	.381		990	17.70		1,007.70	1,125
9019	36" x 36" x 36", 6300 CFM		21	.762		1,975	35.50		2,010.50	2,225
9020	36" x 48" x 36", 6300 CFM		14	1.143		2,575	53		2,628	2,900
9021	36" x 60" x 36", 6300 CFM		13	1.231		3,275	57		3,332	3,675
9022	48" x 48" x 36", 6300 CFM		12	1.333		3,475	62		3,537	3,925
9023	24" x 24" x 60", 6000 CFM		3.60	4.444		1,150	207		1,357	1,550
9026	24" x 30" x 60", 7000 CFM		3.50	4.571		1,275	212		1,487	1,725
9030	24" x 36" x 60", 8000 CFM		3.40	4.706		1,400	219		1,619	1,875
9034	24" x 48" x 60", 11,000 CFM		3.30	4.848		1,200	225		1,425	1,675
9038	36" x 18" x 60", 6300 CFM		3.50	4.571		950	212		1,162	1,375
9042	36" x 36" x 60", 12,000 CFM		3.30	4.848		1,525	225		1,750	2,025
9046	36" x 48" x 60", 17,000 CFM		3.10	5.161		2,000	240		2,240	2,575
9050	36" x 60" x 60", 20,000 CFM		2.80	5.714		2,450	266		2,716	3,100
9054	48" x 48" x 60", 23,000 CFM		2.60	6.154		2,825	286		3,111	3,525
9056	48" x 60" x 60", 23,000 CFM		2.10	7.619		3,525	355		3,880	4,400

23 33 23 – Turning Vanes

23 33 23.13 Air Turning Vanes

		Crew	Daily Output	Labor-Hours	Unit	Material	2011 Bare Costs Labor	Equipment	Total	Total Incl O&P
0010	**AIR TURNING VANES**									
9400	Turning vane components									
9410	Turning vane rail	1 Shee	160	.050	L.F.	.83	2.58		3.41	4.82
9420	Double thick, factory fab. vane		300	.027		1.17	1.38		2.55	3.37
9428	12" high set		170	.047		1.68	2.43		4.11	5.55
9432	14" high set		160	.050		1.96	2.58		4.54	6.05
9434	16" high set		150	.053		2.24	2.75		4.99	6.65
9436	18" high set		144	.056		2.52	2.87		5.39	7.10
9438	20" high set		138	.058		2.80	2.99		5.79	7.60
9440	22" high set		130	.062		3.08	3.18		6.26	8.20
9442	24" high set		124	.065		3.36	3.33		6.69	8.75
9444	26" high set		116	.069		3.64	3.56		7.20	9.40
9446	30" high set		112	.071		4.20	3.69		7.89	10.20

23 33 33 – Duct-Mounting Access Doors

23 33 33.13 Duct Access Doors

		Crew	Daily Output	Labor-Hours	Unit	Material	2011 Bare Costs Labor	Equipment	Total	Total Incl O&P
0010	**DUCT ACCESS DOORS**									
1000	Duct access door, insulated, 6" x 6"	1 Shee	14	.571	Ea.	15	29.50		44.50	61
1020	10" x 10"		11	.727		18.15	37.50		55.65	77
1040	12" x 12"		10	.800		19.40	41.50		60.90	84
1050	12" x 18"		9	.889		34.50	46		80.50	108
1060	16" x 12"		9	.889		27.50	46		73.50	99.50
1070	18" x 18"		8	1		34.50	51.50		86	116
1074	24" x 18"		8	1		41	51.50		92.50	124
1080	24" x 24"		8	1		43.50	51.50		95	126

23 33 46 – Flexible Ducts

23 33 46.10 Flexible Air Ducts

		Crew	Daily Output	Labor-Hours	Unit	Material	2011 Bare Costs Labor	Equipment	Total	Total Incl O&P
0010	**FLEXIBLE AIR DUCTS**	R233100-20								
1280	Add to labor for elevated installation									
1282	of prefabricated (purchased) ductwork									
1283	10' to 15' high						10%			
1284	15' to 20' high						20%			

23 33 46.10 Flexible Air Ducts		Crew	Daily Output	Labor-Hours	Unit	Material	2011 Bare Costs Labor	Equipment	Total	Total Incl O&P
1285	20' to 25' high						25%			
1286	25' to 30' high						35%			
1287	30' to 35' high						40%			
1288	35' to 40' high						50%			
1289	Over 40' high						55%			
1300	Flexible, coated fiberglass fabric on corr. resist. metal helix									
1400	pressure to 12" (WG) UL-181									
1500	Non-insulated, 3" diameter	Q-9	400	.040	L.F.	1.02	1.86		2.88	3.93
1520	4" diameter		360	.044		1.02	2.07		3.09	4.24
1540	5" diameter		320	.050		1.25	2.32		3.57	4.89
1560	6" diameter		280	.057		1.53	2.66		4.19	5.70
1580	7" diameter		240	.067		1.82	3.10		4.92	6.70
1600	8" diameter		200	.080		2.10	3.72		5.82	7.90
1620	9" diameter		180	.089		2.49	4.13		6.62	9
1640	10" diameter		160	.100		2.61	4.65		7.26	9.90
1660	12" diameter		120	.133		3.17	6.20		9.37	12.85
1680	14" diameter		80	.200		3.85	9.30		13.15	18.30
1700	16" diameter		60	.267		5.50	12.40		17.90	25
1800	For adjustable clamps, add				Ea.	.72			.72	.79
1900	Insulated, 1" thick, PE jacket, 3" diameter [G]	Q-9	380	.042	L.F.	2.15	1.96		4.11	5.35
1910	4" diameter [G]		340	.047		2.15	2.19		4.34	5.70
1920	5" diameter [G]		300	.053		2.15	2.48		4.63	6.10
1940	6" diameter [G]		260	.062		2.38	2.86		5.24	6.95
1960	7" diameter [G]		220	.073		2.78	3.38		6.16	8.15
1980	8" diameter [G]		180	.089		3	4.13		7.13	9.55
2000	9" diameter [G]		160	.100		3.40	4.65		8.05	10.80
2020	10" diameter [G]		140	.114		3.51	5.30		8.81	11.90
2040	12" diameter [G]		100	.160		4.36	7.45		11.81	16.05
2060	14" diameter [G]		80	.200		5.25	9.30		14.55	19.85
2080	16" diameter [G]		60	.267		6.55	12.40		18.95	26
2100	18" diameter [G]		45	.356		8.35	16.55		24.90	34
2120	20" diameter [G]	Q-10	65	.369		9.10	17.80		26.90	37
2500	Insulated, heavy duty, coated fiberglass fabric									
2520	4" diameter [G]	Q-9	340	.047	L.F.	3.45	2.19		5.64	7.10
2540	5" diameter [G]		300	.053		3.69	2.48		6.17	7.80
2560	6" diameter [G]		260	.062		4.14	2.86		7	8.90
2580	7" diameter [G]		220	.073		4.59	3.38		7.97	10.15
2600	8" diameter [G]		180	.089		5.20	4.13		9.33	12
2620	9" diameter [G]		160	.100		5.55	4.65		10.20	13.15
2640	10" diameter [G]		140	.114		6.75	5.30		12.05	15.45
2660	12" diameter [G]		100	.160		7.20	7.45		14.65	19.15
2680	14" diameter [G]		80	.200		9.40	9.30		18.70	24.50
2700	16" diameter [G]		60	.267		14.10	12.40		26.50	34.50
2720	18" diameter [G]		45	.356		17.30	16.55		33.85	44
2800	Flexible, aluminum, pressure to 12" (WG) UL-181									
2820	Non-insulated									
2830	3" diameter	Q-9	400	.040	L.F.	1.14	1.86		3	4.06
2831	4" diameter		360	.044		1.36	2.07		3.43	4.62
2832	5" diameter		320	.050		1.75	2.32		4.07	5.45
2833	6" diameter		280	.057		2.23	2.66		4.89	6.45
2834	7" diameter		240	.067		2.68	3.10		5.78	7.65
2835	8" diameter		200	.080		3.07	3.72		6.79	9
2836	9" diameter		180	.089		3.65	4.13		7.78	10.25

23 33 Air Duct Accessories

23 33 46 – Flexible Ducts

23 33 46.10 Flexible Air Ducts

		Crew	Daily Output	Labor-Hours	Unit	Material	2011 Bare Costs Labor	Equipment	Total	Total Incl O&P
2837	10" diameter	Q-9	160	.100	L.F.	4.30	4.65		8.95	11.80
2838	12" diameter		120	.133		5.30	6.20		11.50	15.15
2839	14" diameter		80	.200		6.50	9.30		15.80	21
2840	15" diameter		70	.229		8.25	10.60		18.85	25
2841	16" diameter		60	.267		10.15	12.40		22.55	30
2842	18" diameter		45	.356		11.30	16.55		27.85	37.50
2843	20" diameter	Q-10	65	.369		12.50	17.80		30.30	41
2880	Insulated, 1" thick with 3/4 lb., PE jacket									
2890	3" diameter G	Q-9	380	.042	L.F.	1.97	1.96		3.93	5.15
2891	4" diameter G		340	.047		2.27	2.19		4.46	5.80
2892	5" diameter G		300	.053		3.12	2.48		5.60	7.20
2893	6" diameter G		260	.062		3.45	2.86		6.31	8.15
2894	7" diameter G		220	.073		4.25	3.38		7.63	9.80
2895	8" diameter G		180	.089		4.55	4.13		8.68	11.25
2896	9" diameter G		160	.100		5.15	4.65		9.80	12.70
2897	10" diameter G		140	.114		6.30	5.30		11.60	15
2898	12" diameter G		100	.160		6.95	7.45		14.40	18.90
2899	14" diameter G		80	.200		8.45	9.30		17.75	23.50
2900	15" diameter G		70	.229		10.90	10.60		21.50	28
2901	16" diameter G		60	.267		12.60	12.40		25	32.50
2902	18" diameter G		45	.356		14.70	16.55		31.25	41
3000	Transitions, 3" deep, square to round									
3010	6" x 6" to round	Q-9	44	.364	Ea.	8.50	16.90		25.40	35
3014	6" x 12" to round		42	.381		15.85	17.70		33.55	44.50
3018	8" x 8" to round		40	.400		10.20	18.60		28.80	39
3022	9" x 9" to round		38	.421		10.20	19.55		29.75	40.50
3026	9" x 12" to round		37	.432		20.50	20		40.50	53
3030	9" x 15" to round		36	.444		24	20.50		44.50	57
3034	10" x 10" to round		34	.471		12.45	22		34.45	46.50
3038	10" x 22" to round		32	.500		29	23		52	67
3042	12" x 12" to round		30	.533		12.45	25		37.45	51
3046	12" x 18" to round		28	.571		29	26.50		55.50	72
3050	12" x 24" to round		26	.615		29	28.50		57.50	75.50
3054	15" x 15" to round		25	.640		15.85	30		45.85	62.50
3058	15" x 18" to round		24	.667		31.50	31		62.50	82
3062	18" x 18" to round		22	.727		19.85	34		53.85	73
3066	21" x 21" to round		21	.762		34	35.50		69.50	91
3070	22" x 22" to round		20	.800		44	37		81	105
3074	24" x 24" to round		18	.889		44	41.50		85.50	111

23 33 53 – Duct Liners

23 33 53.10 Duct Liner Board

0010	**DUCT LINER BOARD**									
3340	Board type fiberglass liner, FSK, 1-1/2 lb. density									
3344	1" thick G	Q-14	150	.107	S.F.	.57	4.50		5.07	7.60
3345	1-1/2" thick G		130	.123		.62	5.20		5.82	8.75
3346	2" thick G		120	.133		.73	5.65		6.38	9.50
3348	3" thick G		110	.145		.94	6.15		7.09	10.55
3350	4" thick G		100	.160		1.15	6.75		7.90	11.70
3356	3 lb. density, 1" thick G		150	.107		.77	4.50		5.27	7.80
3358	1-1/2" thick G		130	.123		.97	5.20		6.17	9.10
3360	2" thick G		120	.133		1.18	5.65		6.83	10
3362	2-1/2" thick G		110	.145		1.38	6.15		7.53	11

23 33 Air Duct Accessories

23 33 53 – Duct Liners

23 33 53.10 Duct Liner Board		Crew	Daily Output	Labor-Hours	Unit	Material	2011 Bare Costs Labor	Equipment	Total	Total Incl O&P
3364	3" thick	G	Q-14	100	.160	S.F.	1.59	6.75		8.34	12.20
3366	4" thick	G		90	.178		2	7.50		9.50	13.80
3370	6 lb. density, 1" thick	G		140	.114		1.09	4.82		5.91	8.65
3374	1-1/2" thick	G		120	.133		1.46	5.65		7.11	10.30
3378	2" thick	G		100	.160		1.82	6.75		8.57	12.45
3490	Board type, fiberglass liner, 3 lb. density										
3500	Fire resistant, black pigmented, 1 side										
3520	1" thick	G	Q-14	150	.107	S.F.	.77	4.50		5.27	7.80
3540	1-1/2" thick	G		130	.123		.97	5.20		6.17	9.10
3560	2" thick	G		120	.133		1.18	5.65		6.83	10
3600	FSK vapor barrier										
3620	1" thick	G	Q-14	150	.107	S.F.	.77	4.50		5.27	7.80
3630	1-1/2" thick	G		130	.123		.97	5.20		6.17	9.10
3640	2" thick	G		120	.133		1.18	5.65		6.83	10
3680	No finish										
3700	1" thick	G	Q-14	170	.094	S.F.	.40	3.97		4.37	6.60
3710	1-1/2" thick	G		140	.114		.60	4.82		5.42	8.10
3720	2" thick	G		130	.123		.81	5.20		6.01	8.95
3940	Board type, non-fibrous foam										
3950	Temperature, bacteria and fungi resistant										
3960	1" thick	G	Q-14	150	.107	S.F.	2.25	4.50		6.75	9.45
3970	1-1/2" thick	G		130	.123		3	5.20		8.20	11.35
3980	2" thick	G		120	.133		3.60	5.65		9.25	12.65

23 34 HVAC Fans

23 34 13 – Axial HVAC Fans

23 34 13.10 Axial Flow HVAC Fans

			Crew	Daily Output	Labor-Hours	Unit	Material	2011 Bare Costs Labor	Equipment	Total	Total Incl O&P
0010	**AXIAL FLOW HVAC FANS**	R233100-10									
0020	Air conditioning and process air handling										
0030	Axial flow, compact, low sound, 2.5" S.P.	R233400-10									
0050	3,800 CFM, 5 HP		Q-20	3.40	5.882	Ea.	4,625	278		4,903	5,525
0080	6,400 CFM, 5 HP			2.80	7.143		5,175	335		5,510	6,200
0100	10,500 CFM, 7-1/2 HP			2.40	8.333		6,450	395		6,845	7,700
0120	15,600 CFM, 10 HP			1.60	12.500		8,125	590		8,715	9,825
0140	23,000 CFM, 15 HP			.70	28.571		12,500	1,350		13,850	15,700
0160	28,000 CFM, 20 HP			.40	50		13,900	2,350		16,250	18,900
0500	Axial flow, constant speed										
0505	Direct drive, 1/8" S.P.										
0510	12", 1060 CFM, 1/6 HP		Q-20	3	6.667	Ea.	560	315		875	1,100
0514	12", 2095 CFM, 1/2 HP			3	6.667		660	315		975	1,200
0518	16", 2490 CFM, 1/3 HP			2.80	7.143		660	335		995	1,225
0522	20", 4130 CFM, 3/4 HP			2.60	7.692		785	365		1,150	1,400
0526	22", 4700 CFM, 3/4 HP			2.60	7.692		1,175	365		1,540	1,825
0530	24", 5850 CFM, 1 HP			2.50	8		1,200	380		1,580	1,900
0534	24", 7925 CFM, 1-1/2 HP			2.40	8.333		1,475	395		1,870	2,225
0538	30", 10640 CFM, 2 HP			2.20	9.091		1,900	430		2,330	2,750
0542	30", 14765 CFM, 2-1/2 HP			2.10	9.524		2,225	450		2,675	3,125
0546	36", 16780 CFM, 2 HP			2.10	9.524		2,275	450		2,725	3,200
0550	36", 22920 CFM, 5 HP			1.80	11.111		2,500	525		3,025	3,575
0560	Belt drive, 1/8" S.P.										
0562	15", 2800 CFM, 1/3 HP		Q-20	3.20	6.250	Ea.	880	295		1,175	1,425

23 34 HVAC Fans

23 34 13 – Axial HVAC Fans

23 34 13.10 Axial Flow HVAC Fans

		Crew	Daily Output	Labor-Hours	Unit	Material	2011 Bare Costs Labor	Equipment	Total	Total Incl O&P
0564	15", 3400 CFM, 1/2 HP	Q-20	3	6.667	Ea.	905	315		1,220	1,475
0568	18", 3280 CFM, 1/3 HP		2.80	7.143		965	335		1,300	1,550
0572	18", 3900 CFM, 1/2 HP		2.80	7.143		980	335		1,315	1,575
0576	18", 5250 CFM, 1 HP		2.70	7.407		1,100	350		1,450	1,725
0584	24", 6430 CFM, 1 HP		2.50	8		1,300	380		1,680	2,025
0588	24", 8860 CFM, 2 HP		2.40	8.333		1,350	395		1,745	2,100
0592	30", 9250 CFM, 1 HP		2.20	9.091		1,725	430		2,155	2,550
0596	30", 16900 CFM, 5 HP		2	10		1,875	470		2,345	2,750
0604	36", 14475 CFM, 2 HP		2.30	8.696		2,025	410		2,435	2,850
0608	36", 20080 CFM, 5 HP		1.80	11.111		2,150	525		2,675	3,150
0612	36", 14475 CFM, 7-1/2 HP		1.60	12.500		2,325	590		2,915	3,450
0616	42", 29000 CFM, 7-1/2 HP	▼	1.40	14.286	▼	2,875	675		3,550	4,175
1000	Return air fan									
1010	9200 CFM	Q-20	1.40	14.286	Ea.	3,775	675		4,450	5,175
1020	13,200 CFM		1.30	15.385		5,425	725		6,150	7,075
1030	16,500 CFM		1.20	16.667		6,775	785		7,560	8,625
1040	19,500 CFM		1	20		8,025	945		8,970	10,300
1500	Vaneaxial, low pressure, 2000 CFM, 1/2 HP		3.60	5.556		1,975	262		2,237	2,575
1520	4,000 CFM, 1 HP		3.20	6.250		2,275	295		2,570	2,950
1540	8,000 CFM, 2 HP		2.80	7.143		2,900	335		3,235	3,700
1560	16,000 CFM, 5 HP	▼	2.40	8.333	▼	4,175	395		4,570	5,175

23 34 14 – Blower HVAC Fans

23 34 14.10 Blower Type HVAC Fans

		Crew	Daily Output	Labor-Hours	Unit	Material	2011 Bare Costs Labor	Equipment	Total	Total Incl O&P
0010	**BLOWER TYPE HVAC FANS**									
2000	Blowers, direct drive with motor, complete									
2020	1045 CFM @ .5" S.P., 1/5 HP	Q-20	18	1.111	Ea.	254	52.50		306.50	360
2040	1385 CFM @ .5" S.P., 1/4 HP		18	1.111		256	52.50		308.50	360
2060	1640 CFM @ .5" S.P., 1/3 HP		18	1.111		232	52.50		284.50	335
2080	1760 CFM @ .5" S.P., 1/2 HP	▼	18	1.111	▼	242	52.50		294.50	345
2090	4 speed									
2100	1164 to 1739 CFM @ .5" S.P., 1/3 HP	Q-20	16	1.250	Ea.	249	59		308	365
2120	1467 to 2218 CFM @ 1.0" S.P., 3/4 HP	"	14	1.429	"	288	67.50		355.50	415
2500	Ceiling fan, right angle, extra quiet, 0.10" S.P.									
2520	95 CFM	Q-20	20	1	Ea.	252	47		299	350
2540	210 CFM		19	1.053		298	49.50		347.50	400
2560	385 CFM		18	1.111		380	52.50		432.50	495
2580	885 CFM		16	1.250		745	59		804	910
2600	1,650 CFM		13	1.538		1,025	72.50		1,097.50	1,225
2620	2,960 CFM	▼	11	1.818		1,375	86		1,461	1,625
2640	For wall or roof cap, add	1 Shee	16	.500		252	26		278	315
2660	For straight thru fan, add					10%				
2680	For speed control switch, add	1 Elec	16	.500	▼	137	25		162	189
7500	Utility set, steel construction, pedestal, 1/4" S.P.									
7520	Direct drive, 150 CFM, 1/8 HP	Q-20	6.40	3.125	Ea.	775	148		923	1,075
7540	485 CFM, 1/6 HP		5.80	3.448		980	163		1,143	1,325
7560	1950 CFM, 1/2 HP		4.80	4.167		1,150	197		1,347	1,550
7580	2410 CFM, 3/4 HP		4.40	4.545		2,125	215		2,340	2,650
7600	3328 CFM, 1-1/2 HP	▼	3	6.667	▼	2,350	315		2,665	3,075
7680	V-belt drive, drive cover, 3 phase									
7700	800 CFM, 1/4 HP	Q-20	6	3.333	Ea.	855	157		1,012	1,175
7720	1,300 CFM, 1/3 HP		5	4		895	189		1,084	1,275
7740	2,000 CFM, 1 HP	▼	4.60	4.348		1,050	205		1,255	1,475

23 34 14 – Blower HVAC Fans

23 34 14.10 Blower Type HVAC Fans	Crew	Daily Output	Labor-Hours	Unit	Material	2011 Bare Costs Labor	Equipment	Total	Total Incl O&P
7760	2,900 CFM, 3/4 HP	Q-20	4.20	4.762	Ea.	1,425	225		1,650	1,925
7780	3,600 CFM, 3/4 HP		4	5		1,775	236		2,011	2,300
7800	4,800 CFM, 1 HP		3.50	5.714		2,075	270		2,345	2,700
7820	6,700 CFM, 1-1/2 HP		3	6.667		2,575	315		2,890	3,300
7830	7,500 CFM, 2 HP		2.50	8		3,475	380		3,855	4,400
7840	11,000 CFM, 3 HP		2	10		4,650	470		5,120	5,825
7860	13,000 CFM, 3 HP		1.60	12.500		4,725	590		5,315	6,100
7880	15,000 CFM, 5 HP		1	20		4,900	945		5,845	6,800
7900	17,000 CFM, 7-1/2 HP		.80	25		5,250	1,175		6,425	7,525
7920	20,000 CFM, 7-1/2 HP		.80	25		6,250	1,175		7,425	8,650

23 34 16 – Centrifugal HVAC Fans

23 34 16.10 Centrifugal Type HVAC Fans

		Crew	Daily Output	Labor-Hours	Unit	Material	2011 Bare Costs Labor	Equipment	Total	Total Incl O&P
0010	**CENTRIFUGAL TYPE HVAC FANS**									
0200	In-line centrifugal, supply/exhaust booster									
0220	aluminum wheel/hub, disconnect switch, 1/4" S.P.									
0240	500 CFM, 10" diameter connection	Q-20	3	6.667	Ea.	1,150	315		1,465	1,750
0260	1,380 CFM, 12" diameter connection		2	10		1,225	470		1,695	2,050
0280	1,520 CFM, 16" diameter connection		2	10		1,325	470		1,795	2,150
0300	2,560 CFM, 18" diameter connection		1	20		1,450	945		2,395	3,025
0320	3,480 CFM, 20" diameter connection		.80	25		1,725	1,175		2,900	3,650
0326	5,080 CFM, 20" diameter connection		.75	26.667		1,875	1,250		3,125	3,950
0340	7,500 CFM, 22" diameter connection		.70	28.571		2,225	1,350		3,575	4,475
0350	10,000 CFM, 27" diameter connection		.65	30.769		2,700	1,450		4,150	5,175
3500	Centrifugal, airfoil, motor and drive, complete									
3520	1000 CFM, 1/2 HP	Q-20	2.50	8	Ea.	1,675	380		2,055	2,425
3540	2,000 CFM, 1 HP		2	10		1,875	470		2,345	2,750
3560	4,000 CFM, 3 HP		1.80	11.111		2,375	525		2,900	3,400
3580	8,000 CFM, 7-1/2 HP		1.40	14.286		3,675	675		4,350	5,075
3600	12,000 CFM, 10 HP		1	20		4,425	945		5,370	6,300
4000	Single width, belt drive, not incl. motor, capacities									
4020	at 2000 fpm, 2.5" S.P. for indicated motor									
4040	6900 CFM, 5 HP	Q-9	2.40	6.667	Ea.	3,375	310		3,685	4,200
4060	10,340 CFM, 7-1/2 HP		2.20	7.273		4,700	340		5,040	5,675
4080	15,320 CFM, 10 HP		2	8		5,550	370		5,920	6,650
4100	22,780 CFM, 15 HP		1.80	8.889		8,250	415		8,665	9,700
4120	33,840 CFM, 20 HP		1.60	10		11,000	465		11,465	12,800
4140	41,400 CFM, 25 HP		1.40	11.429		14,500	530		15,030	16,800
4160	50,100 CFM, 30 HP		.80	20		18,500	930		19,430	21,700
4200	Double width wheel, 12,420 CFM, 7.5 HP		2.20	7.273		4,825	340		5,165	5,825
4220	18,620 CFM, 15 HP		2	8		6,925	370		7,295	8,175
4240	27,580 CFM, 20 HP		1.80	8.889		8,475	415		8,890	9,925
4260	40,980 CFM, 25 HP		1.50	10.667		13,300	495		13,795	15,400
4280	60,920 CFM, 40 HP		1	16		17,900	745		18,645	20,700
4300	74,520 CFM, 50 HP		.80	20		21,200	930		22,130	24,700
4320	90,160 CFM, 50 HP		.70	22.857		29,000	1,050		30,050	33,400
4340	110,300 CFM, 60 HP		.50	32		39,200	1,475		40,675	45,400
4360	134,960 CFM, 75 HP		.40	40		52,000	1,850		53,850	60,500
4500	Corrosive fume resistant, plastic									
4600	roof ventilators, centrifugal, V belt drive, motor									
4620	1/4" S.P., 250 CFM, 1/4 HP	Q-20	6	3.333	Ea.	3,975	157		4,132	4,600
4640	895 CFM, 1/3 HP		5	4		4,325	189		4,514	5,025
4660	1630 CFM, 1/2 HP		4	5		5,125	236		5,361	5,975

23 34 HVAC Fans

23 34 16 – Centrifugal HVAC Fans

23 34 16.10 Centrifugal Type HVAC Fans	Crew	Daily Output	Labor-Hours	Unit	Material	2011 Bare Costs Labor	Equipment	Total	Total Incl O&P
4680 2240 CFM, 1 HP	Q-20	3	6.667	Ea.	5,325	315		5,640	6,350
4700 3810 CFM, 2 HP		2	10		5,925	470		6,395	7,225
4710 5000 CFM, 2 HP		1.80	11.111		12,500	525		13,025	14,500
4715 8000 CFM, 5 HP		1.40	14.286		13,900	675		14,575	16,200
4720 11760 CFM, 5 HP		1	20		14,100	945		15,045	16,900
4740 18810 CFM, 10 HP		.70	28.571		14,600	1,350		15,950	18,000
4800 For intermediate capacity, motors may be varied									
4810 For explosion proof motor, add				Ea.	15%				
5000 Utility set, centrifugal, V belt drive, motor									
5020 1/4" S.P., 1200 CFM, 1/4 HP	Q-20	6	3.333	Ea.	1,550	157		1,707	1,950
5040 1520 CFM, 1/3 HP		5	4		1,725	189		1,914	2,175
5060 1850 CFM, 1/2 HP		4	5		1,725	236		1,961	2,250
5080 2180 CFM, 3/4 HP		3	6.667		2,050	315		2,365	2,725
5100 1/2" S.P., 3600 CFM, 1 HP		2	10		2,350	470		2,820	3,300
5120 4250 CFM, 1-1/2 HP		1.60	12.500		2,575	590		3,165	3,750
5140 4800 CFM, 2 HP		1.40	14.286		2,900	675		3,575	4,225
5160 6920 CFM, 5 HP		1.30	15.385		3,600	725		4,325	5,050
5180 7700 CFM, 7-1/2 HP		1.20	16.667		4,650	785		5,435	6,275
5200 For explosion proof motor, add					15%				
5300 Fume exhauster without hose and intake nozzle									
5310 630 CFM, 1-1/2 HP	Q-20	2	10	Ea.	1,025	470		1,495	1,825
5320 Hose extension kit 5'	"	20	1	"	82	47		129	161
5500 Fans, industrial exhauster, for air which may contain granular matl.									
5520 1000 CFM, 1-1/2 HP	Q-20	2.50	8	Ea.	2,500	380		2,880	3,325
5540 2000 CFM, 3 HP		2	10		3,100	470		3,570	4,100
5560 4000 CFM, 7-1/2 HP		1.80	11.111		4,375	525		4,900	5,600
5580 8000 CFM, 15 HP		1.40	14.286		5,650	675		6,325	7,250
5600 12,000 CFM, 30 HP		1	20		9,275	945		10,220	11,600
7000 Roof exhauster, centrifugal, aluminum housing, 12" galvanized									
7020 curb, bird screen, back draft damper, 1/4" S.P.									
7100 Direct drive, 320 CFM, 11" sq. damper	Q-20	7	2.857	Ea.	610	135		745	875
7120 600 CFM, 11" sq. damper		6	3.333		760	157		917	1,075
7140 815 CFM, 13" sq. damper		5	4		760	189		949	1,125
7160 1450 CFM, 13" sq. damper		4.20	4.762		1,100	225		1,325	1,550
7180 2050 CFM, 16" sq. damper		4	5		1,450	236		1,686	1,950
7200 V-belt drive, 1650 CFM, 12" sq. damper		6	3.333		1,100	157		1,257	1,425
7220 2750 CFM, 21" sq. damper		5	4		1,300	189		1,489	1,700
7230 3500 CFM, 21" sq. damper		4.50	4.444		1,475	210		1,685	1,950
7240 4910 CFM, 23" sq. damper		4	5		1,775	236		2,011	2,325
7260 8525 CFM, 28" sq. damper		3	6.667		2,400	315		2,715	3,125
7280 13,760 CFM, 35" sq. damper		2	10		3,225	470		3,695	4,250
7300 20,558 CFM, 43" sq. damper		1	20		6,550	945		7,495	8,650
7320 For 2 speed winding, add					15%				
7340 For explosionproof motor, add					495			495	545
7360 For belt driven, top discharge, add					15%				
7400 Roof mounted kitchen exhaust, aluminum, centrifugal									
7410 Direct drive, 2 speed, temp to 200°F									
7412 1/12 HP, 9-3/4"	Q-20	8	2.500	Ea.	570	118		688	805
7414 1/3 HP, 12-5/8"		7	2.857		700	135		835	975
7416 1/2 HP, 13-1/2"		6	3.333		775	157		932	1,075
7418 3/4 HP, 15"		5	4		1,100	189		1,289	1,500
7424 Belt drive, temp to 250°F									
7426 3/4 HP, 20"	Q-20	5	4	Ea.	2,425	189		2,614	2,950

23 34 16 – Centrifugal HVAC Fans

23 34 16.10 Centrifugal Type HVAC Fans		Crew	Daily Output	Labor-Hours	Unit	Material	2011 Bare Costs Labor	Equipment	Total	Total Incl O&P
7428	1-1/2 HP, 24-1/2"	Q-20	4	5	Ea.	2,675	236		2,911	3,300
7430	1-1/2 HP, 30"	↓	4	5	↓	3,550	236		3,786	4,250
7450	Upblast, propeller, w/BDD, BS, + C									
7454	30,300 CFM @ 3/8" S.P., 5 HP	Q-20	1.60	12.500	Ea.	3,900	590		4,490	5,200
7458	36,000 CFM @ 1/2" S.P., 15 HP	"	1.60	12.500	"	5,550	590		6,140	7,025
8500	Wall exhausters, centrifugal, auto damper, 1/8" S.P.									
8520	Direct drive, 610 CFM, 1/20 HP	Q-20	14	1.429	Ea.	365	67.50		432.50	500
8540	796 CFM, 1/12 HP		13	1.538		735	72.50		807.50	915
8560	822 CFM, 1/6 HP		12	1.667		925	78.50		1,003.50	1,150
8580	1,320 CFM, 1/4 HP		12	1.667		1,100	78.50		1,178.50	1,325
8600	1756 CFM, 1/4 HP		11	1.818		1,100	86		1,186	1,325
8620	1983 CFM, 1/4 HP		10	2		1,125	94.50		1,219.50	1,375
8640	2900 CFM, 1/2 HP		9	2.222		1,175	105		1,280	1,450
8660	3307 CFM, 3/4 HP		8	2.500		1,275	118		1,393	1,575
8670	5940 CFM, 1 HP	↓	7	2.857	↓	340	135		475	580
9500	V-belt drive, 3 phase									
9520	2,800 CFM, 1/4 HP	Q-20	9	2.222	Ea.	1,650	105		1,755	1,975
9540	3,740 CFM, 1/2 HP		8	2.500		1,725	118		1,843	2,075
9560	4400 CFM, 3/4 HP		7	2.857		1,750	135		1,885	2,125
9580	5700 CFM, 1-1/2 HP	↓	6	3.333	↓	1,825	157		1,982	2,225

23 34 23 – HVAC Power Ventilators

23 34 23.10 HVAC Power Circulators and Ventilators

		Crew	Daily Output	Labor-Hours	Unit	Material	2011 Bare Costs Labor	Equipment	Total	Total Incl O&P
0010	**HVAC POWER CIRCULATORS AND VENTILATORS**									
3000	Paddle blade air circulator, 3 speed switch									
3020	42", 5,000 CFM high, 3000 CFM low G	1 Elec	2.40	3.333	Ea.	154	168		322	415
3040	52", 6,500 CFM high, 4000 CFM low G	"	2.20	3.636	"	154	183		337	440
3100	For antique white motor, same cost									
3200	For brass plated motor, same cost									
3300	For light adaptor kit, add G				Ea.	38.50			38.50	42.50
3310	Industrial grade, reversible, 4 blade									
3312	5500 CFM	Q-20	5	4	Ea.	170	189		359	470
3314	7600 CFM		5	4		185	189		374	490
3316	21,015 CFM	↓	4	5	↓	240	236		476	620
6000	Propeller exhaust, wall shutter, 1/4" S.P.									
6020	Direct drive, two speed									
6100	375 CFM, 1/10 HP	Q-20	10	2	Ea.	430	94.50		524.50	615
6120	730 CFM, 1/7 HP		9	2.222		480	105		585	685
6140	1000 CFM, 1/8 HP		8	2.500		665	118		783	915
6160	1890 CFM, 1/4 HP		7	2.857		680	135		815	950
6180	3275 CFM, 1/2 HP		6	3.333		690	157		847	995
6200	4720 CFM, 1 HP	↓	5	4	↓	1,075	189		1,264	1,450
6300	V-belt drive, 3 phase									
6320	6175 CFM, 3/4 HP	Q-20	5	4	Ea.	870	189		1,059	1,250
6340	7500 CFM, 3/4 HP		5	4		960	189		1,149	1,325
6360	10,100 CFM, 1 HP		4.50	4.444		1,050	210		1,260	1,475
6380	14,300 CFM, 1-1/2 HP		4	5		1,200	236		1,436	1,675
6400	19,800 CFM, 2 HP		3	6.667		1,475	315		1,790	2,100
6420	26,250 CFM, 3 HP		2.60	7.692		1,875	365		2,240	2,600
6440	38,500 CFM, 5 HP		2.20	9.091		2,200	430		2,630	3,075
6460	46,000 CFM, 7-1/2 HP		2	10		2,350	470		2,820	3,300
6480	51,500 CFM, 10 HP	↓	1.80	11.111	↓	2,450	525		2,975	3,500
6490	V-belt drive, 115 V, residential, whole house									

23 34 HVAC Fans

23 34 23 – HVAC Power Ventilators

23 34 23.10 HVAC Power Circulators and Ventilators	Crew	Daily Output	Labor-Hours	Unit	Material	2011 Bare Costs Labor	Equipment	Total	Total Incl O&P
6500	Ceiling-wall, 5200 CFM, 1/4 HP, 30" x 30"	1 Shee	6	1.333	Ea.	470	69		539	620
6510	7500 CFM, 1/3 HP, 36" x 36"		5	1.600		490	82.50		572.50	665
6520	10,500 CFM, 1/3 HP, 42" x 42"		5	1.600		510	82.50		592.50	690
6530	13,200 CFM, 1/3 HP, 48" x 48"		4	2		565	103		668	780
6540	15,445 CFM, 1/2 HP, 48" x 48"		4	2		590	103		693	800
6550	17,025 CFM, 1/2 HP, 54" x 54"		4	2		1,125	103		1,228	1,400
6560	For two speed motor, add					20%				
6570	Shutter, automatic, ceiling/wall									
6580	30" x 30"	1 Shee	8	1	Ea.	153	51.50		204.50	246
6590	36" x 36"		8	1		174	51.50		225.50	270
6600	42" x 42"		8	1		218	51.50		269.50	320
6610	48" x 48"		7	1.143		236	59		295	350
6620	54" x 54"		6	1.333		310	69		379	450
6630	Timer, shut off, to 12 hour		20	.400		50	20.50		70.50	86
6650	Residential, bath exhaust, grille, back draft damper									
6660	50 CFM	Q-20	24	.833	Ea.	47.50	39.50		87	112
6670	110 CFM		22	.909		79	43		122	151
6672	180 CFM		22	.909		152	43		195	232
6673	210 CFM		22	.909		204	43		247	289
6674	260 CFM		22	.909		228	43		271	315
6675	300 CFM		22	.909		310	43		353	405
6680	Light combination, squirrel cage, 100 watt, 70 CFM		24	.833		85.50	39.50		125	154
6700	Light/heater combination, ceiling mounted									
6710	70 CFM, 1450 watt	Q-20	24	.833	Ea.	127	39.50		166.50	200
6800	Heater combination, recessed, 70 CFM		24	.833		51	39.50		90.50	116
6820	With 2 infrared bulbs		23	.870		79.50	41		120.50	150
6840	Wall mount, 170 CFM		22	.909		234	43		277	325
6846	Ceiling mount, 180 CFM		22	.909		236	43		279	325
6900	Kitchen exhaust, grille, complete, 160 CFM		22	.909		87.50	43		130.50	161
6910	180 CFM		20	1		74.50	47		121.50	153
6920	270 CFM		18	1.111		134	52.50		186.50	227
6930	350 CFM		16	1.250		105	59		164	204
6940	Residential roof jacks and wall caps									
6944	Wall cap with back draft damper									
6946	3" & 4" dia. round duct	1 Shee	11	.727	Ea.	20.50	37.50		58	80
6948	6" dia. round duct	"	11	.727	"	52.50	37.50		90	115
6958	Roof jack with bird screen and back draft damper									
6960	3" & 4" dia. round duct	1 Shee	11	.727	Ea.	21.50	37.50		59	80.50
6962	3-1/4" x 10" rectangular duct	"	10	.800	"	39	41.50		80.50	105
6980	Transition									
6982	3-1/4" x 10" to 6" dia. round	1 Shee	20	.400	Ea.	23	20.50		43.50	56.50
8020	Attic, roof type									
8030	Aluminum dome, damper & curb									
8040	6" diameter, 300 CFM	1 Elec	16	.500	Ea.	440	25		465	525
8050	7" diameter, 450 CFM		15	.533		480	27		507	570
8060	9" diameter, 900 CFM		14	.571		525	28.50		553.50	625
8080	12" diameter, 1000 CFM (gravity)		10	.800		495	40		535	605
8090	16" diameter, 1500 CFM (gravity)		9	.889		595	44.50		639.50	720
8100	20" diameter, 2500 CFM (gravity)		8	1		730	50.50		780.50	880
8110	26" diameter, 4000 CFM (gravity)		7	1.143		885	57.50		942.50	1,050
8120	32" diameter, 6500 CFM (gravity)		6	1.333		1,225	67		1,292	1,450
8130	38" diameter, 8000 CFM (gravity)		5	1.600		1,800	80.50		1,880.50	2,100
8140	50" diameter, 13,000 CFM (gravity)		4	2		2,625	101		2,726	3,025

23 34 HVAC Fans

23 34 23 – HVAC Power Ventilators

23 34 23.10 HVAC Power Circulators and Ventilators		Crew	Daily Output	Labor-Hours	Unit	Material	2011 Bare Costs Labor	Equipment	Total	Total Incl O&P
8160	Plastic, ABS dome									
8180	1050 CFM	1 Elec	14	.571	Ea.	146	28.50		174.50	203
8200	1600 CFM	"	12	.667	"	218	33.50		251.50	290
8240	Attic, wall type, with shutter, one speed									
8250	12" diameter, 1000 CFM	1 Elec	14	.571	Ea.	345	28.50		373.50	425
8260	14" diameter, 1500 CFM		12	.667		375	33.50		408.50	465
8270	16" diameter, 2000 CFM	↓	9	.889	↓	425	44.50		469.50	535
8290	Whole house, wall type, with shutter, one speed									
8300	30" diameter, 4800 CFM	1 Elec	7	1.143	Ea.	910	57.50		967.50	1,075
8310	36" diameter, 7000 CFM		6	1.333		990	67		1,057	1,200
8320	42" diameter, 10,000 CFM		5	1.600		1,100	80.50		1,180.50	1,350
8330	48" diameter, 16,000 CFM	↓	4	2		1,375	101		1,476	1,675
8340	For two speed, add				↓	83			83	91.50
8350	Whole house, lay-down type, with shutter, one speed									
8360	30" diameter, 4500 CFM	1 Elec	8	1	Ea.	970	50.50		1,020.50	1,150
8370	36" diameter, 6500 CFM		7	1.143		1,050	57.50		1,107.50	1,225
8380	42" diameter, 9000 CFM		6	1.333		1,150	67		1,217	1,350
8390	48" diameter, 12,000 CFM	↓	5	1.600		1,300	80.50		1,380.50	1,550
8440	For two speed, add					62.50			62.50	68.50
8450	For 12 hour timer switch, add	1 Elec	32	.250	↓	62.50	12.60		75.10	87

23 34 33 – Air Curtains

23 34 33.10 Air Barrier Curtains

		Crew	Daily Output	Labor-Hours	Unit	Material	Labor	Equipment	Total	Total Incl O&P
0010	**AIR BARRIER CURTAINS**, Incl. motor starters, transformers,									
0050	door switches & temperature controls									
0100	Shipping and receiving doors, unheated, minimal wind stoppage									
0150	8' high, multiples of 3' wide	2 Shee	6	2.667	L.F.	244	138		382	475
0160	5' wide		10	1.600		235	82.50		317.50	385
0210	10' high, multiples of 4' wide		8	2		194	103		297	370
0250	12' high, 3'-6" wide		7	2.286		210	118		328	410
0260	12' wide		6	2.667		225	138		363	455
0350	16' high, 3'-6" wide		7	2.286		210	118		328	410
0360	12' wide	↓	6	2.667	↓	225	138		363	455
1500	Customer entrance doors, unheated, minimal wind stoppage									
1550	10' high, multiples of 3' wide	2 Shee	6	2.667	L.F.	220	138		358	450
1560	5' wide		10	1.600		235	82.50		317.50	385
1650	Maximum wind stoppage, 12' high, multiples of 4' wide	↓	8	2	↓	350	103		453	540
1700	Heated, minimal wind stoppage, electric heat									
1750	8' high, multiples of 3' wide	2 Shee	6	2.667	L.F.	244	138		382	475
1850	10' high, multiples of 3' wide		6	2.667		220	138		358	450
1860	Multiples of 5' wide	↓	10	1.600		510	82.50		592.50	685
1950	Maximum wind stoppage, steam heat									
1960	12' high, multiples of 4' wide	2 Shee	8	2	L.F.	930	103		1,033	1,175
2000	Walk-in coolers and freezers, ambient air, minimal wind stoppage									
2050	8' high, multiples of 3' wide	2 Shee	6	2.667	L.F.	220	138		358	450
2060	Multiples of 5' wide		10	1.600		510	82.50		592.50	685
2250	Maximum wind stoppage, 12' high, multiples of 3' wide		6	2.667		220	138		358	450
2450	Conveyor openings or service windows, unheated, 5' high		5	3.200		172	165		337	440
2460	Heated, electric, 5' high, 2'-6" wide	↓	5	3.200	↓	460	165		625	755

23 35 Special Exhaust Systems

23 35 16 – Engine Exhaust Systems

23 35 16.10 Engine Exhaust Removal Systems	Crew	Daily Output	Labor-Hours	Unit	Material	2011 Bare Costs Labor	Equipment	Total	Total Incl O&P
0010 **ENGINE EXHAUST REMOVAL SYSTEMS** D3090-320									
0500 Engine exhaust, garage, in-floor system									
0510 Single tube outlet assemblies									
0520 For transite pipe ducting, self-storing tube									
0530 3" tubing adapter plate	1 Shee	16	.500	Ea.	300	26		326	370
0540 4" tubing adapter plate		16	.500		305	26		331	375
0550 5" tubing adapter plate	↓	16	.500	↓	305	26		331	375
0600 For vitrified tile ducting									
0610 3" tubing adapter plate, self-storing tube	1 Shee	16	.500	Ea.	300	26		326	370
0620 4" tubing adapter plate, self-storing tube		16	.500		300	26		326	370
0660 5" tubing adapter plate, self-storing tube	↓	16	.500	↓	300	26		326	370
0800 Two tube outlet assemblies									
0810 For transite pipe ducting, self-storing tube									
0820 3" tubing, dual exhaust adapter plate	1 Shee	16	.500	Ea.	315	26		341	390
0850 For vitrified tile ducting									
0860 3" tubing, dual exhaust, self-storing tube	1 Shee	16	.500	Ea.	315	26		341	390
0870 3" tubing, double outlet, non-storing tubes	"	16	.500	"	315	26		341	390
0900 Accessories for metal tubing, (overhead systems also)									
0910 Adapters, for metal tubing end									
0920 3" tail pipe type				Ea.	62			62	68.50
0930 4" tail pipe type					66.50			66.50	73.50
0940 5" tail pipe type					68			68	74.50
0990 5" diesel stack type					400			400	440
1000 6" diesel stack type				↓	415			415	455
1100 Bullnose (guide) required for in-floor assemblies									
1110 3" tubing size				Ea.	34.50			34.50	38
1120 4" tubing size					37			37	40.50
1130 5" tubing size				↓	39			39	43
1150 Plain rings, for tubing end									
1160 3" tubing size				Ea.	30			30	33
1170 4" tubing size				"	33.50			33.50	36.50
1200 Tubing, galvanized, flexible, (for overhead systems also)									
1210 3" ID				L.F.	14.60			14.60	16.05
1220 4" ID					17.95			17.95	19.75
1230 5" ID					21			21	23.50
1240 6" ID				↓	24.50			24.50	27
1250 Stainless steel, flexible, (for overhead system, also)									
1260 3" ID				L.F.	3.25			3.25	3.58
1270 4" ID					4.45			4.45	4.90
1280 5" ID					5.05			5.05	5.55
1290 6" ID				↓	5.90			5.90	6.45
1500 Engine exhaust, garage, overhead components, for neoprene tubing									
1510 Alternate metal tubing & accessories see above									
1550 Adapters, for neoprene tubing end									
1560 3" tail pipe, adjustable, neoprene				Ea.	60			60	66
1570 3" tail pipe, heavy wall neoprene					66.50			66.50	73.50
1580 4" tail pipe, heavy wall neoprene					128			128	140
1590 5" tail pipe, heavy wall neoprene				↓	135			135	148
1650 Connectors, tubing									
1660 3" interior, aluminum				Ea.	26			26	28.50
1670 4" interior, aluminum					33.50			33.50	36.50
1710 5" interior, neoprene					64.50			64.50	71
1750 3" spiralock, neoprene				↓	30			30	33

23 35 Special Exhaust Systems

23 35 16 – Engine Exhaust Systems

23 35 16.10 Engine Exhaust Removal Systems	Crew	Daily Output	Labor-Hours	Unit	Material	2011 Bare Costs Labor	Equipment	Total	Total Incl O&P
1760	4" spiralock, neoprene				Ea.	33.50			33.50	36.50
1780	Y for 3" ID tubing, neoprene, dual exhaust					251			251	276
1790	Y for 4" ID tubing, aluminum, dual exhaust				↓	251			251	276
1850	Elbows, aluminum, splice into tubing for strap									
1860	3" neoprene tubing size				Ea.	62			62	68.50
1870	4" neoprene tubing size					52			52	57
1900	Flange assemblies, connect tubing to overhead duct				↓	47.50			47.50	52.50
2000	Hardware and accessories									
2020	Cable, galvanized, 1/8" diameter				L.F.	.51			.51	.56
2040	Cleat, tie down cable or rope				Ea.	5.75			5.75	6.35
2060	Pulley					7.95			7.95	8.75
2080	Pulley hook, universal				↓	5.75			5.75	6.35
2100	Rope, nylon, 1/4" diameter				L.F.	.43			.43	.47
2120	Winch, 1" diameter				Ea.	123			123	135
2150	Lifting strap, mounts on neoprene									
2160	3" tubing size				Ea.	29			29	32
2170	4" tubing size					29			29	32
2180	5" tubing size					29			29	32
2190	6" tubing size				↓	29			29	32
2200	Tubing, neoprene, 11' lengths									
2210	3" ID				L.F.	12.35			12.35	13.55
2220	4" ID					20			20	22
2230	5" ID				↓	28			28	31
2500	Engine exhaust, thru-door outlet									
2510	3" tube size	1 Carp	16	.500	Ea.	68	21.50		89.50	108
2530	4" tube size	"	16	.500	"	68	21.50		89.50	108
3000	Tubing, exhaust, flex hose, with									
3010	coupler, damper and tail pipe adapter									
3020	Neoprene									
3040	3" x 20'	1 Shee	6	1.333	Ea.	375	69		444	515
3050	4" x 15'		5.40	1.481		495	76.50		571.50	655
3060	4" x 20'		5	1.600		595	82.50		677.50	780
3070	5" x 15'	↓	4.40	1.818	↓	660	94		754	865
3100	Galvanized									
3110	3" x 20'	1 Shee	6	1.333	Ea.	415	69		484	560
3120	4" x 17'		5.60	1.429		435	74		509	590
3130	4" x 20'		5	1.600		490	82.50		572.50	665
3140	5" x 17'	↓	4.60	1.739	↓	530	90		620	715
8000	Blower, for tailpipe exhaust system									
8010	Direct drive									
8012	495 CFM, 1/3 HP	Q-9	5	3.200	Ea.	1,800	149		1,949	2,200
8014	1445 CFM, 3/4 HP		4	4		2,225	186		2,411	2,725
8016	1840 CFM, 1-1/2 HP	↓	3	5.333	↓	2,650	248		2,898	3,300
8030	Beltdrive									
8032	1400 CFM, 1 HP	Q-9	4	4	Ea.	2,225	186		2,411	2,725
8034	2023 CFM, 1-1/2 HP		3	5.333		3,025	248		3,273	3,700
8036	2750 CFM, 2 HP		2	8		3,600	370		3,970	4,525
8038	4400 CFM, 3 HP		1.80	8.889		5,300	415		5,715	6,450
8040	7060 CFM, 5 HP	↓	1.40	11.429	↓	7,950	530		8,480	9,550

23 35 Special Exhaust Systems

23 35 43 – Welding Fume Elimination Systems

23 35 43.10 Welding Fume Elimination System Components	Crew	Daily Output	Labor-Hours	Unit	Material	2011 Bare Costs Labor	Equipment	Total	Total Incl O&P
0010 **WELDING FUME ELIMINATION SYSTEM COMPONENTS**									
7500 Welding fume elimination accessories for garage exhaust systems									
7600 Cut off (blast gate)									
7610 3" tubing size, 3" x 6" opening	1 Shee	24	.333	Ea.	30	17.20		47.20	59
7620 4" tubing size, 4" x 8" opening		24	.333		31	17.20		48.20	60
7630 5" tubing size, 5" x 10" opening		24	.333		37	17.20		54.20	66.50
7640 6" tubing size		24	.333		39	17.20		56.20	69
7650 8" tubing size		24	.333		54	17.20		71.20	85.50
7700 Hoods, magnetic, with handle & screen									
7710 3" tubing size, 3" x 6" opening	1 Shee	24	.333	Ea.	108	17.20		125.20	145
7720 4" tubing size, 4" x 8" opening		24	.333		108	17.20		125.20	145
7730 5" tubing size, 5" x 10" opening		24	.333		108	17.20		125.20	145

23 36 Air Terminal Units

23 36 13 – Constant-Air-Volume Units

23 36 13.10 Constant Volume Mixing Boxes

	Crew	Daily Output	Labor-Hours	Unit	Material	2011 Bare Costs Labor	Equipment	Total	Total Incl O&P
0010 **CONSTANT VOLUME MIXING BOXES**									
5180 Mixing box, includes electric or pneumatic motor									
5192 Recommend use with silencer, see Line 23 33 19.10 0010									
5200 Constant volume, 150 to 270 CFM	Q-9	12	1.333	Ea.	645	62		707	805
5210 270 to 600 CFM		11	1.455		665	67.50		732.50	830
5230 550 to 1000 CFM		9	1.778		665	82.50		747.50	855
5240 1000 to 1600 CFM		8	2		680	93		773	890
5250 1300 to 1900 CFM		6	2.667		695	124		819	950
5260 550 to 2640 CFM		5.60	2.857		765	133		898	1,050
5270 650 to 3120 CFM		5.20	3.077		815	143		958	1,100

23 36 16 – Variable-Air-Volume Units

23 36 16.10 Variable Volume Mixing Boxes

	Crew	Daily Output	Labor-Hours	Unit	Material	2011 Bare Costs Labor	Equipment	Total	Total Incl O&P
0010 **VARIABLE VOLUME MIXING BOXES**									
5180 Mixing box, includes electric or pneumatic motor									
5192 Recommend use with attenuator, see Line 23 33 19.10 0010									
5500 VAV Cool only, pneumatic, pressure independent 300 to 600 CFM	Q-9	11	1.455	Ea.	370	67.50		437.50	510
5510 500 to 1000 CFM		9	1.778		380	82.50		462.50	545
5520 800 to 1600 CFM		9	1.778		395	82.50		477.50	560
5530 1100 to 2000 CFM		8	2		400	93		493	580
5540 1500 to 3000 CFM		7	2.286		430	106		536	630
5550 2000 to 4000 CFM		6	2.667		435	124		559	660
5560 For electric, w/thermostat, pressure dependent, add					23			23	25.50
5600 VAV Cool & HW coils, damper, actuator and t'stat									
5610 200 CFM	Q-9	11	1.455	Ea.	725	67.50		792.50	900
5620 400 CFM		10	1.600		735	74.50		809.50	915
5630 600 CFM		10	1.600		735	74.50		809.50	915
5640 800 CFM		8	2		755	93		848	970
5650 1000 CFM		8	2		755	93		848	970
5660 1250 CFM		6	2.667		825	124		949	1,100
5670 1500 CFM		6	2.667		825	124		949	1,100
5680 2000 CFM		4	4		900	186		1,086	1,275
5684 3000 CFM		3.80	4.211		980	196		1,176	1,375
5700 VAV Cool only, fan powered, damper, actuator, thermostat									
5710 200 CFM	Q-9	10	1.600	Ea.	995	74.50		1,069.50	1,200

23 36 Air Terminal Units

23 36 16 – Variable-Air-Volume Units

23 36 16.10 Variable Volume Mixing Boxes		Crew	Daily Output	Labor-Hours	Unit	Material	2011 Bare Costs Labor	Equipment	Total	Total Incl O&P
5720	400 CFM	Q-9	9	1.778	Ea.	1,025	82.50		1,107.50	1,250
5730	600 CFM		9	1.778		1,025	82.50		1,107.50	1,250
5740	800 CFM		7	2.286		1,100	106		1,206	1,375
5750	1000 CFM		7	2.286		1,100	106		1,206	1,375
5760	1250 CFM		5	3.200		1,200	149		1,349	1,525
5770	1500 CFM		5	3.200		1,200	149		1,349	1,525
5780	2000 CFM	▼	4	4	▼	1,325	186		1,511	1,725
5800	VAV Fan powr'd, cooling, with HW coils, dampers, actuators, t'stat									
5810	200 CFM	Q-9	10	1.600	Ea.	1,200	74.50		1,274.50	1,425
5820	400 CFM		9	1.778		1,225	82.50		1,307.50	1,475
5830	600 CFM		9	1.778		1,225	82.50		1,307.50	1,475
5840	800 CFM		7	2.286		1,300	106		1,406	1,600
5850	1000 CFM		7	2.286		1,300	106		1,406	1,600
5860	1250 CFM		5	3.200		1,500	149		1,649	1,875
5870	1500 CFM		5	3.200		1,500	149		1,649	1,875
5880	2000 CFM	▼	3.50	4.571	▼	1,600	212		1,812	2,075

23 37 Air Outlets and Inlets

23 37 13 – Diffusers, Registers, and Grilles

23 37 13.10 Diffusers

			Crew	Daily Output	Labor-Hours	Unit	Material	2011 Bare Costs Labor	Equipment	Total	Total Incl O&P
0010	**DIFFUSERS**, Aluminum, opposed blade damper unless noted	R233100-10									
0100	Ceiling, linear, also for sidewall										
0120	2" wide	R233100-30	1 Shee	32	.250	L.F.	44.50	12.90		57.40	68.50
0140	3" wide			30	.267		51.50	13.75		65.25	77.50
0160	4" wide	R233700-60		26	.308		60	15.90		75.90	90
0180	6" wide			24	.333		74	17.20		91.20	108
0200	8" wide			22	.364		86.50	18.80		105.30	124
0220	10" wide			20	.400		99	20.50		119.50	140
0240	12" wide		▼	18	.444	▼	117	23		140	163
0260	For floor or sill application, add						15%				
0500	Perforated, 24" x 24" lay-in panel size, 6" x 6"		1 Shee	16	.500	Ea.	57.50	26		83.50	103
0520	8" x 8"			15	.533		59	27.50		86.50	107
0530	9" x 9"			14	.571		63	29.50		92.50	114
0540	10" x 10"			14	.571		65	29.50		94.50	116
0560	12" x 12"			12	.667		66	34.50		100.50	125
0580	15" x 15"			11	.727		76	37.50		113.50	141
0590	16" x 16"			11	.727		109	37.50		146.50	177
0600	18" x 18"			10	.800		115	41.50		156.50	190
0610	20" x 20"			10	.800		133	41.50		174.50	209
0620	24" x 24"			9	.889		165	46		211	252
1000	Rectangular, 1 to 4 way blow, 6" x 6"			16	.500		46.50	26		72.50	90
1010	8" x 8"			15	.533		55.50	27.50		83	103
1014	9" x 9"			15	.533		61	27.50		88.50	109
1016	10" x 10"			15	.533		75.50	27.50		103	125
1020	12" x 6"			15	.533		67.50	27.50		95	116
1040	12" x 9"			14	.571		73.50	29.50		103	126
1060	12" x 12"			12	.667		78.50	34.50		113	139
1070	14" x 6"			13	.615		73	32		105	129
1074	14" x 14"			12	.667		121	34.50		155.50	185
1080	18" x 12"			11	.727		130	37.50		167.50	200
1120	15" x 15"		▼	10	.800		106	41.50		147.50	180

23 37 13 – Diffusers, Registers, and Grilles

23 37 13.10 Diffusers		Crew	Daily Output	Labor-Hours	Unit	Material	2011 Bare Costs Labor	Equipment	Total	Total Incl O&P
1140	18" x 15"	1 Shee	9	.889	Ea.	152	46		198	238
1150	18" x 18"		9	.889		132	46		178	215
1160	21" x 21"		8	1		210	51.50		261.50	310
1170	24" x 12"		10	.800		160	41.50		201.50	239
1180	24" x 24"		7	1.143		263	59		322	380
1500	Round, butterfly damper, steel, diffuser size, 6" diameter		18	.444		24	23		47	61
1520	8" diameter		16	.500		25.50	26		51.50	67
1540	10" diameter		14	.571		31.50	29.50		61	79.50
1560	12" diameter		12	.667		41.50	34.50		76	97.50
1580	14" diameter		10	.800		51.50	41.50		93	119
1600	18" diameter		9	.889		132	46		178	215
1610	20" diameter		9	.889		141	46		187	225
1620	22" diameter		8	1		154	51.50		205.50	247
1630	24" diameter		8	1		171	51.50		222.50	266
1640	28" diameter		7	1.143		210	59		269	320
1650	32" diameter		6	1.333		235	69		304	365
1700	Round, steel, adjustable core, annular segmented damper, diffuser size									
1704	6" diameter	1 Shee	18	.444	Ea.	80	23		103	123
1708	8" diameter		16	.500		86	26		112	134
1712	10" diameter		14	.571		96.50	29.50		126	151
1716	12" diameter		12	.667		107	34.50		141.50	170
1720	14" diameter		10	.800		119	41.50		160.50	194
1724	18" diameter		9	.889		182	46		228	271
1728	20" diameter		9	.889		211	46		257	300
1732	24" diameter		8	1		330	51.50		381.50	440
1736	30" diameter		7	1.143		570	59		629	720
1740	36" diameter		6	1.333		700	69		769	870
2000	T bar mounting, 24" x 24" lay-in frame, 6" x 6"		16	.500		44	26		70	87.50
2020	8" x 8"		14	.571		46	29.50		75.50	95
2040	12" x 12"		12	.667		54.50	34.50		89	112
2060	16" x 16"		11	.727		73	37.50		110.50	138
2080	18" x 18"		10	.800		81	41.50		122.50	152
2500	Combination supply and return									
2520	21" x 21" supply, 15" x 15" return	Q-9	10	1.600	Ea.	325	74.50		399.50	465
2540	24" x 24" supply, 18" x 18" return		9.50	1.684		405	78.50		483.50	570
2560	27" x 27" supply, 18" x 18" return		9	1.778		500	82.50		582.50	670
2580	30" x 30" supply, 21" x 21" return		8.50	1.882		630	87.50		717.50	825
2600	33" x 33" supply, 24" x 24" return		8	2		680	93		773	885
2620	36" x 36" supply, 24" x 24" return		7.50	2.133		865	99		964	1,100
3000	Baseboard, white enameled steel									
3100	18" long	1 Shee	20	.400	Ea.	13.20	20.50		33.70	45.50
3120	24" long		18	.444		25.50	23		48.50	63
3140	48" long		16	.500		47	26		73	91
3400	For matching return, deduct					10%				
4000	Floor, steel, adjustable pattern									
4100	2" x 10"	1 Shee	34	.235	Ea.	17.65	12.15		29.80	38
4120	2" x 12"		32	.250		19.40	12.90		32.30	41
4140	2" x 14"		30	.267		22	13.75		35.75	45
4200	4" x 10"		28	.286		20.50	14.75		35.25	45
4220	4" x 12"		26	.308		22.50	15.90		38.40	49
4240	4" x 14"		25	.320		25.50	16.55		42.05	53
4260	6" x 10"		26	.308		31	15.90		46.90	58
4280	6" x 12"		24	.333		32	17.20		49.20	61.50

23 37 13.10 Diffusers

		Crew	Daily Output	Labor-Hours	Unit	Material	2011 Bare Costs Labor	Equipment	Total	Total Incl O&P
4300	6" x 14"	1 Shee	22	.364	Ea.	33.50	18.80		52.30	65.50
4500	Linear bar diffuser in wall, sill or ceiling									
4510	2" wide	1 Shee	32	.250	L.F.	36	12.90		48.90	59.50
4512	3" wide		30	.267		39	13.75		52.75	64
4514	4" wide		26	.308		48.50	15.90		64.40	77
4516	6" wide		24	.333		60	17.20		77.20	91.50
4520	10" wide		20	.400		79.50	20.50		100	119
4522	12" wide		18	.444		89.50	23		112.50	133
5000	Sidewall, aluminum, 3 way dispersion									
5100	8" x 4"	1 Shee	24	.333	Ea.	9.70	17.20		26.90	36.50
5110	8" x 6"		24	.333		10.35	17.20		27.55	37.50
5120	10" x 4"		22	.364		10.15	18.80		28.95	39.50
5130	10" x 6"		22	.364		10.90	18.80		29.70	40.50
5140	10" x 8"		20	.400		14.85	20.50		35.35	47.50
5160	12" x 4"		18	.444		10.95	23		33.95	46.50
5170	12" x 6"		17	.471		12.10	24.50		36.60	50.50
5180	12" x 8"		16	.500		15.70	26		41.70	56.50
5200	14" x 4"		15	.533		12.20	27.50		39.70	55
5220	14" x 6"		14	.571		13.45	29.50		42.95	59.50
5240	14" x 8"		13	.615		17.55	32		49.55	67.50
5260	16" x 6"		12.50	.640		19.25	33		52.25	71
6000	For steel diffusers instead of aluminum, deduct					10%				

23 37 13.30 Grilles

		Crew	Daily Output	Labor-Hours	Unit	Material	2011 Bare Costs Labor	Equipment	Total	Total Incl O&P
0010	**GRILLES**									
0020	Aluminum									
0100	Air supply, single deflection, adjustable									
0120	8" x 4"	1 Shee	30	.267	Ea.	19.60	13.75		33.35	42.50
0140	8" x 8"		28	.286		24	14.75		38.75	48.50
0160	10" x 4"		24	.333		21.50	17.20		38.70	50
0180	10" x 10"		23	.348		28.50	17.95		46.45	58.50
0200	12" x 6"		23	.348		25.50	17.95		43.45	55
0220	12" x 12"		22	.364		33.50	18.80		52.30	65.50
0230	14" x 6"		23	.348		28.50	17.95		46.45	58.50
0240	14" x 8"		23	.348		30.50	17.95		48.45	60.50
0260	14" x 14"		22	.364		41	18.80		59.80	74
0270	18" x 8"		23	.348		35	17.95		52.95	65.50
0280	18" x 10"		23	.348		40.50	17.95		58.45	71.50
0300	18" x 18"		21	.381		63	19.70		82.70	99
0320	20" x 12"		22	.364		46	18.80		64.80	79
0340	20" x 20"		21	.381		76	19.70		95.70	114
0350	24" x 8"		20	.400		44	20.50		64.50	79.50
0360	24" x 14"		18	.444		62	23		85	103
0380	24" x 24"		15	.533		107	27.50		134.50	160
0400	30" x 8"		20	.400		56.50	20.50		77	93.50
0420	30" x 10"		19	.421		62.50	22		84.50	102
0440	30" x 12"		18	.444		68.50	23		91.50	110
0460	30" x 16"		17	.471		89	24.50		113.50	135
0480	30" x 18"		17	.471		99.50	24.50		124	147
0500	30" x 30"		14	.571		178	29.50		207.50	240
0520	36" x 12"		17	.471		85	24.50		109.50	131
0540	36" x 16"		16	.500		109	26		135	158
0560	36" x 18"		15	.533		120	27.50		147.50	174

23 37 13 – Diffusers, Registers, and Grilles

23 37 13.30 Grilles		Crew	Daily Output	Labor-Hours	Unit	Material	2011 Bare Costs Labor	Equipment	Total	Total Incl O&P
0570	36" x 20"	1 Shee	15	.533	Ea.	132	27.50		159.50	187
0580	36" x 24"		14	.571		168	29.50		197.50	229
0600	36" x 28"		13	.615		190	32		222	257
0620	36" x 30"		12	.667		230	34.50		264.50	305
0640	36" x 32"		12	.667		248	34.50		282.50	325
0660	36" x 34"		11	.727		275	37.50		312.50	360
0680	36" x 36"		11	.727		299	37.50		336.50	385
0700	For double deflecting, add					70%				
1000	Air return, 6" x 6"	1 Shee	26	.308		17.55	15.90		33.45	43.50
1020	10" x 6"		24	.333		17.55	17.20		34.75	45.50
1040	14" x 6"		23	.348		20.50	17.95		38.45	49.50
1060	10" x 8"		23	.348		20.50	17.95		38.45	49.50
1080	16" x 8"		22	.364		22	18.80		40.80	53
1100	12" x 12"		22	.364		24	18.80		42.80	54.50
1120	24" x 12"		18	.444		33	23		56	70.50
1140	30" x 12"		16	.500		42.50	26		68.50	85.50
1160	14" x 14"		22	.364		28.50	18.80		47.30	59.50
1180	16" x 16"		22	.364		31.50	18.80		50.30	63.50
1200	18" x 18"		21	.381		36.50	19.70		56.20	70
1220	24" x 18"		16	.500		41.50	26		67.50	84.50
1240	36" x 18"		15	.533		63.50	27.50		91	112
1260	24" x 24"		15	.533		48.50	27.50		76	95
1280	36" x 24"		14	.571		71	29.50		100.50	123
1300	48" x 24"		12	.667		122	34.50		156.50	186
1320	48" x 30"		11	.727		157	37.50		194.50	230
1340	36" x 36"		13	.615		94.50	32		126.50	152
1360	48" x 36"		11	.727		195	37.50		232.50	272
1380	48" x 48"		8	1		241	51.50		292.50	345
2000	Door grilles, 12" x 12"		22	.364		54	18.80		72.80	87.50
2020	18" x 12"		22	.364		61	18.80		79.80	96
2040	24" x 12"		18	.444		66.50	23		89.50	108
2060	18" x 18"		18	.444		82	23		105	125
2080	24" x 18"		16	.500		93.50	26		119.50	142
2100	24" x 24"		15	.533		129	27.50		156.50	184
3000	Filter grille with filter, 12" x 12"		24	.333		47	17.20		64.20	77.50
3020	18" x 12"		20	.400		65.50	20.50		86	104
3040	24" x 18"		18	.444		77	23		100	119
3060	24" x 24"		16	.500		89	26		115	137
3080	30" x 24"		14	.571		111	29.50		140.50	167
3100	30" x 30"		13	.615		131	32		163	192
3950	Eggcrate, framed, 6" x 6" opening		26	.308		19.25	15.90		35.15	45
3954	8" x 8" opening		24	.333		20.50	17.20		37.70	48.50
3960	10" x 10" opening		23	.348		22	17.95		39.95	51.50
3970	12" x 12" opening		22	.364		26.50	18.80		45.30	58
3980	14" x 14" opening		22	.364		28	18.80		46.80	59
3984	16" x 16" opening		21	.381		41	19.70		60.70	75
3990	18" x 18" opening		21	.381		47	19.70		66.70	81.50
4020	22" x 22" opening		17	.471		61	24.50		85.50	105
4040	24" x 24" opening		15	.533		77	27.50		104.50	126
4044	28" x 28" opening		14	.571		119	29.50		148.50	176
4048	36" x 36" opening		12	.667		136	34.50		170.50	202
4050	48" x 24" opening		12	.667		173	34.50		207.50	242
4060	Eggcrate, lay-in, T-bar system									

23 37 13 – Diffusers, Registers, and Grilles

23 37 13.30 Grilles

		Crew	Daily Output	Labor-Hours	Unit	Material	2011 Bare Costs Labor	Equipment	Total	Total Incl O&P
4070	48" x 24" sheet	1 Shee	40	.200	Ea.	93	10.35		103.35	118
5000	Transfer grille, vision proof, 8" x 4"		30	.267		36	13.75		49.75	60.50
5020	8" x 6"		28	.286		36	14.75		50.75	62
5040	8" x 8"		26	.308		40.50	15.90		56.40	68.50
5060	10" x 6"		24	.333		37.50	17.20		54.70	67.50
5080	10" x 10"		23	.348		47.50	17.95		65.45	79.50
5090	12" x 6"		23	.348		39	17.95		56.95	70
5100	12" x 10"		22	.364		52	18.80		70.80	85.50
5110	12" x 12"		22	.364		56.50	18.80		75.30	90.50
5120	14" x 10"		22	.364		56.50	18.80		75.30	90.50
5140	16" x 8"		21	.381		56.50	19.70		76.20	92
5160	18" x 12"		20	.400		73.50	20.50		94	112
5170	18" x 18"		21	.381		104	19.70		123.70	144
5180	20" x 12"		20	.400		79	20.50		99.50	118
5200	20" x 20"		19	.421		125	22		147	171
5220	24" x 12"		17	.471		89.50	24.50		114	136
5240	24" x 24"		16	.500		171	26		197	228
5260	30" x 6"		15	.533		79	27.50		106.50	129
5280	30" x 8"		15	.533		92.50	27.50		120	143
5300	30" x 12"		14	.571		115	29.50		144.50	172
5320	30" x 16"		13	.615		137	32		169	199
5340	30" x 20"		12	.667		173	34.50		207.50	242
5360	30" x 24"		11	.727		213	37.50		250.50	291
5380	30" x 30"		10	.800		280	41.50		321.50	375
6000	For steel grilles instead of aluminum in above, deduct					10%				
6200	Plastic, eggcrate, lay-in, T-bar system									
6210	48" x 24" sheet	1 Shee	50	.160	Ea.	23	8.25		31.25	38
6250	Steel door louver									
6270	With fire link, steel only									
6350	12" x 18"	1 Shee	22	.364	Ea.	194	18.80		212.80	243
6410	12" x 24"		21	.381		223	19.70		242.70	275
6470	18" x 18"		20	.400		217	20.50		237.50	270
6530	18" x 24"		19	.421		250	22		272	310
6600	18" x 36"		17	.471		290	24.50		314.50	355
6660	24" x 24"		15	.533		252	27.50		279.50	320
6720	24" x 30"		15	.533		290	27.50		317.50	360
6780	30" x 30"		13	.615		340	32		372	425
6840	30" x 36"		12	.667		395	34.50		429.50	480
6900	36" x 36"		10	.800		475	41.50		516.50	585

23 37 13.60 Registers

		Crew	Daily Output	Labor-Hours	Unit	Material	2011 Bare Costs Labor	Equipment	Total	Total Incl O&P
0010	**REGISTERS**									
0980	Air supply									
1000	Ceiling/wall, O.B. damper, anodized aluminum									
1010	One or two way deflection, adj. curved face bars									
1014	6" x 6"	1 Shee	24	.333	Ea.	34	17.20		51.20	63
1020	8" x 4"		26	.308		34	15.90		49.90	61
1040	8" x 8"		24	.333		42	17.20		59.20	72.50
1060	10" x 6"		20	.400		39	20.50		59.50	73.50
1080	10" x 10"		19	.421		52.50	22		74.50	90.50
1100	12" x 6"		19	.421		39	22		61	76
1120	12" x 12"		18	.444		62	23		85	103
1140	14" x 8"		17	.471		54	24.50		78.50	96

23 37 13.60 Registers		Crew	Daily Output	Labor-Hours	Unit	Material	2011 Bare Costs Labor	Equipment	Total	Total Incl O&P
1160	14" x 14"	1 Shee	18	.444	Ea.	72	23		95	114
1170	16" x 16"		17	.471		80	24.50		104.50	125
1180	18" x 8"		18	.444		60	23		83	101
1200	18" x 18"		17	.471		110	24.50		134.50	158
1220	20" x 4"		19	.421		55	22		77	93.50
1240	20" x 6"		18	.444		55	23		78	95
1260	20" x 8"		18	.444		60.50	23		83.50	101
1280	20" x 20"		17	.471		133	24.50		157.50	184
1290	22" x 22"		15	.533		153	27.50		180.50	211
1300	24" x 4"		17	.471		68	24.50		92.50	112
1320	24" x 6"		16	.500		68	26		94	114
1340	24" x 8"		13	.615		74.50	32		106.50	130
1350	24" x 18"		12	.667		139	34.50		173.50	204
1360	24" x 24"		11	.727		188	37.50		225.50	264
1380	30" x 4"		16	.500		85.50	26		111.50	133
1400	30" x 6"		15	.533		85.50	27.50		113	136
1420	30" x 8"		14	.571		97.50	29.50		127	152
1440	30" x 24"		12	.667		243	34.50		277.50	320
1460	30" x 30"	▼	10	.800	▼	310	41.50		351.50	405
1504	4 way deflection, adjustable curved face bars									
1510	6" x 6"	1 Shee	26	.308	Ea.	40.50	15.90		56.40	68.50
1514	8" x 8"		24	.333		50.50	17.20		67.70	81.50
1518	10" x 10"		19	.421		63	22		85	102
1522	12" x 6"		19	.421		47	22		69	84.50
1526	12" x 12"		18	.444		74	23		97	116
1530	14" x 14"		18	.444		86.50	23		109.50	130
1534	16" x 16"		17	.471		96	24.50		120.50	143
1538	18" x 18"		17	.471		132	24.50		156.50	182
1542	22" x 22"	▼	16	.500	▼	184	26		210	241
1980	One way deflection, adj. vert. or horiz. face bars									
1990	6" x 6"	1 Shee	26	.308	Ea.	35.50	15.90		51.40	63
2000	8" x 4"		26	.308		34.50	15.90		50.40	62
2020	8" x 8"		24	.333		41	17.20		58.20	71.50
2040	10" x 6"		20	.400		41.50	20.50		62	76.50
2060	10" x 10"		19	.421		48.50	22		70.50	86.50
2080	12" x 6"		19	.421		45	22		67	82.50
2100	12" x 12"		18	.444		58.50	23		81.50	98.50
2120	14" x 8"		17	.471		53.50	24.50		78	96
2140	14" x 12"		18	.444		64	23		87	105
2160	14" x 14"		18	.444		72.50	23		95.50	114
2180	16" x 6"		18	.444		53.50	23		76.50	93.50
2200	16" x 12"		18	.444		71	23		94	113
2220	16" x 16"		17	.471		89.50	24.50		114	136
2240	18" x 8"		18	.444		62.50	23		85.50	103
2260	18" x 12"		17	.471		76	24.50		100.50	121
2280	18" x 18"		16	.500		110	26		136	160
2300	20" x 10"		18	.444		74	23		97	116
2320	20" x 16"		16	.500		107	26		133	156
2340	20" x 20"		16	.500		135	26		161	187
2360	24" x 12"		15	.533		93	27.50		120.50	145
2380	24" x 16"		14	.571		124	29.50		153.50	182
2400	24" x 20"		12	.667		155	34.50		189.50	223
2420	24" x 24"	▼	10	.800		190	41.50		231.50	272

23 37 Air Outlets and Inlets

23 37 13 – Diffusers, Registers, and Grilles

23 37 13.60 Registers		Crew	Daily Output	Labor-Hours	Unit	Material	2011 Bare Costs Labor	Equipment	Total	Total Incl O&P
2440	30" x 12"	1 Shee	13	.615	Ea.	121	32		153	181
2460	30" x 16"		13	.615		157	32		189	221
2480	30" x 24"		11	.727		245	37.50		282.50	325
2500	30" x 30"		9	.889		315	46		361	415
2520	36" x 12"		11	.727		150	37.50		187.50	222
2540	36" x 24"		10	.800		300	41.50		341.50	395
2560	36" x 36"		8	1		505	51.50		556.50	635
2600	For 2 way deflect., adj. vert. or horiz. face bars, add					40%				
2700	Above registers in steel instead of aluminum, deduct					10%				
3000	Baseboard, hand adj. damper, enameled steel									
3012	8" x 6"	1 Shee	26	.308	Ea.	14.45	15.90		30.35	40
3020	10" x 6"		24	.333		15.70	17.20		32.90	43.50
3040	12" x 5"		23	.348		17.05	17.95		35	46
3060	12" x 6"		23	.348		17.05	17.95		35	46
3080	12" x 8"		22	.364		24.50	18.80		43.30	55.50
3100	14" x 6"		20	.400		18.50	20.50		39	51.50
4000	Floor, toe operated damper, enameled steel									
4020	4" x 8"	1 Shee	32	.250	Ea.	26.50	12.90		39.40	49
4040	4" x 12"		26	.308		31.50	15.90		47.40	58.50
4060	6" x 8"		28	.286		29	14.75		43.75	54
4080	6" x 14"		22	.364		36.50	18.80		55.30	68.50
4100	8" x 10"		22	.364		32.50	18.80		51.30	64.50
4120	8" x 16"		20	.400		54.50	20.50		75	91
4140	10" x 10"		20	.400		39	20.50		59.50	74
4160	10" x 16"		18	.444		59.50	23		82.50	99.50
4180	12" x 12"		18	.444		48	23		71	87.50
4200	12" x 24"		16	.500		166	26		192	222
4220	14" x 14"		16	.500		137	26		163	190
4240	14" x 20"		15	.533		183	27.50		210.50	243
4300	Spiral pipe supply register									
4310	Aluminum, double deflection, w/damper extractor									
4320	4" x 12", for 6" thru 10" diameter duct	1 Shee	25	.320	Ea.	60	16.55		76.55	91
4330	4" x 18", for 6" thru 10" diameter duct		18	.444		74.50	23		97.50	117
4340	6" x 12", for 8" thru 12" diameter duct		19	.421		65.50	22		87.50	105
4350	6" x 18", for 8" thru 12" diameter duct		18	.444		85	23		108	128
4360	6" x 24", for 8" thru 12" diameter duct		16	.500		105	26		131	155
4370	6" x 30", for 8" thru 12" diameter duct		15	.533		134	27.50		161.50	189
4380	8" x 18", for 10" thru 14" diameter duct		18	.444		91	23		114	135
4390	8" x 24", for 10" thru 14" diameter duct		15	.533		114	27.50		141.50	167
4400	8" x 30", for 10" thru 14" diameter duct		14	.571		151	29.50		180.50	211
4410	10" x 24", for 12" thru 18" diameter duct		13	.615		125	32		157	185
4420	10" x 30", for 12" thru 18" diameter duct		12	.667		165	34.50		199.50	233
4430	10" x 36", for 12" thru 18" diameter duct		11	.727		204	37.50		241.50	282
4980	Air return									
5000	Ceiling or wall, fixed 45° face blades									
5010	Adjustable O.B. damper, anodized aluminum									
5020	4" x 8"	1 Shee	26	.308	Ea.	24	15.90		39.90	50
5040	6" x 8"		24	.333		27	17.20		44.20	56
5060	6" x 10"		19	.421		27	22		49	63
5070	6" x 12"		19	.421		31.50	22		53.50	68
5080	6" x 16"		18	.444		35	23		58	73
5100	8" x 10"		19	.421		31.50	22		53.50	68
5120	8" x 12"		16	.500		33.50	26		59.50	76

23 37 Air Outlets and Inlets

23 37 13 – Diffusers, Registers, and Grilles

23 37 13.60 Registers		Crew	Daily Output	Labor-Hours	Unit	Material	2011 Bare Costs Labor	Equipment	Total	Total Incl O&P
5140	10" x 10"	1 Shee	18	.444	Ea.	33.50	23		56.50	71.50
5160	10" x 16"		17	.471		41.50	24.50		66	82.50
5170	12" x 12"		18	.444		36.50	23		59.50	74.50
5180	12" x 18"		18	.444		47.50	23		70.50	87
5184	12" x 24"		14	.571		54.50	29.50		84	105
5200	12" x 30"		12	.667		72.50	34.50		107	132
5210	12" x 36"		10	.800		99	41.50		140.50	172
5220	16" x 16"		17	.471		50.50	24.50		75	92.50
5240	18" x 18"		16	.500		60	26		86	105
5250	18" x 24"		13	.615		71.50	32		103.50	127
5254	18" x 30"		11	.727		110	37.50		147.50	178
5260	18" x 36"		10	.800		130	41.50		171.50	206
5270	20" x 20"		10	.800		69	41.50		110.50	139
5274	20" x 24"		12	.667		82	34.50		116.50	143
5280	24" x 24"		11	.727		92	37.50		129.50	158
5290	24" x 30"		10	.800		102	41.50		143.50	175
5300	24" x 36"		8	1		144	51.50		195.50	237
5320	24" x 48"		6	1.333		215	69		284	340
5340	30" x 30"	▼	6.50	1.231		169	63.50		232.50	282
5344	30" x 120"	Q-9	6	2.667		725	124		849	985
5360	36" x 36"	1 Shee	6	1.333		370	69		439	515
5370	40" x 32"	"	6	1.333	▼	465	69		534	620
5400	Ceiling or wall, removable-reversible core, single deflection									
5402	Adjustable O.B. damper, radiused frame with border, aluminum									
5410	8" x 4"	1 Shee	26	.308	Ea.	30.50	15.90		46.40	57.50
5414	8" x 6"		24	.333		30.50	17.20		47.70	59.50
5418	10" x 6"		19	.421		39	22		61	76
5422	10" x 10"		18	.444		44.50	23		67.50	83.50
5426	12" x 6"		19	.421		39	22		61	76
5430	12" x 12"		18	.444		50	23		73	89.50
5434	16" x 16"		16	.500		75	26		101	122
5438	18" x 18"		17	.471		94.50	24.50		119	141
5442	20" x 20"		16	.500		114	26		140	164
5446	24" x 12"		15	.533		75	27.50		102.50	124
5450	24" x 18"		13	.615		114	32		146	173
5454	24" x 24"		11	.727		158	37.50		195.50	231
5458	30" x 12"		13	.615		100	32		132	158
5462	30" x 18"		12.50	.640		144	33		177	209
5466	30" x 24"		11	.727		203	37.50		240.50	280
5470	30" x 30"		9	.889		253	46		299	350
5474	36" x 12"		11	.727		125	37.50		162.50	194
5478	36" x 18"		10.50	.762		175	39.50		214.50	252
5482	36" x 24"		10	.800		244	41.50		285.50	330
5486	36" x 30"		9	.889		330	46		376	430
5490	36" x 36"	▼	8	1	▼	420	51.50		471.50	545
6000	For steel construction instead of aluminum, deduct					10%				

23 37 15 – Louvers

23 37 15.40 HVAC Louvers

		Crew	Daily Output	Labor-Hours	Unit	Material	2011 Bare Costs Labor	Equipment	Total	Total Incl O&P
0010	**HVAC LOUVERS**									
0100	Aluminum, extruded, with screen, mill finish									
1002	Brick vent, see also Section 04 05 23.19									
1100	Standard, 4" deep, 8" wide, 5" high	1 Shee	24	.333	Ea.	34.50	17.20		51.70	63.50

23 37 15 – Louvers

23 37 15.40 HVAC Louvers

	23 37 15.40 HVAC Louvers	Crew	Daily Output	Labor-Hours	Unit	Material	2011 Bare Costs Labor	Equipment	Total	Total Incl O&P
1200	Modular, 4" deep, 7-3/4" wide, 5" high	1 Shee	24	.333	Ea.	36	17.20		53.20	65.50
1300	Speed brick, 4" deep, 11-5/8" wide, 3-7/8" high		24	.333		36	17.20		53.20	65.50
1400	Fuel oil brick, 4" deep, 8" wide, 5" high		24	.333		61.50	17.20		78.70	94
2000	Cooling tower and mechanical equip., screens, light weight		40	.200	S.F.	15.40	10.35		25.75	32.50
2020	Standard weight		35	.229		41	11.80		52.80	63
2500	Dual combination, automatic, intake or exhaust		20	.400		56	20.50		76.50	92.50
2520	Manual operation		20	.400		41.50	20.50		62	77
2540	Electric or pneumatic operation		20	.400		41.50	20.50		62	77
2560	Motor, for electric or pneumatic		14	.571	Ea.	480	29.50		509.50	575
3000	Fixed blade, continuous line									
3100	Mullion type, stormproof	1 Shee	28	.286	S.F.	41.50	14.75		56.25	68.50
3200	Stormproof		28	.286		41.50	14.75		56.25	68.50
3300	Vertical line		28	.286		49.50	14.75		64.25	76.50
3500	For damper to use with above, add					50%	30%			
3520	Motor, for damper, electric or pneumatic	1 Shee	14	.571	Ea.	480	29.50		509.50	575
4000	Operating, 45°, manual, electric or pneumatic		24	.333	S.F.	50	17.20		67.20	81
4100	Motor, for electric or pneumatic		14	.571	Ea.	480	29.50		509.50	575
4200	Penthouse, roof		56	.143	S.F.	24.50	7.40		31.90	38
4300	Walls		40	.200		58	10.35		68.35	79.50
5000	Thinline, under 4" thick, fixed blade		40	.200		24	10.35		34.35	42
5010	Finishes, applied by mfr. at additional cost, available in colors									
5020	Prime coat only, add				S.F.	3.30			3.30	3.63
5040	Baked enamel finish coating, add					6.10			6.10	6.70
5060	Anodized finish, add					6.60			6.60	7.25
5080	Duranodic finish, add					12			12	13.20
5100	Fluoropolymer finish coating, add					18.90			18.90	21
9980	For small orders (under 10 pieces), add					25%				

23 37 23 – HVAC Gravity Ventilators

23 37 23.10 HVAC Gravity Air Ventilators

	23 37 23.10 HVAC Gravity Air Ventilators	Crew	Daily Output	Labor-Hours	Unit	Material	2011 Bare Costs Labor	Equipment	Total	Total Incl O&P
0010	**HVAC GRAVITY AIR VENTILATORS**, Incl. base and damper									
1280	Rotary ventilators, wind driven, galvanized									
1300	4" neck diameter	Q-9	20	.800	Ea.	50	37		87	111
1320	5" neck diameter		18	.889		56	41.50		97.50	124
1340	6" neck diameter		16	1		56	46.50		102.50	132
1360	8" neck diameter		14	1.143		58	53		111	145
1380	10" neck diameter		12	1.333		75	62		137	176
1400	12" neck diameter		10	1.600		81	74.50		155.50	202
1420	14" neck diameter		10	1.600		113	74.50		187.50	236
1440	16" neck diameter		9	1.778		136	82.50		218.50	275
1460	18" neck diameter, 1700 CFM		9	1.778		158	82.50		240.50	299
1480	20" neck diameter, 2100 CFM		8	2		176	93		269	335
1500	24" neck diameter, 3,100 CFM		8	2		236	93		329	400
1520	30" neck diameter, 4500 CFM		7	2.286		490	106		596	700
1540	36" neck diameter, 5,500 CFM		6	2.667		695	124		819	950
1600	For aluminum, add					300%				
1620	For stainless steel, add					600%				
1630	For copper, add					600%				
2000	Stationary, gravity, syphon, galvanized									
2100	3" neck diameter, 40 CFM	Q-9	24	.667	Ea.	45.50	31		76.50	97
2120	4" neck diameter, 50 CFM		20	.800		49.50	37		86.50	111
2140	5" neck diameter, 58 CFM		18	.889		50	41.50		91.50	118
2160	6" neck diameter, 66 CFM		16	1		53	46.50		99.50	129

23 37 23 – HVAC Gravity Ventilators

23 37 23.10 HVAC Gravity Air Ventilators	Crew	Daily Output	Labor-Hours	Unit	Material	2011 Bare Costs Labor	Equipment	Total	Total Incl O&P
2180	7" neck diameter, 86 CFM	Q-9	15	1.067	Ea.	54.50	49.50		104	135
2200	8" neck diameter, 110 CFM		14	1.143		55	53		108	141
2220	10" neck diameter, 140 CFM		12	1.333		71	62		133	172
2240	12" neck diameter, 160 CFM		10	1.600		91	74.50		165.50	212
2260	14" neck diameter, 250 CFM		10	1.600		114	74.50		188.50	238
2280	16" neck diameter, 380 CFM		9	1.778		153	82.50		235.50	293
2300	18" neck diameter, 500 CFM		9	1.778		203	82.50		285.50	350
2320	20" neck diameter, 625 CFM		8	2		215	93		308	380
2340	24" neck diameter, 900 CFM		8	2		248	93		341	415
2360	30" neck diameter, 1375 CFM		7	2.286		565	106		671	780
2380	36" neck diameter, 2,000 CFM		6	2.667		775	124		899	1,050
2400	42" neck diameter, 3000 CFM		4	4		1,100	186		1,286	1,475
2410	48" neck diameter, 4000 CFM		3	5.333		1,475	248		1,723	2,000
2500	For aluminum, add					300%				
2520	For stainless steel, add					600%				
2530	For copper, add					600%				
3000	Rotating chimney cap, galvanized, 4" neck diameter	Q-9	20	.800		39	37		76	98.50
3020	5" neck diameter		18	.889		40	41.50		81.50	106
3040	6" neck diameter		16	1		40.50	46.50		87	115
3060	7" neck diameter		15	1.067		44	49.50		93.50	123
3080	8" neck diameter		14	1.143		45.50	53		98.50	131
3100	10" neck diameter		12	1.333		58.50	62		120.50	158
3600	Stationary chimney rain cap, galvanized, 3" neck diameter		24	.667		34	31		65	84.50
3620	4" neck diameter		20	.800		35	37		72	94.50
3640	6" neck diameter		16	1		36.50	46.50		83	111
3680	8" neck diameter		14	1.143		39	53		92	124
3700	10" neck diameter		12	1.333		49.50	62		111.50	148
3720	12" neck diameter		10	1.600		82.50	74.50		157	203
3740	14" neck diameter		10	1.600		88	74.50		162.50	209
3760	16" neck diameter		9	1.778		108	82.50		190.50	244
3770	18" neck diameter		9	1.778		156	82.50		238.50	296
3780	20" neck diameter		8	2		208	93		301	370
3790	24" neck diameter		8	2		243	93		336	410
4200	Stationary mushroom, aluminum, 16" orifice diameter		10	1.600		535	74.50		609.50	700
4220	26" orifice diameter		6.15	2.602		790	121		911	1,050
4230	30" orifice diameter		5.71	2.802		1,150	130		1,280	1,475
4240	38" orifice diameter		5	3.200		1,650	149		1,799	2,050
4250	42" orifice diameter		4.70	3.404		2,200	158		2,358	2,675
4260	50" orifice diameter		4.44	3.604		2,625	168		2,793	3,125
5000	Relief vent									
5500	Rectangular, aluminum, galvanized curb									
5510	intake/exhaust, 0.033" SP									
5580	500 CFM, 12" x 12"	Q-9	8.60	1.860	Ea.	625	86.50		711.50	820
5600	600 CFM, 12" x 16"		8	2		705	93		798	915
5620	750 CFM, 12" x 20"		7.20	2.222		755	103		858	990
5640	1000 CFM, 12" x 24"		6.60	2.424		795	113		908	1,050
5660	1500 CFM, 12" x 36"		5.80	2.759		1,050	128		1,178	1,350
5680	3000 CFM, 20" x 42"		4	4		1,400	186		1,586	1,800
5700	5000 CFM, 20" x 72"		3	5.333		2,000	248		2,248	2,575
5720	6000 CFM, 24" x 72"		2.30	6.957		2,175	325		2,500	2,875
5740	10,000 CFM, 48" x 60"		1.80	8.889		2,750	415		3,165	3,650
5760	12,000 CFM, 48" x 72"		1.60	10		3,250	465		3,715	4,300
5780	13,750 CFM, 60" x 66"		1.40	11.429		3,600	530		4,130	4,750

23 37 Air Outlets and Inlets

23 37 23 – HVAC Gravity Ventilators

23 37 23.10 HVAC Gravity Air Ventilators	Crew	Daily Output	Labor-Hours	Unit	Material	2011 Bare Costs Labor	Equipment	Total	Total Incl O&P
5800	15,000 CFM, 60" x 72"	Q-9	1.30	12.308	Ea.	3,850	570		4,420	5,125
5820	18,000 CFM, 72" x 72"	↓	1.20	13.333	↓	4,375	620		4,995	5,750
5880	Size is throat area, volume is at 500 fpm									
7000	Note: sizes based on exhaust. Intake, with 0.125" SP									
7100	loss, approximately twice listed capacity.									

23 38 Ventilation Hoods

23 38 13 – Commercial-Kitchen Hoods

23 38 13.10 Hood and Ventilation Equipment

		Crew	Daily Output	Labor-Hours	Unit	Material	2011 Bare Costs Labor	Equipment	Total	Total Incl O&P
0010	**HOOD AND VENTILATION EQUIPMENT**									
2970	Exhaust hood, sst, gutter on all sides, 4' x 4' x 2'	1 Carp	1.80	4.444	Ea.	4,525	191		4,716	5,275
2980	4' x 4' x 7'	"	1.60	5	"	7,100	215		7,315	8,150
7800	Vent hood, wall canopy with fire protection	L-3A	9	1.333	L.F.	385	63		448	515
7810	Without fire protection		10	1.200		268	56.50		324.50	380
7820	Island canopy with fire protection		7	1.714		405	81		486	570
7830	Without fire protection		8	1.500		292	71		363	430
7840	Back shelf with fire protection		11	1.091		385	51.50		436.50	500
7850	Without fire protection	↓	12	1	↓	268	47.50		315.50	365
7860	Range hood & CO_2 system, minimum	1 Carp	2.50	3.200	Ea.	2,825	138		2,963	3,300
7870	Maximum	"	1	8		33,800	345		34,145	37,600
7950	Hood fire protection system, minimum	Q-1	3	5.333		4,400	257		4,657	5,225
8050	Maximum	"	1	16	↓	35,300	770		36,070	40,100

23 38 13.20 Kitchen Ventilation

		Crew	Daily Output	Labor-Hours	Unit	Material	2011 Bare Costs Labor	Equipment	Total	Total Incl O&P
0010	**KITCHEN VENTILATION**, Commercial									
1010	Heat reclaim unit, air to air									
1100	Heat pipe exchanger									
1110	Combined supply/exhaust air volume									
1120	2.5 to 6.0 MCFM	Q-10	2.80	8.571	MCFM	6,775	415		7,190	8,075
1130	6 to 16 MCFM		5	4.800		4,150	231		4,381	4,925
1140	16 to 22 MCFM	↓	6	4	↓	3,250	193		3,443	3,875
2010	Packaged ventilation									
2100	Combined supply/exhaust air volume									
2110	With ambient supply									
2120	3 to 7 MCFM	Q-10	5	4.800	MCFM	1,225	231		1,456	1,700
2130	7 to 12 MCFM		7.10	3.380		815	163		978	1,150
2140	12 to 22 MCFM	↓	10.70	2.243	↓	575	108		683	800
2400	With tempered supply									
2410	3 to 7 MCFM	Q-10	5	4.800	MCFM	2,125	231		2,356	2,675
2420	7 to 12 MCFM		7.10	3.380		1,375	163		1,538	1,750
2430	12 to 22 MCFM	↓	10.70	2.243	↓	1,000	108		1,108	1,275
6000	Exhaust hoods									
6100	Centrifugal grease extraction									
6110	Water wash type									
6120	Per foot of length, minimum	Q-10	7.10	3.380	L.F.	1,625	163		1,788	2,025
6130	Maximum	"	7.10	3.380	"	2,775	163		2,938	3,300
6200	Non water wash type									
6210	Per foot of length	Q-10	13.40	1.791	L.F.	1,200	86.50		1,286.50	1,425
6400	Island style ventilator									
6410	Water wash type									
6420	Per foot of length, minimum	Q-10	7.10	3.380	L.F.	2,150	163		2,313	2,625
6430	Maximum	"	7.10	3.380	"	3,600	163		3,763	4,225

23 38 Ventilation Hoods

23 38 13 – Commercial-Kitchen Hoods

23 38 13.20 Kitchen Ventilation	Crew	Daily Output	Labor-Hours	Unit	Material	Labor	Equipment	Total	Total Incl O&P
6460	Non water wash type									
6470	Per foot of length, minimum	Q-10	13.40	1.791	L.F.	935	86.50		1,021.50	1,150
6480	Maximum	"	13.40	1.791	"	1,900	86.50		1,986.50	2,225

23 41 Particulate Air Filtration

23 41 13 – Panel Air Filters

23 41 13.10 Panel Type Air Filters

0010	**PANEL TYPE AIR FILTERS**									
2950	Mechanical media filtration units									
3000	High efficiency type, with frame, non-supported	G				MCFM	45			45	49.50
3100	Supported type	G				"	60			60	66
5500	Throwaway glass or paper media type					Ea.	3.35			3.35	3.69

23 41 16 – Renewable-Media Air Filters

23 41 16.10 Disposable Media Air Filters

		Crew	Daily Output	Labor-Hours	Unit	Material	Labor	Equipment	Total	Total Incl O&P
0010	**DISPOSABLE MEDIA AIR FILTERS**									
5000	Renewable disposable roll				MCFM	19.35			19.35	21.50
5800	Filter, bag type									
5810	90-95% efficiency									
5820	24" x 12" x 29", .75-1.25 MCFM	1 Shee	1.90	4.211	Ea.	51	217		268	385
5830	24" x 24" x 29", 1.5-2.5 MCFM	"	1.90	4.211	"	103	217		320	445
5850	80-85% efficiency									
5860	24" x 12" x 29", .75-1.25 MCFM	1 Shee	1.90	4.211	Ea.	50.50	217		267.50	385
5870	24" x 24" x 29", 1.5-2.5 MCFM	"	1.90	4.211	"	83	217		300	420

23 41 19 – Washable Air Filters

23 41 19.10 Permanent Air Filters

0010	**PERMANENT AIR FILTERS**									
4500	Permanent washable	G			MCFM	20			20	22

23 41 23 – Extended Surface Filters

23 41 23.10 Expanded Surface Filters

0010	**EXPANDED SURFACE FILTERS**									
4000	Medium efficiency, extended surface	G			MCFM	5.50			5.50	6.05

23 41 33 – High-Efficiency Particulate Filtration

23 41 33.10 HEPA Filters

		Crew	Daily Output	Labor-Hours	Unit	Material	Labor	Equipment	Total	Total Incl O&P
0010	**HEPA FILTERS**									
6000	HEPA filter complete w/particle board,									
6010	kraft paper frame, separator material									
6020	95% DOP efficiency									
6030	12" x 12" x 6", 150 CFM	1 Shee	3.70	2.162	Ea.	44	112		156	218
6034	24" x 12" x 6", 375 CFM		1.90	4.211		73.50	217		290.50	410
6038	24" x 18" x 6", 450 CFM		1.90	4.211		90	217		307	430
6042	24" x 24" x 6", 700 CFM		1.90	4.211		101	217		318	440
6046	12" x 12" x 12", 250 CFM		3.70	2.162		74.50	112		186.50	251
6050	24" x 12" x 12", 500 CFM		1.90	4.211		105	217		322	445
6054	24" x 18" x 12", 875 CFM		1.90	4.211		146	217		363	490
6058	24" x 24" x 12", 1000 CFM		1.90	4.211		158	217		375	505
6100	99% DOP efficiency									
6110	12" x 12" x 6", 150 CFM	1 Shee	3.70	2.162	Ea.	46	112		158	220
6114	24" x 12" x 6", 325 CFM		1.90	4.211		76	217		293	415
6118	24" x 18" x 6", 550 CFM		1.90	4.211		93	217		310	430

23 41 Particulate Air Filtration

23 41 33 – High-Efficiency Particulate Filtration

23 41 33.10 HEPA Filters

		Crew	Daily Output	Labor-Hours	Unit	Material	2011 Bare Costs Labor	Equipment	Total	Total Incl O&P
6122	24" x 24" x 6", 775 CFM	1 Shee	1.90	4.211	Ea.	105	217		322	445
6124	24" x 48" x 6", 775 CFM		1.30	6.154		201	320		521	700
6126	12" x 12" x 12", 250 CFM		3.60	2.222		77	115		192	259
6130	24" x 12" x 12", 500 CFM		1.90	4.211		109	217		326	450
6134	24" x 18" x 12", 775 CFM		1.90	4.211		152	217		369	495
6138	24" x 24" x 12", 1100 CFM	▼	1.90	4.211	▼	164	217		381	510
6500	HEPA filter housing, 14 ga. galv. sheet metal									
6510	12" x 12" x 6"	1 Shee	2.50	3.200	Ea.	660	165		825	975
6514	24" x 12" x 6"		2	4		740	207		947	1,125
6518	12" x 12" x 12"		2.40	3.333		660	172		832	985
6522	14" x 12" x 12"		2.30	3.478		660	180		840	995
6526	24" x 18" x 6"		1.90	4.211		940	217		1,157	1,350
6530	24" x 24" x 6"		1.80	4.444		855	230		1,085	1,275
6534	24" x 48" x 6"		1.70	4.706		1,150	243		1,393	1,625
6538	24" x 72" x 6"		1.60	5		1,475	258		1,733	2,025
6542	24" x 18" x 12"		1.80	4.444		940	230		1,170	1,375
6546	24" x 24" x 12"		1.70	4.706		1,050	243		1,293	1,525
6550	24" x 48" x 12"		1.60	5		1,225	258		1,483	1,750
6554	24" x 72" x 12"	▼	1.50	5.333		1,525	275		1,800	2,125
6558	48" x 48" x 6"	Q-9	2.80	5.714		1,700	266		1,966	2,250
6562	48" x 72" x 6"		2.60	6.154		2,025	286		2,311	2,650
6566	48" x 96" x 6"		2.40	6.667		2,575	310		2,885	3,300
6570	48" x 48" x 12"		2.70	5.926		1,700	275		1,975	2,275
6574	48" x 72" x 12"		2.50	6.400		2,025	297		2,322	2,675
6578	48" x 96" x 12"		2.30	6.957		2,575	325		2,900	3,325
6582	114" x 72" x 12"	▼	2	8	▼	3,550	370		3,920	4,450

23 42 Gas-Phase Air Filtration

23 42 13 – Activated-Carbon Air Filtration

23 42 13.10 Charcoal Type Air Filtration

		Crew	Daily Output	Labor-Hours	Unit	Material	Labor	Equipment	Total	Total Incl O&P
0010	**CHARCOAL TYPE AIR FILTRATION**									
0050	Activated charcoal type, full flow				MCFM	600			600	660
0060	Full flow, impregnated media 12" deep					225			225	248
0070	HEPA filter & frame for field erection					300			300	330
0080	HEPA filter-diffuser, ceiling install.				▼	300			300	330

23 42 16 – Chemically-Impregnated Adsorption Air Filtration

23 42 16.10 Chemical Adsorption Air Filtration

		Crew	Daily Output	Labor-Hours	Unit	Material	Labor	Equipment	Total	Total Incl O&P
0010	**CHEMICAL ADSORPTION AIR FILTRATION**									
0500	Chemical media filtration type									
1100	Industrial air fume & odor scrubber unit w/pump & motor									
1110	corrosion resistant PVC construction									
1120	Single pack filter, horizontal type									
1130	500 CFM	Q-9	14	1.143	Ea.	6,225	53		6,278	6,925
1140	1000 CFM		11	1.455		7,300	67.50		7,367.50	8,125
1150	2000 CFM		8	2		8,850	93		8,943	9,875
1160	3000 CFM		7	2.286		10,300	106		10,406	11,500
1170	5000 CFM		5	3.200		13,400	149		13,549	14,900
1180	8000 CFM		4	4		18,700	186		18,886	20,800
1190	12,000 CFM		3	5.333		24,800	248		25,048	27,600
1200	16,000 CFM	▼	2.50	6.400		30,600	297		30,897	34,100

23 42 16 – Chemically-Impregnated Adsorption Air Filtration

23 42 16.10 Chemical Adsorption Air Filtration	Crew	Daily Output	Labor-Hours	Unit	Material	2011 Bare Costs Labor	Equipment	Total	Total Incl O&P
1210	20,000 CFM	Q-9	2	8	Ea.	38,300	370		38,670	42,800
1220	26,000 CFM	↓	1.50	10.667		44,100	495		44,595	49,400
1230	30,000 CFM	Q-10	2	12		50,000	580		50,580	56,000
1240	40,000 CFM	"	1.50	16		65,500	770		66,270	73,000
1250	50,000 CFM	Q-11	2	16		80,000	785		80,785	89,000
1260	55,000 CFM		1.80	17.778		87,000	875		87,875	97,500
1270	60,000 CFM	↓	1.50	21.333	↓	94,000	1,050		95,050	105,000
1300	Double pack filter, horizontal type									
1310	500 CFM	Q-9	10	1.600	Ea.	8,100	74.50		8,174.50	9,000
1320	1000 CFM		8	2		9,700	93		9,793	10,800
1330	2000 CFM		6	2.667		11,700	124		11,824	13,100
1340	3000 CFM		5	3.200		14,000	149		14,149	15,600
1350	5000 CFM		4	4		17,800	186		17,986	19,800
1360	8000 CFM		3	5.333		23,500	248		23,748	26,200
1370	12,000 CFM		2.50	6.400		31,900	297		32,197	35,600
1380	16,000 CFM		2	8		41,600	370		41,970	46,300
1390	20,000 CFM		1.50	10.667		50,500	495		50,995	56,500
1400	26,000 CFM	↓	1	16		58,500	745		59,245	65,000
1410	30,000 CFM	Q-10	1.50	16		68,500	770		69,270	76,500
1420	40,000 CFM	"	1.30	18.462		87,500	890		88,390	98,000
1430	50,000 CFM	Q-11	1.50	21.333		109,500	1,050		110,550	122,000
1440	55,000 CFM		1.30	24.615		120,000	1,200		121,200	134,000
1450	60,000 CFM	↓	1	32	↓	129,500	1,575		131,075	145,000
1500	Single pack filter, vertical type									
1510	500 CFM	Q-9	24	.667	Ea.	5,950	31		5,981	6,600
1520	1000 CFM		18	.889		7,025	41.50		7,066.50	7,800
1530	2000 CFM		12	1.333		8,425	62		8,487	9,350
1540	3000 CFM		9	1.778		9,850	82.50		9,932.50	10,900
1550	5000 CFM		6	2.667		13,100	124		13,224	14,600
1560	8000 CFM		4	4		18,700	186		18,886	20,800
1570	12,000 CFM		3	5.333		25,400	248		25,648	28,400
1580	16,000 CFM		2	8		29,700	370		30,070	33,300
1590	20,000 CFM		1.80	8.889		37,500	415		37,915	41,800
1600	24,000 CFM	↓	1.60	10	↓	41,900	465		42,365	46,800
1650	Double pack filter, vertical type									
1660	500 CFM	Q-9	22	.727	Ea.	7,700	34		7,734	8,525
1670	1000 CFM		16	1		9,200	46.50		9,246.50	10,200
1680	2000 CFM		10	1.600		11,000	74.50		11,074.50	12,300
1690	3000 CFM		7	2.286		13,000	106		13,106	14,500
1700	5000 CFM		5	3.200		17,000	149		17,149	18,900
1710	8000 CFM		3	5.333		23,300	248		23,548	26,000
1720	12,000 CFM		2.50	6.400		31,400	297		31,697	35,100
1730	16,000 CFM		2	8		39,500	370		39,870	44,100
1740	20,000 CFM		1.50	10.667		49,300	495		49,795	55,000
1750	24,000 CFM	↓	1	16		55,000	745		55,745	61,500
1800	Inlet or outlet transition, horizontal									
1810	Single pack to 12,000 CFM, add					4%				
1820	Single pack to 30,000 CFM, add					6%				
1830	Single pack to 60,000 CFM, add					8%				
1840	Double pack to 12,000 CFM, add					3%				
1850	Double pack to 30,000 CFM, add					5%				
1860	Double pack to 60,000 CFM, add					6%				
1870	Inlet or outlet transition, vertical									

23 42 Gas-Phase Air Filtration

23 42 16 – Chemically-Impregnated Adsorption Air Filtration

23 42 16.10 Chemical Adsorption Air Filtration	Crew	Daily Output	Labor-Hours	Unit	Material	2011 Bare Costs Labor	Equipment	Total	Total Incl O&P
1880	Single pack to 5000 CFM, add					2%				
1890	Single pack to 24,000 CFM, add					3%				
1900	Double pack to 24,000 CFM, add					2%				

23 43 Electronic Air Cleaners

23 43 13 – Washable Electronic Air Cleaners

23 43 13.10 Electronic Air Cleaners

		Crew	Daily Output	Labor-Hours	Unit	Material	2011 Bare Costs Labor	Equipment	Total	Total Incl O&P
0010	**ELECTRONIC AIR CLEANERS**									
2000	Electronic air cleaner, duct mounted									
2150	400 – 1000 CFM	1 Shee	2.30	3.478	Ea.	178	180		358	470
2200	1000 – 1400 CFM		2.20	3.636		1,600	188		1,788	2,050
2250	1400 – 2000 CFM		2.10	3.810		1,625	197		1,822	2,075
2260	2000 – 2500 CFM		2	4		530	207		737	895

23 51 Breechings, Chimneys, and Stacks

23 51 13 – Draft Control Devices

23 51 13.13 Draft-Induction Fans

		Crew	Daily Output	Labor-Hours	Unit	Material	2011 Bare Costs Labor	Equipment	Total	Total Incl O&P
0010	**DRAFT-INDUCTION FANS**									
1000	Breeching installation									
1800	Hot gas, 600°F, variable pitch pulley and motor									
1840	6" diam. inlet, 1/4 H.P., 1 phase, 400 CFM	Q-9	6	2.667	Ea.	1,875	124		1,999	2,225
1860	8" diam. inlet, 1/4 H.P., 1phase, 1120 CFM		4	4		2,675	186		2,861	3,225
1870	9" diam. inlet, 3/4 H.P., 1 phase, 1440 CFM		3.60	4.444		3,300	207		3,507	3,925
1880	10" diam. inlet, 3/4 H.P., 1 phase, 2000 CFM		3.30	4.848		3,300	225		3,525	3,975
1900	12" diam. inlet, 3/4 H.P., 3 phase, 2960 CFM		3	5.333		3,650	248		3,898	4,400
1910	14" diam. inlet, 1 H.P., 3 phase, 4160 CFM		2.60	6.154		3,675	286		3,961	4,475
1920	16" diam. inlet, 2 H.P., 3 phase, 5500 CFM		2.30	6.957		4,075	325		4,400	4,975
1950	20" diam. inlet, 3 H.P., 3 phase, 9760 CFM		1.50	10.667		5,850	495		6,345	7,200
1960	22" diam. inlet, 5 H.P., 3 phase, 13,360 CFM		1	16		9,525	745		10,270	11,600
1980	24" diam. inlet, 7-1/2 H.P., 3 phase, 17,760 CFM		.80	20		10,500	930		11,430	12,900
2300	For multi-blade damper at fan inlet, add					20%				
3600	Chimney-top installation									
3700	6" size	1 Shee	8	1	Ea.	810	51.50		861.50	970
3740	8" size		7	1.143		815	59		874	985
3750	10" size		6.50	1.231		1,075	63.50		1,138.50	1,275
3780	13" size		6	1.333		1,100	69		1,169	1,300
3880	For speed control switch, add					62			62	68
3920	For thermal fan control, add					62			62	68
5500	Flue blade style damper for draft control,									
5510	locking quadrant blade									
5550	8" size	Q-9	8	2	Ea.	405	93		498	585
5560	9" size		7.50	2.133		405	99		504	595
5570	10" size		7	2.286		415	106		521	615
5580	12" size		6.50	2.462		420	114		534	640
5590	14" size		6	2.667		430	124		554	660
5600	16" size		5.50	2.909		445	135		580	695
5610	18" size		5	3.200		465	149		614	735
5620	20" size		4.50	3.556		490	165		655	790
5630	22" size		4	4		515	186		701	845

23 51 Breechings, Chimneys, and Stacks

23 51 13 – Draft Control Devices

23 51 13.13 Draft-Induction Fans

		Crew	Daily Output	Labor-Hours	Unit	Material	2011 Bare Costs Labor	Equipment	Total	Total Incl O&P
5640	24" size	Q-9	3.50	4.571	Ea.	545	212		757	915
5650	27" size		3	5.333		570	248		818	1,000
5660	30" size		2.50	6.400		625	297		922	1,125
5670	32" size		2	8		655	370		1,025	1,275
5680	36" size		1.50	10.667		725	495		1,220	1,550

23 51 13.16 Vent Dampers

		Crew	Daily Output	Labor-Hours	Unit	Material	2011 Bare Costs Labor	Equipment	Total	Total Incl O&P
0010	**VENT DAMPERS**									
5000	Vent damper, bi-metal, gas, 3" diameter	Q-9	24	.667	Ea.	44	31		75	95.50
5010	4" diameter		24	.667		44	31		75	95.50
5020	5" diameter		23	.696		44	32.50		76.50	97.50
5030	6" diameter		22	.727		44	34		78	99.50
5040	7" diameter		21	.762		45.50	35.50		81	104
5050	8" diameter		20	.800		49	37		86	110

23 51 13.19 Barometric Dampers

		Crew	Daily Output	Labor-Hours	Unit	Material	2011 Bare Costs Labor	Equipment	Total	Total Incl O&P
0010	**BAROMETRIC DAMPERS**									
1000	Barometric, gas fired system only, 6" size for 5" and 6" pipes	1 Shee	20	.400	Ea.	61	20.50		81.50	98.50
1020	7" size, for 6" and 7" pipes		19	.421		65.50	22		87.50	105
1040	8" size, for 7" and 8" pipes		18	.444		84.50	23		107.50	128
1060	9" size, for 8" and 9" pipes		16	.500		95	26		121	144
2000	All fuel, oil, oil/gas, coal									
2020	10" for 9" and 10" pipes	1 Shee	15	.533	Ea.	142	27.50		169.50	198
2040	12" for 11" and 12" pipes		15	.533		185	27.50		212.50	246
2060	14" for 13" and 14" pipes		14	.571		240	29.50		269.50	310
2080	16" for 15" and 16" pipes		13	.615		335	32		367	420
2100	18" for 17" and 18" pipes		12	.667		445	34.50		479.50	540
2120	20" for 19" and 21" pipes		10	.800		535	41.50		576.50	655
2140	24" for 22" and 25" pipes	Q-9	12	1.333		655	62		717	815
2160	28" for 26" and 30" pipes		10	1.600		815	74.50		889.50	1,000
2180	32" for 31" and 34" pipes		8	2		1,050	93		1,143	1,300
3260	For thermal switch for above, add	1 Shee	24	.333		58.50	17.20		75.70	90.50

23 51 23 – Gas Vents

23 51 23.10 Gas Chimney Vents

		Crew	Daily Output	Labor-Hours	Unit	Material	2011 Bare Costs Labor	Equipment	Total	Total Incl O&P
0010	**GAS CHIMNEY VENTS,** Prefab metal, U.L. listed									
0020	Gas, double wall, galvanized steel									
0080	3" diameter	Q-9	72	.222	V.L.F.	5.55	10.35		15.90	22
0100	4" diameter		68	.235		6.95	10.95		17.90	24
0120	5" diameter		64	.250		8.10	11.60		19.70	26.50
0140	6" diameter		60	.267		9.45	12.40		21.85	29
0160	7" diameter		56	.286		13.85	13.30		27.15	35.50
0180	8" diameter		52	.308		15.45	14.30		29.75	38.50
0200	10" diameter		48	.333		35	15.50		50.50	62
0220	12" diameter		44	.364		47	16.90		63.90	77
0240	14" diameter		42	.381		79	17.70		96.70	114
0260	16" diameter		40	.400		107	18.60		125.60	146
0280	18" diameter		38	.421		138	19.55		157.55	182
0300	20" diameter	Q-10	36	.667		145	32		177	208
0320	22" diameter		34	.706		207	34		241	279
0340	24" diameter		32	.750		255	36		291	335
0360	26" diameter		31	.774		305	37.50		342.50	390
0380	28" diameter		30	.800		325	38.50		363.50	415
0400	30" diameter		28	.857		340	41.50		381.50	440
0420	32" diameter		27	.889		395	43		438	500

23 51 23.10 Gas Chimney Vents		Crew	Daily Output	Labor-Hours	Unit	Material	2011 Bare Costs Labor	Equipment	Total	Total Incl O&P
0440	34" diameter	Q-10	26	.923	V.L.F.	450	44.50		494.50	565
0460	36" diameter		25	.960		475	46.50		521.50	595
0480	38" diameter		24	1		520	48		568	645
0500	40" diameter		23	1.043		580	50.50		630.50	715
0520	42" diameter		22	1.091		610	52.50		662.50	750
0540	44" diameter		21	1.143		670	55		725	825
0560	46" diameter		20	1.200		740	58		798	905
0580	48" diameter		19	1.263		820	61		881	990
0600	For 4", 5" and 6" oval, add					50%				
0650	Gas, double wall, galvanized steel, fittings									
0660	Elbow 45°, 3" diameter	Q-9	36	.444	Ea.	13.80	20.50		34.30	46
0670	4" diameter		34	.471		16.35	22		38.35	51
0680	5" diameter		32	.500		19.25	23		42.25	56
0690	6" diameter		30	.533		23	25		48	63
0700	7" diameter		28	.571		33	26.50		59.50	76.50
0710	8" diameter		26	.615		38	28.50		66.50	85.50
0720	10" diameter		24	.667		102	31		133	159
0730	12" diameter		22	.727		129	34		163	193
0740	14" diameter		21	.762		181	35.50		216.50	253
0750	16" diameter		20	.800		227	37		264	305
0760	18" diameter		19	.842		292	39		331	380
0770	20" diameter	Q-10	18	1.333		345	64.50		409.50	475
0780	22" diameter		17	1.412		530	68		598	690
0790	24" diameter		16	1.500		660	72.50		732.50	835
0800	26" diameter		16	1.500		1,300	72.50		1,372.50	1,525
0810	28" diameter		15	1.600		1,325	77		1,402	1,600
0820	30" diameter		14	1.714		1,400	82.50		1,482.50	1,675
0830	32" diameter		14	1.714		1,575	82.50		1,657.50	1,875
0840	34" diameter		13	1.846		1,750	89		1,839	2,050
0850	36" diameter		12	2		1,775	96.50		1,871.50	2,100
0860	38" diameter		12	2		1,925	96.50		2,021.50	2,275
0870	40" diameter		11	2.182		2,025	105		2,130	2,375
0880	42" diameter		11	2.182		2,150	105		2,255	2,500
0890	44" diameter		10	2.400		2,275	116		2,391	2,675
0900	46" diameter		10	2.400		2,425	116		2,541	2,850
0910	48" diameter		10	2.400		2,525	116		2,641	2,975
0916	Adjustable length									
0918	3" diameter, to 12"	Q-9	36	.444	Ea.	11.35	20.50		31.85	43.50
0920	4" diameter, to 12"		34	.471		13.40	22		35.40	48
0924	6" diameter, to 12"		30	.533		17.25	25		42.25	56.50
0928	8" diameter, to 12"		26	.615		25.50	28.50		54	71.50
0930	10" diameter, to 18"		24	.667		103	31		134	160
0932	12" diameter, to 18"		22	.727		126	34		160	189
0936	16" diameter, to 18"		20	.800		261	37		298	345
0938	18" diameter, to 18"		19	.842		288	39		327	375
0944	24" diameter, to 18"	Q-10	16	1.500		515	72.50		587.50	680
0950	Elbow 90°, adjustable, 3" diameter	Q-9	36	.444		23	20.50		43.50	56.50
0960	4" diameter		34	.471		28	22		50	64
0970	5" diameter		32	.500		33	23		56	71
0980	6" diameter		30	.533		39	25		64	80.50
0990	7" diameter		28	.571		62	26.50		88.50	108
1010	8" diameter		26	.615		66	28.50		94.50	116
1020	Wall thimble, 4 to 7" adjustable, 3" diameter		36	.444		14.60	20.50		35.10	47

23 51 23.10 Gas Chimney Vents		Crew	Daily Output	Labor-Hours	Unit	Material	2011 Bare Costs Labor	Equipment	Total	Total Incl O&P
1022	4" diameter	Q-9	34	.471	Ea.	16.30	22		38.30	51
1024	5" diameter		32	.500		19.65	23		42.65	56.50
1026	6" diameter		30	.533		22.50	25		47.50	62
1028	7" diameter		28	.571		26	26.50		52.50	68.50
1030	8" diameter		26	.615		27	28.50		55.50	73
1040	Roof flashing, 3" diameter		36	.444		25.50	20.50		46	59.50
1050	4" diameter		34	.471		29	22		51	65
1060	5" diameter		32	.500		31	23		54	69
1070	6" diameter		30	.533		32.50	25		57.50	73.50
1080	7" diameter		28	.571		55	26.50		81.50	101
1090	8" diameter		26	.615		56	28.50		84.50	105
1100	10" diameter		24	.667		86.50	31		117.50	143
1110	12" diameter		22	.727		111	34		145	173
1120	14" diameter		20	.800		129	37		166	198
1130	16" diameter		18	.889		160	41.50		201.50	239
1140	18" diameter		16	1		217	46.50		263.50	310
1150	20" diameter	Q-10	18	1.333		281	64.50		345.50	405
1160	22" diameter		14	1.714		355	82.50		437.50	515
1170	24" diameter		12	2		430	96.50		526.50	615
1200	Tee, 3" diameter	Q-9	27	.593		34	27.50		61.50	79
1210	4" diameter		26	.615		35.50	28.50		64	82.50
1220	5" diameter		25	.640		38.50	30		68.50	87.50
1230	6" diameter		24	.667		41.50	31		72.50	92.50
1240	7" diameter		23	.696		45.50	32.50		78	99
1250	8" diameter		22	.727		63	34		97	120
1260	10" diameter		21	.762		184	35.50		219.50	256
1270	12" diameter		20	.800		221	37		258	299
1280	14" diameter		18	.889		360	41.50		401.50	460
1290	16" diameter		16	1		510	46.50		556.50	630
1300	18" diameter		14	1.143		635	53		688	780
1310	20" diameter	Q-10	17	1.412		865	68		933	1,050
1320	22" diameter		13	1.846		1,075	89		1,164	1,300
1330	24" diameter		12	2		1,250	96.50		1,346.50	1,525
1340	26" diameter		12	2		1,775	96.50		1,871.50	2,125
1350	28" diameter		11	2.182		2,000	105		2,105	2,350
1360	30" diameter		10	2.400		2,175	116		2,291	2,575
1370	32" diameter		10	2.400		2,525	116		2,641	2,950
1380	34" diameter		9	2.667		2,550	129		2,679	3,025
1390	36" diameter		9	2.667		3,050	129		3,179	3,550
1400	38" diameter		8	3		3,050	145		3,195	3,575
1410	40" diameter		7	3.429		3,350	165		3,515	3,925
1420	42" diameter		6	4		3,650	193		3,843	4,300
1430	44" diameter		5	4.800		3,950	231		4,181	4,700
1440	46" diameter		5	4.800		4,350	231		4,581	5,125
1450	48" diameter		5	4.800		4,625	231		4,856	5,450
1460	Tee cap, 3" diameter	Q-9	45	.356		2.43	16.55		18.98	27.50
1470	4" diameter		42	.381		2.74	17.70		20.44	30
1480	5" diameter		40	.400		3.61	18.60		22.21	32
1490	6" diameter		37	.432		4.23	20		24.23	35
1500	7" diameter		35	.457		6.85	21.50		28.35	39.50
1510	8" diameter		34	.471		8.05	22		30.05	42
1520	10" diameter		32	.500		16.60	23		39.60	53.50
1530	12" diameter		30	.533		22	25		47	62

23 51 Breechings, Chimneys, and Stacks

23 51 23 – Gas Vents

23 51 23.10 Gas Chimney Vents	Crew	Daily Output	Labor-Hours	Unit	Material	2011 Bare Costs Labor	Equipment	Total	Total Incl O&P
1540	14" diameter	Q-9	28	.571	Ea.	45.50	26.50		72	90
1550	16" diameter		25	.640		53	30		83	103
1560	18" diameter		24	.667		60	31		91	113
1570	20" diameter	Q-10	27	.889		64	43		107	136
1580	22" diameter		22	1.091		73.50	52.50		126	160
1590	24" diameter		21	1.143		82.50	55		137.50	175
1600	26" diameter		20	1.200		143	58		201	245
1610	28" diameter		19	1.263		340	61		401	460
1620	30" diameter		18	1.333		390	64.50		454.50	525
1630	32" diameter		17	1.412		455	68		523	605
1640	34" diameter		16	1.500		455	72.50		527.50	610
1650	36" diameter		15	1.600		470	77		547	630
1660	38" diameter		14	1.714		475	82.50		557.50	650
1670	40" diameter		14	1.714		490	82.50		572.50	665
1680	42" diameter		13	1.846		505	89		594	690
1690	44" diameter		12	2		510	96.50		606.50	705
1700	46" diameter		12	2		520	96.50		616.50	715
1710	48" diameter		11	2.182		525	105		630	735
1750	Top, 3" diameter	Q-9	46	.348		12.80	16.15		28.95	38.50
1760	4" diameter		44	.364		16.85	16.90		33.75	44
1770	5" diameter		42	.381		18.80	17.70		36.50	47.50
1780	6" diameter		40	.400		23.50	18.60		42.10	53.50
1790	7" diameter		38	.421		34.50	19.55		54.05	67.50
1800	8" diameter		36	.444		45	20.50		65.50	80.50
1810	10" diameter		34	.471		114	22		136	158
1820	12" diameter		32	.500		172	23		195	224
1830	14" diameter		30	.533		196	25		221	253
1840	16" diameter		28	.571		355	26.50		381.50	430
1850	18" diameter		26	.615		620	28.50		648.50	730
1860	20" diameter	Q-10	28	.857		690	41.50		731.50	825
1870	22" diameter		22	1.091		940	52.50		992.50	1,100
1880	24" diameter		20	1.200		1,175	58		1,233	1,400
1882	26" diameter		19	1.263		1,775	61		1,836	2,050
1884	28" diameter		18	1.333		2,025	64.50		2,089.50	2,325
1886	30" diameter		17	1.412		2,250	68		2,318	2,600
1888	32" diameter		16	1.500		2,375	72.50		2,447.50	2,725
1890	34" diameter		15	1.600		2,550	77		2,627	2,925
1892	36" diameter		14	1.714		2,725	82.50		2,807.50	3,125
1900	Gas, double wall, galvanized steel, oval									
1904	4" x 1'	Q-9	68	.235	V.L.F.	20.50	10.95		31.45	39
1906	5"		64	.250		57	11.60		68.60	80
1908	5"/6" x 1'		60	.267		57	12.40		69.40	81.50
1910	Oval fittings									
1912	Adjustable length									
1914	4" diameter to 12" lg	Q-9	34	.471	Ea.	21.50	22		43.50	56.50
1916	5" diameter to 12" lg		32	.500		39.50	23		62.50	78
1918	5"/6" diameter to 12"		30	.533		41.50	25		66.50	83.50
1920	Elbow 45°									
1922	4"	Q-9	34	.471	Ea.	30.50	22		52.50	66.50
1924	5"		32	.500		71	23		94	113
1926	5"/6"		30	.533		67	25		92	112
1930	Elbow 45°, flat									
1932	4"	Q-9	34	.471	Ea.	37.50	22		59.50	74

23 51 Breechings, Chimneys, and Stacks

23 51 23 – Gas Vents

23 51 23.10 Gas Chimney Vents		Crew	Daily Output	Labor-Hours	Unit	Material	2011 Bare Costs Labor	Equipment	Total	Total Incl O&P
1934	5"	Q-9	32	.500	Ea.	71	23		94	113
1936	5"/6"		30	.533		71	25		96	116
1940	Top									
1942	4"	Q-9	44	.364	Ea.	28	16.90		44.90	56
1944	5"		42	.381		19.40	17.70		37.10	48.50
1946	5"/6"		40	.400		47	18.60		65.60	80
1950	Adjustable flashing									
1952	4"	Q-9	34	.471	Ea.	16.75	22		38.75	51.50
1954	5"		32	.500		49.50	23		72.50	89.50
1956	5"/6"		30	.533		55.50	25		80.50	98.50
1960	Tee									
1962	4"	Q-9	26	.615	Ea.	55.50	28.50		84	105
1964	5"		25	.640		106	30		136	162
1966	5"/6"		24	.667		108	31		139	165
1970	Tee with short snout									
1972	4"	Q-9	26	.615	Ea.	55.50	28.50		84	105

23 51 26 – All-Fuel Vent Chimneys

23 51 26.10 All-Fuel Vent Chimneys, Press. Tight, Dbl. Wall

		Crew	Daily Output	Labor-Hours	Unit	Material	2011 Bare Costs Labor	Equipment	Total	Total Incl O&P
0010	**ALL-FUEL VENT CHIMNEYS, PRESSURE TIGHT, DOUBLE WALL**									
3200	All fuel, pressure tight, double wall, U.L. listed, 1400°F.									
3210	304 stainless steel liner, aluminized steel outer jacket									
3220	6" diameter	Q-9	60	.267	L.F.	49	12.40		61.40	73
3221	8" diameter		52	.308		60	14.30		74.30	87.50
3222	10" diameter		48	.333		67	15.50		82.50	97.50
3223	12" diameter		44	.364		76	16.90		92.90	110
3224	14" diameter		42	.381		84	17.70		101.70	120
3225	16" diameter		40	.400		93	18.60		111.60	130
3226	18" diameter		38	.421		105	19.55		124.55	146
3227	20" diameter	Q-10	36	.667		120	32		152	181
3228	24" diameter		32	.750		154	36		190	224
3229	28" diameter		30	.800		177	38.50		215.50	254
3230	32" diameter		27	.889		199	43		242	284
3231	36" diameter		25	.960		233	46.50		279.50	325
3232	42" diameter		22	1.091		268	52.50		320.50	375
3233	48" diameter		19	1.263		315	61		376	435
3260	For 316 stainless steel liner add					6%				
3280	All fuel, pressure tight, double wall fittings									
3284	304 stainless steel inner, aluminized steel jacket									
3288	Adjustable 18"/30" section									
3291	5" diameter	Q-9	32	.500	Ea.	126	23		149	174
3292	6" diameter		30	.533		159	25		184	213
3293	8" diameter		26	.615		183	28.50		211.50	245
3294	10" diameter		24	.667		206	31		237	273
3295	12" diameter		22	.727		234	34		268	310
3296	14" diameter		21	.762		270	35.50		305.50	350
3297	16" diameter		20	.800		281	37		318	365
3298	18" diameter		19	.842		345	39		384	440
3299	20" diameter	Q-10	18	1.333		400	64.50		464.50	535
3300	24" diameter		16	1.500		515	72.50		587.50	680
3301	28" diameter		15	1.600		575	77		652	745
3302	32" diameter		14	1.714		675	82.50		757.50	870
3303	36" diameter		12	2		710	96.50		806.50	925

23 51 26.10 All-Fuel Vent Chimneys, Press. Tight, Dbl. Wall		Crew	Daily Output	Labor-Hours	Unit	Material	2011 Bare Costs Labor	Equipment	Total	Total Incl O&P
3304	42" diameter	Q-10	11	2.182	Ea.	775	105		880	1,025
3305	48" diameter	↓	10	2.400	↓	910	116		1,026	1,175
3326	For 316 stainless steel liner, add				↓	15%				
3350	Elbow 90° fixed									
3353	5" diameter	Q-9	32	.500	Ea.	236	23		259	294
3354	6" diameter		30	.533		277	25		302	345
3355	8" diameter		26	.615		305	28.50		333.50	380
3356	10" diameter		24	.667		350	31		381	430
3357	12" diameter		22	.727		395	34		429	485
3358	14" diameter		21	.762		465	35.50		500.50	565
3359	16" diameter		20	.800		520	37		557	625
3360	18" diameter	↓	19	.842		605	39		644	725
3361	20" diameter	Q-10	18	1.333		675	64.50		739.50	835
3362	24" diameter		16	1.500		835	72.50		907.50	1,025
3363	28" diameter		15	1.600		1,050	77		1,127	1,275
3364	32" diameter		14	1.714		1,400	82.50		1,482.50	1,675
3365	36" diameter		12	2		1,650	96.50		1,746.50	1,950
3366	42" diameter		11	2.182		2,500	105		2,605	2,900
3367	48" diameter	↓	10	2.400		3,125	116		3,241	3,625
3380	For 316 stainless steel liner, add				↓	20%				
3400	Elbow 45°									
3404	6" diameter	Q-9	30	.533	Ea.	174	25		199	230
3405	8" diameter		26	.615		198	28.50		226.50	261
3406	10" diameter		24	.667		222	31		253	291
3407	12" diameter		22	.727		253	34		287	330
3408	14" diameter		21	.762		291	35.50		326.50	375
3409	16" diameter		20	.800		325	37		362	415
3410	18" diameter	↓	19	.842		370	39		409	470
3411	20" diameter	Q-10	18	1.333		425	64.50		489.50	565
3412	24" diameter		16	1.500		545	72.50		617.50	710
3413	28" diameter		15	1.600		615	77		692	795
3414	32" diameter		14	1.714		825	82.50		907.50	1,025
3415	36" diameter		12	2		1,175	96.50		1,271.50	1,425
3416	42" diameter		11	2.182		1,575	105		1,680	1,900
3417	48" diameter	↓	10	2.400		2,125	116		2,241	2,525
3430	For 316 stainless steel liner, add				↓	20%				
3450	Tee 90°									
3454	6" diameter	Q-9	24	.667	Ea.	182	31		213	248
3455	8" diameter		22	.727		210	34		244	282
3456	10" diameter		21	.762		246	35.50		281.50	325
3457	12" diameter		20	.800		286	37		323	370
3458	14" diameter		18	.889		340	41.50		381.50	435
3459	16" diameter		16	1		380	46.50		426.50	490
3460	18" diameter	↓	14	1.143		445	53		498	570
3461	20" diameter	Q-10	17	1.412		505	68		573	660
3462	24" diameter		12	2		620	96.50		716.50	830
3463	28" diameter		11	2.182		775	105		880	1,025
3464	32" diameter		10	2.400		1,075	116		1,191	1,350
3465	36" diameter		9	2.667		1,250	129		1,379	1,575
3466	42" diameter		6	4		1,975	193		2,168	2,475
3467	48" diameter	↓	5	4.800		2,525	231		2,756	3,125
3480	For Tee Cap, add					35%	20%			
3500	For 316 stainless steel liner, add				↓	22%				

23 51 26.10 All-Fuel Vent Chimneys, Press. Tight, Dbl. Wall	Crew	Daily Output	Labor-Hours	Unit	Material	2011 Bare Costs Labor	Equipment	Total	Total Incl O&P
3520	Plate support, galvanized									
3524	6" diameter	Q-9	26	.615	Ea.	95	28.50		123.50	148
3525	8" diameter		22	.727		108	34		142	169
3526	10" diameter		20	.800		117	37		154	185
3527	12" diameter		18	.889		122	41.50		163.50	197
3528	14" diameter		17	.941		148	44		192	229
3529	16" diameter		16	1		156	46.50		202.50	243
3530	18" diameter	▼	15	1.067		164	49.50		213.50	256
3531	20" diameter	Q-10	16	1.500		174	72.50		246.50	300
3532	24" diameter		14	1.714		183	82.50		265.50	325
3533	28" diameter		13	1.846		224	89		313	380
3534	32" diameter		12	2		292	96.50		388.50	465
3535	36" diameter		10	2.400		365	116		481	575
3536	42" diameter		9	2.667		645	129		774	905
3537	48" diameter	▼	8	3	▼	775	145		920	1,075
3570	Bellows, lined, 316 stainless steel only									
3574	6" diameter	Q-9	30	.533	Ea.	455	25		480	540
3575	8" diameter		26	.615		495	28.50		523.50	590
3576	10" diameter		24	.667		545	31		576	645
3577	12" diameter		22	.727		605	34		639	715
3578	14" diameter		21	.762		680	35.50		715.50	800
3579	16" diameter		20	.800		750	37		787	885
3580	18" diameter	▼	19	.842		840	39		879	985
3581	20" diameter	Q-10	18	1.333		945	64.50		1,009.50	1,150
3582	24" diameter		16	1.500		1,175	72.50		1,247.50	1,400
3583	28" diameter		15	1.600		1,325	77		1,402	1,600
3584	32" diameter		14	1.714		1,525	82.50		1,607.50	1,800
3585	36" diameter		12	2		1,750	96.50		1,846.50	2,100
3586	42" diameter		11	2.182		2,625	105		2,730	3,050
3587	48" diameter	▼	10	2.400	▼	2,725	116		2,841	3,175
3600	Ventilated roof thimble, 304 stainless steel									
3620	6" diameter	Q-9	26	.615	Ea.	234	28.50		262.50	300
3624	8" diameter		22	.727		250	34		284	325
3625	10" diameter		20	.800		270	37		307	355
3626	12" diameter		18	.889		292	41.50		333.50	385
3627	14" diameter		17	.941		315	44		359	410
3628	16" diameter		16	1		335	46.50		381.50	435
3629	18" diameter	▼	15	1.067		360	49.50		409.50	470
3630	20" diameter	Q-10	16	1.500		385	72.50		457.50	535
3631	24" diameter		14	1.714		435	82.50		517.50	605
3632	28" diameter		13	1.846		500	89		589	685
3633	32" diameter		12	2		560	96.50		656.50	760
3634	36" diameter		10	2.400		640	116		756	880
3635	42" diameter		9	2.667		855	129		984	1,125
3636	48" diameter	▼	8	3	▼	1,075	145		1,220	1,400
3650	For 316 stainless steel, add					10%				
3670	Exit cone, 316 stainless steel only									
3674	6" diameter	Q-9	46	.348	Ea.	185	16.15		201.15	228
3675	8" diameter		42	.381		194	17.70		211.70	240
3676	10" diameter		40	.400		210	18.60		228.60	259
3677	12" diameter		38	.421		219	19.55		238.55	271
3678	14" diameter		37	.432		244	20		264	299
3679	16" diameter	▼	36	.444	▼	268	20.50		288.50	325

23 51 26.10 All-Fuel Vent Chimneys, Press. Tight, Dbl. Wall

		Crew	Daily Output	Labor-Hours	Unit	Material	2011 Bare Costs Labor	Equipment	Total	Total Incl O&P
3680	18" diameter	Q-9	35	.457	Ea.	330	21.50		351.50	395
3681	20" diameter	Q-10	28	.857		395	41.50		436.50	495
3682	24" diameter		26	.923		505	44.50		549.50	625
3683	28" diameter		25	.960		615	46.50		661.50	745
3684	32" diameter		24	1		760	48		808	915
3685	36" diameter		22	1.091		915	52.50		967.50	1,075
3686	42" diameter		21	1.143		1,175	55		1,230	1,350
3687	48" diameter		20	1.200		1,500	58		1,558	1,750
3720	Roof support assembly, incl. 30" pipe sect, 304 stainless steel									
3724	6" diameter	Q-9	25	.640	Ea.	269	30		299	340
3725	8" diameter		21	.762		294	35.50		329.50	380
3726	10" diameter		19	.842		320	39		359	415
3727	12" diameter		17	.941		340	44		384	440
3728	14" diameter		16	1		385	46.50		431.50	490
3729	16" diameter		15	1.067		410	49.50		459.50	525
3730	18" diameter		14	1.143		440	53		493	560
3731	20" diameter	Q-10	15	1.600		465	77		542	630
3732	24" diameter		13	1.846		520	89		609	705
3733	28" diameter		12	2		610	96.50		706.50	820
3734	32" diameter		11	2.182		735	105		840	965
3735	36" diameter		9	2.667		890	129		1,019	1,175
3736	42" diameter		8	3		1,400	145		1,545	1,750
3737	48" diameter		7	3.429		1,725	165		1,890	2,125
3750	For 316 stainless steel, add					20%				
3770	Stack cap, 316 stainless steel only									
3773	5" diameter	Q-9	48	.333	Ea.	176	15.50		191.50	217
3774	6" diameter		46	.348		206	16.15		222.15	252
3775	8" diameter		42	.381		238	17.70		255.70	288
3776	10" diameter		40	.400		280	18.60		298.60	340
3777	12" diameter		38	.421		325	19.55		344.55	390
3778	14" diameter		37	.432		385	20		405	450
3779	16" diameter		36	.444		435	20.50		455.50	510
3780	18" diameter		35	.457		495	21.50		516.50	570
3781	20" diameter	Q-10	28	.857		565	41.50		606.50	685
3782	24" diameter		26	.923		680	44.50		724.50	815
3783	28" diameter		25	.960		740	46.50		786.50	885
3784	32" diameter		24	1		855	48		903	1,025
3785	36" diameter		22	1.091		1,050	52.50		1,102.50	1,225
3786	42" diameter		21	1.143		1,275	55		1,330	1,500
3787	48" diameter		20	1.200		1,500	58		1,558	1,750
3800	Round storm collar									
3806	5" diameter	Q-9	32	.500	Ea.	27.50	23		50.50	65

23 51 26.30 All-Fuel Vent Chimneys, Double Wall, St. Stl.

		Crew	Daily Output	Labor-Hours	Unit	Material	2011 Bare Costs Labor	Equipment	Total	Total Incl O&P
0010	**ALL-FUEL VENT CHIMNEYS, DOUBLE WALL, STAINLESS STEEL**									
7800	All fuel, double wall, stainless steel, 6" diameter	Q-9	60	.267	V.L.F.	58.50	12.40		70.90	83.50
7802	7" diameter		56	.286		76	13.30		89.30	104
7804	8" diameter		52	.308		90	14.30		104.30	121
7806	10" diameter		48	.333		130	15.50		145.50	167
7808	12" diameter		44	.364		174	16.90		190.90	217
7810	14" diameter		42	.381		228	17.70		245.70	278
8000	All fuel, double wall, stainless steel fittings									
8010	Roof support 6" diameter	Q-9	30	.533	Ea.	108	25		133	156

23 51 26 — All-Fuel Vent Chimneys

23 51 26.30 All-Fuel Vent Chimneys, Double Wall, St. Stl.		Crew	Daily Output	Labor-Hours	Unit	Material	2011 Bare Costs Labor	2011 Bare Costs Equipment	Total	Total Incl O&P
8020	7" diameter	Q-9	28	.571	Ea.	121	26.50		147.50	174
8030	8" diameter		26	.615		136	28.50		164.50	193
8040	10" diameter		24	.667		156	31		187	219
8050	12" diameter		22	.727		188	34		222	258
8060	14" diameter		21	.762		238	35.50		273.50	315
8100	Elbow 15°, 6" diameter		30	.533		172	25		197	227
8120	7" diameter		28	.571		194	26.50		220.50	253
8140	8" diameter		26	.615		221	28.50		249.50	287
8160	10" diameter		24	.667		289	31		320	365
8180	12" diameter		22	.727		350	34		384	435
8200	14" diameter		21	.762		420	35.50		455.50	515
8300	Insulated tee with insulated tee cap, 6" diameter		30	.533		162	25		187	217
8340	7" diameter		28	.571		213	26.50		239.50	274
8360	8" diameter		26	.615		240	28.50		268.50	310
8380	10" diameter		24	.667		340	31		371	415
8400	12" diameter		22	.727		475	34		509	570
8420	14" diameter		21	.762		620	35.50		655.50	740
8500	Joist shield, 6" diameter		30	.533		49.50	25		74.50	92
8510	7" diameter		28	.571		53.50	26.50		80	99
8520	8" diameter		26	.615		66	28.50		94.50	117
8530	10" diameter		24	.667		89.50	31		120.50	146
8540	12" diameter		22	.727		111	34		145	174
8550	14" diameter		21	.762		138	35.50		173.50	206
8600	Round top, 6" diameter		30	.533		55	25		80	98
8620	7" diameter		28	.571		75	26.50		101.50	123
8640	8" diameter		26	.615		101	28.50		129.50	155
8660	10" diameter		24	.667		187	31		218	253
8680	12" diameter		22	.727		259	34		293	335
8700	14" diameter		21	.762		345	35.50		380.50	435
8800	Adjustable roof flashing, 6" diameter		30	.533		65.50	25		90.50	110
8820	7" diameter		28	.571		75	26.50		101.50	123
8840	8" diameter		26	.615		81.50	28.50		110	133
8860	10" diameter		24	.667		104	31		135	162
8880	12" diameter		22	.727		135	34		169	199
8900	14" diameter		21	.762	Ea.	168	35.50		203.50	239

23 51 33 — Insulated Sectional Chimneys

23 51 33.10 Prefabricated Insulated Sectional Chimneys

		Crew	Daily Output	Labor-Hours	Unit	Material	Labor	Equipment	Total	Total Incl O&P
0010	**PREFABRICATED INSULATED SECTIONAL CHIMNEYS**									
9000	High temp. (2000°F), steel jacket, acid resistant refractory lining									
9010	11 ga. galvanized jacket, U.L. listed									
9020	Straight section, 48" long, 10" diameter	Q-10	13.30	1.805	Ea.	400	87		487	570
9030	12" diameter		11.20	2.143		440	103		543	635
9040	18" diameter		7.40	3.243		615	156		771	910
9050	24" diameter		4.60	5.217		840	251		1,091	1,300
9060	30" diameter		3.70	6.486		1,075	315		1,390	1,650
9070	36" diameter		2.70	8.889		1,325	430		1,755	2,100
9080	42" diameter	Q-11	3.10	10.323		1,625	510		2,135	2,575
9090	48" diameter		2.70	11.852		1,925	585		2,510	2,975
9100	54" diameter		2.20	14.545		2,325	715		3,040	3,625
9110	60" diameter		2	16		3,050	785		3,835	4,550
9120	Tee section, 10" diameter	Q-10	4.40	5.455		855	263		1,118	1,350
9130	12" diameter		3.70	6.486		900	315		1,215	1,475

23 51 33 – Insulated Sectional Chimneys

23 51 33.10 Prefabricated Insulated Sectional Chimneys		Crew	Daily Output	Labor-Hours	Unit	Material	2011 Bare Costs Labor	Equipment	Total	Total Incl O&P
9140	18" diameter	Q-10	2.40	10	Ea.	1,225	480		1,705	2,075
9150	24" diameter		1.50	16		1,675	770		2,445	3,025
9160	30" diameter		1.20	20		2,350	965		3,315	4,050
9170	36" diameter	▼	.80	30		3,975	1,450		5,425	6,550
9180	42" diameter	Q-11	1	32		5,225	1,575		6,800	8,125
9190	48" diameter		.90	35.556		6,775	1,750		8,525	10,100
9200	54" diameter		.70	45.714		9,900	2,250		12,150	14,300
9210	60" diameter	▼	.60	53.333		12,500	2,625		15,125	17,800
9220	Cleanout pier section, 10" diameter	Q-10	3.50	6.857		655	330		985	1,225
9230	12" diameter		2.50	9.600		690	465		1,155	1,450
9240	18" diameter		1.90	12.632		885	610		1,495	1,900
9250	24" diameter		1.30	18.462		1,125	890		2,015	2,575
9260	30" diameter		1	24		1,375	1,150		2,525	3,250
9270	36" diameter	▼	.75	32		1,625	1,550		3,175	4,100
9280	42" diameter	Q-11	.90	35.556		2,050	1,750		3,800	4,925
9290	48" diameter		.75	42.667		2,475	2,100		4,575	5,900
9300	54" diameter		.60	53.333		2,950	2,625		5,575	7,225
9310	60" diameter	▼	.50	64		3,775	3,150		6,925	8,900
9320	For drain, add					53%				
9330	Elbow, 30° and 45°, 10" diameter	Q-10	6.60	3.636		555	175		730	875
9340	12" diameter		5.60	4.286		600	207		807	965
9350	18" diameter		3.70	6.486		900	315		1,215	1,475
9360	24" diameter		2.30	10.435		1,350	505		1,855	2,225
9370	30" diameter		1.85	12.973		1,850	625		2,475	2,975
9380	36" diameter	▼	1.30	18.462		2,225	890		3,115	3,800
9390	42" diameter	Q-11	1.60	20		3,050	985		4,035	4,875
9400	48" diameter		1.35	23.704		3,900	1,175		5,075	6,050
9410	54" diameter		1.10	29.091		5,050	1,425		6,475	7,725
9420	60" diameter	▼	1	32		6,150	1,575		7,725	9,150
9430	For 60° and 90° elbow, add					112%				
9440	End cap, 10" diameter	Q-10	26	.923		955	44.50		999.50	1,125
9450	12" diameter		22	1.091		1,100	52.50		1,152.50	1,275
9460	18" diameter		15	1.600		1,350	77		1,427	1,600
9470	24" diameter		9	2.667		1,650	129		1,779	2,025
9480	30" diameter		7.30	3.288		2,050	158		2,208	2,500
9490	36" diameter	▼	5	4.800		2,350	231		2,581	2,950
9500	42" diameter	Q-11	6.30	5.079		2,950	250		3,200	3,600
9510	48" diameter		5.30	6.038		3,475	297		3,772	4,250
9520	54" diameter		4.40	7.273		3,975	360		4,335	4,900
9530	60" diameter	▼	3.90	8.205		4,675	405		5,080	5,750
9540	Increaser (1 diameter), 10" diameter	Q-10	6.60	3.636		475	175		650	785
9550	12" diameter		5.60	4.286		505	207		712	865
9560	18" diameter		3.70	6.486		740	315		1,055	1,300
9570	24" diameter		2.30	10.435		1,050	505		1,555	1,925
9580	30" diameter		1.85	12.973		1,325	625		1,950	2,400
9590	36" diameter	▼	1.30	18.462		1,450	890		2,340	2,950
9592	42" diameter	Q-11	1.60	20		1,600	985		2,585	3,250
9594	48" diameter		1.35	23.704		1,925	1,175		3,100	3,875
9596	54" diameter	▼	1.10	29.091		2,325	1,425		3,750	4,725
9600	For expansion joints, add to straight section					7%				
9610	For 1/4" hot rolled steel jacket, add					157%				
9620	For 2950°F very high temperature, add		▼			89%				
9630	26 ga. aluminized jacket, straight section, 48" long									

23 51 Breechings, Chimneys, and Stacks

23 51 33 – Insulated Sectional Chimneys

23 51 33.10 Prefabricated Insulated Sectional Chimneys

		Crew	Daily Output	Labor-Hours	Unit	Material	2011 Bare Costs Labor	Equipment	Total	Total Incl O&P
9640	10" diameter	Q-10	15.30	1.569	V.L.F.	237	75.50		312.50	375
9650	12" diameter		12.90	1.860		252	89.50		341.50	415
9660	18" diameter		8.50	2.824		365	136		501	610
9670	24" diameter		5.30	4.528		520	218		738	900
9680	30" diameter		4.30	5.581		665	269		934	1,150
9690	36" diameter		3.10	7.742		845	375		1,220	1,500
9700	Accessories (all models)									
9710	Guy band, 10" diameter	Q-10	32	.750	Ea.	74	36		110	136
9720	12" diameter		30	.800		78	38.50		116.50	144
9730	18" diameter		26	.923		102	44.50		146.50	180
9740	24" diameter		24	1		149	48		197	236
9750	30" diameter		20	1.200		173	58		231	278
9760	36" diameter		18	1.333		191	64.50		255.50	305
9770	42" diameter	Q-11	22	1.455		345	71.50		416.50	490
9780	48" diameter		20	1.600		380	78.50		458.50	535
9790	54" diameter		16	2		420	98.50		518.50	610
9800	60" diameter		12	2.667		455	131		586	700
9810	Draw band, galv. stl., 11 gauge, 10" diameter	Q-10	32	.750		65	36		101	126
9820	12" diameter		30	.800		69	38.50		107.50	134
9830	18" diameter		26	.923		86	44.50		130.50	162
9840	24" diameter		24	1		110	48		158	194
9850	30" diameter		20	1.200		127	58		185	227
9860	36" diameter		18	1.333		146	64.50		210.50	257
9870	42" diameter	Q-11	22	1.455		196	71.50		267.50	325
9880	48" diameter		20	1.600		251	78.50		329.50	395
9890	54" diameter		16	2		291	98.50		389.50	470
9900	60" diameter		12	2.667		370	131		501	605
9910	Draw band, aluminized stl., 26 gauge, 10" diameter	Q-10	32	.750		18.95	36		54.95	75.50
9920	12" diameter		30	.800		18.95	38.50		57.45	79.50
9930	18" diameter		26	.923		24	44.50		68.50	94
9940	24" diameter		24	1		28	48		76	104
9950	30" diameter		20	1.200		32	58		90	123
9960	36" diameter		18	1.333		43	64.50		107.50	144
9962	42" diameter	Q-11	22	1.455		43	71.50		114.50	155
9964	48" diameter		20	1.600		51	78.50		129.50	175
9966	54" diameter		16	2		51	98.50		149.50	205
9968	60" diameter		12	2.667		56	131		187	260

23 52 Heating Boilers

23 52 13 – Electric Boilers

23 52 13.10 Electric Boilers, ASME

			Crew	Daily Output	Labor-Hours	Unit	Material	2011 Bare Costs Labor	Equipment	Total	Total Incl O&P
0010	**ELECTRIC BOILERS, ASME**, Standard controls and trim	R235000-30									
1000	Steam, 6 KW, 20.5 MBH		Q-19	1.20	20	Ea.	3,625	985		4,610	5,475
1040	9 KW, 30.7 MBH	D3020–102		1.20	20		3,775	985		4,760	5,625
1060	18 KW, 61.4 MBH			1.20	20		3,875	985		4,860	5,750
1080	24 KW, 81.8 MBH	D3020–104		1.10	21.818		4,525	1,075		5,600	6,575
1120	36 KW, 123 MBH			1.10	21.818		5,000	1,075		6,075	7,100
1160	60 KW, 205 MBH			1	24		6,250	1,175		7,425	8,625
1220	112 KW, 382 MBH			.75	32		8,550	1,575		10,125	11,800
1240	148 KW, 505 MBH			.65	36.923		9,000	1,825		10,825	12,600
1260	168 KW, 573 MBH			.60	40		11,800	1,975		13,775	15,900

23 52 13.10 Electric Boilers, ASME	Crew	Daily Output	Labor-Hours	Unit	Material	2011 Bare Costs Labor	Equipment	Total	Total Incl O&P
1280	222 KW, 758 MBH	Q-19	.55	43.636	Ea.	13,500	2,150		15,650	18,100
1300	296 KW, 1010 MBH		.45	53.333		13,600	2,625		16,225	18,900
1320	300 KW, 1023 MBH		.40	60		17,500	2,950		20,450	23,700
1340	370 KW, 1263 MBH		.35	68.571		18,800	3,375		22,175	25,700
1360	444 KW, 1515 MBH		.30	80		20,200	3,950		24,150	28,100
1380	518 KW, 1768 MBH	Q-21	.36	88.889		22,800	4,475		27,275	31,800
1400	592 KW, 2020 MBH		.34	94.118		25,800	4,750		30,550	35,500
1420	666 KW, 2273 MBH		.32	100		27,200	5,050		32,250	37,400
1460	740 KW, 2526 MBH		.28	114		28,500	5,775		34,275	39,900
1480	814 KW, 2778 MBH		.25	128		31,900	6,450		38,350	44,700
1500	962 KW, 3283 MBH		.22	145		32,800	7,350		40,150	46,900
1520	1036 KW, 3536 MBH		.20	160		34,400	8,075		42,475	49,800
1540	1110 KW, 3788 MBH		.19	168		36,100	8,500		44,600	52,500
1560	2070 KW, 7063 MBH		.18	177		50,500	8,975		59,475	69,000
1580	2250 KW, 7677 MBH		.17	188		59,000	9,500		68,500	79,000
1600	2,340 KW, 7984 MBH		.16	200		65,500	10,100		75,600	87,000
2000	Hot water, 7.5 KW, 25.6 MBH	Q-19	1.30	18.462		3,825	910		4,735	5,550
2020	15 KW, 51.2 MBH		1.30	18.462		3,925	910		4,835	5,675
2040	30 KW, 102 MBH		1.20	20		4,075	985		5,060	5,950
2060	45 KW, 164 MBH		1.20	20		4,550	985		5,535	6,475
2070	60 KW, 205 MBH		1.20	20		4,800	985		5,785	6,775
2080	75 KW, 256 MBH		1.10	21.818		5,175	1,075		6,250	7,300
2100	90 KW, 307 MBH		1.10	21.818		5,575	1,075		6,650	7,775
2120	105 KW, 358 MBH		1	24		5,950	1,175		7,125	8,300
2140	120 KW, 410 MBH		.90	26.667		6,250	1,325		7,575	8,825
2160	135 KW, 461 MBH		.75	32		6,775	1,575		8,350	9,800
2180	150 KW, 512 MBH		.65	36.923		7,050	1,825		8,875	10,500
2200	165 KW, 563 MBH		.60	40		7,300	1,975		9,275	11,000
2220	296 KW, 1010 MBH		.55	43.636		11,900	2,150		14,050	16,300
2280	370 KW, 1263 MBH		.40	60		13,200	2,950		16,150	18,900
2300	444 KW, 1515 MBH		.35	68.571		16,000	3,375		19,375	22,600
2340	518 KW, 1768 MBH	Q-21	.44	72.727		17,700	3,675		21,375	25,000
2360	592 KW, 2020 MBH		.43	74.419		19,600	3,750		23,350	27,100
2400	666 KW, 2273 MBH		.40	80		20,900	4,025		24,925	29,000
2420	740 KW, 2526 MBH		.39	82.051		21,400	4,150		25,550	29,800
2440	814 KW, 2778 MBH		.38	84.211		22,100	4,250		26,350	30,600
2460	888 KW, 3031 MBH		.37	86.486		23,300	4,375		27,675	32,100
2480	962 KW, 3283 MBH		.36	88.889		26,100	4,475		30,575	35,400
2500	1036 KW, 3536 MBH		.34	94.118		27,700	4,750		32,450	37,600
2520	1110 KW, 3788 MBH		.33	96.970		30,700	4,900		35,600	41,000
2540	1440 KW, 4915 MBH		.32	100		35,200	5,050		40,250	46,200
2560	1560 KW, 5323 MBH		.31	103		39,600	5,200		44,800	51,500
2580	1680 KW, 5733 MBH		.30	106		42,200	5,375		47,575	54,500
2600	1800 KW, 6143 MBH		.29	110		44,500	5,575		50,075	57,000
2620	1980 KW, 6757 MBH		.28	114		47,500	5,775		53,275	61,000
2640	2100 KW, 7167 MBH		.27	118		53,000	5,975		58,975	67,500
2660	2220 KW, 7576 MBH		.26	123		55,000	6,200		61,200	69,500
2680	2400 KW, 8191 MBH		.25	128		58,000	6,450		64,450	73,000
2700	2610 KW, 8905 MBH		.24	133		62,000	6,725		68,725	78,500
2720	2790 KW, 9519 MBH		.23	139		65,000	7,025		72,025	82,000
2740	2970 KW, 10133 MBH		.21	152		66,500	7,700		74,200	84,500
2760	3150 KW, 10748 MBH		.19	168		73,000	8,500		81,500	92,500
2780	3240 KW, 11055 MBH		.18	177		74,000	8,975		82,975	95,000

23 52 Heating Boilers

23 52 13 – Electric Boilers

23 52 13.10 Electric Boilers, ASME		Crew	Daily Output	Labor-Hours	Unit	Material	2011 Bare Costs Labor	Equipment	Total	Total Incl O&P
2800	3420 KW, 11669 MBH	Q-21	.17	188	Ea.	78,500	9,500		88,000	100,500
2820	3600 KW, 12,283 MBH	▼	.16	200	▼	81,500	10,100		91,600	104,500

23 52 16 – Condensing Boilers

23 52 16.24 Condensing Boilers

	23 52 16.24 Condensing Boilers		Crew	Daily Output	Labor-Hours	Unit	Material	Labor	Equipment	Total	Total Incl O&P
0010	**CONDENSING BOILERS**, Cast iron										
0020	Packaged with standard controls, circulator and trim										
0030	Intermittent (spark) pilot, natural or LP gas										
0040	Hot water, DOE MBH output, (AFUE)										
0100	42 MBH, (84.0%)	G	Q-5	1.80	8.889	Ea.	1,425	435		1,860	2,225
0120	57 MBH, (84.3%)	G		1.60	10		1,600	485		2,085	2,475
0140	85 MBH, (84.0%)	G		1.40	11.429		1,750	555		2,305	2,750
0160	112 MBH, (83.7%)	G	▼	1.20	13.333		1,975	650		2,625	3,125
0180	140 MBH, (83.3%)	G	Q-6	1.60	15		2,250	760		3,010	3,600
0200	167 MBH, (83.0%)	G		1.40	17.143		2,625	865		3,490	4,175
0220	194 MBH, (82.7%)	G	▼	1.20	20	▼	2,775	1,000		3,775	4,550

23 52 19 – Pulse Combustion Boilers

23 52 19.20 Pulse Type Combustion Boilers

	23 52 19.20 Pulse Type Combustion Boilers		Crew	Daily Output	Labor-Hours	Unit	Material	Labor	Equipment	Total	Total Incl O&P
0010	**PULSE TYPE COMBUSTION BOILERS**										
7990	Special feature gas fired boilers										
8000	Pulse combustion, standard controls/trim										
8050	88,000 BTU	G	Q-5	1.40	11.429	Ea.	4,700	555		5,255	6,000
8080	134,000 BTU	G	"	1.20	13.333	"	5,275	650		5,925	6,775

23 52 23 – Cast-Iron Boilers

23 52 23.20 Gas-Fired Boilers

	23 52 23.20 Gas-Fired Boilers		Crew	Daily Output	Labor-Hours	Unit	Material	Labor	Equipment	Total	Total Incl O&P
0010	**GAS-FIRED BOILERS,** Natural or propane, standard controls, packaged	R235000-10									
1000	Cast iron, with insulated jacket										
2000	Steam, gross output, 81 MBH	R235000-20	Q-7	1.40	22.857	Ea.	2,025	1,175		3,200	3,975
2020	102 MBH			1.30	24.615		2,300	1,275		3,575	4,425
2040	122 MBH	R235000-30		1	32		2,475	1,650		4,125	5,150
2060	163 MBH			.90	35.556		2,950	1,825		4,775	5,975
2080	203 MBH	R235000-50		.90	35.556		3,225	1,825		5,050	6,275
2100	240 MBH			.85	37.647		3,225	1,950		5,175	6,450
2120	280 MBH	R235000-70		.80	40		4,350	2,050		6,400	7,875
2140	320 MBH			.70	45.714		4,375	2,350		6,725	8,350
2160	360 MBH	R235000-80		.63	51.200		5,100	2,650		7,750	9,575
2180	400 MBH			.56	56.838		5,400	2,925		8,325	10,300
2200	440 MBH			.51	62.500		5,750	3,225		8,975	11,100
2220	544 MBH			.45	71.588		8,825	3,700		12,525	15,200
2240	765 MBH			.43	74.419		10,600	3,825		14,425	17,400
2260	892 MBH			.38	84.211		12,300	4,350		16,650	20,100
2280	1275 MBH			.34	94.118		15,100	4,850		19,950	23,900
2300	1530 MBH			.32	100		16,100	5,150		21,250	25,500
2320	1,875 MBH			.30	106		19,000	5,500		24,500	29,100
2340	2170 MBH			.26	122		20,400	6,300		26,700	31,800
2360	2675 MBH			.20	163		23,400	8,425		31,825	38,400
2380	3060 MBH			.19	172		24,900	8,875		33,775	40,600
2400	3570 MBH			.18	181		27,600	9,375		36,975	44,300
2420	4207 MBH			.16	205		30,100	10,600		40,700	48,900
2440	4,720 MBH			.15	207		70,000	10,700		80,700	93,000
2460	5660 MBH			.15	220		75,000	11,400		86,400	99,500
2480	6,100 MBH	▼		.13	246		77,500	12,700		90,200	104,000

23 52 23.20 Gas-Fired Boilers

		Crew	Daily Output	Labor-Hours	Unit	Material	2011 Bare Costs Labor	Equipment	Total	Total Incl O&P
2500	6390 MBH	Q-7	.12	266	Ea.	79,500	13,700		93,200	108,000
2520	6680 MBH		.11	290		81,500	15,000		96,500	112,000
2540	6,970 MBH		.10	320		84,500	16,500		101,000	117,500
3000	Hot water, gross output, 80 MBH		1.46	21.918		1,850	1,125		2,975	3,725
3020	100 MBH		1.35	23.704		2,375	1,225		3,600	4,450
3040	122 MBH		1.10	29.091		2,750	1,500		4,250	5,250
3060	163 MBH		1	32		2,925	1,650		4,575	5,650
3080	203 MBH		1	32		3,075	1,650		4,725	5,825
3100	240 MBH		.95	33.684		3,225	1,725		4,950	6,150
3120	280 MBH		.90	35.556		4,300	1,825		6,125	7,450
3140	320 MBH		.80	40		4,375	2,050		6,425	7,900
3160	360 MBH		.71	45.070		4,875	2,325		7,200	8,850
3180	400 MBH		.64	50		5,175	2,575		7,750	9,550
3200	440 MBH		.58	54.983		5,550	2,825		8,375	10,300
3220	544 MBH		.51	62.992		8,550	3,250		11,800	14,300
3240	765 MBH		.46	70.022		10,300	3,600		13,900	16,800
3260	1,088 MBH		.40	80		12,900	4,125		17,025	20,400
3280	1,275 MBH		.36	89.888		14,700	4,625		19,325	23,100
3300	1,530 MBH		.31	104		15,900	5,400		21,300	25,600
3320	2,000 MBH		.26	125		18,700	6,450		25,150	30,100
3340	2,312 MBH		.22	148		21,300	7,625		28,925	34,800
3360	2,856 MBH		.20	160		22,400	8,250		30,650	36,900
3380	3,264 MBH		.18	179		23,900	9,275		33,175	40,100
3400	3,808 MBH		.16	195		26,200	10,100		36,300	43,900
3420	4,488 MBH		.15	210		29,000	10,800		39,800	48,100
3440	4,720 MBH		.15	220		70,000	11,400		81,400	94,000
3460	5,520 MBH		.14	228		75,000	11,800		86,800	100,000
3480	6,100 MBH		.13	250		77,500	12,900		90,400	104,000
3500	6,390 MBH		.11	285		79,500	14,700		94,200	109,500
3520	6,680 MBH		.10	310		81,500	16,000		97,500	113,500
3540	6,970 MBH	▼	.09	359		84,500	18,500		103,000	120,500
7000	For tankless water heater, add					10%				
7050	For additional zone valves up to 312 MBH add			▼		157			157	173

23 52 23.30 Gas/Oil Fired Boilers

		Crew	Daily Output	Labor-Hours	Unit	Material	2011 Bare Costs Labor	Equipment	Total	Total Incl O&P
0010	**GAS/OIL FIRED BOILERS,** Combination with burners and controls, packaged									
1000	Cast iron with insulated jacket									
2000	Steam, gross output, 720 MBH	Q-7	.43	74.074	Ea.	14,400	3,825		18,225	21,600
2020	810 MBH		.38	83.990		14,400	4,325		18,725	22,400
2040	1,084 MBH		.34	93.023		15,900	4,800		20,700	24,600
2060	1,360 MBH		.33	98.160		17,700	5,050		22,750	27,100
2080	1,600 MBH		.30	107		17,700	5,525		23,225	27,600
2100	2,040 MBH		.25	130		24,800	6,700		31,500	37,300
2120	2,450 MBH		.21	156		27,200	8,050		35,250	41,900
2140	2,700 MBH		.19	165		28,600	8,550		37,150	44,300
2160	3,000 MBH		.18	175		30,200	9,050		39,250	46,700
2180	3,270 MBH		.17	183		31,700	9,475		41,175	48,900
2200	3,770 MBH		.17	191		65,000	9,875		74,875	86,000
2220	4,070 MBH		.16	200		68,500	10,300		78,800	91,000
2240	4,650 MBH		.15	210		72,000	10,800		82,800	95,500
2260	5,230 MBH		.14	223		80,000	11,500		91,500	105,000
2280	5,520 MBH		.14	235		87,000	12,100		99,100	114,000
2300	5,810 MBH	▼	.13	248		88,000	12,800		100,800	116,000

23 52 Heating Boilers

23 52 23 – Cast-Iron Boilers

23 52 23.30 Gas/Oil Fired Boilers

		Crew	Daily Output	Labor-Hours	Unit	Material	2011 Bare Costs Labor	Equipment	Total	Total Incl O&P
2320	6,100 MBH	Q-7	.12	260	Ea.	88,500	13,400		101,900	117,500
2340	6,390 MBH		.11	296		93,500	15,300		108,800	125,500
2360	6,680 MBH		.10	320		96,500	16,500		113,000	130,500
2380	6,970 MBH		.09	372		99,000	19,200		118,200	137,500
2900	Hot water, gross output									
2910	200 MBH	Q-6	.62	39.024	Ea.	8,275	1,975		10,250	12,100
2920	300 MBH		.49	49.080		8,275	2,475		10,750	12,800
2930	400 MBH		.41	57.971		9,650	2,925		12,575	15,000
2940	500 MBH		.36	67.039		10,400	3,375		13,775	16,600
3000	584 MBH	Q-7	.44	72.072		11,500	3,725		15,225	18,300
3020	876 MBH		.41	79.012		15,200	4,075		19,275	22,800
3040	1,168 MBH		.31	103		19,400	5,350		24,750	29,300
3060	1,460 MBH		.28	113		23,900	5,825		29,725	35,000
3080	2,044 MBH		.26	122		26,400	6,300		32,700	38,500
3100	2,628 MBH		.21	150		30,800	7,750		38,550	45,500
3120	3,210 MBH		.18	174		38,000	9,000		47,000	55,000
3140	3,796 MBH		.17	186		42,500	9,575		52,075	61,000
3160	4,088 MBH		.16	195		48,300	10,100		58,400	68,000
3180	4,672 MBH		.16	203		52,500	10,500		63,000	73,500
3200	5,256 MBH		.15	217		56,500	11,200		67,700	78,500
3220	6,000 MBH, 179 BHP		.13	256		84,500	13,200		97,700	112,500
3240	7,130 MBH, 213 BHP		.08	385		88,500	19,900		108,400	127,000
3260	9,800 MBH, 286 BHP		.06	533		102,000	27,500		129,500	153,000
3280	10,900 MBH, 325.6 BHP		.05	592		123,500	30,500		154,000	181,500
3290	12,200 MBH, 364.5 BHP		.05	666		135,000	34,400		169,400	200,000
3300	13,500 MBH, 403.3 BHP		.04	727		149,500	37,500		187,000	220,500

23 52 23.40 Oil-Fired Boilers

		Crew	Daily Output	Labor-Hours	Unit	Material	2011 Bare Costs Labor	Equipment	Total	Total Incl O&P
0010	**OIL-FIRED BOILERS,** Standard controls, flame retention burner, packaged									
1000	Cast iron, with insulated flush jacket									
2000	Steam, gross output, 109 MBH	Q-7	1.20	26.667	Ea.	2,050	1,375		3,425	4,300
2020	144 MBH		1.10	29.091		2,325	1,500		3,825	4,775
2040	173 MBH		1	32		2,600	1,650		4,250	5,300
2060	207 MBH		.90	35.556		2,800	1,825		4,625	5,825
2080	236 MBH		.85	37.647		3,350	1,950		5,300	6,575
2100	300 MBH		.70	45.714		3,825	2,350		6,175	7,750
2120	480 MBH		.50	64		4,925	3,300		8,225	10,300
2140	665 MBH		.45	71.111		6,700	3,675		10,375	12,800
2160	794 MBH		.41	78.049		7,925	4,025		11,950	14,700
2180	1,084 MBH		.38	85.106		9,400	4,375		13,775	16,900
2200	1,360 MBH		.33	98.160		10,800	5,050		15,850	19,500
2220	1,600 MBH		.26	122		12,100	6,300		18,400	22,700
2240	2,175 MBH		.24	133		17,300	6,900		24,200	29,400
2260	2,480 MBH		.21	156		18,600	8,050		26,650	32,400
2280	3,000 MBH		.19	170		21,600	8,775		30,375	36,800
2300	3,550 MBH		.17	187		23,900	9,650		33,550	40,700
2320	3,820 MBH		.16	200		34,000	10,300		44,300	52,500
2340	4,360 MBH		.15	214		37,300	11,100		48,400	57,500
2360	4,940 MBH		.14	225		58,500	11,600		70,100	81,500
2380	5,520 MBH		.14	235		68,500	12,100		80,600	93,500
2400	6,100 MBH		.13	256		78,500	13,200		91,700	106,000
2420	6,390 MBH		.11	290		83,000	15,000		98,000	113,500
2440	6,680 MBH		.10	313		85,000	16,200		101,200	117,500

23 52 Heating Boilers

23 52 23 – Cast-Iron Boilers

23 52 23.40 Oil-Fired Boilers

		Crew	Daily Output	Labor-Hours	Unit	Material	2011 Bare Costs Labor	Equipment	Total	Total Incl O&P
2460	6,970 MBH	Q-7	.09	363	Ea.	88,000	18,700		106,700	125,000
3000	Hot water, same price as steam									
4000	For tankless coil in smaller sizes, add				Ea.	15%				

23 52 23.60 Solid-Fuel Boilers

		Crew	Daily Output	Labor-Hours	Unit	Material	2011 Bare Costs Labor	Equipment	Total	Total Incl O&P
0010	**SOLID-FUEL BOILERS**									
3000	Stoker fired (coal) cast iron with flush jacket and									
3400	insulation, steam or water, gross output, 1280 MBH	Q-6	.36	66.667	Ea.	34,800	3,375		38,175	43,300
3420	1460 MBH		.30	80		38,200	4,050		42,250	48,000
3440	1640 MBH		.28	85.714		40,700	4,325		45,025	51,500
3460	1820 MBH		.26	92.308		43,600	4,650		48,250	55,000
3480	2000 MBH		.25	96		46,400	4,850		51,250	58,000
3500	2360 MBH		.23	104		52,500	5,275		57,775	66,000
3540	2725 MBH		.20	120		58,500	6,050		64,550	73,000
3800	2950 MBH	Q-7	.16	200		97,500	10,300		107,800	123,000
3820	3210 MBH		.15	213		102,500	11,000		113,500	129,000
3840	3480 MBH		.14	228		107,000	11,800		118,800	135,500
3860	3745 MBH		.14	228		112,500	11,800		124,300	141,000
3880	4000 MBH		.13	246		117,000	12,700		129,700	147,500
3900	4200 MBH		.13	246		122,000	12,700		134,700	153,000
3920	4400 MBH		.12	266		126,500	13,700		140,200	160,000
3940	4600 MBH		.12	266		131,500	13,700		145,200	165,000

23 52 26 – Steel Boilers

23 52 26.40 Oil-Fired Boilers

		Crew	Daily Output	Labor-Hours	Unit	Material	2011 Bare Costs Labor	Equipment	Total	Total Incl O&P
0010	**OIL-FIRED BOILERS**, Standard controls, flame retention burner									
5000	Steel, with insulated flush jacket									
7000	Hot water, gross output, 103 MBH	Q-6	1.60	15	Ea.	1,725	760		2,485	3,025
7020	122 MBH		1.45	16.506		1,825	835		2,660	3,250
7040	137 MBH		1.36	17.595		1,925	890		2,815	3,450
7060	168 MBH		1.30	18.405		2,025	930		2,955	3,600
7080	225 MBH		1.22	19.704		2,700	995		3,695	4,425
7100	315 MBH		.96	25.105		5,350	1,275		6,625	7,775
7120	420 MBH		.70	34.483		5,625	1,750		7,375	8,775
7140	525 MBH		.57	42.403		6,500	2,150		8,650	10,400
7180	735 MBH		.48	50.104		7,950	2,525		10,475	12,500
7220	1,050 MBH		.37	65.753		13,700	3,325		17,025	20,100
7280	2,310 MBH		.21	114		25,300	5,800		31,100	36,600
7320	3,150 MBH		.13	184		39,300	9,325		48,625	57,000
7340	For tankless coil in steam or hot water, add					7%				

23 52 26.70 Packaged Water Tube Boilers

		Crew	Daily Output	Labor-Hours	Unit	Material	2011 Bare Costs Labor	Equipment	Total	Total Incl O&P
0010	**PACKAGED WATER TUBE BOILERS**									
2000	Packaged water tube, #2 oil, steam or hot water, gross output									
2010	200 MBH	Q-6	1.50	16	Ea.	4,825	810		5,635	6,525
2014	275 MBH		1.40	17.143		5,100	865		5,965	6,900
2018	360 MBH		1.10	21.818		5,350	1,100		6,450	7,525
2022	520 MBH		.65	36.923		6,300	1,875		8,175	9,700
2026	600 MBH		.60	40		6,425	2,025		8,450	10,100
2030	720 MBH		.55	43.636		7,250	2,200		9,450	11,300
2034	960 MBH		.48	50		9,700	2,525		12,225	14,500
2040	1200 MBH	Q-7	.50	64		15,300	3,300		18,600	21,700
2044	1440 MBH		.45	71.111		20,500	3,675		24,175	28,000
2060	1600 MBH		.40	80		24,100	4,125		28,225	32,700
2068	1920 MBH		.35	91.429		26,100	4,700		30,800	35,700

23 52 Heating Boilers

23 52 26 – Steel Boilers

23 52 26.70 Packaged Water Tube Boilers

		Crew	Daily Output	Labor-Hours	Unit	Material	2011 Bare Costs Labor	Equipment	Total	Total Incl O&P
2072	2160 MBH	Q-7	.33	96.970	Ea.	29,000	5,000		34,000	39,400
2080	2400 MBH		.30	106		32,400	5,500		37,900	43,900
2100	3200 MBH		.25	128		37,100	6,600		43,700	50,500
2120	4800 MBH		.20	160		48,400	8,250		56,650	65,500
2200	Gas fired									
2204	200 MBH	Q-6	1.50	16	Ea.	6,025	810		6,835	7,825
2208	275 MBH		1.40	17.143		6,275	865		7,140	8,200
2212	360 MBH		1.10	21.818		6,525	1,100		7,625	8,825
2216	520 MBH		.65	36.923		7,475	1,875		9,350	11,000
2220	600 MBH		.60	40		7,600	2,025		9,625	11,400
2224	720 MBH		.55	43.636		8,375	2,200		10,575	12,500
2228	960 MBH		.48	50		10,600	2,525		13,125	15,400
2232	1220 MBH	Q-7	.50	64		15,400	3,300		18,700	21,900
2236	1440 MBH		.45	71.111		16,000	3,675		19,675	23,100
2240	1680 MBH		.40	80		20,800	4,125		24,925	29,100
2244	1920 MBH		.35	91.429		20,900	4,700		25,600	30,000
2248	2160 MBH		.33	96.970		21,200	5,000		26,200	30,900
2252	2400 MBH		.30	106		23,800	5,500		29,300	34,400

23 52 28 – Swimming Pool Boilers

23 52 28.10 Swimming Pool Heaters

		Crew	Daily Output	Labor-Hours	Unit	Material	2011 Bare Costs Labor	Equipment	Total	Total Incl O&P
0010	**SWIMMING POOL HEATERS**, Not including wiring, external									
0020	piping, base or pad,									
0160	Gas fired, input, 155 MBH	Q-6	1.50	16	Ea.	1,625	810		2,435	2,975
0200	199 MBH		1	24		1,725	1,200		2,925	3,700
0220	250 MBH		.70	34.286		1,875	1,725		3,600	4,625
0240	300 MBH		.60	40		1,950	2,025		3,975	5,175
0260	399 MBH		.50	48		2,200	2,425		4,625	6,050
0280	500 MBH		.40	60		7,325	3,025		10,350	12,600
0300	650 MBH		.35	68.571		7,800	3,475		11,275	13,800
0320	750 MBH		.33	72.727		8,525	3,675		12,200	14,900
0360	990 MBH		.22	109		11,400	5,500		16,900	20,800
0370	1,260 MBH		.21	114		13,400	5,775		19,175	23,400
0380	1,440 MBH		.19	126		14,300	6,375		20,675	25,200
0400	1,800 MBH		.14	171		15,900	8,650		24,550	30,400
0410	2,070 MBH		.13	184		18,800	9,325		28,125	34,600
2000	Electric, 12 KW, 4,800 gallon pool	Q-19	3	8		2,025	395		2,420	2,800
2020	15 KW, 7,200 gallon pool		2.80	8.571		2,175	420		2,595	3,000
2040	24 KW, 9,600 gallon pool		2.40	10		2,375	490		2,865	3,325
2060	30 KW, 12,000 gallon pool		2	12		2,400	590		2,990	3,525
2080	36 KW, 14,400 gallon pool		1.60	15		2,775	740		3,515	4,150
2100	57 KW, 24,000 gallon pool		1.20	20		3,475	985		4,460	5,300
9000	To select pool heater: 12 BTUH x S.F. pool area									
9010	X temperature differential = required output									
9050	For electric, KW = gallons x 2.5 divided by 1000									
9100	For family home type pool, double the									
9110	Rated gallon capacity = 1/2°F rise per hour									

23 52 39 – Fire-Tube Boilers

23 52 39.13 Scotch Marine Boilers

		Crew	Daily Output	Labor-Hours	Unit	Material	2011 Bare Costs Labor	Equipment	Total	Total Incl O&P
0010	**SCOTCH MARINE BOILERS**									
1000	Packaged fire tube, #2 oil, gross output									
1006	15 PSI steam									
1010	1005 MBH, 30 HP	Q-6	.33	72.072	Ea.	33,900	3,650		37,550	42,700

23 52 Heating Boilers

23 52 39 – Fire-Tube Boilers

23 52 39.13 Scotch Marine Boilers		Crew	Daily Output	Labor-Hours	Unit	Material	2011 Bare Costs Labor	Equipment	Total	Total Incl O&P
1014	1675 MBH, 50 HP	Q-6	.26	93.023	Ea.	42,500	4,700		47,200	54,000
1020	3348 MBH, 100 HP	Q-7	.21	152		56,000	7,850		63,850	73,000
1030	5025 MBH, 150 HP		.17	192		70,500	9,925		80,425	93,000
1040	6696 MBH, 200 HP		.14	223		80,500	11,500		92,000	105,500
1050	8375 MBH, 250 HP		.14	231		90,500	11,900		102,400	117,500
1060	10,044 MBH, 300 HP		.13	251		109,500	13,000		122,500	140,000
1080	16,740 MBH, 500 HP		.08	380		148,000	19,600		167,600	192,500
1100	23,435 MBH, 700 HP		.07	484		181,500	25,000		206,500	237,000
1140	To fire #6, and gas, add	↓	.42	76.190	↓	16,100	3,925		20,025	23,700
1180	For duplex package feed system									
1200	To 3348 MBH boiler, add	Q-7	.54	59.259	Ea.	8,700	3,050		11,750	14,100
1220	To 6696 MBH boiler, add		.41	78.049		10,300	4,025		14,325	17,300
1240	To 10,044 MBH boiler, add		.38	84.211		11,800	4,350		16,150	19,500
1260	To 16,740 MBH boiler, add		.28	114		13,500	5,900		19,400	23,600
1280	To 23,435 MBH boiler, add	↓	.25	128	↓	15,900	6,600		22,500	27,400
1300	150 PSI steam									
1310	1005 MBH, 30 HP	Q-6	.33	72.072	Ea.	37,700	3,650		41,350	46,900
1320	1675 MBH, 50 HP	"	.26	93.023		47,900	4,700		52,600	59,500
1330	3350 MBH, 100 HP	Q-7	.21	152		68,500	7,850		76,350	87,000
1340	5025 MBH, 150 HP		.17	192		87,500	9,925		97,425	111,000
1350	6700 MBH, 200 HP		.14	223		90,000	11,500		101,500	116,000
1360	10,050 MBH, 300 HP	↓	.13	251	↓	136,500	13,000		149,500	169,500
1400	Packaged scotch marine, #6 oil									
1410	15 PSI steam									
1420	1675 MBH, 50 HP	Q-6	.25	97.166	Ea.	52,000	4,900		56,900	65,000
1430	3350 MBH, 100 HP	Q-7	.20	159		68,000	8,200		76,200	87,000
1440	5025 MBH, 150 HP		.16	202		83,000	10,400		93,400	107,000
1450	6700 MBH, 200 HP		.14	233		95,000	12,000		107,000	122,500
1460	8375 MBH, 250 HP		.13	244		108,500	12,600		121,100	138,500
1470	10,050 MBH, 300 HP		.12	262		122,500	13,500		136,000	154,500
1480	13,400 MBH, 400 HP	↓	.09	351		150,500	18,100		168,600	192,500
1500	150 PSI steam									
1510	1675 MBH, 50 HP	Q-6	.25	97.166	Ea.	62,000	4,900		66,900	75,500
1520	3350 MBH, 100 HP	Q-7	.20	159		78,000	8,200		86,200	98,000
1530	5025 MBH, 150 HP		.16	202		96,500	10,400		106,900	122,000
1540	6700 MBH, 200 HP		.14	233		106,500	12,000		118,500	135,000
1550	8375 MBH, 250 HP		.13	244		129,000	12,600		141,600	161,000
1560	10,050 MBH, 300 HP		.12	262		143,500	13,500		157,000	178,000
1570	13,400 MBH, 400 HP	↓	.09	351		166,500	18,100		184,600	210,000

23 52 84 – Boiler Blowdown

23 52 84.10 Boiler Blowdown Systems

		Crew	Daily Output	Labor-Hours	Unit	Material	Labor	Equipment	Total	Total Incl O&P
0010	**BOILER BLOWDOWN SYSTEMS**									
1010	Boiler blowdown, auto/manual to 2000 MBH	Q-5	3.75	4.267	Ea.	3,350	208		3,558	4,000
1020	7300 MBH	"	3	5.333	"	4,575	260		4,835	5,425

23 52 88 – Burners

23 52 88.10 Replacement Type Burners

		Crew	Daily Output	Labor-Hours	Unit	Material	Labor	Equipment	Total	Total Incl O&P
0010	**REPLACEMENT TYPE BURNERS**									
0990	Residential, conversion, gas fired, LP or natural									
1000	Gun type, atmospheric input 50 to 225 MBH	Q-1	2.50	6.400	Ea.	720	310		1,030	1,250
1020	100 to 400 MBH		2	8		1,200	385		1,585	1,900
1040	300 to 1000 MBH	↓	1.70	9.412	↓	3,625	455		4,080	4,650
2000	Commercial and industrial, gas/oil, input									

23 52 Heating Boilers

23 52 88 – Burners

23 52 88.10 Replacement Type Burners		Crew	Daily Output	Labor-Hours	Unit	Material	2011 Bare Costs Labor	Equipment	Total	Total Incl O&P
2050	400 MBH	Q-1	1.50	10.667	Ea.	3,400	515		3,915	4,525
2090	670 MBH		1.40	11.429		3,400	550		3,950	4,575
2140	1155 MBH		1.30	12.308		3,800	595		4,395	5,050
2200	1800 MBH		1.20	13.333		3,800	645		4,445	5,125
2260	3000 MBH		1.10	14.545		4,575	700		5,275	6,075
2320	4100 MBH		1	16		4,675	770		5,445	6,300
2500	Impinged jet, rectangular, burner only, 3-1/2" WC,									
2520	Gas fired, input, 420 to 640 MBH	Q-1	2	8	Ea.	950	385		1,335	1,625
2540	560 to 860 MBH		2	8		1,475	385		1,860	2,200
2560	630 to 950 MBH		1.90	8.421		1,650	405		2,055	2,425
2580	950 to 1200 MBH		1.80	8.889		2,200	430		2,630	3,075
2600	1120 to 1700 MBH		1.80	8.889		2,925	430		3,355	3,875
2620	1700 to 1900 MBH		1.70	9.412		3,300	455		3,755	4,300
2640	1400 to 2100 MBH		1.60	10		3,675	485		4,160	4,750
2660	1600 to 2400 MBH		1.60	10		3,800	485		4,285	4,900
2680	1700 to 2500 MBH		1.50	10.667		4,400	515		4,915	5,625
2700	1880 to 2880 MBH		1.40	11.429		4,500	550		5,050	5,775
2720	2260 to 3680 MBH		1.40	11.429		4,975	550		5,525	6,300
2740	2590 to 4320 MBH		1.30	12.308		6,600	595		7,195	8,125
2760	3020 to 5040 MBH		1.20	13.333		7,700	645		8,345	9,425
2780	3450 to 5760 MBH		1.20	13.333		8,800	645		9,445	10,600
2850	Burner pilot		50	.320		24	15.45		39.45	49.50
2860	Thermocouple		50	.320		25	15.45		40.45	50.50
2870	Thermocouple & pilot bracket		45	.356		91.50	17.15		108.65	127
2900	For 7" W.C. pressure, increase MBH 40%									
3000	Flame retention oil fired assembly, input									
3020	.50 to 2.25 GPH	Q-1	2.40	6.667	Ea.	289	320		609	800
3040	2.0 to 5.0 GPH		2	8		345	385		730	955
3060	3.0 to 7.0 GPH		1.80	8.889		620	430		1,050	1,325
3080	6.0 to 12.0 GPH		1.60	10		990	485		1,475	1,800
4600	Gas safety, shut off valve, 3/4" threaded	1 Stpi	20	.400		178	21.50		199.50	229
4610	1" threaded		19	.421		172	23		195	224
4620	1-1/4" threaded		15	.533		194	29		223	256
4630	1-1/2" threaded		13	.615		210	33.50		243.50	281
4640	2" threaded		11	.727		235	39.50		274.50	320
4650	2-1/2" threaded	Q-1	15	1.067		269	51.50		320.50	375
4660	3" threaded		13	1.231		370	59.50		429.50	495
4670	4" flanged		3	5.333		2,625	257		2,882	3,275
4680	6" flanged	Q-2	3	8		5,875	400		6,275	7,075

23 53 Heating Boiler Feedwater Equipment

23 53 24 – Shot Chemical Feeder

23 53 24.10 Chemical Feeder

		Crew	Daily Output	Labor-Hours	Unit	Material	2011 Bare Costs Labor	Equipment	Total	Total Incl O&P
0010	**CHEMICAL FEEDER**									
0200	Shot chem feeder, by pass, in line mount, 125 PSIG, 1.7 gallon	2 Stpi	1.80	8.889	Ea.	298	480		778	1,050
0220	Floor mount, 175 PSIG, 5 gallon		1.24	12.903		350	700		1,050	1,425
0230	12 gallon		.93	17.204		540	930		1,470	2,000
0300	150 lb., ASME, 5 gallon		1.24	12.903		350	700		1,050	1,425
0320	10 gallon		.93	17.204		500	930		1,430	1,950
0400	300 lb., ASME, 5 gallon		1.24	12.903		350	700		1,050	1,425
0420	10 gallon		.93	17.204		500	930		1,430	1,950

23 54 Furnaces

23 54 13 – Electric-Resistance Furnaces

23 54 13.10 Electric Furnaces		Crew	Daily Output	Labor-Hours	Unit	Material	2011 Bare Costs Labor	Equipment	Total	Total Incl O&P
0010	**ELECTRIC FURNACES**, Hot air, blowers, std. controls									
0011	not including gas, oil or flue piping									
1000	Electric, UL listed									
1020	10.2 MBH	Q-20	5	4	Ea.	475	189		664	805
1040	17.1 MBH		4.80	4.167		495	197		692	840
1060	27.3 MBH		4.60	4.348		600	205		805	970
1100	34.1 MBH		4.40	4.545		635	215		850	1,025
1120	51.6 MBH		4.20	4.762		685	225		910	1,100
1140	68.3 MBH		4	5		735	236		971	1,150
1160	85.3 MBH		3.80	5.263		745	249		994	1,200

23 54 16 – Fuel-Fired Furnaces

23 54 16.13 Gas-Fired Furnaces

		Crew	Daily Output	Labor-Hours	Unit	Material	Labor	Equipment	Total	Total Incl O&P
0010	**GAS-FIRED FURNACES**									
3000	Gas, AGA certified, upflow, direct drive models									
3020	45 MBH input	Q-9	4	4	Ea.	575	186		761	915
3040	60 MBH input		3.80	4.211		615	196		811	970
3060	75 MBH input		3.60	4.444		685	207		892	1,075
3100	100 MBH input		3.20	5		690	232		922	1,100
3120	125 MBH input		3	5.333		700	248		948	1,150
3130	150 MBH input		2.80	5.714		705	266		971	1,175
3140	200 MBH input		2.60	6.154		2,425	286		2,711	3,075
3160	300 MBH input		2.30	6.957		2,525	325		2,850	3,300
3180	400 MBH input		2	8		2,825	370		3,195	3,650
3200	Gas fired wall furnace									
3204	Horizontal flow									
3208	7.7 MBH	Q-9	7	2.286	Ea.	470	106		576	680
3212	14 MBH		6.50	2.462		470	114		584	695
3216	24 MBH		5	3.200		495	149		644	770
3220	49 MBH		4	4		705	186		891	1,050
3224	65 MBH		3.60	4.444		895	207		1,102	1,300
3240	Up-flow									
3244	7.7 MBH	Q-9	7	2.286	Ea.	470	106		576	680
3248	14 MBH		6.50	2.462		470	114		584	695
3252	24 MBH		5	3.200		590	149		739	870
3254	35 MBH		4.50	3.556		705	165		870	1,025
3256	49 MBH		4	4		920	186		1,106	1,275
3260	65 MBH		3.60	4.444		1,025	207		1,232	1,425
3500	Gas furnace									
3510	Up-flow									
3514	1200 CFM, 45 MBH	Q-9	4	4	Ea.	825	186		1,011	1,200
3518	75 MBH		3.60	4.444		865	207		1,072	1,250
3522	100 MBH		3.20	5		1,025	232		1,257	1,475
3526	125 MBH		3	5.333		1,125	248		1,373	1,625
3530	150 MBH		2.80	5.714		1,325	266		1,591	1,850
3540	2000 CFM, 75 MBH		3.60	4.444		530	207		737	890
3544	100 MBH		3.20	5		570	232		802	975
3548	120 MBH		3	5.333		600	248		848	1,025
3552	150 MBH		2.80	5.714		615	266		881	1,075
3560	Horizontal flow									
3564	1200 CFM, 45 MBH	Q-9	4	4	Ea.	955	186		1,141	1,325
3568	60 MBH		3.80	4.211		1,000	196		1,196	1,400
3572	70 MBH		3.60	4.444		1,025	207		1,232	1,425

23 54 Furnaces

23 54 16 – Fuel-Fired Furnaces

23 54 16.13 Gas-Fired Furnaces		Crew	Daily Output	Labor-Hours	Unit	Material	2011 Bare Costs Labor	Equipment	Total	Total Incl O&P
3576	2000 CFM, 90 MBH	Q-9	3.30	4.848	Ea.	1,125	225		1,350	1,600
3580	110 MBH	↓	3.10	5.161	↓	1,300	240		1,540	1,825

23 54 16.16 Oil-Fired Furnaces

		Crew	Daily Output	Labor-Hours	Unit	Material	Labor	Equipment	Total	Total Incl O&P
0010	**OIL-FIRED FURNACES**									
6000	Oil, UL listed, atomizing gun type burner									
6020	56 MBH output	Q-9	3.60	4.444	Ea.	1,475	207		1,682	1,925
6030	84 MBH output		3.50	4.571		1,850	212		2,062	2,375
6040	95 MBH output		3.40	4.706		1,875	219		2,094	2,400
6060	134 MBH output		3.20	5		2,175	232		2,407	2,750
6080	151 MBH output		3	5.333		2,250	248		2,498	2,850
6100	200 MBH input		2.60	6.154		2,475	286		2,761	3,150
6120	300 MBH input		2.30	6.957		3,375	325		3,700	4,225
6140	400 MBH input	↓	2	8	↓	3,875	370		4,245	4,825
6200	Up flow									
6210	105 MBH	Q-9	3.40	4.706	Ea.	1,575	219		1,794	2,075
6220	140 MBH		3.30	4.848		1,725	225		1,950	2,225
6230	168 MBH	↓	2.90	5.517	↓	2,425	256		2,681	3,075
6260	Lo-boy style									
6270	105 MBH	Q-9	3.40	4.706	Ea.	1,575	219		1,794	2,075
6280	140 MBH		3.30	4.848		1,725	225		1,950	2,225
6290	189 MBH	↓	2.80	5.714		2,425	266		2,691	3,075
6320	Down flow									
6330	105 MBH	Q-9	3.40	4.706	Ea.	1,525	219		1,744	2,000
6340	140 MBH		3.30	4.848		1,725	225		1,950	2,225
6350	168 MBH	↓	2.90	5.517	↓	2,425	256		2,681	3,075
6380	Horizontal									
6390	105 MBH	Q-9	3.40	4.706	Ea.	1,575	219		1,794	2,075
6400	140 MBH		3.30	4.848		1,725	225		1,950	2,225
6410	189 MBH	↓	2.80	5.714		2,225	266		2,491	2,850

23 54 16.21 Solid Fuel-Fired Furnaces

			Crew	Daily Output	Labor-Hours	Unit	Material	Labor	Equipment	Total	Total Incl O&P
0010	**SOLID FUEL-FIRED FURNACES**										
6020	Wood fired furnaces										
6030	Includes hot water coil, thermostat, and auto draft control										
6040	24" long firebox	G	Q-9	4	4	Ea.	3,975	186		4,161	4,625
6050	30" long firebox	G		3.60	4.444		4,675	207		4,882	5,425
6060	With fireplace glass doors	G	↓	3.20	5	↓	6,000	232		6,232	6,950
6200	Wood/oil fired furnaces, includes two thermostats										
6210	Includes hot water coil and auto draft control										
6240	24" long firebox	G	Q-9	3.40	4.706	Ea.	4,825	219		5,044	5,625
6250	30" long firebox	G		3	5.333		5,400	248		5,648	6,325
6260	With fireplace glass doors	G	↓	2.80	5.714	↓	6,700	266		6,966	7,775
6400	Wood/gas fired furnaces, includes two thermostats										
6410	Includes hot water coil and auto draft control										
6440	24" long firebox	G	Q-9	2.80	5.714	Ea.	5,450	266		5,716	6,400
6450	30" long firebox	G		2.40	6.667		6,150	310		6,460	7,250
6460	With fireplace glass doors	G	↓	2	8	↓	7,425	370		7,795	8,700
6600	Wood/oil/gas fired furnaces, optional accessories										
6610	Hot air plenum		Q-9	16	1	Ea.	93	46.50		139.50	173
6620	Safety heat dump			24	.667		82	31		113	137
6630	Auto air intake			18	.889		132	41.50		173.50	208
6640	Cold air return package		↓	14	1.143		142	53		195	237
6650	Wood fork					↓	44			44	48.50

23 54 Furnaces

23 54 16 – Fuel-Fired Furnaces

23 54 16.21 Solid Fuel-Fired Furnaces		Crew	Daily Output	Labor-Hours	Unit	Material	2011 Bare Costs Labor	Equipment	Total	Total Incl O&P
6700	Wood fired outdoor furnace									
6740	24" long firebox	G	Q-9	3.80	4.211	Ea.	4,200	196		4,396	4,900
6760	Wood fired outdoor furnace, optional accessories									
6770	Chimney section, stainless steel, 6" ID x 3' lg.	G	Q-9	36	.444	Ea.	121	20.50		141.50	164
6780	Chimney cap, stainless steel	G	"	40	.400	"	66	18.60		84.60	101
6800	Wood fired hot water furnace									
6820	Includes 200 gal. hot water storage, thermostat, and auto draft control									
6840	30" long firebox	G	Q-9	2.10	7.619	Ea.	8,275	355		8,630	9,625
6850	Water to air heat exchanger									
6870	Includes mounting kit and blower relay									
6880	140 MBH, 18.75" W x 18.75" L		Q-9	7.50	2.133	Ea.	370	99		469	555
6890	200 MBH, 24" W x 24" L		"	7	2.286	"	525	106		631	740
6900	Water to water heat exchanger									
6940	100 MBH		Q-9	6.50	2.462	Ea.	370	114		484	585
6960	290 MBH		"	6	2.667	"	535	124		659	775
7000	Optional accessories									
7010	Large volume circulation pump, (2 Included)		Q-9	14	1.143	Ea.	242	53		295	345
7020	Air bleed fittings, (package)			24	.667		49	31		80	101
7030	Domestic water preheater			6	2.667		219	124		343	430
7040	Smoke pipe kit			4	4		74	186		260	365

23 54 24 – Furnace Components for Cooling

23 54 24.10 Furnace Components and Combinations		Crew	Daily Output	Labor-Hours	Unit	Material	2011 Bare Costs Labor	Equipment	Total	Total Incl O&P
0010	**FURNACE COMPONENTS AND COMBINATIONS**									
0080	Coils, A/C evaporator, for gas or oil furnaces									
0090	Add-on, with holding charge									
0100	Upflow									
0120	1-1/2 ton cooling	Q-5	4	4	Ea.	267	195		462	585
0130	2 ton cooling		3.70	4.324		298	211		509	645
0140	3 ton cooling		3.30	4.848		460	236		696	855
0150	4 ton cooling		3	5.333		610	260		870	1,050
0160	5 ton cooling		2.70	5.926		635	289		924	1,125
0300	Downflow									
0330	2-1/2 ton cooling	Q-5	3	5.333	Ea.	365	260		625	790
0340	3-1/2 ton cooling		2.60	6.154		560	300		860	1,050
0350	5 ton cooling		2.20	7.273		635	355		990	1,225
0600	Horizontal									
0630	2 ton cooling	Q-5	3.90	4.103	Ea.	460	200		660	805
0640	3 ton cooling		3.50	4.571		470	223		693	845
0650	4 ton cooling		3.20	5		530	244		774	945
0660	5 ton cooling		2.90	5.517		530	269		799	980
2000	Cased evaporator coils for air handlers									
2100	1-1/2 ton cooling	Q-5	4.40	3.636	Ea.	345	177		522	640
2110	2 ton cooling		4.10	3.902		380	190		570	700
2120	2-1/2 ton cooling		3.90	4.103		420	200		620	760
2130	3 ton cooling		3.70	4.324		465	211		676	825
2140	3-1/2 ton cooling		3.50	4.571		515	223		738	895
2150	4 ton cooling		3.20	5		575	244		819	995
2160	5 ton cooling		2.90	5.517		610	269		879	1,075
3010	Air handler, modular									
3100	With cased evaporator cooling coil									
3120	1-1/2 ton cooling	Q-5	3.80	4.211	Ea.	1,175	205		1,380	1,600
3130	2 ton cooling		3.50	4.571		1,200	223		1,423	1,650

23 54 Furnaces

23 54 24 – Furnace Components for Cooling

23 54 24.10 Furnace Components and Combinations

		Crew	Daily Output	Labor-Hours	Unit	Material	2011 Bare Costs Labor	Equipment	Total	Total Incl O&P
3140	2-1/2 ton cooling	Q-5	3.30	4.848	Ea.	1,275	236		1,511	1,750
3150	3 ton cooling		3.10	5.161		1,400	251		1,651	1,900
3160	3-1/2 ton cooling		2.90	5.517		1,475	269		1,744	2,025
3170	4 ton cooling		2.50	6.400		1,625	310		1,935	2,275
3180	5 ton cooling		2.10	7.619		1,850	370		2,220	2,600
3500	With no cooling coil									
3520	1-1/2 ton coil size	Q-5	12	1.333	Ea.	795	65		860	970
3530	2 ton coil size		10	1.600		850	78		928	1,050
3540	2-1/2 ton coil size		10	1.600		870	78		948	1,075
3554	3 ton coil size		9	1.778		990	86.50		1,076.50	1,225
3560	3-1/2 ton coil size		9	1.778		1,075	86.50		1,161.50	1,300
3570	4 ton coil size		8.50	1.882		1,225	91.50		1,316.50	1,475
3580	5 ton coil size		8	2		1,450	97.50		1,547.50	1,750
4000	With heater									
4120	5 kW, 17.1 MBH	Q-5	16	1	Ea.	665	48.50		713.50	810
4130	7.5 kW, 25.6 MBH		15.60	1.026		710	50		760	855
4140	10 kW, 34.2 MBH		15.20	1.053		730	51.50		781.50	875
4150	12.5 KW, 42.7 MBH		14.80	1.081		770	52.50		822.50	925
4160	15 KW, 51.2 MBH		14.40	1.111		800	54		854	960
4170	25 KW, 85.4 MBH		14	1.143		860	55.50		915.50	1,025
4180	30 KW, 102 MBH		13	1.231		980	60		1,040	1,175

23 55 Fuel-Fired Heaters

23 55 13 – Fuel-Fired Duct Heaters

23 55 13.16 Gas-Fired Duct Heaters

		Crew	Daily Output	Labor-Hours	Unit	Material	2011 Bare Costs Labor	Equipment	Total	Total Incl O&P
0010	**GAS-FIRED DUCT HEATERS**, Includes burner, controls, stainless steel									
0020	heat exchanger. Gas fired, electric ignition									
0030	Indoor installation									
0080	100 MBH output	Q-5	5	3.200	Ea.	1,875	156		2,031	2,300
0100	120 MBH output		4	4		2,125	195		2,320	2,650
0130	200 MBH output		2.70	5.926		2,850	289		3,139	3,550
0140	240 MBH output		2.30	6.957		2,975	340		3,315	3,750
0160	280 MBH output		2	8		3,350	390		3,740	4,275
0180	320 MBH output		1.60	10		3,550	485		4,035	4,650
0300	For powered venter and adapter, add					460			460	510
0502	For required flue pipe, see Section 23 51 23.10									
1000	Outdoor installation, with power venter									
1020	75 MBH output	Q-5	4	4	Ea.	2,775	195		2,970	3,350
1040	94 MBH output		4	4		3,125	195		3,320	3,750
1060	120 MBH output		4	4		3,525	195		3,720	4,200
1080	157 MBH output		3.50	4.571		4,100	223		4,323	4,825
1100	187 MBH output		3	5.333		5,050	260		5,310	5,950
1120	225 MBH output		2.50	6.400		5,225	310		5,535	6,225
1140	300 MBH output		1.80	8.889		6,100	435		6,535	7,350
1160	375 MBH output		1.60	10		6,425	485		6,910	7,800
1180	450 MBH output		1.40	11.429		7,450	555		8,005	9,025
1200	600 MBH output		1	16		8,400	780		9,180	10,400
1300	Aluminized exchanger, subtract					15%				
1500	For two stage gas valve, add					190			190	208

23 55 Fuel-Fired Heaters

23 55 33 – Fuel-Fired Unit Heaters

23 55 33.13 Oil-Fired Unit Heaters

		Crew	Daily Output	Labor-Hours	Unit	Material	2011 Bare Costs Labor	Equipment	Total	Total Incl O&P
0010	**OIL-FIRED UNIT HEATERS**, Cabinet, grilles, fan, ctrl, burner, no piping									
6000	Oil fired, suspension mounted, 94 MBH output	Q-5	4	4	Ea.	3,400	195		3,595	4,025
6040	140 MBH output		3	5.333		3,625	260		3,885	4,375
6060	184 MBH output		3	5.333		3,900	260		4,160	4,700

23 55 33.16 Gas-Fired Unit Heaters

		Crew	Daily Output	Labor-Hours	Unit	Material	2011 Bare Costs Labor	Equipment	Total	Total Incl O&P
0010	**GAS-FIRED UNIT HEATERS**, Cabinet, grilles, fan, ctrls., burner, no piping									
0022	thermostat, no piping. For flue see Section 23 51 23.10									
1000	Gas fired, floor mounted									
1100	60 MBH output	Q-5	10	1.600	Ea.	805	78		883	1,000
1120	80 MBH output		9	1.778		815	86.50		901.50	1,025
1140	100 MBH output		8	2		890	97.50		987.50	1,125
1160	120 MBH output		7	2.286		1,050	111		1,161	1,325
1180	180 MBH output		6	2.667		1,275	130		1,405	1,600
1500	Rooftop mounted, power vent, stainless steel exchanger									
1520	75 MBH input	Q-6	4	6	Ea.	4,200	305		4,505	5,075
1540	100 MBH input		3.60	6.667		4,425	335		4,760	5,375
1560	125 MBH input		3.30	7.273		4,625	365		4,990	5,650
1580	150 MBH input		3	8		5,050	405		5,455	6,150
1600	175 MBH input		2.60	9.231		5,200	465		5,665	6,400
1620	225 MBH input		2.30	10.435		6,050	525		6,575	7,425
1640	300 MBH input		1.90	12.632		7,125	640		7,765	8,775
1660	350 MBH input		1.40	17.143		7,825	865		8,690	9,900
1680	450 MBH input		1.20	20		8,250	1,000		9,250	10,600
1720	700 MBH input		.80	30		10,300	1,525		11,825	13,600
1760	1200 MBH input		.30	80		10,700	4,050		14,750	17,700
1900	For aluminized steel exchanger, subtract					10%				
2000	Suspension mounted, propeller fan, 20 MBH output	Q-5	8.50	1.882		985	91.50		1,076.50	1,200
2020	40 MBH output		7.50	2.133		1,075	104		1,179	1,350
2040	60 MBH output		7	2.286		1,200	111		1,311	1,500
2060	80 MBH output		6	2.667		1,225	130		1,355	1,550
2080	100 MBH output		5.50	2.909		1,375	142		1,517	1,700
2100	130 MBH output		5	3.200		1,500	156		1,656	1,875
2120	140 MBH output		4.50	3.556		1,550	173		1,723	1,950
2140	160 MBH output		4	4		1,600	195		1,795	2,050
2160	180 MBH output		3.50	4.571		1,700	223		1,923	2,200
2180	200 MBH output		3	5.333		1,800	260		2,060	2,375
2200	240 MBH output		2.70	5.926		2,075	289		2,364	2,700
2220	280 MBH output		2.30	6.957		2,425	340		2,765	3,175
2240	320 MBH output		2	8		2,650	390		3,040	3,475
2500	For powered venter and adapter, add					355			355	390
3000	Suspension mounted, blower type, 40 MBH output	Q-5	6.80	2.353		605	115		720	835
3020	60 MBH output		6.60	2.424		775	118		893	1,025
3040	84 MBH output		5.80	2.759		810	134		944	1,100
3060	104 MBH output		5.20	3.077		920	150		1,070	1,225
3080	140 MBH output		4.30	3.721		1,025	181		1,206	1,400
3100	180 MBH output		3.30	4.848		1,150	236		1,386	1,600
3120	240 MBH output		2.50	6.400		1,500	310		1,810	2,125
3140	280 MBH output		2	8		1,875	390		2,265	2,650
4000	Suspension mounted, sealed combustion system,									
4020	Aluminized steel exchanger, powered vent									
4040	100 MBH output	Q-5	5	3.200	Ea.	1,800	156		1,956	2,225
4060	120 MBH output		4.70	3.404		2,025	166		2,191	2,475

23 55 Fuel-Fired Heaters

23 55 33 – Fuel-Fired Unit Heaters

23 55 33.16 Gas-Fired Unit Heaters		Crew	Daily Output	Labor-Hours	Unit	Material	2011 Bare Costs Labor	Equipment	Total	Total Incl O&P
4080	160 MBH output	Q-5	3.70	4.324	Ea.	2,225	211		2,436	2,750
4100	200 MBH output		2.90	5.517		2,650	269		2,919	3,325
4120	240 MBH output		2.50	6.400		2,850	310		3,160	3,600
4140	320 MBH output		1.70	9.412		3,400	460		3,860	4,400
5000	Wall furnace, 17.5 MBH output		6	2.667		715	130		845	980
5020	24 MBH output		5	3.200		740	156		896	1,050
5040	35 MBH output		4	4		895	195		1,090	1,275

23 56 Solar Energy Heating Equipment

23 56 16 – Packaged Solar Heating Equipment

23 56 16.40 Solar Heating Systems

			Crew	Daily Output	Labor-Hours	Unit	Material	2011 Bare Costs Labor	Equipment	Total	Total Incl O&P
0010	**SOLAR HEATING SYSTEMS**	R235616-60									
0020	System/Package prices, not including connecting										
0030	pipe, insulation, or special heating/plumbing fixtures										
0152	For solar ultraviolet pipe insulation see Section 22 07 19.10										
0500	Hot water, standard package, low temperature										
0540	1 collector, circulator, fittings, 65 gal. tank	G	Q-1	.50	32	Ea.	1,550	1,550		3,100	4,000
0580	2 collectors, circulator, fittings, 120 gal. tank	G		.40	40		2,250	1,925		4,175	5,350
0620	3 collectors, circulator, fittings, 120 gal. tank	G		.34	47.059		3,075	2,275		5,350	6,775
0700	Medium temperature package										
0720	1 collector, circulator, fittings, 80 gal. tank	G	Q-1	.50	32	Ea.	2,250	1,550		3,800	4,775
0740	2 collectors, circulator, fittings, 120 gal. tank	G		.40	40		3,275	1,925		5,200	6,475
0780	3 collectors, circulator, fittings, 120 gal. tank	G		.30	53.333		4,250	2,575		6,825	8,525
0980	For each additional 120 gal. tank, add	G					1,475			1,475	1,625

23 56 19 – Solar Heating Components

23 56 19.50 Solar Heating Ancillary

			Crew	Daily Output	Labor-Hours	Unit	Material	2011 Bare Costs Labor	Equipment	Total	Total Incl O&P
0010	**SOLAR HEATING ANCILLARY**										
2300	Circulators, air										
2310	Blowers										
2330	100-300 S.F. system, 1/10 HP	G	Q-9	16	1	Ea.	229	46.50		275.50	325
2340	300-500 S.F. system, 1/5 HP	G		15	1.067		299	49.50		348.50	405
2350	Two speed, 100-300 S.F., 1/10 HP	G		14	1.143		136	53		189	231
2400	Reversible fan, 20" diameter, 2 speed	G		18	.889		108	41.50		149.50	182
2550	Booster fan 6" diameter, 120 CFM	G		16	1		35	46.50		81.50	109
2570	6" diameter, 225 CFM	G		16	1		43	46.50		89.50	118
2580	8" diameter, 150 CFM	G		16	1		39.50	46.50		86	114
2590	8" diameter, 310 CFM	G		14	1.143		60.50	53		113.50	147
2600	8" diameter, 425 CFM	G		14	1.143		68	53		121	156
2650	Rheostat	G		32	.500		14.80	23		37.80	51.50
2660	Shutter/damper	G		12	1.333		49	62		111	148
2670	Shutter motor	G		16	1		119	46.50		165.50	202
2800	Circulators, liquid, 1/25 HP, 5.3 GPM	G	Q-1	14	1.143		268	55		323	380
2820	1/20 HP, 17 GPM	G		12	1.333		154	64.50		218.50	266
2850	1/20 HP, 17 GPM, stainless steel	G		12	1.333		243	64.50		307.50	365
2870	1/12 HP, 30 GPM	G		10	1.600		330	77		407	480
3000	Collector panels, air with aluminum absorber plate										
3010	Wall or roof mount										
3040	Flat black, plastic glazing										
3080	4' x 8'	G	Q-9	6	2.667	Ea.	635	124		759	880
3100	4' x 10'	G		5	3.200	"	780	149		929	1,075

23 56 19.50 Solar Heating Ancillary		Crew	Daily Output	Labor-Hours	Unit	Material	2011 Bare Costs Labor	Equipment	Total	Total Incl O&P
3200	Flush roof mount, 10' to 16' x 22" wide	G	Q-9	96	.167	L.F.	450	7.75		457.75	505
3210	Manifold, by L.F. width of collectors	G	↓	160	.100	"	127	4.65		131.65	147
3300	Collector panels, liquid with copper absorber plate										
3320	Black chrome, tempered glass glazing										
3330	Alum. frame, 4' x 8', 5/32" single glazing	G	Q-1	9.50	1.684	Ea.	945	81.50		1,026.50	1,175
3390	Alum. frame, 4' x 10', 5/32" single glazing	G		6	2.667		1,075	129		1,204	1,400
3450	Flat black, alum. frame, 3.5' x 7.5'	G		9	1.778		690	86		776	890
3500	4' x 8'	G		5.50	2.909		865	140		1,005	1,150
3520	4' x 10'	G		10	1.600		970	77		1,047	1,200
3540	4' x 12.5'	G	↓	5	3.200	↓	1,175	154		1,329	1,500
3550	Liquid with fin tube absorber plate										
3560	Alum. frame 4' x 8' tempered glass	G	Q-1	10	1.600	Ea.	495	77		572	660
3580	Liquid with vacuum tubes, 4' x 6'-10"	G		9	1.778		850	86		936	1,075
3600	Liquid, full wetted, plastic, alum. frame, 3' x 10'	G		5	3.200		234	154		388	490
3650	Collector panel mounting, flat roof or ground rack	G		7	2.286	↓	217	110		327	405
3670	Roof clamps	G		70	.229	Set	2.35	11.05		13.40	19.05
3700	Roof strap, teflon	G	1 Plum	205	.039	L.F.	19.95	2.09		22.04	25
3900	Differential controller with two sensors										
3930	Thermostat, hard wired	G	1 Plum	8	1	Ea.	79.50	53.50		133	167
3950	Line cord and receptacle	G		12	.667		525	35.50		560.50	630
4050	Pool valve system	G		2.50	3.200		264	172		436	545
4070	With 12 VAC actuator	G		2	4		276	214		490	625
4080	Pool pump system, 2" pipe size	G		6	1.333		172	71.50		243.50	297
4100	Five station with digital read-out	G	↓	3	2.667	↓	218	143		361	450
4150	Sensors										
4200	Brass plug, 1/2" MPT	G	1 Plum	32	.250	Ea.	17.95	13.40		31.35	40
4210	Brass plug, reversed	G		32	.250		24.50	13.40		37.90	47
4220	Freeze prevention	G		32	.250		22	13.40		35.40	44
4240	Screw attached	G		32	.250		8.35	13.40		21.75	29
4250	Brass, immersion	G		32	.250	↓	26.50	13.40		39.90	49
4300	Heat exchanger										
4330	Fluid to air coil, up flow, 45 MBH	G	Q-1	4	4	Ea.	295	193		488	615
4380	70 MBH	G		3.50	4.571		335	221		556	695
4400	80 MBH	G	↓	3	5.333	↓	445	257		702	875
4580	Fluid to fluid package includes two circulating pumps										
4590	expansion tank, check valve, relief valve										
4600	controller, high temperature cutoff and sensors	G	Q-1	2.50	6.400	Ea.	695	310		1,005	1,225
4650	Heat transfer fluid										
4700	Propylene glycol, inhibited anti-freeze	G	1 Plum	28	.286	Gal.	12.05	15.30		27.35	36.50
4800	Solar storage tanks, knocked down										
4810	Air, galvanized steel clad, double wall, 4" fiberglass										
4820	insulation, 20 Mil PVC lining										
4870	4' high, 4' x 4', = 64 C.F./450 gallons	G	Q-9	2	8	Ea.	2,275	370		2,645	3,050
4880	4' x 8' = 128 C.F./900 gallons	G		1.50	10.667		3,225	495		3,720	4,300
4890	4' x 12' = 190 C.F./1300 gallons	G		1.30	12.308		4,300	570		4,870	5,600
4900	8' x 8' = 250 C.F./1700 gallons	G	↓	1	16		4,800	745		5,545	6,400
5010	6'-3" high, 7' x 7' = 306 C.F./2000 gallons	G	Q-10	1.20	20		11,700	965		12,665	14,400
5020	7' x 10'-6" = 459 C.F./3000 gallons	G		.80	30		14,700	1,450		16,150	18,400
5030	7' x 14' = 613 C.F./4000 gallons	G		.60	40		17,200	1,925		19,125	21,900
5040	10'-6" x 10'-6" = 689 C.F./4500 gallons	G		.50	48		17,800	2,325		20,125	23,000
5050	10'-6" x 14' = 919 C.F./6000 gallons	G		.40	60		21,000	2,900		23,900	27,500
5060	14' x 14' = 1225 C.F./8000 gallons	G	Q-11	.40	80		24,500	3,925		28,425	32,900
5070	14' x 17'-6" = 1531 C.F./10,000 gallons	G	↓	.30	106		27,600	5,250		32,850	38,200

23 56 19.50 Solar Heating Ancillary		Crew	Daily Output	Labor-Hours	Unit	Material	2011 Bare Costs Labor	Equipment	Total	Total Incl O&P
5080	17'-6" x 17'-6" = 1914 C.F./12,500 gallons	G	Q-11	.25	128	Ea.	32,500	6,300		38,800	45,200
5090	17'-6" x 21' = 2297 C.F./15,000 gallons	G		.20	160		36,200	7,875		44,075	51,500
5100	21' x 21' = 2756 C.F./18,000 gallons	G		.18	177		41,500	8,750		50,250	59,000
5120	30 Mil reinforced Chemflex lining,										
5140	4' high, 4' x 4' = 64 C.F./450 gallons	G	Q-9	2	8	Ea.	2,525	370		2,895	3,325
5150	4' x 8' = 128 C.F./900 gallons	G		1.50	10.667		3,825	495		4,320	4,950
5160	4' x 12' = 190 C.F./1300 gallons	G		1.30	12.308		5,050	570		5,620	6,450
5170	8' x 8' = 250 C.F./1700 gallons	G		1	16		5,650	745		6,395	7,325
5190	6'-3" high, 7' x 7' = 306 C.F./2000 gallons	G	Q-10	1.20	20		12,500	965		13,465	15,200
5200	7' x 10'-6" = 459 C.F./3000 gallons	G		.80	30		15,200	1,450		16,650	19,000
5210	7' x 14' = 613 C.F./4000 gallons	G		.60	40		17,700	1,925		19,625	22,400
5220	10'-6" x 10'-6" = 689 C.F./4500 gallons	G		.50	48		18,300	2,325		20,625	23,600
5230	10'-6" x 14' = 919 C.F./6000 gallons	G		.40	60		21,500	2,900		24,400	28,000
5240	14' x 14' = 1225 C.F./8000 gallons	G	Q-11	.40	80		25,100	3,925		29,025	33,700
5250	14' x 17'-6" = 1531 C.F./10,000 gallons	G		.30	106		28,100	5,250		33,350	38,900
5260	17'-6" x 17'-6" = 1914 C.F./12,500 gallons	G		.25	128		33,100	6,300		39,400	45,900
5270	17'-6" x 21' = 2297 C.F./15,000 gallons	G		.20	160		37,300	7,875		45,175	53,000
5280	21' x 21' = 2756 C.F./18,000 gallons	G		.18	177		42,700	8,750		51,450	60,000
5290	30 Mil reinforced Hypalon lining, add						.02%				
7000	Solar control valves and vents										
7050	Air purger, 1" pipe size	G	1 Plum	12	.667	Ea.	43.50	35.50		79	101
7070	Air eliminator, automatic 3/4" size	G		32	.250		28.50	13.40		41.90	51.50
7090	Air vent, automatic, 1/8" fitting	G		32	.250		12.65	13.40		26.05	34
7100	Manual, 1/8" NPT	G		32	.250		2.78	13.40		16.18	23
7120	Backflow preventer, 1/2" pipe size	G		16	.500		58.50	27		85.50	104
7130	3/4" pipe size	G		16	.500		106	27		133	157
7150	Balancing valve, 3/4" pipe size	G		20	.400		43	21.50		64.50	79
7180	Draindown valve, 1/2" copper tube	G		9	.889		202	47.50		249.50	293
7200	Flow control valve, 1/2" pipe size	G		22	.364		107	19.50		126.50	147
7220	Expansion tank, up to 5 gal.	G		32	.250		61.50	13.40		74.90	87.50
7250	Hydronic controller (aquastat)	G		8	1		172	53.50		225.50	269
7400	Pressure gauge, 2" dial	G		32	.250		22	13.40		35.40	44.50
7450	Relief valve, temp. and pressure 3/4" pipe size	G		30	.267		15.60	14.30		29.90	38.50
7500	Solenoid valve, normally closed										
7520	Brass, 3/4" NPT, 24V	G	1 Plum	9	.889	Ea.	109	47.50		156.50	191
7530	1" NPT, 24V	G		9	.889		850	47.50		897.50	1,000
7750	Vacuum relief valve, 3/4" pipe size	G		32	.250		26.50	13.40		39.90	49
7800	Thermometers										
7820	Digital temperature monitoring, 4 locations	G	1 Plum	2.50	3.200	Ea.	128	172		300	395
7900	Upright, 1/2" NPT	G		8	1		24	53.50		77.50	107
7970	Remote probe, 2" dial	G		8	1		30.50	53.50		84	114
7990	Stem, 2" dial, 9" stem	G		16	.500		20	27		47	62
8250	Water storage tank with heat exchanger and electric element										
8270	66 gal. with 2" x 2 lb. density insulation	G	1 Plum	1.60	5	Ea.	970	268		1,238	1,475
8300	80 gal. with 2" x 2 lb. density insulation	G		1.60	5		1,100	268		1,368	1,600
8380	120 gal. with 2" x 2 lb. density insulation	G		1.40	5.714		1,225	305		1,530	1,800
8400	120 gal. with 2" x 2 lb. density insul., 40 S.F. heat coil	G		1.40	5.714		1,575	305		1,880	2,175
8500	Water storage module, plastic										
8600	Tubular, 12" diameter, 4' high	G	1 Carp	48	.167	Ea.	103	7.20		110.20	124
8610	12" diameter, 8' high	G		40	.200		159	8.60		167.60	188
8620	18" diameter, 5' high	G		38	.211		174	9.05		183.05	205
8630	18" diameter, 10' high	G		32	.250		229	10.75		239.75	269
8640	58" diameter, 5' high	G	2 Carp	32	.500		490	21.50		511.50	570

23 56 Solar Energy Heating Equipment

23 56 19 – Solar Heating Components

23 56 19.50 Solar Heating Ancillary		Crew	Daily Output	Labor-Hours	Unit	Material	2011 Bare Costs Labor	Equipment	Total	Total Incl O&P
8650	Cap, 12" diameter	G			Ea.	19			19	21
8660	18" diameter	G			▼	24			24	26.50

23 57 Heat Exchangers for HVAC

23 57 16 – Steam-to-Water Heat Exchangers

23 57 16.10 Shell/Tube Type Steam-to-Water Heat Exch.

		Crew	Daily Output	Labor-Hours	Unit	Material	2011 Bare Costs Labor	Equipment	Total	Total Incl O&P
0010	**SHELL AND TUBE TYPE STEAM-TO-WATER HEAT EXCHANGERS**									
0016	Shell & tube type, 2 or 4 pass, 3/4" O.D. copper tubes,									
0020	C.I. heads, C.I. tube sheet, steel shell									
0100	Hot water 40°F to 180°F, by steam at 10 PSI									
0120	8 GPM	Q-5	6	2.667	Ea.	1,825	130		1,955	2,200
0140	10 GPM		5	3.200		2,750	156		2,906	3,250
0160	40 GPM		4	4		4,250	195		4,445	4,975
0180	64 GPM		2	8		6,525	390		6,915	7,750
0200	96 GPM	▼	1	16		8,725	780		9,505	10,800
0220	120 GPM	Q-6	1.50	16		11,400	810		12,210	13,800
0240	168 GPM		1	24		14,000	1,200		15,200	17,200
0260	240 GPM		.80	30		21,900	1,525		23,425	26,400
0300	600 GPM	▼	.70	34.286	▼	47,100	1,725		48,825	54,500
0500	For bronze head and tube sheet, add					50%				

23 57 19 – Liquid-to-Liquid Heat Exchangers

23 57 19.13 Plate-Type, Liquid-to-Liquid Heat Exchangers

		Crew	Daily Output	Labor-Hours	Unit	Material	2011 Bare Costs Labor	Equipment	Total	Total Incl O&P
0010	**PLATE-TYPE, LIQUID-TO-LIQUID HEAT EXCHANGERS**									
3000	Plate type,									
3100	400 GPM	Q-6	.80	30	Ea.	32,300	1,525		33,825	37,800
3120	800 GPM	"	.50	48		55,500	2,425		57,925	64,500
3140	1200 GPM	Q-7	.34	94.118		82,500	4,850		87,350	98,500
3160	1800 GPM	"	.24	133	▼	110,000	6,875		116,875	131,500

23 57 19.16 Shell-Type, Liquid-to-Liquid Heat Exchangers

		Crew	Daily Output	Labor-Hours	Unit	Material	2011 Bare Costs Labor	Equipment	Total	Total Incl O&P
0010	**SHELL-TYPE, LIQUID-TO-LIQUID HEAT EXCHANGERS**									
1000	Hot water 40°F to 140°F, by water at 200°F									
1020	7 GPM	Q-5	6	2.667	Ea.	2,225	130		2,355	2,650
1040	16 GPM		5	3.200		3,175	156		3,331	3,725
1060	34 GPM		4	4		4,800	195		4,995	5,575
1080	55 GPM		3	5.333		6,975	260		7,235	8,050
1100	74 GPM		1.50	10.667		8,700	520		9,220	10,400
1120	86 GPM	▼	1.40	11.429		11,600	555		12,155	13,600
1140	112 GPM	Q-6	2	12		14,500	605		15,105	16,800
1160	126 GPM		1.80	13.333		18,100	675		18,775	20,900
1180	152 GPM	▼	1	24	▼	25,800	1,200		27,000	30,100

23 61 Refrigerant Compressors

23 61 15 – Rotary Refrigerant Compressors

23 61 15.10 Rotary Compressors

23 61 15.10 Rotary Compressors	Crew	Daily Output	Labor-Hours	Unit	Material	2011 Bare Costs Labor	Equipment	Total	Total Incl O&P
0010 **ROTARY COMPRESSORS**									
0100 Refrigeration, hermetic, switches and protective devices									
0210 1.25 ton	1 Stpi	3	2.667	Ea.	221	144		365	460
0220 1.42 ton		3	2.667		223	144		367	460
0230 1.68 ton		2.80	2.857		243	155		398	500
0240 2.00 ton		2.60	3.077		250	166		416	525
0250 2.37 ton		2.50	3.200		282	173		455	570
0260 2.67 ton		2.40	3.333		296	180		476	595
0270 3.53 ton		2.30	3.478		325	188		513	635
0280 4.43 ton		2.20	3.636		415	197		612	750
0290 5.08 ton		2.10	3.810		455	206		661	810
0300 5.22 ton	▼	2	4		640	216		856	1,025
0310 7.31 ton	Q-5	3	5.333		1,025	260		1,285	1,525
0320 9.95 ton		2.60	6.154		1,200	300		1,500	1,775
0330 11.6 ton		2.50	6.400		1,550	310		1,860	2,175
0340 15.25 ton		2.30	6.957		2,125	340		2,465	2,825
0350 17.7 ton	▼	2.10	7.619	▼	2,275	370		2,645	3,050

23 61 16 – Reciprocating Refrigerant Compressors

23 61 16.10 Reciprocating Compressors

23 61 16.10 Reciprocating Compressors	Crew	Daily Output	Labor-Hours	Unit	Material	2011 Bare Costs Labor	Equipment	Total	Total Incl O&P
0010 **RECIPROCATING COMPRESSORS**									
0990 Refrigeration, recip. hermetic, switches & protective devices									
1000 10 ton	Q-5	1	16	Ea.	8,850	780		9,630	10,900
1100 20 ton	Q-6	.72	33.333		13,400	1,675		15,075	17,200
1200 30 ton		.64	37.500		15,100	1,900		17,000	19,400
1300 40 ton		.44	54.545		16,600	2,750		19,350	22,400
1400 50 ton	▼	.20	120		17,500	6,050		23,550	28,300
1500 75 ton	Q-7	.27	118		19,300	6,100		25,400	30,400
1600 130 ton	"	.21	152	▼	22,100	7,850		29,950	36,000

23 61 19 – Scroll Refrigerant Compressors

23 61 19.10 Scroll Compressors

23 61 19.10 Scroll Compressors	Crew	Daily Output	Labor-Hours	Unit	Material	2011 Bare Costs Labor	Equipment	Total	Total Incl O&P
0010 **SCROLL COMPRESSORS**									
1800 Refrigeration, scroll type									
1810 1.5 ton	1 Stpi	2.70	2.963	Ea.	885	160		1,045	1,225
1820 2.5 ton		2.50	3.200		885	173		1,058	1,225
1830 2.75 ton		2.35	3.404		905	184		1,089	1,275
1840 3 ton		2.30	3.478		940	188		1,128	1,300
1850 3.5 ton		2.25	3.556		965	192		1,157	1,325
1860 4 ton		2.20	3.636		1,075	197		1,272	1,475
1870 5 ton	▼	2.10	3.810		1,175	206		1,381	1,575
1880 6 ton	Q-5	3.70	4.324		1,250	211		1,461	1,700
1900 7.5 ton		3.20	5		2,600	244		2,844	3,225
1910 10 ton		3.10	5.161		2,650	251		2,901	3,275
1920 13 ton		2.90	5.517		3,000	269		3,269	3,700
1960 25 ton	▼	2.20	7.273	▼	5,000	355		5,355	6,000

23 62 Packaged Compressor and Condenser Units

23 62 13 – Packaged Air-Cooled Refrigerant Compressor and Condenser Units

23 62 13.10 Packaged Air-Cooled Refrig. Condensing Units	Crew	Daily Output	Labor-Hours	Unit	Material	2011 Bare Costs Labor	Equipment	Total	Total Incl O&P
0010 **PACKAGED AIR-COOLED REFRIGERANT CONDENSING UNITS**									
0020 Condensing unit									
0030 Air cooled, compressor, standard controls									
0050 1.5 ton	Q-5	2.50	6.400	Ea.	1,125	310		1,435	1,700
0100 2 ton		2.10	7.619		1,350	370		1,720	2,025
0200 2.5 ton		1.70	9.412		1,375	460		1,835	2,175
0300 3 ton		1.30	12.308		1,400	600		2,000	2,450
0350 3.5 ton		1.10	14.545		1,625	710		2,335	2,850
0400 4 ton		.90	17.778		1,825	865		2,690	3,325
0500 5 ton		.60	26.667		2,225	1,300		3,525	4,400
0550 7.5 ton		.55	29.091		3,625	1,425		5,050	6,125
0560 8.5 ton		.53	30.189		4,550	1,475		6,025	7,225
0600 10 ton		.50	32		4,950	1,550		6,500	7,775
0620 12.5 ton		.48	33.333		7,475	1,625		9,100	10,600
0650 15 ton		.40	40		8,125	1,950		10,075	11,900
0700 20 ton	Q-6	.40	60		11,000	3,025		14,025	16,600
0720 25 ton		.35	68.571		15,500	3,475		18,975	22,200
0750 30 ton		.30	80		17,800	4,050		21,850	25,600
0800 40 ton		.20	120		22,500	6,050		28,550	33,900
0840 50 ton		.18	133		26,800	6,725		33,525	39,600
0860 60 ton		.16	150		30,900	7,575		38,475	45,300
0900 70 ton		.14	171		35,300	8,650		43,950	51,500
1000 80 ton		.12	200		39,200	10,100		49,300	58,000
1100 100 ton		.09	266		48,100	13,500		61,600	73,000

23 62 23 – Packaged Water-Cooled Refrigerant Compressor and Condenser Units

23 62 23.10 Packaged Water-Cooled Refrigerant Condensing Units

	Crew	Daily Output	Labor-Hours	Unit	Material	Labor	Equipment	Total	Total Incl O&P
0010 **PACKAGED WATER-COOLED REFRIGERANT CONDENSING UNITS**									
2000 Water cooled, compressor, heat exchanger, controls									
2100 5 ton	Q-5	.70	22.857	Ea.	12,600	1,125		13,725	15,600
2110 10 ton		.60	26.667		14,700	1,300		16,000	18,200
2200 15 ton		.50	32		19,900	1,550		21,450	24,200
2300 20 ton	Q-6	.40	60		23,000	3,025		26,025	29,800
2310 30 ton		.30	80		26,400	4,050		30,450	35,000
2400 40 ton		.20	120		28,700	6,050		34,750	40,700
2410 60 ton		.18	133		42,000	6,725		48,725	56,500
2420 80 ton		.14	171		48,500	8,650		57,150	66,500
2500 100 ton		.11	218		59,500	11,000		70,500	82,000
2510 120 ton		.10	240		95,500	12,100		107,600	123,500

23 63 Refrigerant Condensers

23 63 13 – Air-Cooled Refrigerant Condensers

23 63 13.10 Air-Cooled Refrig. Condensers

	Crew	Daily Output	Labor-Hours	Unit	Material	Labor	Equipment	Total	Total Incl O&P
0010 **AIR-COOLED REFRIG. CONDENSERS**, Ratings for 30°F TD, R-22.									
0080 Air cooled, belt drive, propeller fan									
0220 45 ton	Q-6	.70	34.286	Ea.	9,250	1,725		10,975	12,800
0240 50 ton		.69	34.985		9,975	1,775		11,750	13,600
0260 54 ton		.64	37.795		10,900	1,900		12,800	14,900
0280 59 ton		.58	41.308		12,000	2,075		14,075	16,300
0300 65 ton		.53	45.541		14,300	2,300		16,600	19,100
0320 73 ton		.47	51.173		15,600	2,575		18,175	21,100

23 63 Refrigerant Condensers

23 63 13 – Air-Cooled Refrigerant Condensers

23 63 13.10 Air-Cooled Refrig. Condensers		Crew	Daily Output	Labor-Hours	Unit	Material	2011 Bare Costs Labor	Equipment	Total	Total Incl O&P
0340	81 ton	Q-6	.42	56.738	Ea.	17,200	2,875		20,075	23,300
0360	86 ton		.40	60.302		18,100	3,050		21,150	24,500
0380	88 ton	▼	.39	61.697		19,400	3,125		22,525	26,100
0400	101 ton	Q-7	.45	70.640		24,800	3,650		28,450	32,700
0500	159 ton		.31	102		34,100	5,300		39,400	45,400
0600	228 ton	▼	.22	148	▼	48,000	7,625		55,625	64,500
1500	May be specified single or multi-circuit									
1550	Air cooled, direct drive, propeller fan									
1590	1 ton	Q-5	3.80	4.211	Ea.	685	205		890	1,050
1600	1-1/2 ton		3.60	4.444		840	216		1,056	1,250
1620	2 ton		3.20	5		900	244		1,144	1,350
1630	3 ton		2.40	6.667		1,000	325		1,325	1,575
1640	5 ton		2	8		1,825	390		2,215	2,600
1650	8 ton		1.80	8.889		2,400	435		2,835	3,275
1660	10 ton		1.40	11.429		3,025	555		3,580	4,150
1670	12 ton		1.30	12.308		3,350	600		3,950	4,575
1680	14 ton		1.20	13.333		3,800	650		4,450	5,150
1690	16 ton		1.10	14.545		4,150	710		4,860	5,625
1700	21 ton		1	16		4,850	780		5,630	6,500
1720	26 ton		.84	19.002		5,400	925		6,325	7,325
1740	30 ton	▼	.70	22.792		8,825	1,100		9,925	11,400
1760	41 ton	Q-6	.77	31.008		9,700	1,575		11,275	13,000
1780	52 ton		.66	36.419		13,800	1,850		15,650	18,000
1800	63 ton		.55	44.037		15,300	2,225		17,525	20,100
1820	76 ton		.45	52.980		17,300	2,675		19,975	23,100
1840	86 ton		.40	60		21,400	3,025		24,425	28,100
1860	97 ton	▼	.35	67.989		24,400	3,425		27,825	31,900
1880	105 ton	Q-7	.44	73.563		28,100	3,800		31,900	36,600
1890	118 ton		.39	82.687		30,300	4,250		34,550	39,700
1900	126 ton		.36	88.154		35,200	4,550		39,750	45,500
1910	136 ton		.34	95.238		39,400	4,900		44,300	50,500
1920	142 ton	▼	.32	99.379	▼	40,400	5,125		45,525	52,000

23 63 30 – Lubricants

23 63 30.10 Lubricant Oils

		Crew	Daily Output	Labor-Hours	Unit	Material	Labor	Equipment	Total	Total Incl O&P
0010	**LUBRICANT OILS**									
8000	Oils									
8100	Lubricating									
8120	Oil, lubricating				Oz.	.32			.32	.35
8500	Refrigeration									
8520	Oil, refrigeration				Gal.	38.50			38.50	42.50
8525	Oil, refrigeration				Qt.	9.60			9.60	10.60

23 63 33 – Evaporative Refrigerant Condensers

23 63 33.10 Evaporative Condensers

		Crew	Daily Output	Labor-Hours	Unit	Material	Labor	Equipment	Total	Total Incl O&P
0010	**EVAPORATIVE CONDENSERS**									
3400	Evaporative, copper coil, pump, fan motor									
3440	10 ton	Q-5	.54	29.630	Ea.	8,400	1,450		9,850	11,400
3460	15 ton		.50	32		8,575	1,550		10,125	11,800
3480	20 ton		.47	34.043		8,850	1,650		10,500	12,200
3500	25 ton		.45	35.556		9,050	1,725		10,775	12,500
3520	30 ton	▼	.42	38.095		9,500	1,850		11,350	13,200
3540	40 ton	Q-6	.49	48.980		10,300	2,475		12,775	15,100
3560	50 ton	▼	.39	61.538		11,700	3,100		14,800	17,600

377

23 63 Refrigerant Condensers

23 63 33 – Evaporative Refrigerant Condensers

23 63 33.10 Evaporative Condensers		Crew	Daily Output	Labor-Hours	Unit	Material	2011 Bare Costs Labor	Equipment	Total	Total Incl O&P
3580	65 ton	Q-6	.35	68.571	Ea.	12,500	3,475		15,975	19,000
3600	80 ton		.33	72.727		13,900	3,675		17,575	20,700
3620	90 ton		.29	82.759		15,200	4,175		19,375	23,000
3640	100 ton	Q-7	.36	88.889		16,700	4,575		21,275	25,200
3660	110 ton		.33	96.970		17,500	5,000		22,500	26,800
3680	125 ton		.30	106		19,000	5,500		24,500	29,000
3700	135 ton		.28	114		20,000	5,900		25,900	30,800
3720	150 ton		.25	128		21,700	6,600		28,300	33,800
3740	165 ton		.23	139		23,000	7,175		30,175	36,000
3760	185 ton		.22	145		26,000	7,500		33,500	39,800
3860	For fan damper control, add	Q-5	2	8		790	390		1,180	1,450

23 64 Packaged Water Chillers

23 64 13 – Absorption Water Chillers

23 64 13.13 Direct-Fired Absorption Water Chillers

		Crew	Daily Output	Labor-Hours	Unit	Material	Labor	Equipment	Total	Total Incl O&P
0010	**DIRECT-FIRED ABSORPTION WATER CHILLERS**									
3000	Gas fired, air cooled									
3220	5 ton	Q-5	.60	26.667	Ea.	8,975	1,300		10,275	11,800
3270	10 ton	"	.40	40	"	24,600	1,950		26,550	29,900
4000	Water cooled, duplex									
4130	100 ton	Q-7	.13	246	Ea.	136,000	17,700		148,700	168,500
4140	200 ton		.11	283		183,000	14,600		197,600	223,500
4150	300 ton		.11	299		219,500	15,400		234,900	264,500
4160	400 ton		.10	316		285,500	16,300		301,800	338,500
4170	500 ton		.10	329		359,500	17,000		376,500	420,500
4180	600 ton		.09	340		424,500	17,500		442,000	492,500
4190	700 ton		.09	359		470,500	18,500		489,000	545,000
4200	800 ton		.08	380		568,000	19,600		587,600	654,500
4210	900 ton		.08	400		658,000	20,600		678,600	755,000
4220	1000 ton		.08	421		731,500	21,700		753,200	837,000

23 64 13.16 Indirect-Fired Absorption Water Chillers

		Crew	Daily Output	Labor-Hours	Unit	Material	Labor	Equipment	Total	Total Incl O&P
0010	**INDIRECT-FIRED ABSORPTION WATER CHILLERS**									
0020	Steam or hot water, water cooled									
0050	100 ton	Q-7	.13	240	Ea.	125,000	12,400		137,400	156,000
0100	148 ton		.12	258		200,500	13,300		213,800	240,500
0200	200 ton		.12	275		223,000	14,200		237,200	266,500
0240	250 ton		.11	290		245,500	15,000		260,500	292,500
0300	354 ton		.11	304		306,500	15,700		322,200	361,000
0400	420 ton		.10	323		335,000	16,700		351,700	394,000
0500	665 ton		.09	340		470,500	17,500		488,000	543,500
0600	750 ton		.09	363		550,000	18,700		568,700	633,000
0700	850 ton		.08	385		587,000	19,900		606,900	675,000
0800	955 ton		.08	410		627,500	21,100		648,600	722,000
0900	1125 ton		.08	421		724,500	21,700		746,200	829,500
1000	1250 ton		.07	444		779,000	22,900		801,900	891,000
1100	1465 ton		.07	463		924,000	23,900		947,900	1,051,500
1200	1660 ton		.07	477		1,087,000	24,600		1,111,600	1,232,000
2000	For two stage unit, add					80%	25%			

23 64 Packaged Water Chillers

23 64 16 – Centrifugal Water Chillers

23 64 16.10 Centrifugal Type Water Chillers	Crew	Daily Output	Labor-Hours	Unit	Material	2011 Bare Costs Labor	Equipment	Total	Total Incl O&P
0010 **CENTRIFUGAL TYPE WATER CHILLERS**, With standard controls									
0020 Centrifugal liquid chiller, water cooled									
0030 not including water tower									
0100 2000 ton (twin 1000 ton units)	Q-7	.07	477	Ea.	847,500	24,600		872,100	969,000
0274 Centrifugal, packaged unit, water cooled, not incl. tower									
0280 400 ton	Q-7	.11	283	Ea.	129,500	14,600		144,100	164,500
0282 450 ton		.11	290		190,500	15,000		205,500	232,000
0290 500 ton		.11	296		212,000	15,300		227,300	256,000
0292 550 ton		.11	304		233,000	15,700		248,700	280,000
0300 600 ton		.10	310		254,000	16,000		270,000	303,500
0302 650 ton		.10	320		275,500	16,500		292,000	327,500
0304 700 ton		.10	326		296,500	16,800		313,300	351,500
0306 750 ton		.10	333		318,000	17,200		335,200	375,000
0310 800 ton		.09	340		339,000	17,500		356,500	399,000
0312 850 ton		.09	351		360,000	18,100		378,100	423,000
0316 900 ton		.09	359		381,500	18,500		400,000	447,000
0318 950 ton		.09	363		402,500	18,700		421,200	471,000
0320 1000 ton		.09	372		424,000	19,200		443,200	494,500
0324 1100 ton		.08	385		453,000	19,900		472,900	528,000
0330 1200 ton		.08	395		494,500	20,400		514,900	574,000
0340 1300 ton		.08	410		535,500	21,100		556,600	620,500
0350 1400 ton		.08	421		576,500	21,700		598,200	667,000
0360 1500 ton		.08	426		618,000	22,000		640,000	712,500
0370 1600 ton		.07	438		659,000	22,600		681,600	758,500
0380 1700 ton		.07	450		700,000	23,200		723,200	804,500
0390 1800 ton		.07	457		741,500	23,600		765,100	850,500
0400 1900 ton		.07	470		782,500	24,200		806,700	897,000
0430 2000 ton		.07	492		824,000	25,400		849,400	944,000
0440 2500 ton	▼	.06	524	▼	1,029,500	27,000		1,056,500	1,173,000

23 64 19 – Reciprocating Water Chillers

23 64 19.10 Reciprocating Type Water Chillers

	Crew	Daily Output	Labor-Hours	Unit	Material	2011 Bare Costs Labor	Equipment	Total	Total Incl O&P
0010 **RECIPROCATING TYPE WATER CHILLERS**, With standard controls									
0494 Water chillers, integral air cooled condenser									
0538 60 ton cooling	Q-7	.28	115	Ea.	47,700	5,925		53,625	61,500
0546 70 ton cooling		.27	119		54,500	6,175		60,675	69,000
0552 80 ton cooling		.26	123		59,000	6,350		65,350	74,500
0554 90 ton cooling		.25	125		63,500	6,500		70,000	79,000
0600 100 ton cooling		.25	129		71,500	6,650		78,150	88,500
0620 110 ton cooling		.24	132		76,500	6,825		83,325	94,000
0630 130 ton cooling		.24	135		91,000	6,950		97,950	110,500
0640 150 ton cooling		.23	137		101,000	7,100		108,100	121,500
0650 175 ton cooling		.23	140		113,000	7,275		120,275	135,500
0654 190 ton cooling		.22	144		124,500	7,425		131,925	148,000
0660 210 ton cooling		.22	148		128,000	7,625		135,625	152,500
0662 250 ton cooling		.21	151		148,500	7,825		156,325	175,000
0664 275 ton cooling		.21	156		154,000	8,050		162,050	181,000
0666 300 ton cooling		.20	160		172,000	8,250		180,250	202,000
0668 330 ton cooling		.20	164		201,000	8,450		209,450	233,500
0670 360 ton cooling		.19	168		221,000	8,675		229,675	256,000
0672 390 ton cooling		.19	172		234,000	8,925		242,925	271,000
0674 420 ton cooling	▼	.18	177	▼	244,500	9,150		253,650	282,500
0980 Water cooled, multiple compressor, semi-hermetic, tower not incl.									

23 64 Packaged Water Chillers

23 64 19 – Reciprocating Water Chillers

23 64 19.10 Reciprocating Type Water Chillers

		Crew	Daily Output	Labor-Hours	Unit	Material	2011 Bare Costs Labor	Equipment	Total	Total Incl O&P
1000	15 ton cooling	Q-6	.36	65.934	Ea.	17,500	3,325		20,825	24,300
1020	25 ton cooling	Q-7	.41	78.049		18,900	4,025		22,925	26,800
1040	28 ton cooling		.36	89.888		20,200	4,625		24,825	29,100
1060	35 ton cooling		.31	101		22,600	5,250		27,850	32,700
1080	40 ton cooling		.30	108		30,200	5,575		35,775	41,500
1100	50 ton cooling		.28	113		32,300	5,875		38,175	44,300
1120	60 ton cooling		.25	125		35,400	6,500		41,900	48,600
1130	75 ton cooling		.23	139		48,900	7,175		56,075	64,500
1140	85 ton cooling		.21	151		51,000	7,825		58,825	67,500
1150	95 ton cooling		.19	164		54,500	8,500		63,000	72,500
1160	100 ton cooling		.18	179		58,500	9,275		67,775	78,500
1170	115 ton cooling		.17	190		61,500	9,825		71,325	82,000
1180	125 ton cooling		.16	196		66,500	10,100		76,600	88,000
1200	145 ton cooling		.16	202		77,500	10,400		87,900	101,000
1210	155 ton cooling		.15	210		82,000	10,800		92,800	106,000
1300	Water cooled, single compressor, semi-hermetic, tower not incl.									
1320	40 ton cooling,	Q-7	.30	108	Ea.	16,000	5,575		21,575	25,900
1340	50 ton cooling		.28	113		19,800	5,875		25,675	30,600
1360	60 ton cooling		.25	125		22,000	6,500		28,500	33,900
1451	Water cooled, dual compressors, semi-hermetic, tower not incl.									
1500	80 ton cooling	Q-7	.14	222	Ea.	27,400	11,500		38,900	47,200
1520	100 ton cooling		.14	228		37,000	11,800		48,800	58,000
1540	120 ton cooling		.14	231		44,800	11,900		56,700	67,000
4000	Packaged chiller, remote air cooled condensers not incl.									
4020	15 ton cooling	Q-7	.30	108	Ea.	14,600	5,575		20,175	24,400
4030	20 ton cooling		.28	115		15,900	5,975		21,875	26,400
4040	25 ton cooling		.25	125		18,000	6,500		24,500	29,500
4050	35 ton cooling		.24	133		19,400	6,900		26,300	31,600
4060	40 ton cooling		.22	144		22,700	7,425		30,125	36,000
4070	50 ton cooling		.21	153		26,300	7,925		34,225	40,800
4080	60 ton cooling		.20	164		30,700	8,450		39,150	46,300
4090	75 ton cooling		.18	173		39,300	8,950		48,250	56,500
4100	85 ton cooling		.17	183		41,300	9,475		50,775	59,500
4110	95 ton cooling		.17	193		44,500	10,000		54,500	64,000
4120	105 ton cooling		.16	203		48,000	10,500		58,500	68,500
4130	115 ton cooling		.15	213		52,000	11,000		63,000	73,500
4140	125 ton cooling		.14	223		55,500	11,500		67,000	78,000
4150	145 ton cooling		.14	233		77,500	12,000		89,500	103,500

23 64 23 – Scroll Water Chillers

23 64 23.10 Scroll Water Chillers

		Crew	Daily Output	Labor-Hours	Unit	Material	2011 Bare Costs Labor	Equipment	Total	Total Incl O&P
0010	**SCROLL WATER CHILLERS,** With standard controls									
0490	Packaged w/integral air cooled condenser, 15 ton cool	Q-7	.37	86.486	Ea.	20,400	4,450		24,850	29,200
0500	20 ton cooling		.34	94.118		26,400	4,850		31,250	36,300
0510	25 ton cooling		.34	94.118		27,600	4,850		32,450	37,700
0515	30 ton cooling		.31	101		28,300	5,250		33,550	39,000
0517	35 ton cooling		.31	104		31,800	5,400		37,200	43,100
0520	40 ton cooling		.30	108		34,100	5,575		39,675	45,800
0528	45 ton cooling		.29	109		35,300	5,675		40,975	47,400
0536	50 ton cooling		.28	113		39,000	5,825		44,825	51,500
0680	Scroll water cooled, single compressor, hermetic, tower not incl.									
0700	2 ton cooling	Q-5	.57	28.070	Ea.	3,225	1,375		4,600	5,600
0710	5 ton cooling		.57	28.070		3,850	1,375		5,225	6,300

23 64 Packaged Water Chillers

23 64 23 – Scroll Water Chillers

23 64 23.10 Scroll Water Chillers	Crew	Daily Output	Labor-Hours	Unit	Material	2011 Bare Costs Labor	Equipment	Total	Total Incl O&P
0720	6 ton cooling	Q-5	.42	38.005	Ea.	4,850	1,850		6,700	8,125
0740	8 ton cooling	↓	.31	52.117		5,325	2,550		7,875	9,650
0760	10 ton cooling	Q-6	.36	67.039		6,125	3,375		9,500	11,800
0780	15 ton cooling	"	.33	72.727		9,625	3,675		13,300	16,100
0800	20 ton cooling	Q-7	.38	83.990		10,900	4,325		15,225	18,500
0820	30 ton cooling	"	.33	96.096	↓	12,200	4,950		17,150	20,900

23 64 26 – Rotary-Screw Water Chillers

23 64 26.10 Rotary-Screw Type Water Chillers

		Crew	Daily Output	Labor-Hours	Unit	Material	Labor	Equipment	Total	Total Incl O&P
0010	**ROTARY-SCREW TYPE WATER CHILLERS**, With standard controls									
0110	Screw, liquid chiller, air cooled, insulated evaporator									
0120	130 ton	Q-7	.14	228	Ea.	83,500	11,800		95,300	109,500
0124	160 ton		.13	246		103,000	12,700		115,700	132,000
0128	180 ton		.13	250		115,500	12,900		128,400	146,000
0132	210 ton		.12	258		127,500	13,300		140,800	160,000
0136	270 ton		.12	266		145,500	13,700		159,200	180,500
0140	320 ton	↓	.12	275	↓	183,000	14,200		197,200	222,000
0200	Packaged unit, water cooled, not incl. tower									
0210	80 ton	Q-7	.14	223	Ea.	42,700	11,500		54,200	64,000
0220	100 ton		.14	230		53,000	11,900		64,900	76,000
0230	150 ton		.13	240		70,500	12,400		82,900	96,500
0240	200 ton		.13	251		82,000	13,000		95,000	109,500
0250	250 ton		.12	260		96,500	13,400		109,900	126,000
0260	300 ton		.12	266		111,000	13,700		124,700	142,500
0270	350 ton	↓	.12	275	↓	158,500	14,200		172,700	195,500
1450	Water cooled, tower not included									
1560	135 ton cooling, screw compressors	Q-7	.14	235	Ea.	55,000	12,100		67,100	78,500
1580	150 ton cooling, screw compressors		.13	240		63,500	12,400		75,900	88,500
1620	200 ton cooling, screw compressors		.13	250		87,500	12,900		100,400	115,500
1660	291 ton cooling, screw compressors	↓	.12	260		91,500	13,400		104,900	121,000

23 64 33 – Direct Expansion Water Chillers

23 64 33.10 Direct Expansion Type Water Chillers

		Crew	Daily Output	Labor-Hours	Unit	Material	Labor	Equipment	Total	Total Incl O&P
0010	**DIRECT EXPANSION TYPE WATER CHILLERS**, With standard controls									
8000	Direct expansion, shell and tube type, for built up systems									
8020	1 ton	Q-5	2	8	Ea.	6,675	390		7,065	7,925
8030	5 ton		1.90	8.421		11,100	410		11,510	12,800
8040	10 ton		1.70	9.412		13,700	460		14,160	15,800
8050	20 ton		1.50	10.667		15,700	520		16,220	18,100
8060	30 ton		1	16		20,000	780		20,780	23,200
8070	50 ton	↓	.90	17.778		32,100	865		32,965	36,600
8080	100 ton	Q-6	.90	26.667	↓	57,000	1,350		58,350	65,000

23 65 13.10 Forced-Draft Type Cooling Towers	Crew	Daily Output	Labor-Hours	Unit	Material	2011 Bare Costs Labor	Equipment	Total	Total Incl O&P
0010 **FORCED-DRAFT TYPE COOLING TOWERS**, Packaged units D3030–310									
0070 Galvanized steel									
0080 Induced draft, crossflow									
0100 Vertical, belt drive, 61 tons	Q-6	90	.267	TonAC	171	13.45		184.45	208
0150 100 ton		100	.240.		163	12.10		175.10	198
0200 115 ton		109	.220		153	11.10		164.10	185
0250 131 ton		120	.200		151	10.10		161.10	181
0260 162 ton		132	.182		149	9.20		158.20	178
1000 For higher capacities, use multiples									
1500 Induced air, double flow									
1900 Vertical, gear drive, 167 ton	Q-6	126	.190	TonAC	145	9.60		154.60	173
2000 297 ton		129	.186		90.50	9.40		99.90	114
2100 582 ton		132	.182		73	9.20		82.20	94
2150 849 ton		142	.169		70.50	8.55		79.05	91
2200 1016 ton		150	.160		65.50	8.10		73.60	84
2500 Blow through, centrifugal type									
2510 50 ton	Q-6	2.28	10.526	Ea.	10,900	530		11,430	12,700
2520 75 ton		1.52	15.789		14,000	795		14,795	16,600
2524 100 ton		1.22	19.672		16,600	995		17,595	19,700
2528 125 ton	Q-7	1.31	24.427		19,400	1,250		20,650	23,200
2532 200 ton		.82	39.216		25,300	2,025		27,325	30,900
2536 250 ton		.65	49.231		30,200	2,525		32,725	37,000
2540 300 ton		.54	59.259		34,700	3,050		37,750	42,800
2544 350 ton		.47	68.085		39,200	3,500		42,700	48,400
2548 400 ton		.41	78.049		40,900	4,025		44,925	51,000
2552 450 ton		.36	88.154		49,200	4,550		53,750	61,000
2556 500 ton		.32	100		52,500	5,150		57,650	65,000
2560 550 ton		.30	107		56,500	5,550		62,050	70,500
2564 600 ton		.27	117		59,000	6,050		65,050	74,000
2568 650 ton		.25	127		66,000	6,575		72,575	82,500
2572 700 ton		.23	137		77,000	7,075		84,075	95,500
2576 750 ton		.22	146		83,000	7,575		90,575	102,500
2580 800 ton		.21	152		89,500	7,850		97,350	110,000
2584 850 ton		.20	156		95,500	8,075		103,575	117,000
2588 900 ton		.20	161		101,000	8,325		109,325	123,500
2592 950 ton		.19	170		106,500	8,775		115,275	130,000
2596 1000 ton		.18	180		110,000	9,325		119,325	135,000
2700 Axial fan, induced draft									
2710 50 ton	Q-6	2.28	10.526	Ea.	9,450	530		9,980	11,200
2720 75 ton		1.52	15.789		11,100	795		11,895	13,400
2724 100 ton		1.22	19.672		11,800	995		12,795	14,400
2728 125 ton	Q-7	1.31	24.427		14,500	1,250		15,750	17,800
2732 200 ton		.82	39.216		24,600	2,025		26,625	30,000
2736 250 ton		.65	49.231		27,700	2,525		30,225	34,300
2740 300 ton		.54	59.259		32,300	3,050		35,350	40,200
2744 350 ton		.47	68.085		34,700	3,500		38,200	43,300
2748 400 ton		.41	78.049		39,700	4,025		43,725	49,700
2752 450 ton		.36	88.154		44,100	4,550		48,650	55,500
2756 500 ton		.32	100		46,800	5,150		51,950	59,000
2760 550 ton		.30	107		52,500	5,550		58,050	66,000
2764 600 ton		.27	117		55,000	6,050		61,050	69,500
2768 650 ton		.25	127		60,500	6,575		67,075	76,500
2772 700 ton		.23	137		68,500	7,075		75,575	86,000

23 65 Cooling Towers

23 65 13 – Forced-Draft Cooling Towers

23 65 13.10 Forced-Draft Type Cooling Towers

	23 65 13.10 Forced-Draft Type Cooling Towers	Crew	Daily Output	Labor-Hours	Unit	Material	2011 Bare Costs Labor	Equipment	Total	Total Incl O&P
2776	750 ton	Q-7	.22	146	Ea.	71,000	7,575		78,575	90,000
2780	800 ton		.21	152		74,500	7,850		82,350	93,000
2784	850 ton		.20	156		75,000	8,075		83,075	94,500
2788	900 ton		.20	161		77,500	8,325		85,825	97,500
2792	950 ton		.19	170		82,000	8,775		90,775	103,000
2796	1000 ton		.18	180		86,500	9,325		95,825	109,000
3000	For higher capacities, use multiples									
3500	For pumps and piping, add	Q-6	38	.632	TonAC	85.50	32		117.50	142
4000	For absorption systems, add				"	75%	75%			
4100	Cooling water chemical feeder	Q-5	3	5.333	Ea.	315	260		575	735
5000	Fiberglass tower on galvanized steel support structure									
5010	Draw thru									
5100	100 ton	Q-6	1.40	17.143	Ea.	10,100	865		10,965	12,400
5120	120 ton		1.20	20		11,700	1,000		12,700	14,400
5140	140 ton		1	24		12,700	1,200		13,900	15,800
5160	160 ton		.80	30		14,100	1,525		15,625	17,800
5180	180 ton		.65	36.923		16,000	1,875		17,875	20,400
5200	200 ton		.48	50		18,100	2,525		20,625	23,700
5300	For stainless steel support structure, add					30%				
5360	For higher capacities, use multiples of each size									
6000	Stainless steel									
6010	Induced draft, crossflow, horizontal, belt drive									
6100	57 ton	Q-6	1.50	16	Ea.	24,100	810		24,910	27,700
6120	91 ton		.99	24.242		29,000	1,225		30,225	33,800
6140	111 ton		.43	55.814		37,000	2,825		39,825	44,900
6160	126 ton		.22	109		37,000	5,500		42,500	48,900
6170	Induced draft, crossflow, vertical, gear drive									
6172	167 ton	Q-6	.75	32	Ea.	47,300	1,625		48,925	54,500
6174	297 ton		.43	55.814		52,500	2,825		55,325	62,000
6176	582 ton		.23	104		83,000	5,275		88,275	99,500
6178	849 ton		.17	141		117,500	7,125		124,625	139,500
6180	1016 ton		.15	160		141,000	8,075		149,075	167,500

23 72 Air-to-Air Energy Recovery Equipment

23 72 13 – Heat-Wheel Air-to-Air Energy-Recovery Equipment

23 72 13.10 Heat-Wheel Air-to-Air Energy Recovery Equip.

	23 72 13.10 Heat-Wheel ...		Crew	Daily Output	Labor-Hours	Unit	Material	2011 Bare Costs Labor	Equipment	Total	Total Incl O&P
0010	**HEAT-WHEEL AIR-TO-AIR ENERGY RECOVERY EQUIPMENT**										
0100	Air to air										
4000	Enthalpy recovery wheel										
4010	1000 max CFM	G	Q-9	1.20	13.333	Ea.	6,050	620		6,670	7,575
4020	2000 max CFM	G		1	16		7,075	745		7,820	8,900
4030	4000 max CFM	G		.80	20		8,200	930		9,130	10,400
4040	6000 max CFM	G		.70	22.857		9,575	1,050		10,625	12,100
4050	8000 max CFM	G	Q-10	1	24		10,600	1,150		11,750	13,400
4060	10,000 max CFM	G		.90	26.667		12,700	1,275		13,975	15,900
4070	20,000 max CFM	G		.80	30		22,900	1,450		24,350	27,400
4080	25,000 max CFM	G		.70	34.286		28,000	1,650		29,650	33,300
4090	30,000 max CFM	G		.50	48		31,100	2,325		33,425	37,700
4100	40,000 max CFM	G		.45	53.333		42,900	2,575		45,475	51,000
4110	50,000 max CFM	G		.40	60		49,900	2,900		52,800	59,500

23 72 Air-to-Air Energy Recovery Equipment

23 72 16 – Heat-Pipe Air-To-Air Energy-Recovery Equipment

23 72 16.10 Heat Pipes		Crew	Daily Output	Labor-Hours	Unit	Material	2011 Bare Costs Labor	Equipment	Total	Total Incl O&P
0010	**HEAT PIPES**									
8000	Heat pipe type, glycol, 50% efficient									
8010	100 MBH, 1700 CFM	1 Stpi	.80	10	Ea.	3,550	540		4,090	4,725
8020	160 MBH, 2700 CFM		.60	13.333		4,775	720		5,495	6,325
8030	620 MBH, 4000 CFM	↓	.40	20	↓	7,225	1,075		8,300	9,575

23 73 Indoor Central-Station Air-Handling Units

23 73 13 – Modular Indoor Central-Station Air-Handling Units

23 73 13.10 Air Handling Units

		Crew	Daily Output	Labor-Hours	Unit	Material	2011 Bare Costs Labor	Equipment	Total	Total Incl O&P
0010	**AIR HANDLING UNITS**, Built-Up									
0100	With cooling/heating coil section, filters, mixing box									
0880	Single zone, horizontal/vertical									
0890	Constant volume									
0900	1600 CFM	Q-5	1.20	13.333	Ea.	3,325	650		3,975	4,625
0906	2000 CFM		1.10	14.545		4,150	710		4,860	5,600
0910	3000 CFM		1	16		5,175	780		5,955	6,875
0916	4000 CFM	↓	.95	16.842		6,900	820		7,720	8,825
0920	5000 CFM	Q-6	1.40	17.143		8,625	865		9,490	10,800
0926	6500 CFM		1.30	18.462		11,200	930		12,130	13,700
0930	7500 CFM		1.20	20		12,900	1,000		13,900	15,700
0936	9200 CFM		1.10	21.818		15,900	1,100		17,000	19,200
0940	11,500 CFM		1	24		17,500	1,200		18,700	21,000
0946	13,200 CFM		.90	26.667		20,100	1,350		21,450	24,100
0950	16,500 CFM		.80	30		25,100	1,525		26,625	29,900
0960	19,500 CFM		.70	34.286		29,600	1,725		31,325	35,200
0970	22,000 CFM		.60	40		33,400	2,025		35,425	39,800
0980	27,000 CFM		.50	48		37,300	2,425		39,725	44,600
0990	34,000 CFM		.40	60		47,000	3,025		50,025	56,000
1000	40,000 CFM		.30	80		51,000	4,050		55,050	62,500
1010	47,000 CFM	↓	.20	120		65,000	6,050		71,050	80,500
1140	60,000 CFM	Q-7	.18	177		75,000	9,150		84,150	95,500
1160	75,000 CFM	"	.16	200	↓	93,500	10,300		103,800	118,500
2000	Variable air volume									
2100	75,000 CFM	Q-7	.12	266	Ea.	94,500	13,700		108,200	124,500
2140	100,000 CFM		.11	290		128,500	15,000		143,500	164,000
2160	150,000 CFM	↓	.10	320	↓	197,000	16,500		213,500	241,000
2300	Multi-zone, horizontal or vertical, blow-thru fan									
2310	3000 CFM	Q-5	1	16	Ea.	6,050	780		6,830	7,825
2314	4000 CFM	"	.95	16.842		7,975	820		8,795	9,975
2318	5000 CFM	Q-6	1.40	17.143		9,925	865		10,790	12,200
2322	6500 CFM		1.30	18.462		12,900	930		13,830	15,600
2326	7500 CFM		1.20	20		14,900	1,000		15,900	17,900
2330	9200 CFM		1.10	21.818		18,300	1,100		19,400	21,900
2334	11,500 CFM		1	24		20,200	1,200		21,400	24,000
2336	13,200 CFM		.90	26.667		23,100	1,350		24,450	27,400
2340	16,500 CFM		.80	30		28,800	1,525		30,325	34,000
2344	19,500 CFM		.70	34.286		34,000	1,725		35,725	40,000
2348	22,000 CFM		.60	40		38,400	2,025		40,425	45,200
2352	27,000 CFM		.50	48		42,900	2,425		45,325	51,000
2356	34,000 CFM	↓	.40	60	↓	54,000	3,025		57,025	64,000

23 73 Indoor Central-Station Air-Handling Units

23 73 13 – Modular Indoor Central-Station Air-Handling Units

23 73 13.10 Air Handling Units	Crew	Daily Output	Labor-Hours	Unit	Material	2011 Bare Costs Labor	Equipment	Total	Total Incl O&P
2360	40,000 CFM	Q-6	.30	80	Ea.	58,500	4,050		62,550	70,500
2364	47,000 CFM	↓	.20	120	↓	74,500	6,050		80,550	91,000

23 73 13.20 Air Handling Units, Packaged Indoor Type

		Crew	Daily Output	Labor-Hours	Unit	Material	Labor	Equipment	Total	Total Incl O&P
0010	**AIR HANDLING UNITS, PACKAGED INDOOR TYPE**									
0040	Cooling coils may be chilled water or DX									
0050	Heating coils may be hot water, steam or electric									
1090	Constant volume									
1200	2000 CFM	Q-5	1.10	14.545	Ea.	9,950	710		10,660	12,000
1400	5000 CFM	Q-6	.80	30		14,600	1,525		16,125	18,300
1550	10,000 CFM		.54	44.444		32,400	2,250		34,650	39,000
1670	15,000 CFM		.38	63.158		46,700	3,200		49,900	56,500
1700	20,000 CFM		.31	77.419		58,500	3,900		62,400	70,000
1710	30,000 CFM	↓	.27	88.889	↓	78,500	4,500		83,000	93,000
2300	Variable air volume									
2330	2000 CFM	Q-5	1	16	Ea.	12,100	780		12,880	14,500
2340	5000 CFM	Q-6	.70	34.286		18,000	1,725		19,725	22,400
2350	10,000 CFM		.50	48		34,600	2,425		37,025	41,700
2360	15,000 CFM		.40	60		49,500	3,025		52,525	59,000
2370	20,000 CFM		.29	82.759		64,500	4,175		68,675	77,500
2380	30,000 CFM	↓	.20	120	↓	88,500	6,050		94,550	106,500

23 73 39 – Indoor, Direct Gas-Fired Heating and Ventilating Units

23 73 39.10 Make-Up Air Unit

		Crew	Daily Output	Labor-Hours	Unit	Material	Labor	Equipment	Total	Total Incl O&P
0010	**MAKE-UP AIR UNIT**									
0020	Indoor suspension, natural/LP gas, direct fired,									
0032	standard control. For flue see Section 23 51 23.10									
0040	70°F temperature rise, MBH is input									
0100	75 MBH input	Q-6	3.60	6.667	Ea.	2,400	335		2,735	3,150
0120	100 MBH input		3.40	7.059		2,425	355		2,780	3,200
0140	125 MBH input		3.20	7.500		2,550	380		2,930	3,375
0160	150 MBH input		3	8		2,850	405		3,255	3,725
0180	175 MBH input		2.80	8.571		2,950	435		3,385	3,875
0200	200 MBH input		2.60	9.231		3,050	465		3,515	4,050
0220	225 MBH input		2.40	10		3,300	505		3,805	4,375
0240	250 MBH input		2	12		3,925	605		4,530	5,200
0260	300 MBH input		1.90	12.632		4,025	640		4,665	5,400
0280	350 MBH input		1.75	13.714		4,825	695		5,520	6,325
0300	400 MBH input		1.60	15		7,450	760		8,210	9,300
0320	840 MBH input		1.20	20		10,400	1,000		11,400	13,000
0340	1200 MBH input	↓	.88	27.273		14,900	1,375		16,275	18,500
0600	For discharge louver assembly, add					5%				
0700	For filters, add					10%				
0800	For air shut-off damper section, add				↓	30%				

23 74 Packaged Outdoor HVAC Equipment

23 74 13 – Packaged, Outdoor, Central-Station Air-Handling Units

23 74 13.10 Packaged, Outdoor Type, Central-Station AHU		Crew	Daily Output	Labor-Hours	Unit	Material	2011 Bare Costs Labor	Equipment	Total	Total Incl O&P
0010	**PACKAGED, OUTDOOR TYPE, CENTRAL-STATION AIR-HANDLING UNITS**									
2900	Cooling coils may be chilled water or DX									
2910	Heating coils may be hot water, steam or electric									
3000	Weathertight, ground or rooftop									
3010	Constant volume									
3100	2000 CFM	Q-6	1	24	Ea.	10,100	1,200		11,300	12,900
3140	5000 CFM		.79	30.380		24,500	1,525		26,025	29,300
3150	10,000 CFM		.52	46.154		52,000	2,325		54,325	60,500
3160	15,000 CFM		.41	58.537		76,500	2,950		79,450	88,500
3170	20,000 CFM		.30	80		106,000	4,050		110,050	122,500
3180	30,000 CFM		.24	100		158,000	5,050		163,050	181,500
3200	Variable air volume									
3210	2000 CFM	Q-6	.96	25	Ea.	12,200	1,275		13,475	15,300
3240	5000 CFM		.73	32.877		27,600	1,650		29,250	32,900
3250	10,000 CFM		.50	48		58,500	2,425		60,925	67,500
3260	15,000 CFM		.40	60		87,500	3,025		90,525	100,500
3270	20,000 CFM		.29	82.759		112,000	4,175		116,175	129,500
3280	30,000 CFM		.20	120		158,000	6,050		164,050	183,000

23 74 23 – Packaged, Outdoor, Heating-Only Makeup-Air Units

23 74 23.16 Packaged, Ind.-Fired, In/Outdoor

			Crew	Daily Output	Labor-Hours	Unit	Material	Labor	Equipment	Total	Total Incl O&P
0010	**PACKAGED, IND.-FIRED, IN/OUTDOOR**, Heat-Only Makeup-Air Units										
1000	Rooftop unit, natural gas, gravity vent, S.S. exchanger										
1010	70°F temperature rise, MBH is input										
1020	250 MBH		Q-6	4	6	Ea.	11,400	305		11,705	13,000
1040	400 MBH			3.60	6.667		11,400	335		11,735	13,100
1060	550 MBH			3.30	7.273		12,500	365		12,865	14,300
1080	750 MBH			3	8		13,800	405		14,205	15,800
1100	1000 MBH			2.60	9.231		16,400	465		16,865	18,700
1120	1750 MBH			2.30	10.435		20,600	525		21,125	23,500
1140	2500 MBH			1.90	12.632		26,900	640		27,540	30,600
1160	3250 MBH			1.40	17.143		32,500	865		33,365	37,100
1180	4000 MBH			1.20	20		39,700	1,000		40,700	45,200
1200	6000 MBH			1	24		49,400	1,200		50,600	56,500
1600	For cleanable filters, add						5%				
1700	For electric modulating gas control, add						10%				
8000	Rooftop unit, electric, with curb, controls										
8010	33 kW		Q-5	4	4	Ea.	3,550	195		3,745	4,200
8020	100 kW			3.60	4.444		5,925	216		6,141	6,850
8030	150 kW			3.20	5		6,825	244		7,069	7,875
8040	250 kW			2.80	5.714		8,950	278		9,228	10,200
8050	400 kW			2.20	7.273		13,200	355		13,555	15,000

23 74 33 – Dedicated Outdoor-Air Units

23 74 33.10 Roof Top Air Conditioners

			Crew	Daily Output	Labor-Hours	Unit	Material	Labor	Equipment	Total	Total Incl O&P
0010	**ROOF TOP AIR CONDITIONERS**, Standard controls, curb, economizer										
1000	Single zone, electric cool, gas heat										
1100	3 ton cooling, 60 MBH heating	D3050-150	Q-5	.70	22.857	Ea.	3,600	1,125		4,725	5,600
1120	4 ton cooling, 95 MBH heating	D3050-155		.61	26.403		4,250	1,275		5,525	6,600
1140	5 ton cooling, 112 MBH heating	R236000-10		.56	28.521		4,975	1,400		6,375	7,550
1145	6 ton cooling, 140 MBH heating	R236000-20		.52	30.769		5,825	1,500		7,325	8,650
1150	7.5 ton cooling, 170 MBH heating			.50	32.258		7,650	1,575		9,225	10,800
1160	10 ton cooling, 200 MBH heating		Q-6	.67	35.982		10,600	1,825		12,425	14,300
1170	12.5 ton cooling, 230 MBH heating			.63	37.975		13,400	1,925		15,325	17,600

23 74 33.10 Roof Top Air Conditioners		Crew	Daily Output	Labor-Hours	Unit	Material	2011 Bare Costs Labor	Equipment	Total	Total Incl O&P
1180	15 ton cooling, 270 MBH heating	Q-6	.57	42.032	Ea.	15,900	2,125		18,025	20,700
1190	17.5 ton cooling, 330 MBH heating	↓	.52	45.889		18,800	2,325		21,125	24,200
1200	20 ton cooling, 360 MBH heating	Q-7	.67	47.976		21,900	2,475		24,375	27,800
1210	25 ton cooling, 450 MBH heating		.56	57.554		25,000	2,975		27,975	31,900
1220	30 ton cooling, 540 MBH heating		.47	68.376		25,500	3,525		29,025	33,400
1240	40 ton cooling, 675 MBH heating		.35	91.168		33,400	4,700		38,100	43,700
1260	50 ton cooling, 810 MBH heating		.28	113		39,500	5,875		45,375	52,500
1265	60 ton cooling, 900 MBH heating		.23	136		47,800	7,050		54,850	63,000
1270	80 ton cooling, 1000 MBH heating		.20	160		79,000	8,250		87,250	99,500
1275	90 ton cooling, 1200 MBH heating		.17	188		82,000	9,700		91,700	104,500
1280	100 ton cooling, 1350 MBH heating	↓	.14	228	↓	100,000	11,800		111,800	127,500
1300	Electric cool, electric heat									
1310	5 ton	Q-5	.63	25.600	Ea.	3,750	1,250		5,000	5,975
1314	10 ton	Q-6	.74	32.432		6,825	1,650		8,475	9,975
1318	15 ton	"	.64	37.795		10,300	1,900		12,200	14,200
1322	20 ton	Q-7	.74	43.185		13,700	2,225		15,925	18,300
1326	25 ton		.62	51.696		17,100	2,675		19,775	22,800
1330	30 ton		.52	61.657		20,500	3,175		23,675	27,400
1334	40 ton		.39	82.051		27,300	4,225		31,525	36,400
1338	50 ton		.31	102		32,500	5,275		37,775	43,600
1342	60 ton		.26	123		39,000	6,350		45,350	52,500
1346	75 ton	↓	.21	153	↓	48,700	7,925		56,625	65,500
2000	Multizone, electric cool, gas heat, economizer									
2100	15 ton cooling, 360 MBH heating	Q-7	.61	52.545	Ea.	57,000	2,700		59,700	66,500
2120	20 ton cooling, 360 MBH heating		.53	60.038		61,000	3,100		64,100	72,000
2140	25 ton cooling, 450 MBH heating		.45	71.910		73,500	3,700		77,200	86,000
2160	28 ton cooling, 450 MBH heating		.41	79.012		83,500	4,075		87,575	98,000
2180	30 ton cooling, 540 MBH heating		.37	85.562		94,000	4,400		98,400	109,500
2200	40 ton cooling, 540 MBH heating		.28	113		107,000	5,875		112,875	127,000
2210	50 ton cooling, 540 MBH heating		.23	142		133,500	7,325		140,825	157,500
2220	70 ton cooling, 1500 MBH heating		.16	198		144,500	10,200		154,700	174,500
2240	80 ton cooling, 1500 MBH heating		.14	228		165,000	11,800		176,800	199,000
2260	90 ton cooling, 1500 MBH heating		.13	256		173,000	13,200		186,200	209,500
2280	105 ton cooling, 1500 MBH heating	↓	.11	290		190,500	15,000		205,500	232,000
2400	For hot water heat coil, deduct					5%				
2500	For steam heat coil, deduct					2%				
2600	For electric heat, deduct				↓	3%	5%			
4000	Single zone, electric cool, gas heat, variable air volume									
4100	12.5 ton cooling, 230 MBH heating	Q-6	.55	43.716	Ea.	15,100	2,200		17,300	19,900
4120	18 ton cooling, 330 MBH heating	"	.46	52.747		20,800	2,675		23,475	26,900
4140	25 ton cooling, 450 MBH heating	Q-7	.48	66.116		27,500	3,400		30,900	35,300
4160	40 ton cooling, 675 MBH heating		.31	104		37,800	5,400		43,200	49,700
4180	60 ton cooling, 900 MBH heating		.20	157		55,500	8,125		63,625	73,000
4200	80 ton cooling, 1000 MBH heating	↓	.15	209	↓	88,000	10,800		98,800	112,500
5000	Single zone, electric cool only									
5050	3 ton cooling	Q-5	.88	18.203	Ea.	2,475	885		3,360	4,050
5060	4 ton cooling		.76	21.108		2,875	1,025		3,900	4,675
5070	5 ton cooling		.70	22.792		3,525	1,100		4,625	5,525
5080	6 ton cooling		.65	24.502		3,975	1,200		5,175	6,150
5090	7.5 ton cooling		.62	25.806		5,550	1,250		6,800	7,975
5100	8.5 ton cooling	↓	.59	26.981		6,200	1,325		7,525	8,775
5110	10 ton cooling	Q-6	.83	28.812		7,325	1,450		8,775	10,200
5120	12.5 ton cooling	↓	.80	30	↓	9,350	1,525		10,875	12,600

23 74 Packaged Outdoor HVAC Equipment

23 74 33 – Dedicated Outdoor-Air Units

23 74 33.10 Roof Top Air Conditioners		Crew	Daily Output	Labor-Hours	Unit	Material	2011 Bare Costs Labor	Equipment	Total	Total Incl O&P
5130	15 ton cooling	Q-6	.71	33.613	Ea.	15,100	1,700		16,800	19,200
5140	17.5 ton cooling		.65	36.697		18,600	1,850		20,450	23,300
5150	20 ton cooling	Q-7	.83	38.415		20,700	1,975		22,675	25,800
5160	25 ton cooling		.70	45.977		22,700	2,375		25,075	28,500
5170	30 ton cooling		.59	54.701		24,200	2,825		27,025	30,800
5180	40 ton cooling		.44	73.059		32,500	3,775		36,275	41,400
5400	For low heat, add					7%				
5410	For high heat, add					10%				
6000	Single zone electric cooling with variable volume distribution									
6020	20 ton cooling	Q-7	.72	44.199	Ea.	24,600	2,275		26,875	30,400
6030	25 ton cooling		.61	52.893		28,900	2,725		31,625	35,900
6040	30 ton cooling		.51	62.868		29,700	3,250		32,950	37,500
6050	40 ton cooling		.38	83.990		39,900	4,325		44,225	50,500
6060	50 ton cooling		.31	104		44,200	5,400		49,600	56,500
6070	60 ton cooling		.25	125		52,000	6,500		58,500	67,000
6200	For low heat, add					7%				
6210	For high heat, add					10%				
7000	Multizone, cool/heat, variable volume distribution									
7100	50 ton cooling	Q-7	.20	164	Ea.	99,000	8,450		107,450	121,500
7110	70 ton cooling		.14	228		138,500	11,800		150,300	170,000
7120	90 ton cooling		.11	296		159,000	15,300		174,300	198,000
7130	105 ton cooling		.10	333		178,000	17,200		195,200	221,500
7140	120 ton cooling		.08	380		203,500	19,600		223,100	253,500
7150	140 ton cooling		.07	444		218,000	22,900		240,900	274,000
7300	Penthouse unit, cool/heat, variable volume distribution									
7400	150 ton cooling	Q-7	.10	310	Ea.	356,500	16,000		372,500	416,000
7410	170 ton cooling		.10	320		380,500	16,500		397,000	443,500
7420	200 ton cooling		.10	326		448,000	16,800		464,800	517,500
7430	225 ton cooling		.09	340		490,000	17,500		507,500	565,000
7440	250 ton cooling		.09	359		505,500	18,500		524,000	583,500
7450	270 ton cooling		.09	363		546,000	18,700		564,700	628,500
7460	300 ton cooling		.09	372		607,000	19,200		626,200	696,000
9400	Sub assemblies for assembly systems									
9410	Ductwork, per ton, rooftop 1 zone units	Q-9	.96	16.667	Ton	249	775		1,024	1,450
9420	Ductwork, per ton, rooftop multizone units		.50	32		340	1,475		1,815	2,625
9430	Ductwork, VAV cooling/ton, rooftop multizone, 1 zone/ton		.32	50		460	2,325		2,785	4,025
9440	Ductwork per ton, packaged water & air cooled units		1.26	12.698		70	590		660	965
9450	Ductwork per ton, split system remote condensing units		1.32	12.121		66	565		631	925
9500	Ductwork package for residential units, 1 ton		.80	20	Ea.	340	930		1,270	1,775
9520	2 ton		.40	40		685	1,850		2,535	3,550
9530	3 ton		.27	59.259		1,025	2,750		3,775	5,300
9540	4 ton	Q-10	.30	80		1,375	3,850		5,225	7,325
9550	5 ton		.23	104		1,700	5,025		6,725	9,475
9560	6 ton		.20	120		2,050	5,775		7,825	11,000
9570	7 ton		.17	141		2,400	6,800		9,200	12,900
9580	8 ton		.15	160		2,725	7,700		10,425	14,700

23 76 Evaporative Air-Cooling Equipment

23 76 13 – Direct Evaporative Air Coolers

23 76 13.10 Evaporative Coolers

		Crew	Daily Output	Labor-Hours	Unit	Material	2011 Bare Costs Labor	2011 Bare Costs Equipment	Total	Total Incl O&P
0010	**EVAPORATIVE COOLERS**, (Swamp Coolers) ducted, not incl. duct.									
0100	Side discharge style, capacities at .25" S.P.									
0120	1785 CFM, 1/3 HP, 115 V [G]	Q-9	5	3.200	Ea.	445	149		594	715
0140	2740 CFM, 1/3 HP, 115 V [G]		4.50	3.556		540	165		705	845
0160	3235 CFM, 1/2 HP, 115 V [G]		4	4		555	186		741	890
0180	3615 CFM, 1/2 HP, 230 V [G]		3.60	4.444		685	207		892	1,075
0200	4215 CFM, 3/4 HP, 230 V [G]		3.20	5		710	232		942	1,125
0220	5255 CFM. 1 HP, 115/230 V [G]		3	5.333		1,400	248		1,648	1,925
0240	6090 CFM, 1 HP, 230/460 V [G]		2.80	5.714		1,475	266		1,741	2,025
0260	8300 CFM, 1-1/2 HP, 230/460 V [G]		2.60	6.154		2,150	286		2,436	2,800
0280	8360 CFM, 1-1/2 HP, 230/460 V [G]		2.20	7.273		2,150	340		2,490	2,875
0300	9725 CFM, 2 HP, 230/460 V [G]		1.80	8.889		2,275	415		2,690	3,125
0320	11,715 CFM, 3 HP, 230/460 V [G]		1.40	11.429		2,800	530		3,330	3,875
0340	14,410 CFM, 5 HP, 230/460 V [G]		1	16		2,950	745		3,695	4,375
0400	For two-speed motor, add					5%				
0500	For down discharge style, add					10%				

23 76 26 – Fan Coil Evaporators for Coolers/Freezers

23 76 26.10 Evaporators

		Crew	Daily Output	Labor-Hours	Unit	Material	2011 Bare Costs Labor	2011 Bare Costs Equipment	Total	Total Incl O&P
0010	**EVAPORATORS**, DX coils, remote compressors not included.									
1000	Coolers, reach-in type, above freezing temperatures									
1300	Shallow depth, wall mount, 7 fins per inch, air defrost									
1310	600 BTUH, 8" fan, rust-proof core	Q-5	3.80	4.211	Ea.	640	205		845	1,000
1320	900 BTUH, 8" fan, rust-proof core		3.30	4.848		740	236		976	1,175
1330	1200 BTUH, 8" fan, rust-proof core		3	5.333		835	260		1,095	1,300
1340	1800 BTUH, 8" fan, rust-proof core		2.40	6.667		970	325		1,295	1,550
1350	2500 BTUH, 10" fan, rust-proof core		2.20	7.273		995	355		1,350	1,625
1360	3500 BTUH, 10" fan		2	8		1,125	390		1,515	1,825
1370	4500 BTUH, 10" fan		1.90	8.421		1,175	410		1,585	1,900
1390	For pan drain, add					55			55	60.50
1600	Undercounter refrigerators, ceiling or wall mount,									
1610	8 fins per inch, air defrost, rust-proof core									
1630	800 BTUH, one 6" fan	Q-5	5	3.200	Ea.	288	156		444	550
1640	1300 BTUH, two 6" fans		4	4		350	195		545	675
1650	1700 BTUH, two 6" fans		3.60	4.444		390	216		606	755
2000	Coolers, reach-in and walk-in types, above freezing temperatures									
2600	Two-way discharge, ceiling mount, 150-4100 CFM,									
2610	above 34°F applications, air defrost									
2630	900 BTUH, 7 fins per inch, 8" fan	Q-5	4	4	Ea.	520	195		715	860
2660	2500 BTUH, 7 fins per inch, 10" fan		2.60	6.154		815	300		1,115	1,350
2690	5500 BTUH, 7 fins per inch, 12" fan		1.70	9.412		1,400	460		1,860	2,225
2720	8500 BTUH, 8 fins per inch, 16" fan		1.20	13.333		1,700	650		2,350	2,825
2750	15,000 BTUH, 7 fins per inch, 18" fan		1.10	14.545		2,350	710		3,060	3,625
2770	24,000 BTUH, 7 fins per inch, two 16" fans		1	16		3,525	780		4,305	5,050
2790	30,000 BTUH, 7 fins per inch, two 18" fans		.90	17.778		4,375	865		5,240	6,100
2850	Two-way discharge, low profile, ceiling mount,									
2860	8 fins per inch, 200-570 CFM, air defrost									
2880	800 BTUH, one 6" fan	Q-5	4	4	Ea.	345	195		540	670
2890	1300 BTUH, two 6" fans		3.50	4.571		415	223		638	785
2900	1800 BTUH, two 6" fans		3.30	4.848		435	236		671	825
2910	2700 BTUH, three 6" fans		2.70	5.926		510	289		799	990
3000	Coolers, walk-in type, above freezing temperatures									
3300	General use, ceiling mount, 108-2080 CFM									

23 76 26.10 Evaporators	Crew	Daily Output	Labor-Hours	Unit	Material	2011 Bare Costs Labor	Equipment	Total	Total Incl O&P
3320	600 BTUH, 7 fins per inch, 6" fan	Q-5	5.20	3.077	Ea.	305	150		455	565
3340	1200 BTUH, 7 fins per inch, 8" fan		3.70	4.324		425	211		636	780
3360	1800 BTUH, 7 fins per inch, 10" fan		2.70	5.926		595	289		884	1,075
3380	3500 BTUH, 7 fins per inch, 12" fan		2.40	6.667		875	325		1,200	1,450
3400	5500 BTUH, 8 fins per inch, 12" fan		2	8		975	390		1,365	1,650
3430	8500 BTUH, 8 fins per inch, 16" fan		1.40	11.429		1,450	555		2,005	2,425
3460	15,000 BTUH, 7 fins per inch, two 16" fans		1.10	14.545		2,325	710		3,035	3,625
3640	Low velocity, high latent load, ceiling mount,									
3650	1050-2420 CFM, air defrost, 6 fins per inch									
3670	6700 BTUH, two 10" fans	Q-5	1.30	12.308	Ea.	1,925	600		2,525	3,000
3680	10,000 BTUH, three 10" fans		1	16		2,425	780		3,205	3,850
3690	13,500 BTUH, three 10" fans		1	16		2,775	780		3,555	4,225
3700	18,000 BTUH, four 10" fans		1	16		3,750	780		4,530	5,300
3710	26,500 BTUH, four 10" fans		.80	20		4,375	975		5,350	6,275
3730	For electric defrost, add					17%				
5000	Freezers and coolers, reach-in type, above 34°F									
5030	to sub-freezing temperature range, low latent load,									
5050	air defrost, 7 fins per inch									
5070	1200 BTUH, 8" fan, rustproof core	Q-5	4.40	3.636	Ea.	570	177		747	890
5080	1500 BTUH, 8" fan, rustproof core		4.20	3.810		580	186		766	915
5090	1800 BTUH, 8" fan, rustproof core		3.70	4.324		600	211		811	975
5100	2500 BTUH, 10" fan, rustproof core		2.90	5.517		650	269		919	1,125
5110	3500 BTUH, 10" fan, rustproof core		2.50	6.400		815	310		1,125	1,350
5120	4500 BTUH, 12" fan, rustproof core		2.10	7.619		1,000	370		1,370	1,650
6000	Freezers and coolers, walk-in type									
6050	1960-18,000 CFM, medium profile									
6060	Standard motor, 6 fins per inch, to -30°F, all aluminum									
6080	10,500 BTUH, air defrost, one 18" fan	Q-5	1.40	11.429	Ea.	1,550	555		2,105	2,525
6090	12,500 BTUH, air defrost, one 18" fan		1.40	11.429		1,675	555		2,230	2,650
6100	16,400 BTUH, air defrost, one 18" fan		1.30	12.308		1,675	600		2,275	2,750
6110	20,900 BTUH, air defrost, one 18" fans		1	16		1,975	780		2,755	3,350
6120	27,000 BTUH, air defrost, two 18" fans		1	16		2,925	780		3,705	4,375
6130	32,900 BTUH, air defrost, two 20" fans	Q-6	1.30	18.462		3,250	930		4,180	4,975
6140	39,000 BTUH, air defrost, two 20" fans		1.25	19.200		3,775	970		4,745	5,625
6150	44,100 BTUH, air defrost, three 20" fans		1	24		4,300	1,200		5,500	6,525
6155	53,000 BTUH, air defrost, three 20" fans		1	24		4,300	1,200		5,500	6,525
6160	66,200 BTUH, air defrost, four 20" fans		1	24		4,875	1,200		6,075	7,175
6170	78,000 BTUH, air defrost, four 20" fans		.80	30		6,075	1,525		7,600	8,950
6180	88,200 BTUH, air defrost, four 24" fans		.60	40		6,450	2,025		8,475	10,100
6190	110,000 BTUH, air defrost, four 24" fans		.50	48		7,325	2,425		9,750	11,700
6330	Hot gas defrost, standard motor units, add					35%				
6370	Electric defrost, 230V, standard motor, add					16%				
6410	460V, standard or high capacity motor, add					25%				
6800	Eight fins per inch, increases BTUH 75%									
6810	Standard capacity, add				Ea.	5%				
6830	Hot gas & electric defrost not recommended									
7000	800-4150 CFM, 12" fans									
7010	Four fins per inch									
7030	3400 BTUH, 1 fan	Q-5	2.80	5.714	Ea.	520	278		798	990
7040	4200 BTUH, 1 fan		2.40	6.667		715	325		1,040	1,275
7050	5300 BTUH, 1 fan		1.90	8.421		785	410		1,195	1,475
7060	6800 BTUH, 1 fan		1.70	9.412		935	460		1,395	1,700
7070	8400 BTUH, 2 fans		1.60	10		1,200	485		1,685	2,050

23 76 Evaporative Air-Cooling Equipment

23 76 26 – Fan Coil Evaporators for Coolers/Freezers

23 76 26.10 Evaporators	Crew	Daily Output	Labor-Hours	Unit	Material	2011 Bare Costs Labor	Equipment	Total	Total Incl O&P
7080	10,500 BTUH, 2 fans	Q-5	1.40	11.429	Ea.	1,350	555		1,905	2,300
7090	13,000 BTUH, 3 fans		1.30	12.308		1,850	600		2,450	2,950
7100	17,000 BTUH, 4 fans		1.20	13.333		2,375	650		3,025	3,575
7110	21,500 BTUH, 5 fans		1	16		2,950	780		3,730	4,400
7160	Six fins per inch increases BTUH 35%, add					5%				
7180	Eight fins per inch increases BTUH 53%,									
7190	not recommended for below 35°F, add				Ea.	8%				
7210	For 12°F temperature differential, add					20%				
7230	For adjustable thermostat control, add					132			132	145
7240	For hot gas defrost, except 8 fin models									
7250	on applications below 35°F, add				Ea.	58%				
7260	For electric defrost, except 8 fin models									
7270	on applications below 35°F, add				Ea.	36%				
8000	Freezers, pass-thru door uprights, walk-in storage									
8300	Low temperature, thin profile, electric defrost									
8310	138-1800 CFM, 5 fins per inch									
8320	900 BTUH, one 6" fan	Q-5	3.30	4.848	Ea.	475	236		711	875
8340	1500 BTUH, two 6" fans		2.20	7.273		745	355		1,100	1,350
8360	2600 BTUH, three 6" fans		1.90	8.421		1,000	410		1,410	1,725
8370	3300 BTUH, four 6" fans		1.70	9.412		1,175	460		1,635	1,975
8380	4400 BTUH, five 6" fans		1.40	11.429		1,650	555		2,205	2,625
8400	7000 BTUH, three 10" fans		1.30	12.308		1,850	600		2,450	2,950
8410	8700 BTUH, four 10" fans		1.10	14.545		2,250	710		2,960	3,525
8450	For air defrost, deduct					240			240	264

23 81 Decentralized Unitary HVAC Equipment

23 81 13 – Packaged Terminal Air-Conditioners

23 81 13.10 Packaged Cabinet Type Air-Conditioners

		Crew	Daily Output	Labor-Hours	Unit	Material	2011 Bare Costs Labor	Equipment	Total	Total Incl O&P
0010	**PACKAGED CABINET TYPE AIR-CONDITIONERS**, Cabinet, wall sleeve,									
0100	louver, electric heat, thermostat, manual changeover, 208 V									
0200	6,000 BTUH cooling, 8800 BTU heat	Q-5	6	2.667	Ea.	715	130		845	980
0220	9,000 BTUH cooling, 13,900 BTU heat		5	3.200		810	156		966	1,125
0240	12,000 BTUH cooling, 13,900 BTU heat		4	4		895	195		1,090	1,275
0260	15,000 BTUH cooling, 13,900 BTU heat		3	5.333		985	260		1,245	1,475
0280	18,000 BTUH cooling, 10 KW heat		2.40	6.667		1,725	325		2,050	2,350
0300	24,000 BTUH cooling, 10 KW heat		1.90	8.421		1,750	410		2,160	2,525
0320	30,000 BTUH cooling, 10 KW heat		1.40	11.429		1,975	555		2,530	3,000
0340	36,000 BTUH cooling, 10 KW heat		1.25	12.800		2,075	625		2,700	3,200
0360	42,000 BTUH cooling, 10 KW heat		1	16		2,550	780		3,330	3,975
0380	48,000 BTUH cooling, 10 KW heat		.90	17.778		2,700	865		3,565	4,275
0500	For hot water coil, increase heat by 10%, add					5%	10%			
1000	For steam, increase heat output by 30%, add					8%	10%			

23 81 19 – Self-Contained Air-Conditioners

23 81 19.10 Window Unit Air Conditioners

		Crew	Daily Output	Labor-Hours	Unit	Material	2011 Bare Costs Labor	Equipment	Total	Total Incl O&P
0010	**WINDOW UNIT AIR CONDITIONERS**									
4000	Portable/window, 15 amp 125 V grounded receptacle required									
4060	5000 BTUH	1 Carp	8	1	Ea.	232	43		275	320
4340	6000 BTUH		8	1		278	43		321	370
4480	8000 BTUH		6	1.333		430	57.50		487.50	560
4500	10,000 BTUH		6	1.333		540	57.50		597.50	685

23 81 19.10 Window Unit Air Conditioners	Crew	Daily Output	Labor-Hours	Unit	Material	2011 Bare Costs Labor	Equipment	Total	Total Incl O&P
4520 12,000 BTUH	L-2	8	2	Ea.	660	75.50		735.50	840
4600 Window/thru-the-wall, 15 amp 230 V grounded receptacle required									
4780 17,000 BTUH	L-2	6	2.667	Ea.	890	101		991	1,125
4940 25,000 BTUH		4	4		1,125	151		1,276	1,475
4960 29,000 BTUH		4	4		1,225	151		1,376	1,575

23 81 19.20 Self-Contained Single Package	Crew	Daily Output	Labor-Hours	Unit	Material	2011 Bare Costs Labor	Equipment	Total	Total Incl O&P
0010 **SELF-CONTAINED SINGLE PACKAGE**									
0100 Air cooled, for free blow or duct, not incl. remote condenser									
0110 Constant volume									
0200 3 ton cooling	Q-5	1	16	Ea.	3,475	780		4,255	5,000
0210 4 ton cooling	"	.80	20		3,800	975		4,775	5,625
0220 5 ton cooling	Q-6	1.20	20		4,150	1,000		5,150	6,075
0230 7.5 ton cooling	"	.90	26.667		5,300	1,350		6,650	7,825
0240 10 ton cooling	Q-7	1	32		7,000	1,650		8,650	10,200
0250 15 ton cooling		.95	33.684		9,725	1,725		11,450	13,300
0260 20 ton cooling		.90	35.556		15,600	1,825		17,425	19,900
0270 25 ton cooling		.85	37.647		17,200	1,950		19,150	21,800
0280 30 ton cooling		.80	40		23,600	2,050		25,650	29,100
0300 40 ton cooling		.60	53.333		34,200	2,750		36,950	41,700
0320 50 ton cooling		.50	64		44,500	3,300		47,800	54,000
0340 60 ton cooling	Q-8	.40	80		53,500	4,125	81	57,706	65,000
0380 With hot water heat, variable air volume									
0390 10 ton cooling	Q-7	.96	33.333	Ea.	9,125	1,725		10,850	12,600
0400 20 ton cooling		.86	37.209		19,600	1,925		21,525	24,500
0410 30 ton cooling		.76	42.105		29,700	2,175		31,875	36,000
0420 40 ton cooling		.58	55.172		42,000	2,850		44,850	50,500
0430 50 ton cooling		.48	66.667		54,500	3,425		57,925	65,000
0440 60 ton cooling	Q-8	.38	84.211		66,500	4,350	85	70,935	80,000
0490 For duct mounting no price change									
0500 For steam heating coils, add				Ea.	10%	10%			
0550 Packaged, with electric heat									
0552 Constant volume									
0560 5 ton cooling	Q-6	1.20	20	Ea.	7,375	1,000		8,375	9,600
0570 10 ton cooling	Q-7	1	32		11,900	1,650		13,550	15,600
0580 20 ton cooling		.90	35.556		22,700	1,825		24,525	27,600
0590 30 ton cooling		.80	40		35,500	2,050		37,550	42,100
0600 40 ton cooling		.60	53.333		47,000	2,750		49,750	55,500
0610 50 ton cooling		.50	64		62,000	3,300		65,300	73,500
0700 Variable air volume									
0710 10 ton cooling	Q-7	1	32	Ea.	13,400	1,650		15,050	17,200
0720 20 ton cooling		.90	35.556		25,100	1,825		26,925	30,300
0730 30 ton cooling		.80	40		39,200	2,050		41,250	46,200
0740 40 ton cooling		.60	53.333		51,500	2,750		54,250	60,500
0750 50 ton cooling		.50	64		67,500	3,300		70,800	79,500
0760 60 ton cooling	Q-8	.40	80		77,500	4,125	81	81,706	91,000
1000 Water cooled for free blow or duct, not including tower									
1010 Constant volume									
1100 3 ton cooling	Q-6	1	24	Ea.	3,475	1,200		4,675	5,625
1120 5 ton cooling		1	24		4,550	1,200		5,750	6,825
1130 7.5 ton cooling		.80	30		6,400	1,525		7,925	9,300
1140 10 ton cooling	Q-7	.90	35.556		8,550	1,825		10,375	12,100
1150 15 ton cooling		.85	37.647		14,100	1,950		16,050	18,400

23 81 Decentralized Unitary HVAC Equipment

23 81 19 – Self-Contained Air-Conditioners

23 81 19.20 Self-Contained Single Package

		Crew	Daily Output	Labor-Hours	Unit	Material	2011 Bare Costs Labor	Equipment	Total	Total Incl O&P
1160	20 ton cooling	Q-7	.80	40	Ea.	26,800	2,050		28,850	32,600
1170	25 ton cooling		.75	42.667		31,600	2,200		33,800	38,100
1180	30 ton cooling		.70	45.714		35,500	2,350		37,850	42,600
1200	40 ton cooling		.40	80		45,000	4,125		49,125	56,000
1220	50 ton cooling		.30	106		56,000	5,500		61,500	69,500
1240	60 ton cooling	Q-8	.30	106		65,500	5,500	108	71,108	80,500
1300	For hot water or steam heat coils, add					12%	10%			
2000	Water cooled with electric heat, not including tower									
2010	Constant volume									
2100	5 ton cooling	Q-6	1.20	20	Ea.	7,775	1,000		8,775	10,100
2120	10 ton cooling	Q-7	1	32		13,500	1,650		15,150	17,300
2130	15 ton cooling	"	.90	35.556		21,200	1,825		23,025	26,000
2400	Variable air volume									
2440	10 ton cooling	Q-7	1	32	Ea.	14,900	1,650		16,550	18,900
2450	15 ton cooling	"	.90	35.556	"	23,600	1,825		25,425	28,600
2600	Water cooled, hot water coils, not including tower									
2610	Variable air volume									
2640	10 ton cooling	Q-7	1	32	Ea.	10,800	1,650		12,450	14,400
2650	15 ton cooling	"	.80	40	"	18,000	2,050		20,050	22,900

23 81 23 – Computer-Room Air-Conditioners

23 81 23.10 Computer Room Units

		Crew	Daily Output	Labor-Hours	Unit	Material	2011 Bare Costs Labor	Equipment	Total	Total Incl O&P
0010	**COMPUTER ROOM UNITS**　　D3050-185									
1000	Air cooled, includes remote condenser but not									
1020	interconnecting tubing or refrigerant									
1080	3 ton	Q-5	.50	32	Ea.	17,000	1,550		18,550	21,000
1120	5 ton		.45	35.556		18,100	1,725		19,825	22,600
1160	6 ton		.30	53.333		32,000	2,600		34,600	39,100
1200	8 ton		.27	59.259		33,900	2,875		36,775	41,600
1240	10 ton		.25	64		35,400	3,125		38,525	43,700
1260	12 ton		.24	66.667		36,700	3,250		39,950	45,300
1280	15 ton		.22	72.727		39,000	3,550		42,550	48,200
1290	18 ton		.20	80		44,400	3,900		48,300	54,500
1300	20 ton	Q-6	.26	92.308		46,700	4,650		51,350	58,500
1320	22 ton		.24	100		47,300	5,050		52,350	59,500
1360	30 ton		.21	114		58,000	5,775		63,775	72,500
2200	Chilled water, for connection to									
2220	existing chiller system of adequate capacity									
2260	5 ton	Q-5	.74	21.622	Ea.	13,000	1,050		14,050	15,900
2280	6 ton		.52	30.769		13,100	1,500		14,600	16,600
2300	8 ton		.50	32		13,200	1,550		14,750	16,800
2320	10 ton		.49	32.653		13,300	1,600		14,900	17,000
2330	12 ton		.49	32.990		13,600	1,600		15,200	17,400
2360	15 ton		.48	33.333		13,900	1,625		15,525	17,700
2400	20 ton		.46	34.783		14,800	1,700		16,500	18,800
2440	25 ton	Q-6	.63	38.095		15,800	1,925		17,725	20,200
2460	30 ton		.57	42.105		17,100	2,125		19,225	22,000
2480	35 ton		.52	46.154		17,300	2,325		19,625	22,500
2500	45 ton		.46	52.174		18,300	2,625		20,925	24,000
2520	50 ton		.42	57.143		21,000	2,875		23,875	27,300
2540	60 ton		.38	63.158		22,200	3,200		25,400	29,300
2560	77 ton		.35	68.571		37,600	3,475		41,075	46,600
2580	95 ton		.32	75		43,900	3,800		47,700	54,000

23 81 Decentralized Unitary HVAC Equipment

23 81 23 – Computer-Room Air-Conditioners

23 81 23.10 Computer Room Units

		Crew	Daily Output	Labor-Hours	Unit	Material	2011 Bare Costs Labor	2011 Bare Costs Equipment	Total	Total Incl O&P
4000	Glycol system, complete except for interconnecting tubing									
4060	3 ton	Q-5	.40	40	Ea.	21,300	1,950		23,250	26,300
4100	5 ton		.38	42.105		23,300	2,050		25,350	28,800
4120	6 ton		.25	64		34,100	3,125		37,225	42,300
4140	8 ton		.23	69.565		37,500	3,400		40,900	46,300
4160	10 ton		.21	76.190		39,900	3,700		43,600	49,400
4180	12 ton		.19	84.211		43,200	4,100		47,300	53,500
4200	15 ton	Q-6	.26	92.308		49,400	4,650		54,050	61,500
4240	20 ton		.24	100		54,500	5,050		59,550	67,500
4280	22 ton		.23	104		57,000	5,275		62,275	70,500
4430	30 ton		.22	109		70,500	5,500		76,000	86,000
8000	Water cooled system, not including condenser,									
8020	water supply or cooling tower									
8060	3 ton	Q-5	.62	25.806	Ea.	17,300	1,250		18,550	20,900
8100	5 ton		.54	29.630		18,800	1,450		20,250	22,900
8120	6 ton		.35	45.714		29,300	2,225		31,525	35,500
8140	8 ton		.33	48.485		30,800	2,350		33,150	37,400
8160	10 ton		.31	51.613		31,900	2,525		34,425	38,800
8180	12 ton		.29	55.172		34,600	2,675		37,275	42,000
8200	15 ton		.27	59.259		37,400	2,875		40,275	45,400
8240	20 ton	Q-6	.38	63.158		40,200	3,200		43,400	49,000
8280	22 ton		.35	68.571		42,100	3,475		45,575	51,500
8300	30 ton		.30	80		52,500	4,050		56,550	63,500

23 81 26 – Split-System Air-Conditioners

23 81 26.10 Split Ductless Systems

		Crew	Daily Output	Labor-Hours	Unit	Material	2011 Bare Costs Labor	2011 Bare Costs Equipment	Total	Total Incl O&P
0010	**SPLIT DUCTLESS SYSTEMS**									
0100	Cooling only, single zone									
0110	Wall mount									
0120	3/4 ton cooling	Q-5	2	8	Ea.	690	390		1,080	1,350
0130	1 ton cooling		1.80	8.889		765	435		1,200	1,500
0140	1-1/2 ton cooling		1.60	10		1,200	485		1,685	2,050
0150	2 ton cooling		1.40	11.429		1,375	555		1,930	2,325
1000	Ceiling mount									
1020	2 ton cooling	Q-5	1.40	11.429	Ea.	2,475	555		3,030	3,550
1030	3 ton cooling	"	1.20	13.333	"	3,200	650		3,850	4,475
3000	Multizone									
3010	Wall mount									
3020	2 @ 3/4 ton cooling	Q-5	1.80	8.889	Ea.	2,100	435		2,535	2,975
5000	Cooling/Heating									
5010	Wall mount									
5110	1 ton cooling	Q-5	1.70	9.412	Ea.	765	460		1,225	1,525
5120	1-1/2 ton cooling	"	1.50	10.667	"	1,175	520		1,695	2,075
7000	Accessories for all split ductless systems									
7010	Add for ambient frost control	Q-5	8	2	Ea.	179	97.50		276.50	340
7020	Add for tube/wiring kit, (Line sets)									
7030	15' kit	Q-5	32	.500	Ea.	40	24.50		64.50	80.50
7036	25' kit		28	.571		95	28		123	147
7040	35' kit		24	.667		129	32.50		161.50	191
7050	50' kit		20	.800		167	39		206	242

23 81 Decentralized Unitary HVAC Equipment

23 81 43 – Air-Source Unitary Heat Pumps

23 81 43.10 Air-Source Heat Pumps

23 81 43.10 Air-Source Heat Pumps	Crew	Daily Output	Labor-Hours	Unit	Material	2011 Bare Costs Labor	Equipment	Total	Total Incl O&P
0010	**AIR-SOURCE HEAT PUMPS**, Not including interconnecting tubing									
1000	Air to air, split system, not including curbs, pads, fan coil and ductwork									
1010	For curbs/pads see Section 23 91 00									
1012	Outside condensing unit only, for fan coil see Section 23 82 19.10									
1015	1.5 ton cooling, 7 MBH heat @ 0°F	Q-5	2.40	6.667	Ea.	2,075	325		2,400	2,750
1020	2 ton cooling, 8.5 MBH heat @ 0°F		2	8		2,125	390		2,515	2,900
1030	2.5 ton cooling, 10 MBH heat @ 0°F		1.60	10		2,350	485		2,835	3,300
1040	3 ton cooling, 13 MBH heat @ 0°F		1.20	13.333		2,650	650		3,300	3,900
1050	3.5 ton cooling, 18 MBH heat @ 0°F		1	16		2,850	780		3,630	4,300
1054	4 ton cooling, 24 MBH heat @ 0°F		.80	20		3,150	975		4,125	4,900
1060	5 ton cooling, 27 MBH heat @ 0°F		.50	32		3,425	1,550		4,975	6,100
1080	7.5 ton cooling, 33 MBH heat @ 0°F		.45	35.556		5,925	1,725		7,650	9,075
1100	10 ton cooling, 50 MBH heat @ 0°F	Q-6	.64	37.500		8,250	1,900		10,150	11,900
1120	15 ton cooling, 64 MBH heat @ 0°F		.50	48		12,200	2,425		14,625	17,000
1130	20 ton cooling, 85 MBH heat @ 0°F		.35	68.571		16,000	3,475		19,475	22,800
1140	25 ton cooling, 119 MBH heat @ 0°F		.25	96		19,300	4,850		24,150	28,400
1500	Single package, not including curbs, pads, or plenums									
1502	1/2 ton cooling, supplementary heat not incl.	Q-5	8	2	Ea.	1,975	97.50		2,072.50	2,325
1504	3/4 ton cooling, supplementary heat not incl.		6	2.667		2,225	130		2,355	2,650
1506	1 ton cooling, supplementary heat not incl.		4	4		2,300	195		2,495	2,825
1510	1.5 ton cooling, 5 MBH heat @ 0°F		1.55	10.323		2,550	505		3,055	3,550
1520	2 ton cooling, 6.5 MBH heat @ 0°F		1.50	10.667		2,750	520		3,270	3,800
1540	2.5 ton cooling, 8 MBH heat @ 0°F		1.40	11.429		3,000	555		3,555	4,125
1560	3 ton cooling, 10 MBH heat @ 0°F		1.20	13.333		3,275	650		3,925	4,575
1570	3.5 ton cooling, 11 MBH heat @ 0°F		1	16		3,575	780		4,355	5,125
1580	4 ton cooling, 13 MBH heat @ 0°F		.96	16.667		3,850	810		4,660	5,450
1620	5 ton cooling, 27 MBH heat @ 0°F		.65	24.615		4,150	1,200		5,350	6,375
1640	7.5 ton cooling, 35 MBH heat @ 0°F		.40	40		6,850	1,950		8,800	10,400
1648	10 ton cooling, 45 MBH heat @ 0°F	Q-6	.40	60		9,000	3,025		12,025	14,400
1652	12 ton cooling, 50 MBH heat @ 0°F	"	.36	66.667		10,400	3,375		13,775	16,400
1696	Supplementary electric heat coil incl., except as noted									

23 81 46 – Water-Source Unitary Heat Pumps

23 81 46.10 Water Source Heat Pumps

23 81 46.10 Water Source Heat Pumps	Crew	Daily Output	Labor-Hours	Unit	Material	2011 Bare Costs Labor	Equipment	Total	Total Incl O&P
0010	**WATER SOURCE HEAT PUMPS**, Not incl. connecting tubing or water source									
2000	Water source to air, single package									
2100	1 ton cooling, 13 MBH heat @ 75°F	Q-5	2	8	Ea.	1,700	390		2,090	2,425
2120	1.5 ton cooling, 17 MBH heat @ 75°F		1.80	8.889		1,775	435		2,210	2,625
2140	2 ton cooling, 19 MBH heat @ 75°F		1.70	9.412		1,875	460		2,335	2,750
2160	2.5 ton cooling, 25 MBH heat @ 75°F		1.60	10		1,975	485		2,460	2,900
2180	3 ton cooling, 27 MBH heat @ 75°F		1.40	11.429		2,025	555		2,580	3,075
2190	3.5 ton cooling, 29 MBH heat @ 75°F		1.30	12.308		2,300	600		2,900	3,450
2200	4 ton cooling, 31 MBH heat @ 75°F		1.20	13.333		2,475	650		3,125	3,700
2220	5 ton cooling, 29 MBH heat @ 75°F		.90	17.778		2,825	865		3,690	4,400
2240	7.5 ton cooling, 35 MBH heat @ 75°F		.60	26.667		6,600	1,300		7,900	9,200
2250	8.5 ton cooling, 40 MBH heat @ 75°F		.58	27.586		8,025	1,350		9,375	10,900
2260	10 ton cooling, 50 MBH heat @ 75°F		.53	30.189		8,575	1,475		10,050	11,700
2280	15 ton cooling, 64 MBH heat @ 75°F	Q-6	.47	51.064		14,000	2,575		16,575	19,300
2300	20 ton cooling, 100 MBH heat @ 75°F		.41	58.537		15,300	2,950		18,250	21,200
2310	25 ton cooling, 100 MBH heat @ 75°F		.32	75		20,800	3,800		24,600	28,600
2320	30 ton cooling, 128 MBH heat @ 75°F		.24	102		22,800	5,150		27,950	32,800
2340	40 ton cooling, 200 MBH heat @ 75°F		.21	117		32,200	5,900		38,100	44,200
2360	50 ton cooling, 200 MBH heat @ 75°F		.15	160		36,400	8,075		44,475	52,000

23 81 Decentralized Unitary HVAC Equipment

23 81 46 – Water-Source Unitary Heat Pumps

23 81 46.10 Water Source Heat Pumps	Crew	Daily Output	Labor-Hours	Unit	Material	2011 Bare Costs Labor	Equipment	Total	Total Incl O&P
3960	For supplementary heat coil, add				Ea.	10%				
4000	For increase in capacity thru use									
4020	of solar collector, size boiler at 60%									

23 82 Convection Heating and Cooling Units

23 82 13 – Valance Heating and Cooling Units

23 82 13.16 Valance Units

		Crew	Daily Output	Labor-Hours	Unit	Material	2011 Bare Costs Labor	Equipment	Total	Total Incl O&P
0010	**VALANCE UNITS**									
6000	Valance units, complete with 1/2" cooling coil, enclosure									
6020	2 tube	Q-5	18	.889	L.F.	30.50	43.50		74	98
6040	3 tube		16	1		36.50	48.50		85	113
6060	4 tube		16	1		41.50	48.50		90	118
6080	5 tube		15	1.067		47	52		99	130
6100	6 tube		15	1.067		52	52		104	135
6120	8 tube		14	1.143		75.50	55.50		131	167
6200	For 3/4" cooling coil, add					10%				

23 82 16 – Air Coils

23 82 16.10 Flanged Coils

		Crew	Daily Output	Labor-Hours	Unit	Material	2011 Bare Costs Labor	Equipment	Total	Total Incl O&P
0010	**FLANGED COILS**									
0100	Basic water, DX, or condenser coils									
0110	Copper tubes, alum. fins, galv. end sheets									
0112	H is finned height, L is finned length									
0120	3/8" x .016 tube, .0065 aluminum fins									
0130	2 row, 8 fins per inch									
0140	4" H x 12" L	Q-5	48	.333	Ea.	570	16.25		586.25	650
0150	4" H x 24" L		38.38	.417		590	20.50		610.50	680
0160	4" H x 48" L		19.25	.831		625	40.50		665.50	750
0170	4" H x 72" L		12.80	1.250		810	61		871	980
0180	6" H x 12" L		48	.333		550	16.25		566.25	630
0190	6" H x 24" L		25.60	.625		575	30.50		605.50	675
0200	6" H x 48" L		12.80	1.250		620	61		681	775
0210	6" H x 72" L		8.53	1.876		820	91.50		911.50	1,025
0220	10" H x 12" L		30.73	.521		585	25.50		610.50	685
0230	10" H x 24" L		15.33	1.044		620	51		671	760
0240	10" H x 48" L		7.69	2.081		695	101		796	910
0250	10" H x 72" L		5.12	3.125		920	152		1,072	1,250
0260	12" H x 12" L		25.60	.625		600	30.50		630.50	705
0270	12" H x 24" L		12.80	1.250		640	61		701	795
0280	12" H x 48" L		6.40	2.500		725	122		847	975
0290	12" H x 72" L		4.27	3.747		970	182		1,152	1,350
0300	16" H x 16" L		14.38	1.113		655	54		709	800
0310	16" H x 24" L		9.59	1.668		690	81.50		771.50	880
0320	16" H x 48" L		4.80	3.333		870	162		1,032	1,200
0330	16" H x 72" L		3.20	5		1,075	244		1,319	1,550
0340	20" H x 24" L		7.69	2.081		765	101		866	990
0350	20" H x 48" L		3.84	4.167		975	203		1,178	1,375
0360	20" H x 72" L		2.56	6.250		1,200	305		1,505	1,775
0370	24" H x 24" L		6.40	2.500		825	122		947	1,075
0380	24" H x 48" L		3.20	5		1,075	244		1,319	1,550
0390	24" H x 72" L		2.13	7.512		1,325	365		1,690	2,000

23 82 16 – Air Coils

23 82 16.10 Flanged Coils	Crew	Daily Output	Labor-Hours	Unit	Material	Labor	2011 Bare Costs Equipment	Total	Total Incl O&P
0400 30" H x 28" L	Q-5	4.39	3.645	Ea.	1,025	177		1,202	1,425
0410 30" H x 48" L		2.56	6.250		1,225	305		1,530	1,800
0420 30" H x 72" L		1.71	9.357		1,525	455		1,980	2,350
0430 30" H x 84" L		1.46	10.959		1,650	535		2,185	2,625
0440 34" H x 32" L		3.39	4.720		1,150	230		1,380	1,600
0450 34" H x 48" L		2.26	7.080		1,300	345		1,645	1,975
0460 34" H x 72" L		1.51	10.596		1,625	515		2,140	2,575
0470 34" H x 84" L		1.29	12.403		1,800	605		2,405	2,875
0600 For 10 fins per inch, add					1%				
0610 For 12 fins per inch, add					3%				
0620 For 4 row, add				Ea.	30%	50%			
0630 For 6 row, add					60%	100%			
0640 For 8 row, add					80%	150%			
1100 1/2" x .017 tube, .0065 aluminum fins									
1110 2 row, 8 fins per inch									
1120 5" H x 5" L	Q-5	40	.400	Ea.	565	19.50		584.50	650
1130 5" H x 20" L		37.10	.431		595	21		616	685
1140 5" H x 45" L		16.41	.975		650	47.50		697.50	785
1150 5" H x 90" L		8.21	1.949		950	95		1,045	1,200
1160 7.5" H x 5" L		38.50	.416		545	20		565	630
1170 7.5" H x 20" L		24.62	.650		590	31.50		621.50	695
1180 7.5" H x 45" L		10.94	1.463		660	71		731	830
1190 7.5" H x 90" L		5.46	2.930		1,000	143		1,143	1,325
1200 12.5" H x 10" L		29.49	.543		620	26.50		646.50	720
1210 12.5" H x 20" L		14.75	1.085		665	53		718	810
1220 12.5" H x 45" L		6.55	2.443		775	119		894	1,025
1230 12.5" H x 90" L		3.28	4.878		1,200	238		1,438	1,675
1240 15" H x 15" L		16.41	.975		665	47.50		712.50	805
1250 15" H x 30" L		8.21	1.949		745	95		840	960
1260 15" H x 45" L		5.46	2.930		820	143		963	1,125
1270 15" H x 90" L		2.73	5.861		1,300	285		1,585	1,850
1280 20" H x 20" L		9.22	1.735		760	84.50		844.50	960
1290 20" H x 45" L		4.10	3.902		1,000	190		1,190	1,375
1300 20" H x 90" L		2.05	7.805		1,475	380		1,855	2,200
1310 25" H x 25" L		5.90	2.712		905	132		1,037	1,200
1320 25" H x 45" L		3.28	4.878		1,150	238		1,388	1,625
1330 25" H x 90" L		1.64	9.756		1,700	475		2,175	2,575
1340 30" H x 30" L		4.10	3.902		1,100	190		1,290	1,500
1350 30" H x 45" L		2.73	5.861		1,275	285		1,560	1,825
1360 30" H x 90" L		1.37	11.679		1,900	570		2,470	2,950
1370 35" H x 35" L		3.01	5.316		1,275	259		1,534	1,775
1380 35" H x 45" L		2.34	6.838		1,400	335		1,735	2,025
1390 35" H x 90" L		1.17	13.675		2,100	665		2,765	3,300
1400 42.5" H x 45" L		1.93	8.290		1,625	405		2,030	2,375
1410 42.5" H x 75" L		1.16	13.793		2,150	670		2,820	3,375
1420 42.5" H x 105" L		.83	19.277		2,650	940		3,590	4,325
1600 For 10 fins per inch, add					2%				
1610 For 12 fins per inch, add					5%				
1620 For 4 row, add					35%	50%			
1630 For 6 row, add					65%	100%			
1640 For 8 row, add					90%	150%			
2010 5/8" x .020" tube, .0065 aluminum fins									
2020 2 row, 8 fins per inch									

23 82 16.10 Flanged Coils		Crew	Daily Output	Labor-Hours	Unit	Material	2011 Bare Costs Labor	Equipment	Total	Total Incl O&P
2030	6" H x 6" L	Q-5	38	.421	Ea.	580	20.50		600.50	670
2040	6" H x 24" L		24.96	.641		630	31		661	735
2050	6" H x 48" L		12.48	1.282		690	62.50		752.50	855
2060	6" H x 96" L		6.24	2.564		1,025	125		1,150	1,300
2070	12" H x 12" L		24.96	.641		605	31		636	710
2080	12" H x 24" L		12.48	1.282		660	62.50		722.50	825
2090	12" H x 48" L		6.24	2.564		770	125		895	1,025
2100	12" H x 96" L		3.12	5.128		1,225	250		1,475	1,725
2110	18" H x 24" L		8.32	1.923		780	93.50		873.50	995
2120	18" H x 48" L		4.16	3.846		1,025	187		1,212	1,400
2130	18" H x 96" L		2.08	7.692		1,500	375		1,875	2,200
2140	24" H x 24" L		6.24	2.564		865	125		990	1,150
2150	24" H x 48" L		3.12	5.128		1,150	250		1,400	1,650
2160	24" H x 96" L		1.56	10.256		1,750	500		2,250	2,675
2170	30" H x 30" L		3.99	4.010		1,125	195		1,320	1,550
2180	30" H x 48" L		2.50	6.400		1,350	310		1,660	1,950
2190	30" H x 96" L		1.25	12.800		2,050	625		2,675	3,175
2200	30" H x 120" L		1	16		2,400	780		3,180	3,800
2210	36" H x 36" L		2.77	5.776		1,350	281		1,631	1,925
2220	36" H x 48" L		2.08	7.692		1,500	375		1,875	2,200
2230	36" H x 90" L		1.11	14.414		2,225	700		2,925	3,500
2240	36" H x 120" L		.83	19.277		2,700	940		3,640	4,375
2250	42" H x 48" L		1.78	8.989		1,675	440		2,115	2,475
2260	42" H x 90" L		.95	16.842		2,475	820		3,295	3,950
2270	42" H x 120" L		.71	22.535		3,025	1,100		4,125	4,975
2280	51" H x 48" L		1.47	10.884		1,950	530		2,480	2,950
2290	51" H x 96" L		.73	21.918		3,025	1,075		4,100	4,925
2300	51" H x 120" L	▼	.59	27.119	▼	3,550	1,325		4,875	5,875
2500	For 10 fins per inch, add					3%				
2510	For 12 fins per inch, add					6%				
2520	For 4 row, add				Ea.	40%	50%			
2530	For 6 row, add					80%	100%			
2540	For 8 row, add				▼	110%	150%			
3000	Hot water booster coils									
3010	Copper tubes, alum. fins, galv. end sheets									
3012	H is finned height, L is finned length									
3020	1/2" x .017" tube, .0065 Al fins									
3030	1 row, 10 fins per inch									
3040	5"H x 10"L	Q-5	40	.400	Ea.	265	19.50		284.50	320
3050	5"H x 25"L		33.18	.482		290	23.50		313.50	355
3060	10"H x 10"L		39	.410		292	20		312	350
3070	10"H x 20"L		20.73	.772		315	37.50		352.50	400
3080	10"H x 30"L		13.83	1.157		340	56.50		396.50	455
3090	15"H x 15"L		18.44	.868		330	42.50		372.50	430
3110	15"H x 20"L		13.83	1.157		345	56.50		401.50	465
3120	15"H x 30"L		9.22	1.735		380	84.50		464.50	545
3130	20"H x 20"L		10.37	1.543		380	75		455	525
3140	20"H x 30"L		6.91	2.315		420	113		533	635
3150	20"H x 40"L		5.18	3.089		465	150		615	735
3160	25"H x 25"L		6.64	2.410		435	117		552	655
3170	25"H x 35"L		4.74	3.376		555	164		719	855
3180	25"H x 45"L		3.69	4.336		625	211		836	1,000
3190	30"H x 30"L	▼	4.61	3.471	▼	570	169		739	880

23 82 16.10 Flanged Coils		Crew	Daily Output	Labor-Hours	Unit	Material	2011 Bare Costs Labor	Equipment	Total	Total Incl O&P
3200	30"H x 40"L	Q-5	3.46	4.624	Ea.	655	225		880	1,050
3210	30"H x 50"L		2.76	5.797		730	282		1,012	1,225
3300	For 2 row, add					20%	50%			
3400	1/2" x .020" tube, .008 Al fins									
3410	2 row, 10 fins per inch									
3420	20" H x 27" L	Q-5	7.27	2.201	Ea.	560	107		667	775
3430	20" H x 36" L		5.44	2.941		635	143		778	915
3440	20" H x 45" L		4.35	3.678		790	179		969	1,125
3450	25" H x 45" L		3.49	4.585		910	223		1,133	1,325
3460	25" H x 54" L		2.90	5.517		1,075	269		1,344	1,600
3470	30" H x 54" L		2.42	6.612		1,225	320		1,545	1,825
3480	35" H x 54" L		2.07	7.729		1,375	375		1,750	2,050
3490	35" H x 63" L		1.78	8.989		1,500	440		1,940	2,300
3500	40" H x 72" L		1.36	11.765		1,850	575		2,425	2,875
3510	55" H x 63" L		1.13	14.159		2,175	690		2,865	3,425
3520	55" H x 72" L		.99	16.162		2,400	785		3,185	3,825
3530	55" H x 81" L		.88	18.182		2,650	885		3,535	4,250
3540	75" H x 81" L		.64	25		3,450	1,225		4,675	5,625
3550	75" H x 90" L		.58	27.586		3,775	1,350		5,125	6,150
3560	75" H x 108" L		.48	33.333		4,425	1,625		6,050	7,300
3800	5/8" x .020" tubing, .0065 aluminum fins									
3810	1 row, 10 fins per inch									
3820	6" H x 12" L	Q-5	38	.421	Ea.	296	20.50		316.50	355
3830	6" H x 24" L		28.80	.556		299	27		326	370
3840	6" H x 30" L		23.04	.694		305	34		339	385
3850	12" H x 12" L		28.80	.556		305	27		332	375
3860	12" H x 24" L		14.40	1.111		330	54		384	445
3870	12" H x 30" L		11.52	1.389		345	67.50		412.50	480
3880	18" H x 18" L		12.80	1.250		350	61		411	475
3890	18" H x 24" L		9.60	1.667		370	81		451	525
3900	18" H x 36" L		6.40	2.500		410	122		532	630
3910	24" H x 24" L		7.20	2.222		405	108		513	605
3920	24" H x 30" L		4.80	3.333		535	162		697	830
3930	24" H x 36" L		5.05	3.168		580	154		734	870
3940	30" H x 30" L		4.61	3.471		600	169		769	910
3950	30" H x 36" L		3.85	4.159		650	203		853	1,025
3960	30" H x 42" L		3.29	4.863		705	237		942	1,125
3970	36" H x 36" L		3.20	5		720	244		964	1,150
3980	36" H x 42" L		2.74	5.839		785	284		1,069	1,275
4100	For 2 row, add					30%	50%			
4400	Steam and hot water heating coils									
4410	Copper tubes, alum. fins. galv. end sheets									
4412	H is finned height, L is finned length									
4420	5/8" tubing, to 175 psig working pressure									
4430	1 row, 8 fins per inch									
4440	6"H x 6"L	Q-5	38	.421	Ea.	550	20.50		570.50	635
4450	6"H x 12"L		37	.432		565	21		586	650
4460	6"H x 24"L		24.48	.654		590	32		622	695
4470	6"H x 36"L		16.32	.980		610	48		658	740
4480	6"H x 48"L		12.24	1.307		635	63.50		698.50	795
4490	6"H x 84"L		6.99	2.289		1,225	111		1,336	1,525
4500	6" H x 108" L		5.44	2.941		1,375	143		1,518	1,725
4510	12" H x 12" L		24.48	.654		665	32		697	780

23 82 Convection Heating and Cooling Units

23 82 16 – Air Coils

23 82 16.10 Flanged Coils		Crew	Daily Output	Labor-Hours	Unit	Material	2011 Bare Costs Labor	Equipment	Total	Total Incl O&P
4520	12" H x 24" L	Q-5	12.24	1.307	Ea.	705	63.50		768.50	870
4530	12" H x 36" L		8.16	1.961		750	95.50		845.50	970
4540	12" H x 48" L		6.12	2.614		865	127		992	1,150
4550	12" H x 84" L		3.50	4.571		1,525	223		1,748	2,000
4560	12" H x 108" L		2.72	5.882		1,750	286		2,036	2,350
4570	18" H x 18" L		10.88	1.471		795	71.50		866.50	980
4580	18" H x 24" L		8.16	1.961		825	95.50		920.50	1,050
4590	18" H x 48" L		4.08	3.922		1,025	191		1,216	1,400
4600	18" H x 84" L		2.33	6.867		1,825	335		2,160	2,525
4610	18" H x 108" L		1.81	8.840		2,125	430		2,555	2,975
4620	24" H x 24" L		6.12	2.614		1,000	127		1,127	1,300
4630	24" H x 48" L		3.06	5.229		1,200	255		1,455	1,700
4640	24" H x 84" L		1.75	9.143		2,150	445		2,595	3,025
4650	24" H x 108" L		1.36	11.765		2,500	575		3,075	3,600
4660	36" H x 36" L		2.72	5.882		1,400	286		1,686	1,950
4670	36" H x 48" L		2.04	7.843		1,525	380		1,905	2,250
4680	36" H x 84" L		1.17	13.675		2,750	665		3,415	4,025
4690	36" H x 108" L		.91	17.582		3,250	855		4,105	4,850
4700	48"H x 48"L		1.53	10.458		1,875	510		2,385	2,800
4710	48"H x 84"L		.87	18.391		3,400	895		4,295	5,075
4720	48"H x 108"L		.68	23.529		4,025	1,150		5,175	6,125
4730	48"H x 126"L	▼	.58	27.586	▼	4,300	1,350		5,650	6,725
4910	For 10 fins per inch, add					2%				
4920	For 12 fins per inch, add					5%				
4930	For 2 row, add					30%	50%			
4940	For 3 row, add					50%	100%			
4950	For 4 row, add					80%	123%			
4960	For 230 PSIG heavy duty, add					20%	30%			
6000	Steam, heavy duty, 200 PSIG									
6010	Copper tubes, alum. fins, galv. end sheets									
6012	H is finned height, L is finned length									
6020	1" x .035" tubing, .010 aluminum fins									
6030	1 row, 8 fins per inch									
6040	12"H x 12"L	Q-5	24.48	.654	Ea.	795	32		827	925
6050	12"H x 24"L		12.24	1.307		900	63.50		963.50	1,075
6060	12"H x 36"L		8.16	1.961		1,000	95.50		1,095.50	1,250
6070	12"H x 48"L		6.12	2.614		1,200	127		1,327	1,500
6080	12"H x 84"L		3.50	4.571		1,675	223		1,898	2,150
6090	12"H x 108"L		2.72	5.882		1,975	286		2,261	2,600
6100	18"H x 24"L		8.16	1.961		1,125	95.50		1,220.50	1,375
6110	18"H x 36"L		5.44	2.941		1,350	143		1,493	1,700
6120	18"H x 48"L		4.08	3.922		1,525	191		1,716	1,950
6130	18"H x 84"L		2.33	6.867		2,150	335		2,485	2,850
6140	18"H x 108"L		1.81	8.840		2,575	430		3,005	3,500
6150	24"H x 24"L		6.12	2.614		1,450	127		1,577	1,800
6160	24"H x 36"L		4.08	3.922		1,625	191		1,816	2,050
6170	24"H x 48"L		3.06	5.229		1,825	255		2,080	2,400
6180	24"H x 84"L		1.75	9.143		2,625	445		3,070	3,575
6190	24"H x 108"L		1.36	11.765		3,175	575		3,750	4,350
6200	30"H x 36"L		3.26	4.908		1,875	239		2,114	2,425
6210	30"H x 48"L		2.45	6.531		2,150	320		2,470	2,850
6220	30"H x 84"L		1.40	11.429		3,100	555		3,655	4,250
6230	30"H x 108"L	▼	1.09	14.679	▼	3,775	715		4,490	5,225

23 82 16 – Air Coils

23 82 16.10 Flanged Coils		Crew	Daily Output	Labor-Hours	Unit	Material	2011 Bare Costs Labor	Equipment	Total	Total Incl O&P
6240	36"H x 36"L	Q-5	2.72	5.882	Ea.	2,150	286		2,436	2,800
6250	36"H x 48"L		2.04	7.843		2,475	380		2,855	3,300
6260	36"H x 84"L		1.17	13.675		3,600	665		4,265	4,950
6270	36"H x 108"L		.91	17.582		4,375	855		5,230	6,100
6280	36"H x 120"L		.82	19.512		4,700	950		5,650	6,600
6290	48"H x 48"L		1.53	10.458		4,525	510		5,035	5,725
6300	48"H x 84"L		.87	18.391		4,550	895		5,445	6,350
6310	48"H x 108"L		.68	23.529		5,600	1,150		6,750	7,850
6320	48"H x 120"L	▼	.61	26.230		6,025	1,275		7,300	8,525
6400	For 10 fins per inch, add					4%				
6410	For 12 fins per inch, add				▼	10%				

23 82 16.20 Duct Heaters

		Crew	Daily Output	Labor-Hours	Unit	Material	Labor	Equipment	Total	Total Incl O&P
0010	**DUCT HEATERS**, Electric, 480 V, 3 Ph.									
0020	Finned tubular insert, 500°F									
0100	8" wide x 6" high, 4.0 kW	Q-20	16	1.250	Ea.	705	59		764	865
0120	12" high, 8.0 kW		15	1.333		1,175	63		1,238	1,375
0140	18" high, 12.0 kW		14	1.429		1,650	67.50		1,717.50	1,900
0160	24" high, 16.0 kW		13	1.538		2,100	72.50		2,172.50	2,425
0180	30" high, 20.0 kW		12	1.667		2,575	78.50		2,653.50	2,950
0300	12" wide x 6" high, 6.7 kW		15	1.333		750	63		813	920
0320	12" high, 13.3 kW		14	1.429		1,200	67.50		1,267.50	1,425
0340	18" high, 20.0 kW		13	1.538		1,700	72.50		1,772.50	1,975
0360	24" high, 26.7 kW		12	1.667		2,175	78.50		2,253.50	2,525
0380	30" high, 33.3 kW		11	1.818		2,650	86		2,736	3,050
0500	18" wide x 6" high, 13.3 kW		14	1.429		805	67.50		872.50	985
0520	12" high, 26.7 kW		13	1.538		1,375	72.50		1,447.50	1,625
0540	18" high, 40.0 kW		12	1.667		1,850	78.50		1,928.50	2,150
0560	24" high, 53.3 kW		11	1.818		2,450	86		2,536	2,825
0580	30" high, 66.7 kW		10	2		3,050	94.50		3,144.50	3,525
0700	24" wide x 6" high, 17.8 kW		13	1.538		890	72.50		962.50	1,100
0720	12" high, 35.6 kW		12	1.667		1,500	78.50		1,578.50	1,775
0740	18" high, 53.3 kW		11	1.818		2,100	86		2,186	2,450
0760	24" high, 71.1 kW		10	2		2,725	94.50		2,819.50	3,150
0780	30" high, 88.9 kW		9	2.222		3,325	105		3,430	3,825
0900	30" wide x 6" high, 22.2 kW		12	1.667		940	78.50		1,018.50	1,150
0920	12" high, 44.4 kW		11	1.818		1,575	86		1,661	1,875
0940	18" high, 66.7 kW		10	2		2,250	94.50		2,344.50	2,625
0960	24" high, 88.9 kW		9	2.222		2,750	105		2,855	3,175
0980	30" high, 111.0 kW	▼	8	2.500	▼	3,525	118		3,643	4,050
1400	Note decreased kW available for									
1410	each duct size at same cost									
1420	See line 5000 for modifications and accessories									
2000	Finned tubular flange with insulated									
2020	terminal box, 500°F									
2100	12" wide x 36" high, 54 kW	Q-20	10	2	Ea.	3,475	94.50		3,569.50	3,975
2120	40" high, 60 kW		9	2.222		4,000	105		4,105	4,550
2200	24" wide x 36" high, 118.8 kW		9	2.222		3,925	105		4,030	4,475
2220	40" high, 132 kW		8	2.500		4,675	118		4,793	5,300
2400	36" wide x 8" high, 40 kW		11	1.818		1,800	86		1,886	2,100
2420	16" high, 80 kW		10	2		2,350	94.50		2,444.50	2,750
2440	24" high, 120 kW		9	2.222		3,075	105		3,180	3,550
2460	32" high, 160 kW	▼	8	2.500	▼	3,950	118		4,068	4,525

23 82 Convection Heating and Cooling Units

23 82 16 – Air Coils

23 82 16.20 Duct Heaters

		Crew	Daily Output	Labor-Hours	Unit	Material	2011 Bare Costs Labor	Equipment	Total	Total Incl O&P
2480	36" high, 180 kW	Q-20	7	2.857	Ea.	4,875	135		5,010	5,550
2500	40" high, 200 kW		6	3.333		5,400	157		5,557	6,150
2600	40" wide x 8" high, 45 kW		11	1.818		1,850	86		1,936	2,150
2620	16" high, 90 kW		10	2		2,550	94.50		2,644.50	2,950
2640	24" high, 135 kW		9	2.222		3,225	105		3,330	3,700
2660	32" high, 180 kW		8	2.500		4,300	118		4,418	4,900
2680	36" high, 202.5 kW		7	2.857		4,950	135		5,085	5,650
2700	40" high, 225 kW		6	3.333		5,625	157		5,782	6,425
2800	48" wide x 8" high, 54.8 kW		10	2		1,925	94.50		2,019.50	2,275
2820	16" high, 109.8 kW		9	2.222		2,675	105		2,780	3,100
2840	24" high, 164.4 kW		8	2.500		3,400	118		3,518	3,925
2860	32" high, 219.2 kW		7	2.857		4,550	135		4,685	5,200
2880	36" high, 246.6 kW		6	3.333		5,225	157		5,382	5,975
2900	40" high, 274 kW		5	4		5,925	189		6,114	6,775
3000	56" wide x 8" high, 64 kW		9	2.222		2,200	105		2,305	2,575
3020	16" high, 128 kW		8	2.500		3,050	118		3,168	3,525
3040	24" high, 192 kW		7	2.857		3,700	135		3,835	4,275
3060	32" high, 256 kW		6	3.333		5,175	157		5,332	5,900
3080	36" high, 288kW		5	4		5,925	189		6,114	6,800
3100	40" high, 320kW		4	5		6,550	236		6,786	7,575
3200	64" wide x 8" high, 74kW		8	2.500		2,275	118		2,393	2,675
3220	16" high, 148kW		7	2.857		3,150	135		3,285	3,650
3240	24" high, 222kW		6	3.333		4,000	157		4,157	4,625
3260	32" high, 296kW		5	4		5,350	189		5,539	6,175
3280	36" high, 333kW		4	5		6,375	236		6,611	7,375
3300	40" high, 370kW		3	6.667		7,050	315		7,365	8,225
3800	Note decreased kW available for									
3820	each duct size at same cost									
5000	Duct heater modifications and accessories									
5120	T.C.O. limit auto or manual reset	Q-20	42	.476	Ea.	116	22.50		138.50	161
5140	Thermostat		28	.714		495	33.50		528.50	595
5160	Overheat thermocouple (removable)		7	2.857		700	135		835	975
5180	Fan interlock relay		18	1.111		168	52.50		220.50	264
5200	Air flow switch		20	1		146	47		193	231
5220	Split terminal box cover		100	.200		48	9.45		57.45	67
8000	To obtain BTU multiply kW by 3413									

23 82 19 – Fan Coil Units

23 82 19.10 Fan Coil Air Conditioning

		Crew	Daily Output	Labor-Hours	Unit	Material	2011 Bare Costs Labor	Equipment	Total	Total Incl O&P
0010	**FAN COIL AIR CONDITIONING** Cabinet mounted, filters, controls									
0100	Chilled water, 1/2 ton cooling	Q-5	8	2	Ea.	710	97.50		807.50	925
0110	3/4 ton cooling		7	2.286		735	111		846	975
0120	1 ton cooling		6	2.667		815	130		945	1,100
0140	1.5 ton cooling		5.50	2.909		920	142		1,062	1,200
0150	2 ton cooling		5.25	3.048		1,150	148		1,298	1,500
0160	2.5 ton cooling		5	3.200		1,525	156		1,681	1,900
0180	3 ton cooling		4	4		1,725	195		1,920	2,200
0262	For hot water coil, add					40%	10%			
0300	Console, 2 pipe with electric heat									
0310	1/2 ton cooling	Q-5	8	2	Ea.	1,175	97.50		1,272.50	1,450
0315	3/4 ton cooling		7	2.286		1,300	111		1,411	1,600
0320	1 ton cooling		6	2.667		1,400	130		1,530	1,750
0330	1.5 ton cooling		5.50	2.909		1,725	142		1,867	2,100

23 82 19.10 Fan Coil Air Conditioning

		Crew	Daily Output	Labor-Hours	Unit	Material	2011 Bare Costs Labor	2011 Bare Costs Equipment	Total	Total Incl O&P
0340	2 ton cooling	Q-5	4.70	3.404	Ea.	2,275	166		2,441	2,750
0345	2-1/2 ton cooling		4.70	3.404		3,300	166		3,466	3,875
0350	3 ton cooling		4	4		4,525	195		4,720	5,275
0360	4 ton cooling		4	4		4,700	195		4,895	5,475
0940	Direct expansion, for use w/air cooled condensing unit, 1.5 ton cooling		5	3.200		690	156		846	995
0950	2 ton cooling		4.80	3.333		750	162		912	1,075
0960	2.5 ton cooling		4.40	3.636		790	177		967	1,125
0970	3 ton cooling		3.80	4.211		985	205		1,190	1,375
0980	3.5 ton cooling		3.60	4.444		1,025	216		1,241	1,475
0990	4 ton cooling		3.40	4.706		1,150	229		1,379	1,625
1000	5 ton cooling		3	5.333		1,325	260		1,585	1,850
1020	7.5 ton cooling		3	5.333		2,475	260		2,735	3,125
1040	10 ton cooling	Q-6	2.60	9.231		2,975	465		3,440	3,975
1042	12.5 ton cooling		2.30	10.435		3,875	525		4,400	5,025
1050	15 ton cooling		1.60	15		4,100	760		4,860	5,625
1060	20 ton cooling		.70	34.286		5,575	1,725		7,300	8,725
1070	25 ton cooling		.65	36.923		7,000	1,875		8,875	10,500
1080	30 ton cooling		.60	40		7,325	2,025		9,350	11,100
1500	For hot water coil, add					40%	10%			
1512	For condensing unit add see Section 23 62									
3000	Chilled water, horizontal unit, housing, 2 pipe, fan control, no valves									
3100	1/2 ton cooling	Q-5	8	2	Ea.	1,100	97.50		1,197.50	1,375
3110	1 ton cooling		6	2.667		1,325	130		1,455	1,675
3120	1.5 ton cooling		5.50	2.909		1,600	142		1,742	1,950
3130	2 ton cooling		5.25	3.048		1,925	148		2,073	2,325
3140	3 ton cooling		4	4		2,225	195		2,420	2,750
3150	3.5 ton cooling		3.80	4.211		2,450	205		2,655	2,975
3160	4 ton cooling		3.80	4.211		2,450	205		2,655	2,975
3170	5 ton cooling		3.40	4.706		2,825	229		3,054	3,475
3180	6 ton cooling		3.40	4.706		2,825	229		3,054	3,475
3190	7 ton cooling		2.90	5.517		3,100	269		3,369	3,800
3200	8 ton cooling		2.90	5.517		3,100	269		3,369	3,800
3210	10 ton cooling	Q-6	2.80	8.571		3,750	435		4,185	4,775
3240	For hot water coil, add					40%	10%			
4000	With electric heat, 2 pipe									
4100	1/2 ton cooling	Q-5	8	2	Ea.	1,350	97.50		1,447.50	1,625
4105	3/4 ton cooling		7	2.286		1,500	111		1,611	1,825
4110	1 ton cooling		6	2.667		1,625	130		1,755	2,000
4120	1.5 ton cooling		5.50	2.909		1,825	142		1,967	2,200
4130	2 ton cooling		5.25	3.048		2,300	148		2,448	2,750
4135	2.5 ton cooling		4.80	3.333		3,150	162		3,312	3,725
4140	3 ton cooling		4	4		4,325	195		4,520	5,050
4150	3.5 ton cooling		3.80	4.211		5,150	205		5,355	5,975
4160	4 ton cooling		3.60	4.444		5,975	216		6,191	6,900
4170	5 ton cooling		3.40	4.706		6,775	229		7,004	7,800
4180	6 ton cooling		3.20	5		6,775	244		7,019	7,825
4190	7 ton cooling		2.90	5.517		7,050	269		7,319	8,150
4200	8 ton cooling		2.20	7.273		7,950	355		8,305	9,275
4210	10 ton cooling	Q-6	2.80	8.571		8,450	435		8,885	9,950

23 82 19.20 Heating and Ventilating Units

0010	**HEATING AND VENTILATING UNITS**, Classroom units									
0020	Includes filter, heating/cooling coils, standard controls									

23 82 Convection Heating and Cooling Units

23 82 19 – Fan Coil Units

23 82 19.20 Heating and Ventilating Units	Crew	Daily Output	Labor-Hours	Unit	Material	2011 Bare Costs Labor	Equipment	Total	Total Incl O&P
0080	750 CFM, 2 tons cooling	Q-6	2	12	Ea.	3,775	605		4,380	5,050
0100	1000 CFM, 2-1/2 tons cooling		1.60	15		4,450	760		5,210	6,025
0120	1250 CFM, 3 tons cooling		1.40	17.143		4,600	865		5,465	6,350
0140	1500 CFM, 4 tons cooling		.80	30		4,925	1,525		6,450	7,650
0160	2000 CFM 5 tons cooling		.50	48		5,775	2,425		8,200	9,975
0500	For electric heat, add					35%				
1000	For no cooling, deduct					25%	10%			

23 82 27 – Infrared Units

23 82 27.10 Infrared Type Heating Units

		Crew	Daily Output	Labor-Hours	Unit	Material	Labor	Equipment	Total	Total Incl O&P
0010	**INFRARED TYPE HEATING UNITS**									
0020	Gas fired, unvented, electric ignition, 100% shutoff.									
0030	Piping and wiring not included									
0060	Input, 15 MBH	Q-5	7	2.286	Ea.	505	111		616	720
0100	30 MBH		6	2.667		1,050	130		1,180	1,350
0120	45 MBH		5	3.200		1,500	156		1,656	1,875
0140	50 MBH		4.50	3.556		1,575	173		1,748	1,975
0160	60 MBH		4	4		1,600	195		1,795	2,050
0180	75 MBH		3	5.333		1,625	260		1,885	2,175
0200	90 MBH		2.50	6.400		1,700	310		2,010	2,350
0220	105 MBH		2	8		1,850	390		2,240	2,600
0240	120 MBH		2	8		1,975	390		2,365	2,725
2000	Electric, single or three phase									
2050	6 kW, 20,478 BTU	1 Elec	2.30	3.478	Ea.	475	175		650	780
2100	13.5 KW, 40,956 BTU		2.20	3.636		685	183		868	1,025
2150	24 KW, 81,912 BTU		2	4		1,350	201		1,551	1,800
3000	Oil fired, pump, controls, fusible valve, oil supply tank									
3050	91,000 BTU	Q-5	2.50	6.400	Ea.	6,200	310		6,510	7,275
3080	105,000 BTU		2.25	7.111		6,575	345		6,920	7,750
3110	119,000 BTU		2	8		7,175	390		7,565	8,475

23 82 29 – Radiators

23 82 29.10 Hydronic Heating

		Crew	Daily Output	Labor-Hours	Unit	Material	Labor	Equipment	Total	Total Incl O&P
0010	**HYDRONIC HEATING,** Terminal units, not incl. main supply pipe									
1000	Radiation									
1100	Panel, baseboard, C.I., including supports, no covers	Q-5	46	.348	L.F.	32.50	16.95		49.45	61
3000	Radiators, cast iron									
3100	Free standing or wall hung, 6 tube, 25" high	Q-5	96	.167	Section	36	8.10		44.10	51.50
3150	4 tube 25" high		96	.167		27	8.10		35.10	41.50
3200	4 tube, 19" high		96	.167		25	8.10		33.10	39.50
3250	Adj. brackets, 2 per wall radiator up to 30 sections	1 Stpi	32	.250	Ea.	28	13.55		41.55	50.50
3500	Recessed, 20" high x 5" deep, without grille	Q-5	60	.267	Section	30.50	13		43.50	53
3600	For inlet grille, add				"	2.12			2.12	2.33
9500	To convert SFR to BTU rating: Hot water, 150 x SFR									
9510	Forced hot water, 180 x SFR; steam, 240 x SFR									

23 82 33 – Convectors

23 82 33.10 Convector Units

		Crew	Daily Output	Labor-Hours	Unit	Material	Labor	Equipment	Total	Total Incl O&P
0010	**CONVECTOR UNITS,** Terminal units, not incl. main supply pipe									
2204	Convector, multifin, 2 pipe w/cabinet									
2210	17" H x 24" L	Q-5	10	1.600	Ea.	98.50	78		176.50	224
2214	17" H x 36" L		8.60	1.860		148	90.50		238.50	297
2218	17" H x 48" L		7.40	2.162		197	105		302	375
2222	21" H x 24" L		9	1.778		96.50	86.50		183	235

23 82 33 – Convectors

23 82 33.10 Convector Units		Crew	Daily Output	Labor-Hours	Unit	Material	2011 Bare Costs Labor	Equipment	Total	Total Incl O&P
2226	21" H x 36" L	Q-5	8.20	1.951	Ea.	145	95		240	300
2228	21" H x 48" L	↓	6.80	2.353	↓	193	115		308	385
2240	For knob operated damper, add					140%				
2241	For metal trim strips, add	Q-5	64	.250	Ea.	19.25	12.20		31.45	39
2243	For snap-on inlet grille, add					10%	10%			
2245	For hinged access door, add	Q-5	64	.250	Ea.	35.50	12.20		47.70	57
2246	For air chamber, auto-venting, add	"	58	.276	"	6.35	13.45		19.80	27

23 82 36 – Finned-Tube Radiation Heaters

23 82 36.10 Finned Tube Radiation

		Crew	Daily Output	Labor-Hours	Unit	Material	2011 Bare Costs Labor	Equipment	Total	Total Incl O&P
0010	**FINNED TUBE RADIATION**, Terminal units, not incl. main supply pipe									
1150	Fin tube, wall hung, 14" slope top cover, with damper									
1200	1-1/4" copper tube, 4-1/4" alum. fin	Q-5	38	.421	L.F.	40.50	20.50		61	75.50
1250	1-1/4" steel tube, 4-1/4" steel fin		36	.444		38.50	21.50		60	75
1255	2" steel tube, 4-1/4" steel fin	↓	32	.500	↓	41.50	24.50		66	82
1260	21", two tier, slope top, w/damper									
1262	1-1/4" copper tube, 4-1/4" alum. fin	Q-5	30	.533	L.F.	48.50	26		74.50	92
1264	1-1/4" steel tube, 4-1/4" steel fin		28	.571		46	28		74	92
1266	2" steel tube, 4-1/4" steel fin	↓	25	.640	↓	49	31		80	101
1270	10", flat top, with damper									
1272	1-1/4" copper tube, 4-1/4" alum. fin	Q-5	38	.421	L.F.	28.50	20.50		49	62
1274	1-1/4" steel tube, 4-1/4" steel fin		36	.444		38.50	21.50		60	75
1276	2" steel tube, 4-1/4" steel fin	↓	32	.500	↓	41.50	24.50		66	82
1280	17", two tier, flat top, w/damper									
1282	1-1/4" copper tube, 4-1/4" alum. fin	Q-5	30	.533	L.F.	49	26		75	93
1284	1-1/4" steel tube, 4-1/4" steel fin		28	.571		46	28		74	92
1286	2" steel tube, 4-1/4" steel fin	↓	25	.640	↓	49	31		80	101
1301	Rough in wall hung, steel fin w/supply & balance valves	1 Stpi	.91	8.791	Ea.	261	475		736	995
1310	Baseboard, pkgd, 1/2" copper tube, alum. fin, 7" high	Q-5	60	.267	L.F.	7.55	13		20.55	27.50
1320	3/4" copper tube, alum. fin, 7" high		58	.276		8	13.45		21.45	29
1340	1" copper tube, alum. fin, 8-7/8" high		56	.286		17.60	13.90		31.50	40.50
1360	1-1/4" copper tube, alum. fin, 8-7/8" high		54	.296		26	14.45		40.45	50
1380	1-1/4" IPS steel tube with steel fins	↓	52	.308	↓	26	15		41	51
1381	Rough in baseboard panel & fin tube, supply & balance valves	1 Stpi	1.06	7.547	Ea.	251	410		661	885
1500	Note: fin tube may also require corners, caps, etc.									

23 82 39 – Unit Heaters

23 82 39.13 Cabinet Unit Heaters

		Crew	Daily Output	Labor-Hours	Unit	Material	2011 Bare Costs Labor	Equipment	Total	Total Incl O&P
0010	**CABINET UNIT HEATERS**									
5400	Cabinet, horizontal, hot water, blower type									
5420	20 MBH	Q-5	10	1.600	Ea.	795	78		873	990
5430	60 MBH		7	2.286		1,375	111		1,486	1,700
5440	100 MBH		5	3.200		1,450	156		1,606	1,825
5450	120 MBH	↓	4	4	↓	1,500	195		1,695	1,950

23 82 39.16 Propeller Unit Heaters

		Crew	Daily Output	Labor-Hours	Unit	Material	2011 Bare Costs Labor	Equipment	Total	Total Incl O&P
0010	**PROPELLER UNIT HEATERS**									
3950	Unit heaters, propeller, 115 V 2 psi steam, 60°F entering air									
4000	Horizontal, 12 MBH	Q-5	12	1.333	Ea.	335	65		400	465
4020	28.2 MBH		10	1.600		435	78		513	590
4040	36.5 MBH		8	2		505	97.50		602.50	700
4060	43.9 MBH		8	2		505	97.50		602.50	700
4080	56.5 MBH		7.50	2.133		540	104		644	750
4100	65.6 MBH	↓	7	2.286		555	111		666	775

23 82 39.16 Propeller Unit Heaters		Crew	Daily Output	Labor-Hours	Unit	Material	2011 Bare Costs Labor	2011 Bare Costs Equipment	Total	Total Incl O&P
4120	87.6 MBH	Q-5	6.50	2.462	Ea.	600	120		720	840
4140	96.8 MBH		6	2.667		735	130		865	1,000
4160	133.3 MBH		5	3.200		800	156		956	1,125
4180	157.6 MBH		4	4		980	195		1,175	1,375
4200	197.7 MBH		3	5.333		1,125	260		1,385	1,625
4220	257.2 MBH		2.50	6.400		1,550	310		1,860	2,175
4240	286.9 MBH		2	8		1,600	390		1,990	2,325
4250	326.0 MBH		1.90	8.421		1,825	410		2,235	2,600
4260	364 MBH		1.80	8.889		1,900	435		2,335	2,750
4270	404 MBH	▼	1.60	10		2,500	485		2,985	3,475
4300	For vertical diffuser, add					174			174	191
4310	Vertical flow, 40 MBH	Q-5	11	1.455		500	71		571	655
4314	58.5 MBH		8	2		520	97.50		617.50	715
4318	92 MBH		7	2.286		660	111		771	895
4322	109.7 MBH		6	2.667		780	130		910	1,050
4326	131.0 MBH		4	4		780	195		975	1,150
4330	160.0 MBH		3	5.333		825	260		1,085	1,300
4334	194.0 MBH		2.20	7.273		960	355		1,315	1,575
4338	212.0 MBH		2.10	7.619		1,175	370		1,545	1,850
4342	247.0 MBH		1.96	8.163		1,400	400		1,800	2,150
4346	297.0 MBH	▼	1.80	8.889		1,500	435		1,935	2,300
4350	333.0 MBH	Q-6	1.90	12.632		1,625	640		2,265	2,725
4354	420 MBH, (460 V)		1.80	13.333		2,000	675		2,675	3,200
4358	500 MBH, (460 V)		1.71	14.035		2,650	710		3,360	3,950
4362	570 MBH, (460 V)		1.40	17.143		3,725	865		4,590	5,400
4366	620 MBH, (460 V)		1.30	18.462		4,275	930		5,205	6,100
4370	960 MBH, (460 V)	▼	1.10	21.818	▼	7,200	1,100		8,300	9,550
4400	Unit heaters, propeller, hot water, 115 V, 60 Deg. F entering air									
4404	Horizontal									
4408	6 MBH	Q-5	11	1.455	Ea.	335	71		406	475
4412	12 MBH		11	1.455		335	71		406	475
4416	16 MBH		10	1.600		375	78		453	530
4420	24 MBH		9	1.778		435	86.50		521.50	605
4424	29 MBH		8	2		505	97.50		602.50	700
4428	47 MBH		7	2.286		555	111		666	775
4432	63 MBH		6	2.667		600	130		730	855
4436	81 MBH		5.50	2.909		735	142		877	1,025
4440	90 MBH		5	3.200		800	156		956	1,125
4460	133 MBH		4	4		980	195		1,175	1,375
4464	139 MBH		3.50	4.571		1,125	223		1,348	1,550
4468	198 MBH		2.50	6.400		1,550	310		1,860	2,175
4472	224 MBH		2	8		1,750	390		2,140	2,500
4476	273 MBH	▼	1.50	10.667	▼	1,900	520		2,420	2,875
4500	Vertical									
4504	8 MBH	Q-5	13	1.231	Ea.	500	60		560	640
4508	12 MBH		13	1.231		500	60		560	640
4512	16 MBH		13	1.231		500	60		560	640
4516	30 MBH		12	1.333		500	65		565	645
4520	43 MBH		11	1.455		520	71		591	675
4524	57 MBH		8	2		635	97.50		732.50	845
4528	68 MBH		7	2.286		660	111		771	895
4540	105 MBH		6	2.667		780	130		910	1,050
4544	123 MBH	▼	5	3.200	▼	825	156		981	1,150

23 82 Convection Heating and Cooling Units

23 82 39 – Unit Heaters

23 82 39.16 Propeller Unit Heaters		Crew	Daily Output	Labor-Hours	Unit	Material	2011 Bare Costs Labor	Equipment	Total	Total Incl O&P
4550	140 MBH	Q-5	5	3.200	Ea.	1,125	156		1,281	1,450
4554	156 MBH		4	4		1,175	195		1,370	1,600
4560	210 MBH		4	4		1,500	195		1,695	1,950
4564	223 MBH		3	5.333		1,625	260		1,885	2,175
4568	257 MBH		2	8		1,625	390		2,015	2,350

23 83 Radiant Heating Units

23 83 16 – Radiant-Heating Hydronic Piping

23 83 16.10 Radiant Floor Heating

		Crew	Daily Output	Labor-Hours	Unit	Material	2011 Bare Costs Labor	Equipment	Total	Total Incl O&P
0010	**RADIANT FLOOR HEATING**									
0100	Tubing, PEX (cross-linked polyethylene)									
0110	Oxygen barrier type for systems with ferrous materials									
0120	1/2"	Q-5	800	.020	L.F.	.99	.97		1.96	2.54
0130	3/4"		535	.030		1.40	1.46		2.86	3.71
0140	1"		400	.040		2.19	1.95		4.14	5.30
0200	Non barrier type for ferrous free systems									
0210	1/2"	Q-5	800	.020	L.F.	.57	.97		1.54	2.08
0220	3/4"		535	.030		1.05	1.46		2.51	3.33
0230	1"		400	.040		1.80	1.95		3.75	4.89
1000	Manifolds									
1110	Brass									
1120	With supply and return valves, flow meter, thermometer,									
1122	auto air vent and drain/fill valve.									
1130	1", 2 circuit	Q-5	14	1.143	Ea.	236	55.50		291.50	340
1140	1", 3 circuit		13.50	1.185		269	57.50		326.50	380
1150	1", 4 circuit		13	1.231		292	60		352	410
1154	1", 5 circuit		12.50	1.280		350	62.50		412.50	480
1158	1", 6 circuit		12	1.333		380	65		445	510
1162	1", 7 circuit		11.50	1.391		415	68		483	555
1166	1", 8 circuit		11	1.455		460	71		531	610
1172	1", 9 circuit		10.50	1.524		495	74		569	655
1174	1", 10 circuit		10	1.600		530	78		608	700
1178	1", 11 circuit		9.50	1.684		555	82		637	730
1182	1", 12 circuit		9	1.778		610	86.50		696.50	805
1610	Copper manifold header, (cut to size)									
1620	1" header, 12 – 1/2" sweat outlets	Q-5	3.33	4.805	Ea.	76	234		310	435
1630	1-1/4" header, 12 – 1/2" sweat outlets		3.20	5		88	244		332	460
1640	1-1/4" header, 12 – 3/4" sweat outlets		3	5.333		95	260		355	495
1650	1-1/2" header, 12 – 3/4" sweat outlets		3.10	5.161		114	251		365	500
1660	2" header, 12 – 3/4" sweat outlets		2.90	5.517		168	269		437	585
3000	Valves									
3110	Thermostatic zone valve actuator with end switch	Q-5	40	.400	Ea.	36.50	19.50		56	69
3114	Thermostatic zone valve actuator	"	36	.444	"	81	21.50		102.50	122
3120	Motorized straight zone valve with operator complete									
3130	3/4"	Q-5	35	.457	Ea.	125	22.50		147.50	171
3140	1"		32	.500		137	24.50		161.50	188
3150	1-1/4"		29.60	.541		165	26.50		191.50	222
3500	4 Way mixing valve, manual, brass									
3530	1"	Q-5	13.30	1.203	Ea.	126	58.50		184.50	227
3540	1-1/4"		11.40	1.404		137	68.50		205.50	253
3550	1-1/2"		11	1.455		203	71		274	330

23 83 Radiant Heating Units

23 83 16 – Radiant-Heating Hydronic Piping

23 83 16.10 Radiant Floor Heating		Crew	Daily Output	Labor-Hours	Unit	Material	2011 Bare Costs Labor	Equipment	Total	Total Incl O&P
3560	2"	Q-5	10.60	1.509	Ea.	288	73.50		361.50	425
3800	Mixing valve motor, 4 way for valves, 1" and 1-1/4"		34	.471		310	23		333	375
3810	Mixing valve motor, 4 way for valves, 1-1/2" and 2"	↓	30	.533	↓	355	26		381	430
5000	Radiant floor heating, zone control panel									
5120	4 Zone actuator valve control, expandable	Q-5	20	.800	Ea.	163	39		202	237
5130	6 Zone actuator valve control, expandable		18	.889		199	43.50		242.50	284
6070	Thermal track, straight panel for long continuous runs, 5.333 S.F.		40	.400		16.60	19.50		36.10	47.50
6080	Thermal track, utility panel, for direction reverse at run end, 5.333 S.F.		40	.400		16.60	19.50		36.10	47.50
6090	Combination panel, for direction reverse plus straight run, 5.333 S.F.	↓	40	.400	↓	16.60	19.50		36.10	47.50
7000	PEX tubing fittings									
7100	Compression type									
7116	Coupling									
7120	1/2" x 1/2"	1 Stpi	27	.296	Ea.	6.40	16.05		22.45	31
7124	3/4" x 3/4"	"	23	.348	"	10.10	18.80		28.90	39
7130	Adapter									
7132	1/2" x female sweat 1/2"	1 Stpi	27	.296	Ea.	4.15	16.05		20.20	28.50
7134	1/2" x female sweat 3/4"		26	.308		4.88	16.65		21.53	30.50
7136	5/8" x female sweat 3/4"	↓	24	.333	↓	7	18.05		25.05	34.50
7140	Elbow									
7142	1/2" x female sweat 1/2"	1 Stpi	27	.296	Ea.	6.65	16.05		22.70	31.50
7144	1/2" x female sweat 3/4"		26	.308		7.80	16.65		24.45	33.50
7146	5/8" x female sweat 3/4"	↓	24	.333		8.75	18.05		26.80	36.50
7200	Insert type									
7206	PEX x male NPT									
7210	1/2" x 1/2"	1 Stpi	29	.276	Ea.	3.80	14.90		18.70	26.50
7220	3/4" x 3/4"		27	.296		4.61	16.05		20.66	29
7230	1" x 1"	↓	26	.308	↓	6.20	16.65		22.85	32
7300	PEX coupling									
7310	1/2" x 1/2"	1 Stpi	30	.267	Ea.	1.07	14.45		15.52	22.50
7320	3/4" x 3/4"		29	.276		1.62	14.90		16.52	24.50
7330	1" x 1"	↓	28	.286		3.01	15.45		18.46	26.50
7400	PEX stainless crimp ring									
7410	1/2" x 1/2"	1 Stpi	86	.093	Ea.	.26	5.05		5.31	7.80
7420	3/4" x 3/4"		84	.095		.36	5.15		5.51	8.10
7430	1" x 1"	↓	82	.098	↓	.42	5.30		5.72	8.35

23 83 33 – Electric Radiant Heaters

23 83 33.10 Electric Heating		Crew	Daily Output	Labor-Hours	Unit	Material	2011 Bare Costs Labor	Equipment	Total	Total Incl O&P
0010	**ELECTRIC HEATING**, not incl. conduit or feed wiring									
1100	Rule of thumb: Baseboard units, including control	1 Elec	4.40	1.818	kW	95.50	91.50		187	240
1300	Baseboard heaters, 2' long, 375 watt		8	1	Ea.	40	50.50		90.50	119
1400	3' long, 500 watt		8	1		46	50.50		96.50	125
1600	4' long, 750 watt		6.70	1.194		54	60		114	149
1800	5' long, 935 watt		5.70	1.404		68	70.50		138.50	180
2000	6' long, 1125 watt		5	1.600		74	80.50		154.50	201
2200	7' long, 1310 watt		4.40	1.818		84	91.50		175.50	228
2400	8' long, 1500 watt		4	2		92	101		193	250
2600	9' long, 1680 watt		3.60	2.222		104	112		216	280
2800	10' long, 1875 watt	↓	3.30	2.424	↓	193	122		315	395
2950	Wall heaters with fan, 120 to 277 volt									
3160	Recessed, residential, 750 watt	1 Elec	6	1.333	Ea.	134	67		201	247
3170	1000 watt		6	1.333		134	67		201	247
3180	1250 watt	↓	5	1.600	↓	134	80.50		214.50	266

23 83 Radiant Heating Units

23 83 33 – Electric Radiant Heaters

23 83 33.10 Electric Heating	Crew	Daily Output	Labor-Hours	Unit	Material	2011 Bare Costs Labor	Equipment	Total	Total Incl O&P
3190	1500 watt	1 Elec	4	2	Ea.	134	101		235	296
3210	2000 watt		4	2		134	101		235	296
3230	2500 watt		3.50	2.286		335	115		450	540
3240	3000 watt		3	2.667		495	134		629	740
3250	4000 watt		2.70	2.963		495	149		644	765
3260	Commercial, 750 watt		6	1.333		176	67		243	293
3270	1000 watt		6	1.333		176	67		243	293
3280	1250 watt		5	1.600		176	80.50		256.50	310
3290	1500 watt		4	2		176	101		277	340
3300	2000 watt		4	2		176	101		277	340
3310	2500 watt		3.50	2.286		176	115		291	365
3320	3000 watt		3	2.667		320	134		454	550
3330	4000 watt		2.70	2.963		320	149		469	570
3600	Thermostats, integral		16	.500		30	25		55	70.50
3800	Line voltage, 1 pole		8	1		31	50.50		81.50	109
3810	2 pole		8	1		29.50	50.50		80	107
3820	Low voltage, 1 pole	▼	8	1	▼	34.50	50.50		85	113
4000	Heat trace system, 400 degree									
4020	115 V, 2.5 watts per L.F.	1 Elec	530	.015	L.F.	7.35	.76		8.11	9.20
4030	5 watts per L.F.		530	.015		7.35	.76		8.11	9.20
4050	10 watts per L.F.		530	.015		7.35	.76		8.11	9.20
4060	208 V, 5 watts per L.F.		530	.015		7.35	.76		8.11	9.20
4080	480 V, 8 watts per L.F.	▼	530	.015	▼	7.35	.76		8.11	9.20
4200	Heater raceway									
4260	Heat transfer cement									
4280	1 gallon				Ea.	48			48	53
4300	5 gallon				"	180			180	198
4320	Snap band, clamp									
4340	3/4" pipe size	1 Elec	470	.017	Ea.		.86		.86	1.27
4360	1" pipe size		444	.018			.91		.91	1.34
4380	1-1/4" pipe size		400	.020			1.01		1.01	1.49
4400	1-1/2" pipe size		355	.023			1.13		1.13	1.68
4420	2" pipe size		320	.025			1.26		1.26	1.86
4440	3" pipe size		160	.050			2.52		2.52	3.73
4460	4" pipe size		100	.080			4.02		4.02	5.95
4480	Thermostat NEMA 3R, 22 amp, 0-150 Deg, 10' cap.		8	1		229	50.50		279.50	325
4500	Thermostat NEMA 4X, 25 amp, 40 Deg, 5-1/2' cap.		7	1.143		229	57.50		286.50	335
4520	Thermostat NEMA 4X, 22 amp, 25-325 Deg, 10' cap.		7	1.143		615	57.50		672.50	765
4540	Thermostat NEMA 4X, 22 amp, 15-140 Deg,		6	1.333		520	67		587	670
4580	Thermostat NEMA 4,7,9, 22 amp, 25-325 Deg, 10' cap.		3.60	2.222		760	112		872	1,000
4600	Thermostat NEMA 4,7,9, 22 amp, 15-140 Deg,		3	2.667		755	134		889	1,025
4720	Fiberglass application tape, 36 yard roll		11	.727		71.50	36.50		108	133
5000	Radiant heating ceiling panels, 2' x 4', 500 watt		16	.500		252	25		277	315
5050	750 watt		16	.500		277	25		302	345
5200	For recessed plaster frame, add		32	.250		70.50	12.60		83.10	96
5300	Infrared quartz heaters, 120 volts, 1000 watts		6.70	1.194		222	60		282	335
5350	1500 watt		5	1.600		222	80.50		302.50	365
5400	240 volts, 1500 watt		5	1.600		222	80.50		302.50	365
5450	2000 watt		4	2		222	101		323	395
5500	3000 watt		3	2.667		246	134		380	470
5550	4000 watt		2.60	3.077		246	155		401	500
5570	Modulating control	▼	.80	10	▼	266	505		771	1,050
5600	Unit heaters, heavy duty, with fan & mounting bracket									

23 83 33 – Electric Radiant Heaters

23 83 33.10 Electric Heating		Crew	Daily Output	Labor-Hours	Unit	Material	2011 Bare Costs Labor	Equipment	Total	Total Incl O&P
5650	Single phase, 208-240-277 volt, 3 kW	1 Elec	3.20	2.500	Ea.	405	126		531	630
5750	5 kW		2.40	3.333		415	168		583	705
5800	7 kW		1.90	4.211		640	212		852	1,025
5850	10 kW		1.30	6.154		730	310		1,040	1,275
5950	15 kW		.90	8.889		1,175	445		1,620	1,950
6000	480 volt, 3 kW		3.30	2.424		505	122		627	735
6020	4 kW		3	2.667		450	134		584	690
6040	5 kW		2.60	3.077		460	155		615	735
6060	7 kW		2	4		690	201		891	1,050
6080	10 kW		1.40	5.714		750	287		1,037	1,250
6100	13 kW		1.10	7.273		1,175	365		1,540	1,850
6120	15 kW		1	8		1,175	400		1,575	1,900
6140	20 kW		.90	8.889		1,525	445		1,970	2,325
6300	3 phase, 208-240 volt, 5 kW		2.40	3.333		390	168		558	680
6320	7 kW		1.90	4.211		605	212		817	980
6340	10 kW		1.30	6.154		655	310		965	1,175
6360	15 kW		.90	8.889		1,100	445		1,545	1,850
6380	20 kW		.70	11.429		1,550	575		2,125	2,575
6400	25 kW		.50	16		1,850	805		2,655	3,225
6500	480 volt, 5 kW		2.60	3.077		550	155		705	835
6520	7 kW		2	4		690	201		891	1,050
6540	10 kW		1.40	5.714		730	287		1,017	1,225
6560	13 kW		1.10	7.273		1,175	365		1,540	1,850
6580	15 kW		1	8		1,175	400		1,575	1,900
6600	20 kW		.90	8.889		1,200	445		1,645	1,975
6620	25 kW		.60	13.333		1,850	670		2,520	3,025
6630	30 kW		.70	11.429		2,125	575		2,700	3,200
6640	40 kW		.60	13.333		2,725	670		3,395	4,000
6650	50 kW		.50	16		3,275	805		4,080	4,800
6800	Vertical discharge heaters, with fan									
6820	Single phase, 208-240-277 volt, 10 kW	1 Elec	1.30	6.154	Ea.	675	310		985	1,200
6840	15 kW		.90	8.889		1,125	445		1,570	1,900
6900	3 phase, 208-240 volt, 10 kW		1.30	6.154		655	310		965	1,175
6920	15 kW		.90	8.889		1,100	445		1,545	1,850
6940	20 kW		.70	11.429		1,550	575		2,125	2,575
6960	25 kW		.50	16		1,850	805		2,655	3,225
6980	30 kW		.40	20		2,125	1,000		3,125	3,850
7000	40 kW		.36	22.222		2,725	1,125		3,850	4,650
7020	50 kW		.32	25		3,275	1,250		4,525	5,475
7100	480 volt, 10 kW		1.40	5.714		730	287		1,017	1,225
7120	15 kW		1	8		1,175	400		1,575	1,900
7140	20 kW		.90	8.889		1,525	445		1,970	2,325
7160	25 kW		.60	13.333		1,850	670		2,520	3,025
7180	30 kW		.50	16		2,125	805		2,930	3,550
7200	40 kW		.40	20		2,725	1,000		3,725	4,500
7220	50 kW		.35	22.857		3,275	1,150		4,425	5,300
7900	Cabinet convector heaters, 240 volt									
7920	3' long, 2000 watt	1 Elec	5.30	1.509	Ea.	1,825	76		1,901	2,125
7940	3000 watt		5.30	1.509		1,900	76		1,976	2,200
7960	4000 watt		5.30	1.509		1,975	76		2,051	2,250
7980	6000 watt		4.60	1.739		2,050	87.50		2,137.50	2,375
8000	8000 watt		4.60	1.739		2,175	87.50		2,262.50	2,525
8020	4' long, 4000 watt		4.60	1.739		2,000	87.50		2,087.50	2,325

23 83 Radiant Heating Units

23 83 33 – Electric Radiant Heaters

23 83 33.10 Electric Heating		Crew	Daily Output	Labor-Hours	Unit	Material	2011 Bare Costs Labor	Equipment	Total	Total Incl O&P
8040	6000 watt	1 Elec	4	2	Ea.	2,075	101		2,176	2,425
8060	8000 watt		4	2		2,150	101		2,251	2,500
8080	10,000 watt	↓	4	2	↓	2,150	101		2,251	2,525
8100	Available also in 208 or 277 volt									
8200	Cabinet unit heaters, 120 to 277 volt, 1 pole,									
8220	wall mounted, 2 kW	1 Elec	4.60	1.739	Ea.	1,925	87.50		2,012.50	2,250
8230	3 kW		4.60	1.739		2,000	87.50		2,087.50	2,325
8240	4 kW		4.40	1.818		2,050	91.50		2,141.50	2,375
8250	5 kW		4.40	1.818		2,125	91.50		2,216.50	2,450
8260	6 kW		4.20	1.905		2,125	96		2,221	2,500
8270	8 kW		4	2		2,200	101		2,301	2,575
8280	10 kW		3.80	2.105		2,250	106		2,356	2,625
8290	12 kW		3.50	2.286		2,300	115		2,415	2,700
8300	13.5 kW		2.90	2.759		2,400	139		2,539	2,825
8310	16 kW		2.70	2.963		2,375	149		2,524	2,825
8320	20 kW		2.30	3.478		3,425	175		3,600	4,000
8330	24 kW		1.90	4.211		3,625	212		3,837	4,325
8350	Recessed, 2 kW		4.40	1.818		1,925	91.50		2,016.50	2,250
8370	3 kW		4.40	1.818		2,075	91.50		2,166.50	2,400
8380	4 kW		4.20	1.905		2,125	96		2,221	2,475
8390	5 kW		4.20	1.905		2,175	96		2,271	2,550
8400	6 kW		4	2		2,200	101		2,301	2,575
8410	8 kW		3.80	2.105		2,300	106		2,406	2,700
8420	10 kW		3.50	2.286		2,775	115		2,890	3,225
8430	12 kW		2.90	2.759		2,800	139		2,939	3,275
8440	13.5 kW		2.70	2.963		2,825	149		2,974	3,350
8450	16 kW		2.30	3.478		2,825	175		3,000	3,350
8460	20 kW		1.90	4.211		3,400	212		3,612	4,050
8470	24 kW		1.60	5		3,400	252		3,652	4,100
8490	Ceiling mounted, 2 kW		3.20	2.500		1,925	126		2,051	2,300
8510	3 kW		3.20	2.500		2,025	126		2,151	2,400
8520	4 kW		3	2.667		2,050	134		2,184	2,450
8530	5 kW		3	2.667		2,125	134		2,259	2,525
8540	6 kW		2.80	2.857		2,200	144		2,344	2,650
8550	8 kW		2.40	3.333		2,300	168		2,468	2,800
8560	10 kW		2.20	3.636		2,775	183		2,958	3,325
8570	12 kW		2	4		2,800	201		3,001	3,375
8580	13.5 kW		1.50	5.333		2,825	268		3,093	3,525
8590	16 kW		1.30	6.154		2,825	310		3,135	3,575
8600	20 kW		.90	8.889		3,400	445		3,845	4,375
8610	24 kW	↓	.60	13.333	↓	3,400	670		4,070	4,725
8630	208 to 480 V, 3 pole									
8650	Wall mounted, 2 kW	1 Elec	4.60	1.739	Ea.	2,000	87.50		2,087.50	2,325
8670	3 kW		4.60	1.739		2,075	87.50		2,162.50	2,400
8680	4 kW		4.40	1.818		2,125	91.50		2,216.50	2,450
8690	5 kW		4.40	1.818		2,175	91.50		2,266.50	2,525
8700	6 kW		4.20	1.905		2,200	96		2,296	2,575
8710	8 kW		4	2		2,275	101		2,376	2,650
8720	10 kW		3.80	2.105		2,325	106		2,431	2,700
8730	12 kW		3.50	2.286		2,375	115		2,490	2,775
8740	13.5 kW		2.90	2.759		2,500	139		2,639	2,950
8750	16 kW		2.70	2.963		2,450	149		2,599	2,925
8760	20 kW	↓	2.30	3.478	↓	3,350	175		3,525	3,925

411

23 83 Radiant Heating Units

23 83 33 – Electric Radiant Heaters

23 83 33.10 Electric Heating		Crew	Daily Output	Labor-Hours	Unit	Material	2011 Bare Costs Labor	Equipment	Total	Total Incl O&P
8770	24 kW	1 Elec	1.90	4.211	Ea.	3,400	212		3,612	4,050
8790	Recessed, 2 kW		4.40	1.818		2,000	91.50		2,091.50	2,325
8810	3 kW		4.40	1.818		2,075	91.50		2,166.50	2,400
8820	4 kW		4.20	1.905		3,825	96		3,921	4,350
8830	5 kW		4.20	1.905		2,175	96		2,271	2,550
8840	6 kW		4	2		2,200	101		2,301	2,575
8850	8 kW		3.80	2.105		2,275	106		2,381	2,650
8860	10 kW		3.50	2.286		2,325	115		2,440	2,725
8870	12 kW		2.90	2.759		2,375	139		2,514	2,800
8880	13.5 kW		2.70	2.963		2,375	149		2,524	2,850
8890	16 kW		2.30	3.478		2,450	175		2,625	2,950
8900	20 kW		1.90	4.211		3,350	212		3,562	4,000
8920	24 kW		1.60	5		3,400	252		3,652	4,100
8940	Ceiling mount, 2 kW		3.20	2.500		2,000	126		2,126	2,375
8950	3 kW		3.20	2.500		2,075	126		2,201	2,450
8960	4 kW		3	2.667		2,125	134		2,259	2,525
8970	5 kW		3	2.667		2,175	134		2,309	2,600
8980	6 kW		2.80	2.857		2,200	144		2,344	2,650
8990	8 kW		2.40	3.333		2,275	168		2,443	2,750
9000	10 kW		2.20	3.636		2,325	183		2,508	2,825
9020	13.5 kW		1.50	5.333		2,500	268		2,768	3,150
9030	16 kW		1.30	6.154		2,550	310		2,860	3,250
9040	20 kW		.90	8.889		3,350	445		3,795	4,325
9060	24 kW		.60	13.333		3,400	670		4,070	4,725

23 84 Humidity Control Equipment

23 84 13 – Humidifiers

23 84 13.10 Humidifier Units		Crew	Daily Output	Labor-Hours	Unit	Material	2011 Bare Costs Labor	Equipment	Total	Total Incl O&P
0010	**HUMIDIFIER UNITS**									
0520	Steam, room or duct, filter, regulators, auto. controls, 220 V									
0540	11 lb. per hour	Q-5	6	2.667	Ea.	2,450	130		2,580	2,900
0560	22 lb. per hour		5	3.200		2,700	156		2,856	3,200
0580	33 lb. per hour		4	4		2,775	195		2,970	3,350
0600	50 lb. per hour		4	4		3,400	195		3,595	4,050
0620	100 lb. per hour		3	5.333		4,075	260		4,335	4,875
0640	150 lb. per hour		2.50	6.400		5,400	310		5,710	6,400
0660	200 lb. per hour		2	8		6,675	390		7,065	7,925
0700	With blower									
0720	11 lb. per hour	Q-5	5.50	2.909	Ea.	3,625	142		3,767	4,200
0740	22 lb. per hour		4.75	3.368		3,875	164		4,039	4,500
0760	33 lb. per hour		3.75	4.267		3,925	208		4,133	4,625
0780	50 lb. per hour		3.50	4.571		4,525	223		4,748	5,300
0800	100 lb. per hour		2.75	5.818		5,025	283		5,308	5,950
0820	150 lb. per hour		2	8		7,525	390		7,915	8,850
0840	200 lb. per hour		1.50	10.667		8,750	520		9,270	10,400
1000	Steam for duct installation, manifold with pneum. controls									
1010	3 – 191 lb/hr, 24" x 24"	Q-5	5	3.200	Ea.	1,800	156		1,956	2,200
1020	3 – 191 lb/hr, 48" x 48"		4.60	3.478		3,500	169		3,669	4,100
1030	65 – 334 lb/hr, 36" x 36"		4.80	3.333		3,425	162		3,587	4,025
1040	65 – 334 lb/hr, 72" x 72"		4	4		3,700	195		3,895	4,375
1050	100 – 690 lb/hr, 48" x 48"		4	4		4,800	195		4,995	5,575

412

23 84 Humidity Control Equipment

23 84 13 – Humidifiers

23 84 13.10 Humidifier Units

	23 84 13.10 Humidifier Units	Crew	Daily Output	Labor-Hours	Unit	Material	2011 Bare Costs Labor	Equipment	Total	Total Incl O&P
1060	100 – 690 lb/hr, 84" x 84"	Q-5	3.60	4.444	Ea.	4,950	216		5,166	5,775
1070	220 – 2000 lb/hr, 72" x 72"		3.80	4.211		5,025	205		5,230	5,825
1080	220 – 2000 lb/hr, 144" x 144"		3.40	4.706		5,875	229		6,104	6,800
1090	For electrically controlled unit, add					805			805	885
1092	For additional manifold, add					25%	15%			
1200	Electric steam, self generating, with manifold									
1210	1 – 96 lb/hr, 12" x 72"	Q-5	3.60	4.444	Ea.	7,825	216		8,041	8,950
1220	20 – 168 lb/hr, 12" x 72"		3.60	4.444		13,700	216		13,916	15,400
1290	For room mounting, add		16	1		225	48.50		273.50	320
5000	Furnace type, wheel bypass									
5020	10 GPD	1 Stpi	4	2	Ea.	260	108		368	450
5040	14 GPD		3.80	2.105		274	114		388	470
5060	19 GPD		3.60	2.222		293	120		413	500

23 84 16 – Mechanical Dehumidification Units

23 84 16.10 Dehumidifier Units

		Crew	Daily Output	Labor-Hours	Unit	Material	2011 Bare Costs Labor	Equipment	Total	Total Incl O&P
0010	**DEHUMIDIFIER UNITS**									
6000	Self contained with filters and standard controls									
6040	1.5 lb./hr., 50 CFM	1 Plum	8	1	Ea.	5,300	53.50		5,353.50	5,900
6060	3 lb./hr., 150 CFM	Q-1	12	1.333		5,525	64.50		5,589.50	6,175
6065	6 lb./hr., 150 CFM		9	1.778		9,000	86		9,086	10,100
6070	16 to 20 lb./hr., 600 CFM		5	3.200		18,300	154		18,454	20,400
6080	30 to 40 lb./hr., 1125 CFM		4	4		28,600	193		28,793	31,700
6090	60 to 75 lb./hr., 2250 CFM		3	5.333		37,300	257		37,557	41,400
6100	120 to 155 lb./hr., 4500 CFM		2	8		66,000	385		66,385	73,000
6110	240 to 310 lb./hr., 9000 CFM		1.50	10.667		94,500	515		95,015	105,000
6120	400 to 515 lb./hr., 15,000 CFM	Q-2	1.60	15		128,000	750		128,750	142,000
6130	530 to 690 lb./hr., 20,000 CFM		1.40	17.143		139,000	860		139,860	154,000
6140	800 to 1030 lb./hr., 30,000 CFM		1.20	20		158,000	1,000		159,000	175,500
6150	1060 to 1375 lb./hr., 40,000 CFM		1	24		220,000	1,200		221,200	244,000

23 91 Prefabricated Equipment Supports

23 91 10 – Prefabricated Curbs, Pads and Stands

23 91 10.10 Prefabricated Pads and Stands

		Crew	Daily Output	Labor-Hours	Unit	Material	2011 Bare Costs Labor	Equipment	Total	Total Incl O&P
0010	**PREFABRICATED PADS AND STANDS**									
6000	Pad, fiberglass reinforced concrete with polystyrene foam core									
6050	Condenser, 2" thick, 20" x 38"	1 Shee	8	1	Ea.	24	51.50		75.50	105
6070	24" x 24"		16	.500		19.25	26		45.25	60
6090	24" x 36"		12	.667		31	34.50		65.50	86
6110	24" x 42"		8	1		35.50	51.50		87	118
6150	26" x 36"	Q-9	8	2		33.50	93		126.50	178
6170	28" x 38"	1 Shee	8	1		38	51.50		89.50	120
6190	30" x 30"		12	.667		32.50	34.50		67	87.50
6220	30" x 36"		8	1		38	51.50		89.50	120
6240	30" x 40"	Q-9	8	2		41.50	93		134.50	187
6260	32" x 32"		7	2.286		36	106		142	201
6280	36" x 36"		8	2		44.50	93		137.50	190
6300	36" x 40"		7	2.286		50.50	106		156.50	217
6320	36" x 48"		7	2.286		59	106		165	226
6340	36" x 54"		6	2.667		71.50	124		195.50	266
6360	36" x 60" x 3"		7	2.286		85	106		191	254

23 91 10.10 Prefabricated Pads and Stands	Crew	Daily Output	Labor-Hours	Unit	Material	2011 Bare Costs Labor	Equipment	Total	Total Incl O&P
6400	41" x 47" x 3"	Q-9	6	2.667	Ea.	79	124		203	274
6490	26" round		10	1.600		19	74.50		93.50	133
6550	30" round		9	1.778		25	82.50		107.50	153
6600	36" round	↓	8	2		34.50	93		127.50	179
6700	For 3" thick pad, add				↓	20%	5%			
8800	Stand, corrosion-resistant plastic									
8850	Heat pump, 6" high	1 Shee	32	.250	Ea.	5	12.90		17.90	25
8870	12" high	"	32	.250	"	11.35	12.90		24.25	32

Estimating Tips

26 05 00 Common Work Results for Electrical

- Conduit should be taken off in three main categories—power distribution, branch power, and branch lighting—so the estimator can concentrate on systems and components, therefore making it easier to ensure all items have been accounted for.

- For cost modifications for elevated conduit installation, add the percentages to labor according to the height of installation, and only to the quantities exceeding the different height levels, not to the total conduit quantities.

- Remember that aluminum wiring of equal ampacity is larger in diameter than copper and may require larger conduit.

- If more than three wires at a time are being pulled, deduct percentages from the labor hours of that grouping of wires.

- When taking off grounding systems, identify separately the type and size of wire, and list each unique type of ground connection.

- The estimator should take the weights of materials into consideration when completing a takeoff. Topics to consider include: How will the materials be supported? What methods of support are available? How high will the support structure have to reach? Will the final support structure be able to withstand the total burden? Is the support material included or separate from the fixture, equipment, and material specified?

- Do not overlook the costs for equipment used in the installation. If scaffolding or highlifts are available in the field, contractors may use them in lieu of the proposed ladders and rolling staging.

26 20 00 Low-Voltage Electrical Transmission

- Supports and concrete pads may be shown on drawings for the larger equipment, or the support system may be only a piece of plywood for the back of a panelboard. In either case, it must be included in the costs.

26 40 00 Electrical and Cathodic Protection

- When taking off cathodic protections systems, identify the type and size of cable, and list each unique type of anode connection.

26 50 00 Lighting

- Fixtures should be taken off room by room, using the fixture schedule, specifications, and the ceiling plan. For large concentrations of lighting fixtures in the same area, deduct the percentages from labor hours.

Reference Numbers

Reference numbers are shown in shaded boxes at the beginning of some major classifications. These numbers refer to related items in the Reference Section. The reference information may be an estimating procedure, an alternate pricing method, or technical information.

Note: Not all subdivisions listed here necessarily appear in this publication.

Note: **Trade Service,** *in part, has been used as a reference source for some of the material prices used in Division 26.*

26 05 33.95 Cutting and Drilling	Crew	Daily Output	Labor-Hours	Unit	Material	2011 Bare Costs Labor	Equipment	Total	Total Incl O&P
0010 **CUTTING AND DRILLING**									
0100 Hole drilling to 10' high, concrete wall									
0110 8" thick, 1/2" pipe size	R-31	12	.667	Ea.	5.05	33.50	3.92	42.47	59.50
0120 3/4" pipe size		12	.667		5.05	33.50	3.92	42.47	59.50
0130 1" pipe size		9.50	.842		9.80	42.50	4.95	57.25	79
0140 1-1/4" pipe size		9.50	.842		9.80	42.50	4.95	57.25	79
0150 1-1/2" pipe size		9.50	.842		9.80	42.50	4.95	57.25	79
0160 2" pipe size		4.40	1.818		15.05	91.50	10.70	117.25	163
0170 2-1/2" pipe size		4.40	1.818		15.05	91.50	10.70	117.25	163
0180 3" pipe size		4.40	1.818		15.05	91.50	10.70	117.25	163
0190 3-1/2" pipe size		3.30	2.424		18.90	122	14.25	155.15	218
0200 4" pipe size		3.30	2.424		18.90	122	14.25	155.15	218
0500 12" thick, 1/2" pipe size		9.40	.851		7.80	43	5	55.80	77.50
0520 3/4" pipe size		9.40	.851		7.80	43	5	55.80	77.50
0540 1" pipe size		7.30	1.096		14.15	55	6.45	75.60	104
0560 1-1/4" pipe size		7.30	1.096		14.15	55	6.45	75.60	104
0570 1-1/2" pipe size		7.30	1.096		14.15	55	6.45	75.60	104
0580 2" pipe size		3.60	2.222		20.50	112	13.05	145.55	203
0590 2-1/2" pipe size		3.60	2.222		20.50	112	13.05	145.55	203
0600 3" pipe size		3.60	2.222		20.50	112	13.05	145.55	203
0610 3-1/2" pipe size		2.80	2.857		26	144	16.80	186.80	260
0630 4" pipe size		2.50	3.200		26	161	18.80	205.80	287
0650 16" thick, 1/2" pipe size		7.60	1.053		10.55	53	6.20	69.75	97
0670 3/4" pipe size		7	1.143		10.55	57.50	6.70	74.75	104
0690 1" pipe size		6	1.333		18.45	67	7.85	93.30	129
0710 1-1/4" pipe size		5.50	1.455		18.45	73	8.55	100	138
0730 1-1/2" pipe size		5.50	1.455		18.45	73	8.55	100	138
0750 2" pipe size		3	2.667		26.50	134	15.70	176.20	245
0770 2-1/2" pipe size		2.70	2.963		26.50	149	17.40	192.90	269
0790 3" pipe size		2.50	3.200		26.50	161	18.80	206.30	288
0810 3-1/2" pipe size		2.30	3.478		33	175	20.50	228.50	320
0830 4" pipe size		2	4		33	201	23.50	257.50	360
0850 20" thick, 1/2" pipe size		6.40	1.250		13.30	63	7.35	83.65	116
0870 3/4" pipe size		6	1.333		13.30	67	7.85	88.15	123
0890 1" pipe size		5	1.600		23	80.50	9.40	112.90	154
0910 1-1/4" pipe size		4.80	1.667		23	84	9.80	116.80	160
0930 1-1/2" pipe size		4.60	1.739		23	87.50	10.25	120.75	166
0950 2" pipe size		2.70	2.963		32	149	17.40	198.40	275
0970 2-1/2" pipe size		2.40	3.333		32	168	19.60	219.60	305
0990 3" pipe size		2.20	3.636		32	183	21.50	236.50	330
1010 3-1/2" pipe size		2	4		39.50	201	23.50	264	370
1030 4" pipe size		1.70	4.706		39.50	237	27.50	304	425
1050 24" thick, 1/2" pipe size		5.50	1.455		16.10	73	8.55	97.65	135
1070 3/4" pipe size		5.10	1.569		16.10	79	9.20	104.30	145
1090 1" pipe size		4.30	1.860		27	93.50	10.95	131.45	181
1110 1-1/4" pipe size		4	2		27	101	11.75	139.75	192
1130 1-1/2" pipe size		4	2		27	101	11.75	139.75	192
1150 2" pipe size		2.40	3.333		37.50	168	19.60	225.10	310
1170 2-1/2" pipe size		2.20	3.636		37.50	183	21.50	242	335
1190 3" pipe size		2	4		37.50	201	23.50	262	365
1210 3-1/2" pipe size		1.80	4.444		46.50	224	26	296.50	410
1230 4" pipe size		1.50	5.333		46.50	268	31.50	346	480
1500 Brick wall, 8" thick, 1/2" pipe size		18	.444		5.05	22.50	2.61	30.16	41.50

26 05 33.95 Cutting and Drilling	Crew	Daily Output	Labor-Hours	Unit	Material	2011 Bare Costs Labor	Equipment	Total	Total Incl O&P
1520	3/4" pipe size	R-31	18	.444	Ea.	5.05	22.50	2.61	30.16	41.50
1540	1" pipe size		13.30	.602		9.80	30.50	3.54	43.84	59.50
1560	1-1/4" pipe size		13.30	.602		9.80	30.50	3.54	43.84	59.50
1580	1-1/2" pipe size		13.30	.602		9.80	30.50	3.54	43.84	59.50
1600	2" pipe size		5.70	1.404		15.05	70.50	8.25	93.80	131
1620	2-1/2" pipe size		5.70	1.404		15.05	70.50	8.25	93.80	131
1640	3" pipe size		5.70	1.404		15.05	70.50	8.25	93.80	131
1660	3-1/2" pipe size		4.40	1.818		18.90	91.50	10.70	121.10	168
1680	4" pipe size		4	2		18.90	101	11.75	131.65	183
1700	12" thick, 1/2" pipe size		14.50	.552		7.80	28	3.24	39.04	53
1720	3/4" pipe size		14.50	.552		7.80	28	3.24	39.04	53
1740	1" pipe size		11	.727		14.15	36.50	4.28	54.93	74.50
1760	1-1/4" pipe size		11	.727		14.15	36.50	4.28	54.93	74.50
1780	1-1/2" pipe size		11	.727		14.15	36.50	4.28	54.93	74.50
1800	2" pipe size		5	1.600		20.50	80.50	9.40	110.40	152
1820	2-1/2" pipe size		5	1.600		20.50	80.50	9.40	110.40	152
1840	3" pipe size		5	1.600		20.50	80.50	9.40	110.40	152
1860	3-1/2" pipe size		3.80	2.105		26	106	12.40	144.40	199
1880	4" pipe size		3.30	2.424		26	122	14.25	162.25	225
1900	16" thick, 1/2" pipe size		12.30	.650		10.55	32.50	3.82	46.87	64.50
1920	3/4" pipe size		12.30	.650		10.55	32.50	3.82	46.87	64.50
1940	1" pipe size		9.30	.860		18.45	43.50	5.05	67	90
1960	1-1/4" pipe size		9.30	.860		18.45	43.50	5.05	67	90
1980	1-1/2" pipe size		9.30	.860		18.45	43.50	5.05	67	90
2000	2" pipe size		4.40	1.818		26.50	91.50	10.70	128.70	176
2010	2-1/2" pipe size		4.40	1.818		26.50	91.50	10.70	128.70	176
2030	3" pipe size		4.40	1.818		26.50	91.50	10.70	128.70	176
2050	3-1/2" pipe size		3.30	2.424		33	122	14.25	169.25	233
2070	4" pipe size		3	2.667		33	134	15.70	182.70	252
2090	20" thick, 1/2" pipe size		10.70	.748		13.30	37.50	4.40	55.20	75
2110	3/4" pipe size		10.70	.748		13.30	37.50	4.40	55.20	75
2130	1" pipe size		8	1		23	50.50	5.90	79.40	106
2150	1-1/4" pipe size		8	1		23	50.50	5.90	79.40	106
2170	1-1/2" pipe size		8	1		23	50.50	5.90	79.40	106
2190	2" pipe size		4	2		32	101	11.75	144.75	197
2210	2-1/2" pipe size		4	2		32	101	11.75	144.75	197
2230	3" pipe size		4	2		32	101	11.75	144.75	197
2250	3-1/2" pipe size		3	2.667		39.50	134	15.70	189.20	260
2270	4" pipe size		2.70	2.963		39.50	149	17.40	205.90	284
2290	24" thick, 1/2" pipe size		9.40	.851		16.10	43	5	64.10	86.50
2310	3/4" pipe size		9.40	.851		16.10	43	5	64.10	86.50
2330	1" pipe size		7.10	1.127		27	56.50	6.65	90.15	121
2350	1-1/4" pipe size		7.10	1.127		27	56.50	6.65	90.15	121
2370	1-1/2" pipe size		7.10	1.127		27	56.50	6.65	90.15	121
2390	2" pipe size		3.60	2.222		37.50	112	13.05	162.55	222
2410	2-1/2" pipe size		3.60	2.222		37.50	112	13.05	162.55	222
2430	3" pipe size		3.60	2.222		37.50	112	13.05	162.55	222
2450	3-1/2" pipe size		2.80	2.857		46.50	144	16.80	207.30	283
2470	4" pipe size		2.50	3.200		46.50	161	18.80	226.30	310
3000	Knockouts to 8' high, metal boxes & enclosures									
3020	With hole saw, 1/2" pipe size	1 Elec	53	.151	Ea.		7.60		7.60	11.25
3040	3/4" pipe size		47	.170			8.55		8.55	12.70
3050	1" pipe size		40	.200			10.05		10.05	14.90

26 05 33 – Raceway and Boxes for Electrical Systems

26 05 33.95 Cutting and Drilling

		Crew	Daily Output	Labor-Hours	Unit	Material	2011 Bare Costs Labor	Equipment	Total	Total Incl O&P
3060	1-1/4" pipe size	1 Elec	36	.222	Ea.		11.20		11.20	16.55
3070	1-1/2" pipe size		32	.250			12.60		12.60	18.65
3080	2" pipe size		27	.296			14.90		14.90	22
3090	2-1/2" pipe size		20	.400			20		20	30
4010	3" pipe size		16	.500			25		25	37.50
4030	3-1/2" pipe size		13	.615			31		31	46
4050	4" pipe size		11	.727			36.50		36.50	54
4070	With hand punch set, 1/2" pipe size		40	.200			10.05		10.05	14.90
4090	3/4" pipe size		32	.250			12.60		12.60	18.65
4110	1" pipe size		30	.267			13.40		13.40	19.85
4130	1-1/4" pipe size		28	.286			14.35		14.35	21.50
4150	1-1/2" pipe size		26	.308			15.50		15.50	23
4170	2" pipe size		20	.400			20		20	30
4190	2-1/2" pipe size		17	.471			23.50		23.50	35
4200	3" pipe size		15	.533			27		27	39.50
4220	3-1/2" pipe size		12	.667			33.50		33.50	49.50
4240	4" pipe size		10	.800			40		40	59.50
4260	With hydraulic punch, 1/2" pipe size		44	.182			9.15		9.15	13.55
4280	3/4" pipe size		38	.211			10.60		10.60	15.70
4300	1" pipe size		38	.211			10.60		10.60	15.70
4320	1-1/4" pipe size		38	.211			10.60		10.60	15.70
4340	1-1/2" pipe size		38	.211			10.60		10.60	15.70
4360	2" pipe size		32	.250			12.60		12.60	18.65
4380	2-1/2" pipe size		27	.296			14.90		14.90	22
4400	3" pipe size		23	.348			17.50		17.50	26
4420	3-1/2" pipe size		20	.400			20		20	30
4440	4" pipe size		18	.444			22.50		22.50	33

26 05 80 – Wiring Connections

26 05 80.10 Motor Connections

		Crew	Daily Output	Labor-Hours	Unit	Material	2011 Bare Costs Labor	Equipment	Total	Total Incl O&P
0010	**MOTOR CONNECTIONS**									
0020	Flexible conduit and fittings, 115 volt, 1 phase, up to 1 HP motor	1 Elec	8	1	Ea.	9.70	50.50		60.20	85
0050	2 HP motor		6.50	1.231		9.90	62		71.90	102
0100	3 HP motor		5.50	1.455		15.40	73		88.40	125
0120	230 volt, 10 HP motor, 3 phase		4.20	1.905		16.40	96		112.40	160
0150	15 HP motor		3.30	2.424		28.50	122		150.50	213
0200	25 HP motor		2.70	2.963		34	149		183	258
0400	50 HP motor		2.20	3.636		83	183		266	365
0600	100 HP motor		1.50	5.333		201	268		469	615
1500	460 volt, 5 HP motor, 3 phase		8	1		10.50	50.50		61	86
1520	10 HP motor		8	1		10.50	50.50		61	86
1530	25 HP motor		6	1.333		18.35	67		85.35	120
1540	30 HP motor		6	1.333		18.35	67		85.35	120
1550	40 HP motor		5	1.600		28.50	80.50		109	151
1560	50 HP motor		5	1.600		32	80.50		112.50	154
1570	60 HP motor		3.80	2.105		36	106		142	197
1580	75 HP motor		3.50	2.286		48	115		163	223
1590	100 HP motor		2.50	3.200		71	161		232	315
1600	125 HP motor		2	4		93.50	201		294.50	400
1610	150 HP motor		1.80	4.444		98	224		322	440
1620	200 HP motor		1.50	5.333		154	268		422	565
2005	460 Volt, 5 HP motor, 3 Phase, w/sealtite		8	1		16	50.50		66.50	92
2010	10 HP motor		8	1		16	50.50		66.50	92

26 05 80 – Wiring Connections

26 05 80.10 Motor Connections	Crew	Daily Output	Labor-Hours	Unit	Material	2011 Bare Costs Labor	Equipment	Total	Total Incl O&P
2015	25 HP motor	1 Elec	6	1.333	Ea.	28.50	67		95.50	131
2020	30 HP motor		6	1.333		28.50	67		95.50	131
2025	40 HP motor		5	1.600		53.50	80.50		134	178
2030	50 HP motor		5	1.600		53.50	80.50		134	178
2035	60 HP motor		3.80	2.105		83.50	106		189.50	249
2040	75 HP motor		3.50	2.286		87	115		202	266
2045	100 HP motor		2.50	3.200		117	161		278	365
2055	150 HP motor		1.80	4.444		122	224		346	465
2060	200 HP motor		1.50	5.333		860	268		1,128	1,350

26 05 90 – Residential Applications

26 05 90.10 Residential Wiring

		Crew	Daily Output	Labor-Hours	Unit	Material	2011 Bare Costs Labor	Equipment	Total	Total Incl O&P
0010	**RESIDENTIAL WIRING**									
0020	20' avg. runs and #14/2 wiring incl. unless otherwise noted									
1000	Service & panel, includes 24' SE-AL cable, service eye, meter,									
1010	Socket, panel board, main bkr., ground rod, 15 or 20 amp									
1020	1-pole circuit breakers, and misc. hardware									
1100	100 amp, with 10 branch breakers	1 Elec	1.19	6.723	Ea.	560	340		900	1,125
1110	With PVC conduit and wire		.92	8.696		630	435		1,065	1,350
1120	With RGS conduit and wire		.73	10.959		790	550		1,340	1,675
1150	150 amp, with 14 branch breakers		1.03	7.767		865	390		1,255	1,525
1170	With PVC conduit and wire		.82	9.756		1,000	490		1,490	1,825
1180	With RGS conduit and wire		.67	11.940		1,300	600		1,900	2,350
1200	200 amp, with 18 branch breakers	2 Elec	1.80	8.889		1,150	445		1,595	1,900
1220	With PVC conduit and wire		1.46	10.959		1,275	550		1,825	2,225
1230	With RGS conduit and wire		1.24	12.903		1,700	650		2,350	2,825
1800	Lightning surge suppressor for above services, add	1 Elec	32	.250		55	12.60		67.60	79
2000	Switch devices									
2100	Single pole, 15 amp, Ivory, with a 1-gang box, cover plate,									
2110	Type NM (Romex) cable	1 Elec	17.10	.468	Ea.	10.70	23.50		34.20	47
2120	Type MC (BX) cable		14.30	.559		28.50	28		56.50	73
2130	EMT & wire		5.71	1.401		34	70.50		104.50	141
2150	3-way, #14/3, type NM cable		14.55	.550		14.85	27.50		42.35	57.50
2170	Type MC cable		12.31	.650		38.50	32.50		71	91
2180	EMT & wire		5	1.600		37.50	80.50		118	160
2200	4-way, #14/3, type NM cable		14.55	.550		31.50	27.50		59	76
2220	Type MC cable		12.31	.650		55.50	32.50		88	110
2230	EMT & wire		5	1.600		54.50	80.50		135	179
2250	S.P., 20 amp, #12/2, type NM cable		13.33	.600		18.05	30		48.05	64.50
2270	Type MC cable		11.43	.700		33.50	35		68.50	88.50
2280	EMT & wire		4.85	1.649		42.50	83		125.50	170
2290	S.P. rotary dimmer, 600W, no wiring		17	.471		21	23.50		44.50	58
2300	S.P. rotary dimmer, 600W, type NM cable		14.55	.550		26	27.50		53.50	70
2320	Type MC cable		12.31	.650		44	32.50		76.50	97
2330	EMT & wire		5	1.600		51	80.50		131.50	175
2350	3-way rotary dimmer, type NM cable		13.33	.600		33.50	30		63.50	81
2370	Type MC cable		11.43	.700		51	35		86	109
2380	EMT & wire		4.85	1.649		58	83		141	187
2400	Interval timer wall switch, 20 amp, 1-30 min., #12/2									
2410	Type NM cable	1 Elec	14.55	.550	Ea.	53	27.50		80.50	99.50
2420	Type MC cable		12.31	.650		61	32.50		93.50	116
2430	EMT & wire		5	1.600		77.50	80.50		158	204
2500	Decorator style									

26 05 90.10 Residential Wiring	Crew	Daily Output	Labor-Hours	Unit	Material	2011 Bare Costs Labor	Equipment	Total	Total Incl O&P
2510	S.P., 15 amp, type NM cable	1 Elec	17.10	.468	Ea.	15.55	23.50		39.05	52
2520	Type MC cable		14.30	.559		33.50	28		61.50	78
2530	EMT & wire		5.71	1.401		38.50	70.50		109	147
2550	3-way, #14/3, type NM cable		14.55	.550		19.70	27.50		47.20	62.50
2570	Type MC cable		12.31	.650		43.50	32.50		76	96
2580	EMT & wire		5	1.600		42.50	80.50		123	166
2600	4-way, #14/3, type NM cable		14.55	.550		36.50	27.50		64	81.50
2620	Type MC cable		12.31	.650		60.50	32.50		93	115
2630	EMT & wire		5	1.600		59.50	80.50		140	184
2650	S.P., 20 amp, #12/2, type NM cable		13.33	.600		23	30		53	69.50
2670	Type MC cable		11.43	.700		38	35		73	94
2680	EMT & wire		4.85	1.649		47.50	83		130.50	175
2700	S.P., slide dimmer, type NM cable		17.10	.468		33.50	23.50		57	72
2720	Type MC cable		14.30	.559		51.50	28		79.50	98
2730	EMT & wire		5.71	1.401		58.50	70.50		129	168
2750	S.P., touch dimmer, type NM cable		17.10	.468		28.50	23.50		52	66.50
2770	Type MC cable		14.30	.559		46.50	28		74.50	92.50
2780	EMT & wire		5.71	1.401		53	70.50		123.50	163
2800	3-way touch dimmer, type NM cable		13.33	.600		48	30		78	97
2820	Type MC cable		11.43	.700		65.50	35		100.50	125
2830	EMT & wire	▼	4.85	1.649	▼	72.50	83		155.50	203
3000	Combination devices									
3100	S.P. switch/15 amp recpt., Ivory, 1-gang box, plate									
3110	Type NM cable	1 Elec	11.43	.700	Ea.	22.50	35		57.50	77
3120	Type MC cable		10	.800		40.50	40		80.50	104
3130	EMT & wire		4.40	1.818		47	91.50		138.50	187
3150	S.P. switch/pilot light, type NM cable		11.43	.700		23.50	35		58.50	78
3170	Type MC cable		10	.800		41.50	40		81.50	105
3180	EMT & wire		4.43	1.806		48	91		139	188
3190	2-S.P. switches, 2-#14/2, no wiring		14	.571		8.40	28.50		36.90	51.50
3200	2-S.P. switches, 2-#14/2, type NM cables		10	.800		26.50	40		66.50	89
3220	Type MC cable		8.89	.900		53.50	45.50		99	126
3230	EMT & wire		4.10	1.951		51	98		149	201
3250	3-way switch/15 amp recpt., #14/3, type NM cable		10	.800		30.50	40		70.50	93
3270	Type MC cable		8.89	.900		54.50	45.50		100	127
3280	EMT & wire		4.10	1.951		53.50	98		151.50	204
3300	2-3 way switches, 2-#14/3, type NM cables		8.89	.900		40.50	45.50		86	112
3320	Type MC cable		8	1		79.50	50.50		130	162
3330	EMT & wire		4	2		61	101		162	216
3350	S.P. switch/20 amp recpt., #12/2, type NM cable		10	.800		34.50	40		74.50	97.50
3370	Type MC cable		8.89	.900		42.50	45.50		88	114
3380	EMT & wire	▼	4.10	1.951	▼	59	98		157	210
3400	Decorator style									
3410	S.P. switch/15 amp recpt., type NM cable	1 Elec	11.43	.700	Ea.	27.50	35		62.50	82
3420	Type MC cable		10	.800		45.50	40		85.50	110
3430	EMT & wire		4.40	1.818		52	91.50		143.50	193
3450	S.P. switch/pilot light, type NM cable		11.43	.700		28.50	35		63.50	83
3470	Type MC cable		10	.800		46	40		86	111
3480	EMT & wire		4.40	1.818		53	91.50		144.50	193
3500	2-S.P. switches, 2-#14/2, type NM cables		10	.800		31.50	40		71.50	94
3520	Type MC cable		8.89	.900		58.50	45.50		104	132
3530	EMT & wire		4.10	1.951		56	98		154	207
3550	3-way/15 amp recpt., #14/3, type NM cable	▼	10	.800		35.50	40		75.50	98.50

26 05 90 – Residential Applications

26 05 90.10 Residential Wiring		Crew	Daily Output	Labor-Hours	Unit	Material	2011 Bare Costs Labor	Equipment	Total	Total Incl O&P
3570	Type MC cable	1 Elec	8.89	.900	Ea.	59	45.50		104.50	132
3580	EMT & wire		4.10	1.951		58	98		156	209
3650	2-3 way switches, 2-#14/3, type NM cables		8.89	.900		45.50	45.50		91	117
3670	Type MC cable		8	1		84	50.50		134.50	167
3680	EMT & wire		4	2		65.50	101		166.50	222
3700	S.P. switch/20 amp recpt., #12/2, type NM cable		10	.800		39.50	40		79.50	103
3720	Type MC cable		8.89	.900		47.50	45.50		93	119
3730	EMT & wire		4.10	1.951		64	98		162	216
4000	Receptacle devices									
4010	Duplex outlet, 15 amp recpt., Ivory, 1-gang box, plate									
4015	Type NM cable	1 Elec	14.55	.550	Ea.	9.10	27.50		36.60	51
4020	Type MC cable		12.31	.650		27	32.50		59.50	78
4030	EMT & wire		5.33	1.501		32	75.50		107.50	148
4050	With #12/2, type NM cable		12.31	.650		11.80	32.50		44.30	61.50
4070	Type MC cable		10.67	.750		27	37.50		64.50	86
4080	EMT & wire		4.71	1.699		36	85.50		121.50	167
4100	20 amp recpt., #12/2, type NM cable		12.31	.650		21	32.50		53.50	71.50
4120	Type MC cable		10.67	.750		36.50	37.50		74	96
4130	EMT & wire		4.71	1.699		45.50	85.50		131	177
4140	For GFI see Section 26 05 90.10 line 4300 below									
4150	Decorator style, 15 amp recpt., type NM cable	1 Elec	14.55	.550	Ea.	13.95	27.50		41.45	56.50
4170	Type MC cable		12.31	.650		32	32.50		64.50	83.50
4180	EMT & wire		5.33	1.501		37	75.50		112.50	153
4200	With #12/2, type NM cable		12.31	.650		16.65	32.50		49.15	67
4220	Type MC cable		10.67	.750		32	37.50		69.50	91
4230	EMT & wire		4.71	1.699		41	85.50		126.50	172
4250	20 amp recpt. #12/2, type NM cable		12.31	.650		26	32.50		58.50	77
4270	Type MC cable		10.67	.750		41	37.50		78.50	102
4280	EMT & wire		4.71	1.699		50.50	85.50		136	183
4300	GFI, 15 amp recpt., type NM cable		12.31	.650		43.50	32.50		76	96.50
4320	Type MC cable		10.67	.750		61.50	37.50		99	124
4330	EMT & wire		4.71	1.699		66.50	85.50		152	201
4350	GFI with #12/2, type NM cable		10.67	.750		46.50	37.50		84	107
4370	Type MC cable		9.20	.870		61.50	43.50		105	133
4380	EMT & wire		4.21	1.900		71	95.50		166.50	220
4400	20 amp recpt., #12/2 type NM cable		10.67	.750		48.50	37.50		86	109
4420	Type MC cable		9.20	.870		63.50	43.50		107	135
4430	EMT & wire		4.21	1.900		73	95.50		168.50	222
4500	Weather-proof cover for above receptacles, add		32	.250		5.25	12.60		17.85	24.50
4550	Air conditioner outlet, 20 amp-240 volt recpt.									
4560	30' of #12/2, 2 pole circuit breaker									
4570	Type NM cable	1 Elec	10	.800	Ea.	61	40		101	127
4580	Type MC cable		9	.889		79.50	44.50		124	154
4590	EMT & wire		4	2		85	101		186	243
4600	Decorator style, type NM cable		10	.800		65.50	40		105.50	132
4620	Type MC cable		9	.889		84	44.50		128.50	159
4630	EMT & wire		4	2		89.50	101		190.50	248
4650	Dryer outlet, 30 amp-240 volt recpt., 20' of #10/3									
4660	2 pole circuit breaker									
4670	Type NM cable	1 Elec	6.41	1.248	Ea.	67.50	63		130.50	168
4680	Type MC cable		5.71	1.401		80	70.50		150.50	192
4690	EMT & wire		3.48	2.299		80.50	116		196.50	260
4700	Range outlet, 50 amp-240 volt recpt., 30' of #8/3									

26 05 90 – Residential Applications

26 05 90.10 Residential Wiring		Crew	Daily Output	Labor-Hours	Unit	Material	2011 Bare Costs Labor	Equipment	Total	Total Incl O&P
4710	Type NM cable	1 Elec	4.21	1.900	Ea.	96.50	95.50		192	248
4720	Type MC cable		4	2		165	101		266	330
4730	EMT & wire		2.96	2.703		110	136		246	320
4750	Central vacuum outlet, Type NM cable		6.40	1.250		65	63		128	165
4770	Type MC cable		5.71	1.401		91	70.50		161.50	204
4780	EMT & wire	▼	3.48	2.299	▼	90	116		206	270
4800	30 amp-110 volt locking recpt., #10/2 circ. bkr.									
4810	Type NM cable	1 Elec	6.20	1.290	Ea.	75.50	65		140.50	179
4830	EMT & wire	"	3.20	2.500	"	100	126		226	296
4900	Low voltage outlets									
4910	Telephone recpt., 20' of 4/C phone wire	1 Elec	26	.308	Ea.	11	15.50		26.50	35
4920	TV recpt., 20' of RG59U coax wire, F type connector	"	16	.500	"	19.25	25		44.25	58.50
4950	Door bell chime, transformer, 2 buttons, 60' of bellwire									
4970	Economy model	1 Elec	11.50	.696	Ea.	77.50	35		112.50	138
4980	Custom model		11.50	.696		121	35		156	185
4990	Luxury model, 3 buttons	▼	9.50	.842	▼	325	42.50		367.50	420
6000	Lighting outlets									
6050	Wire only (for fixture), type NM cable	1 Elec	32	.250	Ea.	7.15	12.60		19.75	26.50
6070	Type MC cable		24	.333		17.25	16.75		34	44
6080	EMT & wire		10	.800		21	40		61	82.50
6100	Box (4"), and wire (for fixture), type NM cable		25	.320		17.70	16.10		33.80	43.50
6120	Type MC cable		20	.400		28	20		48	60.50
6130	EMT & wire	▼	11	.727	▼	31.50	36.50		68	89
6200	Fixtures (use with lines 6050 or 6100 above)									
6210	Canopy style, economy grade	1 Elec	40	.200	Ea.	34.50	10.05		44.55	53
6220	Custom grade		40	.200		53	10.05		63.05	73.50
6250	Dining room chandelier, economy grade		19	.421		86	21		107	126
6260	Custom grade		19	.421		253	21		274	310
6270	Luxury grade		15	.533		570	27		597	665
6310	Kitchen fixture (fluorescent), economy grade		30	.267		64	13.40		77.40	90.50
6320	Custom grade		25	.320		175	16.10		191.10	217
6350	Outdoor, wall mounted, economy grade		30	.267		32.50	13.40		45.90	56
6360	Custom grade		30	.267		112	13.40		125.40	143
6370	Luxury grade		25	.320		253	16.10		269.10	300
6410	Outdoor PAR floodlights, 1 lamp, 150 watt		20	.400		35	20		55	68.50
6420	2 lamp, 150 watt each		20	.400		58	20		78	94
6430	For infrared security sensor, add		32	.250		117	12.60		129.60	148
6450	Outdoor, quartz-halogen, 300 watt flood		20	.400		42	20		62	76.50
6600	Recessed downlight, round, pre-wired, 50 or 75 watt trim		30	.267		43.50	13.40		56.90	67.50
6610	With shower light trim		30	.267		53.50	13.40		66.90	79
6620	With wall washer trim		28	.286		64.50	14.35		78.85	92.50
6630	With eye-ball trim	▼	28	.286		64.50	14.35		78.85	92.50
6640	For direct contact with insulation, add					2.30			2.30	2.53
6700	Porcelain lamp holder	1 Elec	40	.200		4.50	10.05		14.55	19.85
6710	With pull switch		40	.200		5.25	10.05		15.30	20.50
6750	Fluorescent strip, 1-20 watt tube, wrap around diffuser, 24"		24	.333		54	16.75		70.75	84.50
6770	2-34 watt tubes, 48"		20	.400		82.50	20		102.50	121
6780	With residential ballast		20	.400		92.50	20		112.50	132
6800	Bathroom heat lamp, 1-250 watt		28	.286		47.50	14.35		61.85	74
6810	2-250 watt lamps	▼	28	.286	▼	73.50	14.35		87.85	103
6820	For timer switch, see Section 26 05 90.10 line 2400									
6900	Outdoor post lamp, incl. post, fixture, 35' of #14/2									
6910	Type NMC cable	1 Elec	3.50	2.286	Ea.	203	115		318	395

26 05 90 – Residential Applications

26 05 90.10 Residential Wiring	Crew	Daily Output	Labor-Hours	Unit	Material	2011 Bare Costs Labor	Equipment	Total	Total Incl O&P
6920 Photo-eye, add	1 Elec	27	.296	Ea.	35	14.90		49.90	60.50
6950 Clock dial time switch, 24 hr., w/enclosure, type NM cable		11.43	.700		67	35		102	126
6970 Type MC cable		11	.727		84.50	36.50		121	147
6980 EMT & wire		4.85	1.649		90	83		173	222
7000 Alarm systems									
7050 Smoke detectors, box, #14/3, type NM cable	1 Elec	14.55	.550	Ea.	35	27.50		62.50	79.50
7070 Type MC cable		12.31	.650		51.50	32.50		84	105
7080 EMT & wire		5	1.600		50.50	80.50		131	175
7090 For relay output to security system, add					13			13	14.30
8000 Residential equipment									
8050 Disposal hook-up, incl. switch, outlet box, 3' of flex									
8060 20 amp-1 pole circ. bkr., and 25' of #12/2									
8070 Type NM cable	1 Elec	10	.800	Ea.	31.50	40		71.50	94
8080 Type MC cable		8	1		48.50	50.50		99	128
8090 EMT & wire		5	1.600		58	80.50		138.50	183
8100 Trash compactor or dishwasher hook-up, incl. outlet box,									
8110 3' of flex, 15 amp-1 pole circ. bkr., and 25' of #14/2									
8120 Type NM cable	1 Elec	10	.800	Ea.	23	40		63	84.50
8130 Type MC cable		8	1		43	50.50		93.50	122
8140 EMT & wire		5	1.600		50	80.50		130.50	174
8150 Hot water sink dispensor hook-up, use line 8100									
8200 Vent/exhaust fan hook-up, type NM cable	1 Elec	32	.250	Ea.	7.15	12.60		19.75	26.50
8220 Type MC cable		24	.333		17.25	16.75		34	44
8230 EMT & wire		10	.800		21	40		61	82.50
8250 Bathroom vent fan, 50 CFM (use with above hook-up)									
8260 Economy model	1 Elec	15	.533	Ea.	25	27		52	67
8270 Low noise model		15	.533		34	27		61	77
8280 Custom model		12	.667		117	33.50		150.50	179
8300 Bathroom or kitchen vent fan, 110 CFM									
8310 Economy model	1 Elec	15	.533	Ea.	69.50	27		96.50	116
8320 Low noise model	"	15	.533	"	86	27		113	134
8350 Paddle fan, variable speed (w/o lights)									
8360 Economy model (AC motor)	1 Elec	10	.800	Ea.	115	40		155	187
8362 With light kit		10	.800		153	40		193	229
8370 Custom model (AC motor)		10	.800		182	40		222	260
8372 With light kit		10	.800		220	40		260	305
8380 Luxury model (DC motor)		8	1		360	50.50		410.50	470
8382 With light kit		8	1		400	50.50		450.50	515
8390 Remote speed switch for above, add		12	.667		30	33.50		63.50	82.50
8500 Whole house exhaust fan, ceiling mount, 36", variable speed									
8510 Remote switch, incl. shutters, 20 amp-1 pole circ. bkr.									
8520 30' of #12/2, type NM cable	1 Elec	4	2	Ea.	1,125	101		1,226	1,400
8530 Type MC cable		3.50	2.286		1,150	115		1,265	1,450
8540 EMT & wire		3	2.667		1,150	134		1,284	1,475
8600 Whirlpool tub hook-up, incl. timer switch, outlet box									
8610 3' of flex, 20 amp-1 pole GFI circ. bkr.									
8620 30' of #12/2, type NM cable	1 Elec	5	1.600	Ea.	128	80.50		208.50	259
8630 Type MC cable		4.20	1.905		139	96		235	295
8640 EMT & wire		3.40	2.353		148	118		266	340
8650 Hot water heater hook-up, incl. 1-2 pole circ. bkr., box;									
8660 3' of flex, 20' of #10/2, type NM cable	1 Elec	5	1.600	Ea.	33	80.50		113.50	155
8670 Type MC cable		4.20	1.905		56	96		152	204
8680 EMT & wire		3.40	2.353		53	118		171	233

26 05 Common Work Results for Electrical

26 05 90 – Residential Applications

26 05 90.10 Residential Wiring	Crew	Daily Output	Labor-Hours	Unit	Material	2011 Bare Costs Labor	Equipment	Total	Total Incl O&P
9000	Heating/air conditioning									
9050	Furnace/boiler hook-up, incl. firestat, local on-off switch									
9060	Emergency switch, and 40' of type NM cable	1 Elec	4	2	Ea.	56.50	101		157.50	211
9070	Type MC cable		3.50	2.286		83.50	115		198.50	262
9080	EMT & wire	↓	1.50	5.333	↓	92.50	268		360.50	495
9100	Air conditioner hook-up, incl. local 60 amp disc. switch									
9110	3' sealtite, 40 amp, 2 pole circuit breaker									
9130	40' of #8/2, type NM cable	1 Elec	3.50	2.286	Ea.	225	115		340	415
9140	Type MC cable		3	2.667		325	134		459	560
9150	EMT & wire	↓	1.30	6.154	↓	259	310		569	745
9200	Heat pump hook-up, 1-40 & 1-100 amp 2 pole circ. bkr.									
9210	Local disconnect switch, 3' sealtite									
9220	40' of #8/2 & 30' of #3/2									
9230	Type NM cable	1 Elec	1.30	6.154	Ea.	570	310		880	1,100
9240	Type MC cable		1.08	7.407		755	375		1,130	1,375
9250	EMT & wire	↓	.94	8.511	↓	620	430		1,050	1,325
9500	Thermostat hook-up, using low voltage wire									
9520	Heating only, 25' of #18-3	1 Elec	24	.333	Ea.	9.70	16.75		26.45	35.50
9530	Heating/cooling, 25' of #18-4	"	20	.400	"	12.25	20		32.25	43.50

26 24 Switchboards and Panelboards

26 24 19 – Motor-Control Centers

26 24 19.40 Motor Starters and Controls

	26 24 19.40 Motor Starters and Controls	Crew	Daily Output	Labor-Hours	Unit	Material	2011 Bare Costs Labor	Equipment	Total	Total Incl O&P
0010	**MOTOR STARTERS AND CONTROLS**									
0050	Magnetic, FVNR, with enclosure and heaters, 480 volt									
0080	2 HP, size 00	1 Elec	3.50	2.286	Ea.	211	115		326	400
0100	5 HP, size 0		2.30	3.478		255	175		430	540
0200	10 HP, size 1	↓	1.60	5		287	252		539	690
0300	25 HP, size 2	2 Elec	2.20	7.273		540	365		905	1,125
0400	50 HP, size 3		1.80	8.889		880	445		1,325	1,625
0500	100 HP, size 4		1.20	13.333		1,950	670		2,620	3,150
0600	200 HP, size 5		.90	17.778		4,550	895		5,445	6,350
0610	400 HP, size 6	↓	.80	20		18,500	1,000		19,500	21,800
0620	NEMA 7, 5 HP, size 0	1 Elec	1.60	5		1,375	252		1,627	1,900
0630	10 HP, size 1	"	1.10	7.273		1,425	365		1,790	2,125
0640	25 HP, size 2	2 Elec	1.80	8.889		2,325	445		2,770	3,200
0650	50 HP, size 3		1.20	13.333		3,475	670		4,145	4,825
0660	100 HP, size 4		.90	17.778		5,625	895		6,520	7,500
0670	200 HP, size 5	↓	.50	32		13,400	1,600		15,000	17,200
0700	Combination, with motor circuit protectors, 5 HP, size 0	1 Elec	1.80	4.444		830	224		1,054	1,250
0800	10 HP, size 1	"	1.30	6.154		860	310		1,170	1,400
0900	25 HP, size 2	2 Elec	2	8		1,200	400		1,600	1,925
1000	50 HP, size 3		1.32	12.121		1,750	610		2,360	2,825
1200	100 HP, size 4	↓	.80	20		3,800	1,000		4,800	5,675
1220	NEMA 7, 5 HP, size 0	1 Elec	1.30	6.154		2,450	310		2,760	3,150
1230	10 HP, size 1	"	1	8		2,525	400		2,925	3,375
1240	25 HP, size 2	2 Elec	1.32	12.121		3,375	610		3,985	4,600
1250	50 HP, size 3		.80	20		5,550	1,000		6,550	7,600
1260	100 HP, size 4		.60	26.667		8,650	1,350		10,000	11,500
1270	200 HP, size 5	↓	.40	40		18,800	2,000		20,800	23,700
1400	Combination, with fused switch, 5 HP, size 0	1 Elec	1.80	4.444	↓	635	224		859	1,025

26 24 Switchboards and Panelboards

26 24 19 – Motor-Control Centers

26 24 19.40 Motor Starters and Controls	Crew	Daily Output	Labor-Hours	Unit	Material	2011 Bare Costs Labor	2011 Bare Costs Equipment	Total	Total Incl O&P
1600	10 HP, size 1	1 Elec	1.30	6.154	Ea.	675	310		985	1,200
1800	25 HP, size 2	2 Elec	2	8		1,100	400		1,500	1,800
2000	50 HP, size 3		1.32	12.121		1,850	610		2,460	2,950
2200	100 HP, size 4		.80	20		3,250	1,000		4,250	5,075
3500	Magnetic FVNR with NEMA 12, enclosure & heaters, 480 volt									
3600	5 HP, size 0	1 Elec	2.20	3.636	Ea.	240	183		423	535
3700	10 HP, size 1	"	1.50	5.333		360	268		628	795
3800	25 HP, size 2	2 Elec	2	8		675	400		1,075	1,350
3900	50 HP, size 3		1.60	10		1,050	505		1,555	1,900
4000	100 HP, size 4		1	16		2,500	805		3,305	3,950
4100	200 HP, size 5		.80	20		5,975	1,000		6,975	8,075
4200	Combination, with motor circuit protectors, 5 HP, size 0	1 Elec	1.70	4.706		795	237		1,032	1,225
4300	10 HP, size 1	"	1.20	6.667		830	335		1,165	1,400
4400	25 HP, size 2	2 Elec	1.80	8.889		1,250	445		1,695	2,000
4500	50 HP, size 3		1.20	13.333		2,025	670		2,695	3,225
4600	100 HP, size 4		.74	21.622		4,550	1,100		5,650	6,625
4700	Combination, with fused switch, 5 HP, size 0	1 Elec	1.70	4.706		770	237		1,007	1,200
4800	10 HP, size 1	"	1.20	6.667		805	335		1,140	1,375
4900	25 HP, size 2	2 Elec	1.80	8.889		1,225	445		1,670	2,000
5000	50 HP, size 3		1.20	13.333		1,950	670		2,620	3,150
5100	100 HP, size 4		.74	21.622		3,975	1,100		5,075	5,975
5200	Factory installed controls, adders to size 0 thru 5									
5300	Start-stop push button	1 Elec	32	.250	Ea.	50.50	12.60		63.10	74
5400	Hand-off-auto-selector switch		32	.250		50.50	12.60		63.10	74
5500	Pilot light		32	.250		94.50	12.60		107.10	123
5600	Start-stop-pilot		32	.250		145	12.60		157.60	178
5700	Auxiliary contact, NO or NC		32	.250		69.50	12.60		82.10	94.50
5800	NO-NC		32	.250		139	12.60		151.60	171

26 29 Low-Voltage Controllers

26 29 13 – Enclosed Controllers

26 29 13.20 Control Stations

0010	CONTROL STATIONS									
0050	NEMA 1, heavy duty, stop/start	1 Elec	8	1	Ea.	137	50.50		187.50	225
0100	Stop/start, pilot light		6.20	1.290		186	65		251	300
0200	Hand/off/automatic		6.20	1.290		101	65		166	207
0400	Stop/start/reverse		5.30	1.509		184	76		260	315
0500	NEMA 7, heavy duty, stop/start		6	1.333		360	67		427	495
0600	Stop/start, pilot light		4	2		435	101		536	630

26 29 23 – Variable-Frequency Motor Controllers

26 29 23.10 Variable Frequency Drives/Adj. Frequency Drives

0010	VARIABLE FREQUENCY DRIVES/ADJ. FREQUENCY DRIVES										
0100	Enclosed (NEMA 1), 460 volt, for 3 HP motor size	G	1 Elec	.80	10	Ea.	1,075	505		1,580	1,925
0110	5 HP motor size	G		.80	10		1,200	505		1,705	2,075
0120	7.5 HP motor size	G		.67	11.940		1,425	600		2,025	2,475
0130	10 HP motor size	G		.67	11.940		1,700	600		2,300	2,775
0140	15 HP motor size	G	2 Elec	.89	17.978		2,025	905		2,930	3,575
0150	20 HP motor size	G		.89	17.978		2,425	905		3,330	4,025
0160	25 HP motor size	G		.67	23.881		3,000	1,200		4,200	5,075
0170	30 HP motor size	G		.67	23.881		3,700	1,200		4,900	5,850

26 29 Low-Voltage Controllers

26 29 23 – Variable-Frequency Motor Controllers

26 29 23.10 Variable Frequency Drives/Adj. Frequency Drives		Crew	Daily Output	Labor-Hours	Unit	Material	2011 Bare Costs Labor	Equipment	Total	Total Incl O&P
0180	40 HP motor size **G**	2 Elec	.67	23.881	Ea.	4,425	1,200		5,625	6,625
0190	50 HP motor size **G**	▼	.53	30.189		5,525	1,525		7,050	8,325
0200	60 HP motor size **G**	R-3	.56	35.714		6,575	1,775	230	8,580	10,100
0210	75 HP motor size **G**		.56	35.714		7,625	1,775	230	9,630	11,300
0220	100 HP motor size **G**		.50	40		8,875	2,000	258	11,133	13,000
0230	125 HP motor size **G**		.50	40		9,875	2,000	258	12,133	14,100
0240	150 HP motor size **G**		.50	40		11,400	2,000	258	13,658	15,700
0250	200 HP motor size **G**	▼	.42	47.619		14,700	2,375	305	17,380	20,100
1100	Custom-engineered, 460 volt, for 3 HP motor size **G**	1 Elec	.56	14.286		2,475	720		3,195	3,775
1110	5 HP motor size **G**		.56	14.286		2,475	720		3,195	3,775
1120	7.5 HP motor size **G**		.47	17.021		2,575	855		3,430	4,125
1130	10 HP motor size **G**	▼	.47	17.021		2,700	855		3,555	4,250
1140	15 HP motor size **G**	2 Elec	.62	25.806		3,375	1,300		4,675	5,650
1150	20 HP motor size **G**		.62	25.806		3,800	1,300		5,100	6,125
1160	25 HP motor size **G**		.47	34.043		4,450	1,700		6,150	7,425
1170	30 HP motor size **G**		.47	34.043		5,575	1,700		7,275	8,650
1180	40 HP motor size **G**		.47	34.043		6,575	1,700		8,275	9,775
1190	50 HP motor size **G**	▼	.37	43.243		7,550	2,175		9,725	11,500
1200	60 HP motor size **G**	R-3	.39	51.282		11,400	2,550	330	14,280	16,600
1210	75 HP motor size **G**		.39	51.282		12,100	2,550	330	14,980	17,400
1220	100 HP motor size **G**		.35	57.143		13,300	2,850	370	16,520	19,200
1230	125 HP motor size **G**		.35	57.143		14,200	2,850	370	17,420	20,200
1240	150 HP motor size **G**		.35	57.143		16,200	2,850	370	19,420	22,500
1250	200 HP motor size **G**	▼	.29	68.966	▼	21,200	3,425	445	25,070	29,000
2000	For complex & special design systems to meet specific									
2010	requirements, obtain quote from vendor.									

26 42 Cathodic Protection

26 42 16 – Passive Cathodic Protection for Underground Storage Tank

26 42 16.50 Cathodic Protection Wiring Methods

		Crew	Daily Output	Labor-Hours	Unit	Material	Labor	Equipment	Total	Total Incl O&P
0010	**CATHODIC PROTECTION WIRING METHODS**									
1000	Anodes, magnesium type, 9 #	R-15	18.50	2.595	Ea.	38.50	128	17.20	183.70	251
1010	17 #		13	3.692		69.50	182	24.50	276	375
1020	32 #		10	4.800		122	237	32	391	520
1030	48 #	▼	7.20	6.667		161	330	44	535	710
1100	Graphite type w/epoxy cap, 3" x 60" (32 #)	R-22	8.40	4.438		123	190		313	420
1110	4" x 80" (68 #)		6	6.213		235	266		501	660
1120	6" x 72" (80 #)		5.20	7.169		1,475	305		1,780	2,075
1130	6" x 36" (45 #)	▼	9.60	3.883		745	166		911	1,075
2000	Rectifiers, silicon type, air cooled, 28 V/10 A	R-19	3.50	5.714		2,100	288		2,388	2,750
2010	20 V/20 A		3.50	5.714		2,200	288		2,488	2,850
2100	Oil immersed, 28 V/10 A		3	6.667		2,975	335		3,310	3,775
2110	20 V/20 A	▼	3	6.667	▼	3,100	335		3,435	3,900
3000	Anode backfill, coke breeze	R-22	3850	.010	Lb.	.18	.41		.59	.82
4000	Cable, HMWPE, No. 8		2.40	15.533	M.L.F.	480	665		1,145	1,525
4010	No. 6		2.40	15.533		655	665		1,320	1,725
4020	No. 4		2.40	15.533		980	665		1,645	2,075
4030	No. 2		2.40	15.533		1,525	665		2,190	2,675
4040	No. 1		2.20	16.945		2,050	725		2,775	3,350
4050	No. 1/0		2.20	16.945		2,575	725		3,300	3,900
4060	No. 2/0	▼	2.20	16.945	▼	4,100	725		4,825	5,575

26 42 Cathodic Protection

26 42 16 – Passive Cathodic Protection for Underground Storage Tank

26 42 16.50 Cathodic Protection Wiring Methods	Crew	Daily Output	Labor-Hours	Unit	Material	2011 Bare Costs Labor	Equipment	Total	Total Incl O&P
4070	No. 4/0	R-22	2	18.640	M.L.F.	4,475	800		5,275	6,125
5000	Test station, 7 terminal box, flush curb type w/lockable cover	R-19	12	1.667	Ea.	70.50	84		154.50	202
5010	Reference cell, 2" dia PVC conduit, cplg., plug, set flush	"	4.80	4.167	"	148	210		358	475

Division Notes

	CREW	DAILY OUTPUT	LABOR-HOURS	UNIT	2011 BARE COSTS				TOTAL INCL O&P
					MAT.	LABOR	EQUIP.	TOTAL	

Estimating Tips

- When estimating material costs for electronic safety and security systems, it is always prudent to obtain manufacturers' quotations for equipment prices and special installation requirements that affect the total cost.

- Fire alarm systems consist of control panels, annunciator panels, battery with rack, charger, and fire alarm actuating and indicating devices. Some fire alarm systems include speakers, telephone lines, door closer controls, and other components. Be careful not to overlook the costs related to installation for these items.

Also be aware of costs for integrated automation instrumentation and terminal devices, control equipment, control wiring, and programming.

- Security equipment includes items such as CCTV, access control, and other detection and identification systems to perform alert and alarm functions. Be sure to consider the costs related to installation for this security equipment, such as for integrated automation instrumentation and terminal devices, control equipment, control wiring, and programming.

Reference Numbers

Reference numbers are shown in shaded boxes at the beginning of some major classifications. These numbers refer to related items in the Reference Section. The reference information may be an estimating procedure, an alternate pricing method, or technical information.

Note: Not all subdivisions listed here necessarily appear in this publication.

Division 28 - Electronic Safety and Security

28 13 Access Control

28 13 53 – Security Access Detection

28 13 53.13 Security Access Metal Detectors

		Crew	Daily Output	Labor-Hours	Unit	Material	2011 Bare Costs Labor	Equipment	Total	Total Incl O&P
0010	**SECURITY ACCESS METAL DETECTORS**									
0240	Metal detector, hand-held, wand type, unit only				Ea.	89			89	98
0250	Metal detector, walk through portal type, single zone	1 Elec	2	4		2,750	201		2,951	3,325
0260	Multi-zone	"	2	4		3,500	201		3,701	4,150

28 13 53.16 Security Access X-Ray Equipment

		Crew	Daily Output	Labor-Hours	Unit	Material	2011 Bare Costs Labor	Equipment	Total	Total Incl O&P
0010	**SECURITY ACCESS X-RAY EQUIPMENT**									
0290	X-ray machine, desk top, for mail/small packages/letters	1 Elec	4	2	Ea.	3,000	101		3,101	3,450
0300	Conveyor type, incl monitor, minimum		2	4		14,000	201		14,201	15,700
0310	Maximum		2	4		25,000	201		25,201	27,800
0320	X-ray machine, large unit, for airports, incl monitor, min	2 Elec	1	16		35,000	805		35,805	39,700
0330	Maximum	"	.50	32		60,000	1,600		61,600	68,500

28 13 53.23 Security Access Explosive Detection Equipment

		Crew	Daily Output	Labor-Hours	Unit	Material	2011 Bare Costs Labor	Equipment	Total	Total Incl O&P
0010	**SECURITY ACCESS EXPLOSIVE DETECTION EQUIPMENT**									
0270	Explosives detector, walk through portal type	1 Elec	2	4	Ea.	3,500	201		3,701	4,150
0280	Hand-held, battery operated				"				25,500	28,100

28 16 Intrusion Detection

28 16 16 – Intrusion Detection Systems Infrastructure

28 16 16.50 Intrusion Detection

		Crew	Daily Output	Labor-Hours	Unit	Material	2011 Bare Costs Labor	Equipment	Total	Total Incl O&P
0010	**INTRUSION DETECTION**, not including wires & conduits									
0100	Burglar alarm, battery operated, mechanical trigger	1 Elec	4	2	Ea.	272	101		373	450
0200	Electrical trigger		4	2		325	101		426	505
0400	For outside key control, add		8	1		82	50.50		132.50	165
0600	For remote signaling circuitry, add		8	1		122	50.50		172.50	210
0800	Card reader, flush type, standard		2.70	2.963		910	149		1,059	1,225
1000	Multi-code		2.70	2.963		1,175	149		1,324	1,525

28 23 Video Surveillance

28 23 23 – Video Surveillance Systems Infrastructure

28 23 23.50 Video Surveillance Equipments

		Crew	Daily Output	Labor-Hours	Unit	Material	2011 Bare Costs Labor	Equipment	Total	Total Incl O&P
0010	**VIDEO SURVEILLANCE EQUIPMENTS**									
0200	Video cameras, wireless, hidden in exit signs, clocks, etc, incl receiver	1 Elec	3	2.667	Ea.	143	134		277	355
0210	Accessories for video recorder, single camera		3	2.667		260	134		394	485
0220	For multiple cameras		3	2.667		1,750	134		1,884	2,125
0230	Video cameras, wireless, for under vehicle searching, complete		2	4		9,975	201		10,176	11,300

28 31 Fire Detection and Alarm

28 31 23 – Fire Detection and Alarm Annunciation Panels and Fire Stations

28 31 23.50 Alarm Panels and Devices

		Crew	Daily Output	Labor-Hours	Unit	Material	2011 Bare Costs Labor	Equipment	Total	Total Incl O&P
0010	**ALARM PANELS AND DEVICES**, not including wires & conduits									
3594	Fire, alarm control panel									
3600	4 zone	2 Elec	2	8	Ea.	390	400		790	1,025
3800	8 zone		1	16		760	805		1,565	2,025
4000	12 zone		.67	23.988		2,300	1,200		3,500	4,300
4020	Alarm device	1 Elec	8	1		229	50.50		279.50	325
4050	Actuating device		8	1		325	50.50		375.50	430
4200	Battery and rack		4	2		395	101		496	585

28 31 Fire Detection and Alarm

28 31 23 – Fire Detection and Alarm Annunciation Panels and Fire Stations

28 31 23.50 Alarm Panels and Devices

		Crew	Daily Output	Labor-Hours	Unit	Material	2011 Bare Costs Labor	2011 Bare Costs Equipment	Total	Total Incl O&P
4400	Automatic charger	1 Elec	8	1	Ea.	560	50.50		610.50	695
4600	Signal bell		8	1		70.50	50.50		121	152
4800	Trouble buzzer or manual station		8	1		80.50	50.50		131	163
5600	Strobe and horn		5.30	1.509		147	76		223	273
5800	Fire alarm horn		6.70	1.194		58.50	60		118.50	154
6000	Door holder, electro-magnetic		4	2		99	101		200	258
6200	Combination holder and closer		3.20	2.500		119	126		245	315
6600	Drill switch		8	1		360	50.50		410.50	470
6800	Master box		2.70	2.963		6,175	149		6,324	7,025
7000	Break glass station		8	1		65	50.50		115.50	146
7800	Remote annunciator, 8 zone lamp		1.80	4.444		201	224		425	550
8000	12 zone lamp	2 Elec	2.60	6.154		325	310		635	815
8200	16 zone lamp	"	2.20	7.273		405	365		770	985

28 31 43 – Fire Detection Sensors

28 31 43.50 Fire and Heat Detectors

		Crew	Daily Output	Labor-Hours	Unit	Material	2011 Bare Costs Labor	2011 Bare Costs Equipment	Total	Total Incl O&P
0010	**FIRE & HEAT DETECTORS**									
5000	Detector, rate of rise	1 Elec	8	1	Ea.	49	50.50		99.50	129
5100	Fixed temperature	"	8	1	"	49	50.50		99.50	129

28 31 46 – Smoke Detection Sensors

28 31 46.50 Smoke Detectors

		Crew	Daily Output	Labor-Hours	Unit	Material	2011 Bare Costs Labor	2011 Bare Costs Equipment	Total	Total Incl O&P
0010	**SMOKE DETECTORS**									
5200	Smoke detector, ceiling type	1 Elec	6.20	1.290	Ea.	126	65		191	235
5400	Duct type	"	3.20	2.500	"	315	126		441	530

28 33 Fuel-Gas Detection and Alarm

28 33 33 – Fuel-Gas Detection Sensors

28 33 33.50 Tank Leak Detection Systems

		Crew	Daily Output	Labor-Hours	Unit	Material	2011 Bare Costs Labor	2011 Bare Costs Equipment	Total	Total Incl O&P
0010	**TANK LEAK DETECTION SYSTEMS** Liquid and vapor									
0100	For hydrocarbons and hazardous liquids/vapors									
0120	Controller, data acquisition, incl. printer, modem, RS232 port									
0140	24 channel, for use with all probes				Ea.	3,500			3,500	3,850
0160	9 channel, for external monitoring				"	770			770	845
0200	Probes									
0210	Well monitoring									
0220	Liquid phase detection				Ea.	400			400	440
0230	Hydrocarbon vapor, fixed position					400			400	440
0240	Hydrocarbon vapor, float mounted					400			400	440
0250	Both liquid and vapor hydrocarbon					400			400	440
0300	Secondary containment, liquid phase									
0310	Pipe trench/manway sump				Ea.	690			690	760
0320	Double wall pipe and manual sump					690			690	760
0330	Double wall fiberglass annular space					272			272	299
0340	Double wall steel tank annular space					282			282	310
0500	Accessories									
0510	Modem, non-dedicated phone line				Ea.	255			255	280
0600	Monitoring, internal									
0610	Automatic tank gauge, incl. overfill				Ea.	965			965	1,075
0620	Product line				"	965			965	1,075

28 33 33.50 Tank Leak Detection Systems	Crew	Daily Output	Labor-Hours	Unit	Material	2011 Bare Costs Labor	Equipment	Total	Total Incl O&P
0700 Monitoring, special									
0710 Cathodic protection				Ea.	605			605	670
0720 Annular space chemical monitor				"	825			825	910

Estimating Tips

31 05 00 Common Work Results for Earthwork

• Estimating the actual cost of performing earthwork requires careful consideration of the variables involved. This includes items such as type of soil, whether water will be encountered, dewatering, whether banks need bracing, disposal of excavated earth, and length of haul to fill or spoil sites, etc. If the project has large quantities of cut or fill, consider raising or lowering the site to reduce costs, while paying close attention to the effect on site drainage and utilities.

• If the project has large quantities of fill, creating a borrow pit on the site can significantly lower the costs.

• It is very important to consider what time of year the project is scheduled for completion. Bad weather can create large cost overruns from dewatering, site repair, and lost productivity from cold weather.

Reference Numbers

Reference numbers are shown in shaded boxes at the beginning of some major classifications. These numbers refer to related items in the Reference Section. The reference information may be an estimating procedure, an alternate pricing method, or technical information.

Note: Not all subdivisions listed here necessarily appear in this publication.

Division 31 – Earthwork

31 05 Common Work Results for Earthwork

31 05 23 – Cement and Concrete for Earthwork

31 05 23.30 Plant Mixed Bituminous Concrete	Crew	Daily Output	Labor-Hours	Unit	Material	2011 Bare Costs Labor	Equipment	Total	Total Incl O&P
0010 **PLANT MIXED BITUMINOUS CONCRETE**									
0020 Asphaltic concrete plant mix (145 lb. per C.F.)				Ton	64			64	70.50
0040 Asphaltic concrete less than 300 tons add trucking costs									
0050 See Section 31 23 23.20 for hauling costs									
0200 All weather patching mix, hot				Ton	65			65	71.50
0250 Cold patch				↓	65			65	71.50
0300 Berm mix				↓	65			65	71.50

31 23 Excavation and Fill

31 23 16 – Excavation

31 23 16.13 Excavating, Trench

	Crew	Daily Output	Labor-Hours	Unit	Material	Labor	Equipment	Total	Total Incl O&P
0010 **EXCAVATING, TRENCH** G1030–805									
0011 Or continuous footing									
0020 Common earth with no sheeting or dewatering included									
0050 1' to 4' deep, 3/8 C.Y. excavator	B-11C	150	.107	B.C.Y.		4.25	2.14	6.39	8.75
0060 1/2 C.Y. excavator	B-11M	200	.080			3.19	1.75	4.94	6.75
0090 4' to 6' deep, 1/2 C.Y. excavator	"	200	.080			3.19	1.75	4.94	6.75
0100 5/8 C.Y. excavator	B-12Q	250	.064			2.59	2.12	4.71	6.25
0110 3/4 C.Y. excavator	B-12F	300	.053			2.16	2.13	4.29	5.60
0300 1/2 C.Y. excavator, truck mounted	B-12J	200	.080			3.23	4.29	7.52	9.60
0500 6' to 10' deep, 3/4 C.Y. excavator	B-12F	225	.071			2.88	2.85	5.73	7.45
0600 1 C.Y. excavator, truck mounted	B-12K	400	.040			1.62	2.50	4.12	5.20
0900 10' to 14' deep, 3/4 C.Y. excavator	B-12F	200	.080			3.23	3.20	6.43	8.40
1000 1-1/2 C.Y. excavator	B-12B	540	.030			1.20	1.75	2.95	3.74
1300 14' to 20' deep, 1 C.Y. excavator	B-12A	320	.050			2.02	2.28	4.30	5.55
1340 20' to 24' deep, 1 C.Y. excavator	"	288	.056			2.25	2.53	4.78	6.20
1352 4' to 6' deep, 1/2 C.Y. excavator w/trench box	B-13H	188	.085			3.44	5.15	8.59	10.90
1354 5/8 C.Y. excavator	"	235	.068			2.75	4.13	6.88	8.70
1356 3/4 C.Y. excavator	B-13G	282	.057			2.29	2.67	4.96	6.40
1362 6' to 10' deep, 3/4 C.Y. excavator w/trench box		212	.075			3.05	3.56	6.61	8.50
1374 10' to 14' deep, 3/4 C.Y. excavator w/trench box	↓	188	.085			3.44	4.01	7.45	9.60
1376 1-1/2 C.Y. excavator	B-13E	508	.032			1.27	2.09	3.36	4.22
1381 14' to 20' deep, 1 C.Y. excavator w/trench box	B-13D	301	.053			2.15	2.80	4.95	6.35
1386 20' to 24' deep, 1 C.Y. excavator w/trench box	"	271	.059			2.39	3.11	5.50	7.05
1400 By hand with pick and shovel 2' to 6' deep, light soil	1 Clab	8	1			34.50		34.50	52.50
1500 Heavy soil	"	4	2	↓		68.50		68.50	105
1700 For tamping backfilled trenches, air tamp, add	A-1G	100	.080	E.C.Y.		2.75	.50	3.25	4.77
1900 Vibrating plate, add	B-18	180	.133	"		4.67	.22	4.89	7.40
2100 Trim sides and bottom for concrete pours, common earth		1500	.016	S.F.		.56	.03	.59	.89
2300 Hardpan	↓	600	.040	"		1.40	.07	1.47	2.22
5020 Loam & Sandy clay with no sheeting or dewatering included									
5050 1' to 4' deep, 3/8 C.Y. tractor loader/backhoe	B-11C	162	.099	B.C.Y.		3.94	1.98	5.92	8.15
5060 1/2 C.Y. excavator	B-11M	216	.074			2.95	1.62	4.57	6.25
5080 4' to 6' deep, 1/2 C.Y. excavator	"	216	.074			2.95	1.62	4.57	6.25
5090 5/8 C.Y. excavator	B-12Q	276	.058			2.34	1.92	4.26	5.65
5100 3/4 C.Y. excavator	B-12F	324	.049			2	1.98	3.98	5.20
5130 1/2 C.Y. excavator, truck mounted	B-12J	216	.074			2.99	3.97	6.96	8.90
5140 6' to 10' deep, 3/4 C.Y. excavator	B-12F	243	.066			2.66	2.64	5.30	6.90
5160 1 C.Y. excavator, truck mounted	B-12K	432	.037			1.50	2.32	3.82	4.81
5190 10' to 14' deep, 3/4 C.Y. excavator	B-12F	216	.074			2.99	2.97	5.96	7.80
5210 1-1/2 C.Y. excavator	B-12B	583	.027			1.11	1.62	2.73	3.47

31 23 Excavation and Fill

31 23 16 - Excavation

31 23 16.13 Excavating, Trench

		Crew	Daily Output	Labor-Hours	Unit	Material	2011 Bare Costs Labor	Equipment	Total	Total Incl O&P
5250	14' to 20' deep, 1 C.Y. excavator	B-12A	346	.046	B.C.Y.		1.87	2.11	3.98	5.15
5300	20' to 24' deep, 1 C.Y. excavator	"	311	.051			2.08	2.35	4.43	5.70
5352	4' to 6' deep, 1/2 C.Y. excavator w/trench box	B-13H	205	.078			3.16	4.74	7.90	9.95
5354	5/8 C.Y. excavator	"	257	.062			2.52	3.78	6.30	7.95
5356	3/4 C.Y. excavator	B-13G	308	.052			2.10	2.45	4.55	5.85
5362	6' to 10' deep, 3/4 C.Y. excavator w/trench box		231	.069			2.80	3.26	6.06	7.80
5370	10' to 14' deep, 3/4 C.Y. excavator w/trench box		205	.078			3.16	3.68	6.84	8.80
5374	1-1/2 C.Y. excavator	B-13E	554	.029			1.17	1.91	3.08	3.87
5382	14' to 20' deep, 1 C.Y. excavator w/trench box	B-13D	329	.049			1.97	2.56	4.53	5.80
5392	20' to 24' deep, 1 C.Y. excavator w/trench box	"	295	.054			2.19	2.86	5.05	6.45
6020	Sand & gravel with no sheeting or dewatering included									
6050	1' to 4' deep, 3/8 C.Y. excavator	B-11C	165	.097	B.C.Y.		3.86	1.94	5.80	8
6060	1/2 C.Y. excavator	B-11M	220	.073			2.90	1.59	4.49	6.15
6080	4' to 6' deep, 1/2 C.Y. excavator	"	220	.073			2.90	1.59	4.49	6.15
6090	5/8 C.Y. excavator	B-12Q	275	.058			2.35	1.93	4.28	5.65
6100	3/4 C.Y. excavator	B-12F	330	.048			1.96	1.94	3.90	5.10
6130	1/2 C.Y. excavator, truck mounted	B-12J	220	.073			2.94	3.90	6.84	8.75
6140	6' to 10' deep, 3/4 C.Y. excavator	B-12F	248	.065			2.61	2.58	5.19	6.80
6160	1 C.Y. excavator, truck mounted	B-12K	440	.036			1.47	2.27	3.74	4.72
6190	10' to 14' deep, 3/4 C.Y. excavator	B-12F	220	.073			2.94	2.91	5.85	7.65
6210	1-1/2 C.Y. excavator	B-12B	594	.027			1.09	1.60	2.69	3.40
6250	14' to 20' deep, 1 C.Y. excavator	B-12A	352	.045			1.84	2.07	3.91	5.05
6300	20' to 24' deep, 1 C.Y. excavator	"	317	.050			2.04	2.30	4.34	5.60
6352	4' to 6' deep, 1/2 C.Y. excavator w/trench box	B-13H	209	.077			3.10	4.65	7.75	9.80
6354	5/8 C.Y. excavator	"	261	.061			2.48	3.72	6.20	7.85
6356	3/4 C.Y. excavator	B-13G	314	.051			2.06	2.40	4.46	5.75
6362	6' to 10' deep, 3/4 C.Y. excavator w/trench box		236	.068			2.74	3.19	5.93	7.65
6370	10' to 14' deep, 3/4 C.Y. excavator w/trench box		209	.077			3.10	3.61	6.71	8.65
6374	1-1/2 C.Y. excavator	B-13E	564	.028			1.15	1.88	3.03	3.80
6382	14' to 20' deep, 1 C.Y. excavator w/trench box	B-13D	334	.048			1.94	2.52	4.46	5.70
6392	20' to 24' deep, 1 C.Y. excavator w/trench box	"	301	.053			2.15	2.80	4.95	6.35
7020	Dense hard clay with no sheeting or dewatering included									
7050	1' to 4' deep, 3/8 C.Y. excavator	B-11C	132	.121	B.C.Y.		4.83	2.43	7.26	9.95
7060	1/2 C.Y. excavator	B-11M	176	.091			3.62	1.99	5.61	7.70
7080	4' to 6' deep, 1/2 C.Y. excavator	"	176	.091			3.62	1.99	5.61	7.70
7090	5/8 C.Y. excavator	B-12Q	220	.073			2.94	2.41	5.35	7.10
7100	3/4 C.Y. excavator	B-12F	264	.061			2.45	2.43	4.88	6.35
7130	1/2 C.Y. excavator, truck mounted	B-12J	176	.091			3.68	4.87	8.55	10.90
7140	6' to 10' deep, 3/4 C.Y. excavator	B-12F	198	.081			3.27	3.23	6.50	8.50
7160	1 C.Y. excavator, truck mounted	B-12K	352	.045			1.84	2.84	4.68	5.90
7190	10' to 14' deep, 3/4 C.Y. excavator	B-12F	176	.091			3.68	3.64	7.32	9.55
7210	1-1/2 C.Y. excavator	B-12B	475	.034			1.36	1.99	3.35	4.25
7250	14' to 20' deep, 1 C.Y. excavator	B-12A	282	.057			2.29	2.59	4.88	6.30
7300	20' to 24' deep, 1 C.Y. excavator	"	254	.063			2.55	2.87	5.42	7

31 23 16.14 Excavating, Utility Trench

		Crew	Daily Output	Labor-Hours	Unit	Material	2011 Bare Costs Labor	Equipment	Total	Total Incl O&P
0010	**EXCAVATING, UTILITY TRENCH**									
0011	Common earth									
0050	Trenching with chain trencher, 12 H.P., operator walking									
0100	4" wide trench, 12" deep	B-53	800	.010	L.F.		.44	.07	.51	.73
0150	18" deep		750	.011			.47	.08	.55	.78
0200	24" deep		700	.011			.50	.08	.58	.83
0300	6" wide trench, 12" deep		650	.012			.54	.09	.63	.90

31 23 16.14 Excavating, Utility Trench

		Crew	Daily Output	Labor-Hours	Unit	Material	2011 Bare Costs Labor	Equipment	Total	Total Incl O&P
0350	18" deep	B-53	600	.013	L.F.		.58	.10	.68	.98
0400	24" deep		550	.015			.63	.10	.73	1.06
0450	36" deep		450	.018			.78	.13	.91	1.30
0600	8" wide trench, 12" deep		475	.017			.73	.12	.85	1.23
0650	18" deep		400	.020			.87	.14	1.01	1.46
0700	24" deep		350	.023			1	.16	1.16	1.67
0750	36" deep	▼	300	.027	▼		1.16	.19	1.35	1.95
1000	Backfill by hand including compaction, add									
1050	4" wide trench, 12" deep	A-1G	800	.010	L.F.		.34	.06	.40	.60
1100	18" deep		530	.015			.52	.09	.61	.90
1150	24" deep		400	.020			.69	.12	.81	1.19
1300	6" wide trench, 12" deep		540	.015			.51	.09	.60	.88
1350	18" deep		405	.020			.68	.12	.80	1.17
1400	24" deep		270	.030			1.02	.18	1.20	1.76
1450	36" deep		180	.044			1.53	.28	1.81	2.64
1600	8" wide trench, 12" deep		400	.020			.69	.12	.81	1.19
1650	18" deep		265	.030			1.04	.19	1.23	1.80
1700	24" deep		200	.040			1.37	.25	1.62	2.38
1750	36" deep	▼	135	.059	▼		2.04	.37	2.41	3.52
2000	Chain trencher, 40 H.P. operator riding									
2050	6" wide trench and backfill, 12" deep	B-54	1200	.007	L.F.		.29	.25	.54	.70
2100	18" deep		1000	.008			.35	.30	.65	.85
2150	24" deep		975	.008			.36	.30	.66	.87
2200	36" deep		900	.009			.39	.33	.72	.94
2250	48" deep		750	.011			.47	.40	.87	1.14
2300	60" deep		650	.012			.54	.46	1	1.30
2400	8" wide trench and backfill, 12" deep		1000	.008			.35	.30	.65	.85
2450	18" deep		950	.008			.37	.31	.68	.89
2500	24" deep		900	.009			.39	.33	.72	.94
2550	36" deep		800	.010			.44	.37	.81	1.06
2600	48" deep		650	.012			.54	.46	1	1.30
2700	12" wide trench and backfill, 12" deep		975	.008			.36	.30	.66	.87
2750	18" deep		860	.009			.41	.35	.76	.99
2800	24" deep		800	.010			.44	.37	.81	1.06
2850	36" deep		725	.011			.48	.41	.89	1.17
3000	16" wide trench and backfill, 12" deep		835	.010			.42	.36	.78	1.01
3050	18" deep		750	.011			.47	.40	.87	1.14
3100	24" deep	▼	700	.011	▼		.50	.42	.92	1.21
3200	Compaction with vibratory plate, add								35%	35%
5100	Hand excavate and trim for pipe bells after trench excavation									
5200	8" pipe	1 Clab	155	.052	L.F.		1.77		1.77	2.72
5300	18" pipe	"	130	.062	"		2.11		2.11	3.24

31 23 16.16 Structural Excavation for Minor Structures

		Crew	Daily Output	Labor-Hours	Unit	Material	2011 Bare Costs Labor	Equipment	Total	Total Incl O&P
0010	**STRUCTURAL EXCAVATION FOR MINOR STRUCTURES**									
0015	Hand, pits to 6' deep, sandy soil	1 Clab	8	1	B.C.Y.		34.50		34.50	52.50
0100	Heavy soil or clay		4	2			68.50		68.50	105
0300	Pits 6' to 12' deep, sandy soil		5	1.600			55		55	84.50
0500	Heavy soil or clay		3	2.667			91.50		91.50	141
0700	Pits 12' to 18' deep, sandy soil		4	2			68.50		68.50	105
0900	Heavy soil or clay	▼	2	4	▼		137		137	211
1500	For wet or muck hand excavation, add to above				%				50%	50%

31 23 Excavation and Fill

31 23 16 – Excavation

31 23 16.30 Drilling and Blasting Rock

	31 23 16.30 Drilling and Blasting Rock	Crew	Daily Output	Labor-Hours	Unit	Material	2011 Bare Costs Labor	Equipment	Total	Total Incl O&P
0010	**DRILLING AND BLASTING ROCK**									
0020	Rock, open face, under 1500 C.Y.	B-47	225	.107	B.C.Y.	2.82	4.06	5.80	12.68	15.65
0100	Over 1500 C.Y.		300	.080		2.82	3.05	4.35	10.22	12.50
0200	Areas where blasting mats are required, under 1500 C.Y.		175	.137		2.82	5.25	7.45	15.52	19.25
0250	Over 1500 C.Y.		250	.096		2.82	3.66	5.20	11.68	14.40
2200	Trenches, up to 1500 C.Y.		22	1.091		8.20	41.50	59.50	109.20	137
2300	Over 1500 C.Y.		26	.923		8.20	35	50	93.20	118

31 23 19 – Dewatering

31 23 19.20 Dewatering Systems

	31 23 19.20 Dewatering Systems	Crew	Daily Output	Labor-Hours	Unit	Material	2011 Bare Costs Labor	Equipment	Total	Total Incl O&P
0010	**DEWATERING SYSTEMS**									
0020	Excavate drainage trench, 2' wide, 2' deep	B-11C	90	.178	C.Y.		7.10	3.56	10.66	14.60
0100	2' wide, 3' deep, with backhoe loader	"	135	.119			4.72	2.37	7.09	9.75
0200	Excavate sump pits by hand, light soil	1 Clab	7.10	1.127			38.50		38.50	59.50
0300	Heavy soil	"	3.50	2.286			78.50		78.50	120
0500	Pumping 8 hr., attended 2 hrs. per day, including 20 L.F.									
0550	of suction hose & 100 L.F. discharge hose									
0600	2" diaphragm pump used for 8 hours	B-10H	4	3	Day		125	17.60	142.60	207
0650	4" diaphragm pump used for 8 hours	B-10I	4	3			125	29.50	154.50	221
0800	8 hrs. attended, 2" diaphragm pump	B-10H	1	12			500	70.50	570.50	835
0900	3" centrifugal pump	B-10J	1	12			500	77.50	577.50	840
1000	4" diaphragm pump	B-10I	1	12			500	117	617	885
1100	6" centrifugal pump	B-10K	1	12			500	315	815	1,100
1300	CMP, incl. excavation 3' deep, 12" diameter	B-6	115	.209	L.F.	8.65	7.80	2.79	19.24	24.50
1400	18" diameter		100	.240	"	14.35	9	3.20	26.55	33
1600	Sump hole construction, incl. excavation and gravel, pit		1250	.019	C.F.	.99	.72	.26	1.97	2.46
1700	With 12" gravel collar, 12" pipe, corrugated, 16 ga.		70	.343	L.F.	23	12.85	4.58	40.43	50
1800	15" pipe, corrugated, 16 ga.		55	.436		29.50	16.35	5.85	51.70	64
1900	18" pipe, corrugated, 16 ga.		50	.480		34.50	17.95	6.40	58.85	72.50
2000	24" pipe, corrugated, 14 ga.		40	.600		41.50	22.50	8	72	88.50
2200	Wood lining, up to 4' x 4', add		300	.080	SFCA	17	2.99	1.07	21.06	24.50
9950	See Section 31 23 19.40 for wellpoints									
9960	See Section 31 23 19.30 for deep well systems									

31 23 19.30 Wells

	31 23 19.30 Wells	Crew	Daily Output	Labor-Hours	Unit	Material	2011 Bare Costs Labor	Equipment	Total	Total Incl O&P
0010	**WELLS**									
0011	For dewatering 10' to 20' deep, 2' diameter									
0020	with steel casing, minimum	B-6	165	.145	V.L.F.	35	5.45	1.94	42.39	49
0050	Average		98	.245		40	9.15	3.27	52.42	61.50
0100	Maximum		49	.490		45	18.35	6.55	69.90	84.50
0300	For dewatering pumps see 01 54 33 in Reference Section									
0500	For domestic water wells, see Section 33 21 13.10									

31 23 19.40 Wellpoints

	31 23 19.40 Wellpoints	Crew	Daily Output	Labor-Hours	Unit	Material	2011 Bare Costs Labor	Equipment	Total	Total Incl O&P
0010	**WELLPOINTS** R312319-90									
0011	For equipment rental, see 01 54 33 in Reference Section									
0100	Installation and removal of single stage system									
0110	Labor only, .75 labor-hours per L.F., minimum	1 Clab	10.70	.748	LF Hdr		25.50		25.50	39.50
0200	2.0 labor-hours per L.F., maximum	"	4	2	"		68.50		68.50	105
0400	Pump operation, 4 @ 6 hr. shifts									
0410	Per 24 hour day	4 Eqlt	1.27	25.197	Day		1,100		1,100	1,650
0500	Per 168 hour week, 160 hr. straight, 8 hr. double time		.18	177	Week		7,750		7,750	11,600
0550	Per 4.3 week month		.04	800	Month		34,900		34,900	52,000
0600	Complete installation, operation, equipment rental, fuel &									

31 23 Excavation and Fill

31 23 19 — Dewatering

31 23 19.40 Wellpoints

		Crew	Daily Output	Labor-Hours	Unit	Material	2011 Bare Costs Labor	2011 Bare Costs Equipment	Total	Total Incl O&P
0610	removal of system with 2" wellpoints 5' O.C.									
0700	100' long header, 6" diameter, first month	4 Eqlt	3.23	9.907	LF Hdr	157	430		587	815
0800	Thereafter, per month		4.13	7.748		125	340		465	645
1000	200' long header, 8" diameter, first month		6	5.333		142	233		375	500
1100	Thereafter, per month		8.39	3.814		70.50	166		236.50	325
1300	500' long header, 8" diameter, first month		10.63	3.010		55	131		186	257
1400	Thereafter, per month		20.91	1.530		39	66.50		105.50	143
1600	1,000' long header, 10" diameter, first month		11.62	2.754		47	120		167	231
1700	Thereafter, per month		41.81	.765		23.50	33.50		57	76
1900	Note: above figures include pumping 168 hrs. per week									
1910	and include the pump operator and one stand-by pump.									

31 23 23 — Fill

31 23 23.13 Backfill

		Crew	Daily Output	Labor-Hours	Unit	Material	2011 Bare Costs Labor	2011 Bare Costs Equipment	Total	Total Incl O&P
0010	**BACKFILL** G1030–805									
0015	By hand, no compaction, light soil	1 Clab	14	.571	L.C.Y.		19.65		19.65	30
0100	Heavy soil		11	.727	"		25		25	38.50
0300	Compaction in 6" layers, hand tamp, add to above		20.60	.388	E.C.Y.		13.35		13.35	20.50
0400	Roller compaction operator walking, add	B-10A	100	.120			5	1.47	6.47	9.15
0500	Air tamp, add	B-9D	190	.211			7.30	1.20	8.50	12.50
0600	Vibrating plate, add	A-1D	60	.133			4.58	.51	5.09	7.60
0800	Compaction in 12" layers, hand tamp, add to above	1 Clab	34	.235			8.10		8.10	12.40
0900	Roller compaction operator walking, add	B-10A	150	.080			3.33	.98	4.31	6.10
1000	Air tamp, add	B-9	285	.140			4.88	.68	5.56	8.25
1100	Vibrating plate, add	A-1E	90	.089			3.05	.43	3.48	5.15
1300	Dozer backfilling, bulk, up to 300' haul, no compaction	B-10B	1200	.010	L.C.Y.		.42	.88	1.30	1.59
1400	Air tamped, add	B-11B	80	.200	E.C.Y.		7.80	3.11	10.91	15.20
1600	Compacting backfill, 6" to 12" lifts, vibrating roller	B-10C	800	.015			.63	1.81	2.44	2.93
1700	Sheepsfoot roller	B-10D	750	.016			.67	2	2.67	3.20
1900	Dozer backfilling, trench, up to 300' haul, no compaction	B-10B	900	.013	L.C.Y.		.56	1.17	1.73	2.12
2000	Air tamped, add	B-11B	80	.200	E.C.Y.		7.80	3.11	10.91	15.20
2200	Compacting backfill, 6" to 12" lifts, vibrating roller	B-10C	700	.017			.71	2.06	2.77	3.35
2300	Sheepsfoot roller	B-10D	650	.018			.77	2.31	3.08	3.70

31 23 23.16 Fill By Borrow and Utility Bedding

		Crew	Daily Output	Labor-Hours	Unit	Material	2011 Bare Costs Labor	2011 Bare Costs Equipment	Total	Total Incl O&P
0010	**FILL BY BORROW AND UTILITY BEDDING**									
0049	Utility bedding, for pipe & conduit, not incl. compaction									
0050	Crushed or screened bank run gravel	B-6	150	.160	L.C.Y.	25	6	2.14	33.14	39
0100	Crushed stone 3/4" to 1/2"		150	.160		38	6	2.14	46.14	53.50
0200	Sand, dead or bank		150	.160		15	6	2.14	23.14	28
0500	Compacting bedding in trench	A-1D	90	.089	E.C.Y.		3.05	.34	3.39	5.05
0600	If material source exceeds 2 miles, add for extra mileage.									
0610	See Section 31 23 23.20 for hauling mileage add.									

31 23 23.17 General Fill

		Crew	Daily Output	Labor-Hours	Unit	Material	2011 Bare Costs Labor	2011 Bare Costs Equipment	Total	Total Incl O&P
0010	**GENERAL FILL**									
0011	Spread dumped material, no compaction									
0020	By dozer, no compaction	B-10B	1000	.012	L.C.Y.		.50	1.05	1.55	1.91
0100	By hand	1 Clab	12	.667	"		23		23	35
0500	Gravel fill, compacted, under floor slabs, 4" deep	B-37	10000	.005	S.F.	.40	.17	.01	.58	.73
0600	6" deep		8600	.006		.60	.20	.02	.82	.99
0700	9" deep		7200	.007		1	.24	.02	1.26	1.49
0800	12" deep		6000	.008		1.40	.29	.02	1.71	2.01
1000	Alternate pricing method, 4" deep		120	.400	E.C.Y.	30	14.50	1.21	45.71	56.50
1100	6" deep		160	.300		30	10.85	.91	41.76	50.50

31 23 Excavation and Fill

31 23 23 – Fill

31 23 23.17 General Fill

		Crew	Daily Output	Labor-Hours	Unit	Material	2011 Bare Costs Labor	Equipment	Total	Total Incl O&P
1200	9" deep	B-37	200	.240	E.C.Y.	30	8.70	.73	39.43	47
1300	12" deep	↓	220	.218	↓	30	7.90	.66	38.56	46
1400	Granular fill				L.C.Y.	20			20	22

31 23 23.20 Hauling

		Crew	Daily Output	Labor-Hours	Unit	Material	2011 Bare Costs Labor	Equipment	Total	Total Incl O&P
0010	**HAULING**									
0011	Excavated or borrow, loose cubic yards									
0012	no loading equipment, including hauling, waiting, loading/dumping									
0013	time per cycle (wait, load, travel, unload or dump & return)									
0014	8 C.Y. truck, 15 MPH ave, cycle 0.5 miles, 10 min. wait/Ld./Uld.	B-34A	320	.025	L.C.Y.		.86	.99	1.85	2.39
0016	cycle 1 mile		272	.029			1.01	1.16	2.17	2.82
0018	cycle 2 miles		208	.038			1.32	1.52	2.84	3.68
0020	cycle 4 miles		144	.056			1.91	2.19	4.10	5.30
0022	cycle 6 miles		112	.071			2.46	2.82	5.28	6.85
0024	cycle 8 miles		88	.091			3.13	3.58	6.71	8.70
0026	20 MPH ave, cycle 0.5 mile		336	.024			.82	.94	1.76	2.27
0028	cycle 1 mile		296	.027			.93	1.07	2	2.58
0030	cycle 2 miles		240	.033			1.15	1.31	2.46	3.19
0032	cycle 4 miles		176	.045			1.56	1.79	3.35	4.34
0034	cycle 6 miles		136	.059			2.02	2.32	4.34	5.60
0036	cycle 8 miles		112	.071			2.46	2.82	5.28	6.85
0044	25 MPH ave, cycle 4 miles		192	.042			1.43	1.64	3.07	3.99
0046	cycle 6 miles		160	.050			1.72	1.97	3.69	4.78
0048	cycle 8 miles		128	.063			2.15	2.46	4.61	6
0050	30 MPH ave, cycle 4 miles		216	.037			1.27	1.46	2.73	3.55
0052	cycle 6 miles		176	.045			1.56	1.79	3.35	4.34
0054	cycle 8 miles		144	.056			1.91	2.19	4.10	5.30
0114	15 MPH ave, cycle 0.5 mile, 15 min. wait/Ld./Uld.		224	.036			1.23	1.41	2.64	3.42
0116	cycle 1 mile		200	.040			1.38	1.58	2.96	3.82
0118	cycle 2 miles		168	.048			1.64	1.88	3.52	4.56
0120	cycle 4 miles		120	.067			2.29	2.63	4.92	6.35
0122	cycle 6 miles		96	.083			2.87	3.29	6.16	7.95
0124	cycle 8 miles		80	.100			3.44	3.94	7.38	9.60
0126	20 MPH ave, cycle 0.5 mile		232	.034			1.19	1.36	2.55	3.30
0128	cycle 1 mile		208	.038			1.32	1.52	2.84	3.68
0130	cycle 2 miles		184	.043			1.50	1.71	3.21	4.16
0132	cycle 4 miles		144	.056			1.91	2.19	4.10	5.30
0134	cycle 6 miles		112	.071			2.46	2.82	5.28	6.85
0136	cycle 8 miles		96	.083			2.87	3.29	6.16	7.95
0144	25 MPH ave, cycle 4 miles		152	.053			1.81	2.08	3.89	5.05
0146	cycle 6 miles		128	.063			2.15	2.46	4.61	6
0148	cycle 8 miles		112	.071			2.46	2.82	5.28	6.85
0150	30 MPH ave, cycle 4 miles		168	.048			1.64	1.88	3.52	4.56
0152	cycle 6 miles		144	.056			1.91	2.19	4.10	5.30
0154	cycle 8 miles		120	.067			2.29	2.63	4.92	6.35
0214	15 MPH ave, cycle 0.5 mile, 20 min wait/Ld./Uld.		176	.045			1.56	1.79	3.35	4.34
0216	cycle 1 mile		160	.050			1.72	1.97	3.69	4.78
0218	cycle 2 miles		136	.059			2.02	2.32	4.34	5.60
0220	cycle 4 miles		104	.077			2.65	3.03	5.68	7.35
0222	cycle 6 miles		88	.091			3.13	3.58	6.71	8.70
0224	cycle 8 miles		72	.111			3.82	4.38	8.20	10.60
0226	20 MPH ave, cycle 0.5 mile		176	.045			1.56	1.79	3.35	4.34
0228	cycle 1 mile	↓	168	.048	↓		1.64	1.88	3.52	4.56

31 23 23.20 Hauling	Crew	Daily Output	Labor-Hours	Unit	Material	2011 Bare Costs Labor	2011 Bare Costs Equipment	Total	Total Incl O&P
0230	cycle 2 miles	B-34A	144	.056	L.C.Y.		1.91	2.19	4.10	5.30
0232	cycle 4 miles		120	.067			2.29	2.63	4.92	6.35
0234	cycle 6 miles		96	.083			2.87	3.29	6.16	7.95
0236	cycle 8 miles		88	.091			3.13	3.58	6.71	8.70
0244	25 MPH ave, cycle 4 miles		128	.063			2.15	2.46	4.61	6
0246	cycle 6 miles		112	.071			2.46	2.82	5.28	6.85
0248	cycle 8 miles		96	.083			2.87	3.29	6.16	7.95
0250	30 MPH ave, cycle 4 miles		136	.059			2.02	2.32	4.34	5.60
0252	cycle 6 miles		120	.067			2.29	2.63	4.92	6.35
0254	cycle 8 miles		104	.077			2.65	3.03	5.68	7.35
0314	15 MPH ave, cycle 0.5 mile, 25 min wait/Ld./Uld.		144	.056			1.91	2.19	4.10	5.30
0316	cycle 1 mile		128	.063			2.15	2.46	4.61	6
0318	cycle 2 miles		112	.071			2.46	2.82	5.28	6.85
0320	cycle 4 miles		96	.083			2.87	3.29	6.16	7.95
0322	cycle 6 miles		80	.100			3.44	3.94	7.38	9.60
0324	cycle 8 miles		64	.125			4.30	4.93	9.23	11.95
0326	20 MPH ave, cycle 0.5 mile		144	.056			1.91	2.19	4.10	5.30
0328	cycle 1 mile		136	.059			2.02	2.32	4.34	5.60
0330	cycle 2 miles		120	.067			2.29	2.63	4.92	6.35
0332	cycle 4 miles		104	.077			2.65	3.03	5.68	7.35
0334	cycle 6 miles		88	.091			3.13	3.58	6.71	8.70
0336	cycle 8 miles		80	.100			3.44	3.94	7.38	9.60
0344	25 MPH ave, cycle 4 miles		112	.071			2.46	2.82	5.28	6.85
0346	cycle 6 miles		96	.083			2.87	3.29	6.16	7.95
0348	cycle 8 miles		88	.091			3.13	3.58	6.71	8.70
0350	30 MPH ave, cycle 4 miles		112	.071			2.46	2.82	5.28	6.85
0352	cycle 6 miles		104	.077			2.65	3.03	5.68	7.35
0354	cycle 8 miles		96	.083			2.87	3.29	6.16	7.95
0414	15 MPH ave, cycle 0.5 mile, 30 min wait/Ld./Uld.		120	.067			2.29	2.63	4.92	6.35
0416	cycle 1 mile		112	.071			2.46	2.82	5.28	6.85
0418	cycle 2 miles		96	.083			2.87	3.29	6.16	7.95
0420	cycle 4 miles		80	.100			3.44	3.94	7.38	9.60
0422	cycle 6 miles		72	.111			3.82	4.38	8.20	10.60
0424	cycle 8 miles		64	.125			4.30	4.93	9.23	11.95
0426	20 MPH ave, cycle 0.5 mile		120	.067			2.29	2.63	4.92	6.35
0428	cycle 1 mile		112	.071			2.46	2.82	5.28	6.85
0430	cycle 2 miles		104	.077			2.65	3.03	5.68	7.35
0432	cycle 4 miles		88	.091			3.13	3.58	6.71	8.70
0434	cycle 6 miles		80	.100			3.44	3.94	7.38	9.60
0436	cycle 8 miles		72	.111			3.82	4.38	8.20	10.60
0444	25 MPH ave, cycle 4 miles		96	.083			2.87	3.29	6.16	7.95
0446	cycle 6 miles		88	.091			3.13	3.58	6.71	8.70
0448	cycle 8 miles		80	.100			3.44	3.94	7.38	9.60
0450	30 MPH ave, cycle 4 miles		96	.083			2.87	3.29	6.16	7.95
0452	cycle 6 miles		88	.091			3.13	3.58	6.71	8.70
0454	cycle 8 miles		80	.100			3.44	3.94	7.38	9.60
0514	15 MPH ave, cycle 0.5 mile, 35 min wait/Ld./Uld.		104	.077			2.65	3.03	5.68	7.35
0516	cycle 1 mile		96	.083			2.87	3.29	6.16	7.95
0518	cycle 2 miles		88	.091			3.13	3.58	6.71	8.70
0520	cycle 4 miles		72	.111			3.82	4.38	8.20	10.60
0522	cycle 6 miles		64	.125			4.30	4.93	9.23	11.95
0524	cycle 8 miles		56	.143			4.91	5.65	10.56	13.65
0526	20 MPH ave, cycle 0.5 mile		104	.077			2.65	3.03	5.68	7.35

31 23 Excavation and Fill

31 23 23 – Fill

31 23 23.20 Hauling		Crew	Daily Output	Labor-Hours	Unit	Material	2011 Bare Costs Labor	2011 Bare Costs Equipment	Total	Total Incl O&P
0528	cycle 1 mile	B-34A	96	.083	L.C.Y.		2.87	3.29	6.16	7.95
0530	cycle 2 miles		96	.083			2.87	3.29	6.16	7.95
0532	cycle 4 miles		80	.100			3.44	3.94	7.38	9.60
0534	cycle 6 miles		72	.111			3.82	4.38	8.20	10.60
0536	cycle 8 miles		64	.125			4.30	4.93	9.23	11.95
0544	25 MPH ave, cycle 4 miles		88	.091			3.13	3.58	6.71	8.70
0546	cycle 6 miles		80	.100			3.44	3.94	7.38	9.60
0548	cycle 8 miles		72	.111			3.82	4.38	8.20	10.60
0550	30 MPH ave, cycle 4 miles		88	.091			3.13	3.58	6.71	8.70
0552	cycle 6 miles		80	.100			3.44	3.94	7.38	9.60
0554	cycle 8 miles		72	.111			3.82	4.38	8.20	10.60
1014	12 C.Y. truck, cycle 0.5 mile, 15 MPH ave, 15 min. wait/Ld./Uld.	B-34B	336	.024			.82	1.51	2.33	2.90
1016	cycle 1 mile		300	.027			.92	1.69	2.61	3.25
1018	cycle 2 miles		252	.032			1.09	2.01	3.10	3.87
1020	cycle 4 miles		180	.044			1.53	2.81	4.34	5.40
1022	cycle 6 miles		144	.056			1.91	3.51	5.42	6.75
1024	cycle 8 miles		120	.067			2.29	4.22	6.51	8.10
1025	cycle 10 miles		96	.083			2.87	5.25	8.12	10.15
1026	20 MPH ave, cycle 0.5 mile		348	.023			.79	1.45	2.24	2.80
1028	cycle 1 mile		312	.026			.88	1.62	2.50	3.12
1030	cycle 2 miles		276	.029			1	1.83	2.83	3.53
1032	cycle 4 miles		216	.037			1.27	2.34	3.61	4.52
1034	cycle 6 miles		168	.048			1.64	3.01	4.65	5.80
1036	cycle 8 miles		144	.056			1.91	3.51	5.42	6.75
1038	cycle 10 miles		120	.067			2.29	4.22	6.51	8.10
1040	25 MPH ave, cycle 4 miles		228	.035			1.21	2.22	3.43	4.27
1042	cycle 6 miles		192	.042			1.43	2.64	4.07	5.10
1044	cycle 8 miles		168	.048			1.64	3.01	4.65	5.80
1046	cycle 10 miles		144	.056			1.91	3.51	5.42	6.75
1050	30 MPH ave, cycle 4 miles		252	.032			1.09	2.01	3.10	3.87
1052	cycle 6 miles		216	.037			1.27	2.34	3.61	4.52
1054	cycle 8 miles		180	.044			1.53	2.81	4.34	5.40
1056	cycle 10 miles		156	.051			1.76	3.24	5	6.25
1060	35 MPH ave, cycle 4 miles		264	.030			1.04	1.92	2.96	3.69
1062	cycle 6 miles		228	.035			1.21	2.22	3.43	4.27
1064	cycle 8 miles		204	.039			1.35	2.48	3.83	4.78
1066	cycle 10 miles		180	.044			1.53	2.81	4.34	5.40
1068	cycle 20 miles		120	.067			2.29	4.22	6.51	8.10
1069	cycle 30 miles		84	.095			3.28	6	9.28	11.65
1070	cycle 40 miles		72	.111			3.82	7.05	10.87	13.55
1072	40 MPH ave, cycle 6 miles		240	.033			1.15	2.11	3.26	4.06
1074	cycle 8 miles		216	.037			1.27	2.34	3.61	4.52
1076	cycle 10 miles		192	.042			1.43	2.64	4.07	5.10
1078	cycle 20 miles		120	.067			2.29	4.22	6.51	8.10
1080	cycle 30 miles		96	.083			2.87	5.25	8.12	10.15
1082	cycle 40 miles		72	.111			3.82	7.05	10.87	13.55
1084	cycle 50 miles		60	.133			4.59	8.45	13.04	16.25
1094	45 MPH ave, cycle 8 miles		216	.037			1.27	2.34	3.61	4.52
1096	cycle 10 miles		204	.039			1.35	2.48	3.83	4.78
1098	cycle 20 miles		132	.061			2.09	3.83	5.92	7.40
1100	cycle 30 miles		108	.074			2.55	4.68	7.23	9
1102	cycle 40 miles		84	.095			3.28	6	9.28	11.65
1104	cycle 50 miles		72	.111			3.82	7.05	10.87	13.55

31 23 23 – Fill

31 23 23.20 Hauling		Crew	Daily Output	Labor-Hours	Unit	Material	2011 Bare Costs		Total	Total Incl O&P
							Labor	Equipment		
1106	50 MPH ave, cycle 10 miles	B-34B	216	.037	L.C.Y.		1.27	2.34	3.61	4.52
1108	cycle 20 miles		144	.056			1.91	3.51	5.42	6.75
1110	cycle 30 miles		108	.074			2.55	4.68	7.23	9
1112	cycle 40 miles		84	.095			3.28	6	9.28	11.65
1114	cycle 50 miles		72	.111			3.82	7.05	10.87	13.55
1214	15 MPH ave, cycle 0.5 mile, 20 min. wait/Ld./Uld.		264	.030			1.04	1.92	2.96	3.69
1216	cycle 1 mile		240	.033			1.15	2.11	3.26	4.06
1218	cycle 2 miles		204	.039			1.35	2.48	3.83	4.78
1220	cycle 4 miles		156	.051			1.76	3.24	5	6.25
1222	cycle 6 miles		132	.061			2.09	3.83	5.92	7.40
1224	cycle 8 miles		108	.074			2.55	4.68	7.23	9
1225	cycle 10 miles		96	.083			2.87	5.25	8.12	10.15
1226	20 MPH ave, cycle 0.5 mile		264	.030			1.04	1.92	2.96	3.69
1228	cycle 1 mile		252	.032			1.09	2.01	3.10	3.87
1230	cycle 2 miles		216	.037			1.27	2.34	3.61	4.52
1232	cycle 4 miles		180	.044			1.53	2.81	4.34	5.40
1234	cycle 6 miles		144	.056			1.91	3.51	5.42	6.75
1236	cycle 8 miles		132	.061			2.09	3.83	5.92	7.40
1238	cycle 10 miles		108	.074			2.55	4.68	7.23	9
1240	25 MPH ave, cycle 4 miles		192	.042			1.43	2.64	4.07	5.10
1242	cycle 6 miles		168	.048			1.64	3.01	4.65	5.80
1244	cycle 8 miles		144	.056			1.91	3.51	5.42	6.75
1246	cycle 10 miles		132	.061			2.09	3.83	5.92	7.40
1250	30 MPH ave, cycle 4 miles		204	.039			1.35	2.48	3.83	4.78
1252	cycle 6 miles		180	.044			1.53	2.81	4.34	5.40
1254	cycle 8 miles		156	.051			1.76	3.24	5	6.25
1256	cycle 10 miles		144	.056			1.91	3.51	5.42	6.75
1260	35 MPH ave, cycle 4 miles		216	.037			1.27	2.34	3.61	4.52
1262	cycle 6 miles		192	.042			1.43	2.64	4.07	5.10
1264	cycle 8 miles		168	.048			1.64	3.01	4.65	5.80
1266	cycle 10 miles		156	.051			1.76	3.24	5	6.25
1268	cycle 20 miles		108	.074			2.55	4.68	7.23	9
1269	cycle 30 miles		72	.111			3.82	7.05	10.87	13.55
1270	cycle 40 miles		60	.133			4.59	8.45	13.04	16.25
1272	40 MPH ave, cycle 6 miles		192	.042			1.43	2.64	4.07	5.10
1274	cycle 8 miles		180	.044			1.53	2.81	4.34	5.40
1276	cycle 10 miles		156	.051			1.76	3.24	5	6.25
1278	cycle 20 miles		108	.074			2.55	4.68	7.23	9
1280	cycle 30 miles		84	.095			3.28	6	9.28	11.65
1282	cycle 40 miles		72	.111			3.82	7.05	10.87	13.55
1284	cycle 50 miles		60	.133			4.59	8.45	13.04	16.25
1294	45 MPH ave, cycle 8 miles		180	.044			1.53	2.81	4.34	5.40
1296	cycle 10 miles		168	.048			1.64	3.01	4.65	5.80
1298	cycle 20 miles		120	.067			2.29	4.22	6.51	8.10
1300	cycle 30 miles		96	.083			2.87	5.25	8.12	10.15
1302	cycle 40 miles		72	.111			3.82	7.05	10.87	13.55
1304	cycle 50 miles		60	.133			4.59	8.45	13.04	16.25
1306	50 MPH ave, cycle 10 miles		180	.044			1.53	2.81	4.34	5.40
1308	cycle 20 miles		132	.061			2.09	3.83	5.92	7.40
1310	cycle 30 miles		96	.083			2.87	5.25	8.12	10.15
1312	cycle 40 miles		84	.095			3.28	6	9.28	11.65
1314	cycle 50 miles		72	.111			3.82	7.05	10.87	13.55
1414	15 MPH ave, cycle 0.5 mile, 25 min. wait/Ld./Uld.		204	.039			1.35	2.48	3.83	4.78

31 23 23.20 Hauling		Crew	Daily Output	Labor-Hours	Unit	Material	2011 Bare Costs		Total	Total Incl O&P
							Labor	Equipment		
1416	cycle 1 mile	B-34B	192	.042	L.C.Y.		1.43	2.64	4.07	5.10
1418	cycle 2 miles		168	.048			1.64	3.01	4.65	5.80
1420	cycle 4 miles		132	.061			2.09	3.83	5.92	7.40
1422	cycle 6 miles		120	.067			2.29	4.22	6.51	8.10
1424	cycle 8 miles		96	.083			2.87	5.25	8.12	10.15
1425	cycle 10 miles		84	.095			3.28	6	9.28	11.65
1426	20 MPH ave, cycle 0.5 mile		216	.037			1.27	2.34	3.61	4.52
1428	cycle 1 mile		204	.039			1.35	2.48	3.83	4.78
1430	cycle 2 miles		180	.044			1.53	2.81	4.34	5.40
1432	cycle 4 miles		156	.051			1.76	3.24	5	6.25
1434	cycle 6 miles		132	.061			2.09	3.83	5.92	7.40
1436	cycle 8 miles		120	.067			2.29	4.22	6.51	8.10
1438	cycle 10 miles		96	.083			2.87	5.25	8.12	10.15
1440	25 MPH ave, cycle 4 miles		168	.048			1.64	3.01	4.65	5.80
1442	cycle 6 miles		144	.056			1.91	3.51	5.42	6.75
1444	cycle 8 miles		132	.061			2.09	3.83	5.92	7.40
1446	cycle 10 miles		108	.074			2.55	4.68	7.23	9
1450	30 MPH ave, cycle 4 miles		168	.048			1.64	3.01	4.65	5.80
1452	cycle 6 miles		156	.051			1.76	3.24	5	6.25
1454	cycle 8 miles		132	.061			2.09	3.83	5.92	7.40
1456	cycle 10 miles		120	.067			2.29	4.22	6.51	8.10
1460	35 MPH ave, cycle 4 miles		180	.044			1.53	2.81	4.34	5.40
1462	cycle 6 miles		156	.051			1.76	3.24	5	6.25
1464	cycle 8 miles		144	.056			1.91	3.51	5.42	6.75
1466	cycle 10 miles		132	.061			2.09	3.83	5.92	7.40
1468	cycle 20 miles		96	.083			2.87	5.25	8.12	10.15
1469	cycle 30 miles		72	.111			3.82	7.05	10.87	13.55
1470	cycle 40 miles		60	.133			4.59	8.45	13.04	16.25
1472	40 MPH ave, cycle 6 miles		168	.048			1.64	3.01	4.65	5.80
1474	cycle 8 miles		156	.051			1.76	3.24	5	6.25
1476	cycle 10 miles		144	.056			1.91	3.51	5.42	6.75
1478	cycle 20 miles		96	.083			2.87	5.25	8.12	10.15
1480	cycle 30 miles		84	.095			3.28	6	9.28	11.65
1482	cycle 40 miles		60	.133			4.59	8.45	13.04	16.25
1484	cycle 50 miles		60	.133			4.59	8.45	13.04	16.25
1494	45 MPH ave, cycle 8 miles		156	.051			1.76	3.24	5	6.25
1496	cycle 10 miles		144	.056			1.91	3.51	5.42	6.75
1498	cycle 20 miles		108	.074			2.55	4.68	7.23	9
1500	cycle 30 miles		84	.095			3.28	6	9.28	11.65
1502	cycle 40 miles		72	.111			3.82	7.05	10.87	13.55
1504	cycle 50 miles		60	.133			4.59	8.45	13.04	16.25
1506	50 MPH ave, cycle 10 miles		156	.051			1.76	3.24	5	6.25
1508	cycle 20 miles		120	.067			2.29	4.22	6.51	8.10
1510	cycle 30 miles		96	.083			2.87	5.25	8.12	10.15
1512	cycle 40 miles		72	.111			3.82	7.05	10.87	13.55
1514	cycle 50 miles		60	.133			4.59	8.45	13.04	16.25
1614	15 MPH, cycle 0.5 mile, 30 min. wait/Ld./Uld.		180	.044			1.53	2.81	4.34	5.40
1616	cycle 1 mile		168	.048			1.64	3.01	4.65	5.80
1618	cycle 2 miles		144	.056			1.91	3.51	5.42	6.75
1620	cycle 4 miles		120	.067			2.29	4.22	6.51	8.10
1622	cycle 6 miles		108	.074			2.55	4.68	7.23	9
1624	cycle 8 miles		84	.095			3.28	6	9.28	11.65
1625	cycle 10 miles		84	.095			3.28	6	9.28	11.65

31 23 23.20 Hauling		Crew	Daily Output	Labor-Hours	Unit	Material	2011 Bare Costs Labor	Equipment	Total	Total Incl O&P
1626	20 MPH ave, cycle 0.5 mile	B-34B	180	.044	L.C.Y.		1.53	2.81	4.34	5.40
1628	cycle 1 mile		168	.048			1.64	3.01	4.65	5.80
1630	cycle 2 miles		156	.051			1.76	3.24	5	6.25
1632	cycle 4 miles		132	.061			2.09	3.83	5.92	7.40
1634	cycle 6 miles		120	.067			2.29	4.22	6.51	8.10
1636	cycle 8 miles		108	.074			2.55	4.68	7.23	9
1638	cycle 10 miles		96	.083			2.87	5.25	8.12	10.15
1640	25 MPH ave, cycle 4 miles		144	.056			1.91	3.51	5.42	6.75
1642	cycle 6 miles		132	.061			2.09	3.83	5.92	7.40
1644	cycle 8 miles		108	.074			2.55	4.68	7.23	9
1646	cycle 10 miles		108	.074			2.55	4.68	7.23	9
1650	30 MPH ave, cycle 4 miles		144	.056			1.91	3.51	5.42	6.75
1652	cycle 6 miles		132	.061			2.09	3.83	5.92	7.40
1654	cycle 8 miles		120	.067			2.29	4.22	6.51	8.10
1656	cycle 10 miles		108	.074			2.55	4.68	7.23	9
1660	35 MPH ave, cycle 4 miles		156	.051			1.76	3.24	5	6.25
1662	cycle 6 miles		144	.056			1.91	3.51	5.42	6.75
1664	cycle 8 miles		132	.061			2.09	3.83	5.92	7.40
1666	cycle 10 miles		120	.067			2.29	4.22	6.51	8.10
1668	cycle 20 miles		84	.095			3.28	6	9.28	11.65
1669	cycle 30 miles		72	.111			3.82	7.05	10.87	13.55
1670	cycle 40 miles		60	.133			4.59	8.45	13.04	16.25
1672	40 MPH, cycle 6 miles		144	.056			1.91	3.51	5.42	6.75
1674	cycle 8 miles		132	.061			2.09	3.83	5.92	7.40
1676	cycle 10 miles		120	.067			2.29	4.22	6.51	8.10
1678	cycle 20 miles		96	.083			2.87	5.25	8.12	10.15
1680	cycle 30 miles		72	.111			3.82	7.05	10.87	13.55
1682	cycle 40 miles		60	.133			4.59	8.45	13.04	16.25
1684	cycle 50 miles		48	.167			5.75	10.55	16.30	20.50
1694	45 MPH ave, cycle 8 miles		144	.056			1.91	3.51	5.42	6.75
1696	cycle 10 miles		132	.061			2.09	3.83	5.92	7.40
1698	cycle 20 miles		96	.083			2.87	5.25	8.12	10.15
1700	cycle 30 miles		84	.095			3.28	6	9.28	11.65
1702	cycle 40 miles		60	.133			4.59	8.45	13.04	16.25
1704	cycle 50 miles		60	.133			4.59	8.45	13.04	16.25
1706	50 MPH ave, cycle 10 miles		132	.061			2.09	3.83	5.92	7.40
1708	cycle 20 miles		108	.074			2.55	4.68	7.23	9
1710	cycle 30 miles		84	.095			3.28	6	9.28	11.65
1712	cycle 40 miles		72	.111			3.82	7.05	10.87	13.55
1714	cycle 50 miles	▼	60	.133	▼		4.59	8.45	13.04	16.25
2000	Hauling, 8 C.Y. truck, small project cost per hour	B-34A	8	1	Hr.		34.50	39.50	74	96
2100	12 C.Y. Truck	B-34B	8	1			34.50	63.50	98	122
2150	16.5 C.Y. Truck	B-34C	8	1			34.50	73	107.50	133
2175	18 C.Y. 8 wheel Truck	B-34I	8	1			34.50	82.50	117	144
2200	20 C.Y. Truck	B-34D	8	1	▼		34.50	75	109.50	135
2300	Grading at dump, or embankment if required, by dozer	B-10B	1000	.012	L.C.Y.		.50	1.05	1.55	1.91
2310	Spotter at fill or cut, if required	1 Clab	8	1	Hr.		34.50		34.50	52.50
9014	18 C.Y. truck, 8 wheels,15 min. wait/Ld./Uld.,15 MPH, cycle 0.5 mi.	B-34I	504	.016	L.C.Y.		.55	1.31	1.86	2.27
9016	cycle 1 mile		450	.018			.61	1.47	2.08	2.55
9018	cycle 2 miles		378	.021			.73	1.75	2.48	3.03
9020	cycle 4 miles		270	.030			1.02	2.45	3.47	4.24
9022	cycle 6 miles		216	.037			1.27	3.06	4.33	5.30
9024	cycle 8 miles	▼	180	.044	▼		1.53	3.67	5.20	6.35

31 23 23 – Fill

31 23 23.20 Hauling		Crew	Daily Output	Labor-Hours	Unit	Material	2011 Bare Costs		Total	Total Incl O&P
							Labor	Equipment		
9025	cycle 10 miles	B-34I	144	.056	L.C.Y.		1.91	4.59	6.50	7.95
9026	20 MPH ave, cycle 0.5 mile		522	.015			.53	1.27	1.80	2.19
9028	cycle 1 mile		468	.017			.59	1.41	2	2.44
9030	cycle 2 miles		414	.019			.66	1.60	2.26	2.77
9032	cycle 4 miles		324	.025			.85	2.04	2.89	3.53
9034	cycle 6 miles		252	.032			1.09	2.62	3.71	4.54
9036	cycle 8 miles		216	.037			1.27	3.06	4.33	5.30
9038	cycle 10 miles		180	.044			1.53	3.67	5.20	6.35
9040	25 MPH ave, cycle 4 miles		342	.023			.80	1.93	2.73	3.35
9042	cycle 6 miles		288	.028			.96	2.29	3.25	3.97
9044	cycle 8 miles		252	.032			1.09	2.62	3.71	4.54
9046	cycle 10 miles		216	.037			1.27	3.06	4.33	5.30
9050	30 MPH ave, cycle 4 miles		378	.021			.73	1.75	2.48	3.03
9052	cycle 6 miles		324	.025			.85	2.04	2.89	3.53
9054	cycle 8 miles		270	.030			1.02	2.45	3.47	4.24
9056	cycle 10 miles		234	.034			1.18	2.82	4	4.90
9060	35 MPH ave, cycle 4 miles		396	.020			.69	1.67	2.36	2.90
9062	cycle 6 miles		342	.023			.80	1.93	2.73	3.35
9064	cycle 8 miles		288	.028			.96	2.29	3.25	3.97
9066	cycle 10 miles		270	.030			1.02	2.45	3.47	4.24
9068	cycle 20 miles		162	.049			1.70	4.08	5.78	7.05
9070	cycle 30 miles		126	.063			2.18	5.25	7.43	9.05
9072	cycle 40 miles		90	.089			3.06	7.35	10.41	12.75
9074	40 MPH ave, cycle 6 miles		360	.022			.76	1.84	2.60	3.18
9076	cycle 8 miles		324	.025			.85	2.04	2.89	3.53
9078	cycle 10 miles		288	.028			.96	2.29	3.25	3.97
9080	cycle 20 miles		180	.044			1.53	3.67	5.20	6.35
9082	cycle 30 miles		144	.056			1.91	4.59	6.50	7.95
9084	cycle 40 miles		108	.074			2.55	6.10	8.65	10.60
9086	cycle 50 miles		90	.089			3.06	7.35	10.41	12.75
9094	45 MPH ave, cycle 8 miles		324	.025			.85	2.04	2.89	3.53
9096	cycle 10 miles		306	.026			.90	2.16	3.06	3.75
9098	cycle 20 miles		198	.040			1.39	3.34	4.73	5.80
9100	cycle 30 miles		144	.056			1.91	4.59	6.50	7.95
9102	cycle 40 miles		126	.063			2.18	5.25	7.43	9.05
9104	cycle 50 miles		108	.074			2.55	6.10	8.65	10.60
9106	50 MPH ave, cycle 10 miles		324	.025			.85	2.04	2.89	3.53
9108	cycle 20 miles		216	.037			1.27	3.06	4.33	5.30
9110	cycle 30 miles		162	.049			1.70	4.08	5.78	7.05
9112	cycle 40 miles		126	.063			2.18	5.25	7.43	9.05
9114	cycle 50 miles		108	.074			2.55	6.10	8.65	10.60
9214	20 min. wait/Ld./Uld.,15 MPH, cycle 0.5 mi.		396	.020			.69	1.67	2.36	2.90
9216	cycle 1 mile		360	.022			.76	1.84	2.60	3.18
9218	cycle 2 miles		306	.026			.90	2.16	3.06	3.75
9220	cycle 4 miles		234	.034			1.18	2.82	4	4.90
9222	cycle 6 miles		198	.040			1.39	3.34	4.73	5.80
9224	cycle 8 miles		162	.049			1.70	4.08	5.78	7.05
9225	cycle 10 miles		144	.056			1.91	4.59	6.50	7.95
9226	20 MPH ave, cycle 0.5 mile		396	.020			.69	1.67	2.36	2.90
9228	cycle 1 mile		378	.021			.73	1.75	2.48	3.03
9230	cycle 2 miles		324	.025			.85	2.04	2.89	3.53
9232	cycle 4 miles		270	.030			1.02	2.45	3.47	4.24
9234	cycle 6 miles		216	.037			1.27	3.06	4.33	5.30

31 23 23 – Fill

31 23 23.20 Hauling		Crew	Daily Output	Labor-Hours	Unit	Material	2011 Bare Costs Labor	Equipment	Total	Total Incl O&P
9236	cycle 8 miles	B-34I	198	.040	L.C.Y.		1.39	3.34	4.73	5.80
9238	cycle 10 miles		162	.049			1.70	4.08	5.78	7.05
9240	25 MPH ave, cycle 4 miles		288	.028			.96	2.29	3.25	3.97
9242	cycle 6 miles		252	.032			1.09	2.62	3.71	4.54
9244	cycle 8 miles		216	.037			1.27	3.06	4.33	5.30
9246	cycle 10 miles		198	.040			1.39	3.34	4.73	5.80
9250	30 MPH ave, cycle 4 miles		306	.026			.90	2.16	3.06	3.75
9252	cycle 6 miles		270	.030			1.02	2.45	3.47	4.24
9254	cycle 8 miles		234	.034			1.18	2.82	4	4.90
9256	cycle 10 miles		216	.037			1.27	3.06	4.33	5.30
9260	35 MPH ave, cycle 4 miles		324	.025			.85	2.04	2.89	3.53
9262	cycle 6 miles		288	.028			.96	2.29	3.25	3.97
9264	cycle 8 miles		252	.032			1.09	2.62	3.71	4.54
9266	cycle 10 miles		234	.034			1.18	2.82	4	4.90
9268	cycle 20 miles		162	.049			1.70	4.08	5.78	7.05
9270	cycle 30 miles		108	.074			2.55	6.10	8.65	10.60
9272	cycle 40 miles		90	.089			3.06	7.35	10.41	12.75
9274	40 MPH ave, cycle 6 miles		288	.028			.96	2.29	3.25	3.97
9276	cycle 8 miles		270	.030			1.02	2.45	3.47	4.24
9278	cycle 10 miles		234	.034			1.18	2.82	4	4.90
9280	cycle 20 miles		162	.049			1.70	4.08	5.78	7.05
9282	cycle 30 miles		126	.063			2.18	5.25	7.43	9.05
9284	cycle 40 miles		108	.074			2.55	6.10	8.65	10.60
9286	cycle 50 miles		90	.089			3.06	7.35	10.41	12.75
9294	45 MPH ave, cycle 8 miles		270	.030			1.02	2.45	3.47	4.24
9296	cycle 10 miles		252	.032			1.09	2.62	3.71	4.54
9298	cycle 20 miles		180	.044			1.53	3.67	5.20	6.35
9300	cycle 30 miles		144	.056			1.91	4.59	6.50	7.95
9302	cycle 40 miles		108	.074			2.55	6.10	8.65	10.60
9304	cycle 50 miles		90	.089			3.06	7.35	10.41	12.75
9306	50 MPH ave, cycle 10 miles		270	.030			1.02	2.45	3.47	4.24
9308	cycle 20 miles		198	.040			1.39	3.34	4.73	5.80
9310	cycle 30 miles		144	.056			1.91	4.59	6.50	7.95
9312	cycle 40 miles		126	.063			2.18	5.25	7.43	9.05
9314	cycle 50 miles		108	.074			2.55	6.10	8.65	10.60
9414	25 min. wait/Ld./Uld.,15 MPH, cycle 0.5 mi.		306	.026			.90	2.16	3.06	3.75
9416	cycle 1 mile		288	.028			.96	2.29	3.25	3.97
9418	cycle 2 miles		252	.032			1.09	2.62	3.71	4.54
9420	cycle 4 miles		198	.040			1.39	3.34	4.73	5.80
9422	cycle 6 miles		180	.044			1.53	3.67	5.20	6.35
9424	cycle 8 miles		144	.056			1.91	4.59	6.50	7.95
9425	cycle 10 miles		126	.063			2.18	5.25	7.43	9.05
9426	20 MPH ave, cycle 0.5 mile		324	.025			.85	2.04	2.89	3.53
9428	cycle 1 mile		306	.026			.90	2.16	3.06	3.75
9430	cycle 2 miles		270	.030			1.02	2.45	3.47	4.24
9432	cycle 4 miles		234	.034			1.18	2.82	4	4.90
9434	cycle 6 miles		198	.040			1.39	3.34	4.73	5.80
9436	cycle 8 miles		180	.044			1.53	3.67	5.20	6.35
9438	cycle 10 miles		144	.056			1.91	4.59	6.50	7.95
9440	25 MPH ave, cycle 4 miles		252	.032			1.09	2.62	3.71	4.54
9442	cycle 6 miles		216	.037			1.27	3.06	4.33	5.30
9444	cycle 8 miles		198	.040			1.39	3.34	4.73	5.80
9446	cycle 10 miles		180	.044			1.53	3.67	5.20	6.35

31 23 Excavation and Fill

31 23 23 – Fill

31 23 23.20 Hauling		Crew	Daily Output	Labor-Hours	Unit	Material	2011 Bare Costs Labor	2011 Bare Costs Equipment	Total	Total Incl O&P
9450	30 MPH ave, cycle 4 miles	B-34I	252	.032	L.C.Y.		1.09	2.62	3.71	4.54
9452	cycle 6 miles		234	.034			1.18	2.82	4	4.90
9454	cycle 8 miles		198	.040			1.39	3.34	4.73	5.80
9456	cycle 10 miles		180	.044			1.53	3.67	5.20	6.35
9460	35 MPH ave, cycle 4 miles		270	.030			1.02	2.45	3.47	4.24
9462	cycle 6 miles		234	.034			1.18	2.82	4	4.90
9464	cycle 8 miles		216	.037			1.27	3.06	4.33	5.30
9466	cycle 10 miles		198	.040			1.39	3.34	4.73	5.80
9468	cycle 20 miles		144	.056			1.91	4.59	6.50	7.95
9470	cycle 30 miles		108	.074			2.55	6.10	8.65	10.60
9472	cycle 40 miles		90	.089			3.06	7.35	10.41	12.75
9474	40 MPH ave, cycle 6 miles		252	.032			1.09	2.62	3.71	4.54
9476	cycle 8 miles		234	.034			1.18	2.82	4	4.90
9478	cycle 10 miles		216	.037			1.27	3.06	4.33	5.30
9480	cycle 20 miles		144	.056			1.91	4.59	6.50	7.95
9482	cycle 30 miles		126	.063			2.18	5.25	7.43	9.05
9484	cycle 40 miles		90	.089			3.06	7.35	10.41	12.75
9486	cycle 50 miles		90	.089			3.06	7.35	10.41	12.75
9494	45 MPH ave, cycle 8 miles		234	.034			1.18	2.82	4	4.90
9496	cycle 10 miles		216	.037			1.27	3.06	4.33	5.30
9498	cycle 20 miles		162	.049			1.70	4.08	5.78	7.05
9500	cycle 30 miles		126	.063			2.18	5.25	7.43	9.05
9502	cycle 40 miles		108	.074			2.55	6.10	8.65	10.60
9504	cycle 50 miles		90	.089			3.06	7.35	10.41	12.75
9506	50 MPH ave, cycle 10 miles		234	.034			1.18	2.82	4	4.90
9508	cycle 20 miles		180	.044			1.53	3.67	5.20	6.35
9510	cycle 30 miles		144	.056			1.91	4.59	6.50	7.95
9512	cycle 40 miles		108	.074			2.55	6.10	8.65	10.60
9514	cycle 50 miles		90	.089			3.06	7.35	10.41	12.75
9614	30 min. wait/Ld./Uld.,15 MPH, cycle 0.5 mi.		270	.030			1.02	2.45	3.47	4.24
9616	cycle 1 mile		252	.032			1.09	2.62	3.71	4.54
9618	cycle 2 miles		216	.037			1.27	3.06	4.33	5.30
9620	cycle 4 miles		180	.044			1.53	3.67	5.20	6.35
9622	cycle 6 miles		162	.049			1.70	4.08	5.78	7.05
9624	cycle 8 miles		126	.063			2.18	5.25	7.43	9.05
9625	cycle 10 miles		126	.063			2.18	5.25	7.43	9.05
9626	20 MPH ave, cycle 0.5 mile		270	.030			1.02	2.45	3.47	4.24
9628	cycle 1 mile		252	.032			1.09	2.62	3.71	4.54
9630	cycle 2 miles		234	.034			1.18	2.82	4	4.90
9632	cycle 4 miles		198	.040			1.39	3.34	4.73	5.80
9634	cycle 6 miles		180	.044			1.53	3.67	5.20	6.35
9636	cycle 8 miles		162	.049			1.70	4.08	5.78	7.05
9638	cycle 10 miles		144	.056			1.91	4.59	6.50	7.95
9640	25 MPH ave, cycle 4 miles		216	.037			1.27	3.06	4.33	5.30
9642	cycle 6 miles		198	.040			1.39	3.34	4.73	5.80
9644	cycle 8 miles		180	.044			1.53	3.67	5.20	6.35
9646	cycle 10 miles		162	.049			1.70	4.08	5.78	7.05
9650	30 MPH ave, cycle 4 miles		216	.037			1.27	3.06	4.33	5.30
9652	cycle 6 miles		198	.040			1.39	3.34	4.73	5.80
9654	cycle 8 miles		180	.044			1.53	3.67	5.20	6.35
9656	cycle 10 miles		162	.049			1.70	4.08	5.78	7.05
9660	35 MPH ave, cycle 4 miles		234	.034			1.18	2.82	4	4.90
9662	cycle 6 miles		216	.037			1.27	3.06	4.33	5.30

31 23 Excavation and Fill

31 23 23 – Fill

31 23 23.20 Hauling

		Crew	Daily Output	Labor-Hours	Unit	Material	2011 Bare Costs Labor	Equipment	Total	Total Incl O&P
9664	cycle 8 miles	B-34I	198	.040	L.C.Y.		1.39	3.34	4.73	5.80
9666	cycle 10 miles		180	.044			1.53	3.67	5.20	6.35
9668	cycle 20 miles		126	.063			2.18	5.25	7.43	9.05
9670	cycle 30 miles		108	.074			2.55	6.10	8.65	10.60
9672	cycle 40 miles		90	.089			3.06	7.35	10.41	12.75
9674	40 MPH ave, cycle 6 miles		216	.037			1.27	3.06	4.33	5.30
9676	cycle 8 miles		198	.040			1.39	3.34	4.73	5.80
9678	cycle 10 miles		180	.044			1.53	3.67	5.20	6.35
9680	cycle 20 miles		144	.056			1.91	4.59	6.50	7.95
9682	cycle 30 miles		108	.074			2.55	6.10	8.65	10.60
9684	cycle 40 miles		90	.089			3.06	7.35	10.41	12.75
9686	cycle 50 miles		72	.111			3.82	9.20	13.02	15.90
9694	45 MPH ave, cycle 8 miles		216	.037			1.27	3.06	4.33	5.30
9696	cycle 10 miles		198	.040			1.39	3.34	4.73	5.80
9698	cycle 20 miles		144	.056			1.91	4.59	6.50	7.95
9700	cycle 30 miles		126	.063			2.18	5.25	7.43	9.05
9702	cycle 40 miles		108	.074			2.55	6.10	8.65	10.60
9704	cycle 50 miles		90	.089			3.06	7.35	10.41	12.75
9706	50 MPH ave, cycle 10 miles		198	.040			1.39	3.34	4.73	5.80
9708	cycle 20 miles		162	.049			1.70	4.08	5.78	7.05
9710	cycle 30 miles		126	.063			2.18	5.25	7.43	9.05
9712	cycle 40 miles		108	.074			2.55	6.10	8.65	10.60
9714	cycle 50 miles		90	.089			3.06	7.35	10.41	12.75

31 41 Shoring

31 41 16 – Sheet Piling

31 41 16.10 Sheet Piling Systems

			Crew	Daily Output	Labor-Hours	Unit	Material	2011 Bare Costs Labor	Equipment	Total	Total Incl O&P
0010	**SHEET PILING SYSTEMS**										
0020	Sheet piling steel, not incl. wales, 22 psf, 15' excav., left in place		B-40	10.81	5.920	Ton	1,550	256	300	2,106	2,425
0100	Drive, extract & salvage	R314116-45		6	10.667		545	460	545	1,550	1,925
0300	20' deep excavation, 27 psf, left in place			12.95	4.942		1,550	213	252	2,015	2,325
0400	Drive, extract & salvage			6.55	9.771		545	420	500	1,465	1,800
1200	15' deep excavation, 22 psf, left in place			983	.065	S.F.	18	2.81	3.32	24.13	28
1300	Drive, extract & salvage			545	.117		6.10	5.05	6	17.15	21
1500	20' deep excavation, 27 psf, left in place			960	.067		22.50	2.88	3.40	28.78	33
1600	Drive, extract & salvage			485	.132		7.90	5.70	6.75	20.35	25
2100	Rent steel sheet piling and wales, first month					Ton	310			310	340
2200	Per added month						31			31	34
2300	Rental piling left in place, add to rental						1,250			1,250	1,375
2500	Wales, connections & struts, 2/3 salvage						495			495	545
3900	Wood, solid sheeting, incl. wales, braces and spacers,	R314116-40									
3910	drive, extract & salvage, 8' deep excavation		B-31	330	.121	S.F.	1.51	4.42	.55	6.48	9.05
4000	10' deep, 50 S.F./hr. in & 150 S.F./hr. out			300	.133		1.55	4.87	.60	7.02	9.80
4100	12' deep, 45 S.F./hr. in & 135 S.F./hr. out			270	.148		1.60	5.40	.67	7.67	10.80
4200	14' deep, 42 S.F./hr. in & 126 S.F./hr. out			250	.160		1.65	5.85	.72	8.22	11.55
4300	16' deep, 40 S.F./hr. in & 120 S.F./hr. out			240	.167		1.70	6.10	.76	8.56	12.05
4400	18' deep, 38 S.F./hr. in & 114 S.F./hr. out			230	.174		1.75	6.35	.79	8.89	12.55
4520	Left in place, 8' deep, 55 S.F./hr.			440	.091		2.72	3.32	.41	6.45	8.55
4540	10' deep, 50 S.F./hr.			400	.100		2.86	3.65	.45	6.96	9.25
4560	12' deep, 45 S.F./hr.			360	.111		3.02	4.05	.50	7.57	10.05
4565	14' deep, 42 S.F./hr.			335	.119		3.20	4.36	.54	8.10	10.80

31 41 Shoring

31 41 16 - Sheet Piling

31 41 16.10 Sheet Piling Systems	Crew	Daily Output	Labor-Hours	Unit	Material	2011 Bare Costs Labor	Equipment	Total	Total Incl O&P
4570 16' deep, 40 S.F./hr.	B-31	320	.125	S.F.	3.40	4.56	.57	8.53	11.35
4700 Alternate pricing, left in place, 8' deep		1.76	22.727	M.B.F.	610	830	103	1,543	2,050
4800 Drive, extract and salvage, 8' deep		1.32	30.303	"	545	1,100	137	1,782	2,450

31 52 Cofferdams

31 52 16 - Timber Cofferdams

31 52 16.10 Cofferdams

		Crew	Daily Output	Labor-Hours	Unit	Material	2011 Bare Costs Labor	Equipment	Total	Total Incl O&P
0010	**COFFERDAMS**									
0011	Incl. mobilization and temporary sheeting									
0080	Soldier beams & lagging H piles with 3" wood sheeting									
0090	horizontal between piles, including removal of wales & braces									
0100	No hydrostatic head, 15' deep, 1 line of braces, minimum	B-50	545	.206	S.F.	6.80	8.45	4.24	19.49	25.50
0200	Maximum		495	.226		7.55	9.30	4.67	21.52	28
1300	No hydrostatic head, left in place, 15' dp., 1 line of braces, min.		635	.176		9.05	7.25	3.64	19.94	25.50
1400	Maximum		575	.195		9.70	8	4.02	21.72	27.50
2350	Lagging only, 3" thick wood between piles 8' O.C., minimum	B-46	400	.120		1.51	4.62	.10	6.23	9
2370	Maximum		250	.192		2.27	7.40	.17	9.84	14.30
2400	Open sheeting no bracing, for trenches to 10' deep, min.		1736	.028		.68	1.06	.02	1.76	2.45
2450	Maximum		1510	.032		.76	1.22	.03	2.01	2.79
2500	Tie-back method, add to open sheeting, add, minimum								20%	20%
2550	Maximum								60%	60%
2700	Tie-backs only, based on tie-backs total length, minimum	B-46	86.80	.553	L.F.	12.55	21.50	.48	34.53	48
2750	Maximum		38.50	1.247	"	22	48	1.08	71.08	101
3500	Tie-backs only, typical average, 25' long		2	24	Ea.	555	925	21	1,501	2,075
3600	35' long		1.58	30.380	"	735	1,175	26.50	1,936.50	2,700

Division Notes

	CREW	DAILY OUTPUT	LABOR-HOURS	UNIT	2011 BARE COSTS				TOTAL INCL O&P
					MAT.	LABOR	EQUIP.	TOTAL	

Estimating Tips

32 01 00 Operations and Maintenance of Exterior Improvements

- Recycling of asphalt pavement is becoming very popular and is an alternative to removal and replacement. It can be a good value engineering proposal if removed pavement can be recycled, either at the project site or at another site that is reasonably close to the project site. Two new sections on repair of flexible and rigid pavement have been added.

32 10 00 Bases, Ballasts, and Paving

- When estimating paving, keep in mind the project schedule. If an asphaltic paving project is in a colder climate and runs through to the spring, consider placing the base course in the autumn and then topping it in the spring, just prior to completion. This could save considerable costs in spring repair. Keep in mind that prices for asphalt and concrete are generally higher in the cold seasons. New lines have been added for paving intersections and bridge approaches.

32 90 00 Planting

- The timing of planting and guarantee specifications often dictate the costs for establishing tree and shrub growth and a stand of grass or ground cover. Establish the work performance schedule to coincide with the local planting season. Maintenance and growth guarantees can add from 20%–100% to the total landscaping cost. The cost to replace trees and shrubs can be as high as 5% of the total cost, depending on the planting zone, soil conditions, and time of year.

Reference Numbers

Reference numbers are shown in shaded boxes at the beginning of some major classifications. These numbers refer to related items in the Reference Section. The reference information may be an estimating procedure, an alternate pricing method, or technical information.

Note: Not all subdivisions listed here necessarily appear in this publication.

32 12 16.13 Plant-Mix Asphalt Paving

		Crew	Daily Output	Labor-Hours	Unit	Material	2011 Bare Costs Labor	Equipment	Total	Total Incl O&P
0010	**PLANT-MIX ASPHALT PAVING**									
0020	And large paved areas with no hauling included									
0025	See Section 31 23 23.20 for hauling costs									
0080	Binder course, 1-1/2" thick	B-25	7725	.011	S.Y.	5.25	.43	.32	6	6.75
0120	2" thick		6345	.014		6.95	.52	.39	7.86	8.85
0130	2-1/2" thick		5620	.016		8.70	.59	.44	9.73	10.95
0160	3" thick		4905	.018		10.45	.67	.51	11.63	13.10
0170	3-1/2" thick		4520	.019		12.20	.73	.55	13.48	15.10
0200	4" thick		4140	.021		13.90	.80	.60	15.30	17.15
0300	Wearing course, 1" thick	B-25B	10575	.009		3.45	.35	.26	4.06	4.61
0340	1-1/2" thick		7725	.012		5.80	.47	.35	6.62	7.50
0380	2" thick		6345	.015		7.80	.58	.43	8.81	9.90
0420	2-1/2" thick		5480	.018		9.60	.67	.49	10.76	12.15
0460	3" thick		4900	.020		11.45	.75	.55	12.75	14.35
0470	3-1/2" thick		4520	.021		13.45	.81	.60	14.86	16.70
0480	4" thick		4140	.023		15.35	.89	.65	16.89	18.95
0500	Open graded friction course	B-25C	5000	.010		3.76	.37	.43	4.56	5.20
0800	Alternate method of figuring paving costs									
0810	Binder course, 1-1/2" thick	B-25	630	.140	Ton	64	5.25	3.94	73.19	83
0811	2" thick		690	.128		64	4.79	3.60	72.39	82
0812	3" thick		800	.110		64	4.13	3.10	71.23	80
0813	4" thick		900	.098		64	3.67	2.76	70.43	79
0850	Wearing course, 1" thick	B-25B	575	.167		65	6.35	4.71	76.06	86.50
0851	1-1/2" thick		630	.152		65	5.80	4.30	75.10	85
0852	2" thick		690	.139		65	5.30	3.92	74.22	84
0853	2-1/2" thick		765	.125		65	4.79	3.54	73.33	82.50
0854	3" thick		800	.120		65	4.58	3.38	72.96	82
1000	Pavement replacement over trench, 2" thick	B-37	90	.533	S.Y.	7.20	19.30	1.62	28.12	39
1050	4" thick		70	.686		14.20	25	2.08	41.28	56
1080	6" thick		55	.873		22.50	31.50	2.64	56.64	76

32 12 16.14 Asphaltic Concrete Paving

		Crew	Daily Output	Labor-Hours	Unit	Material	2011 Bare Costs Labor	Equipment	Total	Total Incl O&P
0011	**ASPHALTIC CONCRETE PAVING**, parking lots & driveways									
0015	No asphalt hauling included									
0018	Use 6.05 C. Y. per inch per M.S.F. for hauling									
0020	6" stone base, 2" binder course, 1" topping	B-25C	9000	.005	S.F.	1.85	.20	.24	2.29	2.62
0025	2" binder course, 2" topping		9000	.005		2.25	.20	.24	2.69	3.05
0030	3" binder course, 2" topping		9000	.005		2.64	.20	.24	3.08	3.48
0035	4" binder course, 2" topping		9000	.005		3.02	.20	.24	3.46	3.90
0040	1.5" binder course, 1" topping		9000	.005		1.66	.20	.24	2.10	2.41
0042	3" binder course, 1" topping		9000	.005		2.24	.20	.24	2.68	3.05
0045	3" binder course, 3" topping		9000	.005		3.03	.20	.24	3.47	3.92
0050	4" binder course, 3" topping		9000	.005		3.42	.20	.24	3.86	4.34
0055	4" binder course, 4" topping		9000	.005		3.81	.20	.24	4.25	4.77
0300	Binder course, 1-1/2" thick		35000	.001		.58	.05	.06	.69	.79
0400	2" thick		25000	.002		.75	.07	.09	.91	1.04
0500	3" thick		15000	.003		1.16	.12	.14	1.42	1.63
0600	4" thick		10800	.004		1.52	.17	.20	1.89	2.16
0800	Sand finish course, 3/4" thick		41000	.001		.31	.04	.05	.40	.47
0900	1" thick		34000	.001		.38	.05	.06	.49	.57
1000	Fill pot holes, hot mix, 2" thick	B-16	4200	.008		.80	.27	.12	1.19	1.42
1100	4" thick		3500	.009		1.17	.32	.14	1.63	1.94
1120	6" thick		3100	.010		1.58	.36	.16	2.10	2.46

32 12 Flexible Paving

32 12 16 – Asphalt Paving

32 12 16.14 Asphaltic Concrete Paving

	32 12 16.14 Asphaltic Concrete Paving	Crew	Daily Output	Labor-Hours	Unit	Material	2011 Bare Costs Labor	Equipment	Total	Total Incl O&P
1140	Cold patch, 2" thick	B-51	3000	.016	S.F.	.81	.55	.06	1.42	1.80
1160	4" thick		2700	.018		1.55	.61	.06	2.22	2.71
1180	6" thick		1900	.025		2.41	.87	.09	3.37	4.09

32 84 Planting Irrigation

32 84 23 – Underground Sprinklers

32 84 23.10 Sprinkler Irrigation System

	32 84 23.10 Sprinkler Irrigation System	Crew	Daily Output	Labor-Hours	Unit	Material	2011 Bare Costs Labor	Equipment	Total	Total Incl O&P
0010	**SPRINKLER IRRIGATION SYSTEM**									
0011	For lawns									
0100	Golf course with fully automatic system	C-17	.05	1600	9 holes	100,000	71,500		171,500	220,000
0200	24' diam. head at 15' O.C. incl. piping, auto oper., minimum	B-20	70	.343	Head	21	13.15		34.15	43
0300	Maximum		40	.600		52	23		75	92.50
0500	60' diam. head at 40' O.C. incl. piping, auto oper., minimum		28	.857		66	33		99	123
0600	Maximum		23	1.043		183	40		223	263
0800	Residential system, custom, 1" supply		2000	.012	S.F.	.32	.46		.78	1.06
0900	1-1/2" supply		1800	.013	"	.37	.51		.88	1.19

453

	CREW	DAILY OUTPUT	LABOR-HOURS	UNIT	2011 BARE COSTS				TOTAL INCL O&P
					MAT.	LABOR	EQUIP.	TOTAL	

Estimating Tips

33 10 00 Water Utilities
33 30 00 Sanitary Sewerage Utilities
33 40 00 Storm Drainage Utilities

- Never assume that the water, sewer, and drainage lines will go in at the early stages of the project. Consider the site access needs before dividing the site in half with open trenches, loose pipe, and machinery obstructions. Always inspect the site to establish that the site drawings are complete. Check off all existing utilities on your drawings as you locate them. Be especially careful with underground utilities because appurtenances are sometimes buried during regrading or repaving operations. If you find any discrepancies, mark up the site plan for further research. Differing site conditions can be very costly if discovered later in the project.

- See also Section 33 01 00 for restoration of pipe where removal/replacement may be undesirable. Use of new types of piping materials can reduce the overall project cost. Owners/design engineers should consider the installing contractor as a valuable source of current information on utility products and local conditions that could lead to significant cost savings.

Reference Numbers

Reference numbers are shown in shaded boxes at the beginning of some major classifications. These numbers refer to related items in the Reference Section. The reference information may be an estimating procedure, an alternate pricing method, or technical information.

Note: Not all subdivisions listed here necessarily appear in this publication.

Note: **Trade Service,** *in part, has been used as a reference source for some of the material prices used in Division 33.*

33 01 Operation and Maintenance of Utilities

33 01 10 – Operation and Maintenance of Water Utilities

33 01 10.10 Corrosion Resistance

		Crew	Daily Output	Labor-Hours	Unit	Material	2011 Bare Costs Labor	Equipment	Total	Total Incl O&P
0010	**CORROSION RESISTANCE**									
0012	Wrap & coat, add to pipe, 4" diameter				L.F.	2.15			2.15	2.37
0020	5" diameter					2.70			2.70	2.97
0040	6" diameter					3.23			3.23	3.55
0060	8" diameter					4.19			4.19	4.61
0080	10" diameter					5.25			5.25	5.80
0100	12" diameter					6.35			6.35	6.95
0120	14" diameter					7.30			7.30	8.05
0140	16" diameter					8.25			8.25	9.10
0160	18" diameter					9.10			9.10	10
0180	20" diameter					10.05			10.05	11.05
0200	24" diameter					12.20			12.20	13.45
0220	Small diameter pipe, 1" diameter, add					1.40			1.40	1.54
0240	2" diameter					1.54			1.54	1.69
0260	2-1/2" diameter					1.69			1.69	1.86
0280	3" diameter					1.69			1.69	1.86
0300	Fittings, field covered, add				S.F.	8.85			8.85	9.75
0500	Coating, bituminous, per diameter inch, 1 coat, add				L.F.	.46			.46	.51
0540	3 coat					1.37			1.37	1.51
0560	Coal tar epoxy, per diameter inch, 1 coat, add					.27			.27	.30
0600	3 coat					.55			.55	.61
1000	Polyethylene H.D. extruded, .025" thk., 1/2" diameter add					.36			.36	.40
1020	3/4" diameter					.38			.38	.42
1040	1" diameter					.43			.43	.47
1060	1-1/4" diameter					.53			.53	.58
1080	1-1/2" diameter					.55			.55	.61
1100	.030" thk., 2" diameter					.59			.59	.65
1120	2-1/2" diameter					.69			.69	.76
1140	.035" thk., 3" diameter					.86			.86	.95
1160	3-1/2" diameter					.97			.97	1.07
1180	4" diameter					1.08			1.08	1.19
1200	.040" thk, 5" diameter					1.48			1.48	1.63
1220	6" diameter					1.56			1.56	1.72
1240	8" diameter					2.02			2.02	2.22
1260	10" diameter					2.47			2.47	2.72
1280	12" diameter					2.97			2.97	3.27
1300	.060" thk., 14" diameter					4.08			4.08	4.49
1320	16" diameter					4.78			4.78	5.25
1340	18" diameter					5.65			5.65	6.25
1360	20" diameter					6.35			6.35	7
1380	Fittings, field wrapped, add				S.F.	6.55			6.55	7.20

33 01 10.20 Pipe Repair

		Crew	Daily Output	Labor-Hours	Unit	Material	2011 Bare Costs Labor	Equipment	Total	Total Incl O&P
0010	**PIPE REPAIR**									
0020	Not including excavation or backfill									
0100	Clamp, stainless steel, lightweight, for steel pipe									
0110	3" long, 1/2" diameter pipe	1 Plum	34	.235	Ea.	13.15	12.60		25.75	33.50
0120	3/4" diameter pipe		32	.250		13.60	13.40		27	35
0130	1" diameter pipe		30	.267		14.35	14.30		28.65	37.50
0140	1-1/4" diameter pipe		28	.286		15.15	15.30		30.45	39.50
0150	1-1/2" diameter pipe		26	.308		15.85	16.50		32.35	42
0160	2" diameter pipe		24	.333		17.35	17.85		35.20	45.50
0170	2-1/2" diameter pipe		23	.348		18.65	18.65		37.30	48.50

33 01 10.20 Pipe Repair	Crew	Daily Output	Labor-Hours	Unit	Material	2011 Bare Costs Labor	Equipment	Total	Total Incl O&P
0180	3" diameter pipe	1 Plum	22	.364	Ea.	20.50	19.50		40	51.50
0190	3-1/2" diameter pipe	↓	21	.381		21.50	20.50		42	54
0200	4" diameter pipe	B-20	56	.429		21.50	16.45		37.95	48.50
0210	5" diameter pipe		53	.453		28	17.40		45.40	57.50
0220	6" diameter pipe		48	.500		29.50	19.20		48.70	62
0230	8" diameter pipe		30	.800		35	30.50		65.50	85.50
0240	10" diameter pipe		28	.857		62.50	33		95.50	119
0250	12" diameter pipe		24	1		66	38.50		104.50	132
0260	14" diameter pipe		22	1.091		218	42		260	305
0270	16" diameter pipe		20	1.200		235	46		281	330
0280	18" diameter pipe		18	1.333		247	51		298	350
0290	20" diameter pipe		16	1.500		261	57.50		318.50	375
0300	24" diameter pipe	↓	14	1.714		290	66		356	420
0360	For 6" long, add					100%	40%			
0370	For 9" long, add					200%	100%			
0380	For 12" long, add					300%	150%			
0390	For 18" long, add				↓	500%	200%			
0400	Pipe Freezing for live repairs of systems 3/8 inch to 6 inch									
0410	Note: Pipe Freezing can also be used to install a valve into a live system									
0420	Pipe Freezing each side 3/8 inch	2 Skwk	8	2	Ea.	500	89		589	685
0425	Pipe Freezing each side 3/8 inch, second location same kit		8	2		20	89		109	158
0430	Pipe Freezing each side 3/4 inch		8	2		500	89		589	685
0435	Pipe Freezing each side 3/4 inch, second location same kit		8	2		20	89		109	158
0440	Pipe Freezing each side 1-1/2 inch		6	2.667		500	118		618	730
0445	Pipe Freezing each side 1-1/2 inch , second location same kit		6	2.667		20	118		138	204
0450	Pipe Freezing each side 2 inch		6	2.667		650	118		768	895
0455	Pipe Freezing each side 2 inch, second location same kit		6	2.667		20	118		138	204
0460	Pipe Freezing each side 2-1/2 inch-3 inch		6	2.667		895	118		1,013	1,175
0465	Pipe Freeze each side 2-1/2 -3 inch, second location same kit		6	2.667		40.50	118		158.50	227
0470	Pipe Freezing each side 4 inch		4	4		1,500	178		1,678	1,925
0475	Pipe Freezing each side 4 inch, second location same kit		4	4		65.50	178		243.50	345
0480	Pipe Freezing each side 5-6 inch		4	4		3,925	178		4,103	4,600
0485	Pipe Freezing each side 5-6 inch, second location same kit	↓	4	4		197	178		375	490
0490	Pipe Frz extra 20 lb CO_2 cylinders (3/8" to 2" - 1 ea, 3" -2 ea)					194			194	213
0500	Pipe Freezing extra 50 lb CO_2 cylinders (4" - 2 ea, 5"-6" -6 ea)				↓	495			495	540
1000	Clamp, stainless steel, with threaded service tap									
1040	Full seal for iron, steel, PVC pipe									
1100	6" long, 2" diameter pipe	1 Plum	17	.471	Ea.	130	25		155	181
1110	2-1/2" diameter pipe		16	.500		133	27		160	186
1120	3" diameter pipe		15.60	.513		145	27.50		172.50	201
1130	3-1/2" diameter pipe	↓	15	.533		150	28.50		178.50	208
1140	4" diameter pipe	B-20	40	.600		156	23		179	207
1150	6" diameter pipe		34	.706		181	27		208	241
1160	8" diameter pipe		21	1.143		213	44		257	305
1170	10" diameter pipe		20	1.200		249	46		295	345
1180	12" diameter pipe	↓	17	1.412		281	54		335	395
1205	For 9" long, add					20%	45%			
1210	For 12" long, add					40%	80%			
1220	For 18" long, add				↓	70%	110%			
1600	Clamp, stainless steel, single section									
1640	Full seal for iron, steel, PVC pipe									
1700	6" long, 2" diameter pipe	1 Plum	17	.471	Ea.	75.50	25		100.50	121
1710	2-1/2" diameter pipe	↓	16	.500	↓	78.50	27		105.50	127

33 01 10.20 **Pipe Repair**		Crew	Daily Output	Labor-Hours	Unit	Material	2011 Bare Costs Labor	Equipment	Total	Total Incl O&P
1720	3" diameter pipe	1 Plum	15.60	.513	Ea.	91	27.50		118.50	141
1730	3-1/2" diameter pipe		15	.533		95.50	28.50		124	148
1740	4" diameter pipe	B-20	40	.600		101	23		124	148
1750	6" diameter pipe		34	.706		126	27		153	181
1760	8" diameter pipe		21	1.143		160	44		204	244
1770	10" diameter pipe		20	1.200		195	46		241	285
1780	12" diameter pipe		17	1.412		227	54		281	335
1800	For 9" long, add					40%	45%			
1810	For 12" long, add					60%	80%			
1820	For 18" long, add					120%	110%			
2000	Clamp, stainless steel, two section									
2040	Full seal, for iron, steel, PVC pipe									
2100	6" long, 4" diameter pipe	B-20	24	1	Ea.	155	38.50		193.50	229
2110	6" diameter pipe		20	1.200		179	46		225	268
2120	8" diameter pipe		13	1.846		203	71		274	330
2130	10" diameter pipe		12	2		213	76.50		289.50	350
2140	12" diameter pipe		10	2.400		273	92		365	440
2200	9" long, 4" diameter pipe		16	1.500		204	57.50		261.50	315
2210	6" diameter pipe		13	1.846		230	71		301	360
2220	8" diameter pipe		9	2.667		256	102		358	440
2230	10" diameter pipe		8	3		340	115		455	550
2240	12" diameter pipe		7	3.429		380	132		512	620
2250	14" diameter pipe		6.40	3.750		635	144		779	920
2260	16" diameter pipe		6	4		715	153		868	1,025
2270	18" diameter pipe		5	4.800		720	184		904	1,075
2280	20" diameter pipe		4.60	5.217		780	200		980	1,175
2290	24" diameter pipe		4	6		1,025	230		1,255	1,475
2320	For 12" long, add to 9"					15%	25%			
2330	For 18" long, add to 9"					70%	55%			
8000	For internal cleaning and inspection, see Section 33 01 30.16									
8100	For pipe testing, see Section 23 05 93.50									

33 01 30.16 **TV Inspection of Sewer Pipelines**		Crew	Daily Output	Labor-Hours	Unit	Material	2011 Bare Costs Labor	Equipment	Total	Total Incl O&P
0010	**TV INSPECTION OF SEWER PIPELINES**									
0100	Pipe internal cleaning & inspection, cleaning, pressure pipe systems									
0120	Pig method, lengths 1000' to 10,000'									
0140	4" diameter thru 24" diameter, minimum				L.F.				2	2.30
0160	Maximum				"				8.75	10.15
6000	Sewage/sanitary systems									
6100	Power rodder with header & cutters									
6110	Mobilization charge, minimum				Total				695	800
6120	Mobilization charge, maximum				"				8,700	10,000
6140	Cleaning 4"-12 " diameter				L.F.				3.70	4.25
6190	14"-24" diameter								4.33	4.98
6240	30" diameter								6.30	7.25
6250	36" diameter								7.35	8.45
6260	48" diameter								8.40	9.65
6270	60" diameter								9.45	10.85
6280	72" diameter								10.50	12.10
9000	Inspection, television camera with video									
9060	up to 500 linear feet				Total				765	880

33 01 Operation and Maintenance of Utilities

33 01 30 – Operation and Maintenance of Sewer Utilities

33 01 30.71 Pipebursting

		Crew	Daily Output	Labor-Hours	Unit	Material	2011 Bare Costs Labor	Equipment	Total	Total Incl O&P
0010	**PIPEBURSTING**									
0011	300' runs, replace with HDPE pipe									
0020	Not including excavation, backfill, shoring, or dewatering									
0100	6" to 15" diameter, minimum				L.F.				100	110
0200	Maximum								200	220
0300	18" to 36" diameter, minimum								200	220
0400	Maximum				↓				410	450
0500	Mobilize and demobilize, minimum				Job				3,000	3,300
0600	Maximum				"				30,000	33,000

33 01 30.74 HDPE Pipe Lining

		Crew	Daily Output	Labor-Hours	Unit	Material	2011 Bare Costs Labor	Equipment	Total	Total Incl O&P
0010	**HDPE PIPE LINING**, excludes cleaning and video inspection									
0020	Pipe relined with one pipe size smaller than original (4" for 6")									
0100	6" diameter, original size	B-6B	600	.080	L.F.	3.64	2.80	1.34	7.78	9.80
0150	8" diameter, original size		600	.080		10.70	2.80	1.34	14.84	17.60
0200	10" diameter, original size		600	.080		12.35	2.80	1.34	16.49	19.40
0250	12" diameter, original size		400	.120		19.15	4.20	2.01	25.36	29.50
0300	14" diameter, original size	↓	400	.120		27	4.20	2.01	33.21	38
0350	16" diameter, original size	B-6C	300	.160		32.50	5.60	4.86	42.96	50
0400	18" diameter, original size	"	300	.160	↓	42	5.60	4.86	52.46	60.50
1000	Pipe HDPE lining, make service line taps	B-6	4	6	Ea.	112	225	80	417	550

33 05 Common Work Results for Utilities

33 05 23 – Trenchless Utility Installation

33 05 23.19 Microtunneling

		Crew	Daily Output	Labor-Hours	Unit	Material	2011 Bare Costs Labor	Equipment	Total	Total Incl O&P
0010	**MICROTUNNELING**									
0011	Not including excavation, backfill, shoring,									
0020	or dewatering, average 50'/day, slurry method									
0100	24" to 48" outside diameter, minimum				L.F.				790	870
0110	Adverse conditions, add				%				50%	50%
1000	Rent microtunneling machine, average monthly lease				Month				88,000	97,000
1010	Operating technician				Day				585	650
1100	Mobilization and demobilization, minimum				Job				38,000	42,000
1110	Maximum				"				400,000	440,000

33 05 23.20 Horizontal Boring

		Crew	Daily Output	Labor-Hours	Unit	Material	2011 Bare Costs Labor	Equipment	Total	Total Incl O&P
0010	**HORIZONTAL BORING**									
0011	Casing only, 100' minimum,									
0020	not incl. jacking pits or dewatering									
0100	Roadwork, 1/2" thick wall, 24" diameter casing	B-42	20	3.200	L.F.	142	124	65	331	420
0200	36" diameter		16	4		214	155	81	450	565
0300	48" diameter		15	4.267		285	165	86.50	536.50	670
0500	Railroad work, 24" diameter		15	4.267		142	165	86.50	393.50	510
0600	36" diameter		14	4.571		214	177	93	484	610
0700	48" diameter	↓	12	5.333		285	206	108	599	755
0900	For ledge, add									20%
1000	Small diameter boring, 3", sandy soil	B-82	900	.018		22	.69	.08	22.77	25.50
1040	Rocky soil	"	500	.032	↓	22	1.25	.15	23.40	26.50
1100	Prepare jacking pits, incl. mobilization & demobilization, minimum				Ea.				3,000	3,450
1101	Maximum				"				20,000	23,000

33 05 Common Work Results for Utilities

33 05 26 – Utility Identification

33 05 26.10 Utility Accessories		Crew	Daily Output	Labor-Hours	Unit	Material	2011 Bare Costs Labor	Equipment	Total	Total Incl O&P
0010	**UTILITY ACCESSORIES**	G1030–805								
0400	Underground tape, detectable, reinforced, alum. foil core, 2"	1 Clab	150	.053	C.L.F.	2.25	1.83		4.08	5.30
0500	6"	"	140	.057	"	6.55	1.96		8.51	10.20

33 11 Water Utility Distribution Piping

33 11 13 – Public Water Utility Distribution Piping

33 11 13.15 Water Supply, Ductile Iron Pipe

		Crew	Daily Output	Labor-Hours	Unit	Material	2011 Bare Costs Labor	Equipment	Total	Total Incl O&P
0010	**WATER SUPPLY, DUCTILE IRON PIPE**	R331113-80								
0020	Not including excavation or backfill									
2000	Pipe, class 50 water piping, 18' lengths									
2020	Mechanical joint, 4" diameter	B-21A	200	.200	L.F.	12.50	8.55	2.17	23.22	29
2040	6" diameter		160	.250		14.60	10.70	2.72	28.02	35
2060	8" diameter		133.33	.300		21	12.80	3.26	37.06	46
2080	10" diameter		114.29	.350		26.50	14.95	3.80	45.25	55.50
2100	12" diameter		105.26	.380		32	16.25	4.13	52.38	64
2120	14" diameter		100	.400		48.50	17.10	4.34	69.94	84.50
2140	16" diameter		72.73	.550		50	23.50	5.95	79.45	97
2160	18" diameter		68.97	.580		53	25	6.30	84.30	103
2170	20" diameter		57.14	.700		56	30	7.60	93.60	115
2180	24" diameter		47.06	.850		61	36.50	9.25	106.75	132
3000	Tyton, push-on joint, 4" diameter		400	.100		10.80	4.27	1.09	16.16	19.55
3020	6" diameter		333.33	.120		15.20	5.15	1.30	21.65	26
3040	8" diameter		200	.200		21.50	8.55	2.17	32.22	39.50
3060	10" diameter		181.82	.220		29.50	9.40	2.39	41.29	49.50
3080	12" diameter		160	.250		31	10.70	2.72	44.42	53
3100	14" diameter		133.33	.300		34	12.80	3.26	50.06	60.50
3120	16" diameter		114.29	.350		37	14.95	3.80	55.75	67
3140	18" diameter		100	.400		41	17.10	4.34	62.44	76
3160	20" diameter		88.89	.450		42.50	19.25	4.89	66.64	81
3180	24" diameter		76.92	.520		44.50	22	5.65	72.15	88.50
8000	Fittings, mechanical joint									
8006	90° bend, 4" diameter	B-20A	16	2	Ea.	248	83.50		331.50	400
8020	6" diameter		12.80	2.500		370	105		475	565
8040	8" diameter		10.67	2.999		530	125		655	775
8060	10" diameter	B-21A	11.43	3.500		810	150	38	998	1,150
8080	12" diameter		10.53	3.799		1,075	162	41.50	1,278.50	1,500
8100	14" diameter		10	4		1,525	171	43.50	1,739.50	1,975
8120	16" diameter		7.27	5.502		1,925	235	60	2,220	2,550
8140	18" diameter		6.90	5.797		2,725	248	63	3,036	3,450
8160	20" diameter		5.71	7.005		3,400	299	76	3,775	4,275
8180	24" diameter		4.70	8.511		5,975	365	92.50	6,432.50	7,200
8200	Wye or tee, 4" diameter	B-20A	10.67	2.999		360	125		485	585
8220	6" diameter		8.53	3.751		530	157		687	820
8240	8" diameter		7.11	4.501		785	188		973	1,150
8260	10" diameter	B-21A	7.62	5.249		1,325	224	57	1,606	1,850
8280	12" diameter		7.02	5.698		1,750	244	62	2,056	2,350
8300	14" diameter		6.67	5.997		1,575	256	65	1,896	2,200
8320	16" diameter		4.85	8.247		2,925	350	89.50	3,364.50	3,850
8340	18" diameter		4.60	8.696		4,100	370	94.50	4,564.50	5,175
8360	20" diameter		3.81	10.499		5,325	450	114	5,889	6,650

33 11 Water Utility Distribution Piping

33 11 13 – Public Water Utility Distribution Piping

33 11 13.15 Water Supply, Ductile Iron Pipe

		Crew	Daily Output	Labor-Hours	Unit	Material	2011 Bare Costs Labor	Equipment	Total	Total Incl O&P
8380	24" diameter	B-21A	3.14	12.739	Ea.	9,525	545	138	10,208	11,500
8398	45° bends, 4" diameter	B-20A	16	2		225	83.50		308.50	375
8400	6" diameter	"	12.80	2.500		320	105		425	510
8410	12" diameter	B-21A	10.53	3.799		915	162	41.50	1,118.50	1,300
8420	16" diameter		7.27	5.502		1,650	235	60	1,945	2,225
8430	20" diameter		5.71	7.005		2,650	299	76	3,025	3,450
8440	24" diameter		4.70	8.511		3,950	365	92.50	4,407.50	5,000
8450	Decreaser, 6" x 4" diameter	B-20A	14.22	2.250		260	94		354	430
8460	8" x 6" diameter	"	11.64	2.749		380	115		495	595
8470	10" x 6" diameter	B-21A	13.33	3.001		455	128	32.50	615.50	730
8480	12" x 6" diameter		12.70	3.150		595	135	34	764	895
8490	16" x 6" diameter		10	4		1,175	171	43.50	1,389.50	1,600
8500	20" x 6" diameter		8.42	4.751		2,600	203	51.50	2,854.50	3,200
8552	For water utility valves see Section 33 12 16									
8600	12" diameter	B-21	3	9.333	Ea.	2,425	370	43	2,838	3,250
8610	14" diameter		2	14		4,075	555	64.50	4,694.50	5,400
8620	16" diameter		2	14		4,775	555	64.50	5,394.50	6,175
8700	Joint restraint, ductile iron mechanical joints									
8710	4" diameter	B-20A	32	1	Ea.	34	42		76	101
8720	6" diameter		25.60	1.250		41.50	52.50		94	125
8730	8" diameter		21.33	1.500		60.50	62.50		123	161
8740	10" diameter		18.28	1.751		86	73		159	206
8750	12" diameter		16.84	1.900		117	79.50		196.50	249
8760	14" diameter		16	2		187	83.50		270.50	330
8770	16" diameter		11.64	2.749		207	115		322	400
8780	18" diameter		11.03	2.901		305	121		426	520
8785	20" diameter		9.14	3.501		375	146		521	630
8790	24" diameter		7.53	4.250		485	178		663	805
9600	Steel sleeve with tap, 4" diameter	B-20	3	8		400	305		705	910
9620	6" diameter		2	12		455	460		915	1,200
9630	8" diameter		2	12		510	460		970	1,275

33 11 13.25 Water Supply, Polyvinyl Chloride Pipe

		Crew	Daily Output	Labor-Hours	Unit	Material	2011 Bare Costs Labor	Equipment	Total	Total Incl O&P
0010	**WATER SUPPLY, POLYVINYL CHLORIDE PIPE** R331113-80									
0020	Not including excavation or backfill, unless specified									
2100	PVC pipe, Class 150, 1-1/2" diameter	Q-1A	750	.013	L.F.	.32	.72		1.04	1.42
2120	2" diameter		686	.015		.49	.79		1.28	1.72
2140	2-1/2" diameter		500	.020		.73	1.08		1.81	2.41
2160	3" diameter	B-20	430	.056		1.11	2.14		3.25	4.50
3010	AWWA C905, PR 100, DR 25									
3030	14" diameter	B-20A	213	.150	L.F.	10.70	6.30		17	21.50
3040	16" diameter		200	.160		13.90	6.70		20.60	25.50
3050	18" diameter		160	.200		21	8.35		29.35	35.50
3060	20" diameter		133	.241		21	10.05		31.05	38
3070	24" diameter		107	.299		29.50	12.50		42	51.50
3080	30" diameter		80	.400		59	16.70		75.70	90.50
3090	36" diameter		80	.400		91	16.70		107.70	126
3100	42" diameter		60	.533		123	22.50		145.50	169
3200	48" diameter		60	.533		161	22.50		183.50	211
4520	Pressure pipe Class 150, SDR 18, AWWA C900, 4" diameter		380	.084		1.76	3.52		5.28	7.25
4530	6" diameter		316	.101		5.15	4.23		9.38	12.05
4540	8" diameter		264	.121		8	5.05		13.05	16.45
4550	10" diameter		220	.145		12.60	6.10		18.70	23

33 11 13.25 Water Supply, Polyvinyl Chloride Pipe	Crew	Daily Output	Labor-Hours	Unit	Material	2011 Bare Costs Labor	Equipment	Total	Total Incl O&P
4560	12" diameter	B-20A	186	.172	L.F.	17.20	7.20		24.40	30
8000	Fittings with rubber gasket									
8003	Class 150, D.R. 18									
8006	90° Bend , 4" diameter	B-20	100	.240	Ea.	54	9.20		63.20	73.50
8020	6" diameter		90	.267		97	10.25		107.25	123
8040	8" diameter		80	.300		185	11.50		196.50	221
8060	10" diameter		50	.480		282	18.40		300.40	340
8080	12" diameter		30	.800		430	30.50		460.50	520
8100	Tee, 4" diameter		90	.267		57	10.25		67.25	78.50
8120	6" diameter		80	.300		154	11.50		165.50	187
8140	8" diameter		70	.343		215	13.15		228.15	256
8160	10" diameter		40	.600		455	23		478	535
8180	12" diameter		20	1.200		635	46		681	770
8200	45° Bend, 4" diameter		100	.240		53.50	9.20		62.70	73
8220	6" diameter		90	.267		96.50	10.25		106.75	122
8240	8" diameter		50	.480		182	18.40		200.40	229
8260	10" diameter		50	.480		286	18.40		304.40	345
8280	12" diameter		30	.800		380	30.50		410.50	465
8300	Reducing tee 6" x 4"		100	.240		131	9.20		140.20	158
8320	8" x 6"		90	.267		266	10.25		276.25	310
8330	10" x 6"		90	.267		310	10.25		320.25	355
8340	10" x 8"		90	.267		330	10.25		340.25	375
8350	12" x 6"		90	.267		415	10.25		425.25	470
8360	12" x 8"		90	.267		460	10.25		470.25	520
8400	Tapped service tee (threaded type) 6" x 6" x 3/4"		100	.240		86.50	9.20		95.70	109
8430	6" x 6" x 1"		90	.267		86.50	10.25		96.75	111
8440	6" x 6" x 1-1/2"		90	.267		86.50	10.25		96.75	111
8450	6" x 6" x 2"		90	.267		86.50	10.25		96.75	111
8460	8" x 8" x 3/4"		90	.267		128	10.25		138.25	156
8470	8" x 8" x 1"		90	.267		128	10.25		138.25	156
8480	8" x 8" x 1-1/2"		90	.267		128	10.25		138.25	156
8490	8" x 8" x 2"		90	.267		128	10.25		138.25	156
8500	Repair coupling 4"		100	.240		91	9.20		100.20	114
8520	6" diameter		90	.267		114	10.25		124.25	142
8540	8" diameter		50	.480		133	18.40		151.40	176
8560	10" diameter		50	.480		185	18.40		203.40	233
8580	12" diameter		50	.480		223	18.40		241.40	275
8600	Plug end 4"		100	.240		34	9.20		43.20	51
8620	6" diameter		90	.267		50	10.25		60.25	70.50
8640	8" diameter		50	.480		76	18.40		94.40	112
8660	10" diameter		50	.480		105	18.40		123.40	145
8680	12" diameter		50	.480		129	18.40		147.40	171
8700	PVC pipe, joint restraint									
8710	4" diameter	B-20A	32	1	Ea.	34.50	42		76.50	101
8720	6" diameter		25.60	1.250		43	52.50		95.50	126
8730	8" diameter		21.33	1.500		61.50	62.50		124	162
8740	10" diameter		18.28	1.751		113	73		186	235
8750	12" diameter		16.84	1.900		119	79.50		198.50	250
8760	14" diameter		16	2		209	83.50		292.50	355
8770	16" diameter		11.64	2.749		272	115		387	475
8780	18" diameter		11.03	2.901		340	121		461	560
8785	20" diameter		9.14	3.501		405	146		551	665
8790	24" diameter		7.53	4.250		470	178		648	785

33 11 13 – Public Water Utility Distribution Piping

33 11 13.35 Water Supply, HDPE	Crew	Daily Output	Labor-Hours	Unit	Material	2011 Bare Costs Labor	2011 Bare Costs Equipment	Total	Total Incl O&P
0010 **WATER SUPPLY, HDPE**									
0011 Butt fusion joints, SDR 21, 40′ lengths not including excavation or backfill									
0100 4″ diameter	B-22A	400	.100	L.F.	3.64	3.92	1.59	9.15	11.70
0200 6″ diameter		380	.105		10.70	4.13	1.67	16.50	19.95
0300 8″ diameter		320	.125		12.35	4.90	1.98	19.23	23
0400 10″ diameter		300	.133		19.15	5.25	2.11	26.51	31.50
0500 12″ diameter		260	.154		27	6.05	2.44	35.49	41.50
0600 14″ diameter	B-22B	220	.182		32.50	7.15	4.84	44.49	52
0700 16″ diameter		180	.222		42	8.70	5.90	56.60	66.50
0800 18″ diameter		140	.286		53.50	11.20	7.60	72.30	84.50
0900 24″ diameter		100	.400		95.50	15.70	10.65	121.85	141
1000 Fittings									
1100 Elbows, 90 degrees									
1200 4″ diameter	B-22A	32	1.250	Ea.	33.50	49	19.85	102.35	134
1300 6″ diameter		28	1.429		104	56	22.50	182.50	225
1400 8″ diameter		24	1.667		220	65.50	26.50	312	370
1500 10″ diameter		18	2.222		350	87	35	472	555
1600 12″ diameter		12	3.333		530	131	53	714	840
1700 14″ diameter	B-22B	9	4.444		635	174	118	927	1,100
1800 16″ diameter		6	6.667		770	261	177	1,208	1,450
1900 18″ diameter		4	10		1,050	390	266	1,706	2,050
2000 24″ diameter		3	13.333		2,125	525	355	3,005	3,500
2100 Tees									
2200 4″ diameter	B-22A	30	1.333	Ea.	41.50	52.50	21	115	149
2300 6″ diameter		26	1.538		123	60.50	24.50	208	254
2400 8″ diameter		22	1.818		310	71.50	29	410.50	480
2500 10″ diameter		15	2.667		405	105	42.50	552.50	650
2600 12″ diameter		10	4		565	157	63.50	785.50	930
2700 14″ diameter	B-22B	8	5		670	196	133	999	1,175
2800 16″ diameter		6	6.667		785	261	177	1,223	1,450
2900 18″ diameter		4	10		1,050	390	266	1,706	2,050
3000 24″ diameter		2	20		2,175	785	530	3,490	4,150

33 12 Water Utility Distribution Equipment

33 12 13 – Water Service Connections

33 12 13.15 Tapping, Crosses and Sleeves

33 12 13.15 Tapping, Crosses and Sleeves	Crew	Daily Output	Labor-Hours	Unit	Material	Labor	Equipment	Total	Total Incl O&P
0010 **TAPPING, CROSSES AND SLEEVES**									
4000 Drill and tap pressurized main (labor only)									
4100 6″ main, 1″ to 2″ service	Q-1	3	5.333	Ea.		257		257	385
4150 8″ main, 1″ to 2″ service	″	2.75	5.818	″		281		281	420
4500 Tap and insert gate valve									
4600 8″ main, 4″ branch	B-21	3.20	8.750	Ea.		345	40.50	385.50	575
4650 6″ branch		2.70	10.370			410	47.50	457.50	680
4700 10″ Main, 4″ branch		2.70	10.370			410	47.50	457.50	680
4750 6″ branch		2.35	11.915			470	55	525	780
4800 12″ main, 6″ branch		2.35	11.915			470	55	525	780
4850 8″ branch		2.35	11.915			470	55	525	780
7020 Crosses, 4″ x 4″		37	.757		1,250	30	3.48	1,283.48	1,425
7030 6″ x 4″		25	1.120		1,050	44.50	5.15	1,099.65	1,250
7040 6″ x 6″		25	1.120		1,050	44.50	5.15	1,099.65	1,250

33 12 13.15 Tapping, Crosses and Sleeves	Crew	Daily Output	Labor-Hours	Unit	Material	2011 Bare Costs		Total	Total Incl O&P
						Labor	Equipment		
7060	8" x 6"	B-21	21	1.333	Ea.	1,300	52.50	6.15	1,358.65	1,500
7080	8" x 8"		21	1.333		1,350	52.50	6.15	1,408.65	1,575
7100	10" x 6"		21	1.333		2,125	52.50	6.15	2,183.65	2,400
7120	10" x 10"		21	1.333		2,275	52.50	6.15	2,333.65	2,575
7140	12" x 6"		18	1.556		2,125	61.50	7.15	2,193.65	2,425
7160	12" x 12"		18	1.556		2,950	61.50	7.15	3,018.65	3,350
7180	14" x 6"		16	1.750		5,750	69	8.05	5,827.05	6,450
7200	14" x 14"		16	1.750		6,100	69	8.05	6,177.05	6,825
7220	16" x 6"		14	2		6,100	79	9.20	6,188.20	6,825
7240	16" x 10"		14	2		6,250	79	9.20	6,338.20	7,000
7260	16" x 16"		14	2		6,575	79	9.20	6,663.20	7,375
7280	18" x 6"		10	2.800		9,225	111	12.90	9,348.90	10,400
7300	18" x 12"		10	2.800		9,300	111	12.90	9,423.90	10,400
7320	18" x 18"		10	2.800		8,500	111	12.90	8,623.90	9,525
7340	20" x 6"		8	3.500		7,475	138	16.10	7,629.10	8,425
7360	20" x 12"		8	3.500		8,000	138	16.10	8,154.10	9,025
7380	20" x 20"		8	3.500		11,700	138	16.10	11,854.10	13,100
7400	24" x 6"		6	4.667		9,650	184	21.50	9,855.50	10,900
7420	24" x 12"		6	4.667		9,775	184	21.50	9,980.50	11,100
7440	24" x 18"		6	4.667		14,600	184	21.50	14,805.50	16,400
7460	24" x 24"		6	4.667		15,200	184	21.50	15,405.50	17,000
7600	Cut-in sleeves with rubber gaskets, 4"		18	1.556		245	61.50	7.15	313.65	370
7620	6"		12	2.333		294	92	10.75	396.75	480
7640	8"		10	2.800		385	111	12.90	508.90	610
7660	10"		10	2.800		490	111	12.90	613.90	725
7680	12"		9	3.111		655	123	14.30	792.30	925
7800	Cut-in valves with rubber gaskets, 4"		18	1.556		355	61.50	7.15	423.65	490
7820	6"		12	2.333		475	92	10.75	577.75	680
7840	8"		10	2.800		735	111	12.90	858.90	995
7860	10"		10	2.800		760	111	12.90	883.90	1,025
7880	12"		9	3.111		905	123	14.30	1,042.30	1,200
7900	Tapping Valve 4 inch, MJ, ductile iron		18	1.556		795	61.50	7.15	863.65	975
7920	6 inch, MJ, ductile iron		12	2.333		890	92	10.75	992.75	1,125
8000	Sleeves with rubber gaskets, 4" x 4"		37	.757		695	30	3.48	728.48	815
8010	6" x 4"		25	1.120		710	44.50	5.15	759.65	855
8020	6" x 6"		25	1.120		710	44.50	5.15	759.65	860
8030	8" x 4"		21	1.333		725	52.50	6.15	783.65	885
8040	8" x 6"		21	1.333		835	52.50	6.15	893.65	1,000
8060	8" x 8"		21	1.333		950	52.50	6.15	1,008.65	1,125
8070	10" x 4"		21	1.333		880	52.50	6.15	938.65	1,050
8080	10" x 6"		21	1.333		1,350	52.50	6.15	1,408.65	1,550
8090	10" x 8"		21	1.333		1,500	52.50	6.15	1,558.65	1,725
8100	10" x 10"		21	1.333		1,600	52.50	6.15	1,658.65	1,850
8110	12" x 4"		18	1.556		1,500	61.50	7.15	1,568.65	1,750
8120	12" x 6"		18	1.556		1,575	61.50	7.15	1,643.65	1,850
8130	12" x 8"		18	1.556		1,700	61.50	7.15	1,768.65	1,950
8135	12" x 10"		18	1.556		1,750	61.50	7.15	1,818.65	2,025
8140	12" x 12"		18	1.556		1,850	61.50	7.15	1,918.65	2,150
8160	14" x 6"		16	1.750		1,575	69	8.05	1,652.05	1,875
8180	14" x 14"		16	1.750		1,725	69	8.05	1,802.05	2,025
8200	16" x 6"		14	2		1,850	79	9.20	1,938.20	2,150
8220	16" x 10"		14	2		2,225	79	9.20	2,313.20	2,575
8240	16" x 16"		14	2		2,900	79	9.20	2,988.20	3,325

33 12 Water Utility Distribution Equipment

33 12 13 – Water Service Connections

33 12 13.15 Tapping, Crosses and Sleeves

		Crew	Daily Output	Labor-Hours	Unit	Material	2011 Bare Costs Labor	Equipment	Total	Total Incl O&P
8260	18" x 6"	B-21	10	2.800	Ea.	1,900	111	12.90	2,023.90	2,275
8280	18" x 12"		10	2.800		2,500	111	12.90	2,623.90	2,925
8300	18" x 18"		10	2.800		3,200	111	12.90	3,323.90	3,700
8320	20" x 6"		8	3.500		2,200	138	16.10	2,354.10	2,650
8340	20" x 12"		8	3.500		3,375	138	16.10	3,529.10	3,925
8360	20" x 20"		8	3.500		3,975	138	16.10	4,129.10	4,600
8380	24" x 6"		6	4.667		1,650	184	21.50	1,855.50	2,125
8400	24" x 12"		6	4.667		2,900	184	21.50	3,105.50	3,500
8420	24" x 18"		6	4.667		3,500	184	21.50	3,705.50	4,150
8440	24" x 24"		6	4.667		5,350	184	21.50	5,555.50	6,200
8800	Hydrant valve box, 6' long	B-20	20	1.200		180	46		226	269
8820	8' long		18	1.333		225	51		276	325
8830	Valve box w/lid 4' deep		14	1.714		130	66		196	244
8840	Valve box and large base w/lid		14	1.714		250	66		316	375

33 12 16 – Water Utility Distribution Valves

33 12 16.10 Valves

		Crew	Daily Output	Labor-Hours	Unit	Material	2011 Bare Costs Labor	Equipment	Total	Total Incl O&P
0010	**VALVES**, water distribution									
0011	See Sections 22 05 23.20 and 22 05 23.60									
3000	Butterfly valves with boxes, cast iron, mech. jt.									
3100	4" diameter	B-6	6	4	Ea.	925	150	53.50	1,128.50	1,300
3140	6" diameter		6	4		975	150	53.50	1,178.50	1,350
3180	8" diameter		6	4		1,250	150	53.50	1,453.50	1,650
3300	10" diameter		6	4		1,825	150	53.50	2,028.50	2,275
3340	12" diameter		6	4		2,550	150	53.50	2,753.50	3,075
3400	14" diameter		4	6		4,325	225	80	4,630	5,175
3440	16" diameter		4	6		5,025	225	80	5,330	5,950
3460	18" diameter		4	6		6,725	225	80	7,030	7,825
3480	20" diameter		4	6		7,750	225	80	8,055	8,950
3500	24" diameter		4	6		10,300	225	80	10,605	11,700
3600	With lever operator									
3610	4" diameter	B-6	6	4	Ea.	795	150	53.50	998.50	1,150
3614	6" diameter		6	4		845	150	53.50	1,048.50	1,225
3616	8" diameter		6	4		1,125	150	53.50	1,328.50	1,500
3618	10" diameter		6	4		1,675	150	53.50	1,878.50	2,125
3620	12" diameter		6	4		2,425	150	53.50	2,628.50	2,925
3622	14" diameter		4	6		4,075	225	80	4,380	4,900
3624	16" diameter		4	6		4,775	225	80	5,080	5,675
3626	18" diameter		4	6		6,475	225	80	6,780	7,550
3628	20" diameter		4	6		7,500	225	80	7,805	8,675
3630	24" diameter		4	6		10,100	225	80	10,405	11,500
3700	Check valves, flanged									
3710	4" diameter	B-6	6	4	Ea.	535	150	53.50	738.50	875
3714	6" diameter		6	4		800	150	53.50	1,003.50	1,175
3716	8" diameter		6	4		1,175	150	53.50	1,378.50	1,550
3718	10" diameter		6	4		2,100	150	53.50	2,303.50	2,575
3720	12" diameter		6	4		3,300	150	53.50	3,503.50	3,900
3722	14" diameter		4	6		4,650	225	80	4,955	5,525
3724	16" diameter		4	6		3,350	225	80	3,655	4,125
3726	18" diameter		4	6		5,500	225	80	5,805	6,475
3728	20" diameter		4	6		10,600	225	80	10,905	12,100
3730	24" diameter		4	6		16,100	225	80	16,405	18,100
3800	Gate valves, C.I., 250 PSI, mechanical joint, w/boxes									

33 12 Water Utility Distribution Equipment

33 12 16 – Water Utility Distribution Valves

33 12 16.10 Valves		Crew	Daily Output	Labor-Hours	Unit	Material	2011 Bare Costs		Total	Total Incl O&P
							Labor	Equipment		
3810	4" diameter	B-6	6	4	Ea.	455	150	53.50	658.50	785
3814	6" diameter		6	4		685	150	53.50	888.50	1,050
3816	8" diameter		6	4		905	150	53.50	1,108.50	1,275
3818	10" diameter		6	4		1,425	150	53.50	1,628.50	1,850
3820	12" diameter		6	4		1,650	150	53.50	1,853.50	2,100
3822	14" diameter		4	6		4,750	225	80	5,055	5,650
3824	16" diameter		4	6		5,050	225	80	5,355	5,975
3826	18" diameter		4	6		8,100	225	80	8,405	9,325
3828	20" diameter		4	6		10,900	225	80	11,205	12,400
3830	24" diameter		4	6		16,400	225	80	16,705	18,400
3831	30" diameter		4	6		40,500	225	80	40,805	45,000
3832	36" diameter	▼	4	6	▼	61,000	225	80	61,305	67,500

33 12 16.20 Valves

33 12 16.20 Valves		Crew	Daily Output	Labor-Hours	Unit	Material	2011 Bare Costs		Total	Total Incl O&P
							Labor	Equipment		
0010	**VALVES**									
0011	Special trim or use									
9000	Valves, gate valve, N.R.S. PIV with post, 4" diameter	B-6	6	4	Ea.	1,300	150	53.50	1,503.50	1,725
9020	6" diameter		6	4		1,550	150	53.50	1,753.50	1,975
9040	8" diameter		6	4		1,750	150	53.50	1,953.50	2,200
9060	10" diameter		6	4		2,275	150	53.50	2,478.50	2,775
9080	12" diameter		6	4		2,500	150	53.50	2,703.50	3,025
9100	14" diameter		6	4		5,475	150	53.50	5,678.50	6,300
9120	O.S.&Y., 4" diameter		6	4		580	150	53.50	783.50	925
9140	6" diameter		6	4		775	150	53.50	978.50	1,150
9160	8" diameter		6	4		1,200	150	53.50	1,403.50	1,600
9180	10" diameter		6	4		1,925	150	53.50	2,128.50	2,400
9200	12" diameter		6	4		2,325	150	53.50	2,528.50	2,825
9220	14" diameter		4	6		3,400	225	80	3,705	4,150
9400	Check valves, rubber disc, 2-1/2" diameter		6	4		435	150	53.50	638.50	765
9420	3" diameter		6	4		435	150	53.50	638.50	765
9440	4" diameter		6	4		535	150	53.50	738.50	875
9480	6" diameter		6	4		800	150	53.50	1,003.50	1,175
9500	8" diameter		6	4		1,175	150	53.50	1,378.50	1,550
9520	10" diameter		6	4		2,100	150	53.50	2,303.50	2,575
9540	12" diameter		6	4		3,300	150	53.50	3,503.50	3,900
9542	14" diameter		4	6		4,650	225	80	4,955	5,525
9700	Detector check valves, reducing, 4" diameter		6	4		1,375	150	53.50	1,578.50	1,775
9720	6" diameter		6	4		1,825	150	53.50	2,028.50	2,275
9740	8" diameter		6	4		2,650	150	53.50	2,853.50	3,200
9760	10" diameter		6	4		4,725	150	53.50	4,928.50	5,475
9800	Galvanized, 4" diameter		6	4		1,925	150	53.50	2,128.50	2,400
9820	6" diameter		6	4		2,625	150	53.50	2,828.50	3,175
9840	8" diameter		6	4		4,125	150	53.50	4,328.50	4,800
9860	10" diameter	▼	6	4	▼	4,475	150	53.50	4,678.50	5,200

33 16 Water Utility Storage Tanks

33 16 13 – Aboveground Water Utility Storage Tanks

33 16 13.13 Steel Water Storage Tanks

		Crew	Daily Output	Labor-Hours	Unit	Material	2011 Bare Costs Labor	Equipment	Total	Total Incl O&P
0010	**STEEL WATER STORAGE TANKS**									
0910	Steel, ground level, ht/diam. less than 1, not incl. fdn., 100,000 gallons				Ea.				178,000	196,000
1000	250,000 gallons								237,000	260,000
1200	500,000 gallons								335,000	368,000
1250	750,000 gallons								432,000	475,000
1300	1,000,000 gallons								530,000	583,000
1500	2,000,000 gallons								921,000	1,013,000
1600	4,000,000 gallons								1,703,000	1,873,000
1800	6,000,000 gallons								2,485,000	2,733,000
1850	8,000,000 gallons								3,266,000	3,593,000
1910	10,000,000 gallons				▼				4,055,000	4,460,000
2100	Steel standpipes, hgt/diam. more than 1, 100' to overflow, no fdn									
2200	500,000 gallons				Ea.				439,000	482,000
2400	750,000 gallons								580,000	638,000
2500	1,000,000 gallons								568,000	625,000
2700	1,500,000 gallons								1,404,000	1,544,000
2800	2,000,000 gallons				▼				1,049,000	1,154,000

33 16 13.16 Prestressed Conc. Aboveground Water Utility Storage Tanks

		Crew	Daily Output	Labor-Hours	Unit	Material	2011 Bare Costs Labor	Equipment	Total	Total Incl O&P
0010	**PRESTRESSED CONC. ABOVEGROUND WATER UTILITY STORAGE TANKS**									
0020	Not including fdn., pipe or pumps, 250,000 gallons				Ea.				270,000	297,000
0100	500,000 gallons								441,000	485,000
0300	1,000,000 gallons								639,000	703,000
0400	2,000,000 gallons								970,000	1,067,000
0600	4,000,000 gallons								1,543,000	1,697,000
0700	6,000,000 gallons								2,050,000	2,255,000
0750	8,000,000 gallons								2,645,000	2,909,000
0800	10,000,000 gallons				▼				3,196,000	3,519,000

33 16 13.23 Plastic-Coated Fabric Pillow Water Tanks

		Crew	Daily Output	Labor-Hours	Unit	Material	2011 Bare Costs Labor	Equipment	Total	Total Incl O&P
0010	**PLASTIC-COATED FABRIC PILLOW WATER TANKS**									
7000	Water tanks, vinyl coated fabric pillow tanks, freestanding, 5,000 gallons	4 Clab	4	8	Ea.	9,500	275		9,775	10,900
7100	Supporting embankment not included, 25,000 gallons	6 Clab	2	24		14,900	825		15,725	17,700
7200	50,000 gallons	8 Clab	1.50	42.667		27,600	1,475		29,075	32,700
7300	100,000 gallons	9 Clab	.90	80		60,000	2,750		62,750	70,000
7400	150,000 gallons		.50	144		82,000	4,950		86,950	97,500
7500	200,000 gallons		.40	180		100,500	6,175		106,675	120,000
7600	250,000 gallons	▼	.30	240	▼	143,000	8,250		151,250	170,000

33 16 19 – Elevated Water Utility Storage Tanks

33 16 19.50 Elevated Water Storage Tanks

		Crew	Daily Output	Labor-Hours	Unit	Material	2011 Bare Costs Labor	Equipment	Total	Total Incl O&P
0010	**ELEVATED WATER STORAGE TANKS**									
0011	Not incl. pipe, pumps or foundation									
3000	Elevated water tanks, 100' to bottom capacity line, incl. painting									
3010	50,000 gallons				Ea.				182,000	200,000
3300	100,000 gallons								273,000	300,000
3400	250,000 gallons								767,000	844,000
3600	500,000 gallons								1,364,000	1,500,000
3700	750,000 gallons								1,655,000	1,820,000
3900	1,000,000 gallons				▼				2,364,000	2,600,000

33 21 Water Supply Wells

33 21 13 – Public Water Supply Wells

33 21 13.10 Wells and Accessories	Crew	Daily Output	Labor-Hours	Unit	Material	2011 Bare Costs Labor	Equipment	Total	Total Incl O&P
0010 **WELLS & ACCESSORIES**									
0011 Domestic									
0100 Drilled, 4" to 6" diameter	B-23	120	.333	L.F.		11.60	22.50	34.10	43
0200 8" diameter	"	95.20	.420	"		14.60	28.50	43.10	53.50
0400 Gravel pack well, 40' deep, incl. gravel & casing, complete									
0500 24" diameter casing x 18" diameter screen	B-23	.13	307	Total	33,800	10,700	20,800	65,300	76,500
0600 36" diameter casing x 18" diameter screen		.12	333	"	35,900	11,600	22,500	70,000	82,000
0800 Observation wells, 1-1/4" riser pipe	↓	163	.245	V.L.F.	17.70	8.55	16.60	42.85	51
0900 For flush Buffalo roadway box, add	1 Skwk	16.60	.482	Ea.	48.50	21.50		70	86
1200 Test well, 2-1/2" diameter, up to 50' deep (15 to 50 GPM)	B-23	1.51	26.490	"	725	920	1,800	3,445	4,175
1300 Over 50' deep, add	"	121.80	.328	L.F.	19.25	11.40	22	52.65	63
1500 Pumps, installed in wells to 100' deep, 4" submersible									
1510 1/2 H.P.	Q-1	3.22	4.969	Ea.	395	240		635	795
1520 3/4 H.P.		2.66	6.015		485	290		775	970
1600 1 H.P.	↓	2.29	6.987		570	335		905	1,125
1700 1-1/2 H.P.	Q-22	1.60	10		810	485	410	1,705	2,050
1800 2 H.P.		1.33	12.030		1,125	580	495	2,200	2,650
1900 3 H.P.		1.14	14.035		1,325	675	575	2,575	3,075
2000 5 H.P.		1.14	14.035		2,275	675	575	3,525	4,125
2050 Remove and install motor only, 4 H.P.		1.14	14.035		2,350	675	575	3,600	4,225
3000 Pump, 6" submersible, 25' to 150' deep, 25 H.P., 249 to 297 GPM		.89	17.978		6,850	865	735	8,450	9,650
3100 25' to 500' deep, 30 H.P., 100 to 300 GPM	↓	.73	21.918	↓	8,425	1,050	900	10,375	11,800
8110 Well screen assembly, stainless steel, 2" diameter	B-23A	273	.088	L.F.	85	3.40	9.50	97.90	109
8120 3" diameter		253	.095		119	3.67	10.25	132.92	148
8130 4" diameter		200	.120		146	4.64	12.95	163.59	182
8140 5" diameter		168	.143		175	5.55	15.40	195.95	218
8150 6" diameter		126	.190		205	7.35	20.50	232.85	260
8160 8" diameter		98.50	.244		273	9.40	26.50	308.90	345
8170 10" diameter		73	.329		340	12.70	35.50	388.20	435
8180 12" diameter		62.50	.384		390	14.85	41.50	446.35	500
8190 14" diameter		54.30	.442		440	17.10	47.50	504.60	565
8200 16" diameter		48.30	.497		490	19.20	53.50	562.70	630
8210 18" diameter		39.20	.612		610	23.50	66	699.50	780
8220 20" diameter		31.20	.769		680	30	83	793	885
8230 24" diameter		23.80	1.008		820	39	109	968	1,075
8240 26" diameter		21	1.143		890	44	123	1,057	1,175
8300 Slotted PVC, 1-1/4" diameter		521	.046		2.81	1.78	4.97	9.56	11.25
8310 1-1/2" diameter		488	.049		3.42	1.90	5.30	10.62	12.50
8320 2" diameter		273	.088		3.40	3.40	9.50	16.30	19.35
8330 3" diameter		253	.095		4.71	3.67	10.25	18.63	22
8340 4" diameter		200	.120		5.85	4.64	12.95	23.44	27.50
8350 5" diameter		168	.143		7.10	5.55	15.40	28.05	33
8360 6" diameter		126	.190		8.70	7.35	20.50	36.55	43.50
8370 8" diameter	↓	98.50	.244		13.20	9.40	26.50	49.10	58
8400 Artificial gravel pack, 2" screen, 6" casing	B-23B	174	.138		3.24	5.35	16.55	25.14	30
8405 8" casing		111	.216		4.16	8.35	26	38.51	46
8410 10" casing		74.50	.322		5.95	12.45	38.50	56.90	68
8415 12" casing		60	.400		8.30	15.45	48	71.75	85.50
8420 14" casing		50.20	.478		9.70	18.50	57.50	85.70	102
8425 16" casing		40.70	.590		12.25	23	71	106.25	126
8430 18" casing		36	.667		14.70	26	80	120.70	143
8435 20" casing		29.50	.814		17.30	31.50	97.50	146.30	174
8440 24" casing	↓	25.70	.934	↓	19.70	36	112	167.70	200

33 21 Water Supply Wells

33 21 13 – Public Water Supply Wells

33 21 13.10 Wells and Accessories

		Crew	Daily Output	Labor-Hours	Unit	Material	2011 Bare Costs Labor	Equipment	Total	Total Incl O&P
8445	26" casing	B-23B	24.60	.976	L.F.	22	37.50	117	176.50	211
8450	30" casing		20	1.200		25	46.50	144	215.50	257
8455	36" casing		16.40	1.463		28.50	56.50	176	261	310
8500	Develop well		8	3	Hr.	297	116	360	773	895
8550	Pump test well		8	3		79	116	360	555	660
8560	Standby well	B-23A	8	3		76.50	116	325	517.50	615
8570	Standby, drill rig		8	3			116	325	441	530
8580	Surface seal well, concrete filled		1	24	Ea.	785	930	2,600	4,315	5,125
8590	Well test pump, install & remove	B-23	1	40			1,400	2,700	4,100	5,100
8600	Well sterilization, chlorine	2 Clab	1	16		135	550		685	995
8610	Well water pressure switch	1 Clab	12	.667		64	23		87	105
9950	See Section 31 23 19.40 for wellpoints									
9960	See Section 31 23 19.30 for drainage wells									

33 21 13.20 Water Supply Wells, Pumps

		Crew	Daily Output	Labor-Hours	Unit	Material	2011 Bare Costs Labor	Equipment	Total	Total Incl O&P
0010	**WATER SUPPLY WELLS, PUMPS**									
0011	With pressure control									
1000	Deep well, jet, 42 gal. galvanized tank									
1040	3/4 HP	1 Plum	.80	10	Ea.	1,450	535		1,985	2,400
3000	Shallow well, jet, 30 gal. galvanized tank									
3040	1/2 HP	1 Plum	2	4	Ea.	1,200	214		1,414	1,650

33 31 Sanitary Utility Sewerage Piping

33 31 13 – Public Sanitary Utility Sewerage Piping

33 31 13.15 Sewage Collection, Concrete Pipe

		Crew	Daily Output	Labor-Hours	Unit	Material	2011 Bare Costs Labor	Equipment	Total	Total Incl O&P
0010	**SEWAGE COLLECTION, CONCRETE PIPE**									
0020	See Section 33 41 13.60 for sewage/drainage collection, concrete pipe									

33 31 13.25 Sewage Collection, Polyvinyl Chloride Pipe

		Crew	Daily Output	Labor-Hours	Unit	Material	2011 Bare Costs Labor	Equipment	Total	Total Incl O&P
0010	**SEWAGE COLLECTION, POLYVINYL CHLORIDE PIPE**									
0020	Not including excavation or backfill									
2000	20' lengths, S.D.R. 35, B&S, 4" diameter	B-20	375	.064	L.F.	1.85	2.46		4.31	5.80
2040	6" diameter		350	.069		3.68	2.63		6.31	8.10
2080	13' lengths , S.D.R. 35, B&S, 8" diameter		335	.072		7.60	2.75		10.35	12.55
2120	10" diameter	B-21	330	.085		11.30	3.35	.39	15.04	18
2160	12" diameter		320	.088		13.05	3.46	.40	16.91	20
2200	15" diameter		240	.117		14.35	4.61	.54	19.50	23.50
4000	Piping, DWV PVC, no exc/bkfill, 10' L, Sch 40, 4" diameter	B-20	375	.064		3.85	2.46		6.31	8
4010	6" diameter		350	.069		7.65	2.63		10.28	12.50
4020	8" diameter		335	.072		13.40	2.75		16.15	18.90

33 41 13 – Public Storm Utility Drainage Piping

33 41 13.40 Piping, Storm Drainage, Corrugated Metal	Crew	Daily Output	Labor-Hours	Unit	Material	2011 Bare Costs Labor	Equipment	Total	Total Incl O&P
0010 **PIPING, STORM DRAINAGE, CORRUGATED METAL**									
0020 Not including excavation or backfill									
2000 Corrugated metal pipe, galvanized									
2020 Bituminous coated with paved invert, 20' lengths									
2040 8" diameter, 16 ga.	B-14	330	.145	L.F.	10.70	5.25	.97	16.92	21
2060 10" diameter, 16 ga.		260	.185		11.15	6.70	1.23	19.08	24
2080 12" diameter, 16 ga.		210	.229		11.50	8.30	1.53	21.33	27
2100 15" diameter, 16 ga.		200	.240		18.75	8.70	1.60	29.05	35.50
2120 18" diameter, 16 ga.		190	.253		19.10	9.15	1.69	29.94	37
2140 24" diameter, 14 ga.		160	.300		22	10.85	2	34.85	43
2160 30" diameter, 14 ga.	B-13	120	.467		26	17.35	6.25	49.60	62
2180 36" diameter, 12 ga.		120	.467		31.50	17.35	6.25	55.10	68
2200 48" diameter, 12 ga.		100	.560		52.50	21	7.50	81	97.50
2220 60" diameter, 10 ga.	B-13B	75	.747		79	28	16.10	123.10	147
2240 72" diameter, 8 ga.	"	45	1.244		90.50	46.50	27	164	200
2500 Galvanized, uncoated, 20' lengths									
2520 8" diameter, 16 ga.	B-14	355	.135	L.F.	8.25	4.90	.90	14.05	17.55
2540 10" diameter, 16 ga.		280	.171		8.75	6.20	1.15	16.10	20.50
2560 12" diameter, 16 ga.		220	.218		9	7.90	1.46	18.36	23.50
2580 15" diameter, 16 ga.		220	.218		13	7.90	1.46	22.36	28
2600 18" diameter, 16 ga.		205	.234		14.75	8.50	1.56	24.81	31
2620 24" diameter, 14 ga.		175	.274		18.75	9.95	1.83	30.53	37.50
2640 30" diameter, 14 ga.	B-13	130	.431		23	16.05	5.75	44.80	56
2660 36" diameter, 12 ga.		130	.431		27.50	16.05	5.75	49.30	61.50
2680 48" diameter, 12 ga.		110	.509		45.50	18.95	6.80	71.25	86.50
2690 60" diameter, 10 ga.	B-13B	78	.718		69	26.50	15.50	111	134
2780 End sections, 8" diameter	B-14	24	2	Ea.	34	72.50	13.35	119.85	163
2785 10" diameter		22	2.182		44.50	79	14.55	138.05	186
2790 12" diameter		35	1.371		78	49.50	9.15	136.65	172
2800 18" diameter		30	1.600		108	58	10.70	176.70	218
2810 24" diameter	B-13	25	2.240		143	83.50	30	256.50	315
2820 30" diameter		25	2.240		246	83.50	30	359.50	430
2825 36" diameter		20	2.800		365	104	37.50	506.50	605
2830 48" diameter		10	5.600		930	208	75	1,213	1,425
2835 60" diameter	B-13B	5	11.200		985	415	242	1,642	1,975
2840 72" diameter	"	4	14		1,575	520	300	2,395	2,850

33 41 13.60 Sewage/Drainage Collection, Concrete Pipe

33 41 13.60 Sewage/Drainage Collection, Concrete Pipe	Crew	Daily Output	Labor-Hours	Unit	Material	2011 Bare Costs Labor	Equipment	Total	Total Incl O&P
0010 **SEWAGE/DRAINAGE COLLECTION, CONCRETE PIPE**									
0020 Not including excavation or backfill									
1000 Non-reinforced pipe, extra strength, B&S or T&G joints									
1010 6" diameter	B-14	265.04	.181	L.F.	6.25	6.55	1.21	14.01	18.25
1020 8" diameter		224	.214		6.90	7.75	1.43	16.08	21
1030 10" diameter		216	.222		7.65	8.05	1.48	17.18	22.50
1040 12" diameter		200	.240		9.35	8.70	1.60	19.65	25.50
1050 15" diameter		180	.267		12.80	9.65	1.78	24.23	31
1060 18" diameter		144	.333		15.10	12.10	2.23	29.43	37.50
1070 21" diameter		112	.429		18.60	15.55	2.86	37.01	47
1080 24" diameter		100	.480		25	17.40	3.21	45.61	57.50
2000 Reinforced culvert, class 3, no gaskets									
2010 12" diameter	B-14	150	.320	L.F.	11.50	11.60	2.14	25.24	32.50
2020 15" diameter		150	.320		15.70	11.60	2.14	29.44	37.50
2030 18" diameter		132	.364		19.70	13.15	2.43	35.28	44

33 41 13 – Public Storm Utility Drainage Piping

33 41 13.60 Sewage/Drainage Collection, Concrete Pipe		Crew	Daily Output	Labor-Hours	Unit	Material	2011 Bare Costs Labor	Equipment	Total	Total Incl O&P
2035	21" diameter	B-14	120	.400	L.F.	23.50	14.50	2.67	40.67	51
2040	24" diameter	↓	100	.480		27.50	17.40	3.21	48.11	60.50
2045	27" diameter	B-13	92	.609		40	22.50	8.15	70.65	87.50
2050	30" diameter		88	.636		48	23.50	8.50	80	98.50
2060	36" diameter	↓	72	.778		58	29	10.40	97.40	119
2070	42" diameter	B-13B	72	.778		76.50	29	16.80	122.30	147
2080	48" diameter		64	.875		95.50	32.50	18.85	146.85	176
2090	60" diameter		48	1.167		123	43.50	25	191.50	229
2100	72" diameter		40	1.400		200	52	30	282	330
2120	84" diameter		32	1.750		288	65	38	391	455
2140	96" diameter	↓	24	2.333		345	87	50.50	482.50	570
2200	With gaskets, class 3, 12" diameter	B-21	168	.167		12.65	6.60	.77	20.02	25
2220	15" diameter		160	.175		17.25	6.90	.81	24.96	30.50
2230	18" diameter		152	.184		21.50	7.30	.85	29.65	36
2240	24" diameter	↓	136	.206		33.50	8.15	.95	42.60	50.50
2260	30" diameter	B-13	88	.636		55	23.50	8.50	87	106
2270	36" diameter	"	72	.778		66.50	29	10.40	105.90	128
2290	48" diameter	B-13B	64	.875		107	32.50	18.85	158.35	189
2310	72" diameter	"	40	1.400		217	52	30	299	350
2330	Flared ends, 6'-1" long, 12" diameter	B-21	190	.147		36.50	5.85	.68	43.03	50
2340	15" diameter		155	.181		40	7.15	.83	47.98	55.50
2400	6'-2" long, 18" diameter		122	.230		51	9.05	1.06	61.11	71.50
2420	24" diameter	↓	88	.318		55	12.60	1.46	69.06	81.50
2440	36" diameter	B-13	60	.933		127	34.50	12.50	174	207
3080	Radius pipe, add to pipe prices, 12" to 60" diameter					50%				
3090	Over 60" diameter, add				↓	20%				
3500	Reinforced elliptical, 8' lengths, C507 class 3									
3520	14" x 23" inside, round equivalent 18" diameter	B-21	82	.341	L.F.	54.50	13.50	1.57	69.57	82
3530	24" x 38" inside, round equivalent 30" diameter	B-13	58	.966		81	36	12.90	129.90	158
3540	29" x 45" inside, round equivalent 36" diameter		52	1.077		97	40	14.40	151.40	184
3550	38" x 60" inside, round equivalent 48" diameter		38	1.474		161	55	19.70	235.70	282
3560	48" x 76" inside, round equivalent 60" diameter		26	2.154		204	80	29	313	380
3570	58" x 91" inside, round equivalent 72" diameter	↓	22	2.545	↓	305	94.50	34	433.50	520
3780	Concrete slotted pipe, class 4 mortar joint									
3800	12" diameter	B-21	168	.167	L.F.	35	6.60	.77	42.37	49.50
3840	18" diameter	"	152	.184	"	40	7.30	.85	48.15	56
3900	Concrete slotted pipe, Class 4 O-ring joint									
3940	12" diameter	B-21	168	.167	L.F.	35	6.60	.77	42.37	49.50
3960	18" diameter	"	152	.184	"	40	7.30	.85	48.15	56

33 42 Culverts

33 42 16 – Concrete Culverts

33 42 16.15 Oval Arch Culverts

		Crew	Daily Output	Labor-Hours	Unit	Material	2011 Bare Costs Labor	Equipment	Total	Total Incl O&P
0010	**OVAL ARCH CULVERTS**									
3000	Corrugated galvanized or aluminum, coated & paved									
3020	17" x 13", 16 ga., 15" equivalent	B-14	200	.240	L.F.	17.45	8.70	1.60	27.75	34
3040	21" x 15", 16 ga., 18" equivalent		150	.320		24.50	11.60	2.14	38.24	47
3060	28" x 20", 14 ga., 24" equivalent		125	.384		28	13.90	2.57	44.47	55
3080	35" x 24", 14 ga., 30" equivalent	↓	100	.480		37	17.40	3.21	57.61	70.50
3100	42" x 29", 12 ga., 36" equivalent	B-13	100	.560		46	21	7.50	74.50	91
3120	49" x 33", 12 ga., 42" equivalent	↓	90	.622	↓	62.50	23	8.30	93.80	113

33 42 Culverts

33 42 16 - Concrete Culverts

33 42 16.15 Oval Arch Culverts

		Crew	Daily Output	Labor-Hours	Unit	Material	2011 Bare Costs Labor	Equipment	Total	Total Incl O&P
3140	57" x 38", 12 ga., 48" equivalent	B-13	75	.747	L.F.	75.50	28	10	113.50	137
3160	Steel, plain oval arch culverts, plain									
3180	17" x 13", 16 ga., 15" equivalent	B-14	225	.213	L.F.	10.25	7.75	1.43	19.43	24.50
3200	21" x 15", 16 ga., 18" equivalent		175	.274		12.40	9.95	1.83	24.18	31
3220	28" x 20", 14 ga., 24" equivalent	↓	150	.320		16.60	11.60	2.14	30.34	38.50
3240	35" x 24", 14 ga., 30" equivalent	B-13	108	.519		21.50	19.30	6.95	47.75	61
3260	42" x 29", 12 ga., 36" equivalent		108	.519		27	19.30	6.95	53.25	67
3280	49" x 33", 12 ga., 42" equivalent		92	.609		31	22.50	8.15	61.65	77.50
3300	57" x 38", 12 ga., 48" equivalent		75	.747	↓	47.50	28	10	85.50	106
3320	End sections, 17" x 13"		22	2.545	Ea.	105	94.50	34	233.50	298
3340	42" x 29"	↓	17	3.294	"	292	123	44	459	555
3360	Multi-plate arch, steel	B-20	1690	.014	Lb.	1.20	.54		1.74	2.16

33 44 Storm Utility Water Drains

33 44 13 - Utility Area Drains

33 44 13.13 Catchbasins

		Crew	Daily Output	Labor-Hours	Unit	Material	2011 Bare Costs Labor	Equipment	Total	Total Incl O&P
0010	**CATCHBASINS**									
0011	Not including footing & excavation									
1600	Frames & covers, C.I., 24" square, 500 lb.	B-6	7.80	3.077	Ea.	345	115	41	501	600
1700	26" D shape, 600 lb.		7	3.429		510	128	46	684	805
1800	Light traffic, 18" diameter, 100 lb.		10	2.400		154	90	32	276	340
1900	24" diameter, 300 lb.		8.70	2.759		236	103	37	376	460
2000	36" diameter, 900 lb.		5.80	4.138		510	155	55	720	855
2100	Heavy traffic, 24" diameter, 400 lb.		7.80	3.077		220	115	41	376	460
2200	36" diameter, 1150 lb.		3	8		830	299	107	1,236	1,500
2300	Mass. State standard, 26" diameter, 475 lb.		7	3.429		250	128	46	424	520
2400	30" diameter, 620 lb.		7	3.429		340	128	46	514	620
2500	Watertight, 24" diameter, 350 lb.		7.80	3.077		235	115	41	391	480
2600	26" diameter, 500 lb.		7	3.429		395	128	46	569	680
2700	32" diameter, 575 lb.	↓	6	4	↓	800	150	53.50	1,003.50	1,175
2800	3 piece cover & frame, 10" deep,									
2900	1200 lb., for heavy equipment	B-6	3	8	Ea.	1,100	299	107	1,506	1,800
3000	Raised for paving 1-1/4" to 2" high									
3100	4 piece expansion ring									
3200	20" to 26" diameter	1 Clab	3	2.667	Ea.	143	91.50		234.50	298
3300	30" to 36" diameter	"	3	2.667	"	199	91.50		290.50	360
3320	Frames and covers, existing, raised for paving, 2", including									
3340	row of brick, concrete collar, up to 12" wide frame	B-6	18	1.333	Ea.	45	50	17.80	112.80	145
3360	20" to 26" wide frame		11	2.182		67.50	81.50	29	178	230
3380	30" to 36" wide frame	↓	9	2.667		83.50	100	35.50	219	283
3400	Inverts, single channel brick	D-1	3	5.333		97	208		305	415
3500	Concrete		5	3.200		75.50	125		200.50	270
3600	Triple channel, brick		2	8		148	310		458	635
3700	Concrete	↓	3	5.333	↓	128	208		336	450

33 46 Subdrainage

33 46 16 – Subdrainage Piping

33 46 16.25 Piping, Subdrainage, Corrugated Metal	Crew	Daily Output	Labor-Hours	Unit	Material	2011 Bare Costs Labor	Equipment	Total	Total Incl O&P
0010 **PIPING, SUBDRAINAGE, CORRUGATED METAL**									
0021 Not including excavation and backfill									
2010 Aluminum, perforated									
2020 6" diameter, 18 ga.	B-14	380	.126	L.F.	7.55	4.58	.84	12.97	16.25
2200 8" diameter, 16 ga.		370	.130		10.90	4.70	.87	16.47	20
2220 10" diameter, 16 ga.		360	.133		13.65	4.83	.89	19.37	23.50
2240 12" diameter, 16 ga.		285	.168		15.30	6.10	1.13	22.53	27.50
2260 18" diameter, 16 ga.		205	.234		23	8.50	1.56	33.06	39.50
3000 Uncoated galvanized, perforated									
3020 6" diameter, 18 ga.	B-20	380	.063	L.F.	7.55	2.42		9.97	12
3200 8" diameter, 16 ga.	"	370	.065		10.90	2.49		13.39	15.80
3220 10" diameter, 16 ga.	B-21	360	.078		16.40	3.07	.36	19.83	23
3240 12" diameter, 16 ga.		285	.098		17.15	3.88	.45	21.48	25.50
3260 18" diameter, 16 ga.		205	.137		26	5.40	.63	32.03	38
4000 Steel, perforated, asphalt coated									
4020 6" diameter 18 ga.	B-20	380	.063	L.F.	7.55	2.42		9.97	12
4030 8" diameter 18 ga.	"	370	.065		10.90	2.49		13.39	15.80
4040 10" diameter 16 ga.	B-21	360	.078		13.25	3.07	.36	16.68	19.70
4050 12" diameter 16 ga.		285	.098		15.10	3.88	.45	19.43	23
4060 18" diameter 16 ga.		205	.137		21	5.40	.63	27.03	32

33 46 16.30 Piping, Subdrainage, Plastic	Crew	Daily Output	Labor-Hours	Unit	Material	2011 Bare Costs Labor	Equipment	Total	Total Incl O&P
0010 **PIPING, SUBDRAINAGE, PLASTIC**									
0020 Not including excavation and backfill									
2100 Perforated PVC, 4" diameter	B-14	314	.153	L.F.	1.85	5.55	1.02	8.42	11.60
2110 6" diameter		300	.160		3.68	5.80	1.07	10.55	14.05
2120 8" diameter		290	.166		6.90	6	1.11	14.01	17.95
2130 10" diameter		280	.171		9.05	6.20	1.15	16.40	20.50
2140 12" diameter		270	.178		13.05	6.45	1.19	20.69	25.50

33 49 Storm Drainage Structures

33 49 13 – Storm Drainage Manholes, Frames, and Covers

33 49 13.10 Storm Drainage Manholes, Frames and Covers	Crew	Daily Output	Labor-Hours	Unit	Material	2011 Bare Costs Labor	Equipment	Total	Total Incl O&P
0010 **STORM DRAINAGE MANHOLES, FRAMES & COVERS**									
0020 Excludes footing, excavation, backfill (See line items for frame & cover)									
0050 Brick, 4' inside diameter, 4' deep	D-1	1	16	Ea.	445	625		1,070	1,425
0100 6' deep		.70	22.857		625	890		1,515	2,000
0150 8' deep		.50	32		805	1,250		2,055	2,750
0200 For depths over 8', add		4	4	V.L.F.	155	156		311	405
0400 Concrete blocks (radial), 4' I.D., 4' deep		1.50	10.667	Ea.	370	415		785	1,025
0500 6' deep		1	16		495	625		1,120	1,475
0600 8' deep		.70	22.857		625	890		1,515	2,000
0700 For depths over 8', add		5.50	2.909	V.L.F.	65.50	113		178.50	242
0800 Concrete, cast in place, 4' x 4', 8" thick, 4' deep	C-14H	2	24	Ea.	430	1,025	11.50	1,466.50	2,075
0900 6' deep		1.50	32		615	1,350	15.35	1,980.35	2,775
1000 8' deep		1	48		895	2,050	23	2,968	4,125
1100 For depths over 8', add		8	6	V.L.F.	98	255	2.88	355.88	500
1110 Precast, 4' I.D., 4' deep	B-22	4.10	7.317	Ea.	690	293	47	1,030	1,250
1120 6' deep		3	10		795	400	64.50	1,259.50	1,550
1130 8' deep		2	15		990	600	97	1,687	2,125
1140 For depths over 8', add		16	1.875	V.L.F.	101	75	12.10	188.10	238
1150 5' I.D., 4' deep	B-6	3	8	Ea.	1,400	299	107	1,806	2,125

33 49 Storm Drainage Structures

33 49 13 – Storm Drainage Manholes, Frames, and Covers

33 49 13.10 Storm Drainage Manholes, Frames and Covers	Crew	Daily Output	Labor-Hours	Unit	Material	2011 Bare Costs Labor	Equipment	Total	Total Incl O&P
1160	6' deep	B-6	2	12	Ea.	1,700	450	160	2,310	2,725
1170	8' deep		1.50	16	↓	1,950	600	214	2,764	3,300
1180	For depths over 8', add		12	2	V.L.F.	180	75	26.50	281.50	340
1190	6' I.D., 4' deep		2	12	Ea.	2,000	450	160	2,610	3,050
1200	6' deep		1.50	16		2,700	600	214	3,514	4,125
1210	8' deep		1	24	↓	3,250	900	320	4,470	5,325
1220	For depths over 8', add	↓	8	3	V.L.F.	252	112	40	404	490
1250	Slab tops, precast, 8" thick									
1300	4' diameter manhole	B-6	8	3	Ea.	234	112	40	386	470
1400	5' diameter manhole		7.50	3.200		400	120	42.50	562.50	670
1500	6' diameter manhole	↓	7	3.429		615	128	46	789	920
3800	Steps, heavyweight cast iron, 7" x 9"	1 Bric	40	.200		16.70	8.65		25.35	31.50
3900	8" x 9"		40	.200		25	8.65		33.65	40.50
3928	12" x 10-1/2"		40	.200		23.50	8.65		32.15	38.50
4000	Standard sizes, galvanized steel		40	.200		22	8.65		30.65	37.50
4100	Aluminum	↓	40	.200	↓	22	8.65		30.65	37.50

33 51 Natural-Gas Distribution

33 51 13 – Natural-Gas Piping

33 51 13.10 Piping, Gas Service and Distribution, P.E.

		Crew	Daily Output	Labor-Hours	Unit	Material	2011 Bare Costs Labor	Equipment	Total	Total Incl O&P
0010	**PIPING, GAS SERVICE AND DISTRIBUTION, POLYETHYLENE**									
0020	Not including excavation or backfill									
1000	60 psi coils, compression coupling @ 100', 1/2" diameter, SDR 11	B-20A	608	.053	L.F.	.75	2.20		2.95	4.15
1010	1" diameter, SDR 11		544	.059		1.95	2.46		4.41	5.85
1040	1-1/4" diameter, SDR 11		544	.059		2.39	2.46		4.85	6.35
1100	2" diameter, SDR 11		488	.066		3.26	2.74		6	7.70
1160	3" diameter, SDR 11	↓	408	.078		4.31	3.28		7.59	9.70
1500	60 PSI 40' joints with coupling, 3" diameter, SDR 11	B-21A	408	.098		5.35	4.19	1.06	10.60	13.35
1540	4" diameter, SDR 11		352	.114		7.75	4.86	1.23	13.84	17.15
1600	6" diameter, SDR 11		328	.122		27	5.20	1.32	33.52	39
1640	8" diameter, SDR 11	↓	272	.147	↓	32.50	6.30	1.60	40.40	46.50

33 52 Liquid Fuel Distribution

33 52 16 – Gasoline Distribution

33 52 16.13 Gasoline Piping

		Crew	Daily Output	Labor-Hours	Unit	Material	2011 Bare Costs Labor	Equipment	Total	Total Incl O&P
0010	**GASOLINE PIPING**									
0020	Primary containment pipe, fiberglass-reinforced									
0030	Plastic pipe 15' & 30' lengths									
0040	2" diameter	Q-6	425	.056	L.F.	6.25	2.85		9.10	11.10
0050	3" diameter		400	.060		6.65	3.03		9.68	11.85
0060	4" diameter	↓	375	.064	↓	8.60	3.23		11.83	14.25
0100	Fittings									
0110	Elbows, 90° & 45°, bell-ends, 2"	Q-6	24	1	Ea.	51.50	50.50		102	132
0120	3" diameter		22	1.091		54	55		109	142
0130	4" diameter		20	1.200		69	60.50		129.50	166
0200	Tees, bell ends, 2"		21	1.143		51	57.50		108.50	143
0210	3" diameter		18	1.333		61	67.50		128.50	168
0220	4" diameter		15	1.600		153	81		234	289
0230	Flanges bell ends, 2"	↓	24	1	↓	36	50.50		86.50	115

33 52 Liquid Fuel Distribution

33 52 16 – Gasoline Distribution

33 52 16.13 Gasoline Piping

		Crew	Daily Output	Labor-Hours	Unit	Material	2011 Bare Costs Labor	2011 Bare Costs Equipment	Total	Total Incl O&P
0240	3" diameter	Q-6	22	1.091	Ea.	47	55		102	134
0250	4" diameter		20	1.200		57.50	60.50		118	154
0260	Sleeve couplings, 2"		21	1.143		21.50	57.50		79	110
0270	3" diameter		18	1.333		25	67.50		92.50	128
0280	4" diameter		15	1.600		29.50	81		110.50	154
0290	Threaded adapters 2"		21	1.143		18	57.50		75.50	106
0300	3" diameter		18	1.333		31.50	67.50		99	136
0310	4" diameter		15	1.600		42.50	81		123.50	168
0320	Reducers, 2"		27	.889		38.50	45		83.50	110
0330	3" diameter		22	1.091		50.50	55		105.50	138
0340	4" diameter	▼	20	1.200	▼	68.50	60.50		129	166
1010	Gas station product line for secondary containment (double wall)									
1100	Fiberglass reinforced plastic pipe 25' lengths									
1120	Pipe, plain end, 3" diameter	Q-6	375	.064	L.F.	13.40	3.23		16.63	19.50
1130	4" diameter		350	.069		15.30	3.46		18.76	22
1140	5" diameter		325	.074		20	3.73		23.73	27.50
1150	6" diameter	▼	300	.080	▼	20.50	4.04		24.54	29
1200	Fittings									
1230	Elbows, 90° & 45°, 3" diameter	Q-6	18	1.333	Ea.	117	67.50		184.50	229
1240	4" diameter		16	1.500		123	76		199	248
1250	5" diameter		14	1.714		282	86.50		368.50	440
1260	6" diameter		12	2		285	101		386	465
1270	Tees, 3" diameter		15	1.600		121	81		202	254
1280	4" diameter		12	2		142	101		243	305
1290	5" diameter		9	2.667		290	135		425	520
1300	6" diameter		6	4		305	202		507	635
1310	Couplings, 3" diameter		18	1.333		95	67.50		162.50	206
1320	4" diameter		16	1.500		125	76		201	251
1330	5" diameter		14	1.714		260	86.50		346.50	415
1340	6" diameter		12	2		270	101		371	450
1350	Cross-over nipples, 3" diameter		18	1.333		16.50	67.50		84	119
1360	4" diameter		16	1.500		19.50	76		95.50	135
1370	5" diameter		14	1.714		25	86.50		111.50	157
1380	6" diameter		12	2		30.50	101		131.50	185
1400	Telescoping, reducers, concentric 4" x 3"		18	1.333		168	67.50		235.50	286
1410	5" x 4"		17	1.412		168	71.50		239.50	291
1420	6" x 5"	▼	16	1.500	▼	168	76		244	298

33 61 Hydronic Energy Distribution

33 61 13 – Underground Hydronic Energy Distribution

33 61 13.20 Pipe Conduit, Prefabricated/Preinsulated

		Crew	Daily Output	Labor-Hours	Unit	Material	2011 Bare Costs Labor	2011 Bare Costs Equipment	Total	Total Incl O&P
0010	**PIPE CONDUIT, PREFABRICATED/PREINSULATED**									
0020	Does not include trenching, fittings or crane.									
0300	For cathodic protection, add 12 to 14%									
0310	of total built-up price (casing plus service pipe)									
0580	Polyurethane insulated system, 250°F. max. temp.									
0620	Black steel service pipe, standard wt., 1/2" insulation									
0660	3/4" diam. pipe size	Q-17	54	.296	L.F.	47.50	14.45	.60	62.55	74
0670	1" diam. pipe size		50	.320		52	15.60	.65	68.25	81
0680	1-1/4" diam. pipe size		47	.340		58	16.60	.69	75.29	90
0690	1-1/2" diam. pipe size	▼	45	.356	▼	63	17.30	.72	81.02	96

33 61 Hydronic Energy Distribution

33 61 13 – Underground Hydronic Energy Distribution

33 61 13.20 Pipe Conduit, Prefabricated/Preinsulated	Crew	Daily Output	Labor-Hours	Unit	Material	2011 Bare Costs Labor	Equipment	Total	Total Incl O&P
0700	2" diam. pipe size	Q-17	42	.381	L.F.	65.50	18.55	.77	84.82	100
0710	2-1/2" diam. pipe size		34	.471		66.50	23	.96	90.46	108
0720	3" diam. pipe size		28	.571		77	28	1.16	106.16	127
0730	4" diam. pipe size		22	.727		97	35.50	1.48	133.98	162
0740	5" diam. pipe size		18	.889		124	43.50	1.80	169.30	202
0750	6" diam. pipe size	Q-18	23	1.043		144	52.50	1.41	197.91	238
0760	8" diam. pipe size		19	1.263		210	64	1.71	275.71	330
0770	10" diam. pipe size		16	1.500		277	76	2.03	355.03	420
0780	12" diam. pipe size		13	1.846		345	93	2.49	440.49	520
0790	14" diam. pipe size		11	2.182		385	110	2.95	497.95	585
0800	16" diam. pipe size		10	2.400		440	121	3.24	564.24	670
0810	18" diam. pipe size		8	3		510	152	4.05	666.05	790
0820	20" diam. pipe size		7	3.429		565	173	4.63	742.63	885
0830	24" diam. pipe size		6	4		690	202	5.40	897.40	1,075
0900	For 1" thick insulation, add					10%				
0940	For 1-1/2" thick insulation, add					13%				
0980	For 2" thick insulation, add					20%				
1500	Gland seal for system, 3/4" diam. pipe size	Q-17	32	.500	Ea.	640	24.50	1.02	665.52	745
1510	1" diam. pipe size		32	.500		640	24.50	1.02	665.52	745
1540	1-1/4" diam. pipe size		30	.533		685	26	1.08	712.08	795
1550	1-1/2" diam. pipe size		30	.533		685	26	1.08	712.08	795
1560	2" diam. pipe size		28	.571		815	28	1.16	844.16	940
1570	2-1/2" diam. pipe size		26	.615		875	30	1.25	906.25	1,000
1580	3" diam. pipe size		24	.667		940	32.50	1.35	973.85	1,075
1590	4" diam. pipe size		22	.727		1,125	35.50	1.48	1,161.98	1,275
1600	5" diam. pipe size		19	.842		1,375	41	1.71	1,417.71	1,575
1610	6" diam. pipe size	Q-18	26	.923		1,450	46.50	1.25	1,497.75	1,675
1620	8" diam. pipe size		25	.960		1,700	48.50	1.30	1,749.80	1,925
1630	10" diam. pipe size		23	1.043		2,050	52.50	1.41	2,103.91	2,325
1640	12" diam. pipe size		21	1.143		2,275	57.50	1.54	2,334.04	2,600
1650	14" diam. pipe size		19	1.263		2,550	64	1.71	2,615.71	2,900
1660	16" diam. pipe size		18	1.333		2,975	67.50	1.80	3,044.30	3,375
1670	18" diam. pipe size		16	1.500		3,175	76	2.03	3,253.03	3,625
1680	20" diam. pipe size		14	1.714		3,625	86.50	2.31	3,713.81	4,100
1690	24" diam. pipe size		12	2		4,025	101	2.70	4,128.70	4,575
2000	Elbow, 45° for system									
2020	3/4" diam. pipe size	Q-17	14	1.143	Ea.	405	55.50	2.32	462.82	530
2040	1" diam. pipe size		13	1.231		415	60	2.50	477.50	545
2050	1-1/4" diam. pipe size		11	1.455		470	71	2.95	543.95	625
2060	1-1/2" diam. pipe size		9	1.778		490	86.50	3.61	580.11	675
2070	2" diam. pipe size		6	2.667		515	130	5.40	650.40	765
2080	2-1/2" diam. pipe size		4	4		555	195	8.10	758.10	910
2090	3" diam. pipe size		3.50	4.571		645	223	9.30	877.30	1,050
2100	4" diam. pipe size		3	5.333		750	260	10.85	1,020.85	1,225
2110	5" diam. pipe size		2.80	5.714		965	278	11.60	1,254.60	1,475
2120	6" diam. pipe size	Q-18	4	6		1,100	305	8.10	1,413.10	1,650
2130	8" diam. pipe size		3	8		1,575	405	10.80	1,990.80	2,375
2140	10" diam. pipe size		2.40	10		2,025	505	13.50	2,543.50	3,000
2150	12" diam. pipe size		2	12		2,650	605	16.20	3,271.20	3,850
2160	14" diam. pipe size		1.80	13.333		3,300	675	18	3,993	4,650
2170	16" diam. pipe size		1.60	15		3,925	760	20.50	4,705.50	5,475
2180	18" diam. pipe size		1.30	18.462		4,925	930	25	5,880	6,850
2190	20" diam. pipe size		1	24		6,200	1,200	32.50	7,432.50	8,650

33 61 Hydronic Energy Distribution

33 61 13 – Underground Hydronic Energy Distribution

33 61 13.20 Pipe Conduit, Prefabricated/Preinsulated

		Crew	Daily Output	Labor-Hours	Unit	Material	2011 Bare Costs Labor	Equipment	Total	Total Incl O&P
2200	24" diam. pipe size	Q-18	.70	34.286	Ea.	7,800	1,725	46.50	9,571.50	11,200
2260	For elbow, 90°, add					25%				
2300	For tee, straight, add					85%	30%			
2340	For tee, reducing, add					170%	30%			
2380	For weldolet, straight, add					50%				
2400	Polyurethane insulation, 1"									
2410	FRP carrier and casing									
2420	4"	Q-5	18	.889	L.F.	51	43.50		94.50	121
2422	6"	Q-6	23	1.043		89	52.50		141.50	177
2424	8"		19	1.263		145	64		209	254
2426	10"		16	1.500		195	76		271	330
2428	12"		13	1.846		255	93		348	420
2430	FRP carrier and PVC casing									
2440	4"	Q-5	18	.889	L.F.	20	43.50		63.50	86.50
2444	8"	Q-6	19	1.263		42	64		106	142
2446	10"		16	1.500		60.50	76		136.50	180
2448	12"		13	1.846		77.50	93		170.50	225
2450	PVC carrier and casing									
2460	4"	Q-1	36	.444	L.F.	10	21.50		31.50	43
2462	6"	"	29	.552		13.95	26.50		40.45	55
2464	8"	Q-2	36	.667		19.45	33.50		52.95	71.50
2466	10"		32	.750		26.50	37.50		64	85.50
2468	12"		31	.774		31	38.50		69.50	92

33 63 Steam Energy Distribution

33 63 13 – Underground Steam and Condensate Distribution Piping

33 63 13.10 Calcium Silicate Insulated System

		Crew	Daily Output	Labor-Hours	Unit	Material	2011 Bare Costs Labor	Equipment	Total	Total Incl O&P
0010	**CALCIUM SILICATE INSULATED SYSTEM**									
0011	High temp. (1200 degrees F)									
2840	Steel casing with protective exterior coating									
2850	6-5/8" diameter	Q-18	52	.462	L.F.	91	23.50	.62	115.12	136
2860	8-5/8" diameter		50	.480		99.50	24	.65	124.15	146
2870	10-3/4" diameter		47	.511		116	26	.69	142.69	167
2880	12-3/4" diameter		44	.545		127	27.50	.74	155.24	181
2890	14" diameter		41	.585		143	29.50	.79	173.29	202
2900	16" diameter		39	.615		153	31	.83	184.83	215
2910	18" diameter		36	.667		169	33.50	.90	203.40	237
2920	20" diameter		34	.706		190	35.50	.95	226.45	263
2930	22" diameter		32	.750		264	38	1.01	303.01	350
2940	24" diameter		29	.828		300	42	1.12	343.12	395
2950	26" diameter		26	.923		345	46.50	1.25	392.75	450
2960	28" diameter		23	1.043		430	52.50	1.41	483.91	550
2970	30" diameter		21	1.143		455	57.50	1.54	514.04	590
2980	32" diameter		19	1.263		510	64	1.71	575.71	660
2990	34" diameter		18	1.333		520	67.50	1.80	589.30	675
3000	36" diameter		16	1.500		555	76	2.03	633.03	725
3040	For multi-pipe casings, add					10%				
3060	For oversize casings, add					2%				
3400	Steel casing gland seal, single pipe									
3420	6-5/8" diameter	Q-18	25	.960	Ea.	1,275	48.50	1.30	1,324.80	1,475
3440	8-5/8" diameter		23	1.043		1,475	52.50	1.41	1,528.91	1,700

33 63 13 – Underground Steam and Condensate Distribution Piping

33 63 13.10 Calcium Silicate Insulated System		Crew	Daily Output	Labor-Hours	Unit	Material	2011 Bare Costs Labor	Equipment	Total	Total Incl O&P
3450	10-3/4" diameter	Q-18	21	1.143	Ea.	1,675	57.50	1.54	1,734.04	1,925
3460	12-3/4" diameter		19	1.263		1,975	64	1.71	2,040.71	2,275
3470	14" diameter		17	1.412		2,175	71.50	1.91	2,248.41	2,475
3480	16" diameter		16	1.500		2,525	76	2.03	2,603.03	2,925
3490	18" diameter		15	1.600		2,825	81	2.16	2,908.16	3,225
3500	20" diameter		13	1.846		3,125	93	2.49	3,220.49	3,575
3510	22" diameter		12	2		3,500	101	2.70	3,603.70	4,025
3520	24" diameter		11	2.182		3,925	110	2.95	4,037.95	4,500
3530	26" diameter		10	2.400		4,500	121	3.24	4,624.24	5,100
3540	28" diameter		9.50	2.526		5,075	128	3.41	5,206.41	5,775
3550	30" diameter		9	2.667		5,175	135	3.60	5,313.60	5,900
3560	32" diameter		8.50	2.824		5,800	143	3.81	5,946.81	6,625
3570	34" diameter		8	3		6,325	152	4.05	6,481.05	7,200
3580	36" diameter	▼	7	3.429		6,725	173	4.63	6,902.63	7,675
3620	For multi-pipe casings, add				▼	5%				
4000	Steel casing anchors, single pipe									
4020	6-5/8" diameter	Q-18	8	3	Ea.	1,125	152	4.05	1,281.05	1,475
4040	8-5/8" diameter		7.50	3.200		1,200	162	4.32	1,366.32	1,550
4050	10-3/4" diameter		7	3.429		1,550	173	4.63	1,727.63	2,000
4060	12-3/4" diameter		6.50	3.692		1,675	186	4.98	1,865.98	2,100
4070	14" diameter		6	4		1,950	202	5.40	2,157.40	2,450
4080	16" diameter		5.50	4.364		2,275	220	5.90	2,500.90	2,825
4090	18" diameter		5	4.800		2,550	242	6.50	2,798.50	3,200
4100	20" diameter		4.50	5.333		2,825	269	7.20	3,101.20	3,500
4110	22" diameter		4	6		3,150	305	8.10	3,463.10	3,900
4120	24" diameter		3.50	6.857		3,425	345	9.25	3,779.25	4,300
4130	26" diameter		3	8		3,875	405	10.80	4,290.80	4,900
4140	28" diameter		2.50	9.600		4,250	485	12.95	4,747.95	5,425
4150	30" diameter		2	12		4,525	605	16.20	5,146.20	5,900
4160	32" diameter		1.50	16		5,425	810	21.50	6,256.50	7,175
4170	34" diameter		1	24		6,075	1,200	32.50	7,307.50	8,500
4180	36" diameter	▼	1	24		6,600	1,200	32.50	7,832.50	9,075
4220	For multi-pipe, add				▼	5%	20%			
4800	Steel casing elbow									
4820	6-5/8" diameter	Q-18	15	1.600	Ea.	1,550	81	2.16	1,633.16	1,850
4830	8-5/8" diameter		15	1.600		1,675	81	2.16	1,758.16	1,950
4850	10-3/4" diameter		14	1.714		2,000	86.50	2.31	2,088.81	2,325
4860	12-3/4" diameter		13	1.846		2,375	93	2.49	2,470.49	2,775
4870	14" diameter		12	2		2,525	101	2.70	2,628.70	2,950
4880	16" diameter		11	2.182		2,775	110	2.95	2,887.95	3,225
4890	18" diameter		10	2.400		3,175	121	3.24	3,299.24	3,650
4900	20" diameter		9	2.667		3,375	135	3.60	3,513.60	3,925
4910	22" diameter		8	3		3,650	152	4.05	3,806.05	4,225
4920	24" diameter		7	3.429		4,025	173	4.63	4,202.63	4,700
4930	26" diameter		6	4		4,375	202	5.40	4,582.40	5,125
4940	28" diameter		5	4.800		4,725	242	6.50	4,973.50	5,575
4950	30" diameter		4	6		4,750	305	8.10	5,063.10	5,675
4960	32" diameter		3	8		5,425	405	10.80	5,840.80	6,575
4970	34" diameter		2	12		5,950	605	16.20	6,571.20	7,450
4980	36" diameter	▼	2	12	▼	6,325	605	16.20	6,946.20	7,900
5500	Black steel service pipe, std. wt., 1" thick insulation									
5510	3/4" diameter pipe size	Q-17	54	.296	L.F.	39.50	14.45	.60	54.55	65.50
5540	1" diameter pipe size	▼	50	.320		41	15.60	.65	57.25	69.50

33 63 13 – Underground Steam and Condensate Distribution Piping

33 63 13.10 Calcium Silicate Insulated System	Crew	Daily Output	Labor-Hours	Unit	Material	2011 Bare Costs Labor	Equipment	Total	Total Incl O&P
5550	1-1/4" diameter pipe size	Q-17	47	.340	L.F.	46	16.60	.69	63.29	77
5560	1-1/2" diameter pipe size		45	.356		50.50	17.30	.72	68.52	82.50
5570	2" diameter pipe size		42	.381		56	18.55	.77	75.32	90
5580	2-1/2" diameter pipe size		34	.471		59	23	.96	82.96	100
5590	3" diameter pipe size		28	.571		67	28	1.16	96.16	117
5600	4" diameter pipe size		22	.727		86	35.50	1.48	122.98	149
5610	5" diameter pipe size		18	.889		116	43.50	1.80	161.30	194
5620	6" diameter pipe size	Q-18	23	1.043		127	52.50	1.41	180.91	219
6000	Black steel service pipe, std. wt., 1-1/2" thick insul.									
6010	3/4" diameter pipe size	Q-17	54	.296	L.F.	40	14.45	.60	55.05	66
6040	1" diameter pipe size		50	.320		45	15.60	.65	61.25	73.50
6050	1-1/4" diameter pipe size		47	.340		50.50	16.60	.69	67.79	81.50
6060	1-1/2" diameter pipe size		45	.356		55	17.30	.72	73.02	87.50
6070	2" diameter pipe size		42	.381		59.50	18.55	.77	78.82	94
6080	2-1/2" diameter pipe size		34	.471		63.50	23	.96	87.46	105
6090	3" diameter pipe size		28	.571		71.50	28	1.16	100.66	121
6100	4" diameter pipe size		22	.727		91	35.50	1.48	127.98	155
6110	5" diameter pipe size		18	.889		116	43.50	1.80	161.30	194
6120	6" diameter pipe size	Q-18	23	1.043		132	52.50	1.41	185.91	225
6130	8" diameter pipe size		19	1.263		190	64	1.71	255.71	305
6140	10" diameter pipe size		16	1.500		248	76	2.03	326.03	390
6150	12" diameter pipe size		13	1.846		296	93	2.49	391.49	465
6190	For 2" thick insulation, add					15%				
6220	For 2-1/2" thick insulation, add					25%				
6260	For 3" thick insulation, add					30%				
6800	Black steel service pipe, ex. hvy. wt., 1" thick insul.									
6820	3/4" diameter pipe size	Q-17	50	.320	L.F.	41.50	15.60	.65	57.75	70
6840	1" diameter pipe size		47	.340		44.50	16.60	.69	61.79	75
6850	1-1/4" diameter pipe size		44	.364		51	17.70	.74	69.44	84
6860	1-1/2" diameter pipe size		42	.381		54.50	18.55	.77	73.82	88.50
6870	2" diameter pipe size		40	.400		57.50	19.50	.81	77.81	93.50
6880	2-1/2" diameter pipe size		31	.516		71.50	25	1.05	97.55	117
6890	3" diameter pipe size		27	.593		80.50	29	1.20	110.70	133
6900	4" diameter pipe size		21	.762		106	37	1.55	144.55	173
6910	5" diameter pipe size		17	.941		148	46	1.91	195.91	234
6920	6" diameter pipe size	Q-18	22	1.091		136	55	1.47	192.47	234
7400	Black steel service pipe, ex. hvy. wt., 1-1/2" thick insul.									
7420	3/4" diameter pipe size	Q-17	50	.320	L.F.	34.50	15.60	.65	50.75	62
7440	1" diameter pipe size		47	.340		39.50	16.60	.69	56.79	69.50
7450	1-1/4" diameter pipe size		44	.364		46	17.70	.74	64.44	78
7460	1-1/2" diameter pipe size		42	.381		50.50	18.55	.77	69.82	84
7470	2" diameter pipe size		40	.400		51	19.50	.81	71.31	86
7480	2-1/2" diameter pipe size		31	.516		60.50	25	1.05	86.55	105
7490	3" diameter pipe size		27	.593		71	29	1.20	101.20	122
7500	4" diameter pipe size		21	.762		92.50	37	1.55	131.05	159
7510	5" diameter pipe size		17	.941		128	46	1.91	175.91	211
7520	6" diameter pipe size	Q-18	22	1.091		141	55	1.47	197.47	239
7530	8" diameter pipe size		18	1.333		211	67.50	1.80	280.30	335
7540	10" diameter pipe size		15	1.600		251	81	2.16	334.16	400
7550	12" diameter pipe size		13	1.846		310	93	2.49	405.49	480
7590	For 2" thick insulation, add					13%				
7640	For 2-1/2" thick insulation, add					18%				
7680	For 3" thick insulation, add					24%				

33 63 13 – Underground Steam and Condensate Distribution Piping

33 63 13.20 Combined Steam Pipe With Condensate Return	Crew	Daily Output	Labor-Hours	Unit	Material	2011 Bare Costs Labor	Equipment	Total	Total Incl O&P
9000 **COMBINED STEAM PIPE WITH CONDENSATE RETURN**									
9010 8" and 4" in 24" case	Q-18	100	.240	L.F.	445	12.10	.32	457.42	510
9020 6" and 3" in 20" case		104	.231		305	11.65	.31	316.96	355
9030 3" and 1-1/2" in 16" case		110	.218		228	11	.29	239.29	267
9040 2" and 1-1/4" in 12-3/4" case		114	.211		198	10.65	.28	208.93	234
9050 1-1/2" and 1-1/4" in 10-3/4" case	↓	116	.207	↓	173	10.45	.28	183.73	206
9100 Steam pipe only (no return)									
9110 6" in 18" case	Q-18	108	.222	L.F.	250	11.20	.30	261.50	292
9120 4" in 14" case		112	.214		198	10.80	.29	209.09	234
9130 3" in 14" case		114	.211		183	10.65	.28	193.93	217
9140 2-1/2" in 12-3/4" case		118	.203		171	10.25	.27	181.52	204
9150 2" in 12-3/4" case	↓	122	.197	↓	163	9.95	.27	173.22	195

Estimating Tips

There are four basic types of pollution: Air, Noise, Water, and Solid.

These systems may be interrelated and care must be taken that the complete systems are estimated. For example, Air Pollution Equipment may include dust and air-entrained particles that have to be collected. The vacuum systems could be noisy, requiring silencers to reduce noise pollution, and the collected solids have to be disposed of to prevent Solid Pollution.

Water Treatment may be accomplished as comparatively small packaged plants providing Chemical, Biological or Thermal Treatment. These are usually priced by GPH or GPD.

Large plants and municipal treatment facilities are sized MGD. These facilities are usually priced in two steps. First, the large concrete structures used as settling ponds, etc.; and second, the equipment required. Storm water does not need as much treatment as sanitary wastewater.

The total cost of the system is greatly affected by the level of purification required by specifications and codes before release.

An additional cost may be encountered if there is a possibility that the water being treated may contain petroleum products, which have to be isolated and disposed of in an environmentally acceptable manner.

Reference Numbers

Reference numbers are shown in shaded boxes at the beginning of some major classifications. These numbers refer to related items in the Reference Section. The reference information may be an estimating procedure, an alternate pricing method, or technical information.

Note: Not all subdivisions listed here necessarily appear in this publication.

44 11 16.10 Dust Collection Systems	Crew	Daily Output	Labor-Hours	Unit	Material	2011 Bare Costs Labor	Equipment	Total	Total Incl O&P
0010 **DUST COLLECTION SYSTEMS** Commercial/Industrial									
0120 Central vacuum units									
0130 Includes stand, filters and motorized shaker									
0200 500 CFM, 10" inlet, 2 HP	Q-20	2.40	8.333	Ea.	3,900	395		4,295	4,900
0220 1000 CFM, 10" inlet, 3 HP		2.20	9.091		4,100	430		4,530	5,175
0240 1500 CFM, 10" inlet, 5 HP		2	10		4,325	470		4,795	5,450
0260 3000 CFM, 13" inlet, 10 HP		1.50	13.333		12,200	630		12,830	14,400
0280 5000 CFM, 16" inlet, 2 @ 10 HP		1	20		12,900	945		13,845	15,600
1000 Vacuum tubing, galvanized									
1100 2-1/8" OD, 16 ga.	Q-9	440	.036	L.F.	2.55	1.69		4.24	5.35
1110 2-1/2" OD, 16 ga.		420	.038		2.93	1.77		4.70	5.90
1120 3" OD, 16 ga.		400	.040		3.35	1.86		5.21	6.50
1130 3-1/2" OD, 16 ga.		380	.042		4.86	1.96		6.82	8.30
1140 4" OD, 16 ga.		360	.044		5.45	2.07		7.52	9.10
1150 5" OD, 14 ga.		320	.050		9.80	2.32		12.12	14.25
1160 6" OD, 14 ga.		280	.057		11.30	2.66		13.96	16.45
1170 8" OD, 14 ga.		200	.080		17.05	3.72		20.77	24.50
1180 10" OD, 12 ga.		160	.100		27	4.65		31.65	36.50
1190 12" OD, 12 ga.		120	.133		29.50	6.20		35.70	42
1200 14" OD, 12 ga.		80	.200		36	9.30		45.30	53.50
1940 Hose, flexible wire reinforced rubber									
1956 3" dia.	Q-9	400	.040	L.F.	7.60	1.86		9.46	11.15
1960 4" dia.		360	.044		8.80	2.07		10.87	12.75
1970 5" dia.		320	.050		11.05	2.32		13.37	15.65
1980 6" dia.		280	.057		11.95	2.66		14.61	17.15
2000 90° Elbow, slip fit									
2110 2-1/8" dia.	Q-9	70	.229	Ea.	10.65	10.60		21.25	28
2120 2-1/2" dia.		65	.246		14.90	11.45		26.35	33.50
2130 3" dia.		60	.267		20	12.40		32.40	41
2140 3-1/2" dia.		55	.291		25	13.50		38.50	48
2150 4" dia.		50	.320		31	14.85		45.85	56.50
2160 5" dia.		45	.356		58.50	16.55		75.05	89.50
2170 6" dia.		40	.400		81	18.60		99.60	117
2180 8" dia.		30	.533		154	25		179	207
2400 45° Elbow, slip fit									
2410 2-1/8" dia.	Q-9	70	.229	Ea.	9.30	10.60		19.90	26.50
2420 2-1/2" dia.		65	.246		13.70	11.45		25.15	32.50
2430 3" dia.		60	.267		16.55	12.40		28.95	37
2440 3-1/2" dia.		55	.291		21	13.50		34.50	43.50
2450 4" dia.		50	.320		27	14.85		41.85	52
2460 5" dia.		45	.356		45.50	16.55		62.05	75
2470 6" dia.		40	.400		61.50	18.60		80.10	96
2480 8" dia.		35	.457		121	21.50		142.50	165
2800 90° TY, slip fit thru 6" dia.									
2810 2-1/8" dia.	Q-9	42	.381	Ea.	21	17.70		38.70	50
2820 2-1/2" dia.		39	.410		27	19.05		46.05	59
2830 3" dia.		36	.444		35.50	20.50		56	70.50
2840 3-1/2" dia.		33	.485		45.50	22.50		68	84
2850 4" dia.		30	.533		65.50	25		90.50	110
2860 5" dia.		27	.593		124	27.50		151.50	179
2870 6" dia.		24	.667		169	31		200	233
2880 8" dia., butt end					360			360	395
2890 10" dia., butt end					650			650	715

44 11 16 – Fugitive Dust Barrier Systems

44 11 16.10 Dust Collection Systems	Crew	Daily Output	Labor-Hours	Unit	Material	2011 Bare Costs Labor	Equipment	Total	Total Incl O&P
2900	12" dia., butt end				Ea.	650			650	715
2910	14" dia., butt end					1,175			1,175	1,300
2920	6" x 4" dia., butt end					103			103	114
2930	8" x 4" dia., butt end					196			196	215
2940	10" x 4" dia., butt end					256			256	282
2950	12" x 4" dia., butt end				▼	294			294	325
3100	90° Elbow, butt end segmented									
3110	8" dia., butt end, segmented				Ea.	154			154	169
3120	10" dia., butt end, segmented					380			380	420
3130	12" dia., butt end, segmented					485			485	530
3140	14" dia., butt end, segmented				▼	605			605	665
3200	45° Elbow, butt end segmented									
3210	8" dia., butt end, segmented				Ea.	121			121	133
3220	10" dia., butt end, segmented					276			276	305
3230	12" dia., butt end, segmented					282			282	310
3240	14" dia., butt end, segmented				▼	345			345	380
3400	All butt end fittings require one coupling per joint.									
3410	Labor for fitting included with couplings.									
3460	Compression coupling, galvanized, neoprene gasket									
3470	2-1/8" dia.	Q-9	44	.364	Ea.	12.35	16.90		29.25	39
3480	2-1/2" dia.		44	.364		12.35	16.90		29.25	39
3490	3" dia.		38	.421		17.80	19.55		37.35	49
3500	3-1/2" dia.		35	.457		20	21.50		41.50	54
3510	4" dia.		33	.485		22	22.50		44.50	58
3520	5" dia.		29	.552		25	25.50		50.50	66.50
3530	6" dia.		26	.615		25	28.50		53.50	71
3540	8" dia.		22	.727		52	34		86	109
3550	10" dia.		20	.800		70.50	37		107.50	134
3560	12" dia.		18	.889		87	41.50		128.50	159
3570	14" dia.	▼	16	1	▼	111	46.50		157.50	193
3800	Air gate valves, galvanized									
3810	2-1/8" dia.	Q-9	30	.533	Ea.	180	25		205	236
3820	2-1/2" dia.		28	.571		199	26.50		225.50	259
3830	3" dia.		26	.615		217	28.50		245.50	283
3840	4" dia.		23	.696		273	32.50		305.50	350
3850	6" dia.	▼	18	.889	▼	370	41.50		411.50	470

44 11 96 – Dry Plate Electrostatic Precipitator Equipment

44 11 96.10 Electrostatic Precipitators	Crew	Daily Output	Labor-Hours	Unit	Material	2011 Bare Costs Labor	Equipment	Total	Total Incl O&P
0010	**ELECTROSTATIC PRECIPITATORS**									
4100	Smoke pollution control									
4110	Exhaust air volume									
4120	1 to 2.4 MCFM	Q-10	2.15	11.163	MCFM	7,975	540		8,515	9,575
4130	2.4 to 4.8 MCFM		2.70	8.889		9,050	430		9,480	10,600
4140	4.8 to 7.2 MCFM		3.60	6.667		9,225	320		9,545	10,600
4150	7.2 to 12 MCFM	▼	5.40	4.444	▼	9,650	214		9,864	10,900
4400	Electrostatic precipitators with built-in exhaust fans									
4410	Smoke pollution control									
4420	Exhaust air volume									
4440	1 to 2.4 MCFM	Q-10	2.15	11.163	MCFM	9,575	540		10,115	11,300
4450	2.4 to 4.8 MCFM		2.70	8.889		7,600	430		8,030	9,000
4460	4.8 to 7.2 MCFM		3.60	6.667		6,500	320		6,820	7,625
4470	7.2 to 12 MCFM	▼	5.40	4.444		5,600	214		5,814	6,500

Division Notes

	CREW	DAILY OUTPUT	LABOR-HOURS	UNIT	2011 BARE COSTS				TOTAL INCL O&P
					MAT.	LABOR	EQUIP.	TOTAL	

Estimating Tips

This new division contains information about water and wastewater equipment and systems, which was formerly located in Division 44.

The main areas of focus are total wastewater treatment plants and components of wastewater treatment plants. In addition, there are assemblies such as sewage treatment lagoons that can be found in some publications under G30 Site Mechnical Utilities.

Also included in this section are oil/water separators for wastewater treatment.

Reference Numbers

Reference numbers are shown in shaded boxes at the beginning of some major classifications. These numbers refer to related items in the Reference Section. The reference information may be an estimating procedure, an alternate pricing method, or technical information.

Note: Not all subdivisions listed here necessarily appear in this publication.

Division 46 - Waste and Wastewater Equip.

46 25 Oil and Grease Separation and Removal Equipment

46 25 16 – API Oil-Water Separators

46 25 16.10 Oil/Water Separators	Crew	Daily Output	Labor-Hours	Unit	Material	2011 Bare Costs Labor	Equipment	Total	Total Incl O&P
0010 **OIL/WATER SEPARATORS**									
0020 Complete system, not including excavation and backfill									
0100 Treated capacity 0.2 cubic feet per second	B-22	1	30	Ea.	9,450	1,200	194	10,844	12,400
0200 0.5 cubic foot per second	B-13	.75	74.667		14,400	2,775	1,000	18,175	21,100
0300 1.0 cubic foot per second		.60	93.333		20,500	3,475	1,250	25,225	29,300
0400 2.4 – 3 cubic feet per second		.30	186		40,500	6,950	2,500	49,950	58,000
0500 11 cubic feet per second		.17	329		45,500	12,300	4,400	62,200	73,500
0600 22 cubic feet per second	B-13	.10	560		68,000	20,800	7,475	96,275	114,500

Assemblies Section

Table of Contents

How to Use the Assemblies Cost Tables

The following is a detailed explanation of a sample Assemblies Cost Table. Most Assembly Tables are separated into three parts: (1) an illustration of the system to be estimated, (2) the components and related costs of a typical system, and (3) the costs for similar systems with dimensional and/or size variations. For costs of the components that comprise these systems, or assemblies, refer to the Unit Price Section. Next to each bold number below is the described item with the appropriate component of the sample entry following in parentheses. In most cases, if the work is to be subcontracted, the general contractor will need to add an additional markup (RSMeans suggests using 10%) to the "Total" figures.

1 System/Line Numbers (D3010 510 1760)

Each Assemblies Cost Line has been assigned a unique identification number based on the UniFormat classification sytem.

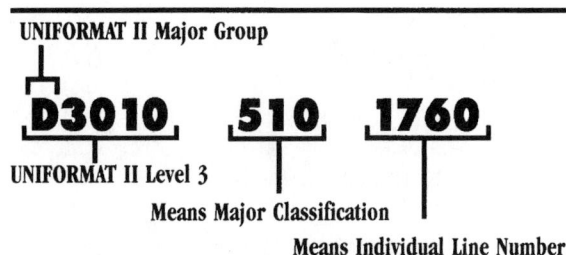

UNIFORMAT II Major Group

D3010   **510**   **1760**

UNIFORMAT II Level 3

Means Major Classification

Means Individual Line Number

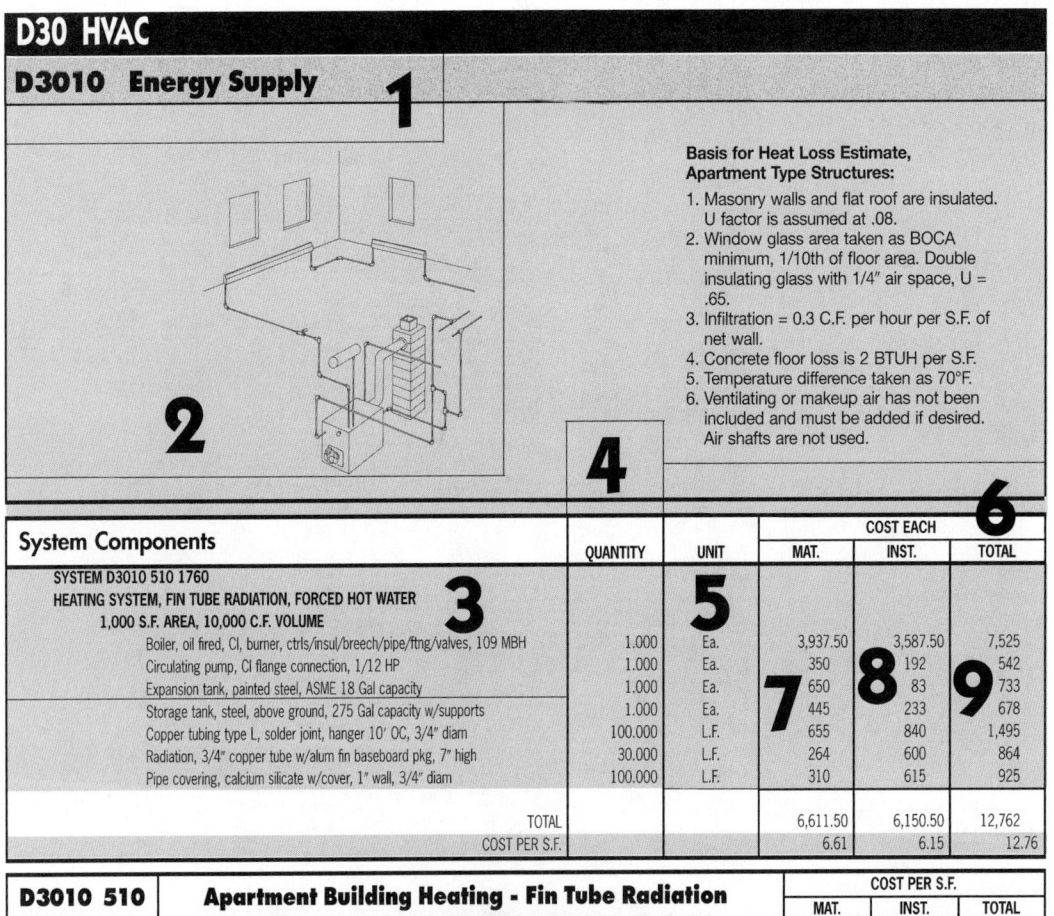

D30 HVAC

D3010 Energy Supply

Basis for Heat Loss Estimate, Apartment Type Structures:

1. Masonry walls and flat roof are insulated. U factor is assumed at .08.
2. Window glass area taken as BOCA minimum, 1/10th of floor area. Double insulating glass with 1/4" air space, U = .65.
3. Infiltration = 0.3 C.F. per hour per S.F. of net wall.
4. Concrete floor loss is 2 BTUH per S.F.
5. Temperature difference taken as 70°F.
6. Ventilating or makeup air has not been included and must be added if desired. Air shafts are not used.

System Components	QUANTITY	UNIT	COST EACH MAT.	COST EACH INST.	COST EACH TOTAL
SYSTEM D3010 510 1760					
HEATING SYSTEM, FIN TUBE RADIATION, FORCED HOT WATER					
1,000 S.F. AREA, 10,000 C.F. VOLUME					
Boiler, oil fired, CI, burner, ctrls/insul/breech/pipe/fttng/valves, 109 MBH	1.000	Ea.	3,937.50	3,587.50	7,525
Circulating pump, CI flange connection, 1/12 HP	1.000	Ea.	350	192	542
Expansion tank, painted steel, ASME 18 Gal capacity	1.000	Ea.	650	83	733
Storage tank, steel, above ground, 275 Gal capacity w/supports	1.000	Ea.	445	233	678
Copper tubing type L, solder joint, hanger 10' OC, 3/4" diam	100.000	L.F.	655	840	1,495
Radiation, 3/4" copper tube w/alum fin baseboard pkg, 7" high	30.000	L.F.	264	600	864
Pipe covering, calcium silicate w/cover, 1" wall, 3/4" diam	100.000	L.F.	310	615	925
TOTAL			6,611.50	6,150.50	12,762
COST PER S.F.			6.61	6.15	12.76

D3010 510	Apartment Building Heating - Fin Tube Radiation		COST PER S.F. MAT.	COST PER S.F. INST.	COST PER S.F. TOTAL
1740	Heating systems, fin tube radiation, forced hot water				
1760	1,000 S.F. area, 10,000 C.F. volume		6.60	6.15	12.75
1800	10,000 S.F. area, 100,000 C.F. volume	R235000 -10	3.09	3.71	6.80
1840	20,000 S.F. area, 200,000 C.F. volume		3.52	4.17	7.69
1880	30,000 S.F. area, 300,000 C.F. volume	R235000 -20	3.38	4.05	7.43
1890					

488

2 Illustration

At the top of most assembly pages are an illustration, a brief description, and the design criteria used to develop the cost.

3 System Components

The components of a typical system are listed separately to show what has been included in the development of the total system price. The table below contains prices for other similar systems with dimensional and/or size variations.

4 Quantity

This is the number of line item units required for one system unit. For example, we assume that it will take 30 linear feet of radiation for the 1,000 S.F. area shown.

5 Unit of Measure for Each Item

The abbreviated designation indicates the unit of measure, as defined by industry standards, upon which the price of the component is based. For example, baseboard radiation is priced by the linear foot. For a complete listing of abbreviations, see the Reference Section.

6 Unit of Measure for Each System (Each)

Costs shown in the three right-hand columns have been adjusted by the component quantity and unit of measure for the entire system. In this example, "Cost Each" is the unit of measure for this system, or assembly.

7 Materials (6,611.50)

This column contains the Materials Cost of each component. These cost figures are bare costs plus 10% for profit.

8 Installation (6,150.50)

Installation includes labor and equipment plus the installing contractor's overhead and profit. Equipment costs are the bare rental costs plus 10% for profit. The labor overhead and profit is defined on the inside back cover of this book.

9 Total (12,762.00)

The figure in this column is the sum of the material and installation costs.

Material Cost	+	Installation Cost	=	Total
$6,611.50	+	$6,150.50	=	$12,762.00

A Substructure

A20 Basement Construction

A2020 Basement Walls

A2020 220	Subdrainage Piping	COST PER L.F.		
		MAT.	INST.	TOTAL
2000	Piping, excavation & backfill excluded, PVC, perforated			
2110	3" diameter	2.04	3.77	5.81
2130	4" diameter	2.04	3.77	5.81
2140	5" diameter	4.05	4.04	8.09
2150	6" diameter	4.05	4.04	8.09
3000	Metal alum. or steel, perforated asphalt coated			
3150	6" diameter	8.30	7.95	16.25
3160	8" diameter	12	8.10	20.10
3170	10" diameter	15	8.35	23.35
3180	12" diameter	16.80	10.55	27.35
3220	18" diameter	25	14.65	39.65
4000	Corrugated HDPE tubing			
4130	4" diameter	.67	.70	1.37
4150	6" diameter	2.16	.94	3.10
4160	8" diameter	2.37	1.20	3.57
4180	12" diameter	7.90	4.15	12.05
4200	15" diameter	8.45	4.71	13.16
4220	18" diameter	11	6.65	17.65

In this closed-loop indirect collection system, fluid with a low freezing temperature, such as propylene glycol, transports heat from the collectors to water storage. The transfer fluid is contained in a closed-loop consisting of collectors, supply and return piping, and a remote heat exchanger. The heat exchanger transfers heat energy from the

fluid in the collector loop to potable water circulated in a storage loop. A typical two-or-three panel system contains 5 to 6 gallons of heat transfer fluid.

When the collectors become approximately 20°F warmer than the storage temperature, a controller activates the circulator on the collector and storage

loops. The circulators will move the fluid and potable water through the heat exchanger until heat collection no longer occurs. At that point, the system shuts down. Since the heat transfer medium is a fluid with a very low freezing temperature, there is no need for it to be drained from the system between periods of collection.

D2020 Domestic Water Distribution

System Components		QUANTITY	UNIT	COST EACH		
				MAT.	INST.	TOTAL
	SYSTEM D2020 265 2760					
	SOLAR, CLOSED LOOP, ADD-ON HOT WATER SYS., EXTERNAL HEAT EXCHANGER					
	3/4″ TUBING, TWO 3′X7′ BLACK CHROME COLLECTORS					
A,B,G,L,K,M	Heat exchanger fluid-fluid pkg incl 2 circulators, expansion tank,					
	Check valve, relief valve, controller, hi temp cutoff, & 2 sensors	1.000	Ea.	765	460	1,225
C	Thermometer, 2″ dial	3.000	Ea.	66	120	186
D, T	Fill & drain valve, brass, 3/4″ connection	1.000	Ea.	11.65	26.50	38.15
E	Air vent, manual, 1/8″ fitting	2.000	Ea.	6.12	40	46.12
F	Air purger	1.000	Ea.	47.50	53.50	101
H	Strainer, Y type, bronze body, 3/4″ IPS	1.000	Ea.	28.50	34	62.50
I	Valve, gate, bronze, NRS, soldered 3/4″ diam	6.000	Ea.	267	192	459
J	Neoprene vent flashing	2.000	Ea.	27.40	64	91.40
N-1, N	Relief valve temp & press, 150 psi 210°F self-closing 3/4″ IPS	1.000	Ea.	17.15	21.50	38.65
O	Pipe covering, urethane, ultraviolet cover, 1″ wall 3/4″ diam	20.000	L.F.	48	114	162
P	Pipe covering, fiberglass, all service jacket, 1″ wall, 3/4″ diam	50.000	L.F.	57	227	284
Q	Collector panel solar energy blk chrome on copper, 1/8″ temp glass 3′x7′	2.000	Ea.	2,100	242	2,342
	Roof clamps for solar energy collector panels	2.000	Set	5.18	32.90	38.08
R	Valve, swing check, bronze, regrinding disc, 3/4″ diam	2.000	Ea.	158	64	222
S	Pressure gauge, 60 psi, 2″ dial	1.000	Ea.	24.50	20	44.50
U	Valve, water tempering, bronze, sweat connections, 3/4″ diam	1.000	Ea.	125	32	157
W-2, V	Tank water storage w/heating element, drain, relief valve, existing	1.000	Ea.			
	Copper tubing type L, solder joint, hanger 10′ OC 3/4″ diam	20.000	L.F.	131	168	299
	Copper tubing, type M, solder joint, hanger 10′ OC 3/4″ diam	70.000	L.F.	350	574	924
	Sensor wire, #22-2 conductor multistranded	.500	C.L.F.	8.25	29.75	38
	Solar energy heat transfer fluid, propylene glycol anti-freeze	6.000	Gal.	79.50	138	217.50
	Wrought copper fittings & solder, 3/4″ diam	76.000	Ea.	350.36	2,546	2,896.36
	TOTAL			4,673.11	5,199.15	9,872.26

D2020 265	Solar, Closed Loop, Add-On Hot Water Systems		COST EACH		
			MAT.	INST.	TOTAL
2550	Solar, closed loop, add-on hot water system, external heat exchanger				
2570	3/8″ tubing, 3 ea. 4′ x 4′-4″ vacuum tube collectors		5,300	4,875	10,175
2580	1/2″ tubing, 4 ea. 4 x 4′-4″ vacuum tube collectors		5,875	5,275	11,150
2600	2 ea. 3′x7′ black chrome collectors	R235616 -60	4,225	4,975	9,200
2620	3 ea. 3′x7′ black chrome collectors		5,275	5,100	10,375
2640	2 ea. 3′x7′ flat black collectors		3,650	4,975	8,625
2660	3 ea. 3′x7′ flat black collectors		4,400	5,125	9,525
2700	3/4″ tubing, 3 ea. 3′x7′ black chrome collectors		5,725	5,325	11,050
2720	3 ea. 3′x7′ flat black absorber plate collectors		4,850	5,350	10,200
2740	2 ea. 4′x9′ flat black w/plastic glazing collectors		4,475	5,375	9,850
2760	2 ea. 3′x7′ black chrome collectors		4,675	5,200	9,875
2780	1″ tubing, 4 ea 2′x9′ plastic absorber & glazing collectors		5,825	6,050	11,875
2800	4 ea. 3′x7′ black chrome absorber collectors		7,825	6,075	13,900
2820	4 ea. 3′x7′ flat black absorber collectors		6,675	6,100	12,775

D2020   Domestic Water Distribution

In the drainback indirect-collection system, the heat transfer fluid is distilled water contained in a loop consisting of collectors, supply and return piping, and an unpressurized holding tank. A large heat exchanger containing incoming potable water is immersed in the holding tank. When a controller activates solar collection, the distilled water is pumped through the collectors and heated and pumped back down to the holding tank. When the temperature differential between the water in the collectors and water in storage is such that collection no longer occurs, the pump turns off and gravity causes the distilled water in the collector loop to drain back to the holding tank. All the loop piping is pitched so that the water can drain out of the collectors and piping and not freeze there. As hot water is needed in the home, incoming water first flows through the holding tank with the immersed heat exchanger and is warmed and then flows through a conventional heater for any supplemental heating that is necessary.

D2020 Domestic Water Distribution

System Components	QUANTITY	UNIT	COST EACH		
			MAT.	INST.	TOTAL
SYSTEM D2020 270 2760					
SOLAR, DRAINBACK, ADD ON, HOT WATER, IMMERSED HEAT EXCHANGER					
3/4″ TUBING, THREE EA 3′X7′ BLACK CHROME COLLECTOR					
A, B Differential controller 2 sensors, thermostat, solar energy system	1.000	Ea.	575	53.50	628.50
C Thermometer 2″ dial	3.000	Ea.	66	120	186
D, T Fill & drain valve, brass, 3/4″ connection	1.000	Ea.	11.65	26.50	38.15
E-1 Automatic air vent 1/8″ fitting	1.000	Ea.	13.95	20	33.95
H Strainer, Y type, bronze body, 3/4″ IPS	1.000	Ea.	28.50	34	62.50
I Valve, gate, bronze, NRS, soldered 3/4″ diam	2.000	Ea.	89	64	153
J Neoprene vent flashing	2.000	Ea.	27.40	64	91.40
L Circulator, solar heated liquid, 1/20 HP	1.000	Ea.	170	96	266
N Relief valve temp. & press. 150 psi 210°F self-closing 3/4″ IPS	1.000	Ea.	17.15	21.50	38.65
O Pipe covering, urethane, ultraviolet cover, 1″ wall, 3/4″ diam	20.000	L.F.	48	114	162
P Pipe covering, fiberglass, all service jacket, 1″ wall, 3/4″ diam	50.000	L.F.	57	227	284
Q Collector panel solar energy blk chrome on copper, 1/8″ temp glas 3′x7′	3.000	Ea.	3,150	363	3,513
Roof clamps for solar energy collector panels	3.000	Set	7.77	49.35	57.12
R Valve, swing check, bronze, regrinding disc, 3/4″ diam	1.000	Ea.	79	32	111
U Valve, water tempering, bronze sweat connections, 3/4″ diam	1.000	Ea.	125	32	157
V Tank, water storage w/heating element, drain, relief valve, existing	1.000	Ea.			
W Tank, water storage immersed heat exchr elec 2″x1/2# insul 120 gal	1.000	Ea.	1,350	455	1,805
X Valve, globe, bronze, rising stem, 3/4″ diam, soldered	3.000	Ea.	336	96	432
Y Flow control valve	1.000	Ea.	118	29	147
Z Valve, ball, bronze, solder 3/4″ diam, solar loop flow control	1.000	Ea.	20.50	32	52.50
Copper tubing, type L, solder joint, hanger 10′ OC 3/4″ diam	20.000	L.F.	131	168	299
Copper tubing, type M, solder joint, hanger 10′ OC 3/4″ diam	70.000	L.F.	350	574	924
Sensor wire, #22-2 conductor, multistranded	.500	C.L.F.	8.25	29.75	38
Wrought copper fittings & solder, 3/4″ diam	76.000	Ea.	350.36	2,546	2,896.36
TOTAL			7,129.53	5,246.60	12,376.13

D2020 270	Solar, Drainback, Hot Water Systems		COST EACH		
			MAT.	INST.	TOTAL
2550	Solar, drainback, hot water, immersed heat exchanger				
2560	3/8″ tubing, 3 ea. 4′ x 4′-4″ vacuum tube collectors	R235616 -60	6,475	4,725	11,200
2580	1/2″ tubing, 4 ea 4′ x 4′-4″ vacuum tube collectors, 80 gal tank		7,075	5,125	12,200
2600	120 gal tank		7,225	5,200	12,425
2640	2 ea. 3′x7′ blk chrome collectors, 80 gal tank		5,425	4,825	10,250
2660	3 ea. 3′x7′ blk chrome collectors, 120 gal tank		6,625	5,025	11,650
2700	2 ea. 3′x7′ flat blk collectors, 120 gal tank		5,000	4,900	9,900
2720	3 ea. 3′x7′ flat blk collectors, 120 gal tank		5,775	5,050	10,825
2760	3/4″ tubing, 3 ea 3′x7′ black chrome collectors, 120 gal tank		7,125	5,250	12,375
2780	3 ea. 3′x7′ flat black absorber collectors, 120 gal tank		6,250	5,275	11,525
2800	2 ea. 4′x9′ flat blk w/plastic glazing collectors 120 gal tank		5,875	5,275	11,150
2840	1″ tubing, 4 ea. 2′x9′ plastic absorber & glazing collectors, 120 gal tank		7,300	5,975	13,275
2860	4 ea. 3′x7′ black chrome absorber collectors, 120 gal tank		9,325	6,000	15,325
2880	4 ea. 3′x7′ flat black absorber collectors, 120 gal tank		8,175	6,025	14,200

In the draindown direct-collection system, incoming domestic water is heated in the collectors. When the controller activates solar collection, domestic water is first heated as it flows through the collectors and is then pumped to storage. When conditions are no longer suitable for heat collection, the pump shuts off and the water in the loop drains down and out of the system by means of solenoid valves and properly pitched piping.

D2020   Domestic Water Distribution

System Components	QUANTITY	UNIT	COST EACH		
			MAT.	INST.	TOTAL
SYSTEM D2020 275 2760					
SOLAR, DRAINDOWN, HOT WATER, DIRECT COLLECTION					
3/4″ TUBING, THREE 3′X7′ BLACK CHROME COLLECTORS					
A, B Differential controller, 2 sensors, thermostat, solar energy system	1.000	Ea.	575	53.50	628.50
A-1 Solenoid valve, solar heating loop, brass, 3/4″ diam, 24 volts	3.000	Ea.	360	213	573
B-1 Solar energy sensor, freeze prevention	1.000	Ea.	24	20	44
C Thermometer, 2″ dial	3.000	Ea.	66	120	186
E-1 Vacuum relief valve, 3/4″ diam	1.000	Ea.	29	20	49
F-1 Air vent, automatic, 1/8″ fitting	1.000	Ea.	13.95	20	33.95
H Strainer, Y type, bronze body, 3/4″ IPS	1.000	Ea.	28.50	34	62.50
I Valve, gate, bronze, NRS, soldered, 3/4″ diam	2.000	Ea.	89	64	153
J Vent flashing neoprene	2.000	Ea.	27.40	64	91.40
K Circulator, solar heated liquid, 1/25 HP	1.000	Ea.	295	82.50	377.50
N Relief valve temp & press 150 psi 210°F self-closing 3/4″ IPS	1.000	Ea.	17.15	21.50	38.65
O Pipe covering, urethane, ultraviolet cover, 1″ wall, 3/4″ diam	20.000	L.F.	48	114	162
P Pipe covering, fiberglass, all service jacket, 1″ wall, 3/4″ diam	50.000	L.F.	57	227	284
Roof clamps for solar energy collector panels	3.000	Set	7.77	49.35	57.12
Q Collector panel solar energy blk chrome on copper, 1/8″ temp glass 3′x7′	3.000	Ea.	3,150	363	3,513
R Valve, swing check, bronze, regrinding disc, 3/4″ diam, soldered	2.000	Ea.	158	64	222
T Drain valve, brass, 3/4″ connection	2.000	Ea.	23.30	53	76.30
U Valve, water tempering, bronze, sweat connections, 3/4″ diam	1.000	Ea.	125	32	157
W-2, W Tank, water storage elec elem 2″x1/2# insul 120 gal	1.000	Ea.	1,350	455	1,805
X Valve, globe, bronze, rising stem, 3/4″ diam, soldered	1.000	Ea.	112	32	144
Copper tubing, type L, solder joints, hangers 10′ OC 3/4″ diam	20.000	L.F.	131	168	299
Copper tubing, type M, solder joints, hangers 10′ OC 3/4″ diam	70.000	L.F.	350	574	924
Sensor wire, #22-2 conductor, multistranded	.500	C.L.F.	8.25	29.75	38
Wrought copper fittings & solder, 3/4″ diam	76.000	Ea.	350.36	2,546	2,896.36
TOTAL			7,395.68	5,419.60	12,815.28

D2020 275	Solar, Draindown, Hot Water Systems		COST EACH		
			MAT.	INST.	TOTAL
2550	Solar, draindown, hot water				
2560	3/8″ tubing, 3 ea. 4′ x 4′-4″ vacuum tube collectors, 80 gal tank		6,450	5,175	11,625
2580	1/2″ tubing, 4 ea. 4′ x 4′-4″ vacuum tube collectors, 80 gal tank	R235616 -60	7,375	5,325	12,700
2600	120 gal tank		7,525	5,375	12,900
2640	2 ea. 3′x7′ black chrome collectors, 80 gal tank		5,725	5,025	10,750
2660	3 ea. 3′x7′ black chrome collectors, 120 gal tank		6,950	5,200	12,150
2700	2 ea. 3′x7′ flat black collectors, 120 gal tank		5,300	5,075	10,375
2720	3 ea. 3′x7′ flat black collectors, 120 gal tank		6,025	5,200	11,225
2760	3/4″ tubing, 3 ea. 3′x7′ black chrome collectors, 120 gal tank		7,400	5,425	12,825
2780	3 ea. 3′x7′ flat collectors, 120 gal tank		6,525	5,450	11,975
2800	2 ea. 4′x9′ flat black & plastic glazing collectors, 120 gal tank		6,150	5,450	11,600
2840	1″ tubing, 4 ea. 2′x9′ plastic absorber & glazing collectors, 120 gal tank		9,925	6,150	16,075
2860	4 ea. 3′x7′ black chrome absorber collectors, 120 gal tank		11,900	6,150	18,050
2880	4 ea. 3′x7′ flat black absorber collectors, 120 gal tank		10,800	6,200	17,000

In the recirculation system, (a direct-collection system), incoming domestic water is heated in the collectors. When the controller activates solar collection, domestic water is heated as it flows through the collectors and then it flows back to storage. When conditions are not suitable for heat collection, the pump shuts off and the flow of the water stops. In this type of system, water remains in the collector loop at all times. A "frost sensor" at the collector activates the circulation of warm water from storage through the collectors when protection from freezing is required.

D2020 Domestic Water Distribution

System Components	QUANTITY	UNIT	COST EACH		
			MAT.	INST.	TOTAL
SYSTEM D2020 280 2820					
SOLAR, RECIRCULATION, HOT WATER					
3/4″ TUBING, TWO 3′X7′ BLACK CHROME COLLECTORS					
A, B Differential controller 2 sensors, thermostat, for solar energy system	1.000	Ea.	575	53.50	628.50
A-1 Solenoid valve, solar heating loop, brass, 3/4″ IPS, 24 volts	2.000	Ea.	240	142	382
B-1 Solar energy sensor freeze prevention	1.000	Ea.	24	20	44
C Thermometer, 2″ dial	3.000	Ea.	66	120	186
D Drain valve, brass, 3/4″ connection	1.000	Ea.	11.65	26.50	38.15
F-1 Air vent, automatic, 1/8″ fitting	1.000	Ea.	13.95	20	33.95
H Strainer, Y type, bronze body, 3/4″ IPS	1.000	Ea.	28.50	34	62.50
I Valve, gate, bronze, 125 lb, soldered 3/4″ diam	3.000	Ea.	133.50	96	229.50
J Vent flashing, neoprene	2.000	Ea.	27.40	64	91.40
L Circulator, solar heated liquid, 1/20 HP	1.000	Ea.	170	96	266
N Relief valve, temp & press 150 psi 210° F self-closing 3/4″ IPS	2.000	Ea.	34.30	43	77.30
O Pipe covering, urethane, ultraviolet cover, 1″ wall, 3/4″ diam	20.000	L.F.	48	114	162
P Pipe covering, fiberglass, all service jacket, 1″ wall 3/4″ diam	50.000	L.F.	57	227	284
Q Collector panel solar energy blk chrome on copper, 1/8″ temp glass 3′x7′	2.000	Ea.	2,100	242	2,342
Roof clamps for solar energy collector panels	2.000	Set	5.18	32.90	38.08
R Valve, swing check, bronze, 125 lb, regrinding disc, soldered 3/4″ diam	2.000	Ea.	158	64	222
U Valve, water tempering, bronze, sweat connections, 3/4″ diam	1.000	Ea.	125	32	157
V Tank, water storage, w/heating element, drain, relief valve, existing	1.000	Ea.			
X Valve, globe, bronze, 125 lb, 3/4″ diam	2.000	Ea.	224	64	288
Copper tubing, type L, solder joints, hangers 10′ OC 3/4″ diam	20.000	L.F.	131	168	299
Copper tubing, type M, solder joints, hangers 10′ OC 3/4″ diam	70.000	L.F.	350	574	924
Wrought copper fittings & solder, 3/4″ diam	76.000	Ea.	350.36	2,546	2,896.36
Sensor wire, #22-2 conductor, multistranded	.500	C.L.F.	8.25	29.75	38
TOTAL			**4,881.09**	**4,808.65**	**9,689.74**

D2020 280	Solar Recirculation, Domestic Hot Water Systems		COST EACH		
			MAT.	INST.	TOTAL
2550	Solar recirculation, hot water				
2560	3/8″ tubing, 3 ea. 4′ x 4′-4″ vacuum tube collectors		5,450	4,475	9,925
2580	1/2″ tubing, 4 ea. 4′ x 4′-4″ vacuum tube collectors	R235616 -60	6,025	4,900	10,925
2640	2 ea. 3′x7′ black chrome collectors		4,400	4,575	8,975
2660	3 ea. 3′x7′ black chrome collectors		5,450	4,725	10,175
2700	2 ea. 3′x7′ flat black collectors		3,800	4,600	8,400
2720	3 ea. 3′x7′ flat black collectors		4,575	4,750	9,325
2760	3/4″ tubing, 3 ea. 3′x7′ black chrome collectors		5,925	4,950	10,875
2780	3 ea. 3′x7′ flat black absorber plate collectors		5,075	4,975	10,050
2800	2 ea. 4′x9′ flat black w/plastic glazing collectors		4,675	4,975	9,650
2820	2 ea. 3′x7′ black chrome collectors		4,875	4,800	9,675
2840	1″ tubing, 4 ea. 2′x9′ black plastic absorber & glazing collectors		7,725	5,675	13,400
2860	4 ea. 3′x7′ black chrome absorber collectors		9,750	5,700	15,450
2880	4 ea. 3′x7′ flat black absorber collectors		8,600	5,725	14,325

The thermosyphon domestic hot water system, a direct collection system, operates under city water pressure and does not require pumps for system operation. An insulated water storage tank is located above the collectors. As the sun heats the collectors, warm water in them rises by means of natural convection; the colder water in the storage tank flows into the collectors by means of gravity. As long as the sun is shining the water continues to flow through the collectors and to become warmer.

To prevent freezing, the system must be drained or the collectors covered with an insulated lid when the temperature drops below 32°F.

D2020 Domestic Water Distribution

System Components	QUANTITY	UNIT	COST EACH		
			MAT.	INST.	TOTAL
SYSTEM D2020 285 0960					
SOLAR, THERMOSYPHON, WATER HEATER					
3/4" TUBING, TWO 3'X7' BLACK CHROME COLLECTORS					
D-1 Framing lumber, fir, 2" x 6" x 8', tank cradle	.008	M.B.F.	4.64	6.76	11.40
F-1 Framing lumber, fir, 2" x 4" x 24', sleepers	.016	M.B.F.	8.72	20.40	29.12
I Valve, gate, bronze, 125 lb, soldered 1/2" diam	2.000	Ea.	77	53	130
J Vent flashing, neoprene	4.000	Ea.	54.80	128	182.80
O Pipe covering, urethane, ultraviolet cover, 1" wall, 1/2" diam	40.000	L.F.	86.40	224	310.40
P Pipe covering fiberglass all service jacket 1" wall 1/2" diam	160.000	L.F.	168	696	864
Q Collector panel solar, blk chrome on copper, 3/16" temp glass 3'-6" x 7.5'	2.000	Ea.	1,520	256	1,776
Y Flow control valve, globe, bronze, 125#, soldered, 1/2" diam	1.000	Ea.	83.50	26.50	110
U Valve, water tempering, bronze, sweat connections, 1/2" diam	1.000	Ea.	108	26.50	134.50
W Tank, water storage, solar energy system, 80 Gal, 2" x 1/2 lb insul	1.000	Ea.	1,200	400	1,600
Copper tubing type L, solder joints, hangers 10' OC 1/2" diam	150.000	L.F.	655.50	1,185	1,840.50
Copper tubing type M, solder joints, hangers 10' OC 1/2" diam	50.000	L.F.	170	380	550
Sensor wire, #22-2 gauge multistranded	.500	C.L.F.	8.25	29.75	38
Wrought copper fittings & solder, 1/2" diam	75.000	Ea.	167.25	2,400	2,567.25
TOTAL			4,312.06	5,831.91	10,143.97

D2020 285	Thermosyphon, Hot Water		COST EACH		
			MAT.	INST.	TOTAL
0960 0970	Solar, thermosyphon, hot water, two collector system	R235616 -60	4,300	5,825	10,125

ATTIC FLOOR

CWS

HWS

This domestic hot water pre-heat system includes heat exchanger with a circulating pump, blower, air-to-water, coil and controls, mounted in the upper collector manifold. Heat from the hot air coming out of the collectors is transferred through the heat exchanger. For each degree of DHW preheating gained, one degree less heating is needed from the fuel fired water heater. The system is simple, inexpensive to operate and can provide a substantial portion of DHW requirements for modest additional cost.

D20 Plumbing

D2020 Domestic Water Distribution

System Components	QUANTITY	UNIT	COST EACH MAT.	COST EACH INST.	COST EACH TOTAL
SYSTEM D2020 290 2560					
SOLAR, HOT WATER, AIR TO WATER HEAT EXCHANGE					
THREE COLLECTORS, OPTICAL BLACK ON ALUMINUM, 10'X2',80 GAL TANK					
A, B Differential controller, 2 sensors, thermostat, solar energy system	1.000	Ea.	575	53.50	628.50
C Thermometer, 2" dial	2.000	Ea.	44	80	124
C-1 Heat exchanger, air to fluid, up flow 70 MBH	1.000	Ea.	365	330	695
E Air vent, manual, for solar energy system 1/8" fitting	1.000	Ea.	3.06	20	23.06
F Air purger	1.000	Ea.	47.50	53.50	101
G Expansion tank	1.000	Ea.	67.50	20	87.50
H Strainer, Y type, bronze body, 1/2" IPS	1.000	Ea.	25	32.50	57.50
I Valve, gate, bronze, 125 lb, NRS, soldered 1/2" diam	5.000	Ea.	192.50	132.50	325
N Relief valve, temp & pressure solar 150 psi 210° F self-closing	1.000	Ea.	17.15	21.50	38.65
P Pipe covering, fiberglass, all service jacket, 1" wall, 1/2" diam	60.000	L.F.	63	261	324
Q Collector panel solar energy, air, black on alum plate, flush mount, 10'x2'	30.000	L.F.	14,850	351	15,201
B-1, R-1 Shutter damper for solar heater circulator	1.000	Ea.	54	93.50	147.50
B-1, R-1 Shutter motor for solar heater circulator	1.000	Ea.	131	70.50	201.50
R Backflow preventer for solar energy system, 1/2" diam	2.000	Ea.	128	80	208
T Drain valve, brass, 3/4" connection	1.000	Ea.	11.65	26.50	38.15
U Valve, water tempering, bronze, sweat connections, 1/2" diam	1.000	Ea.	108	26.50	134.50
W Tank, water storage, solar energy system, 80 Gal, 2" x 2 lb insul	1.000	Ea.	1,200	400	1,600
V Tank, water storage, w/heating element, drain, relief valve, existing	1.000	System			
X Valve, globe, bronze, 125 lb, rising stem, 1/2" diam	1.000	Ea.	83.50	26.50	110
Copper tubing type M, solder joints, hangers 10' OC 1/2" diam	50.000	L.F.	170	380	550
Copper tubing type L, solder joints, hangers 10' OC 1/2" diam	10.000	L.F.	43.70	79	122.70
Wrought copper fittings & solder, 1/2" diam	10.000	Ea.	22.30	320	342.30
Sensor wire, #22-2 conductor multistranded	.500	C.L.F.	8.25	29.75	38
Q-1, Q-2 Ductwork, fiberglass, aluminzed jacket, 1-1/2" thick, 8" diam	32.000	S.F.	184	200	384
Q-3 Manifold for flush mount solar energy collector panels, air	6.000	L.F.	840	42.30	882.30
TOTAL			19,234.11	3,130.05	22,364.16

D2020 290	Solar Hot Water, Air To Water Heat Exchange		COST EACH MAT.	COST EACH INST.	COST EACH TOTAL
2550	Solar hot water, air to water heat exchange				
2560	Three collectors, optical black on aluminum, 10' x 2', 80 Gal tank		19,200	3,125	22,325
2580	Four collectors, optical black on aluminum, 10' x 2', 80 Gal tank	R235616 -60	24,500	3,300	27,800
2600	Four collectors, optical black on aluminum, 10' x 2', 120 Gal tank		24,700	3,350	28,050

In this closed-loop indirect collection system, fluid with a low freezing temperature, such as propylene glycol, transports heat from the collectors to water storage. The transfer fluid is contained in a closed-loop consisting of collectors, supply and return piping, and a heat exchanger immersed in the storage tank. A typical two-or-three panel system contains 5 to 6 gallons of heat transfer fluid.

When the collectors become approximately 20°F warmer than the storage temperature, a controller activates the circulator. The circulator moves the fluid continuously through the collectors until the temperature difference between the collectors and storage is such that heat collection no longer occurs; at that point, the circulator shuts off. Since the heat transfer fluid has a very low freezing temperature, there is no need for it to be drained from the collectors between periods of collection.

D2020 Domestic Water Distribution

System Components	QUANTITY	UNIT	COST EACH MAT.	COST EACH INST.	COST EACH TOTAL
SYSTEM D2020 295 2760					
SOLAR, CLOSED LOOP, HOT WATER SYSTEM, IMMERSED HEAT EXCHANGER					
3/4″ TUBING, THREE 3′ X 7′ BLACK CHROME COLLECTORS					
A, B Differential controller, 2 sensors, thermostat, solar energy system	1.000	Ea.	575	53.50	628.50
C Thermometer 2″ dial	3.000	Ea.	66	120	186
D, T Fill & drain valves, brass, 3/4″ connection	3.000	Ea.	34.95	79.50	114.45
E Air vent, manual, 1/8″ fitting	1.000	Ea.	3.06	20	23.06
F Air purger	1.000	Ea.	47.50	53.50	101
G Expansion tank	1.000	Ea.	67.50	20	87.50
I Valve, gate, bronze, NRS, soldered 3/4″ diam	3.000	Ea.	133.50	96	229.50
J Neoprene vent flashing	2.000	Ea.	27.40	64	91.40
K Circulator, solar heated liquid, 1/25 HP	1.000	Ea.	295	82.50	377.50
N-1, N Relief valve, temp & press 150 psi 210°F self-closing 3/4″ IPS	2.000	Ea.	34.30	43	77.30
O Pipe covering, urethane ultraviolet cover, 1″ wall, 3/4″ diam	20.000	L.F.	48	114	162
P Pipe covering, fiberglass, all service jacket, 1″ wall, 3/4″ diam	50.000	L.F.	57	227	284
Roof clamps for solar energy collector panel	3.000	Set	7.77	49.35	57.12
Q Collector panel solar blk chrome on copper, 1/8″ temp glass, 3′x7′	3.000	Ea.	3,150	363	3,513
R-1 Valve, swing check, bronze, regrinding disc, 3/4″ diam, soldered	1.000	Ea.	79	32	111
S Pressure gauge, 60 psi, 2-1/2″ dial	1.000	Ea.	24.50	20	44.50
U Valve, water tempering, bronze, sweat connections, 3/4″ diam	1.000	Ea.	125	32	157
W-2, W Tank, water storage immersed heat exchr elec elem 2″x2# insul 120 Gal	1.000	Ea.	1,350	455	1,805
X Valve, globe, bronze, rising stem, 3/4″ diam, soldered	1.000	Ea.	112	32	144
Copper tubing type L, solder joint, hanger 10′ OC 3/4″ diam	20.000	L.F.	131	168	299
Copper tubing, type M, solder joint, hanger 10′ OC 3/4″ diam	70.000	L.F.	350	574	924
Sensor wire, #22-2 conductor multistranded	.500	C.L.F.	8.25	29.75	38
Solar energy heat transfer fluid, propylene glycol, anti-freeze	6.000	Gal.	79.50	138	217.50
Wrought copper fittings & solder, 3/4″ diam	76.000	Ea.	350.36	2,546	2,896.36
TOTAL			**7,156.59**	**5,412.10**	**12,568.69**

D2020 295	Solar, Closed Loop, Hot Water Systems		COST EACH MAT.	COST EACH INST.	COST EACH TOTAL
2550	Solar, closed loop, hot water system, immersed heat exchanger				
2560	3/8″ tubing, 3 ea. 4′ x 4′-4″ vacuum tube collectors, 80 gal. tank		6,575	4,925	11,500
2580	1/2″ tubing, 4 ea. 4′ x 4′-4″ vacuum tube collectors, 80 gal. tank		7,175	5,300	12,475
2600	120 gal. tank		7,325	5,375	12,700
2640	2 ea. 3′x7′ black chrome collectors, 80 gal. tank	R235616 -60	5,575	5,075	10,650
2660	120 gal. tank		5,675	5,075	10,750
2700	2 ea. 3′x7′ flat black collectors, 120 gal. tank		5,100	5,075	10,175
2720	3 ea. 3′x7′ flat black collectors, 120 gal. tank		5,850	5,225	11,075
2760	3/4″ tubing, 3 ea. 3′x7′ black chrome collectors, 120 gal. tank		7,150	5,400	12,550
2780	3 ea. 3′x7′ flat black collectors, 120 gal. tank		6,275	5,425	11,700
2800	2 ea. 4′x9′ flat black w/plastic glazing collectors 120 gal. tank		5,900	5,450	11,350
2840	1″ tubing, 4 ea. 2′x9′ plastic absorber & glazing collectors 120 gal. tank		7,150	6,100	13,250
2860	4 ea. 3′x7′ black chrome collectors, 120 gal. tank		9,175	6,125	15,300
2880	4 ea. 3′x7′ flat black absorber collectors, 120 gal. tank		8,000	6,150	14,150

D3010   Energy Supply

Basis for Heat Loss Estimate, Apartment Type Structures:

1. Masonry walls and flat roof are insulated. U factor is assumed at .08.
2. Window glass area taken as BOCA minimum, 1/10th of floor area. Double insulating glass with 1/4″ air space, U = .65.
3. Infiltration = 0.3 C.F. per hour per S.F. of net wall.
4. Concrete floor loss is 2 BTUH per S.F.
5. Temperature difference taken as 70°F.
6. Ventilating or makeup air has not been included and must be added if desired. Air shafts are not used.

System Components	QUANTITY	UNIT	COST EACH		
			MAT.	INST.	TOTAL
SYSTEM D3010 510 1760					
HEATING SYSTEM, FIN TUBE RADIATION, FORCED HOT WATER					
1,000 S.F. AREA, 10,000 C.F. VOLUME					
Boiler, oil fired, CI, burner, ctrls/insul/breech/pipe/fttng/valves, 109 MBH	1.000	Ea.	3,937.50	3,587.50	7,525
Circulating pump, CI flange connection, 1/12 HP	1.000	Ea.	350	192	542
Expansion tank, painted steel, ASME 18 Gal capacity	1.000	Ea.	650	83	733
Storage tank, steel, above ground, 275 Gal capacity w/supports	1.000	Ea.	445	233	678
Copper tubing type L, solder joint, hanger 10′ OC, 3/4″ diam	100.000	L.F.	655	840	1,495
Radiation, 3/4″ copper tube w/alum fin baseboard pkg, 7″ high	30.000	L.F.	264	600	864
Pipe covering, calcium silicate w/cover, 1″ wall, 3/4″ diam	100.000	L.F.	310	615	925
TOTAL			6,611.50	6,150.50	12,762
COST PER S.F.			6.61	6.15	12.76

D3010 510	Apartment Building Heating - Fin Tube Radiation		COST PER S.F.		
			MAT.	INST.	TOTAL
1740	Heating systems, fin tube radiation, forced hot water				
1760	1,000 S.F. area, 10,000 C.F. volume		6.60	6.15	12.75
1800	10,000 S.F. area, 100,000 C.F. volume	R235000 -10	3.09	3.71	6.80
1840	20,000 S.F. area, 200,000 C.F. volume		3.52	4.17	7.69
1880	30,000 S.F. area, 300,000 C.F. volume	R235000 -20	3.38	4.05	7.43
1890					

D3010 Energy Supply

Fin Tube Radiator

Basis for Heat Loss Estimate, Factory or Commercial Type Structures:

1. Walls and flat roof are of lightly insulated concrete block or metal. U factor is assumed at .17.
2. Windows and doors are figured at 1-1/2 S.F. per linear foot of building perimeter. The U factor for flat single glass is 1.13.
3. Infiltration is approximately 5 C.F. per hour per S.F. of wall.
4. Concrete floor loss is 2 BTUH per S.F. of floor.
5. Temperature difference is assumed at 70°F.
6. Ventilation or makeup air has not been included and must be added if desired.

System Components	QUANTITY	UNIT	COST EACH		
			MAT.	INST.	TOTAL
SYSTEM D3010 520 1960					
HEATING SYSTEM, FIN TUBE RADIATION, FORCED HOT WATER					
1,000 S.F. BLDG., ONE FLOOR					
Boiler, oil fired, CI, burner/ctrls/insul/breech/pipe/ftngs/valves, 109 MBH	1.000	Ea.	3,937.50	3,587.50	7,525
Expansion tank, painted steel, ASME 18 Gal capacity	1.000	Ea.	2,450	97	2,547
Storage tank, steel, above ground, 550 Gal capacity w/supports	1.000	Ea.	3,100	430	3,530
Circulating pump, CI flanged, 1/8 HP	1.000	Ea.	590	192	782
Pipe, steel, black, sch. 40, threaded, cplg & hngr 10'OC, 1-1/2" diam.	260.000	L.F.	2,496	3,744	6,240
Pipe covering, calcium silicate w/cover , 1" wall, 1-1/2" diam	260.000	L.F.	819	1,651	2,470
Radiation, steel 1-1/4" tube & 4-1/4" fin w/cover & damper, wall hung	42.000	L.F.	1,785	1,365	3,150
Rough in, steel fin tube radiation w/supply & balance valves	4.000	Set	1,148	2,840	3,988
TOTAL			16,325.50	13,906.50	30,232
COST PER S.F.			16.33	13.91	30.24

D3010 520	Commercial Building Heating - Fin Tube Radiation		COST PER S.F.		
			MAT.	INST.	TOTAL
1940	Heating systems, fin tube radiation, forced hot water				
1960	1,000 S.F. bldg, one floor		16.30	13.90	30.20
2000	10,000 S.F., 100,000 C.F., total two floors		4.48	5.30	9.78
2040	100,000 S.F., 1,000,000 C.F., total three floors	R235000 -10	1.72	2.33	4.05
2080	1,000,000 S.F., 10,000,000 C.F., total five floors		.88	1.23	2.11
2090		R235000 -20			

D3010 Energy Supply

Unit Heater

Basis for Heat Loss Estimate, Factory or Commercial Type Structures:

1. Walls and flat roof are of lightly insulated concrete block or metal. U factor is assumed at .17.
2. Windows and doors are figured at 1-1/2 S.F. per linear foot of building perimeter. The U factor for flat single glass is 1.13.
3. Infiltration is approximately 5 C.F. per hour per S.F. of wall.
4. Concrete floor loss is 2 BTUH per S.F. of floor.
5. Temperature difference is assumed at 70°F.
6. Ventilation or makeup air has not been included and must be added if desired.

System Components	QUANTITY	UNIT	COST EACH		
			MAT.	INST.	TOTAL
SYSTEM D3010 530 1880					
HEATING SYSTEM, TERMINAL UNIT HEATERS, FORCED HOT WATER					
1,000 S.F. BLDG., ONE FLOOR					
Boiler oil fired, CI, burner/ctrls/insul/breech/pipe/ftngs/valves, 109 MBH	1.000	Ea.	3,937.50	3,587.50	7,525
Expansion tank, painted steel, ASME 18 Gal capacity	1.000	Ea.	2,450	97	2,547
Storage tank, steel, above ground, 550 Gal capacity w/supports	1.000	Ea.	3,100	430	3,530
Circulating pump, CI, flanged, 1/8 HP	1.000	Ea.	590	192	782
Pipe, steel, black, schedule 40, threaded, cplg & hngr 10' OC 1-1/2" diam	260.000	L.F.	2,496	3,744	6,240
Pipe covering, calcium silicate w/cover, 1" wall, 1-1/2" diam	260.000	L.F.	819	1,651	2,470
Unit heater, 1 speed propeller, horizontal, 200° EWT, 26.9 MBH	2.000	Ea.	1,110	290	1,400
Unit heater piping hookup with controls	2.000	Set	1,370	2,750	4,120
TOTAL			15,872.50	12,741.50	28,614
COST PER S.F.			15.87	12.74	28.61

D3010 530	Commercial Bldg. Heating - Terminal Unit Heaters	COST PER S.F.		
		MAT.	INST.	TOTAL
1860	Heating systems, terminal unit heaters, forced hot water			
1880	1,000 S.F. bldg., one floor	15.90	12.70	28.60
1920	10,000 S.F. bldg., 100,000 C.F. total two floors	4	4.37	8.37
1960	100,000 S.F. bldg., 1,000,000 C.F. total three floors _R235000 -10_	1.72	2.12	3.84
2000	1,000,000 S.F. bldg., 10,000,000 C.F. total five floors	1.13	1.37	2.50
2010	_R235000 -20_			

D3010 Energy Supply

Many styles of active solar energy systems exist. Those shown on the following page represent the majority of systems now being installed in different regions of the country.

The five active domestic hot water (DHW) systems which follow are typically specified as two or three panel systems with additional variations being the type of glazing and size of the storage tanks. Various combinations have been costed for the user's evaluation and comparison. The basic specifications from which the following systems were developed satisfy the construction detail requirements specified in the HUD Intermediate Minimum Property Standards (IMPS). If these standards are not complied with, the renewable energy system's costs could be significantly lower than shown.

To develop the system's specifications and costs it was necessary to make a number of assumptions about the systems. Certain systems are more appropriate to one climatic region than another or the systems may require modifications to be usable in particular locations. Specific instances in which the systems are not appropriate throughout the country as specified, or in which modification will be needed include the following:

- The freeze protection mechanisms provided in the DHW systems vary greatly. In harsh climates, where freeze is a major concern, a closed-loop indirect collection system may be more appropriate than a direct collection system.
- The thermosyphon water heater system described cannot be used when temperatures drop below 32°F.
- In warm climates it may be necessary to modify the systems installed to prevent overheating.

For each renewable resource (solar) system a schematic diagram and descriptive summary of the system is provided along with a list of all the components priced as part of the system. The costs were developed based on these specifications.

Considerations affecting costs which may increase or decrease beyond the estimates presented here include the following:

- Special structural qualities (allowance for earthquake, future expansion, high winds, and unusual spans or shapes);
- Isolated building site or rough terrain that would affect the transportation of personnel, material, or equipment;
- Unusual climatic conditions during the construction process;
- Substitution of other materials or system components for those used in the system specifications.

In this closed-loop indirect collection system, fluid with a low freezing temperature, propylene glycol, transports heat from the collectors to water storage. The transfer fluid is contained in a closed loop consisting of collectors, supply and return piping, and a heat exchanger immersed in the storage tank.

When the collectors become approximately 20°F warmer than the storage temperature, the controller activates the circulator. The circulator moves the fluid continuously until the temperature difference between fluid in the collectors and storage is such that the collection will no longer occur and then the circulator turns off. Since the heat transfer fluid has a very low freezing temperature, there is no need for it to be drained from the collectors between periods of collection.

D3010 Energy Supply

System Components	QUANTITY	UNIT	COST EACH		
			MAT.	INST.	TOTAL
SYSTEM D3010 650 2750					
SOLAR, CLOSED LOOP, SPACE/HOT WATER					
1″ TUBING, TEN 3′X7′ BLK CHROME ON COPPER ABSORBER COLLECTORS					
A, B Differential controller 2 sensors, thermostat, solar energy system	2.000	Ea.	1,150	107	1,257
C Thermometer, 2″ dial	10.000	Ea.	220	400	620
C-1 Heat exchanger, solar energy system, fluid to air, up flow, 80 MBH	1.000	Ea.	490	385	875
D, T Fill & drain valves, brass, 3/4″ connection	5.000	Ea.	58.25	132.50	190.75
D-1 Fan center	1.000	Ea.	131	132	263
E Air vent, manual, 1/8″ fitting	1.000	Ea.	3.06	20	23.06
E-2 Thermostat, 2 stage for sensing room temperature	1.000	Ea.	160	81	241
F Air purger	2.000	Ea.	95	107	202
F-1 Controller, liquid temperature, solar energy system	1.000	Ea.	105	128	233
G Expansion tank	2.000	Ea.	135	40	175
I Valve, gate, bronze, 125 lb, soldered, 1″ diam	5.000	Ea.	312.50	167.50	480
J Vent flashing, neoprene	2.000	Ea.	27.40	64	91.40
K Circulator, solar heated liquid, 1/25 HP	2.000	Ea.	590	165	755
L Circulator, solar heated liquid, 1/20 HP	1.000	Ea.	170	96	266
N Relief valve, temp & pressure 150 psi 210°F self-closing	4.000	Ea.	68.60	86	154.60
N-1 Relief valve, pressure poppet, bronze, 30 psi, 3/4″ IPS	3.000	Ea.	123	69	192
O Pipe covering, urethane, ultraviolet cover, 1″ wall, 1″ diam	50.000	L.F.	120	285	405
P Pipe covering, fiberglass, all service jacket, 1″ wall, 1″ diam	60.000	L.F.	68.40	272.40	340.80
Q Collector panel solar energy blk chrome on copper 1/8″ temp glass 3′x7′	10.000	Ea.	10,500	1,210	11,710
Roof clamps for solar energy collector panel	10.000	Set	25.90	164.50	190.40
R Valve, swing check, bronze, 125 lb, regrinding disc, 3/4″ & 1″ diam	4.000	Ea.	456	134	590
S Pressure gage, 0-60 psi, for solar energy system	2.000	Ea.	49	40	89
U Valve, water tempering, bronze, sweat connections, 3/4″ diam	1.000	Ea.	125	32	157
W-2, W-1, W Tank, water storage immersed heat xchr elec elem 2″x1/2# ins 120 gal	4.000	Ea.	5,400	1,820	7,220
X Valve, globe, bronze, 125 lb, rising stem, 1″ diam	3.000	Ea.	528	100.50	628.50
Y Valve, flow control	1.000	Ea.	118	29	147
Copper tubing type M, solder joint, hanger 10′ OC, 1″ diam	110.000	L.F.	885.50	1,006.50	1,892
Copper tubing, type L, solder joint, hanger 10′ OC 3/4″ diam	20.000	L.F.	131	168	299
Wrought copper fittings & solder, 3/4″ & 1″ diam	121.000	Ea.	1,488.30	4,840	6,328.30
Sensor, wire, #22-2 conductor, multistranded	.700	C.L.F.	11.55	41.65	53.20
Ductwork, galvanized steel, for heat exchanger	8.000	Lb.	5.84	59.60	65.44
Solar energy heat transfer fluid propylene glycol, anti-freeze	25.000	Gal.	331.25	575	906.25
TOTAL			24,082.55	12,958.15	37,040.70

D3010 650	Solar, Closed Loop, Space/Hot Water Systems		COST EACH		
			MAT.	INST.	TOTAL
2540	Solar, closed loop, space/hot water				
2550	1/2″ tubing, 12 ea. 4′x4′4″ vacuum tube collectors		22,400	12,100	34,500
2600	3/4″ tubing, 12 ea. 4′x4′4″ vacuum tube collectors		23,500	12,400	35,900
2650	10 ea. 3′ x 7′ black chrome absorber collectors	R235616 -60	22,800	12,100	34,900
2700	10 ea. 3′ x 7′ flat black absorber collectors		19,900	12,100	32,000
2750	1″ tubing, 10 ea. 3′ x 7′ black chrome absorber collectors		24,100	13,000	37,100
2800	10 ea. 3′ x 7′ flat black absorber collectors		21,300	13,000	34,300
2850	6 ea. 4′ x 9′ flat black w/plastic glazing collectors		19,400	13,000	32,400
2900	12 ea. 2′ x 9′ plastic absorber and glazing collectors		20,200	13,200	33,400

This draindown pool system uses a differential thermostat similar to those used in solar domestic hot water and space heating applications. To heat the pool, the pool water passes through the conventional pump-filter loop and then flows through the collectors. When collection is not possible, or when the pool temperature is reached, all water drains from the solar loop back to the pool through the existing piping. The modes are controlled by solenoid valves or other automatic valves in conjunction with a vacuum breaker relief valve, which facilitates draindown.

D3010 Energy Supply

System Components	QUANTITY	UNIT	COST EACH		
			MAT.	INST.	TOTAL
SYSTEM D3010 660 2640					
SOLAR SWIMMING POOL HEATER, ROOF MOUNTED COLLECTORS					
TEN 4' X 10' FULLY WETTED UNGLAZED PLASTIC ABSORBERS					
A Differential thermostat/controller, 110V, adj pool pump system	1.000	Ea.	305	320	625
A-1 Solenoid valve, PVC, normally 1 open 1 closed (included)	2.000	Ea.			
B Sensor, thermistor type (included)	2.000	Ea.			
E-1 Valve, vacuum relief	1.000	Ea.	29	20	49
Q Collector panel, solar energy, plastic, liquid full wetted, 4' x 10'	10.000	Ea.	2,570	2,310	4,880
R Valve, ball check, PVC, socket, 1-1/2" diam	1.000	Ea.	101	32	133
Z Valve, ball, PVC, socket, 1-1/2" diam	3.000	Ea.	199.50	96	295.50
Pipe, PVC, sch 40, 1-1/2" diam	80.000	L.F.	504	1,424	1,928
Pipe fittings, PVC sch 40, socket joint, 1-1/2" diam	10.000	Ea.	18.20	320	338.20
Sensor wire, #22-2 conductor, multistranded	.500	C.L.F.	8.25	29.75	38
Roof clamps for solar energy collector panels	10.000	Set	25.90	164.50	190.40
Roof strap, teflon for solar energy collector panels	26.000	L.F.	572	81.12	653.12
TOTAL			4,332.85	4,797.37	9,130.22

D3010 660	Solar Swimming Pool Heater Systems		COST EACH		
			MAT.	INST.	TOTAL
2530	Solar swimming pool heater systems, roof mounted collectors				
2540	10 ea. 3'x7' black chrome absorber, 1/8" temp. glass		12,300	3,700	16,000
2560	10 ea. 4'x8' black chrome absorber, 3/16" temp. glass	R235616 -60	13,800	4,400	18,200
2580	10 ea. 3'8"x6' flat black absorber, 3/16" temp. glass		9,375	3,775	13,150
2600	10 ea. 4'x9' flat black absorber, plastic glazing		11,300	4,575	15,875
2620	10 ea. 2'x9' rubber absorber, plastic glazing		12,700	4,950	17,650
2640	10 ea. 4'x10' fully wetted unglazed plastic absorber		4,325	4,800	9,125
2660	Ground mounted collectors				
2680	10 ea. 3'x7' black chrome absorber, 1/8" temp. glass		12,300	4,250	16,550
2700	10 ea. 4'x8' black chrome absorber, 3/16" temp. glass		13,800	4,975	18,775
2720	10 ea. 3'8"x6' flat blk absorber, 3/16" temp. glass		9,400	4,325	13,725
2740	10 ea. 4'x9' flat blk absorber, plastic glazing		11,300	5,150	16,450
2760	10 ea. 2'x9' rubber absorber, plastic glazing		12,700	5,350	18,050
2780	10 ea. 4'x10' fully wetted unglazed plastic absorber		4,375	5,350	9,725

The complete Solar Air Heating System provides maximum savings of conventional fuel with both space heating and year-round domestic hot water heating. It allows for the home air conditioning to operate simultaneously and independently from the solar domestic water heating in summer. The system's modes of operation are:

Mode 1: The building is heated directly from the collectors with air circulated by the Solar Air Mover.

Mode 2: When heat is not needed in the building, the dampers change within the air mover to circulate the air from the collectors to the rock storage bin.

Mode 3: When heat is not available from the collector array and is available in rock storage, the air mover draws heated air from rock storage and directs it into the building. When heat is not available from the collectors or the rock storage bin, the auxiliary heating unit will provide heat for the building. The size of the collector array is typically 25% the size of the main floor area.

D30 HVAC

D3010 Energy Supply

System Components		QUANTITY	UNIT	COST EACH		
				MAT.	INST.	TOTAL
	SYSTEM D3010 675 1210					
	SOLAR, SPACE/HOT WATER, AIR TO WATER HEAT EXCHANGE					
A, B	Differential controller 2 sensors thermos., solar energy sys liquid loop	1.000	Ea.	575	53.50	628.50
A-1, B	Differential controller 2 sensors 6 station solar energy sys air loop	1.000	Ea.	239	213	452
B-1	Solar energy sensor, freeze prevention	1.000	Ea.	24	20	44
C	Thermometer for solar energy system, 2″ dial	2.000	Ea.	44	80	124
C-1	Heat exchanger, solar energy system, air to fluid, up flow, 70 MBH	1.000	Ea.	365	330	695
D	Drain valve, brass, 3/4″ connection	2.000	Ea.	23.30	53	76.30
E	Air vent, manual, for solar energy system 1/8″ fitting	1.000	Ea.	3.06	20	23.06
E-1	Thermostat, 2 stage for sensing room temperature	1.000	Ea.	160	81	241
F	Air purger	1.000	Ea.	47.50	53.50	101
G	Expansion tank, for solar energy system	1.000	Ea.	67.50	20	87.50
I	Valve, gate, bronze, 125 lb, NRS, soldered 3/4″ diam	2.000	Ea.	89	64	153
K	Circulator, solar heated liquid, 1/25 HP	1.000	Ea.	295	82.50	377.50
N	Relief valve temp & press 150 psi 210°F self-closing, 3/4″ IPS	1.000	Ea.	17.15	21.50	38.65
N-1	Relief valve, pressure, poppet, bronze, 30 psi, 3/4″ IPS	1.000	Ea.	41	23	64
P	Pipe covering, fiberglass, all service jacket, 1″ wall, 3/4″ diam	60.000	L.F.	68.40	272.40	340.80
Q-3	Manifold for flush mount solar energy collector panels	20.000	L.F.	2,800	141	2,941
Q	Collector panel solar energy, air, black on alum. plate, flush mount 10′x2′	100.000	L.F.	49,500	1,170	50,670
R	Valve, swing check, bronze, 125 lb, regrinding disc, 3/4″ diam	2.000	Ea.	158	64	222
S	Pressure gage, 2″ dial, for solar energy system	1.000	Ea.	24.50	20	44.50
U	Valve, water tempering, bronze, sweat connections, 3/4″ diam	1.000	Ea.	125	32	157
W-2, W	Tank, water storage, solar, elec element 2″x1/2# insul, 80 Gal	1.000	Ea.	1,200	400	1,600
X	Valve, globe, bronze, 125 lb, soldered, 3/4″ diam	1.000	Ea.	112	32	144
	Copper tubing type L, solder joints, hangers 10′ OC 3/4″ diam	10.000	L.F.	65.50	84	149.50
	Copper tubing type M, solder joints, hangers 10′ OC 3/4″ diam	60.000	L.F.	300	492	792
	Wrought copper fittings & solder, 3/4″ diam	26.000	Ea.	119.86	871	990.86
	Sensor wire, #22-2 conductor multistranded	1.200	C.L.F.	19.80	71.40	91.20
Q-1	Duct work, rigid fiberglass, rectangular	400.000	S.F.	340	2,000	2,340
	12″ x 10″	8.000	Ea.	188	376	564
R-2	Shutter/damper for solar heater circulator	9.000	Ea.	486	841.50	1,327.50
K-1	Shutter motor for solar heater circulator blower	9.000	Ea.	1,179	634.50	1,813.50
K-1	Fan, solar energy heated air circulator, space & DHW system	1.000	Ea.	1,625	2,250	3,875
Z-1	Tank, solar energy air storage, 6′-3″H 7′x7′ = 306 CF/2000 Gal	1.000	Ea.	12,900	1,450	14,350
Z-2	Crushed stone 1-1/2″	11.000	C.Y.	462	70.73	532.73
C-2	Themometer, remote probe, 2″ dial	1.000	Ea.	33.50	80	113.50
R-1	Solenoid valve	1.000	Ea.	120	71	191
V, R-4, R-3	Furnace, supply diffusers, return grilles, existing	1.000	Ea.	12,900	1,450	14,350
	TOTAL			86,717.07	13,988.53	100,705.60

D3010 675	Air To Water Heat Exchange		COST EACH		
			MAT.	INST.	TOTAL
1210 1220	Solar, air to water heat exchange, for space/hot water heating	R235616 -60	86,500	13,900	100,400

Boiler   Baseboard Radiation

Small Electric Boiler
System Considerations:
1. Terminal units are fin tube baseboard radiation rated at 720 BTU/hr with 200° water temperature or 820 BTU/hr steam.
2. Primary use being for residential or smaller supplementary areas, the floor levels are based on 7-1/2' ceiling heights.
3. All distribution piping is copper for boilers through 205 MBH. All piping for larger systems is steel pipe.

System Components	QUANTITY	UNIT	COST EACH MAT.	INST.	TOTAL
SYSTEM D3020 102 1120					
SMALL HEATING SYSTEM, HYDRONIC, ELECTRIC BOILER					
1,480 S.F., 61 MBH, STEAM, 1 FLOOR					
Boiler, electric steam, std cntrls, trim, ftngs and valves, 18 KW, 61.4 MBH	1.000	Ea.	4,702.50	1,622.50	6,325
Copper tubing type L, solder joint, hanger 10'OC, 1-1/4" diam	160.000	L.F.	2,184	1,768	3,952
Radiation, 3/4" copper tube w/alum fin baseboard pkg 7" high	60.000	L.F.	528	1,200	1,728
Rough in baseboard panel or fin tube with valves & traps	10.000	Set	2,760	6,100	8,860
Pipe covering, calcium silicate w/cover 1" wall 1-1/4" diam	160.000	L.F.	496	1,016	1,512
Low water cut-off, quick hookup, in gage glass tappings	1.000	Ea.	239	40.50	279.50
TOTAL			10,909.50	11,747	22,656.50
COST PER S.F.			7.37	7.94	15.31

D3020 102	Small Heating Systems, Hydronic, Electric Boilers		COST PER S.F. MAT.	INST.	TOTAL
1100	Small heating systems, hydronic, electric boilers				
1120	Steam, 1 floor, 1480 S.F., 61 M.B.H.		7.36	7.91	15.27
1160	3,000 S.F., 123 M.B.H.	R235000 -10	5.75	6.95	12.70
1200	5,000 S.F., 205 M.B.H.		5.05	6.45	11.50
1240	2 floors, 12,400 S.F., 512 M.B.H.	R235000 -20	3.99	6.40	10.39
1280	3 floors, 24,800 S.F., 1023 M.B.H.		4.12	6.30	10.42
1320	34,750 S.F., 1,433 M.B.H.	R235000 -30	3.79	6.10	9.89
1360	Hot water, 1 floor, 1,000 S.F., 41 M.B.H.		10.50	4.40	14.90
1400	2,500 S.F., 103 M.B.H.		8.40	7.95	16.35
1440	2 floors, 4,850 S.F., 205 M.B.H.		8.70	9.50	18.20
1480	3 floors, 9,700 S.F., 410 M.B.H.		9.60	9.85	19.45

D3020 Heat Generating Systems

Boiler

Unit Heater

Large Electric Boiler System Considerations:

1. Terminal units are all unit heaters of the same size. Quantities are varied to accommodate total requirements.
2. All air is circulated through the heaters a minimum of three times per hour.
3. As the capacities are adequate for commercial use, floor levels are based on 10′ ceiling heights.
4. All distribution piping is black steel pipe.

System Components	QUANTITY	UNIT	COST EACH		
			MAT.	INST.	TOTAL
SYSTEM D3020 104 1240					
LARGE HEATING SYSTEM, HYDRONIC, ELECTRIC BOILER					
9,280 S.F., 135 KW, 461 MBH, 1 FLOOR					
Boiler, electric hot water, std ctrls, trim, ftngs, valves, 135 KW, 461 MBH	1.000	Ea.	11,175	3,525	14,700
Expansion tank, painted steel, 60 Gal capacity ASME	1.000	Ea.	3,825	194	4,019
Circulating pump, CI, close cpld, 50 GPM, 2 HP, 2″ pipe conn	1.000	Ea.	2,275	385	2,660
Unit heater, 1 speed propeller, horizontal, 200° EWT, 72.7 MBH	7.000	Ea.	5,670	1,477	7,147
Unit heater piping hookup with controls	7.000	Set	4,795	9,625	14,420
Pipe, steel, black, schedule 40, welded, 2-1/2″ diam	380.000	L.F.	4,047	9,598.80	13,645.80
Pipe covering, calcium silicate w/cover, 1″ wall, 2-1/2″ diam	380.000	L.F.	1,459.20	2,489	3,948.20
TOTAL			33,246.20	27,293.80	60,540
COST PER S.F.			3.58	2.94	6.52

D3020 104	Large Heating Systems, Hydronic, Electric Boilers		COST PER S.F.		
			MAT.	INST.	TOTAL
1230	Large heating systems, hydronic, electric boilers				
1240	9,280 S.F., 135 K.W., 461 M.B.H., 1 floor		3.58	2.94	6.52
1280	14,900 S.F., 240 K.W., 820 M.B.H., 2 floors		4.38	4.74	9.12
1320	18,600 S.F., 296 K.W., 1,010 M.B.H., 3 floors	R235000 -10	4.33	5.15	9.48
1360	26,100 S.F., 420 K.W., 1,432 M.B.H., 4 floors	R235000 -20	4.29	5.05	9.34
1400	39,100 S.F., 666 K.W., 2,273 M.B.H., 4 floors		3.68	4.23	7.91
1440	57,700 S.F., 900 K.W., 3,071 M.B.H., 5 floors	R235000 -30	3.46	4.16	7.62
1480	111,700 S.F., 1,800 K.W., 6,148 M.B.H., 6 floors		3.19	3.60	6.79
1520	149,000 S.F., 2,400 K.W., 8,191 M.B.H., 8 floors		3.16	3.58	6.74
1560	223,300 S.F., 3,600 K.W., 12,283 M.B.H., 14 floors		3.34	4.08	7.42

D3020 Heat Generating Systems

Fossil Fuel Boiler System Considerations:

1. Terminal units are horizontal unit heaters. Quantities are varied to accommodate total heat loss per building.
2. Unit heater selection was determined by their capacity to circulate the building volume a minimum of three times per hour in addition to the BTU output.

3. Systems shown are forced hot water. Steam boilers cost slightly more than hot water boilers. However, this is compensated for by the smaller size or fewer terminal units required with steam.
4. Floor levels are based on 10′ story heights.
5. MBH requirements are gross boiler output.

Unit Heater

System Components	QUANTITY	UNIT	COST EACH		
			MAT.	INST.	TOTAL
SYSTEM D3020 108 1280					
HEATING SYSTEM, HYDRONIC, FOSSIL FUEL, TERMINAL UNIT HEATERS					
CAST IRON BOILER, GAS, 80 MBH, 1,070 S.F. BUILDING					
Boiler, gas, hot water, CI, burner, controls, insulation, breeching, 80 MBH	1.000	Ea.	2,152.50	1,758.75	3,911.25
Pipe, steel, black, schedule 40, threaded, cplg & hngr 10′OC, 2″ diam	200.000	L.F.	2,520	3,600	6,120
Unit heater, 1 speed propeller, horizontal, 200° EWT, 72.7 MBH	2.000	Ea.	1,620	422	2,042
Unit heater piping hookup with controls	2.000	Set	1,370	2,750	4,120
Expansion tank, painted steel, ASME, 18 Gal capacity	1.000	Ea.	2,450	97	2,547
Circulating pump, CI, flange connection, 1/12 HP	1.000	Ea.	350	192	542
Pipe covering, calcium silicate w/cover, 1″ wall, 2″ diam	200.000	L.F.	722	1,310	2,032
TOTAL			11,184.50	10,129.75	21,314.25
COST PER S.F.			10.45	9.47	19.92

D3020 108	Heating Systems, Unit Heaters		COST PER S.F.		
			MAT.	INST.	TOTAL
1260	Heating systems, hydronic, fossil fuel, terminal unit heaters,				
1280	Cast iron boiler, gas, 80 M.B.H., 1,070 S.F. bldg.		10.47	9.44	19.91
1320	163 M.B.H., 2,140 S.F. bldg.		7	6.45	13.45
1360	544 M.B.H., 7,250 S.F. bldg.	R235000 -10	4.63	4.50	9.13
1400	1,088 M.B.H., 14,500 S.F. bldg.		3.91	4.26	8.17
1440	3,264 M.B.H., 43,500 S.F. bldg.	R235000 -20	2.99	3.24	6.23
1480	5,032 M.B.H., 67,100 S.F. bldg.		3.67	3.32	6.99
1520	Oil, 109 M.B.H., 1,420 S.F. bldg.	R235000 -30	10.80	8.45	19.25
1560	235 M.B.H., 3,150 S.F. bldg.		6.45	5.95	12.40
1600	940 M.B.H., 12,500 S.F. bldg.		4.37	3.93	8.30
1640	1,600 M.B.H., 21,300 S.F. bldg.		4.28	3.75	8.03
1680	2,480 M.B.H., 33,100 S.F. bldg.		4.33	3.39	7.72
1720	3,350 M.B.H., 44,500 S.F. bldg.		3.73	3.45	7.18
1760	Coal, 148 M.B.H., 1,975 S.F. bldg.		9.20	5.50	14.70
1800	300 M.B.H., 4,000 S.F. bldg.		6.95	4.32	11.27
1840	2,360 M.B.H., 31,500 S.F. bldg.		4.52	3.44	7.96

D3020 Heat Generating Systems

Fin Tube Radiator

Fossil Fuel Boiler System Considerations:

1. Terminal units are commercial steel fin tube radiation. Quantities are varied to accommodate total heat loss per building.
2. Systems shown are forced hot water. Steam boilers cost slightly more than hot water boilers. However, this is compensated for by the smaller size or fewer terminal units required with steam.
3. Floor levels are based on 10' story heights.
4. MBH requirements are gross boiler output.

System Components	QUANTITY	UNIT	COST EACH		
			MAT.	INST.	TOTAL
SYSTEM D3020 110 3240					
HEATING SYSTEM, HYDRONIC, FOSSIL FUEL, FIN TUBE RADIATION					
CAST IRON BOILER, GAS, 80 MBH, 1,070 S.F. BLDG.					
Boiler, gas, hot water, CI, burner, controls, insulation, breeching, 80 MBH	1.000	Ea.	2,152.50	1,758.75	3,911.25
Pipe, steel, black, schedule 40, threaded, 2″ diam	200.000	L.F.	2,520	3,600	6,120
Radiation, steel 1-1/4″ tube & 4-1/4″ fin w/cover & damper, wall hung	80.000	L.F.	3,400	2,600	6,000
Rough-in, wall hung steel fin radiation with supply & balance valves	8.000	Set	2,296	5,680	7,976
Expansion tank, painted steel, ASME, 18 Gal capacity	1.000	Ea.	2,450	97	2,547
Circulating pump, CI flanged, 1/12 HP	1.000	Ea.	350	192	542
Pipe covering, calcium silicate w/cover, 1″ wall, 2″ diam	200.000	L.F.	722	1,310	2,032
TOTAL			13,890.50	15,237.75	29,128.25
COST PER S.F.			12.98	14.24	27.22

D3020 110	Heating System, Fin Tube Radiation		COST PER S.F.		
			MAT.	INST.	TOTAL
3230	Heating systems, hydronic, fossil fuel, fin tube radiation				
3240	Cast iron boiler, gas, 80 MBH, 1,070 S.F. bldg.		12.99	14.21	27.20
3280	169 M.B.H., 2,140 S.F. bldg.	R235000 -10	8.20	9.05	17.25
3320	544 M.B.H., 7,250 S.F. bldg.		6.45	7.70	14.15
3360	1,088 M.B.H., 14,500 S.F. bldg.	R235000 -20	5.80	7.55	13.35
3400	3,264 M.B.H., 43,500 S.F. bldg.		5.05	6.70	11.75
3440	5,032 M.B.H., 67,100 S.F. bldg.	R235000 -30	5.70	6.80	12.50
3480	Oil, 109 M.B.H., 1,420 S.F. bldg.		15.10	15.55	30.65
3520	235 M.B.H., 3,150 S.F. bldg.		8.25	9.05	17.30
3560	940 M.B.H., 12,500 S.F. bldg.		6.30	7.25	13.55
3600	1,600 M.B.H., 21,300 S.F. bldg.		6.35	7.20	13.55
3640	2,480 M.B.H., 33,100 S.F. bldg.		6.40	6.85	13.25
3680	3,350 M.B.H., 44,500 S.F. bldg.		5.75	6.90	12.65
3720	Coal, 148 M.B.H., 1,975 S.F. bldg.		10.95	8.60	19.55
3760	300 M.B.H., 4,000 S.F. bldg.		8.65	7.40	16.05
3800	2,360 M.B.H., 31,500 S.F. bldg.		6.50	6.85	13.35
4080	Steel boiler, oil, 97 M.B.H., 1,300 S.F. bldg.		12.35	12.75	25.10
4120	315 M.B.H., 4,550 S.F. bldg.		6.05	6.45	12.50
4160	525 M.B.H., 7,000 S.F. bldg.		7.45	7.45	14.90
4200	1,050 M.B.H., 14,000 S.F. bldg.		6.65	7.45	14.10
4240	2,310 M.B.H., 30,800 S.F. bldg.		6.45	6.90	13.35
4280	3,150 M.B.H., 42,000 S.F. bldg.		6.30	7	13.30

D3020 Heat Generating Systems

Pressure Relief Valve — Supply Piping — Circulating Pump — Return Piping — Expansion Tank

Electric Boiler, Hot Water

System Components	QUANTITY	UNIT	COST EACH		
			MAT.	INST.	TOTAL
SYSTEM D3020 126 1010					
BOILER, ELECTRIC, HOT WATER, 15 kW, 52 MBH					
Boilers, electric, ASME, hot water, 15 KW, 52 MBH	1.000	Ea.	4,325	1,350	5,675
Pipe, black steel, Sch 40, threaded, W/coupling & hangers, 10' OC, 3/4" dia	10.000	L.F.	55.20	120.75	175.95
Pipe, black steel, Sch 40, threaded, W/coupling & hangers, 10' OC, 1" dia	20.000	L.F.	145.13	260.15	405.28
Elbow, 90°, black, straight, 3/4" dia.	6.000	Ea.	27.48	273	300.48
Elbow, 90°, black, straight, 1" dia.	6.000	Ea.	50.70	294	344.70
Tee, black, straight, 3/4" dia.	1.000	Ea.	5.30	71	76.30
Tee, black, straight, 1" dia.	2.000	Ea.	18.20	160	178.20
Tee, black, reducing, 1" dia.	2.000	Ea.	29.20	160	189.20
Union, black with brass seat, 3/4" dia.	1.000	Ea.	14.05	49	63.05
Union, black with brass seat, 1" dia.	2.000	Ea.	36.60	107	143.60
Pipe nipples, 3/4" diam.	3.000	Ea.	7.20	15.75	22.95
Pipe nipples, 1" diam.	3.000	Ea.	10.13	18.15	28.28
Valves, bronze, gate, N.R.S., threaded, class 150, 3/4" size	2.000	Ea.	139	64	203
Valves, bronze, gate, N.R.S., threaded, class 150, 1" size	2.000	Ea.	177	67	244
Thermometer, stem type, 9" case, 8" stem, 3/4" NPT	1.000	Ea.	372	46	418
Tank, steel, liquid expansion, ASME, painted, 15 gallon capacity	1.000	Ea.	615	68.50	683.50
Pump, circulating, bronze, flange connection, 3/4" to 1-1/2" size, 1/8 HP	1.000	Ea.	940	192	1,132
Insulation, fiberglass pipe covering, 1" wall, 1" IPS	20.000	L.F.	24.60	95	119.60
TOTAL			6,991.79	3,411.30	10,403.09

D3020 126	Electric Boiler, Hot Water		COST EACH		
			MAT.	INST.	TOTAL
1010	Boiler, electric, hot water, 15 kW, 52 MBH		7,000	3,400	10,400
1020	30 kW, 103 MBH		7,850	3,600	11,450
1030	60 kW, 205 MBH		8,025	3,725	11,750
1040	120 kW, 410 MBH		10,300	4,475	14,775
1050	150 kW, 510 MBH		15,900	8,500	24,400
1060	210 kW, 716 MBH		18,700	9,350	28,050
1070	296 kW, 1010 MBH		24,100	9,625	33,725
1080	444 kW, 1515 MBH		29,100	12,900	42,000
1090	666 kW, 2273 MBH		34,800	14,000	48,800
1100	900 kW, 3071 MBH		43,300	16,400	59,700
1110	1320 kW, 4505 MBH		51,500	17,200	68,700
1120	2100 kW, 7167 MBH		77,500	20,900	98,400
1130	2610 kW, 8905 MBH		96,000	25,200	121,200
1140	3600 kW, 12,283 MBH		117,000	30,200	147,200

D30 HVAC

D3020 Heat Generating Systems

D3020 128	Electric Boiler, Steam	COST EACH		
		MAT.	INST.	TOTAL
1010	Boiler, electric, steam, 18 kW, 61.4 MBH	5,425	3,275	8,700
1020	36 kW, 123 MBH	6,825	3,475	10,300
1030	60 kW, 205 MBH	14,100	7,050	21,150
1040	120 kW, 409 MBH	16,600	7,775	24,375
1050	150 kW, 512 MBH	19,800	11,600	31,400
1060	210 kW, 716 MBH	25,400	12,500	37,900
1070	300 kW, 1023 MBH	29,800	13,700	43,500
1080	510 kW, 1740 MBH	37,300	17,800	55,100
1090	720 kW, 2456 MBH	42,100	18,600	60,700
1100	1080 kW, 3685 MBH	49,500	22,300	71,800
1110	1260 kW, 4300 MBH	52,500	23,700	76,200
1120	1620 kW, 5527 MBH	54,500	26,700	81,200
1130	2070 kW, 7063 MBH	72,500	28,100	100,600
1140	2340 kW, 7984 MBH	94,500	31,900	126,400

D3020 Heat Generating Systems

Cast Iron Boiler, Hot Water, Gas Fired

System Components	QUANTITY	UNIT	COST EACH		
			MAT.	INST.	TOTAL
SYSTEM D3020 130 1010					
BOILER, CAST IRON, GAS, HOT WATER, 100 MBH					
Boilers, gas fired, std controls, CI, insulated, HW, gross output 100 MBH	1.000	Ea.	2,625	1,825	4,450
Pipe, black steel, Sch 40, threaded, W/coupling & hangers, 10' OC, 3/4" dia	20.000	L.F.	108	236.25	344.25
Pipe, black steel, Sch 40, threaded, W/coupling & hangers, 10' OC, 1" dia.	20.000	L.F.	145.13	260.15	405.28
Elbow, 90°, black, straight, 3/4" dia.	9.000	Ea.	41.22	409.50	450.72
Elbow, 90°, black, straight, 1" dia.	6.000	Ea.	50.70	294	344.70
Tee, black, straight, 3/4" dia.	2.000	Ea.	10.60	142	152.60
Tee, black, straight, 1" dia.	2.000	Ea.	18.20	160	178.20
Tee, black, reducing, 1" dia.	2.000	Ea.	29.20	160	189.20
Pipe cap, black, 3/4" dia.	1.000	Ea.	3.80	20	23.80
Union, black with brass seat, 3/4" dia.	2.000	Ea.	28.10	98	126.10
Union, black with brass seat, 1" dia.	2.000	Ea.	36.60	107	143.60
Pipe nipples, black, 3/4" dia.	5.000	Ea.	12	26.25	38.25
Pipe nipples, black, 1" dia.	3.000	Ea.	10.13	18.15	28.28
Valves, bronze, gate, N.R.S., threaded, class 150, 3/4" size	2.000	Ea.	139	64	203
Valves, bronze, gate, N.R.S., threaded, class 150, 1" size	2.000	Ea.	177	67	244
Gas cock, brass, 3/4" size	1.000	Ea.	12.30	29	41.30
Thermometer, stem type, 9" case, 8" stem, 3/4" NPT	2.000	Ea.	372	46	418
Tank, steel, liquid expansion, ASME, painted, 15 gallon capacity	1.000	Ea.	615	68.50	683.50
Pump, circulating, bronze, flange connection, 3/4" to 1-1/2" size, 1/8 HP	1.000	Ea.	940	192	1,132
Vent chimney, all fuel, pressure tight, double wall, SS, 6" dia.	20.000	L.F.	1,080	375	1,455
Vent chimney, elbow, 90° fixed, 6" dia.	2.000	Ea.	610	75	685
Vent chimney, Tee, 6" dia.	2.000	Ea.	402	94	496
Vent chimney, ventilated roof thimble, 6" dia.	1.000	Ea.	257	43.50	300.50
Vent chimney, adjustable roof flashing, 6" dia.	1.000	Ea.	72	37.50	109.50
Vent chimney, stack cap, 6" diameter	1.000	Ea.	227	24.50	251.50
Insulation, fiberglass pipe covering, 1" wall, 1" IPS	20.000	L.F.	24.60	95	119.60
TOTAL			8,046.58	4,967.30	13,013.88

D3020 130	Boiler, Cast Iron, Hot Water, Gas	COST EACH		
		MAT.	INST.	TOTAL
1010	Boiler, cast iron, gas, hot water, 100 MBH	8,050	4,975	13,025
1020	200 MBH	8,975	5,675	14,650
1030	320 MBH	9,700	6,175	15,875
1040	440 MBH	12,800	7,975	20,775
1050	544 MBH	21,300	12,400	33,700
1060	765 MBH	26,300	13,300	39,600

D30 HVAC

D3020 Heat Generating Systems

D3020 130	Boiler, Cast Iron, Hot Water, Gas	COST EACH		
		MAT.	INST.	TOTAL
1070	1088 MBH	29,200	14,200	43,400
1080	1530 MBH	34,000	18,800	52,800
1090	2312 MBH	39,900	22,100	62,000
1100	2856 MBH	46,900	23,900	70,800
1110	3808 MBH	52,000	27,900	79,900
1120	4720 MBH	101,500	31,500	133,000
1130	6100 MBH	110,000	35,000	145,000
1140	6970 MBH	118,000	43,200	161,200

D3020 134	Boiler, Cast Iron, Steam, Gas	COST EACH		
		MAT.	INST.	TOTAL
1010	Boiler, cast iron, gas, steam, 100 MBH	6,525	4,825	11,350
1020	200 MBH	13,500	9,050	22,550
1030	320 MBH	15,300	10,200	25,500
1040	544 MBH	22,900	15,800	38,700
1050	765 MBH	26,000	16,500	42,500
1060	1275 MBH	31,000	18,100	49,100
1070	2675 MBH	42,700	26,400	69,100
1080	3570 MBH	50,000	29,600	79,600
1090	4720 MBH	97,000	31,600	128,600
1100	6100 MBH	107,500	36,800	144,300
1110	6970 MBH	115,500	43,000	158,500

D3020 136	Boiler, Cast Iron, Hot Water, Gas/Oil	COST EACH		
		MAT.	INST.	TOTAL
1010	Boiler, cast iron, gas & oil, hot water, 584 MBH	24,900	13,800	38,700
1020	876 MBH	32,000	14,800	46,800
1030	1168 MBH	36,600	16,700	53,300
1040	1460 MBH	43,000	20,100	63,100
1050	2044 MBH	45,800	20,800	66,600
1060	2628 MBH	57,500	25,100	82,600
1070	3210 MBH	65,500	26,900	92,400
1080	3796 MBH	70,000	27,800	97,800
1090	4672 MBH	81,500	29,200	110,700
1100	5256 MBH	96,000	35,400	131,400
1110	6000 MBH	118,000	36,100	154,100
1120	9800 MBH	148,000	62,000	210,000
1130	12,200 MBH	184,500	72,500	257,000
1140	13,500 MBH	200,500	77,000	277,500

D3020 138	Boiler, Cast Iron, Steam, Gas/Oil	COST EACH		
		MAT.	INST.	TOTAL
1010	Boiler, cast iron, gas & oil, steam, 810 MBH	30,600	17,900	48,500
1020	1360 MBH	34,300	19,100	53,400
1030	2040 MBH	44,600	24,500	69,100
1040	2700 MBH	48,800	27,300	76,100
1050	3270 MBH	55,000	30,400	85,400
1060	3770 MBH	91,500	31,000	122,500
1070	4650 MBH	102,500	35,300	137,800
1080	5520 MBH	119,000	37,200	156,200
1090	6100 MBH	120,500	39,100	159,600
1100	6970 MBH	132,000	47,700	179,700

D3020 Heat Generating Systems

Base Mounted End—Suction Pump

System Components	QUANTITY	UNIT	COST EACH MAT.	COST EACH INST.	COST EACH TOTAL
SYSTEM D3020 330 1010					
PUMP, BASE MTD. WITH MOTOR, END-SUCTION, 2-1/2″ SIZE, 3 HP, TO 150 GPM					
Pump, circulating, CI, base mounted, 2-1/2″ size, 3 HP, to 150 GPM	1.000	Ea.	7,450	640	8,090
Pipe, black steel, Sch. 40, on yoke & roll hangers, 10′ O.C., 2-1/2″ dia	12.000	L.F.	127.80	303.12	430.92
Elbow, 90°, weld joint, steel, 2-1/2″ pipe size	1.000	Ea.	30.50	148.46	178.96
Flange, weld neck, 150 LB, 2-1/2″ pipe size	8.000	Ea.	280	593.84	873.84
Valve, iron body, gate, 125 lb., N.R.S., flanged, 2-1/2″ size	1.000	Ea.	620	231	851
Strainer, Y type, iron body, flanged, 125 lb., 2-1/2″ pipe size	1.000	Ea.	113	233	346
Multipurpose valve, CI body, 2″ size	1.000	Ea.	455	81	536
Expansion joint, flanged spool, 6″ F to F, 2-1/2″ dia	2.000	Ea.	572	188	760
T-O-L, weld joint, socket, 1/4″ pipe size, nozzle	2.000	Ea.	14.70	103.10	117.80
T-O-L, weld joint, socket, 1/2″ pipe size, nozzle	2.000	Ea.	14.70	108.24	122.94
Control gauges, pressure or vacuum, 3-1/2″ diameter dial	2.000	Ea.	79	40	119
Pressure/temperature relief plug, 316 SS, 3/4″ OD, 7-1/2″ insertion	2.000	Ea.	188	40	228
Insulation, fiberglass pipe covering, 1-1/2″ wall, 2-1/2″ IPS	12.000	L.F.	34.56	69.60	104.16
Pump control system	1.000	Ea.	1,225	585	1,810
Pump balancing	1.000	Ea.		284	284
TOTAL			11,204.26	3,648.36	14,852.62

D3020 330	Circulating Pump Systems, End Suction	COST EACH MAT.	COST EACH INST.	COST EACH TOTAL
1010	Pump, base mtd with motor, end-suction, 2-1/2″ size, 3 HP, to 150 GPM	11,200	3,650	14,850
1020	3″ size, 5 HP, to 225 GPM	13,400	4,200	17,600
1030	4″ size, 7-1/2 HP, to 350 GPM	15,800	5,025	20,825
1040	5″ size, 15 HP, to 1000 GPM	21,700	7,650	29,350
1050	6″ size, 25 HP, to 1550 GPM	29,700	9,275	38,975

D3020 340	Circulating Pump Systems, Double Suction	COST EACH MAT.	COST EACH INST.	COST EACH TOTAL
1010	Pump, base mtd w/motor, double suction, 6″ size, 50 HP, to 1200 GPM	19,600	13,400	33,000
1020	8″ size, 75 HP, to 2500 GPM	36,000	18,100	54,100
1030	8″ size, 100 HP, to 3000 GPM	48,300	20,300	68,600
1040	10″ size, 150 HP, to 4000 GPM	74,500	23,000	97,500

D3030 Cooling Generating Systems

Chilled Water Supply & Return Piping

Air Cooled Water Chiller Unit

Roof

Insulate
Return
Fan Coil Unit
Supply — Finish Ceiling

Design Assumptions: The chilled water, air cooled systems priced, utilize reciprocating hermetic compressors and propeller-type condenser fans. Piping with pumps and expansion tanks is included based on a two pipe system. No ducting is included and the fan-coil units are cooling only. Water treatment and balancing are not included. Chilled water piping is insulated. Area distribution is through the use of multiple fan coil units. Fewer but larger fan coil units with duct distribution would be approximately the same S.F. cost.

System Components	QUANTITY	UNIT	COST EACH		
			MAT.	INST.	TOTAL
SYSTEM D3030 110 1200					
PACKAGED CHILLER, AIR COOLED, WITH FAN COIL UNIT					
APARTMENT CORRIDORS, 3,000 S.F., 5.50 TON					
Fan coil air conditioning unit, cabinet mounted & filters chilled water	1.000	Ea.	3,482.70	533.40	4,016.10
Water chiller, air conditioning unit, air cooled	1.000	Ea.	7,975	1,993.75	9,968.75
Chilled water unit coil connections	1.000	Ea.	1,275	1,450	2,725
Chilled water distribution piping	440.000	L.F.	7,832	19,580	27,412
TOTAL			20,564.70	23,557.15	44,121.85
COST PER S.F.			6.85	7.85	14.70

*Cooling requirements would lead to choosing a water cooled unit.

D3030 110	Chilled Water, Air Cooled Condenser Systems	COST PER S.F.		
		MAT.	INST.	TOTAL
1180	Packaged chiller, air cooled, with fan coil unit			
1200	Apartment corridors, 3,000 S.F., 5.50 ton	6.87	7.87	14.74
1240	6,000 S.F., 11.00 ton	5.80	6.30	12.10
1280	10,000 S.F., 18.33 ton	5.15	4.78	9.93
1320	20,000 S.F., 36.66 ton	3.71	3.47	7.18
1360	40,000 S.F., 73.33 ton	4.17	3.51	7.68
1440	Banks and libraries, 3,000 S.F., 12.50 ton	11.05	9.25	20.30
1480	6,000 S.F., 25.00 ton	10.25	7.50	17.75
1520	10,000 S.F., 41.66 ton	7.40	5.25	12.65
1560	20,000 S.F., 83.33 ton	7.55	4.75	12.30
1600	40,000 S.F., 167 ton*			
1680	Bars and taverns, 3,000 S.F., 33.25 ton	18.30	11	29.30
1720	6,000 S.F., 66.50 ton	16.60	8.40	25
1760	10,000 S.F., 110.83 ton	13.40	2.74	16.14
1800	20,000 S.F., 220 ton*			
1840	40,000 S.F., 440 ton*			
1920	Bowling alleys, 3,000 S.F., 17.00 ton	14.15	10.40	24.55
1960	6,000 S.F., 34.00 ton	10	7.30	17.30
2000	10,000 S.F., 56.66 ton	8.85	5.25	14.10
2040	20,000 S.F., 113.33 ton	8.50	4.83	13.33
2080	40,000 S.F., 227 ton*			
2160	Department stores, 3,000 S.F., 8.75 ton	9.75	8.75	18.50
2200	6,000 S.F., 17.50 ton	7.75	6.75	14.50
2240	10,000 S.F., 29.17 ton	5.80	4.90	10.70
2280	20,000 S.F., 58.33 ton	4.97	3.59	8.56
2320	40,000 S.F., 116.66 ton	5.45	3.63	9.08

D30 HVAC

D3030 Cooling Generating Systems

D3030 110	Chilled Water, Air Cooled Condenser Systems	COST PER S.F.		
		MAT.	INST.	TOTAL
2400	Drug stores, 3,000 S.F., 20.00 ton	16.15	10.75	26.90
2440	6,000 S.F., 40.00 ton	11.95	8	19.95
2480	10,000 S.F., 66.66 ton	11.75	6.85	18.60
2520	20,000 S.F., 133.33 ton	10.55	5.10	15.65
2560	40,000 S.F., 267 ton*			
2640	Factories, 2,000 S.F., 10.00 ton	9.60	8.90	18.50
2680	6,000 S.F., 20.00 ton	8.80	7.15	15.95
2720	10,000 S.F., 33.33 ton	6.35	5	11.35
2760	20,000 S.F., 66.66 ton	6.45	4.53	10.98
2800	40,000 S.F., 133.33 ton	5.95	3.74	9.69
2880	Food supermarkets, 3,000 S.F., 8.50 ton	9.60	8.70	18.30
2920	6,000 S.F., 17.00 ton	7.60	6.75	14.35
2960	10,000 S.F., 28.33 ton	5.50	4.77	10.27
3000	20,000 S.F., 56.66 ton	4.77	3.52	8.29
3040	40,000 S.F., 113.33 ton	5.15	3.64	8.79
3120	Medical centers, 3,000 S.F., 7.00 ton	8.55	8.50	17.05
3160	6,000 S.F., 14.00 ton	6.70	6.50	13.20
3200	10,000 S.F., 23.33 ton	6.10	4.99	11.09
3240	20,000 S.F., 46.66 ton	4.47	3.66	8.13
3280	40,000 S.F., 93.33 ton	4.64	3.57	8.21
3360	Offices, 3,000 S.F., 9.50 ton	8.85	8.65	17.50
3400	6,000 S.F., 19.00 ton	8.50	7.05	15.55
3440	10,000 S.F., 31.66 ton	6.15	4.97	11.12
3480	20,000 S.F., 63.33 ton	6.45	4.60	11.05
3520	40,000 S.F., 126.66 ton	5.85	3.75	9.60
3600	Restaurants, 3,000 S.F., 15.00 ton	12.55	9.60	22.15
3640	6,000 S.F., 30.00 ton	9.50	7.30	16.80
3680	10,000 S.F., 50.00 ton	8.40	5.35	13.75
3720	20,000 S.F., 100.00 ton	8.65	5.05	13.70
3760	40,000 S.F., 200 ton*			
3840	Schools and colleges, 3,000 S.F., 11.50 ton	10.50	9.10	19.60
3880	6,000 S.F., 23.00 ton	9.70	7.35	17.05
3920	10,000 S.F., 38.33 ton	6.95	5.15	12.10
3960	20,000 S.F., 76.66 ton	7.35	4.81	12.16
4000	40,000 S.F., 153 ton*			

D3030 Cooling Generating Systems

General: Water cooled chillers are available in the same sizes as air cooled units. They are also available in larger capacities.

Design Assumptions: The chilled water systems with water cooled condenser include reciprocating hermetic compressors, water cooling tower, pumps, piping and expansion tanks and are based on a two pipe system. Chilled water piping is insulated. No ducts are included and fan-coil units are cooling only. Area distribution is through use of multiple fan coil units. Fewer but larger fan coil units with duct distribution would be approximately the same S.F. cost. Water treatment and balancing are not included.

System Components	QUANTITY	UNIT	COST EACH		
			MAT.	INST.	TOTAL
SYSTEM D3030 115 1320					
PACKAGED CHILLER, WATER COOLED, WITH FAN COIL UNIT					
APARTMENT CORRIDORS, 4,000 S.F., 7.33 TON					
Fan coil air conditioner unit, cabinet mounted & filters, chilled water	2.000	Ea.	4,641.70	710.91	5,352.61
Water chiller, water cooled, 1 compressor, hermetic scroll,	1.000	Ea.	5,358.60	3,480.80	8,839.40
Cooling tower, draw thru single flow, belt drive	1.000	Ea.	1,378.04	146.60	1,524.64
Cooling tower pumps & piping	1.000	System	689.02	348.18	1,037.20
Chilled water unit coil connections	2.000	Ea.	2,550	2,900	5,450
Chilled water distribution piping	520.000	L.F.	9,256	23,140	32,396
TOTAL			23,873.36	30,726.49	54,599.85
COST PER S.F.			5.97	7.68	13.65

*Cooling requirements would lead to choosing a water cooled unit.

D3030 115	Chilled Water, Cooling Tower Systems		COST PER S.F.		
			MAT.	INST.	TOTAL
1300	Packaged chiller, water cooled, with fan coil unit				
1320	Apartment corridors, 4,000 S.F., 7.33 ton		5.98	7.68	13.66
1360	6,000 S.F., 11.00 ton	R236000 -20	5	6.65	11.65
1400	10,000 S.F., 18.33 ton		5.10	4.92	10.02
1440	20,000 S.F., 26.66 ton		4.05	3.68	7.73
1480	40,000 S.F., 73.33 ton		3.80	3.79	7.59
1520	60,000 S.F., 110.00 ton		3.97	3.89	7.86
1600	Banks and libraries, 4,000 S.F., 16.66 ton		11	8.45	19.45
1640	6,000 S.F., 25.00 ton		9.10	7.45	16.55
1680	10,000 S.F., 41.66 ton		8.10	5.55	13.65
1720	20,000 S.F., 83.33 ton		6.60	5.30	11.90
1760	40,000 S.F., 166.66 ton		7.30	6.35	13.65
1800	60,000 S.F., 250.00 ton		7.40	6.75	14.15
1880	Bars and taverns, 4,000 S.F., 44.33 ton		19.50	10.70	30.20
1920	6,000 S.F., 66.50 ton		19.30	11.10	30.40
1960	10,000 S.F., 110.83 ton		15.40	8.80	24.20
2000	20,000 S.F., 221.66 ton		18.70	8.85	27.55
2040	40,000 S.F., 440 ton*				
2080	60,000 S.F., 660 ton*				
2160	Bowling alleys, 4,000 S.F., 22.66 ton		12.90	9.25	22.15
2200	6,000 S.F., 34.00 ton		11.30	8.20	19.50
2240	10,000 S.F., 56.66 ton		9.80	6	15.80
2280	20,000 S.F., 113.33 ton		8.50	5.60	14.10
2320	40,000 S.F., 226.66 ton		10.40	6.35	16.75

D30 HVAC

D3030 Cooling Generating Systems

D3030 115	Chilled Water, Cooling Tower Systems	COST PER S.F.		
		MAT.	INST.	TOTAL
2360	60,000 S.F., 340 ton			
2440	Department stores, 4,000 S.F., 11.66 ton	6.85	8.35	15.20
2480	6,000 S.F., 17.50 ton	8.05	6.90	14.95
2520	10,000 S.F., 29.17 ton	6.10	5.15	11.25
2560	20,000 S.F., 58.33 ton	5.10	3.82	8.92
2600	40,000 S.F., 116.66 ton	4.97	3.98	8.95
2640	60,000 S.F., 175.00 ton	6.65	6.15	12.80
2720	Drug stores, 4,000 S.F., 26.66 ton	13.55	9.55	23.10
2760	6,000 S.F., 40.00 ton	12.90	8.20	21.10
2800	10,000 S.F., 66.66 ton	12.20	7.50	19.70
2840	20,000 S.F., 133.33 ton	9.85	5.85	15.70
2880	40,000 S.F., 266.67 ton	10.35	7.20	17.55
2920	60,000 S.F., 400 ton*			
3000	Factories, 4,000 S.F., 13.33 ton	9.40	8.05	17.45
3040	6,000 S.F., 20.00 ton	8.15	6.95	15.10
3080	10,000 S.F., 33.33 ton	6.70	5.30	12
3120	20,000 S.F., 66.66 ton	6.60	4.80	11.40
3160	40,000 S.F., 133.33 ton	5.60	4.13	9.73
3200	60,000 S.F., 200.00 ton	6.60	6.50	13.10
3280	Food supermarkets, 4,000 S.F., 11.33 ton	6.75	8.30	15.05
3320	6,000 S.F., 17.00 ton	7.20	6.75	13.95
3360	10,000 S.F., 28.33 ton	6	5.10	11.10
3400	20,000 S.F., 56.66 ton	5.15	3.84	8.99
3440	40,000 S.F., 113.33 ton	4.93	3.99	8.92
3480	60,000 S.F., 170.00 ton	6.55	6.15	12.70
3560	Medical centers, 4.000 S.F., 9.33 ton	5.80	7.60	13.40
3600	6,000 S.F., 14.00 ton	6.95	6.60	13.55
3640	10,000 S.F., 23.33 ton	5.45	4.95	10.40
3680	20,000 S.F., 46.66 ton	4.60	3.74	8.34
3720	40,000 S.F., 93.33 ton	4.42	3.91	8.33
3760	60,000 S.F., 140.00 ton	5.30	6.25	11.55
3840	Offices, 4,000 S.F., 12.66 ton	9.10	7.95	17.05
3880	6,000 S.F., 19.00 ton	8.10	7.10	15.20
3920	10,000 S.F., 31.66 ton	6.70	5.35	12.05
3960	20,000 S.F., 63.33 ton	6.50	4.79	11.29
4000	40,000 S.F., 126.66 ton	5.95	6.10	12.05
4040	60,000 S.F., 190.00 ton	6.40	6.40	12.80
4120	Restaurants, 4,000 S.F., 20.00 ton	11.50	8.60	20.10
4160	6,000 S.F., 30.00 ton	10.15	7.70	17.85
4200	10,000 S.F., 50.00 ton	8.95	5.75	14.70
4240	20,000 S.F., 100.00 ton	7.90	5.60	13.50
4280	40,000 S.F., 200.00 ton	7.80	6.15	13.95
4320	60,000 S.F., 300.00 ton	8.35	7	15.35
4400	Schools and colleges, 4,000 S.F., 15.33 ton	10.35	8.30	18.65
4440	6,000 S.F., 23.00 ton	8.60	7.30	15.90
4480	10,000 S.F., 38.33 ton	7.65	5.45	13.10
4520	20,000 S.F., 76.66 ton	6.40	5.25	11.65
4560	40,000 S.F., 153.33 ton	6.85	6.20	13.05
4600	60,000 S.F., 230.00 ton	6.80	6.50	13.30

D3030 Cooling Generating Systems

Chilled Water Piping

Reciprocating Chiller, Water Cooled

Condenser Water Piping

System Components	QUANTITY	UNIT	COST EACH		
			MAT.	INST.	TOTAL
SYSTEM D3030 130 1010					
CHILLER, RECIPROCATING, WATER COOLED, STD. CONTROLS, 60 TON					
Water chiller, reciprocating, water cooled, 60 ton cooling	1.000	Ea.	38,900	9,700	48,600
Pipe, black steel, welded, Sch. 40, on yoke & roll hgrs, 10' O.C., 4" dia	40.000	L.F.	694	1,278.40	1,972.40
Elbow, 90°, weld joint, butt, 4" pipe size	8.000	Ea.	424	1,905.20	2,329.20
Flange, welding neck, 150 lb., 4" pipe size	16.000	Ea.	744	1,897.12	2,641.12
Valves, iron body, butterfly, lug type, lever actuator, 4" size	4.000	Ea.	772	924	1,696
Expansion joints, bellows type, flange spool, 6" F to F, 4" dia	4.000	Ea.	1,380	552	1,932
T-O-L socket, weld joint, 1/4" pipe size, nozzle	4.000	Ea.	29.40	206.20	235.60
T-O-L socket, weld joint, 3/4" pipe size, nozzle	4.000	Ea.	34	226.80	260.80
Thermometers, stem type, 9" case, 8" stem, 3/4" NPT	4.000	Ea.	744	92	836
Gauges, pressure or vacuum, 3-1/2" diameter dial	4.000	Ea.	158	80	238
Insulation, fiberglass pipe covering, 1-1/2" wall, 4" IPS	20.000	L.F.	68.40	149	217.40
Chiller balancing	1.000	Ea.		550	550
TOTAL			43,947.80	17,560.72	61,508.52

D3030 130	Reciprocating Chiller, Water Cooled	COST EACH		
		MAT.	INST.	TOTAL
1010	Chiller, reciprocating, water cooled, std. controls, 60 Ton	43,900	17,600	61,500
1020	100 Ton	46,600	26,500	73,100
1030	150 Ton	77,000	29,200	106,200
1040	200 Ton	104,000	30,800	134,800

D3030 135	Reciprocating Chiller, Air Cooled	COST EACH		
		MAT.	INST.	TOTAL
1010	Chiller, reciprocating, air cooled, std. controls, 20 Ton	31,000	10,400	41,400
1020	30 Ton	33,100	11,000	44,100
1030	40 Ton	39,700	11,700	51,400
1040	60 Ton	55,000	13,100	68,100
1050	70 Ton	62,500	13,500	76,000
1060	80 Ton	67,500	13,800	81,300
1070	90 Ton	72,000	14,000	86,000
1080	100 Ton	81,000	14,200	95,200
1090	110 Ton	87,500	15,400	102,900
1100	130 Ton	103,500	15,600	119,100

D30 HVAC

D3030 Cooling Generating Systems

D3030 135	Reciprocating Chiller, Air Cooled	COST EACH		
		MAT.	INST.	TOTAL
1110	150 Ton	114,500	15,800	130,300
1120	175 Ton	128,000	16,000	144,000
1130	210 Ton	144,500	17,500	162,000

D3030 Cooling Generating Systems

Chilled Water Piping

Centrifugal Chiller, Water Cooled, Hermetic

Condenser Water Piping

System Components	QUANTITY	UNIT	COST EACH		
			MAT.	INST.	TOTAL
SYSTEM D3030 140 1010					
CHILLER, CENTRIF., WATER COOLED, PKGD. HERMETIC, STD. CONTROLS, 200 TON					
Water chiller, screw, water cooled, not incl tower, 200 ton	1.000	Ea.	90,000	19,400	109,400
Pipe, black steel, welded, Sch. 40, on yoke & roll hgrs, 10' o.c., 6" dia	40.000	L.F.	1,100	2,039.60	3,139.60
Elbow, 90°, weld joint, 6" pipe size	8.000	Ea.	928	2,937.20	3,865.20
Flange, welding neck, 150 lb., 6" pipe size	16.000	Ea.	1,120	2,921.28	4,041.28
Valves, iron body, butterfly, lug type, lever actuator, 6" size	4.000	Ea.	1,300	1,440	2,740
Expansion joints, bellows type, flange spool, 6" F to F, 6" dia	4.000	Ea.	1,740	684	2,424
T-O-L socket, weld jiont, 1/4" pipe size, nozzle	4.000	Ea.	29.40	206.20	235.60
T-O-L socket, weld joint, 3/4" pipe size, nozzle	4.000	Ea.	34	226.80	260.80
Thermometers, stem type, 9" case, 8" stem, 3/4" NPT	4.000	Ea.	744	92	836
Gauges, pressure or vacuum, 3-1/2" diameter dial	4.000	Ea.	158	80	238
Insulation, fiberglass pipe covering, 1-1/2" wall, 6" IPS	40.000	L.F.	162	380	542
Chiller balancing	1.000	Ea.		550	550
TOTAL			97,315.40	30,957.08	128,272.48

D3030 140	Centrifugal Chiller, Water Cooled	COST EACH		
		MAT.	INST.	TOTAL
1010	Chiller, centrif., water cooled, pkgd. hermetic, std controls, 200 Ton	97,500	31,000	128,500
1020	400 Ton	156,500	42,500	199,000
1030	1,000 Ton	503,000	64,000	567,000
1040	1,500 Ton	737,500	78,000	815,500

Chilled Water Piping
(Condenser Water Piping Similar)

Gas Absorption Chiller, Water Cooled

Gas Piping

System Components	QUANTITY	UNIT	COST EACH		
			MAT.	INST.	TOTAL
SYSTEM D3030 145 1010					
CHILLER, GAS FIRED ABSORPTION, WATER COOLED, STD. CONTROLS, 800 TON					
Absorption water chiller, gas fired, water cooled, 800 ton	1.000	Ea.	625,000	29,300	654,300
Pipe, black steel, weld, Sch. 40, on yoke & roll hangers, 10' O.C., 10" dia	20.000	L.F.	1,120	1,519.80	2,639.80
Pipe, black steel, welded, Sch. 40, 14" dia., (two rod roll type hanger)	44.000	L.F.	4,708	5,340.72	10,048.72
Pipe, black steel, welded, Sch. 40, on yoke & roll hangers, 10' O.C., 4" di	20.000	L.F.	347	639.20	986.20
Elbow, 90°, weld joint, 10" pipe size	4.000	Ea.	1,740	2,427.60	4,167.60
Elbow, 90°, weld joint, 14" pipe size	6.000	Ea.	6,450	5,477.40	11,927.40
Elbow, 90°, weld joint, 4" pipe size	3.000	Ea.	159	714.45	873.45
Flange, welding neck, 150 lb., 10" pipe size	8.000	Ea.	1,592	2,439.60	4,031.60
Flange, welding neck, 150 lb., 14" pipe size	14.000	Ea.	5,810	5,711.30	11,521.30
Flange, welding neck, 150 lb., 4" pipe size	3.000	Ea.	139.50	355.71	495.21
Valves, iron body, butterfly, lug type, gear operated, 10" size	2.000	Ea.	1,940	900	2,840
Valves, iron body, butterfly, lug type, gear operated, 14" size	5.000	Ea.	13,625	3,900	17,525
Valves, semi-steel, lubricated plug valve, flanged, 4" pipe size	1.000	Ea.	470	385	855
Expansion joints, bellows type, flange spool, 6" F to F, 10" dia	2.000	Ea.	1,400	466	1,866
Expansion joints, bellows type, flange spool, 10" F to F, 14" dia	2.000	Ea.	2,400	610	3,010
T-O-L socket, weld joint, 1/4" pipe size, nozzle	4.000	Ea.	29.40	206.20	235.60
T-O-L socket, weld joint, 3/4" pipe size, nozzle	4.000	Ea.	34	226.80	260.80
Thermometers, stem type, 9" case, 8" stem, 3/4" NPT	4.000	Ea.	744	92	836
Gauges, pressure or vacuum, 3-1/2" diameter dial	4.000	Ea.	158	80	238
Insulation, fiberglass pipe covering, 2" wall, 10" IPS	20.000	L.F.	180	298	478
Vent chimney, all fuel, pressure tight, double wall, SS, 12" dia	20.000	L.F.	1,680	510	2,190
Chiller balancing	1.000	Ea.		550	550
TOTAL			669,725.90	62,149.78	731,875.68

D3030 145	Gas Absorption Chiller, Water Cooled	COST EACH		
		MAT.	INST.	TOTAL
1010	Chiller, gas fired absorption, water cooled, std. ctrls, 800 Ton	669,500	62,000	731,500
1020	1000 Ton	877,500	79,500	957,000

D3030 150	Steam Absorption Chiller, Water Cooled	COST EACH		
		MAT.	INST.	TOTAL
1010	Chiller, steam absorption, water cooled, std. ctrls, 750 Ton	665,000	73,500	738,500
1020	955 Ton	775,500	88,000	863,500
1030	1465 Ton	1,124,500	101,000	1,225,500
1040	1660 Ton	1,358,000	114,500	1,472,500

D3030 Cooling Generating Systems

Cold Water Fill

Drain To Waste Piping

Cooling Tower

Condenser Water Piping

System Components	QUANTITY	UNIT	COST EACH		
			MAT.	INST.	TOTAL
SYSTEM D3030 310 1010					
COOLING TOWER, GALVANIZED STEEL, PACKAGED UNIT, DRAW THRU, 60 TON					
Cooling towers, draw thru, single flow, belt drive, 60 tons	1.000	Ea.	11,280	1,200	12,480
Pipe, black steel, welded, Sch. 40, on yoke & roll hangers, 10' OC, 4" dia.	40.000	L.F.	694	1,278.40	1,972.40
Tubing, copper, Type L, couplings & hangers 10' OC, 1-1/2" dia.	20.000	L.F.	346	246	592
Elbow, 90°, copper, cu x cu, 1-1/2" dia.	3.000	Ea.	87	147	234
Elbow, 90°, weld joint, 4" pipe size	6.000	Ea.	318	1,428.90	1,746.90
Flange, welding neck, 150 lb., 4" pipe size	6.000	Ea.	279	711.42	990.42
Valves, bronze, gate, N.R.S., soldered, 125 psi, 1-1/2" size	1.000	Ea.	107	49	156
Valves, iron body, butterfly, lug type, lever actuator, 4" size	2.000	Ea.	386	462	848
Electric heat trace system, 400 degree, 115v, 5 watts per L.F.	160.000	L.F.	1,296	179.20	1,475.20
Insulation, fiberglass pipe covering, , 1-1/2" wall, 4" IPS	40.000	L.F.	136.80	298	434.80
Insulation, fiberglass pipe covering, 1" wall, 1-1/2" IPS	20.000	L.F.	28.80	99.40	128.20
Cooling tower condenser control system	1.000	Ea.	5,325	2,625	7,950
Cooling tower balancing	1.000	Ea.		420	420
TOTAL			20,283.60	9,144.32	29,427.92

D3030 310	Galvanized Draw Through Cooling Tower	COST EACH		
		MAT.	INST.	TOTAL
1010	Cooling tower, galvanized steel, packaged unit, draw thru, 60 Ton	20,300	9,150	29,450
1020	110 Ton	28,400	10,800	39,200
1030	300 Ton	42,000	16,300	58,300
1040	600 Ton	68,500	24,300	92,800
1050	1000 Ton	106,000	35,900	141,900

D3030 320	Fiberglass Draw Through Cooling Tower	COST EACH		
		MAT.	INST.	TOTAL
1010	Cooling tower, fiberglass, packaged unit, draw thru, 100 Ton	20,100	9,250	29,350
1020	120 Ton	22,800	10,400	33,200
1030	140 Ton	26,200	13,900	40,100
1040	160 Ton	35,600	18,400	54,000
1050	180 Ton	51,500	26,600	78,100

D3030 330	Stainless Steel Draw Through Cooling Tower	COST EACH		
		MAT.	INST.	TOTAL
1010	Cooling tower, stainless steel, packaged unit, draw thru, 60 Ton	35,500	9,150	44,650
1020	110 Ton	41,900	10,800	52,700
1030	300 Ton	70,000	16,300	86,300
1040	600 Ton	111,500	24,000	135,500
1050	1000 Ton	189,500	35,900	225,400

D3040 Distribution Systems

Two-Way Valve Piping

Three-Way Valve Piping

Built-Up Air Handling Unit

System Components	QUANTITY	UNIT	COST EACH		
			MAT.	INST.	TOTAL
SYSTEM D3040 106 1010					
AHU, FIELD FAB., BLT. UP, COOL/HEAT COILS, FLTRS., CONST. VOL., 40,000 CFM					
AHU, Built-up, cool/heat coils, filters, mix box, constant volume	1.000	Ea.	56,500	6,025	62,525
Pipe, black steel, welded, Sch. 40, on yoke & roll hgrs, 10' OC, 3" dia.	26.000	L.F.	325	723.58	1,048.58
Pipe, black steel, welded, Sch. 40, on yoke & roll hgrs, 10' OC, 4" dia	20.000	L.F.	347	639.20	986.20
Elbow, 90°, weld joint, 3" pipe size	5.000	Ea.	160	850.50	1,010.50
Elbow, 90°, weld joint, 4" pipe size	4.000	Ea.	212	952.60	1,164.60
Tee, welded, reducing on outlet, 3" by 2-1/2" pipe	3.000	Ea.	390	890.70	1,280.70
Tee, welded, reducing on outlet, 4" by 3" pipe	2.000	Ea.	274	793.80	1,067.80
Reducer, eccentric, weld joint, 3" pipe size	4.000	Ea.	260	593.84	853.84
Reducer, eccentric, weld joint, 4" pipe size	4.000	Ea.	332	791.80	1,123.80
Flange, weld neck, 150 LB, 2-1/2" pipe size	3.000	Ea.	105	222.69	327.69
Flange, weld neck, 150 lb., 3" pipe size	14.000	Ea.	539	1,190.70	1,729.70
Flange, weld neck, 150 lb., 4" pipe size	10.000	Ea.	465	1,185.70	1,650.70
Valves, iron body, butterfly, lug type, lever actuator, 3" size	2.000	Ea.	318	288	606
Valves, iron body, butterfly, lug type, lever actuator, 4" size	2.000	Ea.	386	462	848
Strainer, Y type, iron body, flanged, 125 lb., 3" pipe size	1.000	Ea.	146	258	404
Strainer, Y type, iron body, flanged, 125 lb., 4" pipe size	1.000	Ea.	267	390	657
Valve, electric motor actuated, iron body, 3 way, flanged, 2-1/2" pipe	1.000	Ea.	1,225	315	1,540
Valve, electric motor actuated, iron body, two way, flanged, 3" pipe size	1.000	Ea.	1,275	315	1,590
Expansion joints, bellows type, flange spool, 6" F to F, 3" dia	2.000	Ea.	640	204	844
Expansion joints, bellows type, flange spool, 6" F to F, 4" dia	2.000	Ea.	690	276	966
T-O-L socket, weld joint, 1/4" pipe size, nozzle	4.000	Ea.	29.40	206.20	235.60
T-O-L socket, weld joint, 1/2" pipe size, nozzle	12.000	Ea.	88.20	649.44	737.64
Thermometers, stem type, 9" case, 8" stem, 3/4" NPT	4.000	Ea.	744	92	836
Pressure/temperature relief plug, 316 SS, 3/4"OD, 7-1/2" insertion	4.000	Ea.	376	80	456
Insulation, fiberglass pipe covering, 1-1/2" wall, 3" IPS	26.000	L.F.	78.26	159.90	238.16
Insulation, fiberglass pipe covering, 1-1/2" wall, 4" IPS	20.000	L.F.	68.40	149	217.40
Fan coil control system	1.000	Ea.	1,200	585	1,785
Re-heat coil balancing	2.000	Ea.		240	240
Air conditioner balancing	1.000	Ea.		750	750
TOTAL			67,440.26	20,279.65	87,719.91

D3040 106	Field Fabricated Air Handling Unit	COST EACH		
		MAT.	INST.	TOTAL
1010	AHU, Field fab, blt up, cool/heat coils, fltrs, const vol, 40,000 CFM	67,500	20,300	87,800
1020	60,000 CFM	97,000	31,500	128,500
1030	75,000 CFM	118,000	33,200	151,200

D30 HVAC

D3040 Distribution Systems

D3040 108	Field Fabricated VAV Air Handling Unit	COST EACH		
		MAT.	INST.	TOTAL
1010	AHU, Field fab, blt up, cool/heat coils, fltrs, VAV, 75,000 CFM	119,000	38,100	157,100
1020	100,000 CFM	179,500	50,500	230,000
1030	150,000 CFM	297,000	81,000	378,000

Chilled Water Piping   Central Station Air Handling Unit   Hot Water Piping

System Components	QUANTITY	UNIT	COST EACH		
			MAT.	INST.	TOTAL
SYSTEM D3040 110 1010					
AHU, CENTRAL STATION, COOL/HEAT COILS, CONST. VOL., FLTRS., 2,000 CFM					
Central station air handling unit, chilled water, 2000 CFM	1.000	Ea.	10,900	1,050	11,950
Pipe, black steel, Sch 40, threaded, W/cplgs & hangers, 10' OC, 1" dia.	26.000	L.F.	219.38	393.25	612.63
Pipe, black steel, Sch 40, threaded, W/cplg & hangers, 10' OC, 1-1/4" dia.	20.000	L.F.	215.80	336.70	552.50
Elbow, 90°, black, straight, 1" dia.	4.000	Ea.	33.80	196	229.80
Elbow, 90°, black, straight, 1-1/4" dia.	4.000	Ea.	53.60	210	263.60
Tee, black, reducing, 1" dia.	8.000	Ea.	116.80	640	756.80
Tee, black, reducing, 1-1/4" dia.	8.000	Ea.	204	660	864
Reducer, concentric, black, 1" dia.	4.000	Ea.	42.20	170	212.20
Reducer, concentric, black, 1-1/4" dia.	4.000	Ea.	47.20	178	225.20
Union, black with brass seat, 1" dia.	4.000	Ea.	73.20	214	287.20
Union, black with brass seat, 1-1/4" dia.	2.000	Ea.	53	110	163
Pipe nipple, black, 1" dia	13.000	Ea.	43.88	78.65	122.53
Pipe nipple, black, 1-1/4" dia	12.000	Ea.	49.80	77.70	127.50
Valves, bronze, gate, N.R.S., threaded, class 150, 1" size	2.000	Ea.	177	67	244
Valves, bronze, gate, N.R.S., threaded, class 150, 1-1/4" size	2.000	Ea.	240	85	325
Strainer, Y type, bronze body, screwed, 150 lb., 1" pipe size	1.000	Ea.	34	38	72
Strainer, Y type, bronze body, screwed, 150 lb., 1-1/4" pipe size	1.000	Ea.	68	43	111
Valve, electric motor actuated, brass, 3 way, screwed, 3/4" pipe	1.000	Ea.	325	34.50	359.50
Valve, electric motor actuated, brass, 2 way, screwed, 1" pipe size	1.000	Ea.	345	33.50	378.50
Thermometers, stem type, 9" case, 8" stem, 3/4" NPT	4.000	Ea.	744	92	836
Pressure/temperature relief plug, 316 SS, 3/4"OD, 7-1/2" insertion	4.000	Ea.	376	80	456
Insulation, fiberglass pipe covering, 1" wall, 1" IPS	26.000	L.F.	31.98	123.50	155.48
Insulation, fiberglass pipe covering, 1" wall, 1-1/4" IPS	20.000	L.F.	26.60	99.40	126
Fan coil control system	1.000	Ea.	1,200	585	1,785
Re-heat coil balancing	2.000	Ea.		240	240
Multizone A/C balancing	1.000	Ea.		560	560
TOTAL			15,620.24	6,395.20	22,015.44

D3040 110	Central Station Air Handling Unit	COST EACH		
		MAT.	INST.	TOTAL
1010	AHU, Central station, cool/heat coils, constant vol, fltrs, 2,000 CFM	15,600	6,400	22,000
1020	5,000 CFM	22,100	8,225	30,325
1030	10,000 CFM	44,400	11,500	55,900
1040	15,000 CFM	63,000	15,500	78,500
1050	20,000 CFM	76,000	18,000	94,000

D30 HVAC

D3040 Distribution Systems

D3040 112	Central Station Air Handling Unit, VAV	COST EACH		
		MAT.	INST.	TOTAL
1010	AHU, Central station, cool/heat coils, VAV, fltrs, 5,000 CFM	25,900	8,550	34,450
1020	10,000 CFM	46,900	11,700	58,600
1030	15,000 CFM	66,000	15,300	81,300
1040	20,000 CFM	83,000	18,400	101,400
1050	30,000 CFM	109,500	21,200	130,700

Two-Way Valve Piping

Three-Way Valve Piping

Rooftop Air Handling Unit

System Components	QUANTITY	UNIT	COST EACH		
			MAT.	INST.	TOTAL
SYSTEM D3040 114 1010					
AHU, ROOFTOP, COOL/HEAT COILS, FLTRS., CONST. VOL., 2,000 CFM					
AHU, Built-up, cool/heat coils, filter, mix box, rooftop, constant volume	1.000	Ea.	11,100	1,800	12,900
Pipe, black steel, Sch 40, threaded, W/cplgs & hangers, 10' OC, 1" dia.	26.000	L.F.	219.38	393.25	612.63
Pipe, black steel, Sch 40, threaded, W/cplg & hangers, 10' OC, 1-1/4" dia.	20.000	L.F.	215.80	336.70	552.50
Elbow, 90°, black, straight, 1" dia.	4.000	Ea.	33.80	196	229.80
Elbow, 90°, black, straight, 1-1/4" dia.	4.000	Ea.	53.60	210	263.60
Tee, black, reducing, 1" dia.	8.000	Ea.	116.80	640	756.80
Tee, black, reducing, 1-1/4" dia.	8.000	Ea.	204	660	864
Reducer, black, concentric, 1" dia.	4.000	Ea.	42.20	170	212.20
Reducer, black, concentric, 1-1/4" dia.	4.000	Ea.	47.20	178	225.20
Union, black with brass seat, 1" dia.	4.000	Ea.	73.20	214	287.20
Union, black with brass seat, 1-1/4" dia.	2.000	Ea.	53	110	163
Pipe nipples, 1" dia	13.000	Ea.	43.88	78.65	122.53
Pipe nipples, 1-1/4" dia	12.000	Ea.	49.80	77.70	127.50
Valves, bronze, gate, N.R.S., threaded, class 150, 1" size	2.000	Ea.	177	67	244
Valves, bronze, gate, N.R.S., threaded, class 150, 1-1/4" size	2.000	Ea.	240	85	325
Strainer, Y type, bronze body, screwed, 150 lb., 1" pipe size	1.000	Ea.	34	38	72
Strainer, Y type, bronze body, screwed, 150 lb., 1-1/4" pipe size	1.000	Ea.	68	43	111
Valve, electric motor actuated, brass, 3 way, screwed, 3/4" pipe	1.000	Ea.	325	34.50	359.50
Valve, electric motor actuated, brass, 2 way, screwed, 1" pipe size	1.000	Ea.	345	33.50	378.50
Thermometers, stem type, 9" case, 8" stem, 3/4" NPT	4.000	Ea.	744	92	836
Pressure/temperature relief plug, 316 SS, 3/4"OD, 7-1/2" insertion	4.000	Ea.	376	80	456
Insulation, fiberglass pipe covering, 1" wall, 1" IPS	26.000	L.F.	31.98	123.50	155.48
Insulation, fiberglass pipe covering, 1" wall, 1-1/4" IPS	20.000	L.F.	26.60	99.40	126
Insulation pipe covering .016" aluminum jacket finish, add	45.000	S.F.	55.35	234	289.35
Fan coil control system	1.000	Ea.	1,200	585	1,785
Re-heat coil balancing	2.000	Ea.		240	240
Rooftop unit heat/cool balancing	1.000	Ea.		435	435
TOTAL			15,875.59	7,254.20	23,129.79

D3040 114		Rooftop Air Handling Unit	COST EACH		
			MAT.	INST.	TOTAL
1010	AHU, Rooftop, cool/heat coils, constant vol., fltrs, 2,000 CFM		15,900	7,250	23,150
1020	5,000 CFM		33,100	8,450	41,550
1030	10,000 CFM		66,000	11,700	77,700
1040	15,000 CFM		95,500	15,400	110,900

D3040   Distribution Systems

D3040 114	Rooftop Air Handling Unit	COST EACH		
		MAT.	INST.	TOTAL
1050	20,000 CFM	128,500	18,500	147,000

D3040 116	Rooftop Air Handling Unit, VAV	COST EACH		
		MAT.	INST.	TOTAL
1010	AHU, Rooftop, cool/heat coils, VAV, fltrs, 5,000 CFM	36,500	8,625	45,125
1020	10,000 CFM	73,000	11,800	84,800
1030	15,000 CFM	107,500	15,500	123,000
1040	20,000 CFM	135,000	18,700	153,700
1050	30,000 CFM	186,000	21,500	207,500

D3040 Distribution Systems

Galvanized Steel Duct

Fiberglass Duct Insulation

Flexible Fiberglass Duct

Horizontal Fan Coil Air Conditioning System

System Components	QUANTITY	UNIT	COST EACH		
			MAT.	INST.	TOTAL
SYSTEM D3040 124 1010					
FAN COIL A/C SYSTEM, HORIZONTAL, W/HOUSING, CONTROLS, 2 PIPE, 1/2 TON					
Fan coil A/C, horizontal housing, filters, chilled water, 1/2 ton cooling	1.000	Ea.	1,225	145	1,370
Pipe, black steel, Sch 40, threaded, W/cplgs & hangers, 10' OC, 3/4" dia.	20.000	L.F.	103.20	225.75	328.95
Elbow, 90°, black, straight, 3/4" dia.	6.000	Ea.	27.48	273	300.48
Tee, black, straight, 3/4" dia.	2.000	Ea.	10.60	142	152.60
Union, black with brass seat, 3/4" dia.	2.000	Ea.	28.10	98	126.10
Pipe nipples, 3/4" diam	3.000	Ea.	7.20	15.75	22.95
Valves, bronze, gate, N.R.S., threaded, class 150, 3/4" size	2.000	Ea.	139	64	203
Circuit setter, balance valve, bronze body, threaded, 3/4" pipe size	1.000	Ea.	81	32.50	113.50
Insulation, fiberglass pipe covering, 1" wall, 3/4" IPS	20.000	L.F.	22.80	90.80	113.60
Ductwork, 12" x 8" fabricated, galvanized steel, 12 LF	55.000	Lb.	40.15	409.75	449.90
Insulation, ductwork, blanket type, fiberglass, 1" thk, 1-1/2 LB density	40.000	S.F.	30	119.20	149.20
Diffusers, aluminum, OB damper, ceiling, perf, 24"x24" panel size, 6"x6"	2.000	Ea.	127	78	205
Ductwork, flexible, fiberglass fabric, insulated, 1"thk, PE jacket, 6" dia	16.000	L.F.	41.92	69.28	111.20
Round volume control damper 6" dia.	2.000	Ea.	51	57	108
Fan coil unit balancing	1.000	Ea.		81	81
Re-heat coil balancing	1.000	Ea.		120	120
Diffuser/register, high, balancing	2.000	Ea.		224	224
TOTAL			1,934.45	2,245.03	4,179.48

D3040 118	Fan Coil A/C Unit, Two Pipe	COST EACH		
		MAT.	INST.	TOTAL
1010	Fan coil A/C system, cabinet mounted, controls, 2 pipe, 1/2 Ton	1,200	1,200	2,400
1020	1 Ton	1,450	1,350	2,800
1030	1-1/2 Ton	1,575	1,375	2,950
1040	2 Ton	2,475	1,675	4,150
1050	3 Ton	3,100	1,725	4,825

D3040 120	Fan Coil A/C Unit, Two Pipe, Electric Heat	COST EACH		
		MAT.	INST.	TOTAL
1010	Fan coil A/C system, cabinet mntd, elect. ht, controls, 2 pipe, 1/2 Ton	1,725	1,200	2,925
1020	1 Ton	2,125	1,350	3,475
1030	1-1/2 Ton	2,475	1,375	3,850
1040	2 Ton	3,700	1,700	5,400
1050	3 Ton	6,175	1,725	7,900

D3040 Distribution Systems

D3040 122	Fan Coil A/C Unit, Four Pipe	COST EACH		
		MAT.	INST.	TOTAL
1010	Fan coil A/C system, cabinet mounted, controls, 4 pipe, 1/2 Ton	1,950	2,325	4,275
1020	1 Ton	2,250	2,475	4,725
1030	1-1/2 Ton	2,375	2,500	4,875
1040	2 Ton	2,975	2,600	5,575
1050	3 Ton	4,000	2,800	6,800

D3040 124	Fan Coil A/C, Horizontal, Duct Mount, 2 Pipe	COST EACH		
		MAT.	INST.	TOTAL
1010	Fan coil A/C system, horizontal w/housing, controls, 2 pipe, 1/2 Ton	1,925	2,250	4,175
1020	1 Ton	2,500	3,275	5,775
1030	1-1/2 Ton	3,050	4,225	7,275
1040	2 Ton	4,100	5,100	9,200
1050	3 Ton	4,875	6,725	11,600
1060	3-1/2 Ton	5,100	6,975	12,075
1070	4 Ton	5,225	7,900	13,125
1080	5 Ton	6,125	10,000	16,125
1090	6 Ton	6,225	10,800	17,025
1100	7 Ton	6,275	11,200	17,475
1110	8 Ton	6,275	11,400	17,675
1120	10 Ton	7,325	11,800	19,125

D3040 126	Fan Coil A/C, Horiz., Duct Mount, 2 Pipe, Elec. Ht.	COST EACH		
		MAT.	INST.	TOTAL
1010	Fan coil A/C system, horiz. hsng, elect. ht, ctrls, 2 pipe, 1/2 Ton	2,175	2,250	4,425
1020	1 Ton	2,825	3,275	6,100
1030	1-1/2 Ton	3,300	4,225	7,525
1040	2 Ton	4,525	5,100	9,625
1050	3 Ton	7,175	6,950	14,125
1060	3-1/2 Ton	8,100	6,975	15,075
1070	4 Ton	9,125	7,925	17,050
1080	5 Ton	10,500	10,000	20,500
1090	6 Ton	10,500	10,800	21,300
1100	7 Ton	11,100	11,400	22,500
1110	8 Ton	11,700	11,900	23,600
1120	10 Ton	12,500	11,200	23,700

D3040 128	Fan Coil A/C, Horizontal, Duct Mount, 4 Pipe	COST EACH		
		MAT.	INST.	TOTAL
1010	Fan coil A/C system, horiz. w/cabinet, controls, 4 pipe, 1/2 Ton	2,850	3,375	6,225
1020	1 Ton	3,525	4,525	8,050
1030	1-1/2 Ton	4,150	5,375	9,525
1040	2 Ton	4,925	6,025	10,950
1050	3 Ton	6,000	8,025	14,025
1060	3.5 Ton	6,725	9,800	16,525
1070	4 Ton	6,700	9,900	16,600
1080	5 Ton	8,025	13,600	21,625
1090	6 Ton	8,575	16,100	24,675
1100	7 Ton	9,775	18,100	27,875
1110	8 Ton	9,925	19,100	29,025
1120	10 Ton	12,300	23,400	35,700

D3040 Distribution Systems

Fiberglass Duct Insulation

Galvanized Steel Duct

Flexible Insulated Duct

VAV Terminal

Fan Powered VAV Terminal

System Components	QUANTITY	UNIT	COST EACH		
			MAT.	INST.	TOTAL
SYSTEM D3040 132 1010					
VAV TERMINAL, COOLING ONLY, WITH ACTUATOR/CONTROLS, 200 CFM					
Mixing box variable volume 300 to 600 CFM, cool only	1.000	Ea.	451	112.20	563.20
Ductwork, 12" x 8" fabricated, galvanized steel, 12 LF	55.000	Lb.	40.15	409.75	449.90
Insulation, ductwork, blanket type, fiberglass, 1" thk, 1-1/2 LB density	40.000	S.F.	30	119.20	149.20
Diffusers, aluminum, OB damper, ceiling, perf, 24"x24" panel size, 6"x6"	2.000	Ea.	127	78	205
Ductwork, flexible, fiberglass fabric, insulated, 1"thk, PE jacket, 6" dia	16.000	L.F.	41.92	69.28	111.20
Round volume control damper 6" dia.	2.000	Ea.	51	57	108
VAV box balancing	1.000	Ea.		75	75
Diffuser/register, high, balancing	2.000	Ea.		224	224
TOTAL			741.07	1,144.43	1,885.50

D3040 132	VAV Terminal, Cooling Only	COST EACH		
		MAT.	INST.	TOTAL
1010	VAV Terminal, cooling only, with actuator/controls, 200 CFM	741.07	1,144.43	1,885.40
1020	400 CFM	920	2,025	2,945
1030	600 CFM	1,175	3,000	4,175
1040	800 CFM	1,275	3,625	4,900
1050	1000 CFM	1,450	4,225	5,675
1060	1250 CFM	1,700	5,350	7,050
1070	1500 CFM	1,850	6,325	8,175
1080	2000 CFM	2,425	8,950	11,375

D3040 134	VAV Terminal, Hot Water Reheat	COST EACH		
		MAT.	INST.	TOTAL
1010	VAV Terminal, cooling, HW reheat, with actuator/controls, 200 CFM	1,839.09	2,140.03	3,980.70
1020	400 CFM	2,025	2,900	4,925
1030	600 CFM	2,275	3,975	6,250
1040	800 CFM	2,375	4,650	7,025
1050	1000 CFM	2,550	5,225	7,775
1060	1250 CFM	2,875	6,400	9,275
1070	1500 CFM	3,250	7,600	10,850
1080	2000 CFM	3,875	10,300	14,175

D3040 136	Fan Powered VAV Terminal, Cooling Only	COST EACH		
		MAT.	INST.	TOTAL
1010	VAV Terminal, cooling, fan powrd, with actuator/controls, 200 CFM	1,390.07	1,144.23	2,524.40
1020	400 CFM	1,600	1,875	3,475

D30 HVAC

D3040 Distribution Systems

D3040 136	Fan Powered VAV Terminal, Cooling Only	COST EACH		
		MAT.	INST.	TOTAL
1030	600 CFM	1,850	2,975	4,825
1040	800 CFM	2,025	3,575	5,600
1050	1000 CFM	2,200	4,125	6,325
1060	1250 CFM	2,525	5,425	7,950
1070	1500 CFM	2,650	6,350	9,000
1080	2000 CFM	3,350	9,225	12,575

D3040 138	Fan Powered VAV Terminal, Hot Water Reheat	COST EACH		
		MAT.	INST.	TOTAL
1010	VAV Terminal, cool, HW reht, fan powrd, with actuator/ctrls, 200 CFM	2,350	2,125	4,475
1020	400 CFM	2,575	2,925	5,500
1030	600 CFM	2,825	4,000	6,825
1040	800 CFM	3,000	4,575	7,575
1050	1000 CFM	3,175	5,150	8,325
1060	1250 CFM	3,625	6,450	10,075
1070	1500 CFM	4,000	7,675	11,675
1080	2000 CFM	4,650	10,600	15,250

D3040 Distribution Systems

Axial Flow Centrifugal Fan
Belt Drive

Belt Drive Utility Set

Centrifugal Roof Exhaust Fan
Direct Drive

System Components	QUANTITY	UNIT	COST EACH		
			MAT.	INST.	TOTAL
SYSTEM D3040 220 1010					
FAN SYSTEM, IN-LINE CENTRIFUGAL, 500 CFM					
Fans, in-line centrifugal, supply/exhaust, 500 CFM, 10" dia conn	1.000	Ea.	1,275	475	1,750
Louver, alum., W/screen, damper, mill fin, fxd blade, cont line, stormproof	4.050	S.F.	186.30	91.13	277.43
Motor for louver damper, electric or pneumatic	1.000	Ea.	530	44.50	574.50
Ductwork, 12" x 8" fabricated, galvanized steel, 60 LF	200.000	Lb.	187.20	1,859	2,046.20
Grilles, aluminum, air supply, single deflection, adjustable, 12" x 6"	1.000	Ea.	84	81	165
Fire damper, curtain type, vertical, 12" x 8"	1.000	Ea.	27.50	28.50	56
Duct access door, insulated, 10" x 10"	1.000	Ea.	19.95	57	76.95
Duct access door, insulated, 16" x 12"	1.000	Ea.	30	69.50	99.50
Pressure controller/switch, air flow controller	1.000	Ea.	180	54	234
Fan control switch and mounting base	1.000	Ea.	107	40.50	147.50
Fan balancing	1.000	Ea.		375	375
Diffuser/register balancing	3.000	Ea.		225	225
TOTAL			2,626.95	3,400.13	6,027.08

D3040 220	Centrifugal In Line Fan Systems	COST EACH		
		MAT.	INST.	TOTAL
1010	Fan system, in-line centrifugal, 500 CFM	2,625	3,400	6,025
1020	1300 CFM	3,150	5,875	9,025
1030	1500 CFM	3,525	6,800	10,325
1040	2500 CFM	5,550	20,700	26,250
1050	3500 CFM	6,975	27,900	34,875
1060	5000 CFM	9,200	41,600	50,800
1070	7500 CFM	10,800	42,500	53,300
1080	10,000 CFM	13,100	49,800	62,900

D3040 230	Utility Set Fan Systems	COST EACH		
		MAT.	INST.	TOTAL
1010	Utility fan set system, belt drive, 2000 CFM	2,550	11,000	13,550
1020	3500 CFM	4,325	19,400	23,725
1030	5000 CFM	5,525	27,000	32,525
1040	7500 CFM	7,650	31,300	38,950
1050	10,000 CFM	10,300	41,200	51,500
1060	15,000 CFM	10,900	42,000	52,900
1070	20,000 CFM	12,600	47,600	60,200

D30 HVAC

D3040 Distribution Systems

D3040 240	Roof Exhaust Fan Systems	COST EACH		
		MAT.	INST.	TOTAL
1010	Roof vent. system, centrifugal, alum., galv curb, BDD, 500 CFM	1,200	2,225	3,425
1020	800 CFM	1,450	4,300	5,750
1030	1500 CFM	2,050	6,375	8,425
1040	2750 CFM	3,200	14,400	17,600
1050	3500 CFM	3,975	18,500	22,475
1060	5000 CFM	5,650	30,600	36,250
1070	8500 CFM	7,500	39,400	46,900
1080	13,800 CFM	10,500	57,000	67,500

D3040 Controls & Instrumentation

Commercial/Industrial Dust Collection System

System Components	QUANTITY	UNIT	COST EACH		
			MAT.	INST.	TOTAL
SYSTEM D3040 250 1010					
COMMERCIAL/INDUSTRIAL VACUUM DUST COLLECTION SYSTEM, 500 CFM					
Dust collection central vac unit inc. stand, filter & shaker, 500 CFM 2 HP	1.000	Ea.	4,300	595	4,895
Galvanized tubing, 16 ga., 4" OD	80.000	L.F.	480	249.60	729.60
Galvanized tubing, 90° ell, 4" dia.	3.000	Ea.	102	67.50	169.50
Galvanized tubing, 45° ell, 4" dia.	4.000	Ea.	118	90	208
Galvanized tubing, TY slip fit, 4" dia.	4.000	Ea.	288	150	438
Flexible rubber hose, 4" dia.	32.000	L.F.	308.80	99.84	408.64
Galvanized air gate valve, 4" dia.	4.000	Ea.	1,200	196	1,396
Galvanized compression coupling, 4" dia.	36.000	Ea.	864	1,224	2,088
Pipe hangers, steel, 4" pipe size	16.000	Ea.	66.56	168	234.56
TOTAL			7,727.36	2,839.94	10,567.30

D3040 250	Commercial/Industrial Vacuum Dust Collection	COST EACH		
		MAT.	INST.	TOTAL
1010	Commercial/industrial, vacuum dust collection system, 500 CFM	7,727.36	2,839.94	10,575.90
1020	1000 CFM	10,100	3,225	13,325
1030	1500 CFM	12,700	3,575	16,275
1040	3000 CFM	27,900	5,000	32,900
1050	5000 CFM	35,100	7,525	42,625

D3040 Controls & Instrumentation

Commercial Kitchen Exhaust/Make-up Air Rooftop System

System Components	QUANTITY	UNIT	COST EACH		
			MAT.	INST.	TOTAL
SYSTEM D3040 260 1010					
COMMERCIAL KITCHEN EXHAUST/MAKE-UP AIR SYSTEM, ROOFTOP, GAS, 2000 CFM					
Make-up air rooftop unit, gas, 750 MBH	1.000	Ea.	3,480.80	138.55	3,619.35
Pipe, black steel, Sch 40, thrded, nipples, W/cplngs/hngrs, 10' OC, 1" dia.	20.000	L.F.	148.50	266.20	414.70
Elbow, 90°, black steel, straight, 3/4" dia.	2.000	Ea.	9.16	91	100.16
Elbow, 90°, black steel, straight, 1" dia.	3.000	Ea.	25.35	147	172.35
Tee, black steel, reducing, 1" dia.	1.000	Ea.	14.60	80	94.60
Union, black with brass seat, 1" dia.	1.000	Ea.	18.30	53.50	71.80
Pipe nipples, black, 3/4" dia	2.000	Ea.	9.60	21	30.60
Pipe cap, malleable iron, black, 1" dia.	1.000	Ea.	4.59	21.50	26.09
Gas cock, brass, 1" size	1.000	Ea.	19.45	33.50	52.95
Ductwork, 20" x 14" fabricated, galv. steel, 30 LF	220.000	Lb.	154	1,507	1,661
Ductwork, 18" x 10" fabricated, galv. steel, 12 LF	72.000	Lb.	50.40	493.20	543.60
Ductwork, 30" x 10" fabricated, galv. steel, 30 LF	260.000	Lb.	182	1,781	1,963
Ductwork, 16" x 10" fabricated, galv. steel, 12 LF	70.000	Lb.	49	479.50	528.50
Kitchen ventilation, island style, water wash, Stainless Steel	5.000	L.F.	11,875	1,230	13,105
Control system, Make-up Air Unit/Exhauster	1.000	Ea.	2,650	2,275	4,925
Rooftop unit heat/cool balancing	1.000	Ea.		435	435
Roof fan balancing	1.000	Ea.		250	250
TOTAL			18,690.75	9,302.95	27,993.70

D3040 260	Kitchen Exhaust/Make-Up Air	COST EACH		
		MAT.	INST.	TOTAL
1010	Commercial kitchen exhaust/make-up air system, rooftop, gas, 2000 CFM	18,700	9,300	28,000
1020	3000 CFM	28,000	12,900	40,900
1030	5000 CFM	30,800	15,300	46,100
1040	8000 CFM	46,300	22,700	69,000
1050	12,000 CFM	61,000	26,700	87,700
1060	16,000 CFM	75,500	40,800	116,300

D3040　Distribution Systems

Plate Heat Exchanger　　　　Shell and Tube Heat Exchanger

System Components	QUANTITY	UNIT	COST EACH		
			MAT.	INST.	TOTAL
SYSTEM D3040 620 1010					
SHELL & TUBE HEAT EXCHANGER, 40 GPM					
Heat exchanger, 4 pass, 3/4" O.D. copper tubes, by steam at 10 psi, 40 GPM	1.000	Ea.	4,675	291	4,966
Pipe, black steel, Sch 40, threaded, W/couplings & hangers, 10' OC, 2" dia.	40.000	L.F.	636.30	909	1,545.30
Elbow, 90°, straight, 2" dia.	16.000	Ea.	456	1,024	1,480
Reducer, black steel, concentric, 2" dia.	2.000	Ea.	49	110	159
Valves, bronze, gate, rising stem, threaded, class 150, 2" size	6.000	Ea.	1,152	348	1,500
Valves, bronze, globe, class 150, rising stem, threaded, 2" size	2.000	Ea.	1,020	116	1,136
Strainers, Y type, bronze body, screwed, 150 lb., 2" pipe size	2.000	Ea.	194	99	293
Valve, electric motor actuated, brass, 2 way, screwed, 1-1/2" pipe size	1.000	Ea.	450	49.50	499.50
Union, black with brass seat, 2" dia.	8.000	Ea.	304	544	848
Tee, black, straight, 2" dia.	9.000	Ea.	247.50	945	1,192.50
Tee, black, reducing run and outlet, 2" dia.	4.000	Ea.	196	420	616
Pipe nipple, black, 2" dia.	21.000	Ea.	6.30	9	15.30
Thermometers, stem type, 9" case, 8" stem, 3/4" NPT	2.000	Ea.	372	46	418
Gauges, pressure or vacuum, 3-1/2" diameter dial	2.000	Ea.	79	40	119
Insulation, fiberglass pipe covering, 1" wall, 2" IPS	40.000	L.F.	62	208	270
Heat exchanger control system	1.000	Ea.	2,575	2,050	4,625
Coil balancing	1.000	Ea.		119	119
TOTAL			12,474.10	7,327.50	19,801.60

D3040 610	Heat Exchanger, Plate Type	COST EACH		
		MAT.	INST.	TOTAL
1010	Plate heat exchanger, 400 GPM	49,100	14,700	63,800
1020	800 GPM	78,000	18,700	96,700
1030	1200 GPM	118,000	25,300	143,300
1040	1800 GPM	160,500	31,600	192,100

D3040 620	Heat Exchanger, Shell & Tube	COST EACH		
		MAT.	INST.	TOTAL
1010	Shell & tube heat exchanger, 40 GPM	12,500	7,325	19,825
1020	96 GPM	22,200	12,900	35,100
1030	240 GPM	40,500	17,500	58,000
1040	600 GPM	78,500	24,700	103,200

D3050 Terminal & Package Units

Weather Cap Roof Vent →

Roof Flashing →

Vent Pipe ←

90° Elbow ←

Gas Cock →

Tee ←

Tee Cap ←

Cabinet Unit Heater   **Gas Unit Heater**   **Hydronic Unit Heater**

System Components	QUANTITY	UNIT	COST EACH		
			MAT.	INST.	TOTAL
SYSTEM D3050 120 1010					
SPACE HEATER, SUSPENDED, GAS FIRED, PROPELLER FAN, 20 MBH					
Space heater, propeller fan, 20 MBH output	1.000	Ea.	1,075	137	1,212
Pipe, black steel, Sch 40, threaded, W/coupling & hangers, 10' OC, 3/4" dia	20.000	L.F.	100.80	220.50	321.30
Elbow, 90°, black steel, straight, 1/2" dia.	3.000	Ea.	11.19	127.50	138.69
Elbow, 90°, black steel, straight, 3/4" dia.	3.000	Ea.	13.74	136.50	150.24
Tee, black steel, reducing, 3/4" dia.	1.000	Ea.	8.45	71	79.45
Union, black with brass seat, 3/4" dia.	1.000	Ea.	14.05	49	63.05
Pipe nipples, black, 1/2" dia	2.000	Ea.	8.36	20.30	28.66
Pipe nipples, black, 3/4" dia	2.000	Ea.	4.80	10.50	15.30
Pipe cap, black, 3/4" dia.	1.000	Ea.	3.80	20	23.80
Gas cock, brass, 3/4" size	1.000	Ea.	12.30	29	41.30
Thermostat, 1 set back, electric, timed	1.000	Ea.	33	81	114
Wiring, thermostat hook-up,25' of #18-3	1.000	Ea.	10.65	25	35.65
Vent chimney, prefab metal, U.L. listed, gas, double wall, galv. st, 4" dia	12.000	L.F.	91.80	198.60	290.40
Vent chimney, gas, double wall, galv. steel, elbow 90°, 4" dia.	2.000	Ea.	62	66	128
Vent chimney, gas, double wall, galv. steel, Tee, 4" dia.	1.000	Ea.	39	43.50	82.50
Vent chimney, gas, double wall, galv. steel, T cap, 4" dia.	1.000	Ea.	3.01	27	30.01
Vent chimney, gas, double wall, galv. steel, roof flashing, 4" dia.	1.000	Ea.	32	33	65
Vent chimney, gas, double wall, galv. steel, top, 4" dia.	1.000	Ea.	18.50	25.50	44
TOTAL			1,542.45	1,320.90	2,863.35

D3050 120	Unit Heaters, Gas	COST EACH		
		MAT.	INST.	TOTAL
1010	Space heater, suspended, gas fired, propeller fan, 20 MBH	1,550	1,325	2,875
1020	60 MBH	1,825	1,375	3,200
1030	100 MBH	2,050	1,450	3,500
1040	160 MBH	2,375	1,600	3,975
1050	200 MBH	2,825	1,775	4,600
1060	280 MBH	3,525	1,875	5,400
1070	320 MBH	4,650	2,150	6,800

D30 HVAC

D3050 Terminal & Package Units

D3050 130	Unit Heaters, Hydronic	COST EACH		
		MAT.	INST.	TOTAL
1010	Space heater, suspended, horiz. mount, HW, prop. fan, 20 MBH	1,850	1,400	3,250
1020	60 MBH	2,375	1,575	3,950
1030	100 MBH	2,850	1,725	4,575
1040	150 MBH	3,475	1,900	5,375
1050	200 MBH	4,025	2,150	6,175
1060	300 MBH	4,525	2,525	7,050

D3050 140	Cabinet Unit Heaters, Hydronic	COST EACH		
		MAT.	INST.	TOTAL
1010	Unit heater, cabinet type, horizontal blower, hot water, 20 MBH	2,150	1,375	3,525
1020	60 MBH	3,025	1,525	4,550
1030	100 MBH	3,375	1,675	5,050
1040	120 MBH	3,425	1,725	5,150

D3050 Terminal & Package Units

System Description: Rooftop single zone units are electric cooling and gas heat. Duct systems are low velocity, galvanized steel supply and return. Price variations between sizes are due to several factors. Jumps in the cost of the rooftop unit occur when the manufacturer shifts from the largest capacity unit on a small frame to the smallest capacity on the next larger frame, or changes from one compressor to two. As the unit capacity increases for larger areas the duct distribution grows in proportion. For most applications there is a tradeoff point where it is less expensive and more efficient to utilize smaller units with short simple distribution systems. Larger units also require larger initial supply and return ducts which can create a space problem. Supplemental heat may be desired in colder locations. The table below is based on one unit supplying the area listed. The 10,000 S.F. unit for bars and taverns is not listed because a nominal 110 ton unit would be required and this is above the normal single zone rooftop capacity.

System Components	QUANTITY	UNIT	COST EACH		
			MAT.	INST.	TOTAL
SYSTEM D3050 150 1280					
ROOFTOP, SINGLE ZONE, AIR CONDITIONER					
APARTMENT CORRIDORS, 500 S.F., .92 TON					
Rooftop air conditioner, 1 zone, electric cool, standard controls, curb	1.000	Ea.	1,817	759	2,576
Ductwork package for rooftop single zone units	1.000	System	252.08	1,081	1,333.08
TOTAL			2,069.08	1,840	3,909.08
COST PER S.F.			4.14	3.68	7.82

*Size would suggest multiple units.

D3050 150	Rooftop Single Zone Unit Systems		COST PER S.F.		
			MAT.	INST.	TOTAL
1260	Rooftop, single zone, air conditioner				
1280	Apartment corridors, 500 S.F., .92 ton		4.15	3.70	7.85
1320	1,000 S.F., 1.83 ton	R236000 -20	4.11	3.66	7.77
1360	1500 S.F., 2.75 ton		2.64	3.03	5.67
1400	3,000 S.F., 5.50 ton		2.51	2.91	5.42
1440	5,000 S.F., 9.17 ton		2.63	2.65	5.28
1480	10,000 S.F., 18.33 ton		2.71	2.49	5.20
1560	Banks or libraries, 500 S.F., 2.08 ton		9.35	8.30	17.65
1600	1,000 S.F., 4.17 ton		6	6.90	12.90
1640	1,500 S.F., 6.25 ton		5.70	6.65	12.35
1680	3,000 S.F., 12.50 ton		5.95	6.05	12
1720	5,000 S.F., 20.80 ton		6.15	5.65	11.80
1760	10,000 S.F., 41.67 ton		4.96	5.65	10.61
1840	Bars and taverns, 500 S.F. 5.54 ton		13.65	11.10	24.75
1880	1,000 S.F., 11.08 ton		14.35	9.50	23.85
1920	1,500 S.F., 16.62 ton		14.45	8.85	23.30
1960	3,000 S.F., 33.25 ton		11.90	8.45	20.35
2000	5,000 S.F., 55.42 ton		11.15	8.45	19.60
2040	10,000 S.F., 110.83 ton*				
2080	Bowling alleys, 500 S.F., 2.83 ton		8.15	9.40	17.55
2120	1,000 S.F., 5.67 ton		7.75	9	16.75
2160	1,500 S.F., 8.50 ton		8.10	8.20	16.30
2200	3,000 S.F., 17.00 ton		8.15	7.85	16
2240	5,000 S.F., 28.33 ton		6.85	7.65	14.50
2280	10,000 S.F., 56.67 ton		6.50	7.65	14.15
2360	Department stores, 500 S.F., 1.46 ton		6.55	5.85	12.40

D3050 Terminal & Package Units

D3050 150	Rooftop Single Zone Unit Systems	COST PER S.F.		
		MAT.	INST.	TOTAL
2400	1,000 S.F., 2.92 ton	4.21	4.83	9.04
2440	1,500 S.F., 4.37 ton	3.99	4.63	8.62
2480	3,000 S.F., 8.75 ton	4.18	4.22	8.40
2520	5,000 S.F., 14.58 ton	4.20	4.05	8.25
2560	10,000 S.F., 29.17 ton	3.53	3.94	7.47
2640	Drug stores, 500 S.F., 3.33 ton	9.60	11.05	20.65
2680	1,000 S.F., 6.67 ton	9.15	10.60	19.75
2720	1,500 S.F., 10.00 ton	9.55	9.65	19.20
2760	3,000 S.F., 20.00 ton	9.85	9.05	18.90
2800	5,000 S.F., 33.33 ton	8.05	9	17.05
2840	10,000 S.F., 66.67 ton	7.65	9	16.65
2920	Factories, 500 S.F., 1.67 ton	7.50	6.65	14.15
2960	1,000 S.F., 3.33 ton	4.80	5.50	10.30
3000	1,500 S.F., 5.00 ton	4.56	5.30	9.86
3040	3,000 S.F., 10.00 ton	4.78	4.82	9.60
3080	5,000 S.F., 16.67 ton	4.80	4.63	9.43
3120	10,000 S.F., 33.33 ton	4.03	4.50	8.53
3200	Food supermarkets, 500 S.F., 1.42 ton	6.35	5.70	12.05
3240	1,000 S.F., 2.83 ton	4.05	4.68	8.73
3280	1,500 S.F., 4.25 ton	3.88	4.51	8.39
3320	3,000 S.F., 8.50 ton	4.07	4.10	8.17
3360	5,000 S.F., 14.17 ton	4.08	3.93	8.01
3400	10,000 S.F., 28.33 ton	3.43	3.83	7.26
3480	Medical centers, 500 S.F., 1.17 ton	5.25	4.67	9.92
3520	1,000 S.F., 2.33 ton	5.25	4.66	9.91
3560	1,500 S.F., 3.50 ton	3.37	3.86	7.23
3600	3,000 S.F., 7.00 ton	3.20	3.71	6.91
3640	5,000 S.F., 11.67 ton	3.35	3.37	6.72
3680	10,000 S.F., 23.33 ton	3.45	3.17	6.62
3760	Offices, 500 S.F., 1.58 ton	7.15	6.30	13.45
3800	1,000 S.F., 3.17 ton	4.57	5.25	9.82
3840	1,500 S.F., 4.75 ton	4.34	5.05	9.39
3880	3,000 S.F., 9.50 ton	4.54	4.58	9.12
3920	5,000 S.F., 15.83 ton	4.57	4.39	8.96
3960	10,000 S.F., 31.67 ton	3.84	4.27	8.11
4000	Restaurants, 500 S.F., 2.50 ton	11.25	10	21.25
4040	1,000 S.F., 5.00 ton	6.85	7.95	14.80
4080	1,500 S.F., 7.50 ton	7.15	7.25	14.40
4120	3,000 S.F., 15.00 ton	7.20	6.95	14.15
4160	5,000 S.F., 25.00 ton	7.40	6.80	14.20
4200	10,000 S.F., 50.00 ton	5.70	6.75	12.45
4240	Schools and colleges, 500 S.F., 1.92 ton	8.60	7.65	16.25
4280	1,000 S.F., 3.83 ton	5.55	6.35	11.90
4320	1,500 S.F., 5.75 ton	5.25	6.10	11.35
4360	3,000 S.F., 11.50 ton	5.50	5.55	11.05
4400	5,000 S.F., 19.17 ton	5.65	5.20	10.85
4440	10,000 S.F., 38.33 ton	4.57	5.15	9.72
5000				
5100				
5200				
9000	Components of ductwork packages for above systems, per ton of cooling:			
9010	Ductwork; galvanized steel, 120 pounds			
9020	Insulation; fiberglass 2" thick, FRK faced, 52 SF			
9030	Diffusers; aluminum, 24" x 12", one			
9040	Registers; aluminum, return with dampers, one			

D3050 Terminal & Package Units

Roof — **Roof Top Unit**

Return Ducts —

Insulated Supply Ducts

Finish Ceiling

Return Grille (Typ.) **Supply Diff. (Typ.)**

System Description: Rooftop units are multizone with up to 12 zones, and include electric cooling, gas heat, thermostats, filters, supply and return fans complete. Duct systems are low velocity, galvanized steel supply and return with insulated supplies.

Multizone units cost more per ton of cooling than single zone. However, they offer flexibility where load conditions are varied due to heat generating areas or exposure to radiational heating. For example, perimeter offices on the "sunny side" may require cooling at the same

time "shady side" or central offices may require heating. It is possible to accomplish similar results using duct heaters in branches of the single zone unit. However, heater location could be a problem and total system operating energy efficiency could be lower.

System Components		QUANTITY	UNIT	COST EACH		
				MAT.	INST.	TOTAL
SYSTEM D3050 155 1280						
ROOFTOP, MULTIZONE, AIR CONDITIONER						
APARTMENT CORRIDORS, 3,000 S.F., 5.50 TON						
Rooftop multizone unit, standard controls, curb		1.000	Ea.	27,500	1,782	29,282
Ductwork package for rooftop multizone units		1.000	System	2,062.50	12,375	14,437.50
	TOTAL			29,562.50	14,157	43,719.50
	COST PER S.F.			9.85	4.72	14.57

Note A: Small single zone unit recommended.
Note B: A combination of multizone units recommended.

D3050 155	Rooftop Multizone Unit Systems		COST PER S.F.		
			MAT.	INST.	TOTAL
1240	Rooftop, multizone, air conditioner				
1260	Apartment corridors, 1,500 S.F., 2.75 ton. See Note A.				
1280	3,000 S.F., 5.50 ton	R236000 -20	9.87	4.73	14.60
1320	10,000 S.F., 18.30 ton		6.90	4.54	11.44
1360	15,000 S.F., 27.50 ton		6.70	4.53	11.23
1400	20,000 S.F., 36.70 ton		7	4.53	11.53
1440	25,000 S.F., 45.80 ton		6.55	4.55	11.10
1520	Banks or libraries, 1,500 S.F., 6.25 ton		22.50	10.75	33.25
1560	3,000 S.F., 12.50 ton		18.90	10.50	29.40
1600	10,000 S.F., 41.67 ton		14.85	10.35	25.20
1640	15,000 S.F., 62.50 ton		11.05	10.30	21.35
1680	20,000 S.F., 83.33 ton		11	10.30	21.30
1720	25,000 S.F., 104.00 ton		9.85	10.25	20.10
1800	Bars and taverns, 1,500 S.F., 16.62 ton		48.50	15.45	63.95
1840	3,000 S.F., 33.24 ton		40	14.90	54.90
1880	10,000 S.F., 110.83 ton		24	14.85	38.85
1920	15,000 S.F., 165 ton, See Note B				
1960	20,000 S.F., 220 ton, See Note B				
2000	25,000 S.F., 275 ton, See Note B				
2080	Bowling alleys, 1,500 S.F., 8.50 ton		30.50	14.60	45.10
2120	3,000 S.F., 17.00 ton		25.50	14.30	39.80
2160	10,000 S.F., 56.70 ton		20	14.10	34.10
2200	15,000 S.F., 85.00 ton		15	14	29
2240	20,000 S.F., 113.00 ton		13.40	13.90	27.30
2280	25,000 S.F., 140.00 ton see Note B				
2360	Department stores, 1,500 S.F., 4.37 ton, See Note A.				

D3050 Terminal & Package Units

D3050 155	Rooftop Multizone Unit Systems	COST PER S.F.		
		MAT.	INST.	TOTAL
2400	3,000 S.F., 8.75 ton	15.65	7.50	23.15
2440	10,000 S.F., 29.17 ton	11.10	7.20	18.30
2480	15,000 S.F., 43.75 ton	10.40	7.25	17.65
2520	20,000 S.F., 58.33 ton	7.70	7.20	14.90
2560	25,000 S.F., 72.92 ton	7.70	7.20	14.90
2640	Drug stores, 1,500 S.F., 10.00 ton	36	17.15	53.15
2680	3,000 S.F., 20.00 ton	25	16.55	41.55
2720	10,000 S.F., 66.66 ton	17.65	16.45	34.10
2760	15,000 S.F., 100.00 ton	15.80	16.40	32.20
2800	20,000 S.F., 135 ton, See Note B.			
2840	25,000 S.F., 165 ton, See Note B.			
2920	Factories, 1,500 S.F., 5 ton, See Note A.			
2960	3,000 S.F., 10.00 ton	17.90	8.60	26.50
3000	10,000 S.F., 33.33 ton	12.70	8.25	20.95
3040	15,000 S.F., 50.00 ton	11.90	8.30	20.20
3080	20,000 S.F., 66.66 ton	8.80	8.25	17.05
3120	25,000 S.F., 83.33 ton	8.80	8.25	17.05
3200	Food supermarkets, 1,500 S.F., 4.25 ton, See Note A.			
3240	3,000 S.F., 8.50 ton	15.25	7.30	22.55
3280	10,000 S.F., 28.33 ton	10.35	7	17.35
3320	15,000 S.F., 42.50 ton	10.10	7.05	17.15
3360	20,000 S.F., 56.67 ton	7.50	7	14.50
3400	25,000 S.F., 70.83 ton	7.50	7	14.50
3480	Medical centers, 1,500 S.F., 3.5 ton, See Note A.			
3520	3,000 S.F., 7.00 ton	12.55	6	18.55
3560	10,000 S.F., 23.33 ton	8.40	5.75	14.15
3600	15,000 S.F., 35.00 ton	8.90	5.75	14.65
3640	20,000 S.F., 46.66 ton	8.30	5.80	14.10
3680	25,000 S.F., 58.33 ton	6.15	5.75	11.90
3760	Offices, 1,500 S.F., 4.75 ton, See Note A.			
3800	3,000 S.F., 9.50 ton	17	8.15	25.15
3840	10,000 S.F., 31.66 ton	12.05	7.80	19.85
3880	15,000 S.F., 47.50 ton	11.30	7.90	19.20
3920	20,000 S.F., 63.33 ton	8.40	7.80	16.20
3960	25,000 S.F., 79.16 ton	8.35	7.80	16.15
4000	Restaurants, 1,500 S.F., 7.50 ton	27	12.85	39.85
4040	3,000 S.F., 15.00 ton	22.50	12.60	35.10
4080	10,000 S.F., 50.00 ton	17.80	12.45	30.25
4120	15,000 S.F., 75.00 ton	13.25	12.35	25.60
4160	20,000 S.F., 100.00 ton	11.85	12.30	24.15
4200	25,000 S.F., 125 ton, See Note B.			
4240	Schools and colleges, 1,500 S.F., 5.75 ton	20.50	9.85	30.35
4280	3,000 S.F., 11.50 ton	17.40	9.65	27.05
4320	10,000 S.F., 38.33 ton	13.65	9.55	23.20
4360	15,000 S.F., 57.50 ton	10.15	9.45	19.60
4400	20,000 S.F.,76.66 ton	10.15	9.45	19.60
4440	25,000 S.F., 95.83 ton	9.55	9.45	19
4450				
9000	Components of ductwork packages for above systems, per ton of cooling:			
9010	Ductwork; galvanized steel, 240 pounds			
9020	Insulation; fiberglass, 2" thick, FRK faced, 104 SF			
9030	Diffusers; aluminum, 24" x 12", two			
9040	Registers; aluminum, return with dampers, one			

D3050 Terminal & Package Units

System Description: Self-contained, single package water cooled units include cooling tower, pump, piping allowance. Systems for 1000 S.F. and up include duct and diffusers to provide for even distribution of air. Smaller units distribute air through a supply air plenum, which is integral with the unit.

Returns are not ducted and supplies are not insulated.

Hot water or steam heating coils are included but piping to boiler and the boiler itself is not included.

Where local codes or conditions permit single pass cooling for the smaller units, deduct 10%.

System Components	QUANTITY	UNIT	COST EACH		
			MAT.	INST.	TOTAL
SYSTEM D3050 160 1300					
SELF-CONTAINED, WATER COOLED UNIT					
APARTMENT CORRIDORS, 500 S.F., .92 TON					
Self contained, water cooled, single package air conditioner unit	1.000	Ea.	1,285.20	604.80	1,890
Ductwork package for water or air cooled packaged units	1.000	System	70.84	818.80	889.64
Cooling tower, draw thru single flow, belt drive	1.000	Ea.	172.96	18.40	191.36
Cooling tower pumps & piping	1.000	System	86.48	43.70	130.18
TOTAL			1,615.48	1,485.70	3,101.18
COST PER S.F.			3.23	2.97	6.20

D3050 160	Self-contained, Water Cooled Unit Systems	COST PER S.F.		
		MAT.	INST.	TOTAL
1280	Self-contained, water cooled unit	3.23	2.98	6.21
1300	Apartment corridors, 500 S.F., .92 ton	3.25	2.95	6.20
1320	1,000 S.F., 1.83 ton	3.22	2.97	6.19
1360	3,000 S.F., 5.50 ton	2.68	2.49	5.17
1400	5,000 S.F., 9.17 ton	2.55	2.31	4.86
1440	10,000 S.F., 18.33 ton	3.63	2.07	5.70
1520	Banks or libraries, 500 S.F., 2.08 ton	7	3.02	10.02
1560	1,000 S.F., 4.17 ton	6.10	5.65	11.75
1600	3,000 S.F., 12.50 ton	5.80	5.25	11.05
1640	5,000 S.F., 20.80 ton	8.25	4.68	12.93
1680	10,000 S.F., 41.66 ton	7.15	4.69	11.84
1760	Bars and taverns, 500 S.F., 5.54 ton	15.30	5.10	20.40
1800	1,000 S.F., 11.08 ton	14.90	9	23.90
1840	3,000 S.F., 33.25 ton	19.30	7.10	26.40
1880	5,000 S.F., 55.42 ton	18.20	7.35	25.55
1920	10,000 S.F., 110.00 ton	17.80	7.25	25.05
2000	Bowling alleys, 500 S.F., 2.83 ton	9.55	4.12	13.67
2040	1,000 S.F., 5.66 ton	8.30	7.65	15.95
2080	3,000 S.F., 17.00 ton	11.25	6.40	17.65
2120	5,000 S.F., 28.33 ton	10.15	6.15	16.30
2160	10,000 S.F., 56.66 ton	9.50	6.25	15.75
2200	Department stores, 500 S.F., 1.46 ton	4.91	2.13	7.04
2240	1,000 S.F., 2.92 ton	4.27	3.96	8.23
2280	3,000 S.F., 8.75 ton	4.06	3.67	7.73
2320	5,000 S.F., 14.58 ton	5.75	3.29	9.04
2360	10,000 S.F., 29.17 ton	5.20	3.18	8.38

R236000 -20

D3050 Terminal & Package Units

D3050 160	Self-contained, Water Cooled Unit Systems	COST PER S.F.		
		MAT.	INST.	TOTAL
2440	Drug stores, 500 S.F., 3.33 ton	11.25	4.85	16.10
2480	1,000 S.F., 6.66 ton	9.75	9	18.75
2520	3,000 S.F., 20.00 ton	13.20	7.50	20.70
2560	5,000 S.F., 33.33 ton	13.85	7.40	21.25
2600	10,000 S.F., 66.66 ton	11.20	7.40	18.60
2680	Factories, 500 S.F., 1.66 ton	5.60	2.43	8.03
2720	1,000 S.F. 3.37 ton	4.93	4.56	9.49
2760	3,000 S.F., 10.00 ton	4.65	4.20	8.85
2800	5,000 S.F., 16.66 ton	6.60	3.76	10.36
2840	10,000 S.F., 33.33 ton	6	3.63	9.63
2920	Food supermarkets, 500 S.F., 1.42 ton	4.77	2.06	6.83
2960	1,000 S.F., 2.83 ton	4.99	4.58	9.57
3000	3,000 S.F., 8.50 ton	4.15	3.83	7.98
3040	5,000 S.F., 14.17 ton	3.95	3.56	7.51
3080	10,000 S.F., 28.33 ton	5.10	3.08	8.18
3160	Medical centers, 500 S.F., 1.17 ton	3.93	1.70	5.63
3200	1,000 S.F., 2.33 ton	4.11	3.78	7.89
3240	3,000 S.F., 7.00 ton	3.42	3.16	6.58
3280	5,000 S.F., 11.66 ton	3.25	2.94	6.19
3320	10,000 S.F., 23.33 ton	4.62	2.63	7.25
3400	Offices, 500 S.F., 1.58 ton	5.35	2.30	7.65
3440	1,000 S.F., 3.17 ton	5.60	5.10	10.70
3480	3,000 S.F., 9.50 ton	4.41	3.98	8.39
3520	5,000 S.F., 15.83 ton	6.25	3.57	9.82
3560	10,000 S.F., 31.67 ton	5.70	3.44	9.14
3640	Restaurants, 500 S.F., 2.50 ton	8.40	3.64	12.04
3680	1,000 S.F., 5.00 ton	7.35	6.75	14.10
3720	3,000 S.F., 15.00 ton	6.95	6.30	13.25
3760	5,000 S.F., 25.00 ton	9.90	5.65	15.55
3800	10,000 S.F., 50.00 ton	6.10	5.20	11.30
3880	Schools and colleges, 500 S.F., 1.92 ton	6.45	2.79	9.24
3920	1,000 S.F., 3.83 ton	5.60	5.20	10.80
3960	3,000 S.F., 11.50 ton	5.35	4.82	10.17
4000	5,000 S.F., 19.17 ton	7.60	4.32	11.92
4040	10,000 S.F., 38.33 ton	6.60	4.32	10.92
4050				
9000	Components of ductwork packages for above systems, per ton of cooling:			
9010	Ductwork; galvanized steel, 108 pounds			
9020	Diffusers, aluminum, 24" x 12", two			

D30 HVAC

D3050 Terminal & Package Units

System Description: Self-contained air cooled units with remote air cooled condenser and interconnecting tubing. Systems for 1000 S.F. and up include duct and diffusers. Smaller units distribute air directly.

Returns are not ducted and supplies are not insulated.

Potential savings may be realized by using a single zone rooftop system or through-the-wall unit, especially in the smaller capacities, if the application permits.

Hot water or steam heating coils are included but piping to boiler and the boiler itself is not included.

Condenserless models are available for 15% less where remote refrigerant source is available.

System Components	QUANTITY	UNIT	COST EACH MAT.	COST EACH INST.	COST EACH TOTAL
SYSTEM D3050 165 1320					
SELF-CONTAINED, AIR COOLED UNIT					
APARTMENT CORRIDORS, 500 S.F., .92 TON					
Air cooled, package unit	1.000	Ea.	1,300.50	399.50	1,700
Ductwork package for water or air cooled packaged units	1.000	System	70.84	818.80	889.64
Refrigerant piping	1.000	System	487.60	533.60	1,021.20
Air cooled condenser, direct drive, propeller fan	1.000	Ea.	341	150.35	491.35
TOTAL			2,199.94	1,902.25	4,102.19
COST PER S.F.			4.40	3.80	8.20

D3050 165	Self-contained, Air Cooled Unit Systems		COST PER S.F. MAT.	COST PER S.F. INST.	COST PER S.F. TOTAL
1300	Self-contained, air cooled unit				
1320	Apartment corridors, 500 S.F., .92 ton		4.40	3.80	8.20
1360	1,000 S.F., 1.83 ton	R236000 -20	4.34	3.78	8.12
1400	3,000 S.F., 5.50 ton		3.70	3.51	7.21
1440	5,000 S.F., 9.17 ton		3.28	3.33	6.61
1480	10,000 S.F., 18.33 ton		3.31	3.07	6.38
1560	Banks or libraries, 500 S.F., 2.08 ton		9.50	4.86	14.36
1600	1,000 S.F., 4.17 ton		8.35	8	16.35
1640	3,000 S.F., 12.50 ton		7.45	7.60	15.05
1680	5,000 S.F., 20.80 ton		7.55	6.95	14.50
1720	10,000 S.F., 41.66 ton		7.95	6.85	14.80
1800	Bars and taverns, 500 S.F., 5.54 ton		19.70	10.60	30.30
1840	1,000 S.F., 11.08 ton		19.40	15.25	34.65
1880	3,000 S.F., 33.25 ton		20.50	13.20	33.70
1920	5,000 S.F., 55.42 ton		21	13.20	34.20
1960	10,000 S.F., 110.00 ton		21.50	13.10	34.60
2040	Bowling alleys, 500 S.F., 2.83 ton		13	6.65	19.65
2080	1,000 S.F., 5.66 ton		11.45	10.85	22.30
2120	3,000 S.F., 17.00 ton		10.25	9.50	19.75
2160	5,000 S.F., 28.33 ton		10.65	9.30	19.95
2200	10,000 S.F., 56.66 ton		11.10	9.25	20.35
2240	Department stores, 500 S.F., 1.46 ton		6.70	3.41	10.11
2280	1,000 S.F., 2.92 ton		5.90	5.60	11.50
2320	3,000 S.F., 8.75 ton		5.20	5.30	10.50

D30 HVAC

D3050 Terminal & Package Units

D3050 165	Self-contained, Air Cooled Unit Systems	COST PER S.F.		
		MAT.	INST.	TOTAL
2360	5,000 S.F., 14.58 ton	5.20	5.30	10.50
2400	10,000 S.F., 29.17 ton	5.50	4.78	10.28
2480	Drug stores, 500 S.F., 3.33 ton	15.30	7.80	23.10
2520	1,000 S.F., 6.66 ton	13.40	12.75	26.15
2560	3,000 S.F., 20.00 ton	12.15	11.20	23.35
2600	5,000 S.F., 33.33 ton	12.55	10.90	23.45
2640	10,000 S.F., 66.66 ton	13.10	10.90	24
2720	Factories, 500 S.F., 1.66 ton	7.75	3.94	11.69
2760	1,000 S.F., 3.33 ton	6.80	6.40	13.20
2800	3,000 S.F., 10.00 ton	5.95	6.10	12.05
2840	5,000 S.F., 16.66 ton	6	5.60	11.60
2880	10,000 S.F., 33.33 ton	6.30	5.45	11.75
2960	Food supermarkets, 500 S.F., 1.42 ton	6.50	3.32	9.82
3000	1,000 S.F., 2.83 ton	6.70	5.85	12.55
3040	3,000 S.F., 8.50 ton	5.70	5.45	11.15
3080	5,000 S.F., 14.17 ton	5.05	5.15	10.20
3120	10,000 S.F., 28.33 ton	5.35	4.64	9.99
3200	Medical centers, 500 S.F., 1.17 ton	5.40	2.74	8.14
3240	1,000 S.F., 2.33 ton	5.55	4.82	10.37
3280	3,000 S.F., 7.00 ton	4.72	4.47	9.19
3320	5,000 S.F., 16.66 ton	4.19	4.25	8.44
3360	10,000 S.F., 23.33 ton	4.23	3.91	8.14
3440	Offices, 500 S.F., 1.58 ton	7.30	3.71	11.01
3480	1,000 S.F., 3.16 ton	7.55	6.55	14.10
3520	3,000 S.F., 9.50 ton	5.65	5.80	11.45
3560	5,000 S.F., 15.83 ton	5.75	5.30	11.05
3600	10,000 S.F., 31.66 ton	5.95	5.20	11.15
3680	Restaurants, 500 S.F., 2.50 ton	11.45	5.85	17.30
3720	1,000 S.F., 5.00 ton	10.10	9.60	19.70
3760	3,000 S.F., 15.00 ton	8.95	9.10	18.05
3800	5,000 S.F., 25.00 ton	9.40	8.20	17.60
3840	10,000 S.F., 50.00 ton	9.90	8.15	18.05
3920	Schools and colleges, 500 S.F., 1.92 ton	8.80	4.49	13.29
3960	1,000 S.F., 3.83 ton	7.80	7.35	15.15
4000	3,000 S.F., 11.50 ton	6.85	7	13.85
4040	5,000 S.F., 19.17 ton	6.95	6.40	13.35
4080	10,000 S.F., 38.33 ton	7.30	6.30	13.60
4090				
9000	Components of ductwork packages for above systems, per ton of cooling:			
9010	Ductwork; galvanized steel, 102 pounds			
9020	Diffusers; aluminum, 24" x 12", two			

D30 HVAC

D3050 Terminal & Package Units

Refrigerant Piping — **Air Cooled Condensing Unit** — **Roof** — **Supply Duct**

Fin. Ceiling — **Return Grille** — **DX Air Handling Unit** — **Supply Diffuser**

General: Split systems offer several important advantages which should be evaluated when a selection is to be made. They provide a greater degree of flexibility in component selection which permits an accurate match-up of the proper equipment size and type with the particular needs of the building. This allows for maximum use of modern energy saving concepts in heating and cooling. Outdoor installation of the air cooled condensing unit allows space savings in the building and also isolates the equipment operating sounds from building occupants.

Design Assumptions: The systems below are comprised of a direct expansion air handling unit and air cooled condensing unit with interconnecting copper tubing. Ducts and diffusers are also included for distribution of air. Systems are priced for cooling only. Heat can be added as desired either by putting hot water/steam coils into the air unit or into the duct supplying the particular area of need. Gas fired duct furnaces are also available. Refrigerant liquid line is insulated.

System Components	QUANTITY	UNIT	COST EACH		
			MAT.	INST.	TOTAL
SYSTEM D3050 170 1280					
SPLIT SYSTEM, AIR COOLED CONDENSING UNIT					
APARTMENT CORRIDORS, 1,000 S.F., 1.80 TON					
Fan coil AC unit, cabinet mntd & filters direct expansion air cool	1.000	Ea.	530.70	142.74	673.44
Ductwork package, for split system, remote condensing unit	1.000	System	66.34	777.75	844.09
Refrigeration piping	1.000	System	409.92	887.55	1,297.47
Condensing unit, air cooled, incls compressor & standard controls	1.000	Ea.	1,494.50	567.30	2,061.80
TOTAL			2,501.46	2,375.34	4,876.80
COST PER S.F.			2.50	2.38	4.88

*Cooling requirements would lead to choosing a water cooled unit.

D3050 170	Split Systems With Air Cooled Condensing Units		COST PER S.F.		
			MAT.	INST.	TOTAL
1260	Split system, air cooled condensing unit				
1280	Apartment corridors, 1,000 S.F., 1.83 ton		2.50	2.38	4.88
1320	2,000 S.F., 3.66 ton	R236000 -20	1.94	2.40	4.34
1360	5,000 S.F., 9.17 ton		2.14	3.01	5.15
1400	10,000 S.F., 18.33 ton		2.41	3.27	5.68
1440	20,000 S.F., 36.66 ton		2.65	3.33	5.98
1520	Banks and libraries, 1,000 S.F., 4.17 ton		4.40	5.45	9.85
1560	2,000 S.F., 8.33 ton		4.86	6.80	11.66
1600	5,000 S.F., 20.80 ton		5.50	7.45	12.95
1640	10,000 S.F., 41.66 ton		6	7.55	13.55
1680	20,000 S.F., 83.32 ton		7.15	7.90	15.05
1760	Bars and taverns, 1,000 S.F., 11.08 ton		11.30	10.75	22.05
1800	2,000 S.F., 22.16 ton		14.90	12.85	27.75
1840	5,000 S.F., 55.42 ton		13.15	12.05	25.20
1880	10,000 S.F., 110.84 ton		15.70	12.55	28.25
1920	20,000 S.F., 220 ton*				
2000	Bowling alleys, 1,000 S.F., 5.66 ton		6.10	10.20	16.30
2040	2,000 S.F., 11.33 ton		6.65	9.30	15.95
2080	5,000 S.F., 28.33 ton		7.50	10.10	17.60
2120	10,000 S.F., 56.66 ton		8.20	10.30	18.50
2160	20,000 S.F., 113.32 ton		10.70	11.15	21.85
2320	Department stores, 1,000 S.F., 2.92 ton		3.12	3.75	6.87
2360	2,000 S.F., 5.83 ton		3.14	5.25	8.39
2400	5,000 S.F., 14.58 ton		3.40	4.77	8.17
2440	10,000 S.F., 29.17 ton		3.85	5.20	9.05
2480	20,000 S.F., 58.33 ton		4.22	5.30	9.52

D3050 Terminal & Package Units

D3050 170	Split Systems With Air Cooled Condensing Units	COST PER S.F.		
		MAT.	INST.	TOTAL
2560	Drug stores, 1,000 S.F., 6.66 ton	7.15	12	19.15
2600	2,000 S.F., 13.32 ton	7.80	10.90	18.70
2640	5,000 S.F., 33.33 ton	9.65	12.10	21.75
2680	10,000 S.F., 66.66 ton	10.45	12.65	23.10
2720	20,000 S.F., 133.32 ton*			
2800	Factories, 1,000 S.F., 3.33 ton	3.56	4.28	7.84
2840	2,000 S.F., 6.66 ton	3.59	6	9.59
2880	5,000 S.F., 16.66 ton	4.40	5.95	10.35
2920	10,000 S.F., 33.33 ton	4.81	6.05	10.86
2960	20,000 S.F., 66.66 ton	5.20	6.35	11.55
3040	Food supermarkets, 1,000 S.F., 2.83 ton	3.01	3.63	6.64
3080	2,000 S.F., 5.66 ton	3.05	5.10	8.15
3120	5,000 S.F., 14.66 ton	3.31	4.64	7.95
3160	10,000 S.F., 28.33 ton	3.74	5.05	8.79
3200	20,000 S.F., 56.66 ton	4.10	5.15	9.25
3280	Medical centers, 1,000 S.F., 2.33 ton	2.68	2.94	5.62
3320	2,000 S.F., 4.66 ton	2.51	4.20	6.71
3360	5,000 S.F., 11.66 ton	2.72	3.81	6.53
3400	10,000 S.F., 23.33 ton	3.08	4.16	7.24
3440	20,000 S.F., 46.66 ton	3.38	4.23	7.61
3520	Offices, 1,000 S.F., 3.17 ton	3.38	4.08	7.46
3560	2,000 S.F., 6.33 ton	3.41	5.70	9.11
3600	5,000 S.F., 15.83 ton	3.71	5.20	8.91
3640	10,000 S.F., 31.66 ton	4.18	5.65	9.83
3680	20,000 S.F., 63.32 ton	4.95	6	10.95
3760	Restaurants, 1,000 S.F., 5.00 ton	5.40	9	14.40
3800	2,000 S.F., 10.00 ton	5.85	8.20	14.05
3840	5,000 S.F., 25.00 ton	6.60	8.90	15.50
3880	10,000 S.F., 50.00 ton	7.25	9.05	16.30
3920	20,000 S.F., 100.00 ton	9.45	9.85	19.30
4000	Schools and colleges, 1,000 S.F., 3.83 ton	4.05	5.05	9.10
4040	2,000 S.F., 7.66 ton	4.49	6.30	10.79
4080	5,000 S.F., 19.17 ton	5.05	6.85	11.90
4120	10,000 S.F., 38.33 ton	5.55	6.95	12.50
4160	20,000 S.F., 76.66 ton	6	7.25	13.25
5000				
5100				
5200				
9000	Components of ductwork packages for above systems, per ton of cooling:			
9010	Ductwork; galvanized steel, 102 pounds			
9020	Diffusers; aluminum, 24" x 12", two			

D3050 Terminal & Package Units

Gas
Cock

Roof
Curb

Gas Piping

Rooftop Air Conditioner Unit

System Components	QUANTITY	UNIT	COST EACH		
			MAT.	INST.	TOTAL
SYSTEM D3050 175 1010					
A/C, ROOFTOP, DX COOL, GAS HEAT, CURB, ECONOMIZER, FILTERS., 5 TON					
Roof top A/C, curb, economizer, sgl zone, elec cool, gas ht, 5 ton, 112 MBH	1.000	Ea.	5,475	2,075	7,550
Pipe, black steel, Sch 40, threaded, W/cplgs & hangers, 10' OC, 1" dia	20.000	L.F.	141.75	254.10	395.85
Elbow, 90°, black, straight, 3/4" dia.	3.000	Ea.	13.74	136.50	150.24
Elbow, 90°, black, straight, 1" dia.	3.000	Ea.	25.35	147	172.35
Tee, black, reducing, 1" dia.	1.000	Ea.	14.60	80	94.60
Union, black with brass seat, 1" dia.	1.000	Ea.	18.30	53.50	71.80
Pipe nipple, black, 3/4" dia	2.000	Ea.	9.60	21	30.60
Pipe nipple, black, 1" dia	2.000	Ea.	6.75	12.10	18.85
Cap, black, 1" dia.	1.000	Ea.	4.59	21.50	26.09
Gas cock, brass, 1" size	1.000	Ea.	19.45	33.50	52.95
Control system, pneumatic, Rooftop A/C unit	1.000	Ea.	2,650	2,275	4,925
Rooftop unit heat/cool balancing	1.000	Ea.		435	435
TOTAL			8,379.13	5,544.20	13,923.33

D3050 175	Rooftop Air Conditioner, Const. Volume	COST EACH		
		MAT.	INST.	TOTAL
1010	A/C, Rooftop, DX cool, gas heat, curb, economizer, fltrs, 5 Ton	8,375	5,550	13,925
1020	7-1/2 Ton	11,300	5,775	17,075
1030	12-1/2 Ton	17,700	6,350	24,050
1040	18 Ton	23,700	6,925	30,625
1050	25 Ton	30,500	7,900	38,400
1060	40 Ton	39,800	10,500	50,300

D3050 180	Rooftop Air Conditioner, Variable Air Volume	COST EACH		
		MAT.	INST.	TOTAL
1010	A/C, Rooftop, DX cool, gas heat, curb, ecmizr, fltrs, VAV, 12-1/2 Ton	19,600	6,775	26,375
1020	18 Ton	25,900	7,450	33,350
1030	25 Ton	33,200	8,550	41,750
1040	40 Ton	44,700	11,600	56,300
1050	60 Ton	64,000	15,800	79,800
1060	80 Ton	99,500	19,800	119,300

D30 HVAC

D3050 Terminal & Package Units

Computer rooms impose special requirements on air conditioning systems. A prime requirement is reliability, due to the potential monetary loss that could be incurred by a system failure. A second basic requirement is the tolerance of control with which temperature and humidity are regulated, and dust eliminated. As the air conditioning system reliability is so vital, the additional cost of reserve capacity and redundant components is often justified.

System Descriptions: Computer areas may be environmentally controlled by one of three methods as follows:

1. Self-contained Units
These are units built to higher standards of performance and

reliability. They usually contain alarms and controls to indicate component operation failure, filter change, etc. It should be remembered that these units in the room will occupy space that is relatively expensive to build and that all alterations and service of the equipment will also have to be accomplished within the computer area.

2. Decentralized Air Handling Units In operation these are similar to the self-contained units except that their cooling capability comes from remotely located refrigeration equipment as refrigerant or chilled water. As no compressors or refrigerating equipment are required in the air units, they are smaller and require less service than

self-contained units. An added plus for this type of system occurs if some of the computer components themselves also require chilled water for cooling.

3. Central System Supply Cooling is obtained from a central source which, since it is not located within the computer room, may have excess capacity and permit greater flexibility without interfering with the computer components. System performance criteria must still be met.

Note: The costs shown below do not include an allowance for ductwork or piping.

D3050 185	Computer Room Cooling Units	COST EACH		
		MAT.	INST.	TOTAL
0560	Computer room unit, air cooled, includes remote condenser			
0580	3 ton	18,700	2,325	21,025
0600	5 ton	20,000	2,575	22,575
0620	8 ton	37,300	4,300	41,600
0640	10 ton	39,000	4,650	43,650
0660	15 ton	42,900	5,275	48,175
0680	20 ton	52,000	7,550	59,550
0700	23 ton	64,000	8,625	72,625
0800	Chilled water, for connection to existing chiller system			
0820	5 ton	14,300	1,575	15,875
0840	8 ton	14,500	2,325	16,825
0860	10 ton	14,600	2,375	16,975
0880	15 ton	15,300	2,425	17,725
0900	20 ton	16,300	2,525	18,825
0920	23 ton	17,300	2,875	20,175
1000	Glycol system, complete except for interconnecting tubing			
1020	3 ton	23,400	2,900	26,300
1040	5 ton	25,700	3,050	28,750
1060	8 ton	41,200	5,050	46,250
1080	10 ton	43,800	5,550	49,350
1100	15 ton	54,500	6,950	61,450
1120	20 ton	60,000	7,550	67,550
1140	23 ton	62,500	7,875	70,375
1240	Water cooled, not including condenser water supply or cooling tower			
1260	3 ton	19,000	1,875	20,875
1280	5 ton	20,700	2,150	22,850
1300	8 ton	33,900	3,525	37,425
1320	15 ton	41,100	4,300	45,400
1340	20 ton	44,200	4,750	48,950
1360	23 ton	46,300	5,175	51,475

D3050 201	Packaged A/C, Elec. Ht., Const. Volume	COST EACH		
		MAT.	INST.	TOTAL
1010	A/C Packaged, DX, air cooled, electric heat, constant vol., 5 Ton	8,375	3,950	12,325
1020	10 Ton	13,400	5,450	18,850

D3050 Terminal & Package Units

D3050 201	Packaged A/C, Elec. Ht., Const. Volume	COST EACH		
		MAT.	INST.	TOTAL
1030	20 Ton	25,400	7,150	32,550
1040	30 Ton	39,600	7,950	47,550
1050	40 Ton	52,000	9,125	61,125
1060	50 Ton	69,000	10,200	79,200

D3050 202	Packaged Air Conditioner, Elect. Heat, VAV	COST EACH		
		MAT.	INST.	TOTAL
1010	A/C Packaged, DX, air cooled, electric heat, VAV, 10 Ton	15,200	6,775	21,975
1020	20 Ton	28,300	8,925	37,225
1030	30 Ton	43,900	9,725	53,625
1040	40 Ton	57,500	11,200	68,700
1050	50 Ton	75,500	12,400	87,900
1060	60 Ton	86,000	14,500	100,500

D3050 203	Packaged A/C, Hot Wtr. Heat, Const. Volume	COST EACH		
		MAT.	INST.	TOTAL
1010	A/C Packaged, DX, air cooled, H/W heat, constant vol., 5 Ton	7,200	5,825	13,025
1020	10 Ton	10,700	7,650	18,350
1030	20 Ton	22,400	10,300	32,700
1040	30 Ton	33,200	12,700	45,900
1050	40 Ton	46,400	14,300	60,700
1060	50 Ton	59,500	15,500	75,000

D3050 204	Packaged A/C, Hot Wtr. Heat, VAV	COST EACH		
		MAT.	INST.	TOTAL
1010	A/C Packaged, DX, air cooled, H/W heat, VAV, 10 Ton	12,400	8,850	21,250
1020	20 Ton	25,300	11,600	36,900
1030	30 Ton	37,400	14,300	51,700
1040	40 Ton	51,500	15,500	67,000
1050	50 Ton	66,000	18,000	84,000
1060	60 Ton	79,500	20,100	99,600

D3050 Terminal & Package Units

Condenser Supply Water Piping

Condenser Return Water Piping

Self—Contained Air Conditioner Unit, Water Cooled

System Components	QUANTITY	UNIT	COST EACH		
			MAT.	INST.	TOTAL
SYSTEM D3050 210 1010					
A/C, SELF CONTAINED, SINGLE PKG., WATER COOLED, ELECT. HEAT, 5 TON					
Self-contained, water cooled, elect. heat, not inc tower, 5 ton, const vol	1.000	Ea.	8,550	1,500	10,050
Pipe, black steel, Sch 40, threaded, W/cplg & hangers, 10' OC, 1-1/4" dia.	20.000	L.F.	178.45	278.43	456.88
Elbow, 90°, black, straight, 1-1/4" dia.	6.000	Ea.	80.40	315	395.40
Tee, black, straight, 1-1/4" dia.	2.000	Ea.	29.50	165	194.50
Tee, black, reducing, 1-1/4" dia.	2.000	Ea.	51	165	216
Thermometers, stem type, 9" case, 8" stem, 3/4" NPT	2.000	Ea.	372	46	418
Union, black with brass seat, 1-1/4" dia.	2.000	Ea.	53	110	163
Pipe nipple, black, 1-1/4" dia	3.000	Ea.	12.45	19.43	31.88
Valves, bronze, gate, N.R.S., threaded, class 150, 1-1/4" size	2.000	Ea.	240	85	325
Circuit setter, bal valve, bronze body, threaded, 1-1/4" pipe size	1.000	Ea.	154	43	197
Insulation, fiberglass pipe covering, 1" wall, 1-1/4" IPS	20.000	L.F.	26.60	99.40	126
Control system, pneumatic, A/C Unit with heat	1.000	Ea.	2,650	2,275	4,925
Re-heat coil balancing	1.000	Ea.		120	120
Rooftop unit heat/cool balancing	1.000	Ea.		435	435
TOTAL			12,397.40	5,656.26	18,053.66

D3050 210	AC Unit, Package, Elec. Ht., Water Cooled	COST EACH		
		MAT.	INST.	TOTAL
1010	A/C, Self contained, single pkg., water cooled, elect. heat, 5 Ton	12,400	5,650	18,050
1020	10 Ton	19,200	6,975	26,175
1030	20 Ton	28,000	8,800	36,800
1040	30 Ton	43,900	9,575	53,475
1050	40 Ton	56,500	10,600	67,100
1060	50 Ton	74,000	12,700	86,700

D3050 215	AC Unit, Package, Elec. Ht., Water Cooled, VAV	COST EACH		
		MAT.	INST.	TOTAL
1010	A/C, Self contained, single pkg., water cooled, elect. ht, VAV, 10 Ton	20,800	6,975	27,775
1020	20 Ton	30,600	8,800	39,400
1030	30 Ton	48,000	9,575	57,575
1040	40 Ton	61,500	10,600	72,100
1050	50 Ton	80,000	12,700	92,700
1060	60 Ton	91,500	15,300	106,800

D30 HVAC

D3050 Terminal & Package Units

D3050 220	AC Unit, Package, Hot Water Coil, Water Cooled	COST EACH		
		MAT.	INST.	TOTAL
1010	A/C, Self contn'd, single pkg., water cool, H/W ht, const. vol, 5 Ton	11,000	7,850	18,850
1020	10 Ton	16,100	9,350	25,450
1030	20 Ton	40,200	11,700	51,900
1040	30 Ton	52,000	14,700	66,700
1050	40 Ton	64,000	17,900	81,900
1060	50 Ton	79,000	22,000	101,000

D3050 225	AC Unit, Package, HW Coil, Water Cooled, VAV	COST EACH		
		MAT.	INST.	TOTAL
1010	A/C, Self contn'd, single pkg., water cool, H/W ht, VAV, 10 Ton	17,700	8,800	26,500
1020	20 Ton	27,500	11,400	38,900
1030	30 Ton	41,900	14,100	56,000
1040	40 Ton	56,000	15,400	71,400
1050	50 Ton	71,000	18,100	89,100
1060	60 Ton	84,500	19,500	104,000

D3050 Terminal & Package Units

Heat Pump, Air/Air, Rooftop w/Economizer

System Components	QUANTITY	UNIT	COST EACH		
			MAT.	INST.	TOTAL
SYSTEM D3050 245 1010					
HEAT PUMP, ROOF TOP, AIR/AIR, CURB., ECONOMIZER., SUP. ELECT. HEAT, 10 TON					
Heat pump, air to air, rooftop, curb, economizer, elect ht, 10 ton cooling	1.000	Ea.	55,500	4,525	60,025
Ductwork, 32" x 12" fabricated, galvanized steel, 16 LF	176.000	Lb.	128.48	1,311.20	1,439.68
Insulation, ductwork, blanket type, fiberglass, 1" thk, 1-1/2 LB density	117.000	S.F.	87.75	348.66	436.41
Rooftop unit heat/cool balancing	1.000	Ea.		435	435
TOTAL			55,716.23	6,619.86	62,336.09

D3050 230	Heat Pump, Water Source, Central Station	COST EACH		
		MAT.	INST.	TOTAL
1010	Heat pump, central station, water source, constant vol., 5 Ton	4,300	3,175	7,475
1020	10 Ton	11,200	4,450	15,650
1030	20 Ton	18,900	8,200	27,100
1040	30 Ton	27,300	12,000	39,300
1050	40 Ton	37,600	13,100	50,700
1060	50 Ton	42,700	17,600	60,300

D3050 235	Heat Pump, Water Source, Console	COST EACH		
		MAT.	INST.	TOTAL
1020	Heat pump, console, water source, 1 Ton	2,425	1,725	4,150
1030	1-1/2 Ton	2,550	1,800	4,350
1040	2 Ton	3,275	2,125	5,400
1050	3 Ton	3,450	2,275	5,725
1060	3-1/2 Ton	3,750	2,350	6,100

D3050 240	Heat Pump, Horizontal, Duct Mounted	COST EACH		
		MAT.	INST.	TOTAL
1020	Heat pump, horizontal, ducted, water source, 1 Ton	2,775	3,700	6,475
1030	1-1/2 Ton	3,075	4,650	7,725
1040	2 Ton	3,850	5,550	9,400
1050	3 Ton	4,325	7,275	11,600
1060	3-1/2 Ton	4,650	7,525	12,175

D30 HVAC

D3050 Terminal & Package Units

D3050 245	Heat Pump, Roof Top, Air/Air	COST EACH		
		MAT.	INST.	TOTAL
1010	Heat pump, roof top, air/air, curb, economizer, sup. elect heat, 10 Ton	55,500	6,625	62,125
1020	20 Ton	79,500	11,300	90,800
1030	30 Ton	108,000	15,800	123,800
1040	40 Ton	120,000	18,200	138,200
1050	50 Ton	140,000	20,400	160,400
1060	60 Ton	147,500	23,100	170,600

D3050 Terminal & Package Units

Thru—Wall Heat Pump   Thru—Wall Air Conditioner

System Components	QUANTITY	UNIT	COST EACH		
			MAT.	INST.	TOTAL
SYSTEM D3050 255 1010					
A/C UNIT, THRU-THE-WALL, SUPPLEMENTAL ELECT. HEAT, CABINET, LOUVER, 1/2 TON					
A/C unit, thru-wall, electric heat., cabinet, louver, 1/2 ton	1.000	Ea.	785	194	979
TOTAL			785	194	979

D3050 255	Thru-Wall A/C Unit	COST EACH		
		MAT.	INST.	TOTAL
1010	A/C Unit, thru-the-wall, supplemental elect. heat, cabinet, louver, 1/2 Ton	785	194	979
1020	3/4 Ton	890	233	1,123
1030	1 Ton	980	291	1,271
1040	1-1/2 Ton	1,875	485	2,360
1050	2 Ton	1,925	610	2,535

D3050 260	Thru-Wall Heat Pump	COST EACH		
		MAT.	INST.	TOTAL
1010	Heat pump, thru-the-wall, cabinet, louver, 1/2 Ton	2,175	145	2,320
1020	3/4 Ton	2,450	194	2,644
1030	1 Ton	2,525	291	2,816
1040	Supplemental. elect. heat, 1-1/2 Ton	2,800	750	3,550
1050	2 Ton	3,025	775	3,800

D3090 Fume Exhaust Systems

Roof Exhaust Fan

10" Dia. Fiberglass 90° Elbows

10" Dia. Fiberglass Reinforced Duct

Fume Hood

Fume Hood Exhaust

System Components	QUANTITY	UNIT	COST EACH		
			MAT.	INST.	TOTAL
SYSTEM D3090 310 1010		Ea.			
FUME HOOD EXHAUST SYSTEM, 3' LONG, 1000 CFM					
Fume hood	3.000	L.F.	5,925	1,320	7,245
Ductwork, plastic, FM approved for acid fume & smoke, straight, 10" dia	60.000	L.F.	1,320	795	2,115
Ductwork, plastic, FM approved for acid fume & smoke, elbow 90°, 10" dia	3.000	Ea.	342	432	774
Fan, corrosive fume resistant plastic, 1630 CFM, 1/2 HP	1.000	Ea.	5,625	355	5,980
Fan control switch & mounting base	1.000	Ea.	107	40.50	147.50
Roof fan balancing	1.000	Ea.		250	250
TOTAL			13,319	3,192.50	16,511.50

D3090 310	Fume Hood Exhaust Systems	COST EACH		
		MAT.	INST.	TOTAL
1010	Fume hood exhaust system, 3' long, 1000 CFM	13,300	3,200	16,500
1020	4' long, 2000 CFM	15,900	4,050	19,950
1030	6' long, 3500 CFM	21,700	6,025	27,725
1040	6' long, 5000 CFM	30,100	7,100	37,200
1050	10' long, 8000 CFM	40,500	9,725	50,225

D3090 Other HVAC Systems/Equip

Cast Iron Garage Exhaust System

Dual Exhaust System

System Components	QUANTITY	UNIT	COST EACH		
			MAT.	INST.	TOTAL
SYSTEM D3090 320 1040					
GARAGE, EXHAUST, SINGLE 3″ EXHAUST OUTLET, CARS & LIGHT TRUCKS					
A Outlet top assy. for engine exhaust system with adapters and ftngs, 3″ diam	1.000	Ea.	330	39	369
F Bullnose (guide) for engine exhaust system, 3″ diam	1.000	Ea.	38		38
G Galvanized flexible tubing for engine exhaust system, 3″ diam	8.000	L.F.	128.40		128.40
H Adapter for metal tubing end of engine exhaust system, 3″ tail pipe	1.000	Ea.	68.50		68.50
J Pipe, sewer, cast iron, push-on joint, 8″ diam.	18.000	L.F.	846	711	1,557
24″ deep	18.000	L.F.		30.06	30.06
Backfill utility trench by hand, incl. compaction, 8″ wide 24″ deep	18.000	L.F.		42.84	42.84
Stand for blower, concrete over polystyrene core, 6″ high	1.000	Ea.	5.50	19.55	25.05
AC&V duct spiral reducer 10″x8″	1.000	Ea.	15.75	35	50.75
AC&V duct spiral reducer 12″x10″	1.000	Ea.	17.90	47	64.90
AC&V duct spiral preformed 45° elbow 8″ diam	1.000	Ea.	9.80	40	49.80
AC&V utility fan, belt drive, 3 phase, 2000 CFM, 1 HP	2.000	Ea.	2,350	620	2,970
Safety switch, heavy duty fused, 240V, 3 pole, 30 amp	1.000	Ea.	211	186	397
TOTAL			4,020.85	1,770.45	5,791.30

D3090 320	Garage Exhaust Systems	COST PER BAY		
		MAT.	INST.	TOTAL
1040	Garage, single 3″ exhaust outlet, cars & light trucks, one bay	4,025	1,750	5,775
1060	Additional bays up to seven bays	1,025	475	1,500
1500	4″ outlet, trucks, one bay	4,050	1,750	5,800
1520	Additional bays up to six bays	1,075	475	1,550
1600	5″ outlet, diesel trucks, one bay	4,450	1,750	6,200
1650	Additional single bays up to six	1,575	560	2,135
1700	Two adjoining bays	4,450	1,750	6,200
2000	Dual exhaust, 3″ outlets, pair of adjoining bays	5,050	2,225	7,275
2100	Additional pairs of adjoining bays	1,675	560	2,235

D3090 Other HVAC Systems/Equip

NOTES:
1) ALL HOUSEKEEPING PAD REINFORCING SHOULD TO ACI STANDARDS FOR MINIMUM AREA AND CONCRETE COVERAGE
2) HIGH CONCENTRATED LOADS OR POINT SUPPORTED EQUIPMENT REQUIRE SPECIAL TIE DOWN CONSIDERATIONS NOT COVERED BY THIS DETAIL
3) MINIMUM OF (4) HK PAD TO FLOOR ANCHOR BOLTS REQUIRED FOR EACH HOUSEKEEPING PAD.

System Components	QUANTITY	UNIT	COST EACH		
			MAT.	INST.	TOTAL
SYSTEM D3090 750 1000					
SEISMIC PAD FOR 20-38 SELF-CONTAINED UNIT					
Form equipment pad	28.000	L.F.	26.32	110.32	136.64
Reinforcing in hold down pad, #3 to #7	60.000	Lb.	30	35.40	65.40
Concrete 3000 PSI	.866	C.Y.	94.39		94.39
Place concrete pad	.866	C.Y.		29.58	29.58
Finish concrete equipment pad	47.000	S.F.		34.78	34.78
Drill for 3/8 inch slab anchors	10.000	Ea.	.50	84	84.50
Exisiting slab anchors 3/8 inch	10.000	Ea.	7.40	37.70	45.10
Equipment anchors 3/8 inch for plate/foot	8.000	Ea.	13.28	32.24	45.52
TOTAL			171.89	364.02	535.91

D3090 750	Seismic Concrete Equipment Pads	COST PER EACH		
		MAT.	INST.	TOTAL
0990	Seismic pad for self-contained AC units for plate/foot attachment			
1000	Seismic pad for 20-38 tons Self-contained AC unit without economizer .25g	172	365	537
1010	0.5 g seismic acceleration	172	365	537
1020	0.75 g seismic acceleration	172	365	537
1030	1.0 g seismic acceleration	172	365	537
1040	1.5 g seismic acceleration	175	410	585
1050	2.0 g seismic acceleration	188	565	753
1060	2.5 g seismic acceleration	197	710	907
1100	Seismic pad for 40-80 tons Self contained AC unit without economizer .25g	268	525	793
1110	0.5 g seismic acceleration	268	525	793
1120	0.75 g seismic acceleration	268	525	793
1130	1.0 g seismic acceleration	268	525	793
1140	1.5 g seismic acceleration	291	750	1,041
1150	2.0 g seismic acceleration	340	1,150	1,490

D30 HVAC

D3090 Other HVAC Systems/Equip

D3090 750	Seismic Concrete Equipment Pads	COST PER EACH		
		MAT.	INST.	TOTAL
1160	2.5 g seismic acceleration	395	1,650	2,045
1200	Seismic pad for 80-110 tons Self contained AC unit without economizer .25g	295	570	865
1210	0.5 g seismic acceleration	295	570	865
1220	0.75 g seismic acceleration	295	570	865
1230	1.0 g seismic acceleration	295	570	865
1240	1.5 g seismic acceleration	320	805	1,125
1250	2.0 g seismic acceleration	375	1,250	1,625
1260	2.5 g seismic acceleration	430	1,775	2,205
1990	Seismic pad for self-contained AC units for isolator/snubber bolts			
2000	Seismic pad for 20-38 tons Self contained AC unit without economizer .25g	185	395	580
2010	0.5 g seismic acceleration	185	395	580
2020	0.75 g seismic acceleration	185	395	580
2030	1.0 g seismic acceleration	185	395	580
2040	1.5 g seismic acceleration	188	445	633
2050	2.0 g seismic acceleration	204	605	809
2060	2.5 g seismic acceleration	214	750	964
2100	Seismic pad for 40-80 tons Self contained AC unit without economizer .25g	285	565	850
2110	0.5 g seismic acceleration	285	565	850
2120	0.75 g seismic acceleration	285	565	850
2130	1.0 g seismic acceleration	285	565	850
2140	1.5 g seismic acceleration	310	790	1,100
2150	2.0 g seismic acceleration	365	1,200	1,565
2160	2.5 g seismic acceleration	420	1,725	2,145
2200	Seismic pad for 80-110 tons Self contained AC unit without economizer .25g	310	610	920
2210	0.5 g seismic acceleration	310	610	920
2220	0.75 g seismic acceleration	310	610	920
2230	1.0 g seismic acceleration	310	610	920
2240	1.5 g seismic acceleration	335	845	1,180
2250	2.0 g seismic acceleration	395	1,300	1,695
2260	2.5 g seismic acceleration	455	1,850	2,305
2990	Seismic pad for self-contained AC economizer units for plate/foot attachment			
3000	Seismic pad for 20-38 tons Self contained AC unit economizer .25 g-2.0 g	115	213	328
3010	2.5 g seismic acceleration	122	282	404
3100	Seismic pad for 40-80 tons Self contained AC unit economizer .25-1.5 g	153	282	435
3110	2.0 g seismic acceleration	158	325	483
3120	2.5 g seismic acceleration	164	385	549
3190	Seismic pad for self-contained AC unit economizer for isolator/snubber bolts			
3200	Seismic pad for 20-38 tons Self contained AC unit economizer .25 g-2.0 g	120	225	345
3210	2.5 g seismic acceleration	130	300	430
3300	Seismic pad for 40-80 tons Self contained AC unit economizer .25-1.5 g	163	305	468
3310	2.0 g seismic acceleration	168	350	518
3320	2.5 g seismic acceleration	174	410	584

G1030 Site Earthwork

Trenching Systems are shown on a cost per linear foot basis. The systems include: excavation; backfill and removal of spoil; and compaction for various depths and trench bottom widths. The backfill has been reduced to accommodate a pipe of suitable diameter and bedding.

The slope for trench sides varies from none to 1:1.

The Expanded System Listing shows Trenching Systems that range from 2' to 12' in width. Depths range from 2' to 25'.

System Components	QUANTITY	UNIT	COST PER L.F.		
			EQUIP.	LABOR	TOTAL
SYSTEM G1030 805 1310					
TRENCHING COMMON EARTH, NO SLOPE, 2' WIDE, 2' DP, 3/8 C.Y. BUCKET					
Excavation, trench, hyd. backhoe, track mtd., 3/8 C.Y. bucket	.148	B.C.Y.	.35	.95	1.30
Backfill and load spoil, from stockpile	.153	L.C.Y.	.10	.29	.39
Compaction by vibrating plate, 6" lifts, 4 passes	.118	E.C.Y.	.03	.36	.39
Remove excess spoil, 8 C.Y. dump truck, 2 mile roundtrip	.040	L.C.Y.	.12	.14	.26
TOTAL			.60	1.74	2.34

G1030 805	Trenching Common Earth	COST PER L.F.		
		EQUIP.	LABOR	TOTAL
1310	Trenching, common earth, no slope, 2' wide, 2' deep, 3/8 C.Y. bucket	.60	1.74	2.34
1320	3' deep, 3/8 C.Y. bucket	.85	2.61	3.46
1330	4' deep, 3/8 C.Y. bucket	1.10	3.48	4.58
1340	6' deep, 3/8 C.Y. bucket	1.42	4.54	5.96
1350	8' deep, 1/2 C.Y. bucket	1.88	6	7.88
1360	10' deep, 1 C.Y. bucket	3.19	7.15	10.34
1400	4' wide, 2' deep, 3/8 C.Y. bucket	1.33	3.40	4.73
1410	3' deep, 3/8 C.Y. bucket	1.84	5.15	6.99
1420	4' deep, 1/2 C.Y. bucket	2.12	5.80	7.92
1430	6' deep, 1/2 C.Y. bucket	3.44	9.30	12.74
1440	8' deep, 1/2 C.Y. bucket	5.95	12	17.95
1450	10' deep, 1 C.Y. bucket	7.20	14.90	22.10
1460	12' deep, 1 C.Y. bucket	9.35	19	28.35
1470	15' deep, 1-1/2 C.Y. bucket	8.05	16.95	25
1480	18' deep, 2-1/2 C.Y. bucket	11.15	23.50	34.65
1520	6' wide, 6' deep, 5/8 C.Y. bucket w/trench box	7.90	13.75	21.65
1530	8' deep, 3/4 C.Y. bucket	10.45	18.15	28.60
1540	10' deep, 1 C.Y. bucket	10.15	18.80	28.95
1550	12' deep, 1-1/2 C.Y. bucket	10.90	20	30.90
1560	16' deep, 2-1/2 C.Y. bucket	15.60	25.50	41.10
1570	20' deep, 3-1/2 C.Y. bucket	18.85	30	48.85
1580	24' deep, 3-1/2 C.Y. bucket	22.50	36.50	59
1640	8' wide, 12' deep, 1-1/2 C.Y. bucket w/trench box	15.20	25.50	40.70
1650	15' deep, 1-1/2 C.Y. bucket	19.95	33.50	53.45
1660	18' deep, 2-1/2 C.Y. bucket	22.50	33.50	56
1680	24' deep, 3-1/2 C.Y. bucket	30.50	47	77.50
1730	10' wide, 20' deep, 3-1/2 C.Y. bucket w/trench box	24.50	44.50	69
1740	24' deep, 3-1/2 C.Y. bucket	37	53.50	90.50
1780	12' wide, 20' deep, 3-1/2 C.Y. bucket w/trench box	39	56.50	95.50
1790	25' deep, bucket	48	72	120
1800	1/2 to 1 slope, 2' wide, 2' deep, 3/8 C.Y. bucket	.85	2.61	3.46
1810	3' deep, 3/8 C.Y. bucket	1.42	4.59	6.01

G10 Site Preparation

G1030   Site Earthwork

G1030 805	Trenching Common Earth	COST PER L.F.		
		EQUIP.	LABOR	TOTAL
1820	4' deep, 3/8 C.Y. bucket	2.13	7	9.13
1840	6' deep, 3/8 C.Y. bucket	3.42	11.40	14.82
1860	8' deep, 1/2 C.Y. bucket	5.45	18.20	23.65
1880	10' deep, 1 C.Y. bucket	11	25.50	36.50
2300	4' wide, 2' deep, 3/8 C.Y. bucket	1.59	4.29	5.88
2310	3' deep, 3/8 C.Y. bucket	2.42	7.15	9.57
2320	4' deep, 1/2 C.Y. bucket	3.03	8.85	11.88
2340	6' deep, 1/2 C.Y. bucket	5.80	16.55	22.35
2360	8' deep, 1/2 C.Y. bucket	11.55	24	35.55
2380	10' deep, 1 C.Y. bucket	16	34	50
2400	12' deep, 1 C.Y. bucket	20.50	46	66.50
2430	15' deep, 1-1/2 C.Y. bucket	23	49.50	72.50
2460	18' deep, 2-1/2 C.Y. bucket	42	77.50	119.50
2840	6' wide, 6' deep, 5/8 C.Y. bucket w/trench box	11.70	20.50	32.20
2860	8' deep, 3/4 C.Y. bucket	17	30.50	47.50
2880	10' deep, 1 C.Y. bucket	16.40	31.50	47.90
2900	12' deep, 1-1/2 C.Y. bucket	21	41	62
2940	16' deep, 2-1/2 C.Y. bucket	36	60.50	96.50
2980	20' deep, 3-1/2 C.Y. bucket	47	81	128
3020	24' deep, 3-1/2 C.Y. bucket	67	111	178
3100	8' wide, 12' deep, 1-1/2 C.Y. bucket w/trench box	25.50	46.50	72
3120	15' deep, 1-1/2 C.Y. bucket	37.50	68	105.50
3140	18' deep, 2-1/2 C.Y. bucket	49	80.50	129.50
3180	24' deep, 3-1/2 C.Y. bucket	75	122	197
3270	10' wide, 20' deep, 3-1/2 C.Y. bucket w/trench box	47.50	93.50	141
3280	24' deep, 3-1/2 C.Y. bucket	83	133	216
3370	12' wide, 20' deep, 3-1/2 C.Y. bucket w/trench box	69.50	108	177.50
3380	25' deep, 3-1/2 C.Y. bucket	96.50	155	251.50
3500	1 to 1 slope, 2' wide, 2' deep, 3/8 C.Y. bucket	1.11	3.49	4.60
3520	3' deep, 3/8 C.Y. bucket	3.04	7.80	10.84
3540	4' deep, 3/8 C.Y. bucket	3.15	10.55	13.70
3560	6' deep, 3/8 C.Y. bucket	3.42	11.40	14.82
3580	8' deep, 1/2 C.Y. bucket	6.75	23	29.75
3600	10' deep, 1 C.Y. bucket	18.80	43.50	62.30
3800	4' wide, 2' deep, 3/8 C.Y. bucket	1.84	5.15	6.99
3820	3' deep, 3/8 C.Y. bucket	2.99	9.10	12.09
3840	4' deep, 1/2 C.Y. bucket	3.91	11.90	15.81
3860	6' deep, 1/2 C.Y. bucket	8.15	24	32.15
3880	8' deep, 1/2 C.Y. bucket	17.15	36.50	53.65
3900	10' deep, 1 C.Y. bucket	25	53.50	78.50
3920	12' deep, 1 C.Y. bucket	36.50	77	113.50
3940	15' deep, 1-1/2 C.Y. bucket	37.50	82.50	120
3960	18' deep, 2-1/2 C.Y. bucket	58	107	165
4030	6' wide, 6' deep, 5/8 C.Y. bucket w/trench box	15.40	27.50	42.90
4040	8' deep, 3/4 C.Y. bucket	22.50	38	60.50
4050	10' deep, 1 C.Y. bucket	23.50	46	69.50
4060	12' deep, 1-1/2 C.Y. bucket	32	62.50	94.50
4070	16' deep, 2-1/2 C.Y. bucket	56	95.50	151.50
4080	20' deep, 3-1/2 C.Y. bucket	76	132	208
4090	24' deep, 3-1/2 C.Y. bucket	111	186	297
4500	8' wide, 12' deep, 1-1/2 C.Y. bucket w/trench box	36	68	104
4550	15' deep, 1-1/2 C.Y. bucket	55	102	157
4600	18' deep, 2-1/2 C.Y. bucket	74.50	125	199.50
4650	24' deep, 3-1/2 C.Y. bucket	119	196	315
4800	10' wide, 20' deep, 3-1/2 C.Y. bucket w/trench box	70.50	142	212.50
4850	24' deep, 3-1/2 C.Y. bucket	127	208	335
4950	12' wide, 20' deep, 3-1/2 C.Y. bucket w/trench box	100	160	260
4980	25' deep, 3-1/2 C.Y. bucket	144	235	379

G1030 Site Earthwork

Trenching Systems are shown on a cost per linear foot basis. The systems include: excavation; backfill and removal of spoil; and compaction for various depths and trench bottom widths. The backfill has been reduced to accommodate a pipe of suitable diameter and bedding.

The slope for trench sides varies from none to 1:1.

The Expanded System Listing shows Trenching Systems that range from 2' to 12' in width. Depths range from 2' to 25'.

System Components	QUANTITY	UNIT	COST PER L.F.		
			EQUIP.	LABOR	TOTAL
SYSTEM G1030 806 1310					
TRENCHING LOAM & SANDY CLAY, NO SLOPE, 2' WIDE, 2' DP, 3/8 C.Y. BUCKET					
Excavation, trench, hyd. backhoe, track mtd., 3/8 C.Y. bucket	.148	B.C.Y.	.32	.88	1.20
Backfill and load spoil, from stockpile	.165	L.C.Y.	.11	.31	.42
Compaction by vibrating plate 18" wide, 6" lifts, 4 passes	.118	E.C.Y.	.03	.36	.39
Remove excess spoil, 8 C.Y. dump truck, 2 mile roundtrip	.042	L.C.Y.	.12	.15	.27
TOTAL			.60	1.74	2.34

G1030 806	Trenching Loam & Sandy Clay	COST PER L.F.		
		EQUIP.	LABOR	TOTAL
1310	Trenching, loam & sandy clay, no slope, 2' wide, 2' deep, 3/8 C.Y. bucket	.58	1.70	2.28
1320	3' deep, 3/8 C.Y. bucket	.92	2.80	3.72
1330	4' deep, 3/8 C.Y. bucket	1.07	3.41	4.48
1340	6' deep, 3/8 C.Y. bucket	1.54	4.06	5.60
1350	8' deep, 1/2 C.Y. bucket	2.03	5.35	7.40
1360	10' deep, 1 C.Y. bucket	2.31	5.75	8.05
1400	4' wide, 2' deep, 3/8 C.Y. bucket	1.32	3.33	4.65
1410	3' deep, 3/8 C.Y. bucket	1.82	5.05	6.85
1420	4' deep, 1/2 C.Y. bucket	2.10	5.70	7.80
1430	6' deep, 1/2 C.Y. bucket	3.68	8.35	12.05
1440	8' deep, 1/2 C.Y. bucket	5.75	11.80	17.55
1450	10' deep, 1 C.Y. bucket	5.45	12.10	17.55
1460	12' deep, 1 C.Y. bucket	6.90	15.05	22
1470	15' deep, 1-1/2 C.Y. bucket	8.40	17.55	26
1480	18' deep, 2-1/2 C.Y. bucket	10.40	19.25	29.50
1520	6' wide, 6' deep, 5/8 C.Y. bucket w/trench box	7.60	13.50	21
1530	8' deep, 3/4 C.Y. bucket	10.05	17.80	28
1540	10' deep, 1 C.Y. bucket	9.25	18.05	27.50
1550	12' deep, 1-1/2 C.Y. bucket	10.60	20.50	31
1560	16' deep, 2-1/2 C.Y. bucket	15.10	25.50	40.50
1570	20' deep, 3-1/2 C.Y. bucket	17.30	30	47.50
1580	24' deep, 3-1/2 C.Y. bucket	21.50	37	58.50
1640	8' wide, 12' deep, 1-1/4 C.Y. bucket w/trench box	14.90	25.50	40.50
1650	15' deep, 1-1/2 C.Y. bucket	18.25	32.50	51
1660	18' deep, 2-1/2 C.Y. bucket	23	36.50	59.50
1680	24' deep, 3-1/2 C.Y. bucket	29.50	47.50	77
1730	10' wide, 20' deep, 3-1/2 C.Y. bucket w/trench box	29.50	47.50	77
1740	24' deep, 3-1/2 C.Y. bucket	37	59	96
1780	12' wide, 20' deep, 3-1/2 C.Y. bucket w/trench box	36	56.50	92.50
1790	25' deep, bucket	46.50	73	120
1800	1/2:1 slope, 2' wide, 2' deep, 3/8 C.Y. BK	.82	2.56	3.38
1810	3' deep, 3/8 C.Y. bucket	1.39	4.48	5.85
1820	4' deep, 3/8 C.Y. bucket	2.06	6.85	8.90
1840	6' deep, 3/8 C.Y. bucket	3.70	10.20	13.90

G1030 Site Earthwork

G1030 806	Trenching Loam & Sandy Clay	COST PER L.F.		
		EQUIP.	LABOR	TOTAL
1860	8' deep, 1/2 C.Y. bucket	5.90	16.25	22
1880	10' deep, 1 C.Y. bucket	7.90	20.50	28.50
2300	4' wide, 2' deep, 3/8 C.Y. bucket	1.57	4.20	5.75
2310	3' deep, 3/8 C.Y. bucket	2.37	7	9.35
2320	4' deep, 1/2 C.Y. bucket	2.97	8.70	11.65
2340	6' deep, 1/2 C.Y. bucket	6.20	14.85	21
2360	8' deep, 1/2 C.Y. bucket	11.15	24	35
2380	10' deep, 1 C.Y. bucket	12	28	40
2400	12' deep, 1 C.Y. bucket	20	46	66
2430	15' deep, 1-1/2 C.Y. bucket	24	51.50	75.50
2460	18' deep, 2-1/2 C.Y. bucket	41	78.50	120
2840	6' wide, 6' deep, 5/8 C.Y. bucket w/trench box	11.15	20.50	31.50
2860	8' deep, 3/4 C.Y. bucket	16.30	30	46.50
2880	10' deep, 1 C.Y. bucket	16.55	34	50.50
2900	12' deep, 1-1/2 C.Y. bucket	20.50	41.50	62
2940	16' deep, 2-1/2 C.Y. bucket	34.50	61	95.50
2980	20' deep, 3-1/2 C.Y. bucket	45.50	82.50	128
3020	24' deep, 3-1/2 C.Y. bucket	64.50	113	178
3100	8' wide, 12' deep, 1-1/2 C.Y. bucket w/trench box	25	47	72
3120	15' deep, 1-1/2 C.Y. bucket	34	66	100
3140	18' deep, 2-1/2 C.Y. bucket	47.50	81.50	129
3180	24' deep, 3-1/2 C.Y. bucket	72.50	124	197
3270	10' wide, 20' deep, 3-1/2 C.Y. bucket w/trench box	58	100	158
3280	24' deep, 3-1/2 C.Y. bucket	80	135	215
3320	12' wide, 20' deep, 3-1/2 C.Y. bucket w/trench box	64	109	173
3380	25' deep, 3-1/2 C.Y. bucket w/trench box	87.50	146	234
3500	1:1 slope, 2' wide, 2' deep, 3/8 C.Y. bucket	1.07	3.41	4.48
3520	3' deep, 3/8 C.Y. bucket	1.94	6.40	8.35
3540	4' deep, 3/8 C.Y. bucket	3.05	10.30	13.35
3560	6' deep, 3/8 C.Y. bucket	3.70	10.20	13.90
3580	8' deep, 1/2 C.Y. bucket	9.75	27	37
3600	10' deep, 1 C.Y. bucket	13.45	35	48.50
3800	4' wide, 2' deep, 3/8 C.Y. bucket	1.82	5.05	6.85
3820	3' deep, 3/8 C.Y. bucket	2.93	8.90	11.85
3840	4' deep, 1/2 C.Y. bucket	3.84	11.70	15.55
3860	6' deep, 1/2 C.Y. bucket	8.75	21.50	30.50
3880	8' deep, 1/2 C.Y. bucket	16.60	36	52.50
3900	10' deep, 1 C.Y. bucket	18.55	43.50	62
3920	12' deep, 1 C.Y. bucket	27	61.50	88.50
3940	15' deep, 1-1/2 C.Y. bucket	39	85	124
3960	18' deep, 2-1/2 C.Y. bucket	56.50	108	165
4030	6' wide, 6' deep, 5/8 C.Y. bucket w/trench box	14.65	27.50	42
4040	8' deep, 3/4 C.Y. bucket	22.50	42.50	65
4050	10' deep, 1 C.Y. bucket	24	50	74
4060	12' deep, 1-1/2 C.Y. bucket	31	63	94
4070	16' deep, 2-1/2 C.Y. bucket	54	96.50	151
4080	20' deep, 3-1/2 C.Y. bucket	73.50	134	208
4090	24' deep, 3-1/2 C.Y. bucket	107	189	296
4500	8' wide, 12' deep, 1-1/4 C.Y. bucket w/trench box	35	68	103
4550	15' deep, 1-1/2 C.Y. bucket	50	99	149
4600	18' deep, 2-1/2 C.Y. bucket	72.50	126	199
4650	24' deep, 3-1/2 C.Y. bucket	115	199	315
4800	10' wide, 20' deep, 3-1/2 C.Y. bucket w/trench box	86	152	238
4850	24' deep, 3-1/2 C.Y. bucket	123	211	335
4950	12' wide, 20' deep, 3-1/2 C.Y. bucket w/trench box	92.50	161	254
4980	25' deep, bucket	139	238	375

G1030 Site Earthwork

Trenching Systems are shown on a cost per linear foot basis. The systems include: excavation; backfill and removal of spoil; and compaction for various depths and trench bottom widths. The backfill has been reduced to accommodate a pipe of suitable diameter and bedding.

The slope for trench sides varies from none to 1:1.

The Expanded System Listing shows Trenching Systems that range from 2' to 12' in width. Depths range from 2' to 25'.

System Components		QUANTITY	UNIT	COST PER L.F.		
				EQUIP.	LABOR	TOTAL
SYSTEM G1030 807 1310						
TRENCHING SAND & GRAVEL, NO SLOPE, 2' WIDE, 2' DEEP, 3/8 C.Y. BUCKET						
Excavation, trench, hyd. backhoe, track mtd., 3/8 C.Y. bucket		.148	B.C.Y.	.32	.87	1.19
Backfill and load spoil, from stockpile		.140	L.C.Y.	.09	.26	.35
Compaction by vibrating plate 18" wide, 6" lifts, 4 passes		.118	E.C.Y.	.03	.36	.39
Remove excess spoil, 8 C.Y. dump truck, 2 mile roundtrip		.035	L.C.Y.	.10	.12	.22
	TOTAL			.54	1.61	2.15

G1030 807	Trenching Sand & Gravel	COST PER L.F.		
		EQUIP.	LABOR	TOTAL
1310	Trenching, sand & gravel, no slope, 2' wide, 2' deep, 3/8 C.Y. bucket	.54	1.61	2.15
1320	3' deep, 3/8 C.Y. bucket	.87	2.69	3.56
1330	4' deep, 3/8 C.Y. bucket	.99	3.24	4.23
1340	6' deep, 3/8 C.Y. bucket	1.46	3.87	5.35
1350	8' deep, 1/2 C.Y. bucket	1.94	5.15	7.10
1360	10' deep, 1 C.Y. bucket	2.16	5.45	7.60
1400	4' wide, 2' deep, 3/8 C.Y. bucket	1.21	3.15	4.36
1410	3' deep, 3/8 C.Y. bucket	1.68	4.80	6.50
1420	4' deep, 1/2 C.Y. bucket	1.94	5.40	7.35
1430	6' deep, 1/2 C.Y. bucket	3.49	7.95	11.45
1440	8' deep, 1/2 C.Y. bucket	5.45	11.20	16.65
1450	10' deep, 1 C.Y. bucket	5.15	11.45	16.60
1460	12' deep, 1 C.Y. bucket	6.50	14.30	21
1470	15' deep, 1-1/2 C.Y. bucket	7.95	16.55	24.50
1480	18' deep, 2-1/2 C.Y. bucket	9.85	18.10	28
1520	6' wide, 6' deep, 5/8 C.Y. bucket w/trench box	7.20	12.80	20
1530	8' deep, 3/4 C.Y. bucket	9.55	16.90	26.50
1540	10' deep, 1 C.Y. bucket	8.75	17.05	26
1550	12' deep, 1-1/2 C.Y. bucket	10	19.10	29
1560	16' deep, 2 C.Y. bucket	14.50	24	38.50
1570	20' deep, 3-1/2 C.Y. bucket	16.40	28.50	45
1580	24' deep, 3-1/2 C.Y. bucket	20.50	34.50	55
1640	8' wide, 12' deep, 1-1/2 C.Y. bucket w/trench box	14.10	24	38
1650	15' deep, 1-1/2 C.Y. bucket	17.15	30.50	47.50
1660	18' deep, 2-1/2 C.Y. bucket	22	34.50	56.50
1680	24' deep, 3-1/2 C.Y. bucket	28	44.50	72.50
1730	10' wide, 20' deep, 3-1/2 C.Y. bucket w/trench box	28	44.50	72.50
1740	24' deep, 3-1/2 C.Y. bucket	35	55	90
1780	12' wide, 20' deep, 3-1/2 C.Y. bucket w/trench box	34	53	87
1790	25' deep, 3-1/2 C.Y. bucket	44	68.50	113
1800	1/2:1 slope, 2' wide, 2' deep, 3/8 C.Y. bucket	.78	2.43	3.21
1810	3' deep, 3/8 C.Y. bucket	1.30	4.28	5.60
1820	4' deep, 3/8 C.Y. bucket	1.94	6.55	8.50
1840	6' deep, 3/8 C.Y. bucket	3.54	9.70	13.25

G10 Site Preparation

G1030 Site Earthwork

G1030 807	Trenching Sand & Gravel	COST PER L.F.		
		EQUIP.	LABOR	TOTAL
1860	8' deep, 1/2 C.Y. bucket	5.65	15.55	21
1880	10' deep, 1 C.Y. bucket	7.40	19.20	26.50
2300	4' wide, 2' deep, 3/8 C.Y. bucket	1.45	3.98	5.45
2310	3' deep, 3/8 C.Y. bucket	2.21	6.65	8.85
2320	4' deep, 1/2 C.Y. bucket	2.76	8.25	11
2340	6' deep, 1/2 C.Y. bucket	5.95	14.20	20
2360	8' deep, 1/2 C.Y. bucket	10.60	22.50	33
2380	10' deep, 1 C.Y. bucket	11.40	26.50	38
2400	12' deep, 1 C.Y. bucket	19.20	43.50	62.50
2430	15' deep, 1-1/2 C.Y. bucket	22.50	48.50	71
2460	18' deep, 2-1/2 C.Y. bucket	39	74	113
2840	6' wide, 6' deep, 5/8 C.Y. bucket w/trench box	10.60	19.45	30
2860	8' deep, 3/4 C.Y. bucket	15.50	28.50	44
2880	10' deep, 1 C.Y. bucket	15.70	32	47.50
2900	12' deep, 1-1/2 C.Y. bucket	19.65	39	58.50
2940	16' deep, 2 C.Y. bucket	33.50	57.50	91
2980	20' deep, 3-1/2 C.Y. bucket	43.50	77	121
3020	24' deep, 3-1/2 C.Y. bucket	61.50	106	168
3100	8' wide, 12' deep, 1-1/4 C.Y. bucket w/trench box	24	44.50	68.50
3120	15' deep, 1-1/2 C.Y. bucket	32	62	94
3140	18' deep, 2-1/2 C.Y. bucket	45.50	76.50	122
3180	24' deep, 3-1/2 C.Y. bucket	69	116	185
3270	10' wide, 20' deep, 3-1/2 C.Y. bucket w/trench box	55	93.50	149
3280	24' deep, 3-1/2 C.Y. bucket	76	126	202
3370	12' wide, 20' deep, 3-1/2 C.Y. bucket w/trench box	61	104	165
3380	25' deep, 3-1/2 C.Y. bucket	88.50	146	235
3500	1:1 slope, 2' wide, 2' deep, 3/8 C.Y. bucket	1.66	3.97	5.65
3520	3' deep, 3/8 C.Y. bucket	1.82	6.10	7.90
3540	4' deep, 3/8 C.Y. bucket	2.87	9.80	12.65
3560	6' deep, 3/8 C.Y. bucket	3.55	9.70	13.25
3580	8' deep, 1/2 C.Y. bucket	9.35	26	35.50
3600	10' deep, 1 C.Y. bucket	12.65	33	45.50
3800	4' wide, 2' deep, 3/8 C.Y. bucket	1.68	4.80	6.50
3820	3' deep, 3/8 C.Y. bucket	2.73	8.50	11.25
3840	4' deep, 1/2 C.Y. bucket	3.57	11.10	14.65
3860	6' deep, 1/2 C.Y. bucket	8.35	20.50	29
3880	8' deep, 1/2 C.Y. bucket	15.80	34.50	50.50
3900	10' deep, 1 C.Y. bucket	17.60	41	58.50
3920	12' deep, 1 C.Y. bucket	25.50	58.50	84
3940	15' deep, 1-1/2 C.Y. bucket	37	80.50	118
3960	18' deep, 2-1/2 C.Y. bucket	53.50	102	156
4030	6' wide, 6' deep, 5/8 C.Y. bucket w/trench box	13.95	26	40
4040	8' deep, 3/4 C.Y. bucket	21.50	40.50	62
4050	10' deep, 1 C.Y. bucket	22.50	47	69.50
4060	12' deep, 1-1/2 C.Y. bucket	29.50	59	88.50
4070	16' deep, 2 C.Y. bucket	52	91	143
4080	20' deep, 3-1/2 C.Y. bucket	70	126	196
4090	24' deep, 3-1/2 C.Y. bucket	102	177	279
4500	8' wide, 12' deep, 1-1/2 C.Y. bucket w/trench box	33.50	64.50	98
4550	15' deep, 1-1/2 C.Y. bucket	47	93.50	141
4600	18' deep, 2-1/2 C.Y. bucket	69	119	188
4650	24' deep, 3-1/2 C.Y. bucket	110	187	297
4800	10' wide, 20' deep, 3-1/2 C.Y. bucket w/trench box	81.50	142	224
4850	24' deep, 3-1/2 C.Y. bucket	117	198	315
4950	12' wide, 20' deep, 3-1/2 C.Y. bucket w/trench box	87.50	151	239
4980	25' deep, 3-1/2 C.Y. bucket	133	223	355

G1030 Site Earthwork

The Pipe Bedding System is shown for various pipe diameters. Compacted bank sand is used for pipe bedding and to fill 12″ over the pipe. No backfill is included. Various side slopes are shown to accommodate different soil conditions. Pipe sizes vary from 6″ to 84″ diameter.

System Components	QUANTITY	UNIT	COST PER L.F.		
			MAT.	INST.	TOTAL
SYSTEM G1030 815 1440					
PIPE BEDDING, SIDE SLOPE 0 TO 1, 1′ WIDE, PIPE SIZE 6″ DIAMETER					
Borrow, bank sand, 2 mile haul, machine spread	.086	C.Y.	1.41	.55	1.96
Compaction, vibrating plate	.086	C.Y.		.19	.19
TOTAL			1.41	.74	2.15

G1030 815	Pipe Bedding	COST PER L.F.		
		MAT.	INST.	TOTAL
1440	Pipe bedding, side slope 0 to 1, 1′ wide, pipe size 6″ diameter	1.41	.74	2.15
1460	2′ wide, pipe size 8″ diameter	3.05	1.61	4.66
1480	Pipe size 10″ diameter	3.12	1.64	4.76
1500	Pipe size 12″ diameter	3.18	1.68	4.86
1520	3′ wide, pipe size 14″ diameter	5.15	2.72	7.87
1540	Pipe size 15″ diameter	5.20	2.75	7.95
1560	Pipe size 16″ diameter	5.25	2.78	8.03
1580	Pipe size 18″ diameter	5.35	2.84	8.19
1600	4′ wide, pipe size 20″ diameter	7.60	4.03	11.63
1620	Pipe size 21″ diameter	7.70	4.06	11.76
1640	Pipe size 24″ diameter	7.85	4.14	11.99
1660	Pipe size 30″ diameter	8	4.22	12.22
1680	6′ wide, pipe size 32″ diameter	13.70	7.25	20.95
1700	Pipe size 36″ diameter	14	7.40	21.40
1720	7′ wide, pipe size 48″ diameter	22.50	11.80	34.30
1740	8′ wide, pipe size 60″ diameter	27	14.35	41.35
1760	10′ wide, pipe size 72″ diameter	37.50	19.95	57.45
1780	12′ wide, pipe size 84″ diameter	50	26.50	76.50
2140	Side slope 1/2 to 1, 1′ wide, pipe size 6″ diameter	2.63	1.39	4.02
2160	2′ wide, pipe size 8″ diameter	4.47	2.36	6.83
2180	Pipe size 10″ diameter	4.80	2.53	7.33
2200	Pipe size 12″ diameter	5.10	2.68	7.78
2220	3′ wide, pipe size 14″ diameter	7.35	3.88	11.23
2240	Pipe size 15″ diameter	7.50	3.97	11.47
2260	Pipe size 16″ diameter	7.70	4.08	11.78
2280	Pipe size 18″ diameter	8.10	4.28	12.38
2300	4′ wide, pipe size 20″ diameter	10.70	5.65	16.35
2320	Pipe size 21″ diameter	10.90	5.75	16.65
2340	Pipe size 24″ diameter	11.60	6.15	17.75
2360	Pipe size 30″ diameter	12.90	6.80	19.70
2380	6′ wide, pipe size 32″ diameter	19	10.05	29.05
2400	Pipe size 36″ diameter	20	10.70	30.70
2420	7′ wide, pipe size 48″ diameter	31.50	16.65	48.15
2440	8′ wide, pipe size 60″ diameter	40	21	61
2460	10′ wide, pipe size 72″ diameter	55	29	84
2480	12′ wide, pipe size 84″ diameter	72	38	110
2620	Side slope 1 to 1, 1′ wide, pipe size 6″ diameter	3.85	2.03	5.88
2640	2′ wide, pipe size 8″ diameter	5.94	3.13	9.10

G10 Site Preparation

G1030 Site Earthwork

G1030 815	Pipe Bedding	COST PER L.F.		
		MAT.	INST.	TOTAL
2660	Pipe size 10" diameter	6.45	3.42	9.87
2680	Pipe size 12" diameter	7.05	3.71	10.76
2700	3' wide, pipe size 14" diameter	9.50	5.05	14.55
2720	Pipe size 15" diameter	9.80	5.20	15
2740	Pipe size 16" diameter	10.15	5.35	15.50
2760	Pipe size 18" diameter	10.85	5.70	16.55
2780	4' wide, pipe size 20" diameter	13.75	7.25	21
2800	Pipe size 21" diameter	14.15	7.45	21.60
2820	Pipe size 24" diameter	15.35	8.10	23.45
2840	Pipe size 30" diameter	17.70	9.35	27.05
2860	6' wide, pipe size 32" diameter	24.50	12.85	37.35
2880	Pipe size 36" diameter	26.50	13.90	40.40
2900	7' wide, pipe size 48" diameter	41	21.50	62.50
2920	8' wide, pipe size 60" diameter	53	28	81
2940	10' wide, pipe size 72" diameter	72	38.50	110.50
2960	12' wide, pipe size 84" diameter	94	49.50	143.50

G30 Site Mechanical Utilities

G3010 Water Supply

G3010 110	Water Distribution Piping	COST PER L.F.		
		MAT.	INST.	TOTAL
2000	Piping, excav. & backfill excl., ductile iron class 250, mech. joint			
2130	4" diameter	13.75	15.30	29.05
2150	6" diameter	16.05	19.10	35.15
2160	8" diameter	23	23	46
2170	10" diameter	29	26.50	55.50
2180	12" diameter	35	29	64
2210	16" diameter	55	42	97
2220	18" diameter	58.50	44.50	103
3000	Tyton joint			
3130	4" diameter	11.90	7.65	19.55
3150	6" diameter	16.70	9.20	25.90
3160	8" diameter	24	15.30	39.30
3170	10" diameter	32.50	16.80	49.30
3180	12" diameter	34	19.10	53.10
3210	16" diameter	40.50	26.50	67
3220	18" diameter	45	31	76
3230	20" diameter	49.50	34.50	84
3250	24" diameter	49	39.50	88.50
4000	Copper tubing, type K			
4050	3/4" diameter	8.45	2.88	11.33
4060	1" diameter	11.05	3.60	14.65
4080	1-1/2" diameter	17.90	4.35	22.25
4090	2" diameter	27.50	5	32.50
4110	3" diameter	56.50	8.60	65.10
4130	4" diameter	93.50	12.15	105.65
4150	6" diameter	237	22.50	259.50
5000	Polyvinyl chloride class 160, S.D.R. 26			
5130	1-1/2" diameter	.35	1.07	1.42
5150	2" diameter	.54	1.18	1.72
5160	4" diameter	1.94	5.30	7.24
5170	6" diameter	5.65	6.40	12.05
5180	8" diameter	8.80	7.65	16.45
6000	Polyethylene 160 psi, S.D.R. 7			
6050	3/4" diameter	.66	1.54	2.20
6060	1" diameter	.92	1.66	2.58
6080	1-1/2" diameter	1.20	1.79	2.99
6090	2" diameter	1.79	2.21	4

G30 Site Mechanical Utilities

G3020 Sanitary Sewer

G3020 110	Drainage & Sewage Piping	COST PER L.F.		
		MAT.	INST.	TOTAL
2000	Piping, excavation & backfill excluded, PVC, plain			
2130	4" diameter	2.04	3.77	5.81
2150	6" diameter	4.05	4.04	8.09
2160	8" diameter	8.35	4.22	12.57
2900	Box culvert, precast, 8' long			
3000	6' x 3'	375	33	408
3020	6' x 7'	560	37	597
3040	8' x 3'	550	41	591
3060	8' x 8'	710	46	756
3080	10' x 3'	775	42	817
3100	10' x 8'	990	58	1,048
3120	12' x 3'	990	46	1,036
3140	12' x 8'	1,200	69	1,269
4000	Concrete, nonreinforced			
4150	6" diameter	6.90	11.35	18.25
4160	8" diameter	7.60	13.40	21
4170	10" diameter	8.40	13.95	22.35
4180	12" diameter	10.30	15	25.30
4200	15" diameter	14.10	16.70	30.80
4220	18" diameter	16.60	21	37.60
4250	24" diameter	27.50	30	57.50
4400	Reinforced, no gasket			
4580	12" diameter	12.65	20	32.65
4600	15" diameter	17.25	20	37.25
4620	18" diameter	21.50	22.50	44
4650	24" diameter	30.50	30	60.50
4670	30" diameter	53	45.50	98.50
4680	36" diameter	63.50	55.50	119
4690	42" diameter	84.50	62.50	147
4700	48" diameter	105	70.50	175.50
4720	60" diameter	135	93.50	228.50
4730	72" diameter	220	112	332
4740	84" diameter	315	141	456
4800	With gasket			
4980	12" diameter	13.90	10.90	24.80
5000	15" diameter	19	11.45	30.45
5020	18" diameter	24	12.05	36.05
5050	24" diameter	37	13.50	50.50
5070	30" diameter	60.50	45.50	106
5080	36" diameter	73	55.50	128.50
5090	42" diameter	94.50	56.50	151
5100	48" diameter	118	70.50	188.50
5120	60" diameter	176	89.50	265.50
5130	72" diameter	239	112	351
5140	84" diameter	380	188	568
5700	Corrugated metal, alum. or galv. bit. coated			
5760	8" diameter	11.75	9.10	20.85
5770	10" diameter	12.25	11.55	23.80
5780	12" diameter	12.65	14.35	27
5800	15" diameter	20.50	15	35.50
5820	18" diameter	21	15.80	36.80
5850	24" diameter	24	18.80	42.80
5870	30" diameter	28.50	33.50	62
5880	36" diameter	34.50	33.50	68
5900	48" diameter	57.50	40	97.50
5920	60" diameter	87	60	147
5930	72" diameter	99.50	100	199.50
6000	Plain			

G3020 Sanitary Sewer

G3020 110	Drainage & Sewage Piping	COST PER L.F.		
		MAT.	INST.	TOTAL
6060	8" diameter	9.10	8.45	17.55
6070	10" diameter	9.65	10.75	20.40
6080	12" diameter	9.90	13.65	23.55
6100	15" diameter	14.30	13.65	27.95
6120	18" diameter	16.25	14.65	30.90
6140	24" diameter	20.50	17.15	37.65
6170	30" diameter	25	31	56
6180	36" diameter	30.50	31	61.50
6200	48" diameter	50	36.50	86.50
6220	60" diameter	76	57.50	133.50
6230	72" diameter	165	140	305
6300	Steel or alum. oval arch, coated & paved invert			
6400	15" equivalent diameter	19.20	15	34.20
6420	18" equivalent diameter	27	20	47
6450	24" equivalent diameter	31	24	55
6470	30" equivalent diameter	40.50	30	70.50
6480	36" equivalent diameter	51	40	91
6490	42" equivalent diameter	68.50	44	112.50
6500	48" equivalent diameter	83.50	53.50	137
6600	Plain			
6700	15" equivalent diameter	11.30	13.35	24.65
6720	18" equivalent diameter	13.65	17.15	30.80
6750	24" equivalent diameter	18.25	20	38.25
6770	30" equivalent diameter	24	37	61
6780	36" equivalent diameter	30	37	67
6790	42" equivalent diameter	34	43.50	77.50
6800	48" equivalent diameter	52	53.50	105.50
8000	Polyvinyl chloride SDR 35			
8130	4" diameter	2.04	3.77	5.81
8150	6" diameter	4.05	4.04	8.09
8160	8" diameter	8.35	4.22	12.57
8170	10" diameter	12.45	5.55	18
8180	12" diameter	14.35	5.75	20.10
8200	15" diameter	15.80	7.65	23.45

G3030 Storm Sewer

Manhole   Catch Basin

The Manhole and Catch Basin System includes: excavation with a backhoe; a formed concrete footing; frame and cover; cast iron steps and compacted backfill.

The Expanded System Listing shows manholes that have a 4′, 5′ and 6′ inside diameter riser. Depths range from 4′ to 14′. Construction material shown is either concrete, concrete block, precast concrete, or brick.

System Components	QUANTITY	UNIT	COST PER EACH		
			MAT.	INST.	TOTAL
SYSTEM G3030 210 1920					
MANHOLE/CATCH BASIN, BRICK, 4′ I.D. RISER, 4′ DEEP					
Excavation, hydraulic backhoe, 3/8 C.Y. bucket	14.815	B.C.Y.		110.67	110.67
Trim sides and bottom of excavation	64.000	S.F.		56.96	56.96
Forms in place, manhole base, 4 uses	20.000	SFCA	12.80	97	109.80
Reinforcing in place footings, #4 to #7	.019	Ton	17.86	22.33	40.19
Concrete, 3000 psi	.925	C.Y.	100.83		100.83
Place and vibrate concrete, footing, direct chute	.925	C.Y.		44.80	44.80
Catch basin or MH, brick, 4′ ID, 4′ deep	1.000	Ea.	490	935	1,425
Catch basin or MH steps; heavy galvanized cast iron	1.000	Ea.	18.40	12.95	31.35
Catch basin or MH frame and cover	1.000	Ea.	242	220	462
Fill, granular	12.954	L.C.Y.	284.99		284.99
Backfill, spread with wheeled front end loader	12.954	L.C.Y.		27.46	27.46
Backfill compaction, 12″ lifts, air tamp	12.954	E.C.Y.		106.88	106.88
TOTAL			1,166.88	1,634.05	2,800.93

G3030 210	Manholes & Catch Basins	COST PER EACH		
		MAT.	INST.	TOTAL
1920	Manhole/catch basin, brick, 4′ I.D. riser, 4′ deep	1,175	1,650	2,825
1940	6′ deep	1,650	2,250	3,900
1960	8′ deep	2,200	3,100	5,300
1980	10′ deep	2,775	3,850	6,625
3000	12′ deep	3,650	4,175	7,825
3020	14′ deep	4,600	5,850	10,450
3200	Block, 4′ I.D. riser, 4′ deep	1,075	1,325	2,400
3220	6′ deep	1,500	1,850	3,350
3240	8′ deep	2,000	2,550	4,550
3260	10′ deep	2,375	3,175	5,550
3280	12′ deep	3,050	4,025	7,075
3300	14′ deep	3,800	4,900	8,700
4620	Concrete, cast-in-place, 4′ I.D. riser, 4′ deep	1,150	2,275	3,425
4640	6′ deep	1,650	3,025	4,675
4660	8′ deep	2,300	4,375	6,675
4680	10′ deep	2,750	5,425	8,175
4700	12′ deep	3,475	6,725	10,200
4720	14′ deep	4,325	8,075	12,400
5820	Concrete, precast, 4′ I.D. riser, 4′ deep	1,425	1,200	2,625
5840	6′ deep	1,825	1,600	3,425

G30 Site Mechanical Utilities

G3030 Storm Sewer

G3030 210	Manholes & Catch Basins	COST PER EACH		
		MAT.	INST.	TOTAL
5860	8' deep	2,425	2,250	4,675
5880	10' deep	2,875	2,775	5,650
5900	12' deep	3,625	3,400	7,025
5920	14' deep	4,450	4,350	8,800
6000	5' I.D. riser, 4' deep	2,300	1,350	3,650
6020	6' deep	2,900	1,900	4,800
6040	8' deep	3,575	2,525	6,100
6060	10' deep	4,450	3,200	7,650
6080	12' deep	5,425	4,075	9,500
6100	14' deep	6,525	5,000	11,525
6200	6' I.D. riser, 4' deep	3,075	1,775	4,850
6220	6' deep	4,200	2,375	6,575
6240	8' deep	5,275	3,300	8,575
6260	10' deep	6,375	4,200	10,575
6280	12' deep	7,600	5,275	12,875
6300	14' deep	8,925	6,400	15,325

G30 Site Mechanical Utilities

G3060 Fuel Distribution

G3060 110	Gas Service Piping	COST PER L.F.		
		MAT.	INST.	TOTAL
2000	Piping, excavation & backfill excluded, polyethylene			
2070	1-1/4" diam, SDR 10	2.62	3.71	6.33
2090	2" diam, SDR 11	3.58	4.14	7.72
2110	3" diam, SDR 11.5	4.74	4.95	9.69
2130	4" diam, SDR 11	8.50	8.65	17.15
2150	6" diam, SDR 21	29.50	9.30	38.80
2160	8" diam, SDR 21	35.50	11.20	46.70
3000	Steel, schedule 40, plain end, tarred & wrapped			
3060	1" diameter	7.10	8.25	15.35
3090	2" diameter	11.15	8.85	20
3110	3" diameter	18.45	9.55	28
3130	4" diameter	24	15	39
3140	5" diameter	35	17.40	52.40
3150	6" diameter	42.50	21.50	64
3160	8" diameter	67.50	27.50	95

G3060 Fuel Distribution

Fiberglass Underground Fuel Storage Tank, Double Wall

System Components	QUANTITY	UNIT	COST EACH		
			MAT.	INST.	TOTAL
SYSTEM G3060 310 1010					
STORAGE TANK, FUEL, UNDERGROUND, DOUBLE WALL FIBERGLASS, 600 GAL.		Ea.			
Excavating trench, no sheeting or dewatering, 6'-10 ' D, 3/4 CY hyd backhoe	18.000	Ea.		134.46	134.46
Corrosion resistance wrap & coat, small diam pipe, 1" diam, add	120.000	Ea.	184.80		184.80
Fiberglass-reinforced elbows 90 & 45 degree, 2 inch	7.000	Ea.	395.50	528.50	924
Fiberglass-reinforced threaded adapters, 2 inch	2.000	Ea.	39.60	172	211.60
Remote tank gauging system, 30', 5" pointer travel	1.000	Ea.	5,200	325	5,525
Corrosion resistance wrap & coat, small diam pipe. 4 " diam, add	30.000	Ea.	71.10		71.10
Pea gravel	9.000	Ea.	684	414	1,098
Reinforcing in hold down pad, #3 to #7	120.000	Ea.	60	70.80	130.80
Tank leak detector system, 8 channel, external monitoring	1.000	Ea.	845		845
Tubbing copper, type L, 3/4" diam	120.000	Ea.	786	1,008	1,794
Elbow 90 degree ,wrought cu x cu 3/4 " diam	4.000	Ea.	18.44	134	152.44
Pipe, black steel welded, sch 40, 3" diam	30.000	Ea.	375	834.90	1,209.90
Elbow 90 degree, steel welded butt joint, 3 " diam	3.000	Ea.	96	510.30	606.30
Fiberglass-reinforced plastic pipe, 2 inch	156.000	Ea.	1,068.60	664.56	1,733.16
Fiberglass-reinforced plastic pipe, 3 inch	36.000	Ea.	529.20	173.52	702.72
Fiberglass-reinforced elbows 90 & 45 degree, 3 inch	3.000	Ea.	384	303	687
Tank leak detection system, probes, well monitoring, liquid phase detection	2.000	Ea.	880		880
Double wall fiberglass annular space	1.000	Ea.	299		299
Flange, weld neck, 150 lb, 3 " pipe size	1.000	Ea.	38.50	85.05	123.55
Vent protector/breather, 2" dia.	1.000	Ea.	24.50	20	44.50
Corrosion resistance wrap & coat, small diam pipe, 2" diam, add	36.000	Ea.	60.84		60.84
Concrete hold down pad, 6' x 5' x 8" thick	30.000	Ea.	86.70	41.70	128.40
600 gallon capacity	1.000	Ea.	7,625	480	8,105
Tank, for hold-downs 500-4000 gal., add	1.000	Ea.	485	154	639
Foot valve, single poppet, 3/4" dia	1.000	Ea.	65	36	101
Fuel fill box, locking inner cover, 3" dia.	1.000	Ea.	130	129	259
Fuel oil specialties, valve,ball chk, globe type fusible, 3/4" dia	2.000	Ea.	154	65	219
TOTAL			20,585.78	6,283.79	26,869.57

G3060 310	Fiberglass Fuel Tank, Double Wall	COST EACH		
		MAT.	INST.	TOTAL
1010	Storage tank, fuel, underground, double wall fiberglass, 600 Gal.	20,600	6,275	26,875
1020	2500 Gal.	31,300	8,675	39,975

G3060 Fuel Distribution

G3060 310	Fiberglass Fuel Tank, Double Wall	COST EACH		
		MAT.	INST.	TOTAL
1030	4000 Gal.	37,500	10,700	48,200
1040	6000 Gal.	40,700	12,200	52,900
1050	8000 Gal.	48,300	14,600	62,900
1060	10,000 Gal.	53,500	16,800	70,300
1070	15,000 Gal.	72,000	19,400	91,400
1080	20,000 Gal.	89,000	23,500	112,500
1090	25,000 Gal.	111,500	27,300	138,800
1100	30,000 Gal.	127,000	28,900	155,900

G3060 Fuel Distribution

Steel Underground Fuel Storage Tank, Double Wall

System Components	QUANTITY	UNIT	COST EACH		
			MAT.	INST.	TOTAL
SYSTEM G3060 320 1010					
STORAGE TANK, FUEL, UNDERGROUND, DOUBLE WALL STEEL, 550 GAL.		Ea.			
Fiberglass-reinforced plastic pipe, 2 inch	156.000	Ea.	1,068.60	664.56	1,733.16
Fiberglass-reinforced plastic pipe, 3 inch	36.000	Ea.	529.20	173.52	702.72
Fiberglass-reinforced elbows 90 & 45 degree, 3 inch	3.000	Ea.	384	303	687
Tank leak detection system, probes, well monitoring, liquid phase detection	2.000	Ea.	880		880
Flange, weld neck, 150 lb, 3" pipe size	1.000	Ea.	38.50	85.05	123.55
Vent protector/breather, 2" dia.	1.000	Ea.	24.50	20	44.50
Concrete hold down pad, 6' x 5' x 8" thick	30.000	Ea.	86.70	41.70	128.40
Tanks, for hold-downs, 500-2000 gal, add	1.000	Ea.	207	154	361
Double wall steel tank annular space	1.000	Ea.	310		310
Foot valve, single poppet, 3/4" dia	1.000	Ea.	65	36	101
Fuel fill box, locking inner cover, 3" dia.	1.000	Ea.	130	129	259
Fuel oil specialties, valve, ball chk, globe type fusible, 3/4" dia	2.000	Ea.	154	65	219
Pea gravel	9.000	Ea.	684	414	1,098
Reinforcing in hold down pad, #3 to #7	120.000	Ea.	60	70.80	130.80
Tank, for manways, add	1.000	Ea.	1,650		1,650
Tank leak detector system, 8 channel, external monitoring	1.000	Ea.	845		845
Tubbing copper, type L, 3/4" diam	120.000	Ea.	786	1,008	1,794
Elbow 90 degree, wrought cu x cu 3/4" diam	8.000	Ea.	36.88	268	304.88
Pipe, black steel welded, sch 40, 3" diam	30.000	Ea.	375	834.90	1,209.90
Elbow 90 degree, steel welded butt joint, 3" diam	3.000	Ea.	96	510.30	606.30
Excavating trench, no sheeting or dewatering, 6'-10' D, 3/4 CY hyd backhoe	18.000	Ea.		134.46	134.46
Fiberglass-reinforced elbows 90 & 45 degree, 2 inch	7.000	Ea.	395.50	528.50	924
Fiberglass-reinforced threaded adapters, 2 inch	2.000	Ea.	39.60	172	211.60
500 gallon capacity	1.000	Ea.	4,300	485	4,785
Remote tank gauging system, 30', 5" pointer travel	1.000	Ea.	5,200	325	5,525
TOTAL			18,345.48	6,422.79	24,768.27

G3060 320	Steel Fuel Tank, Double Wall	COST EACH		
		MAT.	INST.	TOTAL
1010	Storage tank, fuel, underground, double wall steel, 500 Gal.	18,300	6,425	24,725
1020	2000 Gal.	25,200	8,400	33,600
1030	4000 Gal.	35,000	10,900	45,900
1040	6000 Gal.	38,900	12,100	51,000
1050	8000 Gal.	44,600	14,800	59,400
1060	10,000 Gal.	49,300	17,000	66,300

G30 Site Mechanical Utilities

G3060 Fuel Distribution

G3060 320	Steel Fuel Tank, Double Wall	COST EACH		
		MAT.	INST.	TOTAL
1070	15,000 Gal.	63,000	19,800	82,800
1080	20,000 Gal.	83,500	23,800	107,300
1090	25,000 Gal.	108,000	27,200	135,200
1100	30,000 Gal.	123,500	29,000	152,500
1110	40,000 Gal.	155,000	36,400	191,400
1120	50,000 Gal.	186,000	43,700	229,700

G3060 325	Steel Tank, Above Ground, Single Wall	COST EACH		
		MAT.	INST.	TOTAL
1010	Storage tank, fuel, above ground, single wall steel, 550 Gal.	12,500	4,450	16,950
1020	2000 Gal.	15,700	4,550	20,250
1030	5000 Gal.	17,500	7,000	24,500
1040	10,000 Gal.	39,800	7,475	47,275
1050	15,000 Gal.	38,100	7,700	45,800
1060	20,000 Gal.	45,600	7,950	53,550
1070	25,000 Gal.	46,500	8,150	54,650
1080	30,000 Gal.	57,500	8,475	65,975

G3060 330	Steel Tank, Above Ground, Double Wall	COST EACH		
		MAT.	INST.	TOTAL
1010	Storage tank, fuel, above ground, double wall steel, 500 Gal.	13,000	4,500	17,500
1020	2000 Gal.	19,600	4,600	24,200
1030	4000 Gal.	27,000	4,700	31,700
1040	6000 Gal.	31,400	7,275	38,675
1050	8000 Gal.	37,100	7,475	44,575
1060	10,000 Gal.	40,100	7,625	47,725
1070	15,000 Gal.	54,500	7,900	62,400
1080	20,000 Gal.	60,000	8,150	68,150
1090	25,000 Gal.	70,000	8,400	78,400
1100	30,000 Gal.	76,000	8,700	84,700

Reference Section

All the reference information is in one section, making it easy to find what you need to know . . . and easy to use the book on a daily basis. This section is visually identified by a vertical gray bar on the page edges.

In this Reference Section, we've included Equipment Rental Costs, a listing of rental and operating costs; Crew Listings, a full listing of all crews and equipment, and their costs; Historical Cost Indexes for cost comparisons over time; City Cost Indexes and Location Factors for adjusting costs to the region you are in; Reference Tables, where you will find explanations, estimating information and procedures, or technical data; Change Orders, information on pricing changes to contract documents; Square Foot Costs that allow you to make a rough estimate for the overall cost of a project; and an explanation of all the Abbreviations in the book.

Table of Contents

Estimating Tips

- This section contains the average costs to rent and operate hundreds of pieces of construction equipment. This is useful information when estimating the time and material requirements of any particular operation in order to establish a unit or total cost. Equipment costs include not only rental, but also operating costs for equipment under normal use.

Rental Costs

- Equipment rental rates are obtained from industry sources throughout North America–contractors, suppliers, dealers, manufacturers, and distributors.
- Rental rates vary throughout the country, with larger cities generally having lower rates. Lease plans for new equipment are available for periods in excess of six months, with a percentage of payments applying toward purchase.
- Monthly rental rates vary from 2% to 5% of the purchase price of the equipment depending on the anticipated life of the equipment and its wearing parts.
- Weekly rental rates are about 1/3 the monthly rates, and daily rental rates are about 1/3 the weekly rate.
- Rental rates can also be treated as reimbursement costs for contractor-owned equipment. Owned equipment costs include depreciation, loan payments, interest, taxes, insurance, storage, and major repairs.

Operating Costs

- The operating costs include parts and labor for routine servicing, such as repair and replacement of pumps, filters and worn lines. Normal operating expendables, such as fuel, lubricants, tires and electricity (where applicable), are also included.
- Extraordinary operating expendables with highly variable wear patterns, such as diamond bits and blades, are excluded. These costs can be found as material costs in the Unit Price section.
- The hourly operating costs listed do not include the operator's wages.

Equipment Cost/Day

- Any power equipment required by a crew is shown in the Crew Listings with a daily cost.
- The daily cost of equipment needed by a crew is based on dividing the weekly rental rate by 5 (number of working days in the week), and then adding the hourly operating cost times 8 (the number of hours in a day). This "Equipment Cost/Day" is shown in the far right column of the Equipment Rental pages.
- If equipment is needed for only one or two days, it is best to develop your own cost by including components for daily rent and hourly operating cost. This is important when the listed Crew for a task does not contain the equipment needed, such as a crane for lifting mechanical heating/cooling equipment up onto a roof.

- If the quantity of work is less than the crew's Daily Output shown for a Unit Price line item that includes a bare unit equipment cost, it is recommended to estimate one day's rental cost and operating cost for equipment shown in the Crew Listing for that line item.

Mobilization/ Demobilization

- The cost to move construction equipment from an equipment yard or rental company to the jobsite and back again is not included in equipment rental costs listed in the Reference section, nor in the bare equipment cost of any Unit Price line item, nor in any equipment costs shown in the Crew listings.
- Mobilization (to the site) and demobilization (from the site) costs can be found in the Unit Price section.
- If a piece of equipment is already at the jobsite, it is not appropriate to utilize mobil./ demob. costs again in an estimate.

01 54 33 | Equipment Rental

			UNIT	HOURLY OPER. COST	RENT PER DAY	RENT PER WEEK	RENT PER MONTH	EQUIPMENT COST/DAY	
10	0010	**CONCRETE EQUIPMENT RENTAL** without operators [R015433 -10]							10
	0200	Bucket, concrete lightweight, 1/2 C.Y.	Ea.	.75	23	69	207	19.80	
	0300	1 C.Y.		.80	28	84	252	23.20	
	0400	1-1/2 C.Y.		1.00	38.50	115	345	31	
	0500	2 C.Y.		1.10	45	135	405	35.80	
	0580	8 C.Y.		5.70	265	795	2,375	204.60	
	0600	Cart, concrete, self-propelled, operator walking, 10 C.F.		2.50	58.50	175	525	55	
	0700	Operator riding, 18 C.F.		4.20	96.50	290	870	91.60	
	0800	Conveyer for concrete, portable, gas, 16" wide, 26' long		9.20	127	380	1,150	149.60	
	0900	46' long		9.55	153	460	1,375	168.40	
	1000	56' long		9.70	162	485	1,450	174.60	
	1100	Core drill, electric, 2-1/2 H.P., 1" to 8" bit diameter		1.40	59.50	179	535	47	
	1150	11 H.P., 8" to 18" cores		5.10	115	345	1,025	109.80	
	1200	Finisher, concrete floor, gas, riding trowel, 96" wide		8.55	148	445	1,325	157.40	
	1300	Gas, walk-behind, 3 blade, 36" trowel		1.40	20.50	62	186	23.60	
	1400	4 blade, 48" trowel		2.70	28	84	252	38.40	
	1500	Float, hand-operated (Bull float) 48" wide		.08	13.65	41	123	8.85	
	1570	Curb builder, 14 H.P., gas, single screw		11.95	253	760	2,275	247.60	
	1590	Double screw		12.60	293	880	2,650	276.80	
	1600	Floor grinder, concrete and terrazzo, electric, 22" path		1.62	102	305	915	73.95	
	1700	Edger, concrete, electric, 7" path		.90	51.50	155	465	38.20	
	1750	Vacuum pick-up system for floor grinders, wet/dry		1.35	81.50	245	735	59.80	
	1800	Mixer, powered, mortar and concrete, gas, 6 C.F., 18 H.P.		6.45	120	360	1,075	123.60	
	1900	10 C.F., 25 H.P.		7.90	147	440	1,325	151.20	
	2000	16 C.F.		8.20	170	510	1,525	167.60	
	2100	Concrete, stationary, tilt drum, 2 C.Y.		6.45	238	715	2,150	194.60	
	2120	Pump, concrete, truck mounted 4" line 80' boom		23.25	925	2,775	8,325	741	
	2140	5" line, 110' boom		30.70	1,250	3,770	11,300	999.60	
	2160	Mud jack, 50 C.F. per hr.		6.10	128	385	1,150	125.80	
	2180	225 C.F. per hr		8.15	147	440	1,325	153.20	
	2190	Shotcrete pump rig, 12 C.Y./hr.		14.15	227	680	2,050	249.20	
	2200	35 C.Y./hr.		16.75	243	730	2,200	280	
	2600	Saw, concrete, manual, gas, 18 H.P.		4.65	43.50	130	390	63.20	
	2650	Self-propelled, gas, 30 H.P.		8.90	105	315	945	134.20	
	2700	Vibrators, concrete, electric, 60 cycle, 2 H.P.		.31	9	27	81	7.90	
	2800	3 H.P.		.43	12	36	108	10.65	
	2900	Gas engine, 5 H.P.		1.30	16.35	49	147	20.20	
	3000	8 H.P.		1.75	15.35	46	138	23.20	
	3050	Vibrating screed, gas engine, 8 H.P.		2.81	71.50	215	645	65.50	
	3120	Concrete transit mixer, 6 x 4, 250 H.P., 8 C.Y., rear discharge		43.95	590	1,765	5,300	704.60	
	3200	Front discharge		51.35	725	2,175	6,525	845.80	
	3300	6 x 6, 285 H.P., 12 C.Y., rear discharge		50.45	685	2,050	6,150	813.60	
	3400	Front discharge	▼	52.65	735	2,200	6,600	861.20	
20	0010	**EARTHWORK EQUIPMENT RENTAL** without operators [R015433 -10]							20
	0040	Aggregate spreader, push type 8' to 12' wide	Ea.	2.25	26	78	234	33.60	
	0045	Tailgate type, 8' wide		2.20	33.50	100	300	37.60	
	0055	Earth auger, truck-mounted, for fence & sign posts, utility poles		11.60	515	1,545	4,625	401.80	
	0060	For borings and monitoring wells		38.45	665	1,995	5,975	706.60	
	0070	Portable, trailer mounted		2.45	32.50	98	294	39.20	
	0075	Truck-mounted, for caissons, water wells		83.90	3,025	9,075	27,200	2,486	
	0080	Horizontal boring machine, 12" to 36" diameter, 45 H.P.		20.85	200	600	1,800	286.80	
	0090	12" to 48" diameter, 65 H.P.		28.90	350	1,055	3,175	442.20	
	0095	Auger, for fence posts, gas engine, hand held		.45	6	18	54	7.20	
	0100	Excavator, diesel hydraulic, crawler mounted, 1/2 C.Y. cap.		16.70	385	1,155	3,475	364.60	
	0120	5/8 C.Y. capacity		24.50	555	1,670	5,000	530	
	0140	3/4 C.Y. capacity		30.55	660	1,980	5,950	640.40	
	0150	1 C.Y. capacity	▼	36.35	730	2,195	6,575	729.80	

01 54 33 | Equipment Rental

		UNIT	HOURLY OPER. COST	RENT PER DAY	RENT PER WEEK	RENT PER MONTH	EQUIPMENT COST/DAY	
20	0200	1-1/2 C.Y. capacity	Ea.	44.55	985	2,955	8,875	947.40	20
	0300	2 C.Y. capacity		58.25	1,275	3,790	11,400	1,224	
	0320	2-1/2 C.Y. capacity		77.80	1,650	4,955	14,900	1,613	
	0325	3-1/2 C.Y. capacity		104.95	2,225	6,655	20,000	2,171	
	0330	4-1/2 C.Y. capacity		128.05	2,750	8,225	24,700	2,669	
	0335	6 C.Y. capacity		158.65	3,050	9,160	27,500	3,101	
	0340	7 C.Y. capacity		160.90	3,150	9,460	28,400	3,179	
	0342	Excavator attachments, bucket thumbs		3.00	245	735	2,200	171	
	0345	Grapples		2.70	213	640	1,925	149.60	
	0347	Hydraulic hammer for boom mounting, 5000 ft lbs		12.65	420	1,255	3,775	352.20	
	0349	11,000 ft lbs		21.50	750	2,250	6,750	622	
	0350	Gradall type, truck mounted, 3 ton @ 15' radius, 5/8 C.Y.		36.80	940	2,815	8,450	857.40	
	0370	1 C.Y. capacity		41.00	1,125	3,365	10,100	1,001	
	0400	Backhoe-loader, 40 to 45 H.P., 5/8 C.Y. capacity		11.20	223	670	2,000	223.60	
	0450	45 H.P. to 60 H.P., 3/4 C.Y. capacity		16.30	315	950	2,850	320.40	
	0460	80 H.P., 1-1/4 C.Y. capacity		17.80	345	1,035	3,100	349.40	
	0470	112 H.P., 1-1/2 C.Y. capacity		30.25	610	1,830	5,500	608	
	0482	Backhoe-loader attachment, compactor, 20,000 lb.		5.40	145	435	1,300	130.20	
	0485	Hydraulic hammer, 750 ft lbs		3.10	103	310	930	86.80	
	0486	Hydraulic hammer, 1200 ft lbs		5.50	190	570	1,700	158	
	0500	Brush chipper, gas engine, 6" cutter head, 35 H.P.		8.10	110	330	990	130.80	
	0550	12" cutter head, 130 H.P.		13.25	185	555	1,675	217	
	0600	15" cutter head, 165 H.P.		19.45	222	665	2,000	288.60	
	0750	Bucket, clamshell, general purpose, 3/8 C.Y.		1.20	38.50	115	345	32.60	
	0800	1/2 C.Y.		1.30	45	135	405	37.40	
	0850	3/4 C.Y.		1.45	56.50	170	510	45.60	
	0900	1 C.Y.		1.50	60	180	540	48	
	0950	1-1/2 C.Y.		2.40	81.50	245	735	68.20	
	1000	2 C.Y.		2.50	91.50	275	825	75	
	1010	Bucket, dragline, medium duty, 1/2 C.Y.		.70	24.50	73	219	20.20	
	1020	3/4 C.Y.		.70	25.50	77	231	21	
	1030	1 C.Y.		.75	27	81	243	22.20	
	1040	1-1/2 C.Y.		1.15	41.50	125	375	34.20	
	1050	2 C.Y.		1.20	46.50	140	420	37.60	
	1070	3 C.Y.		1.85	63.50	190	570	52.80	
	1200	Compactor, manually guided 2-drum vibratory smooth roller, 7.5 H.P.		5.85	167	500	1,500	146.80	
	1250	Rammer/tamper, gas, 8"		2.20	43.50	130	390	43.60	
	1260	15"		2.45	50	150	450	49.60	
	1300	Vibratory plate, gas, 18" plate, 3000 lb. blow		2.10	23.50	70	210	30.80	
	1350	21" plate, 5000 lb. blow		2.60	30.50	91	273	39	
	1370	Curb builder/extruder, 14 H.P., gas, single screw		11.95	253	760	2,275	247.60	
	1390	Double screw		12.60	293	880	2,650	276.80	
	1500	Disc harrow attachment, for tractor		.41	68.50	205	615	44.30	
	1810	Feller buncher, shearing & accumulating trees, 100 H.P.		27.75	570	1,710	5,125	564	
	1860	Grader, self-propelled, 25,000 lb.		22.90	470	1,405	4,225	464.20	
	1910	30,000 lb.		26.35	555	1,660	4,975	542.80	
	1920	40,000 lb.		43.05	1,050	3,155	9,475	975.40	
	1930	55,000 lb.		55.95	1,425	4,275	12,800	1,303	
	1950	Hammer, pavement breaker, self-propelled, diesel, 1000 to 1250 lb.		22.90	360	1,080	3,250	399.20	
	2000	1300 to 1500 lb.		34.35	720	2,160	6,475	706.80	
	2050	Pile driving hammer, steam or air, 4150 ft lbs @ 225 bpm		9.80	500	1,495	4,475	377.40	
	2100	8750 ft lbs @ 145 bpm		11.80	700	2,095	6,275	513.40	
	2150	15,000 ft lbs @ 60 bpm		13.40	845	2,535	7,600	614.20	
	2200	24,450 ft lbs @ 111 bpm		14.40	940	2,820	8,450	679.20	
	2250	Leads, 60' high for pile driving hammers up to 20,000 ft lbs		3.10	82	246	740	74	
	2300	90' high for hammers over 20,000 ft lbs		4.70	149	448	1,350	127.20	
	2350	Diesel type hammer, 22,400 ft lbs		24.80	655	1,970	5,900	592.40	
	2400	41,300 ft lbs		31.25	710	2,135	6,400	677	

01 54 33 | Equipment Rental

		UNIT	HOURLY OPER. COST	RENT PER DAY	RENT PER WEEK	RENT PER MONTH	EQUIPMENT COST/DAY	
2450	141,000 ft lbs	Ea.	47.65	1,225	3,660	11,000	1,113	20
2500	Vib. elec. hammer/extractor, 200 kW diesel generator, 34 H.P.		36.95	680	2,045	6,125	704.60	
2550	80 H.P.		64.95	1,000	3,025	9,075	1,125	
2600	150 H.P.		120.90	1,950	5,865	17,600	2,140	
2700	Extractor, steam or air, 700 ft lbs		17.75	520	1,560	4,675	454	
2750	1000 ft lbs		20.20	635	1,910	5,725	543.60	
2800	Log chipper, up to 22" diameter, 600 H.P.		35.35	455	1,360	4,075	554.80	
2850	Logger, for skidding & stacking logs, 150 H.P.		41.75	850	2,550	7,650	844	
2860	Mulcher, diesel powered, trailer mounted		15.60	223	670	2,000	258.80	
2900	Rake, spring tooth, with tractor		12.66	294	882	2,650	277.70	
3000	Roller, vibratory, tandem, smooth drum, 20 H.P.		7.40	143	430	1,300	145.20	
3050	35 H.P.		9.40	248	745	2,225	224.20	
3100	Towed type vibratory compactor, smooth drum, 50 H.P.		22.45	360	1,075	3,225	394.60	
3150	Sheepsfoot, 50 H.P.		24.75	420	1,255	3,775	449	
3170	Landfill compactor, 220 H.P.		68.75	1,425	4,290	12,900	1,408	
3200	Pneumatic tire roller, 80 H.P.		12.65	345	1,040	3,125	309.20	
3250	120 H.P.		19.15	600	1,800	5,400	513.20	
3300	Sheepsfoot vibratory roller, 240 H.P.		55.10	1,100	3,265	9,800	1,094	
3320	340 H.P.		73.05	1,675	5,025	15,100	1,589	
3350	Smooth drum vibratory roller, 75 H.P.		20.60	555	1,670	5,000	498.80	
3400	125 H.P.		25.60	655	1,960	5,875	596.80	
3410	Rotary mower, brush, 60", with tractor		16.60	268	805	2,425	293.80	
3420	Rototiller, walk-behind, gas, 5 H.P.		2.23	67.50	203	610	58.45	
3422	8 H.P.		3.47	103	310	930	89.75	
3440	Scrapers, towed type, 7 C.Y. capacity		5.25	110	330	990	108	
3450	10 C.Y. capacity		6.00	153	460	1,375	140	
3500	15 C.Y. capacity		6.45	178	535	1,600	158.60	
3525	Self-propelled, single engine, 14 C.Y. capacity		77.50	1,475	4,460	13,400	1,512	
3550	Dual engine, 21 C.Y. capacity		107.80	1,675	5,050	15,200	1,872	
3600	31 C.Y. capacity		152.05	2,600	7,805	23,400	2,777	
3640	44 C.Y. capacity		193.60	3,600	10,780	32,300	3,705	
3650	Elevating type, single engine, 11 C.Y. capacity		54.35	1,100	3,270	9,800	1,089	
3700	22 C.Y. capacity		99.85	2,150	6,440	19,300	2,087	
3710	Screening plant 110 H.P. w/5' x 10' screen		30.50	420	1,260	3,775	496	
3720	5' x 16' screen		32.65	520	1,560	4,675	573.20	
3850	Shovel, crawler-mounted, front-loading, 7 C.Y. capacity		168.30	2,950	8,815	26,400	3,109	
3855	12 C.Y. capacity		225.05	3,750	11,230	33,700	4,046	
3860	Shovel/backhoe bucket, 1/2 C.Y.		2.25	63.50	190	570	56	
3870	3/4 C.Y.		2.30	71.50	215	645	61.40	
3880	1 C.Y.		2.40	80	240	720	67.20	
3890	1-1/2 C.Y.		2.55	95	285	855	77.40	
3910	3 C.Y.		2.85	130	390	1,175	100.80	
3950	Stump chipper, 18" deep, 30 H.P.		7.17	187	560	1,675	169.35	
4110	Dozer, crawler, torque converter, diesel 80 H.P.		21.70	400	1,195	3,575	412.60	
4150	105 H.P.		25.80	500	1,495	4,475	505.40	
4200	140 H.P.		37.35	800	2,400	7,200	778.80	
4260	200 H.P.		50.75	1,075	3,225	9,675	1,051	
4310	300 H.P.		71.80	1,725	5,165	15,500	1,607	
4360	410 H.P.		97.20	2,200	6,635	19,900	2,105	
4370	500 H.P.		130.15	3,000	9,020	27,100	2,845	
4380	700 H.P.		191.35	4,375	13,095	39,300	4,150	
4400	Loader, crawler, torque conv., diesel, 1-1/2 C.Y., 80 H.P.		22.70	475	1,425	4,275	466.60	
4450	1-1/2 to 1-3/4 C.Y., 95 H.P.		25.75	600	1,795	5,375	565	
4510	1-3/4 to 2-1/4 C.Y., 130 H.P.		33.70	840	2,520	7,550	773.60	
4530	2-1/2 to 3-1/4 C.Y., 190 H.P.		46.40	1,050	3,185	9,550	1,008	
4560	3-1/2 to 5 C.Y., 275 H.P.		64.35	1,525	4,585	13,800	1,432	
4610	Front end loader, 4WD, articulated frame, 1 to 1-1/4 C.Y., 70 H.P.		13.35	215	645	1,925	235.80	
4620	1-1/2 to 1-3/4 C.Y., 95 H.P.		17.30	300	905	2,725	319.40	

01 54 33 | Equipment Rental

		UNIT	HOURLY OPER. COST	RENT PER DAY	RENT PER WEEK	RENT PER MONTH	EQUIPMENT COST/DAY	
20	4650	1-3/4 to 2 C.Y., 130 H.P.	Ea.	19.90	370	1,105	3,325	380.20	20
	4710	2-1/2 to 3-1/2 C.Y., 145 H.P.		21.20	380	1,140	3,425	397.60	
	4730	3 to 4-1/2 C.Y., 185 H.P.		28.95	610	1,835	5,500	598.60	
	4760	5-1/4 to 5-3/4 C.Y., 270 H.P.		47.05	930	2,785	8,350	933.40	
	4810	7 to 9 C.Y., 475 H.P.		83.30	1,900	5,725	17,200	1,811	
	4870	9 - 11 C.Y., 620 H.P.		108.70	2,675	8,030	24,100	2,476	
	4880	Skid steer loader, wheeled, 10 C.F., 30 H.P. gas		7.85	148	445	1,325	151.80	
	4890	1 C.Y., 78 H.P., diesel		13.90	250	750	2,250	261.20	
	4892	Skid-steer attachment, auger		.53	88	264	790	57.05	
	4893	Backhoe		.71	119	356	1,075	76.90	
	4894	Broom		.74	123	370	1,100	79.90	
	4895	Forks		.24	40.50	122	365	26.30	
	4896	Grapple		.55	91.50	274	820	59.20	
	4897	Concrete hammer		1.07	178	534	1,600	115.35	
	4898	Tree spade		.78	129	388	1,175	83.85	
	4899	Trencher		.76	127	380	1,150	82.10	
	4900	Trencher, chain, boom type, gas, operator walking, 12 H.P.		3.55	48.50	145	435	57.40	
	4910	Operator riding, 40 H.P.		14.25	305	915	2,750	297	
	5000	Wheel type, diesel, 4' deep, 12" wide		61.50	850	2,545	7,625	1,001	
	5100	6' deep, 20" wide		77.45	1,975	5,910	17,700	1,802	
	5150	Chain type, diesel, 5' deep, 8" wide		27.10	600	1,805	5,425	577.80	
	5200	Diesel, 8' deep, 16" wide		114.45	2,650	7,950	23,900	2,506	
	5202	Rock trencher, wheel type, 6" wide x 18" deep		17.35	325	970	2,900	332.80	
	5206	Chain type, 18" wide x 7' deep		95.95	2,950	8,850	26,600	2,538	
	5210	Tree spade, self-propelled		13.73	267	800	2,400	269.85	
	5250	Truck, dump, 2-axle, 12 ton, 8 C.Y. payload, 220 H.P.		21.80	235	705	2,125	315.40	
	5300	Three axle dump, 16 ton, 12 CY payload, 400 H.P.		38.00	335	1,010	3,025	506	
	5310	Four axle dump, 25 ton, 18 C.Y. payload, 450 H.P.		45.60	495	1,480	4,450	660.80	
	5350	Dump trailer only, rear dump, 16-1/2 C.Y.		5.10	142	425	1,275	125.80	
	5400	20 C.Y.		5.50	160	480	1,450	140	
	5450	Flatbed, single axle, 1-1/2 ton rating		16.05	71.50	215	645	171.40	
	5500	3 ton rating		19.50	102	305	915	217	
	5550	Off highway rear dump, 25 ton capacity		57.25	1,350	4,030	12,100	1,264	
	5600	35 ton capacity		52.80	1,200	3,565	10,700	1,135	
	5610	50 ton capacity		71.20	1,625	4,895	14,700	1,549	
	5620	65 ton capacity		72.90	1,625	4,840	14,500	1,551	
	5630	100 ton capacity		93.85	2,100	6,290	18,900	2,009	
	6000	Vibratory plow, 25 H.P., walking		5.95	63.50	190	570	85.60	
40	0010	**GENERAL EQUIPMENT RENTAL** without operators							40
	0150	Aerial lift, scissor type, to 15' high, 1000 lb. cap., electric [R015433 -10]	Ea.	2.90	56.50	170	510	57.20	
	0160	To 25' high, 2000 lb. capacity		3.30	71.50	215	645	69.40	
	0170	Telescoping boom to 40' high, 500 lb. capacity, gas		15.85	320	955	2,875	317.80	
	0180	To 45' high, 500 lb. capacity		16.95	370	1,105	3,325	356.60	
	0190	To 60' high, 600 lb. capacity		19.20	480	1,445	4,325	442.60	
	0195	Air compressor, portable, 6.5 CFM, electric		.33	14	42	126	11.05	
	0196	Gasoline		.80	21	63	189	19	
	0200	Towed type, gas engine, 60 CFM		10.15	48.50	145	435	110.20	
	0300	160 CFM		11.75	51.50	155	465	125	
	0400	Diesel engine, rotary screw, 250 CFM		12.35	107	320	960	162.80	
	0500	365 CFM		16.40	132	395	1,175	210.20	
	0550	450 CFM		20.70	165	495	1,475	264.60	
	0600	600 CFM		35.95	228	685	2,050	424.60	
	0700	750 CFM		36.10	238	715	2,150	431.80	
	0800	For silenced models, small sizes, add to rent		3%	5%	5%	5%		
	0900	Large sizes, add to rent		5%	7%	7%	7%		
	0930	Air tools, breaker, pavement, 60 lb.	Ea.	.50	9.65	29	87	9.80	
	0940	80 lb.		.50	10.35	31	93	10.20	
	0950	Drills, hand (jackhammer) 65 lb.		.55	17.35	52	156	14.80	

01 54 33 | Equipment Rental

		UNIT	HOURLY OPER. COST	RENT PER DAY	RENT PER WEEK	RENT PER MONTH	EQUIPMENT COST/DAY	
40	0960	Track or wagon, swing boom, 4" drifter	Ea.	49.45	755	2,270	6,800	849.60	**40**
	0970	5" drifter		59.40	815	2,450	7,350	965.20	
	0975	Track mounted quarry drill, 6" diameter drill		82.30	1,250	3,725	11,200	1,403	
	0980	Dust control per drill		.58	18.65	56	168	15.85	
	0990	Hammer, chipping, 12 lb.		.50	24.50	73	219	18.60	
	1000	Hose, air with couplings, 50' long, 3/4" diameter		.03	4.67	14	42	3.05	
	1100	1" diameter		.04	6.35	19	57	4.10	
	1200	1-1/2" diameter		.06	9.65	29	87	6.30	
	1300	2" diameter		.07	12	36	108	7.75	
	1400	2-1/2" diameter		.11	19	57	171	12.30	
	1410	3" diameter		.14	23	69	207	14.90	
	1450	Drill, steel, 7/8" x 2'		.05	8.65	26	78	5.60	
	1460	7/8" x 6'		.05	9	27	81	5.80	
	1520	Moil points		.02	4	12	36	2.55	
	1525	Pneumatic nailer w/accessories		.49	32.50	98	294	23.50	
	1530	Sheeting driver for 60 lb. breaker		.04	6	18	54	3.90	
	1540	For 90 lb. breaker		.12	8	24	72	5.75	
	1550	Spade, 25 lb.		.45	7.35	22	66	8	
	1560	Tamper, single, 35 lb.		.55	36.50	109	325	26.20	
	1570	Triple, 140 lb.		.82	54.50	164	490	39.35	
	1580	Wrenches, impact, air powered, up to 3/4" bolt		.40	10	30	90	9.20	
	1590	Up to 1-1/4" bolt		.50	23.50	71	213	18.20	
	1600	Barricades, barrels, reflectorized, 1 to 50 barrels		.03	4.60	13.80	41.50	3	
	1610	100 to 200 barrels		.02	3.53	10.60	32	2.30	
	1620	Barrels with flashers, 1 to 50 barrels		.03	5.25	15.80	47.50	3.40	
	1630	100 to 200 barrels		.03	4.20	12.60	38	2.75	
	1640	Barrels with steady burn type C lights		.04	7	21	63	4.50	
	1650	Illuminated board, trailer mounted, with generator		.75	122	365	1,100	79	
	1670	Portable barricade, stock, with flashers, 1 to 6 units		.03	5.25	15.80	47.50	3.40	
	1680	25 to 50 units		.03	4.90	14.70	44	3.20	
	1685	Butt fusion machine, wheeled, 1.5 HP electric, 2" - 8" diam. pipe		2.56	167	500	1,500	120.50	
	1690	Tracked, 20 HP diesel, 4"-12" diam. pipe		10.43	490	1,465	4,400	376.45	
	1695	83 HP diesel, 8" - 24" diam. pipe		27.52	975	2,930	8,800	806.15	
	1700	Carts, brick, hand powered, 1000 lb. capacity		.44	72.50	218	655	47.10	
	1800	Gas engine, 1500 lb., 7-1/2' lift		4.04	115	345	1,025	101.30	
	1822	Dehumidifier, medium, 6 lb/hr, 150 CFM		.93	60	180	540	43.45	
	1824	Large, 18 lb/hr, 600 CFM		1.89	122	366	1,100	88.30	
	1830	Distributor, asphalt, trailer mounted, 2000 gal., 38 H.P. diesel		9.60	340	1,020	3,050	280.80	
	1840	3000 gal., 38 H.P. diesel		11.00	370	1,105	3,325	309	
	1850	Drill, rotary hammer, electric		.82	26.50	80	240	22.55	
	1860	Carbide bit, 1-1/2" diameter, add to electric rotary hammer		.02	6.25	18.80	56.50	3.90	
	1865	Rotary, crawler, 250 H.P.		122.20	2,050	6,125	18,400	2,203	
	1870	Emulsion sprayer, 65 gal., 5 H.P. gas engine		2.89	99	297	890	82.50	
	1880	200 gal., 5 H.P. engine		6.85	165	495	1,475	153.80	
	1900	Floor auto-scrubbing machine, walk-behind, 28" path		3.26	213	640	1,925	154.10	
	1930	Floodlight, mercury vapor, or quartz, on tripod, 1000 watt		.27	14	42	126	10.55	
	1940	2000 watt		.50	27	81	243	20.20	
	1950	Floodlights, trailer mounted with generator, 1 - 300 watt light		3.10	73.50	220	660	68.80	
	1960	2 - 1000 watt lights		3.95	98.50	295	885	90.60	
	2000	4 - 300 watt lights		3.75	93.50	280	840	86	
	2005	Foam spray rig, incl. box trailer, compressor, generator, proportioner		23.78	495	1,490	4,475	488.25	
	2020	Forklift, straight mast, 12' lift, 5000 lb., 2 wheel drive, gas		19.80	202	605	1,825	279.40	
	2040	21' lift, 5000 lb., 4 wheel drive, diesel		16.20	248	745	2,225	278.60	
	2050	For rough terrain, 42' lift, 35' reach, 9000 lb., 110 H.P.		22.30	450	1,345	4,025	447.40	
	2060	For plant, 4 ton capacity, 80 H.P., 2 wheel drive, gas		12.05	102	305	915	157.40	
	2080	10 ton capacity, 120 H.P., 2 wheel drive, diesel		17.70	177	530	1,600	247.60	
	2100	Generator, electric, gas engine, 1.5 kW to 3 kW		2.80	11.35	34	102	29.20	
	2200	5 kW		3.55	15.35	46	138	37.60	

01 54 33 | Equipment Rental

		UNIT	HOURLY OPER. COST	RENT PER DAY	RENT PER WEEK	RENT PER MONTH	EQUIPMENT COST/DAY	
40	2300	10 kW	Ea.	6.65	38.50	115	345	76.20	**40**
	2400	25 kW		7.80	88.50	265	795	115.40	
	2500	Diesel engine, 20 kW		9.00	75	225	675	117	
	2600	50 kW		15.85	105	315	945	189.80	
	2700	100 kW		30.25	145	435	1,300	329	
	2800	250 kW		59.65	253	760	2,275	629.20	
	2850	Hammer, hydraulic, for mounting on boom, to 500 ft lbs		2.40	76.50	230	690	65.20	
	2860	1000 ft lbs		4.05	127	380	1,150	108.40	
	2900	Heaters, space, oil or electric, 50 MBH		1.84	8	24	72	19.50	
	3000	100 MBH		3.31	10.65	32	96	32.90	
	3100	300 MBH		10.62	36.50	110	330	106.95	
	3150	500 MBH		21.42	48.50	145	435	200.35	
	3200	Hose, water, suction with coupling, 20' long, 2" diameter		.02	3	9	27	1.95	
	3210	3" diameter		.03	4.67	14	42	3.05	
	3220	4" diameter		.03	5	15	45	3.25	
	3230	6" diameter		.11	18.35	55	165	11.90	
	3240	8" diameter		.20	33.50	100	300	21.60	
	3250	Discharge hose with coupling, 50' long, 2" diameter		.01	1.33	4	12	.90	
	3260	3" diameter		.02	2.67	8	24	1.75	
	3270	4" diameter		.02	3.67	11	33	2.35	
	3280	6" diameter		.06	9.65	29	87	6.30	
	3290	8" diameter		.20	33.50	100	300	21.60	
	3295	Insulation blower		.20	6	18	54	5.20	
	3300	Ladders, extension type, 16' to 36' long		.15	25.50	76	228	16.40	
	3400	40' to 60' long		.22	36	108	325	23.35	
	3405	Lance for cutting concrete		2.17	104	313	940	79.95	
	3407	Lawn mower, rotary, 22", 5 H.P.		2.14	63.50	190	570	55.10	
	3408	48" self propelled		2.88	94.50	283	850	79.65	
	3410	Level, electronic, automatic, with tripod and leveling rod		1.45	96.50	290	870	69.60	
	3430	Laser type, for pipe and sewer line and grade		.76	50.50	152	455	36.50	
	3440	Rotating beam for interior control		1.16	77.50	232	695	55.70	
	3460	Builder's optical transit, with tripod and rod		.09	14.65	44	132	9.50	
	3500	Light towers, towable, with diesel generator, 2000 watt		3.75	93.50	280	840	86	
	3600	4000 watt		3.95	98.50	295	885	90.60	
	3700	Mixer, powered, plaster and mortar, 6 C.F., 7 H.P.		1.85	30	90	270	32.80	
	3800	10 C.F., 9 H.P.		2.00	31.50	95	285	35	
	3850	Nailer, pneumatic		.49	32.50	98	294	23.50	
	3900	Paint sprayers complete, 8 CFM		.73	48.50	145	435	34.85	
	4000	17 CFM		1.35	89.50	269	805	64.60	
	4020	Pavers, bituminous, rubber tires, 8' wide, 50 H.P., diesel		38.00	1,050	3,165	9,500	937	
	4030	10' wide, 150 H.P.		80.45	1,700	5,120	15,400	1,668	
	4050	Crawler, 8' wide, 100 H.P., diesel		78.80	1,800	5,415	16,200	1,713	
	4060	10' wide, 150 H.P.		90.60	2,125	6,405	19,200	2,006	
	4070	Concrete paver, 12' to 24' wide, 250 H.P.		78.20	1,625	4,875	14,600	1,601	
	4080	Placer-spreader-trimmer, 24' wide, 300 H.P.		111.00	2,650	7,975	23,900	2,483	
	4100	Pump, centrifugal gas pump, 1-1/2" diam., 65 GPM		3.50	48.50	145	435	57	
	4200	2" diameter, 130 GPM		4.50	53.50	160	480	68	
	4300	3" diameter, 250 GPM		4.75	55	165	495	71	
	4400	6" diameter, 1500 GPM		22.95	180	540	1,625	291.60	
	4500	Submersible electric pump, 1-1/4" diameter, 55 GPM		.31	16.35	49	147	12.30	
	4600	1-1/2" diameter, 83 GPM		.35	18.65	56	168	14	
	4700	2" diameter, 120 GPM		1.30	23.50	70	210	24.40	
	4800	3" diameter, 300 GPM		2.20	41.50	125	375	42.60	
	4900	4" diameter, 560 GPM		9.50	163	490	1,475	174	
	5000	6" diameter, 1590 GPM		13.95	220	660	1,975	243.60	
	5100	Diaphragm pump, gas, single, 1-1/2" diameter		1.15	49.50	148	445	38.80	
	5200	2" diameter		3.70	61.50	185	555	66.60	
	5300	3" diameter		3.70	61.50	185	555	66.60	

01 54 33 | Equipment Rental

		UNIT	HOURLY OPER. COST	RENT PER DAY	RENT PER WEEK	RENT PER MONTH	EQUIPMENT COST/DAY	
40	5400	Double, 4" diameter	Ea.	5.55	108	325	975	109.40	40
	5450	Pressure washer 5 GPM, 3000 psi		3.85	46.50	140	420	58.80	
	5500	Trash pump, self-priming, gas, 2" diameter		3.70	21	63	189	42.20	
	5600	Diesel, 4" diameter		9.10	58.50	175	525	107.80	
	5650	Diesel, 6" diameter		29.10	130	390	1,175	310.80	
	5655	Grout Pump		14.75	100	300	900	178	
	5700	Salamanders, L.P. gas fired, 100,000 Btu		3.35	16.35	49	147	36.60	
	5705	50,000 Btu		2.50	10.35	31	93	26.20	
	5720	Sandblaster, portable, open top, 3 C.F. capacity		.55	27	81	243	20.60	
	5730	6 C.F. capacity		.90	40	120	360	31.20	
	5740	Accessories for above		.12	20.50	61	183	13.15	
	5750	Sander, floor		.60	20	60	180	16.80	
	5760	Edger		.61	24.50	74	222	19.70	
	5800	Saw, chain, gas engine, 18" long		1.75	19.65	59	177	25.80	
	5900	Hydraulic powered, 36" long		.70	60	180	540	41.60	
	5950	60" long		.70	61.50	185	555	42.60	
	6000	Masonry, table mounted, 14" diameter, 5 H.P.		1.25	56.50	170	510	44	
	6050	Portable cut-off, 8 H.P.		1.90	33	99	297	35	
	6100	Circular, hand held, electric, 7-1/4" diameter		.14	4.67	14	42	3.90	
	6200	12" diameter		.21	7.65	23	69	6.30	
	6250	Wall saw, w/hydraulic power, 10 H.P.		6.50	61.50	185	555	89	
	6275	Shot blaster, walk-behind, 20" wide		4.65	298	895	2,675	216.20	
	6280	Sidewalk broom, walk-behind		2.09	60.50	182	545	53.10	
	6300	Steam cleaner, 100 gallons per hour		3.15	60.50	182	545	61.60	
	6310	200 gallons per hour		4.25	73.50	220	660	78	
	6340	Tar Kettle/Pot, 400 gallons		4.97	81.50	245	735	88.75	
	6350	Torch, cutting, acetylene-oxygen, 150' hose		.50	23.50	70	210	18	
	6360	Hourly operating cost includes tips and gas		10.35				82.80	
	6410	Toilet, portable chemical		.12	19.65	59	177	12.75	
	6420	Recycle flush type		.14	23.50	71	213	15.30	
	6430	Toilet, fresh water flush, garden hose,		.17	27.50	83	249	17.95	
	6440	Hoisted, non-flush, for high rise		.14	23.50	70	210	15.10	
	6450	Toilet, trailers, minimum		.24	40.50	122	365	26.30	
	6460	Maximum		.74	124	372	1,125	80.30	
	6465	Tractor, farm with attachment		15.50	305	910	2,725	306	
	6500	Trailers, platform, flush deck, 2 axle, 25 ton capacity		5.15	117	350	1,050	111.20	
	6600	40 ton capacity		6.65	162	485	1,450	150.20	
	6700	3 axle, 50 ton capacity		7.20	180	540	1,625	165.60	
	6800	75 ton capacity		8.95	237	710	2,125	213.60	
	6810	Trailer mounted cable reel for high voltage line work		5.04	240	720	2,150	184.30	
	6820	Trailer mounted cable tensioning rig		10.01	475	1,430	4,300	366.10	
	6830	Cable pulling rig		67.52	2,675	8,020	24,100	2,144	
	6900	Water tank trailer, engine driven discharge, 5000 gallons		6.65	155	465	1,400	146.20	
	6925	10,000 gallons		9.15	218	655	1,975	204.20	
	6950	Water truck, off highway, 6000 gallons		64.15	860	2,585	7,750	1,030	
	7010	Tram car for high voltage line work, powered, 2 conductor		4.39	130	391	1,175	113.30	
	7020	Transit (builder's level) with tripod		.09	14.65	44	132	9.50	
	7030	Trench box, 3000 lbs. 6' x 8'		.56	93	279	835	60.30	
	7040	7200 lbs. 6' x 20'		1.05	175	525	1,575	113.40	
	7050	8000 lbs., 8' x 16'		1.05	175	525	1,575	113.40	
	7060	9500 lbs., 8' x 20'		1.19	199	596	1,800	128.70	
	7065	11,000 lbs., 8' x 24'		1.25	209	627	1,875	135.40	
	7070	12,000 lbs., 10' x 20'		1.36	227	680	2,050	146.90	
	7100	Truck, pickup, 3/4 ton, 2 wheel drive		8.70	58.50	175	525	104.60	
	7200	4 wheel drive		8.95	75	225	675	116.60	
	7250	Crew carrier, 9 passenger		12.30	90	270	810	152.40	
	7290	Flat bed truck, 20,000 lbs. GVW		13.75	133	400	1,200	190	
	7300	Tractor, 4 x 2, 220 H.P.		19.40	207	620	1,850	279.20	

01 54 33 | Equipment Rental

		UNIT	HOURLY OPER. COST	RENT PER DAY	RENT PER WEEK	RENT PER MONTH	EQUIPMENT COST/DAY	
40	7410	330 H.P.	Ea.	28.45	283	850	2,550	397.60	**40**
	7500	6 x 4, 380 H.P.		32.55	330	995	2,975	459.40	
	7600	450 H.P.		39.85	400	1,200	3,600	558.80	
	7610	Tractor, with A frame, boom and winch, 225 H.P.		22.10	283	850	2,550	346.80	
	7620	Vacuum truck, hazardous material, 2500 gallons		11.35	315	945	2,825	279.80	
	7625	5,000 gallons		13.89	440	1,320	3,950	375.10	
	7650	Vacuum, HEPA, 16 gallon, wet/dry		.36	18	54	162	13.70	
	7655	55 gallon, wet/dry		.59	27	81	243	20.90	
	7660	Water tank, portable		.17	28.50	85.50	257	18.45	
	7690	Sewer/catch basin vacuum, 14 C.Y., 1500 gallons		18.51	660	1,980	5,950	544.10	
	7700	Welder, electric, 200 amp		1.70	17.35	52	156	24	
	7800	300 amp		2.48	21	63	189	32.45	
	7900	Gas engine, 200 amp		10.15	25.50	76	228	96.40	
	8000	300 amp		11.60	27	81	243	109	
	8100	Wheelbarrow, any size		.08	12.65	38	114	8.25	
	8200	Wrecking ball, 4000 lbs.		2.20	73.50	220	660	61.60	
50	0010	**HIGHWAY EQUIPMENT RENTAL** without operators	R015433 -10						**50**
	0050	Asphalt batch plant, portable drum mixer, 100 ton/hr	Ea.	68.35	1,475	4,400	13,200	1,427	
	0060	200 ton/hr		76.45	1,550	4,655	14,000	1,543	
	0070	300 ton/hr		88.65	1,825	5,485	16,500	1,806	
	0100	Backhoe attachment, long stick, up to 185 H.P., 10.5' long		.33	22	66	198	15.85	
	0140	Up to 250 H.P., 12' long		.36	23.50	71	213	17.10	
	0180	Over 250 H.P., 15' long		.49	32.50	97	291	23.30	
	0200	Special dipper arm, up to 100 H.P., 32' long		1.00	66.50	199	595	47.80	
	0240	Over 100 H.P., 33' long		1.24	82.50	248	745	59.50	
	0280	Catch basin/sewer cleaning truck, 3 ton, 9 C.Y., 1000 gal		30.85	415	1,250	3,750	496.80	
	0300	Concrete batch plant, portable, electric, 200 C.Y./hr.		18.20	530	1,585	4,750	462.60	
	0520	Grader/dozer attachment, ripper/scarifier, rear mounted, up to 135 H.P.		3.30	68.50	205	615	67.40	
	0540	Up to 180 H.P.		3.90	86.50	260	780	83.20	
	0580	Up to 250 H.P.		4.30	100	300	900	94.40	
	0700	Pvmt. removal bucket, for hyd. excavator, up to 90 H.P.		1.75	51.50	155	465	45	
	0740	Up to 200 H.P.		1.95	73.50	220	660	59.60	
	0780	Over 200 H.P.		2.10	85	255	765	67.80	
	0900	Aggregate spreader, self-propelled, 187 H.P.		40.05	715	2,140	6,425	748.40	
	1000	Chemical spreader, 3 C.Y.		2.75	45	135	405	49	
	1900	Hammermill, traveling, 250 H.P.		67.55	1,925	5,760	17,300	1,692	
	2000	Horizontal borer, 3" diameter, 13 H.P. gas driven		5.25	56.50	170	510	76	
	2150	Horizontal directional drill, 20,000 lb. thrust, 78 H.P. diesel		26.20	690	2,065	6,200	622.60	
	2160	30,000 lb. thrust, 115 H.P.		32.30	1,050	3,150	9,450	888.40	
	2170	50,000 lb. thrust, 170 H.P.		46.00	1,350	4,030	12,100	1,174	
	2190	Mud trailer for HDD, 1500 gallons, 175 H.P., gas		21.10	162	485	1,450	265.80	
	2200	Hydromulcher, diesel, 3000 gallon, for truck mounting		15.00	263	790	2,375	278	
	2300	Gas, 600 gallon		6.40	103	310	930	113.20	
	2400	Joint & crack cleaner, walk behind, 25 H.P.		2.75	51.50	155	465	53	
	2500	Filler, trailer mounted, 400 gallons, 20 H.P.		7.70	222	665	2,000	194.60	
	3000	Paint striper, self-propelled, double line, 30 H.P.		6.10	165	495	1,475	147.80	
	3200	Post drivers, 6" I-Beam frame, for truck mounting		11.45	440	1,325	3,975	356.60	
	3400	Road sweeper, self-propelled, 8' wide, 90 H.P.		31.10	595	1,780	5,350	604.80	
	3450	Road sweeper, vacuum assisted, 4 C.Y., 220 gallons		51.45	650	1,945	5,825	800.60	
	4000	Road mixer, self-propelled, 130 H.P.		41.45	785	2,350	7,050	801.60	
	4100	310 H.P.		71.70	2,200	6,580	19,700	1,890	
	4220	Cold mix paver, incl. pug mill and bitumen tank, 165 H.P.		90.60	2,150	6,455	19,400	2,016	
	4250	Paver, asphalt, wheel or crawler, 130 H.P., diesel		89.10	2,050	6,180	18,500	1,949	
	4300	Paver, road widener, gas 1' to 6', 67 H.P.		42.15	885	2,650	7,950	867.20	
	4400	Diesel, 2' to 14', 88 H.P.		53.00	1,075	3,215	9,650	1,067	
	4600	Slipform pavers, curb and gutter, 2 track, 75 H.P.		35.60	765	2,300	6,900	744.80	
	4700	4 track, 165 H.P.		44.65	820	2,455	7,375	848.20	
	4800	Median barrier, 215 H.P.		45.40	850	2,550	7,650	873.20	

01 54 33 | Equipment Rental

			UNIT	HOURLY OPER. COST	RENT PER DAY	RENT PER WEEK	RENT PER MONTH	EQUIPMENT COST/DAY	
50	4901	Trailer, low bed, 75 ton capacity	Ea.	9.65	237	710	2,125	219.20	50
	5000	Road planer, walk behind, 10" cutting width, 10 H.P.		2.70	33.50	100	300	41.60	
	5100	Self-propelled, 12" cutting width, 64 H.P.		7.00	115	345	1,025	125	
	5120	Traffic line remover, metal ball blaster, truck mounted, 115 H.P.		47.15	765	2,300	6,900	837.20	
	5140	Grinder, truck mounted, 115 H.P.		50.10	835	2,505	7,525	901.80	
	5160	Walk-behind, 11 H.P.		3.05	55	165	495	57.40	
	5200	Pavement profiler, 4' to 6' wide, 450 H.P.		210.55	3,425	10,285	30,900	3,741	
	5300	8' to 10' wide, 750 H.P.		332.90	4,950	14,815	44,400	5,626	
	5400	Roadway plate, steel, 1" x 8' x 20'		.07	12	36	108	7.75	
	5600	Stabilizer, self-propelled, 150 H.P.		39.35	615	1,850	5,550	684.80	
	5700	310 H.P.		66.85	1,325	3,955	11,900	1,326	
	5800	Striper, thermal, truck mounted 120 gallon paint, 150 H.P.		42.85	520	1,555	4,675	653.80	
	6000	Tar kettle, 330 gallon, trailer mounted		4.58	63.50	190	570	74.65	
	7000	Tunnel locomotive, diesel, 8 to 12 ton		27.45	585	1,755	5,275	570.60	
	7005	Electric, 10 ton		24.75	670	2,005	6,025	599	
	7010	Muck cars, 1/2 C.Y. capacity		1.90	23.50	71	213	29.40	
	7020	1 C.Y. capacity		2.15	32.50	98	294	36.80	
	7030	2 C.Y. capacity		2.25	36.50	110	330	40	
	7040	Side dump, 2 C.Y. capacity		2.45	45	135	405	46.60	
	7050	3 C.Y. capacity		3.30	51.50	155	465	57.40	
	7060	5 C.Y. capacity		4.70	65	195	585	76.60	
	7100	Ventilating blower for tunnel, 7-1/2 H.P.		1.02	51.50	155	465	39.15	
	7110	10 H.P.		1.11	51.50	155	465	39.90	
	7120	20 H.P.		1.67	67.50	202	605	53.75	
	7140	40 H.P.		2.77	96.50	290	870	80.15	
	7160	60 H.P.		4.26	152	455	1,375	125.10	
	7175	75 H.P.		5.51	202	607	1,825	165.50	
	7180	200 H.P.		11.15	305	910	2,725	271.20	
	7800	Windrow loader, elevating		49.40	1,350	4,075	12,200	1,210	
60	0010	**LIFTING AND HOISTING EQUIPMENT RENTAL** without operators [R015433 -10]							60
	0120	Aerial lift truck, 2 person, to 80'	Ea.	24.90	730	2,185	6,550	636.20	
	0140	Boom work platform, 40' snorkel [R015433 -15]		14.25	278	835	2,500	281	
	0150	Crane, flatbed mounted, 3 ton capacity		13.45	197	590	1,775	225.60	
	0200	Crane, climbing, 106' jib, 6000 lb. capacity, 410 fpm [R312316 -45]		29.50	1,525	4,580	13,700	1,152	
	0300	101' jib, 10,250 lb. capacity, 270 fpm		35.60	1,925	5,800	17,400	1,445	
	0500	Tower, static, 130' high, 106' jib, 6200 lb. capacity at 400 fpm		33.05	1,775	5,290	15,900	1,322	
	0600	Crawler mounted, lattice boom, 1/2 C.Y., 15 tons at 12' radius		30.74	685	2,060	6,175	657.90	
	0700	3/4 C.Y., 20 tons at 12' radius		40.99	855	2,570	7,700	841.90	
	0800	1 C.Y., 25 tons at 12' radius		54.65	1,150	3,430	10,300	1,123	
	0900	1-1/2 C.Y., 40 tons at 12' radius		54.65	1,150	3,455	10,400	1,128	
	1000	2 C.Y., 50 tons at 12' radius		58.55	1,350	4,055	12,200	1,279	
	1100	3 C.Y., 75 tons at 12' radius		63.05	1,575	4,710	14,100	1,446	
	1200	100 ton capacity, 60' boom		71.50	1,800	5,405	16,200	1,653	
	1300	165 ton capacity, 60' boom		89.20	2,100	6,325	19,000	1,979	
	1400	200 ton capacity, 70' boom		111.15	2,675	7,995	24,000	2,488	
	1500	350 ton capacity, 80' boom		157.20	4,050	12,160	36,500	3,690	
	1600	Truck mounted, lattice boom, 6 x 4, 20 tons at 10' radius		37.39	1,225	3,680	11,000	1,035	
	1700	25 tons at 10' radius		40.40	1,325	4,000	12,000	1,123	
	1800	8 x 4, 30 tons at 10' radius		43.77	1,425	4,260	12,800	1,202	
	1900	40 tons at 12' radius		46.65	1,475	4,450	13,400	1,263	
	2000	60 tons at 15' radius		52.35	1,575	4,710	14,100	1,361	
	2050	82 tons at 15' radius		58.46	1,675	5,030	15,100	1,474	
	2100	90 tons at 15' radius		65.56	1,825	5,490	16,500	1,622	
	2200	115 tons at 15' radius		73.91	2,050	6,130	18,400	1,817	
	2300	150 tons at 18' radius		76.90	2,150	6,455	19,400	1,906	
	2350	165 tons at 18' radius		86.63	2,275	6,840	20,500	2,061	
	2400	Truck mounted, hydraulic, 12 ton capacity		41.45	540	1,620	4,850	655.60	

01 54 33 | Equipment Rental

		UNIT	HOURLY OPER. COST	RENT PER DAY	RENT PER WEEK	RENT PER MONTH	EQUIPMENT COST/DAY	
60	2500	25 ton capacity	Ea.	43.70	665	1,995	5,975	748.60	**60**
	2550	33 ton capacity		44.25	680	2,045	6,125	763	
	2560	40 ton capacity		57.20	800	2,405	7,225	938.60	
	2600	55 ton capacity		75.80	1,000	3,010	9,025	1,208	
	2700	80 ton capacity		97.75	1,525	4,555	13,700	1,693	
	2720	100 ton capacity		91.65	1,525	4,585	13,800	1,650	
	2740	120 ton capacity		106.15	1,625	4,910	14,700	1,831	
	2760	150 ton capacity		124.30	2,225	6,710	20,100	2,336	
	2800	Self-propelled, 4 x 4, with telescoping boom, 5 ton		14.60	235	705	2,125	257.80	
	2900	12-1/2 ton capacity		25.80	380	1,140	3,425	434.40	
	3000	15 ton capacity		26.45	405	1,210	3,625	453.60	
	3050	20 ton capacity		28.25	475	1,425	4,275	511	
	3100	25 ton capacity		29.80	535	1,610	4,825	560.40	
	3150	40 ton capacity		36.95	605	1,815	5,450	658.60	
	3200	Derricks, guy, 20 ton capacity, 60' boom, 75' mast		23.35	375	1,122	3,375	411.20	
	3300	100' boom, 115' mast		36.76	645	1,930	5,800	680.10	
	3400	Stiffleg, 20 ton capacity, 70' boom, 37' mast		25.72	485	1,460	4,375	497.75	
	3500	100' boom, 47' mast		39.56	775	2,330	7,000	782.50	
	3550	Helicopter, small, lift to 1250 lbs. maximum, w/pilot		95.92	3,025	9,060	27,200	2,579	
	3600	Hoists, chain type, overhead, manual, 3/4 ton		.15	1	3	9	1.80	
	3900	10 ton		.70	7	21	63	9.80	
	4000	Hoist and tower, 5000 lb. cap., portable electric, 40' high		3.89	215	646	1,950	160.30	
	4100	For each added 10' section, add		.10	17	51	153	11	
	4200	Hoist and single tubular tower, 5000 lb. electric, 100' high		5.34	300	902	2,700	223.10	
	4300	For each added 6'-6" section, add		.17	29	87	261	18.75	
	4400	Hoist and double tubular tower, 5000 lb., 100' high		5.79	330	993	2,975	244.90	
	4500	For each added 6'-6" section, add		.19	32.50	97	291	20.90	
	4550	Hoist and tower, mast type, 6000 lb., 100' high		6.14	345	1,030	3,100	255.10	
	4570	For each added 10' section, add		.12	20.50	61	183	13.15	
	4600	Hoist and tower, personnel, electric, 2000 lb., 100' @ 125 fpm		14.36	915	2,740	8,225	662.90	
	4700	3000 lb., 100' @ 200 fpm		16.38	1,025	3,110	9,325	753.05	
	4800	3000 lb., 150' @ 300 fpm		18.23	1,150	3,480	10,400	841.85	
	4900	4000 lb., 100' @ 300 fpm		18.74	1,175	3,550	10,700	859.90	
	5000	6000 lb., 100' @ 275 fpm		19.92	1,250	3,720	11,200	903.35	
	5100	For added heights up to 500', add	L.F.	.01	1.67	5	15	1.10	
	5200	Jacks, hydraulic, 20 ton	Ea.	.05	2	6	18	1.60	
	5500	100 ton		.35	11.65	35	105	9.80	
	6100	Jacks, hydraulic, climbing w/50' jackrods, control console, 30 ton cap.		1.86	124	371	1,125	89.10	
	6150	For each added 10' jackrod section, add		.05	3.33	10	30	2.40	
	6300	50 ton capacity		2.99	199	597	1,800	143.30	
	6350	For each added 10' jackrod section, add		.06	4	12	36	2.90	
	6500	125 ton capacity		7.85	525	1,570	4,700	376.80	
	6550	For each added 10' jackrod section, add		.54	35.50	107	320	25.70	
	6600	Cable jack, 10 ton capacity with 200' cable		1.56	104	311	935	74.70	
	6650	For each added 50' of cable, add		.18	12	36	108	8.65	
70	0010	**WELLPOINT EQUIPMENT RENTAL** without operators							**70**
	0020	Based on 2 months rental							
	0100	Combination jetting & wellpoint pump, 60 H.P. diesel	Ea.	15.74	305	920	2,750	309.90	
	0200	High pressure gas jet pump, 200 H.P., 300 psi	"	36.50	262	786	2,350	449.20	
	0300	Discharge pipe, 8" diameter	L.F.	.01	.50	1.50	4.50	.40	
	0350	12" diameter		.01	.73	2.20	6.60	.50	
	0400	Header pipe, flows up to 150 GPM, 4" diameter		.01	.45	1.35	4.05	.35	
	0500	400 GPM, 6" diameter		.01	.53	1.60	4.80	.40	
	0600	800 GPM, 8" diameter		.01	.73	2.20	6.60	.50	
	0700	1500 GPM, 10" diameter		.01	.77	2.31	6.95	.55	
	0800	2500 GPM, 12" diameter		.02	1.46	4.38	13.15	1.05	
	0900	4500 GPM, 16" diameter		.03	1.87	5.61	16.85	1.35	

R015433 -10

01 54 33 | Equipment Rental

			UNIT	HOURLY OPER. COST	RENT PER DAY	RENT PER WEEK	RENT PER MONTH	EQUIPMENT COST/DAY	
70	0950	For quick coupling aluminum and plastic pipe, add	L.F.	.03	1.93	5.80	17.40	1.40	70
	1100	Wellpoint, 25' long, with fittings & riser pipe, 1-1/2" or 2" diameter	Ea.	.06	3.86	11.58	34.50	2.80	
	1200	Wellpoint pump, diesel powered, 4" suction, 20 H.P.		6.81	177	530	1,600	160.50	
	1300	6" suction, 30 H.P.		9.26	219	658	1,975	205.70	
	1400	8" suction, 40 H.P.		12.51	300	902	2,700	280.50	
	1500	10" suction, 75 H.P.		19.00	350	1,054	3,150	362.80	
	1600	12" suction, 100 H.P.		27.26	560	1,680	5,050	554.10	
	1700	12" suction, 175 H.P.		40.08	615	1,850	5,550	690.65	
80	0010	**MARINE EQUIPMENT RENTAL** without operators R015433 -10							80
	0200	Barge, 400 Ton, 30' wide x 90' long	Ea.	16.05	1,150	3,460	10,400	820.40	
	0240	800 Ton, 45' wide x 90' long		19.50	1,400	4,190	12,600	994	
	2000	Tugboat, diesel, 100 H.P.		25.55	215	645	1,925	333.40	
	2040	250 H.P.		49.40	395	1,185	3,550	632.20	
	2080	380 H.P.		108.20	1,175	3,505	10,500	1,567	

Crews

Crew A-1	Hr.	Daily	Hr.	Daily	Bare Costs	Incl. O&P
1 Building Laborer	$34.35	$274.80	$52.70	$421.60	$34.35	$52.70
1 Concrete saw, gas manual		63.20		69.52	7.90	8.69
8 L.H., Daily Totals		$338.00		$491.12	$42.25	$61.39

Crew A-1A	Hr.	Daily	Hr.	Daily	Bare Costs	Incl. O&P
1 Skilled Worker	$44.40	$355.20	$68.10	$544.80	$44.40	$68.10
1 Shot Blaster, 20"		216.20		237.82	27.02	29.73
8 L.H., Daily Totals		$571.40		$782.62	$71.42	$97.83

Crew A-1B	Hr.	Daily	Hr.	Daily	Bare Costs	Incl. O&P
1 Building Laborer	$34.35	$274.80	$52.70	$421.60	$34.35	$52.70
1 Concrete Saw		134.20		147.62	16.77	18.45
8 L.H., Daily Totals		$409.00		$569.22	$51.13	$71.15

Crew A-1C	Hr.	Daily	Hr.	Daily	Bare Costs	Incl. O&P
1 Building Laborer	$34.35	$274.80	$52.70	$421.60	$34.35	$52.70
1 Chain Saw, gas, 18"		25.80		28.38	3.23	3.55
8 L.H., Daily Totals		$300.60		$449.98	$37.58	$56.25

Crew A-1D	Hr.	Daily	Hr.	Daily	Bare Costs	Incl. O&P
1 Building Laborer	$34.35	$274.80	$52.70	$421.60	$34.35	$52.70
1 Vibrating plate, gas, 18"		30.80		33.88	3.85	4.24
8 L.H., Daily Totals		$305.60		$455.48	$38.20	$56.94

Crew A-1E	Hr.	Daily	Hr.	Daily	Bare Costs	Incl. O&P
1 Building Laborer	$34.35	$274.80	$52.70	$421.60	$34.35	$52.70
1 Vibratory Plate, gas, 21"		39.00		42.90	4.88	5.36
8 L.H., Daily Totals		$313.80		$464.50	$39.23	$58.06

Crew A-1F	Hr.	Daily	Hr.	Daily	Bare Costs	Incl. O&P
1 Building Laborer	$34.35	$274.80	$52.70	$421.60	$34.35	$52.70
1 Rammer/tamper, gas, 8"		43.60		47.96	5.45	6.00
8 L.H., Daily Totals		$318.40		$469.56	$39.80	$58.70

Crew A-1G	Hr.	Daily	Hr.	Daily	Bare Costs	Incl. O&P
1 Building Laborer	$34.35	$274.80	$52.70	$421.60	$34.35	$52.70
1 Rammer/tamper, gas, 15"		49.60		54.56	6.20	6.82
8 L.H., Daily Totals		$324.40		$476.16	$40.55	$59.52

Crew A-1H	Hr.	Daily	Hr.	Daily	Bare Costs	Incl. O&P
1 Building Laborer	$34.35	$274.80	$52.70	$421.60	$34.35	$52.70
1 Exterior Steam Cleaner		61.60		67.76	7.70	8.47
8 L.H., Daily Totals		$336.40		$489.36	$42.05	$61.17

Crew A-1J	Hr.	Daily	Hr.	Daily	Bare Costs	Incl. O&P
1 Building Laborer	$34.35	$274.80	$52.70	$421.60	$34.35	$52.70
1 Cultivator, Walk-Behind, 5 H.P.		58.45		64.30	7.31	8.04
8 L.H., Daily Totals		$333.25		$485.89	$41.66	$60.74

Crew A-1K	Hr.	Daily	Hr.	Daily	Bare Costs	Incl. O&P
1 Building Laborer	$34.35	$274.80	$52.70	$421.60	$34.35	$52.70
1 Cultivator, Walk-Behind, 8 H.P.		89.75		98.72	11.22	12.34
8 L.H., Daily Totals		$364.55		$520.33	$45.57	$65.04

Crew A-1M	Hr.	Daily	Hr.	Daily	Bare Costs	Incl. O&P
1 Building Laborer	$34.35	$274.80	$52.70	$421.60	$34.35	$52.70
1 Snow Blower, Walk-Behind		53.10		58.41	6.64	7.30
8 L.H., Daily Totals		$327.90		$480.01	$40.99	$60.00

Crew A-2	Hr.	Daily	Hr.	Daily	Bare Costs	Incl. O&P
2 Laborers	$34.35	$549.60	$52.70	$843.20	$34.03	$52.05
1 Truck Driver (light)	33.40	267.20	50.75	406.00		
1 Flatbed Truck, Gas, 1.5 Ton		171.40		188.54	7.14	7.86
24 L.H., Daily Totals		$988.20		$1437.74	$41.17	$59.91

Crew A-2A	Hr.	Daily	Hr.	Daily	Bare Costs	Incl. O&P
2 Laborers	$34.35	$549.60	$52.70	$843.20	$34.03	$52.05
1 Truck Driver (light)	33.40	267.20	50.75	406.00		
1 Flatbed Truck, Gas, 1.5 Ton		171.40		188.54		
1 Concrete Saw		134.20		147.62	12.73	14.01
24 L.H., Daily Totals		$1122.40		$1585.36	$46.77	$66.06

Crew A-2B	Hr.	Daily	Hr.	Daily	Bare Costs	Incl. O&P
1 Truck Driver (light)	$33.40	$267.20	$50.75	$406.00	$33.40	$50.75
1 Flatbed Truck, Gas, 1.5 Ton		171.40		188.54	21.43	23.57
8 L.H., Daily Totals		$438.60		$594.54	$54.83	$74.32

Crew A-3A	Hr.	Daily	Hr.	Daily	Bare Costs	Incl. O&P
1 Truck Driver (light)	$33.40	$267.20	$50.75	$406.00	$33.40	$50.75
1 Pickup truck, 4 x 4, 3/4 ton		116.60		128.26	14.57	16.03
8 L.H., Daily Totals		$383.80		$534.26	$47.98	$66.78

Crew A-3B	Hr.	Daily	Hr.	Daily	Bare Costs	Incl. O&P
1 Equip. Oper. (med.)	$45.35	$362.80	$67.75	$542.00	$39.88	$60.00
1 Truck Driver (heav.)	34.40	275.20	52.25	418.00		
1 Dump Truck, 12 C.Y., 400 H.P.		506.00		556.60		
1 F.E. Loader, W.M.,2.5 C.Y.		397.60		437.36	56.48	62.12
16 L.H., Daily Totals		$1541.60		$1953.96	$96.35	$122.12

Crew A-3C	Hr.	Daily	Hr.	Daily	Bare Costs	Incl. O&P
1 Equip. Oper. (light)	$43.60	$348.80	$65.15	$521.20	$43.60	$65.15
1 Loader, Skid Steer, 78 H.P.		261.20		287.32	32.65	35.91
8 L.H., Daily Totals		$610.00		$808.52	$76.25	$101.07

Crew A-3D	Hr.	Daily	Hr.	Daily	Bare Costs	Incl. O&P
1 Truck Driver, Light	$33.40	$267.20	$50.75	$406.00	$33.40	$50.75
1 Pickup truck, 4 x 4, 3/4 ton		116.60		128.26		
1 Flatbed Trailer, 25 Ton		111.20		122.32	28.48	31.32
8 L.H., Daily Totals		$495.00		$656.58	$61.88	$82.07

Crew A-3E	Hr.	Daily	Hr.	Daily	Bare Costs	Incl. O&P
1 Equip. Oper. (crane)	$46.50	$372.00	$69.45	$555.60	$40.45	$60.85
1 Truck Driver (heavy)	34.40	275.20	52.25	418.00		
1 Pickup truck, 4 x 4, 3/4 ton		116.60		128.26	7.29	8.02
16 L.H., Daily Totals		$763.80		$1101.86	$47.74	$68.87

Crew A-3F	Hr.	Daily	Hr.	Daily	Bare Costs	Incl. O&P
1 Equip. Oper. (crane)	$46.50	$372.00	$69.45	$555.60	$40.45	$60.85
1 Truck Driver (heavy)	34.40	275.20	52.25	418.00		
1 Pickup truck, 4 x 4, 3/4 ton		116.60		128.26		
1 Truck Tractor, 6x4, 380 H.P.		459.40		505.34		
1 Lowbed Trailer, 75 Ton		219.20		241.12	49.70	54.67
16 L.H., Daily Totals		$1442.40		$1848.32	$90.15	$115.52

Crew No.	Bare Costs Hr.	Daily	Incl. Subs O&P Hr.	Daily	Cost Per Labor-Hour Bare Costs	Incl. O&P
Crew A-3G	Hr.	Daily	Hr.	Daily	Bare Costs	Incl. O&P
1 Equip. Oper. (crane)	$46.50	$372.00	$69.45	$555.60	$40.45	$60.85
1 Truck Driver (heavy)	34.40	275.20	52.25	418.00		
1 Pickup truck, 4 x 4, 3/4 ton		116.60		128.26		
1 Truck Tractor, 6x4, 450 H.P.		558.80		614.68		
1 Lowbed Trailer, 75 Ton		219.20		241.12	55.91	61.50
16 L.H., Daily Totals		$1541.80		$1957.66	$96.36	$122.35
Crew A-3H	Hr.	Daily	Hr.	Daily	Bare Costs	Incl. O&P
1 Equip. Oper. (crane)	$46.50	$372.00	$69.45	$555.60	$46.50	$69.45
1 Hyd. crane, 12 Ton (daily)		871.60		958.76	108.95	119.85
8 L.H., Daily Totals		$1243.60		$1514.36	$155.45	$189.29
Crew A-3I	Hr.	Daily	Hr.	Daily	Bare Costs	Incl. O&P
1 Equip. Oper. (crane)	$46.50	$372.00	$69.45	$555.60	$46.50	$69.45
1 Hyd. crane, 25 Ton (daily)		1015.00		1116.50	126.88	139.56
8 L.H., Daily Totals		$1387.00		$1672.10	$173.38	$209.01
Crew A-3J	Hr.	Daily	Hr.	Daily	Bare Costs	Incl. O&P
1 Equip. Oper. (crane)	$46.50	$372.00	$69.45	$555.60	$46.50	$69.45
1 Hyd. crane, 40 Ton (daily)		1258.00		1383.80	157.25	172.97
8 L.H., Daily Totals		$1630.00		$1939.40	$203.75	$242.43
Crew A-3K	Hr.	Daily	Hr.	Daily	Bare Costs	Incl. O&P
1 Equip. Oper. (crane)	$46.50	$372.00	$69.45	$555.60	$43.40	$64.83
1 Equip. Oper. Oiler	40.30	322.40	60.20	481.60		
1 Hyd. crane, 55 Ton (daily)		1611.00		1772.10		
1 P/U Truck, 3/4 Ton (daily)		129.60		142.56	108.79	119.67
16 L.H., Daily Totals		$2435.00		$2951.86	$152.19	$184.49
Crew A-3L	Hr.	Daily	Hr.	Daily	Bare Costs	Incl. O&P
1 Equip. Oper. (crane)	$46.50	$372.00	$69.45	$555.60	$43.40	$64.83
1 Equip. Oper. Oiler	40.30	322.40	60.20	481.60		
1 Hyd. crane, 80 Ton (daily)		2302.00		2532.20		
1 P/U Truck, 3/4 Ton (daily)		129.60		142.56	151.97	167.17
16 L.H., Daily Totals		$3126.00		$3711.96	$195.38	$232.00
Crew A-3M	Hr.	Daily	Hr.	Daily	Bare Costs	Incl. O&P
1 Equip. Oper. (crane)	$46.50	$372.00	$69.45	$555.60	$43.40	$64.83
1 Equip. Oper. Oiler	40.30	322.40	60.20	481.60		
1 Hyd. crane, 100 Ton (daily)		2263.00		2489.30		
1 P/U Truck, 3/4 Ton (daily)		129.60		142.56	149.54	164.49
16 L.H., Daily Totals		$3087.00		$3669.06	$192.94	$229.32
Crew A-3N	Hr.	Daily	Hr.	Daily	Bare Costs	Incl. O&P
1 Equip. Oper. (crane)	$46.50	$372.00	$69.45	$555.60	$46.50	$69.45
1 Tower crane (monthly)		987.20		1085.92	123.40	135.74
8 L.H., Daily Totals		$1359.20		$1641.52	$169.90	$205.19
Crew A-3P	Hr.	Daily	Hr.	Daily	Bare Costs	Incl. O&P
1 Equip. Oper., Light	$43.60	$348.80	$65.15	$521.20	$43.60	$65.15
1 A.T. Forklift, 42' lift		447.40		492.14	55.92	61.52
8 L.H., Daily Totals		$796.20		$1013.34	$99.53	$126.67
Crew A-4	Hr.	Daily	Hr.	Daily	Bare Costs	Incl. O&P
2 Carpenters	$43.05	$688.80	$66.05	$1056.80	$41.22	$62.60
1 Painter, Ordinary	37.55	300.40	55.70	445.60		
24 L.H., Daily Totals		$989.20		$1502.40	$41.22	$62.60

Crew No.	Bare Costs Hr.	Daily	Incl. Subs O&P Hr.	Daily	Cost Per Labor-Hour Bare Costs	Incl. O&P
Crew A-5	Hr.	Daily	Hr.	Daily	Bare Costs	Incl. O&P
2 Laborers	$34.35	$549.60	$52.70	$843.20	$34.24	$52.48
.25 Truck Driver (light)	33.40	66.80	50.75	101.50		
.25 Flatbed Truck, Gas, 1.5 Ton		42.85		47.13	2.38	2.62
18 L.H., Daily Totals		$659.25		$991.84	$36.63	$55.10
Crew A-6	Hr.	Daily	Hr.	Daily	Bare Costs	Incl. O&P
1 Instrument Man	$44.40	$355.20	$68.10	$544.80	$43.23	$65.63
1 Rodman/Chainman	42.05	336.40	63.15	505.20		
1 Level, electronic		69.60		76.56	4.35	4.79
16 L.H., Daily Totals		$761.20		$1126.56	$47.58	$70.41
Crew A-7	Hr.	Daily	Hr.	Daily	Bare Costs	Incl. O&P
1 Chief Of Party	$54.45	$435.60	$83.10	$664.80	$46.97	$71.45
1 Instrument Man	44.40	355.20	68.10	544.80		
1 Rodman/Chainman	42.05	336.40	63.15	505.20		
1 Level, electronic		69.60		76.56	2.90	3.19
24 L.H., Daily Totals		$1196.80		$1791.36	$49.87	$74.64
Crew A-8	Hr.	Daily	Hr.	Daily	Bare Costs	Incl. O&P
1 Chief of Party	$54.45	$435.60	$83.10	$664.80	$45.74	$69.38
1 Instrument Man	44.40	355.20	68.10	544.80		
2 Rodmen/Chainmen	42.05	672.80	63.15	1010.40		
1 Level, electronic		69.60		76.56	2.17	2.39
32 L.H., Daily Totals		$1533.20		$2296.56	$47.91	$71.77
Crew A-9	Hr.	Daily	Hr.	Daily	Bare Costs	Incl. O&P
1 Asbestos Foreman	$47.40	$379.20	$73.30	$586.40	$46.96	$72.60
7 Asbestos Workers	46.90	2626.40	72.50	4060.00		
64 L.H., Daily Totals		$3005.60		$4646.40	$46.96	$72.60
Crew A-10A	Hr.	Daily	Hr.	Daily	Bare Costs	Incl. O&P
1 Asbestos Foreman	$47.40	$379.20	$73.30	$586.40	$47.07	$72.77
2 Asbestos Workers	46.90	750.40	72.50	1160.00		
24 L.H., Daily Totals		$1129.60		$1746.40	$47.07	$72.77
Crew A-10B	Hr.	Daily	Hr.	Daily	Bare Costs	Incl. O&P
1 Asbestos Foreman	$47.40	$379.20	$73.30	$586.40	$47.02	$72.70
3 Asbestos Workers	46.90	1125.60	72.50	1740.00		
32 L.H., Daily Totals		$1504.80		$2326.40	$47.02	$72.70
Crew A-10C	Hr.	Daily	Hr.	Daily	Bare Costs	Incl. O&P
3 Asbestos Workers	$46.90	$1125.60	$72.50	$1740.00	$46.90	$72.50
1 Flatbed Truck, Gas, 1.5 Ton		171.40		188.54	7.14	7.86
24 L.H., Daily Totals		$1297.00		$1928.54	$54.04	$80.36
Crew A-10D	Hr.	Daily	Hr.	Daily	Bare Costs	Incl. O&P
2 Asbestos Workers	$46.90	$750.40	$72.50	$1160.00	$45.15	$68.66
1 Equip. Oper. (crane)	46.50	372.00	69.45	555.60		
1 Equip. Oper. Oiler	40.30	322.40	60.20	481.60		
1 Hydraulic Crane, 33 Ton		763.00		839.30	23.84	26.23
32 L.H., Daily Totals		$2207.80		$3036.50	$68.99	$94.89
Crew A-11	Hr.	Daily	Hr.	Daily	Bare Costs	Incl. O&P
1 Asbestos Foreman	$47.40	$379.20	$73.30	$586.40	$46.96	$72.60
7 Asbestos Workers	46.90	2626.40	72.50	4060.00		
2 Chip. Hammers, 12 Lb., Elec.		37.20		40.92	.58	.64
64 L.H., Daily Totals		$3042.80		$4687.32	$47.54	$73.24

Crews

Crew A-12

Crew No.	Bare Costs Hr.	Daily	Incl. Subs O&P Hr.	Daily	Cost Per Labor-Hour Bare Costs	Incl. O&P
1 Asbestos Foreman	$47.40	$379.20	$73.30	$586.40	$46.96	$72.60
7 Asbestos Workers	46.90	2626.40	72.50	4060.00		
1 Trk-mtd vac, 14 CY, 1500 Gal.		544.10		598.51		
1 Flatbed Truck, 20,000 GVW		190.00		209.00	11.47	12.62
64 L.H., Daily Totals		$3739.70		$5453.91	$58.43	$85.22

Crew A-13

Crew No.	Bare Costs Hr.	Daily	Incl. Subs O&P Hr.	Daily	Cost Per Labor-Hour Bare Costs	Incl. O&P
1 Equip. Oper. (light)	$43.60	$348.80	$65.15	$521.20	$43.60	$65.15
1 Trk-mtd vac, 14 CY, 1500 Gal.		544.10		598.51		
1 Flatbed Truck, 20,000 GVW		190.00		209.00	91.76	100.94
8 L.H., Daily Totals		$1082.90		$1328.71	$135.36	$166.09

Crew B-1

Crew No.	Bare Costs Hr.	Daily	Incl. Subs O&P Hr.	Daily	Cost Per Labor-Hour Bare Costs	Incl. O&P
1 Labor Foreman (outside)	$36.35	$290.80	$55.75	$446.00	$35.02	$53.72
2 Laborers	34.35	549.60	52.70	843.20		
24 L.H., Daily Totals		$840.40		$1289.20	$35.02	$53.72

Crew B-1A

Crew No.	Bare Costs Hr.	Daily	Incl. Subs O&P Hr.	Daily	Cost Per Labor-Hour Bare Costs	Incl. O&P
1 Laborer Foreman	$36.35	$290.80	$55.75	$446.00	$35.02	$53.72
2 Laborers	34.35	549.60	52.70	843.20		
2 Cutting Torches		36.00		39.60		
2 Set of Gases		165.60		182.16	8.40	9.24
24 L.H., Daily Totals		$1042.00		$1510.96	$43.42	$62.96

Crew B-1B

Crew No.	Bare Costs Hr.	Daily	Incl. Subs O&P Hr.	Daily	Cost Per Labor-Hour Bare Costs	Incl. O&P
1 Laborer Foreman	$36.35	$290.80	$55.75	$446.00	$37.89	$57.65
2 Laborers	34.35	549.60	52.70	843.20		
1 Equip. Oper. (crane)	46.50	372.00	69.45	555.60		
2 Cutting Torches		36.00		39.60		
2 Set of Gases		165.60		182.16		
1 Hyd. Crane, 12 Ton		655.60		721.16	26.79	29.47
32 L.H., Daily Totals		$2069.60		$2787.72	$64.67	$87.12

Crew B-2

Crew No.	Bare Costs Hr.	Daily	Incl. Subs O&P Hr.	Daily	Cost Per Labor-Hour Bare Costs	Incl. O&P
1 Labor Foreman (outside)	$36.35	$290.80	$55.75	$446.00	$34.75	$53.31
4 Laborers	34.35	1099.20	52.70	1686.40		
40 L.H., Daily Totals		$1390.00		$2132.40	$34.75	$53.31

Crew B-3

Crew No.	Bare Costs Hr.	Daily	Incl. Subs O&P Hr.	Daily	Cost Per Labor-Hour Bare Costs	Incl. O&P
1 Labor Foreman (outside)	$36.35	$290.80	$55.75	$446.00	$36.53	$55.57
2 Laborers	34.35	549.60	52.70	843.20		
1 Equip. Oper. (med.)	45.35	362.80	67.75	542.00		
2 Truck Drivers (heavy)	34.40	550.40	52.25	836.00		
1 Crawler Loader, 3 C.Y.		1008.00		1108.80		
2 Dump Trucks 12 C.Y., 400 H.P.		1012.00		1113.20	42.08	46.29
48 L.H., Daily Totals		$3773.60		$4889.20	$78.62	$101.86

Crew B-3A

Crew No.	Bare Costs Hr.	Daily	Incl. Subs O&P Hr.	Daily	Cost Per Labor-Hour Bare Costs	Incl. O&P
4 Laborers	$34.35	$1099.20	$52.70	$1686.40	$36.55	$55.71
1 Equip. Oper. (med.)	45.35	362.80	67.75	542.00		
1 Hyd. Excavator, 1.5 C.Y.		947.40		1042.14	23.68	26.05
40 L.H., Daily Totals		$2409.40		$3270.54	$60.23	$81.76

Crew B-3B

Crew No.	Bare Costs Hr.	Daily	Incl. Subs O&P Hr.	Daily	Cost Per Labor-Hour Bare Costs	Incl. O&P
2 Laborers	$34.35	$549.60	$52.70	$843.20	$37.11	$56.35
1 Equip. Oper. (med.)	45.35	362.80	67.75	542.00		
1 Truck Driver (heavy)	34.40	275.20	52.25	418.00		
1 Backhoe Loader, 80 H.P.		349.40		384.34		
1 Dump Truck, 12 C.Y., 400 H.P.		506.00		556.60	26.73	29.40
32 L.H., Daily Totals		$2043.00		$2744.14	$63.84	$85.75

Crew B-3C

Crew No.	Bare Costs Hr.	Daily	Incl. Subs O&P Hr.	Daily	Cost Per Labor-Hour Bare Costs	Incl. O&P
3 Laborers	$34.35	$824.40	$52.70	$1264.80	$37.10	$56.46
1 Equip. Oper. (med.)	45.35	362.80	67.75	542.00		
1 Crawler Loader, 4 C.Y.		1432.00		1575.20	44.75	49.23
32 L.H., Daily Totals		$2619.20		$3382.00	$81.85	$105.69

Crew B-4

Crew No.	Bare Costs Hr.	Daily	Incl. Subs O&P Hr.	Daily	Cost Per Labor-Hour Bare Costs	Incl. O&P
1 Labor Foreman (outside)	$36.35	$290.80	$55.75	$446.00	$34.69	$53.13
4 Laborers	34.35	1099.20	52.70	1686.40		
1 Truck Driver (heavy)	34.40	275.20	52.25	418.00		
1 Truck Tractor, 220 H.P.		279.20		307.12		
1 Flatbed Trailer, 40 Ton		150.20		165.22	8.95	9.84
48 L.H., Daily Totals		$2094.60		$3022.74	$43.64	$62.97

Crew B-5

Crew No.	Bare Costs Hr.	Daily	Incl. Subs O&P Hr.	Daily	Cost Per Labor-Hour Bare Costs	Incl. O&P
1 Labor Foreman (outside)	$36.35	$290.80	$55.75	$446.00	$37.78	$57.44
4 Laborers	34.35	1099.20	52.70	1686.40		
2 Equip. Oper. (med.)	45.35	725.60	67.75	1084.00		
1 Air Compressor, 250 cfm		162.80		179.08		
2 Breakers, Pavement, 60 lb.		19.60		21.56		
2 -50' Air Hoses, 1.5"		12.60		13.86		
1 Crawler Loader, 3 C.Y.		1008.00		1108.80	21.48	23.63
56 L.H., Daily Totals		$3318.60		$4539.70	$59.26	$81.07

Crew B-5A

Crew No.	Bare Costs Hr.	Daily	Incl. Subs O&P Hr.	Daily	Cost Per Labor-Hour Bare Costs	Incl. O&P
1 Foreman	$36.35	$290.80	$55.75	$446.00	$37.13	$56.42
6 Laborers	34.35	1648.80	52.70	2529.60		
2 Equip. Oper. (med.)	45.35	725.60	67.75	1084.00		
1 Equip. Oper. (light)	43.60	348.80	65.15	521.20		
2 Truck Drivers (heavy)	34.40	550.40	52.25	836.00		
1 Air Compressor, 365 cfm		210.20		231.22		
2 Breakers, Pavement, 60 lb.		19.60		21.56		
8 -50' Air Hoses, 1"		32.80		36.08		
2 Dump Trucks, 8 C.Y., 220 H.P.		630.80		693.88	9.31	10.24
96 L.H., Daily Totals		$4457.80		$6399.54	$46.44	$66.66

Crew B-5B

Crew No.	Bare Costs Hr.	Daily	Incl. Subs O&P Hr.	Daily	Cost Per Labor-Hour Bare Costs	Incl. O&P
1 Powderman	$44.40	$355.20	$68.10	$544.80	$39.72	$60.06
2 Equip. Oper. (med.)	45.35	725.60	67.75	1084.00		
3 Truck Drivers (heavy)	34.40	825.60	52.25	1254.00		
1 F.E. Loader, W.M.,2.5 C.Y.		397.60		437.36		
3 Dump Trucks, 12 C.Y., 400 H.P.		1518.00		1669.80		
1 Air Compressor, 365 cfm		210.20		231.22	44.29	48.72
48 L.H., Daily Totals		$4032.20		$5221.18	$84.00	$108.77

Crew B-5C

Crew No.	Bare Costs Hr.	Daily	Incl. Subs O&P Hr.	Daily	Cost Per Labor-Hour Bare Costs	Incl. O&P
3 Laborers	$34.35	$824.40	$52.70	$1264.80	$38.00	$57.50
1 Equip. Oper. (med.)	45.35	362.80	67.75	542.00		
2 Truck Drivers (heav.)	34.40	550.40	52.25	836.00		
1 Equip. Oper. (crane)	46.50	372.00	69.45	555.60		
1 Equip. Oper. Oiler	40.30	322.40	60.20	481.60		
2 Dump Trucks, 12 C.Y., 400 H.P.		1012.00		1113.20		
1 Crawler Loader, 4 C.Y.		1432.00		1575.20		
1 S.P. Crane, 4x4, 25 Ton		560.40		616.44	46.94	51.64
64 L.H., Daily Totals		$5436.40		$6984.84	$84.94	$109.14

Crew B-6

Crew No.	Bare Costs Hr.	Daily	Incl. Subs O&P Hr.	Daily	Cost Per Labor-Hour Bare Costs	Incl. O&P
2 Laborers	$34.35	$549.60	$52.70	$843.20	$37.43	$56.85
1 Equip. Oper. (light)	43.60	348.80	65.15	521.20		
1 Backhoe Loader, 48 H.P.		320.40		352.44	13.35	14.69
24 L.H., Daily Totals		$1218.80		$1716.84	$50.78	$71.53

Crew B-6A

Crew No.	Bare Costs Hr.	Bare Costs Daily	Incl. Subs O&P Hr.	Incl. Subs O&P Daily	Cost Per Labor-Hour Bare Costs	Cost Per Labor-Hour Incl. O&P
.5 Labor Foreman (outside)	$36.35	$145.40	$55.75	$223.00	$39.15	$59.33
1 Laborer	34.35	274.80	52.70	421.60		
1 Equip. Oper. (med.)	45.35	362.80	67.75	542.00		
1 Vacuum Trk.,5000 Gal.		375.10		412.61	18.75	20.63
20 L.H., Daily Totals		$1158.10		$1599.21	$57.91	$79.96

Crew B-6B

Crew No.	Hr.	Daily	Hr.	Daily	Bare Costs	Incl. O&P
2 Labor Foremen (out)	$36.35	$581.60	$55.75	$892.00	$35.02	$53.72
4 Laborers	34.35	1099.20	52.70	1686.40		
1 S.P. Crane, 4x4, 5 Ton		257.80		283.58		
1 Flatbed Truck, Gas, 1.5 Ton		171.40		188.54		
1 Butt Fusion Mach., 4"-12" diam.		376.45		414.10	16.78	18.46
48 L.H., Daily Totals		$2486.45		$3464.61	$51.80	$72.18

Crew B-6C

Crew No.	Hr.	Daily	Hr.	Daily	Bare Costs	Incl. O&P
2 Labor Foremen (out)	$36.35	$581.60	$55.75	$892.00	$35.02	$53.72
4 Laborers	34.35	1099.20	52.70	1686.40		
1 S.P. Crane, 4x4, 12 Ton		434.40		477.84		
1 Flatbed Truck, Gas, 3 Ton		217.00		238.70		
1 Butt Fusion Mach., 8"-24" diam.		806.15		886.76	30.37	33.40
48 L.H., Daily Totals		$3138.35		$4181.70	$65.38	$87.12

Crew B-7

Crew No.	Hr.	Daily	Hr.	Daily	Bare Costs	Incl. O&P
1 Labor Foreman (outside)	$36.35	$290.80	$55.75	$446.00	$36.52	$55.72
4 Laborers	34.35	1099.20	52.70	1686.40		
1 Equip. Oper. (med.)	45.35	362.80	67.75	542.00		
1 Brush Chipper, 12", 130 H.P.		217.00		238.70		
1 Crawler Loader, 3 C.Y.		1008.00		1108.80		
2 Chain Saws, gas, 36" Long		83.20		91.52	27.25	29.98
48 L.H., Daily Totals		$3061.00		$4113.42	$63.77	$85.70

Crew B-7A

Crew No.	Hr.	Daily	Hr.	Daily	Bare Costs	Incl. O&P
2 Laborers	$34.35	$549.60	$52.70	$843.20	$37.43	$56.85
1 Equip. Oper. (light)	43.60	348.80	65.15	521.20		
1 Rake w/Tractor		277.70		305.47		
2 Chain Saws, gas, 18"		51.60		56.76	13.72	15.09
24 L.H., Daily Totals		$1227.70		$1726.63	$51.15	$71.94

Crew B-8

Crew No.	Hr.	Daily	Hr.	Daily	Bare Costs	Incl. O&P
1 Labor Foreman (outside)	$36.35	$290.80	$55.75	$446.00	$38.11	$57.67
2 Laborers	34.35	549.60	52.70	843.20		
2 Equip. Oper. (med.)	45.35	725.60	67.75	1084.00		
1 Equip. Oper. Oiler	40.30	322.40	60.20	481.60		
2 Truck Drivers (heavy)	34.40	550.40	52.25	836.00		
1 Hyd. Crane, 25 Ton		748.60		823.46		
1 Crawler Loader, 3 C.Y.		1008.00		1108.80		
2 Dump Trucks, 12 C.Y., 400 H.P.		1012.00		1113.20	43.26	47.59
64 L.H., Daily Totals		$5207.40		$6736.26	$81.37	$105.25

Crew B-9

Crew No.	Hr.	Daily	Hr.	Daily	Bare Costs	Incl. O&P
1 Labor Foreman (outside)	$36.35	$290.80	$55.75	$446.00	$34.75	$53.31
4 Laborers	34.35	1099.20	52.70	1686.40		
1 Air Compressor, 250 cfm		162.80		179.08		
2 Breakers, Pavement, 60 lb.		19.60		21.56		
2 -50' Air Hoses, 1.5"		12.60		13.86	4.88	5.36
40 L.H., Daily Totals		$1585.00		$2346.90	$39.63	$58.67

Crew B-9A

Crew No.	Bare Costs Hr.	Bare Costs Daily	Incl. Subs O&P Hr.	Incl. Subs O&P Daily	Cost Per Labor-Hour Bare Costs	Cost Per Labor-Hour Incl. O&P
2 Laborers	$34.35	$549.60	$52.70	$843.20	$34.37	$52.55
1 Truck Driver (heavy)	34.40	275.20	52.25	418.00		
1 Water Tank Trailer, 5000 Gal.		146.20		160.82		
1 Truck Tractor, 220 H.P.		279.20		307.12		
2 -50' Discharge Hoses, 3"		3.50		3.85	17.87	19.66
24 L.H., Daily Totals		$1253.70		$1732.99	$52.24	$72.21

Crew B-9B

Crew No.	Hr.	Daily	Hr.	Daily	Bare Costs	Incl. O&P
2 Laborers	$34.35	$549.60	$52.70	$843.20	$34.37	$52.55
1 Truck Driver (heavy)	34.40	275.20	52.25	418.00		
2 -50' Discharge Hoses, 3"		3.50		3.85		
1 Water Tank Trailer, 5000 Gal.		146.20		160.82		
1 Truck Tractor, 220 H.P.		279.20		307.12		
1 Pressure Washer		58.80		64.68	20.32	22.35
24 L.H., Daily Totals		$1312.50		$1797.67	$54.69	$74.90

Crew B-9D

Crew No.	Hr.	Daily	Hr.	Daily	Bare Costs	Incl. O&P
1 Labor Foreman (Outside)	$36.35	$290.80	$55.75	$446.00	$34.75	$53.31
4 Common Laborers	34.35	1099.20	52.70	1686.40		
1 Air Compressor, 250 cfm		162.80		179.08		
2 -50' Air Hoses, 1.5"		12.60		13.86		
2 Air Powered Tampers		52.40		57.64	5.70	6.26
40 L.H., Daily Totals		$1617.80		$2382.98	$40.45	$59.57

Crew B-10

Crew No.	Hr.	Daily	Hr.	Daily	Bare Costs	Incl. O&P
1 Equip. Oper. (med.)	$45.35	$362.80	$67.75	$542.00	$41.68	$62.73
.5 Laborer	34.35	137.40	52.70	210.80		
12 L.H., Daily Totals		$500.20		$752.80	$41.68	$62.73

Crew B-10A

Crew No.	Hr.	Daily	Hr.	Daily	Bare Costs	Incl. O&P
1 Equip. Oper. (med.)	$45.35	$362.80	$67.75	$542.00	$41.68	$62.73
.5 Laborer	34.35	137.40	52.70	210.80		
1 Roller, 2-Drum, W.B., 7.5 H.P.		146.80		161.48	12.23	13.46
12 L.H., Daily Totals		$647.00		$914.28	$53.92	$76.19

Crew B-10B

Crew No.	Hr.	Daily	Hr.	Daily	Bare Costs	Incl. O&P
1 Equip. Oper. (med.)	$45.35	$362.80	$67.75	$542.00	$41.68	$62.73
.5 Laborer	34.35	137.40	52.70	210.80		
1 Dozer, 200 H.P.		1051.00		1156.10	87.58	96.34
12 L.H., Daily Totals		$1551.20		$1908.90	$129.27	$159.07

Crew B-10C

Crew No.	Hr.	Daily	Hr.	Daily	Bare Costs	Incl. O&P
1 Equip. Oper. (med.)	$45.35	$362.80	$67.75	$542.00	$41.68	$62.73
.5 Laborer	34.35	137.40	52.70	210.80		
1 Dozer, 200 H.P.		1051.00		1156.10		
1 Vibratory Roller, Towed, 23 Ton		394.60		434.06	120.47	132.51
12 L.H., Daily Totals		$1945.80		$2342.96	$162.15	$195.25

Crew B-10D

Crew No.	Hr.	Daily	Hr.	Daily	Bare Costs	Incl. O&P
1 Equip. Oper. (med.)	$45.35	$362.80	$67.75	$542.00	$41.68	$62.73
.5 Laborer	34.35	137.40	52.70	210.80		
1 Dozer, 200 H.P.		1051.00		1156.10		
1 Sheepsft. Roller, Towed		449.00		493.90	125.00	137.50
12 L.H., Daily Totals		$2000.20		$2402.80	$166.68	$200.23

Crew B-10E

Crew No.	Hr.	Daily	Hr.	Daily	Bare Costs	Incl. O&P
1 Equip. Oper. (med.)	$45.35	$362.80	$67.75	$542.00	$41.68	$62.73
.5 Laborer	34.35	137.40	52.70	210.80		
1 Tandem Roller, 5 Ton		145.20		159.72	12.10	13.31
12 L.H., Daily Totals		$645.40		$912.52	$53.78	$76.04

Crews

Crew B-10F

Crew No.	Bare Costs		Incl. Subs O&P		Cost Per Labor-Hour	
Crew B-10F	Hr.	Daily	Hr.	Daily	Bare Costs	Incl. O&P
1 Equip. Oper. (med.)	$45.35	$362.80	$67.75	$542.00	$41.68	$62.73
.5 Laborer	34.35	137.40	52.70	210.80		
1 Tandem Roller, 10 Ton		224.20		246.62	18.68	20.55
12 L.H., Daily Totals		$724.40		$999.42	$60.37	$83.28

Crew B-10G

Crew No.	Bare Costs		Incl. Subs O&P		Cost Per Labor-Hour	
Crew B-10G	Hr.	Daily	Hr.	Daily	Bare Costs	Incl. O&P
1 Equip. Oper. (med.)	$45.35	$362.80	$67.75	$542.00	$41.68	$62.73
.5 Laborer	34.35	137.40	52.70	210.80		
1 Sheepsft. Roll., 240 H.P.		1094.00		1203.40	91.17	100.28
12 L.H., Daily Totals		$1594.20		$1956.20	$132.85	$163.02

Crew B-10H

Crew No.	Bare Costs		Incl. Subs O&P		Cost Per Labor-Hour	
Crew B-10H	Hr.	Daily	Hr.	Daily	Bare Costs	Incl. O&P
1 Equip. Oper. (med.)	$45.35	$362.80	$67.75	$542.00	$41.68	$62.73
.5 Laborer	34.35	137.40	52.70	210.80		
1 Diaphragm Water Pump, 2"		66.60		73.26		
1 -20' Suction Hose, 2"		1.95		2.15		
2 -50' Discharge Hoses, 2"		1.80		1.98	5.86	6.45
12 L.H., Daily Totals		$570.55		$830.18	$47.55	$69.18

Crew B-10I

Crew No.	Bare Costs		Incl. Subs O&P		Cost Per Labor-Hour	
Crew B-10I	Hr.	Daily	Hr.	Daily	Bare Costs	Incl. O&P
1 Equip. Oper. (med.)	$45.35	$362.80	$67.75	$542.00	$41.68	$62.73
.5 Laborer	34.35	137.40	52.70	210.80		
1 Diaphragm Water Pump, 4"		109.40		120.34		
1 -20' Suction Hose, 4"		3.25		3.58		
2 -50' Discharge Hoses, 4"		4.70		5.17	9.78	10.76
12 L.H., Daily Totals		$617.55		$881.88	$51.46	$73.49

Crew B-10J

Crew No.	Bare Costs		Incl. Subs O&P		Cost Per Labor-Hour	
Crew B-10J	Hr.	Daily	Hr.	Daily	Bare Costs	Incl. O&P
1 Equip. Oper. (med.)	$45.35	$362.80	$67.75	$542.00	$41.68	$62.73
.5 Laborer	34.35	137.40	52.70	210.80		
1 Centrifugal Water Pump, 3"		71.00		78.10		
1 -20' Suction Hose, 3"		3.05		3.36		
2 -50' Discharge Hoses, 3"		3.50		3.85	6.46	7.11
12 L.H., Daily Totals		$577.75		$838.11	$48.15	$69.84

Crew B-10K

Crew No.	Bare Costs		Incl. Subs O&P		Cost Per Labor-Hour	
Crew B-10K	Hr.	Daily	Hr.	Daily	Bare Costs	Incl. O&P
1 Equip. Oper. (med.)	$45.35	$362.80	$67.75	$542.00	$41.68	$62.73
.5 Laborer	34.35	137.40	52.70	210.80		
1 Centr. Water Pump, 6"		291.60		320.76		
1 -20' Suction Hose, 6"		11.90		13.09		
2 -50' Discharge Hoses, 6"		12.60		13.86	26.34	28.98
12 L.H., Daily Totals		$816.30		$1100.51	$68.03	$91.71

Crew B-10L

Crew No.	Bare Costs		Incl. Subs O&P		Cost Per Labor-Hour	
Crew B-10L	Hr.	Daily	Hr.	Daily	Bare Costs	Incl. O&P
1 Equip. Oper. (med.)	$45.35	$362.80	$67.75	$542.00	$41.68	$62.73
.5 Laborer	34.35	137.40	52.70	210.80		
1 Dozer, 80 H.P.		412.60		453.86	34.38	37.82
12 L.H., Daily Totals		$912.80		$1206.66	$76.07	$100.56

Crew B-10M

Crew No.	Bare Costs		Incl. Subs O&P		Cost Per Labor-Hour	
Crew B-10M	Hr.	Daily	Hr.	Daily	Bare Costs	Incl. O&P
1 Equip. Oper. (med.)	$45.35	$362.80	$67.75	$542.00	$41.68	$62.73
.5 Laborer	34.35	137.40	52.70	210.80		
1 Dozer, 300 H.P.		1607.00		1767.70	133.92	147.31
12 L.H., Daily Totals		$2107.20		$2520.50	$175.60	$210.04

Crew B-10N

Crew No.	Bare Costs		Incl. Subs O&P		Cost Per Labor-Hour	
Crew B-10N	Hr.	Daily	Hr.	Daily	Bare Costs	Incl. O&P
1 Equip. Oper. (med.)	$45.35	$362.80	$67.75	$542.00	$41.68	$62.73
.5 Laborer	34.35	137.40	52.70	210.80		
1 F.E. Loader, T.M., 1.5 C.Y		466.60		513.26	38.88	42.77
12 L.H., Daily Totals		$966.80		$1266.06	$80.57	$105.51

Crew B-100

Crew No.	Bare Costs		Incl. Subs O&P		Cost Per Labor-Hour	
Crew B-100	Hr.	Daily	Hr.	Daily	Bare Costs	Incl. O&P
1 Equip. Oper. (med.)	$45.35	$362.80	$67.75	$542.00	$41.68	$62.73
.5 Laborer	34.35	137.40	52.70	210.80		
1 F.E. Loader, T.M., 2.25 C.Y.		773.60		850.96	64.47	70.91
12 L.H., Daily Totals		$1273.80		$1603.76	$106.15	$133.65

Crew B-10P

Crew No.	Bare Costs		Incl. Subs O&P		Cost Per Labor-Hour	
Crew B-10P	Hr.	Daily	Hr.	Daily	Bare Costs	Incl. O&P
1 Equip. Oper. (med.)	$45.35	$362.80	$67.75	$542.00	$41.68	$62.73
.5 Laborer	34.35	137.40	52.70	210.80		
1 Crawler Loader, 3 C.Y.		1008.00		1108.80	84.00	92.40
12 L.H., Daily Totals		$1508.20		$1861.60	$125.68	$155.13

Crew B-10Q

Crew No.	Bare Costs		Incl. Subs O&P		Cost Per Labor-Hour	
Crew B-10Q	Hr.	Daily	Hr.	Daily	Bare Costs	Incl. O&P
1 Equip. Oper. (med.)	$45.35	$362.80	$67.75	$542.00	$41.68	$62.73
.5 Laborer	34.35	137.40	52.70	210.80		
1 Crawler Loader, 4 C.Y.		1432.00		1575.20	119.33	131.27
12 L.H., Daily Totals		$1932.20		$2328.00	$161.02	$194.00

Crew B-10R

Crew No.	Bare Costs		Incl. Subs O&P		Cost Per Labor-Hour	
Crew B-10R	Hr.	Daily	Hr.	Daily	Bare Costs	Incl. O&P
1 Equip. Oper. (med.)	$45.35	$362.80	$67.75	$542.00	$41.68	$62.73
.5 Laborer	34.35	137.40	52.70	210.80		
1 F.E. Loader, W.M., 1 C.Y.		235.80		259.38	19.65	21.61
12 L.H., Daily Totals		$736.00		$1012.18	$61.33	$84.35

Crew B-10S

Crew No.	Bare Costs		Incl. Subs O&P		Cost Per Labor-Hour	
Crew B-10S	Hr.	Daily	Hr.	Daily	Bare Costs	Incl. O&P
1 Equip. Oper. (med.)	$45.35	$362.80	$67.75	$542.00	$41.68	$62.73
.5 Laborer	34.35	137.40	52.70	210.80		
1 F.E. Loader, W.M., 1.5 C.Y.		319.40		351.34	26.62	29.28
12 L.H., Daily Totals		$819.60		$1104.14	$68.30	$92.01

Crew B-10T

Crew No.	Bare Costs		Incl. Subs O&P		Cost Per Labor-Hour	
Crew B-10T	Hr.	Daily	Hr.	Daily	Bare Costs	Incl. O&P
1 Equip. Oper. (med.)	$45.35	$362.80	$67.75	$542.00	$41.68	$62.73
.5 Laborer	34.35	137.40	52.70	210.80		
1 F.E. Loader, W.M.,2.5 C.Y.		397.60		437.36	33.13	36.45
12 L.H., Daily Totals		$897.80		$1190.16	$74.82	$99.18

Crew B-10U

Crew No.	Bare Costs		Incl. Subs O&P		Cost Per Labor-Hour	
Crew B-10U	Hr.	Daily	Hr.	Daily	Bare Costs	Incl. O&P
1 Equip. Oper. (med.)	$45.35	$362.80	$67.75	$542.00	$41.68	$62.73
.5 Laborer	34.35	137.40	52.70	210.80		
1 F.E. Loader, W.M., 5.5 C.Y.		933.40		1026.74	77.78	85.56
12 L.H., Daily Totals		$1433.60		$1779.54	$119.47	$148.29

Crew B-10V

Crew No.	Bare Costs		Incl. Subs O&P		Cost Per Labor-Hour	
Crew B-10V	Hr.	Daily	Hr.	Daily	Bare Costs	Incl. O&P
1 Equip. Oper. (med.)	$45.35	$362.80	$67.75	$542.00	$41.68	$62.73
.5 Laborer	34.35	137.40	52.70	210.80		
1 Dozer, 700 H.P.		4150.00		4565.00	345.83	380.42
12 L.H., Daily Totals		$4650.20		$5317.80	$387.52	$443.15

Crew B-10W

Crew No.	Bare Costs		Incl. Subs O&P		Cost Per Labor-Hour	
Crew B-10W	Hr.	Daily	Hr.	Daily	Bare Costs	Incl. O&P
1 Equip. Oper. (med.)	$45.35	$362.80	$67.75	$542.00	$41.68	$62.73
.5 Laborer	34.35	137.40	52.70	210.80		
1 Dozer, 105 H.P.		505.40		555.94	42.12	46.33
12 L.H., Daily Totals		$1005.60		$1308.74	$83.80	$109.06

Crew B-10X

Crew No.	Bare Costs		Incl. Subs O&P		Cost Per Labor-Hour	
Crew B-10X	Hr.	Daily	Hr.	Daily	Bare Costs	Incl. O&P
1 Equip. Oper. (med.)	$45.35	$362.80	$67.75	$542.00	$41.68	$62.73
.5 Laborer	34.35	137.40	52.70	210.80		
1 Dozer, 410 H.P.		2105.00		2315.50	175.42	192.96
12 L.H., Daily Totals		$2605.20		$3068.30	$217.10	$255.69

Crew No.	Bare Costs		Incl. Subs O&P		Cost Per Labor-Hour	
Crew B-10Y	Hr.	Daily	Hr.	Daily	Bare Costs	Incl. O&P
1 Equip. Oper. (med.)	$45.35	$362.80	$67.75	$542.00	$41.68	$62.73
.5 Laborer	34.35	137.40	52.70	210.80		
1 Vibr. Roller, Towed, 12 Ton		498.80		548.68	41.57	45.72
12 L.H., Daily Totals		$999.00		$1301.48	$83.25	$108.46

Crew No.	Bare Costs		Incl. Subs O&P		Cost Per Labor-Hour	
Crew B-11A	Hr.	Daily	Hr.	Daily	Bare Costs	Incl. O&P
1 Equipment Oper. (med.)	$45.35	$362.80	$67.75	$542.00	$39.85	$60.23
1 Laborer	34.35	274.80	52.70	421.60		
1 Dozer, 200 H.P.		1051.00		1156.10	65.69	72.26
16 L.H., Daily Totals		$1688.60		$2119.70	$105.54	$132.48

Crew No.	Bare Costs		Incl. Subs O&P		Cost Per Labor-Hour	
Crew B-11B	Hr.	Daily	Hr.	Daily	Bare Costs	Incl. O&P
1 Equipment Oper. (light)	$43.60	$348.80	$65.15	$521.20	$38.98	$58.92
1 Laborer	34.35	274.80	52.70	421.60		
1 Air Powered Tamper		26.20		28.82		
1 Air Compressor, 365 cfm		210.20		231.22		
2 -50' Air Hoses, 1.5"		12.60		13.86	15.56	17.12
16 L.H., Daily Totals		$872.60		$1216.70	$54.54	$76.04

Crew No.	Bare Costs		Incl. Subs O&P		Cost Per Labor-Hour	
Crew B-11C	Hr.	Daily	Hr.	Daily	Bare Costs	Incl. O&P
1 Equipment Oper. (med.)	$45.35	$362.80	$67.75	$542.00	$39.85	$60.23
1 Laborer	34.35	274.80	52.70	421.60		
1 Backhoe Loader, 48 H.P.		320.40		352.44	20.02	22.03
16 L.H., Daily Totals		$958.00		$1316.04	$59.88	$82.25

Crew No.	Bare Costs		Incl. Subs O&P		Cost Per Labor-Hour	
Crew B-11J	Hr.	Daily	Hr.	Daily	Bare Costs	Incl. O&P
1 Equipment Oper. (med.)	$45.35	$362.80	$67.75	$542.00	$39.85	$60.23
1 Laborer	34.35	274.80	52.70	421.60		
1 Grader, 30,000 Lbs.		542.80		597.08		
1 Ripper, beam & 1 shank		83.20		91.52	39.13	43.04
16 L.H., Daily Totals		$1263.60		$1652.20	$78.97	$103.26

Crew No.	Bare Costs		Incl. Subs O&P		Cost Per Labor-Hour	
Crew B-11K	Hr.	Daily	Hr.	Daily	Bare Costs	Incl. O&P
1 Equipment Oper. (med.)	$45.35	$362.80	$67.75	$542.00	$39.85	$60.23
1 Laborer	34.35	274.80	52.70	421.60		
1 Trencher, Chain Type, 8' D		2506.00		2756.60	156.63	172.29
16 L.H., Daily Totals		$3143.60		$3720.20	$196.47	$232.51

Crew No.	Bare Costs		Incl. Subs O&P		Cost Per Labor-Hour	
Crew B-11L	Hr.	Daily	Hr.	Daily	Bare Costs	Incl. O&P
1 Equipment Oper. (med.)	$45.35	$362.80	$67.75	$542.00	$39.85	$60.23
1 Laborer	34.35	274.80	52.70	421.60		
1 Grader, 30,000 Lbs.		542.80		597.08	33.92	37.32
16 L.H., Daily Totals		$1180.40		$1560.68	$73.78	$97.54

Crew No.	Bare Costs		Incl. Subs O&P		Cost Per Labor-Hour	
Crew B-11M	Hr.	Daily	Hr.	Daily	Bare Costs	Incl. O&P
1 Equipment Oper. (med.)	$45.35	$362.80	$67.75	$542.00	$39.85	$60.23
1 Laborer	34.35	274.80	52.70	421.60		
1 Backhoe Loader, 80 H.P.		349.40		384.34	21.84	24.02
16 L.H., Daily Totals		$987.00		$1347.94	$61.69	$84.25

Crew No.	Bare Costs		Incl. Subs O&P		Cost Per Labor-Hour	
Crew B-11N	Hr.	Daily	Hr.	Daily	Bare Costs	Incl. O&P
1 Labor Foreman	$36.35	$290.80	$55.75	$446.00	$37.05	$56.08
2 Equipment Operators (med.)	45.35	725.60	67.75	1084.00		
6 Truck Drivers (hvy.)	34.40	1651.20	52.25	2508.00		
1 F.E. Loader, W.M., 5.5 C.Y.		933.40		1026.74		
1 Dozer, 410 H.P.		2105.00		2315.50		
6 Dump Trucks, Off Hwy., 50 Ton		9294.00		10223.40	171.28	188.41
72 L.H., Daily Totals		$15000.00		$17603.64	$208.33	$244.50

Crew No.	Bare Costs		Incl. Subs O&P		Cost Per Labor-Hour	
Crew B-11Q	Hr.	Daily	Hr.	Daily	Bare Costs	Incl. O&P
1 Equipment Operator (med.)	$45.35	$362.80	$67.75	$542.00	$41.68	$62.73
.5 Laborer	34.35	137.40	52.70	210.80		
1 Dozer, 140 H.P.		778.80		856.68	64.90	71.39
12 L.H., Daily Totals		$1279.00		$1609.48	$106.58	$134.12

Crew No.	Bare Costs		Incl. Subs O&P		Cost Per Labor-Hour	
Crew B-11R	Hr.	Daily	Hr.	Daily	Bare Costs	Incl. O&P
1 Equipment Operator (med.)	$45.35	$362.80	$67.75	$542.00	$41.68	$62.73
.5 Laborer	34.35	137.40	52.70	210.80		
1 Dozer, 200 H.P.		1051.00		1156.10	87.58	96.34
12 L.H., Daily Totals		$1551.20		$1908.90	$129.27	$159.07

Crew No.	Bare Costs		Incl. Subs O&P		Cost Per Labor-Hour	
Crew B-11S	Hr.	Daily	Hr.	Daily	Bare Costs	Incl. O&P
1 Equipment Operator (med.)	$45.35	$362.80	$67.75	$542.00	$41.68	$62.73
.5 Laborer	34.35	137.40	52.70	210.80		
1 Dozer, 300 H.P.		1607.00		1767.70		
1 Ripper, beam & 1 shank		83.20		91.52	140.85	154.94
12 L.H., Daily Totals		$2190.40		$2612.02	$182.53	$217.67

Crew No.	Bare Costs		Incl. Subs O&P		Cost Per Labor-Hour	
Crew B-11T	Hr.	Daily	Hr.	Daily	Bare Costs	Incl. O&P
1 Equipment Operator (med.)	$45.35	$362.80	$67.75	$542.00	$41.68	$62.73
.5 Laborer	34.35	137.40	52.70	210.80		
1 Dozer, 410 H.P.		2105.00		2315.50		
1 Ripper, beam & 2 shanks		94.40		103.84	183.28	201.61
12 L.H., Daily Totals		$2699.60		$3172.14	$224.97	$264.35

Crew No.	Bare Costs		Incl. Subs O&P		Cost Per Labor-Hour	
Crew B-11U	Hr.	Daily	Hr.	Daily	Bare Costs	Incl. O&P
1 Equipment Operator (med.)	$45.35	$362.80	$67.75	$542.00	$41.68	$62.73
.5 Laborer	34.35	137.40	52.70	210.80		
1 Dozer, 520 H.P.		2845.00		3129.50	237.08	260.79
12 L.H., Daily Totals		$3345.20		$3882.30	$278.77	$323.52

Crew No.	Bare Costs		Incl. Subs O&P		Cost Per Labor-Hour	
Crew B-11V	Hr.	Daily	Hr.	Daily	Bare Costs	Incl. O&P
3 Laborers	$34.35	$824.40	$52.70	$1264.80	$34.35	$52.70
1 Roller, 2-Drum, W.B., 7.5 H.P.		146.80		161.48	6.12	6.73
24 L.H., Daily Totals		$971.20		$1426.28	$40.47	$59.43

Crew No.	Bare Costs		Incl. Subs O&P		Cost Per Labor-Hour	
Crew B-11W	Hr.	Daily	Hr.	Daily	Bare Costs	Incl. O&P
1 Equipment Operator (med.)	$45.35	$362.80	$67.75	$542.00	$35.31	$53.58
1 Common Laborer	34.35	274.80	52.70	421.60		
10 Truck Drivers (hvy.)	34.40	2752.00	52.25	4180.00		
1 Dozer, 200 H.P.		1051.00		1156.10		
1 Vibratory Roller, Towed, 23 Ton		394.60		434.06		
10 Dump Trucks, 8 C.Y., 220 H.P.		3154.00		3469.40	47.91	52.70
96 L.H., Daily Totals		$7989.20		$10203.16	$83.22	$106.28

Crew No.	Bare Costs		Incl. Subs O&P		Cost Per Labor-Hour	
Crew B-11Y	Hr.	Daily	Hr.	Daily	Bare Costs	Incl. O&P
1 Labor Foreman (Outside)	$36.35	$290.80	$55.75	$446.00	$38.24	$58.06
5 Common Laborers	34.35	1374.00	52.70	2108.00		
3 Equipment Operators (med.)	45.35	1088.40	67.75	1626.00		
1 Dozer, 80 H.P.		412.60		453.86		
2 Roller, 2-Drum, W.B., 7.5 H.P.		293.60		322.96		
4 Vibratory Plates, gas, 21"		156.00		171.60	11.98	13.17
72 L.H., Daily Totals		$3615.40		$5128.42	$50.21	$71.23

Crew No.	Bare Costs		Incl. Subs O&P		Cost Per Labor-Hour	
Crew B-12A	Hr.	Daily	Hr.	Daily	Bare Costs	Incl. O&P
1 Equip. Oper. (crane)	$46.50	$372.00	$69.45	$555.60	$40.42	$61.08
1 Laborer	34.35	274.80	52.70	421.60		
1 Hyd. Excavator, 1 C.Y.		729.80		802.78	45.61	50.17
16 L.H., Daily Totals		$1376.60		$1779.98	$86.04	$111.25

Crews

Crew B-12B

Crew B-12B	Bare Costs Hr.	Bare Costs Daily	Incl. Subs O&P Hr.	Incl. Subs O&P Daily	Cost Per Labor-Hour Bare Costs	Cost Per Labor-Hour Incl. O&P
1 Equip. Oper. (crane)	$46.50	$372.00	$69.45	$555.60	$40.42	$61.08
1 Laborer	34.35	274.80	52.70	421.60		
1 Hyd. Excavator, 1.5 C.Y.		947.40		1042.14	59.21	65.13
16 L.H., Daily Totals		$1594.20		$2019.34	$99.64	$126.21

Crew B-12C

Crew B-12C	Bare Costs Hr.	Bare Costs Daily	Incl. Subs O&P Hr.	Incl. Subs O&P Daily	Cost Per Labor-Hour Bare Costs	Cost Per Labor-Hour Incl. O&P
1 Equip. Oper. (crane)	$46.50	$372.00	$69.45	$555.60	$40.42	$61.08
1 Laborer	34.35	274.80	52.70	421.60		
1 Hyd. Excavator, 2 C.Y.		1224.00		1346.40	76.50	84.15
16 L.H., Daily Totals		$1870.80		$2323.60	$116.93	$145.22

Crew B-12D

Crew B-12D	Bare Costs Hr.	Bare Costs Daily	Incl. Subs O&P Hr.	Incl. Subs O&P Daily	Cost Per Labor-Hour Bare Costs	Cost Per Labor-Hour Incl. O&P
1 Equip. Oper. (crane)	$46.50	$372.00	$69.45	$555.60	$40.42	$61.08
1 Laborer	34.35	274.80	52.70	421.60		
1 Hyd. Excavator, 3.5 C.Y.		2171.00		2388.10	135.69	149.26
16 L.H., Daily Totals		$2817.80		$3365.30	$176.11	$210.33

Crew B-12E

Crew B-12E	Bare Costs Hr.	Bare Costs Daily	Incl. Subs O&P Hr.	Incl. Subs O&P Daily	Cost Per Labor-Hour Bare Costs	Cost Per Labor-Hour Incl. O&P
1 Equip. Oper. (crane)	$46.50	$372.00	$69.45	$555.60	$40.42	$61.08
1 Laborer	34.35	274.80	52.70	421.60		
1 Hyd. Excavator, .5 C.Y.		364.60		401.06	22.79	25.07
16 L.H., Daily Totals		$1011.40		$1378.26	$63.21	$86.14

Crew B-12F

Crew B-12F	Bare Costs Hr.	Bare Costs Daily	Incl. Subs O&P Hr.	Incl. Subs O&P Daily	Cost Per Labor-Hour Bare Costs	Cost Per Labor-Hour Incl. O&P
1 Equip. Oper. (crane)	$46.50	$372.00	$69.45	$555.60	$40.42	$61.08
1 Laborer	34.35	274.80	52.70	421.60		
1 Hyd. Excavator, .75 C.Y.		640.40		704.44	40.02	44.03
16 L.H., Daily Totals		$1287.20		$1681.64	$80.45	$105.10

Crew B-12G

Crew B-12G	Bare Costs Hr.	Bare Costs Daily	Incl. Subs O&P Hr.	Incl. Subs O&P Daily	Cost Per Labor-Hour Bare Costs	Cost Per Labor-Hour Incl. O&P
1 Equip. Oper. (crane)	$46.50	$372.00	$69.45	$555.60	$40.42	$61.08
1 Laborer	34.35	274.80	52.70	421.60		
1 Crawler Crane, 15 Ton		657.90		723.69		
1 Clamshell Bucket, .5 C.Y.		37.40		41.14	43.46	47.80
16 L.H., Daily Totals		$1342.10		$1742.03	$83.88	$108.88

Crew B-12H

Crew B-12H	Bare Costs Hr.	Bare Costs Daily	Incl. Subs O&P Hr.	Incl. Subs O&P Daily	Cost Per Labor-Hour Bare Costs	Cost Per Labor-Hour Incl. O&P
1 Equip. Oper. (crane)	$46.50	$372.00	$69.45	$555.60	$40.42	$61.08
1 Laborer	34.35	274.80	52.70	421.60		
1 Crawler Crane, 25 Ton		1123.00		1235.30		
1 Clamshell Bucket, 1 C.Y.		48.00		52.80	73.19	80.51
16 L.H., Daily Totals		$1817.80		$2265.30	$113.61	$141.58

Crew B-12I

Crew B-12I	Bare Costs Hr.	Bare Costs Daily	Incl. Subs O&P Hr.	Incl. Subs O&P Daily	Cost Per Labor-Hour Bare Costs	Cost Per Labor-Hour Incl. O&P
1 Equip. Oper. (crane)	$46.50	$372.00	$69.45	$555.60	$40.42	$61.08
1 Laborer	34.35	274.80	52.70	421.60		
1 Crawler Crane, 20 Ton		841.90		926.09		
1 Dragline Bucket, .75 C.Y.		21.00		23.10	53.93	59.32
16 L.H., Daily Totals		$1509.70		$1926.39	$94.36	$120.40

Crew B-12J

Crew B-12J	Bare Costs Hr.	Bare Costs Daily	Incl. Subs O&P Hr.	Incl. Subs O&P Daily	Cost Per Labor-Hour Bare Costs	Cost Per Labor-Hour Incl. O&P
1 Equip. Oper. (crane)	$46.50	$372.00	$69.45	$555.60	$40.42	$61.08
1 Laborer	34.35	274.80	52.70	421.60		
1 Gradall, 5/8 C.Y.		857.40		943.14	53.59	58.95
16 L.H., Daily Totals		$1504.20		$1920.34	$94.01	$120.02

Crew B-12K

Crew B-12K	Bare Costs Hr.	Bare Costs Daily	Incl. Subs O&P Hr.	Incl. Subs O&P Daily	Cost Per Labor-Hour Bare Costs	Cost Per Labor-Hour Incl. O&P
1 Equip. Oper. (crane)	$46.50	$372.00	$69.45	$555.60	$40.42	$61.08
1 Laborer	34.35	274.80	52.70	421.60		
1 Gradall, 3 Ton, 1 C.Y.		1001.00		1101.10	62.56	68.82
16 L.H., Daily Totals		$1647.80		$2078.30	$102.99	$129.89

Crew B-12L

Crew B-12L	Bare Costs Hr.	Bare Costs Daily	Incl. Subs O&P Hr.	Incl. Subs O&P Daily	Cost Per Labor-Hour Bare Costs	Cost Per Labor-Hour Incl. O&P
1 Equip. Oper. (crane)	$46.50	$372.00	$69.45	$555.60	$40.42	$61.08
1 Laborer	34.35	274.80	52.70	421.60		
1 Crawler Crane, 15 Ton		657.90		723.69		
1 F.E. Attachment, .5 C.Y.		56.00		61.60	44.62	49.08
16 L.H., Daily Totals		$1360.70		$1762.49	$85.04	$110.16

Crew B-12M

Crew B-12M	Bare Costs Hr.	Bare Costs Daily	Incl. Subs O&P Hr.	Incl. Subs O&P Daily	Cost Per Labor-Hour Bare Costs	Cost Per Labor-Hour Incl. O&P
1 Equip. Oper. (crane)	$46.50	$372.00	$69.45	$555.60	$40.42	$61.08
1 Laborer	34.35	274.80	52.70	421.60		
1 Crawler Crane, 20 Ton		841.90		926.09		
1 F.E. Attachment, .75 C.Y.		61.40		67.54	56.46	62.10
16 L.H., Daily Totals		$1550.10		$1970.83	$96.88	$123.18

Crew B-12N

Crew B-12N	Bare Costs Hr.	Bare Costs Daily	Incl. Subs O&P Hr.	Incl. Subs O&P Daily	Cost Per Labor-Hour Bare Costs	Cost Per Labor-Hour Incl. O&P
1 Equip. Oper. (crane)	$46.50	$372.00	$69.45	$555.60	$40.42	$61.08
1 Laborer	34.35	274.80	52.70	421.60		
1 Crawler Crane, 25 Ton		1123.00		1235.30		
1 F.E. Attachment, 1 C.Y.		67.20		73.92	74.39	81.83
16 L.H., Daily Totals		$1837.00		$2286.42	$114.81	$142.90

Crew B-12O

Crew B-12O	Bare Costs Hr.	Bare Costs Daily	Incl. Subs O&P Hr.	Incl. Subs O&P Daily	Cost Per Labor-Hour Bare Costs	Cost Per Labor-Hour Incl. O&P
1 Equip. Oper. (crane)	$46.50	$372.00	$69.45	$555.60	$40.42	$61.08
1 Laborer	34.35	274.80	52.70	421.60		
1 Crawler Crane, 40 Ton		1128.00		1240.80		
1 F.E. Attachment, 1.5 C.Y.		77.40		85.14	75.34	82.87
16 L.H., Daily Totals		$1852.20		$2303.14	$115.76	$143.95

Crew B-12P

Crew B-12P	Bare Costs Hr.	Bare Costs Daily	Incl. Subs O&P Hr.	Incl. Subs O&P Daily	Cost Per Labor-Hour Bare Costs	Cost Per Labor-Hour Incl. O&P
1 Equip. Oper. (crane)	$46.50	$372.00	$69.45	$555.60	$40.42	$61.08
1 Laborer	34.35	274.80	52.70	421.60		
1 Crawler Crane, 40 Ton		1128.00		1240.80		
1 Dragline Bucket, 1.5 C.Y.		34.20		37.62	72.64	79.90
16 L.H., Daily Totals		$1809.00		$2255.62	$113.06	$140.98

Crew B-12Q

Crew B-12Q	Bare Costs Hr.	Bare Costs Daily	Incl. Subs O&P Hr.	Incl. Subs O&P Daily	Cost Per Labor-Hour Bare Costs	Cost Per Labor-Hour Incl. O&P
1 Equip. Oper. (crane)	$46.50	$372.00	$69.45	$555.60	$40.42	$61.08
1 Laborer	34.35	274.80	52.70	421.60		
1 Hyd. Excavator, 5/8 C.Y.		530.00		583.00	33.13	36.44
16 L.H., Daily Totals		$1176.80		$1560.20	$73.55	$97.51

Crew B-12S

Crew B-12S	Bare Costs Hr.	Bare Costs Daily	Incl. Subs O&P Hr.	Incl. Subs O&P Daily	Cost Per Labor-Hour Bare Costs	Cost Per Labor-Hour Incl. O&P
1 Equip. Oper. (crane)	$46.50	$372.00	$69.45	$555.60	$40.42	$61.08
1 Laborer	34.35	274.80	52.70	421.60		
1 Hyd. Excavator, 2.5 C.Y.		1613.00		1774.30	100.81	110.89
16 L.H., Daily Totals		$2259.80		$2751.50	$141.24	$171.97

Crew B-12T

Crew B-12T	Bare Costs Hr.	Bare Costs Daily	Incl. Subs O&P Hr.	Incl. Subs O&P Daily	Cost Per Labor-Hour Bare Costs	Cost Per Labor-Hour Incl. O&P
1 Equip. Oper. (crane)	$46.50	$372.00	$69.45	$555.60	$40.42	$61.08
1 Laborer	34.35	274.80	52.70	421.60		
1 Crawler Crane, 75 Ton		1446.00		1590.60		
1 F.E. Attachment, 3 C.Y.		100.80		110.88	96.67	106.34
16 L.H., Daily Totals		$2193.60		$2678.68	$137.10	$167.42

615

Crew B-12V

Crew No.	Bare Costs Hr.	Daily	Incl. Subs O&P Hr.	Daily	Cost Per Labor-Hour Bare Costs	Incl. O&P
1 Equip. Oper. (crane)	$46.50	$372.00	$69.45	$555.60	$40.42	$61.08
1 Laborer	34.35	274.80	52.70	421.60		
1 Crawler Crane, 75 Ton		1446.00		1590.60		
1 Dragline Bucket, 3 C.Y.		52.80		58.08	93.67	103.04
16 L.H., Daily Totals		$2145.60		$2625.88	$134.10	$164.12

Crew B-12Y

Crew No.	Bare Costs Hr.	Daily	Incl. Subs O&P Hr.	Daily	Cost Per Labor-Hour Bare Costs	Incl. O&P
1 Equip. Oper. (crane)	$46.50	$372.00	$69.45	$555.60	$38.40	$58.28
2 Laborers	34.35	549.60	52.70	843.20		
1 Hyd. Excavator, 3.5 C.Y.		2171.00		2388.10	90.46	99.50
24 L.H., Daily Totals		$3092.60		$3786.90	$128.86	$157.79

Crew B-12Z

Crew No.	Bare Costs Hr.	Daily	Incl. Subs O&P Hr.	Daily	Cost Per Labor-Hour Bare Costs	Incl. O&P
1 Equip. Oper. (crane)	$46.50	$372.00	$69.45	$555.60	$38.40	$58.28
2 Laborers	34.35	549.60	52.70	843.20		
1 Hyd. Excavator, 2.5 C.Y.		1613.00		1774.30	67.21	73.93
24 L.H., Daily Totals		$2534.60		$3173.10	$105.61	$132.21

Crew B-13

Crew No.	Bare Costs Hr.	Daily	Incl. Subs O&P Hr.	Daily	Cost Per Labor-Hour Bare Costs	Incl. O&P
1 Labor Foreman (outside)	$36.35	$290.80	$55.75	$446.00	$37.22	$56.60
4 Laborers	34.35	1099.20	52.70	1686.40		
1 Equip. Oper. (crane)	46.50	372.00	69.45	555.60		
1 Equip. Oper. Oiler	40.30	322.40	60.20	481.60		
1 Hyd. Crane, 25 Ton		748.60		823.46	13.37	14.70
56 L.H., Daily Totals		$2833.00		$3993.06	$50.59	$71.30

Crew B-13A

Crew No.	Bare Costs Hr.	Daily	Incl. Subs O&P Hr.	Daily	Cost Per Labor-Hour Bare Costs	Incl. O&P
1 Foreman	$36.35	$290.80	$55.75	$446.00	$37.79	$57.31
2 Laborers	34.35	549.60	52.70	843.20		
2 Equipment Operators (med.)	45.35	725.60	67.75	1084.00		
2 Truck Drivers (heavy)	34.40	550.40	52.25	836.00		
1 Crawler Crane, 75 Ton		1446.00		1590.60		
1 Crawler Loader, 4 C.Y.		1432.00		1575.20		
2 Dump Trucks, 8 C.Y., 220 H.P.		630.80		693.88	62.66	68.92
56 L.H., Daily Totals		$5625.20		$7068.88	$100.45	$126.23

Crew B-13B

Crew No.	Bare Costs Hr.	Daily	Incl. Subs O&P Hr.	Daily	Cost Per Labor-Hour Bare Costs	Incl. O&P
1 Labor Foreman (outside)	$36.35	$290.80	$55.75	$446.00	$37.22	$56.60
4 Laborers	34.35	1099.20	52.70	1686.40		
1 Equip. Oper. (crane)	46.50	372.00	69.45	555.60		
1 Equip. Oper. Oiler	40.30	322.40	60.20	481.60		
1 Hyd. Crane, 55 Ton		1208.00		1328.80	21.57	23.73
56 L.H., Daily Totals		$3292.40		$4498.40	$58.79	$80.33

Crew B-13C

Crew No.	Bare Costs Hr.	Daily	Incl. Subs O&P Hr.	Daily	Cost Per Labor-Hour Bare Costs	Incl. O&P
1 Labor Foreman (outside)	$36.35	$290.80	$55.75	$446.00	$37.22	$56.60
4 Laborers	34.35	1099.20	52.70	1686.40		
1 Equip. Oper. (crane)	46.50	372.00	69.45	555.60		
1 Equip. Oper. Oiler	40.30	322.40	60.20	481.60		
1 Crawler Crane, 100 Ton		1653.00		1818.30	29.52	32.47
56 L.H., Daily Totals		$3737.40		$4987.90	$66.74	$89.07

Crew B-13D

Crew No.	Bare Costs Hr.	Daily	Incl. Subs O&P Hr.	Daily	Cost Per Labor-Hour Bare Costs	Incl. O&P
1 Laborer	$34.35	$274.80	$52.70	$421.60	$40.42	$61.08
1 Equip. Oper. (crane)	46.50	372.00	69.45	555.60		
1 Hyd. Excavator, 1 C.Y.		729.80		802.78		
1 Trench Box		113.40		124.74	52.70	57.97
16 L.H., Daily Totals		$1490.00		$1904.72	$93.13	$119.05

Crew B-13E

Crew No.	Bare Costs Hr.	Daily	Incl. Subs O&P Hr.	Daily	Cost Per Labor-Hour Bare Costs	Incl. O&P
1 Laborer	$34.35	$274.80	$52.70	$421.60	$40.42	$61.08
1 Equip. Oper. (crane)	46.50	372.00	69.45	555.60		
1 Hyd. Excavator, 1.5 C.Y.		947.40		1042.14		
1 Trench Box		113.40		124.74	66.30	72.93
16 L.H., Daily Totals		$1707.60		$2144.08	$106.72	$134.01

Crew B-13F

Crew No.	Bare Costs Hr.	Daily	Incl. Subs O&P Hr.	Daily	Cost Per Labor-Hour Bare Costs	Incl. O&P
1 Laborer	$34.35	$274.80	$52.70	$421.60	$40.42	$61.08
1 Equip. Oper. (crane)	46.50	372.00	69.45	555.60		
1 Hyd. Excavator, 3.5 C.Y.		2171.00		2388.10		
1 Trench Box		113.40		124.74	142.78	157.05
16 L.H., Daily Totals		$2931.20		$3490.04	$183.20	$218.13

Crew B-13G

Crew No.	Bare Costs Hr.	Daily	Incl. Subs O&P Hr.	Daily	Cost Per Labor-Hour Bare Costs	Incl. O&P
1 Laborer	$34.35	$274.80	$52.70	$421.60	$40.42	$61.08
1 Equip. Oper. (crane)	46.50	372.00	69.45	555.60		
1 Hyd. Excavator, .75 C.Y.		640.40		704.44		
1 Trench Box		113.40		124.74	47.11	51.82
16 L.H., Daily Totals		$1400.60		$1806.38	$87.54	$112.90

Crew B-13H

Crew No.	Bare Costs Hr.	Daily	Incl. Subs O&P Hr.	Daily	Cost Per Labor-Hour Bare Costs	Incl. O&P
1 Laborer	$34.35	$274.80	$52.70	$421.60	$40.42	$61.08
1 Equip. Oper. (crane)	46.50	372.00	69.45	555.60		
1 Gradall, 5/8 C.Y.		857.40		943.14		
1 Trench Box		113.40		124.74	60.67	66.74
16 L.H., Daily Totals		$1617.60		$2045.08	$101.10	$127.82

Crew B-13I

Crew No.	Bare Costs Hr.	Daily	Incl. Subs O&P Hr.	Daily	Cost Per Labor-Hour Bare Costs	Incl. O&P
1 Laborer	$34.35	$274.80	$52.70	$421.60	$40.42	$61.08
1 Equip. Oper. (crane)	46.50	372.00	69.45	555.60		
1 Gradall, 3 Ton, 1 C.Y.		1001.00		1101.10		
1 Trench Box		113.40		124.74	69.65	76.61
16 L.H., Daily Totals		$1761.20		$2203.04	$110.08	$137.69

Crew B-13J

Crew No.	Bare Costs Hr.	Daily	Incl. Subs O&P Hr.	Daily	Cost Per Labor-Hour Bare Costs	Incl. O&P
1 Laborer	$34.35	$274.80	$52.70	$421.60	$40.42	$61.08
1 Equip. Oper. (crane)	46.50	372.00	69.45	555.60		
1 Hyd. Excavator, 2.5 C.Y.		1613.00		1774.30		
1 Trench Box		113.40		124.74	107.90	118.69
16 L.H., Daily Totals		$2373.20		$2876.24	$148.32	$179.76

Crew B-14

Crew No.	Bare Costs Hr.	Daily	Incl. Subs O&P Hr.	Daily	Cost Per Labor-Hour Bare Costs	Incl. O&P
1 Labor Foreman (outside)	$36.35	$290.80	$55.75	$446.00	$36.23	$55.28
4 Laborers	34.35	1099.20	52.70	1686.40		
1 Equip. Oper. (light)	43.60	348.80	65.15	521.20		
1 Backhoe Loader, 48 H.P.		320.40		352.44	6.67	7.34
48 L.H., Daily Totals		$2059.20		$3006.04	$42.90	$62.63

Crew B-14A

Crew No.	Bare Costs Hr.	Daily	Incl. Subs O&P Hr.	Daily	Cost Per Labor-Hour Bare Costs	Incl. O&P
1 Equip. Oper. (crane)	$46.50	$372.00	$69.45	$555.60	$42.45	$63.87
.5 Laborer	34.35	137.40	52.70	210.80		
1 Hyd. Excavator, 4.5 C.Y.		2669.00		2935.90	222.42	244.66
12 L.H., Daily Totals		$3178.40		$3702.30	$264.87	$308.52

Crew B-14B

Crew No.	Bare Costs Hr.	Daily	Incl. Subs O&P Hr.	Daily	Cost Per Labor-Hour Bare Costs	Incl. O&P
1 Equip. Oper. (crane)	$46.50	$372.00	$69.45	$555.60	$42.45	$63.87
.5 Laborer	34.35	137.40	52.70	210.80		
1 Hyd. Excavator, 6 C.Y.		3101.00		3411.10	258.42	284.26
12 L.H., Daily Totals		$3610.40		$4177.50	$300.87	$348.13

Crews

Crew No.	Bare Costs		Incl. Subs O&P		Cost Per Labor-Hour	

Crew B-14C	Hr.	Daily	Hr.	Daily	Bare Costs	Incl. O&P
1 Equip. Oper. (crane)	$46.50	$372.00	$69.45	$555.60	$42.45	$63.87
.5 Laborer	34.35	137.40	52.70	210.80		
1 Hyd. Excavator, 7 C.Y.		3179.00		3496.90	264.92	291.41
12 L.H., Daily Totals		$3688.40		$4263.30	$307.37	$355.27

Crew B-14F	Hr.	Daily	Hr.	Daily	Bare Costs	Incl. O&P
1 Equip. Oper. (crane)	$46.50	$372.00	$69.45	$555.60	$42.45	$63.87
.5 Laborer	34.35	137.40	52.70	210.80		
1 Hyd. Shovel, 7 C.Y.		3109.00		3419.90	259.08	284.99
12 L.H., Daily Totals		$3618.40		$4186.30	$301.53	$348.86

Crew B-14G	Hr.	Daily	Hr.	Daily	Bare Costs	Incl. O&P
1 Equip. Oper. (crane)	$46.50	$372.00	$69.45	$555.60	$42.45	$63.87
.5 Laborer	34.35	137.40	52.70	210.80		
1 Hyd. Shovel, 12 C.Y.		4046.00		4450.60	337.17	370.88
12 L.H., Daily Totals		$4555.40		$5217.00	$379.62	$434.75

Crew B-14J	Hr.	Daily	Hr.	Daily	Bare Costs	Incl. O&P
1 Equip. Oper. (med.)	$45.35	$362.80	$67.75	$542.00	$41.68	$62.73
.5 Laborer	34.35	137.40	52.70	210.80		
1 F.E. Loader, 8 C.Y.		1811.00		1992.10	150.92	166.01
12 L.H., Daily Totals		$2311.20		$2744.90	$192.60	$228.74

Crew B-14K	Hr.	Daily	Hr.	Daily	Bare Costs	Incl. O&P
1 Equip. Oper. (med.)	$45.35	$362.80	$67.75	$542.00	$41.68	$62.73
.5 Laborer	34.35	137.40	52.70	210.80		
1 F.E. Loader, 10 C.Y.		2476.00		2723.60	206.33	226.97
12 L.H., Daily Totals		$2976.20		$3476.40	$248.02	$289.70

Crew B-15	Hr.	Daily	Hr.	Daily	Bare Costs	Incl. O&P
1 Equipment Oper. (med.)	$45.35	$362.80	$67.75	$542.00	$37.52	$56.74
.5 Laborer	34.35	137.40	52.70	210.80		
2 Truck Drivers (heavy)	34.40	550.40	52.25	836.00		
2 Dump Trucks, 12 C.Y., 400 H.P.		1012.00		1113.20		
1 Dozer, 200 H.P.		1051.00		1156.10	73.68	81.05
28 L.H., Daily Totals		$3113.60		$3858.10	$111.20	$137.79

Crew B-16	Hr.	Daily	Hr.	Daily	Bare Costs	Incl. O&P
1 Labor Foreman (outside)	$36.35	$290.80	$55.75	$446.00	$34.86	$53.35
2 Laborers	34.35	549.60	52.70	843.20		
1 Truck Driver (heavy)	34.40	275.20	52.25	418.00		
1 Dump Truck, 12 C.Y., 400 H.P.		506.00		556.60	15.81	17.39
32 L.H., Daily Totals		$1621.60		$2263.80	$50.67	$70.74

Crew B-17	Hr.	Daily	Hr.	Daily	Bare Costs	Incl. O&P
2 Laborers	$34.35	$549.60	$52.70	$843.20	$36.67	$55.70
1 Equip. Oper. (light)	43.60	348.80	65.15	521.20		
1 Truck Driver (heavy)	34.40	275.20	52.25	418.00		
1 Backhoe Loader, 48 H.P.		320.40		352.44		
1 Dump Truck, 8 C.Y., 220 H.P.		315.40		346.94	19.87	21.86
32 L.H., Daily Totals		$1809.40		$2481.78	$56.54	$77.56

Crew B-17A	Hr.	Daily	Hr.	Daily	Bare Costs	Incl. O&P
2 Laborer Foremen	$36.35	$581.60	$55.75	$892.00	$36.96	$56.70
6 Laborers	34.35	1648.80	52.70	2529.60		
1 Skilled Worker Foreman	46.40	371.20	71.20	569.60		
1 Skilled Worker	44.40	355.20	68.10	544.80		
80 L.H., Daily Totals		$2956.80		$4536.00	$36.96	$56.70

Crew B-18	Hr.	Daily	Hr.	Daily	Bare Costs	Incl. O&P
1 Labor Foreman (outside)	$36.35	$290.80	$55.75	$446.00	$35.02	$53.72
2 Laborers	34.35	549.60	52.70	843.20		
1 Vibratory Plate, gas, 21"		39.00		42.90	1.63	1.79
24 L.H., Daily Totals		$879.40		$1332.10	$36.64	$55.50

Crew B-19	Hr.	Daily	Hr.	Daily	Bare Costs	Incl. O&P
1 Pile Driver Foreman	$44.00	$352.00	$70.55	$564.40	$43.16	$67.38
4 Pile Drivers	42.00	1344.00	67.35	2155.20		
2 Equip. Oper. (crane)	46.50	744.00	69.45	1111.20		
1 Equip. Oper. Oiler	40.30	322.40	60.20	481.60		
1 Crawler Crane, 40 Ton		1128.00		1240.80		
1 Lead, 90' high		127.20		139.92		
1 Hammer, Diesel, 22k ft-lb		592.40		651.64	28.87	31.76
64 L.H., Daily Totals		$4610.00		$6344.76	$72.03	$99.14

Crew B-19A	Hr.	Daily	Hr.	Daily	Bare Costs	Incl. O&P
1 Pile Driver Foreman	$44.00	$352.00	$70.55	$564.40	$43.16	$67.38
4 Pile Drivers	42.00	1344.00	67.35	2155.20		
2 Equip. Oper. (crane)	46.50	744.00	69.45	1111.20		
1 Equip. Oper. Oiler	40.30	322.40	60.20	481.60		
1 Crawler Crane, 75 Ton		1446.00		1590.60		
1 Lead, 90' high		127.20		139.92		
1 Hammer, Diesel, 41k ft-lb		677.00		744.70	35.16	38.68
64 L.H., Daily Totals		$5012.60		$6787.62	$78.32	$106.06

Crew B-20	Hr.	Daily	Hr.	Daily	Bare Costs	Incl. O&P
1 Labor Foreman (out)	$36.35	$290.80	$55.75	$446.00	$38.37	$58.85
1 Skilled Worker	44.40	355.20	68.10	544.80		
1 Laborer	34.35	274.80	52.70	421.60		
24 L.H., Daily Totals		$920.80		$1412.40	$38.37	$58.85

Crew B-20A	Hr.	Daily	Hr.	Daily	Bare Costs	Incl. O&P
1 Labor Foreman	$36.35	$290.80	$55.75	$446.00	$41.80	$63.13
1 Laborer	34.35	274.80	52.70	421.60		
1 Plumber	53.60	428.80	80.00	640.00		
1 Plumber Apprentice	42.90	343.20	64.05	512.40		
32 L.H., Daily Totals		$1337.60		$2020.00	$41.80	$63.13

Crew B-21	Hr.	Daily	Hr.	Daily	Bare Costs	Incl. O&P
1 Labor Foreman (out)	$36.35	$290.80	$55.75	$446.00	$39.53	$60.36
1 Skilled Worker	44.40	355.20	68.10	544.80		
1 Laborer	34.35	274.80	52.70	421.60		
.5 Equip. Oper. (crane)	46.50	186.00	69.45	277.80		
.5 S.P. Crane, 4x4, 5 Ton		128.90		141.79	4.60	5.06
28 L.H., Daily Totals		$1235.70		$1831.99	$44.13	$65.43

Crew B-21A	Hr.	Daily	Hr.	Daily	Bare Costs	Incl. O&P
1 Labor Foreman	$36.35	$290.80	$55.75	$446.00	$42.74	$64.39
1 Laborer	34.35	274.80	52.70	421.60		
1 Plumber	53.60	428.80	80.00	640.00		
1 Plumber Apprentice	42.90	343.20	64.05	512.40		
1 Equip. Oper. (crane)	46.50	372.00	69.45	555.60		
1 S.P. Crane, 4x4, 12 Ton		434.40		477.84	10.86	11.95
40 L.H., Daily Totals		$2144.00		$3053.44	$53.60	$76.34

Crew B-21B	Hr.	Daily	Hr.	Daily	Bare Costs	Incl. O&P
1 Laborer Foreman	$36.35	$290.80	$55.75	$446.00	$37.18	$56.66
3 Laborers	34.35	824.40	52.70	1264.80		
1 Equip. Oper. (crane)	46.50	372.00	69.45	555.60		
1 Hyd. Crane, 12 Ton		655.60		721.16	16.39	18.03
40 L.H., Daily Totals		$2142.80		$2987.56	$53.57	$74.69

Crews

Crew B-21C

Crew No.	Bare Costs Hr.	Daily	Incl. Subs O&P Hr.	Daily	Cost Per Labor-Hour Bare Costs	Incl. O&P
1 Laborer Foreman	$36.35	$290.80	$55.75	$446.00	$37.22	$56.60
4 Laborers	34.35	1099.20	52.70	1686.40		
1 Equip. Oper. (crane)	46.50	372.00	69.45	555.60		
1 Equip. Oper. Oiler	40.30	322.40	60.20	481.60		
2 Cutting Torches		36.00		39.60		
2 Set of Gases		165.60		182.16		
1 Lattice Boom Crane, 90 Ton		1622.00		1784.20	32.56	35.82
56 L.H., Daily Totals		$3908.00		$5175.56	$69.79	$92.42

Crew B-22

	Hr.	Daily	Hr.	Daily	Bare Costs	Incl. O&P
1 Labor Foreman (outside)	$36.35	$290.80	$55.75	$446.00	$39.99	$60.97
1 Skilled Worker	44.40	355.20	68.10	544.80		
1 Laborer	34.35	274.80	52.70	421.60		
.75 Equip. Oper. (crane)	46.50	279.00	69.45	416.70		
.75 S.P. Crane, 4x4, 5 Ton		193.35		212.69	6.45	7.09
30 L.H., Daily Totals		$1393.15		$2041.79	$46.44	$68.06

Crew B-22A

	Hr.	Daily	Hr.	Daily	Bare Costs	Incl. O&P
1 Labor Foreman (outside)	$36.35	$290.80	$55.75	$446.00	$39.19	$59.74
1 Skilled Worker	44.40	355.20	68.10	544.80		
2 Laborers	34.35	549.60	52.70	843.20		
1 Equipment Oper. (crane)	46.50	372.00	69.45	555.60		
1 S.P. Crane, 4x4, 5 Ton		257.80		283.58		
1 Butt Fusion Mach., 4"-12" diam.		376.45		414.10	15.86	17.44
40 L.H., Daily Totals		$2201.85		$3087.28	$55.05	$77.18

Crew B-22B

	Hr.	Daily	Hr.	Daily	Bare Costs	Incl. O&P
1 Labor Foreman (outside)	$36.35	$290.80	$55.75	$446.00	$39.19	$59.74
1 Skilled Worker	44.40	355.20	68.10	544.80		
2 Laborers	34.35	549.60	52.70	843.20		
1 Equip. Oper. (crane)	46.50	372.00	69.45	555.60		
1 S.P. Crane, 4x4, 5 Ton		257.80		283.58		
1 Butt Fusion Mach., 8"-24" diam.		806.15		886.76	26.60	29.26
40 L.H., Daily Totals		$2631.55		$3559.95	$65.79	$89.00

Crew B-22C

	Hr.	Daily	Hr.	Daily	Bare Costs	Incl. O&P
1 Skilled Worker	$44.40	$355.20	$68.10	$544.80	$39.38	$60.40
1 Laborer	34.35	274.80	52.70	421.60		
1 Butt Fusion Mach., 2"-8" diam.		120.50		132.55	7.53	8.28
16 L.H., Daily Totals		$750.50		$1098.95	$46.91	$68.68

Crew B-23

	Hr.	Daily	Hr.	Daily	Bare Costs	Incl. O&P
1 Labor Foreman (outside)	$36.35	$290.80	$55.75	$446.00	$34.75	$53.31
4 Laborers	34.35	1099.20	52.70	1686.40		
1 Drill Rig, Truck-Mounted		2486.00		2734.60		
1 Flatbed Truck, Gas, 3 Ton		217.00		238.70	67.58	74.33
40 L.H., Daily Totals		$4093.00		$5105.70	$102.33	$127.64

Crew B-23A

	Hr.	Daily	Hr.	Daily	Bare Costs	Incl. O&P
1 Labor Foreman (outside)	$36.35	$290.80	$55.75	$446.00	$38.68	$58.73
1 Laborer	34.35	274.80	52.70	421.60		
1 Equip. Operator (med.)	45.35	362.80	67.75	542.00		
1 Drill Rig, Truck-Mounted		2486.00		2734.60		
1 Pickup Truck, 3/4 Ton		104.60		115.06	107.94	118.74
24 L.H., Daily Totals		$3519.00		$4259.26	$146.63	$177.47

Crew B-23B

	Hr.	Daily	Hr.	Daily	Bare Costs	Incl. O&P
1 Labor Foreman (outside)	$36.35	$290.80	$55.75	$446.00	$38.68	$58.73
1 Laborer	34.35	274.80	52.70	421.60		
1 Equip. Operator (med.)	45.35	362.80	67.75	542.00		
1 Drill Rig, Truck-Mounted		2486.00		2734.60		
1 Pickup Truck, 3/4 Ton		104.60		115.06		
1 Centr. Water Pump, 6"		291.60		320.76	120.09	132.10
24 L.H., Daily Totals		$3810.60		$4580.02	$158.78	$190.83

Crew B-24

	Hr.	Daily	Hr.	Daily	Bare Costs	Incl. O&P
1 Cement Finisher	$40.85	$326.80	$59.50	$476.00	$39.42	$59.42
1 Laborer	34.35	274.80	52.70	421.60		
1 Carpenter	43.05	344.40	66.05	528.40		
24 L.H., Daily Totals		$946.00		$1426.00	$39.42	$59.42

Crew B-25

	Hr.	Daily	Hr.	Daily	Bare Costs	Incl. O&P
1 Labor Foreman	$36.35	$290.80	$55.75	$446.00	$37.53	$57.08
7 Laborers	34.35	1923.60	52.70	2951.20		
3 Equip. Oper. (med.)	45.35	1088.40	67.75	1626.00		
1 Asphalt Paver, 130 H.P.		1949.00		2143.90		
1 Tandem Roller, 10 Ton		224.20		246.62		
1 Roller, Pneum. Whl, 12 Ton		309.20		340.12	28.21	31.03
88 L.H., Daily Totals		$5785.20		$7753.84	$65.74	$88.11

Crew B-25B

	Hr.	Daily	Hr.	Daily	Bare Costs	Incl. O&P
1 Labor Foreman	$36.35	$290.80	$55.75	$446.00	$38.18	$57.97
7 Laborers	34.35	1923.60	52.70	2951.20		
4 Equip. Oper. (med.)	45.35	1451.20	67.75	2168.00		
1 Asphalt Paver, 130 H.P.		1949.00		2143.90		
2 Tandem Rollers, 10 Ton		448.40		493.24		
1 Roller, Pneum. Whl, 12 Ton		309.20		340.12	28.19	31.01
96 L.H., Daily Totals		$6372.20		$8542.46	$66.38	$88.98

Crew B-25C

	Hr.	Daily	Hr.	Daily	Bare Costs	Incl. O&P
1 Labor Foreman	$36.35	$290.80	$55.75	$446.00	$38.35	$58.23
3 Laborers	34.35	824.40	52.70	1264.80		
2 Equip. Oper. (med.)	45.35	725.60	67.75	1084.00		
1 Asphalt Paver, 130 H.P.		1949.00		2143.90		
1 Tandem Roller, 10 Ton		224.20		246.62	45.27	49.80
48 L.H., Daily Totals		$4014.00		$5185.32	$83.63	$108.03

Crew B-26

	Hr.	Daily	Hr.	Daily	Bare Costs	Incl. O&P
1 Labor Foreman (outside)	$36.35	$290.80	$55.75	$446.00	$38.40	$58.57
6 Laborers	34.35	1648.80	52.70	2529.60		
2 Equip. Oper. (med.)	45.35	725.60	67.75	1084.00		
1 Rodman (reinf.)	48.40	387.20	77.35	618.80		
1 Cement Finisher	40.85	326.80	59.50	476.00		
1 Grader, 30,000 Lbs.		542.80		597.08		
1 Paving Mach. & Equip.		2483.00		2731.30	34.38	37.82
88 L.H., Daily Totals		$6405.00		$8482.78	$72.78	$96.40

Crew B-26A

	Hr.	Daily	Hr.	Daily	Bare Costs	Incl. O&P
1 Labor Foreman (outside)	$36.35	$290.80	$55.75	$446.00	$38.40	$58.57
6 Laborers	34.35	1648.80	52.70	2529.60		
2 Equip. Oper. (med.)	45.35	725.60	67.75	1084.00		
1 Rodman (reinf.)	48.40	387.20	77.35	618.80		
1 Cement Finisher	40.85	326.80	59.50	476.00		
1 Grader, 30,000 Lbs.		542.80		597.08		
1 Paving Mach. & Equip.		2483.00		2731.30		
1 Concrete Saw		134.20		147.62	35.91	39.50
88 L.H., Daily Totals		$6539.20		$8630.40	$74.31	$98.07

Crew B-26B

	Bare Costs Hr.	Daily	Incl. Subs O&P Hr.	Daily	Bare Costs	Incl. O&P
1 Labor Foreman (outside)	$36.35	$290.80	$55.75	$446.00	$38.98	$59.34
6 Laborers	34.35	1648.80	52.70	2529.60		
3 Equip. Oper. (med.)	45.35	1088.40	67.75	1626.00		
1 Rodman (reinf.)	48.40	387.20	77.35	618.80		
1 Cement Finisher	40.85	326.80	59.50	476.00		
1 Grader, 30,000 Lbs.		542.80		597.08		
1 Paving Mach. & Equip.		2483.00		2731.30		
1 Concrete Pump, 110' Boom		999.60		1099.56	41.93	46.12
96 L.H., Daily Totals		$7767.40		$10124.34	$80.91	$105.46

Crew B-27

	Bare Costs Hr.	Daily	Incl. Subs O&P Hr.	Daily	Bare Costs	Incl. O&P
1 Labor Foreman (outside)	$36.35	$290.80	$55.75	$446.00	$34.85	$53.46
3 Laborers	34.35	824.40	52.70	1264.80		
1 Berm Machine		276.80		304.48	8.65	9.52
32 L.H., Daily Totals		$1392.00		$2015.28	$43.50	$62.98

Crew B-28

	Bare Costs Hr.	Daily	Incl. Subs O&P Hr.	Daily	Bare Costs	Incl. O&P
2 Carpenters	$43.05	$688.80	$66.05	$1056.80	$40.15	$61.60
1 Laborer	34.35	274.80	52.70	421.60		
24 L.H., Daily Totals		$963.60		$1478.40	$40.15	$61.60

Crew B-29

	Bare Costs Hr.	Daily	Incl. Subs O&P Hr.	Daily	Bare Costs	Incl. O&P
1 Labor Foreman (outside)	$36.35	$290.80	$55.75	$446.00	$37.22	$56.60
4 Laborers	34.35	1099.20	52.70	1686.40		
1 Equip. Oper. (crane)	46.50	372.00	69.45	555.60		
1 Equip. Oper. Oiler	40.30	322.40	60.20	481.60		
1 Gradall, 5/8 C.Y.		857.40		943.14	15.31	16.84
56 L.H., Daily Totals		$2941.80		$4112.74	$52.53	$73.44

Crew B-30

	Bare Costs Hr.	Daily	Incl. Subs O&P Hr.	Daily	Bare Costs	Incl. O&P
1 Equip. Oper. (med.)	$45.35	$362.80	$67.75	$542.00	$38.05	$57.42
2 Truck Drivers (heavy)	34.40	550.40	52.25	836.00		
1 Hyd. Excavator, 1.5 C.Y.		947.40		1042.14		
2 Dump Trucks, 12 C.Y., 400 H.P.		1012.00		1113.20	81.64	89.81
24 L.H., Daily Totals		$2872.60		$3533.34	$119.69	$147.22

Crew B-31

	Bare Costs Hr.	Daily	Incl. Subs O&P Hr.	Daily	Bare Costs	Incl. O&P
1 Labor Foreman (outside)	$36.35	$290.80	$55.75	$446.00	$36.49	$55.98
3 Laborers	34.35	824.40	52.70	1264.80		
1 Carpenter	43.05	344.40	66.05	528.40		
1 Air Compressor, 250 cfm		162.80		179.08		
1 Sheeting Driver		5.75		6.33		
2 -50' Air Hoses, 1.5"		12.60		13.86	4.53	4.98
40 L.H., Daily Totals		$1640.75		$2438.47	$41.02	$60.96

Crew B-32

	Bare Costs Hr.	Daily	Incl. Subs O&P Hr.	Daily	Bare Costs	Incl. O&P
1 Laborer	$34.35	$274.80	$52.70	$421.60	$42.60	$63.99
3 Equip. Oper. (med.)	45.35	1088.40	67.75	1626.00		
1 Grader, 30,000 Lbs.		542.80		597.08		
1 Tandem Roller, 10 Ton		224.20		246.62		
1 Dozer, 200 H.P.		1051.00		1156.10	56.81	62.49
32 L.H., Daily Totals		$3181.20		$4047.40	$99.41	$126.48

Crew B-32A

	Bare Costs Hr.	Daily	Incl. Subs O&P Hr.	Daily	Bare Costs	Incl. O&P
1 Laborer	$34.35	$274.80	$52.70	$421.60	$41.68	$62.73
2 Equip. Oper. (med.)	45.35	725.60	67.75	1084.00		
1 Grader, 30,000 Lbs.		542.80		597.08		
1 Roller, Vibratory, 25 Ton		596.80		656.48	47.48	52.23
24 L.H., Daily Totals		$2140.00		$2759.16	$89.17	$114.97

Crew B-32B

	Bare Costs Hr.	Daily	Incl. Subs O&P Hr.	Daily	Bare Costs	Incl. O&P
1 Laborer	$34.35	$274.80	$52.70	$421.60	$41.68	$62.73
2 Equip. Oper. (med.)	45.35	725.60	67.75	1084.00		
1 Dozer, 200 H.P.		1051.00		1156.10		
1 Roller, Vibratory, 25 Ton		596.80		656.48	68.66	75.52
24 L.H., Daily Totals		$2648.20		$3318.18	$110.34	$138.26

Crew B-32C

	Bare Costs Hr.	Daily	Incl. Subs O&P Hr.	Daily	Bare Costs	Incl. O&P
1 Labor Foreman	$36.35	$290.80	$55.75	$446.00	$40.18	$60.73
2 Laborers	34.35	549.60	52.70	843.20		
3 Equip. Oper. (med.)	45.35	1088.40	67.75	1626.00		
1 Grader, 30,000 Lbs.		542.80		597.08		
1 Tandem Roller, 10 Ton		224.20		246.62		
1 Dozer, 200 H.P.		1051.00		1156.10	37.88	41.66
48 L.H., Daily Totals		$3746.80		$4915.00	$78.06	$102.40

Crew B-33A

	Bare Costs Hr.	Daily	Incl. Subs O&P Hr.	Daily	Bare Costs	Incl. O&P
1 Equip. Oper. (med.)	$45.35	$362.80	$67.75	$542.00	$42.21	$63.45
.5 Laborer	34.35	137.40	52.70	210.80		
.25 Equip. Oper. (med.)	45.35	90.70	67.75	135.50		
1 Scraper, Towed, 7 C.Y.		108.00		118.80		
1.25 Dozers, 300 H.P.		2008.75		2209.63	151.20	166.32
14 L.H., Daily Totals		$2707.65		$3216.72	$193.40	$229.77

Crew B-33B

	Bare Costs Hr.	Daily	Incl. Subs O&P Hr.	Daily	Bare Costs	Incl. O&P
1 Equip. Oper. (med.)	$45.35	$362.80	$67.75	$542.00	$42.21	$63.45
.5 Laborer	34.35	137.40	52.70	210.80		
.25 Equip. Oper. (med.)	45.35	90.70	67.75	135.50		
1 Scraper, Towed, 10 C.Y.		140.00		154.00		
1.25 Dozers, 300 H.P.		2008.75		2209.63	153.48	168.83
14 L.H., Daily Totals		$2739.65		$3251.93	$195.69	$232.28

Crew B-33C

	Bare Costs Hr.	Daily	Incl. Subs O&P Hr.	Daily	Bare Costs	Incl. O&P
1 Equip. Oper. (med.)	$45.35	$362.80	$67.75	$542.00	$42.21	$63.45
.5 Laborer	34.35	137.40	52.70	210.80		
.25 Equip. Oper. (med.)	45.35	90.70	67.75	135.50		
1 Scraper, Towed, 15 C.Y.		158.60		174.46		
1.25 Dozers, 300 H.P.		2008.75		2209.63	154.81	170.29
14 L.H., Daily Totals		$2758.25		$3272.39	$197.02	$233.74

Crew B-33D

	Bare Costs Hr.	Daily	Incl. Subs O&P Hr.	Daily	Bare Costs	Incl. O&P
1 Equip. Oper. (med.)	$45.35	$362.80	$67.75	$542.00	$42.21	$63.45
.5 Laborer	34.35	137.40	52.70	210.80		
.25 Equip. Oper. (med.)	45.35	90.70	67.75	135.50		
1 S.P. Scraper, 14 C.Y.		1512.00		1663.20		
.25 Dozer, 300 H.P.		401.75		441.93	136.70	150.37
14 L.H., Daily Totals		$2504.65		$2993.43	$178.90	$213.82

Crew B-33E

	Bare Costs Hr.	Daily	Incl. Subs O&P Hr.	Daily	Bare Costs	Incl. O&P
1 Equip. Oper. (med.)	$45.35	$362.80	$67.75	$542.00	$42.21	$63.45
.5 Laborer	34.35	137.40	52.70	210.80		
.25 Equip. Oper. (med.)	45.35	90.70	67.75	135.50		
1 S.P. Scraper, 21 C.Y.		1872.00		2059.20		
.25 Dozer, 300 H.P.		401.75		441.93	162.41	178.65
14 L.H., Daily Totals		$2864.65		$3389.43	$204.62	$242.10

Crew No.	Bare Costs Hr.	Daily	Incl. Subs O&P Hr.	Daily	Cost Per Labor-Hour Bare Costs	Incl. O&P
Crew B-33F	Hr.	Daily	Hr.	Daily	Bare Costs	Incl. O&P
1 Equip. Oper. (med.)	$45.35	$362.80	$67.75	$542.00	$42.21	$63.45
.5 Laborer	34.35	137.40	52.70	210.80		
.25 Equip. Oper. (med.)	45.35	90.70	67.75	135.50		
1 Elev. Scraper, 11 C.Y.		1089.00		1197.90		
.25 Dozer, 300 H.P.		401.75		441.93	106.48	117.13
14 L.H., Daily Totals		$2081.65		$2528.13	$148.69	$180.58
Crew B-33G	Hr.	Daily	Hr.	Daily	Bare Costs	Incl. O&P
1 Equip. Oper. (med.)	$45.35	$362.80	$67.75	$542.00	$42.21	$63.45
.5 Laborer	34.35	137.40	52.70	210.80		
.25 Equip. Oper. (med.)	45.35	90.70	67.75	135.50		
1 Elev. Scraper, 22 C.Y.		2087.00		2295.70		
.25 Dozer, 300 H.P.		401.75		441.93	177.77	195.54
14 L.H., Daily Totals		$3079.65		$3625.93	$219.97	$258.99
Crew B-33H	Hr.	Daily	Hr.	Daily	Bare Costs	Incl. O&P
.5 Laborer	$34.35	$137.40	$52.70	$210.80	$42.21	$63.45
1 Equipment Operator (med.)	45.35	362.80	67.75	542.00		
.25 Equipment Operator (med.)	45.35	90.70	67.75	135.50		
1 S.P. Scraper, 44 C.Y.		3705.00		4075.50		
.25 Dozer, 410 H.P.		526.25		578.88	302.23	332.46
14 L.H., Daily Totals		$4822.15		$5542.68	$344.44	$395.91
Crew B-33J	Hr.	Daily	Hr.	Daily	Bare Costs	Incl. O&P
1 Equipment Operator (med.)	$45.35	$362.80	$67.75	$542.00	$45.35	$67.75
1 S.P. Scraper, 14 C.Y.		1512.00		1663.20	189.00	207.90
8 L.H., Daily Totals		$1874.80		$2205.20	$234.35	$275.65
Crew B-33K	Hr.	Daily	Hr.	Daily	Bare Costs	Incl. O&P
1 Equipment Operator (med.)	$45.35	$362.80	$67.75	$542.00	$42.21	$63.45
.25 Equipment Operator (med.)	45.35	90.70	67.75	135.50		
.5 Laborer	34.35	137.40	52.70	210.80		
1 S.P. Scraper, 31 C.Y.		2777.00		3054.70		
.25 Dozer, 410 H.P.		526.25		578.88	235.95	259.54
14 L.H., Daily Totals		$3894.15		$4521.88	$278.15	$322.99
Crew B-34A	Hr.	Daily	Hr.	Daily	Bare Costs	Incl. O&P
1 Truck Driver (heavy)	$34.40	$275.20	$52.25	$418.00	$34.40	$52.25
1 Dump Truck, 8 C.Y., 220 H.P.		315.40		346.94	39.42	43.37
8 L.H., Daily Totals		$590.60		$764.94	$73.83	$95.62
Crew B-34B	Hr.	Daily	Hr.	Daily	Bare Costs	Incl. O&P
1 Truck Driver (heavy)	$34.40	$275.20	$52.25	$418.00	$34.40	$52.25
1 Dump Truck, 12 C.Y., 400 H.P.		506.00		556.60	63.25	69.58
8 L.H., Daily Totals		$781.20		$974.60	$97.65	$121.83
Crew B-34C	Hr.	Daily	Hr.	Daily	Bare Costs	Incl. O&P
1 Truck Driver (heavy)	$34.40	$275.20	$52.25	$418.00	$34.40	$52.25
1 Truck Tractor, 6x4, 380 H.P.		459.40		505.34		
1 Dump Trailer, 16.5 C.Y.		125.80		138.38	73.15	80.47
8 L.H., Daily Totals		$860.40		$1061.72	$107.55	$132.72
Crew B-34D	Hr.	Daily	Hr.	Daily	Bare Costs	Incl. O&P
1 Truck Driver (heavy)	$34.40	$275.20	$52.25	$418.00	$34.40	$52.25
1 Truck Tractor, 6x4, 380 H.P.		459.40		505.34		
1 Dump Trailer, 20 C.Y.		140.00		154.00	74.92	82.42
8 L.H., Daily Totals		$874.60		$1077.34	$109.33	$134.67

Crew No.	Bare Costs Hr.	Daily	Incl. Subs O&P Hr.	Daily	Cost Per Labor-Hour Bare Costs	Incl. O&P
Crew B-34E	Hr.	Daily	Hr.	Daily	Bare Costs	Incl. O&P
1 Truck Driver (heavy)	$34.40	$275.20	$52.25	$418.00	$34.40	$52.25
1 Dump Truck, Off Hwy., 25 Ton		1264.00		1390.40	158.00	173.80
8 L.H., Daily Totals		$1539.20		$1808.40	$192.40	$226.05
Crew B-34F	Hr.	Daily	Hr.	Daily	Bare Costs	Incl. O&P
1 Truck Driver (heavy)	$34.40	$275.20	$52.25	$418.00	$34.40	$52.25
1 Dump Truck, Off Hwy., 35 Ton		1135.00		1248.50	141.88	156.06
8 L.H., Daily Totals		$1410.20		$1666.50	$176.28	$208.31
Crew B-34G	Hr.	Daily	Hr.	Daily	Bare Costs	Incl. O&P
1 Truck Driver (heavy)	$34.40	$275.20	$52.25	$418.00	$34.40	$52.25
1 Dump Truck, Off Hwy., 50 Ton		1549.00		1703.90	193.63	212.99
8 L.H., Daily Totals		$1824.20		$2121.90	$228.03	$265.24
Crew B-34H	Hr.	Daily	Hr.	Daily	Bare Costs	Incl. O&P
1 Truck Driver (heavy)	$34.40	$275.20	$52.25	$418.00	$34.40	$52.25
1 Dump Truck, Off Hwy., 65 Ton		1551.00		1706.10	193.88	213.26
8 L.H., Daily Totals		$1826.20		$2124.10	$228.28	$265.51
Crew B-34I	Hr.	Daily	Hr.	Daily	Bare Costs	Incl. O&P
1 Truck Driver (heavy)	$34.40	$275.20	$52.25	$418.00	$34.40	$52.25
1 Dump Truck, 18 C.Y., 450 H.P.		660.80		726.88	82.60	90.86
8 L.H., Daily Totals		$936.00		$1144.88	$117.00	$143.11
Crew B-34J	Hr.	Daily	Hr.	Daily	Bare Costs	Incl. O&P
1 Truck Driver (heavy)	$34.40	$275.20	$52.25	$418.00	$34.40	$52.25
1 Dump Truck, Off Hwy., 100 Ton		2009.00		2209.90	251.13	276.24
8 L.H., Daily Totals		$2284.20		$2627.90	$285.52	$328.49
Crew B-34K	Hr.	Daily	Hr.	Daily	Bare Costs	Incl. O&P
1 Truck Driver (heavy)	$34.40	$275.20	$52.25	$418.00	$34.40	$52.25
1 Truck Tractor, 6x4, 450 H.P.		558.80		614.68		
1 Lowbed Trailer, 75 Ton		219.20		241.12	97.25	106.97
8 L.H., Daily Totals		$1053.20		$1273.80	$131.65	$159.22
Crew B-34L	Hr.	Daily	Hr.	Daily	Bare Costs	Incl. O&P
1 Equip. Oper. (light)	$43.60	$348.80	$65.15	$521.20	$43.60	$65.15
1 Flatbed Truck, Gas, 1.5 Ton		171.40		188.54	21.43	23.57
8 L.H., Daily Totals		$520.20		$709.74	$65.03	$88.72
Crew B-34M	Hr.	Daily	Hr.	Daily	Bare Costs	Incl. O&P
1 Equip. Oper. (light)	$43.60	$348.80	$65.15	$521.20	$43.60	$65.15
1 Flatbed Truck, Gas, 3 Ton		217.00		238.70	27.13	29.84
8 L.H., Daily Totals		$565.80		$759.90	$70.72	$94.99
Crew B-34N	Hr.	Daily	Hr.	Daily	Bare Costs	Incl. O&P
1 Truck Driver (heavy)	$34.40	$275.20	$52.25	$418.00	$34.40	$52.25
1 Dump Truck, 8 C.Y., 220 H.P.		315.40		346.94		
1 Flatbed Trailer, 40 Ton		150.20		165.22	58.20	64.02
8 L.H., Daily Totals		$740.80		$930.16	$92.60	$116.27
Crew B-34P	Hr.	Daily	Hr.	Daily	Bare Costs	Incl. O&P
1 Pipe Fitter	$54.10	$432.80	$80.75	$646.00	$44.28	$66.42
1 Truck Driver (light)	33.40	267.20	50.75	406.00		
1 Equip. Oper. (med.)	45.35	362.80	67.75	542.00		
1 Flatbed Truck, Gas, 3 Ton		217.00		238.70		
1 Backhoe Loader, 48 H.P.		320.40		352.44	22.39	24.63
24 L.H., Daily Totals		$1600.20		$2185.14	$66.67	$91.05

| Crew No. | Bare Costs | | Incl. Subs O&P | | Cost Per Labor-Hour | |

Crew B-34Q

Crew B-34Q	Hr.	Daily	Hr.	Daily	Bare Costs	Incl. O&P
1 Pipe Fitter	$54.10	$432.80	$80.75	$646.00	$44.67	$66.98
1 Truck Driver (light)	33.40	267.20	50.75	406.00		
1 Eqip. Oper. (crane)	46.50	372.00	69.45	555.60		
1 Flatbed Trailer, 25 Ton		111.20		122.32		
1 Dump Truck, 8 C.Y., 220 H.P.		315.40		346.94		
1 Hyd. Crane, 25 Ton		748.60		823.46	48.97	53.86
24 L.H., Daily Totals		$2247.20		$2900.32	$93.63	$120.85

Crew B-34R	Hr.	Daily	Hr.	Daily	Bare Costs	Incl. O&P
1 Pipe Fitter	$54.10	$432.80	$80.75	$646.00	$44.67	$66.98
1 Truck Driver (light)	33.40	267.20	50.75	406.00		
1 Eqip. Oper. (crane)	46.50	372.00	69.45	555.60		
1 Flatbed Trailer, 25 Ton		111.20		122.32		
1 Dump Truck, 8 C.Y., 220 H.P.		315.40		346.94		
1 Hyd. Crane, 25 Ton		748.60		823.46		
1 Hyd. Excavator, 1 C.Y.		729.80		802.78	79.38	87.31
24 L.H., Daily Totals		$2977.00		$3703.10	$124.04	$154.30

Crew B-34S	Hr.	Daily	Hr.	Daily	Bare Costs	Incl. O&P
2 Pipe Fitters	$54.10	$865.60	$80.75	$1292.00	$47.27	$70.80
1 Truck Driver (heavy)	34.40	275.20	52.25	418.00		
1 Eqip. Oper. (crane)	46.50	372.00	69.45	555.60		
1 Flatbed Trailer, 40 Ton		150.20		165.22		
1 Truck Tractor, 6x4, 380 H.P.		459.40		505.34		
1 Hyd. Crane, 80 Ton		1693.00		1862.30		
1 Hyd. Excavator, 2 C.Y.		1224.00		1346.40	110.21	121.23
32 L.H., Daily Totals		$5039.40		$6144.86	$157.48	$192.03

Crew B-34T	Hr.	Daily	Hr.	Daily	Bare Costs	Incl. O&P
2 Pipe Fitters	$54.10	$865.60	$80.75	$1292.00	$47.27	$70.80
1 Truck Driver (heavy)	34.40	275.20	52.25	418.00		
1 Eqip. Oper. (crane)	46.50	372.00	69.45	555.60		
1 Flatbed Trailer, 40 Ton		150.20		165.22		
1 Truck Tractor, 6x4, 380 H.P.		459.40		505.34		
1 Hyd. Crane, 80 Ton		1693.00		1862.30	71.96	79.15
32 L.H., Daily Totals		$3815.40		$4798.46	$119.23	$149.95

Crew B-35	Hr.	Daily	Hr.	Daily	Bare Costs	Incl. O&P
1 Laborer Foreman (out)	$36.35	$290.80	$55.75	$446.00	$42.58	$64.37
1 Skilled Worker	44.40	355.20	68.10	544.80		
1 Welder (plumber)	53.60	428.80	80.00	640.00		
1 Laborer	34.35	274.80	52.70	421.60		
1 Equip. Oper. (crane)	46.50	372.00	69.45	555.60		
1 Equip. Oper. Oiler	40.30	322.40	60.20	481.60		
1 Welder, electric, 300 amp		32.45		35.70		
1 Hyd. Excavator, .75 C.Y.		640.40		704.44	14.02	15.42
48 L.H., Daily Totals		$2716.85		$3829.74	$56.60	$79.79

Crew B-35A	Hr.	Daily	Hr.	Daily	Bare Costs	Incl. O&P
1 Laborer Foreman (out)	$36.35	$290.80	$55.75	$446.00	$41.41	$62.70
2 Laborers	34.35	549.60	52.70	843.20		
1 Skilled Worker	44.40	355.20	68.10	544.80		
1 Welder (plumber)	53.60	428.80	80.00	640.00		
1 Equip. Oper. (crane)	46.50	372.00	69.45	555.60		
1 Equip. Oper. Oiler	40.30	322.40	60.20	481.60		
1 Welder, gas engine, 300 amp		109.00		119.90		
1 Crawler Crane, 75 Ton		1446.00		1590.60	27.77	30.54
56 L.H., Daily Totals		$3873.80		$5221.70	$69.17	$93.24

Crew B-36	Hr.	Daily	Hr.	Daily	Bare Costs	Incl. O&P
1 Labor Foreman (outside)	$36.35	$290.80	$55.75	$446.00	$39.15	$59.33
2 Laborers	34.35	549.60	52.70	843.20		
2 Equip. Oper. (med.)	45.35	725.60	67.75	1084.00		
1 Dozer, 200 H.P.		1051.00		1156.10		
1 Aggregate Spreader		33.60		36.96		
1 Tandem Roller, 10 Ton		224.20		246.62	32.72	35.99
40 L.H., Daily Totals		$2874.80		$3812.88	$71.87	$95.32

Crew B-36A	Hr.	Daily	Hr.	Daily	Bare Costs	Incl. O&P
1 Labor Foreman (outside)	$36.35	$290.80	$55.75	$446.00	$40.92	$61.74
2 Laborers	34.35	549.60	52.70	843.20		
4 Equip. Oper. (med.)	45.35	1451.20	67.75	2168.00		
1 Dozer, 200 H.P.		1051.00		1156.10		
1 Aggregate Spreader		33.60		36.96		
1 Tandem Roller, 10 Ton		224.20		246.62		
1 Roller, Pneum. Whl, 12 Ton		309.20		340.12	28.89	31.78
56 L.H., Daily Totals		$3909.60		$5237.00	$69.81	$93.52

Crew B-36B	Hr.	Daily	Hr.	Daily	Bare Costs	Incl. O&P
1 Labor Foreman (outside)	$36.35	$290.80	$55.75	$446.00	$40.11	$60.55
2 Laborers	34.35	549.60	52.70	843.20		
4 Equip. Oper. (med.)	45.35	1451.20	67.75	2168.00		
1 Truck Driver, Heavy	34.40	275.20	52.25	418.00		
1 Grader, 30,000 Lbs.		542.80		597.08		
1 F.E. Loader, crl, 1.5 C.Y.		565.00		621.50		
1 Dozer, 300 H.P.		1607.00		1767.70		
1 Roller, Vibratory, 25 Ton		596.80		656.48		
1 Truck Tractor, 6x4, 450 H.P.		558.80		614.68		
1 Water Tank Trailer, 5000 Gal.		146.20		160.82	62.76	69.04
64 L.H., Daily Totals		$6583.40		$8293.46	$102.87	$129.59

Crew B-36C	Hr.	Daily	Hr.	Daily	Bare Costs	Incl. O&P
1 Labor Foreman (outside)	$36.35	$290.80	$55.75	$446.00	$41.36	$62.25
3 Equip. Oper. (med.)	45.35	1088.40	67.75	1626.00		
1 Truck Driver, Heavy	34.40	275.20	52.25	418.00		
1 Grader, 30,000 Lbs.		542.80		597.08		
1 Dozer, 300 H.P.		1607.00		1767.70		
1 Roller, Vibratory, 25 Ton		596.80		656.48		
1 Truck Tractor, 6x4, 450 H.P.		558.80		614.68		
1 Water Tank Trailer, 5000 Gal.		146.20		160.82	86.29	94.92
40 L.H., Daily Totals		$5106.00		$6286.76	$127.65	$157.17

Crew B-36D	Hr.	Daily	Hr.	Daily	Bare Costs	Incl. O&P
1 Labor Foreman (outside)	$36.35	$290.80	$55.75	$446.00	$43.10	$64.75
3 Equip. Oper. (medium)	45.35	1088.40	67.75	1626.00		
1 Grader, 30,000 Lbs.		542.80		597.08		
1 Dozer, 300 H.P.		1607.00		1767.70		
1 Roller, Vibratory, 25 Ton		596.80		656.48	85.83	94.41
32 L.H., Daily Totals		$4125.80		$5093.26	$128.93	$159.16

Crew B-37	Hr.	Daily	Hr.	Daily	Bare Costs	Incl. O&P
1 Labor Foreman (outside)	$36.35	$290.80	$55.75	$446.00	$36.23	$55.28
4 Laborers	34.35	1099.20	52.70	1686.40		
1 Equip. Oper. (light)	43.60	348.80	65.15	521.20		
1 Tandem Roller, 5 Ton		145.20		159.72	3.02	3.33
48 L.H., Daily Totals		$1884.00		$2813.32	$39.25	$58.61

Crew No.	Bare Costs		Incl. Subs O&P		Cost Per Labor-Hour	

Crew B-37A	Hr.	Daily	Hr.	Daily	Bare Costs	Incl. O&P
2 Laborers	$34.35	$549.60	$52.70	$843.20	$34.03	$52.05
1 Truck Driver (light)	33.40	267.20	50.75	406.00		
1 Flatbed Truck, Gas, 1.5 Ton		171.40		188.54		
1 Tar Kettle, T.M.		74.65		82.11	10.25	11.28
24 L.H., Daily Totals		$1062.85		$1519.86	$44.29	$63.33

Crew B-37B	Hr.	Daily	Hr.	Daily	Bare Costs	Incl. O&P
3 Laborers	$34.35	$824.40	$52.70	$1264.80	$34.11	$52.21
1 Truck Driver (light)	33.40	267.20	50.75	406.00		
1 Flatbed Truck, Gas, 1.5 Ton		171.40		188.54		
1 Tar Kettle, T.M.		74.65		82.11	7.69	8.46
32 L.H., Daily Totals		$1337.65		$1941.45	$41.80	$60.67

Crew B-37C	Hr.	Daily	Hr.	Daily	Bare Costs	Incl. O&P
2 Laborers	$34.35	$549.60	$52.70	$843.20	$33.88	$51.73
2 Truck Drivers (light)	33.40	534.40	50.75	812.00		
2 Flatbed Truck, Gas, 1.5 Ton		342.80		377.08		
1 Tar Kettle, T.M.		74.65		82.11	13.05	14.35
32 L.H., Daily Totals		$1501.45		$2114.40	$46.92	$66.07

Crew B-37D	Hr.	Daily	Hr.	Daily	Bare Costs	Incl. O&P
1 Laborer	$34.35	$274.80	$52.70	$421.60	$33.88	$51.73
1 Truck Driver (light)	33.40	267.20	50.75	406.00		
1 Pickup Truck, 3/4 Ton		104.60		115.06	6.54	7.19
16 L.H., Daily Totals		$646.60		$942.66	$40.41	$58.92

Crew B-37E	Hr.	Daily	Hr.	Daily	Bare Costs	Incl. O&P
3 Laborers	$34.35	$824.40	$52.70	$1264.80	$36.97	$56.07
1 Equip. Oper. (light)	43.60	348.80	65.15	521.20		
1 Equip. Oper. (medium)	45.35	362.80	67.75	542.00		
2 Truck Drivers (light)	33.40	534.40	50.75	812.00		
4 Barrel w/ Flasher		13.60		14.96		
1 Concrete Saw		134.20		147.62		
1 Rotary Hammer Drill		22.55		24.81		
1 Hammer Drill Bit		3.90		4.29		
1 Loader, Skid Steer, 30 H.P., gas		151.80		166.98		
1 Conc. Hammer Attach.		115.35		126.89		
1 Vibrating plate, gas, 18"		30.80		33.88		
2 Flatbed Truck, Gas, 1.5 Ton		342.80		377.08	14.55	16.01
56 L.H., Daily Totals		$2885.40		$4036.50	$51.52	$72.08

Crew B-37F	Hr.	Daily	Hr.	Daily	Bare Costs	Incl. O&P
3 Laborers	$34.35	$824.40	$52.70	$1264.80	$34.11	$52.21
1 Truck Driver (light)	33.40	267.20	50.75	406.00		
4 Barrel w/ Flasher		13.60		14.96		
1 Concrete Mixer, 10 C.F.		151.20		166.32		
1 Air Compressor, 60 cfm		110.20		121.22		
1 -50' Air Hose, 3/4"		3.05		3.36		
1 Spade (chipper)		8.00		8.80		
1 Flatbed Truck, Gas, 1.5 Ton		171.40		188.54	14.30	15.72
32 L.H., Daily Totals		$1549.05		$2173.99	$48.41	$67.94

Crew B-38	Hr.	Daily	Hr.	Daily	Bare Costs	Incl. O&P
1 Labor Foreman (outside)	$36.35	$290.80	$55.75	$446.00	$38.80	$58.81
2 Laborers	34.35	549.60	52.70	843.20		
1 Equip. Oper. (light)	43.60	348.80	65.15	521.20		
1 Equip. Oper. (med.)	45.35	362.80	67.75	542.00		
1 Backhoe Loader, 48 H.P.		320.40		352.44		
1 Hyd.Hammer, (1200 lb.)		158.00		173.80		
1 F.E. Loader, W.M., 4 C.Y.		598.60		658.46		
1 Pvmt. Rem. Bucket		59.60		65.56	28.41	31.26
40 L.H., Daily Totals		$2688.60		$3602.66	$67.22	$90.07

Crew B-39	Hr.	Daily	Hr.	Daily	Bare Costs	Incl. O&P
1 Labor Foreman (outside)	$36.35	$290.80	$55.75	$446.00	$36.23	$55.28
4 Laborers	34.35	1099.20	52.70	1686.40		
1 Equip. Oper. (light)	43.60	348.80	65.15	521.20		
1 Air Compressor, 250 cfm		162.80		179.08		
2 Breakers, Pavement, 60 lb.		19.60		21.56		
2 -50' Air Hoses, 1.5"		12.60		13.86	4.06	4.47
48 L.H., Daily Totals		$1933.80		$2868.10	$40.29	$59.75

Crew B-40	Hr.	Daily	Hr.	Daily	Bare Costs	Incl. O&P
1 Pile Driver Foreman (out)	$44.00	$352.00	$70.55	$564.40	$43.16	$67.38
4 Pile Drivers	42.00	1344.00	67.35	2155.20		
2 Equip. Oper. (crane)	46.50	744.00	69.45	1111.20		
1 Equip. Oper. Oiler	40.30	322.40	60.20	481.60		
1 Crawler Crane, 40 Ton		1128.00		1240.80		
1 Vibratory Hammer & Gen.		2140.00		2354.00	51.06	56.17
64 L.H., Daily Totals		$6030.40		$7907.20	$94.22	$123.55

Crew B-40B	Hr.	Daily	Hr.	Daily	Bare Costs	Incl. O&P
1 Laborer Foreman	$36.35	$290.80	$55.75	$446.00	$37.70	$57.25
3 Laborers	34.35	824.40	52.70	1264.80		
1 Equip. Oper. (crane)	46.50	372.00	69.45	555.60		
1 Equip. Oper. Oiler	40.30	322.40	60.20	481.60		
1 Lattice Boom Crane, 40 Ton		1263.00		1389.30	26.31	28.94
48 L.H., Daily Totals		$3072.60		$4137.30	$64.01	$86.19

Crew B-41	Hr.	Daily	Hr.	Daily	Bare Costs	Incl. O&P
1 Labor Foreman (outside)	$36.35	$290.80	$55.75	$446.00	$35.54	$54.36
4 Laborers	34.35	1099.20	52.70	1686.40		
.25 Equip. Oper. (crane)	46.50	93.00	69.45	138.90		
.25 Equip. Oper. Oiler	40.30	80.60	60.20	120.40		
.25 Crawler Crane, 40 Ton		282.00		310.20	6.41	7.05
44 L.H., Daily Totals		$1845.60		$2701.90	$41.95	$61.41

Crew B-42	Hr.	Daily	Hr.	Daily	Bare Costs	Incl. O&P
1 Labor Foreman (outside)	$36.35	$290.80	$55.75	$446.00	$38.64	$60.24
4 Laborers	34.35	1099.20	52.70	1686.40		
1 Equip. Oper. (crane)	46.50	372.00	69.45	555.60		
1 Equip. Oper. Oiler	40.30	322.40	60.20	481.60		
1 Welder	48.55	388.40	85.75	686.00		
1 Hyd. Crane, 25 Ton		748.60		823.46		
1 Welder, gas engine, 300 amp		109.00		119.90		
1 Horz. Boring Csg. Mch.		442.20		486.42	20.31	22.34
64 L.H., Daily Totals		$3772.60		$5285.38	$58.95	$82.58

Crews

Crew No.	Hr.	Daily	Hr.	Daily	Bare Costs	Incl. O&P
Crew B-43						
1 Labor Foreman (outside)	$36.35	$290.80	$55.75	$446.00	$37.70	$57.25
3 Laborers	34.35	824.40	52.70	1264.80		
1 Equip. Oper. (crane)	46.50	372.00	69.45	555.60		
1 Equip. Oper. Oiler	40.30	322.40	60.20	481.60		
1 Drill Rig, Truck-Mounted		2486.00		2734.60	51.79	56.97
48 L.H., Daily Totals		$4295.60		$5482.60	$89.49	$114.22
Crew B-44						
1 Pile Driver Foreman	$44.00	$352.00	$70.55	$564.40	$42.42	$66.44
4 Pile Drivers	42.00	1344.00	67.35	2155.20		
2 Equip. Oper. (crane)	46.50	744.00	69.45	1111.20		
1 Laborer	34.35	274.80	52.70	421.60		
1 Crawler Crane, 40 Ton		1128.00		1240.80		
1 Lead, 60' high		74.00		81.40		
1 Hammer, diesel, 15K ft.-lbs.		614.20		675.62	28.38	31.22
64 L.H., Daily Totals		$4531.00		$6250.22	$70.80	$97.66
Crew B-45						
1 Equip. Oper. (med.)	$45.35	$362.80	$67.75	$542.00	$39.88	$60.00
1 Truck Driver (heavy)	34.40	275.20	52.25	418.00		
1 Dist. Tanker, 3000 Gallon		309.00		339.90		
1 Truck Tractor, 6x4, 380 H.P.		459.40		505.34	48.02	52.83
16 L.H., Daily Totals		$1406.40		$1805.24	$87.90	$112.83
Crew B-46						
1 Pile Driver Foreman	$44.00	$352.00	$70.55	$564.40	$38.51	$60.56
2 Pile Drivers	42.00	672.00	67.35	1077.60		
3 Laborers	34.35	824.40	52.70	1264.80		
1 Chain Saw, gas, 36" Long		41.60		45.76	.87	.95
48 L.H., Daily Totals		$1890.00		$2952.56	$39.38	$61.51
Crew B-47						
1 Blast Foreman	$36.35	$290.80	$55.75	$446.00	$38.10	$57.87
1 Driller	34.35	274.80	52.70	421.60		
1 Equip. Oper. (light)	43.60	348.80	65.15	521.20		
1 Air Track Drill, 4"		849.60		934.56		
1 Air Compressor, 600 cfm		424.60		467.06		
2 -50' Air Hoses, 3"		29.80		32.78	54.33	59.77
24 L.H., Daily Totals		$2218.40		$2823.20	$92.43	$117.63
Crew B-47A						
1 Drilling Foreman	$36.35	$290.80	$55.75	$446.00	$41.05	$61.80
1 Equip. Oper. (heavy)	46.50	372.00	69.45	555.60		
1 Oiler	40.30	322.40	60.20	481.60		
1 Air Track Drill, 5"		965.20		1061.72	40.22	44.24
24 L.H., Daily Totals		$1950.40		$2544.92	$81.27	$106.04
Crew B-47C						
1 Laborer	$34.35	$274.80	$52.70	$421.60	$38.98	$58.92
1 Equip. Oper. (light)	43.60	348.80	65.15	521.20		
1 Air Compressor, 750 cfm		431.80		474.98		
2 -50' Air Hoses, 3"		29.80		32.78		
1 Air Track Drill, 4"		849.60		934.56	81.95	90.14
16 L.H., Daily Totals		$1934.80		$2385.12	$120.93	$149.07
Crew B-47E						
1 Laborer Foreman	$36.35	$290.80	$55.75	$446.00	$34.85	$53.46
3 Laborers	34.35	824.40	52.70	1264.80		
1 Flatbed Truck, Gas, 3 Ton		217.00		238.70	6.78	7.46
32 L.H., Daily Totals		$1332.20		$1949.50	$41.63	$60.92
Crew B-47G						
1 Laborer Foreman	$36.35	$290.80	$55.75	$446.00	$37.16	$56.58
2 Laborers	34.35	549.60	52.70	843.20		
1 Equip. Oper. (light)	43.60	348.80	65.15	521.20		
1 Air Track Drill, 4"		849.60		934.56		
1 Air Compressor, 600 cfm		424.60		467.06		
2 -50' Air Hoses, 3"		29.80		32.78		
1 Gunnite Pump Rig		178.00		195.80	46.31	50.94
32 L.H., Daily Totals		$2671.20		$3440.60	$83.47	$107.52
Crew B-47H						
1 Skilled Worker Foreman	$46.40	$371.20	$71.20	$569.60	$44.90	$68.88
3 Skilled Workers	44.40	1065.60	68.10	1634.40		
1 Flatbed Truck, Gas, 3 Ton		217.00		238.70	6.78	7.46
32 L.H., Daily Totals		$1653.80		$2442.70	$51.68	$76.33
Crew B-48						
1 Labor Foreman (outside)	$36.35	$290.80	$55.75	$446.00	$38.54	$58.38
3 Laborers	34.35	824.40	52.70	1264.80		
1 Equip. Oper. (crane)	46.50	372.00	69.45	555.60		
1 Equip. Oper. Oiler	40.30	322.40	60.20	481.60		
1 Equip. Oper. (light)	43.60	348.80	65.15	521.20		
1 Centr. Water Pump, 6"		291.60		320.76		
1 -20' Suction Hose, 6"		11.90		13.09		
1 -50' Discharge Hose, 6"		6.30		6.93		
1 Drill Rig, Truck-Mounted		2486.00		2734.60	49.92	54.92
56 L.H., Daily Totals		$4954.20		$6344.58	$88.47	$113.30
Crew B-49						
1 Labor Foreman (outside)	$36.35	$290.80	$55.75	$446.00	$40.05	$61.18
3 Laborers	34.35	824.40	52.70	1264.80		
2 Equip. Oper. (crane)	46.50	744.00	69.45	1111.20		
2 Equip. Oper. Oilers	40.30	644.80	60.20	963.20		
1 Equip. Oper. (light)	43.60	348.80	65.15	521.20		
2 Pile Drivers	42.00	672.00	67.35	1077.60		
1 Hyd. Crane, 25 Ton		748.60		823.46		
1 Centr. Water Pump, 6"		291.60		320.76		
1 -20' Suction Hose, 6"		11.90		13.09		
1 -50' Discharge Hose, 6"		6.30		6.93		
1 Drill Rig, Truck-Mounted		2486.00		2734.60	40.28	44.31
88 L.H., Daily Totals		$7069.20		$9282.84	$80.33	$105.49
Crew B-50						
2 Pile Driver Foremen	$44.00	$704.00	$70.55	$1128.80	$41.17	$64.46
6 Pile Drivers	42.00	2016.00	67.35	3232.80		
2 Equip. Oper. (crane)	46.50	744.00	69.45	1111.20		
1 Equip. Oper. Oiler	40.30	322.40	60.20	481.60		
3 Laborers	34.35	824.40	52.70	1264.80		
1 Crawler Crane, 40 Ton		1128.00		1240.80		
1 Lead, 60' high		74.00		81.40		
1 Hammer, diesel, 15K ft.-lbs.		614.20		675.62		
1 Air Compressor, 600 cfm		424.60		467.06		
2 -50' Air Hoses, 3"		29.80		32.78		
1 Chain Saw, gas, 36" Long		41.60		45.76	20.64	22.71
112 L.H., Daily Totals		$6923.00		$9762.62	$61.81	$87.17
Crew B-51						
1 Labor Foreman (outside)	$36.35	$290.80	$55.75	$446.00	$34.52	$52.88
4 Laborers	34.35	1099.20	52.70	1686.40		
1 Truck Driver (light)	33.40	267.20	50.75	406.00		
1 Flatbed Truck, Gas, 1.5 Ton		171.40		188.54	3.57	3.93
48 L.H., Daily Totals		$1828.60		$2726.94	$38.10	$56.81

Crew No.	Bare Costs		Incl. Subs O&P		Cost Per Labor-Hour	
Crew B-52	Hr.	Daily	Hr.	Daily	Bare Costs	Incl. O&P
1 Carpenter Foreman	$45.05	$360.40	$69.10	$552.80	$39.84	$60.76
1 Carpenter	43.05	344.40	66.05	528.40		
3 Laborers	34.35	824.40	52.70	1264.80		
1 Cement Finisher	40.85	326.80	59.50	476.00		
.5 Rodman (reinf.)	48.40	193.60	77.35	309.40		
.5 Equip. Oper. (med.)	45.35	181.40	67.75	271.00		
.5 Crawler Loader, 3 C.Y.		504.00		554.40	9.00	9.90
56 L.H., Daily Totals		$2735.00		$3956.80	$48.84	$70.66

Crew No.					Bare Costs	Incl. O&P
Crew B-53	Hr.	Daily	Hr.	Daily		
1 Equip. Oper. (light)	$43.60	$348.80	$65.15	$521.20	$43.60	$65.15
1 Trencher, Chain, 12 H.P.		57.40		63.14	7.17	7.89
8 L.H., Daily Totals		$406.20		$584.34	$50.77	$73.04

Crew No.					Bare Costs	Incl. O&P
Crew B-54	Hr.	Daily	Hr.	Daily		
1 Equip. Oper. (light)	$43.60	$348.80	$65.15	$521.20	$43.60	$65.15
1 Trencher, Chain, 40 H.P.		297.00		326.70	37.13	40.84
8 L.H., Daily Totals		$645.80		$847.90	$80.72	$105.99

Crew No.					Bare Costs	Incl. O&P
Crew B-54A	Hr.	Daily	Hr.	Daily		
.17 Labor Foreman (outside)	$36.35	$49.44	$55.75	$75.82	$44.04	$66.01
1 Equipment Operator (med.)	45.35	362.80	67.75	542.00		
1 Wheel Trencher, 67 H.P.		1001.00		1101.10	106.94	117.64
9.36 L.H., Daily Totals		$1413.24		$1718.92	$150.99	$183.65

Crew No.					Bare Costs	Incl. O&P
Crew B-54B	Hr.	Daily	Hr.	Daily		
.25 Labor Foreman (outside)	$36.35	$72.70	$55.75	$111.50	$43.55	$65.35
1 Equipment Operator (med.)	45.35	362.80	67.75	542.00		
1 Wheel Trencher, 150 H.P.		1802.00		1982.20	180.20	198.22
10 L.H., Daily Totals		$2237.50		$2635.70	$223.75	$263.57

Crew No.					Bare Costs	Incl. O&P
Crew B-54C	Hr.	Daily	Hr.	Daily		
1 Laborer	$34.35	$274.80	$52.70	$421.60	$39.85	$60.23
1 Equipment Operator (med.)	45.35	362.80	67.75	542.00		
1 Wheel Trencher, 67 H.P.		1001.00		1101.10	62.56	68.82
16 L.H., Daily Totals		$1638.60		$2064.70	$102.41	$129.04

Crew No.					Bare Costs	Incl. O&P
Crew B-54D	Hr.	Daily	Hr.	Daily		
1 Laborer	$34.35	$274.80	$52.70	$421.60	$39.85	$60.23
1 Equipment Operator (med.)	45.35	362.80	67.75	542.00		
1 Rock trencher, 6" width		332.80		366.08	20.80	22.88
16 L.H., Daily Totals		$970.40		$1329.68	$60.65	$83.11

Crew No.					Bare Costs	Incl. O&P
Crew B-54E	Hr.	Daily	Hr.	Daily		
1 Laborer	$34.35	$274.80	$52.70	$421.60	$39.85	$60.23
1 Equipment Operator (med.)	45.35	362.80	67.75	542.00		
1 Rock trencher, 18" width		2538.00		2791.80	158.63	174.49
16 L.H., Daily Totals		$3175.60		$3755.40	$198.47	$234.71

Crew No.					Bare Costs	Incl. O&P
Crew B-55	Hr.	Daily	Hr.	Daily		
2 Laborers	$34.35	$549.60	$52.70	$843.20	$34.03	$52.05
1 Truck Driver (light)	33.40	267.20	50.75	406.00		
1 Truck-mounted earth auger		706.60		777.26		
1 Flatbed Truck, Gas, 3 Ton		217.00		238.70	38.48	42.33
24 L.H., Daily Totals		$1740.40		$2265.16	$72.52	$94.38

Crew No.	Bare Costs		Incl. Subs O&P		Cost Per Labor-Hour	
Crew B-56	Hr.	Daily	Hr.	Daily	Bare Costs	Incl. O&P
1 Laborer	$34.35	$274.80	$52.70	$421.60	$38.98	$58.92
1 Equip. Oper. (light)	43.60	348.80	65.15	521.20		
1 Air Track Drill, 4"		849.60		934.56		
1 Air Compressor, 600 cfm		424.60		467.06		
1 -50' Air Hose, 3"		14.90		16.39	80.57	88.63
16 L.H., Daily Totals		$1912.70		$2360.81	$119.54	$147.55

Crew No.					Bare Costs	Incl. O&P
Crew B-57	Hr.	Daily	Hr.	Daily		
1 Labor Foreman (outside)	$36.35	$290.80	$55.75	$446.00	$39.24	$59.33
2 Laborers	34.35	549.60	52.70	843.20		
1 Equip. Oper. (crane)	46.50	372.00	69.45	555.60		
1 Equip. Oper. (light)	43.60	348.80	65.15	521.20		
1 Equip. Oper. Oiler	40.30	322.40	60.20	481.60		
1 Crawler Crane, 25 Ton		1123.00		1235.30		
1 Clamshell Bucket, 1 C.Y.		48.00		52.80		
1 Centr. Water Pump, 6"		291.60		320.76		
1 -20' Suction Hose, 6"		11.90		13.09		
20 -50' Discharge Hoses, 6"		126.00		138.60	33.34	36.68
48 L.H., Daily Totals		$3484.10		$4608.15	$72.59	$96.00

Crew No.					Bare Costs	Incl. O&P
Crew B-58	Hr.	Daily	Hr.	Daily		
2 Laborers	$34.35	$549.60	$52.70	$843.20	$37.43	$56.85
1 Equip. Oper. (light)	43.60	348.80	65.15	521.20		
1 Backhoe Loader, 48 H.P.		320.40		352.44		
1 Small Helicopter, w/pilot		2579.00		2836.90	120.81	132.89
24 L.H., Daily Totals		$3797.80		$4553.74	$158.24	$189.74

Crew No.					Bare Costs	Incl. O&P
Crew B-59	Hr.	Daily	Hr.	Daily		
1 Truck Driver (heavy)	$34.40	$275.20	$52.25	$418.00	$34.40	$52.25
1 Truck Tractor, 220 H.P.		279.20		307.12		
1 Water Tank Trailer, 5000 Gal.		146.20		160.82	53.17	58.49
8 L.H., Daily Totals		$700.60		$885.94	$87.58	$110.74

Crew No.					Bare Costs	Incl. O&P
Crew B-59A	Hr.	Daily	Hr.	Daily		
2 Laborers	$34.35	$549.60	$52.70	$843.20	$34.37	$52.55
1 Truck Driver (heavy)	34.40	275.20	52.25	418.00		
1 Water Tank Trailer, 5000 Gal.		146.20		160.82		
1 Truck Tractor, 220 H.P.		279.20		307.12	17.73	19.50
24 L.H., Daily Totals		$1250.20		$1729.14	$52.09	$72.05

Crew No.					Bare Costs	Incl. O&P
Crew B-60	Hr.	Daily	Hr.	Daily		
1 Labor Foreman (outside)	$36.35	$290.80	$55.75	$446.00	$39.86	$60.16
2 Laborers	34.35	549.60	52.70	843.20		
1 Equip. Oper. (crane)	46.50	372.00	69.45	555.60		
2 Equip. Oper. (light)	43.60	697.60	65.15	1042.40		
1 Equip. Oper. Oiler	40.30	322.40	60.20	481.60		
1 Crawler Crane, 40 Ton		1128.00		1240.80		
1 Lead, 60' high		74.00		81.40		
1 Hammer, diesel, 15K ft.-lbs.		614.20		675.62		
1 Backhoe Loader, 48 H.P.		320.40		352.44	38.15	41.97
56 L.H., Daily Totals		$4369.00		$5719.06	$78.02	$102.13

Crew No.					Bare Costs	Incl. O&P
Crew B-61	Hr.	Daily	Hr.	Daily		
1 Labor Foreman (outside)	$36.35	$290.80	$55.75	$446.00	$36.60	$55.80
3 Laborers	34.35	824.40	52.70	1264.80		
1 Equip. Oper. (light)	43.60	348.80	65.15	521.20		
1 Cement Mixer, 2 C.Y.		194.60		214.06		
1 Air Compressor, 160 cfm		125.00		137.50	7.99	8.79
40 L.H., Daily Totals		$1783.60		$2583.56	$44.59	$64.59

Crews

Crew B-62

Crew B-62	Hr.	Daily	Hr.	Daily	Bare Costs	Incl. O&P
2 Laborers	$34.35	$549.60	$52.70	$843.20	$37.43	$56.85
1 Equip. Oper. (light)	43.60	348.80	65.15	521.20		
1 Loader, Skid Steer, 30 H.P., gas		151.80		166.98	6.33	6.96
24 L.H., Daily Totals		$1050.20		$1531.38	$43.76	$63.81

Crew B-63

Crew B-63	Hr.	Daily	Hr.	Daily	Bare Costs	Incl. O&P
4 Laborers	$34.35	$1099.20	$52.70	$1686.40	$36.20	$55.19
1 Equip. Oper. (light)	43.60	348.80	65.15	521.20		
1 Loader, Skid Steer, 30 H.P., gas		151.80		166.98	3.79	4.17
40 L.H., Daily Totals		$1599.80		$2374.58	$39.99	$59.36

Crew B-64

Crew B-64	Hr.	Daily	Hr.	Daily	Bare Costs	Incl. O&P
1 Laborer	$34.35	$274.80	$52.70	$421.60	$33.88	$51.73
1 Truck Driver (light)	33.40	267.20	50.75	406.00		
1 Power Mulcher (small)		130.80		143.88		
1 Flatbed Truck, Gas, 1.5 Ton		171.40		188.54	18.89	20.78
16 L.H., Daily Totals		$844.20		$1160.02	$52.76	$72.50

Crew B-65

Crew B-65	Hr.	Daily	Hr.	Daily	Bare Costs	Incl. O&P
1 Laborer	$34.35	$274.80	$52.70	$421.60	$33.88	$51.73
1 Truck Driver (light)	33.40	267.20	50.75	406.00		
1 Power Mulcher (large)		258.80		284.68		
1 Flatbed Truck, Gas, 1.5 Ton		171.40		188.54	26.89	29.58
16 L.H., Daily Totals		$972.20		$1300.82	$60.76	$81.30

Crew B-66

Crew B-66	Hr.	Daily	Hr.	Daily	Bare Costs	Incl. O&P
1 Equip. Oper. (light)	$43.60	$348.80	$65.15	$521.20	$43.60	$65.15
1 Loader-Backhoe		223.60		245.96	27.95	30.75
8 L.H., Daily Totals		$572.40		$767.16	$71.55	$95.89

Crew B-67

Crew B-67	Hr.	Daily	Hr.	Daily	Bare Costs	Incl. O&P
1 Millwright	$45.35	$362.80	$66.05	$528.40	$44.48	$65.60
1 Equip. Oper. (light)	43.60	348.80	65.15	521.20		
1 Forklift, R/T, 4,000 Lb.		278.60		306.46	17.41	19.15
16 L.H., Daily Totals		$990.20		$1356.06	$61.89	$84.75

Crew B-68

Crew B-68	Hr.	Daily	Hr.	Daily	Bare Costs	Incl. O&P
2 Millwrights	$45.35	$725.60	$66.05	$1056.80	$44.77	$65.75
1 Equip. Oper. (light)	43.60	348.80	65.15	521.20		
1 Forklift, R/T, 4,000 Lb.		278.60		306.46	11.61	12.77
24 L.H., Daily Totals		$1353.00		$1884.46	$56.38	$78.52

Crew B-69

Crew B-69	Hr.	Daily	Hr.	Daily	Bare Costs	Incl. O&P
1 Labor Foreman (outside)	$36.35	$290.80	$55.75	$446.00	$37.70	$57.25
3 Laborers	34.35	824.40	52.70	1264.80		
1 Equip Oper. (crane)	46.50	372.00	69.45	555.60		
1 Equip Oper. Oiler	40.30	322.40	60.20	481.60		
1 Hyd. Crane, 80 Ton		1693.00		1862.30	35.27	38.80
48 L.H., Daily Totals		$3502.60		$4610.30	$72.97	$96.05

Crew B-69A

Crew B-69A	Hr.	Daily	Hr.	Daily	Bare Costs	Incl. O&P
1 Labor Foreman	$36.35	$290.80	$55.75	$446.00	$37.60	$56.85
3 Laborers	34.35	824.40	52.70	1264.80		
1 Equip. Oper. (med.)	45.35	362.80	67.75	542.00		
1 Concrete Finisher	40.85	326.80	59.50	476.00		
1 Curb/Gutter Paver, 2-Track		744.80		819.28	15.52	17.07
48 L.H., Daily Totals		$2549.60		$3548.08	$53.12	$73.92

Crew B-69B

Crew B-69B	Hr.	Daily	Hr.	Daily	Bare Costs	Incl. O&P
1 Labor Foreman	$36.35	$290.80	$55.75	$446.00	$37.60	$56.85
3 Laborers	34.35	824.40	52.70	1264.80		
1 Equip. Oper. (med.)	45.35	362.80	67.75	542.00		
1 Cement Finisher	40.85	326.80	59.50	476.00		
1 Curb/Gutter Paver, 4-Track		848.20		933.02	17.67	19.44
48 L.H., Daily Totals		$2653.00		$3661.82	$55.27	$76.29

Crew B-70

Crew B-70	Hr.	Daily	Hr.	Daily	Bare Costs	Incl. O&P
1 Labor Foreman (outside)	$36.35	$290.80	$55.75	$446.00	$39.35	$59.59
3 Laborers	34.35	824.40	52.70	1264.80		
3 Equip. Oper. (med.)	45.35	1088.40	67.75	1626.00		
1 Grader, 30,000 Lbs.		542.80		597.08		
1 Ripper, beam & 1 shank		83.20		91.52		
1 Road Sweeper, S.P., 8' wide		604.80		665.28		
1 F.E. Loader, W.M. 1.5 C.Y.		319.40		351.34	27.68	30.45
56 L.H., Daily Totals		$3753.80		$5042.02	$67.03	$90.04

Crew B-70A

Crew B-70A	Hr.	Daily	Hr.	Daily	Bare Costs	Incl. O&P
1 Laborer	$34.35	$274.80	$52.70	$421.60	$43.15	$64.74
4 Equip. Oper. (med.)	45.35	1451.20	67.75	2168.00		
1 Grader, 40,000 Lbs.		975.40		1072.94		
1 F.E. Loader, W.M.,2.5 C.Y.		397.60		437.36		
1 Dozer, 80 H.P.		412.60		453.86		
1 Roller, Pneum. Whl, 12 Ton		309.20		340.12	52.37	57.61
40 L.H., Daily Totals		$3820.80		$4893.88	$95.52	$122.35

Crew B-71

Crew B-71	Hr.	Daily	Hr.	Daily	Bare Costs	Incl. O&P
1 Labor Foreman (outside)	$36.35	$290.80	$55.75	$446.00	$39.35	$59.59
3 Laborers	34.35	824.40	52.70	1264.80		
3 Equip. Oper. (med.)	45.35	1088.40	67.75	1626.00		
1 Pvmt. Profiler, 750 H.P.		5626.00		6188.60		
1 Road Sweeper, S.P., 8' wide		604.80		665.28		
1 F.E. Loader, W.M., 1.5 C.Y.		319.40		351.34	116.97	128.66
56 L.H., Daily Totals		$8753.80		$10542.02	$156.32	$188.25

Crew B-72

Crew B-72	Hr.	Daily	Hr.	Daily	Bare Costs	Incl. O&P
1 Labor Foreman (outside)	$36.35	$290.80	$55.75	$446.00	$40.10	$60.61
3 Laborers	34.35	824.40	52.70	1264.80		
4 Equip. Oper. (med.)	45.35	1451.20	67.75	2168.00		
1 Pvmt. Profiler, 750 H.P.		5626.00		6188.60		
1 Hammermill, 250 H.P.		1692.00		1861.20		
1 Windrow Loader		1210.00		1331.00		
1 Mix Paver 165 H.P.		2016.00		2217.60		
1 Roller, Pneum. Whl, 12 Ton		309.20		340.12	169.58	186.54
64 L.H., Daily Totals		$13419.60		$15817.32	$209.68	$247.15

Crew B-73

Crew B-73	Hr.	Daily	Hr.	Daily	Bare Costs	Incl. O&P
1 Labor Foreman (outside)	$36.35	$290.80	$55.75	$446.00	$41.48	$62.49
2 Laborers	34.35	549.60	52.70	843.20		
5 Equip. Oper. (med.)	45.35	1814.00	67.75	2710.00		
1 Road Mixer, 310 H.P.		1890.00		2079.00		
1 Tandem Roller, 10 Ton		224.20		246.62		
1 Hammermill, 250 H.P.		1692.00		1861.20		
1 Grader, 30,000 Lbs.		542.80		597.08		
.5 F.E. Loader, W.M., 1.5 C.Y.		159.70		175.67		
.5 Truck Tractor, 220 H.P.		139.60		153.56		
.5 Water Tank Trailer, 5000 Gal.		73.10		80.41	73.77	81.15
64 L.H., Daily Totals		$7375.80		$9192.74	$115.25	$143.64

| Crew No. | Bare Costs | | Incl. Subs O&P | | Cost Per Labor-Hour | |

Crew B-74

	Hr.	Daily	Hr.	Daily	Bare Costs	Incl. O&P
1 Labor Foreman (outside)	$36.35	$290.80	$55.75	$446.00	$40.11	$60.49
1 Laborer	34.35	274.80	52.70	421.60		
4 Equip. Oper. (med.)	45.35	1451.20	67.75	2168.00		
2 Truck Drivers (heavy)	34.40	550.40	52.25	836.00		
1 Grader, 30,000 Lbs.		542.80		597.08		
1 Ripper, beam & 1 shank		83.20		91.52		
2 Stabilizers, 310 H.P.		2652.00		2917.20		
1 Flatbed Truck, Gas, 3 Ton		217.00		238.70		
1 Chem. Spreader, Towed		49.00		53.90		
1 Roller, Vibratory, 25 Ton		596.80		656.48		
1 Water Tank Trailer, 5000 Gal.		146.20		160.82		
1 Truck Tractor, 220 H.P.		279.20		307.12	71.35	78.48
64 L.H., Daily Totals		$7133.40		$8894.42	$111.46	$138.98

Crew B-75

	Hr.	Daily	Hr.	Daily	Bare Costs	Incl. O&P
1 Labor Foreman (outside)	$36.35	$290.80	$55.75	$446.00	$40.93	$61.67
1 Laborer	34.35	274.80	52.70	421.60		
4 Equip. Oper. (med.)	45.35	1451.20	67.75	2168.00		
1 Truck Driver (heavy)	34.40	275.20	52.25	418.00		
1 Grader, 30,000 Lbs.		542.80		597.08		
1 Ripper, beam & 1 shank		83.20		91.52		
2 Stabilizers, 310 H.P.		2652.00		2917.20		
1 Dist. Tanker, 3000 Gallon		309.00		339.90		
1 Truck Tractor, 6x4, 380 H.P.		459.40		505.34		
1 Roller, Vibratory, 25 Ton		596.80		656.48	82.91	91.21
56 L.H., Daily Totals		$6935.20		$8561.12	$123.84	$152.88

Crew B-76

	Hr.	Daily	Hr.	Daily	Bare Costs	Incl. O&P
1 Dock Builder Foreman	$44.00	$352.00	$70.55	$564.40	$43.03	$67.38
5 Dock Builders	42.00	1680.00	67.35	2694.00		
2 Equip. Oper. (crane)	46.50	744.00	69.45	1111.20		
1 Equip. Oper. Oiler	40.30	322.40	60.20	481.60		
1 Crawler Crane, 50 Ton		1279.00		1406.90		
1 Barge, 400 Ton		820.40		902.44		
1 Hammer, diesel, 15K ft.-lbs.		614.20		675.62		
1 Lead, 60' high		74.00		81.40		
1 Air Compressor, 600 cfm		424.60		467.06		
2 -50' Air Hoses, 3"		29.80		32.78	45.03	49.53
72 L.H., Daily Totals		$6340.40		$8417.40	$88.06	$116.91

Crew B-76A

	Hr.	Daily	Hr.	Daily	Bare Costs	Incl. O&P
1 Laborer Foreman	$36.35	$290.80	$55.75	$446.00	$36.86	$56.11
5 Laborers	34.35	1374.00	52.70	2108.00		
1 Equip. Oper. (crane)	46.50	372.00	69.45	555.60		
1 Equip. Oper. Oiler	40.30	322.40	60.20	481.60		
1 Crawler Crane, 50 Ton		1279.00		1406.90		
1 Barge, 400 Ton		820.40		902.44	32.80	36.08
64 L.H., Daily Totals		$4458.60		$5900.54	$69.67	$92.20

Crew B-77

	Hr.	Daily	Hr.	Daily	Bare Costs	Incl. O&P
1 Labor Foreman	$36.35	$290.80	$55.75	$446.00	$34.56	$52.92
3 Laborers	34.35	824.40	52.70	1264.80		
1 Truck Driver (light)	33.40	267.20	50.75	406.00		
1 Crack Cleaner, 25 H.P.		53.00		58.30		
1 Crack Filler, Trailer Mtd.		194.60		214.06		
1 Flatbed Truck, Gas, 3 Ton		217.00		238.70	11.62	12.78
40 L.H., Daily Totals		$1847.00		$2627.86	$46.17	$65.70

Crew B-78

	Hr.	Daily	Hr.	Daily	Bare Costs	Incl. O&P
1 Labor Foreman	$36.35	$290.80	$55.75	$446.00	$34.52	$52.88
4 Laborers	34.35	1099.20	52.70	1686.40		
1 Truck Driver (light)	33.40	267.20	50.75	406.00		
1 Paint Striper, S.P.		147.80		162.58		
1 Flatbed Truck, Gas, 3 Ton		217.00		238.70		
1 Pickup Truck, 3/4 Ton		104.60		115.06	9.78	10.76
48 L.H., Daily Totals		$2126.60		$3054.74	$44.30	$63.64

Crew B-78A

	Hr.	Daily	Hr.	Daily	Bare Costs	Incl. O&P
1 Equip. Oper. (light)	$43.60	$348.80	$65.15	$521.20	$43.60	$65.15
1 Line Rem. (metal balls) 115 H.P.		837.20		920.92	104.65	115.11
8 L.H., Daily Totals		$1186.00		$1442.12	$148.25	$180.26

Crew B-78B

	Hr.	Daily	Hr.	Daily	Bare Costs	Incl. O&P
2 Laborers	$34.35	$549.60	$52.70	$843.20	$35.38	$54.08
.25 Equip. Oper. (light)	43.60	87.20	65.15	130.30		
1 Pickup Truck, 3/4 Ton		104.60		115.06		
1 Line Rem., 11 H.P., walk behind		57.40		63.14		
.25 Road Sweeper, S.P., 8' wide		151.20		166.32	17.40	19.14
18 L.H., Daily Totals		$950.00		$1318.02	$52.78	$73.22

Crew B-79

	Hr.	Daily	Hr.	Daily	Bare Costs	Incl. O&P
1 Labor Foreman	$36.35	$290.80	$55.75	$446.00	$34.56	$52.92
3 Laborers	34.35	824.40	52.70	1264.80		
1 Truck Driver (light)	33.40	267.20	50.75	406.00		
1 Thermo. Striper, T.M.		653.80		719.18		
1 Flatbed Truck, Gas, 3 Ton		217.00		238.70		
2 Pickup Trucks, 3/4 Ton		209.20		230.12	27.00	29.70
40 L.H., Daily Totals		$2462.40		$3304.80	$61.56	$82.62

Crew B-79A

	Hr.	Daily	Hr.	Daily	Bare Costs	Incl. O&P
1.5 Equip. Oper. (light)	$43.60	$523.20	$65.15	$781.80	$43.60	$65.15
.5 Line Remov. (grinder) 115 H.P.		450.90		495.99		
1 Line Rem. (metal balls) 115 H.P.		837.20		920.92	107.34	118.08
12 L.H., Daily Totals		$1811.30		$2198.71	$150.94	$183.23

Crew B-80

	Hr.	Daily	Hr.	Daily	Bare Costs	Incl. O&P
1 Labor Foreman	$36.35	$290.80	$55.75	$446.00	$36.92	$56.09
1 Laborer	34.35	274.80	52.70	421.60		
1 Truck Driver (light)	33.40	267.20	50.75	406.00		
1 Equip. Oper. (light)	43.60	348.80	65.15	521.20		
1 Flatbed Truck, Gas, 3 Ton		217.00		238.70		
1 Earth Auger, Truck-Mtd.		401.80		441.98	19.34	21.27
32 L.H., Daily Totals		$1800.40		$2475.48	$56.26	$77.36

Crew B-80A

	Hr.	Daily	Hr.	Daily	Bare Costs	Incl. O&P
3 Laborers	$34.35	$824.40	$52.70	$1264.80	$34.35	$52.70
1 Flatbed Truck, Gas, 3 Ton		217.00		238.70	9.04	9.95
24 L.H., Daily Totals		$1041.40		$1503.50	$43.39	$62.65

Crew B-80B

	Hr.	Daily	Hr.	Daily	Bare Costs	Incl. O&P
3 Laborers	$34.35	$824.40	$52.70	$1264.80	$36.66	$55.81
1 Equip. Oper. (light)	43.60	348.80	65.15	521.20		
1 Crane, Flatbed Mounted, 3 Ton		225.60		248.16	7.05	7.75
32 L.H., Daily Totals		$1398.80		$2034.16	$43.71	$63.57

Crew No.	Bare Costs		Incl. Subs O&P		Cost Per Labor-Hour	
Crew B-80C	Hr.	Daily	Hr.	Daily	Bare Costs	Incl. O&P
2 Laborers	$34.35	$549.60	$52.70	$843.20	$34.03	$52.05
1 Truck Driver (light)	33.40	267.20	50.75	406.00		
1 Flatbed Truck, Gas, 1.5 Ton		171.40		188.54		
1 Manual fence post auger, gas		7.20		7.92	7.44	8.19
24 L.H., Daily Totals		$995.40		$1445.66	$41.48	$60.24

Crew No.	Bare Costs		Incl. Subs O&P		Cost Per Labor-Hour	
Crew B-81	Hr.	Daily	Hr.	Daily	Bare Costs	Incl. O&P
1 Laborer	$34.35	$274.80	$52.70	$421.60	$38.03	$57.57
1 Equip. Oper. (med.)	45.35	362.80	67.75	542.00		
1 Truck Driver (heavy)	34.40	275.20	52.25	418.00		
1 Hydromulcher, T.M., 3000 Gal.		278.00		305.80		
1 Truck Tractor, 220 H.P.		279.20		307.12	23.22	25.54
24 L.H., Daily Totals		$1470.00		$1994.52	$61.25	$83.11

Crew B-81A	Hr.	Daily	Hr.	Daily	Bare Costs	Incl. O&P
1 Laborer	$34.35	$274.80	$52.70	$421.60	$33.88	$51.73
1 Truck Driver (light)	33.40	267.20	50.75	406.00		
1 Hydromulcher, T.M., 600 Gal.		113.20		124.52		
1 Flatbed Truck, Gas, 3 Ton		217.00		238.70	20.64	22.70
16 L.H., Daily Totals		$872.20		$1190.82	$54.51	$74.43

Crew B-82	Hr.	Daily	Hr.	Daily	Bare Costs	Incl. O&P
1 Laborer	$34.35	$274.80	$52.70	$421.60	$38.98	$58.92
1 Equip. Oper. (light)	43.60	348.80	65.15	521.20		
1 Horiz. Borer, 6 H.P.		76.00		83.60	4.75	5.22
16 L.H., Daily Totals		$699.60		$1026.40	$43.73	$64.15

Crew B-82A	Hr.	Daily	Hr.	Daily	Bare Costs	Incl. O&P
1 Laborer	$34.35	$274.80	$52.70	$421.60	$38.98	$58.92
1 Equip. Oper. (light)	43.60	348.80	65.15	521.20		
1 Flatbed Truck, Gas, 3 Ton		217.00		238.70		
1 Flatbed Trailer, 25 Ton		111.20		122.32		
1 Horiz. Dir. Drill, 20k lb. thrust		622.60		684.86	59.42	65.37
16 L.H., Daily Totals		$1574.40		$1988.68	$98.40	$124.29

Crew B-82B	Hr.	Daily	Hr.	Daily	Bare Costs	Incl. O&P
2 Laborers	$34.35	$549.60	$52.70	$843.20	$37.43	$56.85
1 Equip. Oper. (light)	43.60	348.80	65.15	521.20		
1 Flatbed Truck, Gas, 3 Ton		217.00		238.70		
1 Flatbed Trailer, 25 Ton		111.20		122.32		
1 Horiz. Dir. Drill, 30k lb. thrust		888.40		977.24	50.69	55.76
24 L.H., Daily Totals		$2115.00		$2702.66	$88.13	$112.61

Crew B-82C	Hr.	Daily	Hr.	Daily	Bare Costs	Incl. O&P
2 Laborers	$34.35	$549.60	$52.70	$843.20	$37.43	$56.85
1 Equip. Oper. (light)	43.60	348.80	65.15	521.20		
1 Flatbed Truck, Gas, 3 Ton		217.00		238.70		
1 Flatbed Trailer, 25 Ton		111.20		122.32		
1 Horiz. Dir. Drill, 50k lb. thrust		1174.00		1291.40	62.59	68.85
24 L.H., Daily Totals		$2400.60		$3016.82	$100.03	$125.70

Crew B-82D	Hr.	Daily	Hr.	Daily	Bare Costs	Incl. O&P
1 Equip. Oper. (light)	$43.60	$348.80	$65.15	$521.20	$43.60	$65.15
1 Mud Trailer for HDD, 1500 gallon		265.80		292.38	33.23	36.55
8 L.H., Daily Totals		$614.60		$813.58	$76.83	$101.70

Crew No.	Bare Costs		Incl. Subs O&P		Cost Per Labor-Hour	
Crew B-83	Hr.	Daily	Hr.	Daily	Bare Costs	Incl. O&P
1 Tugboat Captain	$45.35	$362.80	$67.75	$542.00	$39.85	$60.23
1 Tugboat Hand	34.35	274.80	52.70	421.60		
1 Tugboat, 250 H.P.		632.20		695.42	39.51	43.46
16 L.H., Daily Totals		$1269.80		$1659.02	$79.36	$103.69

Crew B-84	Hr.	Daily	Hr.	Daily	Bare Costs	Incl. O&P
1 Equip. Oper. (med.)	$45.35	$362.80	$67.75	$542.00	$45.35	$67.75
1 Rotary Mower/Tractor		293.80		323.18	36.73	40.40
8 L.H., Daily Totals		$656.60		$865.18	$82.08	$108.15

Crew B-85	Hr.	Daily	Hr.	Daily	Bare Costs	Incl. O&P
3 Laborers	$34.35	$824.40	$52.70	$1264.80	$36.56	$55.62
1 Equip. Oper. (med.)	45.35	362.80	67.75	542.00		
1 Truck Driver (heavy)	34.40	275.20	52.25	418.00		
1 Aerial Lift Truck, 80'		636.20		699.82		
1 Brush Chipper, 12", 130 H.P.		217.00		238.70		
1 Pruning Saw, Rotary		6.30		6.93	21.49	23.64
40 L.H., Daily Totals		$2321.90		$3170.25	$58.05	$79.26

Crew B-86	Hr.	Daily	Hr.	Daily	Bare Costs	Incl. O&P
1 Equip. Oper. (med.)	$45.35	$362.80	$67.75	$542.00	$45.35	$67.75
1 Stump Chipper, S.P.		169.35		186.29	21.17	23.29
8 L.H., Daily Totals		$532.15		$728.28	$66.52	$91.04

Crew B-86A	Hr.	Daily	Hr.	Daily	Bare Costs	Incl. O&P
1 Equip. Oper. (med.)	$45.35	$362.80	$67.75	$542.00	$45.35	$67.75
1 Grader, 30,000 Lbs.		542.80		597.08	67.85	74.64
8 L.H., Daily Totals		$905.60		$1139.08	$113.20	$142.38

Crew B-86B	Hr.	Daily	Hr.	Daily	Bare Costs	Incl. O&P
1 Equip. Oper. (med.)	$45.35	$362.80	$67.75	$542.00	$45.35	$67.75
1 Dozer, 200 H.P.		1051.00		1156.10	131.38	144.51
8 L.H., Daily Totals		$1413.80		$1698.10	$176.72	$212.26

Crew B-87	Hr.	Daily	Hr.	Daily	Bare Costs	Incl. O&P
1 Laborer	$34.35	$274.80	$52.70	$421.60	$43.15	$64.74
4 Equip. Oper. (med.)	45.35	1451.20	67.75	2168.00		
2 Feller Bunchers, 100 H.P.		1128.00		1240.80		
1 Log Chipper, 22" Tree		554.80		610.28		
1 Dozer, 105 H.P.		505.40		555.94		
1 Chain Saw, gas, 36" Long		41.60		45.76	55.74	61.32
40 L.H., Daily Totals		$3955.80		$5042.38	$98.89	$126.06

Crew B-88	Hr.	Daily	Hr.	Daily	Bare Costs	Incl. O&P
1 Laborer	$34.35	$274.80	$52.70	$421.60	$43.78	$65.60
6 Equip. Oper. (med.)	45.35	2176.80	67.75	3252.00		
2 Feller Bunchers, 100 H.P.		1128.00		1240.80		
1 Log Chipper, 22" Tree		554.80		610.28		
2 Log Skidders, 50 H.P.		1688.00		1856.80		
1 Dozer, 105 H.P.		505.40		555.94		
1 Chain Saw, gas, 36" Long		41.60		45.76	69.96	76.96
56 L.H., Daily Totals		$6369.40		$7983.18	$113.74	$142.56

Crew B-89	Hr.	Daily	Hr.	Daily	Bare Costs	Incl. O&P
1 Equip. Oper. (light)	$43.60	$348.80	$65.15	$521.20	$38.50	$57.95
1 Truck Driver (light)	33.40	267.20	50.75	406.00		
1 Flatbed Truck, Gas, 3 Ton		217.00		238.70		
1 Concrete Saw		134.20		147.62		
1 Water Tank, 65 Gal.		18.45		20.30	23.10	25.41
16 L.H., Daily Totals		$985.65		$1333.82	$61.60	$83.36

Crew No.	Bare Costs		Incl. Subs O&P		Cost Per Labor-Hour	

Crew B-89A

	Hr.	Daily	Hr.	Daily	Bare Costs	Incl. O&P
1 Skilled Worker	$44.40	$355.20	$68.10	$544.80	$39.38	$60.40
1 Laborer	34.35	274.80	52.70	421.60		
1 Core Drill (large)		109.80		120.78	6.86	7.55
16 L.H., Daily Totals		$739.80		$1087.18	$46.24	$67.95

Crew B-89B

	Hr.	Daily	Hr.	Daily	Bare Costs	Incl. O&P
1 Equip. Oper. (light)	$43.60	$348.80	$65.15	$521.20	$38.50	$57.95
1 Truck Driver, Light	33.40	267.20	50.75	406.00		
1 Wall Saw, Hydraulic, 10 H.P.		89.00		97.90		
1 Generator, Diesel, 100 kW		329.00		361.90		
1 Water Tank, 65 Gal.		18.45		20.30		
1 Flatbed Truck, Gas, 3 Ton		217.00		238.70	40.84	44.92
16 L.H., Daily Totals		$1269.45		$1645.99	$79.34	$102.87

Crew B-90

	Hr.	Daily	Hr.	Daily	Bare Costs	Incl. O&P
1 Labor Foreman (outside)	$36.35	$290.80	$55.75	$446.00	$36.92	$56.08
3 Laborers	34.35	824.40	52.70	1264.80		
2 Equip. Oper. (light)	43.60	697.60	65.15	1042.40		
2 Truck Drivers (heavy)	34.40	550.40	52.25	836.00		
1 Road Mixer, 310 H.P.		1890.00		2079.00		
1 Dist. Truck, 2000 Gal.		280.80		308.88	33.92	37.31
64 L.H., Daily Totals		$4534.00		$5977.08	$70.84	$93.39

Crew B-90A

	Hr.	Daily	Hr.	Daily	Bare Costs	Incl. O&P
1 Labor Foreman	$36.35	$290.80	$55.75	$446.00	$40.92	$61.74
2 Laborers	34.35	549.60	52.70	843.20		
4 Equip. Oper. (med.)	45.35	1451.20	67.75	2168.00		
2 Graders, 30,000 Lbs.		1085.60		1194.16		
1 Tandem Roller, 10 Ton		224.20		246.62		
1 Roller, Pneum. Whl, 12 Ton		309.20		340.12	28.91	31.80
56 L.H., Daily Totals		$3910.60		$5238.10	$69.83	$93.54

Crew B-90B

	Hr.	Daily	Hr.	Daily	Bare Costs	Incl. O&P
1 Labor Foreman	$36.35	$290.80	$55.75	$446.00	$40.18	$60.73
2 Laborers	34.35	549.60	52.70	843.20		
3 Equip. Oper. (med.)	45.35	1088.40	67.75	1626.00		
1 Tandem Roller, 10 Ton		224.20		246.62		
1 Roller, Pneum. Whl, 12 Ton		309.20		340.12		
1 Road Mixer, 310 H.P.		1890.00		2079.00	50.49	55.54
48 L.H., Daily Totals		$4352.20		$5580.94	$90.67	$116.27

Crew B-91

	Hr.	Daily	Hr.	Daily	Bare Costs	Incl. O&P
1 Labor Foreman (outside)	$36.35	$290.80	$55.75	$446.00	$40.11	$60.55
2 Laborers	34.35	549.60	52.70	843.20		
4 Equip. Oper. (med.)	45.35	1451.20	67.75	2168.00		
1 Truck Driver (heavy)	34.40	275.20	52.25	418.00		
1 Dist. Tanker, 3000 Gallon		309.00		339.90		
1 Truck Tractor, 6x4, 380 H.P.		459.40		505.34		
1 Aggreg. Spreader, S.P.		748.40		823.24		
1 Roller, Pneum. Whl, 12 Ton		309.20		340.12		
1 Tandem Roller, 10 Ton		224.20		246.62	32.03	35.24
64 L.H., Daily Totals		$4617.00		$6130.42	$72.14	$95.79

Crew B-92

	Hr.	Daily	Hr.	Daily	Bare Costs	Incl. O&P
1 Labor Foreman (outside)	$36.35	$290.80	$55.75	$446.00	$34.85	$53.46
3 Laborers	34.35	824.40	52.70	1264.80		
1 Crack Cleaner, 25 H.P.		53.00		58.30		
1 Air Compressor, 60 cfm		110.20		121.22		
1 Tar Kettle, T.M.		74.65		82.11		
1 Flatbed Truck, Gas, 3 Ton		217.00		238.70	14.21	15.64
32 L.H., Daily Totals		$1570.05		$2211.14	$49.06	$69.10

Crew B-93

	Hr.	Daily	Hr.	Daily	Bare Costs	Incl. O&P
1 Equip. Oper. (med.)	$45.35	$362.80	$67.75	$542.00	$45.35	$67.75
1 Feller Buncher, 100 H.P.		564.00		620.40	70.50	77.55
8 L.H., Daily Totals		$926.80		$1162.40	$115.85	$145.30

Crew B-94A

	Hr.	Daily	Hr.	Daily	Bare Costs	Incl. O&P
1 Laborer	$34.35	$274.80	$52.70	$421.60	$34.35	$52.70
1 Diaphragm Water Pump, 2"		66.60		73.26		
1 -20' Suction Hose, 2"		1.95		2.15		
2 -50' Discharge Hoses, 2"		1.80		1.98	8.79	9.67
8 L.H., Daily Totals		$345.15		$498.99	$43.14	$62.37

Crew B-94B

	Hr.	Daily	Hr.	Daily	Bare Costs	Incl. O&P
1 Laborer	$34.35	$274.80	$52.70	$421.60	$34.35	$52.70
1 Diaphragm Water Pump, 4"		109.40		120.34		
1 -20' Suction Hose, 4"		3.25		3.58		
2 -50' Discharge Hoses, 4"		4.70		5.17	14.67	16.14
8 L.H., Daily Totals		$392.15		$550.68	$49.02	$68.84

Crew B-94C

	Hr.	Daily	Hr.	Daily	Bare Costs	Incl. O&P
1 Laborer	$34.35	$274.80	$52.70	$421.60	$34.35	$52.70
1 Centrifugal Water Pump, 3"		71.00		78.10		
1 -20' Suction Hose, 3"		3.05		3.36		
2 -50' Discharge Hoses, 3"		3.50		3.85	9.69	10.66
8 L.H., Daily Totals		$352.35		$506.90	$44.04	$63.36

Crew B-94D

	Hr.	Daily	Hr.	Daily	Bare Costs	Incl. O&P
1 Laborer	$34.35	$274.80	$52.70	$421.60	$34.35	$52.70
1 Centr. Water Pump, 6"		291.60		320.76		
1 -20' Suction Hose, 6"		11.90		13.09		
2 -50' Discharge Hoses, 6"		12.60		13.86	39.51	43.46
8 L.H., Daily Totals		$590.90		$769.31	$73.86	$96.16

Crew C-1

	Hr.	Daily	Hr.	Daily	Bare Costs	Incl. O&P
3 Carpenters	$43.05	$1033.20	$66.05	$1585.20	$40.88	$62.71
1 Laborer	34.35	274.80	52.70	421.60		
32 L.H., Daily Totals		$1308.00		$2006.80	$40.88	$62.71

Crew C-2

	Hr.	Daily	Hr.	Daily	Bare Costs	Incl. O&P
1 Carpenter Foreman (out)	$45.05	$360.40	$69.10	$552.80	$41.93	$64.33
4 Carpenters	43.05	1377.60	66.05	2113.60		
1 Laborer	34.35	274.80	52.70	421.60		
48 L.H., Daily Totals		$2012.80		$3088.00	$41.93	$64.33

Crew C-2A

	Hr.	Daily	Hr.	Daily	Bare Costs	Incl. O&P
1 Carpenter Foreman (out)	$45.05	$360.40	$69.10	$552.80	$41.57	$63.24
3 Carpenters	43.05	1033.20	66.05	1585.20		
1 Cement Finisher	40.85	326.80	59.50	476.00		
1 Laborer	34.35	274.80	52.70	421.60		
48 L.H., Daily Totals		$1995.20		$3035.60	$41.57	$63.24

Crew C-3

	Hr.	Daily	Hr.	Daily	Bare Costs	Incl. O&P
1 Rodman Foreman	$50.40	$403.20	$80.55	$644.40	$44.54	$70.06
4 Rodmen (reinf.)	48.40	1548.80	77.35	2475.20		
1 Equip. Oper. (light)	43.60	348.80	65.15	521.20		
2 Laborers	34.35	549.60	52.70	843.20		
3 Stressing Equipment		29.40		32.34		
.5 Grouting Equipment		76.60		84.26	1.66	1.82
64 L.H., Daily Totals		$2956.40		$4600.60	$46.19	$71.88

Crews

Crew C-4	Hr.	Daily	Hr.	Daily	Bare Costs	Incl. O&P
1 Rodman Foreman	$50.40	$403.20	$80.55	$644.40	$48.90	$78.15
3 Rodmen (reinf.)	48.40	1161.60	77.35	1856.40		
3 Stressing Equipment		29.40		32.34	.92	1.01
32 L.H., Daily Totals		$1594.20		$2533.14	$49.82	$79.16

Crew C-4A	Hr.	Daily	Hr.	Daily	Bare Costs	Incl. O&P
2 Rodmen (reinf.)	$48.40	$774.40	$77.35	$1237.60	$48.40	$77.35
4 Stressing Equipment		39.20		43.12	2.45	2.69
16 L.H., Daily Totals		$813.60		$1280.72	$50.85	$80.05

Crew C-5	Hr.	Daily	Hr.	Daily	Bare Costs	Incl. O&P
1 Rodman Foreman	$50.40	$403.20	$80.55	$644.40	$47.26	$74.23
4 Rodmen (reinf.)	48.40	1548.80	77.35	2475.20		
1 Equip. Oper. (crane)	46.50	372.00	69.45	555.60		
1 Equip. Oper. Oiler	40.30	322.40	60.20	481.60		
1 Hyd. Crane, 25 Ton		748.60		823.46	13.37	14.70
56 L.H., Daily Totals		$3395.00		$4980.26	$60.63	$88.93

Crew C-6	Hr.	Daily	Hr.	Daily	Bare Costs	Incl. O&P
1 Labor Foreman (outside)	$36.35	$290.80	$55.75	$446.00	$35.77	$54.34
4 Laborers	34.35	1099.20	52.70	1686.40		
1 Cement Finisher	40.85	326.80	59.50	476.00		
2 Gas Engine Vibrators		46.40		51.04	.97	1.06
48 L.H., Daily Totals		$1763.20		$2659.44	$36.73	$55.41

Crew C-7	Hr.	Daily	Hr.	Daily	Bare Costs	Incl. O&P
1 Labor Foreman (outside)	$36.35	$290.80	$55.75	$446.00	$37.18	$56.30
5 Laborers	34.35	1374.00	52.70	2108.00		
1 Cement Finisher	40.85	326.80	59.50	476.00		
1 Equip. Oper. (med.)	45.35	362.80	67.75	542.00		
1 Equip. Oper. (oiler)	40.30	322.40	60.20	481.60		
2 Gas Engine Vibrators		46.40		51.04		
1 Concrete Bucket, 1 C.Y.		23.20		25.52		
1 Hyd. Crane, 55 Ton		1208.00		1328.80	17.74	19.52
72 L.H., Daily Totals		$3954.40		$5458.96	$54.92	$75.82

Crew C-7A	Hr.	Daily	Hr.	Daily	Bare Costs	Incl. O&P
1 Labor Foreman (outside)	$36.35	$290.80	$55.75	$446.00	$34.61	$52.97
5 Laborers	34.35	1374.00	52.70	2108.00		
2 Truck Drivers (Heavy)	34.40	550.40	52.25	836.00		
2 Conc. Transit Mixers		1722.40		1894.64	26.91	29.60
64 L.H., Daily Totals		$3937.60		$5284.64	$61.52	$82.57

Crew C-7B	Hr.	Daily	Hr.	Daily	Bare Costs	Incl. O&P
1 Labor Foreman (outside)	$36.35	$290.80	$55.75	$446.00	$36.86	$56.11
5 Laborers	34.35	1374.00	52.70	2108.00		
1 Equipment Oper. (crane)	46.50	372.00	69.45	555.60		
1 Equipment Oiler	40.30	322.40	60.20	481.60		
1 Conc. Bucket, 2 C.Y.		35.80		39.38		
1 Lattice Boom Crane, 165 Ton		2061.00		2267.10	32.76	36.04
64 L.H., Daily Totals		$4456.00		$5897.68	$69.63	$92.15

Crew C-7C	Hr.	Daily	Hr.	Daily	Bare Costs	Incl. O&P
1 Labor Foreman (outside)	$36.35	$290.80	$55.75	$446.00	$37.35	$56.84
5 Laborers	34.35	1374.00	52.70	2108.00		
2 Equipment Operators (med.)	45.35	725.60	67.75	1084.00		
2 F.E. Loaders, W.M., 4 C.Y.		1197.20		1316.92	18.71	20.58
64 L.H., Daily Totals		$3587.60		$4954.92	$56.06	$77.42

Crew C-7D	Hr.	Daily	Hr.	Daily	Bare Costs	Incl. O&P
1 Labor Foreman (outside)	$36.35	$290.80	$55.75	$446.00	$36.21	$55.29
5 Laborers	34.35	1374.00	52.70	2108.00		
1 Equip. Oper. (med.)	45.35	362.80	67.75	542.00		
1 Concrete Conveyer		174.60		192.06	3.12	3.43
56 L.H., Daily Totals		$2202.20		$3288.06	$39.33	$58.72

Crew C-8	Hr.	Daily	Hr.	Daily	Bare Costs	Incl. O&P
1 Labor Foreman (outside)	$36.35	$290.80	$55.75	$446.00	$38.06	$57.23
3 Laborers	34.35	824.40	52.70	1264.80		
2 Cement Finishers	40.85	653.60	59.50	952.00		
1 Equip. Oper. (med.)	45.35	362.80	67.75	542.00		
1 Concrete Pump (small)		741.00		815.10	13.23	14.56
56 L.H., Daily Totals		$2872.60		$4019.90	$51.30	$71.78

Crew C-8A	Hr.	Daily	Hr.	Daily	Bare Costs	Incl. O&P
1 Labor Foreman (outside)	$36.35	$290.80	$55.75	$446.00	$36.85	$55.48
3 Laborers	34.35	824.40	52.70	1264.80		
2 Cement Finishers	40.85	653.60	59.50	952.00		
48 L.H., Daily Totals		$1768.80		$2662.80	$36.85	$55.48

Crew C-8B	Hr.	Daily	Hr.	Daily	Bare Costs	Incl. O&P
1 Labor Foreman (outside)	$36.35	$290.80	$55.75	$446.00	$36.95	$56.32
3 Laborers	34.35	824.40	52.70	1264.80		
1 Equip. Oper. (med.)	45.35	362.80	67.75	542.00		
1 Vibrating Power Screed		65.50		72.05		
1 Roller, Vibratory, 25 Ton		596.80		656.48		
1 Dozer, 200 H.P.		1051.00		1156.10	42.83	47.12
40 L.H., Daily Totals		$3191.30		$4137.43	$79.78	$103.44

Crew C-8C	Hr.	Daily	Hr.	Daily	Bare Costs	Incl. O&P
1 Labor Foreman (outside)	$36.35	$290.80	$55.75	$446.00	$37.60	$56.85
3 Laborers	34.35	824.40	52.70	1264.80		
1 Cement Finisher	40.85	326.80	59.50	476.00		
1 Equip. Oper. (med.)	45.35	362.80	67.75	542.00		
1 Shotcrete Rig, 12 C.Y./hr		249.20		274.12		
1 Air Compressor, 160 cfm		125.00		137.50		
4 -50' Air Hoses, 1"		16.40		18.04		
4 -50' Air Hoses, 2"		31.00		34.10	8.78	9.66
48 L.H., Daily Totals		$2226.40		$3192.56	$46.38	$66.51

Crew C-8D	Hr.	Daily	Hr.	Daily	Bare Costs	Incl. O&P
1 Labor Foreman (outside)	$36.35	$290.80	$55.75	$446.00	$38.79	$58.27
1 Laborer	34.35	274.80	52.70	421.60		
1 Cement Finisher	40.85	326.80	59.50	476.00		
1 Equipment Oper. (light)	43.60	348.80	65.15	521.20		
1 Air Compressor, 250 cfm		162.80		179.08		
2 -50' Air Hoses, 1"		8.20		9.02	5.34	5.88
32 L.H., Daily Totals		$1412.20		$2052.90	$44.13	$64.15

Crew C-8E	Hr.	Daily	Hr.	Daily	Bare Costs	Incl. O&P
1 Labor Foreman (outside)	$36.35	$290.80	$55.75	$446.00	$37.31	$56.42
3 Laborers	34.35	824.40	52.70	1264.80		
1 Cement Finisher	40.85	326.80	59.50	476.00		
1 Equipment Oper. (light)	43.60	348.80	65.15	521.20		
1 Shotcrete Rig, 35 C.Y./hr		280.00		308.00		
1 Air Compressor, 250 cfm		162.80		179.08		
4 -50' Air Hoses, 1"		16.40		18.04		
4 -50' Air Hoses, 2"		31.00		34.10	10.21	11.23
48 L.H., Daily Totals		$2281.00		$3247.22	$47.52	$67.65

Crew No.	Bare Costs Hr.	Daily	Incl. Subs O&P Hr.	Daily	Cost Per Labor-Hour Bare Costs	Incl. O&P
Crew C-10						
1 Laborer	$34.35	$274.80	$52.70	$421.60	$38.68	$57.23
2 Cement Finishers	40.85	653.60	59.50	952.00		
24 L.H., Daily Totals		$928.40		$1373.60	$38.68	$57.23
Crew C-10B						
3 Laborers	$34.35	$824.40	$52.70	$1264.80	$36.95	$55.42
2 Cement Finishers	40.85	653.60	59.50	952.00		
1 Concrete Mixer, 10 C.F.		151.20		166.32		
2 Trowels, 48" Walk-Behind		76.80		84.48	5.70	6.27
40 L.H., Daily Totals		$1706.00		$2467.60	$42.65	$61.69
Crew C-10C						
1 Laborer	$34.35	$274.80	$52.70	$421.60	$38.68	$57.23
2 Cement Finishers	40.85	653.60	59.50	952.00		
1 Trowel, 48" Walk-Behind		38.40		42.24	1.60	1.76
24 L.H., Daily Totals		$966.80		$1415.84	$40.28	$58.99
Crew C-10D						
1 Laborer	$34.35	$274.80	$52.70	$421.60	$38.68	$57.23
2 Cement Finishers	40.85	653.60	59.50	952.00		
1 Vibrating Power Screed		65.50		72.05		
1 Trowel, 48" Walk-Behind		38.40		42.24	4.33	4.76
24 L.H., Daily Totals		$1032.30		$1487.89	$43.01	$62.00
Crew C-10E						
1 Laborer	$34.35	$274.80	$52.70	$421.60	$38.68	$57.23
2 Cement Finishers	40.85	653.60	59.50	952.00		
1 Vibrating Power Screed		65.50		72.05		
1 Cement Trowel, 96" Ride-On		157.40		173.14	9.29	10.22
24 L.H., Daily Totals		$1151.30		$1618.79	$47.97	$67.45
Crew C-11						
1 Struc. Steel Foreman	$50.55	$404.40	$89.25	$714.00	$47.63	$81.49
6 Struc. Steel Workers	48.55	2330.40	85.75	4116.00		
1 Equip. Oper. (crane)	46.50	372.00	69.45	555.60		
1 Equip. Oper. Oiler	40.30	322.40	60.20	481.60		
1 Lattice Boom Crane, 150 Ton		1906.00		2096.60	26.47	29.12
72 L.H., Daily Totals		$5335.20		$7963.80	$74.10	$110.61
Crew C-12						
1 Carpenter Foreman (out)	$45.05	$360.40	$69.10	$552.80	$42.51	$64.90
3 Carpenters	43.05	1033.20	66.05	1585.20		
1 Laborer	34.35	274.80	52.70	421.60		
1 Equip. Oper. (crane)	46.50	372.00	69.45	555.60		
1 Hyd. Crane, 12 Ton		655.60		721.16	13.66	15.02
48 L.H., Daily Totals		$2696.00		$3836.36	$56.17	$79.92
Crew C-13						
1 Struc. Steel Worker	$48.55	$388.40	$85.75	$686.00	$46.72	$79.18
1 Welder	48.55	388.40	85.75	686.00		
1 Carpenter	43.05	344.40	66.05	528.40		
1 Welder, gas engine, 300 amp		109.00		119.90	4.54	5.00
24 L.H., Daily Totals		$1230.20		$2020.30	$51.26	$84.18
Crew C-14						
1 Carpenter Foreman (out)	$45.05	$360.40	$69.10	$552.80	$42.21	$64.90
5 Carpenters	43.05	1722.00	66.05	2642.00		
4 Laborers	34.35	1099.20	52.70	1686.40		
4 Rodmen (reinf.)	48.40	1548.80	77.35	2475.20		
2 Cement Finishers	40.85	653.60	59.50	952.00		
1 Equip. Oper. (crane)	46.50	372.00	69.45	555.60		
1 Equip. Oper. Oiler	40.30	322.40	60.20	481.60		
1 Hyd. Crane, 80 Ton		1693.00		1862.30	11.76	12.93
144 L.H., Daily Totals		$7771.40		$11207.90	$53.97	$77.83
Crew C-14A						
1 Carpenter Foreman (out)	$45.05	$360.40	$69.10	$552.80	$43.29	$66.72
16 Carpenters	43.05	5510.40	66.05	8454.40		
4 Rodmen (reinf.)	48.40	1548.80	77.35	2475.20		
2 Laborers	34.35	549.60	52.70	843.20		
1 Cement Finisher	40.85	326.80	59.50	476.00		
1 Equip. Oper. (med.)	45.35	362.80	67.75	542.00		
1 Gas Engine Vibrator		23.20		25.52		
1 Concrete Pump (small)		741.00		815.10	3.82	4.20
200 L.H., Daily Totals		$9423.00		$14184.22	$47.12	$70.92
Crew C-14B						
1 Carpenter Foreman (out)	$45.05	$360.40	$69.10	$552.80	$43.20	$66.44
16 Carpenters	43.05	5510.40	66.05	8454.40		
4 Rodmen (reinf.)	48.40	1548.80	77.35	2475.20		
2 Laborers	34.35	549.60	52.70	843.20		
2 Cement Finishers	40.85	653.60	59.50	952.00		
1 Equip. Oper. (med.)	45.35	362.80	67.75	542.00		
1 Gas Engine Vibrator		23.20		25.52		
1 Concrete Pump (small)		741.00		815.10	3.67	4.04
208 L.H., Daily Totals		$9749.80		$14660.22	$46.87	$70.48
Crew C-14C						
1 Carpenter Foreman (out)	$45.05	$360.40	$69.10	$552.80	$41.31	$63.60
6 Carpenters	43.05	2066.40	66.05	3170.40		
2 Rodmen (reinf.)	48.40	774.40	77.35	1237.60		
4 Laborers	34.35	1099.20	52.70	1686.40		
1 Cement Finisher	40.85	326.80	59.50	476.00		
1 Gas Engine Vibrator		23.20		25.52	.21	.23
112 L.H., Daily Totals		$4650.40		$7148.72	$41.52	$63.83
Crew C-14D						
1 Carpenter Foreman (out)	$45.05	$360.40	$69.10	$552.80	$42.87	$65.81
18 Carpenters	43.05	6199.20	66.05	9511.20		
2 Rodmen (reinf.)	48.40	774.40	77.35	1237.60		
2 Laborers	34.35	549.60	52.70	843.20		
1 Cement Finisher	40.85	326.80	59.50	476.00		
1 Equip. Oper. (med.)	45.35	362.80	67.75	542.00		
1 Gas Engine Vibrator		23.20		25.52		
1 Concrete Pump (small)		741.00		815.10	3.82	4.20
200 L.H., Daily Totals		$9337.40		$14003.42	$46.69	$70.02
Crew C-14E						
1 Carpenter Foreman (out)	$45.05	$360.40	$69.10	$552.80	$42.60	$66.20
2 Carpenters	43.05	688.80	66.05	1056.80		
4 Rodmen (reinf.)	48.40	1548.80	77.35	2475.20		
3 Laborers	34.35	824.40	52.70	1264.80		
1 Cement Finisher	40.85	326.80	59.50	476.00		
1 Gas Engine Vibrator		23.20		25.52	.26	.29
88 L.H., Daily Totals		$3772.40		$5851.12	$42.87	$66.49

Crews

Crew No.	Bare Costs		Incl. Subs O&P		Cost Per Labor-Hour	

Crew C-14F

	Hr.	Daily	Hr.	Daily	Bare Costs	Incl. O&P
1 Laborer Foreman (out)	$36.35	$290.80	$55.75	$446.00	$38.91	$57.57
2 Laborers	34.35	549.60	52.70	843.20		
6 Cement Finishers	40.85	1960.80	59.50	2856.00		
1 Gas Engine Vibrator		23.20		25.52	.32	.35
72 L.H., Daily Totals		$2824.40		$4170.72	$39.23	$57.93

Crew C-14G

	Hr.	Daily	Hr.	Daily	Bare Costs	Incl. O&P
1 Laborer Foreman (out)	$36.35	$290.80	$55.75	$446.00	$38.35	$57.02
2 Laborers	34.35	549.60	52.70	843.20		
4 Cement Finishers	40.85	1307.20	59.50	1904.00		
1 Gas Engine Vibrator		23.20		25.52	.41	.46
56 L.H., Daily Totals		$2170.80		$3218.72	$38.76	$57.48

Crew C-14H

	Hr.	Daily	Hr.	Daily	Bare Costs	Incl. O&P
1 Carpenter Foreman (out)	$45.05	$360.40	$69.10	$552.80	$42.46	$65.13
2 Carpenters	43.05	688.80	66.05	1056.80		
1 Rodman (reinf.)	48.40	387.20	77.35	618.80		
1 Laborer	34.35	274.80	52.70	421.60		
1 Cement Finisher	40.85	326.80	59.50	476.00		
1 Gas Engine Vibrator		23.20		25.52	.48	.53
48 L.H., Daily Totals		$2061.20		$3151.52	$42.94	$65.66

Crew C-14L

	Hr.	Daily	Hr.	Daily	Bare Costs	Incl. O&P
1 Carpenter Foreman (out)	$45.05	$360.40	$69.10	$552.80	$40.13	$61.31
6 Carpenters	43.05	2066.40	66.05	3170.40		
4 Laborers	34.35	1099.20	52.70	1686.40		
1 Cement Finisher	40.85	326.80	59.50	476.00		
1 Gas Engine Vibrator		23.20		25.52	.24	.27
96 L.H., Daily Totals		$3876.00		$5911.12	$40.38	$61.57

Crew C-15

	Hr.	Daily	Hr.	Daily	Bare Costs	Incl. O&P
1 Carpenter Foreman (out)	$45.05	$360.40	$69.10	$552.80	$40.48	$61.74
2 Carpenters	43.05	688.80	66.05	1056.80		
3 Laborers	34.35	824.40	52.70	1264.80		
2 Cement Finishers	40.85	653.60	59.50	952.00		
1 Rodman (reinf.)	48.40	387.20	77.35	618.80		
72 L.H., Daily Totals		$2914.40		$4445.20	$40.48	$61.74

Crew C-16

	Hr.	Daily	Hr.	Daily	Bare Costs	Incl. O&P
1 Labor Foreman (outside)	$36.35	$290.80	$55.75	$446.00	$38.06	$57.23
3 Laborers	34.35	824.40	52.70	1264.80		
2 Cement Finishers	40.85	653.60	59.50	952.00		
1 Equip. Oper. (med.)	45.35	362.80	67.75	542.00		
1 Gunite Pump Rig		178.00		195.80		
2 -50' Air Hoses, 3/4"		6.10		6.71		
2 -50' Air Hoses, 2"		15.50		17.05	3.56	3.92
56 L.H., Daily Totals		$2331.20		$3424.36	$41.63	$61.15

Crew C-17

	Hr.	Daily	Hr.	Daily	Bare Costs	Incl. O&P
2 Skilled Worker Foremen	$46.40	$742.40	$71.20	$1139.20	$44.80	$68.72
8 Skilled Workers	44.40	2841.60	68.10	4358.40		
80 L.H., Daily Totals		$3584.00		$5497.60	$44.80	$68.72

Crew C-17A

	Hr.	Daily	Hr.	Daily	Bare Costs	Incl. O&P
2 Skilled Worker Foremen	$46.40	$742.40	$71.20	$1139.20	$44.82	$68.73
8 Skilled Workers	44.40	2841.60	68.10	4358.40		
.125 Equip. Oper. (crane)	46.50	46.50	69.45	69.45		
.125 Hyd. Crane, 80 Ton		211.63		232.79	2.61	2.87
81 L.H., Daily Totals		$3842.13		$5799.84	$47.43	$71.60

Crew C-17B

	Hr.	Daily	Hr.	Daily	Bare Costs	Incl. O&P
2 Skilled Worker Foremen	$46.40	$742.40	$71.20	$1139.20	$44.84	$68.74
8 Skilled Workers	44.40	2841.60	68.10	4358.40		
.25 Equip. Oper. (crane)	46.50	93.00	69.45	138.90		
.25 Hyd. Crane, 80 Ton		423.25		465.57		
.25 Trowel, 48" Walk-Behind		9.60		10.56	5.28	5.81
82 L.H., Daily Totals		$4109.85		$6112.64	$50.12	$74.54

Crew C-17C

	Hr.	Daily	Hr.	Daily	Bare Costs	Incl. O&P
2 Skilled Worker Foremen	$46.40	$742.40	$71.20	$1139.20	$44.86	$68.75
8 Skilled Workers	44.40	2841.60	68.10	4358.40		
.375 Equip. Oper. (crane)	46.50	139.50	69.45	208.35		
.375 Hyd. Crane, 80 Ton		634.88		698.36	7.65	8.41
83 L.H., Daily Totals		$4358.38		$6404.31	$52.51	$77.16

Crew C-17D

	Hr.	Daily	Hr.	Daily	Bare Costs	Incl. O&P
2 Skilled Worker Foremen	$46.40	$742.40	$71.20	$1139.20	$44.88	$68.75
8 Skilled Workers	44.40	2841.60	68.10	4358.40		
.5 Equip. Oper. (crane)	46.50	186.00	69.45	277.80		
.5 Hyd. Crane, 80 Ton		846.50		931.15	10.08	11.09
84 L.H., Daily Totals		$4616.50		$6706.55	$54.96	$79.84

Crew C-17E

	Hr.	Daily	Hr.	Daily	Bare Costs	Incl. O&P
2 Skilled Worker Foremen	$46.40	$742.40	$71.20	$1139.20	$44.80	$68.72
8 Skilled Workers	44.40	2841.60	68.10	4358.40		
1 Hyd. Jack with Rods		89.10		98.01	1.11	1.23
80 L.H., Daily Totals		$3673.10		$5595.61	$45.91	$69.95

Crew C-18

	Hr.	Daily	Hr.	Daily	Bare Costs	Incl. O&P
.125 Labor Foreman (out)	$36.35	$36.35	$55.75	$55.75	$34.57	$53.04
1 Laborer	34.35	274.80	52.70	421.60		
1 Concrete Cart, 10 C.F.		55.00		60.50	6.11	6.72
9 L.H., Daily Totals		$366.15		$537.85	$40.68	$59.76

Crew C-19

	Hr.	Daily	Hr.	Daily	Bare Costs	Incl. O&P
.125 Labor Foreman (out)	$36.35	$36.35	$55.75	$55.75	$34.57	$53.04
1 Laborer	34.35	274.80	52.70	421.60		
1 Concrete Cart, 18 C.F.		91.60		100.76	10.18	11.20
9 L.H., Daily Totals		$402.75		$578.11	$44.75	$64.23

Crew C-20

	Hr.	Daily	Hr.	Daily	Bare Costs	Incl. O&P
1 Labor Foreman (outside)	$36.35	$290.80	$55.75	$446.00	$36.79	$55.81
5 Laborers	34.35	1374.00	52.70	2108.00		
1 Cement Finisher	40.85	326.80	59.50	476.00		
1 Equip. Oper. (med.)	45.35	362.80	67.75	542.00		
2 Gas Engine Vibrators		46.40		51.04		
1 Concrete Pump (small)		741.00		815.10	12.30	13.53
64 L.H., Daily Totals		$3141.80		$4438.14	$49.09	$69.35

Crew C-21

	Hr.	Daily	Hr.	Daily	Bare Costs	Incl. O&P
1 Labor Foreman (outside)	$36.35	$290.80	$55.75	$446.00	$36.79	$55.81
5 Laborers	34.35	1374.00	52.70	2108.00		
1 Cement Finisher	40.85	326.80	59.50	476.00		
1 Equip. Oper. (med.)	45.35	362.80	67.75	542.00		
2 Gas Engine Vibrators		46.40		51.04		
1 Concrete Conveyer		174.60		192.06	3.45	3.80
64 L.H., Daily Totals		$2575.40		$3815.10	$40.24	$59.61

Crew No.	Bare Costs		Incl. Subs O&P		Cost Per Labor-Hour	

Crew C-22

Crew C-22	Hr.	Daily	Hr.	Daily	Bare Costs	Incl. O&P
1 Rodman Foreman	$50.40	$403.20	$80.55	$644.40	$48.54	$77.36
4 Rodmen (reinf.)	48.40	1548.80	77.35	2475.20		
.125 Equip. Oper. (crane)	46.50	46.50	69.45	69.45		
.125 Equip. Oper. Oiler	40.30	40.30	60.20	60.20		
.125 Hyd. Crane, 25 Ton		93.58		102.93	2.23	2.45
42 L.H., Daily Totals		$2132.38		$3352.18	$50.77	$79.81

Crew C-23

Crew C-23	Hr.	Daily	Hr.	Daily	Bare Costs	Incl. O&P
2 Skilled Worker Foremen	$46.40	$742.40	$71.20	$1139.20	$44.60	$68.06
6 Skilled Workers	44.40	2131.20	68.10	3268.80		
1 Equip. Oper. (crane)	46.50	372.00	69.45	555.60		
1 Equip. Oper. Oiler	40.30	322.40	60.20	481.60		
1 Lattice Boom Crane, 90 Ton		1622.00		1784.20	20.27	22.30
80 L.H., Daily Totals		$5190.00		$7229.40	$64.88	$90.37

Crew C-23A

Crew C-23A	Hr.	Daily	Hr.	Daily	Bare Costs	Incl. O&P
1 Labor Foreman (outside)	$36.35	$290.80	$55.75	$446.00	$38.37	$58.16
2 Laborers	34.35	549.60	52.70	843.20		
1 Equip. Oper. (crane)	46.50	372.00	69.45	555.60		
1 Equip. Oper. Oiler	40.30	322.40	60.20	481.60		
1 Crawler Crane, 100 Ton		1653.00		1818.30		
3 Conc. Buckets, 8 C.Y.		613.80		675.18	56.67	62.34
40 L.H., Daily Totals		$3801.60		$4819.88	$95.04	$120.50

Crew C-24

Crew C-24	Hr.	Daily	Hr.	Daily	Bare Costs	Incl. O&P
2 Skilled Worker Foremen	$46.40	$742.40	$71.20	$1139.20	$44.60	$68.06
6 Skilled Workers	44.40	2131.20	68.10	3268.80		
1 Equip. Oper. (crane)	46.50	372.00	69.45	555.60		
1 Equip. Oper. Oiler	40.30	322.40	60.20	481.60		
1 Lattice Boom Crane, 150 Ton		1906.00		2096.60	23.82	26.21
80 L.H., Daily Totals		$5474.00		$7541.80	$68.42	$94.27

Crew C-25

Crew C-25	Hr.	Daily	Hr.	Daily	Bare Costs	Incl. O&P
2 Rodmen (reinf.)	$48.40	$774.40	$77.35	$1237.60	$37.75	$61.23
2 Rodmen Helpers	27.10	433.60	45.10	721.60		
32 L.H., Daily Totals		$1208.00		$1959.20	$37.75	$61.23

Crew C-27

Crew C-27	Hr.	Daily	Hr.	Daily	Bare Costs	Incl. O&P
2 Cement Finishers	$40.85	$653.60	$59.50	$952.00	$40.85	$59.50
1 Concrete Saw		134.20		147.62	8.39	9.23
16 L.H., Daily Totals		$787.80		$1099.62	$49.24	$68.73

Crew C-28

Crew C-28	Hr.	Daily	Hr.	Daily	Bare Costs	Incl. O&P
1 Cement Finisher	$40.85	$326.80	$59.50	$476.00	$40.85	$59.50
1 Portable Air Compressor, Gas		19.00		20.90	2.38	2.61
8 L.H., Daily Totals		$345.80		$496.90	$43.23	$62.11

Crew C-29

Crew C-29	Hr.	Daily	Hr.	Daily	Bare Costs	Incl. O&P
1 Laborer	$34.35	$274.80	$52.70	$421.60	$34.35	$52.70
1 Pressure Washer		58.80		64.68	7.35	8.09
8 L.H., Daily Totals		$333.60		$486.28	$41.70	$60.78

Crew D-1

Crew D-1	Hr.	Daily	Hr.	Daily	Bare Costs	Incl. O&P
1 Bricklayer	$43.15	$345.20	$64.75	$518.00	$38.98	$58.48
1 Bricklayer Helper	34.80	278.40	52.20	417.60		
16 L.H., Daily Totals		$623.60		$935.60	$38.98	$58.48

Crew D-2

Crew D-2	Hr.	Daily	Hr.	Daily	Bare Costs	Incl. O&P
3 Bricklayers	$43.15	$1035.60	$64.75	$1554.00	$40.10	$60.30
2 Bricklayer Helpers	34.80	556.80	52.20	835.20		
.5 Carpenter	43.05	172.20	66.05	264.20		
44 L.H., Daily Totals		$1764.60		$2653.40	$40.10	$60.30

Crew D-3

Crew D-3	Hr.	Daily	Hr.	Daily	Bare Costs	Incl. O&P
3 Bricklayers	$43.15	$1035.60	$64.75	$1554.00	$39.96	$60.03
2 Bricklayer Helpers	34.80	556.80	52.20	835.20		
.25 Carpenter	43.05	86.10	66.05	132.10		
42 L.H., Daily Totals		$1678.50		$2521.30	$39.96	$60.03

Crew D-4

Crew D-4	Hr.	Daily	Hr.	Daily	Bare Costs	Incl. O&P
1 Bricklayer	$43.15	$345.20	$64.75	$518.00	$39.09	$58.58
2 Bricklayer Helpers	34.80	556.80	52.20	835.20		
1 Equip. Oper. (light)	43.60	348.80	65.15	521.20		
1 Grout Pump, 50 C.F./hr.		125.80		138.38	3.93	4.32
32 L.H., Daily Totals		$1376.60		$2012.78	$43.02	$62.90

Crew D-5

Crew D-5	Hr.	Daily	Hr.	Daily	Bare Costs	Incl. O&P
1 Bricklayer	$43.15	$345.20	$64.75	$518.00	$43.15	$64.75
8 L.H., Daily Totals		$345.20		$518.00	$43.15	$64.75

Crew D-6

Crew D-6	Hr.	Daily	Hr.	Daily	Bare Costs	Incl. O&P
3 Bricklayers	$43.15	$1035.60	$64.75	$1554.00	$39.14	$58.78
3 Bricklayer Helpers	34.80	835.20	52.20	1252.80		
.25 Carpenter	43.05	86.10	66.05	132.10		
50 L.H., Daily Totals		$1956.90		$2938.90	$39.14	$58.78

Crew D-7

Crew D-7	Hr.	Daily	Hr.	Daily	Bare Costs	Incl. O&P
1 Tile Layer	$40.45	$323.60	$59.00	$472.00	$36.15	$52.73
1 Tile Layer Helper	31.85	254.80	46.45	371.60		
16 L.H., Daily Totals		$578.40		$843.60	$36.15	$52.73

Crew D-8

Crew D-8	Hr.	Daily	Hr.	Daily	Bare Costs	Incl. O&P
3 Bricklayers	$43.15	$1035.60	$64.75	$1554.00	$39.81	$59.73
2 Bricklayer Helpers	34.80	556.80	52.20	835.20		
40 L.H., Daily Totals		$1592.40		$2389.20	$39.81	$59.73

Crew D-9

Crew D-9	Hr.	Daily	Hr.	Daily	Bare Costs	Incl. O&P
3 Bricklayers	$43.15	$1035.60	$64.75	$1554.00	$38.98	$58.48
3 Bricklayer Helpers	34.80	835.20	52.20	1252.80		
48 L.H., Daily Totals		$1870.80		$2806.80	$38.98	$58.48

Crew D-10

Crew D-10	Hr.	Daily	Hr.	Daily	Bare Costs	Incl. O&P
1 Bricklayer Foreman	$45.15	$361.20	$67.75	$542.00	$42.40	$63.54
1 Bricklayer	43.15	345.20	64.75	518.00		
1 Bricklayer Helper	34.80	278.40	52.20	417.60		
1 Equip. Oper. (crane)	46.50	372.00	69.45	555.60		
1 S.P. Crane, 4x4, 12 Ton		434.40		477.84	13.57	14.93
32 L.H., Daily Totals		$1791.20		$2511.04	$55.98	$78.47

Crew D-11

Crew D-11	Hr.	Daily	Hr.	Daily	Bare Costs	Incl. O&P
1 Bricklayer Foreman	$45.15	$361.20	$67.75	$542.00	$41.03	$61.57
1 Bricklayer	43.15	345.20	64.75	518.00		
1 Bricklayer Helper	34.80	278.40	52.20	417.60		
24 L.H., Daily Totals		$984.80		$1477.60	$41.03	$61.57

Crews

Crew No.	Bare Costs		Incl. Subs O&P		Cost Per Labor-Hour	

Crew D-12	Hr.	Daily	Hr.	Daily	Bare Costs	Incl. O&P
1 Bricklayer Foreman	$45.15	$361.20	$67.75	$542.00	$39.48	$59.23
1 Bricklayer	43.15	345.20	64.75	518.00		
2 Bricklayer Helpers	34.80	556.80	52.20	835.20		
32 L.H., Daily Totals		$1263.20		$1895.20	$39.48	$59.23

Crew D-13	Hr.	Daily	Hr.	Daily	Bare Costs	Incl. O&P
1 Bricklayer Foreman	$45.15	$361.20	$67.75	$542.00	$41.24	$62.07
1 Bricklayer	43.15	345.20	64.75	518.00		
2 Bricklayer Helpers	34.80	556.80	52.20	835.20		
1 Carpenter	43.05	344.40	66.05	528.40		
1 Equip. Oper. (crane)	46.50	372.00	69.45	555.60		
1 S.P. Crane, 4x4, 12 Ton		434.40		477.84	9.05	9.96
48 L.H., Daily Totals		$2414.00		$3457.04	$50.29	$72.02

Crew E-1	Hr.	Daily	Hr.	Daily	Bare Costs	Incl. O&P
1 Welder Foreman	$50.55	$404.40	$89.25	$714.00	$47.57	$80.05
1 Welder	48.55	388.40	85.75	686.00		
1 Equip. Oper. (light)	43.60	348.80	65.15	521.20		
1 Welder, gas engine, 300 amp		109.00		119.90	4.54	5.00
24 L.H., Daily Totals		$1250.60		$2041.10	$52.11	$85.05

Crew E-2	Hr.	Daily	Hr.	Daily	Bare Costs	Incl. O&P
1 Struc. Steel Foreman	$50.55	$404.40	$89.25	$714.00	$47.36	$80.27
4 Struc. Steel Workers	48.55	1553.60	85.75	2744.00		
1 Equip. Oper. (crane)	46.50	372.00	69.45	555.60		
1 Equip. Oper. Oiler	40.30	322.40	60.20	481.60		
1 Lattice Boom Crane, 90 Ton		1622.00		1784.20	28.96	31.86
56 L.H., Daily Totals		$4274.40		$6279.40	$76.33	$112.13

Crew E-3	Hr.	Daily	Hr.	Daily	Bare Costs	Incl. O&P
1 Struc. Steel Foreman	$50.55	$404.40	$89.25	$714.00	$49.22	$86.92
1 Struc. Steel Worker	48.55	388.40	85.75	686.00		
1 Welder	48.55	388.40	85.75	686.00		
1 Welder, gas engine, 300 amp		109.00		119.90	4.54	5.00
24 L.H., Daily Totals		$1290.20		$2205.90	$53.76	$91.91

Crew E-4	Hr.	Daily	Hr.	Daily	Bare Costs	Incl. O&P
1 Struc. Steel Foreman	$50.55	$404.40	$89.25	$714.00	$49.05	$86.63
3 Struc. Steel Workers	48.55	1165.20	85.75	2058.00		
1 Welder, gas engine, 300 amp		109.00		119.90	3.41	3.75
32 L.H., Daily Totals		$1678.60		$2891.90	$52.46	$90.37

Crew E-5	Hr.	Daily	Hr.	Daily	Bare Costs	Incl. O&P
2 Struc. Steel Foremen	$50.55	$808.80	$89.25	$1428.00	$47.92	$82.27
5 Struc. Steel Workers	48.55	1942.00	85.75	3430.00		
1 Equip. Oper. (crane)	46.50	372.00	69.45	555.60		
1 Welder	48.55	388.40	85.75	686.00		
1 Equip. Oper. Oiler	40.30	322.40	60.20	481.60		
1 Lattice Boom Crane, 90 Ton		1622.00		1784.20		
1 Welder, gas engine, 300 amp		109.00		119.90	21.64	23.80
80 L.H., Daily Totals		$5564.60		$8485.30	$69.56	$106.07

Crew E-6	Hr.	Daily	Hr.	Daily	Bare Costs	Incl. O&P
3 Struc. Steel Foremen	$50.55	$1213.20	$89.25	$2142.00	$47.97	$82.50
9 Struc. Steel Workers	48.55	3495.60	85.75	6174.00		
1 Equip. Oper. (crane)	46.50	372.00	69.45	555.60		
1 Welder	48.55	388.40	85.75	686.00		
1 Equip. Oper. Oiler	40.30	322.40	60.20	481.60		
1 Equip. Oper. (light)	43.60	348.80	65.15	521.20		
1 Lattice Boom Crane, 90 Ton		1622.00		1784.20		
1 Welder, gas engine, 300 amp		109.00		119.90		
1 Air Compressor, 160 cfm		125.00		137.50		
2 Impact Wrenches		36.40		40.04	14.78	16.26
128 L.H., Daily Totals		$8032.80		$12642.04	$62.76	$98.77

Crew E-7	Hr.	Daily	Hr.	Daily	Bare Costs	Incl. O&P
1 Struc. Steel Foreman	$50.55	$404.40	$89.25	$714.00	$47.92	$82.27
4 Struc. Steel Workers	48.55	1553.60	85.75	2744.00		
1 Equip. Oper. (crane)	46.50	372.00	69.45	555.60		
1 Equip. Oper. Oiler	40.30	322.40	60.20	481.60		
1 Welder Foreman	50.55	404.40	89.25	714.00		
2 Welders	48.55	776.80	85.75	1372.00		
1 Lattice Boom Crane, 90 Ton		1622.00		1784.20		
2 Welders, gas engine, 300 amp		218.00		239.80	23.00	25.30
80 L.H., Daily Totals		$5673.60		$8605.20	$70.92	$107.57

Crew E-8	Hr.	Daily	Hr.	Daily	Bare Costs	Incl. O&P
1 Struc. Steel Foreman	$50.55	$404.40	$89.25	$714.00	$47.68	$81.48
4 Struc. Steel Workers	48.55	1553.60	85.75	2744.00		
1 Welder Foreman	50.55	404.40	89.25	714.00		
4 Welders	48.55	1553.60	85.75	2744.00		
1 Equip. Oper. (crane)	46.50	372.00	69.45	555.60		
1 Equip. Oper. Oiler	40.30	322.40	60.20	481.60		
1 Equip. Oper. (light)	43.60	348.80	65.15	521.20		
1 Lattice Boom Crane, 90 Ton		1622.00		1784.20		
4 Welders, gas engine, 300 amp		436.00		479.60	19.79	21.77
104 L.H., Daily Totals		$7017.20		$10738.20	$67.47	$103.25

Crew E-9	Hr.	Daily	Hr.	Daily	Bare Costs	Incl. O&P
2 Struc. Steel Foremen	$50.55	$808.80	$89.25	$1428.00	$47.97	$82.50
5 Struc. Steel Workers	48.55	1942.00	85.75	3430.00		
1 Welder Foreman	50.55	404.40	89.25	714.00		
5 Welders	48.55	1942.00	85.75	3430.00		
1 Equip. Oper. (crane)	46.50	372.00	69.45	555.60		
1 Equip. Oper. Oiler	40.30	322.40	60.20	481.60		
1 Equip. Oper. (light)	43.60	348.80	65.15	521.20		
1 Lattice Boom Crane, 90 Ton		1622.00		1784.20		
5 Welders, gas engine, 300 amp		545.00		599.50	16.93	18.62
128 L.H., Daily Totals		$8307.40		$12944.10	$64.90	$101.13

Crew E-10	Hr.	Daily	Hr.	Daily	Bare Costs	Incl. O&P
1 Welder Foreman	$50.55	$404.40	$89.25	$714.00	$49.55	$87.50
1 Welder	48.55	388.40	85.75	686.00		
1 Welder, gas engine, 300 amp		109.00		119.90		
1 Flatbed Truck, Gas, 3 Ton		217.00		238.70	20.38	22.41
16 L.H., Daily Totals		$1118.80		$1758.60	$69.92	$109.91

Crew E-11	Hr.	Daily	Hr.	Daily	Bare Costs	Incl. O&P
2 Painters, Struc. Steel	$38.65	$618.40	$69.15	$1106.40	$38.81	$64.04
1 Building Laborer	34.35	274.80	52.70	421.60		
1 Equip. Oper. (light)	43.60	348.80	65.15	521.20		
1 Air Compressor, 250 cfm		162.80		179.08		
1 Sandblaster, portable, 3 C.F.		20.60		22.66		
1 Set Sand Blasting Accessories		13.15		14.47	6.14	6.76
32 L.H., Daily Totals		$1438.55		$2265.41	$44.95	$70.79

Crew E-12	Bare Costs Hr.	Daily	Incl. Subs O&P Hr.	Daily	Cost Per Labor-Hour Bare Costs	Incl. O&P
1 Welder Foreman	$50.55	$404.40	$89.25	$714.00	$47.08	$77.20
1 Equip. Oper. (light)	43.60	348.80	65.15	521.20		
1 Welder, gas engine, 300 amp		109.00		119.90	6.81	7.49
16 L.H., Daily Totals		$862.20		$1355.10	$53.89	$84.69

Crew E-13	Bare Costs Hr.	Daily	Incl. Subs O&P Hr.	Daily	Cost Per Labor-Hour Bare Costs	Incl. O&P
1 Welder Foreman	$50.55	$404.40	$89.25	$714.00	$48.23	$81.22
.5 Equip. Oper. (light)	43.60	174.40	65.15	260.60		
1 Welder, gas engine, 300 amp		109.00		119.90	9.08	9.99
12 L.H., Daily Totals		$687.80		$1094.50	$57.32	$91.21

Crew E-14	Bare Costs Hr.	Daily	Incl. Subs O&P Hr.	Daily	Cost Per Labor-Hour Bare Costs	Incl. O&P
1 Welder Foreman	$50.55	$404.40	$89.25	$714.00	$50.55	$89.25
1 Welder, gas engine, 300 amp		109.00		119.90	13.63	14.99
8 L.H., Daily Totals		$513.40		$833.90	$64.17	$104.24

Crew E-16	Bare Costs Hr.	Daily	Incl. Subs O&P Hr.	Daily	Cost Per Labor-Hour Bare Costs	Incl. O&P
1 Welder Foreman	$50.55	$404.40	$89.25	$714.00	$49.55	$87.50
1 Welder	48.55	388.40	85.75	686.00		
1 Welder, gas engine, 300 amp		109.00		119.90	6.81	7.49
16 L.H., Daily Totals		$901.80		$1519.90	$56.36	$94.99

Crew E-17	Bare Costs Hr.	Daily	Incl. Subs O&P Hr.	Daily	Cost Per Labor-Hour Bare Costs	Incl. O&P
1 Structural Steel Foreman	$50.55	$404.40	$89.25	$714.00	$49.55	$87.50
1 Structural Steel Worker	48.55	388.40	85.75	686.00		
16 L.H., Daily Totals		$792.80		$1400.00	$49.55	$87.50

Crew E-18	Bare Costs Hr.	Daily	Incl. Subs O&P Hr.	Daily	Cost Per Labor-Hour Bare Costs	Incl. O&P
1 Structural Steel Foreman	$50.55	$404.40	$89.25	$714.00	$48.31	$82.85
3 Structural Steel Workers	48.55	1165.20	85.75	2058.00		
1 Equipment Operator (med.)	45.35	362.80	67.75	542.00		
1 Lattice Boom Crane, 20 Ton		1035.00		1138.50	25.88	28.46
40 L.H., Daily Totals		$2967.40		$4452.50	$74.19	$111.31

Crew E-19	Bare Costs Hr.	Daily	Incl. Subs O&P Hr.	Daily	Cost Per Labor-Hour Bare Costs	Incl. O&P
1 Structural Steel Worker	$48.55	$388.40	$85.75	$686.00	$47.57	$80.05
1 Structural Steel Foreman	50.55	404.40	89.25	714.00		
1 Equip. Oper. (light)	43.60	348.80	65.15	521.20		
1 Lattice Boom Crane, 20 Ton		1035.00		1138.50	43.13	47.44
24 L.H., Daily Totals		$2176.60		$3059.70	$90.69	$127.49

Crew E-20	Bare Costs Hr.	Daily	Incl. Subs O&P Hr.	Daily	Cost Per Labor-Hour Bare Costs	Incl. O&P
1 Structural Steel Foreman	$50.55	$404.40	$89.25	$714.00	$47.51	$80.96
5 Structural Steel Workers	48.55	1942.00	85.75	3430.00		
1 Equip. Oper. (crane)	46.50	372.00	69.45	555.60		
1 Oiler	40.30	322.40	60.20	481.60		
1 Lattice Boom Crane, 40 Ton		1263.00		1389.30	19.73	21.71
64 L.H., Daily Totals		$4303.80		$6570.50	$67.25	$102.66

Crew E-22	Bare Costs Hr.	Daily	Incl. Subs O&P Hr.	Daily	Cost Per Labor-Hour Bare Costs	Incl. O&P
1 Skilled Worker Foreman	$46.40	$371.20	$71.20	$569.60	$45.07	$69.13
2 Skilled Workers	44.40	710.40	68.10	1089.60		
24 L.H., Daily Totals		$1081.60		$1659.20	$45.07	$69.13

Crew E-24	Bare Costs Hr.	Daily	Incl. Subs O&P Hr.	Daily	Cost Per Labor-Hour Bare Costs	Incl. O&P
3 Structural Steel Workers	$48.55	$1165.20	$85.75	$2058.00	$47.75	$81.25
1 Equipment Operator (med.)	45.35	362.80	67.75	542.00		
1 Hyd. Crane, 25 Ton		748.60		823.46	23.39	25.73
32 L.H., Daily Totals		$2276.60		$3423.46	$71.14	$106.98

Crew E-25	Bare Costs Hr.	Daily	Incl. Subs O&P Hr.	Daily	Cost Per Labor-Hour Bare Costs	Incl. O&P
1 Welder Foreman	$50.55	$404.40	$89.25	$714.00	$50.55	$89.25
1 Cutting Torch		18.00		19.80		
1 Set of Gases		82.80		91.08	12.60	13.86
8 L.H., Daily Totals		$505.20		$824.88	$63.15	$103.11

Crew F-3	Bare Costs Hr.	Daily	Incl. Subs O&P Hr.	Daily	Cost Per Labor-Hour Bare Costs	Incl. O&P
4 Carpenters	$43.05	$1377.60	$66.05	$2113.60	$43.74	$66.73
1 Equip. Oper. (crane)	46.50	372.00	69.45	555.60		
1 Hyd. Crane, 12 Ton		655.60		721.16	16.39	18.03
40 L.H., Daily Totals		$2405.20		$3390.36	$60.13	$84.76

Crew F-4	Bare Costs Hr.	Daily	Incl. Subs O&P Hr.	Daily	Cost Per Labor-Hour Bare Costs	Incl. O&P
4 Carpenters	$43.05	$1377.60	$66.05	$2113.60	$43.17	$65.64
1 Equip. Oper. (crane)	46.50	372.00	69.45	555.60		
1 Equip. Oper. Oiler	40.30	322.40	60.20	481.60		
1 Hyd. Crane, 55 Ton		1208.00		1328.80	25.17	27.68
48 L.H., Daily Totals		$3280.00		$4479.60	$68.33	$93.33

Crew F-5	Bare Costs Hr.	Daily	Incl. Subs O&P Hr.	Daily	Cost Per Labor-Hour Bare Costs	Incl. O&P
1 Carpenter Foreman	$45.05	$360.40	$69.10	$552.80	$43.55	$66.81
3 Carpenters	43.05	1033.20	66.05	1585.20		
32 L.H., Daily Totals		$1393.60		$2138.00	$43.55	$66.81

Crew F-6	Bare Costs Hr.	Daily	Incl. Subs O&P Hr.	Daily	Cost Per Labor-Hour Bare Costs	Incl. O&P
2 Carpenters	$43.05	$688.80	$66.05	$1056.80	$40.26	$61.39
2 Building Laborers	34.35	549.60	52.70	843.20		
1 Equip. Oper. (crane)	46.50	372.00	69.45	555.60		
1 Hyd. Crane, 12 Ton		655.60		721.16	16.39	18.03
40 L.H., Daily Totals		$2266.00		$3176.76	$56.65	$79.42

Crew F-7	Bare Costs Hr.	Daily	Incl. Subs O&P Hr.	Daily	Cost Per Labor-Hour Bare Costs	Incl. O&P
2 Carpenters	$43.05	$688.80	$66.05	$1056.80	$38.70	$59.38
2 Building Laborers	34.35	549.60	52.70	843.20		
32 L.H., Daily Totals		$1238.40		$1900.00	$38.70	$59.38

Crew G-1	Bare Costs Hr.	Daily	Incl. Subs O&P Hr.	Daily	Cost Per Labor-Hour Bare Costs	Incl. O&P
1 Roofer Foreman	$38.55	$308.40	$64.20	$513.60	$34.14	$56.83
4 Roofers, Composition	36.55	1169.60	60.85	1947.20		
2 Roofer Helpers	27.10	433.60	45.10	721.60		
1 Application Equipment		168.40		185.24		
1 Tar Kettle/Pot		88.75		97.63		
1 Crew Truck		152.40		167.64	7.31	8.04
56 L.H., Daily Totals		$2321.15		$3632.91	$41.45	$64.87

Crew G-2	Bare Costs Hr.	Daily	Incl. Subs O&P Hr.	Daily	Cost Per Labor-Hour Bare Costs	Incl. O&P
1 Plasterer	$39.40	$315.20	$58.60	$468.80	$36.27	$54.47
1 Plasterer Helper	35.05	280.40	52.10	416.80		
1 Building Laborer	34.35	274.80	52.70	421.60		
1 Grout Pump, 50 C.F./hr		125.80		138.38	5.24	5.77
24 L.H., Daily Totals		$996.20		$1445.58	$41.51	$60.23

Crew G-2A	Bare Costs Hr.	Daily	Incl. Subs O&P Hr.	Daily	Cost Per Labor-Hour Bare Costs	Incl. O&P
1 Roofer, composition	$36.55	$292.40	$60.85	$486.80	$32.67	$52.88
1 Roofer Helper	27.10	216.80	45.10	360.80		
1 Building Laborer	34.35	274.80	52.70	421.60		
1 Foam spray rig, trailer-mtd.		488.25		537.08		
1 Pickup Truck, 3/4 Ton		104.60		115.06	24.70	27.17
24 L.H., Daily Totals		$1376.85		$1921.34	$57.37	$80.06

Crew G-3	Hr.	Daily	Hr.	Daily	Bare Costs	Incl. O&P
2 Sheet Metal Workers	$51.65	$826.40	$78.10	$1249.60	$43.00	$65.40
2 Building Laborers	34.35	549.60	52.70	843.20		
32 L.H., Daily Totals		$1376.00		$2092.80	$43.00	$65.40

Crew G-4	Hr.	Daily	Hr.	Daily	Bare Costs	Incl. O&P
1 Labor Foreman (outside)	$36.35	$290.80	$55.75	$446.00	$35.02	$53.72
2 Building Laborers	34.35	549.60	52.70	843.20		
1 Flatbed Truck, Gas, 1.5 Ton		171.40		188.54		
1 Air Compressor, 160 cfm		125.00		137.50	12.35	13.59
24 L.H., Daily Totals		$1136.80		$1615.24	$47.37	$67.30

Crew G-5	Hr.	Daily	Hr.	Daily	Bare Costs	Incl. O&P
1 Roofer Foreman	$38.55	$308.40	$64.20	$513.60	$33.17	$55.22
2 Roofers, Composition	36.55	584.80	60.85	973.60		
2 Roofer Helpers	27.10	433.60	45.10	721.60		
1 Application Equipment		168.40		185.24	4.21	4.63
40 L.H., Daily Totals		$1495.20		$2394.04	$37.38	$59.85

Crew G-6A	Hr.	Daily	Hr.	Daily	Bare Costs	Incl. O&P
2 Roofers Composition	$36.55	$584.80	$60.85	$973.60	$36.55	$60.85
1 Small Compressor, Electric		11.05		12.15		
2 Pneumatic Nailers		47.00		51.70	3.63	3.99
16 L.H., Daily Totals		$642.85		$1037.45	$40.18	$64.84

Crew G-7	Hr.	Daily	Hr.	Daily	Bare Costs	Incl. O&P
1 Carpenter	$43.05	$344.40	$66.05	$528.40	$43.05	$66.05
1 Small Compressor, Electric		11.05		12.15		
1 Pneumatic Nailer		23.50		25.85	4.32	4.75
8 L.H., Daily Totals		$378.95		$566.40	$47.37	$70.80

Crew H-1	Hr.	Daily	Hr.	Daily	Bare Costs	Incl. O&P
2 Glaziers	$42.05	$672.80	$63.15	$1010.40	$45.30	$74.45
2 Struc. Steel Workers	48.55	776.80	85.75	1372.00		
32 L.H., Daily Totals		$1449.60		$2382.40	$45.30	$74.45

Crew H-2	Hr.	Daily	Hr.	Daily	Bare Costs	Incl. O&P
2 Glaziers	$42.05	$672.80	$63.15	$1010.40	$39.48	$59.67
1 Building Laborer	34.35	274.80	52.70	421.60		
24 L.H., Daily Totals		$947.60		$1432.00	$39.48	$59.67

Crew H-3	Hr.	Daily	Hr.	Daily	Bare Costs	Incl. O&P
1 Glazier	$42.05	$336.40	$63.15	$505.20	$37.35	$56.52
1 Helper	32.65	261.20	49.90	399.20		
16 L.H., Daily Totals		$597.60		$904.40	$37.35	$56.52

Crew H-4	Hr.	Daily	Hr.	Daily	Bare Costs	Incl. O&P
1 Carpenter	$43.05	$344.40	$66.05	$528.40	$40.34	$61.28
1 Carpenter Helper	32.65	261.20	49.90	399.20		
.5 Electrician	50.30	201.20	74.50	298.00		
20 L.H., Daily Totals		$806.80		$1225.60	$40.34	$61.28

Crew J-1	Hr.	Daily	Hr.	Daily	Bare Costs	Incl. O&P
3 Plasterers	$39.40	$945.60	$58.60	$1406.40	$37.66	$56.00
2 Plasterer Helpers	35.05	560.80	52.10	833.60		
1 Mixing Machine, 6 C.F.		123.60		135.96	3.09	3.40
40 L.H., Daily Totals		$1630.00		$2375.96	$40.75	$59.40

Crew J-2	Hr.	Daily	Hr.	Daily	Bare Costs	Incl. O&P
3 Plasterers	$39.40	$945.60	$58.60	$1406.40	$37.81	$56.04
2 Plasterer Helpers	35.05	560.80	52.10	833.60		
1 Lather	38.55	308.40	56.25	450.00		
1 Mixing Machine, 6 C.F.		123.60		135.96	2.58	2.83
48 L.H., Daily Totals		$1938.40		$2825.96	$40.38	$58.87

Crew J-3	Hr.	Daily	Hr.	Daily	Bare Costs	Incl. O&P
1 Terrazzo Worker	$40.10	$320.80	$58.45	$467.60	$36.80	$53.65
1 Terrazzo Helper	33.50	268.00	48.85	390.80		
1 Floor Grinder, 22" path		73.95		81.34		
1 Terrazzo Mixer		167.60		184.36	15.10	16.61
16 L.H., Daily Totals		$830.35		$1124.11	$51.90	$70.26

Crew J-4	Hr.	Daily	Hr.	Daily	Bare Costs	Incl. O&P
2 Cement Finishers	$40.85	$653.60	$59.50	$952.00	$38.68	$57.23
1 Laborer	34.35	274.80	52.70	421.60		
1 Floor Grinder, 22" path		73.95		81.34		
1 Floor Edger, 7" path		38.20		42.02		
1 Vacuum Pick-Up System		59.80		65.78	7.16	7.88
24 L.H., Daily Totals		$1100.35		$1562.74	$45.85	$65.11

Crew J-4A	Hr.	Daily	Hr.	Daily	Bare Costs	Incl. O&P
2 Cement Finishers	$40.85	$653.60	$59.50	$952.00	$37.60	$56.10
2 Laborers	34.35	549.60	52.70	843.20		
1 Floor Grinder, 22" path		73.95		81.34		
1 Floor Edger, 7" path		38.20		42.02		
1 Vacuum Pick-Up System		59.80		65.78		
1 Floor Auto Scrubber		154.10		169.51	10.19	11.21
32 L.H., Daily Totals		$1529.25		$2153.86	$47.79	$67.31

Crew J-4B	Hr.	Daily	Hr.	Daily	Bare Costs	Incl. O&P
1 Laborer	$34.35	$274.80	$52.70	$421.60	$34.35	$52.70
1 Floor Auto Scrubber		154.10		169.51	19.26	21.19
8 L.H., Daily Totals		$428.90		$591.11	$53.61	$73.89

Crew K-1	Hr.	Daily	Hr.	Daily	Bare Costs	Incl. O&P
1 Carpenter	$43.05	$344.40	$66.05	$528.40	$38.23	$58.40
1 Truck Driver (light)	33.40	267.20	50.75	406.00		
1 Flatbed Truck, Gas, 3 Ton		217.00		238.70	13.56	14.92
16 L.H., Daily Totals		$828.60		$1173.10	$51.79	$73.32

Crew K-2	Hr.	Daily	Hr.	Daily	Bare Costs	Incl. O&P
1 Struc. Steel Foreman	$50.55	$404.40	$89.25	$714.00	$44.17	$75.25
1 Struc. Steel Worker	48.55	388.40	85.75	686.00		
1 Truck Driver (light)	33.40	267.20	50.75	406.00		
1 Flatbed Truck, Gas, 3 Ton		217.00		238.70	9.04	9.95
24 L.H., Daily Totals		$1277.00		$2044.70	$53.21	$85.20

Crew L-1	Hr.	Daily	Hr.	Daily	Bare Costs	Incl. O&P
1 Electrician	$50.30	$402.40	$74.50	$596.00	$51.95	$77.25
1 Plumber	53.60	428.80	80.00	640.00		
16 L.H., Daily Totals		$831.20		$1236.00	$51.95	$77.25

Crew L-2	Hr.	Daily	Hr.	Daily	Bare Costs	Incl. O&P
1 Carpenter	$43.05	$344.40	$66.05	$528.40	$37.85	$57.98
1 Carpenter Helper	32.65	261.20	49.90	399.20		
16 L.H., Daily Totals		$605.60		$927.60	$37.85	$57.98

Crew No.	Bare Costs Hr.	Daily	Incl. Subs O&P Hr.	Daily	Cost Per Labor-Hour Bare Costs	Incl. O&P
Crew L-3	Hr.	Daily	Hr.	Daily	Bare Costs	Incl. O&P
1 Carpenter	$43.05	$344.40	$66.05	$528.40	$47.01	$71.17
.5 Electrician	50.30	201.20	74.50	298.00		
.5 Sheet Metal Worker	51.65	206.60	78.10	312.40		
16 L.H., Daily Totals		$752.20		$1138.80	$47.01	$71.17
Crew L-3A	Hr.	Daily	Hr.	Daily	Bare Costs	Incl. O&P
1 Carpenter Foreman (outside)	$45.05	$360.40	$69.10	$552.80	$47.25	$72.10
.5 Sheet Metal Worker	51.65	206.60	78.10	312.40		
12 L.H., Daily Totals		$567.00		$865.20	$47.25	$72.10
Crew L-4	Hr.	Daily	Hr.	Daily	Bare Costs	Incl. O&P
2 Skilled Workers	$44.40	$710.40	$68.10	$1089.60	$40.48	$62.03
1 Helper	32.65	261.20	49.90	399.20		
24 L.H., Daily Totals		$971.60		$1488.80	$40.48	$62.03
Crew L-5	Hr.	Daily	Hr.	Daily	Bare Costs	Incl. O&P
1 Struc. Steel Foreman	$50.55	$404.40	$89.25	$714.00	$48.54	$83.92
5 Struc. Steel Workers	48.55	1942.00	85.75	3430.00		
1 Equip. Oper. (crane)	46.50	372.00	69.45	555.60		
1 Hyd. Crane, 25 Ton		748.60		823.46	13.37	14.70
56 L.H., Daily Totals		$3467.00		$5523.06	$61.91	$98.63
Crew L-5A	Hr.	Daily	Hr.	Daily	Bare Costs	Incl. O&P
1 Structural Steel Foreman	$50.55	$404.40	$89.25	$714.00	$48.54	$82.55
2 Structural Steel Workers	48.55	776.80	85.75	1372.00		
1 Equip. Oper. (crane)	46.50	372.00	69.45	555.60		
1 S.P. Crane, 4x4, 25 Ton		560.40		616.44	17.51	19.26
32 L.H., Daily Totals		$2113.60		$3258.04	$66.05	$101.81
Crew L-5B	Hr.	Daily	Hr.	Daily	Bare Costs	Incl. O&P
1 Structural Steel Foreman	$50.55	$404.40	$89.25	$714.00	$49.25	$77.88
2 Structural Steel Workers	48.55	776.80	85.75	1372.00		
2 Electricians	50.30	804.80	74.50	1192.00		
2 Steamfitters/Pipefitters	54.10	865.60	80.75	1292.00		
1 Equip. Oper. (crane)	46.50	372.00	69.45	555.60		
1 Equip. Oper. Oiler	40.30	322.40	60.20	481.60		
1 Hyd. Crane, 80 Ton		1693.00		1862.30	23.51	25.87
72 L.H., Daily Totals		$5239.00		$7469.50	$72.76	$103.74
Crew L-6	Hr.	Daily	Hr.	Daily	Bare Costs	Incl. O&P
1 Plumber	$53.60	$428.80	$80.00	$640.00	$52.50	$78.17
.5 Electrician	50.30	201.20	74.50	298.00		
12 L.H., Daily Totals		$630.00		$938.00	$52.50	$78.17
Crew L-7	Hr.	Daily	Hr.	Daily	Bare Costs	Incl. O&P
2 Carpenters	$43.05	$688.80	$66.05	$1056.80	$41.60	$63.44
1 Building Laborer	34.35	274.80	52.70	421.60		
.5 Electrician	50.30	201.20	74.50	298.00		
28 L.H., Daily Totals		$1164.80		$1776.40	$41.60	$63.44
Crew L-8	Hr.	Daily	Hr.	Daily	Bare Costs	Incl. O&P
2 Carpenters	$43.05	$688.80	$66.05	$1056.80	$45.16	$68.84
.5 Plumber	53.60	214.40	80.00	320.00		
20 L.H., Daily Totals		$903.20		$1376.80	$45.16	$68.84

Crew No.	Bare Costs Hr.	Daily	Incl. Subs O&P Hr.	Daily	Cost Per Labor-Hour Bare Costs	Incl. O&P
Crew L-9	Hr.	Daily	Hr.	Daily	Bare Costs	Incl. O&P
1 Labor Foreman (inside)	$34.85	$278.80	$53.45	$427.60	$39.39	$62.63
2 Building Laborers	34.35	549.60	52.70	843.20		
1 Struc. Steel Worker	48.55	388.40	85.75	686.00		
.5 Electrician	50.30	201.20	74.50	298.00		
36 L.H., Daily Totals		$1418.00		$2254.80	$39.39	$62.63
Crew L-10	Hr.	Daily	Hr.	Daily	Bare Costs	Incl. O&P
1 Structural Steel Foreman	$50.55	$404.40	$89.25	$714.00	$48.53	$81.48
1 Structural Steel Worker	48.55	388.40	85.75	686.00		
1 Equip. Oper. (crane)	46.50	372.00	69.45	555.60		
1 Hyd. Crane, 12 Ton		655.60		721.16	27.32	30.05
24 L.H., Daily Totals		$1820.40		$2676.76	$75.85	$111.53
Crew L-11	Hr.	Daily	Hr.	Daily	Bare Costs	Incl. O&P
2 Wreckers	$34.35	$549.60	$57.80	$924.80	$39.70	$62.55
1 Equip. Oper. (crane)	46.50	372.00	69.45	555.60		
1 Equip. Oper. (light)	43.60	348.80	65.15	521.20		
1 Hyd. Excavator, 2.5 C.Y.		1613.00		1774.30		
1 Loader, Skid Steer, 78 H.P.		261.20		287.32	58.57	64.43
32 L.H., Daily Totals		$3144.60		$4063.22	$98.27	$126.98
Crew M-1	Hr.	Daily	Hr.	Daily	Bare Costs	Incl. O&P
3 Elevator Constructors	$67.40	$1617.60	$100.00	$2400.00	$64.03	$95.00
1 Elevator Apprentice	53.90	431.20	80.00	640.00		
5 Hand Tools		49.00		53.90	1.53	1.68
32 L.H., Daily Totals		$2097.80		$3093.90	$65.56	$96.68
Crew M-3	Hr.	Daily	Hr.	Daily	Bare Costs	Incl. O&P
1 Electrician Foreman (out)	$52.30	$418.40	$77.45	$619.60	$51.60	$76.96
1 Common Laborer	34.35	274.80	52.70	421.60		
.25 Equipment Operator, Med.	45.35	90.70	67.75	135.50		
1 Elevator Constructor	67.40	539.20	100.00	800.00		
1 Elevator Apprentice	53.90	431.20	80.00	640.00		
.25 S.P. Crane, 4x4, 20 Ton		127.75		140.53	3.76	4.13
34 L.H., Daily Totals		$1882.05		$2757.22	$55.35	$81.09
Crew M-4	Hr.	Daily	Hr.	Daily	Bare Costs	Incl. O&P
1 Electrician Foreman (out)	$52.30	$418.40	$77.45	$619.60	$51.03	$76.13
1 Common Laborer	34.35	274.80	52.70	421.60		
.25 Equipment Operator, Crane	46.50	93.00	69.45	138.90		
.25 Equipment Operator, Oiler	40.30	80.60	60.20	120.40		
1 Elevator Constructor	67.40	539.20	100.00	800.00		
1 Elevator Apprentice	53.90	431.20	80.00	640.00		
.25 S.P. Crane, 4x4, 40 Ton		164.65		181.12	4.57	5.03
36 L.H., Daily Totals		$2001.85		$2921.61	$55.61	$81.16
Crew Q-1	Hr.	Daily	Hr.	Daily	Bare Costs	Incl. O&P
1 Plumber	$53.60	$428.80	$80.00	$640.00	$48.25	$72.03
1 Plumber Apprentice	42.90	343.20	64.05	512.40		
16 L.H., Daily Totals		$772.00		$1152.40	$48.25	$72.03
Crew Q-1A	Hr.	Daily	Hr.	Daily	Bare Costs	Incl. O&P
.25 Plumber Foreman (out)	$55.60	$111.20	$83.00	$166.00	$54.00	$80.60
1 Plumber	53.60	428.80	80.00	640.00		
10 L.H., Daily Totals		$540.00		$806.00	$54.00	$80.60

Crew Q-1C

	Bare Costs Hr.	Daily	Incl. Subs O&P Hr.	Daily	Cost Per Labor-Hour Bare Costs	Incl. O&P
1 Plumber	$53.60	$428.80	$80.00	$640.00	$47.28	$70.60
1 Plumber Apprentice	42.90	343.20	64.05	512.40		
1 Equip. Oper. (medium)	45.35	362.80	67.75	542.00		
1 Trencher, Chain Type, 8' D		2506.00		2756.60	104.42	114.86
24 L.H., Daily Totals		$3640.80		$4451.00	$151.70	$185.46

Crew Q-2

	Bare Costs Hr.	Daily	Incl. Subs O&P Hr.	Daily	Cost Per Labor-Hour Bare Costs	Incl. O&P
2 Plumbers	$53.60	$857.60	$80.00	$1280.00	$50.03	$74.68
1 Plumber Apprentice	42.90	343.20	64.05	512.40		
24 L.H., Daily Totals		$1200.80		$1792.40	$50.03	$74.68

Crew Q-3

	Bare Costs Hr.	Daily	Incl. Subs O&P Hr.	Daily	Cost Per Labor-Hour Bare Costs	Incl. O&P
1 Plumber Foreman (inside)	$54.10	$432.80	$80.75	$646.00	$51.05	$76.20
2 Plumbers	53.60	857.60	80.00	1280.00		
1 Plumber Apprentice	42.90	343.20	64.05	512.40		
32 L.H., Daily Totals		$1633.60		$2438.40	$51.05	$76.20

Crew Q-4

	Bare Costs Hr.	Daily	Incl. Subs O&P Hr.	Daily	Cost Per Labor-Hour Bare Costs	Incl. O&P
1 Plumber Foreman (inside)	$54.10	$432.80	$80.75	$646.00	$51.05	$76.20
1 Plumber	53.60	428.80	80.00	640.00		
1 Welder (plumber)	53.60	428.80	80.00	640.00		
1 Plumber Apprentice	42.90	343.20	64.05	512.40		
1 Welder, electric, 300 amp		32.45		35.70	1.01	1.12
32 L.H., Daily Totals		$1666.05		$2474.09	$52.06	$77.32

Crew Q-5

	Bare Costs Hr.	Daily	Incl. Subs O&P Hr.	Daily	Cost Per Labor-Hour Bare Costs	Incl. O&P
1 Steamfitter	$54.10	$432.80	$80.75	$646.00	$48.70	$72.70
1 Steamfitter Apprentice	43.30	346.40	64.65	517.20		
16 L.H., Daily Totals		$779.20		$1163.20	$48.70	$72.70

Crew Q-6

	Bare Costs Hr.	Daily	Incl. Subs O&P Hr.	Daily	Cost Per Labor-Hour Bare Costs	Incl. O&P
2 Steamfitters	$54.10	$865.60	$80.75	$1292.00	$50.50	$75.38
1 Steamfitter Apprentice	43.30	346.40	64.65	517.20		
24 L.H., Daily Totals		$1212.00		$1809.20	$50.50	$75.38

Crew Q-7

	Bare Costs Hr.	Daily	Incl. Subs O&P Hr.	Daily	Cost Per Labor-Hour Bare Costs	Incl. O&P
1 Steamfitter Foreman (inside)	$54.60	$436.80	$81.50	$652.00	$51.52	$76.91
2 Steamfitters	54.10	865.60	80.75	1292.00		
1 Steamfitter Apprentice	43.30	346.40	64.65	517.20		
32 L.H., Daily Totals		$1648.80		$2461.20	$51.52	$76.91

Crew Q-8

	Bare Costs Hr.	Daily	Incl. Subs O&P Hr.	Daily	Cost Per Labor-Hour Bare Costs	Incl. O&P
1 Steamfitter Foreman (inside)	$54.60	$436.80	$81.50	$652.00	$51.52	$76.91
1 Steamfitter	54.10	432.80	80.75	646.00		
1 Welder (steamfitter)	54.10	432.80	80.75	646.00		
1 Steamfitter Apprentice	43.30	346.40	64.65	517.20		
1 Welder, electric, 300 amp		32.45		35.70	1.01	1.12
32 L.H., Daily Totals		$1681.25		$2496.90	$52.54	$78.03

Crew Q-9

	Bare Costs Hr.	Daily	Incl. Subs O&P Hr.	Daily	Cost Per Labor-Hour Bare Costs	Incl. O&P
1 Sheet Metal Worker	$51.65	$413.20	$78.10	$624.80	$46.48	$70.28
1 Sheet Metal Apprentice	41.30	330.40	62.45	499.60		
16 L.H., Daily Totals		$743.60		$1124.40	$46.48	$70.28

Crew Q-10

	Bare Costs Hr.	Daily	Incl. Subs O&P Hr.	Daily	Cost Per Labor-Hour Bare Costs	Incl. O&P
2 Sheet Metal Workers	$51.65	$826.40	$78.10	$1249.60	$48.20	$72.88
1 Sheet Metal Apprentice	41.30	330.40	62.45	499.60		
24 L.H., Daily Totals		$1156.80		$1749.20	$48.20	$72.88

Crew Q-11

	Bare Costs Hr.	Daily	Incl. Subs O&P Hr.	Daily	Cost Per Labor-Hour Bare Costs	Incl. O&P
1 Sheet Metal Foreman (inside)	$52.15	$417.20	$78.85	$630.80	$49.19	$74.38
2 Sheet Metal Workers	51.65	826.40	78.10	1249.60		
1 Sheet Metal Apprentice	41.30	330.40	62.45	499.60		
32 L.H., Daily Totals		$1574.00		$2380.00	$49.19	$74.38

Crew Q-12

	Bare Costs Hr.	Daily	Incl. Subs O&P Hr.	Daily	Cost Per Labor-Hour Bare Costs	Incl. O&P
1 Sprinkler Installer	$53.10	$424.80	$79.35	$634.80	$47.80	$71.42
1 Sprinkler Apprentice	42.50	340.00	63.50	508.00		
16 L.H., Daily Totals		$764.80		$1142.80	$47.80	$71.42

Crew Q-13

	Bare Costs Hr.	Daily	Incl. Subs O&P Hr.	Daily	Cost Per Labor-Hour Bare Costs	Incl. O&P
1 Sprinkler Foreman (inside)	$53.60	$428.80	$80.10	$640.80	$50.58	$75.58
2 Sprinkler Installers	53.10	849.60	79.35	1269.60		
1 Sprinkler Apprentice	42.50	340.00	63.50	508.00		
32 L.H., Daily Totals		$1618.40		$2418.40	$50.58	$75.58

Crew Q-14

	Bare Costs Hr.	Daily	Incl. Subs O&P Hr.	Daily	Cost Per Labor-Hour Bare Costs	Incl. O&P
1 Asbestos Worker	$46.90	$375.20	$72.50	$580.00	$42.20	$65.25
1 Asbestos Apprentice	37.50	300.00	58.00	464.00		
16 L.H., Daily Totals		$675.20		$1044.00	$42.20	$65.25

Crew Q-15

	Bare Costs Hr.	Daily	Incl. Subs O&P Hr.	Daily	Cost Per Labor-Hour Bare Costs	Incl. O&P
1 Plumber	$53.60	$428.80	$80.00	$640.00	$48.25	$72.03
1 Plumber Apprentice	42.90	343.20	64.05	512.40		
1 Welder, electric, 300 amp		32.45		35.70	2.03	2.23
16 L.H., Daily Totals		$804.45		$1188.10	$50.28	$74.26

Crew Q-16

	Bare Costs Hr.	Daily	Incl. Subs O&P Hr.	Daily	Cost Per Labor-Hour Bare Costs	Incl. O&P
2 Plumbers	$53.60	$857.60	$80.00	$1280.00	$50.03	$74.68
1 Plumber Apprentice	42.90	343.20	64.05	512.40		
1 Welder, electric, 300 amp		32.45		35.70	1.35	1.49
24 L.H., Daily Totals		$1233.25		$1828.10	$51.39	$76.17

Crew Q-17

	Bare Costs Hr.	Daily	Incl. Subs O&P Hr.	Daily	Cost Per Labor-Hour Bare Costs	Incl. O&P
1 Steamfitter	$54.10	$432.80	$80.75	$646.00	$48.70	$72.70
1 Steamfitter Apprentice	43.30	346.40	64.65	517.20		
1 Welder, electric, 300 amp		32.45		35.70	2.03	2.23
16 L.H., Daily Totals		$811.65		$1198.90	$50.73	$74.93

Crew Q-17A

	Bare Costs Hr.	Daily	Incl. Subs O&P Hr.	Daily	Cost Per Labor-Hour Bare Costs	Incl. O&P
1 Steamfitter	$54.10	$432.80	$80.75	$646.00	$47.97	$71.62
1 Steamfitter Apprentice	43.30	346.40	64.65	517.20		
1 Equip. Oper. (crane)	46.50	372.00	69.45	555.60		
1 Hyd. Crane, 12 Ton		655.60		721.16		
1 Welder, electric, 300 amp		32.45		35.70	28.67	31.54
24 L.H., Daily Totals		$1839.25		$2475.66	$76.64	$103.15

Crew Q-18

	Bare Costs Hr.	Daily	Incl. Subs O&P Hr.	Daily	Cost Per Labor-Hour Bare Costs	Incl. O&P
2 Steamfitters	$54.10	$865.60	$80.75	$1292.00	$50.50	$75.38
1 Steamfitter Apprentice	43.30	346.40	64.65	517.20		
1 Welder, electric, 300 amp		32.45		35.70	1.35	1.49
24 L.H., Daily Totals		$1244.45		$1844.90	$51.85	$76.87

Crew Q-19

	Bare Costs Hr.	Daily	Incl. Subs O&P Hr.	Daily	Cost Per Labor-Hour Bare Costs	Incl. O&P
1 Steamfitter	$54.10	$432.80	$80.75	$646.00	$49.23	$73.30
1 Steamfitter Apprentice	43.30	346.40	64.65	517.20		
1 Electrician	50.30	402.40	74.50	596.00		
24 L.H., Daily Totals		$1181.60		$1759.20	$49.23	$73.30

Crew Q-20

Crew No.	Hr.	Daily	Hr.	Daily	Bare Costs	Incl. O&P
1 Sheet Metal Worker	$51.65	$413.20	$78.10	$624.80	$47.24	$71.12
1 Sheet Metal Apprentice	41.30	330.40	62.45	499.60		
.5 Electrician	50.30	201.20	74.50	298.00		
20 L.H., Daily Totals		$944.80		$1422.40	$47.24	$71.12

Crew Q-21

Crew No.	Hr.	Daily	Hr.	Daily	Bare Costs	Incl. O&P
2 Steamfitters	$54.10	$865.60	$80.75	$1292.00	$50.45	$75.16
1 Steamfitter Apprentice	43.30	346.40	64.65	517.20		
1 Electrician	50.30	402.40	74.50	596.00		
32 L.H., Daily Totals		$1614.40		$2405.20	$50.45	$75.16

Crew Q-22

Crew No.	Hr.	Daily	Hr.	Daily	Bare Costs	Incl. O&P
1 Plumber	$53.60	$428.80	$80.00	$640.00	$48.25	$72.03
1 Plumber Apprentice	42.90	343.20	64.05	512.40		
1 Hyd. Crane, 12 Ton		655.60		721.16	40.98	45.07
16 L.H., Daily Totals		$1427.60		$1873.56	$89.22	$117.10

Crew Q-22A

Crew No.	Hr.	Daily	Hr.	Daily	Bare Costs	Incl. O&P
1 Plumber	$53.60	$428.80	$80.00	$640.00	$44.34	$66.55
1 Plumber Apprentice	42.90	343.20	64.05	512.40		
1 Laborer	34.35	274.80	52.70	421.60		
1 Equip. Oper. (crane)	46.50	372.00	69.45	555.60		
1 Hyd. Crane, 12 Ton		655.60		721.16	20.49	22.54
32 L.H., Daily Totals		$2074.40		$2850.76	$64.83	$89.09

Crew Q-23

Crew No.	Hr.	Daily	Hr.	Daily	Bare Costs	Incl. O&P
1 Plumber Foreman	$55.60	$444.80	$83.00	$664.00	$51.52	$76.92
1 Plumber	53.60	428.80	80.00	640.00		
1 Equip. Oper. (med.)	45.35	362.80	67.75	542.00		
1 Lattice Boom Crane, 20 Ton		1035.00		1138.50	43.13	47.44
24 L.H., Daily Totals		$2271.40		$2984.50	$94.64	$124.35

Crew R-1

Crew No.	Hr.	Daily	Hr.	Daily	Bare Costs	Incl. O&P
1 Electrician Foreman	$50.80	$406.40	$75.25	$602.00	$44.50	$66.42
3 Electricians	50.30	1207.20	74.50	1788.00		
2 Helpers	32.65	522.40	49.90	798.40		
48 L.H., Daily Totals		$2136.00		$3188.40	$44.50	$66.42

Crew R-1A

Crew No.	Hr.	Daily	Hr.	Daily	Bare Costs	Incl. O&P
1 Electrician	$50.30	$402.40	$74.50	$596.00	$41.48	$62.20
1 Helper	32.65	261.20	49.90	399.20		
16 L.H., Daily Totals		$663.60		$995.20	$41.48	$62.20

Crew R-2

Crew No.	Hr.	Daily	Hr.	Daily	Bare Costs	Incl. O&P
1 Electrician Foreman	$50.80	$406.40	$75.25	$602.00	$44.79	$66.86
3 Electricians	50.30	1207.20	74.50	1788.00		
2 Helpers	32.65	522.40	49.90	798.40		
1 Equip. Oper. (crane)	46.50	372.00	69.45	555.60		
1 S.P. Crane, 4x4, 5 Ton		257.80		283.58	4.60	5.06
56 L.H., Daily Totals		$2765.80		$4027.58	$49.39	$71.92

Crew R-3

Crew No.	Hr.	Daily	Hr.	Daily	Bare Costs	Incl. O&P
1 Electrician Foreman	$50.80	$406.40	$75.25	$602.00	$49.74	$73.79
1 Electrician	50.30	402.40	74.50	596.00		
.5 Equip. Oper. (crane)	46.50	186.00	69.45	277.80		
.5 S.P. Crane, 4x4, 5 Ton		128.90		141.79	6.45	7.09
20 L.H., Daily Totals		$1123.70		$1617.59	$56.19	$80.88

Crew R-4

Crew No.	Hr.	Daily	Hr.	Daily	Bare Costs	Incl. O&P
1 Struc. Steel Foreman	$50.55	$404.40	$89.25	$714.00	$49.30	$84.20
3 Struc. Steel Workers	48.55	1165.20	85.75	2058.00		
1 Electrician	50.30	402.40	74.50	596.00		
1 Welder, gas engine, 300 amp		109.00		119.90	2.73	3.00
40 L.H., Daily Totals		$2081.00		$3487.90	$52.02	$87.20

Crew R-5

Crew No.	Hr.	Daily	Hr.	Daily	Bare Costs	Incl. O&P
1 Electrician Foreman	$50.80	$406.40	$75.25	$602.00	$43.93	$65.62
4 Electrician Linemen	50.30	1609.60	74.50	2384.00		
2 Electrician Operators	50.30	804.80	74.50	1192.00		
4 Electrician Groundmen	32.65	1044.80	49.90	1596.80		
1 Crew Truck		152.40		167.64		
1 Flatbed Truck, 20,000 GVW		190.00		209.00		
1 Pickup Truck, 3/4 Ton		104.60		115.06		
.2 Hyd. Crane, 55 Ton		241.60		265.76		
.2 Hyd. Crane, 12 Ton		131.12		144.23		
.2 Earth Auger, Truck-Mtd.		80.36		88.40		
1 Tractor w/Winch		346.80		381.48	14.17	15.59
88 L.H., Daily Totals		$5112.48		$7146.37	$58.10	$81.21

Crew R-6

Crew No.	Hr.	Daily	Hr.	Daily	Bare Costs	Incl. O&P
1 Electrician Foreman	$50.80	$406.40	$75.25	$602.00	$43.93	$65.62
4 Electrician Linemen	50.30	1609.60	74.50	2384.00		
2 Electrician Operators	50.30	804.80	74.50	1192.00		
4 Electrician Groundmen	32.65	1044.80	49.90	1596.80		
1 Crew Truck		152.40		167.64		
1 Flatbed Truck, 20,000 GVW		190.00		209.00		
1 Pickup Truck, 3/4 Ton		104.60		115.06		
.2 Hyd. Crane, 55 Ton		241.60		265.76		
.2 Hyd. Crane, 12 Ton		131.12		144.23		
.2 Earth Auger, Truck-Mtd.		80.36		88.40		
1 Tractor w/Winch		346.80		381.48		
3 Cable Trailers		552.90		608.19		
.5 Tensioning Rig		183.05		201.35		
.5 Cable Pulling Rig		1072.00		1179.20	34.71	38.19
88 L.H., Daily Totals		$6920.43		$9135.11	$78.64	$103.81

Crew R-7

Crew No.	Hr.	Daily	Hr.	Daily	Bare Costs	Incl. O&P
1 Electrician Foreman	$50.80	$406.40	$75.25	$602.00	$35.67	$54.13
5 Electrician Groundmen	32.65	1306.00	49.90	1996.00		
1 Crew Truck		152.40		167.64	3.17	3.49
48 L.H., Daily Totals		$1864.80		$2765.64	$38.85	$57.62

Crew R-8

Crew No.	Hr.	Daily	Hr.	Daily	Bare Costs	Incl. O&P
1 Electrician Foreman	$50.80	$406.40	$75.25	$602.00	$44.50	$66.42
3 Electrician Linemen	50.30	1207.20	74.50	1788.00		
2 Electrician Groundmen	32.65	522.40	49.90	798.40		
1 Pickup Truck, 3/4 Ton		104.60		115.06		
1 Crew Truck		152.40		167.64	5.35	5.89
48 L.H., Daily Totals		$2393.00		$3471.10	$49.85	$72.31

Crew R-9

Crew No.	Hr.	Daily	Hr.	Daily	Bare Costs	Incl. O&P
1 Electrician Foreman	$50.80	$406.40	$75.25	$602.00	$41.54	$62.29
1 Electrician Lineman	50.30	402.40	74.50	596.00		
2 Electrician Operators	50.30	804.80	74.50	1192.00		
4 Electrician Groundmen	32.65	1044.80	49.90	1596.80		
1 Pickup Truck, 3/4 Ton		104.60		115.06		
1 Crew Truck		152.40		167.64	4.02	4.42
64 L.H., Daily Totals		$2915.40		$4269.50	$45.55	$66.71

Crew No.	Bare Costs		Incl. Subs O&P		Cost Per Labor-Hour	

Crew R-10	Hr.	Daily	Hr.	Daily	Bare Costs	Incl. O&P
1 Electrician Foreman	$50.80	$406.40	$75.25	$602.00	$47.44	$70.53
4 Electrician Linemen	50.30	1609.60	74.50	2384.00		
1 Electrician Groundman	32.65	261.20	49.90	399.20		
1 Crew Truck		152.40		167.64		
3 Tram Cars		339.90		373.89	10.26	11.28
48 L.H., Daily Totals		$2769.50		$3926.73	$57.70	$81.81

Crew R-11	Hr.	Daily	Hr.	Daily	Bare Costs	Incl. O&P
1 Electrician Foreman	$50.80	$406.40	$75.25	$602.00	$47.55	$70.77
4 Electricians	50.30	1609.60	74.50	2384.00		
1 Equip. Oper. (crane)	46.50	372.00	69.45	555.60		
1 Common Laborer	34.35	274.80	52.70	421.60		
1 Crew Truck		152.40		167.64		
1 Hyd. Crane, 12 Ton		655.60		721.16	14.43	15.87
56 L.H., Daily Totals		$3470.80		$4852.00	$61.98	$86.64

Crew R-12	Hr.	Daily	Hr.	Daily	Bare Costs	Incl. O&P
1 Carpenter Foreman	$43.55	$348.40	$66.80	$534.40	$40.64	$63.21
4 Carpenters	43.05	1377.60	66.05	2113.60		
4 Common Laborers	34.35	1099.20	52.70	1686.40		
1 Equip. Oper. (med.)	45.35	362.80	67.75	542.00		
1 Steel Worker	48.55	388.40	85.75	686.00		
1 Dozer, 200 H.P.		1051.00		1156.10		
1 Pickup Truck, 3/4 Ton		104.60		115.06	13.13	14.45
88 L.H., Daily Totals		$4732.00		$6833.56	$53.77	$77.65

Crew R-13	Hr.	Daily	Hr.	Daily	Bare Costs	Incl. O&P
1 Electrician Foreman	$50.80	$406.40	$75.25	$602.00	$48.31	$71.68
3 Electricians	50.30	1207.20	74.50	1788.00		
.25 Equip. Oper. (crane)	46.50	93.00	69.45	138.90		
1 Equipment Oiler	40.30	322.40	60.20	481.60		
.25 Hydraulic Crane, 33 Ton		190.75		209.82	4.54	5.00
42 L.H., Daily Totals		$2219.75		$3220.32	$52.85	$76.67

Crew R-15	Hr.	Daily	Hr.	Daily	Bare Costs	Incl. O&P
1 Electrician Foreman	$50.80	$406.40	$75.25	$602.00	$49.27	$73.07
4 Electricians	50.30	1609.60	74.50	2384.00		
1 Equipment Oper. (light)	43.60	348.80	65.15	521.20		
1 Aerial Lift Truck		317.80		349.58	6.62	7.28
48 L.H., Daily Totals		$2682.60		$3856.78	$55.89	$80.35

Crew R-18	Hr.	Daily	Hr.	Daily	Bare Costs	Incl. O&P
.25 Electrician Foreman	$50.80	$101.60	$75.25	$150.50	$39.48	$59.42
1 Electrician	50.30	402.40	74.50	596.00		
2 Helpers	32.65	522.40	49.90	798.40		
26 L.H., Daily Totals		$1026.40		$1544.90	$39.48	$59.42

Crew R-19	Hr.	Daily	Hr.	Daily	Bare Costs	Incl. O&P
.5 Electrician Foreman	$50.80	$203.20	$75.25	$301.00	$50.40	$74.65
2 Electricians	50.30	804.80	74.50	1192.00		
20 L.H., Daily Totals		$1008.00		$1493.00	$50.40	$74.65

Crew R-21	Hr.	Daily	Hr.	Daily	Bare Costs	Incl. O&P
1 Electrician Foreman	$50.80	$406.40	$75.25	$602.00	$50.30	$74.52
3 Electricians	50.30	1207.20	74.50	1788.00		
.1 Equip. Oper. (med.)	45.35	36.28	67.75	54.20		
.1 S.P. Crane, 4x4, 25 Ton		56.04		61.64	1.71	1.88
32.8 L.H., Daily Totals		$1705.92		$2505.84	$52.01	$76.40

Crew R-22	Hr.	Daily	Hr.	Daily	Bare Costs	Incl. O&P
.66 Electrician Foreman	$50.80	$268.22	$75.25	$397.32	$42.80	$64.05
2 Helpers	32.65	522.40	49.90	798.40		
2 Electricians	50.30	804.80	74.50	1192.00		
37.28 L.H., Daily Totals		$1595.42		$2387.72	$42.80	$64.05

Crew R-30	Hr.	Daily	Hr.	Daily	Bare Costs	Incl. O&P
.25 Electrician Foreman (out)	$52.30	$104.60	$77.45	$154.90	$40.64	$61.31
1 Electrician	50.30	402.40	74.50	596.00		
2 Laborers, (Semi-Skilled)	34.35	549.60	52.70	843.20		
26 L.H., Daily Totals		$1056.60		$1594.10	$40.64	$61.31

Crew R-31	Hr.	Daily	Hr.	Daily	Bare Costs	Incl. O&P
1 Electrician	$50.30	$402.40	$74.50	$596.00	$50.30	$74.50
1 Core Drill, Electric, 2.5 H.P.		47.00		51.70	5.88	6.46
8 L.H., Daily Totals		$449.40		$647.70	$56.17	$80.96

Crew W-41E	Hr.	Daily	Hr.	Daily	Bare Costs	Incl. O&P
.5 Plumber Foreman (out)	$55.60	$222.40	$83.00	$332.00	$46.30	$69.68
1 Plumber	53.60	428.80	80.00	640.00		
1 Laborer	34.35	274.80	52.70	421.60		
20 L.H., Daily Totals		$926.00		$1393.60	$46.30	$69.68

Historical Cost Indexes

The table below lists both the RSMeans Historical Cost Index based on Jan. 1, 1993 = 100 as well as the computed value of an index based on Jan. 1, 2011 costs. Since the Jan. 1, 2011 figure is estimated, space is left to write in the actual index figures as they become available through either the quarterly *Means Construction Cost Indexes* or as printed in the *Engineering News-Record*. To compute the actual index based on Jan. 1, 2011 = 100, divide the Historical Cost Index for a particular year by the actual Jan. 1, 2011 Construction Cost Index. Space has been left to advance the index figures as the year progresses.

Year	Historical Cost Index Jan. 1, 1993 = 100		Current Index Based on Jan. 1, 2011 = 100		Year	Historical Cost Index Jan. 1, 1993 = 100	Current Index Based on Jan. 1, 2011 = 100		Year	Historical Cost Index Jan. 1, 1993 = 100	Current Index Based on Jan. 1, 2011 = 100	
	Est.	Actual	Est.	Actual		Actual	Est.	Actual		Actual	Est.	Actual
Oct 2011*					July 1996	110.2	59.6		July 1978	53.5	28.9	
July 2011*					1995	107.6	58.2		1977	49.5	26.8	
April 2011*					1994	104.4	56.4		1976	46.9	25.4	
Jan 2011*	185.0		100.0	100.0	1993	101.7	55.0		1975	44.8	24.2	
July 2010		183.5	99.2		1992	99.4	53.7		1974	41.4	22.4	
2009		180.1	97.4		1991	96.8	52.3		1973	37.7	20.4	
2008		180.4	97.5		1990	94.3	51.0		1972	34.8	18.8	
2007		169.4	91.6		1989	92.1	49.8		1971	32.1	17.4	
2006		162.0	87.6		1988	89.9	48.6		1970	28.7	15.5	
2005		151.6	81.9		1987	87.7	47.4		1969	26.9	14.5	
2004		143.7	77.7		1986	84.2	45.5		1968	24.9	13.5	
2003		132.0	71.4		1985	82.6	44.7		1967	23.5	12.7	
2002		128.7	69.6		1984	82.0	44.3		1966	22.7	12.3	
2001		125.1	67.6		1983	80.2	43.3		1965	21.7	11.7	
2000		120.9	65.4		1982	76.1	41.2		1964	21.2	11.5	
1999		117.6	63.6		1981	70.0	37.8		1963	20.7	11.2	
1998		115.1	62.2		1980	62.9	34.0		1962	20.2	10.9	
1997		112.8	61.0		1979	57.8	31.2		1961	19.8	10.7	

Adjustments to Costs

The Historical Cost Index can be used to convert National Average building costs at a particular time to the approximate building costs for some other time.

Example:

Estimate and compare construction costs for different years in the same city.

To estimate the National Average construction cost of a building in 1970, knowing that it cost $900,000 in 2011:

INDEX in 1970 = 28.7

INDEX in 2011 = 185.0

Note: The City Cost Indexes for Canada can be used to convert U.S. National averages to local costs in Canadian dollars.

*Historical Cost Index updates and other resources are provided on the following website.
http://www.reedconstructiondata.com/rsmeans/chgnotice/842339

Time Adjustment using the Historical Cost Indexes:

$$\frac{\text{Index for Year A}}{\text{Index for Year B}} \times \text{Cost in Year B} = \text{Cost in Year A}$$

$$\frac{\text{INDEX 1970}}{\text{INDEX 2011}} \times \text{Cost 2011} = \text{Cost 1970}$$

$$\frac{28.7}{185.0} \times \$900,000 = .155 \times \$900,000 = \$139,500$$

The construction cost of the building in 1970 is $139,500.

How to Use the City Cost Indexes

What you should know before you begin

RSMeans City Cost Indexes (CCI) are an extremely useful tool to use when you want to compare costs from city to city and region to region.

This publication contains average construction cost indexes for 731 U.S. and Canadian cities covering over 930 three-digit zip code locations, as listed directly under each city.

Keep in mind that a City Cost Index number is a percentage ratio of a specific city's cost to the national average cost of the same item at a stated time period.

In other words, these index figures represent relative construction factors (or, if you prefer, multipliers) for Material and Installation costs, as well as the weighted average for Total In Place costs for each CSI MasterFormat division. Installation costs include both labor and equipment rental costs. When estimating equipment rental rates only, for a specific location, use 015433 CONTRACTOR EQUIPMENT index.

The 30 City Average Index is the average of 30 major U.S. cities and serves as a National Average.

Index figures for both material and installation are based on the 30 major city average of 100 and represent the cost relationship as of July 1, 2010. The index for each division is computed from representative material and labor quantities for that division. The weighted average for each city is a weighted total of the components listed above it, but does not include relative productivity between trades or cities.

As changes occur in local material prices, labor rates, and equipment rental rates, (including fuel costs) the impact of these changes should be accurately measured by the change in the City Cost Index for each particular city (as compared to the 30 City Average).

Therefore, if you know (or have estimated) building costs in one city today, you can easily convert those costs to expected building costs in another city.

In addition, by using the Historical Cost Index, you can easily convert National Average building costs at a particular time to the approximate building costs for some other time. The City Cost Indexes can then be applied to calculate the costs for a particular city.

Quick Calculations

Location Adjustment Using the City Cost Indexes:

$$\frac{\text{Index for City A}}{\text{Index for City B}} \times \text{Cost in City B} = \text{Cost in City A}$$

Time Adjustment for the National Average Using the Historical Cost Index:

$$\frac{\text{Index for Year A}}{\text{Index for Year B}} \times \text{Cost in Year B} = \text{Cost in Year A}$$

Adjustment from the National Average:

$$\frac{\text{Index for City A}}{100} \times \text{National Average Cost} = \text{Cost in City A}$$

Since each of the other RSMeans publications contains many different items, any *one* item multiplied by the particular city index may give incorrect results. However, the larger the number of items compiled, the closer the results should be to actual costs for that particular city.

The City Cost Indexes for Canadian cities are calculated using Canadian material and equipment prices and labor rates, in Canadian dollars. Therefore, indexes for Canadian cities can be used to convert U.S. National Average prices to local costs in Canadian dollars.

How to use this section

1. Compare costs from city to city.

In using the RSMeans Indexes, remember that an index number is not a fixed number but a ratio: It's a percentage ratio of a building component's cost at any stated time to the National Average cost of that same component at the same time period. Put in the form of an equation:

$$\frac{\text{Specific City Cost}}{\text{National Average Cost}} \times 100 = \text{City Index Number}$$

Therefore, when making cost comparisons between cities, do not subtract one city's index number from the index number of another city and read the result as a percentage difference. Instead, divide one city's index number by that of the other city. The resulting number may then be used as a multiplier to calculate cost differences from city to city.

The formula used to find cost differences between cities for the purpose of comparison is as follows:

$$\frac{\text{City A Index}}{\text{City B Index}} \times \text{City B Cost (Known)} = \text{City A Cost (Unknown)}$$

In addition, you can use RSMeans CCI to calculate and compare costs division by division between cities using the same basic formula. (Just be sure that you're comparing similar divisions.)

2. Compare a specific city's construction costs with the National Average.

When you're studying construction location feasibility, it's advisable to compare a prospective project's cost index with an index of the National Average cost.

For example, divide the weighted average index of construction costs of a specific city by that of the 30 City Average, which = 100.

$$\frac{\text{City Index}}{100} = \% \text{ of National Average}$$

As a result, you get a ratio that indicates the relative cost of construction in that city in comparison with the National Average.

3. Convert U.S. National Average to actual costs in Canadian City.

$$\frac{\text{Index for Canadian City}}{100} \times \text{National Average Cost} = \text{Cost in Canadian City in \$ CAN}$$

4. Adjust construction cost data based on a National Average.

When you use a source of construction cost data which is based on a National Average (such as RSMeans cost data publications), it is necessary to adjust those costs to a specific location.

$$\frac{\text{City Index}}{100} \times \frac{\text{"Book" Cost Based on}}{\text{National Average Costs}} = \frac{\text{City Cost}}{\text{(Unknown)}}$$

5. When applying the City Cost Indexes to demolition projects, use the appropriate division installation index. For example, for removal of existing doors and windows, use Division 8 (Openings) index.

What you might like to know about how we developed the Indexes

The information presented in the CCI is organized according to the Construction Specifications Institute (CSI) MasterFormat 2010.

To create a reliable index, RSMeans researched the building type most often constructed in the United States and Canada. Because it was concluded that no one type of building completely represented the building construction industry, nine different types of buildings were combined to create a composite model.

The exact material, labor, and equipment quantities are based on detailed analysis of these nine building types, then each quantity is weighted in proportion to expected usage. These various material items, labor hours, and equipment rental rates are thus combined to form a composite building representing as closely as possible the actual usage of materials, labor and equipment used in the North American Building Construction Industry.

The following structures were chosen to make up that composite model:

1. Factory, 1 story
2. Office, 2–4 story
3. Store, Retail
4. Town Hall, 2–3 story
5. High School, 2–3 story
6. Hospital, 4–8 story
7. Garage, Parking
8. Apartment, 1–3 story
9. Hotel/Motel, 2–3 story

For the purposes of ensuring the timeliness of the data, the components of the index for the composite model have been streamlined. They currently consist of:

- specific quantities of 66 commonly used construction materials;
- specific labor-hours for 21 building construction trades; and
- specific days of equipment rental for 6 types of construction equipment (normally used to install the 66 material items by the 21 trades.) Fuel costs and routine maintenance costs are included in the equipment cost.

A sophisticated computer program handles the updating of all costs for each city on a quarterly basis. Material and equipment price quotations are gathered quarterly from 731 cities in the United States and Canada. These prices and the latest negotiated labor wage rates for 21 different building trades are used to compile the quarterly update of the City Cost Index.

The 30 major U.S. cities used to calculate the National Average are:

Atlanta, GA	Memphis, TN
Baltimore, MD	Milwaukee, WI
Boston, MA	Minneapolis, MN
Buffalo, NY	Nashville, TN
Chicago, IL	New Orleans, LA
Cincinnati, OH	New York, NY
Cleveland, OH	Philadelphia, PA
Columbus, OH	Phoenix, AZ
Dallas, TX	Pittsburgh, PA
Denver, CO	St. Louis, MO
Detroit, MI	San Antonio, TX
Houston, TX	San Diego, CA
Indianapolis, IN	San Francisco, CA
Kansas City, MO	Seattle, WA
Los Angeles, CA	Washington, DC

What the CCI does not indicate

The weighted average for each city is a total of the divisional components weighted to reflect typical usage, but it does not include the productivity variations between trades or cities.

In addition, the CCI does not take into consideration factors such as the following:

- managerial efficiency
- competitive conditions
- automation
- restrictive union practices
- unique local requirements
- regional variations due to specific building codes

City Cost Indexes

DIVISION		UNITED STATES 30 CITY AVERAGE			ANNISTON 362			BIRMINGHAM 350 - 352			BUTLER 369			DECATUR 356			DOTHAN 363		
		MAT.	INST.	TOTAL	MAT.	INST.	TOTAL	MAT.	INST.	TOTAL	MAT.	INST.	TOTAL	MAT.	INST.	TOTAL	MAT.	INST.	TOTAL
									ALABAMA											
015433	CONTRACTOR EQUIPMENT		100.0	100.0		100.8	100.8		100.9	100.9		97.8	97.8		100.8	100.8		97.8	97.8
0241, 31 - 34	SITE & INFRASTRUCTURE, DEMOLITION	100.0	100.0	100.0	87.3	93.5	91.6	85.2	94.1	91.4	101.5	87.2	91.6	80.7	92.1	88.5	99.0	87.1	90.8
0310	Concrete Forming & Accessories	100.0	100.0	100.0	92.6	50.9	56.4	92.8	73.9	76.4	89.1	46.1	51.7	96.8	49.0	55.3	97.9	45.5	52.4
0320	Concrete Reinforcing	100.0	100.0	100.0	88.9	84.6	86.7	88.9	84.9	86.9	93.9	49.5	71.1	88.9	79.3	84.0	93.9	48.9	70.8
0330	Cast-in-Place Concrete	100.0	100.0	100.0	92.1	52.5	77.0	98.3	67.9	86.7	89.9	58.2	77.7	91.6	66.1	81.9	89.9	48.8	74.1
03	CONCRETE	100.0	100.0	100.0	91.5	59.4	76.2	90.8	74.9	83.2	92.1	52.7	73.3	87.6	62.2	75.5	91.6	49.1	71.3
04	MASONRY	100.0	100.0	100.0	85.7	72.1	77.5	88.1	80.9	83.7	89.5	52.4	67.0	86.6	55.4	67.7	90.7	42.9	61.7
05	METALS	100.0	100.0	100.0	93.6	92.1	93.1	94.5	93.9	94.3	92.6	79.2	88.0	95.8	88.5	93.3	92.7	77.9	87.6
06	WOOD, PLASTICS & COMPOSITES	100.0	100.0	100.0	94.5	46.2	66.8	96.4	72.5	82.7	89.6	45.9	64.5	99.5	46.5	69.1	101.1	45.9	69.4
07	THERMAL & MOISTURE PROTECTION	100.0	100.0	100.0	91.9	56.8	77.9	93.9	83.4	89.8	91.9	60.3	79.4	91.7	62.1	79.9	91.9	52.6	76.3
08	OPENINGS	100.0	100.0	100.0	93.9	52.1	83.3	97.6	74.8	91.8	94.0	47.7	82.3	97.9	53.6	86.7	94.0	47.7	82.3
0920	Plaster & Gypsum Board	100.0	100.0	100.0	89.7	45.1	58.3	98.4	72.2	79.9	87.5	44.8	57.4	94.6	45.4	60.0	97.7	44.8	60.4
0950, 0980	Ceilings & Acoustic Treatment	100.0	100.0	100.0	89.7	45.1	60.2	95.6	72.2	80.0	89.7	44.8	59.9	94.4	45.4	61.9	89.7	44.8	59.9
0960	Flooring	100.0	100.0	100.0	103.2	42.5	85.8	105.9	63.7	93.8	109.1	59.2	94.8	105.9	53.5	90.8	116.1	29.1	91.1
0970, 0990	Wall Finishes & Painting/Coating	100.0	100.0	100.0	96.5	33.9	58.5	96.5	59.8	74.3	96.5	57.4	72.8	96.5	59.9	74.3	96.5	57.4	72.8
09	FINISHES	100.0	100.0	100.0	96.4	46.3	68.8	99.5	70.0	83.2	100.0	49.2	72.0	98.2	50.1	71.7	103.1	43.2	70.0
COVERS	DIVS. 10 - 14, 25, 28, 41, 43, 44, 46	100.0	100.0	100.0	100.0	76.6	95.1	100.0	88.5	97.6	100.0	46.1	88.8	100.0	48.2	89.2	100.0	46.2	88.8
21, 22, 23	FIRE SUPPRESSION, PLUMBING & HVAC	100.0	100.0	100.0	99.0	60.8	83.1	99.8	67.4	86.3	96.9	36.5	71.7	99.8	44.7	76.9	96.9	36.4	71.7
26, 27, 3370	ELECTRICAL, COMMUNICATIONS & UTIL.	100.0	100.0	100.0	96.4	62.2	78.2	105.6	62.7	82.8	98.6	43.6	69.4	98.2	66.2	81.2	97.3	62.3	78.7
MF2010	WEIGHTED AVERAGE	100.0	100.0	100.0	95.1	65.5	81.9	97.0	75.6	87.4	95.7	52.6	76.4	95.8	60.8	80.1	95.8	52.6	76.5

DIVISION		EVERGREEN 364			GADSDEN 359			HUNTSVILLE 357 - 358			JASPER 355			MOBILE 365 - 366			MONTGOMERY 360 - 361		
		MAT.	INST.	TOTAL	MAT.	INST.	TOTAL	MAT.	INST.	TOTAL	MAT.	INST.	TOTAL	MAT.	INST.	TOTAL	MAT.	INST.	TOTAL
									ALABAMA											
015433	CONTRACTOR EQUIPMENT		97.8	97.8		100.8	100.8		100.8	100.8		100.8	100.8		97.8	97.8		97.8	97.8
0241, 31 - 34	SITE & INFRASTRUCTURE, DEMOLITION	101.9	87.4	91.9	86.0	93.0	90.8	80.5	93.1	89.2	85.5	92.8	90.5	93.6	88.2	89.9	92.3	88.1	89.4
0310	Concrete Forming & Accessories	85.6	47.9	52.8	89.1	44.6	50.5	96.8	65.6	69.7	94.3	37.1	44.6	96.9	58.0	63.1	94.7	46.7	53.1
0320	Concrete Reinforcing	94.0	49.5	71.1	94.1	83.7	88.8	88.9	79.8	84.2	88.9	83.7	86.2	91.8	83.0	87.2	91.8	83.6	87.5
0330	Cast-in-Place Concrete	89.9	61.0	78.8	91.6	51.6	76.3	89.2	64.4	79.7	101.5	46.3	80.4	94.2	70.7	85.2	96.2	49.2	78.2
03	CONCRETE	92.3	54.4	74.2	91.3	56.2	74.5	86.5	69.1	78.1	94.9	51.0	73.9	89.4	68.5	79.4	90.2	56.3	74.0
04	MASONRY	89.5	54.4	67.0	84.8	46.4	61.5	87.7	61.6	71.9	81.2	44.5	59.0	88.4	56.9	69.3	86.8	42.5	59.9
05	METALS	92.6	79.2	88.0	93.7	91.2	92.8	95.8	90.1	93.8	93.6	90.5	92.5	94.7	92.2	93.8	94.5	90.8	93.2
06	WOOD, PLASTICS & COMPOSITES	85.7	45.9	62.9	90.2	41.9	62.5	99.5	66.5	80.6	96.4	34.1	60.7	99.6	57.3	75.4	96.0	45.9	67.3
07	THERMAL & MOISTURE PROTECTION	91.9	61.2	79.7	91.8	67.0	82.0	91.6	75.9	85.3	91.8	48.6	74.6	91.6	74.5	84.8	91.1	65.8	81.1
08	OPENINGS	94.0	47.7	82.3	93.9	50.9	83.0	97.9	64.4	89.4	93.9	49.8	82.7	97.9	61.9	88.8	97.9	55.4	87.1
0920	Plaster & Gypsum Board	85.8	44.8	56.9	87.1	40.7	54.4	94.6	66.0	74.5	90.2	32.7	49.7	94.6	56.5	67.8	98.4	44.8	60.6
0950, 0980	Ceilings & Acoustic Treatment	89.7	44.8	59.9	89.7	40.7	57.2	96.5	66.0	76.3	89.7	32.7	51.9	96.5	56.5	70.0	95.6	44.8	61.9
0960	Flooring	106.5	59.2	92.9	101.4	47.5	85.9	105.9	57.1	91.9	104.1	42.5	86.4	115.1	59.2	99.0	113.5	31.3	89.9
0970, 0990	Wall Finishes & Painting/Coating	96.5	57.4	72.8	96.5	65.4	77.6	96.5	62.9	76.1	96.5	41.5	63.2	100.0	58.4	74.8	96.5	57.4	72.8
09	FINISHES	99.1	50.4	72.3	95.4	45.7	68.0	98.7	63.4	79.2	96.5	37.1	63.7	103.4	57.2	77.9	102.7	43.5	70.0
COVERS	DIVS. 10 - 14, 25, 28, 41, 43, 44, 46	100.0	47.8	89.2	100.0	79.9	95.8	100.0	84.8	96.8	100.0	44.2	88.4	100.0	81.1	96.1	100.0	79.8	95.8
21, 22, 23	FIRE SUPPRESSION, PLUMBING & HVAC	96.9	39.0	72.8	101.4	39.4	75.6	99.8	56.8	81.9	101.4	57.7	83.2	99.8	63.6	84.7	99.9	37.2	73.8
26, 27, 3370	ELECTRICAL, COMMUNICATIONS & UTIL.	95.7	43.6	68.0	98.2	62.7	79.3	99.4	66.2	*81.8	97.7	62.7	79.1	99.4	62.0	79.5	100.5	66.1	82.2
MF2010	WEIGHTED AVERAGE	95.3	54.1	76.9	95.6	58.2	78.9	95.8	69.0	83.8	96.0	58.2	79.1	96.8	67.9	83.9	96.8	57.2	79.1

DIVISION		PHENIX CITY 368			SELMA 367			TUSCALOOSA 354			ANCHORAGE 995 - 996			FAIRBANKS 997			JUNEAU 998		
		MAT.	INST.	TOTAL	MAT.	INST.	TOTAL	MAT.	INST.	TOTAL	MAT.	INST.	TOTAL	MAT.	INST.	TOTAL	MAT.	INST.	TOTAL
				ALABAMA									ALASKA								
015433	CONTRACTOR EQUIPMENT		97.8	97.8		97.8	97.8		100.8	100.8		115.9	115.9		115.9	115.9		115.9	115.9
0241, 31 - 34	SITE & INFRASTRUCTURE, DEMOLITION	105.9	88.1	93.6	98.8	88.0	91.4	80.9	93.2	89.4	129.3	131.2	130.6	121.2	131.2	128.1	129.0	131.2	130.5
0310	Concrete Forming & Accessories	92.5	41.4	48.2	90.3	45.9	51.8	96.7	50.5	56.6	137.3	118.9	121.3	140.5	118.9	121.8	136.5	118.7	121.1
0320	Concrete Reinforcing	93.8	66.1	79.6	93.9	83.4	88.5	88.9	83.7	86.2	153.8	107.1	129.8	148.1	107.1	127.0	115.7	107.1	111.3
0330	Cast-in-Place Concrete	89.9	49.8	74.5	89.9	49.0	74.2	92.9	51.6	77.1	144.8	117.2	134.2	136.9	117.5	129.5	144.7	117.1	134.1
03	CONCRETE	94.6	50.8	73.7	91.1	55.8	74.2	88.3	58.8	74.2	132.9	115.4	124.5	120.4	115.5	118.1	123.5	115.3	119.6
04	MASONRY	89.4	44.3	62.1	92.8	41.7	61.8	86.7	49.6	64.2	200.7	120.0	151.8	203.5	120.0	152.9	185.9	120.0	146.0
05	METALS	92.6	84.2	89.7	92.6	89.7	91.6	94.9	91.1	93.6	122.1	101.5	115.0	119.1	101.6	113.1	119.1	101.4	113.0
06	WOOD, PLASTICS & COMPOSITES	94.2	38.3	62.1	91.2	45.9	65.2	99.5	48.1	70.0	112.6	118.7	116.1	117.6	118.7	118.2	112.6	118.7	116.1
07	THERMAL & MOISTURE PROTECTION	92.2	65.7	81.7	91.8	55.3	77.3	91.7	69.3	82.8	192.8	116.0	162.3	181.6	114.8	155.0	183.0	116.0	156.4
08	OPENINGS	93.9	47.0	82.1	93.9	55.4	84.2	97.9	61.5	88.7	120.1	114.0	118.5	116.6	114.0	116.0	115.9	114.0	115.4
0920	Plaster & Gypsum Board	91.1	37.0	53.0	88.9	44.8	57.8	94.6	47.0	61.1	146.7	119.0	127.2	153.1	119.0	129.1	134.6	119.0	123.6
0950, 0980	Ceilings & Acoustic Treatment	89.7	37.0	54.7	89.7	44.8	59.9	96.5	47.0	63.7	140.9	119.0	126.4	126.0	119.0	121.4	120.3	119.0	119.5
0960	Flooring	111.7	31.3	88.6	109.8	30.1	86.9	105.9	39.6	86.9	162.0	125.2	151.4	152.3	125.2	144.5	152.7	125.2	144.8
0970, 0990	Wall Finishes & Painting/Coating	96.5	57.4	72.8	96.5	57.4	72.8	96.5	49.4	68.0	150.8	117.3	130.5	149.8	122.3	133.1	150.8	117.3	130.5
09	FINISHES	101.7	39.3	67.3	100.1	43.3	68.8	98.7	47.1	70.2	152.6	120.4	134.8	145.2	120.9	131.8	142.4	120.4	130.2
COVERS	DIVS. 10 - 14, 25, 28, 41, 43, 44, 46	100.0	79.3	95.7	100.0	46.1	88.8	100.0	81.8	96.2	100.0	112.3	102.6	100.0	112.3	102.6	100.0	112.3	102.6
21, 22, 23	FIRE SUPPRESSION, PLUMBING & HVAC	96.9	37.6	72.2	96.9	36.7	71.8	99.8	37.0	73.7	100.4	105.9	102.7	100.3	108.0	103.5	100.4	96.9	98.9
26, 27, 3370	ELECTRICAL, COMMUNICATIONS & UTIL.	98.0	74.2	85.4	97.0	43.6	68.6	98.8	62.7	79.6	141.8	116.3	128.2	147.2	116.3	130.8	137.5	116.3	126.3
MF2010	WEIGHTED AVERAGE	96.2	56.3	78.4	95.5	52.5	76.3	95.8	59.2	79.4	127.2	114.5	121.5	124.4	115.0	120.2	122.8	112.6	118.3

ALASKA / ARIZONA

DIVISION		KETCHIKAN 999			CHAMBERS 865			FLAGSTAFF 860			GLOBE 855			KINGMAN 864			MESA/TEMPE 852		
		MAT.	INST.	TOTAL	MAT.	INST.	TOTAL	MAT.	INST.	TOTAL	MAT.	INST.	TOTAL	MAT.	INST.	TOTAL	MAT.	INST.	TOTAL
015433	CONTRACTOR EQUIPMENT		115.9	115.9		94.8	94.8		94.8	94.8		97.8	97.8		94.8	94.8		97.8	97.8
0241, 31 - 34	SITE & INFRASTRUCTURE, DEMOLITION	180.9	131.2	146.6	61.0	97.5	86.2	77.3	99.3	92.5	88.8	100.9	97.2	60.9	99.5	87.5	80.3	102.5	95.6
0310	Concrete Forming & Accessories	132.2	118.7	120.5	96.6	46.4	53.0	101.7	69.1	73.4	98.1	46.2	53.1	94.9	68.6	72.1	100.8	68.5	72.8
0320	Concrete Reinforcing	116.3	107.1	111.5	96.5	79.2	87.6	96.4	81.3	88.6	102.4	79.1	90.4	96.7	81.3	88.8	103.1	81.3	91.8
0330	Cast-in-Place Concrete	277.0	117.1	215.8	92.7	55.2	78.3	92.8	75.7	86.3	91.2	55.5	77.5	92.4	73.6	85.2	91.9	73.9	85.0
03	CONCRETE	205.7	115.3	162.4	96.9	56.0	77.3	116.0	73.6	95.7	109.6	56.0	84.0	96.5	72.7	85.1	101.0	72.8	87.5
04	MASONRY	212.0	120.0	156.3	93.1	42.1	62.2	93.1	63.7	75.3	107.8	42.1	67.9	93.1	62.6	74.6	107.9	62.6	80.5
05	METALS	119.3	101.4	113.2	94.6	69.0	85.8	95.1	75.0	88.2	92.4	69.3	84.5	95.2	74.9	88.2	92.7	75.3	86.7
06	WOOD, PLASTICS & COMPOSITES	108.8	118.7	114.5	97.1	44.3	66.8	102.9	69.8	83.9	91.7	44.4	64.5	93.1	69.8	79.7	94.9	69.9	80.6
07	THERMAL & MOISTURE PROTECTION	184.5	116.0	157.3	96.6	51.9	78.8	98.5	66.7	85.8	104.5	49.7	82.7	96.6	62.8	83.2	103.6	63.8	87.8
08	OPENINGS	118.3	114.0	117.2	102.7	51.8	89.8	102.8	72.1	95.1	97.4	51.9	85.9	102.9	72.1	95.1	97.4	72.2	91.0
0920	Plaster & Gypsum Board	141.1	119.0	125.6	94.9	42.6	58.1	99.2	68.9	77.8	101.8	42.6	60.1	89.0	68.9	74.8	104.4	68.9	79.4
0950, 0980	Ceilings & Acoustic Treatment	120.7	119.0	119.6	122.0	42.6	69.4	122.9	68.9	87.1	113.0	42.6	66.3	122.9	68.9	87.1	113.0	68.9	83.7
0960	Flooring	152.3	125.2	144.5	92.8	41.8	78.1	95.1	42.0	79.8	98.8	41.8	82.4	91.9	56.7	81.8	100.2	52.3	86.4
0970, 0990	Wall Finishes & Painting/Coating	149.8	117.3	130.1	91.5	39.7	60.1	91.5	58.9	71.8	96.5	39.7	62.1	91.5	58.9	71.8	96.5	58.9	73.7
09	FINISHES	147.6	120.4	132.6	99.5	43.6	68.7	102.7	62.4	80.5	100.2	43.7	69.1	98.7	64.7	80.0	100.0	63.9	80.1
COVERS	DIVS. 10 - 14, 25, 28, 41, 43, 44, 46	100.0	112.3	102.6	100.0	77.5	95.3	100.0	82.7	96.4	100.0	77.8	95.4	100.0	82.4	96.3	100.0	82.7	96.4
21, 22, 23	FIRE SUPPRESSION, PLUMBING & HVAC	98.9	96.9	98.1	96.8	73.1	86.9	100.2	84.5	93.7	94.9	68.0	83.7	96.8	83.9	91.4	100.3	76.0	90.2
26, 27, 3370	ELECTRICAL, COMMUNICATIONS & UTIL.	147.1	116.3	130.8	101.4	70.0	84.7	100.2	63.6	80.7	97.0	65.1	80.1	101.4	63.6	81.3	93.5	63.6	77.6
MF2010	WEIGHTED AVERAGE	136.9	112.6	126.0	96.8	63.5	81.9	100.7	74.4	88.9	98.5	62.0	82.2	96.7	74.2	86.7	98.2	72.8	86.8

ARIZONA / ARKANSAS

DIVISION		PHOENIX 850,853			PRESCOTT 863			SHOW LOW 859			TUCSON 856 - 857			BATESVILLE 725			CAMDEN 717		
		MAT.	INST.	TOTAL	MAT.	INST.	TOTAL	MAT.	INST.	TOTAL	MAT.	INST.	TOTAL	MAT.	INST.	TOTAL	MAT.	INST.	TOTAL
015433	CONTRACTOR EQUIPMENT		98.4	98.4		94.8	94.8		97.8	97.8		97.8	97.8		87.8	87.8		87.8	87.8
0241, 31 - 34	SITE & INFRASTRUCTURE, DEMOLITION	80.7	103.0	96.1	66.6	97.5	87.9	90.8	100.9	97.8	76.3	102.7	94.5	73.8	84.1	80.9	74.6	83.7	80.9
0310	Concrete Forming & Accessories	101.7	71.5	75.5	98.1	46.2	53.0	104.5	46.5	54.2	101.1	71.1	75.0	83.6	48.6	53.2	83.4	32.9	39.6
0320	Concrete Reinforcing	101.5	81.4	91.1	96.4	79.4	87.6	103.1	79.3	90.8	85.1	81.3	83.1	99.3	65.5	81.9	100.6	65.3	82.4
0330	Cast-in-Place Concrete	92.0	76.1	85.9	92.7	55.1	78.3	91.2	55.6	77.6	94.6	76.0	87.5	81.3	47.8	68.5	83.4	40.1	66.8
03	CONCRETE	100.7	74.9	88.3	102.1	55.9	80.0	111.8	56.3	85.2	99.6	74.6	87.6	84.0	52.2	68.8	85.7	42.6	65.1
04	MASONRY	95.0	64.3	76.4	93.1	47.4	65.4	107.8	43.6	68.9	93.0	63.1	74.9	105.1	44.0	68.1	113.4	33.9	66.2
05	METALS	94.3	76.3	88.1	95.1	69.2	86.2	92.2	69.8	84.5	93.5	75.3	87.3	97.5	67.0	87.0	97.6	66.4	86.9
06	WOOD, PLASTICS & COMPOSITES	95.8	72.8	82.6	98.4	44.3	67.3	99.2	44.4	67.8	95.2	72.8	82.3	86.7	49.3	65.2	86.4	32.1	55.3
07	THERMAL & MOISTURE PROTECTION	103.3	67.0	88.8	97.1	51.2	78.9	104.7	52.1	83.8	104.8	64.1	88.6	95.8	46.8	76.3	95.6	37.9	72.7
08	OPENINGS	98.6	73.7	92.3	102.8	51.8	89.9	96.4	51.9	85.1	93.7	73.8	88.7	95.9	47.8	83.7	92.0	42.1	79.3
0920	Plaster & Gypsum Board	107.0	71.8	82.2	95.1	42.6	58.2	107.1	42.6	61.7	111.0	71.8	83.4	75.4	48.3	56.3	75.4	30.6	43.8
0950, 0980	Ceilings & Acoustic Treatment	120.5	71.8	88.2	121.0	42.6	69.1	113.0	42.6	66.3	113.9	71.8	86.0	86.6	48.3	61.2	86.6	30.6	49.5
0960	Flooring	100.4	54.6	87.3	93.7	41.8	78.8	101.9	47.6	86.3	91.0	42.0	76.9	101.4	62.4	90.2	101.3	41.8	84.2
0970, 0990	Wall Finishes & Painting/Coating	96.5	59.1	73.8	91.5	39.7	60.1	96.5	39.7	62.1	97.0	58.9	73.9	97.1	44.3	65.1	97.1	44.5	65.2
09	FINISHES	102.0	66.6	82.5	100.1	43.6	68.9	102.0	44.7	70.4	98.1	64.2	79.4	88.1	50.4	67.3	88.1	35.7	59.2
COVERS	DIVS. 10 - 14, 25, 28, 41, 43, 44, 46	100.0	83.4	96.5	100.0	77.5	95.3	100.0	77.8	95.4	100.0	83.3	96.5	100.0	44.0	88.4	100.0	39.5	87.4
21, 22, 23	FIRE SUPPRESSION, PLUMBING & HVAC	100.2	84.6	93.7	100.2	72.8	88.8	94.9	73.2	85.8	100.2	72.0	88.5	94.7	54.9	78.1	94.7	52.1	77.0
26, 27, 3370	ELECTRICAL, COMMUNICATIONS & UTIL.	100.4	64.9	81.5	99.9	63.5	80.6	93.9	65.0	78.5	95.9	60.2	76.9	100.2	64.7	81.3	96.1	61.7	77.8
MF2010	WEIGHTED AVERAGE	98.6	75.9	88.5	98.4	63.0	82.6	98.6	63.5	82.9	96.9	72.0	85.8	94.3	56.9	77.6	94.2	50.7	74.8

ARKANSAS

DIVISION		FAYETTEVILLE 727			FORT SMITH 729			HARRISON 726			HOT SPRINGS 719			JONESBORO 724			LITTLE ROCK 720 - 722		
		MAT.	INST.	TOTAL	MAT.	INST.	TOTAL	MAT.	INST.	TOTAL	MAT.	INST.	TOTAL	MAT.	INST.	TOTAL	MAT.	INST.	TOTAL
015433	CONTRACTOR EQUIPMENT		87.8	87.8		87.8	87.8		87.8	87.8		87.8	87.8		106.7	106.7		87.8	87.8
0241, 31 - 34	SITE & INFRASTRUCTURE, DEMOLITION	73.2	85.7	81.8	77.9	85.6	83.2	78.5	84.1	82.4	78.0	85.1	82.9	98.8	99.4	99.2	85.3	85.6	85.5
0310	Concrete Forming & Accessories	79.5	41.6	46.6	97.2	59.4	64.4	87.7	48.5	53.7	81.0	38.5	44.1	87.3	52.1	56.8	89.6	63.9	67.3
0320	Concrete Reinforcing	99.3	70.4	84.4	100.4	71.2	85.4	98.8	65.4	81.6	98.7	65.5	81.6	96.1	67.8	81.5	100.6	66.2	82.9
0330	Cast-in-Place Concrete	81.3	52.3	70.2	93.0	67.7	83.3	90.2	45.7	73.2	85.3	40.6	68.2	88.6	58.3	77.0	93.4	67.8	83.6
03	CONCRETE	83.8	51.6	68.4	92.0	64.9	79.0	91.2	51.5	72.2	89.0	45.3	68.1	88.7	58.6	74.3	91.7	66.0	79.4
04	MASONRY	94.5	44.7	64.3	99.8	53.1	71.5	105.5	42.1	67.1	84.5	36.2	55.2	95.9	45.3	65.2	96.6	53.1	70.2
05	METALS	97.5	69.1	87.8	99.9	70.9	89.9	98.7	66.7	87.6	97.5	67.0	87.0	94.4	80.5	89.6	96.1	69.3	86.8
06	WOOD, PLASTICS & COMPOSITES	82.3	38.4	57.1	102.6	61.9	79.3	91.9	49.3	67.5	83.6	38.6	57.8	90.1	52.3	68.5	97.2	67.8	80.4
07	THERMAL & MOISTURE PROTECTION	96.6	48.4	77.4	96.9	56.7	81.0	96.1	45.4	76.0	95.8	40.2	73.7	100.4	52.6	80.7	96.0	57.4	80.7
08	OPENINGS	95.9	47.3	83.5	96.7	58.1	87.0	96.7	48.5	84.5	91.9	44.0	79.8	98.8	56.5	88.1	96.7	61.4	87.7
0920	Plaster & Gypsum Board	72.3	37.1	47.5	81.4	61.3	67.2	80.0	48.3	57.7	73.7	37.3	48.1	91.9	51.1	63.2	86.8	67.4	73.2
0950, 0980	Ceilings & Acoustic Treatment	86.6	37.1	53.8	90.3	61.3	71.1	90.3	48.3	62.5	86.6	37.3	53.9	90.9	51.1	64.5	92.2	67.4	75.8
0960	Flooring	98.4	62.4	88.0	109.0	64.2	96.1	104.1	62.4	92.1	100.0	62.4	89.2	72.6	56.4	68.0	110.3	64.2	97.1
0970, 0990	Wall Finishes & Painting/Coating	97.1	31.6	57.4	97.1	57.9	73.4	97.1	44.3	65.1	97.1	50.9	69.1	85.9	52.1	65.4	97.1	59.5	74.3
09	FINISHES	86.9	43.1	62.7	92.1	60.4	74.6	90.6	50.4	68.4	87.9	43.2	63.3	85.2	52.3	67.1	93.8	64.1	77.4
COVERS	DIVS. 10 - 14, 25, 28, 41, 43, 44, 46	100.0	54.4	90.5	100.0	77.3	95.3	100.0	54.9	90.6	100.0	40.6	87.7	100.0	50.3	89.7	100.0	78.0	95.4
21, 22, 23	FIRE SUPPRESSION, PLUMBING & HVAC	94.8	53.4	77.5	100.2	52.7	80.4	94.7	52.5	77.2	94.7	51.6	76.8	100.4	55.8	81.8	100.1	60.6	83.7
26, 27, 3370	ELECTRICAL, COMMUNICATIONS & UTIL.	93.1	55.6	73.2	97.1	70.1	82.7	98.5	40.3	67.5	98.6	68.6	82.7	103.9	64.8	83.1	101.6	71.2	85.5
MF2010	WEIGHTED AVERAGE	92.9	54.9	75.9	97.0	63.4	82.0	95.7	53.0	76.6	93.4	53.6	75.6	96.7	61.6	81.0	97.0	65.9	83.1

ARKANSAS / CALIFORNIA

DIVISION		PINE BLUFF 716			RUSSELLVILLE 728			TEXARKANA 718			WEST MEMPHIS 723			ALHAMBRA 917-918			ANAHEIM 928		
		MAT.	INST.	TOTAL	MAT.	INST.	TOTAL	MAT.	INST.	TOTAL	MAT.	INST.	TOTAL	MAT.	INST.	TOTAL	MAT.	INST.	TOTAL
015433	CONTRACTOR EQUIPMENT		87.8	87.8		87.8	87.8		88.5	88.5		106.7	106.7		99.1	99.1		101.6	101.6
0241, 31 - 34	SITE & INFRASTRUCTURE, DEMOLITION	79.7	85.6	83.8	75.1	84.0	81.2	93.3	86.1	88.3	106.1	99.4	101.5	99.6	109.4	106.4	98.7	109.2	105.9
0310	Concrete Forming & Accessories	80.7	63.8	66.0	84.2	57.7	61.2	86.7	43.4	49.1	92.2	52.3	57.6	121.7	119.9	120.2	105.6	123.1	120.7
0320	Concrete Reinforcing	100.5	66.2	82.9	100.0	65.3	82.1	100.1	65.5	82.3	96.1	67.8	81.5	120.7	119.9	120.3	99.1	119.9	109.8
0330	Cast-in-Place Concrete	85.3	67.8	78.6	85.2	47.8	70.9	93.0	45.6	74.9	92.9	58.4	79.7	90.3	119.0	101.3	93.7	121.6	104.3
03	CONCRETE	89.7	65.9	78.4	87.2	56.2	72.4	88.6	49.2	69.8	95.5	58.7	77.9	103.7	118.6	110.8	101.9	120.9	111.0
04	MASONRY	122.1	53.1	80.3	99.1	40.3	63.5	99.8	35.8	61.0	81.4	45.3	59.5	113.5	116.9	115.6	82.6	115.1	102.3
05	METALS	98.4	69.1	88.3	97.5	66.5	86.9	91.7	67.0	83.2	93.5	80.7	89.1	86.6	102.1	91.9	102.7	103.1	102.8
06	WOOD, PLASTICS & COMPOSITES	83.1	67.8	74.4	88.0	61.9	73.0	91.4	45.5	65.1	95.9	52.3	70.9	93.3	119.2	108.1	93.6	123.1	110.5
07	THERMAL & MOISTURE PROTECTION	95.9	57.4	80.6	96.8	46.6	76.9	96.6	47.1	77.0	100.8	52.5	81.6	102.5	115.3	107.6	102.6	119.4	109.3
08	OPENINGS	91.9	61.4	84.2	95.9	57.4	86.1	97.2	48.5	84.9	98.8	56.5	88.1	91.0	119.0	98.1	103.2	123.3	108.3
0920	Plaster & Gypsum Board	73.2	67.4	69.1	75.4	61.3	65.5	78.3	44.4	54.4	94.6	51.1	64.0	108.3	119.8	116.4	99.1	119.9	109.8
0950, 0980	Ceilings & Acoustic Treatment	86.6	67.4	73.9	86.6	61.3	69.8	92.2	44.4	60.5	88.8	51.1	63.8	98.6	119.8	112.7	111.4	123.7	119.5
0960	Flooring	99.8	64.2	89.6	101.1	62.4	90.0	102.1	53.9	88.3	75.1	56.4	69.7	98.7	105.9	100.7	110.3	112.4	110.9
0970, 0990	Wall Finishes & Painting/Coating	97.1	59.5	74.3	97.1	37.7	61.1	97.1	32.6	58.0	85.9	52.1	65.4	105.9	109.0	107.8	102.5	109.0	106.4
09	FINISHES	87.9	64.1	74.7	88.2	57.1	71.0	91.0	43.5	64.8	86.5	52.3	67.7	99.4	116.3	108.7	105.9	119.7	113.5
COVERS	DIVS. 10 - 14, 25, 28, 41, 43, 44, 46	100.0	78.0	95.4	100.0	45.4	88.7	100.0	36.2	86.7	100.0	50.3	89.7	100.0	109.5	102.0	100.0	110.3	102.1
21, 22, 23	FIRE SUPPRESSION, PLUMBING & HVAC	100.2	55.2	81.4	94.8	54.7	78.1	100.2	58.6	82.9	95.0	67.8	83.7	94.8	107.8	100.2	100.1	112.6	105.3
26, 27, 3370	ELECTRICAL, COMMUNICATIONS & UTIL.	96.3	71.2	83.0	97.0	48.5	71.2	98.5	39.1	66.9	105.8	61.6	82.3	116.1	114.5	115.3	93.8	107.7	101.2
MF2010	WEIGHTED AVERAGE	96.7	64.8	82.4	94.2	56.1	77.2	95.9	51.9	76.3	95.9	63.7	81.5	98.6	112.7	104.9	99.9	113.7	106.1

CALIFORNIA

DIVISION		BAKERSFIELD 932-933			BERKELEY 947			EUREKA 955			FRESNO 936-938			INGLEWOOD 903-905			LONG BEACH 906-908		
		MAT.	INST.	TOTAL	MAT.	INST.	TOTAL	MAT.	INST.	TOTAL	MAT.	INST.	TOTAL	MAT.	INST.	TOTAL	MAT.	INST.	TOTAL
015433	CONTRACTOR EQUIPMENT		99.4	99.4		98.4	98.4		99.1	99.1		99.4	99.4		98.8	98.8		98.8	98.8
0241, 31 - 34	SITE & INFRASTRUCTURE, DEMOLITION	101.1	106.6	104.9	127.7	104.4	111.6	112.6	103.7	106.5	104.1	106.5	105.7	91.0	106.2	101.5	98.4	106.2	103.8
0310	Concrete Forming & Accessories	103.7	122.6	120.1	119.8	146.4	142.9	115.6	129.3	127.5	101.6	130.2	126.4	109.9	119.9	118.6	105.1	119.9	118.0
0320	Concrete Reinforcing	111.6	119.8	115.8	93.0	121.1	107.5	108.3	120.6	114.7	94.5	120.3	107.8	114.3	119.9	117.2	113.3	119.9	116.7
0330	Cast-in-Place Concrete	95.4	120.6	105.1	126.7	143.6	133.2	101.6	116.6	107.3	98.7	117.3	105.8	89.7	119.8	101.2	102.1	119.8	108.9
03	CONCRETE	102.2	120.4	110.9	115.4	138.8	126.6	112.8	122.0	117.2	103.5	122.7	112.7	99.3	119.0	108.8	109.4	119.0	114.0
04	MASONRY	102.5	114.5	109.8	139.6	131.5	134.7	108.3	130.1	121.5	105.8	121.8	115.5	75.0	117.0	100.5	83.6	117.0	103.9
05	METALS	98.0	102.3	99.5	105.9	103.7	105.1	102.6	101.7	102.3	102.6	103.0	102.7	93.4	103.5	96.9	93.3	103.6	96.8
06	WOOD, PLASTICS & COMPOSITES	87.8	123.2	108.1	115.3	151.1	135.8	106.6	132.6	121.5	99.0	132.6	118.3	92.1	119.1	107.6	85.8	119.1	104.9
07	THERMAL & MOISTURE PROTECTION	101.3	112.0	105.5	115.6	136.1	123.7	106.7	117.0	110.8	96.1	115.6	103.8	101.4	116.7	107.5	101.7	117.0	107.7
08	OPENINGS	100.7	118.3	105.2	101.2	140.0	111.0	102.0	112.8	104.7	102.5	123.4	107.8	92.8	119.0	99.4	92.8	119.0	99.4
0920	Plaster & Gypsum Board	99.3	123.7	116.5	110.5	152.1	139.8	112.6	133.4	127.3	101.9	133.4	124.1	104.9	119.8	115.4	100.5	119.8	114.1
0950, 0980	Ceilings & Acoustic Treatment	112.9	123.7	120.0	116.4	152.1	140.0	117.0	133.4	127.9	114.5	133.4	127.0	98.9	119.8	112.7	98.9	119.8	112.7
0960	Flooring	110.6	98.3	107.0	119.8	128.0	122.2	115.0	123.1	117.3	117.1	141.0	124.0	100.6	105.9	102.1	97.7	105.9	100.1
0970, 0990	Wall Finishes & Painting/Coating	104.4	95.2	98.8	106.2	143.2	128.7	103.7	49.8	71.0	120.9	107.8	113.0	99.3	109.0	105.2	99.3	109.0	105.2
09	FINISHES	107.2	116.0	112.0	115.3	144.1	131.2	112.3	121.7	117.5	111.5	131.2	122.4	101.1	116.3	109.5	100.2	116.3	109.1
COVERS	DIVS. 10 - 14, 25, 28, 41, 43, 44, 46	100.0	110.5	102.2	100.0	125.5	105.3	100.0	122.1	104.6	100.0	122.0	104.6	100.0	109.4	102.0	100.0	109.4	102.0
21, 22, 23	FIRE SUPPRESSION, PLUMBING & HVAC	100.1	106.0	102.6	94.9	145.0	115.7	94.7	105.8	99.4	100.2	113.7	105.8	94.8	107.8	100.2	94.8	107.8	100.2
26, 27, 3370	ELECTRICAL, COMMUNICATIONS & UTIL.	105.8	103.4	104.5	104.0	138.6	122.4	101.7	119.4	111.1	93.2	99.4	96.5	106.2	114.5	110.6	105.9	114.5	110.5
MF2010	WEIGHTED AVERAGE	101.4	110.4	105.4	106.8	133.6	118.8	103.1	115.2	108.5	101.7	115.3	107.8	96.1	112.6	103.5	97.8	112.6	104.5

CALIFORNIA

DIVISION		LOS ANGELES 900-902			MARYSVILLE 959			MODESTO 953			MOJAVE 935			OAKLAND 946			OXNARD 930		
		MAT.	INST.	TOTAL	MAT.	INST.	TOTAL	MAT.	INST.	TOTAL	MAT.	INST.	TOTAL	MAT.	INST.	TOTAL	MAT.	INST.	TOTAL
015433	CONTRACTOR EQUIPMENT		100.5	100.5		99.1	99.1		99.1	99.1		99.4	99.4		98.4	98.4		98.2	98.2
0241, 31 - 34	SITE & INFRASTRUCTURE, DEMOLITION	97.5	108.6	105.2	108.5	105.9	106.7	102.5	106.0	104.9	97.5	106.6	103.8	134.5	104.3	113.7	104.3	104.6	104.6
0310	Concrete Forming & Accessories	107.2	123.0	120.9	105.9	130.1	126.9	102.5	130.3	126.6	114.8	117.3	116.9	109.0	146.2	141.3	106.4	123.0	120.8
0320	Concrete Reinforcing	114.3	120.0	117.2	108.3	120.1	114.4	112.3	120.4	116.4	113.7	119.7	116.8	95.1	121.1	108.5	111.6	119.8	115.8
0330	Cast-in-Place Concrete	97.7	120.2	106.3	113.5	117.4	115.0	101.6	117.5	107.7	85.4	120.5	98.9	120.1	121.3	120.5	99.4	120.9	107.6
03	CONCRETE	105.9	120.5	112.9	114.6	122.6	118.5	105.3	122.8	113.7	95.6	117.9	106.3	115.2	131.1	122.8	104.4	120.6	112.2
04	MASONRY	88.9	117.8	106.4	109.3	118.1	114.6	107.1	122.6	116.5	103.9	113.8	109.9	147.9	131.5	138.0	107.6	113.2	111.0
05	METALS	100.7	103.9	101.8	102.0	103.0	102.4	99.2	103.6	100.7	99.6	102.0	100.5	100.1	103.6	101.3	97.8	102.6	99.5
06	WOOD, PLASTICS & COMPOSITES	88.6	122.9	108.3	94.1	132.6	116.2	90.2	132.6	114.5	99.2	116.2	109.0	102.6	151.1	130.4	94.0	123.2	110.7
07	THERMAL & MOISTURE PROTECTION	101.4	119.0	108.4	106.3	118.1	111.0	105.9	117.1	110.4	102.1	110.9	105.6	113.1	136.1	122.2	103.7	100.7	101.9
08	OPENINGS	99.1	123.2	105.2	101.2	124.3	107.0	100.0	124.5	106.2	96.6	113.2	100.8	101.3	140.0	111.1	99.6	123.3	105.6
0920	Plaster & Gypsum Board	104.9	123.7	118.1	104.8	133.4	125.0	108.0	133.4	125.9	115.5	116.5	116.2	103.2	152.1	137.6	107.5	123.7	118.9
0950, 0980	Ceilings & Acoustic Treatment	108.2	123.7	118.5	116.0	133.4	127.6	111.4	133.4	126.0	116.0	116.5	116.3	119.5	152.1	141.1	116.6	123.7	121.3
0960	Flooring	99.1	112.4	102.9	110.0	111.5	110.4	110.6	117.7	112.6	117.3	98.3	111.8	114.3	128.0	118.2	109.0	112.4	109.9
0970, 0990	Wall Finishes & Painting/Coating	99.3	109.0	105.2	103.7	117.7	112.2	103.7	118.5	112.7	103.7	94.6	98.2	106.2	143.2	128.7	103.7	100.7	101.9
09	FINISHES	102.9	119.6	112.1	109.2	127.2	119.1	108.1	128.4	119.3	111.8	111.7	111.7	114.0	144.1	130.6	108.7	118.9	114.3
COVERS	DIVS. 10 - 14, 25, 28, 41, 43, 44, 46	100.0	109.9	102.1	100.0	122.0	104.6	100.0	122.1	104.6	100.0	109.7	102.0	100.0	125.5	105.3	100.0	110.5	102.2
21, 22, 23	FIRE SUPPRESSION, PLUMBING & HVAC	100.1	106.3	105.3	94.7	107.5	100.0	100.1	113.7	105.8	94.7	101.3	97.5	100.3	144.9	118.9	100.1	112.6	105.3
26, 27, 3370	ELECTRICAL, COMMUNICATIONS & UTIL.	102.8	118.2	111.0	97.8	103.9	101.0	100.6	109.9	105.6	92.0	100.4	96.5	103.1	146.7	126.3	97.7	109.8	104.1
MF2010	WEIGHTED AVERAGE	100.6	115.4	107.2	102.4	113.8	107.5	101.9	116.6	108.5	98.2	107.7	102.4	107.5	133.7	119.2	101.4	113.1	106.6

CALIFORNIA

DIVISION		PALM SPRINGS 922			PALO ALTO 943			PASADENA 910 - 912			REDDING 960			RICHMOND 948			RIVERSIDE 925		
		MAT.	INST.	TOTAL	MAT.	INST.	TOTAL	MAT.	INST.	TOTAL	MAT.	INST.	TOTAL	MAT.	INST.	TOTAL	MAT.	INST.	TOTAL
015433	CONTRACTOR EQUIPMENT		100.3	100.3		98.4	98.4		99.1	99.1		99.1	99.1		98.4	98.4		100.3	100.3
0241, 31 - 34	SITE & INFRASTRUCTURE, DEMOLITION	90.2	107.2	101.9	123.1	104.3	110.1	96.4	109.4	105.4	108.3	105.9	106.7	134.2	104.3	113.6	96.9	107.2	104.0
0310	Concrete Forming & Accessories	102.2	120.1	117.7	107.1	139.7	135.4	111.1	119.9	118.8	105.7	130.1	126.9	122.2	145.8	142.6	105.9	123.0	120.7
0320	Concrete Reinforcing	113.9	119.8	116.9	93.0	120.8	107.3	121.7	119.9	120.8	108.3	120.1	114.4	93.0	120.8	107.3	110.6	119.8	115.4
0330	Cast-in-Place Concrete	89.3	121.5	101.6	107.1	121.0	112.4	85.7	119.0	98.5	112.6	117.4	114.4	123.2	121.1	122.4	97.1	121.6	106.5
03	CONCRETE	96.8	119.5	107.7	103.7	128.1	115.4	99.2	118.6	108.5	113.6	122.6	117.9	117.2	130.8	123.7	103.1	120.9	111.6
04	MASONRY	80.4	114.0	100.7	116.6	127.6	123.3	97.8	116.9	109.4	109.3	118.1	114.6	139.3	127.6	132.2	81.3	114.0	101.1
05	METALS	103.3	102.7	103.1	97.4	102.8	99.3	86.6	102.1	91.9	102.4	103.0	102.6	97.5	102.8	99.3	102.9	102.9	102.9
06	WOOD, PLASTICS & COMPOSITES	88.3	119.3	106.1	100.0	142.9	124.6	80.5	119.2	102.7	93.8	132.6	116.1	119.0	151.1	137.4	93.6	123.1	110.5
07	THERMAL & MOISTURE PROTECTION	102.4	116.5	108.0	112.7	129.5	119.4	102.2	115.3	107.4	106.3	118.1	111.0	113.2	132.5	120.9	102.9	118.3	109.0
08	OPENINGS	98.7	119.1	103.8	101.3	134.6	109.7	91.0	119.1	98.1	102.2	124.3	107.8	101.3	139.0	110.8	101.7	123.3	107.2
0920	Plaster & Gypsum Board	102.1	119.8	114.6	101.0	143.6	131.0	101.3	119.8	114.3	105.1	133.4	125.0	112.5	152.1	140.4	107.0	123.7	118.7
0950, 0980	Ceilings & Acoustic Treatment	109.5	119.8	116.3	117.4	143.6	134.8	98.6	119.8	112.7	122.2	133.4	129.6	117.4	152.1	140.4	114.7	123.7	120.7
0960	Flooring	112.5	105.9	110.6	113.3	121.9	115.8	93.2	105.9	96.8	110.0	111.5	110.4	121.8	121.9	121.8	114.4	112.4	113.8
0970, 0990	Wall Finishes & Painting/Coating	100.6	119.8	112.2	106.2	138.0	125.5	105.9	109.0	107.8	103.7	117.7	112.2	106.2	138.0	125.5	100.6	109.0	105.7
09	FINISHES	104.7	117.6	111.8	112.1	137.7	126.2	96.7	116.3	107.5	110.4	127.2	119.7	117.0	142.5	131.0	107.4	119.7	114.2
COVERS	DIVS. 10 - 14, 25, 28, 41, 43, 44, 46	100.0	109.8	102.0	100.0	124.6	105.1	100.0	109.5	102.0	100.0	122.0	104.6	100.0	125.5	105.3	100.0	110.3	102.1
21, 22, 23	FIRE SUPPRESSION, PLUMBING & HVAC	94.7	107.8	100.2	94.9	138.9	113.2	94.8	107.8	100.2	100.1	107.5	103.2	94.9	135.4	111.8	100.1	112.6	105.3
26, 27, 3370	ELECTRICAL, COMMUNICATIONS & UTIL.	97.8	106.7	102.5	103.0	140.8	123.1	112.6	114.5	113.6	101.2	103.9	102.6	103.6	133.0	119.2	93.9	106.7	100.7
MF2010	WEIGHTED AVERAGE	97.5	111.5	103.8	102.3	129.4	114.4	96.5	112.7	103.7	104.1	113.8	108.4	106.1	128.9	116.3	100.0	113.3	105.9

CALIFORNIA

DIVISION		SACRAMENTO 942,956 - 958			SALINAS 939			SAN BERNARDINO 923 - 924			SAN DIEGO 919 - 921			SAN FRANCISCO 940 - 941			SAN JOSE 951		
		MAT.	INST.	TOTAL	MAT.	INST.	TOTAL	MAT.	INST.	TOTAL	MAT.	INST.	TOTAL	MAT.	INST.	TOTAL	MAT.	INST.	TOTAL
015433	CONTRACTOR EQUIPMENT		98.0	98.0		99.4	99.4		100.3	100.3		99.1	99.1		107.7	107.7		99.8	99.8
0241, 31 - 34	SITE & INFRASTRUCTURE, DEMOLITION	105.8	109.9	108.6	119.0	106.6	110.5	75.9	107.1	97.5	102.2	103.1	102.8	136.9	110.1	118.4	138.8	100.8	112.6
0310	Concrete Forming & Accessories	107.1	132.2	128.9	109.1	133.6	130.3	109.3	120.0	118.6	108.5	111.9	111.5	108.5	146.9	141.8	108.4	146.0	141.0
0320	Concrete Reinforcing	88.0	120.4	104.7	112.3	120.8	116.7	110.6	119.8	115.3	117.9	119.7	118.8	108.3	121.4	115.0	98.7	121.1	110.2
0330	Cast-in-Place Concrete	101.6	118.1	107.9	98.2	117.8	105.7	67.1	121.4	87.9	93.9	107.6	99.2	123.2	122.6	122.9	118.1	120.8	119.1
03	CONCRETE	104.7	123.5	113.7	114.2	124.5	119.1	76.8	119.5	97.2	104.3	111.3	107.6	118.6	132.4	125.2	112.0	131.3	121.2
04	MASONRY	118.6	122.6	121.0	103.5	128.0	118.4	88.2	114.0	103.8	89.7	111.1	102.7	148.4	134.7	140.1	137.9	131.5	134.1
05	METALS	95.7	98.3	96.6	102.2	104.7	103.1	102.8	102.6	102.8	102.2	103.5	102.6	106.5	111.1	108.1	97.7	109.3	101.7
06	WOOD, PLASTICS & COMPOSITES	96.2	135.0	118.4	98.3	135.9	119.9	97.2	119.3	109.9	93.4	109.6	102.7	102.6	151.2	130.5	101.6	150.9	129.9
07	THERMAL & MOISTURE PROTECTION	119.3	120.7	119.9	102.7	125.2	111.7	101.6	116.5	107.5	108.2	104.6	106.8	115.3	137.5	124.1	102.2	137.3	116.2
08	OPENINGS	113.7	125.8	116.8	101.0	131.8	108.8	98.7	119.1	103.9	98.8	112.8	102.3	105.7	140.1	114.4	92.9	139.9	104.8
0920	Plaster & Gypsum Board	97.4	135.7	124.3	108.8	136.8	128.5	108.4	119.8	116.4	108.7	109.7	109.4	105.6	152.1	138.3	102.5	152.1	137.4
0950, 0980	Ceilings & Acoustic Treatment	117.4	135.7	129.5	116.0	136.8	129.8	111.1	119.8	117.0	106.1	109.7	108.5	125.8	152.1	143.2	108.2	152.1	137.3
0960	Flooring	113.5	111.7	113.0	111.3	128.0	116.1	116.6	105.9	113.5	101.5	112.4	104.6	114.3	128.0	118.2	102.8	128.0	110.0
0970, 0990	Wall Finishes & Painting/Coating	103.6	118.5	112.7	104.8	143.2	128.1	100.6	106.1	103.9	102.6	109.0	106.5	106.2	153.2	134.7	104.2	143.2	127.9
09	FINISHES	110.4	128.8	120.5	111.3	134.8	124.3	105.8	116.1	111.4	102.4	111.4	107.4	115.8	145.4	132.1	106.6	144.0	127.2
COVERS	DIVS. 10 - 14, 25, 28, 41, 43, 44, 46	100.0	122.8	104.7	100.0	122.5	104.7	100.0	109.8	102.0	100.0	108.4	101.8	100.0	126.0	105.4	100.0	125.1	105.2
21, 22, 23	FIRE SUPPRESSION, PLUMBING & HVAC	100.1	114.9	106.3	94.7	120.6	105.5	94.7	107.8	100.2	100.1	111.7	105.0	100.2	165.7	127.5	100.1	144.4	118.6
26, 27, 3370	ELECTRICAL, COMMUNICATIONS & UTIL.	98.5	111.4	105.3	93.2	128.6	112.0	97.9	105.8	102.1	99.1	100.3	99.8	103.2	152.8	129.6	104.0	149.7	128.3
MF2010	WEIGHTED AVERAGE	103.9	117.2	109.8	102.0	123.0	111.4	95.2	111.2	102.3	100.6	108.2	104.0	109.6	140.8	123.6	104.6	134.3	117.9

CALIFORNIA

DIVISION		SAN LUIS OBISPO 934			SAN MATEO 944			SAN RAFAEL 949			SANTA ANA 926 - 927			SANTA BARBARA 931			SANTA CRUZ 950		
		MAT.	INST.	TOTAL	MAT.	INST.	TOTAL	MAT.	INST.	TOTAL	MAT.	INST.	TOTAL	MAT.	INST.	TOTAL	MAT.	INST.	TOTAL
015433	CONTRACTOR EQUIPMENT		99.4	99.4		98.4	98.4		98.5	98.5		100.3	100.3		99.4	99.4		99.8	99.8
0241, 31 - 34	SITE & INFRASTRUCTURE, DEMOLITION	110.4	106.6	107.8	131.3	104.3	112.7	120.4	110.0	113.2	88.5	107.1	101.4	104.3	106.6	105.9	137.9	100.7	112.2
0310	Concrete Forming & Accessories	116.4	119.8	119.3	112.8	145.8	141.5	115.5	145.6	141.7	109.6	120.0	118.6	107.4	122.9	120.9	108.4	133.7	130.4
0320	Concrete Reinforcing	113.7	119.7	116.8	93.0	121.0	107.4	93.6	121.0	107.7	114.4	119.8	117.2	111.6	119.8	115.8	122.3	120.8	121.5
0330	Cast-in-Place Concrete	105.5	120.6	111.3	119.3	121.1	120.0	138.9	120.1	131.7	85.7	121.5	99.4	99.1	120.7	107.3	117.3	119.7	118.2
03	CONCRETE	112.4	119.1	115.6	113.4	130.9	121.8	136.7	130.3	133.6	94.4	119.5	106.4	104.3	120.5	112.1	115.1	125.4	120.0
04	MASONRY	105.5	113.8	110.5	139.0	130.8	134.0	113.9	130.8	124.1	77.4	114.3	99.8	104.1	113.8	110.0	142.1	128.1	133.7
05	METALS	100.3	102.2	100.9	97.3	103.1	99.3	98.9	100.2	99.3	102.9	102.7	102.8	98.2	102.5	99.7	104.6	108.2	105.8
06	WOOD, PLASTICS & COMPOSITES	101.5	119.4	111.8	107.7	151.1	132.6	104.5	150.8	131.1	98.8	119.3	110.6	94.0	123.2	110.7	101.6	136.0	121.3
07	THERMAL & MOISTURE PROTECTION	102.9	113.6	107.1	113.0	132.5	120.8	117.6	132.2	123.4	102.7	116.6	108.2	102.3	114.0	106.9	102.1	127.9	112.4
08	OPENINGS	98.8	115.0	102.9	101.2	139.0	110.8	111.0	138.9	118.0	97.9	119.1	103.3	100.7	123.3	106.4	94.1	131.8	103.7
0920	Plaster & Gypsum Board	117.2	119.8	119.0	106.8	152.1	138.6	109.2	152.1	139.4	109.3	119.8	116.7	107.5	123.7	118.9	108.3	136.8	128.3
0950, 0980	Ceilings & Acoustic Treatment	116.0	119.8	118.5	117.4	152.1	140.4	123.9	152.1	142.6	111.4	119.8	117.0	116.6	123.7	121.3	113.2	136.8	128.8
0960	Flooring	118.3	98.3	112.6	116.3	121.9	117.9	127.4	121.9	125.8	117.2	105.9	113.9	110.6	98.3	107.0	107.3	128.0	113.2
0970, 0990	Wall Finishes & Painting/Coating	103.7	100.7	101.9	106.2	138.0	125.5	102.3	138.0	124.0	100.6	106.1	103.9	103.7	100.7	101.90	104.5	143.2	128.0
09	FINISHES	113.4	113.4	113.4	114.4	142.1	129.7	116.4	142.3	130.7	107.3	116.1	112.2	109.4	116.5	113.3	109.7	134.9	123.6
COVERS	DIVS. 10 - 14, 25, 28, 41, 43, 44, 46	100.0	120.7	104.3	100.0	125.5	105.3	100.0	124.8	105.2	100.0	109.8	102.0	100.0	110.5	102.2	100.0	122.9	104.7
21, 22, 23	FIRE SUPPRESSION, PLUMBING & HVAC	94.7	107.8	100.2	94.9	138.5	113.0	94.8	149.0	117.4	94.7	107.8	100.2	100.1	112.6	105.3	100.1	120.7	108.7
26, 27, 3370	ELECTRICAL, COMMUNICATIONS & UTIL.	92.0	98.7	95.6	103.0	144.2	124.9	99.9	118.8	109.9	97.9	107.2	102.8	90.9	110.5	101.3	102.8	128.6	116.5
MF2010	WEIGHTED AVERAGE	101.2	109.8	105.0	105.1	131.4	116.9	107.4	130.3	117.6	97.2	111.4	103.6	100.6	113.1	106.2	106.4	123.1	113.9

City Cost Indexes

		SANTA ROSA 954			STOCKTON 952			SUSANVILLE 961			VALLEJO 945			VAN NUYS 913 - 916			ALAMOSA 811		
DIVISION		MAT.	INST.	TOTAL	MAT.	INST.	TOTAL	MAT.	INST.	TOTAL	MAT.	INST.	TOTAL	MAT.	INST.	TOTAL	MAT.	INST.	TOTAL
015433	CONTRACTOR EQUIPMENT		99.6	99.6		99.1	99.1		99.1	99.1		98.5	98.5		99.1	99.1		96.4	96.4
0241, 31 - 34	SITE & INFRASTRUCTURE, DEMOLITION	103.1	106.1	105.1	102.2	106.0	104.8	115.8	105.9	109.0	105.0	110.0	108.4	114.3	109.4	111.0	122.3	92.7	101.9
0310	Concrete Forming & Accessories	103.1	145.5	139.9	106.3	130.3	127.1	106.9	130.1	127.0	106.2	145.1	139.9	117.2	119.9	119.6	106.0	72.8	77.2
0320	Concrete Reinforcing	109.3	121.2	115.4	112.3	120.4	116.4	108.3	120.1	114.4	94.7	121.2	108.3	121.7	119.9	120.8	104.4	79.6	91.6
0330	Cast-in-Place Concrete	111.5	119.1	114.4	99.0	117.5	106.1	102.4	117.4	108.2	110.6	119.3	113.9	90.4	119.0	101.3	103.4	80.6	94.7
03	CONCRETE	114.2	130.3	121.9	104.3	122.8	113.2	114.9	122.6	118.6	110.1	129.8	119.5	113.2	118.6	115.8	116.0	77.0	97.4
04	MASONRY	106.8	133.4	122.9	107.2	122.6	116.6	108.2	118.1	114.2	84.5	133.3	114.1	113.5	116.9	115.6	126.1	77.2	96.5
05	METALS	103.3	106.4	104.4	99.4	103.6	100.8	101.5	103.0	102.0	98.8	100.4	99.4	85.6	102.1	91.3	96.5	82.3	91.6
06	WOOD, PLASTICS & COMPOSITES	89.7	150.6	124.6	95.6	132.6	116.8	95.7	132.6	116.9	93.4	150.8	126.3	88.4	119.2	106.0	95.7	73.4	82.9
07	THERMAL & MOISTURE PROTECTION	103.1	134.9	115.8	106.0	118.5	110.9	106.8	118.1	111.3	115.2	134.6	122.9	103.1	115.3	108.0	106.3	81.2	96.3
08	OPENINGS	99.6	139.7	109.8	100.0	124.5	106.2	102.1	124.3	107.7	113.3	139.8	120.1	90.8	119.0	98.0	93.9	78.9	90.1
0920	Plaster & Gypsum Board	103.5	152.1	137.7	108.0	133.4	125.9	106.2	133.4	125.3	102.6	152.1	137.4	106.0	119.8	115.7	85.5	72.3	76.2
0950, 0980	Ceilings & Acoustic Treatment	111.4	152.1	138.3	116.6	133.4	127.7	116.0	133.4	127.6	126.0	152.1	143.3	96.8	119.8	111.4	118.2	72.3	87.8
0960	Flooring	113.1	121.4	115.5	110.6	117.7	112.6	110.6	111.5	110.8	122.3	128.0	123.9	96.4	105.9	99.1	112.3	57.9	96.6
0970, 0990	Wall Finishes & Painting/Coating	100.6	143.2	126.5	103.7	118.5	112.7	103.7	117.7	112.2	103.3	143.2	127.5	105.9	109.0	107.8	107.9	25.4	57.8
09	FINISHES	106.5	142.4	126.3	109.2	128.4	119.8	110.4	127.2	119.7	112.8	143.6	129.8	99.2	116.3	108.6	108.6	64.7	84.4
COVERS	DIVS. 10 - 14, 25, 28, 41, 43, 44, 46	100.0	124.0	105.0	100.0	122.1	104.6	100.0	122.0	104.6	100.0	124.4	105.1	100.0	109.5	102.0	100.0	89.7	97.9
21, 22, 23	FIRE SUPPRESSION, PLUMBING & HVAC	94.7	164.9	123.9	100.1	113.7	105.8	94.7	107.5	100.0	100.2	129.0	112.2	94.8	107.8	100.2	94.6	69.2	84.1
26, 27, 3370	ELECTRICAL, COMMUNICATIONS & UTIL.	98.2	123.5	111.7	100.6	114.7	108.1	101.6	103.9	102.8	95.6	128.3	113.0	112.6	114.5	113.6	96.0	76.7	85.7
MF2010	WEIGHTED AVERAGE	101.7	134.9	116.5	101.9	117.3	108.8	103.1	113.8	107.9	102.9	127.8	114.0	99.7	112.7	105.5	102.1	76.1	90.5

CALIFORNIA (Santa Rosa 954, Stockton 952, Susanville 961, Vallejo 945, Van Nuys 913-916); COLORADO (Alamosa 811)

		BOULDER 803			COLORADO SPRINGS 808 - 809			DENVER 800 - 802			DURANGO 813			FORT COLLINS 805			FORT MORGAN 807		
DIVISION		MAT.	INST.	TOTAL	MAT.	INST.	TOTAL	MAT.	INST.	TOTAL	MAT.	INST.	TOTAL	MAT.	INST.	TOTAL	MAT.	INST.	TOTAL
015433	CONTRACTOR EQUIPMENT		96.2	96.2		94.8	94.8		99.2	99.2		96.4	96.4		96.2	96.2		96.2	96.2
0241, 31 - 34	SITE & INFRASTRUCTURE, DEMOLITION	88.7	97.2	94.6	90.0	94.8	93.3	89.1	102.9	98.6	116.5	92.7	100.1	99.6	96.4	97.4	90.9	96.8	94.9
0310	Concrete Forming & Accessories	105.5	81.6	84.7	96.1	81.8	83.7	101.6	82.0	84.6	111.6	72.8	77.9	102.7	78.0	81.2	105.8	81.0	84.3
0320	Concrete Reinforcing	96.3	79.7	87.7	95.5	84.8	90.0	95.5	84.8	90.0	104.4	79.5	91.6	96.3	78.4	87.1	96.5	79.6	87.8
0330	Cast-in-Place Concrete	105.4	80.8	96.0	108.3	81.4	98.0	102.5	81.1	94.3	118.9	80.6	104.2	119.0	75.5	102.3	103.4	79.6	94.3
03	CONCRETE	103.3	81.2	92.7	106.6	82.5	95.0	100.9	82.4	92.1	119.1	77.0	98.9	114.1	77.4	96.5	101.8	80.4	91.5
04	MASONRY	91.6	77.8	83.2	93.4	77.8	83.9	95.3	77.8	84.7	113.6	77.2	91.6	108.0	49.1	72.3	106.3	77.4	88.8
05	METALS	98.9	85.7	94.3	102.0	88.3	97.3	104.5	88.3	98.9	96.5	82.3	91.6	100.1	80.6	93.4	98.6	82.3	93.0
06	WOOD, PLASTICS & COMPOSITES	99.8	83.9	90.7	90.3	84.1	86.7	97.7	84.0	89.8	103.9	73.4	86.4	97.5	83.9	89.7	100.1	83.9	90.8
07	THERMAL & MOISTURE PROTECTION	103.5	79.4	94.0	104.3	83.2	95.9	102.7	77.5	92.7	106.2	81.2	96.2	104.0	66.0	88.9	103.5	82.5	95.1
08	OPENINGS	98.0	84.6	94.6	102.5	86.2	98.4	102.6	86.1	98.4	101.7	78.9	95.9	98.0	84.6	94.6	98.0	84.6	94.6
0920	Plaster & Gypsum Board	108.6	83.7	91.0	88.7	83.7	85.1	104.7	83.7	89.9	97.8	72.3	79.9	102.9	83.7	89.3	109.5	83.7	91.3
0950, 0980	Ceilings & Acoustic Treatment	106.7	83.7	91.4	114.2	83.7	93.9	113.2	83.7	93.7	118.2	72.3	87.8	106.7	83.7	91.4	106.7	83.7	91.4
0960	Flooring	109.9	86.6	103.2	101.3	72.1	92.9	106.4	89.6	101.6	117.0	57.9	100.0	106.3	57.9	92.4	110.4	57.9	95.3
0970, 0990	Wall Finishes & Painting/Coating	103.6	71.7	84.2	103.3	43.6	67.1	103.6	71.7	84.2	107.9	25.4	57.8	103.6	43.3	67.0	103.6	57.7	75.7
09	FINISHES	106.6	81.8	92.9	103.3	76.1	88.3	106.4	82.4	93.2	110.7	64.7	85.3	105.4	71.0	86.4	106.8	74.4	88.9
COVERS	DIVS. 10 - 14, 25, 28, 41, 43, 44, 46	100.0	89.9	97.9	100.0	90.3	98.0	100.0	90.0	97.9	100.0	89.7	97.9	100.0	87.4	97.4	100.0	90.0	97.9
21, 22, 23	FIRE SUPPRESSION, PLUMBING & HVAC	94.6	81.1	89.0	100.1	75.9	90.0	99.9	82.1	92.5	94.6	66.9	83.1	100.0	77.3	90.5	94.6	81.0	89.0
26, 27, 3370	ELECTRICAL, COMMUNICATIONS & UTIL.	98.3	85.8	91.7	102.2	86.5	93.9	103.9	85.8	94.3	95.3	74.5	84.3	98.3	85.8	91.6	98.7	85.8	91.8
MF2010	WEIGHTED AVERAGE	98.5	83.7	91.8	101.3	82.4	92.8	101.4	84.9	94.0	102.6	75.3	90.4	102.3	77.1	91.0	99.1	82.3	91.6

COLORADO (Boulder 803, Colorado Springs 808-809, Denver 800-802, Durango 813, Fort Collins 805, Fort Morgan 807)

		GLENWOOD SPRINGS 816			GOLDEN 804			GRAND JUNCTION 815			GREELEY 806			MONTROSE 814			PUEBLO 810		
DIVISION		MAT.	INST.	TOTAL	MAT.	INST.	TOTAL	MAT.	INST.	TOTAL	MAT.	INST.	TOTAL	MAT.	INST.	TOTAL	MAT.	INST.	TOTAL
015433	CONTRACTOR EQUIPMENT		99.1	99.1		96.2	96.2		99.1	99.1		96.2	96.2		97.7	97.7		96.4	96.4
0241, 31 - 34	SITE & INFRASTRUCTURE, DEMOLITION	130.5	99.6	109.2	100.7	97.1	98.3	115.8	99.3	104.4	87.3	95.2	92.8	124.6	95.8	104.7	108.1	94.5	98.7
0310	Concrete Forming & Accessories	103.2	80.8	83.7	98.6	81.1	83.4	110.3	80.3	84.3	100.9	78.8	81.7	102.8	80.5	83.4	107.9	82.0	85.5
0320	Concrete Reinforcing	103.3	79.5	91.1	96.5	79.6	87.8	103.6	79.5	91.2	96.3	78.2	86.9	103.1	79.5	91.0	99.9	84.8	92.1
0330	Cast-in-Place Concrete	103.4	79.9	94.4	103.5	79.7	94.4	114.5	79.2	101.0	99.4	52.4	81.4	103.4	79.9	94.4	102.6	82.5	94.9
03	CONCRETE	121.1	80.4	101.6	111.3	80.4	96.5	115.3	79.9	98.3	98.4	69.9	84.7	112.3	80.2	96.9	105.1	83.0	94.5
04	MASONRY	100.9	77.4	86.7	109.0	77.5	89.9	133.6	77.2	99.4	102.1	36.8	62.5	106.7	77.2	88.9	97.3	77.8	85.5
05	METALS	96.2	82.8	91.6	98.8	82.6	93.2	97.9	81.5	92.2	100.1	79.5	93.0	95.5	81.8	90.8	99.4	89.1	95.9
06	WOOD, PLASTICS & COMPOSITES	90.8	84.1	86.9	92.7	83.9	87.7	101.4	84.1	91.4	94.9	90.2	92.2	92.1	84.2	87.6	98.1	84.4	90.2
07	THERMAL & MOISTURE PROTECTION	106.1	82.3	96.6	104.4	77.3	93.7	105.0	71.9	91.8	103.4	57.5	85.2	106.3	82.3	96.8	104.6	83.1	96.0
08	OPENINGS	100.9	84.7	96.8	98.0	84.6	94.6	101.7	84.7	97.4	98.0	88.0	95.5	101.8	84.8	97.5	95.7	86.3	93.3
0920	Plaster & Gypsum Board	112.5	83.7	92.2	100.2	83.7	88.5	124.9	83.7	95.9	101.1	90.1	93.4	83.7	83.7	83.7	90.5	83.7	85.7
0950, 0980	Ceilings & Acoustic Treatment	117.3	83.7	95.0	106.7	83.7	91.4	117.3	83.7	95.0	106.7	90.1	95.7	118.2	83.7	95.3	128.5	83.7	98.8
0960	Flooring	111.4	53.3	94.7	104.1	57.9	90.8	116.4	57.9	99.6	105.2	57.9	91.6	114.5	47.9	95.4	113.2	89.6	106.4
0970, 0990	Wall Finishes & Painting/Coating	107.8	57.7	77.4	103.6	71.7	84.2	107.8	71.7	85.9	103.6	27.0	57.1	107.9	25.4	57.8	107.9	40.4	66.9
09	FINISHES	111.7	73.7	90.8	105.2	76.7	89.5	112.9	76.1	92.6	104.1	70.7	85.7	109.0	69.0	87.0	109.9	79.2	93.0
COVERS	DIVS. 10 - 14, 25, 28, 41, 43, 44, 46	100.0	90.2	98.0	100.0	90.0	97.9	100.0	90.1	98.0	100.0	85.1	96.9	100.0	90.5	98.0	100.0	91.0	98.1
21, 22, 23	FIRE SUPPRESSION, PLUMBING & HVAC	94.6	66.7	83.0	94.6	80.7	88.8	99.9	65.5	85.6	100.0	72.8	88.7	94.6	66.7	83.0	99.9	69.4	87.2
26, 27, 3370	ELECTRICAL, COMMUNICATIONS & UTIL.	92.5	74.5	83.0	98.7	85.8	91.8	95.0	58.5	75.6	98.3	85.8	91.6	95.0	58.5	75.6	96.0	76.8	85.8
MF2010	WEIGHTED AVERAGE	102.1	78.0	91.3	100.6	82.5	92.5	104.7	75.2	91.5	99.6	73.6	88.0	101.2	74.7	89.4	100.8	80.2	91.6

COLORADO (Glenwood Springs 816, Golden 804, Grand Junction 815, Greeley 806, Montrose 814, Pueblo 810)

City Cost Indexes

COLORADO / CONNECTICUT

	DIVISION	COLORADO SALIDA 812			CONNECTICUT BRIDGEPORT 066			BRISTOL 060			HARTFORD 061			MERIDEN 064			NEW BRITAIN 060		
		MAT.	INST.	TOTAL	MAT.	INST.	TOTAL	MAT.	INST.	TOTAL	MAT.	INST.	TOTAL	MAT.	INST.	TOTAL	MAT.	INST.	TOTAL
015433	CONTRACTOR EQUIPMENT		97.7	97.7		100.3	100.3		100.3	100.3		100.3	100.3		100.8	100.8		100.3	100.3
0241, 31 - 34	SITE & INFRASTRUCTURE, DEMOLITION	116.2	96.1	102.3	103.5	103.8	103.7	102.7	103.8	103.5	96.7	103.8	101.6	101.0	104.6	103.5	102.9	103.8	103.5
0310	Concrete Forming & Accessories	110.2	80.7	84.6	96.5	118.3	115.4	96.5	121.8	118.5	94.5	121.8	118.2	96.2	121.8	118.4	96.8	121.8	118.5
0320	Concrete Reinforcing	102.9	79.5	90.9	108.1	127.0	117.8	108.1	127.0	117.8	108.1	127.0	117.8	108.1	127.0	117.8	108.1	127.0	117.8
0330	Cast-in-Place Concrete	118.4	79.9	103.7	101.4	124.5	110.3	95.0	124.5	106.3	100.7	124.5	109.8	91.4	124.5	104.1	96.6	124.5	107.3
03	CONCRETE	114.2	80.3	98.0	109.2	121.9	115.3	106.0	123.5	114.4	108.7	123.5	115.8	104.2	123.5	113.4	106.8	123.5	114.8
04	MASONRY	135.2	77.2	100.1	108.3	127.9	120.2	100.2	127.9	117.0	100.7	127.9	117.2	99.9	127.9	116.9	102.2	127.9	117.8
05	METALS	95.1	82.0	90.6	99.8	124.3	108.3	99.8	124.3	108.2	104.5	124.3	111.3	97.1	124.3	106.4	96.2	124.3	105.9
06	WOOD, PLASTICS & COMPOSITES	99.1	84.2	90.6	98.1	116.2	108.5	98.1	121.0	111.3	96.6	121.0	110.6	98.1	121.0	111.3	98.1	121.0	111.3
07	THERMAL & MOISTURE PROTECTION	105.1	82.3	96.1	96.4	125.4	107.9	96.5	123.6	107.3	97.7	123.6	108.0	96.5	123.6	107.2	96.5	123.6	107.3
08	OPENINGS	94.0	84.8	91.6	105.6	127.2	111.1	105.6	129.9	111.8	106.3	129.9	112.3	108.5	129.9	113.9	105.6	129.9	111.8
0920	Plaster & Gypsum Board	85.0	83.7	84.1	99.4	116.0	111.1	99.4	121.0	114.6	97.5	121.0	114.0	102.9	121.0	115.1	99.4	121.0	114.6
0950, 0980	Ceilings & Acoustic Treatment	118.2	83.7	95.3	99.4	116.0	110.4	99.4	121.0	113.7	97.5	121.0	113.1	102.9	121.0	114.9	99.4	121.0	113.7
0960	Flooring	119.5	47.9	99.0	95.8	120.8	103.0	95.8	120.8	103.0	95.8	120.8	103.0	95.8	120.8	103.0	95.8	120.8	103.0
0970, 0990	Wall Finishes & Painting/Coating	107.9	25.4	57.8	92.3	114.5	105.8	92.3	116.9	107.2	92.3	116.9	107.2	92.3	116.9	107.2	92.3	116.9	107.2
09	FINISHES	109.7	69.0	87.3	100.1	118.0	110.0	100.1	121.1	111.7	98.6	121.1	111.0	101.1	121.1	112.1	100.1	121.1	111.7
COVERS	DIVS. 10 - 14, 25, 28, 41, 43, 44, 46	100.0	90.6	98.1	100.0	108.0	101.7	100.0	108.6	101.8	100.0	108.6	101.8	100.0	108.6	101.8	100.0	108.6	101.8
21, 22, 23	FIRE SUPPRESSION, PLUMBING & HVAC	94.6	69.2	84.0	100.2	112.8	105.4	100.2	112.8	105.4	100.2	112.8	105.5	94.8	112.8	102.3	100.2	112.8	105.4
26, 27, 3370	ELECTRICAL, COMMUNICATIONS & UTIL.	95.3	76.7	85.4	100.8	105.9	103.5	100.8	110.4	105.9	99.7	110.4	105.4	100.8	109.5	105.4	100.9	110.4	105.9
MF2010	WEIGHTED AVERAGE	102.0	77.8	91.2	102.2	116.4	108.6	101.4	117.8	108.7	102.1	117.8	109.1	99.7	117.7	107.8	101.0	117.8	108.5

CONNECTICUT

	DIVISION	NEW HAVEN 065			NEW LONDON 063			NORWALK 068			STAMFORD 069			WATERBURY 067			WILLIMANTIC 062		
		MAT.	INST.	TOTAL	MAT.	INST.	TOTAL	MAT.	INST.	TOTAL	MAT.	INST.	TOTAL	MAT.	INST.	TOTAL	MAT.	INST.	TOTAL
015433	CONTRACTOR EQUIPMENT		100.8	100.8		100.8	100.8		100.3	100.3		100.3	100.3		100.3	100.3		100.3	100.3
0241, 31 - 34	SITE & INFRASTRUCTURE, DEMOLITION	102.7	104.6	104.0	95.4	104.6	101.7	103.3	103.8	103.7	103.9	103.8	103.9	103.2	103.8	103.6	103.4	103.8	103.7
0310	Concrete Forming & Accessories	96.2	121.8	118.4	96.2	121.8	118.4	96.5	118.3	115.4	96.4	118.6	115.7	96.5	121.8	118.5	96.5	121.8	118.5
0320	Concrete Reinforcing	108.1	127.0	117.8	84.8	127.0	106.5	108.1	127.1	117.9	108.1	127.1	117.9	108.1	127.0	117.8	108.1	127.0	117.8
0330	Cast-in-Place Concrete	98.3	124.5	108.3	83.7	124.5	99.3	99.8	125.9	109.8	101.4	126.0	110.9	101.4	124.5	110.3	94.7	124.5	106.1
03	CONCRETE	123.0	123.5	123.2	93.7	123.5	108.0	108.4	122.4	115.1	109.2	122.6	115.6	109.2	123.5	116.0	105.9	123.5	114.3
04	MASONRY	100.5	127.9	117.1	98.9	127.9	116.5	99.9	129.5	117.8	100.7	129.5	118.2	100.7	127.9	117.2	100.2	127.9	117.0
05	METALS	96.4	124.3	106.0	96.1	124.3	105.8	99.8	124.4	108.3	99.8	124.8	108.4	99.8	124.3	108.2	99.6	124.3	108.1
06	WOOD, PLASTICS & COMPOSITES	98.1	121.0	111.3	98.1	121.0	111.3	98.1	116.2	108.5	98.1	116.2	108.5	98.1	121.0	111.3	98.1	121.0	111.3
07	THERMAL & MOISTURE PROTECTION	96.6	123.6	107.3	96.4	123.6	107.2	96.5	126.1	108.3	96.5	126.1	108.2	96.5	123.6	107.3	96.7	123.6	107.4
08	OPENINGS	105.6	129.9	111.8	108.9	129.9	114.2	105.6	127.2	111.1	105.6	127.2	111.1	105.6	129.9	111.8	108.9	129.9	114.2
0920	Plaster & Gypsum Board	99.4	121.0	114.6	99.4	121.0	114.6	99.4	116.0	111.1	99.4	116.0	111.1	99.4	121.0	114.6	99.4	121.0	114.6
0950, 0980	Ceilings & Acoustic Treatment	99.4	121.0	113.7	97.3	121.0	113.0	99.4	116.0	110.4	99.4	116.0	110.4	99.4	121.0	113.7	97.3	121.0	113.0
0960	Flooring	95.8	120.8	103.0	95.8	120.8	103.0	95.8	120.8	103.0	95.8	120.8	103.0	95.8	120.8	103.0	95.8	120.8	103.0
0970, 0990	Wall Finishes & Painting/Coating	92.3	114.5	105.8	92.3	116.9	107.2	92.3	114.5	105.8	92.3	114.5	105.8	92.3	116.9	107.2	92.3	116.9	107.2
09	FINISHES	100.2	120.9	111.6	99.1	121.1	111.3	100.1	118.0	110.0	100.2	118.0	110.0	100.0	121.1	111.6	99.8	121.1	111.6
COVERS	DIVS. 10 - 14, 25, 28, 41, 43, 44, 46	100.0	108.6	101.8	100.0	108.6	101.8	100.0	108.0	101.7	100.0	108.2	101.7	100.0	108.6	101.8	100.0	108.6	101.8
21, 22, 23	FIRE SUPPRESSION, PLUMBING & HVAC	100.2	112.8	105.4	94.8	112.8	102.3	100.2	112.8	105.4	100.2	112.9	105.5	100.2	112.8	105.4	100.2	112.8	105.4
26, 27, 3370	ELECTRICAL, COMMUNICATIONS & UTIL.	100.8	109.5	105.4	97.2	109.5	103.7	100.8	104.4	102.7	100.8	151.0	127.5	100.2	109.8	105.3	100.8	110.4	105.9
MF2010	WEIGHTED AVERAGE	102.9	117.7	109.5	97.7	117.7	106.6	101.6	116.5	108.3	101.8	123.0	111.3	101.7	117.7	108.9	101.6	117.8	108.9

D.C. / DELAWARE / FLORIDA

	DIVISION	D.C. WASHINGTON 200 - 205			DELAWARE DOVER 199			NEWARK 197			WILMINGTON 198			FLORIDA DAYTONA BEACH 321			FORT LAUDERDALE 333		
		MAT.	INST.	TOTAL	MAT.	INST.	TOTAL	MAT.	INST.	TOTAL	MAT.	INST.	TOTAL	MAT.	INST.	TOTAL	MAT.	INST.	TOTAL
015433	CONTRACTOR EQUIPMENT		103.6	103.6		117.7	117.7		117.7	117.7		117.8	117.8		97.8	97.8		90.5	90.5
0241, 31 - 34	SITE & INFRASTRUCTURE, DEMOLITION	113.8	92.3	99.0	94.0	112.2	106.6	94.2	112.2	106.6	85.3	112.5	104.0	113.2	89.3	96.7	98.7	77.3	83.9
0310	Concrete Forming & Accessories	98.7	80.3	82.7	100.2	99.8	99.9	102.4	99.8	100.1	103.3	99.8	100.3	97.5	73.6	76.8	96.2	74.3	77.2
0320	Concrete Reinforcing	94.6	89.6	92.0	92.7	103.6	98.3	93.5	103.6	98.7	93.5	103.6	98.7	91.8	78.2	84.8	91.8	77.8	84.6
0330	Cast-in-Place Concrete	129.7	87.2	113.5	92.1	102.1	95.9	82.7	102.1	90.1	89.2	102.1	94.2	91.4	75.5	85.3	95.9	82.9	90.9
03	CONCRETE	116.2	85.7	101.6	97.1	102.3	99.6	92.7	102.3	97.3	96.0	102.3	99.0	88.1	76.2	82.4	90.3	79.0	84.9
04	MASONRY	102.2	81.9	89.9	104.4	97.6	100.3	102.7	97.6	99.6	108.7	97.6	102.0	87.1	70.9	77.3	87.4	73.8	79.1
05	METALS	93.5	107.2	98.2	104.6	118.0	109.2	104.6	118.0	109.2	101.6	118.0	107.2	96.1	93.0	95.1	96.0	93.9	95.3
06	WOOD, PLASTICS & COMPOSITES	96.8	79.2	86.7	100.4	98.1	99.1	103.4	98.1	100.3	105.8	98.1	101.4	100.5	75.1	86.0	96.3	72.9	82.9
07	THERMAL & MOISTURE PROTECTION	99.1	84.9	93.4	97.5	113.9	104.0	97.4	113.9	104.0	97.9	113.9	104.3	93.8	79.4	88.1	93.8	87.3	91.2
08	OPENINGS	104.2	88.8	100.3	95.8	109.0	99.2	95.8	109.0	99.2	95.6	109.0	99.0	100.5	71.9	93.2	97.9	70.9	91.1
0920	Plaster & Gypsum Board	104.3	78.6	86.2	105.2	97.9	100.1	106.6	97.9	100.5	107.1	97.9	100.6	94.6	74.9	80.7	93.7	72.6	78.8
0950, 0980	Ceilings & Acoustic Treatment	107.6	78.6	88.4	109.0	97.9	101.6	109.0	97.9	101.6	104.2	97.9	100.0	96.5	74.9	82.2	96.5	72.6	80.7
0960	Flooring	120.1	96.6	113.4	90.3	106.2	94.8	90.5	106.2	95.0	90.1	106.2	94.7	119.4	77.8	107.4	119.4	73.2	106.1
0970, 0990	Wall Finishes & Painting/Coating	121.0	88.5	101.3	94.7	101.6	98.9	94.7	101.6	98.9	94.7	101.6	98.9	111.3	79.8	92.2	105.9	75.6	87.5
09	FINISHES	108.4	83.5	94.7	99.8	100.3	100.1	100.3	100.3	100.3	99.1	100.3	99.8	107.6	74.8	89.5	104.9	73.5	87.6
COVERS	DIVS. 10 - 14, 25, 28, 41, 43, 44, 46	100.0	98.1	99.6	100.0	96.5	99.3	100.0	96.5	99.3	100.0	96.5	99.3	100.0	81.2	96.1	100.0	88.0	97.5
21, 22, 23	FIRE SUPPRESSION, PLUMBING & HVAC	100.2	94.9	98.0	99.9	116.1	106.6	100.1	116.1	106.8	100.1	116.1	106.8	100.1	80.7	91.1	99.8	69.4	87.1
26, 27, 3370	ELECTRICAL, COMMUNICATIONS & UTIL.	101.4	100.2	100.7	101.2	112.5	107.2	99.6	112.5	106.4	103.5	112.5	108.3	98.1	59.6	77.6	98.1	75.9	86.3
MF2010	WEIGHTED AVERAGE	102.8	91.9	97.9	99.9	108.6	103.8	99.3	108.6	103.4	99.6	108.6	103.6	98.0	76.4	88.3	97.3	76.7	88.1

FLORIDA

DIVISION		FORT MYERS 339,341			GAINESVILLE 326,344			JACKSONVILLE 320,322			LAKELAND 338			MELBOURNE 329			MIAMI 330 - 332,340		
		MAT.	INST.	TOTAL	MAT.	INST.	TOTAL	MAT.	INST.	TOTAL	MAT.	INST.	TOTAL	MAT.	INST.	TOTAL	MAT.	INST.	TOTAL
015433	CONTRACTOR EQUIPMENT		97.8	97.8		97.8	97.8		97.8	97.8		97.8	97.8		97.8	97.8		90.5	90.5
0241, 31 - 34	SITE & INFRASTRUCTURE, DEMOLITION	110.5	88.6	95.4	123.5	88.3	99.2	113.2	88.6	96.3	112.6	88.8	96.2	121.3	88.6	98.7	100.5	77.1	84.3
0310	Concrete Forming & Accessories	91.5	81.1	82.5	92.4	58.6	63.1	97.3	59.0	64.0	87.9	81.8	82.6	93.7	75.4	77.8	100.7	74.8	78.2
0320	Concrete Reinforcing	92.8	100.1	96.6	97.4	68.6	82.6	91.8	68.7	79.9	95.1	101.0	98.1	92.8	78.2	85.3	91.8	77.9	84.6
0330	Cast-in-Place Concrete	100.0	73.1	89.7	105.0	66.5	90.2	92.3	72.8	84.9	102.3	74.5	91.6	110.2	78.1	97.9	97.8	84.0	92.5
03	CONCRETE	91.2	82.9	87.2	98.8	64.7	82.5	88.5	67.0	78.2	93.1	83.8	88.6	99.6	77.9	89.2	91.5	79.5	85.8
04	MASONRY	80.7	66.9	72.3	100.2	66.0	79.5	87.0	66.1	74.3	95.7	81.9	87.3	84.0	75.3	78.8	85.9	77.7	80.9
05	METALS	98.1	101.5	99.3	95.1	87.7	92.5	94.7	88.0	92.4	98.0	102.3	99.5	104.0	93.1	100.3	100.7	92.7	98.0
06	WOOD, PLASTICS & COMPOSITES	93.2	84.1	88.0	94.4	56.1	72.5	100.5	56.1	75.1	88.9	84.1	86.1	96.0	75.1	84.0	102.3	72.9	85.4
07	THERMAL & MOISTURE PROTECTION	93.7	86.3	90.7	94.1	65.4	82.7	94.0	66.1	82.9	93.6	91.3	92.7	94.2	82.0	89.4	102.4	81.1	93.9
08	OPENINGS	99.0	80.6	94.4	98.4	56.7	87.9	100.5	60.1	90.2	99.0	81.3	94.5	99.6	75.4	93.4	97.9	70.9	91.1
0920	Plaster & Gypsum Board	88.9	84.1	85.5	89.7	55.3	65.5	94.6	55.3	66.9	86.6	84.1	84.9	89.7	74.9	79.3	92.0	72.6	78.4
0950, 0980	Ceilings & Acoustic Treatment	89.7	84.1	86.0	89.7	55.3	66.9	96.5	55.3	69.2	89.7	84.1	86.0	91.8	74.9	80.6	93.7	72.6	79.7
0960	Flooring	115.7	58.2	99.2	116.2	45.6	95.9	119.4	68.2	104.7	112.9	59.7	97.6	116.4	77.8	105.3	127.0	77.1	112.7
0970, 0990	Wall Finishes & Painting/Coating	111.3	65.4	83.5	111.3	65.4	83.5	111.3	66.7	84.3	111.3	65.4	83.5	111.3	99.0	103.8	105.9	75.6	87.5
09	FINISHES	104.2	75.5	88.3	105.7	55.8	78.2	107.7	60.5	81.6	103.2	75.8	88.1	105.9	78.1	90.5	106.5	74.7	89.0
COVERS	DIVS. 10 - 14, 25, 28, 41, 43, 44, 46	100.0	81.5	96.2	100.0	79.4	95.7	100.0	77.4	95.3	100.0	81.5	96.2	100.0	82.8	96.4	100.0	88.6	97.6
21, 22, 23	FIRE SUPPRESSION, PLUMBING & HVAC	96.9	59.7	81.4	98.9	69.1	86.5	99.8	66.7	86.0	96.9	88.7	93.5	99.8	81.0	92.0	99.8	73.0	88.6
26, 27, 3370	ELECTRICAL, COMMUNICATIONS & UTIL.	100.6	66.9	82.7	98.4	77.9	87.5	97.8	67.6	81.7	98.4	66.8	81.6	99.4	69.5	83.5	103.9	78.5	90.4
MF2010	WEIGHTED AVERAGE	97.3	75.4	87.5	99.5	70.6	86.6	97.8	69.7	85.2	98.0	83.5	91.5	100.4	79.6	91.1	99.1	78.1	89.7

FLORIDA

DIVISION		ORLANDO 327 - 328,347			PANAMA CITY 324			PENSACOLA 325			SARASOTA 342			ST. PETERSBURG 337			TALLAHASSEE 323		
		MAT.	INST.	TOTAL	MAT.	INST.	TOTAL	MAT.	INST.	TOTAL	MAT.	INST.	TOTAL	MAT.	INST.	TOTAL	MAT.	INST.	TOTAL
015433	CONTRACTOR EQUIPMENT		97.8	97.8		97.8	97.8		97.8	97.8		97.8	97.8		97.8	97.8		97.8	97.8
0241, 31 - 34	SITE & INFRASTRUCTURE, DEMOLITION	113.8	88.5	96.3	127.5	87.5	99.9	127.0	88.0	100.1	114.7	88.6	96.6	114.0	88.1	96.1	110.0	87.7	94.6
0310	Concrete Forming & Accessories	97.1	78.0	80.6	96.6	47.1	53.7	94.2	56.1	61.1	97.5	81.3	83.4	95.5	55.1	60.4	99.4	47.7	54.5
0320	Concrete Reinforcing	91.8	77.5	84.4	95.8	78.2	86.8	98.3	78.8	88.3	91.8	100.9	96.5	95.1	86.4	90.6	91.8	68.5	79.8
0330	Cast-in-Place Concrete	114.6	75.4	99.6	97.0	60.2	82.9	119.8	70.3	100.9	104.6	74.2	93.0	103.4	68.5	90.1	99.8	59.8	84.5
03	CONCRETE	100.9	78.0	89.9	96.2	59.2	78.5	107.1	66.8	87.8	94.7	83.4	89.3	94.5	67.1	81.4	92.4	57.5	75.7
04	MASONRY	90.4	70.9	78.6	91.4	51.0	66.9	109.7	58.7	78.8	88.1	81.9	84.3	133.0	52.8	84.4	89.8	58.1	70.6
05	METALS	101.6	92.3	98.4	95.9	89.7	93.8	97.1	91.6	95.2	98.7	101.7	99.8	99.0	94.1	97.3	89.9	87.4	89.1
06	WOOD, PLASTICS & COMPOSITES	100.1	81.7	89.5	99.3	45.2	68.3	97.4	55.9	73.6	100.5	84.1	91.1	97.8	54.3	72.9	97.9	44.7	67.4
07	THERMAL & MOISTURE PROTECTION	95.0	80.0	89.0	94.3	60.8	81.0	94.2	67.0	83.4	93.8	91.3	92.8	93.8	62.1	81.2	99.7	77.0	90.7
08	OPENINGS	100.5	74.4	93.8	98.0	50.3	85.9	97.9	61.6	88.7	100.4	80.3	95.3	99.0	64.2	90.2	100.9	50.9	88.3
0920	Plaster & Gypsum Board	105.4	81.6	88.7	92.4	44.1	58.4	92.9	55.0	66.3	94.6	84.1	87.2	91.5	53.4	64.7	110.8	43.6	63.5
0950, 0980	Ceilings & Acoustic Treatment	98.4	81.6	87.3	91.8	44.1	60.2	91.8	55.0	67.4	94.4	84.1	87.6	91.8	53.4	66.4	98.4	43.6	62.0
0960	Flooring	119.6	77.8	107.6	118.9	45.4	97.8	113.6	66.5	100.1	119.4	61.1	102.6	118.0	58.3	100.8	119.4	65.5	103.9
0970, 0990	Wall Finishes & Painting/Coating	111.3	75.6	89.6	111.3	67.7	84.9	111.3	67.7	84.9	111.3	65.4	83.5	111.3	65.4	83.5	111.3	67.7	84.9
09	FINISHES	109.1	78.2	92.1	107.7	48.1	74.8	105.8	58.5	79.7	107.2	76.0	90.0	105.9	55.8	78.3	109.4	51.8	77.6
COVERS	DIVS. 10 - 14, 25, 28, 41, 43, 44, 46	100.0	81.9	96.2	100.0	51.5	89.9	100.0	51.6	89.9	100.0	81.5	96.2	100.0	62.4	92.2	100.0	73.8	94.6
21, 22, 23	FIRE SUPPRESSION, PLUMBING & HVAC	99.7	62.0	84.0	99.8	41.3	75.4	99.8	57.5	82.2	99.7	60.2	83.3	99.8	58.6	82.6	99.8	42.7	76.0
26, 27, 3370	ELECTRICAL, COMMUNICATIONS & UTIL.	104.1	63.6	82.6	96.7	58.5	76.4	100.9	58.4	78.3	98.1	66.8	81.5	98.4	66.8	81.6	104.1	58.6	79.9
MF2010	WEIGHTED AVERAGE	101.2	74.2	89.1	99.2	57.5	80.6	101.9	65.1	85.4	99.2	77.3	89.4	101.3	66.1	85.6	98.6	59.7	81.2

FLORIDA / GEORGIA

DIVISION		TAMPA 335 - 336,346			WEST PALM BEACH 334,349			ALBANY 317,398			ATHENS 306			ATLANTA 300 - 303,399			AUGUSTA 308 - 309		
		MAT.	INST.	TOTAL	MAT.	INST.	TOTAL	MAT.	INST.	TOTAL	MAT.	INST.	TOTAL	MAT.	INST.	TOTAL	MAT.	INST.	TOTAL
015433	CONTRACTOR EQUIPMENT		97.8	97.8		90.5	90.5		91.3	91.3		92.7	92.7		93.3	93.3		92.7	92.7
0241, 31 - 34	SITE & INFRASTRUCTURE, DEMOLITION	114.4	88.7	96.7	95.6	77.4	83.0	98.9	78.5	84.8	101.2	92.3	95.0	97.2	94.1	95.1	93.8	92.0	92.6
0310	Concrete Forming & Accessories	98.3	81.9	84.1	99.5	73.9	77.3	96.8	46.6	53.2	94.8	44.3	51.0	98.2	74.1	77.3	95.7	68.4	72.0
0320	Concrete Reinforcing	91.8	101.0	96.5	94.4	77.6	85.7	91.4	84.5	87.8	99.4	84.0	91.4	98.8	85.9	92.2	99.7	76.6	87.8
0330	Cast-in-Place Concrete	101.1	74.5	90.9	91.2	79.7	86.8	97.6	48.4	78.8	110.0	53.6	88.4	110.0	72.3	95.5	103.9	50.4	83.4
03	CONCRETE	93.0	83.8	88.6	87.3	77.7	82.7	91.0	55.9	74.2	101.0	55.8	79.3	98.9	75.8	87.9	94.5	64.1	79.9
04	MASONRY	88.3	81.9	84.4	86.9	71.9	77.9	89.7	45.6	63.0	74.6	58.1	64.6	88.0	69.2	76.6	88.2	42.2	60.3
05	METALS	98.0	102.4	99.5	95.2	93.4	94.6	95.4	87.3	92.6	91.1	75.5	85.8	92.0	79.6	87.8	90.8	73.0	84.7
06	WOOD, PLASTICS & COMPOSITES	101.8	84.1	91.7	101.1	72.9	84.9	99.5	41.1	66.0	89.3	38.0	59.9	93.3	76.1	83.4	90.7	74.3	81.3
07	THERMAL & MOISTURE PROTECTION	94.0	91.3	92.9	93.6	75.6	86.4	91.5	61.7	79.7	94.5	54.3	78.6	94.4	74.7	86.6	94.1	58.3	79.9
08	OPENINGS	100.4	83.8	96.2	97.0	70.9	90.4	97.9	47.4	85.1	91.9	46.1	80.3	97.8	72.4	91.3	91.9	65.7	85.3
0920	Plaster & Gypsum Board	94.6	84.1	87.2	97.3	72.6	79.9	94.3	39.8	56.0	100.5	36.5	55.5	102.7	75.8	83.7	101.4	73.9	82.0
0950, 0980	Ceilings & Acoustic Treatment	96.5	84.1	88.3	89.7	72.6	78.4	95.6	39.8	58.6	100.6	36.5	58.1	100.6	75.8	84.1	101.6	73.9	83.2
0960	Flooring	119.4	59.7	102.2	121.8	70.2	107.0	119.4	37.9	96.0	89.5	56.9	80.1	90.9	68.9	84.6	89.9	41.5	76.0
0970, 0990	Wall Finishes & Painting/Coating	111.3	65.4	83.5	105.9	75.6	87.5	105.9	55.5	75.3	99.0	37.7	61.8	99.0	86.4	91.3	99.0	39.4	62.8
09	FINISHES	107.7	75.8	90.1	104.3	72.9	87.0	104.7	43.9	71.2	96.1	44.3	67.5	96.3	74.4	84.2	95.8	61.0	76.6
COVERS	DIVS. 10 - 14, 25, 28, 41, 43, 44, 46	100.0	82.1	96.3	100.0	88.0	97.5	100.0	80.2	95.9	100.0	43.5	88.3	100.0	84.7	96.8	100.0	55.4	90.7
21, 22, 23	FIRE SUPPRESSION, PLUMBING & HVAC	99.8	88.7	95.2	96.9	68.8	85.2	99.8	69.8	87.3	94.6	74.5	86.2	99.9	74.4	88.3	99.8	67.2	86.3
26, 27, 3370	ELECTRICAL, COMMUNICATIONS & UTIL.	98.2	66.8	81.5	99.4	75.9	86.9	93.5	59.8	75.6	100.2	74.2	86.4	99.5	77.1	87.6	100.8	67.1	82.9
MF2010	WEIGHTED AVERAGE	99.0	83.6	92.1	96.0	75.7	86.9	96.9	62.0	81.3	94.9	64.9	81.5	97.2	76.8	88.1	95.9	65.5	82.3

DIVISION		GEORGIA																	
		COLUMBUS 318 - 319			DALTON 307			GAINESVILLE 305			MACON 310 - 312			SAVANNAH 313 - 314			STATESBORO 304		
		MAT.	INST.	TOTAL	MAT.	INST.	TOTAL	MAT.	INST.	TOTAL	MAT.	INST.	TOTAL	MAT.	INST.	TOTAL	MAT.	INST.	TOTAL
015433	CONTRACTOR EQUIPMENT		91.3	91.3		104.7	104.7		92.7	92.7		102.0	102.0		92.3	92.3		92.2	92.2
0241, 31 - 34	SITE & INFRASTRUCTURE, DEMOLITION	98.8	78.7	84.9	102.0	96.2	98.0	100.8	92.3	94.9	99.4	93.9	95.6	99.8	79.4	85.7	102.5	76.4	84.5
0310	Concrete Forming & Accessories	96.7	58.2	63.3	88.2	49.9	54.9	97.9	46.1	53.0	96.0	55.3	60.7	96.8	51.6	57.6	82.8	43.2	48.4
0320	Concrete Reinforcing	91.8	85.3	88.4	98.9	78.9	88.6	99.2	84.0	91.3	92.9	84.6	88.6	92.7	76.8	84.5	98.6	44.6	70.8
0330	Cast-in-Place Concrete	97.2	57.1	81.9	106.8	47.3	84.0	115.6	53.4	91.8	96.0	50.7	78.6	107.0	49.8	85.1	109.8	62.7	91.7
03	CONCRETE	90.9	64.3	78.1	99.8	56.1	78.9	103.1	56.6	80.8	90.4	60.7	76.1	95.9	57.2	77.4	100.4	51.9	77.2
04	MASONRY	89.2	59.7	71.3	75.5	34.6	50.7	83.6	58.5	68.4	101.6	40.3	64.4	88.6	52.7	66.8	77.4	42.5	56.2
05	METALS	94.9	89.1	92.9	91.9	85.3	89.6	90.4	75.2	85.1	91.2	88.1	90.2	91.6	84.6	89.2	95.4	72.4	87.5
06	WOOD, PLASTICS & COMPOSITES	99.5	56.5	74.8	73.3	51.4	60.7	93.1	41.0	63.2	109.0	55.5	78.3	115.1	48.5	76.9	66.9	42.6	53.0
07	THERMAL & MOISTURE PROTECTION	91.5	67.0	81.8	96.4	52.5	78.9	94.5	57.4	79.7	90.1	63.4	79.5	93.7	58.4	79.7	95.1	50.8	77.5
08	OPENINGS	97.9	57.7	87.7	93.6	47.9	82.0	91.9	43.2	79.5	96.1	56.2	86.0	99.1	49.4	86.5	94.3	34.1	79.1
0920	Plaster & Gypsum Board	94.3	55.7	67.1	85.0	50.4	60.6	102.7	39.6	58.3	102.9	54.6	68.9	93.3	47.4	61.0	85.7	41.3	54.5
0950, 0980	Ceilings & Acoustic Treatment	95.6	55.7	69.1	115.0	50.4	72.2	100.6	39.6	60.2	90.2	54.6	66.6	98.4	47.4	64.6	110.1	41.3	64.5
0960	Flooring	119.4	53.3	100.4	90.3	37.9	75.2	90.9	37.9	75.6	92.9	38.3	77.2	119.9	49.0	99.5	106.9	47.4	89.8
0970, 0990	Wall Finishes & Painting/Coating	105.9	68.9	83.4	89.7	57.5	70.2	99.0	37.7	61.8	108.2	55.5	76.2	107.3	58.4	77.6	97.0	35.3	59.6
09	FINISHES	104.6	57.4	78.6	105.9	47.8	73.9	96.5	41.6	66.2	91.8	51.0	69.3	105.5	50.9	75.4	108.4	42.4	72.0
COVERS	DIVS. 10 - 14, 25, 28, 41, 43, 44, 46	100.0	82.0	96.3	100.0	24.7	84.4	100.0	43.8	88.3	100.0	79.9	95.8	100.0	54.3	90.5	100.0	40.4	87.6
21, 22, 23	FIRE SUPPRESSION, PLUMBING & HVAC	99.8	51.0	79.5	94.6	59.0	79.8	94.6	74.5	86.2	99.8	67.8	86.5	99.8	60.5	83.4	95.0	55.5	78.5
26, 27, 3370	ELECTRICAL, COMMUNICATIONS & UTIL.	93.7	74.3	83.4	109.9	73.5	90.6	100.2	74.2	86.4	92.7	63.4	77.1	97.7	58.7	77.0	100.3	58.7	78.2
MF2010	WEIGHTED AVERAGE	96.8	65.3	82.7	96.8	60.5	80.6	95.5	64.7	81.7	95.6	65.0	81.9	97.7	60.8	81.2	96.8	54.1	77.7

DIVISION		GEORGIA						HAWAII									IDAHO		
		VALDOSTA 316			WAYCROSS 315			HILO 967			HONOLULU 968			STATES & POSS., GUAM 969			BOISE 836 - 837		
		MAT.	INST.	TOTAL	MAT.	INST.	TOTAL	MAT.	INST.	TOTAL	MAT.	INST.	TOTAL	MAT.	INST.	TOTAL	MAT.	INST.	TOTAL
015433	CONTRACTOR EQUIPMENT		91.3	91.3		91.3	91.3		99.9	99.9		99.9	99.9		165.5	165.5		100.6	100.6
0241, 31 - 34	SITE & INFRASTRUCTURE, DEMOLITION	108.8	78.5	87.9	105.7	77.2	86.1	127.4	108.0	114.0	129.9	108.0	114.8	169.4	105.6	125.4	74.1	99.4	91.6
0310	Concrete Forming & Accessories	86.3	45.7	51.1	88.2	67.8	70.5	108.6	134.5	131.1	115.7	134.5	132.0	110.8	68.4	74.0	100.7	75.0	78.4
0320	Concrete Reinforcing	93.6	72.5	82.8	93.6	78.4	85.8	112.7	120.1	116.5	111.6	120.1	116.0	196.9	32.5	112.2	98.5	76.0	86.9
0330	Cast-in-Place Concrete	95.6	52.4	79.0	108.0	51.0	86.2	189.4	125.4	164.9	159.8	125.4	146.6	164.2	110.7	143.7	93.6	76.4	87.0
03	CONCRETE	94.4	54.7	75.4	98.8	64.9	82.6	149.9	127.5	139.2	135.3	127.5	131.6	149.7	76.5	114.7	101.1	75.8	89.0
04	MASONRY	94.8	52.2	69.0	95.7	42.4	63.4	126.2	125.4	125.7	118.0	125.4	122.5	179.4	46.9	99.1	122.6	63.9	87.0
05	METALS	94.6	83.5	90.7	93.6	80.2	89.0	109.4	107.8	108.8	130.1	107.8	122.4	140.9	78.5	119.4	100.4	75.5	91.8
06	WOOD, PLASTICS & COMPOSITES	86.8	40.1	60.0	89.1	75.7	81.4	97.4	137.6	120.4	108.2	137.6	125.0	109.3	72.5	88.2	90.4	77.3	82.9
07	THERMAL & MOISTURE PROTECTION	91.7	64.8	81.0	91.5	52.9	76.1	105.8	123.9	113.0	115.7	123.9	118.9	121.8	69.3	100.9	96.2	68.0	85.0
08	OPENINGS	93.3	43.5	80.7	93.3	60.4	84.9	99.5	131.3	107.6	104.7	131.3	111.5	104.3	58.1	92.6	94.4	71.9	88.7
0920	Plaster & Gypsum Board	85.8	38.8	52.7	87.1	75.5	78.9	108.0	138.4	129.5	139.1	138.4	138.6	207.0	60.5	103.9	90.8	76.5	80.7
0950, 0980	Ceilings & Acoustic Treatment	91.8	38.8	56.7	89.7	75.5	80.3	113.2	138.4	130.0	118.5	138.4	131.7	224.9	60.5	115.9	122.2	76.5	91.9
0960	Flooring	112.0	49.1	94.0	113.4	31.6	89.9	134.9	131.0	133.8	158.6	131.0	150.7	158.3	49.0	126.9	97.3	67.1	88.6
0970, 0990	Wall Finishes & Painting/Coating	105.9	55.5	75.3	105.9	42.9	67.7	98.4	144.0	126.1	102.7	144.0	127.8	103.7	38.7	64.2	98.5	42.5	64.5
09	FINISHES	101.6	45.7	70.8	101.3	59.5	78.2	117.4	135.9	127.6	130.2	135.9	133.4	191.5	63.0	120.7	101.9	71.1	84.9
COVERS	DIVS. 10 - 14, 25, 28, 41, 43, 44, 46	100.0	53.4	90.3	100.0	44.0	88.4	100.0	116.5	103.4	100.0	116.5	103.4	100.0	82.7	96.4	100.0	64.5	92.6
21, 22, 23	FIRE SUPPRESSION, PLUMBING & HVAC	99.8	69.7	87.3	96.3	59.4	80.9	100.1	108.6	103.7	100.2	108.6	103.7	102.3	40.9	76.7	100.0	67.2	86.3
26, 27, 3370	ELECTRICAL, COMMUNICATIONS & UTIL.	91.6	58.2	73.8	96.2	61.8	77.9	108.4	126.1	117.9	111.1	126.1	119.1	160.4	44.7	98.9	95.5	74.6	84.4
MF2010	WEIGHTED AVERAGE	96.7	61.2	80.9	96.7	61.8	81.1	111.9	120.6	115.8	115.1	120.6	117.5	133.3	61.7	101.3	99.5	73.3	87.8

DIVISION		IDAHO															ILLINOIS		
		COEUR D'ALENE 838			IDAHO FALLS 834			LEWISTON 835			POCATELLO 832			TWIN FALLS 833			BLOOMINGTON 617		
		MAT.	INST.	TOTAL	MAT.	INST.	TOTAL	MAT.	INST.	TOTAL	MAT.	INST.	TOTAL	MAT.	INST.	TOTAL	MAT.	INST.	TOTAL
015433	CONTRACTOR EQUIPMENT		94.6	94.6		100.6	100.6		94.6	94.6		100.6	100.6		100.6	100.6		101.0	101.0
0241, 31 - 34	SITE & INFRASTRUCTURE, DEMOLITION	74.0	94.7	88.3	72.6	98.5	90.5	79.9	95.2	90.5	75.6	99.4	92.0	81.6	97.6	92.7	98.8	96.8	97.4
0310	Concrete Forming & Accessories	117.8	58.5	66.3	95.2	42.1	49.1	122.5	60.0	68.3	100.9	74.7	78.2	101.7	30.1	39.6	88.1	118.8	114.8
0320	Concrete Reinforcing	105.8	94.9	100.1	100.4	74.9	87.3	105.8	95.1	100.3	98.9	76.1	87.1	100.7	75.2	87.5	99.6	104.8	102.2
0330	Cast-in-Place Concrete	101.2	88.7	96.4	89.1	53.0	75.3	105.2	90.0	99.3	96.2	76.3	88.6	98.7	40.3	76.4	99.7	114.5	105.4
03	CONCRETE	108.2	76.2	92.9	93.3	53.0	74.0	112.0	77.3	95.4	100.5	75.6	88.6	108.1	43.3	77.1	96.7	114.6	105.3
04	MASONRY	124.1	81.7	98.4	117.8	35.3	67.8	124.5	84.6	100.3	119.9	65.8	87.1	122.7	33.3	68.5	113.8	119.1	117.0
05	METALS	94.8	84.7	91.3	108.1	72.8	95.9	94.2	85.8	91.3	108.1	75.3	96.8	108.2	72.2	95.8	95.0	110.1	100.2
06	WOOD, PLASTICS & COMPOSITES	95.6	51.6	70.3	84.5	40.6	59.3	100.3	51.6	72.4	90.4	77.3	82.9	91.2	28.8	55.4	84.1	117.8	103.4
07	THERMAL & MOISTURE PROTECTION	148.4	74.6	119.1	95.8	45.4	75.8	148.6	75.3	119.5	96.2	62.5	82.8	97.0	39.4	74.1	98.4	113.3	104.3
08	OPENINGS	114.3	57.1	99.8	98.7	46.8	85.6	114.3	59.8	100.5	95.2	66.5	88.0	98.8	38.2	83.4	90.3	105.5	94.1
0920	Plaster & Gypsum Board	155.8	50.1	81.4	83.9	38.7	52.1	156.7	50.1	81.6	87.9	76.5	79.9	88.3	26.5	44.8	91.2	118.2	110.2
0950, 0980	Ceilings & Acoustic Treatment	160.9	50.1	87.5	119.1	38.7	65.8	160.9	50.1	87.5	128.5	76.5	94.0	122.0	26.5	58.7	89.5	118.2	108.5
0960	Flooring	136.5	48.5	111.2	97.1	45.5	82.3	139.5	90.3	125.3	100.4	67.1	90.8	101.6	45.5	85.5	95.1	113.2	100.3
0970, 0990	Wall Finishes & Painting/Coating	117.4	63.2	84.5	98.5	34.9	59.9	117.4	63.2	84.5	98.4	44.2	65.5	98.5	26.1	54.5	96.2	122.3	112.0
09	FINISHES	172.0	55.6	107.8	99.9	41.3	67.6	173.1	64.4	113.2	103.8	71.3	85.9	103.5	31.4	63.8	94.9	118.8	108.1
COVERS	DIVS. 10 - 14, 25, 28, 41, 43, 44, 46	100.0	53.9	90.4	100.0	40.7	87.7	100.0	67.3	93.2	100.0	64.5	92.6	100.0	36.6	86.8	100.0	105.2	101.1
21, 22, 23	FIRE SUPPRESSION, PLUMBING & HVAC	99.8	77.4	90.5	100.7	60.8	84.1	100.8	78.8	91.6	99.9	67.2	86.3	99.9	56.9	82.0	94.7	95.5	94.6
26, 27, 3370	ELECTRICAL, COMMUNICATIONS & UTIL.	88.0	79.7	83.6	87.7	74.6	80.7	85.9	82.5	84.1	92.9	74.6	83.2	89.2	64.3	76.0	93.2	95.5	94.5
MF2010	WEIGHTED AVERAGE	108.5	75.4	93.7	99.0	59.0	81.1	109.3	78.3	95.4	100.4	73.0	88.2	101.7	53.1	80.0	96.1	108.7	101.7

ILLINOIS

DIVISION		CARBONDALE 629			CENTRALIA 628			CHAMPAIGN 618 - 619			CHICAGO 606 - 608			DECATUR 625			EAST ST. LOUIS 620 - 622		
		MAT.	INST.	TOTAL	MAT.	INST.	TOTAL	MAT.	INST.	TOTAL	MAT.	INST.	TOTAL	MAT.	INST.	TOTAL	MAT.	INST.	TOTAL
015433	CONTRACTOR EQUIPMENT		107.4	107.4		107.4	107.4		101.7	101.7		93.4	93.4		101.7	101.7		107.4	107.4
0241, 31 - 34	SITE & INFRASTRUCTURE, DEMOLITION	107.1	97.0	100.1	107.4	98.3	101.1	108.5	97.5	100.9	115.1	94.0	100.6	92.2	97.5	95.9	109.2	98.4	101.7
0310	Concrete Forming & Accessories	96.5	103.7	102.7	98.3	111.4	109.7	94.0	116.6	113.6	100.3	152.2	145.3	98.2	116.0	113.7	94.1	111.7	109.3
0320	Concrete Reinforcing	98.5	91.8	95.0	98.5	106.1	102.4	99.6	101.9	100.8	98.8	157.4	129.0	96.5	101.7	99.2	98.4	106.1	102.4
0330	Cast-in-Place Concrete	93.3	101.4	96.4	93.8	115.1	102.0	115.7	110.7	113.8	112.8	147.0	125.9	101.9	112.1	105.8	95.4	115.3	103.0
03	CONCRETE	87.8	101.5	94.3	88.2	112.3	99.8	109.3	111.7	110.5	104.4	150.3	126.3	98.0	111.9	104.7	89.2	112.5	100.4
04	MASONRY	76.0	103.0	92.4	76.1	117.2	101.1	136.9	114.1	123.1	96.0	154.3	131.3	72.8	113.2	97.3	76.2	117.3	101.1
05	METALS	94.3	110.4	99.8	94.3	119.0	102.8	95.0	107.1	99.2	95.6	133.4	108.6	97.9	107.4	101.1	95.5	119.4	103.7
06	WOOD, PLASTICS & COMPOSITES	98.0	102.2	100.4	100.5	108.8	105.3	90.7	116.0	105.2	100.2	149.9	128.7	97.9	116.0	108.3	95.3	108.8	103.1
07	THERMAL & MOISTURE PROTECTION	93.7	100.1	96.2	93.7	113.0	101.4	99.0	112.5	104.4	102.4	144.6	119.2	99.1	108.1	102.7	93.7	112.5	101.2
08	OPENINGS	86.4	107.1	91.7	86.4	114.4	93.5	91.0	111.2	96.1	100.9	152.9	114.1	96.8	111.1	100.4	86.5	114.4	93.6
0920	Plaster & Gypsum Board	95.6	102.2	100.3	97.0	109.0	105.4	94.3	116.4	109.9	96.3	151.3	135.0	98.5	116.4	111.1	94.3	109.0	104.7
0950, 0980	Ceilings & Acoustic Treatment	89.5	102.2	97.9	89.5	109.0	102.4	89.5	116.4	107.3	104.8	151.3	135.6	95.1	116.4	109.2	89.5	109.0	102.4
0960	Flooring	113.4	126.9	117.3	114.5	119.6	116.0	98.2	126.9	106.5	93.1	149.9	109.4	101.8	114.0	105.3	112.2	119.6	114.4
0970, 0990	Wall Finishes & Painting/Coating	104.0	98.4	100.6	104.0	108.0	106.4	96.2	114.7	107.4	94.9	156.2	132.1	96.2	110.7	105.0	104.0	105.8	105.1
09	FINISHES	98.4	106.5	102.8	98.9	112.3	106.3	97.0	118.8	109.0	98.0	152.8	128.2	99.0	116.0	108.4	97.9	112.3	105.9
COVERS	DIVS. 10 - 14, 25, 28, 41, 43, 44, 46	100.0	104.0	100.8	100.0	106.9	101.4	100.0	104.1	100.9	100.0	125.5	105.3	100.0	103.8	100.8	100.0	106.9	101.4
21, 22, 23	FIRE SUPPRESSION, PLUMBING & HVAC	94.6	98.9	96.4	94.6	93.3	94.1	94.7	104.4	98.7	100.1	136.5	115.3	100.1	100.3	100.2	100.0	95.4	98.1
26, 27, 3370	ELECTRICAL, COMMUNICATIONS & UTIL.	92.0	100.6	96.6	93.5	103.9	99.0	96.8	95.8	96.3	95.3	134.8	116.3	96.3	91.3	93.6	93.0	103.9	98.8
MF2010	WEIGHTED AVERAGE	92.9	102.3	97.1	93.2	106.9	99.3	99.8	107.3	103.1	99.7	138.9	117.2	97.0	105.3	100.7	94.6	107.5	100.4

ILLINOIS

DIVISION		EFFINGHAM 624			GALESBURG 614			JOLIET 604			KANKAKEE 609			LA SALLE 613			NORTH SUBURBAN 600 - 603		
		MAT.	INST.	TOTAL	MAT.	INST.	TOTAL	MAT.	INST.	TOTAL	MAT.	INST.	TOTAL	MAT.	INST.	TOTAL	MAT.	INST.	TOTAL
015433	CONTRACTOR EQUIPMENT		101.7	101.7		101.0	101.0		91.5	91.5		91.5	91.5		101.0	101.0		91.5	91.5
0241, 31 - 34	SITE & INFRASTRUCTURE, DEMOLITION	100.1	97.1	98.0	101.4	96.7	98.1	114.8	93.2	99.9	107.9	92.3	97.2	100.8	97.8	98.7	113.8	92.5	99.1
0310	Concrete Forming & Accessories	103.1	110.2	109.2	93.9	114.5	111.8	102.4	154.7	147.8	96.0	130.8	126.2	107.2	119.9	118.2	101.8	148.3	142.1
0320	Concrete Reinforcing	99.6	97.0	98.3	99.1	104.5	101.9	98.8	145.2	122.7	99.5	138.3	119.5	99.2	135.4	117.9	98.8	152.0	126.2
0330	Cast-in-Place Concrete	101.6	107.0	103.6	102.7	105.5	103.8	112.7	144.7	125.0	105.1	127.3	113.6	102.6	117.5	108.3	112.8	142.1	124.0
03	CONCRETE	98.8	106.7	102.6	99.5	109.6	104.3	104.5	148.2	125.4	98.4	130.3	113.6	100.2	122.1	110.7	104.5	145.7	124.2
04	MASONRY	80.2	106.7	96.2	114.0	116.2	115.3	98.7	149.5	129.5	95.5	129.0	115.8	114.0	119.6	117.4	96.0	145.6	126.0
05	METALS	94.9	104.5	98.3	95.0	109.1	99.9	93.3	126.3	104.7	93.3	121.8	103.1	95.1	127.6	106.3	94.6	128.6	106.3
06	WOOD, PLASTICS & COMPOSITES	100.3	111.1	106.5	90.6	112.8	103.4	101.8	155.4	132.6	94.8	130.5	115.3	104.7	118.1	112.4	100.2	149.4	128.4
07	THERMAL & MOISTURE PROTECTION	98.7	106.7	101.8	98.5	107.4	102.1	102.2	141.5	117.8	101.4	130.1	112.8	98.7	114.4	104.9	102.7	135.8	115.8
08	OPENINGS	90.8	107.2	94.9	90.3	106.5	94.4	98.9	152.6	112.5	90.4	137.4	102.3	90.3	121.6	98.2	98.9	151.2	112.2
0920	Plaster & Gypsum Board	98.7	111.3	107.6	94.3	113.1	107.6	93.2	157.0	138.1	90.6	131.4	119.3	101.0	118.6	113.4	96.3	150.8	134.7
0950, 0980	Ceilings & Acoustic Treatment	89.5	111.3	103.9	89.5	113.1	105.2	104.8	157.0	139.4	104.8	131.4	122.4	89.5	118.6	108.8	104.8	150.8	135.3
0960	Flooring	103.0	126.9	109.9	98.1	113.2	102.5	92.7	139.3	106.1	89.8	139.3	104.1	105.2	117.4	108.7	93.1	139.3	106.4
0970, 0990	Wall Finishes & Painting/Coating	96.2	103.5	100.6	96.2	95.9	96.0	93.0	146.1	125.3	93.0	122.3	110.8	96.2	122.3	112.0	94.9	146.1	126.0
09	FINISHES	98.3	113.6	106.8	96.3	113.0	105.5	97.4	152.3	127.7	95.7	129.4	114.3	99.2	118.6	109.9	97.9	147.7	125.4
COVERS	DIVS. 10 - 14, 25, 28, 41, 43, 44, 46	100.0	74.6	94.7	100.0	104.3	100.9	100.0	124.8	105.2	100.0	116.3	103.4	100.0	104.4	100.9	100.0	121.2	104.4
21, 22, 23	FIRE SUPPRESSION, PLUMBING & HVAC	94.7	100.9	97.3	94.7	101.9	97.7	100.2	136.2	115.2	94.8	122.1	106.2	94.7	122.9	106.4	100.1	123.0	109.6
26, 27, 3370	ELECTRICAL, COMMUNICATIONS & UTIL.	94.0	100.6	97.5	94.2	89.6	91.8	94.6	138.5	117.9	89.2	137.6	114.9	91.1	137.6	115.8	94.4	121.2	108.6
MF2010	WEIGHTED AVERAGE	95.1	103.6	98.9	96.8	104.8	100.4	99.2	137.7	116.4	95.2	125.3	108.7	96.9	121.4	107.8	99.3	131.0	113.4

ILLINOIS

DIVISION		PEORIA 615 - 616			QUINCY 623			ROCK ISLAND 612			ROCKFORD 610 - 611			SOUTH SUBURBAN 605			SPRINGFIELD 626 - 627		
		MAT.	INST.	TOTAL	MAT.	INST.	TOTAL	MAT.	INST.	TOTAL	MAT.	INST.	TOTAL	MAT.	INST.	TOTAL	MAT.	INST.	TOTAL
015433	CONTRACTOR EQUIPMENT		101.0	101.0		101.7	101.7		101.0	101.0		101.0	101.0		91.5	91.5		101.7	101.7
0241, 31 - 34	SITE & INFRASTRUCTURE, DEMOLITION	101.5	96.6	98.1	98.8	97.0	97.6	99.4	96.1	97.1	100.5	97.9	98.7	113.8	92.5	99.1	100.2	97.6	98.4
0310	Concrete Forming & Accessories	96.4	115.0	112.5	100.9	112.5	110.9	95.4	100.3	99.6	101.1	129.5	125.8	101.8	148.3	142.1	99.2	116.5	114.3
0320	Concrete Reinforcing	96.5	104.6	100.7	99.2	101.3	100.3	99.1	101.3	100.2	91.2	140.2	116.4	98.8	152.0	126.2	96.5	102.4	99.5
0330	Cast-in-Place Concrete	99.6	114.8	105.4	101.7	98.4	100.5	100.5	98.1	99.6	102.0	116.4	107.5	112.8	142.1	124.0	91.6	108.4	98.1
03	CONCRETE	96.7	113.0	104.5	98.5	105.6	101.9	97.6	100.1	98.8	97.4	126.9	111.5	104.5	145.7	124.2	92.9	111.0	101.6
04	MASONRY	114.1	115.4	114.9	99.9	100.7	100.4	113.8	97.6	104.0	88.9	139.1	119.3	96.0	145.6	126.0	77.6	114.3	99.9
05	METALS	97.9	110.0	102.0	95.0	106.2	98.8	95.1	106.9	99.2	97.8	129.2	108.6	94.6	128.6	106.3	100.2	107.7	102.8
06	WOOD, PLASTICS & COMPOSITES	97.9	112.8	106.5	97.7	116.0	108.2	92.1	99.1	96.1	97.9	127.5	114.8	100.2	149.4	128.4	98.9	116.0	108.7
07	THERMAL & MOISTURE PROTECTION	99.2	111.5	104.1	98.7	103.2	100.5	98.5	98.1	98.4	101.7	129.6	112.8	102.7	135.8	115.8	98.9	111.6	103.9
08	OPENINGS	96.9	110.5	100.4	91.6	111.1	96.6	90.3	98.4	92.3	96.9	130.6	105.5	98.9	151.2	112.2	97.9	111.3	101.3
0920	Plaster & Gypsum Board	98.5	113.1	108.8	97.0	116.4	110.7	94.3	99.0	97.6	98.5	128.2	119.4	96.3	150.8	134.7	100.1	116.4	111.6
0950, 0980	Ceilings & Acoustic Treatment	95.1	113.1	107.0	89.5	116.4	107.3	89.5	99.0	95.8	95.1	128.2	117.0	104.8	150.8	135.3	94.2	116.4	108.9
0960	Flooring	101.8	113.2	105.1	101.8	104.9	102.7	99.3	95.5	98.2	101.8	117.4	106.3	93.1	139.3	106.4	101.8	108.4	103.7
0970, 0990	Wall Finishes & Painting/Coating	96.2	122.3	112.0	96.2	105.6	101.9	96.2	99.0	97.9	96.2	129.6	116.4	94.9	146.1	126.0	96.2	110.2	104.7
09	FINISHES	99.1	115.7	108.3	97.7	111.8	105.4	96.6	99.2	98.0	99.2	127.6	114.8	97.9	147.7	125.4	98.9	115.1	107.8
COVERS	DIVS. 10 - 14, 25, 28, 41, 43, 44, 46	100.0	104.3	100.9	100.0	76.8	95.2	100.0	89.2	99.6	100.0	96.8	99.3	100.0	121.2	104.4	100.0	104.2	100.9
21, 22, 23	FIRE SUPPRESSION, PLUMBING & HVAC	100.1	104.2	101.8	94.7	96.3	95.3	94.7	98.4	96.3	100.1	114.3	106.0	100.1	123.0	109.6	99.9	102.5	101.0
26, 27, 3370	ELECTRICAL, COMMUNICATIONS & UTIL.	95.4	95.7	95.5	91.2	83.4	87.0	85.7	94.5	91.5	95.7	127.1	112.4	94.4	121.2	108.6	98.9	94.4	96.5
MF2010	WEIGHTED AVERAGE	99.2	107.3	102.8	95.8	99.6	97.5	95.7	99.0	97.2	98.1	122.7	109.1	99.3	131.0	113.4	97.5	106.2	101.4

INDIANA

DIVISION		ANDERSON 460 MAT.	INST.	TOTAL	BLOOMINGTON 474 MAT.	INST.	TOTAL	COLUMBUS 472 MAT.	INST.	TOTAL	EVANSVILLE 476 - 477 MAT.	INST.	TOTAL	FORT WAYNE 467 - 468 MAT.	INST.	TOTAL	GARY 463 - 464 MAT.	INST.	TOTAL
015433	CONTRACTOR EQUIPMENT		96.3	96.3		87.3	87.3		87.3	87.3		117.4	117.4		96.3	96.3		96.3	96.3
0241, 31 - 34	SITE & INFRASTRUCTURE, DEMOLITION	89.3	94.5	92.9	77.4	94.7	89.4	74.4	94.5	88.3	82.2	125.1	111.8	90.6	94.4	93.2	89.9	98.1	95.5
0310	Concrete Forming & Accessories	93.4	79.7	81.5	100.1	80.2	82.8	94.5	77.9	80.1	94.0	79.8	81.7	92.0	73.8	76.2	93.5	111.5	109.1
0320	Concrete Reinforcing	93.9	80.4	86.9	90.0	80.1	84.9	90.4	80.1	85.1	98.6	77.7	87.8	93.9	79.0	86.2	93.9	104.8	99.5
0330	Cast-in-Place Concrete	101.6	80.4	93.5	98.2	76.8	90.0	97.7	75.3	89.1	93.8	86.7	91.1	108.0	76.4	95.9	106.1	108.8	107.2
03	CONCRETE	94.1	80.6	87.6	100.8	78.8	90.3	100.2	77.3	89.2	101.0	82.0	91.9	97.2	76.3	87.2	96.4	109.2	102.5
04	MASONRY	84.2	82.2	83.0	88.4	78.5	82.4	88.3	78.4	82.3	84.2	83.2	83.6	87.7	78.9	82.4	85.5	112.7	102.0
05	METALS	93.4	88.3	91.7	94.5	77.4	88.6	94.6	76.6	88.4	88.5	84.6	87.2	93.4	87.4	91.4	93.4	105.6	97.6
06	WOOD, PLASTICS & COMPOSITES	109.6	79.7	92.4	120.4	80.1	97.2	115.1	77.1	93.3	98.9	78.0	86.9	109.3	72.6	88.3	107.0	110.7	109.1
07	THERMAL & MOISTURE PROTECTION	99.4	79.5	91.5	94.2	81.4	89.1	93.9	81.6	89.0	98.1	84.0	92.5	99.2	81.6	92.2	98.2	109.2	102.6
08	OPENINGS	97.3	80.6	93.1	105.3	80.9	99.1	100.5	79.2	95.1	97.3	79.0	92.7	97.3	73.8	91.4	97.3	114.0	101.5
0920	Plaster & Gypsum Board	99.5	79.5	85.4	94.0	80.1	84.2	90.0	77.0	80.9	88.5	76.8	80.2	98.6	72.2	80.1	92.0	111.5	105.7
0950, 0980	Ceilings & Acoustic Treatment	94.3	79.5	84.5	80.5	80.1	80.2	80.5	77.0	78.2	85.6	76.8	79.8	94.3	72.2	79.7	94.3	111.5	105.7
0960	Flooring	97.0	82.0	92.7	104.8	79.2	97.5	99.8	79.2	93.9	98.6	85.2	94.7	97.0	79.2	91.9	97.0	122.5	104.4
0970, 0990	Wall Finishes & Painting/Coating	102.9	71.5	83.9	92.2	85.1	87.9	92.2	85.1	87.9	97.6	84.5	89.7	102.9	76.0	86.6	102.9	118.6	112.4
09	FINISHES	97.6	79.3	87.5	96.3	80.5	87.6	94.3	78.7	85.7	94.8	81.3	87.3	97.3	75.0	85.0	96.6	114.4	106.4
COVERS	DIVS. 10 - 14, 25, 28, 41, 43, 44, 46	100.0	91.9	98.3	100.0	89.8	97.9	100.0	89.5	97.8	100.0	77.8	95.4	100.0	90.3	98.0	100.0	106.4	101.3
21, 22, 23	FIRE SUPPRESSION, PLUMBING & HVAC	99.9	75.7	89.8	99.7	79.4	91.2	94.3	79.9	88.3	99.9	80.9	92.0	99.9	76.0	90.0	99.9	105.9	102.4
26, 27, 3370	ELECTRICAL, COMMUNICATIONS & UTIL.	85.2	88.2	86.8	100.0	83.8	91.4	99.1	88.1	93.3	96.1	86.3	90.9	85.9	75.7	80.5	97.5	107.4	102.8
MF2010	WEIGHTED AVERAGE	95.3	82.8	89.7	98.1	81.5	90.7	95.9	81.6	89.5	95.9	86.0	91.5	95.9	79.2	88.5	96.7	108.1	101.8

INDIANA

DIVISION		INDIANAPOLIS 461 - 462 MAT.	INST.	TOTAL	KOKOMO 469 MAT.	INST.	TOTAL	LAFAYETTE 479 MAT.	INST.	TOTAL	LAWRENCEBURG 470 MAT.	INST.	TOTAL	MUNCIE 473 MAT.	INST.	TOTAL	NEW ALBANY 471 MAT.	INST.	TOTAL
015433	CONTRACTOR EQUIPMENT		91.8	91.8		96.3	96.3		87.3	87.3		106.1	106.1		96.1	96.1		95.8	95.8
0241, 31 - 34	SITE & INFRASTRUCTURE, DEMOLITION	88.8	97.8	95.0	86.6	94.5	92.0	75.4	94.5	88.6	73.4	111.9	100.0	77.5	94.9	89.5	69.6	98.0	89.2
0310	Concrete Forming & Accessories	94.1	86.5	87.5	96.5	76.2	78.9	92.4	81.2	82.6	91.0	74.9	77.0	92.6	79.5	81.2	88.2	72.7	74.8
0320	Concrete Reinforcing	93.3	83.6	88.3	85.0	80.2	82.5	90.0	80.3	85.0	89.3	79.5	84.2	99.6	80.3	89.7	90.7	77.7	84.0
0330	Cast-in-Place Concrete	97.7	82.9	92.1	100.6	83.3	94.0	98.3	82.7	92.3	91.9	74.3	85.2	103.2	79.6	94.1	94.9	73.4	86.7
03	CONCRETE	95.7	84.4	90.3	91.1	80.0	85.8	100.4	81.3	91.2	93.6	76.2	85.2	99.9	80.2	90.5	98.7	74.3	87.0
04	MASONRY	91.7	84.6	87.4	83.9	80.0	81.6	93.5	82.0	86.6	73.7	72.0	72.7	90.2	82.2	85.3	80.6	71.5	75.1
05	METALS	95.6	80.6	90.4	89.9	88.0	89.3	93.0	77.2	87.6	89.9	86.2	88.6	96.4	88.5	93.7	91.7	81.9	88.3
06	WOOD, PLASTICS & COMPOSITES	106.3	86.9	95.2	112.7	74.7	90.9	112.4	81.1	94.4	97.7	74.0	84.1	114.7	79.5	94.5	99.4	73.1	84.3
07	THERMAL & MOISTURE PROTECTION	97.4	84.7	92.4	99.1	78.2	90.7	93.8	83.1	89.6	97.4	78.2	89.8	96.8	79.5	89.9	85.9	71.1	80.0
08	OPENINGS	105.3	85.4	100.3	91.6	77.9	88.2	98.8	81.4	94.4	98.8	76.6	93.1	100.1	80.5	95.1	96.8	75.8	91.5
0920	Plaster & Gypsum Board	97.2	86.8	89.9	104.8	74.4	83.4	87.2	81.2	83.0	74.0	73.8	73.8	88.5	79.5	82.2	86.8	72.6	76.8
0950, 0980	Ceilings & Acoustic Treatment	98.0	86.8	90.6	94.3	74.4	81.1	76.8	81.2	79.7	91.3	73.8	79.7	81.5	79.5	80.2	85.6	72.6	77.0
0960	Flooring	96.8	88.2	94.4	101.4	89.6	98.0	98.7	89.9	96.2	73.4	88.2	77.7	98.3	82.0	93.6	96.7	66.4	88.0
0970, 0990	Wall Finishes & Painting/Coating	102.9	85.1	92.1	102.9	77.2	87.3	92.2	82.3	86.2	93.2	68.6	78.3	92.2	71.5	79.6	97.6	69.5	80.6
09	FINISHES	98.4	87.0	92.1	99.3	78.5	87.9	92.9	82.9	87.4	86.6	77.0	81.3	93.6	79.1	85.6	94.3	71.3	81.6
COVERS	DIVS. 10 - 14, 25, 28, 41, 43, 44, 46	100.0	94.0	98.8	100.0	89.7	97.9	100.0	91.7	98.3	100.0	47.1	89.0	100.0	91.4	98.2	100.0	46.6	88.9
21, 22, 23	FIRE SUPPRESSION, PLUMBING & HVAC	99.9	81.3	92.1	94.5	80.0	88.5	94.3	79.7	88.2	95.0	75.8	87.0	99.7	75.7	89.7	94.5	76.1	86.8
26, 27, 3370	ELECTRICAL, COMMUNICATIONS & UTIL.	99.1	88.2	93.3	90.1	79.0	84.2	98.5	84.5	91.0	93.8	76.1	84.4	91.5	80.8	85.8	94.3	78.0	85.6
MF2010	WEIGHTED AVERAGE	98.3	85.7	92.7	93.1	81.7	88.0	95.6	82.8	89.9	92.5	78.9	86.4	96.8	81.7	90.1	93.6	76.4	85.9

INDIANA / IOWA

DIVISION		SOUTH BEND 465 - 466 MAT.	INST.	TOTAL	TERRE HAUTE 478 MAT.	INST.	TOTAL	WASHINGTON 475 MAT.	INST.	TOTAL	BURLINGTON 526 MAT.	INST.	TOTAL	CARROLL 514 MAT.	INST.	TOTAL	CEDAR RAPIDS 522 - 524 MAT.	INST.	TOTAL
015433	CONTRACTOR EQUIPMENT		105.0	105.0		117.4	117.4		117.4	117.4		99.8	99.8		99.8	99.8		96.4	96.4
0241, 31 - 34	SITE & INFRASTRUCTURE, DEMOLITION	91.5	94.9	93.9	83.9	125.3	112.5	84.1	123.2	111.1	98.3	96.7	97.2	86.2	96.8	93.5	100.3	96.1	97.4
0310	Concrete Forming & Accessories	95.9	78.0	80.4	94.9	81.6	83.3	95.8	78.7	80.9	95.4	78.8	81.0	85.0	53.8	57.9	100.2	83.4	85.7
0320	Concrete Reinforcing	93.9	77.0	85.2	98.6	80.5	89.3	91.2	52.5	71.2	95.9	83.6	89.6	96.7	83.0	89.7	96.7	86.1	91.3
0330	Cast-in-Place Concrete	98.8	80.5	91.8	90.8	88.6	89.9	98.8	88.4	94.8	109.1	57.9	89.5	109.1	63.0	91.5	109.4	82.4	99.1
03	CONCRETE	89.0	80.0	84.7	103.5	84.0	94.1	109.2	77.1	93.8	104.0	73.3	89.3	103.0	63.8	84.3	104.1	84.2	94.6
04	MASONRY	84.0	81.1	82.2	91.4	82.2	85.8	84.4	84.5	84.4	103.2	69.8	83.0	105.6	76.6	88.0	109.6	82.1	92.9
05	METALS	93.4	99.2	95.4	89.1	87.1	88.4	83.8	70.5	79.2	89.2	92.5	90.3	89.2	92.1	90.2	91.6	94.6	92.6
06	WOOD, PLASTICS & COMPOSITES	106.8	77.4	89.9	100.9	81.1	89.5	101.3	77.9	87.9	97.4	79.6	87.2	85.5	48.6	64.4	103.0	82.4	91.2
07	THERMAL & MOISTURE PROTECTION	93.8	83.6	89.8	98.2	83.1	92.2	98.3	84.2	92.7	105.7	78.4	94.9	106.1	72.9	92.9	106.6	82.1	96.9
08	OPENINGS	90.6	77.6	87.3	97.8	81.4	93.7	94.4	69.0	87.9	90.7	72.6	86.1	95.9	57.3	86.1	96.8	83.4	93.4
0920	Plaster & Gypsum Board	90.8	77.1	81.2	88.5	79.9	82.5	88.3	76.7	80.1	99.2	79.0	85.0	94.3	47.2	61.1	103.3	82.2	88.4
0950, 0980	Ceilings & Acoustic Treatment	88.7	77.1	81.0	85.6	79.9	81.8	80.0	76.7	77.8	104.7	79.0	87.7	104.7	47.2	66.5	108.4	82.2	91.0
0960	Flooring	95.4	84.1	92.1	98.6	88.2	95.6	99.6	80.7	94.2	116.1	39.9	94.2	109.0	35.1	87.7	134.7	82.7	119.8
0970, 0990	Wall Finishes & Painting/Coating	102.4	85.7	92.3	97.6	87.0	91.2	97.6	85.1	90.0	101.4	84.5	91.1	101.4	80.0	88.4	104.3	73.9	85.9
09	FINISHES	94.8	79.9	86.6	94.8	83.3	88.5	94.2	80.4	86.6	107.0	72.2	87.8	102.4	51.2	74.2	113.9	82.2	96.4
COVERS	DIVS. 10 - 14, 25, 28, 41, 43, 44, 46	100.0	73.1	94.4	100.0	94.5	98.9	100.0	77.7	95.4	100.0	90.6	98.1	100.0	72.4	94.3	100.0	91.7	98.3
21, 22, 23	FIRE SUPPRESSION, PLUMBING & HVAC	99.8	78.0	90.7	99.9	80.0	91.6	94.5	80.1	88.5	94.8	79.7	88.5	94.8	72.7	85.6	100.2	83.8	93.4
26, 27, 3370	ELECTRICAL, COMMUNICATIONS & UTIL.	102.7	86.0	93.8	94.0	83.8	88.6	94.6	86.3	90.2	98.6	73.4	85.2	99.4	84.2	91.3	96.0	82.7	89.0
MF2010	WEIGHTED AVERAGE	95.3	83.3	89.9	96.6	86.7	92.2	94.6	83.3	89.5	97.5	78.6	89.1	97.3	73.6	86.7	100.5	85.5	93.8

IOWA

DIVISION		COUNCIL BLUFFS 515			CRESTON 508			DAVENPORT 527 - 528			DECORAH 521			DES MOINES 500 - 503,509			DUBUQUE 520		
		MAT.	INST.	TOTAL	MAT.	INST.	TOTAL	MAT.	INST.	TOTAL	MAT.	INST.	TOTAL	MAT.	INST.	TOTAL	MAT.	INST.	TOTAL
015433	CONTRACTOR EQUIPMENT		96.1	96.1		99.8	99.8		99.8	99.8		99.8	99.8		101.5	101.5		95.1	95.1
0241, 31 - 34	SITE & INFRASTRUCTURE, DEMOLITION	108.5	92.7	97.6	86.2	96.4	93.2	97.9	99.1	98.7	96.8	96.2	96.4	87.1	100.1	96.1	98.1	93.2	94.7
0310	Concrete Forming & Accessories	84.5	73.5	75.0	85.5	70.2	72.2	99.8	89.9	91.2	93.3	47.4	53.5	101.1	80.1	82.8	85.6	72.4	74.2
0320	Concrete Reinforcing	98.6	79.1	88.5	96.1	83.1	89.4	96.7	97.2	97.0	95.9	82.6	89.1	96.7	83.3	89.8	95.3	86.0	90.5
0330	Cast-in-Place Concrete	113.8	78.4	100.2	108.8	67.2	92.9	105.4	88.9	99.1	106.2	59.5	88.3	91.7	92.9	92.2	107.1	90.0	100.6
03	CONCRETE	106.8	77.1	92.5	102.8	72.5	88.3	102.0	91.5	97.0	101.9	59.7	81.7	94.1	85.8	90.1	100.9	81.8	91.8
04	MASONRY	110.9	77.5	90.7	105.6	83.8	92.4	106.0	87.7	94.9	125.3	72.7	93.4	99.8	84.9	90.8	110.6	78.3	91.0
05	METALS	96.7	90.9	94.7	89.1	92.4	90.2	91.6	102.4	95.3	89.3	90.2	89.6	91.7	94.7	92.8	90.2	94.0	91.5
06	WOOD, PLASTICS & COMPOSITES	84.5	73.0	77.9	86.0	66.9	75.0	103.0	89.0	94.9	94.7	40.1	63.4	102.8	77.8	88.4	86.1	68.6	76.1
07	THERMAL & MOISTURE PROTECTION	106.1	71.1	92.2	106.8	80.6	96.4	106.1	87.7	98.8	105.9	55.8	86.0	107.5	81.7	97.2	106.4	73.6	93.4
08	OPENINGS	95.9	78.2	91.4	108.5	67.5	98.1	96.8	90.8	95.3	94.2	52.6	83.7	96.8	83.8	93.5	95.9	77.7	91.3
0920	Plaster & Gypsum Board	94.3	72.4	78.9	94.3	66.0	74.4	103.3	88.7	93.0	97.8	38.4	56.0	95.4	77.2	82.6	94.7	68.0	75.9
0950, 0980	Ceilings & Acoustic Treatment	104.7	72.4	83.3	104.7	66.0	79.0	108.4	88.7	95.3	104.7	38.4	60.7	105.6	77.2	86.7	104.7	68.0	80.3
0960	Flooring	108.0	85.2	101.5	109.2	35.1	87.9	118.9	87.7	110.0	115.6	49.8	96.7	118.6	85.3	109.0	123.6	80.5	111.2
0970, 0990	Wall Finishes & Painting/Coating	98.4	69.0	80.5	101.4	80.0	88.4	101.4	90.5	94.7	101.4	35.4	61.4	101.4	84.2	91.0	103.3	68.6	82.3
09	FINISHES	103.6	74.9	87.8	102.5	63.8	81.1	109.1	89.4	98.2	106.6	44.8	72.5	106.0	81.0	92.2	108.5	72.6	88.7
COVERS	DIVS. 10 - 14, 25, 28, 41, 43, 44, 46	100.0	84.8	96.8	100.0	77.0	95.2	100.0	93.7	98.7	100.0	85.0	96.9	100.0	91.6	98.3	100.0	89.6	97.8
21, 22, 23	FIRE SUPPRESSION, PLUMBING & HVAC	100.2	76.3	90.2	94.8	82.6	89.7	100.2	91.7	96.7	94.8	75.0	86.5	100.0	83.2	93.0	100.2	78.2	91.0
26, 27, 3370	ELECTRICAL, COMMUNICATIONS & UTIL.	101.7	83.8	92.2	93.8	84.2	88.7	93.8	88.5	91.0	96.0	48.1	70.6	107.0	84.2	94.9	100.2	80.8	89.9
MF2010	WEIGHTED AVERAGE	101.4	80.3	91.9	98.0	80.2	90.0	99.4	92.0	96.1	98.4	66.7	84.3	98.9	86.2	93.2	99.7	81.2	91.4

IOWA

DIVISION		FORT DODGE 505			MASON CITY 504			OTTUMWA 525			SHENANDOAH 516			SIBLEY 512			SIOUX CITY 510 - 511		
		MAT.	INST.	TOTAL	MAT.	INST.	TOTAL	MAT.	INST.	TOTAL	MAT.	INST.	TOTAL	MAT.	INST.	TOTAL	MAT.	INST.	TOTAL
015433	CONTRACTOR EQUIPMENT		99.8	99.8		99.8	99.8		95.1	95.1		96.1	96.1		99.8	99.8		99.8	99.8
0241, 31 - 34	SITE & INFRASTRUCTURE, DEMOLITION	93.7	95.1	94.7	93.7	96.3	95.5	99.0	91.3	93.7	106.7	91.7	96.3	108.9	95.0	99.3	110.0	95.9	100.2
0310	Concrete Forming & Accessories	86.0	48.5	53.4	90.0	48.2	53.7	91.6	60.6	64.7	85.9	60.6	64.0	86.3	40.4	46.5	100.2	67.2	71.6
0320	Concrete Reinforcing	96.1	65.4	80.3	95.9	82.5	89.0	95.9	83.9	89.7	98.6	69.7	83.7	98.6	65.3	81.5	96.7	65.7	80.7
0330	Cast-in-Place Concrete	102.1	46.6	80.8	102.1	60.2	86.1	109.8	57.3	89.7	110.0	63.3	92.1	107.8	47.1	84.5	108.5	58.3	89.3
03	CONCRETE	98.0	52.5	76.2	98.3	60.2	80.1	103.7	65.1	85.2	104.1	64.4	85.1	103.0	49.0	77.2	103.4	64.9	84.9
04	MASONRY	103.8	40.9	65.7	117.9	71.5	89.7	106.9	60.7	78.9	110.5	77.5	90.5	131.1	41.2	76.6	103.0	57.6	75.5
05	METALS	89.2	81.7	86.6	89.2	90.6	89.7	89.1	92.1	90.1	95.7	85.5	92.2	89.4	80.4	86.3	91.6	83.2	88.7
06	WOOD, PLASTICS & COMPOSITES	86.5	48.9	65.0	90.6	40.1	61.6	92.5	62.1	75.0	86.1	57.2	69.6	86.8	38.8	59.3	103.0	67.3	82.5
07	THERMAL & MOISTURE PROTECTION	106.2	60.7	88.1	105.7	66.3	90.0	106.5	65.8	90.4	105.4	69.3	91.1	105.7	51.4	84.2	106.1	66.4	90.4
08	OPENINGS	101.1	51.4	88.5	91.7	52.5	81.8	95.8	67.6	88.7	85.6	58.1	78.6	92.0	45.9	80.3	96.8	63.2	88.3
0920	Plaster & Gypsum Board	94.7	47.5	61.5	95.2	38.4	55.2	96.1	61.2	71.5	94.7	56.3	67.6	94.7	37.0	54.1	103.3	66.4	77.3
0950, 0980	Ceilings & Acoustic Treatment	104.7	47.5	66.7	104.7	38.4	60.7	104.7	61.2	75.9	104.7	56.3	72.6	104.7	37.0	59.8	108.4	66.4	80.5
0960	Flooring	111.0	49.8	93.4	113.8	49.8	95.4	127.4	53.0	106.0	108.9	36.2	88.0	110.0	35.9	88.7	119.3	56.7	101.3
0970, 0990	Wall Finishes & Painting/Coating	101.4	63.3	78.3	101.4	32.4	59.5	103.3	82.9	90.9	98.4	69.0	80.5	101.4	63.3	78.3	101.8	63.3	78.5
09	FINISHES	104.6	49.3	74.1	105.5	44.6	71.9	109.9	60.6	82.7	103.8	55.8	77.3	106.4	40.6	70.1	111.0	64.7	85.5
COVERS	DIVS. 10 - 14, 25, 28, 41, 43, 44, 46	100.0	80.7	96.0	100.0	85.4	97.0	100.0	83.1	96.5	100.0	73.7	94.5	100.0	79.7	95.8	100.0	87.4	97.4
21, 22, 23	FIRE SUPPRESSION, PLUMBING & HVAC	94.8	68.4	83.8	94.8	76.1	87.0	94.8	74.2	86.2	94.8	73.8	86.0	94.8	64.2	82.0	100.2	70.9	88.0
26, 27, 3370	ELECTRICAL, COMMUNICATIONS & UTIL.	101.2	45.5	71.6	100.2	60.4	79.0	98.4	69.4	83.0	96.0	83.8	89.5	96.0	45.0	68.9	96.0	76.0	85.4
MF2010	WEIGHTED AVERAGE	97.7	60.6	81.1	97.5	68.9	84.7	98.4	71.9	86.6	97.9	73.6	87.1	98.9	57.3	80.3	100.1	71.9	87.5

DIVISION		IOWA						KANSAS											
		SPENCER 513			WATERLOO 506 - 507			BELLEVILLE 669			COLBY 677			DODGE CITY 678			EMPORIA 668		
		MAT.	INST.	TOTAL	MAT.	INST.	TOTAL	MAT.	INST.	TOTAL	MAT.	INST.	TOTAL	MAT.	INST.	TOTAL	MAT.	INST.	TOTAL
015433	CONTRACTOR EQUIPMENT		99.8	99.8		99.8	99.8		102.7	102.7		102.7	102.7		102.7	102.7		101.0	101.0
0241, 31 - 34	SITE & INFRASTRUCTURE, DEMOLITION	109.0	95.0	99.3	98.4	96.3	96.9	116.8	93.7	100.9	114.1	93.7	100.0	116.3	93.3	100.5	107.6	91.5	96.5
0310	Concrete Forming & Accessories	91.5	40.4	47.2	100.6	50.7	57.3	96.9	57.8	63.0	103.1	62.4	67.8	96.4	62.3	66.8	88.7	70.0	72.4
0320	Concrete Reinforcing	98.6	65.3	81.5	96.7	85.8	91.1	104.5	56.4	79.7	107.1	56.4	81.0	104.5	56.5	79.8	103.0	57.0	79.3
0330	Cast-in-Place Concrete	107.8	47.1	84.5	109.4	57.4	89.5	120.8	58.8	97.1	116.2	58.8	94.2	118.2	58.4	95.3	116.8	49.0	90.8
03	CONCRETE	103.3	49.0	77.3	104.1	61.1	83.5	116.9	59.2	89.3	114.1	61.2	88.8	115.3	61.0	89.3	109.6	61.4	86.5
04	MASONRY	131.1	41.2	76.6	104.5	77.5	88.2	97.9	59.7	74.8	99.3	59.7	75.3	106.3	62.3	79.6	103.6	70.2	83.4
05	METALS	89.3	80.4	86.3	91.6	93.5	92.2	93.2	79.0	88.3	93.5	79.0	88.5	95.1	77.9	89.2	92.9	80.3	88.5
06	WOOD, PLASTICS & COMPOSITES	92.3	38.8	61.6	103.5	40.1	67.1	92.3	56.6	71.8	99.2	62.8	78.3	91.7	62.8	75.1	84.3	72.4	77.5
07	THERMAL & MOISTURE PROTECTION	106.6	51.4	84.7	105.9	73.5	93.0	99.9	60.2	84.1	99.9	60.8	84.4	99.9	61.8	84.7	98.5	75.6	89.4
08	OPENINGS	105.2	45.9	90.2	92.6	55.2	83.1	96.6	51.6	85.2	96.7	55.0	86.2	96.7	55.0	86.1	94.3	60.2	85.6
0920	Plaster & Gypsum Board	96.1	37.0	54.5	103.3	38.4	57.6	96.5	55.2	67.4	101.0	61.6	73.3	95.6	61.6	71.7	92.5	71.5	77.7
0950, 0980	Ceilings & Acoustic Treatment	104.7	37.0	59.8	108.4	38.4	62.0	89.5	55.2	66.7	89.5	61.6	71.0	89.5	61.6	71.0	89.5	71.5	77.5
0960	Flooring	113.3	35.9	91.1	120.7	79.7	108.9	101.4	41.4	84.2	105.1	41.4	86.8	101.2	41.4	84.0	96.3	38.5	79.7
0970, 0990	Wall Finishes & Painting/Coating	101.4	63.3	78.3	101.8	83.7	90.8	96.2	41.9	63.2	96.2	41.9	63.2	96.2	41.9	63.2	96.2	41.9	63.2
09	FINISHES	107.5	40.6	70.6	109.6	57.4	80.8	98.9	52.8	73.5	100.4	56.5	76.2	98.7	56.5	75.4	95.9	61.5	76.9
COVERS	DIVS. 10 - 14, 25, 28, 41, 43, 44, 46	100.0	79.7	95.8	100.0	86.8	97.3	100.0	44.9	88.6	100.0	45.7	88.7	100.0	45.7	88.7	100.0	44.6	88.5
21, 22, 23	FIRE SUPPRESSION, PLUMBING & HVAC	94.8	64.2	82.0	100.2	83.1	93.1	94.7	72.4	85.4	94.7	69.0	84.0	100.1	69.0	87.1	94.7	72.7	85.5
26, 27, 3370	ELECTRICAL, COMMUNICATIONS & UTIL.	97.9	45.0	69.7	96.0	60.4	77.1	98.8	68.7	82.8	98.8	68.7	82.8	95.3	75.3	84.7	95.7	75.2	84.8
MF2010	WEIGHTED AVERAGE	100.6	57.3	81.3	99.4	73.4	87.8	99.6	66.5	84.8	99.5	66.8	84.9	101.0	67.8	86.2	97.8	70.8	85.7

KANSAS

DIVISION		FORT SCOTT 667			HAYS 676			HUTCHINSON 675			INDEPENDENCE 673			KANSAS CITY 660 - 662			LIBERAL 679		
		MAT.	INST.	TOTAL	MAT.	INST.	TOTAL	MAT.	INST.	TOTAL	MAT.	INST.	TOTAL	MAT.	INST.	TOTAL	MAT.	INST.	TOTAL
015433	CONTRACTOR EQUIPMENT		101.9	101.9		102.7	102.7		102.7	102.7		102.7	102.7		99.5	99.5		102.7	102.7
0241, 31 - 34	SITE & INFRASTRUCTURE, DEMOLITION	103.9	92.0	95.7	120.5	93.7	102.0	96.5	93.8	94.7	117.9	93.9	101.3	96.3	92.2	93.4	120.4	93.8	102.1
0310	Concrete Forming & Accessories	103.7	82.5	85.3	100.6	62.4	67.5	91.1	62.2	66.1	111.5	71.0	76.3	100.1	99.3	99.4	96.9	62.3	66.8
0320	Concrete Reinforcing	102.3	90.4	96.2	104.5	56.4	79.7	104.5	56.4	79.7	103.9	61.4	82.0	99.3	95.8	97.5	106.0	56.4	80.5
0330	Cast-in-Place Concrete	108.3	57.5	88.8	91.6	58.8	79.0	84.8	48.4	70.9	118.7	48.6	91.9	92.8	98.5	95.0	91.6	48.4	75.1
03	CONCRETE	104.5	76.1	90.9	104.5	61.2	83.8	88.0	57.6	73.5	116.5	62.5	90.7	96.0	98.6	97.2	106.5	57.6	83.1
04	MASONRY	104.3	60.7	77.9	107.0	59.7	78.3	98.6	61.9	76.4	96.4	64.6	77.1	106.8	99.6	102.5	104.1	61.9	78.5
05	METALS	92.8	92.0	92.5	93.1	79.0	88.2	92.9	77.8	87.7	92.8	80.4	88.6	100.6	99.9	100.4	93.4	77.8	88.0
06	WOOD, PLASTICS & COMPOSITES	101.2	90.8	95.2	96.5	62.8	77.2	86.8	62.8	73.0	108.8	73.9	88.8	97.4	99.8	98.8	92.3	62.8	75.4
07	THERMAL & MOISTURE PROTECTION	99.2	72.1	88.5	100.2	60.8	84.5	98.9	61.7	84.1	99.9	74.2	89.7	98.6	99.3	98.9	100.3	61.7	85.0
08	OPENINGS	94.3	82.2	91.2	96.6	55.0	86.1	96.6	55.0	86.1	94.3	62.2	86.1	95.7	92.6	94.9	96.7	55.0	86.1
0920	Plaster & Gypsum Board	99.2	90.4	93.0	99.2	61.6	72.7	94.3	61.6	71.3	107.6	73.0	83.2	91.2	99.7	97.2	96.5	61.6	71.9
0950, 0980	Ceilings & Acoustic Treatment	89.5	90.4	90.1	89.5	61.6	71.0	89.5	61.6	71.0	89.5	73.0	78.6	89.5	99.7	96.3	89.5	61.6	71.0
0960	Flooring	112.4	41.6	92.1	103.9	41.4	85.9	98.1	41.4	81.8	109.8	41.4	90.1	90.0	99.2	92.7	101.4	41.4	84.2
0970, 0990	Wall Finishes & Painting/Coating	98.1	41.9	64.0	96.2	41.9	63.2	96.2	41.9	63.2	96.2	41.9	63.2	104.0	72.9	85.1	96.2	41.9	63.2
09	FINISHES	101.7	72.1	85.3	100.4	56.5	76.2	95.9	56.5	74.2	102.7	63.0	80.8	96.2	95.6	95.9	99.6	56.5	75.8
COVERS	DIVS. 10 - 14, 25, 28, 41, 43, 44, 46	100.0	51.4	89.9	100.0	45.7	88.7	100.0	45.7	88.7	100.0	47.0	89.0	100.0	68.8	93.5	100.0	45.7	88.7
21, 22, 23	FIRE SUPPRESSION, PLUMBING & HVAC	94.7	70.7	84.7	94.7	69.0	84.0	94.7	69.0	84.0	94.7	71.0	84.8	100.0	101.3	100.6	94.7	69.0	84.0
26, 27, 3370	ELECTRICAL, COMMUNICATIONS & UTIL.	94.9	75.2	84.4	97.6	68.7	82.3	92.0	68.7	79.6	94.5	75.1	84.2	101.4	98.0	99.6	95.3	71.1	82.5
MF2010	WEIGHTED AVERAGE	97.6	75.2	87.6	98.7	66.8	84.4	94.5	66.4	82.0	99.2	70.6	86.4	99.2	97.3	98.3	98.5	66.8	84.3

KANSAS / KENTUCKY

DIVISION		SALINA 674			TOPEKA 664 - 666			WICHITA 670 - 672			ASHLAND 411 - 412			BOWLING GREEN 421 - 422			CAMPTON 413 - 414		
		MAT.	INST.	TOTAL	MAT.	INST.	TOTAL	MAT.	INST.	TOTAL	MAT.	INST.	TOTAL	MAT.	INST.	TOTAL	MAT.	INST.	TOTAL
015433	CONTRACTOR EQUIPMENT		102.7	102.7		101.0	101.0		102.7	102.7		98.4	98.4		95.8	95.8		102.3	102.3
0241, 31 - 34	SITE & INFRASTRUCTURE, DEMOLITION	105.8	93.8	97.5	97.8	90.5	92.8	97.6	92.4	94.0	106.2	87.0	93.0	69.6	98.8	89.8	78.9	99.0	92.8
0310	Concrete Forming & Accessories	93.1	58.2	62.8	100.3	43.5	51.1	99.4	47.4	54.3	88.7	102.1	100.4	84.6	81.1	81.5	89.3	48.6	54.0
0320	Concrete Reinforcing	103.9	65.3	84.0	96.5	94.7	95.6	96.5	77.0	86.4	92.1	100.7	96.6	89.4	81.4	85.2	90.2	65.1	77.3
0330	Cast-in-Place Concrete	102.6	52.1	83.2	92.4	48.1	75.4	90.2	51.2	75.3	86.3	102.2	92.4	85.6	101.7	91.8	95.5	60.0	81.9
03	CONCRETE	102.2	58.7	81.4	93.3	56.4	75.7	92.2	55.7	74.7	94.5	102.4	98.3	92.7	88.3	90.6	96.8	56.6	77.5
04	MASONRY	121.3	59.7	84.0	100.6	58.0	74.8	94.4	54.3	70.1	89.9	105.3	99.3	91.3	74.1	80.9	89.8	40.7	60.0
05	METALS	94.9	82.8	90.7	100.2	95.8	98.7	100.2	84.7	94.9	90.9	108.2	96.3	92.5	83.2	89.3	91.7	74.1	85.6
06	WOOD, PLASTICS & COMPOSITES	88.2	56.6	70.1	93.5	40.3	63.0	95.0	46.1	66.9	80.5	101.5	92.5	94.6	80.5	86.5	93.2	47.1	66.8
07	THERMAL & MOISTURE PROTECTION	99.4	60.2	83.8	100.4	64.2	86.1	98.8	55.4	81.5	90.4	98.8	93.7	85.8	82.5	84.5	98.5	51.5	79.8
08	OPENINGS	96.6	52.5	85.4	96.9	56.8	86.8	96.9	54.1	86.1	95.7	96.7	95.9	96.8	76.2	91.6	98.5	56.9	87.9
0920	Plaster & Gypsum Board	94.3	55.2	66.7	93.4	38.5	54.7	93.2	44.4	58.8	60.6	101.6	89.5	82.8	80.2	80.9	82.8	44.9	56.1
0950, 0980	Ceilings & Acoustic Treatment	89.5	55.2	66.7	90.4	38.5	56.0	89.5	44.4	59.6	80.4	101.6	94.5	85.6	80.2	82.0	85.6	44.9	58.6
0960	Flooring	99.4	65.4	89.6	102.8	77.8	95.6	101.6	66.6	91.5	78.5	105.9	86.4	94.6	80.1	90.4	96.1	43.6	81.0
0970, 0990	Wall Finishes & Painting/Coating	96.2	41.9	63.2	96.2	56.6	72.1	96.2	51.6	69.1	99.8	90.7	94.3	99.8	71.3	81.7	97.6	37.2	61.0
09	FINISHES	97.2	57.5	75.3	97.6	49.5	71.1	97.2	50.8	71.6	82.7	101.9	93.3	93.1	79.9	85.8	93.7	45.3	67.0
COVERS	DIVS. 10 - 14, 25, 28, 41, 43, 44, 46	100.0	56.0	90.9	100.0	52.7	90.2	100.0	52.0	90.0	100.0	102.6	100.6	100.0	63.2	92.3	100.0	54.5	90.5
21, 22, 23	FIRE SUPPRESSION, PLUMBING & HVAC	100.1	69.1	87.2	100.1	58.5	82.8	99.9	65.1	85.4	94.4	90.0	92.6	99.9	83.0	92.9	94.6	66.3	82.8
26, 27, 3370	ELECTRICAL, COMMUNICATIONS & UTIL.	95.0	75.3	84.5	105.4	75.2	89.3	102.1	75.3	87.8	91.8	99.4	95.8	94.7	82.5	88.2	91.7	71.8	81.1
MF2010	WEIGHTED AVERAGE	99.7	68.0	85.5	99.2	65.2	84.0	98.3	65.1	83.4	93.2	98.5	95.6	94.7	82.8	89.4	94.4	62.6	80.2

KENTUCKY

DIVISION		CORBIN 407 - 409			COVINGTON 410			ELIZABETHTOWN 427			FRANKFORT 406			HAZARD 417 - 418			HENDERSON 424		
		MAT.	INST.	TOTAL	MAT.	INST.	TOTAL	MAT.	INST.	TOTAL	MAT.	INST.	TOTAL	MAT.	INST.	TOTAL	MAT.	INST.	TOTAL
015433	CONTRACTOR EQUIPMENT		102.3	102.3		106.1	106.1		95.8	95.8		102.3	102.3		102.3	102.3		117.4	117.4
0241, 31 - 34	SITE & INFRASTRUCTURE, DEMOLITION	76.4	99.0	92.0	74.7	114.2	102.0	65.0	98.7	88.2	80.9	99.9	94.0	76.9	100.6	93.3	73.6	125.7	109.5
0310	Concrete Forming & Accessories	86.5	48.2	53.3	85.3	98.4	96.7	80.2	80.1	80.1	96.6	59.0	64.0	86.5	42.0	47.9	92.3	77.9	79.8
0320	Concrete Reinforcing	89.8	65.0	77.1	88.8	97.3	93.2	89.8	91.5	90.7	91.1	79.7	85.2	90.7	65.1	77.8	89.5	88.7	89.1
0330	Cast-in-Place Concrete	90.4	59.8	78.7	91.4	96.3	93.3	77.4	75.3	76.6	87.4	66.9	79.6	91.9	87.6	90.2	75.7	90.6	81.4
03	CONCRETE	92.0	56.3	74.9	95.1	97.4	96.2	85.1	80.8	83.0	91.4	66.3	79.4	93.6	63.1	79.0	89.8	84.5	87.2
04	MASONRY	89.5	41.4	60.4	103.2	108.8	106.6	76.0	80.1	78.5	87.1	73.2	78.6	86.7	50.1	64.6	95.3	93.1	94.0
05	METALS	91.6	73.7	85.4	89.9	94.1	91.3	91.6	87.9	90.3	98.4	83.3	93.2	91.7	73.9	85.6	83.6	89.0	85.4
06	WOOD, PLASTICS & COMPOSITES	90.2	47.1	65.5	91.1	91.4	91.3	89.7	80.5	84.4	102.0	56.9	76.2	90.2	34.6	58.3	96.9	73.7	83.6
07	THERMAL & MOISTURE PROTECTION	98.3	52.6	80.1	97.6	99.6	98.4	85.5	79.3	83.0	98.1	70.0	87.0	98.4	51.0	79.5	97.8	92.4	95.7
08	OPENINGS	97.7	49.6	85.5	99.9	90.2	97.4	96.9	83.8	93.6	97.0	67.8	89.6	98.9	40.6	84.1	94.9	79.3	90.9
0920	Plaster & Gypsum Board	81.4	44.9	55.7	70.1	91.7	85.3	81.0	80.2	80.4	93.1	55.0	66.3	81.4	32.0	46.6	85.2	72.3	76.1
0950, 0980	Ceilings & Acoustic Treatment	85.6	44.9	58.6	88.5	91.7	90.6	85.6	80.2	82.0	87.5	55.0	66.0	85.6	32.0	50.1	80.0	72.3	74.9
0960	Flooring	94.4	43.6	79.8	71.0	88.9	76.1	92.0	70.8	85.9	98.9	57.5	87.0	94.4	44.3	80.0	97.8	90.9	95.8
0970, 0990	Wall Finishes & Painting/Coating	97.6	37.2	61.0	93.2	83.1	87.1	97.6	78.7	86.1	97.6	50.6	69.1	97.6	30.5	56.9	97.6	95.8	96.5
09	FINISHES	92.8	45.3	66.6	85.0	94.9	90.5	91.7	78.0	84.1	96.2	56.7	74.5	92.9	39.7	63.6	92.5	81.6	86.5
COVERS	DIVS. 10 - 14, 25, 28, 41, 43, 44, 46	100.0	49.1	89.4	100.0	105.0	101.0	100.0	93.4	98.6	100.0	67.0	93.1	100.0	56.2	90.9	100.0	70.6	93.9
21, 22, 23	FIRE SUPPRESSION, PLUMBING & HVAC	94.5	65.8	82.5	95.0	93.0	94.2	94.9	81.5	89.3	100.0	81.5	92.3	94.6	43.1	73.2	94.9	76.3	87.1
26, 27, 3370	ELECTRICAL, COMMUNICATIONS & UTIL.	92.0	71.8	81.2	96.1	80.5	87.8	91.5	82.5	86.7	103.1	94.6	98.6	91.7	54.5	71.9	94.0	82.4	87.8
MF2010	WEIGHTED AVERAGE	93.5	62.0	79.4	94.4	96.0	95.1	91.0	83.4	87.6	97.1	74.3	86.9	93.7	55.7	76.7	92.3	86.3	89.6

KENTUCKY

	DIVISION	LEXINGTON 403 - 405			LOUISVILLE 400 - 402			OWENSBORO 423			PADUCAH 420			PIKEVILLE 415 - 416			SOMERSET 425 - 426		
		MAT.	INST.	TOTAL	MAT.	INST.	TOTAL	MAT.	INST.	TOTAL	MAT.	INST.	TOTAL	MAT.	INST.	TOTAL	MAT.	INST.	TOTAL
015433	CONTRACTOR EQUIPMENT		102.3	102.3		95.8	95.8		117.4	117.4		117.4	117.4		98.4	98.4		102.3	102.3
0241, 31 - 34	SITE & INFRASTRUCTURE, DEMOLITION	78.0	101.1	93.9	67.8	98.7	89.1	82.2	125.3	112.0	76.2	125.1	109.9	116.2	85.6	95.1	70.7	99.5	90.6
0310	Concrete Forming & Accessories	97.6	74.2	77.3	93.0	80.1	81.8	90.9	69.8	72.6	88.8	80.3	81.4	96.8	67.4	71.3	87.2	46.7	52.1
0320	Concrete Reinforcing	98.6	91.0	94.7	98.6	91.5	95.0	89.5	90.3	89.9	90.0	85.5	87.7	92.6	89.6	91.1	89.8	78.7	84.1
0330	Cast-in-Place Concrete	92.5	84.1	89.3	94.4	75.3	87.1	88.3	97.4	91.8	80.6	88.7	83.7	94.9	92.4	94.0	75.7	101.9	85.7
03	CONCRETE	95.1	81.1	88.4	95.8	80.8	88.6	101.5	83.5	92.9	94.1	84.3	89.4	107.6	81.4	95.1	80.4	72.5	76.6
04	MASONRY	88.3	63.7	73.4	89.4	80.1	83.8	88.2	69.3	76.7	91.2	90.5	90.8	88.0	103.0	97.1	82.2	60.8	69.2
05	METALS	94.0	88.3	92.0	100.7	87.9	96.3	85.1	88.3	86.2	82.0	86.9	83.7	90.8	99.1	93.6	91.6	80.3	87.7
06	WOOD, PLASTICS & COMPOSITES	104.6	74.6	87.4	102.0	80.5	89.7	95.0	66.2	78.5	92.7	79.4	85.1	89.4	59.7	72.3	90.8	39.9	61.6
07	THERMAL & MOISTURE PROTECTION	98.5	84.6	93.0	92.5	79.3	87.3	98.2	74.7	88.8	97.9	82.4	91.7	91.0	69.5	82.5	97.8	57.0	81.6
08	OPENINGS	97.8	77.9	92.8	97.8	83.8	94.3	94.9	76.7	90.3	94.0	78.5	90.1	96.5	68.4	89.4	97.7	56.3	87.3
0920	Plaster & Gypsum Board	91.4	73.2	78.6	84.4	80.2	81.4	83.9	64.5	70.2	82.5	78.2	79.5	64.2	58.5	60.2	81.9	37.5	50.6
0950, 0980	Ceilings & Acoustic Treatment	91.3	73.2	79.3	87.5	80.2	82.6	80.0	64.5	69.7	80.0	78.2	78.8	80.4	58.5	65.9	85.6	37.5	53.7
0960	Flooring	99.3	58.6	87.6	97.8	70.8	90.0	97.0	80.1	92.2	95.8	62.7	86.3	82.9	51.8	74.0	94.6	43.6	79.9
0970, 0990	Wall Finishes & Painting/Coating	97.6	67.6	79.4	97.6	78.7	86.1	97.6	95.6	96.4	97.6	80.6	87.3	99.8	87.4	92.3	97.6	51.3	69.5
09	FINISHES	96.7	70.0	82.0	94.1	78.0	85.2	92.7	73.6	82.2	91.8	77.0	83.6	85.3	65.4	74.3	92.3	44.8	66.1
COVERS	DIVS. 10 - 14, 25, 28, 41, 43, 44, 46	100.0	93.4	98.6	100.0	93.4	98.6	100.0	79.8	95.8	100.0	67.4	93.2	100.0	59.0	91.5	100.0	58.3	91.3
21, 22, 23	FIRE SUPPRESSION, PLUMBING & HVAC	99.9	62.9	84.5	100.0	81.5	92.3	99.9	66.6	86.0	94.9	82.0	89.5	94.4	76.1	86.8	94.9	62.4	81.4
26, 27, 3370	ELECTRICAL, COMMUNICATIONS & UTIL.	95.2	71.3	82.5	106.2	82.5	93.6	94.1	82.4	87.9	96.8	82.3	89.0	95.2	79.6	86.9	92.2	71.2	81.0
MF2010	WEIGHTED AVERAGE	96.3	75.3	86.9	97.7	83.4	91.3	95.1	80.2	88.4	92.6	86.0	89.7	95.7	80.3	88.8	91.6	66.6	80.4

LOUISIANA

	DIVISION	ALEXANDRIA 713 - 714			BATON ROUGE 707 - 708			HAMMOND 704			LAFAYETTE 705			LAKE CHARLES 706			MONROE 712		
		MAT.	INST.	TOTAL	MAT.	INST.	TOTAL	MAT.	INST.	TOTAL	MAT.	INST.	TOTAL	MAT.	INST.	TOTAL	MAT.	INST.	TOTAL
015433	CONTRACTOR EQUIPMENT		88.5	88.5		88.3	88.3		89.0	89.0		89.0	89.0		88.3	88.3		88.5	88.5
0241, 31 - 34	SITE & INFRASTRUCTURE, DEMOLITION	100.0	85.4	90.0	108.0	85.1	92.2	109.7	86.4	93.6	110.3	86.3	93.7	110.9	84.9	93.0	100.0	84.8	89.5
0310	Concrete Forming & Accessories	82.6	36.8	42.9	94.2	63.1	67.2	79.1	49.3	53.2	95.6	54.1	59.6	97.1	58.4	63.5	82.1	37.4	43.3
0320	Concrete Reinforcing	102.0	58.3	79.5	100.6	60.1	79.7	99.2	61.9	80.0	100.6	59.7	79.5	100.6	59.1	79.2	100.8	58.1	78.8
0330	Cast-in-Place Concrete	97.2	39.4	75.1	95.1	58.4	81.0	97.9	45.2	77.7	97.4	45.7	77.6	102.5	67.6	89.2	97.2	56.6	81.6
03	CONCRETE	94.0	43.1	69.6	92.4	61.4	77.6	92.7	51.2	72.8	93.7	53.0	74.2	96.3	62.3	80.0	93.8	49.1	72.4
04	MASONRY	117.5	48.8	75.9	105.9	50.6	72.4	105.5	52.3	73.3	105.5	43.9	68.2	104.4	57.9	76.2	113.0	46.7	72.8
05	METALS	90.2	71.7	83.9	102.7	70.0	91.5	98.5	70.5	88.8	97.5	69.3	87.8	97.5	70.2	88.1	90.2	69.7	83.1
06	WOOD, PLASTICS & COMPOSITES	86.3	35.8	57.3	100.7	69.3	82.7	85.8	50.9	65.8	107.1	58.2	79.0	105.2	62.4	80.6	85.7	36.1	57.2
07	THERMAL & MOISTURE PROTECTION	97.0	47.9	77.5	95.7	57.2	80.4	97.5	56.8	81.3	98.0	52.0	79.7	97.3	59.2	82.1	97.0	50.7	78.6
08	OPENINGS	99.0	42.3	84.6	102.2	60.8	91.7	97.7	55.4	87.0	102.2	54.8	90.2	102.2	57.9	91.0	98.9	46.3	85.6
0920	Plaster & Gypsum Board	74.5	34.4	46.2	94.6	68.7	76.4	88.1	49.8	61.2	98.2	57.3	69.4	98.2	61.6	72.4	74.0	34.7	46.4
0950, 0980	Ceilings & Acoustic Treatment	89.4	34.4	52.9	103.7	68.7	80.5	100.7	49.8	67.0	98.9	57.3	71.3	99.9	61.6	74.5	89.4	34.7	53.2
0960	Flooring	100.0	64.7	89.9	103.9	64.7	92.6	94.8	48.9	81.6	104.1	36.6	84.7	104.1	77.0	96.3	99.6	60.3	88.3
0970, 0990	Wall Finishes & Painting/Coating	97.1	34.3	59.0	100.7	39.3	63.4	100.7	52.0	71.1	100.7	39.3	63.4	100.7	36.2	61.6	97.1	41.0	63.1
09	FINISHES	89.8	40.8	62.8	98.9	61.7	78.4	94.8	49.4	69.8	98.4	49.2	71.2	98.6	59.8	77.2	89.6	41.0	62.8
COVERS	DIVS. 10 - 14, 25, 28, 41, 43, 44, 46	100.0	52.7	90.2	100.0	79.2	95.7	100.0	50.6	89.7	100.0	77.6	95.4	100.0	78.5	95.5	100.0	48.3	89.3
21, 22, 23	FIRE SUPPRESSION, PLUMBING & HVAC	100.2	52.3	80.3	100.0	60.4	83.5	94.7	46.9	74.8	100.1	60.6	83.7	100.1	61.1	83.9	100.2	51.3	79.8
26, 27, 3370	ELECTRICAL, COMMUNICATIONS & UTIL.	95.2	57.7	75.3	114.6	62.7	87.0	100.3	57.8	77.7	101.7	68.4	84.0	101.2	68.4	83.8	97.3	58.9	76.9
MF2010	WEIGHTED AVERAGE	97.2	53.8	77.8	101.5	63.6	84.5	97.4	56.1	78.9	99.7	60.5	82.2	100.0	65.0	84.3	97.2	54.3	78.0

	DIVISION	LOUISIANA NEW ORLEANS 700 - 701			SHREVEPORT 710 - 711			THIBODAUX 703			MAINE AUGUSTA 043			BANGOR 044			BATH 045		
		MAT.	INST.	TOTAL	MAT.	INST.	TOTAL	MAT.	INST.	TOTAL	MAT.	INST.	TOTAL	MAT.	INST.	TOTAL	MAT.	INST.	TOTAL
015433	CONTRACTOR EQUIPMENT		89.2	89.2		88.5	88.5		89.0	89.0		100.3	100.3		100.3	100.3		100.3	100.3
0241, 31 - 34	SITE & INFRASTRUCTURE, DEMOLITION	113.7	88.0	95.9	102.6	84.7	90.2	112.4	86.3	94.4	82.5	100.6	95.0	84.7	98.4	94.1	83.1	100.6	95.2
0310	Concrete Forming & Accessories	98.9	64.1	68.7	96.7	39.5	47.1	91.7	62.9	66.7	95.2	99.3	98.7	90.8	89.7	89.8	87.0	99.2	97.6
0320	Concrete Reinforcing	100.6	62.2	80.8	100.4	58.6	78.9	99.2	61.9	80.0	88.3	99.8	94.3	88.0	99.9	94.1	87.0	99.7	93.5
0330	Cast-in-Place Concrete	101.5	71.0	89.8	96.7	45.3	77.0	105.1	48.4	83.4	89.7	62.6	79.3	78.6	107.0	89.5	78.6	62.5	72.5
03	CONCRETE	95.9	66.7	81.9	93.8	46.2	71.0	98.0	58.3	79.0	105.1	86.2	96.1	98.8	96.8	97.9	98.9	86.1	92.8
04	MASONRY	105.7	61.8	79.1	102.7	48.4	69.8	133.7	49.5	82.7	99.2	61.2	76.2	113.7	61.0	81.7	120.3	58.1	82.6
05	METALS	110.6	73.9	98.0	87.0	69.7	81.1	98.5	70.7	88.9	87.4	83.9	86.2	87.0	80.0	84.6	85.4	83.6	84.8
06	WOOD, PLASTICS & COMPOSITES	101.6	64.1	80.1	101.9	39.2	65.9	94.2	68.5	79.5	96.7	110.4	104.6	91.6	96.8	94.6	87.1	110.4	100.5
07	THERMAL & MOISTURE PROTECTION	97.7	62.2	83.6	96.2	49.6	77.7	97.7	55.2	80.8	96.1	64.8	83.7	96.6	66.2	84.6	96.6	64.0	83.7
08	OPENINGS	103.0	64.9	93.3	96.7	44.4	83.5	103.2	64.9	93.5	102.3	90.0	99.2	102.3	82.4	97.2	102.2	89.7	99.1
0920	Plaster & Gypsum Board	96.4	63.4	73.2	81.4	37.9	50.8	92.1	67.9	75.1	96.8	110.0	106.1	94.2	96.0	95.5	91.5	110.0	104.6
0950, 0980	Ceilings & Acoustic Treatment	98.9	63.4	75.4	90.3	37.9	55.6	100.7	67.9	79.0	91.9	110.0	103.9	91.9	96.0	94.7	89.8	110.0	103.2
0960	Flooring	104.5	64.7	93.1	108.4	64.7	95.9	101.2	46.6	85.5	95.8	58.4	85.1	93.8	64.9	85.5	92.0	58.4	82.3
0970, 0990	Wall Finishes & Painting/Coating	102.1	65.9	80.1	97.1	34.3	59.0	102.1	52.0	71.7	92.3	37.1	58.8	92.3	38.6	59.7	92.3	37.2	58.8
09	FINISHES	98.8	64.3	79.8	93.5	42.8	65.6	97.5	59.3	76.5	96.4	86.8	91.1	95.8	80.7	87.5	94.5	86.8	90.2
COVERS	DIVS. 10 - 14, 25, 28, 41, 43, 44, 46	100.0	82.5	96.4	100.0	67.2	93.2	100.0	79.7	95.8	100.0	96.7	99.3	100.0	106.3	101.3	100.0	96.7	99.3
21, 22, 23	FIRE SUPPRESSION, PLUMBING & HVAC	100.0	66.0	85.8	100.0	56.3	81.8	94.7	61.0	80.7	100.0	66.8	86.2	100.2	62.7	84.6	94.8	66.8	83.2
26, 27, 3370	ELECTRICAL, COMMUNICATIONS & UTIL.	101.7	72.1	86.0	103.0	68.0	84.4	98.7	70.8	83.9	99.0	79.9	88.8	98.6	79.9	88.7	96.4	79.9	87.6
MF2010	WEIGHTED AVERAGE	102.0	69.2	87.3	96.9	57.1	79.1	100.3	64.3	84.2	97.9	80.0	89.9	97.9	79.0	89.4	96.3	79.6	88.8

MAINE

DIVISION		HOULTON 047 MAT.	INST.	TOTAL	KITTERY 039 MAT.	INST.	TOTAL	LEWISTON 042 MAT.	INST.	TOTAL	MACHIAS 046 MAT.	INST.	TOTAL	PORTLAND 040-041 MAT.	INST.	TOTAL	ROCKLAND 048 MAT.	INST.	TOTAL
015433	CONTRACTOR EQUIPMENT		100.3	100.3		100.3	100.3		100.3	100.3		100.3	100.3		100.3	100.3		100.3	100.3
0241, 31 - 34	SITE & INFRASTRUCTURE, DEMOLITION	84.8	100.5	95.6	79.3	100.7	94.1	81.9	98.4	93.3	84.2	100.5	95.4	80.2	98.4	92.7	80.7	100.5	94.4
0310	Concrete Forming & Accessories	93.8	98.3	97.7	87.5	99.7	98.1	95.5	89.7	90.5	91.3	98.3	97.4	94.4	89.7	90.3	92.2	98.7	97.8
0320	Concrete Reinforcing	88.0	99.8	94.1	86.5	99.8	93.3	108.1	99.9	103.9	88.0	99.8	94.1	108.1	99.9	103.9	88.0	99.9	94.1
0330	Cast-in-Place Concrete	78.6	60.7	71.8	78.6	63.5	72.8	80.2	107.0	90.5	78.6	60.8	71.8	99.5	107.0	102.3	80.3	61.9	73.2
03	CONCRETE	99.8	85.2	92.8	94.0	86.7	90.5	98.5	96.8	97.7	99.3	85.2	92.5	108.1	96.8	102.7	96.1	85.7	91.1
04	MASONRY	97.0	59.2	74.1	110.3	60.8	80.3	97.6	61.0	75.4	97.0	59.5	74.3	97.7	61.0	75.5	91.4	61.0	73.0
05	METALS	85.6	83.8	85.0	85.3	83.8	84.8	90.1	80.0	86.6	85.6	83.7	85.0	91.4	80.0	87.5	85.5	83.9	84.9
06	WOOD, PLASTICS & COMPOSITES	95.5	110.4	104.0	87.5	110.4	100.6	97.4	96.8	97.1	92.5	110.4	102.8	96.6	96.8	96.7	93.4	110.4	103.1
07	THERMAL & MOISTURE PROTECTION	96.7	62.3	83.0	96.3	65.0	83.9	96.5	66.2	84.5	96.7	62.3	83.0	99.6	66.2	86.3	96.4	63.2	83.2
08	OPENINGS	102.3	88.0	98.7	102.2	89.3	98.9	105.6	82.4	99.7	102.3	88.0	98.7	105.6	82.4	99.7	102.3	88.0	98.7
0920	Plaster & Gypsum Board	96.8	110.0	106.1	91.5	110.0	104.6	100.2	96.0	97.3	95.0	110.0	105.6	101.0	96.0	97.5	95.0	110.0	105.6
0950, 0980	Ceilings & Acoustic Treatment	89.8	110.0	103.2	89.8	110.0	103.2	99.4	96.0	97.2	89.8	110.0	103.2	96.6	96.0	96.2	89.8	110.0	103.2
0960	Flooring	94.9	54.7	83.3	92.1	60.9	83.1	96.5	64.9	87.4	94.1	54.6	82.8	95.8	64.9	87.0	94.3	54.7	82.9
0970, 0990	Wall Finishes & Painting/Coating	92.3	38.6	59.7	92.3	37.2	58.8	92.3	38.6	59.7	92.3	38.6	59.7	92.3	38.6	59.7	92.3	38.6	59.7
09	FINISHES	96.0	85.8	90.4	94.1	87.5	90.5	98.5	80.7	88.7	95.5	85.8	90.2	97.4	80.7	88.2	95.2	86.0	90.1
COVERS	DIVS. 10 - 14, 25, 28, 41, 43, 44, 46	100.0	96.0	99.2	100.0	98.4	99.7	100.0	106.3	101.3	100.0	90.8	98.1	100.0	106.3	101.3	100.0	96.3	99.2
21, 22, 23	FIRE SUPPRESSION, PLUMBING & HVAC	94.8	65.7	82.7	94.8	67.4	83.4	100.2	62.7	84.6	94.8	65.8	82.7	100.0	62.7	84.5	94.8	66.2	82.9
26, 27, 3370	ELECTRICAL, COMMUNICATIONS & UTIL.	100.9	79.9	89.7	96.4	79.9	87.6	100.9	79.9	89.7	100.9	79.9	89.7	104.5	79.9	91.4	100.8	79.9	89.7
MF2010	WEIGHTED AVERAGE	95.9	79.1	88.4	95.0	80.2	88.4	98.2	79.0	89.6	95.8	78.9	88.2	99.8	79.0	90.5	94.9	79.5	88.0

MAINE / MARYLAND

DIVISION		MAINE WATERVILLE 049 MAT.	INST.	TOTAL	ANNAPOLIS 214 MAT.	INST.	TOTAL	BALTIMORE 210-212 MAT.	INST.	TOTAL	COLLEGE PARK 207-208 MAT.	INST.	TOTAL	CUMBERLAND 215 MAT.	INST.	TOTAL	EASTON 216 MAT.	INST.	TOTAL
015433	CONTRACTOR EQUIPMENT		100.3	100.3		99.1	99.1		102.6	102.6		103.5	103.5		99.1	99.1		99.1	99.1
0241, 31 - 34	SITE & INFRASTRUCTURE, DEMOLITION	84.8	100.6	95.7	98.4	90.4	92.9	98.8	94.4	95.8	111.1	91.4	97.5	90.3	89.7	89.9	97.1	85.9	89.4
0310	Concrete Forming & Accessories	86.5	99.3	97.6	92.5	51.6	57.0	95.2	76.8	79.2	86.7	72.4	74.3	86.6	78.7	79.7	84.7	29.7	37.0
0320	Concrete Reinforcing	88.0	99.8	94.1	96.5	83.3	89.7	96.5	83.4	89.7	94.6	81.9	88.1	85.2	73.4	79.1	84.5	17.2	49.9
0330	Cast-in-Place Concrete	78.6	62.6	72.5	128.8	77.7	109.2	114.7	78.5	100.8	130.9	78.3	110.8	102.4	79.8	93.7	113.7	39.7	85.3
03	CONCRETE	100.6	86.2	93.7	115.6	68.1	92.9	108.6	79.6	94.7	116.2	77.4	97.6	97.0	79.0	88.4	105.6	32.8	70.7
04	MASONRY	107.4	61.2	79.4	96.1	69.8	80.2	97.3	69.8	80.7	116.4	70.8	88.8	95.0	78.8	85.2	109.1	31.3	61.9
05	METALS	85.5	83.9	85.0	92.3	95.0	93.2	93.4	95.8	94.2	81.3	96.1	86.4	90.0	90.9	90.3	90.3	59.0	79.5
06	WOOD, PLASTICS & COMPOSITES	86.6	110.4	100.2	93.5	44.3	65.3	96.3	78.2	85.9	82.3	72.7	76.8	86.7	78.0	81.7	84.3	29.7	53.0
07	THERMAL & MOISTURE PROTECTION	96.7	64.8	84.1	94.9	76.6	87.6	95.2	80.4	89.3	99.4	78.4	91.0	94.6	77.4	87.8	94.7	37.1	71.8
08	OPENINGS	102.3	90.0	99.2	95.4	64.7	87.6	97.7	83.0	94.0	98.3	76.5	92.8	96.6	78.4	92.0	94.6	26.8	77.4
0920	Plaster & Gypsum Board	91.5	110.0	104.6	96.9	43.0	59.0	98.5	77.8	83.9	95.4	71.9	78.9	93.2	77.7	82.2	91.8	27.9	46.8
0950, 0980	Ceilings & Acoustic Treatment	89.8	110.0	103.2	93.2	43.0	60.0	94.2	77.8	83.3	98.3	71.9	80.8	94.2	77.7	83.2	94.2	27.9	50.3
0960	Flooring	91.8	58.4	82.2	94.2	81.6	90.6	94.2	81.6	90.6	111.9	54.5	95.4	90.1	82.9	88.0	89.3	17.2	68.6
0970, 0990	Wall Finishes & Painting/Coating	92.3	37.1	58.8	98.5	85.3	90.5	98.5	85.3	90.5	121.0	80.4	96.4	98.5	79.0	86.6	98.5	21.9	52.0
09	FINISHES	94.6	86.8	90.3	94.9	58.1	74.6	95.5	78.1	85.9	102.8	69.3	84.3	93.0	79.4	85.5	95.1	27.1	56.7
COVERS	DIVS. 10 - 14, 25, 28, 41, 43, 44, 46	100.0	96.7	99.3	100.0	83.0	96.5	100.0	87.2	97.3	100.0	57.8	91.2	100.0	89.5	97.8	100.0	39.1	87.3
21, 22, 23	FIRE SUPPRESSION, PLUMBING & HVAC	94.8	66.8	83.2	100.0	85.7	94.1	100.1	85.7	94.1	94.9	88.1	92.0	94.6	71.3	84.9	94.6	27.8	66.8
26, 27, 3370	ELECTRICAL, COMMUNICATIONS & UTIL.	100.9	79.9	89.7	96.4	91.2	93.6	97.3	93.7	95.4	100.9	97.2	98.9	94.9	82.2	88.2	94.4	62.5	77.4
MF2010	WEIGHTED AVERAGE	96.3	80.0	89.0	99.1	78.5	89.9	98.9	84.8	92.6	99.2	83.0	92.0	94.7	80.1	88.1	96.4	41.9	72.0

MARYLAND / MASSACHUSETTS

DIVISION		ELKTON 219 MAT.	INST.	TOTAL	HAGERSTOWN 217 MAT.	INST.	TOTAL	SALISBURY 218 MAT.	INST.	TOTAL	SILVER SPRING 209 MAT.	INST.	TOTAL	WALDORF 206 MAT.	INST.	TOTAL	BOSTON 020-022, 024 MAT.	INST.	TOTAL
015433	CONTRACTOR EQUIPMENT		99.1	99.1		99.1	99.1		99.1	99.1		96.2	96.2		96.2	96.2		106.1	106.1
0241, 31 - 34	SITE & INFRASTRUCTURE, DEMOLITION	84.6	86.3	85.8	88.3	90.8	90.0	97.1	85.5	89.1	97.6	84.0	88.2	105.0	83.1	89.9	90.3	107.9	102.4
0310	Concrete Forming & Accessories	90.3	79.8	81.2	85.8	78.5	79.4	97.6	28.2	37.4	95.6	52.3	58.0	102.4	45.7	53.2	99.3	140.8	135.3
0320	Concrete Reinforcing	84.5	50.4	67.0	85.2	73.3	79.1	84.5	50.4	67.0	93.6	76.9	85.0	94.2	50.4	71.7	107.0	147.5	127.8
0330	Cast-in-Place Concrete	92.0	45.0	74.0	97.5	79.7	90.7	113.7	38.5	84.9	134.0	60.5	105.9	150.2	53.9	113.3	101.4	151.4	120.6
03	CONCRETE	89.0	62.8	76.4	93.0	78.8	86.2	106.4	37.3	73.3	114.7	61.4	89.2	126.6	50.9	90.4	112.3	144.7	127.8
04	MASONRY	95.3	34.1	58.2	101.0	78.8	87.5	108.7	33.1	62.9	115.5	61.3	82.7	98.7	57.9	74.0	116.6	159.3	142.5
05	METALS	90.3	67.5	82.4	90.1	90.6	90.3	90.3	60.6	80.1	85.3	90.8	87.2	85.3	74.1	81.4	102.0	128.6	111.2
06	WOOD, PLASTICS & COMPOSITES	91.0	94.4	93.0	85.5	78.1	81.2	100.2	26.3	57.8	88.5	51.2	67.1	95.7	45.9	67.1	100.7	140.2	123.4
07	THERMAL & MOISTURE PROTECTION	94.3	46.2	75.2	94.3	71.2	85.1	94.9	54.2	78.7	101.7	74.2	90.8	102.2	69.9	89.4	96.8	150.0	117.9
08	OPENINGS	94.6	67.4	87.7	94.6	77.2	90.2	94.9	22.5	76.5	91.4	63.7	84.4	92.2	44.5	80.2	101.6	142.8	112.0
0920	Plaster & Gypsum Board	95.4	94.6	94.8	91.8	77.8	81.9	101.1	24.5	47.2	101.2	50.6	65.5	104.7	45.1	62.7	103.1	140.8	129.7
0950, 0980	Ceilings & Acoustic Treatment	94.2	94.6	94.4	95.2	77.8	83.7	94.2	24.5	48.0	107.6	50.6	69.8	107.6	45.1	66.2	98.4	140.8	126.5
0960	Flooring	91.9	51.7	80.4	89.8	82.9	87.8	95.7	51.7	83.1	119.0	54.5	100.5	123.2	51.7	102.7	96.3	176.2	119.3
0970, 0990	Wall Finishes & Painting/Coating	98.5	80.4	87.5	98.5	33.5	59.0	98.5	24.7	53.7	127.9	80.4	99.0	127.9	57.6	85.3	94.9	157.1	132.6
09	FINISHES	93.5	77.8	84.8	92.8	74.4	82.7	96.1	31.0	60.2	102.7	54.1	75.9	104.8	47.1	73.0	98.4	149.6	126.7
COVERS	DIVS. 10 - 14, 25, 28, 41, 43, 44, 46	100.0	47.5	89.1	100.0	89.6	97.8	100.0	38.3	87.2	100.0	50.1	89.6	100.0	41.0	87.7	100.0	120.3	104.2
21, 22, 23	FIRE SUPPRESSION, PLUMBING & HVAC	94.6	73.5	85.9	100.0	85.0	93.8	94.6	63.5	81.7	94.9	82.7	89.8	94.9	80.3	88.8	100.2	129.0	112.2
26, 27, 3370	ELECTRICAL, COMMUNICATIONS & UTIL.	96.3	53.5	73.5	94.7	82.2	88.1	93.1	62.5	76.8	98.1	97.2	97.6	95.3	97.2	96.3	100.0	134.8	118.5
MF2010	WEIGHTED AVERAGE	93.6	64.9	80.7	95.5	82.2	89.5	96.8	51.0	76.3	98.3	74.5	87.7	99.1	68.5	85.4	102.3	136.8	117.7

MASSACHUSETTS

DIVISION		BROCKTON 023			BUZZARDS BAY 025			FALL RIVER 027			FITCHBURG 014			FRAMINGHAM 017			GREENFIELD 013		
		MAT.	INST.	TOTAL	MAT.	INST.	TOTAL	MAT.	INST.	TOTAL	MAT.	INST.	TOTAL	MAT.	INST.	TOTAL	MAT.	INST.	TOTAL
015433	CONTRACTOR EQUIPMENT		102.2	102.2		102.2	102.2		103.2	103.2		100.3	100.3		101.4	101.4		100.3	100.3
0241, 31 - 34	SITE & INFRASTRUCTURE, DEMOLITION	87.5	104.3	99.1	79.0	104.2	96.4	86.6	104.4	98.8	80.5	104.2	96.8	77.3	103.8	95.6	83.7	102.7	96.8
0310	Concrete Forming & Accessories	99.7	129.4	125.5	97.5	128.7	124.6	99.7	129.0	125.2	91.9	120.5	116.7	98.5	129.2	125.1	90.4	109.8	107.2
0320	Concrete Reinforcing	108.1	147.4	128.3	86.7	123.1	105.4	108.1	123.1	115.8	86.7	142.7	115.6	86.7	147.0	117.8	90.3	121.9	106.6
0330	Cast-in-Place Concrete	96.6	151.3	117.5	80.2	151.0	107.3	93.4	151.6	115.7	84.1	145.3	107.5	84.1	139.1	105.2	86.5	123.2	100.5
03	CONCRETE	109.6	139.4	123.8	92.5	134.4	112.6	108.0	134.8	120.8	90.6	132.4	110.6	93.7	135.1	113.5	94.6	116.0	104.9
04	MASONRY	112.6	155.6	138.6	103.2	155.6	134.9	113.0	155.5	138.8	99.5	149.3	129.7	106.1	149.4	132.4	103.7	123.4	115.6
05	METALS	98.7	126.2	108.1	93.5	115.7	101.1	98.7	116.1	104.7	93.4	120.8	102.8	93.4	125.3	104.4	95.6	107.8	99.8
06	WOOD, PLASTICS & COMPOSITES	99.6	127.6	115.6	96.7	127.6	114.4	99.6	127.8	115.8	93.2	116.5	106.6	99.2	127.3	115.3	91.4	108.4	101.1
07	THERMAL & MOISTURE PROTECTION	96.7	145.8	116.2	96.1	139.7	113.4	96.7	138.9	113.5	96.2	136.6	112.3	96.2	141.7	114.3	96.2	117.8	104.8
08	OPENINGS	99.6	133.1	108.1	94.9	122.4	101.9	99.6	122.6	105.4	104.2	126.0	109.7	95.4	132.9	104.9	104.4	112.1	106.3
0920	Plaster & Gypsum Board	97.3	127.7	118.7	93.0	127.7	117.4	97.3	127.7	118.7	94.2	116.3	109.8	97.3	127.7	118.7	95.4	108.0	104.2
0950, 0980	Ceilings & Acoustic Treatment	100.5	127.7	118.5	90.9	127.7	115.3	100.5	127.7	118.5	89.8	116.3	107.4	89.8	127.7	114.9	97.3	108.0	104.4
0960	Flooring	96.8	176.2	119.7	95.5	176.2	118.7	96.6	176.2	119.5	93.8	176.2	117.5	95.3	176.2	118.6	93.0	135.4	105.2
0970, 0990	Wall Finishes & Painting/Coating	94.4	142.1	123.3	94.4	142.1	123.3	94.4	142.1	123.3	92.3	142.1	122.5	93.4	142.1	122.9	92.3	111.7	104.1
09	FINISHES	98.4	139.6	121.1	94.8	139.6	119.5	98.3	139.8	121.2	94.1	133.1	115.6	94.8	139.4	119.4	96.0	114.7	106.3
COVERS	DIVS. 10 - 14, 25, 28, 41, 43, 44, 46	100.0	117.5	103.6	100.0	117.5	103.6	100.0	118.1	103.8	100.0	92.9	98.5	100.0	117.0	103.5	100.0	105.6	101.2
21, 22, 23	FIRE SUPPRESSION, PLUMBING & HVAC	100.2	110.3	104.4	94.8	110.0	101.1	100.2	110.0	104.3	95.6	112.5	102.7	95.6	123.2	107.1	95.6	103.5	98.9
26, 27, 3370	ELECTRICAL, COMMUNICATIONS & UTIL.	99.1	101.0	100.1	96.0	101.0	98.6	99.1	101.0	100.1	101.4	104.0	102.8	97.5	126.9	113.1	101.4	95.0	98.0
MF2010	WEIGHTED AVERAGE	100.9	124.5	111.5	95.0	122.2	107.1	100.7	122.3	110.4	96.2	121.1	107.3	95.6	129.4	110.7	97.4	108.6	102.4

MASSACHUSETTS

DIVISION		HYANNIS 026			LAWRENCE 019			LOWELL 018			NEW BEDFORD 027			PITTSFIELD 012			SPRINGFIELD 010 - 011		
		MAT.	INST.	TOTAL	MAT.	INST.	TOTAL	MAT.	INST.	TOTAL	MAT.	INST.	TOTAL	MAT.	INST.	TOTAL	MAT.	INST.	TOTAL
015433	CONTRACTOR EQUIPMENT		102.2	102.2		102.2	102.2		100.3	100.3		103.2	103.2		100.3	100.3		100.3	100.3
0241, 31 - 34	SITE & INFRASTRUCTURE, DEMOLITION	84.4	104.2	98.0	88.2	104.3	99.3	87.3	104.2	99.0	85.2	104.4	98.4	88.3	102.6	98.2	87.8	102.9	98.2
0310	Concrete Forming & Accessories	92.1	128.7	123.9	99.5	129.5	125.5	96.5	129.6	125.2	99.7	129.0	125.2	96.5	108.4	106.8	96.7	110.4	108.6
0320	Concrete Reinforcing	86.7	123.1	105.4	107.2	137.3	122.7	108.1	137.0	123.0	108.1	123.1	115.8	89.6	106.2	98.2	108.1	121.9	115.2
0330	Cast-in-Place Concrete	88.1	151.0	112.2	97.4	148.2	116.8	88.5	148.2	111.4	82.3	151.6	108.8	96.6	121.4	106.1	92.1	124.2	104.4
03	CONCRETE	99.6	134.4	116.3	109.8	136.5	122.6	100.3	136.3	117.5	102.4	134.8	117.9	101.6	111.8	106.5	102.1	116.7	109.1
04	MASONRY	111.5	155.6	138.2	111.7	155.6	138.3	98.2	149.3	129.2	111.4	155.5	138.1	98.9	120.3	111.9	98.5	125.1	114.6
05	METALS	95.0	115.7	102.1	96.0	122.1	105.0	96.0	119.4	104.0	98.7	116.1	104.7	95.8	101.2	97.6	98.6	107.8	101.8
06	WOOD, PLASTICS & COMPOSITES	90.7	127.6	111.8	99.6	127.6	115.6	99.0	127.6	115.4	99.6	127.8	115.8	99.0	108.4	104.4	99.0	108.4	104.4
07	THERMAL & MOISTURE PROTECTION	96.4	139.7	113.6	96.7	145.4	116.0	96.5	143.4	115.1	96.7	138.9	113.5	96.5	116.5	104.5	96.5	118.6	105.3
08	OPENINGS	95.6	122.4	102.4	99.6	130.4	107.4	105.6	130.4	111.9	99.6	122.6	105.4	105.6	108.0	106.2	105.6	112.1	107.3
0920	Plaster & Gypsum Board	89.5	127.7	116.4	100.2	127.7	119.6	100.2	127.7	119.6	97.3	127.7	118.7	100.2	108.0	105.7	100.2	108.0	105.7
0950, 0980	Ceilings & Acoustic Treatment	93.0	127.7	116.0	99.4	127.7	118.2	99.4	127.7	118.2	100.5	127.7	118.5	99.4	108.0	105.1	99.4	108.0	105.1
0960	Flooring	93.4	176.2	117.2	95.8	176.2	118.9	95.8	176.2	118.9	96.6	176.2	119.5	96.0	135.4	107.4	95.7	135.4	107.2
0970, 0990	Wall Finishes & Painting/Coating	94.4	142.1	123.3	92.4	142.1	122.5	92.3	142.1	122.5	94.4	142.1	123.3	92.3	111.7	104.1	93.8	111.7	104.7
09	FINISHES	94.6	139.6	119.4	98.1	139.6	120.9	98.0	139.6	120.9	98.2	139.8	121.1	98.1	114.0	106.8	98.1	115.1	107.5
COVERS	DIVS. 10 - 14, 25, 28, 41, 43, 44, 46	100.0	117.5	103.6	100.0	117.6	103.7	100.0	117.6	103.7	100.0	118.1	103.8	100.0	104.5	100.9	100.0	106.2	101.3
21, 22, 23	FIRE SUPPRESSION, PLUMBING & HVAC	100.2	110.0	104.3	100.2	113.0	105.5	100.2	123.2	109.8	100.2	110.0	104.3	100.2	96.6	98.7	100.2	104.4	102.0
26, 27, 3370	ELECTRICAL, COMMUNICATIONS & UTIL.	96.5	101.0	98.9	100.3	126.9	114.4	100.9	126.8	114.7	99.9	101.0	100.5	100.9	95.0	97.7	100.9	96.9	98.8
MF2010	WEIGHTED AVERAGE	98.0	122.2	108.8	100.6	127.8	112.8	99.4	129.0	112.6	100.0	122.3	110.0	99.6	105.3	102.2	100.0	109.5	104.2

MASSACHUSETTS / MICHIGAN

DIVISION		WORCESTER (MA) 015 - 016			ANN ARBOR 481			BATTLE CREEK 490			BAY CITY 487			DEARBORN 481			DETROIT 482		
		MAT.	INST.	TOTAL	MAT.	INST.	TOTAL	MAT.	INST.	TOTAL	MAT.	INST.	TOTAL	MAT.	INST.	TOTAL	MAT.	INST.	TOTAL
015433	CONTRACTOR EQUIPMENT		100.3	100.3		110.9	110.9		103.6	103.6		110.9	110.9		110.9	110.9		98.5	98.5
0241, 31 - 34	SITE & INFRASTRUCTURE, DEMOLITION	87.7	104.2	99.1	73.4	96.1	89.1	82.5	87.9	86.2	66.2	94.7	85.8	73.2	96.1	89.0	86.5	97.7	94.2
0310	Concrete Forming & Accessories	97.0	124.7	121.1	98.7	104.0	103.3	98.2	91.4	92.3	98.7	89.0	90.3	98.5	112.1	110.3	99.2	112.2	110.5
0320	Concrete Reinforcing	108.1	146.3	127.8	100.8	125.5	113.6	98.6	92.5	95.5	100.8	124.7	113.1	100.8	125.7	113.6	100.1	125.7	113.3
0330	Cast-in-Place Concrete	91.6	145.4	112.2	82.3	115.3	94.9	92.4	104.8	97.2	78.7	95.2	85.0	80.5	113.3	93.0	88.3	113.3	97.9
03	CONCRETE	101.9	135.0	117.7	90.3	112.3	100.9	95.0	95.5	95.3	88.5	98.6	93.4	89.4	115.3	101.8	93.3	114.2	103.3
04	MASONRY	98.0	149.3	129.1	96.8	110.4	105.0	95.3	88.1	90.9	96.4	89.6	92.3	96.6	113.7	107.0	95.8	113.7	106.7
05	METALS	98.7	122.4	106.8	93.1	120.2	102.4	92.9	87.7	91.1	93.7	116.9	101.7	93.1	120.6	102.6	97.0	102.8	99.0
06	WOOD, PLASTICS & COMPOSITES	99.4	122.0	112.4	100.4	101.7	101.2	100.9	91.6	95.6	100.4	88.3	93.5	100.4	112.2	107.2	100.8	112.2	107.3
07	THERMAL & MOISTURE PROTECTION	96.5	137.2	112.7	100.2	111.5	104.7	94.3	89.1	92.2	98.0	97.9	98.0	98.8	116.7	105.9	97.3	116.7	105.0
08	OPENINGS	105.6	129.9	111.8	92.1	105.6	95.5	92.0	85.5	90.4	92.1	95.9	93.1	92.1	110.6	96.8	93.6	111.3	98.1
0920	Plaster & Gypsum Board	100.2	122.0	115.5	99.1	100.7	100.3	90.1	87.4	88.2	99.1	86.9	90.5	99.1	111.5	107.8	99.1	111.5	107.8
0950, 0980	Ceilings & Acoustic Treatment	99.4	122.0	114.4	94.3	100.7	98.6	91.3	87.4	88.7	95.3	86.9	89.7	94.3	111.5	105.7	95.3	111.5	106.0
0960	Flooring	95.8	170.2	117.2	89.7	118.1	97.9	98.2	97.1	97.9	89.6	81.2	87.2	89.4	118.9	97.8	89.6	118.9	98.0
0970, 0990	Wall Finishes & Painting/Coating	92.3	142.1	122.5	93.0	104.2	99.8	97.6	82.8	88.6	93.0	85.1	88.2	93.0	104.3	99.90	94.5	104.3	100.5
09	FINISHES	98.0	135.1	118.4	92.6	106.2	100.1	96.1	91.6	93.6	92.3	86.4	89.1	92.4	112.7	103.6	93.8	112.7	104.2
COVERS	DIVS. 10 - 14, 25, 28, 41, 43, 44, 46	100.0	109.9	102.1	100.0	106.1	101.3	100.0	106.8	101.4	100.0	100.2	100.0	100.0	107.5	101.6	100.0	107.5	101.6
21, 22, 23	FIRE SUPPRESSION, PLUMBING & HVAC	100.2	112.6	105.3	100.0	102.1	100.9	99.9	92.0	96.6	100.0	87.5	94.8	100.0	107.8	103.3	100.1	111.2	104.7
26, 27, 3370	ELECTRICAL, COMMUNICATIONS & UTIL.	100.9	104.8	103.0	96.9	110.3	104.1	94.7	85.2	89.7	95.9	90.5	93.0	96.9	111.6	104.7	97.9	111.5	105.2
MF2010	WEIGHTED AVERAGE	100.0	122.7	110.1	95.2	107.7	100.8	95.8	90.4	93.4	94.6	93.9	94.3	95.0	111.2	102.2	96.7	110.2	102.8

657

MICHIGAN

DIVISION		FLINT 484-485			GAYLORD 497			GRAND RAPIDS 493,495			IRON MOUNTAIN 498-499			JACKSON 492			KALAMAZOO 491		
		MAT.	INST.	TOTAL	MAT.	INST.	TOTAL	MAT.	INST.	TOTAL	MAT.	INST.	TOTAL	MAT.	INST.	TOTAL	MAT.	INST.	TOTAL
015433	CONTRACTOR EQUIPMENT		110.9	110.9		105.9	105.9		103.6	103.6		94.4	94.4		105.9	105.9		103.6	103.6
0241, 31 - 34	SITE & INFRASTRUCTURE, DEMOLITION	64.2	95.4	85.8	78.2	86.1	83.7	83.9	87.7	86.6	85.4	94.5	91.7	98.0	87.4	90.7	82.8	87.9	86.3
0310	Concrete Forming & Accessories	101.6	90.9	92.3	95.4	56.9	62.0	98.4	80.8	83.1	87.3	88.0	87.9	92.9	88.5	89.1	98.2	91.1	92.0
0320	Concrete Reinforcing	100.8	125.0	113.3	91.9	94.1	93.1	98.6	91.8	95.1	91.7	95.8	93.8	89.3	124.8	107.6	98.6	92.4	95.4
0330	Cast-in-Place Concrete	82.9	97.1	88.3	92.2	72.5	84.6	98.4	96.9	97.8	109.1	91.6	102.4	92.0	95.5	93.4	94.3	104.7	98.3
03	CONCRETE	90.8	100.2	95.3	91.4	70.8	81.6	98.1	87.9	93.2	101.1	90.8	96.2	86.5	98.4	92.2	98.3	95.3	96.9
04	MASONRY	96.8	98.0	97.5	105.6	84.1	92.6	91.2	81.8	85.5	92.1	90.8	91.3	86.3	94.7	91.4	94.0	88.1	90.4
05	METALS	93.1	117.7	101.6	94.0	101.7	96.6	94.0	85.1	90.9	93.4	91.5	92.7	94.2	115.2	101.4	92.9	87.3	91.0
06	WOOD, PLASTICS & COMPOSITES	103.6	88.6	95.0	93.3	49.8	68.4	98.1	80.1	87.8	87.9	88.0	88.0	92.1	86.9	89.1	100.9	91.6	95.6
07	THERMAL & MOISTURE PROTECTION	98.0	101.5	99.4	92.8	71.9	84.5	96.1	69.6	85.6	96.0	85.8	92.0	92.2	98.1	94.5	94.3	89.0	92.2
08	OPENINGS	92.1	96.1	93.1	94.0	55.0	84.2	95.7	70.7	89.4	99.3	74.9	93.1	93.0	88.3	91.8	92.0	85.5	90.4
0920	Plaster & Gypsum Board	100.9	87.2	91.2	89.3	47.0	59.5	90.7	75.6	80.0	59.5	88.1	79.6	87.8	85.2	85.9	90.1	87.4	88.2
0950, 0980	Ceilings & Acoustic Treatment	94.3	87.2	89.6	88.5	47.0	61.0	89.4	75.6	80.2	89.2	88.1	88.5	89.4	85.2	86.6	91.3	87.4	88.7
0960	Flooring	89.6	83.9	87.9	91.1	50.9	79.5	98.4	35.8	80.4	112.3	97.7	108.1	89.6	70.0	84.0	98.2	76.5	91.9
0970, 0990	Wall Finishes & Painting/Coating	93.0	92.0	92.4	93.2	46.5	64.9	97.6	34.7	59.5	114.3	58.3	80.3	93.2	77.3	83.5	97.6	82.8	88.6
09	FINISHES	91.9	88.8	90.2	96.1	52.6	72.1	95.9	67.0	80.0	99.6	86.7	92.5	97.2	83.0	89.4	96.1	87.5	91.4
COVERS	DIVS. 10 - 14, 25, 28, 41, 43, 44, 46	100.0	101.5	100.3	100.0	95.1	99.0	100.0	105.2	101.1	100.0	97.3	99.4	100.0	101.4	100.3	100.0	106.8	101.4
21, 22, 23	FIRE SUPPRESSION, PLUMBING & HVAC	100.0	96.1	98.4	94.9	82.6	89.7	99.9	80.6	91.8	94.8	87.4	91.7	94.9	93.1	94.1	99.9	83.0	92.9
26, 27, 3370	ELECTRICAL, COMMUNICATIONS & UTIL.	96.9	97.0	97.0	92.4	50.0	69.8	103.1	49.9	74.8	99.4	87.2	92.9	96.8	110.3	104.0	94.6	88.8	91.5
MF2010	WEIGHTED AVERAGE	94.9	98.2	96.4	94.5	73.4	85.1	97.3	76.8	88.1	96.7	88.8	93.2	94.0	96.8	95.2	96.1	88.4	92.7

MICHIGAN / MINNESOTA

DIVISION		LANSING 488-489			MUSKEGON 494			ROYAL OAK 480,483			SAGINAW 486			TRAVERSE CITY 496			BEMIDJI 566		
		MAT.	INST.	TOTAL	MAT.	INST.	TOTAL	MAT.	INST.	TOTAL	MAT.	INST.	TOTAL	MAT.	INST.	TOTAL	MAT.	INST.	TOTAL
015433	CONTRACTOR EQUIPMENT		110.9	110.9		103.6	103.6		95.8	95.8		110.9	110.9		94.4	94.4		98.6	98.6
0241, 31 - 34	SITE & INFRASTRUCTURE, DEMOLITION	81.0	94.8	90.5	80.5	87.8	85.5	77.0	95.5	89.8	67.0	94.7	86.1	72.8	94.0	87.4	95.7	98.4	97.6
0310	Concrete Forming & Accessories	102.2	89.9	91.6	98.6	89.6	90.8	94.9	101.2	100.3	98.7	88.9	90.2	87.3	72.3	74.3	87.4	96.1	95.0
0320	Concrete Reinforcing	100.8	124.8	113.2	99.3	91.8	95.4	91.4	122.4	107.3	100.8	124.7	113.1	93.1	91.6	92.3	98.7	110.4	104.7
0330	Cast-in-Place Concrete	87.4	94.9	90.3	92.1	97.8	94.3	72.0	113.0	87.7	81.3	95.2	86.6	85.1	64.5	77.3	103.6	109.5	105.9
03	CONCRETE	93.1	99.0	95.9	93.4	92.2	92.8	77.9	108.4	92.5	89.8	98.6	94.0	83.1	73.9	78.7	99.1	104.4	101.6
04	MASONRY	90.5	95.2	93.3	92.5	77.3	83.3	91.9	113.5	105.0	98.3	89.6	93.0	90.3	84.0	86.5	103.6	107.6	106.0
05	METALS	91.7	117.1	100.4	90.6	86.1	89.1	96.1	99.1	97.2	93.2	116.8	101.3	93.3	89.3	91.9	92.5	123.9	103.3
06	WOOD, PLASTICS & COMPOSITES	103.0	88.4	94.6	97.0	90.4	93.2	95.8	97.8	96.9	96.1	88.3	91.6	87.9	71.0	78.2	73.2	93.0	84.6
07	THERMAL & MOISTURE PROTECTION	99.2	94.1	97.2	93.5	78.9	87.7	96.7	114.6	103.8	98.8	97.9	98.5	95.2	72.5	86.2	107.4	99.7	104.3
08	OPENINGS	92.1	96.0	93.1	91.3	89.0	90.7	92.2	102.5	94.8	90.0	95.9	91.5	99.2	66.2	90.9	96.6	113.3	100.8
0920	Plaster & Gypsum Board	91.9	87.0	88.5	76.9	86.1	83.4	95.9	96.7	96.4	99.1	86.9	90.5	59.5	70.6	67.3	100.6	93.1	95.3
0950, 0980	Ceilings & Acoustic Treatment	94.3	87.0	89.4	92.2	86.1	88.2	93.5	96.7	95.6	94.3	86.9	89.4	89.2	70.6	76.8	132.5	93.1	106.3
0960	Flooring	99.1	83.9	94.8	97.4	65.9	88.3	87.5	117.4	96.1	89.7	81.2	87.3	112.2	50.9	94.6	112.6	127.6	116.9
0970, 0990	Wall Finishes & Painting/Coating	103.9	85.9	93.0	96.9	83.1	88.5	94.5	104.2	100.4	93.0	85.1	88.2	114.3	46.5	73.2	98.4	102.4	100.8
09	FINISHES	96.0	87.7	91.4	93.8	84.5	88.6	91.6	103.9	98.4	92.2	86.4	89.0	98.6	64.8	80.0	111.4	102.1	106.3
COVERS	DIVS. 10 - 14, 25, 28, 41, 43, 44, 46	100.0	101.1	100.2	100.0	106.4	101.3	100.0	104.5	100.9	100.0	100.2	100.0	100.0	93.6	98.7	100.0	99.4	99.9
21, 22, 23	FIRE SUPPRESSION, PLUMBING & HVAC	99.9	93.8	97.3	99.8	85.0	93.6	95.0	104.9	99.1	100.0	87.2	94.7	94.8	82.1	89.5	94.9	90.1	92.9
26, 27, 3370	ELECTRICAL, COMMUNICATIONS & UTIL.	99.5	95.0	97.1	95.0	78.8	86.4	99.3	96.1	97.6	94.5	92.1	93.2	94.1	78.3	85.7	102.4	101.4	101.9
MF2010	WEIGHTED AVERAGE	95.7	96.5	96.1	94.7	85.3	90.5	92.8	103.6	97.6	94.5	94.0	94.3	93.4	79.4	87.2	98.5	102.2	100.2

MINNESOTA

DIVISION		BRAINERD 564			DETROIT LAKES 565			DULUTH 556-558			MANKATO 560			MINNEAPOLIS 553-555			ROCHESTER 559		
		MAT.	INST.	TOTAL	MAT.	INST.	TOTAL	MAT.	INST.	TOTAL	MAT.	INST.	TOTAL	MAT.	INST.	TOTAL	MAT.	INST.	TOTAL
015433	CONTRACTOR EQUIPMENT		101.5	101.5		98.6	98.6		101.0	101.0		101.5	101.5		104.7	104.7		101.0	101.0
0241, 31 - 34	SITE & INFRASTRUCTURE, DEMOLITION	93.8	103.5	100.5	93.7	98.9	97.3	96.5	101.9	100.2	90.1	104.2	99.8	99.1	107.5	104.9	98.0	101.8	100.6
0310	Concrete Forming & Accessories	90.7	99.1	98.0	84.8	99.0	97.1	100.6	118.1	115.8	97.4	107.0	105.7	102.6	132.1	128.2	103.3	113.0	111.7
0320	Concrete Reinforcing	97.6	110.6	104.3	98.7	110.4	104.7	93.1	110.8	102.2	97.4	121.2	109.7	93.3	121.6	107.9	93.1	121.3	107.6
0330	Cast-in-Place Concrete	112.6	114.6	113.4	100.6	113.9	105.7	105.0	109.4	106.7	103.7	113.7	107.5	117.5	122.1	119.3	115.2	105.7	111.6
03	CONCRETE	103.4	107.5	105.4	96.5	107.2	101.6	102.0	114.3	107.9	98.5	112.8	105.4	110.3	127.0	118.3	107.3	112.9	110.0
04	MASONRY	127.8	120.2	123.2	127.6	115.4	120.2	116.3	119.4	118.2	115.4	117.0	116.4	115.7	131.5	125.3	114.9	119.6	117.7
05	METALS	93.6	124.2	104.1	92.4	123.8	103.2	101.0	126.1	109.6	93.4	130.3	106.1	103.9	133.7	114.2	100.9	131.9	111.6
06	WOOD, PLASTICS & COMPOSITES	94.4	92.8	93.5	70.7	93.0	83.5	109.8	119.0	115.1	102.4	104.9	103.8	113.0	130.6	123.1	113.0	111.9	112.3
07	THERMAL & MOISTURE PROTECTION	107.0	113.6	109.6	107.2	113.9	109.9	107.9	119.4	112.5	107.3	104.8	106.3	107.7	133.6	118.0	108.0	106.5	107.4
08	OPENINGS	84.8	113.1	92.0	96.6	113.3	100.8	93.7	122.6	101.1	89.5	123.7	98.2	96.8	137.6	107.1	93.7	127.4	102.3
0920	Plaster & Gypsum Board	85.4	93.1	90.8	99.7	93.1	95.0	92.3	120.0	111.8	89.4	105.6	100.8	100.0	131.9	122.4	99.4	112.6	108.7
0950, 0980	Ceilings & Acoustic Treatment	65.5	93.1	83.8	132.5	93.1	106.3	88.3	120.0	109.3	65.5	105.6	92.0	90.2	131.9	117.8	88.3	112.6	104.5
0960	Flooring	111.4	127.6	116.0	111.2	127.6	115.9	108.3	129.0	114.2	113.5	127.6	117.6	106.1	127.6	112.3	108.5	127.6	114.0
0970, 0990	Wall Finishes & Painting/Coating	92.7	102.4	98.6	98.4	102.4	100.8	89.0	110.0	101.7	104.0	105.4	104.8	97.7	130.8	117.8	92.3	105.4	100.2
09	FINISHES	94.6	103.9	99.7	110.7	104.0	107.0	96.2	119.9	109.3	96.2	110.6	104.1	97.6	131.6	116.4	97.4	115.1	107.2
COVERS	DIVS. 10 - 14, 25, 28, 41, 43, 44, 46	100.0	101.7	100.4	100.0	102.1	100.4	100.0	101.0	100.2	100.0	101.9	100.4	100.0	111.1	102.3	100.0	104.9	101.0
21, 22, 23	FIRE SUPPRESSION, PLUMBING & HVAC	94.3	94.1	94.2	94.9	93.8	94.5	100.0	99.7	99.9	94.3	92.9	93.7	99.9	124.0	109.9	99.9	100.3	100.1
26, 27, 3370	ELECTRICAL, COMMUNICATIONS & UTIL.	99.8	111.2	105.9	102.2	70.7	85.5	100.0	111.2	106.0	106.9	91.0	98.4	98.9	115.6	107.8	98.0	91.0	94.2
MF2010	WEIGHTED AVERAGE	97.6	107.3	101.9	99.3	100.7	99.9	100.6	112.2	105.8	97.6	106.4	101.5	102.3	125.0	112.5	101.0	109.2	104.7

		MINNESOTA																MISSISSIPPI		
DIVISION		SAINT PAUL			ST. CLOUD			THIEF RIVER FALLS			WILLMAR			WINDOM			BILOXI		
		550 - 551			563			567			562			561			395		
		MAT.	INST.	TOTAL	MAT.	INST.	TOTAL	MAT.	INST.	TOTAL	MAT.	INST.	TOTAL	MAT.	INST.	TOTAL	MAT.	INST.	TOTAL
015433	CONTRACTOR EQUIPMENT		101.0	101.0		101.5	101.5		98.6	98.6		101.5	101.5		101.5	101.5		98.3	98.3
0241, 31 - 34	SITE & INFRASTRUCTURE, DEMOLITION	98.8	102.6	101.5	88.1	105.5	100.1	94.4	98.4	97.1	88.0	103.2	98.5	81.3	101.9	95.5	100.9	88.5	92.4
0310	Concrete Forming & Accessories	97.4	131.8	127.3	88.3	127.6	122.4	87.9	95.5	94.5	88.1	98.3	97.0	91.9	89.8	90.1	95.5	57.3	62.3
0320	Concrete Reinforcing	89.9	121.6	106.2	97.6	121.4	109.9	99.1	110.3	104.9	97.2	121.1	109.5	97.2	120.1	109.0	91.8	60.6	75.7
0330	Cast-in-Place Concrete	117.5	121.5	119.0	99.3	121.1	107.7	102.7	93.6	99.2	100.9	91.4	97.3	87.3	88.3	87.7	106.2	56.1	87.0
03	CONCRETE	112.0	126.7	119.0	94.3	124.6	108.8	97.7	98.7	98.2	94.3	101.3	97.7	84.8	89.5	87.0	95.3	59.1	78.0
04	MASONRY	125.3	131.5	129.1	112.2	129.3	122.6	103.6	107.6	106.0	116.7	120.2	118.8	127.2	92.1	105.9	89.1	46.7	63.4
05	METALS	100.4	133.4	111.7	94.2	131.6	107.1	92.6	123.2	103.1	93.3	129.1	105.7	93.3	126.6	104.7	92.7	84.0	89.7
06	WOOD, PLASTICS & COMPOSITES	106.1	130.5	120.1	91.9	125.0	110.9	74.1	93.0	84.9	91.7	93.0	92.4	95.5	90.1	92.4	99.8	60.9	77.5
07	THERMAL & MOISTURE PROTECTION	109.2	133.4	118.8	107.1	124.9	114.2	108.3	102.6	106.0	107.0	112.0	109.0	106.9	87.2	99.1	91.6	58.0	78.2
08	OPENINGS	91.3	137.5	103.0	89.5	134.6	100.9	96.6	113.3	100.8	87.1	98.0	89.9	91.3	96.4	92.6	97.9	61.8	88.8
0920	Plaster & Gypsum Board	89.6	131.9	119.3	85.0	126.3	114.1	100.1	93.1	95.2	85.0	93.3	90.9	85.4	90.3	88.8	98.6	60.2	71.5
0950, 0980	Ceilings & Acoustic Treatment	86.5	131.9	116.6	65.5	126.3	105.8	132.5	93.1	106.3	65.5	93.3	83.9	65.5	90.3	81.9	96.5	60.2	72.4
0960	Flooring	102.4	127.6	109.7	107.2	127.6	113.0	112.2	127.6	116.6	109.2	127.6	114.5	112.0	127.6	116.5	119.4	55.8	101.1
0970, 0990	Wall Finishes & Painting/Coating	97.7	128.2	116.2	104.0	130.8	120.3	98.4	102.4	100.8	98.4	102.4	100.8	98.4	105.4	102.6	105.9	46.8	70.0
09	FINISHES	94.5	131.2	114.8	93.6	128.3	112.7	111.2	102.1	106.2	93.8	101.5	98.0	94.1	96.7	95.5	105.4	56.1	78.2
COVERS	DIVS. 10 - 14, 25, 28, 41, 43, 44, 46	100.0	110.8	102.2	100.0	108.7	101.8	100.0	99.3	99.9	100.0	101.7	100.3	100.0	96.5	99.3	100.0	62.0	92.1
21, 22, 23	FIRE SUPPRESSION, PLUMBING & HVAC	99.9	116.8	106.9	99.7	123.9	109.8	94.9	89.7	92.8	94.3	112.6	101.9	94.3	87.3	91.4	99.8	51.3	79.6
26, 27, 3370	ELECTRICAL, COMMUNICATIONS & UTIL.	98.5	115.6	107.6	99.8	115.6	108.2	99.3	70.7	84.1	99.8	92.5	95.9	106.9	90.9	98.4	98.7	59.4	77.8
MF2010	WEIGHTED AVERAGE	101.6	123.0	111.2	97.3	123.2	108.9	98.1	97.1	97.7	95.9	107.2	100.9	96.2	95.3	95.8	97.6	60.8	81.2

		MISSISSIPPI																		
DIVISION		CLARKSDALE			COLUMBUS			GREENVILLE			GREENWOOD			JACKSON			LAUREL		
		386			397			387			389			390 - 392			394		
		MAT.	INST.	TOTAL	MAT.	INST.	TOTAL	MAT.	INST.	TOTAL	MAT.	INST.	TOTAL	MAT.	INST.	TOTAL	MAT.	INST.	TOTAL
015433	CONTRACTOR EQUIPMENT		98.3	98.3		98.3	98.3		98.3	98.3		98.3	98.3		98.3	98.3		98.3	98.3
0241, 31 - 34	SITE & INFRASTRUCTURE, DEMOLITION	98.6	88.3	91.5	99.7	88.6	92.1	104.6	88.6	93.6	101.6	88.0	92.3	96.2	88.6	90.9	106.0	88.4	93.9
0310	Concrete Forming & Accessories	86.4	53.4	57.8	84.5	55.6	59.4	83.0	69.2	71.0	95.4	53.4	59.0	94.8	62.3	66.6	84.5	57.4	61.0
0320	Concrete Reinforcing	98.9	44.8	71.0	98.3	45.9	71.3	99.5	65.0	81.7	98.9	56.6	77.1	91.8	58.6	74.7	98.9	42.1	69.7
0330	Cast-in-Place Concrete	103.6	57.0	85.7	108.3	62.5	90.8	106.7	59.3	88.5	111.4	56.6	90.4	90.4	60.1	78.8	105.9	58.8	87.9
03	CONCRETE	93.5	54.7	74.9	96.9	57.7	78.2	98.2	66.3	82.9	99.6	56.6	79.0	87.3	62.3	75.3	98.7	56.6	78.6
04	MASONRY	88.9	48.6	64.5	111.8	55.8	77.8	130.4	52.2	83.0	89.5	48.5	64.6	89.8	52.2	67.0	108.3	52.0	74.2
05	METALS	89.8	76.4	85.1	89.6	77.3	85.4	90.8	86.1	89.2	89.8	77.9	85.7	92.6	83.4	89.5	89.7	76.2	85.1
06	WOOD, PLASTICS & COMPOSITES	87.5	55.6	69.2	85.5	56.6	68.9	83.9	74.0	78.2	99.8	55.6	74.4	98.4	64.7	79.1	86.3	59.2	70.8
07	THERMAL & MOISTURE PROTECTION	91.4	50.6	75.2	91.5	55.2	77.1	91.7	64.2	80.8	91.7	50.5	75.4	93.4	63.2	81.4	91.7	57.0	77.9
08	OPENINGS	97.2	53.4	86.1	97.2	53.5	86.1	97.2	66.4	89.4	97.2	54.8	86.5	98.3	59.6	88.5	93.8	54.4	83.8
0920	Plaster & Gypsum Board	87.1	54.7	64.3	86.6	55.8	64.9	85.8	73.7	77.3	97.3	54.7	67.3	92.7	64.1	72.6	85.8	58.5	66.6
0950, 0980	Ceilings & Acoustic Treatment	89.7	54.7	66.5	89.7	55.8	67.2	91.8	73.7	79.8	89.7	54.7	66.5	93.7	64.1	74.1	89.7	58.5	69.0
0960	Flooring	113.0	55.8	96.6	111.8	62.6	97.7	110.9	55.8	95.1	119.4	55.8	101.1	119.8	55.8	101.4	109.9	55.8	94.4
0970, 0990	Wall Finishes & Painting/Coating	105.9	49.6	71.8	105.9	57.0	76.3	105.9	70.9	84.7	105.9	49.6	71.8	105.9	70.9	84.7	105.9	65.6	81.4
09	FINISHES	100.5	53.8	74.7	100.3	57.4	76.6	100.8	67.8	82.6	104.0	53.8	76.3	104.4	62.4	81.2	100.2	58.6	77.2
COVERS	DIVS. 10 - 14, 25, 28, 41, 43, 44, 46	100.0	62.4	92.2	100.0	63.6	92.4	100.0	65.5	92.8	100.0	62.3	92.2	100.0	64.4	92.6	100.0	41.6	87.9
21, 22, 23	FIRE SUPPRESSION, PLUMBING & HVAC	97.5	45.3	75.8	96.9	41.2	73.7	99.8	52.2	80.0	97.5	43.2	74.9	99.8	57.8	82.3	96.9	50.3	77.5
26, 27, 3370	ELECTRICAL, COMMUNICATIONS & UTIL.	98.0	50.3	72.6	95.9	53.3	73.3	98.0	68.0	82.1	98.0	47.2	70.9	101.9	68.0	83.9	97.6	58.9	77.0
MF2010	WEIGHTED AVERAGE	95.7	56.3	78.1	97.0	57.7	79.4	99.3	66.1	84.5	97.0	55.8	78.6	96.9	65.3	82.8	97.0	59.3	80.2

		MISSISSIPPI									MISSOURI									
DIVISION		MCCOMB			MERIDIAN			TUPELO			BOWLING GREEN			CAPE GIRARDEAU			CHILLICOTHE		
		396			393			388			633			637			646		
		MAT.	INST.	TOTAL	MAT.	INST.	TOTAL	MAT.	INST.	TOTAL	MAT.	INST.	TOTAL	MAT.	INST.	TOTAL	MAT.	INST.	TOTAL
015433	CONTRACTOR EQUIPMENT		98.3	98.3		98.3	98.3		98.3	98.3		106.5	106.5		106.5	106.5		101.6	101.6
0241, 31 - 34	SITE & INFRASTRUCTURE, DEMOLITION	92.3	88.2	89.5	96.2	88.9	91.1	96.1	88.2	90.7	92.4	93.4	93.1	93.6	93.4	93.4	98.8	90.4	93.0
0310	Concrete Forming & Accessories	84.5	55.3	59.2	81.9	68.6	70.4	83.5	55.6	59.3	96.2	89.0	90.0	89.7	76.6	78.3	91.2	91.9	91.8
0320	Concrete Reinforcing	99.5	44.2	71.0	98.3	58.7	77.9	96.8	56.7	76.1	103.7	98.4	101.0	105.0	84.6	94.5	97.9	101.7	99.9
0330	Cast-in-Place Concrete	94.0	56.4	79.6	100.6	62.7	86.1	103.6	58.5	86.3	94.0	77.2	87.6	93.0	84.3	89.7	94.2	66.7	83.7
03	CONCRETE	87.1	55.1	71.8	91.2	66.0	79.2	93.0	58.2	76.3	94.5	87.9	91.4	93.7	82.3	88.2	99.7	85.5	92.9
04	MASONRY	113.3	47.9	73.7	88.7	56.5	69.2	119.9	51.7	78.6	126.0	98.0	109.1	123.1	79.0	96.4	99.7	92.8	95.5
05	METALS	89.9	74.5	84.6	90.8	83.6	88.3	89.7	78.2	85.7	96.5	112.5	102.0	97.6	106.0	100.5	92.0	97.2	93.8
06	WOOD, PLASTICS & COMPOSITES	85.5	58.6	70.1	82.8	70.7	75.9	84.5	56.6	68.5	93.5	89.3	91.1	86.5	73.8	79.2	85.7	98.1	92.9
07	THERMAL & MOISTURE PROTECTION	91.2	49.7	74.7	91.3	65.7	81.1	91.4	56.8	77.7	99.0	100.4	99.6	98.9	82.5	92.4	97.6	86.2	93.1
08	OPENINGS	97.2	55.2	86.6	97.2	68.9	90.0	97.2	54.7	86.4	99.6	84.8	95.8	99.5	75.8	93.5	99.6	86.2	88.7
0920	Plaster & Gypsum Board	86.6	57.8	66.4	85.8	70.3	74.9	85.8	55.8	64.7	94.8	89.0	90.7	92.1	73.0	78.7	98.0	97.9	97.9
0950, 0980	Ceilings & Acoustic Treatment	89.7	57.8	68.6	91.8	70.3	77.6	89.7	55.8	67.2	85.9	89.0	87.9	85.9	73.0	77.3	86.8	97.9	94.2
0960	Flooring	111.8	55.8	95.7	109.9	55.8	94.3	111.3	55.8	95.3	97.2	96.6	97.0	93.6	88.8	92.2	94.7	45.4	80.6
0970, 0990	Wall Finishes & Painting/Coating	105.9	45.5	69.3	105.9	82.0	91.4	105.9	63.2	80.0	100.9	105.6	103.8	100.9	74.3	84.8	95.1	48.3	66.7
09	FINISHES	99.7	54.9	75.0	99.7	68.3	82.4	99.8	56.7	76.0	95.8	91.1	93.2	94.4	76.7	84.6	96.0	81.0	87.7
COVERS	DIVS. 10 - 14, 25, 28, 41, 43, 44, 46	100.0	66.2	93.0	100.0	66.6	93.1	100.0	63.6	92.4	100.0	92.9	98.5	100.0	93.9	98.7	100.0	57.5	91.2
21, 22, 23	FIRE SUPPRESSION, PLUMBING & HVAC	96.9	37.9	72.3	99.8	60.1	83.3	97.8	47.5	76.9	94.5	88.7	92.1	99.9	97.1	98.7	94.9	45.0	74.1
26, 27, 3370	ELECTRICAL, COMMUNICATIONS & UTIL.	94.3	53.7	72.7	97.6	60.5	77.9	97.7	53.5	74.2	99.2	84.7	91.5	99.2	110.2	105.0	92.5	35.7	62.3
MF2010	WEIGHTED AVERAGE	95.4	55.3	77.5	95.9	67.0	83.0	97.1	58.8	80.0	98.1	92.2	95.5	99.2	91.3	95.7	95.2	71.1	84.4

659

City Cost Indexes

MISSOURI

DIVISION		COLUMBIA 652 MAT.	INST.	TOTAL	FLAT RIVER 636 MAT.	INST.	TOTAL	HANNIBAL 634 MAT.	INST.	TOTAL	HARRISONVILLE 647 MAT.	INST.	TOTAL	JEFFERSON CITY 650-651 MAT.	INST.	TOTAL	JOPLIN 648 MAT.	INST.	TOTAL
015433	CONTRACTOR EQUIPMENT		107.4	107.4		106.5	106.5		106.5	106.5		101.6	101.6		107.4	107.4		105.0	105.0
0241, 31-34	SITE & INFRASTRUCTURE, DEMOLITION	107.4	95.7	99.3	95.2	93.5	94.0	90.0	93.2	92.2	90.6	95.5	94.0	103.5	95.6	98.0	98.0	97.8	97.8
0310	Concrete Forming & Accessories	89.4	81.1	82.2	101.9	89.0	90.7	94.6	79.0	81.0	88.4	106.1	103.7	99.3	81.0	83.4	102.6	75.0	78.7
0320	Concrete Reinforcing	100.3	113.4	107.0	105.0	107.4	106.2	103.1	98.3	100.6	97.5	113.2	105.6	99.0	108.4	103.9	101.0	73.4	86.8
0330	Cast-in-Place Concrete	91.2	84.3	88.6	97.1	96.5	96.9	89.0	76.7	84.3	96.6	103.4	99.2	91.8	84.3	88.6	102.1	67.0	88.6
03	CONCRETE	86.3	89.7	88.0	97.5	96.3	96.9	90.7	83.3	87.2	96.2	106.7	101.2	87.8	88.7	88.2	99.5	72.9	86.7
04	MASONRY	133.3	87.4	105.5	123.6	80.5	97.5	117.1	98.0	105.5	94.5	104.6	100.6	96.6	87.4	91.0	92.6	86.6	89.0
05	METALS	92.9	119.5	102.1	96.4	115.7	103.1	96.5	111.3	101.6	92.4	112.7	99.4	92.3	117.3	100.9	95.1	88.8	92.9
06	WOOD, PLASTICS & COMPOSITES	90.7	78.8	83.9	100.6	89.3	94.2	91.3	76.8	83.0	82.3	106.0	95.9	101.2	78.8	88.3	96.5	76.1	84.8
07	THERMAL & MOISTURE PROTECTION	93.8	87.6	91.4	99.2	95.7	97.8	98.9	94.1	97.0	96.9	106.5	100.7	93.9	87.6	91.4	97.3	73.2	87.7
08	OPENINGS	91.3	95.3	92.3	99.5	87.5	96.5	99.6	78.0	94.1	89.9	108.2	94.5	88.6	94.0	89.9	90.8	73.8	86.5
0920	Plaster & Gypsum Board	93.9	78.1	82.8	100.1	89.0	92.3	92.6	76.0	80.9	91.2	106.0	101.7	98.9	78.1	84.3	105.1	75.2	84.0
0950, 0980	Ceilings & Acoustic Treatment	89.5	78.1	81.9	85.9	89.0	87.9	85.9	76.0	79.3	86.8	106.0	99.6	88.5	78.1	81.6	88.7	75.2	79.7
0960	Flooring	109.4	96.7	105.8	100.4	88.8	97.0	96.3	96.6	96.4	90.3	104.9	94.5	117.0	96.7	111.2	119.9	42.6	97.7
0970, 0990	Wall Finishes & Painting/Coating	104.0	81.1	90.1	100.9	63.9	78.5	100.9	105.6	103.8	99.8	107.9	104.7	104.0	81.1	90.1	94.6	37.8	60.1
09	FINISHES	96.9	82.6	89.0	97.6	85.1	90.7	95.1	83.7	88.8	93.6	106.0	100.4	99.8	82.6	90.3	102.8	65.6	82.3
COVERS	DIVS. 10-14, 25, 28, 41, 43, 44, 46	100.0	95.6	99.1	100.0	96.2	99.2	100.0	56.6	91.0	100.0	99.9	100.0	100.0	95.6	99.1	100.0	64.1	92.5
21, 22, 23	FIRE SUPPRESSION, PLUMBING & HVAC	100.0	97.4	98.9	94.5	88.7	92.1	94.5	88.7	92.0	94.8	103.5	98.5	100.0	91.6	96.5	100.3	49.1	79.0
26, 27, 3370	ELECTRICAL, COMMUNICATIONS & UTIL.	94.3	89.9	91.9	104.2	110.2	107.4	97.8	84.7	90.8	99.7	103.8	101.9	99.1	89.9	94.2	90.2	70.5	79.8
MF2010	WEIGHTED AVERAGE	97.4	93.7	95.7	99.1	94.7	97.1	96.9	88.8	93.3	94.8	104.8	99.3	95.9	92.1	94.2	97.1	71.4	85.6

MISSOURI

DIVISION		KANSAS CITY 640-641 MAT.	INST.	TOTAL	KIRKSVILLE 635 MAT.	INST.	TOTAL	POPLAR BLUFF 639 MAT.	INST.	TOTAL	ROLLA 654-655 MAT.	INST.	TOTAL	SEDALIA 653 MAT.	INST.	TOTAL	SIKESTON 638 MAT.	INST.	TOTAL
015433	CONTRACTOR EQUIPMENT		102.8	102.8		98.9	98.9		101.1	101.1		107.4	107.4		99.5	99.5		101.1	101.1
0241, 31-34	SITE & INFRASTRUCTURE, DEMOLITION	92.0	97.7	96.0	91.6	89.8	90.4	78.1	93.0	88.4	107.0	93.8	97.9	98.3	92.1	94.0	81.6	93.2	89.6
0310	Concrete Forming & Accessories	101.3	106.9	106.2	87.2	79.6	80.6	87.6	74.1	75.9	96.4	92.4	92.9	93.6	81.5	83.1	88.5	75.2	76.9
0320	Concrete Reinforcing	96.0	114.1	105.3	103.9	87.3	95.4	107.1	84.6	95.5	100.8	68.9	84.4	99.3	106.4	102.9	106.4	84.5	95.1
0330	Cast-in-Place Concrete	94.9	107.8	99.8	97.0	76.6	89.2	74.6	81.5	77.3	93.3	87.5	91.1	97.5	85.1	92.7	79.8	83.9	81.3
03	CONCRETE	95.5	108.8	101.8	107.6	80.9	94.9	84.0	79.7	81.9	88.1	87.6	87.9	99.7	88.3	94.2	87.6	80.9	84.4
04	MASONRY	96.5	108.4	103.7	129.6	89.3	105.2	121.5	73.1	92.2	110.0	80.5	92.1	116.1	87.8	99.0	121.2	77.9	95.0
05	METALS	102.7	113.7	106.5	96.1	97.6	96.6	96.7	97.7	97.0	92.2	103.6	96.2	90.9	108.2	96.8	97.0	96.9	97.0
06	WOOD, PLASTICS & COMPOSITES	95.8	106.4	101.9	79.6	78.8	79.1	78.8	73.8	76.0	97.8	95.2	96.3	90.8	78.8	83.9	80.2	73.8	76.6
07	THERMAL & MOISTURE PROTECTION	97.0	109.5	101.9	104.9	87.3	97.9	103.6	82.3	95.1	94.0	92.8	93.5	99.2	94.8	97.4	103.8	80.7	94.6
08	OPENINGS	96.9	108.8	99.9	103.9	76.5	96.9	104.9	75.1	97.3	91.3	89.1	90.7	94.7	88.9	93.2	104.9	68.5	95.7
0920	Plaster & Gypsum Board	100.9	106.5	104.8	87.2	78.1	80.8	87.7	73.0	77.3	95.6	95.0	95.2	88.5	78.1	81.2	89.5	73.0	77.9
0950, 0980	Ceilings & Acoustic Treatment	94.3	106.5	102.4	84.0	78.1	80.1	85.9	73.0	77.3	89.5	95.0	93.2	89.5	78.1	81.9	85.9	73.0	77.3
0960	Flooring	95.9	108.8	99.6	74.8	96.7	81.1	88.7	82.2	86.8	113.3	96.7	108.6	86.8	97.9	90.0	89.1	82.2	87.1
0970, 0990	Wall Finishes & Painting/Coating	99.8	115.5	109.4	96.3	81.1	87.0	95.8	73.7	82.4	104.0	88.2	94.4	104.0	107.9	106.4	95.8	58.9	73.4
09	FINISHES	97.9	108.0	103.5	94.3	81.7	87.3	94.4	74.1	83.2	98.3	92.2	94.9	95.2	86.1	90.2	95.0	73.0	82.8
COVERS	DIVS. 10-14, 25, 28, 41, 43, 44, 46	100.0	101.3	100.3	100.0	54.9	90.6	100.0	91.8	98.3	100.0	59.1	91.5	100.0	93.1	98.6	100.0	93.5	98.7
21, 22, 23	FIRE SUPPRESSION, PLUMBING & HVAC	100.2	108.7	103.7	94.5	88.1	91.8	94.5	93.5	94.1	94.6	89.0	92.3	94.6	98.2	96.1	94.5	86.4	91.1
26, 27, 3370	ELECTRICAL, COMMUNICATIONS & UTIL.	101.3	104.2	102.8	98.0	84.6	90.9	98.3	110.2	104.6	92.6	89.8	91.1	94.0	103.8	99.2	97.3	110.2	104.1
MF2010	WEIGHTED AVERAGE	99.0	107.3	102.7	100.1	85.4	93.5	96.6	88.4	92.9	95.0	89.5	92.5	96.5	94.6	95.7	97.1	87.1	92.6

MISSOURI / MONTANA

DIVISION		SPRINGFIELD 656-658 MAT.	INST.	TOTAL	ST. JOSEPH 644-645 MAT.	INST.	TOTAL	ST. LOUIS 630-631 MAT.	INST.	TOTAL	BILLINGS 590-591 MAT.	INST.	TOTAL	BUTTE 597 MAT.	INST.	TOTAL	GREAT FALLS 594 MAT.	INST.	TOTAL
015433	CONTRACTOR EQUIPMENT		101.9	101.9		101.6	101.6		107.6	107.6		98.9	98.9		98.6	98.6		98.6	98.6
0241, 31-34	SITE & INFRASTRUCTURE, DEMOLITION	100.1	93.7	95.6	93.5	93.7	93.6	93.4	97.2	96.1	95.9	96.8	96.5	106.4	96.1	99.3	110.4	96.8	101.0
0310	Concrete Forming & Accessories	101.7	75.7	79.1	101.2	89.4	91.0	101.9	102.7	102.6	98.0	67.3	71.3	87.2	65.3	68.2	97.8	68.8	72.6
0320	Concrete Reinforcing	96.5	112.9	104.9	94.8	112.8	104.1	96.0	109.2	102.8	96.6	81.5	88.8	104.9	81.5	92.8	96.7	81.5	88.8
0330	Cast-in-Place Concrete	99.1	73.9	89.4	94.8	104.5	98.5	93.0	108.3	98.9	113.0	67.6	95.6	124.3	69.3	103.3	131.4	65.6	106.2
03	CONCRETE	96.8	83.0	90.2	95.3	99.7	97.4	93.1	106.7	99.6	106.0	71.0	89.3	109.7	70.8	91.1	115.0	71.1	94.0
04	MASONRY	90.0	86.3	87.8	96.2	96.0	96.1	103.7	110.5	107.8	130.4	71.4	94.7	125.4	73.2	93.8	130.1	75.6	97.0
05	METALS	96.9	106.0	100.0	98.9	111.6	103.3	102.5	119.6	108.4	110.0	90.8	103.4	102.9	90.8	98.7	106.4	90.8	101.0
06	WOOD, PLASTICS & COMPOSITES	98.1	74.9	84.8	96.5	86.9	91.0	100.5	100.7	100.6	98.0	67.6	80.6	87.4	64.9	74.5	99.1	67.6	81.0
07	THERMAL & MOISTURE PROTECTION	98.0	81.6	91.5	97.4	97.0	97.3	99.0	107.1	102.2	106.3	70.2	92.0	106.2	70.1	91.9	106.7	69.0	91.7
08	OPENINGS	96.9	82.1	93.2	95.5	98.0	96.1	98.9	108.3	101.3	95.5	67.6	88.4	93.5	65.5	86.4	96.6	66.9	89.3
0920	Plaster & Gypsum Board	97.0	74.0	80.8	106.4	86.4	92.3	101.0	100.7	100.8	94.6	67.0	75.1	95.3	64.2	73.4	103.3	67.0	77.7
0950, 0980	Ceilings & Acoustic Treatment	89.5	74.0	79.2	93.4	86.4	88.7	90.5	100.7	97.3	102.1	67.0	78.8	106.5	64.2	78.5	108.4	67.0	80.9
0960	Flooring	111.2	42.6	91.5	99.1	97.9	98.7	100.1	100.1	100.1	111.9	74.2	101.1	109.1	78.4	100.3	117.8	78.4	106.4
0970, 0990	Wall Finishes & Painting/Coating	98.1	54.0	71.4	95.1	90.7	92.4	100.9	109.0	105.8	100.0	50.4	69.9	98.4	55.2	72.2	98.4	52.4	70.5
09	FINISHES	100.7	66.6	81.9	99.1	90.4	94.3	98.5	102.2	100.5	104.3	66.3	83.3	104.1	66.2	83.2	108.4	68.3	86.3
COVERS	DIVS. 10-14, 25, 28, 41, 43, 44, 46	100.0	93.7	98.7	100.0	97.1	99.4	100.0	102.2	100.5	100.0	77.7	95.4	100.0	77.5	95.3	100.0	79.2	95.7
21, 22, 23	FIRE SUPPRESSION, PLUMBING & HVAC	100.1	72.3	88.5	100.3	89.2	95.7	99.9	103.3	101.3	100.1	73.1	88.8	100.2	70.0	87.6	100.2	75.2	89.8
26, 27, 3370	ELECTRICAL, COMMUNICATIONS & UTIL.	98.7	74.0	85.5	99.7	78.1	88.2	101.9	110.6	106.5	94.7	78.3	86.0	102.7	74.1	87.5	94.8	75.1	84.3
MF2010	WEIGHTED AVERAGE	98.2	80.9	90.5	98.3	93.2	96.0	99.4	106.6	102.6	103.2	75.9	91.0	103.2	74.6	90.4	104.7	76.5	92.1

MONTANA

DIVISION		HAVRE 595			HELENA 596			KALISPELL 599			MILES CITY 593			MISSOULA 598			WOLF POINT 592		
		MAT.	INST.	TOTAL	MAT.	INST.	TOTAL	MAT.	INST.	TOTAL	MAT.	INST.	TOTAL	MAT.	INST.	TOTAL	MAT.	INST.	TOTAL
015433	CONTRACTOR EQUIPMENT		98.6	98.6		98.6	98.6		98.6	98.6		98.6	98.6		98.6	98.6		98.6	98.6
0241, 31 - 34	SITE & INFRASTRUCTURE, DEMOLITION	115.9	96.5	102.5	99.1	96.0	96.9	95.6	95.7	95.7	103.2	96.1	98.3	86.4	95.9	93.0	124.4	96.2	105.0
0310	Concrete Forming & Accessories	81.3	58.4	61.4	98.0	68.0	72.0	89.8	56.3	60.7	96.5	56.8	62.1	89.8	63.6	67.1	90.3	51.8	56.8
0320	Concrete Reinforcing	105.7	65.3	84.9	100.0	65.0	82.0	107.7	81.2	94.0	105.2	65.3	84.7	106.6	81.5	93.7	106.8	65.3	85.4
0330	Cast-in-Place Concrete	133.9	64.2	107.2	102.0	64.6	87.6	107.9	61.7	90.0	118.2	63.8	97.4	91.6	63.8	80.9	132.4	64.2	106.3
03	CONCRETE	118.4	62.8	91.8	100.7	67.2	84.6	98.9	64.1	82.2	106.3	62.0	85.1	86.8	68.1	77.8	122.3	59.9	92.4
04	MASONRY	126.4	73.6	94.4	121.2	70.0	90.2	124.3	68.3	90.4	132.3	68.7	93.7	150.9	71.8	102.9	133.5	69.4	94.7
05	METALS	99.1	82.6	93.5	105.5	82.0	97.4	99.0	89.2	95.6	98.3	82.5	92.9	99.6	89.9	96.2	98.4	82.5	92.9
06	WOOD, PLASTICS & COMPOSITES	80.0	55.2	65.8	99.1	67.6	81.1	90.3	53.5	69.2	96.6	55.7	73.1	90.3	63.7	75.1	89.4	48.5	65.9
07	THERMAL & MOISTURE PROTECTION	106.5	66.8	90.7	106.5	71.0	92.4	105.8	67.7	90.7	106.1	64.9	89.8	105.5	69.5	91.2	107.0	63.9	89.9
08	OPENINGS	93.6	55.6	83.9	96.3	61.2	87.4	93.5	53.3	84.9	93.0	55.8	83.6	93.5	64.6	86.2	93.0	51.7	82.6
0920	Plaster & Gypsum Board	90.4	54.2	64.9	95.4	67.0	75.4	95.7	52.4	65.2	102.3	54.6	68.7	95.7	62.9	72.6	97.4	47.2	62.1
0950, 0980	Ceilings & Acoustic Treatment	106.5	54.2	71.8	105.6	67.0	80.0	106.5	52.4	70.6	104.7	54.6	71.5	106.5	62.9	77.6	104.7	47.2	66.6
0960	Flooring	105.7	78.4	97.8	117.8	62.4	101.9	111.3	78.4	101.8	117.5	74.2	105.1	111.3	78.4	101.8	113.0	74.2	101.9
0970, 0990	Wall Finishes & Painting/Coating	98.4	52.4	70.5	98.4	46.6	66.9	98.4	40.3	63.1	98.4	48.3	68.0	98.4	51.2	69.8	98.4	52.4	70.5
09	FINISHES	103.3	60.5	79.7	106.3	64.9	83.5	104.2	57.7	78.6	106.9	58.4	80.1	103.5	64.7	82.1	106.5	54.7	77.9
COVERS	DIVS. 10 - 14, 25, 28, 41, 43, 44, 46	100.0	77.1	95.2	100.0	79.2	95.7	100.0	76.1	95.0	100.0	75.4	94.9	100.0	76.9	95.2	100.0	74.8	94.8
21, 22, 23	FIRE SUPPRESSION, PLUMBING & HVAC	94.8	73.8	86.0	100.1	71.5	88.2	94.8	69.7	84.3	94.8	71.2	85.0	100.2	69.1	87.2	94.8	71.6	85.1
26, 27, 3370	ELECTRICAL, COMMUNICATIONS & UTIL.	94.8	67.4	80.2	101.0	75.1	87.2	99.3	73.5	85.6	94.8	70.2	81.7	100.4	73.5	86.1	94.8	65.4	79.2
MF2010	WEIGHTED AVERAGE	101.9	71.3	88.2	102.5	73.2	89.4	99.4	71.3	86.8	100.6	70.2	87.0	100.5	73.4	88.4	103.2	68.6	87.7

NEBRASKA

DIVISION		ALLIANCE 693			COLUMBUS 686			GRAND ISLAND 688			HASTINGS 689			LINCOLN 683 - 685			MCCOOK 690		
		MAT.	INST.	TOTAL	MAT.	INST.	TOTAL	MAT.	INST.	TOTAL	MAT.	INST.	TOTAL	MAT.	INST.	TOTAL	MAT.	INST.	TOTAL
015433	CONTRACTOR EQUIPMENT		96.8	96.8		101.0	101.0		101.0	101.0		101.0	101.0		101.0	101.0		101.0	101.0
0241, 31 - 34	SITE & INFRASTRUCTURE, DEMOLITION	100.8	97.4	98.5	99.4	90.8	93.5	103.9	92.0	95.7	103.0	91.2	94.9	91.3	92.0	91.8	103.9	91.0	95.0
0310	Concrete Forming & Accessories	90.4	57.5	61.8	98.8	60.4	65.5	98.4	77.7	80.4	101.6	75.9	79.3	103.2	61.6	67.1	96.1	57.8	62.9
0320	Concrete Reinforcing	114.4	48.3	80.4	105.6	47.1	75.5	105.1	76.6	90.4	105.1	47.8	75.6	96.5	76.1	86.0	107.1	76.2	91.2
0330	Cast-in-Place Concrete	110.2	59.4	90.7	112.3	56.2	90.8	118.9	71.2	100.6	118.9	65.2	98.4	88.8	73.7	83.0	118.9	59.1	96.0
03	CONCRETE	118.3	56.9	88.9	104.6	57.5	82.1	109.3	75.9	93.3	109.5	61.6	89.5	91.7	69.5	81.1	109.4	62.8	87.1
04	MASONRY	110.8	68.3	85.1	115.9	73.2	90.0	108.9	80.8	91.8	117.8	90.5	101.3	97.2	71.2	81.4	106.0	68.4	83.2
05	METALS	98.5	65.2	87.0	91.1	74.9	85.5	92.9	87.0	90.9	93.4	76.5	87.6	97.9	86.1	93.8	93.7	84.9	90.6
06	WOOD, PLASTICS & COMPOSITES	84.3	56.6	68.4	95.5	61.6	76.0	94.7	78.5	85.4	98.4	78.5	87.0	97.6	57.0	74.3	92.7	56.7	72.1
07	THERMAL & MOISTURE PROTECTION	104.9	68.9	90.6	98.8	65.8	85.7	98.9	76.6	90.0	98.9	88.4	94.7	98.4	71.8	87.9	98.9	66.0	85.8
08	OPENINGS	90.9	52.7	81.2	90.6	57.3	82.2	90.7	74.2	86.5	90.7	66.5	84.5	96.1	58.4	86.6	90.7	59.9	82.9
0920	Plaster & Gypsum Board	86.4	55.3	64.5	95.2	60.4	70.7	94.3	77.8	82.7	96.5	77.8	83.3	96.4	55.6	67.7	94.3	55.3	66.8
0950, 0980	Ceilings & Acoustic Treatment	93.9	55.3	68.3	89.5	60.4	70.2	89.5	77.8	81.7	89.5	77.8	81.7	92.3	55.6	68.0	89.5	55.3	66.8
0960	Flooring	101.4	86.7	97.2	100.4	106.0	102.0	100.2	113.3	104.0	101.6	106.0	102.9	101.8	88.3	97.9	99.2	86.7	95.6
0970, 0990	Wall Finishes & Painting/Coating	172.0	59.0	103.4	96.2	68.6	79.4	96.2	72.8	82.4	96.2	68.6	79.4	96.2	83.8	88.7	96.2	53.8	70.6
09	FINISHES	100.6	62.2	79.4	97.6	68.6	81.6	97.7	82.9	89.6	98.4	80.1	88.3	98.5	68.0	81.7	97.4	61.3	77.5
COVERS	DIVS. 10 - 14, 25, 28, 41, 43, 44, 46	100.0	70.5	93.9	100.0	80.1	95.9	100.0	85.7	97.0	100.0	84.3	96.7	100.0	83.2	96.5	100.0	69.7	93.7
21, 22, 23	FIRE SUPPRESSION, PLUMBING & HVAC	94.8	60.8	80.7	94.7	74.0	86.1	100.1	80.1	91.7	94.7	77.2	87.4	99.9	80.0	91.7	94.7	74.5	86.3
26, 27, 3370	ELECTRICAL, COMMUNICATIONS & UTIL.	91.2	75.3	82.7	90.8	83.7	87.0	89.4	68.8	78.4	88.7	90.7	89.8	100.4	68.8	83.6	93.5	75.3	83.8
MF2010	WEIGHTED AVERAGE	99.6	66.8	85.0	96.7	72.9	86.1	98.5	79.8	90.1	97.7	80.7	90.1	97.7	74.9	87.5	97.5	71.9	86.0

NEBRASKA / NEVADA

DIVISION		NORFOLK 687			NORTH PLATTE 691			OMAHA 680 - 681			VALENTINE 692			CARSON CITY 897			ELKO 898		
		MAT.	INST.	TOTAL	MAT.	INST.	TOTAL	MAT.	INST.	TOTAL	MAT.	INST.	TOTAL	MAT.	INST.	TOTAL	MAT.	INST.	TOTAL
015433	CONTRACTOR EQUIPMENT		91.8	91.8		101.0	101.0		91.8	91.8		94.6	94.6		100.6	100.6		100.6	100.6
0241, 31 - 34	SITE & INFRASTRUCTURE, DEMOLITION	81.4	90.2	87.5	104.8	91.2	95.4	81.4	90.9	88.0	86.9	94.5	92.2	65.9	102.3	91.0	54.3	102.0	92.2
0310	Concrete Forming & Accessories	84.6	75.0	76.3	98.3	75.7	78.7	93.5	74.6	77.1	86.0	57.0	60.9	101.5	87.1	89.0	108.4	84.6	87.8
0320	Concrete Reinforcing	105.8	49.8	77.0	106.5	77.3	91.5	101.1	76.6	88.5	107.2	46.8	76.1	103.6	119.3	111.7	105.3	93.2	99.1
0330	Cast-in-Place Concrete	113.4	64.4	94.7	118.9	66.7	98.9	95.3	80.4	89.6	105.1	56.7	86.6	98.2	86.4	93.7	99.1	84.0	93.3
03	CONCRETE	103.5	66.9	86.0	109.5	73.5	92.3	94.3	77.2	86.1	105.9	55.7	81.9	102.3	93.1	97.9	100.0	86.2	93.4
04	MASONRY	122.7	71.0	91.4	93.1	90.5	91.6	102.7	81.6	89.9	105.8	68.4	83.1	116.9	75.8	92.0	119.3	68.5	88.5
05	METALS	94.3	68.1	85.3	93.1	86.9	91.0	97.2	79.1	91.0	104.3	66.3	91.2	94.7	104.2	98.0	93.7	91.4	92.9
06	WOOD, PLASTICS & COMPOSITES	79.1	78.0	78.4	94.6	78.5	85.4	89.2	74.1	80.5	79.0	56.2	65.9	90.2	87.0	88.3	99.6	86.3	92.0
07	THERMAL & MOISTURE PROTECTION	98.3	75.1	89.0	98.8	82.1	92.2	94.5	82.0	89.5	99.0	68.8	87.0	99.7	85.8	94.2	101.0	76.4	91.2
08	OPENINGS	92.3	67.0	85.9	89.9	73.5	85.8	99.0	73.9	92.6	92.4	53.9	82.7	95.2	106.4	98.0	98.0	83.9	94.4
0920	Plaster & Gypsum Board	94.7	77.8	82.8	94.3	77.8	82.7	99.3	73.8	81.4	96.7	55.3	67.5	94.0	86.4	88.7	98.9	85.8	89.6
0950, 0980	Ceilings & Acoustic Treatment	104.9	77.8	86.9	89.5	77.8	81.7	105.8	73.8	84.6	107.4	55.3	72.8	119.1	86.4	97.5	122.0	85.8	98.0
0960	Flooring	126.9	106.0	120.9	100.1	106.0	101.8	131.6	86.6	118.7	129.4	84.3	116.5	101.7	82.7	96.3	107.1	54.0	91.8
0970, 0990	Wall Finishes & Painting/Coating	170.4	68.6	108.6	96.2	68.6	79.4	170.4	70.4	109.7	170.4	71.1	110.2	99.5	85.0	90.7	98.5	85.0	90.3
09	FINISHES	118.9	80.1	97.5	97.7	80.1	88.0	121.1	76.1	96.3	120.9	62.6	88.8	103.2	85.3	93.3	104.6	79.1	90.5
COVERS	DIVS. 10 - 14, 25, 28, 41, 43, 44, 46	100.0	83.0	96.5	100.0	73.4	94.5	100.0	84.0	96.7	100.0	67.6	93.3	100.0	98.3	99.6	100.0	69.3	93.6
21, 22, 23	FIRE SUPPRESSION, PLUMBING & HVAC	94.5	77.1	87.2	100.1	75.4	89.8	99.9	79.9	91.5	94.4	58.8	79.6	99.9	79.9	91.5	97.4	76.5	88.7
26, 27, 3370	ELECTRICAL, COMMUNICATIONS & UTIL.	89.7	84.3	86.8	91.5	83.7	87.3	100.1	84.6	91.9	88.5	94.0	91.4	95.8	100.6	98.3	93.7	100.6	97.8
MF2010	WEIGHTED AVERAGE	98.4	76.4	88.6	97.8	80.9	90.3	99.8	80.6	91.2	99.5	68.7	85.7	98.6	90.8	95.1	97.7	84.4	91.8

City Cost Indexes

DIVISION		NEVADA								NEW HAMPSHIRE								
		ELY			LAS VEGAS			RENO			CHARLESTON			CLAREMONT			CONCORD		
		893			889 - 891			894 - 895			036			037			032 - 033		
		MAT.	INST.	TOTAL	MAT.	INST.	TOTAL	MAT.	INST.	TOTAL	MAT.	INST.	TOTAL	MAT.	INST.	TOTAL	MAT.	INST.	TOTAL
015433	CONTRACTOR EQUIPMENT		100.6	100.6		100.6	100.6		100.6	100.6		100.3	100.3		100.3	100.3		100.3	100.3
0241, 31 - 34	SITE & INFRASTRUCTURE, DEMOLITION	58.9	102.5	89.0	60.7	104.9	91.2	58.8	102.3	88.8	81.6	96.4	91.8	75.8	96.4	90.0	86.8	98.3	94.8
0310	Concrete Forming & Accessories	102.0	106.9	106.2	103.7	117.3	115.5	98.8	87.0	88.6	85.3	44.2	49.7	90.9	44.2	50.7	92.4	86.2	87.0
0320	Concrete Reinforcing	104.1	86.3	94.9	96.0	127.3	112.1	98.4	126.0	112.6	86.5	47.4	66.4	86.5	47.4	66.4	86.5	90.8	88.7
0330	Cast-in-Place Concrete	106.4	106.2	106.3	103.0	114.1	107.3	112.5	86.3	102.5	94.1	50.6	77.4	86.5	50.6	72.7	102.4	88.0	96.9
03	CONCRETE	108.5	102.5	105.6	103.7	117.6	110.3	108.5	94.3	101.7	103.9	47.9	77.1	95.2	47.9	72.5	106.3	87.6	97.3
04	MASONRY	124.4	80.8	98.0	111.7	109.1	110.1	118.4	75.8	92.6	94.0	38.6	60.4	93.5	38.6	60.2	96.4	81.9	87.6
05	METALS	93.7	91.5	92.9	100.8	112.1	104.7	95.1	106.9	99.2	92.7	61.6	82.0	92.7	61.6	82.0	93.5	87.3	91.4
06	WOOD, PLASTICS & COMPOSITES	91.0	108.7	101.1	90.2	117.5	105.8	86.6	87.0	86.8	85.9	45.8	62.9	91.9	45.8	65.5	95.7	89.8	92.3
07	THERMAL & MOISTURE PROTECTION	101.7	88.7	96.5	115.2	102.8	110.3	101.2	85.8	95.1	95.9	41.8	74.4	95.8	41.8	74.3	94.7	84.9	90.8
08	OPENINGS	97.9	94.1	96.9	96.8	124.8	103.9	95.5	101.8	97.1	102.1	41.4	86.7	103.4	41.4	87.7	103.4	81.0	97.7
0920	Plaster & Gypsum Board	94.5	108.8	104.6	94.8	117.9	111.0	87.9	86.4	86.9	91.5	43.5	57.7	93.3	43.5	58.3	95.9	88.8	90.9
0950, 0980	Ceilings & Acoustic Treatment	122.0	108.8	113.3	132.3	117.9	122.7	128.5	86.4	100.6	89.8	43.5	59.1	89.8	43.5	59.1	88.9	88.8	88.8
0960	Flooring	104.6	60.4	91.9	95.7	104.3	98.2	101.3	82.7	96.0	90.7	33.9	74.4	93.1	33.9	76.1	90.8	94.6	91.9
0970, 0990	Wall Finishes & Painting/Coating	98.5	120.8	112.0	101.1	124.7	115.4	98.5	85.0	90.3	92.3	49.2	66.2	92.3	49.2	66.2	92.3	98.6	96.2
09	FINISHES	103.9	100.6	102.1	103.4	116.0	110.4	103.3	85.3	93.4	92.8	43.4	65.5	93.2	43.4	65.7	93.9	89.9	91.7
COVERS	DIVS. 10 - 14, 25, 28, 41, 43, 44, 46	100.0	69.4	93.6	100.0	96.9	99.4	100.0	98.3	99.6	100.0	39.7	87.5	100.0	39.7	87.5	100.0	63.8	92.5
21, 22, 23	FIRE SUPPRESSION, PLUMBING & HVAC	97.4	107.2	101.5	100.0	110.0	104.2	99.9	79.9	91.6	94.8	40.2	72.0	94.8	40.2	72.1	100.0	83.2	93.0
26, 27, 3370	ELECTRICAL, COMMUNICATIONS & UTIL.	94.0	111.3	103.2	98.1	119.1	109.2	94.4	100.6	97.7	98.0	34.9	64.4	98.0	34.9	64.4	98.9	60.7	78.6
MF2010	WEIGHTED AVERAGE	99.1	99.6	99.3	100.2	112.8	105.9	99.2	91.1	95.6	96.5	47.6	74.6	95.4	47.6	74.0	98.7	82.5	91.5

DIVISION		NEW HAMPSHIRE															NEW JERSEY		
		KEENE			LITTLETON			MANCHESTER			NASHUA			PORTSMOUTH			ATLANTIC CITY		
		034			035			031			030			038			082,084		
		MAT.	INST.	TOTAL	MAT.	INST.	TOTAL	MAT.	INST.	TOTAL	MAT.	INST.	TOTAL	MAT.	INST.	TOTAL	MAT.	INST.	TOTAL
015433	CONTRACTOR EQUIPMENT		100.3	100.3		100.3	100.3		100.3	100.3		100.3	100.3		100.3	100.3		98.4	98.4
0241, 31 - 34	SITE & INFRASTRUCTURE, DEMOLITION	88.6	96.5	94.1	75.9	96.1	89.9	85.1	98.3	94.3	89.3	98.3	95.6	83.5	99.7	94.7	92.1	103.6	100.0
0310	Concrete Forming & Accessories	89.6	42.3	48.5	99.7	47.2	54.1	94.8	86.8	87.8	96.7	86.8	88.1	86.6	90.7	90.2	104.5	126.8	123.8
0320	Concrete Reinforcing	86.5	56.4	71.0	87.3	34.3	60.0	108.1	90.9	99.3	108.1	90.9	99.3	86.5	91.0	88.8	82.4	116.4	99.9
0330	Cast-in-Place Concrete	94.5	52.1	78.3	84.8	52.8	72.6	105.3	110.8	107.4	89.5	110.8	97.6	84.9	117.0	97.2	80.2	130.2	99.4
03	CONCRETE	103.5	49.2	77.5	94.5	47.4	71.9	111.1	95.6	103.7	103.2	95.6	99.6	94.6	99.6	97.0	96.6	124.9	110.1
04	MASONRY	97.7	40.8	63.2	105.7	44.3	68.5	99.5	81.9	88.9	99.8	81.9	88.6	94.6	91.4	92.6	101.4	127.0	116.9
05	METALS	93.4	65.9	83.9	93.4	56.2	80.6	98.6	87.9	94.9	98.6	87.9	94.9	95.0	90.9	93.6	93.2	104.8	97.2
06	WOOD, PLASTICS & COMPOSITES	90.3	40.9	62.0	100.6	46.8	69.8	95.7	89.8	92.3	99.0	89.8	93.7	87.1	89.8	88.6	109.1	126.8	119.2
07	THERMAL & MOISTURE PROTECTION	96.3	43.8	75.5	95.9	47.7	76.7	95.9	88.4	92.9	96.7	88.4	93.4	96.3	112.0	102.5	96.4	126.6	108.4
08	OPENINGS	100.6	46.2	86.8	104.5	39.2	87.9	105.6	81.0	99.4	105.6	81.0	99.4	106.2	76.3	98.6	103.0	124.1	108.4
0920	Plaster & Gypsum Board	92.4	38.4	54.4	105.7	44.5	62.6	103.7	88.8	93.2	100.2	88.8	92.2	91.5	88.8	89.6	104.3	126.9	120.2
0950, 0980	Ceilings & Acoustic Treatment	89.8	38.4	55.8	89.8	44.5	59.8	95.7	88.8	91.1	99.4	88.8	92.4	91.9	88.8	89.9	89.8	126.9	114.4
0960	Flooring	92.6	55.6	82.0	101.6	33.9	82.2	96.9	94.6	96.2	95.8	94.6	95.5	91.2	94.6	92.2	99.5	134.3	109.5
0970, 0990	Wall Finishes & Painting/Coating	92.3	49.2	66.2	92.3	64.5	75.5	92.3	98.6	96.2	92.3	98.6	96.2	92.3	98.6	96.2	92.3	123.7	111.3
09	FINISHES	94.7	45.2	67.4	97.2	46.8	69.4	97.7	89.9	93.4	98.5	89.9	93.8	94.1	92.1	93.0	98.6	128.4	115.1
COVERS	DIVS. 10 - 14, 25, 28, 41, 43, 44, 46	100.0	54.5	90.6	100.0	41.5	87.8	100.0	63.9	92.5	100.0	63.9	92.5	100.0	67.1	93.2	100.0	112.7	102.6
21, 22, 23	FIRE SUPPRESSION, PLUMBING & HVAC	94.8	43.5	73.4	94.8	67.1	83.3	100.0	83.2	93.0	100.2	83.2	93.1	100.2	88.0	95.1	99.8	123.7	109.7
26, 27, 3370	ELECTRICAL, COMMUNICATIONS & UTIL.	98.0	34.9	64.4	99.3	60.6	78.7	99.6	77.7	87.9	100.8	77.7	88.5	98.6	77.7	87.5	95.5	135.1	116.6
MF2010	WEIGHTED AVERAGE	97.0	49.9	76.0	96.7	57.4	79.1	100.8	86.1	94.2	100.2	86.1	93.9	97.7	89.9	94.2	98.1	122.7	109.1

DIVISION		NEW JERSEY																	
		CAMDEN			DOVER			ELIZABETH			HACKENSACK			JERSEY CITY			LONG BRANCH		
		081			078			072			076			073			077		
		MAT.	INST.	TOTAL	MAT.	INST.	TOTAL	MAT.	INST.	TOTAL	MAT.	INST.	TOTAL	MAT.	INST.	TOTAL	MAT.	INST.	TOTAL
015433	CONTRACTOR EQUIPMENT		98.4	98.4		100.3	100.3		100.3	100.3		100.3	100.3		98.4	98.4		98.0	98.0
0241, 31 - 34	SITE & INFRASTRUCTURE, DEMOLITION	92.8	104.1	100.6	103.9	104.7	104.5	108.0	104.7	105.7	104.6	105.2	105.0	92.8	105.2	101.3	97.8	105.0	102.8
0310	Concrete Forming & Accessories	96.7	126.9	122.9	93.2	127.5	123.0	104.0	127.5	124.4	93.2	127.5	123.0	96.7	127.6	123.5	97.2	127.3	123.3
0320	Concrete Reinforcing	108.1	118.2	113.3	83.3	125.3	104.9	83.3	125.3	104.9	83.3	125.3	104.9	108.1	125.3	117.0	83.3	125.3	104.9
0330	Cast-in-Place Concrete	77.8	132.0	98.6	99.8	129.2	111.0	85.7	129.2	102.4	97.5	129.2	109.6	77.8	129.3	97.5	86.5	130.1	103.2
03	CONCRETE	97.4	125.9	111.0	104.6	126.7	115.2	99.9	126.8	112.8	102.6	126.8	114.2	97.4	126.6	111.4	102.3	126.8	114.0
04	MASONRY	92.5	130.3	115.4	96.9	127.5	115.5	112.5	127.5	121.6	100.7	127.5	117.0	89.7	127.5	112.6	105.2	127.0	118.4
05	METALS	98.5	105.2	100.8	93.2	112.3	99.8	94.7	112.3	100.8	93.2	112.4	99.8	98.5	110.0	102.5	93.3	109.8	99.0
06	WOOD, PLASTICS & COMPOSITES	99.0	126.8	114.9	98.4	126.8	114.7	112.3	126.8	120.6	98.4	126.8	114.7	99.0	126.8	114.7	100.1	126.7	115.4
07	THERMAL & MOISTURE PROTECTION	96.3	126.9	108.4	96.7	127.4	108.9	96.9	127.4	109.0	96.5	124.9	107.8	96.2	127.4	108.6	96.4	126.8	108.5
08	OPENINGS	105.6	124.1	110.3	108.6	125.0	112.8	106.6	125.0	111.2	105.8	125.0	110.7	105.6	125.0	110.5	101.4	125.0	107.4
0920	Plaster & Gypsum Board	100.2	126.9	119.0	96.8	126.9	118.0	104.8	126.9	120.3	96.8	126.9	118.0	100.2	126.9	119.0	99.0	126.9	118.6
0950, 0980	Ceilings & Acoustic Treatment	99.4	126.9	117.6	89.8	126.9	114.4	91.9	126.9	115.1	89.8	126.9	114.4	99.4	126.9	117.6	89.8	126.9	114.4
0960	Flooring	95.8	134.3	106.9	94.8	165.7	115.1	99.9	165.7	118.8	94.8	165.7	115.1	95.8	165.7	115.9	96.1	165.7	116.1
0970, 0990	Wall Finishes & Painting/Coating	92.3	123.7	111.3	92.2	125.7	112.5	92.2	125.7	112.5	92.2	125.7	112.5	92.3	125.7	112.6	92.3	123.7	111.3
09	FINISHES	99.0	128.4	115.2	96.5	134.1	117.2	99.9	134.1	118.8	96.3	134.1	117.2	98.9	134.1	118.4	97.4	133.7	117.5
COVERS	DIVS. 10 - 14, 25, 28, 41, 43, 44, 46	100.0	112.8	102.7	100.0	114.6	103.0	100.0	114.6	103.0	100.0	114.6	103.0	100.0	114.6	103.0	100.0	112.6	102.6
21, 22, 23	FIRE SUPPRESSION, PLUMBING & HVAC	100.2	123.7	110.0	99.8	127.1	111.2	100.2	127.1	111.6	99.8	127.1	111.2	100.2	127.1	111.4	99.8	127.5	111.4
26, 27, 3370	ELECTRICAL, COMMUNICATIONS & UTIL.	101.2	139.9	121.8	96.9	135.9	117.7	97.7	135.9	118.0	96.9	135.6	117.5	102.3	135.5	120.0	96.6	127.9	113.2
MF2010	WEIGHTED AVERAGE	99.4	124.0	110.4	99.6	125.5	111.2	100.6	125.6	111.8	99.3	125.4	111.0	99.3	125.3	110.9	98.9	124.1	110.2

City Cost Indexes

NEW JERSEY

	DIVISION	NEW BRUNSWICK 088 - 089			NEWARK 070 - 071			PATERSON 074 - 075			POINT PLEASANT 087			SUMMIT 079			TRENTON 085 - 086		
		MAT.	INST.	TOTAL	MAT.	INST.	TOTAL	MAT.	INST.	TOTAL	MAT.	INST.	TOTAL	MAT.	INST.	TOTAL	MAT.	INST.	TOTAL
015433	CONTRACTOR EQUIPMENT		98.0	98.0		100.3	100.3		100.3	100.3		98.0	98.0		100.3	100.3		98.0	98.0
0241, 31 - 34	SITE & INFRASTRUCTURE, DEMOLITION	105.4	105.1	105.2	108.2	104.7	105.8	106.1	105.2	105.5	107.0	105.0	105.6	105.5	104.7	105.0	90.3	105.0	100.5
0310	Concrete Forming & Accessories	99.5	127.5	123.8	93.2	127.5	123.0	95.3	127.5	123.3	94.6	127.1	122.8	95.9	127.5	123.3	94.0	116.5	113.5
0320	Concrete Reinforcing	83.3	125.3	104.9	108.1	125.3	117.0	108.1	125.3	117.0	83.3	125.2	104.9	83.3	125.3	104.9	108.1	120.6	114.6
0330	Cast-in-Place Concrete	99.1	130.4	111.1	103.4	129.2	113.3	99.2	129.2	110.7	99.1	130.0	111.0	82.8	129.2	100.6	95.5	130.1	108.8
03	CONCRETE	114.8	127.0	120.7	110.0	126.8	118.0	108.0	126.8	117.0	114.5	126.6	120.3	96.8	126.8	111.1	106.1	121.1	113.3
04	MASONRY	99.0	127.5	116.3	102.1	127.5	117.5	97.3	127.5	115.6	90.0	127.0	112.4	99.3	127.5	116.4	88.9	127.0	112.0
05	METALS	93.2	109.9	99.0	98.5	112.3	103.3	93.8	112.4	100.2	93.2	109.6	98.9	93.2	112.3	99.8	93.7	107.4	98.4
06	WOOD, PLASTICS & COMPOSITES	103.2	126.7	116.7	97.7	126.8	114.4	100.8	126.8	115.7	96.7	126.7	113.9	101.9	126.8	116.2	96.1	112.2	105.4
07	THERMAL & MOISTURE PROTECTION	96.5	127.0	108.6	96.4	127.4	108.8	96.8	124.9	107.9	96.6	126.8	108.6	97.1	127.4	109.2	92.6	124.9	105.4
08	OPENINGS	97.4	125.0	104.4	111.7	125.0	115.0	111.7	125.0	115.0	99.5	127.1	106.5	113.9	125.0	116.7	106.2	115.6	108.6
0920	Plaster & Gypsum Board	101.2	126.9	119.3	97.9	126.9	118.3	100.2	126.9	119.0	96.8	126.9	118.0	99.0	126.9	118.6	98.1	112.0	107.9
0950, 0980	Ceilings & Acoustic Treatment	89.8	126.9	114.4	95.7	126.9	116.4	99.4	126.9	117.6	89.8	126.9	114.4	89.8	126.9	114.4	96.6	112.0	106.8
0960	Flooring	97.4	165.7	117.0	96.0	165.7	116.1	95.8	165.7	115.9	94.8	134.3	106.1	96.1	165.7	116.1	96.2	165.7	116.2
0970, 0990	Wall Finishes & Painting/Coating	92.3	125.7	112.6	92.2	125.7	112.5	92.2	125.7	112.5	92.3	123.7	111.3	92.2	125.7	112.5	92.3	123.7	111.3
09	FINISHES	98.9	134.1	118.3	97.8	134.1	117.9	99.1	134.1	118.4	97.6	128.4	114.6	97.5	134.1	117.7	97.7	125.2	112.9
COVERS	DIVS. 10 - 14, 25, 28, 41, 43, 44, 46	100.0	114.5	103.0	100.0	114.6	103.0	100.0	114.6	103.0	100.0	108.1	101.7	100.0	114.6	103.0	100.0	111.0	102.3
21, 22, 23	FIRE SUPPRESSION, PLUMBING & HVAC	99.8	127.8	111.4	100.3	127.6	111.7	100.2	127.1	111.4	99.8	127.4	111.3	99.8	127.6	111.4	100.3	127.5	111.6
26, 27, 3370	ELECTRICAL, COMMUNICATIONS & UTIL.	96.3	132.8	115.7	100.6	135.6	119.2	102.3	135.9	120.2	95.5	127.9	112.7	97.7	135.9	118.0	102.4	137.7	121.2
MF2010	WEIGHTED AVERAGE	100.1	125.0	111.2	102.3	125.6	112.7	101.4	125.5	112.2	99.6	123.3	110.2	99.6	125.6	111.2	99.4	122.7	109.8

	DIVISION	NEW JERSEY VINELAND 080,083			NEW MEXICO ALBUQUERQUE 870 - 872			CARRIZOZO 883			CLOVIS 881			FARMINGTON 874			GALLUP 873		
		MAT.	INST.	TOTAL	MAT.	INST.	TOTAL	MAT.	INST.	TOTAL	MAT.	INST.	TOTAL	MAT.	INST.	TOTAL	MAT.	INST.	TOTAL
015433	CONTRACTOR EQUIPMENT		98.4	98.4		113.0	113.0		113.0	113.0		113.0	113.0		113.0	113.0		113.0	113.0
0241, 31 - 34	SITE & INFRASTRUCTURE, DEMOLITION	96.6	104.1	101.8	75.1	108.5	98.1	94.7	108.5	104.2	83.6	108.4	100.7	81.1	108.5	100.0	88.4	108.5	102.2
0310	Concrete Forming & Accessories	92.3	126.9	122.3	99.2	67.2	71.5	99.2	67.2	71.5	99.2	67.1	71.3	99.2	67.2	71.5	99.2	67.2	71.5
0320	Concrete Reinforcing	82.4	116.4	99.9	99.1	72.2	85.2	106.2	72.2	88.7	107.4	72.1	89.2	108.2	72.2	89.7	103.6	72.2	87.4
0330	Cast-in-Place Concrete	86.5	130.3	103.3	105.2	75.3	93.8	97.9	75.3	89.3	97.9	75.3	89.2	106.2	75.3	94.4	99.9	75.3	90.5
03	CONCRETE	101.8	124.9	112.9	105.0	72.0	89.2	119.2	72.0	96.6	107.9	71.9	90.7	108.6	72.0	91.1	114.5	72.0	94.2
04	MASONRY	89.2	127.0	112.1	107.9	64.3	81.5	104.8	64.3	80.3	104.8	64.3	80.3	114.8	64.3	84.2	104.9	64.3	80.3
05	METALS	93.1	104.9	97.2	101.0	89.4	97.0	97.8	89.4	94.9	97.5	89.2	94.6	98.8	89.4	95.5	97.8	89.4	94.9
06	WOOD, PLASTICS & COMPOSITES	93.9	126.8	112.7	90.3	67.6	77.3	90.4	67.6	77.3	90.4	67.6	77.3	90.4	67.6	77.3	90.4	67.6	77.3
07	THERMAL & MOISTURE PROTECTION	96.2	126.6	108.2	100.7	75.0	90.5	102.5	75.0	91.6	101.2	75.0	90.8	101.0	75.0	90.7	102.1	75.0	91.3
08	OPENINGS	99.0	124.1	105.4	95.3	70.2	88.9	94.1	70.2	88.0	94.3	70.2	88.2	98.3	70.2	91.2	98.3	70.2	91.2
0920	Plaster & Gypsum Board	95.0	126.9	117.5	93.2	66.2	74.2	85.0	66.2	71.8	85.0	66.2	71.8	85.0	66.2	71.8	85.0	66.2	71.8
0950, 0980	Ceilings & Acoustic Treatment	89.8	126.9	114.4	122.4	66.2	85.2	118.2	66.2	83.7	118.2	66.2	83.7	118.2	66.2	83.7	118.2	66.2	83.7
0960	Flooring	94.0	134.3	105.6	99.2	69.7	90.7	101.3	69.7	92.2	101.3	69.7	92.2	101.3	69.7	92.2	101.3	69.7	92.2
0970, 0990	Wall Finishes & Painting/Coating	92.3	123.7	111.3	103.1	52.8	72.6	98.5	52.8	70.8	98.5	52.8	70.8	98.5	52.8	70.8	98.5	52.8	70.8
09	FINISHES	96.3	128.4	114.0	102.8	65.9	82.5	103.2	65.9	82.6	102.0	65.9	82.1	101.6	65.9	81.9	102.7	65.9	82.4
COVERS	DIVS. 10 - 14, 25, 28, 41, 43, 44, 46	100.0	112.8	102.7	100.0	83.3	96.5	100.0	83.3	96.5	100.0	83.3	96.5	100.0	83.3	96.5	100.0	83.3	96.5
21, 22, 23	FIRE SUPPRESSION, PLUMBING & HVAC	99.8	123.7	109.7	100.0	69.8	87.4	96.9	69.8	85.6	96.9	69.4	85.4	99.9	69.8	87.4	96.7	69.8	85.5
26, 27, 3370	ELECTRICAL, COMMUNICATIONS & UTIL.	95.5	135.1	116.6	89.7	72.6	80.6	89.7	72.6	80.6	87.1	72.6	79.4	87.8	72.6	79.7	86.9	72.6	79.3
MF2010	WEIGHTED AVERAGE	97.5	122.8	108.8	99.1	75.0	88.3	100.0	75.0	88.8	97.9	74.9	87.6	99.8	75.0	88.7	99.3	75.0	88.5

NEW MEXICO

	DIVISION	LAS CRUCES 880			LAS VEGAS 877			ROSWELL 882			SANTA FE 875			SOCORRO 878			TRUTH/CONSEQUENCES 879		
		MAT.	INST.	TOTAL	MAT.	INST.	TOTAL	MAT.	INST.	TOTAL	MAT.	INST.	TOTAL	MAT.	INST.	TOTAL	MAT.	INST.	TOTAL
015433	CONTRACTOR EQUIPMENT		87.9	87.9		113.0	113.0		113.0	113.0		113.0	113.0		113.0	113.0		87.9	87.9
0241, 31 - 34	SITE & INFRASTRUCTURE, DEMOLITION	85.8	86.7	86.4	80.7	108.5	99.9	85.4	108.5	101.3	79.0	108.5	99.3	77.3	108.5	98.8	96.6	86.7	89.8
0310	Concrete Forming & Accessories	96.1	66.0	70.0	99.2	67.2	71.5	99.2	67.2	71.5	98.4	67.2	71.4	99.2	67.2	71.5	96.1	66.0	70.0
0320	Concrete Reinforcing	102.4	72.0	86.7	105.3	72.2	88.2	107.4	72.2	89.3	106.2	72.2	88.7	107.4	72.2	89.3	101.2	72.0	86.2
0330	Cast-in-Place Concrete	92.4	67.4	82.8	103.3	75.3	92.6	97.9	75.3	89.2	107.6	75.3	95.2	101.2	75.3	91.3	111.0	67.5	94.3
03	CONCRETE	86.6	68.4	77.9	105.9	72.0	89.7	108.7	72.0	91.1	107.2	72.0	90.3	104.5	72.0	89.0	99.0	68.4	84.4
04	MASONRY	100.8	64.0	78.5	105.2	64.3	80.4	115.5	64.3	84.5	109.1	64.3	82.0	105.1	64.3	80.4	102.8	64.0	79.3
05	METALS	98.7	82.6	93.1	97.6	89.4	94.8	98.7	89.4	95.5	98.7	89.4	95.5	97.8	89.4	94.9	97.4	82.7	92.3
06	WOOD, PLASTICS & COMPOSITES	81.6	66.4	72.9	90.4	67.6	77.3	90.4	67.6	77.3	89.9	67.6	77.1	90.4	67.6	77.3	81.6	66.4	72.9
07	THERMAL & MOISTURE PROTECTION	87.1	70.3	80.4	100.6	75.0	90.5	101.2	75.0	90.8	100.3	75.0	90.3	100.6	75.0	90.3	87.7	70.3	80.8
08	OPENINGS	88.3	69.5	83.6	94.3	70.2	88.2	94.1	70.2	88.0	94.2	70.2	88.1	94.1	70.2	88.0	88.2	69.5	83.5
0920	Plaster & Gypsum Board	87.4	66.2	72.5	85.0	66.2	71.8	85.0	66.2	71.8	92.9	66.2	74.1	85.0	66.2	71.8	87.4	66.2	72.5
0950, 0980	Ceilings & Acoustic Treatment	108.8	66.2	80.6	118.2	66.2	83.7	118.2	66.2	83.7	119.1	66.2	84.0	118.2	66.2	83.7	108.8	66.2	80.6
0960	Flooring	133.7	69.7	115.3	101.3	69.7	92.2	101.3	69.7	92.2	101.3	69.7	92.2	101.3	69.7	92.2	133.7	69.7	115.3
0970, 0990	Wall Finishes & Painting/Coating	90.2	52.8	67.5	98.5	52.8	70.8	98.5	52.8	70.8	98.5	52.8	70.8	98.5	52.8	70.8	90.2	52.8	67.5
09	FINISHES	115.9	65.0	87.8	101.4	65.9	81.8	102.1	65.9	82.1	102.8	65.9	82.4	101.4	65.9	81.8	116.5	65.0	88.1
COVERS	DIVS. 10 - 14, 25, 28, 41, 43, 44, 46	100.0	80.4	95.9	100.0	83.3	96.5	100.0	83.3	96.5	100.0	83.3	96.5	100.0	83.3	96.5	100.0	80.4	95.9
21, 22, 23	FIRE SUPPRESSION, PLUMBING & HVAC	100.3	67.5	87.5	96.7	69.8	85.5	99.9	69.8	87.4	99.9	69.8	87.4	99.9	69.8	87.4	96.6	69.5	85.3
26, 27, 3370	ELECTRICAL, COMMUNICATIONS & UTIL.	88.3	67.8	77.4	89.7	72.6	80.6	88.6	72.6	80.1	100.9	72.6	85.4	87.5	72.6	79.6	91.4	72.6	81.4
MF2010	WEIGHTED AVERAGE	96.3	70.9	85.0	97.7	75.0	87.6	99.6	75.0	88.6	100.1	75.0	88.9	97.2	75.0	87.3	97.5	71.6	85.9

		NEW MEXICO			NEW YORK													
		TUCUMCARI			ALBANY			BINGHAMTON			BRONX			BROOKLYN			BUFFALO		
	DIVISION	884			120 - 122			137 - 139			104			112			140 - 142		
		MAT.	INST.	TOTAL	MAT.	INST.	TOTAL	MAT.	INST.	TOTAL	MAT.	INST.	TOTAL	MAT.	INST.	TOTAL	MAT.	INST.	TOTAL
015433	CONTRACTOR EQUIPMENT		113.0	113.0		114.2	114.2		114.5	114.5		114.5	114.5		115.4	115.4		96.1	96.1
0241, 31 - 34	SITE & INFRASTRUCTURE, DEMOLITION	83.3	108.4	100.6	70.4	107.4	95.9	94.4	91.9	92.7	123.4	125.4	124.8	121.6	129.3	126.9	94.9	96.9	96.3
0310	Concrete Forming & Accessories	99.2	67.1	71.3	99.0	97.4	97.6	103.6	84.4	86.9	104.2	171.6	162.7	110.2	171.2	163.1	100.2	112.4	110.8
0320	Concrete Reinforcing	105.3	72.1	88.2	97.1	90.0	93.5	96.2	90.3	93.2	97.1	182.2	141.0	97.5	182.2	141.1	100.1	100.3	100.2
0330	Cast-in-Place Concrete	97.9	75.3	89.2	76.6	104.5	87.3	98.3	96.7	97.7	91.8	159.3	117.7	100.3	157.6	122.3	104.3	117.7	109.5
03	CONCRETE	107.1	71.9	90.3	96.0	99.1	97.5	98.0	91.4	94.8	99.8	167.6	132.3	110.7	166.8	137.6	109.1	111.2	110.1
04	MASONRY	117.0	64.3	85.1	92.0	103.1	98.7	107.2	91.4	97.6	94.4	165.1	137.2	117.4	165.0	146.3	107.8	116.4	113.0
05	METALS	97.5	89.2	94.6	100.0	104.1	101.4	96.6	113.6	102.5	102.2	143.1	116.2	104.5	140.8	117.0	101.7	94.2	99.1
06	WOOD, PLASTICS & COMPOSITES	90.4	67.6	77.3	99.1	96.2	97.4	109.6	83.0	94.3	103.3	175.2	144.6	111.3	174.9	147.8	103.1	112.3	108.4
07	THERMAL & MOISTURE PROTECTION	101.1	75.0	90.7	89.9	95.5	92.1	104.9	84.8	97.0	107.7	159.2	128.1	107.5	158.2	127.6	103.3	107.9	105.1
08	OPENINGS	94.0	70.2	88.0	94.5	88.6	93.0	89.6	79.9	87.1	87.3	171.6	108.6	89.4	170.0	109.8	90.6	100.4	93.1
0920	Plaster & Gypsum Board	85.0	66.2	71.8	103.6	95.8	98.1	114.4	82.0	91.6	109.4	177.1	157.1	112.0	177.1	157.8	100.8	112.4	109.0
0950, 0980	Ceilings & Acoustic Treatment	118.2	66.2	83.7	97.8	95.8	96.5	100.6	82.0	88.3	88.4	177.1	147.2	97.0	177.1	150.1	101.1	112.4	108.6
0960	Flooring	101.3	69.7	92.2	88.8	104.8	93.4	100.5	90.4	97.6	99.4	175.0	121.1	107.5	175.0	126.9	95.8	115.3	101.4
0970, 0990	Wall Finishes & Painting/Coating	98.5	52.8	70.8	85.8	80.1	82.4	87.9	83.5	85.3	99.8	156.0	133.9	116.0	156.0	140.3	94.5	111.6	104.9
09	FINISHES	101.9	65.9	82.0	95.7	97.1	96.4	98.0	85.1	90.9	104.9	171.5	141.6	114.4	171.2	145.7	97.3	113.7	106.3
COVERS	DIVS. 10 - 14, 25, 28, 41, 43, 44, 46	100.0	83.3	96.5	100.0	93.9	98.7	100.0	94.6	98.9	100.0	127.5	105.7	100.0	126.6	105.5	100.0	105.9	101.2
21, 22, 23	FIRE SUPPRESSION, PLUMBING & HVAC	96.9	69.4	85.4	100.1	94.2	97.6	100.5	82.5	93.0	100.3	161.5	125.8	99.8	161.4	125.4	99.9	92.6	96.9
26, 27, 3370	ELECTRICAL, COMMUNICATIONS & UTIL.	89.7	72.6	80.6	102.9	93.6	98.0	101.4	92.2	96.5	97.0	172.8	137.3	100.8	172.8	139.1	99.9	98.7	99.3
MF2010	WEIGHTED AVERAGE	98.6	74.9	88.0	97.2	97.9	97.5	98.8	90.2	95.0	99.9	160.3	126.9	104.0	160.1	129.1	100.5	102.9	101.6

		NEW YORK																
		ELMIRA			FAR ROCKAWAY			FLUSHING			GLENS FALLS			HICKSVILLE			JAMAICA		
	DIVISION	148 - 149			116			113			128			115,117,118			114		
		MAT.	INST.	TOTAL	MAT.	INST.	TOTAL	MAT.	INST.	TOTAL	MAT.	INST.	TOTAL	MAT.	INST.	TOTAL	MAT.	INST.	TOTAL
015433	CONTRACTOR EQUIPMENT		115.9	115.9		115.4	115.4		115.4	115.4		114.2	114.2		115.4	115.4		115.4	115.4
0241, 31 - 34	SITE & INFRASTRUCTURE, DEMOLITION	90.8	93.3	92.5	125.7	129.3	128.2	125.7	129.3	128.2	62.6	106.7	93.0	114.0	128.1	123.8	119.5	129.3	126.2
0310	Concrete Forming & Accessories	85.2	87.1	86.8	96.6	171.2	161.3	100.5	171.2	161.8	87.5	88.6	88.5	93.3	142.7	136.2	100.5	171.2	161.8
0320	Concrete Reinforcing	101.4	87.2	94.1	97.5	182.2	141.1	99.2	182.2	141.9	96.8	90.0	93.3	97.5	182.0	141.0	97.5	182.2	141.1
0330	Cast-in-Place Concrete	97.5	96.7	97.2	108.5	157.6	127.3	108.5	157.6	127.3	74.8	102.1	85.2	92.1	153.1	115.5	100.3	157.6	122.3
03	CONCRETE	102.6	92.2	97.6	117.2	166.8	140.9	117.7	166.8	141.2	88.4	94.4	91.3	102.2	152.4	126.3	110.1	166.8	137.3
04	MASONRY	104.1	93.0	97.3	121.0	165.0	147.7	114.8	165.0	145.2	92.9	98.9	96.5	111.2	158.9	140.1	119.3	165.0	147.0
05	METALS	96.9	114.7	103.0	104.5	140.8	117.0	104.5	140.8	117.0	96.0	103.8	98.7	106.1	137.6	116.9	104.5	140.8	117.0
06	WOOD, PLASTICS & COMPOSITES	86.1	86.3	86.2	94.9	174.9	140.8	99.7	174.9	142.8	85.7	86.8	86.3	91.2	140.8	119.7	99.7	174.9	142.8
07	THERMAL & MOISTURE PROTECTION	103.9	86.1	96.9	107.4	158.2	127.6	107.5	158.2	127.6	89.4	92.0	90.4	107.1	150.0	124.1	107.3	158.2	127.5
08	OPENINGS	91.6	80.9	88.9	87.7	170.0	108.6	87.7	170.0	108.6	88.1	83.5	86.9	87.7	151.6	103.9	87.7	170.0	108.6
0920	Plaster & Gypsum Board	94.7	85.6	88.3	101.1	177.1	154.6	103.8	177.1	155.4	97.0	86.1	89.3	100.2	142.0	129.6	103.8	177.1	155.4
0950, 0980	Ceilings & Acoustic Treatment	97.3	85.6	89.6	86.7	177.1	146.6	86.7	177.1	146.6	95.1	86.1	89.1	85.6	142.0	123.0	86.7	177.1	146.6
0960	Flooring	85.7	90.4	87.1	102.3	175.0	123.2	103.8	175.0	124.3	80.8	104.8	87.7	102.1	150.7	115.4	103.8	175.0	124.3
0970, 0990	Wall Finishes & Painting/Coating	93.0	79.5	84.8	116.0	156.0	140.3	116.0	156.0	140.3	85.8	83.6	84.5	116.0	156.0	140.3	116.0	156.0	140.3
09	FINISHES	92.4	86.7	89.3	109.9	171.2	143.7	110.7	171.2	144.1	91.3	90.9	91.1	108.2	144.7	128.3	110.2	171.2	143.8
COVERS	DIVS. 10 - 14, 25, 28, 41, 43, 44, 46	100.0	91.1	98.2	100.0	126.6	105.5	100.0	126.6	105.5	100.0	85.6	97.0	100.0	120.5	104.3	100.0	126.6	105.5
21, 22, 23	FIRE SUPPRESSION, PLUMBING & HVAC	94.7	85.4	90.9	94.4	161.4	122.3	94.4	161.4	122.3	95.0	86.5	91.5	99.8	146.9	119.4	94.4	161.4	122.3
26, 27, 3370	ELECTRICAL, COMMUNICATIONS & UTIL.	98.1	95.5	96.7	109.0	172.8	142.9	109.0	172.8	142.9	96.3	90.4	93.2	100.1	144.0	123.5	99.0	172.8	138.2
MF2010	WEIGHTED AVERAGE	96.9	92.0	94.7	103.9	160.1	129.1	103.8	160.1	129.0	92.6	93.2	92.8	101.7	145.2	121.2	101.9	160.1	127.9

		NEW YORK																
		JAMESTOWN			KINGSTON			LONG ISLAND CITY			MONTICELLO			MOUNT VERNON			NEW ROCHELLE		
	DIVISION	147			124			111			127			105			108		
		MAT.	INST.	TOTAL	MAT.	INST.	TOTAL	MAT.	INST.	TOTAL	MAT.	INST.	TOTAL	MAT.	INST.	TOTAL	MAT.	INST.	TOTAL
015433	CONTRACTOR EQUIPMENT		92.9	92.9		115.4	115.4		115.4	115.4		115.4	115.4		114.5	114.5		114.5	114.5
0241, 31 - 34	SITE & INFRASTRUCTURE, DEMOLITION	92.1	92.6	92.4	121.3	125.7	124.3	122.9	129.3	127.3	116.5	125.7	122.9	133.0	120.1	124.1	131.5	120.1	123.6
0310	Concrete Forming & Accessories	85.2	80.0	80.7	89.3	109.8	107.1	104.7	171.2	162.4	96.2	109.8	108.0	94.8	131.9	127.0	109.4	131.9	128.9
0320	Concrete Reinforcing	101.5	74.6	87.7	97.1	133.3	115.8	97.5	182.2	141.1	96.4	133.3	115.4	95.9	181.1	139.8	96.0	181.1	139.8
0330	Cast-in-Place Concrete	101.2	95.6	99.1	101.0	133.0	113.3	103.5	157.6	124.2	94.6	133.0	109.3	102.6	132.9	114.2	102.6	132.9	114.2
03	CONCRETE	105.9	84.4	95.6	111.1	121.8	116.2	113.4	166.8	138.9	105.7	121.8	113.4	110.1	140.5	124.6	109.3	140.5	124.3
04	MASONRY	113.1	91.0	99.7	108.4	136.2	125.2	114.3	165.0	145.0	101.7	136.2	122.6	100.0	130.1	118.2	100.0	130.1	118.2
05	METALS	94.2	80.9	89.6	104.5	114.5	107.9	104.5	140.8	117.0	104.5	114.6	107.9	101.9	134.0	112.9	102.2	134.0	113.1
06	WOOD, PLASTICS & COMPOSITES	84.6	76.6	80.0	87.0	104.9	97.2	105.3	174.9	145.2	95.4	104.9	100.8	92.6	135.5	117.2	110.6	135.5	124.8
07	THERMAL & MOISTURE PROTECTION	103.5	84.9	96.1	109.1	135.7	119.7	107.4	158.2	127.6	108.8	135.7	119.5	108.6	137.5	120.1	108.6	137.5	120.1
08	OPENINGS	91.5	74.7	87.2	93.0	121.3	100.2	87.7	170.0	108.6	87.8	121.3	96.3	87.3	148.7	102.8	87.3	148.7	102.9
0920	Plaster & Gypsum Board	89.3	75.7	79.7	95.8	105.0	102.2	107.3	177.1	156.4	99.3	105.0	103.3	103.5	136.1	126.5	116.8	136.1	130.4
0950, 0980	Ceilings & Acoustic Treatment	94.5	75.7	82.0	80.1	105.0	96.6	86.7	177.1	146.6	80.1	105.0	96.6	86.5	136.1	119.4	86.5	136.1	119.4
0960	Flooring	88.6	93.1	89.9	94.7	76.6	89.5	105.4	175.0	125.4	97.4	76.6	91.4	91.3	173.8	115.0	98.4	173.8	120.1
0970, 0990	Wall Finishes & Painting/Coating	94.5	79.9	85.6	113.6	105.0	108.4	116.0	156.0	140.3	113.6	103.9	107.7	97.8	156.0	133.1	97.8	156.0	133.1
09	FINISHES	92.2	81.8	86.5	105.1	101.0	102.8	111.3	171.2	144.3	105.9	100.9	103.1	102.3	141.9	124.1	105.7	141.9	125.7
COVERS	DIVS. 10 - 14, 25, 28, 41, 43, 44, 46	100.0	93.2	98.6	100.0	109.4	101.9	100.0	126.6	105.5	100.0	109.4	101.9	100.0	115.3	103.2	100.0	115.3	103.2
21, 22, 23	FIRE SUPPRESSION, PLUMBING & HVAC	94.6	83.9	90.2	94.9	117.4	104.3	99.8	161.4	125.4	94.9	119.9	105.3	95.0	120.5	105.6	95.0	120.5	105.6
26, 27, 3370	ELECTRICAL, COMMUNICATIONS & UTIL.	97.0	81.8	88.9	97.9	118.2	108.7	99.5	172.8	138.5	97.9	118.2	108.7	95.2	153.6	126.3	95.2	153.6	126.3
MF2010	WEIGHTED AVERAGE	97.2	84.4	91.5	101.6	118.7	109.2	103.6	160.1	128.9	100.0	119.2	108.6	100.0	134.3	115.3	100.3	134.3	115.5

City Cost Indexes

NEW YORK

DIVISION		NEW YORK 100 - 102			NIAGARA FALLS 143			PLATTSBURGH 129			POUGHKEEPSIE 125 - 126			QUEENS 110			RIVERHEAD 119		
		MAT.	INST.	TOTAL	MAT.	INST.	TOTAL	MAT.	INST.	TOTAL	MAT.	INST.	TOTAL	MAT.	INST.	TOTAL	MAT.	INST.	TOTAL
015433	CONTRACTOR EQUIPMENT		114.8	114.8		92.9	92.9		99.0	99.0		115.4	115.4		115.4	115.4		115.4	115.4
0241, 31 - 34	SITE & INFRASTRUCTURE, DEMOLITION	133.1	126.2	128.4	94.2	93.4	93.6	87.3	102.5	97.8	117.3	125.6	123.0	117.6	129.3	125.6	115.1	128.1	124.1
0310	Concrete Forming & Accessories	108.7	187.6	177.1	85.2	108.1	105.0	91.9	82.4	83.6	89.3	156.6	147.7	93.5	171.2	160.9	97.7	142.6	136.7
0320	Concrete Reinforcing	102.8	182.3	143.7	100.1	92.7	96.3	101.2	89.4	95.1	97.1	133.6	115.9	99.2	182.2	141.9	99.3	179.0	140.4
0330	Cast-in-Place Concrete	103.4	169.5	128.7	104.7	102.4	103.8	91.5	94.3	92.6	97.8	131.5	110.7	95.3	157.6	119.2	93.6	153.1	116.4
03	CONCRETE	110.6	178.2	142.9	108.4	102.5	105.6	102.3	87.6	95.3	108.2	142.2	124.4	105.6	166.8	134.9	102.9	151.9	126.3
04	MASONRY	105.7	173.9	147.0	121.1	111.7	115.4	88.9	88.3	88.5	101.7	132.2	120.2	108.6	165.0	142.8	116.8	158.9	142.3
05	METALS	116.5	143.3	125.7	96.9	88.7	94.1	100.5	83.4	94.6	104.5	116.5	108.6	104.5	140.8	117.0	106.5	136.5	116.8
06	WOOD, PLASTICS & COMPOSITES	107.8	192.2	156.2	84.6	107.4	97.7	93.5	80.5	86.1	87.0	168.4	133.7	91.4	174.9	139.3	96.7	140.8	122.0
07	THERMAL & MOISTURE PROTECTION	107.7	165.0	130.4	103.6	102.1	103.0	102.8	87.1	96.6	109.1	141.9	122.1	107.1	158.2	127.4	107.7	150.0	124.5
08	OPENINGS	92.6	180.7	114.9	91.5	95.4	92.5	97.3	79.4	92.8	93.0	152.3	108.0	87.7	170.0	108.6	87.7	150.9	103.7
0920	Plaster & Gypsum Board	117.5	194.6	171.8	89.3	107.4	102.0	119.9	79.2	91.2	95.8	170.4	148.3	100.2	177.1	154.4	102.3	142.0	130.2
0950, 0980	Ceilings & Acoustic Treatment	108.1	194.6	165.4	94.5	107.4	103.0	109.8	79.2	89.5	80.1	170.4	140.0	86.7	177.1	146.6	86.6	142.0	123.3
0960	Flooring	100.7	175.0	122.0	88.4	107.6	94.0	104.3	100.7	103.3	94.7	147.1	109.8	101.2	175.0	122.4	102.3	150.7	116.2
0970, 0990	Wall Finishes & Painting/Coating	99.8	156.0	133.9	94.5	102.5	99.3	110.6	77.0	90.2	113.6	103.9	107.7	116.0	156.0	140.3	116.0	156.0	140.3
09	FINISHES	111.2	183.6	151.2	92.3	108.2	101.1	101.9	85.2	92.7	104.9	151.7	130.7	108.6	171.2	143.1	108.9	144.7	128.6
COVERS	DIVS. 10 - 14, 25, 28, 41, 43, 44, 46	100.0	137.6	107.8	100.0	100.8	100.2	100.0	51.6	89.9	100.0	115.3	103.2	100.0	126.6	105.5	100.0	120.5	104.3
21, 22, 23	FIRE SUPPRESSION, PLUMBING & HVAC	100.2	166.0	127.6	94.6	92.4	93.7	94.9	81.4	89.2	94.9	119.7	105.2	99.8	161.4	125.4	99.9	145.6	119.0
26, 27, 3370	ELECTRICAL, COMMUNICATIONS & UTIL.	105.6	173.1	141.5	95.4	90.6	92.8	94.6	84.6	89.3	97.9	118.2	108.7	100.1	172.8	138.8	101.9	144.0	124.3
MF2010	WEIGHTED AVERAGE	106.0	166.3	133.0	98.3	98.0	98.2	97.5	85.0	91.9	100.7	130.5	114.0	101.9	160.1	127.9	102.5	144.7	121.4

NEW YORK

DIVISION		ROCHESTER 144 - 146			SCHENECTADY 123			STATEN ISLAND 103			SUFFERN 109			SYRACUSE 130 - 132			UTICA 133 - 135		
		MAT.	INST.	TOTAL	MAT.	INST.	TOTAL	MAT.	INST.	TOTAL	MAT.	INST.	TOTAL	MAT.	INST.	TOTAL	MAT.	INST.	TOTAL
015433	CONTRACTOR EQUIPMENT		116.6	116.6		114.2	114.2		114.5	114.5		114.5	114.5		114.2	114.2		114.2	114.2
0241, 31 - 34	SITE & INFRASTRUCTURE, DEMOLITION	71.8	108.5	97.1	72.0	107.3	96.4	138.0	125.4	129.3	127.8	121.1	123.2	93.4	106.7	102.6	70.2	104.4	93.8
0310	Concrete Forming & Accessories	101.5	93.7	94.7	103.4	97.0	97.8	94.3	171.6	161.4	103.2	125.5	122.6	102.4	89.1	90.9	104.0	82.2	85.1
0320	Concrete Reinforcing	101.0	90.6	95.7	95.8	90.0	92.8	97.1	182.2	141.0	96.0	133.6	115.3	97.1	90.7	93.8	97.1	88.3	92.6
0330	Cast-in-Place Concrete	97.9	99.0	98.3	90.3	103.9	95.5	102.6	159.3	124.3	99.5	129.8	111.1	91.3	96.9	93.4	83.5	89.3	85.7
03	CONCRETE	116.3	95.8	106.5	103.0	98.7	100.9	111.9	167.6	138.6	106.2	127.9	116.6	101.6	92.9	97.5	99.8	86.8	93.6
04	MASONRY	104.2	98.4	100.7	91.1	102.0	97.7	108.2	165.1	142.7	100.1	125.6	115.6	98.5	100.1	99.5	89.8	87.8	88.6
05	METALS	98.2	105.4	100.7	100.1	104.1	101.5	100.1	143.1	114.9	100.1	118.5	106.4	100.0	103.2	101.1	97.9	102.0	99.3
06	WOOD, PLASTICS & COMPOSITES	100.0	93.7	96.4	105.8	96.2	100.3	91.1	175.2	139.4	102.9	131.1	119.1	105.8	87.4	95.2	105.8	84.2	93.4
07	THERMAL & MOISTURE PROTECTION	98.7	94.6	97.1	90.5	95.0	92.3	108.0	159.2	128.3	108.5	134.2	118.7	101.1	96.2	99.2	90.3	90.7	90.5
08	OPENINGS	93.8	87.3	92.2	94.5	88.6	93.0	87.3	171.6	108.6	87.3	137.3	100.0	92.7	80.1	89.5	94.5	79.9	90.8
0920	Plaster & Gypsum Board	106.2	93.5	97.3	108.6	95.8	99.6	103.6	177.1	155.3	108.4	131.7	124.8	108.6	86.8	93.3	108.6	83.5	90.9
0950, 0980	Ceilings & Acoustic Treatment	102.1	93.5	96.4	100.6	95.8	97.4	88.4	177.1	147.2	86.5	131.7	116.5	100.6	86.8	91.5	100.6	83.5	89.3
0960	Flooring	84.6	104.0	90.2	88.2	104.8	93.0	95.4	175.0	118.3	94.6	150.7	110.7	90.1	90.5	90.2	88.2	91.2	89.1
0970, 0990	Wall Finishes & Painting/Coating	94.3	97.6	96.3	85.8	80.1	82.4	99.8	156.0	133.9	97.8	112.3	106.6	92.5	95.4	94.3	85.8	95.4	91.6
09	FINISHES	98.5	96.3	97.3	96.7	96.8	96.8	104.5	171.5	141.4	103.4	126.3	116.0	98.5	89.8	93.7	96.8	85.2	90.4
COVERS	DIVS. 10 - 14, 25, 28, 41, 43, 44, 46	100.0	94.5	98.8	100.0	93.5	98.7	100.0	127.5	105.7	100.0	113.2	102.7	100.0	95.7	99.1	100.0	87.2	97.3
21, 22, 23	FIRE SUPPRESSION, PLUMBING & HVAC	99.9	89.1	95.4	100.3	93.6	97.5	100.3	161.5	125.8	95.0	114.4	103.1	100.3	86.7	94.6	100.3	80.2	91.9
26, 27, 3370	ELECTRICAL, COMMUNICATIONS & UTIL.	100.1	91.6	95.6	101.5	93.6	97.3	97.0	172.8	137.3	103.7	126.4	115.8	101.5	92.4	96.6	99.6	92.4	95.8
MF2010	WEIGHTED AVERAGE	100.3	95.6	98.2	98.2	97.5	97.9	102.1	160.3	128.1	100.1	123.1	110.4	99.4	93.5	96.7	97.1	88.7	93.4

NEW YORK / NORTH CAROLINA

DIVISION		WATERTOWN 136			WHITE PLAINS 106			YONKERS 107			ASHEVILLE 287 - 288			CHARLOTTE 281 - 282			DURHAM 277		
		MAT.	INST.	TOTAL	MAT.	INST.	TOTAL	MAT.	INST.	TOTAL	MAT.	INST.	TOTAL	MAT.	INST.	TOTAL	MAT.	INST.	TOTAL
015433	CONTRACTOR EQUIPMENT		114.2	114.2		114.5	114.5		114.5	114.5		94.9	94.9		94.9	94.9		100.4	100.4
0241, 31 - 34	SITE & INFRASTRUCTURE, DEMOLITION	78.2	106.8	97.9	123.0	120.1	121.0	132.5	119.9	123.8	109.5	74.8	85.6	110.7	74.9	86.0	109.8	83.6	91.7
0310	Concrete Forming & Accessories	89.2	92.0	91.6	108.1	131.9	128.8	108.4	131.9	128.8	92.0	38.1	45.3	96.5	40.0	47.4	94.6	38.7	46.1
0320	Concrete Reinforcing	97.7	90.7	94.1	96.0	181.1	139.8	99.8	181.2	141.7	90.2	41.6	65.2	90.6	38.8	63.9	91.6	48.7	69.5
0330	Cast-in-Place Concrete	97.2	96.7	97.0	91.1	132.9	107.1	101.9	132.9	113.8	113.2	45.8	87.4	119.5	47.7	92.0	115.1	46.0	88.6
03	CONCRETE	113.0	94.1	104.0	99.3	140.5	119.0	109.5	140.5	124.3	107.1	43.3	76.6	109.9	44.3	78.5	107.7	45.0	77.7
04	MASONRY	90.9	101.0	97.0	99.5	130.1	118.0	105.3	130.1	120.3	85.2	39.0	57.2	91.1	42.7	61.8	90.5	35.1	56.9
05	METALS	98.0	103.2	99.8	101.7	134.0	112.8	111.9	134.0	119.5	84.7	72.7	80.6	88.6	71.8	82.8	97.4	75.9	90.0
06	WOOD, PLASTICS & COMPOSITES	87.5	91.4	89.7	108.9	135.5	124.1	108.7	135.5	124.1	93.5	37.6	61.5	99.1	39.7	65.0	96.2	39.1	63.4
07	THERMAL & MOISTURE PROTECTION	90.6	97.0	93.1	108.3	137.5	119.9	108.6	137.5	120.3	98.2	42.1	75.9	95.7	43.6	75.0	98.9	44.0	77.1
08	OPENINGS	94.5	86.0	92.4	87.4	148.7	102.9	90.5	148.7	105.2	94.5	37.4	80.0	98.6	38.8	83.5	98.6	41.3	84.1
0920	Plaster & Gypsum Board	99.3	90.9	93.4	111.9	136.1	129.0	117.0	136.1	130.5	99.7	35.6	54.6	100.0	37.8	56.2	104.5	37.1	57.0
0950, 0980	Ceilings & Acoustic Treatment	100.6	90.9	94.2	86.5	136.1	119.4	106.2	136.1	126.0	92.0	35.6	54.6	94.8	37.8	57.0	95.7	37.1	56.8
0960	Flooring	81.3	92.0	84.4	96.7	173.8	118.8	96.3	173.8	118.5	106.2	44.0	88.3	109.3	44.2	90.6	109.9	44.0	91.0
0970, 0990	Wall Finishes & Painting/Coating	85.8	86.5	86.2	97.8	156.0	133.1	97.8	156.0	133.1	114.7	34.5	66.0	114.7	39.4	69.0	114.7	36.1	67.0
09	FINISHES	94.3	91.4	92.7	103.8	141.9	124.8	109.3	141.9	127.2	100.7	38.6	66.4	101.8	40.5	68.0	103.2	39.0	67.8
COVERS	DIVS. 10 - 14, 25, 28, 41, 43, 44, 46	100.0	96.1	99.2	100.0	115.3	103.2	100.0	120.4	104.2	100.0	66.9	93.1	100.0	67.4	93.2	100.0	71.6	94.1
21, 22, 23	FIRE SUPPRESSION, PLUMBING & HVAC	100.3	81.8	92.6	100.5	120.5	108.8	100.5	120.5	108.8	100.0	37.8	74.3	100.0	37.9	74.2	100.4	36.6	73.8
26, 27, 3370	ELECTRICAL, COMMUNICATIONS & UTIL.	101.5	87.5	94.1	95.2	153.6	126.3	103.8	153.6	130.3	97.1	37.4	65.3	99.5	48.6	72.4	97.8	47.3	71.0
MF2010	WEIGHTED AVERAGE	98.9	92.6	96.1	99.9	134.3	115.3	104.8	134.5	118.0	97.4	46.0	74.4	99.2	48.4	76.5	100.3	48.4	77.1

City Cost Indexes

NORTH CAROLINA

DIVISION		ELIZABETH CITY 279			FAYETTEVILLE 283			GASTONIA 280			GREENSBORO 270,272 - 274			HICKORY 286			KINSTON 285		
		MAT.	INST.	TOTAL	MAT.	INST.	TOTAL	MAT.	INST.	TOTAL	MAT.	INST.	TOTAL	MAT.	INST.	TOTAL	MAT.	INST.	TOTAL
015433	CONTRACTOR EQUIPMENT		104.7	104.7		100.4	100.4		94.9	94.9		100.4	100.4		100.4	100.4		100.4	100.4
0241, 31 - 34	SITE & INFRASTRUCTURE, DEMOLITION	115.3	84.8	94.2	108.5	83.7	91.4	109.2	74.8	85.4	109.7	83.7	91.8	108.9	83.7	91.5	107.6	83.5	91.0
0310	Concrete Forming & Accessories	81.4	33.6	39.9	91.7	39.4	46.4	98.3	39.0	46.9	94.3	39.4	46.6	88.4	37.2	44.0	85.0	31.5	38.6
0320	Concrete Reinforcing	89.7	40.5	64.4	93.7	48.8	70.6	90.6	41.6	65.3	90.6	57.1	73.3	90.2	18.9	53.5	89.7	18.3	53.0
0330	Cast-in-Place Concrete	115.3	37.3	85.4	118.7	49.0	92.0	110.7	53.1	88.6	114.2	47.3	88.6	113.2	46.9	87.8	109.2	37.5	81.8
03	CONCRETE	108.1	38.2	74.6	109.3	46.4	79.2	105.6	46.2	77.2	107.1	47.3	78.5	106.9	39.2	74.5	103.5	33.3	69.9
04	MASONRY	102.9	26.3	56.4	88.5	40.0	59.1	89.7	36.9	57.7	88.4	35.8	56.5	74.9	40.2	53.9	81.0	28.2	49.0
05	METALS	86.7	72.2	81.7	101.1	76.0	92.5	85.6	72.7	81.2	91.7	79.1	87.4	84.7	66.1	78.3	83.8	65.0	77.3
06	WOOD, PLASTICS & COMPOSITES	80.5	34.8	54.3	92.5	39.0	61.8	101.8	39.0	65.8	95.9	39.0	63.2	88.3	37.4	59.1	85.0	31.7	54.5
07	THERMAL & MOISTURE PROTECTION	98.2	33.1	72.3	98.1	42.6	76.0	98.4	42.3	76.1	98.7	41.7	76.0	98.5	41.0	75.6	98.4	31.8	71.9
08	OPENINGS	95.5	32.7	79.6	94.6	41.3	81.1	98.6	38.9	83.5	98.6	43.6	84.7	94.6	30.3	78.3	94.7	27.3	77.6
0920	Plaster & Gypsum Board	96.3	32.0	51.0	104.2	37.1	56.9	106.5	37.0	57.6	106.5	37.0	57.6	99.7	35.4	54.4	99.4	29.5	50.2
0950, 0980	Ceilings & Acoustic Treatment	95.7	32.0	53.5	92.9	37.1	55.9	95.7	37.0	56.8	95.7	37.0	56.8	92.0	35.4	54.4	95.7	29.5	51.8
0960	Flooring	100.3	23.9	78.3	106.3	44.0	88.4	109.9	44.0	91.0	109.9	40.7	90.0	106.0	33.9	85.3	103.1	19.4	79.0
0970, 0990	Wall Finishes & Painting/Coating	114.7	44.5	72.1	114.7	34.5	66.0	114.7	39.5	69.1	114.7	33.5	65.5	114.7	33.5	65.5	114.7	30.5	63.6
09	FINISHES	99.7	32.8	62.8	101.4	39.4	67.2	103.3	39.8	68.3	103.4	38.7	67.7	100.8	35.9	65.0	100.5	29.1	61.1
COVERS	DIVS. 10 - 14, 25, 28, 41, 43, 44, 46	100.0	75.5	94.9	100.0	72.4	94.3	100.0	66.8	93.1	100.0	67.1	93.2	100.0	66.8	93.1	100.0	70.5	93.9
21, 22, 23	FIRE SUPPRESSION, PLUMBING & HVAC	94.9	29.6	67.7	100.2	37.7	74.2	100.3	37.4	74.1	100.3	37.9	74.3	94.9	31.1	68.3	94.9	25.1	65.8
26, 27, 3370	ELECTRICAL, COMMUNICATIONS & UTIL.	97.6	29.7	61.5	95.7	51.6	72.2	96.4	34.7	63.6	96.8	37.6	65.3	94.4	31.4	60.9	94.1	49.0	70.1
MF2010	WEIGHTED AVERAGE	97.5	41.0	72.3	100.0	50.0	77.6	98.2	46.0	74.8	99.2	47.9	76.2	95.2	42.8	71.8	94.9	40.6	70.6

NORTH CAROLINA / NORTH DAKOTA

DIVISION		MURPHY 289			RALEIGH 275 - 276			ROCKY MOUNT 278			WILMINGTON 284			WINSTON-SALEM 271			BISMARCK 585		
		MAT.	INST.	TOTAL	MAT.	INST.	TOTAL	MAT.	INST.	TOTAL	MAT.	INST.	TOTAL	MAT.	INST.	TOTAL	MAT.	INST.	TOTAL
015433	CONTRACTOR EQUIPMENT		94.9	94.9		100.4	100.4		100.4	100.4		94.9	94.9		100.4	100.4		98.6	98.6
0241, 31 - 34	SITE & INFRASTRUCTURE, DEMOLITION	111.5	74.5	86.0	110.2	83.8	92.0	113.2	83.4	92.6	110.8	74.8	85.9	110.0	83.7	91.8	97.2	97.2	97.2
0310	Concrete Forming & Accessories	98.8	24.3	34.1	94.8	40.2	47.4	87.3	29.3	36.9	93.2	38.7	45.9	95.7	38.4	45.9	95.7	43.1	50.1
0320	Concrete Reinforcing	89.8	18.2	52.9	90.6	48.8	69.1	89.7	26.7	57.3	90.9	47.0	68.3	90.6	38.6	63.8	104.9	86.9	95.6
0330	Cast-in-Place Concrete	117.1	35.5	85.9	122.4	52.4	95.6	112.8	34.7	82.9	112.8	47.0	87.6	117.1	44.7	89.4	99.2	49.9	80.3
03	CONCRETE	110.4	29.4	71.6	111.3	47.9	80.9	108.7	32.8	72.3	106.9	45.0	77.3	108.6	42.5	77.0	99.9	55.1	78.4
04	MASONRY	77.5	26.0	46.3	89.6	40.9	60.1	81.3	26.3	48.0	75.2	37.0	52.1	88.6	36.8	57.2	107.5	60.9	79.3
05	METALS	82.8	64.5	76.5	89.6	75.9	84.9	85.9	65.3	78.8	84.6	75.0	81.3	89.6	71.3	83.3	96.3	86.7	93.0
06	WOOD, PLASTICS & COMPOSITES	102.5	22.4	56.5	96.2	39.7	63.8	87.3	29.4	54.1	95.3	38.4	62.6	95.9	38.3	62.9	82.3	37.6	56.7
07	THERMAL & MOISTURE PROTECTION	98.4	31.1	71.6	97.9	42.9	76.0	98.7	29.1	71.1	98.2	42.0	75.9	98.7	41.3	75.9	106.3	53.5	85.3
08	OPENINGS	94.5	24.9	76.9	95.4	41.4	81.7	94.7	26.0	77.3	94.7	39.9	80.8	98.6	38.2	83.3	96.9	49.4	84.8
0920	Plaster & Gypsum Board	105.4	19.9	45.2	100.0	37.7	56.1	98.6	27.1	48.3	101.3	36.4	55.6	106.5	36.3	57.1	95.9	36.0	53.7
0950, 0980	Ceilings & Acoustic Treatment	92.0	19.9	44.2	94.8	37.7	57.0	92.9	27.1	49.3	92.9	36.4	55.4	95.7	36.3	56.4	139.9	36.0	71.0
0960	Flooring	110.2	24.8	85.7	109.9	44.0	91.0	104.4	22.5	80.9	107.0	44.0	88.9	109.9	44.0	91.0	118.2	76.2	106.2
0970, 0990	Wall Finishes & Painting/Coating	114.7	18.5	56.3	114.7	38.0	68.2	114.7	32.4	64.8	114.7	33.4	65.4	114.7	35.2	66.5	98.4	31.7	57.9
09	FINISHES	102.8	23.1	58.8	102.2	40.4	68.1	100.8	28.0	60.6	101.3	38.8	66.8	103.4	38.9	67.9	114.3	46.1	76.7
COVERS	DIVS. 10 - 14, 25, 28, 41, 43, 44, 46	100.0	64.0	92.5	100.0	72.7	94.3	100.0	69.5	93.7	100.0	72.1	94.2	100.0	66.8	93.1	100.0	81.1	96.1
21, 22, 23	FIRE SUPPRESSION, PLUMBING & HVAC	94.9	18.8	63.2	100.1	36.1	73.5	94.9	32.7	69.0	100.3	37.5	74.2	100.3	36.5	73.7	100.2	56.4	82.0
26, 27, 3370	ELECTRICAL, COMMUNICATIONS & UTIL.	98.1	26.2	59.9	103.5	37.3	68.2	99.9	35.8	65.8	97.3	36.1	64.7	96.8	40.9	67.1	100.4	66.0	82.1
MF2010	WEIGHTED AVERAGE	96.2	33.4	68.1	99.6	48.1	76.6	96.6	39.8	71.2	96.9	46.3	74.3	99.1	46.5	75.6	100.8	63.0	83.9

NORTH DAKOTA

DIVISION		DEVILS LAKE 583			DICKINSON 586			FARGO 580 - 581			GRAND FORKS 582			JAMESTOWN 584			MINOT 587		
		MAT.	INST.	TOTAL	MAT.	INST.	TOTAL	MAT.	INST.	TOTAL	MAT.	INST.	TOTAL	MAT.	INST.	TOTAL	MAT.	INST.	TOTAL
015433	CONTRACTOR EQUIPMENT		98.6	98.6		98.6	98.6		98.6	98.6		98.6	98.6		98.6	98.6		98.6	98.6
0241, 31 - 34	SITE & INFRASTRUCTURE, DEMOLITION	106.9	95.1	98.7	116.7	93.7	100.9	97.8	97.2	97.4	111.2	93.7	99.1	105.7	93.7	97.4	108.2	97.2	100.6
0310	Concrete Forming & Accessories	99.1	37.5	45.6	89.6	36.7	43.7	96.9	43.8	50.9	93.0	36.7	44.2	90.7	36.4	43.6	89.4	61.9	65.5
0320	Concrete Reinforcing	104.9	87.2	95.8	105.8	42.7	73.3	96.7	87.0	91.7	103.3	87.1	94.9	105.5	50.5	77.2	106.8	87.2	96.7
0330	Cast-in-Place Concrete	120.4	48.4	92.9	109.0	46.9	85.2	98.3	51.8	80.5	109.0	46.9	85.2	118.9	46.7	91.3	108.9	48.9	85.9
03	CONCRETE	113.7	52.1	84.2	113.4	42.7	79.5	108.5	56.1	83.4	110.2	51.0	81.9	112.3	44.0	79.6	108.7	63.2	86.9
04	MASONRY	116.3	66.7	86.2	119.4	60.3	83.6	109.4	60.9	80.0	110.6	66.2	83.7	130.1	35.4	72.7	110.0	65.6	83.1
05	METALS	96.3	85.8	92.7	96.2	64.7	85.4	98.7	87.1	94.7	96.3	81.1	91.1	96.3	66.0	85.8	96.6	87.4	93.4
06	WOOD, PLASTICS & COMPOSITES	85.7	34.6	56.4	85.1	34.6	56.1	82.2	38.1	56.9	79.2	34.6	53.6	77.2	34.6	52.8	75.3	62.7	68.1
07	THERMAL & MOISTURE PROTECTION	107.1	51.6	85.1	107.7	49.1	84.4	104.7	54.6	84.8	107.4	51.6	85.2	107.0	43.5	81.8	107.1	57.3	87.4
08	OPENINGS	96.9	43.1	83.3	96.9	32.5	80.6	96.8	49.7	84.9	96.9	43.1	83.3	96.9	34.7	81.1	97.0	62.9	88.4
0920	Plaster & Gypsum Board	113.6	32.9	56.8	104.7	32.9	54.2	95.9	36.6	54.1	106.5	32.9	54.7	106.0	32.9	54.6	104.7	61.8	74.5
0950, 0980	Ceilings & Acoustic Treatment	140.9	32.9	69.3	140.9	32.9	69.3	139.9	36.6	71.4	140.9	32.9	69.3	140.9	32.9	69.3	140.9	61.8	88.5
0960	Flooring	122.5	37.9	98.2	114.0	37.9	92.2	117.9	76.2	105.9	116.4	37.9	93.9	115.1	37.9	92.9	113.8	84.9	105.5
0970, 0990	Wall Finishes & Painting/Coating	98.4	24.2	53.4	98.4	24.2	53.4	98.4	75.3	84.4	98.4	24.2	53.4	98.4	24.2	53.4	98.4	26.5	54.8
09	FINISHES	118.6	34.6	72.3	116.0	34.6	71.1	114.3	51.3	79.6	116.3	34.6	71.3	115.4	34.6	70.8	115.1	62.1	85.8
COVERS	DIVS. 10 - 14, 25, 28, 41, 43, 44, 46	100.0	35.9	86.7	100.0	36.1	86.7	100.0	81.1	96.1	100.0	36.0	86.7	100.0	78.9	95.6	100.0	84.0	96.7
21, 22, 23	FIRE SUPPRESSION, PLUMBING & HVAC	94.9	56.1	78.7	94.9	69.6	84.4	100.2	59.4	83.2	100.3	36.4	73.7	94.9	38.4	71.4	100.3	56.4	82.0
26, 27, 3370	ELECTRICAL, COMMUNICATIONS & UTIL.	94.1	38.8	64.7	103.5	66.5	83.8	102.1	64.0	81.9	97.9	58.1	76.7	94.1	38.8	64.7	101.3	82.1	91.1
MF2010	WEIGHTED AVERAGE	101.7	56.0	81.3	102.8	58.2	82.9	102.4	64.3	85.4	102.6	53.8	80.8	101.9	46.8	77.3	102.5	70.0	88.0

City Cost Indexes

		NORTH DAKOTA			OHIO															
	DIVISION	WILLISTON			AKRON			ATHENS			CANTON			CHILLICOTHE			CINCINNATI		
		588			442 - 443			457			446 - 447			456			451 - 452		
		MAT.	INST.	TOTAL	MAT.	INST.	TOTAL	MAT.	INST.	TOTAL	MAT.	INST.	TOTAL	MAT.	INST.	TOTAL	MAT.	INST.	TOTAL
015433	CONTRACTOR EQUIPMENT		98.6	98.6		96.8	96.8		91.3	91.3		96.8	96.8		101.6	101.6		101.4	101.4
0241, 31 - 34	SITE & INFRASTRUCTURE, DEMOLITION	109.3	93.7	98.6	94.1	103.8	100.8	93.1	92.2	92.5	94.2	103.6	100.7	81.9	106.0	98.5	79.2	105.8	97.5
0310	Concrete Forming & Accessories	94.5	36.7	44.4	101.2	93.9	94.9	93.2	90.1	90.5	101.2	82.3	84.8	94.9	94.7	94.7	96.6	83.7	85.4
0320	Concrete Reinforcing	107.9	42.7	74.3	99.1	91.0	94.9	97.2	85.6	91.2	99.1	76.3	87.4	94.0	81.8	87.7	99.6	84.4	91.8
0330	Cast-in-Place Concrete	109.0	46.9	85.2	100.5	100.4	100.5	98.8	96.7	98.0	101.5	96.0	99.4	89.7	90.0	89.8	82.6	88.3	84.8
03	CONCRETE	110.1	42.7	77.9	98.8	94.8	96.9	98.4	90.5	94.6	99.3	85.3	92.6	93.0	90.5	91.8	88.0	85.5	86.8
04	MASONRY	104.5	60.3	77.7	95.5	97.6	96.8	72.4	91.0	83.7	96.2	84.1	88.9	80.5	93.5	88.4	80.1	87.9	84.8
05	METALS	96.4	64.7	85.5	93.0	80.2	88.6	100.8	70.2	90.3	93.0	73.5	86.3	93.3	84.8	90.4	95.7	86.0	92.4
06	WOOD, PLASTICS & COMPOSITES	80.4	34.6	54.1	99.9	93.4	96.2	81.1	92.3	87.5	100.4	82.7	90.2	93.3	96.3	95.0	95.6	81.7	87.6
07	THERMAL & MOISTURE PROTECTION	107.3	49.1	84.2	107.8	98.5	104.1	101.4	93.4	98.2	108.9	92.0	102.2	102.6	90.9	97.9	100.9	92.2	97.5
08	OPENINGS	97.0	32.5	80.6	111.6	92.8	106.8	101.3	82.6	96.5	105.4	79.1	98.7	92.6	87.8	91.4	101.1	81.8	96.2
0920	Plaster & Gypsum Board	106.5	32.9	54.7	92.0	92.9	92.6	91.8	91.7	91.8	93.4	81.8	85.2	93.3	96.4	95.5	95.3	81.4	85.5
0950, 0980	Ceilings & Acoustic Treatment	140.9	32.9	69.3	91.5	92.9	92.4	97.3	91.7	93.6	91.5	81.8	85.1	91.3	96.4	94.7	92.2	81.4	85.0
0960	Flooring	117.6	37.9	94.7	101.1	92.6	98.6	129.8	97.6	120.5	101.2	76.2	94.0	105.8	98.2	103.6	106.8	91.2	102.3
0970, 0990	Wall Finishes & Painting/Coating	98.4	24.2	53.4	101.3	101.2	101.2	105.9	44.9	68.9	101.3	83.6	90.6	102.9	86.6	93.0	102.9	85.1	92.1
09	FINISHES	116.6	34.6	71.4	99.5	94.5	96.8	102.0	87.9	94.2	99.8	81.3	89.6	99.0	95.5	97.0	99.4	84.8	91.4
COVERS	DIVS. 10 - 14, 25, 28, 41, 43, 44, 46	100.0	36.1	86.7	100.0	81.4	96.1	100.0	58.0	91.3	100.0	61.3	91.9	100.0	77.7	95.4	100.0	92.5	98.5
21, 22, 23	FIRE SUPPRESSION, PLUMBING & HVAC	94.9	69.6	84.4	99.9	89.5	95.5	94.6	56.0	78.5	99.9	74.7	89.4	95.0	82.2	89.7	99.9	82.6	92.7
26, 27, 3370	ELECTRICAL, COMMUNICATIONS & UTIL.	98.3	66.5	81.4	99.9	92.5	96.0	98.3	98.0	98.1	99.1	88.8	93.6	98.4	79.5	88.3	96.9	80.6	88.2
MF2010	WEIGHTED AVERAGE	101.0	58.2	81.9	99.8	92.6	96.6	97.0	81.0	89.8	99.2	82.5	91.8	94.5	88.5	91.8	96.0	86.3	91.7

		OHIO																		
	DIVISION	CLEVELAND			COLUMBUS			DAYTON			HAMILTON			LIMA			LORAIN		
		441			430 - 432			453 - 454			450			458			440		
		MAT.	INST.	TOTAL	MAT.	INST.	TOTAL	MAT.	INST.	TOTAL	MAT.	INST.	TOTAL	MAT.	INST.	TOTAL	MAT.	INST.	TOTAL
015433	CONTRACTOR EQUIPMENT		97.1	97.1		95.6	95.6		96.4	96.4		101.6	101.6		93.8	93.8		96.8	96.8
0241, 31 - 34	SITE & INFRASTRUCTURE, DEMOLITION	94.1	104.6	101.4	92.3	100.7	98.1	78.3	105.6	97.2	78.2	105.9	97.3	87.8	93.9	92.0	93.5	103.3	100.3
0310	Concrete Forming & Accessories	101.2	99.3	99.6	99.3	88.9	90.3	96.6	83.1	84.9	96.6	83.2	85.0	93.2	86.9	87.7	101.2	89.2	90.8
0320	Concrete Reinforcing	99.6	91.3	95.4	99.6	84.2	91.7	99.6	80.4	89.7	99.6	84.4	91.8	97.2	80.5	88.6	99.1	91.1	95.0
0330	Cast-in-Place Concrete	98.6	108.0	102.2	88.2	93.0	90.0	76.8	89.4	81.6	82.3	87.8	84.4	90.8	98.8	93.9	95.7	102.6	98.4
03	CONCRETE	97.9	100.0	98.9	93.4	89.0	91.3	85.1	84.4	84.8	87.9	85.1	86.5	91.9	89.5	90.7	96.3	93.5	95.0
04	MASONRY	100.3	104.6	102.9	91.0	97.1	94.7	79.7	85.0	82.9	80.0	86.9	84.2	98.8	84.5	90.1	91.8	97.5	95.3
05	METALS	94.7	83.9	91.0	94.7	81.2	90.1	94.8	77.3	88.8	95.0	85.9	91.8	100.8	80.4	93.8	93.6	81.8	89.6
06	WOOD, PLASTICS & COMPOSITES	98.9	97.0	97.8	103.4	86.7	93.8	96.9	82.1	88.4	95.6	81.7	87.6	81.0	87.0	84.5	99.9	87.2	92.6
07	THERMAL & MOISTURE PROTECTION	106.5	108.3	107.3	101.0	96.4	99.2	105.6	90.4	99.6	102.7	92.0	98.4	100.9	96.7	99.2	108.8	101.4	105.9
08	OPENINGS	101.3	94.8	99.6	103.5	83.2	98.4	101.1	81.0	96.0	98.5	81.8	94.2	101.3	81.1	96.2	105.4	89.4	101.4
0920	Plaster & Gypsum Board	91.1	96.6	95.0	97.1	86.3	89.5	95.3	81.8	85.8	95.3	81.4	85.5	91.8	86.3	87.9	92.0	86.5	88.1
0950, 0980	Ceilings & Acoustic Treatment	89.6	96.6	94.2	88.1	86.3	86.9	93.2	81.8	85.7	92.2	81.4	85.0	96.3	86.3	89.6	91.5	86.5	88.2
0960	Flooring	100.9	103.4	101.6	95.5	90.8	94.2	109.7	81.2	101.5	106.8	91.2	102.3	120.9	89.7	117.7	101.0	105.4	102.5
0970, 0990	Wall Finishes & Painting/Coating	101.3	106.6	104.5	97.8	93.3	95.1	102.9	80.1	89.1	102.9	81.4	89.9	105.9	85.9	93.8	101.3	106.6	104.5
09	FINISHES	99.1	100.7	100.0	97.5	89.5	93.1	100.5	82.2	90.4	99.4	84.2	91.0	101.2	87.2	93.5	99.6	93.8	96.4
COVERS	DIVS. 10 - 14, 25, 28, 41, 43, 44, 46	100.0	98.6	99.7	100.0	95.0	99.0	100.0	92.1	98.4	100.0	92.1	98.4	100.0	79.9	95.8	100.0	95.4	99.0
21, 22, 23	FIRE SUPPRESSION, PLUMBING & HVAC	99.8	99.3	99.6	99.9	90.7	96.1	100.6	78.9	91.6	100.4	82.1	92.8	94.6	85.2	90.7	99.9	84.8	93.6
26, 27, 3370	ELECTRICAL, COMMUNICATIONS & UTIL.	99.4	108.9	104.5	99.3	86.7	92.6	95.6	78.8	86.6	95.9	65.1	79.5	98.7	76.0	86.6	99.2	91.1	94.9
MF2010	WEIGHTED AVERAGE	99.0	100.5	99.7	97.9	90.3	94.5	95.8	83.6	90.3	95.7	83.8	90.4	97.4	85.0	91.9	98.7	91.5	95.5

		OHIO																		
	DIVISION	MANSFIELD			MARION			SPRINGFIELD			STEUBENVILLE			TOLEDO			YOUNGSTOWN		
		448 - 449			433			455			439			434 - 436			444 - 445		
		MAT.	INST.	TOTAL	MAT.	INST.	TOTAL	MAT.	INST.	TOTAL	MAT.	INST.	TOTAL	MAT.	INST.	TOTAL	MAT.	INST.	TOTAL
015433	CONTRACTOR EQUIPMENT		96.8	96.8		95.3	95.3		96.4	96.4		100.0	100.0		98.2	98.2		96.8	96.8
0241, 31 - 34	SITE & INFRASTRUCTURE, DEMOLITION	90.5	104.3	100.0	88.6	99.3	96.0	78.6	104.2	96.3	134.3	109.0	116.8	91.5	100.2	97.5	94.0	104.2	101.0
0310	Concrete Forming & Accessories	91.3	87.9	88.4	96.1	75.4	78.2	96.6	86.0	87.4	98.7	87.8	89.2	99.3	96.8	97.2	101.2	84.1	86.4
0320	Concrete Reinforcing	90.2	76.9	83.3	91.8	84.1	87.9	99.6	81.4	90.3	89.4	86.6	87.9	99.6	91.0	95.2	99.1	86.4	92.5
0330	Cast-in-Place Concrete	93.1	91.8	92.6	80.6	89.5	84.0	78.9	89.3	82.9	87.5	94.6	90.2	88.2	104.7	94.5	99.6	98.6	99.2
03	CONCRETE	91.1	86.6	88.9	85.7	81.7	83.8	86.1	85.9	86.0	89.6	89.3	89.4	93.4	98.0	95.6	98.3	88.9	93.8
04	MASONRY	94.6	97.2	96.2	92.9	96.1	94.8	79.8	85.0	83.0	79.8	92.5	87.5	99.6	101.0	100.5	95.8	90.0	92.3
05	METALS	93.8	75.6	87.5	93.9	77.6	88.3	94.8	77.5	88.9	90.6	79.0	86.6	94.4	86.8	91.8	93.0	78.4	88.0
06	WOOD, PLASTICS & COMPOSITES	88.4	87.2	87.7	99.6	72.9	84.3	98.3	86.4	91.5	95.4	87.0	90.6	103.4	96.7	99.5	99.9	82.3	89.8
07	THERMAL & MOISTURE PROTECTION	108.3	96.8	103.7	100.6	84.4	94.2	105.5	90.8	99.7	112.4	94.5	105.3	102.7	105.4	103.8	109.0	91.5	102.1
08	OPENINGS	104.5	80.2	98.3	97.5	74.3	91.6	98.9	81.7	94.5	98.3	84.1	94.7	101.3	92.8	99.1	105.4	84.2	100.0
0920	Plaster & Gypsum Board	85.2	86.5	86.1	94.8	72.1	78.8	95.3	86.3	88.9	95.4	86.1	88.8	97.1	96.5	96.7	92.0	81.5	84.6
0950, 0980	Ceilings & Acoustic Treatment	92.4	86.5	88.5	88.1	72.1	77.5	93.2	86.3	88.6	84.9	86.1	85.7	88.1	96.5	93.7	91.5	81.5	84.8
0960	Flooring	96.6	104.2	98.8	94.0	73.7	88.1	109.7	81.2	101.5	122.5	84.1	111.4	94.8	100.2	96.4	101.2	87.8	97.4
0970, 0990	Wall Finishes & Painting/Coating	101.3	93.1	96.3	97.8	51.1	69.5	102.9	80.1	89.1	113.1	91.4	99.9	97.8	105.1	102.2	101.3	90.9	95.0
09	FINISHES	97.3	91.7	94.2	96.4	72.2	83.0	100.5	84.7	91.8	115.8	87.2	100.0	97.3	98.6	98.0	99.7	85.0	91.6
COVERS	DIVS. 10 - 14, 25, 28, 41, 43, 44, 46	100.0	83.3	96.5	100.0	58.3	91.3	100.0	92.6	98.5	100.0	83.5	96.6	100.0	82.4	96.3	100.0	78.7	95.6
21, 22, 23	FIRE SUPPRESSION, PLUMBING & HVAC	94.5	83.6	89.9	94.6	80.7	88.8	100.6	78.9	91.6	94.8	87.1	91.6	99.9	92.7	96.9	99.9	90.8	96.1
26, 27, 3370	ELECTRICAL, COMMUNICATIONS & UTIL.	96.5	78.9	87.1	92.7	74.3	82.9	95.6	76.5	85.4	87.3	98.2	93.1	99.6	97.6	98.5	99.2	80.0	89.0
MF2010	WEIGHTED AVERAGE	96.2	87.0	92.1	94.1	80.8	88.2	95.7	83.8	90.4	96.4	90.5	93.8	98.1	95.9	97.1	99.1	87.5	93.9

667

Table 1

DIVISION		OHIO ZANESVILLE 437 - 438			OKLAHOMA ARDMORE 734			OKLAHOMA CLINTON 736			OKLAHOMA DURANT 747			OKLAHOMA ENID 737			OKLAHOMA GUYMON 739		
		MAT.	INST.	TOTAL	MAT.	INST.	TOTAL	MAT.	INST.	TOTAL	MAT.	INST.	TOTAL	MAT.	INST.	TOTAL	MAT.	INST.	TOTAL
015433	CONTRACTOR EQUIPMENT		95.3	95.3		81.0	81.0		80.1	80.1		80.1	80.1		80.1	80.1		80.1	80.1
0241, 31 - 34	SITE & INFRASTRUCTURE, DEMOLITION	91.3	100.9	97.9	103.2	90.3	94.3	104.7	88.9	93.8	97.7	88.6	91.4	106.0	88.9	94.2	109.5	87.4	94.3
0310	Concrete Forming & Accessories	93.2	75.4	77.7	91.7	40.3	47.1	90.6	34.9	42.2	83.9	40.3	46.1	93.8	34.2	42.1	97.3	22.2	32.1
0320	Concrete Reinforcing	91.3	86.1	88.6	100.3	78.9	89.3	100.8	78.9	89.5	101.2	62.5	81.3	100.2	78.9	89.2	100.8	27.7	63.1
0330	Cast-in-Place Concrete	84.9	88.7	86.3	99.2	43.4	77.9	95.9	43.4	75.8	92.7	43.3	73.8	95.9	45.8	76.7	95.9	31.0	71.1
03	CONCRETE	89.1	81.8	85.6	95.7	49.2	73.5	95.0	46.8	71.9	90.4	46.0	69.1	95.5	47.3	72.4	98.1	27.0	64.1
04	MASONRY	89.9	84.0	86.3	96.8	57.0	72.7	125.2	57.0	83.9	95.4	65.0	77.0	106.0	57.0	76.3	99.9	22.6	53.1
05	METALS	95.2	79.9	89.9	91.5	69.8	84.1	91.6	69.7	84.1	91.6	63.1	81.8	93.1	69.7	85.1	92.1	39.8	74.1
06	WOOD, PLASTICS & COMPOSITES	95.8	72.9	82.7	97.2	38.9	63.8	96.2	31.6	59.1	88.1	38.9	59.9	99.9	30.7	60.2	103.6	20.3	55.8
07	THERMAL & MOISTURE PROTECTION	100.7	87.7	95.5	97.0	60.1	82.3	97.1	59.3	82.1	96.8	57.9	81.3	97.2	59.2	82.1	97.5	28.2	70.0
08	OPENINGS	97.5	76.3	92.1	95.1	49.6	83.5	95.1	45.6	82.6	95.1	45.5	82.5	95.1	45.0	82.4	95.2	21.7	76.6
0920	Plaster & Gypsum Board	92.2	72.1	78.0	78.4	37.8	49.8	77.9	30.2	44.3	73.5	37.8	48.4	79.3	29.3	44.1	79.8	18.6	36.7
0950, 0980	Ceilings & Acoustic Treatment	88.1	72.1	77.5	82.8	37.8	53.0	82.8	30.2	47.9	82.8	37.8	53.0	82.8	29.3	47.4	84.7	18.6	40.9
0960	Flooring	92.2	77.6	88.0	105.9	45.6	88.6	104.5	43.5	87.0	100.7	65.4	90.6	106.6	43.5	88.5	108.6	25.7	84.8
0970, 0990	Wall Finishes & Painting/Coating	97.8	72.8	82.7	97.1	55.2	71.7	97.1	55.2	71.7	97.1	55.2	71.7	97.1	55.2	71.7	97.1	19.7	50.1
09	FINISHES	95.8	74.7	84.2	91.3	41.0	63.6	91.0	36.3	60.9	88.8	45.2	64.7	91.9	35.8	60.9	93.4	22.2	54.1
COVERS	DIVS. 10 - 14, 25, 28, 41, 43, 44, 46	100.0	73.7	94.5	100.0	75.3	94.9	100.0	74.5	94.7	100.0	75.4	94.9	100.0	74.4	94.7	100.0	72.3	94.2
21, 22, 23	FIRE SUPPRESSION, PLUMBING & HVAC	94.6	84.9	90.6	94.8	62.7	81.4	94.8	62.7	81.4	94.8	62.8	81.5	100.2	62.7	84.6	94.8	23.6	65.1
26, 27, 3370	ELECTRICAL, COMMUNICATIONS & UTIL.	92.8	69.5	80.4	95.6	70.0	82.0	96.7	70.0	82.5	98.6	52.8	74.2	96.7	70.0	82.5	98.6	16.7	55.0
MF2010	WEIGHTED AVERAGE	94.6	81.1	88.6	95.2	61.0	79.9	96.7	59.6	80.2	94.3	58.5	78.3	97.4	59.6	80.5	96.5	31.1	67.3

Table 2

DIVISION		OKLAHOMA LAWTON 735			OKLAHOMA MCALESTER 745			OKLAHOMA MIAMI 743			OKLAHOMA MUSKOGEE 744			OKLAHOMA OKLAHOMA CITY 730 - 731			OKLAHOMA PONCA CITY 746		
		MAT.	INST.	TOTAL	MAT.	INST.	TOTAL	MAT.	INST.	TOTAL	MAT.	INST.	TOTAL	MAT.	INST.	TOTAL	MAT.	INST.	TOTAL
015433	CONTRACTOR EQUIPMENT		81.0	81.0		80.1	80.1		88.5	88.5		88.5	88.5		81.2	81.2		80.1	80.1
0241, 31 - 34	SITE & INFRASTRUCTURE, DEMOLITION	101.5	90.3	93.8	90.5	89.0	89.4	92.3	86.2	88.1	92.3	84.8	87.1	99.9	90.8	93.7	98.3	88.9	91.8
0310	Concrete Forming & Accessories	97.1	46.1	52.9	82.0	41.6	47.0	94.0	70.7	73.8	98.2	30.6	39.6	96.2	40.0	47.4	89.8	40.7	47.2
0320	Concrete Reinforcing	100.4	78.9	89.4	100.8	41.6	70.3	99.3	79.0	88.8	100.2	31.6	64.9	100.4	78.9	89.3	100.2	78.9	89.2
0330	Cast-in-Place Concrete	92.7	45.9	74.8	81.3	45.9	67.7	85.2	45.2	69.9	86.2	35.9	66.9	93.9	48.6	76.5	95.2	38.5	73.5
03	CONCRETE	91.8	52.6	73.1	81.3	43.5	63.2	85.6	64.1	75.3	87.3	33.9	61.7	92.4	50.8	72.5	92.4	47.7	71.5
04	MASONRY	99.8	57.0	73.9	114.6	57.6	80.1	98.1	58.2	73.9	117.4	46.2	74.3	99.3	58.8	74.8	90.7	57.6	70.6
05	METALS	96.7	69.8	87.5	91.5	53.5	78.4	91.5	82.3	88.3	93.1	55.9	80.3	98.6	69.8	88.7	91.5	69.5	83.9
06	WOOD, PLASTICS & COMPOSITES	102.6	46.7	70.6	85.7	42.2	60.7	99.8	79.2	88.0	104.0	30.4	61.8	101.9	37.6	65.0	95.5	39.5	63.4
07	THERMAL & MOISTURE PROTECTION	97.0	60.9	82.7	96.4	56.7	80.6	96.8	61.3	82.7	96.8	41.8	75.0	93.7	60.7	80.6	96.9	59.3	82.0
08	OPENINGS	96.7	53.7	85.8	95.1	40.9	81.4	95.1	71.9	89.2	95.0	29.5	78.4	96.7	48.7	84.6	95.1	49.9	83.6
0920	Plaster & Gypsum Board	81.4	45.8	56.3	72.2	41.2	50.3	79.3	79.1	79.2	81.4	28.8	44.4	83.4	36.4	50.3	77.9	38.4	50.1
0950, 0980	Ceilings & Acoustic Treatment	90.3	45.8	60.8	82.8	41.2	55.2	82.8	79.1	80.4	90.3	28.8	49.6	90.3	36.4	54.6	82.8	38.4	53.4
0960	Flooring	109.0	43.5	90.2	99.6	43.5	83.5	107.4	65.4	95.4	110.0	42.0	90.5	109.0	43.5	90.2	104.1	43.5	86.7
0970, 0990	Wall Finishes & Painting/Coating	97.1	55.2	71.7	97.1	40.7	62.9	97.1	83.4	88.8	97.1	29.6	56.1	97.1	55.2	71.7	97.1	55.2	71.7
09	FINISHES	93.9	45.2	67.0	87.7	41.1	62.0	90.7	72.2	80.5	93.4	32.5	59.8	93.8	40.2	64.3	90.4	41.1	63.3
COVERS	DIVS. 10 - 14, 25, 28, 41, 43, 44, 46	100.0	76.2	95.1	100.0	75.9	95.0	100.0	80.7	96.0	100.0	74.0	94.6	100.0	75.8	95.0	100.0	75.5	94.9
21, 22, 23	FIRE SUPPRESSION, PLUMBING & HVAC	100.2	62.7	84.6	94.8	30.8	68.2	94.8	61.6	81.0	100.2	23.9	68.4	100.0	63.6	84.9	94.8	61.2	80.8
26, 27, 3370	ELECTRICAL, COMMUNICATIONS & UTIL.	98.6	70.0	83.4	96.8	67.8	81.4	98.5	67.9	82.2	96.3	28.6	60.3	104.0	70.0	85.9	96.3	64.0	79.1
MF2010	WEIGHTED AVERAGE	97.6	62.3	81.8	93.7	51.2	74.7	94.0	69.0	82.8	96.7	39.6	71.2	98.2	61.5	81.8	94.3	59.6	78.8

Table 3

DIVISION		OKLAHOMA POTEAU 749			OKLAHOMA SHAWNEE 748			OKLAHOMA TULSA 740 - 741			OKLAHOMA WOODWARD 738			OREGON BEND 977			OREGON EUGENE 974		
		MAT.	INST.	TOTAL	MAT.	INST.	TOTAL	MAT.	INST.	TOTAL	MAT.	INST.	TOTAL	MAT.	INST.	TOTAL	MAT.	INST.	TOTAL
015433	CONTRACTOR EQUIPMENT		87.8	87.8		80.1	80.1		88.5	88.5		80.1	80.1		99.4	99.4		99.4	99.4
0241, 31 - 34	SITE & INFRASTRUCTURE, DEMOLITION	76.2	84.8	82.2	101.6	88.8	92.8	98.6	86.2	90.0	105.1	88.9	93.9	115.3	104.6	107.9	103.3	104.6	104.2
0310	Concrete Forming & Accessories	87.8	43.0	48.9	83.8	39.3	45.2	98.1	39.8	47.5	90.7	34.8	42.2	109.7	95.3	97.2	106.4	95.2	96.7
0320	Concrete Reinforcing	101.3	78.9	89.8	100.2	45.7	72.1	100.4	78.8	89.3	100.2	78.9	89.2	104.3	102.9	103.6	108.8	102.9	105.8
0330	Cast-in-Place Concrete	85.2	44.7	69.7	98.3	42.9	77.1	94.0	44.7	75.1	95.9	43.4	75.8	100.2	106.2	102.5	97.0	106.2	100.5
03	CONCRETE	87.6	51.6	70.4	94.2	42.3	69.3	92.5	50.2	72.2	95.3	46.7	72.0	111.8	100.3	106.3	102.8	100.2	101.6
04	MASONRY	99.1	57.6	74.0	115.9	56.4	79.8	98.9	58.8	74.6	93.8	57.0	71.5	109.9	101.5	104.8	107.4	101.5	103.8
05	METALS	91.5	81.9	88.2	91.4	55.3	79.0	96.4	80.9	91.1	91.7	69.7	84.1	88.7	97.9	91.9	89.4	97.8	92.3
06	WOOD, PLASTICS & COMPOSITES	92.6	42.4	63.8	87.9	38.7	59.7	103.2	38.7	66.2	96.4	31.5	59.1	98.0	93.1	95.1	93.9	93.1	93.4
07	THERMAL & MOISTURE PROTECTION	96.9	57.0	81.0	96.9	59.9	82.2	96.8	55.4	80.4	97.2	59.3	82.1	106.3	94.2	101.5	105.6	91.2	99.9
08	OPENINGS	95.1	52.0	84.2	95.0	38.1	80.6	96.7	48.9	84.6	95.1	45.6	82.6	97.5	100.6	98.3	97.9	100.6	98.6
0920	Plaster & Gypsum Board	76.6	41.2	51.7	73.5	37.5	48.2	81.4	37.4	50.4	78.5	30.1	44.4	106.9	92.7	96.9	104.6	92.7	96.2
0950, 0980	Ceilings & Acoustic Treatment	82.8	41.2	55.2	82.8	37.5	52.8	90.3	37.4	55.3	84.7	30.1	48.5	101.1	92.7	95.5	102.1	92.7	95.8
0960	Flooring	103.4	65.4	92.5	100.6	34.6	81.7	108.8	45.2	90.5	104.5	45.6	87.6	110.6	100.7	107.8	108.8	100.7	106.5
0970, 0990	Wall Finishes & Painting/Coating	97.1	55.2	71.7	97.1	36.7	60.5	97.1	43.6	64.7	97.1	55.2	71.7	108.1	70.5	85.3	108.1	70.5	85.3
09	FINISHES	88.3	47.2	65.6	89.0	36.7	60.2	93.2	40.0	63.9	91.6	36.6	61.3	107.4	93.1	99.5	105.5	93.1	98.7
COVERS	DIVS. 10 - 14, 25, 28, 41, 43, 44, 46	100.0	76.2	95.0	100.0	75.0	94.8	100.0	76.2	95.0	100.0	74.4	94.7	100.0	100.3	100.1	100.0	100.3	100.1
21, 22, 23	FIRE SUPPRESSION, PLUMBING & HVAC	94.8	61.3	80.8	94.8	52.0	81.1	100.2	48.4	78.6	94.8	62.7	81.4	94.7	98.7	96.4	100.1	98.7	99.5
26, 27, 3370	ELECTRICAL, COMMUNICATIONS & UTIL.	96.4	67.9	81.3	98.7	70.0	83.5	98.6	37.3	66.0	98.5	70.0	83.3	102.4	95.2	98.6	100.9	95.2	97.9
MF2010	WEIGHTED AVERAGE	93.3	62.3	79.5	96.0	57.3	78.7	97.4	54.2	78.1	95.4	59.7	79.4	100.3	98.4	99.5	99.8	98.3	99.1

City Cost Indexes

OREGON

DIVISION		KLAMATH FALLS 976 MAT.	INST.	TOTAL	MEDFORD 975 MAT.	INST.	TOTAL	PENDLETON 978 MAT.	INST.	TOTAL	PORTLAND 970-972 MAT.	INST.	TOTAL	SALEM 973 MAT.	INST.	TOTAL	VALE 979 MAT.	INST.	TOTAL
015433	CONTRACTOR EQUIPMENT		99.4	99.4		99.4	99.4		96.7	96.7		99.4	99.4		99.4	99.4		96.7	96.7
0241, 31 - 34	SITE & INFRASTRUCTURE, DEMOLITION	120.0	104.6	109.3	112.1	104.6	106.9	110.5	98.0	101.8	106.2	104.6	105.1	98.4	104.6	102.7	97.0	97.9	97.6
0310	Concrete Forming & Accessories	103.2	95.1	96.2	102.4	95.1	96.1	102.9	95.5	96.5	107.5	95.4	97.0	105.2	95.3	96.6	108.9	94.5	96.4
0320	Concrete Reinforcing	104.3	102.9	103.6	106.1	102.9	104.4	103.6	103.1	103.3	109.6	103.1	106.2	109.7	103.1	106.3	101.0	102.9	102.0
0330	Cast-in-Place Concrete	100.2	106.2	102.5	100.2	106.2	102.5	100.9	107.6	103.5	99.7	106.3	102.2	88.9	106.2	95.5	79.5	107.2	90.1
03	CONCRETE	114.8	100.2	107.8	109.2	100.2	104.9	94.6	100.9	97.6	104.3	100.4	102.4	98.8	100.4	99.5	79.2	100.3	89.3
04	MASONRY	124.5	101.5	110.6	104.3	101.5	102.6	114.1	103.7	107.8	108.0	103.6	105.3	112.3	103.6	107.0	112.4	103.7	107.1
05	METALS	88.7	97.7	91.8	89.0	97.7	92.0	94.5	98.6	95.9	90.2	98.2	93.0	89.8	98.1	92.6	94.4	97.6	95.5
06	WOOD, PLASTICS & COMPOSITES	89.2	93.1	91.4	88.1	93.1	90.9	91.0	93.2	92.2	94.8	93.1	93.8	91.1	93.1	92.2	99.2	93.2	95.8
07	THERMAL & MOISTURE PROTECTION	106.5	92.7	101.0	106.2	92.7	100.8	98.0	93.5	96.2	105.6	97.3	102.3	103.5	94.8	100.1	97.4	91.1	94.9
08	OPENINGS	97.5	100.6	98.3	100.5	100.6	100.5	93.1	100.7	95.0	95.5	100.6	96.8	97.3	100.6	98.2	93.1	93.2	93.1
0920	Plaster & Gypsum Board	102.1	92.7	95.5	101.2	92.7	95.2	82.4	92.7	89.6	104.9	92.7	96.3	99.2	92.7	94.6	89.1	92.7	91.6
0950, 0980	Ceilings & Acoustic Treatment	109.5	92.7	98.3	114.7	92.7	100.1	69.1	92.7	84.7	104.4	92.7	96.6	104.4	92.7	96.6	69.1	92.7	84.7
0960	Flooring	107.3	100.7	105.4	106.6	100.7	104.9	74.9	100.7	82.3	106.3	100.7	104.7	109.0	100.7	106.6	76.8	100.7	83.7
0970, 0990	Wall Finishes & Painting/Coating	108.1	62.6	80.5	108.1	62.6	80.5	96.2	70.5	80.6	107.7	70.5	85.1	108.1	70.5	85.3	96.2	70.5	80.6
09	FINISHES	108.1	92.3	99.4	108.0	92.3	99.3	73.1	93.2	84.2	105.3	93.1	98.6	104.9	93.1	98.4	73.4	93.2	84.3
COVERS	DIVS. 10 - 14, 25, 28, 41, 43, 44, 46	100.0	100.3	100.1	100.0	100.3	100.1	100.0	98.8	99.8	100.0	100.4	100.1	100.0	100.3	100.1	100.0	100.6	100.1
21, 22, 23	FIRE SUPPRESSION, PLUMBING & HVAC	94.7	98.7	96.4	100.1	98.7	99.5	96.1	108.3	101.2	100.1	98.7	99.6	100.1	98.7	99.6	96.1	75.5	87.5
26, 27, 3370	ELECTRICAL, COMMUNICATIONS & UTIL.	101.0	84.7	92.4	104.8	84.7	94.1	92.6	96.4	94.6	101.1	102.4	101.8	106.3	95.2	100.4	92.6	78.4	85.0
MF2010	WEIGHTED AVERAGE	101.5	96.7	99.3	101.5	96.7	99.3	94.9	100.3	97.3	100.0	99.7	99.9	99.8	98.6	99.3	92.6	90.4	91.6

PENNSYLVANIA

DIVISION		ALLENTOWN 181 MAT.	INST.	TOTAL	ALTOONA 166 MAT.	INST.	TOTAL	BEDFORD 155 MAT.	INST.	TOTAL	BRADFORD 167 MAT.	INST.	TOTAL	BUTLER 160 MAT.	INST.	TOTAL	CHAMBERSBURG 172 MAT.	INST.	TOTAL
015433	CONTRACTOR EQUIPMENT		114.2	114.2		114.2	114.2		111.3	111.3		114.2	114.2		114.2	114.2		113.5	113.5
0241, 31 - 34	SITE & INFRASTRUCTURE, DEMOLITION	92.3	105.9	101.7	96.3	105.9	103.0	112.5	102.8	105.8	91.8	105.3	101.1	87.5	107.3	101.2	86.2	103.5	98.1
0310	Concrete Forming & Accessories	101.8	114.3	112.6	87.2	79.4	80.4	87.6	81.8	82.6	89.4	79.7	81.0	88.7	93.7	93.0	86.7	80.9	81.7
0320	Concrete Reinforcing	97.1	113.3	105.5	94.2	91.5	92.8	95.9	80.5	88.0	96.2	91.5	93.8	94.8	103.7	99.4	94.6	99.9	97.3
0330	Cast-in-Place Concrete	82.7	102.5	90.3	92.2	86.6	90.0	106.3	74.0	93.9	88.1	89.0	88.4	81.4	97.6	87.6	89.4	73.7	83.4
03	CONCRETE	96.2	110.8	103.2	91.2	85.6	88.6	101.9	80.3	91.5	98.1	86.6	92.6	82.9	98.1	90.2	104.2	83.6	94.3
04	MASONRY	95.3	101.1	98.8	96.9	69.6	80.3	100.6	85.7	91.6	94.1	83.1	87.4	99.0	96.9	97.7	95.2	89.5	91.7
05	METALS	100.2	124.7	108.6	94.1	111.9	100.2	97.1	104.7	99.7	98.0	111.3	102.6	93.8	119.1	102.5	98.0	114.3	103.6
06	WOOD, PLASTICS & COMPOSITES	105.1	118.0	112.5	83.3	81.3	82.2	87.8	81.2	84.0	90.2	77.5	82.9	84.9	91.2	88.5	83.8	81.3	86.6
07	THERMAL & MOISTURE PROTECTION	101.1	117.1	107.4	100.5	91.4	96.9	105.5	89.0	98.9	101.1	90.2	96.8	100.2	100.2	100.2	100.2	75.9	90.5
08	OPENINGS	92.7	116.9	98.8	87.3	85.9	86.9	93.9	82.9	91.1	93.1	87.6	91.7	87.3	101.4	90.8	89.5	82.8	87.8
0920	Plaster & Gypsum Board	106.5	118.3	114.8	97.8	80.5	85.6	98.3	80.5	85.7	99.2	76.6	83.3	98.3	90.7	92.9	103.2	80.5	87.2
0950, 0980	Ceilings & Acoustic Treatment	93.1	118.3	109.8	96.9	80.5	86.0	98.7	80.5	86.6	95.1	76.6	82.9	97.9	90.7	93.1	95.1	80.5	85.4
0960	Flooring	90.1	97.0	92.1	84.0	93.0	86.6	89.4	93.1	90.5	84.2	97.3	87.9	84.8	85.6	85.0	87.8	48.8	76.6
0970, 0990	Wall Finishes & Painting/Coating	92.5	76.0	82.5	88.6	105.7	99.0	95.0	72.8	81.5	92.5	94.2	93.6	88.6	105.7	99.0	92.5	62.2	74.1
09	FINISHES	96.7	107.6	102.7	95.1	83.9	88.9	96.9	81.5	88.4	94.5	82.7	88.0	95.0	92.2	93.5	95.1	72.6	82.6
COVERS	DIVS. 10 - 14, 25, 28, 41, 43, 44, 46	100.0	105.2	101.1	100.0	95.8	99.1	100.0	99.0	99.8	100.0	98.9	99.8	100.0	103.1	101.1	100.0	95.9	99.1
21, 22, 23	FIRE SUPPRESSION, PLUMBING & HVAC	100.3	110.3	104.5	99.8	82.9	92.8	94.7	87.6	91.8	94.9	89.2	92.5	94.5	94.5	94.5	95.0	89.0	92.5
26, 27, 3370	ELECTRICAL, COMMUNICATIONS & UTIL.	100.7	96.8	98.6	90.0	105.4	98.2	93.6	105.4	99.9	93.6	105.4	99.8	90.7	104.2	97.9	92.1	84.4	88.0
MF2010	WEIGHTED AVERAGE	98.3	108.6	102.9	95.0	90.5	93.0	97.6	91.1	94.7	95.9	93.1	94.7	92.6	100.3	96.0	96.1	88.6	92.7

PENNSYLVANIA

DIVISION		DOYLESTOWN 189 MAT.	INST.	TOTAL	DUBOIS 158 MAT.	INST.	TOTAL	ERIE 164-165 MAT.	INST.	TOTAL	GREENSBURG 156 MAT.	INST.	TOTAL	HARRISBURG 170-171 MAT.	INST.	TOTAL	HAZLETON 182 MAT.	INST.	TOTAL
015433	CONTRACTOR EQUIPMENT		94.4	94.4		111.3	111.3		114.2	114.2		111.3	111.3		113.5	113.5		114.2	114.2
0241, 31 - 34	SITE & INFRASTRUCTURE, DEMOLITION	107.2	91.6	96.4	118.5	102.9	107.8	92.8	106.5	102.3	107.9	104.9	105.8	82.2	104.8	97.8	85.6	106.1	99.8
0310	Concrete Forming & Accessories	86.1	131.0	125.1	87.1	82.9	83.4	101.3	87.3	89.1	93.7	93.6	93.6	94.7	86.0	87.1	83.9	87.2	86.7
0320	Concrete Reinforcing	94.0	141.0	118.2	95.4	91.8	93.5	96.2	93.4	94.8	95.4	103.7	99.6	97.1	102.0	99.6	94.3	108.0	101.3
0330	Cast-in-Place Concrete	78.2	91.1	83.1	102.4	92.3	98.5	90.6	85.5	88.6	98.6	97.2	98.1	87.1	94.3	89.9	78.2	90.9	83.1
03	CONCRETE	91.9	119.0	104.9	103.0	89.1	96.3	90.0	89.1	89.6	96.7	97.9	97.3	96.1	93.3	94.7	88.7	93.8	91.1
04	MASONRY	98.5	130.7	118.0	100.6	86.6	92.1	87.6	91.2	89.7	110.9	96.9	102.4	95.1	89.6	91.7	107.4	96.1	100.6
05	METALS	97.7	129.7	108.7	97.1	110.8	101.8	94.3	112.2	100.5	97.0	117.7	104.2	102.3	118.6	107.9	99.9	121.2	107.2
06	WOOD, PLASTICS & COMPOSITES	85.4	133.2	112.8	86.7	81.2	83.5	101.1	85.7	92.3	94.4	91.2	92.5	98.9	85.0	90.9	84.0	85.0	84.5
07	THERMAL & MOISTURE PROTECTION	99.4	134.3	113.3	105.7	94.7	101.3	100.8	93.2	97.8	105.3	99.8	103.1	107.0	108.0	107.4	100.7	106.4	102.9
08	OPENINGS	94.3	142.9	106.6	93.9	89.6	92.8	87.4	88.4	87.7	93.9	101.4	96.0	92.7	91.6	92.4	92.8	93.2	93.2
0920	Plaster & Gypsum Board	96.2	133.9	122.8	97.0	80.5	85.4	106.5	85.1	91.4	99.4	90.7	93.3	103.3	84.3	89.9	96.8	84.3	88.2
0950, 0980	Ceilings & Acoustic Treatment	92.3	133.9	119.9	98.7	80.5	86.6	93.1	85.1	87.8	97.8	90.7	93.1	96.9	84.3	88.5	94.2	84.3	87.6
0960	Flooring	74.4	134.5	91.7	89.1	99.2	92.0	89.5	97.0	91.6	92.6	71.9	86.6	90.7	92.6	91.3	81.6	95.7	85.6
0970, 0990	Wall Finishes & Painting/Coating	92.0	74.3	81.3	95.0	105.7	101.5	97.0	89.4	92.4	95.0	105.7	101.5	92.5	87.8	89.7	92.5	94.6	93.8
09	FINISHES	87.5	124.8	108.1	97.2	87.5	91.9	97.4	88.9	92.7	97.4	90.6	93.6	95.8	86.9	90.9	92.7	87.4	89.8
COVERS	DIVS. 10 - 14, 25, 28, 41, 43, 44, 46	100.0	76.9	95.2	100.0	99.0	99.8	100.0	100.9	100.2	100.0	103.0	100.6	100.0	96.3	99.2	100.0	99.8	100.0
21, 22, 23	FIRE SUPPRESSION, PLUMBING & HVAC	94.5	127.3	108.1	94.7	87.9	91.9	99.8	92.2	96.6	94.7	92.1	93.7	100.1	89.8	95.8	94.9	98.3	96.4
26, 27, 3370	ELECTRICAL, COMMUNICATIONS & UTIL.	91.3	120.5	107.6	94.3	105.4	100.2	92.0	86.0	88.8	94.3	105.4	100.2	97.3	84.5	90.5	94.5	87.6	90.8
MF2010	WEIGHTED AVERAGE	95.1	121.9	107.1	98.0	94.2	96.3	94.8	93.5	94.2	97.5	99.4	98.3	98.0	93.8	96.1	95.5	97.3	96.3

City Cost Indexes

PENNSYLVANIA

DIVISION		INDIANA 157 MAT.	INST.	TOTAL	JOHNSTOWN 159 MAT.	INST.	TOTAL	KITTANNING 162 MAT.	INST.	TOTAL	LANCASTER 175-176 MAT.	INST.	TOTAL	LEHIGH VALLEY 180 MAT.	INST.	TOTAL	MONTROSE 188 MAT.	INST.	TOTAL
015433	CONTRACTOR EQUIPMENT		111.3	111.3		111.3	111.3		114.2	114.2		113.5	113.5		114.2	114.2		114.2	114.2
0241, 31 - 34	SITE & INFRASTRUCTURE, DEMOLITION	105.8	103.7	104.4	113.2	104.1	106.9	90.2	107.3	102.0	78.2	104.8	96.6	89.7	105.9	100.9	88.6	104.1	99.3
0310	Concrete Forming & Accessories	88.2	86.0	86.3	87.1	83.1	83.6	88.7	93.8	93.1	88.6	85.7	86.1	95.4	114.1	111.7	84.7	86.3	86.1
0320	Concrete Reinforcing	94.6	103.7	99.3	95.9	103.5	99.8	94.8	103.8	99.4	94.3	101.9	98.2	94.3	113.3	104.1	98.8	107.8	103.4
0330	Cast-in-Place Concrete	96.7	97.1	96.8	107.2	91.8	101.3	84.6	97.7	89.6	76.1	94.5	83.1	84.6	102.4	91.4	83.0	89.3	85.4
03	CONCRETE	94.4	94.4	94.4	102.7	91.3	97.2	85.5	98.2	91.6	91.4	93.2	92.3	95.2	110.7	102.6	93.8	92.6	93.2
04	MASONRY	96.8	98.5	97.9	97.7	86.1	90.7	103.1	98.5	100.3	100.7	90.2	94.3	95.3	101.1	98.8	95.3	95.8	95.6
05	METALS	97.2	117.5	104.2	97.1	116.3	103.7	93.9	119.3	102.7	98.0	118.1	104.9	99.8	124.3	108.3	98.1	116.3	104.4
06	WOOD, PLASTICS & COMPOSITES	88.6	81.2	84.4	86.7	81.2	83.5	84.9	91.2	88.5	96.4	85.0	89.9	97.2	118.0	109.1	84.8	85.3	85.1
07	THERMAL & MOISTURE PROTECTION	105.2	97.8	102.3	105.5	93.2	100.6	100.3	100.6	100.4	99.8	97.1	98.7	101.0	105.6	102.8	100.7	94.4	98.2
08	OPENINGS	93.9	93.2	93.7	93.9	89.5	92.8	87.3	98.7	90.2	89.5	95.6	91.1	93.4	116.9	99.4	89.7	93.0	90.5
0920	Plaster & Gypsum Board	98.8	80.5	85.9	96.7	80.5	85.3	98.3	90.7	92.9	105.4	84.3	90.5	101.2	118.3	113.3	97.0	84.6	88.3
0950, 0980	Ceilings & Acoustic Treatment	98.7	80.5	86.6	97.8	80.5	86.3	97.9	90.7	93.1	95.1	84.3	87.9	94.2	118.3	110.2	95.1	84.6	88.2
0960	Flooring	89.8	99.2	92.5	89.1	93.1	90.3	84.8	103.3	90.1	88.8	92.6	89.9	86.9	97.0	89.8	82.2	61.2	76.2
0970, 0990	Wall Finishes & Painting/Coating	95.0	105.7	101.5	95.0	105.7	101.5	88.6	105.7	99.0	92.5	61.5	73.7	92.5	71.5	79.8	92.5	98.4	96.1
09	FINISHES	96.6	88.8	92.3	96.5	86.4	90.9	95.1	95.4	95.3	95.0	84.1	89.0	95.2	106.0	101.1	93.4	82.5	87.4
COVERS	DIVS. 10 - 14, 25, 28, 41, 43, 44, 46	100.0	101.8	100.4	100.0	99.2	99.8	100.0	103.1	100.6	100.0	96.3	99.2	100.0	105.4	101.1	100.0	96.5	99.3
21, 22, 23	FIRE SUPPRESSION, PLUMBING & HVAC	94.7	92.1	93.6	94.7	88.2	92.0	94.5	96.7	95.4	95.0	90.0	92.9	94.9	110.3	101.3	94.9	97.8	96.1
26, 27, 3370	ELECTRICAL, COMMUNICATIONS & UTIL.	94.3	105.4	100.2	94.3	105.4	100.2	90.0	105.4	98.2	93.6	47.8	69.2	94.5	141.5	119.5	93.6	95.1	94.4
MF2010	WEIGHTED AVERAGE	96.3	98.2	97.1	97.6	94.9	96.4	93.2	101.4	96.9	94.7	88.3	91.8	96.0	114.2	104.2	94.9	96.3	95.5

PENNSYLVANIA

DIVISION		NEW CASTLE 161 MAT.	INST.	TOTAL	NORRISTOWN 194 MAT.	INST.	TOTAL	OIL CITY 163 MAT.	INST.	TOTAL	PHILADELPHIA 190-191 MAT.	INST.	TOTAL	PITTSBURGH 150-152 MAT.	INST.	TOTAL	POTTSVILLE 179 MAT.	INST.	TOTAL
015433	CONTRACTOR EQUIPMENT		114.2	114.2		97.9	97.9		114.2	114.2		97.3	97.3		112.6	112.6		113.5	113.5
0241, 31 - 34	SITE & INFRASTRUCTURE, DEMOLITION	88.0	107.3	101.3	88.0	98.9	95.6	86.4	105.5	99.6	94.4	98.2	97.0	111.4	107.1	108.4	81.1	104.9	97.5
0310	Concrete Forming & Accessories	88.7	94.0	93.3	88.6	133.4	127.5	88.7	81.0	82.0	104.8	137.6	133.3	101.1	94.9	95.7	80.6	86.7	85.9
0320	Concrete Reinforcing	93.6	93.4	93.5	92.2	148.5	121.2	94.8	93.2	94.0	95.1	148.5	122.6	96.5	104.0	100.3	93.6	101.9	97.8
0330	Cast-in-Place Concrete	82.2	96.7	87.8	78.1	131.3	98.5	79.9	94.6	85.5	95.8	131.3	109.4	102.4	97.4	100.5	81.0	78.8	80.2
03	CONCRETE	83.3	96.0	89.4	87.5	135.2	110.3	81.7	89.4	85.4	99.7	137.1	117.6	100.1	98.6	99.3	95.4	88.3	92.0
04	MASONRY	98.3	96.9	97.4	110.0	130.8	122.6	99.2	90.5	94.0	96.4	131.0	117.4	92.0	101.9	98.0	99.3	99.7	99.5
05	METALS	93.9	113.8	100.8	100.4	135.2	112.4	93.9	112.2	100.2	104.3	135.3	115.0	98.6	118.4	105.4	98.2	118.2	105.1
06	WOOD, PLASTICS & COMPOSITES	84.9	92.5	89.3	86.8	133.1	113.4	84.9	77.5	80.7	106.0	138.7	124.7	104.1	92.4	97.4	86.7	85.0	85.7
07	THERMAL & MOISTURE PROTECTION	100.2	97.3	99.0	97.1	134.8	112.1	100.1	93.4	97.4	97.9	135.4	112.8	105.5	101.3	103.8	99.9	105.6	102.2
08	OPENINGS	87.3	96.0	89.5	89.7	145.0	103.7	87.3	80.0	85.4	98.2	148.0	110.8	96.8	102.2	98.1	89.6	91.0	89.9
0920	Plaster & Gypsum Board	98.3	92.0	93.8	89.7	133.9	120.8	98.3	76.6	83.0	98.1	139.7	127.4	106.7	92.0	96.4	99.7	84.3	88.8
0950, 0980	Ceilings & Acoustic Treatment	97.9	92.0	94.0	104.3	133.9	123.9	97.9	76.6	83.8	104.3	139.7	127.7	98.7	92.0	94.3	95.1	84.3	87.9
0960	Flooring	84.8	60.7	77.9	85.2	140.6	101.1	84.8	99.2	88.9	92.8	140.6	106.5	96.2	101.0	97.6	85.0	92.6	87.2
0970, 0990	Wall Finishes & Painting/Coating	88.6	105.7	99.0	96.2	145.4	126.0	88.6	105.7	99.0	96.2	142.3	124.2	95.0	114.8	107.0	92.5	99.1	96.5
09	FINISHES	95.0	88.4	91.4	95.3	135.8	117.7	94.8	85.1	89.5	98.9	138.8	120.9	99.7	97.0	98.2	93.4	88.7	90.8
COVERS	DIVS. 10 - 14, 25, 28, 41, 43, 44, 46	100.0	103.2	100.7	100.0	119.3	104.0	100.0	100.1	100.0	100.0	121.5	104.5	100.0	103.1	100.7	100.0	97.1	99.4
21, 22, 23	FIRE SUPPRESSION, PLUMBING & HVAC	94.5	94.3	94.4	94.6	128.7	108.8	94.5	91.8	93.4	100.0	129.0	112.1	100.1	100.4	100.3	95.0	97.6	96.0
26, 27, 3370	ELECTRICAL, COMMUNICATIONS & UTIL.	90.7	93.3	92.1	93.4	141.8	119.1	92.7	104.2	98.8	98.3	141.9	121.5	96.9	105.4	101.4	91.6	89.6	90.5
MF2010	WEIGHTED AVERAGE	92.6	97.2	94.7	95.1	131.2	111.3	92.6	94.9	93.6	99.7	132.2	114.3	99.3	102.9	100.9	94.6	95.6	95.1

PENNSYLVANIA

DIVISION		READING 195-196 MAT.	INST.	TOTAL	SCRANTON 184-185 MAT.	INST.	TOTAL	STATE COLLEGE 168 MAT.	INST.	TOTAL	STROUDSBURG 183 MAT.	INST.	TOTAL	SUNBURY 178 MAT.	INST.	TOTAL	UNIONTOWN 154 MAT.	INST.	TOTAL
015433	CONTRACTOR EQUIPMENT		117.7	117.7		114.2	114.2		113.5	113.5		114.2	114.2		114.2	114.2		111.3	111.3
0241, 31 - 34	SITE & INFRASTRUCTURE, DEMOLITION	92.0	111.1	105.2	92.7	106.2	102.0	83.3	104.8	98.2	87.3	104.2	99.0	91.4	105.9	101.4	106.4	104.9	105.4
0310	Concrete Forming & Accessories	104.3	87.0	89.3	101.9	87.7	89.6	85.8	79.3	80.2	90.0	87.4	87.7	93.3	85.8	86.8	81.7	93.6	92.0
0320	Concrete Reinforcing	93.5	102.1	97.9	97.1	108.0	102.7	95.4	92.0	93.7	97.4	120.3	109.2	96.2	100.2	98.2	95.4	103.7	99.6
0330	Cast-in-Place Concrete	70.0	95.8	79.9	86.4	91.1	88.2	83.3	69.4	78.0	81.4	75.4	79.1	88.6	94.2	90.7	96.7	97.2	96.9
03	CONCRETE	86.4	94.3	90.2	98.0	94.1	96.2	98.4	79.9	89.5	92.6	90.8	91.7	98.8	92.8	95.9	94.1	97.9	95.9
04	MASONRY	98.3	92.9	95.0	95.6	96.2	96.0	100.8	80.6	88.5	91.7	100.7	97.1	94.1	87.3	90.0	113.1	96.9	103.2
05	METALS	100.7	119.3	107.1	102.4	121.5	108.9	97.8	112.7	102.9	99.9	121.8	107.5	97.9	117.4	104.6	96.9	117.7	104.0
06	WOOD, PLASTICS & COMPOSITES	106.1	84.6	93.8	105.1	85.3	93.8	92.5	81.3	86.1	91.1	85.3	87.8	94.7	85.0	89.1	80.2	91.2	86.5
07	THERMAL & MOISTURE PROTECTION	97.5	109.8	102.4	101.0	97.9	99.8	100.0	92.3	96.9	100.8	87.0	95.4	101.0	104.6	102.4	105.1	99.8	103.0
08	OPENINGS	95.6	95.4	95.5	92.7	93.0	92.8	89.5	85.9	88.6	93.4	89.7	92.5	89.7	90.4	89.9	93.8	101.4	95.8
0920	Plaster & Gypsum Board	102.6	84.0	89.5	108.6	84.6	91.7	101.4	80.5	86.6	98.9	84.6	88.9	100.0	84.3	88.9	93.6	90.7	91.6
0950, 0980	Ceilings & Acoustic Treatment	93.9	84.0	87.3	100.6	84.6	90.0	93.2	80.5	84.8	92.3	84.6	87.2	93.2	84.3	87.3	97.8	90.7	93.1
0960	Flooring	90.5	92.6	91.1	90.1	107.6	95.2	87.4	103.0	91.9	84.6	54.2	75.9	85.9	92.6	87.8	86.5	99.2	90.1
0970, 0990	Wall Finishes & Painting/Coating	94.7	99.1	97.4	92.5	98.4	96.1	92.5	105.7	100.5	92.5	69.5	78.6	92.5	99.1	96.5	95.0	105.7	101.5
09	FINISHES	96.5	88.0	91.8	98.5	91.8	94.8	94.1	85.9	89.6	93.7	78.9	85.5	94.7	88.1	91.1	94.7	94.4	94.6
COVERS	DIVS. 10 - 14, 25, 28, 41, 43, 44, 46	100.0	99.0	99.8	100.0	99.9	100.0	100.0	92.6	98.5	100.0	67.2	93.2	100.0	96.3	99.2	100.0	103.0	100.6
21, 22, 23	FIRE SUPPRESSION, PLUMBING & HVAC	100.1	106.9	102.9	100.3	98.4	99.5	95.0	84.7	90.7	94.9	100.1	97.1	94.9	89.6	92.7	94.7	92.1	93.7
26, 27, 3370	ELECTRICAL, COMMUNICATIONS & UTIL.	100.5	89.6	94.7	100.7	95.1	97.7	92.6	105.4	99.4	94.5	141.5	119.5	92.0	88.5	90.1	90.9	105.4	98.6
MF2010	WEIGHTED AVERAGE	97.4	99.6	98.4	99.0	98.7	98.9	95.5	91.4	93.7	95.3	102.3	98.4	95.5	94.0	94.8	96.6	99.9	98.0

PENNSYLVANIA

DIVISION		WASHINGTON 153			WELLSBORO 169			WESTCHESTER 193			WILKES-BARRE 186 - 187			WILLIAMSPORT 177			YORK 173 - 174		
		MAT.	INST.	TOTAL	MAT.	INST.	TOTAL	MAT.	INST.	TOTAL	MAT.	INST.	TOTAL	MAT.	INST.	TOTAL	MAT.	INST.	TOTAL
015433	CONTRACTOR EQUIPMENT		111.3	111.3		114.2	114.2		97.9	97.9		114.2	114.2		114.2	114.2		113.5	113.5
0241, 31 - 34	SITE & INFRASTRUCTURE, DEMOLITION	106.4	104.8	105.3	95.9	104.0	101.5	93.7	96.7	95.8	85.4	106.1	99.7	82.6	104.7	97.8	81.5	104.8	97.6
0310	Concrete Forming & Accessories	88.4	93.7	93.0	88.7	84.2	84.8	95.3	132.1	127.2	92.5	87.6	88.3	90.0	62.9	66.5	83.9	86.2	85.9
0320	Concrete Reinforcing	95.4	103.8	99.7	95.4	107.6	101.7	91.3	114.6	103.3	96.2	108.0	102.3	95.4	69.4	82.0	96.2	102.0	99.2
0330	Cast-in-Place Concrete	96.7	97.3	96.9	87.3	87.0	87.2	86.5	130.7	103.8	78.2	90.9	83.1	75.0	75.3	75.1	81.4	94.7	86.5
03	CONCRETE	94.5	98.0	96.2	100.8	90.9	96.1	95.1	127.9	110.8	89.5	94.0	91.7	86.3	70.3	78.6	96.6	93.5	95.1
04	MASONRY	97.3	98.5	98.0	100.8	89.9	94.2	103.7	130.8	120.1	108.1	95.8	100.7	86.2	91.5	89.4	95.1	90.2	92.1
05	METALS	96.8	117.9	104.1	97.9	116.4	104.3	100.4	118.6	106.7	98.1	121.6	106.2	98.0	101.9	99.3	99.7	118.6	106.2
06	WOOD, PLASTICS & COMPOSITES	88.6	91.2	90.1	89.6	85.0	86.9	94.5	133.1	116.6	93.7	85.2	88.8	91.1	55.3	70.6	90.5	85.0	87.3
07	THERMAL & MOISTURE PROTECTION	105.2	100.2	103.3	101.3	92.1	97.7	97.4	134.4	112.1	100.6	106.3	102.9	100.5	102.7	101.4	99.9	108.3	103.2
08	OPENINGS	93.8	101.4	95.8	93.1	92.8	93.0	89.7	129.4	99.7	89.7	97.5	91.7	89.7	57.9	81.6	89.5	91.6	90.1
0920	Plaster & Gypsum Board	98.5	90.7	93.0	98.7	84.3	88.5	91.9	133.9	121.5	100.1	84.5	89.1	99.7	53.8	67.3	101.5	84.3	89.3
0950, 0980	Ceilings & Acoustic Treatment	97.8	90.7	93.1	93.2	84.3	87.3	104.3	133.9	123.9	95.1	84.5	88.1	95.1	53.8	67.7	94.1	84.3	87.6
0960	Flooring	89.9	99.2	92.6	83.9	53.7	75.2	88.6	140.6	103.5	85.6	95.7	88.5	84.6	52.5	75.4	86.4	92.6	88.2
0970, 0990	Wall Finishes & Painting/Coating	95.0	105.7	101.5	92.5	98.4	96.1	96.2	145.4	126.0	92.5	98.4	96.1	92.5	98.4	96.1	92.5	87.8	89.7
09	FINISHES	96.4	94.8	95.5	94.4	79.7	86.3	97.1	133.7	117.3	94.5	89.3	91.6	94.0	63.9	77.4	93.9	87.0	90.1
COVERS	DIVS. 10 - 14, 25, 28, 41, 43, 44, 46	100.0	103.0	100.6	100.0	92.8	98.5	100.0	119.2	104.0	100.0	99.7	99.9	100.0	93.1	98.6	100.0	96.5	99.3
21, 22, 23	FIRE SUPPRESSION, PLUMBING & HVAC	94.7	95.5	95.1	94.9	89.0	92.5	94.6	127.2	108.1	94.9	98.2	96.3	94.9	89.7	92.8	100.3	90.1	96.0
26, 27, 3370	ELECTRICAL, COMMUNICATIONS & UTIL.	93.6	105.4	99.9	93.6	79.9	86.3	93.3	123.4	109.3	94.5	87.6	90.8	92.5	50.6	70.2	93.6	84.5	88.8
MF2010	WEIGHTED AVERAGE	96.2	100.8	98.3	96.7	91.0	94.2	96.1	124.7	108.9	95.2	97.7	96.3	93.3	79.7	87.2	96.5	94.0	95.4

DIVISION		PUERTO RICO SAN JUAN 009			RHODE ISLAND NEWPORT 028			PROVIDENCE 029			SOUTH CAROLINA AIKEN 298			BEAUFORT 299			CHARLESTON 294		
		MAT.	INST.	TOTAL	MAT.	INST.	TOTAL	MAT.	INST.	TOTAL	MAT.	INST.	TOTAL	MAT.	INST.	TOTAL	MAT.	INST.	TOTAL
015433	CONTRACTOR EQUIPMENT		89.6	89.6		101.9	101.9		101.9	101.9		100.0	100.0		100.0	100.0		100.0	100.0
0241, 31 - 34	SITE & INFRASTRUCTURE, DEMOLITION	125.4	90.5	101.3	82.5	103.5	97.0	83.8	103.5	97.4	127.9	86.0	99.0	122.3	84.2	96.0	103.5	83.8	89.9
0310	Concrete Forming & Accessories	92.1	17.6	27.4	99.7	115.6	113.5	98.4	115.6	113.3	96.0	71.0	74.3	95.0	40.1	47.4	94.1	41.9	48.8
0320	Concrete Reinforcing	194.4	12.4	100.7	108.1	114.4	111.3	108.1	114.4	111.3	91.7	68.8	79.9	90.8	27.7	58.3	90.6	59.8	74.7
0330	Cast-in-Place Concrete	101.6	30.6	74.4	78.6	120.3	94.5	96.5	120.3	105.6	89.3	72.9	83.0	89.3	49.5	74.1	104.8	50.2	83.9
03	CONCRETE	112.8	22.0	69.3	100.5	116.5	108.2	109.5	116.5	112.8	109.8	72.3	91.8	106.9	43.0	76.3	102.4	49.8	77.2
04	MASONRY	88.9	16.6	45.1	105.1	123.8	116.4	108.5	123.8	117.7	79.5	63.9	70.0	95.2	35.5	59.0	96.5	39.7	62.1
05	METALS	112.9	33.5	85.6	98.6	109.6	102.4	98.6	109.6	102.4	83.6	86.4	84.5	83.6	69.7	78.8	85.4	77.5	82.7
06	WOOD, PLASTICS & COMPOSITES	92.5	17.3	49.4	99.5	115.2	108.5	98.6	115.2	108.1	98.8	72.3	83.6	97.2	40.9	64.9	95.9	40.9	64.3
07	THERMAL & MOISTURE PROTECTION	125.6	21.1	84.0	96.6	115.4	104.1	94.3	115.4	102.7	99.5	69.8	87.7	99.2	38.5	75.1	98.3	43.1	76.4
08	OPENINGS	151.4	14.7	116.8	99.6	114.2	103.3	100.2	114.2	103.8	94.6	68.2	87.9	94.6	39.6	80.7	98.6	44.6	84.9
0920	Plaster & Gypsum Board	116.2	14.8	44.8	96.8	114.9	109.6	94.7	114.9	109.0	103.7	71.3	80.9	107.4	39.0	59.2	109.4	39.0	59.8
0950, 0980	Ceilings & Acoustic Treatment	234.7	14.8	88.9	96.5	114.9	108.7	95.8	114.9	108.5	92.0	71.3	78.3	95.7	39.0	58.1	95.7	39.0	58.1
0960	Flooring	200.5	13.9	146.9	96.6	131.6	106.7	96.3	131.6	106.4	108.7	69.9	97.5	110.4	54.1	94.2	109.9	61.5	96.0
0970, 0990	Wall Finishes & Painting/Coating	197.9	12.9	87.2	94.4	123.4	112.0	94.4	123.4	112.0	114.7	73.4	89.6	114.7	37.3	67.8	114.7	42.2	70.7
09	FINISHES	204.5	16.6	100.8	97.5	119.8	109.8	96.8	119.8	109.5	105.1	71.1	86.3	106.3	42.8	71.3	104.2	45.1	71.6
COVERS	DIVS. 10 - 14, 25, 28, 41, 43, 44, 46	100.0	17.1	82.8	100.0	103.2	100.7	100.0	103.2	100.7	100.0	79.4	95.7	100.0	48.2	89.2	100.0	72.5	94.3
21, 22, 23	FIRE SUPPRESSION, PLUMBING & HVAC	103.4	13.6	66.0	100.2	108.0	103.4	100.0	108.0	103.3	94.9	65.6	82.7	94.9	25.9	66.2	100.3	40.8	75.5
26, 27, 3370	ELECTRICAL, COMMUNICATIONS & UTIL.	128.5	12.9	67.0	99.9	101.2	100.6	98.3	101.2	99.8	94.4	69.4	81.1	98.2	32.9	63.5	96.6	96.4	96.5
MF2010	WEIGHTED AVERAGE	121.8	24.0	78.1	99.3	111.5	104.7	100.3	111.5	105.3	96.7	71.9	85.6	97.5	43.0	73.1	98.0	58.3	80.2

DIVISION		SOUTH CAROLINA COLUMBIA 290 - 292			FLORENCE 295			GREENVILLE 296			ROCK HILL 297			SPARTANBURG 293			SOUTH DAKOTA ABERDEEN 574		
		MAT.	INST.	TOTAL	MAT.	INST.	TOTAL	MAT.	INST.	TOTAL	MAT.	INST.	TOTAL	MAT.	INST.	TOTAL	MAT.	INST.	TOTAL
015433	CONTRACTOR EQUIPMENT		100.0	100.0		100.0	100.0		100.0	100.0		100.0	100.0		100.0	100.0		98.6	98.6
0241, 31 - 34	SITE & INFRASTRUCTURE, DEMOLITION	101.3	83.7	89.2	115.0	83.8	93.4	109.3	83.4	91.4	107.1	83.3	90.6	109.1	83.4	91.4	96.1	94.6	95.0
0310	Concrete Forming & Accessories	95.8	41.0	48.2	82.3	41.1	46.6	93.5	41.0	47.9	92.1	38.6	45.7	96.6	41.0	48.3	96.3	39.7	47.2
0320	Concrete Reinforcing	90.6	59.6	74.6	90.3	59.7	74.5	90.2	47.6	68.3	91.0	47.4	68.6	90.2	47.6	68.3	103.4	41.0	71.3
0330	Cast-in-Place Concrete	107.6	50.8	85.8	89.3	50.0	74.2	89.3	49.9	74.2	89.3	44.8	72.2	89.3	49.9	74.2	103.8	46.2	81.8
03	CONCRETE	103.9	49.5	77.9	101.3	49.3	76.4	99.7	47.0	74.5	97.8	44.1	72.1	99.9	47.0	74.6	102.7	43.6	74.4
04	MASONRY	90.8	37.6	58.5	79.5	39.7	55.4	77.3	39.7	54.5	103.0	34.8	61.7	79.5	39.7	55.4	111.6	57.9	79.1
05	METALS	85.4	76.7	82.4	84.4	77.0	81.9	84.4	72.6	80.3	83.6	72.1	79.6	84.4	72.6	80.3	102.4	66.2	90.0
06	WOOD, PLASTICS & COMPOSITES	98.3	40.4	65.1	81.7	40.4	58.0	95.6	40.4	63.9	93.8	38.4	62.1	99.8	40.4	65.7	97.2	38.9	63.8
07	THERMAL & MOISTURE PROTECTION	97.2	41.8	75.2	98.6	43.1	76.6	98.5	43.1	76.5	98.4	40.6	75.5	98.6	43.1	76.5	106.1	51.0	84.2
08	OPENINGS	98.6	44.4	84.9	94.6	44.4	81.9	94.6	41.5	81.1	94.6	40.6	80.9	94.6	41.5	81.1	92.7	39.7	79.3
0920	Plaster & Gypsum Board	100.0	38.5	56.7	95.5	38.5	55.4	101.9	38.5	57.2	101.0	36.4	55.5	105.0	38.5	58.2	105.4	37.4	57.5
0950, 0980	Ceilings & Acoustic Treatment	94.8	38.5	57.5	92.9	38.5	56.8	92.0	38.5	56.5	92.0	36.4	55.2	92.0	38.5	56.5	108.1	37.4	61.2
0960	Flooring	106.5	46.1	89.1	100.8	46.1	85.1	107.4	60.4	93.9	106.4	47.0	89.3	108.9	60.4	95.0	118.6	53.2	99.8
0970, 0990	Wall Finishes & Painting/Coating	112.3	42.2	69.7	114.7	42.2	70.7	114.7	42.2	70.7	114.7	42.2	70.7	114.7	42.2	70.7	98.4	40.0	62.9
09	FINISHES	101.1	41.7	68.4	100.6	42.2	68.4	102.6	44.7	70.7	101.9	40.2	67.9	103.5	44.7	71.0	108.3	41.8	71.6
COVERS	DIVS. 10 - 14, 25, 28, 41, 43, 44, 46	100.0	72.2	94.2	100.0	72.2	94.2	100.0	72.2	94.2	100.0	71.4	94.0	100.0	72.2	94.2	100.0	45.2	88.6
21, 22, 23	FIRE SUPPRESSION, PLUMBING & HVAC	100.0	36.1	73.4	100.3	36.1	73.6	100.3	34.6	73.0	94.9	33.7	69.4	100.3	34.6	73.0	100.1	43.7	76.6
26, 27, 3370	ELECTRICAL, COMMUNICATIONS & UTIL.	98.2	49.3	72.2	94.4	46.9	69.1	96.6	50.7	72.2	96.6	50.7	72.2	96.6	50.7	72.2	98.4	55.5	75.6
MF2010	WEIGHTED AVERAGE	97.6	50.0	76.3	96.2	50.0	75.5	96.2	49.6	75.3	95.7	47.7	74.3	96.4	49.6	75.5	101.2	52.8	79.6

SOUTH DAKOTA

DIVISION		MITCHELL 573			MOBRIDGE 576			PIERRE 575			RAPID CITY 577			SIOUX FALLS 570-571			WATERTOWN 572		
		MAT.	INST.	TOTAL	MAT.	INST.	TOTAL	MAT.	INST.	TOTAL	MAT.	INST.	TOTAL	MAT.	INST.	TOTAL	MAT.	INST.	TOTAL
015433	CONTRACTOR EQUIPMENT		98.6	98.6		98.6	98.6		98.6	98.6		98.6	98.6		99.7	99.7		98.6	98.6
0241, 31 - 34	SITE & INFRASTRUCTURE, DEMOLITION	93.1	94.6	94.1	93.0	94.6	94.1	91.3	94.6	93.6	94.1	94.7	94.5	91.9	96.3	95.0	93.0	94.6	94.1
0310	Concrete Forming & Accessories	95.6	40.0	47.4	87.7	39.8	46.2	95.1	40.8	48.0	102.4	38.8	47.2	95.7	43.4	50.4	84.8	39.6	45.6
0320	Concrete Reinforcing	102.8	49.9	75.5	105.5	41.0	72.3	102.9	60.0	80.8	96.7	60.1	77.8	96.7	60.3	77.9	100.0	41.1	69.7
0330	Cast-in-Place Concrete	100.9	44.9	79.5	100.9	46.2	80.0	87.1	44.5	70.8	100.1	45.1	79.1	92.6	45.5	74.6	100.9	46.2	80.0
03	CONCRETE	100.4	44.8	73.8	100.3	43.6	73.2	93.5	47.1	71.3	99.6	46.5	74.2	94.2	48.7	72.4	99.4	43.6	72.7
04	MASONRY	99.8	55.7	73.1	109.0	57.9	78.1	107.1	56.0	76.1	109.0	57.1	77.5	104.6	57.8	76.2	136.1	60.2	90.1
05	METALS	101.4	66.3	89.4	101.5	66.1	89.3	102.4	75.0	93.0	104.3	75.5	94.4	104.5	75.8	94.6	101.4	66.7	89.5
06	WOOD, PLASTICS & COMPOSITES	96.2	39.5	63.7	86.9	39.2	59.6	95.1	39.8	63.4	100.3	36.0	63.4	95.9	42.5	65.3	84.0	38.9	58.1
07	THERMAL & MOISTURE PROTECTION	105.9	48.7	83.2	106.0	51.1	84.2	105.9	49.0	83.3	106.4	50.6	84.2	108.2	52.0	85.9	105.8	51.6	84.3
08	OPENINGS	91.2	40.3	78.3	94.1	39.3	80.2	95.8	45.8	83.2	96.8	43.8	84.5	97.1	47.3	84.5	91.2	39.4	78.1
0920	Plaster & Gypsum Board	103.8	38.0	57.5	97.9	37.7	55.5	95.9	38.3	55.3	104.6	34.4	55.2	102.5	41.1	59.3	96.2	37.4	54.8
0950, 0980	Ceilings & Acoustic Treatment	105.3	38.0	60.7	108.1	37.7	61.4	104.3	38.3	60.5	109.9	34.4	59.9	107.1	41.1	63.4	105.3	37.4	60.3
0960	Flooring	118.1	53.2	99.4	112.4	53.2	95.4	117.8	37.9	94.8	117.8	76.0	105.8	117.8	78.9	106.6	110.7	53.2	94.2
0970, 0990	Wall Finishes & Painting/Coating	98.4	43.7	65.2	98.4	44.8	65.9	98.4	48.0	67.8	98.4	48.0	67.8	98.4	48.0	67.8	98.4	40.0	62.9
09	FINISHES	107.2	42.5	71.5	105.4	42.5	70.7	105.8	40.2	69.6	108.3	45.8	73.9	107.3	50.0	75.7	104.1	41.8	69.7
COVERS	DIVS. 10 - 14, 25, 28, 41, 43, 44, 46	100.0	41.9	87.9	100.0	45.2	88.6	100.0	76.9	95.2	100.0	76.8	95.2	100.0	77.3	95.3	100.0	45.2	88.6
21, 22, 23	FIRE SUPPRESSION, PLUMBING & HVAC	94.7	43.3	73.3	94.7	43.8	73.5	100.0	43.4	76.4	100.1	43.9	76.7	100.1	42.2	76.0	94.7	43.8	73.5
26, 27, 3370	ELECTRICAL, COMMUNICATIONS & UTIL.	96.7	47.7	70.7	98.4	48.7	72.0	98.4	53.2	74.4	94.8	53.2	72.7	100.7	64.5	81.4	95.8	48.7	70.7
MF2010	WEIGHTED AVERAGE	98.3	51.6	77.4	99.1	52.0	78.0	99.7	54.5	79.5	101.0	55.3	80.6	100.6	57.9	81.5	99.7	52.2	78.4

TENNESSEE

DIVISION		CHATTANOOGA 373-374			COLUMBIA 384			COOKEVILLE 385			JACKSON 383			JOHNSON CITY 376			KNOXVILLE 377-379		
		MAT.	INST.	TOTAL	MAT.	INST.	TOTAL	MAT.	INST.	TOTAL	MAT.	INST.	TOTAL	MAT.	INST.	TOTAL	MAT.	INST.	TOTAL
015433	CONTRACTOR EQUIPMENT		103.8	103.8		97.8	97.8		97.8	97.8		104.0	104.0		97.7	97.7		97.7	97.7
0241, 31 - 34	SITE & INFRASTRUCTURE, DEMOLITION	101.5	98.4	99.3	89.5	86.9	87.7	95.5	86.3	89.2	98.0	96.9	97.2	110.0	87.1	94.2	87.7	87.6	87.6
0310	Concrete Forming & Accessories	97.2	58.2	63.4	83.5	62.3	65.1	83.7	38.0	44.0	90.2	48.3	53.8	84.5	41.5	47.1	96.0	44.6	51.4
0320	Concrete Reinforcing	83.5	63.1	73.0	86.6	63.1	74.5	86.6	63.0	74.5	86.6	59.9	72.8	84.1	61.9	72.7	83.5	62.0	72.4
0330	Cast-in-Place Concrete	97.6	63.3	84.5	92.3	54.0	77.7	104.6	45.6	82.1	102.2	52.2	83.0	78.5	64.0	72.9	91.7	50.8	76.0
03	CONCRETE	89.4	62.6	76.5	90.3	61.2	76.4	99.8	47.5	74.8	92.3	53.7	73.8	93.2	55.1	75.0	86.5	52.0	70.0
04	MASONRY	98.3	52.8	70.7	110.6	57.8	78.6	105.2	47.2	70.1	110.9	50.0	74.0	109.8	47.6	72.1	75.4	55.5	63.4
05	METALS	96.4	87.7	93.4	90.7	88.2	89.8	90.8	87.2	89.5	92.9	86.2	90.6	93.7	86.6	91.2	97.1	86.8	93.5
06	WOOD, PLASTICS & COMPOSITES	109.3	60.0	81.0	74.1	64.2	68.4	74.3	36.4	52.6	90.7	49.2	66.9	81.4	40.0	57.6	95.4	40.0	63.6
07	THERMAL & MOISTURE PROTECTION	93.7	60.5	80.5	89.2	60.0	77.6	89.6	49.0	73.5	91.4	52.0	75.7	89.3	56.4	76.2	87.5	60.0	76.6
08	OPENINGS	101.6	58.3	90.7	92.2	58.6	83.7	92.2	44.7	80.1	101.4	53.7	89.4	97.1	47.8	84.6	94.0	47.9	82.3
0920	Plaster & Gypsum Board	78.4	59.4	65.0	83.3	63.6	69.4	83.3	35.0	49.3	85.8	48.1	59.3	96.2	38.7	55.7	104.4	38.7	58.1
0950, 0980	Ceilings & Acoustic Treatment	97.4	59.4	72.2	84.1	63.6	70.5	84.1	35.0	51.5	93.3	48.1	63.4	93.3	38.7	57.1	94.2	38.7	57.4
0960	Flooring	100.0	59.8	88.5	94.0	21.7	73.2	94.1	61.9	84.9	93.0	43.1	78.7	96.2	42.9	80.9	101.2	57.7	88.7
0970, 0990	Wall Finishes & Painting/Coating	103.1	56.8	75.0	94.6	35.9	59.0	94.6	39.3	61.0	95.9	53.4	70.1	100.6	45.3	67.1	95.2	46.7	67.3
09	FINISHES	96.4	58.5	75.5	96.0	52.5	72.0	96.6	41.3	66.1	95.9	47.1	69.0	99.8	41.0	67.4	92.6	46.7	67.3
COVERS	DIVS. 10 - 14, 25, 28, 41, 43, 44, 46	100.0	45.8	88.7	100.0	52.4	90.1	100.0	45.6	88.7	100.0	47.9	89.2	100.0	48.1	89.2	100.0	50.9	89.8
21, 22, 23	FIRE SUPPRESSION, PLUMBING & HVAC	100.0	63.1	84.7	97.1	76.1	88.3	97.1	70.7	86.1	99.9	70.5	87.7	99.8	57.6	82.2	99.8	66.5	85.9
26, 27, 3370	ELECTRICAL, COMMUNICATIONS & UTIL.	103.3	70.8	86.0	94.8	62.3	77.5	96.6	62.3	78.4	101.7	63.5	81.4	93.0	50.3	70.3	99.3	58.7	77.7
MF2010	WEIGHTED AVERAGE	98.2	66.9	84.2	94.9	67.3	82.6	96.2	60.3	80.1	98.2	63.8	82.8	97.6	57.4	79.6	94.5	61.7	79.9

TENNESSEE / TEXAS

DIVISION		MCKENZIE 382			MEMPHIS 375,380-381			NASHVILLE 370-372			ABILENE 795-796			AMARILLO 790-791			AUSTIN 786-787		
		MAT.	INST.	TOTAL	MAT.	INST.	TOTAL	MAT.	INST.	TOTAL	MAT.	INST.	TOTAL	MAT.	INST.	TOTAL	MAT.	INST.	TOTAL
015433	CONTRACTOR EQUIPMENT		97.8	97.8		101.7	101.7		104.8	104.8		88.5	88.5		88.5	88.5		87.9	87.9
0241, 31 - 34	SITE & INFRASTRUCTURE, DEMOLITION	95.1	86.6	89.3	89.6	92.6	91.7	97.2	99.9	99.0	99.0	85.8	89.9	99.1	86.9	90.7	95.2	86.0	88.8
0310	Concrete Forming & Accessories	91.3	41.0	47.6	96.2	61.5	66.1	97.4	65.2	69.5	95.6	40.9	48.2	96.2	55.2	60.7	89.7	54.4	59.1
0320	Concrete Reinforcing	86.7	63.8	74.9	89.7	64.5	76.7	89.7	63.5	76.2	97.5	52.6	74.3	97.5	50.8	73.4	97.7	48.6	72.4
0330	Cast-in-Place Concrete	102.4	61.0	86.6	88.3	63.7	78.9	89.2	67.9	81.1	97.5	49.1	79.0	96.6	56.1	81.1	84.0	49.7	70.8
03	CONCRETE	98.4	54.3	77.3	85.9	64.4	75.6	86.5	67.2	77.3	93.7	47.1	71.4	93.3	55.5	75.2	82.4	52.4	68.0
04	MASONRY	109.2	49.9	73.2	91.1	65.1	75.3	80.9	60.5	68.5	102.7	58.8	76.1	102.7	59.6	76.5	96.1	48.6	67.3
05	METALS	90.8	88.5	90.0	98.1	90.0	95.3	98.1	89.2	95.0	99.9	69.0	89.3	99.9	68.8	89.2	95.0	64.2	84.4
06	WOOD, PLASTICS & COMPOSITES	83.2	39.2	58.0	100.8	62.5	78.8	106.1	66.4	83.3	99.7	39.4	65.1	98.1	57.7	75.0	94.1	57.8	73.3
07	THERMAL & MOISTURE PROTECTION	89.6	50.2	74.0	88.4	65.6	79.3	92.2	63.3	80.7	96.9	51.6	78.9	97.0	54.8	80.2	98.2	51.6	73.0
08	OPENINGS	92.2	46.7	80.7	100.7	62.9	91.1	99.5	65.2	90.8	93.1	42.9	80.4	93.1	54.8	83.4	98.4	55.0	87.3
0920	Plaster & Gypsum Board	87.3	37.9	52.5	96.2	61.8	72.0	90.8	65.8	73.2	81.4	38.2	50.9	82.9	57.0	64.7	81.3	57.1	64.2
0950, 0980	Ceilings & Acoustic Treatment	84.1	37.9	53.5	95.5	61.8	73.1	92.3	65.8	74.7	90.3	38.2	55.7	90.3	57.0	68.2	84.3	57.1	66.2
0960	Flooring	97.3	41.6	81.3	96.5	38.1	79.7	102.3	67.9	92.4	109.0	77.9	100.1	108.8	77.9	99.9	90.8	41.8	76.7
0970, 0990	Wall Finishes & Painting/Coating	94.6	53.4	69.6	96.7	56.4	72.3	102.6	70.4	83.0	95.7	52.9	69.7	95.7	52.2	69.3	93.0	43.0	62.7
09	FINISHES	97.9	41.1	66.6	96.6	56.1	74.3	98.4	66.1	80.6	93.2	48.3	68.5	93.5	59.2	74.6	85.4	50.8	66.3
COVERS	DIVS. 10 - 14, 25, 28, 41, 43, 44, 46	100.0	30.3	85.5	100.0	80.8	96.0	100.0	81.6	96.2	100.0	75.7	94.9	100.0	73.9	94.6	100.0	77.6	95.3
21, 22, 23	FIRE SUPPRESSION, PLUMBING & HVAC	97.1	70.8	86.2	99.9	71.2	87.9	99.8	75.9	89.8	100.2	46.2	77.7	100.0	52.3	80.2	100.1	56.5	82.0
26, 27, 3370	ELECTRICAL, COMMUNICATIONS & UTIL.	96.4	68.3	81.4	100.3	66.4	82.3	100.9	63.3	80.9	98.8	51.6	73.7	101.5	68.5	84.0	100.7	65.9	82.2
MF2010	WEIGHTED AVERAGE	96.3	62.1	81.1	96.6	70.3	84.9	96.7	72.7	85.9	97.9	54.8	78.6	98.1	61.9	81.9	95.1	59.4	79.1

City Cost Indexes

TEXAS

DIVISION		BEAUMONT 776 - 777			BROWNWOOD 768			BRYAN 778			CHILDRESS 792			CORPUS CHRISTI 783 - 784			DALLAS 752 - 753		
		MAT.	INST.	TOTAL	MAT.	INST.	TOTAL	MAT.	INST.	TOTAL	MAT.	INST.	TOTAL	MAT.	INST.	TOTAL	MAT.	INST.	TOTAL
015433	CONTRACTOR EQUIPMENT		89.5	89.5		88.5	88.5		89.5	89.5		88.5	88.5		95.4	95.4		98.1	98.1
0241, 31 - 34	SITE & INFRASTRUCTURE, DEMOLITION	93.2	86.0	88.2	105.6	85.2	91.5	84.2	87.6	86.6	110.7	85.7	93.4	131.5	81.7	97.1	116.2	87.3	96.2
0310	Concrete Forming & Accessories	102.8	51.1	57.9	94.1	34.7	42.6	81.7	42.3	47.5	94.1	62.8	66.9	96.2	38.7	46.3	94.8	65.1	69.1
0320	Concrete Reinforcing	97.4	43.3	69.5	97.8	49.8	73.1	99.9	48.7	73.5	97.6	50.0	73.1	97.0	48.0	71.8	99.1	52.2	75.0
0330	Cast-in-Place Concrete	92.9	53.8	78.0	105.1	44.1	81.8	74.4	67.2	71.7	100.0	50.7	81.1	93.1	46.4	75.3	101.5	57.0	84.4
03	CONCRETE	95.2	51.5	74.3	101.0	41.9	72.7	79.7	53.2	67.0	101.8	56.7	80.2	89.5	45.0	68.2	97.5	61.2	80.1
04	MASONRY	106.7	59.8	78.3	132.1	51.3	83.1	144.8	58.4	92.5	107.5	52.6	74.2	85.1	50.3	64.0	103.3	59.6	76.8
05	METALS	98.0	65.4	86.8	97.5	62.4	85.4	97.5	68.0	87.3	97.3	64.4	86.0	94.5	75.4	88.0	100.9	80.0	93.7
06	WOOD, PLASTICS & COMPOSITES	112.5	51.3	77.4	98.3	30.8	59.6	78.5	35.5	53.9	99.1	68.0	81.3	113.6	38.0	70.2	98.4	68.1	81.0
07	THERMAL & MOISTURE PROTECTION	103.9	57.6	85.5	97.1	46.1	76.8	96.4	57.6	81.0	97.5	50.8	79.0	96.2	46.4	76.4	90.2	64.4	80.0
08	OPENINGS	96.5	47.8	84.2	89.0	38.2	76.1	97.9	43.6	84.1	90.1	57.0	81.7	105.5	39.3	88.7	101.9	59.0	91.0
0920	Plaster & Gypsum Board	92.1	50.4	62.7	80.3	29.2	44.4	80.0	34.2	47.8	80.3	67.6	71.4	86.0	36.6	51.2	95.1	67.6	75.7
0950, 0980	Ceilings & Acoustic Treatment	102.1	50.4	67.8	86.6	29.2	48.6	95.2	34.2	54.8	86.6	67.6	74.0	84.3	36.6	52.2	93.1	67.6	76.2
0960	Flooring	107.2	75.3	98.0	107.9	47.2	90.5	80.0	55.7	73.0	106.9	43.2	88.6	103.3	66.6	92.8	99.8	50.4	85.6
0970, 0990	Wall Finishes & Painting/Coating	92.3	51.2	67.4	95.7	30.8	56.3	88.1	63.8	73.3	95.7	40.7	62.3	105.3	60.8	78.3	103.8	54.3	73.7
09	FINISHES	90.2	55.3	70.9	92.6	35.8	61.3	78.4	45.3	60.1	93.0	57.8	73.6	93.2	45.7	67.0	99.0	61.6	78.4
COVERS	DIVS. 10 - 14, 25, 28, 41, 43, 44, 46	100.0	79.2	95.7	100.0	40.4	87.6	100.0	78.4	95.5	100.0	76.1	95.0	100.0	76.5	95.1	100.0	81.8	96.2
21, 22, 23	FIRE SUPPRESSION, PLUMBING & HVAC	100.0	60.6	83.6	94.8	49.5	75.9	94.6	67.2	83.2	94.8	53.9	77.8	100.0	41.0	75.5	99.9	64.5	85.1
26, 27, 3370	ELECTRICAL, COMMUNICATIONS & UTIL.	97.7	74.2	85.2	94.2	46.4	68.8	95.2	67.0	80.2	98.7	46.0	70.7	95.0	55.9	74.2	97.4	69.1	82.3
MF2010	WEIGHTED AVERAGE	98.2	62.9	82.4	98.0	49.6	76.3	95.1	62.1	80.3	97.5	58.1	79.9	97.6	52.8	77.6	99.9	67.5	85.4

TEXAS

DIVISION		DEL RIO 788			DENTON 762			EASTLAND 764			EL PASO 798 - 799,885			FORT WORTH 760 - 761			GALVESTON 775		
		MAT.	INST.	TOTAL	MAT.	INST.	TOTAL	MAT.	INST.	TOTAL	MAT.	INST.	TOTAL	MAT.	INST.	TOTAL	MAT.	INST.	TOTAL
015433	CONTRACTOR EQUIPMENT		87.9	87.9		94.7	94.7		88.5	88.5		88.5	88.5		88.5	88.5		98.8	98.8
0241, 31 - 34	SITE & INFRASTRUCTURE, DEMOLITION	115.1	84.5	94.0	105.7	78.4	86.8	108.6	85.2	92.4	96.3	85.2	88.6	101.0	86.2	90.8	116.8	85.5	95.2
0310	Concrete Forming & Accessories	90.8	38.4	45.3	99.5	34.1	42.8	94.7	37.7	45.3	94.9	45.8	52.3	91.8	64.5	68.1	90.6	63.1	66.8
0320	Concrete Reinforcing	97.7	45.0	70.5	99.2	51.1	74.4	98.0	49.8	73.2	97.5	47.9	71.9	97.5	52.1	74.1	99.3	65.0	81.6
0330	Cast-in-Place Concrete	100.2	43.9	78.7	81.3	46.0	67.8	111.1	44.8	85.7	91.1	42.0	72.3	100.3	52.7	82.1	98.8	68.3	87.1
03	CONCRETE	105.3	42.5	75.3	79.8	43.4	62.4	105.6	43.5	75.9	90.4	45.8	69.1	94.9	58.7	77.6	96.8	66.8	82.4
04	MASONRY	100.0	53.0	71.5	140.7	49.7	85.5	101.2	51.1	70.9	100.4	48.9	69.2	96.8	59.6	74.2	101.4	58.5	75.4
05	METALS	93.4	59.3	81.7	97.1	76.0	89.9	97.3	62.8	85.4	99.7	63.3	87.2	99.0	80.3	93.4	99.0	88.0	95.2
06	WOOD, PLASTICS & COMPOSITES	93.5	36.5	60.8	108.1	33.7	65.4	105.2	34.6	64.7	99.2	48.5	70.1	100.7	68.0	81.9	95.8	63.1	77.1
07	THERMAL & MOISTURE PROTECTION	93.0	45.6	74.1	95.5	45.9	75.8	97.4	45.3	76.7	93.1	54.4	77.7	97.8	54.8	80.7	95.7	66.2	83.9
08	OPENINGS	99.2	34.8	82.9	103.5	39.0	87.2	78.0	40.3	68.5	93.1	44.5	80.8	87.9	58.9	80.6	102.8	62.9	92.7
0920	Plaster & Gypsum Board	80.9	35.2	48.7	83.8	32.1	47.4	80.3	33.2	47.1	85.9	47.4	58.8	83.9	67.6	72.4	87.0	62.5	69.8
0950, 0980	Ceilings & Acoustic Treatment	80.5	35.2	50.5	91.3	32.1	52.1	86.6	33.2	51.2	90.3	47.4	61.9	92.2	67.6	75.9	98.0	62.5	74.5
0960	Flooring	87.9	36.0	73.0	100.5	45.2	84.6	140.7	47.2	113.9	109.0	65.9	96.6	142.0	43.0	113.6	93.8	55.7	82.8
0970, 0990	Wall Finishes & Painting/Coating	93.0	27.1	53.0	105.5	31.2	60.4	97.1	32.2	57.7	95.7	36.8	60.0	97.1	53.8	70.9	99.5	49.5	69.2
09	FINISHES	85.6	36.4	58.5	89.1	35.2	59.4	102.9	38.2	67.2	93.4	48.8	68.8	104.2	60.0	79.8	88.3	59.9	72.7
COVERS	DIVS. 10 - 14, 25, 28, 41, 43, 44, 46	100.0	40.1	87.6	100.0	38.9	87.3	100.0	41.6	87.9	100.0	77.4	95.3	100.0	81.4	96.1	100.0	83.4	96.6
21, 22, 23	FIRE SUPPRESSION, PLUMBING & HVAC	94.6	47.3	74.9	94.8	40.7	72.3	94.8	49.8	76.1	100.0	36.2	73.5	100.0	58.9	82.9	94.6	67.2	83.2
26, 27, 3370	ELECTRICAL, COMMUNICATIONS & UTIL.	96.8	31.9	62.3	96.6	46.5	70.0	94.1	46.4	68.7	97.7	54.4	74.7	97.7	63.8	79.7	97.1	68.7	82.0
MF2010	WEIGHTED AVERAGE	97.0	47.0	74.6	97.3	48.4	75.4	96.8	50.3	76.0	97.0	51.7	76.8	97.7	63.6	82.4	97.6	69.2	84.9

TEXAS

DIVISION		GIDDINGS 789			GREENVILLE 754			HOUSTON 770 - 772			HUNTSVILLE 773			LAREDO 780			LONGVIEW 756		
		MAT.	INST.	TOTAL	MAT.	INST.	TOTAL	MAT.	INST.	TOTAL	MAT.	INST.	TOTAL	MAT.	INST.	TOTAL	MAT.	INST.	TOTAL
015433	CONTRACTOR EQUIPMENT		87.9	87.9		95.7	95.7		98.6	98.6		89.5	89.5		87.9	87.9		90.3	90.3
0241, 31 - 34	SITE & INFRASTRUCTURE, DEMOLITION	101.3	84.8	89.9	108.6	82.1	90.3	114.5	85.2	94.3	101.9	85.6	90.7	95.1	85.7	88.6	103.6	89.4	93.8
0310	Concrete Forming & Accessories	89.0	62.8	66.8	87.2	28.6	36.4	92.5	63.3	67.1	88.3	58.3	62.3	90.8	38.4	46.3	84.3	29.5	36.8
0320	Concrete Reinforcing	98.3	46.9	71.8	99.6	26.0	61.7	99.1	65.5	81.8	100.2	49.0	73.8	97.7	47.9	72.0	98.6	23.7	60.0
0330	Cast-in-Place Concrete	85.0	44.4	69.4	92.4	38.2	71.7	95.8	68.3	85.3	102.4	49.1	82.0	71.8	60.0	67.2	107.4	35.8	80.0
03	CONCRETE	85.8	52.2	69.7	90.1	33.5	63.0	94.5	66.9	81.3	102.8	54.1	79.5	80.9	48.9	65.6	105.6	31.9	70.3
04	MASONRY	108.1	53.1	74.7	164.1	48.4	94.0	101.2	68.7	81.5	144.2	44.8	83.9	93.8	48.5	66.3	159.5	44.6	89.9
05	METALS	92.9	61.7	82.1	98.1	63.5	86.2	102.0	88.9	97.5	97.4	63.9	85.9	96.0	63.5	84.8	91.8	50.6	77.6
06	WOOD, PLASTICS & COMPOSITES	92.5	63.1	75.6	89.7	27.7	54.2	98.1	63.1	78.0	87.0	61.7	72.5	93.5	37.9	61.6	83.9	30.0	53.0
07	THERMAL & MOISTURE PROTECTION	93.6	49.9	76.2	90.3	41.1	70.7	95.3	68.5	84.6	97.4	47.9	77.7	92.0	49.4	75.1	91.5	36.4	69.6
08	OPENINGS	98.4	51.2	86.5	100.7	27.3	82.1	105.0	62.9	94.3	97.8	50.3	85.8	99.0	40.0	84.1	91.6	27.3	75.4
0920	Plaster & Gypsum Board	79.6	62.5	67.6	89.6	26.1	44.9	89.6	62.5	70.5	84.4	61.1	68.0	82.0	36.6	50.0	87.3	28.5	45.9
0950, 0980	Ceilings & Acoustic Treatment	80.5	62.5	68.6	89.4	26.1	47.4	102.7	62.5	76.1	95.2	61.1	72.6	84.3	36.6	52.7	85.6	28.5	47.8
0960	Flooring	88.0	36.0	73.0	95.7	49.6	82.5	95.0	55.7	83.7	83.5	36.0	69.9	87.8	66.6	81.7	101.1	36.4	82.5
0970, 0990	Wall Finishes & Painting/Coating	93.0	32.2	56.1	103.8	26.5	56.9	99.5	63.8	77.8	88.1	33.0	54.7	93.0	57.3	71.3	95.6	30.9	56.3
09	FINISHES	84.4	52.9	67.0	95.7	31.9	60.5	95.5	61.5	76.8	81.1	52.6	65.4	84.8	45.2	63.0	101.2	30.9	62.4
COVERS	DIVS. 10 - 14, 25, 28, 41, 43, 44, 46	100.0	46.6	88.9	100.0	37.1	86.9	100.0	83.4	96.6	100.0	47.0	89.0	100.0	74.3	94.7	100.0	23.4	84.1
21, 22, 23	FIRE SUPPRESSION, PLUMBING & HVAC	94.6	61.5	80.9	94.6	46.7	74.6	99.9	67.4	86.4	94.6	62.4	81.2	100.0	39.7	74.9	94.5	29.5	67.4
26, 27, 3370	ELECTRICAL, COMMUNICATIONS & UTIL.	93.3	46.0	68.2	93.7	28.7	59.1	99.4	67.0	82.2	95.2	46.0	69.0	96.9	63.6	79.2	94.5	46.6	69.0
MF2010	WEIGHTED AVERAGE	94.1	57.0	77.5	99.6	43.7	74.6	100.0	70.4	86.8	98.6	56.7	79.9	94.7	53.1	76.1	99.7	40.7	73.4

TEXAS

DIVISION		LUBBOCK 793-794 MAT.	INST.	TOTAL	LUFKIN 759 MAT.	INST.	TOTAL	MCALLEN 785 MAT.	INST.	TOTAL	MCKINNEY 750 MAT.	INST.	TOTAL	MIDLAND 797 MAT.	INST.	TOTAL	ODESSA 797 MAT.	INST.	TOTAL
015433	CONTRACTOR EQUIPMENT		97.2	97.2		90.3	90.3		95.4	95.4		95.7	95.7		97.2	97.2		88.5	88.5
0241, 31 - 34	SITE & INFRASTRUCTURE, DEMOLITION	125.6	84.1	97.0	98.0	90.5	92.8	136.2	81.6	98.5	104.2	82.7	89.4	129.4	84.1	98.1	99.2	86.1	90.2
0310	Concrete Forming & Accessories	93.6	53.6	58.9	87.3	59.9	63.5	95.6	36.3	44.1	86.4	38.3	44.7	97.0	40.3	47.8	95.6	40.2	47.5
0320	Concrete Reinforcing	98.7	52.6	74.9	100.2	61.7	80.4	97.2	47.9	71.8	99.6	51.1	74.6	99.7	51.9	75.1	97.5	51.8	74.0
0330	Cast-in-Place Concrete	97.6	49.4	79.2	96.1	45.6	76.8	101.2	44.7	79.6	86.6	46.8	71.4	103.8	46.6	81.9	97.5	45.6	77.6
03	CONCRETE	92.5	53.6	73.9	97.7	55.9	77.7	94.9	43.3	70.2	85.5	45.4	66.3	97.0	46.7	72.9	93.7	45.5	70.6
04	MASONRY	101.9	57.1	74.7	121.2	44.8	74.9	100.1	50.3	69.9	176.8	51.0	100.6	120.1	44.6	74.3	102.7	44.5	67.4
05	METALS	103.2	80.9	95.5	98.2	67.5	87.7	93.3	74.9	87.0	98.0	75.0	90.1	101.3	79.7	93.9	99.2	67.5	88.3
06	WOOD, PLASTICS & COMPOSITES	98.9	57.9	75.3	92.1	63.0	75.4	112.2	35.5	68.2	88.4	38.5	59.7	103.3	40.0	67.0	99.7	39.9	65.4
07	THERMAL & MOISTURE PROTECTION	87.7	53.8	74.2	91.3	47.6	73.9	96.2	44.7	75.7	90.2	47.9	73.4	87.9	45.2	70.9	96.9	44.4	76.1
08	OPENINGS	103.1	51.7	90.1	74.8	55.6	69.9	103.8	38.4	87.3	100.7	41.4	85.7	102.0	41.6	86.7	93.1	41.6	80.0
0920	Plaster & Gypsum Board	81.9	57.0	64.4	87.0	62.5	69.8	86.0	33.9	49.3	88.3	37.1	52.3	83.3	38.6	51.8	81.4	38.6	51.3
0950, 0980	Ceilings & Acoustic Treatment	92.2	57.0	68.9	80.0	62.5	68.4	84.3	33.9	50.9	89.4	37.1	54.7	89.4	38.6	55.7	90.3	38.6	56.1
0960	Flooring	101.5	40.8	84.1	135.2	39.6	107.7	102.8	66.6	92.4	95.4	36.9	78.6	103.0	50.4	87.9	109.0	50.4	92.2
0970, 0990	Wall Finishes & Painting/Coating	106.0	52.2	73.3	95.6	39.8	61.7	105.3	30.4	59.9	103.8	43.4	67.2	106.0	52.2	73.3	95.7	52.2	69.3
09	FINISHES	95.8	51.2	71.2	109.8	54.8	79.5	93.5	40.8	64.4	95.1	38.6	63.9	96.3	42.7	66.7	93.2	42.6	65.3
COVERS	DIVS. 10 - 14, 25, 28, 41, 43, 44, 46	100.0	77.5	95.3	100.0	77.4	95.3	100.0	74.2	94.6	100.0	43.0	88.2	100.0	77.1	95.2	100.0	76.8	95.2
21, 22, 23	FIRE SUPPRESSION, PLUMBING & HVAC	99.8	45.3	77.1	94.5	62.5	81.2	94.6	31.1	68.2	94.6	44.4	73.7	94.4	45.5	74.0	100.2	45.4	77.4
26, 27, 3370	ELECTRICAL, COMMUNICATIONS & UTIL.	97.4	52.6	73.5	96.0	46.0	69.4	94.1	36.3	63.3	93.8	48.0	69.4	97.4	42.8	68.3	98.9	42.7	69.0
MF2010	WEIGHTED AVERAGE	99.6	57.5	80.8	96.7	59.1	79.9	97.5	46.9	74.9	99.5	50.8	77.7	99.7	51.9	78.3	97.8	50.7	76.8

TEXAS

DIVISION		PALESTINE 758 MAT.	INST.	TOTAL	SAN ANGELO 769 MAT.	INST.	TOTAL	SAN ANTONIO 781-782 MAT.	INST.	TOTAL	TEMPLE 765 MAT.	INST.	TOTAL	TEXARKANA 755 MAT.	INST.	TOTAL	TYLER 757 MAT.	INST.	TOTAL
015433	CONTRACTOR EQUIPMENT		90.3	90.3		88.5	88.5		90.4	90.4		88.5	88.5		90.3	90.3		90.3	90.3
0241, 31 - 34	SITE & INFRASTRUCTURE, DEMOLITION	105.7	90.3	95.1	101.6	86.2	91.0	94.8	90.3	91.7	89.3	84.9	86.3	91.5	90.2	90.6	103.2	90.3	94.3
0310	Concrete Forming & Accessories	79.5	59.0	61.8	94.4	38.0	45.5	90.9	57.0	61.5	97.3	35.8	43.9	93.5	35.8	43.4	88.8	34.6	41.8
0320	Concrete Reinforcing	97.8	49.8	73.1	97.7	51.8	74.0	104.6	47.4	75.2	97.8	49.5	72.9	97.7	51.8	74.1	98.6	52.0	74.6
0330	Cast-in-Place Concrete	87.8	44.1	71.1	99.2	43.7	77.9	70.4	68.1	69.5	81.3	49.3	69.1	88.5	44.9	71.8	105.5	43.8	81.9
03	CONCRETE	98.5	52.7	76.6	96.3	43.7	71.1	81.3	59.7	71.0	82.4	44.2	64.1	90.8	43.2	68.0	105.0	42.4	75.0
04	MASONRY	116.5	51.5	77.1	128.6	43.9	77.2	93.7	60.5	73.6	140.1	49.8	85.4	181.5	47.4	100.2	170.5	50.3	97.7
05	METALS	98.0	62.3	85.7	97.7	64.5	86.3	96.7	66.2	86.2	97.4	63.4	85.7	91.7	66.8	83.1	97.8	67.3	87.3
06	WOOD, PLASTICS & COMPOSITES	82.9	63.0	71.5	98.6	38.1	63.9	93.5	55.2	71.6	107.6	34.6	65.7	94.4	34.1	59.8	93.5	32.2	58.3
07	THERMAL & MOISTURE PROTECTION	91.7	48.7	74.6	96.9	44.0	75.9	91.7	64.1	80.7	96.5	45.7	76.3	91.1	49.4	74.5	91.6	53.9	76.6
08	OPENINGS	74.7	52.6	69.1	89.0	40.6	76.8	100.9	52.9	88.8	75.9	36.9	66.0	91.6	39.2	78.3	74.7	38.2	65.4
0920	Plaster & Gypsum Board	82.2	62.5	68.3	80.3	36.8	49.7	82.0	54.4	62.5	80.3	33.2	47.1	91.3	32.7	50.1	87.0	30.7	47.4
0950, 0980	Ceilings & Acoustic Treatment	80.0	62.5	68.4	86.6	36.8	53.6	84.3	54.4	64.5	86.6	33.2	51.2	85.6	32.7	50.6	80.0	30.7	47.3
0960	Flooring	126.9	61.2	108.0	108.0	36.0	87.3	87.8	66.8	81.7	143.3	43.3	114.6	108.7	61.6	95.2	136.9	38.4	108.6
0970, 0990	Wall Finishes & Painting/Coating	95.6	32.2	57.1	95.7	52.2	69.3	93.0	57.3	71.3	97.1	35.4	59.7	96.5	41.5	62.8	95.6	41.5	62.8
09	FINISHES	107.5	57.4	79.9	92.4	38.3	62.6	84.9	58.0	70.0	102.2	36.6	66.1	103.2	40.1	68.4	110.7	34.3	68.6
COVERS	DIVS. 10 - 14, 25, 28, 41, 43, 44, 46	100.0	44.2	88.4	100.0	74.2	94.6	100.0	80.6	96.0	100.0	38.4	87.2	100.0	74.3	94.7	100.0	74.0	94.6
21, 22, 23	FIRE SUPPRESSION, PLUMBING & HVAC	94.5	59.0	79.7	94.8	44.6	73.9	99.9	66.4	86.0	94.8	35.7	70.2	94.5	35.2	69.8	94.5	59.8	80.1
26, 27, 3370	ELECTRICAL, COMMUNICATIONS & UTIL.	91.6	46.0	67.4	98.7	48.6	72.1	97.3	63.6	79.4	95.5	68.8	81.3	95.8	64.8	79.3	94.5	55.1	73.5
MF2010	WEIGHTED AVERAGE	96.1	57.4	78.8	97.6	50.1	76.3	95.0	65.1	81.6	95.3	50.0	75.1	99.1	51.5	77.8	100.2	54.9	80.0

TEXAS / UTAH

DIVISION		VICTORIA 779 MAT.	INST.	TOTAL	WACO 766-767 MAT.	INST.	TOTAL	WAXAHACHIE 751 MAT.	INST.	TOTAL	WHARTON 774 MAT.	INST.	TOTAL	WICHITA FALLS 763 MAT.	INST.	TOTAL	LOGAN 843 MAT.	INST.	TOTAL
015433	CONTRACTOR EQUIPMENT		97.6	97.6		88.5	88.5		95.7	95.7		98.8	98.8		88.5	88.5		99.8	99.8
0241, 31 - 34	SITE & INFRASTRUCTURE, DEMOLITION	122.3	80.7	93.6	97.8	86.2	89.8	106.6	83.2	90.5	129.0	83.6	97.7	98.5	85.8	89.7	85.2	99.5	95.1
0310	Concrete Forming & Accessories	90.9	37.1	44.2	96.0	41.3	48.5	86.4	60.4	63.8	85.6	38.6	44.8	96.0	39.2	46.7	105.4	61.4	67.2
0320	Concrete Reinforcing	95.4	31.6	62.6	97.5	48.6	72.3	99.6	51.1	74.7	99.3	53.9	75.9	97.5	51.8	73.9	98.9	77.0	87.6
0330	Cast-in-Place Concrete	111.0	42.2	84.7	88.0	53.9	74.9	91.4	42.6	72.7	114.1	52.7	90.6	94.0	46.2	75.7	89.2	73.9	83.4
03	CONCRETE	104.5	40.0	73.6	88.9	48.1	69.4	89.3	53.9	72.4	108.6	48.3	79.7	91.9	45.2	69.6	108.0	69.1	89.4
04	MASONRY	118.0	36.2	68.4	99.4	57.0	73.7	164.7	51.0	95.8	102.6	47.2	69.0	99.9	58.8	75.0	107.3	63.4	80.7
05	METALS	97.6	71.3	88.5	99.8	66.4	88.3	98.1	77.0	90.9	99.0	78.6	92.0	99.9	69.0	89.2	100.4	77.6	92.6
06	WOOD, PLASTICS & COMPOSITES	99.7	37.9	64.2	106.0	37.1	66.5	88.4	68.1	76.7	89.7	34.4	58.0	106.0	37.0	66.4	82.2	59.7	69.3
07	THERMAL & MOISTURE PROTECTION	99.2	41.1	76.1	97.4	49.2	78.3	90.3	48.0	73.4	96.0	51.2	78.2	87.9	42.1	76.3	99.8	65.1	86.0
08	OPENINGS	102.6	37.1	86.0	87.9	37.9	75.3	100.7	57.4	89.8	102.8	39.0	86.7	87.9	42.1	76.3	100.0	58.7	82.1
0920	Plaster & Gypsum Board	85.0	36.5	50.9	81.4	35.8	49.2	88.6	67.6	73.8	82.6	32.9	47.6	81.4	35.6	49.2	85.3	58.3	66.3
0950, 0980	Ceilings & Acoustic Treatment	99.0	36.5	57.6	90.3	35.8	54.2	90.3	67.6	75.3	98.0	32.9	54.9	90.3	35.6	54.1	119.1	58.3	78.8
0960	Flooring	92.9	39.6	77.6	142.0	46.8	114.7	95.4	39.6	79.3	91.1	47.2	78.5	142.8	77.9	124.2	101.3	50.4	86.7
0970, 0990	Wall Finishes & Painting/Coating	99.6	39.8	63.3	97.1	35.4	59.7	103.8	39.8	65.0	99.5	33.0	59.2	99.6	52.2	70.8	98.5	63.3	77.1
09	FINISHES	86.4	37.9	59.7	103.3	40.5	68.7	95.6	55.7	73.6	87.9	38.6	60.7	103.7	46.8	72.3	102.8	58.9	78.6
COVERS	DIVS. 10 - 14, 25, 28, 41, 43, 44, 46	100.0	39.6	87.4	100.0	77.9	95.4	100.0	43.6	88.3	100.0	41.9	87.9	100.0	71.2	94.0	100.0	59.1	91.5
21, 22, 23	FIRE SUPPRESSION, PLUMBING & HVAC	94.6	33.5	69.2	100.2	52.4	80.3	94.6	43.4	73.3	94.6	62.5	81.2	100.2	47.7	78.3	99.9	70.2	87.6
26, 27, 3370	ELECTRICAL, COMMUNICATIONS & UTIL.	102.6	42.1	70.5	98.9	69.1	83.0	93.8	69.1	80.6	101.5	46.5	72.2	100.7	59.1	78.6	94.4	75.3	84.2
MF2010	WEIGHTED AVERAGE	99.9	44.4	75.1	97.5	57.1	79.4	99.4	58.2	81.0	99.8	54.8	79.7	98.1	55.5	79.1	99.5	70.7	86.6

DIVISION		UTAH												VERMONT					
		OGDEN			PRICE			PROVO			SALT LAKE CITY			BELLOWS FALLS			BENNINGTON		
		842,844			845			846 - 847			840 - 841			051			052		
		MAT.	INST.	TOTAL	MAT.	INST.	TOTAL	MAT.	INST.	TOTAL	MAT.	INST.	TOTAL	MAT.	INST.	TOTAL	MAT.	INST.	TOTAL
015433	CONTRACTOR EQUIPMENT		99.8	99.8		98.7	98.7		98.7	98.7		99.7	99.7		100.3	100.3		100.3	100.3
0241, 31 - 34	SITE & INFRASTRUCTURE, DEMOLITION	74.6	99.5	91.8	83.1	97.8	93.2	81.9	97.8	92.8	74.4	99.4	91.7	78.4	98.2	92.1	77.7	98.0	91.7
0310	Concrete Forming & Accessories	105.5	61.4	67.3	107.5	52.1	59.4	106.8	61.3	67.3	108.0	61.3	67.5	96.5	49.7	55.9	94.4	54.1	59.4
0320	Concrete Reinforcing	98.5	77.0	87.5	106.0	76.9	91.0	106.9	77.0	91.5	100.8	77.0	88.6	86.5	64.6	75.2	86.5	64.6	75.2
0330	Cast-in-Place Concrete	90.6	73.9	84.2	89.3	63.1	79.3	89.3	73.8	83.4	99.0	73.8	89.4	91.2	72.4	84.0	91.2	69.0	82.7
03	CONCRETE	97.9	69.1	84.2	109.2	61.3	86.3	107.7	69.1	89.2	117.2	69.1	94.2	99.9	60.9	81.2	99.7	61.7	81.5
04	MASONRY	100.9	63.4	78.1	112.7	63.4	82.8	112.8	63.4	82.8	114.3	63.4	83.4	97.5	61.9	75.9	106.6	61.8	79.4
05	METALS	100.9	77.6	92.9	97.7	76.9	90.5	98.7	77.4	91.4	105.6	77.4	95.9	93.0	69.4	84.8	92.9	68.4	84.5
06	WOOD, PLASTICS & COMPOSITES	82.2	59.7	69.3	84.9	48.1	63.8	83.5	59.7	69.8	83.8	59.7	70.0	99.0	39.1	64.7	96.2	45.7	67.2
07	THERMAL & MOISTURE PROTECTION	98.6	65.1	85.3	101.6	60.7	85.4	101.6	65.1	87.1	105.4	65.1	89.4	95.9	60.3	81.8	95.8	55.3	79.7
08	OPENINGS	90.0	58.7	82.1	94.6	51.3	83.6	94.6	58.7	85.5	91.9	58.7	83.5	102.2	41.9	86.9	102.2	44.2	87.5
0920	Plaster & Gypsum Board	85.3	58.3	66.3	87.5	46.4	58.5	85.7	58.3	66.4	88.2	58.3	67.2	98.1	36.6	54.8	95.9	43.5	59.0
0950, 0980	Ceilings & Acoustic Treatment	119.1	58.3	78.8	119.1	46.4	70.9	119.1	58.3	78.8	112.9	58.3	76.7	89.8	36.6	54.6	89.8	43.5	59.1
0960	Flooring	99.1	50.4	85.1	102.5	37.5	83.8	102.1	50.4	87.3	102.5	50.4	87.5	95.8	32.2	77.6	94.7	32.2	76.7
0970, 0990	Wall Finishes & Painting/Coating	98.5	63.3	77.1	98.5	40.4	63.2	98.5	51.7	70.1	101.4	51.7	71.2	92.3	28.7	53.7	92.3	28.7	53.7
09	FINISHES	100.9	58.9	77.7	103.9	46.9	72.5	103.5	57.6	78.2	101.7	57.6	77.4	94.8	42.8	66.1	94.2	46.7	68.0
COVERS	DIVS. 10 - 14, 25, 28, 41, 43, 44, 46	100.0	59.1	91.5	100.0	51.3	89.9	100.0	59.1	91.5	100.0	59.1	91.5	100.0	47.3	89.1	100.0	47.9	89.2
21, 22, 23	FIRE SUPPRESSION, PLUMBING & HVAC	99.9	70.2	87.6	97.2	70.4	86.1	99.9	70.2	87.5	100.1	70.2	87.6	94.8	96.9	95.7	94.8	91.5	93.4
26, 27, 3370	ELECTRICAL, COMMUNICATIONS & UTIL.	94.7	75.3	84.4	100.3	75.3	87.0	95.1	64.9	79.1	97.6	64.9	80.2	100.6	96.9	98.7	100.6	66.3	82.4
MF2010	WEIGHTED AVERAGE	97.5	70.7	85.5	100.0	67.1	85.3	100.0	68.9	86.1	102.0	69.1	87.3	96.6	73.7	86.4	97.0	68.9	84.4

DIVISION		VERMONT																	
		BRATTLEBORO			BURLINGTON			GUILDHALL			MONTPELIER			RUTLAND			ST. JOHNSBURY		
		053			054			059			056			057			058		
		MAT.	INST.	TOTAL	MAT.	INST.	TOTAL	MAT.	INST.	TOTAL	MAT.	INST.	TOTAL	MAT.	INST.	TOTAL	MAT.	INST.	TOTAL
015433	CONTRACTOR EQUIPMENT		100.3	100.3		100.3	100.3		100.3	100.3		100.3	100.3		100.3	100.3		100.3	100.3
0241, 31 - 34	SITE & INFRASTRUCTURE, DEMOLITION	79.1	98.2	92.3	80.2	96.9	91.7	77.5	96.2	90.4	81.2	96.9	92.0	80.9	96.9	91.9	77.6	96.4	90.6
0310	Concrete Forming & Accessories	96.7	49.5	55.8	93.9	52.3	57.8	95.0	44.6	51.3	99.1	52.3	58.5	97.0	52.3	58.2	94.0	44.7	51.2
0320	Concrete Reinforcing	85.6	64.6	74.8	108.1	81.2	94.3	87.3	38.2	62.0	86.1	81.2	83.6	108.1	81.2	94.3	85.6	43.6	64.0
0330	Cast-in-Place Concrete	94.1	69.3	84.6	99.5	101.2	100.1	88.4	55.5	75.8	105.3	101.2	103.7	89.4	101.2	93.9	88.4	55.8	75.9
03	CONCRETE	102.3	59.8	81.9	108.1	75.0	92.2	97.3	47.9	73.7	108.1	75.0	92.3	103.2	75.0	89.7	97.0	49.0	74.0
04	MASONRY	105.7	61.9	79.1	109.1	50.1	73.3	106.0	39.9	65.9	92.6	50.1	66.9	87.9	50.1	65.0	131.4	39.9	76.0
05	METALS	92.9	69.2	84.8	100.4	77.9	92.7	93.0	57.9	80.9	92.9	77.9	87.8	98.6	77.9	91.5	93.0	58.3	81.0
06	WOOD, PLASTICS & COMPOSITES	99.3	39.1	64.8	94.2	51.7	69.8	95.7	45.7	67.0	98.4	51.7	71.7	99.4	51.7	72.1	92.0	45.7	65.5
07	THERMAL & MOISTURE PROTECTION	95.9	54.2	79.3	95.8	60.1	81.6	95.7	41.7	74.3	95.3	60.6	81.5	96.0	60.6	81.9	95.7	41.7	74.2
08	OPENINGS	102.2	41.9	86.9	105.6	52.8	92.3	102.2	38.9	86.2	102.2	52.8	89.7	105.6	52.8	92.3	102.2	38.9	86.2
0920	Plaster & Gypsum Board	98.1	36.6	54.8	99.7	49.6	64.5	103.9	43.5	61.3	107.4	49.6	66.7	98.9	49.6	64.2	107.4	43.5	62.4
0950, 0980	Ceilings & Acoustic Treatment	89.8	36.6	54.6	91.9	49.6	63.9	89.8	43.5	59.1	88.9	49.6	62.9	94.7	49.6	64.8	89.8	43.5	59.1
0960	Flooring	96.0	32.2	77.6	96.0	58.3	85.2	99.0	32.2	79.8	102.5	58.3	89.8	95.8	58.3	85.0	102.5	32.2	82.3
0970, 0990	Wall Finishes & Painting/Coating	92.3	28.7	53.7	92.3	33.4	56.6	92.3	22.3	49.8	92.3	33.4	56.6	92.3	33.4	56.6	92.3	23.4	50.5
09	FINISHES	94.9	42.8	66.2	95.6	50.2	70.6	96.3	40.1	65.3	97.8	50.2	71.5	96.0	50.2	70.7	97.8	40.2	66.0
COVERS	DIVS. 10 - 14, 25, 28, 41, 43, 44, 46	100.0	47.3	89.1	100.0	73.1	94.4	100.0	39.7	87.5	100.0	73.1	94.4	100.0	73.1	94.4	100.0	39.7	87.5
21, 22, 23	FIRE SUPPRESSION, PLUMBING & HVAC	94.8	94.7	94.8	100.0	63.5	84.8	94.8	61.7	81.0	94.6	63.5	81.6	100.2	63.5	84.9	94.8	61.7	81.0
26, 27, 3370	ELECTRICAL, COMMUNICATIONS & UTIL.	100.6	96.9	98.7	99.8	66.3	82.0	100.6	66.3	82.4	101.2	66.3	82.6	100.7	66.3	82.4	100.6	66.3	82.4
MF2010	WEIGHTED AVERAGE	97.4	72.9	86.5	100.9	66.1	85.4	96.9	55.7	78.5	97.7	66.2	83.6	99.2	66.2	84.4	98.3	55.9	79.3

DIVISION		VERMONT			VIRGINIA														
		WHITE RIVER JCT.			ALEXANDRIA			ARLINGTON			BRISTOL			CHARLOTTESVILLE			CULPEPER		
		050			223			222			242			229			227		
		MAT.	INST.	TOTAL	MAT.	INST.	TOTAL	MAT.	INST.	TOTAL	MAT.	INST.	TOTAL	MAT.	INST.	TOTAL	MAT.	INST.	TOTAL
015433	CONTRACTOR EQUIPMENT		100.3	100.3		101.7	101.7		100.4	100.4		100.4	100.4		104.8	104.8		100.4	100.4
0241, 31 - 34	SITE & INFRASTRUCTURE, DEMOLITION	81.4	96.3	91.7	120.9	88.1	98.3	132.3	84.8	99.5	114.9	84.4	93.9	120.2	86.5	97.0	119.1	85.2	95.7
0310	Concrete Forming & Accessories	91.9	42.9	49.4	89.4	74.7	76.6	88.9	71.6	73.9	85.2	45.0	50.3	83.7	51.2	55.5	81.0	70.3	71.7
0320	Concrete Reinforcing	86.5	38.2	61.6	80.5	81.0	80.8	90.9	72.6	81.5	90.9	69.5	79.9	90.3	70.5	80.1	90.9	66.7	78.4
0330	Cast-in-Place Concrete	94.1	55.2	79.2	117.1	81.5	103.5	114.0	77.3	99.9	113.5	49.8	89.1	118.1	58.1	95.1	116.7	67.8	98.0
03	CONCRETE	104.3	47.1	76.9	110.0	79.2	95.3	114.4	74.4	95.3	110.6	53.1	83.1	111.6	58.9	86.4	108.3	69.6	89.8
04	MASONRY	118.4	38.6	70.0	91.8	71.6	79.5	105.7	66.3	81.8	93.9	42.5	62.7	119.7	49.1	76.9	107.6	63.2	80.7
05	METALS	93.0	58.0	80.9	93.0	95.2	93.7	91.7	82.3	88.5	90.5	86.2	89.0	90.8	88.6	90.1	90.9	81.2	87.5
06	WOOD, PLASTICS & COMPOSITES	93.2	44.0	65.0	95.0	73.8	82.8	91.9	73.8	81.5	84.6	43.2	60.9	83.0	49.3	63.7	81.9	73.8	77.3
07	THERMAL & MOISTURE PROTECTION	96.0	36.7	72.4	97.2	80.5	90.6	99.1	69.1	87.2	98.7	49.1	79.0	98.3	54.7	81.0	98.5	63.2	84.5
08	OPENINGS	102.2	40.2	86.5	98.6	75.7	92.8	96.5	69.6	89.7	99.9	49.4	87.1	97.9	54.3	86.9	98.3	65.1	89.9
0920	Plaster & Gypsum Board	94.2	41.6	57.2	106.5	72.8	82.8	103.0	72.8	81.8	98.4	41.4	58.2	97.9	46.9	62.0	98.6	72.8	80.5
0950, 0980	Ceilings & Acoustic Treatment	89.8	41.6	57.9	95.7	72.8	80.6	92.9	72.8	79.6	92.0	41.4	58.4	92.0	46.9	62.1	92.9	72.8	79.6
0960	Flooring	93.8	32.2	76.1	109.9	85.7	102.9	108.1	63.9	95.4	104.2	39.5	85.6	102.7	46.6	86.6	102.7	63.9	91.5
0970, 0990	Wall Finishes & Painting/Coating	92.3	19.5	48.2	129.1	108.3	116.5	129.1	82.8	101.0	114.7	44.8	72.3	114.7	63.3	83.5	129.1	56.3	84.9
09	FINISHES	94.0	38.6	63.4	104.4	79.8	90.8	104.2	71.9	86.4	100.5	43.1	68.9	100.1	50.8	72.9	100.8	68.8	83.2
COVERS	DIVS. 10 - 14, 25, 28, 41, 43, 44, 46	100.0	39.2	87.4	100.0	88.2	97.5	100.0	72.5	94.3	100.0	75.9	95.0	100.0	75.8	95.0	100.0	61.6	92.0
21, 22, 23	FIRE SUPPRESSION, PLUMBING & HVAC	94.8	61.3	80.9	100.3	89.6	95.9	100.3	86.6	94.6	94.9	55.0	78.3	94.9	58.2	79.6	94.9	64.3	82.2
26, 27, 3370	ELECTRICAL, COMMUNICATIONS & UTIL.	100.6	66.3	82.4	96.8	96.9	96.9	94.2	96.7	95.5	96.5	38.4	65.6	96.4	70.1	82.4	99.2	56.6	76.5
MF2010	WEIGHTED AVERAGE	98.3	55.1	79.0	100.2	85.5	93.6	101.2	80.1	91.8	98.4	55.1	79.0	99.8	63.4	83.5	99.1	67.8	85.1

VIRGINIA

DIVISION		FAIRFAX 220 - 221			FARMVILLE 239			FREDERICKSBURG 224 - 225			GRUNDY 246			HARRISONBURG 228			LYNCHBURG 245		
		MAT.	INST.	TOTAL	MAT.	INST.	TOTAL	MAT.	INST.	TOTAL	MAT.	INST.	TOTAL	MAT.	INST.	TOTAL	MAT.	INST.	TOTAL
015433	CONTRACTOR EQUIPMENT		100.4	100.4		104.8	104.8		100.4	100.4		100.4	100.4		100.4	100.4		100.4	100.4
0241, 31 - 34	SITE & INFRASTRUCTURE, DEMOLITION	131.4	85.7	99.8	115.9	85.8	95.1	118.5	85.4	95.6	112.5	83.2	92.3	128.0	84.5	98.0	113.4	84.6	93.5
0310	Concrete Forming & Accessories	83.6	72.1	73.7	95.0	46.6	53.0	83.6	50.9	55.2	88.1	38.8	45.3	79.9	45.1	49.7	85.2	64.5	67.2
0320	Concrete Reinforcing	90.9	80.9	85.7	90.9	53.7	71.7	91.6	80.0	85.6	89.7	54.3	71.4	90.9	48.9	69.2	90.3	70.2	79.9
0330	Cast-in-Place Concrete	114.0	78.3	100.3	118.1	56.1	94.3	115.7	54.7	92.3	113.5	50.1	89.3	114.0	61.3	93.8	113.5	61.2	93.5
03	CONCRETE	114.0	77.1	96.4	110.5	52.8	82.9	107.9	59.4	84.7	109.3	47.2	79.5	111.8	53.4	83.8	109.2	65.8	88.4
04	MASONRY	105.7	66.6	82.0	102.6	48.3	69.7	106.6	47.0	70.5	97.1	51.1	69.2	103.9	53.2	73.2	111.8	55.4	77.6
05	METALS	90.9	96.1	92.7	90.8	78.0	86.4	90.9	92.0	91.3	90.5	71.7	84.1	90.8	84.5	88.6	90.7	87.6	89.6
06	WOOD, PLASTICS & COMPOSITES	84.6	73.8	78.4	97.0	47.6	68.7	84.6	48.6	64.0	87.5	37.9	59.0	80.8	42.2	58.6	84.6	67.9	75.0
07	THERMAL & MOISTURE PROTECTION	99.1	69.5	87.3	98.3	49.5	78.9	98.5	52.7	80.3	98.6	48.8	78.9	98.9	62.8	84.6	98.5	56.9	82.0
08	OPENINGS	96.5	77.0	91.6	98.2	46.4	85.1	97.9	56.3	87.4	99.9	37.1	84.0	98.3	52.9	86.8	98.2	62.8	89.2
0920	Plaster & Gypsum Board	98.6	72.8	80.5	105.4	45.2	63.0	98.6	46.9	62.2	98.8	35.9	54.5	97.9	40.3	57.3	98.4	66.8	76.1
0950, 0980	Ceilings & Acoustic Treatment	92.9	72.8	79.6	92.0	45.2	61.0	92.9	46.9	62.4	92.0	35.9	54.8	92.0	40.3	57.7	92.0	66.8	75.3
0960	Flooring	104.5	66.8	93.7	109.9	63.7	96.6	104.5	78.2	96.9	105.7	35.6	85.6	102.3	78.2	95.4	104.2	44.1	86.9
0970, 0990	Wall Finishes & Painting/Coating	129.1	82.8	101.0	114.7	41.7	70.4	129.1	69.1	92.7	114.7	37.8	68.1	129.1	54.6	83.9	114.7	44.8	72.3
09	FINISHES	102.6	72.4	85.9	102.8	49.5	73.4	101.3	56.8	76.8	100.9	37.8	66.1	101.2	51.1	73.6	100.3	59.0	77.5
COVERS	DIVS. 10 - 14, 25, 28, 41, 43, 44, 46	100.0	72.5	94.3	100.0	50.8	89.8	100.0	76.7	95.2	100.0	48.6	89.3	100.0	75.6	94.9	100.0	79.5	95.7
21, 22, 23	FIRE SUPPRESSION, PLUMBING & HVAC	94.9	86.9	91.6	94.9	37.5	71.0	94.9	80.4	88.9	94.9	42.1	72.9	94.9	54.9	78.3	94.9	57.7	79.4
26, 27, 3370	ELECTRICAL, COMMUNICATIONS & UTIL.	97.7	96.7	97.2	93.4	50.9	70.8	94.4	93.2	93.8	96.5	48.4	70.9	96.7	84.8	90.4	97.6	57.2	76.1
MF2010	WEIGHTED AVERAGE	99.9	82.3	92.0	98.7	53.0	78.3	98.5	72.2	86.7	98.3	50.3	76.9	99.3	64.0	83.5	99.0	64.6	83.6

VIRGINIA

DIVISION		NEWPORT NEWS 236			NORFOLK 233 - 235			PETERSBURG 238			PORTSMOUTH 237			PULASKI 243			RICHMOND 230 - 232		
		MAT.	INST.	TOTAL	MAT.	INST.	TOTAL	MAT.	INST.	TOTAL	MAT.	INST.	TOTAL	MAT.	INST.	TOTAL	MAT.	INST.	TOTAL
015433	CONTRACTOR EQUIPMENT		104.8	104.8		105.4	105.4		104.8	104.8		104.7	104.7		100.4	100.4		104.8	104.8
0241, 31 - 34	SITE & INFRASTRUCTURE, DEMOLITION	114.0	87.2	95.5	112.1	88.2	95.6	118.7	87.5	97.2	112.3	86.6	94.5	111.8	83.2	92.1	111.3	87.5	94.9
0310	Concrete Forming & Accessories	94.3	67.4	71.0	95.8	67.5	71.3	88.1	60.8	64.4	84.3	55.4	59.3	88.1	39.0	45.5	93.8	60.8	65.2
0320	Concrete Reinforcing	90.6	70.9	80.4	90.6	70.9	80.4	90.3	71.2	80.4	90.3	70.8	80.3	89.7	54.3	71.5	90.6	71.2	80.6
0330	Cast-in-Place Concrete	114.7	61.4	94.3	125.4	61.5	100.9	121.9	56.2	96.7	113.6	61.0	93.5	113.5	50.2	89.3	110.6	56.2	89.8
03	CONCRETE	107.4	67.3	88.2	112.8	67.3	91.1	113.4	62.6	89.1	106.2	61.7	84.9	109.3	47.3	79.6	105.3	62.6	84.9
04	MASONRY	97.4	55.3	71.8	103.4	55.3	74.2	111.1	56.8	78.2	102.8	55.1	73.9	90.6	51.1	66.7	95.1	56.8	71.9
05	METALS	93.0	89.6	91.8	92.0	89.7	91.2	90.9	90.1	90.6	92.0	88.1	90.6	90.6	71.9	84.2	95.1	90.1	93.4
06	WOOD, PLASTICS & COMPOSITES	95.9	71.8	82.1	98.0	71.8	82.9	88.0	63.4	73.9	84.3	55.7	67.9	87.5	37.9	59.0	95.0	63.4	76.9
07	THERMAL & MOISTURE PROTECTION	98.3	57.3	82.0	95.9	57.3	80.6	98.3	57.6	82.2	98.3	55.0	81.1	98.6	48.8	78.9	97.1	57.6	81.4
08	OPENINGS	98.6	65.5	90.2	98.6	65.5	90.2	97.9	62.0	88.8	98.7	56.8	88.1	100.0	37.1	84.0	98.6	62.0	89.3
0920	Plaster & Gypsum Board	106.5	70.1	80.9	100.2	70.1	79.0	99.1	61.5	72.6	98.5	53.5	66.8	98.8	35.9	54.5	109.7	61.5	75.8
0950, 0980	Ceilings & Acoustic Treatment	95.7	70.1	78.7	95.7	70.1	78.7	92.9	61.5	72.1	95.7	53.5	67.7	92.0	35.9	54.8	98.5	61.5	74.0
0960	Flooring	109.9	53.4	93.7	109.3	53.4	93.2	105.2	75.2	96.6	101.8	67.5	92.0	105.7	35.6	85.6	109.3	75.2	99.5
0970, 0990	Wall Finishes & Painting/Coating	114.7	45.3	72.6	114.7	69.3	87.1	114.7	69.3	87.1	114.7	69.3	87.1	114.7	37.8	68.1	114.7	63.9	81.8
09	FINISHES	103.5	62.9	81.1	102.1	65.6	82.0	101.1	63.9	80.6	100.2	58.5	77.2	100.9	37.8	66.1	103.8	63.9	81.8
COVERS	DIVS. 10 - 14, 25, 28, 41, 43, 44, 46	100.0	81.3	96.1	100.0	81.3	96.1	100.0	76.1	95.0	100.0	79.5	95.7	100.0	48.6	89.3	100.0	76.1	95.0
21, 22, 23	FIRE SUPPRESSION, PLUMBING & HVAC	100.3	64.2	85.3	100.1	63.5	84.9	94.9	56.3	78.8	100.3	63.5	85.0	94.9	42.4	73.0	100.1	56.3	81.9
26, 27, 3370	ELECTRICAL, COMMUNICATIONS & UTIL.	96.5	59.3	76.7	101.7	61.2	80.1	96.7	70.1	82.6	94.6	61.1	76.8	96.5	52.3	73.0	101.3	70.1	84.7
MF2010	WEIGHTED AVERAGE	100.0	67.6	85.5	101.0	68.2	86.3	99.6	66.8	84.9	99.4	65.5	84.2	98.0	50.9	77.0	100.2	66.8	85.3

VIRGINIA / WASHINGTON

DIVISION		ROANOKE 240 - 241			STAUNTON 244			WINCHESTER 226			CLARKSTON 994			EVERETT 982			OLYMPIA 985		
		MAT.	INST.	TOTAL	MAT.	INST.	TOTAL	MAT.	INST.	TOTAL	MAT.	INST.	TOTAL	MAT.	INST.	TOTAL	MAT.	INST.	TOTAL
015433	CONTRACTOR EQUIPMENT		100.4	100.4		104.8	104.8		100.4	100.4		91.3	91.3		103.9	103.9		103.9	103.9
0241, 31 - 34	SITE & INFRASTRUCTURE, DEMOLITION	111.5	84.5	92.9	116.5	86.3	95.7	126.5	84.6	97.6	99.2	90.6	93.3	94.5	113.8	107.8	95.8	113.8	108.2
0310	Concrete Forming & Accessories	94.1	64.2	68.2	87.7	53.5	58.0	82.2	50.2	54.4	122.4	71.5	78.2	109.1	101.4	102.4	99.0	101.6	101.2
0320	Concrete Reinforcing	90.6	70.2	80.1	90.3	52.7	70.9	90.3	85.1	87.6	110.2	85.8	97.6	115.1	102.0	108.3	115.7	102.0	108.7
0330	Cast-in-Place Concrete	129.0	61.1	103.0	118.1	54.3	93.6	114.0	57.4	92.3	102.6	83.7	95.4	103.8	108.3	105.6	94.5	108.5	99.9
03	CONCRETE	114.5	65.6	91.1	110.7	55.3	84.2	111.1	60.8	87.0	100.7	78.5	90.1	99.6	103.3	101.4	95.5	103.5	99.3
04	MASONRY	98.4	57.1	73.3	108.1	55.5	76.2	101.5	51.0	70.9	108.3	80.5	91.5	132.1	101.4	113.5	121.0	101.5	109.2
05	METALS	92.8	87.4	90.9	90.8	83.0	88.1	90.9	91.2	91.0	89.0	81.2	86.3	101.8	92.4	98.6	97.0	92.5	95.4
06	WOOD, PLASTICS & COMPOSITES	95.9	67.9	79.9	87.5	54.0	68.3	83.2	48.6	63.4	96.8	69.3	81.0	105.4	101.2	103.0	92.0	101.2	97.3
07	THERMAL & MOISTURE PROTECTION	98.3	56.0	81.5	98.3	53.4	80.5	99.0	60.1	83.5	152.5	77.5	122.7	101.8	99.6	100.9	101.4	98.5	100.3
08	OPENINGS	98.6	62.8	89.5	98.2	52.0	86.5	100.0	57.6	89.3	106.6	72.4	97.9	102.3	100.7	101.9	102.2	100.7	101.8
0920	Plaster & Gypsum Board	106.5	66.8	78.6	98.8	51.8	65.7	98.6	46.9	62.2	134.2	68.2	87.7	119.5	101.2	106.6	108.7	101.2	103.4
0950, 0980	Ceilings & Acoustic Treatment	95.7	66.8	76.6	92.0	51.8	65.3	92.9	46.9	62.4	103.9	68.2	80.2	113.3	101.2	105.3	107.5	101.2	103.3
0960	Flooring	109.9	47.6	92.0	105.1	41.7	86.9	103.8	78.2	96.4	104.4	49.8	88.7	117.4	96.7	111.4	110.2	96.9	106.4
0970, 0990	Wall Finishes & Painting/Coating	114.7	44.8	72.3	114.7	31.2	64.1	129.1	48.1	79.9	108.3	63.0	80.8	104.7	76.4	87.6	104.7	85.9	93.3
09	FINISHES	103.3	59.8	79.3	100.6	49.5	72.4	101.8	54.4	75.7	118.5	65.5	89.3	110.8	97.9	103.7	106.3	99.0	102.3
COVERS	DIVS. 10 - 14, 25, 28, 41, 43, 44, 46	100.0	79.4	95.7	100.0	79.1	95.7	100.0	57.1	91.1	100.0	83.8	96.6	100.0	100.3	100.1	100.0	101.7	100.3
21, 22, 23	FIRE SUPPRESSION, PLUMBING & HVAC	100.3	48.5	78.8	94.9	56.7	79.0	94.9	69.0	84.1	95.0	76.2	87.2	100.1	92.8	97.1	99.9	92.9	97.0
26, 27, 3370	ELECTRICAL, COMMUNICATIONS & UTIL.	96.5	41.3	67.1	95.2	44.9	68.4	94.9	46.6	69.2	101.8	94.2	97.8	104.4	84.1	93.6	105.8	95.8	100.5
MF2010	WEIGHTED AVERAGE	100.7	60.7	82.8	98.9	59.1	81.1	99.2	63.2	83.1	101.9	79.8	92.0	103.4	97.1	100.6	101.3	98.9	100.2

City Cost Indexes

WASHINGTON

DIVISION		RICHLAND 993 MAT.	INST.	TOTAL	SEATTLE 980-981,987 MAT.	INST.	TOTAL	SPOKANE 990-992 MAT.	INST.	TOTAL	TACOMA 983-984 MAT.	INST.	TOTAL	VANCOUVER 986 MAT.	INST.	TOTAL	WENATCHEE 988 MAT.	INST.	TOTAL
015433	CONTRACTOR EQUIPMENT		91.3	91.3		103.7	103.7		91.3	91.3		103.9	103.9		98.2	98.2		103.9	103.9
0241, 31 - 34	SITE & INFRASTRUCTURE, DEMOLITION	100.2	90.8	93.7	98.5	111.9	107.7	99.5	90.7	93.5	97.6	113.8	108.8	108.7	100.5	103.0	108.6	111.8	110.8
0310	Concrete Forming & Accessories	122.6	81.2	86.6	100.7	102.2	102.0	127.0	80.7	86.8	100.7	101.6	101.4	101.6	87.5	89.4	102.1	79.2	82.2
0320	Concrete Reinforcing	105.7	85.8	95.5	113.6	102.1	107.7	106.5	85.8	95.8	113.6	102.0	107.6	114.4	101.7	108.0	114.5	86.3	100.0
0330	Cast-in-Place Concrete	102.9	85.3	96.2	109.0	108.7	108.9	106.9	85.1	98.6	106.8	108.5	107.4	119.3	101.5	112.4	109.0	80.3	98.0
03	CONCRETE	100.3	83.3	92.2	102.6	104.0	103.2	102.7	83.1	93.3	101.5	103.5	102.4	111.8	94.9	103.7	109.0	80.9	95.6
04	MASONRY	108.1	83.0	92.9	125.5	101.5	111.0	108.8	83.0	93.1	125.4	101.5	110.9	126.5	96.1	108.1	128.5	82.9	100.9
05	METALS	89.4	81.7	86.8	103.7	94.5	100.5	91.8	81.3	88.2	103.7	92.5	99.8	101.0	93.0	98.3	101.0	82.8	94.7
06	WOOD, PLASTICS & COMPOSITES	97.0	80.4	87.5	96.5	101.2	99.2	105.6	80.4	91.2	95.2	101.2	98.7	87.5	85.2	86.2	96.3	78.2	85.9
07	THERMAL & MOISTURE PROTECTION	153.7	79.8	124.3	101.6	99.7	100.8	150.1	80.7	122.5	101.5	98.5	100.3	101.8	91.0	97.5	102.0	81.1	93.7
08	OPENINGS	109.1	74.0	100.2	104.6	100.7	103.6	109.7	74.0	100.6	102.9	100.7	102.3	100.5	91.1	98.1	102.5	72.8	94.9
0920	Plaster & Gypsum Board	134.2	79.7	95.8	112.8	101.2	104.6	134.5	79.7	95.9	115.6	101.2	105.5	113.6	85.0	93.5	117.4	77.4	89.3
0950, 0980	Ceilings & Acoustic Treatment	109.1	79.7	89.6	115.9	101.2	106.1	106.0	79.7	88.5	116.1	101.2	106.2	111.5	85.0	94.0	109.3	77.4	88.2
0960	Flooring	104.6	84.6	98.8	110.1	104.8	108.6	104.7	90.1	100.5	110.6	96.9	106.7	116.7	107.9	114.1	113.6	66.2	100.0
0970, 0990	Wall Finishes & Painting/Coating	108.3	63.0	80.8	104.7	85.9	93.3	108.3	63.0	80.8	104.7	85.9	93.3	109.6	66.6	83.6	104.7	63.0	79.4
09	FINISHES	119.9	79.7	97.7	108.7	100.6	104.2	119.4	80.8	98.1	109.1	99.0	103.5	107.3	88.7	97.0	109.8	74.6	90.4
COVERS	DIVS. 10 - 14, 25, 28, 41, 43, 44, 46	100.0	69.2	93.6	100.0	101.7	100.4	100.0	69.1	93.6	100.0	101.7	100.3	100.0	65.7	92.9	100.0	85.3	96.9
21, 22, 23	FIRE SUPPRESSION, PLUMBING & HVAC	100.5	102.2	101.2	100.0	118.3	107.6	100.4	77.5	90.8	100.1	92.9	97.1	100.2	94.1	97.7	94.7	84.4	90.4
26, 27, 3370	ELECTRICAL, COMMUNICATIONS & UTIL.	99.5	95.1	97.2	104.2	110.2	107.4	96.5	80.8	88.1	104.2	95.8	99.7	109.5	100.1	104.5	105.1	81.9	92.8
MF2010	WEIGHTED AVERAGE	103.4	88.0	96.5	103.8	106.6	105.0	103.7	80.8	93.5	103.5	98.9	101.4	104.8	93.8	99.9	103.3	83.7	94.6

WASHINGTON / WEST VIRGINIA

DIVISION		YAKIMA 989 MAT.	INST.	TOTAL	BECKLEY 258-259 MAT.	INST.	TOTAL	BLUEFIELD 247-248 MAT.	INST.	TOTAL	BUCKHANNON 262 MAT.	INST.	TOTAL	CHARLESTON 250-253 MAT.	INST.	TOTAL	CLARKSBURG 263-264 MAT.	INST.	TOTAL
015433	CONTRACTOR EQUIPMENT		103.9	103.9		100.4	100.4		100.4	100.4		100.4	100.4		100.4	100.4		100.4	100.4
0241, 31 - 34	SITE & INFRASTRUCTURE, DEMOLITION	100.4	112.4	108.7	106.1	86.7	92.7	105.8	86.7	92.6	112.6	86.4	94.5	103.7	88.3	93.1	113.2	86.4	94.7
0310	Concrete Forming & Accessories	101.0	96.7	97.2	83.8	88.3	87.7	85.1	87.4	87.1	84.5	85.5	85.4	99.4	90.2	91.4	82.3	85.5	85.1
0320	Concrete Reinforcing	114.1	86.4	99.8	89.3	85.3	87.2	89.3	67.1	77.9	89.8	84.3	87.0	90.6	85.5	88.0	89.8	84.9	87.3
0330	Cast-in-Place Concrete	114.1	84.9	102.9	111.0	102.7	107.9	111.0	102.4	107.7	110.7	101.1	107.0	102.7	102.9	102.8	121.3	101.1	113.6
03	CONCRETE	106.5	90.2	98.7	104.7	93.3	99.2	104.8	89.4	97.4	107.9	91.3	100.0	101.6	94.2	98.1	112.7	91.4	102.5
04	MASONRY	118.3	82.8	96.8	97.5	88.9	92.3	92.3	88.9	90.2	103.7	84.7	92.2	94.0	90.8	92.1	106.2	84.7	93.2
05	METALS	101.7	82.5	95.1	90.8	96.0	92.6	90.8	88.7	90.1	90.9	96.5	92.9	93.0	97.2	94.5	90.9	96.7	92.9
06	WOOD, PLASTICS & COMPOSITES	95.7	101.2	98.9	84.6	89.8	87.6	86.3	89.8	88.3	85.6	85.2	85.4	102.0	89.8	95.0	82.4	85.2	84.0
07	THERMAL & MOISTURE PROTECTION	101.6	82.7	94.1	98.4	87.1	93.9	98.4	87.1	93.9	98.7	86.9	94.0	95.7	87.9	92.6	98.6	86.9	94.0
08	OPENINGS	102.4	81.8	97.2	99.3	82.9	95.1	100.2	78.7	94.7	100.2	80.1	95.1	99.9	82.9	95.5	100.2	82.1	95.6
0920	Plaster & Gypsum Board	115.7	101.2	105.5	96.1	89.3	91.3	96.9	89.3	91.6	97.5	84.6	88.4	99.7	89.3	92.4	95.3	84.6	87.7
0950, 0980	Ceilings & Acoustic Treatment	111.4	101.2	104.6	90.1	89.3	89.6	90.1	89.3	89.6	92.0	84.6	87.1	93.8	89.3	90.8	92.0	84.6	87.1
0960	Flooring	111.8	55.0	95.4	101.2	106.0	102.6	101.8	106.0	103.0	101.5	100.2	101.1	109.9	106.0	108.8	100.3	100.2	100.2
0970, 0990	Wall Finishes & Painting/Coating	104.7	63.0	79.4	114.7	37.7	68.0	114.7	37.7	68.0	114.7	85.5	97.0	114.7	84.9	96.7	114.7	85.5	97.0
09	FINISHES	108.7	85.9	96.1	98.3	87.0	92.1	98.6	87.0	92.2	99.5	88.1	93.2	101.6	92.7	96.7	98.8	88.1	92.9
COVERS	DIVS. 10 - 14, 25, 28, 41, 43, 44, 46	100.0	98.7	99.7	100.0	59.3	91.5	100.0	59.2	91.5	100.0	97.7	99.5	100.0	99.3	99.9	100.0	97.7	99.5
21, 22, 23	FIRE SUPPRESSION, PLUMBING & HVAC	100.1	102.3	101.0	94.9	82.0	89.5	94.9	56.1	78.8	94.9	89.1	92.5	100.1	84.0	93.4	94.9	88.9	92.4
26, 27, 3370	ELECTRICAL, COMMUNICATIONS & UTIL.	107.5	95.1	100.9	92.4	89.4	90.8	95.3	70.4	82.0	96.8	93.5	95.0	99.8	89.5	94.3	96.8	93.5	95.0
MF2010	WEIGHTED AVERAGE	103.7	93.1	99.0	97.0	87.2	92.6	97.1	77.7	88.4	98.5	89.7	94.6	99.0	90.1	95.0	99.2	89.7	94.9

WEST VIRGINIA

DIVISION		GASSAWAY 266 MAT.	INST.	TOTAL	HUNTINGTON 255-257 MAT.	INST.	TOTAL	LEWISBURG 249 MAT.	INST.	TOTAL	MARTINSBURG 254 MAT.	INST.	TOTAL	MORGANTOWN 265 MAT.	INST.	TOTAL	PARKERSBURG 261 MAT.	INST.	TOTAL
015433	CONTRACTOR EQUIPMENT		100.4	100.4		100.4	100.4		100.4	100.4		100.4	100.4		100.4	100.4		100.4	100.4
0241, 31 - 34	SITE & INFRASTRUCTURE, DEMOLITION	109.7	86.9	93.9	110.5	89.1	95.8	123.1	86.7	98.0	110.0	84.5	92.4	106.9	86.6	92.9	115.4	88.2	96.7
0310	Concrete Forming & Accessories	84.0	85.7	85.5	95.3	92.5	92.9	82.6	87.9	87.2	83.9	78.4	79.1	82.6	85.2	84.9	86.5	87.7	87.5
0320	Concrete Reinforcing	89.8	85.7	87.6	90.6	85.6	88.0	90.8	67.1	78.2	89.3	72.4	80.6	89.8	84.8	87.2	89.3	84.2	86.7
0330	Cast-in-Place Concrete	115.9	101.2	110.3	121.1	103.8	114.5	111.1	102.6	107.8	116.3	92.2	107.1	110.7	101.0	107.0	113.2	94.0	105.8
03	CONCRETE	108.5	91.6	100.5	110.6	95.6	103.4	114.7	89.7	102.7	108.8	82.6	96.2	104.5	91.3	98.2	110.1	89.9	100.4
04	MASONRY	107.3	87.7	95.4	97.8	92.4	94.5	96.5	88.9	91.9	99.3	72.1	82.8	126.4	84.7	101.1	82.4	86.6	84.9
05	METALS	90.9	97.2	93.1	93.1	97.4	94.5	90.9	89.2	90.3	91.1	84.5	88.8	91.0	96.4	92.9	91.7	96.5	93.3
06	WOOD, PLASTICS & COMPOSITES	84.7	85.2	85.0	95.9	91.9	93.6	82.7	89.8	86.8	84.6	79.9	81.9	82.7	85.2	84.1	85.9	86.8	86.4
07	THERMAL & MOISTURE PROTECTION	98.4	87.7	94.2	98.5	88.8	94.6	99.3	87.1	94.4	98.6	71.6	87.9	98.5	86.9	93.9	98.3	87.2	93.9
08	OPENINGS	98.1	80.4	93.6	98.6	84.1	94.9	100.2	78.7	94.7	101.6	67.7	93.0	101.6	82.1	96.7	99.2	81.0	94.6
0920	Plaster & Gypsum Board	96.6	84.6	88.1	105.4	91.5	95.7	95.3	89.3	91.1	96.6	79.1	84.3	95.3	84.6	87.7	98.4	86.2	89.8
0950, 0980	Ceilings & Acoustic Treatment	92.0	84.6	87.1	92.0	91.5	91.7	92.0	89.3	90.2	92.0	84.6	87.1	92.0	84.6	87.1	92.0	86.2	88.2
0960	Flooring	101.3	106.0	102.6	109.7	106.0	108.6	100.4	106.0	102.0	101.2	56.7	88.4	100.4	100.2	100.3	104.7	97.4	102.6
0970, 0990	Wall Finishes & Painting/Coating	114.7	84.9	96.7	114.7	84.9	96.7	114.7	37.7	68.0	114.7	44.5	72.1	114.7	85.5	97.0	114.7	85.0	96.7
09	FINISHES	99.0	89.2	93.6	102.3	94.4	97.9	100.0	87.0	92.8	99.0	71.4	83.8	98.4	88.1	92.7	100.7	89.1	94.3
COVERS	DIVS. 10 - 14, 25, 28, 41, 43, 44, 46	100.0	97.7	99.5	100.0	100.2	100.0	100.0	59.2	91.5	100.0	57.3	91.1	100.0	68.5	93.4	100.0	98.8	99.8
21, 22, 23	FIRE SUPPRESSION, PLUMBING & HVAC	94.9	89.9	92.8	100.3	83.3	93.3	94.9	81.8	89.5	94.9	84.5	90.6	94.9	88.8	92.4	100.2	86.8	94.6
26, 27, 3370	ELECTRICAL, COMMUNICATIONS & UTIL.	96.8	89.5	92.9	96.5	95.9	96.2	92.4	70.4	80.7	98.8	82.1	89.9	97.0	93.5	95.1	97.0	92.2	94.4
MF2010	WEIGHTED AVERAGE	98.4	89.9	94.6	100.2	91.6	96.3	98.9	83.2	91.9	98.6	79.1	89.9	99.2	88.8	94.6	99.1	89.4	94.8

WEST VIRGINIA / WISCONSIN

DIVISION		PETERSBURG 268 MAT.	INST.	TOTAL	ROMNEY 267 MAT.	INST.	TOTAL	WHEELING 260 MAT.	INST.	TOTAL	BELOIT 535 MAT.	INST.	TOTAL	EAU CLAIRE 547 MAT.	INST.	TOTAL	GREEN BAY 541-543 MAT.	INST.	TOTAL
015433	CONTRACTOR EQUIPMENT		100.4	100.4		100.4	100.4		100.4	100.4		101.0	101.0		100.7	100.7		98.6	98.6
0241, 31 - 34	SITE & INFRASTRUCTURE, DEMOLITION	106.1	87.0	92.9	109.1	87.0	93.9	116.2	87.3	96.2	94.2	106.3	102.5	95.9	103.1	100.9	99.9	99.4	99.5
0310	Concrete Forming & Accessories	85.4	80.6	81.2	82.0	80.8	80.9	88.1	85.7	86.0	96.4	96.7	96.7	98.3	98.5	98.4	102.9	97.8	98.4
0320	Concrete Reinforcing	89.3	73.3	81.0	89.8	70.6	79.9	88.7	84.9	86.7	98.3	118.7	108.8	95.2	103.3	99.4	93.5	89.7	91.5
0330	Cast-in-Place Concrete	110.7	95.7	104.9	115.9	95.7	108.2	113.2	101.2	108.6	101.9	95.2	99.3	101.6	96.0	99.5	105.1	100.7	103.4
03	CONCRETE	104.6	85.2	95.3	108.4	84.8	97.1	110.1	91.5	101.2	99.5	100.5	100.0	101.0	98.8	99.9	104.0	97.5	100.9
04	MASONRY	99.3	84.7	90.5	96.9	84.0	89.0	106.9	84.7	93.5	104.7	95.6	100.1	98.3	100.1	98.3	129.4	99.3	111.2
05	METALS	91.0	91.6	91.2	91.1	90.8	91.0	91.8	96.8	93.5	94.0	109.5	99.3	95.5	104.3	98.5	97.7	98.2	97.9
06	WOOD, PLASTICS & COMPOSITES	86.5	79.8	82.7	81.9	79.8	80.7	87.5	85.2	86.2	102.8	99.2	100.7	108.7	99.3	103.3	109.0	99.3	103.4
07	THERMAL & MOISTURE PROTECTION	98.5	71.7	87.9	98.6	72.0	88.0	98.7	87.4	94.2	105.2	93.1	100.4	105.2	86.7	97.9	106.7	87.0	98.9
08	OPENINGS	101.6	74.7	94.8	101.5	74.1	94.6	100.1	82.1	95.6	100.0	107.0	101.8	98.2	97.9	98.1	96.6	96.6	96.6
0920	Plaster & Gypsum Board	97.5	79.1	84.5	94.8	79.1	83.7	98.8	84.6	88.8	95.7	99.5	98.4	100.4	99.5	99.8	94.5	99.5	98.0
0950, 0980	Ceilings & Acoustic Treatment	92.0	79.1	83.4	92.0	79.1	83.4	92.0	84.6	87.1	83.7	99.5	94.2	97.2	99.5	98.7	89.8	99.5	96.3
0960	Flooring	102.3	100.2	101.7	100.1	98.7	99.7	105.7	100.2	104.1	107.3	118.6	110.5	98.7	114.3	103.2	118.8	114.3	117.5
0970, 0990	Wall Finishes & Painting/Coating	114.7	84.9	96.7	114.7	84.9	96.7	114.7	85.5	97.0	95.9	94.5	95.0		93.1	91.9	100.4	82.0	89.2
09	FINISHES	99.2	84.9	91.3	98.5	84.6	90.8	101.0	88.1	93.9	100.6	101.3	101.0	99.0	101.2	100.2	104.0	99.8	101.7
COVERS	DIVS. 10 - 14, 25, 28, 41, 43, 44, 46	100.0	67.8	93.3	100.0	97.1	99.4	100.0	99.4	99.9	100.0	92.9	98.5	100.0	95.3	99.0	100.0	95.0	99.0
21, 22, 23	FIRE SUPPRESSION, PLUMBING & HVAC	94.9	89.9	92.8	94.9	89.8	92.8	100.3	88.9	95.6	99.9	91.1	96.2	100.1	86.7	94.5	100.3	84.8	93.9
26, 27, 3370	ELECTRICAL, COMMUNICATIONS & UTIL.	100.5	82.2	90.8	99.7	82.2	90.4	93.9	93.5	93.7	99.1	89.4	94.0	102.1	92.2	96.8	96.1	84.6	90.0
MF2010	WEIGHTED AVERAGE	98.2	85.0	92.3	98.4	85.6	92.7	100.3	89.9	95.7	99.3	97.8	98.6	99.3	96.1	97.9	101.7	93.3	98.0

WISCONSIN

DIVISION		KENOSHA 531 MAT.	INST.	TOTAL	LA CROSSE 546 MAT.	INST.	TOTAL	LANCASTER 538 MAT.	INST.	TOTAL	MADISON 537 MAT.	INST.	TOTAL	MILWAUKEE 530,532 MAT.	INST.	TOTAL	NEW RICHMOND 540 MAT.	INST.	TOTAL
015433	CONTRACTOR EQUIPMENT		99.1	99.1		100.7	100.7		101.0	101.0		101.0	101.0		89.0	89.0		101.5	101.5
0241, 31 - 34	SITE & INFRASTRUCTURE, DEMOLITION	100.6	103.8	102.8	89.1	103.2	98.8	93.9	106.6	102.7	89.9	106.7	101.5	95.0	96.7	96.2	90.9	103.8	99.8
0310	Concrete Forming & Accessories	99.8	105.5	104.8	87.1	98.6	97.1	95.8	97.6	97.4	95.8	99.4	98.9	98.3	117.3	114.8	94.5	105.4	104.0
0320	Concrete Reinforcing	98.1	101.9	100.1	95.0	92.0	93.4	99.6	91.8	95.6	98.3	92.2	95.2	98.3	102.3	100.4	92.5	103.2	98.0
0330	Cast-in-Place Concrete	110.9	99.8	106.7	91.4	93.4	92.1	101.3	99.0	100.4	93.2	99.7	95.7	99.6	111.8	104.2	105.8	82.9	97.0
03	CONCRETE	104.2	102.9	103.6	92.1	95.8	93.9	99.2	97.1	98.2	95.1	98.2	96.6	98.6	111.7	104.9	98.9	97.4	98.2
04	MASONRY	101.9	109.4	106.4	94.8	102.1	99.2	104.7	101.8	102.9	102.0	104.4	103.4	104.6	120.8	114.4	123.8	96.4	107.2
05	METALS	94.7	103.3	97.6	95.4	98.9	96.6	91.5	95.9	93.0	96.7	98.0	97.1	97.5	96.8	97.2	95.5	103.3	98.2
06	WOOD, PLASTICS & COMPOSITES	102.5	104.9	103.9	95.7	99.3	97.7	102.1	99.2	100.4	100.1	99.2	99.6	104.4	117.3	111.8	98.3	111.5	105.8
07	THERMAL & MOISTURE PROTECTION	105.0	98.1	102.3	104.8	86.3	97.4	105.0	81.2	95.6	101.8	92.5	98.1	103.2	113.8	107.4	107.2	93.9	101.9
08	OPENINGS	95.4	106.5	98.3	98.2	85.6	95.0	95.3	86.4	93.1	100.0	99.5	99.9	102.1	113.2	104.9	86.7	104.5	91.2
0920	Plaster & Gypsum Board	80.9	105.4	98.1	95.0	99.5	98.2	94.6	99.5	98.1	97.8	99.5	99.0	96.7	118.0	111.7	86.1	112.3	104.6
0950, 0980	Ceilings & Acoustic Treatment	83.7	105.4	98.1	95.3	99.5	98.1	79.9	99.5	92.9	87.6	99.5	95.5	87.4	118.0	107.7	64.5	112.3	96.2
0960	Flooring	125.9	113.2	122.2	91.3	114.3	97.9	106.4	107.1	106.6	101.3	107.1	103.0	110.3	120.4	113.2	111.2	114.3	112.1
0970, 0990	Wall Finishes & Painting/Coating	104.3	105.9	105.2	90.0	74.0	80.3	95.9	71.5	81.1	95.9	99.5	98.0	97.8	116.2	108.9	104.0	93.1	97.4
09	FINISHES	105.2	107.5	106.4	95.2	99.6	97.6	99.4	96.2	97.7	99.5	101.2	100.4	102.6	118.4	111.3	95.0	105.7	100.9
COVERS	DIVS. 10 - 14, 25, 28, 41, 43, 44, 46	100.0	100.3	100.1	100.0	96.0	99.2	100.0	52.8	90.2	100.0	95.9	99.1	100.0	103.9	100.8	100.0	93.0	98.5
21, 22, 23	FIRE SUPPRESSION, PLUMBING & HVAC	100.1	97.0	98.8	100.1	87.5	94.9	94.5	87.3	91.5	99.8	96.8	98.6	99.9	104.0	101.6	94.3	84.2	90.1
26, 27, 3370	ELECTRICAL, COMMUNICATIONS & UTIL.	99.6	101.5	100.6	102.4	92.4	97.1	98.9	92.4	95.4	102.1	96.2	98.9	100.6	100.3	100.4	100.0	92.4	96.0
MF2010	WEIGHTED AVERAGE	100.0	102.7	101.2	97.6	94.9	96.4	97.0	93.2	95.3	98.9	99.1	99.0	100.2	107.6	103.5	97.3	96.2	96.8

WISCONSIN

DIVISION		OSHKOSH 549 MAT.	INST.	TOTAL	PORTAGE 539 MAT.	INST.	TOTAL	RACINE 534 MAT.	INST.	TOTAL	RHINELANDER 545 MAT.	INST.	TOTAL	SUPERIOR 548 MAT.	INST.	TOTAL	WAUSAU 544 MAT.	INST.	TOTAL
015433	CONTRACTOR EQUIPMENT		98.6	98.6		101.0	101.0		101.0	101.0		98.6	98.6		101.5	101.5		98.6	98.6
0241, 31 - 34	SITE & INFRASTRUCTURE, DEMOLITION	91.3	98.3	96.1	84.0	106.7	99.7	94.0	107.3	103.2	105.1	99.5	101.2	87.3	104.1	98.9	86.7	99.4	95.5
0310	Concrete Forming & Accessories	88.6	86.5	86.8	89.9	98.5	97.4	96.7	105.5	104.4	86.6	97.7	96.2	93.1	94.7	94.5	88.2	97.7	96.5
0320	Concrete Reinforcing	93.6	78.9	86.0	99.7	91.9	95.7	98.3	101.9	100.1	93.7	94.1	93.9	92.5	94.3	93.6	93.7	92.0	92.8
0330	Cast-in-Place Concrete	97.5	90.5	94.8	87.0	101.3	92.5	100.0	99.6	99.8	110.5	101.4	107.0	99.7	99.3	99.5	90.8	100.7	94.6
03	CONCRETE	93.6	86.9	90.4	87.1	98.3	92.5	98.5	102.8	100.6	106.2	98.5	102.5	93.2	96.6	94.8	88.4	97.9	93.0
04	MASONRY	109.1	89.9	97.4	103.6	101.7	102.5	104.6	109.3	107.5	126.6	101.3	111.3	123.1	106.3	112.9	108.6	99.3	103.0
05	METALS	95.7	90.8	94.1	92.2	96.8	93.8	95.7	103.3	98.3	95.6	99.2	96.9	96.5	100.8	98.0	95.5	99.1	96.7
06	WOOD, PLASTICS & COMPOSITES	92.1	88.1	89.8	94.5	99.2	97.2	103.0	104.9	104.1	89.8	99.3	95.2	96.8	93.3	94.8	91.6	99.3	96.0
07	THERMAL & MOISTURE PROTECTION	105.9	79.9	95.5	104.5	91.0	99.1	105.3	97.7	102.3	106.7	85.3	98.2	106.9	97.0	103.0	105.7	86.0	97.9
08	OPENINGS	92.0	84.9	90.2	95.5	93.5	95.0	100.0	106.5	101.6	92.0	88.6	91.2	86.7	93.3	88.3	92.2	97.2	93.5
0920	Plaster & Gypsum Board	84.4	88.0	86.9	89.1	99.5	96.4	95.7	105.4	102.5	83.9	99.5	94.9	85.9	93.6	91.3	84.4	99.5	95.0
0950, 0980	Ceilings & Acoustic Treatment	89.8	88.0	88.6	80.9	99.5	93.2	83.7	105.4	98.1	89.8	99.5	96.3	65.5	93.6	84.1	89.8	99.5	96.3
0960	Flooring	109.2	100.6	106.7	102.1	113.7	105.4	107.3	113.2	109.0	108.2	114.3	109.9	112.7	123.6	115.8	109.0	114.3	110.5
0970, 0990	Wall Finishes & Painting/Coating	98.4	69.8	81.0	95.9	71.5	81.1	95.8	111.0	105.0	98.4	65.0	78.1	92.7	102.5	98.6	98.4	82.0	88.4
09	FINISHES	99.2	86.9	92.4	96.9	98.3	97.7	100.6	108.0	104.7	100.0	97.4	98.6	94.5	100.8	98.0	98.8	99.8	99.3
COVERS	DIVS. 10 - 14, 25, 28, 41, 43, 44, 46	100.0	66.4	93.0	100.0	68.4	93.4	100.0	100.3	100.1	100.0	58.9	91.5	100.0	92.6	98.5	100.0	95.0	99.0
21, 22, 23	FIRE SUPPRESSION, PLUMBING & HVAC	94.9	76.3	87.2	94.5	92.3	93.6	99.9	97.0	98.7	94.9	86.8	91.5	94.3	91.8	93.2	94.9	85.5	91.0
26, 27, 3370	ELECTRICAL, COMMUNICATIONS & UTIL.	100.7	77.8	88.5	103.0	95.7	99.1	98.9	99.4	99.2	100.0	83.2	91.0	105.3	98.9	101.9	101.9	83.2	91.9
MF2010	WEIGHTED AVERAGE	97.0	84.1	91.2	95.4	96.3	95.8	99.4	102.8	100.9	99.8	92.2	96.4	97.1	98.1	97.6	96.2	93.4	95.0

WYOMING

DIVISION		CASPER 826			CHEYENNE 820			NEWCASTLE 827			RAWLINS 823			RIVERTON 825			ROCK SPRINGS 829 - 831		
		MAT.	INST.	TOTAL	MAT.	INST.	TOTAL	MAT.	INST.	TOTAL	MAT.	INST.	TOTAL	MAT.	INST.	TOTAL	MAT.	INST.	TOTAL
015433	CONTRACTOR EQUIPMENT		100.6	100.6		100.6	100.6		100.6	100.6		100.6	100.6		100.6	100.6		100.6	100.6
0241, 31 - 34	SITE & INFRASTRUCTURE, DEMOLITION	89.4	98.5	95.7	82.6	98.6	93.6	75.8	97.4	90.7	89.1	97.4	94.8	82.9	97.4	92.9	79.0	97.4	91.7
0310	Concrete Forming & Accessories	103.0	42.0	50.1	104.7	57.1	63.4	95.9	44.7	51.5	99.8	44.7	52.0	94.8	44.7	51.3	101.5	44.7	52.2
0320	Concrete Reinforcing	105.3	65.1	84.6	99.4	65.9	82.1	106.8	65.7	85.6	106.4	65.7	85.5	107.4	65.7	85.9	107.4	67.3	86.8
0330	Cast-in-Place Concrete	96.4	73.3	87.6	94.7	73.5	86.6	95.7	52.9	79.3	95.8	52.9	79.4	95.8	52.9	79.4	95.7	52.9	79.3
03	CONCRETE	101.7	58.0	80.8	106.0	64.9	83.5	101.1	52.3	77.7	114.2	52.3	84.6	109.0	52.3	81.8	101.6	52.6	78.1
04	MASONRY	101.5	36.6	62.2	98.3	50.5	69.4	95.1	45.3	64.9	95.1	45.3	64.9	95.1	45.3	64.9	153.6	45.3	88.0
05	METALS	98.1	68.7	88.0	99.3	70.3	89.3	94.4	68.1	85.3	94.4	68.1	85.4	94.5	68.1	85.4	95.3	68.7	86.1
06	WOOD, PLASTICS & COMPOSITES	94.3	39.2	62.7	96.0	59.0	74.8	86.3	45.7	63.0	90.3	45.7	64.7	85.3	45.7	62.6	93.7	45.7	66.2
07	THERMAL & MOISTURE PROTECTION	98.0	48.4	78.3	99.6	54.2	81.5	101.0	47.5	79.7	102.7	47.5	80.8	102.0	47.5	80.4	101.1	47.5	79.8
08	OPENINGS	94.7	44.1	81.9	96.8	54.9	86.2	101.1	47.6	87.5	100.7	47.6	87.3	100.9	47.6	87.4	101.5	48.0	87.9
0920	Plaster & Gypsum Board	101.1	37.2	56.1	92.6	57.7	68.0	86.6	43.9	56.6	87.9	43.9	57.0	86.6	43.9	56.6	96.3	43.9	59.4
0950, 0980	Ceilings & Acoustic Treatment	124.9	37.2	66.8	116.1	57.7	77.4	119.3	43.9	69.3	119.3	43.9	69.3	119.3	43.9	69.3	119.3	43.9	69.3
0960	Flooring	100.3	34.9	81.5	104.7	56.0	90.7	98.5	49.3	84.3	101.3	49.3	86.3	97.8	49.3	83.8	103.4	49.3	87.9
0970, 0990	Wall Finishes & Painting/Coating	98.5	46.3	66.8	101.8	46.3	68.2	97.7	29.7	56.4	97.7	29.7	56.4	97.7	29.7	56.4	97.7	29.7	56.4
09	FINISHES	106.9	39.8	69.9	106.2	55.8	78.4	101.0	43.2	69.1	103.6	43.2	70.3	101.8	43.2	69.5	103.7	43.2	70.3
COVERS	DIVS. 10 - 14, 25, 28, 41, 43, 44, 46	100.0	77.1	95.2	100.0	79.4	95.7	100.0	59.3	91.5	100.0	59.3	91.5	100.0	59.3	91.5	100.0	59.3	91.5
21, 22, 23	FIRE SUPPRESSION, PLUMBING & HVAC	100.0	56.2	81.8	100.0	56.3	81.8	97.2	53.9	79.2	97.2	53.9	79.2	97.2	53.9	79.2	99.9	53.9	80.8
26, 27, 3370	ELECTRICAL, COMMUNICATIONS & UTIL.	98.9	68.3	82.6	97.3	74.4	85.2	95.2	68.3	80.9	95.2	68.3	80.9	95.2	68.3	80.9	93.3	68.3	80.0
MF2010	WEIGHTED AVERAGE	99.5	58.5	81.2	99.2	64.7	83.8	97.3	58.0	79.7	99.6	58.0	81.0	98.6	58.0	80.4	101.5	58.1	82.1

DIVISION		WYOMING												CANADA					
		SHERIDAN 828			WHEATLAND 822			WORLAND 824			YELLOWSTONE NAT'L PA 821			BARRIE, ONTARIO			BATHURST, NEW BRUNSWICK		
		MAT.	INST.	TOTAL	MAT.	INST.	TOTAL	MAT.	INST.	TOTAL	MAT.	INST.	TOTAL	MAT.	INST.	TOTAL	MAT.	INST.	TOTAL
015433	CONTRACTOR EQUIPMENT		100.6	100.6		100.6	100.6		100.6	100.6		100.6	100.6		101.6	101.6		101.4	101.4
0241, 31 - 34	SITE & INFRASTRUCTURE, DEMOLITION	83.0	98.6	93.7	79.8	97.5	92.0	77.5	97.4	91.2	77.5	97.8	91.5	115.0	101.5	105.7	94.2	97.3	96.3
0310	Concrete Forming & Accessories	102.6	47.1	54.4	97.8	45.1	52.1	97.9	44.7	51.7	97.9	47.0	53.7	123.1	95.7	99.3	102.2	68.0	72.5
0320	Concrete Reinforcing	107.4	65.8	86.0	106.8	65.7	85.6	107.4	65.7	85.9	109.3	66.8	87.4	180.1	89.0	133.2	148.4	60.0	102.9
0330	Cast-in-Place Concrete	99.0	73.4	89.2	99.9	53.5	82.1	95.7	52.9	79.3	95.7	56.4	80.6	167.6	94.5	139.6	138.2	65.1	110.2
03	CONCRETE	109.1	60.4	85.8	106.0	52.6	80.4	101.3	52.3	77.9	101.6	54.7	79.1	152.8	94.1	124.7	135.5	66.1	102.3
04	MASONRY	95.4	49.9	67.8	95.4	46.2	65.6	95.1	45.3	64.9	95.2	51.4	68.6	168.4	104.3	129.5	158.7	69.0	104.4
05	METALS	98.1	70.0	88.4	94.4	68.1	85.3	94.5	68.1	85.4	95.2	68.5	86.0	107.4	93.5	102.6	104.7	73.4	94.0
06	WOOD, PLASTICS & COMPOSITES	97.2	45.7	67.7	88.0	45.7	63.7	88.0	45.7	63.8	88.0	45.7	63.8	118.4	94.2	104.5	94.9	68.3	79.6
07	THERMAL & MOISTURE PROTECTION	102.1	52.6	82.4	101.3	48.5	80.3	101.0	47.5	79.7	100.6	50.7	80.8	111.4	94.9	104.8	104.8	66.1	89.4
08	OPENINGS	101.8	48.6	88.3	99.6	47.6	86.5	101.3	47.6	87.7	94.1	47.9	82.4	91.8	92.1	91.9	86.9	59.5	80.0
0920	Plaster & Gypsum Board	111.4	43.9	63.9	86.6	43.9	56.6	86.6	43.9	56.6	86.9	43.9	56.6	157.5	94.0	112.8	163.1	67.2	95.6
0950, 0980	Ceilings & Acoustic Treatment	122.4	43.9	70.4	119.3	43.9	69.3	119.3	43.9	69.3	120.2	43.9	69.6	95.2	94.0	94.4	96.2	67.2	77.0
0960	Flooring	101.3	49.3	86.4	99.8	49.3	85.3	99.8	49.3	85.3	99.8	49.3	85.3	132.0	102.7	123.5	115.2	49.0	96.2
0970, 0990	Wall Finishes & Painting/Coating	99.9	46.3	67.4	97.7	29.7	56.4	97.7	29.7	56.4	97.7	29.7	56.4	112.2	95.1	101.8	111.4	54.6	76.9
09	FINISHES	108.5	46.2	74.2	101.8	43.4	69.6	101.4	43.2	69.3	101.7	44.7	70.2	118.8	97.2	106.9	112.9	63.2	85.5
COVERS	DIVS. 10 - 14, 25, 28, 41, 43, 44, 46	100.0	77.9	95.4	100.0	59.6	91.6	100.0	59.3	91.5	100.0	61.4	92.0	140.0	82.1	128.0	140.0	71.9	125.9
21, 22, 23	FIRE SUPPRESSION, PLUMBING & HVAC	97.2	56.2	80.2	97.2	54.3	79.4	97.2	53.9	79.2	97.2	57.0	80.5	102.4	107.3	104.5	102.6	75.4	91.3
26, 27, 3370	ELECTRICAL, COMMUNICATIONS & UTIL.	97.7	68.3	82.1	95.2	74.4	84.1	95.2	68.3	80.9	94.1	68.3	80.4	120.1	96.6	107.6	121.9	66.0	92.2
MF2010	WEIGHTED AVERAGE	100.1	61.5	82.8	98.0	59.2	80.6	97.5	58.0	79.8	96.8	60.0	80.4	118.9	98.8	109.9	114.2	71.2	94.9

CANADA

DIVISION		BRANDON, MANITOBA			BRANTFORD, ONTARIO			BRIDGEWATER, NOVA SCOTIA			CALGARY, ALBERTA			CAP-DE-LA-MADELEINE, QUEBEC			CHARLESBOURG, QUEBEC		
		MAT.	INST.	TOTAL	MAT.	INST.	TOTAL	MAT.	INST.	TOTAL	MAT.	INST.	TOTAL	MAT.	INST.	TOTAL	MAT.	INST.	TOTAL
015433	CONTRACTOR EQUIPMENT		103.8	103.8		101.6	101.6		101.2	101.2		106.6	106.6		102.0	102.0		102.0	102.0
0241, 31 - 34	SITE & INFRASTRUCTURE, DEMOLITION	110.6	100.3	103.5	112.7	101.8	105.2	96.9	99.0	98.4	125.5	105.6	111.8	92.5	100.7	98.2	92.5	100.7	98.2
0310	Concrete Forming & Accessories	122.4	77.4	83.4	122.9	103.7	106.2	96.4	76.5	79.2	127.5	98.4	102.2	128.1	94.1	98.6	128.1	94.1	98.6
0320	Concrete Reinforcing	161.6	56.5	107.5	175.0	87.6	130.0	148.4	49.6	97.5	148.3	73.6	109.8	148.4	83.7	115.1	148.4	83.7	115.1
0330	Cast-in-Place Concrete	151.1	81.0	124.2	163.5	117.6	145.9	168.0	76.2	132.9	200.4	108.8	165.3	131.9	103.9	121.2	131.9	103.9	121.2
03	CONCRETE	135.2	75.3	106.5	150.0	105.2	128.6	148.0	72.0	111.6	167.1	97.3	133.7	130.6	95.5	113.8	130.6	95.5	113.8
04	MASONRY	162.8	71.9	107.7	164.7	109.1	131.0	159.7	76.5	109.3	215.0	92.7	140.9	159.7	92.6	119.0	159.7	92.6	119.0
05	METALS	107.6	79.0	97.8	106.8	94.2	102.5	105.9	76.3	95.7	148.0	89.8	128.0	105.0	90.5	100.0	105.0	90.5	100.0
06	WOOD, PLASTICS & COMPOSITES	114.0	78.4	93.6	116.3	102.3	108.3	87.1	75.9	80.7	114.5	98.1	105.1	127.7	94.0	108.4	127.7	94.0	108.4
07	THERMAL & MOISTURE PROTECTION	105.0	76.3	93.6	111.3	100.9	107.2	106.3	74.4	93.6	123.3	95.8	112.4	105.2	95.7	101.4	105.2	95.7	101.4
08	OPENINGS	92.9	69.0	86.8	91.7	98.7	93.4	83.5	69.0	79.8	92.9	87.5	91.5	92.9	86.3	91.2	92.9	86.3	91.2
0920	Plaster & Gypsum Board	130.2	77.4	93.0	157.5	102.4	118.7	156.2	75.1	99.1	173.7	97.6	120.1	188.5	93.7	121.7	188.5	93.7	121.7
0950, 0980	Ceilings & Acoustic Treatment	95.2	77.4	83.4	95.2	102.4	100.0	95.2	75.1	81.9	142.4	97.6	112.7	95.2	93.7	94.2	95.2	93.7	94.2
0960	Flooring	131.9	73.0	115.0	131.9	102.7	123.5	111.6	69.9	99.6	134.2	96.2	123.3	131.9	103.4	123.7	131.9	103.4	123.7
0970, 0990	Wall Finishes & Painting/Coating	111.4	62.1	81.5	111.4	104.4	107.2	111.4	67.9	85.0	111.3	112.0	111.7	111.4	97.1	102.7	111.4	97.1	102.7
09	FINISHES	114.6	75.3	92.9	117.5	104.0	110.1	110.5	74.6	90.7	131.9	99.9	114.3	120.3	96.4	107.2	120.3	96.4	107.2
COVERS	DIVS. 10 - 14, 25, 28, 41, 43, 44, 46	140.0	74.4	126.4	140.0	84.5	128.5	140.0	73.3	126.1	140.0	105.4	132.8	140.0	94.0	130.5	140.0	94.0	130.5
21, 22, 23	FIRE SUPPRESSION, PLUMBING & HVAC	102.6	90.9	97.7	102.6	110.6	105.9	102.6	91.2	97.9	100.4	94.7	98.0	103.2	98.4	101.2	103.2	98.4	101.2
26, 27, 3370	ELECTRICAL, COMMUNICATIONS & UTIL.	121.8	74.3	96.6	119.7	96.0	107.1	125.8	68.7	95.4	127.0	86.0	105.2	119.0	77.5	96.9	119.0	77.5	96.9
MF2010	WEIGHTED AVERAGE	116.1	80.3	100.1	118.1	103.0	111.3	115.8	79.1	99.4	130.9	94.8	114.8	114.9	93.0	105.1	114.9	93.0	105.1

City Cost Indexes

CANADA

DIVISION		CHARLOTTETOWN, PRINCE EDWARD ISLAND			CHICOUTIMI, QUEBEC			CORNER BROOK, NEWFOUNDLAND			CORNWALL, ONTARIO			DALHOUSIE, NEW BRUNSWICK			DARTMOUTH, NOVA SCOTIA		
		MAT.	INST.	TOTAL	MAT.	INST.	TOTAL	MAT.	INST.	TOTAL	MAT.	INST.	TOTAL	MAT.	INST.	TOTAL	MAT.	INST.	TOTAL
015433	CONTRACTOR EQUIPMENT		101.1	101.1		102.0	102.0		102.4	102.4		101.6	101.6		101.4	101.4		101.2	101.2
0241, 31 - 34	SITE & INFRASTRUCTURE, DEMOLITION	104.7	96.4	99.0	92.0	99.4	97.1	112.7	97.7	102.4	110.7	101.3	104.2	94.1	97.3	96.3	104.6	99.0	100.7
0310	Concrete Forming & Accessories	96.6	62.1	66.7	128.1	91.7	96.5	103.8	65.8	70.8	120.9	95.8	99.2	102.2	68.2	72.7	96.4	76.5	79.2
0320	Concrete Reinforcing	158.7	49.6	102.5	111.9	93.5	102.4	148.4	51.5	98.5	175.0	87.3	129.8	148.4	60.0	102.9	155.0	49.6	100.7
0330	Cast-in-Place Concrete	159.3	64.3	122.9	130.1	99.0	118.2	172.3	74.5	134.8	147.0	107.1	131.7	138.2	65.2	110.2	169.8	76.2	134.0
03	CONCRETE	153.7	61.2	109.4	124.3	94.5	110.1	162.4	66.9	116.7	141.6	98.1	120.8	135.5	66.2	102.3	149.9	72.0	112.6
04	MASONRY	160.1	66.6	103.4	160.2	92.6	119.2	160.0	68.2	104.4	163.4	100.0	125.0	156.6	69.0	103.5	171.2	76.5	113.8
05	METALS	105.9	69.1	93.2	104.5	91.6	100.1	108.1	74.7	96.6	106.8	92.8	101.9	104.7	73.6	94.0	108.0	76.3	97.1
06	WOOD, PLASTICS & COMPOSITES	87.4	61.6	72.6	127.7	91.7	107.1	95.9	64.8	78.1	115.1	95.1	103.7	94.9	68.3	79.6	87.1	75.9	80.7
07	THERMAL & MOISTURE PROTECTION	105.8	63.4	88.9	105.1	94.8	101.0	108.3	65.7	91.4	111.1	95.1	104.7	108.8	66.1	91.8	106.3	74.4	93.6
08	OPENINGS	95.2	53.1	84.5	92.4	79.4	89.1	99.2	60.8	89.5	92.9	91.8	92.6	86.9	59.5	80.0	83.5	69.0	79.8
0920	Plaster & Gypsum Board	160.3	60.4	90.0	188.5	91.3	120.1	163.1	63.6	93.0	228.3	95.0	134.4	163.1	67.2	95.6	158.8	75.1	99.8
0950, 0980	Ceilings & Acoustic Treatment	108.3	60.4	76.6	95.2	91.3	92.6	96.2	63.6	74.6	98.0	95.0	96.0	96.2	67.2	77.0	104.6	75.1	85.0
0960	Flooring	111.6	65.3	98.3	131.9	103.4	123.7	115.9	59.5	99.7	131.9	101.2	123.1	115.2	74.9	103.7	111.6	69.9	99.6
0970, 0990	Wall Finishes & Painting/Coating	111.4	45.5	71.4	111.4	103.2	106.4	111.4	65.4	83.5	111.4	97.3	102.8	111.4	54.6	76.9	111.4	67.9	85.0
09	FINISHES	114.7	61.3	85.3	120.3	95.5	106.6	114.1	64.6	86.8	126.5	97.3	110.4	112.9	68.4	88.4	112.8	74.6	91.7
COVERS	DIVS. 10 - 14, 25, 28, 41, 43, 44, 46	140.0	71.4	125.7	140.0	93.4	130.3	140.0	72.2	125.9	140.0	81.7	127.9	140.0	71.9	125.8	140.0	73.3	126.1
21, 22, 23	FIRE SUPPRESSION, PLUMBING & HVAC	102.6	68.9	88.6	102.6	97.8	100.6	102.6	77.6	92.2	103.2	108.2	105.3	102.6	75.4	91.3	102.6	91.2	97.9
26, 27, 3370	ELECTRICAL, COMMUNICATIONS & UTIL.	119.0	55.9	85.4	118.3	84.8	100.5	118.9	62.2	88.8	120.4	97.1	108.0	122.1	62.4	90.4	124.0	68.7	94.6
MF2010	WEIGHTED AVERAGE	117.6	66.4	94.7	113.9	93.2	104.7	119.6	71.5	98.1	118.0	99.1	109.5	114.2	71.4	95.1	117.2	79.1	100.1

CANADA

DIVISION		EDMONTON, ALBERTA			FORT MCMURRAY, ALBERTA			FREDERICTON, NEW BRUNSWICK			GATINEAU, QUEBEC			GRANBY, QUEBEC			HALIFAX, NOVA SCOTIA		
		MAT.	INST.	TOTAL	MAT.	INST.	TOTAL	MAT.	INST.	TOTAL	MAT.	INST.	TOTAL	MAT.	INST.	TOTAL	MAT.	INST.	TOTAL
015433	CONTRACTOR EQUIPMENT		106.6	106.6		103.8	103.8		101.4	101.4		102.0	102.0		102.0	102.0		101.2	101.2
0241, 31 - 34	SITE & INFRASTRUCTURE, DEMOLITION	150.7	105.6	119.6	118.6	102.4	107.4	93.7	97.3	96.2	92.3	100.7	98.1	92.9	100.7	98.3	97.2	98.8	98.3
0310	Concrete Forming & Accessories	129.8	98.4	102.5	121.1	98.6	101.5	120.3	68.6	75.4	128.1	94.0	98.5	128.1	93.9	98.4	96.6	79.0	81.3
0320	Concrete Reinforcing	148.3	73.6	109.8	161.6	73.6	116.3	149.3	60.2	103.4	157.1	83.7	119.3	157.1	83.6	119.3	155.0	62.1	107.2
0330	Cast-in-Place Concrete	208.9	108.8	170.5	217.2	108.0	175.4	131.5	65.2	106.1	130.1	103.9	120.1	134.4	103.8	122.7	171.2	82.2	137.1
03	CONCRETE	171.5	97.3	136.0	174.9	97.2	137.7	133.6	66.4	101.4	130.9	95.4	113.9	133.1	95.4	115.1	150.6	77.5	115.6
04	MASONRY	202.5	92.7	136.0	204.3	92.7	136.6	177.3	70.7	112.7	159.6	92.6	119.0	160.0	92.6	119.1	172.8	84.2	119.1
05	METALS	148.6	89.8	128.4	132.1	89.9	117.6	104.6	74.2	94.2	105.0	90.3	100.0	105.0	90.3	100.0	119.0	81.8	106.2
06	WOOD, PLASTICS & COMPOSITES	111.0	98.1	103.6	110.8	98.1	103.5	116.9	68.3	89.0	127.7	94.0	108.4	127.7	94.0	108.4	87.3	77.9	81.9
07	THERMAL & MOISTURE PROTECTION	124.4	95.8	113.0	118.8	95.2	109.4	109.0	67.2	92.4	105.1	94.1	100.7	105.1	94.1	100.7	106.8	78.5	95.6
08	OPENINGS	92.9	87.5	91.5	92.9	87.5	91.5	86.8	58.3	79.6	92.9	81.3	89.9	92.9	81.3	89.9	83.5	72.8	80.8
0920	Plaster & Gypsum Board	164.5	97.6	117.4	156.8	97.7	115.2	181.0	67.2	100.9	147.7	93.7	109.7	150.6	93.7	110.5	159.2	77.1	101.4
0950, 0980	Ceilings & Acoustic Treatment	156.8	97.6	117.6	104.6	97.7	100.0	101.8	67.2	78.9	95.2	93.7	94.2	95.2	93.7	94.2	104.6	77.1	86.4
0960	Flooring	135.3	96.2	124.1	131.9	96.2	121.6	127.5	78.2	113.3	131.9	103.4	123.7	131.9	103.4	123.7	111.6	91.8	105.9
0970, 0990	Wall Finishes & Painting/Coating	111.4	112.0	111.8	111.4	101.1	105.1	111.4	69.6	86.0	111.4	97.1	102.7	111.4	97.1	102.7	111.4	81.5	93.2
09	FINISHES	136.0	99.9	116.1	120.9	98.7	108.6	119.9	70.7	92.8	115.5	96.4	105.0	115.8	96.4	105.1	112.8	81.9	95.8
COVERS	DIVS. 10 - 14, 25, 28, 41, 43, 44, 46	140.0	105.4	132.8	140.0	105.2	132.8	140.0	71.9	125.9	140.0	94.0	130.5	140.0	94.0	130.5	140.0	74.0	126.3
21, 22, 23	FIRE SUPPRESSION, PLUMBING & HVAC	100.4	94.7	98.0	103.0	106.8	104.6	102.6	85.7	95.6	103.2	98.4	101.2	102.6	98.4	100.8	99.7	79.8	91.4
26, 27, 3370	ELECTRICAL, COMMUNICATIONS & UTIL.	117.8	86.0	100.9	113.0	91.2	101.4	132.0	82.3	105.5	119.0	77.5	96.9	119.6	77.5	97.2	124.9	81.5	101.8
MF2010	WEIGHTED AVERAGE	131.1	94.8	114.9	127.0	97.7	113.9	116.8	76.8	98.9	114.6	92.7	104.8	114.8	92.7	104.9	118.1	81.7	101.8

CANADA

DIVISION		HAMILTON, ONTARIO			HULL, QUEBEC			JOLIETTE, QUEBEC			KAMLOOPS, BRITISH COLUMBIA			KINGSTON, ONTARIO			KITCHENER, ONTARIO		
		MAT.	INST.	TOTAL	MAT.	INST.	TOTAL	MAT.	INST.	TOTAL	MAT.	INST.	TOTAL	MAT.	INST.	TOTAL	MAT.	INST.	TOTAL
015433	CONTRACTOR EQUIPMENT		108.5	108.5		102.0	102.0		102.0	102.0		105.3	105.3		103.9	103.9		103.8	103.8
0241, 31 - 34	SITE & INFRASTRUCTURE, DEMOLITION	113.4	113.3	113.3	92.3	100.7	98.1	93.0	100.7	98.3	116.6	104.3	108.1	110.7	105.1	106.8	100.3	107.3	103.7
0310	Concrete Forming & Accessories	130.9	96.4	100.9	128.1	94.0	98.5	128.1	94.1	98.6	123.3	97.1	100.6	121.0	95.9	99.2	117.3	87.5	91.4
0320	Concrete Reinforcing	167.9	87.2	126.4	157.1	83.7	119.3	148.4	83.7	115.1	116.6	81.5	98.5	175.0	87.3	129.8	110.5	87.1	98.5
0330	Cast-in-Place Concrete	151.9	98.1	131.3	130.1	103.9	120.1	135.5	103.9	123.4	117.3	107.9	113.7	147.0	107.0	131.7	140.7	88.8	120.9
03	CONCRETE	143.6	95.3	120.5	130.9	95.4	113.9	132.4	95.5	114.7	138.4	97.8	119.0	141.6	98.1	120.8	122.5	88.1	106.1
04	MASONRY	180.7	102.7	133.4	159.6	92.6	119.0	160.0	92.6	119.2	166.6	101.0	126.8	169.9	100.0	127.5	161.4	94.6	120.9
05	METALS	126.0	93.2	114.7	105.0	90.3	100.0	105.0	90.5	100.0	107.4	90.4	101.6	108.5	92.7	103.1	115.4	93.0	107.7
06	WOOD, PLASTICS & COMPOSITES	115.6	95.5	104.1	127.7	94.0	108.4	127.7	94.0	108.4	98.6	95.5	96.8	115.1	95.3	103.7	109.6	86.7	96.5
07	THERMAL & MOISTURE PROTECTION	121.8	97.6	112.2	105.1	95.7	101.4	105.1	95.7	101.4	120.5	92.4	109.3	111.1	96.3	105.2	110.7	92.6	103.5
08	OPENINGS	92.9	93.8	93.1	92.9	81.3	89.9	92.9	86.3	91.2	88.7	91.2	89.3	92.9	91.5	92.5	84.0	87.2	84.8
0920	Plaster & Gypsum Board	193.9	95.4	124.5	147.7	93.7	109.7	188.5	93.7	121.7	137.8	94.9	107.6	232.2	95.1	135.7	155.2	86.3	106.7
0950, 0980	Ceilings & Acoustic Treatment	124.1	95.4	105.0	95.2	93.7	94.2	95.2	93.7	94.2	95.2	94.9	95.0	112.1	95.1	100.8	104.6	86.3	92.5
0960	Flooring	134.2	106.2	126.2	131.9	103.4	123.7	131.9	103.4	123.7	129.8	57.8	109.1	131.9	101.2	123.1	127.9	106.2	121.7
0970, 0990	Wall Finishes & Painting/Coating	111.4	103.1	106.3	111.4	97.1	102.7	111.4	97.1	102.7	111.4	88.9	97.7	111.4	89.9	98.3	111.4	93.0	100.2
09	FINISHES	129.2	99.1	112.6	115.5	96.4	105.0	120.3	96.4	107.2	116.7	89.4	101.7	130.0	96.5	111.5	117.2	91.2	102.9
COVERS	DIVS. 10 - 14, 25, 28, 41, 43, 44, 46	140.0	108.0	133.4	140.0	94.0	130.5	140.0	94.0	130.5	140.0	105.5	132.8	140.0	81.7	127.9	140.0	104.8	132.7
21, 22, 23	FIRE SUPPRESSION, PLUMBING & HVAC	100.5	86.0	94.5	102.6	98.4	100.9	102.6	98.4	100.9	102.6	102.0	102.4	103.2	108.4	105.3	99.7	91.2	96.1
26, 27, 3370	ELECTRICAL, COMMUNICATIONS & UTIL.	127.5	89.4	107.2	121.1	77.5	97.9	119.6	77.5	97.2	123.5	87.5	104.3	120.4	95.7	107.2	120.9	86.9	102.8
MF2010	WEIGHTED AVERAGE	122.6	95.4	110.5	114.6	92.7	104.8	115.1	93.0	105.2	117.2	96.2	107.8	118.9	99.2	110.1	114.0	92.1	104.2

City Cost Indexes

CANADA

DIVISION		LAVAL, QUEBEC			LETHBRIDGE, ALBERTA			LLOYDMINSTER, ALBERTA			LONDON, ONTARIO			MEDICINE HAT, ALBERTA			MONCTON, NEW BRUNSWICK		
		MAT.	INST.	TOTAL	MAT.	INST.	TOTAL	MAT.	INST.	TOTAL	MAT.	INST.	TOTAL	MAT.	INST.	TOTAL	MAT.	INST.	TOTAL
015433	CONTRACTOR EQUIPMENT		102.0	102.0		103.8	103.8		103.8	103.8		104.0	104.0		103.8	103.8		101.4	101.4
0241, 31 - 34	SITE & INFRASTRUCTURE, DEMOLITION	92.8	100.7	98.3	110.6	103.1	105.4	110.5	102.4	104.9	113.3	105.5	107.9	109.2	102.5	104.6	93.6	97.5	96.3
0310	Concrete Forming & Accessories	128.2	94.0	98.5	122.3	98.6	101.8	120.6	88.1	92.4	131.0	90.2	95.6	122.2	88.1	92.6	102.2	68.5	73.0
0320	Concrete Reinforcing	157.1	83.7	119.3	161.6	73.6	116.3	161.6	73.6	116.3	146.7	86.0	115.4	161.6	73.6	116.3	148.4	63.2	104.5
0330	Cast-in-Place Concrete	134.4	103.9	122.7	162.9	108.0	141.9	151.1	104.0	133.1	149.2	95.9	128.8	151.1	104.0	133.0	133.1	65.4	107.2
03	CONCRETE	133.1	95.4	115.1	147.7	97.2	123.5	141.7	91.2	117.5	139.2	91.5	116.4	141.8	91.1	117.5	133.0	67.2	101.5
04	MASONRY	159.9	92.6	119.1	177.7	92.7	126.2	159.9	85.7	114.9	182.1	99.8	132.2	159.9	85.7	114.9	158.3	69.1	104.2
05	METALS	105.1	90.3	100.0	125.5	89.9	113.3	107.6	89.8	101.5	123.6	92.4	112.9	107.6	89.7	101.5	104.7	79.4	96.0
06	WOOD, PLASTICS & COMPOSITES	127.8	94.0	108.4	114.1	98.1	104.9	110.8	87.6	97.5	115.6	88.3	99.9	114.1	87.6	98.9	94.9	68.3	79.6
07	THERMAL & MOISTURE PROTECTION	105.7	95.7	101.7	116.2	95.2	107.9	113.4	90.5	104.3	122.6	95.3	111.8	119.4	90.5	107.9	108.8	67.6	92.4
08	OPENINGS	92.9	81.3	89.9	92.9	87.5	91.5	92.9	81.8	90.1	94.0	88.7	92.7	92.9	81.8	90.1	86.9	62.8	80.8
0920	Plaster & Gypsum Board	151.1	93.7	110.6	138.8	97.7	109.9	133.8	86.9	100.8	197.3	88.0	120.3	136.5	86.9	101.5	163.1	67.2	95.6
0950, 0980	Ceilings & Acoustic Treatment	95.2	93.7	94.2	103.7	97.7	99.7	95.2	86.9	89.7	136.3	88.0	104.3	95.2	86.9	89.7	96.2	67.2	77.0
0960	Flooring	131.9	103.4	123.7	131.9	96.2	121.6	131.9	96.2	121.6	134.4	106.2	126.3	131.9	96.2	121.6	115.2	74.9	103.7
0970, 0990	Wall Finishes & Painting/Coating	111.4	97.1	102.7	111.3	110.3	110.7	111.4	86.0	96.0	111.4	99.8	104.3	111.3	86.0	95.9	111.4	54.6	76.9
09	FINISHES	115.9	96.4	105.2	117.6	99.7	107.7	115.3	89.0	100.8	132.3	94.0	111.2	115.4	89.0	100.9	112.9	68.4	88.4
COVERS	DIVS. 10 - 14, 25, 28, 41, 43, 44, 46	140.0	94.0	130.5	140.0	105.2	132.8	140.0	101.5	132.0	140.0	106.6	133.1	140.0	101.5	132.0	140.0	71.9	125.9
21, 22, 23	FIRE SUPPRESSION, PLUMBING & HVAC	99.7	98.4	99.1	102.8	103.2	103.0	103.2	103.2	103.2	100.4	83.1	93.2	102.6	99.6	101.4	102.6	75.5	91.3
26, 27, 3370	ELECTRICAL, COMMUNICATIONS & UTIL.	120.8	77.5	97.8	114.5	91.2	102.1	112.0	91.2	100.9	119.5	86.9	102.2	112.0	91.2	100.9	127.2	66.0	94.7
MF2010	WEIGHTED AVERAGE	114.3	92.7	104.6	120.9	97.1	110.3	116.2	93.6	106.1	121.4	91.9	108.2	116.3	92.8	105.8	114.5	72.8	95.8

CANADA

DIVISION		MONTREAL, QUEBEC			MOOSE JAW, SASKATCHEWAN			NEW GLASGOW, NOVA SCOTIA			NEWCASTLE, NEW BRUNSWICK			NORTH BAY, ONTARIO			OSHAWA, ONTARIO		
		MAT.	INST.	TOTAL	MAT.	INST.	TOTAL	MAT.	INST.	TOTAL	MAT.	INST.	TOTAL	MAT.	INST.	TOTAL	MAT.	INST.	TOTAL
015433	CONTRACTOR EQUIPMENT		103.6	103.6		100.0	100.0		101.2	101.2		101.4	101.4		101.6	101.6		103.8	103.8
0241, 31 - 34	SITE & INFRASTRUCTURE, DEMOLITION	99.6	99.7	99.7	110.3	96.7	100.9	97.1	99.0	98.4	94.2	97.3	96.3	112.7	100.8	104.5	112.6	105.1	107.4
0310	Concrete Forming & Accessories	136.0	92.0	97.8	106.1	64.7	70.1	96.4	76.5	79.2	102.2	68.3	72.7	122.9	93.1	97.0	123.1	95.3	99.0
0320	Concrete Reinforcing	153.4	93.6	122.6	114.0	65.7	89.1	148.4	49.6	97.5	148.4	60.0	102.9	175.0	86.8	129.6	175.0	89.6	131.0
0330	Cast-in-Place Concrete	180.9	100.4	150.1	146.6	75.1	119.2	169.8	76.2	134.0	138.2	65.2	110.2	163.5	92.4	136.3	162.8	93.3	136.2
03	CONCRETE	156.4	95.1	127.1	125.2	69.0	98.3	148.9	72.0	112.1	135.5	66.2	102.3	150.0	91.8	122.1	149.7	93.6	122.8
04	MASONRY	167.3	92.6	122.1	158.5	68.4	103.9	159.8	76.5	109.3	158.7	69.0	104.4	164.7	95.6	122.8	164.3	100.4	125.6
05	METALS	122.2	92.1	111.8	103.9	76.1	94.3	105.9	76.3	95.7	104.7	73.6	94.0	106.8	92.4	101.8	106.8	94.7	102.7
06	WOOD, PLASTICS & COMPOSITES	127.8	92.0	107.3	96.4	63.1	77.3	87.1	75.9	80.7	94.9	68.3	79.6	116.3	93.6	103.3	116.5	94.2	103.7
07	THERMAL & MOISTURE PROTECTION	118.6	95.3	109.3	104.6	68.2	90.1	106.3	74.4	93.6	108.8	66.1	91.8	111.3	91.2	103.3	111.7	93.0	104.3
08	OPENINGS	92.9	81.4	90.0	87.7	60.4	80.8	83.5	69.0	79.8	86.9	59.5	80.0	91.7	89.1	91.0	91.7	93.6	92.2
0920	Plaster & Gypsum Board	168.2	91.3	114.1	130.8	61.9	82.3	156.2	75.1	99.1	163.1	67.2	95.6	157.5	93.4	112.4	159.0	94.0	113.3
0950, 0980	Ceilings & Acoustic Treatment	113.8	91.3	98.9	95.2	61.9	73.2	95.2	75.1	81.9	96.2	67.2	77.0	95.2	93.4	94.0	99.0	94.0	95.7
0960	Flooring	133.6	103.4	124.9	120.6	65.7	104.9	111.6	69.9	99.6	115.2	74.9	103.7	131.9	101.2	123.1	131.9	109.0	125.3
0970, 0990	Wall Finishes & Painting/Coating	111.4	103.2	106.4	111.4	70.7	86.7	111.4	67.9	85.0	111.4	54.6	76.9	111.4	96.5	102.4	111.4	107.2	108.8
09	FINISHES	122.6	95.7	107.7	111.4	65.2	85.9	110.5	74.6	90.7	112.9	68.4	88.4	117.5	95.4	105.3	118.4	99.1	107.7
COVERS	DIVS. 10 - 14, 25, 28, 41, 43, 44, 46	140.0	94.0	130.4	140.0	71.1	125.7	140.0	73.3	126.1	140.0	71.9	125.9	140.0	80.2	127.6	140.0	107.5	133.3
21, 22, 23	FIRE SUPPRESSION, PLUMBING & HVAC	100.4	97.9	99.4	103.2	82.2	94.5	102.6	91.2	97.9	102.6	75.4	91.3	102.6	106.1	104.1	99.7	108.8	103.5
26, 27, 3370	ELECTRICAL, COMMUNICATIONS & UTIL.	128.3	84.8	105.1	122.1	67.0	92.8	119.3	68.7	92.4	121.1	66.0	91.8	119.7	97.0	107.6	122.2	96.6	108.6
MF2010	WEIGHTED AVERAGE	122.0	93.5	109.3	113.4	73.6	95.6	115.3	79.1	96.1	114.2	71.9	95.3	118.1	96.7	108.5	117.7	100.0	109.8

CANADA

DIVISION		OTTAWA, ONTARIO			OWEN SOUND, ONTARIO			PETERBOROUGH, ONTARIO			PORTAGE LA PRAIRIE, MANITOBA			PRINCE ALBERT, SASKATCHEWAN			PRINCE GEORGE, BRITISH COLUMBIA		
		MAT.	INST.	TOTAL	MAT.	INST.	TOTAL	MAT.	INST.	TOTAL	MAT.	INST.	TOTAL	MAT.	INST.	TOTAL	MAT.	INST.	TOTAL
015433	CONTRACTOR EQUIPMENT		103.8	103.8		101.6	101.6		101.6	101.6		103.8	103.8		100.0	100.0		105.3	105.3
0241, 31 - 34	SITE & INFRASTRUCTURE, DEMOLITION	111.4	105.4	107.2	115.0	101.3	105.5	112.7	101.2	104.8	111.0	100.2	103.6	105.0	96.9	99.4	120.3	104.3	109.3
0310	Concrete Forming & Accessories	131.9	92.8	97.9	123.1	91.5	95.7	122.9	94.2	98.0	122.4	76.9	82.9	106.1	64.4	70.0	111.4	91.5	94.1
0320	Concrete Reinforcing	160.4	85.9	122.0	180.1	89.0	133.2	175.0	87.4	129.9	161.6	56.5	107.5	119.2	65.6	91.6	116.6	81.5	98.5
0330	Cast-in-Place Concrete	167.3	95.1	139.6	167.6	88.1	137.1	163.5	94.2	137.0	151.1	80.3	124.0	132.8	75.0	110.7	147.1	107.9	132.1
03	CONCRETE	150.3	92.4	122.6	152.8	90.0	122.7	150.0	93.0	122.7	135.2	74.8	106.3	119.0	68.9	95.0	152.6	95.3	125.2
04	MASONRY	181.1	99.8	131.8	168.4	101.7	127.9	164.7	102.8	127.2	162.8	70.8	107.0	157.5	68.4	103.5	169.0	101.0	127.8
05	METALS	124.4	92.3	113.4	107.4	93.3	102.5	106.8	92.9	102.0	107.6	79.0	97.8	103.9	75.9	94.3	107.4	90.5	101.6
06	WOOD, PLASTICS & COMPOSITES	115.4	92.0	102.0	118.4	90.1	102.2	116.3	92.2	102.5	114.0	78.4	93.6	96.4	63.1	77.3	98.6	87.7	92.3
07	THERMAL & MOISTURE PROTECTION	124.8	94.8	112.9	111.4	92.0	103.7	111.3	97.1	105.7	105.0	75.8	93.4	104.5	66.9	89.6	114.7	91.6	105.5
08	OPENINGS	92.9	90.7	92.3	91.8	88.4	91.0	91.7	91.0	91.5	92.9	69.0	86.8	86.3	60.4	79.7	88.7	87.0	88.3
0920	Plaster & Gypsum Board	234.1	91.8	133.9	157.5	89.8	109.8	157.5	91.9	111.3	130.2	77.4	93.0	130.8	61.9	82.3	134.8	86.9	101.1
0950, 0980	Ceilings & Acoustic Treatment	133.5	91.8	105.8	95.2	89.8	91.6	95.2	91.9	93.0	95.2	77.4	83.4	95.2	61.9	73.2	95.2	86.9	89.7
0960	Flooring	134.2	101.2	124.7	132.0	102.7	123.5	131.9	101.2	123.1	131.9	73.0	115.0	120.6	65.7	104.9	126.8	79.2	113.1
0970, 0990	Wall Finishes & Painting/Coating	111.4	94.6	101.2	112.2	95.1	101.8	111.4	99.0	103.8	111.4	62.1	81.5	111.4	60.3	80.4	111.4	88.9	97.7
09	FINISHES	135.4	94.7	113.0	118.8	94.1	105.2	117.5	96.1	105.7	114.6	75.1	92.8	111.4	64.0	85.2	115.5	88.5	100.6
COVERS	DIVS. 10 - 14, 25, 28, 41, 43, 44, 46	140.0	104.4	132.6	140.0	80.7	127.7	140.0	81.9	127.9	140.0	73.9	126.3	140.0	71.1	125.7	140.0	104.7	132.7
21, 22, 23	FIRE SUPPRESSION, PLUMBING & HVAC	100.5	87.6	93.4	102.4	106.0	103.9	102.6	109.9	105.7	102.6	90.3	97.5	103.2	74.1	91.1	102.6	102.1	102.4
26, 27, 3370	ELECTRICAL, COMMUNICATIONS & UTIL.	119.5	87.8	102.7	121.6	95.7	107.8	119.7	96.6	107.4	121.8	64.1	91.1	122.1	67.0	92.8	119.8	87.5	102.6
MF2010	WEIGHTED AVERAGE	123.0	92.3	109.3	119.1	96.8	109.1	118.1	98.8	109.5	116.1	78.5	99.3	112.3	71.6	94.1	118.5	95.4	108.2

CANADA

DIVISION		QUEBEC, QUEBEC			RED DEER, ALBERTA			REGINA, SASKATCHEWAN			RIMOUSKI, QUEBEC			ROUYN-NORANDA, QUEBEC			SAINT HYACINTHE, QUEBEC		
		MAT.	INST.	TOTAL	MAT.	INST.	TOTAL	MAT.	INST.	TOTAL	MAT.	INST.	TOTAL	MAT.	INST.	TOTAL	MAT.	INST.	TOTAL
015433	CONTRACTOR EQUIPMENT		104.0	104.0		103.8	103.8		100.0	100.0		102.0	102.0		102.0	102.0		102.0	102.0
0241, 31 - 34	SITE & INFRASTRUCTURE, DEMOLITION	97.4	99.8	99.0	109.2	102.5	104.6	111.6	98.4	102.5	92.8	99.4	97.4	92.3	100.7	98.1	92.8	100.7	98.3
0310	Concrete Forming & Accessories	137.5	92.2	98.2	139.4	88.1	94.9	106.3	90.0	92.2	128.1	91.7	96.5	128.1	94.0	98.5	128.1	94.0	98.5
0320	Concrete Reinforcing	145.4	93.6	118.7	161.6	73.6	116.3	126.2	79.0	101.8	111.9	93.5	102.4	157.1	83.7	119.3	157.1	83.7	119.3
0330	Cast-in-Place Concrete	148.2	100.8	130.0	151.1	104.0	133.0	156.2	95.4	132.9	136.9	99.0	122.4	130.1	103.9	120.1	134.4	103.9	122.7
03	CONCRETE	138.9	95.4	118.1	142.8	91.1	118.1	131.8	89.7	111.7	127.7	94.5	111.8	130.9	95.4	113.9	133.1	95.4	115.1
04	MASONRY	174.5	92.6	124.9	159.9	85.7	114.9	165.6	93.1	121.7	159.4	92.6	118.9	159.6	92.6	119.0	159.9	92.6	119.1
05	METALS	123.8	92.4	113.0	107.6	89.7	101.5	104.8	84.2	97.7	104.5	91.6	100.1	105.0	90.3	100.0	105.0	90.3	100.0
06	WOOD, PLASTICS & COMPOSITES	128.6	92.0	107.6	114.1	87.6	98.9	96.6	90.8	93.3	127.7	91.7	107.1	127.7	94.0	108.4	127.7	94.0	108.4
07	THERMAL & MOISTURE PROTECTION	120.8	95.5	110.7	129.1	90.5	113.8	106.1	85.9	98.1	105.1	94.8	101.0	105.1	95.7	101.4	105.4	95.7	101.6
08	OPENINGS	92.9	89.6	92.1	92.9	81.8	90.1	87.7	79.2	85.6	92.4	79.4	89.1	92.9	81.3	89.9	92.9	81.3	89.9
0920	Plaster & Gypsum Board	196.6	91.3	122.5	136.5	86.9	101.5	166.4	90.5	113.0	188.0	91.3	119.9	147.2	93.7	109.5	150.1	93.7	110.4
0950, 0980	Ceilings & Acoustic Treatment	119.1	91.3	100.7	95.2	86.9	89.7	116.8	90.5	99.4	93.4	91.3	92.0	93.4	93.7	93.6	93.4	93.7	93.6
0960	Flooring	134.2	103.4	125.3	134.2	96.2	123.3	120.6	68.0	105.5	131.9	103.4	123.7	131.9	103.4	123.7	131.9	103.4	123.7
0970, 0990	Wall Finishes & Painting/Coating	111.8	103.2	106.6	111.3	86.0	95.9	111.4	89.2	97.9	111.4	103.2	106.4	111.4	97.1	102.7	111.4	97.1	102.7
09	FINISHES	127.5	95.7	110.0	116.1	89.0	101.2	120.5	86.4	101.7	119.9	95.5	106.4	115.0	96.4	104.8	115.4	96.4	104.9
COVERS	DIVS. 10 - 14, 25, 28, 41, 43, 44, 46	140.0	94.1	130.5	140.0	101.5	132.0	140.0	77.9	127.1	140.0	93.4	130.3	140.0	94.0	130.5	140.0	94.0	130.5
21, 22, 23	FIRE SUPPRESSION, PLUMBING & HVAC	100.4	97.9	99.4	102.6	99.6	101.4	99.8	84.2	93.3	102.6	97.8	100.6	102.6	98.4	100.9	98.4	98.4	98.4
26, 27, 3370	ELECTRICAL, COMMUNICATIONS & UTIL.	119.6	84.8	101.1	112.0	91.2	100.9	123.6	89.2	105.3	119.6	84.8	101.1	119.6	77.5	97.2	120.4	77.5	97.6
MF2010	WEIGHTED AVERAGE	120.1	94.0	108.4	116.8	92.8	106.1	114.9	87.7	102.7	114.4	93.2	104.9	114.5	92.7	104.7	113.9	92.7	104.4

CANADA

DIVISION		SAINT JOHN, NEW BRUNSWICK			SARNIA, ONTARIO			SASKATOON, SASKATCHEWAN			SAULT STE MARIE, ONTARIO			SHERBROOKE, QUEBEC			SOREL, QUEBEC		
		MAT.	INST.	TOTAL	MAT.	INST.	TOTAL	MAT.	INST.	TOTAL	MAT.	INST.	TOTAL	MAT.	INST.	TOTAL	MAT.	INST.	TOTAL
015433	CONTRACTOR EQUIPMENT		101.4	101.4		101.6	101.6		100.0	100.0		101.6	101.6		102.0	102.0		102.0	102.0
0241, 31 - 34	SITE & INFRASTRUCTURE, DEMOLITION	94.4	98.3	97.1	111.2	101.3	104.4	106.1	98.4	100.8	101.2	100.9	101.0	92.9	100.7	98.3	93.0	100.7	98.3
0310	Concrete Forming & Accessories	120.3	65.2	72.5	121.7	101.9	104.6	106.3	90.0	92.2	111.4	89.9	92.8	128.1	94.0	98.5	128.1	94.1	98.6
0320	Concrete Reinforcing	148.4	63.4	104.6	123.5	88.8	105.6	119.2	79.0	98.5	112.4	87.0	99.3	157.1	83.7	119.3	148.4	83.7	115.1
0330	Cast-in-Place Concrete	136.1	65.2	109.0	150.8	108.7	134.7	142.0	95.4	124.1	135.5	92.5	119.0	134.4	103.9	122.7	135.5	103.9	123.4
03	CONCRETE	135.8	65.7	102.3	136.0	101.6	119.5	123.7	89.7	107.4	119.8	90.4	105.7	133.1	95.4	115.1	132.4	95.5	114.7
04	MASONRY	176.7	71.3	112.8	175.2	105.8	133.1	165.3	93.1	121.6	161.2	97.2	122.4	159.9	92.6	119.1	160.0	92.6	119.1
05	METALS	104.6	80.0	96.2	106.7	93.3	102.1	104.9	84.2	97.7	105.9	91.5	101.0	105.0	90.3	100.0	105.0	90.5	100.0
06	WOOD, PLASTICS & COMPOSITES	116.9	63.7	86.4	115.4	101.2	107.2	94.9	90.8	92.6	103.0	88.9	94.9	127.7	94.0	108.4	127.7	94.0	108.4
07	THERMAL & MOISTURE PROTECTION	109.0	66.5	92.1	111.3	100.9	107.2	105.2	85.9	97.5	110.2	92.7	103.2	105.2	95.7	101.4	105.1	95.7	101.4
08	OPENINGS	86.8	59.1	79.8	94.0	95.2	94.3	86.8	79.2	84.9	84.4	87.2	85.1	92.9	81.3	89.9	92.9	86.3	91.2
0920	Plaster & Gypsum Board	181.0	62.5	97.6	179.8	101.2	124.5	146.0	90.5	106.9	141.9	88.6	104.4	150.1	93.7	110.4	188.0	93.7	121.6
0950, 0980	Ceilings & Acoustic Treatment	101.8	62.5	75.7	99.0	101.2	100.5	116.8	90.5	99.4	95.2	88.6	90.8	93.4	93.7	93.6	93.4	93.7	93.6
0960	Flooring	127.5	74.9	112.4	131.9	110.7	125.8	120.6	68.0	105.5	123.8	104.7	118.3	131.9	103.4	123.7	131.9	103.4	123.7
0970, 0990	Wall Finishes & Painting/Coating	111.4	73.2	88.2	111.4	111.9	111.7	111.4	89.2	97.9	111.4	103.6	106.7	111.4	97.1	102.7	111.4	97.1	102.7
09	FINISHES	119.9	67.6	91.1	121.0	105.1	112.2	117.8	86.4	100.5	112.4	94.0	102.3	115.4	96.4	104.9	119.9	96.4	107.0
COVERS	DIVS. 10 - 14, 25, 28, 41, 43, 44, 46	140.0	71.3	125.7	140.0	83.5	128.3	140.0	77.9	127.1	140.0	106.0	132.9	140.0	94.0	130.5	140.0	94.0	130.5
21, 22, 23	FIRE SUPPRESSION, PLUMBING & HVAC	102.6	80.8	93.5	102.6	116.5	108.4	99.7	84.2	93.3	102.6	102.4	102.5	103.2	98.4	101.2	102.6	98.4	100.9
26, 27, 3370	ELECTRICAL, COMMUNICATIONS & UTIL.	130.6	86.4	107.1	122.9	99.0	110.2	123.9	89.2	105.4	121.5	97.0	108.4	119.7	77.5	97.2	119.6	77.5	97.2
MF2010	WEIGHTED AVERAGE	116.9	76.5	98.8	117.8	103.6	111.4	113.4	87.7	101.9	112.7	96.3	105.4	114.9	92.7	105.0	115.1	93.0	105.2

CANADA

DIVISION		ST CATHARINES, ONTARIO			ST JEROME, QUEBEC			ST JOHNS, NEWFOUNDLAND			SUDBURY, ONTARIO			SUMMERSIDE, PRINCE EDWARD ISLAND			SYDNEY, NOVA SCOTIA		
		MAT.	INST.	TOTAL	MAT.	INST.	TOTAL	MAT.	INST.	TOTAL	MAT.	INST.	TOTAL	MAT.	INST.	TOTAL	MAT.	INST.	TOTAL
015433	CONTRACTOR EQUIPMENT		101.6	101.6		102.0	102.0		102.4	102.4		101.6	101.6		101.1	101.1		101.2	101.2
0241, 31 - 34	SITE & INFRASTRUCTURE, DEMOLITION	101.1	102.0	101.8	92.3	100.7	98.1	113.1	99.2	103.5	101.2	101.5	101.4	105.0	96.4	99.1	92.3	99.0	96.9
0310	Concrete Forming & Accessories	115.2	96.1	98.6	128.1	94.0	98.5	103.8	74.0	78.0	111.4	89.9	92.8	96.6	62.2	66.8	96.4	76.5	79.2
0320	Concrete Reinforcing	111.5	87.2	98.9	157.1	83.7	119.3	164.6	65.0	113.3	112.4	85.6	98.6	148.4	49.7	97.5	148.4	49.6	97.5
0330	Cast-in-Place Concrete	134.4	97.9	120.4	130.1	103.9	120.1	172.3	86.9	139.6	135.5	94.2	119.7	161.1	64.3	124.1	130.9	76.2	109.9
03	CONCRETE	119.3	95.0	107.7	130.9	95.4	113.9	164.7	77.3	122.9	119.8	90.8	105.9	153.1	61.3	109.2	129.4	72.0	101.9
04	MASONRY	160.8	102.7	125.6	159.6	92.6	119.0	162.9	79.0	112.0	160.9	97.7	122.6	159.1	66.6	103.0	156.7	76.5	108.1
05	METALS	105.9	92.8	101.4	105.0	90.3	100.0	108.1	79.8	98.3	105.9	92.4	101.2	105.9	69.3	93.3	105.9	76.3	95.7
06	WOOD, PLASTICS & COMPOSITES	107.1	95.5	100.5	127.7	94.0	108.4	96.0	72.7	82.6	103.0	88.9	94.9	87.4	61.6	72.6	87.1	75.9	80.7
07	THERMAL & MOISTURE PROTECTION	110.7	98.3	105.8	105.1	95.7	101.4	109.7	76.8	96.6	110.2	92.7	103.2	105.7	64.9	89.5	106.3	74.4	93.6
08	OPENINGS	83.4	92.3	85.7	92.9	81.3	89.9	99.2	67.5	91.2	84.4	87.2	85.1	95.2	53.1	84.5	83.5	69.0	79.8
0920	Plaster & Gypsum Board	136.9	95.4	107.7	147.2	93.7	109.5	165.6	71.7	99.5	141.9	88.6	104.4	156.6	60.4	88.9	156.2	75.1	99.1
0950, 0980	Ceilings & Acoustic Treatment	99.0	95.4	96.6	93.4	93.7	93.6	103.7	71.7	82.5	95.2	88.6	90.8	95.2	60.4	72.2	95.2	75.1	81.9
0960	Flooring	126.4	102.7	119.6	131.9	103.4	123.7	115.9	61.5	100.3	123.8	104.7	118.3	111.6	65.3	98.3	111.6	69.9	99.6
0970, 0990	Wall Finishes & Painting/Coating	111.4	103.1	106.3	111.4	97.1	102.7	111.4	76.3	90.1	111.4	94.0	100.9	111.4	45.5	71.4	111.4	67.9	85.0
09	FINISHES	113.3	98.5	105.2	115.0	96.4	104.8	116.0	71.9	91.7	112.4	93.1	101.7	111.5	61.3	83.8	110.5	74.6	90.7
COVERS	DIVS. 10 - 14, 25, 28, 41, 43, 44, 46	140.0	81.8	127.9	140.0	94.0	130.5	140.0	74.7	126.4	140.0	106.1	133.0	140.0	71.4	125.7	140.0	73.3	126.1
21, 22, 23	FIRE SUPPRESSION, PLUMBING & HVAC	99.7	84.8	93.5	102.6	98.4	100.9	99.9	72.9	88.7	102.6	87.9	96.5	102.6	68.9	88.6	102.6	91.2	97.9
26, 27, 3370	ELECTRICAL, COMMUNICATIONS & UTIL.	122.7	87.9	104.2	120.3	77.5	97.6	119.7	73.8	95.3	119.7	87.9	102.8	118.2	55.9	85.1	119.3	68.7	92.4
MF2010	WEIGHTED AVERAGE	112.1	93.1	103.6	114.5	92.7	104.8	119.7	76.9	100.6	112.5	92.1	103.4	117.1	66.4	94.5	112.6	79.1	97.6

City Cost Indexes

CANADA

DIVISION		THUNDER BAY, ONTARIO			TIMMINS, ONTARIO			TORONTO, ONTARIO			TROIS RIVIERES, QUEBEC			TRURO, NOVA SCOTIA			VANCOUVER, BRITISH COLUMBIA		
		MAT.	INST.	TOTAL	MAT.	INST.	TOTAL	MAT.	INST.	TOTAL	MAT.	INST.	TOTAL	MAT.	INST.	TOTAL	MAT.	INST.	TOTAL
015433	CONTRACTOR EQUIPMENT		101.6	101.6		101.6	101.6		104.0	104.0		102.0	102.0		101.2	101.2		112.1	112.1
0241, 31 - 34	SITE & INFRASTRUCTURE, DEMOLITION	106.5	102.0	103.4	112.7	100.8	104.5	138.2	106.1	116.1	93.0	100.7	98.3	97.1	99.0	98.4	120.5	108.6	112.3
0310	Concrete Forming & Accessories	123.0	93.6	97.5	122.9	93.1	97.0	132.3	101.7	105.7	128.1	94.1	98.6	96.4	76.6	79.2	157.2	80.9	93.9
0320	Concrete Reinforcing	99.7	86.6	92.9	175.0	86.8	129.6	167.3	87.8	126.4	148.4	83.7	115.1	148.4	49.6	97.5	157.2	80.9	117.9
0330	Cast-in-Place Concrete	148.0	96.3	128.2	163.5	92.4	136.3	153.2	117.9	139.7	135.5	103.9	123.4	169.8	76.2	134.0	162.8	103.4	140.1
03	CONCRETE	128.3	93.3	111.5	150.0	91.8	122.1	144.3	104.4	125.2	132.4	95.5	114.7	148.9	72.0	112.1	153.2	92.6	124.2
04	MASONRY	161.7	101.9	125.4	164.7	95.6	122.8	194.1	107.3	141.5	160.0	92.6	119.1	159.8	76.5	109.3	176.7	95.9	127.7
05	METALS	105.7	91.7	100.9	106.8	92.4	101.8	126.1	93.8	115.0	105.0	90.5	100.0	105.9	76.3	95.7	162.8	91.8	138.3
06	WOOD, PLASTICS & COMPOSITES	116.5	92.3	102.6	116.3	93.6	103.3	116.5	100.2	107.2	127.7	94.0	108.4	87.1	75.9	80.7	114.9	87.0	98.9
07	THERMAL & MOISTURE PROTECTION	110.9	95.0	104.6	111.3	91.2	103.3	122.7	102.5	114.7	105.1	95.7	101.4	106.3	74.4	93.6	133.0	90.6	116.2
08	OPENINGS	82.4	89.9	84.3	91.7	89.1	91.0	91.7	98.4	93.4	92.9	86.3	91.2	83.5	69.0	79.8	94.6	86.0	92.4
0920	Plaster & Gypsum Board	172.5	92.0	115.8	157.5	93.4	112.4	180.3	100.2	123.9	188.0	93.7	121.6	156.2	75.1	99.1	155.1	86.1	106.5
0950, 0980	Ceilings & Acoustic Treatment	95.2	92.0	93.1	95.2	93.4	94.0	131.7	100.2	110.8	93.4	93.7	93.6	95.2	75.1	81.9	148.7	86.1	107.2
0960	Flooring	131.9	110.0	125.6	131.9	101.2	123.1	135.3	112.6	128.8	131.9	103.4	123.7	111.6	69.9	99.6	134.6	99.7	124.5
0970, 0990	Wall Finishes & Painting/Coating	111.4	91.4	99.3	111.4	96.5	102.4	113.4	107.2	109.6	111.4	97.1	102.7	111.4	67.9	85.0	111.3	83.1	94.2
09	FINISHES	118.9	96.7	106.6	117.5	95.4	105.3	131.6	104.4	116.6	119.9	96.4	107.0	110.5	74.6	90.7	130.8	89.9	108.2
COVERS	DIVS. 10 - 14, 25, 28, 41, 43, 44, 46	140.0	81.6	127.9	140.0	80.2	127.6	140.0	109.8	133.7	140.0	94.0	130.5	140.0	73.3	126.1	140.0	103.3	132.4
21, 22, 23	FIRE SUPPRESSION, PLUMBING & HVAC	99.7	84.6	93.4	102.6	106.1	104.1	100.4	99.9	100.2	102.6	98.4	100.9	102.6	91.2	97.9	100.4	86.6	94.7
26, 27, 3370	ELECTRICAL, COMMUNICATIONS & UTIL.	120.9	86.3	102.5	121.5	97.0	108.4	121.1	90.1	104.6	119.6	77.5	97.2	119.3	68.7	92.4	121.8	79.4	99.3
MF2010	WEIGHTED AVERAGE	113.6	91.9	103.9	118.3	96.7	108.6	123.6	100.7	113.4	115.1	93.0	105.2	115.3	79.1	99.1	129.1	90.7	111.9

CANADA

DIVISION		VICTORIA, BRITISH COLUMBIA			WHITEHORSE, YUKON			WINDSOR, ONTARIO			WINNIPEG, MANITOBA			YARMOUTH, NOVA SCOTIA			YELLOWKNIFE, NWT		
		MAT.	INST.	TOTAL	MAT.	INST.	TOTAL	MAT.	INST.	TOTAL	MAT.	INST.	TOTAL	MAT.	INST.	TOTAL	MAT.	INST.	TOTAL
015433	CONTRACTOR EQUIPMENT		108.6	108.6		101.4	101.4		101.6	101.6		105.9	105.9		101.2	101.2		101.2	101.2
0241, 31 - 34	SITE & INFRASTRUCTURE, DEMOLITION	120.3	108.1	111.9	106.7	97.5	100.3	96.2	101.9	100.1	119.0	101.7	107.1	96.9	99.0	98.4	128.7	102.2	110.4
0310	Concrete Forming & Accessories	111.4	88.3	91.3	106.2	63.6	69.2	123.0	91.9	96.0	131.5	71.2	79.2	96.4	76.5	79.2	106.3	87.2	89.7
0320	Concrete Reinforcing	116.6	80.8	98.2	119.2	63.7	90.6	109.2	85.9	97.2	148.3	59.0	102.3	148.4	49.6	97.5	119.2	67.6	92.6
0330	Cast-in-Place Concrete	147.1	102.7	130.1	146.6	74.2	118.8	137.7	97.2	122.2	190.9	76.8	147.2	168.0	76.2	132.9	243.2	94.7	186.4
03	CONCRETE	152.6	91.9	123.6	126.0	67.9	98.2	121.1	92.7	107.5	162.5	71.3	118.9	148.0	72.0	111.6	174.5	86.2	132.3
04	MASONRY	168.1	95.9	124.4	157.9	66.7	102.6	161.0	100.7	124.4	180.3	72.4	114.9	159.7	76.5	109.3	166.4	80.9	114.6
05	METALS	107.4	87.6	100.6	105.1	75.6	95.0	105.8	92.2	101.1	156.2	75.6	128.5	105.9	76.3	95.7	106.0	81.8	97.7
06	WOOD, PLASTICS & COMPOSITES	98.6	86.9	91.9	96.4	62.2	76.8	116.5	90.8	101.7	114.5	72.1	90.1	87.1	75.9	80.7	96.4	88.9	92.1
07	THERMAL & MOISTURE PROTECTION	114.7	90.4	105.1	104.5	65.9	89.1	110.7	94.0	104.1	116.0	73.6	99.1	106.3	74.4	93.6	107.4	84.7	98.4
08	OPENINGS	88.7	82.0	87.0	86.3	59.0	79.4	82.1	89.6	84.0	92.9	63.9	85.5	83.5	69.0	79.8	86.3	74.4	83.3
0920	Plaster & Gypsum Board	134.8	86.1	100.5	119.2	60.9	78.1	160.8	90.5	111.3	164.3	70.7	98.4	156.2	75.1	99.1	169.1	88.5	112.3
0950, 0980	Ceilings & Acoustic Treatment	95.2	86.1	89.2	95.2	60.9	72.5	95.2	90.5	92.1	136.4	70.7	92.9	95.2	75.1	81.9	128.0	88.5	101.8
0960	Flooring	126.8	79.2	113.1	120.6	63.7	104.3	131.9	107.1	124.8	133.0	75.5	116.5	111.6	69.9	99.6	120.6	97.8	114.1
0970, 0990	Wall Finishes & Painting/Coating	111.4	83.1	94.2	111.4	59.3	79.8	111.4	95.9	102.0	111.4	52.7	75.8	111.4	67.9	85.0	111.4	86.4	96.2
09	FINISHES	115.5	86.3	99.4	110.0	62.9	84.0	117.1	95.4	105.1	128.0	70.4	96.3	110.5	74.6	90.7	122.9	88.5	103.9
COVERS	DIVS. 10 - 14, 25, 28, 41, 43, 44, 46	140.0	77.2	127.0	140.0	70.3	125.5	140.0	80.9	127.7	140.0	73.0	126.1	140.0	73.3	126.1	140.0	75.0	126.5
21, 22, 23	FIRE SUPPRESSION, PLUMBING & HVAC	102.6	86.6	95.9	102.6	80.8	93.5	99.7	84.2	93.3	100.5	69.5	87.6	102.6	91.2	97.9	102.9	103.2	103.1
26, 27, 3370	ELECTRICAL, COMMUNICATIONS & UTIL.	120.9	79.4	98.8	122.5	66.4	92.7	125.6	87.9	105.5	125.6	69.8	95.9	119.3	68.7	92.4	122.8	92.5	106.7
MF2010	WEIGHTED AVERAGE	118.5	88.7	105.2	113.2	72.4	95.0	112.7	91.6	103.3	128.8	73.5	104.1	115.1	79.1	99.0	121.5	90.5	107.7

683

Location Factors

Costs shown in RSMeans cost data publications are based on national averages for materials and installation. To adjust these costs to a specific location, simply multiply the base cost by the factor and divide by 100 for that city. The data is arranged alphabetically by state and postal zip code numbers. For a city not listed, use the factor for a nearby city with similar economic characteristics.

STATE/ZIP	CITY	MAT.	INST.	TOTAL
ALABAMA				
350-352	Birmingham	97.0	75.6	87.4
354	Tuscaloosa	95.8	59.2	79.4
355	Jasper	96.0	58.2	79.1
356	Decatur	95.8	60.8	80.1
357-358	Huntsville	95.8	69.0	83.8
359	Gadsden	95.6	58.2	78.9
360-361	Montgomery	96.8	57.2	79.1
362	Anniston	95.1	65.5	81.9
363	Dothan	95.8	52.6	76.5
364	Evergreen	95.3	54.1	76.9
365-366	Mobile	96.8	67.9	83.9
367	Selma	95.5	52.5	76.3
368	Phenix City	96.2	56.3	78.4
369	Butler	95.7	52.6	76.4
ALASKA				
995-996	Anchorage	127.2	114.5	121.5
997	Fairbanks	124.4	115.0	120.2
998	Juneau	122.8	112.6	118.3
999	Ketchikan	136.9	112.6	126.0
ARIZONA				
850,853	Phoenix	98.6	75.9	88.5
851,852	Mesa/Tempe	98.2	72.8	86.8
855	Globe	98.5	62.0	82.2
856-857	Tucson	96.9	72.0	85.8
859	Show Low	98.6	63.5	82.9
860	Flagstaff	100.7	74.4	88.9
863	Prescott	98.4	63.0	82.6
864	Kingman	96.7	74.2	86.7
865	Chambers	96.8	63.5	81.9
ARKANSAS				
716	Pine Bluff	96.7	64.8	82.4
717	Camden	94.2	50.7	74.8
718	Texarkana	95.9	51.9	76.3
719	Hot Springs	93.4	53.6	75.6
720-722	Little Rock	97.0	65.9	83.1
723	West Memphis	95.9	63.7	81.5
724	Jonesboro	96.7	61.6	81.0
725	Batesville	94.3	56.9	77.6
726	Harrison	95.7	53.0	76.6
727	Fayetteville	92.9	54.9	75.9
728	Russellville	94.2	56.1	77.2
729	Fort Smith	97.0	63.4	82.0
CALIFORNIA				
900-902	Los Angeles	100.6	115.4	107.2
903-905	Inglewood	96.1	112.6	103.5
906-908	Long Beach	97.8	112.6	104.5
910-912	Pasadena	96.5	112.7	103.7
913-916	Van Nuys	99.7	112.7	105.5
917-918	Alhambra	98.6	112.7	104.9
919-921	San Diego	100.6	108.2	104.0
922	Palm Springs	97.5	111.5	103.8
923-924	San Bernardino	95.2	111.2	102.3
925	Riverside	100.0	113.3	105.9
926-927	Santa Ana	97.2	111.4	103.6
928	Anaheim	99.9	113.7	106.1
930	Oxnard	101.4	113.1	106.6
931	Santa Barbara	100.6	113.1	106.2
932-933	Bakersfield	101.4	110.4	105.4
934	San Luis Obispo	101.2	109.8	105.0
935	Mojave	98.2	107.7	102.4
936-938	Fresno	101.7	115.3	107.8
939	Salinas	102.0	123.0	111.4
940-941	San Francisco	109.6	140.8	123.6
942,956-958	Sacramento	103.9	117.2	109.8
943	Palo Alto	102.3	129.4	114.4
944	San Mateo	105.1	131.4	116.9
945	Vallejo	102.9	127.8	114.0
946	Oakland	107.5	133.7	119.2
947	Berkeley	106.8	133.6	118.8
948	Richmond	106.1	128.9	116.3
949	San Rafael	107.4	130.3	117.6
950	Santa Cruz	106.4	123.1	113.9

STATE/ZIP	CITY	MAT.	INST.	TOTAL
CALIFORNIA (CONT'D)				
951	San Jose	104.6	134.3	117.9
952	Stockton	101.9	117.3	108.8
953	Modesto	101.9	116.6	108.5
954	Santa Rosa	101.7	134.9	116.5
955	Eureka	103.1	115.2	108.5
959	Marysville	102.4	113.8	107.5
960	Redding	104.1	113.8	108.4
961	Susanville	103.1	113.8	107.9
COLORADO				
800-802	Denver	101.4	84.9	94.0
803	Boulder	98.5	83.7	91.8
804	Golden	100.6	82.5	92.5
805	Fort Collins	102.3	77.1	91.0
806	Greeley	99.6	73.6	88.0
807	Fort Morgan	99.1	82.3	91.6
808-809	Colorado Springs	101.3	82.4	92.8
810	Pueblo	100.8	80.2	91.6
811	Alamosa	102.1	76.1	90.5
812	Salida	102.0	77.8	91.2
813	Durango	102.6	75.3	90.4
814	Montrose	102.1	74.7	89.4
815	Grand Junction	104.7	75.2	91.5
816	Glenwood Springs	102.1	78.0	91.3
CONNECTICUT				
060	New Britain	101.0	117.8	108.5
061	Hartford	102.1	117.8	109.1
062	Willimantic	101.6	117.8	108.9
063	New London	97.7	117.7	106.6
064	Meriden	99.7	117.7	107.8
065	New Haven	102.9	117.7	109.5
066	Bridgeport	102.2	116.4	108.6
067	Waterbury	101.7	117.7	108.9
068	Norwalk	101.6	116.5	108.3
069	Stamford	101.8	123.0	111.3
D.C.				
200-205	Washington	102.8	91.9	97.9
DELAWARE				
197	Newark	99.3	108.6	103.4
198	Wilmington	99.6	108.6	103.6
199	Dover	99.9	108.6	103.8
FLORIDA				
320,322	Jacksonville	97.8	69.7	85.2
321	Daytona Beach	98.0	76.4	88.3
323	Tallahassee	98.6	59.7	81.2
324	Panama City	99.2	57.5	80.6
325	Pensacola	101.9	65.1	85.4
326,344	Gainesville	99.5	70.6	86.6
327-328,347	Orlando	101.2	74.2	89.1
329	Melbourne	100.4	79.6	91.1
330-332,340	Miami	99.1	78.1	89.7
333	Fort Lauderdale	97.3	76.7	88.1
334,349	West Palm Beach	96.0	75.7	86.9
335-336,346	Tampa	99.0	83.6	92.1
337	St. Petersburg	101.3	66.1	85.6
338	Lakeland	98.0	83.5	91.5
339,341	Fort Myers	97.3	75.4	87.5
342	Sarasota	99.2	77.3	89.4
GEORGIA				
300-303,399	Atlanta	97.2	76.8	88.1
304	Statesboro	96.8	54.1	77.7
305	Gainesville	95.5	64.7	81.7
306	Athens	94.9	64.9	81.5
307	Dalton	96.8	60.5	80.6
308-309	Augusta	95.9	65.5	82.3
310-312	Macon	95.6	65.0	81.9
313-314	Savannah	97.7	60.8	81.2
315	Waycross	96.7	61.8	81.1
316	Valdosta	96.7	61.2	80.9
317,398	Albany	96.9	62.0	81.3
318-319	Columbus	96.8	65.3	82.7

Location Factors

STATE/ZIP	CITY	MAT.	INST.	TOTAL
HAWAII				
967	Hilo	111.9	120.6	115.8
968	Honolulu	115.1	120.6	117.5
STATES & POSS.				
969	Guam	133.3	61.7	101.3
IDAHO				
832	Pocatello	100.4	73.0	88.2
833	Twin Falls	101.7	53.1	80.0
834	Idaho Falls	99.0	59.0	81.1
835	Lewiston	109.3	78.3	95.4
836-837	Boise	99.5	73.3	87.8
838	Coeur d'Alene	108.5	75.4	93.7
ILLINOIS				
600-603	North Suburban	99.3	131.0	113.4
604	Joliet	99.2	137.7	116.4
605	South Suburban	99.3	131.0	113.4
606-608	Chicago	99.7	138.9	117.2
609	Kankakee	95.2	125.3	108.7
610-611	Rockford	98.1	122.7	109.1
612	Rock Island	95.7	99.0	97.2
613	La Salle	96.9	121.4	107.8
614	Galesburg	96.8	104.8	100.4
615-616	Peoria	99.2	107.3	102.8
617	Bloomington	96.1	108.7	101.7
618-619	Champaign	99.8	107.3	103.1
620-622	East St. Louis	94.6	107.5	100.4
623	Quincy	95.8	99.6	97.5
624	Effingham	95.1	103.6	98.9
625	Decatur	97.0	105.3	100.7
626-627	Springfield	97.5	106.2	101.4
628	Centralia	93.2	106.9	99.3
629	Carbondale	92.9	102.3	97.1
INDIANA				
460	Anderson	95.3	82.8	89.7
461-462	Indianapolis	98.3	85.7	92.7
463-464	Gary	96.7	108.1	101.8
465-466	South Bend	95.3	83.3	89.9
467-468	Fort Wayne	95.9	79.2	88.5
469	Kokomo	93.1	81.7	88.0
470	Lawrenceburg	92.5	78.9	86.4
471	New Albany	93.6	76.4	85.9
472	Columbus	95.9	81.6	89.5
473	Muncie	96.8	81.7	90.1
474	Bloomington	98.1	81.5	90.7
475	Washington	94.6	83.3	89.5
476-477	Evansville	95.9	86.0	91.5
478	Terre Haute	96.6	86.7	92.2
479	Lafayette	95.6	82.8	89.9
IOWA				
500-503,509	Des Moines	98.9	86.2	93.2
504	Mason City	97.5	68.9	84.7
505	Fort Dodge	97.7	60.6	81.1
506-507	Waterloo	99.4	73.4	87.8
508	Creston	98.0	80.2	90.0
510-511	Sioux City	100.1	71.9	87.5
512	Sibley	98.9	57.3	80.3
513	Spencer	100.6	57.3	81.3
514	Carroll	97.3	73.6	86.7
515	Council Bluffs	101.4	80.3	91.9
516	Shenandoah	97.9	73.6	87.1
520	Dubuque	99.7	81.2	91.4
521	Decorah	98.4	66.7	84.3
522-524	Cedar Rapids	100.5	85.5	93.8
525	Ottumwa	98.4	71.9	86.6
526	Burlington	97.5	78.6	89.1
527-528	Davenport	99.4	92.0	96.1
KANSAS				
660-662	Kansas City	99.2	97.3	98.3
664-666	Topeka	99.2	65.2	84.0
667	Fort Scott	97.6	75.2	87.6
668	Emporia	97.8	70.8	85.7
669	Belleville	99.6	66.5	84.8
670-672	Wichita	98.3	65.1	83.4
673	Independence	99.2	70.6	86.4
674	Salina	99.7	68.0	85.5
675	Hutchinson	94.5	66.4	82.0
676	Hays	98.7	66.8	84.4
677	Colby	99.5	66.8	84.9

STATE/ZIP	CITY	MAT.	INST.	TOTAL
KANSAS (CONT'D)				
678	Dodge City	101.0	67.8	86.2
679	Liberal	98.5	66.8	84.3
KENTUCKY				
400-402	Louisville	97.7	83.4	91.3
403-405	Lexington	96.3	75.3	86.9
406	Frankfort	97.1	74.3	86.9
407-409	Corbin	93.5	62.0	79.4
410	Covington	94.4	96.0	95.1
411-412	Ashland	93.2	98.5	95.6
413-414	Campton	94.4	62.6	80.2
415-416	Pikeville	95.7	80.3	88.8
417-418	Hazard	93.7	55.7	76.7
420	Paducah	92.6	86.0	89.7
421-422	Bowling Green	94.7	82.8	89.4
423	Owensboro	95.1	80.2	88.4
424	Henderson	92.3	86.3	89.6
425-426	Somerset	91.6	66.6	80.4
427	Elizabethtown	91.0	83.4	87.6
LOUISIANA				
700-701	New Orleans	102.0	69.2	87.3
703	Thibodaux	100.3	64.3	84.2
704	Hammond	97.4	56.1	78.9
705	Lafayette	99.7	60.5	82.2
706	Lake Charles	100.0	65.0	84.3
707-708	Baton Rouge	101.5	63.6	84.5
710-711	Shreveport	96.9	57.1	79.1
712	Monroe	97.2	54.3	78.0
713-714	Alexandria	97.2	53.8	77.8
MAINE				
039	Kittery	95.0	80.2	88.4
040-041	Portland	99.8	79.0	90.5
042	Lewiston	98.2	79.0	89.6
043	Augusta	97.9	80.0	89.9
044	Bangor	97.9	79.0	89.4
045	Bath	96.3	79.6	88.8
046	Machias	95.8	78.9	88.2
047	Houlton	95.9	79.1	88.4
048	Rockland	94.9	79.5	88.0
049	Waterville	96.3	80.0	89.0
MARYLAND				
206	Waldorf	99.1	68.5	85.4
207-208	College Park	99.2	83.0	92.0
209	Silver Spring	98.3	74.5	87.7
210-212	Baltimore	98.9	84.8	92.6
214	Annapolis	99.1	78.5	89.9
215	Cumberland	94.7	80.1	88.1
216	Easton	96.4	41.9	72.0
217	Hagerstown	95.5	82.2	89.5
218	Salisbury	96.8	51.0	76.3
219	Elkton	93.6	64.9	80.7
MASSACHUSETTS				
010-011	Springfield	100.0	109.5	104.2
012	Pittsfield	99.6	105.3	102.2
013	Greenfield	97.4	108.6	102.4
014	Fitchburg	96.2	121.1	107.3
015-016	Worcester	100.0	122.7	110.1
017	Framingham	95.6	129.4	110.7
018	Lowell	99.4	129.0	112.6
019	Lawrence	100.6	127.8	112.8
020-022, 024	Boston	102.3	136.8	117.7
023	Brockton	100.9	124.5	111.5
025	Buzzards Bay	95.0	122.2	107.1
026	Hyannis	98.0	122.2	108.8
027	New Bedford	100.0	122.3	110.0
MICHIGAN				
480,483	Royal Oak	92.8	103.6	97.6
481	Ann Arbor	95.2	107.7	100.8
482	Detroit	96.7	110.2	102.8
484-485	Flint	94.9	98.2	96.4
486	Saginaw	94.5	94.0	94.3
487	Bay City	94.6	93.9	94.3
488-489	Lansing	95.7	96.5	96.1
490	Battle Creek	95.8	90.4	93.4
491	Kalamazoo	96.1	88.4	92.7
492	Jackson	94.0	96.8	95.2
493,495	Grand Rapids	97.3	76.8	88.1
494	Muskegon	94.7	85.3	90.5

Location Factors

STATE/ZIP	CITY	MAT.	INST.	TOTAL
MICHIGAN (CONT'D)				
496	Traverse City	93.4	79.4	87.2
497	Gaylord	94.5	73.4	85.1
498-499	Iron mountain	96.7	88.8	93.2
MINNESOTA				
550-551	Saint Paul	101.6	123.0	111.2
553-555	Minneapolis	102.3	125.0	112.5
556-558	Duluth	100.6	112.2	105.8
MINNESOTA (CONT'd)				
559	Rochester	101.0	109.2	104.7
560	Mankato	97.6	106.4	101.5
561	Windom	96.2	95.3	95.8
562	Willmar	95.9	107.2	100.9
563	St. Cloud	97.3	123.2	108.9
564	Brainerd	97.6	107.3	101.9
565	Detroit Lakes	99.3	100.7	99.9
566	Bemidji	98.5	102.2	100.2
567	Thief River Falls	98.1	97.1	97.7
MISSISSIPPI				
386	Clarksdale	95.7	56.3	78.1
387	Greenville	99.3	66.1	84.5
388	Tupelo	97.1	58.8	80.0
389	Greenwood	97.0	55.8	78.6
390-392	Jackson	96.9	65.3	82.8
393	Meridian	95.9	67.0	83.0
394	Laurel	97.0	59.3	80.2
395	Biloxi	97.6	60.8	81.2
396	Mccomb	95.4	55.3	77.5
397	Columbus	97.0	57.7	79.4
MISSOURI				
630-631	St. Louis	99.4	106.6	102.6
633	Bowling Green	98.1	92.2	95.5
634	Hannibal	96.9	88.8	93.3
635	Kirksville	100.1	85.4	93.5
636	Flat River	99.1	94.7	97.1
637	Cape Girardeau	99.2	91.3	95.7
638	Sikeston	97.1	87.1	92.6
639	Poplar Bluff	96.6	88.4	92.9
640-641	Kansas City	99.0	107.3	102.7
644-645	St. Joseph	98.3	93.2	96.0
646	Chillicothe	95.2	71.1	84.4
647	Harrisonville	94.8	104.8	99.3
648	Joplin	97.1	71.4	85.6
650-651	Jefferson City	95.9	92.1	94.2
652	Columbia	97.4	93.7	95.7
653	Sedalia	96.5	94.6	95.7
654-655	Rolla	95.0	89.5	92.5
656-658	Springfield	98.2	80.9	90.5
MONTANA				
590-591	Billings	103.2	75.9	91.0
592	Wolf Point	103.2	68.6	87.7
593	Miles City	100.6	70.2	87.0
594	Great Falls	104.7	76.5	92.1
595	Havre	101.9	71.3	88.2
596	Helena	102.5	73.2	89.4
597	Butte	103.2	74.6	90.4
598	Missoula	100.5	73.4	88.4
599	Kalispell	99.4	71.3	86.8
NEBRASKA				
680-681	Omaha	99.8	80.6	91.2
683-685	Lincoln	97.7	74.9	87.5
686	Columbus	96.7	72.9	86.1
687	Norfolk	98.4	76.4	88.6
688	Grand Island	98.5	79.8	90.1
689	Hastings	97.7	80.7	90.1
690	Mccook	97.5	71.9	86.0
691	North Platte	97.8	80.9	90.3
692	Valentine	99.5	68.7	85.7
693	Alliance	99.6	66.8	85.0
NEVADA				
889-891	Las Vegas	100.2	112.8	105.9
893	Ely	99.1	99.6	99.3
894-895	Reno	99.2	91.1	95.6
897	Carson City	98.6	90.8	95.1
898	Elko	97.7	84.4	91.8
NEW HAMPSHIRE				
030	Nashua	100.2	86.1	93.9
031	Manchester	100.8	86.1	94.2

STATE/ZIP	CITY	MAT.	INST.	TOTAL
NEW HAMPSHIRE (CONT'D)				
032-033	Concord	98.7	82.5	91.5
034	Keene	97.0	49.9	76.0
035	Littleton	96.7	57.4	79.1
036	Charleston	96.5	47.6	74.6
037	Claremont	95.4	47.6	74.0
038	Portsmouth	97.7	89.9	94.2
NEW JERSEY				
070-071	Newark	102.3	125.6	112.7
072	Elizabeth	100.6	125.6	111.8
073	Jersey City	99.3	125.3	110.9
074-075	Paterson	101.4	125.5	112.2
076	Hackensack	99.3	125.4	111.0
077	Long Branch	98.9	124.1	110.2
078	Dover	99.6	125.5	111.2
079	Summit	99.6	125.6	111.2
080,083	Vineland	97.5	122.8	108.8
081	Camden	99.4	124.0	110.4
082,084	Atlantic City	98.1	122.7	109.1
085-086	Trenton	99.4	122.7	109.8
087	Point Pleasant	99.6	123.3	110.2
088-089	New Brunswick	100.1	125.0	111.2
NEW MEXICO				
870-872	Albuquerque	99.1	75.0	88.3
873	Gallup	99.3	75.0	88.5
874	Farmington	99.8	75.0	88.7
875	Santa Fe	100.1	75.0	88.9
877	Las Vegas	97.7	75.0	87.6
878	Socorro	97.2	75.0	87.3
879	Truth/Consequences	97.5	71.6	85.9
880	Las Cruces	96.3	70.9	85.0
881	Clovis	97.9	74.9	87.6
882	Roswell	99.6	75.0	88.6
883	Carrizozo	100.0	75.0	88.8
884	Tucumcari	98.6	74.9	88.0
NEW YORK				
100-102	New York	106.0	166.3	133.0
103	Staten Island	102.1	160.3	128.1
104	Bronx	99.9	160.3	126.9
105	Mount Vernon	100.0	134.3	115.3
106	White Plains	99.9	134.3	115.3
107	Yonkers	104.8	134.5	118.0
108	New Rochelle	100.3	134.3	115.5
109	Suffern	100.1	123.1	110.4
110	Queens	101.9	160.1	127.9
111	Long Island City	103.6	160.1	128.9
112	Brooklyn	104.0	160.1	129.1
113	Flushing	103.8	160.1	129.0
114	Jamaica	101.9	160.1	127.9
115,117,118	Hicksville	101.7	145.2	121.2
116	Far Rockaway	103.9	160.1	129.1
119	Riverhead	102.5	144.7	121.4
120-122	Albany	97.2	97.9	97.5
123	Schenectady	98.2	97.5	97.9
124	Kingston	101.6	118.7	109.2
125-126	Poughkeepsie	100.7	130.5	114.0
127	Monticello	100.0	119.2	108.6
128	Glens Falls	92.6	93.2	92.8
129	Plattsburgh	97.5	85.0	91.9
130-132	Syracuse	99.4	93.5	96.7
133-135	Utica	97.1	88.7	93.4
136	Watertown	98.9	92.6	96.1
137-139	Binghamton	98.8	90.2	95.0
140-142	Buffalo	100.5	102.9	101.6
143	Niagara Falls	98.3	98.0	98.2
144-146	Rochester	100.3	95.6	98.2
147	Jamestown	97.2	84.4	91.5
148-149	Elmira	96.9	92.0	94.7
NORTH CAROLINA				
270,272-274	Greensboro	99.2	47.9	76.2
271	Winston-Salem	99.1	46.5	75.6
275-276	Raleigh	99.6	48.1	76.6
277	Durham	100.3	48.4	77.1
278	Rocky Mount	96.6	39.8	71.2
279	Elizabeth City	97.5	41.0	72.3
280	Gastonia	98.2	46.0	74.8
281-282	Charlotte	99.2	48.4	76.5
283	Fayetteville	100.0	50.0	77.6
284	Wilmington	96.9	46.3	74.3
285	Kinston	94.9	40.6	70.6

Location Factors

STATE/ZIP	CITY	MAT.	INST.	TOTAL
NORTH CAROLINA (CONT'D)				
286	Hickory	95.2	42.8	71.8
287-288	Asheville	97.4	46.0	74.4
289	Murphy	96.2	33.4	68.1
NORTH DAKOTA				
580-581	Fargo	102.4	64.3	85.4
582	Grand Forks	102.6	53.8	80.8
583	Devils Lake	101.7	56.0	81.3
584	Jamestown	101.9	46.8	77.3
585	Bismarck	100.8	63.0	83.9
586	Dickinson	102.8	58.2	82.9
587	Minot	102.5	70.0	88.0
588	Williston	101.0	58.2	81.9
OHIO				
430-432	Columbus	97.9	90.3	94.5
433	Marion	94.1	80.8	88.2
434-436	Toledo	98.1	95.9	97.1
437-438	Zanesville	94.6	81.1	88.6
439	Steubenville	96.4	90.5	93.8
440	Lorain	98.7	91.5	95.5
441	Cleveland	99.0	100.5	99.7
442-443	Akron	99.8	92.6	96.6
444-445	Youngstown	99.1	87.5	93.9
446-447	Canton	99.2	82.5	91.8
448-449	Mansfield	96.2	87.0	92.1
450	Hamilton	95.7	83.8	90.4
451-452	Cincinnati	96.0	86.3	91.7
453-454	Dayton	95.8	83.6	90.3
455	Springfield	95.7	83.8	90.4
456	Chillicothe	94.5	88.5	91.8
457	Athens	97.0	81.0	89.8
458	Lima	97.4	85.0	91.9
OKLAHOMA				
730-731	Oklahoma City	98.2	61.5	81.8
734	Ardmore	95.2	61.0	79.9
735	Lawton	97.6	62.3	81.8
736	Clinton	96.7	59.6	80.2
737	Enid	97.4	59.6	80.5
738	Woodward	95.4	59.7	79.4
739	Guymon	96.5	31.1	67.3
740-741	Tulsa	97.4	54.2	78.1
743	Miami	94.0	69.0	82.8
744	Muskogee	96.7	39.6	71.2
745	Mcalester	93.7	51.2	74.7
746	Ponca City	94.3	59.6	78.8
747	Durant	94.3	58.5	78.3
748	Shawnee	96.0	57.3	78.7
749	Poteau	93.3	62.3	79.5
OREGON				
970-972	Portland	100.0	99.7	99.9
973	Salem	99.8	98.6	99.3
974	Eugene	99.8	98.3	99.1
975	Medford	101.5	96.7	99.3
976	Klamath Falls	101.5	96.7	99.3
977	Bend	100.3	98.4	99.5
978	Pendleton	94.9	100.3	97.3
979	Vale	92.6	90.4	91.6
PENNSYLVANIA				
150-152	Pittsburgh	99.3	102.9	100.9
153	Washington	96.2	100.8	98.3
154	Uniontown	96.6	99.9	98.0
155	Bedford	97.6	91.1	94.7
156	Greensburg	97.5	99.4	98.3
157	Indiana	96.3	98.2	97.1
158	Dubois	98.0	94.2	96.3
159	Johnstown	97.6	94.9	96.4
160	Butler	92.6	100.3	96.0
161	New Castle	92.6	97.2	94.7
162	Kittanning	93.2	101.4	96.9
163	Oil City	92.6	94.9	93.6
164-165	Erie	94.8	93.5	94.2
166	Altoona	95.0	90.5	93.0
167	Bradford	95.9	93.1	94.7
168	State College	95.5	91.4	93.7
169	Wellsboro	96.7	91.0	94.2
170-171	Harrisburg	98.0	93.8	96.1
172	Chambersburg	96.1	88.6	92.7
173-174	York	96.5	94.0	95.4
175-176	Lancaster	94.7	88.3	91.8

STATE/ZIP	CITY	MAT.	INST.	TOTAL
PENNSYLVANIA (CONT'D)				
177	Williamsport	93.3	79.7	87.2
178	Sunbury	95.5	94.0	94.8
179	Pottsville	94.6	95.6	95.1
180	Lehigh Valley	96.0	114.2	104.2
181	Allentown	98.3	108.6	102.9
182	Hazleton	95.5	97.3	96.3
183	Stroudsburg	95.3	102.3	98.4
184-185	Scranton	99.0	98.7	98.9
186-187	Wilkes-Barre	95.2	97.7	96.3
188	Montrose	94.9	96.3	95.5
189	Doylestown	95.1	121.9	107.1
190-191	Philadelphia	99.7	132.2	114.3
193	Westchester	96.1	124.7	108.9
194	Norristown	95.1	131.2	111.3
195-196	Reading	97.4	99.6	98.4
PUERTO RICO				
009	San Juan	121.8	24.0	78.1
RHODE ISLAND				
028	Newport	99.3	111.5	104.7
029	Providence	100.3	111.5	105.3
SOUTH CAROLINA				
290-292	Columbia	97.6	50.0	76.3
293	Spartanburg	96.4	49.6	75.5
294	Charleston	98.0	58.3	80.2
295	Florence	96.2	50.0	75.5
296	Greenville	96.2	49.6	75.3
297	Rock Hill	95.7	47.7	74.3
298	Aiken	96.7	71.9	85.6
299	Beaufort	97.5	43.0	73.1
SOUTH DAKOTA				
570-571	Sioux Falls	100.6	57.9	81.5
572	Watertown	99.7	52.2	78.4
573	Mitchell	98.3	51.6	77.4
574	Aberdeen	101.2	52.8	79.6
575	Pierre	99.7	54.5	79.5
576	Mobridge	99.1	52.0	78.0
577	Rapid City	101.0	55.3	80.6
TENNESSEE				
370-372	Nashville	96.7	72.7	85.9
373-374	Chattanooga	98.2	66.9	84.2
375,380-381	Memphis	96.6	70.3	84.9
376	Johnson City	97.6	57.4	79.6
377-379	Knoxville	94.5	61.7	79.9
382	Mckenzie	96.3	62.1	81.1
383	Jackson	98.2	63.8	82.8
384	Columbia	94.9	67.3	82.6
385	Cookeville	96.2	60.3	80.1
TEXAS				
750	Mckinney	99.5	50.8	77.7
751	Waxahackie	99.4	58.2	81.0
752-753	Dallas	99.9	67.5	85.4
754	Greenville	99.6	43.7	74.6
755	Texarkana	99.1	51.5	77.8
756	Longview	99.7	40.7	73.4
757	Tyler	100.2	54.9	80.0
758	Palestine	96.1	57.4	78.8
759	Lufkin	96.7	59.1	79.9
760-761	Fort Worth	97.7	63.6	82.4
762	Denton	97.3	48.4	75.4
763	Wichita Falls	98.1	55.5	79.1
764	Eastland	96.8	50.3	76.0
765	Temple	95.3	50.0	75.1
766-767	Waco	97.5	57.1	79.4
768	Brownwood	98.0	49.6	76.3
769	San Angelo	97.6	50.1	76.3
770-772	Houston	100.0	70.4	86.8
773	Huntsville	98.6	56.7	79.9
774	Wharton	99.8	54.8	79.7
775	Galveston	97.6	69.2	84.9
776-777	Beaumont	98.2	62.9	82.4
778	Bryan	95.1	62.1	80.3
779	Victoria	99.9	44.4	75.1
780	Laredo	94.7	53.1	76.1
781-782	San Antonio	95.0	65.1	81.6
783-784	Corpus Christi	97.6	52.8	77.6
785	Mc Allen	97.5	46.9	74.9
786-787	Austin	95.1	59.4	79.1

Location Factors

STATE/ZIP	CITY	MAT.	INST.	TOTAL
TEXAS (CONT'D)				
788	Del Rio	97.0	47.0	74.6
789	Giddings	94.1	57.0	77.5
790-791	Amarillo	98.1	61.9	81.9
792	Childress	97.5	58.1	79.9
793-794	Lubbock	99.6	57.5	80.8
795-796	Abilene	97.9	54.8	78.6
797	Midland	99.7	51.9	78.3
798-799,885	El Paso	97.0	51.7	76.8
UTAH				
840-841	Salt Lake City	102.0	69.1	87.3
842,844	Ogden	97.5	70.7	85.5
843	Logan	99.5	70.7	86.6
845	Price	100.0	67.1	85.3
846-847	Provo	100.0	68.9	86.1
VERMONT				
050	White River Jct.	98.3	55.1	79.0
051	Bellows Falls	96.6	73.7	86.4
052	Bennington	97.0	68.9	84.4
053	Brattleboro	97.4	72.9	86.5
054	Burlington	100.9	66.1	85.4
056	Montpelier	97.7	66.2	83.6
057	Rutland	99.2	66.2	84.4
058	St. Johnsbury	98.3	55.9	79.3
059	Guildhall	96.9	55.7	78.5
VIRGINIA				
220-221	Fairfax	99.9	82.3	92.0
222	Arlington	101.2	80.1	91.8
223	Alexandria	100.2	85.5	93.6
224-225	Fredericksburg	98.5	72.2	86.7
226	Winchester	99.2	63.2	83.1
227	Culpeper	99.1	67.8	85.1
228	Harrisonburg	99.3	64.0	83.5
229	Charlottesville	99.8	63.4	83.5
230-232	Richmond	100.2	66.8	85.3
233-235	Norfolk	101.0	68.2	86.3
236	Newport News	100.0	67.6	85.5
237	Portsmouth	99.4	65.5	84.2
238	Petersburg	99.6	66.8	84.9
239	Farmville	98.7	53.0	78.3
240-241	Roanoke	100.7	60.7	82.8
242	Bristol	98.4	55.1	79.0
243	Pulaski	98.0	50.9	77.0
244	Staunton	98.9	59.1	81.1
245	Lynchburg	99.0	64.6	83.6
246	Grundy	98.3	50.3	76.9
WASHINGTON				
980-981,987	Seattle	103.8	106.6	105.0
982	Everett	103.4	97.1	100.6
983-984	Tacoma	103.5	98.9	101.4
985	Olympia	101.3	98.9	100.2
986	Vancouver	104.8	93.8	99.9
988	Wenatchee	103.3	83.7	94.6
989	Yakima	103.7	93.1	99.0
990-992	Spokane	103.7	80.8	93.5
993	Richland	103.4	88.0	96.5
994	Clarkston	101.9	79.8	92.0
WEST VIRGINIA				
247-248	Bluefield	97.1	77.7	88.4
249	Lewisburg	98.9	83.2	91.9
250-253	Charleston	99.0	90.1	95.0
254	Martinsburg	98.6	79.1	89.9
255-257	Huntington	100.2	91.6	96.3
258-259	Beckley	97.0	87.2	92.6
260	Wheeling	100.3	89.9	95.7
261	Parkersburg	99.1	89.4	94.8
262	Buckhannon	98.5	89.7	94.6
263-264	Clarksburg	99.2	89.7	94.9
265	Morgantown	99.2	88.8	94.6
266	Gassaway	98.4	89.9	94.6
267	Romney	98.4	85.6	92.7
268	Petersburg	98.2	85.0	92.3
WISCONSIN				
530,532	Milwaukee	100.2	107.6	103.5
531	Kenosha	100.0	102.7	101.2
534	Racine	99.4	102.8	100.9
535	Beloit	99.3	97.8	98.6
537	Madison	98.9	99.1	99.0

STATE/ZIP	CITY	MAT.	INST.	TOTAL
WISCONSIN (CONT'D)				
538	Lancaster	97.0	93.2	95.3
539	Portage	95.4	96.3	95.8
540	New Richmond	97.3	96.2	96.8
541-543	Green Bay	101.7	93.3	98.0
544	Wausau	96.2	93.4	95.0
545	Rhinelander	99.8	92.2	96.4
546	La Crosse	97.6	94.9	96.4
547	Eau Claire	99.3	96.1	97.9
548	Superior	97.1	98.1	97.6
549	Oshkosh	97.0	84.1	91.2
WYOMING				
820	Cheyenne	99.2	64.7	83.8
821	Yellowstone Nat'l Park	96.8	60.0	80.4
822	Wheatland	98.0	59.2	80.6
823	Rawlins	99.6	58.0	81.0
824	Worland	97.5	58.0	79.8
825	Riverton	98.6	58.0	80.4
826	Casper	99.5	58.5	81.2
827	Newcastle	97.3	58.0	79.7
828	Sheridan	100.1	61.5	82.8
829-831	Rock Springs	101.5	58.1	82.1
CANADIAN FACTORS (reflect Canadian currency)				
ALBERTA				
	Calgary	130.9	94.8	114.8
	Edmonton	131.1	94.8	114.9
	Fort McMurray	127.0	97.7	113.9
	Lethbridge	120.9	97.1	110.3
	Lloydminster	116.2	93.6	106.1
	Medicine Hat	116.3	92.8	105.8
	Red Deer	116.8	92.8	106.1
BRITISH COLUMBIA				
	Kamloops	117.2	96.2	107.8
	Prince George	118.5	95.4	108.2
	Vancouver	129.1	90.7	111.9
	Victoria	118.5	88.7	105.2
MANITOBA				
	Brandon	116.1	80.3	100.1
	Portage la Prairie	116.1	78.5	99.3
	Winnipeg	128.8	73.5	104.1
NEW BRUNSWICK				
	Bathurst	114.2	71.2	94.9
	Dalhousie	114.2	71.4	95.1
	Fredericton	116.8	76.8	98.9
	Moncton	114.5	72.8	95.8
	Newcastle	114.2	71.9	95.3
	St. John	116.9	76.5	98.8
NEWFOUNDLAND				
	Corner Brook	119.6	71.5	98.1
	St Johns	119.7	76.9	100.6
NORTHWEST TERRITORIES				
	Yellowknife	121.5	90.5	107.7
NOVA SCOTIA				
	Bridgewater	115.8	79.1	99.4
	Dartmouth	117.2	79.1	100.1
	Halifax	118.1	81.7	101.8
	New Glasgow	115.3	79.1	99.1
	Sydney	112.6	79.1	97.6
	Truro	115.3	79.1	99.1
	Yarmouth	115.1	79.1	99.0
ONTARIO				
	Barrie	118.9	98.8	109.9
	Brantford	118.1	103.0	111.3
	Cornwall	118.0	99.1	109.5
	Hamilton	122.6	95.4	110.5
	Kingston	118.9	99.2	110.1
	Kitchener	114.0	92.1	104.2
	London	121.4	91.9	108.2
	North Bay	118.1	96.7	108.5
	Oshawa	117.7	100.0	109.8
	Ottawa	123.0	92.3	109.3
	Owen Sound	119.1	96.8	109.1
	Peterborough	118.1	98.8	109.5
	Sarnia	117.8	103.6	111.4

Location Factors

STATE/ZIP	CITY	MAT.	INST.	TOTAL
ONTARIO (CONT'D)	Sault Ste Marie	112.7	96.3	105.4
	St. Catharines	112.1	93.1	103.6
	Sudbury	112.5	92.1	103.4
	Thunder Bay	113.6	91.9	103.9
	Timmins	118.3	96.7	108.6
	Toronto	123.6	100.7	113.4
	Windsor	112.7	91.6	103.3
PRINCE EDWARD ISLAND				
	Charlottetown	117.6	66.4	94.7
	Summerside	117.1	66.4	94.5
QUEBEC				
	Cap-de-la-Madeleine	114.9	93.0	105.1
	Charlesbourg	114.9	93.0	105.1
	Chicoutimi	113.9	93.2	104.7
	Gatineau	114.6	92.7	104.8
	Granby	114.8	92.7	104.9
	Hull	114.6	92.7	104.8
	Joliette	115.1	93.0	105.2
	Laval	114.3	92.7	104.6
	Montreal	122.0	93.5	109.3
	Quebec	120.1	94.0	108.4
	Rimouski	114.4	93.2	104.9
	Rouyn-Noranda	114.5	92.7	104.7
	Saint Hyacinthe	113.9	92.7	104.4
	Sherbrooke	114.9	92.7	105.0
	Sorel	115.1	93.0	105.2
	St Jerome	114.5	92.7	104.8
	Trois Rivieres	115.1	93.0	105.2
SASKATCHEWAN				
	Moose Jaw	113.4	73.6	95.6
	Prince Albert	112.3	71.6	94.1
	Regina	114.9	87.7	102.7
	Saskatoon	113.4	87.7	101.9
YUKON				
	Whitehorse	113.2	72.4	95.0

R011105-05 Tips for Accurate Estimating

1. Use pre-printed or columnar forms for orderly sequence of dimensions and locations and for recording telephone quotations.

2. Use only the front side of each paper or form except for certain pre-printed summary forms.

3. Be consistent in listing dimensions: For example, length x width x height. This helps in rechecking to ensure that, the total length of partitions is appropriate for the building area.

4. Use printed (rather than measured) dimensions where given.

5. Add up multiple printed dimensions for a single entry where possible.

6. Measure all other dimensions carefully.

7. Use each set of dimensions to calculate multiple related quantities.

8. Convert foot and inch measurements to decimal feet when listing. Memorize decimal equivalents to .01 parts of a foot (1/8″ equals approximately .01′).

9. Do not "round off" quantities until the final summary.

10. Mark drawings with different colors as items are taken off.

11. Keep similar items together, different items separate.

12. Identify location and drawing numbers to aid in future checking for completeness.

13. Measure or list everything on the drawings or mentioned in the specifications.

14. It may be necessary to list items not called for to make the job complete.

15. Be alert for: Notes on plans such as N.T.S. (not to scale); changes in scale throughout the drawings; reduced size drawings; discrepancies between the specifications and the drawings.

16. Develop a consistent pattern of performing an estimate.
   For example:
   a. Start the quantity takeoff at the lower floor and move to the next higher floor.
   b. Proceed from the main section of the building to the wings.
   c. Proceed from south to north or vice versa, clockwise or counterclockwise.
   d. Take off floor plan quantities first, elevations next, then detail drawings.

17. List all gross dimensions that can be either used again for different quantities, or used as a rough check of other quantities for verification (exterior perimeter, gross floor area, individual floor areas, etc.).

18. Utilize design symmetry or repetition (repetitive floors, repetitive wings, symmetrical design around a center line, similar room layouts, etc.). Note: Extreme caution is needed here so as not to omit or duplicate an area.

19. Do not convert units until the final total is obtained. For instance, when estimating concrete work, keep all units to the nearest cubic foot, then summarize and convert to cubic yards.

20. When figuring alternatives, it is best to total all items involved in the basic system, then total all items involved in the alternates. Therefore you work with positive numbers in all cases. When adds and deducts are used, it is often confusing whether to add or subtract a portion of an item; especially on a complicated or involved alternate.

R011105-10 Unit Gross Area Requirements

The figures in the table below indicate typical ranges in square feet as a function of the "occupant" unit. This table is best used in the preliminary design stages to help determine the probable size requirement for the total project.

Building Type	Unit	Gross Area in S.F.		
		1/4	Median	3/4
Apartments	Unit	660	860	1,100
Auditorium & Play Theaters	Seat	18	25	38
Bowling Alleys	Lane		940	
Churches & Synagogues	Seat	20	28	39
Dormitories	Bed	200	230	275
Fraternity & Sorority Houses	Bed	220	315	370
Garages, Parking	Car	325	355	385
Hospitals	Bed	685	850	1,075
Hotels	Rental Unit	475	600	710
Housing for the elderly	Unit	515	635	755
Housing, Public	Unit	700	875	1,030
Ice Skating Rinks	Total	27,000	30,000	36,000
Motels	Rental Unit	360	465	620
Nursing Homes	Bed	290	350	450
Restaurants	Seat	23	29	39
Schools, Elementary	Pupil	65	77	90
Junior High & Middle		85	110	129
Senior High		102	130	145
Vocational		110	135	195
Shooting Ranges	Point		450	
Theaters & Movies	Seat		15	

R011105-20   Floor Area Ratios

Table below lists commonly used gross to net area and net to gross area ratios expressed in % for various building types.

Building Type	Gross to Net Ratio	Net to Gross Ratio	Building Type	Gross to Net Ratio	Net to Gross Ratio
Apartment	156	64	School Buildings (campus type)		
Bank	140	72	Administrative	150	67
Church	142	70	Auditorium	142	70
Courthouse	162	61	Biology	161	62
Department Store	123	81	Chemistry	170	59
Garage	118	85	Classroom	152	66
Hospital	183	55	Dining Hall	138	72
Hotel	158	63	Dormitory	154	65
Laboratory	171	58	Engineering	164	61
Library	132	76	Fraternity	160	63
Office	135	75	Gymnasium	142	70
Restaurant	141	70	Science	167	60
Warehouse	108	93	Service	120	83
			Student Union	172	59

The gross area of a building is the total floor area based on outside dimensions.

The net area of a building is the usable floor area for the function intended and excludes such items as stairways, corridors and mechanical rooms. In the case of a commercial building, it might be considered as the "leasable area."

R011105-30   Occupancy Determinations

Description		S.F. Required per Person		
		BOCA	SBC	UBC
Assembly Areas	Fixed Seats	**	6	7
	Movable Seats		15	15
	Concentrated	7		
	Unconcentrated	15		
	Standing Space	3		
Educational	Unclassified			
	Classrooms	20	40	20
	Shop Areas	50	100	50
Institutional	Unclassified		125	
	In-Patient Areas	240		
	Sleeping Areas	120		
Mercantile	Basement	30	30	20
	Ground Floor	30	30	30
	Upper Floors	60	60	50
Office		100	100	100

BOCA=Building Officials & Code Administrators
SBC=Southern Building Code
UBC=Uniform Building Code

** The occupancy load for assembly area with fixed seats shall be determined by the number of fixed seats installed.

R011110-10   Architectural Fees

Tabulated below are typical percentage fees by project size, for good professional architectural service. Fees may vary from those listed depending upon degree of design difficulty and economic conditions in any particular area.

Rates can be interpolated horizontally and vertically. Various portions of the same project requiring different rates should be adjusted proportionately. For alterations, add 50% to the fee for the first $500,000 of project cost and add 25% to the fee for project cost over $500,000.

Architectural fees tabulated below include Structural, Mechanical and Electrical Engineering Fees. They do not include the fees for special consultants such as kitchen planning, security, acoustical, interior design, etc.

Civil Engineering fees are included in the Architectural fee for project sites requiring minimal design such as city sites. However, separate Civil Engineering fees must be added when utility connections require design, drainage calculations are needed, stepped foundations are required, or provisions are required to protect adjacent wetlands.

Building Types	Total Project Size in Thousands of Dollars						
	100	250	500	1,000	5,000	10,000	50,000
Factories, garages, warehouses, repetitive housing	9.0%	8.0%	7.0%	6.2%	5.3%	4.9%	4.5%
Apartments, banks, schools, libraries, offices, municipal buildings	12.2	12.3	9.2	8.0	7.0	6.6	6.2
Churches, hospitals, homes, laboratories, museums, research	15.0	13.6	12.7	11.9	9.5	8.8	8.0
Memorials, monumental work, decorative furnishings	—	16.0	14.5	13.1	10.0	9.0	8.3

R011110-30   Engineering Fees

Typical **Structural Engineering Fees** based on type of construction and total project size. These fees are included in Architectural Fees.

Type of Construction	Total Project Size (in thousands of dollars)			
	$500	$500-$1,000	$1,000-$5,000	Over $5000
Industrial buildings, factories & warehouses	Technical payroll times 2.0 to 2.5	1.60%	1.25%	1.00%
Hotels, apartments, offices, dormitories, hospitals, public buildings, food stores		2.00%	1.70%	1.20%
Museums, banks, churches and cathedrals		2.00%	1.75%	1.25%
Thin shells, prestressed concrete, earthquake resistive		2.00%	1.75%	1.50%
Parking ramps, auditoriums, stadiums, convention halls, hangars & boiler houses		2.50%	2.00%	1.75%
Special buildings, major alterations, underpinning & future expansion		Add to above 0.5%	Add to above 0.5%	Add to above 0.5%

For complex reinforced concrete or unusually complicated structures, add 20% to 50%.

Typical **Mechanical and Electrical Engineering Fees** are based on the size of the subcontract. The fee structure for both are shown below. These fees are included in Architectural Fees.

Type of Construction	Subcontract Size							
	$25,000	$50,000	$100,000	$225,000	$350,000	$500,000	$750,000	$1,000,000
Simple structures	6.4%	5.7%	4.8%	4.5%	4.4%	4.3%	4.2%	4.1%
Intermediate structures	8.0	7.3	6.5	5.6	5.1	5.0	4.9	4.8
Complex structures	10.1	9.0	9.0	8.0	7.5	7.5	7.0	7.0

For renovations, add 15% to 25% to applicable fee.

R012153-10   Repair and Remodeling

Cost figures are based on new construction utilizing the most cost-effective combination of labor, equipment and material with the work scheduled in proper sequence to allow the various trades to accomplish their work in an efficient manner.

The costs for repair and remodeling work must be modified due to the following factors that may be present in any given repair and remodeling project.

1. Equipment usage curtailment due to the physical limitations of the project, with only hand-operated equipment being used.

2. Increased requirement for shoring and bracing to hold up the building while structural changes are being made and to allow for temporary storage of construction materials on above-grade floors.

3. Material handling becomes more costly due to having to move within the confines of an enclosed building. For multi-story construction, low capacity elevators and stairwells may be the only access to the upper floors.

4. Large amount of cutting and patching and attempting to match the existing construction is required. It is often more economical to remove entire walls rather than create many new door and window openings. This sort of trade-off has to be carefully analyzed.

5. Cost of protection of completed work is increased since the usual sequence of construction usually cannot be accomplished.

6. Economies of scale usually associated with new construction may not be present. If small quantities of components must be custom fabricated due to job requirements, unit costs will naturally increase. Also, if only small work areas are available at a given time, job scheduling between trades becomes difficult and subcontractor quotations may reflect the excessive start-up and shut-down phases of the job.

7. Work may have to be done on other than normal shifts and may have to be done around an existing production facility which has to stay in production during the course of the repair and remodeling.

8. Dust and noise protection of adjoining non-construction areas can involve substantial special protection and alter usual construction methods.

9. Job may be delayed due to unexpected conditions discovered during demolition or removal. These delays ultimately increase construction costs.

10. Piping and ductwork runs may not be as simple as for new construction. Wiring may have to be snaked through walls and floors.

11. Matching "existing construction" may be impossible because materials may no longer be manufactured. Substitutions may be expensive.

12. Weather protection of existing structure requires additional temporary structures to protect building at openings.

13. On small projects, because of local conditions, it may be necessary to pay a tradesman for a minimum of four hours for a task that is completed in one hour.

All of the above areas can contribute to increased costs for a repair and remodeling project. Each of the above factors should be considered in the planning, bidding and construction stage in order to minimize the increased costs associated with repair and remodeling jobs.

R012909-80   Sales Tax by State

State sales tax on materials is tabulated below (5 states have no sales tax). Many states allow local jurisdictions, such as a county or city, to levy additional sales tax.

Some projects may be sales tax exempt, particularly those constructed with public funds.

State	Tax (%)	State	Tax (%)	State	Tax (%)	State	Tax (%)
Alabama	4	Illinois	6.25	Montana	0	Rhode Island	7
Alaska	0	Indiana	7	Nebraska	5.5	South Carolina	6
Arizona	5.6	Iowa	6	Nevada	6.85	South Dakota	4
Arkansas	6	Kansas	5.3	New Hampshire	0	Tennessee	7
California	8.25	Kentucky	6	New Jersey	7	Texas	6.25
Colorado	2.9	Louisiana	4	New Mexico	5	Utah	4.7
Connecticut	6	Maine	5	New York	4	Vermont	6
Delaware	0	Maryland	6	North Carolina	5.75	Virginia	5
District of Columbia	6	Massachusetts	6.25	North Dakota	5	Washington	6.5
Florida	6	Michigan	6	Ohio	5.5	West Virginia	6
Georgia	4	Minnesota	6.875	Oklahoma	4.5	Wisconsin	5
Hawaii	4	Mississippi	7	Oregon	0	Wyoming	4
Idaho	6	Missouri	4.225	Pennsylvania	6	Average	5.04 %

Sales Tax by Province (Canada)

GST - a value-added tax, which the government imposes on most goods and services provided in or imported into Canada. PST - a retail sales tax, which five of the provinces impose on the price of most goods and some

services. QST - a value-added tax, similar to the federal GST, which Quebec imposes. HST - Three provinces have combined their retail sales tax with the federal GST into one harmonized tax.

Province	PST (%)	QST (%)	GST(%)	HST(%)
Alberta	0	0	5	0
British Columbia	0	0	0	12
Manitoba	7	0	5	0
New Brunswick	0	0	0	13
Newfoundland	0	0	0	13
Northwest Territories	0	0	5	0
Nova Scotia	0	0	0	15
Ontario	0	0	0	13
Prince Edward Island	10	0	5	0
Quebec	0	8.5	5	0
Saskatchewan	5	0	5	0
Yukon	0	0	5	0

R012909-85   Unemployment Taxes and Social Security Taxes

State Unemployment Tax rates vary not only from state to state, but also with the experience rating of the contractor. The Federal Unemployment Tax rate is 6.2% of the first $7,000 of wages. This is reduced by a credit of up to 5.4% for timely payment to the state. The minimum Federal Unemployment Tax is 0.8% after all credits.

Social Security (FICA) for 2011 is estimated at time of publication to be 7.65% of wages up to $106,800.

R012909-90   Overtime

One way to improve the completion date of a project or eliminate negative float from a schedule is to compress activity duration times. This can be achieved by increasing the crew size or working overtime with the proposed crew.

To determine the costs of working overtime to compress activity duration times, consider the following examples. Below is an overtime efficiency and cost chart based on a five, six, or seven day week with an eight through twelve hour day. Payroll percentage increases for time and one half and double time are shown for the various working days.

Days per Week	Hours per Day	Production Efficiency					Payroll Cost Factors	
		1st Week	2nd Week	3rd Week	4th Week	Average 4 Weeks	@ 1-1/2 Times	@ 2 Times
5	8	100%	100%	100%	100%	100 %	100 %	100 %
	9	100	100	95	90	96.25	105.6	111.1
	10	100	95	90	85	91.25	110.0	120.0
	11	95	90	75	65	81.25	113.6	127.3
	12	90	85	70	60	76.25	116.7	133.3
6	8	100	100	95	90	96.25	108.3	116.7
	9	100	95	90	85	92.50	113.0	125.9
	10	95	90	85	80	87.50	116.7	133.3
	11	95	85	70	65	78.75	119.7	139.4
	12	90	80	65	60	73.75	122.2	144.4
7	8	100	95	85	75	88.75	114.3	128.6
	9	95	90	80	70	83.75	118.3	136.5
	10	90	85	75	65	78.75	121.4	142.9
	11	85	80	65	60	72.50	124.0	148.1
	12	85	75	60	55	68.75	126.2	152.4

R013113-40 Builder's Risk Insurance

Builder's Risk Insurance is insurance on a building during construction. Premiums are paid by the owner or the contractor. Blasting, collapse and underground insurance would raise total insurance costs above those listed. Floater policy for materials delivered to the job runs $.75 to $1.25 per $100 value. Contractor equipment insurance runs $.50 to $1.50 per $100 value. Insurance for miscellaneous tools to $1,500 value runs from $3.00 to $7.50 per $100 value.

Tabulated below are New England Builder's Risk insurance rates in dollars per $100 value for $1,000 deductible. For $25,000 deductible, rates can be reduced 13% to 34%. On contracts over $1,000,000, rates may be lower than those tabulated. Policies are written annually for the total completed value in place. For "all risk" insurance (excluding flood, earthquake and certain other perils) add $.025 to total rates below.

Coverage	Frame Construction (Class 1)			Brick Construction (Class 4)			Fire Resistive (Class 6)		
	Range		Average	Range		Average	Range		Average
Fire Insurance	$.350 to $.850		$.600	$.158 to $.189		$.174	$.052 to $.080		$.070
Extended Coverage	.115 to .200		.158	.080 to .105		.101	.081 to .105		.100
Vandalism	.012 to .016		.014	.008 to .011		.011	.008 to .011		.010
Total Annual Rate	$.477 to $1.066		$.772	$.246 to $.305		$.286	$.141 to $.196		$.180

R013113-50 General Contractor's Overhead

There are two distinct types of overhead on a construction project: Project Overhead and Main Office Overhead. Project Overhead includes those costs at a construction site not directly associated with the installation of construction materials. Examples of Project Overhead costs include the following:

1. Superintendent
2. Construction office and storage trailers
3. Temporary sanitary facilities
4. Temporary utilities
5. Security fencing
6. Photographs
7. Clean up
8. Performance and payment bonds

The above Project Overhead items are also referred to as General Requirements and therefore are estimated in Division 1. Division 1 is the first division listed in the CSI MasterFormat but it is usually the last division estimated. The sum of the costs in Divisions 1 through 49 is referred to as the sum of the direct costs.

All construction projects also include indirect costs. The primary components of indirect costs are the contractor's Main Office Overhead and profit. The amount of the Main Office Overhead expense varies depending on the following:

1. Owner's compensation
2. Project managers and estimator's wages
3. Clerical support wages
4. Office rent and utilities
5. Corporate legal and accounting costs
6. Advertising
7. Automobile expenses
8. Association dues
9. Travel and entertainment expenses

These costs are usually calculated as a percentage of annual sales volume. This percentage can range from 35% for a small contractor doing less than $500,000 to 5% for a large contractor with sales in excess of $100 million.

R013113-60 Workers' Compensation Insurance Rates by Trade

The table below tabulates the national averages for Workers' Compensation insurance rates by trade and type of building. The average "Insurance Rate" is multiplied by the "% of Building Cost" for each trade. This produces the "Workers' Compensation Cost" by % of total labor cost, to be added for each trade by building type to determine the weighted average Workers' Compensation rate for the building types analyzed.

Trade	Insurance Rate (% Labor Cost) Range		Average	% of Building Cost Office Bldgs.	Schools & Apts.	Mfg.	Workers' Compensation Office Bldgs.	Schools & Apts.	Mfg.
Excavation, Grading, etc.	3.5 % to	18.2%	9.1%	4.8%	4.9%	4.5%	0.44%	0.45%	0.41%
Piles & Foundations	4.8 to	51.8	18.0	7.1	5.2	8.7	1.28	0.94	1.57
Concrete	4.7 to	28.9	13.3	5.0	14.8	3.7	0.67	1.97	0.49
Masonry	4 to	28.5	12.7	6.9	7.5	1.9	0.88	0.95	0.24
Structural Steel	4.8 to	155.9	36.3	10.7	3.9	17.6	3.88	1.42	6.39
Miscellaneous & Ornamental Metals	4 to	21.2	9.9	2.8	4.0	3.6	0.28	0.40	0.36
Carpentry & Millwork	4.8 to	53.2	16.1	3.7	4.0	0.5	0.60	0.64	0.08
Metal or Composition Siding	4.8 to	40	14.5	2.3	0.3	4.3	0.33	0.04	0.62
Roofing	4.8 to	77.1	29.2	2.3	2.6	3.1	0.67	0.76	0.91
Doors & Hardware	2.5 to	40	10.9	0.9	1.4	0.4	0.10	0.15	0.04
Sash & Glazing	4.8 to	34.6	12.9	3.5	4.0	1.0	0.45	0.52	0.13
Lath & Plaster	3.1 to	29.9	11.4	3.3	6.9	0.8	0.38	0.79	0.09
Tile, Marble & Floors	2.8 to	18.9	8.5	2.6	3.0	0.5	0.22	0.26	0.04
Acoustical Ceilings	3.1 to	24.5	8.6	2.4	0.2	0.3	0.21	0.02	0.03
Painting	4.3 to	29.6	11.1	1.5	1.6	1.6	0.17	0.18	0.18
Interior Partitions	4.8 to	53.2	16.1	3.9	4.3	4.4	0.63	0.69	0.71
Miscellaneous Items	2.6 to	109.2	14.7	5.2	3.7	9.7	0.77	0.55	1.43
Elevators	1.9 to	11.9	6.1	2.1	1.1	2.2	0.13	0.07	0.13
Sprinklers	2.2 to	17.2	7.1	0.5	—	2.0	0.04	—	0.14
Plumbing	2.6 to	17.5	7.0	4.9	7.2	5.2	0.34	0.50	0.36
Heat., Vent., Air Conditioning	3.8 to	22.7	8.9	13.5	11.0	12.9	1.20	0.98	1.15
Electrical	2.4 to	11.7	5.8	10.1	8.4	11.1	0.59	0.49	0.64
Total	1.9 % to	155.9%	—	100.0%	100.0%	100.0%	14.26%	12.77%	16.14%
		Overall Weighted Average	14.39%						

Workers' Compensation Insurance Rates by States

The table below lists the weighted average Workers' Compensation base rate for each state with a factor comparing this with the national average of 14.1%.

State	Weighted Average	Factor	State	Weighted Average	Factor	State	Weighted Average	Factor
Alabama	20.0%	142	Kentucky	16.9%	120	North Dakota	11.2%	79
Alaska	16.7	118	Louisiana	28.5	202	Ohio	8.6	61
Arizona	8.9	63	Maine	13.3	94	Oklahoma	13.7	97
Arkansas	11.3	80	Maryland	15.1	107	Oregon	12.4	88
California	28.0	199	Massachusetts	12.7	90	Pennsylvania	16.1	114
Colorado	7.7	55	Michigan	17.7	126	Rhode Island	11.7	83
Connecticut	21.3	151	Minnesota	26.2	186	South Carolina	20.0	142
Delaware	9.6	68	Mississippi	14.9	106	South Dakota	17.3	123
District of Columbia	11.1	79	Missouri	18.4	130	Tennessee	13.5	96
Florida	10.7	76	Montana	14.9	106	Texas	11.3	80
Georgia	27.1	192	Nebraska	19.4	138	Utah	7.8	55
Hawaii	9.0	64	Nevada	10.2	72	Vermont	14.8	105
Idaho	10.1	72	New Hampshire	20.1	143	Virginia	10.3	73
Illinois	22.9	162	New Jersey	12.8	91	Washington	9.6	68
Indiana	5.7	40	New Mexico	17.7	126	West Virginia	13.0	92
Iowa	10.9	77	New York	6.1	43	Wisconsin	15.3	109
Kansas	8.4	60	North Carolina	17.5	124	Wyoming	5.3	38
			Weighted Average for U.S. is	14.4% of payroll = 100%				

Rates in the following table are the base or manual costs per $100 of payroll for Workers' Compensation in each state. Rates are usually applied to straight time wages only and not to premium time wages and bonuses.

The weighted average skilled worker rate for 35 trades is 14.1%. For bidding purposes, apply the full value of Workers' Compensation directly to total labor costs, or if labor is 38%, materials 42% and overhead and profit 20% of total cost, carry 38/80 x 14.1% =6.7% of cost (before overhead and profit) into overhead. Rates vary not only from state to state but also with the experience rating of the contractor.

Rates are the most current available at the time of publication.

R013113-60 Workers' Compensation Insurance Rates by Trade and State (cont.)

State	Carpentry — 3 stories or less 5651	Carpentry — interior cab. work 5437	Carpentry — general 5403	Concrete Work — NOC 5213	Concrete Work — flat (flr, sdwk.) 5221	Electrical Wiring — inside 5190	Excavation — earth NOC 6217	Excavation — rock 6217	Glaziers 5462	Insulation Work 5479	Lathing 5443	Masonry 5022	Painting & Decorating 5474	Pile Driving 6003	Plastering 5480	Plumbing 5183	Roofing 5551	Sheet Metal Work (HVAC) 5538	Steel Erection — door & sash 5102	Steel Erection — inter., ornam. 5102	Steel Erection — structure 5040	Steel Erection — NOC 5057	Tile Work — (interior ceramic) 5348	Waterproofing 9014	Wrecking 5701
AL	25.95	16.64	27.92	13.20	9.70	9.29	11.19	11.19	22.26	17.03	10.11	25.50	17.74	21.90	15.41	8.77	47.76	9.10	9.40	9.40	45.83	21.35	15.04	7.80	45.83
AK	10.17	10.72	14.12	11.68	10.00	6.69	10.27	10.27	33.82	21.22	8.87	19.14	13.19	34.01	17.99	7.07	29.77	8.15	7.49	7.49	22.50	34.08	5.76	5.77	22.50
AZ	9.58	6.05	15.80	7.78	4.09	3.84	4.98	4.98	6.16	11.90	3.88	7.81	7.11	9.50	5.20	4.77	17.80	6.44	7.96	7.96	20.43	11.14	2.90	3.33	20.43
AR	10.25	6.32	13.03	7.90	6.10	5.13	7.72	7.72	8.76	9.71	5.62	8.38	8.83	12.60	11.45	5.58	21.23	5.31	6.19	6.19	27.01	32.27	6.75	3.52	27.01
CA	40.00	40.00	40.00	17.64	17.64	11.71	18.18	18.18	22.67	18.58	17.85	28.46	21.99	20.82	29.93	17.51	69.24	22.67	21.16	21.16	33.84	20.64	10.83	21.99	20.69
CO	7.64	4.80	6.96	6.20	4.69	2.91	5.75	5.75	5.42	9.91	3.93	7.92	5.68	8.44	5.60	4.11	14.33	3.85	5.41	5.41	24.62	9.37	4.39	3.11	9.37
CT	15.80	17.74	21.69	25.96	13.53	7.98	13.75	13.75	19.29	19.38	14.94	26.57	19.57	21.58	16.34	10.58	44.27	13.02	15.01	15.01	52.47	21.56	12.35	6.03	52.47
DE	9.85	9.85	7.39	7.74	6.33	3.39	5.87	5.87	8.38	7.39	8.38	8.64	10.01	11.74	8.38	5.02	17.66	6.50	8.74	8.74	17.91	8.74	5.92	8.64	17.91
DC	9.38	9.50	8.30	11.23	8.01	5.31	7.58	7.58	13.48	6.63	5.70	9.54	5.84	13.94	9.73	7.23	16.66	7.74	13.51	13.51	19.78	13.05	11.01	3.80	19.78
FL	9.80	6.83	10.73	10.69	5.31	4.72	5.78	5.78	8.37	7.60	4.04	8.78	7.50	38.24	12.58	4.75	17.85	6.49	8.06	8.06	20.73	10.19	4.07	4.15	20.73
GA	35.73	21.35	25.92	17.90	15.31	10.45	16.30	16.30	19.14	22.15	14.61	24.16	24.99	26.75	24.86	12.18	63.28	17.45	20.26	20.26	71.30	39.53	13.97	8.92	71.30
HI	7.56	7.43	13.92	8.59	4.90	4.26	5.38	5.38	8.46	6.34	6.06	9.00	7.28	11.84	10.47	4.00	12.79	4.33	6.28	6.28	29.40	10.64	5.80	4.77	29.40
ID	9.90	6.33	10.07	10.29	5.89	3.41	6.44	6.44	8.75	7.70	7.95	6.62	7.75	9.61	13.76	3.96	23.71	6.82	7.17	7.17	27.50	8.29	7.39	4.49	22.50
IL	23.76	14.07	22.20	28.86	11.72	9.02	10.65	10.65	18.69	19.31	15.58	20.36	12.84	27.34	20.79	11.11	35.16	13.05	18.56	18.56	84.61	24.43	16.31	5.51	84.61
IN	11.17	4.02	6.52	4.66	3.27	2.82	4.33	4.33	5.93	6.16	3.07	3.96	4.60	5.13	3.12	2.57	9.84	4.52	3.95	3.95	13.05	5.36	2.93	2.36	13.05
IA	13.13	9.09	11.58	14.33	7.45	4.42	7.61	7.61	8.93	6.14	5.23	8.37	6.58	9.12	8.83	5.86	20.63	4.44	5.40	5.40	33.37	19.27	7.35	3.98	33.37
KS	14.43	6.21	8.57	6.75	5.51	3.76	4.69	4.69	7.85	6.43	3.93	6.66	6.88	8.42	4.99	4.62	13.89	4.59	4.56	4.56	24.54	15.33	5.61	3.86	15.33
KY	12.63	12.30	24.88	12.96	7.51	5.36	11.65	11.65	16.68	13.04	8.83	7.70	11.00	21.98	12.55	5.87	34.41	11.42	10.07	10.07	60.00	17.23	18.88	4.22	60.00
LA	22.02	24.93	53.17	26.43	15.61	9.88	17.49	17.49	20.14	21.43	24.51	28.33	29.58	31.25	22.37	8.64	77.12	21.20	18.98	18.98	51.76	24.21	13.77	13.78	66.41
ME	14.70	10.85	21.69	14.37	9.79	5.55	9.56	9.56	13.86	11.69	6.39	11.50	12.62	13.12	11.21	8.38	21.49	8.22	6.97	6.97	30.84	16.36	5.67	4.87	30.84
MD	16.12	12.09	13.45	17.65	5.24	5.74	7.23	7.23	11.72	13.83	4.53	10.17	7.24	22.67	11.81	6.40	29.04	12.17	8.52	8.52	56.70	24.19	7.39	5.42	24.19
MA	7.50	5.93	11.92	21.45	6.62	3.17	4.17	4.17	8.93	8.94	6.53	11.62	5.01	14.99	4.89	3.88	30.35	5.74	6.76	6.76	44.61	39.97	6.45	2.50	29.46
MI	20.23	9.63	20.23	17.30	8.33	4.58	11.76	11.76	12.90	10.98	13.59	14.95	13.12	51.77	10.05	6.78	29.95	11.08	10.17	10.17	51.77	13.99	9.52	6.17	51.77
MN	21.58	21.93	31.55	10.75	14.50	5.25	11.40	11.40	34.58	20.50	12.03	17.40	14.38	28.68	12.03	9.30	59.58	18.78	11.58	11.58	155.90	9.80	13.50	6.93	9.80
MS	15.14	10.81	14.94	10.42	6.74	5.60	9.95	9.95	10.94	9.28	6.27	14.58	12.36	26.40	13.13	8.38	36.90	7.23	13.97	13.97	37.66	12.49	9.54	4.69	37.66
MO	18.48	13.84	14.93	21.56	11.45	8.02	12.16	12.16	11.41	15.50	10.00	16.97	12.30	21.47	15.98	11.55	34.76	9.09	12.35	12.35	34.33	46.40	12.73	6.91	34.33
MT	12.71	9.67	25.69	13.47	11.92	6.12	13.11	13.11	9.86	28.05	7.55	10.94	7.34	18.44	11.08	8.28	37.59	8.86	8.53	8.53	20.44	10.38	10.20	8.69	10.38
NE	22.48	16.10	22.18	21.90	12.55	8.83	14.33	14.33	13.63	21.55	9.85	18.65	13.40	17.78	15.03	10.45	37.18	11.60	13.93	13.93	45.53	36.08	9.05	6.43	64.50
NV	11.97	7.52	11.36	8.77	7.79	5.12	8.77	8.77	8.23	8.26	4.07	6.55	7.54	9.58	6.66	6.13	14.83	9.39	7.46	7.46	24.81	16.86	4.41	4.95	18.14
NH	19.09	11.63	18.72	25.24	11.74	6.65	11.22	11.22	10.69	13.49	8.48	21.03	25.47	16.39	10.11	9.96	43.20	9.01	12.97	12.97	79.22	23.21	11.47	6.64	79.22
NJ	14.86	9.38	14.81	14.44	9.32	4.32	8.00	8.00	7.69	10.21	10.65	12.88	10.30	16.15	10.65	5.67	38.90	6.72	9.23	9.23	16.02	16.09	7.69	5.87	16.52
NM	19.58	8.17	18.99	13.33	10.11	8.08	8.28	8.28	18.17	11.75	6.68	15.36	11.43	18.20	14.39	8.14	37.47	9.92	12.79	12.79	54.49	47.45	7.66	6.66	54.49
NY	5.97	2.54	5.59	6.71	5.39	2.37	3.49	3.49	15.45	2.98	5.00	6.56	4.25	6.03	3.16	3.16	12.06	4.13	5.04	5.04	10.02	7.90	2.79	2.53	4.13
NC	15.50	11.55	18.63	17.12	8.14	9.96	13.11	13.11	11.87	12.57	9.34	11.05	11.85	16.82	13.63	8.92	33.88	11.87	11.97	11.97	70.40	20.69	8.12	5.41	70.40
ND	10.56	10.56	10.56	5.17	5.17	3.66	4.44	4.44	10.56	10.56	7.94	6.35	6.94	14.58	7.94	5.56	20.76	5.56	14.58	14.58	14.58	19.14	10.56	20.76	9.31
OH	8.30	4.39	6.71	6.58	5.27	3.78	4.99	4.99	7.82	8.20	23.30	7.55	7.86	7.63	5.54	4.36	17.26	4.87	6.38	6.38	12.99	11.61	6.16	5.65	7.74
OK	11.94	8.70	12.16	13.07	6.43	6.00	10.98	10.98	12.76	12.68	8.07	9.29	7.99	15.81	16.64	7.06	21.53	9.05	11.30	11.30	36.18	23.97	6.37	6.21	36.18
OR	16.95	7.57	14.30	10.92	7.82	5.37	9.61	9.61	11.51	11.15	7.46	12.31	11.60	14.60	9.99	5.10	25.58	6.63	7.09	7.09	21.15	21.87	7.63	5.93	21.15
PA	16.22	16.22	13.12	16.61	12.01	6.79	9.95	9.95	12.89	13.12	12.89	14.10	14.78	17.69	12.89	8.58	32.97	8.95	16.30	16.30	24.57	15.11	9.36	14.00	24.57
RI	10.79	8.37	10.96	13.14	10.29	5.15	7.50	7.50	13.33	17.64	6.04	11.62	9.27	27.21	9.50	4.42	14.82	7.75	5.68	5.68	20.92	14.98	5.62	4.72	14.98
SC	21.43	18.13	22.49	16.89	9.59	11.45	11.78	11.78	16.99	18.58	11.06	13.93	15.50	20.61	15.75	12.36	50.08	13.06	13.91	13.91	41.22	32.53	10.50	6.33	41.22
SD	21.47	9.62	21.86	25.23	7.35	5.33	9.54	9.54	15.06	13.87	8.26	11.62	10.45	20.56	11.87	10.92	26.74	12.70	8.86	8.86	59.15	31.01	7.69	5.99	59.15
TN	14.09	10.66	11.66	12.50	6.18	6.09	12.14	12.14	12.52	9.27	8.02	12.10	9.94	24.60	10.80	6.06	25.65	8.80	8.58	8.58	25.35	25.53	8.11	4.08	25.35
TX	9.17	8.39	9.17	9.00	7.06	5.76	7.88	7.88	8.56	11.97	5.74	10.31	8.18	16.38	8.18	5.88	19.42	12.05	7.87	7.87	33.20	11.24	5.53	6.19	8.70
UT	11.11	6.27	8.58	8.13	6.04	2.96	6.70	6.70	7.35	5.44	3.32	7.07	6.19	8.25	4.53	3.79	21.11	5.03	5.21	5.21	14.21	11.11	4.24	3.59	18.87
VT	12.73	12.09	17.18	11.43	8.18	5.46	11.10	11.10	15.78	14.64	7.03	16.92	9.46	13.49	8.82	8.68	25.30	8.45	9.53	9.53	40.51	30.71	9.66	6.41	40.51
VA	10.44	8.34	8.71	9.83	5.37	4.39	7.46	7.46	8.80	7.36	8.95	7.50	8.31	10.07	7.92	4.93	23.55	6.37	6.66	6.66	27.15	20.56	4.98	3.09	27.15
WA	7.58	7.58	7.58	7.65	7.65	2.92	6.45	6.45	12.60	7.04	7.58	9.40	12.79	17.67	9.14	4.28	16.55	4.09	7.25	7.25	7.25	7.61	6.59	16.55	7.25
WV	14.41	10.17	14.80	12.02	6.15	6.42	8.69	8.69	11.48	9.73	8.06	10.16	10.75	17.16	10.53	6.79	24.97	8.18	7.97	7.97	34.22	7.91	8.31	3.83	17.17
WI	11.71	9.86	17.41	11.55	8.51	6.39	8.17	8.17	11.44	13.60	5.38	14.52	12.34	24.28	10.17	6.03	33.37	8.83	10.63	10.63	20.55	48.12	17.76	4.94	20.55
WY	4.77	4.77	4.77	4.77	4.77	4.77	4.77	4.77	4.77	4.77	4.77	4.77	4.77	4.77	4.77	4.77	4.77	4.77	4.77	4.77	4.77	4.77	4.77	4.77	4.77
AVG.	14.52	10.93	16.07	13.33	8.35	5.81	9.10	9.10	12.89	12.30	8.59	12.66	11.07	18.00	11.43	7.04	29.20	8.94	9.86	9.86	36.30	20.12	8.45	6.43	30.85

R013113-60 Workers' Compensation (cont.) (Canada in Canadian dollars)

Province		Alberta	British Columbia	Manitoba	Ontario	New Brunswick	Newfndld. & Labrador	Northwest Territories	Nova Scotia	Prince Edward Island	Quebec	Saskat-chewan	Yukon
Carpentry—3 stories or less	Rate	6.42	3.64	4.23	4.35	5.20	10.67	3.54	6.91	5.61	13.18	4.07	9.22
	Code	42143	721028	40102	723	238130	4226	4-41	4226	401	80110	1226	202
Carpentry—interior cab. work	Rate	2.43	2.62	4.23	4.35	4.64	4.29	3.54	4.73	3.44	13.18	2.41	9.22
	Code	42133	721021	40102	723	238350	4279	4-41	4274	402	80110	B1127	202
CARPENTRY—general	Rate	6.42	3.64	4.23	4.35	5.20	4.29	3.54	6.91	5.61	13.18	4.07	9.22
	Code	42143	721028	40102	723	238130	4299	4-41	4226	401	80110	12.02	202
CONCRETE WORK—NOC	Rate	4.11	3.87	6.79	16.50	5.20	10.67	3.54	4.85	5.61	14.82	4.80	7.12
	Code	42104	721010	40110	748	238110	4224	4-41	4224	401	80100	B1314	203
CONCRETE WORK—flat (flr. sidewalk)	Rate	4.11	3.87	6.79	16.50	5.20	10.67	3.54	4.85	5.61	14.82	4.80	7.12
	Code	42104	721010	40110	748	238110	4224	4-41	4224	401	80100	B1314	203
ELECTRICAL Wiring—inside	Rate	1.97	1.41	1.96	3.25	2.54	2.76	2.75	2.62	3.44	5.25	2.41	5.51
	Code	42124	721019	40203	704	238210	4261	4-46	4261	402	80170	B1105	206
EXCAVATION—earth NOC	Rate	2.36	3.25	4.52	4.68	5.20	3.67	3.79	2.85	2.82	8.43	2.42	5.51
	Code	40604	721031	40706	711	238190	4214	4-43	4214	404	80030	R1106	207
EXCAVATION—rock	Rate	2.36	3.25	4.52	4.68	5.20	3.67	3.79	2.85	2.82	8.43	2.42	5.51
	Code	40604	721031	40706	711	238190	4214	4-43	4214	404	80030	R1106	207
GLAZIERS	Rate	2.75	3.05	4.23	9.25	5.20	5.40	3.54	5.76	3.44	12.63	4.80	5.51
	Code	42121	715020	40109	751	238150	4233	4-41	4233	402	80150	B1304	212
INSULATION WORK	Rate	2.04	6.91	4.23	9.25	4.64	5.40	3.54	5.76	5.61	13.18	4.07	9.22
	Code	42184	721029	40102	751	238310	4234	4-41	4234	401	80110	B1207	202
LATHING	Rate	4.15	6.27	4.23	4.35	4.64	4.29	3.54	4.73	3.44	13.18	4.80	9.22
	Code	42135	721042	40102	723	238390	4279	4-41	4271	402	80110	B1316	202
MASONRY	Rate	4.11	3.50	4.23	11.15	5.20	5.40	3.54	5.76	5.61	14.82	4.80	9.22
	Code	42102	721037	40102	741	238140	4231	4-41	4231	401	80100	B1318	202
PAINTING & DECORATING	Rate	3.39	3.36	3.49	6.75	4.64	4.29	3.54	4.73	3.44	13.18	4.07	9.22
	Code	42111	721041	40105	719	238320	4275	4-41	4275	402	80110	B1201	202
PILE DRIVING	Rate	4.04	3.82	4.52	6.34	5.20	5.67	3.79	5.19	5.61	8.43	4.80	9.22
	Code	42159	722004	40706	732	238190	4129	4-43	4221	401	80030	B1310	202
PLASTERING	Rate	4.15	6.27	4.46	6.75	4.64	4.29	3.54	4.73	3.44	13.18	4.07	9.22
	Code	42135	721042	40108	719	238390	4271	4-41	4271	402	80110	B1221	202
PLUMBING	Rate	1.97	1.90	3.09	3.98	2.54	2.77	2.75	2.76	3.44	6.24	2.41	5.51
	Code	42122	721043	40204	707	238220	4241	4-46	4241	402	80160	B1101	214
ROOFING	Rate	6.59	6.18	7.23	13.30	5.20	10.67	3.54	7.45	5.61	19.78	4.80	9.22
	Code	42118	721036	40403	728	238160	4236	4-41	4236	401	80130	B1320	202
SHEET METAL WORK (HVAC)	Rate	1.97	1.90	7.23	3.98	2.54	2.77	2.75	2.76	3.44	6.24	2.41	5.51
	Code	42117	721043	40402	707	238220	4244	4-46	4244	402	80160	B1107	208
STEEL ERECTION—door & sash	Rate	2.04	12.66	10.32	16.50	5.20	10.67	3.54	5.76	5.61	22.38	4.80	9.22
	Code	42106	722005	40502	748	238120	4227	4-41	4227	401	80080	B1322	202
STEEL ERECTION—inter., ornam.	Rate	2.04	12.66	10.32	16.50	5.20	10.67	3.54	5.76	5.61	22.38	4.80	9.22
	Code	42106	722005	40502	748	238120	4227	4-41	4227	401	80080	B1322	202
STEEL ERECTION—structure	Rate	2.04	12.66	10.32	16.50	5.20	10.67	3.54	5.76	5.61	22.38	4.80	9.22
	Code	42106	722005	40502	748	238120	4227	4-41	4227	401	80080	B1322	202
STEEL ERECTION—NOC	Rate	2.04	12.66	10.32	16.50	5.20	10.67	3.54	5.76	5.61	22.38	4.80	9.22
	Code	42106	722005	40502	748	238120	4227	4-41	4227	401	80080	B1322	202
TILE WORK—inter. (ceramic)	Rate	3.24	3.62	2.00	6.75	4.64	4.29	3.54	4.73	3.44	13.18	4.80	9.22
	Code	42113	721054	40103	719	238340	4276	4-41	4276	402	80110	B1301	202
WATERPROOFING	Rate	3.39	3.35	4.23	4.35	5.20	4.29	3.54	5.76	3.44	19.78	4.07	9.22
	Code	42139	721016	40102	723	238190	4299	4-41	4239	402	80130	B1217	202
WRECKING	Rate	2.36	4.44	7.34	16.50	5.20	3.67	3.79	2.85	5.61	13.18	2.42	9.22
	Code	40604	721005	40106	748	238190	4211	4-43	4211	401	80110	R1125	202

R013113-80   Performance Bond

This table shows the cost of a Performance Bond for a construction job scheduled to be completed in 12 months. Add 1% of the premium cost per month for jobs requiring more than 12 months to complete. The rates are "standard" rates offered to contractors that the bonding company considers financially sound and capable of doing the work. Preferred rates are offered by some bonding companies based upon financial strength of the contractor. Actual rates vary from contractor to contractor and from bonding company to bonding company. Contractors should prequalify through a bonding agency before submitting a bid on a contract that requires a bond.

Contract Amount		Building Construction Class B Projects			Highways & Bridges					
					Class A New Construction			Class A-1 Highway Resurfacing		
First	$ 100,000 bid	\$25.00 per M			\$15.00 per M			\$9.40 per M		
Next	400,000 bid	$ 2,500	plus \$15.00	per M	$ 1,500	plus \$10.00	per M	$ 940	plus \$7.20	per M
Next	2,000,000 bid	8,500	plus 10.00	per M	5,500	plus 7.00	per M	3,820	plus 5.00	per M
Next	2,500,000 bid	28,500	plus 7.50	per M	19,500	plus 5.50	per M	15,820	plus 4.50	per M
Next	2,500,000 bid	47,250	plus 7.00	per M	33,250	plus 5.00	per M	28,320	plus 4.50	per M
Over	7,500,000 bid	64,750	plus 6.00	per M	45,750	plus 4.50	per M	39,570	plus 4.00	per M

R015423-10   Steel Tubular Scaffolding

On new construction, tubular scaffolding is efficient up to 60' high or five stories. Above this it is usually better to use a hung scaffolding if construction permits. Swing scaffolding operations may interfere with tenants. In this case, the tubular is more practical at all heights.

In repairing or cleaning the front of an existing building the cost of tubular scaffolding per S.F. of building front increases as the height increases above the first tier. The first tier cost is relatively high due to leveling and alignment.

The minimum efficient crew for erecting and dismantling is three workers. They can set up and remove 18 frame sections per day up to 5 stories high. For 6 to 12 stories high, a crew of four is most efficient. Use two or more on top and two on the bottom for handing up or hoisting. They can also set up and remove 18 frame sections per day. At 7' horizontal spacing, this will run about 800 S.F. per day of erecting and dismantling. Time for placing and removing planks must be added to the above. A crew of three can place and remove 72 planks per day up to 5 stories. For over 5 stories, a crew of four can place and remove 80 planks per day.

The table below shows the number of pieces required to erect tubular steel scaffolding for 1000 S.F. of building frontage. This area is made up of a scaffolding system that is 12 frames (11 bays) long by 2 frames high.

For jobs under twenty-five frames, add 50% to rental cost. Rental rates will be lower for jobs over three months duration. Large quantities for long periods can reduce rental rates by 20%.

Description of Component	Number of Pieces for 1000 S.F. of Building Front	Unit
5' Wide Standard Frame, 6'-4" High	24	Ea.
Leveling Jack & Plate	24	
Cross Brace	44	
Side Arm Bracket, 21"	12	
Guardrail Post	12	
Guardrail, 7' section	22	
Stairway Section	2	
Stairway Starter Bar	1	
Stairway Inside Handrail	2	
Stairway Outside Handrail	2	
Walk-Thru Frame Guardrail	2	

Scaffolding is often used as falsework over 15' high during construction of cast-in-place concrete beams and slabs. Two foot wide scaffolding is generally used for heavy beam construction. The span between frames depends upon the load to be carried with a maximum span of 5'.

Heavy duty shoring frames with a capacity of 10,000#/leg can be spaced up to 10' O.C. depending upon form support design and loading.

Scaffolding used as horizontal shoring requires less than half the material required with conventional shoring.

On new construction, erection is done by carpenters.

Rolling towers supporting horizontal shores can reduce labor and speed the job. For maintenance work, catwalks with spans up to 70' can be supported by the rolling towers.

R015433-10   Contractor Equipment

Rental Rates shown elsewhere in the book pertain to late model high quality machines in excellent working condition, rented from equipment dealers. Rental rates from contractors may be substantially lower than the rental rates from equipment dealers depending upon economic conditions; for older, less productive machines, reduce rates by a maximum of 15%. Any overtime must be added to the base rates. For shift work, rates are lower. Usual rule of thumb is 150% of one shift rate for two shifts; 200% for three shifts.

For periods of less than one week, operated equipment is usually more economical to rent than renting bare equipment and hiring an operator.

Costs to move equipment to a job site (mobilization) or from a job site (demobilization) are not included in rental rates, nor in any Equipment costs on any Unit Price line items or crew listings. These costs can be found elsewhere. If a piece of equipment is already at a job site, it is not appropriate to utilize mob/demob costs in an estimate again.

Rental rates vary throughout the country with larger cities generally having lower rates. Lease plans for new equipment are available for periods in excess of six months with a percentage of payments applying toward purchase.

Rental rates can also be treated as reimbursement costs for contractor-owned equipment. Owned equipment costs include depreciation, loan payments, interest, taxes, insurance, storage, and major repairs.

Monthly rental rates vary from 2% to 5% of the cost of the equipment depending on the anticipated life of the equipment and its wearing parts. Weekly rates are about 1/3 the monthly rates and daily rental rates about 1/3 the weekly rate.

The hourly operating costs for each piece of equipment include costs to the user such as fuel, oil, lubrication, normal expendables for the equipment, and a percentage of mechanic's wages chargeable to maintenance. The hourly operating costs listed do not include the operator's wages.

The daily cost for equipment used in the standard crews is figured by dividing the weekly rate by five, then adding eight times the hourly operating cost to give the total daily equipment cost, not including the operator. This figure is in the right hand column of the Equipment listings under Equipment Cost/Day.

Pile Driving rates shown for pile hammer and extractor do not include leads, crane, boiler or compressor. Vibratory pile driving requires an added field specialist during set-up and pile driving operation for the electric model. The hydraulic model requires a field specialist for set-up only. Up to 125 reuses of sheet piling are possible using vibratory drivers. For normal conditions, crane capacity for hammer type and size are as follows.

Crane Capacity	Hammer Type and Size		
	Air or Steam	Diesel	Vibratory
25 ton	to 8,750 ft.-lb.		70 H.P.
40 ton	15,000 ft.-lb.	to 32,000 ft.-lb.	170 H.P.
60 ton	25,000 ft.-lb.		300 H.P.
100 ton		112,000 ft.-lb.	

Cranes should be specified for the job by size, building and site characteristics, availability, performance characteristics, and duration of time required.

Backhoes & Shovels rent for about the same as equivalent size cranes but maintenance and operating expense is higher. Crane operators rate must be adjusted for high boom heights. Average adjustments: for 150' boom add 2% per hour; over 185', add 4% per hour; over 210', add 6% per hour; over 250', add 8% per hour and over 295', add 12% per hour.

Tower Cranes of the climbing or static type have jibs from 50' to 200' and capacities at maximum reach range from 4,000 to 14,000 pounds. Lifting capacities increase up to maximum load as the hook radius decreases.

Typical rental rates, based on purchase price are about 2% to 3% per month.

Erection and dismantling runs between 500 and 2000 labor hours. Climbing operation takes 10 labor hours per 20' climb. Crane dead time is about 5 hours per 40' climb. If crane is bolted to side of the building add cost of ties and extra mast sections. Climbing cranes have from 80' to 180' of mast while static cranes have 80' to 800' of mast.

Truck Cranes can be converted to tower cranes by using tower attachments. Mast heights over 400' have been used.

A single 100' high material **Hoist and Tower** can be erected and dismantled in about 400 labor hours; a double 100' high hoist and tower in about 600 labor hours. Erection times for additional heights are 3 and 4 labor hours

per vertical foot respectively up to 150', and 4 to 5 labor hours per vertical foot over 150' high. A 40' high portable Buck hoist takes about 160 labor hours to erect and dismantle. Additional heights take 2 labor hours per vertical foot to 80' and 3 labor hours per vertical foot for the next 100'. Most material hoists do not meet local code requirements for carrying personnel.

A 150' high **Personnel Hoist** requires about 500 to 800 labor hours to erect and dismantle. Budget erection time at 5 labor hours per vertical foot for all trades. Local code requirements or labor scarcity requiring overtime can add up to 50% to any of the above erection costs.

Earthmoving Equipment: The selection of earthmoving equipment depends upon the type and quantity of material, moisture content, haul distance, haul road, time available, and equipment available. Short haul cut and fill operations may require dozers only, while another operation may require excavators, a fleet of trucks, and spreading and compaction equipment. Stockpiled material and granular material are easily excavated with front end loaders. Scrapers are most economically used with hauls between 300' and 1-1/2 miles if adequate haul roads can be maintained. Shovels are often used for blasted rock and any material where a vertical face of 8' or more can be excavated. Special conditions may dictate the use of draglines, clamshells, or backhoes. Spreading and compaction equipment must be matched to the soil characteristics, the compaction required and the rate the fill is being supplied.

R024119-10   Demolition Defined

Whole Building Demolition - Demolition of the whole building with no concern for any particular building element, component, or material type being demolished. This type of demolition is accomplished with large pieces of construction equipment that break up the structure, load it into trucks and haul it to a disposal site, but disposal or dump fees are not included. Demolition of below-grade foundation elements, such as footings, foundation walls, grade beams, slabs on grade, etc., is not included. Certain mechanical equipment containing flammable liquids or ozone-depleting refrigerants, electric lighting elements, communication equipment components, and other building elements may contain hazardous waste, and must be removed, either selectively or carefully, as hazardous waste before the building can be demolished.

Foundation Demolition - Demolition of below-grade foundation footings, foundation walls, grade beams, and slabs on grade. This type of demolition is accomplished by hand or pneumatic hand tools, and does not include saw cutting, or handling, loading, hauling, or disposal of the debris.

Gutting - Removal of building interior finishes and electrical/mechanical systems down to the load-bearing and sub-floor elements of the rough building frame, with no concern for any particular building element, component, or material type being demolished. This type of demolition is accomplished by hand or pneumatic hand tools, and includes loading into trucks, but not hauling, disposal or dump fees, scaffolding, or shoring. Certain mechanical equipment containing flammable liquids or ozone-depleting refrigerants, electric lighting elements, communication equipment components, and other building elements may contain hazardous waste, and must be removed, either selectively or carefully, as hazardous waste, before the building is gutted.

Selective Demolition - Demolition of a selected building element, component, or finish, with some concern for surrounding or adjacent elements, components, or finishes (see the first Subdivision (s) at the beginning of appropriate Divisions). This type of demolition is accomplished by hand or pneumatic hand tools, and does not include handling, loading,

storing, hauling, or disposal of the debris, scaffolding, or shoring. "Gutting" methods may be used in order to save time, but damage that is caused to surrounding or adjacent elements, components, or finishes may have to be repaired at a later time.

Careful Removal - Removal of a piece of service equipment, building element or component, or material type, with great concern for both the removed item and surrounding or adjacent elements, components or finishes. The purpose of careful removal may be to protect the removed item for later re-use, preserve a higher salvage value of the removed item, or replace an item while taking care to protect surrounding or adjacent elements, components, connections, or finishes from cosmetic and/or structural damage. An approximation of the time required to perform this type of removal is 1/3 to 1/2 the time it would take to install a new item of like kind (see Reference Number R220105-10). This type of removal is accomplished by hand or pneumatic hand tools, and does not include loading, hauling, or storing the removed item, scaffolding, shoring, or lifting equipment.

Cutout Demolition - Demolition of a small quantity of floor, wall, roof, or other assembly, with concern for the appearance and structural integrity of the surrounding materials. This type of demolition is accomplished by hand or pneumatic hand tools, and does not include saw cutting, handling, loading, hauling, or disposal of debris, scaffolding, or shoring.

Rubbish Handling - Work activities that involve handling, loading or hauling of debris. Generally, the cost of rubbish handling must be added to the cost of all types of demolition, with the exception of whole building demolition.

Minor Site Demolition - Demolition of site elements outside the footprint of a building. This type of demolition is accomplished by hand or pneumatic hand tools, or with larger pieces of construction equipment, and may include loading a removed item onto a truck (check the Crew for equipment used). It does not include saw cutting, hauling or disposal of debris, and, sometimes, handling or loading.

R024119-20   Dumpsters

Dumpster rental costs on construction sites are presented in two ways.

The cost per week rental includes the delivery of the dumpster; its pulling or emptying once per week, and its final removal. The assumption is made that the dumpster contractor could choose to empty a dumpster by simply bringing in an empty unit and removing the full one. These costs also include the disposal of the materials in the dumpster.

The Alternate Pricing can be used when actual planned conditions are not approximated by the weekly numbers. For example, these lines can be used when a dumpster is needed for 4 weeks and will need to be emptied 2 or 3 times per week. Conversely the Alternate Pricing lines can be used when a dumpster will be rented for several weeks or months but needs to be emptied only a few times over this period.

Existing Conditions   R0265 Underground Storage Tank Removal

R026510-20   Underground Storage Tank Removal

Underground Storage Tank Removal can be divided into two categories: Non-Leaking and Leaking. Prior to removing an underground storage tank, tests should be made, with the proper authorities present, to determine whether a tank has been leaking or the surrounding soil has been contaminated.

To safely remove Liquid Underground Storage Tanks:
1. Excavate to the top of the tank.
2. Disconnect all piping.
3. Open all tank vents and access ports.

4. Remove all liquids and/or sludge.
5. Purge the tank with an inert gas.
6. Provide access to the inside of the tank and clean out the interior using proper personal protective equipment (PPE).
7. Excavate soil surrounding the tank using proper PPE for on-site personnel.
8. Pull and properly dispose of the tank.
9. Clean up the site of all contaminated material.
10. Install new tanks or close the excavation.

Existing Conditions — R0282 Asbestos Remediation

R028213-20 Asbestos Removal Process

Asbestos removal is accomplished by a specialty contractor who understands the federal and state regulations regarding the handling and disposal of the material. The process of asbestos removal is divided into many individual steps. An accurate estimate can be calculated only after all the steps have been priced.

The steps are generally as follows:

1. Obtain an asbestos abatement plan from an industrial hygienist.
2. Monitor the air quality in and around the removal area and along the path of travel between the removal area and transport area. This establishes the background contamination.
3. Construct a two part decontamination chamber at entrance to removal area.
4. Install a HEPA filter to create a negative pressure in the removal area.
5. Install wall, floor and ceiling protection as required by the plan, usually 2 layers of fireproof 6 mil polyethylene.
6. Industrial hygienist visually inspects work area to verify compliance with plan.
7. Provide temporary supports for conduit and piping affected by the removal process.
8. Proceed with asbestos removal and bagging process. Monitor air quality as described in Step #2. Discontinue operations when contaminate levels exceed applicable standards.
9. Document the legal disposal of materials in accordance with EPA standards.
10. Thoroughly clean removal area including all ledges, crevices and surfaces.
11. Post abatement inspection by industrial hygienist to verify plan compliance.
12. Provide a certificate from a licensed industrial hygienist attesting that contaminate levels are within acceptable standards before returning area to regular use.

Thermal & Moist. Protec. — R0784 Firestopping

R078413-30 Firestopping

Firestopping is the sealing of structural, mechanical, electrical and other penetrations through fire-rated assemblies. The basic components of firestop systems are safing insulation and firestop sealant on both sides of wall penetrations and the top side of floor penetrations.

Pipe penetrations are assumed to be through concrete, grout, or joint compound and can be sleeved or unsleeved. Costs for the penetrations and sleeves are not included. An annular space of 1″ is assumed. Escutcheons are not included.

Metallic pipe is assumed to be copper, aluminum, cast iron or similar metallic material. Insulated metallic pipe is assumed to be covered with a thermal insulating jacket of varying thickness and materials.

Non-metallic pipe is assumed to be PVC, CPVC, FR Polypropylene or similar plastic piping material. Intumescent firestop sealant or wrap strips are included. Collars on both sides of wall penetrations and a sheet metal plate on the underside of floor penetrations are included.

Ductwork is assumed to be sheet metal, stainless steel or similar metallic material. Duct penetrations are assumed to be through concrete, grout or joint compound. Costs for penetrations and sleeves are not included. An annular space of 1/2″ is assumed.

Multi-trade openings include costs for sheet metal forms, firestop mortar, wrap strips, collars and sealants as necessary.

Structural penetrations joints are assumed to be 1/2″ or less. CMU walls are assumed to be within 1-1/2″ of metal deck. Drywall walls are assumed to be tight to the underside of metal decking.

Metal panel, glass or curtain wall systems include a spandrel area of 5′ filled with mineral wool foil-faced insulation. Fasteners and stiffeners are included.

Finishes — R0991 Painting

R099100-10 Painting Estimating Techniques

Proper estimating methodology is needed to obtain an accurate painting estimate. There is no known reliable shortcut or square foot method. The following steps should be followed:

- List all surfaces to be painted, with an accurate quantity (area) of each. Items having similar surface condition, finish, application method and accessibility may be grouped together.
- List all the tasks required for each surface to be painted, including surface preparation, masking, and protection of adjacent surfaces. Surface preparation may include minor repairs, washing, sanding and puttying.
- Select the proper Means line for each task. Review and consider all adjustments to labor and materials for type of paint and location of work. Apply the height adjustment carefully. For instance, when applying the adjustment for work over 8' high to a wall that is 12' high, apply the adjustment only to the area between 8' and 12' high, and not to the entire wall.

When applying more than one percent (%) adjustment, apply each to the base cost of the data, rather than applying one percentage adjustment on top of the other.

When estimating the cost of painting walls and ceilings remember to add the brushwork for all cut-ins at inside corners and around windows and doors as a LF measure. One linear foot of cut-in with brush equals one square foot of painting.

All items for spray painting include the labor for roll-back.

Deduct for openings greater than 100 SF, or openings that extend from floor to ceiling and are greater than 5' wide. Do not deduct small openings.

The cost of brushes, rollers, ladders and spray equipment are considered to be part of a painting contractor's overhead, and should not be added to the estimate. The cost of rented equipment such as scaffolding and swing staging should be added to the estimate.

R220105-10 Demolition (Selective vs. Removal for Replacement)

Demolition can be divided into two basic categories.

One type of demolition involves the removal of material with no concern for its replacement. The labor-hours to estimate this work are found under "Selective Demolition" in the Fire Protection, Plumbing and HVAC Divisions. It is selective in that individual items or all the material installed as a system or trade grouping such as plumbing or heating systems are removed. This may be accomplished by the easiest way possible, such as sawing, torch cutting, or sledge hammer as well as simple unbolting.

The second type of demolition is the removal of some item for repair or replacement. This removal may involve careful draining, opening of unions,

disconnecting and tagging of electrical connections, capping of pipes/ducts to prevent entry of debris or leakage of the material contained as well as transport of the item away from its in-place location to a truck/dumpster. An approximation of the time required to accomplish this type of demolition is to use half of the time indicated as necessary to install a new unit. For example; installation of a new pump might be listed as requiring 6 labor-hours so if we had to estimate the removal of the old pump we would allow an additional 3 hours for a total of 9 hours. That is, the complete replacement of a defective pump with a new pump would be estimated to take 9 labor-hours.

Plumbing | **R2205 Common Work Results for Plumbing**

R220523-80 Valve Materials

VALVE MATERIALS

Bronze:
Bronze is one of the oldest materials used to make valves. It is most commonly used in hot and cold water systems and other non-corrosive services. It is often used as a seating surface in larger iron body valves to ensure tight closure.

Carbon Steel:
Carbon steel is a high strength material. Therefore, valves made from this metal are used in higher pressure services, such as steam lines up to 600 psi at 850° F. Many steel valves are available with butt-weld ends for economy and are generally used in high pressure steam service as well as other higher pressure non-corrosive services.

Forged Steel:
Valves from tough carbon steel are used in service up to 2000 psi and temperatures up to 1000° F in Gate, Globe and Check valves.

Iron:
Valves are normally used in medium to large pipe lines to control non-corrosive fluid and gases, where pressures do not exceed 250 psi at 450° F or 500 psi cold water, oil or gas.

Stainless Steel:
Developed steel alloys can be used in over 90% corrosive services.

Plastic PVC:
This is used in a great variety of valves generally in high corrosive service with lower temperatures and pressures.

VALVE SERVICE PRESSURES

Pressure ratings on valves provide an indication of the safe operating pressure for a valve at some elevated temperature. This temperature is dependent upon the materials used and the fabrication of the valve. When specific data is not available, a good "rule-of-thumb" to follow is the temperature of saturated steam on the primary rating indicated on the valve body. Example: The valve has the number 150S printed on the side indicating 150 psi and hence, a maximum operating temperature of 367° F (temperature of saturated steam and 150 psi).

DEFINITIONS
1. "WOG" – Water, oil, gas (cold working pressures).
2. "SWP" – Steam working pressure.
3. 100% area (full port) – means the area through the valve is equal to or greater than the area of standard pipe.
4. "Standard Opening" – means that the area through the valve is less than the area of standard pipe and therefore these valves should be used only where restriction of flow is unimportant.
5. "Round Port" – means the valve has a full round opening through the plug and body, of the same size and area as standard pipe.
6. "Rectangular Port" – valves have rectangular shaped ports through the plug body. The area of the port is either equal to 100% of the area of standard pipe, or restricted (standard opening). In either case it is clearly marked.
7. "ANSI" – American National Standards Institute.

R220523-90 Valve Selection Considerations

INTRODUCTION: In any piping application, valve performance is critical. Valves should be selected to give the best performance at the lowest cost.

The following is a list of performance characteristics generally expected of valves.
1. Stopping flow or starting it.
2. Throttling flow (Modulation).
3. Flow direction changing.
4. Checking backflow (Permitting flow in only one direction).
5. Relieving or regulating pressure.

In order to properly select the right valve, some facts must be determined.

A. What liquid or gas will flow through the valve?
B. Does the fluid contain suspended particles?
C. Does the fluid remain in liquid form at all times?

D. Which metals does fluid corrode?
E. What are the pressure and temperature limits? (As temperature and pressure rise, so will the price of the valve.)
F. Is there constant line pressure?
G. Is the valve merely an on-off valve?
H. Will checking of backflow be required?
I. Will the valve operate frequently or infrequently?

Valves are classified by design type into such classifications as Gate, Globe, Angle, Check, Ball, Butterfly and Plug. They are also classified by end connection, stem, pressure restrictions and material such as bronze, cast iron, etc. Each valve has a specific use. A quality valve used correctly will provide a lifetime of trouble-free service, but a high quality valve installed in the wrong service may require frequent attention.

STEM TYPES
(OS & Y)—Rising Stem-Outside Screw and Yoke

Offers a visual indication of whether the valve is open or closed. Recommended where high temperatures, corrosives, and solids in the line might cause damage to inside-valve stem threads. The stem threads are engaged by the yoke bushing so the stem rises through the hand wheel as it is turned.

(R.S.)—Rising Stem-Inside Screw

Adequate clearance for operation must be provided because both the hand wheel and the stem rise.
The valve wedge position is indicated by the position of the stem and hand wheel.

(N.R.S.)—Non-Rising Stem-Inside Screw

A minimum clearance is required for operating this type of valve. Excessive wear or damage to stem threads inside the valve may be caused by heat, corrosion, and solids. Because the hand wheel and stem do not rise, wedge position cannot be visually determined.

VALVE TYPES
Gate Valves

Provide full flow, minute pressure drop, minimum turbulence and minimum fluid trapped in the line.
They are normally used where operation is infrequent.

Globe Valves

Globe valves are designed for throttling and/or frequent operation with positive shut-off. Particular attention must be paid to the several types of seating materials available to avoid unnecessary wear. The seats must be compatible with the fluid in service and may be composition or metal. The configuration of the Globe valve opening causes turbulence which results in increased resistance. Most bronze Globe valves are rising stem-inside screw, but they are also available on O.S. & Y.

Angle Valves

The fundamental difference between the Angle valve and the Globe valve is the fluid flow through the Angle valve. It makes a 90° turn and offers less resistance to flow than the Globe valve while replacing an elbow. An Angle valve thus reduces the number of joints and installation time.

Check Valves

Check valves are designed to prevent backflow by automatically seating when the direction of fluid is reversed.

Swing Check valves are generally installed with Gate-valves, as they provide comparable full flow. Usually recommended for lines where flow velocities are low and should not be used on lines with pulsating flow. Recommended for horizontal installation, or in vertical lines only where flow is upward.

Lift Check Valves

These are commonly used with Globe and Angle valves since they have similar diaphragm seating arrangements and are recommended for preventing backflow of steam, air, gas and water, and on vapor lines with high flow velocities. For horizontal lines, horizontal lift checks should be used and vertical lift checks for vertical lines.

Ball Valves

Ball valves are light and easily installed, yet because of modern elastomeric seats, provide tight closure. Flow is controlled by rotating up to 90° a drilled ball which fits tightly against resilient seals. This ball seats with flow in either direction, and valve handle indicates the degree of opening. Recommended for frequent operation readily adaptable to automation, ideal for installation where space is limited.

Butterfly Valves

Butterfly valves provide bubble-tight closure with excellent throttling characteristics. They can be used for full-open, closed and for throttling applications.

The Butterfly valve consists of a disc within the valve body which is controlled by a shaft. In its closed position, the valve disc seals against a resilient seat. The disc position throughout the full 90° rotation is visually indicated by the position of the operator.

A Butterfly valve is only a fraction of the weight of a Gate valve and requires no gaskets between flanges in most cases. Recommended for frequent operation and adaptable to automation where space is limited.

Wafer and Lug type bodies when installed between two pipe flanges, can be easily removed from the line. The pressure of the bolted flanges holds the valve in place.
Locating lugs makes installation easier.

Plug Valves

Lubricated plug valves, because of the wide range of service to which they are adapted, may be classified as all purpose valves. They can be safely used at all pressure and vacuums, and at all temperatures up to the limits of available lubricants. They are the most satisfactory valves for the handling of gritty suspensions and many other destructive, erosive, corrosive and chemical solutions.

R221113-50   Pipe Material Considerations

1. Malleable fittings should be used for gas service.
2. Malleable fittings are used where there are stresses/strains due to expansion and vibration.
3. Cast fittings may be broken as an aid to disassembling of heating lines frozen by long use, temperature and minerals.
4. Cast iron pipe is extensively used for underground and submerged service.
5. Type M (light wall) copper tubing is available in hard temper only and is used for nonpressure and less severe applications than K and L.
6. Type L (medium wall) copper tubing, available hard or soft for interior service.
7. Type K (heavy wall) copper tubing, available in hard or soft temper for use where conditions are severe. For underground and interior service.
8. Hard drawn tubing requires fewer hangers or supports but should not be bent. Silver brazed fittings are recommended, however soft solder is normally used.
9. Type DMV (very light wall) copper tubing designed for drainage, waste and vent plus other non-critical pressure services.

Domestic/Imported Pipe and Fittings Cost

The prices shown in this publication for steel/cast iron pipe and steel, cast iron, malleable iron fittings are based on domestic production sold at the normal trade discounts. The above listed items of foreign manufacture may be available at prices of 1/3 to 1/2 those shown. Some imported items after minor machining or finishing operations are being sold as domestic to further complicate the system.

Caution: Most pipe prices in this book also include a coupling and pipe hangers which for the larger sizes can add significantly to the per foot cost and should be taken into account when comparing "book cost" with quoted supplier's cost.

R221113-70 Piping to 10' High

When taking off pipe, it is important to identify the different material types and joining procedures, as well as distances between supports and components required for proper support.

During the takeoff, measure through all fittings. Do not subtract the lengths of the fittings, valves, or strainers, etc. This added length plus the final rounding of the totals will compensate for nipples and waste.

When rounding off totals always increase the actual amount to correspond with manufacturer's shipping lengths.

A. Both red brass and yellow brass pipe are normally furnished in 12' lengths, plain end. The Unit Price section includes in the linear foot costs two field threads and one coupling per 10' length. A carbon steel clevis type hanger assembly every 10' is also prorated into the linear foot costs, including both material and labor.

B. Cast iron soil pipe is furnished in either 5' or 10' lengths. For pricing purposes, the Unit Price section features 10' lengths with a joint and a carbon steel clevis hanger assembly every 5' prorated into the per foot costs of both material and labor.

Three methods of joining are considered: lead and oakum poured joints or push-on gasket type joints for the bell and spigot pipe and a joint clamp for the no-hub soil pipe. The labor and material costs for each of these individual joining procedures are also prorated into the linear costs per foot.

C. Copper tubing covers types K, L, M, and DWV which are furnished in 20' lengths. Means pricing data is based on a tubing cut each length and a coupling and two soft soldered joints every 10'. A carbon steel, clevis type hanger assembly every 10' is also prorated into the per foot costs. The prices for refrigeration tubing are for materials only. Labor for full lengths may be based on the type L labor but short cut measures in tight areas can increase the installation labor-hours from 20 to 40%.

D. Corrosion-resistant piping does not lend itself to one particular standard of hanging or support assembly due to its diversity of application and placement. The several varieties of corrosion-resistant piping do not include any material or labor costs for hanger assemblies (See the Unit Price section for appropriate selection).

E. Glass pipe is furnished in standard lengths either 5' or 10' long, beaded on one end. Special orders for diverse lengths beaded on both ends are also available. For pricing purposes, R.S. Means features 10' lengths with a coupling and a carbon steel band hanger assembly every 10' prorated into the per foot linear costs.

Glass pipe is also available with conical ends and standard lengths ranging from 6" through 3' in 6" increments, then up to 10' in 12" increments. Special lengths can be customized for particular installation requirements.

For pricing purposes, Means has based the labor and material pricing on 10' lengths. Included in these costs per linear foot are the prorated costs for a flanged assembly every 10' consisting of two flanges, a gasket, two insertable seals, and the required number of bolts and nuts. A carbon steel band hanger assembly based on 10' center lines has also been prorated into the costs per foot for labor and materials.

F. Plastic pipe of several compositions and joining methods are considered. Fiberglass reinforced pipe (FRP) is priced, based on 10' lengths (20' lengths are also available), with coupling and epoxy joints every 10'. FRP is furnished in both "General Service" and "High Strength." A carbon steel clevis hanger assembly, 3 for every 10', is built into the prorated labor and material costs on a per foot basis.

The PVC and CPVC pipe schedules 40, 80 and 120, plus SDR ratings are all based on 20' lengths with a coupling installed every 10', as well as a carbon steel clevis hanger assembly every 3'. The PVC and

ABS type DWV piping is based on 10' lengths with solvent weld couplings every 10', and with carbon steel clevis hanger assemblies, 3 for every 10'. The rest of the plastic piping in this section is based on flexible 100' coils and does not include any coupling or supports.

This section ends with PVC drain and sewer piping based on 10' lengths with bell and spigot ends and 0-ring type, push-on joints.

G. Stainless steel piping includes both weld end and threaded piping, both in the type 304 and 316 specification and in the following schedules, 5, 10, 40, 80, and 160. Although this piping is usually furnished in 20' lengths, this cost grouping has a joint (either heli-arc butt-welded or threads and coupling) every 10'. A carbon steel clevis type hanger assembly is also included at 10' intervals and prorated into the linear foot costs.

H. Carbon steel pipe includes both black and galvanized. This section encompasses schedules 40 (standard) and 80 (extra heavy).

Several common methods of joining steel pipe — such as thread and coupled, butt welded, and flanged (150 lb. weld neck flanges) are also included.

For estimating purposes, it is assumed that the piping is purchased in 20' lengths and that a compatible joint is made up every 10'. These joints are prorated into the labor and material costs per linear foot. The following hanger and support assemblies every 10' are also included: carbon steel clevis for the T & C pipe, and single rod roll type for both the welded and flanged piping. All of these hangers are oversized to accommodate pipe insulation 3/4" thick through 5" pipe size and 1-1/2" thick from 6" through 12" pipe size.

I. Grooved joint steel pipe is priced both black and galvanized, in schedules 10, 40, and 80, furnished in 20' lengths. This section describes two joining methods: cut groove and roll groove. The schedule 10 piping is roll-grooved, while the heavier schedules are cut-grooved. The labor and material costs are prorated into per linear foot prices, including a coupled joint every 10', as well as a carbon steel clevis hanger assembly.

Notes:

The pipe hanger assemblies mentioned in the preceding paragraphs include the described hanger; appropriately sized steel, box-type insert and nut; plus 18" of threaded hanger rod.

C clamps are used when the pipe is to be supported from steel shapes rather than anchored in the slab. C clamps are slightly less costly than inserts. However, to save time in estimating, it is advisable to use the given line number cost, rather than substituting a C clamp for the insert.

Add to piping labor for elevated installation:

10' to 14.5' high	10%	30' to 34.5' high	40%
15' to 19.5' high	20%	35' to 39.5' high	50%
20' to 24.5' high	25%	Over 40' and higher	55%
25' to 29.5' high	35%		

When using the percentage adds for elevated piping installations as shown above, bear in mind that the given heights are for the pipe supports, even though the insert, anchor, or clamp may be several feet higher than the pipe itself.

An allowance has been included in the piping installation time for testing and minor tightening of leaking joints, fittings, stuffing boxes, packing glands, etc. For extraordinary test requirements such as x-rays, prolonged pressure or demonstration tests, a percentage of the piping labor, based on the estimator's experience, must be added to the labor total. A testing service specializing in weld x-rays should be consulted for pricing if it is an estimate requirement. Equipment installation time includes start-up with associated adjustments.

R230500-10 Subcontractors

On the unit cost pages of the R.S. Means Cost Data books, the last column is entitled "Total Incl. O&P". This is normally the cost of the installing contractor. In the HVAC Division, this is the cost of the mechanical contractor. If the particular work being estimated is to be performed by a sub to the mechanical contractor, the mechanical's profit and handling charge (usually 10%) is added to the total of the last column.

R233100-10 Loudness Levels for Moving Air thru Fans, Diffusers, Register, Etc. (Measured in Sones)

	Recommended Loudness Levels (Sones)										
Area	Very Quiet	Quiet	Noisy	Area	Very Quiet	Quiet	Noisy	Area	Very Quiet	Quiet	Noisy
Auditoriums				**Hotels**				**Offices**			
Auditorium lobbies	3	4	6	Banquet Rooms	1.5	3	6	Conference rooms	1	1.7	3
Concert and opera halls	0.8	1	1.5	Individual rooms, suites	1	2	4	Drafting	2	4	8
Courtrooms	2	3	4	Kitchens and laundries	4	7	10	General open offices	2	4	8
Lecture halls	1.5	2	3	Lobbies	2	4	8	Halls and corridors	2.5	5	10
Movie theaters	1.5	2	3	**Indoor Sports**				Professional offices	1.5	3	6
Churches and Schools				Bowling alleys	3	4	6	Reception room	1.5	3	6
Kitchens	4	6	8	Gymnasiums	2	4	6	Tabulation & computation	3	6	12
Laboratories	2	4	6	Swimming pools	4	7	10	**Public Buildings**			
Libraries	1.5	2	3	**Manufacturing Areas**				Banks	2	4	6
Recreation halls	2	4	8	Assembly lines	5	12	30	Court houses	2.5	4	6
Sanctuaries	1	1.7	3	Foreman's office	3	5	8	Museums	2	3	4
Schools and classrooms	1.5	2.5	4	Foundries	10	20	40	Planetariums	1.5	2	3
Hospital and Clinics				General storage	5	10	20	Post offices	2.5	4	6
Laboratories	2	4	6	Heavy machinery	10	25	60	Public libraries	1.5	2	4
Lobbies, waiting rooms	2	4	6	Light machinery	5	12	30	Waiting rooms	3	5	8
Halls and corridors	2	4	6	Tool maintenance	4	7	10	**Restaurants**			
Operating rooms	1.5	2.5	4	**Stores**				Cafeterias	3	6	10
Private rooms	1	1.7	3	Department stores	3	6	8	Night clubs	2.5	4	6
Wards	1.5	2.5	4	Supermarkets	4	7	10	Restaurants	2	4	8

R233100-20 Ductwork

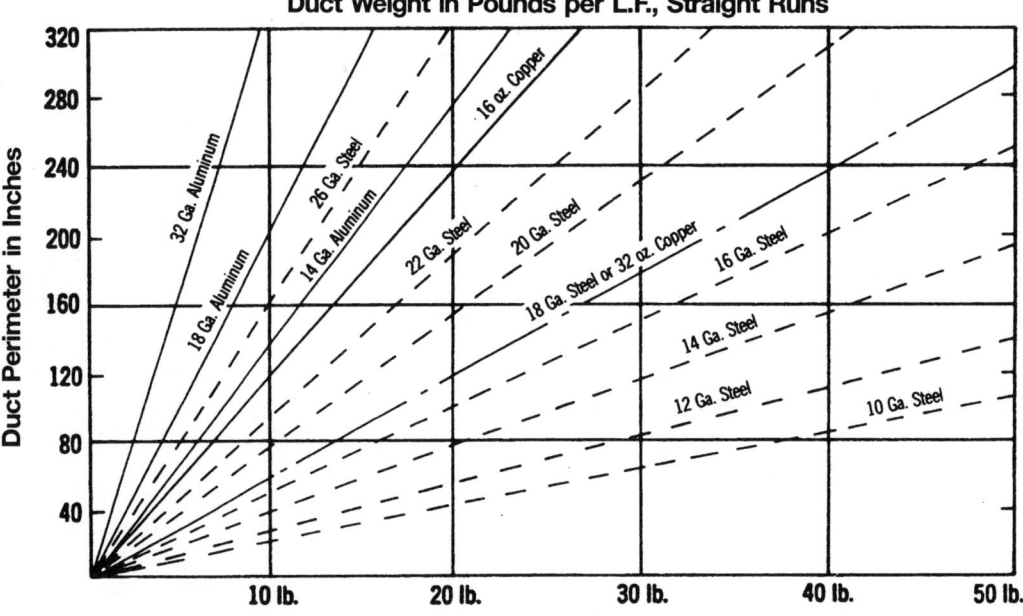

Add to the above for fittings; 90° elbow is 3 L.F.; 45° elbow is 2.5 L.F.; offset is 4 L.F.; transition offset is 6 L.F.; square-to-round transition is 4 L.F.; 90° reducing elbow is 5 L.F. For bracing and waste, add 20% to aluminum and copper, 15% to steel.

R233100-30 Duct Fabrication/Installation

The labor cost for fabricated sheet metal duct includes both the cost of fabrication and installation of the duct. The split is approximately 40% for fabrication, 60% for installation. It is for this reason that the percentage add for elevated installation of fabricated duct is less than the percentage add for prefabricated/preformed duct.

Example: assume a piece of duct cost $100 installed (labor only)

$$\text{Sheet Metal Fabrication} = 40\% = \$40$$
$$\text{Installation} = 60\% = \$60$$

The add for elevated installation is:

$$\frac{\text{Based on total labor}}{\text{(fabrication & installation)}} = \$100 \times 6\% = \$6.00$$

$$\frac{\text{Based on installation cost only}}{\text{(Material purchased prefabricated)}} = \$60 \times 10\% = \$6.00$$

The $6.00 markup (10' to 15' high) is the same.

R233100-40 Sheet Metal Calculator (Weight in Lb./Ft. of Length)

Gauge	26	24	22	20	18	16	Gauge	26	24	22	20	18	16
Wt.-Lb./S.F.	.906	1.156	1.406	1.656	2.156	2.656	Wt.-Lb./S.F.	.906	1.156	1.406	1.656	2.156	2.656
SMACNA Max. Dimension – Long Side		30″	54″	84″	85″ Up		SMACNA Max. Dimension – Long Side		30″	54″	84″	85″ Up	
Sum-2 sides							**Sum-2 Sides**						
2	.3	.40	.50	.60	.80	.90	56	9.3	12.0	14.0	16.2	21.3	25.2
3	.5	.65	.80	.90	1.1	1.4	57	9.5	12.3	14.3	16.5	21.7	25.7
4	.7	.85	1.0	1.2	1.5	1.8	58	9.7	12.5	14.5	16.8	22.0	26.1
5	.8	1.1	1.3	1.5	1.9	2.3	59	9.8	12.7	14.8	17.1	22.4	26.6
6	1.0	1.3	1.5	1.7	2.3	2.7	60	10.0	12.9	15.0	17.4	22.8	27.0
7	1.2	1.5	1.8	2.0	2.7	3.2	61	10.2	13.1	15.3	17.7	23.2	27.5
8	1.3	1.7	2.0	2.3	3.0	3.6	62	10.3	13.3	15.5	18.0	23.6	27.9
9	1.5	1.9	2.3	2.6	3.4	4.1	63	10.5	13.5	15.8	18.3	24.0	28.4
10	1.7	2.2	2.5	2.9	3.8	4.5	64	10.7	13.7	16.0	18.6	24.3	28.8
11	1.8	2.4	2.8	3.2	4.2	5.0	65	10.8	13.9	16.3	18.9	24.7	29.3
12	2.0	2.6	3.0	3.5	4.6	5.4	66	11.0	14.1	16.5	19.1	25.1	29.7
13	2.2	2.8	3.3	3.8	4.9	5.9	67	11.2	14.3	16.8	19.4	25.5	30.2
14	2.3	3.0	3.5	4.1	5.3	6.3	68	11.3	14.6	17.0	19.7	25.8	30.6
15	2.5	3.2	3.8	4.4	5.7	6.8	69	11.5	14.8	17.3	20.0	26.2	31.1
16	2.7	3.4	4.0	4.6	6.1	7.2	70	11.7	15.0	17.5	20.3	26.6	31.5
17	2.8	3.7	4.3	4.9	6.5	7.7	71	11.8	15.2	17.8	20.6	27.0	32.0
18	3.0	3.9	4.5	5.2	6.8	8.1	72	12.0	15.4	18.0	20.9	27.4	32.4
19	3.2	4.1	4.8	5.5	7.2	8.6	73	12.2	15.6	18.3	21.2	27.7	32.9
20	3.3	4.3	5.0	5.8	7.6	9.0	74	12.3	15.8	18.5	21.5	28.1	33.3
21	3.5	4.5	5.3	6.1	8.0	9.5	75	12.5	16.1	18.8	21.8	28.5	33.8
22	3.7	4.7	5.5	6.4	8.4	9.9	76	12.7	16.3	19.0	22.0	28.9	34.2
23	3.8	5.0	5.8	6.7	8.7	10.4	77	12.8	16.5	19.3	22.3	29.3	34.7
24	4.0	5.2	6.0	7.0	9.1	10.8	78	13.0	16.7	19.5	22.6	29.6	35.1
25	4.2	5.4	6.3	7.3	9.5	11.3	79	13.2	16.9	19.8	22.9	30.0	35.6
26	4.3	5.6	6.5	7.5	9.9	11.7	80	13.3	17.1	20.0	23.2	30.4	36.0
27	4.5	5.8	6.8	7.8	10.3	12.2	81	13.5	17.3	20.3	23.5	30.8	36.5
28	4.7	6.0	7.0	8.1	10.6	12.6	82	13.7	17.5	20.5	23.8	31.2	36.9
29	4.8	6.2	7.3	8.4	11.0	13.1	83	13.8	17.8	20.8	24.1	31.5	37.4
30	5.0	6.5	7.5	8.7	11.4	13.5	84	14.0	18.0	21.0	24.4	31.9	37.8
31	5.2	6.7	7.8	9.0	11.8	14.0	85	14.2	18.2	21.3	24.7	32.3	38.3
32	5.3	6.9	8.0	9.3	12.2	14.4	86	14.3	18.4	21.5	24.9	32.7	38.7
33	5.5	7.1	8.3	9.6	12.5	14.9	87	14.5	18.6	21.8	25.2	33.1	39.2
34	5.7	7.3	8.5	9.9	12.9	15.3	88	14.7	18.8	22.0	25.5	33.4	39.6
35	5.8	7.5	8.8	10.2	13.3	15.8	89	14.8	19.0	22.3	25.8	33.8	40.1
36	6.0	7.8	9.0	10.4	13.7	16.2	90	15.0	19.3	22.5	26.1	34.2	40.5
37	6.2	8.0	9.3	10.7	14.1	16.7	91	15.2	19.5	22.8	26.4	34.6	41.0
38	6.3	8.2	9.5	11.0	14.4	17.1	92	15.3	19.7	23.0	26.7	35.0	41.4
39	6.5	8.4	9.8	11.3	14.8	17.6	93	15.5	19.9	23.3	27.0	35.3	41.9
40	6.7	8.6	10.0	11.6	15.2	18.0	94	15.7	20.1	23.5	27.3	35.7	42.3
41	6.8	8.8	10.3	11.9	15.6	18.5	95	15.8	20.3	23.8	27.6	36.1	42.8
42	7.0	9.0	10.5	12.2	16.0	18.9	96	16.0	20.5	24.0	27.8	36.5	43.2
43	7.2	9.2	10.8	12.5	16.3	19.4	97	16.2	20.8	24.3	28.1	36.9	43.7
44	7.3	9.5	11.0	12.8	16.7	19.8	98	16.3	21.0	24.5	28.4	37.2	44.1
45	7.5	9.7	11.3	13.1	17.1	20.3	99	16.5	21.2	24.8	28.7	37.6	44.6
46	7.7	9.9	11.5	13.3	17.5	20.7	100	16.7	21.4	25.0	29.0	38.0	45.0
47	7.8	10.1	11.8	13.6	17.9	21.2	101	16.8	21.6	25.3	29.3	38.4	45.5
48	8.0	10.3	12.0	13.9	18.2	21.6	102	17.0	21.8	25.5	29.6	38.8	45.9
49	8.2	10.5	12.3	14.2	18.6	22.1	103	17.2	22.0	25.8	29.9	39.1	46.4
50	8.3	10.7	12.5	14.5	19.0	22.5	104	17.3	22.3	26.0	30.2	39.5	46.8
51	8.5	11.0	12.8	14.8	19.4	23.0	105	17.5	22.5	26.3	30.5	39.9	47.3
52	8.7	11.2	13.0	15.1	19.8	23.4	106	17.7	22.7	26.5	30.7	40.3	47.7
53	8.8	11.4	13.3	15.4	20.1	23.9	107	17.8	22.9	26.8	31.0	40.7	48.2
54	9.0	11.6	13.5	15.7	20.5	24.3	108	18.0	23.1	27.0	31.3	41.0	48.6
55	9.2	11.8	13.8	16.0	20.9	24.8	109	18.2	23.3	27.3	31.6	41.4	49.1
							110	18.3	23.5	27.5	31.9	41.8	49.5

Example: If duct is 34″ x 20″ x 15′ long, 34″ is greater than 30″ maximum, for 24 ga. so must be 22 ga. 34″ + 20″ = 54″ going across from 54″ find 13.5 lb. per foot. 13.5 x 15′ = 202.5 lbs. For S.F. of surface area

202.5 ÷ 1.406 = 144 S.F.
Note: Figures include an allowance for scrap.

Heating, Ventilating & A.C. | R2331 HVAC Ducts & Casings

R233100-50 Ductwork Packages (per Ton of Cooling)

System	Sheet Metal	Insulation	Diffusers	Return Register
Roof Top Unit Single Zone	120 Lbs.	52 S.F.	1	1
Roof Top Unit Multizone	240 Lbs.	104 S.F.	2	1
Self-contained Air or Water Cooled	108 Lbs.	—	2	—
Split System Air Cooled	102 Lbs.	—	2	—

Systems reflect most common usage.
Refer to system graphics for duct layout.

Heating, Ventilating & A.C. | R2334 HVAC Fans

R233400-10 Recommended Ventilation Air Changes

Table below lists range of time in minutes per change for various types of facilities.

Assembly Halls	2-10	Dance Halls	2-10	Laundries	1-3
Auditoriums	2-10	Dining Rooms	3-10	Markets	2-10
Bakeries	2-3	Dry Cleaners	1-5	Offices	2-10
Banks	3-10	Factories	2-5	Pool Rooms	2-5
Bars	2-5	Garages	2-10	Recreation Rooms	2-10
Beauty Parlors	2-5	Generator Rooms	2-5	Sales Rooms	2-10
Boiler Rooms	1-5	Gymnasiums	2-10	Theaters	2-8
Bowling Alleys	2-10	Kitchens-Hospitals	2-5	Toilets	2-5
Churches	5-10	Kitchens-Restaurant	1-3	Transformer Rooms	1-5

CFM air required for changes = Volume of room in cubic feet ÷ Minutes per change.

Heating, Ventilating & A.C. | R2337 Air Outlets & Inlets

R233700-60 Diffuser Evaluation

CFM = V × An × K where V = Outlet velocity in feet per minute. An = Neck area in square feet and K = Diffuser delivery factor. An undersized diffuser for a desired CFM will produce a high velocity and noise level. When air moves past people at a velocity in excess of 25 FPM, an annoying draft is felt. An oversized diffuser will result in low velocity with poor mixing. Consideration must be given to avoid vertical stratification or horizontal areas of stagnation.

Heating, Ventilating & A.C. R2350 Central Heating Equipment

R235000-10 Heating Systems

Heating Systems

The basic function of a heating system is to bring an enclosed volume up to a desired temperature and then maintain that temperature within a reasonable range. To accomplish this, the selected system must have sufficient capacity to offset transmission losses resulting from the temperature difference on the interior and exterior of the enclosing walls in addition to losses due to cold air infiltration through cracks, crevices and around doors and windows. The amount of heat to be furnished is dependent upon the building size, construction, temperature difference, air leakage, use, shape, orientation and exposure. Air circulation is also an important consideration. Circulation will prevent stratification which could result in heat losses through uneven temperatures at various levels. For example, the most

efficient use of unit heaters can usually be achieved by circulating the space volume through the total number of units once every 20 minutes or 3 times an hour. This general rule must, of course, be adapted for special cases such as large buildings with low ratios of heat transmitting surface to cubical volume. The type of occupancy of a building will have considerable bearing on the number of heat transmitting units and the location selected. It is axiomatic, however, that the basis of any successful heating system is to provide the maximum amount of heat at the points of maximum heat loss such as exposed walls, windows, and doors. Large roof areas, wind direction, and wide doorways create problems of excessive heat loss and require special consideration and treatment.

Heat Transmission

Heat transfer is an important parameter to be considered during selection of the exterior wall style, material and window area. A high rate of transfer will permit greater heat loss during the wintertime with the resultant increase in heating energy costs and a greater rate of heat gain in the summer with proportionally greater cooling cost. Several terms are used to describe various aspects of heat transfer. However, for general estimating purposes this book lists U values for systems of construction materials. U is the "overall heat transfer coefficient." It is defined as the heat flow per hour through one square foot when the temperature difference in the air on either side of the structure wall, roof, ceiling or floor is one degree Fahrenheit. The structural segment may be a single homogeneous material or a composite.

Total heat transfer is found using the following equation:

$Q = AU(T_2 - T_1)$ where
Q = Heat flow, BTU per hour
A = Area, square feet
U = Overall heat transfer coefficient
$(T_2 - T_1)$ = Difference in temperature of air on each side of the construction component. (Also abbreviated TD)

Note that heat can flow through all surfaces of any building and this flow is in addition to heat gain or loss due to ventilation, infiltration and generation (appliances, machinery, people).

R235000-20 Heating Approximations for Quick Estimating

Oil Piping & Boiler Room Piping

Small System . . . . 20 to 30% of Boiler

Complex System

with Pumps, Headers, Etc. . . . . 80 to 110% of Boiler

Breeching With Insulation:

Small . . . . 10 to 15% of Boiler

Large . . . . 15 to 25% of Boiler

Coils: . . . . 15 to 30% of Containing Unit

Balancing (Independent) . . . . 1/2% of H.V.A.C. Estimate

Quality/Complexity Adjustment: For all heating installations add these adjustments to the estimate to more closely allow for the equipment and conditions of the particular job under consideration.

Economy installation, add . . . . 0 to 5% of System

Good quality, medium complexity, add . . . . 5 to 15% of System

Above average quality and complexity, add . . . . 15 to 25% of System

Reference Tables

R235000-30 The Basics of a Heating System

The function of a heating system is to achieve and maintain a desired temperature in a room or building by replacing the amount of heat being dissipated. There are four kinds of heating systems: hot-water, steam, warm-air and electric resistance. Each has certain essential and similar elements with the exception of electric resistance heating.

The basic elements of a heating system are:

A. A **combustion chamber** in which fuel is burned and heat transferred to a conveying medium.

B. The **"fluid"** used for conveying the heat (water, steam or air).

C. **Conductors** or pipes for transporting the fluid to specific desired locations.

D. A means of disseminating the heat, sometimes called **terminal units.**

A. The **combustion chamber** in a furnace heats air which is then distributed. This is called a warm-air system.

The combustion chamber in a boiler heats water which is either distributed as hot water or steam and this is termed a hydronic system.

The maximum allowable working pressures are limited by ASME "Code for Heating Boilers" to 15 PSI for steam and 160 PSI for hot water heating boilers, with a maximum temperature limitation of 250° F. Hot water boilers are generally rated for a working pressure of 30 PSI. High pressure boilers are governed by the ASME "Code for Power Boilers" which is used almost universally for boilers operating over 15 PSIG. High pressure boilers used for a combination of heating/process loads are usually designed for 150 PSIG.

Boiler ratings are usually indicated as either Gross or Net Output. The Gross Load is equal to the Net Load plus a piping and pickup allowance. When this allowance cannot be determined, divide the gross output rating by 1.25 for a value equal to or greater than the net heat loss requirement of the building.

B. Of the three **fluids** used, steam carries the greatest amount of heat per unit volume. This is due to the fact that it gives up its latent heat of vaporization at a temperature considerably above room temperature. Another advantage is that the pressure to produce a positive circulation is readily available. Piping conducts the steam to terminal units and returns condensate to the boiler.

The **steam system** is well adapted to large buildings because of its positive circulation, its comparatively economical installation and its ability to deliver large quantities of heat. Nearly all large office buildings, stores, hotels, and industrial buildings are so heated, in addition to many residences.

Hot water, when used as the heat carrying fluid, gives up a portion of its sensible heat and then returns to the boiler or heating apparatus for reheating. As the heat conveyed by each pound of water is about one-fiftieth of the heat conveyed by a pound of steam, it is necessary to circulate about fifty times as much water as steam by weight (although only one-thirtieth as much by volume). The hot water system is usually, although not necessarily, designed to operate at temperatures below that of the ordinary steam system and so the amount of heat transfer surface must be correspondingly greater. A temperature of 190° F to 200° F is normally the maximum. Circulation in small buildings may depend on the difference in density between hot water and the cool water returning to the boiler; circulating pumps are normally used to maintain a desired rate of flow. Pumps permit a greater degree of flexibility and better control.

In **warm-air** furnace systems, cool air is taken from one or more points in the building, passed over the combustion chamber and flue gas passages and then distributed through a duct system. A disadvantage of this system is that the ducts take up much more building volume than steam or hot water pipes. Advantages of this system are the relative ease with which humidification can be accomplished by the evaporation of water as the air circulates through the heater, and the lack of need for expensive disseminating units as the warm air simply becomes part of the interior atmosphere of the building.

C. Conductors (pipes and ducts) have been lightly treated in the discussion of conveying fluids. For more detailed information such as sizing and distribution methods, the reader is referred to technical publications such as the American Society of Heating, Refrigerating and Air-Conditioning Engineers "Handbook of Fundamentals."

D. Terminal units come in an almost infinite variety of sizes and styles, but the basic principles of operation are very limited. As previously mentioned, warm-air systems require only a simple register or diffuser to mix heated air with that present in the room. Special application items such as radiant coils and infrared heaters are available to meet particular conditions but are not usually considered for general heating needs. Most heating is accomplished by having air flow over coils or pipes containing the heat transporting medium (steam, hot-water, electricity). These units, while varied, may be separated into two general types, (1) radiator/convectors and (2) unit heaters.

Radiator/convectors may be cast, fin-tube or pipe assemblies. They may be direct, indirect, exposed, concealed or mounted within a cabinet enclosure, upright or baseboard style. These units are often collectively referred to as "radiatiors" or "radiation" although none gives off heat either entirely by radiation or by convection but rather a combination of both. The air flows over the units as a gravity "current." It is necessary to have one or more heat-emitting units in each room. The most efficient placement is low along an outside wall or under a window to counteract the cold coming into the room and achieve an even distribution.

In contrast to radiator/convectors which operate most effectively against the walls of smaller rooms, **unit heaters** utilize a fan to move air over heating coils and are very effective in locations of relatively large volume. Unit heaters, while usually suspended overhead, may be floor mounted. They also may take in fresh outside air for ventilation. The heat distributed by unit heaters may be from a remote source and conveyed by a fluid or it may be from the combustion of fuel in each individual heater. In the latter case the only piping required would be for fuel, however, a vent for the products of combustion would be necessary.

The following list gives may of the advantages of unit heaters for applications other than office or residential:

a. Large capacity so smaller number of units are required, **b.** Piping system simplified, **c.** Space saved where they are located overhead out of the way, **d.** Rapid heating directed where needed with effective wide distribution, **e.** Difference between floor and ceiling temperature reduced, **f.** Circulation of air obtained, and ventilation with introduction of fresh air possible, **g.** Heat output flexible and easily controlled.

R235000-50 Factor for Determining Heat Loss for Various Types of Buildings

General: While the most accurate estimates of heating requirements would naturally be based on detailed information about the building being considered, it is possible to arrive at a reasonable approximation using the following procedure:

1. Calculate the cubic volume of the room or building.
2. Select the appropriate factor from Table 1 below. Note that the factors apply only to inside temperatures listed in the first column and to 0° F outside temperature.

3. If the building has bad north and west exposures, multiply the heat loss factor by 1.1.
4. If the outside design temperature is other than 0° F, multiply the factor from Table 1 by the factor from Table 2.
5. Multiply the cubic volume by the factor selected from Table 1. This will give the estimated BTUH heat loss which must be made up to maintain inside temperature.

Table 1 — Building Type	Conditions	Qualifications	Loss Factor*
Factories & Industrial Plants General Office Areas at 70° F	One Story	Skylight in Roof	6.2
		No Skylight in Roof	5.7
	Multiple Story	Two Story	4.6
		Three Story	4.3
		Four Story	4.1
		Five Story	3.9
		Six Story	3.6
	All Walls Exposed	Flat Roof	6.9
		Heated Space Above	5.2
	One Long Warm Common Wall	Flat Roof	6.3
		Heated Space Above	4.7
	Warm Common Walls on Both Long Sides	Flat Roof	5.8
		Heated Space Above	4.1
Warehouses at 60° F	All Walls Exposed	Skylights in Roof	5.5
		No Skylight in Roof	5.1
		Heated Space Above	4.0
	One Long Warm Common Wall	Skylight in Roof	5.0
		No Skylight in Roof	4.9
		Heated Space Above	3.4
	Warm Common Walls on Both Long Sides	Skylight in Roof	4.7
		No Skylight in Roof	4.4
		Heated Space Above	3.0

*Note: This table tends to be conservative particularly for new buildings designed for minimum energy consumption.

Table 2 — Outside Design Temperature Correction Factor (for Degrees Fahrenheit)									
Outside Design Temperature	50	40	30	20	10	0	-10	-20	-30
Correction Factor	.29	.43	.57	.72	.86	1.00	1.14	1.28	1.43

R235000-70 Transmission of Heat

R235000-70 and R235000-80 provide a way to calculate heat transmission of various construction materials from their U values and the TD (Temperature Difference).

1. From the Exterior Enclosure Division or elsewhere, determine U values for the construction desired.
2. Determine the coldest design temperature. The difference between this temperature and the desired interior temperature is the TD (temperature difference).

3. Enter R235000-70 or R235000-80 at correct U Value. Cross horizontally to the intersection with appropriate TD. Read transmission per square foot from bottom of figure.
4. Multiply this value of BTU per hour transmission per square foot of area by the total surface area of that type of construction.

R235000-80 Transmission of Heat (Low Rate)

R235616-60   Solar Heating (Space and Hot Water)

Collectors should face as close to due South as possible, however, variations of up to 20 degrees on either side of true South are acceptable. Local climate and collector type may influence the choice between east or west deviations. Obviously they should be located so they are not shaded from the sun's rays. Incline collectors at a slope of latitude minus 5 degrees for domestic hot water and latitude plus 15 degrees for space heating.

Flat plate collectors consist of a number of components as follows: Insulation to reduce heat loss through the bottom and sides of the collector. The enclosure which contains all the components in this assembly is usually weatherproof and prevents dust, wind and water from coming in contact with the absorber plate. The cover plate usually consists of one or more layers of a variety of glass or plastic and reduces the reradiation by creating an air space which traps the heat between the cover and the absorber plates.

The absorber plate must have a good thermal bond with the fluid passages. The absorber plate is usually metallic and treated with a surface coating which improves absorptivity. Black or dark paints or selective coatings are used for this purpose, and the design of this passage and plate combination helps determine a solar system's effectiveness.

Heat transfer fluid passage tubes are attached above and below or integral with an absorber plate for the purpose of transferring thermal energy from the absorber plate to a heat transfer medium. The heat exchanger is a device for transferring thermal energy from one fluid to another.

Piping and storage tanks should be well insulated to minimize heat losses.

Size domestic water heating storage tanks to hold 20 gallons of water per user, minimum, plus 10 gallons per dishwasher or washing machine. For domestic water heating an optimum collector size is approximately 3/4 square foot of area per gallon of water storage. For space heating of residences and small commercial applications the collector is commonly sized between 30% and 50% of the internal floor area. For space heating of large commercial applications, collector areas less than 30% of the internal floor area can still provide significant heat reductions.

A supplementary heat source is recommended for Northern states for December through February.

The solar energy transmission per square foot of collector surface varies greatly with the material used. Initial cost, heat transmittance and useful life are obviously interrelated.

R236000-10   Air Conditioning

General: The purpose of air conditioning is to control the environment of a space so that comfort is provided for the occupants and/or conditions are suitable for the processes or equipment contained therein. The several items which should be evaluated to define system objectives are:

Temperature Control
Humidity Control
Cleanliness
Odor, smoke and fumes
Ventilation

Efforts to control the above parameters must also include consideration of the degree or tolerance of variation, the noise level introduced, the velocity of air motion and the energy requirements to accomplish the desired results.

The variation in **temperature** and **humidity** is a function of the sensor and the controller. The controller reacts to a signal from the sensor and produces the appropriate suitable response in either the terminal unit, the conductor of the transporting medium (air, steam, chilled water, etc.), or the source (boiler, evaporating coils, etc.).

The **noise level** is a by-product of the energy supplied to moving components of the system. Those items which usually contribute the most noise are pumps, blowers, fans, compressors and diffusers. The level of noise can be partially controlled through use of vibration pads, isolators, proper sizing, shields, baffles and sound absorbing liners.

Some **air motion** is necessary to prevent stagnation and stratification. The maximum acceptable velocity varies with the degree of heating or cooling

which is taking place. Most people feel air moving past them at velocities in excess of 25 FPM as an annoying draft. However, velocities up to 45 FPM may be acceptable in certain cases. Ventilation, expressed as air changes per hour and percentage of fresh air, is usually an item regulated by local codes.

Selection of the system to be used for a particular application is usually a trade-off. In some cases the building size, style, or room available for mechanical use limits the range of possibilities. Prime factors influencing the decision are first cost and total life (operating, maintenance and replacement costs). The accuracy with which each parameter is determined will be an important measure of the reliability of the decision and subsequent satisfactory operation of the installed system.

Heat delivery may be desired from an air conditioning system. Heating capability usually is added as follows: A gas fired burner or hot water/steam/electric coils may be added to the air handling unit directly and heat all air equally. For limited or localized heat requirements the water/steam/electric coils may be inserted into the duct branch supplying the cold areas. Gas fired duct furnaces are also available.

Note: When water or steam coils are used the cost of the piping and boiler must also be added. For a rough estimate use the cost per square foot of the appropriate sized hydronic system with unit heaters. This will provide a cost for the boiler and piping, and the unit heaters of the system would equate to the approximate cost of the heating coils.

R236000-20 Air Conditioning Requirements

BTUs per hour per S.F. of floor area and S.F. per ton of air conditioning.

Type of Building	BTU/Hr per S.F.	S.F. per Ton	Type of Building	BTU/Hr per S.F.	S.F. per Ton	Type of Building	BTU/Hr per S.F.	S.F. per Ton
Apartments, Individual	26	450	Dormitory, Rooms	40	300	Libraries	50	240
Corridors	22	550	Corridors	30	400	Low Rise Office, Exterior	38	320
Auditoriums & Theaters	40	300/18*	Dress Shops	43	280	Interior	33	360
Banks	50	240	Drug Stores	80	150	Medical Centers	28	425
Barber Shops	48	250	Factories	40	300	Motels	28	425
Bars & Taverns	133	90	High Rise Office—Ext. Rms.	46	263	Office (small suite)	43	280
Beauty Parlors	66	180	Interior Rooms	37	325	Post Office, Individual Office	42	285
Bowling Alleys	68	175	Hospitals, Core	43	280	Central Area	46	260
Churches	36	330/20*	Perimeter	46	260	Residences	20	600
Cocktail Lounges	68	175	Hotel, Guest Rooms	44	275	Restaurants	60	200
Computer Rooms	141	85	Corridors	30	400	Schools & Colleges	46	260
Dental Offices	52	230	Public Spaces	55	220	Shoe Stores	55	220
Dept. Stores, Basement	34	350	Industrial Plants, Offices	38	320	Shop'g. Ctrs., Supermarkets	34	350
Main Floor	40	300	General Offices	34	350	Retail Stores	48	250
Upper Floor	30	400	Plant Areas	40	300	Specialty	60	200

*Persons per ton
12,000 BTU = 1 ton of air conditioning

R236000-30 Psychrometric Table

Dewpoint or Saturation Temperature (F)

Relative humidity (%)	32	35	40	45	50	55	60	65	70	75	80	85	90	95	100
100	32	35	40	45	50	55	60	65	70	75	80	85	90	95	100
90	30	33	37	42	47	52	57	62	67	72	77	82	87	92	97
80	27	30	34	39	44	49	54	58	64	68	73	78	83	88	93
70	24	27	31	36	40	45	50	55	60	64	69	74	79	84	88
60	20	24	28	32	36	41	46	51	55	60	65	69	74	79	83
50	16	20	24	28	33	36	41	46	50	55	60	64	69	73	78
40	12	15	18	23	27	31	35	40	45	49	53	58	62	67	71
30	8	10	14	18	21	25	29	33	37	42	46	50	54	59	62
20	6	7	8	9	13	16	20	24	28	31	35	40	43	48	52
10	4	4	5	5	6	8	9	10	13	17	20	24	27	30	34
	32	35	40	45	50	55	60	65	70	75	80	85	90	95	100

Dry bulb temperature (F)

This table shows the relationship between RELATIVE HUMIDITY, DRY BULB TEMPERATURE AND DEWPOINT.

As an example, assume that the thermometer in a room reads 75° F, and we know that the relative humidity is 50%. The chart shows the dewpoint temperature to be 55° F. That is, any surface colder than 55° F will "sweat" or collect condensing moisture. This surface could be the outside of an uninsulated chilled water pipe in the summertime, or the inside surface of a wall or deck in the wintertime. After determining the extreme ambient parameters, the table at the left is useful in determining which surfaces need insulation or vapor barrier protection.

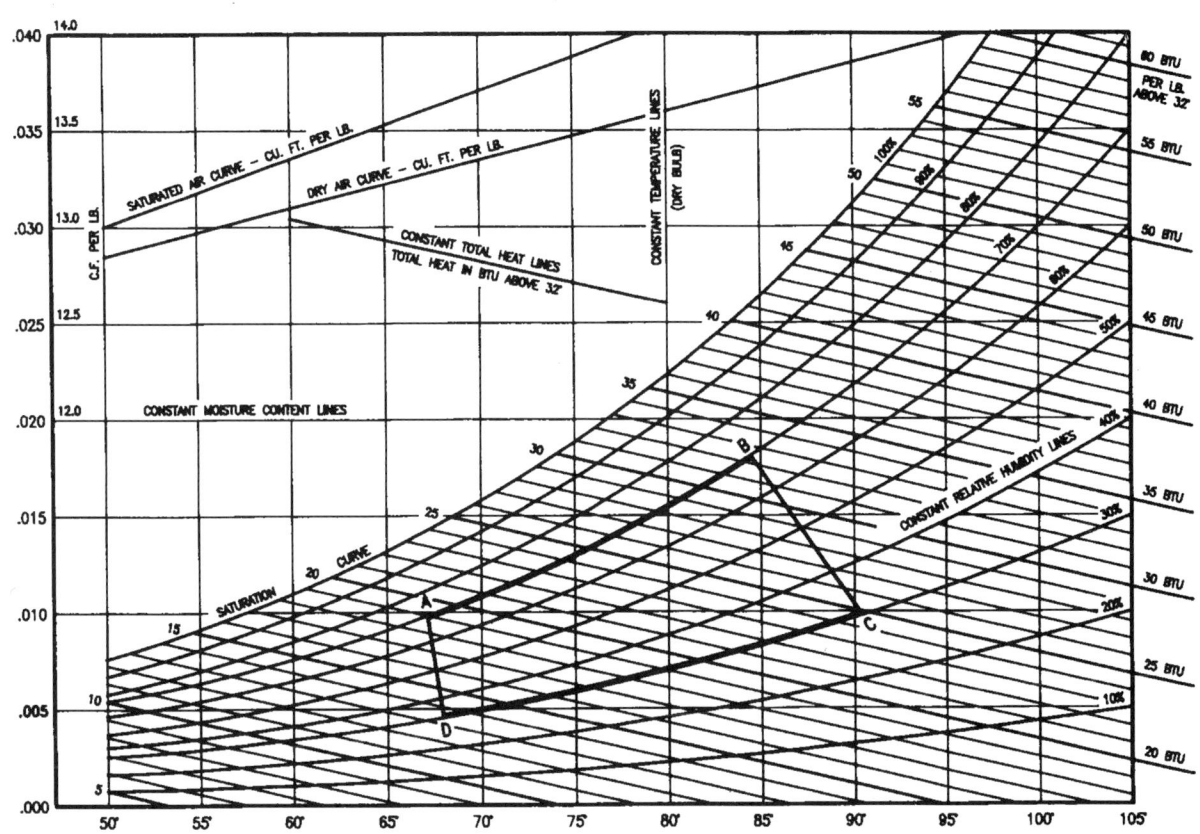

TEMPERATURE DEGREES FAHRENHEIT
TOTAL PRESSURE = 14.696 LB. PER SQ. IN. ABS.

Psychrometric chart showing different variables based on one pound of dry air. Space marked A B C D is temperature-humidity range which is most comfortable for majority of people.

R236000-90 Quality/Complexity Adjustment for Air Conditioning Systems

Economy installation, add .... 0 to 5%

Good quality, medium complexity, add .... 5 to 15%

Above average quality and complexity, add .... 15 to 25%

Add the above adjustments to the estimate to more closely allow for the equipment and conditions of the particular job under consideration.

Fig. R238313-11

R238313-10 Heat Trace Systems

Before you can determine the cost of a HEAT TRACE installation, the method of attachment must be established. There are (4) common methods:

1. Cable is simply attached to the pipe with polyester tape every 12'.
2. Cable is attached with a continuous cover of 2" wide aluminum tape.
3. Cable is attached with factory extruded heat transfer cement and covered with metallic raceway with clips every 10'.
4. Cable is attached between layers of pipe insulation using either clips or polyester tape.

Example: Components for method 3 must include:
   A. Heat trace cable by voltage and watts per linear foot.
   B. Heat transfer cement, 1 gallon per 60 linear feet of cover.
   C. Metallic raceway by size and type.
   D. Raceway clips by size of pipe.

When taking off linear foot lengths of cable add the following for each valve in the system. (E)

In all of the above methods each component of the system must be priced individually.

SCREWED OR WELDED VALVE:			FLANGED VALVE:			BUTTERFLY VALVES:		
1/2"	=	6"	1/2"	=	1'-0"	1/2"	=	0'
3/4"	=	9"	3/4"	=	1'-6"	3/4"	=	0'
1"	=	1'-0"	1"	=	2'-0"	1"	=	1'-0"
1-1/2"	=	1'-6"	1-1/2"	=	2'-6"	1-1/2"	=	1'-6"
2"	=	2'	2"	=	2'-6"	2"	=	2'-0"
2-1/2"	=	2'-6"	2-1/2"	=	3'-0"	2-1/2"	=	2'-6"
3"	=	2'-6"	3"	=	3'-6"	3"	=	2'-6"
4"	=	4'-0"	4"	=	4'-0"	4"	=	3'-0"
6"	=	7'-0"	6"	=	8'-0"	6"	=	3'-6"
8"	=	9'-6"	8"	=	11'-0"	8"	=	4'-0"
10"	=	12'-6"	10"	=	14'-0"	10"	=	4'-0"
12"	=	15'-0"	12"	=	16'-6"	12"	=	5'-0"
14"	=	18'-0"	14"	=	19'-6"	14"	=	5'-6"
16"	=	21'-6"	16"	=	23'-0"	16"	=	6'-0"
18"	=	25'-6"	18"	=	27'-0"	18"	=	6'-6"
20"	=	28'-6"	20"	=	30'-0"	20"	=	7'-0"
24"	=	34'-0"	24"	=	36'-0"	24"	=	8'-0"
30"	=	40'-0"	30"	=	42'-0"	30"	=	10'-0"

R238313-10   Heat Trace Systems (cont.)

Add the following quantities of heat transfer cement to linear foot totals for each valve:

Nominal Valve Size	Gallons of Cement per Valve
1/2″	0.14
3/4″	0.21
1″	0.29
1-1/2″	0.36
2″	0.43
2-1/2″	0.70
3″	0.71
4″	1.00
6″	1.43
8″	1.48
10″	1.50
12″	1.60
14″	1.75
16″	2.00
18″	2.25
20″	2.50
24″	3.00
30″	3.75

The following must be added to the list of components to accurately price HEAT TRACE systems:

1. Expediter fitting and clamp fasteners (F)
2. Junction box and nipple connected to expediter fitting (G)
3. Field installed terminal blocks within junction box
4. Ground lugs
5. Piping from power source to expediter fitting
6. Controls
7. Thermostats
8. Branch wiring
9. Cable splices
10. End of cable terminations
11. Branch piping fittings and boxes

Deduct the following percentages from labor if cable lengths in the same area exceed:

150′ to 250′	10%	351′ to 500′	20%
251′ to 350′	15%	Over 500′	25%

Add the following percentages to labor for elevated installations:

15′ to 20′ high	10%	31′ to 35′ high	40%
21′ to 25′ high	20%	36′ to 40′ high	50%
26′ to 30′ high	30%	Over 40′ high	60%

R238313-20   Spiral-Wrapped Heat Trace Cable (Pitch Table)

In order to increase the amount of heat, occasionally heat trace cable is wrapped in a spiral fashion around a pipe; increasing the number of feet of heater cable per linear foot of pipe.

Engineers first determine the heat loss per foot of pipe (based on the insulating material, its thickness, and the temperature differential across it). A ratio is then calculated by the formula:

$$\text{Feet of Heat Trace per Foot of Pipe} = \frac{\text{Watts/Foot of Heat Loss}}{\text{Watts/Foot of the Cable}}$$

The linear distance between wraps (pitch) is then taken from a chart or table. Generally, the pitch is listed on a drawing leaving the estimator to calculate the total length of heat tape required. An approximation may be taken from this table.

Feet of Heat Trace Per Foot of Pipe																
	Nominal Pipe Size in Inches															
Pitch In Inches	1	1¼	1½	2	2½	3	4	6	8	10	12	14	16	18	20	24
3.5	1.80															
4	1.65															
5	1.46	1.60	1.80													
6	1.34	1.45	1.55	1.75												
7	1.25	1.35	1.43	1.57	1.75											
8	1.20	1.28	1.34	1.45	1.60	1.80										
9	1.16	1.23	1.28	1.37	1.51	1.68										
10	1.13	1.19	1.24	1.32	1.44	1.57	1.82									
15	1.06	1.08	1.10	1.15	1.21	1.29	1.42	1.78								
20	1.04	1.05	1.06	1.08	1.13	1.17	1.25	1.49	1.73							
25		1.04	1.04	1.06	1.08	1.11	1.17	1.33	1.51	1.72						
30				1.04	1.06	1.07	1.12	1.24	1.37	1.54	1.70	1.80				
35					1.05	1.06	1.09	1.17	1.28	1.42	1.54	1.64	1.78			
40						1.05	1.07	1.14	1.22	1.33	1.44	1.52	1.64	1.75		
50							1.05	1.09	1.15	1.22	1.29	1.35	1.44	1.53	1.64	1.83
60								1.06	1.11	1.16	1.21	1.25	1.31	1.39	1.46	1.62
70								1.05	1.08	1.12	1.17	1.19	1.24	1.30	1.35	1.47
80									1.06	1.09	1.13	1.15	1.19	1.24	1.28	1.38
90									1.04	1.06	1.10	1.13	1.16	1.19	1.23	1.32
100										1.05	1.08	1.10	1.13	1.15	1.19	1.23

Note: Common practice would normally limit the lower end of the table to 5% of additional heat and above 80% an engineer would likely opt for two (2) parallel cables.

R312319-90   Wellpoints

A single stage wellpoint system is usually limited to dewatering an average 15' depth below normal ground water level. Multi-stage systems are employed for greater depth with the pumping equipment installed only at the lowest header level. Ejectors with unlimited lift capacity can be economical when two or more stages of wellpoints can be replaced or when horizontal clearance is restricted, such as in deep trenches or tunneling projects, and where low water flows are expected. Wellpoints are usually spaced on 2-1/2' to 10' centers along a header pipe. Wellpoint spacing, header size, and pump size are all determined by the expected flow as dictated by soil conditions.

In almost all soils encountered in wellpoint dewatering, the wellpoints may be jetted into place. Cemented soils and stiff clays may require sand wicks about 12" in diameter around each wellpoint to increase efficiency and eliminate weeping into the excavation. These sand wicks require 1/2 to 3 C.Y. of washed filter sand and are installed by using a 12" diameter steel casing and hole puncher jetted into the ground 2' deeper than the wellpoint. Rock may require predrilled holes.

Labor required for the complete installation and removal of a single stage wellpoint system is in the range of 3/4 to 2 labor-hours per linear foot of header, depending upon jetting conditions, wellpoint spacing, etc.

Continuous pumping is necessary except in some free draining soil where temporary flooding is permissible (as in trenches which are backfilled after each day's work). Good practice requires provision of a stand-by pump during the continuous pumping operation.

Systems for continuous trenching below the water table should be installed three to four times the length of expected daily progress to ensure uninterrupted digging, and header pipe size should not be changed during the job.

For pervious free draining soils, deep wells in place of wellpoints may be economical because of lower installation and maintenance costs. Daily production ranges between two to three wells per day, for 25' to 40' depths, to one well per day for depths over 50'.

Detailed analysis and estimating for any dewatering problem is available at no cost from wellpoint manufacturers. Major firms will quote "sufficient equipment" quotes or their affiliates offer lump sum proposals to cover complete dewatering responsibility.

Description for 200' System with 8" Header		Quantities
Equipment & Material	Wellpoints 25' long, 2" diameter @ 5' O.C.	40 Each
	Header pipe, 8" diameter	200 L.F.
	Discharge pipe, 8" diameter	100 L.F.
	8" valves	3 Each
	Combination jetting & wellpoint pump (standby)	1 Each
	Wellpoint pump, 8" diameter	1 Each
	Transportation to and from site	1 Day
	Fuel for 30 days x 60 gal./day	1800 Gallons
	Lubricants for 30 days x 16 lbs./day	480 Lbs.
	Sand for points	40 C.Y.
Labor	Technician to supervise installation	1 Week
	Labor for installation and removal of system	300 Labor-hours
	4 Operators straight time 40 hrs./wk. for 4.33 wks.	693 Hrs.
	4 Operators overtime 2 hrs./wk. for 4.33 wks.	35 Hrs.

Earthwork   R3141 Shoring

R314116-40   Wood Sheet Piling

Wood sheet piling may be used for depths to 20' where there is no ground water. If moderate ground water is encountered Tongue & Groove sheeting will help to keep it out. When considerable ground water is present, steel sheeting must be used.

For estimating purposes on trench excavation, sizes are as follows:

Depth	Sheeting	Wales	Braces	B.F. per S.F.
To 8'	3 x 12's	6 x 8's, 2 line	6 x 8's, @ 10'	4.0 @ 8'
8' x 12'	3 x 12's	10 x 10's, 2 line	10 x 10's, @ 9'	5.0 average
12' to 20'	3 x 12's	12 x 12's, 3 line	12 x 12's, @ 8'	7.0 average

Sheeting to be toed in at least 2' depending upon soil conditions. A five person crew with an air compressor and sheeting driver can drive and brace 440 SF/day at 8' deep, 360 SF/day at 12' deep, and 320 SF/day at 16' deep.

For normal soils, piling can be pulled in 1/3 the time to install. Pulling difficulty increases with the time in the ground. Production can be increased by high pressure jetting.

R314116-45   Steel Sheet Piling

Limiting weights are 22 to 38#/S.F. of wall surface with 27#/S.F. average for usual types and sizes. (Weights of piles themselves are from 30.7#/L.F. to 57#/L.F. but they are 15" to 21" wide.) Lightweight sections 12" to 28" wide from 3 ga. to 12 ga. thick are also available for shallow excavations. Piles may be driven two at a time with an impact or vibratory hammer (use vibratory to pull) hung from a crane without leads. A reasonable estimate of the life of steel sheet piling is 10 uses with up to 125 uses possible if a vibratory hammer is used. Used piling costs from 50% to 80% of new piling depending on location and market conditions. Sheet piling and H piles can be rented for about 30% of the delivered mill price for the first month and 5% per month thereafter. Allow 1 labor-hour per pile for cleaning and trimming after driving. These costs increase with depth and hydrostatic head. Vibratory drivers are faster in wet granular soils and are excellent for pile extraction. Pulling difficulty increases with the time in the ground and may cost more than driving. It is often economical to abandon the sheet piling, especially if it can be used as the outer wall form. Allow about 1/3 additional length or more for toeing into ground. Add bracing, waler and strut costs. Waler costs can equal the cost per ton of sheeting.

Utilities   R3311 Water Utility Distribution Piping

R331113-80   Piping Designations

There are several systems currently in use to describe pipe and fittings. The following paragraphs will help to identify and clarify classifications of piping systems used for water distribution.

Piping may be classified by schedule. Piping schedules include 5S, 10S, 10, 20, 30, Standard, 40, 60, Extra Strong, 80, 100, 120, 140, 160 and Double Extra Strong. These schedules are dependent upon the pipe wall thickness. The wall thickness of a particular schedule may vary with pipe size.

Ductile iron pipe for water distribution is classified by Pressure Classes such as Class 150, 200, 250, 300 and 350. These classes are actually the rated water working pressure of the pipe in pounds per square inch (psi). The pipe in these pressure classes is designed to withstand the rated water working pressure plus a surge allowance of 100 psi.

The American Water Works Association (AWWA) provides standards for various types of **plastic pipe**. C-900 is the specification for polyvinyl chloride (PVC) piping used for water distribution in sizes ranging from 4" through 12". C-901 is the specification for polyethylene (PE) pressure pipe, tubing and fittings used for water distribution in sizes ranging from 1/2" through 3". C-905 is the specification for PVC piping sizes 14" and greater.

PVC pressure-rated pipe is identified using the standard dimensional ratio (SDR) method. This method is defined by the American Society for Testing and Materials (ASTM) Standard D 2241. This pipe is available in SDR numbers 64, 41, 32.5, 26, 21, 17, and 13.5. Pipe with an SDR of 64 will have the thinnest wall while pipe with an SDR of 13.5 will have the thickest wall. When the pressure rating (PR) of a pipe is given in psi, it is based on a line supplying water at 73 degrees F.

The National Sanitation Foundation (NSF) seal of approval is applied to products that can be used with potable water. These products have been tested to ANSI/NSF Standard 14.

Valves and strainers are classified by American National Standards Institute (ANSI) Classes. These Classes are 125, 150, 200, 250, 300, 400, 600, 900, 1500 and 2500. Within each class there is an operating pressure range dependent upon temperature. Design parameters should be compared to the appropriate material dependent, pressure-temperature rating chart for accurate valve selection.

Change Orders

Change Order Considerations

A Change Order is a written document, usually prepared by the design professional, and signed by the owner, the architect/engineer and the contractor. A change order states the agreement of the parties to: an addition, deletion, or revision in the work; an adjustment in the contract sum, if any; or an adjustment in the contract time, if any. Change orders, or "extras" in the construction process occur after execution of the construction contract and impact architects/engineers, contractors and owners.

Change orders that are properly recognized and managed can ensure orderly, professional and profitable progress for all who are involved in the project. There are many causes for change orders and change order requests. In all cases, change orders or change order requests should be addressed promptly and in a precise and prescribed manner. The following paragraphs include information regarding change order pricing and procedures.

The Causes of Change Orders

Reasons for issuing change orders include:

- Unforeseen field conditions that require a change in the work
- Correction of design discrepancies, errors or omissions in the contract documents
- Owner-requested changes, either by design criteria, scope of work, or project objectives
- Completion date changes for reasons unrelated to the construction process
- Changes in building code interpretations, or other public authority requirements that require a change in the work
- Changes in availability of existing or new materials and products

Procedures

Properly written contract documents must include the correct change order procedures for all parties—owners, design professionals and contractors—to follow in order to avoid costly delays and litigation.

Being "in the right" is not always a sufficient or acceptable defense. The contract provisions requiring notification and documentation must be adhered to within a defined or reasonable time frame.

The appropriate method of handling change orders is by a written proposal and acceptance by all parties involved. Prior to starting work on a project, all parties should identify their authorized agents who may sign and accept change orders, as well as any limits placed on their authority.

Time may be a critical factor when the need for a change arises. For such cases, the contractor might be directed to proceed on a "time and materials" basis, rather than wait for all paperwork to be processed—a delay that could impede progress. In this situation, the contractor must still follow the prescribed change order procedures, including but not limited to, notification and documentation.

All forms used for change orders should be dated and signed by the proper authority. Lack of documentation can be very costly, especially if legal judgments are to be made and if certain field personnel are no longer available. For time and material change orders, the contractor should keep accurate daily records of all labor and material allocated to the change. Forms that can be used to document change order work are available in *Means Forms for Building Construction Professionals.*

Owners or awarding authorities who do considerable and continual building construction (such as the federal government) realize the inevitability of change orders for numerous reasons, both predictable and unpredictable. As a result, the federal government, the American Institute of Architects (AIA), the Engineers Joint Contract Documents Committee (EJCDC) and other contractor, legal and technical organizations have developed standards and procedures to be followed by all parties to achieve contract continuance and timely completion, while being financially fair to all concerned.

In addition to the change order standards put forth by industry associations, there are also many books available on the subject.

Pricing Change Orders

When pricing change orders, regardless of their cause, the most significant factor is when the change occurs. The need for a change may be perceived in the field or requested by the architect/engineer *before* any of the actual installation has begun, or may evolve or appear *during* construction when the item of work in question is partially installed. In the latter cases, the original sequence of construction is disrupted, along with all contiguous and supporting systems. Change orders cause the greatest impact when they occur *after* the installation has been completed and must be uncovered, or even replaced. Post-completion changes may be caused by necessary design changes, product failure, or changes in the owner's requirements that are not discovered until the building or the systems begin to function.

Specified procedures of notification and record keeping must be adhered to and enforced regardless of the stage of construction: *before, during,* or *after* installation. Some bidding documents anticipate change orders by requiring that unit prices including overhead and profit percentages—for additional as well as deductible changes—be listed. Generally these unit prices do not fully take into account the ripple effect, or impact on other trades, and should be used for general guidance only.

When pricing change orders, it is important to classify the time frame in which the change occurs. There are two basic time frames for change orders: *pre-installation change orders,* which occur before the start of construction, and *post-installation change orders,* which involve reworking after the original installation. Change orders that occur between these stages may be priced according to the extent of work completed using a combination of techniques developed for pricing *pre-* and *post-installation* changes.

The following factors are the basis for a check list to use when preparing a change order estimate.

Factors To Consider When Pricing Change Orders

As an estimator begins to prepare a change order, the following questions should be reviewed to determine their impact on the final price.

General

- Is the change order work *pre-installation* or *post-installation?*

  Change order work costs vary according to how much of the installation has been completed. Once workers have the project scoped in their mind, even though they have not started, it can be difficult to refocus.

Consequently they may spend more than the normal amount of time understanding the change. Also, modifications to work in place such as trimming or refitting usually take more time than was initially estimated. The greater the amount of work in place, the more reluctant workers are to change it. Psychologically they may resent the change and as a result the rework takes longer than normal. Post-installation change order estimates must include demolition of existing work as required to accomplish the change. If the work is performed at a later time, additional obstacles such as building finishes may be present which must be protected. Regardless of whether the change occurs pre-installation or post-installation, attempt to isolate the identifiable factors and price them separately. For example, add shipping costs that may be required pre-installation or any demolition required post-installation. Then analyze the potential impact on productivity of psychological and/or learning curve factors and adjust the output rates accordingly. One approach is to break down the typical workday into segments and quantify the impact on each segment. The following chart may be useful as a guide:

	Activities (Productivity) Expressed as Percentages of a Workday		
Task	**Means Mechanical Cost Data (for New Construction)**	**Pre-Installation Change Orders**	**Post-Installation Change Orders**
1. Study plans	3%	6%	6%
2. Material procurement	3%	3%	3%
3. Receiving and storing	3%	3%	3%
4. Mobilization	5%	5%	5%
5. Site movement	5%	5%	8%
6. Layout and marking	8%	10%	12%
7. Actual installation	64%	59%	54%
8. Clean-up	3%	3%	3%
9. Breaks—non-productive	6%	6%	6%
Total	100%	100%	100%

Change Order Installation Efficiency

The labor-hours expressed (for new construction) are based on average installation time, using an efficiency level of approximately 60-65%. For change order situations, adjustments to this efficiency level should reflect the daily labor-hour allocation for that particular occurrence.

If any of the specific percentages expressed in the above chart do not apply to a particular project situation, then those percentage points should be reallocated to the appropriate task(s). Example: Using data for new construction, assume there is no new material being utilized. The percentages for Tasks 2 and 3 would therefore be reallocated to other tasks. If the time required for Tasks 2 and 3 can now be applied to installation, we can add the time allocated for *Material Procurement* and *Receiving and Storing* to the *Actual Installation* time for new construction, thereby increasing the Actual Installation percentage.

This chart shows that, due to reduced productivity, labor costs will be higher than those for new construction by 5% to 15% for pre-installation change orders and by 15% to 25% for post-installation change orders. Each job and change order is unique and must be examined individually. Many factors, covered elsewhere in this section, can each have a significant impact on productivity and change order costs. All such factors should be considered in every case.

- Will the change substantially delay the original completion date?

A significant change in the project may cause the original completion date to be extended. The extended schedule may subject the contractor to new wage rates dictated by relevant labor contracts. Project supervision and other project overhead must also be extended beyond the original completion date. The schedule extension may also put installation into a new weather season. For example, underground piping scheduled for October installation was delayed until January. As a result, frost penetrated the trench area, thereby changing the degree of difficulty of the task. Changes and delays may have a ripple effect throughout the project. This effect must be analyzed and negotiated with the owner.

- What is the net effect of a deduct change order?

In most cases, change orders resulting in a deduction or credit reflect only bare costs. The contractor may retain the overhead and profit based on the original bid.

Materials

- Will you have to pay more or less for the new material, required by the change order, than you paid for the original purchase?

The same material prices or discounts will usually apply to materials purchased for change orders as new construction. In some instances, however, the contractor may forfeit the advantages of competitive pricing for change orders. Consider the following example:

A contractor purchased over $20,000 worth of fan coil units for an installation, and obtained the maximum discount. Some time later it was determined the project required an additional matching unit. The contractor has to purchase this unit from the original supplier to ensure a match. The supplier at this time may not discount the unit because of the small quantity, and the fact that he is no longer in a competitive situation. The impact of quantity on purchase can add between 0% and 25% to material prices and/or subcontractor quotes.

- If materials have been ordered or delivered to the job site, will they be subject to a cancellation charge or restocking fee?

Check with the supplier to determine if ordered materials are subject to a cancellation charge. Delivered materials not used as result of a change order may be subject to a restocking fee if returned to the supplier. Common restocking charges run between 20% and 40%. Also, delivery charges to return the goods to the supplier must be added.

Labor

- How efficient is the existing crew at the actual installation?

Is the same crew that performed the initial work going to do the change order? Possibly the change consists of the installation of a unit identical to one already installed; therefore the change should take less time. Be sure to consider this potential productivity increase and modify the productivity rates accordingly.

- If the crew size is increased, what impact will that have on supervision requirements?

Under most bargaining agreements or management practices, there is a point at which a working foreman is replaced by a nonworking foreman. This replacement increases project overhead by adding a nonproductive worker. If additional workers are added to accelerate the project or to perform changes while maintaining the schedule, be sure to add additional supervision time if warranted. Calculate the hours involved and the additional cost directly if possible.

- What are the other impacts of increased crew size?

The larger the crew, the greater the potential for productivity to decrease. Some of the factors that cause this productivity loss are: overcrowding (producing restrictive conditions in the working space), and possibly a shortage of any special tools and equipment required. Such factors affect not only the crew working on the elements directly involved in the change order, but other crews whose movement may also be hampered.

As the crew increases, check its basic composition for changes by the addition or deletion of apprentices or nonworking foreman and quantify the potential effects of equipment shortages or other logistical factors.

- As new crews, unfamiliar with the project, are brought onto the site, how long will it take them to become oriented to the project requirements?

The orientation time for a new crew to become 100% effective varies with the site and type of project. Orientation is easiest at a new construction site, and most difficult at existing, very restrictive renovation sites. The type of work also affects orientation time. When all elements of the work are exposed, such as concrete or masonry work, orientation is decreased. When the work is concealed or less visible, such as existing electrical systems, orientation takes longer. Usually orientation can be accomplished in one day or less. Costs for added orientation should be itemized and added to the total estimated cost.

- How much actual production can be gained by working overtime?

Short term overtime can be used effectively to accomplish more work in a day. However, as overtime is scheduled to run beyond several weeks, studies have shown marked decreases in output. The following chart shows the effect of long term overtime on worker efficiency. If the anticipated change requires extended overtime to keep the job on schedule, these factors can be used as a guide to predict the impact on time and cost. Add project overhead, particularly supervision, that may also be incurred.

Days per Week	Hours per Day	Production Efficiency					Payroll Cost Factors	
		1 Week	2 Weeks	3 Weeks	4 Weeks	Average 4 Weeks	@ 1-1/2 Times	@ 2 Times
5	8	100%	100%	100%	100%	100%	100%	100%
	9	100	100	95	90	96.25	105.6	111.1
	10	100	95	90	85	91.25	110.0	120.0
	11	95	90	75	65	81.25	113.6	127.3
	12	90	85	70	60	76.25	116.7	133.3
6	8	100	100	95	90	96.25	108.3	116.7
	9	100	95	90	85	92.50	113.0	125.9
	10	95	90	85	80	87.50	116.7	133.3
	11	95	85	70	65	78.75	119.7	139.4
	12	90	80	65	60	73.75	122.2	144.4
7	8	100	95	85	75	88.75	114.3	128.6
	9	95	90	80	70	83.75	118.3	136.5
	10	90	85	75	65	78.75	121.4	142.9
	11	85	80	65	60	72.50	124.0	148.1
	12	85	75	60	55	68.75	126.2	152.4

Effects of Overtime

Caution: Under many labor agreements, Sundays and holidays are paid at a higher premium than the normal overtime rate.

The use of long-term overtime is counterproductive on almost any construction job; that is, the longer the period of overtime, the lower the actual production rate. Numerous studies have been conducted, and while they have resulted in slightly different numbers, all reach the same conclusion. The figure above tabulates the effects of overtime work on efficiency.

As illustrated, there can be a difference between the *actual* payroll cost per hour and the *effective* cost per hour for overtime work. This is due to the reduced production efficiency with the increase in weekly hours beyond 40. This difference between actual and effective cost results from overtime work over a prolonged period. Short-term overtime work does not result in as great a reduction in efficiency, and in such cases, effective cost may not vary significantly from the actual payroll cost. As the total hours per week are increased on a regular basis, more time is lost because of fatigue, lowered morale, and an increased accident rate.

As an example, assume a project where workers are working 6 days a week, 10 hours per day. From the figure above (based on productivity studies), the average effective productive hours over a four-week period are:

$$0.875 \times 60 = 52.5$$

Equipment

- What equipment is required to complete the change order?

  Change orders may require extending the rental period of equipment already on the job site, or the addition of special equipment brought in to accomplish the change work. In either case, the additional rental charges and operator labor charges must be added.

Depending upon the locale and day of week, overtime hours may be paid at time and a half or double time. For time and a half, the overall (average) *actual* payroll cost (including regular and overtime hours) is determined as follows:

$$\frac{40 \text{ reg. hrs.} + (20 \text{ overtime hrs.} \times 1.5)}{60 \text{ hrs.}} = 1.167$$

Based on 60 hours, the payroll cost per hour will be 116.7% of the normal rate at 40 hours per week. However, because the effective production (efficiency) for 60 hours is reduced to the equivalent of 52.5 hours, the effective cost of overtime is calculated as follows:

For time and a half:

$$\frac{40 \text{ reg. hrs.} + (20 \text{ overtime hrs.} \times 1.5)}{52.5 \text{ hrs.}} = 1.33$$

Installed cost will be 133% of the normal rate (for labor).

Thus, when figuring overtime, the actual cost per unit of work will be higher than the apparent overtime payroll dollar increase, due to the reduced productivity of the longer workweek. These efficiency calculations are true only for those cost factors determined by hours worked. Costs that are applied weekly or monthly, such as equipment rentals, will not be similarly affected.

Summary

The preceding considerations and others you deem appropriate should be analyzed and applied to a change order estimate. The impact of each should be quantified and listed on the estimate to form an audit trail.

Change orders that are properly identified, documented, and managed help to ensure the orderly, professional and profitable progress of the work. They also minimize potential claims or disputes at the end of the project.

Estimating Tips

- The cost figures in this Square Foot Cost section were derived from approximately 11,200 projects contained in the RSMeans database of completed construction projects. They include the contractor's overhead and profit, but do not generally include architectural fees or land costs. The figures have been adjusted to January of the current year. New projects are added to our files each year, and outdated projects are discarded. For this reason, certain costs may not show a uniform annual progression. In no case are all subdivisions of a project listed.

- These projects were located throughout the U.S. and reflect a tremendous variation in square foot (S.F.) and cubic foot (C.F.) costs. This is due to differences, not only in labor and material costs, but also in individual owners' requirements. For instance, a bank in a large city would have different features than one in a rural area. This is true of all the different types of buildings analyzed. Therefore, caution should be exercised when using these Square Foot costs. For example, for court houses, costs in the database are local court house costs and will not apply to the larger, more elaborate federal court houses. As a general rule, the projects in the 1/4 column do not include any site work or equipment, while the projects in the 3/4 column may include both equipment and site work. The median figures do not generally include site work.

- None of the figures "go with" any others. All individual cost items were computed and tabulated separately. Thus, the sum of the median figures for Plumbing, HVAC and Electrical will not normally total up to the total Mechanical and Electrical costs arrived at by separate analysis and tabulation of the projects.

- Each building was analyzed as to total and component costs and percentages. The figures were arranged in ascending order with the results tabulated as shown. The 1/4 column shows that 25% of the projects had lower costs and 75% had higher. The 3/4 column shows that 75% of the projects had lower costs and 25% had higher. The median column shows that 50% of the projects had lower costs and 50% had higher.

- There are two times when square foot costs are useful. The first is in the conceptual stage when no details are available. Then square foot costs make a useful starting point. The second is after the bids are in and the costs can be worked back into their appropriate units for information purposes. As soon as details become available in the project design, the square foot approach should be discontinued and the project priced as to its particular components. When more precision is required, or for estimating the replacement cost of specific buildings, the current edition of *RSMeans Square Foot Costs* should be used.

- In using the figures in this section, it is recommended that the median column be used for preliminary figures if no additional information is available. The median figures, when multiplied by the total city construction cost index figures (see City Cost Indexes) and then multiplied by the project size modifier at the end of this section, should present a fairly accurate base figure, which would then have to be adjusted in view of the estimator's experience, local economic conditions, code requirements, and the owner's particular requirements. There is no need to factor the percentage figures, as these should remain constant from city to city. All tabulations mentioning air conditioning had at least partial air conditioning.

- The editors of this book would greatly appreciate receiving cost figures on one or more of your recent projects, which would then be included in the averages for next year. All cost figures received will be kept confidential, except that they will be averaged with other similar projects to arrive at Square Foot cost figures for next year's book. See the last page of the book for details and the discount available for submitting one or more of your projects.

		50 17 00 | S.F. Costs	UNIT	UNIT COSTS			% OF TOTAL			
				1/4	MEDIAN	3/4	1/4	MEDIAN	3/4	
01	0010	**APARTMENTS Low Rise (1 to 3 story)**	S.F.	67	84	112				01
	0020	Total project cost	C.F.	6	7.95	9.80				
	0100	Site work	S.F.	5.70	7.80	13.70	6.05%	10.55%	14.05%	
	0500	Masonry		1.31	3.05	5.30	1.54%	3.67%	6.35%	
	1500	Finishes		7.10	9.70	12	9.05%	10.75%	12.85%	
	1800	Equipment		2.18	3.31	4.92	2.73%	4.03%	5.95%	
	2720	Plumbing		5.20	6.70	8.50	6.65%	8.95%	10.05%	
	2770	Heating, ventilating, air conditioning		3.31	4.08	6	4.20%	5.60%	7.60%	
	2900	Electrical		3.86	5.15	6.90	5.20%	6.65%	8.40%	
	3100	Total: Mechanical & Electrical		13.40	17.05	21.50	15.90%	18.05%	23%	
	9000	Per apartment unit, total cost	Apt.	62,000	95,000	140,000				
	9500	Total: Mechanical & Electrical	"	11,800	18,500	24,200				
02	0010	**APARTMENTS Mid Rise (4 to 7 story)**	S.F.	88.50	107	132				02
	0020	Total project costs	C.F.	6.90	9.55	13.05				
	0100	Site work	S.F.	3.54	7	12.60	5.25%	6.70%	9.15%	
	0500	Masonry		5.90	8.10	11.10	5.10%	7.25%	10.50%	
	1500	Finishes		11.15	14.50	18.30	10.55%	13.45%	17.70%	
	1800	Equipment		2.57	3.85	5.05	2.54%	3.48%	4.31%	
	2500	Conveying equipment		1.91	2.43	2.94	1.94%	2.27%	2.69%	
	2720	Plumbing		5.20	8.30	8.80	5.70%	7.20%	8.95%	
	2900	Electrical		5.85	7.90	9.60	6.65%	7.20%	8.95%	
	3100	Total: Mechanical & Electrical		18.70	23.50	28.50	18.50%	21%	23%	
	9000	Per apartment unit, total cost	Apt.	100,000	118,000	195,500				
	9500	Total: Mechanical & Electrical	"	18,900	21,900	27,600				
03	0010	**APARTMENTS High Rise (8 to 24 story)**	S.F.	100	116	139				03
	0020	Total project costs	C.F.	9.75	11.35	14.50				
	0100	Site work	S.F.	3.64	5.90	8.25	2.58%	4.84%	6.15%	
	0500	Masonry		5.80	10.55	13.15	4.74%	9.65%	11.05%	
	1500	Finishes		11.15	13.90	16.45	9.75%	11.80%	13.70%	
	1800	Equipment		3.23	3.97	5.25	2.78%	3.49%	4.35%	
	2500	Conveying equipment		2.28	3.46	4.71	2.23%	2.78%	3.37%	
	2720	Plumbing		7.40	8.70	12.15	6.80%	7.20%	10.45%	
	2900	Electrical		6.90	8.70	11.75	6.45%	7.65%	8.80%	
	3100	Total: Mechanical & Electrical		20.50	26.50	31.50	17.95%	22.50%	24.50%	
	9000	Per apartment unit, total cost	Apt.	104,500	115,000	159,500				
	9500	Total: Mechanical & Electrical	"	22,600	25,800	27,300				
04	0010	**AUDITORIUMS**	S.F.	104	142	204				04
	0020	Total project costs	C.F.	6.55	9.10	13.05				
	2720	Plumbing	S.F.	6.30	9.15	11.10	5.85%	7.20%	8.70%	
	2900	Electrical		8.05	11.75	19.30	6.80%	9.05%	11.35%	
	3100	Total: Mechanical & Electrical		50.50	65.50	90.50	24.50%	29%	31.50%	
05	0010	**AUTOMOTIVE SALES**	S.F.	77	105	130				05
	0020	Total project costs	C.F.	5.10	6.10	7.90				
	2720	Plumbing	S.F.	3.52	6.10	6.65	2.89%	6.05%	6.50%	
	2770	Heating, ventilating, air conditioning		5.45	8.30	8.95	4.61%	10%	10.35%	
	2900	Electrical		6.20	9.75	13.25	7.25%	8.80%	12.15%	
	3100	Total: Mechanical & Electrical		19.50	27.50	33.50	18.90%	20.50%	22%	
06	0010	**BANKS**	S.F.	152	189	239				06
	0020	Total project costs	C.F.	10.80	14.70	19.40				
	0100	Site work	S.F.	17.35	26.50	38.50	7.85%	12.95%	16.95%	
	0500	Masonry		8.15	15.60	27.50	3.36%	6.95%	10.05%	
	1500	Finishes		14	20.50	25.50	5.85%	8.60%	11.55%	
	1800	Equipment		5.70	12.55	25.50	1%	5.55%	10.50%	
	2720	Plumbing		4.74	6.80	9.90	2.82%	3.90%	4.93%	
	2770	Heating, ventilating, air conditioning		9	12.05	16.05	4.86%	7.15%	8.50%	
	2900	Electrical		14.40	19.15	25	8.20%	10.20%	12.20%	
	3100	Total: Mechanical & Electrical		34.50	46.50	55.50	16.25%	19.45%	23%	
	3500	See also division 11 16 00 & 11 17 00								

		50 17 00 | S.F. Costs	UNIT	UNIT COSTS			% OF TOTAL			
				1/4	MEDIAN	3/4	1/4	MEDIAN	3/4	
13	0010	**CHURCHES**	S.F.	102	129	168				**13**
	0020	Total project costs	C.F.	6.30	8	10.50				
	1800	Equipment	S.F.	1.22	2.91	6.20	.95%	2.11%	4.31%	
	2720	Plumbing		3.97	5.55	8.15	3.51%	4.96%	6.25%	
	2770	Heating, ventilating, air conditioning		9.30	12.10	17.10	7.50%	10%	12%	
	2900	Electrical		8.60	11.80	15.85	7.30%	8.80%	10.95%	
	3100	Total: Mechanical & Electrical	↓	27	35.50	47.50	18.30%	22%	25%	
	3500	See also division 11 91 00								
15	0010	**CLUBS, COUNTRY**	S.F.	109	132	166				**15**
	0020	Total project costs	C.F.	8.80	10.75	14.85				
	2720	Plumbing	S.F.	6.60	9.80	22.50	5.60%	7.90%	10%	
	2900	Electrical		8.60	11.80	15.35	7%	8.95%	11%	
	3100	Total: Mechanical & Electrical	↓	45.50	57	60.50	19%	26.50%	29.50%	
17	0010	**CLUBS, SOCIAL Fraternal**	S.F.	87	125	168				**17**
	0020	Total project costs	C.F.	5.45	8.25	9.85				
	2720	Plumbing	S.F.	5.50	6.85	10.35	5.60%	6.90%	8.55%	
	2770	Heating, ventilating, air conditioning		7.90	9.55	12.30	8.20%	9.25%	14.40%	
	2900	Electrical		6.55	10.80	12.35	6.50%	9.50%	10.55%	
	3100	Total: Mechanical & Electrical	↓	34	36.50	46.50	21%	23%	23.50%	
18	0010	**CLUBS, Y.M.C.A.**	S.F.	114	146	181				**18**
	0020	Total project costs	C.F.	5.05	8.45	12.60				
	2720	Plumbing	S.F.	6.90	13.80	15.45	5.65%	7.60%	10.85%	
	2900	Electrical		8.95	10.95	16.40	6.25%	7.95%	9.25%	
	3100	Total: Mechanical & Electrical	↓	33.50	40.50	68	20.50%	22.50%	28.50%	
19	0010	**COLLEGES Classrooms & Administration**	S.F.	111	151	201				**19**
	0020	Total project costs	C.F.	8.10	11.75	18.10				
	0500	Masonry	S.F.	8.10	15.30	18.80	5.10%	7.45%	10.50%	
	2720	Plumbing		5.65	11.55	21.50	5.10%	6.60%	8.95%	
	2900	Electrical		9.20	14	19.45	7.80%	9.85%	12%	
	3100	Total: Mechanical & Electrical	↓	36.50	51.50	69	23.50%	28%	32%	
21	0010	**COLLEGES Science, Engineering, Laboratories**	S.F.	206	241	279				**21**
	0020	Total project costs	C.F.	11.80	17.20	19.55				
	1800	Equipment	S.F.	11.45	26	28.50	2%	6.45%	12.65%	
	2900	Electrical		16.95	24	37	7.10%	9.40%	12.10%	
	3100	Total: Mechanical & Electrical	↓	63	74.50	116	28.50%	31.50%	41%	
	3500	See also division 11 53 00								
23	0010	**COLLEGES Student Unions**	S.F.	131	178	215				**23**
	0020	Total project costs	C.F.	7.30	9.60	11.85				
	3100	Total: Mechanical & Electrical	S.F.	49.50	53.50	63	23.50%	26%	29%	
25	0010	**COMMUNITY CENTERS**	S.F.	108	134	181				**25**
	0020	Total project costs	C.F.	7.05	10.10	13.10				
	1800	Equipment	S.F.	2.62	4.44	7.05	1.50%	3.04%	5.45%	
	2720	Plumbing		5.15	9	12.25	4.85%	7%	8.95%	
	2770	Heating, ventilating, air conditioning		8.20	12.05	17.15	6.80%	10.35%	12.90%	
	2900	Electrical		8.85	11.80	17	7.20%	8.95%	10.45%	
	3100	Total: Mechanical & Electrical	↓	30.50	38	55	20%	25%	31%	
28	0010	**COURT HOUSES**	S.F.	156	184	254				**28**
	0020	Total project costs	C.F.	12	14.35	18.05				
	2720	Plumbing	S.F.	7.45	10.40	14.90	5.95%	7.45%	8.20%	
	2900	Electrical		16.60	18.45	27	8.90%	10.45%	11.85%	
	3100	Total: Mechanical & Electrical	↓	46.50	61	66.50	22.50%	27.50%	30%	
30	0010	**DEPARTMENT STORES**	S.F.	58	78	98.50				**30**
	0020	Total project costs	C.F.	3.10	4.01	5.45				
	2720	Plumbing	S.F.	1.80	2.27	3.45	1.82%	4.21%	5.90%	
	2770	Heating, ventilating, air conditioning	↓	5.25	8.10	12.20	8.20%	9.10%	14.80%	

		50 17 00 | S.F. Costs	UNIT	UNIT COSTS			% OF TOTAL			
				1/4	MEDIAN	3/4	1/4	MEDIAN	3/4	
30	2900	Electrical	S.F.	6.65	9.10	10.75	9.05%	12.15%	14.95%	30
	3100	Total: Mechanical & Electrical	↓	11.65	14.95	26	13.20%	21.50%	50%	
31	0010	**DORMITORIES Low Rise (1 to 3 story)**	S.F.	110	143	190				31
	0020	Total project costs	C.F.	6.20	10	15				
	2720	Plumbing	S.F.	6.60	8.80	11.15	8.05%	9%	9.65%	
	2770	Heating, ventilating, air conditioning		6.95	8.35	11.10	4.61%	8.05%	10%	
	2900	Electrical		7.25	11.05	15.10	6.40%	8.65%	9.50%	
	3100	Total: Mechanical & Electrical	↓	38	40.50	63	22%	25%	27%	
	9000	Per bed, total cost	Bed	46,400	51,500	110,500				
32	0010	**DORMITORIES Mid Rise (4 to 8 story)**	S.F.	135	176	217				32
	0020	Total project costs	C.F.	14.85	16.30	19.50				
	2900	Electrical	S.F.	14.30	16.25	22	8.20%	10.20%	11.95%	
	3100	Total: Mechanical & Electrical	"	40	80	81	25.50%	34.50%	37.50%	
	9000	Per bed, total cost	Bed	19,200	43,700	253,000				
34	0010	**FACTORIES**	S.F.	51	76	117				34
	0020	Total project costs	C.F.	3.27	4.87	8.10				
	0100	Site work	S.F.	5.80	10.60	16.75	6.95%	11.45%	17.95%	
	2720	Plumbing		2.74	5.10	8.45	3.73%	6.05%	8.10%	
	2770	Heating, ventilating, air conditioning		5.35	7.65	10.35	5.25%	8.45%	11.35%	
	2900	Electrical		6.35	10	15.30	8.10%	10.50%	14.20%	
	3100	Total: Mechanical & Electrical	↓	18.10	27.50	36.50	21%	28.50%	35.50%	
36	0010	**FIRE STATIONS**	S.F.	101	139	187				36
	0020	Total project costs	C.F.	5.90	8.10	10.80				
	0500	Masonry	S.F.	14.95	26.50	35	8.35%	11.55%	16.15%	
	1140	Roofing		3.29	8.90	10.15	1.90%	4.94%	5.05%	
	1580	Painting		2.56	3.82	3.91	1.37%	1.57%	2.07%	
	1800	Equipment		1.26	2.41	4.45	.62%	1.86%	3.54%	
	2720	Plumbing		5.65	9.10	13.15	5.85%	7.35%	9.45%	
	2770	Heating, ventilating, air conditioning		5.60	9.05	14	5.15%	7.40%	9.40%	
	2900	Electrical		7.20	12.65	17.15	6.80%	8.60%	10.60%	
	3100	Total: Mechanical & Electrical	↓	37	47.50	53.50	18.40%	23%	26%	
37	0010	**FRATERNITY HOUSES & Sorority Houses**	S.F.	101	130	178				37
	0020	Total project costs	C.F.	10.05	10.50	12.65				
	2720	Plumbing	S.F.	7.65	8.75	16	6.80%	8%	10.85%	
	2900	Electrical		6.65	14.40	17.65	6.60%	9.90%	10.65%	
	3100	Total: Mechanical & Electrical	↓	7.85	25.50	30.50		15.10%	15.90%	
38	0010	**FUNERAL HOMES**	S.F.	107	145	264				38
	0020	Total project costs	C.F.	10.90	12.10	23.50				
	2900	Electrical	S.F.	4.71	8.65	9.45	3.58%	4.44%	5.95%	
	3100	Total: Mechanical & Electrical	↓	16.75	24.50	34	12.90%	12.90%	12.90%	
39	0010	**GARAGES, COMMERCIAL (Service)**	S.F.	60.50	93.50	129				39
	0020	Total project costs	C.F.	3.97	5.85	8.50				
	1800	Equipment	S.F.	3.40	7.65	11.90	2.21%	4.62%	6.80%	
	2720	Plumbing		4.18	6.45	11.75	5.45%	7.85%	10.65%	
	2730	Heating & ventilating		5.50	7.30	9.85	5.25%	6.85%	8.20%	
	2900	Electrical		5.75	8.75	12.60	7.15%	9.25%	10.85%	
	3100	Total: Mechanical & Electrical	↓	12.65	24.50	36	12.35%	17.40%	26%	
40	0010	**GARAGES, MUNICIPAL (Repair)**	S.F.	94	119	169				40
	0020	Total project costs	C.F.	5.55	7.05	12.10				
	0500	Masonry	S.F.	8.30	16.25	25.50	5.60%	9.15%	12.50%	
	2720	Plumbing		3.98	7.65	14.40	3.59%	6.70%	7.95%	
	2730	Heating & ventilating		6.80	9.85	19	6.15%	7.45%	13.50%	
	2900	Electrical		6.55	10.50	14.95	6.65%	9.25%	11.15%	
	3100	Total: Mechanical & Electrical	↓	30.50	42	61	21.50%	23%	28.50%	

50 17 00 | S.F. Costs		UNIT	UNIT COSTS			% OF TOTAL		
			1/4	MEDIAN	3/4	1/4	MEDIAN	3/4
41 0010	**GARAGES, PARKING**	S.F.	34.50	50.50	86.50				**41**
0020	Total project costs	C.F.	3.24	4.40	6.40			
2720	Plumbing	S.F.	.98	1.52	2.34	1.72%	2.70%	3.85%
2900	Electrical		1.89	2.32	3.65	4.33%	5.20%	6.30%
3100	Total: Mechanical & Electrical	↓	3.87	5.40	6.70	7%	8.90%	11.05%
3200								
9000	Per car, total cost	Car	14,600	18,300	23,300			
43 0010	**GYMNASIUMS**	S.F.	96	128	173				**43**
0020	Total project costs	C.F.	4.79	6.50	8			
1800	Equipment	S.F.	2.28	4.28	8.20	1.98%	3.30%	6.70%
2720	Plumbing		6.05	7.20	9.30	4.65%	6.40%	7.75%
2770	Heating, ventilating, air conditioning		6.55	9.95	19.95	5.15%	9.05%	11.10%
2900	Electrical		7.40	10.05	13.25	6.60%	8.30%	10.30%
3100	Total: Mechanical & Electrical	↓	27	36.50	43.50	19.60%	23.50%	28%
3500	See also division 11 67 00							
46 0010	**HOSPITALS**	S.F.	184	230	315				**46**
0020	Total project costs	C.F.	13.95	17.40	25			
1800	Equipment	S.F.	4.68	9	15.50	.80%	2.53%	4.80%
2720	Plumbing		15.85	22.50	29	7.60%	9.10%	10.85%
2770	Heating, ventilating, air conditioning		23.50	30	41.50	7.80%	12.95%	16.65%
2900	Electrical		20.50	27.50	39	9.90%	11.75%	14.10%
3100	Total: Mechanical & Electrical	↓	59.50	82.50	123	28%	33.50%	37%
9000	Per bed or person, total cost	Bed	213,500	294,000	338,500			
9900	See also division 11 71 00							
48 0010	**HOUSING For the Elderly**	S.F.	90.50	115	141				**48**
0020	Total project costs	C.F.	6.45	8.95	11.45			
0100	Site work	S.F.	6.30	9.80	14.35	5.05%	7.90%	12.10%
0500	Masonry		2.76	10.30	15.05	1.30%	6.05%	11%
1800	Equipment		2.19	3.02	4.80	1.88%	3.23%	4.43%
2510	Conveying systems		2.20	2.96	4.02	1.78%	2.20%	2.81%
2720	Plumbing		6.75	8.60	10.80	8.15%	9.55%	10.50%
2730	Heating, ventilating, air conditioning		3.45	4.89	7.30	3.30%	5.60%	7.25%
2900	Electrical		6.75	9.15	11.75	7.30%	8.50%	10.25%
3100	Total: Mechanical & Electrical	↓	24.50	29	37	18.10%	22.50%	29%
9000	Per rental unit, total cost	Unit	84,500	98,500	110,000			
9500	Total: Mechanical & Electrical	"	18,800	21,600	25,200			
50 0010	**HOUSING Public (Low Rise)**	S.F.	76	106	138				**50**
0020	Total project costs	C.F.	6.80	8.45	10.50			
0100	Site work	S.F.	9.70	13.95	22.50	8.35%	11.75%	16.50%
1800	Equipment		2.07	3.38	5.15	2.26%	3.03%	4.24%
2720	Plumbing		5.50	7.25	9.20	7.15%	9.05%	11.60%
2730	Heating, ventilating, air conditioning		2.76	5.35	5.85	4.26%	6.05%	6.45%
2900	Electrical		4.61	6.85	9.50	5.10%	6.55%	8.25%
3100	Total: Mechanical & Electrical	↓	22	28	31.50	14.50%	17.55%	26.50%
9000	Per apartment, total cost	Apt.	83,500	95,000	119,500			
9500	Total: Mechanical & Electrical	"	17,800	22,000	24,400			
51 0010	**ICE SKATING RINKS**	S.F.	65	152	167				**51**
0020	Total project costs	C.F.	4.79	4.90	5.65			
2720	Plumbing	S.F.	2.43	4.56	4.66	3.12%	3.23%	5.65%
2900	Electrical		6.95	10.70	11.30	6.30%	10.15%	15.05%
3100	Total: Mechanical & Electrical	↓	11.55	16.35	20	18.95%	18.95%	18.95%
52 0010	**JAILS**	S.F.	198	256	330				**52**
0020	Total project costs	C.F.	17.85	25	29			
1800	Equipment	S.F.	7.75	23	39	2.77%	5.55%	10.35%
2720	Plumbing		19.10	25.50	34	7%	8.90%	13.35%
2770	Heating, ventilating, air conditioning		17.90	24	46	7.50%	9.45%	17.75%
2900	Electrical	↓	21	28	35	8.80%	11.55%	14.95%

		**50 17 00	S.F. Costs**	UNIT	UNIT COSTS			% OF TOTAL			
				1/4	MEDIAN	3/4	1/4	MEDIAN	3/4	
52	3100	Total: Mechanical & Electrical	S.F.	54	98.50	117	27.50%	30%	34%	52
53	0010	**LIBRARIES**	S.F.	127	162	209				53
	0020	Total project costs	C.F.	8.55	10.75	13.70				
	0500	Masonry	S.F.	10	17.50	29.50	5.80%	7.60%	11.60%	
	1800	Equipment		1.71	4.60	6.95	.37%	1.50%	4.07%	
	2720	Plumbing		4.63	6.75	9.10	3.38%	4.60%	5.70%	
	2770	Heating, ventilating, air conditioning		10.25	17.35	22.50	7.80%	10.95%	12.80%	
	2900	Electrical		12.85	16.65	21.50	8.35%	10.30%	11.95%	
	3100	Total: Mechanical & Electrical		37.50	48	59.50	20.50%	23%	26.50%	
54	0010	**LIVING, ASSISTED**	S.F.	116	137	161				54
	0020	Total project costs	C.F.	9.75	11.40	12.95				
	0500	Masonry	S.F.	3.41	4.06	4.77	2.36%	3.16%	3.86%	
	1800	Equipment		2.63	3.06	3.92	2.12%	2.47%	2.87%	
	2720	Plumbing		9.70	13	13.45	6.05%	8.15%	10.60%	
	2770	Heating, ventilating, air conditioning		11.50	12.05	13.20	7.95%	9.35%	9.70%	
	2900	Electrical		11.35	12.55	14.50	9%	10%	10.70%	
	3100	Total: Mechanical & Electrical		31.50	37	43	25%	28.50%	31.50%	
55	0010	**MEDICAL CLINICS**	S.F.	118	145	184				55
	0020	Total project costs	C.F.	8.65	11.20	14.85				
	1800	Equipment	S.F.	3.19	6.70	10.40	1.05%	2.94%	6.35%	
	2720	Plumbing		7.85	11.05	14.75	6.15%	8.40%	10.10%	
	2770	Heating, ventilating, air conditioning		9.35	12.25	18	6.65%	8.85%	11.35%	
	2900	Electrical		10.15	14.40	18.80	8.10%	10%	12.25%	
	3100	Total: Mechanical & Electrical		32.50	44	60	22.50%	27%	33.50%	
	3500	See also division 11 71 00								
57	0010	**MEDICAL OFFICES**	S.F.	111	137	168				57
	0020	Total project costs	C.F.	8.30	11.20	15.15				
	1800	Equipment	S.F.	3.66	7.25	10.30	.70%	5.10%	7.05%	
	2720	Plumbing		6.15	9.45	12.75	5.60%	6.80%	8.50%	
	2770	Heating, ventilating, air conditioning		7.40	10.70	14.10	6.10%	8%	9.70%	
	2900	Electrical		9.05	12.85	17.90	7.60%	9.80%	11.40%	
	3100	Total: Mechanical & Electrical		24	34.50	50	19.30%	22.50%	28.50%	
59	0010	**MOTELS**	S.F.	70	101	132				59
	0020	Total project costs	C.F.	6.25	8.35	13.65				
	2720	Plumbing	S.F.	7.10	9.05	10.75	9.45%	10.60%	12.55%	
	2770	Heating, ventilating, air conditioning		4.32	6.45	11.55	5.60%	5.60%	10%	
	2900	Electrical		6.60	8.35	10.40	7.45%	9.05%	10.45%	
	3100	Total: Mechanical & Electrical		21	28	48.50	18.50%	24%	25.50%	
	5000									
	9000	Per rental unit, total cost	Unit	35,600	67,500	73,000				
	9500	Total: Mechanical & Electrical	"	6,950	10,500	12,200				
60	0010	**NURSING HOMES**	S.F.	110	142	173				60
	0020	Total project costs	C.F.	8.60	10.75	14.70				
	1800	Equipment	S.F.	3.46	4.58	7.65	2.02%	3.62%	4.99%	
	2720	Plumbing		9.40	14.25	17.15	8.75%	10.10%	12.70%	
	2770	Heating, ventilating, air conditioning		9.90	15.05	19.90	9.70%	11.45%	11.80%	
	2900	Electrical		10.85	13.55	18.45	9.40%	10.55%	12.45%	
	3100	Total: Mechanical & Electrical		26	36.50	60.50	26%	29.50%	30.50%	
	9000	Per bed or person, total cost	Bed	48,700	61,000	78,500				
61	0010	**OFFICES Low Rise (1 to 4 story)**	S.F.	92.50	120	156				61
	0020	Total project costs	C.F.	6.60	9.10	12				
	0100	Site work	S.F.	7.40	13.15	19.35	6.20%	9.70%	13.55%	
	0500	Masonry		3.58	7.35	13.15	2.62%	5.45%	8.45%	
	1800	Equipment		.98	1.92	5.25	.60%	1.50%	3.50%	
	2720	Plumbing		3.31	5.10	7.50	3.66%	4.50%	6.10%	
	2770	Heating, ventilating, air conditioning		7.30	10.20	14.90	7.20%	10.30%	11.70%	
	2900	Electrical		7.55	10.75	15.25	7.45%	9.65%	11.35%	

		50 17 00 | S.F. Costs	UNIT	UNIT COSTS			% OF TOTAL			
				1/4	MEDIAN	3/4	1/4	MEDIAN	3/4	
61	3100	Total: Mechanical & Electrical	S.F.	21	28.50	42.50	18.20%	22%	27%	61
62	0010	**OFFICES Mid Rise (5 to 10 story)**	S.F.	98	119	161				62
	0020	Total project costs	C.F.	6.95	8.85	12.55				
	2720	Plumbing	S.F.	2.96	4.58	6.60	2.83%	3.74%	4.50%	
	2770	Heating, ventilating, air conditioning		7.45	10.60	16.95	7.65%	9.40%	11%	
	2900	Electrical		7.25	9.40	14.05	6.35%	7.80%	10%	
	3100	Total: Mechanical & Electrical		18.80	24.50	46	18.95%	21%	27.50%	
63	0010	**OFFICES High Rise (11 to 20 story)**	S.F.	120	152	187				63
	0020	Total project costs	C.F.	8.40	10.50	15.05				
	2900	Electrical	S.F.	7.30	8.90	13.20	5.80%	7.85%	10.50%	
	3100	Total: Mechanical & Electrical		23.50	31.50	53	16.90%	23.50%	34%	
64	0010	**POLICE STATIONS**	S.F.	144	189	239				64
	0020	Total project costs	C.F.	11.50	14.05	19.25				
	0500	Masonry	S.F.	13.55	24	30	6.70%	9.10%	11.35%	
	1800	Equipment		2.29	10.05	15.95	.98%	3.35%	6.70%	
	2720	Plumbing		8.05	16.05	20	5.65%	6.90%	10.75%	
	2770	Heating, ventilating, air conditioning		12.60	16.75	23.50	5.85%	10.55%	11.70%	
	2900	Electrical		15.75	22.50	30	9.80%	11.85%	14.80%	
	3100	Total: Mechanical & Electrical		51.50	62	84	25%	31.50%	32.50%	
65	0010	**POST OFFICES**	S.F.	114	141	179				65
	0020	Total project costs	C.F.	6.85	8.65	9.85				
	2720	Plumbing	S.F.	5.15	6.35	8	4.24%	5.30%	5.60%	
	2770	Heating, ventilating, air conditioning		8	9.90	11	6.65%	7.15%	9.35%	
	2900	Electrical		9.40	13.20	15.70	7.25%	9%	11%	
	3100	Total: Mechanical & Electrical		27	35.50	40	16.25%	18.80%	22%	
66	0010	**POWER PLANTS**	S.F.	790	1,050	1,925				66
	0020	Total project costs	C.F.	21.50	47.50	101				
	2900	Electrical	S.F.	56	118	176	9.30%	12.75%	21.50%	
	8100	Total: Mechanical & Electrical		139	450	1,025	32.50%	32.50%	52.50%	
67	0010	**RELIGIOUS EDUCATION**	S.F.	91.50	120	147				67
	0020	Total project costs	C.F.	5.10	7.30	9.10				
	2720	Plumbing	S.F.	3.83	5.45	7.70	4.40%	5.30%	7.10%	
	2770	Heating, ventilating, air conditioning		9.70	10.95	15.50	10.05%	11.45%	12.35%	
	2900	Electrical		7.25	10.25	13.55	7.60%	9.05%	10.35%	
	3100	Total: Mechanical & Electrical		29.50	39	47.50	22%	23%	27%	
69	0010	**RESEARCH Laboratories & Facilities**	S.F.	137	197	287				69
	0020	Total project costs	C.F.	10.35	20	24				
	1800	Equipment	S.F.	6.10	12	29.50	.94%	4.58%	9.10%	
	2720	Plumbing		13.75	17.60	28.50	6.15%	8.30%	10.80%	
	2770	Heating, ventilating, air conditioning		12.35	41.50	49	7.25%	16.50%	17.50%	
	2900	Electrical		16.05	26.50	43.50	9.45%	11.15%	15.40%	
	3100	Total: Mechanical & Electrical		52.50	91.50	132	31.50%	37.50%	42%	
70	0010	**RESTAURANTS**	S.F.	133	172	223				70
	0020	Total project costs	C.F.	11.20	14.70	19.25				
	1800	Equipment	S.F.	8.60	21.50	32	6.10%	13%	15.65%	
	2720	Plumbing		10.55	12.80	16.80	6.10%	8.15%	9%	
	2770	Heating, ventilating, air conditioning		13.40	18.50	22.50	9.20%	12%	12.40%	
	2900	Electrical		14.10	17.35	22.50	8.35%	10.55%	11.55%	
	3100	Total: Mechanical & Electrical		43.50	46.50	60.50	21%	24.50%	29.50%	
	9000	Per seat unit, total cost	Seat	4,875	6,500	7,700				
	9500	Total: Mechanical & Electrical	"	1,225	1,625	1,925				
72	0010	**RETAIL STORES**	S.F.	62	83.50	111				72
	0020	Total project costs	C.F.	4.21	6	8.35				
	2720	Plumbing	S.F.	2.25	3.76	6.40	3.26%	4.60%	6.80%	
	2770	Heating, ventilating, air conditioning		4.87	6.65	10	6.75%	8.75%	10.15%	
	2900	Electrical		5.60	7.65	11.05	7.25%	9.90%	11.60%	
	3100	Total: Mechanical & Electrical		14.90	19.05	26.50	17.05%	21%	23.50%	

				UNIT COSTS			% OF TOTAL			
	50 17 00 | S.F. Costs		UNIT	1/4	MEDIAN	3/4	1/4	MEDIAN	3/4	
74	0010	**SCHOOLS Elementary**	S.F.	101	125	154				74
	0020	Total project costs	C.F.	6.65	8.50	11				
	0500	Masonry	S.F.	9.10	15.60	23	5.45%	10.65%	14.10%	
	1800	Equipment		2.77	4.73	8.70	1.90%	3.20%	4.70%	
	2720	Plumbing		5.85	8.25	11	5.70%	7.15%	9.35%	
	2730	Heating, ventilating, air conditioning		8.75	13.90	19.45	8.15%	10.80%	14.90%	
	2900	Electrical		9.55	12.65	16.10	8.40%	10.05%	11.70%	
	3100	Total: Mechanical & Electrical	↓	34.50	43	53	25%	27.50%	30%	
	9000	Per pupil, total cost	Ea.	11,600	17,300	38,700				
	9500	Total: Mechanical & Electrical	"	3,275	4,150	10,500				
76	0010	**SCHOOLS Junior High & Middle**	S.F.	105	128	157				76
	0020	Total project costs	C.F.	6.65	8.60	9.65				
	0500	Masonry	S.F.	13.40	17.60	20.50	8.60%	11.60%	14.35%	
	1800	Equipment		3.34	5.40	8.10	1.79%	3.09%	4.86%	
	2720	Plumbing		6.10	7.50	9.30	5.30%	6.80%	7.25%	
	2770	Heating, ventilating, air conditioning		12.15	14.80	26	8.90%	11.55%	14.20%	
	2900	Electrical		10.25	12.35	15.95	7.90%	9.35%	10.60%	
	3100	Total: Mechanical & Electrical	↓	33.50	43	53	23.50%	27%	29.50%	
	9000	Per pupil, total cost	Ea.	13,300	17,400	23,400				
78	0010	**SCHOOLS Senior High**	S.F.	108	133	167				78
	0020	Total project costs	C.F.	6.55	9.65	15.55				
	1800	Equipment	S.F.	2.86	6.75	9.50	1.88%	2.91%	4.56%	
	2720	Plumbing		6.10	9.15	16.75	5.60%	6.90%	8.30%	
	2770	Heating, ventilating, air conditioning		12.45	14.25	27	8.95%	11.60%	15%	
	2900	Electrical		11	14.25	21.50	8.65%	10.20%	12.25%	
	3100	Total: Mechanical & Electrical	↓	36	42.50	71	23.50%	26.50%	28.50%	
	9000	Per pupil, total cost	Ea.	10,300	20,900	26,100				
80	0010	**SCHOOLS Vocational**	S.F.	88	128	158				80
	0020	Total project costs	C.F.	5.45	7.85	10.85				
	0500	Masonry	S.F.	5.15	12.75	19.50	3.20%	4.61%	10.95%	
	1800	Equipment		2.75	6.85	9.55	1.24%	3.10%	4.26%	
	2720	Plumbing		5.65	8.40	12.35	5.40%	6.90%	8.55%	
	2770	Heating, ventilating, air conditioning		7.90	14.65	24.50	8.60%	11.90%	14.65%	
	2900	Electrical		9.15	12.55	17.15	8.45%	11%	13.20%	
	3100	Total: Mechanical & Electrical	↓	34.50	35	60.50	27.50%	29.50%	31%	
	9000	Per pupil, total cost	Ea.	12,300	32,800	49,000				
83	0010	**SPORTS ARENAS**	S.F.	77	103	158				83
	0020	Total project costs	C.F.	4.18	7.50	9.65				
	2720	Plumbing	S.F.	4.46	6.75	14.30	4.35%	6.35%	9.40%	
	2770	Heating, ventilating, air conditioning		9.60	11.35	15.75	8.80%	10.20%	13.55%	
	2900	Electrical		8	10.90	14.05	8.60%	9.90%	12.25%	
	3100	Total: Mechanical & Electrical	↓	19.95	35	47	21.50%	25%	27.50%	
85	0010	**SUPERMARKETS**	S.F.	71	82.50	96.50				85
	0020	Total project costs	C.F.	3.96	4.79	7.25				
	2720	Plumbing	S.F.	3.97	5	5.80	5.40%	6%	7.45%	
	2770	Heating, ventilating, air conditioning		5.85	7.75	9.45	8.60%	8.65%	9.60%	
	2900	Electrical		8.90	10.25	12.10	10.40%	12.45%	13.60%	
	3100	Total: Mechanical & Electrical	↓	23	25	32.50	20.50%	26.50%	31%	
86	0010	**SWIMMING POOLS**	S.F.	115	193	410				86
	0020	Total project costs	C.F.	9.25	11.50	12.55				
	2720	Plumbing	S.F.	10.65	12.15	16.45	4.80%	9.70%	20.50%	
	2900	Electrical		8.65	14	20.50	5.75%	6.95%	7.60%	
	3100	Total: Mechanical & Electrical	↓	21	54.50	73	11.15%	14.10%	23.50%	
87	0010	**TELEPHONE EXCHANGES**	S.F.	153	224	284				87
	0020	Total project costs	C.F.	9.50	15.25	21				
	2720	Plumbing	S.F.	6.45	9.95	14.60	4.52%	5.80%	6.90%	
	2770	Heating, ventilating, air conditioning	↓	14.95	30	37.50	11.80%	16.05%	18.40%	

		50 17 00 | S.F. Costs	UNIT	UNIT COSTS			% OF TOTAL			
				1/4	MEDIAN	3/4	1/4	MEDIAN	3/4	
87	2900	Electrical	S.F.	15.55	24.50	43.50	10.90%	14%	17.85%	87
	3100	Total: Mechanical & Electrical	↓	46	87	123	29.50%	33.50%	44.50%	
91	0010	**THEATERS**	S.F.	96	120	182				91
	0020	Total project costs	C.F.	4.44	6.55	9.65				
	2720	Plumbing	S.F.	3.20	3.47	14.20	2.92%	4.70%	6.80%	
	2770	Heating, ventilating, air conditioning		9.35	11.30	14	8%	12.25%	13.40%	
	2900	Electrical		8.40	11.35	23	8.05%	9.35%	12.25%	
	3100	Total: Mechanical & Electrical	↓	21.50	29	34.50	21.50%	25.50%	27.50%	
94	0010	**TOWN HALLS City Halls & Municipal Buildings**	S.F.	107	136	178				94
	0020	Total project costs	C.F.	9.80	11.75	16.50				
	2720	Plumbing	S.F.	4.48	8.40	15.40	4.31%	5.95%	7.95%	
	2770	Heating, ventilating, air conditioning		8.10	16.10	23.50	7.05%	9.05%	13.45%	
	2900	Electrical		10.20	14.55	19.85	8.05%	9.45%	11.65%	
	3100	Total: Mechanical & Electrical	↓	35.50	45	68.50	22%	26.50%	31%	
97	0010	**WAREHOUSES & Storage Buildings**	S.F.	40	59.50	85.50				97
	0020	Total project costs	C.F.	2.10	3.28	5.45				
	0100	Site work	S.F.	4.13	8.20	12.40	6.05%	12.95%	19.85%	
	0500	Masonry		2.41	5.70	12.30	3.73%	7.40%	12.30%	
	1800	Equipment		.64	1.39	7.80	.91%	1.82%	5.55%	
	2720	Plumbing		1.34	2.39	4.48	2.90%	4.80%	6.55%	
	2730	Heating, ventilating, air conditioning		1.53	4.30	5.75	2.41%	5%	8.90%	
	2900	Electrical		2.37	4.46	7.40	5.15%	7.20%	10.10%	
	3100	Total: Mechanical & Electrical	↓	6.60	10.15	20	12.75%	18.90%	26%	
99	0010	**WAREHOUSE & OFFICES Combination**	S.F.	49	65.50	90				99
	0020	Total project costs	C.F.	2.52	3.66	5.40				
	1800	Equipment	S.F.	.86	1.65	2.46	.52%	1.20%	2.40%	
	2720	Plumbing		1.90	3.37	4.93	3.74%	4.76%	6.30%	
	2770	Heating, ventilating, air conditioning		3	4.69	6.55	5%	5.65%	10.05%	
	2900	Electrical		3.30	4.91	7.75	5.75%	8%	10%	
	3100	Total: Mechanical & Electrical	↓	9.20	14.20	22.50	14.40%	19.95%	24.50%	

Square Foot Project Size Modifier

One factor that affects the S.F. cost of a particular building is the size. In general, for buildings built to the same specifications in the same locality, the larger building will have the lower S.F. cost. This is due mainly to the decreasing contribution of the exterior walls plus the economy of scale usually achievable in larger buildings. The Area Conversion Scale shown below will give a factor to convert costs for the typical size building to an adjusted cost for the particular project.

The Square Foot Base Size lists the median costs, most typical project size in our accumulated data, and the range in size of the projects.

The Size Factor for your project is determined by dividing your project area in S.F. by the typical project size for the particular Building Type. With this factor, enter the Area Conversion Scale at the appropriate Size Factor and determine the appropriate cost multiplier for your building size.

Example: Determine the cost per S.F. for a 100,000 S.F. Mid-rise apartment building.

$$\frac{\text{Proposed building area} = 100,000 \text{ S.F.}}{\text{Typical size from below} = 50,000 \text{ S.F.}} = 2.00$$

Enter Area Conversion scale at 2.0, intersect curve, read horizontally the appropriate cost multiplier of .94. Size adjusted cost becomes .94 x $107.00 = $101.00 based on national average costs.

Note: For Size Factors less than .50, the Cost Multiplier is 1.1
For Size Factors greater than 3.5, the Cost Multiplier is .90

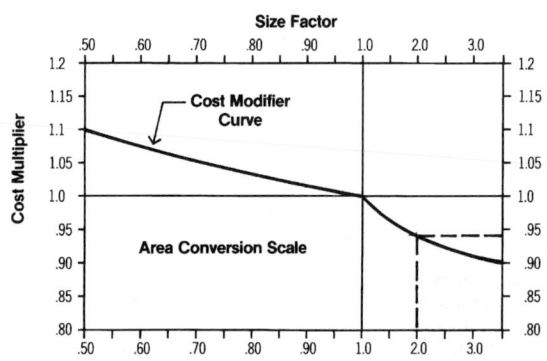

Square Foot Base Size

Building Type	Median Cost per S.F.	Typical Size Gross S.F.	Typical Range Gross S.F.	Building Type	Median Cost per S.F.	Typical Size Gross S.F.	Typical Range Gross S.F.
Apartments, Low Rise	$ 84.00	21,000	9,700 - 37,200	Jails	$ 256.00	40,000	5,500 - 145,000
Apartments, Mid Rise	107.00	50,000	32,000 - 100,000	Libraries	162.00	12,000	7,000 - 31,000
Apartments, High Rise	116.00	145,000	95,000 - 600,000	Living, Assisted	137.00	32,300	23,500 - 50,300
Auditoriums	142.00	25,000	7,600 - 39,000	Medical Clinics	145.00	7,200	4,200 - 15,700
Auto Sales	105.00	20,000	10,800 - 28,600	Medical Offices	137.00	6,000	4,000 - 15,000
Banks	189.00	4,200	2,500 - 7,500	Motels	101.00	40,000	15,800 - 120,000
Churches	129.00	17,000	2,000 - 42,000	Nursing Homes	142.00	23,000	15,000 - 37,000
Clubs, Country	132.00	6,500	4,500 - 15,000	Offices, Low Rise	120.00	20,000	5,000 - 80,000
Clubs, Social	125.00	10,000	6,000 - 13,500	Offices, Mid Rise	119.00	120,000	20,000 - 300,000
Clubs, YMCA	146.00	28,300	12,800 - 39,400	Offices, High Rise	152.00	260,000	120,000 - 800,000
Colleges (Class)	151.00	50,000	15,000 - 150,000	Police Stations	189.00	10,500	4,000 - 19,000
Colleges (Science Lab)	241.00	45,600	16,600 - 80,000	Post Offices	141.00	12,400	6,800 - 30,000
College (Student Union)	178.00	33,400	16,000 - 85,000	Power Plants	1050.00	7,500	1,000 - 20,000
Community Center	134.00	9,400	5,300 - 16,700	Religious Education	120.00	9,000	6,000 - 12,000
Court Houses	184.00	32,400	17,800 - 106,000	Research	197.00	19,000	6,300 - 45,000
Dept. Stores	78.00	90,000	44,000 - 122,000	Restaurants	172.00	4,400	2,800 - 6,000
Dormitories, Low Rise	143.00	25,000	10,000 - 95,000	Retail Stores	83.50	7,200	4,000 - 17,600
Dormitories, Mid Rise	176.00	85,000	20,000 - 200,000	Schools, Elementary	125.00	41,000	24,500 - 55,000
Factories	76.00	26,400	12,900 - 50,000	Schools, Jr. High	128.00	92,000	52,000 - 119,000
Fire Stations	139.00	5,800	4,000 - 8,700	Schools, Sr. High	133.00	101,000	50,500 - 175,000
Fraternity Houses	130.00	12,500	8,200 - 14,800	Schools, Vocational	128.00	37,000	20,500 - 82,000
Funeral Homes	145.00	10,000	4,000 - 20,000	Sports Arenas	103.00	15,000	5,000 - 40,000
Garages, Commercial	93.50	9,300	5,000 - 13,600	Supermarkets	82.50	44,000	12,000 - 60,000
Garages, Municipal	119.00	8,300	4,500 - 12,600	Swimming Pools	193.00	20,000	10,000 - 32,000
Garages, Parking	50.50	163,000	76,400 - 225,300	Telephone Exchange	224.00	4,500	1,200 - 10,600
Gymnasiums	128.00	19,200	11,600 - 41,000	Theaters	120.00	10,500	8,800 - 17,500
Hospitals	230.00	55,000	27,200 - 125,000	Town Halls	136.00	10,800	4,800 - 23,400
House (Elderly)	115.00	37,000	21,000 - 66,000	Warehouses	59.50	25,000	8,000 - 72,000
Housing (Public)	106.00	36,000	14,400 - 74,400	Warehouse & Office	65.50	25,000	8,000 - 72,000
Ice Rinks	152.00	29,000	27,200 - 33,600				

A	Area Square Feet; Ampere	BTUH	BTU per Hour	Cwt.	100 Pounds
ABS	Acrylonitrile Butadiene Stryrene; Asbestos Bonded Steel	B.U.R.	Built-up Roofing	C.W.X.	Cool White Deluxe
A.C.	Alternating Current; Air-Conditioning; Asbestos Cement; Plywood Grade A & C	BX	Interlocked Armored Cable	C.Y.	Cubic Yard (27 cubic feet)
		°C	degree centigrade	C.Y./Hr.	Cubic Yard per Hour
		c	Conductivity, Copper Sweat	Cyl.	Cylinder
		C	Hundred; Centigrade	d	Penny (nail size)
A.C.I.	American Concrete Institute	C/C	Center to Center, Cedar on Cedar	D	Deep; Depth; Discharge
AD	Plywood, Grade A & D	C-C	Center to Center	Dis., Disch.	Discharge
Addit.	Additional	Cab.	Cabinet	Db.	Decibel
Adj.	Adjustable	Cair.	Air Tool Laborer	Dbl.	Double
af	Audio-frequency	Calc	Calculated	DC	Direct Current
A.G.A.	American Gas Association	Cap.	Capacity	DDC	Direct Digital Control
Agg.	Aggregate	Carp.	Carpenter	Demob.	Demobilization
A.H.	Ampere Hours	C.B.	Circuit Breaker	d.f.u.	Drainage Fixture Units
A hr.	Ampere-hour	C.C.A.	Chromate Copper Arsenate	D.H.	Double Hung
A.H.U.	Air Handling Unit	C.C.F.	Hundred Cubic Feet	DHW	Domestic Hot Water
A.I.A.	American Institute of Architects	cd	Candela	DI	Ductile Iron
AIC	Ampere Interrupting Capacity	cd/sf	Candela per Square Foot	Diag.	Diagonal
Allow.	Allowance	CD	Grade of Plywood Face & Back	Diam., Dia	Diameter
alt.	Altitude	CDX	Plywood, Grade C & D, exterior glue	Distrib.	Distribution
Alum.	Aluminum			Div.	Division
a.m.	Ante Meridiem	Cefi.	Cement Finisher	Dk.	Deck
Amp.	Ampere	Cem.	Cement	D.L.	Dead Load; Diesel
Anod.	Anodized	CF	Hundred Feet	DLH	Deep Long Span Bar Joist
Approx.	Approximate	C.F.	Cubic Feet	Do.	Ditto
Apt.	Apartment	CFM	Cubic Feet per Minute	Dp.	Depth
Asb.	Asbestos	c.g.	Center of Gravity	D.P.S.T.	Double Pole, Single Throw
A.S.B.C.	American Standard Building Code	CHW	Chilled Water; Commercial Hot Water	Dr.	Drive
Asbe.	Asbestos Worker			Drink.	Drinking
ASCE.	American Society of Civil Engineers	C.I.	Cast Iron	D.S.	Double Strength
A.S.H.R.A.E.	American Society of Heating, Refrig. & AC Engineers	C.I.P.	Cast in Place	D.S.A.	Double Strength A Grade
		Circ.	Circuit	D.S.B.	Double Strength B Grade
A.S.M.E.	American Society of Mechanical Engineers	C.L.	Carload Lot	Dty.	Duty
		Clab.	Common Laborer	DWV	Drain Waste Vent
A.S.T.M.	American Society for Testing and Materials	Clam	Common maintenance laborer	DX	Deluxe White, Direct Expansion
		C.L.F.	Hundred Linear Feet	dyn	Dyne
Attchmt.	Attachment	CLF	Current Limiting Fuse	e	Eccentricity
Avg., Ave.	Average	CLP	Cross Linked Polyethylene	E	Equipment Only; East
A.W.G.	American Wire Gauge	cm	Centimeter	Ea.	Each
AWWA	American Water Works Assoc.	CMP	Corr. Metal Pipe	E.B.	Encased Burial
Bbl.	Barrel	C.M.U.	Concrete Masonry Unit	Econ.	Economy
B&B	Grade B and Better; Balled & Burlapped	CN	Change Notice	E.C.Y	Embankment Cubic Yards
		Col.	Column	EDP	Electronic Data Processing
B.&S.	Bell and Spigot	CO₂	Carbon Dioxide	EIFS	Exterior Insulation Finish System
B.&W.	Black and White	Comb.	Combination	E.D.R.	Equiv. Direct Radiation
b.c.c.	Body-centered Cubic	Compr.	Compressor	Eq.	Equation
B.C.Y.	Bank Cubic Yards	Conc.	Concrete	EL	elevation
BE	Bevel End	Cont.	Continuous; Continued, Container	Elec.	Electrician; Electrical
B.F.	Board Feet	Corr.	Corrugated	Elev.	Elevator; Elevating
Bg. cem.	Bag of Cement	Cos	Cosine	EMT	Electrical Metallic Conduit; Thin Wall Conduit
BHP	Boiler Horsepower; Brake Horsepower	Cot	Cotangent		
		Cov.	Cover	Eng.	Engine, Engineered
B.I.	Black Iron	C/P	Cedar on Paneling	EPDM	Ethylene Propylene Diene Monomer
Bit., Bitum.	Bituminous	CPA	Control Point Adjustment		
Bit., Conc.	Bituminous Concrete	Cplg.	Coupling	EPS	Expanded Polystyrene
Bk.	Backed	C.P.M.	Critical Path Method	Eqhv.	Equip. Oper., Heavy
Bkrs.	Breakers	CPVC	Chlorinated Polyvinyl Chloride	Eqlt.	Equip. Oper., Light
Bldg.	Building	C.Pr.	Hundred Pair	Eqmd.	Equip. Oper., Medium
Blk.	Block	CRC	Cold Rolled Channel	Eqmm.	Equip. Oper., Master Mechanic
Bm.	Beam	Creos.	Creosote	Eqol.	Equip. Oper., Oilers
Boil.	Boilermaker	Crpt.	Carpet & Linoleum Layer	Equip.	Equipment
B.P.M.	Blows per Minute	CRT	Cathode-ray Tube	ERW	Electric Resistance Welded
BR	Bedroom	CS	Carbon Steel, Constant Shear Bar Joist	E.S.	Energy Saver
Brg.	Bearing			Est.	Estimated
Brhe.	Bricklayer Helper	Csc	Cosecant	esu	Electrostatic Units
Bric.	Bricklayer	C.S.F.	Hundred Square Feet	E.W.	Each Way
Brk.	Brick	CSI	Construction Specifications Institute	EWT	Entering Water Temperature
Brng.	Bearing			Excav.	Excavation
Brs.	Brass	C.T.	Current Transformer	Exp.	Expansion, Exposure
Brz.	Bronze	CTS	Copper Tube Size	Ext.	Exterior
Bsn.	Basin	Cu	Copper, Cubic	Extru.	Extrusion
Btr.	Better	Cu. Ft.	Cubic Foot	f.	Fiber stress
Btu	British Thermal Unit	cw	Continuous Wave	F	Fahrenheit; Female; Fill
		C.W.	Cool White; Cold Water	Fab.	Fabricated

Abbreviations

FBGS	Fiberglass	H.P.	Horsepower; High Pressure	LE	Lead Equivalent
F.C.	Footcandles	H.P.F.	High Power Factor	LED	Light Emitting Diode
f.c.c.	Face-centered Cubic	Hr.	Hour	L.F.	Linear Foot
f'c.	Compressive Stress in Concrete;	Hrs./Day	Hours per Day	L.F. Nose	Linear Foot of Stair Nosing
	Extreme Compressive Stress	HSC	High Short Circuit	L.F. Rsr	Linear Foot of Stair Riser
F.E.	Front End	Ht.	Height	Lg.	Long; Length; Large
FEP	Fluorinated Ethylene Propylene	Htg.	Heating	L & H	Light and Heat
	(Teflon)	Htrs.	Heaters	LH	Long Span Bar Joist
F.G.	Flat Grain	HVAC	Heating, Ventilation & Air-	L.H.	Labor Hours
F.H.A.	Federal Housing Administration		Conditioning	L.L.	Live Load
Fig.	Figure	Hvy.	Heavy	L.L.D.	Lamp Lumen Depreciation
Fin.	Finished	HW	Hot Water	lm	Lumen
Fixt.	Fixture	Hyd.;Hydr.	Hydraulic	lm/sf	Lumen per Square Foot
Fl. Oz.	Fluid Ounces	Hz.	Hertz (cycles)	lm/W	Lumen per Watt
Flr.	Floor	I.	Moment of Inertia	L.O.A.	Length Over All
F.M.	Frequency Modulation;	IBC	International Building Code	log	Logarithm
	Factory Mutual	I.C.	Interrupting Capacity	L-O-L	Lateralolet
Fmg.	Framing	ID	Inside Diameter	long.	longitude
Fdn.	Foundation	I.D.	Inside Dimension; Identification	L.P.	Liquefied Petroleum; Low Pressure
Fori.	Foreman, Inside	I.F.	Inside Frosted	L.P.F.	Low Power Factor
Foro.	Foreman, Outside	I.M.C.	Intermediate Metal Conduit	LR	Long Radius
Fount.	Fountain	In.	Inch	L.S.	Lump Sum
fpm	Feet per Minute	Incan.	Incandescent	Lt.	Light
FPT	Female Pipe Thread	Incl.	Included; Including	Lt. Ga.	Light Gauge
Fr.	Frame	Int.	Interior	L.T.L.	Less than Truckload Lot
F.R.	Fire Rating	Inst.	Installation	Lt. Wt.	Lightweight
FRK	Foil Reinforced Kraft	Insul.	Insulation/Insulated	L.V.	Low Voltage
FRP	Fiberglass Reinforced Plastic	I.P.	Iron Pipe	M	Thousand; Material; Male;
FS	Forged Steel	I.P.S.	Iron Pipe Size		Light Wall Copper Tubing
FSC	Cast Body; Cast Switch Box	I.P.T.	Iron Pipe Threaded	M²CA	Meters Squared Contact Area
Ft.	Foot; Feet	I.W.	Indirect Waste	m/hr.; M.H.	Man-hour
Ftng.	Fitting	J	Joule	mA	Milliampere
Ftg.	Footing	J.I.C.	Joint Industrial Council	Mach.	Machine
Ft lb.	Foot Pound	K	Thousand; Thousand Pounds;	Mag. Str.	Magnetic Starter
Furn.	Furniture		Heavy Wall Copper Tubing, Kelvin	Maint.	Maintenance
FVNR	Full Voltage Non-Reversing	K.A.H.	Thousand Amp. Hours	Marb.	Marble Setter
FXM	Female by Male	KCMIL	Thousand Circular Mils	Mat; Mat'l.	Material
Fy.	Minimum Yield Stress of Steel	KD	Knock Down	Max.	Maximum
g	Gram	K.D.A.T.	Kiln Dried After Treatment	MBF	Thousand Board Feet
G	Gauss	kg	Kilogram	MBH	Thousand BTU's per hr.
Ga.	Gauge	kG	Kilogauss	MC	Metal Clad Cable
Gal., gal.	Gallon	kgf	Kilogram Force	M.C.F.	Thousand Cubic Feet
gpm, GPM	Gallon per Minute	kHz	Kilohertz	M.C.F.M.	Thousand Cubic Feet per Minute
Galv.	Galvanized	Kip.	1000 Pounds	M.C.M.	Thousand Circular Mils
Gen.	General	KJ	Kiljoule	M.C.P.	Motor Circuit Protector
G.F.I.	Ground Fault Interrupter	K.L.	Effective Length Factor	MD	Medium Duty
Glaz.	Glazier	K.L.F.	Kips per Linear Foot	M.D.O.	Medium Density Overlaid
GPD	Gallons per Day	Km	Kilometer	Med.	Medium
GPH	Gallons per Hour	K.S.F.	Kips per Square Foot	MF	Thousand Feet
GPM	Gallons per Minute	K.S.I.	Kips per Square Inch	M.F.B.M.	Thousand Feet Board Measure
GR	Grade	kV	Kilovolt	Mfg.	Manufacturing
Gran.	Granular	kVA	Kilovolt Ampere	Mfrs.	Manufacturers
Grnd.	Ground	K.V.A.R.	Kilovar (Reactance)	mg	Milligram
H	High Henry	KW	Kilowatt	MGD	Million Gallons per Day
H.C.	High Capacity	KWh	Kilowatt-hour	MGPH	Thousand Gallons per Hour
H.D.	Heavy Duty; High Density	L	Labor Only; Length; Long;	MH, M.H.	Manhole; Metal Halide; Man-Hour
H.D.O.	High Density Overlaid		Medium Wall Copper Tubing	MHz	Megahertz
H.D.P.E.	high density polyethelene	Lab.	Labor	Mi.	Mile
Hdr.	Header	lat	Latitude	MI	Malleable Iron; Mineral Insulated
Hdwe.	Hardware	Lath.	Lather	mm	Millimeter
Help.	Helper Average	Lav.	Lavatory	Mill.	Millwright
HEPA	High Efficiency Particulate Air	lb.; #	Pound	Min., min.	Minimum, minute
	Filter	L.B.	Load Bearing; L Conduit Body	Misc.	Miscellaneous
Hg	Mercury	L. & E.	Labor & Equipment	ml	Milliliter, Mainline
HIC	High Interrupting Capacity	lb./hr.	Pounds per Hour	M.L.F.	Thousand Linear Feet
HM	Hollow Metal	lb./L.F.	Pounds per Linear Foot	Mo.	Month
HMWPE	high molecular weight	lbf/sq.in.	Pound-force per Square Inch	Mobil.	Mobilization
	polyethylene	L.C.L.	Less than Carload Lot	Mog.	Mogul Base
H.O.	High Output	L.C.Y.	Loose Cubic Yard	MPH	Miles per Hour
Horiz.	Horizontal	Ld.	Load	MPT	Male Pipe Thread

MRT	Mile Round Trip	Pl.	Plate	S.F.C.A.	Square Foot Contact Area
ms	Millisecond	Plah.	Plasterer Helper	S.F. Flr.	Square Foot of Floor
M.S.F.	Thousand Square Feet	Plas.	Plasterer	S.F.G.	Square Foot of Ground
Mstz.	Mosaic & Terrazzo Worker	Pluh.	Plumbers Helper	S.F. Hor.	Square Foot Horizontal
M.S.Y.	Thousand Square Yards	Plum.	Plumber	S.F.R.	Square Feet of Radiation
Mtd., mtd.	Mounted	Ply.	Plywood	S.F. Shlf.	Square Foot of Shelf
Mthe.	Mosaic & Terrazzo Helper	p.m.	Post Meridiem	S4S	Surface 4 Sides
Mtng.	Mounting	Pntd.	Painted	Shee.	Sheet Metal Worker
Mult.	Multi; Multiply	Pord.	Painter, Ordinary	Sin.	Sine
M.V.A.	Million Volt Amperes	pp	Pages	Skwk.	Skilled Worker
M.V.A.R.	Million Volt Amperes Reactance	PP, PPL	Polypropylene	SL	Saran Lined
MV	Megavolt	P.P.M.	Parts per Million	S.L.	Slimline
MW	Megawatt	Pr.	Pair	Sldr.	Solder
MXM	Male by Male	P.E.S.B.	Pre-engineered Steel Building	SLH	Super Long Span Bar Joist
MYD	Thousand Yards	Prefab.	Prefabricated	S.N.	Solid Neutral
N	Natural; North	Prefin.	Prefinished	S-O-L	Socketolet
nA	Nanoampere	Prop.	Propelled	sp	Standpipe
NA	Not Available; Not Applicable	PSF, psf	Pounds per Square Foot	S.P.	Static Pressure; Single Pole; Self-
N.B.C.	National Building Code	PSI, psi	Pounds per Square Inch		Propelled
NC	Normally Closed	PSIG	Pounds per Square Inch Gauge	Spri.	Sprinkler Installer
N.E.M.A.	National Electrical Manufacturers	PSP	Plastic Sewer Pipe	spwg	Static Pressure Water Gauge
	Assoc.	Pspr.	Painter, Spray	S.P.D.T.	Single Pole, Double Throw
NEHB	Bolted Circuit Breaker to 600V.	Psst.	Painter, Structural Steel	SPF	Spruce Pine Fir
N.L.B.	Non-Load-Bearing	P.T.	Potential Transformer	S.P.S.T.	Single Pole, Single Throw
NM	Non-Metallic Cable	P. & T.	Pressure & Temperature	SPT	Standard Pipe Thread
nm	Nanometer	Ptd.	Painted	Sq.	Square; 100 Square Feet
No.	Number	Ptns.	Partitions	Sq. Hd.	Square Head
NO	Normally Open	Pu	Ultimate Load	Sq. In.	Square Inch
N.O.C.	Not Otherwise Classified	PVC	Polyvinyl Chloride	S.S.	Single Strength; Stainless Steel
Nose.	Nosing	Pvmt.	Pavement	S.S.B.	Single Strength B Grade
N.P.T.	National Pipe Thread	Pwr.	Power	sst, ss	Stainless Steel
NQOD	Combination Plug-on/Bolt on	Q	Quantity Heat Flow	Sswk.	Structural Steel Worker
	Circuit Breaker to 240V.	Qt.	Quart	Sswl.	Structural Steel Welder
N.R.C.	Noise Reduction Coefficient/	Quan., Qty.	Quantity	St.;Stl.	Steel
	Nuclear Regulator Commission	Q.C.	Quick Coupling	S.T.C.	Sound Transmission Coefficient
N.R.S.	Non Rising Stem	r	Radius of Gyration	Std.	Standard
ns	Nanosecond	R	Resistance	Stg.	Staging
nW	Nanowatt	R.C.P.	Reinforced Concrete Pipe	STK	Select Tight Knot
OB	Opposing Blade	Rect.	Rectangle	STP	Standard Temperature & Pressure
OC	On Center	Reg.	Regular	Stpi.	Steamfitter, Pipefitter
OD	Outside Diameter	Reinf.	Reinforced	Str.	Strength; Starter; Straight
O.D.	Outside Dimension	Req'd.	Required	Strd.	Stranded
ODS	Overhead Distribution System	Res.	Resistant	Struct.	Structural
O.G.	Ogee	Resi.	Residential	Sty.	Story
O.H.	Overhead	Rgh.	Rough	Subj.	Subject
O&P	Overhead and Profit	RGS	Rigid Galvanized Steel	Subs.	Subcontractors
Oper.	Operator	R.H.W.	Rubber, Heat & Water Resistant;	Surf.	Surface
Opng.	Opening		Residential Hot Water	Sw.	Switch
Orna.	Ornamental	rms	Root Mean Square	Swbd.	Switchboard
OSB	Oriented Strand Board	Rnd.	Round	S.Y.	Square Yard
O.S.&Y.	Outside Screw and Yoke	Rodm.	Rodman	Syn.	Synthetic
Ovhd.	Overhead	Rofc.	Roofer, Composition	S.Y.P.	Southern Yellow Pine
OWG	Oil, Water or Gas	Rofp.	Roofer, Precast	Sys.	System
Oz.	Ounce	Rohe.	Roofer Helpers (Composition)	t.	Thickness
P.	Pole; Applied Load; Projection	Rots.	Roofer, Tile & Slate	T	Temperature; Ton
p.	Page	R.O.W.	Right of Way	Tan	Tangent
Pape.	Paperhanger	RPM	Revolutions per Minute	T.C.	Terra Cotta
P.A.P.R.	Powered Air Purifying Respirator	R.S.	Rapid Start	T & C	Threaded and Coupled
PAR	Parabolic Reflector	Rsr	Riser	T.D.	Temperature Difference
Pc., Pcs.	Piece, Pieces	RT	Round Trip	Tdd	Telecommunications Device for
P.C.	Portland Cement; Power Connector	S.	Suction; Single Entrance; South		the Deaf
P.C.F.	Pounds per Cubic Foot	SC	Screw Cover	T.E.M.	Transmission Electron Microscopy
P.C.M.	Phase Contrast Microscopy	SCFM	Standard Cubic Feet per Minute	TFE	Tetrafluoroethylene (Teflon)
P.E.	Professional Engineer;	Scaf.	Scaffold	T. & G.	Tongue & Groove;
	Porcelain Enamel;	Sch., Sched.	Schedule		Tar & Gravel
	Polyethylene; Plain End	S.C.R.	Modular Brick	Th., Thk.	Thick
Perf.	Perforated	S.D.	Sound Deadening	Thn.	Thin
PEX	Cross linked polyethylene	S.D.R.	Standard Dimension Ratio	Thrded	Threaded
Ph.	Phase	S.E.	Surfaced Edge	Tilf.	Tile Layer, Floor
P.I.	Pressure Injected	Sel.	Select	Tilh.	Tile Layer, Helper
Pile.	Pile Driver	S.E.R., S.E.U.	Service Entrance Cable	THHN	Nylon Jacketed Wire
Pkg.	Package	S.F.	Square Foot	THW.	Insulated Strand Wire

Abbreviations

THWN	Nylon Jacketed Wire	USP	United States Primed	Wrck.	Wrecker
T.L.	Truckload	UTP	Unshielded Twisted Pair	W.S.P.	Water, Steam, Petroleum
T.M.	Track Mounted	V	Volt	WT., Wt.	Weight
Tot.	Total	V.A.	Volt Amperes	WWF	Welded Wire Fabric
T-O-L	Threadolet	V.C.T.	Vinyl Composition Tile	XFER	Transfer
T.S.	Trigger Start	VAV	Variable Air Volume	XFMR	Transformer
Tr.	Trade	VC	Veneer Core	XHD	Extra Heavy Duty
Transf.	Transformer	Vent.	Ventilation	XHHW,	Cross-Linked Polyethylene Wire
Trhv.	Truck Driver, Heavy	Vert.	Vertical	XLPE	Insulation
Trlr	Trailer	V.F.	Vinyl Faced	XLP	Cross-linked Polyethylene
Trlt.	Truck Driver, Light	V.G.	Vertical Grain	Y	Wye
TTY	Teletypewriter	V.H.F.	Very High Frequency	yd	Yard
TV	Television	VHO	Very High Output	yr	Year
T.W.	Thermoplastic Water Resistant	Vib.	Vibrating	Δ	Delta
	Wire	V.L.F.	Vertical Linear Foot	%	Percent
UCI	Uniform Construction Index	Vol.	Volume	~	Approximately
UF	Underground Feeder	VRP	Vinyl Reinforced Polyester	Ø	Phase; diameter
UGND	Underground Feeder	W	Wire; Watt; Wide; West	@	At
U.H.F.	Ultra High Frequency	w/	With	#	Pound; Number
U.I.	United Inch	W.C.	Water Column; Water Closct	<	Less Than
U.L.	Underwriters Laboratory	W.F.	Wide Flange	>	Greater Than
Uld.	unloading	W.G.	Water Gauge	Z	zone
Unfin.	Unfinished	Wldg.	Welding		
URD	Underground Residential	W. Mile	Wire Mile		
	Distribution	W-O-L	Weldolet		
US	United States	W.R.	Water Resistant		

Index

Index

Index

Index

Index

Reed Construction Data/RSMeans—
a tradition of excellence in construction cost information and services since 1942

For more information visit the RSMeans website at www.rsmeans.com

Book Selection Guide

The following table provides definitive information on the content of each cost data publication. The number of lines of data provided in each unit price or assemblies division, as well as the number of reference tables and crews, is listed for each book. The presence of other elements such as equipment rental costs, historical cost indexes, city cost indexes, square foot models, or cross-referenced indexes is also indicated. You can use the table to help select the RSMeans book that has the quantity and type of information you most need in your work.

Unit Cost Divisions	Building Construction	Mechanical	Electrical	Commercial Renovation	Square Foot	Site Work Landsc.	Green Building	Interior	Concrete Masonry	Open Shop	Heavy Construction	Light Commercial	Facilities Construction	Plumbing	Residential
1	571	392	401	492		509	39	317	455	570	499	235	985	402	172
2	746	266	85	685		967	168	364	221	743	706	440	1163	290	242
3	1526	216	206	887		1329	859	304	1833	1526	1468	375	1552	183	359
4	898	24	0	698		700	182	573	1133	874	593	490	1145	0	405
5	1916	263	167	1024		870	1803	1025	833	1907	1117	895	1910	339	727
6	1855	67	69	1442		450	537	1161	300	1834	381	1544	1466	22	2056
7	1437	209	125	1465		534	766	544	485	1435	26	1143	1497	218	896
8	2093	61	10	2233		297	1146	1731	676	2018	0	1622	2450	0	1397
9	1812	72	26	1627		278	441	1910	386	1774	15	1542	2064	54	1354
10	988	17	10	608		221	22	803	159	988	29	490	1074	232	221
11	1012	208	165	484		126	54	862	28	997	0	231	1026	169	110
12	562	0	2	318		248	150	1706	12	551	19	364	1756	23	317
13	738	119	111	247		349	128	255	69	737	261	80	748	74	79
14	280	36	0	229		28	0	259	0	279	0	12	299	21	6
21	78	0	16	30		0	0	250	0	78	0	60	368	373	0
22	1141	7183	155	1105		1510	1015	621	20	1103	1676	834	7114	8929	632
23	1186	7077	608	799		157	894	677	38	1106	110	661	5154	1872	416
26	1249	496	9916	1006		746	555	1068	55	1237	523	1095	9852	441	560
27	67	0	261	34		9	0	59	0	67	35	48	263	0	4
28	96	58	124	71		0	21	78	0	99	0	40	139	44	25
31	1477	735	608	798		2794	38	7	1208	1421	2865	600	1540	660	606
32	759	54	0	840		4195	343	359	274	730	1710	380	1611	166	429
33	502	1010	465	201		1950	41	0	225	501	1873	113	1536	1181	111
34	109	0	20	4		144	0	0	31	64	122	0	129	0	0
35	18	0	0	0		166	0	0	0	18	281	0	83	0	0
41	52	0	0	24		7	0	22	0	52	30	0	59	14	0
44	75	83	0	0		0	0	0	0	0	0	0	75	75	0
46	23	6	0	0		112	99	0	0	23	112	0	23	23	0
48	0	0	0	0		0	19	0	0	0	0	0	0	0	0
Totals	23266	18652	13550	17351		18696	9320	14955	8441	22732	14451	13294	47081	15805	11094

Assem Div	Building Construction	Mechanical	Electrical	Commercial Renovation	Square Foot	Site Work Landscape	Assemblies	Green Building	Interior	Concrete Masonry	Heavy Construction	Light Commercial	Facilities Construction	Plumbing	Asm Div	Residential
A		15	0	188	150	575	596	0	0	534	569	154	24	0	1	374
B		0	0	825	2465	0	5562	56	329	1911	368	2025	143	0	2	217
C		0	0	647	866	0	1236	0	1561	146	0	766	248	0	3	588
D		1064	811	712	1793	0	2423	233	765	0	0	1270	1026	1083	4	853
E		0	0	86	260	0	297	0	5	0	0	257	5	0	5	392
F		0	0	0	123	0	124	0	0	0	0	123	3	0	6	358
G		527	294	317	202	2673	668	0	0	535	1189	200	293	685	7	300
															8	760
															9	80
															10	0
															11	0
															12	0
Totals		1606	1105	2775	5859	3248	10906	289	2660	3126	2126	4795	1742	1768		3922

Reference Section	Building Construction Costs	Mechanical	Electrical	Commercial Renovation	Square Foot	Site Work Landscape	Assem.	Green Building	Interior	Concrete Masonry	Open Shop	Heavy Construction	Light Commercial	Facilities Construction	Plumbing	Resi.
Reference Tables	yes	yes	yes	yes	no	yes	yes	yes	yes	yes	yes	yes	yes	yes	yes	yes
Models					107								46			28
Crews	505	505	505	483		505		505	505	505	481	505	481	483	505	481
Equipment Rental Costs	yes	yes	yes	yes		yes		yes	yes	yes	yes	yes	yes	yes	yes	yes
Historical Cost Indexes	yes	yes	yes	yes	yes	yes	yes	yes	yes	yes	yes	yes	yes	yes	yes	no
City Cost Indexes	yes	yes	yes	yes	yes	yes	yes	yes	yes	yes	yes	yes	yes	yes	yes	yes

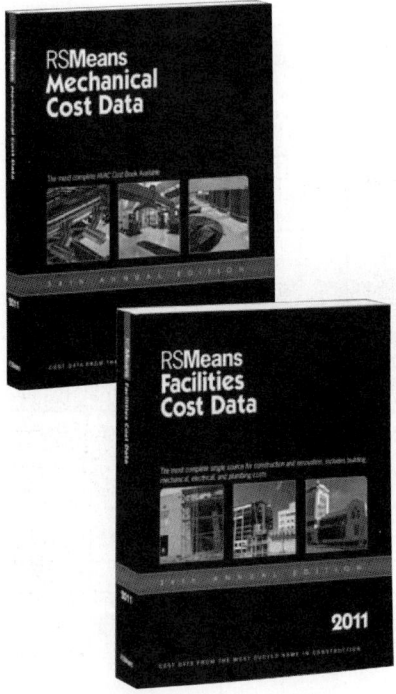

For more information
visit the RSMeans website
at www.rsmeans.com

Annual Cost Guides

RSMeans Square Foot Costs 2011

Accurate and Easy To Use!

- **Updated price information** based on nationwide figures from suppliers, estimators, labor experts, and contractors

- "How-to-Use" sections, with **clear examples** of commercial, residential, industrial, and institutional structures

- Realistic graphics, offering true-to-life illustrations of building projects

- Extensive information on using square foot cost data, including sample estimates and alternate pricing methods

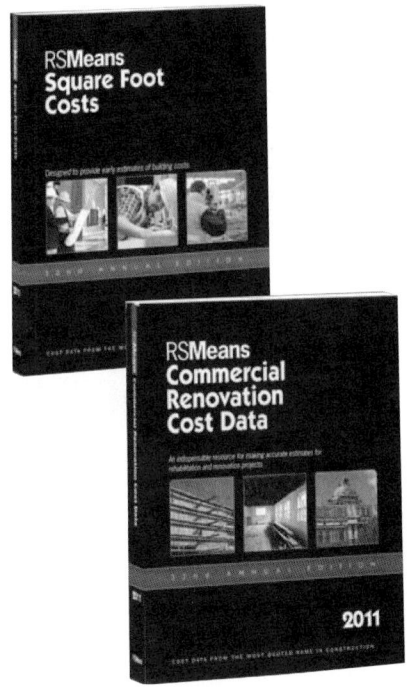

$184.95 per copy
Available Oct. 2010
Catalog no. 60051

RSMeans Commercial Renovation Cost Data 2011

Commercial/Multifamily Residential

Use this valuable tool to estimate commercial and multifamily residential renovation and remodeling.

Includes: New costs for hundreds of unique methods, materials, and conditions that only come up in repair and remodeling, PLUS:

- Unit costs for more than 15,000 construction components
- Installed costs for more than 90 assemblies
- More than 930 "on-site" localization factors for the U.S. and Canada

Unit prices organized according to the latest MasterFormat!

$144.95 per copy
Available Oct. 2010
Catalog no. 60041

RSMeans Electrical Cost Data 2011

Pricing information for every part of electrical cost planning. More than 13,000 unit and systems costs with design tables; clear specifications and drawings; engineering guides; illustrated estimating procedures; complete labor-hour and materials costs for better scheduling and procurement; and the latest electrical products and construction methods.

- A variety of special electrical systems including cathodic protection

- Costs for maintenance, demolition, HVAC/mechanical, specialties, equipment, and more

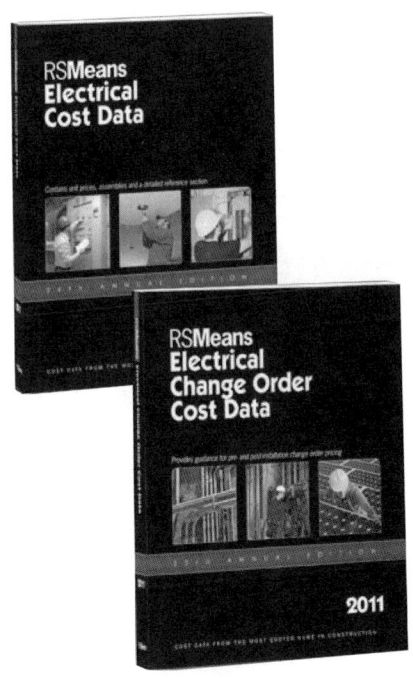

Unit prices organized according to the latest MasterFormat!

$174.95 per copy
Available Oct. 2010
Catalog no. 60031

RSMeans Electrical Change Order Cost Data 2011

RSMeans Electrical Change Order Cost Data provides you with electrical unit prices exclusively for pricing change orders—based on the recent, direct experience of contractors and suppliers. Analyze and check your own change order estimates against the experience others have had doing the same work. It also covers productivity analysis and change order cost justifications. With useful information for calculating the effects of change orders and dealing with their administration.

$169.95 per copy
Available Dec. 2010
Catalog no. 60231

RSMeans Assemblies Cost Data 2011

RSMeans Assemblies Cost Data takes the guesswork out of preliminary or conceptual estimates. Now you don't have to try to calculate the assembled cost by working up individual component costs. We've done all the work for you.

Presents detailed illustrations, descriptions, specifications, and costs for every conceivable building assembly—240 types in all—arranged in the easy-to-use UNIFORMAT II system. Each illustrated "assembled" cost includes a complete grouping of materials and associated installation costs, including the installing contractor's overhead and profit.

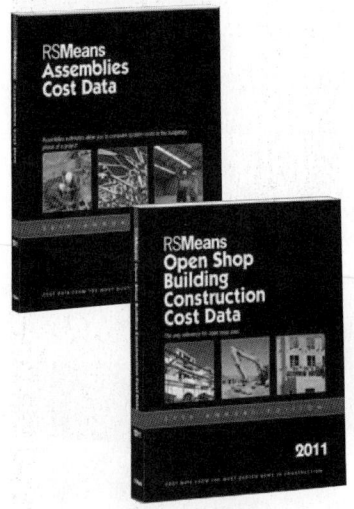

$285.95 per copy
Available Sept. 2010
Catalog no. 60061

RSMeans Open Shop Building Construction Cost Data 2011

The latest costs for accurate budgeting and estimating of new commercial and residential construction. . . renovation work. . . change orders. . . cost engineering.

RSMeans Open Shop "BCCD" will assist you to:
• Develop benchmark prices for change orders
• Plug gaps in preliminary estimates and budgets
• Estimate complex projects
• Substantiate invoices on contracts
• Price ADA-related renovations

Unit prices organized according to the latest MasterFormat!

$134.95 per copy
Available Dec. 2010
Catalog no. 60151

RSMeans Residential Cost Data 2011

Contains square foot costs for 30 basic home models with the look of today, plus hundreds of custom additions and modifications you can quote right off the page. With costs for the 100 residential systems you're most likely to use in the year ahead. Complete with blank estimating forms, sample estimates, and step-by-step instructions.

Now contains line items for cultured stone and brick, PVC trim lumber, and TPO roofing.

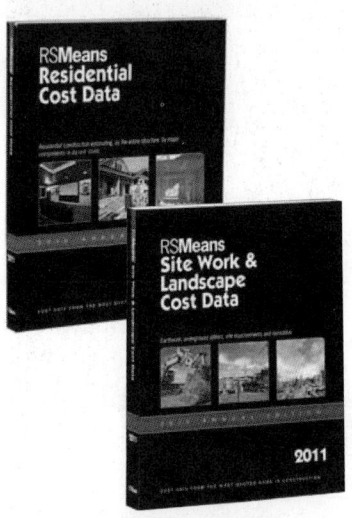

Unit prices organized according to the latest MasterFormat!

$109.95 per copy
Available Oct. 2010
Catalog no. 60171

RSMeans Site Work & Landscape Cost Data 2011

Includes unit and assemblies costs for earthwork, sewerage, piped utilities, site improvements, drainage, paving, trees & shrubs, street openings/ repairs, underground tanks, and more. Contains 57 tables of assemblies costs for accurate conceptual estimates.

Includes:
• Estimating for infrastructure improvements
• Environmentally-oriented construction
• ADA-mandated handicapped access
• Hazardous waste line items

$169.95 per copy
Available Dec. 2010
Catalog no. 60281

RSMeans Facilities Maintenance & Repair Cost Data 2011

RSMeans Facilities Maintenance & Repair Cost Data gives you a complete system to manage and plan your facility repair and maintenance costs and budget efficiently. Guidelines for auditing a facility and developing an annual maintenance plan. Budgeting is included, along with reference tables on cost and management, and information on frequency and productivity of maintenance operations.

The only nationally recognized source of maintenance and repair costs. Developed in cooperation with the Civil Engineering Research Laboratory (CERL) of the Army Corps of Engineers.

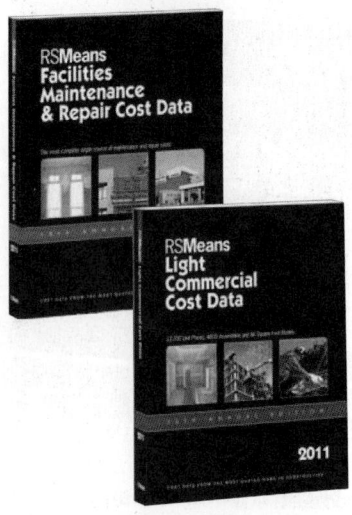

$389.95 per copy
Available Nov. 2010
Catalog no. 60301

RSMeans Light Commercial Cost Data 2011

Specifically addresses the light commercial market, which is a specialized niche in the construction industry. Aids you, the owner/designer/contractor, in preparing all types of estimates—from budgets to detailed bids. Includes new advances in methods and materials.

Assemblies Section allows you to evaluate alternatives in the early stages of design/planning.

Over 11,000 unit costs ensure that you have the prices you need. . . when you need them.

Unit prices organized according to the latest MasterFormat !

$119.95 per copy
Available Nov. 2010
Catalog no. 60181

For more information
visit the RSMeans website
at www.rsmeans.com

Annual Cost Guides

RSMeans Concrete & Masonry Cost Data 2011

Provides you with cost facts for virtually all concrete/masonry estimating needs, from complicated formwork to various sizes and face finishes of brick and block—all in great detail. The comprehensive Unit Price Section contains more than 8,000 selected entries. Also contains an Assemblies [Cost] Section, and a detailed Reference Section that supplements the cost data.

Unit prices organized according to the latest MasterFormat!

$159.95 per copy
Available Dec. 2010
Catalog no. 60111

RSMeans Labor Rates for the Construction Industry 2011

Complete information for estimating labor costs, making comparisons, and negotiating wage rates by trade for more than 300 U.S. and Canadian cities. With 46 construction trades listed by local union number in each city, and historical wage rates included for comparison. Each city chart lists the county and is alphabetically arranged with handy visual flip tabs for quick reference.

$389.95 per copy
Available Dec. 2010
Catalog no. 60121

RSMeans Construction Cost Indexes 2011

What materials and labor costs will change unexpectedly this year? By how much?

- Breakdowns for 318 major cities
- National averages for 30 key cities
- Expanded five major city indexes
- Historical construction cost indexes

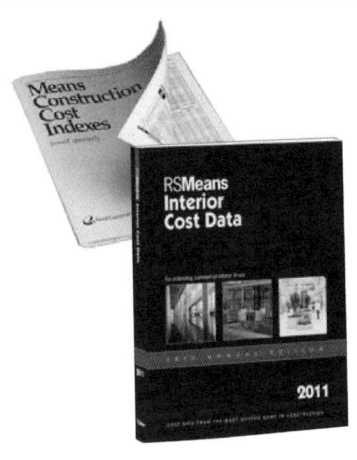

$362.00 per year (subscription)
Catalog no. 50141

$90.50 individual quarters
Catalog no. 60141 A,B,C,D

RSMeans Interior Cost Data 2011

Provides you with prices and guidance needed to make accurate interior work estimates. Contains costs on materials, equipment, hardware, custom installations, furnishings, and labor costs. . . for new and remodel commercial and industrial interior construction, including updated information on office furnishings, and reference information.

Unit prices organized according to the latest MasterFormat!

$179.95 per copy
Available Nov. 2010
Catalog no. 60091

RSMeans Heavy Construction Cost Data 2011

A comprehensive guide to heavy construction costs. Includes costs for highly specialized projects such as tunnels, dams, highways, airports, and waterways. Information on labor rates, equipment, and materials costs is included. Features unit price costs, systems costs, and numerous reference tables for costs and design.

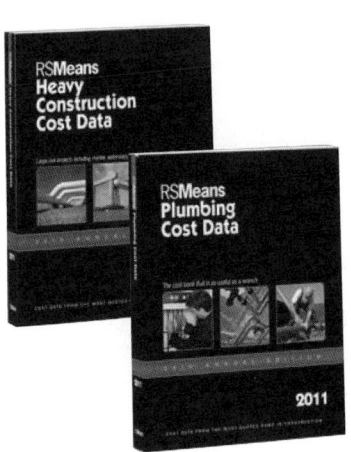

$174.95 per copy
Available Dec. 2010
Catalog no. 60161

Unit prices organized according to the latest MasterFormat!

RSMeans Plumbing Cost Data 2011

Comprehensive unit prices and assemblies for plumbing, irrigation systems, commercial and residential fire protection, point-of-use water heaters, and the latest approved materials. This publication and its companion, *RSMeans Mechanical Cost Data*, provide full-range cost estimating coverage for all the mechanical trades.

Now contains updated costs for potable water and radiant heat systems and high efficiency fixtures.

$174.95 per copy
Available Oct. 2010
Catalog no. 60211

Reference Books

For more information visit the RSMeans website at www.rsmeans.com

Value Engineering: Practical Applications

For Design, Construction, Maintenance & Operations

by Alphonse Dell'Isola, PE

A tool for immediate application—for engineers, architects, facility managers, owners, and contractors. Includes making the case for VE—the management briefing; integrating VE into planning, budgeting, and design; conducting life cycle costing; using VE methodology in design review and consultant selection; case studies; VE workbook; and a life cycle costing program on disk.

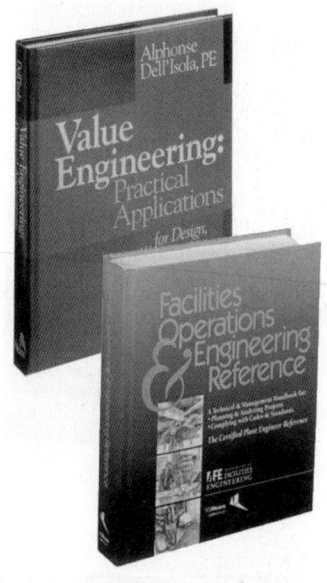

$79.95 per copy
Over 450 pages, illustrated, softcover
Catalog no. 67319A

Facilities Operations & Engineering Reference

by the Association for Facilities Engineering and RSMeans

An all-in-one technical reference for planning and managing facility projects and solving day-to-day operations problems. Selected as the official certified plant engineer reference, this handbook covers financial analysis, maintenance, HVAC and energy efficiency, and more.

$54.98 per copy
Over 700 pages, illustrated, hardcover
Catalog no. 67318

The Building Professional's Guide to Contract Documents

3rd Edition

by Waller S. Poage, AIA, CSI, CVS

A comprehensive reference for owners, design professionals, contractors, and students

- Structure your documents for maximum efficiency.
- Effectively communicate construction requirements.
- Understand the roles and responsibilities of construction professionals.
- Improve methods of project delivery.

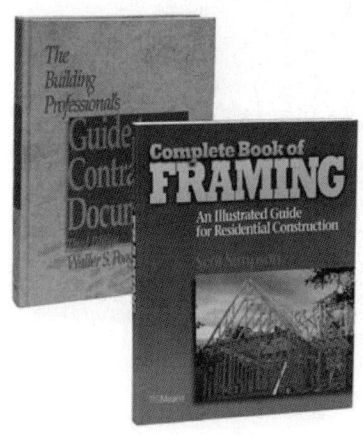

$32.48 per copy, 400 pages
Diagrams and construction forms, hardcover
Catalog no. 67261A

Complete Book of Framing

by Scot Simpson

This straightforward, easy-to-learn method will help framers, carpenters, and handy homeowners build their skills in rough carpentry and framing. Shows how to frame all the parts of a house: floors, walls, roofs, door & window openings, and stairs—with hundreds of color photographs and drawings that show every detail.

The book gives beginners all the basics they need to go from zero framing knowledge to a journeyman level.

$29.95 per copy
352 pages, softcover
Catalog no. 67353

Cost Planning & Estimating for Facilities Maintenance

In this unique book, a team of facilities management authorities shares their expertise on:

- Evaluating and budgeting maintenance operations
- Maintaining and repairing key building components
- Applying *RSMeans Facilities Maintenance & Repair Cost Data* to your estimating

Covers special maintenance requirements of the ten major building types.

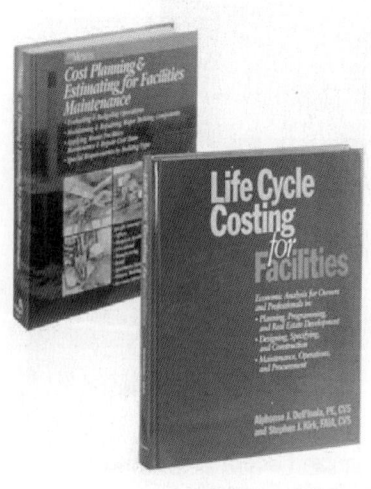

$89.95 per copy
Over 475 pages, hardcover
Catalog no. 67314

Life Cycle Costing for Facilities

by Alphonse Dell'Isola and Dr. Steven Kirk

Guidance for achieving higher quality design and construction projects at lower costs! Cost-cutting efforts often sacrifice quality to yield the cheapest product. Life cycle costing enables building designers and owners to achieve both. The authors of this book show how LCC can work for a variety of projects — from roads to HVAC upgrades to different types of buildings.

$99.95 per copy
450 pages, hardcover
Catalog no. 67341

Interior Home Improvement Costs 9th Edition

Updated estimates for the most popular remodeling and repair projects—from small, do-it-yourself jobs to major renovations and new construction. Includes: kitchens & baths; new living space from your attic, basement, or garage; new floors, paint, and wallpaper; tearing out or building new walls; closets, stairs, and fireplaces; new energy-saving improvements, home theaters, and more!

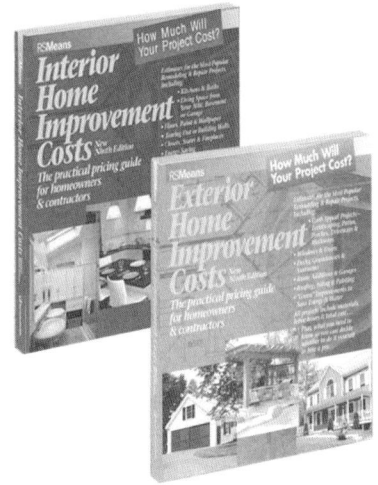

$24.95 per copy
250 pages, illustrated, softcover
Catalog no. 67308E

Exterior Home Improvement Costs 9th Edition

Updated estimates for the most popular remodeling and repair projects—from small, do-it-yourself jobs, to major renovations and new construction. Includes: curb appeal projects—landscaping, patios, porches, driveways, and walkways; new windows and doors; decks, greenhouses, and sunrooms; room additions and garages; roofing, siding, and painting; "green" improvements to save energy & water.

$24.95 per copy
Over 275 pages, illustrated, softcover
Catalog no. 67309E

Builder's Essentials: Plan Reading & Material Takeoff

For Residential and Light Commercial Construction

by Wayne J. DelPico

A valuable tool for understanding plans and specs, and accurately calculating material quantities. Step-by-step instructions and takeoff procedures based on a full set of working drawings.

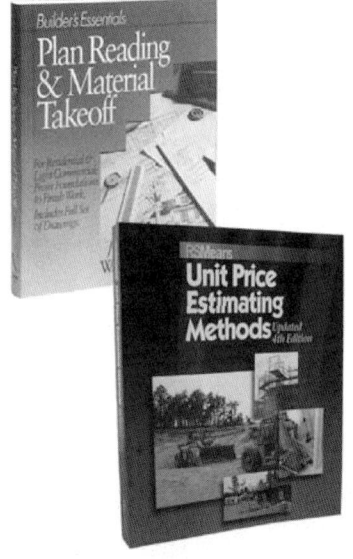

$35.95 per copy
Over 420 pages, softcover
Catalog no. 67307

Means Unit Price Estimating Methods

New 4th Edition

This new edition includes up-to-date cost data and estimating examples, updated to reflect changes to the CSI numbering system and new features of RSMeans cost data. It describes the most productive, universally accepted ways to estimate, and uses checklists and forms to illustrate shortcuts and timesavers. A model estimate demonstrates procedures. A new chapter explores computer estimating alternatives.

$59.95 per copy
Over 350 pages, illustrated, softcover
Catalog no. 67303B

Concrete Repair and Maintenance Illustrated

by Peter Emmons

Hundreds of illustrations show users how to analyze, repair, clean, and maintain concrete structures for optimal performance and cost effectiveness. From parking garages to roads and bridges to structural concrete, this comprehensive book describes the causes, effects, and remedies for concrete wear and failure. Invaluable for planning jobs, selecting materials, and training employees, this book is a must-have for concrete specialists, general contractors, facility managers, civil and structural engineers, and architects.

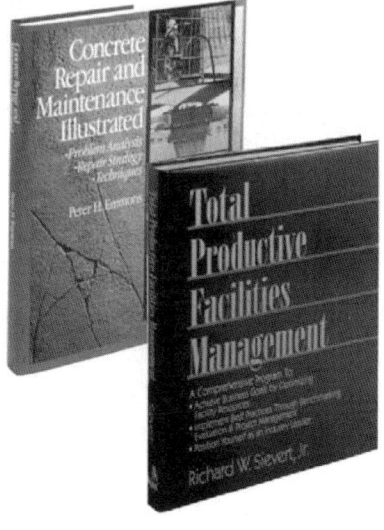

$34.98 per copy
300 pages, illustrated, softcover
Catalog no. 67146

Total Productive Facilities Management

by Richard W. Sievert, Jr.

Today, facilities are viewed as strategic resources. . . elevating the facility manager to the role of asset manager supporting the organization's overall business goals. Now, Richard Sievert Jr., in this well-articulated guidebook, sets forth a new operational standard for the facility manager's emerging role. . . a comprehensive program for managing facilities as a true profit center.

$29.98 per copy
275 pages, softcover
Catalog no. 67321

Reference Books

For more information
visit the RSMeans website
at www.rsmeans.com

Means Illustrated Construction
Dictionary Condensed, 2nd Edition

The best portable dictionary for office or field use—an essential tool for contractors, architects, insurance and real estate personnel, facility managers, homeowners, and anyone who needs quick, clear definitions for construction terms. The second edition has been further enhanced with updates and hundreds of new terms and illustrations . . . in keeping with the most recent developments in the construction industry.

Now with a quick-reference Spanish section. Includes tools and equipment, materials, tasks, and more!

$59.95 per copy
Over 500 pages, softcover
Catalog no. 67282A

Repair & Remodeling
Estimating 4th Edition
By Edward B. Wetherill and RSMeans

This important reference focuses on the unique problems of estimating renovations of existing structures, and helps you determine the true costs of remodeling through careful evaluation of architectural details and a site visit.

New section on disaster restoration costs.

$69.95 per copy
Over 450 pages, illustrated, hardcover
Catalog no. 67265B

Facilities Planning &
Relocation
by David D. Owen

A complete system for planning space needs and managing relocations. Includes step-by-step manual, over 50 forms, and extensive reference section on materials and furnishings.

New lower price and user-friendly format.

$49.98 per copy
Over 450 pages, softcover
Catalog no. 67301

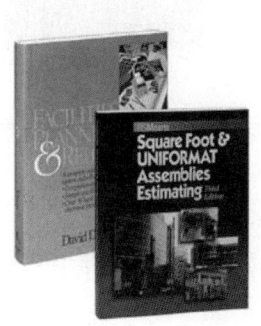

Square Foot &
UNIFORMAT Assemblies
Estimating Methods 3rd Edition

Develop realistic square foot and assemblies costs for budgeting and construction funding. The new edition features updated guidance on square foot and assemblies estimating using UNIFORMAT II. An essential reference for anyone who performs conceptual estimates.

$69.95 per copy
Over 300 pages, illustrated, softcover
Catalog no. 67145B

Electrical Estimating
Methods 3rd Edition

Expanded edition includes sample estimates and cost information in keeping with the latest version of the CSI MasterFormat and UNIFORMAT II. Complete coverage of fiber optic and uninterruptible power supply electrical systems, broken down by components, and explained in detail. Includes a new chapter on computerized estimating methods. A practical companion to *RSMeans Electrical Cost Data.*

$64.95 per copy
Over 325 pages, hardcover
Catalog no. 67230B

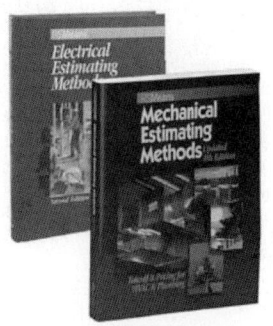

Mechanical
Estimating Methods 4th Edition

Completely updated, this guide assists you in making a review of plans, specs, and bid packages, with suggestions for takeoff procedures, listings, substitutions, and pre-bid scheduling for all components of HVAC. Includes suggestions for budgeting labor and equipment usage. Compares materials and construction methods to allow you to select the best option.

$64.95 per copy
Over 350 pages, illustrated, softcover
Catalog no. 67294B

Construction Business
Management
by Nick Ganaway

Only 43% of construction firms stay in business after four years. Make sure your company thrives with valuable guidance from a pro with 25 years of success as a commercial contractor. Find out what it takes to build all aspects of a business that is profitable, enjoyable, and enduring. With a bonus chapter on retail construction.

$49.95 per copy
200 pages, softcover
Catalog no. 67352

Project Scheduling &
Management for Construction
3rd Edition
by David R. Pierce, Jr.

A comprehensive yet easy-to-follow guide to construction project scheduling and control—from vital project management principles through the latest scheduling, tracking, and controlling techniques. The author is a leading authority on scheduling, with years of field and teaching experience at leading academic institutions. Spend a few hours with this book and come away with a solid understanding of this essential management topic.

$64.95 per copy
Over 300 pages, illustrated, hardcover
Catalog no. 67247B

For more information
visit the RSMeans website
at www.rsmeans.com

Reference Books

The Practice of Cost Segregation Analysis

by Bruce A. Desrosiers and Wayne J. DelPico

This expert guide walks you through the practice of cost segregation analysis, which enables property owners to defer taxes and benefit from "accelerated cost recovery" through depreciation deductions on assets that are properly identified and classified.

With a glossary of terms, sample cost segregation estimates for various building types, key information resources, and updates via a dedicated website, this book is a critical resource for anyone involved in cost segregation analysis.

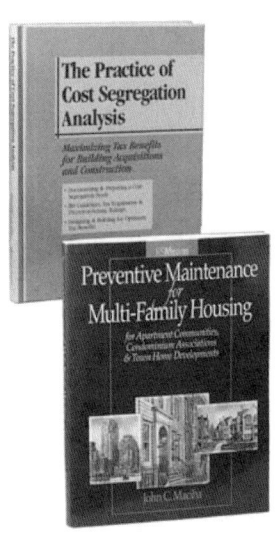

$99.95 per copy
Over 225 pages
Catalog no. 67345

Preventive Maintenance for Multi-Family Housing

by John C. Maciha

Prepared by one of the nation's leading experts on multi-family housing.

This complete PM system for apartment and condominium communities features expert guidance, checklists for buildings and grounds maintenance tasks and their frequencies, a reusable wall chart to track maintenance, and a dedicated website featuring customizable electronic forms. A must-have for anyone involved with multi-family housing maintenance and upkeep.

$89.95 per copy
225 pages
Catalog no. 67346

How to Estimate with Means Data & CostWorks

New 3rd Edition

by RSMeans and Saleh A. Mubarak, Ph.D.

New 3rd Edition—fully updated with new chapters, plus new CD with updated *CostWorks* cost data and MasterFormat organization. Includes all major construction items—with more than 300 exercises and two sets of plans that show how to estimate for a broad range of construction items and systems—including general conditions and equipment costs.

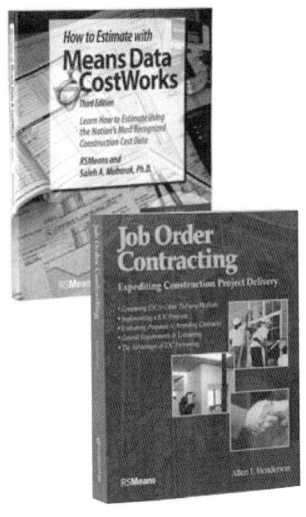

$59.95 per copy
272 pages, softcover
Includes CostWorks CD
Catalog no. 67324B

Job Order Contracting

Expediting Construction Project Delivery

by Allen Henderson

Expert guidance to help you implement JOC—fast becoming the preferred project delivery method for repair and renovation, minor new construction, and maintenance projects in the public sector and in many states and municipalities. The author, a leading JOC expert and practitioner, shows how to:

• Establish a JOC program
• Evaluate proposals and award contracts
• Handle general requirements and estimating
• Partner for maximum benefits

$89.95 per copy
192 pages, illustrated, hardcover
Catalog no. 67348

Estimating Building Costs

For the Residential & Light Commercial Contractor

by Wayne J. DelPico

Step-by-step estimating methods for residential and light commercial contractors. Includes a detailed look at every construction specialty—explaining all the components, takeoff units, and labor needed for well-organized, complete estimates. Covers correctly interpreting plans and specifications, and developing accurate and complete labor and material costs.

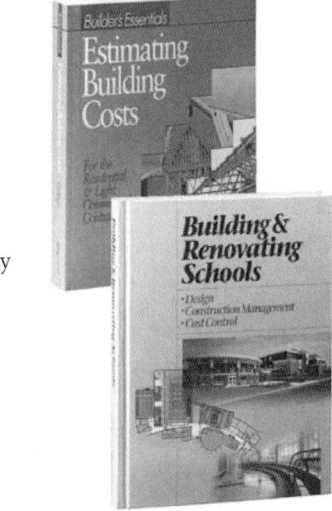

$29.95 per copy
Over 400 pages, illustrated, softcover
Catalog no. 67343

Building & Renovating Schools

This all-inclusive guide covers every step of the school construction process—from initial planning, needs assessment, and design, right through moving into the new facility. A must-have resource for anyone concerned with new school construction or renovation. With square foot cost models for elementary, middle, and high school facilities, and real-life case studies of recently completed school projects.

The contributors to this book—architects, construction project managers, contractors, and estimators who specialize in school construction—provide start-to-finish, expert guidance on the process.

$69.98 per copy
Over 425 pages, hardcover
Catalog no. 67342

Reference Books

Reference Books

Reference Books

Reference Books

The Homeowner's Guide to Mold By Michael Pugliese

Expert guidance to protect your health and your home.

Mold, whether caused by leaks, humidity or flooding, is a real health and financial issue—for homeowners and contractors. This full-color book explains:
- Construction and maintenance practices to prevent mold
- How to inspect for and remove mold
- Mold remediation procedures and costs
- What to do after a flood
- How to deal with insurance companies

$21.95 per copy
144 pages, softcover
Catalog no. 67344

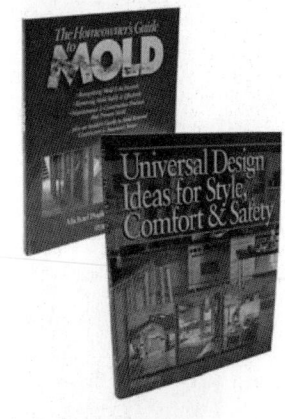

Universal Design Ideas for Style, Comfort & Safety
by RSMeans and Lexicon Consulting, Inc.

Incorporating universal design when building or remodeling helps people of any age and physical ability more fully and safely enjoy their living spaces. This book shows how universal design can be artfully blended into the most attractive homes. It discusses specialized products like adjustable countertops and chair lifts, as well as simple ways to enhance a home's safety and comfort. With color photos and expert guidance, every area of the home is covered. Includes budget estimates that give an idea how much projects will cost.

$21.95
160 pages, illustrated, softcover
Catalog no. 67354

Landscape Estimating Methods
5th Edition

Answers questions about preparing competitive landscape construction estimates, with up-to-date cost estimates and the new MasterFormat classification system. Expanded and revised to address the latest materials and methods, including new coverage on approaches to green building. Includes:
- Step-by-step explanation of the estimating process
- Sample forms and worksheets that save time and prevent errors

$64.95 per copy
Over 350 pages, softcover
Catalog no. 67295C

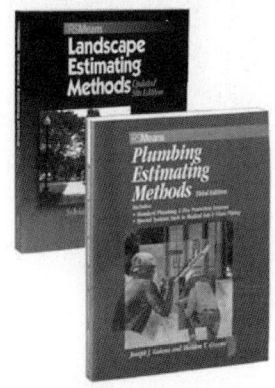

Plumbing Estimating Methods 3rd Edition
by Joseph Galeno and Sheldon Greene

Updated and revised! This practical guide walks you through a plumbing estimate, from basic materials and installation methods through change order analysis. *Plumbing Estimating Methods* covers residential, commercial, industrial, and medical systems, and features sample takeoff and estimate forms and detailed illustrations of systems and components.

$29.98 per copy
330+ pages, softcover
Catalog no. 67283B

Understanding & Negotiating Construction Contracts
by Kit Werremeyer

Take advantage of the author's 30 years' experience in small-to-large (including international) construction projects. Learn how to identify, understand, and evaluate high risk terms and conditions typically found in all construction contracts—then negotiate to lower or eliminate the risk, improve terms of payment, and reduce exposure to claims and disputes. The author avoids "legalese" and gives real-life examples from actual projects.

$69.95 per copy
300 pages, softcover
Catalog no. 67350

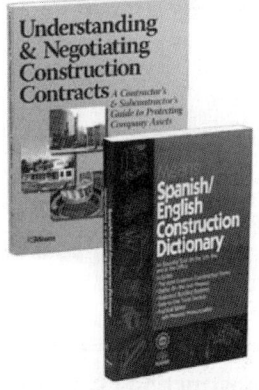

Spanish/English Construction Dictionary 2nd Edition
by RSMeans and the International Code Council

This expanded edition features thousands of the most common words and useful phrases in the construction industry with easy-to-follow pronunciations in both Spanish and English. Over 800 new terms, phrases, and illustrations have been added. It also features a new stand-alone "Safety & Emergencies" section, with colored pages for quick access.

Unique to this dictionary are the systems illustrations showing the relationship of components in the most common building systems for all major trades.

$23.95 per copy
Over 400 pages
Catalog no. 67327A

How Your House Works
by Charlie Wing

A must-have reference for every homeowner, handyman, and contractor—for repair, remodeling, and new construction. This book uncovers the mysteries behind just about every major appliance and building element in your house—from electrical, heating and AC, to plumbing, framing, foundations and appliances. Clear, "exploded" drawings show exactly how things should be put together and how they function—what to check if they don't work, and what you can do that might save you having to call in a professional.

$21.95 per copy
160 pages, softcover
Catalog no. 67351

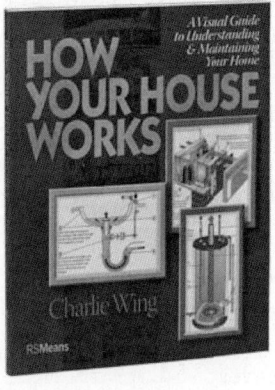

Professional Development

Means CostWorks® CD Training

This one-day course helps users become more familiar with the functionality of *Means CostWorks* program. Each menu, icon, screen, and function found in the program is explained in depth. Time is devoted to hands-on estimating exercises.

Some of what you'll learn:
- Searching the database utilizing all navigation methods
- Exporting RSMeans Data to your preferred spreadsheet format
- Viewing crews, assembly components, and much more!
- Automatically regionalizing the database

You are required to bring your own laptop computer for this course.

When you register for this course you will receive an outline for your laptop requirements.

Also offering web training for CostWorks CD!

Unit Price Estimating

This interactive *two-day* seminar teaches attendees how to interpret project information and process it into final, detailed estimates with the greatest accuracy level.

The most important credential an estimator can take to the job is the ability to visualize construction, and estimate accurately.

Some of what you'll learn:
- Interpreting the design in terms of cost
- The most detailed, time-tested methodology for accurate pricing
- Key cost drivers—material, labor, equipment, staging, and subcontracts
- Understanding direct and indirect costs for accurate job cost accounting and change order management

Who should attend: Corporate and government estimators and purchasers, architects, engineers… and others needing to produce accurate project estimates.

Square Foot & Assemblies Estimating

This *two-day* course teaches attendees how to quickly deliver accurate square foot estimates using limited budget and design information.

Some of what you'll learn:
- How square foot costing gets the estimate done faster
- Taking advantage of a "systems" or "assemblies" format
- The RSMeans "building assemblies/square foot cost approach"
- How to create a reliable preliminary systems estimate using bare-bones design information

Who should attend: Facilities managers, facilities engineers, estimators, planners, developers, construction finance professionals… and others needing to make quick, accurate construction cost estimates at commercial, government, educational, and medical facilities.

Facilities Maintenance & Repair Estimating

This *two-day* course teaches attendees how to plan, budget, and estimate the cost of ongoing and preventive maintenance and repair for existing buildings and grounds.

Some of what you'll learn:
- The most financially favorable maintenance, repair, and replacement scheduling and estimating
- Auditing and value engineering facilities
- Preventive planning and facilities upgrading
- Determining both in-house and contract-out service costs
- Annual, asset-protecting M&R plan

Who should attend: Facility managers, maintenance supervisors, buildings and grounds superintendents, plant managers, planners, estimators… and others involved in facilities planning and budgeting.

Mechanical & Electrical Estimating

This *two-day* course teaches attendees how to prepare more accurate and complete mechanical/electrical estimates, avoiding the pitfalls of omission and double-counting, while understanding the composition and rationale within the RSMeans Mechanical/Electrical database.

Some of what you'll learn:
- The unique way mechanical and electrical systems are interrelated
- M&E estimates–conceptual, planning, budgeting, and bidding stages
- Order of magnitude, square foot, assemblies, and unit price estimating
- Comparative cost analysis of equipment and design alternatives

Who should attend: Architects, engineers, facilities managers, mechanical and electrical contractors… and others needing a highly reliable method for developing, understanding, and evaluating mechanical and electrical contracts.

Green Building: Planning & Construction

In this *two-day* course, learn about tailoring a building and its placement on the site to the local climate, site conditions, culture, and community. Includes information to reduce resource consumption and augment resource supply.

Some of what you'll learn:
- Green technologies, materials, systems, and standards
- Modeling tools and how to use them
- Cost vs. value of green products over their life cycle
- Low-cost green strategies, and economic incentives and funding
- Health, comfort, and productivity

Who should attend: Contractors, project managers, building owners, building officials, healthcare and insurance professionals.

Practical Project Management for Construction Professionals

In this *two-day* course, acquire the essential knowledge and develop the skills to effectively and efficiently execute the day-to-day responsibilities of the construction project manager.

Covers:
- General conditions of the construction contract
- Contract modifications: change orders and construction change directives
- Negotiations with subcontractors and vendors
- Effective writing: notification and communications
- Dispute resolution: claims and liens

Who should attend: Architects, engineers, owner's representatives, project managers.

Facilities Construction Estimating

In this *two-day* course, professionals working in facilities management can get help with their daily challenges to establish budgets for all phase of a project

Some of what you'll learn:
- Determine the full scope of a project
- Identify the scope of risks & opportunities
- Creative solutions to estimating issues
- Organizing estimates for presentation & discussion
- Special techniques for repair/remodel and maintenance projects
- Negotiating project change orders

Who should attend: Facility managers, engineers, contractors, facility trades-people, planners & project managers.

Professional Development

Assessing Scope of Work for Facility Construction Estimating

This *two-day* practical training program addresses the vital importance of understanding the SCOPE of projects in order to produce accurate cost estimates for facility repair and remodeling.

Some of what you'll learn:
- Discussions of site visits, plans/specs, record drawings of facilities, and site-specific lists
- Review of CSI divisions, including means, methods, materials, and the challenges of scoping each topic
- Exercises in SCOPE identification and SCOPE writing for accurate estimating of projects
- Hands-on exercises that require SCOPE, take-off, and pricing

Who should attend: Corporate and government estimators, planners, facility managers, and others who need to produce accurate project estimates.

Site Work & Heavy Construction Estimating

This *two-day* course teaches attendees how to estimate earthwork, site utilities, foundations, and site improvements, using the assemblies and the unit price methods.

Some of what you'll learn:
- Basic site work and heavy construction estimating skills
- Estimating foundations, utilities, earthwork, and site improvements
- Correct equipment usage, quality control, and site investigation for estimating purposes

Who should attend: Project managers, design engineers, estimators, and contractors doing site work and heavy construction.

2011 RSMeans Seminar Schedule

Note: call for exact dates and details.

Location	Dates
Las Vegas, NV	March
Washington, DC	April
Denver, CO	May
San Francisco, CA	June
Washington, DC	September
Dallas, TX	September
Las Vegas, NV	October
Atlantic City, NJ	October
Orlando, FL	November
San Diego, CA	December

2011 Regional Seminar Schedule

Location	Dates	Location	Dates
Seattle, WA	January and August	St. Louis, MO	August
Dallas/Ft. Worth, TX	January	Houston, TX	October
Austin, TX	February	Colorado Springs, CO	October
Anchorage, AK	March and September	Baltimore, MD	November
Oklahoma City, OK	March	Sacramento, CA	November
Phoenix, AZ	March	San Antonio, TX	December
Aiea, HI	May	Raleigh, NC	December
Kansas City, MO	May		
Bethesda, MD	June		
Columbus, GA	June		
El Segundo, CA	August		

1-800-334-3509, ext. 5115

**For more information
visit the RSMeans website
at www.rsmeans.com**

Professional Development

Registration Information

Register early... Save up to $100! Register 30 days before the start date of a seminar and save $100 off your total fee. *Note: This discount can be applied only once per order. It cannot be applied to team discount registrations or any other special offer.*

How to register Register by phone today! The RSMeans toll-free number for making reservations is **1-800-334-3509, ext. 5115.**

Individual seminar registration fee - $935. *Means CostWorks®* **training registration fee - $375.** To register by mail, complete the registration form and return, with your full fee, to: RSMeans Seminars, 700 Longwater Drive, Norwell, MA 02061.

Government pricing All federal government employees save off the regular seminar price. Other promotional discounts cannot be combined with the government discount.

Team discount program for two to four seminar registrations. Call for pricing: 1-800-334-3509, ext. 5115

Multiple course discounts When signing up for two or more courses, call for pricing.

Refund policy Cancellations will be accepted up to ten business days prior to the seminar start. There are no refunds for cancellations received later than ten working days prior to the first day of the seminar. A $150 processing fee will be applied for all cancellations. Written notice of cancellation is required. Substitutions can be made at any time before the session starts. **No-shows are subject to the full seminar fee.**

AACE approved courses Many seminars described and offered here have been approved for 14 hours (1.4 recertification credits) of credit by the AACE International Certification Board toward meeting the continuing education requirements for recertification as a Certified Cost Engineer/Certified Cost Consultant.

AIA Continuing Education We are registered with the AIA Continuing Education System (AIA/CES) and are committed to developing quality learning activities in accordance with the CES criteria. Many seminars meet the AIA/CES criteria for Quality Level 2. AIA members may receive (14) learning units (LUs) for each two-day RSMeans course.

Daily course schedule The first day of each seminar session begins at 8:30 a.m. and ends at 4:30 p.m. The second day begins at 8:00 a.m. and ends at 4:00 p.m. Participants are urged to bring a hand-held calculator, since many actual problems will be worked out in each session.

Continental breakfast Your registration includes the cost of a continental breakfast, and a morning and afternoon refreshment break. These informal segments allow you to discuss topics of mutual interest with other seminar attendees. (You are free to make your own lunch and dinner arrangements.)

Hotel/transportation arrangements RSMeans arranges to hold a block of rooms at most host hotels. To take advantage of special group rates when making your reservation, be sure to mention that you are attending the RSMeans seminar. You are, of course, free to stay at the lodging place of your choice. (**Hotel reservations and transportation arrangements should be made directly by seminar attendees.**)

Important Class sizes are limited, so please register as soon as possible.

Note: Pricing subject to change.

Registration Form   **ADDS-1000**

Call 1-800-334-3509, ext. 5115 to register or FAX this form 1-800-632-6732. Visit our website: www.rsmeans.com

Please register the following people for the RSMeans construction seminars as shown here. We understand that we must make our own hotel reservations if overnight stays are necessary.

☐ Full payment of $_____enclosed.

☐ Bill me

Please print name of registrant(s)

(To appear on certificate of completion)

_____

_____

_____

_____

P.O. #: _____
    GOVERNMENT AGENCIES MUST SUPPLY PURCHASE ORDER NUMBER OR
    TRAINING FORM.

Firm name_____

Address_____

City/State/Zip_____

Telephone no._____ Fax no._____

E-mail address_____

Charge registration(s) to: ☐ MasterCard ☐ VISA ☐ American Express

Account no._____Exp. date_____

Cardholder's signature_____

Seminar name_____

Seminar City _____

Please mail check to: RSMeans Seminars, 700 Longwater Drive, Norwell, MA 02061 USA

New Titles

The Gypsum Construction Handbook by USG

An invaluable reference of construction procedures for gypsum drywall, cement board, veneer plaster, and conventional plaster. This new edition includes the newest product developments, installation methods, fire- and sound-rated construction information, illustrated framing-to-finish application instructions, estimating and planning information, and more. Great for architects and engineers; contractors, builders, and dealers; apprentices and training programs; building inspectors and code officials; and anyone interested in gypsum construction. Features information on tools and safety practices, a glossary of construction terms, and a list of important agencies and associations. Also available in Spanish.

**$29.95
Over 575 pages, illustrated, softcover
Catalog no. 67357A**

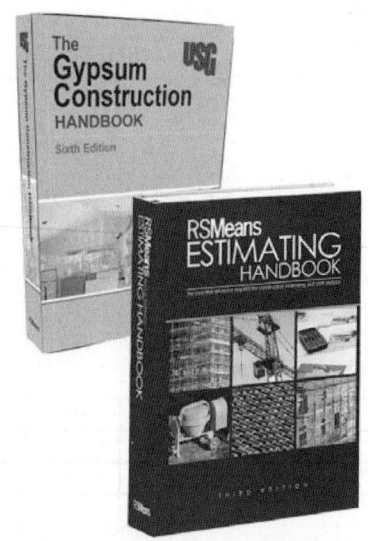

RSMeans Estimating Handbook 3rd Edition

Widely used in the industry for tasks ranging from routine estimates to special cost analysis projects, this handbook has been completely updated and reorganized with new and expanded technical information.

RSMeans Estimating Handbook will help construction professionals:

- Evaluate architectural plans and specifications
- Prepare accurate quantity takeoffs
- Compare design alternatives and costs
- Perform value engineering
- Double-check estimates and quotes
- Estimate change orders

**$99.95 per copy
Over 900 pages, hardcover
Catalog No. 67276B**

Means Illustrated Construction Dictionary
Unabridged 4th Edition, with CD-ROM

Long regarded as the industry's finest, *Means Illustrated Construction Dictionary* is now even better. With nearly 20,000 terms and more than 1,400 illustrations and photos, it is the clear choice for the most comprehensive and current information. The companion CD-ROM that comes with this new edition adds extra features such as: larger graphics and expanded definitions.

**$99.95 per copy
Over 790 pages, illustrated, hardcover
Catalog no. 67292B**

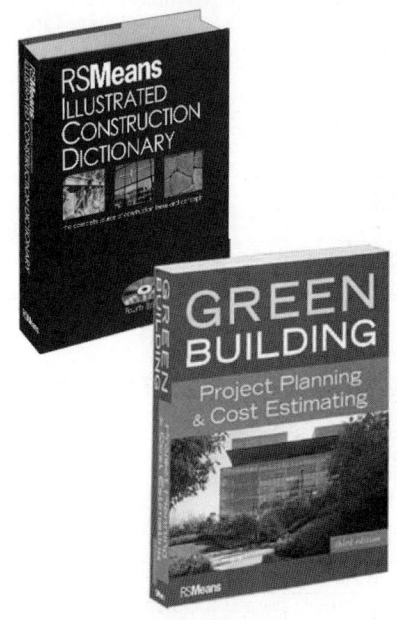

Green Building: Project Planning & Cost Estimating 3rd Edition

Since the widely read first edition of this book, green building has gone from a growing trend to a major force in design and construction.

This new edition has been updated with the latest in green building technologies, design concepts, standards, and costs. Full-color with all new case studies—plus a new chapter on commercial real estate.

**$99.95 per copy
Over 450 pages, softcover
Catalog no. 67338B**

Green Home Improvement
by Daniel D. Chiras, PhD

With energy costs rising and environmental awareness increasing, many people are looking to make their homes greener. This book makes it easy, with 65 projects and actual costs and projected savings that will help homeowners prioritize their green improvements.

Projects range from simple water savers that cost only a few dollars, to bigger-ticket items such as efficient appliances and HVAC systems, green roofing, flooring, landscaping, and much more. With color photos and cost estimates, each project compares options and describes the work involved, the benefits, and the savings.

**$24.95
320 pages, illustrated, softcover
Catalog no. 67355**

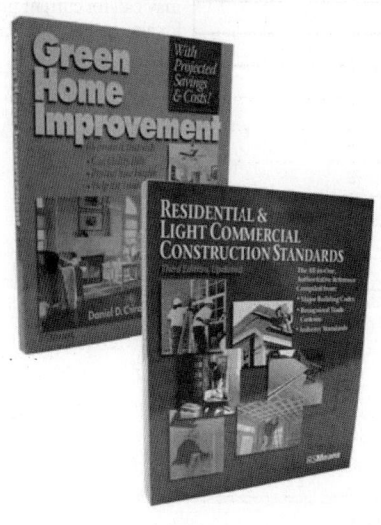

Residential & Light Commercial Construction Standards 3rd Edition, Updated
by RSMeans and contributing authors

This book provides authoritative requirements and recommendations compiled from leading professional associations, industry publications, and building code organizations. This all-in-one reference helps establish a standard for workmanship, quickly resolve disputes, and avoid defect claims. Updated third edition includes new coverage of green building, seismic, hurricane, and mold-resistant construction.

**$59.95
Over 550 pages, illustrated, softcover
Catalog no. 67322B**

2011 Order Form

Qty.	Book no.	COST ESTIMATING BOOKS	Unit Price	Total
	60061	Assemblies Cost Data 2011	$285.95	
	60011	Building Construction Cost Data 2011	174.95	
	60111	Concrete & Masonry Cost Data 2011	154.95	
	50141	Construction Cost Indexes 2011 (subscription)	362.00	
	60141A	Construction Cost Index–January 2011	90.50	
	60141B	Construction Cost Index–April 2011	90.50	
	60141C	Construction Cost Index–July 2011	90.50	
	60141D	Construction Cost Index–October 2011	90.50	
	60341	Contr. Pricing Guide: Resid. R & R Costs 2011	39.95	
	60231	Electrical Change Order Cost Data 2011	169.95	
	60031	Electrical Cost Data 2011	174.95	
	60201	Facilities Construction Cost Data 2011	439.95	
	60301	Facilities Maintenance & Repair Cost Data 2011	389.95	
	60551	Green Building Cost Data 2011	144.95	
	60161	Heavy Construction Cost Data 2011	174.95	
	60091	Interior Cost Data 2011	179.95	
	60121	Labor Rates for the Const. Industry 2011	389.95	
	60181	Light Commercial Cost Data 2011	119.95	
	60021	Mechanical Cost Data 2011	174.95	
	60151	Open Shop Building Const. Cost Data 2011	134.95	
	60211	Plumbing Cost Data 2011	174.95	
	60041	Commercial Renovation Cost Data 2011	144.95	
	60171	Residential Cost Data 2011	109.95	
	60281	Site Work & Landscape Cost Data 2011	169.95	
	60051	Square Foot Costs 2011	184.95	
	62010	Yardsticks for Costing (2010)	174.95	
	62011	Yardsticks for Costing (2011)	174.95	
		REFERENCE BOOKS		
	67329	Bldrs Essentials: Best Bus. Practices for Bldrs	29.95	
	67298A	Bldrs Essentials: Framing/Carpentry 2nd Ed.	15.98	
	67298AS	Bldrs Essentials: Framing/Carpentry Spanish	15.98	
	67307	Bldrs Essentials: Plan Reading & Takeoff	35.95	
	67342	Building & Renovating Schools	69.98	
	67339	Building Security: Strategies & Costs	44.98	
	67353	Complete Book of Framing	29.95	
	67146	Concrete Repair & Maintenance Illustrated	34.98	
	67352	Construction Business Management	49.95	
	67314	Cost Planning & Est. for Facil. Maint.	89.95	
	67328A	Designing & Building with the IBC, 2nd Ed.	99.95	
	67230B	Electrical Estimating Methods, 3rd Ed.	64.95	
	64777A	Environmental Remediation Est. Methods, 2nd Ed.	49.98	
	67343	Estimating Bldg. Costs for Resi. & Lt. Comm.	29.95	
	67276B	Estimating Handbook, 3rd Ed.	$99.95	
	67318	Facilities Operations & Engineering Reference	54.98	
	67301	Facilities Planning & Relocation	49.98	
	67338B	Green Building: Proj. Planning & Cost Est., 3rd Ed.	99.95	
	67355	Green Home Improvement	24.95	
	67357A	The Gypsum Construction Handbook	29.95	
	67357S	The Gypsum Construction Handbook (Spanish)	29.95	
	67323	Historic Preservation: Proj. Planning & Est.	49.98	
	67308E	Home Improvement Costs–Int. Projects, 9th Ed.	24.95	

Qty.	Book no.	REFERENCE BOOKS (Cont.)	Unit Price	Total
	67309E	Home Improvement Costs–Ext. Projects, 9th Ed.	24.95	
	67344	Homeowner's Guide to Mold	21.95	
	67324B	How to Est.w/Means Data & CostWorks, 3rd Ed.	59.95	
	67351	How Your House Works	21.95	
	67282A	Illustrated Const. Dictionary, Condensed, 2nd Ed.	59.95	
	67292B	Illustrated Const. Dictionary, w/CD-ROM, 4th Ed.	99.95	
	67348	Job Order Contracting	89.95	
	67295C	Landscape Estimating Methods, 5th Ed.	64.95	
	67341	Life Cycle Costing for Facilities	99.95	
	67294B	Mechanical Estimating Methods, 4th Ed.	64.95	
	67283B	Plumbing Estimating Methods, 3rd Ed.	29.98	
	67345	Practice of Cost Segregation Analysis	99.95	
	67337	Preventive Maint. for Higher Education Facilities	149.95	
	67346	Preventive Maint. for Multi-Family Housing	89.95	
	67326	Preventive Maint. Guidelines for School Facil.	149.95	
	67247B	Project Scheduling & Management for Constr. 3rd Ed.	64.95	
	67265B	Repair & Remodeling Estimating Methods, 4th Ed.	69.95	
	67322B	Resi. & Light Commercial Const. Stds., 3rd Ed.	59.95	
	67327A	Spanish/English Construction Dictionary, 2nd Ed.	23.95	
	67145B	Sq. Ft. & Assem. Estimating Methods, 3rd Ed.	69.95	
	67321	Total Productive Facilities Management	29.98	
	67350	Understanding and Negotiating Const. Contracts	69.95	
	67303B	Unit Price Estimating Methods, 4th Ed.	59.95	
	67354	Universal Design	21.95	
	67319A	Value Engineering: Practical Applications	79.95	

MA residents add 6.25% state sales tax	
Shipping & Handling**	
Total (U.S. Funds)*	

Prices are subject to change and are for U.S. delivery only. *Canadian customers may call for current prices. **Shipping & handling charges: Add 7% of total order for check and credit card payments. Add 9% of total order for invoiced orders.

Send order to: **ADDV-1000**

Name (please print) _____

Company _____

☐ **Company**

☐ **Home   Address** _____

_____

City/State/Zip _____

Phone # _____   P.O. # _____

(Must accompany all orders being billed)

RSMeans Project Cost Report

By filling out this report, your project data will contribute to the database that supports the RSMeans Project Cost Square Foot Data. When you fill out this form, RSMeans will provide a $30 discount off one of the RSMeans products advertised in the preceding pages. Please complete the form including all items where you have cost data, and all the items marked (✓).

$30.00 Discount per product for each report you submit.

Project Description (No remodeling projects, please.)

✓ Building Use (Office, School...) _____

✓ Address (City, State) _____

_____

✓ Total Building Area (SF) _____

✓ Ground Floor (SF) _____

✓ Frame (Wood, Steel...) _____

✓ Exterior Wall (Brick, Tilt-up...) _____

✓ Basement: (check one) ☐ Full ☐ Partial ☐ None

✓ Number Stories _____

✓ Floor-to-Floor Height _____

✓ Volume (C.F.) _____

% Air Conditioned _____ Tons _____

Total Project Cost $ _____

Owner _____

Architect _____

General Contractor _____

✓ Bid Date _____

✓ Typical Bay Size _____

✓ Capacity _____

✓ Labor Force: ____ % Union   ____ % Non-Union

✓ Project Description (Circle one number in each line)

1. Economy 2. Average 3. Custom 4. Luxury
1. Square 2. Rectangular 3. Irregular 4. Very Irregular

Comments _____

_____

A ✓ General Conditions	$
B ✓ Site Work	$
C ✓ Concrete	$
D ✓ Masonry	$
E ✓ Metals	$
F ✓ Wood & Plastics	$
G ✓ Thermal & Moisture Protection	$
GR Roofing & Flashing	$
H ✓ Doors and Windows	$
J ✓ Finishes	$
JP Painting & Wall Covering	$

K ✓ Specialties	$
L ✓ Equipment	$
M ✓ Furnishings	$
N ✓ Special Construction	$
P ✓ Conveying Systems	$
Q ✓ Mechanical	$
QP Plumbing	$
QB HVAC	$
R ✓ Electrical	$
S ✓ Mech./Elec. Combined	$

Please specify the RSMeans product you wish to receive. Complete the address information.

Product Name _____

Product Number _____

Your Name _____

Title _____

Company _____

☐ Company
☐ Home   Street Address _____

City, State, Zip _____

Return by mail or Fax 888-492-6770.

Method of Payment:

Credit Card # _____

Expiration Date _____

Check _____

Purchase Order _____

RSMeans
a division of Reed Construction Data
Square Foot Costs Department
700 Longwater Drive
Norwell, MA 02061

DISCARD